EVOLUTIONARY NEUROSCIENCE

EVOLUTIONARY NEUROSCIENCE

SECOND EDITION

Edited by

JON H KAAS

ELSEVIER

ACADEMIC PRESS
An imprint of Elsevier

Academic Press is an imprint of Elsevier
125 London Wall, London EC2Y 5AS, United Kingdom
525 B Street, Suite 1650, San Diego, CA 92101, United States
50 Hampshire Street, 5th Floor, Cambridge, MA 02139, United States
The Boulevard, Langford Lane, Kidlington, Oxford OX5 1GB, United Kingdom

Library of Congress Cataloging-in-Publication Data
A catalog record for this book is available from the Library of Congress

British Library Cataloguing-in-Publication Data
A catalogue record for this book is available from the British Library

ISBN: 978-0-12-820584-6

For information on all Academic Press publications visit our website at
https://www.elsevier.com/books-and-journals

Publisher: Nikki Levy
Acquisitions Editor: Natalie Farra
Editorial Project Manager: Samantha Allard
Production Project Manager: Sruthi Satheesh
Cover Designer: Matthew Limbert

Typeset by TNQ Technologies

Contents

3

Early Mammals and Subsequent Adaptations

15. Consistencies and Variances in the Anatomical Organization of Aspects of the Mammalian Brain stem

P.R. MANGER

16. Comparative Anatomy of Glial Cells in Mammals

A. REICHENBACH AND A. BRINGMANN

17. The Monotreme Nervous System

K.W.S. ASHWELL

18. Evolution of Flight and Echolocation in Bats

S.J. STERBING-D'ANGELO AND C.F. MOSS

19. Carnivoran Brains: Effects of Sociality on Inter- and Intraspecific Comparisons of Regional Brain Volumes

S.T. SAKAI AND B.M. ARSZNOV

4

Primates

20. The Phylogeny of Primates

J.G. FLEAGLE AND E.R. SEIFFERT

21. The Expansion of the Cortical Sheet in Primates

S. MAYER AND A.R. KRIEGSTEIN

5
Evolution of Human Brains

36. On the Evolution of the Frontal Eye Field: Comparisons of Monkeys, Apes, and Humans

J.D. SCHALL, W. ZINKE, J.D. COSMAN, M.S. SCHALL, M. PARÉ AND
P. POUGET

37. The Evolution of Auditory Cortex in Humans

J.P. RAUSCHECKER

38. Language Evolution

C. BOECKX

39. The Search for Human Cognitive Specializations

H. CLARK BARRETT

Contributors

J.S. Albert   University of Louisiana, Lafayette, LA, United States

B.M. Arsznov   Minnesota State University Mankato, Mankato, MN, United States

K.W.S. Ashwell   The University of New South Wales, Sydney, NSW, Australia

A.M. Balanoff   Johns Hopkins University, Baltimore, MD, United States; Johns Hopkins University School of Medicine, Baltimore, MD, United States

M.K.L. Baldwin   University of California, Davis, Davis, CA, United States; Monash University, Clayton, VIC, Australia

H. Clark Barrett   UCLA, Los Angeles, CA, United States

G.S. Bever   Johns Hopkins University, Baltimore, MD, United States; Johns Hopkins University School of Medicine, Baltimore, MD, United States

O.R.P. Bininda-Emonds   Carl von Ossietzky University Oldenburg, Oldenburg, Germany

C. Boeckx   ICREA, Universitat de Barcelona, Barcelona, Spain

B. Bogin   Loughborough University, Loughborough, United Kingdom

J.A. Bourne   University of California, Davis, Davis, CA, United States; Monash University, Clayton, VIC, Australia

E.K. Boyle   George Washington University, Washington, DC, United States

A. Bringmann   University of Leipzig Faculty of Medicine, Leipzig, Germany

E. Bruner   Centro Nacional de Investigación sobre la Evolución Humana, Burgos, Spain

R.L. Buckner   Harvard Medical School, Boston, MA, United States; Broad Institute, Cambridge, MA, United States; Harvard University, Cambridge, MA, United States; Massachusetts General Hospital, Boston, MA, United States

E. Candal   Universidade de Santiago de Compostela, Santiago de Compostela, Spain

T.A. Chaplin   Monash University, Melbourne, VIC, Australia

J.D. Cosman   Vanderbilt University, Nashville, TN, United States; Queen's University, Kingston, ON, Canada; Université Pierre et Marie Curie, Institut du Cerveau et de la Moelle épinière, Paris, France

B.L. Finlay   Cornell University, Ithaca, NY, United States

J.G. Fleagle   Stony Brook University, Stony Brook, NY, United States

W.A. Freiwald   National Institutes of Health, Bethesda, MD, United States; University of Rochester, Rochester, NY, United States; The Rockefeller University, New York, NY, United States

A.B. Goldring   University of California, Davis Center for Neuroscience, Davis, CA, United States

A. González   University Complutense of Madrid, Madrid, Spain

O. Güntürkün   Ruhr-University Bochum, Bochum, Germany

A.-M. Hartmann   Carl von Ossietzky University Oldenburg, Oldenburg, Germany

S. Herculano-Houzel   Vanderbilt University, Nashville, TN, United States

L.Z. Holland   University of California at San Diego, La Jolla, CA, United States

J.H. Kaas   Vanderbilt University, Nashville, TN, United States

A.R. Kriegstein   University of California, San Francisco, San Francisco, CA, United States

F.M. Krienen   Harvard Medical School, Boston, MA, United States; Broad Institute, Cambridge, MA, United States; Harvard University, Cambridge, MA, United States; Massachusetts General Hospital, Boston, MA, United States

L.A. Krubitzer   University of California, Davis Center for Neuroscience, Davis, CA, United States

G. Laurent   Max Planck Institute for Brain Research, Frankfurt, Hessen, Germany

D.A. Leopold   National Institutes of Health, Bethesda, MD, United States; University of Rochester, Rochester, NY, United States; The Rockefeller University, New York, NY, United States

J.M. López   University Complutense of Madrid, Madrid, Spain

P.R. Manger   University of the Witwatersrand, Johannesburg, South Africa

S. Mayer   University of California, San Francisco, San Francisco, CA, United States

M. Megías   University of Vigo, Vigo, Spain

J.F. Mitchell   National Institutes of Health, Bethesda, MD, United States; University of Rochester, Rochester, NY, United States; The Rockefeller University, New York, NY, United States

N. Moreno   University Complutense of Madrid, Madrid, Spain

R. Morona　University Complutense of Madrid, Madrid, Spain

C.F. Moss　Johns Hopkins University, Baltimore, MD, United States

R.K. Naumann　Max Planck Institute for Brain Research, Frankfurt, Hessen, Germany

J.M.P. Pakan　University of Edinburgh, Edinburgh, United Kingdom

M. Paré　Vanderbilt University, Nashville, TN, United States; Queen's University, Kingston, ON, Canada; Université Pierre et Marie Curie, Institut du Cerveau et de la Moelle épinière, Paris, France

M.A. Pombal　University of Vigo, Vigo, Spain

S. Pose-Méndez　Universidade de Santiago de Compostela, Santiago de Compostela, Spain

P. Pouget　Vanderbilt University, Nashville, TN, United States; Queen's University, Kingston, ON, Canada; Université Pierre et Marie Curie, Institut du Cerveau et de la Moelle épinière, Paris, France

T.M. Preuss　Emory University, Atlanta, GA, United States

H.-X. Qi　Vanderbilt University, Nashville, TN, United States

I. Quintana-Urzainqui　Universidade de Santiago de Compostela, Santiago de Compostela, Spain

J.P. Rauschecker　Georgetown University Medical Center, Washington, DC, United States; TUM, Munich, Germany

A. Reichenbach　University of Leipzig Faculty of Medicine, Leipzig, Germany

I. Rodríguez-Moldes　Universidade de Santiago de Compostela, Santiago de Compostela, Spain

M.G.P. Rosa　Monash University, Melbourne, VIC, Australia

T.B. Rowe　The University of Texas at Austin, Austin, TX, United States

S.T. Sakai　Michigan State University, East Lansing, MI, United States

G.N. Santos-Durán　Universidade de Santiago de Compostela, Santiago de Compostela, Spain

J.D. Schall　Vanderbilt University, Nashville, TN, United States; Queen's University, Kingston, ON, Canada; Université Pierre et Marie Curie, Institut du Cerveau et de la Moelle épinière, Paris, France

M.S. Schall　Vanderbilt University, Nashville, TN, United States; Queen's University, Kingston, ON, Canada; Université Pierre et Marie Curie, Institut du Cerveau et de la Moelle épinière, Paris, France

E.R. Seiffert　University of Southern California, Los Angeles, CA, United States

M. Stacho　Ruhr-University Bochum, Bochum, Germany

I. Stepniewska　Vanderbilt University, Nashville, TN, United States

S.J. Sterbing-D'Angelo　University of Maryland, College Park, MD, United States; Johns Hopkins University, Baltimore, MD, United States

G.F. Striedter　University of California, Irvine, CA, United States

F. Ströckens　Ruhr-University Bochum, Bochum, Germany

R. Uchiyama　Cornell University, Ithaca, NY, United States

C. Varea　Universidad Autónoma de Madrid, Madrid, Spain

S.P. Wise　Olschefskie Institute for the Neurobiology of Knowledge, Potomac, MD, United States

B. Wood　George Washington University, Washington, DC, United States

Douglas Wylie　Department of Biological Sciences, University of Alberta, Edmonton, AB, Canada

Kara E. Yopak　University of North Carolina Wilmington, Department of Biology and Marine Biology, Wilmington, NC, United States

H.-H. Yu　Monash University, Melbourne, VIC, Australia

W. Zinke　Vanderbilt University, Nashville, TN, United States; Queen's University, Kingston, ON, Canada; Université Pierre et Marie Curie, Institut du Cerveau et de la Moelle épinière, Paris, France

PART 1

History, Concepts, and Theory

CHAPTER

1

A History of Ideas in Evolutionary Neuroscience

G.F. Striedter
University of California, Irvine, CA, United States

Glossary

allometry — The notion that changes in the size of an object (e.g., the body or the brain) entail predictable changes in the proportional sizes of its components. In contrast, isometric scaling involves no changes in an object's proportions.

convergence — The independent evolution of similar structures or functions from non-homologous ancestral precursors.

developmental constraint — The notion that the mechanisms of development bias the production of phenotypic variants that natural selection can act on.

encephalization — Brain size relative to what one would expect in an organism of the same type (i.e., species or other taxonomic group) and body size. Synonym: relative brain size.

heterochrony — Phylogenetic changes in the relative timing of developmental events or in the relative rates of developmental processes.

homology — The relationship between two or more characters that were continuously present since their origin in a shared ancestor. For a more detailed definition, especially for neural characters, see Striedter (1999).

mosaic evolution — The notion that, as brains evolve, individual brain regions may change in size independently of one another. In contrast, concerted evolution indicates that brain regions must change their size in concert with one another.

The field of evolutionary neuroscience is more than 100 years old, and it has deep pre-evolutionary roots. Because that illustrious history has been reviewed repeatedly (Northcutt, 2001; Striedter, 2005) and is treated piecemeal in several articles of this book, I shall not review it fully. Instead, I will discuss a selection of the field's historically most important ideas and how they fit into the larger context of evolutionary theory. I also emphasize ideas that are, or were, controversial. Specifically, I present the field's central ideas in contrast pairs, such as 'common plan versus diversity' and 'natural selection versus constraints'. This approach scrambles the chronology of theoretical developments but helps to disentangle the diverse strands of thought that currently characterize evolutionary neuroscience. It also helps to clarify which future directions are likely to be most fruitful for the field.

1.1 Common Plan versus Diversity

One of the most famous battles of ideas in comparative biology was that between Etienne Geoffroy St. Hilaire and George Cuvier over the existence, or not, of a common plan of construction (or Bauplan) for animals (Appel, 1987). Geoffroy was of the opinion, previously developed by Buffon (1753), that all animals are built according to a single plan or archetype, but Cuvier, France's most illustrious morphologist, recognized at least four different types. Their disagreement erupted into the public sphere when Geoffroy in 1830 endorsed the view that the ventral nerve cord of invertebrates is directly comparable (today we say 'homologous') to the spinal cord of vertebrates. Cuvier responded that Geoffroy was speculating far beyond the available data, and he reasserted publicly that the major types of animals could not be linked by intermediate forms or topological transformations. This Cuvier–Geoffroy debate was followed closely by comparative biologists all across Europe, who were already flirting with the idea of biological evolution or, as they called it, the transmutation of species. If Cuvier was right, then evolution was impossible. On the other hand, some of Geoffroy's hypotheses (e.g., his proposal that insect legs correspond to vertebrate ribs) did seem a trifle fanciful. Thus, the Cuvier–Geoffroy debate embodied much of the ambivalence surrounding evolution in the first half of the nineteenth century.

After Darwin offered a plausible mechanism for the transmutation of species, namely, natural selection (Darwin, 1859), the idea of biological evolution took hold

3

and, by extension, Geoffroy's ideas gained currency. Innumerable homologies were sought and, frequently, revealed (Russel, 1916). Most impressive was the discovery of extensive molecular homologies between species that span the metazoan family tree (Schmidt-Rhaesa, 2003). It was striking, for example, to discover that many of the genes critical for early brain development are homologous between insects and vertebrates (Sprecher and Reichert, 2003). Indeed, the invertebrate and vertebrate genes are sometimes functionally interchangeable (Halder et al., 1995; deRobertis and Sasai, 1996). Those discoveries supported Geoffroy's view that all animals were built according to a common plan, which could now be understood to be a common genetic blueprint or 'program' (Gehring, 1996). Indeed, many biologists proceeded to search for molecular genetic homologies that could reveal previously unimagined morphological homologies (Janies and DeSalle, 1999). Geoffroy would have been thrilled. There are, however, problems with the view that animals are all alike.

The most serious problem, in my view, is that homologous genes may sometimes be involved in the development of adult structures that are clearly not homologous (Striedter and Northcutt, 1991). For example, insect wings and vertebrate nervous systems both depend on *hedgehog* function for normal development, but this does not make neural tubes and insect wings homologous (Baguñã and Garcia-Fernandez, 2003). Instead, findings such as this suggest that evolution tends to work with highly conserved 'master genes' (Gehring, 1996) or, more accurately, tightly knit assemblies of crucial genes (Nilsson, 2004), which it occasionally reshuffles by altering their upstream regulatory elements and/or downstream targets. Evolution is a terrific tinkerer that manages to create novelty from conserved elements. This conclusion echoes Geoffroy's arguments insofar as it acknowledges that "Nature works constantly with the same materials" (Geoffroy, 1807), but it does not mesh with the view that evolution built all animals according to a single plan. What we have, then, is at least a partial rapprochement of the positions held by Cuvier and Geoffroy: adult organisms do conform to several different body plans, but they are built by shuffling repeatedly a highly conserved set of genes (Raff, 1996). Therefore, a crucial question for research is how evolutionary changes in networks of developmentally important genes influence adult structure and function.

Implicit in the preceding discussion has been the idea that adult species differences arise because of evolutionary changes in development (Garstang, 1922). This idea is commonly accepted now, but, back in the nineteenth century, Haeckel (1889) used to promote its polar opposite, namely, the notion that phylogeny creates ontogeny (see Gould, 1977). Haeckel also promoted the idea that all vertebrates pass through a highly conserved phylotypic stage of embryonic development (Slack et al., 1993). Studies have, however, challenged the phylotypic stage idea by showing that the major groups of vertebrates can be distinguished at all stages of embryogenesis (Richardson et al., 1997). An intriguing aspect of that early embryonic variability is that it consists mainly of differences in the timing of developmental processes (Richardson, 1999). Little is known about the genes that generate those changes in developmental timing (also known as heterochrony), but some of them, at least, are likely to be fairly well conserved across species (Pasquinelli and Ruvkun, 2002). More importantly, the notion that adult diversity is based on evolution changing the temporal relationships of conserved processes represents another reconciliation of Cuvier's insistence on adult diversity with Geoffroy's belief in a common plan. Thus, the field of evolutionary developmental biology (evo-devo for short) has overcome the once so prominent dichotomy between conservation and diversity. Its major challenge now is to discover the mechanistic details of how conserved genes and processes are able to produce such diverse adult animals.

Evo-devo thinking has also invaded neuroscience, but evo-devo neurobiology still emphasizes conservation over diversity. For example, we now have extensive evidence that all vertebrate brains are amazingly similar at very early stages of development (Puelles et al., 2000; Puelles and Rubenstein, 2003). However, we still know very little about how and why brain development diverges in the various vertebrate groups after that early, highly conserved stage or period. Looking beyond vertebrates, we find that insect brain development involves at least some genes that are homologous to genes with similar functions in vertebrates (Sprecher and Reichert, 2003). This is remarkable but does not prove that insects and vertebrates are built according to a common plan — if by that we mean that the various parts of adult insect brains all have vertebrate homologues. For example, the finding that several conserved genes, notably *Pax6*, are critical to eye development in both invertebrates and vertebrates, does not indicate that all those eyes are built according to a common plan. The crucial question, which we are just beginning to explore, is how the conserved genes are tinkered with (reshuffled, co-opted, or redeployed) to produce very different adult eyes (Zuber et al., 2003; Nilsson, 2004). This, then, seems to be the future of evo-devo neurobiology: to discover how highly conserved developmental genes and processes are used to different ends in different species. As I have discussed, this research program has ancient roots, but it is just now becoming clear.

1.2 *Scala Naturae* versus Phylogenetic Bush

The idea of evolution proceeding along some kind of scale from simple to complex also has pre-evolutionary roots. Aristotle, for example, ordered animals according to the degree of perfection of their eggs (see Gould, 1977). Later religious thinkers then described an elaborate scale of nature, or *scala naturae*, with inanimate materials on its bottom rung and archangels and God at the other extreme. The early evolutionists, such as Lamarck, transformed this static concept of a *scala naturae* into a dynamic phylogenetic scale that organisms ascended as they evolved. Darwin himself had doubts about arranging species on a scale, but most of his followers had no such qualms (Bowler, 1988). Even today, the phylogenetic scale is taught in many schools and it persists in medicine and academia. For example, the National Institutes of Health's (NIH) guide for institutional animal care and use still recommends that researchers, whenever possible, should work with "species lower on the phylogenetic scale" (Pitts, 2002, p. 97). On the other hand, most contemporary evolutionists have pronounced as dead both the *scala naturae* and its postevolutionary cousin, the phylogenetic scale (Hodos and Campbell, 1969). What do those modern evolutionists cite as the scales' cause of death?

One fatal flaw in the idea that species evolve along a single scale is that, as we now know, evolution made at least some species simpler than their ancestors. Salamanders, for example, are much simpler, especially in brain anatomy (Roth *et al.*, 1993), than one would expect from their phylogenetic position. Even more dramatically, the simplest of all animals, the placozoans, are now thought to have evolved from far more complicated ancestors (Collins, 1998). As more and more molecular data are used to reconstruct phylogenies, it is becoming apparent that such secondary simplification of entire animals has occurred far more frequently than scientists had previously believed (Jenner, 2004) — perhaps because they were so enamored of the phylogenetic scale. A second major problem with *scala naturae* thinking is that the order of species within the scale depends on which organismal features we consider. For example, many fishes would rank higher than mammals if we based our scale on skull complexity, which was reduced dramatically as early mammals evolved (Sidor, 2001). Similarly, dolphins rank high if we look only at brain size, but relatively low if we consider neocortical complexity, which was reduced as the toothed whales evolved (Morgane and Jacobs, 1972). Most people tacitly agree that 'higher animals' are warm-blooded, social, curious, and generally like us, but once we try to be more objective, the single 'chain of being' (Lovejoy,

1936) fractionates into a multitude of different chains, none of which has any special claim to being true.

This multiple-chains idea becomes self-evident once we have grasped that species phylogenies are just like human family trees; they are neither ladders, nor trees with just a single trunk, but bushes or tumbleweeds (Striedter, 2004) with branches growing in divergent directions. Within a given branch, or lineage, complexity may have increased at some points in time and decreased at others, but even if complexity increased more frequently than it decreased, the over- all phylogeny would fail to yield a single scale, because complexity tends to increase divergently in different lineages. For example, bats, honeybees, and hummingbirds are all incredibly complex, compared to their last common ancestor, but they are each complex in different ways. Of course, we can pick one parameter and build a scale for that — we can, for instance, compare the ability of bats, honeybees, and hummingbirds to see ultraviolet (UV) radiation — but different parameters might well yield different scales. Simply put, changes that occurred divergently in different lineages will not, in general, produce a single overarching scale. This insight is old hat to evolutionary biologists, but news to many neuroscientists (Hodos and Campbell, 1969). In part, therefore, the persistence of *scala naturae* thinking in the neurosciences reflects a lack of proper training in contemporary evolutionary theory. In addition, I suspect that human minds possess a natural tendency for ordering disparate items linearly. Such a bias would be useful in many contexts, but it would make it difficult to comprehend (without training) the divergent nature of phylogeny.

Although *scala naturae* thinking persists in neuroscience generally, evolutionary neuroscientists have labored to expunge its ghost. For example, a consortium of 28 comparative neurobiologists revised the nomenclature of avian brains to replace the terms *neo*striatum, *archi*striatum, and *paleo*striatum — which suggested that brains evolved by the sequential addition of new brain regions — with terms devoid of *scala naturae* overtones (Reiner *et al.*, 2004a, 2004b; Jarvis *et al.*, 2005). Some of the replacement names are terms that were already used for brain regions in other vertebrates; they reflect our current understanding of homologies. However, some of the new terms — e.g., nidipallium and arcopallium — are novel and intended to apply exclusively to birds. These novel names were coined because bird brains, particularly bird forebrains, have diverged so much from those of other vertebrates (including reptiles) that strict one-to-one homologies are difficult, if not impossible, to draw for several regions (Striedter, 1998, 1999). Thus, the revised terminology reflects a new consensus view that avian brains did not evolve by the sequential addition of new brain areas, yet also reminds us that bird brains are full of features that evolved quite independently of those that feature in mammalian phylogeny. In other words, the new terminology avoids *scala naturae* overtones and, instead, combines the notion of a common plan with that of divergent complexity.

As comparative neurobiologists reject the notion of a *scala naturae*, they stand to lose a central part of their traditional justification for working on nonhuman brains. No longer can they argue that research on other brains must be useful because nonhuman brains are always simpler, and therefore easier to comprehend, than human brains. Instead, they must admit that some nonhuman brains are stunningly complex and, more importantly, that their phylogenetic paths toward complexity diverged from the primate trajectory. That is, complex bird, fish, or insect brains are not mere steps along the path to human brains, but the outcome of divergent phylogenies (see "Evolution of the Nervous System in Fishes," "Do Birds and Reptiles Possess Homologues of Mammalian Visual, Somatosensory, and Motor Cortices?," "Evolution of Color Vision and Visual Pigments in Invertebrates"). Does this suggest that research on nonhuman brains should cease to be funded? I do not think so, but the justification for working on nonhuman brains ought to be tweaked. One obvious alternative justification is that all brains are likely to share some features, especially if they come from close relatives. Another good justification for research on nonhuman brains is that, compared to human brains, the former are much more amenable to physiological and anatomical research. This line of justification assumes that the model differs from the target system only in those respects that make the model easier to study, and not in the respects that are modeled — an assumption that sometimes fails. It now appears, for example, that the auditory system of owls, which was generally regarded as an ideal model for sound localization in vertebrates, exhibits some highly specialized features (McAlpine and Grothe, 2003). This finding, at first glance, suggests that research on bird brains is wasteful, but this is a simplistic view. Research on the owl's auditory system has taught us much about how neurons compute behaviorally relevant information and it serves as an invaluable reference against which we can compare sound processing in other species, including humans. Furthermore, some differences between a model and its target can lead to surprising discoveries. Much might be gained, for example, from studying why some nonhuman brains are far more capable than primate brains of repairing themselves (Kirsche and Kirsche, 1964). Thus, model systems research can be useful even if the model is imprecise. A third, less

frequently discussed, justification for examining the brains of diverse species is that comparative research can bring to light convergent similarities, which in turn might reveal some principles of brain design. For example, the discovery that olfactory systems in both vertebrates and many different invertebrates exhibit distinctive glomeruli strongly suggests that those glomeruli are needed for some critical aspects of odorant detection and analysis (Strausfeld and Hildebrand, 1999).

Therefore, research on nonhuman brains need not be justified in terms of a presumed phylogenetic scale. Instead, comparative neurobiology is valuable because (1) all brains are likely to share some features, (2) nonhuman brains are more amenable to some types of research, and (3) the study of diverse nonhuman brains can lead to the discovery of design rules for brains. Historically, only the first of these alternatives has been widely discussed, but all are logically sound, and none depend on the existence of a *scala naturae*.

1.3 Relative Size versus Absolute Size

The most obvious difference between species is that they differ enormously in size. Because life began with tiny organisms, evolutionary increases in body size must have outnumbered or outpaced the decreases. This is true of organisms generally, but it also holds for several individual lineages, including mammals and, within mammals, primates (Stanley, 1973; Alroy, 1998). The most fascinating aspect of those changes in body size is that they involved much more than the isometric scaling up or down of the ancestral condition; they involved allometric changes in the proportions of body parts and physiologic processes. For example, skeletal mass increases disproportionately with increasing body size, whereas heart rate decreases. Countless studies — on both vertebrates and invertebrates — have documented these allometries and explored their functional implications (Calder, 1984; Schmidt-Nielsen, 1984).

Much less is known about the causes of allometry. Studies on allometry in insects showed that some scaling relationships are readily modifiable by natural or artificial selection (see Emlen and Nijhout, 2000; Frankino et al., 2005). This finding suggests that even tight scaling laws are not immutable, which would explain why many traits scale differently (e.g., with different exponents) in different taxonomic groups (Pagel and Harvey, 1989). A very different, more theoretical line of research has shown that numerous allometries, specifically those with power law exponents that are multiples of 1/4, may have evolved because the optimal means of delivering metabolic energy to cells is through an hierarchically branching, fractal network of vessels whose termini (e.g., capillaries) are body size-invariant (West et al., 1997; Savage et al., 2004; West and Brown, 2005). This theory is mathematically complex and still controversial (Kozlowski and Konarzewski, 2004; Brown et al., 2005; Hoppeler and Weibel, 2005), but it is elegant. Furthermore, because the theory of West et al. is based in part on the assumption that natural selection optimizes phenotypes, it is consistent with the aforementioned finding that allometries are modifiable by selection. However, West et al.'s (1997) theory cannot explain (or does not yet explain) why some organs, such as the brain, scale with exponents that are not multiples of 1/4. Nor can it easily explain taxonomic differences in scaling exponents. Thus, the causal — physiological and/or developmental — bases of allometry are coming into focus but remain, for now, mysterious.

Brain scaling, in particular, remains quite poorly understood (see "Principles of Brain Scaling," "Scaling the Brain and Its Connections," "How to Build a Bigger Brain; Cellular Scaling Rules for Rodent Brains"). The discovery that brains become proportionately smaller with increasing body size dates back to the late eighteenth century (Haller, 1762; Cuvier, 1805–1845). Since then, numerous studies have documented brain allometry in all the major groups of vertebrates (Deacon, 1990a; van Dongen, 1998) and even some invertebrates (Julian and Gronenberg, 2002; Mares et al., 2005). Generally speaking, those studies confirmed that in double logarithmic plots of brain size versus body size, the data points for different species within a given lineage tend to form a reasonably straight line, indicating the existence of a simple power law. The slope of those best-fit lines are almost always less than 1, which reflects the aforementioned fact that brains generally become proportionately smaller with increasing body size. The large body of work on brain–body scaling further revealed that data points for different taxonomic groups often form lines with similar slopes but different y intercepts. These differences in y intercepts are known as differences in relative brain size or encephalization. They seriously complicate efforts to draw a single allometric line for any large taxonomic group (Pagel and Harvey, 1989), but they allow us to identify evolutionary changes in relative brain size among some smaller taxonomic groups. For example, they allow us to determine that relative brain size increased with the origin of mammals, with the origin of primates, several times within primates, with the origin of the genus Homo, and, last but not least, with the emergence of Homo sapiens (see "Primate Brain Evolution in Phylogenetic Context," "The Hominin Fossil Record and the Emergence of the Modern Human Central Nervous System," "The Evolution of Human Brain and Body Growth Patterns").

Overall, such phylogenetic analyses suggest that, among vertebrates, relative brain size increased more frequently than it decreased (Striedter, 2005).

Enormous effort has gone into determining the functional significance of evolutionary changes in brain–body scaling. Darwin, for example, had argued that relative brain size is related to "higher cognitive powers" (Darwin, 1871), but defining those powers and comparing them across species has proven difficult (Macphail, 1982). Consequently, most subsequent investigators shied away from the notion of general intelligence, or 'biological intelligence' (Jerison, 1973), and focused instead on more specific forms of higher cognition. Parker and Gibson (1977), for example, proposed that a species' degree of encephalization is related to its capacity for extracting nutritious fruits and nuts from their protective shells. Several authors have stressed correlations between brain size and 'social intelligence' (Byrne and Whiten, 1988; Dunbar, 1998; Reader and Laland, 2002). Collectively, these studies reinforced the sense that relative brain size is, somehow, related to some forms of intelligence. However, relative brain size also correlates with several other attributes, such as longevity, home-range size, diet, and metabolic rate (for a review, see van Dongen, 1998). The latter correlations, with diet and metabolism, have received particularly lavish attention (Martin, 1981; McNab, 1989; Aiello and Wheeler, 1995). Paradoxically, the discovery of so many correlations has led some evolutionary neuroscientists to despair: there are too many correlates of relative brain size, and many of them come and go, depending on which taxonomic group is being examined and which statistical methods are used for the analyses (e.g., Bennet and Harvey, 1985; Iwaniuk et al., 1999; Deaner et al., 2000; Beauchamp and Fernández-Juricic, 2004; Jones and MacLarnon, 2004; Martin et al., 2005). Too many contested hypotheses, too little certitude.

There is also not much clarity on why brains scale so predictably with body size. Early workers argued that brains generally scale against body size with a power law exponent close to 2/3 because the brain's sensory and motor functions were related to the body's surface area, which presumably scales with that same exponent (Snell, 1891; Jerison, 1973). According to this view, brain sizes in excess of that predicted by the 2/3 power law are due to increases in the brain's nonsomatic, cognitive regions. This would explain the correlations between relative brain size and some forms of intelligence. Unfortunately, there are two major problems with this view. First, brain–body scaling exponents often differ substantially from 2/3 (van Dongen, 1998; Nealen and Ricklefs, 2001). The second problem is that the brain's more cognitive regions also scale predictably with body size (Fox and Wilczynski, 1986), undermining the assumption that brains are divisible into regions that scale with body size and regions that do not. Therefore, the excess neuron hypothesis (Striedter, 2005) is dead. In searching for an alternative, some have suggested that brain–body allometry is linked to the scaling of metabolic rates. This hypothesis is based on the observation that, in at least some taxonomic groups, brain size and basal metabolic rate scale against body size with similar exponents (Martin, 1981; Mink *et al.*, 1981). However, other studies have shown that the correlation between brain size and metabolism is not tight, once the mutual correlation with body size is factored out (McNab, 1989). This correlational slack presumably arises because species differ in how much of the body's total energy supply they deliver to the brain (Aiello and Wheeler, 1995; Kaufman, 2003), but this just underscores that relative brain size is not so tightly linked to metabolic rate.

Overall, the lack of clarity on what causes brains to scale predictably with body size, and how to interpret deviations from the scaling trends, has caused interest in relative brain size to fade. Increasingly, evolutionary neuroscientists have turned away from relative brain size and asked, instead, how the size of individual brain regions correlates with various behavioural parameters (Harvey and Krebs, 1990; see Brain Size in Primates as a Function of Behavioral Innovation, Mosaic Evolution of Brain Structure in Mammals). This shift in research strategy makes sense, because, after all, the brain is functionally heterogeneous. However, even studies that focus on correlations between single brain areas and specific behaviors — some refer to them as neuroecological studies — are controversial because: (1) the behavioral parameters are difficult to quantify and/or define (Bolhuis and Macphail, 2001), (2) neuronal structure–function relationships are complex and often poorly understood, (3) it is difficult to decide *a priori* whether one should correlate behavioral parameters against a region's absolute size, its proportional size, or its size relative to expectations (Striedter, 2005), and (4) the methods for establishing statistically significant correlations in phylogenetic data remain debatable (Felsenstein, 1985; Garland *et al.*, 1992; Smith, 1994; Martin *et al.*, 2005). Brave neuroscientists are continuing to tackle those problems, but the larger problem of how to deal with relative brain size — how to find its causes and its functional significance — is fading from view. Perhaps we need a new approach to understanding relative brain size — perhaps one that is linked more directly to the physiological and geometric properties of brains (West and Brown, 2005) — but this novel direction is not yet apparent.

As interest in relative brain size waned, interest in absolute brain size waxed, mainly because many of the brain's internal structural and functional features turn out to scale predictably with absolute brain size. Best studied is the phenomenon of size-related shifts in brain region proportions (Sacher, 1970; Finlay and Darlington, 1995). In mammals, for example, the neo-cortex becomes disproportionately large as absolute brain size increases, whereas most other regions become disproportionately small. A second interesting scaling law is that a brain's degree of structural complexity tends to increase with absolute brain size. Within the neocortex, for example, the number of distinct areas increases predictably with neocortex size (Changizi and Shimojo, 2005). A third fascinating aspect of brain scaling is that the amount of white matter within mammalian brains scales allometrically with absolute brain size (Ringo, 1991; Zhang and Sejnowski, 2000). This connectional allometry, taken together with the fact that synapse size and density are relatively size-invariant, indicates that brains become less densely interconnected, on average, as they increase in size (Stevens, 1989; Deacon, 1990a, 1990b; Striedter, 2005; see Scaling the Brain and Its Connections). All of this signifies that brains change structurally in many ways as they vary in absolute size. Many of those changes have clear functional implications. For example, it has been suggested that, as hominid brains increased in size, the axons inter-connecting the two cerebral hemispheres became so sparse and long that the hemispheres became less capable of interacting functionally, which led to an increase in functional asymmetry (Ringo *et al.*, 1994; see Cortical Commissural Connections in Primates, The Evolution of Hemispheric Specializations of the Human Brain). Considerations such as these suggest that absolute brain size is a much better predictor of brain function than relative brain size, at least among close relatives (Striedter, 2005).

In retrospect, we can say that evolutionary neuroscientists historically have overemphasized relative brain size. As Dunbar (2006) put it, comparative neurobiologists have too long been "dragooned into worrying about relativizing brain size by a very peculiar view that body size must be the default determinant of brain volume." Can we explain this undue emphasis? Partly, evolutionary neuroscientists may have worried that focusing on absolute brain size and linking it to higher cognitive powers would force us to conclude that whales and elephants, with their enormous brains, are smarter than humans (see "Cetacean Brain Evolution," "Evolution of the Elephant Brain: A Paradox between Brain Size and Cognitive Behavior"). This is a valid concern, for few would doubt that humans are — or at least can be — the most intelligent creatures on earth. However, whales and elephants are behaviorally complex, and humans may well be special because they are unique in possessing symbolic language (Macphail, 1982). Furthermore, it seems to me that large whales, with large brains, are more intelligent (both socially and in their hunting strategies) than dolphins or small whales. This hypothesis remains to be tested, but it points to a strategy for reconciling absolute and relative brain size: among close relatives, comparisons of absolute brain size are most informative, but in comparisons of distant relatives (e.g., whales and humans), relative brain size is a more potent variable (Striedter, 2005). This view is consistent with the finding that, among primates, social group size correlates more strongly with absolute brain size than with relative brain size (Kudo and Dunbar, 2001; Striedter, 2005). It also serves as a productive counterweight to the field's traditional, almost exclusive emphasis on relative brain size.

1.4 Natural Selection versus Developmental Constraints

Darwin's theory of natural selection entails two main components, namely, that (1) organisms produce offspring with at least some heritable variation and (2) that organisms generally produce more offspring than their environment is able to sustain. Given those two components, some variants are bound to be fitter than others in the sense that their offspring are more likely to survive and produce offspring. This difference, in turn, will cause the heritable traits of the fitter variants to spread in the population. Given this, Darwin's most "dangerous idea" (Dennett, 1995), one can explain an organism's attributes in terms of the selective pressures that promoted their spread and, hence, their current existence. An enormous number of such adaptational explanations have been proposed. Many stress that natural selection optimized features for specific functions; others emphasize that natural selection tends to produce optimal compromises between competing functions and/or costs (Maynard Smith, 1982). Generally speaking, the explanatory power of these adaptational explanations derives solely from natural selection's second step, the sorting of offspring. Generation of the variants that are sorted is usually assumed to be random and, hence, irrelevant to explanations of the phenotype. This 'adaptationist paradigm' (Gould and Lewontin, 1979) has dominated evolutionary theory for most of its history.

In the 1970s and 1980s, however, the adaptationist paradigm was challenged by authors who stressed that the variants available to natural selection may not really be random (Gould and Lewontin, 1979; Alberch, 1982; Maynard Smith et al., 1985). Central to those challenges was the idea that, even if mutations are random at the genetic level, those random genetic mutations are channeled, or filtered, through mechanisms of development that favor the emergence of some phenotypes. Some structures may be impossible

for embryos to develop; others are likely to emerge (Alberch, 1982). If this is true, then natural selection chooses not among a random selection of phenotypes but from a structured set that is determined, or at least biased, by the mechanisms of development. This idea is important, because it suggests that development constrains the power of natural selection to set the course of evolutionary change. It threatens natural selection's widely assumed omnipotence. Some authors carried this threat so far as to exhort biologists to halt their search for adaptive scenarios and to research, instead, the 'generative' mechanisms of development (Goodwin, 1984). Fortunately, most evolutionary biologists today seek a more balanced rapprochement of embryology and evolutionary biology (Gilbert et al., 1996; Wagner and Laubichler, 2004).

Specifically, evo-devo biologists today tend to accept the concept that natural selection is the most prominent determinant of who thrives and who dies, no matter how constrained development might be. They also tend to stress that development itself is subject to descent with modification — i.e., evolution — which means that even fairly tight constraints can change. Therefore, explanations couched in terms of natural selection are not antithetical to those involving developmental constraints, but complementary (Striedter, 2005). Still, the synthesis of natural selection and developmental constraints remains uncertain in one key respect: what if the mechanisms of development were shaped by natural selection to produce variants that are much fitter than one would expect by chance? Then the distinction between the generative and selective components of natural selection (see above) would blur. The developmental production of variants would no longer be random with respect to a species' ecology. This hypothesis, which was pushed furthest by Riedl (1977), is interesting and potentially profound, but not yet supported by much evidence.

Brains were historically considered to be shaped by natural selection, unencumbered by developmental constraints. In general, the size and structure of both entire brains and individual brain regions were thought to be optimized. Jerison (1973, p. 8), made this idea explicit when he wrote that "the importance of a function in the life of each species will be reflected by the absolute amount of neural tissue of that function in each species." How development produced that fine-tuning was never specified. Presumably, the idea was that genetic mutations could vary the size and structure of individual brain regions freely, leading to steady improvements in fitness until an optimum was reached. Little thought was given to the possibility that brains might be constrained in how they could evolve. However, a few authors proposed that trophic dependencies between interconnected brain regions might cause entire circuits or systems to change size in unison rather than piece-meal (Katz and Lasek, 1978). Such 'epigenetic cascades' (Wilczynski, 1984) might channel evolution (Katz *et al.*, 1981), but they would not constrain natural selection, because the cascades help to optimize functional brain systems by matching the size of interconnected neuronal populations. That is, epigenetic cascades act not against, but in conjunction with, the optimizing power of natural selection; they are not classical constraints, which may explain why they have rarely been discussed (Finlay *et al.*, 1987).

The idea of brains evolving under a restrictive developmental rule was proclaimed forcefully by Finlay and Darlington (1995). Their argument was founded on the observation that the various major brain regions in mammals scale against absolute brain size with different allometric slopes (Sacher, 1970; Gould, 1975; Jerison, 1989). Although this finding was well established at the time, it had not been explained; it was a scaling rule without a cause. Finlay and Darlington's major contribution was to propose that the height of a region's allometric slope was related to the region's date of birth (i.e., the time at which the region's precursor cells cease to divide), with late-born regions tending to become disproportionately large with increasing brain size. Why does this relationship exist? Finlay and Darlington (1995) showed that their late-equals- large rule emerges naturally if neurogenetic schedules (i.e., the schedules of what regions are born when) are stretched as brains increase in size and compressed when they shrink. This insight, in turn, prompted Finlay and Darlington to hypothesize that brain evolution is constrained to stretch or compress neurogenetic schedules and cannot, in general, delay or advance the birth of individual regions. In other words, even if evolution 'wanted' to increase the size of only one brain region, it would be 'forced' to change also the size of many other brain regions. Thus, Finlay and Darlington argued that development constrains brains to evolve concertedly, rather than mosaically.

Finlay and Darlington's developmental constraint hypothesis has been challenged by various authors, who all pointed out that brains do sometimes evolve mosaically (Barton and Harvey, 2000; Clark *et al.*, 2001; de Winter and Oxnard, 2001; Iwaniuk *et al.*, 2004; Safi and Dechmann, 2005). In addition, Barton (2001) has argued that correlations between region size and absolute brain size are due to functional requirements, rather than developmental constraints. Specifically, Barton (2001, p. 281) reported that the sizes of interconnected brain regions in what he called a functional system exhibited "significantly correlated evolution after taking variation in a range of other structures and overall brain size into account." Finlay *et al.* (2001) countered that such system-specific evolution may indeed occur, particularly for the so-called limbic system (see also Barton *et al.*, 2003), but that this does not negate the existence of developmental constraints. In a review of this debate, I concluded that most of it may be resolved by arguing that instances of mosaic (and/or system-specific) evolution occur against a background of concerted, developmentally constrained evolution (Striedter, 2005; see "Mosaic Evolution of Brain Structure in Mammals"). Both Finlay and Barton seem open to this kind of rapprochement (Finlay *et al.*, 2001; Barton, 2006).

The debate on mosaic versus concerted evolution highlights how little we know about the evolution of neural development or, for that matter, about the role that natural selection played in shaping brains. The developmental data used to support Finlay *et al.*'s (2001) hypothesis came from just 15 species and were collected by several different laboratories, using diverse methodologies. Moreover, the data are limited to dates of neurogenesis. We know virtually nothing about species differences (or similarities) in how large brain regions are prior to neurogenesis, how quickly the regions grow, or how much cell death they endure. Data on these other, relatively neglected aspects of brain development might reveal additional constraints, and they might clarify how regions can evolve mosaically even if neurogenetic schedules are conserved.

Similarly lacking are data on natural selection and the brain. Although several analyses have shown that the size of some brain regions (relative to absolute brain size) correlates with aspects of a species' behavior or ecology (e.g., Clark *et al.*, 2001; de Winter and Oxnard, 2001; Iwaniuk *et al.*, 2004), such correlations are only indirect evidence for natural selection. More direct data are difficult to gather, because direct demonstrations of natural selection at work require measurements of heritability and fitness functions. As it is, we know so little about how selection acts on brains that debates on its potency are bound to erupt. Clearly, more studies must be performed before we can reach firm conclusions about which aspects of brain development and evolution are tightly constrained and which are subject to specific selective pressures.

1.5 One Law, Many Laws, or None

Is human history explicable in terms of general principles or laws? This question has been debated extensively. Some scholars insist that history is based largely on a few major laws, playing out against a background of far less important noise. Others argue, instead, that history is so full of contingencies (or accidents) that general or universal laws are blown to bits. I am not competent to review this debate but find myself most sympathetic to the intermediate position taken by

Hempel (1942) in his call for a nomological—deductive approach to history. Basically, Hempel argued that historical events can be explained only by reference to various general (deterministic or probabilistic) laws that causally link preceding events or conditions to the event being explained. For example, an account of why an automotive radiator cracked during a frost would involve both historical contingencies and general laws relating temperature to pressure (Hempel, 1942). Similarly, events in human history can be explained by "showing that the event in question was not 'a matter of chance', but was to be expected in view of certain antecedent or simultaneous conditions" (Hempel, 1942) and the operation of several, often implicitly assumed, general laws. This nomological—deductive methodology waxes and wanes in popularity (Kincaid, 1996; McIntyre, 1996), but it seems logical in principle. Naturally, one may debate whether human behavior is predictable enough to yield the kind of laws that are needed for nomological—deductive explanations (Beed and Beed, 2000).

Evolutionary biologists have likewise debated the role of general laws in explaining the past, which in their realm is phylogeny. Some have argued that natural selection is a universal law that can be used to explain the emergence of many, if not most, biological features. Others have countered that natural selection is a mathematical truth, rather than an empirically determined law (Sober, 2000). More importantly, many biologists have pointed out that the results of natural selection are not highly predictable. Gould (1989) made this argument when he declared that rewinding the tape of life on earth and playing it again would not lead to a repeat performance. Biological history is full of accidents, of happenstance. Therefore, Gould argued, evolutionary explanations must be crafted one event at a time, without recourse to general laws. On the other hand, Gould did grant that evolution is constrained by diverse physical principles, by rules of construction and good design, and by some scaling rules (Gould, 1986, 1989). In his view, "the question of questions boils down to the placement of the boundary between predictability under invariant law and the multifarious possibilities of historical contingency" (Gould, 1989, p. 290). Gould placed this boundary "so high that almost every interesting event of life's history falls into the realm of contingency" (Gould, 1989, p. 290). This appears to be an extreme position, for many other evolutionary biologists place that same boundary lower. They tend to be far more impressed than Gould by the degree of convergent evolution in the history of life (Carroll, 2001; Willmer, 2003). They look, for example, at the convergent similarities of eyes in vertebrates and octopi and conclude that some design rules for eyes exist. In sum, disagreements

persist about the placement of Gould's boundary between predictability and contingency, but most biologists accept that evolutionary explanations must involve at least some causal laws (Bock, 1999).

Given this context, it is not surprising that neuroscientists are conflicted about the importance of general laws for explaining the evolutionary history of brains. Marsh (1886) had proposed that brains consistently increase in size over evolutionary time, but later authors vehemently disagreed (see Jerison, 1973; Buchholtz and Seyfarth, 1999). Personally, I think that Marsh did have a point, for brain and body size have both increased, at least on average, in several vertebrate lineages (see Striedter, 2005). Still, Marsh's laws were merely descriptions of phylogenetic trends, not causal laws. The first explicitly causal law of brain evolution was Ariëns Kappers' (1921) law of neurobiotaxis, which states that cell groups in evolution tend to move toward their principal inputs. Unfortunately for Ariëns Kappers, later studies showed that cell groups do not move quite so predictably and called into question some of the mechanisms that supposedly produced neurobiotaxis. The next major putative law of brain evolution was Ebbesson's (1980) parcellation principle, which states that brains become more complex by the division of ancestrally uniform cell groups into daughter aggregates that selectively lose some of their ancestral connections. This principle was strenuously criticized by most comparative neuroanatomists, mainly because its empirical foundation was shaky (see Ebbesson, 1984). Although a weak version of Ebbesson's theory, stating merely that brains become less densely connected as they increase in size, is probably defensible (Deacon, 1990a; Striedter, 2005), the strong version of Ebbesson's original idea has failed the test of time: plenty of data now show that brains evolve not only by the loss of connections, but also by creating novel projections.

Confronted with this abundance of failed brain evolution laws, most evolutionary neuroscientists have emphasized only a single, undisputed regularity of brain evolution, namely, that numerous aspects of brain structure and function are highly conserved across species. Specifically, they focused, á la Geoffroy St. Hilaire, on the existence of common plans of construction and highlighted molecular homologies between invertebrates and vertebrates (see above). This has been productive. It is important to note, however, that the principle of phylogenetic conservation predicts stability and does not deal explicitly with change. Is brain phylogeny subject to just a single law, which states that brains change little over time? Or are there also laws of evolutionary change in brains? I affirmed the second possibility (Striedter, 2005), but laws of evolutionary change in brains are no doubt difficult to find. C. J. Herrick, a founding father of evolutionary neuroscience, put it well:

> Most scientific research has been directed to the discovery of the uniformities of nature and the codification of these in a system of generalizations. This must be done before the changes can be interpreted. The time has come to devote more attention to the processes and mechanisms of these changes... but it is much more difficult to find and describe the mechanisms of ... [the] apparently miraculous production of novelties than it is to discover the mechanical principles of those repetitive processes that yield uniform products (Herrick, 1956, p. 43).

The last few years have seen an uptick in the number of studies that address evolutionary change and novelty in brains (Aboitiz, 1995; Catania et al., 1999; Rosa and Tweedale, 2005), and modern research on brain scaling and developmental constraints (see above) has advanced our understanding of the regularities that lurk within brain variability. In addition, a rapidly increasing number of studies is beginning to reveal genomic changes that are probably linked to changes in brain size and/or structure (e.g., Dorus et al., 2004; Mekel-Bobrov et al., 2005). Therefore, the time Herrick discussed, when evolutionary change becomes a focus of analysis (see also Gans, 1969), is probably at hand.

Thus, I envision a future in which most evolutionary neuroscientists will embrace many different laws, some dealing with constancy and some with change. A few philosophers of science (e.g., Beatty, 1995) might decry such a vision, because they think that any natural law deserving of its name must apply universally, in all contexts and without room for other, countervailing laws. I have no training in philosophy, but think that all scientific laws apply only in specified domains and given assumptions (Striedter, 2005). In the real world, particularly in the complex world of biological systems, most laws or principles are sometimes excepted. This does not make them useless but, instead, prompts us to ask what causes the observed exceptional cases (West and Brown, 2005). If we understand the causal basis of our laws, then the exceptions should, with further work, become explicable. In other words, I think that evolutionary neuroscientists can fruitfully avail themselves of Hempel's nomological–deductive approach to history. To some extent, they always have.

1.6 Conclusions and Prospects

In summary, the history of evolutionary neuroscience features some serious missteps, such as the idea that brains evolved in a phylogenetic series and Ariëns Kappers' law of neurobiotaxis, but it also reveals considerable progress. The scala naturae has ceased to guide the research of evolutionary neuroscientists and the idea of

neurobiotaxis has quietly disappeared. The once stagnant field of brain allometry is showing signs of revival, largely because of new statistical techniques and a new emphasis on absolute brain size. The debate about concerted versus mosaic evolution persists, but directions for rapprochement are emerging. In general, the field has flirted with a broad variety of theoretical ideas and found some of them wanting and others promising. In terms of theory, the field is still quite young, but it is poised to mature now.

Predicting directions of growth for any science is problematic, but I believe that most future developments in evolutionary neuroscience will parallel developments in other, non-neural domains of evolutionary biology. After all, the history of evolutionary neuroscience is full of ideas that originated in non-neural areas of biology. For example, the methodology of phylogenetic reconstruction or cladistics (which I did not discuss in this article but have treated elsewhere; see Striedter, 2005) was originally developed by an entomologist (Hennig, 1950; see also Northcutt, 2001). Similarly, evolutionary developmental biology was burgeoning before it turned to brains (Hall, 1999). Therefore, I think it likely that the future of evolutionary neuroscience has already begun in some non-neural field. Maybe molecular genetics, with its new emphasis on evolutionary change (Dorus et al., 2004), will soon take center stage. Maybe the excitement about linking physiological allometries to metabolic parameters (West and Brown, 2005) will infect some mathematically inclined evolutionary neuroscientists. Or perhaps the next big thing in evolutionary neuroscience will be microevolutionary studies that integrate across the behavioral, physiological, and molecular levels (Lim et al., 2004). Maybe the future lies with computational studies that model in silico how changes in neuronal circuitry impact behavior (e.g., Treves, 2003). It is hoped that all of these new directions – and more – will bloom. If so, the field is headed for exciting times.

On the other hand, evolutionary neuroscientists are still struggling to make their findings relevant to other neuroscientists, other biologists, and other taxpayers (see "Relevance of Understanding Brain Evolution"). It may be interesting to contemplate the evolution of our brains, or even the brains of other animals, but can that knowledge be applied? Does understanding how or why a brain evolved help to decipher how that same brain works or, if it does not work, how it can be repaired? Are advances in evolutionary neuroscience likely to advance some general aspects of evolutionary theory? All of these questions remain underexplored (see Bullock, 1990).

Near the end of the nineteenth century, Jackson (1958) attempted to apply evolutionary ideas to clinical neurology, but his efforts failed. It has been pointed out that some species are far more capable than others at regenerating damaged brain regions (e.g., Kirsche and Kirsche, 1964) and that nonhuman apes tend not to suffer from neurodegenerative diseases such as Alzheimer's (Erwin, 2001). Such species differences in brain vulnerability and healing capacity might well help us elucidate some disease etiologies or lead to novel therapies. Unfortunately, this research strategy has not yet succeeded. Thus far, evolutionary neuroscience's most important contribution has been the discovery that human brains differ substantially from other brains, particularly nonprimate brains, which means that cross-species extrapolations must be conducted cautiously (Preuss, 1995). This is an important message, but it can be construed as negative in tone. Hopefully, the future holds more positive discoveries.

Work on justifying evolutionary science is especially important in the United States, where anti-evolutionary sentiment is on the rise. Many conservative Christians believe that evolution is a dangerous, insidious idea because it makes life meaningless (Dennett, 1995). Add to this fear the notion that our thoughts and feelings are mere products of our brains (e.g., Dennett, 1991) and evolutionary neuroscience seems like a serious threat to God's supremacy. Although this line of argument is well entrenched, Darwin and most of his immediate followers were hardly atheists (Young, 1985). Instead, they either distinguished clearly between God's words and God's works, as Francis Bacon put it, or argued that God's creative act was limited to setting up the laws that control history. Either way, God was seen as quite compatible with evolutionary theory. Moreover, Darwin's view of life need not produce a meaningless void. Instead, it helps to clarify our relationships with other humans, other species, and our environment. Those relationships, in turn, give meaning to our lives, just as linguistic relationships give meaning to our words. Thus, Darwin knew — and we would do well to recall — that evolutionary biology can be useful even if it yields no direct medical or technological applications. Even Huxley (1863), who was a very pragmatic Darwinian and coined the word 'agnostic', knew that the uniquely human quest to comprehend our place in nature is not driven by mere curiosity or technological imperatives, but by a profound need to understand ourselves, our purpose, our existence. Within that larger and enduring enterprise, evolutionary neuroscience will continue to play a crucial role. This essay was originally published in 2007 as a chapter in "Evolution of Nervous Systems: A Comprehensive Reference," published by Elsevier. For a more recent treatment of ideas central to evolutionary neurobiology, see "Brains Through Time: A Natural History of Vertebrates" by G.F. Striedter and R.G. Northcutt, to be published by Oxford University Press in 2020.

References

Aboitiz, F., 1995. Homology in the evolution of the cerebral hemispheres: the case of reptilian dorsal ventricular ridge and its possible correspondence with mammalian neocortex. J. Hirnforsch. 36, 461–472.

Aiello, L.C., Wheeler, P., 1995. The expensive-tissue hypothesis. Curr. Anthropol. 36, 199–221.

Alberch, P., 1982. Developmental constraints in evolutionary processes. In: Bonner, J.T. (Ed.), Evolution and Development. Springer, pp. 313–332.

Alroy, J., 1998. Cope's rule and the dynamics of body mass evolution in North American fossil mammals. Science 280, 731–734.

Appel, T.A., 1987. The Cuvier—Geoffroy Debate. Oxford University Press.

Ariëns Kappers', C.U., 1921. On structural laws in the nervous system: the principles of neurobiotaxis. Brain 44, 125–149.

Baguña, J., Garcia-Fernandez, J., 2003. Evo-devo: the long and winding road. Int. J. Dev. Biol. 47, 705–713.

Barton, R.A., 2001. The coordinated structure of mosaic brain evolution. Behav. Brain Sci. 24, 281–282.

Barton, R.A., 2006. Neuroscientists need to be evolutionarily challenged. Behav. Brain Sci. 29, 13–14.

Barton, R.A., Harvey, P.H., 2000. Mosaic evolution of brain structure in mammals. Nature 405, 1055–1058.

Barton, R.A., Aggleton, J.P., Grenyer, R., 2003. Evolutionary coherence of the mammalian amygdala. Proc. R. Soc. Lond. B 270, 539–543.

Beatty, J., 1995. The evolutionary contingency thesis. In: Wolters, G., Lennox, J.G. (Eds.), Concepts, Theories, and Rationality in the Biological Sciences. University of Pittsburgh Press, pp. 45–81.

Beauchamp, G., Fernández-Juricic, E., 2004. Is there a relationship between forebrain size and group size in birds? Evol. Ecol. Res. 6, 833–842.

Beed, C., Beed, C., 2000. Is the case for social science laws strengthening? J. Theor. Soc. Behav. 30, 131.

Bennet, P.M., Harvey, P.H., 1985. Relative brain size and ecology in birds. J. Zool. Lond. A 207, 151–169.

Bock, W., 1999. Functional and evolutionary explanations in morphology. Neth. J. Zool. 49, 45–65.

Bolhuis, J.J., Macphail, E.M., 2001. A critique of the neuroecology of learning and memory. Trends Cogn. Sci. 5, 426–433.

Bowler, J.P., 1988. The Non-darwinian Revolution: Reinterpreting a Historical Myth. Johns Hopkins University Press.

Brown, J.H., West, G.B., Enquist, B.J., 2005. Yes, West, Brown and Enquist's model of allometric scaling is both mathematically correct and biologically relevant. Funct. Ecol. 19, 735–738.

Buchholtz, E.A., Seyfarth, E.-A., 1999. The gospel of the fossil brain: Tilly Edinger and the science of paleoneurology. Brain Res. Bull. 48, 351–361.

Buffon, G.L. L.C.d., 1753. Histoire Naturelle, générale et particuliére vol. 4.

Bullock, T.H., 1990. Goals of neuroethology: roots, rules, and relevance go beyond stories of special adaptations. Bioscience 40, 244–248.

Byrne, R.W., Whiten, A. (Eds.), 1988. Machiavellian Intelligence: Social Expertise and the Evolution of Intellect in Monkeys, Apes, and Humans. Clarendon.

Calder III, W.A., 1984. Size, Function, and Life History. Harvard University Press.

Carroll, S.B., 2001. Chance and necessity: the evolution of morphological complexity and diversity. Nature 409, 1102–1109.

Catania, K.C., Northcutt, R.G., Kaas, J.H., 1999. The development of a biological novelty: a different way to make appendages as revealed in the snout of the star-nosed mole Condylura cristata. J. Exp. Biol. 202, 2719–2726.

Changizi, M.A., Shimojo, S., 2005. Parcellation and area—area connectivity as a function of neocortex size. Brain Behav. Evol. 66, 88–98.

Clark, D.A., Mitra, P.P., Wang, S.-H., 2001. Scalable architecture in mammalian brains. Nature 411, 189–193.

Collins, A.G., 1998. Evaluating multiple alternative hypotheses for the origin of Bilateria: an analysis of 18S rRNA molecular evidence. Proc. Natl. Acad. Sci. U.S.A. 95, 15458–15463.

Cuvier, G., 1805–1845. Lecons D'anatomie Comparée.

Darwin, C.R., 1859. On the origin of species by means of natural selection. Murray.

Darwin, C., 1871. The descent of man, and selection in relation to Sex. Murray.

de Winter, W., Oxnard, C.E., 2001. Evolutionary radiations and convergences in the structural organization of mammalian brains. Nature 409, 710–714.

Deacon, T.W., 1990a. Fallacies of progression in theories of brain-size evolution. Int. J. Primatol. 11, 193–236.

Deacon, T.W., 1990b. Rethinking mammalian brain evolution. Am. Zool. 30, 629–705.

Deaner, R.O., Nunn, C.L., van Schaik, C.P., 2000. Comparative tests of primate cognition: different scaling methods produce different results. Brain Behav. Evol. 55, 44–52.

Dennett, D.C., 1991. Consciousness Explained. Little, Brown.

Dennett, D.C., 1995. Darwin's Dangerous Idea: Evolution and the Meanings of Life. Simon and Schuster.

deRobertis, E.M., Sasai, Y., 1996. A common plan for dorsoventral patterning in Bilateria. Nature 380, 37–40.

Dorus, S., Vallender, E.J., Evans, P.D., et al., 2004. Accelerated evolution of nervous system genes in the origin of Homo sapiens. Cell 119, 1027–1040.

Dunbar, R.I.M., 1998. The social brain hypothesis. Evol. Anthropol. 6, 178–190.

Dunbar, R.I.M., 2006. Putting humans in their proper place. Behav. Brain Sci. 29, 15–16.

Ebbesson, S.O.E., 1980. The parcellation theory and its relation to interspecific variability in brain organization, evolutionary and ontogenetic development, and neuronal plasticity. Cell Tissue Res. 213, 179–212.

Ebbesson, S.O.E., 1984. Evolution and ontogeny of neural circuits. Behav. Brain Sci. 7, 321–366.

Emlen, D.J., Nijhout, H.F., 2000. The development and evolution of exaggerated morphologies in insects. Annu. Rev. Entomol. 45, 661–708.

Erwin, J., 2001. Ageing apes. New Sci. Mar. 17, 61.

Felsenstein, J., 1985. Phylogenies and the comparative method. Am. Nat. 125, 1–15.

Finlay, B.L., Darlington, R.B., 1995. Linked regularities in the development and evolution of mammalian brains. Science 268, 1578–1584.

Finlay, B.L., Wikler, K.C., Sengelaug, D.R., 1987. Regressive events in brain development and scenarios for vertebrate brain evolution. Brain Behav. Evol 30, 102–117.

Finlay, B.L., Darlington, R.B., Nicastro, N., 2001. Developmental structure in brain evolution. Behav. Brain Sci. 24, 263–308.

Fox, J.H., Wilczynski, W., 1986. Allometry of major CNS divisions: towards a reevaluation of somatic brain–body scaling. Brain Behav. Evol. 28, 157–169.

Frankino, W.A., Zwaan, B.J., Stern, D.L., Brakefield, P.M., 2005. Natural selection and developmental constraints. Science 307, 718–720.

Gans, C., 1969. Discussion: some questions and problems in morphological comparison. Ann. NY Acad. Sci. 167, 506–513.

Garland, T.J., Harvey, P.H., Ives, A.R., 1992. Procedures for the analysis of comparative data using phylogenetically independent contrasts. Syst. Biol. 41, 18–32.

Garstang, W., 1922. The theory of recapitulation: a critical restatement of the biogenetic law. Zool. J. Linn. Soc. Lond. 35, 81–101.

Gehring, W.J., 1996. The master control gene for morphogenesis and evolution of the eye. Genes Cells 1, 11–15.

Geoffroy, S.-H.E., 1807. Considérations sur les pièces de la tête osseuse de animaux vertebrés, et particulièrmenr sur celles du crêne des oiseaux. Ann. Mus. Hist. Nat. 10, 342–343.

Gilbert, S.F., Opitz, J.M., Raff, A.R., 1996. Resynthesizing evolutionary and developmental biology. Dev. Biol. 173, 357–372.

Goodwin, B.C., 1984. Changing from an evolutionary to a generative paradigm in biology. In: Pollard, J.W. (Ed.), Evolutionary Theory: Paths into the Future. Wiley, pp. 99–120.

Gould, S.J., 1975. Allometry in primates, with emphasis on scaling and the evolution of the brain. Contrib. Primatol. 5, 244–292.

Gould, S.J., 1977. Ontogeny and Phylogeny. Harvard University Press.

Gould, S.J., 1986. Evolution and the triumph of homology, or why history matters. Am. Sci. Jan./Feb. 60–69.

Gould, S.J., 1989. Wonderful Life: The Burgess Shale and the Nature of History. Norton.

Gould, S.J., Lewontin, R.C., 1979. The spandrels of San Marco and the Panglossian paradigm: a critique of the adaptationist program. Proc. R. Soc. Lond. 205, 581–598.

Haeckel, E., 1889. Natü Rliche Schö Pfungsgeschichte. Georg Reimer.

Halder, G., Callaerts, P., Gehring, W.J., 1995. Induction of ectopic eyes by targeted expression of the eyeless gene in Drosophila. Science 267, 1788–1792.

Hall, B.K., 1999. Evolutionary Developmental Biology. Kluwer.

Haller, A.v., 1762. Elementa Physiologiae Corporis Humani. Sumptibus M.-M. Bousquet et Sociorum.

Harvey, P.H., Krebs, J.H., 1990. Comparing brains. Science 249, 140–146.

Hempel, C.G., 1942. The function of general laws in history. J. Philos. 39, 35–48.

Hennig, W., 1950. Grundzü ge einer Theorie der Phylogenetischen Systematik. Deutscher Zentralverlag.

Herrick, C.J., 1956. The Evolution of Human Nature. University of Texas Press.

Hodos, W., Campbell, C.B.G., 1969. Scala naturae: why there is no theory in comparative psychology. Psychol. Rev. 76, 337–350.

Hoppeler, H., Weibel, E.R., 2005. Scaling functions to body size: theories and facts. J. Exp. Biol. 208, 1573–1574.

Huxley, T.H., 1863. Man's Place in Nature. University of Michigan Press.

Iwaniuk, A.N., Pellis, S.M., Whishaw, I.Q., 1999. Brain size is not correlated with forelimb dexterity in fissiped carnivores (Carnivora): a comparative test of the principle of proper mass. Brain Behav. Evol. 54, 167–180.

Iwaniuk, A.N., Dean, C., Nelson, J.E., 2004. A mosaic pattern characterizes the evolution of the avian brain. Proc. R. Soc. Lond. B 271, S148–S151.

Jackson, J.H., 1958. Evolution and dissolution of the nervous system. In: Taylor, J. (Ed.), Selected Writings of John Hughlings Jackson. Staples Press, pp. 45–75.

Janies, D., DeSalle, R., 1999. Development, evolution, and corroboration. Anat. Rec. 257, 6–14.

Jarvis, E.D., Gü ntü rkü n, O., Bruce, L., et al., 2005. Avian brains and a new understanding of vertebrate brain evolution. Nat. Rev. Neurosci. 6, 151–159.

Jenner, R.A., 2004. When molecules and morphology clash: reconciling conflicting phylogenies of the Metazoa by considering secondary character loss. Evol. Dev. 6, 372–378.

Jerison, H.J., 1973. Evolution of the Brain and Intelligence. Academic Press.

Jerison, H.J., 1989. Brain Size and the Evolution of Mind. American Museum of Natural History.

Jones, K.E., MacLarnon, A.M., 2004. Affording larger brains: Testing hypotheses of mammalian brain evolution on bats. Am. Nat. 164, E20–E31.

Julian, G.E., Gronenberg, W., 2002. Reduction of brain volume correlates with behavioral changes in queen ants. Brain Behav. Evol. 60, 152–164.

Katz, M.J., Lasek, R.L., 1978. Evolution of the nervous system: role of ontogenetic buffer mechanisms in the evolution of matching populations. Proc. Natl. Acad. Sci. U.S.A. 75, 1349–1352.

Katz, M.J., Lasek, R.L., Kaiserman-Abramof, I.R., 1981. Ontophyletics of the nervous system: eyeless mutants illustrate how ontogenetic buffer mechanisms channel evolution. Proc. Natl. Acad. Sci. U.S.A. 78, 397–401.

Kaufman, J.A., 2003. On the expensive tissue hypothesis: independent support from highly encephalized fish. Curr. Anthropol. 5, 705–707.

Kincaid, H., 1996. Philosophical Foundation of the Social Sciences: Analyzing Controversies in Social Research. Cambridge University Press.

Kirsche, K., Kirsche, W., 1964. Experimental study on the influence of olfactory nerve regeneration on forebrain regeneration of *Ambystoma mexicanum*. J. Hirnforsch. 7, 315–333.

Kozlowski, J., Konarzewski, M., 2004. Is West, Brown and Enquist's model of allometric scaling mathematically correct and biologically relevant? Funct. Ecol. 18, 283–289.

Kudo, H., Dunbar, R.I.M., 2001. Neocortex size and social network size in primates. Anim. Behav. 62, 711–722.

Lim, M.M., Wang, Z., Olzá bal, D.E., Ren, X., Terwilliger, E.F., Young, L.J., 2004. Enhanced partner preference in a promiscuous species by manipulating the expression of a single gene. Nature 429, 754–757.

Lovejoy, A.O., 1936. The Great Chain of Being: A Study of the History of an Idea. Harvard University Press.

Macphail, E.M., 1982. Brain and Intelligence in Vertebrates. Clarendon.

Mares, S., Ash, L., Gronenberg, W., 2005. Brain allometry in bumblebee and honey bee workers. Brain Behav. Evol. 66, 50–61.

Marsh, O.C., 1886. Dinocerata: A Monograph of an Extinct Order of Gigantic Mammals. Government Printing Office.

Martin, R.D., 1981. Relative brain size and basal metabolic rate in terrestrial vertebrates. Nature 293, 57–60.

Martin, R.D., Genoud, M., Hemelrijk, C.K., 2005. Problems of allometric scaling analysis: Examples from mammalian reproductive biology. J. Exp. Biol. 208, 1731–1747.

Maynard Smith, J., 1982. Evolution and the Theory of Games. Cambridge University Press.

Maynard Smith, J., Burian, R., Kauffman, S., et al., 1985. Developmental constraints and evolution. Quart. Rev. Biol. 60, 265–287.

McAlpine, D., Grothe, B., 2003. Sound localization and delay lines – do mammals fit the model? Trends Neurosci. 26, 347–350.

McIntyre, L., 1996. Laws and Explanations in the Social Sciences: Defending a Science of Human Behavior. Westview Press.

McNab, B.K., 1989. Brain size and its relation to the rate of metabolism in mammals. Am. Nat. 133, 157–167.

Mekel-Bobrov, N., Gilbert, S.L., Evans, P.D., et al., 2005. Ongoing adaptive evolution of ASPM, a brain size determinant in *Homo sapiens*. Science 309, 1720–1722.

Mink, J.W., Blumenshine, R.J., Adams, D.B., 1981. Ratio of central nervous system to body metabolism in vertebrates: its constancy and functional basis. Am. J. Physiol. 241, R203–R212.

Morgane, P.J., Jacobs, L.F., 1972. Comparative anatomy of the cetacean nervous system. In: Harrison, R.J. (Ed.), Functional Anatomy of Marine Mammals. Academic Press, pp. 117–244.

Nealen, P.M., Ricklefs, R.E., 2001. Early diversification of the avian brain: body relationship. J. Zool. Lond. 253, 391–404.

Nilsson, D.-E., 2004. Eye evolution: a question of genetic promiscuity. Curr. Opin. Neurobiol. 14, 407–414.

Northcutt, R.G., 2001. Changing views of brain evolution. Brain Res. Bull. 55, 663–674.

Pagel, M.D., Harvey, P.H., 1989. Taxonomic differences in the scaling of brain on body weight among mammals. Science 244, 1589–1593.

Parker, S.T., Gibson, K.R., 1977. Object manipulation, tool use, and sensorimotor intelligence as feeding adaptations in cebus monkeys and great apes. J. Hum. Evol. 6, 623–641.

Pasquinelli, A.E., Ruvkun, G., 2002. Control of developmental timing by microRNAs and their targets. Annu. Rev. Cell Dev. Biol. 18, 495–513.

Pitts, M. (Ed.), 2002. Institutional Animal Care and Use Committee Guidebook. Office of Laboratory Animal Welfare, National Institutes of Health.

Preuss, T.M., 1995. The argument from animals to humans in cognitive neuroscience. In: Gazzaniga, M.S. (Ed.), The Cognitive Neurosciences. MIT Press, pp. 1227–1241.

Puelles, L., Rubenstein, J.L.R., 2003. Forebrain gene expression domains and the evolving prosomeric model. Trends Neurosci. 26, 469–476.

Puelles, L., Kuwana, E., Puelles, E., et al., 2000. Pallial and subpallial derivatives in the embryonic chick and mouse telencephalon, traced by the expression of the genes *Dlx-2, Emx-1, Nkx-2.1, Pax-6* and *Tbr-1*. J. Comp. Neurol. 424, 409–438.

Raff, R.A., 1996. The Shape of Life: Genes, Development, and the Evolution of Animal Form. University of Chicago Press.

Reader, S.M., Laland, K.N., 2002. Social intelligence, innovation, and enhanced brain size in primates. Proc. Natl. Acad. Sci. U.S.A. 99, 4436–4441.

Reiner, A., Perkel, D.J., Bruce, L.L., et al., 2004a. The avian brain nomenclature forum: terminology for a new century in comparative neuroanatomy. J. Comp. Neurol. 473, E1–E6.

Reiner, A., Perkel, D.J., Bruce, L.L., et al., 2004b. Revised nomenclature for avian telencephalon and some related brainstem nuclei. J. Comp. Neurol. 473, 377–414.

Richardson, M.K., 1999. Vertebrate evolution: the developmental origins of adult variation. Bioessays 21, 604–613.

Richardson, M.K., Hanken, J., Gooneratne, M.L., et al., 1997. There is no highly conserved embryonic stage in vertebrates: implications for current theories of evolution and development. Anat. Embryol. 196, 91–106.

Riedl, R., 1977. A systems-analytical approach to macroevolutionary phenomena. Quart. Rev. Biol. 52, 351–370.

Ringo, J.L., 1991. Neuronal interconnection as a function of brain size. Brain Behav. Evol. 38, 1–6.

Ringo, J.L., Doty, R.W., Demeter, S., Simard, P.Y., 1994. Time is of the essence: a conjecture that hemispheric specialization arises from interhemispheric conduction delay. Cereb. Cortex 4, 331–343.

Rosa, M.P.G., Tweedale, R., 2005. Brain maps, great and small: Lessons from comparative studies of primate visual cortical organization. Philos. Trans. R. Soc. Lond. B 360, 665–691.

Roth, G., Nishikawa, K.C., Naujoks-Manteurrel, C., Schmidt, A., Wake, D.B., 1993. Paedomorphosis and simplification in the nervous system of salamanders. Brain Behav. Evol. 42, 137–170.

Russel, E.S., 1916. Form and Function: A Contribution to the History of Animal Morphology. University of Chicago Press.

Sacher, G.A., 1970. Allometric and factorial analysis of brain structure in insectivores and primates. In: Noback, C.R., Montagna, W. (Eds.), The Primate Brain. Appleton Century Crofts, pp. 109–135.

Safi, K., Dechmann, K.N., 2005. Adaptation of brain regions to habitat complexity: a comparative analysis in bats (Chicoptera). Proc. R. Soc. Lond. B Biol. Sci. 272, 179–186.

Savage, V.M., Gillooly, J.F., Woodruff, W.H., et al., 2004. The predominance of quarter-power scaling in biology. Funct. Ecol. 18, 257–282.

Schmidt-Nielsen, K., 1984. Scaling: Why Animal Size Is So Important. Cambridge University Press.

Schmidt-Rhaesa, A., 2003. Old trees, new trees – is there any progress? Zoology 106, 291–301.

Sidor, C.A., 2001. Simplification as a trend in synapsid cranial evolution. Evolution 55, 1419–1442.

Slack, J.M.W., Holland, P.W.H., Graham, C.F., 1993. The zootype and the phylotypic stage. Nature 361, 490–492.

Smith, R.J., 1994. Degrees of freedom in interspecific allometry: an adjustment for the effects of phylogenetic constraint. Am. J. Phys. Anthropol. 93, 95–107.

Snell, O., 1891. Die Abhä ngigkeit des Hirngewichtes von dem Körpergewicht und den geistigen Fä higkeiten. Arch. Psychiat. Nervenkr. 23, 436–446.

Sober, E., 2000. Philosophy of Biology. Westview Press.

Sprecher, S.G., Reichert, H., 2003. The urbilaterian brain: developmental insights into the evolutionary origin of the brain in insects and vertebrates. Arthropod Struct. Dev. 32, 141–156.

Stanley, S.M., 1973. An explanation for Cope's rule. Evolution 27, 1–26.

Stevens, C.F., 1989. How cortical interconnectedness varies with network size. Neural Comput. 1, 473–479.

Strausfeld, N.J., Hildebrand, J.G., 1999. Olfactory systems: common design, uncommon origins? Curr. Opin. Neurobiol. 9, 634–639.

Striedter, G.F., 1998. Progress in the study of brain evolution: from speculative theories to testable hypotheses. Anat. Rec. 253, 105–112.

Striedter, G.F., 1999. Homology in the nervous system: of characters, embryology and levels of analysis. In: Bock, G.R., Cardew, G. (Eds.), Novartis Foundation Symposium, vol. 222. Wiley, pp. 158–172.

Striedter, G.F., 2004. Brain evolution. In: Paxinos, G., May, J.K. (Eds.), The Human Nervous System, second ed. Elsevier, pp. 3–21.

Striedter, G.F., 2005. Principles of Brain Evolution. Sinauer Associates.

Striedter, G.F., Northcutt, R.G., 1991. Biological hierarchies and the concept of homology. Brain Behav. Evol. 38, 177–189.

Treves, A., 2003. Computational constraints that may have favoured the lamination of sensory cortex. J. Comp. Neurosci. 14, 271–282.

van Dongen, B.A.M., 1998. Brain size in vertebrates. In: Nieuwenhuys, R., Ten Donkelaar, H.J., Nicholson, C. (Eds.), The Central Nervous System of Vertebrates. Springer, pp. 2099–2134.

Wagner, G.P., Laubichler, M.D., 2004. Rupert Riedl and the re-synthesis of evolutionary and developmental biology: body plans and evolvability. J. Exp. Zool. Part B 302, 92–102.

West, G.B., Brown, J.H., 2005. The origin of allometric scaling laws in biology from genomes to ecosystems: towards a quantitative unifying theory of biological structure and organization. J. Exp. Biol. 208, 1575–1592.

West, G.B., Brown, J.H., Enquist, B.J., 1997. A general model for the origin of allometric scaling laws in biology. Science 276, 122–126.

Wilczynski, W., 1984. Central nervous systems subserving a homoplasous periphery. Am. Zool. 24, 755–763.

Willmer, P., 2003. Convergence and homoplasy in the evolution of organismal form. In: Mü ller, G.B., Newman, S.A. (Eds.), Origination of Organismal Form. MIT Press, pp. 33–49.

Young, R.M., 1985. Darwin's Metaphor: Nature's Place in Victorian Culture. Cambridge University Press.

Zhang, K., Sejnowski, T.J., 2000. A universal scaling law between gray matter and white matter of cerebral cortex. Proc. Natl. Acad. Sci. U.S.A. 97, 5621–5626.

Zuber, M.E., Gestri, G., Viczan, A.S., Barsacchi, G., Harris, W.A., 2003. Specification of the vertebrate eye by a network of eye field transcription factors. Development 130, 5155–5167.

Further Reading

Butler, A.B., Hodos, W., 2005. Comparative Vertebrate Neuroanatomy, second ed. Wiley-Liss.

Deacon, T.W., 1990. Rethinking mammalian brain evolution. Am. Zool. 30, 629–705.

Finlay, B.L., Darlington, R.B., Nicastro, N., 2001. Developmental structure in brain evolution. Behav. Brain Sci. 24, 263–308.

Jerison, H.J., 1973. Evolution of the Brain and Intelligence. Academic Press.

Nieuwenhuys, R., 1994. Comparative neuroanatomy: place, principles, practice and programme. Eur. J. Morphol. 32, 142–155.

Northcutt, R.G., 2001. Changing views of brain evolution. Brain Res. Bull. 55, 663–674.

Striedter, G.F., 2005. Principles of Brain Evolution. Sinauer Associates.

Striedter, G., 2006. Precis and multiple book review of 'principles of brain evolution'. Behav. Brain Sci. 29, 1–36.

2

Phylogenetic Character Reconstruction

J.S. Albert

University of Louisiana, Lafayette, LA, United States

Glossary

adaptation A feature or phenotype or trait that evolved to serve a particular function or purpose.

anagenesis The origin of evolutionary novelties within a species lineage by changes in gene allele frequencies by the processes of natural selection and/or neutral genetic drift.

character polarity The temporal direction of change between alternative (primitive and derived) states of a character.

character state reconstruction The process of estimating the ancestral or primitive condition of a character at a given node (branching point) in a phylogenetic tree.

clade A complete branch of the tree of life. A monophyletic group.

cladogenesis The origin of daughter species by the splitting of ancestral species; may or may not occur under the influence of natural selection.

cladogram A branching tree-shaped diagram used to summarize comparative (interspecific) data on phenotypes or gene sequences. In contrast to a phylogeny, a cladogram has no time dimension.

comparative method The study of differences between species.

continuous trait A quantitatively defined feature with no easily distinguished boundaries between phenotypes (e.g., size, cell counts, and gene expression levels).

convergence Similarity of structure or function due to independent evolution from different ancestral conditions.

discrete trait A qualitatively defined feature with only a few distinct phenotypes (e.g., polymorphism; presence vs. absence).

homology Similarity of structure or function due to phylogeny (common ancestry).

homoplasy Similarity of structure or function due to convergence, parallelism orreversal.

monophyletic A systematic category that includes an ancestor and all of its descendents; a complete branch of the tree of life; a 'natural' taxon; a clade.

node An internal branching point in a phylogenetic tree.

optimization Methods for estimating ancestral trait values on a tree. Commonly used optimization criteria are: maximum parsimony (MP) which minimizes the amount of trait change, and maximum likelihood (ML) which maximizes the likelihood of a trait at a node given likelihood values for trait evolution.

parallelism Similarity of structure or function due to independent evolution from a common ancestral condition.

paraphyletic A systematic category that includes an ancestor and some but not all of its descendents (e.g., 'invertebrates', 'agnathans', 'fish', and 'reptiles' (sans birds)).

parsimony A principle of scientific inquiry that one should not increase, beyond what is necessary, the number of entities required to explain anything.

phenotypic evolution Change in the developmental program descendents inherit from their ancestors.

phylogenetic character A homologous feature or phenotype or trait of an organism or group of organisms.

phylogenetic systematic A method for reconstructing evolutionary trees in which taxa are grouped exclusively on the presence of shared derived features.

phylogenetic tree Genealogical map of interrelation-ships among species, with a measure of relative or absolute time on one axis. Also called a tree of life or a phylogeny.

phylogeny The evolutionary history of a species or group of species that results from anagenesis and cladogenesis.

polyphyletic A systematic category that includes taxa from multiple phylogenetic origins (e.g., 'homeothermia' consisting of birds and mammals).

reversal Change from a derived character state back to a more primitive state; an atavism. Includes evolutionary losses (e.g., snakes which have 'lost' their paired limbs).

synapomorphy A shared, derived character used as a hypothesis of homology.

taxon A species or monophyletic group of species (plural taxa).

trait evolution The sequence of changes of a feature or phenotype on a phylogeny.

2.1 Introduction to Character State Reconstruction and Evolution

Comparisons among the features of living organisms have played a prominent role in the biological sciences at least since the time of Aristotle. The comparative approach takes advantage of the enormous diversity of organismal form and function to study basic biological processes of physiology, embryology, neurology, and behavior. This approach has given rise to the widespread use of certain species as model systems, based on what has become known as the August Krogh Principle: "For many problems there is an animal on which it can be most conveniently studied" (Krebs, 1975).

From an evolutionary perspective, interspecific (between species) comparisons allow for the systematic study of organismal design. Rensch (1959) conceived of phylogeny as being composed of two distinct sets of processes: anagenesis, the origin of phenotypic novelties within an evolving species lineage (from the Greek *ana* ¼ up þ *genesis* ¼ origin), and cladogenesis, the origin of new species from lineage splitting (speciation) (from the Greek *clado* ¼ branch). Anagenetic changes arise within a population by the forces of natural selection and genetic drift. Cladogenesis may or may not arise from these population-level processes, and in fact many (or perhaps most?) species on Earth are thought to have their origins from geographical (allopatric) speciation under the influence of landscape and geological processes (Mayr, 1963; Coyne and Orr, 1989).

Because species descend from common ancestors in a hierarchical fashion (i.e., from a branching, tree-like process of speciation) closely related species tend to resemble each other more than they do more distantly related species. Patterns in the diversification of phenotypes have therefore been described as mosaic evolution, in which different species inherit distinct combinations of traits depending on the position of that species in the tree of life (McKinney and McNamara, 1990). Under this view, character evolution is regarded as a process of historical transformation from a primitive to a derived state, and study of this process necessarily presumes knowledge of primitive or ancestral conditions. In other words, because character evolution is perceived as trait change on a tree, it is necessary to estimate 'ancestral trait values'.

Direct observations of ancient phenotypes may be taken from fossils, which provide unique information on entirely extinct groups of organisms, and are usually associated with stratigraphic information pertaining to relative and absolute geological ages (Benton, 1993). Nonetheless, the fossil record has many well-known shortcomings, including the famously incomplete levels of preservation, and usually very limited information about the nature of soft tissues such as nerves and brains

(but see Edinger, 1941; Stensiö, 1963). Paleontological information on ancient physiological and behavioral traits is even more scanty (but see Jerison, 1976; MacLeod and Rose, 1993; Rogers, 2005).

Recent years have seen great advances in the formulation of comparative methods to estimate or infer ancestral phenotypes from extant (living) species (Garland *et al.*, 1992, 1999; Martins, 2000). These methods use patterns in the mosaic of traits present among species in the context of an explicit hypothesis of interrelationships. These methods also address new topics, such as whether rates of phenotypic evolution have differed among lineages (clades), the circumstances in which a phenotype first evolved, the selective and developmental mechanisms underlying the origin of new phenotypes, and the evolutionary lability of phenotypes (Albert *et al.*, 1998; Blomberg *et al.*, 2003; Blackledge and Gillespie, 2004).

In this article, I summarize the major recent developments in phylogenetically based methods of studying character evolution, with the goals of explaining both the strengths and weaknesses of alternative methods. Most of the empirical examples cited are among animals with the most complex central nervous systems (e.g., vertebrates) in which neurological and behavioral evolution has been (arguably) most extensively studied. A major goal of this article is to highlight some of the most exciting new developments in the study of character evolution now being explored in this fascinating area of comparative neurobiology.

2.2 Basic Concepts

2.2.1 Homology: Similarity Due to Common Ancestry

All methods of ancestral character state reconstruction make explicit assumptions about the homology of the traits under study. In comparative biology the term 'homology' refers to similarity in form or function arising from common ancestry. In other words, homologous features among organisms can be traced to a single evolutionary origin. In the language of Garstang (1922), a homologous trait is a unique historical change in the developmental program of an evolving lineage. Homologous similarities may be observed in any aspect of the heritable phenotype, from properties of genetic sequences (e.g., base composition and gene order), through aspects of development, including cellular, tissue, and organismal phenotypes, to aspects of behavior that emerge from the organization of the nervous system. Homology in behavioral traits has been examined in a number of taxa, and in a variety of contexts (de

Queiroz and Wimberger, 1993; Wimberger and de Queiroz, 1996; Blomberg *et al.*, 2003). Taxa are individual branches of the tree of life, and may include species or groups of species that share a common ancestor (the latter are also referred to as clades or monophyletic groups).

It is important to note that developmental, structural, positional, compositional, and functional features of phenotypes are all useful in proposing hypotheses of homology. Yet by the evolutionary definition employed above, only features that can be traced to a common ancestor in an explicitly phylogenetic context are regarded as homologues. Because phylogenies are the product of comparative analyses using many traits, it is in fact congruence in the phylogenetic distribution of characters that serves as the ultimate criterion for homology. By this criterion homologous characters are said to have passed the test of congruence. In other words, congruence in the phylogenetic distribution of numerous character states is regarded to be the ultimate evidence for homology (Patterson, 1982; see Primate Brain Evolution in Phylogenetic Context, Electric Fish, Electric Organ Discharges, and Electroreception).

2.2.2 Homoplasy: Convergence, Parallelism, and Reversal

All other forms of phenotypic similarity that arise during the course of evolution are referred to collectively as homoplasy (similarity due to causes other than homology). Homoplastic characters may arise from several sources: convergence due to similar functional pressures and natural selection, parallel (independent) evolution to a common structure or function from organisms with similar genetic and developmental backgrounds, or convergent reversal to a common ancestral (plesiomorphic) condition. Some well-known examples of convergent evolution in the nervous system include: image-forming eyes of cephalopod mollusks (e.g., squids and octopods) and vertebrates (Packard, 1972), and the evolution of G-protein-coupled receptors as odorant receptors in many animal phyla (Eisthen, 2002). Examples of parallel evolution in the nervous system of vertebrates have been summarized in several recent reviews (Nishikawa, 2002; Zakon, 2002). These include: electric communication in mormyriform (African) and gymnotiform (South American) electric fishes (Albert and Crampton, 2005), prey capture among frogs (Nishikawa, 1999), sound localization among owls (Grothe *et al.*, 2005), and thermoreception in snakes (Hartline, 1988; Molenaar, 1992).

Reversals are among the most common forms of homoplasy, and are often the most difficult to detect even in the context of a resolved phylogenetic hypothesis of relationships (Cunningham, 1999). The reason for this is the phenotypes of some reversals may be quite literally identical, as in the case of convergent loss of structures (e.g., the derived loss of paired limbs in snakes and limbless lizards).

2.2.3 Character State Polarity

A central task of ancestral character state reconstruction is determining the direction or polarity of evolutionary change between alternative states of a character. The ancestral state is referred to as plesiomorphic or primitive, and the descendent state is referred to as apomorphic or derived. Establishing the polarity of a character state transformation is critical to understanding the functional significance of that event. Phenotypes determined to be primitive simply mean they precede the derived state in time and are not necessarily functionally inferior. It is often, although by no means always, the case that characters evolve from more simple to more complex states, or from the absence of a particular state to the presence of that state.

There are several methods in use to determine character state polarity. The most widely used method is the so-called outgroup criterion, which employs conditions observed in members of clades other than the clade in which the derived state is present. The basic idea of the outgroup criterion is that for a given character with two or more states within a group, the state occurring in related groups is assumed to represent the plesiomorphic state. In other words, the outgroup criterion states that if one character is found in both ingroup and outgroup, this character is then postulated to be the ancestral state (plesiomorphic). Of course, it is always possible that a given outgroup exhibits an independently derived state of a given character, which is why the condition in several outgroup taxa is regarded as a more reliable test of the plesiomorphic condition.

2.2.4 Character or Trait Data

Methods for estimating ancestral character states and analyzing phenotypic evolution may treat trait data either as continuous (quantitative) or discrete (qualitative) (Zelditch *et al.*, 1995; Rohlf, 1998; Wiens, 2001). Continuously distributed trait values have no easily distinguished boundaries between phenotypes. Examples of continuous traits include the sizes of brains and brain regions (e.g., nuclei), the number of cells in a brain region, pigment intensity, amplitude or timing of communication signals, and the amount of gene expression in a tissue. Continuous phenotypic variation typically reflects the additive effects of alleles at multiple loci and is frequently also influenced by environmental factors. Patterns of intraspecific (within species) continuous variation are often analyzed using parametric statistics, including such devices as the population mean and standard deviation. Methods for the analysis of interspecific (between species) continuous traits are useful for assessing the quantitative relationships among variables to address questions regarding, for example, the trade-offs and constraints among correlated traits.

Discontinuous traits have only a few distinct phenotypes. In many cases alternative alleles generate phenotypes that differ from each other in discrete steps, such that each phenotype can be clearly distinguished from the others. Many classes of phenotypic data are inherently discrete, such as meristic counts (e.g., number of body segments, rhombomeres, and cortical visual maps), and genetic polymorphisms (e.g., left- vs. right-handedness). Nucleotide bases at a locus are discrete states of a character. The presence (or absence) of derived traits on a phylogenetic tree also constitutes a class of discrete phenotypes. Such derived traits that underlie or explain subsequent evolutionary events are referred to as key innovations. Some widely cited examples of putative key innovations in the comparative neurosciences include arthropod cephalic tagmosis (Strausfeld, 1998), cephalopod eyes (Hanlon and Messenger, 1996), craniate neural crest (Northcutt and Gans, 1983), and ray-finned fish genome duplication (Taylor *et al.*, 2003; Postlethwait *et al.*, 2004). Each of these novelties is thought to have been critical in the diversification of the taxon in which it originated.

2.2.5 Adaptation

One of the most widely applied uses of ancestral character state reconstruction is in the study of adaptation. The word adaptation is derived from the Latin *ad* (to, toward) and *aptus* (a fit), and is used to imply a feature or phenotype that evolved to serve a particular function or purpose. For example, the function or purpose of an animal central nervous system is to coordinate sensory information and motor output patterns; that is to say, a centralized brain is an adaptation for sensory-motor coordination. Adaptation is therefore used both as a noun to describe the features that arose because of natural selection, and as a verb, the process of natural selection through which the features originated. In an evolutionary context, an adaptation is not only a static description of the match between form and function, but is also an explanation for the origin of that relationship (Russell, 1916).

It is important to distinguish among several distinct uses of the word 'adaptation' in the biological sciences. A physiological adaptation is an organismal response to a particular stress: if you heat up from the sun you may respond by moving into the shade (a behavioral adaptation), or you may respond by sweating (a physiological adaptation). In an evolutionary context, adaptation is also a change in response to a certain problem, but the change is genetic. Evolutionary adaptations that result from the process of natural selection usually take place over periods of time considerably longer than physiological timescales. Traits are referred to as adaptations only when they evolved as the solutions for a specific problem; that is, for a particular function or purpose. A physiological response can itself be an adaptation in the evolutionary sense.

In reconstructing ancestral phenotypes it is important to bear in mind the primitive condition may be more or less variable than the conditions observed in living species. In some cases physiological or developmental plasticity is itself an evolutionary (genetic) specialization that permits organisms to adapt physiologically or behaviorally. For example, many species are characterized as eurytopic, or tolerant of a wide variety of habitats. Other species are stenotopic, or adapted to a narrow range of habitats. Similarly, individual characters may be more or less variable within a species, and this variability may itself be subject to evolutionary change. Flexible phenotypes may be more adaptive in a variable environment and stereotyped phenotypes more adaptive in a stable environment (van Buskirk, 2002).

2.2.6 Phylogenetic Trees

Implicit in all phylogenetic methods for studying character evolution is a tree-shaped branching diagram, alternatively called a dendrogram, cladogram, phenogram, or tree, depending on the methods used to construct the diagram, and the information content it is intended to convey. It is important to note that each of the many alterative methods for building trees that are currently available was designed to communicate different kinds of information. The methods grouped formally as 'phylogenetic systematics' (cladistics) exclusively use derived similarities (synapomorphies) to hypothesize genealogical relationships. This is to be contrasted with phenetic methods which use measures of overall similarity to group taxa, including both primitive and derived aspects of similarity. Cladistic methods generate branched diagrams referred to as cladograms, which should be viewed as summary diagrams depicting the branching pattern most consistent with a given data set (morphological or molecular). It is important to distinguish raw cladograms from phylogenetic trees; there is no time dimension to a cladogram *per se,* and the branch lengths are simply proportional to the minimum number of steps required to map all the character states

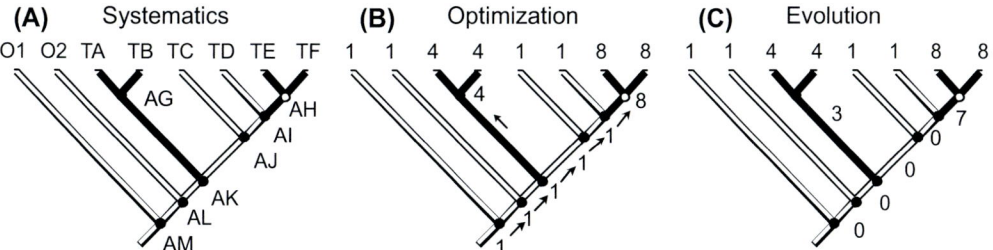

FIGURE 2.1   Summary of the comparative approach for inferring phenotypic evolution. a, Phylogenetic systematics (i.e., tree building): reconstruction of genealogical interrelationships among taxa (extant and/or fossil) using morphological and/or molecular sequence data. Taxa are species or clades (monophyletic groups of species): phylogeny includes six ingroup terminal taxa (TA−TF) and two outgroup taxa (O1 and O2). b, Character state optimization at internal nodes (branching points or hypothesized speciation events). Observed trait values at tips of the tree. Seven internal tree nodes represented by ancestral taxa (AG−AM) with trait values estimated by linear parsimony. c, Evolution: tracing the history of phenotypic changes along branches of the tree. Numbers indicate absolute amount of trait change on the branch.

onto that tree. A robust phylogenetic tree is usually the result of several or many phylogenetic analyses. The geological time frames associated with branching events are usually estimated from external paleontological, molecular, and biogeographic sources of information.

Figure 2.1 provides a conceptual overview for how phylogenetic trees may be used to study phenotypic evolution. All comparative approaches begin by assuming (or building) a hypothesis of genealogical interrelationships among the taxa of interest. There are many methods, even whole philosophies, of tree building, and the reader is referred to Page and Holmes (1998) for an introduction to this literature. Phylogenetic methods are then used to optimize character states at internal nodes of the tree; these nodes or branching points are hypothesized speciation events. Comparisons of trait

values at ancestral and descendant nodes of the tree allow the history of phenotypic changes to be traced. The distribution of these phenotypic changes (also known as steps or transformations) can then be assessed, qualitatively or quantitatively, depending on the types of data examined and the analytical methods employed.

A tree-shaped branching diagram conveys two kinds of information (whether they are intended or not): the tree topology, or the sequential order in which the taxa branch from one another, and the lengths of the individual branches (Figure 2.2). These two aspects of a tree correspond to the cladogenesis and the anagenesis of Rensch (1959). The tree topology (branching order) is reconstructed from the distribution of shared−derived traits among taxa. The traits examined may be

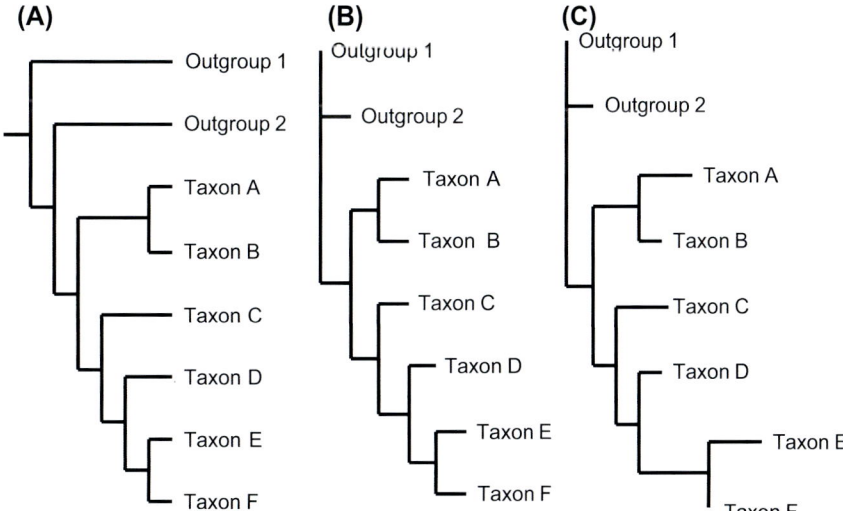

FIGURE 2.2   Alternative branch length models. a, Molecular clock: all terminal taxa equidistant from root to from an ultrametric tree. b, Equal branch lengths: all character evolution (anagenesis) occurs at branching events, as in punctuated equilibrium. c, Empirical: branch lengths proportional to amount of character evolution and/or geological ages determined from fossils. Note: tree topology is transitive; branch lengths are not.

morphological novelties or nucleotide substitutions. Branch lengths may be reconstructed from one or more sources of information, including alternative models (or modes) of character evolution, or from empirical data. Under models of constant (or near constant) evolution (e.g., molecular clocks), all terminal taxa are treated as equidistant from the root (or base) of the tree. Terminal taxa are those at the tips of the tree, as opposed to ancestral taxa at internal nodes (branching points) within the tree. Under models of punctuated equilibrium, all (or most) character evolution occurs at branching points (nodes), and all branches are therefore of equal (or almost equal) length. Branch lengths derived from empirical data sets may be treated as proportional to the amount of character state change on that particular tree topology, or from stochastic models of evolution assuming that DNA nucleotide substitutions occur at an equal rate (Sanderson, 2002). The constant evolution and punctuated equilibrium models represent extremes of branch-length heterogeneity, between which branch lengths derived from empirical data sets usually fall. Branch lengths for clades with known fossilized members can also be estimated from the geological age of these fossils (Benton *et al.*, 2000; Near and Sanderson, 2004). Calibrations based on molecular sequence divergence or fossil data can take one of two forms: assignment of a fixed age to a node, or enforcement of a minimum or maximum age constraint on a node. The latter option is generally a better reflection of the information content of fossil evidence.

It is important to recognize an analytical difference in the two kinds of information represented in a phylogeny: whereas the tree topology is transitive, the branch lengths are not. In the language of formal logic, 'transitive' means that a relationship necessarily holds across (i.e., it transcends) the particularity of data sets. In the case of phylogenetic trees, the branching order derived from analysis of one data set is expected to predict the branching order of independent data sets (e.g., those derived from different genes, genes and morphology, osteology and neurology). Branch lengths, however, are intransitive, meaning the branch length values derived from one data set are not expected to predict those of other data sets. The reason for this is that we believe there has been a single phylogenetic history of life; a unique sequence of speciation events that gave rise to the species richness of the modern world. This single history underlies the evolution of all aspects of organismal phenotypes. There are, however, no such expectations of homogeneity in the rates of phenotypic (or gene sequence) evolution; in fact, the differential effects of directional and stabilizing selection on different phenotypes may be expected to result in longer or shorter branches for some traits than others.

2.3 Methods

2.3.1 Parsimony Optimization of Discrete Traits

The principle of parsimony (i.e., Occam's razor) is widely used in the natural sciences as a method for selecting from among numerous alternative hypotheses. The principle of parsimony underlies all scientific modeling and theory building. The basic idea is that one should not increase, beyond what is necessary, the number of entities required to explain anything. In this context, parsimony means that simpler hypotheses are preferable to more complicated ones. It is not generally meant to imply that Nature itself is simple, but rather that we as observers should prefer the most simple explanations.

Maximum parsimony (MP) is a character-based method used in phylogenetic systematics to reconstruct phylogenetic trees by minimizing the total number of evolutionary transformations (steps) required to explain a given set of data. In other words, MP minimizes the total tree length. The steps may be nucleotide base or amino acid substitutions for sequence data, or gain and loss events for restriction site and morphological data. MP may also be used to infer ancestral states of a character within a phylogenetic tree (this is discussed in the following).

2.3.2 Binary and Multistate Characters

Discrete characters may be characterized as either binary (coded into two mutually exclusive alternative states) or as multistate (a transformation series of three or more discrete states). The alternative states of a binary character are generally (although not necessarily) explicit hypotheses of the primitive and derived (advanced) states of a single evolutionary transformation event, such as the origin (or loss) of a novel feature. A multistate character is a more complex intellectual device with many more interpretations of meaning. Multistate characters may be presented as many stages of a long-term phylogenetic trend (e.g., larger relative brain size, larger body size) or as independent alternative trends from a common ancestral plan (e.g., large brains evolving from enlargement of the cerebellum in chondrichthyans vs. the telencephalon in mammals). An ordered transformation series models a preconceived phylogenetic sequence of changes, such that in the series 1–2–3, state 3 is only permitted to be derived from state 2. In an unordered transformation series, state 3 may be derived from either of states 1 or 2. Following a similar logic, reversals (e.g., from 2 to 1) may be allowed, penalized, or prohibited, depending on the preconceptions of the investigator. Of course, building *a priori* conceptions of order or reversibility into an analysis of character state change precludes the use of that analysis as an independent test of those assumptions. To summarize this section, treating all characters as unpolarized and unordered means that all transitions among states are regarded as equally probable.

2.3.3 Squared-Change and Linear Parsimony

There are two general types of MP widely used in tracing the evolution of continuous traits: squared-change parsimony and linear parsimony. Squared-change algorithms (Rogers, 1984) seek to minimize the amount of squared change along each branch across the entire tree simultaneously, using a formula in which the cost of a change from state x to y is $(x-y)^2$. Squared-change parsimony assigns a single ancestral value to each internal node to minimize the sum of squares change over the tree (Maddison, 1991). When using squared-change parsimony, the absolute amount of evolution over the whole tree is not necessarily minimized,

and some degree of change is forced along most branches. Linear parsimony reconstructs ancestral node values by minimizing total changes (Figure 2.3). Linear-parsimony algorithms (Kluge and Farris, 1969) seek to minimize the total amount of evolution and consider only the three nearest nodes when calculating the ancestral character states. In linear parsimony the cost of a change from x to y is $|x-y|$. The result of this local optimization is that changes are inferred on very few or single branches. Linear parsimony therefore permits the accurate reconstruction of discontinuous events, or of large changes in trait values on a tree. Although evolutionary change is often thought of as gradual, large changes on a tree may result from a variety of real biological processes, not the least of which is the extinction of taxa with intermediate trait values (Butler and Losos, 1997).

2.3.4 Maximum Likelihood and Bayesian Optimization

Maximum likelihood (ML) methods for tracing character evolution select ancestral trait values with highest likelihood on a given phylogenetic hypothesis given a model of trait evolution (defined by user). Bayesian analysis (BA) selects the ancestral trait value with the highest posterior probability, given the probabilities of priors (external evidence) and assumptions of trait

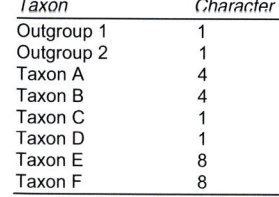

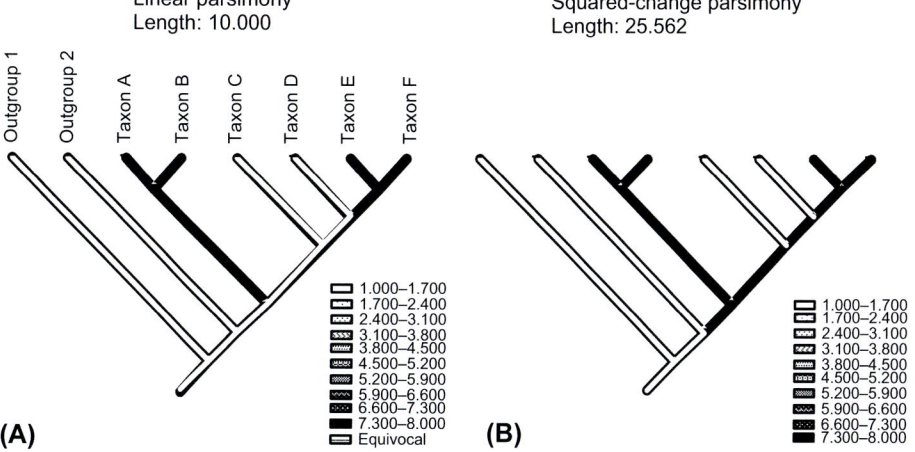

FIGURE 2.3 Alternative methods for estimating ancestral character states. a, Linear parsimony. b, Squared-change parsimony. Character state data by taxon reported in the table.

evolution (defined by user). Because they are model-based approaches, ML and BA optimization methods are more commonly used in the analysis of gene sequence data, using explicit models of changes between nucleotide bases (Liò and Goldman, 1998; Sullivan et al., 1999). ML has been used in the analysis of continuous character evolution where the models may vary from very simple (e.g., Brownian motion) to quite complex; there is a large literature regarding methods to test the validity of using particular models (Diaz-Uriarte and Garland, 1996; Oakley, 2003).

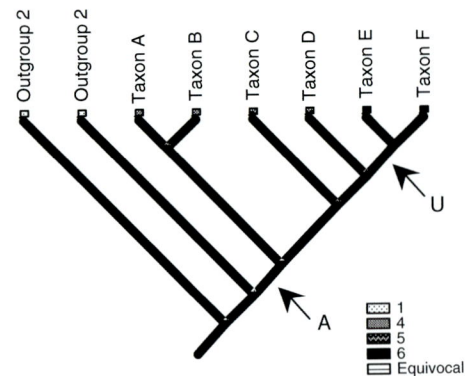

FIGURE 2.4   Ambiguous (A) vs. unambiguous (U) optimizations.

2.3.5 Which Optimization Approach to Use?

Empirical studies using simulated data sets and those derived from evolution in a test tube have concluded that model-driven approaches like ML and BA give more accurate results than MP when the modeled parameters (i.e., likelihood or probability of nucleotide substitutions) are known, but can be positively misleading when the parameters are unknown (Hillis et al., 1992; Oakley and Cunningham, 2000). MP often provides less resolution (more interior tree nodes reconstructed with ambiguous states), than ML or BA methods, which usually give very precise estimates with high confidence levels even under circumstances in which available data are insufficient to the task. In this regard, MP methods are regarded as more conservative, with lower risk of recovering false positives (Webster and Purvis, 2002).

Most studies on the evolution of neural characters use MP approaches because, unlike molecular sequence data, it is not straightforward how to pose or parametrize models on the evolution of complex phenotypes. Continuously varying aspects of neural features, like the size or shape of structures, have been modeled as simple Brownian motion or random walk processes, under the assumptions that the trait has not experienced selection and that there are no constraints on variance through time (Butler and King, 2004). Whether or not the assumptions of Brownian motion or any other specific model are satisfied by real neural or behavioral data is almost completely unknown.

A general conclusion reached by a number of review studies is that, under most circumstances faced by comparative morphologists, linear parsimony is the most conservative method for reconstructing ancestral trait values (Losos, 1999). Unlike squared-change parsimony, linear parsimony does not average out change over the interior nodes of a tree, but rather permits discontinuous changes along a branch. This has the advantageous effect of not forcing gradual trait evolution on the tree, and also of not forcing unnecessary trait reversals (Figure 2.3). A methodological advantage of linear over squared-change parsimony is that it permits the reconstruction of ambiguous ancestral character state reconstructions (Figure 2.4). This is a desirable property in cases where the available data are in fact insufficient to resolve the trait value at a specified internal nodes (Cunningham, 1999). A methodological disadvantage of linear parsimony is that, computationally, it requires a completely resolved tree topology in which all branching events are divided into only two daughter clades. Unfortunately, fully resolved trees are unusual in most studies with many (>30) species. By contrast, squared-change parsimony can be calculated on a tree with unresolved multichotomies (also called polytomies), and therefore often becomes the method of choice by default. One alternative to using squared-change parsimony when faced with an incompletely resolved tree is to use linear parsimony on numerous (100, 1000) arbitrarily resolved trees, and then report statistics (e.g., minimum and maximum) of the trait values obtained. Software for this procedure is available in the freely available Mesquite software package (see 'Relevant Website').

2.3.6 Correlative Comparative Methods

Ordinary least-squares regression allows one to investigate relationships between two variables in order to ask if change in one of these variables is associated with change in the other. One may ask, for example, how is variation in brain size related to body size, ecological role (predator vs. prey), climate, life history mode, or locomotion (Albert et al., 2000; Safi and Dechmann, 2005). The least-squares fitting procedure is commonly used in data analysis in comparative studies, and conventional regression analysis has been one of the main tools available to comparative neurobiology and ecological physiology to study form–function relationships and adaptation (Garland and Carter, 1994). However, it is now widely recognized that interspecific observations generally do not comprise independent

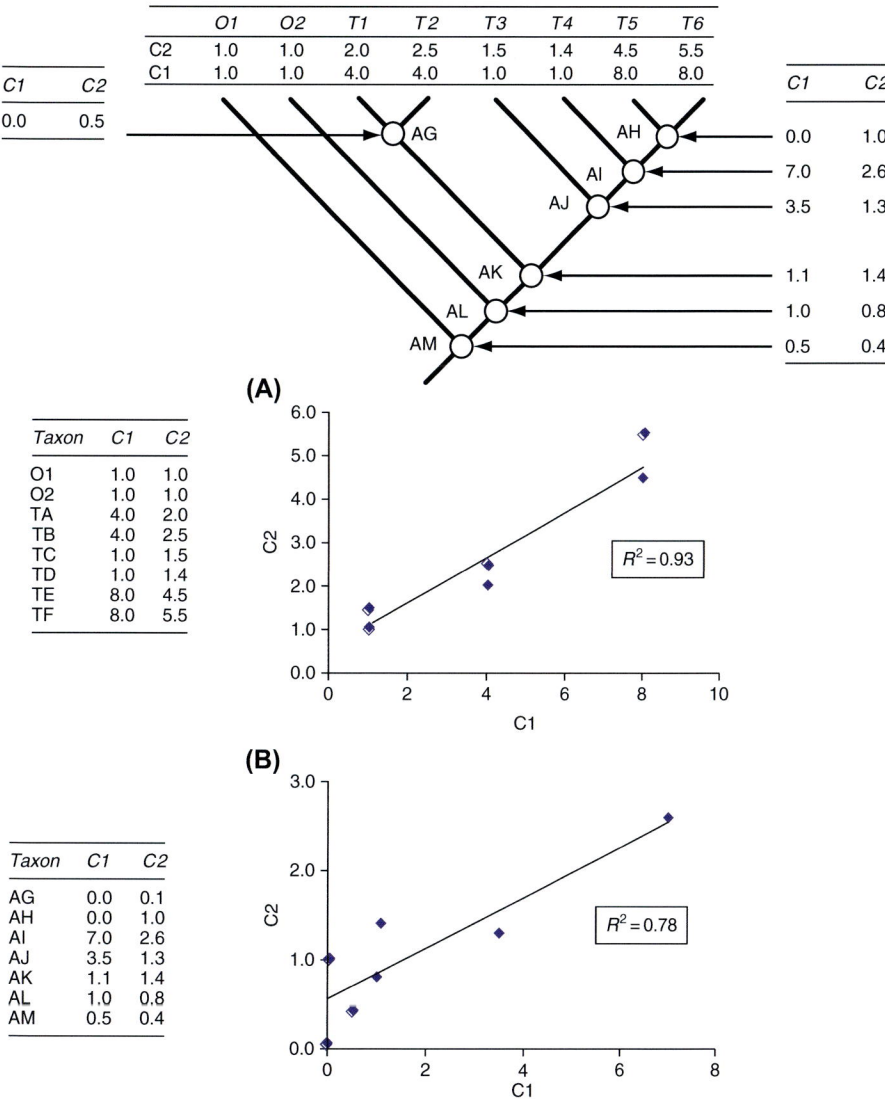

	O1	O2	T1	T2	T3	T4	T5	T6
C2	1.0	1.0	2.0	2.5	1.5	1.4	4.5	5.5
C1	1.0	1.0	4.0	4.0	1.0	1.0	8.0	8.0

C1	C2
0.0	0.5

C1	C2
0.0	1.0
7.0	2.6
3.5	1.3
1.1	1.4
1.0	0.8
0.5	0.4

(A)

Taxon	C1	C2
O1	1.0	1.0
O2	1.0	1.0
TA	4.0	2.0
TB	4.0	2.5
TC	1.0	1.5
TD	1.0	1.4
TE	8.0	4.5
TF	8.0	5.5

$R^2 = 0.93$

(B)

Taxon	C1	C2
AG	0.0	0.1
AH	0.0	1.0
AI	7.0	2.6
AJ	3.5	1.3
AK	1.1	1.4
AL	1.0	0.8
AM	0.5	0.4

$R^2 = 0.78$

FIGURE 2.5   Comparison of conventional and phylogenetic regression analyses. Phylogeny of six terminal taxa (TA−TF) and two outgroup taxa (O1 and O2), represented by two continuously distributed characters (C1 and C2). Tree topology determined from data other than characters 1 and 2, and branch lengths treated as equal. Seven internal tree nodes represented by ancestral taxa (AG−AM) with ancestral trait values estimated by least-squares parsimony. a, Conventional regression of trait values from terminal taxa. b, Phylogenetic regression of trait values at internal tree nodes using the method of independent contrasts. Note that by removing psuedoreplicates, the phylogenetic regression compares fewer taxa, has fewer degrees of freedom, and has a lower correlation coefficient (R^2 value) than does the conventional regression. The phylogenetic regression, therefore, provides a more conservative quantitative measure of correlated evolution between the two traits.

and identically distributed data points, thus violating fundamental assumptions of conventional parametric statistics (Felsenstein, 1985, 1988; Pagel and Harvey, 1989; Harvey and Pagel, 1991).

Phylogenetically based statistical methods allow traditional topics in comparative neuroanatomy and physiology to be addressed with greater rigor, including the form of allometric relationships among traits and whether phenotypes vary predictably in relation to behavior, ecology, or environmental characteristics (Brooks and McLennan, 1991; Frumhoff and Reeve, 1994; Losos, 1996). In a conventional regression analysis

the data points represent terminal taxa. In a phylogenetic regression the data points represent sistertaxon comparisons (Grafen, 1989). These two methods are compared in Figure 2.5, in which identical data are analyzed using conventional and phylogenetic regression methods. The phylogeny of Figure 2.5 includes six terminal taxa (TA−TF) and two outgroup taxa (O1 and O2), which are represented by two continuously distributed characters (C1 and C2). The tree topology has been determined from data other than characters 1 and 2, and the branch lengths are treated as equal (under a model of punctuated equilibrium). There are seven internal tree nodes represented by ancestral taxa (AG—AM) with trait values estimated by least-square parsimony. By removing psuedorepilcates, the phylogenetic regression compares fewer taxa, has fewer degrees of freedom, and has a lower correlation coefficient (R^2 value) than does the conventional regression. The phylogenetic regression, therefore, provides a better quantitative measure of correlated evolution between the two traits, and is a more conservative measure of the strength of adaptive pressures.

Relationships between brain size and the volume of frontal and visual cortices in mammals have recently been studied using the methods of phylogenetic regression analysis (Bush and Allman, 2004a, 2004b). These studies found that size has a profound effect on the structure of the brain, and that many brain structures scale allometrically; that is, their relative size changes systematically as a function of brain size. They also conclude that the three- dimensional shape of visual maps in anthropoid primates is significantly longer and narrower than in strepsirrhine primates. Using conventional regression analyses, von Bonin (1947) showed that frontal cortex hyperscales with brain size, and humans have "precisely the frontal lobe which [we deserve] by virtue of the overall size of [our] brain." These are, of course, precisely the qualitative conclusions arrived at by Bush and Allman using analysis of phylogenetic regressions. In fact, many studies reviewing the uses of phylogenetic methods for reconstructing ancestral states conclude that all methods will recover a very strong historical signal (Losos, 1999).

2.4 Limitations of Methods

The accuracy of ancestral reconstructions has been investigated by comparisons with known phylogenies (e.g., viruses, computer simulations; Oakley and Cunningham, 2000). It is well known that all phylogenetically based methods perform poorly when taxon sampling is low and when rates of evolution in the character of interest are unequal among branches of the tree (Garland *et al.*, 1993; Sullivan *et al.*, 1999; Hillis *et al.*, 2003). Further, all methods for studying character evolution on a tree make certain assumptions about the capacity of trees to faithfully record the actual history of character change. These include the assumptions that: phenotypic diversification results largely from speciation and that the effects of extinction have not erased the signal, that taxon sampling faithfully represent the history of diversification, and that genealogical history is largely or entirely bifurcating (vs. multifurcating or converging). Of course, all methods assume we know the 'true' (or 'nearly true') tree topology. In addition, each of the optimization methods makes assumptions

about critical parameters, including branch lengths, models of character evolution, absolute rates of evolution, homogeneity (vs. heterogeneity) of evolutionary rates, reversibility (or the lack thereof), and the orderedness (or unorderedness) of multistate characters.

The accuracy of ancestral trait reconstruction also depends strongly on parameter estimation (e.g., tree topology, branch lengths, and models of trait evolution). ML and BA perform well when model assumptions match real parameters. ML and BA are positively misleading when model assumptions are violated. MP is more conservative, recovering fewer false positives than ML and BA when biological parameters are not known. Squared-change parsimony, ML, and BA minimize large changes, spreading evolution over the internal tree branches. Linear parsimony permits reconstructions at ancestral nodes with no change, and permits ambiguous reconstructions. 'Independent contrasts' assumes that selection operates in the origin but not maintenance of derived traits. Both conventional and phylogenetic correlations of interspecific character data make assumptions about critical parameters. These assumptions are often of unknown validity, and in some cases are known to be incorrect. Conventional statistics assume that each terminal taxon (tips of the tree) may be treated as independent sample of the relationship under investigation. This means that the character value (phenotype) observed in that taxon evolved independently (without inheritance) from the values in other taxa in the analysis. In an evolutionary context, this is equivalent to assuming that trait values result primarily from stabilizing selection in each species that acts to maintain trait values, rather than from directional selection at the origin of the trait in an ancestral species (Hansen, 1997). In other words, conventional statistics assume traits to be highly labile and without significant phylogenetic inertia. Phylogenetic correlations make converse assumptions, that trait values are due largely or entirely to directional selection at the origin of a feature and that the influence of stabilizing selection is negligible. Phylogenetic correlations also must make particular assumptions about branch lengths and models of trait evolution.

2.5 Conclusions

As in all aspects of historical inquiry, the study of character evolution is exceptionally sensitive to the amount of information that has actually survived up to the present. The reality of neural evolution was in most cases almost certainly very complex, and may be reliably regarded to have included vastly more numbers of independent transformations than has been recorded in the distribution of phenotypes preserved among living species. The signature of many historical events has been overwritten by reversals and convergences, or eliminated altogether by extinctions. Paleontologists estimate that more than 99% of all species that have ever lived are now extinct (Rosenzweig, 1995). This figure, of course, includes higher taxa (e.g., trilobites, placoderms, plesiosaurs) that are now entirely extinct, bringing up the aggregate percentage of extinction for all taxa. The proportion of living species that persists within certain targeted taxa may be much higher (e.g., Lake Victoria cichlid fishes). Nevertheless, in comparative studies of neural, physiological, or behavioral phenotypes, it is rare to have information on all extant species. Whether it is from extinction or incomplete surveys, taxon sampling remains one of the greatest sources of error in phylogenetic estimates of character evolution (Sullivan *et al.*, 1999; Zwickl and Hillis, 2002).

Despite all these reservations, we must continue to estimate ancestral traits in order to study phenotypic evolution. None of the methods reviewed in this article should be regarded as a magic bullet, but rather there are advantages and disadvantages of each method as they are applied under different circumstances. All the methods reviewed here have proved to be useful tools in the phylogenetic toolbox. As in other aspects of science, it is important to make our assumptions explicit, and to use reasonable assumptions. Further, as in other aspects of evolutionary biology, critical insights into the evolution of neural characters will come from a better understanding of the biology of the phenotypes themselves, and the organisms in which they have evolved.

References

Albert, J.S., Crampton, W.G.R., 2005. Electroreception and electrogenesis. In: Evans, D.H., Claiborne, J.B. (Eds.), The Physiology of Fishes, third ed. CRC Press, pp. 431–172.

Albert, J.S., Lannoo, M.J., Yuri, T., 1998. Testing hypotheses of neural evolution in gymnotiform electric fishes using phylogenetic character data. Evolution 52, 1760–1780.

Albert, J.S., Froese, R., Paulay, D., 2000. The Brains Table. In: Froese, R., Paulay, D. (Eds.), FishBase 2000, Concepts, Design and Data Sources. ICLARM, pp. 234–237.

Benton, M.J., 1993. The Fossil Record, vol. 2. Chapman and Hall, p. 845.

Benton, M.J., Wills, M., Hitchin, R., 2000. Quality of the fossil record through time. Nature 403, 534–538.

Blackledge, T.A., Gillespie, R.G., 2004. Convergent evolution of behavior in an adaptive radiation of Hawaiian web-building spiders. Proc. Natl. Acad. Sci. U.S.A. 101, 16228–16233.

Blomberg, S.P., Garland, T., Ives, A.R., 2003. Testing for phylogenetic signal in comparative data: behavioral traits are more labile. Evolution 57, 171–745.

Brooks, D.R., McLennan, D.A., 1991. Phylogeny, Ecology, and Behavior. University of Chicago Press.

Bush, E.C., Allman, J.M., 2004a. Three-dimensional structure and evolution of primate primary visual cortex. Anat. Rec. Part A 281, 1088–1094.

Bush, E.C., Allman, J.M., 2004b. The scaling of frontal cortex in primates and carnivores. Proc. Natl. Acad. Sci. U.S.A. 101 (11), 3962–3966.

Butler, M.A., King, A.A., 2004. Phylogenetic comparative analysis: a modeling approach for adaptive evolution. Am. Nat. 164 (6), 683–695.

Butler, M.A., Losos, J.B., 1997. Testing for unequal amounts of evolution in a continuous character on different branches of a phylogenetic tree using linear and squared-change parsimony: an example using Lesser Antillean *Anolis* lizards. Evolution 51 (5), 1623–1635.

Coyne, J.A., Orr, H.A., 1989. Two Rules of Speciation. In: Otte, D., Endler, J. (Eds.), Speciation and its Consequences. Sinauer Associates, pp. 180–207.

Cunningham, C.W., 1999. Some limitations of ancestral character-state testing evolutionary hypotheses. Syst. Biol. 48 (3), 665–674.

de Queiroz, A., Wimberger, P.H., 1993. The usefulness of behavior for phylogeny estimation: levels of homoplasy in behavioral and morphological characters. Evolution 47, 46–60.

Diaz-Uriarte, R., Garland, T., 1996. Testing hypotheses of correlated evolution using phylogenetically independent contrasts: sensitivity to deviations from Brownian motion. Syst. Biol. 45 (1), 27–47.

Edinger, T., 1941. The brain of *Pterodactylus*. Am. J. Sci. 239 (9), 665–682.

Eisthen, H.L., 2002. Why are olfactory systems of different animals so similar? Brain Behav. Evol. 59, 273–293.

Felsenstein, J., 1985. Phylogenies and the comparative method. Am. Nat. 125, 1–15.

Felsenstein, J., 1988. Phylogenies and quantitative characters. Annu. Rev. Ecol. Syst. 19, 445–471.

Frumhoff, P.C., Reeve, H.K., 1994. Using phylogenies to test hypotheses of adaptation: a critique of some current proposals. Evolution 48, 172–180.

Garland Jr., T., Carter, P.A., 1994. Evolutionary physiology. Annu. Rev. Physiol. 56, 579–621.

Garland Jr., T., Harvey, P.H., Ives, A.R., 1992. Procedures for the analysis of comparative data using phylogenetically independent contrasts. Syst. Biol. 41, 18–32.

Garland Jr., T., Dickerman, A.W., Janis, C.M., Jones, J.A., 1993. Phylogenetic analysis of covariance by computer simulation. Syst. Biol. 42, 265–292.

Garland Jr., T., Midford, P.E., Ives, A.R., 1999. An introduction to phylogenetically based statistical methods, with a new method for confidence intervals on ancestral values. Am. Zool. 39, 347–388.

Garstang, W., 1922. The theory of recapitulation: a critical restatement of the biogenetic law. J. Linn. Soc. Zool. 35, 81–101.

Grafen, A., 1989. The phylogenetic regression. Philos. Trans. R. Soc. Lond. 326, 119–157.

Grothe, B., Carr, C.E., Cassedy, J., Fritzsch, B., Köppl, C., 2005. Brain pathway evolution and neural processing patterns – parallel evolution? In: Manley, G.A., Popper, A.N., Fay, R.R. (Eds.), Evolution of the Vertebrate Auditory System: Springer Handbook of Auditory Research. Springer.

Hanlon, R.T., Messenger, J.B., 1996. Cephalopod Behaviour. Cambridge University Press, p. 232.

Hansen, T.F., 1997. Stabilizing selection and the comparative analysis of adaptation. Evolution 51, 1341–1351.

Hartline, P.H., 1988. Thermoreception in snakes. Prog. Brain Res. 74, 297–312.

Harvey, P.H., Pagel, M.D., 1991. The Comparative Method in Evolutionary Biology. Oxford University Press.

Hillis, D.M., Bull, J.J., White, M.E., Badgett, M.R., Molineux, I.J., 1992. Experimental phylogenetics: Generation of a known phylogeny. Science 255, 589–592.

Hillis, D.M., Pollock, D.D., Mcguire, J.A., Zwickl, D.J., 2003. Is sparse taxon sampling a problem for phylogenetic inference? Syst. Biol. 52 (1), 124–126.

Jerison, H.J., 1976. Paleoneurology and evolution of mind. Sci. Am. 234, 90.

Kluge, A.G., Farris, S.J., 1969. Quantitative phyletics and the evolution of anurans. Syst. Zool. 18, 1–32.

Krebs, H.A., 1975. The August Krogh principle: for many problems there is an animal on which it can be most conveniently studied. J. Exp. Zool. 194 (1), 221–226.

Liò, P., Goldman, N., 1998. Models of molecular evolution and phylogeny. Genome Res. 8 (12), 1233–1244.

Losos, J.B., 1996. Phylogenies and comparative biology, stage II: testing causal hypotheses derived from phylogenies with data drawn from extant taxa. Syst. Biol. 45, 259–260.

Losos, J.B., 1999. Uncertainty in the reconstruction of ancestral character states and limitations on the use of phylogenetic comparative methods. Anim. Behav. 58, 1319–1324.

MacLeod, N., Rose, K.D., 1993. Inferring locomotor behavior in Paleogene mammals via eigenshape analysis. Am. J. Sci. 293-A, 300–355.

Maddison, W.P., 1991. Squared-change parsimony reconstructions of ancestral states for continuous valued characters on a phylogenetic tree. Syst. Zool. 40, 304–314.

Martins, E.P., 2000. Adaptation and the comparative method. Trends Ecol. Evol. 15, 295–299.

Mayr, E., 1963. Populations, Species, and Evolution. Harvard University Press.

McKinney, M.L., McNamara, K.J., 1990. Heterochrony: The Evolution of Ontegeny. Plenum.

Molenaar, G.J., 1992. Infrared sensitivity of snakes. In: Gans, C., Ulinski, P.S. (Eds.), Biology of the Reptilia, vol. 17. University of Chicago Press, pp. 367–453.

Near, T.J., Sanderson, M.J., 2004. Assessing the quality of molecular divergence time estimates by fossil calibrations and fossil-based model selection. Philos. Trans. R. Soc. Lond. B 359, 1477–1483.

Nishikawa, K.C., 1999. Neuromuscular control of prey capture in frogs. Philos. Trans. R. Soc. Lond. B 354, 941–954.

Nishikawa, K.C., 2002. Evolutionary convergence in nervous systems: insights from comparative phylogenetic studies. Brain Behav. Evol. 59, 240–249.

Northcutt, R.G., Gans, C., 1983. The genesis of neural crest and epidermal placodes: a reinterpretation of vertebrate origins. Q. Rev. Biol. 58, 1–28.

Oakley, T.H., 2003. Maximum likelihood models of trait evolution. Comments Theor. Biol. 8, 1–17.

Oakley, T.H., Cunningham, C.W., 2000. Independent contrasts succeed where ancestor reconstruction fails in a known bacteriophage phylogeny. Evolution 54 (2), 397–405.

Packard, A., 1972. Cephalopods and fish – the limits of convergence. Biol. Rev. 47, 241–307.

Page, R.D.M., Holmes, E.C., 1998. Molecular Evolution: A Phylogenetic Approach. Blackwell.

Pagel, M.D., Harvey, P.H., 1989. Taxonomic differences in the sealing of brain on body weight among mammals. Science 244 (4912), 1589–1593.

Patterson, C., 1982. Morphological characters and homology. In: Joysey, K.A., Friday, A.E. (Eds.), Problems in Phylogenetic Reconstruction. Academic Press, pp. 21–74.

Postlethwait, J., Amores, A., Cresko, W., Singer, A., Yan, Y.L., 2004. Subfunction partitioning, the teleost radiation and the annotation of the human genome. Trends Genet. 10, 481–490.

Rensch, B., 1959. Evolution above the Species Level. Columbia University Press, p. 419.

Rogers, J.S., 1984. Deriving phylogenetic trees from allele frequencies. Syst. Zool. 33, 52–63.

Rogers, S.W., 2005. Reconstructing the behaviors of extinct species: an excursion into comparative paleoneurology. Am. J. Med. Genet. Part A 134, 349.

Rohlf, F.J., 1998. On applications of geometric morphometrics to studies of ontogeny and phylogeny. Syst. Biol. 47, 147–158.

Rosenzweig, M.L., 1995. Species Diversity in Space and Time. Cambridge University Press.

Russell, E.S., 1916. Form and function: a Contribution to the history of animal morphology. Murray. 1982. Reprint, with a New Introduction by G. V. Lauder. University of Chicago Press.

Safi, K., Dechmann, D.K.N., 2005. Adaptation of brain regions to habitat complexity: a comparative analysis in bats (Chiroptera). Proc. R. Soc. Biol. Sci. 272 (1559), 179–186.

Sanderson, M.J., 2002. Estimating absolute rates of molecular evolution and divergence times: a penalized likelihood approach. Mol. Biol. Evol. 19, 101–109.

Stensio, E.A., 1963. The brain and cranial nerves in fossil, lower craniate vertebrates. Skrifter Norske Videnskaps-Akademi 1 Oslo I. Mat.-Naturv. Klasse. Ny Serie 13, 1–120.

Strausfeld, N.J., 1998. Crustacean – insect relationships: the use of brain characters to derive phylogeny amongst segmented invertebrates. Brain Behav. Evol. 52, 186–206.

Sullivan, J., Swofford, D.L., Naylor, G.J.P., 1999. The effect of taxon sampling on estimating rate heterogeneity parameters of maximum-likelihood models. Mol. Biol. Evol. 16 (10), 1347–1356.

Taylor, J.S., Braasch, I., Frickey, T., Meyer, A., Van de Peer, Y., 2003. Genome duplication, a trait shared by 22,000 species of ray-finned fish. Genome Res. 13 (3), 382–390.

van Buskirk, J., 2002. A comparative test of the adaptive plasticity hypothesis: relationships between habitat and phenotype in anuran larvae. Am. Nat. 160 (1), 87–102.

von Bonin, G., 1947. Assoc. Res. Nervous Mental Dis. 27, 67–83.

Webster, A.J., Purvis, A., 2002. Testing the accuracy of methods for reconstructing ancestral states of continuous characters. Proc. R. Soc. Lond. B 269, 143–150.

Wiens, J.J., 2001. Character analysis in morphological phylogenetics: problems and solutions. Syst. Biol. 50, 689–699.

Wimberger, P.H., de Queiroz, A., 1996. Comparing behavioral and morphological characters as indicators of phylogeny. In: Martins, E.P. (Ed.), Phylogenies and the Comparative Method in Animal Behavior. Oxford University Press, pp. 206–233.

Zakon, H.H., 2002. Convergent evolution on the molecular level. Brain Behav. Evol. 59, 250–261.

Zelditch, M.L., Fink, W.L., Swiderski, D.L., 1995. Morphometrics, homology, and phylogenetics: Quantified characters as synapomorphies. Syst. Biol. 44, 179–189.

Zwickl, D.J., Hillis, D.M., 2002. Increased taxon sampling greatly reduces phylogenetic error. Syst. Biol. 51 (4), 588–598.

Further Reading

Felsenstein, J., 1985. Phylogenies and the comparative method. Am. Nat. 125, 1–15.

Garland Jr., T., Carter, P.A., 1994. Evolutionary physiology. Annu. Rev. Physiol. 56, 579–621.

Garland Jr., T., Midford, P.E., Ives, A.R., 1999. An introduction to phylogenetically based statistical methods, with a new method for confidence intervals on ancestral values. Am. Zool. 39, 347–388.

Harvey, P.H., Pagel, M.D., 1991. The Comparative Method in Evolutionary Biology. Oxford University Press.

Losos, J.B., 1999. Uncertainty in the reconstruction of ancestral character states and limitations on the use of phylogenetic comparative methods. Anim. Behav. 58, 1319–1324.

Relevant Website

http://mesquiteproject.org – A modular system of evolutionary analysis. Version 1.06, Maddison, W. P., and Maddison, D. R., 2005.

3

The Role of Endocasts in the Study of Brain Evolution

A.M. Balanoff, G.S. Bever*

Johns Hopkins University, Baltimore, MD, United States; Johns Hopkins University School of Medicine, Baltimore, MD, United States

3.1 Introduction

The chordate brain began its remarkable evolutionary history when changes in an increasingly complex and genetically controlled developmental network produced an anterior expansion of the deuterostome dorsal hollow nerve cord. This initial neuroectodermal enlargement is most faithfully conserved among extant chordates in the lancelets (amphioxus) (Butler, 2000; Lacalli, 2008)—a lineage that diverged from our own ancestral line well over 500 million years ago (Yue et al., 2014). The antiquity of the chordate brain thus exceeds, and considerably so, the cranial skeleton that in the majority of chordate groups provides support and protection to the brain and related sensory organs. The precocious appearance of the brain relative to its supportive skeletal framework in evolutionary history is paralleled during chordate ontogeny. Here, the brain begins to emerge during secondary neurulation and then likely serves an underappreciated developmental role as an early signaling center and organizer of the neighboring head mesenchyme—directing cells that will eventually contribute to a wide diversity of adult cranial structures, including the skull (eg, Hu et al., 2015). This signaling relationship between brain and neighboring head mesenchyme certainly moves in both directions (see Creuzet, 2009a,b).

The spatial integration that occurs in ontogeny and phylogeny between the brain and head skeleton creates an interfacing surface, preserved in many vertebrate fossils, that can be estimated and studied as an endocast. The goal of this chapter is to explore the logical framework within which endocasts can be combined with data from modern neuroscience to provide a more complete evolutionary account of the vertebrate brain. We begin the exploration with an explicit description of the inferential model in which fossils inform macroevolutionary patterns and the phylogenetic terminology used to communicate these patterns. Inconsistent application of this terminology by neuroscientists and paleontologists alike is an unnecessary trend that creates confusion and impedes whatever understanding might be achieved through a clear integration of data from these seemingly disparate, but actually complementary, fields. We will then outline what an endocast does, and does not, represent anatomically and thus what lines of neuroanatomical investigation endocast data can potentially support. Examples will be drawn heavily from the reptile side of the amniote tree and especially from the long evolutionary stem lineage that produced modern birds. For reasons that hopefully will become clear during the course of this chapter, the deep history of the avian lineage represents well the challenges, opportunities, and potentialities that endocasts present to the greater neuroscience community.

*Both the authors contributed equally.

30

3. The Role of Endocasts in the Study of Brain Evolution

3.1.1 Crown, Stem, and the Heuristic Potential of Fossil Endocasts

Considered against the backdrop of Life's more than four billion year history (Bell et al., 2015), the extant biota provides us with a rather impoverished view of taxonomic, and probably, process diversity. At the same time, we rely heavily on living species both for our understanding of detailed biological functions and as a framework for establishing broad, macroevolutionary patterns. We tend to perceive extant forms as islands of either actuated or potential insight floating in a rather murky sea of extinction. These islands are not floating freely, of course, but are instead tethered to each other within the historical edifice of the evolutionary process. The logical framework provided by this edifice is what gives us hope that we may meaningfully clarify the deep biological history which extinction has muddled. The involved model of inference, and the beneficial role that fossils play within that model, is perhaps best demonstrated using an explicit crown-stem distinction (Fig. 3.1).

A crown clade is a monophyletic group whose definition is drawn from the ancestral divergence of two lineages, both of which retain at least some extant members (Hennig, 1966; Budd and Jensen, 2000; Gauthier and De Queiroz, 2001). The crown clade of mammals (Mammalia; Rowe, 1987), for example, is defined as the most recent common ancestor of monotremes and therians and all of that ancestor's descendants. The extant sister taxon to crown mammals is the crown clade of reptiles

(Reptilia; Gauthier et al., 1988), which is defined based on the ancestral split between the lineages that would eventually produce the modern radiations of turtles, lepidosaurs (modern lizards, snakes, and tuatara), crocodilians, and birds. The most exclusive clade that includes both Mammalia and Reptilia is the crown clade of amniotes (Amniota; Gauthier et al., 1988). More inclusive crown clades in our own ancestral line as chordates include (Fig. 3.2; see also Rowe, 2004): Tetrapoda (Amniota + Amphibia), Choanata [Tetrapoda + Dipnoi (lungfish)], Sarcopterygii [Choanata + Actinistia (coelacanths)], Osteichthyes [Sarcopterygii + Actinopterygians (ray-finned fish)], Gnathostomata [Osteichthyes + Chondrichthyes (sharks and rays)], Vertebrata [Gnathostomata + Cyclostomata (lampreys and hagfish)], Olfactores [Vertebrata + Urochordata (tuincates or sea squirts)], and Chordata [Olfactores + Cephalochordata (lancelets or amphioxus)].

All crown clades, from the most inclusive to the most exclusive, are separated from each other by some expanse of evolutionary time, which is circumscribed by their collective stem lineages (Fig. 3.1). These stem lineages are somewhat of an evolutionary black box because, by definition, they lack any modern representatives (outside of the crown clade). Our understanding of the evolutionary transformations that populate these stems is based largely on a combination of: (1) the empirical observations that we make on the extant taxa within the associated crown clades—observations that we cannot make directly on stem taxa and (2) phylogenetically justified inferences (Farris, 1983; Bryant and

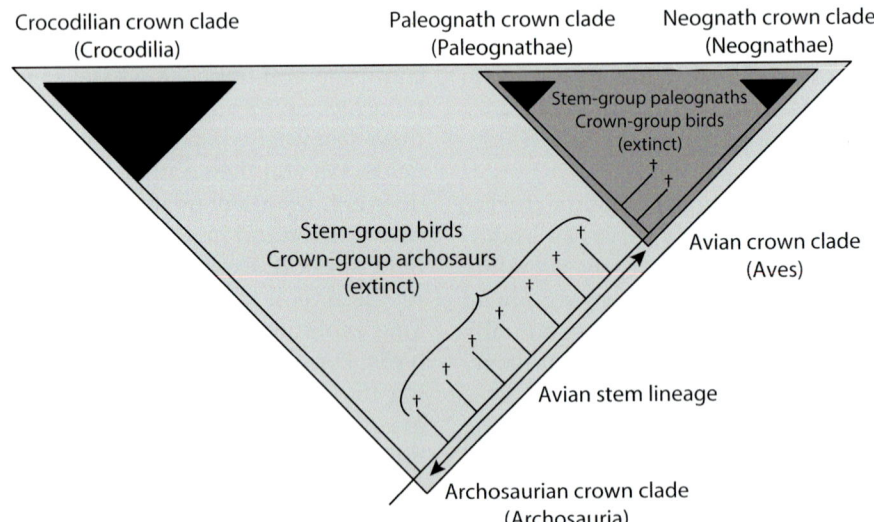

FIGURE 3.1 Tree depicting the phylogenetic concept of crown and stem as it applies to birds. The concept is critical because crown clades define the inferential framework lying at the heart of all comparative biology and within which fossils (including fossil endocasts) play a key role (see text). The extant sister taxon to the avian crown clade (Aves) is the crown clade Crocodilia, and together Crocodilia and Aves define crown-clade Archosauria. The avian stem lineage includes all those extinct forms (including pterosaurs and all nonavian dinosaurs) that are more closely related to Aves than Crocodilia. The crown-stem distinction is a nested concept, so that, for example, stem-group birds are also crown-clade archosaurs.

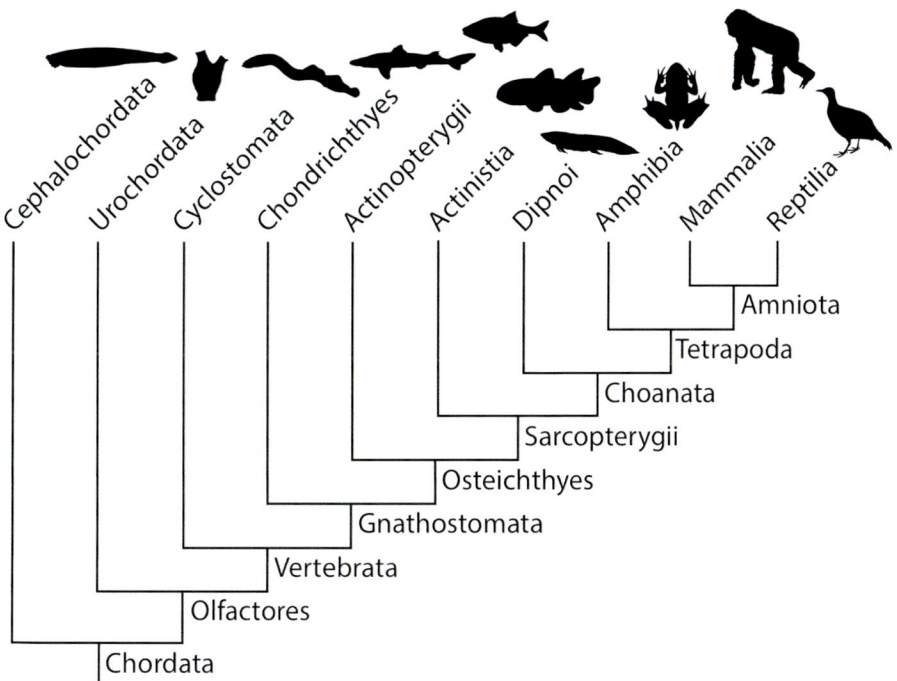

FIGURE 3.2   Phylogenetic relationships of the major chordate crown clades discussed in this chapter. For an expanded discussion of the tree topology, see Rowe (2004), Benton et al. (2015), and Gee (2018).

Russell, 1992; De Queiroz and Gauthier, 1992; Witmer, 1995). This model is rich in explanatory power and forms the basis of all comparative biology, including all the biomedical research that is grounded in the study of model organisms. The problem is that as these phylogenetic stems come to represent more and more time and are inhabited by an increasing number of evolutionary transformations, the model itself grows less and less heuristically powerful. In other words, as the stems lengthen, the explanatory power of the inferential model tends to diminish.

Recognizing the inverse tendencies of this relationship is important, especially when we consider the incredible spans of evolutionary time represented by some stem lineages—including many of those attracting great interest from comparative neuroscientists. The phylogenetic stem of our own human crown clade, for example, is currently estimated at approximately 6 million years (Dos Reis et al., 2012; Benton et al., 2015), whereas that of crown Mammalia is approximately 150 million years (O'Leary et al., 2013; Luo et al., 2015), and that of crown-group birds (Aves; Gauthier, 1986) exceeds 150 million years (Prum et al., 2015). The stems associated with the crown clades informing the earliest history of the vertebrate brain (ie, cephalochordates, cyclostomes, chondrichthyans) may exceed 200 million years (Kuraku and Kuratani, 2006; Chen et al., 2012; Hedges et al., 2015).

The inherent difficulty of inferring details across such long stem lineages can be eased through an effective utilization of the fossil record. Fossils, and only fossils, afford windows (in the form of semaphoronts; Hennig, 1966) into these otherwise empirically opaque histories. Fossils can enlighten, either directly or indirectly, the nature and timing of evolutionary transformations and thus help to "break up" long phylogenetic stems (Fig. 3.1). The most obvious beneficiary of integrating vertebrate fossils into comparative studies is our understanding of transformations within the more readily fossilized bony skeleton. The skeleton enjoys, however, at least some form of correlative relationship with most other anatomical systems, including the brain (see the following section). Establishing the strength of these correlations is critical to maximizing the explanatory potential of fossils for macroevolutionary patterns.

The crown-stem distinction is a nested concept wherein every individual fossil falls along a single phylogenetic stem, but at the same time, is nested within a series of more inclusive crown clades. *Ardipithecus* is a fossil form on the stem of the human crown clade (White et al., 2009) but one that is nested within the crown clades of Catarrhini, Primates, Placentalia, Theria, Mammalia, and so forth. *Archaeopteryx* is a stem bird that lies relatively near, but still outside, the radiation of crown-clade avians (Gauthier, 1986; Turner et al., 2012), but *Archaeopteryx* is a crown-clade archosaur, reptile, and amniote. Fossils, like extant forms, express a combination of primitive (plesiomorphic) and derived (apomorphic) morphologies, not all of which are going to reflect the ancestral phenotype of their most closely related crown clade. A well-supported understanding of a fossil's phylogenetic position is crucial to maximizing its potential for informing the ancestral series of stem transformations (accrual of apomorphies along the stem), which make its associated crown clade unique compared to those of other lineages. Even when this position enjoys widespread consensus, however, the inferential role of the fossil can be confused when the employed terminology and taxonomy mean different things to different researchers.

Paleontologists and neuroscientists often employ "crown" and "stem" in ways that differ significantly from the usage advocated here. Paleontologists are known to apply these terms to wholly extinct groups, especially when these lineages include a subclade that is especially distinct morphologically (eg, crown and stem sauropterygians; Rieppel, 1994). Neuroscientists often describe extant groups lying outside some clade of interest as "stem." For example, Corfield et al. (2015) recently referred to turtles as stem reptiles in a study whose focus was the neuroanatomy of crown-clade birds. It is not that these usages are incorrect—the meanings intended by their authors may be effectively conveyed, especially within their respective research communities. Different usages, however, do confuse the inferential roles of crown and stem and thus hinder meaningful integration of what are complementary datasets.

3.2 Assessing the Anatomical Identity of Endocasts

In the strictest anatomical sense, endocasts are representations (casts) of any enclosed, three-dimensional (3D) space. The term usually is reserved for those cavities defined by the cranial skeleton (eg, nasal capsule, semicircular canals and vestibule, bony sinuses, neurovascular canals) and especially the endocranial cavity housing the brain (Figs. 3.3 and 3.4). It is in this latter, most-restricted, sense that we will confine our usage in this chapter.

3.2.1 Endocranial Cavity as Brain Proxy

The brain is not isolated within the endocranial cavity but shares this space with a variable number of intimately associated structures (see the following section). This anatomical reality creates a differential relationship between the brain and cavity walls that varies widely between vertebrate lineages and has important

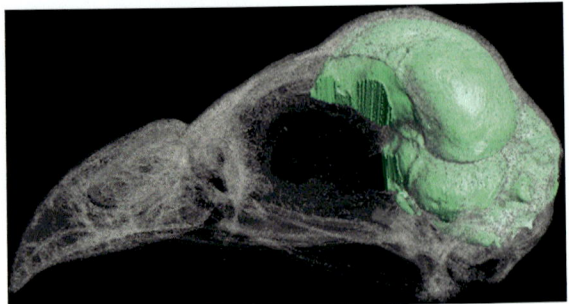

FIGURE 3.3 Digital skull and endocast of the extant red-crested turaco, *Tauraco erythrolophus* (AMNH 27414). The data are derived from a high-resolution X-ray computed tomography scan of specimen. The skull is rendered transparent to show the positional relationship of the endocranial cavity and its endocast to other cranial features. Crown-clade birds express an apomorphically high brain-to-endocranial cavity index, making them particularly conducive for studies that integrate endocasts with other forms of neurological data. AMNH, American Museum of Natural History, New York.

implications for endocast studies. Although this variable relationship has long been recognized (Edinger, 1951; Hopson, 1979; Witmer et al., 2008), it was only recently named the brain-to-endocranial cavity (BEC) index (Balanoff et al., 2016a). High BEC values can be expected to produce an endocast that reflects brain volume and morphology with high fidelity, whereas low BEC values are associated with a more cylindrical endocast bearing less resemblance to the actual brain (Figs. 3.4 and 3.5).

A phylogenetically inclusive survey of empirical BEC values for vertebrates does not yet exist. The current absence of such a valuable contribution likely reflects, at least in part, the labor-intensive nature of assessing this relationship using traditional histological sections (Corfield et al., 2012). Modern visualization techniques and scanning technology are rapidly transcending these logistical issues, so we can expect to learn much of this pattern in the near future (Clement et al., 2015; Gignac and Kley, 2014). Based on qualitative evaluation of cross-sectional anatomy, we can predict that the ancestral crown-clade vertebrate had a low BEC index and that a relatively low value is broadly conserved even within tetrapods. For example, the extant coelacanth, *Latimeria chalumnae*, and the basking shark, *Cetorhinus maximus*, reportedly fill less than 1% of their endocranial cavities (Millot and Anthony, 1965; Kruska, 1988), although these exceedingly low numbers are likely due in part to desiccation of neural tissue in alcohol-preserved specimens. Within crown-group reptiles, an index of 0.33 is reported in the snapping turtle, *Chelydra serpentina* (Humphrey, 1894), whereas the BEC index of the tuatara, *Sphenodon punctatus* is 0.5 (Dendy, 1910). The brain of *Alligator mississippiensis* occupies from 32% to 68% of the endocranial cavity, with the lowest BEC values being found in the largest specimens

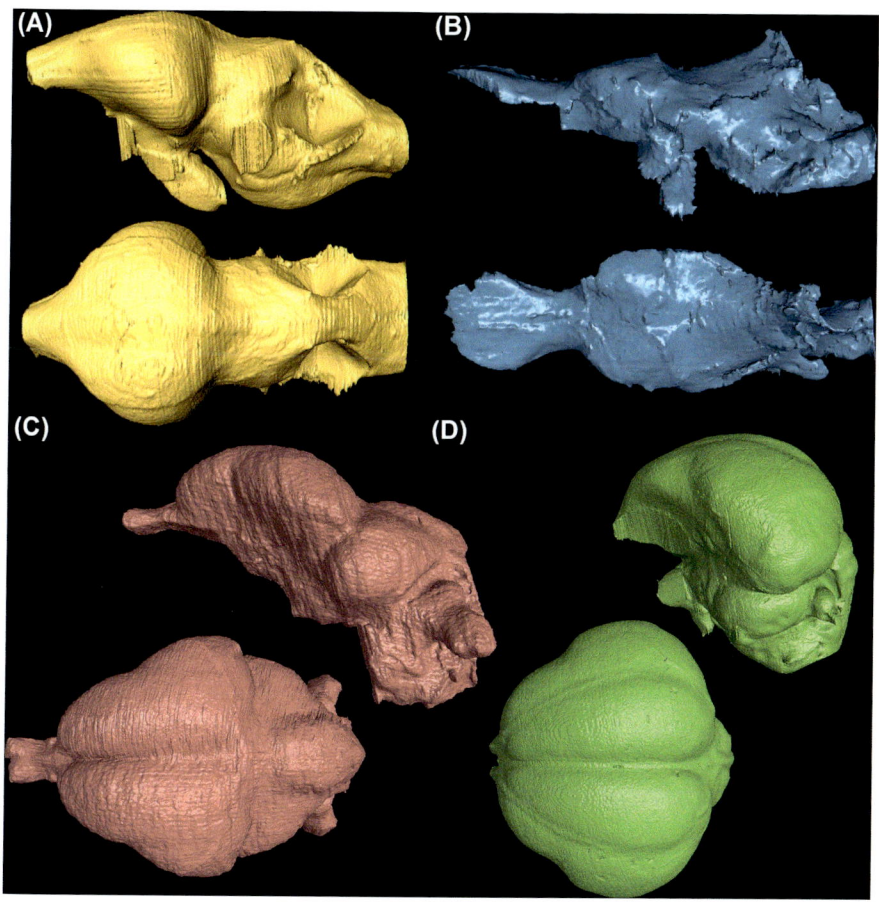

FIGURE 3.4 Digitally rendered endocasts from representative crown-clade archosaurs. (A) *Paleosuchus trigonatus* (AMNH 137175), extant crown-clade crocodilian; (B) *Alioramus altai* (IGM 100/1844), an extinct tyrannosaurid dinosaur (avian stem-group); (C) extinct unnamed troodontid (IGM 100/1126), paravian maniraptoran (avian stem-group); and (D) *Colius striatus* (AMNH 12378), extant crown-clade avian. AMNH, American Museum of Natural History, New York; IGM, Institute of Geology, Mongolian Academy of Sciences, Ulan Bator.

(Hurlburt et al., 2013). A number of vertebrate lineages have independently acquired a high BEC index under a variety of evolutionary and morphological contexts (Coates, 1999; Northcutt, 2002; see Balanoff et al., 2016a). A brief discussion of the anatomical factors that influence BEC values should increase our understanding of endocasts and their neuroanatomical implications.

It is likely no surprise that encephalized taxa—those expressing a high ratio of brain volume to body mass—also tend to enjoy large BEC values. Mammalia, particularly our own branch of the primate tree, is one such taxon (Jerison, 1975; Northcutt, 2002; Isler et al., 2008). Birds are another lineage of highly encephalized amniotes, with encephalization values (see later discussion) rivaling those of most mammals (Figs. 3.3 and 3.4). The BEC values of birds may actually exceed those of mammals (Iwaniuk and Nelson, 2002), making birds a particularly attractive group for integrating endocasts with other forms of neuroscience data.

Moving outside of tetrapods, we find interesting relationships between BEC indices and encephalization.

Chondrichthyans express a relatively high level of encephalization (Northcutt, 1977, 2002), exceeding that of most tetrapods, but retain a plesiomorphically low BEC index (Kruska, 1988). Actinopterygians, a hugely diverse radiation of osteichthyan fish include numerous lineages that express high BEC indices in the absence of significant encephalization (Bjerring, 1991; Northcutt, 2002; Giles and Friedman, 2014). These bony fish demonstrate that the two logical pathways of increasing the BEC index—enlarging the brain within the endocranial cavity or shrinking the endocranial cavity around the brain—are both represented in the evolutionary history of vertebrates.

3.2.2 What Anatomical Structures Share the Endocranial Cavity With the Brain and Thus Lower Brain-to-Endocranial Cavity Values?

The external surface of the brain and the deep surface of the bony and/or cartilaginous endocranial cavity are

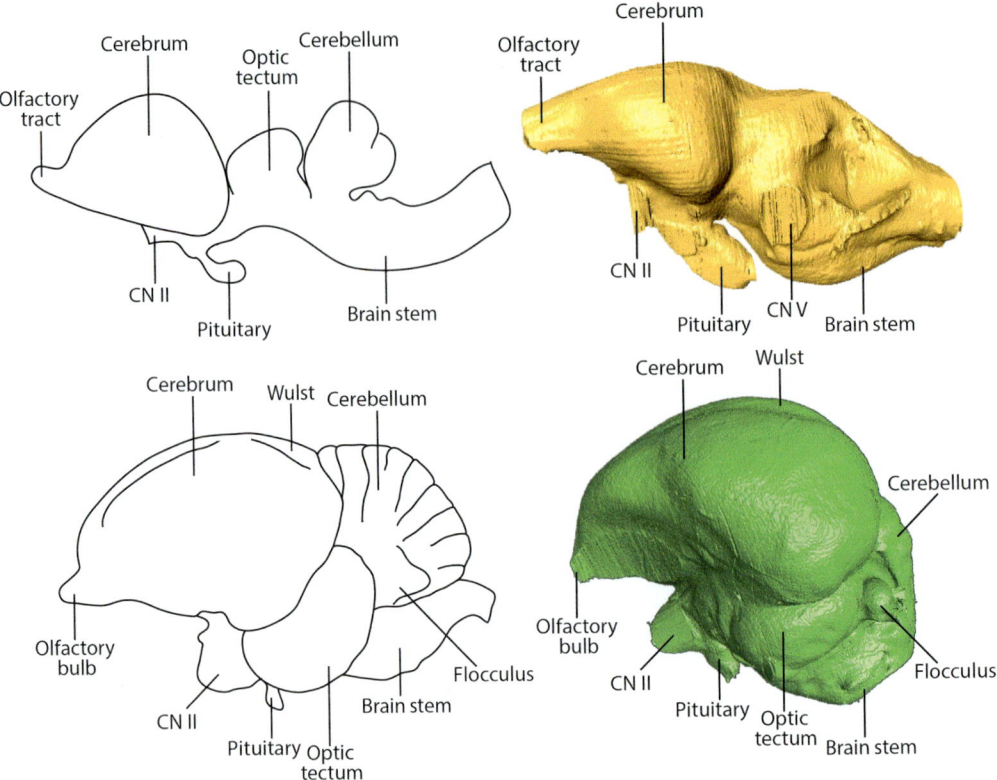

FIGURE 3.5   Comparison between the brain (*line drawing*) and endocast (*digital rendering*) for the crown-group crocodilian *Paleosuchus trigonatus* and the crown-clade bird *Colius striatus* (in left lateral view). Note the strong brain-endocast fidelity and marked cerebral expansion of the bird relative to the crocodilian. This expansion was the primary driver of encephalization along the avian stem-group and likely responsible for the conversion from a more linear, anterior—posterior arrangement of the neuroanatomical regions (as expressed in the crocodilian) to the more s-shaped neuroarchitecture of crown-clade birds and their closest stem relatives. *CN*, cranial nerve.

not in direct contact but rather are consistently buffered by meningeal tissue. These tissues appear to have a rather complicated evolutionary history of differentiation across chordates that is far from well understood. It does seem clear that the plesiomorphic condition is a single, undifferentiated layer known as the primitive menix (Bjerring, 1991; Coates, 1999; Butler and Hodos, 2005). This layer divided, probably somewhere along the tetrapod stem lineage, to form a secondary menix (endomenix) and a more superficial layer of dura mater (Brocklehurst, 1979; Butler and Hodos, 2005). The secondary menix subsequently differentiated to form an internal pia and intermediate arachnoid layer, with cerebrospinal fluid filling the intervening, subarachnoid space. This differentiation is sometimes considered a convergently derived feature of crown-clade mammals and birds and possibly correlated with a homeothermic physiology (see Starck, 1979; Gauthier et al., 1988). At the same time, a third meningeal layer, often explicitly identified as the arachnoid layer, has been described for a diverse assemblage of tetrapods, including crocodilians (Hurlburt, 2014), turtles (Humphrey,

1894), and amphibians (Francis, 1934; Kuhlenbeck, 1973; Brocklehurst, 1979; Joven et al., 2013). It seems unlikely that if three layers are present across crown-clade Tetrapoda, these layers are not homologous (the null hypothesis is homology, with the burden of proof lying with a convergent origin; Hennig, 1966). Homology would not, however, negate the possibility that crown-clade mammals and birds do convergently share a derived transformation of some form, and this apomorphy is causally related to the maintenance of an elevated body temperature. The general conclusion that the transformational history of these tissues across Chordata is in need of further study seems a safe one.

The homology of the individual meningeal layers is of less importance for the stated goals of this chapter than their collective thickness. A systematic survey of vertebrate meningeal thickness has not been attempted, but clade-level variation certainly does exist. Crown-clade birds, for example, possess exceptionally thin meninges relative to other vertebrates (Iwaniuk and Nelson, 2002), helping to explain why avian BEC values may even exceed those of mammals (see previously mentioned). The relatively thick meninges of mammals sometimes obscure surface features of the brain such as cerebral gyri and sulci.

Dural infolding creates a system of intracranial sinuses that help to drain the metabolically active brain of its large quantity of venous blood (Fig. 3.6). These volumetric requirements can translate to relatively large individual sinuses that significantly influence endocast morphology and thus reduce the size and shape correspondence between endocast and brain (lowering BEC values). The vertebrate occipital sinus, for example, extends posteriorly along the sagittal midline of the cerebellum, separating that region from the deep surface of the overlying neurocranial roof. In crown-clade birds, the occipital sinus is rather narrow mediolaterally—likely a derived condition, considering the broadened occipital sinus of crocodilians and other nonavian reptiles (Goodrich, 1930). The phylogenetic stem of birds is witness to an inflation of the occipital sinus that produced a prominent dural peak projecting posterodorsally over the hindbrain and the cerebellar region of the endocast (Fig. 3.4B). This peak has a somewhat complicated history along the avian stem. It probably reaches its greatest development in the tyrannosaurids (Fig. 3.4B) and is variably present in the crownward and more birdlike, deinonychosaurs (Norell et al., 2009; Witmer and Ridgely, 2009; Bever et al., 2011, 2013). The functional significance of the dural peak is unclear, but the structure is not present in any avialan, including *Archaeopteryx* and the entire diversity of the crown radiation.

The occipital sinus of crown-group avians also exhibits interesting variation. For example, diving birds, such as penguins and loons, have a distinctively wide occipital sinus that obscures the details of their cerebellar folia on an endocast (Ksepka et al., 2012; Tambussi et al., 2015). The functional driver of this correlation between diving behavior and a wide occipital sinus is unclear; but if such a correlation does exist, then endocasts could be used in combination with other skeletal signatures (apomorphically thick limb-bone cortices, reduced pneumaticity, reduced sternum; Ksepka et al., 2006; Hinić-Frlog and Motani, 2010; Smith and Clarke, 2013) to identify this behavioral ecology in fossils and thus trace it through deep time. In a similar fashion, the endocast signatures of the sphenotemporal and rostral petrosal sinuses (Fig. 3.6) conceal details of the midbrain in crown and stem avians but at the same time serve as landmarks that allow the optic tectum to be identified and traced in the fossil record (Sedlmayr, 2002; see later discussion). The use of vasculature and other overlying structures to inform our understanding of the size and shape of brain regions is an area of active research and will surely become increasingly refined (Morhardt et al., 2012).

Although their influence on the size and shape of an endocast is not typically large, certain cranial nerves and arteries do course through the endocranial cavity and thus diminish BEC values (Fig. 3.6). The trigeminal ganglion, for example, is an endocranial structure that resides within a subtle excavation of the neurocranial wall—an excavation that is visible on the resultant endocast (Fig. 3.6). The complex of intracranial arteries that extend through the endocranial cavity, such as the internal carotid canals, circle of Willis, basilar arteries (Midtgård, 1984; Baumel, 1993), do not lie within their own excavations of the endocranial wall and thus do not have a specific identity on the endocast; but because they are present and lie between the brain and floor of the neurocranium, they are collectively contributing in a small way to endocast volume and morphology.

Another important predictor of brain-endocast correspondence is the structural architecture of the braincase itself. In many vertebrate lineages, the endocranial cavity is poorly delineated in bone, often reflecting a failure of the orbital cartilages to ossify. This is generally not a large problem for either crown- or stem-group avians, where these anterior cartilages ossify extensively as the laterosphenoid and orbitosphenoid bones. These bones largely close the rostral end of the endocranial cavity, leaving only well-delineated paths for olfactory tracts, cranial nerves II–V, and their associated vasculature. In contrast, in many crown-clade turtles the orbital cartilages and the medial wall of the inner ear both fail to

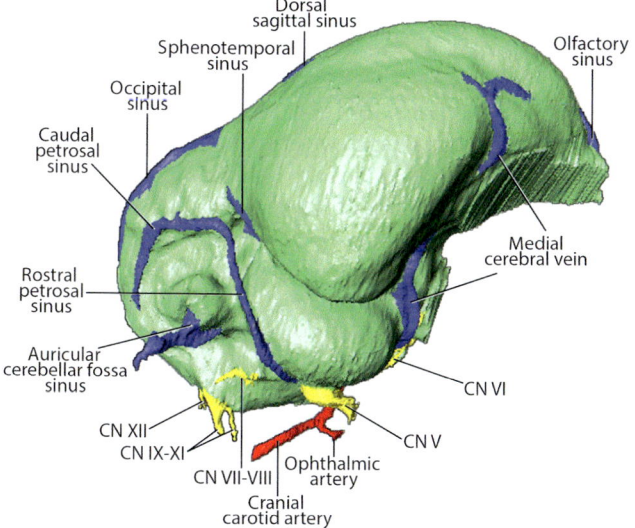

FIGURE 3.6 Digital endocast of the crown-group bird, *Tauraco erythrolophus*, this time showing selected sinuses of the dural venous system (*blue*), intracranial arteries (*red*), and cranial nerves (CN) (*yellow*). All intracranial structures that are not part of the brain proper effectively lower the brain-to-endocranial cavity index and decrease fidelity between brain and endocast. That stated, these "other" features often convey important neuroanatomical information.

ossify. This is a derived condition within reptiles that does not extend across the entire turtle stem lineage (Gaffney, 1990; Bhullar and Bever, 2009; Bever et al., 2015), but it is certainly a structural plan that presents an obstacle to assessing brain evolution in turtles using endocasts (Gaffney, 1977; Carabajal et al., 2013).

3.2.3 Partial Endocasts

The BEC index, defined as the percentage of the endocranial cavity filled by the brain, is an important metric for assessing the explanatory potential of endocasts whose distribution across the vertebrate tree is not well understood. The index, however, does not take into account the anatomical reality that the spatial relationship between the brain and the endocranial wall is not uniform but may vary widely between neuroanatomical regions. The cerebrum of crocodilians and early dinosaurs, for example, appears to fill most of its portion of the endocranial cavity, whereas the hindbrain of these taxa remains poorly defined (Fig. 3.4; Osmólska, 2004; Evans, 2005). The olfactory bulbs comprise another region of the nonavian dinosaur brain that appears to enjoy a strong correspondence with its expression on the endocast when compared to that of other regions. The fidelity of the olfactory bulb and cavity has even been used to investigate the evolution of olfactory acuity among nonavian dinosaurs (Zelenitsky et al., 2009, 2011; Sales and Schultz, 2014).

An elegant demonstration of the variable relationships between brain and endocast is found in a description of the brain and endocranial cavity of the Australian lungfish, *Neoceratodus forsteri* (Clement et al., 2015). The study maps the distance between the brain and endocranial wall and communicates the observed variation using a color gradient (analogous to a heat map). Their data demonstrate that the brain of this important lineage is a close fit with the endocast, except in the areas of the diencephalon and the hindbrain/anterior spinal cord. Such distances could easily be converted to 3D shape data and analyzed morphometrically to provide a clade-level perspective of these spatial relationships and their evolutionary history. The larger point here is that even if the overall BEC index of a taxon is relatively low, an endocast can still convey important morphological details for at least some brain regions.

Although endocasts are likely to reflect varying levels of neuroanatomical resolution for different brain regions, partial endocasts derived from specimens whose neurocranial anatomy is incomplete may serve important roles in constructing and testing evolutionary hypotheses. The vast majority of fossil endocast studies, not surprisingly, are based on well-preserved specimens whose braincases, and thus endocasts, are complete (or nearly so). Most fossils, however, are to varying degrees fragmentary. Simply excluding the majority of the fossil record from evolutionary analyses may leave significant phylogenetic gaps in our sampling, which can easily compromise the explanatory power of our results and interpretations (Heath et al., 2008).

Partial endocasts may be constructed from even a single bony element. The utility of such fragmentary endocasts is largely question dependent, but it is certainly governed by a combination of the neuroanatomical detail provided by that element and the phylogenetic resolution it provides. A fossil must be placed within a phylogenetic tree with some level of statistical support before the anatomical data provided by that fossil can meaningfully inform a tree-based evolutionary hypothesis (see Fig. 3.7). Considering that the number of phylogenetically informative features expressed by a specimen can be expected to decrease as a specimen becomes less and less complete, it must be recognized that fragmentary specimens are not likely to support a taxonomically exclusive identification (eg, to the species or "genus" level) (Bever, 2005; Bell et al., 2010; Bever et al., 2009). It should also be recognized that a fossil may meaningfully inform a macroevolutionary pattern without meeting the requirements of such refined taxonomic resolution. The question being pursued will generally dictate the required level of resolution. The most important thing is that both paleontologists and neuroscientists are aware of where on the tree a fossil endocast can be placed with some confidence and whether that position can support the hypothesis of interest.

Birds provide a helpful example of how partial endocasts might inform the evolutionary history of a profoundly interesting and important neuroanatomical structure. The avian Wulst is a thickening of the cerebral hyperpallium that functions as a processing center for a diverse array of sensory input (Medina and Reiner, 2000; Jarvis, 2009; Jarvis et al., 2013; Reiner et al., 2005; Butler et al., 2011). Anteriorly, the Wulst organizes somatosensory data from several regions, including the body feathers, beak, and feet (Wild, 1987; Funke, 1989; Wild et al., 2008; Cunningham et al., 2013). Visual input is processed largely in the structure's posterior aspect (Reiner et al., 2005). Because a Wulst has yet to be identified in any other extant reptile clade, we can infer that its evolutionary origin lies somewhere along the avian stem lineage.

Birds are an extant lineage of theropod dinosaurs (Ostrom, 1976; Gauthier, 1986). The realization that not all dinosaurs (Fig. 3.7) went extinct at the end of the Cretaceous sparked an exciting and fruitful reconsideration of the modern avian body plan (see Dingus and Rowe, 1997). Features we formerly associated exclusively with birds—feathers, wings, wishbones (furcula),

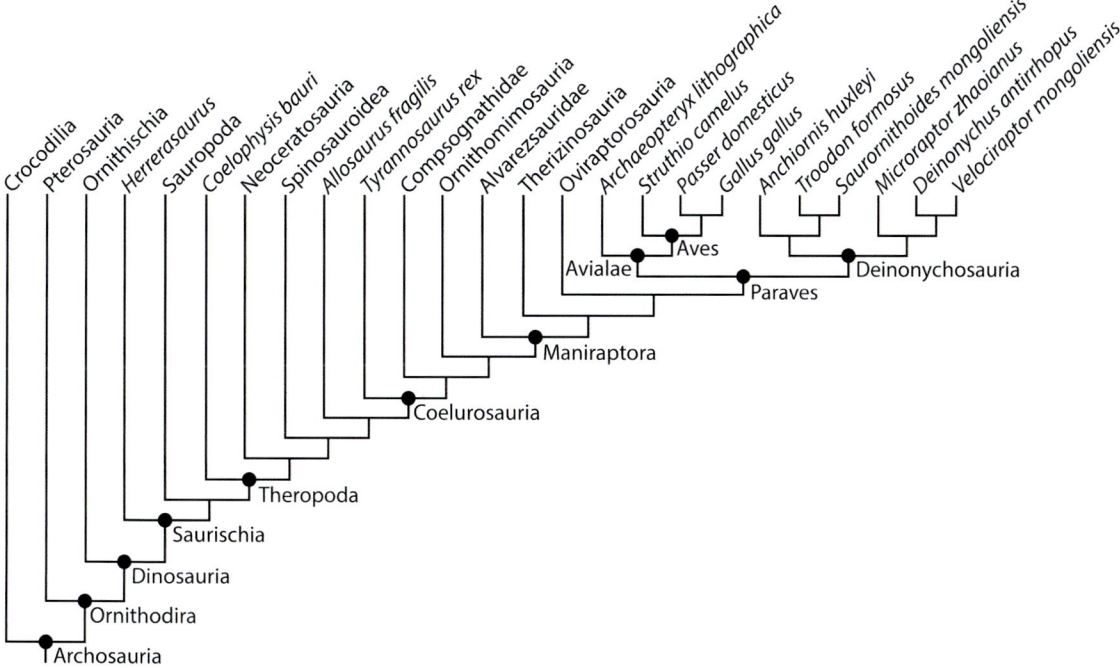

FIGURE 3.7 Phylogenetic tree of crown-clade Archosauria showing selected lineages of the hugely diverse phylogenetic stem of birds. The majority of this stem is nested within the Dinosauria, although the stem also includes nondinosaurian lineages such as the highly encephalized and volant pterosaurs. See Smith et al. (2007), Nesbitt (2011) and Turner et al. (2012) for details of tree topology.

pneumatization of the skeleton, egg-brooding behaviors, and almost undoubtedly a homeothermic physiology—now find their origins deep within nonavian dinosaurs (Ostrom, 1976; Norell et al., 1995; Dong and Currie, 1996; Xu and Norell, 2004; Norell and Xu, 2005; Turner et al., 2007; Nesbitt et al., 2009). Some form of functional bipedalism, a derived feature of birds among extant reptiles, may actually have originated along the archosaur stem lineage, with the obligatory quadrupedal gait of modern crocodilians being secondarily evolved (Hutchinson, 2006; Gauthier et al., 2011). Based on this pattern, we might well expect to find that the apomorphic thickening of the hyperpallium into a proper avian Wulst also occurred well outside of the crown-group radiation of birds.

The first century or so of endocast research, however, produced no records of a Wulst outside of the crown clade (Marsh, 1880; Edinger, 1951; Osborn, 1916; Milner and Walsh, 2009; Witmer and Ridgely, 2009; Balanoff et al., 2014; Walsh et al., 2016. This empirical absence suggests the Wulst evolved right at, or just before, the origin of the avian crown group, providing modern birds with a competitive advantage not enjoyed by their dinosaurian counterparts on the stem (Fig. 3.7). The Wulst presumably allowed crown birds to exploit the vast array of aerial and nonaerial ecologies and to achieve the impressive levels of taxonomic diversity that they currently exhibit (Milner and Walsh, 2009).

This hypothesis has been twice challenged. The first time was in a Ph.D. thesis (Franzosa, 2004) based on the endocast of a troodontid deinonychosaur from the Cretaceous of Mongolia that has yet to be fully described and named. The second challenge was a tentative identification of a Wulst-like structure in the endocast of the early avialan *Archaeopteryx lithographica* (Fig. 3.7; Balanoff et al., 2013). In both cases, the evidence consists of a ridge running along the deep surface of the frontal bone in an area of the endocranial cavity that corresponds to the dorsolateral surface of the cerebrum. The potential Wulst of the Mongolian troodontid has yet to receive much attention or scrutiny, but this will surely come once its description is formally published. The potential Wulst of *Archaeopteryx* was recently re-examined using newly acquired synchrotron data (Beyrand et al., 2019) that are of higher resolution than those available to Balanoff et al., (2013). These data reveal that the ridge-in-question is actually a fracture of the frontal bone, which indirectly supports the traditional hypothesis that the Wulst is a feature unique to the crown clade.

Establishing homology between the aforementioned cerebral structure in the Mongolian troodontid and the avian Wulst requires demonstrating, to some reasonably acceptable level, a continuity of Wulst expression along the backbone of the avian stem lineage—between the ancestral crown-clade bird and that bird's most recent common ancestor with with other paravians. Another, less direct, way of establishing a reasonable probability that the cerebral architecture of the troodontid shares some sort of Wulst with the crown condition is to establish that a Wulst-like structure is more widespread among the nonavian, but extremely birdlike, coelurosaurian theropod dinosaurs (Fig. 3.7). Unfortunately, well-preserved braincases are not abundant in either area of the tree. Most avialan specimens, outside of the crown radiation, are either highly fragmentary or exhibit severe compressional distortion (O'Connor and Chiappe, 2011; Turner et al., 2012). The fossil record of nonavialan coelurosaurs is more promising, but even here, most of the known braincases exhibit some significant level of damage (especially crownward of tyrannosaurs, Fig. 3.7; eg, Makovicky et al., 2003; Balanoff et al., 2009, 2014; Bever et al., 2011, 2013; Turner et al., 2012).

A specimen of the ornithomimosaur *Gallimimus bullatus* (IGM 100/133) provides a nice example of our intended point. The braincase of this specimen is poorly preserved overall, especially because its floor and lateral walls are broken and distorted. In contrast, the dermal roof of the endocranial cavity is nicely preserved and yields a partial digital endocast that appears to closely approximate the morphology of the brain's dorsal surface (Fig. 3.8). The cerebral portion of this surface bears no signature of the Wulst. Therefore, the controversial record of the Wulst-like structure in *Archaeopteryx* remains the only one outside of the crown clade, at least for now. The ability to identify the Wulst using a single bone (the frontal)—as opposed to an entire braincase—greatly increases the probability that the fossil record will inform the potentially complex evolutionary origin of this characteristic feature of avian neuroanatomy.

3.3 Endocast Contributions to Comparative Neuroscience

Endocasts are windows into the deep history of neuroanatomy and as such provide modern neuroscience a more complete appreciation of: (1) the brain's evolutionary potential (by allowing sampling of extinct lineages) and (2) the origins of modern neurological disparity. Endocast research, however, remains very much in its descriptive phase, with the vast majority of studies focused on the neuroanatomical details of a single fossil taxon (eg, Brochu, 2000; Franzosa and Rowe, 2005; Kundrát, 2007; Bever et al., 2011; Carabajal and Succar, 2013; Balanoff et al., 2014) or perhaps the evolutionary patterns within a taxonomically restricted clade (eg, Witmer and Ridgely, 2009; Balanoff et al., 2010; Ksepka et al., 2012; Carril et al., 2016). These descriptive studies are certainly advancing our general understanding of the brain's evolutionary potential as well as its constraints. For example, the huge hypophysial complex of sauropod dinosaurs informs the relationship between gigantism and the pituitary gland (Edinger, 1942; Balanoff et al., 2010).

In terms of modern neurological disparity, there have been exceedingly few attempts, outside of Harry Jerison's groundbreaking contributions, to generate and test endocast-based hypotheses that target the extensive stem lineages of major vertebrate crown clades and the origin of those clades' often highly derived neuroanatomical configurations. Perhaps the best example is Rowe et al. (2011). This study determined that the brain of crown-clade mammals is the product of at least three significant pulses of encephalization that are strongly tied to enhanced olfaction, tactile sensitivity, and neuromuscular coordination. In this section, we will explore the analytical role of endocasts in broad-based

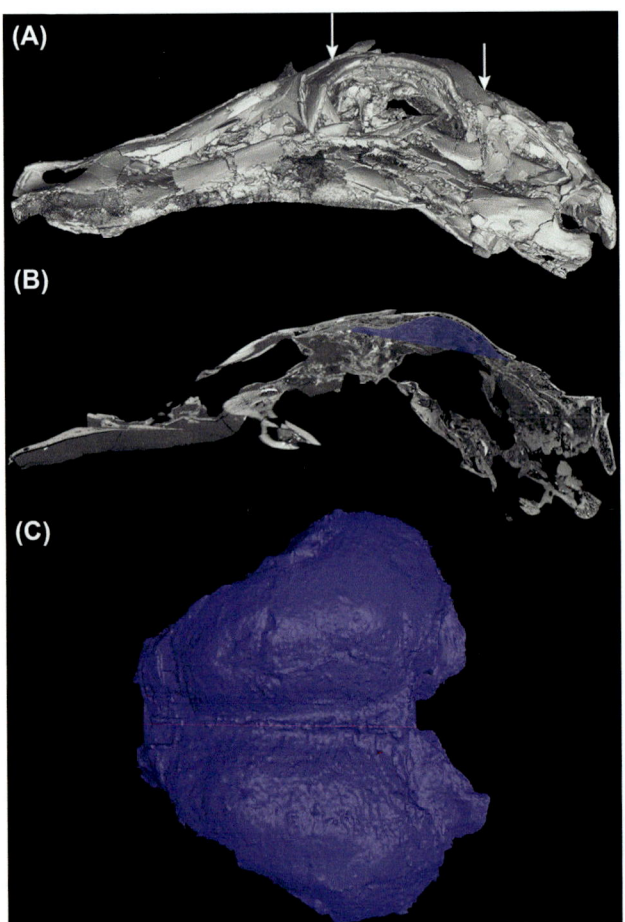

FIGURE 3.8 (A) Digital 3D rendering of the skull of the extinct ornithomimid dinosaur *Gallimimus bullatus* (IGM 100/1133) with *arrows* indicating the anterior and posterior limits of the endocast. (B) Sagittal CT slice revealing the position of a partial cerebral endocast (*blue*). (C) Partial endocast generated from the frontal and parietal bones showing the dorsal surface of the cerebrum. The rendering of this surface exhibits no trace of an avian-like Wulst. IGM, Institute of Geology, Mongolian Academy of Sciences, Ulan Bator.

comparative neuroscience using our relatively scanty understanding of neuroanatomical evolution along the avian stem lineage as a general organizational tool. The origin of the highly encephalized, s-shaped brain of crown-clade birds is almost undoubtedly the product of numerous, but perhaps highly correlated, transformations (Figs. 3.4 and 3.5). While recognizing their probable nonindependence, we are going to divide and discuss these transformations under the individual headings of comparative morphology, encephalization, and correlative change.

3.3.1 Comparative Morphology

Descriptive neuroanatomy is perhaps the most intuitive use of endocasts, especially in reference to their potential for illuminating transformations otherwise concealed within stem lineages (eg, Alonso et al., 2004; Franzosa, 2004; Sampson and Witmer, 2007; Ashwell and Scofield, 2008; Witmer and Ridgely, 2009; Balanoff et al., 2010, 2013, 2014, 2016b; Bever et al., 2011, 2013; Lautenschlager et al., 2012; Ksepka et al., 2012; Kawabe et al., 2013a,b, 2015; Lautenschlager and Hübner, 2013) and/or inferring behavior in the fossil record (Hopson, 1977, 1979; Brochu, 2000; Witmer et al., 2003; Sereno et al., 2007; Witmer and Ridgely, 2009; Walsh and Milner, 2011; Walsh et al., 2013; Marek et al., 2015). Given the relatively strong correspondence between their brain and endocast (Fig. 3.5), it is perhaps not surprising that mammals, birds, and their respective stem lineages have enjoyed the majority of research attention from endocast workers (see Balanoff et al., 2016a for a more complete list). Attention to those vertebrate lineages lying phylogenetically outside of Amniota has steadily increased in recent years. Many of these anamniote data are concentrated in the chondrichthyan crown and stem (Schaeffer, 1981; Maisey, 2004, 2005, 2007, 2011; Pradel et al., 2009; Lane, 2010; Pradel, 2010), but endocast research on actinopterygians (Coates, 1999; Hamel and Poplin, 2008; Giles and Friedman, 2014; Giles et al., 2015a), dipnoan sarcopterygians (Säve-Söderbergh, 1952; Campbell and Barwick, 1982; Challands, 2015; Clément and Ahlberg, 2010; Clement and Ahlberg, 2014; Clement et al., 2015), and both stem (Romer, 1937; Stensiö, 1963; Jarvik, 1972; Chang, 1982; Johanson et al., 2003; Snitting, 2008; Lu et al., 2012; Holland, 2014) and crown tetrapods (eg, stem and crown amphibians; Romer and Edinger, 1942; Maddin et al., 2013) are all steadily increasing.

The comparative biology of early vertebrate fossils remains cryptic and contentious in many respects, including their phylogenetic relationships with extant lineages (Brazeau, 2009; Davis et al., 2012; Brazeau and

Friedman, 2014; Giles et al., 2015b). That stated, this area of the tree has produced a surprising wealth of endocast data (eg, Stensiö, 1925; Janvier, 1981, 1985, 1996, 2008; Chen et al., 1999; Gai et al., 2011). These endocasts are often lacking in surface details, but they still provide important insights into the general shape and extent of individual neuroanatomical regions, as well as the size and distribution of cranial nerves and vessels during what was a critical early history of endocranial evolution (Janvier, 2008). The most recent endocast data flowing from this part of tree are being facilitated by significant improvements to high-resolution CT and increased access to synchrotron X-ray tomographic data. Synchrotron data have proven especially helpful for delineating the lightly ossified and mineralized cartilaginous skeletons of important early vertebrate and gnathostome taxa from the highly indurated matrix in which they tend to be encased (Tafforeau et al., 2006; Sutton, 2008; Cunningham et al., 2014; Dupret et al., 2014; Rahman and Smith, 2014). Access to the endocranial details of these early lineages holds great promise for elucidating the basal transformations that continue to influence patterns of neuroanatomical evolution across the vertebrate tree.

Returning to the reptile side of the amniote radiation and the expansive evolutionary history of the avian stem lineage (Fig. 3.7), we find compelling transformations that help fill empirical gaps in our understanding of archosaur neuroanatomy and thus shed light on the highly derived structural organization characterizing the crown-avian brain. For example, the plesiomorphic condition for reptiles is an elongate brain whose major neuroanatomical regions are arranged in an almost linear, anterior–posterior fashion. This morphology is conserved across much of the reptile crown clade, including the extant sister taxon of birds—the crocodilians (Fig. 3.5; see Hopson, 1979; Butler and Hodos, 2005; George and Holliday, 2013; Ngwenya et al., 2013). Birds, in contrast, exhibit a derived neural architecture in which the expanded cerebrum and cerebellum are pushed into broad contact, obscuring the pineal gland and third ventricle from dorsal view and displacing the optic tectum into a more ventrolateral position (Cohen and Karten, 1974; Hopson, 1979). This major neuroanatomical rearrangement gives the modern bird brain a sinusoidal shape (characterized by two distinct flexure zones: the cephalic flexure between the fore- and midbrain and the pontine flexure within the brain stem between the mid- and hindbrain; Hopson, 1979) (Fig. 3.5). The rearrangement also creates a significant morphological gap between the conditions expressed in the avian and crocodilian crown clades. The origin of this disparity can be inferred to lie along the avian stem lineage, although the timing, tempo, and nature of those rearrangements will remain out-of-the-reach of comparative neuroscience unless the fossil record can help illuminate this deep history.

A linear arrangement of the major neuroanatomical regions is expressed in both ornithischian dinosaurs and the early divergences within Saurischia (sauropodomorphs and early theropods) suggesting that the basic, plesiomorphic architecture of the reptile brain is conserved along the backbone of the avian stem lineage (Galton, 1985, 1988; Giffin, 1989; Galton and Knoll, 2006; Zhou et al., 2007; Evans et al., 2009; Carabajal and Succar, 2013; Lauters et al., 2013; Cruzado-Caballero et al., 2015). There are derivations on this basic plan nested within the major extinct radiations of the avian stem (see Balanoff et al., 2010; Miyashita et al., 2011; Carabajal, 2012; Lautenschlager and Hübner, 2013). Perhaps the most dramatic example of an apomorphic rearrangement of this linear brain is found among the pterosaurs. Pterosaurs lie completely outside of Dinosauria phylogenetically, making them one of the earliest divergences along the avian stem lineage (Fig. 3.7; Nesbitt, 2011). Just as pterosaurs evolved the ability to perform powered flight independently of birds, they also evolved a highly derived neural architecture that includes a birdlike expansion of the cerebrum and overall s-shaped morphology (see Witmer et al., 2003). The independent acquisition of these avian features suggests a highly positive correlation, and possibly causative relationship, between the origin of archosaurian powered flight and the neural expansion and geometric rearrangement of the brain expressed in these two groups (but see later discussion).

The derived transformation responsible for the modern s-shaped brain does not appear in the dinosaurian fossil record until deep within the history of theropods, at the origin of Maniraptora (Figs. 3.4 and 3.7). The earliest history of the maniraptoran sinusoidal brain may be marked by a taxonomically sporadic (ie, homoplastic) pattern of expression (AM Balanoff, personal observation), but the variability in the system was apparently reduced by the origin of Paraves where the morphology is consistently expressed (Fig. 3.4C and D; Alonso et al., 2004; Balanoff et al., 2009, 2013, 2014; Norell et al., 2009; Witmer and Ridgely, 2009; Walsh et al., 2016). Although a relatively late acquisition, the s-shaped brain is still at least 160 million years old (ie, appearing 90 million years prior to the origin of the avian crown clade; Turner et al., 2012; Brusatte et al., 2014; Prum et al., 2015).

3.3.2 Encephalization

Endocasts provide a close approximation of volumetric brain size in groups with a high BEC index (Haight and Nelson, 1987; De Miguel and Henneberg, 1998; Iwaniuk and Nelson, 2002; Isler et al., 2008) and a maximum brain size in those with low BEC values. Endocranial volumes, including those from fossil taxa, have therefore been employed in numerous analyses to assess encephalization dynamics through time (eg, Jerison, 1969, 1973, 1977; Hopson, 1977, 1979; Larsson et al., 2000; Alonso et al., 2004; Franzosa, 2004; Ashwell and Scofield, 2008; Lautenschlager et al., 2012; Balanoff et al., 2013). Encephalization, in its most simplistic conceptualization, is brain size relative to body size (Jerison, 1977) and has been thought, with some amount of controversy, to correlate with cognitive ability or other measures of "intelligence" such as innovation rate—the rate at which novel behaviors or techniques are acquired (Jerison, 1977; see Healy and Rowe, 2007; Lefebvre and Sol, 2008; Overington et al., 2009; Lefebvre, 2013 for overviews).

The most commonly employed metric is the encephalization quotient (EQ). The EQ was first proposed by Jerison (1973) as the ratio between actual brain size (described as either a mass or volume estimate) and expected brain size and was designed to remove the allometric effects of body size on brain-size assessments. Body size remains an inherent factor in the index given that expected brain size reflects the correlative relationship between brain and body size for a given taxonomic group (Jerison, 1973). Expected brain-size estimations must therefore be calculated for individual clades without the expectation that this relationship will remain constant for any area of phylogenetic tree space. EQ is subject to the same general sources of variability and estimation error that underlie all evolvable traits, thus making its meaningful assessment in extinct taxa difficult (see Hurlburt, 1996; Hurlburt et al., 2013). Perhaps the most daunting limitation of the EQ is the notorious difficulty of estimating body mass in fossils. Alleviating these difficulties is an active area of research, which is not surprising considering that body mass is an important variable for a wide assortment of morphological, physiological, and paleoecological considerations (eg, Christiansen and Fariña, 2004; Finarelli and Flynn, 2006; De Esteban-Trivigno et al., 2008; Campione and Evans, 2012; Field et al., 2013). An important step forward is the recognition that every skeletal measurement correlates with body mass at some level and can be used as long as the associated error is explicitly considered (Field et al., 2013).

Attempts have been made to refine the efficacy of using living taxa for estimating the expected brain size within extinct stem taxa. Hurlburt et al. (2013) used variation in the BEC index of extant crocodilians to predict EQ at the base of Dinosauria. These advances may be rendered moot as the utility of the EQ is now being seriously questioned. A myriad of other cognitive proxies, including residual brain size, residual cerebral size, and absolute brain size, may all outperform EQ—at least for certain taxonomic groups (Deaner et al., 2007; Lefebvre and Sol, 2008; Lefebvre, 2013). Each of these estimators can be derived from volumetric measurements thus increasing the explanatory potential of endocasts. Furthermore, digital technologies have facilitated the partitioning of endocasts into functional neuroanatomical regions (Walsh and Milner, 2011, 2013; Balanoff et al., 2013, 2016b; Kawabe et al., 2013b). Such partitioning allows us to pursue more specific questions and makes us less reliant on both total brain values and the vagaries of body size estimation. EQ is still used, but its days in the analytical sun may well be numbered (see Healy and Rowe, 2007; Deaner et al., 2007).

Our understanding of avian encephalization is still largely grounded in the seminal endocast studies of Jerison (1968, 1969, 1973) and Hopson (1977, 1979). These studies examined encephalization as the relationship between endocranial volume and body mass, and in doing so found distinct differences between birds and a paraphyletic reptile group. This reptile assemblage included crown-group turtles, lepidosaurs, and crocodilians, as well as a small series of ornithischian, sauropod, and nonavialan theropod dinosaurs. *Archaeopteryx* was recovered as uniquely transitional between the "reptile" and avian morphospaces suggesting that a pulse of encephalization accompanied the origin of flight.

These studies assumed that "dinosaurs" conserved the relatively low BEC index of *Sphenodon* (0.5), which is undoubtedly not the case, especially as sampling moves progressively crownward (Osmólska, 2004; Evans, 2005; Hurlburt et al., 2013). Combine this point with the fact that these influential analyses could not consider the rich diversity of recently discovered fossils (Fig. 3.7), and one might suspect that the birdlike qualities of the *Archaeopteryx* brain may well characterize a wider array of dinosaurian forms (see Fig. 3.4C and D). If a larger, more avian-like brain is present in other theropods, then its relationship with the origin of flight becomes less clear.

The hypotheses of Professors Jerison and Hopson went uncontested for nearly a quarter century. This period of stagnation reflects the highly comprehensive nature of these early studies, even though the data available during this period were largely restricted to naturally occurring endocasts—a restriction that placed severe limitations on sampling and hypothesis testing. The appearance and rapid spread of nondestructive digital technologies for building and studying endocasts both loosened these sampling constraints and increased the rigor of endocast-based comparative analyses. Digital methods allow endocasts to be generated in a more standardized way (eg, with or without such structures as cranial neurovasculature) that increases the accuracy and precision of their shape and volumetric assessment (see Balanoff et al., 2016a).

The comparative studies emerging as part of the digital era of endocast research have, in many respects, confirmed the early hypothesis that crown-clade birds express apomorphically large brains and that *Archaeopteryx* is an important transitional form. Its degree of encephalization is intermediate between those of modern birds and most reptile clades—including most of the nonavian dinosaur groups (eg, Larsson et al., 2000; Alonso et al., 2004; but see Hurlburt et al., 2013, which found the EQ of *Archaeopteryx* to overlap with the lower range of encephalization expressed by crown-clade birds). The Larsson et al. study drew the important conclusion that encephalization along the avian stem was driven largely by cerebral expansion and that this expansion began deep within theropod history (at least by the time the tyrannosaurs diverged from the backbone of the avian stem lineage; Fig. 3.7). From this conclusion, they made the correlative prediction that more dinosaurs enjoyed encephalization levels comparable to that of *Archaeopteryx* (Fig. 3.9); unfortunately, they did not have the data to test their own hypothesis.

The crownward part of theropod history is just now beginning to be sampled with some density (eg, Lautenschlager et al., 2012; Balanoff et al., 2013, 2016b). The patterns emerging from these studies support the predictions of Larsson et al. (2000) (Fig. 3.9). The data of Balanoff et al. (2013) clearly indicate that cerebral expansion was the primary driver of increasing endocranial volume along the avian stem lineage. These data also suggest that this expansion was not continuous but appeared in pulses (Balanoff et al., 2013). More work is required to determine how much of this pattern reflects evolutionary reality as opposed to sampling artifact (Balanoff et al., 2016b). The limiting factor to testing such patterns is no longer access to the anatomy but rather the time and effort of researchers. This new reality bodes well for the future of encephalization studies that utilize the fossil record.

3.3.3 Correlative Change

One of the most important yet formidable tasks of the comparative neuroscientist is to address questions on how the evolution of the brain and its gross morphology relates to the historical patterns of other biological systems. For example, how does the size and shape of a particular brain region correspond to the functions that the region serves (eg, Iwaniuk et al., 1999, 2000)? How does encephalization impact the morphology and functional anatomy of other cranial modules (eg, Lieberman et al., 2008)? Does the brain play a significant role in patterning other cranial tissues during development through the production of signaling proteins, and how has this developmental dynamic evolved to impact cranial diversity across the tree (eg, Marcucio et al., 2011)?

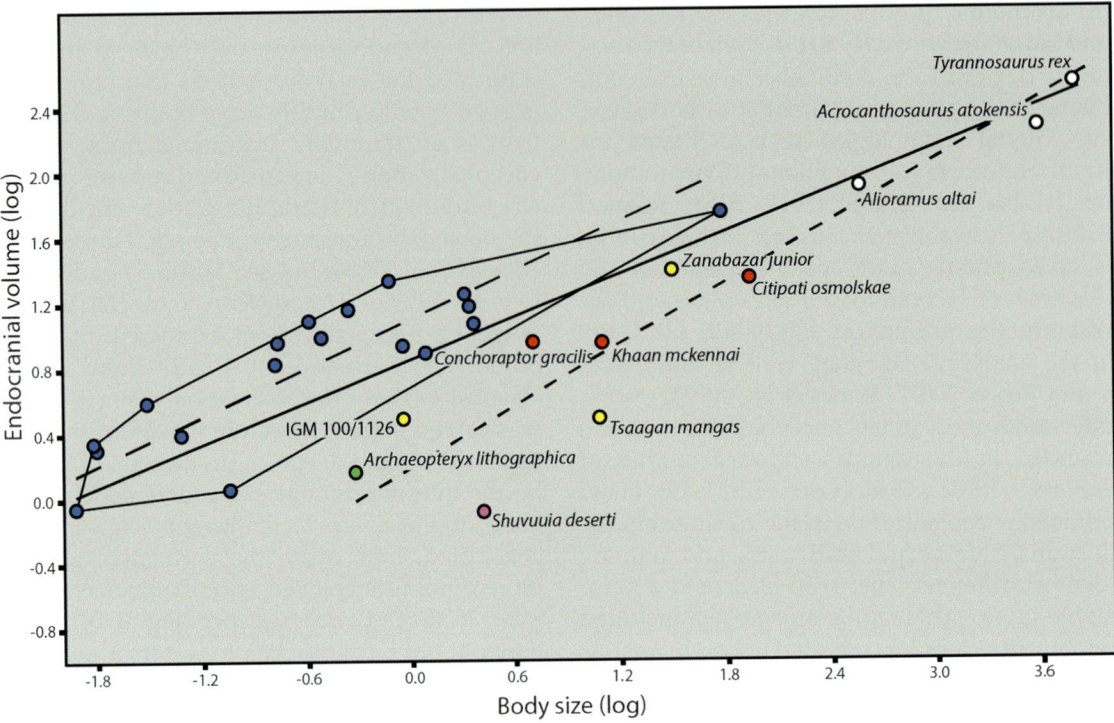

FIGURE 3.9 Bivariate plot of log-transformed body-mass data. Body mass (kg) plotted against total endocranial volume (cm³). Colors indicate crown-clade birds (*blue*), nonmaniraptoran theropods (*white*), *Shuvuuia deserti* (*purple*), oviraptorosaurs (*red*), deinonychosaurs (*yellow*), and *Archaeopteryx lithographica* (*green*). Reduced major-axis regression line for entire sample (*solid line*), crown birds (*large dashes*), and nonavian theropods (*small dashes*). Crown-clade birds display apomorphically high endocranial and cerebral volumes with respect to body size, but *Archaeopteryx* falls within the distribution of other nonavian maniraptoran dinosaurs. *Balanoff, A.M., Bever, G.S., Rowe, T.B., Norell, M.A., 2013. Evolutionary origins of the avian brain. Nature (Nature Publishing Group) 501 (7465), 93–96. https://doi.org/10.1038/nature12424.*

How does the transformational history of the brain correspond to other macroevolutionary patterns of taxonomic diversity (eg, Flinn et al., 2005)?

Here the ability of fossil endocasts to serve as empirical windows into the history of a phylogenetic stem may be especially valuable. To take full advantage of the insights that endocasts can provide and to understand the broader implications of brain evolution, we have to move away from treating endocast data as abstract shapes and sizes. This integrative process is certainly underway, and there are many examples that we could provide to demonstrate this point. To explore a few of them, let us once again return to avian cerebral expansion.

Cerebral hyperinflation is the primary driver of encephalization in the deep history of birds and likely initiated the geometric conversion of the plesiomorphic linear brain to the apomorphic sinusoidal brain (Fig. 3.5; Balanoff et al., 2013, 2016b). Such a dramatic neuroanatomical reconfiguration did not evolve in an anatomical vacuum, but is correlated, at least in part, to an equally impressive transformation of cranial morphology as a whole. Expansion of the cranial vault and overlying dermal roof, enlargement and repositioning of the orbits, and shortening of the facial skeleton (Brusatte et al., 2012; Bhullar et al., 2012) are all stem transformations within

maniraptoran dinosaurs that we now associate with the avian cranial form. This form is distinctive when compared to that of adult crocodilians and early dinosaurs, yet remarkably similar to the juveniles of those same taxa (Bever and Norell, 2009; Bhullar et al., 2012). It now appears likely that these birdlike features of the cranial skeleton are the products of multiple, global shifts in developmental timing (paedomorphosis) (Bhullar et al., 2012), at least some of which are temporally congruent with the neuroanatomical transformations that produced the avian s-shaped brain (Balanoff et al., 2013).

Phylogenetic congruence raises the question of whether the cranial and neuroanatomical transformations are linked in some deeper, perhaps causative manner. Among extant birds, variation in orbit size and shape is correlated with dramatic shifts in endocranial morphology, including a more globular brain and braincase (Kawabe et al., 2013b). A large endocranial volume relative to the length of the cranial base has also been associated with a more globular brain (Marugán-Lobón and Buscalioni, 2003). Whether these correlations reflect geometrically imposed constraints on morphospace or perhaps are the phenotypic products of some shared, deeply conserved linkage in their developmental pathways is as yet unclear. Certainly, these types of broadly integrative questions are of critical importance to the morphologist, paleontologist, and comparative neuroscientist and cannot be answered without input from each discipline (Rowe, 1996; Rowe et al., 2011; Werneburg et al., 2014).

Another correlative relationship that may be informed by the fossil record is the one uniting neuro-anatomy and behavior. For example, the discovery that the brain of *Archaeopteryx* corresponds closely in both shape and size to that of at least some other nonavian theropod dinosaurs (Fig. 3.9; Balanoff et al., 2013) suggests that we may have to rethink the relationship between the origins of avian encephalization and flight. If we accept that *Archaeopteryx* could fly in some capacity (Gatesy and Dial, 1996; Burgers and Chiappe, 1999; Wang et al., 2012), then we must also logically accept that the brain of *Archaeopteryx* had the processing power and cognitive range necessary to support that behavior in an amniote (these neural requirements may differ significantly from those that flight incurs, for example, in arthropods). If the brain of *Archaeopteryx* was "flight ready," then volumetrically speaking, so were the brains of these other apparently nonvolant dinosaurs (Fig. 3.9). This point, in turn, supports the conclusion that avian encephalization was an exaptation for flight rather than an adaptation that coevolved with this highly successful avian locomotory and ecological strategy (Balanoff et al., 2013, 2016b). In fact, flight explains almost none of the observed variation in total endocranial volume for birds and their stem relatives (Balanoff et al., 2016b).

There is a theoretical basis for employing brain size (ie, the amount of neural tissue present), and the size of its constituent regions, as a proxy for processing power and behavioral potential. The hypothesis is formally termed the Principle of Proper Mass (Jerison, 1973), and it is often invoked in studies that attempt to use size measurements drawn from endocasts to support conclusions regarding activity of function and behavior in extinct taxa (Jerison, 1973; Witmer et al., 2003; Zelenitsky et al., 2009, 2011; Lautenschlager et al., 2012; Sales and Schultz, 2014). Even if we set aside for a moment the difficulties of measuring cognitive ability, the relationship between the size of a neural region and the intensity of its use is almost assuredly more nuanced than is apparent from our interpretations (see Healy and Rowe, 2007). The shape of a neural region, for example, appears to be an important arbiter of processing speed and thus should also be explicitly considered when attempting to relate endocasts to behavior. This point was demonstrated elegantly in primates whose globular-shaped brain serves to decrease average connection length. A shorter connection length in turn facilitates more efficient processing and faster communication (see Sepulcre et al., 2010; Smaers and Soligo, 2013). Neuronal density may

outcompete both size and shape as a reflection of a region's intensity of use (see Olkowicz et al., 2016). Although such considerations may complicate our ability to use volumetric data from endocasts for inferring behavior in the fossil record, they also open new pathways for constructing and testing evolutionary and functional hypotheses—thus contributing to the integrative potential of endocast research.

3.4 Concluding Remarks

The study of endocasts was pioneered well over a century ago (eg, Marsh, 1880; Osborn, 1916; Edinger, 1925, 1951, 1975). For most of this history, it is difficult to argue that endocast data had more than a subtle influence on paleobiology or comparative neuroscience. The muted impact of such interesting data largely reflects the severe logistical constraints that once plagued the discipline. A reliance on the chance discoveries of natural endocasts, the suboptimal resolution of latex or other physically generated endocasts, and the destructive nature of traditional serial sectioning meant that the sampling benefits afforded the comparative biologist by the fossil record could not be fully realized for endocasts. These difficulties have now been largely overcome through digital technologies (Carlson et al., 2003; Witmer et al., 2008; Balanoff et al., 2016a). Endocasts are actively being constructed for the best specimens in all the major vertebrate clades. Including an endocast with the description of a newly discovered fossil skull is rapidly becoming the standard operating procedure.

The limiting factor in this golden age of endocast research is not whether the fossil record will produce the data we need to adequately document neuroanatomical transformations in deep time. What ultimately limits research is the degree to which we can correlate endocast morphology with the 3D complexity and modularity of the brain itself (Boire and Baron, 1994; Barton and Harvey, 2000; Iwaniuk et al., 2004; Iwaniuk and Hurd, 2005; Smaers and Soligo, 2013). Attempts to rigorously partition the endocast into neurologically meaningful volumetric regions are a step in the right direction. Although the quantitative relationship between the bony landmarks and the neuroanatomy require verification in a broader sample of lineages, the approach appears useful in tracing the modular nature of the brain in deep time (Balanoff et al., 2013, 2016b).

Volumes will continue to be an important aspect of comparative endocast and brain studies. There are compelling recent data, however, indicating that internal architectural rearrangements may be a more efficient mechanism for increasing processing efficacy and speed than simply increasing the volume of a neuroanatomical region (Sepulcre et al., 2010; Smaers and Soligo, 2013). If this is the case, then shape variation may be as important a predictor of that internal architecture and its neurophysiological implications as volume—a hypothesis that is especially intriguing given the remarkable level of shape variation that vertebrate endocasts appear to possess. The analytical potential of establishing these types of correlative relationships make the study of endocasts fertile ground for the increasingly sophisticated morphometric tools of shape analysis (Kawabe et al., 2013b; Klingenberg and Marugán-Lobón, 2013).

The structural and functional complexity of the brain is always going to exceed the volumes, shapes, and surface features available from an endocast. The point that we hope to convey here is that, just as every skeletal metric has some relationship with body size, each of the complex, 3D components of the brain bear some correlative relationship with the surface of the brain and with the morphology of the endocast. These correlations will not always be strong; certainly some of the brain's architectural and functional nuances will remain beyond the reach of the endocast. But, by gaining an understanding of where these strengths and weakness lie, we will comprehend the range and depth of questions that endocast data can and cannot support. We are willing to wager that this process of discovery will reveal that endocasts have a greater potential for broadly interesting, significant discoveries than most of us might currently conclude.

References

Alonso, P.D., Milner, A.C., Ketcham, R.A., Cookson, M.J., Rowe, T.B., 2004. The avian nature of the brain and inner ear of *Archaeopteryx*. Nature 430, 666–669.

Ashwell, K.W.S., Scofield, R.P., 2008. Big birds and their brains: paleoneurology of the New Zealand moa. Brain Behav. Evol. 71, 151–166.

Balanoff, A.M., Xu, X., Kobayashi, Y., Matsufune, Y., Norell, M.A., 2009. Cranial osteology of the theropod dinosaur *Incisivosaurus gauthieri* (Theropoda: Oviraptorosauria). Am. Mus. Novit. 3651, 1–35.

Balanoff, A.M., Bever, G.S., Ikejiri, T., 2010. The braincase of *Apatosaurus* (Dinosauria: sauropoda) based on computed tomography of a new specimen with comments on variation and evolution in sauropod neuroanatomy. Am. Mus. Novit. 3677, 1–32.

Balanoff, A.M., Bever, G.S., Rowe, T.B., Norell, M.A., 2013. Evolutionary origins of the avian brain. Nature 501, 93–96.

Balanoff, A.M., Bever, G.S., Norell, M.A., 2014. Reconsidering the avian nature of the oviraptorosaur brain (Dinosauria: Theropoda). PLoS One 9, e113559.

Balanoff, A.M., Bever, G.S., Colbert, M.W., et al., 2016a. Best practices for digitally constructing endocranial casts: examples from birds and their dinosaurian relatives. J. Anat. 229, 173–190. https://doi.org/10.1111/joa.12378.

Balanoff, A.M., Smaers, J.B., Turner, A.H., 2016b. Brain modularity across the theropod-bird transition: testing the influence of flight on neuroanatomical variation. J. Anat. 229, 204–214. https://doi.org/10.1111/joa.12403.

Barton, R.A., Harvey, P.H., 2000. Mosaic evolution of brain structure in mammals. Nature 405, 1055–1058.

Baumel, J.J., 1993. Osteologia. In: Baumel, J.J., King, A.S., Breazile, J.E., Evans, H.E., Vanden Berge, J.C. (Eds.), Handbook of Avian Anatomy: Nomina Anatomica Avium, second ed., pp. 45–132.

Bell, C.J., Gauthier, J.A., Bever, G.S., 2010. Covert biases, circularity, and apomorphies: a critical look at the North American Quaternary herpetofaunal stability hypothesis. Quat. Int. 217, 30–36.

Bell, E.A., Boehnke, P., Harrison, T.M., Mao, W.L., 2015. Potentially biogenic carbon preserved in a 4.1 billion-year-old zircon. Proc. Natl. Acad. Sci. U.S.A. 112, 14518–14521.

Benton, M.J., Donoghue, P., Asher, R.J., 2015. Constraints on the timescale of animal evolutionary history. Palaeontol. Electron. 18.1.1FC 1–106.

Bever, G.S., Norell, M.A., 2009. The perinate skull of *Byronosaurus* (Troodontidae) with observations on the cranial ontogeny of paravian theropods. Am. Mus. Novit. 3657, 1–51.

Bever, G.S., Macrini, T.E., Jass, C.N., 2009. A natural endocranial cast of a fossil proboscidean with comments on the evolution of elephant neuroanatomy and the scientific value of 'no data' specimens. In: Farley, G.H., Choate, J.R. (Eds.), Unlocking the Unknown: Papers Honoring Dr. Richard J. Zakrzewski. Fort Hays Studies, Fort Hays, KS, pp. 11–22. Special Issue 1.

Bever, G.S., Brusatte, S.L., Balanoff, A.M., Norell, M.A., 2011. Variation, variability, and the origin of the avian endocranium: insights from the anatomy of *Alioramus altai* (Theropoda: Tyrannosauroidea). PLoS One 6, e23393.

Bever, G.S., Brusatte, S.L., Carr, T.D., et al., 2013. The braincase anatomy of the late Cretaceous dinosaur *Alioramus* (Theropoda: Tyrannosauroidea). Bull. Am. Mus. Nat. Hist. 376, 1–72.

Bever, G.S., Lyson, T.R., Field, D.J., Bhullar, B.-A.S., 2015. Evolutionary origin of the turtle skull. Nature 525, 239–242.

Bever, G.S., 2005. Variation in the ilium of North American *Bufo* (Lissamphibia; Anura) and its implications for species-level identification of fragmentary anuran fossils. J. Vertebr. Paleontol. 25, 548–560.

Beyrand, V., Voeten, D.F.A.E., Bureš, S., et al., 2019. Multiphase progenetic development shaped the brain of flying archosaurs. Sci. Rep. 9, 10807.

Bhullar, B.-A.S., Bever, G.S., 2009. An archosaur-like laterosphenoid in early turtles (Reptilia: Pantestudines). Breviora 518, 1–11.

Bhullar, B.-A.S., Marugán-Lobón, J., Racimo, F., et al., 2012. Birds have paedomorphic dinosaur skulls. Nature 487, 223–226.

Bjerring, H.C., 1991. Two intracranial ligaments supporting the brain of the brachiopterygian fish *Polypterus senegalus*. Acta Zool 72, 41–47.

Boire, D., Baron, G., 1994. Allometric comparison of brain and main brain subdivisions in birds. J. Hirnforsch. 35, 49–66.

Brazeau, M.D., Friedman, M., 2014. The characters of Palaeozoic jawed vertebrates. Zool. J. Linn. Soc. 170, 779–821.

Brazeau, M.D., 2009. The braincase and jaws of a Devonian "acanthodian" and modern gnathostome origins. Nature 457, 305–308.

Brochu, C.A., 2000. A digitally-rendered endocast for *Tyrannosaurus rex*. J. Vertebr. Paleontol. 20, 1–6.

Brocklehurst, G., 1979. The significance of the evolution of the cerebrospinal fluid system. Ann. R. Coll. Surg. Engl. 61, 349–356.

Brusatte, S.L., Sakamoto, M., Montanari, S., Harcourt Smith, W.E.H., 2012. The evolution of cranial form and function in theropod dinosaurs: insights from geometric morphometrics. J. Evol. Biol. 25, 365–377.

Brusatte, S.L., Lloyd, G.T., Wang, S.C., Norell, M.A., 2014. Gradual assembly of avian body plan culminated in rapid rates of evolution across the dinosaur-bird transition. Curr. Biol. 24, 2386–2392.

Bryant, H.N., Russell, A.P., 1992. The role of phylogenetic analysis in the inference of unpreserved attributes of extinct taxa. Philos. Trans. R. Soc. Lond. B 337, 405–418.

Budd, G.E., Jensen, S., 2000. A critical reappraisal of the fossil record of the bilaterian phyla. Biol. Rev. 75, 253–295.

Burgers, P., Chiappe, L.M., 1999. The wing of *Archaeopteryx* as a primary thrust generator. Nature 399, 60–62.

Butler, A.B., Hodos, W., 2005. Comparative Vertebrate Anatomy: Evolution and Adaptation, second ed. John Wiley & Sons, Hoboken.

Butler, A.B., Reiner, A., Karten, H.J., 2011. Evolution of the amniote pallium and the origins of mammalian neocortex. Ann. N.Y. Acad. Sci. 1225, 14–27.

Butler, A.B., 2000. Chordate evolution and the origin of craniates: an old brain in a new head. Anat. Rec. 261, 111–125.

Campbell, K., Barwick, R.E., 1982. The neurocranium of the primitive dipnoan *Dipnorhynchus sussmilchi* (Etheridge). J. Vertebr. Paleontol. 2, 286–327.

Campione, N.E., Evans, D.C., 2012. A universal scaling relationship between body mass and proximal limb bone dimensions in quadrupedal terrestrial tetrapods. BMC Biol. 10, 60.

Carabajal, A.P., Succar, C., 2013. The endocranial morphology and inner ear of the abelisaurid theropod *Aucasaurus garridoi*. Acta Palaeontol. Pol. 60, 141–144.

Carabajal, A.P., Sterli, J., Müller, J., Hilger, A., 2013. Neuroanatomy of the marine Jurassic turtle *Plesiochelys etalloni* (Testudinata, Plesiochelyidae). PLoS One 8, e69264.

Carabajal, A.P., 2012. Neuroanatomy of titanosaurid dinosaurs from the Upper Cretaceous of Patagonia, with comments on endocranial variability within sauropoda. Anat. Rec. Adv. Integr. Anat. Evol. Biol. 295, 2141–2156.

Carlson, W.D., Rowe, T., Ketcham, R.A., Colbert, M.W., 2003. Applications of high-resolution X-ray computed tomography in petrology, meteoritics and palaeontology. Geol. Soc. Lond. Spec. Publ. 215, 7–22.

Carril, J., Tambussi, C.P., Degrange, F.J., Benitez Saldivar, M.J., Picasso, M.B.J., 2016. Comparative brain morphology of Neotropical parrots (Aves, Psittaciformes) inferred from virtual 3D endocasts. J. Anat. 229, 239–251. https://doi.org/10.1111/joa.12325.

Challands, T.J., 2015. The cranial endocast of the Middle Devonian dipnoan *Dipterus valenciennesi* and a fossilized dipnoan otoconial mass. Pap. Palaeontol. 1, 289–317.

Chang, M.M., 1982. The Braincase of Youngolepis, a Lower Devonian Crossopterygian from Yunnan, South-Western China (Ph.D. dissertation). University of Stockholm and Section of Palaeozoology, Swedish Museum of Natural History, p. 113.

Chen, J.-Y., Huang, D.-Y, Li, C.-W., 1999. An early Cambrian craniate-like chordate. Nature 402, 518–522.

Chen, M., Zou, M., Yang, L., He, S., 2012. Basal jawed vertebrate phylogenomics using transcriptomic data from Solexa sequencing. PLoS One 7, e36256.

Christiansen, P., Fariña, R.A., 2004. Mass prediction in theropod dinosaurs. Hist. Biol. 16, 85–92.

Clément, G., Ahlberg, P.E., 2010. The endocranial anatomy of the early sarcopterygian *Powichthys* from Spitsbergen, based on CT scanning. In: Elliott, D.K., Maisey, J.G., Yu, X., Miao, D. (Eds.), Morphology, Phylogeny and Paleobiogeography of Fossil Fishes: Honoring Meemann Chang. Dr. Friedrich Pfeil, Munich, pp. 363–377.

Clement, A.M., Ahlberg, P.E., 2014. The first virtual cranial endocast of a lungfish (Sarcopterygii: Dipnoi). PLoS One 9, e113898.

Clement, A.M., Nysjö, J., Strand, R., Ahlberg, P.E., 2015. Brain–endocast relationship in the Australian lungfish, *Neoceratodus forsteri*, elucidated from tomographic data (Sarcopterygii: Dipnoi). PLoS One 10, e0141277.

Coates, M.I., 1999. Endocranial preservation of a Carboniferous actinopterygian from Lancashire, U.K., and the interrelationships of primitive actinopterygians. Philos. Trans. R. Soc. Biol. Sci. 354, 435–462.

Cohen, D.J., Karten, H.J., 1974. The structural organization of avian brain: an overview. In: Goodman, I.J., Schein, M.W. (Eds.), Birds: Brain and Behavior. Academic Press, New York, pp. 29–76.

Corfield, J.R., Wild, J.M., Parsons, S., Kubke, M.F., 2012. Morphometric analysis of telencephalic structure in a variety of neognath and paleognath bird species reveals regional differences associated with specific behavioral traits. Brain Behav. Evol. 80, 181–195.

Corfield, J.R., Kolominsky, J., Marin, G.J., et al., 2015. Zebrin II expression in the cerebellum of a paleognathous bird, the Chilean Tinamou (*Nothoprocta perdicaria*). Brain Behav. Evol. 85, 94–106.

Creuzet, S.E., 2009. Neural crest contribution to forebrain development. Semin. Cell Dev. Biol. 20, 751–759.

Creuzet, S.E., 2009. Regulation of pre-otic brain development by the cephalic neural crest. Proc. Natl. Acad. Sci. U.S.A. 106, 15774–15779.

Cruzado-Caballero, P., Fortuny, J., Llacer, S., Canudo, J.I., 2015. Paleoneuroanatomy of the European lambeosaurine dinosaur *Arenysaurus ardevoli*. PeerJ 3, e802.

Cunningham, S.J., Corfield, J.R., Iwaniuk, A.N., et al., 2013. The anatomy of the bill tip of Kiwi and associated somatosensory regions of the brain: comparisons with shorebirds. PLoS One 8, e80036.

Cunningham, J.A., Rahman, I.A., Lautenschlager, S., Rayfield, E.J., Donoghue, P.C.J., 2014. A virtual world of paleontology. Trends Ecol. Evol. 29, 347–357.

Davis, S.P., Finarelli, J.A., Coates, M.I., 2012. *Acanthodes* and shark-like conditions in the last common ancestor of modern gnathostomes. Nature 486, 247–250.

De Esteban-Trivigno, S., Mendoza, M., De Renzi, M., 2008. Body mass estimation in Xenarthra: a predictive equation suitable for all quadrupedal terrestrial placentals? J. Morphol. 269, 1276–1293.

De Miguel, C., Henneberg, M., 1998. Encephalization of the koala, *Phascolarctos cinereus*. Aust. Mammal. 20, 315–320.

De Queiroz, K., Gauthier, J., 1992. Phylogenetic taxonomy. Annu. Rev. Ecol. Syst. 23, 449–480.

Deaner, R.O., Isler, K., Burkart, J., van Schaik, C., 2007. Overall brain size, and not encephalization quotient, best predicts cognitive ability across non-human primates. Brain Behav. Evol. 70, 115–124.

Dendy, A., 1910. On the structure, development and morphological interpretation of the pineal organs and adjacent parts of the brain in the tuatara (*Sphenodon punctatus*). Philos. Trans. R. Soc. Lond. B 201, 226–331.

Dingus, L., Rowe, T., 1997. The Mistaken Extinction. W. H. Freeman and Company, New York.

Dong, Z.-M., Currie, P.J., 1996. On the discovery of an oviraptorid skeleton on a nest of eggs at Bayan Mandahu, Inner Mongolia, People's Republic of China. Can. J. Earth Sci. 33, 631–636.

Dos Reis, M., Inoue, J., Hasegawa, M., et al., 2012. Phylogenomic datasets provide both precision and accuracy in estimating the timescale of placental mammal phylogeny. Proc. R. Soc. Biol. Sci. 279, 3491–3500.

Dupret, V., Sanchez, S., Goujet, D., Tafforeau, P., Ahlberg, P.E., 2014. A primitive placoderm sheds light on the origin of the jawed vertebrate face. Nature 507, 500–503.

Edinger, T., 1925. Die *Archaeopteryx*. Nat. Mus. 55, 491–496.

Edinger, T., 1942. The pituitary body in giant animals fossil and living: a survey and a suggestion. Q. Rev. Biol. 17, 31–45.

Edinger, T., 1951. The brains of the Odontognathae. Evolution 5, 6–24.

Edinger, T., 1975. Paleoneurology 1804–1966. An annotated bibliography (with a foreword by Bryan Patterson). Adv. Anat. Embryol. Cell Biol. 49, 1–258.

Evans, D.C., Ridgely, R., Witmer, L.M., 2009. Endocranial anatomy of lambeosaurine hadrosaurids (Dinosauria: Ornithischia): a sensorineural perspective on cranial crest function. Anat. Rec. Adv. Integr. Anat. Evol. Biol. 292, 1315–1337.

Evans, D.C., 2005. New evidence on brain-endocranial cavity relationships in ornithischian dinosaurs. Acta Palaeontol. Pol. 50, 617–622.

Farris, J.S., 1983. The logical basis of phylogenetic analysis. In: Platnick, N.I., Funk, V.A. (Eds.), Advances in Cladistics: Proceedings of the Second Meeting of the Willi Hennig Society, vol. 2. Columbia University Press, New York, pp. 1–47.

Field, D.J., Lynner, C., Brown, C., Darroch, S.A.F., 2013. Skeletal correlates for body mass estimation in modern and fossil flying birds. PLoS One 8, e82000.

Finarelli, J.A., Flynn, J.J., 2006. Ancestral state reconstruction of body size in the caniformia (Carnivora, Mammalia): the effects of incorporating data from the fossil record. Syst. Biol. 55, 301–313.

Flinn, M.V., Geary, D.C., Ward, C.V., 2005. Ecological dominance, social competition, and coalitionary arms races: why humans evolved extraordinary intelligence. Evol. Hum. Behav. 26, 10–46.

Francis, E.T.B., 1934. The Anatomy of the Salamander. Society for the Study of Amphibians and Reptiles, Ithaca reprint 2002.

Franzosa, J., Rowe, T., 2005. Cranial endocast of the Cretaceous theropod dinosaur Acrocanthosaurus atokensis. J. Vertebr. Paleontol. 25, 859–864.

Franzosa, J.W., 2004. Evolution of the Brain in Theropoda (Dinosauria) (Unpublished Ph.D. dissertation). University of Texas at Austin, 357 p.

Funke, K., 1989. Somatosensory areas in the telencephalon of the pigeon. Exp. Brain Res. 76, 603–619.

Gaffney, E.S., 1977. An endocranial cast of the side-necked turtle, Bothremys, with a new reconstruction of the palate. Am. Mus. Novit. 2639, 1–12.

Gaffney, E.S., 1990. Comparative osteology of the Triassic turtle Proganochelys. Bull. Am. Mus. Nat. Hist. 194, 1–263.

Gai, Z., Donoghue, P.C.J., Zhu, M., Janvier, P., Stampanoni, M., 2011. Fossil jawless fish from China foreshadows early jawed vertebrate anatomy. Nature 476, 324–327.

Galton, P.M., Knoll, F., 2006. A saurischian dinosaur braincase from the middle Jurassic (Bathonian) near Oxford, England: from the theropod Megalosaurus or the sauropod Cetiosaurus? Geol. Mag. 143, 905–921.

Galton, P.M., 1985. Cranial anatomy of the prosauropod dinosaur Plateosaurus from the Knollenmergel (middle Keuper, Upper Triassic) of Germany. Geol. Palaeontol. 19, 119–159.

Galton, P.M., 1988. Skull bones and endocranial casts of stegosaurian dinosaur Kentrosaurus Hennig, 1915 from Upper Jurassic of Tanzania, east Africa. Geol. Palaeontol. 22, 123–143.

Gatesy, S.M., Dial, K.P., 1996. From frond to fan: Archaeopteryx and the evolution of short-tailed birds. Evolution 50, 2037–2048.

Gauthier, J., De Queiroz, K., 2001. Feathered dinosaurs, flying dinosaurs, crown dinosaurs, and the name "Aves". In: Gauthier, J., Gall, L.F. (Eds.), New Perspectives on the Origin and Evolution of Birds: Proceedings of the International Symposium in Honor of John H. Ostrom. . Peabody Museum of Natural History, New Haven, pp. 7–41.

Gauthier, J., Kluge, A.G., Rowe, T., 1988. Amniote phylogeny and the importance of fossils. Cladistics 4, 105–209.

Gauthier, J.A., Nesbitt, S.J., Schachner, E.R., Bever, G.S., Joyce, W.G., 2011. The bipedal stem crocodilian Poposaurus gracilis: inferring function in fossils and innovation in archosaur locomotion. Bull. - Peabody Mus. Nat. Hist. 52, 107–126.

Gauthier, J., 1986. Saurischian monophyly and the origin of birds. Mem. Calif. Acad. Sci. 8, 1–55.

Gee, H., 2018. Across the bridge: Understanding the origin of the vertebrates. The University of Chicago Press, Chicago.

George, I.D., Holliday, C.M., 2013. Trigeminal nerve morphology in Alligator mississippiensis and its significance for crocodyliform facial sensation and evolution. Anat. Rec. Adv. Integr. Anat. Evol. Biol. 296, 670–680.

Giffin, E.B., 1989. Pachycephalosaurus paleoneurology (Archosauria: Ornithischia). J. Vertebr. Paleontol. 9, 67–77.

Gignac, P.M., Kley, N.J., 2014. Iodine-enhanced micro-CT imaging: methodological refinements for the study of the soft-tissue anatomy of post-embryonic vertebrates. J. Exp. Zool. Part B Mol. Dev. Evol. 322, 166–176.

Giles, S., Friedman, M., 2014. Virtual reconstruction of endocast anatomy in early ray-finned fishes (Osteichthyes, Actinopterygii). J. Paleontol. 88, 636–651.

Giles, S., Darras, L., Clément, G., Blieck, A., Friedman, M., 2015. An exceptionally preserved Late Devonian actinopterygian provides a new model for primitive cranial anatomy in ray-finned fishes. Proc. R. Soc. Biol. Sci. 282, 20151485.

Giles, S., Friedman, M., Brazeau, M.D., 2015. Osteichthyan-like cranial conditions in an Early Devonian stem gnathostome. Nature 520, 82–85.

Goodrich, E.S., 1930. Studies on the Structure and Development of Vertebrates. The University of Chicago Press, Chicago reprint 1986.

Haight, J.R., Nelson, J.E., 1987. A brain that doesn't fit its skull: a comparative study of the brain and endocranium of the koala, Phascolarctos cinereus (Marsupialia: Phascolarctidae). In: Archer, M. (Ed.), Possums and Opossums: Studies in Evolution. Surrey Beatty & Sons, Sydney, pp. 331–352.

Hamel, M.-H., Poplin, C., 2008. The braincase anatomy of Lawrenciella schaefferi, actinopterygian from the Upper Carboniferous of Kansas (USA). J. Vertebr. Paleontol. 28, 989–1006.

Healy, S.D., Rowe, C., 2007. A critique of comparative studies of brain size. Proc. R. Soc. Biol. Sci. 274, 453–464.

Heath, T.A., Hedtke, S.M., Hillis, D.M., 2008. Taxon sampling and the accuracy of phylogenetic analyses. J. Syst. Evol. 46, 239–257.

Hedges, S.B., Marin, J., Suleski, M., Paymer, M., 2015. Tree of life reveals clock-like speciation and diversification. Mol. Biol. Evol. 32, 835–845.

Hennig, W., 1966. Phylogenetic Systematics. University of Illinois Press, Chicago reprint 1999.

Hinić-Frlog, S., Motani, R., 2010. Relationship between osteology and aquatic locomotion in birds: determining modes of locomotion in extinct Ornithurae. J. Evol. Biol. 23, 372–385.

Holland, T., 2014. The endocranial anatomy of Gogonasus andrewsae Long, 1985 revealed through micro CT-scanning. Earth Environ. Sci. Trans. R. Soc. Edinb. 105, 9–34.

Hopson, J.A., 1977. Relative brain size and behavior in archosaurian reptiles. Annu. Rev. Ecol. Syst. 8, 429–448.

Hopson, J.A., 1979. Paleoneurology. In: Gans, C. (Ed.), Biology of the Reptilia, vol. 9. Academic Press, New York, pp. 39–146.

Hu, D., Young, N.M., Xu, Q., et al., 2015. Signals from the brain induce variation in avian facial shape. Dev. Dyn. 244, 1133–1143.

Humphrey, O.D., 1894. On the brain of the snapping turtle (Chelydra serpentina). J. Comp. Physiol. 4, 73–116.

Hurlburt, G.R., Ridgely, R.C., Witmer, L.M., 2013. Relative size of brain and cerebrum in tyrannosaurid dinosaurs: an analysis using brain-endocast quantitative relationships in extant alligators. In: Parrish, J.M., Molnar, R.E., Currie, P.J., Koppelhus, E.B. (Eds.), Tyrannosaurid Paleobiology. Indiana University Press, Bloomington, pp. 134–154.

Hurlburt, G.R., 1996. Relative Brain Size in Recent and Fossil Amniotes: Determination and Interpretation (Unpublished Ph.D. dissertation). University of Toronto, 250 p.

Hurlburt, G., 2014. First report of an arachnoid mater in a non-avian reptile, Alligator mississippiensis. Society of Vertebrate Paleontology 74th Meeting. Meeting Program and Abstracts, Berlin, p. 151.

Hutchinson, J.R., 2006. The evolution of locomotion in archosaurs. Comptes Rendus Palevol 5, 519–530.

Isler, K., Kirk, E.C., Miller, J.M.A., et al., 2008. Endocranial volumes of primate species: scaling analysis using a comprehensive and reliable data set. J. Hum. Evol. 55, 967–978.

Iwaniuk, A.N., Hurd, P.L., 2005. The evolution of cerebrotypes in birds. Brain Behav. Evol. 65, 215–230.

Iwaniuk, A.N., Nelson, J.E., 2002. Can endocranial volume be used as an estimate of brain size in birds? Can. J. Zool. 80, 16–23.

Iwaniuk, A.N., Pellis, S.M., Whishaw, I.Q., 1999. Brain size is not correlated with forelimb dexterity in fissiped carnivores (Carnivora): a comparative test of the principle of proper mass. Brain Behav. Evol. 54, 167–180.

Iwaniuk, A.N., Nelson, J.E., Whishaw, I.Q., 2000. The relationships between brain regions and forelimb dexterity in marsupials (Marsupialia): a comparative test of the principle of proper mass. Aust. J. Zool. 48, 99–110.

Iwaniuk, A.N., Dean, K.M., Nelson, J.E., 2004. A mosaic pattern characterizes the evolution of the avian brain. Proc. R. Soc. Biol. Sci. 271, S148–S151.

Janvier, P., 1981. Norselaspis glacialis ng, n. sp. et les relations phylogénétiques entre les Kiaeraspidiens (Osteostraci) du Dévonien inférieur du Spitsberg. Palaeovertebrata 11, 19–131.

Janvier, P., 1985. Les Céphalaspides du Spitsberg. Cahiers de Paléontologie, Centre national de la Recherche scientifique, Paris.

Janvier, P., 1996. Early Vertebrates. Oxford University Press, Oxford.

Janvier, P., 2008. The brain in the early fossil jawless vertebrates: evolutionary information from an empty nutshell. Brain Res. Bull. 75, 314–318.

Jarvik, E., 1972. Middle and Upper Devonian Porolepiformes from East Greenland with special reference to Glyptolepis groenlandica n. sp. and a discussion on the structure of the head in the Porolepiformes. In: Meddelelser Om Grønland Udgivne Af Kommissionen for Videnskabelige Undersøgelser I Grønlandvol, vol. 187, pp. 1–307.

Jarvis, E.D., Yu, J., Rivas, M.V., et al., 2013. Global view of the functional molecular organization of the avian cerebrum: mirror images and functional columns. J. Comp. Physiol. 521, 3614–3665.

Jarvis, E.D., 2009. Evolution of the pallium in birds and reptiles. In: Binder, M.D., Nobutaka, H., Windhorst, U. (Eds.), New Encyclopedia of Neuroscience. Springer-Verlag, Berlin, pp. 1390–1400.

Jerison, H.J., 1968. Brain evolution and Archaeopteryx. Nature 219, 1381–1382.

Jerison, H.J., 1969. Brain evolution and dinosaur brains. Am. Nat. 103, 575–588.

Jerison, H.J., 1973. Evolution of the Brain and Intelligence. Academic Press, New York.

Jerison, H.J., 1975. Fossil evidence of the evolution of the human brain. Annu. Rev. Anthropol. 4, 27–58.

Jerison, H.J., 1977. The theory of encephalization. Ann. N.Y. Acad. Sci. 299, 146–160.

Johanson, Z., Ahlberg, P., Ritchie, A., 2003. The braincase and palate of the tetrapodomorph sarcopterygian Mandageria fairfaxi: morphological variability near the fish–tetrapod transition. Palaeontology 46, 271–293.

Joven, A., Morona, R., González, A., Moreno, N., 2013. Spatiotemporal patterns of Pax3, Pax6 and Pax7 expression in the developing brain of a urodele amphibian. Pleurodeles waltl. J. Comp. Physiol. A 521, 3913–3953.

Kawabe, S., Ando, T., Endo, H., 2013. Enigmatic affinity in the brain morphology between plotopterids and penguins, with a comprehensive comparison among water birds. Zool. J. Linn. Soc. 170, 467–493.

Kawabe, S., Shimokawa, T., Miki, H., Matsuda, S., Endo, H., 2013. Variation in avian brain shape: relationship with size and orbital shape. J. Anat. 223, 495–508.

Kawabe, S., Matsuda, S., Tsunekawa, N., Endo, H., 2015. Ontogenetic shape change in the chicken brain: implications for paleontology. PLoS One 10, e0129939.

Klingenberg, C.P., Marugán-Lobón, J., 2013. Evolutionary covariation in geometric morphometric data: analyzing integration, modularity, and allometry in a phylogenetic context. Syst. Biol. 62, 591–610.

Kruska, D.C.T., 1988. The brain of the basking shark (Cetorhinus maximus). Brain Behav. Evol. 32, 353–363.

Ksepka, D.T., Bertelli, S., Giannini, N.P., 2006. The phylogeny of the living and fossil Sphenisciformes (penguins). Cladistics 22, 412–441.

Ksepka, D.T., Balanoff, A.M., Walsh, S., Revan, A., Ho, A., 2012. Evolution of the brain and sensory organs in Sphenisciformes: new data from the stem penguin Paraptenodytes antarcticus. Zool. J. Linn. Soc. 166, 202–219.

Kuhlenbeck, H., 1973. The central nervous system of vertebrates. Part II. In: Overall Morphologic Pattern, vol. 3. S. Karger, Basel.

Kundrát, M., 2007. Avian-like attributes of a virtual brain model of the oviraptorid theropod Conchoraptor gracilis. Naturwissenschaften 94, 499–504.

Kuraku, S., Kuratani, S., 2006. Time scale for cyclostome evolution inferred with a phylogenetic diagnosis of hagfish and lamprey cDNA sequences. Zool. Sci. 23, 1053–1064.

Lacalli, T.C., 2008. Basic features of the ancestral chordate brain: a protochordate perspective. Brain Res. Bull. 75, 319–323.

Lane, J.A., 2010. Morphology of the braincase in the Cretaceous hybodont shark Tribodus limae (Chondrichthyes: Elasmobranchii), based on CT scanning. Am. Mus. Novit. 3681, 1–70.

Larsson, H.C.E., Sereno, P.C., Wilson, J.A., 2000. Forebrain enlargement among nonavian theropod dinosaurs. J. Vertebr. Paleontol. 20, 615–618.

Lautenschlager, S., Hübner, T., 2013. Ontogenetic trajectories in the ornithischian endocranium. J. Evol. Biol. 26, 2044–2050.

Lautenschlager, S., Rayfield, E.J., Altangerel, P., Zanno, L.E., Witmer, L.M., 2012. The endocranial anatomy of Therizinosauria and its implications for sensory and cognitive function. PLoS One 7, e52289.

Lauters, P., Vercauteren, M., Bolotsky, Y.L., Godefroit, P., 2013. Cranial endocast of the lambeosaurine hadrosaurid Amurosaurus riabinini from the Amur Region, Russia. PLoS One 8, e78899.

Lefebvre, L., Sol, D., 2008. Brains, lifestyles and cognition: are there general trends? Brain Behav. Evol. 72, 135–144.

Lefebvre, L., 2013. Brains, innovations, tools and cultural transmission in birds, non-human primates, and fossil hominins. Front. Neurosci. 7, 1–10.

Lieberman, D.E., Hallgrímsson, B., Liu, W., Parsons, T.E., Jamniczky, H.A., 2008. Spatial packing, cranial base angulation, and craniofacial shape variation in the mammalian skull: testing a new model using mice. J. Anat. 212, 720–735.

Lu, J., Zhu, M., Long, J.A., et al., 2012. The earliest known stem-tetrapod from the Lower Devonian of China. Nat. Commun. 3, 1–7.

Luo, Z.-X., Gatesy, S.M., Jenkins Jr., F.A., Amaral, W.W., Shubin, N.H., 2015. Mandibular and dental characteristics of Late Triassic mammaliaform Haramiyavia and their ramifications for basal mammal evolution. Proc. Natl. Acad. Sci. U.S.A. 112, e7101–7109.

Maddin, H.C., Venczel, M., Gardner, J.D., Rage, J.-C., 2013. Micro-computed tomography study of a three-dimensionally preserved neurocranium of Albanerpeton (Lissamphibia, Albanerpetontidae) from the Pliocene of Hungary. J. Vertebr. Paleontol. 33, 568–587.

Maisey, J.G., 2004. Morphology of the braincase in the broadnose sevengill shark Notorynchus (Elasmobranchii, Hexanchiformes), based on CT scanning. Am. Mus. Novit. 3429, 1–52.

Maisey, J.G., 2005. Braincase of the Upper Devonian shark Cladodoides wildungensis (Chondrichthyes, Elasmobranchii), with observations on the braincase in early chondrichthyans. Bull. Am. Mus. Nat. Hist. 288, 1–103.

Maisey, J.G., 2007. The braincase in Paleozoic symmoriiform and cladoselachian sharks. Bull. Am. Mus. Nat. Hist. 307, 1–122.

Maisey, J.G., 2011. The braincase of the middle Triassic shark Acronemus tuberculatus (Bassani, 1886). Palaeontology 54, 417–428.

Makovicky, P., Norell, M.A., Clark, J.M., Rowe, T.B., 2003. Osteology and relationships of *Byronosaurus jaffei* (Theropoda: Troodontidae). Am. Mus. Novit. 3402, 1−32.

Marcucio, R.S., Young, N.M., Hu, D., Hallgrímsson, B., 2011. Mechanisms that underlie co-variation of the brain and face. Genesis 49, 177−189.

Marek, R.D., Moon, B.C., Williams, M., Benton, M.J., 2015. The skull and endocranium of a Lower Jurassic ichthyosaur based on digital reconstructions. Palaeontology 58, 723−742.

Marsh, O.C., 1880. Odontornithes: a monograph on the extinct toothed birds of North America. Rep. Geol. Explor. Fortieth Parallel 7, 1−201.

Marugán-Lobón, J., Buscalioni, Á.D., 2003. Disparity and geometry of the skull in Archosauria (Reptilia: Diapsida). Biol. J. Linn. Soc. 80, 67−88.

Medina, L., Reiner, A., 2000. Do birds possess homologues of mammalian primary visual, somatosensory and motor cortices? Trends Neurosci. 23, 1−12.

Midtgård, U., 1984. The blood vascular system in the head of the herring gull (*Larus argentatus*). J. Morphol. 179, 135−152.

Millot, J., Anthony, J., 1965. Anatomy de Latimeria chalumnae. In: Système nerveux et organes de sens, vol. II. Éditions du Centre National de la Recherche Scientifique, Paris.

Milner, A.C., Walsh, S.A., 2009. Avian brain evolution: new data from Palaeogene birds (Lower Eocene) from England. Zool. J. Linn. Soc. 155, 198−219.

Miyashita, T., Arbour, V.M., Witmer, L.M., Currie, P.J., 2011. The internal cranial morphology of an armoured dinosaur *Euoplocephalus* corroborated by X-ray computed tomographic reconstruction. J. Anat. 219, 661−675.

Morhardt, A.C., Ridgley, R.C., Witmer, L.M., 2012. From endocast to brain: assessing brain size and structure in extinct archosaurs using gross anatomical brain region approximation (GABRA). J. Vertebr. Paleontol. 32 (Suppl. l.), 145.

Nesbitt, S.J., Turner, A.H., Spaulding, M., Conrad, J.L., Norell, M.A., 2009. The theropod furcula. J. Morphol. 270, 856−879.

Nesbitt, S.J., 2011. The early evolution of archosaurs: relationships and the origin of major clades. Bull. Am. Mus. Nat. Hist. 352, 1−292.

Ngwenya, A., Patzke, N., Spocter, M.A., et al., 2013. The continuously growing central nervous system of the Nile Crocodile (*Crocodylus niloticus*). Anat. Rec. Adv. Integr. Anat. Evol. Biol. 296, 1489−1500.

Norell, M.A., Xu, X., 2005. Feathered dinosaurs. Annu. Rev. Earth Planet Sci. 33, 277−299.

Norell, M.A., Clark, J.M., Chiappe, L.M., Dashzeveg, D., 1995. A nesting dinosaur. Nature 378, 774−776.

Norell, M.A., Makovicky, P.J., Bever, G.S., Balanoff, A.M., 2009. A review of the Mongolian Cretaceous dinosaur *Saurornithoides* (Troodontidae: Theropoda). Am. Mus. Novit. 3654, 1−63.

Northcutt, R.G., 1977. Elasmobranch central nervous system organization and its possible evolutionary significance. Am. Zool. 17, 411−429.

Northcutt, R.G., 2002. Understanding vertebrate brain evolution. Integr. Comp. Biol. 42, 743−756.

O'Connor, J.K., Chiappe, L.M., 2011. A revision of enantiornithine (Aves: Ornithothoraces) skull morphology. J. Syst. Palaeontol. 9, 135−157.

O'Leary, M.A., Bloch, J.I., Flynn, J.J., et al., 2013. The placental mammal ancestor and the post−K-Pg radiation of placentals. Science 339, 662−667.

Olkowicz, S., Kocourek, M., Lučan, R.K., et al., 2016. Birds have primate-like numbers of neurons in the forebrain. Proc. Natl. Acad. Sci. U.S.A. 113, 7255−7260.

Osborn, H.F., 1916. Crania of *Tyrannosaurus* and *Allosaurus*. Mem. Am. Mus. Nat. Hist. 1, 1−97.

Osmólska, H., 2004. Brief report: evidence on relation of brain to endocranial cavity in oviraptorid dinosaurs. Acta Palaeontol. Pol. 49, 321−324.

Ostrom, J.H., 1976. *Archaeopteryx* and the origin of birds. Biol. J. Linn. Soc. 8, 91−182.

Overington, S.E., Morand-Ferron, J., Boogert, N.J., Lefebvre, L., 2009. Technical innovations drive the relationship between innovativeness and residual brain size in birds. Anim. Behav. 78, 1001−1010.

Pradel, A., Langer, M., Maisey, J.G., et al., 2009. Skull and brain of a 300-million-year-old chimaeroid fish revealed by synchrotron holotomography. Proc. Natl. Acad. Sci. U.S.A. 106, 5224−5228.

Pradel, A., 2010. Skull and brain anatomy of late Carboniferous Sibyrhynchidae (Chondrichthyes, Iniopterygia) from Kansas and Oklahoma (USA). Geodiversitas 32, 595−661.

Prum, R.O., Berv, J.S., Dornburg, A., et al., 2015. A comprehensive phylogeny of birds (Aves) using targeted next-generation DNA sequencing. Nature 526, 569−573.

Rahman, I.A., Smith, S.Y., 2014. Virtual paleontology: computer-aided analysis of fossil form and function. J. Paleontol. 88, 633−635.

Reiner, A., Yamamoto, K., Karten, H.J., 2005. Organization and evolution of the avian forebrain. Anat. Rec. Adv. Integr. Anat. Evol. Biol. 287A, 1080−1102.

Rieppel, O., 1994. Osteology of *Simosaurus gaillardoti* and the relationships of stem-group *Sauropterygia*. Fieldiana Geol. 1462, 1−85.

Romer, A.S., Edinger, T., 1942. Endocranial casts and brains of living and fossil Amphibia. J. Comp. Neurol. 77, 355−389.

Romer, A.S., 1937. The braincase of the Carboniferous crossopterygian *Megalichthys nitidus*. Bull. Mus. Comp. Zool. 82, 1−73.

Rowe, T.B., Macrini, T.E., Luo, Z.X., 2011. Fossil evidence on origin of the mammalian brain. Science 332, 955−957.

Rowe, T., 1987. Definition and diagnosis in the phylogenetic system. Syst. Biol. 36, 208−211.

Rowe, T., 1996. Coevolution of the mammalian middle ear and neocortex. Science 273, 651−654.

Rowe, T.B., 2004. Chordate phylogeny and development. In: Cracraft, J., Donoghue, M.J. (Eds.), Assembling the Tree of Life, pp. 384−409.

Sales, M.A.F., Schultz, C.L., 2014. Paleoneurology of *Teyumabaita sulcognathus* (Diapsida: Archosauromorpha) and the sense of smell in rhynchosaurs. Palaeontol. Electron. 17, 15A. palaeo-electronica. org/content/2014/705-olfaction-in-rhynchosaurs.

Sampson, S.D., Witmer, L.M., 2007. Craniofacial anatomy of *Majungasaurus crenatissimus* (Theropoda: Abelisauridae) from the late Cretaceous of Madagascar. J. Vertebr. Paleontol. 27, 32−104.

Säve-Söderbergh, G., 1952. On the skull of *Chirodipterus wildungensis* gross, an Upper Devonian dipnoan from Wildungen. Kunglinga Svenska Vetenskapsakademiens Handlingar 4, 1−29.

Schaeffer, B., 1981. The xenacanth shark neurocranium, with comments on elasmobranch monophyly. Bull. Am. Mus. Nat. Hist. 169, 1−66.

Sedlmayr, J.C., 2002. Anatomy, Evolution, and Functional Significance of Cephalic Vasculature in Archosauria (Unpublished Ph.D. dissertation). Ohio University, 398 p.

Sepulcre, J., Liu, H., Talukdar, T., et al., 2010. The organization of local and distant functional connectivity in the human brain. PLoS Comput. Biol. 6, e1000808.

Sereno, P.C., Wilson, J.A., Witmer, L.M., et al., 2007. Structural extremes in a Cretaceous dinosaur. PLoS One 2, e1230.

Smaers, J.B., Soligo, C., 2013. Brain reorganization, not relative brain size, primarily characterizes anthropoid brain evolution. Proc. R. Soc. Biol. Sci. 280, 20130269.

Smith, N.A., Clarke, J.A., 2013. Osteological histology of the Pan-Alcidae (Aves, Charadriiformes): correlates of wing-propelled diving and flightlessness. Anat. Rec. Adv. Integr. Anat. Evol. Biol. 297, 188−199.

Smith, N.D., Makovicky, P.J., Hammer, W.R., Currie, P.J., 2007. Osteology of *Cryolophosaurus ellioti* (Dinosauria: Theropoda) from the early Jurassic of Antarctica and implications for early theropod evolution. Zool. J. Linn. Soc. 151, 377−421.

Snitting, D., 2008. A redescription of the anatomy of the late Devonian *Spodichthys buetleri* Jarvik, 1985 (Sarcopterygii, Tetrapodomorpha) from east Greenland. J. Vertebr. Paleontol. 28, 637−655.

Starck, D., 1979. Cranio-cerebral relations in recent reptiles. In: Gans, C. (Ed.), Biology of the Reptilia, vol. 9. Academic Press, New York, pp. 1—38.

Stensiö, E.A., 1925. On the head of the macropetalichthyids with certain remarks on the head of the other arthrodires. Field Mus. Nat. Hist. Publ. Geol. Ser. 232, 87—197.

Stensiö, E.A., 1963. The brain and the cranial nerves in fossil lower craniate vertebrates. Skrifter utgitt av Det Norske Videnskaps-Akademi 13, 1—120.

Sutton, M.D., 2008. Tomographic techniques for the study of exceptionally preserved fossils. Proc. R. Soc. Biol. Sci. 275, 1587—1593.

Tafforeau, P., Boistel, R., Boller, E., et al., 2006. Applications of X-ray synchrotron microtomography for non-destructive 3D studies of paleontological specimens. Appl. Phys. A 83, 195—202.

Tambussi, C.P., Degrange, F.J., Ksepka, D.T., 2015. Endocranial anatomy of Antarctic Eocene stem penguins: implications for sensory system evolution in Sphenisciformes (Aves). J. Vertebr. Paleontol. 35, e981635.

Turner, A.H., Makovicky, P.J., Norell, M.A., 2007. Feather quill knobs in the dinosaur *Velociraptor*. Science 317, 1721.

Turner, A.H., Makovicky, P.J., Norell, M.A., 2012. A review of dromaeosaurid systematics and paravian phylogeny. Bull. Am. Mus. Nat. Hist. 371, 1—206.

Walsh, S., Milner, A., 2011. *Halcyornis toliapicus* (Aves: lower Eocene, England) indicates advanced neuromorphology in Mesozoic Neornithes. J. Syst. Palaeontol. 9, 173—181.

Walsh, S.A., Iwaniuk, A.N., Knoll, M.A., et al., 2013. Avian cerebellar floccular fossa size is not a proxy for flying ability in birds. PLoS One 8, e67176.

Walsh, S.A., Milner, A.C., Bourdon, E., 2016. A reappraisal of *Cerebavis cenomanica* (Aves, Ornithurae), from Melovatka, Russia. J. Anat. 229, 215—227. https://doi.org/10.1111/joa.12406.

Wang, X., Nudds, R.L., Palmer, C., Dyke, G.J., 2012. Size scaling and stiffness of avian primary feathers: implications for the flight of Mesozoic birds. J. Evol. Biol. 25, 547—555.

Werneburg, I., Morimoto, N., Zollikofer, C.P.E., et al., 2014. Mammalian skull heterochrony reveals modular evolution and a link between cranial development and brain size. Nat. Commun. 5, 1—9.

White, T.D., Asfaw, B., Beyene, Y., et al., 2009. *Ardipithecus ramidus* and the paleobiology of early hominids. Science 326, 64—86.

Wild, J.M., Kubke, M.F., Peña, J.L., 2008. A pathway for predation in the brain of the barn owl (*Tyto alba*): projections of the gracile nucleus to the "claw area" of the rostral Wulst via the dorsal thalamus. J. Comp. Neurol. 509, 156—166.

Wild, J.M., 1987. The avian somatosensory system: connections of regions of body representation in the forebrain of the pigeon. Brain Res. 412, 205—223.

Witmer, L.M., Ridgely, R.C., 2009. New insights into the brain, braincase, and ear region of tyrannosaurs (Dinosauria, Theropoda), with implications for sensory organization and behavior. Anat. Rec. Adv. Integr. Anat. Evol. Biol. 292, 1266—1296.

Witmer, L.M., Chatterjee, S., Franzosa, J., Rowe, T., 2003. Neuroanatomy of flying reptiles and implications for flight, posture and behaviour. Nature 425, 950—953.

Witmer, L.M., Ridgely, R.C., Dufeau, D.L., Semones, M.C., 2008. Using CT to peer into the past: 3D visualization of the brain and ear regions of birds, crocodiles, and nonavian dinosaurs. In: Endo, H., Frey, R. (Eds.), Anatomical Imaging: Towards a New Morphology. Springer-Verlag Tokyo, Tokyo, pp. 67—88.

Witmer, L.M., 1995. The extant phylogenetic bracket and the importance of reconstructing soft tissues in fossils. In: Thomason, J.J. (Ed.), Functional Morphology in Vertebrate Paleontology. Cambridge University Press, New York, pp. 19—33.

Xu, X., Norell, M.A., 2004. A new troodontid dinosaur from China with avian-like sleeping posture. Nature 431, 838—841.

Yue, J.X., Yu, J.K., Putnam, N.H., Holland, L.Z., 2014. The transcriptome of an Amphioxus, *Asymmetron lucayanum*, from the Bahamas: a window into chordate evolution. Genome Biol. Evol. 6, 2681—2696.

Zelenitsky, D.K., Therrien, F., Kobayashi, Y., 2009. Olfactory acuity in theropods: palaeobiological and evolutionary implications. Proc. R. Soc. Biol. Sci. 276, 667—673.

Zelenitsky, D.K., Therrien, F., Ridgely, R.C., McGee, A.R., Witmer, L.M., 2011. Evolution of olfaction in non-avian theropod dinosaurs and birds. Proc. R. Soc. Biol. Sci. 278, 3625—3634.

Zhou, C.-F., Gao, K.-Q., Fox, R.C., Du, X.-K., 2007. Endocranial morphology of psittacosaurus (Dinosauria: Ceratopsia) based on CT scans of new fossils from the Lower Cretaceous, China. Palaeoworld 16, 285—293.

4

Invertebrate Origins of Vertebrate Nervous Systems

L.Z. Holland

University of California at San Diego, La Jolla, CA, United States

4.1 Introduction

The origins of the vertebrate central nervous system (CNS) have long been sought by comparisons with their nearest invertebrate relatives, cephalochordates (amphioxus), and urochordates (tunicates). Together, these three groups of deuterostomes form the Phylum Chordata, so called because they all have a dorsal hollow nerve cord and a notochord. Although tunicates are the sister group of vertebrates, and cephalochordates are basal to them, the focus has been on comparisons between amphioxus and vertebrates, since tunicate nervous systems are probably secondarily simplified. These comparisons have led to a reasonable picture of the CNS of the ancestral chordate. It has been far more problematic to gain an understanding of the nervous system of an ancestral deuterostome. The other deuterostome phylum, the Ambulacraria, consisting of hemichordates and echinoderms (Fig. 4.1), does not have a clear counterpart of the chordate CNS. Hemichordates have both dorsal and ventral nerve cords with relatively few neurons; the relationship, if any, between these nerve cords and those of chordates is quite controversial. Echinoderms are little help as the available evidence indicates that their nerve cords probably arose de novo. Theories of the origin of chordate nerve cords from more distant invertebrate ancestors abound. Current ones consider vertebrate nerve cords to have perhaps evolved from that of a protostome ancestor (annelid or arthropod), or from the nerve net of a cnidarian, or from the nerve cords of an acoel flatworm, currently united with nemertodermatids and xenoturbellids into the phylum Xenacoelomorpha and placed basal to the Bilateria (Cannon et al., 2016; Rouse et al., 2016). The present review first considers the evolution of vertebrate nervous systems from those of a chordate ancestor and then briefly discusses theories concerning the more phylogenetically distant ancestors of chordate nervous systems.

4.2 Correspondence of Major Brain Regions in Amphioxus and Vertebrates

Cephalochordates are evolving very slowly, even more-so than vertebrates. Therefore, although they split from vertebrates over 520 mya, they appear to have retained many features of the common ancestor they shared with vertebrates. In contrast, vertebrates, having undergone two whole-genome duplications (three in the case of teleost fish), have developed a number of new structures (eg, image-forming eyes, limbs, a far more complex brain). Evidence is mounting that the amphioxus nervous system with an estimated 20 000 neurons in a large adult (Nicol and Meinertzhagen, 1991) is a close approximation of that of the last common ancestor of amphioxus and vertebrates. Data from both detailed neuroanatomy and gene expression patterns in amphioxus and vertebrates indicate that the ancestral chordate probably had a brain consisting of a diencephalic forebrain, small midbrain, hindbrain, and spinal cord. Serial transmission electron microscopy (TEM) has allowed 3-D reconstructions of the brain of a late larval amphioxus (Wicht and Lacalli, 2005; Lacalli, 1996, 2002, 2003; Lacalli and Kelly, 1999). Combined with tracing of neural tracts, studies on early embryonic patterning and identification of specific types of neurons by expression of neuropeptides and/or their synthetic enzymes, this has allowed direct comparisons with the larval lamprey brain, generally considered the simplest brain of extant vertebrates.

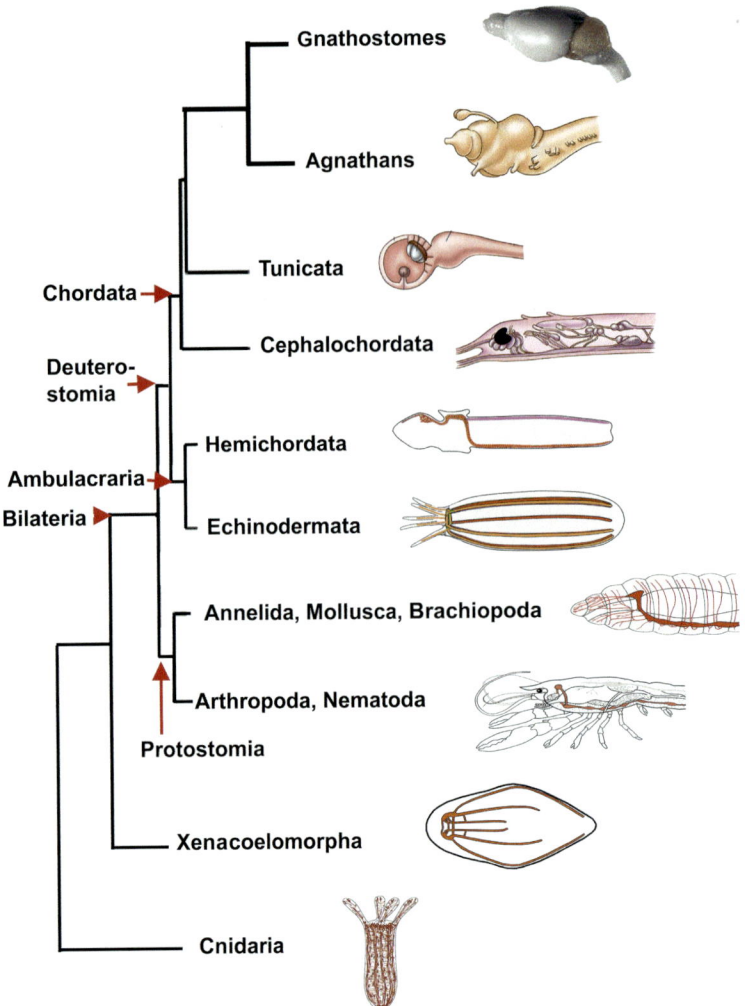

FIGURE 4.1   Phylogenetic tree with nervous systems of representative organisms. These include for gnathostomes, the mouse brain; for agnathans, the lamprey brain; for tunicates, the central nervous system (CNS) of a larval ascidian; for cephalochordate, the CNS of an amphioxus. For the
remaining organisms, the nervous systems (in *orange*) are shown within diagrams of the bodies of representative organisms: for hemichordates, an
enteropneust; for echinoderms, a holothurian; for the annelid, mollusk and brachiopod group, an annelid is shown; for arthropods and nematodes, a
lobster; for Xenacoelomorpha, an acoel flatworm; and for cnidarian, a sea anemone. *Cephalochordate, tunicate, and agnathan diagrams from Feinberg, T.E.,
Mallatt, J., 2013. The evolutionary and genetic origins of consciousness in the Cambrian period over 500 million years ago. Front. Psychol. 4, Article 667; Mouse
brain from the Mouse Brain Library (http://www.mbl.org/atlas232/atlas232_frame.html); annelid brain from Fig. 2.7 Chapt. 2. Basic Plan of the Nervous System,
http://slideplayer.com/slide/686562/, Elsevier; crayfish diagram from http://crescentok.com/staff/jaskew/isr/biology/biolab46b.htm; holothurian diagram from Fig. 4.1
in Mashanov, V.S., et al., 2009. The central nervous system of sea cucumbers (Echinodermata: Holothuroidea) shows positive immunostaining for a chordate glial
secretion. Front. Zool. 6, Article 11; and acoel nervous system from Achatz, G., Martinez, P., 2012. The nervous system of* Isodiametra pulchra *(Acoela) with a
discussion on the neuroanatomy of the Xenacoelomorpha and its evolutionary implications. Front. Zool. 9, 27.*

4.2.1 Anatomy of the Amphioxus Central Nervous System

The amphioxus nerve cord has an anterior swelling,
the cerebral vesicle, which in larvae, narrows abruptly
at its posterior end (Fig. 4.2). This constriction disappears at metamorphosis with the diameter of the CNS
simply increasing steadily from the anterior tip through
at least the equivalent of the hindbrain. In the absence of
constrictions along the developing nerve cord, the
muscular somites, which extend to the anterior tip of

the CNS, serve as markers for anterior/posterior (A/P)
position in the CNS. At the anterior tip of the CNS is a
photoreceptor, which has been called the "frontal eye"
(Figs. 4.2–4.4). It consists of a pigment spot and four
rows of photoreceptor neurons. Posterior to the frontal
eye are ventral balance cells with knob-shaped cilia
and the infundibular organ that secretes an extracellular
fiber similar to Reissner's fiber in vertebrates (Fig. 4.4).
Dorsally there is the lamellar body, a ciliary photoreceptor homologous to the pineal eye of vertebrates
(Figs. 4.3 and 4.4). Numerous large microvillar

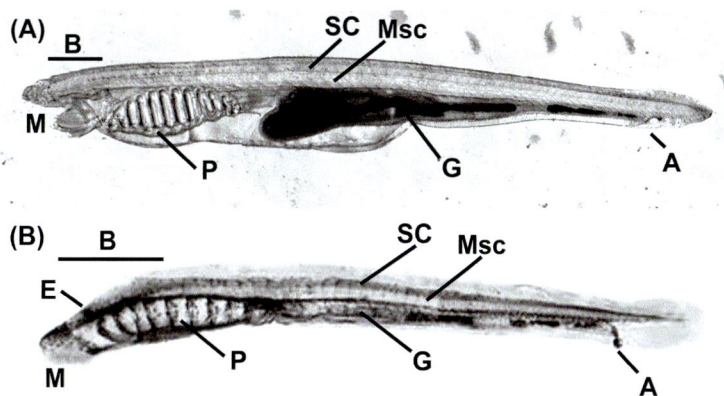

FIGURE 4.2   The body plans of (A) a young juvenile amphioxus (*Branchiostoma floridae*) and (B) a larval lamprey (ammocoete) (*Lampetra planeri*) are very similar, except the lamprey has a much larger brain (*B*), indicated by the bar, and paired eyes (*E*). Anterior to left: *A*, anus; *G*, gut; *M*, mouth; *Msc*, paraxial muscles; *P*, pharynx perforated by gill slits; *SC*, spinal cord.

photoreceptors (Joseph cells) occur immediately posterior to the lamellar body; ventrally at the same level is a segmental series of motor neurons, which extend posteriorly into the CNS. Finally, starting about the level of somite four, there is a series of microvillar photoreceptors each with a pigment cup (the organs of Hesse) along the entire remainder of the nerve cord (Figs. 4.2 and 4.3). These photoreceptors contain melanopsin with a maximum absorbance in the blue (del Pilar Gomez et al., 2009).

4.2.2   Initial Patterning of the Amphioxus Central Nervous System Is Comparable to That in Vertebrates

In spite of being very small, the amphioxus cerebral vesicle is considered to be homologous to the vertebrate diencephalic forebrain plus a small midbrain while the remainder of the nerve cord is comparable to the vertebrate hindbrain and spinal cord (Wicht and Lacalli, 2005). The genes mediating A/P patterning of the CNS as well as those mediating dorsoventral patterning are similarly expressed in amphioxus and vertebrates. As in vertebrates, *Otx* genes are expressed anteriorly in the amphioxus CNS, and the hindbrain has nested expression of *Hox* genes, although in vertebrates, the *Hox2* domain extends anterior to that of *Hox1*, whereas in amphioxus, *Hox1* is expressed anterior to *Hox2* (Fig. 4.5). In addition, in both amphioxus and vertebrates, genes patterning the dorsoventral axis of the CNS are expressed in similar patterns. In both, *Hedgehog* (*Hh*) is expressed in the floor plate of the CNS, although in vertebrates, *Hh* expression extends more anteriorly than in amphioxus, while *Msx* is expressed dorsally, *Gsh* laterally and *Nkx2.2* ventrally (Fig. 4.6).

4.2.3   Amphioxus Has Homologs of the Vertebrate Anterior Neural Ridge, Zona Limitans Intrathalamica, and Midbrain/Hindbrain Boundary

Amphioxus also has homologs of three organizing centers in the vertebrate brain—the anterior neural ridge (ANR), the zona limitans intrathalamica (ZLI), and the midbrain/hindbrain boundary (MHB) (Vieira et al., 2010; Fig. 4.5). Evidence for the presence of these three organizing centers comes chiefly from gene expression. The critical experiment to demonstrate that these regions function as organizers is to ectopically transplant tissue from these regions and determine that the transplants induce neighboring tissue to adopt a fate it would not normally have. Unfortunately, because amphioxus embryos are much, much smaller than vertebrate embryos, such transplantations are not feasible. Therefore, whether these three regions function as organizers in amphioxus embryos is difficult, if not impossible, to determine.

The ANR is at the anterior tip of the vertebrate CNS. In vertebrates, *Distalless* (*Dlx*) is expressed around the anterior edge of the neural plate, while *Fgf8* and *Bf1* (*FoxG1*) are expressed at the anterior tip of the CNS (Vieira et al., 2010; Eagleson and Dempewolf, 2002; Cajal et al., 2011). Gene expression at the anterior tip of the amphioxus CNS is similar. *Dlx* is expressed in the anteriormost portion of the neural plate and in adjacent ectoderm, *Fgf8* throughout the cerebral vesicle, and *Bf1* in the rostral nerves plus in a few cells at the anterior tip of the CNS (Holland et al., 1996; Bertrand et al., 2011; Toresson et al., 1998; Fig. 4.5).

In the vertebrate brain, the ZLI occurs about the midpoint of the diencephalon at the boundary between anterior expression of *Fezf* and posterior expression of

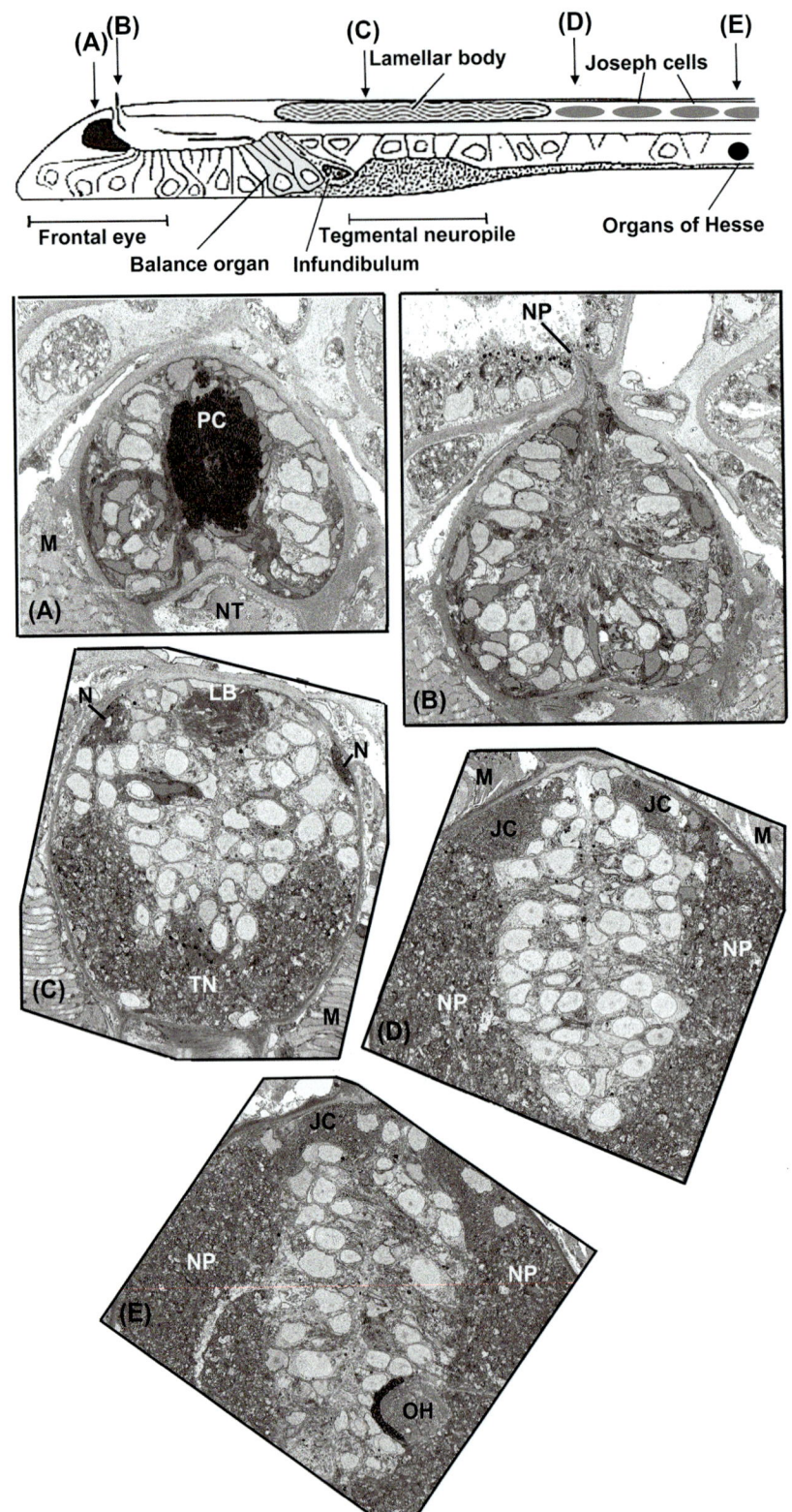

FIGURE 4.3 Cross sections from serial block-face scanning electron microscopy indicating the major features of a juvenile amphioxus nerve cord. *A–E* in diagram indicate levels of sections. (A) Section through a pigment cell (PC) of the frontal eye. (B) Section through the neuropore (NP). Photoreceptor cells are ventral. (C) Section through the lamellar body (LB). The anterodorsal nerves (N) have fused with the nerve cord. The tegmental neuropile (TN) is prominent. (D) Section through two Joseph cells (JC). The neuropile (NP) has extended laterally around the cell nuclei. (E) Section through the most anterior organ of Hesse (OH). *M*, muscle; *NT*, notochord.

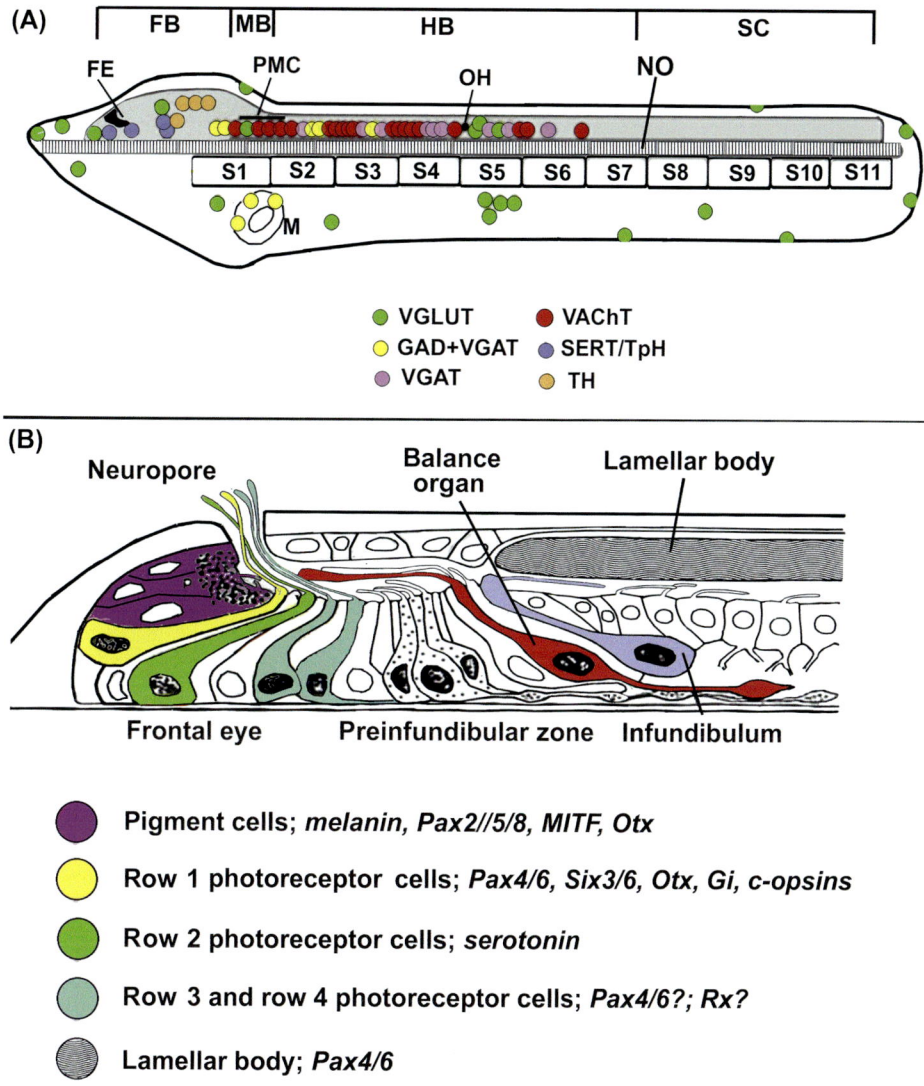

FIGURE 4.4   Diagrams of the anterior portion of the amphioxus central nervous system (CNS) showing major cell types and gene expression. (A) An early larva, anterior to the left indicating cells expressing neuropeptides. *VGLUT*, vesicular glutamate transporter; *GAD*, glutamic acid decarboxylase; *VGAT*, vesicular GABA/glycine transporter; *VACht*, vesicular acetylcholine transporter; *SERT*, serotonin transporter; *TpH*, tryptophan hydroxylase; *TH*, tyrosine hydroxylase. *FE*, frontal eye; *PMC*, primary motor center; *OH*, organ of Hesse (microvillar photoreceptor); *NO*, notochord; *FB*, forebrain; *MB*, midbrain; *HB*, hindbrain; *SC*, spinal cord; *M*, mouth; *NT*, notochord; *S0−S11*, somites. (After Candiani, S., et al., 2012. A neurochemical map of the developing amphioxus nervous system. BMC Neurosci. 13, 59.) (B) The anterior portion of the forebrain of a mid-late larvae of the amphioxus CNS showing gene expression in photoreceptor cells. *(Data according to Vopalensky, et al., 2012. Molecular analysis of the amphioxus frontal eye unravels the evolutionary origin of the retina and pigment cells of the vertebrate eye. Proc. Natl. Acad. Sci. U.S.A. 109, 15383−15383.)*

Irx. Several genes, including *Shh, Wnt 8b, Wnt2a, Fgf8*, and *Nkx2.2*, are expressed at high levels at this boundary (Martinez-Ferre and Martinez, 2012; Fig. 4.5). In amphioxus, the *Fezf* domain abuts that of *Irx* at the middle of the cerebral vesicle (Irimia et al., 2010). *Wnt8* is expressed in this region as is *Nkx2.2* (Holland and Holland, 2000; Schubert et al., 2000). Therefore, amphioxus has much of the genetic machinery for specifying the ZLI. However, *Hh*, which appears to be critical for the organizing properties of the ZLI (Vieira et al., 2010), is not comparably expressed in the diencephalon in amphioxus (Shimeld, 1999). Therefore, while amphioxus has part of the genetic

mechanism, it is missing some critical parts, which evidently evolved later.

The MHB in both vertebrate and amphioxus embryos is marked as the boundary between *Otx* expression in the forebrain and midbrain and *Gbx* expression in the hindbrain (Vieira et al., 2010; Castro et al., 2006; Fig. 4.5). In vertebrates at least, *Gbx* and *Otx* mutually repress one another, and this repression positions the MHB. However, other markers of the vertebrate MHB are not expressed exactly the same in amphioxus. In vertebrates, *Fgf8/17/18* and *Pax2/5/8* genes are upregulated at the MHB, and *Wnt1* and *En* are also expressed there

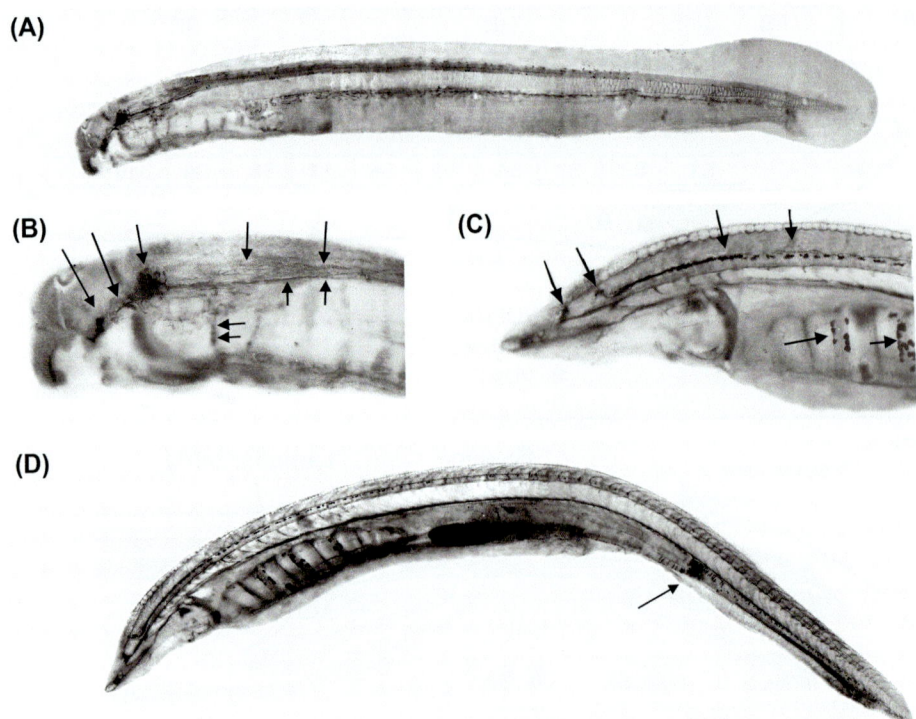

FIGURE 4.5 Serotonin expression in a young ammocoete larva of a lamprey (*Lampetra planeri*) (A, B) is comparable to that in a juvenile amphioxus (*Branchiostoma floridae*). (B) In the lamprey, there are three clusters of serotonergic neurons in the forebrain, midbrain, and anterior hindbrain, which send axons posteriorly in the central nervous system (CNS) (*arrows*). There are also serotonergic neurons associated with the pharyngeal bars between the gill slits. (C) In amphioxus, there are two clusters of serotonergic neurons, one near the frontal eye and another in the anterior hindbrain (*arrows*). These neurons send processes posteriorly in the CNS (*arrows*). Other serotonergic neurons extend posteriorly in the CNS. Numerous serotonergic cells are also associated with the gill bars (*arrows*). Inset: *arrows* show some of pigment cells associated with the organs of Hesse, microvillar photoreceptors that occur from the midpoint of the hindbrain and posteriorly to the end of the CNS. (D) Whole mount of the larvae in (C) showing additional serotonergic cells in the hindgut (*arrow*). The sizes of the ammocoete and the juvenile amphioxus are comparable at about 5 mm in length.

in a stripe. The combined action of all these genes confers organizer properties on the MHB (reviewed in Vieira et al., 2010). Thus, when transplanted into the hindbrain, the MHB will induce neighboring cells to adopt a midbrain identity and when transplanted into the midbrain, the MHB induces neighboring cells to adopt a hindbrain identity (Dworkin and Jane, 2013). However, in amphioxus, while *Pax2/5/8* is expressed from the *Otx/Gbx* boundary posteriorly down the nerve cord, and *Fgf8/17/18* is expressed in the entire cerebral vesicle anterior to that boundary, neither gene is strongly upregulated at the MHB. Moreover, *Wnt1* and *engrailed* (*En*) are not expressed at all at the *Otx/Gbx* boundary in amphioxus (Holland et al., 1997). Therefore, it seems that amphioxus has the initial mechanism for specifying and positioning the MHB but has evolved little of the mechanism that confers organizer properties on it in vertebrates. Since at least *engrailed* and *FGF8* are expressed at the MHB in lamprey as in other vertebrates, organizer properties probably evolved at the MHB at the base of the vertebrates (Holland et al., 1993; Rétaux and Kano, 2010).

Additional evidence that amphioxus has a midbrain homolog is from microanatomy. Because the vertebrate midbrain receives input from the paired eyes and, in amphioxus, the posterior third of the cerebral vesicle receives input from the frontal eye, Lacalli suggested that amphioxus has a small midbrain homolog. However, he could find no clear evidence for an amphioxus homolog of the vertebrate optic tectum (Lacalli, 1996, 2002). The idea that amphioxus has a small midbrain fits with the domains of *Otx* and *Gbx* abutting each other at the posterior limit of the cerebral vesicle (Castro et al., 2006) just as Otx and Gbx domains abut one another at the vertebrate MHB (Fig. 4.5). Another line of evidence that the posterior portion of the amphioxus cerebral vesicle is equivalent to the vertebrate midbrain is that motor neurons extend from this region through the *Otx/Gbx* boundary into the hindbrain equivalent (Lacalli and Kelly, 1999; Bardet et al., 2005; Jackman et al., 2000). Similarly, in vertebrates, motor neurons, marked by expression of *Islet*, extend from the midbrain past the MHB and into the hindbrain (Jackman et al., 2000).

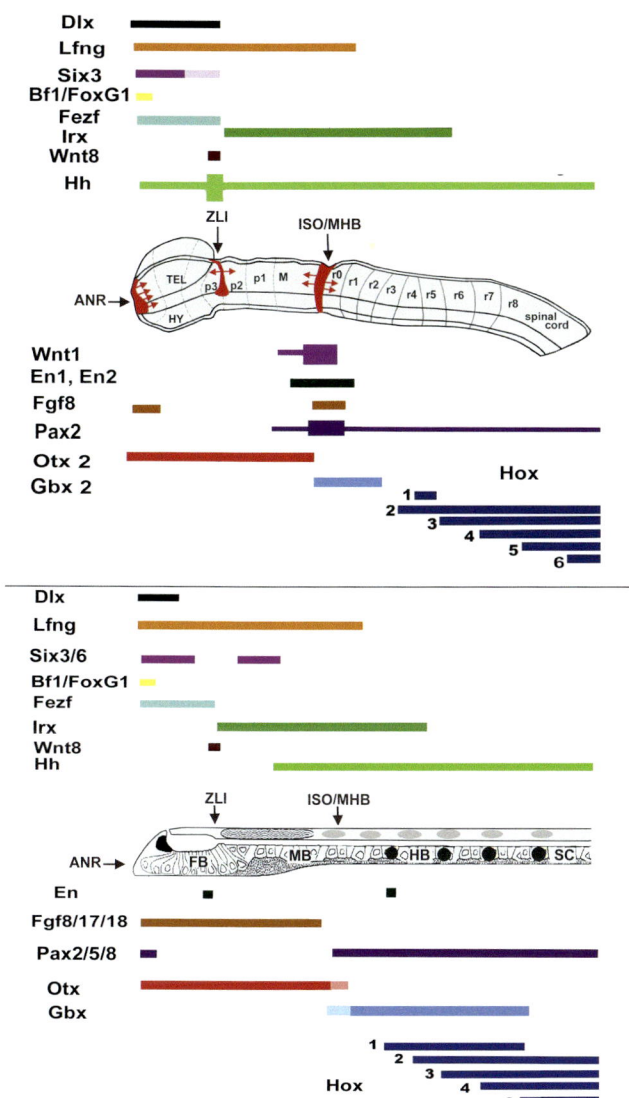

FIGURE 4.6 Comparison of anterior/posterior domains of gene expression in the central nervous system of a generic vertebrate (top) and amphioxus (bottom). In both, *Fezf* and *Irx* domains abut at the ZLI (zona limitans intrathalamica) and *Otx* and *Gbx* domains abut at the MHB (midbrain/hindbrain boundary). Gene expression at the ANR (anterior neural ridge) is similar in the two organisms as well. *ISO*, isthmic organizer; *FB*, forebrain; *MB* and *M*, midbrain; *HB*, hindbrain; *SC*, spinal.

However, in spite of these data, it has been argued that amphioxus completely lacks a midbrain—the *Otx/Gbx* boundary being the boundary between forebrain and hindbrain. This conclusion is based on comparing larval lampreys and amphioxus. In the larval lamprey, the optic nerves extend only into the posterior prosencephalon, not into the midbrain. However, later as the image-forming eye develops in association with metamorphosis, the optic nerves do extend into the midbrain (Suzuki et al., 2015). In addition, the posterior limit of *Pax 6* in the larval lamprey is at the

posterior end of the forebrain (prosencephalon), while in the amphioxus cerebral vesicle, the posterior limit of *Pax6* coincides with the boundary between *Otx* and *Gbx* (Suzuki et al., 2015; Glardon et al., 1998; Murakami et al., 2001; Derobert et al., 2002). Therefore, it was concluded that the cerebral vesicle of amphioxus probably equaled the prosencephalon of the lamprey. Deciding whether amphioxus truly has a midbrain would require additional data. Unfortunately, there are no gene markers that are specific for the vertebrate midbrain that could be used to infer homologies with amphioxus.

4.2.4 Neuropeptide Expression Helps Reveal Homologies Between the Amphioxus and Vertebrate Brains

The expression pattern of neuropeptides, including serotonin (5-hydroxytryptamine), in amphioxus and agnathans underscores that amphioxus really is a reasonable approximation of the ancestral chordate (Figs. 4.4A and 4.7). As Fig. 4.7 shows, the larval lamprey has large serotonergic cells associated with the eye, a second small cluster close to the MHB and a more posterior cluster in the anterior hindbrain. These clusters of serotonergic neurons send processes posteriorly in the nerve cord. Detailed descriptions of serotonergic cells in a larval lamprey can be found in references (Antri et al., 2006; Barreiro-Iglesias et al., 2009). Amphioxus has just two clusters of serotonergic neurons in the brain. Like the lamprey, amphioxus has an anterior cluster of serotonergic neurons associated with the frontal eye (the row 2 cells) and a more posterior cluster which probably corresponds to the cluster of serotonergic neurons in the anterior hindbrain of the larval lamprey. The neurons in these clusters send processes posteriorly in the nerve cord, in which there are additional serotonergic neurons. There are also large serotonergic cells associated with the gill bars in amphioxus and in the lamprey as well as clusters of serotonergic cells in the hindgut (Fig. 4.7; Holland and Holland, 1993).

Although a detailed neurochemical map of embryonic and early larval nerve cord in amphioxus has been constructed for several neuropeptides and/or their related enzymes and transporters [glutamate (GAD, glutamic acid decarboxylase), vGLUT (vesicular glutamate transporter), serotonin GABA, glycine (VGAT, vesicular GABA/glycine transporter), acetylcholine (VaChT, vesicular acetylcholine receptor)] (Fig. 4.4), it is often difficult to identify the precise counterparts of these cells in vertebrates because the vertebrate brain is multilayered and contains far more neurons than that of amphioxus (Candiani et al.,

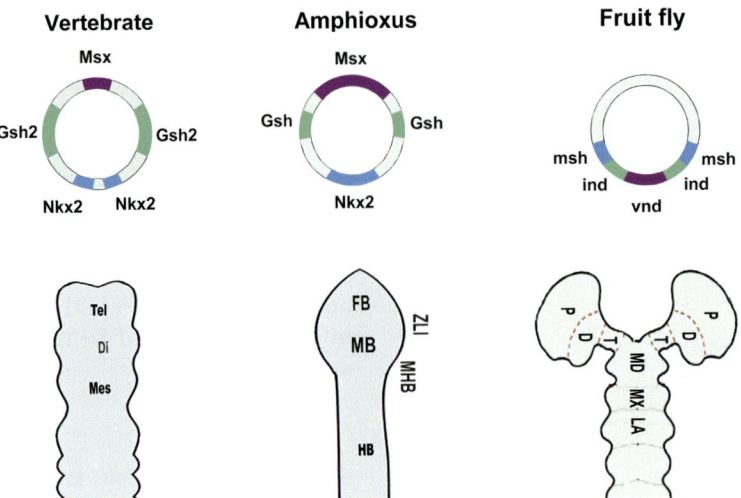

FIGURE 4.7   Dorsoventral gene expression in the central nervous system (CNS) is comparable between vertebrates, amphioxus, and the fruit fly (*Drosophila melanogaster*). Top: Msx (formerly Homeobox 7) is expressed dorsally in the vertebrate and amphioxus brains, and its homolog *vnd* is expressed ventrally in the *Drosophila* CNS. *Gsh* is expressed medially in both the vertebrate and amphioxus CNS, and its homolog *ind* is also expressed medially in the *Drosophila* CNS. *Nkx2* is expressed ventrally in both the amphioxus and vertebrate CNS, while its homolog *msh* is expressed at the lateral edges of the CNS in *Drosophila*. Bottom: schematic views of the anterior portion of the CNS in vertebrates (left), amphioxus (middle), and *Drosophila* (right). *Tel*, telencephalon; *Di*, diencephalon; *Mes*, mesencephalon (midbrain); *FB*, forebrain; *MB*, midbrain; *HB*, hindbrain; *ZLI*, zona limitans intrathalamica; *MHB*, midbrain/hindbrain boundary; *P*, protocerebrum; *D*, deuterocerebrum; *T*, tritocerebrum; *MD*, mandibular segment; *Mx*, maxillary segment; *LA*, labial segment.

2012). Even so, some homologies can be inferred. For example, vGLUT, which is specific for glutamatergic neurons, is expressed in a few cells of the anterior ectoderm of early amphioxus embryos, which has been suggested on the basis of *Pax6* expression to be homologous to the olfactory epithelium of vertebrates (Glardon et al., 1997). In the lamprey, vGLUT is initially expressed in the olfactory epithelium (Villar-Cerviño et al., 2010, 2011), strengthening the proposed homology of the olfactory epithelium of vertebrates and the anterior ectoderm of amphioxus embryos. However, vGLUT is also expressed in cells associated with the frontal eye of amphioxus as well as in some cells in the hindbrain just posterior to the first organ of Hesse to develop (Candiani et al., 2012). Later, additional cells in the cerebral vesicle also express vGLUT. In the lamprey, vGLUT becomes expressed in several cranial ganglia, in the pineal, and in many other cells in the brain and spinal cord. Amphioxus lacks cranial ganglia, and the lamellar body, the pineal homolog, does not express vGLUT. Whether any of the other specific cells in the amphioxus CNS that express vGLUT are homologous to those expressing vGLUT in the lamprey brain is likewise unclear as a precise correlation of most neurons in amphioxus with those in vertebrates has not yet been achieved.

GABAergic neurons are segmentally distributed in the amphioxus CNS from the middle of the cerebral vesicle to just posterior to the most anterior organ of Hesse in the hindbrain (Candiani et al., 2012; Fig. 4.4A). In contrast, in the larval lamprey, there are three segmentally arranged clusters of GABAergic neurons in the olfactory bulbs of the telencephalon, the preoptic area of the hypothalamus, and the midbrain (Reed et al., 2002); expression in the hindbrain and spinal cord was not examined. In addition, for several of these neuropeptides, labeling has been done on sections of adult amphioxus (Anadon et al., 1998), but clear diagrams putting these labeled cells into the broader context of the whole animal have not been produced. Therefore, it is difficult to correlate these sections with specific structures in vertebrates, let alone those of amphioxus.

VAChT is expressed in segmentally arranged neurons in the amphioxus forebrain, midbrain, hindbrain, and spinal cord; expression in the last three regions is largely in motor neurons (Candiani et al., 2008, 2012; Fig. 4.4A). It is also expressed in cells associated with the anterior-most organ of Hesse. Adult lampreys have very many cholinergic neurons, including motor neurons and cells associated with the retina (Pombal et al., 2001). Even though it is difficult to correlate specific neurons in the amphioxus brain with those of the lamprey, if the distribution of serotonergic and motor neurons is any indication, the patterns of some other neuronal subtypes are likely to be similar in the two species with additional subtypes being added as the agnathan brain expanded and evolved some new structures.

4.2.5 Evolution of Eyes

Much of the effort to find commonalities between the amphioxus and vertebrate CNS has focused on the evolution of the vertebrate paired eyes. Based on the fine structure, Lacalli (1996) argued for homology of the amphioxus frontal eye and the vertebrate retina. This conclusion has been supported by evidence from both gene expression and labeling of nerve tracts in amphioxus and larval lampreys. The amphioxus frontal eye is quite simple. There are four rows of ciliated photoreceptor cells in the frontal eye. The cilia extend out of the neuropore (Fig. 4.4B). The row 1 cells express two different c-opsins and a G-protein alpha subunit, which appears to be involved in transduction of the photoreceptor signal (Vopalensky et al., 2012). These photoreceptor cells, like cells in the eyes of many phyla, also express *Pax4/6*, *Otx*, and *Six3/6*. The row 2 photoreceptor cells are serotonergic neurons, which extend cell processes posteriorly to the tegmental neuropile, which is just posterior and ventral to the lamellar body (Suzuki et al., 2015; Holland and Holland, 1993; Vopalensky et al., 2012). It is not clear what genes are expressed in the cells of rows 3 and 4.

Except for the trajectory of the neuronal projections from the anterior photoreceptor, there are no structures in the posterior cerebral vesicle of amphioxus that have clear counterparts in the midbrain of vertebrates. The most striking cells in this region are the large, microvillar photoreceptors, the melanopsin-containing Joseph cells, which appear to be unique to amphioxus (Fig. 4.2). The ipRGCs of the vertebrate retina, which are non-image-forming photoreceptors, contain melanopsin and have been suggested to be evolutionarily related to the Joseph cells. However, the Joseph cells are not located near the frontal eye. Perhaps some of the deep brain photoreceptors, which express a wide variety of opsins and are located in the hypothalamus of nonmammalian vertebrates (António et al., 2012), are homologous to the Joseph cells. However, more work would be needed to see if there are real parallels between deep brain photoreceptors in vertebrates and the Joseph cells of amphioxus (Fernandes et al., 2013; Pulido et al., 2012).

4.3 What Structures Did the Vertebrate Brain Invent?

Major neural structures that agnathans possess that amphioxus seems to lack include a telencephalon, neural crest, and placodes. More than anything, the telencephalon is critical in the evolution of vertebrates because it develops into the pallium or cerebrum, the seat of cognition, and other complex functions in vertebrates. Because the amphioxus frontal eye is at the anterior tip of the CNS, it is assumed that the anteriormost portion of the amphioxus CNS is equivalent to the vertebrate diencephalon. *Bf-1* (*Fox G1*) is expressed in cells associated with the frontal eye in amphioxus, and, although it is expressed in the vertebrate telencephalon, it is also expressed in part of the optic stalk and eyes (Toresson et al., 1998). Therefore, its expression at the tip of the amphioxus nerve cord probably reflects homology of the amphioxus frontal eye and the vertebrate eyes and does not indicate that amphioxus has a telencephalic homolog. Unfortunately, there are no gene markers that are absolutely telencephalic specific that could provide additional evidence.

4.3.1 Neural Crest

In vertebrates, pluripotent neural crest cells at the edges of the neural plate migrate away from it into the pharyngeal arches and other tissues (Fig. 4.8A and B). They develop into many cell types including the cartilage of the head, parts of the cranial ganglia, cells of the adrenal medulla, and pigment cells among others. The relationship between amphioxus and vertebrates in regard to neural crest is relatively clear. Amphioxus does not have definitive neural crest, but it has a migrating cell population at the edges of the neural plate (Holland et al., 1996). Before the neural plate rounds up, the ectoderm on either side of it migrates as sheets over the neural plate to fuse in the dorsal midline. However, this migrating cell population, unlike vertebrate neural crest, remains ectodermal and never breaks up into individual cells.

Although amphioxus lacks neural crest, most of the genetic machinery for specification of the neural plate and its edges is comparable to that in vertebrates (Figs. 4.8C–D and 4.9; Meulemans and Bronner-Fraser, 2004). In amphioxus, the migrating edges of the lateral ectoderm express *distalless* (*Dlx*) as does presumptive neural crest in vertebrates (Holland et al., 1996). In both amphioxus and vertebrates, there is a high level of BMP2/4 signaling in ventral and lateral ectoderm and much lower levels in the neural plate. Upregulating BMP signaling at the blastula stage in amphioxus eliminates the neural plate and other dorsal structures, while upregulating BMP at the early gastrula stage increases the number of ectodermal sensory cells and alters their position (Yu et al., 2007; Lu et al., 2012). Similarly, upregulation of BMP in early development inhibits the formation of dorsal structures in *Xenopus* (Jones et al., 1996). However, while most genes that specify neural crest identity are expressed at the edges of the neural plate in amphioxus, some are not. In addition, most genes

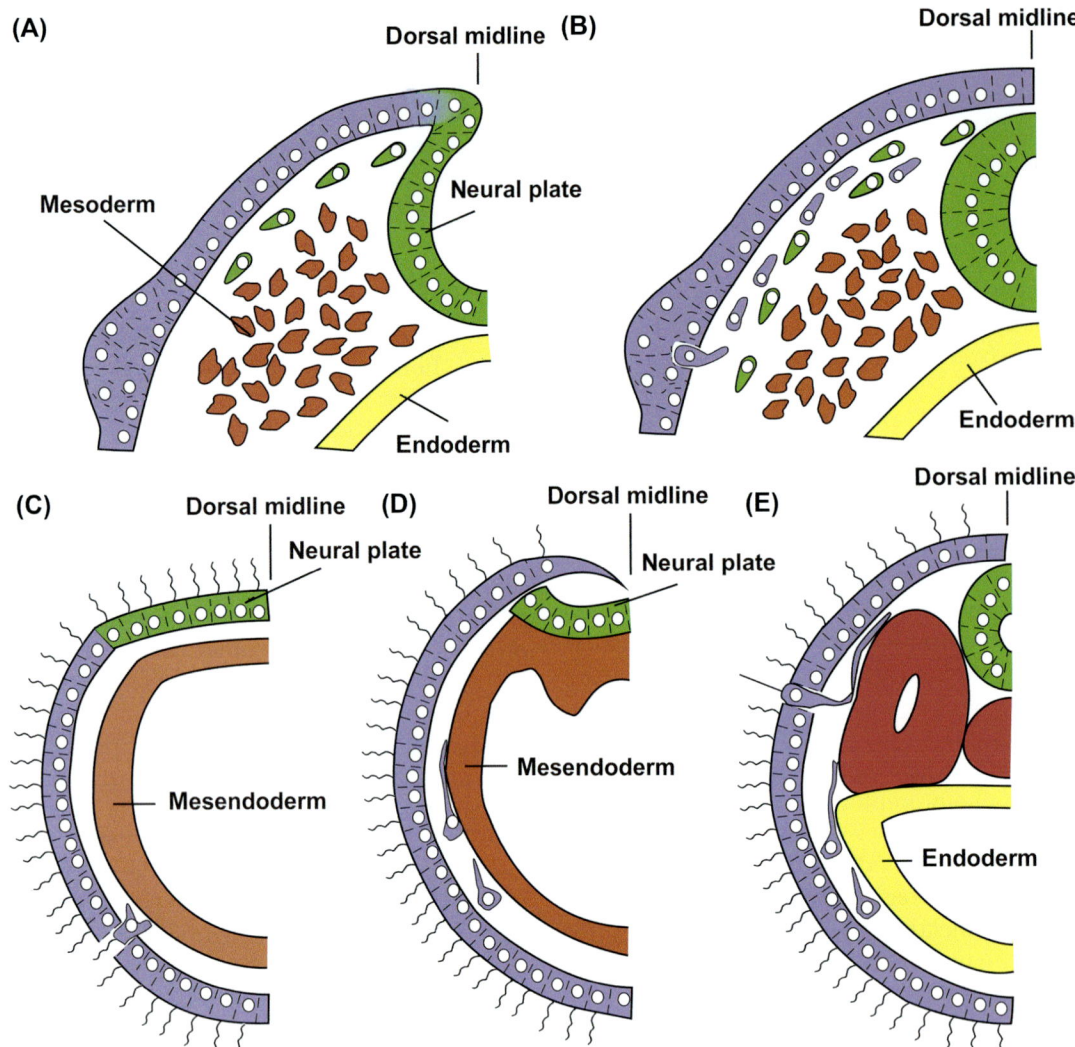

FIGURE 4.8   Migration of neuroblasts from neurogenic placodes in the chick (A, B) resembles the migration of ectodermal sensory cells in amphioxus (C–E). (A) In the chick, neural crest cells (*green*) migrate ventrally from the dorsal edges of the neural tube as it nears closure. (B) Neuroblasts (*blue*) from neurogenic placodes follow the track of the neural crest cells and migrate dorsally to contribute to cranial ganglia. (C) In amphioxus, ectodermal sensory cells, which are generated in a 30-degree arc of ventral ectoderm, lose their cilia and sink beneath the ectoderm. (D) They migrate dorsally underneath the ectoderm and (E) generate axons, which grow into the central nervous system, develop a specialized cilium and reinsert into the ectoderm. Neurulation in amphioxus differs from that in all vertebrates in that the ectoderm adjacent to the presumptive neural plate detaches from it and migrates over it as sheets of ectoderm (D, E). Once the sheets of ectoderm have fused in the dorsal midline, the neural plate (*green*) rounds up to form a neural tube (E). *From Holland, L.Z., 2009. Chordate roots of the vertebrate nervous system: expanding the molecular toolkit. Nat. Rev. Neurosci. 10, 736–746.*

involved in neural crest migration and differentiation in vertebrates are not expressed at the edges of the neural plate in amphioxus (Meulemans and Bronner-Fraser, 2004). Key among the neural crest genes that are not similarly expressed in amphioxus are *FoxD3* and *SoxE*. Experimental data have shown that subsequent to the whole-genome duplications that occurred at the base of the vertebrates, one of the five duplicates of the ancestral *FoxD* gene, *FoxD3*, gained new regulatory elements that allowed it to be expressed at the edges of the neural plate as well as new amino acid sequences at the amino terminal that allowed the FoxD3 protein to induce expression

of downstream neural crest genes (Ono et al., 2014; Yu et al., 2008). This is an excellent example of how cooption of genes for new functions after genome duplications has involved not only the invention of new regulatory elements but also the invention of new amino acid sequences. A second example of the evolution of new regulatory elements is provided by *SoxE*. The single amphioxus *SoxE* gene is not expressed at the edges of the neural plate, while some of the vertebrate duplicates (eg, *Sox9*, *Sox10*) are expressed there (Fig. 4.9). The regulatory region of amphioxus *SoxE*, like that of amphioxus *FoxD*, directs expression to the domains in the vertebrate

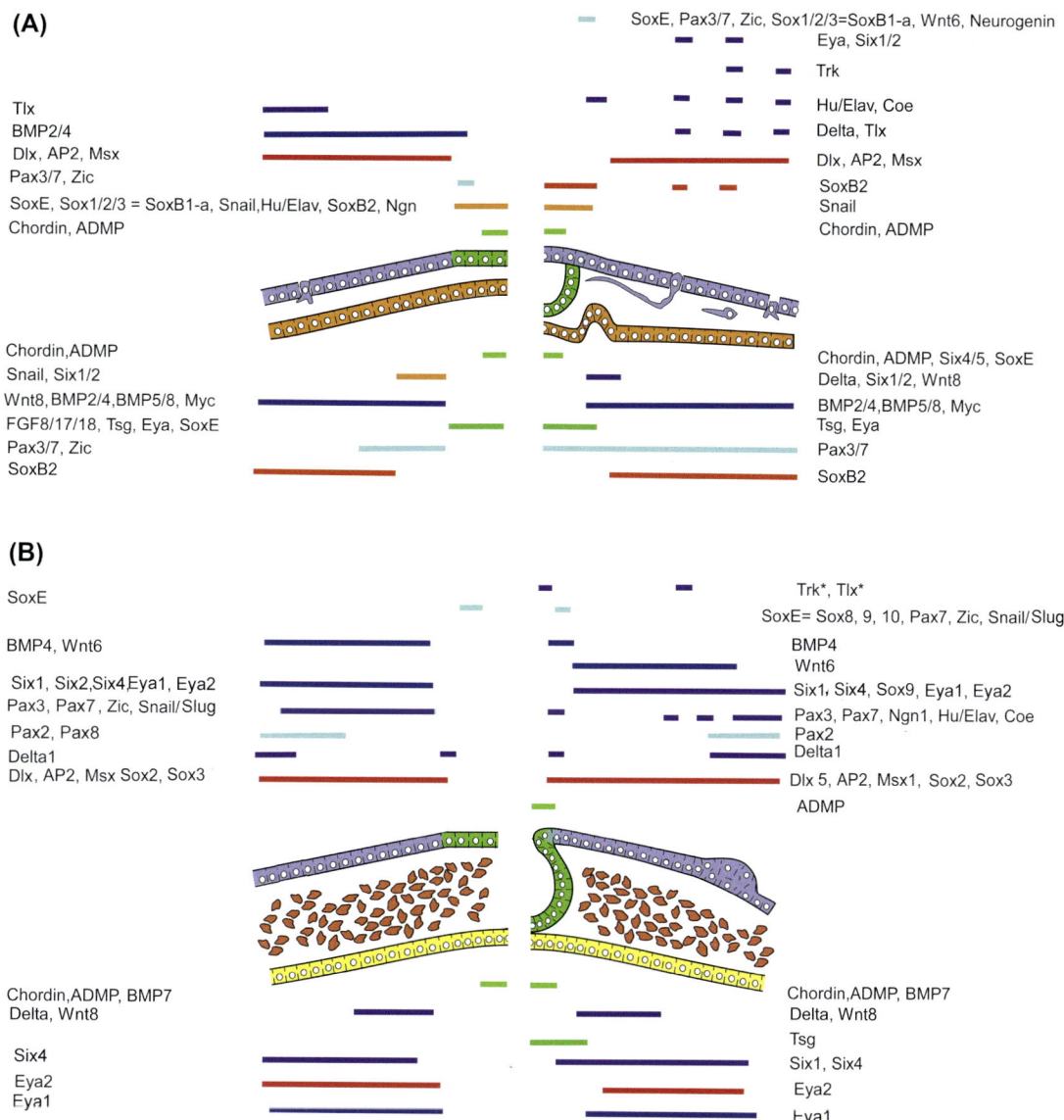

FIGURE 4.9 Gene expression in the neural plate border region, neural tube, and underlying mesendoderm in amphioxus (A) is largely like that in a jawed vertebrate (B). Ectodermal gene expression is shown above and mesodermal and endodermal gene expression is shown below each diagram. (A) Amphioxus embryos at the early neurula (left) and mid-neurula (right). Embryos are shown in cross section about the midpoint along the anterior–posterior axis and are split ventrally along the midline and flattened out. In the early neurula, *Sox1/2/3* is expressed in the entire neural plate, *neurogenin (Ngn)* in the posterior third of the neural plate, *Pax3/7* and *Zic* at the edges of the neural plate, and *Tlx* in the ventral ectoderm giving rise to ectodermal sensory cells. *BMP* is weakly expressed throughout the ectoderm except the neural plate. The BMP antagonists *chordin* and *ADMP* are expressed in the center of the neural plate and in the underlying mesendoderm. At the mid-neurula stage, the migrating and definitive ectodermal sensory cells express a number of genes, including *Eya, Six1/2, Trk, Hu/Elav, Coe, Delta, Tlx,* and *SoxB2*. Expression genes initially broadly expressed in the early neural plate (i.e., *SoxE, Sox1/2/3, Snail, Hu/Elav*) become restricted to the edges of the neural plate together with *Wnt6* and *Pax3/7*. (B) A consensus of gene expression in the neural plate border region and underlying mesoderm in a generic vertebrate. Cross sections through the head of an embryo at the neural plate stage (left) and late neurula before the onset of migration of neural crest and placodal derivatives (right). Because *Trk* and *Tlx* (asterisks) are not expressed in placodal derivatives before their migration, their expression at a later stage is shown. Tissues ventral to the panplacodal region are not shown. Domains along the anterior–posterior extent of the panplacodal region are summed. Expression domains are chiefly based on the chick and frog, but some from the mouse, zebrafish, and lamprey have been included where expression of homologs in the chick and frog is unclear. *From Holland, L.Z., 2009. Chordate roots of the vertebrate nervous system: expanding the molecular toolkit. Nat. Rev. Neurosci. 10, 736–746.*

that are comparable to those that normally express the gene in amphioxus, but not to the edges of the neural plate (Jandzik et al., 2015). It has not yet been determined if the vertebrate SoxE homologs have also gained new

amino acid sequences critical for specification of neural crest. It is likely that more such examples of how duplicate genes acquire new functions will come to light in the future.

4.3.2 Placodes

In vertebrates, lateral line and otic placodes contain hair cells, which are mechanosensory cells with microvilli arranged in a row of descending height, often covered with extracellular material termed a cupula. A cilium is often present as well. It is generally agreed that sensory neurons deriving from vertebrate placodes probably evolved from ancestral ectodermal sensory cells. Cephalochordates have several kinds of ectodermal sensory cells that vary in the morphology of microvilli and cilia. However, neither in amphioxus nor in tunicates, which are the sister group of vertebrates, is the morphology of the ectodermal sensory cells comparable to that of vertebrate hair cells. Moreover, in amphioxus, these cells are not organized into clusters as vertebrates. What the ectodermal sensory cells in amphioxus have in common with cells in neurogenic placodes in vertebrates is that the neuronal precursors of both leave the ectoderm and migrate dorsally underneath it to reach their final destinations (Begbie, 2008; Fig. 4.8). In amphioxus, the first cohort of cells is generated in the ventral midline of the gastrula, leaves the ectoderm, migrates dorsally just underneath the ectoderm, reinserts their apices into the lateral ectoderm, and sends axons into the CNS. In vertebrates, the neurogenic placodes are in the region just outside the neural plate, but the cells also migrate dorsally beneath the ectoderm to their final locations. Most of the placodal neurons in vertebrates do not send axons into the CNS but synapse with interneurons. In amphioxus, the reverse is true. Most of the ectodermal sensory cells send axons to the CNS. An extensive review of vertebrate placodes and their possible evolutionary origins is given in (Patthey et al., 2014).

The genes expressed by ectodermal sensory cells in amphioxus and vertebrates are generally comparable (Holland, 2005). These include *Six1/2*, *Six4/5*, *Eya1/2*, *Tlx* (*Hox11*), *SoxB*, *Trk* (a tyrosine kinase receptor), *Delta*, and *Hu/Elavi*. All these genes are expressed in the panplacodal region and/or placodal neuroblasts in vertebrates as well as in ectodermal sensory cells in amphioxus. Similarly, *Islet*, *ERR* (*estrogen-related receptor*) and an *Irx* gene are expressed in particular placodes in vertebrates and some sensory cells in amphioxus. However, there are also differences. *Dach*, which typically cooperates with *Six* and *Eya* genes, the neural differentiation gene *neurogenin*, *Pax2/5/8* and *Pax3/7* are expressed in vertebrate placodes, but apparently not in amphioxus sensory cells. These comparisons suggest that a conserved gene network initiates the development of ectodermal sensory cells in both amphioxus and vertebrates, but that the more downstream components of the network are divergent. Thus, the evolution of

vertebrate placodes like that of neural crest seems to have involved the conservation of early-expressed genes plus the cooption of genes for late functions.

4.4 What About Tunicates?

Comparisons of the CNSs of amphioxus and vertebrates, which are both evolving slowly, have given a fair picture of the CNS of the ancestral chordate. Tunicates, as the sister group of vertebrates, ought to help give a picture of their common ancestor with vertebrates. Unfortunately, tunicates are evolving much more rapidly than other chordates and have reduced genomes (the genome of the ascidian *Ciona intestinalis* genome is about 170 mb, that of the appendicularian *Oikopleura dioica* is about 70 mb, while those of cephalochordates are ~520 mb). Therefore, the picture painted by tunicates is very blurry.

Two of the five major tunicate groups have nerve cords: ascidians have one only in the larva (Fig. 4.1) while appendicularians have a nerve cord in both larvae and adults. Although the CNS of ascidian larvae has received more attention than that of appendicularians, the latter nerve cord is probably more representative of that in the ancestral tunicate as it contains neurons in the tail nerve cord, whereas that of ascidians does not. Compared to the estimated 20 000 neurons in the adult amphioxus nerve cord, those of tunicates are greatly reduced (Fig. 4.1). The ascidian *C. intestinalis* has about 330 cells in the larval CNS (Meinertzhagen et al., 2004) of which about 80 are neurons (Nicol and Meinertzhagen, 1991). In contrast, the nerve cord of the appendicularian *O. dioica* has fewer cells, about 80–90 in the cerebral ganglion and about 30 in the caudal ganglion of which about 29 are neurons (Soviknes et al., 2005). Likely motor neurons are located in the tail nerve cord and the caudal ganglion (Soviknes et al., 2007). Although motor neurons are probably homologous throughout the chordates, it is very difficult to infer homologies of other neurons because there are so few of them in tunicates and because they appear to have nothing apparently comparable to such structures as the pineal. Ascidian tadpoles and appendicularians do have melanin-containing cells in the CNS that mediate photoreception and gravity sensing, but their homologies with amphioxus and vertebrate photoreceptor cells are unclear.

In spite of having greatly reduced nervous systems, tunicates have some features suggesting that the CNS in the common ancestor of tunicates and vertebrates had the beginnings of neural crest and a midbrain/hindbrain organizer. Evidence for part of the gene network involved in making vertebrate neural crest is that *Twist*, a mesodermally expressed gene, when expressed in the *Ciona* neural plate, will cause some cells to migrate away from the CNS (Abitua et al., 2012). *Twist* is expressed in migrating neural crest in vertebrates, but not in amphioxus, where it is only expressed in mesoderm (Yasui et al., 1998). Therefore, like *FoxD3* and *SoxE* genes, *Twist* may have acquired new regulatory elements at the base of the vertebrates, allowing it to be expressed in migrating neural crest, precisely when the Twist protein acquired the ability to induce neural crest to migrate is uncertain. Whether amphioxus *Twist* can induce neural cells to migrate has not been studied. In addition, in another species of ascidian, cells in the vicinity of the nerve cord were shown to migrate and differentiation into pigment cells (Jeffery et al., 2004). However, the migration of these cells comes after the neural tube has formed, and it is unclear whether they come from the neural tube itself or from nearby tissue. Better evidence for a relative of vertebrate neural crest in tunicates is that two cells on each side of the embryo, which are precursors of the "bipolar tail neurons," move out from the closed neural tube and then migrate anteriorly on either side of it. They remain outside the neural tube and ultimate synapse with motor neurons. It was suggested that these cells may be homologous to vertebrate neurons in the dorsal root ganglia. Even so, these cells lack the pleuropotency of vertebrate neural crest cells, which evidently evolved at the base of the vertebrates.

It has also been suggested that the tunicate CNS has a homolog of the vertebrate MHB since *Pax2/5/8* is expressed in a few cells in the neck region between the sensory vesicle and the visceral ganglion rather than throughout tail nerve cord as in amphioxus. However, an examination of expression domains of additional genes raised the possibility that ascidians lack a midbrain (Ikuta and Saiga, 2007). The situation remains uncertain in part because tunicates lost the *Otx* gene. Moreover, *Pax2/5/8* genes do not appear to be expressed in the CNS in *Oikopleura* (Bassham et al., 2008). Tunicates have clearly lost much of what their ancestor had, in terms of both structures and genes, making it quite difficult, if not impossible, to reconstruct this long extinct ancestor.

4.5 The Roots of the Chordate Nervous System

Although the Ambulacraria (echinoderms and hemichordates) are the sister group of chordates, they are of limited help in clarifying where the chordate CNS came from. Most agree that echinoderm nervous systems originated de novo within the Echinodermata (reviewed in Holland, 2015a). There is no differentiation into a brain, and genes such as *Hox* genes, which are characteristically expressed in the CNSs of chordates,

are not expressed in the nerve cords of adult echinoderms or in the nervous tissues of the larvae (Popodi et al., 1994; Kikuchi et al., 2015). Hemichordates have both dorsal and ventral nerve cords, but they lack a brain (Fig. 4.1). All possibilities for homologies between the hemichordate and chordate nervous systems have been proposed: the dorsal nerve cord is homologous to the chordate nerve cord, the ventral nerve cord is homologous to the chordate nerve cord, neither the dorsal nor ventral nerve cords is homologous to the chordate nerve cord. The nerve cord in the collar region does undergo a sort of neurulation and continues posteriorly as the dorsal nerve cord and anteriorly as nerves that branch out in the proboscis ectoderm. The ventral nerve cord in the trunk is connected to the collar nerve cord by the circumesophageal tract (Fig. 4.1).

Bullock (1944) proposed that the nerve cords of hemichordates were composed of axons and did not contain nerve cell bodies. Therefore, for many years the "real" hemichordate nervous system was thought to be the diffuse nerve net in the general ectoderm (Lowe et al., 2003). Consequently, several studies focused on comparing patterning in the ectodermal nerve net in a direct developing hemichordate, *Saccoglossus kowalevskii* to that in the vertebrate CNS (Lowe et al., 2003; Aronowicz and Lowe, 2006). However, more recently, morphological studies have shown that the nerve cords do contain some neuronal cell bodies (Nomaksteinsky et al., 2009; Kaul-Strehlow et al., 2015) and that the larval nervous system, whether in direct or indirect developing hemichordates, disappears as the adult nerve cords develop (Cunningham and Casey, 2014; Miyamoto et al., 2010). One of these studies proposed that the dorsal nerve cord is evolutionarily related to the chordate nerve cord (Miyamoto and Wada, 2013), while another equivocated as to whether the dorsal or the ventral nerve cord was equivalent to the chordate nerve cord (Nomaksteinsky et al., 2009). As there are relatively few cell bodies in the hemichordate nerve cords, it is uncertain whether one or the other represents a step along the way to development of a more complex nerve cord in cephalochordates or whether the deuterostome ancestor had a more complex nerve cord that was secondarily reduced in the ancestral hemichordate. Sorely needed are studies of gene expression during development of the adult nerve cords, which would allow comparisons with chordate nerve cords.

4.6 Where Did the Chordate Central Nervous System Come From?

The controversial placement of the phylum Xenacoelomorpha (xenoturbellids, nemertodermatids, and acoel flatworms) in the deuterostomes as the sister group of

Ambulacraria (Philippe et al., 2011) created additional problems for understanding the origins of chordate nervous systems. *Xenoturbella* has a diffuse nerve net (Raikova et al., 2000), while nemertodermatids have a basiepithelial nervous system with some axon tracts, and acoels have anterior concentrations of neurons, sometimes called a brain, with up to six tracts of axons extending posteriorly (Perea-Atienza et al., 2015; Fig. 4.1). Thus, if the chordate CNS evolved from a nervous system like any one of these, it becomes very hard to explain similarities between protostome and chordate nervous systems (see later discussion). However, molecular phylogenetic analyses with large sets of nuclear genes have recently firmly placed Xenacoelomorpha at the base of the bilaterians (Cannon et al., 2016; Rouse et al., 2016). This has revived ideas of what the basal bilaterian nervous system was like and whether or not it had a longitudinal nerve cord with a brain, which gave rise to nerve cords in both protostomes and deuterostomes.

In one proposed scenario, the nerve net of an ancestral cnidarian evolved into the nervous system of an ancestral bilaterian that consisted of an anterior concentration of neurons with several nerve cords as in acoel flatworms (Semmler et al., 2010). An animal like this then evolved into a wormlike organism with either a single ventral nerve cord or paired ventral nerve cords and an anterior brain, which may have been tripartite like the brains of arthropods and chordates. This organism then gave rise to deuterostomes on the one hand and to protostomes on the other. The entire CNS was then replaced by the pentameral nerve cords in echinoderms, while in hemichordates the ancestral brain may have become somewhat degenerate and splayed over the proboscis ectoderm. At some point during evolution of deuterostomes, there was a dorsal–ventral inversion so that the ventral nerve cord became dorsal. The CNS with anterior brain was retained in chordates.

The evidence for this scenario comes largely from similarities in gene expression and to some extent from morphological data. Comparisons of the expression patterns of genes that mediate A/P and mediolateral patterning in the CNSs of protostomes and deuterostomes are consistent with the idea that the bilaterian ancestor probably had a longitudinal nerve cord with a brain. For example, *Otd*, the *Drosophila Otx* homolog, is expressed throughout the anteriormost regions of the *Drosophila* brain (the protocerebrum and deutocerebrum), while *unpg* (homologous to *Gbx*) is expressed in the tritocerebrum, subesophageal ganglion, and ventral nerve cord (Hirth et al., 2003). The domains of *Otd* and *umpg* abut between the deutocerebrum and tritocerebrum. This resembles the patterns of *Otx2* and *Gbx2*, which abut at the MHB in vertebrates (Hirth et al., 2003; Inoue et al., 2012) and in amphioxus (Castro et al., 2006; Fig. 4.5). Therefore, it was proposed that

the three sections of the *Drosophila* brain— protocerebrum, deutocerebrum, and tritocerebrum (Hirth and Reichert, 1999; Fig. 4.6)—are homologous to the forebrain, midbrain, and hindbrain of vertebrates (Reichert, 2005). Also, in *D. melanogaster*, the *earmuff* gene (homologous to *Fezf*) is expressed in the anterior part of the larval brain with a posterior boundary at the protocerebrum/deutocerebrum boundary just anterior to the domain of *mirror*, homologous to *Irx* (Pfeiffer et al., 2008). This may be comparable to the domains of *Fezf* and *Irx1* abutting at the ZLI in vertebrates and in the middle of the forebrain in amphioxus (Irimia et al., 2010; Hirata et al., 2006; Fig. 4.5). Other correspondences in gene expression are the *Pax2/5/8* domain just anterior to the deutocerebrum/tritocerebrum boundary in *D. melanogaster*, and expression of the three vertebrate *Pax2/5/8* genes at the MHB and that of the *D. melanogaster Hox1* gene, *labial*, in a stripe at the posterior end of the tritocerebrum compared to *Hoxb1* in a stripe in the vertebrate hindbrain (Hirth et al., 2003; Püschel et al., 1992). In addition, homologs of genes that pattern the *Drosophila* nerve cord mediolaterally are expressed in comparable domains in the vertebrate CNS. Thus, *msh*, *ind*, and *vnd* are expressed in abutting domains from lateral to medial example in the *D. melanogaster* neuroectoderm (reviewed in Bailly et al., 2012; Urbach and Technau, 2008) while their homologs in chordates are expressed in comparable domains—*msh* in the roof plate, *Gsh1*, homologous to *ind* in the alar plate, and *Nkx2.2*, homologous to *vnd* ventrally in the basal plate (Fig. 4.6).

Adding weight to the idea of a bilaterian ancestor with a longitudinal nerve cord and brain are data from the annelid *Platynereis dumerilii* (Tomer et al., 2010; Denes et al., 2007; reviewed in Strausfeld, 2010). Although annelid brains are quite variable, they have some features similar to those of arthropods, including mushroom bodies, which process olfactory information. Based on comparisons of patterns of gene expression, homology between the mushroom bodies and the pallium of the vertebrate brain has been proposed (Tomer et al., 2010; Denes et al., 2007). Comparable expression in the brains of vertebrates and *P. dumerilii* includes *Bf-1* (*FoxG1*) in the anterior part of the telencephalon and the pallium and in the tip of the annelid brain, *Wnt5/8* in the pallium and in the annelid mushroom bodies adjacent to the domain of *Hh* (Tomer et al., 2010), as well as *Six3* and *Otx* domains anteriorly in the CNS of both organisms (Steinmetz et al., 2010). Similarly, *Six3/6* and *Otx* are expressed anteriorly in acoel flatworms (Sikes and Bely, 2010; Hejnol and Martindale, 2008; Achatz and Martinez, 2012), and in *D. melanogaster*. Consequently, the annelid cerebral ganglion has been homologized with the insect protocerebrum. In *P. dumerilii*, the posterior limit of *Otx* abuts the anterior limit of *Gbx* in the first larval segment, and the anterior limits of *Hox1* and *Hox4* are in the second and third larval segments (Steinmetz et al., 2011). Additional evidence is that the patterns of *Hox* genes, *Otx*, *Fezf* etc., are quite similar in protostome and chordate nerve cords, while medial/lateral gene expression is not only similar in the CNSs of *D. melanogaster* and chordates, it is similar in *P. dumerilii* as well (Arendt et al., 2008). These comparisons support a single origin of the CNS in the bilaterian ancestor, although it is never possible to rule out convergent evolution. More recently, more detailed patterns of gene expression have revealed similarities between gene expression in the anterior regions of nervous systems from acoels, lampreys, amphibians, reptiles, mammals, annelids, spiders, and other arthropods (Wolff and Strausfeld, 2015).

The main obstacle to concluding that the ancestral bilaterian had a highly sophisticated brain is the invertebrate deuterostomes, especially cephalochordates. Although the aggregate evidence suggests that the ancestral bilaterian probably had a longitudinal nerve cord with a brain, proposed homologies of specific structures such as mushroom bodies of protostomes and the pallium of the vertebrate brain are problematic. With Xenacoelomorpha no longer in the deuterostomes, the ancestral deuterostome may well have been cephalochordate-like. If so, the protostome mushroom bodies and vertebrate pallium are unlikely to be homologous. The cephalochordate brain lacks a telencephalon; as the pallium is a telencephalic structure, the cephalochordate brain also lacks a pallium. It has nothing resembling the mushroom bodies of arthropods and annelids, or the olfactory bulbs of vertebrate brains, although the rostral nerves of amphioxus do innervate the anterior ectoderm, which contains the organs of de Quatrefages, thought to be mechanosensory cells involved in detection of changes in pressure (Baatrup, 1982). Did amphioxus lose the equivalent of the telencephalon? Loss of structures cannot be proven beyond a reasonable doubt. However, arguing against such a loss is the first that the morphology of modern amphioxus closely resembles that of fossil chordates from the Cambrian such as *Pikaea* and is similar to Cambrian fossils such as *Haikouella*, which appear to have paired eyes and at least a dipartite brain perhaps with olfactory bulbs, and are thought to be closer to the vertebrates (Mallatt and Chen, 2003). It has been argued that the optic tectum was the "original center of multisensory conscious perception" and was followed in evolution of vertebrates by a shift of conscious perception to the dorsal pallium (Feinberg and Mallatt, 2013). Thus, amphioxus, which does not have a telencephalon, presumably lacks conscious perception, while jawless vertebrates, which have a telencephalon, do have conscious perception, which evidently evolved together with true placodes and neural crest at the base of the vertebrates. The second reason why amphioxus probably did not lose a telencephalon is that cephalochordate genomes are evolving even more slowly than those of vertebrates and have conserved a large amount of synteny with vertebrate genomes in spite of the two rounds of whole-genome evolution that occurred in vertebrates. Therefore, when a structure is missing in amphioxus, it seems most likely that it was never there. If this is true, then similar gene expression in the vertebrate pallium and protostome mushroom bodies would represent convergent evolution. Convergent evolution of a complex brain capable of cognition has happened at least once. Cephalopods have evolved large brains capable of learning and paired image-forming eyes (Mather and Kuba, 2013). Moreover, not only do cephalopod brains express neuropeptides such as serotonin common to most nervous systems (Wollesen et al., 2010), they are also patterned by genes such as *Otx* and *Pax6* that pattern the simple brains of other invertebrates as well as those of vertebrates, although their patterns may not correspond precisely to those in chordate brains (Buresi et al., 2012; Navet et al., 2009). If such convergence could happen once, it could happen twice.

It is unfortunate that really clear intermediates between a protostome nerve cord and the amphioxus nerve cord are lacking. The anatomy of the nerve cords of amphioxus and protostomes is really too different for any structural homologies to be evident. Moreover, it is difficult to decide whether the apparently brainless hemichordates have lost a brain or never had one. As noted above, hemichordates have a dorsal nerve cord in the collar and trunk and a ventral nerve cord in the trunk. The two nerve cords are connected by a peripharyngeal nerve ring. Although hemichordates lack a brain, the proboscis ectoderm is highly neural with nerve tracts extending from the collar nerve cord (Nomakstcinsky et al., 2009; Fig. 4.1). The larval nervous systems in both indirect and direct developing hemichordates do not seem to carry over into the adult to any extent (Cunningham and Casey, 2014; Miyamoto et al., 2010). Therefore, while evidence from gene expression indicates that chordate nervous systems probably evolved from that in an ancestral bilaterian, in the absence of intermediates between protostome and chordate brains, it remains uncertain to what extent similarities between, for example, the vertebrate pallium and protostome mushroom bodies represent inheritance from a common ancestor or convergent evolution.

4.7 Where Did the Ancestral Bilaterian Brain Come From?

It is generally agreed that bilaterian nervous systems probably evolved from the ectodermal nerve net in a cnidarian ancestor. Several of the regional patterning genes are expressed similarly in cnidarians and bilaterians (Kelava et al., 2015; Fig. 4.10). At the planula stage, *Six3/6*, *FoxQ2a*, and *irx* are expressed at the aboral end of the planula of *Nematostella vectensis*, a sea anemone, as they are at the anterior end of the early embryos of amphioxus (Sinigaglia et al., 2013; Kozmik et al., 2007; Yu et al., 2003; Kaltenbach et al., 2009; Fig. 4.10). Moreover, *Wnt/β-catenin* expression appears to be similar, being concentrated around the blastopore in amphioxus (Onai et al., 2009, 2012; Holland et al., 2005) and around the blastopore in *Nematostella* (Watanabe et al., 2014). In addition, genes that specify neuronal identity, including *Delta/Notch*, BHLH transcription factors (especially *achaete/scute* and *atonal*), *Hu/Elav*, *SoxB1*, and *SoxB2*, are similarly expressed in developing nerve cells in cnidarians and chordates (Kelava et al., 2015). What is different, and probably critical for evolution of a CNS, is the function of BMPs. There is an ancient role for BMP in patterning the axis perpendicular to the A/P axis (reviewed in Bier and De Robertis, 2015). In cnidarians, this axis is termed the "directive axis." In bilaterians, it is the dorsal/ventral (D/V) axis. In *Nematostella*, *BMP2/4* (*dpp*), *BMP5-8*, and the BMP antagonist *chordin* are coexpressed at one side of the blastopore at the gastrula stage (Watanabe et al., 2014; Matus et al., 2006; Fig. 4.9) However, in spite of being

coexpressed, the maximal level of BMP signaling appears to be on the side of *Nematostella* opposite that of chordin as in most bilaterians (Genikhovich et al., 2015). In later larvae (planula), the domain of *BMPs* expands to the aboral end of the embryo, ultimately being expressed throughout the inner layer or gastrodermis (Genikhovich et al., 2015; Finnerty et al., 2004; Saina et al., 2009). Upregulation of BMP signaling represses chordin expression, indicating that these proteins oppose one another in *Nematostella* as they do in bilaterians. Manipulation of BMP signaling shows that it affects the development of the mesenteries (Genikhovich et al., 2015). However, unlike amphioxus, where excess BMP4 protein applied at the gastrula stage increases the number of ectodermal sensory cells (Lu et al., 2012), development of at least some neuronal subtypes in *Nematostella* is not affected by addition of exogenous BMP protein to early embryos (Watanabe et al., 2014). However, when applied at the planula stage exogenous BMP2 suppresses expression of transcription factors involved in neurogenesis (Watanabe et al., 2014). Thus, the role of BMP versus chordin in axial patterning appears to have been established in cnidarians, but, its function(s) in neurogenesis are not entirely conserved with bilaterians.

Whether the role of BMPs in specification of neural versus nonneural (or less neural) ectoderm evolved just once in basal bilateria and was lost in some bilaterian lineages or evolved more than once is unclear. *Drosophila* embryos mutant for *dpp* show a dorsal expansion of the marker of neurogenic ectoderm, *SoxN* (homologous to chordate *SoxB* genes), which is expressed in the developing ventral nervous system (Crémazy et al., 2000). In flies, graded levels of BMP appear to regulate the partitioning of the neuroectoderm into mediolateral domains expressing *Vnd* in the center, *Ind* in the middle, and *Msh* in the most lateral two rows of neuroblasts. In the annelid *Platynereis*, the effects of up- and downregulating BMP on neural development and the mediolateral arrangement of the *Vnd*-, *Ind*-, and *Msh*-expressing neurons are similar to that in both flies and vertebrates (Denes et al., 2007; Bier and De Robertis, 2015). This is compelling evidence that in spite of large morphological differences between the CNSs of flies, annelids, amphioxus, and vertebrates, a CNS evolved just once in the ancestral bilaterian.

However, in the over 520 million years since phyla separated, some roles of BMP have changed. In *Platynereis*, for example, excess BMP does not affect the number or position of *Hu/Elav*-expressing neurons as it does in amphioxus (Lu et al., 2012). Moreover, in the leech *Helobdella*, which has fixed cell lineages, the role of BMP in D/V patterning has been modified (Weisblat and Kuo, 2014). In the direct developing hemichordate *Saccoglossus*, BMP is expressed dorsally as in protostomes and

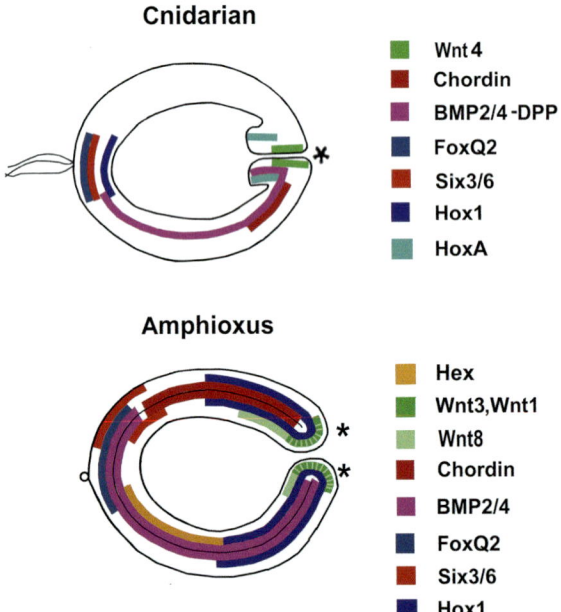

Cnidarian

- Wnt 4
- Chordin
- BMP2/4 -DPP
- FoxQ2
- Six3/6
- Hox1
- HoxA

Amphioxus

- Hex
- Wnt3,Wnt1
- Wnt8
- Chordin
- BMP2/4
- FoxQ2
- Six3/6
- Hox1

FIGURE 4.10 Developmental gene expression during the gastrula stage of a sea anemone (*Nematostella vectensis*) (top) has much in common with that in the chordate amphioxus (*Branchiostoma floridae*) (bottom). In *N. vectensis*, *BMP2/4/Dpp* and its antagonist *chordin* are coexpressed, but *BMP2/4/dpp* appears to exert its effect on the opposite side of the embryo. Asterisks, blastopore. In *N. vectensis*, there is an apical tuft of cilia that is lacking in amphioxus where the second polar body is retained at the anterior end of the embryo.

mediates D/V patterning; however, altering levels of BMPs does not change the number or distribution of ecto-dermal sensory cells as it does in amphioxus (Lowe et al., 2006). Unfortunately, there are no studies examining expression of A/P and mediolateral patterning genes in the developing nerve cords of hemichordates, which might help distinguish whether either of the hemichordate nerve cords is homologous to that of chordates.

4.8 Prevailing Scenarios for Evolution of the Central Nervous System

Several theories have been proposed to explain the evolution of nerve cords of bilaterians (including both protostomes and deuterostomes) from the ciliary bands of either an echinoderm larvalike adult (the auricularia theory) or from the nerve net in something like a cnidarian. These have been discussed extensively (for example, Holland, 2015a,b,c; Holland et al., 2013, 2015; Lacalli, 2010, 2008, 2005; Nielsen, 2015). The present review will consider only two of theories, both of which are closely related to the trochea or amphistome theory (Fig. 4.11 as articulated by Sedgwick and others (Sedgwick, 1884). It proposes that the blastopore of a cnidarian-like ancestor of the bilaterians elongated to form a slit; the mouth evolved from one end of it, the anus from the other, and the tissue in between fused to become the nerve cord(s) (Fig. 4.10). The two theories

differ slightly in that one, articulated by Nielsen (2015) relies chiefly on morphology while that of Marlow et al. (Arendt et al., 2016) relies heavily on shared gene expression between cnidarians, protostomes, and deuterostomes.

In both versions, bilaterians have pelagic larvae with ciliary bands like those of many protostomes and ambulacrarians among the deuterostomes. However, Marlow et al. (Arendt et al., 2016) proposed that, as a cnidarian-like ancestor gave rise to bilaterians, the "blastoporal nervous system" (ie, the neurogenic region around the blastopore which in cnidarians develops into the single gut opening), elongated and evolved into the longitudinal nerve cord while the neurogenic region at the end of the cnidarian opposite the gut opening [apical nervous system (ANS)] evolved into the bilaterian brain. This ANS derives from tissue including the apical organ. In contrast, Nielsen (2015) regards the mouth as evolving from the anterior part of this ancestral slitlike blastopore and the nervous tissue evolving from tissue just posterior to the apical organ. He also regards a trochophore larva as being present in the ancestral chordate.

The evidence for the theory of Marlow et al (Arendt et al., 2016). is shared gene expression between the corresponding regions in both cnidarians and various bilaterians. As conserved expression around the blastopore region, Marlow et al. cite *Brachyury*, *FoxA*, and *hedgehog (Hh)*, which encodes a secreted

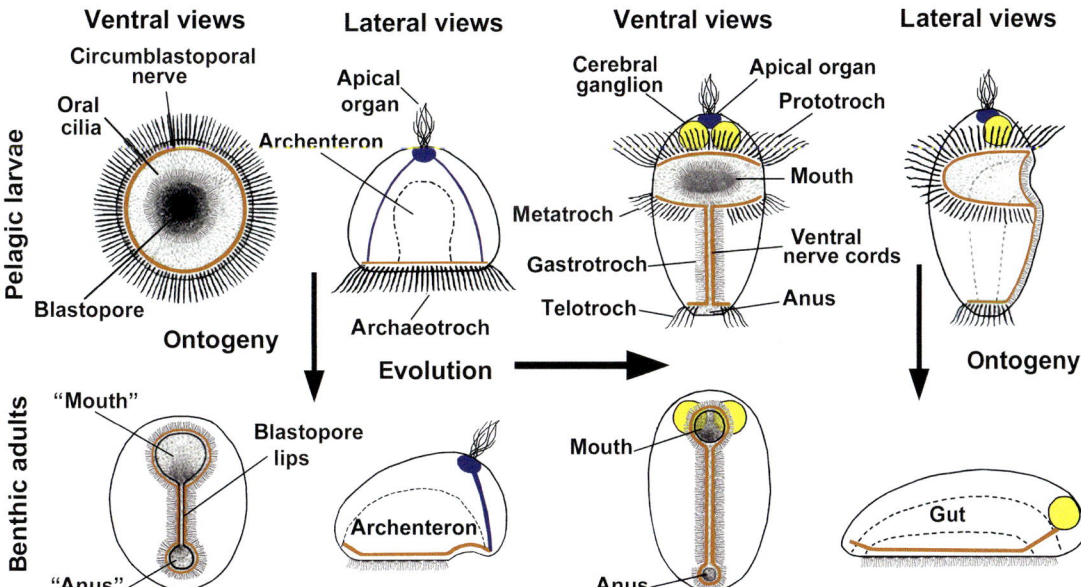

FIGURE 4.11   The trochea theory. According to this theory, a pelagic larva with a ciliated band around the blastopore gives rise to a benthic adult. This larva has an apical tuft of cilia emerging from a neurogenic apical organ. The blastopore constricts in the center to form a slit with mouth open anteriorly and anus open posteriorly. The ciliated region around the slit becomes a ventral nerve cord. During evolution, the pelagic larva transformed into a trochophore-like larva with ciliary bands looping around the mouth, paired ventral nerve cords, and a ciliary band (the telotroch) around the anus. Adjacent the apical organ is a cerebral ganglion. The adult of this form loses the apical organ and the cerebral ganglion becomes the brain.   *After Nielsen, C., 2012. How to make a protostome. Invert. Syst. 26, 25–40.*

signaling protein. Surprisingly, however, the chordate they choose for comparison to annelids and cnidarians is the frog, which has a derived mode of early development. Because frog eggs are very yolky, gastrulation in these animals occurs by involution of tissue over the dorsal blastopore lip. A better species for comparison would be the most basal chordate, amphioxus, which has relatively little yolk, and gastrulates by invagination with virtually no involution over the blastopore lip (Zhang et al., 1997). *Brachyury*, which is involved in invagination/morphogenesis is expressed around the blastopore in amphioxus, echinoderms, and vertebrates (Yamada et al., 2010). *HNF3β* (*FoxA*) is typically an endodermal marker and is expressed in mesendoderm around the blastopore in amphioxus and later in the endoderm and floor plate of the CNS (Shimeld, 1997; Terazawa and Satoh, 1997). In sea urchins, *FoxA* is also expressed around the blastopore (Oliveri et al., 2006). However, *Hh* is not expressed around the blastopore in either *Nematostella*, where it is expressed at the gastrula stage in tissue within the blastopore that will form part of the pharynx (Matus et al., 2008), or in amphioxus, where it is expressed in a wide range of tissues including the notochord and endoderm at the early gastrula stage (Shimeld, 1999). It is also not expressed around the blastopore in the sea urchin (Hara and Katow, 2005). Even so, it is likely that the expression domains of *Brachyury*, *FoxA* and some other genes, but not *Hh*, do indicate an evolutionary relationship between the blastopore in cnidarians and deuterostomes, but not necessarily between the blastopore and a CNS. In fact, in deuterostomes, contrary to what Marlow et al. assert, the tissue around the blastopore in deuterostomes is decidedly not neurogenic. In amphioxus, the floor plate of the neural tube is not derived from blastoporal tissue as Marlow et al. propose for vertebrates. Instead, the entire neural tube, notochord, and somites from the anterior tip of the amphioxus embryo to the middle of the hindbrain form from dorsal ectoderm and mesendoderm at the gastrula through mid-neurula stages. The remainder of these tissues form from the tailbud, which is derived largely from the dorsal blastopore lip.

Moreover, in deuterostomes, neurogenic genes are not expressed in similar patterns as in cnidarians. For example, in the early cnidarian planula, the neurogenic gene *SoxB2* is expressed in ectoderm around the single gut opening as well as in cells in both germ layers at the opposite end of the embryo (Kelava et al., 2015). *Hu/elav* is initially expressed opposite the gut opening at the early planula, and only at the late planula becomes expressed around the gut opening (Kelava et al., 2015). In *Xenopus*, *Soxb* genes are expressed in the neural plate (Nitta et al., 2006), while *Elav* is expressed both in the

neural plate and in preplacodal ectoderm (Perron et al., 1999). They are not expressed around the blastopore. Moreover, in the starfish gastrula, the neural marker *Soxb1* is expressed in the entire ectoderm *except* around the blastopore while *Hu/elav* is expressed in cells associated with the ciliary bands (Yankura et al., 2013). Similarly, in amphioxus, none of the three *SoxB* genes is expressed around the blastopore. *Soxb1a* and *Soxb1b* are expressed throughout the neural plate and underlying mesendoderm; *Soxb1c* and *Soxb2* are expressed in the posterior portion of the neural plate; *Soxb1c* is also expressed in the anterior portion of the neural plate and in ectodermal sensory cells (Meulemans and Bronner-Fraser, 2007; Holland et al., 2000). The disparate expression of neurogenic genes in cnidarians and deuterostome embryos simply does not support the idea that the deuterostome nerve cord derived from tissue around an ancestral blastopore.

Whether either of these theories or any one of numerous others such as those evolving the chordate nerve cords from the ciliated bands of an adult form rather like the auricularia larva of echinoderms (Garstang, 1894a) is correct is difficult to evaluate. Gene expression is of little help. Unfortunately, pieces of gene networks are often coopted for patterning different structures, and lack of homologous genes patterning similar structures in different animals does not mean that they are not homologous. For example, it has never been argued that guts in the larvae of starfish and sea urchins are not homologous. Even so, Hinman and Davidson (Hinman et al., 2007) found that only the core part of the gene regulatory network patterning these guts is conserved. Upstream and downstream parts of the network have diverged. Another caveat is that pieces of gene networks can be used for patterning nonhomologous structures. For example, those involved in formation of vertebrate limbs such as *Fgf8* and *Shh* are also involved in patterning the CNS (Capdevila and Belmonte, 2001). Another example is that *Engrailed*, *hedgehog*, and Wnt signaling sometimes, but not always, go together. For example, while Wnt and hedgehog signaling cooperate in establishing boundaries in the vertebrate brain such as the ZLI, all three genes interact in dorsal—ventral patterning of the chick limb (Logan et al., 1997). It would require a very large amount of dissection of gene networks to determine if there are truly any collections of genes that are only expressed in homologous structures in distantly related animals. Even then, if two structures did not express the entire collection, they still might be homologous (Wagner, 2007).

It would be wonderful if there were some clear intermediates between the cnidarian nerve net and the deuterostome CNS. Unfortunately, xenacoelomorphs are of little help in reconstructing intermediates. Xeno-turbellids lack a CNS. Acoels do have an anterior concentration of neurons, sometimes termed a "brain" and up to six nerve cords (Fig. 4.1). In the absence of any good intermediates between chordates and protostomes, just about anything is possible.

4.9 Conclusion

The evolution of the vertebrate CNS from that of an ancestral chordate can be inferred from comparisons between the brains of the invertebrate chordate amphioxus and the jawless vertebrate, the lamprey. Because amphioxus is evolving very slowly and its genome has retained considerable synteny with vertebrate genomes, it is unlikely that structures the amphioxus brain lacks but are present in the lamprey and other vertebrates represent losses in the amphioxus lineage. These structures include a telencephalon and neural crest. Although the amphioxus brain is divided into the same major regions as the vertebrate brain, the three major organizing centers (ANR, ZLI, and MHB) do not appear to have all of the genetic mechanisms that confer organizer properties. To what extent these were present in the tunicates, which are the sister group of vertebrates, is unclear since tunicates are evolving rapidly and have reduced their genomes and simplified their body plans.

It is more problematic to infer the CNS in the common deuterostome ancestor. It could have been like that in a modern amphioxus or much simpler like one of the nerve cords in a hemichordate. The nervous systems of echinoderms do not appear to be homologous to those of chordates. However, whether or not the hemichordate nerve cords are homologous to those in chordates is uncertain; more work needs to be done on the genetic mechanisms patterning the hemichordate nerve cords.

The considerable similarities in A/P and mediolateral patterning of the nerve cords between chordates and protostomes together with some anatomical similarities have led to proposals that the ancestral bilaterian had a longitudinal nerve cord with a fairly sophisticated brain. However, there is an ongoing debate as to whether the similarities between protostome and deuterostome brains represent convergent evolution of homologies. This debate is not likely to be resolved soon. Several scenarios have been proposed for evolution of the ancestral bilaterian CNS from the nerve net in a cnidarian-like ancestor. Because there are some similarities in axial patterning of the cnidarian nerve net and the bilaterian CNS, it seems likely that such a nerve net is ancestral to bilaterian nerve cords. However, it is still a matter of debate whether the CNS of bilaterians derived from a nervous system similar to that in an acoel flatworm, which in turn arose from a nerve net like that in

cnidarians. In the absence of clear intermediates, the debate as to the origins of bilaterian nerve cords is likely to continue.

Acknowledgment

The research of LZH is partially supported by the U.S.A. National Science Foundation, Grant no. IOS 1353688.

References

Antri, M., Cyr, A., Auclair, F., Dubuc, R., 2006. Ontogeny of 5-HT neurons in the brainstem of the lamprey, *Petromyzon marinus*. J. Comp. Neurol. 495, 788—800.

Anadon, R., Fatima, A., Rodriquez-Moldes, I., 1998. Distribution of GABA immunoreactivity in the central and peripheral nervous system of amphioxus (*Branchiostoma lanceolatum* Pallas). J. Comp. Neurol. 401, 293—307.

António, M., Fernandes, A.M., Fero, K., Arrenberg, A.B., Bergeron, S.A., Driever, W., Burgess, H.A., 2012. Deep brain photoreceptors control light-seeking behavior in zebrafish larvae. Curr. Biol. 22, 2042—2047.

Abitua, P.B., Wagner, E., Navarrete, I.A., Levine, M., 2012. Identification of a rudimentary neural crest in a non-vertebrate chordate. Nature 492, 104—107.

Aronowicz, J., Lowe, C.J., 2006. *Hox* gene expression in the hemichordate *Saccoglossus kowalevskii* and the evolution of deuterostome nervous systems. Integr. Comp. Biol. 46, 890—901.

Achatz, J.G., Martinez, P., 2012. The nervous system of *Isodiametra pulchra* (Acoela) with a discussion on the neuroanatomy of the Xenacoelomorpha and its evolutionary implications. Front. Zool. 9, 27.

Arendt, D., Denes, A.S., Jékely, G., Tessmar-Raible, K., 2008. The evolution of nervous system centralization. Philos. Trans. R. Soc. Biol. Sci. 363, 1523—1528.

Arendt, D., Tosches, M.A., Marlow, H., 2016. From nerve net to nerve ring, nerve cord and brain — evolution of the nervous system. Nat. Rev. Neurosci. 17, 61—72.

Bertrand, S., Camasses, A., Somorjai, I.M.L., Belgacem, M.R., Chabrol, O., Escande, M.L., Pontarotti, P., Escriva, H., 2011. Amphioxus FGF signaling predicts the acquisition of vertebrate morphological traits. Proc. Natl. Acad. Sci. U.S.A. 108, 9160—9165.

Bardet, P.L., Schubert, M., Horard, B., Holland, L.Z., Laudet, V., Holland, N.D., Vanacker, J.M., 2005. Expression of estrogen-receptor related receptors in amphioxus and zebrafish: implications for the evolution of posterior brain segmentation at the invertebrate-to-vertebrate transition. Evol. Dev. 7, 223—233.

Barreiro-Iglesias, A., Aldegunde, M., Anadón, R., Rodicio, M.C., 2009. Extensive presence of serotonergic cells and fibers in the peripheral nervous system of lampreys. J. Comp. Neurol. 512, 478—499.

Begbie, J., 2008. Migration of neuroblasts from neurogenic placodes. Dev. Neurosci. 30, 33—35.

Bassham, S., Canestro, C., Postelthwait, J.H., 2008. Evolution of developmental roles of *Pax2/5/8* paralogs after independent duplication in urochordate and vertebrate lineages. BMC Biol. 3, 17.

Bullock, T.H., 1944. The giant nerve fiber system in balanoglossids. J. Comp. Neurol. 80, 355—367.

Bailly, X., Reichert, H., Hartenstein, V., 2012. The urbilaterian brain revisited: novel insights into old questions from new flatworm clades. Dev. Genes Evol. 1—9.

Baatrup, E., 1982. On the structure of the Corpuscles of de Quatrefages (*Branchiostoma lanceolatum* (P)). Acta Zool. Stockh. 63, 39—44.

Buresi, A., Baratte, S., Da Silva, C., Bonnaud, L., 2012. *Orthodenticle/otx* ortholog expression in the anterior brain and eyes of *Sepia officinalis* (Mollusca, Cephalopoda). Gene Expr. Patterns 12, 109—116.

Bier, E., De Robertis, E.M., 2015. BMP gradients: a paradigm for morphogen-mediated developmental patterning. Science 348, aaa5838.

Cannon, J.T., Vellutini III, B.C., J.S., Ronquist, F., Jondelius, U., Hejnol, A., 2016. Xenacoelomorpha is the sister group to Nephrozoa. Nature 530, 89–93.

Cajal, M., Lawson, K.A., Hill, B., Moreau, A., Rao, J., Ross, A., Collignon, J., Camus, A., 2011. Clonal and molecular analysis of the prospective anterior neural boundary in the mouse embryo. Development 139, 423–436.

Castro, L.C.F., Rasmussen, S.L.K., Holland, P.W.H., Holland, N.D., Holland, L.Z., 2006. A *Gbx* homeobox gene in amphioxus: insights into ancestry of the ANTP class and evolution of the midbrain/hindbrain boundary. Dev. Biol. 295, 40–51.

Candiani, S., Moronti, L., Ramoino, P., Schubert, M., Pestarino, M., 2012. A neurochemical map of the developing amphioxus nervous system. BMC Neurosci. 13, 59.

Candiani, S., Lacalli, T.C., Parodi, M., Oliveri, D., Pestarino, M., 2008. The cholinergic gene locus in amphioxus: molecular characterization and developmental expression patterns. Dev. Dyn. 237, 1399–1411.

Cunningham, D., Casey, E.S., 2014. Spatiotemporal development of the embryonic nervous system of *Saccoglossus kowalevskii*. Dev. Biol. 386, 252–263.

Crémazy, F., Berta, P., Girard, F., 2000. *Sox Neuro*, a new *Drosophila Sox* gene expressed in the developing central nervous system. Mech. Dev. 93, 215–219.

Dworkin, S., Jane, S., 2013. Novel mechanisms that pattern and shape the midbrain-hindbrain boundary. Cell. Mol. Life Sci. 70, 3365–3374.

Derobert, Y., Baratte, B., Lepage, M., Mazan, S., 2002. *Pax6* expression patterns in *Lampetra fluviatilis* and *Scyliorhinus canicula* embryos suggest highly conserved roles in the early regionalization of the vertebrate brain. Brain Res. Bull. 57, 277–280.

Denes, A.S., Jékely, G., Steinmetz, P.R.H., Raible, F., Snyman, H., Prud'homme, B., Ferrier, D.E.K., Balavoine, G., Arendt, D., 2007. Molecular architecture of annelid nerve cord supports common origin of nervous system centralization in Bilateria. Cell 129, 277–288.

Eagleson, G.W., Dempewolf, R.D., 2002. The role of the anterior neural ridge and Fgf-8 in early forebrain patterning and regionalization in *Xenopus laevis*. Comp. Biochem. Physiol. Part B Biochem. Mol. Biol. 132, 179–189.

Fernandes, A.M., Fero, K., Driever, W., Burgess, H.A., 2013. Enlightening the brain: linking deep brain photoreception with behavior and physiology. Bioessays 35, 775–779.

Feinberg, T.E., Mallatt, J., 2013. The evolutionary and genetic origins of consciousness in the Cambrian Period over 500 million years ago. Front. Psychol. 4.

Finnerty, J.R., Pang, K., Burton, P., Paulson, D., Martindale, M.Q., 2004. Origins of bilateral symmetry: *Hox* and *Dpp* expression in a sea anemone. Science 304, 1335–1337.

Glardon, S., Holland, L.Z., Gehring, W.J., Holland, N.D., 1998. Isolation and developmental expression of the amphioxus *Pax-6* gene (*AmphiPax-6*): insights into eye and photoreceptor evolution. Development 125, 2701–2710.

Glardon, S., Callaerts, P., Halder, G., Gehring, W.J., 1997. Conservation of *Pax-6* in a lower chordate, the ascidian *Phallusia mammillata*. Development 124, 817–825.

Genikhovich, G., Fried, P., Prunster, M.M., Schinko, J.B., Gilles, A.F., Fredman, D., Meier, K., Iber, D., Technau, U., 2015. Axis patterning by BMPs: cnidarian network reveals evolutionary constraints. Cell Rep. 10, 1646–1654.

Garstang, W., 1894. Preliminary note on a new theory of the phylogeny of the Chordata. Zool. Anz. 17, 122–125.

Holland, N.D., Panganiban, G., Henyey, E.L., Holland, L.Z., 1996. Sequence and developmental expression of *AmphiDll*, an amphioxus *Distal-less* gene transcribed in the ectoderm, epidermis and nervous system: insights into evolution of craniate forebrain and neural crest. Development 122, 2911–2920.

Holland, L.Z., Holland, N.D., 2000. Developmental expression of *AmphiWnt1*, an amphioxus gene in the *Wnt1/wingless* subfamily. Dev. Genes Evol. 210, 522–524.

Holland, L.Z., Kene, M., Williams, N.A., Holland, N.D., 1997. Sequence and embryonic expression of the amphioxus *engrailed* gene (*AmphiEn*): the metameric pattern of transcription resembles that of its segment-polarity homolog in *Drosophila*. Development 124, 1723–1732.

Holland, N.D., Holland, L.Z., Honma, Y., Fujii, T., 1993. *Engrailed* expression during development of a Lamprey, *Lampetra japonica*: a possible clue to homologies between agnathan and gnathostome muscles of the mandibular arch. Dev. Growth Differ. 35, 153–160.

Holland, N.D., Holland, L.Z., 1993. Serotonin-containing cells in the nervous system and other tissues during ontogeny of a lancelet, *Branchiostoma floridae*. Acta Zool. Stockh. 74, 195–204.

Holland, L.Z., 2005. Non-neural ectoderm is really neural: evolution of developmental patterning mechanisms in non-neural ectoderm of chordates and the problem of sensory cell homologies. J. Exp. Zool. 304B, 304–323.

Holland, L.Z., 2015. Evolution of basal deuterostome nervous systems. J. Exp. Biol. 218, 637–645.

Holland, L.Z., 2015. The origin and evolution of chordate nervous systems. Philos. Trans. R. Soc. Lond. B Biol. Sci. 370.

Holland, N.D., 2015. Nervous systems and scenarios for the invertebrate-to-vertebrate transition. Philos. Trans. R. Soc. Lond. B Biol. Sci. 371.

Hirth, F., Kammermeier, L., Frei, E., Walldorf, U., Noll, M., Reichert, H., 2003. An urbilaterian origin of the tripartite brain: developmental genetic insights from *Drosophila*. Development 130, 2365–2373.

Hirth, F., Reichert, H., 1999. Conserved genetic programs in insect and mammalian brain development. Bioessays 21, 677–684.

Hirata, T., Nakazawa, M., Muraoka, O., Nakayama, R., Suda, Y., Hibi, M., 2006. Zinc-finger genes *Fez* and *Fez-like* function in the establishment of diencephalon subdivisions. Development 133, 3993–4004.

Hejnol, A., Martindale, M.Q., 2008. Acoel development indicates the independent evolution of the bilaterian mouth and anus. Nature 456, 382–386.

Holland, L.Z., Panfilio, K.A., Chastain, R., Schubert, M., Holland, N.D., 2005. Nuclear b-catenin promotes non-neural ectoderm and posterior cell fates in amphioxus embryos. Dev. Dyn. 233, 1430–1443.

Holland, L.Z., Carvalho, J.E., Escriva, H., Laudet, V., Schubert, M., Shimeld, S.M., Yu, J.K., 2013. Evolution of bilaterian central nervous systems: a single origin? EvoDevo 4, 27.

Holland, N.D., Holland, L.Z., Holland, P.W.H., 2015. Scenarios for the making of vertebrates. Nature 520, 450–455.

Hara, Y., Katow, H., 2005. Expression of *hedgehog* in small micromere descendants during early embryogenesis in the sea urchin, *Hemicentrotus pulcherrimus*. Gene Expr. Patterns 5, 503–510.

Holland, L.Z., Schubert, M., Holland, N.D., Neuman, T., 2000. Evolutionary conservation of the presumptive neural plate markers *AmphiSox1/2/3* and *AmphiNeurogenin* in the invertebrate chordate amphioxus. Dev. Biol. 226, 18–33.

Hinman, V.F., Nguyen, A., Davidson, E.H., 2007. Caught in the evolutionary act: precise cis-regulatory basis of difference in the organization of gene networks of sea stars and sea urchins. Dev. Biol. 312, 584–595.

Irimia, M., Pineiro, C., Maeso, I., Gomez-Skarmeta, J.L., Casares, F., Garcia-Fernandez, J., 2010. Conserved developmental expression of *Fezf* in chordates and *Drosophila* and the origin of the *Zona Limitans Intrathalamica* (ZLI) brain organizer. EvoDevo 1 (7), 10.

Ikuta, T., Saiga, S., 2007. Dynamic change in the expression of developmental genes in the ascidian central nervous system: revisit to the tripartite model and the origin of the midbrain-hindbrain boundary region. Dev. Biol. 312, 631–643.

Inoue, F., Kurokawa, D., Takahashi, M., Aizawa, S., 2012. *Gbx2* directly restricts *Otx2* expression to forebrain and midbrain, competing with Class III POU factors. Mol. Cell. Biol. 32, 2618–2627.

Jackman, W.R., Langeland, J.A., Kimmel, C.B., 2000. Islet reveals segmentation in the amphioxus hindbrain homolog. Dev. Biol. 230, 16–26.

Jones, C.M., Dale, L., Hogan, B.L., Wright, C.V., Smith, J.C., 1996. Bone morphogenetic protein-4 (BMP-4) acts during gastrula stages to cause ventralization of *Xenopus* embryos. Development 122, 1545–1554.

Jandzik, D., Garnett, A.T., Square, T.A., Cattell, M.V., Yu, J.-K., Medeiros, D.M., 2015. Evolution of the new vertebrate head by co-option of an ancient chordate skeletal tissue. Nature 518, 534–537.

Jeffery, W.R., Strickler, A.G., Yamamoto, Y., 2004. Migratory neural crest-like cells form body pigmentation in a urochordate embryo. Nature 43, 696–699.

Kikuchi, M., Omori, A., Kurokawa, D., Akasaka, K., 2015. Patterning of anteroposterior body axis displayed in the expression of *Hox* genes in sea cucumber *Apostichopus japonicus*. Dev. Genes Evol. 225, 275–286.

Kaul-Strehlow, S., Urata, M., Minokawa, T., Stach, T., Wanninger, A., 2015. Neurogenesis in directly and indirectly developing enteropneusts: of nets and cords. Org. Divers. Evol. (in press).

Kelava, I., Rentzsch, F., Technau, U., 2015. Evolution of eumetazoan nervous systems: insights from cnidarians. Philos. Trans. R. Soc. Lond. B Biol. Sci. 370.

Kozmik, Z., Holland, N.D., Kreslova, J., Oliveri, D., Schubert, S., Jonasova, K., Holland, L.Z., Pestarino, M., Benes, V., Candiani, S., 2007. *Pax-Six-Eya-Dach* network during amphioxus development: conservation in vitro but context specificity in vivo. Dev. Biol. 306, 143–159.

Kaltenbach, S.L., Holland, L.Z., Holland, N.D., Koop, D., 2009. Developmental expression of the three *iroquois* genes of amphioxus (*BfIrxA*, *BfIrxB*, and *BfIrxC*) with special attention to the gastrula organizer and anteroposterior boundaries in the central nervous system. Gene Expr. Patterns 9, 329–334.

Lacalli, T.C., 1996. Frontal eye circuitry, rostral sensory pathways, and brain organization in amphioxus larvae: evidence from 3D reconstructions. Philos. Trans. R. Soc. B 351, 243–263.

Lacalli, T.C., 2002. Sensory pathways in amphioxus larvae 1. Constituent fibre of the rostral and anterodorsal nerves, their targets and evolutionary significance. Acta Zool. Stockh. 83, 149–166.

Lacalli, T.C., 2003. Ventral neurons in the anterior nerve cord of amphioxus larvae. II. Further data on the pacemaker circuit. J. Morphol. 257, 212–218.

Lacalli, T.C., Kelly, S.J., 1999. Somatic motoneurones in amphioxus larvae: cell types, cell position and innervation patterns. Acta Zool. Stockh. 80, 113–124.

Lu, T.-M., Luo, Y.-J., Yu, J.-K., 2012. BMP and Delta/Notch signaling control the development of amphioxus epidermal sensory neurons: insights into the evolution of the peripheral sensory system. Development 139, 2020–2030.

Lowe, C.J., Wu, M., Salic, A., Evans, S.L., Lander, E., Stange-Thomann, N., Gruber, C.E., Gerhart, J., Kirschner, M., 2003. Anteroposterior patterning in hemichordates and the origins of the chordate nervous system. Cell 113, 853–865.

Lowe, C.J., Terasaki, M., Wu, M., Freeman, R.M., Runft, L., Kwan, K., Haigo, S., Aronowicz, J., Lander, E., Gruber, C., Smith, M., Kirschner, M., Gerhart, J., 2006. Dorsoventral patterning in

hemichordates: insights into early chordate evolution. Public Libr. Sci. Biol. 4, 1603–1619.

Lacalli, T.C., 2010. The emergence of the chordate body plan: some puzzles and problems. Acta Zool 91, 4–10.

Lacalli, T.C., 2008. Basic features of the ancestral chordate brain: a protochordate perspective. Brain Res. Bull. 75, 319–323.

Lacalli, T.C., 2005. Protochordate body plan and the evolutionary role of larvae: old controversies resolved? Can. J. Zool. 83, 216–224.

Martinez-Ferre, A., Martinez, S., 2012. Molecular regionalization of the diencephalon. Front. Neurosci. 6.

Murakami, Y., Oagsawara, M., Sugahara, F., Shigeki, H., Satoh, N., Kuratani, S., 2001. Identification and expression of the lamprey *Pax6* gene: evolutionary origin of the segmented brain of vertebrates. Development 128, 3521–3531.

Meulemans, D., Bronner-Fraser, M., 2004. Gene-regulatory interactions in neural crest evolution and development. Dev. Cell 7, 291–299.

Meinertzhagen, I.A., Lemaire, P., Okamura, Y., 2004. The neurobiology of the ascidian tadpole larva: recent developments in an ancient chordate. Ann. Rev. Neurosci. 27, 453–485.

Miyamoto, N., Nakajima, Y., Wada, H., Saito, Y., 2010. Development of the nervous system in the acorn worm *Balanoglossus simodensis*: insights into nervous system evolution. Evol. Dev. 12, 416–424.

Miyamoto, N., Wada, H., 2013. Hemichordate neurulation and the origin of the neural tube. Nat. Commun. 4.

Mallatt, J., Chen, J-y., 2003. Fossil sister group of craniates: predicted and found. J. Morphol. 258, 1–31.

Mather, J.A., Kuba, M.J., 2013. The cephalopod specialties: complex nervous system, learning, and cognition. Can. J. Zool. 91, 431–449.

Matus, D.Q., Pang, K., Marlow, H., Dunn, C.W., Thomsen, G.H., Martindale, M.Q., 2006. Molecular evidence for deep evolutionary roots of bilaterality in animal development. Proc. Natl. Acad. Sci. U.S.A. 103, 11195–11200.

Matus, D.Q., Magie, C.R., Pang, K., Martindale, M.Q., Thomsen, G.H., 2008. The *Hedgehog* gene family of the cnidarian, *Nematostella vectensis*, and implications for understanding metazoan Hedgehog pathway evolution. Dev. Biol. 313, 501–518.

Meulemans, D., Bronner-Fraser, M., 2007. The amphioxus *SoxB* family: implications for the evolution of vertebrate placodes. Int. J. Biol. Sci. 3, 356–364.

Nicol, D., Meinertzhagen, I.A., 1991. Cell counts and maps in the larval central nervous system of the ascidian *Ciona intestinalis* (L.). J. Comp. Neurol. 309, 415–429.

Nomaksteinsky, M., Dufour, H.D., Chettoug, Z., Lowe, C.J., Martindale, M.Q., Brunet, J.F., 2009. Centralization of the deuterostome nervous system predates chordates. Curr. Biol. 19, 1264–1269.

Navet, S., Andouche, A., Baratte, S., Bonnaud, L., 2009. *Shh* and *Pax6* have unconventional expression patterns in embryonic morphogenesis in *Sepia officinalis* (Cephalopoda). Gene Expr. Patterns 9, 461–467.

Nielsen, C., 2015. Larval nervous systems: true larval and precocious adult. J. Exp. Biol. 218, 629–636.

Nielsen, C., 2015. Evolution of deuterostomy – and the origin of the chordates. Biol. Rev. epub. ahead of print.

Nitta, K.R., Takahashi, S., Haramoto, Y., Fukuda, M., Onuma, Y., Asashima, M., 2006. Expression of *Sox1* during *Xenopus* early embryogenesis. Biochem. Biophys. Res. Commun. 351, 287–293.

Ono, H., Kozmik, Z., Yu, J.-K., Wada, H., 2014. A novel N-terminal motif is responsible for the evolution of neural crest-specific gene-regulatory activity in vertebrate FoxD3. Dev. Biol. 385, 396–404.

Onai, T., Akira, T., Setiamarga, D.H.E., Holland, L.Z., 2012. Essential role of *Dkk3* for head formation by inhibiting Wnt/β-catenin and Nodal/Vg1 signaling pathways in the basal chordate amphioxus. Evol. Dev. 14, 338–350.

Onai, T., Lin, H.-C., Schubert, M., Koop, D., Osborne, P.W., Alvarez, S., Alvarez, R., Holland, N.D., Holland, L.Z., 2009. Retinoic acid and Wnt/beta-catenin have complementary roles in anterior/posterior patterning embryos of the basal chordate amphioxus. Dev. Biol. 332, 223–233.

Oliveri, P., Walton, K.D., Davidson, E.H., McClay, D.R., 2006. Repression of mesodermal fate by foxa, a key endoderm regulator of the sea urchin embryo. Development 133, 4173–4181.

del Pilar Gomez, M., Angueyra, J.M., Nasi, E., 2009. Light-transduction in melanopsin-expressing photoreceptors of amphioxus. Proc. Natl. Acad. Sci. U.S.A. 106, 9081–9086.

Pombal, M.A., Marín, O., González, A., 2001. Distribution of choline acetyltransferase-immunoreactive structures in the lamprey brain. J. Comp. Neurol. 431, 105–126.

Pulido, C., Malagon, G., Ferrer, C., Chen, J.-K., Angueyra, J.M., Nasi, E., Gomez, M.del P., 2012. The light-sensitive conductance of melanopsin-expressing Joseph and Hesse cells in amphioxus. J. Gen. Physiol. 139, 19–30.

Patthey, C., Schlosser, G., Shimeld, S.M., 2014. The evolutionary history of vertebrate cranial placodes – I: cell type evolution. Dev. Biol. 389.

Popodi, E., Andrews, M., Raff, R., 1994. Evolution of body plans–using homeobox genes to examine the development of the radial CNS of echinoderms. Dev. Biol. 163, 540.

Philippe, H., Brinkmann, H., Copley, R., Moroz, L.L., Nakano, H., Poustka, A.J., Wallberg, A., Peterson, K.J., Telford, M.J., 2011. Acoelomorph flatworms are deuterostomes related to Xenoturbella. Nature 470, 255–258.

Perea-Atienza, E., Gavilán, B., Chiodin, M., Abril, J.F., Hoff, K.J., Poustka, A.J., Martinez, P., 2015. The nervous system of Xenacoelomorpha: a genomic perspective. J. Exp. Biol. 218, 618–628.

Pfeiffer, B.D., Jenett, A., Hammonds, A.S., Ngo, T.-T.B., Misra, S., Murphy, C., Scully, A., Carlson, J.W., Wan, K.H., Laverty, T.R., Mungall, C., Svirskas, R., Kadonaga, J.T., Doe, C.Q., Eisen, M.B., Celniker, S.E., Rubin, G.M., 2008. Tools for neuroanatomy and neurogenetics in Drosophila. Proc. Natl. Acad. Sci. U.S.A. 105, 9715–9720.

Püschel, A.W., Westerfield, M., Dressler, G.R., 1992. Comparative analysis of Pax-2 protein distributions during neurulation in mice and zebrafish. Mech. Dev. 38, 197–208.

Perron, M., Furrer, M.-P., Wegnez, M., Théodore, L., 1999. Xenopus elav-like genes are differentially expressed during neurogenesis. Mech. Dev. 84, 139–142.

Rouse, G.W., Wilson, N.G., Carvajal, I.J., Vrijenhoek, R.C., 2016. New deep-sea Xenoturbella and the placement of Xenacoelomorpha. Nature 530, 94–97.

Rétaux, S., Kano, S., 2010. Midline signaling and evolution of the forebrain in chordates: a focus on the lamprey Hedgehog case. Integr. Comp. Biol. 50, 98–109.

Reed, K.L., MacIntyre, J.K., Tobet, S.A., Trudeau, V.L., MacEachern, L., Rubin, B.S., Sower, S.A., 2002. The spatial relationship of γ-aminobutyric acid (GABA) neurons and gonadotropin-releasing hormone (GnRH) neurons in larval and adult sea lamprey, Petromyzon marinus. Brain Behav. Evol. 60, 1–12.

Raikova, O.I., Reuter, M., Jondelius, U., Gustafsson, M.K.S., 2000. An immunocytochemical and ultrastructural study of the nervous and muscular systems of Xenoturbella westbladi (Bilateria inc. sed.). Zoomorphology 120, 107–118.

Reichert, H., 2005. A tripartite organization of the urbilaterian brain: developmental genetic evidence from Drosophila. Brain Res. Bull. 66, 491–494.

Schubert, M., Holland, L.Z., Panopoulou, G.D., Lehrach, H., Holland, N.D., 2000. Characterization of amphioxus AmphiWnt8: insights into the evolution of patterning of the embryonic dorsoventral axis. Evol. Dev. 2, 85–92.

Shimeld, S.M., 1999. The evolution of the hedgehog gene family in chordates: insights from amphioxus hedgehog. Dev. Genes Evol. 209, 40–47.

Suzuki, D.G., Murakami, Y., Escriva, H., Wada, H., 2015. A comparative examination of neural circuit and brain patterning

between the lamprey and amphioxus reveals the evolutionary origin of the vertebrate visual center. J. Comp. Neurol. 523, 251–261.

Soviknes, A.M., Chourrout, D., Glover, J.C., 2005. Development of putative GABAergic neurons in the appendicularian urochordate Oikopleura dioica. J. Comp. Neurol. 490, 12–28.

Soviknes, A.M., Chourrout, D., Glover, J.C., 2007. Development of the caudal nerve cord, motoneurons, and muscle innervation in the appendicularian urochordate Oikopleura dioica. J. Comp. Neurol. 503, 224–243.

Semmler, H., Chiodin, M., Bailly, X., Martinez, P., Wanninger, A., 2010. Steps towards a centralized nervous system in basal bilaterians: insights from neurogenesis of the acoel Symsagittifera roscoffensis. Dev. Growth Differ. 52, 701–713.

Strausfeld, N.J., 2010. Brain homology: Dohrn of a new era? Brain Behav. Evol. 76, 165–167.

Steinmetz, P.R., Urbach, R., Posnien, N., Eriksson, J., Kostyuchenko, C.R.P., Brena, C., Guy, K., Akam, M., Bucher, G., Arendt, D., 2010. Six3 demarcates the anterior-most developing brain region in bilaterian animals. EvoDevo 1, 14.

Sikes, J.M., Bely, A.E., 2010. Making heads from tails: development of a reversed anterior–posterior axis during budding in an acoel. Dev. Biol. 338, 86–97.

Steinmetz, P.R.H., Kostyuchenko, R.P., Fischer, A., Arendt, D., 2011. The segmental pattern of otx, gbx, and Hox genes in the annelid Platynereis dumerilii. Evol. Dev. 13, 72–79.

Sinigaglia, C., Busengdal, H., Leclère, L., Technau, U., Rentzsch, F., 2013. The bilaterian head patterning gene six3/6 controls aboral domain development in a cnidarian. PLoS Biol. 11, e1001488.

Saina, M., Genikhovich, G., Renfer, E., Technau, U., 2009. BMPs and Chordin regulate patterning of the directive axis in a sea anemone. Proc. Natl. Acad. Sci. U.S.A. 106, 18592–18597.

Sedgwick, A., 1884. On the origin of metameric segmentation and some other morphological questions. Q. J. Microsc. Sci. 24, 43–82 (+ pl. II–III).

Shimeld, S.M., 1997. Characterisation of amphioxus HNF-3 genes: conserved expression in the notochord and floor plate. Dev. Biol. 183, 74–85.

Toresson, H., Martinez-Barbera, J.P., Beardsley, A., Caubit, X., Krauss, S., 1998. Conservation of BF-1 expression in amphioxus and zebrafish suggests evolutionary ancestry of anterior cell types that contribute to the vertebrate telencephalon. Dev. Genes Evol. 208, 431–439.

Tomer, R., Denes, A.S., Tessmar-Raible, K., Arendt, D., 2010. Profiling by image registration reveals common origin of annelid mushroom bodies and vertebrate pallium. Cell 142, 800–809.

Terazawa, K., Satoh, N., 1997. Formation of the chordamesoderm in the amphioxus embryo: analysis with Brachyury and fork head/HNF-3 genes. Dev. Genes Evol. 207, 1–11.

Urbach, R., Technau, G.M., 2008. In: Technau, G.M. (Ed.), Dorsoventral Patterning of the Brain: A Comparative Approach, vol. 628. Springer, New York, pp. 42–56.

Vieira, C., Pombero, A., García-Lopez, R., Gimeno, L., Echevarria, D., Martínez, S., 2010. Molecular mechanisms controlling brain development: an overview of neuroepithelial secondary organizers. Int. J. Dev. Biol. 54, 7–20.

Villar-Cerviño, V., Rocancourt, C., Menuet, A., Da Silva, C., Wincker, P., Anadón, R., Mazan, S., Rodicio, M.C., 2010. A vesicular glutamate transporter in lampreys: cDNA cloning and early expression in the nervous system. J. Chem. Neuroanat. 40, 71–81.

Villar-Cerviño, V., Barreiro-Iglesias, A., Mazan, S., Rodicio, M.C., Anadón, R., 2011. Glutamatergic neuronal populations in the forebrain of the sea lamprey, Petromyzon marinus: an in situ hybridization and immunocytochemical study. J. Comp. Neurol. 519, 1712–1735.

Vopalensky, P., Pergner, J., Liegertova, M., Benito-Gutierrez, E., Arendt, D., Kozmik, Z., 2012. Molecular analysis of the amphioxus frontal eye unravels the evolutionary origin of the retina and pigment cells of the vertebrate eye. Proc. Natl. Acad. Sci. U.S.A. 109, 15383–15388.

Wicht, H., Lacalli, T.C., 2005. The nervous system of amphioxus: structure, development, and evolutionary significance. Can. J. Zool. 83, 122–150.

Wolff, G.H., Strausfeld, N.J., 2015. Genealogical correspondence of a forebrain centre implies an executive brain in the protostome–deuterostome bilaterian ancestor. Philos. Trans. R. Soc. Lond. B Biol. Sci. 371.

Wollesen, T., Degnan, B.M., Wanninger, A., 2010. Expression of serotonin (5-HT) during CNS development of the cephalopod mollusk. *Idiosepius notoides*. Cell Tissue Res. 342, 161–178.

Watanabe, H., Kuhn, A., Fushiki, M., Agata, K., Özbek, S., Fujisawa, T., Holstein, T.W., 2014. Sequential actions of β-catenin and Bmp pattern the oral nerve net in *Nematostella vectensis*. Nat. Commun. 5.

Weisblat, D.A., Kuo, D.-H., 2014. Developmental biology of the leech *Helobdella*. Int. J. Dev. Biol. 58, 429–443.

Wagner, G.P., 2007. The developmental genetics of homology. Nat. Rev. Genet. 8, 473–479.

Yu, J.K., Satou, Y., Holland, N.D., Shin-I, T., Kohara, Y., Satoh, N., Bronner-Fraser, M., Holland, L.Z., 2007. Axial patterning in cephalochordates and the evolution of the organizer. Nature 445, 613–617.

Yu, J.-K., Meulemans, D., McKeown, S.J., Bronner-Fraser, M., 2008. Insights from the amphioxus genome on the origin of vertebrate neural crest. Genome Res. 18, 1127–1132.

Yasui, K., Zhang, S.C., Uemura, M., Aizawa, S., Ueki, A., 1998. Expression of a twist-related gene, *Bbtwist*, during the development of a lancelet species and its relation to cephalochordate anterior structures. Dev. Biol. 195, 49–59.

Yu, J.K., Holland, N.D., Holland, L.Z., 2003. *AmphiFoxQ2*, a novel winged helix/forkhead gene, exclusively marks the anterior end of the amphioxus embryo. Dev. Genes Evol. 213, 102–105.

Yamada, A., Martindale, M.Q., Fukui, A., Tochinai, S., 2010. Highly conserved functions of the *Brachyury* gene on morphogenetic movements: insight from the early-diverging phylum Ctenophora. Dev. Biol. 339, 212–222.

Yankura, K.A., Koechlein, C.S., Cryan, A.F., Cheatle, A., Hinman, V.F., 2013. Gene regulatory network for neurogenesis in a sea star embryo connects broad neural specification and localized patterning. Proc. Natl. Acad. Sci. U.S.A. 110, 8591–8596.

Zhang, S.C., Holland, N.D., Holland, L.Z., 1997. Topographic changes in nascent and early mesoderm in amphioxus embryos studied by DiI labeling and by in situ hybridization for a *Brachyury* gene. Dev. Genes Evol. 206, 532–535.

The Brains of Fish, Amphibians, Reptiles, and Birds

5

The Nervous Systems of Jawless Vertebrates

M.A. Pombal, M. Megías

University of Vigo, Vigo, Spain

Abbreviations

1—5 Columns 1—5 of the dV of Nishizawa et al. (1988)
5-HT 5-Hydroxytryptamine (serotonin)
III Oculomotor nucleus
IIIn Oculomotor nerve
IIIv Third ventricle
In Olfactory nerve
IV Trochlear motor nucleus
IVv Fourth ventricle
IX Glossopharyngeal motor nucleus
IXn Glossopharyngeal nerve
V Trigeminal motor nucleus
VI Abducent motor nucleus
VII Facial motor nucleus
VIn Abducent nerve
VIIn Facial nerve
VIIIn Octaval nerve
Vn Trigeminal motor root
Vs Trigeminal sensory root
X Vagal motor nucleus
ABB Alar—basal boundary
AHy Alar hypothalamus
alln Anterior lateral line nerve
aon Anterior octavomotor nucleus
ARN Anterior rhombencephalic reticular nucleus
BDA Biotinylated dextran amines
BHy Basal hypothalamus
ca Cerebral aqueduct
CE Cerebral hemisphere
ch Optic chiasm
ChAT Choline acetyl transferase
CR Calretinin
CSF-c Cerebrospinal fluid—contacting
cvl Commissure vestibulolateralis
D Diencephalon
dn Dorsal nucleus of the OLA
DTsh Dorsal thalamus (subhabenular part)
dV Descending trigeminal tract
eall Electroreceptive anterior lateral line
ell Electroreceptive lateral line
fr Fasciculus retroflexus
GABA Gamma aminobutyric acid
GC Griseum centrale
H Habenula

HB Hindbrain
hc Habenular commissure
hp1 Hypothalamic prosomere 1 (peduncular or caudal hypothalamus)
hp2 Hypothalamic prosomere 2 (prepeduncular or rostral hypothalamus)
HRP Horseradish peroxidase
Hy Hypothalamus
I1 Isthmic reticular cell 1
ibc Interbulbar commissure
inf Infundibulum
IP Interpeduncular nucleus
Is Isthmus
LGE Lateral ganglionic eminence
lv Lateral ventricle
M1—3 Müller cell 1—3
M5 M5 nucleus of Schöber
mall Mechanoreceptive anterior lateral line
mar Mamillary recess
MB Midbrain
MGE Medial ganglionic eminence
mlf Medial longitudinal fascicle
mll Mechanoreceptive lateral line
mn Medial nucleus of the OLA
mr Mesencephalic recess
MRA Mesencephalic reticular area
MRN Medial rhombencephalic reticular nucleus
Mth Mauthner neuron
mv Mesencephalic ventricle
NA Nucleus anterior
NCm Medial nucleus of the central prosencephalic complex
NCvl Ventrolateral nucleus of the central prosencephalic complex
ND Nucleus diffusus
NE Nucleus externus
NH Neurohypophysis
NMLF Nucleus of the medial longitudinal fascicle
not Notochord
NPC Nucleus of the posterior commissure
NPY Neuropeptide Y
NT Nucleus triangularis
NTP Nucleus of the tuberculum posterior
NTPl Lateral nucleus of the tuberculum posterior
NTPm Medial nucleus of the tuberculum posterior
nTPOC Nucleus of the tract of the postoptic commissure
OB Olfactory bulb
oe Olfactory epithelium

og Olfactory glomeruli
OLA Octavolateral area
OT Optic tectum
ot Optic tract
P Pineal organ
p1–p3 Prosomeres 1–3
P1–P5 Pallial layers 1–5
Pa Paraventricular area
Pal Pallium
pc Posterior commissure
pch Choroid plexus
plln Posterior lateral line nerve
PO Preoptic area
POe External nucleus of the preoptic area
POim Intermedial nucleus of the preoptic area
PP Parapineal organ
PR Central prosencephalic complex
pr Preoptic recess
PRN Posterior rhombencephalic reticular nucleus
PT Pretectum
PTh Prethalamus
PVO Periventricular hypothalamic organ
r1–r11 Rhombomeres 1–11
RMN Reticular mesencephalic nucleus
RPa Rostral paraventricular nucleus
RS Reticulospinal cells
SC Spinal cord
SE Septum
SPa Subparaventricular area
SPal Subpallium
ST Striatum
T Telencephalon
Teg Mesencephalic tegmentum
Th Thalamus
TH Tyrosine hydroxylase
TM Tuberomamillary nucleus
tv Impar telencephalic ventricle
vn Ventral nucleus of the OLA
VT Ventral tegmentum
zl Zona limitans intrathalamica

Glossary

Agnathan Animals lacking jaws and refers to the jawless fish including lampreys and hagfishes.

Anadromous (antonym of catadromous) Refers to those species that are born in freshwater, then migrate to the sea where they reach an adult prespawning size, and finally return to freshwater for the spawning season.

Bauplan Model of anatomical organization shared among many members of a certain taxon, in this case by the brain of vertebrates.

Cephalochordates Clade of invertebrate chordates commonly known as lancelets or amphioxus.

Chordates Animals with a dorsal neural tube, a notochord, and a perforated pharynx including urochordates, cephalochordates, and vertebrates.

Craniates Animals with a brain in a well-defined head and the usual set of complex sense organs and include hagfishes, lampreys, and gnathostomes.

Cyclostomes It means round-mouthed vertebrates and refers to the clade of hagfishes and lampreys.

Genoarchitecture Use of gene expression patterns as topographic markers revealing cytoarchitectonic domains and borders during development.

Gnathostomes Clade including living jawed vertebrates (cartilaginous fish, bony fish, and tetrapods).

Neuromeres Segments or transversal units of the developing neural tube including prosomeres (secondary prosencephalon and diencephalon), mesomeres (midbrain), and rhombomeres (hindbrain).

Neuromeric model It assumes transverse (neuromeres) as well as longitudinal units (roof, alar, basal, and floor plates) along the entire anteroposterior neural tube axis and that their arrangement is guided by selective regulatory gene expression that allows for regionalized developmental processes.

Urochordates Clade of invertebrate chordates commonly known as tunicates that includes ascidians, thaliaceans, and appendicularians.

Vertebrates As used here, the chordate clade comprising the last common ancestor of the agnathans and gnathostomes and all its descendants, both living and extant. Some authors exclude hagfishes, thus differentiating vertebrates from craniates.

5.1 Introduction

Urochordates, cephalochordates, and vertebrates, which include agnathans and gnathostomes, are the three members of the phylum Chordata. The term agnathan includes all jawless fish and comprises two extant groups: lampreys (Hyperoartia, Petromyzontidae, or Petromyzontids) and hagfishes (Hyperotreti, Myxinoidea, or Myxinoids), collectively known as cyclostomes (Fig. 5.1). A comparison of their brains in a phylogenetic context should be of help to reconstruct the bauplan of the vertebrate brain. The split between agnathans and gnathostomes occurred about 500 million years ago, with lamprey and hagfish lineages diverging shortly after that split (Kuraku and Kuratani, 2006), thus having a very long and independent evolutionary history; even so, both lineages of agnathans are essentially conservative, and their brains have a large number of similarities between each other. There are also many similar features between agnathan and gnathostome brains; therefore, those characters shared by lampreys, hagfishes, and gnathostomes are considered plesiomorphic for vertebrates. On the other hand, the brains of the two groups of agnathan vertebrates also show some striking differences. Our purpose here is to highlight the main similarities and differences between lamprey and hagfish brains, which should be of help to reconstruct the morphotype of the agnathan/vertebrate ancestor.

The lifestyle of lampreys and hagfishes shows remarkable differences with different needs throughout development, which might be also reflected in the evolution of their brains. In this sense, lampreys have a unique development with a relatively short embryonic period followed by an extremely long and blind larval period

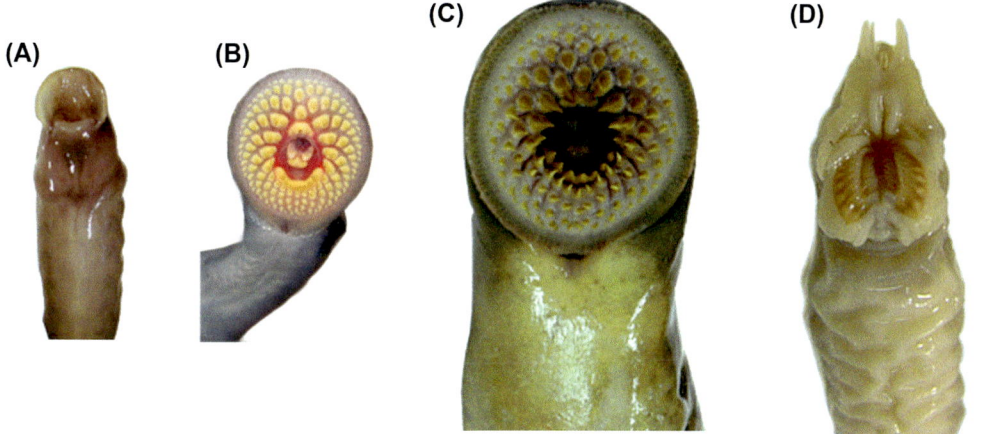

FIGURE 5.1   Ventral views of the rostral portion of the body of a larval (A), a postmetamorphic (B) and an adult (C) sea lamprey *Petromyzon marinus*, and an adult hagfish *Myxine glutinosa* (D). Note the horny teeth in the mouth of the last three specimens (B–D). For abbreviations see list. Scale bars = 5 mm.

that can last for more than 5 years, ie, 60–80% of their entire life span (Beamish and Potter, 1975). During this period, larvae (also known as ammocoetes) behave as burrowing filter-feeders and their body eventually reaches about 12–15 cm in length. Then, they undergo a drastic metamorphosis in a relatively short period of time (comprising between 3 and 6 months) to become young adults. In some species (22 out of 40), known as "brook lampreys", the young adults never feed again, reaching the breeding condition in the next few months and dying shortly thereafter. The remaining lamprey species (18 out of 40) are called "parasitic species" because they feed on the blood and muscle tissues of living preys (usually fish) in rivers, lakes, or seas. This parasitic life can last from a few months to several years, during which the largest anadromous species can reach an adult pre-spawning size of around 80 cm in body length (see Richardson and Wright, 2003; Hardisty, 2006). In contrast, the ~30 known species of hagfishes are strictly marine with no tolerance for low salinities, and they undergo direct development, with no larval stage (Gorbman, 1997). Concerning their reproduction, it is thought that all extant jawless species have external fertilization.

Agnathans have a long, cylindrical, and scaleless eel-like body and are the only fishes without paired fins. Their skeleton is cartilaginous, and they have a prominent gelatinous notochord that persists throughout life and extends almost the entire length of the animal. They develop horny teeth in the mouth, which in hagfishes is surrounded by six or eight barbels (Fig. 5.1D). Lampreys have 7 pairs of gill openings, whereas hagfishes have between 5 and 15 pairs of gills, depending on the species but also showing variations even within a single species. When present, the eyes of myxinoids are small (about 1 mm in diameter) and degenerate. Supposedly they respond only to changes in light intensity. In contrast, the eyes of lampreys are well developed and possess a lens and six extraocular muscles, but no intrinsic eye muscles for accommodation. The two groups possess a single nasal orifice (monorhinic), the nasohypophyseal opening, which is located at the tip of the snout (hagfishes) or on the top of the head (lampreys); behind this opening, lampreys have a well-developed pineal complex, which is not present in hagfishes.

It should be noted that data on embryonic or juvenile hagfishes are limited due to the difficulties of obtaining fertilized eggs. In addition, little is known about how hagfishes feed, grow, or sexually mature (Martini, 1998). The Kuratani laboratory, however, has successfully obtained a few fertilized eggs of the inshore hagfish, *Eptatretus burgeri* (Ota et al., 2007, 2011; Ota and Kuratani, 2008; Oisi et al., 2013a,b; Sugahara et al., 2016), which certainly makes it possible to perform developmental experiments to test if the brain regionalization observed in gnathostomes, specifically determined by differential gene expression patterns (genoarchitecture), is also shared by extant agnathans. Notably, some of the experimental results with those hagfish embryos have already revealed that they share a common body plan with lampreys and gnathostomes (Ota et al., 2007, 2011; Oisi et al., 2013a,b; Sugahara et al., 2016).

5.2 General Aspects of the Agnathan Central Nervous System Morphology and Development

The main rostrocaudal compartments of the gnathostome central nervous system (CNS), namely the secondary prosencephalon—including telencephalon and hypothalamus—diencephalon, mesencephalon, rhombencephalon, and spinal cord, can be also readily identified in the agnathan CNS (Fig. 5.2). Agnathans generally have

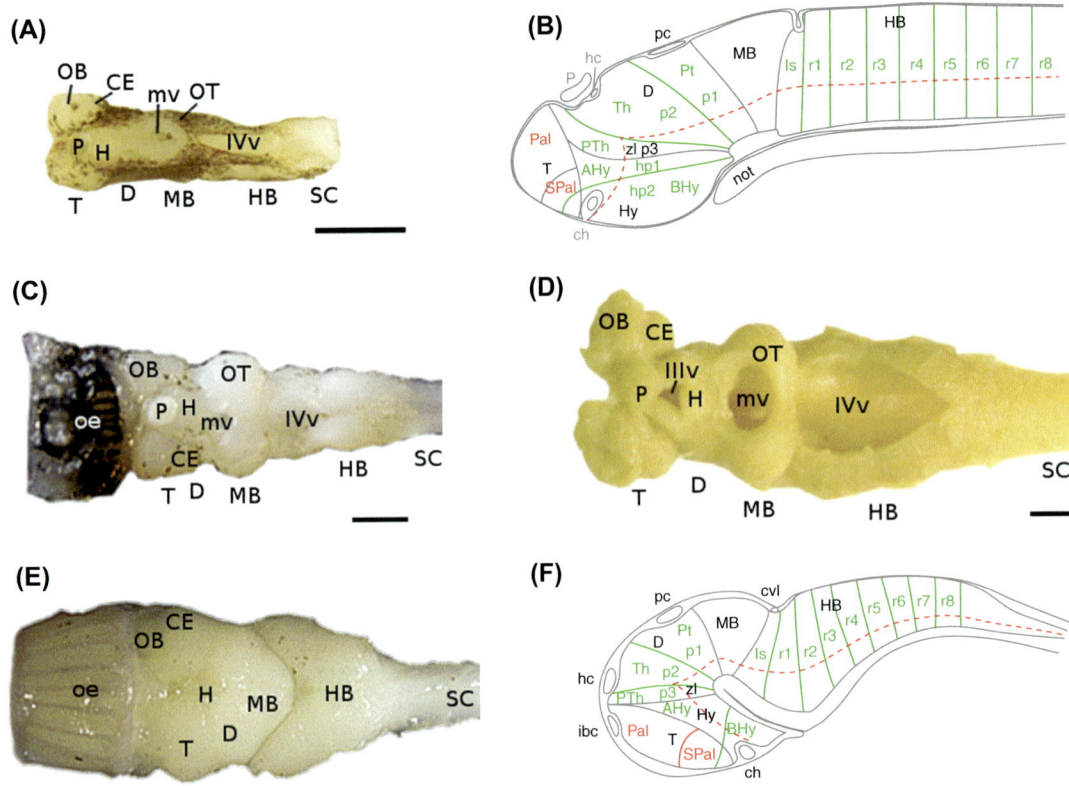

FIGURE 5.2   Main subdivisions of the agnathan brain. (A, C–E) Dorsal views of the brain of a larval (A), a postmetamorphic (C) and an adult (D) sea lamprey *Petromyzon marinus*, and an adult (E) hagfish *Myxine glutinosa*. The olfactory epithelium is also shown in C and E. (B, F) Schematic drawings of lateral views of the brain of a sea lamprey prolarva (B; 29 dpf) and a hagfish, *Eptatretus burgeri* embryo (F; stage 53) showing the main subdivisions according to the updated neuromeric model. The alar–basal boundary is marked by a *dashed red line*. The interprosomeric boundaries of the forebrain and the hindbrain are marked by *green lines* and the pallial–subpallial boundary by an *orange line*. Rostral is to the left. For abbreviations see list. Scale bars = 1 mm.

the lowest ratio of brain to body weight among vertebrates, with hagfishes having brains that are two to three times larger than those of lampreys of the same body size (Platel and Delfini, 1986; Northcutt, 1995). The brain of hagfishes is unusually compact and appears to be compressed along the rostrocaudal axis. Their ventricular system, which is well developed early in development, undergoes a progressive obliteration during ontogeny, becoming reduced to a system of narrow canals and vestigial structures in adults (Fig. 5.3A and B; Wicht and Tusch, 1998); as a consequence, they have no choroid plexuses. In contrast, lampreys possess a well-developed ventricular system with well-developed choroid plexuses, including a true mesencephalic ventricle with the corresponding choroid plexus at the dorsal midline (Fig. 5.3C). Both lampreys and hagfishes lack a true cerebellum, and hagfishes also lack pineal and parapineal organs. The agnathan CNS is surrounded by the meninx primitive, which consists of two layers enveloped by loose connective tissue containing fat cells.

Concerning the sensory systems, hagfishes are characterized by an atrophy of the eyes, which have no lens or extraocular muscles. Their acustico-lateralis system is reduced (there is no evidence for electroreception in hagfishes), but there is a hypertrophy of the olfactory apparatus in these animals (Wicht, 1996; reviewed in Miyashita and Coates, 2015). Lampreys have well-developed eyes as well as olfactory and acustico-lateralis systems. The labyrinth (inner ear) of lampreys is unique in having only two of the three semicircular canals found in gnathostomes and a single well-differentiated macula that can be readily divided into three parts, probably homologous to the utriculus, sacculus, and lagena of gnathostomes (Lowenstein et al., 1968; Maklad et al., 2014). In hagfishes, the labyrinth consists of a single torus-shaped structure containing a single macula that is not as well-differentiated as the macula of lampreys (Jørgensen, 1998). Two dilatations of the torus (anterior and posterior) represent the ampullae, each provided with a ring-shaped crista that lacks a cupula (Lowenstein and Thornhill, 1970).

The neural tube of hagfishes is formed by invagination of the neural plate and fusion of the neural folds at the dorsal midline. In contrast, neurulation in lampreys and teleosts involves the formation of a neural keel, in which a lumen originates secondarily by

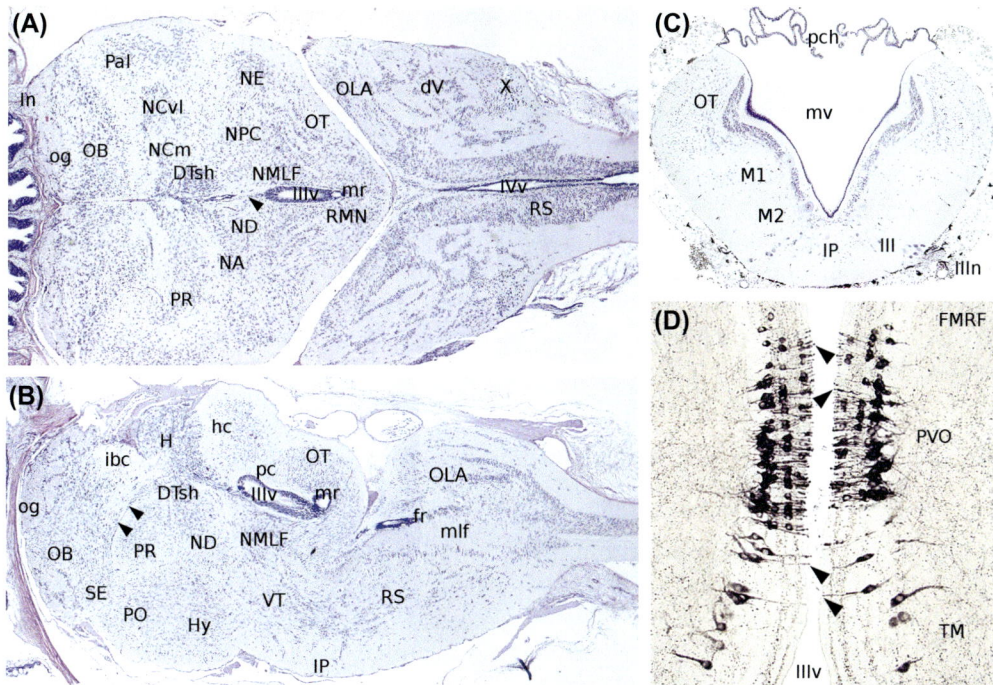

FIGURE 5.3   (A) Photomicrographs of a horizontal (A) and a sagittal (B) sections of the adult brain of the hagfish *Myxine glutinosa*, and a transverse section (C) through the mesencephalon of an adult lamprey *Lampetra fluviatilis* (hematoxylin-stained) showing the general topography. Note the narrow ventricular cavities in A and B (*M. glutinosa*), and the ample mesencephalic ventricle dorsally covered by a prominent choroid plexus in C (adult *L. fluviatilis*). The *black arrowheads* in A and B point to vestiges of ventricular structures. Rostral is to the left in A and B. (D) FMRF-like immunoreactive cells in the periventricular hypothalamic organ and the tuberomamillary nucleus of an adult *Petromyzon marinus*. Note the numerous cells immunoreactive for this peptide that directly contact the third ventricle (*black arrowheads*). For abbreviations see list.

cavitation. At the histological level, in hagfishes most neurons migrate away from the ventricular zone, whereas in lampreys the majority of the cell bodies are grouped in only a few rows at the periventricular level, though some of them are also laterally displaced and arranged in irregularly scattered clusters, particularly in the telencephalon (Fig. 5.3A–C; Pombal and Megías, 2011). At the cellular level, it is remarkable that all of the axons in agnathan brains are unmyelinated (there are no oligodendroglia), as in cephalochordates. This seems to be an ancestral chordate condition (see Nieuwenhuys and Nicholson, 1998). In the brain of hagfishes astroglia-like cells are found in great number (Wicht et al., 1994), whereas the glial population of the lamprey CNS is largely dominated by ependymal cells (Wasowicz et al., 1994). It should be noted that both hagfishes and lampreys possess a large number of cerebrospinal fluid–contacting (CSF-c) cells all along the CNS (Dávid et al., 2003; reviewed in Joly et al., 2007), some of which were labeled with molecular markers such as tyrosine hydroxylase, serotonin, opsins, GABA, or FMRF, particularly in lampreys (Fig. 5.3D; Wicht and Northcutt, 1994; Nieuwenhuys and Nicholson, 1998; Wicht and Nieuwenhuys, 1998; Pombal and Puelles, 1999). The presence of many CSF-c cells appears to reflect the ancestral condition for vertebrates.

Lampreys have a very long larval period of several years during which they gradually increase in size from less than 1 cm at hatching to around 18 cm before metamorphosis, though there is an astonishing variation in growth rates (reviewed in Hardisty and Potter, 1971; Hardisty, 2006). However, at early developmental stages there is a considerable enlargement of the neural tube, with numerous mitotic figures along its rostrocaudal extent that are exclusively located in the proximity of the ventricular lining (reviewed in Nieuwenhuys and Nicholson, 1998; Villar-Cheda et al., 2006; Pombal et al., 2011). During the transition from ammocoetes to young adults, there is an initial increase of the proliferating activity which progressively decreases and finally disappears in late metamorphic stages (Villar-Cheda et al., 2006). Therefore, the remarkable increase in size of the lamprey brain during the metamorphic period (see Fig. 5.2A and B) is most likely due to the enlargement of preexisting neurons, glial cells, and their processes, as suggested by Rovainen (1982). In any case, some areas of the brain, notably the optic tectum, undergo a considerable enlargement (Fig. 5.2A, C, and D). The increase in volume of the optic tectum occurs at the same time as the differentiation of vertebrate-like eyes and appears to be related to the transition from larvae relying predominantly on nonvisual and chemosensory signals to adults relying on both visual and olfactory cues (Salas et al., 2015). The early axonal scaffold, which extends along the boundaries of the expression domains of several transcription factors and appears to guide the later-developing neurons, is also present in the lamprey's developing brain (Kuratani et al., 1998; Barreiro-Iglesias et al., 2008). Importantly, its morphology is comparable to that of the gnathostomes, suggesting that the developmental framework underlying the basic neuronal circuits was already established in the common ancestor of vertebrates (reviewed in Murakami and Watanabe, 2009).

5.3 Forebrain (Secondary Prosencephalon and Diencephalon)

The lamprey brain has been analyzed within the framework of the segmental model initially proposed for the prosencephalon (Pombal and Puelles, 1999) and more recently updated (Pombal et al., 2009; see also Martínez-de-la-Torre et al., 2011; Pombal and Megías, 2011). Therefore, most comments on the lamprey brain will follow the subdivisions and nomenclature proposed by this model, whereas those concerning hagfishes will follow a classic cytoarchitectonic analysis with additional comments on their neuromeric subdivisions in some cases (see Fig. 5.2B and F). As mentioned above, the vast majority of knowledge available on the development and gene expression of the jawless fish brain concerns lampreys due to the difficulties in obtaining fertilized hagfish embryos (see Ota and Kuratani, 2006, 2008; Kuratani and Ota, 2008; Miyashita and Coates, 2015 for review). According to the prosomeric model (Puelles and Rubenstein, 1993, 2003, 2015), the lamprey forebrain (prosencephalon) comprises a rostral portion, the secondary prosencephalon, and a caudal portion, the diencephalon. The secondary prosencephalon is constituted by the telencephalon, which is located dorsally and includes the pallium and the subpallium (with the preoptic region as a portion of the subpallium), and the hypothalamus, which is located ventrally and bears alar and basal subdivisions (Pombal et al., 2009; Pombal and Megías, 2011). Although hagfishes have not yet been analyzed in a segmental context, available data indicate that similar subdivisions are also present in these animals (see below; Fig. 5.2B and F).

5.3.1 Secondary Prosencephalon (Telencephalon and Hypothalamus)

In lampreys, according to the prosomeric model, the dorsocaudal boundary of the secondary prosencephalon is now considered immediately caudal to the interbulbar commissure, whereas the ventrocaudal boundary is located in the apex of the cephalic flexure (Figs. 5.2B

and 5.4A; Pombal et al., 2009, 2011). The same criteria can be applied to the hagfish brain (Figs. 5.2F and 5.4B; Sugahara et al., 2016).

5.3.1.1 Telencephalon

The lamprey telencephalon consists of two lateral evaginations, the olfactory bulbs and the telencephalic (cerebral) hemispheres, and the telencephalon medium (the unevaginated portion; Nieuwenhuys and Nicholson, 1998). Moreover, an intrahypothalamic boundary has recently been described, dividing the hypothalamus into a rostral (prepeduncular) part and a caudal (peduncular) part. This boundary can be prolongated dorsally to reach the rostralmost part of the telencephalic roof plate, just behind the anterior commissure, thus separating the preoptic area from the rest of the subpallium (Figs. 5.2B and 5.4A; Pombal et al., 2009; Pombal and Megías, 2011). The general subdivisions of the hagfish telencephalon resemble those of lampreys, but more data are needed to test if both groups share similar subdivisions at hypothalamic level (compare Fig. 5.4A and B).

The morphological boundaries between neuromeres can be traced based on the expression patterns of regulatory genes involved in the early patterning of brain compartments, which are thought to be highly conserved through evolution. Gene expression patterns are also useful for establishing dorsoventral or rostrocaudal subdivisions in specific neuromeres. Therefore, reference to the expression of some characteristic gene markers will be made when available. In this sense, the telencephalon of hagfish and lamprey embryos has been identified as a domain expressing Fox, Emx, and Pax6 orthologues (Fig. 5.4A and B; Murakami et al., 2001; Derobert et al., 2002; Uchida et al., 2003; Osório et al., 2005; Tank et al., 2009; Sugahara et al., 2011, 2016). In lampreys, the evaginated portion of the telencephalic hemispheres corresponds to the pallium, whereas the subpallium remains in the unevaginated part of the prosencephalon (the telencephalon impar). Transcripts of Lhx2/9 were detected in the entire telencephalon of these animals (Osório et al., 2005), whereas the pallio-subpallial boundary is defined by dorsoventral expression of Pax6 and GliA (in the dorsal pallial division) and Dlx1/6 and GshA (in the ventral subpallial division), with additional EmxA and Lhx1/5 domains included in the dorsal telencephalon (Fig. 5.4A; Murakami et al., 2001; Myojin et al., 2001; Neidert et al., 2001; Shigetani et al., 2002; Osório et al., 2005; Tank et al., 2009; Kuraku et al., 2010; Sugahara et al., 2011; reviewed in Sugahara et al., 2013). In addition, Fgf8/17 expression appears to be restricted to the anterior part of the ventral telencephalon which corresponds to the presumptive subpallium (Uchida et al., 2003; Guérin et al., 2009; Sugahara et al., 2011). Moreover, a subpallial Nkx2.1-expressing subdomain has been identified in lampreys by the expression of two of the three lamprey

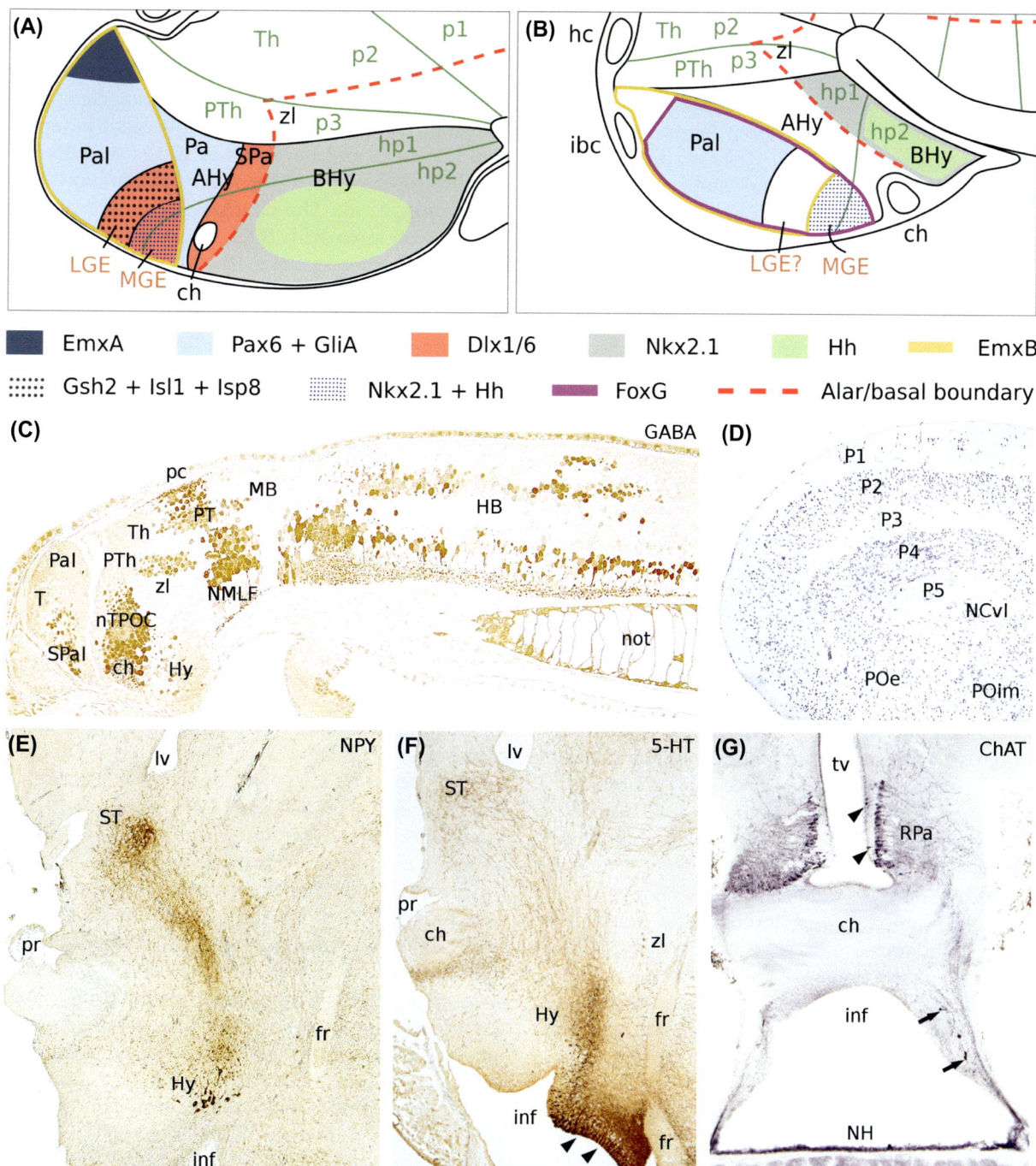

FIGURE 5.4 Schematic drawings of lateral views of the secondary prosencephalon of a lamprey *Petromyzon marinus* prolarvae (A) and a hagfish *Eptatretus burgeri* embryo (B) showing the main subdivisions according to the updated prosomeric model, as well as the expression of some marker genes (combined from various sources, see text). The alar–basal boundary is marked by a *dashed red line* and the interprosomeric boundary of the secondary prosencephalon (hp1/hp2) is marked by a *green line*. (C) Sagittal section of a lamprey *P. marinus* prolarvae showing the distribution of gamma aminobutyric acid (GABA) immunoreactive cells and fibers. (D) Transverse section of the brain of an adult *Myxine glutinosa* showing the pallial layers. (E, F) Sagittal sections through the adult brain of *Lampetra fluviatilis* illustrating the neuropeptide Y (NPY) and serotonin (5-HT) innervation of the lamprey striatum, respectively. Note also the presence of a few serotonergic cells in the zona limitans intrathalamica in F. (G) Photomicrograph of a transverse section showing choline acetyl transferase (ChAT) immunoreactive cells in the rostral paraventricular nucleus of an adult *P. marinus* with most of their processes directed lateroventrally to innervate the neurohypophysis. Note the presence of CSF-c apical processes in some immunopositive cells (*black arrowheads*), as well as the presence of various axonal swellings in the tuberal region (*black arrows*). Rostral is to the left in A–C, E, and F. For abbreviations see list.

orthologues of the gnathostome Nkx2.1 and Nkx2.4 genes, namely Nkx2.1/2.4B and Nkx2.1/2.4C (Sugahara et al., 2016). The latter authors also reported an Nkx2.1 expression subdomain in the hagfish *E. burgeri*, including a more restricted area expressing Hh2 (Fig. 5.4B; Sugahara et al., 2016). These expression patterns define the territory of the embryonic medial ganglionic eminence in gnathostomes (Moreno et al., 2009; Quintana-Urzainqui et al., 2012), thus solving one of the key questions concerning the regionalization of the telencephalon in jawless vertebrates. Moreover, markers of the lateral ganglionic eminence such as Gsh2, Isl1, and Sp8 (Hébert and Fishell, 2008; Moreno et al., 2009) are also expressed in the lamprey ventral telencephalon (Fig. 5.4A; Sugahara et al., 2011; reviewed in Sugahara et al., 2013), which indicates the presence of a domain homologous to the lateral ganglionic eminence of gnathostomes in these animals. Based on the expression of Nkx2.1 and Hh2 (Sugahara et al., 2016), the subpallium of hagfishes appears as a bipartite entity; therefore, as suggested by these authors and pending an equivalent molecular analysis, it appears likely that the subdomain intercalated between the pallium (Pax6-positive) and the MGE corresponds to the LGE of hagfishes (Fig. 5.4B). Taken together, it appears safe to conclude that a telencephalon including pallial and subpallial subdivisions, as well as lateral and medial eminences, was already present in the ancestor of vertebrates before the dichotomy between the gnathostome and agnathan lineages (Sugahara et al., 2011, 2016).

Although initially described in both larval and adult lampreys (see Nieuwenhuys and Nicholson, 1998 for review), a terminal nerve connecting the nasal region with the preoptico-hypothalamic forebrain is apparently lacking in both lampreys (Eisthen and Northcutt, 1996; Médina et al., 2009; reviewed in Pombal and Megías, 2019) and hagfishes (Wicht and Northcutt, 1992b; Wicht and Nieuwenhuys, 1998). Compelling evidence, therefore, suggests that this nerve was acquired in the ancestor of jawed vertebrates.

In general, the organization of the agnathan telencephalon is still an issue of debate, indicating that easy comparison between gnathostome and agnathan structures is not always possible. The hagfish telencephalon is well developed with the olfactory bulb and cerebral hemisphere being much larger than in the lamprey (compare Fig. 5.2D and E). The olfactory bulbs are particularly well developed in both myxinoids and petromyzontids, and their anatomical organization is similar to that of jawed vertebrates (Heier, 1948; Schnitzlein, 1982; Iwahori et al., 1987, 1998; reviewed in Ren et al., 2009), indicating that the olfactory system arose early in vertebrate evolution. It is noteworthy that the olfactory bulbs of agnathans are larger than the cerebral hemispheres illustrating the relevance of this sensory modality in these animals. Both lampreys and hagfishes are monorhinic and possess a single median olfactory organ which is associated with the hypophysis and branchial pharynx. However, it is not clear whether this single median organ was convergently derived by both lampreys and hagfishes or if it is primitive for all vertebrates; the latter is the more parsimonious hypothesis (Braun, 1996). The olfactory bulbs of lampreys are hollow structures with their ventricles connected with the median diencephalic (third) ventricle through a common interventricular foramen (or foramen of Monro). In hagfishes, the olfactory bulbs receive numerous olfactory nerve bundles (about 10 per side). Contrary to the condition in gnathostomes, hagfishes and lampreys display a massive secondary olfactory projection covering most of the pallium (Northcutt and Puzdrowski, 1988; Polenova and Vesselkin, 1993; Wicht and Northcutt, 1993; Northcutt and Wicht, 1997).

According to the prosomeric model, the dorsal pallial region of lampreys comprises a lateral and a ventral portion, and a pallial extended amygdala (Pombal et al., 2009; Martínez-de-la-Torre et al., 2011; Pombal and Megías, 2011). The classical primordium hippocampi of Johnston (1912) (see also Heier, 1948; Schöber, 1964) or medial pallium of Northcutt and Puzdrowski (1988) and Pombal and Puelles (1999) has been included inside the prethalamic eminence by the updated prosomeric model (Pombal et al., 2009). Concerning the classic subhippocampal lobe of lampreys (Heier, 1948; Schöber, 1964; Nieuwenhuys and Nicholson, 1998), there are different interpretations in the literature: some authors identified this structure as a dorsal pallium homologue (Northcutt and Puzdrowski, 1988), whereas others reinterpreted it as an unevaginated part of the pallial amygdala (Pombal et al., 2009; Pombal and Megías, 2011) due to the lack of immunoreactivity for the DLL protein (Martínez-de-la-Torre et al., 2011). More recently, Stephenson-Jones et al. (2011) considered this structure, at least in part, as the putative pallidum of lampreys based on immunohistochemical, tract-tracing, and electrophysiological data.

An interesting issue in lampreys concerns to the origin of the numerous GABAergic cells observed at pallial levels (Fig. 5.4C; Pombal and Puelles, 1999; Meléndez-Ferro et al., 2001; Robertson et al., 2007). Detailed immunohistochemical studies on the early development of the GABAergic system (Meléndez-Ferro et al., 2002; Pombal et al., 2011), together with the recent identification of an MGE in these animals (Sugahara et al., 2016), suggest they are born in the subpallium and subsequently migrate to the pallium.

The pallium is much more developed in hagfishes than in lampreys, displaying a cortical arrangement with five alternating layers of white and gray matter covering most of the telencephalon (Fig. 5.4D; reviewed in Wicht, 1996; Ronan and Northcutt, 1998;

Wicht and Nieuwenhuys, 1998). However, the identification of specific pallial subregions is extremely difficult because of the broad pallial cell migration and a significant reduction of the ventricular system. Although some pallial connections are well known in hagfishes, it was not possible to recognize ventral, lateral, dorsal, and medial components with certainty (Ronan and Northcutt, 1998; Wicht and Nieuwenhuys, 1998). The presence of an extensive secondary olfactory projection led some authors to propose that the entire pallium could be homologous to the lateral pallium of jawed vertebrates (see Butler and Hodos, 2005). In addition to the olfactory input, the pallium in agnathans, as in most jawed fishes, has interhemispheric and thalamic connections (Polenova and Vesselkin, 1993; Wicht and Northcutt, 1993, 1998; Northcutt and Wicht, 1997; Suryanarayana et al., 2017). Therefore, if present, one would expect the agnathan dorsal pallium area to be a relatively small entity in accordance with the low development of the dorsal thalamus in these animals. The pallium in both hagfishes and lampreys projects to the thalamus and optic tectum, indicating that these connections are phylogenetically primitive.

Concerning the subpallium, while septal, striatal, and preoptic subregions have been identified in both hagfishes and lampreys (Wicht and Northcutt, 1992a, 1994; Pombal et al., 1997a,b; Wicht and Nieuwenhuys, 1998), a pallidum-like structure has only been reported in lampreys (Stephenson-Jones et al., 2011). It should be noted, however, that the areas identified as septum and striatum in hagfishes do not show some of the typical characteristics of these regions in other vertebrates (Wicht and Northcutt, 1994). In contrast, the lamprey striatum is quite well characterized (Pombal et al., 1997a,b; Ericsson et al., 2011, 2013; Robertson et al., 2012). It consists of cholinergic cells and cells immunoreactive for GABA, substance P, choline acetyl transferase, as well as of numerous fibers immunoreactive for substance P, dopamine, enkephalin, serotonin, or neuropeptide Y, which are mainly detected in its periventricular neuropil (Fig. 5.4E and F; Pombal et al., 1997a,b, 2001). The identification of additional basal ganglia components in lampreys, together with the analysis of their connections, cellular properties, and transmitters, led to the conclusion that the structure and function of these nuclei were already established in the common ancestor of vertebrates and, therefore, have been highly conserved throughout vertebrate phylogeny (see Grillner et al., 2013 for review).

Although its homology with other vertebrates remains unclear, the preoptic area of hagfishes seems to be highly differentiated and comprises several nuclei: dorsal, externus, intermediate, and internus or periventricular (Fig. 5.4D; Wicht and Northcutt, 1992a; Ronan and Northcutt, 1998). In lampreys, on the other hand, a median preoptic nucleus, a septocommissural preoptic area, and lateral and medial preoptic nuclei have been proposed (Pombal and Puelles, 1999; Pombal et al., 2009).

5.3.1.2 Hypothalamus

The hagfish hypothalamus shows little differentiation and is almost limited to some periventricular cells and a lateral nucleus infundibularis (Wicht and Northcutt, 1992a; Ronan and Northcutt, 1998; Wicht and Nieuwenhuys, 1998); however, according to the prosomeric model, part of the preoptic nuclei could eventually be considered as part of the alar-derived hypothalamic portion, as reported in lampreys (Pombal et al., 2009). In contrast, the lamprey hypothalamus is much more developed, resembles that of gnathostomes in many features, and has been analyzed in detail within the framework of the prosomeric model (Fig. 5.4A; Pombal and Puelles, 1999; Pombal et al., 2009). In this scenario, the alar-derived portion of the prepeduncular hypothalamus (included in the new hp2 prosomere) contains the rostral paraventricular nucleus, the suprachiasmatic and epichiasmatic nuclei, and the anterior hypothalamic area. The basal-derived portion consists of the anterobasal nucleus, nucleus of the tract of the postoptic commissure, tuberal and tuberomamillary nuclei, the rostral part of the periventricular hypothalamic organ (this organ is now interpreted as a longitudinal entity), and the mamillary area. The caudal portion or peduncular hypothalamus, which is part of the new hp1 prosomere, includes the alar-derived caudal paraventricular nucleus and the posterior entopeduncular nucleus, whereas the basal portion comprises the putative posterior hypothalamus area, the caudal part of the periventricular hypothalamic organ, and the retromamillary area.

Similar to jawed vertebrates, the alar portion of the hypothalamus, only defined in lampreys, expresses Pax6 and GliA (Murakami et al., 2001; Derobert et al., 2002; Uchida et al., 2003; Osório et al., 2005; Sugahara et al., 2011) and is constituted by a dorsal paraventricular domain (GABA-negative) and a ventral subparaventricular domain (GABA-positive) expressing Dlx1/6 (Fig. 5.4A and C; Murakami et al., 2001; Myojin et al., 2001; Neidert et al., 2001; Shigetani et al., 2002; Pombal et al., 2009, 2011; Kuraku et al., 2010; Martínez-de-la-Torre et al., 2011). The tuberal hypothalamus of both lampreys and hagfishes is characterized by the coexpression of Hh and Nkx2.1 (Fig. 5.4A and B; Murakami et al., 2001; Ogasawara et al., 2001; Uchida et al., 2003; Osório et al., 2005; Kano et al., 2010; Sugahara et al., 2011, 2016; Oisi et al., 2013b). According to the prosomeric model, in hagfishes the territory located between the telencephalic and tuberal domains corresponds to the alar hypothalamus (see Fig. 5.4B). In addition, and similar to lampreys and gnathostomes, we propose a caudorostral subdivision of the secondary prosencephalon in these animals (hp1, hp2; see Fig. 5.4B), with the caudal portion of the basal hypothalamus expressing Nkx2.1 and the rostral portion coexpressing Nkx2.1 and Hh (Sugahara et al., 2016).

The agnathan hypothalamus shares the presence of several neurochemical markers such as GABA, serotonin, tyrosine hydroxylase, dopamine, gonadotropin-releasing hormone, vasotocin, FMRF, enkephalin, or neuropeptide Y (Figs. 5.3D, 5.4E, and F; Nieuwenhuys and Nicholson, 1998; Wicht and Nieuwenhuys, 1998; Pombal and Puelles, 1999). Apart from the dopaminergic cells, the lamprey alar hypothalamus contains a large number of neurosecretory cells, which project to the infundibular region and contribute to the hypothalamohypophyseal neurosecretory system (see Fig. 5.4G; Nieuwenhuys and Nicholson, 1998; Pombal and Puelles, 1999), indicating that the hypothalamohypophyseal neurosecretory system is well developed in these animals. Although the pituitary morphology in hagfishes is simpler than in lampreys, they appear to have neurohypophyseal and hypothalamic hormones similar to those of other vertebrates (reviewed in Nozaki, 2013). One of the most interesting nuclei inside the basal portion of the lamprey hypothalamus is the periventricular hypothalamic organ. The presence of this nucleus was recognized based on immunohistochemical data and the large number of CSF-c neurons at the ventricular level (Fig. 5.3D; Pombal and Puelles, 1999).

In the prosencephalon of hagfishes there is a prominent nucleus identified as central prosencephalic complex, which appears to originate from the fusion of both telencephalic and diencephalic components (Fig. 5.3A and B). Based on its connections, it was tentatively homologized with portions of the medial pallium, ventral thalamus, and prethalamic eminence of other vertebrates (Amemiya and Northcutt, 1996). In any case, its topographical disposition, per se, represents a hagfish autapomorphy (Khonsari et al., 2009).

5.3.2 Diencephalon

As in other vertebrates, the lamprey diencephalon can be subdivided into three segmental units: p1, p2, and p3 (Figs. 5.2B and 5.5A). The alar territory of p1 includes the pretectal region and comprises a rostral precommissural part, an intermediate juxtacommissural pretectal domain, and a caudal commissural portion, which is covered by the fibers of the posterior commissure. The alar portion of p2 corresponds to the thalamus, the habenula, and the pineal and parapineal organs, with the latter connected to the left habenula Finally, the thalamus and the habenula are separated from the prethalamus and the prethalamic eminence, respectively, by the zona limitans intrathalamica. Some data obtained in developing hagfishes (Sugahara et al., 2016) indicate that at least the three major diencephalic

subdivisions (p1, p2, and p3) are shared with other vertebrates (Figs. 5.2F and 5.5B).

As in gnathostomes, the lamprey diencephalon is defined by the combination of Pax6 and Otx expression (Fig. 5.5A; Ueki et al., 1998; Tomsa and Langeland, 1999; Murakami et al., 2001; Osório et al., 2005; Guérin et al., 2009; reviewed in Murakami and Watanabe, 2009; Suda et al., 2009). Moreover, the pretectal territory (p1) is identified by the overlapping expression of Lhx15 and Pax3/7, the thalamus (p2) by the expression of Lhx2/9, and the prethalamus by the overlapping expression of Lhx1/5 and Dlx1/6 (Fig. 5.5A; Ueki et al., 1998; Tomsa and Langeland, 1999; Murakami et al., 2001; Myojin et al., 2001; Neidert et al., 2001; Shigetani et al., 2002; Osório et al., 2005, 2006; Guérin et al., 2009; Tank et al., 2009; Kuraku et al., 2010; Sugahara et al., 2011; reviewed in Murakami and Watanabe, 2009; Suda et al., 2009). In addition to Pax6 and in contrast to lampreys, the hagfish pretectum also expresses EmxB (Fig. 5.5B; Myojin et al., 2001; Tank et al., 2009; Sugahara et al., 2016). Furthermore, the agnathan zona limitans intrathalamica, located between the prethalamus (p3) and thalamus (p2), expresses Hh2 in hagfishes and HhA and HhB in lampreys. It clearly marks the position of the alar/basal boundary at this level (Fig. 5.5A and B; Uchida et al., 2003; Murakami et al., 2005; Osório et al., 2005; Kano et al., 2010; Sugahara et al., 2011, 2016). In addition, the lamprey zona limitans intrathalamica expresses Lhx15, Fgf (fibroblast growth factor), and Pitx2 (Boorman and Shimeld, 2002; Uchida et al., 2003; Osório et al., 2005; Guérin et al., 2009).

5.3.2.1 Prethalamus

In lampreys, the alar portion of p3 is constituted by the prethalamic eminence and the prethalamus, with the later corresponding to the "old" ventral thalamus (Pombal et al., 2009; Pombal and Megías, 2011). As mentioned previously, the medial pallium is now interpreted by these authors as part of the prethalamic eminence. Moreover, a dorsal nucleus of Bellonci, a medial ventral geniculate nucleus, and a ventral zona incerta were tentatively identified inside the prethalamus (Pombal and Puelles, 1999). It should be noted that a territory homologous to the diencephalic motor region was identified and characterized in the lamprey prethalamus (El Manira et al., 1997; Ménard and Grillner, 2008).

5.3.2.2 Thalamus

In most vertebrates, the epithalamus is formed by the habenula and the epiphysis or pineal complex. The agnathan habenula, which is smaller in hagfishes than in lampreys, is highly asymmetric with the right habenular nucleus being larger than the left one, as it occurs in most fishes (Figs. 5.3B and 5.5C). This unilateral hypertrophy of the habenula accounts for the marked

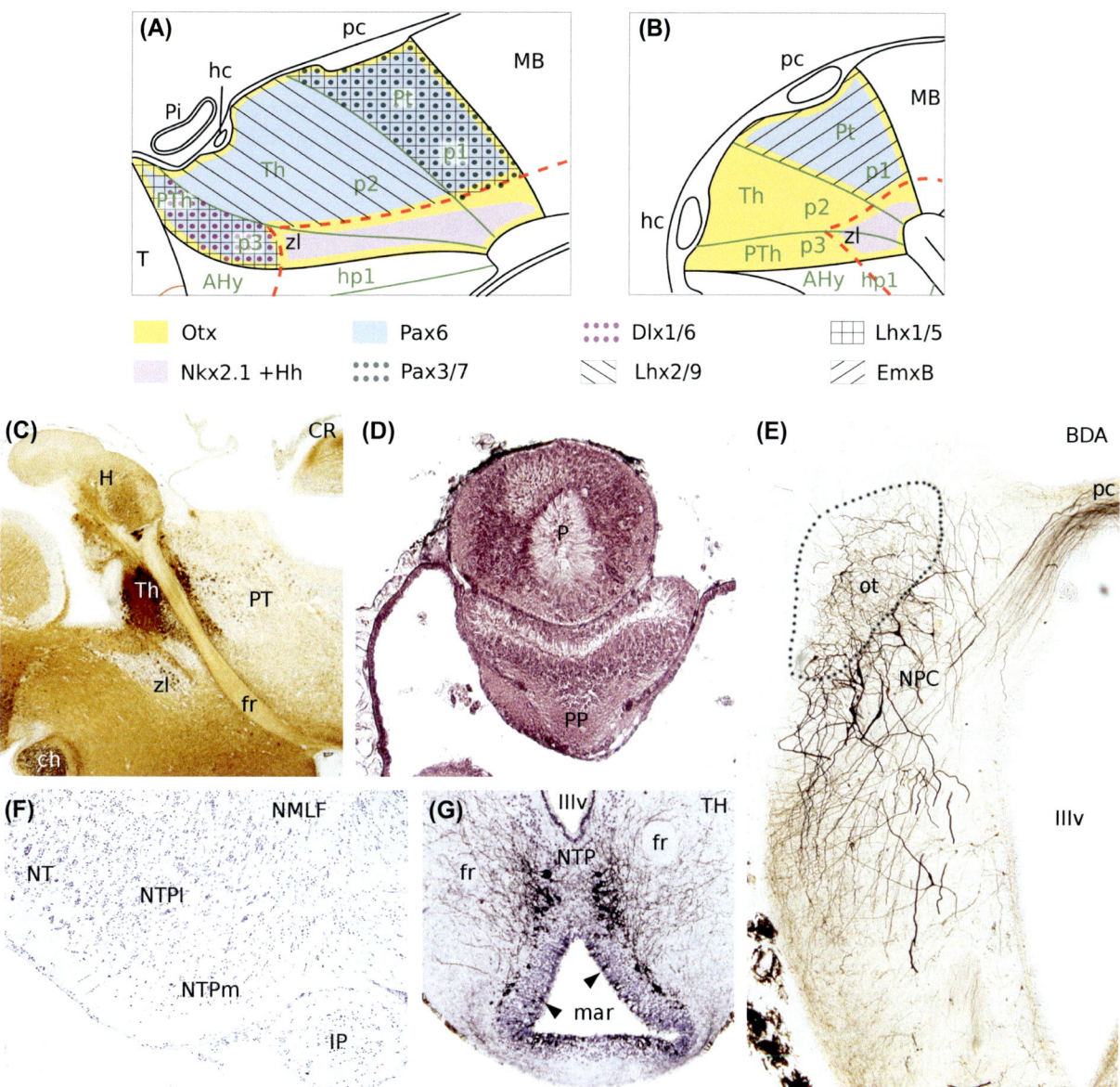

FIGURE 5.5 Schematic drawings of lateral views of the diencephalon of a lamprey *Petromyzon marinus* prolarvae (A) and a hagfish, *Eptatretus burgeri* embryo (B) showing the main subdivisions according to the updated prosomeric model, as well as the expression of some marker genes (combined from various sources, see text). The alar—basal boundary is marked by a *dashed red line* and the interprosomeric boundaries of the diencephalon (p1/p2/p3) are marked by *green lines*. Representative sagittal (C) and transverse (D—G) sections through the diencephalon of adult *Lampetra fluviatilis* (C, D, G), *P. marinus* (E), and *Myxine glutinosa* (F). (C) Sagittal section showing calretinin (CR) immunoreactive cells in the thalamus (p2) and the zona limitans intrathalamica (zl). Note the trajectory of the fasciculus retroflexus. (D) Bodian-stained transverse section of the pineal and parapineal organs. (E) Retrogradely labeled cells of the nucleus of the posterior commissure with numerous apical dendrites distributing inside the optic tract fibers (surrounded by a *dotted line*). (F) Hematoxylin-stained section through the posterior tubercular area and the nucleus of the medial longitudinal fascicle of *M. glutinosa*. Note that the rostral portion of the rhombencephalic interpeduncular nucleus reaches this level. (G) Tyrosine hydroxylase (TH) immunoreactive cells in the mamillary nucleus and the nucleus of the tuberculum posterior. Rostral is to the left in A—C. For abbreviations see list.

asymmetry of its major efferent component, the fasciculus retroflexus or habenulointerpeduncular tract (Fig. 5.5C; Nieuwenhuys, 1977). The position of the prominent habenular commissure and the trajectory of the fasciculus retroflexus in the caudal thalamus have been used as anatomical landmarks for the caudal boundary of p2 (Fig. 5.5C, see also Fig. 5.4E and F; Pombal and Puelles, 1999; Pombal et al., 1999, 2009; Pombal and Megías, 2011). This anatomical landmark is also present in hagfishes (Fig. 5.3B). Based on connectivity and molecular expression (Yáñez and Anadón, 1994; Stephenson-Jones et al., 2012b), it was shown that the distinction between medial and lateral habenula in mammals was already present in lampreys.

Although hagfishes do not show any trace of a pineal organ, the lamprey pineal complex is well developed and consists of two photosensitive organs, the pineal and parapineal (also known as parietal organ) (Fig. 5.5D). The pineal stalk reaches the most caudal part of the epithalamus, just in front of the posterior commissure, whereas the diffuse parapineal tract is associated with the left habenula. Both organs have extensive connections with the forebrain and midbrain (Puzdrowski and Northcutt, 1989; Yáñez et al., 1993, 1999; Pombal et al., 1999). The pineal gland is a source of melatonin for most vertebrates and is involved in the control of photoneuroendocrine functions.

The hagfish thalamus *sensu lato* consists of the following nuclei: anterior, internal, external, and subhabenular thalamic nuclei; triangular thalamic nucleus; intracommissural thalamic nucleus; paracommissural thalamic nucleus; and diffuse nucleus of the thalamus (Wicht and Northcutt, 1992a; Ronan and Northcutt, 1998; Wicht and Nieuwenhuys, 1998). Until now, however, no attempt to interpret the hagfish diencephalon on the framework of the prosomeric model has been performed, which precludes the ascription of these nuclei to either the prethalamus or the thalamus. In any case, the agnathan thalamus is poorly differentiated and is characteristically identified, as in most vertebrates, by its extensive immunoreactivity for calretinin (Fig. 5.5C; Pombal et al., 1999). Apart from its connections with pallial areas (Polenova and Vesselkin, 1993; Northcutt and Wicht, 1997; Suryanarayana et al., 2017), the lamprey thalamus is also connected with the striatum, the hypothalamus, the pretectum, the optic tectum, the mesencephalic tegmentum, and the dorsal isthmal gray (Polenova and Vesselkin, 1993; Northcutt and Wicht, 1997; Robertson et al., 2006; de Arriba and Pombal, 2007; Ericcson et al., 2013). Experimental evidence on the thalamic connections in hagfishes is very similar to that reported for lampreys (Kusunoki and Amemiya, 1983; Wicht and Northcutt, 1990, 1998).

5.3.2.3 Pretectum

The major components inside the alar portion of the agnathan pretectum (p1) are the pretectal nuclei and the nuclei of the posterior commissure (Wicht and Northcutt, 1992a; Pombal and Puelles, 1999). They receive significant visual projections and are involved in the modulation of motor activity (Fig. 5.5E; Ooka-Souda et al., 1995; El Manira et al., 1997; Zompa and Dubuc, 1998; Capantini et al., 2017). Just below the posterior commissure lies the subcommissural organ (Tsuneki, 1986; Joly et al., 2007). This organ is paired in lampreys and its secretory cells produce soluble proteins as well as an insoluble and ever-growing aggregate

named Reissner fiber, which extends caudally along the brain stem ventricular system and the whole length of the central canal of the spinal cord. Notably, this highly conserved gland is present throughout evolution of chordates, from amphioxus to humans (reviewed in Guerra et al., 2015).

5.3.2.4 Basal Diencephalon

The basal portion of the agnathan diencephalon differentiates into the nucleus of the medial longitudinal fascicle in the subpretectal tegmentum and the nucleus of the tuberculum posterior in the basal portion of p2 and p3 (Figs. 5.4C and 5.5F; Wicht and Northcutt, 1992a; Ronan and Northcutt, 1998; Wicht and Nieuwenhuys, 1998; Pombal and Puelles, 1999). As in other vertebrates, the former has ipsilateral projections to the spinal cord, whereas some of the catecholaminergic cells located inside the latter provide the dopaminergic innervation of the lamprey striatum (Fig. 5.5G; Pombal et al., 1997b; Pérez-Fernández et al., 2014). A similar group of TH-positive cells projecting to the telencephalon was reported in the posterior tubercular area of hagfishes (Wicht and Northcutt, 1994); therefore, as in lampreys, this cell population is probably homologous to the substantia nigra pars compacta of amniotes.

5.4 Midbrain (Mesencephalon)

As in gnathostomes, the agnathan mesencephalon is a *Pax6*-negative region whose caudal boundary is defined by the caudal edge of Otx2 expression (Fig. 5.6A and B; Ueki et al., 1998; Tomsa and Langeland, 1999; Murakami et al., 2001; Guérin et al., 2009; Murakami and Watanabe, 2009; Suda et al., 2009; Sugahara et al., 2016). The focal expression of Fgf8/17 at the midbrain—hindbrain boundary of lampreys (Shigetani et al., 2002; Uchida et al., 2003; Murakami et al., 2004; Guérin et al., 2009; Sugahara et al., 2011; Jandzik et al., 2014) and hagfishes (Oisi et al., 2013b; Sugahara et al., 2016) also marks the midbrain's caudal boundary (Fig. 5.6A and B).

The most conspicuous feature of the lamprey mesencephalon is its well-developed ventricle, which is covered by an extensive choroid plexus (Fig. 5.3C). This feature appears to represent a derived character not shared by other vertebrates. The hagfish mesencephalon is small, and only part of the adult dorsal tectum is visible externally (Fig. 5.2E). Contrary to gnathostomes, both lampreys and hagfishes lack a mesencephalic trigeminal nucleus (Rovainen, 1979; Koyama et al., 1987; Ronan, 1988; Anadón et al., 1989; reviewed in Pombal and Megías, 2019), although a group of rhombencephalic cells were suggested as a potential homologue of this nucleus in lampreys (see later discussion).

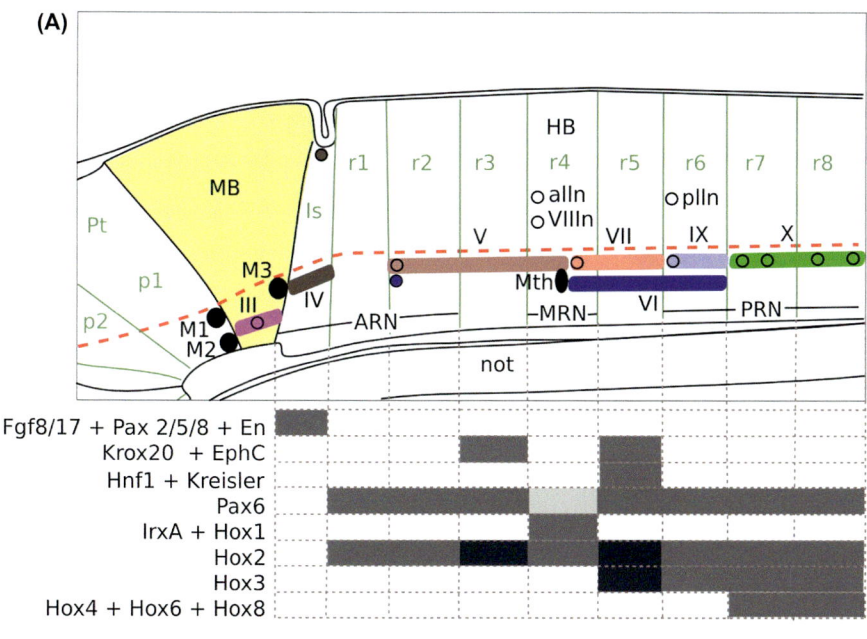

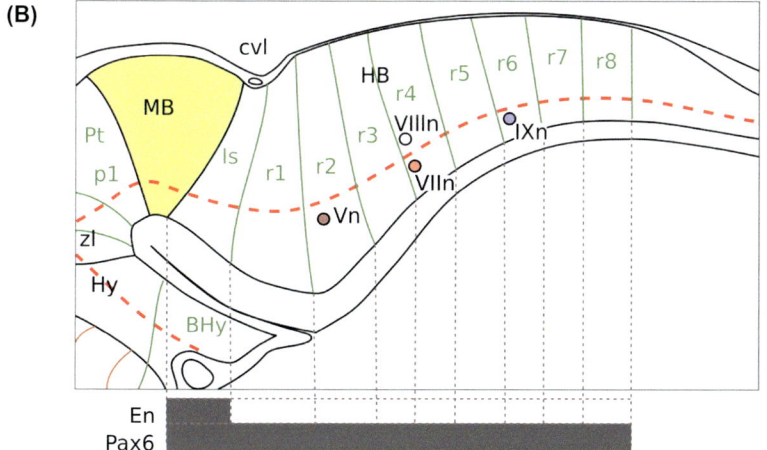

FIGURE 5.6 Schematic drawings of lateral views of the brain stem of a lamprey *Petromyzon marinus* prolarvae (A) and a hagfish *Eptatretus burgeri* embryo (B) showing the hypothetical main rhombomeric subdivisions as well as the expression of some marker genes (combined from various sources, see text). The *yellow color* of the midbrain represents Otx expression. The alar–basal boundary is marked by a *dashed red line* and the interprosomeric boundaries of the hindbrain (r1 to r8) are marked by *green lines*. (A) In lampreys, the relative position of the somatomotor (III, IV, and VI) and branchiomotor (V, VII, IX, and X) nuclei and some of the large reticular cells (M1, M2, M3, and Mth; *full black circles*) was reproduced from Pombal and Puelles (1999) and Murakami et al. (2004). Also shown are the location of the different nerve roots (*black-colored circles*), including the octaval nerve and the anterior and posterior lateral line nerves (*black circles*), as well as the rostrocaudal distribution of the three components of the rhombencephalic reticular formation. (B) The rhombomeric distribution of the V, VII, VIII, and IX nerves in hagfishes was reproduced from Oisi et al. (2013b). Rostral is to the left. For abbreviations see list.

In all vertebrates, the dorsally located optic tectum (superior colliculus of amniotes) is a layered structure subdivided into a varying number of alternating cellular and fibrous laminae or fields. Eight tectal strata (stratum ependymale, stratum cellulare periventriculare, stratum fibrosum periventriculare, stratum cellulare et fibrosum internum, stratum fibrosum centrale, stratum cellulare et fibrosum externun, stratum opticum, and stratum marginale) can be identified in the optic tectum of lampreys (Iwahori et al., 1999; de Arriba and Pombal, 2007), whereas the optic tectum of hagfishes generally contains only four tectal strata (stratum ependymale, stratum periventriculare, stratum cellulare et fibrosum, and stratum marginale) (Fig. 5.7A; Iwahori et al., 1996). These differences are likely related to the fact that the entire visual system is highly reduced in hagfishes. It should be also noted that, with some exceptions, the available data

on tectal lamination clearly show an increase in the number of tectal strata and number of cell types from hagfishes to amniotes (Northcutt, 2002). Although the optic tract fibers are the most important afferent system to the agnathan optic tectum (Fig. 5.7B), many other inputs have also been described (Amemiya, 1983; Robertson et al., 2006; de Arriba and Pombal, 2007).

Most retinopetal cells in lampreys are distributed across three mesencephalic populations: M5 nucleus of Schöber, mesencephalic reticular area (MRA), and the ventrolateral optic tectum (Fig. 5.7B). However, all three cell groups originate from a single location in the mesencephalic tegmentum (Rodicio et al., 1995). More recently, a reinterpretation of the distribution of the retinopetal cells on a prosomeric context revealed that a portion of them is located in the caudal diencephalon (Pombal and Megías, 2019). In myxinoids,

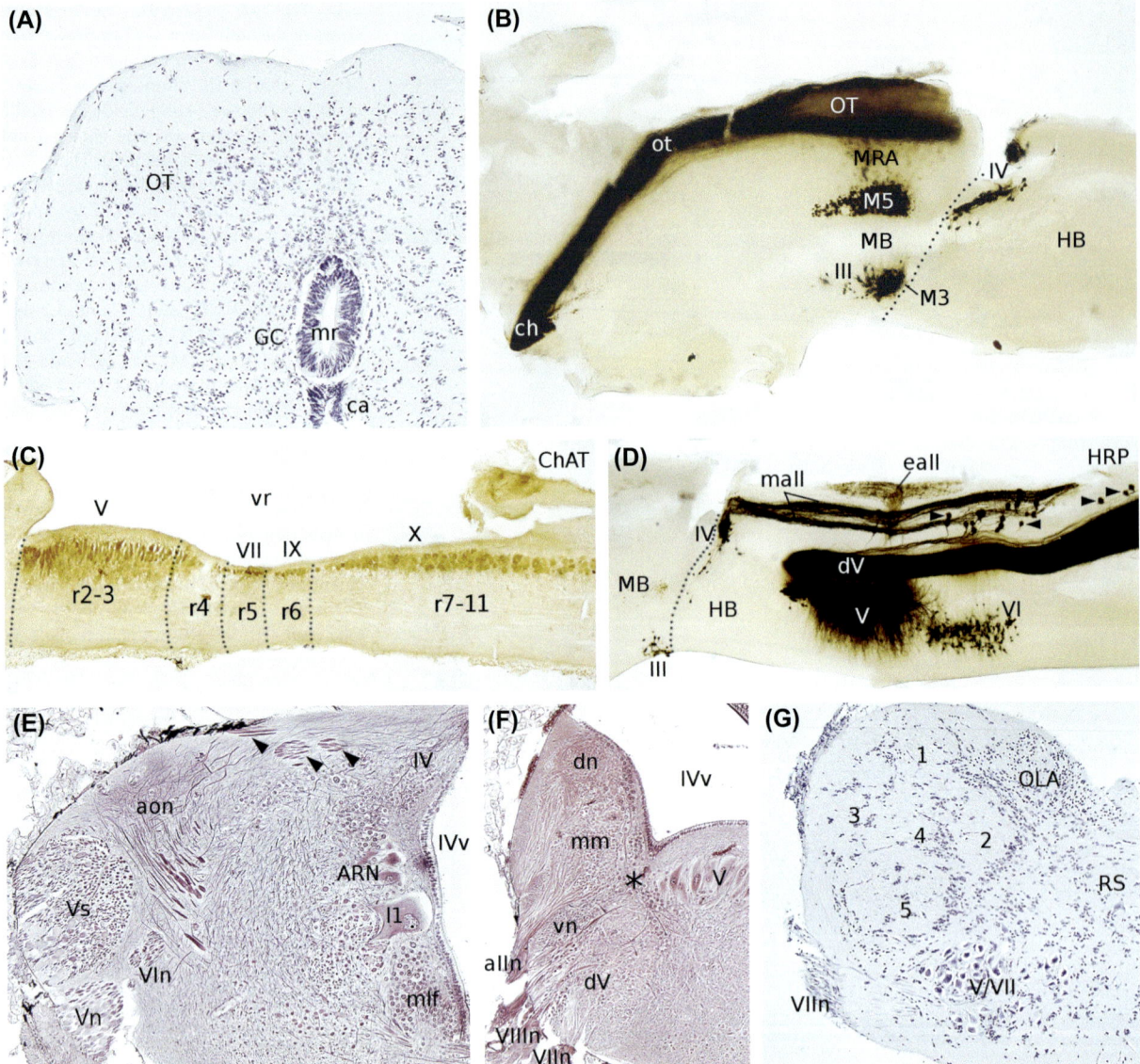

FIGURE 5.7   (A). Hematoxylin-stained section through the mesencephalon of an adult Myxine glutinosa. (B) Photomicrograph of a flattened hemibrain of a 116 mm *Petromyzon marinus* larva showing the labeling observed after horseradish peroxidase (HRP) injections into the contralateral eye orbit. Note the dorsal location of the trochlear motoneurons. (C) Photomicrograph of a sagittal section through the hindbrain of an adult *P. marinus* showing the ChAT immunoreactive neurons of the branchiomotor nuclei (V, VII, IX, and X) and their hypothetical rhombomeric distribution according to Murakami et al. (2004). (D) Lateral view of a flattened hemibrain of a 116 mm *P. marinus* larva showing the labeled structures after injection of HRP into the ipsilateral eye orbit. The *black arrowheads* point to rhombencephalic dorsal cells. (E, F) Bodian-stained transverse sections through the rostral hindbrain (E) and through the trigeminal motor nucleus (F) of an adult *Lampetra fluviatilis*. The *black arrowheads* in E point to the trochlear nerve fascicules and the *asterisk* in F indicates the location of the solitary tract. (G) Hematoxylin-stained transverse section showing the subdivisions of the sensory trigeminal nucleus in an adult *Myxine glutinosa*. The *dotted lines* in B and D represent the midbrain/hindbrain boundary. Rostral is to the left in B–D. For abbreviations see list.

retinopetal neurons have been described only in the adult Pacific hagfish (*Eptatretus stouti*; Wicht and Northcutt, 1990). Although described by these authors as being located in the rostral mesencephalon and compared to the mesencephalic M5 and MRA of lampreys (see Repérant et al., 2006 for review), these cells were initially identified as the nucleus of the posterior commissure (Wicht and Northcutt, 1990), which suggests that they might be diencephalic. Additionally,

some FMRF-amide-like immunoreactive cells projecting to the retina were reported dorsal to the optic chiasm in adult *Lampetra fluviatilis*, but they do not appear to be homologous to the olfactoretinal system of gnathostomes (Médina et al., 2009). As in myxinoids there is no evidence for the presence of an olfactoretinal system (Wicht and Nieuwenhuys, 1998), this system appears to be a synapomorphic characteristic of gnathostomes (see Repérant et al., 2007).

The mesencephalic tegmentum in lampreys includes the oculomotor nucleus (Figs. 5.3C, 5.7B, and D; Fritzsch et al., 1990; Fritzsch and Northcutt, 1993a; Pombal et al., 1994a). Hagfishes have no extraocular muscles and, therefore, lack the corresponding motor nuclei (Ronan and Northcutt, 1998; Wicht and Nieuwenhuys, 1998). The oculomotor nucleus of lampreys is unique, however, in that it innervates only three extraocular muscles (dorsal rectus contralaterally, and rostral rectus and rostral oblique ipsilaterally), whereas it innervates four eye muscles in jawed vertebrates (Fritzsch et al., 1990). Recently, a periventricular nucleus in the ventral mesencephalon and rostral rhombencephalon of lampreys was considered homologous to the periaqueductal grey of mammals and the *griseum centrale* of zebrafish based on its connections and the expression of dopamine D1 and D2 receptors (Olson et al., 2017).

5.5 Hindbrain (Rhombencephalon)

The hindbrain in agnathans is the largest part of the brain and, in hagfishes, it expands rostrally into two horns that embrace the mesencephalon (Figs. 5.2E and 5.3A). Although the rhombencephalon is considered the most conservative portion of the brain, it shows some important differences between agnathans and gnathostomes. As in other vertebrates, during agnathan development some ventricular ridges divide the rhombencephalon into transverse units, the rhombomeres, showing ventricular outpouchings and corresponding outer bulges. In this context, eight rhombomeres were identified in the hindbrain of developing lampreys (Nieuwenhuys and Nicholson, 1998), whereas only six were clearly recognized in a stage 45 *E. burgeri* embryo (Oisi et al., 2013b). It is expected, however, that hagfishes possess a similar number of rhombomeres. Although the presence of pseudorhombomeres (Cambronero and Puelles, 2000; Marín et al., 2008) in agnathans has not yet been investigated, they are likely to exist at least in lampreys due to the large rostrocaudal extension of the vagal nucleus in these animals (Fig. 5.7C; see Pombal and Megías, 2019). The use of additional molecular markers appears necessary to solve this conjecture.

In lampreys, as in gnathostomes, rhombomeric identities are specified by a number of transcription factor–encoding genes (see Murakami et al., 2004, 2005; Takio et al., 2004, 2007; Murakami and Watanabe, 2009; Parker et al., 2014; Miyashita et al., 2019). For instance, Gbx is a marker of the hindbrain in lampreys (Takio et al., 2007) and gnathostomes. The boundary between the midbrain and the hindbrain (MHB; Is or r0), which functions as an organizing center of morphogenic anteroposterior patterning, expresses Fgf8, En, and Pax2/5/8 (Fig. 5.6A; Murakami et al., 2001, 2004; McCauley and Bronner-Fraser, 2002; Shigetani et al., 2002; Uchida

et al., 2003; Takio et al., 2007; Matsuura et al., 2008; Guérin et al., 2009; Hammond et al., 2009; Sugahara et al., 2011, 2016). Moreover, gene cognates of Krox20 and Eph are expressed in r3 and r5 (Suzuki et al., 2015), with r5 expressing also Hnf1 (Jiménez-Guri and Pujades, 2011) and Kreisler (Parker et al., 2014), whereas r4 expresses IrxA (Jiménez-Guri and Pujades, 2011) and very low levels of Pax6 (Fig. 5.6A; Murakami et al., 2001, 2005; Derobert et al., 2002). Furthermore, expression of the various hox genes also shows a segmental distribution in lampreys (see Fig. 5.6A), although some minor differences were observed between Kuratani's lab (Murakami et al., 2004, 2005; Takio et al., 2004, 2007) and that of Parker et al. (2014). Taken together, these examples indicate that the developmental program to generate rhombomeres was already present in the common ancestor of lampreys and gnathostomes.

In addition, the recognition of rhombomeres in lampreys allowed the ascription of reticulospinal and branchiomotor neurons to specific rhombomeres (Fig. 5.6A; Murakami et al., 2004, 2005). In this scenario, the motor nuclei in lampreys follow a segmental pattern similar to that reported for gnathostomes, with the exception of trigeminal (V) and facial (VII) motor neurons, which are not in register with rhombomere boundaries (Figs. 5.6A and 5.7C; Fritzsch, 1998a; Murakami et al., 2004; Murakami and Kuratani, 2008). Another example is the location of the Mauthner cell, a huge neuron with contralateral projection down to the spinal cord involved in escape behavior responses; this cell is only present in aquatic organisms and develops inside r4 (Fig. 5.6A; Murakami et al., 2004; Pombal and Megías, 2019). Therefore, homologies between lamprey and gnathostome reticulospinal neurons may be established based on their rhombomeric distribution. The analysis of the large rostrocaudal populations of the reticular formation, which usually cover more than one rhombomere, would probably benefit when taking into account their specific rhombomeric distribution. A similar segmental pattern is probably present in the hindbrain of hagfishes which, apparently, do not have neuron homologues to Müller and Mauthner cells (Wicht and Nieuwenhuys, 1998; but see Kishida et al., 1986). However, to date only the expression of Pax6 in the hindbrain and the expression of En in the MHB have been described in hagfishes (Fig. 5.6B; Oisi et al., 2013b; Sugahara et al., 2016).

The presence of a cerebellum in the brain of hagfishes and lampreys has long been a matter of debate since there are no obvious cerebellar structures in these animals. Lately, most authors have concluded that agnathans do not have a "true" cerebellum (Pombal and Megías, 2011, 2019). The absence of a rhombic lip-derived cerebellum in lampreys had been assumed on the basis of the lack of Pax6 expression in the dorsal portion of r1 (Murakami et al., 2005). Recent studies, however, demonstrated the presence of a second lamprey Pax6 (Pax6B) gene that is expressed in dorsal r1, although more ventrally than in gnathostomes, together with other rhombic lip markers, such as Atoh1, Wnt1, and Ptf1a (Sugahara et al., 2016). The same authors also demonstrated a Pax6- and Atoh1-positive rhombic lip in the inshore hagfish *E. burgeri* (Sugahara et al., 2016). Altogether, as suggested by Sugahara et al. (2016), these results indicate that certain genetic backgrounds needed for the acquisition of the proper cerebellum of gnathostomes were already present before the splitting of agnathans and gnathostomes. However, the presence of a rhombic lip-like structure expressing these markers appears not to be sufficient for the elaboration of a gnathostome cerebellum because neither Purkinje cells nor cerebellar nuclei have been found in lampreys and hagfishes (Lannoo and Hawkes, 1997; Nieuwenhuys and Nicholson, 1998; Ronan and Northcutt, 1998; Northcutt, 2002). In addition, the main nuclei derived from the rhombic lip that are related to the gnathostome cerebellum, such as the pontine nuclei and the olivary complex or the mesencephalic red nucleus, have not been identified in agnathans (Pombal and Megías, 2011). Their absence likely correlates with the absence of a cerebellum in these animals.

Although the isthmus is not especially developed, a number of nuclei related to the control of locomotion have been described in lampreys and homologized to their gnathostome counterparts. These nuclei include (1) a dopaminergic locus coeruleus projecting to the striatum (Pombal et al., 1997b); (2) a cholinergic mesencephalic locomotor region that, according to the present topological analysis, is in fact located in the isthmus region (Pombal et al., 2001); (3) a nucleus isthmi characterized by its reciprocal connections with the optic tectum and the cholinergic nature of its cells (Pombal et al., 2001; de Arriba and Pombal, 2007); and (4) a substantia nigra pars reticulata (Stephenson-Jones et al., 2012a). Of these nuclei, only a locus coeruleus was tentatively identified in hagfishes (Wicht and Northcutt, 1998).

5.5.1 Somatomotor Zone

In lampreys, the somatomotor neurons of the trochlear nucleus are located at the isthmus level, with their axons leaving the brain dorsally, as in gnathostomes (Figs. 5.6A, 5.7B, D, and E). This nucleus contributes fibers to both the ipsilateral and the contralateral trochlear nerve and innervates the caudal oblique extraocular muscle (Fritzsch and Sonntag, 1988; Fritzsch et al., 1990; Rodicio et al., 1992; Fritzsch and Northcutt, 1993a; Pombal et al., 1994a; Pombal and Megías, 2019). Curiously, in lampreys this nucleus is located exceptionally far dorsal in the so-called velum medullar anterior (Fig. 5.7B, D, and E). The unique dorsal location of the

trochlear motoneurons does not necessarily mean that this nucleus arises from the alar plate; instead, its dorsal location might result from a dorsalization of the basal plate at this level due to differences in the expression of gene markers and the absence of a true cerebellum.

More caudal in the lamprey hindbrain lie the somatic motoneurons of the abducent nucleus, some of which are laterally displaced (Figs. 5.6A and 5.7D; Fritzsch et al., 1990; Fritzsch and Northcutt, 1993a,b; Pombal et al., 1994a; Fritzsch, 1998b; Pombal and Megías, 2019). Surprisingly, the axons of these motoneurons in lampreys do not follow the typical ventral trajectory found in gnathostomes; instead, they course rostrally in parallel with the axons of the trigeminal motoneurons and finally exit the brain at the same level as the trigeminal nerve (in r2). There they exit in a dorsolateral position, as do the visceromotor axons (Fig. 5.7E; Fritzsch and Northcutt, 1993a,b; Pombal et al., 1994a; Pombal and Megías, 2019). As extant hagfishes have no ocular motoneurons, we cannot know whether the dorsal location of the trochlear motoneurons and the rostrodorsal exit of the abducent nerve in lampreys represent primitive features for vertebrates (Fritzsch, 1998a).

Apart from the close relation of the ocular motor nuclei with the medial longitudinal fascicle in all vertebrates, in lampreys the trochlear motoneurons appear to receive direct contacts from octaval and lateral line afferents, whereas the abducent motoneurons send dendrites toward the octaval and descending trigeminal afferents (Fritzsch and Sonntag, 1988; Fritzsch et al., 1990; Rodicio et al., 1992; Pombal et al., 1994a; Fritzsch, 1998b; Pombal and Megías, 2019).

5.5.2 Visceromotor Zone

The agnathan visceromotor zone extends from the isthmus to the region of the obex in the caudal rhombencephalon, where it is continuous with the spinal gray matter (Nieuwenhuys and Nicholson, 1998; Wicht and Nieuwenhuys, 1998; Pombal and Megías, 2019). Most of the motoneurons of this column appear to correspond to the branchiomotor nuclei. The branchial motor column in agnathans is almost continuous from the level of the trigeminal roots to the caudalmost rhombencephalon and comprises the trigeminal (V), facial (VII), glossopharyngeal (IX), and vagal (X) motor nuclei (Figs. 5.6A and 7C). In lampreys their perikaria are located adjacent to the ventricular surface, as they are in gnathostomes, whereas in hagfishes they have migrated to the periphery and occupy a submeningeal position (Kishida et al., 1986; Matsuda et al., 1991; Wicht and Nieuwenhuys, 1998). In both lampreys and hagfishes, the trigeminal motor nucleus is highly developed and consists of large neurons that in lampreys produce a ventricular elevation known as eminentia trigemini (Fig. 5.7C and F).

Although the presence of a visceral component was suggested in both hagfishes and lampreys, particularly in the vagal nerve (Fritzsch and Northcutt, 1993a; Pombal et al., 2001); thus far there is no clear experimental evidence for a segregated visceral component; therefore, if present, visceral motoneurons share perikaryal position and axonal course with branchial motoneurons (Fritzsch and Northcutt, 1993a).

In lampreys the trigeminal nerve roots are located in r2, the facial root in r4, the glossopharyngeal root in r6, and those of the vagal nerve in r7 and r8 (Fig. 5.6A; Fritzsch, 1998a; Pombal and Megías, 2019). As in lampreys, the trigeminal, facial, and glossopharyngeal nerve roots of hagfishes are located in r2, r4, and r6, respectively. However, the vagal roots in hagfishes appear to fuse with the glossopharyngeal nerve to form a cranial nerve complex (Ronan and Northcutt, 1998; Oisi et al., 2013b). Hagfishes also exhibit a conspicuous nucleus identified as nucleus A of Kusunoki (Kishida et al., 1986) lateral to the visceromotor zone at the level of the trigeminal motor nucleus. Apart from its high content of acetyl cholinesterase and its immunoreactivity for several peptides (Kusunoki et al., 1982; Wicht and Nieuwenhuys, 1998), nothing is known about the connections and functions of this nucleus.

5.5.3 Octavolateral System

Sensory information provided by the inner ear is quite similar between lampreys and gnathostomes, but much more limited in hagfishes. The increased complexity of the lamprey inner ear could be associated with the origin of the oculomotor system, which is not present in hagfishes (Braun, 1996). The lateral line system is present in the agnathans and other anamniotes, but lost entirely in amniotes, which may be related to their shift into a terrestrial ecology. In lampreys, this system is very similar to that in anamniotic vertebrates, comprising several lines of mechanoreceptive neuromasts distributed on the surface of the skin over the head and trunk, as well as electroreceptive organs (reviewed in Braun, 1996; Northcutt, 2002). The lateral line system of hagfishes, which is present in members of the family Eptatretidae but absent in the family Myxinidae, is simpler than in lampreys; those species without a lateral line may have secondarily lost it in association with their adoption of a burrowing lifestyle, which makes the lateral line system less useful (Braun, 1996). Although hagfishes do not have electroreceptors (Gibbs, 2004), the presence of ampullary electroreceptors in lampreys and many jawed anamniotes suggests that they were present in primitive vertebrates and lost in the hagfishes (Northcutt, 2002).

Interestingly, lampreys possess a unique photoreceptive system constituted by multivillous receptor cells in the skin, whose stimulation induces photokinesis in both ammocoetes and adult animals (see Braun, 1996; Suzuki and Grillner, 2018 for review). As these receptor cells are innervated by fibers of the lateral line nerve, this photoreceptive system appears to be a new octavolateral modality evolved within the lamprey lineage.

The lamprey octavolateral area lies in the alar plate and comprises three dorsoventral subdivisions: dorsal and medial nuclei receiving electroreceptive and mechanoreceptive information, respectively, via the anterior and posterior lateral line nerves, and a ventral nucleus receiving inner ear information via the octaval (VIII) nerve (Fig. 5.7D and F; Ronan and Northcutt, 1987; Koyama et al., 1989, 1990; González and Anadón, 1992, 1994; reviewed in Pombal and Megías, 2019). Contrary to other vertebrates, lampreys have contralateral, as well as ipsilateral, primary mechanosensory lateral line projections to the medial nucleus (Ronan and Northcutt, 1987; Koyama et al., 1990; González and Anadón, 1992). In eptatretid hagfishes, the anterior and posterior lateral line nerves carry sensory fibers that terminate in the medial nucleus of the octavolateral area (Kishida et al., 1987; Braun and Northcutt, 1997). In addition, the anterior lateral line nerve apparently contains cutaneous sensory fibers projecting to the trigeminal sensory nucleus (Kishida et al., 1987; Nishizawa et al., 1988; Braun and Northcutt, 1997).

After entering the hindbrain, the octaval primary sensory fibers in lampreys bifurcate to form an ascending branch that reaches the hindbrain's rostral end and a descending branch that courses caudally and reaches the rostral spinal cord. In the hindbrain, these fibers contact three nuclei, the anterior, intermediate, and posterior octavomotor nuclei (Fig. 5.7E; Koyama et al., 1989; González and Anadón, 1994; Pombal et al., 1994b; Pombal and Megías, 2019). The primary vestibular projections in hagfishes distribute along the ventral nucleus of the octavolateral area (Figs. 5.3A and 5.7G; Amemiya et al., 1985). As hagfishes have no electroreceptive lateral line system, their dorsal nucleus receives ascending spinal projections through the dorsal column pathway (Ronan and Northcutt, 1990), which hampers its homologization. Contrary to lampreys, hagfishes do not have efferent innervation of the labyrinth (Amemiya et al., 1985; Fritzsch et al., 1989; Koyama et al., 1989; González and Anadón, 1994).

5.5.4 General Somatosensory Zone

The sensory complex of the trigeminal (V) nerve shows a somatotopic organization in agnathans. In hagfishes, it is arranged into several fiber bundles surrounded by the neurons of the huge sensory nucleus; in lampreys, it forms a single fiber bundle, with the sensory neurons randomly interspersed between the descending sensory fibers (Fig. 5.7D–G; Kishida et al., 1986; Koyama et al., 1987; Nishizawa et al., 1988; reviewed in Wicht, 1996; Nieuwenhuys and Nicholson, 1998; Wicht and Nieuwenhuys, 1998; Pombal and Megías, 2019). Importantly, the pattern of trigeminal primary central projections differs between lampreys and hagfishes. In lampreys, as in gnathostomes, the maxillary branch (V2) projects dorsally, the ophthalmic branch (V1) projects ventrally, and the mandibular branch (V3) projects laterally (Koyama et al., 1987). In contrast, the somatotopy in hagfishes appears to be inverted as compared to lampreys and gnathostomes, with V1 projecting dorsally, V3 ventrally, and V2 projecting between V1 and V3 (Koyama et al., 1987; Nishizawa et al., 1988). The somatotopic representation of trigeminal sensory fibers in mice has been linked to rhombomeric compartments, and Hoxa2 is involved in the establishment of those connections (Oury et al., 2006). Curiously, in lampreys the nucleus of the descending trigeminal tract also expresses Hoxa2 (Murakami and Kuratani, 2008), which led these authors to suggest that the basic pattern for sensory trigeminal topographical projections is already present in the lamprey.

In contrast to jawed vertebrates, agnathans have neither a mesencephalic trigeminal nucleus nor an ascending primary trigeminal tract. In addition, both groups of animals possess large intramedullary sensory cells in the hindbrain that have been named dorsal cells (Fig. 5.7D; Rovainen, 1979; Koyama et al., 1987; Ronan, 1988; Anadón et al., 1989; reviewed in Pombal and Megías, 2019). As the intramedullary dorsal cells in the hindbrain of lampreys and hagfishes are not present in the gnathostome hindbrain, they were tentatively homologized with the mesencephalic trigeminal nucleus of gnathostomes—which is located in the mesencephalon but absent in agnathans—because they are the only intraencephalic cells having peripheral processes running off through the trigeminal nerve (Nieuwenhuys, 1977; Anadón et al., 1989).

It should be noted that the trigeminal nerve in hagfishes is not entirely somatosensory because it also carries chemoreceptive information from the so-called Schreiner organs (see below) (Nishizawa et al., 1988; Braun and Northcutt, 1997; Braun, 1998). In addition, collaterals of the glossopharyngeal and vagal nerves project to the sensory nucleus of the trigeminal nerve (Kishida et al., 1986). More caudally in the hindbrain, the general somatosensory zone is constituted by the dorsal column nuclei, which receive ascending projections from the spinal cord (Ronan and Northcutt, 1990; Dubuc et al., 1993; Wicht and Nieuwenhuys, 1998).

5.5.5 Viscerosensory Zone

Hagfishes and lampreys possess solitary chemosensory cells that are distributed over the surface of the skin and innervated by either cranial or spinal nerves, as in most anamniote vertebrates (Finger, 1997). Like gnathostomes, lampreys also have special chemosensory organs in the oropharyngeal region (reviewed in Barreiro-Iglesias et al., 2010). Although these organs have been homologized to gnathostome taste buds, they differ from the gnathostome taste buds in two aspects: (1) The sensory cells in lampreys are ciliated rather than microvillar and the innervating fibers do not directly contact the receptor cells (Braun, 1996); therefore, as indicated by Braun (1996), these lamprey features are either primitive for taste buds or have evolved secondarily in lampreys. (2) Although some authors consider that lampreys lack the facial nerve innervation (see Barreiro-Iglesias et al., 2010), it appears that, as in gnathostomes, taste buds in lampreys are innervated by branches of the facial (VII), glossopharyngeal (IX), and vagal (X) cranial nerves that project to a common region of the hindbrain (the solitary complex) which is probably homologous in all vertebrates (Fritzsch and Northcutt, 1993b). Tracing studies in lampreys labeled a few fibers in each of the three nerves (VII, IX, and X), which is in line with the low number of gustatory organs in these animals (Fritzsch and Northcutt, 1993b; Koyama, 2005). Inside the hindbrain, these fibers course along the dorsomedial edge of the descending trigeminal tract, forming an ascending branch reaching the hindbrain's rostral end and a descending branch reaching the rostral spinal cord, as is the case for gustatory fibers in other vertebrates (Fig. 5.7F; Fritzsch and Northcutt, 1993b; Koyama, 2005; reviewed in Pombal and Megías, 2019).

In contrast, hagfishes have a different type of chemosensory organs identified as Schreiner organs (Braun, 1996, 1998; Northcutt, 2004). Although resembling taste buds, they are not considered homologous structures; they appear to be a specialization of hagfishes (Braun, 1998). Interestingly, Schreiner organs occur in large numbers within the oropharynx and throughout the epidermis and are primarily innervated by trigeminal and spinal nerves (Braun, 1996, 1998). In light of the large number of Schreiner organs in hagfishes and their extensive representation in the brain, they must be an important sensory modality (Braun, 1996). Contrary to other vertebrates, the rhombencephalic viscerosensory zone in hagfishes is located lateral and ventral (Matsuda et al., 1991), which may be related to the unusual pattern of cell migration in the whole brain of these animals. The presence in agnathans of a general viscerosensory system may have been related to a shift in function of the pharynx from filter feeding to predation, as suggested by Northcutt and Gans (1983).

5.6 Conclusions and Perspectives

Agnathans are animals of outstanding phylogenetic interest not only for understanding the agnathan developmental plan but also the early evolution of the vertebrate brain. By the end of the 20th century, most knowledge on the organization of the agnathan CNS relied largely on experimental data obtained on adult lamprey brains. Based on that information, a segmental subdivision of the lamprey prosencephalon was proposed inspired by the prosomeric model initially reported for tetrapods (Puelles and Rubenstein, 1993, 2003; Pombal and Puelles, 1999; Pombal et al., 2009). In addition, the vast majority of developmental studies in agnathans were performed in lampreys, as access to hagfish fertilized embryos was severely limited (Gorbman, 1997; Ota et al., 2006; reviewed in Miyashita et al., 2019). Currently, the most relevant homologies for brain compartments proposed on the framework of the prosomeric model are based on the expression patterns of developmental regulatory genes in lamprey embryos, as well as in the pioneering studies from the Kuratani laboratory on hagfish embryos. An outstanding example refers to the identification of a putative medial ganglionic eminence in both lampreys and hagfishes (Sugahara et al., 2016). The data reviewed herein, in particular those related to the expression of developmental genes, clearly show that the organization of the agnathan brain is more similar to that of gnathostome vertebrates than previously thought and suggest that the basic structure of the vertebrate brain was achieved before the agnathan/gnathostome divergence. The sequencing of the lamprey genome (Smith et al., 2013), together with the recent success in obtaining and raising hagfish embryos (Ota et al., 2007, 2011; Ota and Kuratani, 2008; Oisi et al., 2013a,b; Sugahara et al., 2016), opens new and promising scenarios on the agnathan comparative embryology, which will eventually provide relevant data on the developmental events underlying the emergence of the vertebrate CNS bauplan.

References

Amemiya, F., 1983. Afferent connections to the tectum mesencephali in the hagfish, *Eptatretus burgeri*: an HRP study. J. Hirnforsch. 24, 225–236.

Amemiya, F., Northcutt, R.G., 1996. Afferent and efferent connections of the central prosencephalic nucleus in the Pacific hagfish. Brain Behav. Evol. 47, 149–155.

Amemiya, F., Kishida, R., Goris, R.C., Onishi, H., Kusunoki, T., 1985. Primary vestibular projections in the hagfish, *Eptatretus burgeri*. Brain Res. 337, 73–79.

Anadón, R., De Miguel, E., González-Fuentes, M.J., Rodicio, C., 1989. HRP study of the central components of the trigeminal nerve in the larval sea lamprey: organization and homology of the primary medullary and spinal nucleus of the trigeminus. J. Comp. Neurol. 283, 602–610.

Barreiro-Iglesias, A., Villar-Cheda, B., Abalo, X.M., Anadón, R., Rodicio, M.C., 2008. The early scaffold of axon tracts in the brain of a primitive vertebrate, the sea lamprey. Brain Res. Bull. 75, 42–52.

Barreiro-Iglesias, A., Anadón, R., Rodicio, M.C., 2010. The gustatory system of lampreys. Brain Behav. Evol. 75, 241–250.

Beamish, F.W.H., Potter, I.C., 1975. The biology of the anadromous sea lamprey (*Petromyzon marinus*) in New Brunswick. J. Zool. Lond. 177, 57–72.

Boorman, C.J., Shimeld, S.M., 2002. Cloning and expression of a Pitx homeobox gene from the lamprey, a jawless vertebrate. Dev. Genes Evol. 212, 349–353.

Braun, C.B., 1996. The sensory biology of the living jawless fishes: a phylogenetic assessment. Brain Behav. Evol. 48, 262–276.

Braun, C.B., 1998. Schreiner organs: a new craniate chemosensory modality in hagfishes. J. Comp. Neurol. 392, 135–163.

Braun, C.B., Northcutt, R.G., 1997. The lateral line system of hagfishes (Craniata: Myxinoidea). Acta Zool 78, 247–268.

Butler, A., Hodos, W., 2005. Comparative Vertebrate Neuroanatomy. Wiley, New Jersey.

Cambronero, F., Puelles, L., 2000. Rostrocaudal nuclear relationships in the avian medulla oblongata: a fate map with quail chick chimeras. J. Comp. Neurol. 427, 522–545.

Capantini, L., von Twickel, A., Robertson, B., Grillner, S., 2017. The pretectal connectome in lamprey. J. Comp. Neurol. 525, 753–772.

Dávid, C., Frank, C.L., Lukáts, A., Szél, A., Vígh, B., 2003. Cerebrospinal fluid contacting neurons in the reduced brain ventricular system of the Atlantic hagfish, *Myxine glutinosa*. Acta Biol. Hung. 54, 35–44.

de Arriba, M., del, C., Pombal, M.A., 2007. Afferent connections of the optic tectum in lampreys: an experimental study. Brain Behav. Evol. 69, 37–68.

Derobert, Y., Baratte, B., Lepage, M., Mazan, S., 2002. Pax6 expression patterns in *Lampetra fluviatilis* and *Scyliorhinus canicula* embryos suggest highly conserved roles in the early regionalization of the vertebrate brain. Brain Res. Bull. 57, 277–280.

Dubuc, R., Bongianni, F., Ohta, Y., Grillner, S., 1993. Anatomical and physiological study of brainstem nuclei relaying dorsal column inputs in lampreys. J. Comp. Neurol. 327, 260–270.

Eisthen, H.L., Northcutt, R.G., 1996. Silver lampreys (*Ichthyomyzon unicuspis*) lack a gonadotropin-releasing hormone- and FMRFamide-immunoreactive terminal nerve. J. Comp. Neurol. 370, 159–172.

El Manira, A., Pombal, M.A., Grillner, S., 1997. Diencephalic projection to reticulospinal neurons involved in the initiation of locomotion in adult lampreys *Lampetra fluviatilis*. J. Comp. Neurol. 389, 603–616.

Ericsson, J., Silberberg, G., Robertson, B., Wikström, M.A., Grillner, S., 2011. Striatal cellular properties conserved from lampreys to mammals. J. Physiol. 589, 2979–2992.

Ericsson, J., Stephenson-Jones, M., Kardamakis, A., et al., 2013. Evolutionarily conserved differences in pallial and thalamic short-term synaptic plasticity in striatum. J. Physiol. 591, 859–874.

Finger, T.E., 1997. Evolution of taste and solitary chemoreceptor cell systems. Brain Behav. Evol. 50, 234–243.

Fritzsch, B., 1998a. Of mice and genes: evolution of vertebrate brain development. Brain Behav. Evol. 52, 207–217.

Fritzsch, B., 1998b. Evolution of the vestibulo-ocular system. Otolaryngol. Head Neck Surg. 119, 182–192.

Fritzsch, B., Northcutt, R.G., 1993a. Origin and migration of trochlear, oculomotor and abducent motor neurons in *Petromyzon marinus* L. Brain Res. Dev. Brain Res. 74, 122–126.

Fritzsch, B., Northcutt, R.G., 1993b. Cranial and spinal nerve organization in amphioxus and lampreys: evidence for an ancestral craniate pattern. Acta Anat. 148, 96–109.

Fritzsch, B., Sonntag, R., 1988. The trochlear motoneurons of lampreys (*Lampetra fluviatilis*): location, morphology and numbers as revealed with horseradish peroxidase. Cell Tissue Res. 252, 223–229.

Fritzsch, B., Dubuc, R., Ohta, Y., Grillner, S., 1989. Efferents to the labyrinth of the river lamprey (*Lampetra fluviatilis*) as revealed with retrograde tracing techniques. Neurosci. Lett. 96, 241–246.

Fritzsch, B., Sonntag, R., Dubuc, R., Ohta, Y., Grillner, S., 1990. Organization of the six motor nuclei innervating the ocular muscles in lamprey. J. Comp. Neurol. 294, 491–506.

Gibbs, M.A., 2004. Lateral line receptors: where do they come from developmentally and where is our research going? Brain Behav. Evol. 64, 163–181.

González, M.J., Anadón, R., 1992. Primary projections of the lateral line nerves in larval sea lamprey, *Petromyzon marinus* L. An HRP study. J. Hirnforsch. 33, 185–194.

González, M.J., Anadón, R., 1994. Central projections of the octaval nerve in larval lamprey: an HRP study. J. Hirnforsch. 35, 181–189.

Gorbman, A., 1997. Hagfish development. Zool. Sci. 14, 375–390.

Grillner, S., Robertson, B., Stephenson-Jones, M., 2013. The evolutionary origin of the vertebrate basal ganglia and its role in action selection. J. Physiol. 591, 5425–5431.

Guérin, A., d'Aubenton-Carafa, Y., Marrakchi, E., et al., 2009. Neurodevelopment genes in lampreys reveal trends for forebrain evolution in craniates. PLoS One 4, e5374.

Guerra, M.M., González, C., Caprile, T., et al., 2015. Understanding how the subcommissural organ and other periventricular secretory structures contribute via the cerebrospinal fluid to neurogenesis. Front. Cell. Neurosci. 9, 480.

Hammond, K.L., Baxendale, S., McCauley, D.W., Ingham, P.W., Whitfield, T.T., 2009. Expression of patched, prdm1 and engrailed in the lamprey somite reveals conserved responses to Hedgehog signaling. Evol. Dev. 11, 27–40.

Hardisty, M.W., 2006. Lampreys: Life without Jaws. Forrest Text, Cardigan, UK.

Hardisty, M.W., Potter, I.C., 1971. The behaviour, ecology and growth of larval lampreys. In: Hardisty, M.W., Potter, I.C. (Eds.), The Biology of Lampreys, vol. 1. Academic Press, London, pp. 85–125.

Hébert, J.M., Fishell, G., 2008. The genetics of early telencephalon patterning: some assembly required. Nat. Rev. Neurosci. 9, 678–685.

Heier, P., 1948. Fundamental principles in the structure of the brain. A study of the brain of *Petromyzon fluviatilis*. Acta Anat. S VI 1–213.

Iwahori, N., Kiyota, E., Nakamura, K., 1987. A Golgi study on the olfactory bulb in the lamprey, *Lampetra japonica*. Neurosci. Res. 5, 126–139.

Iwahori, N., Nakamura, K., Tsuda, A., 1996. Neuronal organization of the optic tectum in the hagfish, *Eptatretus burgeri*: a Golgi study. Anat. Embryol. 193, 271–279.

Iwahori, N., Baba, J., Kawawaki, T., 1998. Neuronal organization of the olfactory bulb in the hagfish, *Eptatretus burgeri*: a Golgi study. J. Hirnforsch. 39, 161–173.

Iwahori, N., Kawawaki, T., Baba, J., 1999. Neuronal organization of the optic tectum in the river lamprey, *Lampetra japonica*: a Golgi study. J. Hirnforsch. 39, 409–424.

Jandzik, D., Hawkins, M.B., Cattell, M.V., et al., 2014. Roles for FGF in lamprey pharyngeal pouch formation and skeletogenesis highlight ancestral functions in the vertebrate head. Development 141, 629–638.

Jiménez-Guri, E., Pujades, C., 2011. An ancient mechanism of hindbrain patterning has been conserved in vertebrate evolution. Evol. Dev. 13, 38–46.

Johnston, J.B., 1912. The telencephalon in cyclostomes. J. Comp. Neurol. 22, 341–404.

Joly, J.S., Osorio, J., Alunni, A., et al., 2007. Windows of the brain: towards a developmental biology of circumventricular and other neurohemal organs. Semin. Cell Dev. Biol. 18, 512–524.

Jørgensen, J.M., 1998. Structure of the hagfish inner ear. In: Jorgensen, J.M., Lomholt, J.P., Weber, R.E., Malte, H. (Eds.), The Biology of Hagfishes. Chapman and Hall, London, pp. 557–563.

Kano, S., Xiao, J., Osório, J., et al., 2010. Two lamprey hedgehog genes share non-coding regulatory sequences and expression patterns with gnathostome hedgehogs. PLoS One 5, e13332.

Khonsari, R.H., Li, B., Vernier, P., Northcutt, R.G., Janvier, P., 2009. Agnathan brain anatomy and craniate phylogeny. Acta Zool 90, 52–68.

Kishida, R., Onishi, H., Nishizawa, H., et al., 1986. Organization of the trigeminal and facial motor nuclei in the hagfish, *Eptatretus burgeri*: a retrograde HRP study. Brain Res. 385, 263–272.

Kishida, R., Goris, R.C., Nishizawa, H., et al., 1987. Primary neurons of the lateral line nerves and their central projections in hagfishes. J. Comp. Neurol. 264, 303–310.

Koyama, H., 2005. Organization of the sensory and motor nuclei of the glossopharyngeal and vagal nerves in lampreys. Zool. Sci. 22, 469–476.

Koyama, H., Kishida, R., Goris, R.C., Kusunoki, T., 1987. Organization of sensory and motor nuclei of the trigeminal nerve in lampreys. J. Comp. Neurol. 264, 437–448.

Koyama, H., Kishida, R., Goris, R.C., Kusunoki, T., 1989. Afferent and efferent projections of the VIIIth cranial nerve in the lamprey *Lampetra japonica*. J. Comp. Neurol. 280, 663–671.

Koyama, H., Kishida, R., Goris, R.C., Kusunoki, T., 1990. Organization of the primary projections of the lateral line nerves in the lamprey *Lampetra japonica*. J. Comp. Neurol. 295, 277–289.

Kuraku, S., Kuratani, S., 2006. Time scale for cyclostome evolution inferred with a phylogenetic diagnosis of hagfish and lamprey cDNA sequences. Zool. Sci. 23, 1053–1064.

Kuraku, S., Takio, Y., Sugahara, F., Takechi, M., Kuratani, S., 2010. Evolution of oropharyngeal patterning mechanisms involving Dlx and endothelins in vertebrates. Dev. Biol. 341, 315–323.

Kuratani, S., Ota, K.G., 2008. Hagfish (cyclostomata, vertebrata): searching for the ancestral developmental plan of vertebrates. Bioessays 30, 167–172.

Kuratani, S., Horigome, N., Ueki, T., Aizawa, S., Hirano, S., 1998. Stereotyped axonal bundle formation and neuromeric patterns in embryos of a cyclostome, *Lampetra japonica*. J. Comp. Neurol. 391, 99–114.

Kusunoki, T., Amemiya, F., 1983. Retinal projections in the hagfish, *Eptatretus burgeri*. Brain Res. 262, 295–298.

Kusunoki, T., Kadota, T., Kishida, R., 1982. Chemoarchitectonics of the brain stem of the hagfish, *Eptatretus burgeri*, with special reference to the primordial cerebellum. J. Hirnforsch. 23, 109–119.

Lannoo, M.J., Hawkes, R., 1997. A search for primitive Purkinje cells: zebrin II expression in sea lampreys (*Petromyzon marinus*). Neurosci. Lett. 237, 53–55.

Lara-Ramírez, R., Patthey, C., Shimeld, S.M., 2015. Characterization of two neurogenin genes from the brook lamprey *Lampetra planeri* and their expression in the lamprey nervous system. Dev. Dyn. 244, 1096–1108.

Lowenstein, O., Thornhill, R.A., 1970. The labyrinth of *Myxine*: anatomy, ultrastructure and electrophysiology. Proc. R. Soc. Lond. B 176, 21–42.

Lowenstein, O., Osborne, M.P., Thornhill, R.A., 1968. The anatomy and ultrastructure of the labyrinth of the lamprey (*Lampetra fluviatilis* L.). Proc. R. Soc. Lond. B Biol. Sci. 170, 113–134.

Maklad, A., Reed, C., Johnson, N.S., Fritzsch, B., 2014. Anatomy of the lamprey ear: morphological evidence for occurrence of horizontal semicircular ducts in the labyrinth of *Petromyzon marinus*. J. Anat. 224, 432–446.

Marín, F., Aroca, P., Puelles, L., 2008. Hox gene colinear expression in the avian medulla oblongata is correlated with pseudorhombomeric domains. Dev. Biol. 323, 230–247.

Martínez-de-la-Torre, M., Pombal, M.A., Puelles, L., 2011. Distal-less-like protein distribution in the larval lamprey forebrain. Neuroscience 178, 270–284.

Martini, F., 1998. The ecology of hagfishes. In: Jorgensen, J.M., Lomholt, J.P., Weber, R.E., Malte, H. (Eds.), The Biology of Hagfishes. Chapman and Hall, London, pp. 57–77.

Matsuda, H., Goris, R.C., Kishida, R., 1991. Afferent and efferent projections of the glossopharyngeal-vagal nerve in the hagfish. J. Comp. Neurol. 311, 520–530.

Matsuura, M., Nishihara, H., Onimaru, K., et al., 2008. Identification of four Engrailed genes in the Japanese lamprey, *Lethenteron japonicum*. Dev. Dyn. 237, 1581–1589.

McCauley, D.W., Bronner-Fraser, M., 2002. Conservation of Pax gene expression in ectodermal placodes of the lamprey. Gene 287, 129–139.

Médina, M., Repérant, J., Ward, R., et al., 2009. Preoptic FMRF-amide-like immunoreactive projections to the retina in the lamprey (*Lampetra fluviatilis*). Brain Res. 1273, 58–65.

Meléndez-Ferro, M., Pérez-Costas, E., Rodríguez-Muñoz, R., et al., 2001. GABA immunoreactivity in the olfactory bulbs of the adult sea lamprey *Petromyzon marinus* L. Brain Res. 893, 253–260.

Meléndez-Ferro, M., Pérez-Costas, E., Villar-Cheda, B., et al., 2002. Ontogeny of gamma-aminobutyric acid-immunoreactive neuronal populations in the forebrain and midbrain of the sea lamprey. J. Comp. Neurol. 446, 360–376.

Ménard, A., Grillner, S., 2008. Diencephalic locomotor region in the lamprey—afferents and efferent control. J. Neurophysiol. 100, 1343–1353.

Miyashita, T., Coates, M.I., 2015. Hagfish embryology: staging table and relevance to the evolution and development of vertebrates. In: Edwards, S.L., Goss, G.G. (Eds.), Hagfish Biology. CRC Press, New York, pp. 95–128.

Miyashita, T., Green, S.A., Bronner, M.E., 2019. Comparative development of cyclostomes. In: Johanson, Z., Underwood, C., Richter, M. (Eds.), Evolution and Development of Fishes. Cambridge University Press, Cambridge, pp. 30–58.

Moreno, N., González, A., Rétaux, S., 2009. Development and evolution of the subpallium. Semin. Cell Dev. Biol. 20, 735–743.

Murakami, Y., Kuratani, S., 2008. Brain segmentation and trigeminal projections in the lamprey; with reference to vertebrate brain evolution. Brain Res. Bull. 75, 218–224.

Murakami, Y., Watanabe, A., 2009. Development of the central and peripheral nervous systems in the lamprey. Dev. Growth Diff. 51, 197–205.

Murakami, Y., Ogasawara, M., Sugahara, F., et al., 2001. Identification and expression of the lamprey Pax6 gene: evolutionary origin of the segmented brain of vertebrates. Development 128, 3521–3531.

Murakami, Y., Pasqualetti, M., Takio, Y., et al., 2004. Segmental development of reticulospinal and branchiomotor neurons in lamprey: insights into the evolution of the vertebrate hindbrain. Development 131, 983–995.

Murakami, Y., Uchida, K., Rijli, F.M., Kuratani, S., 2005. Evolution of the brain developmental plan: insights from agnathans. Dev. Biol. 280, 249–259.

Myojin, M., Ueki, T., Sugahara, F., et al., 2001. Isolation of Dlx and Emx gene cognates in an agnathan species, *Lampetra japonica*, and their expression patterns during embryonic and larval development: conserved and diversified regulatory patterns of homeobox genes in vertebrate head evolution. J. Exp. Zool. 291, 68–84.

Neidert, A.H., Virupannavar, V., Hooker, G.W., Langeland, J.A., 2001. Lamprey Dlx genes and early vertebrate evolution. Proc. Natl. Acad. Sci. U.S.A. 98, 1665–1670.

Nieuwenhuys, R., 1977. The brain of the lamprey in a comparative perspective. Ann. N.Y. Acad. Sci. 299, 97–145.

Nieuwenhuys, R., Nicholson, C., 1998. Lampreys, Petromyzontidae. In: Nieuwenhuys, R., ten Donkelaar, H.J., Nicholson, C. (Eds.), The Central Nervous System of Vertebrates, vol. 1. Springer-Verlag, Berlin, Heidelberg, pp. 397–495.

Nishizawa, H., Kishida, R., Kadota, T., Goris, R.C., 1988. Somatotopic organization of the primary sensory trigeminal neurons in the hagfish, *Eptatretus burgeri*. J. Comp. Neurol. 267, 281–295.

Northcutt, R.G., 1995. The forebrain of gnathostomes: in search of a morphotype. Brain Behav. Evol. 46, 275–318.

Northcutt, R.G., 2002. Understanding vertebrate brain evolution. Integr. Comp. Biol. 42, 743–756.

Northcutt, R.G., 2004. Taste buds: development and evolution. Brain Behav. Evol. 64, 198–206.

Northcutt, R.G., Gans, C., 1983. The genesis of neural crest and epidermal placodes: a reinterpretation of vertebrate origins. Q. Rev. Biol. 58, 1–28.

Northcutt, R.G., Puzdrowski, R.L., 1988. Projections of the olfactory bulb and nervus terminalis in the silver lamprey. Brain Behav. Evol. 32, 96–107.

Northcutt, R.G., Wicht, H., 1997. Afferent and efferent connections of the lateral and medial pallia of the silver lamprey. Brain Behav. Evol. 49, 1–19.

Nozaki, M., 2013. Hypothalamic-pituitary-gonadal endocrine system in the hagfish. Front. Endocrinol. 4, 200.

Ogasawara, M., Shigetani, Y., Suzuki, S., Kuratani, S., Satoh, N., 2001. Expression of thyroid transcription factor-1 (TTF-1) gene in the ventral forebrain and endostyle of the agnathan vertebrate, *Lampetra japonica*. Genesis 30, 51–58.

Oisi, Y., Ota, K.G., Fujimoto, S., Kuratani, S., 2013a. Development of the chondrocranium in hagfishes, with special reference to the early evolution of vertebrates. Zool. Sci. 30, 944–961.

Oisi, Y., Ota, K.G., Kuraku, S., Fujimoto, S., Kuratani, S., 2013b. Craniofacial development of hagfishes and the evolution of vertebrates. Nature 493, 175–180.

Olson, I., Suryanarayana, S.M., Robertson, B., Grillner, S., 2017. Griseum centrale, a homologue of the periaqueductal gray in the lamprey. IBRO Rep 2, 24–30.

Ooka-Souda, S., Kadota, T., Kabasawa, H., Takeuchi, H., 1995. A possible retinal information route to the circadian pacemaker through pretectal areas in the hagfish, *Eptatretus burgeri*. Neurosci. Lett. 192, 201–204.

Osório, J., Rétaux, S., 2008. The lamprey in evolutionary studies. Dev. Genes Evol. 218, 221–235.

Osório, J., Mazan, S., Rétaux, S., 2005. Organisation of the lamprey (*Lampetra fluviatilis*) embryonic brain: insights from LIM-homeodomain, Pax and hedgehog genes. Dev. Biol. 288, 100–112.

Osório, J., Megías, M., Pombal, M.A., Rétaux, S., 2006. Dynamic expression of the LIM-homeodomain gene Lhx15 through larval brain development of the sea lamprey (*Petromyzon marinus*). Gene Expr. Patterns 6, 873–878.

Ota, K.G., Kuratani, S., 2006. The history of scientific endeavours towards understanding hagfish embryology. Zool. Sci. 23, 403–418.

Ota, K.G., Kuratani, S., 2008. Developmental biology of hagfishes, with a report on newly obtained embryos of the Japanese inshore hagfish, *Eptatretus burgeri*. Zool. Sci. 25, 999–1011.

Ota, K.G., Kuraku, S., Kuratani, S., 2007. Hagfish embryology with reference to the evolution of the neural crest. Nature 446, 672–675.

Ota, K.G., Fujimoto, S., Oisi, Y., Kuratani, S., 2011. Identification of vertebra-like elements and their possible differentiation from sclerotomes in the hagfish. Nat. Commun. 2, 373.

Oury, F., Murakami, Y., Renaud, J.S., et al., 2006. Hoxa2- and rhombomere-dependent development of the mouse facial somatosensory map. Science 313, 1408–1413.

Parker, H.J., Bronner, M.E., Krumlauf, R., 2014. A Hox regulatory network of hindbrain segmentation is conserved to the base of vertebrates. Nature 514, 490–493.

Pérez-Fernández, J., Stephenson-Jones, M., Suryanarayana, S.M., Robertson, B., Grillner, S., 2014. Evolutionarily conserved organization of the dopaminergic system in lamprey: SNc/VTA afferent and efferent connectivity and D2 receptor expression. J. Comp. Neurol. 522, 3775—3794.

Platel, R., Delfini, C., 1986. Encephalization of the marine lamprey, *Petromyzon marinus* (L.). Quantitative analysis of the principle brain subdivisions. J. Hirnforsch. 27, 279—293.

Polenova, O.A., Vesselkin, N.P., 1993. Olfactory and nonolfactory projections in the river lamprey (*Lampetra fluviatilis*) telencephalon. J. Hirnforsch. 34, 261—279.

Pombal, M.A., Megías, M., 2011. Functional morphology of the brains of agnathans. In: Farrel, A.P. (Ed.), Encyclopedia of Fish Physiology: From Genome to Environment, vol. 1. Elsevier, The Netherlands, pp. 16—25.

Pombal, M.A., Megías, M., 2019. Development and functional organization of the cranial nerves in lampreys. Anat. Rec. (Hoboken) 302, 512—539.

Pombal, M.A., Puelles, L., 1999. Prosomeric map of the lamprey forebrain based on calretinin immunocytochemistry, Nissl stain, and ancillary markers. J. Comp. Neurol. 414, 391—422.

Pombal, M.A., Rodicio, M.C., Anadón, R., 1994a. Development and organization of the ocular motor nuclei in the larval sea lamprey, *Petromyzon marinus* L.: an HRP study. J. Comp. Neurol. 341, 393—406.

Pombal, M.A., Rodicio, M.C., Anadón, R., 1994b. The anatomy of the vestibule-ocular system in lampreys. In: Delgado-García, J.M., Vidal, P.-P., Godaux, E. (Eds.), Information Processing Underlying Gaze Control. Pergamon Press, Oxford, pp. 1—11.

Pombal, M.A., El Manira, A., Grillner, S., 1997a. Organization of the lamprey striatum — transmitters and projections. Brain Res. 766, 249—254.

Pombal, M.A., El Manira, A., Grillner, S., 1997b. Afferents of the lamprey striatum with special reference to the dopaminergic system: a combined tracing and immunohistochemical study. J. Comp. Neurol. 386, 71—91.

Pombal, M.A., Yáñez, J., Marín, O., González, A., Anadón, R., 1999. Cholinergic and GABAergic neuronal elements in the pineal organ of lampreys, and tract-tracing observations of differential connections of pinealofugal neurons. Cell Tissue Res. 295, 215—223.

Pombal, M.A., Marín, O., González, A., 2001. Distribution of choline acetyltransferase-immunoreactive structures in the lamprey brain. J. Comp. Neurol. 431, 105—126.

Pombal, M.A., Megías, M., Bardet, S.M., Puelles, L., 2009. New and old thoughts on the segmental organization of the forebrain in lampreys. Brain Behav. Evol. 74, 7—19.

Pombal, M.A., Álvarez-Otero, R., Pérez-Fernández, J., Solveira, C., Megías, M., 2011. Development and organization of the lamprey telencephalon with special reference to the GABAergic system. Front. Neuroanat. 5, 20.

Puelles, L., Rubenstein, J.L., 1993. Expression patterns of homeobox and other putative regulatory genes in the embryonic mouse forebrain suggest a neuromeric organization. Trends Neurosci. 16, 472—479.

Puelles, L., Rubenstein, J.L., 2003. Forebrain gene expression domains and the evolving prosomeric model. Trends Neurosci. 26, 469—476.

Puelles, L., Rubenstein, J.L., 2015. A new scenario of hypothalamic organization: rationale of new hypotheses introduced in the updated prosomeric model. Front. Neuroanat. 9, 27.

Puzdrowski, R.L., Northcutt, R.G., 1989. Central projections of the pineal complex in the silver lamprey *Ichthyomyzon unicuspis*. Cell Tissue Res. 255, 269—274.

Quintana-Urzainqui, I., Sueiro, C., Carrera, I., et al., 2012. Contributions of developmental studies in the dogfish *Scyliorhinus canicula* to the brain anatomy of elasmobranchs: insights on the basal ganglia. Brain Behav. Evol. 80, 127—141.

Ren, X., Chang, S., Laframboise, A., et al., 2009. Projections from the accessory olfactory organ into the medial region of the olfactory bulb in the sea lamprey (*Petromyzon marinus*): a novel vertebrate sensory structure? J. Comp. Neurol. 516, 105—116.

Repérant, J., Ward, R., Miceli, D., et al., 2006. The centrifugal visual system of vertebrates: a comparative analysis of its functional anatomical organization. Brain Res. Rev. 52, 1—57.

Repérant, J., Médina, M., Ward, R., et al., 2007. The evolution of the centrifugal visual system of vertebrates. A cladistic analysis and new hypotheses. Brain Res. Rev. 53, 161—197.

Richardson, M.K., Wright, G.M., 2003. Developmental transformations in a normal series of embryos of the sea lamprey *Petromyzon marinus* (*Linnaeus*). J. Morphol. 257, 348—363.

Robertson, B., Saitoh, K., Ménard, A., Grillner, S., 2006. Afferents of the lamprey optic tectum with special reference to the GABA input: combined tracing and immunohistochemical study. J. Comp. Neurol. 499, 106—119.

Robertson, B., Auclair, F., Ménard, A., Grillner, S., Dubuc, R., 2007. GABA distribution in lamprey is phylogenetically conserved. J. Comp. Neurol. 503, 47—63.

Robertson, B., Huerta-Ocampo, I., Ericsson, J., et al., 2012. The dopamine D2 receptor gene in lamprey, its expression in the striatum and cellular effects of D2 receptor activation. PLoS One 7, e35642.

Rodicio, M.C., de Miguel, E., Pombal, M.A., Anadón, R., 1992. The origin of the trochlear motoneurons in the sea lamprey, *Petromyzon marinus*, L. A HRP study. Neurosci. Lett. 138, 19—22.

Rodicio, M.C., Pombal, M.A., Anadón, R., 1995. Early development and organization of the retinopetal system in the larval sea lamprey, *Petromyzon marinus* L. An HRP study. Anat. Embryol. 192, 517—526.

Ronan, M., 1988. Anatomical and physiological evidence for electroreception in larval lampreys. Brain Res. 448, 173—177.

Ronan, M., Northcutt, R.G., 1987. Primary projections of the lateral line nerves in adult lampreys. Brain Behav. Evol. 30, 62—81.

Ronan, M., Northcutt, R.G., 1990. Projections ascending from the spinal cord to the brain in petromyzontid and myxinoid agnathans. J. Comp. Neurol. 291, 491—508.

Ronan, M., Northcutt, R.G., 1998. The central nervous system of hagfishes. In: Jorgensen, J.M., Lomholt, J.P., Weber, R.E., Malte, H. (Eds.), The Biology of Hagfishes. Chapman and Hall, London, pp. 452—479.

Rovainen, C.M., 1979. Neurobiology of lampreys. Physiol. Rev. 59, 1007—1077.

Rovainen, C.M., 1982. Neurophysiology. In: Hardisty, M.W., Potter, I.C. (Eds.), The Biology of Lampreys, vol. 4. Academic Press, London, pp. 1—136.

Salas, C.A., Yopak, K.E., Warrington, R.E., et al., 2015. Ontogenetic shifts in brain scaling reflect behavioral changes in the life cycle of the pouched lamprey *Geotria australis*. Front. Neurosci. 9, 251.

Schnitzlein, H.N., 1982. Cyclostomes. In: Crosby, E.C., Schnitzlein, H.N. (Eds.), Comparative Correlative Neuroanatomy of the Vertebrate Telencephalon. Macmillan Press, New York, pp. 4—26.

Schöber, W., 1964. Vergleichend-anatomische Untercuchungen am Gehirn der Larven und Adulten Tiere von *Lampetra fluviatilis* (Linné, 1758) und *Lampetra planeri* (Bloch, 1874). J. Hirnforsch. 7, 107—209.

Shigetani, Y., Sugahara, F., Kawakami, Y., et al., 2002. Heterotopic shift of epithelial-mesenchymal interactions in vertebrate jaw evolution. Science 296, 1316—1319.

Smith, J.J., Kuraku, S., Holt, C., et al., 2013. Sequencing of the sea lamprey (*Petromyzon marinus*) genome provides insights into vertebrate evolution. Nat. Genet. 45, 415—421.

Stephenson-Jones, M., Samuelsson, E., Ericsson, J., Robertson, B., Grillner, S., 2011. Evolutionary conservation of the basal ganglia as a common vertebrate mechanism for action selection. Curr. Biol. 21, 1081—1091.

Stephenson-Jones, M., Ericsson, J., Robertson, B., Grillner, S., 2012a. Evolution of the basal ganglia: dual-output pathways conserved throughout vertebrate phylogeny. J. Comp. Neurol. 520, 2957–2973.

Stephenson-Jones, M., Floros, O., Robertson, B., Grillner, S., 2012b. Evolutionary conservation of the habenular nuclei and their circuitry controlling the dopamine and 5-hydroxytryptophan (5-HT) systems. Proc. Natl. Acad. Sci. U.S.A. 109, E164–E173.

Suda, Y., Kurokawa, D., Takeuchi, M., et al., 2009. Evolution of Otx paralogue usages in early patterning of the vertebrate head. Dev. Biol. 352, 282–295.

Sugahara, F., Aota, S., Kuraku, S., et al., 2011. Involvement of Hedgehog and FGF signalling in the lamprey telencephalon: evolution of regionalization and dorsoventral patterning of the vertebrate forebrain. Development 138, 1217–1226.

Sugahara, F., Murakami, Y., Adachi, N., Kuratani, S., 2013. Evolution of the regionalization and patterning of the vertebrate telencephalon: what can we learn from cyclostomes? Curr. Opin. Genet. Dev. 23, 475–483.

Sugahara, F., Pascual-Anaya, J., Oisi, Y., et al., 2016. Evidence from cyclostomes for complex regionalization of the ancestral vertebrate brain. Nature 531, 97–100.

Suryanarayana, S.M., Robertson, B., Wallén, P., Grillner, S., 2017. The lamprey pallium provides a blueprint of the mammalian layered cortex. Curr. Biol. 27, 3264–3277.

Suzuki, D.G., Grillner, S., 2018. The stepwise development of the lamprey visual system and its evolutionary implications. Biol. Rev. Camb. Philos. Soc. 93, 1461–1477.

Suzuki, D.G., Murakami, Y., Yamazaki, Y., Wada, H., 2015. Expression patterns of Eph genes in the "dual visual development" of the lamprey and their significance in the evolution of vision in vertebrates. Evol. Dev. 17, 139–147.

Takio, Y., Pasqualetti, M., Kuraku, S., et al., 2004. Evolutionary biology: lamprey Hox genes and the evolution of jaws. Nature 429, 1 p. following 262.

Takio, Y., Kuraku, S., Murakami, Y., et al., 2007. Hox gene expression patterns in Lethenteron japonicum embryos-insights into the evolution of the vertebrate Hox code. Dev. Biol. 308, 606–620.

Tank, E.M., Dekker, R.G., Beauchamp, K., et al., 2009. Patterns and consequences of vertebrate Emx gene duplications. Evol. Dev. 11, 343–353.

Tomsa, J.M., Langeland, J.A., 1999. Otx expression during lamprey embryogenesis provides insights into the evolution of the vertebrate head and jaw. Dev. Biol. 207, 26–37.

Tsuneki, K., 1986. A survey of occurrence of about seventeen circumventricular organs in brains of various vertebrates with special reference to lower groups. J. Hirnforsch. 27, 441–470.

Uchida, K., Murakami, Y., Kuraku, S., Hirano, S., Kuratani, S., 2003. Development of the adenohypophysis in the lamprey: evolution of epigenetic patterning programs in organogenesis. J. Exp. Zool. B Mol. Dev. Evol. 300, 32–47.

Ueki, T., Kuratani, S., Hirano, S., Aizawa, S., 1998. Otx cognates in a lamprey, Lampetra japonica. Dev. Genes Evol. 208, 223–228.

Villar-Cheda, B., Pérez-Costas, E., Meléndez-Ferro, M., et al., 2006. Cell proliferation in the forebrain and midbrain of the sea lamprey. J. Comp. Neurol. 494, 986–1006.

Wasowicz, M., Pierre, J., Repérant, J., et al., 1994. Immunoreactivity to glial fibrillary acid protein (GFAP) in the brain and spinal cord of the lamprey (Lampetra fluviatilis). J. Hirnforsch. 35, 71–78.

Wicht, H., 1996. The brains of lampreys and hagfishes: characteristics, characters, and comparisons. Brain Behav. Evol. 48, 248–261.

Wicht, H., Nieuwenhuys, R., 1998. Hagfishes (Myxinoidea). In: Nieuwenhuys, R., ten Donkelaar, H.J., Nicholson, C. (Eds.), The Central Nervous System of Vertebrates, vol. 1. Springer-Verlag, Berlin, Heidelberg, pp. 497–545.

Wicht, H., Northcutt, R.G., 1990. Retinofugal and retinopetal projections in the Pacific hagfish, Eptatretus stouti (Myxinoidea). Brain Behav. Evol. 36, 315–328.

Wicht, H., Northcutt, R.G., 1992a. The forebrain of the Pacific hagfish: a cladistic reconstruction of the ancestral craniate forebrain. Brain Behav. Evol. 40, 25–64.

Wicht, H., Northcutt, R.G., 1992b. FMRFamide-like immunoreactivity in the brain of the Pacific hagfish, Eptatretus stouti (Myxinoidea). Cell Tissue Res. 270, 443–449.

Wicht, H., Northcutt, R.G., 1993. Secondary olfactory projections and pallial topography in the Pacific hagfish, Eptatretus stouti. J. Comp. Neurol. 337, 529–542.

Wicht, H., Northcutt, R.G., 1994. An immunohistochemical study of the telencephalon and the diencephalon in a Myxinoid jawless fish, the Pacific hagfish, Eptatretus stouti. Brain Behav. Evol. 43, 140–161.

Wicht, H., Northcutt, R.G., 1998. Telencephalic connections in the Pacific hagfish (Eptatretus stouti), with special reference to the thalamopallial system. J. Comp. Neurol. 395, 245–260.

Wicht, H., Tusch, U., 1998. Ontogeny of the head and nervous system of myxinoids. In: Jorgensen, J.M., Lomholt, J.P., Weber, R.E., Malte, H. (Eds.), The Biology of Hagfishes. Chapman and Hall, London, pp. 431–451.

Wicht, H., Derouiche, A., Korf, H.W., 1994. An immunocytochemical investigation of glial morphology in the Pacific hagfish: radial and astrocyte-like glia have the same phylogenetic age. J. Neurocytol. 23, 565–576.

Yáñez, J., Anadón, R., 1994. Afferent and efferent connections of the habenula in the larval sea lamprey (Petromyzon marinus L.): an experimental study. J. Comp. Neurol. 345, 148–160.

Yáñez, J., Anadón, R., Holmqvist, B.I., Ekström, P., 1993. Neural projections of the pineal organ in the larval sea lamprey (Petromyzon marinus L.) revealed by indocarbocyanine dye tracing. Neurosci. Lett. 164, 213–216.

Yáñez, J., Pombal, M.A., Anadón, R., 1999. Afferent and efferent connections of the parapineal organ in lampreys: a tract tracing and immunocytochemical study. J. Comp. Neurol. 403, 171–189.

Zompa, I.C., Dubuc, R., 1998. Diencephalic and mesencephalic projections to rhombencephalic reticular nuclei in lampreys. Brain Res. 802, 27–54.

The Brains of Cartilaginous Fishes

I. Rodríguez-Moldes, G.N. Santos-Durán, S. Pose-Méndez,
I. Quintana-Urzainqui, E. Candal

Universidade de Santiago de Compostela, Santiago de Compostela, Spain

6.1 Introduction

The group of cartilaginous fishes, or chondrichthyans, includes two major radiations that diverged over 400 million years ago: Holocephala (chimaeras), which have nonarticulated jaws, and Elasmobranchs (sharks, skates, and rays), which possess articulated jaws. Together, they form a monophyletic group that represents the out-group to all other living jawed vertebrates (gnathostomes) and the sister group of osteichthyans (which comprises actinopterygians, or ray-finned fishes, and sarcopterygians, which in turn include the lobe-finned fishes and tetrapods). This strategic phylogenetic position as one of the most basal extant vertebrates with jaws makes cartilaginous fishes a key group for understanding vertebrate evolution. On one hand, comparisons between cartilaginous fishes and jawless vertebrates (agnathans), such as lampreys, are essential for recognizing changes that occurred in the agnathan—gnathostome transition. On the other hand, comparisons between chondrichthyans and osteichthyans are crucial to understanding conserved traits that define the ancestral state of any particular character in the gnathostome group. Furthermore, whole-genome analysis of holocephalans showed that their genome has evolved significantly slower than that of the "living fossil" coelacanth and other bony vertebrates (Venkatesh et al., 2014), suggesting that chondrichthyans are closer to the gnathostome ancestral pattern than are osteichthyans.

In spite of their relevance as basal gnathostomes, cartilaginous fishes have been largely overlooked or barely considered in many reviews of vertebrate evolution, particularly in those concerned with nervous system evolution. The scarcity of developmental information about chondrichthyans may explain in part why they were not traditionally contemplated in comparative analyses of the nervous system.

The use of model organisms (such as mouse, chicken, *Xenopus*, and zebra fish) has yielded an incredibly large amount of data that have helped to identify conserved developmental processes and some derived traits that might be indicative of evolutionary trends. These developmental model animals were mainly chosen for their technical advantages, such as easy maintenance or short generation time, rather than their ability to represent phylogenetic diversity. Bearing in mind that fishes form the largest vertebrate group, that there exist great differences between cartilaginous and bony fishes (ray-finned fishes or actinopterygians, and lobe-finned fishes), and that bony fishes, but not cartilaginous fishes, went through an extra duplication of their genomes, the widespread use of a single species, the actinopterygian zebra fish, as the canonical model to represent all fishes, is constraining the possibility of gaining knowledge about the diversity of fish brains. Thus, to deepen our understanding of vertebrate brain evolution and fish diversity in the context of neuroscience, studies in at least some cartilaginous fishes are crucial.

A growing amount of data on the nervous system of chondrichthyans comes from studies (mainly developmental ones) on an emergent model organism, the catshark (also known as lesser-spotted dogfish) *Scyliorhinus canicula*. This species belongs to the family *Scyliorhinidae*, genus *Scyliorhinus*, order *Charchariniformes*, and superorder *Galeomorphii*. The catshark has been used in evo-devo studies mainly because it is relatively easy to maintain in captivity and to have access to its eggs and embryos, which are reasonably easy to manipulate for functional experiments, and because its protracted development makes it advantageous for studying all developmental events in detail (reviewed in Coolen et al., 2009). In addition, a useful series of normal developmental stages for this species is already available (Ballard et al., 1993), and the sequencing of its genome is nearly completed (Coolen et al., 2007, 2009). The value of *S. canicula* as a model organism for developmental event has been thoroughly demonstrated (Sauka-Spengler et al., 2004; O'Neill et al., 2007; Ferrando et al., 2010; Oulion et al., 2011; Adachi et al., 2012; Gillis et al., 2012; Debiais-Thibaud et al., 2013, 2015; Compagnucci et al., 2013; Godard and Mazan, 2013; Godard et al., 2014; Enault et al., 2015). These studies have provided much information about the main events taking place during development in cartilaginous fishes, leading to a better understanding of the similarities to, and differences from, other model organisms.

6.2 Neuroecology and Brain Size in Chondrichthyans

While a common brain plan exists in gnathostomes (see later), broad variability in brain morphology and organization (ie, in the relative size or complexity of major brain structures) has been documented in cartilaginous fishes (Yopak et al., 2010; Collin, 2012). Indeed, this group comprises species with extreme neuromorphological diversity, highly developed sensory systems, learning and adaptive capabilities, and complex social behavioral repertoires (see Yopak, 2012 and references therein).

This variability in brain morphology has been related to evolution (phylogeny) but also to ecology encountered during their life history, which in turn influences morphological adaptations, sensory specializations, and behavior (Yopak et al., 2010; Collin, 2012; Yopak, 2012). Indeed, cartilaginous fishes cover a wide range of lifestyles including species living solely on the bottom (benthic), species living near the bottom and occasionally resting on the substratum (benthopelagic), and species living primarily in the water column (pelagic). They also cover a range of habitats including deep sea, demersal, or reef-associated, coastal, and oceanic species (for a review, see Yopak, 2012).

Brain size and brain allometry (changes in the size of brain parts relative to total brain size) have been extensively studied in a representative number of species of cartilaginous fishes in the last decades to determine the influence of both phylogeny and ecology in brain development and morphology (reviewed in Yopak, 2012). These studies highlighted some interesting results regarding brain morphology in this group.

First, while chondrichthyans were considered in classical neuroanatomical studies as primitive animals with small brains mainly dedicated to olfaction (Nieuwenhuys, 1967), studies on brain size and encephalization (a larger than expected brain size for a given body size) have shown that cartilaginous fishes are a highly encephalized group that possesses brain to body mass

ratios that were comparable to all major groups of gnathostomes including birds and mammals (Northcutt, 1989; Yopak, 2012). Indeed, cartilaginous fishes are said to have larger brains at similar body sizes than teleosts (Striedter, 2005).

Second, the allometric scaling of major brain components (telencephalon, diencephalon, mesencephalon, cerebellum, and medulla oblongata) in cartilaginous fishes was found to be unexpectedly similar to that reported in mammals, such that the relative size of brain components can be predicted fairly well from total brain size (with the olfactory bulbs being statistically independent from the rest of the brain; Yopak et al., 2010; Yopak, 2012). As in mammals, the volumes of the telencephalon and cerebellum increase gradually and disproportionately with brain size. Interestingly, cerebellar foliation is related to both absolute and relative cerebellar size (Yopak et al., 2010), and it has been reported that the size and complexity of the chondrichthyan cerebellum are correlated with ecological and behavioral parameters (see later).

Third, similar patterns of brain organization (ie, commonalities in the relative development of major brain components termed cerebrotypes; Clark et al., 2001; Iwaniuk and Hurd, 2005) exist in species of cartilaginous fishes that share lifestyle characteristics such as feeding strategy and habitat, irrespective of phylogenetic grouping (Yopak, 2012). For example, larger brains with well-developed telencephala and highly foliated cerebella are common to species that occupy complex reef or oceanic habitats which may be linked to increased sociality, social intelligence, and cognitive capability. In contrast, the smallest brains with a relatively reduced telencephalon and a smooth cerebellar corpus are found in benthic and benthopelagic demersal species (Yopak et al., 2010; Yopak, 2012). The extent to which different brain morphologies in cartilaginous fishes have evolved as a result of ecology constraints deserves further analyses.

6.3 Evolutionary Changes in Brain Development

6.3.1 Comparisons in Evo-devo

One recurrent problem when comparing developmental processes among different species is that descriptions generally refer to embryonic or postnatal developmental stages that, in turn, refer to body size, days after fertilization or hatching, or to developmental features that are specific to the particular species or clade. This problem becomes pronounced when comparing specific structures, because different structures within an animal mature at different rates (eg, the cerebellum can be mature when the telencephalon is not), and because of changes in the order of these developmental events among related species, a phenomenon that has been termed heterochrony by developmental biologists (Klingerberg, 1998).

An important tool for dealing with this problem is the identification of the phylotypic stage, defined as the point in the development when an animal most closely resembles other species (Richardson, 2012). However, as noted earlier, the lack of synchrony in organ development within an animal and across species often makes it difficult to define a phylotypic stage that truly fits all taxonomic groups within a phylum. Instead, the phylotypic stage in vertebrates is generally identified as an extended period of development that covers several stages (Richardson, 2012). Still, it helps with specifying the course of development of specific structures in terms of landmarks or milestones, rather than chronological age. Importantly, the choice of these landmarks depends on the structure being analyzed; ideally, they can be readily identified in, and compared with, other animal models.

When studying brain regionalization, two main criteria have been used to identify developmental landmarks: morphological and molecular ones. The morphological criterion is founded on the idea that two adjacent neural areas that have different morphogenetic properties come to be separated by morphological boundaries. In some cases, the boundary between two areas is marked by an external constriction and an internal ridge (Larsen et al., 2001). However, these landmarks are transient and may become covert during development, requiring experimental demonstration. Moreover, ridges and bulges may vary across species (Puelles and Rubenstein, 2003). In contrast, the molecular criterion involves gene expression patterns. Genes with distinct temporal and spatial gradients in the developing brains are often functionally relevant in the sense that they control brain regionalization (Kudo et al., 2007). Many studies have shown that gene expression boundaries not only coincide with morphological boundaries, but also are often causally related to them.

6.3.2 Main Stages of Catshark Brain Development

Which morphological landmarks should one select if one wants to analyze the development of the entire brain in any given species? As stated earlier, landmarks should be broadly identifiable across species. Some developmental changes occurring during histogenesis of the neural tube serve as milestones that characterize neural tube development in all vertebrates. Specifically, the walls of the early neural tube consist of a pseudostratified proliferative neuroepithelium (Sauer, 1935). At early stages of development, every cell in this neuroepithelium proliferates and is attached to both the ventricular and pial surfaces. Initially, these progenitor cells divide symmetrically to give off more proliferating cells, thus increasing the thickness of the neuroepithelium. Later, asymmetrical divisions of the proliferating neural cells will give rise to postmitotic neuroblasts, which migrate away from the ventricular zone (VZ) to form a second layer called the intermediate zone (IZ). The IZ becomes progressively thicker as more neuroblasts are added to it; moreover, postmitotic neuroblasts differentiate into neurons and glia in this layer. The neurons then send axons to the pial surface, thereby creating a cell-poor marginal zone (MZ). Based on this three-zone pattern, three developmental periods have been established during brain development in *S. canicula* (Rodríguez-Moldes, 2009; Rodríguez-Moldes et al., 2011). The first period lasts from stage 17 to stage 26 (according to Ballard et al., 1993) and is characterized by the absence of layering in the walls of the developing brain, which consist only of a pseudostratified neuroepithelium that progressively thickens during the period. During the second period (stages 27–31), the three-zone layering of the brain walls becomes progressively apparent. During the third period, extending from stage 32 to hatching, the brain walls present a regionalized, mature-like architecture. The same developmental periods can be defined in the retina, an extension of the brain, although it develops more slowly.

The next question to ask is how one should choose among alternative molecular landmarks. A 2012 work on the phylotypic stage of nematodes showed that it corresponds with a rapid burst of transcriptional activity enriched for developmental genes, a feature characteristic of the phylotypic stage also in other animals (reviewed in Richardson, 2012). Moreover, developmental genes are evolutionarily conserved, and so are their expression patterns along the anteroposterior and dorsoventral axes (Pasini and Wilkinson, 2002). While these molecular codes are initially dynamic, they progress to a more stable pattern that can be compared across species and thus can serve as a phylotypic stage. After this period of relative stability, the spatial expression pattern of these genes may change again, due to downregulation, massive cell migration, or uneven growth rates. In *S. canicula*, these three periods coincide quite well with those defined by morphological criteria. That is, the gene expression patterns are dynamic up to stage 27, become stable between stages 27 and 31, and become dynamic again after stage 32, when important morphological changes may blur the original gene expression boundaries.

Using these criteria we have compared three developmental periods defined in the sharks to the timing of development in mouse, chick, and zebra fish embryos (see Table 6.1). We have taken into account the degree of differentiation of the brain walls as well as other

TABLE 6.1　Equivalence among developmental stages of main model organisms based on key process of brain development

	Fertilization to neurulation	First period Neuroepithelial neural walls	Second period Three-layered neural walls	Third period Regionalized mature-like
Catshark				
Stages	S1–S16	S17–S26	S27–S31	S32–S34
Percentage of development	10%	14%	20%	56%
Main features	Cleavage (S1–S7) Gastrulation (S11–S14) Neural folds closure (S16)	Three primary brain vesicles (S17) Medial longitudinal fascicle (S22–23) Tract of the posterior commissure (S24–25) Pharyngeal clefts C1–C6 open (S26)	Cerebellar plate becomes evident (S27)	Conspicuous cerebellar body and auricles (S32)
Mouse				
Stages (dpc)	E0.5–E8.5	E8.8–E10.5	E10.5–E14.5	E14.5–birth
Percentage of development	42.5%	10%	20%	27.5%
Main features	Early fusion of neural folds (8dpc)	Three primary brain vesicles (E9) Medial longitudinal fascicle (E9.5) Tract of the posterior commissure (E10) Four branchial pouches (E10.5)	Cerebellar plate is evident (E11)	Conspicuous cerebellar body and auricles (prenatal)
Chick				
Stages	HH1–HH8	HH9–HH18	HH19–HH30	HH31–HH45
Percentage of development	6%	8%	21%	65%
Main features	Neural folds meet at the level of the midbrain (HH8)	Three primary brain vesicles (HH10) Medial longitudinal fascicle (HH14) Tract of the posterior commissure (HH17) Four visceral arches visible (HH18)	Cerebellar plate (HH23)	Conspicuous cerebellar body and auricles (prenatal)
Zebra fish				
Stages	10 hpf	11–24 hpf	24–48 hpf	48–72 hpf
Percentage of development	14%	20%	33%	33%
Main features	Neural keel (10 hpf)	Three primary brain vesicles (16 hpf) Medial longitudinal fascicle (18 dpf) Tract of the posterior commissure (20 hpf)	Cerebellar plate (24hpf)	Cerebellar body (72hpf)

Data after Ross, L.S., Parrett, T., Easter, S.S., 1992. Axonogenesis and morphogenesis in the embryonic zebrafish brain. J. Neurosci. 71, 467–482; Kimmel, C.B., Ballard, W.W., Kimmel, S.R., Ullmann, B., Schilling, T.F., 1995. Stages of embryonic development of the zebrafish. Dev. Dyn. 203, 253–310; Mastick, G.S., Easter, S.S., 1996. Initial organization of neurons and tracts in the embryonic mouse fore and midbrain. Dev. Biol. 173, 79–94; Schubert, F.R., Lumsden, A., 2005. Transcriptional control of early tract formation in the embryonic chick midbrain. Development 132, 1785–1793; Rodríguez-Moldes, I., Carrera, I., Pose-Méndez, S., Quintana-Urzainqui, I., Candal, E., Anadón, R., Mazan, S., Ferreiro-Galve, S., 2011. Regionalization of the shark hindbrain: a survey of an ancestral organization. Front. Neuroanat. 5, 1–14; Wullimann, M.F., Mueller, T., Distel, M., Babaryka, A., Grothe, B., Köster, R.W., 2011. The long adventurous journey of rhombic lip cells in jawed vertebrates: a comparative developmental analysis. Front. Neuroanat. 5, 27; Pose-Méndez, S., Candal, E., Mazan, S., Rodríguez-Moldes, I., 2016b. Morphogenesis of the cerebellum and cerebellum-related structures of Scyliorhinus canicula: insights on the ground pattern of the cerebellar ontogeny. Brain Struct. Funct. 221, 1691–1717.

developmental milestones, such as the formation of the three primary brain vesicles and the development of major axon tracts, such as medial longitudinal fascicle (the first longitudinal tract to form) and the tract of the post-optic commissure. These three periods constitute a useful framework for comparative studies in vertebrate development, because they can be unequivocally identified in diverse vertebrates.

6.4 Regionalization of the Chondrichthyan Brain Based on Developmental, Genoarchitectonic, and Neurochemical Evidence

In the early developing brain of cartilaginous fishes, as in other vertebrates, two transversal constrictions subdivide the anterior neural tube into three main vesicles that represent the fundamental anteroposterior (ie,

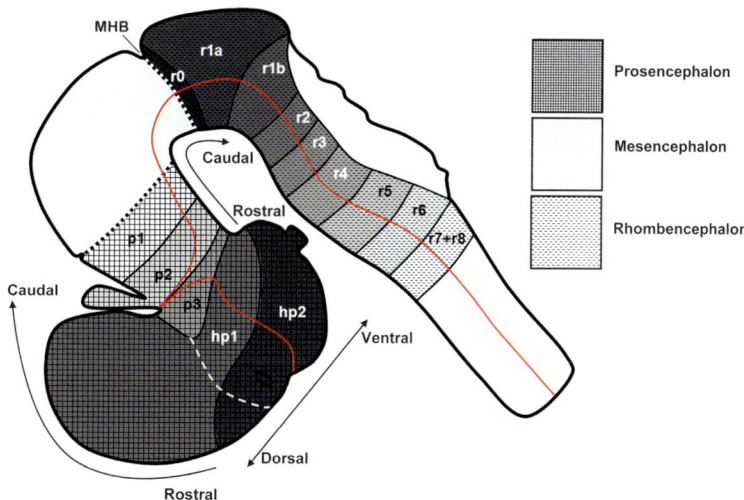

FIGURE 6.1 Schematic drawing of the brain in a *Scyliorhinus canicula* embryo showing the neuromeric subdivisions according to the segmental model. The prosencephalon is subdivided into the primary prosencephalon (containing the prosomeres p1, p2, and p3) and the secondary prosencephalon (containing the prosomeres hp1 and hp2). The mesencephalon presents one mesomere. The rhombencephalon is subdivided into one pseudorhombomere (r0) and eight rhombomeres (the first one containing the subrhombomeres 1a and 1b). *hp, p,* prosomeres; *MHB,* midbrain—hindbrain boundary; *r,* rhombomeres.

rostrocaudal) subdivisions of the vertebrate brain (Fig. 6.1): prosencephalon (forebrain), mesencephalon (midbrain), and rhombencephalon (hindbrain). Moreover, longitudinal zones (roof plate, alar plate, basal plate, and floor plate) subdivide the chondrichthyan brain dorsoventrally. Although these longitudinal divisions are evident in the spinal cord and rhombencephalon, they are not always recognizable at mesencephalic and prosencephalic levels.

We interpret the segmental subdivision of the chondrichthyan brain, including transverse and longitudinal domains, according to the neuromeric model, which integrates gene expression patterns with morphological and ontogenic data to understand vertebrate brain organization. This model has proven useful in comparative neuroanatomy because it facilitates the recognition of changes in development, which are the mechanistic source of adult neuroanatomical diversity. The recognition that the longitudinal axis of the forebrain is bent with respect to the midbrain and hindbrain lies at the heart of the model and defines its primary anteroposterior and dorsoventral divisions (Puelles, 2009; Puelles and Rubenstein, 2015). Specifically, the model establishes that these vesicles are subdivided into genetically specified segments called neuromeres that are more specifically named prosomeres, mesomeres, or rhombomeres, depending on the region considered. These neuromeres are developmentally independent compartments with specific fates. Importantly, the topological relationships among expression patterns and certain morphological landmarks are conserved despite the various morphogenetic deformations of the developing neural tube.

The regionalization process involves a complex series of spatiotemporal interactions among developmental genes that define embryonic territories, which ultimately form adult structures. Genoarchitectonics, the study of combinatorial expression patterns of key developmental genes, has proved to be a useful approach for defining the embryonic origin of specific regions of the adult brain (Puelles and Ferrán, 2012). Moreover, since the network of genes specifying a structure's embryonic precursor tends to be highly conserved, the genoarchitectonic approach also facilitates the search for homologies by comparing gene expression patterns across vertebrates. Under this rationale, we present an overview of the regional organization of the chondrichthyan brain mainly based on updated genoarchitectonic, together with neurochemical and, in a lesser extent, hodological data obtained on the developing brain of *S. canicula*. Data are presented in an anterior—posterior fashion, with special attention to those brain structures on which genoarchitectonic data are available, namely telencephalon, hypothalamus, and cerebellum.

6.4.1 Prosencephalon

The prosencephalon corresponds to the neural tube's most anterior vesicle. As development proceeds, it gives rise to two transverse protosegments: the secondary prosencephalon (rostrally) and the diencephalon or primary prosencephalon (caudally). The secondary prosencephalon is divided into two prosomeres: a rostral one (hp2) that includes the nonevaginated telencephalon or preoptic area (dorsally) and the terminal hypothalamus (ventrally); and a caudal one (hp1) that includes the eva-ginated telencephalon (dorsally) and the peduncular hypothalamus (ventrally). The diencephalon is subdivided from caudal to rostral into prosomeres p1, p2, and p3, which contain the synencephalon (p1), thalamus (alar plate of p2), and prethalamus (alar plate of p3) (Fig. 6.1).

6.4.1.1 Telencephalon

The telencephalon in chondrichthyans, as in most vertebrates, develops by evagination. This morphogenetic process involves an enlargement of the central lumen of the anterior neural tube to form the telencephalic ventricles, followed by an outward expansion of the telencephalic walls (Fig. 6.2). Developmental studies in chondrichthyans have elucidated a telencephalic evagination process that is similar to that observed in tetrapods, lobe-finned fishes, and agnatha (Quintana-Urzainqui et al., 2015), but very different from the eversion process that happens in ray-finned fishes, in which part of the roof of the neural tube becomes thin, elongates, and bends outward (Fig. 6.2), reversing the topology of telencephalic structures in this fish group (see Vernier, this volume). Therefore cartilaginous fishes display lateral ventricles and telencephalic structures more easily comparable with tetrapods, which make them an advantageous fish model for comparative and evolutionary studies of the telencephalon.

The telencephalon of chondrichthyans, as in other vertebrates, consists of two main histogenetic subdivisions: the pallium (dorsally) and the subpallium (ventrally) that includes the preoptic area. In general, the basic telencephalic territories appear to be specified in chondrichthyans by similar sets of genes as in other gnathostomes. Particularly, the catshark pallium and subpallium express equivalent sets of genes as other vertebrates (Fig. 6.3), and they are subdivided in similar ways (Quintana-Urzainqui et al., 2012). The position of

the pallial—subpallial boundary is defined by the expression of Pax6 and *ScTbr1* in the pallium, and *ScDlx2* and GAD in the subpallium (see PSB in Fig. 6.3A).

6.4.1.1.1 Pallium

In the shark pallium, at least four different histogenetic territories (medial, dorsal, lateral, and ventral pallia) can be defined by the differential expression of *ScPax6*, *ScTbr1*, *ScEmx1*, and *ScLhx9* (Fig. 6.3B). *ScPax6* is expressed in the VZ of the whole pallium while the other markers are more selectively distributed (Fig. 6.3B). Thus, the medial pallium (MP) expresses all four of them: ventricular *ScPax6*; subventricular *ScTbr1* and *ScEmx1*; and *ScLhx9* throughout the entire territory. The dorsal pallial (DP) territory expresses *ScPax6* both in the VZ and MZ but not *ScTbr1*, *ScLhx9*, and *ScEmx1*. The lateral pallium (LP) is characterized by the expression of *ScEmx1*, *ScTbr1*, and *ScLhx9* in the MZ, whereas the ventral pallium (VP) expresses both *ScTbr1* and *ScLhx9* (but not *ScEmx1*) in a relatively large area directly adjacent to subpallial territories. Such subdivisions can be reasonably proposed as the shark homologues of those described in the pallium of all tetrapods (reviewed in Medina and Abellán, 2009) and also in bony fishes (Wullimann and Mueller, 2004; Ganz et al., 2012). In lampreys, representatives of agnathans, only two pallial regions have been unequivocally recognized (Sugahara et al., 2011). This is indicative of a big evolutionary leap in the complexity of pallial areas during the transition from jawless to jawed vertebrates implicating the emergence of new genetic programs for the specification of new territories.

The shark pallium presents, however, some remarkable particularities with respect to other gnathostomes. One good example is the expression of *Pax6*. Even though in the catshark *Pax6* is expressed throughout the VZ of the pallium, as in other vertebrates, the gradient described in mammals (lateral-high, medial-low and anterior-high, and posterior-low) is not evident in sharks. Furthermore, a specific sector in the shark pallium strongly expresses Pax6 in a postmitotic population of MZ cells (Quintana-Urzainqui et al., 2012), as shown in Fig. 6.3B. No other vertebrate has been reported to date to express *Pax6* in postmitotic pallial cells. The biological meaning of this difference and its evolutionary implications remain to be elucidated. Moreover, this intriguing pallial sector is the best candidate to represent the dorsal pallium of sharks (Fig. 6.3B), that is, the presumptive homologue of mammalian neocortex, since it seems to be the target region of GABAergic neurons migrating tangentially from the subpallium and expresses Reelin in its MZ (Quintana-Urzainqui et al., 2015), just as specific cells in the developing mammalian cortex do (Cajal—Retzius cells). Although the embryonic

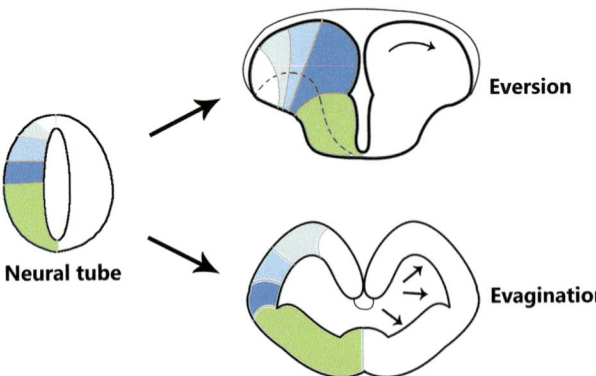

FIGURE 6.2 Schematic representation of the two main morphogenetic processes of telencephalic development in vertebrates: eversion (ray-finned fishes) and evagination (agnathans, chondrichthyans, and sarcopterygians, which include lobe-finned fishes and tetrapods).

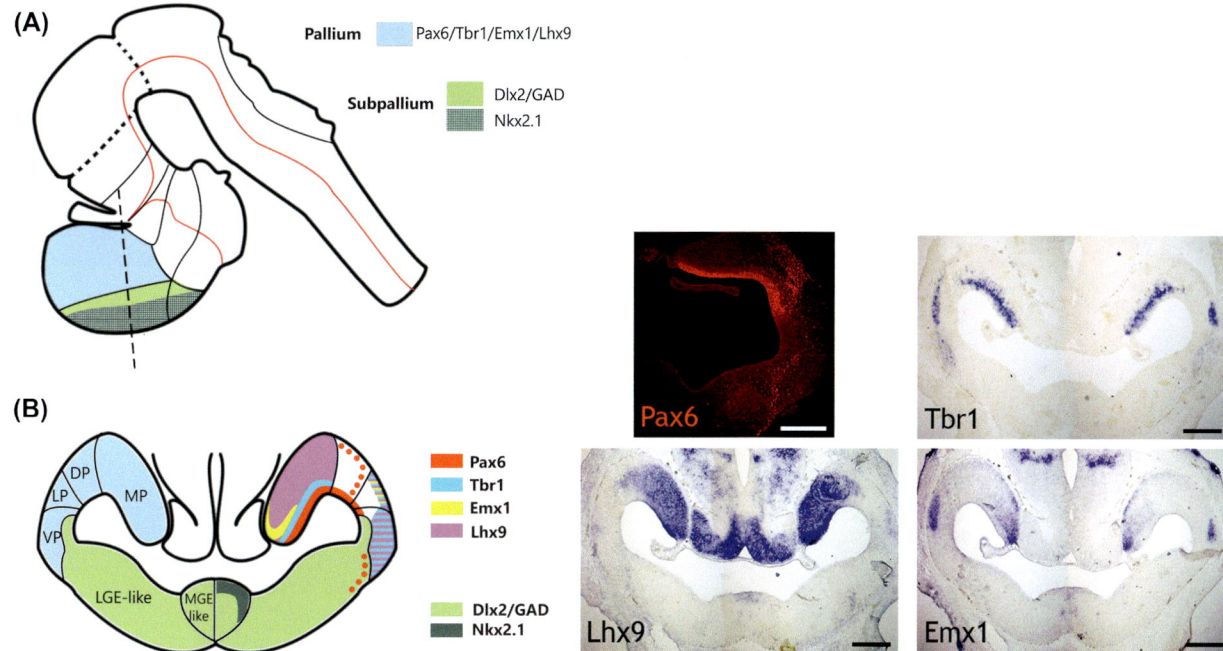

FIGURE 6.3 Main territories proposed in the developing telencephalon of *Scyliorhinus canicula* based on comparative gene expression analysis represented in sagittal (A) and transverse (B) views. The pallium typically expresses *ScPax6*, *ScTbr1*, *ScLhx1*, and *ScLhx9* while the subpallium is characterized by the expression of *ScDlx2*, GAD, and *ScNkx2.1*. DP, dorsal pallium; *LGE-like*, lateral ganglionic eminence-like; *LP*, lateral pallium; *MGE-like*, medial ganglionic eminence-like; *MP*, medial pallium; *PSB*, pallial–subpallial boundary; *VP*, ventral pallium.

homology of this structure with the dorsal pallium seems to be robust, its adult derivative in cartilaginous fishes is far from being a cortexlike structure. The so-called pallium dorsalis in the adult is a nonlayered structure and is proportionally smaller than other pallial areas in the shark telencephalon [ie, the area periventriculairs pallialis (APP) or the olfactory bulbs]. In the adult, the APP appears to be the recipient of secondary olfactory connections (Smeets et al., 1983; Smeets, 1998; Yáñez et al., 2011), but there is no much further information about its function. During late development, both the APP and the olfactory bulbs grow enormously and form incipient layered structures expressing *ScDlx2* (inner layer) and *ScTbr1* (outer layer) (Quintana-Urzainqui et al., 2015), which are typically expressed in developing GABAergic and glutamatergic neurons, respectively. This suggests that the most complex structures and candidates for being the main information processing centers in the shark *S. canicula* telencephalon are not homologues of neocortex and that the event that led to the big expansion of the dorsal pallium happened later in vertebrate evolution.

As mentioned earlier, some of the structures in the mature pallium of chondrichthyans, as in other gnathostomes, are formed by a mixture of excitatory (glutamatergic) and inhibitory (GABAergic) elements. For this to happen, GABAergic cells that were born in subpallial areas migrate toward the pallium where they integrate

to modulate the activity of locally born glutamatergic neurons. This kind of migration, called "tangential migration" (Wichterle et al., 2001), was formerly considered an emergent property of amniotes, but has been reported also in amphibians (Moreno et al., 2008). Evidence that this kind of migration also occurs in sharks is provided by GABAergic markers (Carrera et al., 2008a; Quintana-Urzainqui et al., 2015) and some pilot tracing assays (Fig. 6.4). In these experiments, the tracer neurobiotin was placed in the subpallial territory and subsequently detected in neurons occupying pallial positions (arrows in Fig. 6.4). Although telencephalic tangential migration seems to be a conserved process among gnathostomes, to date we have no evidence of such migratory routes in lampreys (Sugahara et al., 2011). These facts provide further evidence of the evolutionary leap in telencephalic complexity between agnathans and gnathostomes.

6.4.1.1.2 Subpallium

The subpallium seems to be more conserved in terms of gene expression, adult anatomy, and functionality than the pallium. In fact, all vertebrates studied so far form subpallium-derived basal ganglia-like structures (Medina et al., 2014). In mammals, the most prominent components of the subpallium are the embryonic precursors of the subpallial basal ganglia, namely the lateral and medial ganglionic eminences (LGE and MGE),

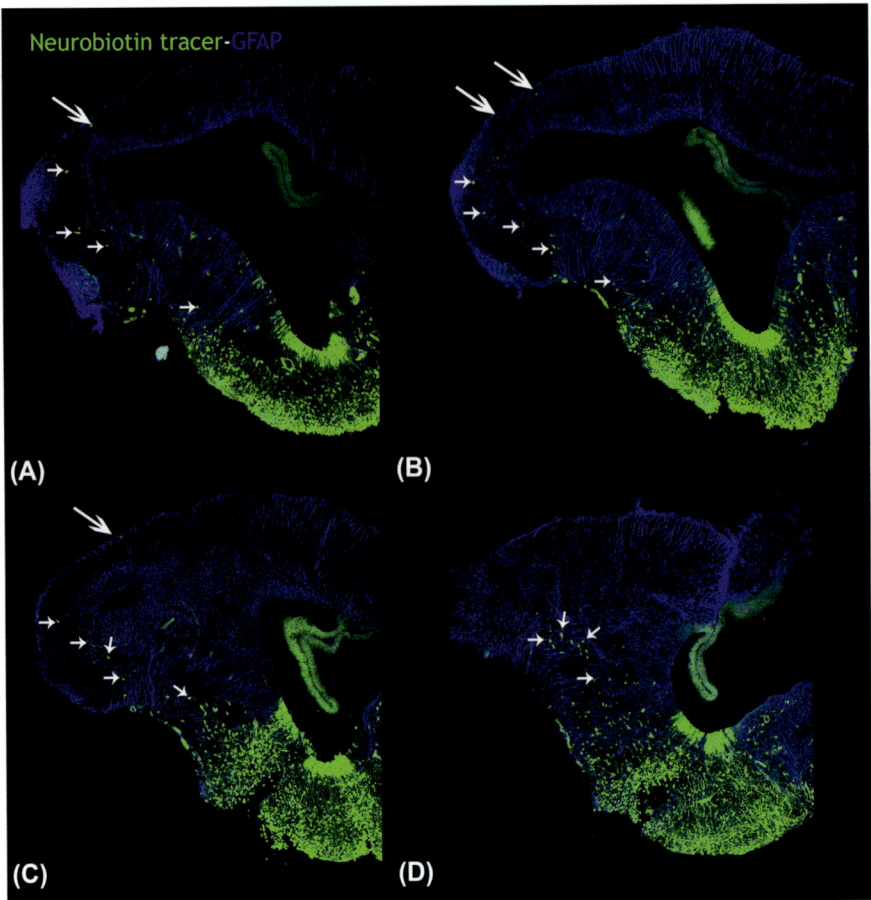

FIGURE 6.4   Neuron-labeling experiments in the telencephalon of stage-31 catshark embryos reveal possible migrations of subpallial cells as seen in serial sections from dorsal (A) to ventral (D) telencephalon. After applying neurobiotin to the subpallium and a short in vitro incubation period, some labeled (*green*) cells appeared outside the place of application, occupying positions compatible with the tangential migratory routes described by Quintana-Urzainqui et al. (2015) (*short white arrows*). Labeled cells were even detected in very dorsal pallial areas (*long arrows*). GFAP (glial fibrillary acid protein) immunofluorescence is shown in *blue*.

which give rise to the striatum and pallidum, respectively. The LGE typically expresses *Dlx2* while the MGE expresses both *Dlx2* and *Nkx2.1* (reviewed in Medina, 2008; in Moreno et al., 2009).

In *S. canicula*, the ventral telencephalon presents two main prominences: one in the midline and one located at a more lateral position. The differential expression analysis of *ScDlx2* and *ScNkx2.1* led to the identification of the medial prominence as the catshark homologue structure of the MGE (MGE-like in Fig. 6.3B), since it expresses both genes. The rest of the subpallial territory was identified as the homologue of the LGE (LGE-like in Fig. 6.3B) since it is *ScDlx2* positive and *ScNkx2.1* negative (Quintana-Urzainqui et al., 2012). It is interesting to note that in mammals the MGE comprises a bilateral pair of structures, whereas the equivalent territory in sharks is formed by just one structure located in the midline. The evolutionary or functional implications of this difference remain to be clarified.

The differential expression of *ScDlx2*, *ScNkx2.1*, and *ScShh* in the catshark telencephalon has allowed us to identify the preoptic area (PO-like), the third histogenetic subdomain in the catshark subpallium, which spreads from the anterior commissure to the *ScOtp*-expressing hypothalamus (Fig. 6.5; Quintana-Urzainqui et al., 2012). The embryonic PO gives rise to cell masses of the adult preoptic area (Fig. 6.6), which in chondrichthyans likely include a nucleus associated with the anterior commissure but not necessarily the peptidergic cell groups of the clasical hypothalamohypophyseal system.

The nested expression of *Dlx*, *Nkx2.1*, and *Shh* orthologs in the subpallium seems to be a trait common to jawed vertebrates related to the existence of an MGE-like structure (*Nkx2.1*-expressing) and its adult derivative, the pallidum (Flames et al., 2007; Bardet et al., 2008; Moreno et al., 2012; Domínguez et al., 2013). The expression of such genes in the subpallium of basal

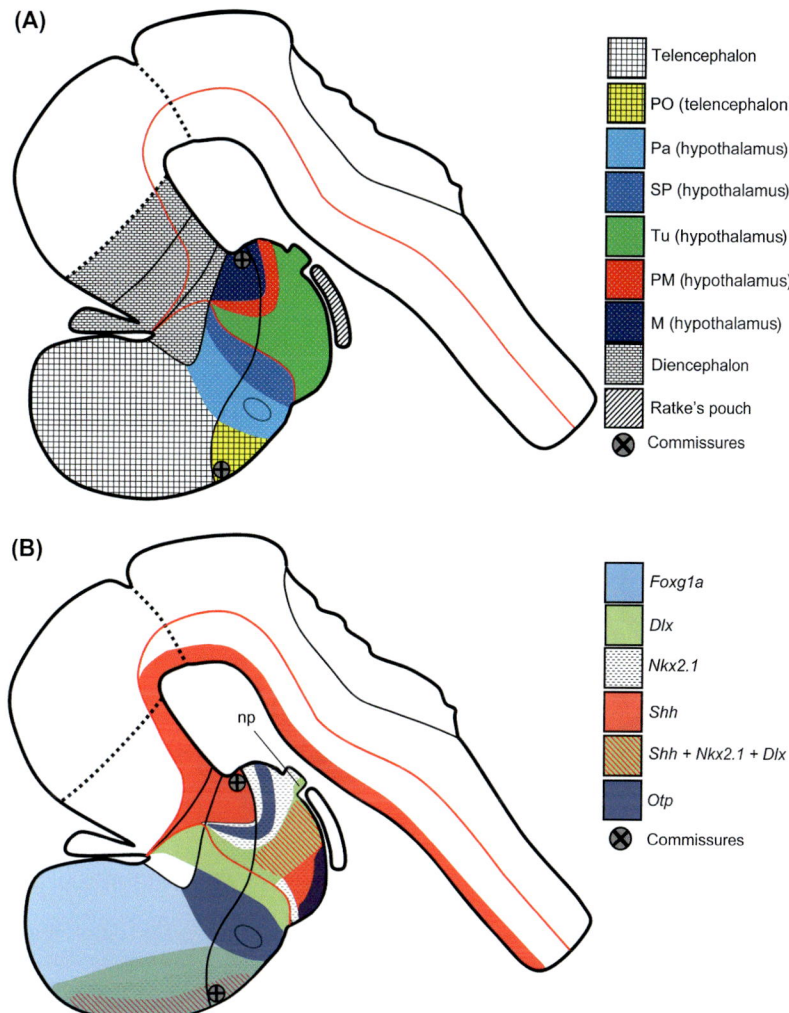

FIGURE 6.5 (A) Regionalization of embryonic hypothalamus and surrounding territories of *Scyliorhinus canicula*. Color codes are related to the adult cell groups and structures depicted in Fig. 6.6A. (B) Expression patterns of the indicated genes in the embryonic hypothalamus and adjacent prosencephalic territories. *np*, neurohypophysis primordium

gnathostomes has been undoubtedly demonstrated in studies on catshark species (Quintana-Urzainqui et al., 2012: Santos-Durán et al., 2015; Sugahara et al., 2016). Moreover, functional studies have shown that in sharks, as in osteichthyans, Shh signaling is required for the initiation of *ScNkx2.1* expression in the prosencephalon (Santos-Durán et al., 2015). Whether pallidal (MGE-like) and PO territories are evolutionary novelties that emerged during the transition from agnathans to gnathostomes is under discussion because a subpallial *Shh* and *Nkx2.1*-positive domain could not be demonstrated in lampreys (Osório et al., 2005; Sugahara et al., 2011, 2013, 2016), but it exists in a hagfish (Sugahara et al., 2016).

6.4.1.2 *Hypothalamus*

The hypothalamus is a functionally very important region located under the telencephalon and rostral to the primary prosencephalon. The hypothalamus of cartilaginous fishes contains highly conserved structures, such as the neurosecretory cells of the classical hypothalamohypophyseal system. However, it also presents some structures that are only found in certain groups of fishes, such as the hypothalamic inferior lobes or the saccus vasculosus. Thus, chondrichthyans are a good model to study both conserved and novel traits.

As for other territories, similarities and differences of the chondrichthyan adult hypothalamus are better understood through the embryonic structure of the hypothalamus. Five longitudinal, dorsoventrally arranged domains can be identified (Fig. 6.5A): the paraventricular (Pa) and subparaventricular (SPa) areas in the alar hypothalamus, and the tuberal (Tu), perimammillar (PM), and mammillar (M) regions in the basal hypothalamus. The combined expression of transcription factors or signaling molecules, such as *ScOtp, ScDlx, ScNkx2.1,*

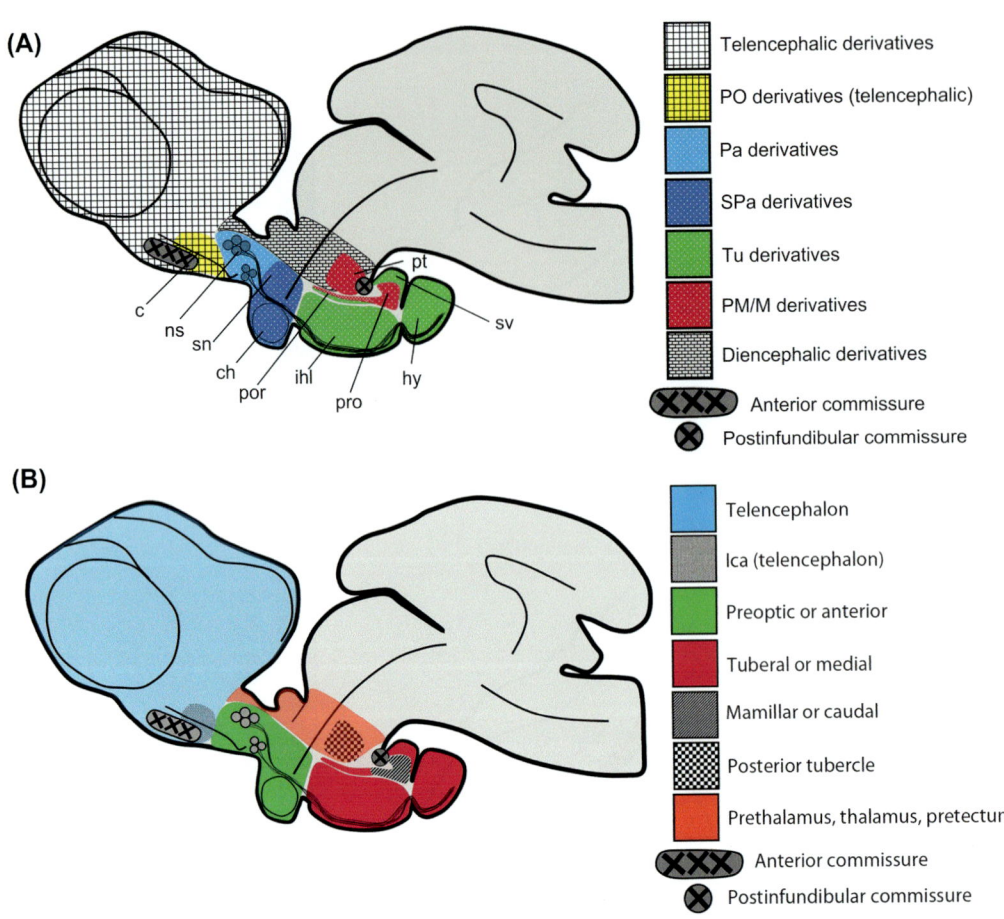

FIGURE 6.6   (A) Correspondence between adult *Scyliorhinus canicula* hypothalamic territories (and adjacent regions) and embryonic domains based on the current prosomeric model. (B) Adult structures or cell groups of the hypothalamus and surrounding territories according the classical interpretation of the chondrichthyan brain regionalization (see Smeets et al., 1983; Smeets, 1998). *c*, commissural cell groups; *ch*, optic chiasm; *hy*, hypophysis (neural lobe); *Ica*, interstitial nucleus of the anterior commissure; *ihl*, inferior hypothalamic lobes; *ns*, neurosecretory cell groups; *por*, paraventricular organ; *pro*, posterior recess organ; *t*, posterior tuberculum; *sn*, suprachiasmatic nucleus; *sv*, saccus vasculosus.

and *ScShh* (Santos-Durán et al., 2015), revealed that these domains are homologous among vertebrates. They also clarify the adult correlation with embryonic domains (Fig. 6.6A).

Otp is a transcription factor fundamental for the development of neurosecretory cell types of the classical hypothalamohypophyseal system (Del Giacco et al., 2008). In the catshark, it is expressed in the embryonic paraventricular region (Pa; Fig. 6.5) which corresponds to part of the adult anterior hypothalamus (Fig. 6.6). This adult region has been classically referred to as preoptic region (Fig. 6.6; Smeets et al., 1983; Smeets, 1998).

The expression of *Dlx* genes has been involved in the development of GABAergic phenotypes (Anderson et al., 1999). In the developing hypothalamus of catshark, *ScDlx* genes are expressed in the embryonic sub-paraventricular region (SPa; Fig. 6.5A), a region rich in inhibitory cells (Ferreiro-Galve et al., 2008), just as they are in other vertebrates (Medina, 2008). This territory gives rise to cell groups of the adult suprachiasmatic

region corresponding to the remaining adult anterior hypothalamus (Fig. 6.6). Moreover, in chondrichthyans and other vertebrates this region is also characterized by the presence of TH-positive cells (Carrera et al., 2012; Domínguez et al., 2015), a fact likely related to the expression of *ScOtp*, since this marker is also involved in the development of the catecholaminergic phenotype (Del Giacco et al., 2008).

The combined expression of *ScNkx2.1* + *ScDlx* + *ScShh* + *ScOtp* defines a complex territory referred to as the embryonic tuberal region (Tu; Fig. 6.5). This domain gives rise to the tuberal part of the adult hypothalamus in elasmobranchs (Tu; Fig. 6.6A) and other vertebrates. The embryonic Tu has a complex histogenetic organization within which different subdomains can be sketched. A subdomain expressing *ScOtp* has been associated with the development of peptidergic phenotypes, as in other parts of the hypothalamus (Del Giacco et al., 2008). In the catshark, it gives rise to cells of the infundibular walls, which probably correspond to the arcuate nucleus of other vertebrates (Joly et al., 2007; Bardet et al., 2008; Del Giacco et al., 2008; Manoli and Driever, 2014). The remaining Tu expresses *ScDlx* and *ScShh* genes, which correlates with the distribution of GABAergic phenotypes in that region (Ferreiro-Galve et al., 2008; Santos-Durán et al., 2015). Furthermore, the embryonic expression of *ScDlx* in the catshark Tu corresponds to the expanded lateral walls of the infundibular hypothalamus that give rise to the inferior hypothalamic lobes (ihl; Fig. 6.6A) which are characteristic tuberal structures involved in feeding behavior only present in the tuberal hypothalamus of gnathostome fishes. Furthermore, the Tu also gives rise to the saccus vasculosus (Figs. 6.5 and 6.6), a circumventricular organ of enigmatic function, probably involved in photoperiodic functions (Nakane et al., 2013). It develops adjacent to the evagination of the neurointermediate lobe of the hypophysis (van de Kamer and Shuurmans, 1953; Sueiro et al., 2007), but is also closely associated with the posterior recess of the adult mammillary hypothalamus (pro; Fig. 6.6B). The Tu also gives rise to the paraventricular organ (por; Fig. 6.6A), a circumventricular specialization of the adult hypothalamus in fishes.

Another embryonic territory characterized by *ScNkx2.1* + *ScOtp* coexpression seems to be involved in the development of peptidergic and catecholaminergic cells of the caudal hypothalamus of the catshark (Santos-Durán et al., 2015), as reported in other fishes (Joly et al., 2007; Del Giacco et al., 2008; Vernier and Wulliman, 2008). This histogenetic domain is referred as the perimammillar region (PM; Fig. 6.5) and also produces cells closely associated with the paraventricular organ and posterior recess (or mammillary) organ (por and pro, respectively; Fig. 6.6A). Both organs have characteristically folded walls formed by a high density of cerebrospinal fluid-contacting neurons of catecholaminergic, serotoninergic, and peptidergic nature (Rodríguez-Moldes and Anadón, 1987; Meurling and Rodríguez, 1990; Molist et al., 1993; Sueiro et al., 2007; Carrera et al., 2008b, 2012)

The mammillary region corresponds to the histogenetic territory where *ScNkx2.1* and *ScShh* genes are expressed (M; Fig. 6.5). Two subdomains have been identified based on the expression of these genes: the *ScNkx2.1*-expressing subdomain, which gives rise to derivatives of the posterior or mammillary recess walls (it corresponds to the adult mammillar hypothalamus, as shown in Fig. 6.6B; see also Smeets et al., 1983; Smeets, 1998), and the subdomain expressing *ScShh*, which defines the retromammillar region (RM). Taking into account that the rostralmost portion of the catecholaminergic population of the posterior tuberculum nucleus, previously described as part of the diencephalon proper (basically a derivative of the basal part of the prosomere 3: Carrera et al., 2012), is located

in this RM *ScShh*-expressing domain, a hypothalamic origin of the shark rostral posterior tuberculum had been proposed (Santos-Durán et al., 2015), which is in agreement with the updated prosomeric framework (Puelles et al., 2012; Puelles and Rubenstein, 2015). The posterior tuberculum had been previously attributed to the hypothalamus or to the diencephalon on the basis of different types of evidence (Vernier and Wullimann, 2008).

Beyond similarities, the differences existing in these homologous domains among vertebrates could account for morphological divergences observed in adults. In the tuberal hypothalamus of the catshark, the expression of *ScShh* is enhanced in the Tu compared with that of other vertebrates. This may be related to the dilated walls of the inferior hypothalamic lobes, which are not present in jawless vertebrates and tetrapods. Along the same line, the lack of *ScShh* in the PM and part of M in the catshark correlates with the development of the posterior recess organ, the caudal expansion of the paraventricular organ (Rodríguez-Moldes, 2011), and the presence of a saccus vasculosus (Figs. 6.5 and 6.6), suggesting involvement in their development. The role of *Shh* in shaping hypothalamic morphogenesis has been demonstrated in mice mutant for *Shh* (Szabó et al., 2009), whose phenotype exhibits many similarities with the organization of the catshark hypothalamus. However, a direct causal role remains to be tested in chondrichthyans, offering an exciting landscape for future experimental work concerning the development of the discussed specializations.

6.4.1.3 Diencephalon

Studies in the catshark based on the comparative analysis of the expression pattern of the *ScPax6* gene and selected neurochemical markers allowed us to recognize in the diencephalon of chondrichthyans a segmental organization with three prosomeres (Ferreiro-Galve et al., 2008) equivalent to those described in other vertebrates. This segmental pattern is clearly evident in the alar parts of the prosomeres that, from rostral to caudal, correspond to the prethalamus (p3), thalamus (p2), and pretectum (p1). The segmental pattern is well reflected in the distribution of certain cell populations and in the existence of boundaries, such as the zona limitans intrathalamica, a signaling center responsible of the formation of prethalamus (rostrally) and thalamus (caudally). A transverse band of GABAergic cells and a stripe of Shh-expressing cells allow us to distinguish unambiguously such a boundary in catshark embryos (Carrera et al., 2006; Ferreiro-Galve et al., 2008; Rodríguez-Moldes, 2009; Santos-Durán et al., 2015).

The distribution of certain markers, especially in the alar diencephalon, reflects a clear segmental pattern.

Particularly helpful is the distribution of Pax6-expressing cells, which occupy the alar parts of prosomeres 3 (prethalamus) and 1 (pretectum), but are absent from the thalamus (alar part of p2) (Ferreiro-Galve et al., 2008; Rodríguez-Moldes, 2009). Similar alternation is observed in the distribution of glycinergic cells, which are absent from the thalamus but present in prethalamus and pretectum (Anadón et al., 2013). Interestingly, the absence of glycinergic cells in the thalamus may be a peculiarity of chondrichthyans, as they exist in chondrosteans and cyclostomes (Villar-Cerviño et al., 2008; Adrio et al., 2011). The thalamus, but not the adjacent alar territories, contains a conspicuous population of calretinin-positive cells equivalent to those of tetrapods (Milán and Puelles, 2000). Prethalamus and thalamus represent relay centers for ascending sensory pathways to the telencephalon, whereas the pretectum receives retinal fibers.

As in other vertebrates, the dorsal part of p2 in chondrichthyans exhibits more or less marked left–right asymmetries. Two relevant structures are formed here: the paired habenula, a markedly asymmetric structure, and the pineal organ or epiphysis, a well-developed unpaired tubular structure without any sign of anatomical asymmetry. The pineal organ is present in all vertebrates excepting some electric rays. The catshark appears to be unique among anamniotes in possessing an extensive GABAergic innervation in the pineal organ (Carrera et al., 2006). Significantly, chondrichthyans lack a parapineal organ, a conspicuous asymmetric structure (in connectivity and position) in lampreys and teleosts which is closely related to the pineal organ (together, they form the pineal complex) and has a role in the establishment of the proper habenular asymmetry (Gamse et al., 2003). Comparison between catsharks and lampreys has revealed conserved molecular asymmetries between the left and right developing habenulae (even though the largest habenula is on the left in catsharks and on the right in lampreys) regardless of the presence or absence of a parapineal organ (Lagadec et al., 2015).

A segmental organization is less apparent in the basal part of the diencephalon, mainly because many cell populations extended over more than one prosomere. This is the case for catecholaminergic cells, which form a column that extends along the basal diencephalon from the posterior tuberculum (whose rostralmost part occupies caudal hypothalamic territories, see the previous sections) to the mesencephalic tegmentum (Carrera et al., 2012) (see the following sections). A conspicuous nucleus that gives rise to a major longitudinal tract, the nucleus of the medial longitudinal fascicle, is located in the basal part of p1, as in other vertebrates. It contains abundant calretinin-positive cells, which are thus a reliable marker of p1 and its adult derivative.

6.4.2 Mesencephalon

During early brain regionalization, the anatomical distinction between the midbrain and hindbrain is not conspicuous. However, genoarchitectonic analysis at early stages has allowed us to identify the caudal limit of the expression domain of ScOtx2 as the caudal limit of the midbrain (Fig. 6.7). Taking into account the gene expression patterns of these and other genes involved in the brain regionalization, the midbrain may be defined in catshark, at least in the tegmentum, as the territory where the expression domains of ScOtx2 and ScEn2 genes are mostly overlapping.

The mesencephalic vesicle, or midbrain, is a single segment that comprises the tectum (alar or dorsal) and the tegmentum (basal or ventral). A longitudinal stripe of Pax6-expressing cells periventricularly located and observed from early embryos to postnatal stages has been related to the position of the alar–basal boundary along the midbrain and the adjacent caudal diencephalon (Ferreiro-Galve et al., 2008; Rodríguez-Moldes, 2009). The isthmic organizer (IsO), a secondary organizer that controls the formation of the optic tectum rostrally and the cerebellum caudally (Figs. 6.1 and 6.7), has been identified at the midbrain–hindbrain boundary (MHB) in catshark (Pose-Méndez et al., 2016a). The sharp limit between the caudal extension of alar prosencephalic Pax6-expressing cells and the Pax6-negative optic tectum marks the diencephalic–mesencephalic boundary (Ferreiro-Galve et al., 2008; Rodríguez-Moldes, 2009). In the tegmentum, the midbrain's caudal limit lies at the level of the oculomotor

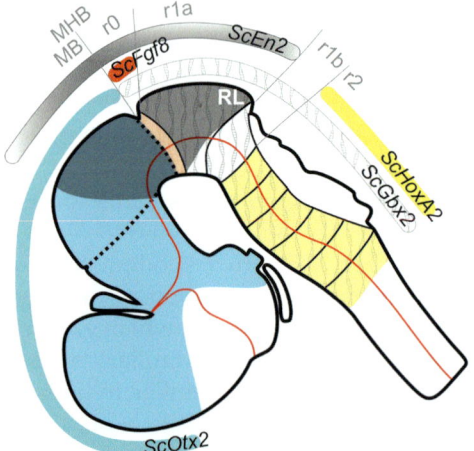

FIGURE 6.7   Rostral hindbrain genoarchitecture. Schematic lateral view of the brain in a *Scyliorhinus canicula* embryo showing the segmentation of the rostral hindbrain (midbrain–hindbrain, rhombomere 0–rhombomere 1, and rhombomere 1a–1b boundaries) as revealed by the expression pattern of the genes *ScOtx2, ScGbx2, ScFgf8, ScEn2,* and *ScHoxA2. MB,* midbrain; *MHB,* midbrain–hindbrain boundary; *r,* rhombomere; *RL,* rhombic lip.

nerve root, which roughly coincides with the sharp separation between the caudalmost tyrosine hydroxylase-immunoreactive (TH-ir) cells of the ventral tegmental area and the rostralmost serotonergic-immunoreactive (5-HT-ir) cells of the reticular formation (Carrera et al., 2008b, 2012; Ferreiro-Galve et al., 2008). By the time the adult structure of the chondrichthyan brain has formed, the anatomical landmarks of the alar midbrain territory are the caudal posterior commissure (rostrally) and the isthmic constriction and the anterior medullary velum (caudally; see Pose-Méndez et al., 2016b). Moreover, commissural tracts containing catecholaminergic fibers (Carrera et al., 2012) mark the forebrain—midbrain boundary and the limit between optic tectum and cerebellum.

6.4.2.1 Optic Tectum

The alar plate of the midbrain gives rise to the optic tectum, a layered structure containing at least five strata (Smeets et al., 1983; Manso and Anadón, 1991a,b, 1993), with the outer layers being the main recipient of visual inputs (Northcutt, 1979; Repérant et al., 1986), while the deep tectal layers receive nonvisual information. The outer cell layer contains abundant glycinergic cells (Anadón et al., 2013) and cells containing calcitonin gene—related peptide (CGRP) (Molist et al., 1995), a neuromodulator in cholinergic and some noncholinergic pathways. In deep layers, such neurons are scarce (glycinergic) or absent (CGRP). The CGRP-positive cells have been considered intrinsic neurons because no CGRP tectal efferent tracts were observed (Molist et al., 1995). The glycinergic cells and fibers may modulate tectal circuits, such as the inhibition of retinal inputs (Anadón et al., 2013). Cholinergic cells were not reported in the tectal walls of chondrichthyans (except those of the trigeminal mesencephalic nucleus, see later), but abundant fibers of cholinergic nature (choline acetyltransferase immunoreactive) were described in the outer tectal layers, consistent with the existence of a second-order cholinergic visuotectal circuit (Anadón et al., 2000).

Monoaminergic cells have not been reported in the chondrichthyan tectum, but the tectum is differentially innervated by serotoninergic and catecholaminergic fibers, with the serotoninergic fibers being more abundant in the outer and intermediate layers (5-HT ir, Carrera et al., 2008b) while the catecholaminergic fibers (tyrosine hydroxylase immunoreactive, Carrera et al., 2012) are mainly restricted to the periventricular region. Developmental studies have revealed that the arrival of catecholaminergic and serotoninergic fibers in the tectum takes place during late development (prehatchings), long after the differentiation of retinal ganglion cells and their retinofugal projections, which in catshark happens by stage 29 (Ferreiro-Galve et al., 2008). These results indicate that monoamines are not involved in the early maturation of the tectum. In contrast, abundant cells containing glutamic acid decarboxylase (GAD)—the enzyme responsible for the synthesis of γ-aminobutyric acid (GABA)—appear in the early optic tectum prior to the completion of the layering process (stage 26) (Carrera et al., 2006; Rodríguez-Moldes et al., 2011). These findings suggest that, in these early shark embryos, GABA may act as a neurotrophic or signaling factor that controls tectal development by promoting synaptogenesis, growth, and differentiation of neurons.

The internal cell layer of the optic tectum also contains the large primary sensory neurons of the trigeminal mesencephalic nucleus. The cholinergic nature of these cells has been reported in *S. canicula* (Anadón et al., 2000).

6.4.2.2 Tegmentum

The midbrain basal plate, or tegmentum, is caudally continuous with the basal plate of the hindbrain. In chondrichthyans, as in all vertebrates, the midbrain tegmentum contains the most rostral motor nucleus, the nucleus of the oculomotor (III) nerve. This nucleus in *S. canicula* is cholinergic (Anadón et al., 2000) and immunoreactive for CGRP (Molist et al., 1995).

As in sarcopterygians (lungfishes, amphibians, reptiles, birds, and mammals), the midbrain tegmentum in elasmobranchs contains catecholaminergic groups of the ventrotegmental area and substantia nigra (VTA/SN). These nuclei are not present in the midbrain of other fish groups, such as actinopterygians and cyclostomes. Developmental studies in catshark revealed that, although catecholaminergic cells of the VTA/SN form a continuous column with the catecholaminergic population of the posterior tuberculum, the appearance of the catecholaminergic midbrain population is delayed with respect to that of the posterior tuberculum supporting an independent origin for the two clusters (Carrera et al., 2012). Cholinergic and CGRP-ir cells have been described in the VTA/SN of *S. canicula* (Molist et al., 1995; Anadón et al., 2000). Lateral to the SN lies the red nucleus, which contains glutamatergic and calretinin-positive cells and has been identified as a mesencephalic precerebellar nucleus (Pose-Méndez et al., 2014). Other precerebellar mesencephalic groups of cells containing calretinin-positive and glutamatergic cells have been identified in *S. canicula* (Pose-Méndez et al., 2014). Numerous perikarya positive for calretinin are found in the midbrain tegmentum from embryonic stage 29 onward (Rodríguez-Moldes et al., 2011) while abundant GABAergic cells are seen in the midbrain tegmentum even earlier (stage 26, Carrera et al., 2006).

Two other important populations have been characterized in the midbrain tegmentum of *S. canicula*. They are the large population of glycinergic cells in the region of the lateral tegmental nucleus, probably involved in both sensory and descending input processing in this region (Anadón et al., 2013), and the Edinger–Westphal (EW) nucleus, which is a motor nucleus located dorsolateral to the oculomotor nucleus that contains cholinergic, glycinergic, and peptidergic (substance P–like positive) cells (Rodríguez-Moldes et al., 1993; Anadón et al., 2000, 2013). This nucleus gives rise to the cholinergic visceromotor fibers coursing along nerve III that innervate the iris and ciliary body of the eye (Anadón et al., 2000), and it is involved in pineal neural circuits that could be related with the photic control of brain functions (Mandado et al., 2001). In addition, an enigmatic gonadotropin-immunoreactive (GnRHir) nucleus has been described in the midbrain of elasmobranchs, including *S. canicula* (Wright and Demski, 1991; Mandado et al., 2001). This nucleus also seems to be involved in secondary pineal pathways that may allow environmental photic levels to influence neural activity in widespread brain areas (Mandado et al., 2001).

6.4.3 Rhombencephalon

The organization of the hindbrain is very similar among chondrichthyans, except for the existence of specializations as the electric lobes in electric rays. Functionally, it represents the brain region where the information from different motor and sensory systems, including electroreception, is processed (New, 2001; Wueringer, 2012). As in other vertebrates, the chondrichthyan hindbrain exhibits a segmental organization that is clearly recognized during development by a series of prominent ventricular outpouchings or ridges separated by constrictions that divide the rhombencephalon in transverse units, or rhombomeres.

The analysis of the expression pattern of genes related to the segmentation of the anterior hindbrain in different developmental stages of *S. canicula* (Pose-Méndez et al., 2016a) has allowed the recognition of rostral hindbrain subdivisions and boundaries in this species, including the MHB and rhombomere 0 (isthmus or isthmic territory), as represented in Fig. 6.7.

The MHB is determined by the abutting expression domains of the gene *ScOtx2* (expressed rostrally in the forebrain and midbrain) and the gene *ScGbx2* (expressed caudally in the hindbrain). Just caudal to MHB lies rhombomere 0 (r0), which corresponds to the isthmus or isthmic territory and is identified by the expression of *ScFgf8*. The genes *ScGbx2* and *ScEn2* are also expressed in r0. The cholinergic cells of the trochlear motor nucleus and calretinin-positive cells of the interpeduncular nucleus differentiate early in the basal part of r0 (Rodríguez-Moldes et al., 2011).

A striking feature of the chondrichthyan rhombencephalon is the large size of rhombomere 1 (r1), which exhibits an internal constriction (potentially misperceived as an interneuromeric boundary). The subdivision of the catshark r1 into subrhombomeres 1a and 1b (r1a–r1b) has been demonstrated on the basis of *ScEn2* expression in the rostral half of r1. The r1 territory is noticeably larger than a normal rhombomere also in other vertebrates, and special features including compartmentation in subrhombomeres have been reported in chick and mouse (Aroca and Puelles, 2005; Moreno-Bravo et al., 2014; Green et al., 2014). Other characteristic rhombencephalic structures differentiate early within r1, such as the catecholaminergic cells of the locus coeruleus (dorsal part of the basal region of r1) and the serotoninergic cells of the rostralmost reticular cells of the superior reticular formation (ventral part of the basal region of r1).

The more caudal rhombomeres (r2–r8) exhibit a rather regular segmental organization (Figs. 6.1 and 6.7). The location of the exit of the trigeminal nerve root is worth mentioning. Classically, the segmental organization of the chondrichthyan rhombencephalon was considered peculiar because the exit of this nerve root was related to r3 instead of r2 as in other vertebrates (Kuratani and Horigome, 2000; Gilland and Baker, 2005). However, analyses in early *S. canicula* embryos of the expression of *ScHoxA2*, the only *Hox* gene expressed in rhombomere 2, and *ScWnt8*, which specifies rhombomere 4, demonstrated that the trigeminal nerve root actually lies in r2 (Rodríguez-Moldes et al., 2011). Such observations reveal the evolutionary conservation of the segmental rhombencephalic entities in which the position of nerve roots is fixed during the formation of rhombomeres. In contrast to previous proposals (Kuratani and Horigome, 2000), no active movement of nerve roots across the interrhombomeric boundaries takes place.

There are several studies on development of rhombencephalic structures that help to better understand the adult structure of the rhombencephalon by identifying the hindbrain subdivisions. In adults, the rhombomeres are no longer visible, but the segmental hindbrain organization remains in some hindbrain centers, such as the (cholinergic) motor nuclei (Anadón et al., 2000) and the reticular formation, as revealed by the distribution of conspicuous serotoninergic cell groups (Carrera et al., 2008b). In the most caudal hindbrain, the inferior olive also projects to the cerebellum. This precerebellar nucleus contains calretinin- and glutamate-immunoreactive cells, and it is probably involved in the timing and learning of movements (Álvarez-Otero et al., 1993; Pose-Méndez et al., 2014).

6.4.3.1 Cerebellum

The cerebellum is a remarkable rhombencephalic structure that is relatively large in chondrichthyans. There is a huge variability in cerebellum size and shape among chondrichthyan groups, mainly due to different degrees of cerebellar cortex folding, ranging from bilobed and quite smooth to highly folded. These differences may be due to phylogenetic constraints (for instance, ancient groups appear to have lower foliation indices: Yopak et al., 2007), functional specialization of the cerebellum linked to ecological adaptations (benthic species present characteristic smooth cerebellum: Lisney et al., 2008; the reef-associated species have highly foliated cerebella: Yopak et al., 2010; Yopak, 2012), enhanced cognitive capabilities, or behavioral parameters (such as habitat dimensionality, activity levels, and agile prey capture: Yopak and Frank, 2007; Yopak and Montgomery, 2008; Yopak et al., 2010). The possibility that these differences may be due to variation in the developmental program of the cerebellum has also been pointed out (Puzdrowski and Gruber, 2009).

Two main divisions are present in the cerebellum of chondrichthyans: the cerebellar body or spinocerebellum, and the cerebellar auricle (subdivided into upper and lower leaves) or vestibulocerebellum, which are homologous to the cerebellar vermis and floculonodular lobe of mammals, respectively. The spinocerebellum is involved in the integration of sensory and premotor sources and in the timing and learning of movements (New, 2001; Montgomery and Bodznick, 2010) while the vestibulocerebellum is mainly involved in the electroreception and mechanoreception (Schmidt and Bodznick, 1987; Bodznick et al., 1999). There are also other hindbrain structures that bear similarities to the cerebellum, known as cerebellum-like structures. They include the medial and dorsal octaval nuclei in cartilaginous fishes (Montgomery et al., 2012). Their resemblance has raised the question of whether the cerebellum itself is a cerebellum-like structure (Devor, 2000; Bell, 2002; Montgomery et al., 2012). A 2016 computational study described an adaptive filter model that shows that the cerebellum-like structures cancel predictable components of the signal generated by own movements (Bratby et al., 2016).

The combined analysis of cerebellar morphogenesis and the expression pattern of isthmus-related genes has contributed to the identification of the rhombomeric origin of the cerebellar body and upper and lower auricular leaves (Pose-Méndez et al., 2016b). As shown in Fig. 6.8, the *ScEn2* expression domains delimit different functional domains in the cerebellum, interpreted as the subrhombomeres r1a (cerebellar body) and r1b (cerebellar auricle), hence making the posterolateral fissure (a fissure flanked by the cerebellar body and the auricles) a consistent anatomical landmark. More caudally, the rostral boundary of *ScHoxA2* in late embryos still delimits the r1–r2 boundary caudal to the cerebellum, suggesting the auricles (both upper and lower leaves) originate from r1 and that no part of the cerebellum develops from r2 (Pose-Méndez et al., 2016b), as suggested also in other vertebrates (Butts et al., 2014).

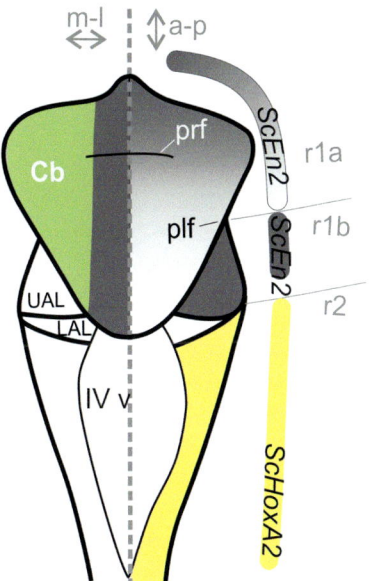

FIGURE 6.8 Compartmentalization of the cerebellum. Schematic dorsal view of the embryonic *Scyliorhinus canicula* hindbrain, showing how the cerebellar body and auricles become distinct compartments along the median–lateral (left), and anteroposterior (right) axes based on the expression patterns of Aldo-C and *ScEn2*. Median–lateral compartmentalization is defined by a single *ScEn2*-expressing paramedian band (*dark gray*) and a single lateral band positive for Aldo-C (*green*). In the anteroposterior axis, the differential *ScEn2* expression in r1 and its correlation with functional domains (cerebellar body from subrhombomere 1a and cerebellar auricles from subrhombomere 1b) are shown. In the cerebellar body, the *ScEn2* expression is intense in the rostral half and decreases gradually in the caudal half. The strong *ScEn2* expression in the area adjacent to the posterolateral fissure correlates with the upper auricular leaf. The auricles originate rostrally to the r1–r2 boundary identified as the anterior limit of *ScHoxA2* expression (*yellow*). a, anterior; *Cb*, cerebellar body; *IVv*, fourth ventricle; *l*, lateral; *LAL*, lower auricular leaf; *m*, medial; *p*, posterior; *plf*, posterolateral fissure; *prf*, primary transverse fissure; *r*, rhombomere; *UAL*, upper auricular leaf.

The aforementioned studies on the segmental identity of the cerebellar primordium (Pose-Méndez et al., 2016a) have helped to gain insight into the evolutionary origin of the cerebellum, a structure that emerged during the agnathan/gnathostome transition and has been maintained as a unique trait of all jawed vertebrates. Because only vertebrates present the whole set of isthmus-related genes at the IsO-like signaling center, differences in their expression patterns between catsharks and lampreys may reveal evolutionary changes that happened at the agnatha–gnathostome transition and, in turn, explain why the isthmic organizer induces the formation of a true cerebellum only in jawed vertebrates (Pose-Méndez et al., 2016a). Features observed in the catshark but not in agnathans, such as the downregulation of *Irx3* at the MHB and r0, and the expression of the gene *Lhx9* in r1 (Fig. 6.9), have been proposed as significant changes in the gene network at the agnathan–

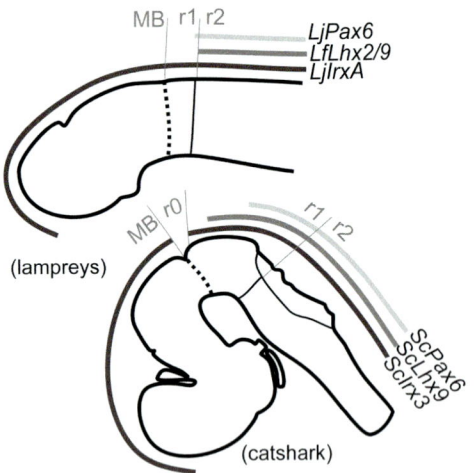

FIGURE 6.9 Schematic drawings of the brain of a lamprey and a catshark embryo from lateral view, showing differences in the expression domains of *Irx*, *Lhx*, and *Pax6* genes in the rostral hindbrain. *MB*, midbrain; *r*, rhombomere.

gnathostome transition related to the emergence of cerebellum (Pose-Méndez et al., 2016a). Although the expression of *Pax6* in the dorsal part (rhombic lip) of r1 in catshark but not in lampreys (Murakami et al., 2001; Kuratani et al., 2002; Murakami et al., 2005; Rodríguez-Moldes et al., 2008; Murakami and Watanabe, 2009) has been directly related to the emergence of cerebellum, 2016 observations of *Pax6* expression in the rhombic lip of hagfish revealed that the regionalization of the rhombic lip by Pax6, which happens early than previously thought, is not sufficient for the differentiation of a morphologically distinct cerebellum from r1 (Sugahara et al., 2016). Additional fortuitous changes during the agnathan–gnathostome transition, such as genetic duplications (eg, the appearance of new *Irx* and *Lhx* isoforms) and/or changes in developmental programs (Wagner, 2008; Montgomery et al., 2012), or even differences in the threshold and temporal dynamics of expression of certain genes (more details in Pose-Méndez et al., 2016a), could all be involved in the emergence of the cerebellum.

In spite of the differences in size and shape among the cerebella of elasmobranchs, cerebellar morphogenesis appears to be similar among shark species and rather similar to that of other groups (Nieuwenhuys, 1967; Larsell, 1967; Pose-Méndez et al., 2016b). Taking catshark development as a reference, different periods of the cerebellar development have been recognized (Pose-Méndez et al., 2016b). (1) An early period (stages 20–29), when the primordium of the cerebellum or rhombic lip emerges from the rostral hindbrain and the cerebellar plate forms. (2) An intermediate or transition period (stages 30 and 31), during which the cerebellar body and auricles acquire their mature-like

shape and the cerebellar peduncle begins to define the cerebellar nucleus (the main cerebellar output, considered homologous to the deep cerebellar nuclei of mammals) (Álvarez-Otero et al., 1996). (3) The late period (stage 32 onward), in which the cerebellum grows and matures and develops the main peculiarity of the chondrichthyan cerebellum, namely the distribution of the granular layer into paramedian eminences. Once the patterning along the anteroposterior axis cerebellum is well defined, layering, ingrowth of afferent projections, and median–lateral compartmentalization take place (Pose-Méndez et al., 2014, 2016b). In the catshark, as in other cartilaginous and bony fishes, but differing from the condition in mammals, there is no multibanding or zebralike pattern along the median–lateral axis of the cerebellar cortex. Instead, the cerebellum shows a simple compartmentalization (Fig. 6.8), defined by a single paramedian band positive for *ScEn2* expression and a single/continuous lateral band positive for Aldo-C (a marker of Purkinje cells); this pattern has been suggested to be the basal state of cerebellar compartmentalization (Pose-Méndez et al., 2016b). The folding of the cerebellar cortex takes place during late development as revealed by the magnetic resonance imaging (MRI) study performed in the whale shark *Rhincodon typus*, a plankton-feeder species with one of the largest highly foliated cerebella among chondrichthyans (Yopak and Frank, 2007).

The cell types of the cerebellar cortex have been well characterized in adults (Anadón et al., 2000, 2009, 2013; Rodríguez-Moldes et al., 2008) and are similar to those in the mammalian cerebellum. The outer, molecular layer contains stellate cells, which have been reported to be calretinin-ir. Below the molecular layer lies the Purkinje cell layer (aldolase-C-ir) and a fibrous layer that contains the axons of Purkinje cells as well as cerebellar afferents. Fiber systems projecting to the cerebellar cortex include the GABAergic, cholinergic, catecholaminergic, and serotonergic systems. The characteristic granule layer of chondrichthyans, grouped into the two paramedian eminences (GL in Fig. 6.10A and B), contains both granule (Pax6-ir) and Golgi cells (GABA, ChAT or cholinergic, glycine, and calretinin-ir) as observed in late embryos, juveniles, and adults of catshark (Anadón et al., 2000, 2009, 2013; Rodríguez-Moldes et al., 2008). Cartilaginous fishes differ from ray-finned fishes in having glycinergic cells in their cerebellum (Anadón et al., 2013). The cerebellar nucleus contains small GABAergic cells in the subventricular part and large glutamate-immunoreactive cells in the central or medial region (Álvarez-Otero et al., 1996), as well as some cholinergic and glycinergic cells (Anadón et al., 2000, 2013). Such a nucleus is not present in actinopterygians and cyclostomes.

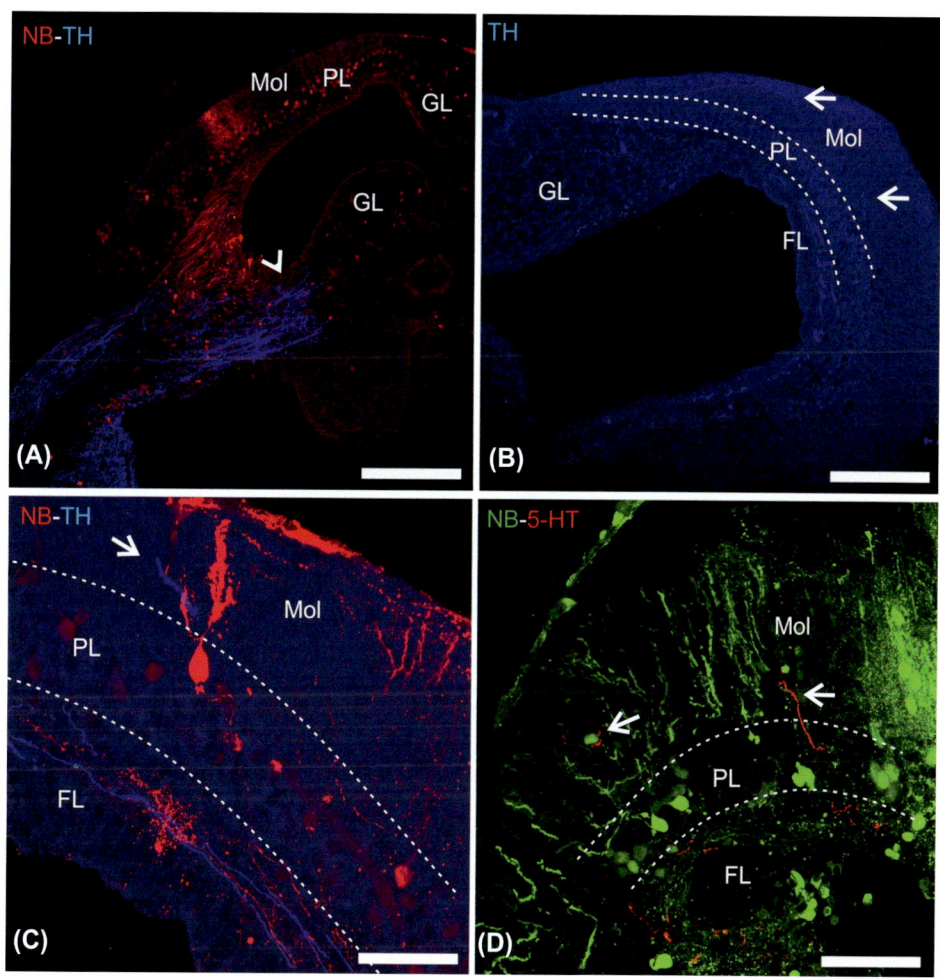

FIGURE 6.10 Serotonergic and catecholaminergic afferents in the cartshark cerebellar cortex. (A–D). Transverse sections of the cerebellar cortex showing afferent TH-ir (*blue* in A–C) and 5-HT-ir (*red* in D) fibers ending in different layers of the cerebellar cortex. The neurobiotin (NB) labeling (*red* in A and C, and *green* in D) helps to distinguish the cerebellar cell layers and was obtained by retrograde labeling and/or diffusion of the tracer close to the tracer application site in the cerebellar cortex. *Arrows* indicate fibers in the molecular layer. *Arrowhead* indicates some fibers entering the granular layer. FL, fibrous layer; GL, granular layer; Mol, molecular layer; PL, Purkinje layer. Scale bars: 300 μm (a), 150 μm (b), 75 μm (d), and 50 μm (C).

Projections from the cerebellar nucleus and extracerebellar nuclei or precerebellar nuclei have been characterized in developing catshark (Pose-Méndez et al., 2014), including the climbing fibers ending on the perikarya of Purkinje cells (not in the dendrites like in mammals), and mossy fibers ending in the granular layer. Additional inputs to the cerebellar cortex are serotoninergic and catecholaminergic fibers that end diffusely in the molecular layer (Fig. 6.10) probably representing modulatory fibers. At least some of the catecholaminergic fibers may correspond to cerebellar projection from the locus coeruleus (Pose-Méndez et al., 2014). The cerebellar projecting nuclei or precerebellar nuclei in the hindbrain are originated, like the cerebellum, from the rhombic lips, and reach their final destination following specific and intricate migratory pathways (Watanabe and Murakami, 2009). Some evidence of migrating

neuroblasts, apparently following migration routes of rhombic lip derivatives equivalent to that described in other jawed vertebrates, was found in *S. canicula* (Pose-Méndez et al., 2014). Those neuroblasts appear related to two precerebellar nuclei, namely reticular cells scattered through the hindbrain that may have migrated tangentially in a marginal position (1 in Fig. 6.11), crossed the midline (2 in Fig. 6.11), or migrated radially (3 in Fig. 6.11) and, second, cells of the inferior olive that migrated tangentially in a submarginal position (4 in Fig. 6.11). These findings suggest that the migratory routes of cerebellum-projecting cells originating in the rhombic lips appeared very early in gnathostome evolution.

The differentiation of cells projecting to the cerebellum is largely concurrent with the differentiation of the cerebellum's efferent projections (axons of Purkinje cells) and

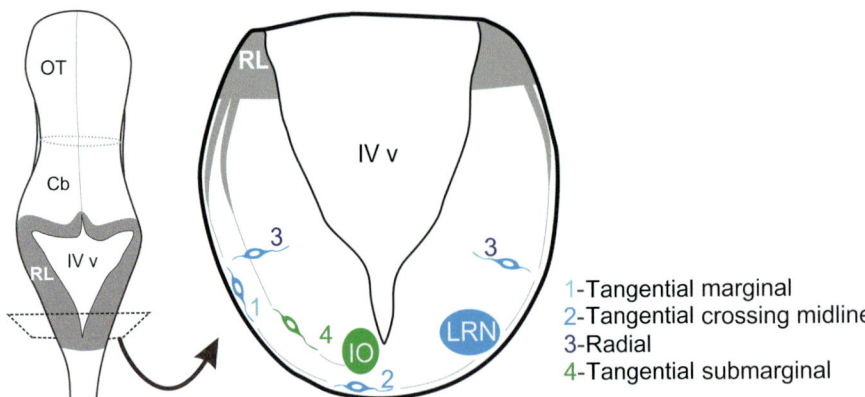

1-Tangential marginal
2-Tangential crossing midline
3-Radial
4-Tangential submarginal

FIGURE 6.11   Rhombic lip derivatives. Schematic drawing of a transverse section of the caudal hindbrain (at the level indicated in the scheme) in *Scyliorhinus canicula*, showing migrating neuroblasts that originated in the rhombic lips. The numbers indicate neuroblasts migrating: tangentially marginally (1), crossing the midline (2), radially migrating (3), and tangentially submarginally (4). *Cb*, cerebellum; *IO*, inferior olive; *IVv*, fourth ventricle; *LRN*, lateral reticular nucleus; *OT*, optic tectum; *RL*, rhombic lip.

the completion of cerebellar cortex layering (Rodríguez-Moldes et al., 2008; Pose-Méndez et al., 2014). The orderly arrival of several afferents, for example, the arrival of spinal cord afferent before the arrival of inferior olive projections, is highly conserved throughout evolution (Pose-Méndez et al., 2014). The mature cerebellar body in chondrichthyans receives inputs from all major brain areas and the spinal cord, as reported in diverse species using various experimental procedures (the thornback guitarfish *Platyrhinoidis triseriata*: Fiebig, 1988; the stingray *Dasyatis sabina*: Puzdrowski and Gruber, 2009; the catshark *S. canicula*: Pose-Méndez et al., 2014, 2016b; see also Fig. 6.12). Once the information from the afferents is processed, the Purkinje cells connect with the cerebellar nucleus, which is the main output of the cerebellum. Three main efferent pathways from the cerebellar nucleus have been described: a cerebellobulbar tract, the brachium conjunctivum, and a cerebellomotor tract (reviewed in New, 2001). The connections of the vestibulocerebellum, or cerebellar auricle, were described in the little skate. It receives afferents mainly from the octavolateral, vestibular systems, funiculus nucleus, nucleus of the medial longitudinal fasciculus, and the spinal cord. Furthermore, it sends projections to some of those centers, as well as to the cerebellar body (Schmidt and Bodznick, 1987).

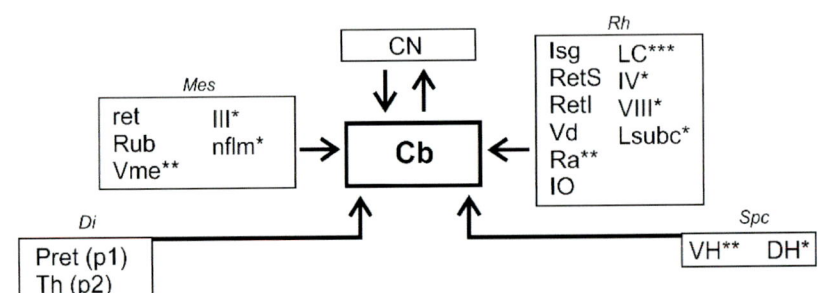

FIGURE 6.12   Connections of the cerebellum in elasmobranchs. Cerebellar afferents from precerebellar nuclei and efferents of the cerebellar body in three elasmobranch species (the lesser spotted dogfish and two batoids, the thornback guitarfish, and the Atlantic stingray). The names of several precerebellar nuclei in batoids (considered homologous to reticular cells in the lesser spotted dogfish) were omitted for the sake of simplicity. The *asterisks* indicate precerebellar nuclei: (*) present in the thornback guitarfish and/or the Atlantic stingray, but absent in the catshark, (**) present in the catshark, but absent in batoids, and (***) present in the catshark and thornback guitarfish, but absent in Atlantic stingray. *Cb*, cerebellum; *CN*, cerebellar nucleus; *DH*, dorsal horn; *Di*, diencephalon; *III*, oculomotor nucleus; *IO*, inferior olive; *Isg*, isthmic group; *IV*, trochlear nucleus; *LC*, locus coeruleus; *Lsubc*, locus subcoeruleus; *Mes*, mesencephalon; *nflm*, nucleus of the medial longitudinal fascicle; *p*, prosomere; *Pret*, pretectum; *Ra*, raphe nucleus; *ret*, reticular cells; *RetI*, inferior reticular formation; *RetS*, superior reticular formation; *Rh*, rhombencephalon; *Rub*, nucleus ruber; *Spc*, spinal cord; *Th*, thalamus; *Vd*, descending trigeminal root nucleus; *VH*, ventral horn; *VIII*, octaval nucleus; *Vme*, trigeminal mesencephalic nucleus. *Data were taken from Fiebig, E., 1988. Connections of the corpus cerebelli in the thornback guitarfish,* Platyrhinoidis triseriata *(Elasmobranchii): a study with WGA-HRP and extracellular granule cell recording. J. Comp. Neurol. 268, 567–583; Puzdrowski, R.L., Gruber, S., 2009. Morphologic features of the cerebellum of the Atlantic stingray, and their possible evolutionary significance. Integr. Zool. 4, 110–122; Pose-Méndez, S., Candal, E., Mazan, S., Rodríguez-Moldes, I., 2016b. Morphogenesis of the cerebellum and cerebellum-related structures of* Scyliorhinus canicula: *insights on the ground pattern of the cerebellar ontogeny. Brain Struct. Funct. 221, 1691–1717.*

Acknowledgment

Supported by Ministerio de Economía y Competitividad-FEDER (BFU2014-58631)

References

Adachi, N., Takechi, M., Hirai, T., Kuratani, S., 2012. Development of the head and trunk mesoderm in the dogfish, *Scyliorhinus torazame*: II. Comparison of gene expression between the head mesoderm and somites with reference to the origin of the vertebrate head. Evol. Dev. 14, 257–276.

Adrio, F., Rodríguez-Moldes, I., Anadón, R., 2011. Distribution of glycine immunoreactivity in the brain of the Siberian sturgeon (*Acipenser baeri*): comparison with gamma-aminobutyric acid. J. Comp. Neurol. 519, 1115–1142.

Álvarez-Otero, R., Pérez, S.E., Rodríguez, M.A., Anadón, R., 1996. Organisation of the cerebellar nucleus of the dogfish, *Scyliorhinus canicula* L.: a light microscopic, immunocytochemical, and ultrastructural study. J. Comp. Neurol. 368, 487–502.

Álvarez-Otero, R., Regueira, S.D., Anadón, R., 1993. New structural aspects of the synaptic contacts on Purkinje cells in an elasmobranch cerebellum. J. Anat. 182, 13–21.

Anadón, R., Molist, P., Rodríguez-Moldes, I., López, J.M., Quintela, I., Cerviño, M.C., Barja, P., González, A., 2000. Distribution of choline acetyltransferase immunoreactivity in the brain of an elasmobranch, the lesser spotted dogfish (*Scyliorhinus canicula*). J. Comp. Neurol. 420, 139–170.

Anadón, R., Ferreiro-Galve, S., Sueiro, C., Graña, P., Carrera, I., Yáñez, J., Rodríguez-Moldes, I., 2009. Calretinin-immunoreactive systems in the cerebellum and cerebellum-related lateral-line medullary nuclei of an elasmobranch, *Scyliorhinus canicula*. J. Chem. Neuroanat. 37, 46–54.

Anadón, R., Rodríguez-Moldes, I., Adrio, F., 2013. Glycine-immunoreactive neurons in the brain of a shark (*Scyliorhinus canicula* L.). J. Comp. Neurol. 521, 3057–3082.

Anderson, S., Mione, M., Yun, K., Rubenstein, J.L., 1999. Differential origins of neocortical projection and local circuit neurons: role of Dlx genes in neocortical interneuronogenesis. Cerebellar Cortex 9, 646–654.

Aroca, P., Puelles, L., 2005. Postulated boundaries and differential fate in the developing rostral hindbrain. Brain Res. Rev. 49, 179–190.

Ballard, W.W., Mellinger, J., Lechenault, H., 1993. A series of normal stages for development of *Scyliorhinus canicula*, the lesser spotted dogfish (*Chondrichthyes: Scyliorhinidae*). J. Exp. Zool. 267, 318–336.

Bardet, S.M., Ferrán, J.L.E., Sánchez-Arrones, L., Puelles, L., 2010. Ontogenetic expression of sonic hedgehog in the chicken subpallium. Front. Neuroanat. 4, 28.

Bardet, S.M., Martínez-de-la-Torre, M., Northcutt, R.G., Rubenstein, J.L.R., Puelles, L., 2008. Conserved pattern of OTP-positive cells in the paraventricular nucleus and other hypothalamic sites of tetrapods. Brain Res. Bull. 75, 231–235.

Bell, C.C., 2002. Evolution of cerebellum-like structures. Brain Behav. Evol. 59, 312–326.

Bodznick, D., Montgomery, J.C., Carey, M., 1999. Adaptive mechanisms in the elasmobranch hindbrain. J. Exp. Biol. 202, 1357–1364.

Bratby, R., Sneyd, J., Montgomery, J., 2016. Computational architecture of the granular layer of cerebellum-like structures. Cerebellum. PMID: 26801651.

Butts, T., Green, M.J., Wingate, R.J.T., 2014. Development of the cerebellum: simple steps to make a little brain. Development 141, 4031–4041.

Carrera, I., Sueiro, C., Molist, P., Holstein, G.R., Martinelli, G.P., Rodríguez-Moldes, I., Anadón, R., 2006. GABAergic system of the pineal

organ of an elasmobranch (*Scyliorhinus canicula*): a developmental immunocytochemical study. Cell Tissue Res. 323, 273–281.

Carrera, I., Ferreiro-Galve, S., Sueiro, C., Anadón, R., Rodríguez-Moldes, I., 2008. Tangentially migrating GABAergic cells of subpallial origin invade massively the pallium in developing sharks. Brain Res. Bull. 75, 405–409.

Carrera, I., Molist, P., Anadón, R., Rodríguez-Moldes, I., 2008. Development of the serotoninergic system in the central nervous system of a shark, the lesser spotted dogfish *Scyliorhinus canicula*. J. Comp. Neurol. 511, 804–831.

Carrera, I., Anadón, R., Rodríguez-Moldes, I., 2012. Development of tyrosine hydroxylase-immunoreactive cell populations and fiber pathways in the brain of the dogfish *Scyliorhinus canicula*: new perspectives on the evolution of the vertebrate catecholaminergic system. J. Comp. Neurol. 520, 3574–3603.

Clark, D.A., Mitra, P.P., Wang, S.S.-H., 2001. Scalable architecture in mammalian brains. Nature 411, 189–193.

Collin, S.P., 2012. The neuroecology of cartilaginous fishes: sensory strategies for survival. Brain Behav. Evol. 80, 80–96.

Compagnucci, C., Debiais-Thibaud, M., Coolen, M., Fish, J., Griffin, J.N., Bertocchini, F., Minoux, M., Rijli, F.M., Borday-Birraux, V., Casane, D., Mazan, S., Depew, M.J., 2013. Pattern and polarity in the development and evolution of the gnathostome jaw: both conservation and heterotopy in the branchial arches of the shark, *Scyliorhinus canicula*. Dev. Biol. 377, 428–448.

Coolen, M., Sauka-Spengler, T., Nicolle, D., Le-Mentec, C., Lallemand, Y., Da Silva, C., Plouhinec, J.L., Robert, B., Wincker, P., Shi, D.L., Mazan, S., 2007. Molecular characterization of the early dogfish embryo: evolutionary implications on the mechanisms of early head and antero-posterior specification in jawed vertebrates. PLoS One 2, e374.

Coolen, M., Menuet, A., Chassoux, D., Compagnucci, C., Henry, S., Lévèque, L., Da Silva, C., Gavory, F., Samain, S., Wincker, P., Thermes, C., D'Aubenton-Carafa, Y., Rodriguez-Moldes, I., Naylor, G., Depew, M., Sourdaine, P., Mazan, S., 2009. The dogfish *Scyliorhinus canicula*, a reference in jawed vertebrates. In: Behringer, R.R., Johnson, A.D., Krumlauf, R.E. (Eds.), Emerging Model Organisms, A Laboratory Manual, vol. 1. Cold Spring Harbor Laboratory Press, Cold Spring Harbor, NY, pp. 431–446.

Debiais-Thibaud, M., Chiori, R., Enault, S., Oulion, S., Germon, I., Martinand-Mari, C., Casane, D., Borday-Birraux, V., 2015. Tooth and scalemorphogenesis in shark: an alternative process to the mammalian enamel knot system. Biomed. Cent. Evol. Biol. 15, 292.

Debiais-Thibaud, M., Metcalfe, C.J., Pollack, J., Germon, I., Ekker, M., Depew, M., Laurenti, P., Borday-Birraux, V., Casane, D., 2013. Heterogeneous conservation of *Dlx* paralog co-expression in jawed vertebrates. PLoS One 8, e68182.

Devor, A., 2000. Is the cerebellum like cerebellar-like structures? Brain Res. Rev. 34, 149–156.

Del Giacco, L., Pistocchi, A., Cotelli, F., Fortunato, A.E., Sordino, P., 2008. A peek inside the neurosecretory brain through Orthopedia lenses. Dev. Dyn. 237, 2295–2303.

Domínguez, L., González, A., Moreno, N., 2015. Patterns of hypothalamic regionalization in amphibians and reptiles: common traits revealed by a genoarchitectonic approach. Front. Neuroanat. 9, 3.

Domínguez, L., Morona, R., González, A., Moreno, N., 2013. Characterization of the hypothalamus of *Xenopus laevis* during development. I. The alar regions. J. Comp. Neurol. 521, 725–759.

Enault, S., Muñoz, D.N., Silva, W.A.F., Borday-Birraux, V., Bonade, M., Oulion, S., Ventéo, S., Marcellini, S., Debiais-Thibaud, M., 2015. Molecular footprinting of skeletal tissues in the catshark *Scyliorhinus canicula* and the clawed frog *Xenopus tropicalis* identifies conserved and derived feature of vertebrate calcification. Front. Genet. 6, 283.

Ferrando, S., Gallus, L., Gambardella, C., Ghigliotti, L., Ravera, S., Vallarino, M., Vacchi, M., Tagliafierro, G., 2010. Cell proliferation and apoptosis in the olfactory epithelium of the shark *Scyliorhinus canicula*. J. Chem. Neuroanat. 40, 293–300.

Ferreiro-Galve, S., Carrera, I., Candal, E., Villar-Cheda, B., Anadón, R., Mazan, S., Rodríguez-Moldes, I., 2008. The segmental organization of the developing shark brain based on neurochemical markers, with special attention to the prosencephalon. Brain Res. Bull. 75, 236–240.

Fiebig, E., 1988. Connections of the corpus cerebelli in the thornback guitarfish, *Platyrhinoidis triseriata* (Elasmobranchii): a study with WGA-HRP and extracellular granule cell recording. J. Comp. Neurol. 268, 567–583.

Flames, N., Pla, R., Gelman, D.M., Rubenstein, J.L., Puelles, L., Marín, O., 2007. Delineation of multiple subpallial progenitor domains by the combinatorial expression of transcriptional codes. J. Neurosci. 27, 9682–9695.

Gamse, J.T., Thisse, C., Thisse, B., Halpern, M.E., 2003. The parapineal mediates left-right asymmetry in the zebrafish diencephalon. Development 130, 1059–1068.

Ganz, J., Kaslin, J., Freudenreich, D., Machate, A., Geffarth, M., Brand, M., 2012. Subdivisions of the adult zebrafish subpallium by molecular marker analysis. J. Comp. Neurol. 520, 633–655.

Gilland, E., Baker, R., 2005. Evolutionary patterns of cranial nerve efferent nuclei in vertebrates. Brain Behav. Evol. 66, 234–254.

Gillis, J.A., Dahn, R.D., Shubin, N.H., 2009. Shared developmental mechanisms pattern the vertebrate gill arch and paired fin skeletons. Proc. Natl. Acad. Sci. 106, 5720–5724.

Gillis, J.A., Modrell, M.S., Baker, C.V.H., 2012. A timeline of pharyngeal endoskeletal condensation in the shark, *Scyliorhinus canicula*, and the paddlefish, *Polyodon spathula*. J. Appl. Ichthyol. 28, 341–345.

Godard, B.G., Mazan, S., 2013. Early patterning in a chondrichthyan model, the small spotted dogfish: towards the gnathostome ancestral state. J. Anat. 222, 56–66.

Godard, B.G., Coolen, M., Le Panse, S., Gombault, A., Ferreiro-Galve, S., Laguerre, L., Lagadec, R., Wincker, P., Poulain, J., Da Silva, C., Kuraku, S., Carre, W., Boutet, A., Mazan, S., 2014. Mechanisms of endoderm formation in a cartilaginous fish reveal ancestral and homoplastic traits in jawed vertebrates. Biol. Open 3, 1098–1107.

Green, M.J., Myat, A.M., Emmenegger, B.A., Wechsler-Reya, R.J., Wilson, L.J., Wingate, R.J., 2014. Independently specified Atoh1 domains define novel developmental compartments in rhombomere 1. Development 141, 389–398.

Iwaniuk, A.N., Hurd, P.L., 2005. The evolution of cerebrotypes in birds. Brain Behav. Evol. 65, 215–230.

Joly, J.S., Osório, J., Alunni, A., Auger, H., Kano, S., Rétaux, S., 2007. Windows of the brain: towards a developmental biology of circumventricular and other neurohemal organs. Semin. Cell Dev. Biol. 18, 512–524.

Kimmel, C.B., Ballard, W.W., Kimmel, S.R., Ullmann, B., Schilling, T.F., 1995. Stages of embryonic development of the zebrafish. Dev. Dyn. 203, 253–310.

Klingenberg, C.P., 1998. Heterochrony and allometry: the analysis of evolutionary change in ontogeny. Biol. Rev. Camb. Philos. Soc. 73, 79–123.

Kudo, L.C., Karsten, S.L., Chen, J., Levitt, P., Geschwind, D.H., 2007. Genetic analysis of anterior posterior expression gradients in the developing mammalian forebrain. Cerebellar Cortex 17, 2108–2122.

Kuratani, S., Horigome, N., 2000. Developmental morphology of branchiomeric nerves in a cat shark, *Scyliorhinus torazame*, with special reference to rhombomeres, cephalic mesoderm, and distribution patterns of cephalic crest cells. Zool. Sci. 17, 893–909.

Kuratani, S., Kuraku, S., Murakami, Y., 2002. Lamprey as an evo-devo model: lessons from comparative embryology and molecular phylogenetics. Genesis 34, 175–183.

Lagadec, R., Laguerre, L., Menuet, A., Amara, A., Rocancourt, C., Péricard, P., Godard, B.G., Rodicio, M.C., Rodriguez-Moldes, I., Mayeur, H., Rougemont, Q., Mazan, S., Boutet, A., 2015. The ancestral role of nodal signalling in breaking L/R symmetry in the vertebrate forebrain. Nat. Commun. 6, 6686.

Larsell, O., 1967. The Comparative Anatomy and Histology of the Cerebellum from Myxinoids through Birds. University of Minnesota Press, Minneapolis.

Larsen, C.W., Zelster, L.M., Lumsden, A., 2001. Boundary formation and compartition in the aviandiencephalon. J. Neurosci. 21, 4699–4711.

Lisney, T., Yopak, K., Montgomery, J., Collin, S., 2008. Variation in brain organization and cerebellar foliation in chondrichthyans: batoids. Brain Behav. Evol. 72, 262–282.

Mandado, M., Molist, P., Anadón, R., Yáñez, J., 2001. A DiI-tracing study of the neural connections of the pineal organ in two elasmobranchs (*Scyliorhinus canicula* and *Raja montagui*) suggests a pineal projection to the midbrain GnRH-immunoreactive nucleus. Cell Tissue Res. 303, 391–401.

Manoli, M., Driever, W., 2014. Nkx2.1 and Nkx2.4 genes function partially redundant during development of the zebrafish hypothalamus, preoptic region, and pallidum. Front. Neuroanat. 8, 1–16.

Manso, M.J., Anadón, R., 1991. The optic tectum of the dogfish *Scyliorhinus canicula* L.: a Golgi study. J. Comp. Neurol. 307, 335–349.

Manso, M.J., Anadón, R., 1991. Specialized presynaptic dendrites in the stratum cellulare externum of the optic tectum of an elasmobranch. *Scyliorhinus canicula* L. Neurosci. Lett. 129, 291–293.

Manso, M.J., Anadón, R., 1993. Distribution of types of neurons in the optic tectum of the small-spotted dogfish, *Scyliorhinus canicula* L. A Golgi study. Brain Behav. Evol. 41, 82–87.

Mastick, G.S., Easter, S.S., 1996. Initial organization of neurons and tracts in the embryonic mouse fore and midbrain. Dev. Biol. 173, 79–94.

Medina, L., 2008. Evolution and embryological development of the forebrain. In: Binder, M.D., Hirokawa, N., Windhorst, U. (Eds.), Encyclopedia of Neuroscience. Springer-Verlag, Berlin, pp. 1172–1192.

Medina, L., Abellán, A., 2009. Development and evolution of the pallium. Semin. Cell Dev. Biol. 20, 698–711.

Medina, L., Abellán, A., Vicario, A., Desfilis, E., 2014. Evolutionary and developmental contributions for understanding the organization of the basal ganglia. Brain Behav. Evol. 83, 112–125.

Meurling, P., Rodríguez, E.M., 1990. The paraventricular and posterior recess organs of elasmobranchs: a system of cerebrospinal fluid-contacting neurons containing immunoreactive serotonin and somatostatin. Cell Tissue Res. 259, 463–473.

Milán, F.J., Puelles, L., 2000. Patterns of calretinin, calbindin, and tyrosine hydroxylase expression are consistent with the prosomeric map of the frog diencephalon. J. Comp. Neurol. 419, 96–121.

Molist, P., Rodríguez-Moldes, I., Anadón, R., 1993. Organization of catecholaminergic systems in the hypothalamus of two elasmobranch species, *Raja undulata* and *Scyliorhinus canicula*. A histofluorescence and immunohistochemical study. Brain Behav. Evol. 41, 290–302.

Molist, P., Rodriguez-Moldes, I., Batten, T.F., Anadon, R., 1995. Distribution of calcitonin gene-related peptide-like immunoreactivity in the brain of the small-spotted dogfish, *Scyliorhinus canicula* L. J. Comp. Neurol. 352, 335–350.

Montgomery, J., Bodznick, D., 2010. Functional origins of the vertebrate cerebellum from a sensory processing antecedent. Curr. Zool. 56, 277–284.

Montgomery, J.C., Bodznick, D., Yopak, K.E., 2012. The cerebellum and cerebellum-like structures of cartilaginous fishes. Brain Behav. Evol. 80, 152–165.

Moreno, N., Domínguez, L., Morona, R., González, A., 2012. Subdivisions of the turtle *Pseudemys scripta* hypothalamus based on the expression of regulatory genes and neuronal markers. J. Comp. Neurol. 520, 453–478.

Moreno, N., González, A., Rétaux, S., 2008. Evidences for tangential migrations in *Xenopus* telencephalon: developmental patterns and cell tracking experiments. Dev. Neurobiol. 68, 504–520.

Moreno, N., González, A., Rétaux, S., 2009. Development and evolution of the subpallium. Semin. Cell Dev. Biol. 20, 735–743.

Moreno-Bravo, J.A., Pérez-Balaguer, A., Martínez-López, J.E., Aroca, P., Puelles, L., Martínez, S., Puelles, E., 2014. Role of Shh in the development of molecularly characterized tegmental nuclei in mouse rhombomere 1. Brain Struct. Funct. 219, 777–792.

Murakami, Y., Ogasawara, M., Sugahara, F., Hirano, S., Satoh, N., Kuratani, S., 2001. Identification and expression of the lamprey Pax6 gene: evolutionary origin of the segmented brain of vertebrates. Development 128, 3521–3531.

Murakami, Y., Uchida, K., Rijli, F.M., Kuratani, S., 2005. Evolution of the brain developmental plan: insights from agnathans. Dev. Biol. 280, 249–259.

Murakami, Y., Watanabe, A., 2009. Development of the central and peripheral nervous systems in the lamprey. Dev. Growth Differ. 51, 197–205.

Nakane, Y., Ikegami, K., Iigo, M., Ono, H., Takeda, K., Takahashi, D., Uesaka, M., Kimijima, M., Hashimoto, R., Arai, N., Suga, T., Kosuge, K., Abe, T., Maeda, R., Senga, T., Amiya, N., Azuma, T., Amano, M., Abe, H., Yamamoto, N., Yoshimura, T., 2013. The saccus vasculosus of fish is a sensor of seasonal changes in day length. Nat. Commun. 4, 2108.

New, J.G., 2001. Comparative neurobiology of the elasmobranch cerebellum: theme and variations on a sensorimotor interface. Environ. Biol. Fishes 60, 93–108.

Nieuwenhuys, R., 1967. Comparative anatomy of olfactory centres and tracts. In: Zotterman, Y. (Ed.), Progress in Brain Research. Elsevier, Amsterdam, pp. 1–64.

Northcutt, R.G., 1979. Retinofugal pathways in fetal and adult spiny dogfish, *Squalus acanthias*. Brain Res. 162, 219–230.

Northcutt, R.G., 1989. Brain variation and phylogenetic trends in elasmobranch fishes. J. Exp. Zool. Suppl. 2, 83–100.

Northcutt, R.G., 2002. Understanding vertebrate brain evolution. Integr. Comp. Biol. 42, 743–756.

O'Neill, P., McCole, R.B., Baker, C.V., 2007. A molecular analysis of neurogenic placode and cranial sensory ganglion development in the shark, *Scyliorhinus canicula*. Dev. Biol. 304, 156–181.

Osório, J., Mazan, S., Rétaux, S., 2005. Organisation of the lamprey (*Lampetra fluviatilis*) embryonic brain: insights from LIM-homeodomain, Pax and hedgehog genes. Dev. Biol. 288, 100–112.

Oulion, S., Borday-Birraux, V., Deviais-Thibaud, M., Mazan, S., Laurenti, P., Casane, D., 2011. Evolution of repeated structures along the body axis of jawed vertebrates, insights from the *Scyliorhinus canicula* Hox code. Evol. Dev. 13, 247–259.

Pasini, A., Wilkinson, D.G., 2002. Stabilizing the regionalisation of the developing vertebrate central nervous system. Bioessays 24, 427–438.

Pose-Méndez, S., Candal, E., Adrio, F., Rodríguez-Moldes, I., 2014. Development of the cerebellar afferent system in the shark *Scyliorhinus canicula*: insights into the basal organization of precerebellar nuclei in gnathostomes. J. Comp. Neurol. 522, 131–168.

Pose-Méndez, S., Candal, E., Mazan, S., Rodríguez-Moldes, I., 2016. Genoarchitecture of the rostral hindbrain of a shark: basis for understanding the emergence of the cerebellum at the agnathan-gnathostome transition. Brain Struct. Funct. 221, 1321–1335.

Pose-Méndez, S., Candal, E., Mazan, S., Rodríguez-Moldes, I., 2016. Morphogenesis of the cerebellum and cerebellum-related structures of *Scyliorhinus canicula*: insights on the ground pattern of the cerebellar ontogeny. Brain Struct. Funct. 221, 1691–1717.

Puelles, L., 2009. Forebrain development: prosomere model. In: Lemke, G. (Ed.), Developmental Neurobiology. Academic Press, London, pp. 315–319.

Puelles, L., Ferrán, J.L., 2012. Concept of neural genoarchitecture and its genomic fundament. Front. Neuroanat. 6, 47.

Puelles, L., Rubenstein, J.L.R., 2003. Forebrain gene expression domains and the evolving prosomeric model. Trends Neurosci. 26, 469–476.

Puelles, L., Rubenstein, J.L.R., 2015. A new scenario of hypothalamic organization: rationale of new hypotheses introduced in the updated prosomeric model. Front. Neuroanat. 9, 27.

Puelles, L., Martínez, S., Martínez-de-la-Torre, M., Rubenstein, J., 2012. Hypothalamus. In: Watson, C., Paxinos, G., Puelles, L. (Eds.), The Mouse Nervous System. Elsevier Academic Press, San Diego, pp. 221–312.

Puzdrowski, R.L., Gruber, S., 2009. Morphologic features of the cerebellum of the Atlantic stingray, and their possible evolutionary significance. Integr. Zool. 4, 110–122.

Quintana-Urzainqui, I., Rodríguez-Moldes, I., Mazan, S., Candal, E., 2015. Tangential migratory pathways of subpallial origin in the embryonic telencephalon of sharks: evolutionary implications. Brain Struct. Funct. 220, 2905–2926.

Quintana-Urzainqui, I., Sueiro, C., Carrera, I., Ferreiro-Galve, S., Santos-Durán, G., Pose-Méndez, S., Mazan, S., Candal, E., Rodríguez-Moldes, I., 2012. Contributions of developmental studies in the dogfish *Scyliorhinus canicula* to the brain anatomy of elasmobranchs: insights on the basal ganglia. Brain Behav. Evol. 80, 127–141.

Repérant, J., Miceli, D., Rio, J.P., Peyrichoux, J., Pierre, J., Kirpitchnikova, E., 1986. The anatomical organization of retinal projections in the shark *Scyliorhinus canicula* with special reference to the evolution of the selachian primary visual system. Brain Res. 396, 227–248.

Richardson, M.K., 2012. A phylotypic stage for all animals? Dev. Cell 22, 903–904.

Rodríguez-Moldes, I., 2009. A developmental approach to forebrain organization in elasmobranchs: new perspectives on the regionalization of the telencephalon. Brain Behav. Evol. 74, 20–29.

Rodríguez-Moldes, I., 2011. Functional morphology of the brains of cartilaginous fishes. In: Farrell, A. (Ed.), Encyclopedia of Fish Physiology: From Genome to Environment, vol. 1. Academic Press, San Diego, pp. 26–36.

Rodríguez-Moldes, I., Anadón, R., 1987. Aminergic neurons in the hypothalamus of the dogfish, *Scyliorhinus canicula* L. (Elasmobranch). A histofluorescence study. J. Hirnforschung 28, 685–693.

Rodríguez-Moldes, I., Carrera, I., Pose-Méndez, S., Quintana-Urzainqui, I., Candal, E., Anadón, R., Mazan, S., Ferreiro-Galve, S., 2011. Regionalization of the shark hindbrain: a survey of an ancestral organization. Front. Neuroanat. 5, 1–14.

Rodríguez-Moldes, I., Ferreiro-Galve, S., Carrera, I., Sueiro, C., Candal, E., Mazan, S., Anadón, R., 2008. Development of the cerebellar body in sharks: spatiotemporal relations of *Pax6*-expression, cell proliferation and differentiation. Neurosci. Lett. 432, 105–110.

Rodríguez-Moldes, I., Manso, M.J., Becerra, M., Molist, P., Anadón, R., 1993. Distribution of substance P-like immunoreactivity in the brain of the elasmobranch *Scyliorhinus canicula*. J. Comp. Neurol. 335, 228–244.

Ross, L.S., Parrett, T., Easter, S.S., 1992. Axonogenesis and morphogenesis in the embryonic zebrafish brain. J. Neurosci. 71, 467–482.

Santos-Durán, G.N., Menuet, A., Lagadec, R., Mayeur, H., Ferreiro-Galve, S., Mazan, S., Rodríguez-Moldes, I., Candal, E., 2015. Prosomeric organization of the hypothalamus in an elasmobranch, the catshark *Scyliorhinus canicula*. Front. Neuroanat. 9, 37.

Sauer, F.C., 1935. Mitosis in the neural tube. J. Comp. Neurol. 62, 377–405.

Sauka-Spengler, T., Plouhinec, J.-L., Mazan, S., 2004. Gastrulation in a chondrichthyan, the dogfish *Scyliorhinus canicula*. In: Stern, C.D. (Ed.), Gastrulation from Cells to Embryo. Cold Spring Harbor Laboratory Press, New York, pp. 151—156 (Chapter 11).

Schmidt, A.W., Bodznick, D., 1987. Afferent and efferent connections of the vestibulolateral cerebellum of the little skate, *Raja erinacea*. Brain Behav. Evol. 30, 282—302.

Schubert, F.R., Lumsden, A., 2005. Transcriptional control of early tract formation in the embryonic chick midbrain. Development 132, 1785—1793.

Smeets, W.J.A.J., Nieuwenhuys, R., Roberts, B.L., 1983. The Central Nervous System of Cartilaginous Fishes. Structure and Functional Correlations. Springer-Verlag, Berlin.

Smeets, W., 1998. Cartilaginous fishes. In: Nieuwenhuys, R., Donkelaar, H.J., Nicholson, C. (Eds.), The Central Nervous System of Vertebrates. Springer-Verlag, Berlin, pp. 551—564.

Striedter, G.F., 2005. Principles of Brain Evolution. Sinauer Associates, Inc, Sunderland, MA.

Striedter, G.F., Northcutt, R.G., 2006. Head size constrains forebrain development and evolution in ray-finned fishes. Evol. Dev. 8, 215—222.

Sueiro, C., Carrera, I., Ferreiro, S., Molist, P., Adrio, F., Anadón, R., Rodríguez-Moldes, I., 2007. New insights on saccus vasculosus evolution: a developmental and immunohistochemical study in elasmobranchs. Brain Behav. Evol. 70, 187—204.

Sugahara, F., Aota, S., Kuraku, S., Murakami, Y., Takio-Ogawa, Y., Hirano, S., Kuratani, S., 2011. Involvement of Hedgehog and FGF signalling in the lamprey telencephalon: evolution of regionalization and dorsoventral patterning of the vertebrate forebrain. Development 138, 1217—1226.

Sugahara, F., Murakami, Y., Adachi, N., Kuratani, S., 2013. Evolution of the regionalization and patterning of the vertebrate telencephalon: what can we learn from cyclostomes? Curr. Opin. Genet. Dev. 23, 475—483.

Sugahara, F., Pascual-Anaya, J., Oisi, Y., Kuraku, S., Aota, S., Adachi, N., Takagi, W., Hirai, T., Sato, N., Murakami, Y., Kuratani, S., 2016. Evidence from cyclostomes for complex regionalization of the ancestral vertebrate brain. Nature 531, 97—100.

Szabó, N.-E., Zhao, T., Cankaya, M., Theil, T., Zhou, X., Álvarez-Bolado, G., 2009. Role of neuroepithelial Sonic hedgehog in hypothalamic patterning. J. Neurosci. 29, 6989—7002.

van de Kamer, J.C., Shuurmans, A.J., 1953. Development and structure of the saccus vasculosus of *Scylliorhinus caniculus* (L.). J. Embryol. Exp. Morphol. 1, 85—96.

Venkatesh, B., Lee, A.P., Ravi, V., Maurya, A., Lian, M.M., Swann, J.B., Ohta, Y., Flajnik, M.F., Sutoh, Y., Kasahara, M., Hoon, S., Gangu, V., Roy, S.W., Irimia, M., Korzh, V., Kondrychyn, I., Lim, Z.W., Tay, B.H., Tohari, S., Kong, K.W., Ho, S., Lorente-Galdos, B., Quilez, J., Marques-Bonet, T., Raney, B.J., Ingham, P.W., Tay, A., Hillier, L.W., Minx, P., Boehm, T., Wilson, R.K., Brenner, S., Warren, W.C., 2014. Elephant shark genome provides unique insights into gnathostome evolution. Nature 505, 174—179.

Vernier, P., Wullimann, M., 2008. Evolution of the posterior tuberculum and preglomerular nuclear complex. In: Binder, M.D., Hirokawa, N., Windhorst, U. (Eds.), Encyclopedia of Neuroscience. Springer-Verlag, Berlin, pp. 1404—1413.

Villar-Cerviño, V., Barreiro-Iglesias, A., Anadón, R., Rodicio, M.C., 2008. Distribution of glycine immunoreactivity in the brain of adult sea lamprey (*Petromyzon marinus*). Comparison with gamma-aminobutyric acid. J. Comp. Neurol. 507, 1441—1463.

Wagner, A., 2008. Gene duplications, robustness and evolutionary innovations. Bioessays 30, 367—373.

Watanabe, H., Murakami, F., 2009. Real time analysis of pontine neurons during initial stages of nucleogenesis. Neurosci. Res. 64, 20—29.

Wichterle, H., Turnbull, D.H., Nery, S., Fishell, G., Alvarez-Buylla, A., 2001. In utero fate mapping reveals distinct migratory pathways and fates of neurons born in the mammalian basal forebrain. Development 128, 3759—3771.

Wright, D.E., Demski, L.S., 1991. Gonadotropin hormone-releasinghormone (GnRH) immunoreactivity in the mesencephalon of sharks and rays. J. Comp. Neurol. 307, 49—56.

Wueringer, B.E., 2012. Electroreception in elasmobranchs: sawfish as a case study. Brain Behav. Evol. 80, 97—107.

Wullimann, M.F., Mueller, T., 2004. Teleostean and mammalian forebrains contrasted: evidence from genes to behavior. J. Comp. Neurol. 475, 143—162.

Wullimann, M.F., Mueller, T., Distel, M., Babaryka, A., Grothe, B., Köster, R.W., 2011. The long adventurous journey of rhombic lip cells in jawed vertebrates: a comparative developmental analysis. Front. Neuroanat. 5, 27.

Yáñez, J., Folgueira, M., Köhler, E., Martínez, C., Anadón, R., 2011. Connections of the terminal nerve and the olfactory system in two galeomorph sharks: an experimental study using a carbocyanine dye. J. Comp. Neurol. 519, 3202—3217.

Yopak, K.E., 2012. Neuroecology of cartilaginous fishes: the functional implications of brain scaling. J. Fish Biol. 80, 1968—2023.

Yopak, K.E., Frank, L.R., 2007. Variation in cerebellar foliation in cartilaginous fishes: ecological and behavioral considerations. Brain Behav. Evol. 70, 210—225.

Yopak, K.E., Montgomery, J.C., 2008. Brain organization and specialization in deep-seachondrichthyans. Brain Behav. Evol. 71, 287—304.

Yopak, K.E., Lisney, T.J., Collin, S.P., Montgomery, J.C., 2007. Variation in brain organization and cerebellar foliation in chondrichthyans: sharks and holocephalans. Brain Behav. Evol. 69, 280—300.

Yopak, K.E., Lisney, T.J., Darlington, R.B., Collin, S.P., Montgomery, J.C., Finlay, B.L., 2010. A conserved pattern of brain scaling from sharks to primates. Proc. Natl. Acad. Sci. U.S.A. 107, 12946—12951.

Further Reading

Beccari, L., Marco-Ferreres, R., Bovolenta, P., 2013. The logic of gene regulatory networks in early vertebrate forebrain patterning. Mech. Dev. 130, 95—111.

Butler, A.B., 2008. Brain evolution. In: Binder, M.D., Hirokawa, N., Windhorst, U. (Eds.), Encyclopedia of Neuroscience. Springer-Verlag, Berlin, pp. 462—472.

Butler, A.B., Hodos, W. (Eds.), 1996. Comparative Vertebrate Neuroanatomy: Evolution and Adaptation. J. W. Sons, Hoboken, NJ.

Hidalgo-Sánchez, M., Millet, S., Bloch-Gallego, E., Alvarado-Mallart, R.M., 2005. Specification of the meso-isthmo-cerebellar region: the *Otx2/Gbx2* boundary. Brain Res. Rev. 49, 134—149.

Hoffmann, M.H., 1999. Nervous system. In: Hamlett, W.C. (Ed.), Sharks, Skates and Rays. The Biology of Elasmobranch Fishes. The Johns Hopkins University Press, Baltimore and London.

Medina, L., Bupesh, M., Abellán, A., 2011. Contribution of genoarchitecture to understanding forebrain evolution and development, with particular emphasis on the amygdala. Brain Behav. Evol. 78, 216—236.

Moreno, N., González, A., 2011. The non-evaginated secondary prosencephalon of vertebrates. Front. Neuroanat. 5, 12.

Müller, G.B., 2007. Evo-devo: extending the evolutionary synthesis. Nat. Rev. Genet. 8, 943—949.

Nieuwenhuys, R., 2011. The structural, functional and molecular organization of the brainstem. Front. Neuroanat. 5, 33.

Puelles, L., Medina, L., 2002. Field homology as a way to reconcile genetic and developmental variability with adult homology. Brain Res. Bull. 57, 243–255.

Quintana-Urzainqui, I., Anadón, R., Candal, E., Rodríguez-Moldes, I., 2014. Development of the terminal nerve system in the shark *Scyliorhinus canicula*. Brain Behav. Evol. 84, 277–287.

Rétaux, S., Kano, S., 2010. Midline signaling and evolution of the forebrain in chordates: a focus on the lamprey Hedgehog case. Integr. Comp. Biol. 50, 98–109.

Ryu, S., Mahler, J., Acampora, D., Holzschuh, J., Erhardt, S., Omodei, D., Simeone, A., Driever, W., 2007. Orthopedia homeodomain protein is essential for diencephalic dopaminergic neuron development. Curr. Biol. 17, 873–880.

7

The Organization of the Central Nervous System of Amphibians

A. González, J.M. López, R. Morona, N. Moreno

University Complutense of Madrid, Madrid, Spain

Abbreviations

A Anterior nucleus of the thalamus
ABB Alar-basal boundary
ac Anterior commissure
Acc Nucleus accumbens
Ad/Av Anterodorsal/anteroventral tegmental nuclei
all Anterior lateral line
AMY Amygdala
AOB Accessory olfactory bulb
aot Accessory olfactory tract
ap Alar plate
BDA Biotinylated dextran amine
bp Basal plate
BST Bed nucleus of the stria terminalis
C Central nucleus of the thalamus
CB Calbindin
Cb Cerebellum
cc Central canal
CeA Central amygdala
Cg Rhombencephalic central gray
ChAT Choline acetyltransferase
CR Calretinin
cTh Caudal thalamus
dh Dorsal horn of the spinal cord
Dien Diencephalon
DMN Dorsal medullary nucleus
DP Dorsal pallium
DPA Dorsal pallidum
E Epiphysis
eaf External arcuate fibers
EW Edinger Westphal nucleus
fr Fasciculus retroflexus
GT Griseum tectale
H Hypothalamus
Hb Habenula
hc Habenular commissure
Hd Dorsal habenular nucleus
hp1,2 Hypothalamic prosomeres 1 and 2
HRP Horseradish peroxidase
Hv Ventral habenular nucleus
iaf Internal arcuate fibers
IC Intercalated cell column

IML Intermediolateral cell column
Is Isthmic nucleus
LA Lateral amygdala
La Lateral anterior nucleus of the thalamus
Lc Locus coeruleus
LDT Laterodorsal tegmental nucleus
lfb Lateral forebrain bundle
LGE Lateral ganglionic eminence
lln Lateral line nucleus
lot Lateral olfactory tract
LP Lateral pallium
Ls Lateral septum
M Mamillary region
MeA Medial amygdala
Mes Mesencephalon
MesV Mesencephalic trigeminal nucleus
mfb Medial forebrain bundle
MGE Medial ganglionic eminence
MOB Main olfactory bulb
MP Medial pallium
Ms Medial septum
NADPHd NADPH-diaphorase
nIV Trochlear nerve
Nsol Nucleus of the solitary tract
nV Trigeminal nerve
nVIII Octaval nerve
oc Optic chiasm
OT Optic tectum
ot Optic tract
P.w Pleurodeles waltl
p1−3 Prosomeres 1−3
p3b Basal part of p3
PA Pallidum
Pa Paraventricular area
PB Parabrachial nucleus
Pdi Posterodorsal tegmental nucleus, isthmic part
PO Preoptic area
PPa Peduncular part of the paraventricular area
PPN Pedunculopontine tegmental nucleus
PSPa Peduncular part of the subparaventricular area
PT Pretectum
PTh Prethalamus
PThE Prethalamic eminence

r0–8 Rhombomeres 0–8
RF Reticular formation
Rhom Rhombencephalon
Ri Inferior reticular nucleus
Rm Median reticular nucleus
RM Retromamillary region
rTh Rostral thalamus
RTu Retrotuberal hypothalamic area
SC Suprachiasmatic nucleus
sgr Stratum granulare (cerebellum)
Smn Somatomotor neurons of spinal cord
smol Stratum moleculare (cerebellum)
SN Substantia nigra
sol Solitary tract
sP Stratum Purkinje (cerebellum)
SPa Subparaventricular area
spc Spinal cord
st Stria terminalis
Str Striatum
Sv Ventral septum
T.c Typhlonectes compressicauda
Tc Commissural nucleus of the torus semicircularis
Tegm Tegmentum
Tel Telencephalon
Th Thalamus
TH Tyrosine hydroxylase
tht Thalamotelencephalic tract
Tl Laminar nucleus of the torus semicircularis
Tmg Magnocellular nucleus of the torus semicircularis
Tor Torus semicircularis
Tp Principal nucleus of the torus semicircularis
TPa Terminal part of the paraventricular area
TRDA Texas red dextran amine
TSPa Terminal part of the subparaventricular area
Tu Tuberal hypothalamus
v Ventricle
Vc Ventral prethalamic nucleus, caudal part
Vd Descending trigeminal tract
VP Ventral pallium
VPA Ventral pallidum
Vr Ventral prethalamic nucleus, rostral part
VTA/SN Ventral tegmental area/substantia nigra complex
X.l Xenopus laevis
Zip Periventricular nucleus of the zona incerta
III Oculomotor nucleus
IV Trochlear nucleus
Vm Trigeminal motor nucleus
VIa Accessory abducens nucleus
VIm Main abducens nucleus
VIIm Facial motor nucleus
VIIId Dorsal part of octaval area
VIIIv Ventral part of octaval area
IXm Glossopharyngeal motor nucleus
Xm Vagal motor nucleus
XI Nucleus of the accessory nerve
XII Hypoglossal nucleus

7.1 Living Amphibians and Phylogenetic Relationships

Extant amphibians (or Lissamphibia) are one of the most diverse radiations of terrestrial vertebrates. They constitute a class of anamniote vertebrates divided into three orders with markedly distinct morphologies and life styles: *Anura* (or *Salentia*), frogs and toads that have short, tailless bodies with long, powerful hind limbs mainly adapted for jumping locomotion; *Caudata* (or *Urodela*), salamanders and newts that have retained a long, muscular body and tail, and locomote by undulating the body; and *Gymnophiona* (or *Apoda*), caecilians that display several features that facilitate their burrowing mode of life, including wormlike elongated bodies, degenerate eyes, external segmentation, a lack of limbs, and a solidly ossified skull. The AmphibiaWeb database (see Relevant Website) currently lists 7501 amphibian species (January 14, 2016), of which 6613 are Anura (anurans), 682 are Caudata (urodeles), and 206 are Gymnophiona (gymnophionans). Both the fossil record and molecular clock estimates suggest the origin of modern amphibians occurred in the late Paleozoic (Carboniferous) period, the oldest fossils being early Permian. However, the origin and phylogenetic relationships of amphibians (particularly among the three orders) have long been hotly debated. Early studies of mitochondrial DNA and nuclear ribosomal DNA sequences supported a close relationship between salamanders and caecilians (a group called Procera) (Feller and Hedges, 1998; Zhang et al., 2003). However, more recent analyses based on large databases of both nuclear and mitochondrial genes, or a combination of both, established frogs and salamanders as sister groups, together forming the Batrachia clade, and caecilians being more distantly relatives. Furthermore, these studies support the monophyly of living amphibians with respect to the other living groups of tetrapods, ie, the amniotes comprising reptiles, birds, and mammals (San Mauro, 2010; San Mauro et al., 2014; Pyron and Wiens, 2011).

7.2 Amphibian Brains, General Features, and Methods of Study

Traditionally amphibians have played a prominent role as experimental animals in brain research for comparative studies. This was prompted by the exceptional position of these vertebrates as the only group of tetrapods that are anamniotes and show striking changes during postembryonic development, known as metamorphosis, that prepare an aquatic larva for a terrestrial existence. They represent a key model for understanding the anamniote/amniote transition, as they share features with other tetrapods (amniotes) and also with fishes (anamniotes). In the context of this transition, the colonization of land by tetrapod ancestors is presumably one of the evolutionary events that entailed significant neural changes.

Early studies in comparative neuroanatomy selected the brain of the tiger salamander, *Ambystoma tigrinum*, as a model in which the basic traits of brain organization of amniotes could be already recognized (see Herrick, 1948). After this classic work, most data have been gathered for anurans, which are usually regarded as being more complex than salamanders, and many of their brain systems have now been thoroughly studied (see Llinás and Precht, 1976). In contrast, the brain of gymnophionans has received much less attention. This is probably due to the difficulty of obtaining these animals, which are restricted to tropical and subtropical regions and have a secretive fossorial lifestyle. However, any study that attempts to clarify the amphibian condition and assess primitive and derived features of the amphibian brain must include results obtained in species from all three orders.

Amphibians have relatively simple brains but the simplified arrangement of neurons and fiber systems in their brains seems to contain all major structural features of amniote brains. All main regions of the vertebrate brain can be easily recognized in amphibians, although some show restricted development and others have been expanded as evolutionary specializations. Amphibian brains are elongated with only minor flexures, except for gymnophionans, which show a sharp bending of the longitudinal axis in the brain stem (Fig. 7.1).

In the forebrain, the telencephalic hemispheres are mainly tubular with wide lateral ventricles; they are particularly large in gymnophionans, extending caudally and covering part of the brain stem. The amphibian telencephalon exhibits well-developed main and accessory olfactory bulbs at the rostral pole. The diencephalon constitutes the caudal part of the forebrain and appears as a narrow cylinder in which the third ventricle is covered dorsally by a choroid plexus and conspicuous habenular ganglia. The midbrain is especially large in anurans, relative to the other amphibian groups, due to pronounced expansion of the optic tectum (OT) and torus semicircularis, which are homologous to the mammalian superior and inferior colliculus, respectively. In contrast, the midbrain is strikingly small in gymnophionans, probably as a secondary simplification that reflects their being almost blind. The principal areas of the hindbrain—the cerebellum and the medulla—are relatively conserved across amphibian groups in their size and general appearance. The cerebellum covers the rostral part of the rhombencephalon (medulla) and is curiously small in amphibians; it is not clear whether this is a primitive trait or the result of a secondary reduction. In gymnophionans the cerebellum was reported to be absent or unrecognizable (Kuhlenbeck, 1922). The medulla lies caudally and is thicker in anurans than in urodeles and gymnophionans, in which the fourth ventricle is strikingly wide and a large choroid plexus covers most of the dorsal rhombencephalon. The usual motor and sensory areas characteristic of all tetrapods are recognized in amphibians. Of interest, the lateral line system is present in amphibians during development and is retained in most adult amphibians with an aquatic lifestyle. The specialized sensory area related to the lateral line system is well developed in the dorsal part of the rhombencephalon throughout its length. The spinal cord extends along the whole length of the vertebral canal in urodeles and gymnophionans, whereas in anurans it extends no more than halfway down the trunk. Distinct brachial (cervical) and lumbar enlargements, which represent the regions of innervation of the extremities, are found in both anurans and urodeles.

Early studies on the microscopic organization on the amphibian brain—based on classical histological techniques such as Nissl, Weigert, reduced silver, and Golgi preparations—noted that the degree of cell migration

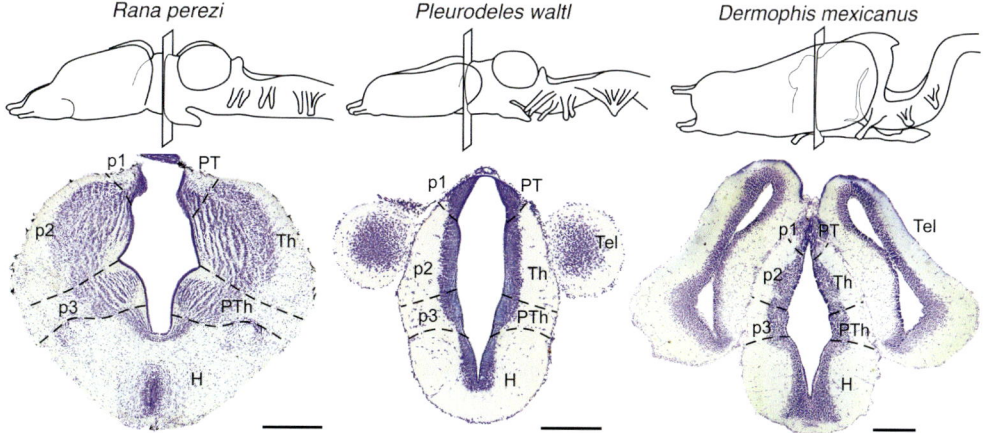

FIGURE 7.1   Schematic drawings of the brains of representative species of the three orders of amphibians: *Rana perezi*, anuran; *Pleurodeles waltl*, urodele; *Dermophis mexicanus*, gymnophionan. Below, transverse sections from each species at the levels indicate in the schemes, showing in Nissl staining the low degree of cell migration from the ventricle. Note the considerable higher cell migration and formation of nuclei in the case of anurans. Scale bars = 500 μm.

from the periventricular zone is reduced and that most neurons are crowded close to the ventricle in a dense central gray layer; in contrast, the outer white zone contains few well-defined fiber tracts. However, migrated nuclei and neuronal lamination are found in several regions of anuran brains, including the thalamus. By contrast, the identification of individual thalamic nuclei in urodeles and gymnophionans is complicated by the restricted radial migration of neurons (Fig. 7.1). Nonetheless, careful studies by Roth and coworkers (see Dicke and Roth, 2007) showed that even the salamander brain possesses virtually all the anatomical and functional properties found in tetrapods. They proposed that the urodele (and gymnophionan) central nervous system is characterized by a secondary simplification, which gives the impression that the brains are more primitive than their phylogenetic position as tetrapods implies.

With the achievement of new technical approaches to the study of the amphibian brain (mainly tract-tracing techniques based on anterograde and retrograde transport of macromolecules), it could be established that the poorly differentiated cytoarchitecture corresponds in most areas with a pattern of organization in which outer white zone contains well-defined tracts (Fig. 7.2A) and the periventricularly located neurons extend their dendrites outward, where they spread widely in the overlying fiber zone (Fig. 7.2B). Furthermore, the use of combined techniques based on immunohistochemistry and axonal tract-tracing revealed that these brains, despite their apparently undifferentiated organization, display distinct groups of cells projecting to well-defined targets and demonstrated that the intrinsic organization within the nonsegregated cell layer shows many common features shared by all tetrapods. Our knowledge about the precise neuroanatomy of most brain regions increased enormously when detailed analysis of their chemoarchitecture became available. Thus, chemically distinct subpopulations of neurons were identified within the periventricular layer (Fig. 7.2C), new neuronal groups were revealed on the basis of the neurotransmitter they use, and complex neurochemical systems (such as catecholaminergic, serotonergic, cholinergic, etc.) led to the recognition that the brain of amphibians is more complex than previously thought and shares all major basic features with those of amniotes.

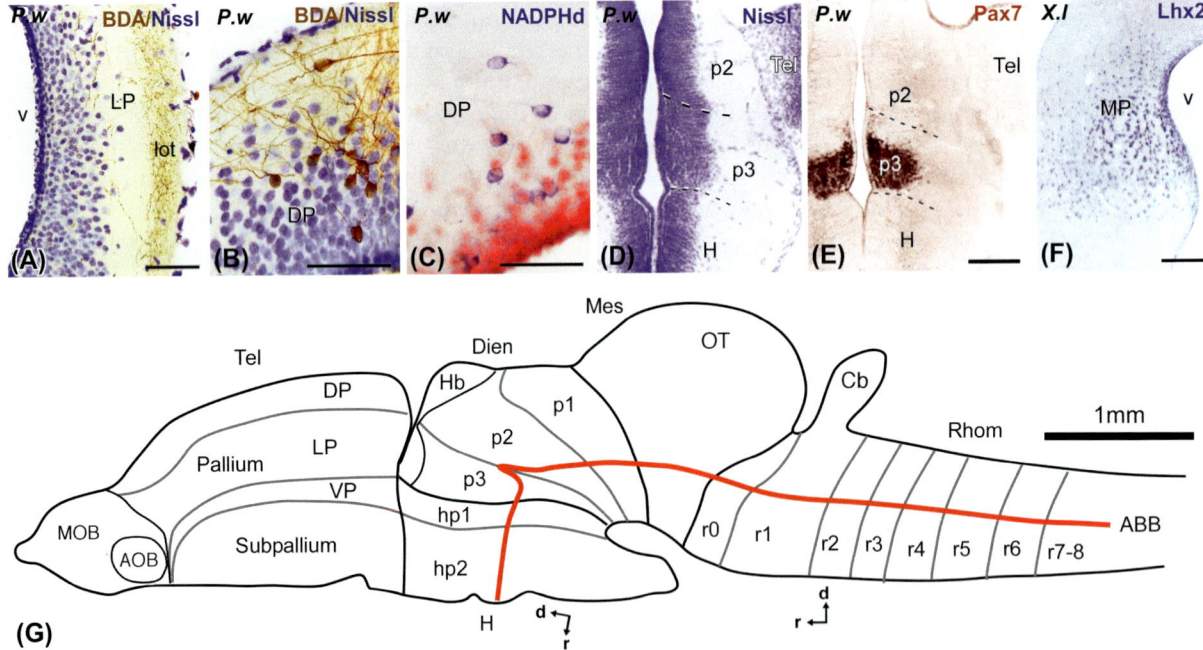

FIGURE 7.2   A series of photomicrographs of transverse sections through the brain of the urodele *Pleurodeles waltl* (A—E) and the anuran *Xenopus laevis* (F), stained with various techniques. (A) Nissl staining with cresyl violet in a detail of the lateral pallium together with BDA-labeled fibers from the olfactory bulb in the lateral olfactory tract, located in the peripheral fiber layer. (B) Dorsal pallium in a section in which some neurons were retrogradely labeled with BDA from the medial pallium, showing the primary morphology of their cell bodies in the periventricular cell layer and their dendrites extending toward the superficial fiber layer. (C) Detail of putative nitrergic (NADPHd-positive) cells in the dorsal pallium. (D, E) Photomicrographs of the same transverse section showing Nissl staining (D) and specific Pax7 (E) immunolabeling; the labeling for Pax7 allows specific cell groups to be recognized within the otherwise homogeneous periventricular cell layer of the prethalamus (p3). (F) Lhx2-expressing cells in the medial pallium of the adult *Xenopus* telencephalon. *(D and E: modified from Joven, A., Morona, R., Gonzalez, A., Moreno, N., 2013. Expression patterns of Pax6 and Pax7 in the adult brain of a urodele amphibian, Pleurodeles waltl. J. Comp. Neurol. 521, 2088—2124.) (G) Diagram of a lateral view of an anuran brain, showing its main subdivisions according to the current model of brain segmentation. For abbreviations, see list. Scale bars = 100 μm (A—C, F), 200 μm (D, E).*

A new impetus to the study of amphibian brains was provided by techniques that revealed regions where particular sets of genes are expressed during development, which led to the concept of *genoarchitecture* (Puelles and Ferran, 2012). Moreover, much information was gathered with the localization of transcription factors that are expressed in particular regions during development, but continue to be expressed in the adult, displaying regionally restricted expression domains with their borders corresponding precisely with morphological landmarks. Thus, in the brain, distinct neuronal populations can be identified within the periventricular layer where cell migration is very limited (Fig. 7.2D and E). More generally, particular groups of cells can be identified by the genes they express, which helps the comparison with counterpart cell groups in other vertebrates (Fig. 7.2F).

In recent years, neuroanatomical studies have usually interpreted the results obtained by different techniques according to the current neuromeric (segmental) model of brain organization (Fig. 7.2G; Puelles and Rubenstein, 1993, 2003; Marín and Puelles, 1995; Puelles et al., 1996; Fritzsch, 1998; Cambronero and Puelles, 2000; Díaz et al., 2000; Straka et al., 2006). This paradigm for comparing brain regions, based on transversal and longitudinal subdivisions of the brain, serves to clarify, in many cases, the principal criterion for the determination of homology, ie, similarity in position (Nieuwenhuys, 1998). This model is frequently used in comparative studies because common patterns of brain regionalization and the number and organization of the brain segments are constant features in all vertebrates. It should be emphasized that, following this approach, regions that express similar genes during development and share a topological location in different vertebrates are probably homologous (Fritzsch, 1998; Puelles, 2001; Puelles and Rubenstein, 2003). Furthermore, the distribution of neurotransmitters (or neurotransmitter-related molecules) in comparable locations and shared hodological characteristics can frequently support homologies (Nieuwenhuys, 1998; Fritzsch and Glover, 2007).

As the anterior neural tube closes, the embryonic brain subdivides into three primary vesicles: the rhombencephalon, the mesencephalon, and the prosencephalon, from caudal to rostral. The rhombencephalon consists of seven to eight rhombomeres (r1—r8) and the mesencephalon of one mesomere (a small second segment has been proposed in the caudal midbrain of mice; Puelles et al., 2012). The primary prosencephalon subdivides into two major components, the caudal diencephalon and the rostral secondary prosencephalon. The secondary prosencephalon gives rise to the hypothalamus ventrally, the eye vesicles dorsolaterally, and the telencephalic vesicles and preoptic area (PO) dorsally (Puelles, 2013). It should be pointed out that the model considers the cephalic flexure that bends the longitudinal brain axis and, therefore, the hypothalamus is topologically rostral (not ventral) to the diencephalon, and the telencephalon develops from the dorsal (alar) part of the hypothalamus (Fig. 7.2G).

Previous compilations of the accumulated knowledge on the organization of the brain of amphibians have concentrated mainly on classical techniques available at the time (Herrick, 1948; Linás and Precht, 1976). Subsequent surveys have summarized elegantly the abundant results obtained primarily with modern cytoarchitectural and experimental tract-tracing techniques (ten Donkelaar, 1998a,b; Dicke and Roth, 2007). Therefore, and due to space constraints, in the present chapter we will deal primarily with the regional anatomy of the amphibian brain, focusing mainly on the chemoarchitecture and genoarchitecture, which serve to define anatomical subdivisions. The combination of developmental transcription factor expression patterns and immunohistochemistry is a powerful tool for comparing brain territories among vertebrates. The resulting data are interpreted within the framework provided by the neuromeric model of brain organization. Where appropriate, aspects of connectivity (hodology) and function will also be described.

7.3 Forebrain

In recent years our understanding of the organization of the forebrain has dramatically changed, in part as a consequence of the large number of morphological, chemoarchitectonic, embryological, and, primarily, genoarchitectonic data. In particular, the forebrain in amphibians has gained importance in evolutionary studies because it shows many features of chemoarchitecture, hodology, and expression patterns for developmental genes that are readily comparable to those in amniotes (Marín et al., 1998b,c; Brox et al., 2004; Endepols et al., 2004; Moreno et al., 2004, 2005a; Moreno and González, 2006, 2007b, 2011; Laberge and Roth, 2007; Domínguez et al., 2015). We will consider below the main regions of the forebrain, as interpreted in the current prosomeric model (Puelles and Rubenstein, 2003, 2015).

7.3.1 Telencephalon

The telencephalon includes the olfactory bulbs in the rostral pole of the hemispheres. These are formed by a dorsal pallial region and a ventral subpallial region, which includes medial septal and lateral striatal components. In addition, the PO is interpreted as a nonevaginated telencephalic region. It should be pointed out that during early embryogenesis the telencephalon develops from the most anterior part of the alar neural plate. Subsequent multiple and complex morphogenetic events relocate it, leaving the subpallium in an apparent "ventral" forebrain position, as opposite to the "dorsal" position of the pallium (Fig. 7.2G).

7.3.1.1 Olfactory Bulbs

Olfactory information reaches the brain via the olfactory nerve. The olfactory organs are particularly well developed in gymnophionans, and, on each side, distinct dorsal and ventral olfactory nerves reach the bulbs (Fig. 7.1). The olfactory bulbs of amphibians evaginate from the rostral telencephalic wall and secondarily form the main (MOB) and the accessory (AOB) bulbs. The AOB is a specialized region in the caudal ventrolateral wall of the olfactory bulb and is the sole recipient of vomeronasal nerve terminals (Northcutt and Kicliter, 1980). The neuronal populations of both the MOB and AOB are distributed in the concentrically organized glomerular (external), mitral cell (intermediate), and granule cell (internal) layers. Most cells are interneurons (local, intrabulbar cells), and only the large mitral cells are projecting neurons whose axons constitute the secondary olfactory pathway leaving the bulbs (Scalia et al., 1991). Although the bulbs are formed from the pallium, studies in *Xenopus* have proposed that, as in amniotes, olfactory bulb interneurons originate in subpallial regions (probably in the striatal region) and migrate rostrally as the bulbs develop (Brox et al., 2003). In contrast, the olfactory projecting cells are clearly born in the pallium (Moreno et al., 2003). The patterns of expression for several transcription factors, such as Tbr1, Eomes, and Lhx5 in mitral cells (Moreno et al., 2003, 2004) and Pax6 and Dll3 in the interneurons (Moreno et al., 2008c; Joven et al., 2013a,b; Bandín et al., 2014), support this interpretation.

7.3.1.2 Pallium

Classically, the pallium of amphibians has been subdivided into three principal regions: the lateral, dorsal, and medial pallial fields (Northcutt and Kicliter, 1980; Neary, 1990). These divisions have generally been compared, respectively, to the piriform, general, and hippocampal cortices of mammals and to the lateral, dorsal, and medial cortices of reptiles (Bruce and Neary, 1995). Interestingly, embryological approaches in amniotes established a fourth pallial division, the ventral pallium, a genetically and functionally definable histogenetic unit (Puelles et al., 2000). Subsequently, the existence of a ventral pallial region was corroborated for amphibians (Brox et al., 2004; Moreno et al., 2004).

In amphibians, the medial pallium is hypertrophied relative to other pallial regions and shows a higher degree of cell migration (Figs. 7.2F, 7.3A and C). At least

in frogs, it appears to be a multimodal center receiving higher-order olfactory projections from telencephalic centers and auditory, somatosensory, and visual inputs primarily via the anterior thalamic nuclei. Considering the full connectivity revealed in several tracing studies, it was proposed that the medial pallium is homologous to the subiculum and Ammon's horn of the mammalian hippocampus (Northcutt and Ronan, 1992; González and López, 2002; Westhoff and Roth, 2002). This notion has recently been supported by the pattern of expression of LIM-homeodomain genes observed in the pallium of amphibians, because the Lhx2-expressing cells distributed within the medial pallium of developing and adult *Xenopus* (Fig. 7.2F; Moreno et al., 2004) are readily comparable to the *Lhx2*-expressing cells detected in the developing and postnatal Ammon's horn and dentate gyrus of the mouse hippocampus (Rétaux et al., 1999; Zhao et al., 1999).

The lateral pallium forms the thin laterodorsal wall of the hemisphere (Fig. 7.3A and C) and is characterized by a scarce cell migration from the periventricular layer in all amphibians. In its external fiber zone courses the lateral olfactory tract originated in the main olfactory bulbs (Fig. 7.2A; see also Fig. 7.6). The direct olfactory connection is a shared feature of the lateral pallium in tetrapods and its counterparts in fishes. Dorsal and ventral divisions within the amphibian lateral pallium had been noted (Neary, 1990) on the basis of immunohistochemical staining for different neurochemicals, such as nitric oxide synthase (Fig. 7.3C). The ventral division corresponded well with the "intermediate zone" between the lateral pallium and the subpallium described in all tetrapods on the basis of the expression of different homeodomain genes (Smith-Fernández et al., 1998) and identified later as the ventral pallium (Puelles et al., 2000). Before the identification of the ventral pallium, the part of the lateral pallium above the striatum was described with different names (striatopallial transition area of Marín et al., 1997a,b) and was demonstrated to be a distinct telencephalic entity with specific connectivity (olfactory and vomeronasal inputs) and chemoarchitecture (Marín et al., 1997a,b; 1998a,b; Moreno and González, 2004). All these studies clearly differentiate the current ventral pallium from the lateral pallium (dorsally) and the striatum (ventrally). The lateral part of the amygdala was first located within the ventral pallium of anurans on chemoarchitectural grounds (Marín et al., 1998a). In agreement with this and based on GAD67 and Dll4 expression, Brox et al. (2003) identified the lateral amygdala (LA) of *Xenopus* in the ventral portion of the ventral pallium (see Section 1.08.3.1.3.2). The distinct expression patterns of LIM-hd genes, in particular Lhx2/9, serve to identify this pallial region (Fig. 7.3D). Apart from the ventropallial component of the basolateral amygdala, other structures in the

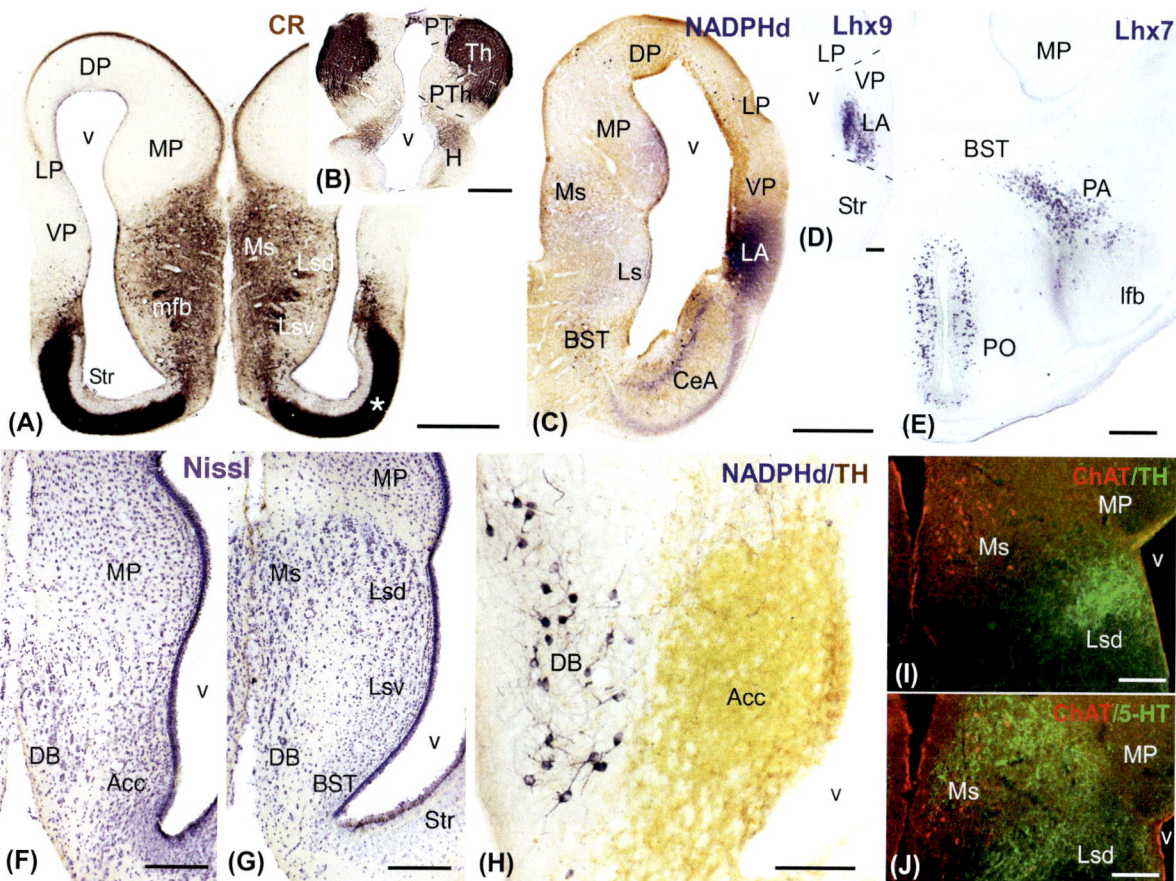

FIGURE 7.3 Photomicrographs of transverse sections through the forebrain of the anurans *Xenopus laevis* (A—E) and *Rana perezi* (F—J): (A) CR immunolabeled cells and fibers at mid-telencephalic levels prevail in the subpallium, especially in the striatal neuropil (*asterisk*), which is the terminal zone for axons from the numerous CR-positive thalamic cells that form the thalamostriatal projections. (B) Intense CR staining in the thalamus. (C) NADPHd staining shows distinct patterns of labeling in the lateral and central amygdala; the densely labeled lateral amygdala region corresponds to the ventral pallial region that specifically expresses the transcription factor Lhx9 during development (D, larval stage 55). (E) Photomicrograph showing Lhx7-expressing cells at caudal telencephalic levels in the pallidum, anterior preoptic area, and BST. *(E: modified from Moreno, N., Bachy, I., Retaux, S., Gonzalez, A., 2004. LIM-homeodomain genes as developmental and adult genetic markers of* Xenopus *forebrain functional subdivisions. J. Comp. Neurol. 472, 52—72.) (F, G) General Nissl staining of the septal region at rostral and mid-telencephalic levels, respectively. (H) Distinct TH immunoreactive plexus in the nucleus accumbens and NADPHd-reactive cells in the area diagonal of Broca at rostral telencephalic levels. (I) Double immunolabeling for ChAT/TH, showing intense TH innervation of the lateral septum, with only scattered fibers reaching the ChAT-ir cell group in the medial septum. (J) Serotonergic (5-HT) fibers in the dorsal septum, showing intense innervation of the lateral and medial septal regions. (I and J: modified from Sánchez-Camacho, C., López, J. M., González, A., 2006. The basal forebrain cholinergic system of the anuran amphibian* Rana perezi: *evidence for a shared organization pattern with amniotes. J. Comp. Neurol. 494, 961—975.) For abbreviations, see list. Scale bars = 500 μm (A—C, E), 200 μm (F, G), 100 μm (D, I, J), 50 μm (H).*

telencephalon of reptiles, birds, and mammals (eg, the posterior dorsal ventricular ridge, the LA, or the claustrum) have been proposed to be derivatives of the ventral pallium (see Martínez-García et al., 2007).

Finally, the narrow zone between the medial and lateral pallial regions was classically named dorsal pallium, which was considered the field homolog of the mammalian neocortex within the amphibian telencephalon. However, the distinct presence in amphibians of dorsal pallial territories (characterized by absence of olfactory input typical for lateral pallium and sensory input functionally separable from that to medial pallium) was difficult to assess. Actually, the existence in amphibians of a dorsal pallium that

would be homolog of the neocortex (isocortex) was questioned (Bruce and Neary, 1995). Detailed studies on its connectivity suggested that the dorsal pallium represents an integrative-associative limbic territory (Westhoff and Roth, 2002). Nevertheless, hodological data have strengthened the comparison with the mammalian cortex (Laberge and Roth, 2007; Roth et al., 2007).

7.3.1.3 Subpallium

The subpallium of amphibian is the major basal subdivision of the telencephalon (Fig. 7.3A, C, and E) and, as in amniotes, it contains the centers that form the basal ganglia, most parts of the amygdaloid complex

(AC), the septum, and the preoptic region (ie, the non-evaginated part of the telencephalon). In amphibians, two main regions in the developing subpallium have been identified as the lateral (striatal) and medial (pallidal) ganglionic eminences (LGE and MGE, respectively). They express different sets of genes that allow their identification and their separation from pallial areas (Fig. 7.4; see below).

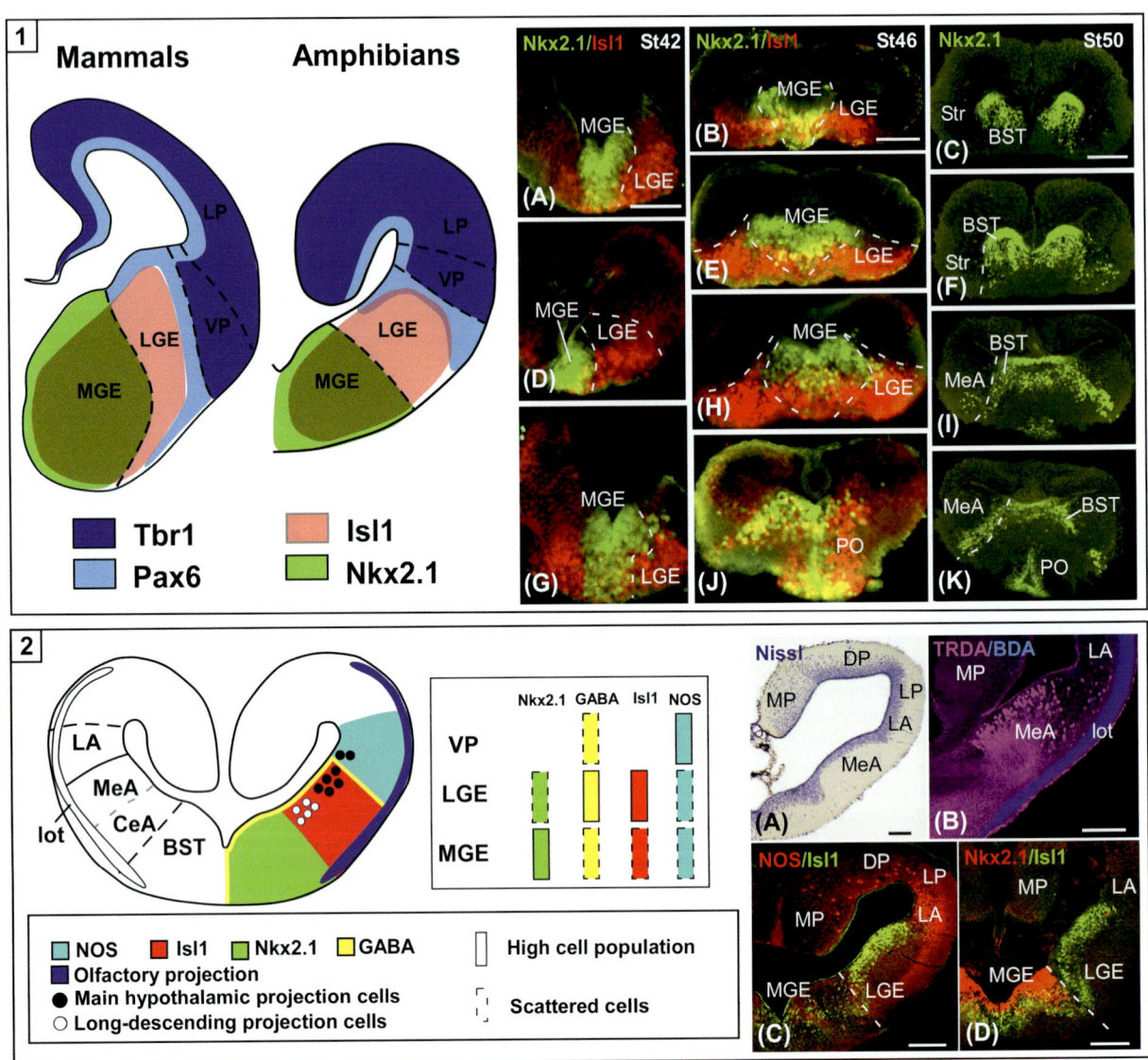

FIGURE 7.4   Panel 1: The schematic drawings correspond to transverse sections through the developing telencephalon of a mouse and a larval urodele amphibian, indicating the regional distribution of the main transcription factors used for identifying striatal (LGE) and pallidal (MGE) regions and their boundaries with neighboring areas. The set of photomicrographs of transverse sections through the telencephalon of *Xenopus laevis* shows double immunofluorescence for Nkx2.1/Isl1 at stages 42 (A, D, G) and 46 (B, E, H, J), as well as Nkx2.1 staining at stage 50 (C, F, I, K). These images show the derivatives of the LGE and MGE. For each stage, the sections are ordered from rostral to caudal levels. In the double-immunostained sections, the use of red or green fluorochromes is indicated for each case. For abbreviations, see list. Scale bars = 100 μm in A (applies to A, D, G); 100 μm in B (applies to B, E, H, J); 100 μm in C (applies to C, F, I, K). *(Modified from Moreno, N., Morona, R., Lopez, J.M., Domínguez, L., Joven, A., Bandín, S., Gonzalez, A., 2012. Characterization of the bed nucleus of the stria terminalis (BST) in the forebrain of anuran amphibians. J. Comp. Neurol. 520, 330–363.) Panel 2: Summary diagram of a transverse section through the caudal telencephalon of Pleurodeles, showing the expression patterns of the main markers used to identify the amygdaloid territories. The photomicrographs show for Pleurodeles: (A) Nissl staining with cresyl violet at caudal levels of the telencephalic hemisphere, illustrating its general appearance and the rough demarcation of its main regions; (B) caudal telencephalon, showing olfactory tract fibers after biotinylated dextran amine (BDA, shown in blue) injection into the olfactory bulb and cells and fibers labeled in the lateral telencephalon after texas red dextran amine (TRDA, shown in magenta) injection into the ventral hypothalamus. (A and B: modified from Moreno, N., Gonzalez, A., 2007. Regionalization of the telencephalon in urodele amphibians and its bearing on the identification of the amygdaloid complex. Front. Neuroanat. 1, 1. https://doi.org/10.3389/neuro.05/001.2007.) (C and D) are two comparable sections double-stained for Isl1 and NOS or Nkx2.1 to illustrate differences between the derivatives of the LGE and MGE. For abbreviations, see list. Scale bars = 200 μm.*

7.3.1.3.1 Basal Ganglia

The clarification of basal ganglia organization in amphibians strongly contributed to a new perspective on the evolution of basal ganglia structures of vertebrates in general, abandoning the previous idea that the fundamental organization of the basal ganglia in vertebrates arose with the appearance of amniotes (Marín et al., 1998a,b,c; Smeets et al., 2000). The application of modern tracing techniques in combination with immunohistochemistry demonstrated that elementary basal ganglia structures were already present in the brain of amphibians and that they were organized according to a general plan shared today by all extant tetrapods. As shown in Fig. 7.5, this pattern includes dorsal (striatum and dorsal pallidum) and ventral (nucleus accumbens and ventral pallidum) striatopallidal systems, reciprocal connections between the striatopallidal complex and the diencephalic and mesencephalic basal plate (striatonigral and nigrostriatal projections), and descending pathways from the striatopallidal system to the midbrain tectum and reticular formation. The connectional similarities are paralleled by similarities in the distribution of chemical markers of striatal and pallidal structures such as dopamine, substance P, and enkephalin. In addition, tracing studies (see Fig. 7.8H) and the specific immunolocalization of calretinin, a calcium-binding protein (Fig. 7.3A and B), served to highlight robust projections from the middle and posterior portions of the thalamus to the striatum. Actually, the thalamostriatal pathway relaying multisensory information is the predominant ascending thalamic connection to the amphibian telencephalon (Roth et al., 2003). However, no evidence was found for separate representations for each sensory system or for a topographically preserved projection from any thalamic nucleus. In addition, the scarce palliostriatal projections in amphibians reveal that functional emphasis of basal ganglia is rather on descending motor control than on reentrant circuits to the pallium.

Further characteristics of the basal ganglia were demonstrated in amphibians, such as a complex neuronal system containing projection cells and several sets of interneurons that share neurochemical and developmental features with amniotes (Moreno et al., 2009; López et al., 2010). Of particular interest was the analysis of the dopaminergic innervation, which is one of the most important features of the organization of the basal ganglia in tetrapods. The location of dorsal and ventral striatal components, ie, the striatum proper and the nucleus accumbens, was identified immunohistochemically primarily by means of antibodies against dopamine and tyrosine hydroxylase, the rate limiting enzyme in catecholamine synthesis (see Smeets and González, 2000). It was found that the striatum proper and the nucleus accumbens occupy a large portion of

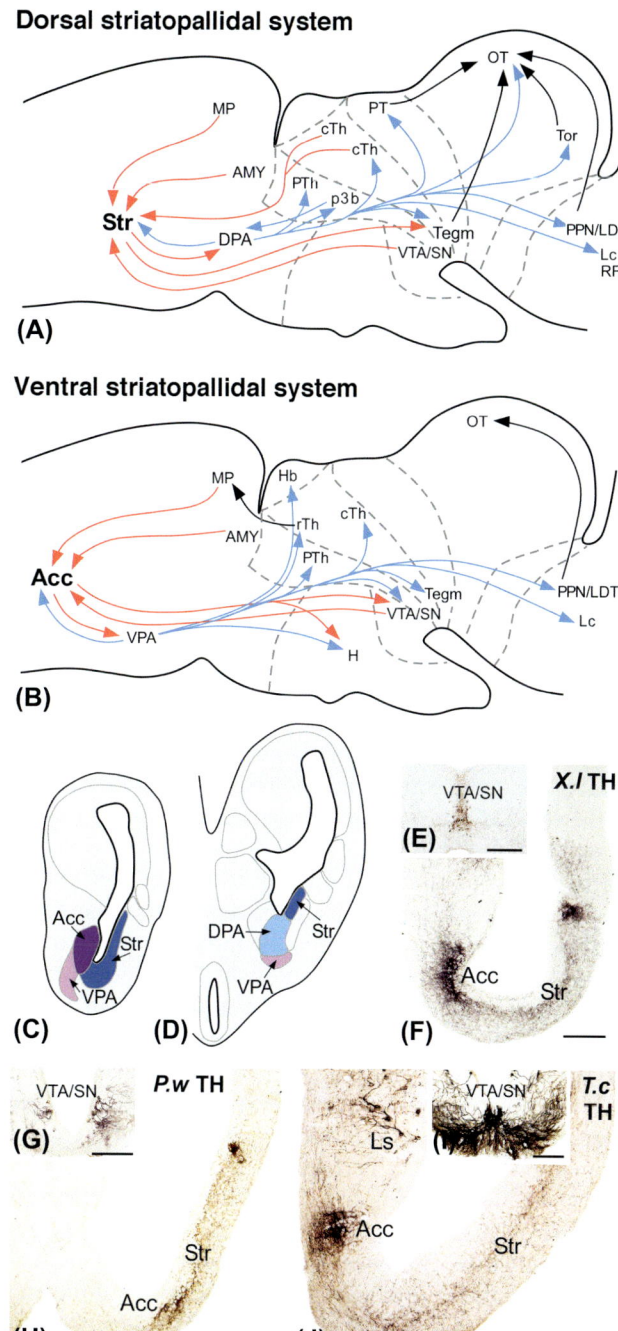

FIGURE 7.5 Schematic drawings that show the main connections of the dorsal (A) and ventral (B) striatopallidal systems, as deduced from studies in amphibians. The two drawings of transverse sections (C, rostral; D, caudal) illustrate the relative position of striatal and pallidal structures for anurans. *(A−D: modified from Marín, O., Smeets, W.J.A.J., González, A., 1998. Evolution of the basal ganglia in tetrapods: a new perspective based on recent studies in amphibians. TINS 21, 487−494.) The photomicrographs show the patterns of dopaminergic (TH positive) innervation of the striatal regions in transverse sections, together with the corresponding cell groups of origin for an anuran (E, F), a urodele (G, H), and a gymnophionan (I, J) (ie, Xenopus laevis, Pleurodeles waltl, and Typhlonectes compressicauda, respectively). For abbreviations, see list. Scale bars = 200 μm.*

the ventrolateral and ventromedial walls, respectively, of the telencephalic hemispheres (Fig. 7.5F, H, and J). The nucleus accumbens is restricted to the rostral one-third of the telencephalic wall and is clearly distinguished from the lateral adjacent striatum by marked differences in catecholamine and neuropeptide immunoreactivities (Marín et al., 1998a). It is worth mentioning that the nucleus accumbens in anurans and gymnophionans is strongly innervated with dopaminergic fibers and lies in the ventromedial part of the hemisphere (Fig. 7.5F and J), but a comparable nucleus accumbens in urodeles lies within the ventrolateral telencephalon (Fig. 7.5H).

Essential differences in the localization and development of the dopaminergic cell groups that project to the basal forebrain were thought to exist between amniotes and anamniotes (Parent, 1986). In amniotes, the basal ganglia receive a strong dopaminergic input primarily from the substantia nigra (SN) pars compacta and the ventral tegmental area (VTA), classically considered to be located in the mesencephalic tegmentum. In contrast, the dopaminergic neurons projecting to the basal forebrain of amphibians constitute a continuous field along the rostrocaudal axis of the diencephalic-mesencephalic basal plate, extending from the retromammillary region to the level of the exit of the oculomotor nerve (Marín et al., 1997d). In urodeles the dopaminergic cells are scarce and form bilateral groups throughout they extend in the diencephalon and mesencephalon, whereas a midline cell group extends into the mesencephalon in anurans and, more conspicuously, gymnophionans (Fig. 7.5E, G, and I). The apparent differences in topography of these cell groups in amniotes and amphibians have traditionally constituted a strong argument against their homology. However, when a segmental approach is applied to the localization of the dopaminergic cell groups in the brain of vertebrates, a different conclusion can be reached (Smeets and González, 2000). It turns out that the VTA and SN cell groups in amniotes are not restricted to the mesencephalic tegmentum but extend into diencephalic (prosomeres p1–p3) and isthmic segments (Marín et al., 1998b,c). In amphibians, the dopaminergic cell field is very similar to the paramedian VTA/SN cell complex of amniotes but lacks a laterally migrated SN and retrorubral field, as well as the isthmic portion of the complex. Thus, the analysis of the development of the VTA/SN cell groups framed in the segmental model gives further support to the notion that the organization of the dopaminergic innervation of the basal ganglia of tetrapods is highly conserved.

Recent developmental studies in amphibians have provided support to the evolutionary constancy of striatal and pallidal components on the basis of comparable molecular specification (genoarchitectonics) of the counterparts of the LGE and MGE (González et al., 2014). By means of immunohistochemistry and in situ hybridization, the localization of the territories expressing Dlx, Nkx2.1, Pax6, GAD67, and Isl1, among others, clearly determined the extent of the components of the basal ganglia in amphibians (González et al., 2002a; Brox et al., 2003; Moreno et al., 2008a, 2009; Joven et al., 2013a,b), readily comparable to their counterparts in amniotes (Fig. 7.4). Thus, Nkx2.1 is expressed from early developmental stages in the caudomedial ventricular zone of the hemisphere, the equivalent to the MGE (Fig. 7.4A–K). The territory expressing Nkx2.1 was originally described as "pallidal," but part of the septum and the AC (the bed nucleus of the stria terminalis, BST) also derive from this Nkx2.1-positive ventricular zone (Moreno et al., 2009, 2012). The expression of Lhx7 reinforced the pallidal entity in this caudal region (Fig. 7.3E; Moreno et al., 2004). In turn, Isl1 and Dlx2 expression served to clarify the dorsal and rostral extents of the striatal (*nonpallidal*) component of the subpallium. The Tbr1 expression in the pallium abuts the expression of the subpallial markers and, rostrally in the hemisphere, only the striatal part of the basal ganglia was molecularly identified being the pallidum a caudal structure in the hemispheres.

Similar results were obtained in urodele amphibians, not only during development but also in the adult (Fig. 7.4C and D). The region clearly comparable to the MGE was revealed by Nkx2.1 expression in the ventricular zone, with Isl1-expressing cells located only in the mantle zone, whereas ventricular zone Isl1 expression highlights the extent of the LGE homolog, showing a sharp boundary with the pallium where Isl1 is not expressed (Moreno and González, 2007b). Moreover, in the adult brain, Pax6 is characteristically expressed by a population of scattered cells in the striatum, but not the pallidum (Joven et al., 2013a,b). The localization of the homologs of the LGE and MGE in amphibians is consistent with our previous proposal based on connectivity and chemoarchitecture (Marín et al., 1998b,c). Noteworthy, the scattered cells expressing GAD67 (GABAergic cells) in the pallium and Nkx2.1 in the striatum were demonstrated to originate in the MGE and migrate tangentially toward these regions in a manner readily comparable to the situation found in amniotes (Moreno et al., 2008b).

7.3.1.3.2 Amygdaloid Complex

For amphibians, and more precisely anurans, the amygdala was first regarded as a single entity in the caudal telencephalon that receives vomeronasal information (Scalia, 1972, 1976). Subsequent studies hypothesized putative medial and lateral components (Northcutt and Kicliter, 1980; Neary, 1990). Even later, a new subdivision of the amygdaloid complex (AC) in anurans was proposed on the basis of chemoarchitecture, and new regions were tentatively named according to their putative counterparts in amniotes (Bruce and Neary, 1995; Marín et al., 1998a). The complex was thus thought to consist of a tier of nuclei organized in several adjacent dorsoventral layers that are arranged in parallel along the telencephalic hemisphere (Fig. 7.6). Hodological techniques and expression pattern of different telencephalic genes were used to clarify the homology between distinct amygdaloid components in anurans and amniotes (Brox et al., 2003, 2004; Moreno and González, 2003, 2004, 2005a,b, 2006; Moreno et al., 2004).

The current view of the AC in tetrapods considers pallial and subpallial components. Thus, the "pallial

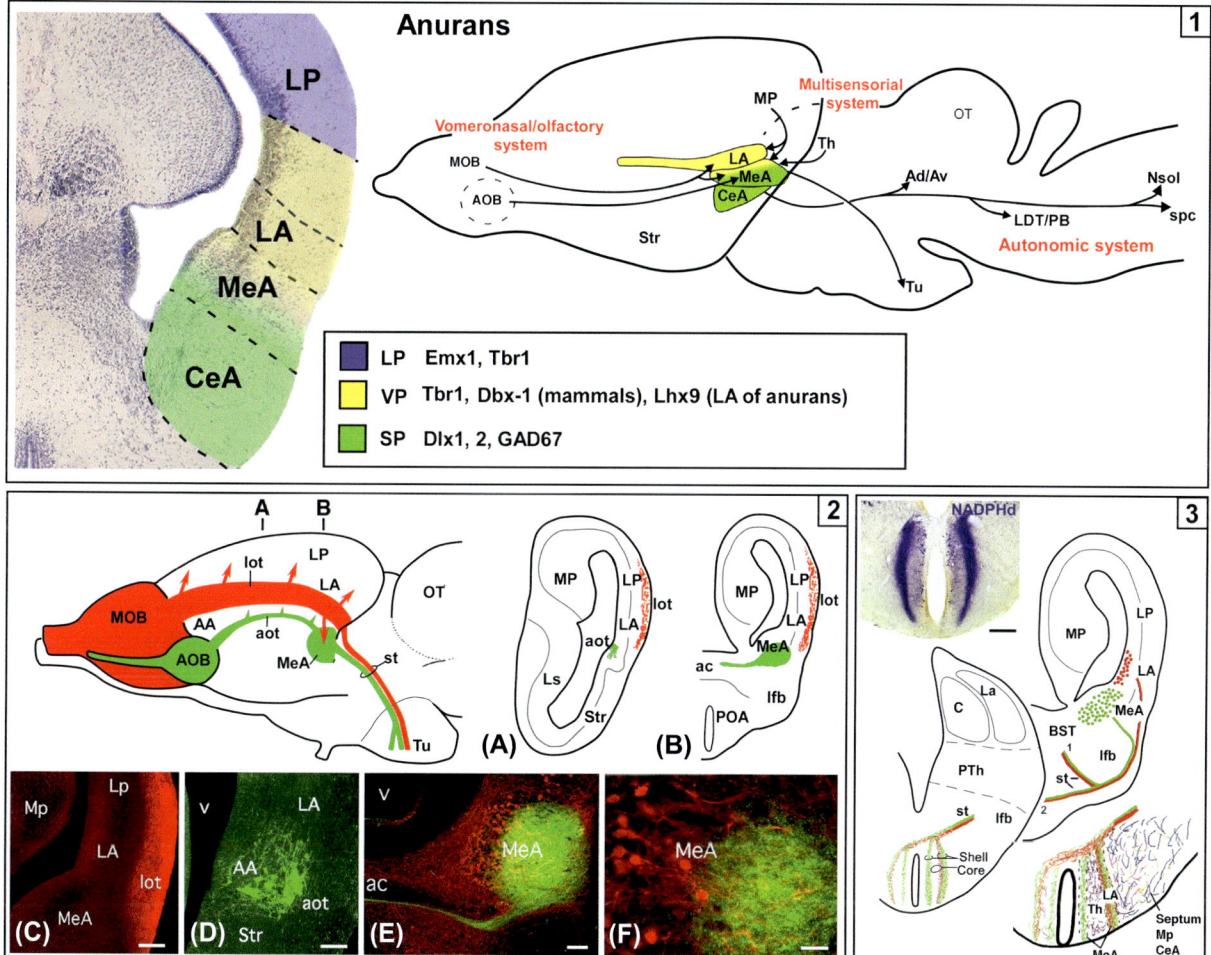

FIGURE 7.6 Panel 1: Schematic drawings illustrating in a transverse section the relative position and main markers of the amygdaloid subdivisions in amphibians. The right side illustrates the connections and main amygdaloid functional systems, as inferred from a comparative analysis of the amygdaloid organization in amphibians. *(Modified from Moreno, N., Gonzalez, A., 2007. Evolution of the amygdaloid complex in vertebrates, with special reference to the anamnio-amniotic transition. J. Anat. 211, 151–163.) Panel 2: Schematic drawing that shows in a lateral view of the anuran brain the efferent olfactory and vomeronasal connections to the lateral and medial amygdala through the lateral and accessory olfactory tracts, respectively. The subsequent projection to the tuberal hypothalamus via the stria terminalis is also indicated. (A and B) are two schematic drawings of transverse sections, at the indicated levels, representing the olfactory and vomeronasal projections through the lateral hemisphere. The photomicrographs are from representative transverse sections: (C) shows the labeled lateral olfactory tract fibers after dextran amine tracer injection into the main olfactory bulb, (D) shows the labeled accessory olfactory tract fibers after dextran amine tracer injection into the accessory olfactory bulb, (E) corresponds to a double-labeled section showing simultaneously the terminal neuropil formed by axons from the accessory olfactory bulb (green) and the labeled cells in the MeA that project to the hypothalamus (red), (F) is a high magnification of the case presented in E. Scale bars = 100 μm (C–E), 50 μm (F). (Modified from Moreno, N., González, A., 2006. The common organization of the amygdaloid complex in tetrapods: new concepts based on developmental, hodological and neurochemical data in anuran amphibians. Prog. Neurobiol. 78, 61–90; Moreno, N., Gonzalez, A., 2003. Hodological characterization of the medial amygdala in anuran amphibians. J. Comp. Neurol. 466, 389–408.) Panel 3: Summary diagram of two transverse sections through the anuran forebrain showing in color code the topographic pattern of projections from the medial (green) and lateral (red) amygdala to the hypothalamus. Large dots represent cell bodies, whereas dashes and small dots represent fibers. The projections from the amygdala pass through the stria terminalis and split into a rostral branch 1 toward the medial telencephalon and a caudal branch 2 to the hypothalamus, where they terminate as distinctly segregated bands. The photomicrograph illustrates the projection to the hypothalamus from the lateral amygdala, as revealed by its NADPHd reactivity. The schema in the lower right depicts the detailed organization of the terminal fields in the hypothalamus, including afferents from the septum, medial pallium, central amygdala, and thalamus. Scale bar = 200 μm.*

amygdala" is composed by derivatives of the lateral and ventral pallium, whereas the "subpallial amygdala" is formed by derivatives of the lateral and medial ganglionic eminences (Puelles et al., 2000; Moreno and González, 2006; Martínez-García et al., 2007). This dual origin makes the AC a histogenetic complex area, and the intense morphogenetic and migratory process during development increase the complexity of the adult structure in all tetrapods (Puelles et al., 2000).

The different components of the AC have been identified mainly in anurans. This complex includes the LA, the medial amygdala (MeA) with a rostral extension named anterior amygdala (AA), and the central amygdala (CeA; Marín et al., 1998a,b). The new terms "medial" and "lateral" are based not on a strict topographical criterion but on a comparison with the names given in amniotes. Thus, the MeA corresponds to the classically named "amygdala pars lateralis," and the new name was given on the basis of its chemoarchitecture and connectivity (Marín et al., 1998a; Moreno and González, 2003). The CeA represents the caudal continuation of the striatum, whereas the LA is located in regions previously included within the ventral lateral pallium (Northcutt and Kicliter, 1980; Neary, 1990; Bruce and Neary, 1995; Marín et al., 1998a,b). These three main components of the AC belong to three different functional systems, namely: vomeronasal (MeA), olfactory/multisensorial (LA), and autonomic (CeA) components (Fig. 7.6).

The MeA is a large structure forming a C-shaped cell population at the level of the postcommissural PO; it corresponds to the cortical and the medial amygdaloid nuclei defined by Scalia et al. (1991). It possesses a narrow rostral part at the border of the ventral pallium, the AA. Whether the MeA is a pallial or subpallial derivative remains uncertain. During development and in the adult, the anuran MeA contains intermingled populations of neurons that likely have different embryonic origins (Fig. 7.6). Scattered Lhx7- and Lhx5/9-expressing cells could have migrated from the MGE and the ventral pallium (Moreno et al., 2004). In addition, the MeA expresses several subpallial markers, including GAD67, Dll4, and Nkx2.1 (Brox et al., 2003; González et al., 2002a).

The MeA receives the bulk of the vomeronasal information relayed from the AOB through the accessory olfactory tract, which courses parallel to the AA and in anurans runs inside the periventricular cellular layer (Fig. 7.6A and D). In salamanders, this vomeronasal tract occupies a superficial position at the ventral part of the lateral olfactory tract (Fig. 7.4B). A large important vomeronasal terminal field is formed at caudal levels in the MeA (Fig. 7.6B and E), which constitutes the main secondary vomeronasal center in the anuran brain, although it also receives an olfactory input to its rostral part (Moreno and González, 2003). Therefore, a certain degree of convergence between olfactory and vomeronasal information exists. In turn, the

MeA influences the animal behavioral response to these chemical stimuli by connections with the hypothalamus. These could be demonstrated in double labeling experiments in which different dextran amines were injected in the hypothalamus and the AOB (Fig. 7.6E and F). This projection courses in the caudal branch of the stria terminalis and terminates bilaterally in the tuberal hypothalamus (Fig. 7.6).

The LA is formed by a rather large cell population that extends from levels just caudal to the olfactory bulb to the level of the anterior commissure that largely resembles the ventral division of the lateral pallium described in anurans (Northcutt and Kicliter, 1980). Of note, in previous studies the term "striatopallial transition area" was proposed to indicate the distinct hodological features of this region (Marín et al., 1997a,b). In this region, the LA was proposed on the basis of distinct neurochemical staining, in particular for nitric oxide synthase revealed with the NADPH-diaphorase (NADPHd) technique (Fig. 7.3C). The data on the expression of the pallial gene Emx1 and the subpallial gene Dll4 in the telencephalon of *Xenopus laevis* (Papalopulu and Kinter, 1993; Smith-Fernández et al., 1998; Bachy et al., 2002; Brox et al., 2003, 2004) indicated that the ventral division of the lateral pallium contains a substantially lower signal for these genes than the surrounding territories. Furthermore, the distinct identity of this region was highlighted by the specific expression of the territorial markers Lhx5 and Lhx9 (Fig. 7.3D; Moreno et al., 2004), which together with the low expression of GAD67, Dll4, and Emx1 (Brox et al., 2003) establish the LA as the main pallial field of the amygdala in *Xenopus*.

Hodological analysis revealed that the LA receives diverse kinds of sensory information, either through direct or indirect pathways. An important direct olfactory projection from the main olfactory bulb to the LA consists mainly of axodendritic contacts onto long dendrites of LA neurons that extend into the lateral olfactory tract (Fig. 7.6). This olfactory information is then relayed to the ventral hypothalamus via the caudal branch of the stria terminalis, and distinct fields are formed by MeA or LA terminals, pointing to a highly topographic organization of afferent fibers in the hypothalamus. This projection is also easily observed with the NADPHd technique (Fig. 7.6). The LA also receives inputs from the caudal portion of the medial pallium (Moreno and González, 2004), which receives olfactory, auditory, somatosensory, and visual inputs (Northcutt and Ronan, 1992; Westhoff and Roth, 2002). Therefore, this multimodal information might reach the LA.

A population of neurons in the anterior and central nuclei of the thalamus projects to the LA, but the projections from the central nucleus are scarce. These thalamic nuclei receive multimodal sensory information that in turn may be relayed to amygdaloid centers (Moreno and González, 2006). All these results strongly suggest that the LA in anurans is a multimodal area in the ventral pallium since, directly or indirectly, olfactory, visual, auditory, somatosensory, vomeronasal, and gustatory information can reach the LA. Taking all these results into consideration, the anuran LA appears as a multimodal area in the ventral pallium that shares features with the amygdaloid ventropallial derivatives of the basolateral complex of amniotes (Moreno and González, 2004). Nevertheless, for some authors, the absence of a conspicuous dorsal thalamic input raises doubts about its homology with the mammalian basolateral complex (Laberge et al., 2006; see Dicke and Roth, 2007). However, if we take into account the gene expression patterns during development, the connections with the hypothalamus, the afferents (although limited) from the thalamus and parabrachial area and a cholinergic projection from the subpallium, we consider that the LA as defined here is a likely homolog of the ventropallial derivatives of the mammalian basolateral complex.

The CeA is an elongated cell population in the caudal part of the hemisphere in close apposition to the caudal aspect of the striatal cell plate (Moreno and González, 2005a). High levels of Dll4 and GAD67 expression, combined with low levels of Nkx2.1 and Lhx7 expression both during development and in the adult, indicate that the CeA has a striatal origin. Developmental analysis in *Xenopus* showed that the CeA originates from the lateral ganglionic eminence and that Nkx2.1- and Lhx7-expressing neurons in the CeA probably migrated from the anuran medial ganglionic eminence (González et al., 2002a; Moreno et al., 2004).

The CeA is the main component of the amygdaloid autonomic system, possessing important connections with brain stem centers, such as the parabrachial nucleus and the nucleus of the solitary tract (Fig. 7.6; Moreno and González, 2005a). A misinterpretation of the striatum's caudal boundary led some authors to consider some CeA connections as striatal (or pallidal) (ten Donkelaar, 1982; Marín et al., 1997a; Endepols et al., 2004). Thus, important connections of the CeA are with the midbrain tegmentum and the torus semicircularis, the parabrachial region, the solitary nucleus, and reticular areas (Moreno and González, 2005a). This supports the notion that the anuran CeA is involved in mediating many autonomic, somatic, endocrine, and behavioral responses through pathways that are similar to those described in amniotes.

As mentioned previously, the main projection from the AC carries vomeronasal, olfactory and, to a lesser extent, multimodal information to the hypothalamus, where its terminal fields are organized in bands around a central core of afferents from other sources (Fig. 7.6; Moreno and González, 2005b). The axons of this projection form the caudal branch of the stria terminalis. A short rostral branch if the stria terminalis courses

through territories of the CeA, forming the BST (Marín et al., 1998a), nucleus accumbens, and the ventral septum, as it does in mammals and reptiles. Developmental gene expression patterns also identified the anuran BST as a subpallial, nonstriatal territory. The BST shows Nkx2.1 and Lhx7 expression and contains an Islet1-positive cell subpopulation derived from the LGE. Immunohistochemistry for diverse peptides and neurotransmitters revealed that the distinct chemoarchitecture of the BST is strongly conserved among tetrapods. In addition, connections were detected between the BST and the CeA, septal territories, medial pallium, PO, lateral hypothalamus, thalamus, and prethalamus. The BST also receives dopaminergic projections from the VTA and is connected with the laterodorsal tegmental nucleus (LDT) and the rostral raphe in the brain stem. All these data suggest that the anuran BST shares many features with its counterpart in amniotes and belongs to a continuum of basal regions comparable to parts of the extended amygdala (Moreno et al., 2012).

In summary, the AC of anurans and probably all the amphibians is molecularly and structurally heterogeneous, just as the AC of amniotes (Puelles et al., 2000), and comprises ventral pallial, striatal, and pallidal (subpallial) components, which can be "extended" rostrally.

7.3.1.3.3 Septum and Preoptic Area

The septal region of amphibians occupies the medial hemispheric wall and is clearly separated dorsally from the medial pallium by a cell-free zone (zona limitans); ventrolaterally it is continuous with basal ganglia structures, lacking a clear boundary. It is homologous to the septal region, as a whole, in mammals (Northcutt and Kicliter, 1980). The septum is particularly well developed in anurans and gymnophionans but is reduced in urodeles. Within the septal complex of anurans, medial and lateral nuclei were traditionally recognized on the basis of cytoarchitecture (Northcutt and Kicliter, 1980). A less clearly defined central nucleus was also considered (Endepols et al., 2005, 2006; Roden et al., 2005). In addition, the nucleus of the diagonal band of Broca was identified as a component of the medial septum (Northcutt and Kicliter, 1980). Other nuclei previously considered as belonging to the septal complex, due to their specific chemoarchitecture, connections, and development, are now regarded as distinct entities of the subpallium, as is the case of the nucleus accumbens (Fig. 7.3H) or the BST, which were discussed earlier.

The large lateral septum was subdivided into dorsolateral and ventrolateral septal nuclei on the basis of distinct chemoarchitectural features (Fig. 7.3A, G, I, and J; Sánchez-Camacho et al., 2003; Endepols et al., 2006). In particular, the presence of a strong catecholaminergic innervation in the lateral nucleus is a feature of the septal region of amniotes, characterized by pericellular baskets of dopaminergic terminals surrounding unstained cell bodies (see Smeets and González, 2000). Catecholaminergic fibers are densely distributed in the dorsolateral septal nucleus (Fig. 7.5I), where they overlap with a strong enkephalinergic and substance P innervation and with a group of enkephalinergic neurons; however, the characteristic pericellular baskets have not been observed in anurans or urodeles (González and Smeets, 1991; González et al., 1993b). Strikingly, gymnophionans possess a stronger dopaminergic innervation, including pericellular baskets as in amniotes (Fig. 7.5J; González and Smeets, 1994). The combination of immunohistochemical staining and retrograde tracing with dextran amines demonstrated that the catecholaminergic innervation of the anuran septum arises in the prethalamus (former ventral thalamus), posterior tubercle, locus coeruleus, and nucleus of the solitary tract (Sánchez-Camacho et al., 2003). In addition, detailed analysis of the septal connectivity revealed that the medial septum receives direct input primarily from regions of the olfactory bulb and from all other limbic structures of the telencephalon (amygdala and nucleus accumbens), as well as from the medial pallium. The anterior nucleus of the thalamus provides the main ascending input to all subnuclei of the anuran septum, which can be interpreted as a limbic/associative pathway (Roden et al., 2005). Other connections of the septum arise in the anterior preoptic nucleus, suprachiasmatic nucleus, ventral hypothalamic nuclei, and the raphe. Therefore, through extensive reciprocal interconnections with limbic forebrain areas and, to a lesser extent, with mesencephalic, lower brain stem, and spinal cord regions, the septum is believed to have important functions in behavioral, autonomic, and endocrine mechanisms.

An important feature of the septum is the presence of a conspicuous cholinergic cell population extending through the diagonal band, medial septal nucleus, BST, and pallidal regions (Fig. 7.5I and J). This population represents the basal forebrain cholinergic system (BFCS), which is readily identified across all tetrapods (Semba, 2004). Hodological studies demonstrated that cholinergic neurons of the anuran medial septal complex contribute to a homolog of the mammalian (medial) septohippocampal system, suggesting that in anurans this projection might participate in learning and memory processes, just as in mammals (González and López, 2002). The cholinergic, catecholaminergic, and serotonergic inputs to the BFCS also show strikingly similarities between anurans and mammals (Sánchez-Camacho et al., 2006).

In terms of development, the septum in vertebrates was classically regarded as a subpallial region, but recent fate-mapping data in the context of molecularly defined longitudinal zones have suggested a partially pallial or roof plate–related nature (see Medina, 2008). In adult *Xenopus*, Pax6 is expressed in the dorsal portion

of the septum (Moreno et al., 2008d), just as in mammals (Puelles et al., 2000). Because Pax6 is generally expressed in the pallial but not the subpallial ventricular zone, the Pax6-expressing cells in the dorsal septum are probably pallial in origin. Therefore, the dorsal septum in *Xenopus* could be considered a subpallial structure with immigrant pallial cells. The developmental expression patterns of xEmx1, Dll4, GAD6, LIM-hd genes (including Isl1), and Nkx2.1 (González et al., 2002a; Brox et al., 2003, 2004; Moreno et al., 2004, 2008a) have further shown that the dorsal septum in anurans is histogenetically complex area and may contain derivatives from several progenitor areas. In addition, numerous neurons were shown to migrate extensively within the septum, as is the case for the Nkx2.1-expressing cells that migrate dorsally out of the MGE (Moreno et al., 2008b). Therefore, the detection of putative immigrant cells (Pax6-expressing cells) and MGE-derived cells expressing Nkx2.1 in the dorsal septum provides evidence that the mixed origin of the septal complex is a conserved feature across tetrapods.

The PO is located immediately in front of the preoptic recess and constitutes the nonevaginated part of the telencephalon. It was classically considered part of the hypothalamus in the rostralmost part of the diencephalon (for review, see Butler and Hodos, 2005). However, this view has changed in recent years and the PO is currently considered part of the telencephalon, due to its topological position in the neural plate and its genetic specification (Medina and Abellán, 2009; Moreno et al., 2009; Sánchez-Arrones et al., 2009; Roth et al., 2010). In anurans, a recent molecular analysis of the PO supports its telencephalic nature (Domínguez et al., 2013). In both amniotes and *Xenopus*, the PO is part of the Dlx-expressing subpallium that also expresses Nkx2.1 (Puelles et al., 2000; Moreno et al., 2009). In addition, the PO in anurans expresses Shh, as it does in other tetrapods (Domínguez et al., 2010, 2013, 2015). In *Xenopus* the ventricular zone of the PO is uniquely defined by the expression of Shh and Nkx2.1; outside of the ventricular lining, a stripe of Nkx2.1- and Lhx7-expressing cells (Fig. 7.3E) is bordered laterally by cells expressing Isl1 (Fig. 7.7B) and Dll4. This pattern suggests putative roles of Shh and Nkx2.1 in PO patterning and cell proliferation, and Dll4 and Isl1 in cell differentiation. This pattern of gene expression during development produces a highly heterogeneous chemoarchitecture in the PO of adult amphibians, giving rise to GABAergic, dopaminergic, serotonergic, and nitrergic cells (among others), just as it does in other vertebrates.

Hodological analysis in anurans showed that the PO is substantially connected with the subpallium, especially with the lateral and medial septal regions and striatum. In addition, distinct connections between the

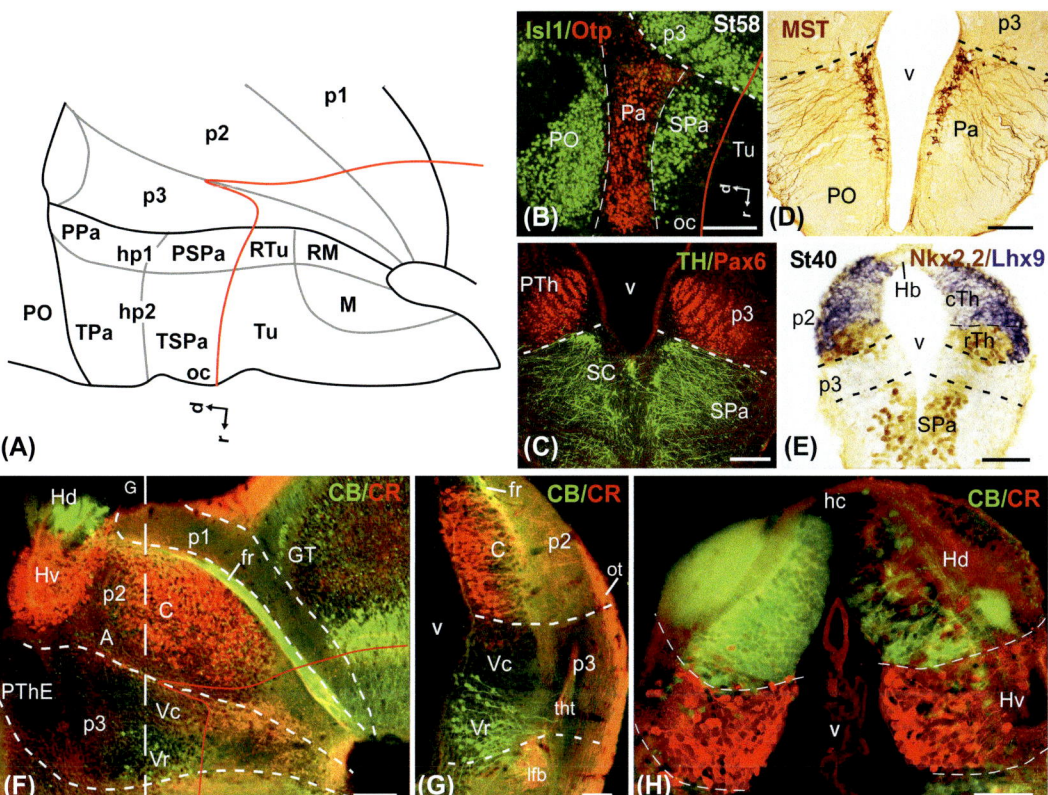

FIGURE 7.7   (A) Schematic drawing of a sagittal view of the hypothalamus, showing the alar–basal boundary (*red*) and the main subdivisions of the hypothalamus alar and basal regions, which are continuous with the alar and basal regions of the diencephalon. Note that the brain flexure causes the alar–basal boundary to bend almost 90 degrees. (B) Photomicrograph of a sagittal section through the main alar hypothalamic subdivisions of *Xenopus* at developmental stage 58, showing the boundaries as revealed by the expression of Isl1 and Otp. *(B: modified from Domiíguez, L., Morona, R., Gonzalez, A., Moreno, N., 2013. Characterization of the hypothalamus of* Xenopus laevis *during development. I. The alar regions. J. Comp. Neurol. 521, 725–759.)* (C) *Transverse section through the adult* Xenopus *prethalamus and hypothalamus, showing their precise boundary as highlighted by the combination of TH labeling in SPa and Pax6 in PTh. (C: modified from Bandín, S., Morona, R., Lopez, J.M., Moreno, N., Gonzalez, A., 2014. Immunohistochemical analysis of Pax6 and Pax7 expression in the CNS of adult* Xenopus laevis. *J. Chem. Neuroanat. https://doi.org/10.1016/j.jchemneu.2014.03.006:24-41.) (D) Transverse section through the hypothalamic Pa, showing the magnocellular group containing mesotocin (MST) in the gymnophionan* Typhlonectes compressicauda. *(E) Transverse section trough the diencephalon and hypothalamus at embryonic stage 40 of* Xenopus laevis, *showing the distinction between rostral and caudal thalamic regions. (E: modified from Bandín, S., Morona, R., Gonzalez, A., 2015. Prepatterning and patterning of the thalamus along embryonic development of* Xenopus laevis. *Front. Neuroanat. 9, 107. https://doi.org/10.3389/fnana.2015.00107.) (F, G) Photomicrographs of doubly labeled sagittal and transverse sections, respectively, through the diencephalon of the urodele* Ambystoma tigrinum, *showing simultaneously the distinct CB- and CR-labeled structures in p2 and p3 (the level of the transverse section G is indicated in F). (H) Transverse section through the epithalamus of* Pleurodeles waltl, *showing the asymmetrical distribution of CB and CR in the habenula. (F–H: modified from Morona, R., Gonzalez, A., 2008. Calbindin-D28k and calretinin expression in the forebrain of anuran and urodele amphibians: further support for newly identified subdivisions. J. Comp. Neurol. 511, 187–220.) For abbreviations, see list. Scale bars = 200 μm (B–D, F, G), 50 μm (E), 100 μm (H).*

PO and the medial and CeA, as well as the BST, have been corroborated (Domínguez et al., 2013). It is known that the preoptic region in mammals regulates reproduction and that this close relationship with the subpallium is related to the control of sexual behavior, since the initiation and coordination of this behavior involve several subpallial nuclei, such as the BST, the lateral and medial septum, and the MeA (reviewed in Canteras, 2011). Therefore, it seems that the pattern of connections between the PO and the subpallium is highly conserved in vertebrate evolution. The anuran PO is also thought to be a key way station in the process by which acoustic stimulation affects the neuroendocrine system

(Wilczynski et al., 1993). According to recent results, and in line with previous findings, auditory information may access the PO through the striatum or through the thalamic nuclei that receive auditory information from the brain stem centers (Allison and Wilczynski, 1991; Brahic and Kelly, 2003; Domínguez et al., 2013). In addition, the PO plays a central role in the activation of calling in terrestrial frogs. However, tracing studies reveal no direct connections between the PO and vocal circuitry. Therefore, this control over vocalization would be indirect and operate via the ventral striatum or the rostral raphe, as previously proposed in *Xenopus* (Brahic and Kelley, 2003).

7.3.2 Hypothalamus

The hypothalamus of amphibians has been thoroughly studied because of its role in the regulation of the endocrine system, being a bridge between the autonomic system and the limbic prosencephalon and thus controlling behavioral responses of the animal, such as reproductive and parental behavior, temperature regulation, territory management, or biological rhythms (for functional and most hodological features of the hypothalamus in amphibians, see ten Donkelaar, 1998a,b).

In the prosomeric model the hypothalamus is "rostral" to the diencephalon (Puelles and Rubenstein, 2003, 2015). Due to the pronounced brain flexure, the rostrocaudal brain axis is bent almost 90 degrees (Figs. 7.2G and 7.7A), which long caused the hypothalamus to be misinterpreted as a "ventral" diencephalic structure (for more detail, see chapter by Moreno et al. in this volume). The hypothalamus is formed by two prosomeres, called hypothalamo-telencephalic prosomeres 1 and 2 (hp1, hp2), or peduncular and terminal segments, respectively (Fig. 7.7A). The hypothalamus includes alar and basal components, each with two main subdivisions that extend through the two prosomeres. The current nomenclature refers to paraventricular and subparaventricular subdivisions in the alar portion, and tuberal and mamillary regions in the basal portion. Furthermore, the topologically most rostral part of the hypothalamus has been identified as a distinct acroterminal region, responsible of the development of structures such as the lamina terminalis or the optic chiasm in the alar hypothalamus, and the median eminence, infundibulum, and the neurohypophysis in the basal part.

Detailed developmental analyses in amphibians have recognized many common patterns of expression for key transcription factors that allowed the identification of the main hypothalamic subdivision. However, no clear intrahypothalamic boundary was observed to distinguish between the two hypothalamic segments, and the acroterminal region has not been genoarchitectonically characterized. We will give here a brief summary of the four main hypothalamic regions, on the basis of recent studies in *Xenopus* (Domínguez et al., 2013, 2014, 2015).

7.3.2.1 Alar Regions

The alar hypothalamus has a caudal boundary with the prethalamic eminence and the prethalamus. Adjacent and topologically ventral to the telencephalic PO lies the paraventricular region (Pa) of the hypothalamus, which was previously named supraoptoparaventricular area because in mammals it includes the supraoptic and the paraventricular nuclei (see Moreno and González, 2011). This most dorsal hypothalamic region is defined developmentally by the expression of orthopedia (Otp) and the lack of Dlx, Shh, Isl1, and Nkx2.1 expression

(Fig. 7.7B). Additional subdivisions have been identified, including a distinct Nkx2.2-expressing area in the rostral domain (Domínguez et al., 2013). This region is the source of most hypothalamic neuroendocrine cells, including the conspicuous arginine vasotocin and mesotocin containing nuclei that have been described in all groups of amphibians (Fig. 7.7D; González and Smeets, 1992, 1997; Smeets and González, 2001). Furthermore, Otp expression has been correlated in *Xenopus* with the emergence of distinct neuroendocrine cells types, such as mesotocin- or somatostatin-containing cells (Domínguez et al., 2013).

The ventral part of the alar territory corresponds to the subparaventricular region (SPa), which is characterized by the expression of Distal-less (Dll) and Isl1, and the lack of Otp expression (Fig. 7.7B; Domínguez et al., 2013). It has also been subdivided on the basis of Nkx2.2 expression in the rostral subdomain (Fig. 7.7E), along with the expression of members of the LIM-hd family, such as Lhx1 and Lhx7 (Domínguez et al., 2013; Moreno et al., 2004). This region gives rise to the suprachiasmatic nucleus, in which important GABAergic cell populations have been described (Brox et al., 2003), suggesting that Dll genes (orthologs of the Dlx genes found in mammals) are involved in the specification of the GABAergic phenotype also in *Xenopus*. In addition, strikingly abundant dopaminergic cells are found in the suprachiasmatic nucleus (Fig. 7.7C; González and Smeets, 1991; González et al., 1993b). The close relationship between these dopaminergic cells and Dll4 expression (Domínguez et al., 2013) suggests that Dlx transcription factors may regulate dopaminergic differentiation in the suprachiasmatic nucleus of amphibians, as reported in mammals (Andrews et al., 2003).

7.3.2.2 Basal Regions

The basal hypothalamus is bordered caudally by the basal plate of the diencephalic prosomere 3. Rostrodorsally it consists of the tuberal region (Tu), which is characterized by the expression of Shh and Nkx2.1 (Moreno and González, 2011; Domínguez et al., 2014). In addition, Dll4 and Isl1 are expressed in the whole tuberal domain, and subdivisions are defined by the exclusive expression of Nkx2.2 caudally and Otp rostrally (Domínguez et al., 2014). Furthermore, Otp expression in this region has been linked to the presence of somatostatin-positive cells (Domínguez et al., 2014), as it has been in mammals (Díaz et al., 2015).

Caudoventrally in the basal hypothalamus lies the mamillary region (Ma), which, in contrast to the tuberal region, expresses Nkx2.1 and Otp but not Isl1 (Domínguez et al., 2014). Additionally, the expression of Shh and Nkx2.1 distinguishes the rostral mamillary area (Nkx2.1+/Shh−), from the caudal retromamillar area (Nkx2.1−/Shh+). For more detailed information on connectivity and functional implications of each hypothalamic region in a comparative framework, see the chapter by Moreno et al. in this volume.

7.3.3 Diencephalon

The diencephalon was classically described as being formed by the four basic longitudinal zones of the epithalamus, dorsal thalamus, ventral thalamus, and hypothalamus (Herrick, 1910). This organization was proposed with consideration of the forebrain axis as parallel to the diencephalic sulci, without recognition of the cephalic flexure and the resulting axial curvature. However, a number of authors did contemplate the flexure and recognized the existence of transverse segments in the diencephalon (eg, Bergquist, 1932; Bergquist and Kallen, 1954). Recent molecular data on the development of the diencephalon could not be explained by the columnar model, leading to the recognition of three transverse segments in the diencephalon named (from caudal to rostral) prosomeres 1–3 (p1–p3) in the current model of forebrain organization (Puelles and Rubenstein, 2003; for more information, see also 1.20, The Optic Tectum: A Structure Evolved for Stimulus Selection). These three prosomeres contain in their large alar regions the pretectum (p1), the thalamus (previously known as the dorsal thalamus) plus the habenula or epithalamus (p2), and the prethalamus (previously known as the ventral thalamus) plus the prethalamic eminence (p3); the basal plate in the three prosomeres is reduced and constitutes the narrow diencephalic tegmentum between the mesencephalon and the hypothalamus (Fig. 7.2G). Puelles et al. (1996) proposed this more detailed morphological framework for studying the diencephalon of amphibians, and it was largely adopted in subsequent studies. Indeed, analysis of the regional distribution for markers such as catecholamines, calcium-binding proteins, and NADPHd highlights the segmental organization of the diencephalon in amphibians (Figs. 7.7F and 7.8A and B). Moreover, molecular specification and developmental patterning in the diencephalon have been shown to be highly conserved between *Xenopus* and amniotes, including the formation of the zona limitans intrathalamica (*zli*), which is a central diencephalic organizer that appears after neurulation in the alar plate between p3 and p2 (Domínguez et al., 2010; Morona et al., 2011a; Bandín et al., 2015). We comment here on the three diencephalic prosomeres of amphibians from rostral to caudal.

7.3.3.1 *Prosomere p3*

The prethalamus is contained in the rostral diencephalic segment p3 and, in its dorsal part, develops the prethalamic eminence (in previous literature called thalamic eminence). The main nuclei described in this region are the "ventromedial and ventrolateral" nuclei (the term ventral was still maintained in spite of their location in the alar plate), which in anurans consist mainly of several laminae of neurons parallel to the ventricle, but are indistinct in urodeles and gymnophionans (Fig. 7.1). The nucleus of Bellonci, the ventral geniculate nucleus, the suprapeduncular nucleus (above the lateral forebrain bundle), and the zona incerta (characterized by its dopaminergic cells) also belong to the prethalamus. Further subdivisions of these main nuclei can be observed with specific markers. Even in the poorly differentiated prethalamus of urodeles and gymnophionans, a rostral (calbindin positive) part can be distinguished from the caudal part (Fig. 7.7F and G). The nucleus of Bellonci and the geniculate nuclei are known to receive retinal inputs, whereas the ventromedial and ventrolateral nuclei receive somatosensory inputs from the spinal cord and dorsal column nuclei, what is in striking contrast with amniotes where these inputs reach directly the thalamus. However, ventral thalamic nuclei in anurans in turn project to the anterior part of the thalamus (Roth et al., 2003; Westhoff et al., 2004). In the basal part of p3 develops the posterior tubercle, which is well developed in most fishes and amphibians. In terms of development, Tbr1 expression in the prethalamic eminence distinguishes it from the rest of the alar p3, and general prethalamic markers during development through the adult comprise Dlx, Pax6 and Pax7, Isl1 and Lhx1/5 (Figs. 7.2E, 7.7B, 7.8C, 7.9G and H). Moreover, the influence from the zli, which in amphibians as in amniotes expresses Shh, modifies the spatiotemporal gene expressions during prethalamic patterning, as, for instance, the downregulation of Pax6 close to the zli (caudal prethalamus) or the expression in this region of Nkx2.2 (Domínguez et al., 2011; Bandín et al., 2015).

7.3.3.2 *Prosomere p2*

The mid-diencephalic segment p2 is the largest of the diencephalic prosomeres in amphibians and houses the thalamus and epithalamus in the alar plate (Figs. 7.1, 7.2G, 7.7F–H, 7.8A and B). Significantly, a large expansion and differentiation of thalamic areas occurs as a derived feature in anurans (Fig. 7.1), in parallel with the expansion of the medial parts of the telencephalic hemispheres (Wilczynski, 2009). The habenula of amphibians is the main component of the epithalamus, together with the pineal complex. In all species studied, the habenula is a large formation in the caudodorsal part of p2 that shows specific chemoarchitecture. In many amphibians, the right and left habenulae are asymmetric in size and staining patterns, as evidenced by calcium-binding proteins that in urodeles discern the dorsal nucleus (calbindin positive) from the ventral nucleus (calretinin positive) (Fig. 7.7F and H). The habenular dorsal nucleus in most species contains cholinergic

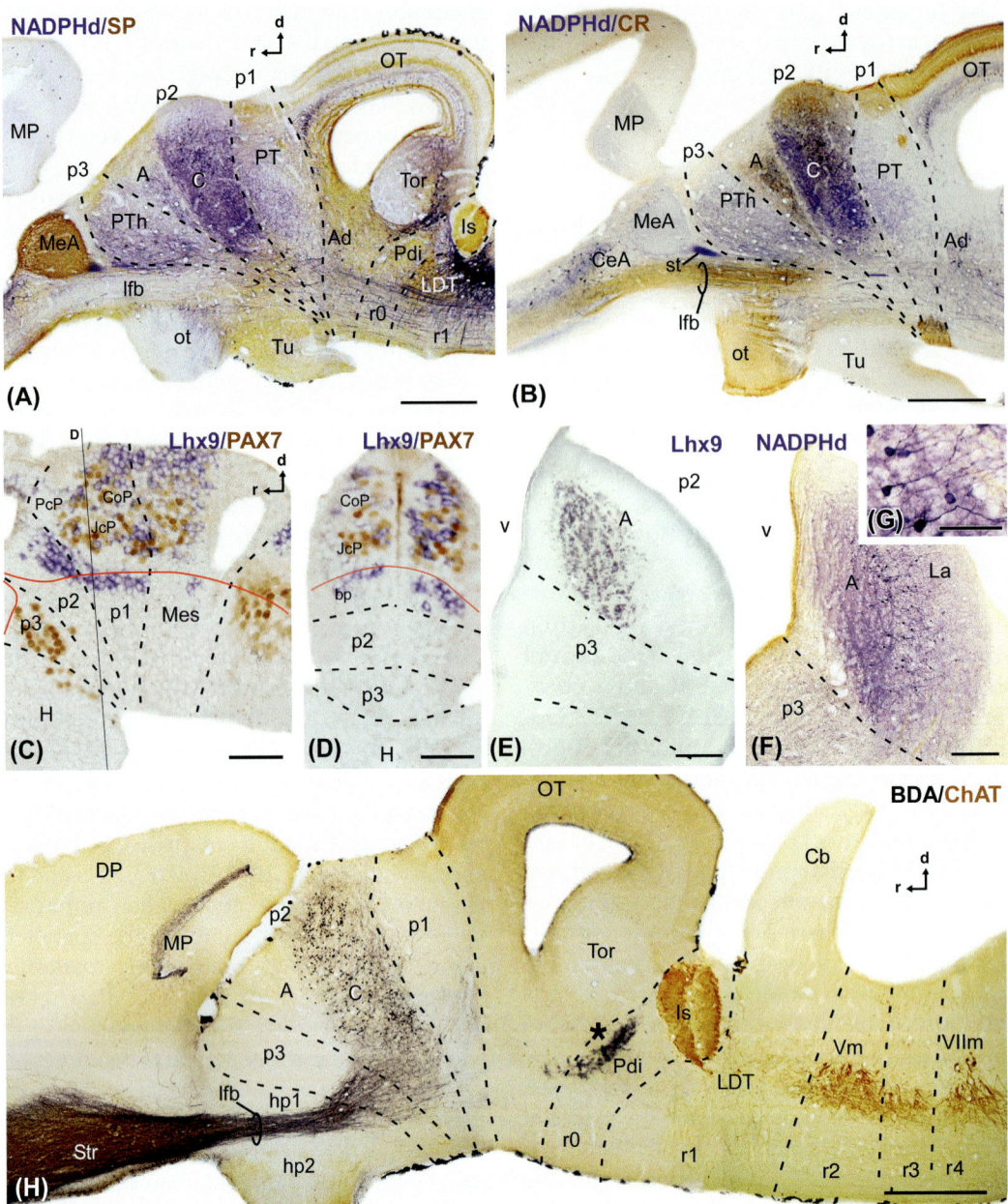

FIGURE 7.8   (A, B) Sagittal sections (rostral is to the left) through the diencephalon and adjacent regions of *Rana perezi* showing the region-alization revealed by NADPHd histochemistry in combination with staining for SP (A) and CR (B). (C, D) Photomicrographs of doubly labeled sagittal and transverse sections, respectively, of *Xenopus laevis* embryos at stage 40, showing in blue the in situ hybridization reaction for Lhx9 and in brown the immunohistochemistry for Pax7 (the level of the transverse section D is indicated in C).   *(C and D: modified from Morona, R., Ferrán, J., Puelles, L., Gonzalez, A., 2011. Embryonic genoarchitecture of pretectum in* Xenopus laevis: *a conserved pattern in tetrapods. J. Comp. Neurol. 519, 1024−1050.) (E, F) Transverse sections through comparable levels of the thalamus of* Xenopus laevis *showing specific in situ hybridization labeling for Lhx9 and NADPHd-reactive cells and fibers, as indicated. (G) Higher magnification of NADPHd-positive cells in the anterior nucleus. (H) Sagittal section through the brain of* Xenopus laevis; *the tracer biotinylated dextran amine (BDA) was injected in the striatum, resulting in retrogradely labeled cells in the caudal thalamus and anterogradely labeled terminal fields in the mesencephalic and isthmal tegmentum (asterisk); the section was immunostained for ChAT, causing labeling in the isthmic nucleus and in motor nuclei of the rostral rhombencephalon. For abbreviations, see list. Scale bars = 250 μm (A, B), 100 μm (C, D), 200 μm (E, F), 50 μm (G), 500 μm (H).*

and nitrergic cells, and the fasciculus retroflexus that originates mainly in the dorsal nucleus courses ventrally in p2 along the p2/p1 boundary toward the interpedun-cular region in the upper rhombencephalon (Fig. 7.7F).

The habenula receives fibers from the stria medularis, which courses through the prethalamic eminence. Within the thalamus of anurans, periventricular (medial) and superficial (lateral) nuclei were discerned,

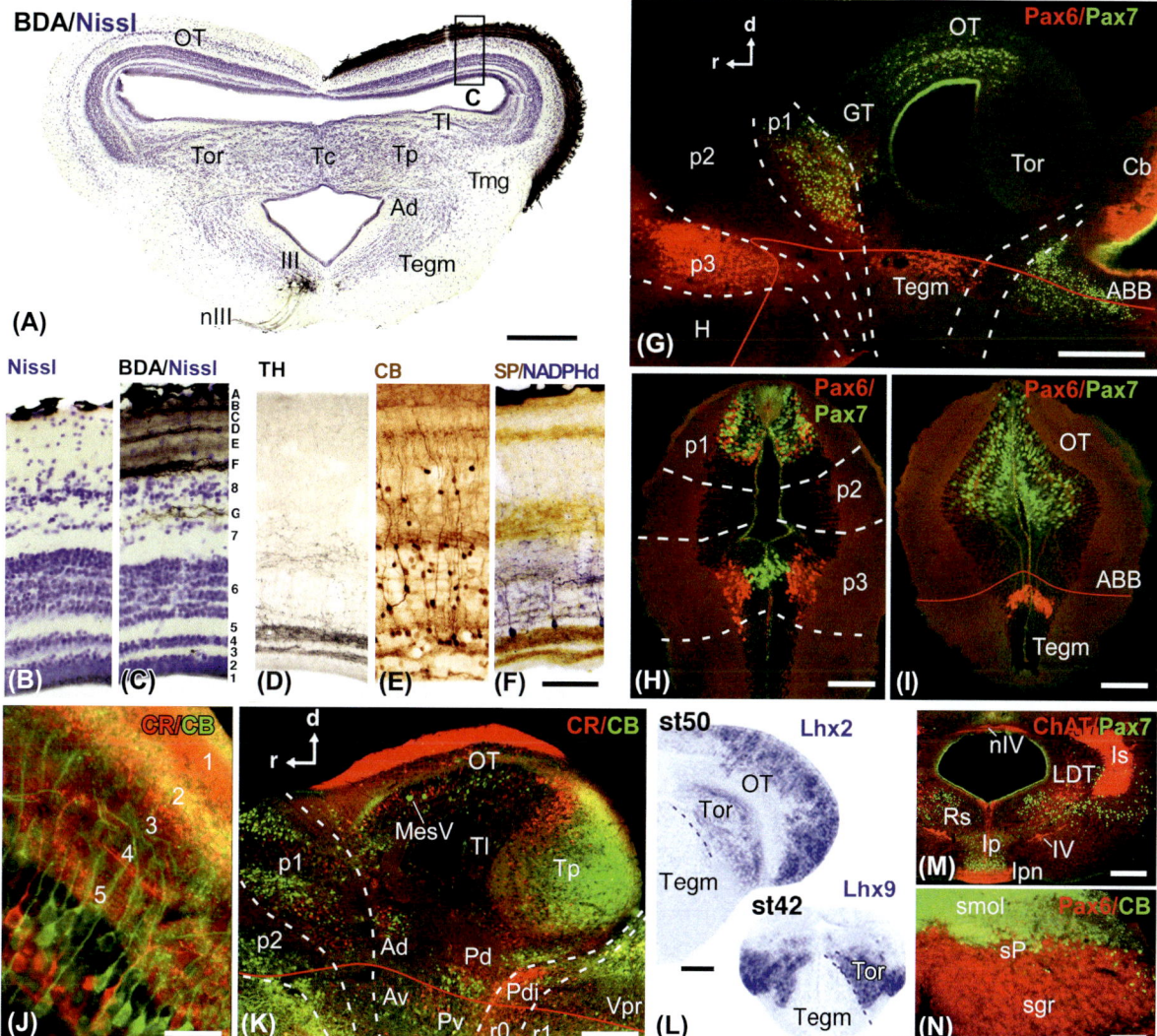

FIGURE 7.9 (A) Transverse section of the frog (*Rana perezi*) midbrain, showing the large optic tectum, the torus semicircularis, and tegmentum. Dark staining in the superficial layers of the right tectum corresponds to the retinal projection, labeled after biotinylated dextran amine (BDA) injection into the left eye. In addition, the left oculomotor nerve was injected, and retrograde-labeled oculomotor neurons are observed in the ipsilateral tegmentum. (B–F) the layered organization of the tectum is shown in Nissl staining (B) and in Nissl staining combined with the labeled retinal projection (C; the numbers and letters used to name the tectal layers follow Kuljis and Karten, 1982; Lázár, 1984); specific layers are occupied by catecholaminergic fibers immunoreactive for TH (D), by cells and fibers containing CB (E), and by a combination of NADPHd-positive cells and SP-positive fibers (F), as indicated. (G) Sagittal section through the brain of a young *Xenopus*, showing the regional distribution of the transcription factors Pax6 and Pax7, revealed with combined immunohistofluorescence. *(G: modified from Bandín, S., Morona, R., Lopez, J.M., Moreno, N., Gonzalez, A., 2014. Immunohistochemical analysis of Pax6 and Pax7 expression in the CNS of adult* Xenopus laevis. *J. Chem. Neuroanat. https://doi.org/10.1016/j.jchemneu.2014.03.006:24-41.) (H, I) Transverse sections through the diencephalon (H) and mesencephalon (I) of* Pleurodeles waltl, *showing the specific labeling of Pax6 and Pax7. (H and I: modified from Joven, A., Morona, R., Gonzalez, A., Moreno, N., 2013. Expression patterns of Pax6 and Pax7 in the adult brain of a urodele amphibian,* Pleurodeles waltl. *J. Comp. Neurol. 521, 2088–2124.) (J) Distinct CB and CR containing cell populations in the tectum of the urodele. (K) Low power image of a lateral sagittal section showing the relationship between CB- and CR-labeled structures in the mesencephalon of* Xenopus. *(J and K: modified from Morona, R., Gonzalez, A., 2009. Immunohistochemical localization of calbindin-D28k and calretinin in the brainstem of anuran and urodele amphibians. J. Comp. Neurol. 515, 503–537.) (L) Transverse sections of* Xenopus *through the mesencephalon showing Lhx2 and Lhx9 in the torus semicircularis and optic tectum in embryos (stage 42) and larvae (stage 50). (L: modified from Moreno, N., Bachy, I., Retaux, S., Gonzalez, A., 2004. LIM-homeodomain genes as developmental and adult genetic markers of* Xenopus *forebrain functional subdivisions. J. Comp. Neurol. 472, 52–72.) (M) Transverse section through r0 and r1 in* Xenopus, *showing the main cholinergic nuclei in relation to Pax7-positive cells. (N) Detail of the corpus cerebelli in* Xenopus *showing distinct staining for Pax6 and calbindin in the three layers. (M and N: modified from Bandín, S., Morona, R., Lopez, J.M., Moreno, N., Gonzalez, A., 2014. Immunohistochemical analysis of Pax6 and Pax7 expression in the CNS of adult* Xenopus laevis. *J. Chem. Neuroanat. https://doi.org/10.1016/j.jchemneu.2014.03.006:24-41.) For abbreviations, see list. Scale bars = 500 μm (A, G), 100 μm (B–F, J, L–N), 200 μm (H, I, K).*

due to a pattern of cell migration that is unusual for amphibians (Puelles et al., 1996). The anterior and central thalamic nuclei are located medially in the thalamus and relatively larger (Figs. 7.7F, G, 7.8A, B, and E–G). Calretinin expression is intense in the central nucleus and in part of the anterior nucleus, whereas calbindin is only present in part of the anterior nucleus (Figs. 7.7F, G and 7.8B; Morona and González, 2008; Morona et al., 2011b). In addition, NADPHd reactivity allows distinction between the central and anterior nuclei (Fig. 7.8A and B). A third, small periventricular cell population is located in rostral p2, just caudal to the p3/p2 boundary, and named intergeniculate leaflet (Puelles et al., 1996); it contains abundant GABAergic cells (Brox et al., 2003). The superficial nuclei of the anuran thalamus are the lateral anterior, lateral posteroventral, and posterodorsal nuclei. Many of their neurons are NADPHd reactive and extend apical dendrites toward the superficial optic neuropil (Fig. 7.8F and G). The thalamus receives multiple types of sensory information, either directly from the brain stem or via the prethalamus, but it seems that none of the nuclei is clearly dedicated to just a single sensory modality. In turn, ascending thalamic projections course mainly from the anterior nucleus toward the medial telencephalic wall (nucleus accumbens, septum, and medial pallium), and from the central nucleus toward the striatum (Fig. 7.8H). Only sparse projections were observed to reach the ventral pallium (the LA, described above). The presence of massive thalamic projections to the subpallium, instead of the pallium, contrasts with the presence of extensive thalamocortical projections in amniotes (Butler, 2009).

Recently, the developmental genoarchitecture of the thalamus has been characterized in *Xenopus* (Bandín et al., 2015). During the patterning phase of the thalamic development, two molecularly distinct progenitor domains are formed, primarily in response to Shh secreted from the *zli* (and basal plate), and inductive influences similar to those described in mammals where suggested for amphibians. A small rostral region occupies the rostroventral part of the thalamus (rostral thalamus, rTh) and seems to be formed under the combined influence of high levels of Shh secreted from the *zli* and the basal plate. The caudodorsal part of the thalamus (caudal thalamus, cTh) is a much larger region and, due to the decreasing concentration gradient of Shh from the zli, is gradually exposed to lower amounts of Shh. The high concentration of Shh that reaches the rTh induces the progenitor cells in this region to express Nkx2.2 (Fig. 7.7E), which ultimately leads to the GABAergic phenotype of thalamic neurons. In turn, progressively less Shh in the cTh induces expression of diverse genes, such as Gli1/2, Ngn1/2, Lhx9 (Figs. 7.7E and 7.8E), Dbx1, and Gbx2, and ultimately leads to the differentiation of the large

population of glutamatergic thalamic neurons. All these data show that the molecular characteristics observed during embryonic development in the thalamus of the anuran *Xenopus* (an anamniote) share many features with those described during thalamic development in amniotes (Martínez-Ferre and Martínez, 2012). The observation of similar features also in zebrafish (Scholpp and Lumsden, 2010; Hagemann and Scholpp, 2012) strengthens the idea of a basic organization of the thalamus across all vertebrates.

7.3.3.3 Prosomere p1

In amphibians, the pretectal region was classically considered to be part of the pars dorsalis thalami (Herrick, 1948), and in most cases it was regarded as a mere transition zone between thalamus and midbrain. However, the physiology of the pretectum has been thoroughly investigated in amphibians, mainly in relation to visuomotor and prey-catching behavior, showing considerable complexity (Ewert et al., 1999). In a cytoarchitectonic analysis of the nuclear organization of the anuran diencephalon, Neary and Northcutt (1983) included most of the pretectal region, as currently conceived, within the large "posterior" nucleus in the thalamus. According to the prosomeric model, p1 corresponds to the caudal diencephalic segment and the pretectum forms the alar part, just caudal to the thalamus and rostral to the mesencephalic tectum. Puelles et al. (1996), adapting the schema of diencephalic neuromeres, revived the segmental concept of the anuran diencephalon and described the pretectum as a neuromeric alar region in which three parts—precommissural (PcP), juxtacommissural (JcP), and commissural (CoP)—can be distinguished. This tripartite schema, subsequently corroborated in amniotes by gene-mapping studies (Ferran et al., 2007, 2008, 2009), was followed by chemoarchitectonic studies in adult *Xenopus* (Milán and Puelles, 2000; Brox et al., 2003; Morona and González, 2008). Detailed molecular characterization of pretectal subdivisions in *Xenopus* embryos (Morona et al., 2011a) indicates that the observed domains can be readily compared with those reported for the domestic chick and mouse, providing evidence for a conserved pattern of pretectal domains and subdomains that is shared by amniotes and amphibians. In particular, a large mixed population of Pax6 and Pax7 cells was localized in the dorsal part of p1. While the caudal PcP is virtually devoid of Pax6 and Pax7 cells, such cells are abundant in the other two parts (Fig. 7.9G and H), and Pax7 expression in combination with other numerous markers during development (Fig. 7.8C and D) was the main tool used for establishing the delineation of the three pretectal territories. (For more detailed information on comparative aspects of the pretectum; see chapter by Moreno et al. in this volume).

7.4 Midbrain

The midbrain or mesencephalon typically is markedly wedge-shaped in the sagittal plane, so that its ventral portion (tegmentum) is narrow, whereas the dorsal part (OT and torus semicircularis; homologs of the superior and inferior colliculi, respectively) is broad (Figs. 7.2G, 7.8A, B, H, and 7.9G). Although this is the general pattern for amphibians, urodeles and, most significantly, gymnophionans have a reduced midbrain, mainly in the dorsal part (Fig. 7.1).

7.4.1 Optic Tectum

The OT receives its name because it represents the main retinorecipient structure of the brain, although it is regarded as an integrative center since it also receives acoustic and somatosensory information. In anurans, the OT is a large bilateral structure that cytoarchitectonically shows a striking layered arrangement of alternating cell and fiber laminae (Fig. 7.9A). Combining classical Nissl staining techniques, tract tracing, and immunohistochemical detection of many makers, nine layers were described and numbered 1 to 9 from the ventricle to the pial surface. Distinct fiber sublaminae, named A to G from the pia to deeper zones of the superficial OT, were also discerned (Fig. 7.9B—F; Kuljis and Karten, 1982; Lázár, 1984). A very heterogeneous population of neurons immunoreactive for different markers has been described in diverse anuran species. For instance, distinct tectal neurons contain calbindin, calretinin (Fig. 7.9E and L), or NADPHd (Figs. 7.8A, B and 7.9F). Of note, the tectum holds the large mesencephalic trigeminal neurons, located mainly at the rostral part of the OT, and a subpopulation of these neurons has been shown to contain calbindin (Figs. 7.9K and 7.10M).

The OT of urodeles and gymnophionans is simpler in structure. It is characterized by a wide, deep cell layer surrounded by a superficial cell-free neuropil. Within the deep cell layer, several cell populations can be distinguished by immunohistochemistry, and the neuropil has been divided into five layers, numbered from the pial surface 1 through 5 (Fig. 7.9J). The superficial half of the neuropil contains the retinal fiber terminations, and the inner half of the neuropil contains somatosensory afferents. The poorly developed appearance of the urodele OT has been explained by the reduction (or absence) of the migration due to paedomorphosis (Roth et al., 1993), but the functional organization of the tectum is essentially the same in frogs and salamanders (Dicke and Roth, 2009).

Visual perception and visuomotor functions are primarily integrated in the OT. Retinotectal fibers are almost entirely crossed in most anurans and terminate in superficial layers (Fig. 7.9A—C) where they form a visuotopic map, ie, adjacent points in visual space are represented at adjacent points in the OT. The nucleus isthmi, a cholinergic cell group in the rostral rhombencephalon (Figs. 7.8H and 7.9M; see below), influence tectal function via strong reciprocal connections with the ipsilateral OT in all species studied so far. Nucleus isthmi project in topographic order also to the contralateral tectum in amphibians. Also the visuotopic map formed by the isthmotectal fibers is in alignment with the retinotectal visuotopic map (Dudkin and Gruberg, 2009). Other direct afferent inputs to the OT arise from all main divisions of the brain except the telencephalon (which has indirect projections to the OT). Thus tectal function may be controlled by the pretectum, thalamus, prethalamus, suprachiasmatic nucleus, mesencephalic tegmentum, rhombencephalic reticular formation, and the cervical spinal cord. In turn, ascending projections from the OT terminate in the rostral part of the thalamus (anterior nucleus) and the pretectum. Descending projections include the contralateral tectum, nucleus isthmi, regions of the tegmentum, some rhombencephalic nuclei, and the rostral spinal cord (see, Saidel, 2009). Functionally, the OT of amphibians is known to play a major role in the control of visually elicited orienting movements, such as prey-catching and avoidance behaviors (for review, see Ewert, 1997). The basal ganglia of amphibians may modulate these visuomotor behaviors by several indirect routes, passing through the pretectum, tegmentum, and torus semicircularis (Fig. 7.5A and B; Marín et al., 1997c).

Various developmental genes are expressed in gradients across the superior colliculus in mice (see Puelles et al., 2012), and some of them have also been detected in amphibians. For example, Pax7 is early expressed in the OT of anuran and urodele amphibians, maintaining their expression into adulthood (Fig. 7.9G and I); its spatiotemporal expression during development serves to guide the incoming retinal axons (Bandín et al., 2013; Joven et al., 2013a,b). Of note, the rostral boundary of the midbrain coincides with the caudal limit of the alar p1 domain, which expresses Pax6 (Fig. 7.9G). Genes of the LIM-hd family have also been investigated in the developing anuran OT (Moreno et al., 2005); Lhx2 expression is especially intense during development (Fig. 7.9L).

7.4.2 Torus Semicircularis

The torus semicircularis is a derivative of the alar plate that forms large bilateral thickenings in the caudal dorsal part of the mesencephalon that in *Xenopus* protrude from the lateral and caudal walls into the ventricular cavity (Fig. 7.9G and K). In *Rana* it is specially enlarged and fused across the midline (Figs. 7.8A, H

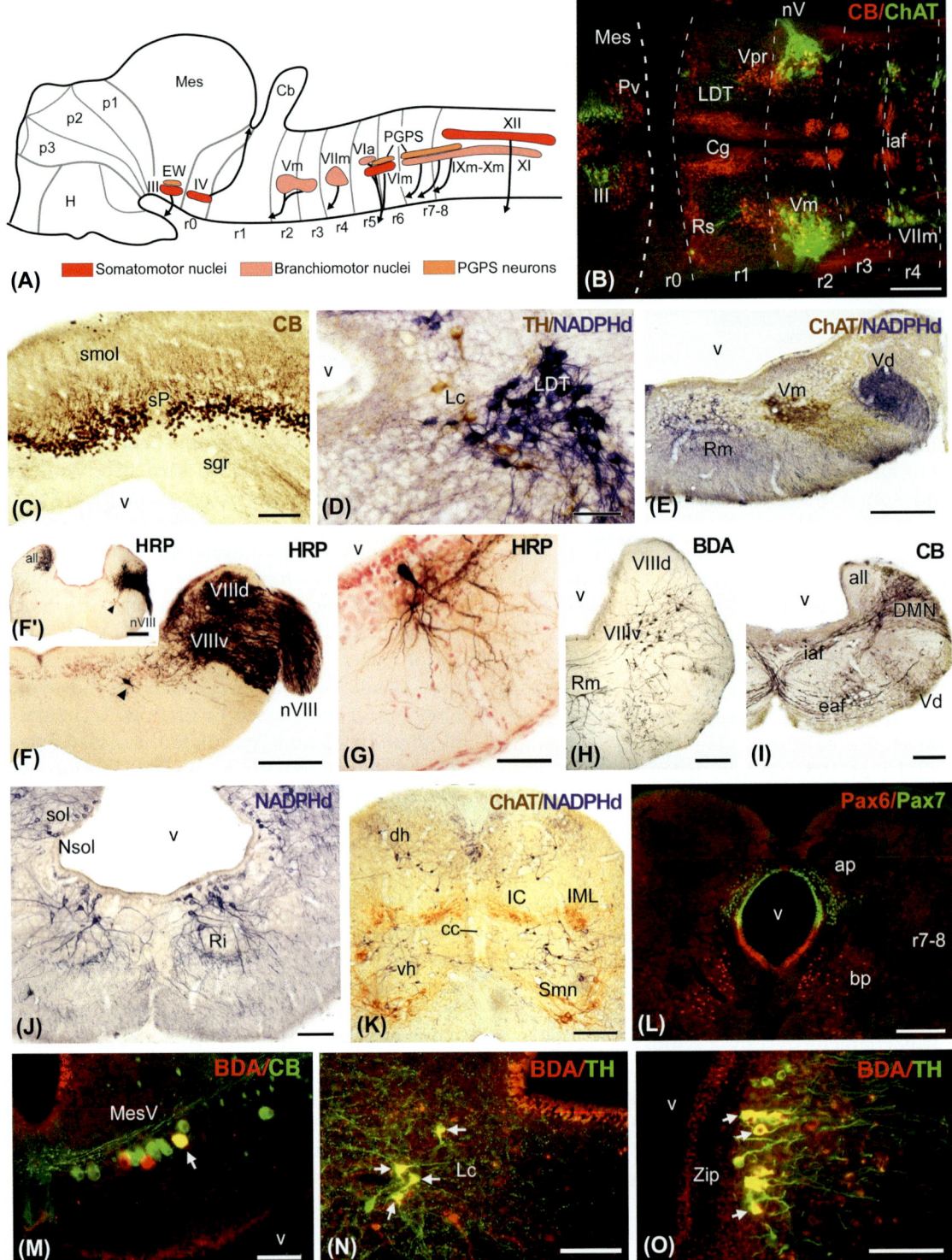

FIGURE 7.10 (A) Schematic drawing of a sagittal section through the brain of an anuran, showing the position of the motor nuclei of the cranial nerves in relation to the segmental domains, as revealed with ChAT immunohistochemistry. *(A: modified from Marín, O., Smeets, W.J.A.J., González, A., 1997. Distribution of choline acetyltransferase immunoreactivity in the brain of anuran (Rana perezi, Xenopus laevis) and urodele (Pleurodeles waltl) amphibians. J. Comp. Neurol. 382, 499–534.) (B) Double-labeled horizontal section through the brain stem of* Xenopus laevis *showing cholinergic nuclei (ChAT positive,* green*) and CB-containing cell groups (*red*) in relation to the rhombomeres. (B: modified from Morona, R., Gonzalez, A., 2009. Immunohistochemical localization of calbindin-D28k and calretinin in the brainstem of anuran and urodele amphibians. J. Comp. Neurol. 515, 503–537.) (C) Transverse section through the cerebellum of* Rana perezi *showing CB immunoreactivity in the Purkinje cells. (D) Transverse section through the locus coeruleus (TH positive) and laterodorsal tegmental nucleus (NADPHd positive) in r1 of* Xenopus laevis*. (E) Transverse section through r2 of* Rana perezi

and 7.9A). From the ventricle to the lateral surface, three different nuclei have been distinguished: the laminar nucleus (Tl) is formed by several alternate cell and fiber layers and located beneath the ventricular lining; the principal nucleus (Tp), the largest of the three, lies ventral to the Tl and is made up of a large population of small cells; the magnocellular nucleus (Tmg) is situated ventrolaterally to the Tp and contains large neurons with long dendrites that are primarily oriented caudodorsally. A commissural nucleus (Tc) can be distinguished in *Rana* but not in *Xenopus* (Fig. 7.9A). Several molecular markers have been used to unravel the toral chemoarchitecture (Endepols et al., 2000), including the calcium-binding proteins (Fig. 7.9K; Morona and González, 2009). In urodeles and gymnophionans, distinct toral nuclei cannot be distinguished.

The torus semicircularis is the main auditory and lateral line center of the amphibian brain stem and is known to be homologous to the mammalian inferior colliculus (ten Donkelaar, 1998a,b; Endepols et al., 2000; Wilczynski and Endepols, 2007). In ranid frogs, auditory input from the dorsal medullary nucleus and the superior olivary nucleus terminates in the magnocellular laminar and, most extensively, principal nuclei of the torus. Notably, the principal nucleus is tonotopically organized and plays a role in spectral information (Feng et al., 1990). In general, low frequencies are represented in a central core, and more peripheral neurons are gradually sensitive to higher frequencies. Multisensory interaction occurs in the torus semicircularis of *Xenopus* (Zittlau et al., 1985) but, in general, auditory input arising directly from the dorsal medullary nucleus is present only in the laminar nucleus, whereas the principal and magnocellular nuclei receive their input from the lateral line nucleus of the medulla. Ascending efferents from all toral nuclei topographically innervate

different dorsal thalamic nuclei. Manteuffel and Naujoks-Manteuffel (1990) interpreted the torus semicircularis of urodeles as a dorsal tegmental area processing auditory and vibratory signals (Manteuffel and Naujoks-Manteuffel, 1990), but the whole rostrocaudal extension of the torus semicircularis receives electrosensory and mechanosensory information, whereas fibers of the acoustic/vestibular system terminate mainly in the caudalmost third of the torus.

During development, the rostral boundary between the torus and the OT is marked by the sharp boundary of Pax7 expression in the tectal ventricular lining (Fig. 7.9G). Several LIM-hd gene members have been shown to be involved in the development of the torus semicircularis; Lhx9 expression is especially intense in *Xenopus* during development and in the adult (Fig. 7.9L; see Moreno et al., 2005).

7.4.3 Mesencephalic Tegmentum

The classical column formed by the anterodorsal (Ad) and posterodorsal (Pd) tegmental nuclei occupies the central stratum of the dorsal tegmentum within the alar territory. In turn, the ventral part of the tegmentum, containing the anteroventral (Av) and posteroventral (Pv) nuclei, is shorter than the alar bands because of the mesencephalon's wedge shape (Fig. 7.9G and K). This profound difference in the development of dorsal and ventral mesencephalic regions has caused an inaccurate identification of the isthmo-mesencephalic boundary in many studies. The obliqueness of this boundary becomes evident in sagittal sections (Palmgren, 1921), but the analysis of traditional "transverse" sections led most authors to consider regions of the isthmus and rostral hindbrain tegmentum as within the mesencephalic tegmentum

showing the cholinergic trigeminal motor nucleus (ChAT positive) in relation to the NADPHd-positive cells and fibers in the reticular formation and descending trigeminal tract. (F) Transverse section through the rhombencephalon of the anuran Discoglossus pictus in which horseradish peroxidase (HRP) was applied to the right VIIIth nerve, showing the dorsal alar plate completely filled by labeled afferent fibers (arrowhead points to a retrograde-labeled cell); the inset (F') shows a transverse section through the rhombencephalon of Xenopus laevis from an experiment in which HRP was applied to the left anterior lateral line nerve and the right VIIIth nerve; the distinct terminal fields formed by the afferent fibers in the dorsal alar plate are labeled (arrowhead points to retrograde-labeled cells). (G) Transverse section through the rhombencephalon of Pleurodeles waltl showing retrogradely labeled neurons of the octavolateralis efferent nucleus labeled after HRP application to the posterior lateral line nerve. (F, F', and G: modified from González, A., Meredith, G.E., Roberts, B.L., 1993. Choline acetyltransferase immunoreactive neurons innervating labyrinthine and lateral line sense organs in amphibians. J. Comp. Neurol. 332, 258–268.) (H) Retrograde-labeled cells in the ventral octaval nucleus and reticular formation after biotinylated dextran amines (BDA) application to the spinal cord of Rana perezi. (I) Transverse section through the rhombencephalon of Xenopus laevis at levels just caudal to the VIIIth nerve entrance showing CB-immunoreactive cells and fibers. (I: modified from Morona, R., Gonzalez, A., 2009. Immunohistochemical localization of calbindin-D28k and calretinin in the brainstem of anuran and urodele amphibians. J. Comp. Neurol. 515, 503–537.) (J) Transverse section at caudal rhombencephalic levels showing NADPHd-reactive reticular neurons in the gymnophionan Dermophis mexicanus. (K) Spinal cord transverse section showing the location of cholinergic neurons in thoracic segment 4 of Rana perezi. (L) Differential dorsoventral location of Pax6 and Pax7 in the ventricular and mantle zones at levels close to the obex in Xenopus laevis. (L: modified from Bandín, S., Morona, R., Lopez, J.M., Moreno, N., Gonzalez, A., 2014. Immunohistochemical analysis of Pax6 and Pax7 expression in the CNS of adult Xenopus laevis. J. Chem. Neuroanat. https://doi.org/10.1016/j.jchemneu.2014.03.006:24-41.) (M–O) Transverse sections showing retrogradely labeled neurons (red) following BDA injections in the spinal cord; they were visualized in combination with immunohistochemistry for CB and TH, as indicated. Doubly labeled neurons appear in a yellowish color (arrows) in the rostral mesencephalic tectum (M) and locus coeruleus (N) of Xenopus laevis, and in the periventricular nucleus of the zona incerta in the gymnophionan Dermophis mexicanus (O). (N and O: modified from Sánchez-Camacho, C., Marín, O., Smeets, W.J.A.J., ten Donkelaar, H.J., González, A., 2001. Descending supraspinal pathways in amphibians. II. Distribution and origin of the catecholaminergic innervation of the spinal cord. J. Comp. Neurol. 434, 209–232.) For abbreviations, see list. Scale bars = 500 μm (B, K), 200 μm (C, E–F', H, I), 100 μm (G, J, L–O), 50 μm (D).

(see ten Donkelaar, 1998a,b). Specific staining, such as that of calbindin and calretinin, highlights the rostral limit of the mesencephalic tegmentum and the sharp and oblique isthmo-mesencephalic border. Thus, it was corroborated that the Pd and Pv nuclei, previously considered in the mesencephalon, are mainly located in the isthmus. However, it was still observed that rostral and caudal groups exist within the mesencephalic tegmentum and, therefore, the terms Pd and Pv have been kept (Fig. 7.9K; Morona and González, 2009).

In amphibians, like in all tetrapods, the ventral tegmentum includes three main groups: the red nucleus, the oculomotor nucleus, and the catecholaminergic A9—A10 group (SN and VTA). The red nucleus of amniotes is a major component of the rostral portion of the basal mesencephalon (Díaz et al., 2000) and is characterized by its contralateral projections to the cerebellum and spinal cord. The red nucleus has been described in all three groups of amphibians on the basis of topography and descending projections to the spinal cord (Sánchez-Camacho et al., 2001a). The oculomotor nucleus occupies a ventral position throughout the rostrocaudal extent of the tegmentum (Fig. 7.9A), and the Edinger—Westphal nucleus has been identified in anurans. The cholinergic oculomotor neurons are more numerous in anurans than in urodeles and gymnophionans, as evidenced by ChAT immunohistochemistry (Marín et al., 1997e; Gonzalez et al., 2002b). As in mammals, species differences in the oculomotor nucleus are revealed by immunostaining for calcium-binding proteins (Fig. 7.10B; Morona and González, 2009). The ventromedial tegmental zone is occupied by a dopaminergic cell group that extends from the retromamillary region to the level of the exit of the oculomotor nerve (González and Smeets, 1994). As we have already described in relation to the basal ganglia, this group is the homolog of the A9—A10 group; its mesencephalic part is particularly well developed in gymnophionans (Fig. 7.5E, G, and I).

In the rodent midbrain, Lhx1/5 and Lhx2/9 show complementary patterns during development. In *Xenopus*, Lhx1/5 are generally expressed in basal tegmental territories, such as the Av nucleus, whereas Lhx2/9 expression appears to be restricted to the Ad nucleus and the torus semicircularis (Moreno et al., 2005). This profuse expression does not extend into the ventral intermediate area, both during development and in the adult. Other gene expression patterns revealed to be conserved in the tegmentum between amphibians and mammals include a longitudinal band of Pax6 cells that extends beneath the alar/basal boundary (Fig. 7.9G and I) and also expresses Nkx2.2 (Domínguez et al., 2011). Interestingly, the orphan nuclear receptor Nurr1 in mammals contributes to the differentiation of mesencephalic dopaminergic neurons in both the VTA and SN (see Ang, 2006). Its zebrafish homolog (NR4A2) reportedly contributes to the differentiation of the dopaminergic cells in the posterior tubercle and is not expressed in dopaminergic cells of the midbrain tegmentum, supporting the hypothesis that the dopaminergic cells of the posterior tubercle are homologous to the dopaminergic mesencephalic nuclei of mammals (Blin et al., 2008). In amphibians, recent studies suggest that Nurr1 is likewise important for the differentiation of the continuous column of dopaminergic cells that extends from the retromamillary region, through the posterior tubercle, into the mesencephalic tegmentum (Velázquez-Ulloa et al., 2011).

7.5 Hindbrain

The hindbrain is a large brain region rostrally continuous with the midbrain and caudally continuous with the spinal cord. We here considered the hindbrain to be formed by the rostral segment of the isthmus, or rhombomere 0 (r0), by rhombomeres 1—7 (r1—r7), and by the long rhombomere r8, which is not clearly defined and probably represents more than one segment. In this context the cerebellum is an outgrowth of the dorsal parts of r1 (the rhombic lip region). This segmentation can be inferred by examining the organization of the patterned motor nuclei (Marín et al., 1997e; Straka et al., 2006). A schematic representation of the topography of the cholinergic cell groups in relation to the segmental domains in the brain stem of anurans is presented in Fig. 7.10A. Only anurans retain the segmental pattern from embryonic stages through the adult, whereas in adult urodeles and gymnophionans the motor nuclei frequently extend across segment boundaries, due to the caudal migration of branchiomotor neurons (Marín et al., 1997e; Fritzsch, 1998).

The fundamental morphological pattern of the hindbrain during development, as revealed by modern molecular studies, is strikingly conserved in vertebrates (see Nieuwenhuys, 2011). This has been corroborated in *Xenopus* with studies on the hindbrain expression of Hox genes, retinoid acid, LIM-hd, and several other genes (Kolm et al., 1997; Moreno et al., 2005). It is beyond the scope of this chapter to deal with the general pattern of hindbrain specification and development. In the following sections we describe only the main components of the hindbrain, pointing out peculiarities found in amphibians.

7.5.1 Rostral Hindbrain (r0—r1)

The structure classically referred to as the isthmus is here considered to be a single segment, named rhombomere 0 (r0). This segment is severely curved behind the oblique isthmo-mesencephalic boundary, due to the excessive expansion of the alar mesencephalon and the bend of the longitudinal axis just rostral to this level in amphibians

(Figs. 7.2G, 7.8A, H, 7.9G, K, and 7.10A). Horizontal sections reveal that this segment is thinner medially and wider ventrolaterally (Fig. 7.10B). In the alar region, r0 possesses the isthmic nucleus (nucleus isthmi), better developed in anurans (especially *Rana*) than in other amphibians. In anurans, nucleus isthmi stand out as a discrete area with a high density of cholinergic cells demarcating much of its outer edge (or cortex) and a lower density of cells in a core medullary area (Figs. 7.8A, H, and 7.9M). In urodele and gymnophionan amphibians, this nucleus is not clearly distinguishable from adjacent tegmental tissue but easily recognized with ChAT staining (Marín et al., 1997e; Gonzalez et al., 2002b). As mentioned earlier in this chapter, the isthmic nucleus possesses strong reciprocal connections with the OT, and its cholinergic projections to the superficial tectal layers can influence the tectum's response to retinal axons that also reach these layers (Marín and González, 1999).

A second characteristic cell group in r0 is the trochlear nucleus, which lies in the isthmic tegmentum, caudal to the mesencephalic oculomotor nucleus. Like other tegmental regions, the trochlear nucleus in amphibians has frequently been considered mesencephalic, in disregard of the oblique plane of the isthmo-mesencephalic boundary (see ten Donkelaar, 1998a,b). The neurons of this nucleus are easily recognized by ChAT immunohistochemistry and by the striking course of their axons, which reach the dorsal part of r0 and form the crossed trochlear nerve (Figs. 7.9M and 7.10A). This nucleus is reduced in urodeles and in gymnophionans, which have reduced eyes and probably lost some eye muscles. Indeed, in *Dermophis*, the reduction of the extraocular musculature led to a reduction of the oculomotor and trochlear nuclei, as indicated by a very small number of ChAT-ir cells (Gonzalez et al., 2002b).

Rhombomere r1 is by far the largest hindbrain segment. Our interpretation of its boundaries and components follows that of Aroca and Puelles (2005) and several previous studies in amphibians (Marín et al., 1997e; Straka et al., 2006). Many r1 derivatives have been described and are distinct enough to have received an anatomical name. Those derivatives include cholinergic cell groups, the locus coeruleus, the parabrachial complex, the main sensory trigeminal nucleus, part of the rostral octavolateral column, and lateral parts of the cerebellum. Moreover, the basal plate is characterized by the absence of cranial nerve motoneurons and the presence of the interpeduncular nucleus.

The cerebellum is a rostral derivative of the alar r1−r2 region known as the cerebellar plate and rostral rhombic lip. Purkinje cells are generated from the r1 cerebellar plate, and cells coming from the rostral rhombic lip migrate rostrally, forming the external granular layer (Wullimann et al., 2011). The cerebellum in amphibians is relatively small in comparison to the total size of the brain. A single transverse plate, or corpus cerebelli, covers the rostral part of the fourth ventricle, and lateral auriculae exist in most species, which are particularly well developed in urodeles and gymnophionans. The trilayered structure of the cerebellar laminae of amniotes is easily recognized in the corpus cerebelli, and all main cell types have been described (ten Donkelaar, 1998a,b). In addition, the relatively simple cerebellum of anurans already possesses the basic pattern of cerebellar connections observed in terrestrial vertebrates. Thus, primary and secondary vestibulocerebellar projections appear to dominate the cerebellar input, but trigeminocerebellar, spinocerebellar, and even olivocerebellar projections have also been demonstrated. As regards cerebellar efferents, both ipsilateral and contralateral, cerebellovestibular projections have been demonstrated. Immunohistochemical studies have helped researchers recognize distinct cell types in the cerebellum of amphibians. Calbindin is a good marker for Purkinje cells in *Rana* (Figs. 7.9N and 7.10C). In addition, Pax6 immunohistochemistry labels the granule cell layer both in embryos and in adults, whereas Pax7 is expressed in the ventricular lining (Fig. 7.9G and N) of both urodeles and anurans (Joven et al., 2013a,b; Bandín et al., 2014). A role for Pax6 in the regulation of migration and patterning of rhombic lip progenitors seems to be a feature that amphibians share with other vertebrates (Wullimann et al., 2011).

Two important alar derivatives have been identified in r1 of amphibians on the basis of their specific neurochemistry and hodology: the locus coeruleus and the LDT. In the three groups of amphibians, the locus coeruleus is characterized by a small and rather disperse group of noradrenergic cell bodies in r1 that has ascending and descending projections to the telencephalon and spinal cord, respectively (Fig. 7.10D and N; Marín et al., 1996, 1997d; Sánchez-Camacho et al., 2001b, 2003). Intermingled with the noradrenergic cells, and extending somewhat more laterodorsally, is the LDT, a cholinergic cell group located ventromedial to the isthmic nucleus (Fig. 7.9M). An additional, more dispersed cholinergic cell population, located slightly more rostrally, was named pedunculopontine tegmental nucleus (PPN in Fig. 7.5A and B); it has been described in anurans but was not recognized in urodeles and gymnophionans. The conspicuous cells of LDT in all three groups of amphibians contain nitric oxide, as demonstrated by immunohistochemistry and NADPHd staining (Fig. 7.10D; Muñoz et al., 1996b; González et al., 1996, 2002c). This cell group is probably homologous to its mammalian namesake because of its widespread connections with telencephalic, midbrain, and spinal cord regions (Marín and González, 1999; Sánchez-Camacho et al., 2006; López et al., 2007).

The most prominent cell population in the basal part of r1 is the interpeduncular nucleus, which forms a long cell column above the neuropil of the axons in the fasciculus retroflexus, easily observed with immunostaining for calbindin and ChAT (Fig. 7.9M). In *Xenopus* and *Pleurodeles*, cells that express Pax7 are only found in the ventricular zone of the alar plate in p1, but progressively through larval development abundant scattered Pax7-labeled cells extend across the basal plate and concentrated in the interpeduncular nucleus (Fig. 7.9M). This observation is very similar to what has been reported in mammals and birds (Lorente-Cánovas et al., 2012), strongly supporting the idea that tangential migrations from alar to basal plate locations occur in amphibians as well as amniotes (Bandín et al., 2013; Joven et al., 2013a,b).

7.5.2 Caudal Hindbrain (r2−r8)

The caudal hindbrain comprises the transverse subdivisions of rhombomeres 2−8 (r2−r8). Due to the patterned, segment-related localization of the cranial nerve motor nuclei, double labeling for ChAT is a good tool for identifying the position of other markers (Figs. 7.8H, 7.10A, B, and E), especially in urodeles and gymnophionans where the neuronal cell bodies form a continuous periventricular gray. The alar plate nuclei in the caudal hindbrain receive and process sensory information, whereas the basal plate harbors cholinergic motoneurons and the transmitter-rich reticular formation. In this part of the hindbrain of amphibians originate all of the mixed cranial nerves (the trigeminal, facial, glossopharyngeus, and vagus nerves), one to three somatic sensory nerves (vestibulocochlear and lateral line nerves), the one cranial nerve with only branchial motoneurons (the accessory nerve), and two of the four cranial nerves with only somatic motoneurons (the abducens and hypoglossus nerves).

General somatosensory information from the face reaches the alar plate of amphibians via the trigeminal, facial, and vagal nerves, whereas special visceral afferents carrying taste information enter the alar plate via the facial, glossopharyngeus, and vagus nerves. The principal and descending trigeminal nuclei and the nucleus of the solitary tracts are the main targets for these sensory fibers. In the dorsal part of the alar plate of amphibians are located the acoustic, vestibular, and lateral line nuclei (area octavolateralis) receiving special somatic afferents. An anterior lateral line nerve innervates mecanoreceptors (neuromasts) and electroreceptors (ampullary organs) located in the head region in most larval and adult urodeles and gymnophionans that possess the lateral line system. Only mechanoreceptors form the lateral line in larval anurans and in the adults of those species that retain an aquatic lifestyle after metamorphosis (*Xenopus*

and *Pipa*). The neuromasts in the trunk are innervated by the posterior lateral line nerve.

In species that do not possess a lateral line system in adulthood, the dorsal alar plate is dedicated only to acoustic and vestibular information, received via the dorsal and ventral branches of the VIIIth nerve that terminate in dorsal and ventral fields (VIIId and VIIIv, in Fig. 7.10F). In contrast, in anurans that retain the lateral line as adults, the neurons in the dorsalmost part of the area octavolateralis receive lateral line information. The octaval terminal fields are located more ventrally (Fig. 7.10F′), in an acoustic dorsal medullary nucleus that strongly projects to the torus semicircularis through external and internal arcuate fibers that form the lateral lemniscus (labeled for calbindin in Fig. 7.10I; Will et al., 1985). Anterograde labeling of afferent fibers from the different organs of the labyrinth in *Rana* has revealed several vestibular nuclei that receive specific sensory inputs from semicircular canal or the otolith organs (Matesz, 1979; Birinyi et al., 2001), as well as dorsally located cochlear and saccular nuclei receiving auditory input (Matesz, 1979). The rostrocaudal extent of these nuclei (from r1 to r7) was assessed by tracing vestibulocerebellar, vestibuloocular, and vestibulospinal projections (Fig. 7.10H). Their boundaries correlate well with genetically defined hindbrain segments that are highly conserved throughout vertebrate evolution (Straka and Baker, 2013).

In amphibians with a complete (electrosensory and mechanosensory) lateral line system, the area octavolateralis consists of three sensory columns: the dorsal column receives electrosensory fibers from the anterior lateral line, the intermediate column is devoted to lateral line mechanoreception, and the ventral column is related to octaval fibers (Fritzsch et al., 1984). Vestibular and auditory subdivisions in urodeles were not distinguished in the octaval (ventral) zone (Fritzsch, 1988).

Much information on the efferent innervation of hair cell sense organs in the ear and lateral line systems has come from research on amphibians. These studies have shown that the efferent innervation uses acetylcholine as the transmitter and has facilitatory as well as inhibitory impacts at the periphery. The efferent cells were identified by retrograde tracing studies (Fig. 7.10F, F′, and G) and corroborated in the hindbrain tegmentum as cholinergic neurons whose axons branch extensively and even innervate sense organs of different modalities (Will, 1982; Fritzsch and Wahnschalffe, 1987). The octavolateral efferent neurons seem to be located in a single, loosely organized nucleus of a few cells that overlaps or lies close to the facial motor nucleus (González et al., 1993a).

Within the basal plate of the hindbrain, the cell bodies of individual cranial nerves have a precise relationship to specific rhombomeres in all vertebrates (Chambers et al., 2009). Segmental organization of the brain stem motor nuclei (Fig. 7.10A) has been analyzed in the three groups of amphibians by means of ChAT immunohisto-chemistry (Marín et al., 1997e; Gonzalez et al., 2002b). The somatomotor nuclei occupy the ventromedial portion of several segments. Similar to the oculomotor and trochlear nuclei described in the mesencephalon and isthmus, the somatomotor neurons of the VIth nerve (main abducens nucleus, VIm) are located in r5 and the motor nucleus of nerve XII is located in r8.

The branchiomotor nuclei constitute a cell column, which extends along the hindbrain and is interrupted at some levels by more or less distinct gaps. The Vth nerve motoneurons form a single nucleus located within r2 and r3 (Fig. 7.10A, B, and E), with the trigeminal root in r2. The organization and topographical relationships of the facial motor nucleus is one of the most striking differences among amphibians. In urodeles and gymnophiones, the facial nerve exits in r4, but the facial motor nucleus lies caudally within r6. Facial axons course initially medially and are arranged as a thick fascicle dorsal to the fasciculus longitudinalis medialis; the fascicle then turns 90 degrees rostralward and continues to the level of the facial nerve roots in r4. There, the fibers seem to genuflect (genu facia-lis) and course laterally to exit the brain. This situation contrasts with the condition in anurans, which the facial motor nucleus lies in front of the nerve root in r4, just caudal to the trigeminal motor nucleus (Fig. 7.10A). As previously mentioned, the octavolateral efferent cells are a unique population of neurons, apparently derived from facial branchial motoneurons. Like facial motoneu-rons, they derive from r4 but segregate as a result of dif-ferential migration in many, but not all, vertebrates. Like many tetrapods, anurans and urodeles possess a retractor bulbi muscle, which serves to retract the eyeball into the orbit, thus allowing the nictitating membrane to become drawn over the cornea. While the main abducens nucleus innervates the lateral rectus muscle, branchiomotor neu-rons of the accessory abducens nucleus (spanning in r5–6) innervate mainly the retractor bulbi muscle (Gonzá-lez and Muñoz, 1987). In some caecilians, several muscle groups underwent modifications in shape, place of inser-tion, and innervation, and parts of the extrinsic muscles are modified to support the special sensory tentacle (Wake, 1993). In some species the lateral rectus muscle may be lost, as seems to be the case of *Dermophis*, in which a VIm nucleus was not found. However, all gymno-phionans have a retractor tentaculi muscle identified as homologous to the retractor bulbi that is innervated by the nerve abducens (Wake, 1993). Therefore, the robust abducens nucleus found in the lateral reticular zone in the rhombencephalon of *Dermophis* corresponds with the accessory nucleus of anurans and urodeles (Gonzalez et al., 2002b).

In anurans, the glossopharyngeal motor nucleus be-gins at the caudal pole of the nucleus abducens and

proceeds along r7 and r8, forming a single column with the vagal motor cells (*Xenopus*), or with a small gap be-tween the two entities (ranid frogs). Caudally, this col-umn overlaps with the accessory nerve nucleus. In urodeles and gymnophionans, the motor nuclei of the glossopharyngeous and vagus were identified as a cell column where the glossopharyngeal neurons were rostral and more closely located to the ventricle than the vagal neurons. It appears that spinal nerves 1 and 2 are compo-nents of the hypoglossal nerve in all caecilians and, in addition, some species possess an occipital nerve that contributes to the hypoglossal nerve (Wake, 1993). ChAT-positive cells, which are topographically associ-ated to the III, VIa, and IXm–Xm nuclei of amphibians most likely represent the preganglionic parasympathetic column reported in amniotes (rhombencephalic pregan-glionic parasympathetic cells (PGPS) and Edinger-Westphal (EW) nucleus in Fig. 7.10A).

Finally, the reticular formation of amphibians was described on the basis of cytoarchitecture as a rather un-differentiated cell mass throughout the rhombenceph-alon divided into inferior, middle, and superior reticular nuclei (see ten Donkelaar, 1998a,b). Subse-quently, the localization of many neurochemicals (such as enkephalin, substance P, somatostatin, nitric oxide, monoamines, and calcium-binding proteins) revealed that the reticular formation is organized complexly and similar to the reticular formation in amniotes (Muñoz et al., 1996b; Stuesse et al., 2001; Morona and González, 2009). The reticular formation receives the bulk of the ascending spinal pathways (Muñoz et al., 1997) and is extensively afferented by the OT. In turn, the large reticular cells characteristic of all amphibians (Fig. 7.10H and J) are the origin for most descending spi-nal pathways (Sánchez-Camacho et al., 2001a). Strik-ingly, up to 10 distinct populations of serotonergic cells in the raphe column were identified on the basis of their location and cellular morphology, which contribute se-rotonin to many brain regions (Zhao and Debski, 2005).

7.6 Spinal Cord

The spinal cord has been a relatively stable region of the central nervous system during vertebrate evolution (Hodos, 2011). The major changes have occurred during the transition from aquatic locomotion, primarily pro-pulsion by body and tail undulation, to rhythmic move-ments of limbs that support the weight of the body in air (Wagner and Chiu, 2003). The new musculature of the limbs that appeared in amphibians led to a high devel-opment of the gray matter columns and to the formation of enlargements in the cervical and lumbar regions. The spinal cord of anuran amphibians has long been used as an important model system for the study of a variety of spinal mechanisms, because of its relatively simple organization. By means of modern neuroanatomical techniques, it was demonstrated that most connections and neurochemical characteristics in the amphibian spinal cord are readily comparable to those of amniotes. Thus, it was demonstrated that the spinal cord of amphibians possesses cholinergic somatomotor and sympathetic neurons, as well as a large and heterogeneous population of interneurons, distinctly marked by a variety of neurochemicals (Fig. 7.10K; Adli et al., 1999; Muñoz et al., 2000; Morona et al., 2006). However, the spinal gray is not differentiated into well-defined cell groups in amphibians and has, therefore, been divided into fields rather than laminae (ten Donkelaar, 1998a).

The complex complement of ascending and descending connections between the brain and the spinal cord has been thoroughly investigated with tract-tracing techniques in representatives of all three amphibian orders (see Muñoz et al., 1994, 1995, 1997; Sánchez-Camacho et al., 2001a; González and ten Donkelaar, 2007). The three main ascending sensory channels, each with largely separate targets, are (1) ascending projections via the dorsal funiculus, including primary and nonprimary projections that mainly terminate somatotopically in the dorsal column nucleus at obex levels; (2) projections ascending via the dorsolateral funiculus to innervate a region considered the amphibian homolog of the lateral cervical nucleus of mammals (Muñoz et al., 1996a), and via fibers ventral to the descending trigeminal tract to terminate in the solitary tract nucleus, the reticular formation, the descending trigeminal nucleus, and the medial aspect of the ventral octaval nucleus; (3) ascending spinal projections via the ventrolateral funiculi that innervate various parts of the reticular formation, the octavolateral area, the cerebellum, the torus semicircularis, the midbrain tegmentum and, sparsely, the OT. Beyond the midbrain, various thalamic and particularly prethalamic nuclei and the posterior tubercle are innervated by this third ascending sensory channel. The presence of these three channels appears to be a shared character in the central nervous system of both amniotes and anamniotes.

The largest descending pathways in amphibians arise in the vestibular part of the octavolateral area and the rhombencephalic reticular formation. Ipsilateral and contralateral spinal projections from the vestibular nuclei are present in all species studied (Fig. 7.10H). Another descending spinal pathway originates from the large Mauthner cells in the contralateral r4 of the hindbrain in urodeles; such a projection has not been described in adult anurans and gymnophionans. A specific serotonergic raphespinal system arising in the caudal portion of the raphe column is found in all amphibians studied (Tan and Miletic, 1990; Sánchez-Camacho et al., 2001a; González and ten Donkelaar,

2007). Similarly, cerebellospinal projections, primarily arising in a cerebellar nucleus located laterally in the cerebellar peduncle, have been demonstrated in all amphibians studied so far. At rostral levels in the hindbrain, amphibians possess cells projecting to the spinal cord from the superior reticular nucleus and from two specific centers: the LDT (cholinergic/nitrergic; López et al., 2007) and the locus coeruleus (noradrenergic; Fig. 7.10N; Sánchez-Camacho et al., 2001b). Spinal projections from the mesencephalon include predominantly contralateral tectospinal projections to the brachial cord and projections from the mesencephalic trigeminal nucleus in anurans (Fig. 7.10M). Amphibians possess spinal projections from the torus semicircularis, mainly from its laminar nucleus, and rather extensive tegmentospinal projections arise in the red nucleus and the interstitial nucleus of the medial longitudinal fascicle. Pretectospinal projections exist primarily in anurans, and descending pathways from the posterior tubercle and prethalamus (from the ventromedial nucleus and dopaminergic periventricular nucleus of the zona incerta; Fig. 7.10O) were demonstrated in the three orders. The suprachiasmatic nucleus, and the magnocellular and parvocellular groups of the Pa of the hypothalamus, together with neurons of the CeA, are the most rostral centers with direct descending projections to the spinal cord.

Finally, the gene expression patterns that lead to the development of the distinct zones and cell types have been conserved in the spinal cord of amphibians. For example, the finding that Pax6 and Pax7 expression distinguish dorsal and ventral regions in the ventricular zone has been corroborated in urodeles and anurans (Fig. 7.10L). This expression is maintained in adult urodeles, but not in *Xenopus* (Joven et al., 2013a,b; Bandín et al., 2014). Interestingly, adult urodeles and anuran larvae (for example, *Xenopus*) can regenerate their spinal cord after injury, and this capacity in the axolotl (urodele amphibian, *Ambystoma mexicanum*) has been related to Pax expression (Tanaka and Ferretti, 2009). Moreover, *Xenopus* can reestablish nerve tracts after spinal cord transection and achieve functional recovery, though this ability is restricted to the larval stages and lost at the end of metamorphosis, when Pax6 and Pax7 expression in the ventricular zone of the spinal cord is lost.

Acknowledgments

This survey largely relies on the extensive work accomplished over the years by our group and with former collaborators to whom we dedicate this chapter: Drs. Margarita Muñoz, Hans J. ten Donkelaar, Wilhelmus J. Smeets, Alberto Muñoz, Óscar Marín, Cristina Sánchez-Camacho, Sylvie Rétaux, Laura Domínguez, Alberto Joven, and Sandra Bandín. This work has been supported by the Spanish Ministry of Economy and Competitiveness (grants: BFU2012-31687 and BFU2015-66041-P).

References

Adli, D.S., Stuesse, S.L., Cruce, W.L., 1999. Immunohistochemical distribution of enkephalin, substance P, and somatostatin in the brainstem of the leopard frog, *Rana pipiens*. J. Comp. Neurol. 404, 387–407.

Allison, J.D., Wilczynski, W., 1991. Thalamic and midbrain auditory projections to the preoptic area and ventral hypothalamus in the green treefrog (*Hyla cinerea*). Brain Behav. Evol. 38, 322–331.

Andrews, G.L., Yun, K., Rubenstein, J.L., Mastick, G.S., 2003. Dlx transcription factors regulate differentiation of dopaminergic neurons of the ventral thalamus. Mol. Cell. Neurosci. 23, 107–120.

Ang, S.L., 2006. Transcriptional control of midbrain dopaminergic neuron development. Development 133, 3499–3506.

Aroca, P., Puelles, L., 2005. Postulated boundaries and differential fate in the developing rostral hindbrain. Brain Res. Brain Res. Rev. 49, 179–190.

Bachy, I., Berthon, J., Rétaux, S., 2002. Defining pallial and subpallial compartments in the developing *Xenopus* forebrain. Mech. Dev. 17, 163–172.

Bandín, S., Morona, R., Moreno, N., González, A., 2013. Regional expression of Pax7 in the brain of *Xenopus laevis* during embryonic and larval development. Front. Neuroanat. 7, 48. https://doi.org/10.3389/fnana.2013.00048.

Bandín, S., Morona, R., López, J.M., Moreno, N., González, A., 2014. Immunohistochemical analysis of Pax6 and Pax7 expression in the CNS of adult *Xenopus laevis*. J. Chem. Neuroanat. 57–58, 24–41.

Bandín, S., Morona, R., González, A., 2015. Prepatterning and patterning of the thalamus along embryonic development of *Xenopus laevis*. Front. Neuroanat. 9, 107. https://doi.org/10.3389/fnana.2015.00107.

Bergquist, H., 1932. Zur morphologie des Zwischenhirns bei niederen vertebraten. Acta Zool. (Stockholm) 13, 57–303.

Bergquist, H., Kallen, B., 1954. Notes on the early histogenesis and morphogenesis of the central nervous system in vertebrates. J. Comp. Neurol. 100, 627–659.

Birinyi, A., Straka, H., Matesz, C., Dieringer, N., 2001. Location of dye-coupled second order and efferent vestibular neurons labeled from individual semicircular canal otolith organs in the frog. Brain Res. 921, 44–59.

Blin, M., Norton, W., Bally-Cuif, L., Vernier, P., 2008. NR4A2 controls the differentiation of selective dopaminergic nuclei in the zebrafish brain. Mol. Cell. Neurosci. 39, 592–604.

Brahic, C.J., Kelley, D.B., 2003. Vocal circuitry in *Xenopus laevis*: telencephalon to laryngeal motor neurons. J. Comp. Neurol. 464, 115–130.

Brox, A., Puelles, L., Ferreiro, B., Medina, L., 2003. Expression of the genes GAD67 and Distal-less-4 in the forebrain of *Xenopus laevis* confirms a common pattern in tetrapods. J. Comp. Neurol. 461, 370–393.

Brox, A., Puelles, L., Ferreiro, B., Medina, L., 2004. Expression of the genes Emx1, Tbr1, and Eomes (Tbr2) in the telencephalon of *Xenopus laevis* confirms the existence of a ventral pallial division in all tetrapods. J. Comp. Neurol. 474, 562–577.

Bruce, L.L., Neary, T.J., 1995. The limbic system of tetrapods: a comparative analysis of cortical and amygdalar populations. Brain Behav. Evol. 46, 224–234.

Butler, A.B., 2009. Evolution of brain: at invertebrate–vertebrate transition. In: Binder, M.D., Hirokawa, N., Windhorst, U. (Eds.), Encyclopedia of Neuroscience. Springer-Verlag GMBH, Berlin Heidelberg, pp. 1236–1240.

Butler, A.B., Hodos, W., 2005. Comparative Vertebrate Neuroanatomy. Wiley, New Jersey.

Cambronero, F., Puelles, L., 2000. Rostrocaudal nuclear relationships in the avian medulla oblongata: a fate map with quail chick chimeras. J. Comp. Neurol. 427, 522–545.

Canteras, N.S., 2011. Hypothalamic goal-directed behaviours—Ingressive, reproductive and defensive. In: Watson, C., Paxinos, G., Puelles, L. (Eds.), The Mouse Nervous System. Academic Press-Elsevier, New York, pp. 539–563.

Chambers, D., Wilson, L.J., Alfonsi, F., Hunter, E., Saxena, U., Blanc, E., Lumsden, A., 2009. Rhombomere-specific analysis reveals the repertoire of genetic cues expressed across the developing hindbrain. Neural Dev. 4, 6. https://doi.org/10.1186/1749-8104-4-6.

Díaz, C., Yanes, C., Trujillo, C.M., Puelles, L., 2000. Cytoarchitectonic subdivisions in the subtectal midbrain of the lizard *Gallotia galloti*. J. Neurocytol. 29, 569–593.

Díaz, C., Morales-Delgado, N., Puelles, L., 2015. Ontogenesis of peptidergic neurons within the genoarchitectonic map of the mouse hypothalamus. Front. Neuroanat. 8, 162. https://doi.org/10.3389/fnana.2014.00162.

Dicke, U., Roth, G., 2007. Evolution of the amphibian nervous system. In: Kaas, J.H. (Ed.), Evolution of Nervous Systems, vol. II. Academic Press, Oxford, pp. 61–124.

Dicke, U., Roth, G., 2009. Evolution of the visual system in amphibians. In: Binder, M.D., Hirokawa, N., Windhorst, U. (Eds.), Encyclopedia of Neuroscience. Springer-Verlag GMBH, Berlin Heidelberg, pp. 1455–1459.

Domínguez, L., González, A., Moreno, N., 2010. Sonic hedgehog expression during *Xenopus laevis* forebrain development. Brain Res. 1347, 19–32.

Domínguez, L., González, A., Moreno, N., 2011. Ontogenetic distribution of the transcription factor Nkx2.2 in the developing forebrain of *Xenopus laevis*. Front. Neuroanat. 5, 11. https://doi.org/10.3389/fnana.2011.00011.

Domínguez, L., Morona, R., González, A., Moreno, N., 2013. Characterization of the hypothalamus of *Xenopus laevis* during development. I. The alar regions. J. Comp. Neurol. 521, 725–759.

Domínguez, L., González, A., Moreno, N., 2014. Characterization of the hypothalamus of *Xenopus laevis* during development. II. The basal regions. J. Comp. Neurol. 522, 1102–1131.

Domínguez, L., González, A., Moreno, N., 2015. Patterns of hypothalamic regionalization in amphibians and reptiles: common traits revealed by a genoarchitectonic approach. Front. Neuroanat. 9, 3. https://doi.org/10.3389/fnana.2015.00003.

Dudkin, E., Gruberg, E., 2009. Evolution of nucleus isthmi. In: Binder, M.D., Hirokawa, N., Windhorst, U. (Eds.), Encyclopedia of Neuroscience. Springer-Verlag GMBH, Berlin Heidelberg, pp. 1258–1262.

Endepols, H., Walkowiak, W., Luksch, H., 2000. Chemoarchitecture of the anuran auditory midbrain. Brain Res. Brain Res. Rev. 33, 179–198.

Endepols, H., Roden, K., Luksch, H., Dicke, U., Walkowiak, W., 2004. Dorsal striatopallidal system in anurans. J. Comp. Neurol. 468, 299–310.

Endepols, H., Roden, K., Walkowiak, W., 2005. Hodological characterization of the septum in anuran amphibians: II. Efferent connections. J. Comp. Neurol. 483, 437–457.

Endepols, H., Mühlenbrock-Lenter, S., Roth, G., Walkowiak, W., 2006. The septal complex of the fire-bellied toad *Bombina orientalis*: chemoarchitecture. J. Chem. Neuroanat. 31, 59–76.

Ewert, J.-P., 1997. Neural correlates of key stimulus and releasing mechanism: a case study and two concepts. Trends Neurosci. 20, 332–339.

Ewert, J.-P., Buxbaum-Conradi, H., Glagow, M., Röttgen, A., Schürg-Pfeiffer, E., Schwippert, W.W., 1999. Forebrain and midbrain structures in prey-catching behaviour of toads: stimulus-response mediating circuits and their modulating loops. Eur. J. Morphol. 37, 172–176.

Feller, A.E., Hedges, S.B., 1998. Molecular evidence for the early history of living amphibians. Mol. Phylogenet. Evol. 9, 509–516.

Feng, A.S., Hall, J.C., Gooler, D.M., 1990. Neural basis of sound pattern recognition in anurans. Prog. Neurobiol. 34, 313–329.

Ferran, J.L., Sánchez-Arrones, L., Sandoval, J.E., Puelles, L., 2007. A model of early molecular regionalization in the chicken embryonic pretectum. J. Comp. Neurol. 505, 379—403.

Ferran, J.L., Sánchez-Arrones, L., Bardet, S.M., Sandoval, J.E., Martínez-de-la-Torre, M., Puelles, L., 2008. Early pretectal gene expression pattern shows a conserved anteroposterior tripartition in mouse and chicken. Brain Res. Bull. 75, 295—298.

Ferran, J.L., de Oliveira, E.D., Merchan, P., et al., 2009. Genoarchitectonic profile of developing nuclear groups in the chicken pretectum. J. Comp. Neurol. 517, 405—451.

Fritzsch, B., 1988. The lateral-line and inner-ear afferents in larval and adult urodeles. Brain Behav. Evol. 31, 325—348.

Fritzsch, B., 1998. Of mice and genes: evolution of vertebrate brain development. Brain Behav. Evol. 52, 207—217.

Fritzsch, B., Glover, J.C., 2007. Evolution of the deuterostome Central nervous system: an intercalation of developmental patterning processes with cellular specification processes. In: Kaas, J.H. (Ed.), Evolution of Nervous Systems, vol. II. Academic Press, Oxford, pp. 1—24.

Fritzsch, B., Wahnschaffe, U., 1987. Electron microscopical evidence for common inner ear and lateral line efferents in urodeles. Neurosci. Lett. 81, 48—52.

Fritzsch, B., Nikundiwe, A.M., Will, U., 1984. Projection patterns of lateral-line afferents in anurans: a comparative HRP study. J. Comp. Neurol. 229, 451—469.

González, A., López, J.M., 2002. A forerunner of septohippocampal cholinergic system is present in amphibians. Neurosci. Lett. 327, 111—114.

González, A., Muñoz, M., 1987. Distribution and morphology of abducens motoneurons innervating the lateral rectus and retractor bulbi muscles in the frog Rana ridibunda. Neurosci. Lett. 79, 29—34.

González, A., Smeets, W.J.A.J., 1991. Comparative analysis of dopamine and tyrosine hydroxylase immunoreactivities in the brain of two amphibians, the anuran Rana ridibunda and the urodele Pleurodeles waltlii. J. Comp. Neurol. 303, 457—477.

González, A., Smeets, W.J.A.J., 1992. Comparative analysis of the vasotocinergic and mesotocinergic cells and fibers in the brain of two amphibians, the anuran Rana ridibunda and the urodele Pleurodeles waltlii. J. Comp. Neurol. 315, 53—73.

González, A., Smeets, W.J.A.J., 1994. Distribution of tyrosine hydroxylase immunoreactivity in the brain of Typhlonectes compressicauda (Amphibia, Gymnophiona): further assessment of primitive and derived traits of amphibian catecholamine systems. J. Chem. Neuroana 8, 19—32.

González, A., Smeets, W.J.A.J., 1997. Distribution of vasotocin- and mesotocin-like immunoreactivities in the brain of Typhlonectes compressicauda (Amphibia, Gymnophiona): further assessment of primitive and derived traits of amphibian neuropeptidergic systems. Cell Tissue Res. 287, 305—314.

González, A., ten Donkelaar, H.J., 2007. Comparative analysis of descending supraspinal projections in amphibians. In: Becker, C.G., Becker, T. (Eds.), Model Organisms in Central Nervous System Regeneration. Wiley-VCH Verlag GmbH, Weinheim, pp. 121—140.

González, A., Meredith, G.E., Roberts, B.L., 1993. Choline acetyltransferase immunoreactive neurons innervating labyrinthine and lateral line sense organs in amphibians. J. Comp. Neurol. 332, 258—268.

González, A., Tuinhof, R., Smeets, W.J.A.J., 1993. Distribution of tyrosine hydroxylase and dopamine immunoreactivities in the brain of the South African clawed frog Xenopus laevis. Anat. Embryol. 187, 193—201.

González, A., Muñoz, A., Muñoz, M., Marín, O., Arévalo, R., Porteros, A., Alonso, J.R., 1996. Nitric oxide synthase in the brain of a urodele amphibian (Pleurodeles waltl) and its relation to catecholaminergic neuronal structures. Brain Res. 727, 49—64.

González, A., López, J.M., Sánchez-Camacho, C., Marín, O., 2002. Regional expression of the homeobox gene NKX2.1 defines pallidal and interneuronal populations in the basal ganglia of amphibians. Neuroscience 114, 567—575.

Gonzalez, A., Lopez, J.M., Sanchez-Camacho, C., Marin, O., 2002. Localization of choline acetyltransferase (ChAT) immunoreactivity in the brain of a caecilian amphibian, Dermophis mexicanus (Amphibia: Gymnophiona). J. Comp. Neurol. 448, 249—267.

González, A., Moreno, N., López, J.M., 2002. Distribution of NADPH-diaphorase/nitric oxide synthase in the brain of the caecilian Dermophis mexicanus (Amphibia: Gymnophiona): comparative aspects in amphibians. Brain Behav. Evol. 60, 80—100.

González, A., Morona, R., Moreno, N., Bandín, S., López, J.M., 2014. Identification of striatal and pallidal regions in the subpallium of anamniotes. Brain Behav. Evol. 83, 93—103.

Hagemann, A.I.H., Scholpp, S., 2012. The tale of the three brothers — Shh, Wnt and Fgf Turing development of the thalamus. Front. Neurosci. 6, 76. https://doi.org/10.3389/fnins.2012.00076.

Herrick, C.J., 1910. The morphology of the forebrain in Amphibia and Reptilia. J. Comp. Neurol. 20, 413—547.

Herrick, C.J., 1948. The Brain of the Tiger Salamander Ambystoma tigrinum. The University of Chicago Press, Chicago.

Hodos, W., 2011. Evolution of the spinal cord. In: Binder, M.D., Hirokawa, N., Windhorst, U. (Eds.), Encyclopedia of Neuroscience. Springer-Verlag GMBH, Berlin Heidelberg, pp. 1419—1421.

Joven, A., Morona, R., González, A., Moreno, N., 2013. Expression patterns of Pax6 and Pax7 in the adult brain of a urodele amphibian. Pleurodeles waltl. J. Comp. Neurol. 521, 2088—2124.

Joven, A., Morona, R., González, A., Moreno, N., 2013. Spatiotemporal patterns of Pax3, Pax6, and Pax7 expression in the developing brain of a urodele amphibian. Pleurodeles waltl. J. Comp. Neurol. 521, 3913—3953.

Kolm, P.J., Apekin, V., Sive, H., 1997. Xenopus hindbrain patterning requires retinoid signaling. Dev. Biol. 192, 1—16.

Kuhlenbeck, H., 1922. Zur Morphologie des Gymnophionengehirns. Jen. Med. Nat. 58, 453—484.

Kuljis, R.O., Karten, H.J., 1982. Laminar organization of peptidelike immunoreactivity in the anuran optic tectum. J. Comp. Neurol. 212, 188—201.

Laberge, F., Roth, G., 2007. Organization of the sensory input to the telencephalon in the fire-bellied toad, Bombina orientalis. J. Comp. Neurol. 502, 55—74.

Laberge, F., Mühlenbrock-Lenter, S., Grunwald, W., Roth, G., 2006. Evolution of the amygdala: new insights from studies in amphibians. Brain Behav. Evol. 67, 177—187.

Lázár, G., 1984. Structure and connections of the frog optic tectum. In: Venegas, H. (Ed.), Comparative Neurology of the Optic Tectum. Plenum, New York, pp. 185—210.

Llinás, R., Precht, W., 1976. Frog Neurobiology. Springer-Verlag, Berlin Heidelberg New York.

López, J.M., Morona, R., Moreno, N., Domíguez, L., González, A., 2007. Origins of spinal cholinergic pathways in amphibians demonstrated by retrograde transport and choline acetyltransferase immunohistochemistry. Neurosci. Lett. 425, 73—77.

Lopez, J.M., Morona, R., Gonzalez, A., 2010. Immunohistochemical localization of DARPP-32 in the brain and spinal cord of anuran amphibians and its relation with the catecholaminergic system. J. Chem. Neuroanat. 40, 325—338.

Lorente-Cánovas, B., Marín, F., Corral-San-Miguel, R., Hidalgo-Sánchez, M., Ferran, J.L., Puelles, L., Aroca, P., 2012. Multiple origins, migratory paths and molecular profiles of cells populating the avian interpeduncular nucleus. Dev. Biol. 361, 12—26.

Manteuffel, G., Naujoks-Manteuffel, C., 1990. Anatomical connections and electrophysiological properties of toral and dorsal tegmental neurons in the terrestrial urodele Salamandra salamandra. J. Hirnforsch. 31, 65—76.

Marín, O., González, A., 1999. Origin of tectal cholinergic projections in amphibians. A combined study of choline acetyltransferase immunohistochemistry and retrograde transport of dextran amines. Vis. Neurosci. 16, 271—283.

Marín, F., Puelles, L., 1995. Morphological fate of rhombomeres in quail/chick chimeras: a segmental analysis of hindbrain nuclei. Eur. J. Neurosci. 7, 1714—1738.

Marín, O., Smeets, W.J.A.J., González, A., 1996. Do amphibians have a true locus coeruleus? Neuroreport 7, 1447—1451.

Marín, O., González, A., Smeets, W.J.A.J., 1997. Basal ganglia organization in amphibians: afferent connections to the striatum and the nucleus accumbens. J. Comp. Neurol. 378, 16—49.

Marín, O., González, A., Smeets, W.J.A.J., 1997. Basal ganglia organization in amphibians: efferent connections of the striatum and the nucleus accumbens. J. Comp. Neurol. 380, 23—50.

Marín, O., González, A., Smeets, W.J.A.J., 1997. Anatomical substrate of amphibian basal ganglia involvement in visuomotor behaviour. Eur. J. Neurosci. 9, 2100—2109.

Marín, O., Smeets, W.J.A.J., González, A., 1997. Basal ganglia organization in amphibians: catecholaminergic innervation of the striatum and the nucleus accumbens. J. Comp. Neurol. 378, 50—69.

Marín, O., Smeets, W.J.A.J., González, A., 1997. Distribution of choline acetyltransferase immunoreactivity in the brain of anuran (*Rana perezi, Xenopus laevis*) and urodele (*Pleurodeles waltl*) amphibians. J. Comp. Neurol. 382, 499—534.

Marín, O., Smeets, W.J.A.J., González, A., 1998. Basal ganglia organization in amphibians: chemoarchitecture. J. Comp. Neurol. 392, 285—312.

Marín, O., Smeets, W.J.A.J., González, A., 1998. Basal ganglia organization in amphibians: evidence for a common pattern in tetrapods. Prog. Neurobiol. 55, 363—397.

Marín, O., Smeets, W.J.A.J., González, A., 1998. Evolution of the basal ganglia in tetrapods: a new perspective based on recent studies in amphibians. Trends Neurosci. 21, 487—494.

Martínez-Ferre, A., Martínez, S., 2012. Molecular regionalization of the diencephalon. Front. Neurosci. 6, 73. https://doi.org/10.3389/fnins.2012.00073.

Martínez-García, F., Novejarque, A., Lanuza, E., 2007. Evolution of the amygdala in vertebrates. In: Kaas, J.H. (Ed.), Evolution of Nervous Systems, vol. II. Academic Press, Oxford, pp. 255—334.

Matesz, C., 1979. Central projection of the VIIIth cranial nerve in the frog. Neuroscience 4, 2061—2071.

Medina, L., 2008. Evolution and embryological development of forebrain. In: Binder, M.D., Hirokawa, N., Windhorst, U. (Eds.), Encyclopedia of Neuroscience. Springer-Verlag GMBH, Berlin Heidelberg, pp. 1172—1192.

Medina, L., Abellán, A., 2009. Development and evolution of the pallium. Semin. Cell Dev. Biol. 20, 698—711.

Milán, F.J., Puelles, L., 2000. Patterns of calretinin, calbindin, and tyrosine-hydroxylase expression are consistent with the prosomeric map of the frog diencephalon. J. Comp. Neurol. 419, 96—121.

Moreno, N., González, A., 2003. Hodological characterization of the medial amygdala in anuran amphibians. J. Comp. Neurol. 466, 389—408.

Moreno, N., González, A., 2004. Localization and connectivity of the lateral amygdala in anuran amphibians. J. Comp. Neurol. 479, 130—148.

Moreno, N., González, A., 2005. The central amygdala in anuran amphibians: neurochemical organization and connectivity. J. Comp. Neurol. 489, 69—91.

Moreno, N., González, A., 2005. Forebrain projections to the hypothalamus are topographically organized in anurans: conservative traits as compared with amniotes. Eur. J. Neurosci. 21, 1895—1910.

Moreno, N., González, A., 2006. The common organization of the amygdaloid complex in tetrapods: new concepts based on developmental, hodological and neurochemical data in anuran amphibians. Prog. Neurobiol. 78, 61—90.

Moreno, N., González, A., 2007. Evolution of the amygdaloid complex in vertebrates, with special reference to the anamnio-amniotic transition. J. Anat. 211, 151—163.

Moreno, N., González, A., 2007. Regionalization of the telencephalon in urodele amphibians and its bearing on the identification of the amygdaloid complex. Front. Neuroanat. 1, 1. https://doi.org/10.3389/neuro.05/001.2007.

Moreno, N., González, A., 2011. The non-evaginated secondary prosencephalon of vertebrates. Front. Neuroanat. 5, 12. https://doi.org/10.3389/fnana.2011.00012.

Moreno, N., Bachy, I., Rétaux, S., González, A., 2003. Pallial origin of mitral cells in the olfactory bulbs of Xenopus. Neuroreport 14, 2355—2358.

Moreno, N., Bachy, I., Rétaux, S., González, A., 2004. LIM-homeodomain genes as developmental and adult genetic markers of *Xenopus* forebrain functional subdivisions. J. Comp. Neurol. 472, 52—72.

Moreno, N., Bachy, I., Rétaux, S., González, A., 2005. LIM-homeodomain genes as territory markers in the brainstem of adult and developing *Xenopus laevis*. J. Comp. Neurol. 485, 240—254.

Moreno, N., Dominguez, L., Retaux, S., Gonzalez, A., 2008. Islet1 as a marker of subdivisions and cell types in the developing forebrain of Xenopus. Neuroscience 154, 1423—1439.

Moreno, N., González, A., Rétaux, S., 2008. Evidences for tangential migrations in *Xenopus* telencephalon: developmental patterns and cell tracking experiments. Dev. Neurobiol. 68, 504—520.

Moreno, N., Morona, R., Lopez, J.M., Domínguez, L., Muñoz, M., González, A., 2008. Anuran olfactory bulb organization: embryology, neurochemistry and hodology. Brain Res. Bull. 75, 241—245.

Moreno, N., Retaúx, S., González, A., 2008. Spatio-temporal expression of Pax6 in *Xenopus* forebrain. Brain Res. 1239, 92—99.

Moreno, N., González, A., Rétaux, S., 2009. Development and evolution of the subpallium. Semin. Cell Dev. Biol. 20, 735—743.

Moreno, N., Morona, R., Lopez, J.M., Domíguez, L., Joven, A., Bandín, S., Gonzalez, A., 2012. Characterization of the bed nucleus of the stria terminalis (BST) in the forebrain of anuran amphibians. J. Comp. Neurol. 520, 330—363.

Morona, R., González, A., 2008. Calbindin-D28k and calretinin expression in the forebrain of anuran and urodele amphibians: further support for newly identified subdivisions. J. Comp. Neurol. 511, 187—220.

Morona, R., González, A., 2009. Immunohistochemical localization of calbindin-D28k and calretinin in the brainstem of anuran and urodele amphibians. J. Comp. Neurol. 515, 503—537.

Morona, R., Moreno, N., López, J.M., González, A., 2006. Immunohistochemical localization of calbindin-D28k and calretinin in the spinal cord of *Xenopus laevis*. J. Comp. Neurol. 494, 763—783.

Morona, R., Ferran, J.L., Puelles, L., González, A., 2011. Embryonic genoarchitecture of the pretectum in *Xenopus laevis*: a conserved pattern in tetrapods. J. Comp. Neurol. 519, 1024—1050.

Morona, R., López, J.M., González, A., 2011. Localization of calbindin-D28k and calretinin in the brain of *Dermophis mexicanus* (Amphibia: Gymnophiona) and its bearing on the interpretation of newly recognized neuroanatomical regions. Brain Behav. Evol. 77, 231—269.

Muñoz, A., Muñoz, M., González, A., ten Donkelaar, H.J., 1994. Spinothalamic projections in amphibians as revealed with anterograde tracing techniques. Neurosci. Lett. 171, 81—84.

Muñoz, A., Muñoz, M., González, A., ten Donkelaar, H.J., 1995. Anuran dorsal column nucleus: organization, immunohistochemical characterization, and fiber connections in *Rana perezi* and *Xenopus laevis*. J. Comp. Neurol. 363, 197—220.

Muñoz, A., Muñoz, M., González, A., ten Donkelaar, H.J., 1996. Evidence for an anuran homologue of the mammalian spinocervicothalamic system: an *in vitro* tract-tracing study in *Xenopus laevis*. Eur. J. Neurosci. 8, 1390—1400.

Muñoz, M., Muñoz, A., Marín, O., Alonso, J.R., Arévalo, R., Porteros, A., González, A., 1996. Topographical distribution of NADPH-diaphorase activity in the central nervous system of the frog, *Rana perezi*. J. Comp. Neurol. 367, 54–69.

Muñoz, A., Muñoz, M., González, A., ten Donkelaar, H.J., 1997. Spinal ascending pathways in amphibians: cells of origin and main targets. J. Comp. Neurol. 378, 205–228.

Muñoz, M., Marín, O., González, A., 2000. Localization of NADPH diaphorase/nitric oxide synthase and choline acetyltransferase in the spinal cord of the frog, *Rana perezi*. J. Comp. Neurol. 419, 451–470.

Neary, T.J., 1990. The Pallium of Anuran Amphibians. In: Jones, E.G., Peters, A. (Eds.), Cerebral Cortex, Comparative Structure and Evolution of Cerebral Cortex, Part 1, vol. 8A. Plenum Press, New York, pp. 107–138.

Neary, T.J., Northcutt, R.G., 1983. Nuclear organization of the bullfrog diencephalon. J. Comp. Neurol. 213, 262–278.

Nieuwenhuys, R., 1998. Comparative neuroanatomy. Place, principles and programme. In: Nieuwenhuys, R., ten Donkelaar, H.J., Nicholson, C. (Eds.), The Central Nervous System of Vertebrates. Springer, London, pp. 273–327.

Nieuwenhuys, R., 2011. The structural, functional, and molecular organization of the brainstem. Front. Neuroanat. 5, 33. https://doi.org/10.3389/fnana.2011.00033.

Northcutt, R.G., Kicliter, E., 1980. Organization of the amphibian telencephalon. In: Ebbesson, S.O.E. (Ed.), Comparative Neurology of the Telencephalon. Plenum, New York, pp. 203–255.

Northcutt, R.G., Ronan, M., 1992. Afferent and efferent connections of the bullfrog medial pallium. Brain Behav. Evol. 40, 1–16.

Palmgren, A., 1921. Embryological and morphological studies on the midbrain and cerebellum of vertebrates. Acta Zool. (Stockholm) 2, 1–94.

Papalopulu, N., Kinter, C., 1993. *Xenopus* Distal-less related homeobox genes are expressed in the developing forebrain and are induced by planar signals. Development 117, 961–975.

Parent, A., 1986. Comparative Neurobiology of the Basal Ganglia. Wiley, New York.

Puelles, L., 2001. Brain segmentation and forebrain development in amniotes. Brain Res. Bull. 55, 695–710.

Puelles, L., 2013. Plan of the developing vertebrate nervous system: relating embryology to the adult nervous system (prosomere model, overview of brain organization). In: Rubenstein, J.R., Rakic, P. (Eds.), Patterning and Cell Type Specification in the Developing CNS and PNS: Comprehensive Developmental Neuroscience, vol. 1. Academic Press, Elsevier, Amsterdam, Boston, pp. 187–209.

Puelles, L., Ferran, J.L., 2012. Concept of neural genoarchitecture and its genomic fundament. Front. Neuroanat. 6, 47. https://doi.org/10.3389/fnana.2012.00047.

Puelles, L., Rubenstein, J.L., 1993. Expression patterns of homeobox and other putative regulatory genes in the embryonic mouse forebrain suggest a neuromeric organization. Trends Neurosci. 16, 472–479.

Puelles, L., Rubenstein, J.L., 2003. Forebrain gene expression domains and the evolving prosomeric model. Trends Neurosci. 26, 469–476.

Puelles, L., Rubenstein, J.L., 2015. A new scenario of hypothalamic organization: rationale of new hypotheses introduced in the updated prosomeric model. Front. Neuroanat. 9, 27. https://doi.org/10.3389/fnana.2015.00027.

Puelles, L., Milán, F.J., Martinez-de-la-Torre, M., 1996. A segmental map of architectonic subdivisions in the diencephalon of the frog *Rana perezi*: acetylcholinesterase-histochemical observations. Brain Behav. Evol. 47, 279–310.

Puelles, L., Kuwana, E., Puelles, E., Bulfone, A., Shimamura, K., Keleher, J., Smiga, S., Rubenstein, J.L., 2000. Pallial and subpallial derivatives in the embryonic chick and mouse telencephalon, traced by the expression of the genes Dlx2, Emx1, Nkx2.1, Pax6 and Tbr1. J. Comp. Neurol. 424, 409–438.

Puelles, E., Martínez-de-la-Torre, M., Watson, C., Puelles, L., 2012. Midbrain. In: Watson, C., Paxinos, G., Puelles, L. (Eds.), The Mouse Nervous System. Academic Press-Elsevier, New York, pp. 337–359.

Pyron, R.A., Wiens, J.J., 2011. A large-scale phylogeny of Amphibia that includes over 2,800 species, and a revised classification of extant frogs, salamanders, and caecilians. Mol. Phylogenet. Evol. 61, 543–583.

Rétaux, S., Rogard, M., Bach, I., Failli, V., Besson, M.J., 1999. Lhx9: a novel LIM-homeodomain gene expressed in the developing forebrain. J. Neurosci. 19, 783–793.

Roden, K., Endepols, H., Walkowiak, W., 2005. Hodological characterization of the septum in anuran amphibians: I. Afferent connections. J. Comp. Neurol. 483, 415–436.

Roth, G., Naujoks-Manteuffel, C., Nishikawa, K., Schmidt, A., Wake, D.B., 1993. The salamander nervous system as a secondarily simplified, paedomorphic system. Brain Behav. Evol. 42, 137–170.

Roth, G., Grunwald, W., Dicke, U., 2003. Morphology, axonal projection pattern, and responses to optic nerve stimulation of thalamic neurons in the fire-bellied toad *Bombina orientalis*. J. Comp. Neurol. 461, 91–110.

Roth, G., Laberge, F., Mühlenbrock-Lenter, S., Grunwald, W., 2007. Organization of the pallium in the fire-bellied toad *Bombina orientalis*. I. Morphology and axonal projection pattern of neurons revealed by intracellular biocytin labeling. J. Comp. Neurol. 501, 443–464.

Roth, M., Bonev, B., Lindsay, J., Lea, R., Panagiotaki, N., Houart, C., Papalopulu, N., 2010. FoxG1 TLE2 act cooperatively to regulate ventral telencephalon formation. Development 137, 1553–1562.

Saidel, W.M., 2009. Evolution of the optic tectum in anamniotes. In: Binder, M.D., Hirokawa, N., Windhorst, U. (Eds.), Encyclopedia of Neuroscience. Springer-Verlag GMBH, Berlin Heidelberg, pp. 1380–1387.

San Mauro, D., 2010. A multilocus timescale for the origin of extant amphibians. Mol. Phylogenet. Evol. 56, 554–561.

San Mauro, D., Gower, D.J., Müller, H., Loader, S.P., Zardoya, R., Nussbaum, R.A., Wilkinson, M., 2014. Life-history evolution and mitogenomic phylogeny of caecilian amphibians. Mol. Phylogenet. Evol. 73, 177–189.

Sánchez-Arrones, L., Ferran, J.L., Rodríguez-Gallardo, L., Puelles, L., 2009. Incipient forebrain boundaries traced by differential gene expression and fate mapping in the chick neural plate. Dev. Biol. 335, 43–65.

Sánchez-Camacho, C., Marín, O., ten Donkelaar, H.J., González, A., 2001. Descending supraspinal pathways in amphibians. I. A dextran amine tracing study of their cells of origin. J. Comp. Neurol. 434, 186–208.

Sánchez-Camacho, C., Marín, O., Smeets, W.J.A.J., ten Donkelaar, H.J., González, A., 2001. Descending supraspinal pathways in amphibians. II. Distribution and origin of the catecholaminergic innervation of the spinal cord. J. Comp. Neurol. 434, 209–232.

Sánchez-Camacho, C., Peña, J.J., González, A., 2003. Catecholaminergic innervation of the septum in the frog: a combined immunohistochemical and tract-tracing study. J. Comp. Neurol. 455, 310–323.

Sánchez-Camacho, C., López, J.M., González, A., 2006. The basal forebrain cholinergic system of the anuran amphibian *Rana perezi*: evidence for a shared organization pattern with amniotes. J. Comp. Neurol. 494, 961–975.

Scalia, F., 1972. The projection of the accessory olfactory bulb in the frog. Brain Res. 36, 409–411.

Scalia, F., 1976. Structure of the olfactory accesory systems. In: Llinás, R., Precht, W. (Eds.), Frog Neurobiology. Springer-Verlag, Berlin, pp. 213–233.

Scalia, F., Gallousis, G., Roca, S., 1991. Differential projections of the main and accessory olfactory bulb in the frog. J. Comp. Neurol. 305, 443–461.

Scholpp, S., Lumsden, A., 2010. Building a bridal chamber: development of the thalamus. Trends Neurosci. 33, 373–380.

Semba, K., 2004. Phylogenetic and ontogenetic aspects of the basal forebrain cholinergic neurons and their innervation of the cerebral cortex. Prog. Brain Res. 145, 3–43.

Smeets, W.J.A.J., González, A., 2000. Catecholamine systems in the brain of vertebrates: new perspectives through a comparative approach. Brain Res. Rev. 33, 308–379.

Smeets, W.J.A.J., González, A., 2001. Vasotocin and mesotocin in the brains of amphibians: state of the art. Microsc. Res. Tech. 54, 125–136.

Smeets, W.J.A.J., Marín, O., González, A., 2000. Evolution of the basal ganglia: new perspectives through a comparative approach. J. Anat. 196, 501–517.

Smith-Fernández, A., Pieau, C., Repérant, J., Boncinelli, E., Wassef, M., 1998. Expression of the *Emx-1* and *Dlx-1* homeobox genes define three molecularly distinct domains in the telencephalon of mouse, chick, turtle and frog embryos: implications for the evolution of telencephalic subdivisions in amniotes. Development 125, 2099–2111.

Straka, H., Baker, R., 2013. Vestibular blueprint in early vertebrates. Front. Neural Circuits 7, 182. https://doi.org/10.3389/fncir.2013.00182.

Straka, H., Baker, R., Gilland, E., 2006. Preservation of segmental hindbrain organization in adult frogs. J. Comp. Neurol. 494, 228–245.

Stuesse, S.L., Adli, D.S., Cruce, W.L., 2001. Immunohistochemical distribution of enkephalin, substance P, and somatostatin in the brainstem of the leopard frog, *Rana pipiens*. Microsc. Res. Tech. 54, 229–245.

Tan, H.J., Miletic, V., 1990. Bulbospinal serotoninergic pathways in the frog *Rana pipiens*. J. Comp. Neurol. 292, 291–302.

Tanaka, E.M., Ferretti, P., 2009. Considering the evolution of regeneration in the central nervous system. Nat. Rev. Neurosci. 10, 713–723.

ten Donkelaar, H.J., 1998. Urodeles. In: Nieuwenhuys, R., ten Donkelaar, H.J., Nicholson, C. (Eds.), The Central Nervous System of Vertebrates. Springer, Berlin, pp. 1045–1150.

ten Donkelaar, H.J., 1998. Anurans. In: Nieuwenhuys, R., ten Donkelaar, H.J., Nicholson, C. (Eds.), The Central Nervous System of Vertebrates. Springer, London, pp. 1151–1314.

ten Donkelaar, H.J., 1982. Organization of descending pathways to the spinal cord in amphibians and reptiles. Prog. Brain Res. 57, 25–67.

Velázquez-Ulloa, N.A., Spitzer, N.C., Dulcis, D., 2011. Contexts for dopamine specification by calcium spike activity in the CNS. J. Neurosci. 31, 78–88.

Wagner, G.P., Chiu, C., 2003. Genetic and epigenetic factors in the origin of the tetrapod limb. In: Muller, G.B., Newman, S.A. (Eds.), Origination of Organismal Form. MIT Press, Cambridge, MA, pp. 265–285.

Wake, M.H., 1993. Evolutionary diversification of cranial and spinal nerves and their targets in the gymnophione amphibians. Acta Anat. 148, 160–168.

Westhoff, G., Roth, G., 2002. Morphology and projection pattern of medial and dorsal pallial neurons in the frog *Discoglossus pictus* and the salamander *Plethodon jordani*. J. Comp. Neurol. 445, 97–121.

Westhoff, G., Roth, G., Straka, H., 2004. Topographic representation of vestibular and somatosensory signals in the anuran thalamus. Neuroscience 124, 669–683.

Wilczynski, W., 2009. Evolution of brain in amphibians. In: Binder, M.D., Hirokawa, N., Windhorst, U. (Eds.), Encyclopedia of Neuroscience. Springer-Verlag GMBH, Berlin Heidelberg, pp. 1301–1304.

Wilczynski, W., Endepols, H., 2007. Central auditory pathways in anuran amphibians: the anatomical basis of hearing and sound communication. In: Narins, P.M., Feng, A.S., Fay, R.R., Popper, A.N. (Eds.), Hearing and Sound Communication in Amphibians. Springer, Berlin, pp. 221–249.

Wilczynski, W., Northcutt, R.G., 1983. Connections of the bullfrog striatum: afferent organization. J. Comp. Neurol. 214, 321–332.

Wilczynski, W., Northcutt, R.G., 1983. Connections of the bullfrog striatum: efferent projections. J. Comp. Neurol. 214, 333–343.

Wilczynski, W., Allison, J.D., Marler, C.A., 1993. Sensory pathways linking social and environmental cues to endocrine control regions of amphibian forebrains. Brain Behav. Evol. 42, 252–264.

Will, U., 1982. Efferent neurons of the lateral-line system and the VIII cranial nerve in the brainstem of anurans. A comparative study using retrograde tracer methods. Cell Tissue Res. 225, 673–685.

Will, U., Luhede, G., Görner, P., 1985. The area octavo-lateralis in *Xenopus laevis*. II. Second order projections and cytoarchitecture. Cell Tissue Res. 239, 163–175.

Wullimann, M.F., Mueller, T., Distel, M., Babaryka, A., Grothe, B., Köster, R.W., 2011. The long adventurous journey of rhombic lip cells in jawed vertebrates: a comparative developmental analysis. Front. Neuroanat. 5, 27. https://doi.org/10.3389/fnana.2011.00027.

Zhang, P., Chen, Y.Q., Zhou, H., Wang, X.L., Qu, L.H., 2003. The complete mitochondrial genome of a relic salamander, *Ranodon sibiricus* (Amphibia: Caudata) and implications for amphibian phylogeny. Mol. Phylogenet. Evol. 28, 620–626.

Zhao, B., Debski, E.A., 2005. Serotonergic reticular formation cells in *Rana pipiens*: categorization, development, and tectal projections. J. Comp. Neurol. 487, 441–456.

Zhao, Y., Sheng, H., Amini, R., Grinberg, A., Lee, E., Huang, S., Taira, M., Westphal, H., 1999. Control of hippocampal morphogenesis and neuronal differentiation by the LIM homeobox gene *Lhx5*. Science 284, 1155–1158.

Zittlau, K.E., Claas, B., Münz, H., Gorner, P., 1985. Multisensory interaction in the torus semicircularis of the clawed toad *Xenopus laevis*. Neurosci. Lett. 60, 77–81.

Relevant Website

http://www.amphibiaweb.org—AmphibiaWeb Provides Information on Amphibian Declines, Natural History, Conservation, and Taxonomy (accessed on 18.05.16.)

8

The Brains of Reptiles and Birds

O. Güntürkün, M. Stacho, F. Ströckens

Ruhr-University Bochum, Bochum, Germany

Abbreviations

AA Arcopallium anterior
Ac Nucleus accumbens
AD Arcopallium dorsale
ADVR Anterior dorsal ventricular ridge
AFP Anterior forebrain pathway
AI Arcopallium intermedium
AIvm Arcopallium intermedium pars ventromedialis
AM Arcopallium mediale
AOB Accessory olfactory bulb
APH Area parahippocampalis
AV Arcopallium ventrale
AVT Area ventralis tegmentalis
Bas Nucleus basorostralis palii
BO Bulbus olfactorius
BSTL Bed nucleus of the stria terminalis, lateral part
CDL Area corticoidea dorsolateralis
CG Nucleus cuneatus and gracilis
CM Caudomedial mesopallium
CPi Cortex piriformis
CPP Cortex prepiriformis
CTB Crossed tectobulbar pathway
D Nucleus of Darkschewitsch
DA Tractus dorsoarcopallialis
DCN Dorsal column nuclei
DIP Nucleus dorsointermedius posterior thalami
DLP Nucleus dorsolateralis posterior thalami
DLM Nucleus dorsolateralis medialis thalami
DM Dorsal medial nucleus of the midbrain
DVR Dorsal ventricular ridge
Ed Entopallium dorsale
Ee Entopallium externum
Ei Entopallium internum
EION Ectopic isthmooptic neurons
Ep Entopallial belt
Ev Entopallium ventrale
Field L1
Field L2
Field L2a
Field L3
GCt Substantia grisea centralis
GLd N. geniculatus lateralis pars dorsalis
GP Globus pallidus
HA Hyperpallium apicale

HD Hyperpallium densocellulare
HI Hyperpallium intercalatum
HL Hyperpallium laterale
HOM Tractus occipitomesencephalicus pars hypothalami
Hp Hippocampus
Hp-DM Dorsomedial nucleus of the hippocampus
Hp-VM Ventromedial nucleus of hippocampus
HVC Letter-based name
Hypoth Hypothalamus
IC Inferior colliculus
ICo Nucleus intercollicularis
IHA Nucleus interstitialis hyperpallii apicalis
INP Nucleus intrapeduncularis INP
Imc Nucleus isthmi pars magnocellularis
INL Inner nuclear layer
IPL Inner plexiform layer
ION N. isthmoopticus
Ipc N. isthmi pars parvocellularis
IS N. of interstitialis Cajal
ITP Ipsilateral tectopontine−tectoreticular pathway
L2/3, L4, L5 Cortical layer 2/3, 4, 5
LFS Lamina frontalis superior
LL Nucleus lemniscus laminaris
LMAN Lateral magnocellular nucleus of anterior nidopallium
LoC Locus coeruleus
MC Mesopallium caudale
MD Mesopallium dorsale
MFV Mesopallium frontoventrale
MLD Nucleus mesencephalicus lateralis pars dorsalis
MM Mesopallium mediale
MOB Main olfactory bulb
MSt Medial striatum
MVex Mesopallium ventrale externum
MVL Mesopallium ventrolaterale
NA N. angularis
NCL Nidopallium caudolaterale
NCM Nidopallium caudomediale
NCVl Nidopallium caudoventrale pars lateralis
NDB N. diagonalis Broca
NFL Nidopallium frontolaterale
NFT Nidopallium frontotrigeminale
NFM Nidopallium frontomediale
NI Nidopallium intermedium
NIf Nucleus interface
NIMl Nidopallium intermedium mediale pars lateralis
NL N. laminaris

NM N. magnocellularis
NMm Nidopallium mediale pars medialis
NIL Nidopallium intermedium laterale
NSTL Nucleus of the stria terminalis
nXIIts Tracheosyringeal part of the nucleus hypoglossus
OS Nucleus olivaris superior
Ov Nucleus ovoidalis
Ov shell Shell of the nucleus ovoidalis
PMI Nucleus paramedianus internus thalami
PoA Nucleus posterioris amygdalopallii
PPC Nucleus principalis precommissuralis
Preopt Preoptic area
R Rhombencephalic tegmental field
RA Robust nucleus of the arcopallium
Re Nucleus reuniens
SNpc Substantia nigra pars parvocellularis
SL Septum laterale
SIu Nucleus isthmi pars semilunaris
SM Septum mediale
SMP Posterior song motor pathway
SPO Nucleus semilunaris parovoidalis
SQ Spinal quotient
SRt Nucleus subrotundus
StL Striatum laterale
StM Striatum mediale
TnA N. taeniae of the amygdala
TO Tectum opticum
TTD Nucleus of the tractus descendens nervi trigemini
TuO Tuberculum olfactorium
Uva Nucleus uvaeformis
VNO Vomeronasal organ
VP Ventral pallidum

Nothing in neuroscience makes sense, except in the light of behavior.

8.1 The Phylogeny of Reptiles and Birds

About 340 million years ago, a group of vertebrates developed the ability to reproduce on land. This evolutionary breakthrough became possible through major changes in the structure of the egg that evolved a fibrous shell membrane (the amnion) that permits sufficient gas exchange but still protects the embryo from drying out. At the same time, the adult forms of these animals started to have keratin-based dry skin with which they protected themselves against the absence of moisture in most areas of land. These changes granted them the ability to move away from coastal areas, even for reproduction. This group of animals would later be called reptile-like amphibians or reptiliomorphs, and we are their descendants.

Slowly, reptiliomorphs became more and more adapted to life on land and spread across the vast territories of our planet's continents. By 312 million years ago, in the late Carboniferous geological period, these changes had finally resulted in the emergence of the first true amniotes, defined as a group of animals characterized by the possession of an egg with sophisticated extraembryonic membranes (Benton and Donoghue, 2006).

The word amnion in classic Greek described a dish in which the blood of sacrificed animals was caught. In Latin it means "membrane around a fetus"—a meaning that resonates better with the critical morphological feature of the amniote egg. Amniotes are a monophyletic group that consists of mammals, reptiles, and birds. Classically they were subdivided on the basis of the number of openings ("apses") on the sides of their skulls. In turtles these openings are missing, which is why they are called "anapsids"—a condition that was often understood as a signature of basal amniotes. Other amniote groups have one ("synapsids") or two ("diapsids") openings on each side (ten Donkelaar, 1998). Since synapsids have one opening more than anapsids, they were thought to represent the first group that diverged from the ancestral line. They constituted the protomammals and later became today's modern mammals. Their single opening is on the ventral part of each side of their skulls. Subsequently, a group of animals developed a second pair of openings at a more dorsal skull position. These animals are called diapsids and are constituted by crocodilians, birds, tuataras, lizards, and snakes.

This kind of evolutionary scenario frames mammals (synapsids) between turtles (anapsids) on the one

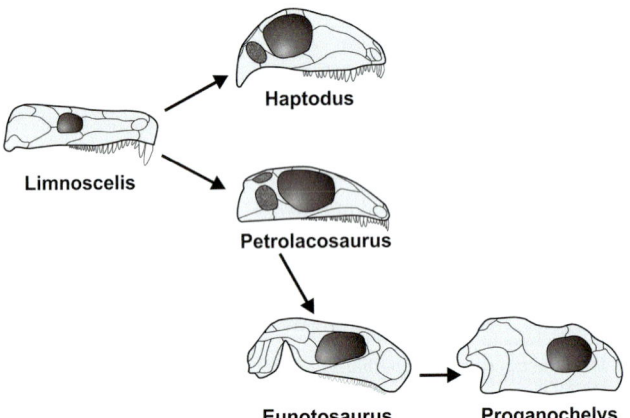

FIGURE 8.1 Generalized phylogeny of amniote skulls. *Arrows* do not imply biological descent but represent transformations in the fenestrae of skulls. *Limnoscelis* is a reptile-like amphibian from the early Permian that retained the anapsid condition of the amniote stem. *Haptodus* is a protomammal from the Carboniferous/Permian transition that shows the synapsid condition with lower temporal fenestrae (*gray/white* speckled area). *Petrolacosaurus* is a reptile from Carbon with a classic diapsid skull. *Eunotosaurus* represents a transitional form in turtle evolution from the late middle Permian with lower temporal fenestrae that are open ventrally and thus look like a prominent invagination. In the juvenile form an upper temporal fenestra is also present. *Proganochelys* is an uncontroversial stem turtle from the late Triassic that shows the classic anapsid condition. Based on the condition in *Eunotosaurus*, the anapsid state of turtles is not a basal but a derived condition. *Modified from Carroll, R.L., 1988. Vertebrate Paleontology and Evolution. W.H. Freeman, New York; Bever, G.S., Lyson, T.R., Field, D.J., Bhullar, B.A.S., 2015. Evolutionary origin of the turtle skull. Nature 525, 239—242.*

hand and diapsid reptiles on the other (ten Donkelaar, 1998). This view on the phylogenetic positioning of turtles seriously eroded in the beginning of the 2000s. Three novel hypotheses emerged. The first hypothesis saw turtles as the extant sister group to crocodiles and birds (Hugall et al., 2007); the second assumed that turtles are the sister group of the lizard—tuatara clade (Lyson et al., 2010), while the third hypothesis placed turtles inside diapsids (Shaffer et al., 2013). Major breakthroughs in gene sequence data (Wang et al., 2013), miRNA analyses (Field et al., 2014), and morphological discoveries (Bever et al., 2015) have largely clarified this issue. Careful analyses on *Eunotosaurus africanus*, a member of an extinct genus of close relatives of turtles from the Middle Permian, have shown that today's anapsid turtles are in fact previous diapsids that became anapsids secondarily (Bever et al., 2015). Thus, turtles started phylogenetically with two openings on each side of the skull and then lost them, giving the appearance of them as being a basal clade (Fig. 8.1).

These and further discoveries now enable a much more concise view on the phylogeny of reptiles and birds. These two groups comprise the sauropsids. In fact, as descendants from dinosaurs, birds could be called "flying reptiles" (Striedter, 2005). However, based on a cladistics analysis of shared derived traits, reptiles are not a monophyletic evolutionary group since it is impossible to define a single common ancestor that includes all reptiles but excludes all nonreptiles such as birds (Fig. 8.2). Aligators and crocodiles, for example, are actually more closely related to birds than to other reptilian lineages (Shine, 2013). Thus, it makes sense to combine sauropsids in one chapter when talking about their brains. Together, these two classes of vertebrates represent more than 18 000 species that live in all major ecosystems of our planet. If we aim to understand the deeper structure of our own brain, we have to study both mammalian and sauropsid brains. Only then can we identify the phylogenetic past and the variations and constancies among amniote brains of which we inherited the primate version.

8.2 Reptilian and Avian Brains in Numbers

8.2.1 Brain Size and Cognition: A Difficult Relation

It is often claimed that brain size is a predictor of an animal's cognitive abilities. This idea can be traced back to Aristotle, who wrote in his text peri zôôn moriôn

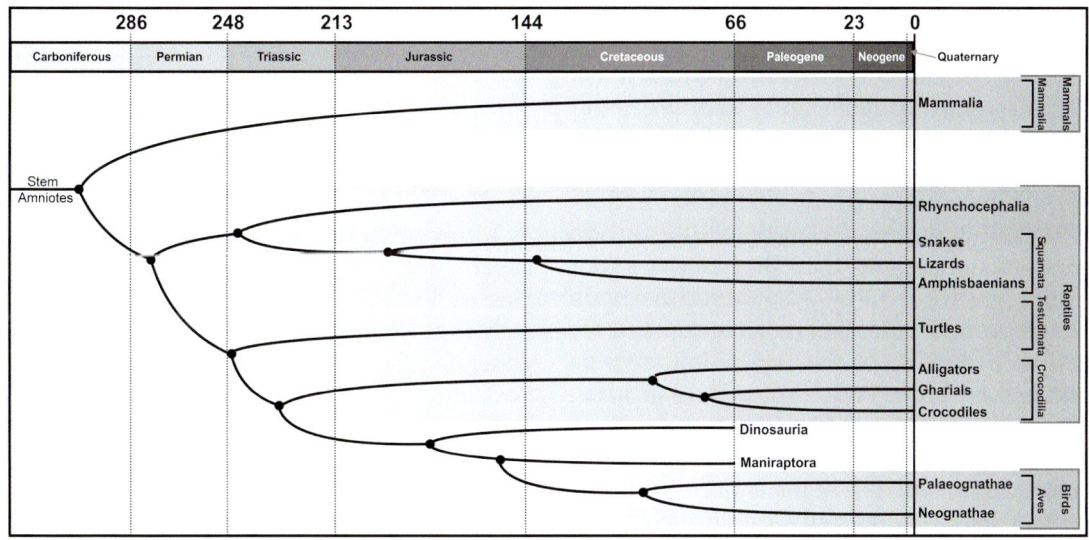

FIGURE 8.2 Genealogical tree of amniotes. The phylogeny shows the amniote radiation along with the time points of the last common ancestors for a given clade. Numbers across the top depict the time before present in millions of years. Geological eras are also shown along the top. Dinosaurs and Maniraptora are shown as extinct relatives of modern birds. *Based on information from Evans, S.E., 2003. At the feet of the dinosaurs: the early history and radiation of lizards. Biol. Rev. 78, 513–551; Green, R.E., Braun, E.L., Armstrong, J., Earl, D., Nguyen, N., Hickey, G., Vandewege, M.W., St John, J.A., Capella-Gutiérrez, S., Castoe, T.A., Kern, C., Fujita, M.K., Opazo, J.C., Jurka, J., Kojima, K.K., Caballero, J., Hubley, R.M., Smit, A.F., Platt, R.N., Lavoie, C.A., Ramakodi, M.P., Finger Jr. J.W., Suh, A., Isberg, S.R., Miles, L., Chong, A.Y., Jaratlerdsiri, W., Gongora, J., Moran, C., Iriarte, A., McCormack, J., Burgess, S.C., Edwards, S.V., Lyons, E., Williams, C., Breen, M., Howard, J.T., Gresham, C.R., Peterson, D.G., Schmitz, J., Pollock, D.D., Haussler, D., Triplett, E.W., Zhang, G., Irie, N., Jarvis, E.D., Brochu, C.A., Schmidt, C.J., McCarthy, F.M., Faircloth, B.C., Hoffmann, F.G., Glenn, T.C., Gabaldón, T., Paten, B., Ray, D.A., 2014. Three crocodilian genomes reveal ancestral patterns of evolution among archosaurs. Science 346, 1254449; Xu, X., Zhou, Z., Dudley, R., Mackem, S., Chuong, C.M., Erickson, G.M., Varricchio, D.J., 2014. An integrative approach to understand bird origins. Science 346, 1253293; Brusatte, S.L., O'Connor, J.K., Jarvis, E.D., 2015. The origin and diversification of birds. Curr. Biol. 25, R888–R898; Prum, R.O., Berv, J.S., Dornburg, A., Field, D.J., Townsend, J.P., Lemmon, E.M., Lemmon, A.R., 2015. A comprehensive phylogeny of birds (Aves) using targeted next-generation DNA sequencing. Nature 526, 569–573.*

(Greek, "On the Parts of Animals"): "Of all animals, man has the largest brain in proportion to his size" (Jerison, 1977). Based on this statement and rightfully assuming that humans possess the highest cognitive abilities of all species, one could conclude that high cognitive abilities or "intelligence" is based purely on the size of the brain. What would that mean for reptiles and birds? In comparison to several mammalian species, reptiles and birds have very small brains, in many cases even in relation to their body mass. Crocodilians, which represent the largest living reptiles (with Nile and saltwater crocodiles sometimes weighing more than 700 kg; Northcutt, 2012), possess brains that weigh only 10−20 g (Northcutt, 2012; Ngwenya et al., 2013, 2016). Paleognathous birds, such as emus and ostriches, with body weights from 60 kg (in emus) to 200 kg (in ostriches) have the largest avian brains, weighing 20−27 g (Peng et al., 2010; Olkowicz et al., 2016). Compared to mammals with approximately the same body mass (eg, horses, sheep, or chimpanzees), these reptilian and avian brain sizes are relatively small, both in terms of absolute size and in relation to body size (from here on called relative brain size) (Roth and Dicke, 2005; Northcutt, 2012). Taking humans into account, with their average body mass of 70 kg and an average brain size of 1450 g (Roth and Dicke, 2005; Herculano-Houzel, 2012), the prospects for higher cognitive abilities in reptiles and birds would seem rather dire, if one assumes that those abilities depend solely on brain size.

Fortunately, the assumption that absolute or relative brain sizes have a causal relationship to complex cognition in vertebrates has come under fire. Sperm Whales and Killer Whales possess the highest absolute brain size in the vertebrate class, reaching up to 9000 g (Roth and Dicke, 2005; Ridgway and Hanson, 2014). But do we have reasons to assume that they surpass our human-typical cognitive abilities? Also for relative brain size, it is not the primate order that ranks on top, but the small, mole-like mammals of the order Eulipotyphla that have the highest brain/body ratios. The European pygmy shrew with a body weight of 4.7 g and a brain weight of 0.1 g has a brain/body ratio of 0.021, which is higher than the one found in humans (Jerison, 1977).

In line with these findings and despite their small brains, reptiles and especially some bird species possess highly complex cognitive abilities. Recent studies demonstrated that some reptilian species are capable of social learning (Wilkinson et al., 2010), problem solving behavior, and rapid associative and reversal learning (Leal and Powell, 2012). It has also been argued that modern reptiles might even have evolved a form of consciousness, possibly independently of consciousness in recent bird and mammalian species (Northcutt, 2012). For birds, a plethora of studies have shown that species from the corvid and parrot orders show cognitive abilities that are on par with those of nonhuman primates when it comes to tool and metatool use (Hunt, 1996; Taylor et al., 2007; Bird and Emery, 2009; Auersperg et al., 2011), mirror-self-recognition (Prior et al., 2008), causal reasoning (Emery and Clayton, 2004; Taylor et al., 2009; Mikolasch et al., 2011; Pepperberg et al., 2013), future planning (Clayton et al., 2003), and imagination (Emery and Clayton, 2001, 2004; for a review, see Güntürkün and Bugnyar, 2016).

Due to the discrepancies between absolute and relative brain mass measures on one side and cognitive abilities on the other, diverse measures were developed to come up with a satisfying correlation between body size, brain size, and cognitive abilities in vertebrates. Proposed measures were, for example, the encephalization quotient (Jerison and Barlow, 1985), brain region relative to total brain size (Krebs et al., 1989), or the use of brain surface instead of brain volume (Sultan, 2002). However, all these attempts were criticized as not being able to explain convincingly the distribution of higher cognitive abilities in vertebrates (Healy and Rowe, 2007).

A recent and more promising approach suggests that neuron numbers per telencephalic volume could explain cognitive skills of a species (Herculano-Houzel, 2011a). Along that line, a scaling analysis of how many neurons are gained as brain volume increases in a given order may also shed light on the cognitive abilities of corvid and parrot species (Olkowicz et al., 2016). This approach will be discussed more thoroughly in the next section.

8.2.2 Brain Sizes in Reptilian and Avian Species

Although brain size alone may not predict cognitive capabilities of a given vertebrate species or taxon, analysis of this rather simple measure allows valuable insights in the evolution of the nervous system. In general, brain mass correlates with body mass over all vertebrates (Martin, 1981), leading to the assumption that bigger bodies need bigger brains (but see below and Ngwenya et al., 2016). However, there are striking differences in relative brain size between vertebrate classes. On average, relative brain sizes are 10 times smaller in reptiles and ray-finned fishes than in birds and mammals, with the latter having rather similar relative brain sizes (Martin, 1981; van Dongen, 1998; Northcutt, 2012). This seems also to be the case for extinct dinosaur species which had, based on endocranial volume measures, relative brain sizes similar to those of modern crocodiles (Jerison, 1973; van Dongen, 1998). In recent reptiles, brain sizes range from 0.03 g in tiny lizard species, over 0.5 g in the tuatara and 1.1 g in varanid species, to 20 g in crocodiles (van Dongen, 1998; Northcutt, 2012). Crocodiles also represent a noteworthy special case in terms of brain/body ratios. The body of Nile crocodiles (*Crocodylus niloticus*) shows a continuous growth over their lifetime. Ngwenya et al. analyzed brain size of Nile crocodiles at different ages with body weights ranging from 90 g to 90 kg. They found that this 10-fold increase in body weight was only accompanied by a 1.8-fold increase in brain size (Ngwenya et al., 2013). Thus, at least within this species, the correlation between body and brain size is not as fixed as had been assumed for vertebrate species in general. Snake species represent another interesting case when comparing relative brain sizes in reptiles, since they seem to have smaller brain/body ratios than the other analyzed reptilian clades and lie below the reptilian regression line (Northcutt, 2012). The reason for this is unclear but could be due to the elongation of their body, since elongated vertebrates tend to have on average smaller brains (van Dongen, 1998).

In contrast to reptiles, for which relatively few studies on brain allometry have been published, extensive research has been done on brain scaling in birds. Since it would be a futile attempt to cover all these findings within the boundaries of this book chapter, we will only cover a small fraction of the data here. However, the interested reader can find more information in Martin, 1981; Armstrong and Bergeron, 1985; Rehkämper et al., 1991a; Iwaniuk et al., 2004 and the Chapter 1.18, Functional Correlates of Brain and Brain Region Sizes in Nonmammalian Vertebrates by Andrew Iwaniuk within this volume.

In birds, brain sizes range from 0.22 g in hummingbirds, over 2 g in pigeons, to 14 g in Keas and ravens and 27 g in ostriches (Rehkämper et al., 1991b; Peng et al., 2010; Olkowicz et al., 2016). Especially noteworthy is that parrots and Passeriformes (perching birds)

generally have higher relative brain sizes than *Palaeognathae* (eg, ostriches; but see Corfield et al., 2008; on kiwis) and *Galloanserae* species (eg, chicken, Rehkämper et al., 1991a; Olkowicz et al., 2016). Thus, birds of the Neoaves clade, which evolved approximately 90 million years ago (Prum et al., 2015), tend to have bigger relative brain sizes than their more basal relatives. These basal avians may represent a recent example for the transition from smaller brained reptiles to bigger brained modern bird species. Domestication of birds (eg, in chicken, ducks, and geese) leads to an opposite trend with a strong reduction in relative brain size in comparison to their wild relatives based on an increase in body size but also in a reduction in absolute brain volume which can reach up to a loss of up to 20% (Ebinger and Löhmer, 1987; Rehkämper et al., 1991a). Examples for brain to body ratios are depicted in Fig. 8.3 for reptiles (adapted from van Dongen, 1998; Northcutt, 2012). Note, however, that the data are restricted to few reptilian species with rather big brains of which many are lizards. This likely reflects a publication bias.

There is a rich literature on comparisons of individual species of vertebrate classes with respect to relative brain sizes (for review, see van Dongen, 1998; Northcutt, 2012). Although some overlap between classes exists, these analyses mostly suggest that during the transition from reptiles to birds and from reptiles to mammals, brain size increased massively. This increase in brain volume was, however, not uniform for all brain areas. When comparing the size of specific brain area in relation to the size of the whole brain, it is mainly the forebrain that increased dramatically. In frogs (*Rana catesbeiana*), the telencephalon constitutes only 22% of the total brain volume, while in reptiles, telencephalic values range from 29% in snakes (*Nerodia sipedon*) over

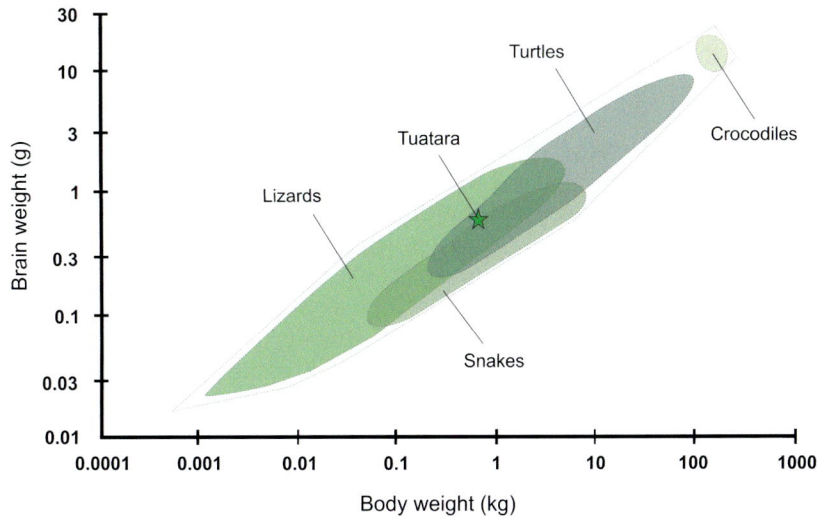

FIGURE 8.3 Brain weight in relation to body weight for the reptilian class. The *solid line* of the convex polygon encloses the data for all reptiles, while the *dotted lines* enclose the different reptilian clades. The tuatara, as the only recent member of the Sphenodontia, is indicated by a *star*. Note the lower brain-body ratios in snakes in comparison to the other reptilian taxa. *Figure modified from Northcutt, R.G., 2012. Variation in reptilian brains and cognition. Brain Behav. Evol. 82, 45–54.*

36% in the tuatara (*Sphenodon punctatus*) and 42% in warans (*Varanus bengalensis*) to 45% in crocodiles (*Caiman crocodilus*, Northcutt, 2012). In birds, the telencephalon constitutes an even bigger portion of the whole brain. The telencephalon takes up 43% of the whole brain in emus (Olkowicz et al., 2016; again, see Corfield et al., 2008 on kiwis, which seem to represent a special case within *Palaeognathae*) and 51% in chicken (Northcutt, 2012). In Neoaves, proportional telencephalon volume is even larger. Among parrots, the telencephalon comprises 68% of the total brain volume in budgerigars, 73% in African Grey parrots, and 77% in Indian ringed parrots (derived from Iwaniuk et al., 2004; see also Olkowicz et al., 2016 and below). Among Passeriformes, the telencephalon constitutes 67% of the entire brain in house sparrows, 68% in Eurasian jays, and 74% in hooded crows (derived from Rehkämper et al., 1991a).

Within the telencephalon, especially the pallium experienced a hypertrophy in both absolute size and in relation to the remaining telencephalon. A recent study showed that in birds, an increase in overall brain size is driven mainly by an increase in pallial volume (Sayol et al., 2016). These results even suggest that relative brain size can be used as a proxy for relative pallium size in comparative studies. In amphibians, the pallium takes only 52% of the total telencephalon volume, increasing to 70% in lizards and 85% in crocodiles and basal birds (Northcutt, 2012). Among Neoaves, the pallium of parrot species comprises 78% of the telencephalic volume in budgerigars, 86% in African Grey parrots, and 83% in Indian ringed parrots (data derived from Iwaniuk et al., 2004; Iwaniuk and Hurd, 2005). Within Passeriformes, the pallium constitutes 90% of the telencephalon in house sparrows, 86% in Eurasian jays, and 88% in hooded crows (data derived from Rehkämper et al., 1991a). This increase in proportional pallial volume probably enabled specific bird species to develop cognitive abilities that are beyond the capabilities of reptilians and bird species with smaller pallial structures. Indeed, several studies have shown that the sizes of certain pallial subdivisions, such as the meso- and nidopallium, correlate with some specific domains of higher cognition, such as innovation rate or tool use (Timmermanns et al., 2000; Lefebvre et al., 2002, 2013; Mehlhorn et al., 2010; Lefebvre et al., 2013).

8.2.3 Neuron Numbers and Scaling Rules

As mentioned above, pure allometric measures of brain sizes alone seem to be insufficient to explain cognitive capabilities of a species (Healy and Rowe, 2007). In response to this problem, a new approach was designed which is based on the number of neurons in a given brain or brain structure. The idea is quite simple: since neurons represent the smallest processing unit of a brain, a higher number of these units would increase information processing capacity (Roth and Dicke, 2005). Originally, it was assumed that neuron numbers scale with a common function of brain size across species (Haug, 1987), but studies during the last decade in mammals have shown that this is utterly wrong. These studies showed a great variety in the cellular composition of different mammalian brains (Herculano-Houzel et al., 2005, 2011b; Gabi et al., 2010; Sarko et al., 2009; Neves et al., 2014). For example, the cerebral cortex of the African elephant is twice as large as that of humans, but has only a third of the number of neurons (Herculano-Houzel et al., 2014a). These studies also revealed that brains of different mammalian orders gain neurons with different scaling rules as brain size increases (Herculano-Houzel et al., 2011b). Within the mammalian class, primates have the most favorable scaling rule of about 1:1 (Herculano-Houzel et al., 2007; Herculano-Houzel, 2009). Thus, their neurons numbers increase directly proportional to the increase of brain weight.

Data on neuron numbers in birds and especially reptiles are unfortunately scarce at the moment. In reptiles, only one study in Nile crocodiles has been conducted so far (Ngwenya et al., 2016). It found that the brains of these animals contain 80.5 million neurons. This corresponds to an overall neuron density of ~25 000 neurons/mg, but these neurons are not evenly distributed over the brain. As in mammals, a disproportionate number of neurons are allocated to the cerebellum (~40%), which shows a neuron density of ~168 000 neurons/mg. Roughly 27% of all neurons in Nile crocodiles are situated in the telencephalon which is similar to the percentage of neurons found in the mammalian cortex (Herculano-Houzel, 2009). Neuron density in the telencephalon (18 500 neurons/mg) is much lower than in the cerebellum, but on average still higher than in the brain stem and spinal cord. The remaining neurons are found in the brain stem and the olfactory bulb (which was analyzed separately from the telencephalon), with the biggest contributor being the mesencephalon, likely because of the cell dense optic tectum. Although the general distribution of neurons in the crocodile brain resembles that found in mammals, neuron density in the whole brain is much lower than in mammals (Herculano-Houzel, 2009). A further interesting finding of Ngwenya et al. (2016) was that these neuron numbers only change marginally during the growth of the animal. As mentioned above, Nile crocodiles grow constantly over their lifetime. However, while there was a 1000-fold increase in body size, neuron numbers increased by only 2.8-fold in the brain and 5.3-fold in the spinal cord. It was suggested that bigger bodies do not necessarily require more neurons to maintain functionality but rather bigger neurons and axons to cope with the increasing distance to the innervation targets (Ngwenya et al., 2016).

Due to a recent publication, more data on neuron numbers are now available for birds. Olkowicz et al. (2016) analyzed the cellular composition of the brains in 28 avian species and found astonishing results. Although the brains of birds are rather small in comparison to mammals, neuron numbers are twice as high as in a primate with the same brain size and up to four times higher in comparison to rodents with a same sized brain (see Fig. 8.4A). Neuron numbers ranged from 136

million in zebra finches, over 310 million in pigeons and 697 million in monk parakeets, to 2.2 billion in ravens and 3.1 billion in macaws (see Fig. 8.4B). With the exception of the analyzed basal birds (chicken: 78 000 neurons/mg, emu: 61 000 neurons/mg), neuron densities are therefore higher in birds than in the analyzed mammalian species (eg, 275 000 neurons/mg in zebra finches, 148 000 neurons/mg in pigeons, 203 000 neurons/mg in monk parakeets, 154 000 neurons/mg in ravens, and 151 000 neurons/mg in macaws).

Although the overall distribution of brain mass across the major brain components is similar between mammals and birds (eg, the telencephalon occupies 72% of the brain in songbirds and 74% in primates), the distribution of neurons is vastly different. While in mammals the majority of neurons are found in the cerebellum (Herculano-Houzel, 2009), 38—62% of all neurons in songbirds and 53—78% of all neurons in parrot species are found in the telencephalon. If the striatum is excluded, to allow a better comparison to the mammalian cortex, 33—55% (songbirds) and 46—61% (parrots) of all neurons in the brain are found in the pallium. In the human brain, only 19% of all neurons are found in the cortex, although it takes up 82% of the brain mass. Thus, even though parrots and songbirds are already outnumbering mammalian species with comparable brain sizes regarding neuron numbers in the whole brain, this advantage gets even further pronounced when only comparing pallial neurons. For example, the cortex (dorsal pallium) of a macaque monkey weights 69.83 g and contains 1.7 billion neurons, whereas the pallium of the blue and yellow macaw weighs one-fifth of that but holds a whopping 1.9 billion neurons.

When comparing neuron numbers between avian species, it becomes apparent that neuron numbers in songbirds and parrots scale similarly with brain weight (see Fig. 8.4B). Thus, a parrot brain contains roughly the same number of neurons as the brain of a Passeriformes species with the same brain weight. Also, in both orders, brain mass gain is faster than neuron gain, leading to lower neuron densities in bigger brained species. In contrast, pigeons, chickens, and emus have relatively low neuronal densities. Given their proportionally lower brain and telencephalic size (see above), their telencephalon contains far fewer neurons than that of a similar sized parrot or songbird brain. As Olkowicz et al. (2016) noted, a chicken brain is 50 times bigger than that of a great tit, but both contain approximately the same number of neurons. Unfortunately, scaling rules for orders outside the Passera clade are currently unavailable, since data from the Columbiformes, Galliformes, and Casuariiformes orders come only from single species.

Still, the obtained data on neuron numbers in combination with the allometric data gathered over decades of research deliver some important evidence on how specific bird species were possibly able to develop cognitive

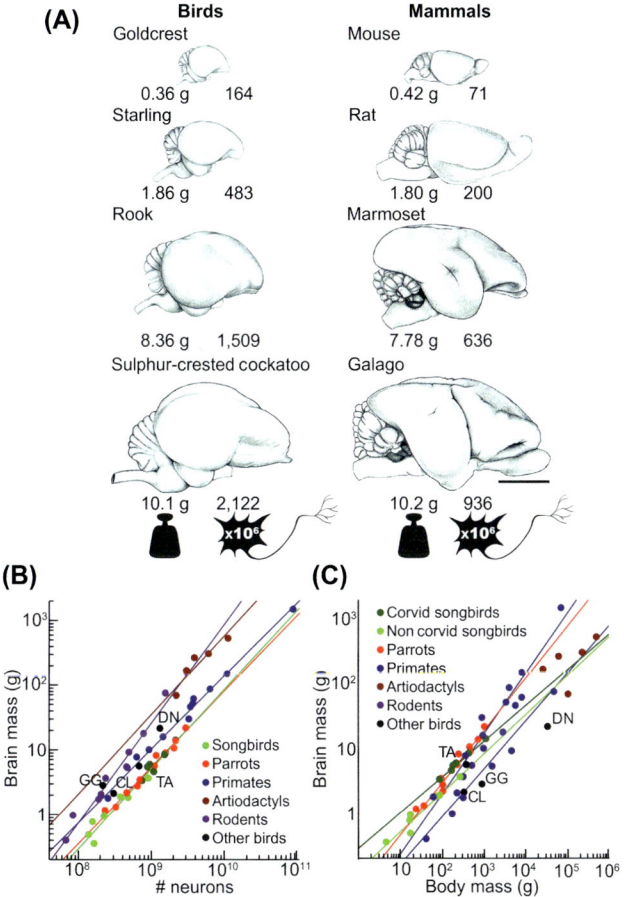

FIGURE 8.4 Neuron numbers and brain weights of selected avian species. (A) Comparison of absolute neuron numbers in four avian species with neuron numbers of four mammalian species with similarly large brains. Neuron numbers in birds are more than twice as high, even when the comparison is done with primate species (eg, rook vs marmoset or sulphur-crested cockatoo vs galago). In (B) neuron numbers in relation to brain mass is depicted for selected avian species in comparison to data from three mammalian orders. (C) shows brain mass in relation to body mass for the same species. *CL, Columba livia* (pigeon); *DN, Dromaius novaehollandiae* (emu); *GG, Gallus gallus* (chicken); *TA, Tyto alba* (barn owl). *Figure adapted from Olkowicz, S., Kocourek, M., Lučan, R.K., Porteš, M., Fitch, W.T., Herculano-Houzel, S., Němec, P., June 13, 2016. Birds have primate-like numbers of neurons in the forebrain. Proc. Natl. Acad. Sci. U.S.A. pii:201517131. [Epub], with permission of the authors.*

abilities which rival those of primate species (Güntürkün and Bugnyar, 2016), while other bird species could not. (1) Songbirds and parrots possess neuronal scaling rules which endow them with neuronal densities surpassing those of primates. (2) Songbirds and parrots developed a proportionately bigger telencephalon with a proportionately bigger pallium than other bird species. (3) Within songbirds, the corvid species possess the most developed cognitive abilities and also the biggest brains. Combining these points implies that corvid species have an absolutely larger number of neurons in their pallium than other bird species; they also have more pallial neurons than a five times bigger primate brain. Thus the processing capacity of the corvid pallium, based on the absolute neuron numbers, is likely to be higher than it is in other bird species and, for that matter, in many primates.

8.3 The Structures of the Reptilian and the Avian Brain

From an embryological point of view, the nervous system of vertebrates is divided into the spinal cord and the three primary brain vesicles, rhombencephalon, mesencephalon, and prosencephalon (Nieuwenhuys, 1998). In the adult form, the transition between the spinal cord and the rhombencephalon is the area between the first cervical spinal root and the exit of the vagal nerve. Despite this clear cut definition, no sharp morphological boundary is discernable; instead, spinal anatomy slowly transforms into the structural constituents of the rhombencephalon. Further anterior, the rhombencephalon borders with the mesencephalon and the cerebellar commissure, with the exit and decussation of the trochlear nerve serving as boundary landmarks. The rhombencephalon and mesencephalon jointly constitute the brain stem. Rostral to the mesencephalon is the prosencephalon with its diencephalic and more rostrally situated telencephalic components. These and further structures are components of the *bauplan* of the vertebrate brain and as such are obviously present both in reptiles and birds. To review all relevant anatomical details of these structures would be a futile attempt for the present treatise, especially since the three-volume book on the central nervous system of vertebrates serves as landmark publication for such a purpose (Nieuwenhuys et al., 1998). Instead, only those components and systems of brain entities will be presented for which specific and relevant adaptations were discovered in some reptile or bird taxa. They will be presented and discussed, moving from caudal to rostral entities.

8.3.1 The Sauropsid Spinal Cord

8.3.1.1 Reptilian and Avian Spinal Cords: Invariant Organization Despite Variances of Behavior

There are no standardized subdivisions of the reptilian spinal cord that are comparable to those of mammals or birds. In fact, there is no vertebrate class with such divergent spinal organization patterns as reptiles. In limbless forms like snakes, the number of spinal segments varies widely, reaching more than 400 in some species (ten Donkelaar, 1998). Snakes rely exclusively on their axial muscles for locomotion and move by large lateral undulations of the body. This is radically different from limbed amniotes like rats in which limbs are crucial for locomotion while axial muscles play only a secondary role. Despite these important differences, the motor neuron pools of rats and the limbless Florida water snake are astonishingly similar (Fetcho, 1986). Thus, even though the details of the arrangements of muscles differ, and the roles of the muscles in locomotion are likely to be very different, the arrangements of the motor pools in the two animals are located in comparable positions of the motor column. The same kind of observation was reported by Ryan et al. (1998) who labeled the motor neuron pools of seven homologous forelimb muscles in mice (*Mus musculus*) and iguanas (*Iguana iguana*) and discovered a similar topography despite dissimilar locomotion patterns. These data on reptiles and mammals suggest that species-typical differences in the locomotor mechanics are accomplished without any dramatic reorganization of the spinal motor column.

This conclusion is supported when studying birds—a group of animals that have developed flapping flight and thus undertook a major change in the concerted action of frontal limb muscles. Goslow et al. (2000) analyzed the spinal topography of motor neurons that innervate key muscles for flight in the European starling and found a pattern that is highly comparable to that seen in nonavian tetrapods. These data indicate that a massive evolutionary change of motor patterns can occur without a corresponding topological reorganization of the corresponding motor column. The evolutionary changes in motor patterns that accompanied the evolution of birds are probably involved alterations in synaptic input from supraspinal sources, not alterations in the topology of the motor columns. This similarity of the spinal motor pool organization among amniotes is in marked contrast to the spinal organization in anamniotes. The transition from anamniotes to amniotes goes along with a breakup of the myomeres into discrete muscles and a subdivision of the spinal motor column into discrete, topographically arranged motor pools serving the individual muscles (Fetcho, 1987).

Dinosaurs were not only the largest reptiles but also the largest animals that ever roamed the land. Their spinal organization as revealed from fossil data provides some clues about their movement patterns. A simple predictor of limb size and extent of limb use is the spinal quotient (SQ), which expresses the enlargement of the spinal limb levels relative to interlimb levels. SQ is lowest in snakes and high in dinosaurs with manipulative forearms (Giffin, 1990). In some dinosaurs, the volume of the lumbar vertebral canal even exceeds the volume of their endocranial cavity (Romer, 1966). Some of this inflation could result from the glycogen body in the lumbosacral region that is sometimes wrongly associated with a "sacral brain"—a myth according to which dinosaurs had a second brain in the spinal cord that compensated for their tiny endocranial nervous system. Studies in birds may help to clarify the true function of the lumbosacral expansion, as outlined in the next section.

8.3.1.2 *The Mystery and the Sobering Reality of the Sacral Brain*

We associate birds with the ability to fly. But they can also walk and this kind of locomotion produces a special challenge: the legs of birds are inserted caudal to the center of gravity, and thus their bipedal walking pattern needs special control of balance. This is even more important when perching on swaying branches. Strikingly, as many farmers know, beheaded chickens can walk and fly for a short while keeping balance. Consequently, scientists had suggested since long that birds should have an extralabyrinthine sense of equilibrium in their abdomen (Mittelstaedt, 1964; Delius and Vollrath, 1973). Subsequent studies suggested that the peculiar glycogen body in the lumbosacral spinal cord might represent such a sense organ (Grimm et al., 1997; Fig. 8.5). The discovery of canals in the lumbosacral region which look similar to the semicircular canals in the inner ear led to the suggestion that some of the specializations in the lumbosacral region may function as a sense organ of equilibrium which is involved in the control of hind limbs (Necker, 1999).

In the avian lumbosacral cord the local vertebrae are fused and tightly connected to the pelvic girdle (Baumel and Witmer, 1993). In addition, the vertebral canal is enlarged considerably. Importantly, this enlargement is not due to an increase of neuronal tissue, but due to the presence of a glycogen body that is embedded in a dorsal groove of the spinal cord (Fig. 8.6). The cord itself is firmly attached by ligaments

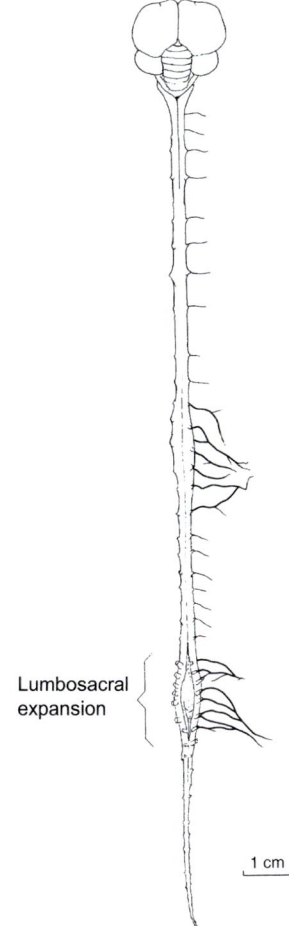

Lumbosacral
expansion

1 cm

FIGURE 8.5   The spinal cord of the pigeon with the lumbosacral enlargement. *Reproduced from Dubbeldam, J.L., 1998. Birds. In: Niewenhuys, R., Ten Donkelaar, H.J., Nicholson, C. (Eds.), The Central Nervous System of Vertebrates. Springer, Berlin, pp. 1525—1636, with permission.*

to the vertebra. Necker (1999) discovered semicircular canal-like structures in the lumbosacral cord and proposed that these specializations could channel cerebrospinal fluid during body movements toward a specialized group of neurons (Necker, 1999). These neurons are equipped with mechanoreceptors (Necker, 2002) and are located in an accessory lobe at the ventrolateral end of the ventral horns (Schroeder and Murray, 1987). The activity of these neurons is transmitted to the cerebellum via paragriseal cells which are at the origin of a ventral spinocerebellar pathway (Necker, 2005a,b). Every time the bird takes a turn, the fluid near the lobes move by inertia to the opposite direction of the turn, thereby activating mechanoreceptors of neurons in the accessory lobe (Fig. 8.6A). Roll and pitch movements could thus be detected by an intraspinal sensory system involved in the control of posture and locomotion on the ground. Indeed, behavioral studies showed that these kinds of movements are less balanced during walking in animals where the

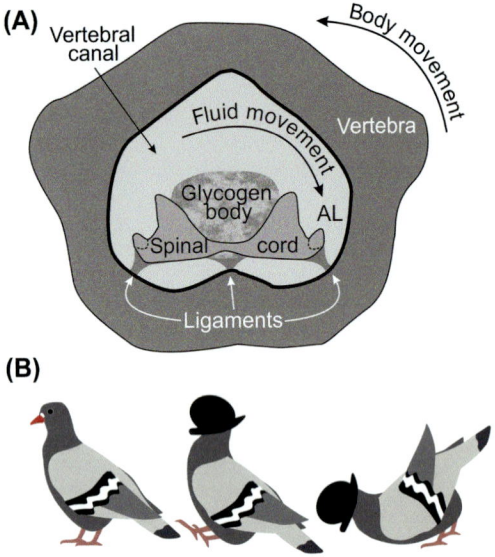

control of equilibrium: the vestibular organ during flight and the lumbosacral system during walking (Necker, 2006).

It seems likely that a similar lumbosacral system existed also in dinosaurs. Control of equilibrium in theropoda was probably at least as complex as in birds since they had often even longer necks than birds. This further decreases the usefulness of a cranial vestibular system for maintaining balance while walking. In addition, some theropoda could grow to enormous sizes. Thus, a vestibular-like system that is close to tail and hind legs is conceivably a faster sensory system—even outside the lineage of modern birds. Still, it is not a "second brain."

8.3.2 Mesencephalon

Moving from medial to lateral, the midbrain consists of the central gray, tegmentum, and tectum. The third ventricle is located in the center of the midbrain, but possesses laterally protruding extensions that are called tectal ventricles. In sauropsids, the "tectum acusticum" is located ventral to the tectal ventricle. In reptiles it is usually called torus semicircularis, and in birds it is nucleus mesencephalicus lateralis dorsalis. The tectum opticum has a position dorsal and lateral to the tectal ventricle. Especially in birds the optic tectum is so extraordinarily enlarged, that it bulges out laterally and is sometimes called the visual lobe (Butler and Hodos, 2005; Fig. 8.7).

In reptiles the optic tectum has six primary layers. Tectal lamination is much more complex in birds with at least 15 different tectal layers being easily identifiable. Despite these differences between reptiles and birds, the tectum possesses the same general organization and harbors highly similar input and output systems (Reiner, 1994). The major common

FIGURE 8.6 The lumbosacral spinal equilibrium system in birds. (A) is a section through the lumbosacral spinal cord, showing the glycogen body in the dorsal groove of the spinal cord. The ventrolateral extensions of the ventral horns constitute the accessory lobes (AL) that are able to detect movements of the cerebrospinal fluid. The cerebrospinal fluid moves in inverse direction to body turns, thereby activating mechanoreceptors in AL. (B) depicts the walking posture of pigeons with punctured lumbosacral cavities. When the birds can see, their gait is mostly normal (left); when blinded by a hood, they constantly tip over (middle and right). *Modified from Necker, R., 2006. Specializations in the lumbosacral vertebral canal and spinal cord of birds: evidence of a function as a sense organ which is involved in the control of walking. J. Comp. Physiol. A 192, 439–448.*

lumbosacral cavity was punctured, whereas flight was normal (Necker et al., 2000). Especially when these lesioned animals were blinded with a hood, they constantly tipped over while walking (Fig. 8.6B). Thus, two different sense organs are involved in the

FIGURE 8.7 (A) Brain of a Nile crocodile. (B) Brain of a pigeon. The brains are not to scale, and the optic lobes are framed. (C) Frontal section through the midbrain of a pigeon. Note the highly laminated optic tectum. The isthmic nuclei are outlined. *Imc*, n. isthmi pars magnocellularis; *ION*, n. isthmoopticus; *Ipc*, n. isthmi pars parvocellularis; *SLu*, n. semilunaris. Crocodile brain. *Courtesy of Mehdi Behroozi.*

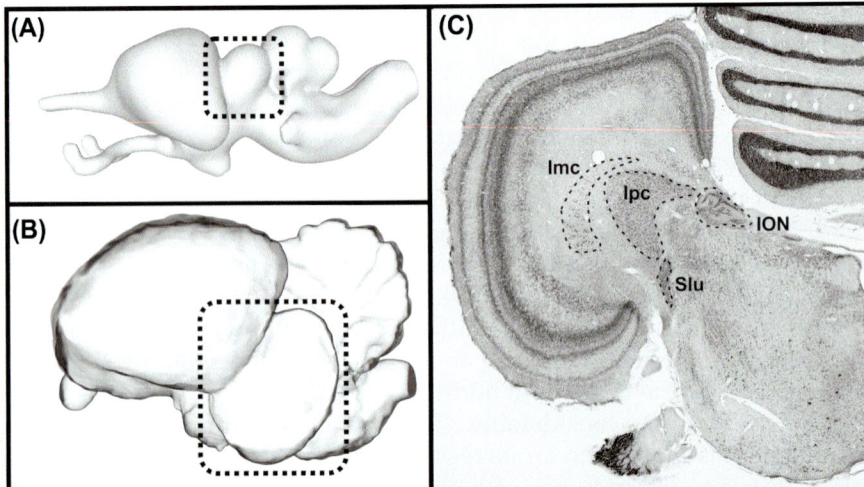

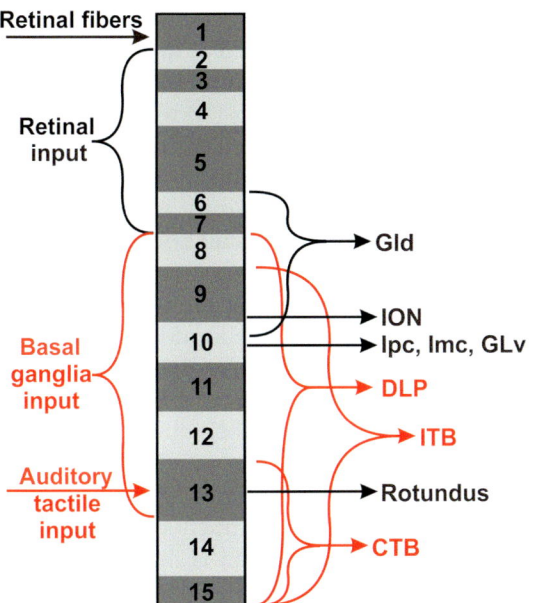

FIGURE 8.8 Major input (left) and output systems (right) of the avian optic tectum. Monomodal visual pathways are shown in *black*; multimodal nonvisual sensory and motor systems are shown in *red*. *CTB*, crossed tectobulbar pathway; *DLP*, n. dorsolateralis posterior thalami (multimodal ascending nucleus); *GLd*, n. geniculatus lateralis pars dorsalis (main ascending visual nucleus of the thalamofugal system); *Imc*, n. isthmi pars magnocellularis; *ION*, n. isthmoopticus; *Ipc*, n. isthmi pars parvocellularis (three isthmic nuclei); *ITP*, ipsilateral tectopontine-tectoreticular pathway. *Modified from Luksch, H., 2003. Cytoarchitecture of the avian optic tectum: neuronal substrate for cellular computation. Rev. Neurosci. 14, 85–106.*

organizational principles are (1) the retinal input from the contralateral eye enters via the most superficial input layer in a topographically organized manner; (2) ascending visual output arises from the intermediate and deeper layers; (3) descending projections to motor areas also arise from intermediate and deeper layers; (4) input from nonvisual sensory pathways terminates mostly in deeper layers (Reiner, 1994; Luksch, 2003; Hellmann et al., 2004; Fig. 8.8). In Section 8.3.2.3, we will take a more detailed look on this pattern when discussing the avian tectum.

8.3.2.1 The Infrared System of Snakes: Seeing the Heat

Being warm-blooded, mammals and birds have a lot of advantages in terms of mobility in the cold, but under certain circumstances, tables are turned: Even in total darkness, their higher body temperature can give their position away. To exploit this information, predators need infrared vision. Two groups of snakes, the Boidae (eg, the Boa constrictor) and the Viperidae (eg, common rattlesnakes and pythons) have evolved infrared vision and can use it to find prey, detect predators, and find warm places to rest.

In rattlesnakes the thermal sensor is a facial pit located on the lateral surface of the head between the external nose cavity and the eye (Fig. 8.9). This pit consists of an open anterior chamber that is closed at the back by a thin membrane that contains sensory receptors. The receptors consist of free nerve endings that are sensitive to radiant heat (Goris and Terashima, 1976; von Düring and Miller, 1979). The pit resembles a pinhole camera for thermal stimuli and, indeed, snakes display directional sensitivity in their thermal responses (Kohl et al., 2012). The thermal receptors can respond to changes as small as 0.001°C in thermal energy (Stanford and Hartline, 1984; Gracheva et al., 2010). The pit organs in rattlesnakes are innervated by fibers of the ophthalmic and the maxillary branches of the trigeminal nerve.

After entering the brain stem, the sensory trigeminal fibers divide into two projection streams. One serves the same purpose as the trigeminal input in all further vertebrates. The second branch, however, conveys thermal information and terminates in the n. descendens lateralis trigemini, which then projects to the n. reticularis caloris of the medulla (Stanford et al., 1981). From there, projections reach the deep layers of the contralateral optic tectum (Kardong and Berkhoudt, 1999). In the tectum, infrared information merges with visual information to create bimodal visual-thermal neurons (Hartline et al., 1978). Some of these neurons respond only to simultaneous bimodal stimulation while others respond to only one modality and are inhibited when simultaneously stimulated by the second modality. These cross-modality interactions could be relevant to disambiguate warm-blooded prey (simultaneous stimulation by visual and infrared input) from cold visual objects that represent nonliving objects (Newman and Hartline, 1981).

A further critical cue for identifying living objects is motion. Behavioral studies show that blindfolded rattlesnakes predominantly respond to moving infrared stimuli (Ebert and Westhoff, 2006). Indeed, slowly moving objects elicit only weak or no responses in tectal units that respond to infrared cues, while increasing object speed increases spike rate (Kaldenbach et al., 2016). This could imply that slow or even stationary objects may not be detected by the infrared system of snakes at all. Indeed, rattlesnakes are ambush predators that wait for prey. Immobile objects are mostly irrelevant as a food resource and do not stimulate the infrared receptors. Thus, the infrared sensory system as represented in the tectum can disambiguate infrared signals from thermal clutter. Rattlesnakes also use their infrared system to seek warm places for thermoregulation (Krochmal and Bakken, 2003). However, when doing so, snakes perform scanning head movements and thus create a

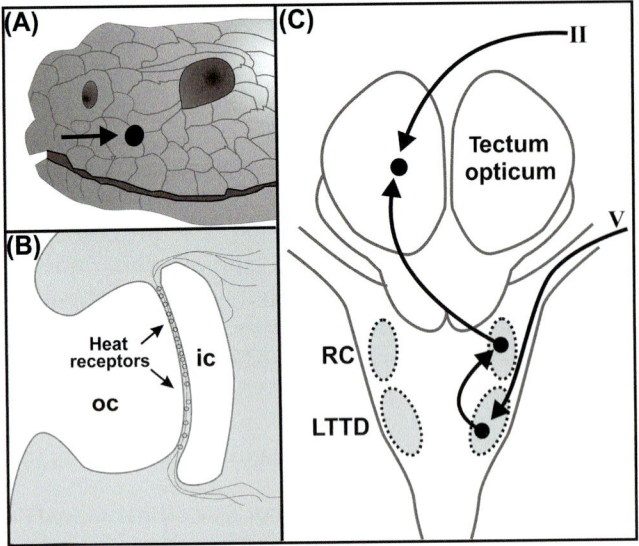

FIGURE 8.9 The infrared system in rattlesnakes. (A) The location of the facial pit containing thermal receptors is indicated by the *arrow*. (B) Cross section of the facial pit, showing the thermal receptors along the membrane suspended between the pit's outer and inner chambers (oc and ic, respectively). Fibers entering the membrane stem from the ophthalmic and maxillary branches of the trigeminal nerve. (C) Schematic dorsal view of the rhombencephalon and mesencephalon. The trigeminal nerve (V) innervates the n. descendens lateralis trigemini (LTTD), which then projects to the n. reticularis caloris of the medulla oblongata (RC). Neurons of the RC project to the optic tectum, where infrared information is merged with incoming visual input from the second cranial nerve (II).

relative movement between warm objects and the receptors (Ebert and Westhoff, 2006).

8.3.2.2 The Centrifugal Visual System: What the Brain Tells the Eye

The optic tectum has topographically organized reciprocal connections with the nucleus isthmi, a complex of several cytoarchitectonically distinguishable nuclei at the mesorhombencephalic border (Yan and Wang, 1986; Güntürkün and Remy, 1990; Wang et al., 2006; Faunes et al., 2013; Fig. 8.7). The isthmic complex is present in most vertebrates (eg, Künzle and Schnyder, 1984) but is most highly differentiated in birds (Wang, 2003), in which it comprises nucleus isthmi pars parvocellularis (Ipc), pars magnocellularis (Imc), pars semilunaris (SLu), and nucleus isthmoopticus (ION). All these structures receive ipsilateral tectal input (Güntürkün, 1987). It is the ION that gives rise to a conspicuous centrifugal projection to the contralateral retina that is present in practically all vertebrates, that is but extremely expanded and differentiated in granivorous birds.

Santiago Ramón y Cajal (1889), the founder of Neuroscience, was the first who discovered in birds axons that project from the central nervous system to the retina. A few years later, Adolf Wallenberg (1898) discovered the

ION as the midbrain nucleus from which these fibers originate. Cajal (1889) suspected that such a system might modulate the retinal input according to expectations generated in the brain. It was long disputed whether such a system is a specialization of birds or is found also in other vertebrates, possibly including humans. Since these early studies, it has become well established that centrifugal visual fibers exist in all classes of vertebrates. Most likely, such a system has evolved multiple times within the vertebrate lineage, with at least eight distinct subsystems located in very different regions of the neuraxis (Repérant et al., 2006, 2007). And yes, centrifugal visual fibers also exist in humans, although they typically number no more than a few dozen (Repérant and Gallego, 1976). The diversity of centrifugal visual systems in vertebrates probably matches the diversity of their functions. In the following sections, the centrifugal system is outlined for reptiles and birds. The emphasis will be the avian centrifugal system since it is the most advanced retinopetal visual pathway of vertebrates and could serve as a model system on how and what the brain tells the eye.

8.3.2.2.1 The Centrifugal Visual System of Reptiles

After tracer injections into the retina of several turtle species, 10–60 retrogradely labeled neurons have been observed in the area of the isthmic region (Haverkamp and Eldred, 1998; Repérant et al., 2006). These centrifugal fibers make extensive collateral branches before penetrating and synapsing in the retina's inner plexiform layer (IPL) (Weiler, 1985). In lizards the situation is very similar (Repérant et al., 2006), although in some species a second source of centrifugal neurons is found in the ventral thalamus (El Hassni et al., 1997). Snakes possess several hundred centrifugal visual neurons, but their centrifugal neurons are found bilaterally in the basal telencephalon, the lateral preoptic area, and the ventral thalamus (Hoogland and Welker, 1981; Repérant et al., 2006). Crocodiles possess between 4000 and 6000 centrifugal visual neurons, depending on the species (Kruger and Maxwell, 1969; Médina et al., 2004). These neurons are mostly located in the isthmic region but can also be found in other tegmental areas. They may be part of a loop that starts with the retinotectal projection and then proceeds via the isthmic nucleus back to the retina (Ferguson et al., 1978).

8.3.2.2.2 The Centrifugal Visual System of Birds

The centrifugal visual system of birds originates in two different mesencephalic cell groups: the isthmooptic nucleus (ION), a folded bilaminate structure in the dorsolateral midbrain tegmentum, and the nucleus of the ectopic isthmooptic neurons (EION), a loosely scattered array of cells with reticular appearance surrounding the ION (Wolf-Oberhollenzer, 1987; Fig. 8.7). Both

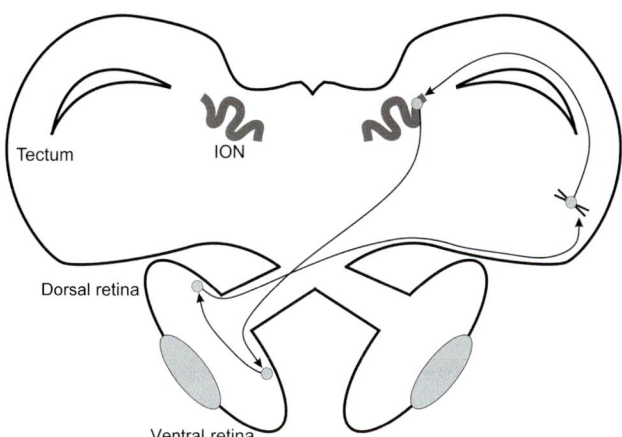

FIGURE 8.10 General organization of the avian centrifugal visual system. Ganglion cells of the dorsal retina (mostly) project to tectal neurons at the border of layer 9/10. These neurons project topographically to ION neurons that then project to association amacrine cells in the ventral retina. Association amacrine cells project to dorsal retina, thereby closing the loop.

structures are part of a closed loop consisting of a projection from the retinal ganglion cells to the contralateral tectum, the efferents of which project to both the ipsilateral ION and the EION, whence back projections lead to the contralateral retina (Güntürkün, 2000). All projections within this system seem to be topographically organized (Li et al., 1998; Fig. 8.10). Weidner et al. (1987) discovered important differences in this system between raptors and ground-feeding birds. In seed or fruit-eating birds, the ION is always large, well differentiated, and laminated. In raptors, the ION is small, poorly differentiated, and reticular in appearance. Thus, the centrifugal system seems to play a specific role in ground-feeding birds that are subject to predation by various animals, including birds of prey. As will be argued later, this condition is possibly relevant to understand the function of the centrifugal system.

In pigeons and chicks, cell bodies of tecto-ION neurons are located at the border of layers 9 and 10 of the tectum, reach up to layer 2 with their dendrites, and can thus pick up direct retinal input (Woodson et al., 1991). This input stems mostly, but not exclusively from the dorsal retina (lower visual field). The tecto-ION neurons project topographically onto the ipsilateral ION. The ION consists of a highly convoluted lamina in which two perikaryal layers are separated by a neuropil in which the dendrites from opposing layers ramify toward the middle of the two layers (Güntürkün, 1987). Afferent axons of tecto-ION neurons pass through this dendritic field and synapse topographically on small dendritic appendages and spines, providing virtually all excitatory synapses in the ION (Cowan, 1970; Angaut and Repérant, 1978). Additionally large numbers of inhibitory synapses on ION dendrites are found which partly originate from a

small number of GABAergic neurons within the ION (Miceli et al., 1995). It is likely that these inhibitory neurons are key to ION function since, as pointed out by Uchiyama (Uchiyama et al., 1998; Uchiyama, 1999) the ION network shows a strong winner-take-all competition which possibly allows the selection of the most salient stimulus. Axons from ION cells proceed, together with those from the EION, to the contralateral retina. The number of efferent axons within the optic nerve is supposed to be about 12 000 in the pigeon, of which the ION contributes about 10 000 (Weidner et al., 1987). Since the tecto-ION and the tecto-EION pathway also consist of about 12 000 neurons, a 1:1 ratio of tectal and centrifugal neurons is likely (Woodson et al., 1991). The centrifugal axons terminate near the IPL, bordering the inner nuclear layer (INL) in the horizontal and ventral retina, barely penetrating the red field that serves frontal binocular vision (Lindstrom et al., 2009). They are composed of two distinct types, with divergent degrees of topographic localizations. Fibers from the ION are called "convergent" and give rise to a single restricted type of terminal fiber, which forms a dense pericellular nest covering the perikaryon of a single association amacrine cell (Uchijama and Ito, 1993; Uchiyama et al., 1995; Lindstrom et al., 2010). Association amacrines have long intraretinal axons, are mainly located in the horizontal plus ventral retina, and project dorsally (Catsicas et al., 1987; Uchiyama et al., 2004). Thus, ION fibers receive input from the dorsal retina (lower visual field), project back to the ventral retina (upper visual field), and are then connected via intraretinal association fibers to the dorsal retina (lower visual field). Axons originating from EION are called "divergent" and give rise to several terminal branches, each constituting an extensive and highly branched arbor (Fritzsch et al., 1990; Woodson et al., 1995).

Electrophysiological data are only available for the ION. Most ION cells have their receptive fields in the inferior anterior visual field and are thus related to the upper posterior parts of the retina (Hayes and Holden, 1983; Catsicas et al., 1987; Uchiyama et al., 2004). Miles (1972) and Holden and Powell (1972) demonstrated that a large number of ION units show a preference for moving shadowlike target movements in the anterior visual field and habituate rapidly to repetitive stimulations. This finding suggests a role in the analysis of transient and dynamic features of the visual environment. In a very sophisticated study, Li et al. (1998) demonstrated that retina, tectum, and ION form a closed loop of topographic excitations. In other words, the same ganglion cells in the dorsal retina that provide input to the ION via the tectum receive feedback from those same ION neurons.

Based on these data, several authors tried to establish the functional importance of the ION and EION in behavioral studies (Rogers and Miles, 1972; Shortess and Klose, 1977; Knipling, 1978; Hahmann and Güntürkün, 1992). Usually, bilateral centrifugal lesions only caused mild or no deficits in visual discrimination experiments. However, Rogers and Miles (1972; but see Hahmann and Güntürkün, 1992) demonstrated profound deficits in the detection of suddenly occurring moving stimuli, suggesting that the centrifugal system may play a role in detecting moving objects under dim light conditions. Recently, Wilson and Lindstrom (2011) formulated a new functional hypothesis that rests on the assumption that the ION system can only be understood if the strange intraretinal projection from the ventral to the dorsal retina is taken into account. They propose that the ION acts as an early warning system that allows the presence of a moving shadow on the ground to trigger a rapid and parallel search of the regions of sky most likely to contain an aerial predator. This dual search could be the function of the intraretinal projection that links the ventral retina (looking into the sky) to the dorsal retina (scanning shadows on the ground). Once

an association between shadow and object is established, the system could link these two stimuli via positive feedback and continue to track shadow and object together. This hypothesis could explain why the centrifugal system is so well developed in granivorous and ground-feeding birds. Bobwhite quail has an annual probability of mortality of 63% from aerial predators (Cox et al., 2004). Thus, any extremely fast neural system that tracks approaching birds of prey and their shadows in parallel could save lives.

8.3.2.3 Projections of the Optic Tectum: From Retinotopy to Functionotopy

Retinal projections to the tectum are retinotopically organized in most vertebrates (Remy and Güntürkün, 1991; Reiner et al., 1996; Dunlop et al., 2007). Retinal fibers and their tectal target cells are then segregated in different intratectal parallel streams, which project to diverse areas along the neuraxis (Reiner, 1994; Güntürkün, 2000; Marín et al., 2003; see also Section 8.2.2). In pigeons, about 90% of retinal ganglion cells project to the

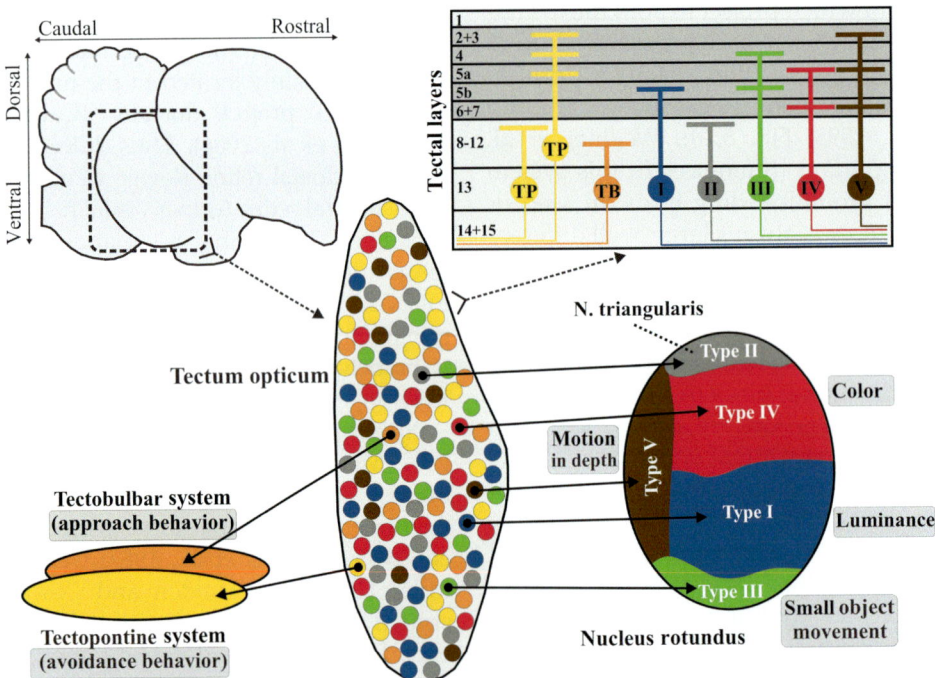

FIGURE 8.11   Schematic outline of the tectal mosaic hypothesis of Hellmann et al. (2014). It is proposed that multiple cell types with diverse visual inputs at about the same location of the tectal map project onto diverse thalamic and rhombencephalic areas. These projections are both retinotopically organized and functionally specific. A pigeon brain with a highlighted optic tectum is represented at upper left. An "unfolded" tectum with a two-dimensional map of the tectal surface is shown in the center. Small circles in different colors represent cells on the tectal map that have descending projections within the tectopontine (TP, in *yellow*) and the tectobulbar systems (TB, in *orange*) or ascending ones within the tectorotundal projection (cell types I–V). Each of these cell types projects to an area with the same color code. It is yet unclear if retinotopy is preserved within such projections. At upper right, a schematic cross section of the optic tectum is shown, with retinorecipient layers 2–7 depicted in *dark gray*. Tectal neurons with descending projections to the brain stem and ascending projections to the rotundus are shown with their main dendritic bifurcations in the superficial tectal strata allowing a cell type–specific organization of visual input. The complete arrangement explains how functionally specific tectal projections arise from a retinotopically arranged organization. *Modified from Hellmann, B., Güntürkün, O., Manns, M., 2004. The tectal mosaic: organization of the descending tectal projections in comparison to the ascending tectofugal pathway in the pigeon. J. Comp. Neurol. 472, 395–410.*

tectum (Remy and Güntürkün, 1991). The outer retinorecipient layers of the tectum are characterized by a precise retinotopic representation with narrowly tuned receptive fields of less than 1 degree (Jassik-Gerschenfeld and Hardy, 1984; Luksch, 2003). However, receptive field widths gradually increase toward layer 13 (Jassik-Gerschenfeld and Guichard, 1972; Frost and DiFranco, 1976), which contains neurons with very large dendritic trees that have characteristic "bottlebrush endings" in specific upper tectal laminae (Luksch et al., 1998). These layer 13 cells are looming sensitive (Frost and Nakayama, 1983; Wu et al., 2005) and project to the diencephalic nucleus rotundus (Rt) (Luksch et al., 1998; Hellmann and Güntürkün, 2001; Marín et al., 2003; Hu et al., 2003). Retinotopic place coding seems to be absent within Rt, since each point of the tectum is connected to nearly the entire rotundus and its dorsal cap, the nucleus triangularis (Benowitz and Karten, 1976; Ngo et al., 1994; Hellmann and Güntürkün, 1999). Instead of retinotopy, a new function-based segregation seems to take place in the thalamus, as electrophysiological data could demonstrate separate rotundal domains in which mainly color, luminance, motion, or looming are processed (Wang and Frost, 1992; Wang et al., 1993). Behavioral data support this view since restricted rotundal lesions affect performance in only specific aspects of visual analysis (Laverghetta and Shimizu, 1999). In contrast to the tectorotundal connection, the rotundoentopallial projection (Benowitz and Karten, 1976; Fredes et al., 2010) and subsequent secondary and tertiary connections within the forebrain are again organized topographically (Benowitz and Karten, 1976; Husband and Shimizu, 1999), suggesting rotundal functional segregation to be carried onto the forebrain.

How is a retinotopically organized tectal system transformed into a functionotopically organized rotundal system (Karten et al., 1997)? According to Hellmann and Güntürkün (2001) and Marín et al. (2003), the transformation is achieved by five morphologically distinct types of tectal layer 13 cells (types I–V) that together establish the tectorotundal system. Each population is characterized by (1) its location on the tectal map, (2) the depth and size of its soma within layer 13, (3) its specific input from tectal laminae 3–11, and (4) its projection onto separate subregions of the rotundal system (Marín et al., 2003; Fig. 8.11).

1. Since the tectum is retinotopically organized, lamina 13 cells sample retinal input mainly from a specific retinotopic area (Gonzalez-Cabrera et al., 2016). However, there are different tectorotundal neuron types and they are differently distributed across the tectal map. For example, type I neurons are four times more common in ventral tectum (representing the frontal, binocular field of view) (Remy and Güntürkün, 1991; Hellmann and Güntürkün, 1999). In contrast, type V neurons are twice as common in the dorsal tectum and transmit information from the dorsal field of view (Hellmann and Güntürkün, 2001).

2. Each tectorotundal neuron type is characterized by a unique combination of soma size and position within the depth of layer 13 (Karten et al., 1997; Hellmann and Güntürkün, 2001; Marín et al., 2003). Thus, different projectional tectofugal streams have different morphologies and positions within the tectum.

3. Retinal ganglion cells can be subdivided according to morphological and physiological criteria into different classes, each of which subserves a different function (Ehrlich et al., 1987; Karten et al., 1990; Mpodozis et al., 1995). These different ganglion cell types terminate in a spatially segregated manner within tectal layers 2–7 (Yamagata and Sanes, 1995; Karten et al., 1997; Repérant and Angaut, 1977; Gonzalez-Cabrera et al., 2016). Therefore, retinorecipient laminae differ in their visual input. Since the tectorotundal neurons sample retinal input from different tectal laminae, they probably process different aspects of vision.

4. Tectorotundal cell type I projects to ventral and central rotundus and probably code for changes in luminance (Wang et al., 1993; Hellmann and Güntürkün, 2001). Fibers of type III neurons terminate in the most ventral rotundus, where the cells strongly respond to moving occlusion edges and very small moving objects, with either excitatory or inhibitory responses (Wang et al., 1993). Axonal projections of type IV neurons ramify within a relatively small area of the dorsal rotundus, which was shown to be highly sensitive for color and/or luminance variations of visual stimuli (Wang et al., 1993). Electrophysiological work revealed the caudal rotundus (termination of type V neurons) to be specialized to three dimensional motion analyses (Wang et al., 1993) with some of these neurons especially computing time to collision for looming stimuli (Wang and Frost, 1992; Sun and Frost, 1998).

Hellmann et al. (2004) showed that the mosaiclike architecture of the ascending tectal projections applies also to the descending fibers that target motor and premotor centers in the mes- and rhombencephalon. As for Rt, the descending motor systems are functionally segregated in pigeons: The crossed tectobulbar tract is involved in approach and orientation toward desired objects, whereas the ipsilateral tectopontine pathway guides movements away from aversive stimuli (Ingle, 1983; Ellard and Goodale, 1988; Dean et al., 1988). The ascending tectal projections to Rt originate mainly from the ventral tectum, representing the frontal inferior field of view. In contrast, the descending tectal projections overrepresent the upper field of view (Hellmann et al., 2004). Thus, the principle of a retinotopic-to-functionotopic transformation seems to apply also for the descending tectal projections. Interestingly, some looming-sensitive layer 13 neurons that project to the Rt also have descending projections to the pons (Wu et al., 2005). Therefore, looming information can directly initiate avoidance behaviors in an animal facing an impending collision. These data support the concept that the tectum is arranged as a mosaic of multiple cell types with diverse input functions at the same location on the tectal map. By a transformation from retinotopic to functionotopic coding, tectal projections onto diverse areas become both retinotopically organized and functionally specific (Fig. 8.11). It is not yet known if retinotopy is preserved in the different functionotopic zones.

8.3.3 Telencephalon

Most reptilian brains only partly fill the cranial cavity. As a result, the shape of reptilian skulls is only slightly influenced by the form and structure of the brain (Starck, 1979). This is especially true for marine turtles, tuataras, and most lizards (ten Donkelaar, 1998). Brains that are much smaller than their skull are also found in many theropod dinosaurs (Witmer and Ridgely, 2009). Exceptions among reptiles are the snakes, in which the space between the brain and the cranial wall is quite narrow. In birds, the brain completely fills the cranial cavity and shapes the skull—a condition already observed in the extinct ancestors of modern birds (Balanoff et al., 2013).

The sauropsid telencephalon consists of paired evaginated hemispheres. Astonishingly, this seemingly inconspicuous brain structure has ignited many heated discussions among comparative neuroanatomists and was subject to major conceptual changes. Based on classic anatomical studies at the turn of the 19th to the 20th century, several leading scholars, including Ludwig Edinger in Germany and Cornelius Ubbo Ariëns Kappers in the Netherlands, decided that the telencephalon of amniotes had gradually expanded by the successive addition of new parts. Comparative neuroanatomists of this time thus followed the ancient concept of *scala naturae* according to which organisms are organized in stepwise increases of complexity and "souls." This conception, as translated into vertebrate comparative neuroanatomy, assumed that the telencephalon of ancestral jawless vertebrates was the starting point and as such related entirely to olfaction. With the advent of jawed fishes, the globus pallidus was added, followed by the addition of the striatum in amphibians. Reptiles

added a three-layered cortex while birds dramatically expanded the volume of their telencephalon by increasing the size of their striatum. With the emergence of mammals, a novel brain entity started to dominate the outer rind of the telencephalon: a six-layered cortex. Since the cortex was seen as the newest addition to the vertebrate telencephalon, it was called "neocortex." As outlined below, the neocortex derives from just one of the four pallial components of the telencephalon. Ventral to these four pallial components are the subpallial basal ganglia. Taken together, reptiles were seen as having developed at least a primitive forerunner of the cortex, while birds had nothing comparable but expanded their basal ganglia instead (Edinger et al., 1903; Ariëns-Kappers et al., 1936).

It was the Swedish neuroanatomist and embryologist Bengt Källén (1962) who departed from this view and proposed that parts of the avian telencephalon are of pallial nature. The strongest shift in general understanding, however, came with the seminal work of the American neuroanatomist Harvey Karten (2015) in pigeons that started in the 1960s and sparked new insights into the organization of the avian forebrain. These and many further studies finally resulted in the Duke Avian Nomenclature Forum of 2002 (Reiner et al., 2004a; Jarvis et al., 2005). Based on an overwhelming body of data from genetics, neurochemistry, anatomy, and physiology, a consortium of neuroscientists at the conference concluded that most of the large dorsal territory of the avian cerebrum is pallial. This pallial territory was seen as homologous to regions of the mammalian brain that includes neocortex, hippocampus, claustrum, and pallial amygdala. The smaller ventral part of the avian cerebrum was identified as subpallial and highly comparable with its mammalian counterpart in all developmental and anatomical details. Thus, bird brains are not dominated by striatum. But how much of the avian pallium is homologous to neocortex? These and further questions are hotly debated (Puelles et al., 2000, 2016; Pfenning et al., 2014; Montiel et al., 2016) and will be discussed by Luis Puelles in this book. We will leave homology questions mostly out of our scope but will concentrate instead on the functional anatomy of the sauropsid forebrain. We begin by briefly charting the overall territory of the telencephalon.

The telencephalon of tetrapods can be divided into a pallial and a subpallial sector. The pallial entity has been subdivided into medial, dorsal, lateral, and ventral components based on cellular migratory patterns, cytoarchitecture, gene expression, and connectivity (Holmgren, 1922; Puelles et al., 2000; Nomura et al., 2008; Montiel et al., 2016). The lateral pallium comprises the olfactory cortex, the medial pallium the hippocampal complex, and the dorsal pallium the neocortex. In reptiles, some parts of the ventral pallium possibly constitute the dorsal ventricular ridge (DVR) along with diverse nuclei of the amygdaloid complex (Northcutt, 2013). Birds do not have a three-layered cortex as seen in reptiles. However, they have a component called "wulst" that is typical for the avian class and is located in the dorsal or dorsofrontal part of the pallium (Striedter, 2005).

When taking a histological frontal section, the division between a pallial and a subpallial component is easily discernable using markers for acetylcholinesterase or dopamine (Reiner et al., 1998). In all vertebrates the basal ganglia play a prominent role in the control of movement patterns. However, neuropsychological studies increasingly make it likely that the selection of future actions is possibly a much more important role (Graybiel and Grafton, 2015). It is likely that this is true not only for humans but also for all sauropsids.

8.3.3.1 The Sauropsid Basal Ganglia

Before reviewing the organization of the avian and reptilian basal ganglia, some more general comments on the functions of these ancient structures of the vertebrate brain seem to be in order. From the earliest days of neurological analyses on, the basal ganglia were seen as a central entity of action generation (Ferrier, 1876). Indeed, neurological disorders that affect the basal ganglia always also affect behavioral output in terms of either a lack of movement (hypokinesia) or a production of undesired movements (hyperkinesia) (Mink, 2003). These observations substantiated the view that the role of the basal ganglia is to produce and control movements. In the last decades this view eroded and gave way to the opinion that the main function of the basal ganglia is to select contextually appropriate actions among many alternatives (Yin, 2016). Indeed, this idea fits with the network structure of corticostriatopallidal-thalamic loops in which the bandwidth of cortically selected behavioral options is successively reduced until only one planned action survives the competition and is then successively produced (Humphries and Prescott, 2010). Parallel to this conceptual shift, the learning of habits and skills at the level of the striatum moved into the focus of scientific attention (Graybiel and Grafton, 2015). While learning such actions requires reward in the beginning, they become increasingly automaticed and are produced without any further reinforcement. Learning psychologists were the first to point out that new behavioral sequences are sensitive to omission of reward in the beginning of acquisition, but become increasingly independent of reward later on, when they are established as habits (Dickinson, 1985). In humans, fMRI studies demonstrate that those subjects who reduce activity patterns in the prefrontal cortex (PFC) sooner are the ones who learn sequential actions faster (Bassett et al., 2015). Concomitantly, striatal units seem to "bracket" an action, thus

firing in the beginning and the end of a habit, as if encapsulating a behavioral unit (Barnes et al., 2005). It is conceivable that the basal ganglia help to store and subsequently select sequences of habit units, depending on the ensuing situation (Graybiel and Grafton, 2015). This short prologema to the basal ganglia will be important toward the end of this section when discussing how the basal ganglia changed during the phylogeny of vertebrates. Now let us turn to the anatomy of the subpallium in general and the basal ganglia in particular.

The subpallium can be subdivided anatomically and developmentally into five entities: (1) the dorsal somatomotor basal ganglia; (2) the ventral viscerolimbic basal ganglia; (3) the extended amygdala; (4) the basal telencephalic cholinergic and noncholinergic corticopetal systems; (5) the septum and septum-associated neuroendocrine systems. In the following we will review the basal ganglia only. Unfortunately, our current knowledge of this system in birds far outweighs what we know from reptiles. A succinct overview of the avian subpallium is provided by Kuenzel et al. (2011).

The sauropsid striatum is considered to be homologous to the mammalian caudate/putamen (Reiner et al., 2004a; Kuenzel et al., 2011). It consists of lateral and medial somatomotor as well as viscerolimbic components. The principal wiring pattern of these components is similar but differs with respect to some connections so that some striatal territories receive more somatic or more viscerolimbic input than others. The vast majority of striatal neurons are GABAergic projection neurons with spiny dendrites. About half of these neurons show a colocalization of GABA with enkephalin (ENK). The other half has a colocalization of substance P (SP) with dynorphin. Both cell types project outside of the striatum. The striatum is rich with cholinergic fibers due to the abundance of local projections of cholinergic interneurons. A further characteristic of the striatum is the dense dopaminergic innervation from the tegmental dopaminergic cell groups.

The striatum shows a winner-take-all dynamic, resulting in a narrowing of many diverse activity fields into a few, or even into just one (Ponzi, 2008). Activity patterns mostly arise from glutamatergic pallial input (Csillag et al., 1997). Indeed, most of the pallium projects onto the striatum. This is true for reptilian cortex/wulst, DVR and for amygdala and hippocampus (Reiner et al., 1998). Descending fibers from the pallium can arise as axon collaterals from cells that project mainly to other pallial areas or from neurons with long projections to subtelencephalic targets (Veenman et al., 1995). The bottom line is that most pallial activity patterns reach the striatum and are then subject to competition with other striatal activity foci elicited by other pallial inputs. A second source of striatal afferents are the dorsomedial group of thalamic nuclei (Reiner, 2002). If the pallial input is blocked or entirely abolished, the tested mammals and birds are able to move and feed and even learn operant tasks, albeit after quite some recovery time (Bjursten et al., 1976; Cerutti and Ferrari, 1995). This is an important insight: the thalamic input into the basal ganglia is sufficient to generate largely normal behavior in amniotes.

The third major input to the striatum consists of the dopaminergic axons from several clusters of tegmental dopamine neurons: the area ventralis tegmentalis (AVT), the substantia nigra pars compacta (SNc), and the retrorubral field. While the AVT input dominates in the ventral striatum, the main source of dopaminergic input to the avian striatum arises from the SNc (Durstewitz et al., 1999b). Autoradiographic D1 receptor binding studies demonstrated that D1 receptors are most abundant in the bird striatum (Schnabel et al., 1997; Stewart et al., 1996). Consequently, the striatal parts of the basal ganglia also exhibit very dense labeling for DARPP-32 (Durstewitz et al., 1998; Schnabel et al., 1997), which is a dopamine- and cAMP-regulated phosphoprotein that acts as a "third messenger" in the D1 receptor stimulation cascade (Berger et al., 1990; Hemmings et al., 1995). The density of D2 receptors in the striatum seems also to be high in birds (Dietl and Palacios, 1988; Stewart et al., 1996).

SP- and ENK-positive GABAergic striatal neurons project densely to the pallidum, which contains sparsely packed large GABAergic neurons; dopaminergic inputs are less abundant. This pattern is seen in both birds and reptiles (Anderson and Reiner, 1990; Brauth, 1984). The second major descending projection from the striatum leads to the GABAergic neurons of the substantia nigra pars reticulata (SNr). Again, both in reptiles and birds, these seem to arise mainly from the SP+ GABAergic neurons. A smaller ENK+ input to dopaminergic tegmental neurons also exists, especially in snakes (Reiner et al., 1998).

Both the pallidum and the SNr project to a small nucleus in the avian dorsal thalamus − the ventrointermediate thalamic area (VIA) (Medina et al., 1997). VIA projects to the most rostral part of the wulst, which serves motor functions and is the source of the avian "corticospinal tract" (see Section 8.3.3.3.1). Through this pathway the striatum is able to modulate the wulst. The organization of this circuit strongly suggests a homology to the mammalian loop from the globus pallidus interna to the thalamus and thence back to the cortex.

A further pallidal projection leads to the thalamic nucleus of the ansa lenticularis (ALa) which projects back by a glutamatergic pathway to the pallidum and the SNr (Jiao et al., 2000). The ALa is homologous to the subthalamic nucleus of mammals that receives input from the external pallidal segment [globus pallidus pars externus (GPe)] and constitutes a component of the

indirect pathway of the basal ganglia. It is presumed that this circuit promotes suppression of unwanted movement patterns (Jiao et al., 2000).

The pallidum of birds also projects to a small number of dorsomedial thalamic nuclei that project back to the striatum as well as several pallial areas (Medina and Reiner, 1997; Veenman et al., 1997). Through this pathway the basal ganglia can modulate processes of the DVR and contribute to aspects of action selection. An important pallidal projection of birds, turtles, crocodiles, and lacertid lizards leads to the pretectum. In birds this GABAergic nucleus is called the nucleus spiriformis lateralis (SpL); in reptiles it is called the dorsal nucleus of the posterior commissure (nDCP) (Reiner et al., 1982a,b, 1998; Medina and Smeets, 1991). SpL and nDCP project to the deeper layers of the tectum, including those that project to premotor cell groups of the hindbrain (see Section 8.3.2.3) (Reiner et al., 1982a). Given the importance of the motor output pathway via the tectum in sauropsids, this projection could provide a major route by which the basal ganglia can influence movements in birds and reptiles (Fig. 8.12). A comparable basal ganglia-pretectotectal pathway seems to be absent in some lizard groups and snakes (Russchen and Jonker, 1988).

The general pattern of this system seems to be ancient and can be traced back to the earliest anamniotes (Grillner and Robertson, 2015), although the loop back to the cortex/pallium seems to be lacking in these animals (Wullimann, 2014). However, the relative contributions of the components can differ between taxa. In anamniotes, the pallium and subpallium are quite small and possibly have less influence on the overall behavior of the animals (Reiner, 2002).

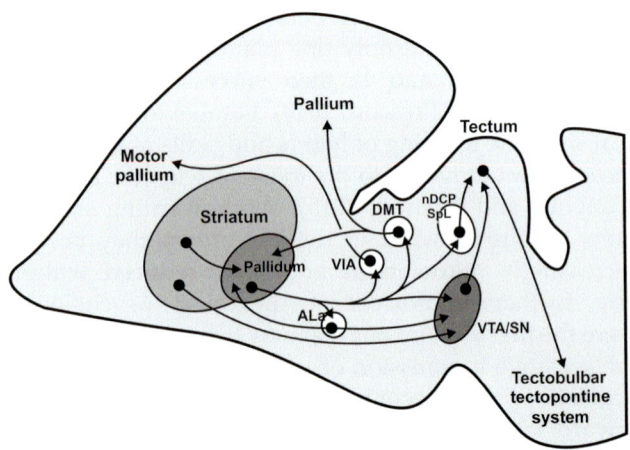

FIGURE 8.12 Highly simplified schema of the sauropsid basal ganglia system. Only a subset of the connections is shown. Abbreviations are given in the list of abbreviations. *Modified from Reiner A., Medina L., Veenman C., 1998. Structural and functional evolution of the basal ganglia in vertebrates. Brain Res. Rev. 28, 235–285.*

8.3.3.2 *The Reptilian Pallium*

The reptilian pallium can be easily discerned from the septum and other subpallial structures by several major anatomical landmarks that are already visible from the outside. Medially, a longitudinal sulcus that is easily visible on the telencephalon marks the border between the medial cortex and the septum. Laterally, a groove divides the pallium from the subpallium. Frontally, the olfactory bulbs are distinctly visible. In most reptiles the bulbs assume a position quite distant from the rest of the telencephalon and are connected by long and slender olfactory tracts; these can be very long in adult crocodiles. In turtles, however, the olfactory bulbs are sessile and frontally abut the more caudal telencephalon. In reptiles the pallium is constituted by the cerebral cortex dorsal to the lateral ventricle and the DVR. We present these two entities separately.

8.3.3.2.1 The Reptilian Dorsal Cortex

Different from birds, reptiles and mammals possess a true multilayered cortex (Ulinski, 1990). The word "cortex" stems from the Latin word "bark" or "skin" but within the neurosciences, it defines a laminated gray matter that harbors multiple layers. To our knowledge, this definition goes back to the 19th century. Leuret and Gratiolet already use it in their famous two-volume book on comparative neuroanatomy (1839, 1857) but possibly, the meaning of "cortex" as referring to a laminated gray matter mantle is even older, given that cortical lamination was first described in 1776 by Francesco Gennari in the human visual cortex. Although cortex is by definition always laminated, there are different numbers of layers that can comprise a cortex. Only the mammalian neocortex that covers the bulk of the cerebral hemispheres has six layers. The human hippocampus (archicortex) has three or four laminae (depending on hippocampal area) while the piriform cortex (CPi) (paleocortex) has three layers.

The reptilian cortex also has three layers. The outer layer 1 is called the superficial plexiform or molecular layer and contains only few scattered interneurons. Afferent axons from the lateral forebrain bundle travel through this layer and fan out in a seemingly nontopographic manner (Naumann et al., 2015). These axons make numerous en-passant synapses on both layer 1 interneurons and distal dendrites of layer 2 principal neurons (Smith et al., 1980). The intermediate layer 2 is called cellular layer and forms a continuous and densely packed sheet of principal neurons, a much smaller number of interneurons as well as afferent and local axons. Layer 3 is the deep plexiform or subcellular layer and is only loosely packed with interneurons; it contains a large number of basal dendrites of principal neurons as well as corticofugal and local axons. A distinct bundle of unmyelinated fibers called alveus is situated deep to layer 3 in most cortical areas (Ulinski, 1990; ten Donkelaar, 1998). These cortical areas show distinct cytoarchitectonic differences that could point to computational specializations.

According to Fournier et al. (2015) and Naumann et al. (2015), the reptilian dorsal cortex strongly resembles the three-layered mammalian CPi. Based on this comparison, and substantiated by some studies in reptiles, they propose a couple of highly interesting functional interpretations of the local dynamics of the turtle dorsal cortex (Fig. 8.13). The starting point of these local dynamics is the system of afferent fibers that reach the dorsal cortex via the dorsal forebrain bundle from sensory thalamus. These axons run across layer 1 and synapse both on inhibitory interneurons of layer 1 as well as on distal dendrites of principal neurons of layer 2 (Kriegstein and Connors, 1986). Interneurons receive more afferent input than the principal cells and thus provide a massive feed-forward inhibition to principal neurons (Fournier et al., 2015). In addition, principal cells activate layer 2 interneurons and receive feedback inhibition. Principal neurons also provide recurrent excitation to other pyramidal neurons via the associational intracortical connections (Fournier et al., 2015). As a result, sensory stimulation evokes strong inhibition, combined with sparse coding properties of principal neurons (Mancilla et al., 1998).

What kinds of computational properties would such a network have? Based on the similarity to the CPi, it is

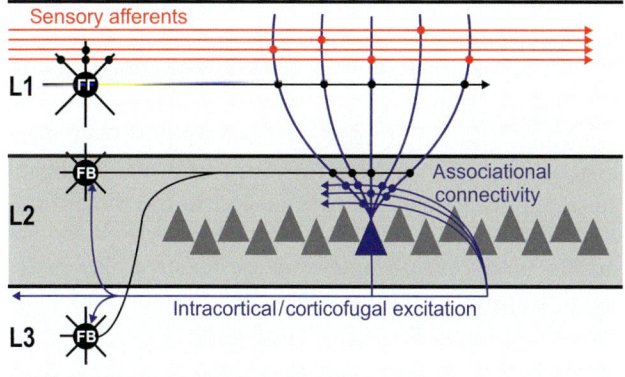

FIGURE 8.13 A schematic overview of the principal wiring pattern of the turtle cortex. Sensory afferents enter layer 1 (L1) and loosely contact apical dendrites of principal neurons that have their somata in layer 2. Inhibitory interneurons of the feedforward type (FF) also receive massive input from sensory afferents and synapse onto the dendrites of principal neurons. These principal neurons receive recurrent excitation from other principal neurons (associational connectivity) and feedback inhibition (FB) from interneurons of layers 2 and 3 that are excited by principal neuron output along the intracortical/corticofugal path. *Modified from Fournier, J., Müller, C.M., Laurent, G., 2015. Looking for the roots of cortical sensory computation in three-layered cortices. Curr. Opin. Neurobiol. 31, 119–126.*

conceivable that coding should not occur by means of topographic maps but by nontopographically organized ensembles of neurons (Fournier et al., 2015; Naumann et al., 2015). Indeed, attempts to discover topography of the visual field in the visual aspects of the turtle cortex were not successful (Mazurskaya, 1973). If this holds, the implication would be both important and unexpected. First, it would be important since it would speak against a functional similarity in the coding properties of mammalian isocortex and reptilian dorsal cortex. Second, it would be unexpected since a nontopographical coding is conceivable for a stimulus-like odor but would be surprising for vision, in which an inherent topographic order is physically provided. In any case, the described attempts to use the architectural similarities between the mammalian CPi and the turtle dorsal cortex as point of explorative departure provides a novel and rich approach to understand the functional properties of the reptilian pallium.

8.3.3.2.2 The Reptilian Dorsal Ventricular Ridge

The DVR is a massive nuclear brain structure that is positioned below and lateral to the lateral ventricle. It is usually divided into a much larger anterior (ADVR) and a smaller posterior (PDVR) component. The ADVR receives thalamic sensory input via the lateral forebrain bundle and can be further subdivided into three longitudinal slabs that receive different types of sensory input (ten Donkelaar, 1998). From lateral to medial, these slabs represent the termination areas of visual, somatosensory, and auditory pathways. They will be discussed in greater detail in Sections 8.4.1–8.4.3.

As outlined in Section 8.3.2.1, rattlesnakes and some other snake species possess a thermal pit below their nasal cavity with which they can detect the heat landscape in front of them. This information is fed into the deep layers of the optic tectum via a branch of the ascending trigeminal system. Some of the deep tectal neurons respond to both infrared and visual input. These bimodal tectal neurons project to the thalamic Rt which projects to the lateral sector of the ADVR (Berson and Hartline, 1988). Single units in this area also evince visual/infrared response properties (Berson and Hartline, 1988). These findings demonstrate two interesting principles. First, despite a large variability in body structures and ecological specializations, sauropsids share a common basic neural bauplan that is evident even in pathways that transport idiosyncratic and highly specialized "unusual" sensory information. In the case of rattlesnakes, infrared sensing is processed along a pathway that exists from snakes to birds across all sauropsids. Second, this commonality is achieved by the incorporation of these special sensory senses into pathways of the "classic" senses. For infrared vision, the trigeminal system merges with the tectofugal visual pathway to transport vision from deep blue to infrared.

The ADVR projects massively to the ipsilateral striatum and to the PDVR (Ulinski, 1978). In addition, ADVR projects via the hippocampal and/or anterior commissures (Bruce and Butler, 1984) to the contralateral hemisphere. Although pallial commissural systems in reptiles are rather small, most of the reptilian pallium is interhemispherically connected (Northcutt, 1981). The PDVR does not receive sensory thalamopallial afferents and resembles the mammalian amygdaloid complex in several respects.

The DVR shows, especially in its anterior component, several specializations that differentiate snakes, lizards, and turtles from crocodiles and birds. In the first group of animals, the ADVR shows a lamination-like pattern in the vicinity of the lateral ventricle. This "first pattern" (ten Donkolaar, 1998) includes a cell-poor zone immediately under the ventricular surface; the second zone consists of clusters of neuronal somata that create a ribbonlike structure along the ventricular border; and the third zone consists of scattered individual neurons. In contrast, the "second pattern" of the DVR in crocodiles and birds lacks the first and second zones, such that the third zone (scattered individual neurons) directly abuts the ventricle.

The first pattern of snakes, lizards, and turtles resembles to some extent a laminar pattern. And indeed, it is conceivable, that is, the remainder of a lamination also encompasses the DVR. This possibility becomes likely when studying the organization of the DVR of the tuatara *S. punctatus*, which is endemic to New Zealand and is the sole survivor of a distinct order, the Rhynchocephalia. Their closest living relatives are the squamates (lizards and snakes). Reiner and Northcutt (2000) demonstrated that in tuatara the distinction between a cortex and the DVR does not exist. Instead, the trilaminar cortex seems to extend into the entire DVR and harbors the termination areas of the ascending visual tectofugal and the ascending auditory pathway. This finding has potentially important implications. If the trilaminar DVR of tuatara is the primitive condition for reptiles, then major aspects of pallial organization in ancestral sauropsids might have been similar to the dorsal cortex of today's reptiles. With the exception of the tuatara, the various sauropsid lineages then must have changed their laminated ancestral DVR into a nuclear arrangement during evolution. A nuclear arrangement, rather than lamination, would then be the derived architecture.

8.3.3.3 *The Small World of the Avian Pallium*

Bird brains are large, and most of their volume consists of the pallium. As nicely outlined by Striedter (2005), the brain of a 1 kg macaw weighs 20 g more than that of a 200 kg alligator and about as much as that of 1000 kg megamouth shark. As outlined in Section

8.2.3, these and further novel insights allow a fresh look at the question of why corvids and parrots are able to master complex cognitive tasks similar to primates. But even if avian brains are large, what is their internal organization? Overall, avian brains have a very large DVR that hardly leaves space for the lateral ventricle. However, bird brains lack an obvious homolog of the reptilian cortex. Instead, birds have a structure that was named "wulst" (ie, "bulge") by German neuroanatomists of the 19th century. The wulst assumes the same position within the telencephalon as the reptilian cortex and resembles mammalian neocortex and reptilian cortex in embryology, genetics, and topology (Puelles et al., 2000; Reiner et al., 2004a; Jarvis et al., 2005). Despite these similarities, the fact remains that the wulst in its architecture is unique to birds. One might assume that brains with so many unique anatomical features produce unique behaviors but, astoundingly, this is not the case. Instead of major differences in behavior and cognition, we find important similarities, at least for cognition in birds and mammals.

Recent years has a surge of comparative studies on "higher" cognitive abilities such as aspects of impulsive control, inferential reasoning, planning ahead, perspective taking, and role understanding. It has been argued that these skills, often subsummed under the term "complex" cognition, form a cognitive tool kit comparable to that of mammals (Emery and Clayton, 2004). Although also reptilian cognition should not be underestimated, nothing at the level and scope of bird cognition has

been reported for this animal group so far (Burghardt, 2013). Critiques have pointed out that most studies on bird cognition have tested these animals in narrowly defined domains with few paradigms that are mostly related to food hoarding (Penn and Povinelli, 2007; Shettleworth, 2010). Using such paradigms, food-caching scrub jays and ravens could show a cognitive prowess that can be interpreted as an indication for corvids having mental capacities that are on par with those of great apes (Clayton and Dickinson, 1998; Bugnyar and Heinrich, 2005; Raby et al., 2007; Prior et al., 2008). On the other hand, the corvid results may be seen as a special adaptation to the very context of food caching. The birds' mental capacities are thus thought to be highly domain specific and not directly comparable with the flexibly used skills of primates (Seed et al., 2009). There are, however, a large number of recent studies that indicate that such a criticism is too restrictive: corvids show various primate-typical behaviors such as alliance formation, third-party intervention, postconflict reconciliation, and consolation (Bugnyar, 2013), and they excel in a variety of experimental tasks and contexts other than caching (Prior et al., 2008; Güntürkün and Bugnyar, 2016). This interpretation becomes even more convincing when parrots are included in the analysis (Pepperberg, 1999; Mikolasch et al., 2011). Even the lowly pigeon can perform noteworthy feats of cognition, such as long-term recollection (Fagot and Cook, 2006), transitive inference reasoning (von Fersen et al., 1990), complex pattern recognition (Yamazaki et al., 2007),

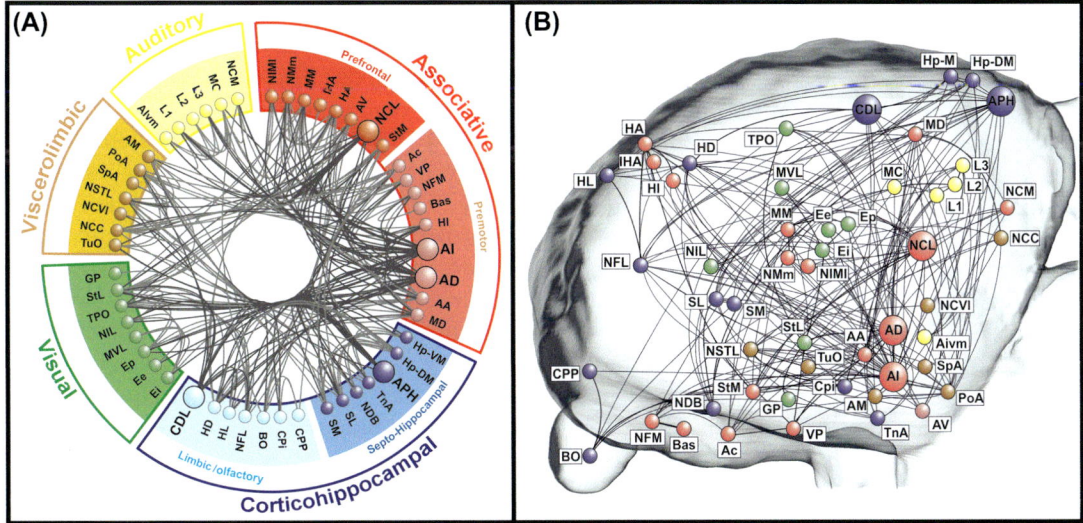

FIGURE 8.14 The connectome of the pigeon telencephalon. (A) Network analysis of the matrix of connections revealed five top-level modules; the associative and the corticohippocampal modules each consist of two lower level modules. Connections to and from hub nodes are shown in a slightly darker color. (B) Sagittal view of the pigeon forebrain with all nodes and their connections. Nodes are colored according to module level membership. Note that the modules are spatially distributed rather than restricted. Color codes: *red*, associative; *blue*, corticohippocampal; *green*, visual; *brown*, viscerolimbic; *yellow*, auditory. Abbreviations are given in the list of abbreviations. *From Shanahan, M., Bingman, V., Shimizu, T., Wild, M., Güntürkün, O., 2013. The large-scale network organization of the avian forebrain: a connectivity matrix and theoretical analysis. Front. Comput. Neurosci. 7, 89, with permission.*

and optimal choice (Herbransen and Schroeder, 2010). But some birds not only reach the same levels of cognitive capacity as mammals but also display identical details of their cognitive architecture as visible in fine-grained analyses of the way they represent objects, categories, or relations (magpies: Pollok et al., 2000; pigeons: Scarf et al., 2011). These similarities of cognitive organization are astounding given that the telencephalon of birds and mammals exhibits a very different anatomical organization.

Looking carefully, it is obvious that the differences apply to the overall organization of the telencephalon but are quite small when it comes to the connectivity of the ascending sensory pathways, associative forebrain areas, and subpallial structures (Reiner et al., 2005; Güntürkün and Bugnyar, 2016]. Thus, avian and mammalian forebrains might have similar connectivities despite a radically different overall organization. These similarities in connectivity might drive similarities in behavior. Indeed, it is a futile enterprise to try to understand cognitive functions of a brain without analyzing information flow within its neural network. To analyze the overall connectivity and possible information flow of the avian telencephalon, Shanahan et al. (2013) compiled a large-scale "wiring diagram" for the pigeon and analyzed it with the mathematical tools of graph theory. Combining more than four decades of tracer studies, they constructed a structural "connectome" of the pigeon telencephalon.

This work revealed, first, that the pigeon pallium is a small-world network. In such a network, neighboring nodes have tight links with each other, but most nodes can be reached from every other node by a small number of steps (Watts and Strogatz, 1998). These properties are achieved by a dense local connectivity (high level of clustering) that is combined with a much smaller number of connections that randomly reach out to far-distant nodes (random graph that creates a short path between two distant nodes). In social networks, this effect is known from the finding that people may live in close-knit societies, but still everybody is linked to any stranger in the world by an astonishingly short chain of acquaintances (Fig. 8.14).

Second, the connectome analysis revealed that the pigeon telencephalon comprises a number of distinct modules, defined as clusters of nodes with dense connections between each other but sparse connections with the nodes of other modules. Remarkably, the pigeon modules were found to be functionally analogous to those of humans. The largest pigeon module is the associative module, which consists of prefrontal and premotor submodules that likely mediate higher cognition. The second largest module is the corticohippocampal module, which includes septohippocampal and limbic/olfactory submodules. They integrate multimodal information that is used, for example, in hippocampus-based spatial orientation and navigation. The visual module represents the tectofugal forebrain areas, including their primary, associative, and descending (motor) components. The tectofugal pathway constitutes the dominant visual system in pigeons, while the auditory module represents subdivisions of the primary auditory fields along with their secondary, associative, and premotor structures. Thus, most of the modules of the pigeon's telencephalon are functionally and/or anatomically comparable to modules that are revealed when network analysis is carried out on human or nonhuman mammalian brains (van den Heuvel et al., 2016). Interestingly, while the top-level modules of mammalian brains are anatomically localized, those of the pigeon brain are more anatomically distributed. So, similar connectome patterns do not necessarily resemble each other in spatial organization.

Third, the pigeon telencephalon has a central connective core, and the hub nodes that comprise this core are functionally analogous to hub nodes in the primate brain's connectome core. What does that mean? Hubs are nodes with a large number of connections to other nodes. They serve functions similar to major international airports: if one flies from a small local airport to another small local airport far away, flights are always routed via connection flights through major airports (hubs). A collection of such hubs within the brain constitutes the functional "backbone" of neural information flow. This neural backbone in pigeon and primate brains consists of very similar hubs. So, if the topologically central connective core of the primate brain plays an important role in high-level cognition (Shanahan, 2012), the required connectional infrastructure seems also to be present in birds. This finding is even more exciting when we realize that the prefrontal-like area of birds and the PFC of primates are not homologues but functionally analogues. Thus, these two structures do not derive from a common ancestral structure but represent the outcomes of two completely independent and convergent evolutionary trajectories. The fact that these two structures constitute such highly similar topological centralities of their respective connectomes suggests the following: if two neural structures of different animals share the same function, they may also share the same connectivity blueprint.

8.3.3.3.1 The Avian Wulst

As outlined above, the avian wulst is a likely candidate for homology with mammalian neocortex and reptilian dorsal cortex, although its internal structure is different (Reiner et al., 2004a; Jarvis et al., 2005). The wulst has three functional zones. Starting from anterior, a very small portion of the most anterior tip of the wulst is motoric. From here, the tractus septomencephalicus descends, like the mammalian tractus corticospinalis, to the cervical spinal cord and terminates predominantly contralaterally in the medial part of the base of the dorsal horn of the upper six to seven cervical segments (Wild and Williams, 2000). More posterior is a slightly larger zone that is somatosensory. In barn owls, this area contains a small protuberance that contains the representation of the contralateral claw (Wild et al., 2008). Even more posterior is the wulst's largest zone, which is visual.

This visual wulst bears some resemblance to the mammalian primary visual cortex. This could be due to a one-to-one homology of the visual wulst to the visual cortex as backed by similarities of chemoarchitecture, afferent inputs from the thalamic n. geniculatus pars dorsolateralis (GLd), output pathways to thalamic and midbrain structures, genetic markers, as well as topological position (Karten et al., 1973; Reiner et al., 2004a; Güntürkün and Karten, 1991). However, the four layerlike areas of the wulst are not truly comparable to laminae of the reptilian dorsal cortex or the mammalian neocortex. Instead, they are called "pseudolayers" by Medina and Reiner (2000), since they have some of the properties of cortical layers; however, they lack pyramidal neurons with translaminar dendritic trees. Thus, the avian visual (and nonvisual) wulst shares many similarities with neocortex but also displays some unique derived features.

8.3.3.3.2 The Avian Dorsal Ventricular Ridge

The avian dorsal ventricular ridge is organized into four major subdivisions that are distinct from each other in terms of gene expression, connectivity, and physiology. The nomenclature conference (Reiner et al., 2004a) renamed them as nidopallium, mesopallium, arcopallium, and pallial amygdala. The nidopallium contains zones that are the termination fields of the ascending auditory, visual tectofugal, and trigeminal thalamopallial pathways. We will discuss the internal connectivity of this area in Section 8.3.3.3.2.3. The mesopallium is an associative pallial field that receives neither ascending sensory input nor harbors descending extratelencephalic output systems. The arcopallium and the amygdala are topologically closely intertwined. While the arcopallium is a premotor area, pallial amygdalar nuclei are limbic in nature. Various one-to-one homologies have been proposed before and after the nomenclature conference for these four subdivisions and various parts of the mammalian pallium, including cerebral cortex (Bruce and Neary, 1995; Puelles et al., 2000, 2016; Dugas-Ford et al., 2012; Belgard et al., 2013; Jarvis et al., 2013; Pfenning et al., 2014). This is an active area of research and debate, and it is far from settled. Güntürkün and Bugnyar (2016) reviewed the different standpoints on this matter and Luis Puelles is providing

a comprehensive account on possible homologies in this volume. We will only discuss two specific aspects of the avian dorsal ventricular ridge; the arcopallium/amygdala dichotomy and the laminar organization of the thalamopallial termination zone in the nidopallium.

8.3.3.3.2.1 The Avian Premotor Arcopallium and the Pallial Amygdala
According to the nomenclature used in the atlas of the pigeon brain (Karten and Hodos, 1967), the highly complex region in the ventral part of the posterolateral telencephalon was called archistriatum. The complexity of this area results from its histological and connectional heterogeneity that includes both premotor and limbic features (Zeier and Karten, 1971). In the new nomenclature, the premotor components are called arcopallium while the remaining portions are assumed to constitute the avian pallial amygdala (Reiner et al., 2004a). We first talk about the arcopallium as a central premotor constituent of the avian brain in which sensory input patterns are translated into action signals.

8.3.3.3.2.2 The Arcopallium as a Premotor Center of the Avian Dorsal Ventricular Ridge
The arcopallium consists of the arcopallium anterius (AA), the arcopallium dorsale (AD), the arcopallium intermedium (AI), and the arcopallium mediale (AM) (Reiner et al., 2004a). The connections and functions of the first three components deviate clearly from an amygdaloid pattern. The situation is far less settled for the AM, and this structure could be more comparable to the anterior part of the medial amygdala of mammals, where olfactory and vomeronasal inputs overlap (Abellán et al., 2013). We therefore discuss only the roles of AA, AD, and AI in sensorimotor transfer. The term "arcopallium" is used as an umbrella name for these three substructures.

The arcopallium receives different sensory information from other pallial entities and projects to brain stem motor systems. A good example for the role of the arcopallium in sensorimotor transformation is the trigeminal system and its role in ingestive behavior. Tactile input from the beak is conveyed to the arcopallium from the frontal trigeminal nidopallium (NFT) which receives input from the n. basorostralis pallii (Bas) (Wild et al., 1985; Schall et al., 1986; Letzner et al., 2016). A second trigeminal pathway runs through the mesopallium frontoventrale (MFV), which is reciprocally connected to Bas and NFT (Atoji and Wild, 2012). As demonstrated by Letzner et al. (2016), both NFT and MFV are interhemispherically connected via arcopallial projections. From the arcopallium, descending fibers reach the medial and the lateral components of the striatum (MSt and LSt), as well as the ventral pallidum (VP) (Veenman et al., 1995; Dubbeldam et al.,

1997; Cohen et al., 1998; Kröner and Güntürkün, 1999). In addition, the arcopallium projects to the deep layers of the optic tectum, from where descending axons reach the rhombencephalic motor fields (Zeier and Karten, 1971; Dubbeldam et al., 1997; Cohen et al., 1998; Hellmann et al., 2004). More medially, OM fibers terminate in the substantia grisea centralis (GCt) and fan out into the medial and lateral aspects of the mesencephalic and medullary reticular formation, as well as to the n. reticularis pontis caudalis and n. reticularis parvocellularis (Zeier and Karten, 1971; Dubbeldam et al., 1997). The picture emerging from this overview is that of a classic sensorimotor pathway. Indeed, perturbations along this pathway affect both sensory and motor aspects of pecking, grasping, and feeding in the pigeon (Wild et al., 1985; Jäger, 1993).

Another example of the arcopallium's sensorimotor role is the ability of birds (especially owls) to respond with a fast gaze shift to novel stimuli. The most important sensory cues for gaze shifts are provided by the visual and the auditory senses. In birds, visual information is conveyed from the retina to the telencephalon via two main visual pathways; the tectofugal and the thalamofugal system (see Sections 8.4.1.1 and 8.4.1.2). Tectofugal projections terminate in the entopallial core of the DVR, which then projects to a penumbra of associative visual areas (Husband and Shimizu, 1999; Krützfeldt and Wild, 2005; Stacho et al., 2016). Nearly all of these areas project to the arcopallium (Husband and Shimizu, 1999; Kröner and Güntürkün, 1999; Letzner et al., 2016) via the tractus dorsoarcopallialis (DA). Thalamofugal visual projections terminate in the avian wulst and then also project to the arcopallium via associative visual areas (Bagnoli and Burkhalter, 1983; Shimizu and Karten, 1990; Shimizu et al., 1995; Kröner and Güntürkün, 1999). The same pattern is observed for the avian auditory pathway (Wild et al., 1993; Cohen et al., 1998; Kröner and Güntürkün, 1999; Letzner et al., 2016). Thus, the arcopallium integrates associative visual and auditory information via associative and modality-specific areas of the avian pallium. As outlined later, some of this input is used to orient toward new targets in the surrounding.

Knudsen et al. (1995) reported that microstimulations of the anterior and intermediate arcopallium of the barn owl evoke orienting movements of the eyes and head. The saccade directions were place-coded in the arcopallium; implying that single arcopallial neurons evinced a receptive field in 2D space. Anatomical studies show that subfields of the arcopallium have downsweeping projections to the brain stem (summarized above for the trigeminal system). Spatial attention guided by the arcopallium incorporates auditory and visual input and is translated into gaze shifts via the projections to the deep layers of the optic tectum (Knudsen and

Knudsen, 1996; Cohen et al., 1998). Indeed, Winkowski and Knudsen (2006) demonstrated that electrical microstimulation in critical areas of the owl's arcopallium can regulate the gain of tectal auditory responses in an attention-like manner. When the arcopallial circuit was stimulated, tectal responses to the place code of the auditory stimuli (the point in space from where the sound emanates) were enhanced, and spatial selectivity was sharpened. At the same time, auditory inputs from other locations were suppressed in the midbrain map.

Taken together, these and many more studies demonstrate that the arcopallial subfields receive associative sensory input and generate premotor outputs to subpallial areas. The studies in barn owls make it clear that the arcopallium's role in this pathway is not that of a simple relay, but that of an integrator of various sensory signals that are associated, based on past experience, with a certain response. According to Shanahan et al. (2013), the AI is among the most important hubs of the avian telencephalon that controls information flow of a large number of forebrain structures. Fig. 8.15 schematically summarizes what we know about the projections of the three arcopallial subfields AA, AI, and AD. It is visible from this schema that the importance of the arcopallium as a hub may have been underestimated by Shanahan et al. (2013) because they limited their analysis to the telencephalon.

8.3.3.3.2.2.1 *The Avian Pallial Amygdala*   The avian pallial amygdala consists of the posterior pallial amygdala (PoA), the nucleus taeniae of the amygdala (TnA), and a component that is usually called the AM but may be a component of the avian pallial amygdala (Reiner et al., 2004a; Atoji et al., 2006; Abellán et al., 2013). According to Letzner et al. (2016), only the ventral part of the PoA, which projects to the hypothalamus, has homotopic interhemispheric connections with the contralateral PoA. Birds also have a subpallial amygdala that was reviewed by Kuenzel et al. (2011) but will not be discussed here.

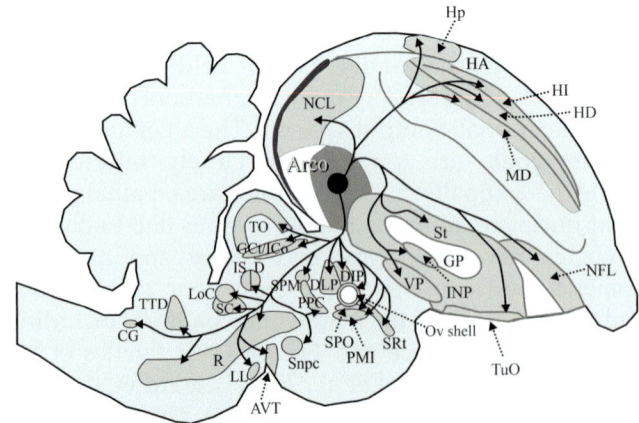

FIGURE 8.15 Projections of the avian arcopallium. The arcopallium as defined here includes the subareas AA, AD, and AI. Abbreviations are listed in the list of abbreviations.

Within the telencephalon, projections of the avian pallial amygdala reach major aspects of the basal ganglia. Both PoA and TnA project to tuberculum olfactorium (TuO), VP, nucleus of the stria terminalis (NSTL), and preoptic nuclei (Veenman et al., 1995; Kröner and Güntürkün, 1999; Cheng et al., 1999). Additionally, PoA projects to the n. accumbens (Ac) and the MSt (Veenman et al., 1995). Within the pallium, axons of the amygdala reach the hippocampal complex (Casini et al., 1986; Cheng et al., 1999; Atoji et al., 2002; Shanahan et al., 2013). Further intratelencephalic projections of TnA target the septum mediale and the nidopallium caudolaterale (NCL) (Cheng et al., 1999). Telencephalic projections to the nuclei of the avian pallial amygdala originate mainly from the hippocampal complex (Casini et al., 1986; Cheng et al., 1999; Atoji et al., 2002), septum laterale, accumbens, NSTL, and bulbus olfactorius (Reiner and Karten, 1985; Cheng et al., 1999; Patzke et al., 2011). Projections descend from the avian pallial amygdala via the tractus occipitomesencephalicus pars hypothalami (HOM) and terminate in hypothalamic subfields (Zeier and Karten, 1971; Dubbeldam et al., 1997; Kröner and Güntürkün, 1999; Cheng et al., 1999), the locus coeruleus (LoC), substantia nigra pars parvocellularis (SNpc), and AVT (Kröner and Güntürkün, 1999; Cheng et al., 1999).

This network resembles that of the mammalian amygdala and places the avian pallial amygdala into the core of a system of various limbic, multimodal, and memory-related structures with which actions can be modulated according to emotional processes. Accordingly, Kingsbury et al. (2015) demonstrated that vasoactive intestinal peptide (VIP) peptide—containing neurons in the AM showed increased transcriptional activity in response to and correlated with nest building activity in zebra finches. Schubloom and Woolley (2016) found that immediate early gene expression in the TnA of female zebra finches was related to the degree of individual preferences for their mate's courtship song. Testosterone levels in TnA also differ relative to breeding or nonbreeding seasons in swamp sparrows (Heimovics et al., 2016). In crows, dominance relationships develop in dyadic encounters. During such social interactions, neural activity levels in TnA correlate with aggressive and submissive behaviors (Nishizawa et al., 2011). Accordingly, lesions of TnA in zebra finches alter the interaction of lesioned males with sexually accessible females only when another male is present (Ikebuchi et al., 2009). These and many more studies demonstrate that the nuclei of the avian pallial amygdala are part of a limbic network that controls emotional behavior during social interactions that include sexual and agonistic components. Fig. 8.16 schematically summarizes the projections of subnuclei of the avian pallial amygdala.

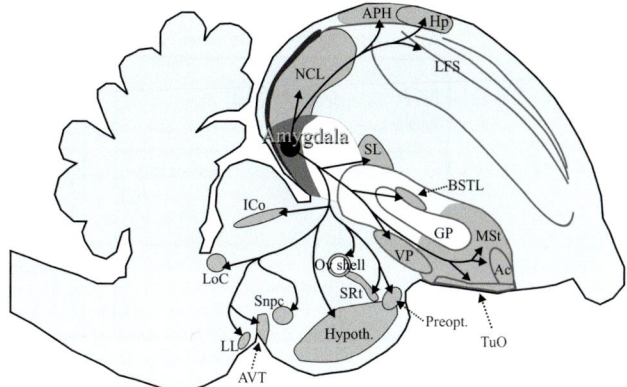

FIGURE 8.16 Projections of the avian pallial amygdala. Efferents of the n. taeniae of the pallial amygdala (TnA) and the n. of the posterior pallial amygdala (PoA) are shown. Abbreviations are given in the list of abbreviations.

8.3.3.3.2.3 Layers in a Nonlaminated Forebrain

At the turn of the 19th to the 20th century, comparative neuroanatomists were sure that they had discovered the core feature that distinguishes mammalian from nonmammalian brains: the six-layered cerebral cortex. Reptiles with their three-layered cortex seemed to possess at least a forerunner of the mammalian cerebral cortex, but birds seemed to possess only the CDL (area corticoidea dorsolateralis), a small and paper-thin three-layered structure in the dorsolateral pallium. The situation changed after the turn to the 21st century. Dugas-Ford et al. (2012) discovered that gene expression patterns of mammalian cortical neurons from granular (layer IV) and infragranular (layer V) laminae corresponded to those of avian pallial clusters that receive sensory thalamic input ("granular") or have descending projections to subpallial targets ("infragranular"). Interestingly, these results spanned both DVR and wulst and incorporated them into a common pattern. Subsequent studies even suggested that most of the avian pallial clusters may be homologous to certain cortical layers, such that most of the avian pallium would have a hidden laminated architecture (Chen et al., 2013; Jarvis et al., 2013; but see Montiel et al., 2016).

Genetic expression patterns are a great tool, but when it comes to the demonstration of a layered organization, local connectivity data are needed. This is what Wang et al. (2010) and Ahumada-Galleguillos et al. (2015) demonstrated in the auditory and the visual tectofugal thalamopallial termination zones of the nidopallium. Using in vitro tracing, they demonstrated three main layerlike entities that can be further subdivided into several sublayers. In this arrangement, neuronal clusters and axonal columns are oriented orthogonally to the layers. The neurons in the sensory recipient laminae are reciprocally connected with the cells in the

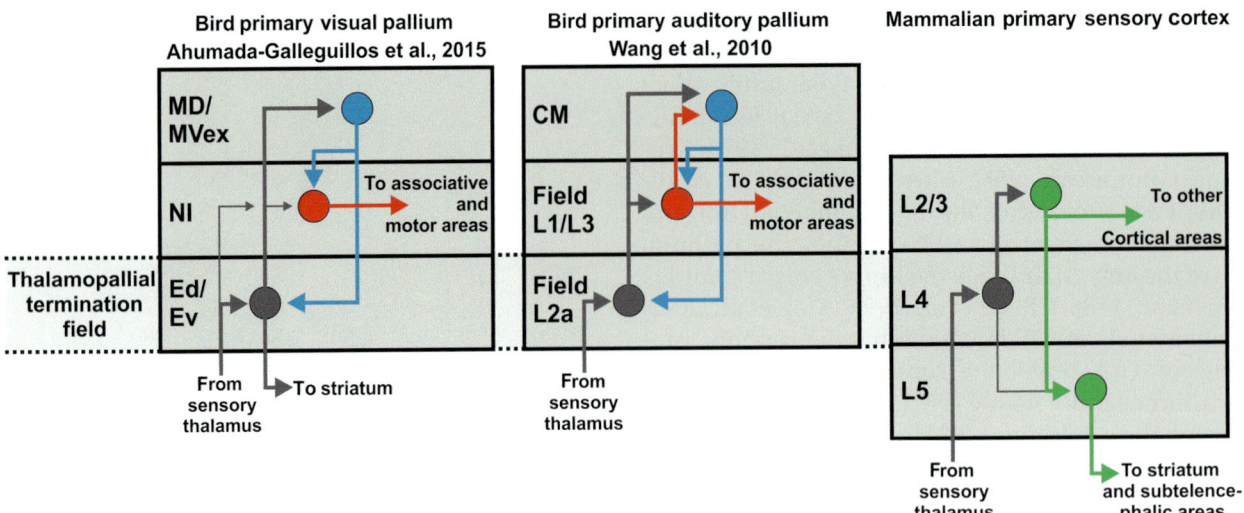

FIGURE 8.17 Overview of the connectivity patterns in the "layered" primary tectofugal visual and primary auditory bird pallium as well as the mammalian primary sensory cortex. For the bird data, some layers were collapsed into one. The cortex schema is shifted vertically so that the thalamopallial projections are aligned; only the main connections are shown. *Thin lines* represent weaker connections. The left two panels represent results from in vitro tracing experiments. The *horizontal arrow* that leads to associative and motor areas depicts connections that are known from the literature (Shanahan et al., 2013) but for which we do not know if they originate from the depicted cell types.

topographically overlaying nidopallial and mesopallial columns. In addition, columns have horizontal projections to associative and motor structures (Fig. 8.17). The entry point to this system is the thalamorecipient layer, which shares genetic expression profiles and morphological features with the cortical granular layer IV (Dugas-Ford et al., 2012; Chen et al., 2013; Belgard et al., 2013). To some extent, this avian circuitry resembles the cortical canonical circuit that is defined by repetitive topographic interlaminar circuits (Douglas and Martin, 2007). These neocortical circuits are the heart of the computational properties that characterize cortical dynamics. Looking carefully at Fig. 8.17, it becomes obvious that mammalian and avian pallial layers are similar, but not identical. If their similarity is due to evolutionary convergence, a laminated forebrain based on repetitive columnar interlaminar circuits could represent a computational necessity for flexible sensorimotor integration. In principle, however, a more mundane interpretation is possible: cascades of interconnected pallial territories around a primary sensory cortical field are also found in cerebral cortex, where diverse associative cortical areas are arranged around a primary sensory cortical field. Thus, the avian pattern could simply reflect sequences of sensory integration along adjacent fields. However, the precise orthogonal arrangement of the cellular columns, combined with the cortical lamina-specific genetic expression patterns, makes the hypothesis of the "invisibly layered" bird pallium attractive. Still, whether birds indeed possess a cortical layerlike organization in the DVR remains an open question.

8.3.3.3.2.4 *The Avian "Prefrontal Cortex"*   In Section 8.3.3.3 we had outlined that avian cognition is not inferior to mammalian cognition. Corvids and parrots even reach the same achievements in complex cognitive tasks as primates, despite having much smaller forebrains (see Section 8.2.2). However, birds are on par with mammals not only with respect to the level of cognitive abilities, but also with respect to the functional details of the cognitive mechanisms (Pollock et al., 2000; Scarf et al., 2011). This is especially true for a cluster of cognitive abilities that are subsumed under the umbrella term "executive functions"—a circumscribed cluster of cognitive functions (working memory, behavioral inhibition of an imminent action, timing, goal shifting, etc.) that reflect the ability to spontaneously generate efficient strategies and schedule future behavior when relying on self-directed task-specific planning. Birds show similar executive functions as mammals (Laude et al., 2016; Castro and Wasserman, 2016). But since birds do not have a cortex, how do they generate their executive functions? The PFC of primates is a large part of the frontal portion of the neocortex. Neither the avian DVR nor the wulst has any entity that even remotely resembles the PFC. Comparative neuroanatomists are happy to accept that common function can emerge in different taxa as a result of convergent evolution, but they then often expect that this process is accompanied by the convergent evolution of similarities in brain architecture. This is, at least at the first glance, not the case for NCL and PFC. The view that similarities of function require similarities of anatomy is the classic trap in which neuroanatomists often step: they know that a certain structure generates a certain

function (structure → function). Erroneously they then conclude that a certain function can only be generated by one kind of structure (function → structure). The analysis of the avian "prefrontal cortex" demonstrates the fallacy of this logic.

Classic neuropsychological studies had demonstrated that lesions of the PFC in humans result in prominent deficits in all aspects of executive functions (Taylor et al., 1986). Subsequent neurobiological advances provided means for mechanism-driven instead of phenomena-driven explanations. For example, a detailed analysis on the biophysical effects of dopamine release within the PFC showed that some of the observed deficits, ranging from working memory to planning, may be the result of a single system failure (Seamans and Yang, 2004; Durstewitz and Seamans, 2008). This is exemplified by working memory tasks, in which the subject has to hold information online until using it for some future actions. The problem in a working memory task is twofold: first, a neuronal/mental trace of a stimulus has to be held over time although the physical representation (the perceived stimulus) is no longer present; second, the neuronal trace has to be shielded against other neuronal processes that result from currently interfering stimuli. Delay activity in the PFC of human and nonhuman primates indeed persists during working memory tasks even if interfering stimuli intervene between the presentation of the sample and the target stimulus (Puig et al., 2014). Durstewitz et al. (1999a) proposed in a biophysically realistic model that dopamine can, via D1 receptor stimulation, selectively increase the firing rate of prefrontal neurons that hold information during a delay period. It thereby also increases inhibitory feedback and thus reduces activity of the "background" neurons. In this manner, dopaminergic effects may act to stabilize current delay activity in a PFC network. Thus, the model offered a mechanistic explanation for the cellular firing properties of PFC neurons or the behavioral deficits observed after blockade or after supranormal stimulation of dopamine receptors in the PFC. Armed with such a mechanistic explanation of a core feature of executive functions in PFC, we now can turn our attention to birds to look if their working memory capacity is realized by similar mechanisms.

In 1982, Mogensen and Divac lesioned an area in the caudolateral aspect of the pigeon's nidopallium and tested the animals in a delayed alternation task. In this task, the animal has to choose one of two keys to obtain reward. After a delay period, it has to select the other key and so forth. The problem is the delay: the subject has to keep in working memory its last choice to be able to select in the subsequent task the next key. Mogensen and Divac (1982) demonstrated

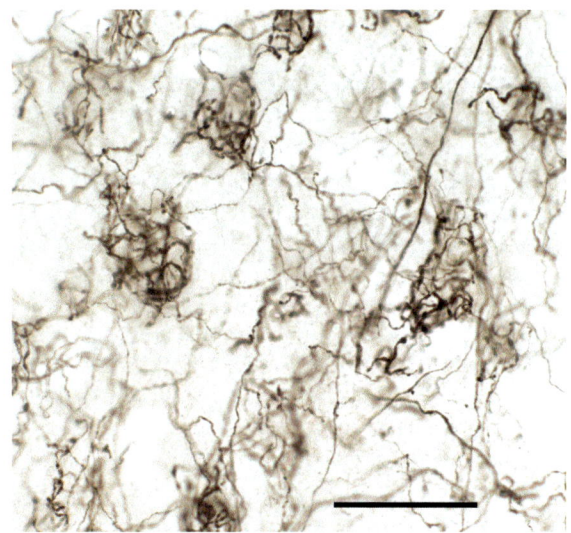

FIGURE 8.18 Tyrosine hydroxylase-positive, presumably dopaminergic, fibers in the pigeon NCL. Note occasional swelling on axons that are probably synapses en-passant. Baskets that tightly wrap around singe neurons are clearly visible. Bar = 50 μm.

that birds with lesions in the NCL displayed deficits in this classic test for executive functions. In subsequent studies, the team also showed that the NCL is densely innervated by catecholaminergic fibers of possibly dopaminergic nature (Divac et al., 1985; Divac and Mogenson, 1985). They concluded that the NCL could be a functional equivalent to the mammalian PFC. Later on, Waldmann and Güntürkün (1993) showed that the NCL is innervated by dopaminergic axons from the SNc and the AVT. Interestingly, these axons either innervate NCL neurons by boutons-en-passant or create dense baskets with which they coil around a soma and possibly bring this neuron under tight dopaminergic control (Wynne and Güntürkün, 1995; Fig. 8.18). NCL neurons within these baskets are never GABAergic interneurons but principal cells that are activated by a D1 receptor cascade (Durstewitz et al., 1998). Some of the principal cells in these baskets are readily elicited by weak excitatory inputs, yet produce a sustained response to a prolonged input—a pattern that favors the function to retain information of their input for a short time (Kröner et al., 2002). Indeed, neurons in the mammalian PFC show enhancement in their firing rate during the delay component of working memory tasks, often also accompanied by brief gamma bursts that possibly gate access to, and prevent sensory interference with, working memory (Lundqvist et al., 2016). Neurons with similar delay activities to those recorded from primate PFC have been observed in the pigeon's NCL during delay tasks (Kalt et al., 1999; Diekamp et al., 2002; Veit et al., 2014). Karakuyu et al. (2007) could show that dopamine in NCL is specifically released during the delay period of working memory tasks and could thus stabilize sustained activity patterns of delay neurons against interference. Since dopamine release in NCL follows (like in PFC) a volume transmission mode, it can affect extended aspects of the network that is currently involved in executing the delay task (Bast et al., 2002). Consequently, locally antagonizing or agonizing D1 receptors in NCL decreases or increases working memory performance, respectively (Herold et al., 2008). These receptors are also massively expressed in NCL when pigeons are subject to cognitive training with working memory tasks (Herold et al., 2012).

Thus, both mammals and birds seem to realize the working memory aspect of their executive functions within their PFC/NCL using mechanisms that are highly similar (Güntürkün, 2005). Astonishingly, these similarities range from the molecular up to the behavioral level. What about executive functions beyond working memory? NCL lesions or local pharmacological alterations of NCL activity patterns do not affect perceptual or motor processes (Gagliardo et al., 1996; Güntürkün, 1997), but they interfere with behavioral inhibition (Güntürkün, 1997; Hartmann and Güntürkün, 1998), self-scheduling along time domains (Kalenscher et al., 2003), response selection (Lissek and Güntürkün, 2004), context integration (Lissek and Güntürkün, 2005), goal shifting (Diekamp et al., 2000), and control of extinction learning (Lissek and Güntürkün, 2003; Lengersdorf et al., 2014). Furthermore, NCL neurons encode cognitive operations like decision-making (Lengersdorf et al., 2014; Veit et al., 2015), rule tracking (Veit and Nieder, 2013), encoding of subjective values (Kalenscher et al., 2005), and the association of outcomes to actions (Starosta et al., 2013). Thus, the full extent of executive functions is encoded at the level of both PFC and NCL.

Thus far, we have not discussed the neuroanatomy of the NCL. In mammals, the PFC is recognized as a hub that connects various sensory, motor, and associative systems (van den Heuvel et al., 2016). Like the PFC, the NCL is also a center of multimodal integration and connects the higher-order sensory input from trigeminal, somatosensory, visual (tecto- and thalamofugal), and olfactory systems and links them to limbic and premotor structures (Leutgeb et al., 1996; Kröner and Güntürkün, 1999; Güntürkün, 2012; Fig. 8.19). Consequently, NCL neurons can integrate and process relevant cues, irrespective of their modality (Moll and Nieder, 2015). Thus, identical to the PFC, the avian NCL is a convergence zone between the ascending sensory and the descending motor systems (Kirsch et al., 2008). Here, all sensory modalities overlap and connect to premotor areas of the arcopallium. However, NCL and PFC are not in all aspects identical to each

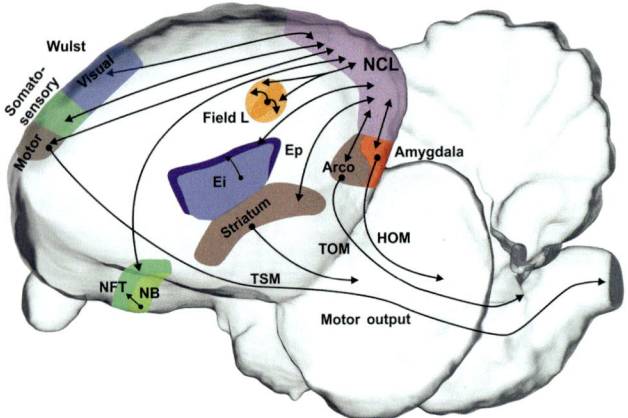

FIGURE 8.19   The NCL is a hub that integrates diverse sensory pathways and links them to limbic and motor structures. Auditory (*orange*), visual (*blue*; thalamofugal, tectofugal), and somatosensory (*green*; somatosensory, trigeminal) regions have reciprocal connections with NCL via their association fields. *Pigeon brain modified from Güntürkün, O., Verhoye, M., De Groof, G., Van der Linden, A., 2013. A 3-dimensional digital atlas of the ascending sensory and the descending motor systems in the pigeon brain. Brain Struct. Funct. 281, 269–281. For abbreviations see list of abbreviations.*

other. The most important difference is the lack of a thalamic input from the mediodorsal thalamic nucleus in birds (Kröner and Güntürkün, 1999). Instead, the n. dorsolateralis posterior thalami (DLP) innervates the NCL (Güntürkün and Kröner, 1999). The DLP integrates multimodal input and is probably homologous either to the intralaminar or the posterior thalamic nuclei in mammals (Korzeniewska and Güntürkün, 1990; Veenman et al., 1997). However, DLP lesions cause deficits that are comparable to lesions of the mammalian nucleus mediodorsalis (Güntürkün, 1997). A second difference to the mammalian prefrontal system is the fact that NCL neurons exhibit high firing rates and are selective for highly familiar stimuli (Veit et al., 2015). This is dissimilar to the PFC but resembles primate association cortices posterior to PFC. Thus, NCL and PFC are highly similar but not identical in all aspects. Despite these similarities, NCL and PFC are certainly not homologous. While NCL is located in the most posterior end of the telencephalon, PFC is at the cortical rostral pole. Thus, it is difficult to conceive how this topological transformation should occur during evolution from a common homologous structure. Also some genetic expression patterns contradict the idea of a homology of NCL and PFC (Puelles et al., 2016). Thus, nonhomologous brain areas converged over the course of 300 million years into mammalian and avian prefrontal structures that serve highly similar functions. In doing so, both areas gained the ability to generate the same cognitive functions using similar cellular properties.

8.4 Functional Systems

8.4.1 Ascending Visual Systems

As pointed out by Butler and Hodos (2005), the dorsal thalamus of anamniotes can be divided into (1) a rostral lemnothalamic component that receives direct retinal and, in some cases, other sensory lemniscal projections; and (2) a caudal collothalamic component that receives its input mostly from the midbrain roof. The lemnothalamus receives its sensory input without an extra synapse in the midbrain roof. Accordingly, the collothalamus receives sensory afferents via a tectal relay. The visual system of sauropsids is characterized by two parallel ascending systems, a lemnothalamic and a collothalamic visual pathway. Especially in avian neuroscience, the terms "lemnothalamic" and "collothalamic" never gained broad acceptance. Instead, scientists generally use the terms "thalamofugal" and "tectofugal visual pathway," respectively. Since avian neuroscience is often used as a reference benchmark for studies on reptiles, most scientists working on the reptilian visual system also refer to thalamofugal and tectofugal systems. To avoid any confusion, we therefore also use these terms.

8.4.1.1 The Thalamofugal Visual Pathway in Reptiles and Birds

As true for practically all aspects of the central nervous system, we know much more about birds than about reptiles. We therefore will first discuss birds before turning our attention to reptiles.

The thalamofugal pathway in birds consists of the retinal projection onto the n. geniculatus lateralis pars dorsalis (GLd) and the bilateral projection of the GLd onto the wulst in the anterodorsal forebrain (Güntürkün, 2000). Due to its anatomical, physiological, and functional properties, the avian thalamofugal pathway probably corresponds to the mammalian geniculostriate system (Shimizu and Karten, 1993).

While the tectofugal pathway receives afferents from the complete extent of the retina, the retinal location of ganglion cells projecting onto the GLd differs in various species. In birds of prey, ganglion cells in the temporal retina subserving frontal vision project primarily onto the GLd (Bravo and Pettigrew, 1981). Consequently, many neurons in the visual wulst of owls, kestrels, and vultures possess binocular visual fields and detect retinal disparity (Pettigrew, 1979; Porciatti et al., 1990). In pigeons, however, efferents to the GLd originate mainly from ganglion cells outside the superiotemporal retina (Remy and Güntürkün, 1991). The paucity of afferents from this retinal field should render the pigeons' thalamofugal pathway largely "laterally oriented," an assumption supported by electrophysiological (Miceli

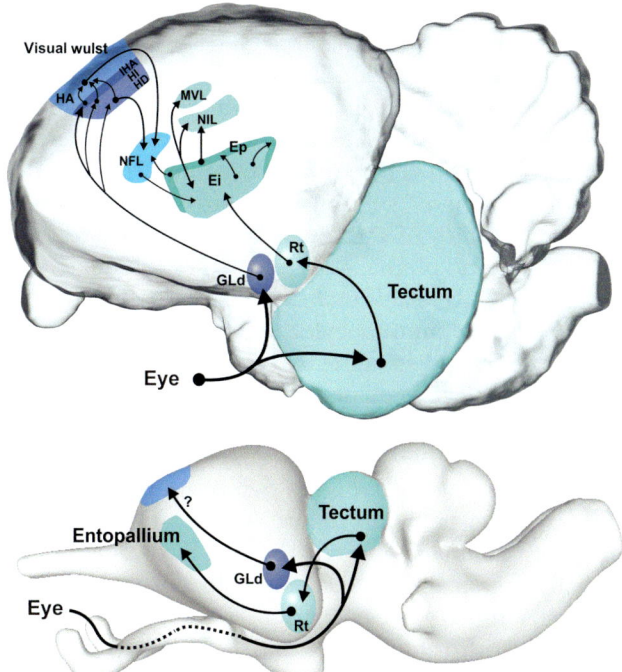

FIGURE 8.20 Ascending visual pathways of the tectofugal (turquoise) and the thalamofugal pathways (*blue*) in the pigeon (above) and the Nile crocodile (below). The projection area of the GLd has been studied in turtles but not yet in crocodiles. The two brains are not drawn to scale. For abbreviations see list of abbreviations. *Pigeon brain modified from Güntürkün, O., Verhoye, M., De Groof, G., Van der Linden, A., 2013. A 3-dimensional digital atlas of the ascending sensory and the descending motor systems in the pigeon brain. Brain Struct. Funct. 281, 269–281.*

et al., 1979) and imaging results (De Groof et al., 2013). Rotundus- and GLd lesions in pigeons also selectively interfere with acuity in the frontal and lateral visual field, respectively (Güntürkün and Hahmann, 1999). Similarly, wulst lesions result in the lateral but not frontal visual far-field deficits (Buszynski and Bingman, 2004). This "lateral orientation" of the pigeon's thalamofugal system is very likely the reason for the virtual absence of behavioral deficits in a variety of discrimination tasks after GLd or wulst lesions in which frontally placed pecking keys were used (Güntürkün, 1991). Thus, in pigeons, frontal and lateral visual acuity performances seem to depend on tecto- and thalamofugal mechanisms, respectively.

The GLd consists of six components, of which four are retinorecipient and project onto the visual wulst (Güntürkün and Karten, 1991; Heyers et al., 2007). Liu et al. (2008) found "distance-to-collision" neurons in the pigeon's GLd that fire briskly at a certain distance when a large surface moves toward the animal. These GLd neurons nicely complement the "time-to-collision" neurons found within the tectofugal system (Xiao et al., 2006).

The projection of the GLd to the wulst is bilateral and topographically organized (Miceli et al., 1990; Fig. 8.20).

In owls with their more frontally oriented eyes, the proportion of ipsi- and contralateral GLd → wulst is about equal (Bagnoli et al., 1990). The visual wulst is organized from dorsal to ventral in four laminae: hyperpallium apicale (HA), interstitial nucleus of HA (IHA), hyperpallium intercalatum (HI), and hyperpallium densocellulare (HD). These subdivisions are based on the cytoarchitectonics of the wulst and do not reflect the full complexity of the structure, since Shimizu and Karten. (1990) were able to distinguish at least eight subdivisions using immunocytochemical techniques. The granular IHA and to some extent also lateral HD and HI are the major recipients of the cholinergic and colecystokinergic GLd input (Watanabe et al., 1983; Güntürkün and Karten, 1991).

Electrophysiological studies demonstrate similarities between the visual wulst of birds of prey and the striate cortex of mammals. The visual wulst of owls is retinotopically organized and contains both simple and complex cells tuned to basic visual parameters such as orientation, direction, and end-stopping (Pettigrew, 1979; Nieder and Wagner, 1999). As in the mammalian primary visual cortex, visual wulst neurons of owls signal the local orientation of features within moving object (Baron et al., 2007). In the visual wulst of further birds of prey, most neurons are primarily concerned with binocular visual processing, are selectively tuned to stereoscopic depth cues, and have small receptive fields that subtend about 1 degree of visual space (Pettigrew and Konishi, 1976; Pettigrew, 1979; Wagner and Frost, 1993). Wulst cells are also clustered into functional domains with orientation pinwheels analogous to those found in cat and monkey V1 (Liu and Pettigrew, 2003). The owl visual wulst also shows cellular correlates of binocular interaction (Pettigrew and Konishi, 1976; Pettigrew, 1979; Nieder and Wagner, 2001) and of illusory contours (Nieder and Wagner, 1999). Thus, in many aspects, the wulst of the barn owl is equivalent to mammalian primary visual cortex. But is the similarity a result of homology or of convergent evolution? It is currently impossible to decide this question, but studies in pigeons make it likely that at least some of these physiological characteristics result from convergence. A study on the neuronal population dynamics of the pigeons' visual wulst captured with voltage-sensitive dye imaging revealed a different kind of dynamic than what was observed in owls. In pigeons, analysis of the imaged spatiotemporal activation patterns revealed no clustered orientation or maplike arrangements as typically found in the wulst of owls and in the primary visual cortices of many mammalian species (Ng et al., 2010). A similar conclusion was also drawn by Bischof et al. (2016): using optical imaging of intrinsic signals, electrophysiological recordings, and retrograde tracers, they discovered that the visual wulst of zebra finches consists of three visual

field representations, each receiving input from distinct subdivisions of the GLd in both hemispheres. No foveal magnification was evident in any of the subdivisions. Bischof et al. (2016) did discover some similarities to the mammalian design but also several features that seem unique to birds.

Astonishingly, the avian thalamofugal system serves two parallel functions. On the one hand, it is a classic visual pathway that transmits object vision from the eyes to the forebrain. On the other, it also has a key role in magnetic compass perception (Mouritsen et al., 2016).

Some avian species migrate over thousands of kilometers, while other just home over a lengthy valley back to their loft. Especially during long flights, but to some extent also during smaller voyages, global cues like those from a compass are very important. Indeed, many bird species have a magnetic compass that was first discovered in European Robins (Wiltschko and Wiltschko, 1972). The avian magnetic compass is an inclination compass, which detects the angle between the magnetic field lines and the Earth's gravity but not their polarity. Consequently, birds do not discriminate North from South, but poleward from equatorward (Wiltschko and Wiltschko, 1995).

How do birds sense the Earth's magnetic field, and where in the brain is magnetic compass information processed? Sensing magnetic fields as weak as that of the Earth is a tall task. Presently, a magnetic compass that is based on a light-dependent, radical-pair-based, chemical compass mechanism is the best candidate (Ritz et al., 2000). The primary sensory molecules appear to be cryptochrome proteins (Mouritsen et al., 2004). Indeed, retinal neurons contain at least four different cryptochromes (Liedvogel and Mouritsen, 2010).

If the avian magnetic compass is light dependent, covering the eyes should abolish compass perception. Indeed, magnetic compass sensing is lost especially when the right eye is covered (Witschko et al., 2002). Although the lateralization of magnetic compass vision is a matter of heated disputes (Hein et al., 2011; Wiltschko et al., 2011), it is clear that vision is required to sense the Earth's magnetic field orientation. This vision is also in need of high-frequency visual input, possibly because the low frequency compass input cannot be disambiguated from ordinary object vision (Stapput et al., 2010). Which parts of the avian brain process magnetic compass information? A forebrain area named "Cluster N" in the visual hyperpallium is by far the most active part of the brain when night-migratory songbirds use magnetic compass information for orientation behavior (Mouritsen et al., 2005). Activation of Cluster N disappears when the eyes are covered (Liedvogel et al., 2007), and neuronal tracing showed that Cluster N is a small part of the visual wulst, which receives its input from the eyes via the thalamofugal visual pathway (Heyers et al., 2007). When Cluster N is inactivated, night-migratory songbirds cannot use their magnetic compass anymore, whereas their sun and star compasses remain functional (Zapka et al., 2009). Since Cluster N is part of the thalamofugal visual pathway, this is very strong evidence that the magnetic compass is light dependent that the primary sensors are in the eyes and that birds perceive magnetic compass information as a visual impression.

To some extent, magnetic compass perception resembles infrared vision in snakes. In both cases, a "classic" sensory pathway is used to transmit a different kind of signal. The result is a change at the sensory input level but not an alteration in the pathway. The tectofugal visual pathway of snakes stays the same, at least from the tectum on, but now incorporates thermal information superimposed on object vision. In birds, magnetic compass information is also superimposed on object vision (Ritz et al., 2000). The thalamofugal pathway stays the same, but now includes a special field, Cluster N, within the visual wulst.

Now let us discuss the thalamofugal system in reptiles. Retinal ganglion cells of all reptilian species project contra- or bilaterally onto the GLd (Ulinski and Nautiyal, 1988; Derobert et al., 1999). The GLd in turtles subsequently projects onto the ipsilateral visual cortex (Mulligan and Ulinski, 1990). In addition, visual cortex has projections back onto both GLd and the optic tectum, as also is the case in birds (Hall et al., 1977; Güntürkün, 2000; Fig. 8.20). This pattern does not apply to all reptiles, however. In lizards, Lohman and van Woerden-Werkley (1978) demonstrated that GLd projects to striatum but not to cortex.

The functional organization of thalamocortical projections in turtles is not resolved. According to Mazurskaya (1973), visual cortical neurons respond to small visual stimuli from everywhere in the visual field. This would imply an absence of retinotopy. In contrast, Mulligan and Ulinski (1990) describe a topological projection from GLd to visual cortex, albeit with multiple boutons-en-passant. They conclude that there is an orderly representation of the rostral–caudal axis of the ipsilateral dorsal lateral geniculate complex within the visual cortex of turtles that is combined with a convergence of inputs from neurons located along a given dorsal–ventral dimension. This would result in topography along one dimension, but not along the other. As outlined in Section 8.3.3.2.1, it is conceivable that place coding in the visual cortex of turtles does not occur by means of topographic maps but by nontopographically organized ensembles of neurons (Fournier et al., 2015; Naumann et al., 2015).

Extensive lesions of the forebrain pathway severely impair the ability of turtles to relearn visual discrimination that they had acquired before surgery (Reiner and Powers, 1980). This is different, when only the visual cortex has been damaged. In this case, deficits are very subtle (Bass et al., 1973).

8.4.1.2 The Tectofugal Visual Pathway in Birds and Reptiles

In all sauropsids, optic nerve axons decussate virtually completely in the optic chiasma and then terminate in diverse areas of the midbrain and thalamus. In birds, the largest contingent of optic axons synapses in the optic tectum. The exact proportion is difficult to estimate but according to the data of Bravo and Pettigrew (1981) in barn owls and Remy and Güntürkün (1991) in pigeons, 75–95% of ganglion cells have axons leading to the tectum in these bird species. With regard to these numbers, the burrowing owl, *Speotyto cunicularia*, is an exception. This bird relies heavily on its thalamofugal pathway and consequently has less than 50% tectally projecting ganglion cells (Bravo and Pettigrew, 1981).

In Section 8.3.2.3 we briefly outlined the organization of the visual input to the optic tectum in birds. Only few things should be added here: in birds, retinal axons only innervate the superficial layers 2–7 and reach their highest synaptic density in layer 5 (Hayes and Webster, 1985). The retinal projection onto the tectum is strictly topographically organized in all species studied, with the inferior retina projecting to the dorsal tectum while the posterior tectum is reached by the nasal retina (Clarke and Whitteridge, 1976; Frost et al., 1990a; Remy and Güntürkün, 1991). The tectal representation of the foveae or the areas of enhanced vision are considerably expanded (Clarke and Whitteridge, 1976; Frost et al., 1990a). Single-unit recordings in the optic tectum demonstrate that the visual receptive fields of neurons in the superficial layers are small (0.5–4 degrees) but increase to up to 150 degrees in deeper laminae (Jassik-Gerschenfeld et al., 1975; Frost et al., 1981). It is possible that these numbers have to be downsized a bit when more objective measures of determining receptive field borders are used (Verhaal and Luksch, 2013). However, the principal pattern of an increasing receptive field size in deeper layers is valid across studies spanning four decades. According to Verhaal and Luksch (2013), about 10% of tectal neurons are luminance sensitive.

Tectal cells also respond selectively to the spatial frequency of drifting sine-wave gratings, with most neurons having their optima between 0.45 and 0.6 c/degree (Jassik-Gerschenfeld and Hardy, 1979). Most of these cells are more selective to spatial frequencies than they are to single bar stimuli (Jassik-Gerschenfeld and Hardy, 1980). Birds therefore appear to be able to perform Fourier analysis of patterns in visual space at the level of the tectum. Indeed, Neuenschwander and Varela (1993) demonstrate visually triggered gamma oscillations in the pigeon's tectum. This oscillatory activity has characteristics similar to those reported in the mammalian neocortex in the context of synchronization of unit responses as a putative physiological basis of perceptual binding (Yu et al., 2008).

In all birds (and possibly in all amniotes), the dominant brain structure for shifting visual attention of the animal toward relevant stimuli is the optic tectum (Luksch, 2003). The tectum works like a saliency map where the relative salience of stimuli is processed and compared with other objects (Dutta and Gutfreund, 2014). Novel or moving objects are potentially important. Consequently, most tectal cells are movement sensitive and play an important role in figure–ground segregation through discontinuities in velocity (Jassik-Gerschenfeld and Guichard, 1972; Frost et al., 1990b; Verhaal and Luksch, 2016).

One type of layer 13 neuron with projections to the Rt has very large, circular dendritic fields that span up to 2 mm and extend into retinorecipient tectal layer 5b. These retinorecipient "bottlebrush" ending-neurons respond with rhythmic bursts (chattering) to depolarizing current injections (Luksch et al., 2001). Such high-frequency bursts have been observed in response to small moving spots in deep tectal neurons of pigeons with burst frequency linearly increasing with stimulus speed (Troje and Frost, 1998). These neurons respond best to fast motion and also show strong directional selectivity. They may be ideal for detecting movement and novelty and subsequently initiating an orienting response (Verhaal and Luksch, 2016).

Indeed, Marín et al. (2007) demonstrated that tectal responses that are triggered by a salient moving stimulus are swiftly transmitted to the layer 13 neurons that then project to Rt (Güntürkün et al., 1998). Marín et al. (2012) showed that tectally initiated visual responses from the isthmic nucleus Ipc send phase-locked feedback signals to the tectum and thus select which afferent activity propagates to the different subdivisions of the Rt and entopallium. The entopallium further projects to multiple visual associative areas including the nidopallium frontolaterale (NFL), mesopallium ventrolaterale (MVL), and nidopallium intermediale pars lateralis (NIL) (Husband and Shimizu, 1999; Krützfeld and Wild, 2005). Stacho et al. (2016) demonstrated that visual stimulus repetition in pigeons results in a reduction of cellular responses in these associative visual regions, just as single-unit recordings revealed reduced activity after repeated or prolonged visual stimulation throughout the primate visual system (Müller et al., 1999). It is likely that this effect reflects a learning-related buildup of stimulus familiarity and represents selective stimulus memory with subsequent response sharpening (Tartaglia et al., 2015). If this interpretation holds, these associative visual telencephalic areas would be part of a distributed visual memory system of birds (Fig. 8.20).

The situation in reptiles is highly similar to that in birds, although far less is known. Since the majority of studies were conducted in crocodilian species, we will first review these experiments. As in birds, retinal fibers in crocodile's project massively to the contralateral tectum end terminate in the upper six layers (Derobert et al., 1999). The pattern is extremely similar to birds with the exception that the first two plexiform tectal layers in crocodiles are practically fused and narrow. Neurons from deep tectal layer (corresponding to the avian tectal layer 13) project bilaterally onto the thalamic Rt (Pritz, 1980). Like in birds, also the Rt of crocodiles can be subdivided anatomically, although no functional data on different cellular properties of these thalamic constituents exist (Pritz, 1997; Pritz and Siadati, 1999). Telencephalic projections of Rt assemble ventromedially and ascend within the dorsal peduncle of the lateral forebrain bundle. At more anterior levels of the telencephalon, these axons turn dorsally and terminate massively in the dorsolateral part of anterior DVR (Pritz, 1975). The termination area corresponds to area G of Rose (1923), is rich in succinate dehydrogenase (Pritz and Northcutt, 1977), and is probably homologous to the avian entopallium (Fig. 8.20).

The situation in other reptiles is comparable. In turtles, a tectofugal pathway very similar to the one described in crocodiles has been discovered (Balaban and Ulinski, 1981). In lizards, large multipolar neurons of the deep tectal layer stratum griseum centrale project toward the Rt (Dávila et al., 2002). The ascending projections of the Rt make synaptic contacts in the striatum and synapse in dorsolateral and ventromedial region of ADVR and the amygdaloid complex (Guirado et al., 2000).

8.4.2 Ascending Somatosensory Systems

Both in reptiles and birds, an important part of the spinal projections terminates in the dorsal column nuclei (DCN) of the caudal rhombencephalon and transmit nonfacial tactile information from the limbs and the trunk. The DCN refers to the gracile and the cuneate nucleus. Both in reptiles (Pritz and Stritzel, 1994b) and birds (Necker, 1991), the spinal input is constituted by direct projections of the dorsal root ganglia to the DCN and a further pathway that involves at least one synapse in the spinal cord before terminating in the DCN. Previously, these nuclei were seen to be a derived system that only exists in amniotes (Hayle, 1973). However, more recent studies could clearly demonstrate a comparable system in frogs (Muñoz et al., 1997; Hiramoto and Cline, 2009).

In birds, the upper cervical spinal segments and the DCN project via the medial lemniscus to the inferior olive, then to the deep tectal layers and finally to the n. intercollicularis (ICo) (Wild, 1989, 1995; Luksch, 2003).

In budgerigars, spinal efferents also reach a small rhombencephalic nucleus which projects directly to the n. basalis prosencephali (Bas) in the telencephalon. Thus, in budgerigars, the Bas has both a head (from the trigeminal input; see below) and a body representation (Wild et al., 1997). In pigeons, no such projection has been demonstrated. Although the situation in reptiles is less clear, spinal and DCN projections to the central nucleus of the torus semicircularis of the midbrain were observed in various species (crocodiles: Ebbesson and Goodman, 1981; Pritz and Stritzel, 1989; turtles: Künzle and Woodson, 1982).

Spinal segments and the DCN of birds project to two main thalamic targets, the DLP and the n. dorsalis intermedius ventralis anterior (DIVA) (Funke, 1989; Korzeniewska and Güntürkün, 1990; Wild et al., 2008). No projection from the ICo to these thalamic targets is reported in this species (Wild, 1987; Korzeniewska and Güntürkün, 1990). In *Caiman*, spinal projections also terminate in a thalamic target, the medialis complex (Pritz and Nortcutt, 1980). Different from pigeons, also the crocodilian torus semicircularis projects to this thalamic nucleus (Pritz and Stritzel, 1990).

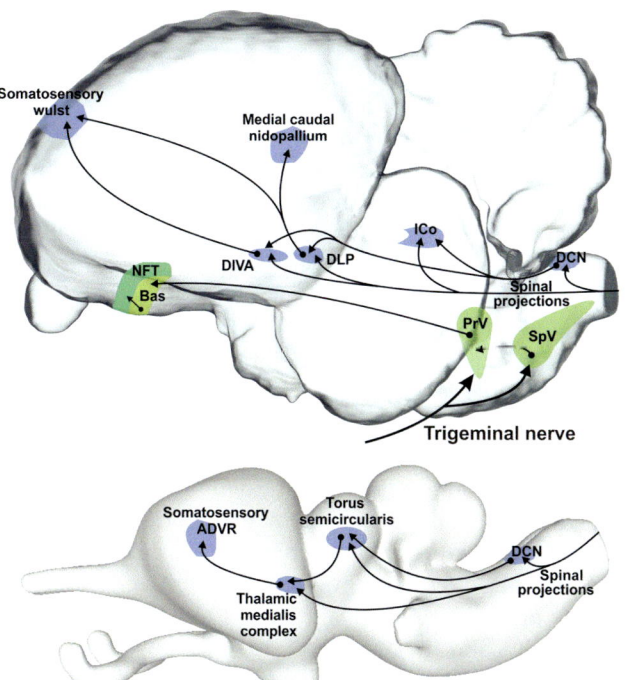

FIGURE 8.21 Ascending somesthetic pathways in birds (above) and crocodiles (below). In budgerigars also a spinal projection via a rhombencephali link to the n. basalis prosencephali (Bas) was demonstrated (not shown here for the pigeon brain). Central trigeminal projections in crocodiles are unknown. For abbreviations, see list of abbreviations. *Pigeon brain modified from Güntürkün, O., Verhoye, M., De Groof, G., Van der Linden, A., 2013. A 3-dimensional digital atlas of the ascending sensory and the descending motor systems in the pigeon brain. Brain Struct. Funct. 281, 269–281. For abbreviations, see list of abbreviations.*

In pigeons, the thalamic somatosensory nuclei DLP and DIVA have different ipsilateral projections to the telencephalon. DLP projects to a somatosensory area in the medial caudal nidopallium and to the somatosensory wulst (Wild, 1987). The main thalamic projection, however, ascends from DIVA and terminates in the rostral somatosensory part of the wulst (Wild et al., 2008). In crocodiles and turtles, the main thalamic somatosensory nuclei (medialis complex in crocodiles, n. caudalis in turtles) project to the central part of the ADVR (Balaban and Ulinski, 1981; Pritz and Stritzel, 1994a; Fig. 8.21). The reptilian medialis complex and n. caudalis are probably comparable to the avian DLP. This accords with their thalamic topography and their projection pattern that is restricted to the DVR. This would imply that the DIVA-wulst projection could be a derived feature of avian evolution, although it resembles the mammalian somatosensory projection in almost all aspects of its features.

A second major source of somatosensory information stems from the head area and is transmitted via the trigeminal system as well as the sensory components of the facial and the glossopharyngeal nerves (Necker et al., 2000). Although crocodiles seem to be extremely sensitive to even slightest touches on their heavily armored jaws (Leitch and Catania, 2012), our knowledge on the trigeminal system in reptiles is extremely limited. The following account is therefore centered on birds.

The somata of the trigeminal nerve in pigeons are located in the trigeminal ganglion gasseri of which the central root enters the brain stem and terminates in the n. principalis nervi trigemini (PrV) and the spinal sensory nucleus of the trigeminal nerve (SpV) (Wild and Zeigler, 1996). In the mallard duck, SpV shows bilateral intratrigeminal projections to the ventral component of PrV as well as ipsilateral projections into various cerebellar lobes (Arends et al., 1984). In addition, a descending part of the trigeminal tract extends caudally to the upper spinal cervical segments and terminates in the n. cuneatus externus (Dubbeldam and Karten, 1978). The only known projection of PrV to higher brain centers is a direct connection via the quintofrontal tract to the n. basalis prosencephali (Bas) in the rostrocaudal telencephalon (Wild et al., 1985; Schall et al., 1986). Bas projects via the nidopallium frontotrigeminale (NFT) to arcopallial substructures and the NCL (Mouritsen et al., 2016). This pathway was outlined in Section 8.3.3.3.2.1 (Fig. 8.21).

The ophthalmic branch of the trigeminal nerve in birds (representing the upper beak) possibly also mediates magnetoreception. Surgical ablation of the ophthalmic branch results in deficits in the detection of magnetic field changes (Mora et al., 2004) or decreases of magnetically induced neural responses of SpV and PrV (Heyers et al., 2010). It is conceivable that the trigeminal system of birds carries positional magnetic information because migratory birds can only compensate for a 1000 km displacement if the ophthalmic nerve remains intact (Kishkinev et al., 2013).

8.4.3 The Olfactory System

Olfaction is among the most ancient sensory systems and still plays a key role in a variety of behaviors that range from feeding to mating. Broadly speaking, the olfactory system comprises two distinct components: the main olfactory system, which is responsible for the sense of smell, and the vomeronasal system, which guides pheromone-based communications. Both systems are extremely sensitive and are, in some species, capable of discriminating between distinct odors of extremely low concentrations. Once chemical molecules bind to receptors cells in the olfactory epithelium, this information is transmitted via the olfactory nerves to the main olfactory bulb (MOB) and, in some species, to the accessory olfactory bulb (AOB) of the vomeronasal system. The MOB exists in nearly all vertebrates, but the AOB first appears in amphibians and is present in reptiles and mammals; it is absent in birds (Hayden and Teeling, 2014).

8.4.3.1 The Olfactory System of Birds

Birds possibly do not have a vomeronasal system. Their MOB projects via the lateral olfactory tract to the CPi, the prepiriform cortex (CPP), the HD, the anterior olfactory nucleus, the TnA, and some perihippocampal structures. Via the intermediate olfactory tract, the olfactory bulb also reaches the medial septum (SM), the TuO and, by crossing the midline, the contralateral bulb (Reiner and Karten, 1985; Patzke et al., 2001; Atoji and Wild, 2014). CPi and CPP are interacted with the visual system and limbic structures (Atoji and Wild, 2014).

Birds were historically considered microsmatic or even anosmic, but their behavior and their neuroanatomy tell a different story (Caro et al., 2015). When pigeons home over previously unexplored areas, they rely on an olfactory map (Wallraff, 2005). The critical role of olfaction in pigeon navigation was first discovered by Papi et al. (1971), who observed that anosmic pigeons were unable to home. He proposed that pigeons acquire an olfactory map by associating the odors carried by the winds at the home area with the directions from which they blow. Once at the release site, they recognize the local odors and determine the direction of displacement. Since then, a large number of studies could firmly establish the role of olfaction in avian navigation (Gagliardo, 2013). The relevance of smell for navigation is also reflected in the neuroanatomy of birds. Olfactory bulbs are spectacularly enlarged in birds known to use olfactory cues for navigation and foraging

such as seabirds (Wallraff, 2005). Manipulation of the olfactory system such as plugging the nostrils (Gagliardo et al., 2007), anaesthetizing the olfactory mucosa (Wallraff, 1988), transecting the olfactory nerve (Papi et al., 1971; Gagliardo et al., 2009), or ablating the CPi (Papi and Casini, 1990) generates remarkable and lateralized disruptions of initial orientation and homing performance in pigeons (Gagliardo, 2013). In addition, when homing pigeons are released at an unfamiliar location, their CPi is much more active, compared to a release at a familiar site. These results implicate the CPi of pigeons in the processing of olfactory map cues over uncharted territories when lying home (Patzke et al., 2010).

8.4.3.2 The Olfactory System of Reptiles

In reptiles, the main olfactory pathway and the vomeronasal system were investigated in quite a number of species, although the majority of the studied species focus on lizards and snakes (Reiner and Karten, 1985; Lanuza and Halpern, 1998; Martinez-Marcos et al., 2002). These studies suggest that snakes especially live in an olfactory world. As outlined later, the olfactory system constitutes a major part of their brain.

The main olfactory system is an open-ended detector of airborne odorants since it is able to represent endless combinations of compounds. The vomeronasal system is different. It evolved for detection of biologically relevant chemical cues (pheromones) that are mostly related to ingestive, sexual, or agonistic interactions. The vomeronasal organs (VNOs) are paired chemosensory organs in the anterior roof of the mouth that reach their highest development in squamate reptiles, and especially in snakes (Burghardt, 1993). Since vomeronasal olfaction serves a different behavioral role than the main olfactory system, their neural substrates differ as well (Martinez-Marcos et al., 2002). While olfaction can be mainly achieved during normal respiration, the vomeronasal system is activated by specific sequences of behavior. In snakes, chemical compounds are gathered in the environment by the tongue and are delivered to the VNOs and the main olfactory system with tongue-flicks. But tongue-flicks are not only about smell; a second kind of newly discovered tongue-flick is optimized for tasting objects on the ground (Daghfous et al., 2012).

The MOBs in snakes project through the lateral, the intermediate, and the medial olfactory tracts to the full extent of the lateral cortex as well as to the external and the ventral anterior amygdala. In addition, olfactory fibers reach the olfactory tubercle, the olfactory gray, and the dorsomedial retrobulbar formation (Lanuza and Halpern, 1998). Interestingly, these structures project back to the bulb, creating a closed loop within the main olfactory system.

The vomeronasal epithelium relays chemosensory information to the AOB, which in turn projects through the accessory olfactory tract to secondary vomeronasal-recipient areas such as the medial amygdala and, especially, to the nucleus sphericus (Lanuza and Halpern, 1998; Martinez-Marcos et al., 2002). The latter structure occupies a very large fraction of the telencephalon, thus testifying to the relevance of vomeronasal input for snakes. N. sphericus projects to the rostral dorsal cortex, the rostral lateral cortex, the olfactostriatum of the rostral basal telencephalon, the ventromedial hypothalamic nucleus, and several amygdaloid nuclei olfactostriatum in the basal telencephalon (Halpern, 1992; Lohman and Smeets, 1993; Lanuza and Halpern, 1997). Minor projections of the AOB also lead to the nucleus of the accessory olfactory tract.

As for the main olfactory system, the vomeronasal pathway features reciprocal projections between the olfactory and its target structures. However, the structures that receive vomeronasal input also have projections to the hypoglossal nucleus which controls the tongue-flicks (Martinez-Marcos et al., 2005). Especially the medial amygdala, which receives both olfactory and vomeronasal afferents, has projections to the hypoglossus via the lateral hypothalamic nucleus. Thus, the olfactory brain of snakes directly feeds back to the recipient sensory areas and controls the tongue with which odorant molecules are gathered (Martinez-Marcos et al., 2001).

8.4.4 Ascending Auditory Systems

The subtelencephalic auditory pathways were outlined in great detail in Chapter 1.14, Evolutionary Trends in Hearing in Nonmammalian Vertebrates by Catherine Carr in this volume. We therefore will only shortly summarize the main auditory brain stem components in birds and will contrast them with those of reptiles. Subsequently, we will review the telencephalic components of the auditory system in birds in some detail, thereby emphasizing both anatomy and function.

In birds, the fibers of the nervus octavus enter the medulla oblongata and split into two branches that terminate in the n. magnocellularis (NM) and the n. angularis (NA). Neurons of NM project bilaterally to n. laminaris (NL), which thus is the first neural entity that integrates input from both ears and is involved in processing interaural time differences (Young and Rubel, 1983; Necker et al., 2000). It seems that NM afferents to NL constitute delay lines, such that NL neurons can act as coincidence detectors, thereby creating an ordered map of interaural time differences (Vergne et al., 2009). Both NA and NL project bilaterally to the n. olivaris superior (OS) which projects back in inhibitory manner to NM and NL to increase the acuity of temporal integration (Burger et al., 2005). Besides these descending projections, OS, NA, and NL project in ascending direction to the n. mesencephalicus lateralis pars dorsalis (MLD) of

the midbrain as well as to diverse subnuclei of the lateral lemniscus (LL; Arends and Zeigler, 1986). Since MLD is believed to be homologous to the mammalian inferior colliculus (IC), especially scientists working on the owl auditory system prefer to use the term IC when referring to the avian MLD (Wagner et al., 2003). The subnuclei of the lateral lemniscus have differential projections, with one component projecting to the forebrain Bas, thereby bypassing the thalamus, while other branches terminate in MLD (Wild, 1987). The ventral component of LL projects to MLD as well as the thalamic relay nuclei n. ovoidalis (Ov) and n. uvaeformis (Uva) (Wild et al., 2010). Uva projects via the pallial n. interface (NIf) to the HVC in songbirds and thus plays a key role in the auditory input into the song system (Mooney, 2014). The midbrain MLD projects ipsilaterally to Ov from where projections ascend ipsilaterally to field L of the telencephalon (Wild et al., 1993). Field L has been divided into three laminae (Ll, L2, L3), and it is L2 where the fibers from the Ov mainly terminate (Carr, 1992; Fig. 8.22).

In reptiles, the auditory nerve also projects topographically to the reptilian version of NM and NA (Burger et al., 2005; Vergne et al., 2009). The functional organization of these cochlear nuclei seems to be very

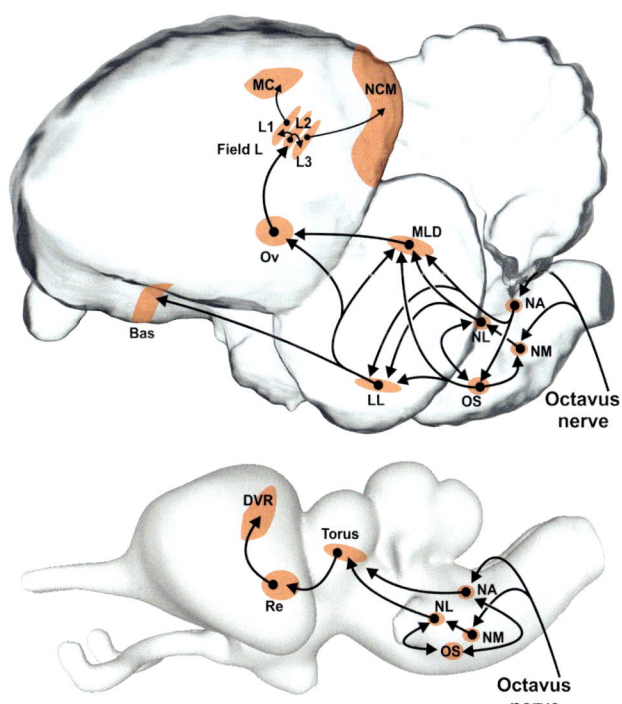

FIGURE 8.22 Ascending auditory pathways in birds (above) and crocodiles (below). The avian auditory pathways are depicted in pigeons, which is not a song bird. Therefore song system–specific structures were omitted. *Pigeon brain modified from Güntürkün, O., Verhoye, M., De Groof, G., Van der Linden, A., 2013. A 3-dimensional digital atlas of the ascending sensory and the descending motor systems in the pigeon brain. Brain Struct. Funct. 281, 269–281.*

similar to that of birds (Manley, 1970). As in birds, the reptilian NL receives afferents from NM and possibly also plays a role in sound localization (Vergne et al., 2009). In caiman, NA and NL project to OS, which then backprojects (Strutz, 1981). Again, this connectivity pattern resembles that of birds and thus could imply that also in reptiles OS projections sharpen auditory temporal integration.

In crocodilians, NA and NL project bilaterally to the torus semicircularis (ten Donkelaar, 1998). This structure shows a clear tonotopic frequency organization (Manley, 1971). According to Pritz (1974a), the central nucleus of the torus semicircularis projects to the core of the thalamic n. reuniens (Re) in caiman. It is likely that the reptilian reuniens and the avian ovoidalis are homologous thalamic auditory relay nuclei. Both structures show a clear tonotopy. Pritz (1974b) also demonstrated that the central core of Re projects to a caudomedial region of the ipsilateral DVR. This is highly similar both in terms of connectivity as well as in terms of topography to the projection of Ov to field L in birds. The auditory caudomedial area in the crocodile DVR shows a similar tonotopic organization as found in the torus semicircularis and in the cochlear nuclei (Weisbach and Schwartz-kopff, 1967). Thus, the frequency-specific projections from the auditory medulla to the dorsal thalamus and thence to the forebrain are well conserved in birds and reptiles (Vergne et al., 2009; Fig. 8.22).

Auditory processes beyond the termination area in the caudomedial DVR were not studied yet in reptiles, but were extensively analyzed for field L in birds. Projections of the Ov mostly terminate in field 2, but only sparsely terminate in the adjacent L1 and L3 (Wild et al., 1993). L2 projects upon L1 and L3, while L1 projects to the caudal mesopallium. Efferents of L3 terminate in the nidopallium caudomediale (NCM) and HVC in songbirds (Reiner et al., 2004) (see Section 8.4.5). It is possible that the HVC of songbirds represents an auditory specialization that derives from the NCL (Feenders et al., 2008). Consequently, in pigeons, L3 projects to NCM and NCL. Axons from NCL terminate in the arcopallium (Kröner and Güntürkün, 1999).

Single-unit recordings in L2 reveal rather simple V-shaped tuning curves with inhibitory side bands (Leppelsack, 1974). Other cells in the entire field L-complex show broad responsiveness to stimuli such as bird calls, pure tones, or white noise (Prather, 2013). Recordings or immediate early gene studies from NCM already evince a high auditory selectivity, with some cells in songbirds being specialized for song of the bird's own species or songs of other species (Phan et al., 2006; Stripling et al., 2001). Most importantly, NCM seems to serve as an acoustic memory (Moorman et al., 2011). This is visible in the ability of NCM neurons to progressively reduce their activity to repeated presentation of the same song, but to then immediately be very active when being presented with presentation of a new song (Prather, 2013). NCM is one of the critical gateways between the ascending auditory pathways and the song system that is outlined in Section 8.4.5.

In birds, a subcomponent of the n. lemniscus lateralis (LL) of the midbrain has a direct projection to the n. basalis prosencephali (Bas) in the frontoventral telencephalon (Schall et al., 1986). The Bas is also the termination area of the trigeminal system (see Section 8.4.2). Why should a trigeminal area receive auditory input? Imagine that you chew a nut. You will sense the haptic component of the nut via your trigeminal system. But you will also hear the cracking sound of the nut via bone-conducted hearing. Thus, auditory input always accompanies eating as a vital fast feedback pathway. This is true for both biting and pecking. Accordingly, Schall and Delius (1986) could show that the characteristics of evoked potentials from Bas make a bone as well as a cochlea-mediated sound input likely. Schall et al. (1986) demonstrated that Bas receives also direct input from the medullary nucleus vestibularis superior. Thus, Bas has a trigeminal, an auditory, and a vestibular input which all bypass the thalamus. As shown by Schall (1987) in pigeons with multiunit recordings, Bas neurons evince a specific directional sensitivity to rotatory vestibular stimulation that results from pitch motions of the head in the downward direction. This is exactly the head motion that occurs during pecking! Taken together, the auditory projection to Bas is possibly part of a sensory system that is highly specialized to represent the relevant sensory properties to guide pecking in birds (Fig. 8.22).

8.4.5 The Avian Song System

All birds vocalize, but only some birds sing. The trick about birdsong is that it has to be learned during early ontogeny (or during each season in some species). Vocal learning is a rare trait that only few animal groups possess, among them humans (Wilbrecht and Nottebohm, 2003). Possibly, vocal learning evolved independently multiple times in different vertebrate species (see Petkov and Jarvis, 2012 for various evolutionary scenarios). Among primates, vocal learning is well developed only in humans but not in nonhuman primates (Egnor and Hauser, 2004; Fischer et al., 2015). There are only a handful of other mammals that are vocal learners. These include marine mammals such as cetaceans (King et al., 2013; Janik, 2014) and pinnipeds (Reichmuth and Casey, 2014) and some terrestrial mammals including bats (Boughman, 1998; Knörnschild, 2014) and elephants (Poole et al., 2005). However, the most numerous vocal learners are three groups of birds—songbirds (Nottebohm and Liu, 2010), hummingbirds (Gaunt et al., 1994; Araya-Salas and Wright, 2013), and parrots (Berg et al., 2012). Learned vocalization can be used to address and label individuals, attract females, repel rivals, and define territory (Notte-bohm and Liu, 2010; Berg et al., 2012; King and Janik, 2013; Janik, 2014; Knörnschild, 2014). It should be noted that there is no simple dichotomy between vocal learners and nonlearners, as some vocal nonlearners possess at least a limited form of vocal learning (Saranathan et al., 2007; Arriaga et al., 2012; Petkov and Jarvis, 2012).

The learning of song in songbirds has many parallels with the human speech acquisition (Doupe and Kuhl, 1999). Both have a critical phase in early life during which they can acquire new vocalizations much easier. Song learning and speech acquisition start with a purely sensory phase followed by a motor (sensory-motor) phase during which vocalizations are produced. In both species, auditory feedback is essential for proper learning. It is interesting to note that although mammals and birds possess different, nonhomologous vocal or-gans, the underlying physical mechanisms of vocaliza-tions produced by these vocal organs might be largely the same in both species (Elemans et al., 2015).

Juvenile songbirds memorize the song of a tutor bird during a sensory period and form an internal representa-tion of its song (Brainard and Doupe, 2002; Konishi, 2010). The learning process is facilitated by an auditory predisposition for conspecific sounds which is likely genetically determined (Wheatcroft and Qvarnström, 2015). Later in the sensorimotor phase, birds start to pro-duce their own vocalizations. The auditory feedback of these developing vocalizations that the individual bird produces is compared to the tutor song template. These early initial vocalizations are called subsong and resemble babbling in humans. The subsong gradually de-velops into plastic song which already incorporates some recognizable elements from the tutor's song. This song is further refined until it reaches its final, crystallized form (Brainard and Doupe, 2002; Bolhuis et al., 2010).

There are several hypotheses on the evolution of vocal learning (Nottebohm and Liu, 2010; Nowicki and Searcy, 2014). One interesting possibility is that vocal learning evolved due to a preexisting sensory bias of fe-males for complex sounds which can be explained by stimulus-specific habituation mechanisms (Searcy, 1992). Since it may be easier to produce more complex songs through learning rather than innate motor pro-grams (Nowicki and Searcy, 2014), sexual selection based on the preference of females for complex songs might have promoted the emergence of vocal learning in males, at least in some species (Soma and Garamszegi, 2011; Woodgate et al., 2011, 2012). In turn, the complexity of male's song seems to have become an indicator for the bird's fitness (Nowicki et al., 1998, 2002; Woodgate et al., 2012). According to one hypothesis, a well-developed song repertoire in males indicates quality of the individual because of the temporal coincidence of song learning and developmental stress (Nowicki et al., 1998, 2000, 2002). Thus, if an individual manages to ac-quire complex songs despite stressful factors during the developmental period, such as limited nutrition, it probably possesses a stress-resistant genotype and more robust phenotype. Accordingly, song repertoire size was shown to correlate with survival of offspring (Woodgate et al., 2012) and learning performance in a foraging task (Boogert et al., 2008). The latter indicates that song complexity may signal to the female, a male's cognitive capacities, which in turn correlates with parental, foraging, and predator-avoidance skills, as well as with territory quality (Searcy, 1992; Nottebohm and Liu, 2010). Therefore, it is likely that one of the main advantages of vocal learning in songbirds was the expansion of the vocal repertoire which then increased mating success (Nowicki and Searcy, 2014).

The neurobiology of vocalization has been extensively studied in songbirds. Because they have a specialized "song system" of cell groups that are easily identifiable, songbirds represent a suitable animal model to investi-gate neurobiology of language, learning, and memory as well as neuronal plasticity and neurogenesis (Jarvis, 2004; Doupe et al., 2005; Bolhuis et al., 2010; Barnea and Pravosudov, 2011; Moorman et al., 2011).

Two specialized neuronal pathways within the song sys-tem have been implicated in vocalization (for reviews, see Brainard and Doupe, 2002; Jarvis, 2004; Bolhuis and Gahr, 2006; Bolhuis et al., 2010; Moorman et al., 2011; Fig. 8.23). These pathways are the posterior song motor pathway (SMP) and the anterior forebrain pathway (AFP). Both originate in the HVC, a song system nucleus in the dorsal aspect of the caudal nidopallium (HVC is its full, letter-based name). However, the two pathways origi-nate from distinct neuronal populations. The HVC neurons that give rise to the SMP project to the robust nucleus of the arcopallium (RA), which in turn projects to the dorsal medial nucleus of the midbrain (DM), to the tracheosyrin-geal part of the nucleus hypoglossus (nXIIts), and to some respiratory brain stem nuclei (Wild, 1997). The nXIIts inner-vates the muscles of the syrinx, the vocal organ of song-birds. The HVC neurons of the AFP project to the AreaX in the medial striatum. The striatal medium spiny neurons in AreaX project to pallidal-like neurons of the AreaX which in turn project to the dorsal lateral nucleus of the medial thalamus (DLM; Carrillo and Doupe, 2004; Kuenzel et al., 2011). DLM projects back to the telencephalic lateral magnocellular nucleus of anterior nidopallium (LMAN). Finally, LMAN projects back to AreaX and also intercon-nects AFP and SMP via its projections to RA.

The SMP generates and coordinates the activity of syringeal and respiratory muscles and is important for song production and certain aspects of song learning (Nottebohm et al., 1976; Wild, 1997; Brainard and

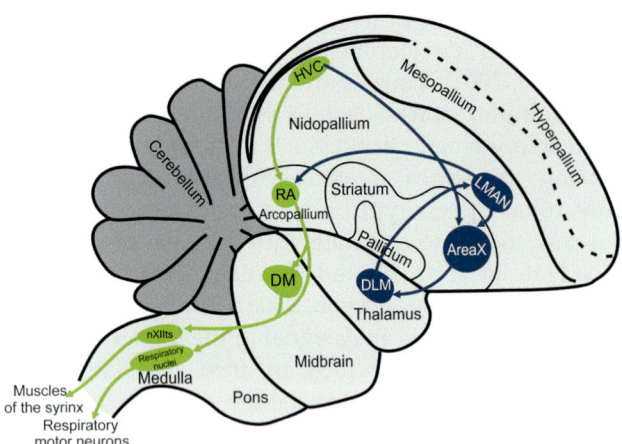

FIGURE 8.23 The figure shows the anterior forebrain pathway (*blue*) and the posterior song motor pathway (*green*) of the song control system in songbirds. Neurons in the HVC project either to the robust nucleus of the arcopallium (RA) or to AreaX in the medial striatum. The pallidal-like neurons of AreaX project to the dorsal lateral nucleus of the medial thalamus (DLM) which projects to the lateral magnocellular nucleus of anterior nidopallium (LMAN). LMAN projects back to AreaX and also connects the two pathways via its projection to RA. RA projects directly and indirectly, via the dorsal medial nucleus of the midbrain (DM), to the tracheosyringeal part of the nucleus hypoglossus (nXIIts) and to some respiratory nuclei in the brain stem. These nuclei innervate syringeal muscles and respiratory motor neurons in the spinal cord, respectively. *Reproduced from Moorman, S., Mello, C.V., Bolhuis, J.J., 2011. From songs to synapses: molecular mechanisms of birdsong memory. Molecular mechanisms of auditory learning in songbirds involve immediate early genes, including zenk and arc, the ERK/MAPK pathway and synapsins. Bioessays 33, 377–385.*

Doupe, 2002; Bolhuis et al., 2010). Consequently, lesions of HVC or RA practically abolish complex vocalizations without impacting the bird's disposition to sing (Nottebohm et al., 1976; Aronov et al., 2008). Interestingly, although the axons of HVC neurons start to grow and reach the dorsal border of RA already during the sensory phase of song development, they do not enter this nucleus until the onset of the motor phase (Mooney and Rao, 1994; Nowicki et al., 1998). Thus, the developmental time point of the SMP underlines its importance for song production. Electrophysiological investigations revealed that the population of RA-projecting HVC neurons seems to represent the temporal sequence of song syllables (Vu et al., 1994; Yu and Margoliash, 1996; Hahnloser et al., 2002). This sequence is then conveyed to the myotopic map in RA (Vicario, 1991; Hanloser et al., 2002), where neurons exhibit temporally precise and structured patterns of burst activity associated with specific notes (Vu et al., 1994; Yu and Margoliash, 1996). Thus, the plasticity of HVC-RA synapses is possibly a key component in the production of learned complex vocalizations (Mooney, 1992; Hahnloser et al., 2002). In addition, the study of Day et al. (2008) suggests that the HVC might control song plasticity during sensorimotor learning.

The AFP is necessary for song learning and adult song plasticity, and it might be involved in memorization of the tutor song (Bolhuis et al., 2010; Bolhuis and Moorman, 2015). Functional connections within this pathway are already established during the sensory phase, considerably earlier than those of SMP (Mooney and Rao, 1994; Nowicki et al., 1998). Lesions within this pathway in juvenile birds have clearly deteriorating effects on the learned song (Scharff and Nottebohm, 1991). However, the consequences for song development differ between AreaX and LMAN lesions (Scharff and Nottebohm, 1991). Lesions of AreaX produce an abnormal song with more variability in terms of notes, intervals, and syllable sequence. On the other hand, lesions of LMAN significantly reduce the number of notes used by the birds. Thus, this and other studies indicate that the LMAN seems to induce variability in the song of juvenile birds necessary for them to acquire the crystallized birdsong by trial-and-error learning (Ölveczky et al., 2005). In contrast, adult lesions of AreaX or LMAN do not alter the birdsong indicating the main role of AFP in song learning rather than adult song production (Scharff and Nottebohm, 1991; Aronov et al., 2008). LMAN and its projection to RA seem especially relevant for producing subsong (Aronov et al., 2008). Neurons in LMAN exhibit premotor activity related to onset (or offset) of syllables of subsong, and inactivation of LMAN entirely eliminates subsong production (Aronov et al., 2008). Although the exact role of the AFP in song learning is still not exactly understood (Bolhuis and Moorman, 2015), the presence of auditory neurons responsive to bird's own song (BOS) within this pathway (Doupe and Konishi, 1991) and the connections to the SMP indicate that AFP provides auditory feedback about the bird's own vocal outcome to match the BOS to the tutor song memory (Doupe, 1993). Such a role of AFP in modification of the own vocalizations by means of auditory feedback is supported by the fact that LMAN lesions prevent song deficits which normally develop after deafening in birds with an intact LMAN (Brainard and Doupe, 2000).

As stated previously, songbirds learn their song from a conspecific tutor and later adjust their own song according to the memorized tutor song (Brainard and Doupe, 2002). Which brain areas are involved in the storage of the tutor song has been the subject of extensive research (for reviews, see Bolhuis and Gahr, 2006; Bolhuis and Moorman, 2015). Although several studies suggested that the AFP might contain the neuronal substrate for the memory of the tutor song, other evidence suggests that the secondary auditory areas NCM and CM are the sites for birdsong storage (Bolhuis and Gahr, 2006; Gobes and Bolhuis, 2007; Bolhuis and Moorman, 2015). In particular, NCM may store the tutor song in male zebra finches, while CM may be relevant for the memory of the father's song in females (Bolhuis and Moorman, 2015). However, the picture is perhaps not that simple. The tutor song is most probably stored in a distributed brain network that also involves the SMP (Roberts et al., 2012; Roberts and Mooney, 2013). In an elegant study, Roberts et al. (2012) demonstrated that the HVC plays a crucial role in encoding the tutor song on a very precise timescale during the sensory phase. They manipulated the activity in HVC of a juvenile zebra finch while listening to the tutor song. When HVC activity was disrupted during the utterance of a specific syllable in the tutor song motif, the bird developed poor copies of the manipulated syllable while producing accurate copies of syllables flanking the target syllable. Thus, the auditory system, as well as several structures of the song system, seems to be involved in the internal representation of the tutor song. Furthermore, differences between species may also exist (Prather et al., 2010; Roberts and Mooney, 2013).

As mentioned above, human speech and birdsong share many features (Doupe and Kuhl, 1999; Elemans et al., 2015). Moreover, the neurobiology of human speech production and birdsong is strikingly similar in numerous respects (Jarvis, 2004; Simonyan et al., 2012). Humans seem to have evolved a specific region in the primary motor cortex that projects monosynaptically to the nucleus ambiguus, which innervates the muscles of the larynx, the vocal organ of humans (Jarvis, 2004; Simonyan and Horwitz, 2011; Simonyan et al., 2012). This region is called the laryngeal motor cortex (LMC) and has not been identified in the primary motor cortex of nonhuman primates (Simonyan and Horwitz, 2011). In nonhuman primates, only the premotor cortex contains a laryngeal region, and it lacks direct projections to laryngeal motor neurons (Simonyan and Horwitz, 2011). It has been suggested that the direct projection from LMC to nucleus ambiguus in humans is crucial for producing human speech (Simonyan et al., 2012). It is also reminiscent of the RA-nXIIts projection in songbirds. These observations indicate that a direct projection from the primary motor areas to the neurons controlling vocal organ muscles is a prerequisite for complex vocalizations (Petkov and Jarvis, 2012). Consistent with this idea, a recent study found that ultrasonic sound production in mice shares several features with the birdsong and involves a direct projection from the primary motor cortex to nucleus ambiguus (Arriaga et al., 2012). The fact that this projection is weak in mice led the authors to suggest that the strength of such projection is proportional to the complexity of vocalizations produced by an animal.

Together with premotor cortical areas, the human LMC is part of a corticobasal ganglia-thalamocortical loop that is comparable to AFP in songbirds (Jarvis, 2004; Simonyan and Horwitz, 2011; Simonyan et al.,

2012). Premotor cortical areas in mammals project to the striatum and from there to motor nuclei of the thalamus via the internal part of the globus pallidus (Reiner et al., 1998; Jarvis, 2004; Simonyan et al., 2012). The thalamus then closes the loop via its projection back to the cortex. This loop is similar to the LMAN-AreaX-DLM-LMAN loop described earlier (Jarvis, 2004; Kuenzel et al., 2011; Simonyan et al., 2012).

These similarities between distantly related species indicate that the neuronal correlates of vocal learning and production may have evolved as specializations of preexisting system present in ancestral amniote (or even vertebrate) brains. The motor theory of vocal learning origin proposes that brain systems for vocal learning and production evolved from ancestral motor system (Feenders et al., 2008; for alternative theories, see Petkov and Jarvis, 2012). This could have happened, for instance, by duplication of whole pathways (Chakraborty and Jarvis, 2015) or by strengthening of existing projections in vocal learners which are sparse or absent in nonvocal learners (Arriaga et al., 2012; Petkov and Jarvis, 2012). Feenders et al. (2008) used molecular imaging to map brain activity of vocal learning and nonlearning birds during body movements. They found movement-associated activity in comparable regions in both vocal learners and nonlearners. Most interestingly, movement-activated areas in vocal learners were adjacent to song system nuclei. The song system nuclei were activated by singing but not by body movements. The findings of this study therefore provide intriguing evidence for the notion that song system in birds emerged from the existing motor system and subsequently specialized for vocal control.

Further support for the motor theory comes from anatomical, electrophysiological, and pharmacological—behavioral studies on motor sequence execution in vocal nonlearning species including pigeons and chickens. Pigeons performing a sequence learning task require brain regions that are similar to the song system nuclei of songbirds (Helduser and Güntürkün, 2012; Helduser et al., 2013). Especially crucial for correct sequence execution are the nidopallium intermedium medialis pars laterale (NIMl), which is comparable to LMAN in songbirds in terms of topology and connectivity (Kröner and Güntürkün, 1999), and the NCL. Furthermore, HVC is adjacent to NCL in songbirds, and the NCL in pigeons contains separate populations of neurons projecting to medial striatum and arcopallium, just as HVC does in songbirds. These findings are consistent with the motor theory.

As mentioned above, the AreaX in the medial striatum of songbirds contains both spiny striatal neurons and aspiny pallidal-like neurons (Kuenzel et al., 2011). However, unlike mammalian medium spiny neurons, the spiny striatal neurons of AreaX do not appear to project outside of the striatum (Reiner et al., 2004b). Rather, these neurons project to a small population of large aspiny neurons which are the output neurons of the AreaX projecting to DLM (Farries et al., 2005a). These neurons show pallidal-like morphology and physiology and express the pallidal marker LANT6 (Farries and Perkel, 2002; Reiner et al., 2004b). The spiny striatal neurons of AreaX seem to express SP, the cotransmitter also presents in mammalian striatal neurons belonging to the direct pathway of basal ganglia (Reiner et al., 2004b). Thus, AreaX consists of both striatal and pallidal components and is possibly part of the direct pathway of the basal ganglia in songbirds (Faries and Perkel, 2002; Carrillo and Doupe, 2004). However, Carillo and Doupe (2004) suggest that functionally, AreaX may contain both the direct and the indirect pathway. This idea was further supported by anatomical and electrophysiological data showing monosynaptic pallial excitatory projections to pallidal-like output neurons as well as a connection between pallidal-like neurons lacking thalamic efferents and pallidal-like output neurons (Farries et al., 2005a). Both of these findings describe pathways that are not anatomically identical to the indirect pathway of mammals but nevertheless elicit an effect opposite to that of the direct pathway.

Farries et al. (2005b) investigated neurons in the striatum of domestic chickens, a vocal nonlearner distantly related to songbirds. Although striatal neurons in chickens exhibit a high diversity in their electrophysiological properties, Farries et al. (2005b) identified aspiny neurons that exhibited properties akin to pallidal neurons. This indicates that mixing of striatal and pallidal features within the striatum might be common to all birds and that AreaX in songbirds might be a specialized subset of these neurons, thus supporting the motor theory of vocal learning origin.

The above mentioned similarities in the pathways and their function raise the question whether similar specialized molecular regulatory mechanisms are responsible for the development and control of vocal behavior in different species. The transcription factor FoxP2 has received particular attention because of its association with the developmental verbal dyspraxia (a speech disorder) in humans and song deficits in songbirds (Bolhuis et al., 2010; Wohlgemuth et al., 2014). In recent years, considerable progress has been made toward mechanistic explanations of FoxP2 function in songbirds (Wohlgemuth et al., 2014). The evidence points to a role of the FoxP2 in the development and proper function of the circuitry required for sensorimotor learning. In a recent study, Pfenning et al. (2014) applied a computational algorithm to analyze a large gene expression database for vocal learning and vocal nonlearning birds and primates. They identified striking similarities between songbird's RA and the human

lateral motor cortex as well as between AreaX and a part of the human striatum activated during speech. The relationships of HVC and LMAN to human brain areas were weaker and had the highest correlation values with the Wernicke and Broca area, respectively. Importantly, none of these relationships were found in vocal nonlearners. Although it is not clear whether such similarities reflect the molecular machinery for the development of vocal learning circuits or whether they are the consequence of these circuits, these data nevertheless indicate that convergent behavioral and anatomical traits of vocal learners are associated with convergent molecular mechanisms.

8.5 Conclusion

There are more than 17 000 different sauropsid species, and they inhabit all major ecosystems of our planet (Shine, 2013). The phylogenetic heterogeneity of this group of animals is mirrored in the diversity of brain organizations of which a small fragment was outlined in this chapter. Comparative neuroscientists traditionally follow one of two scientific traditions to reveal the commonalities and the differences of these sauropsid brains. One tradition primarily analyzes brain anatomy and its ancestral relationships. The other tradition is interested in animal behavior and tries to map functions onto neural entities. Ideally, these two approaches should result in overlapping results. As we have seen, this is by far not always the case. Why?

Comparative neuroanatomists compare not only adult brains but often also expression patterns of genes that are involved in brain development. And they do this in the context of a topological framework of structures. This allows both to detect homologies between brain areas and to reconstruct the changes of brain components during evolution. This approach is quite successful in reconstructing the evolution of the brain, but it has only limited predictive power with regard to behavior. Current studies in the area of evolutionary developmental biology (evo-devo) would not easily expect that nearly identical arrangements of spinal motor pools generate the undulation of snakes, the walking pattern of rodents, or the flight of birds (Section 8.3.1.1). Similarly, a prefrontal-like area in the most posterior corner of the avian ventral pallium (Section 8.3.3.3.2.4) or the existence of an avian song system within the DVR that shows astonishing similarities to the human cortical language circuit comes as a surprise (Section 8.4.5).

The functional approach to comparative neuroscience has its own merits and problems. It is able to identify similar functional circuits in the brains of different animals, but it often fails to provide a strong hypothesis on the evolutionary background of such systems. Functional analyses may support evo-devo conclusions on ancestral conditions in some cases, but they are not useful in establishing conclusions on homologies on their own. Thus, comparative neuroanatomy can reconstruct the phylogeny of brains but often falls short in predicting behavior. Behavioral neuroscience provides insights about function—structure mappings but is mostly unable to conclude on homologies and so to reconstruct ancestral conditions.

A key factor that can explain the differential strengths and weaknesses of these two approaches is neural connectivity. Scientists in the evo-devo field usually try to avoid resting their conclusions on connectivity analyses since axonal pathways are often structured by genes that control late functional maturation and are under less tight evolutionary constraint. In addition, neural connectivity always goes through cycles of massive early maturational overproduction and late maturational pruning. During overproduction, many aberrant connections to nontarget areas are produced that are later eliminated when they do not contribute to proper functioning. But this is possibly exactly the way how new connections can be established rather quickly during evolution when animals are under selection for new perceptual, cognitive, or motor abilities. The seemingly aberrant connections can then contribute to new abilities, thereby increasing the fitness of the individual. This is due to the fact that the function of a neuron is largely determined by its input and far less by its location in the brain. So, if a visual neuron in the tectum of a rattlesnake starts receiving trigeminal thermal input, it will process infrared information in addition to classic vision (Section 8.3.2.1). If a light-dependent molecule in the photoreceptors of birds gains the ability to alter its function relative to the Earth's magnetic field lines, the respective visual pathway starts to see the position of the pole overlayed on object vision (Section 8.4.1.1). If auditory information is funneled to a trigeminal forebrain area, birds start combining the tactile and the auditory feedback of their pecking movements (Section 8.4.1.2). These kinds of changes can happen independently multiple times in evolution since they possibly require only few neural alterations to gain functionality for the individual.

Other neural functions depend on a large number of interwoven circuits to be functional. The mammalian PFC and its control over executive functions is a good example. In such cases, comparative neuroanatomists assumed that a certain macroanatomy is required to enable prefrontal functions. The discovery of a prefrontal-like area in the nonlaminated posterior DVR of birds shows that similar complex functions can be generated in brains with quite a different macroanatomy (Section 8.3.3.3.2.4). The same can be said for the avian telencephalic connectome in comparison to the respective connectomes of mammals (Section 8.3.3.3). Such examples reveal the degree of independence that functional circuits can have from their macroanatomical

framework. At this point it is important to make clear which aspect of macroanatomy we are talking about. The avian prefrontal-like area NCL is not positioned within a laminated dorsal pallium, but still it produces executive functions. But for its functionality, the NCL requires a certain connectivity pattern and, most importantly, input from the dopaminergic system that acts via D1 receptors. This dopaminergic D1 cascade could constitute a "deep homology" between mammalian and avian prefrontal structures that is independent from the overall macroanatomy (Shubin et al., 2009).

In other cases, we have seen that different neural computations can result from a highly comparable macroanatomy. For example, the visual cortex of turtles is homologous to the mammalian visual cortex and holds with its three-layered organization at least some of the critical macroanatomical features of the mammalian cortex. But, as outlined in Section 8.3.3.2.1, the visual cortex of turtles shows a visual representation that does not even remotely resemble the mammalian condition. Thus, even with a similar macroanatomy, local computations can differ substantially. In other cases, however, lamination evolved independently in nonhomologous locations: Sensory areas in the avian DVR show a laminated connectivity pattern with columnar arrangements like found in the sensory cortices of mammals (Section 8.3.3.3.2.3). Similar observations were recently reported from the fish dorsolateral pallium (Trinh et al., 2016).

Taken together, this chapter shows that both studies on homology and studies on function deliver important insights. We cannot replace one of them for the other since these two strands of inquiries often result in very different findings, with both of them telling a part of the truth. Without proper evo-devo-based analyses of homologies, we lose the framework to correctly interpret what has changed in which line of animals during convergent evolution. But a sole analysis of homologies falls far too short to explain the myriads of fascinating observations on sauropsid brains of which some were recapitulated here. We have to appreciate that the ability to properly respond to sensory inputs was the driving force for the evolution of brains. Comparative neuroscience that ignores this most fundamental aspect of brain evolution is prone to neglect the most vital part of its studies.

Nothing in neuroscience makes sense, except in the light of behavior.

References

Abellán, A., Desfilis, E., Medina, L., 2013. The olfactory amygdala in amniotes: an evo-devo approach. Anat. Rec. 296, 1317–1332.

Ahumada-Galleguillos, P., Fernández, M., Marin, G.J., Letelier, J.C., Mpodozis, J., 2015. Anatomical organization of the visual dorsal ventricular ridge in the chick (Gallus gallus): layers and columns in the avian pallium. J. Comp. Neurol. 523, 2618–2636.

Anderson, K.D., Reiner, A., 1990. Extensive co-occurrence of substance P and dynorphin in striatal projection neurons: an evolutionarily conserved feature of basal ganglia organization. J. Comp. Neurol. 295, 339−369.

Angaut, P., Repérant, J., 1978. A light and electron microscopic study of the nucleus isthmo-optic in the pigeon. Arch. Anat. Mikrosk. 67, 63−78.

Araya-Salas, M., Wright, T., 2013. Open-ended song learning in a hummingbird. Biol. Lett. 9, 20130625.

Arends, J.J.A., Zeigler, H.P., 1986. Anatomical identification of an auditory pathway from a nucleus of the lateral lemniscal system to the frontal telencephalon (nucleus basalis) of the pigeon. Brain Res. 398, 375−381.

Arends, J.J., Woelders-Block, A., Dubbeldam, J.L., 1984. The efferent connections of the nuclei of the descending trigeminal tract in the mallard (Anas platyrhynchos L.). Neuroscience 13, 797−817.

Ariëns-Kappers, C.U., Huber, G.C., Crosby, E., 1936. The Comparative Anatomy of the Nervous System of Vertebrates, Including Man. Hafner Press, New York.

Armstrong, E., Bergeron, R., 1985. Relative brain size and metabolism in birds. Brain Behav. Evol. 26, 141−153.

Aronov, D., Andalman, A.S., Fee, M.S., 2008. A specialized forebrain circuit for vocal babbling in the juvenile songbird. Science (New York, N.Y.) 320, 630−634.

Arriaga, G., Zhou, E.P., Jarvis, E.D., 2012. Of mice, birds, and men: the mouse ultrasonic song system has some features similar to humans and song-learning birds. PLoS One 7, e46610.

Atoji, Y., Wild, J.M., 2012. Afferent and efferent projections of the mesopallium in the pigeon (Columba livia). J. Comp. Neurol. 520, 717−741.

Atoji, Y., Wild, J.M., 2014. Efferent and afferent connections of the olfactory bulb and prepiriform cortex in the pigeon (Columba livia). J. Comp. Neurol. 522, 1728−1752.

Atoji, Y., Wild, J.M., Yamamoto, Y., Suzuki, Y., 2002. Intratelencephalic connections of the hippocampus in pigeons (Columba livia). J. Comp. Neurol. 447, 177−199.

Atoji, Y., Saito, S., Wild, J.M., 2006. Fiber connections of the compact division of the posterior pallial amygdala and lateral part of the bed nucleus of the stria terminalis in the pigeon (Columba livia). J. Comp. Neurol. 499, 161−182.

Auersperg, A.M., von Bayern, A.M., Gajdon, G.K., Huber, L., Kacelnik, A., 2011. Flexibility in problem solving and tool use of kea and New Caledonian crows in a multi access box paradigm. PLoS One 6, e20231.

Bagnoli, P., Burkhalter, A., 1983. Organization of afferent projections to wulst in pigeon. J. Comp. Neurol. 214, 103−113.

Bagnoli, P., Fontanesi, G., Casini, G., Porciatti, V., 1990. Binocularity in the little owl, Athene noctua. I. Anatomical investigation of the thalamo-Wulst pathway. Brain Behav. Evol. 35, 31−39.

Balaban, C.D., Ulinski, P.S., 1981. Organization of thalamic afferents to anterior dorsal ventricular ridge in turtles. I. Projections of thalamic nuclei. J. Comp. Neurol. 200, 95−129.

Balanoff, A.M., Bever, G.S., Rowe, T.B., Norell, M.A., 2013. Evolutionary origins of the avian brain. Nature 501, 93−96.

Barnea, A., Pravosudov, V., 2011. Birds as a model to study adult neurogenesis: bridging evolutionary, comparative and neuroethological approaches. Eur. J. Neurosci. 34, 884−907.

Barnes, T.D., Kubota, Y., Hu, D., Jin, D.Z., Graybiel, A.M., 2005. Activity of striatal neurons reflects dynamic encoding and recoding of procedural memories. Nature 437, 1158−1161.

Baron, J., Pinto, L., Dias, M.O., Lima, B., Neuenschwander, S., 2007. Directional responses of visual wulst neurones to grating and plaid patterns in the awake owl. Eur. J. Neurosci. 26, 1950−1968.

Bass, A.H., Pritz, M.B., Northcutt, R.G., 1973. Effects of telencephalic and tectal ablations on visual behavior in the side-necked turtle, Podocnemis unifilis. Brain Res. 55, 455−460.

Bassett, D.S., Yang, M., Wymbs, N.F., Grafton, S.T., 2015. Learning induced autonomy of sensorimotor systems. Nat. Neurosci. 18, 744−751.

Bast, T., Diekamp, D., Thiel, C., Schwarting, R.K.W., Güntürkün, O., 2002. Microdialysis in the 'Prefrontal Cortex' and the striatum of pigeons (Columba livia): evidence for dopaminergic volume transmission in the avian associative forebrain. J. Comp. Neurol. 446, 58−67.

Baumel, J.J., Witmer, L.M., 1993. Osteology. In: Baumel, J.J. (Ed.), Handbook of Avian Anatomy: Nomina Anatomica Avium. Massachusetts Nuttall Ornithology Club, Cambridge, pp. 45−132.

Belgard, T.G., Montiel, J.F., Wang, W.Z., García-Moreno, F., Margulies, E.H., Ponting, C.P., Molnár, Z., 2013. Adult pallium transcriptomes surprise in not reflecting predicted homologies across diverse chicken and mouse pallial sectors. Proc. Natl. Acad. Sci. U.S.A. 110, 13150−13155.

Benowitz, L.I., Karten, H.J., 1976. Organization of tectofugal visual pathway in pigeon: retrograde transport study. J. Comp. Neurol. 167, 503−520.

Benton, M.J., Donoghue, P.C.J., 2006. Palaeontological evidence to date the tree of life. Mol. Biol. Evol. 24, 26−53.

Berg, K.S., Delgado, S., Cortopassi, K.A., Beissinger, S.R., Bradbury, J.W., 2012. Vertical transmission of learned signatures in a wild parrot. Proc. R. Soc. Biol. Sci. 279, 585−591.

Berger, B., Febvret, A., Greengard, P., Goldman-Rakic, P.S., 1990. DARPP-32, a phosphoprotein enriched in dopaminoceptive neurons bearing dopamine D1 receptors: distribution in the cerebral cortex of the newborn and adult rhesus monkey. J. Comp. Neurol. 299, 327−348.

Berson, D.M., Hartline, P.H., 1988. A tecto-rotundo-telencephalic pathway in the rattlesnake: evidence for a forebrain representation of the infrared sense. J. Neurosci. 8, 1074−1088.

Bever, G.S., Lyson, T.R., Field, D.J., Bhullar, B.A.S., 2015. Evolutionary origin of the turtle skull. Nature 525, 239−242.

Bird, C.D., Emery, N.J., 2009. Rooks use stones to raise the water level to reach a floating worm. Curr. Biol. 19, 1410−1414.

Bischof, H.J., Eckmeier, D., Keary, N., Löwel, S., Mayer, U., Michael, N., 2016. Multiple visual field representations in the visual wulst of a laterally eyed bird, the zebra finch (Taeniopygia guttata). PLoS One 11 (5), e0154927.

Bjursten, L.M., Norrsell, K., Norrsell, U., 1976. Behavioural repertory of cats without cerebral cortex from infancy. Exp. Brain Res. 25, 115−130.

Bolhuis, J.J., Gahr, M., 2006. Neural mechanisms of birdsong memory. Nat. Rev. Neurosci. 7, 347−357.

Bolhuis, J.J., Moorman, S., 2015. Birdsong memory and the brain: in search of the template. Neurosci. Biobehav. Rev. 50, 41−55.

Bolhuis, J.J., Okanoya, K., Scharff, C., 2010. Twitter evolution: converging mechanisms in birdsong and human speech. Nat. Rev. Neurosci. 11, 747−759.

Boogert, N.J., Giraldeau, L.-A., Lefebvre, L., 2008. Song complexity correlates with learning ability in zebra finch males. Anim. Behav. 76, 1735−1741.

Boughman, J.W., 1998. Vocal learning by greater spear-nosed bats. Proc. Biol. Sci. R. Soc. 265, 227−233.

Brainard, M.S., Doupe, A.J., 2000. Interruption of a basal ganglia-forebrain circuit prevents plasticity of learned vocalizations. Nature 404, 762−766.

Brainard, M.S., Doupe, A.J., 2002. What songbirds teach us about learning. Nature 417, 351−358.

Brauth, S.E., 1984. Enkephalin-like immunoreactivity within the telencephalon of the reptile Caiman crocodilus. Neuroscience 11, 345−358.

Bravo, H., Pettigrew, J.D., 1981. The distribution of neurons projecting from the retina and visual cortex to the thalamus and tectum opticum of the barn owl, Tyto alba, and the burrowing owl, Speotyto cunicularia. J. Comp. Neurol. 199, 419−441.

Bruce, L.L., Butler, A.B., 1984. Telencephalic connections in lizards. I. Projections to cortex. J. Comp. Neurol. 229, 585–601.

Bruce, L.L., Neary, T.J., 1995. The limbic system of tetrapods: a comparative analysis of cortical and amygdalar populations. Brain Behav. Evol. 46, 224–234.

Brusatte, S.L., O'Connor, J.K., Jarvis, E.D., 2015. The origin and diversification of birds. Curr. Biol. 25, R888–R898.

Budzynski, C.A., Bingman, V.P., 2004. Participation of the thalamofugal visual pathway in a coarse pattern discrimination task in an open arena. Behav. Brain Res. 153, 543–556.

Bugnyar, T., 2013. Social cognition in ravens. Comp. Cogn. Behav. Rev. 8, 1–12.

Bugnyar, T., Heinrich, B., 2005. Ravens, *Corvus corax*, differentiate between knowledgeable and ignorant competitors. Proc. R. Soc. B 272, 1641–1646.

Burger, R.M., Cramer, K.S., Pfeiffer, J.D., Rubel, E.W., 2005. Avian superior olivary nucleus provides divergent inhibitory input to parallel auditory pathways. J. Comp. Neurol. 481, 6–18.

Burghardt, G.M., 1993. The comparative imperative: genetics and ontogeny of chemoreceptive prey responses in natricine snakes. Brain Behav. Evol. 41, 138–146.

Burghardt, G.M., 2013. Environmental enrichment and cognitive complexity in reptiles and amphibians: concepts, review, and implications. Appl. Anim. Behav. Sci. 147, 286–298.

Butler, A.B., Hodos, W., 2005. Comparative Vertebrate Neuroanatomy: Evolution and Adaptation, second ed. Wiley-Liss, New York.

Cajal, S.R., 1889. Sur la morphologie et les connexions des éléments de la rétine des oiseaux. Anat. Anz. 4, 111–128.

Caro, S.P., Balthazart, J., Bonadonna, F., 2015. The perfume of reproduction in birds: chemosignalling in avian social life. Horm. Behav. 68, 25–42.

Carr, C.E., 1992. Evolution of the central auditory system in reptiles and birds. In: Webster, D.B., Fay, R.R., Popper, A.N. (Eds.), The Evolutionary Biology of Hearing. Springer, Heidelberg, pp. 511–543.

Carrillo, G.D., Doupe, A.J., 2004. Is the songbird Area X striatal, pallidal, or both? An anatomical study. J. Comp. Neurol. 473, 415–437.

Carroll, R.L., 1988. Vertebrate Paleontology and Evolution. W.H. Freeman, New York.

Casini, G., Bingman, V.P., Bagnoli, P., 1986. Connections of pigeon dorsomedial forebrain studied with HRP and 3H-proline. J. Comp. Neurol. 245, 454–470.

Castro, L., Wasserman, E.A., 2016. Executive control and task switching in pigeons. Cognition 146, 121–135.

Catsicas, S., Catsicas, M., Clarke, P.G.H., 1987. Long-distance intraretinal connections in birds. Nature 326, 186–187.

Cerutti, S.M., Ferrari, E.A., 1995. Operant discrimination learning in detelencephalated pigeons (*Columba livia*). Braz. J. Med. Biol. Res. 28, 1089–1095.

Chakraborty, M., Jarvis, E.D., 2015. Brain evolution by brain pathway duplication. Philos. Trans. R. Soc. Lond. Ser. B Biol. Sci. 370.

Chen, C.C., Winkler, C.M., Pfenning, A.R., Jarvis, E.D., 2013. Molecular profiling of the developing avian telencephalon: regional timing and brain subdivision continuities. J. Comp. Neurol. 521, 3666–36701.

Cheng, M.-F., Chaiken, M., Zuo, M., Miller, H., 1999. Nucleus taeniae of the amygdala of birds: anatomical and sunctional studies in ring doves (Streptopelia risoria) and European starlings (Sturnus vulgaris). Brain Behav. Evol. 53, 243–270.

Clarke, P.G.H., Whitteridge, D., 1976. Projection of retina, including red area onto optic tectum of pigeon. Q. J. Exp. Psychol. 61, 351–358.

Clayton, N.S., Dickinson, A., 1998. Episodic-like memory during cache recovery by scrub-jays. Nature 395, 272–278.

Clayton, N.S., Bussey, T.J., Dickinson, A., 2003. Can animals recall the past and plan for the future? Nat. Rev. Neurosci. 4, 685–691.

Cohen, Y.E., Miller, G.L., Knudsen, E.I., 1998. Forebrain pathway for auditory space processing in the barn owl. J. Neurophysiol. 79, 891–902.

Corfield, J.R., Wild, J.M., Hauber, M.E., Parsons, S., Kubke, M.F., 2008. Evolution of brain size in the Palaeognath lineage, with an emphasis on New Zealand ratites. Brain Behav. Evol. 71, 87–99.

Cowan, W.M., 1970. Centrifugal fibres to avian retina. Br. Med. Bull. 26, 112–119.

Cox, S.A., Peoples, A.D., DeMaso, S.J., Lusk, J.J., Guthery, F.S., 2004. Survival and cause-specific mortality of northern Bobwhites in western Oklahoma. J. Wildl. Manag. 69, 663–671.

Csillag, A., Székely, A.D., Stewart, M.G., 1997. Synaptic terminals immunolabelled against glutamate in the lobus parolfactorius of domestic chicks (*Gallus domesticus*) in relation to afferents from the archistriatum. Brain Res. 750, 171–179.

Daghfous, G., Smargiassi, M., Libourel, P.A., Wattiez, R., Bels, V., 2012. The function of oscillatory tongue-flicks in snakes: insights from kinematics of tongue-flicking in the banded water snake (*Nerodia fasciata*). Chem. Senses 37, 883–896.

Dávila, J.C., Andreu, M.J., Real, M.A., Puelles, L., Guirado, S., 2002. Mesencephalic and diencephalic afferent connections to the thalamic nucleus rotundus in the lizard, *Psammodromus algirus*. Eur. J. Neurosci. 16, 267–282.

Day, N.F., Kinnischtzke, A.K., Adam, M., Nick, T.A., 2008. Top-down regulation of plasticity in the birdsong system: "premotor" activity in the nucleus HVC predicts song variability better than it predicts song features. J. Neurophysiol. 100, 2956–2965.

Dean, P., Redgrave, P., Mitchell, I.J., 1988. Organization of efferent projections from superior colliculus to brainstem in rat: evidence for functional output channels. Prog. Brain Res. 75, 27–36.

Delius, J.D., Vollrath, W., 1973. Rotation compensating reflexes independent of the labyrinth. Neurosensory correlates in pigeons. J. Comp. Physiol. 83, 123–134.

Derobert, Y., Médina, M., Rio, J.P., Ward, R., Repérant, J., Marchand, M.J., Miceli, D., 1999. Retinal projections in two crocodilian species, *Caiman crocodilus* and *Crocodylus niloticus*. Anat. Embryol. 200, 175–191.

Dickinson, A., 1985. Actions and habits: the development of behavioural autonomy. Philos. Trans. R. Soc. Lond. B Biol. Sci. 308, 67–78.

Diekamp, B., Kalt, T., Ruhm, A., Koch, M., Güntürkün, O., 2000. Impairment in a discrimination reversal task after D1-receptor blockade in the pigeon 'prefrontal cortex'. Behav. Neurosci. 114, 1145–1155.

Diekamp, B., Kalt, T., Güntürkün, O., 2002. Working memory neurons in pigeons. J. Neurosci. 22 (RC210), 1–5.

Dietl, M.M., Palacios, J.M., 1988. Neurotransmitter receptors in the avian brain. I. Dopamine receptors. Brain Res. 439, 354–359.

Divac, I., Mogensen, J., 1985. Prefrontal cortex in pigeon catecholamine histo fluorescence. Neuroscience 15, 677–682.

Divac, J., Mogensen, J., Björklund, A., 1985. The prefrontal "cortex" in the pigeon. Biochemical evidence. Brain Res. 332, 365–368.

van Dongen, P.A.M., 1998. Brain size in vertebrates. In: Nieuwenhuys, R., ten Donkelaar, H.J., Nicholson, C. (Eds.), The Central Nervous System of Vertebrates, vol. 3. Springer, Berlin, pp. 2099–2134.

ten Donkelaar, H.J., 1998. Reptiles. In: Niewenhuys, R., ten Donkelaar, H.J., Nicholson, C. (Eds.), The Central Nervous System of Vertebrates. Springer, Berlin, pp. 1315–1524.

Douglas, R.J., Martin, K.A.C., 2007. Mapping the matrix: the ways of neocortex. Neuron 56, 226–238.

Doupe, A.J., 1993. A neural circuit specialized for vocal learning. Curr. Opin. Neurobiol. 3, 104–111.

Doupe, A.J., Konishi, M., 1991. Song-selective auditory circuits in the vocal control system of the zebra finch. Proc. Natl. Acad. Sci. U.S.A. 88, 11339–11343.

Doupe, A.J., Kuhl, P.K., 1999. Birdsong and human speech: common themes and mechanisms. Annu. Rev. Neurosci. 22, 567–631.

Doupe, A.J., Perkel, D.J., Reiner, A., Stern, E.A., 2005. Birdbrains could teach basal ganglia research a new song. Trends Neurosci. 28, 353–363.

Dubbeldam, J.L., 1998. Birds. In: Niewenhuys, R., Ten Donkelaar, H.J., Nicholson, C. (Eds.), The Central Nervous System of Vertebrates. Springer, Berlin, pp. 1525–1636.

Dubbeldam, J.L., Karten, H.J., 1978. The trigeminal system in the pigeon (Columba livia). I. Projections of the Gasserian ganglion. J. Comp. Neurol. 180, 661–678.

Dubbeldam, J.L., den Boer-Visser, A.M., Bout, R.G., 1997. Organization and efferent connections of the archistriatum of the mallard, Anas platyrhynchos L.: an anterograde and retrograde tracing study. J. Comp. Neurol. 388, 632–657.

Dugas-Ford, J., Rowell, J.J., Ragsdale, C.W., 2012. Cell-type homologies and the origins of the neocortex. Proc. Natl. Acad. Sci. U.S.A. 109, 16974–16979.

Dunlop, S.A., Tee, L.B., Goossens, M.A., Stirling, R.V., Hool, L., Rodger, J., Beazley, L.D., 2007. Regenerating optic axons restore topography after incomplete optic nerve injury. J. Comp. Neurol. 505, 46–57.

von Düring, M., Miller, M.R., 1979. Sensory nerve endings of the skin and deeper structures. In: Gans, C., Northcutt, R.G., Ulinski, P. (Eds.), Biology of Reptilia. Neurology, vol. 9. Academic Press, London, pp. 407–441.

Durstewitz, D., Seamans, J.K., 2008. The dual-state theory of prefrontal cortex dopamine function with relevance to catechol-o-methyl-transferase genotypes and schizophrenia. Biol. Psychiatry 64, 739–749.

Durstewitz, D., Kröner, S., Hemmings Jr., H.C., Güntürkün, O., 1998. The dopaminergic innervation of the pigeon telencephalon: distribution of DARPP-32 and coocurrence with glutamate decarboxylase and tyrosine hydroxylase. Neuroscience 83, 763–779.

Durstewitz, D., Kelc, M., Güntürkün, O., 1999. A neurocomputational theory of the dopaminergic modulation of working memory functions. J. Neurosci. 19, 2807–2822.

Durstewitz, D., Kröner, S., Güntürkün, O., 1999. The dopaminergic innervation of the avian telencephalon. Prog. Neurobiol. 59, 161–195.

Dutta, A., Gutfreund, Y., 2014. Saliency mapping in the optic tectum and its relationship to habituation. Front. Integr. Neurosci. 8, 1.

Ebbesson, S.O.E., Goodman, D.C., 1981. Organization of ascending spinal projections in Caiman crocodilus. Cell Tissue Res. 215, 383–395.

Ebert, J., Westhoff, G., 2006. Behavioural examination of the infrared sensitivity of rattlesnakes (Crotalus atrox). J. Comp. Physiol. 192, 941–947.

Ebinger, P., Löhmer, R., 1987. A volumetric comparison of brains between greylag geese (Anser anser L.) and domestic geese. J. Hirnforschund 28, 291–299.

Edinger, L., Wallenberg, A., Holmes, G.M., 1903. Untersuchungen über die vergleichende Anatomie des Gehirns: 3, das Vorderhirn der Vögel, vol. 20. Abhandlungen der Senkenbergischen Gesellschaft, Frankfurt am Main, pp. 343–426.

Egnor, S.R., Hauser, M.D., 2004. A paradox in the evolution of primate vocal learning. Trends Neurosci. 27, 649–654.

Ehrlich, D., Keyser, K.T., Karten, H.J., 1987. Distribution of substance P-like immunoreactive retinal ganglion cells and their pattern of termination in the optic tectum of chick (Gallus gallus). J. Comp. Neurol. 266, 220–233.

Elemans, C., Rasmussen, J.H., Herbst, C.T., Düring, D.N., Zollinger, S.A., Brumm, H., Srivastava, K., Svane, N., Ding, M., Larsen, O.N., Sober, S.J., Švec, J.G., 2015. Universal mechanisms of sound production and control in birds and mammals. Nat. Commun. 6, 8978.

Ellard, C.G., Goodale, M.A., 1988. A functional analysis of the collicular output pathways: a dissociation of deficits following lesions of the dorsal tegmental decussation and the ipsilateral collicular efferent bundle in the Mongolian gerbil. Exp. Brain Res. 71, 307–319.

Emery, N.J., Clayton, N.S., 2001. Effects of experience and social context on prospective caching strategies by scrub jays. Nature 414, 443–446.

Emery, N.J., Clayton, N.S., 2004. The mentality of crows: convergent evolution of intelligence in corvids and apes. Science 306, 1903–1907.

Evans, S.E., 2003. At the feet of the dinosaurs: the early history and radiation of lizards. Biol. Rev. 78, 513–551.

Fagot, J., Cook, R.G., 2006. Evidence for large long-term memory capacities in baboons and pigeons and its implications for learning and the evolution of cognition. Proc. Natl. Acad. Sci. U.S.A. 103, 17564–17567.

Faries, M.A., Perkel, D.J., 2002. A telencephalic nucleus essential for song learning contains neurons with physiological characteristics of both striatum and globus pallidus. J. Neurosci. 22, 3776–3787.

Farries, M.A., Ding, L., Perkel, D.J., 2005. Evidence for "direct" and "indirect" pathways through the song system basal ganglia. J. Comp. Neurol. 484, 93–104.

Farries, M.A., Meitzen, J., Perkel, D.J., 2005. Electrophysiological properties of neurons in the basal ganglia of the domestic chick: conservation and divergence in the evolution of the avian basal ganglia. J. Neurophysiol. 94, 454–467.

Faunes, M., Fernandez, S., Gutierrez-Ibanez, C., Iwaniuk, A.N., Wylie, D.R., Mpodozis, J., Karten, H.J., Marin, G., 2013. Laminar segregation of GABAergic neurons in the avian nucleus isthmi pars magnocellularis: a retrograde tracer and comparative study. J. Comp. Neurol. 8, 1727–1742.

Feenders, G., Liedvogel, M., Rivas, M., Zapka, M., Horita, H., Hara, E., Wada, K., Mouritsen, H., Jarvis, E.D., 2008. Molecular mapping of movement-associated areas in the avian brain: a motor theory for vocal learning origin. PLoS One 3, e1768.

Ferguson, J.L., Mulvanny, P.J., Brauth, S.E., 1978. Distribution of neurons projecting to the retina of Caiman crocodilus. Brain Behav. Evol. 15, 294–306.

Ferrier, D., 1876. The Functions of the Brain. GP Putnam's Sons, New York.

von Fersen, L., Wynne, C.D., Delius, J.D., Staddon, J.E., 1990. Deductive reasoning in pigeons. Naturwissenschaften 77, 548–549.

Fetcho, J.R., 1986. The organization of the motoneurons innervating the axial musculature of vertebrates. II. Florida water snakes (Nerodia fasciata pictiventris). J. Comp. Neurol. 249, 551–563.

Fetcho, J.R., 1987. A review of the organization and evolution of motoneurons innervating the axial musculature of vertebrates. Brain Res. Rev. 12, 243–280.

Field, D.J., Gauthier, J.A., King, B.L., Pisani, D., Lyson, T.R., Peterson, K.J., 2014. Toward consilience in reptile phylogeny: miRNAs support an archosaur, not lepidosaur, affinity for turtles. Evol. Dev. 16, 189–196.

Fischer, J., Wheeler, B.C., Higham, J.P., 2015. Is there any evidence for vocal learning in chimpanzee food calls? Curr. Biol. 25, R1028.

Fournier, J., Müller, C.M., Laurent, G., 2015. Looking for the roots of cortical sensory computation in three-layered cortices. Curr. Opin. Neurobiol. 31, 119–126.

Fredes, F., Tapia, S., Letelier, J.C., Marin, G., Mpodozis, J., 2010. Topographic arrangement of the rotundo-entopallial projection in the pigeon (Columba livia). J. Comp. Neurol. 518, 4342–4361.

Fritzsch, B., Crapon DeCaprona, M.-D., Clarke, P.G.H., 1990. Development of two morphological types of retinopetal fibers in chick embryos, as shown by the diffusion along axons of a carbocyanine dye in the fixed retina. J. Comp. Neurol. 301, 1–17.

Frost, B.J., DiFranco, D.E., 1976. Motion characteristics of single units in the pigeon optic tectum. Vis. Res. 16, 1229–1234.

Frost, B.J., Nakayama, K., 1983. Single visual neurons code opposing motion independent of direction. Science 220, 744–745.

Frost, B.J., Scilley, P.L., Wong, S.C.P., 1981. Moving background patterns reveal double-opponency of directionally specific pigeon tectal neurons. Exp. Brain Res. 43, 173–185.

Frost, B.J., Wise, L.Z., Morgan, B., Bird, D., 1990. Retinotopic representation of the bifoveate eye of the kestrel (Falco sparverius) on the optic tectum. Vis. Neurosci. 5, 231–239.

Frost, B.J., Wylie, D.R., Wang, Y.-C., 1990. The processing of object and self-motion in the tectofugal and accessory optic pathways of birds. Vis. Res. 30, 1677–1688.

Funke, K., 1989. Somatosensory areas in the telencephalon of the pigeon. II. Spinal pathways and afferent connections. Exp. Brain Res. 76, 620–638.

Gabi, M., Collins, C.E., Wong, P., Torres, L.B., Kaas, J.H., Herculano-Houzel, S., 2010. Cellular scaling rules for the brains of an extended number of primate species. Brain Behav. Evol. 76, 32–44.

Gagliardo, A., 2013. Forty years of olfactory navigation in birds. J. Exp. Biol. 216, 2165–2171.

Gagliardo, A., Bonadonna, F., Divac, I., 1996. Behavioural effects of ablations of the presumed "prefrontal cortex" or the corticoid in pigeons. Behav. Brain Res. 78, 155–162.

Gagliardo, A., Pecchia, T., Savini, M., Odetti, F., Ioalè, P., Vallortigara, G., 2007. Olfactory lateralization in homing pigeons: initial orientation of birds receiving a unilateral olfactory input. Eur. J. Neurosci. 25, 1511–1516.

Gagliardo, A., Ioalè, P., Savini, M., Wild, M., 2009. Navigational abilities of adult and experienced homing pigeons deprived of olfactory or trigeminally mediated magnetic information. J. Exp. Biol. 212, 3119–3124.

Gaunt, S.L., Baptista, L.F., Sánchez, J.E., Hernandez, D., 1994. Song learning as evidenced from song sharing in two hummingbird species (Colibri coruscans and C. thalassinus). Auk 111, 87–103.

Gerfen, C.R., 1992. The neostriatal mosaic: multiple levels of compartmental organization in the basal ganglia. Annu. Rev. Neurosci. 15, 285–320.

Giffin, E.B., 1990. Gross spinal anatomy and limb use in living and fossil reptiles. Paleobiology 16, 448–458.

Gobes, S.M., Bolhuis, J.J., 2007. Birdsong memory: a neural dissociation between song recognition and production. Curr. Biol. 17, 789–793.

Gonzalez-Cabrera, C., Garrido-Charad, F., Mpodozis, J., Bolam, P., Marín, G.J., 2016. Axon terminals from the nucleus isthmi pars parvocellularis control the ascending retinotectofugal output through direct synaptic contact with tectal ganglion cell dendrites. J. Comp. Neurol. 524, 362–379.

Goris, R.C., Terashima, S., 1976. The structure and function of the infrared receptors of snakes. Prog. Brain Res. 43, 159–170.

Goslow Jr., G.E., Wilson, D., Poore, S.O., 2000. Neuromuscular correlates to the evolution of flapping flight in birds. Brain Behav. Evol. 55, 85–99.

Gracheva, E.O., Ingolia, N.T., Kelly, Y.M., Cordero-Morales, J.F., Hollopeter, G., Chesler, A.T., Sánchez, E.E., Perez, J.C., Weissman, J.S., Julius, D., 2010. Molecular basis of infrared detection by snakes. Nature 464, 1006–1011.

Graybiel, A.M., Grafton, S.T., 2015. The striatum: where skills and habits meet. Cold Spring Harb. Perspect. Biol. 7, a021691.

Green, R.E., Braun, E.L., Armstrong, J., Earl, D., Nguyen, N., Hickey, G., Vandewege, M.W., St John, J.A., Capella-Gutiérrez, S., Castoe, T.A., Kern, C., Fujita, M.K., Opazo, J.C., Jurka, J., Kojima, K.K., Caballero, J., Hubley, R.M., Smit, A.F., Platt, R.N., Lavoie, C.A., Ramakodi, M.P., Finger Jr., J.W., Suh, A., Isberg, S.R., Miles, L., Chong, A.Y., Jaratlerdsiri, W., Gongora, J., Moran, C., Iriarte, A., McCormack, J., Burgess, S.C., Edwards, S.V., Lyons, E., Williams, C., Breen, M., Howard, J.T.,

Gresham, C.R., Peterson, D.G., Schmitz, J., Pollock, D.D., Haussler, D., Triplett, E.W., Zhang, G., Irie, N., Jarvis, E.D., Brochu, C.A., Schmidt, C.J., McCarthy, F.M., Faircloth, B.C., Hoffmann, F.G., Glenn, T.C., Gabaldón, T., Paten, B., Ray, D.A., 2014. Three crocodilian genomes reveal ancestral patterns of evolution among archosaurs. Science 346, 1254449.

Grillner, S., Robertson, B., 2015. The basal ganglia downstream control of brainstem motor centres — an evolutionary conserved strategy. Curr. Opin. Neurobiol. 33, 47–52.

Grimm, F., Reese, M., Mittelstaedt, H., 1997. Extravestibuläre Rezeptoren zur Wahrnehmung der Richtung der Schwerkraft bei der Taube (Columba livia, Gmel. 1789). Verhandlungsbericht des 38. Int. Symp. Erkrank. Zoo Wildtiere 38, 97–101.

De Groof, G., Jonckers, E., Güntürkün, O., Denolf, P., Van Auderkerke, J., van der Linden, A., 2013. Functional MRI and functional connectivity of the visual system of awake pigeons. Behav. Brain Res. 239, 43–50.

Guirado, S., Dávila, J.C., Real, M.A., Medina, L., 2000. Light and electron microscopic evidence for projections from the thalamic nucleus rotundus to targets in the basal ganglia, the dorsal ventricular ridge, and the amygdaloid complex in a lizard. J. Comp. Neurol. 424, 216–232.

Güntürkün, O., 1987. A Golgi study of the isthmic nuclei in the pigeon (Columba livia). Cell Tissue Res. 248, 439–448.

Güntürkün, O., 1991. The functional organization of the avian visual system. In: Andrew, R.J. (Ed.), Neural and Behavioural Plasticity: The Use of the Domestic Chick as a Model. Oxford University Press, Oxford, pp. 92–105.

Güntürkün, O., 1997. Cognitive impairments after lesions of the neostriatum caudolaterale and its thalamic afferent: functional similarities to the mammalian prefrontal system? J. Brain Res. 38, 133–143.

Güntürkün, O., 2000. Sensory physiology: vision. In: Whittow, G.C. (Ed.), Sturkie's Avian Physiology. Academic Press, Orlando, pp. 1–19.

Güntürkün, O., 2005. The avian 'prefrontal cortex' and cognition. Curr. Opin. Neurobiol. 15, 686–693.

Güntürkün, O., 2012. The convergent evolution of neural substrates for cognition. Psychol. Res. 76, 212–219.

Güntürkün, O., Bugnyar, T., 2016. Cognition without cortex. Trends Cogn. Sci. 20, 291–303.

Güntürkün, O., Hahmann, U., 1999. Functional subdivisions of the ascending visual pathways in the pigeon. Behav. Brain Res. 98, 193–201.

Güntürkün, O., Karten, H.J., 1991. An immunocytochemical analysis of the lateral geniculate complex in the pigeon (Columba livia). J. Comp. Neurol. 314, 1–29.

Güntürkün, O., Kröner, S., 1999. A polysensory pathway to the forebrain of the pigeon: the ascending projections of the n. dorsolateralis posterior thalami (DLP). Eur. J. Morphol. 37, 124–128.

Güntürkün, O., Remy, M., 1990. The topographical projection of the n. isthmi pars parvocellularis (Ipc) onto the tectum opticum in the pigeon. Neurosci. Lett. 111, 18–22.

Güntürkün, O., Hellmann, B., Melsbach, G., Prior, H., 1998. Asymmetries of representation in the visual system of pigeons. Neuroreport 9, 4127–4130.

Güntürkün, O., Verhoye, M., De Groof, G., Van der Linden, A., 2013. A 3-dimensional digital atlas of the ascending sensory and the descending motor systems in the pigeon brain. Brain Struct. Funct. 281, 269–281.

Hahmann, U., Güntürkün, O., 1992. Visual discrimination deficits after lesions of the centrifugal visual system in pigeons. Vis. Neurosci. 9, 225–234.

Hahnloser, R.H.R., Kozhevnikov, A.A., Fee, M.S., 2002. An ultra-sparse code underlies the generation of neural sequences in a songbird. Nature 419, 65–70.

Hall, J.A., Foster, R.E., Ebner, F.F., Hall, W.C., 1977. Visual cortex in a reptile, the turtle (*Pseudemys scripta* and *Chrysemys picta*). Brain Res. 130, 197–216.

Halpern, M., 1987. The organization and function of the vomeronasal system. Annu. Rev. Neurosci. 10, 325–362.

Halpern, M., 1992. Nasal chemical senses in reptiles: structure and function. In: Gans, C., Crews, D. (Eds.), Biology of the Reptilia: Hormones, Brain and Behavior, vol. 18. University of Chicago Press, Chicago, pp. 423–523.

Hartline, P.H., Kass, L., Loop, M.S., 1978. Merging of modalities in the optic tectum: infrared and visual integration in rattlesnakes. Science 199, 1225–1229.

Hartmann, B., Güntürkün, O., 1998. Selective deficits in reversal learning after neostriatum caudolaterale lesions in pigeons – possible behavioral equivalencies to the mammalian prefrontal system. Behav. Brain Res. 96, 125–133.

El Hassni, M., Repérant, J., Ward, R., Bennis, M., 1997. The retinopetal visual system in the chameleon (*Chamaeleo chameleon*). J. Brain Res. 38, 453–457.

Haug, H., 1987. Brain sizes, surfaces, and neuronal sizes of the cortex cerebri: a stereological investigation of man and his variability and a comparison with some mammals (primates, whales, marsupials, insectivores, and one elephant). Am. J. Anat. 180, 126–142.

Haverkamp, S., Eldred, W.D., 1998. Localization of the origin of retinal efferents in the turtle brain and the involvement of nitric oxide synthase. J. Comp. Neurol. 393, 185–195.

Hayden, S., Teeling, E.C., 2014. The molecular biology of vertebrate olfaction. Anat. Rec. 297, 2216–2226.

Hayes, B.P., Holden, A.L., 1983. The distribution of centrifugal terminals in the pigeon retina. Exp. Brain Res. 49, 189–197.

Hayes, B.P., Webster, K.E., 1985. Cytoarchitectural fields and retinal termination: an axonal transport study of laminar organization in the avian optic tectum. Neuroscience 16, 641–657.

Hayle, T.H., 1973. A comparative study of spinal projections to the brain (except cerebellum) in three classes of poikilothermic vertebrates. J. Comp. Neurol. 149, 463–476.

Healy, S.D., Rowe, C., 2007. A critique of comparative studies of brain size. Proc. R. Soc. Biol. Sci. 274, 453–464.

Heimovics, S.A., Prior, N.H., Ma, C., Soma, K.K., February, 2016. Rapid effects of an aggressive interaction on dehydroepiandrosterone, testosterone and oestradiol levels in the male song sparrow brain: a seasonal comparison. J. Neuroendocrinol. 28 (2). https://doi.org/10.1111/jne.12345.

Hein, C.M., Engels, S., Kishkinev, D., Mouritsen, H., 2011. Robins have a magnetic compass in both eyes. Nature 471, E11–E12.

Helduser, S., Güntürkün, O., 2012. Neural substrates for serial reaction time tasks in pigeons. Behav. Brain Res. 230, 132–143.

Helduser, S., Cheng, S., Güntürkün, O., 2013. Identification of two forebrain structures that mediate execution of memorized sequences in the pigeon. J. Neurophysiol. 109, 958–968.

Hellmann, B., Güntürkün, O., 1999. Visual-field-specific heterogeneity within the tecto-rundal projection of the pigeon. Eur. J. Neurosci. 11, 2635–2650.

Hellmann, B., Güntürkün, O., 2001. The structural organization of parallel information processing within the tectofugal visual system of the pigeon. J. Comp. Neurol. 429, 94–112.

Hellmann, B., Manns, M., Güntürkün, O., 2001. The nucleus isthmi, pars semilunaris as a key component of the tectofugal visual system in the pigeon. J. Comp. Neurol. 2001 (436), 153–166.

Hellmann, B., Güntürkün, O., Manns, M., 2004. The tectal mosaic: organization of the descending tectal projections in comparison to the ascending tectofugal pathway in the pigeon. J. Comp. Neurol. 472, 395–410.

Hemmings Jr., H.C., Nairn, A.C., Bibb, J.A., Greengard, P., 1995. Signal transduction in the striatum: DARPP-32, a molecular integrator of multiple signaling pathways. In: Ariano, M.A., Surmeier, D.J.,

Landes, R.G. (Eds.), Molecular and Cellular Mechanisms of Neostriatal Function. Austin, pp. 283–297.

Herbransen, W.T., Schroeder, J., 2010. Are birds smarter than mathematicians? Pigeons (*Columba livia*) perform optimally on a version of the Monty Hall Dilemma. J. Comp. Psychol. 124, 1–13.

Herculano-Houzel, S., 2009. The human brain in numbers: a linearly scaled-up primate brain. Front. Hum. Neurosci. 3, 31.

Herculano-Houzel, S., 2011. Brains matter, bodies maybe not: the case for examining neuron numbers irrespective of body size. Annu. N.Y. Acad. Sci. 1225, 191–199.

Herculano-Houzel, S., 2011. Not all brains are made the same: new views on brain scaling in evolution. Brain Behav. Evol. 78, 22–36.

Herculano-Houzel, S., 2012. The remarkable, yet not extraordinary, human brain as a scaled-up primate brain and its associated cost. Proc. Natl. Acad. Sci. U.S.A. 109 (Suppl. 1), 10661–10668.

Herculano-Houzel, S., Lent, R., 2005. Isotropic fractionator: a simple, rapid method for the quantification of total cell and neuron numbers in the brain. J. Neurosci. 25, 2518–2521.

Herculano-Houzel, S., Collins, C.E., Wong, P., Kaas, J.H., 2007. Cellular scaling rules for primate brains. Proc. Natl. Acad. Sci. U.S.A. 104, 3562–3567.

Herculano-Houzel, S., Avelino, K., Neves, K., Porfírio, J., Messeder, D., Mattos Feijó, L., Manger, P., 2014. The elephant brain in numbers. Front. Neuroanat. 8, 46.

Herculano-Houzel, S., Manger, P.R., Kaas, J.H., 2014. Brain scaling in mammalian evolution as a consequence of concerted and mosaic changes in numbers of neurons and average neuronal cell size. Front. Neuroanat. 8, 77.

Herold, C., Diekamp, B., Güntürkün, O., 2008. Stimulation of dopamine D1 receptors in the avian fronto-striatal system adjusts daily cognitive fluctuations. Behav. Brain Res. 194, 223–229.

Herold, C., Joshi, I., Hollmann, M., Güntürkün, O., 2012. Prolonged cognitive training increases D5 receptor expression in the avian prefrontal cortex. PLoS One 7, e36484.

van den Heuvel, M.P., Bullmore, E.T., Sporns, O., 2016. Comparative connectomics. Trends Cogn. Sci. 20, 345–361.

Heyers, D., Manns, M., Luksch, H., Güntürkün, O., Mouritsen, H., 2007. A visual pathway links brain structures active during magnetic compass orientation. PLoS One e937.

Heyers, D., Zapka, M., Hoffmeister, M., Wild, J.M., Mouritsen, H., 2010. Magnetic field changes activate the trigeminal brainstem complex in a migratory bird. Proc. Natl. Acad. Sci. U.S.A. 107, 9394–9399.

Hiramoto, M., Cline, H.T., 2009. Convergence of multisensory inputs in *Xenopus* tadpole tectum. Dev. Neurobiol. 69, 959–971.

Holden, A.L., Powell, T.P.S., 1972. The functional organization of the isthmo-optic nucleus in the pigeon. J. Physiol. 233, 419–447.

Holmgren, N., 1922. Points of view concerning forebrain morphology in lower vertebrates. J. Comp. Neurol. 34, 391–459.

Hoogland, P.V., Welker, E., 1981. Telencephalic projections to the eye in *Python reticulatus*. Brain Res. 213, 173–176.

Hu, M., Naito, J., Chen, Y., Ohmori, Y., Fukuta, K., 2003. Afferent and efferent connections of the nucleus rotundus demonstrated by WGA-HRP in the chick. Anat. Histol. Embryol. 32, 335–340.

Hugall, A.F., Foster, R., Lee, M.S.Y., 2007. Calibration choice, rate smoothing, and the pattern of tetrapod diversification according to the long nuclear gene RAG-1. Syst. Biol. 56, 543–563.

Humphries, M.D., Prescott, T.J., 2010. The ventral basal ganglia, a selection mechanism at the crossroads of space, strategy, and reward. Prog. Neurobiol. 90, 385–417.

Hunt, G.R., 1996. Manufacture and use of hook-tools by New Caledonian crows. Nature 379, 249–251.

Husband, S.A., Shimizu, T., 1999. Efferent projections of the ectostriatum in the pigeon (*Columba livia*). J. Comp. Neurol. 406, 329–345.

Ikebuchi, M., Hasegawa, T., Bischof, H.J., 2009. Amygdala and sociosexual behavior in male zebra finches. Brain Behav. Evol. 74, 250–257.

Ingle, D., 1983. Brain mechanisms of visual localization by frogs and toads. In: Ewert, J.P., Capranica, R.R., Ingle, D. (Eds.), Advances in Vertebrate Neuroethology. Plenum Press, New York, pp. 177–226.

Iwaniuk, A.N., Hurd, P.L., 2005. The evolution of cerebrotypes in birds. Brain Behav. Evol. 65, 215–230.

Iwaniuk, A.N., Dean, K.M., Nelson, J.E., 2004. Interspecific allometry of the brain and brain regions in parrots (psittaciformes): comparisons with other birds and primates. Brain Behav. Evol. 65, 40–59.

Jäger, R., 1993. Lateral forebrain lesions affect pecking accuracy in the pigeon. Behav. Process. 28, 181–188.

Janik, V.M., 2014. Cetacean vocal learning and communication. Curr. Opin. Neurobiol. 28, 60–65.

Jarvis, E.D., 2004. Learned birdsong and the neurobiology of human language. Annu. N.Y. Acad. Sci. 1016, 749–777.

Jarvis, E.D., Güntürkün, O., Bruce, L., Csillag, A., Karten, H.J., Kuenzel, W., Medina, L., Paxinos, G., Perkel, D.J., Shimizu, T., Striedter, G., Wild, M., Ball, G.F., Dugas-Ford, J., Durand, S., Hough, G., Husband, S., Kubikova, L., Lee, D., Mello, C.V., Powers, A., Siang, C., Smulders, T.V., Wada, K., White, S.A., Yamamoto, K., Yu, J., Reiner, A., Butler, A.B., 2005. Avian brains and a new understanding of vertebrate brain evolution. Nat. Rev. Neurosci. 6, 151–159.

Jarvis, E.D., Yu, J., Rivas, M.V., Horita, H., Feenders, G., Whitney, O., Jarvis, S.C., Jarvis, E.R., Kubikova, L., Puck, A.E., Siang-Bakshi, C., Martin, S., McElroy, M., Hara, E., Howard, J., Pfenning, A., Mouritsen, H., Chen, C.C., Wada, K., 2013. Global view of the functional molecular organization of the avian cerebrum: mirror images and functional columns. J. Comp. Neurol. 521, 3614–3665.

Jassik-Gerschenfeld, D., Guichard, J., 1972. Visual receptive fields of single cells in the pigeon's optic tectum. Brain Res. 40, 303–317.

Jassik-Gerschenfeld, D., Hardy, O., 1979. Single neuron responses to moving sine-wave gratings in the pigeon optic tectum. Vis. Res. 19, 993–999.

Jassik-Gerschenfeld, D., Hardy, O., 1980. Single-cell responses to bar width and to sine-wave grating frequency in the pigeon. Vis. Res. 21, 745–747.

Jassik-Gerschenfeld, D., Hardy, O., 1984. The avian optic tectum: neurophysiology and behavioural correlations. In: Vanegas, H. (Ed.), Comparative Neurology of the Optic Tectum. Plenum Press, New York, pp. 649–686.

Jassik-Gerschenfeld, D., Guichard, J., Tessier, Y., 1975. Localization of directionally selective and movement sensitive cells in the optic tectum of the pigeon. Vis. Res. 15, 1037–1038.

Jerison, H., 1973. Evolution of the Brain and Intelligence. Academic Press, New York.

Jerison, H.J., 1977. The theory of encephalization. Annu. N.Y. Acad. Sci. 299, 146–160.

Jerison, H.J., Barlow, H.B., 1985. Animal intelligence as encephalization [and discussion]. Philos. Trans. R. Soc. Lond. B Biol. Sci. 308, 21–35.

Jiang, Z.D., King, A.J., Moore, D.R., 1996. Topographic organization of projection from the parabigeminal nucleus to the superior colliculus in the ferret revealed with fluorescent latex microspheres. Brain Res. 12, 217–232.

Jiao, Y., Medina, L., Veenman, C.L., Toledo, C., Puelles, L., Reiner, A., 2000. Identification of the anterior nucleus of the ansa lenticularis in birds as the homologue of the mammalian subthalamic nucleus. J. Neurosci. 20, 6998–7010.

Kaldenbach, F., Bleckmann, H., Kohl, T., 2016. Responses of infrared-sensitive tectal units of the pit viper Crotalus atrox to moving objects. J. Comp. Physiol. 202, 389–398.

Kalenscher, T., Diekamp, B., Güntürkün, O., 2003. A neural network for choice behaviour in a concurrent fixed interval schedule. Eur. J. Neurosci. 18, 2627–2637.

Kalenscher, T., Windmann, S., Rose, J., Diekamp, B., Güntürkün, O., Colombo, M., 2005. Single units in the pigeon brain integrate reward amount and time-to-reward in an impulsive choice task. Curr. Biol. 15, 594–602.

Källén, B., 1962. II. Embryogenesis of brain nuclei in the chick telencephalon. Ergeb. Anat. Entwicklungsgesch. 36, 62–82.

Kalt, T., Diekamp, B., Güntürkün, O., 1999. Single unit activity during a Go/NoGo task in the 'prefrontal cortex' of pigeons. Brain Res. 839, 263–278.

Karakuyu, D., Herold, C., Güntürkün, O., Diekamp, B., 2007. Differential increase of extracellular dopamine and serotonin in 'prefrontal cortex' and striatum of pigeons during working memory. Eur. J. Neurosci. 26, 2293–2302.

Kardong, K.V., Berkhoudt, H., 1999. Rattlesnake hunting behavior: correlations between plasticity of predatory performance and neuroanatomy. Brain Behav. Evol. 53, 20–28.

Karten, H.J., 2015. Vertebrate brains and evolutionary connectomics: on the origins of the mammalian 'neocortex'. Philos. Trans. R. Soc. Lond. B Biol. Sci. 370 (1684).

Karten, H.J., Hodos, W.J.H., 1967. A Stereotaxic Atlas of the Brain of the Pigeon (Columba livia). Johns Hopkins Press, Baltimore, MD.

Karten, H.J., Hodos, W., Nauta, W.J., Revzin, A.M., 1973. Neural connections of the "visual Wulst" of the avian telencephalon. Experimental studies in the pigeon (Columba livia) and owl (Speotyto cunicularia). J. Comp. Neurol. 150, 253–278.

Karten, H.J., Keyser, K.T., Brecha, N.C., 1990. Biochemical and morphological heterogeneity of retinal ganglion cells. In: Cohen, B., Bodis-Wollner, I. (Eds.), Vision and the Brain. Raven Press, New York, pp. 19–33.

Karten, H.J., Cox, K., Mpodozis, J., 1997. Two distinct populations of tectal neurons have unique connections within the retinotectorotundal pathway of the pigeon (Columba livia). J. Comp. Neurol. 387, 449–465.

King, S.L., Janik, V.M., 2013. Bottlenose dolphins can use learned vocal labels to address each other. Proc. Natl. Acad. Sci. U.S.A. 110, 13216–13221.

King, S.L., Sayigh, L.S., Wells, R.S., Fellner, W., Janik, V.M., 2013. Vocal copying of individually distinctive signature whistles in bottlenose dolphins. Proc. R. Soc. Biol. Sci. 280, 20130053.

Kingsbury, M.A., Jan, N., Klatt, J.D., Goodson, J.L., 2015. Nesting behavior is associated with VIP expression and VIP-Fos colocalization in a network-wide manner. Horm. Behav. 69, 68–81.

Kirsch, J., Güntürkün, O., Rose, J., 2008. Insight without cortex: lessons from the avian brain. Conscousnes Cogn 17, 475–483.

Kishkinev, D., Chernetsov, N., Heyers, D., Mouritsen, H., 2013. Migratory reed warblers need intact trigeminal nerves to correct for a 1000 km eastward displacement. PLoS One 8, e65847.

Knipling, R.R., 1978. No deficit in near-field visual acuity of pigeons after transection of the isthmo-optic tract. Brain Res. 22, 813–816.

Knörnschild, M., 2014. Vocal production learning in bats. Curr. Opin. Neurobiol. 28, 80–85.

Knudsen, E.I., Knudsen, P.F., 1996. Contribution of the forebrain archistriatal gaze fields to auditory orienting behaviour in the barn owl. Exp. Brain Res. 108, 23–32.

Knudsen, E.I., Cohen, Y.E., Masino, T., 1995. Characterization of a forebrain gaze field in the archistriatum of the barn owl: microstimulation and anatomical connections. J. Neurosci. 15, 5139–5151.

Kohl, T., Colayori, S.E., Westhoff, G., Bakken, G.S., Young, B.A., 2012. Directional sensitivity in the thermal response of the facial pit in western diamondback rattlesnakes (Crotalus atrox). J. Exp. Biol. 215, 2630–2636.

Konishi, M., 2010. From central pattern generator to sensory template in the evolution of birdsong. Brain Lang. 115, 18–20.

Korzeniewska, E., Güntürkün, O., 1990. Sensory properties and afferents of the n. dorsolateralis posterior thalami (DLP) of the pigeon. J. Comp. Neurol. 292, 457–479.

Krebs, J.R., Sherry, D.F., Healy, S.D., Perry, V.H., Vaccarino, A.L., 1989. Hippocampal specialization of food-storing birds. Proc. Natl. Acad. Sci. U.S.A. 86, 1388—1392.

Kriegstein, R., Connors, B.W., 1986. Cellular physiology of the turtle visual cortex: synaptic properties and intrinsic circuitry. J. Neurosci. 6, 178—191.

Krochmal, A.R., Bakken, G.S., 2003. Thermoregulation in the pits: use of thermal radiation for retreat site selection by rattlesnakes. J. Exp. Biol. 206, 2539—2545.

Kröner, S., Güntürkün, O., 1999. Afferent and efferent connections of the caudolateral neostriatum in the pigeon (Columba livia): a retro- and anterograde pathway tracing study. J. Comp. Neurol. 407, 228—260.

Kröner, S., Gottmann, K., Hatt, H., Güntürkün, O., 2002. Cell types within the neostriatum caudolaterale of the chick: intrinisic electrophysiological and anatomical properties. Neuroscience 110, 473—495.

Kruger, L., Maxwell, D.S., 1969. Wallerian degeneration in the optic nerve of a reptile: an electron microscopic study. Am. J. Anat. 125, 247—269.

Krützfeldt, N.O.E., Wild, J.M., 2005. Definition and novel connections of the entopallium in the pigeon (Columba livia). J. Comp. Neurol. 490, 40—56.

Kuenzel, W.J., Medina, L., Csillag, A., Perkel, D.J., Reiner, A., 2011. The avian subpallium: new insights into structural and functional subdivisions occupying the lateral subpallial wall and their embryological origins. Brain Res. 1424, 67—101.

Künzle, H., Schnyder, H., 1984. The isthmus-tegmentum complex in the turtle and rat: a comparative analysis of its interconnections with the optic tectum. Exp. Brain Res. 3, 509—522.

Künzle, H., Woodson, W., 1982. Mesodiencephalic and other target regions of ascending spinal projections in the turtle, Pseudemys scripta elegans. J. Comp. Neurol. 212, 349—364.

Lanuza, E., Halpern, M., 1997. Afferent and efferent connections of the nucleus sphericus in the snake Thamnophis sirtalis: convergence of olfactory and vomeronasal information in the lateral cortex and the amygdala. J. Comp. Neurol. 385, 627—640.

Lanuza, E., Halpern, M., 1998. Efferents and centrifugal afferents of the main and accessory olfactory bulbs in the snake Thamnophis sirtalis. Brain Behav. Evol. 51, 1—22.

Laude, J.R., Pattison, K.F., Rayburn-Reeves, R.M., Michler, D.M., Zentall, T.R., 2016. Who are the real bird brains? Qualitative differences in behavioral flexibility between dogs (Canis familiaris) and pigeons (Columba livia). Anim. Cogn. 19, 163—169.

Laverghetta, A.V., Shimizu, T., 1999. Visual discrimination in the pigeon (Columba livia): effects of selective lesions of the nucleus rotundus. Neuroreport 10, 981—985.

Leal, M., Powell, B.J., 2012. Behavioural flexibility and problem-solving in a tropical lizard. Biol. Lett. 8, 28—30.

Lefebvre, L., Nicolakakis, N., Boire, D., 2002. Tools and brains in birds. Behaviour 139, 939—973.

Lefebvre, L., Reader, S.M., Sol, D., 2013. Innovating innovation rate and its relationship with brains, ecology and general intelligence. Brain Behav. Evol. 81, 143—145.

Leitch, D.B., Catania, K.C., 2012. Structure, innervation and response properties of integumentary sensory organs in crocodilians. J. Exp. Biol. 215, 4217—4230.

Lengersdorf, D., Stüttgen, M.C., Uengoer, M., Güntürkün, O., 2014. Transient inactivation of the pigeon hippocampus or the nidopallium caudolaterale during extinction learning impairs extinction retrieval in an appetitive conditioning paradigm. Behav. Brain Res. 265, 93—100.

Leppelsack, H.-J., 1974. FunktioneUe Eigenschaften der Hörbahn im Feld L des Neostriatum caudale des Staren (Sturnus vulgaris L., Aves). J. Comp. Physiol. 88, 271—320.

Letzner, S., Simon, A., Güntürkün, O., 2016. The connectivity and the neurochemistry of the commissura anterior of the pigeon (Columba livia). J. Comp. Neurol. 524, 343—361.

Leuret, F., Gratiolet, L.P., 1839. Anatomie comparée du système nerveux, considéré dans ses rapports avec l'intelligence, vol. 1. Baillière, Paris.

Leuret, F., Gratiolet, L.P., 1857. Anatomie comparée du système nerveux, considéré dans ses rapports avec l'intelligence. In: Atlas de 32 planches dessignées d'après nature et gravées, vol. 2. Baillière, Paris.

Leutgeb, S., Husband, S., Riters, L.V., Shimizu, T., Bingman, V.P., 1996. Telencephalic afferents to the caudolateral neostriatum of the pigeon. Brain Res. 730, 173—181.

Li, J.L., Xiao, Q., Fu, Y.X., Wang, S.R., 1998. Centrifugal innervation modulates visual activity of tectal cells in pigeons. Vis. Neurosci. 15, 411—415.

Liedvogel, M., Mouritsen, H., 2010. Cryptochromes — a potential magnetoreceptor: what do we know and what do we want to know? J. R. Soc. Interface 7, S147—S162.

Liedvogel, M., Feenders, G., Wada, K., Troje, N.F., Jarvis, E.D., Mouritsen, H., 2007. Lateralized activation of Cluster N in the brains of migratory songbirds. Eur. J. Neurosci. 25, 1166—1173.

Lindstrom, S.H., Nacsa, N., Blankenship, T., Fitzgerald, P.G., Weller, C., Vaney, D.I., Wilson, M., 2009. Distribution and structure of efferent synapses in the chicken retina. Vis. Neurosci. 26, 215—226.

Lindstrom, S.H., Azizi, N., Weller, C., Wilson, M., 2010. Retinal input to efferent target amacrine cells in the avian retina. Vis. Neurosci. 27, 103—118.

Lissek, S., Güntürkün, O., 2003. Dissociation of extinction and behavioral disinhibition — the role of NMDA receptors in the pigeon associative forebrain during extinction. J. Neurosci. 23, 8119—8124.

Lissek, S., Güntürkün, O., 2004. Maintenance in working memory or response selection — functions of NMDA receptors in the pigeon 'prefrontal cortex'. Behav. Brain Res. 153, 497—506.

Lissek, S., Güntürkün, O., 2005. Out of context — NMDA receptor antagonism in the avian "prefrontal cortex" impairs context processing in a conditional discrimination task. Behav. Neurosci. 119, 797—805.

Liu, G.B., Pettigrew, J.D., 2003. Orientation mosaic in barn owl's visual Wulst revealed by optical imaging: comparison with cat and monkey striate and extra-striate areas. Brain Res. 961, 153—158.

Liu, R.F., Niu, Y.Q., Wang, S.R., 2008. Thalamic neurons in the pigeon compute distance-to-collision of an approaching surface. Brain Behav. Evol. 72, 37—47.

Lohman, A.H., Smeets, W.J., 1993. Overview of the main and accessory olfactory bulb projections in reptiles. Brain Behav. Evol. 41, 147—155.

Lohman, A.H., van Woerden-Verkley, I., 1978. Ascending connections to the forebrain in the Tegu lizard. J. Comp. Neurol. 182, 555—574.

Luksch, H., 2003. Cytoarchitecture of the avian optic tectum: neuronal substrate for cellular computation. Rev. Neurosci. 14, 85—106.

Luksch, H., Cox, K., Karten, H.J., 1998. Bottlebrush dendritic endings and large dendritic fields: motion-detecting neurons in the tectofugal pathway. J. Comp. Neurol. 396, 399—414.

Luksch, H., Karten, H.J., Kleinfeld, D., Wessel, R., 2001. Chattering and differential signal processing in identified motion sensitive neurons of parallel visual pathways in chick tectum. J. Neurosci. 21, 6440—6446.

Lundqvist, M., Rose, J., Herman, P., Brincat, S.L., Buschman, T.J., Miller, E.K., 2016. Gamma and beta bursts underlie working memory. Neuron 90, 152—164.

Lyson, T.R., Bever, G.S., Bhullar, B.A., Joyce, W.G., Gauthier, J.A., 2010. Transitional fossils and the origin of turtles. Biol. Lett. 6, 830—833.

Mancilla, J.G., Fowler, M., Ulinski, P.S., 1998. Responses of regular spiking and fast spiking cells in turtle visual cortex to light flashes. Vis. Neurosci. 15, 979—993.

Manley, G.A., 1970. Frequency sensitivity of auditory neurons in the Caiman cochlear nucleus. Z. Vgl. Physiol. 66, 251—266.

Manley, J., 1971. Single unit studies in the midbrain auditory area of Caiman. Z. Vgl. Physiol. 71, 255–261.

Marín, G., Letelier, J.C., Henny, P., Sentis, E., Farfán, G., Fredes, F., Pohl, N., Karten, H.J., Mpodozis, J., 2003. Spatial organization of the pigeon tectorotundal pathway: an interdigitating topographic arrangement. J. Comp. Neurol. 458, 361–380.

Marín, G., Salas, C., Sentis, E., Rojas, X., Letelier, J.C., Mpodozis, J., 2007. A cholinergic gating mechanism controlled by competitive interactions in the optic tectum of the pigeon. J. Neurosci. 27, 8112–8121.

Marín, G.J., Durán, E., Morales, C., González-Cabrera, C., Sentis, E., Mpodozis, J., Letelier, J.C., 2012. Attentional capture? Synchronized feedback signals from the isthmi boost retinal signals to higher visual areas. J. Neurosci. 32, 1110–1122.

Martin, R.D., 1981. Relative brain size and basal metabolic rate in terrestrial vertebrates. Nature 293, 57–60.

Martínez-Marcos, A., Ubeda-Bañón, I., Halpern, M., 2001. Neural substrates for tongue-flicking behavior in snakes. J. Comp. Neurol. 432, 75–87.

Martinez-Marcos, A., Lanuza, E., Halpern, M., 2002. Neural substrates for processing chemosensory information in snakes. Brain Res. Bull. 57, 543–546.

Martinez-Marcos, A., Ubeda-Bañon, I., Lanuza, E., Halpern, M., 2005. Efferent connections of the "olfactostriatum": a specialized vomeronasal structure within the basal ganglia of snakes. J. Chem. Neuroanat. 29, 217–226.

Mazurskaya, P.Z., 1973. Organization of receptive fields in the forebrain of *Emys orbicularis*. Neurosci. Behav. Physiol. 6, 311–318.

Medina, L., Reiner, A., 1997. The efferent projections of the dorsal and ventral pallidal parts of the pigeon basal ganglia, studied with biotinylated dextran. Neuroscience 81, 773–802.

Medina, L., Reiner, A., 2000. Do birds possess homologues of mammalian primary visual, somatosensory and motor cortices? Trends Neurosci. 23, 1–12.

Medina, L., Smeets, W.J.A.J., 1991. Comparative aspects of the basal ganglia-tectal pathways in reptiles. J. Comp. Neurol. 308, 614–629.

Medina, L., Veenman, C.L., Reiner, A., 1997. Evidence for a possible avian dorsal thalamic region comparable to the mammalian ventral anterior, ventral lateral, and oral ventroposterolateral nuclei. J. Comp. Neurol. 384, 86–108.

Médina, M., Repérant, J., Ward, R., Miceli, D., 2004. Centrifugal visual system of *Crocodylus niloticus*: a hodological, histochemical, and immunocytochemical study. J. Comp. Neurol. 468, 65–85.

Mehlhorn, J., Hunt, G.R., Gray, R.D., Rehkämper, G., Güntürkün, O., 2010. Tool-making new caledonian crows have large associative brain areas. Brain Behav. Evol. 75, 63–70.

Miceli, D., Gioanni, H., Repérant, J., Peyrichoux, J., 1979. The avian visual Wulst: I. An anatomical study of afferent and efferent pathways. II. An electrophysiological study of the functional properties of single neurons. In: Granda, A.M., Maxwell, J.H. (Eds.), Neural Mechanisms of Behavior in Birds. Plenum Press, New York.

Miceli, D., Marchand, L., Repérant, J., Rio, J.-P., 1990. Projections of the dorsolateral anterior complex and adjacent thalamic nuclei upon the visual Wulst in the pigeon. Brain Res. 518, 317–323.

Miceli, D., Repérant, J., Rio, J.-R., Medina, M., 1995. GABA immunoreactivity in the nucleus isthmo-opticus of the centrifugal visual system in the pigeon: a light and electron microscopic study. Vis. Neurosci. 12, 425–441.

Mikolasch, S., Kotrschal, K., Schloegl, C., 2011. African grey parrots (*Psittacus erithacus*) use inference by exclusion to find hidden food. Biol. Lett. 7, 875–877.

Miles, F.A., 1972. Centrifugal control of the avian retina. I. Receptive field properties of retinal ganglion cells. Brain Res. 48, 65–92.

Mink, J.W., 2003. The basal ganglia and involuntary movements: impaired inhibition of competing motor patterns. Arch. Neurol. 60, 1365–1368.

Mittelstaedt, H., 1964. Basic control patterns of orientational homeostasis. Symp. Soc. Exp. Biol. 18, 365–385.

Mogensen, J., Divac, I., 1982. The prefrontal "cortex" in the pigeon. Behavioral evidence. Brain Behav. Evol. 21, 60–66.

Moll, F.W., Nieder, A., 2015. Cross-modal associative mnemonic signals in crow endbrain neurons. Curr. Biol. 25, 2196–2201.

Montiel, J.F., Vasistha, N.A., Garcia-Moreno, F., Molnar, Z., 2016. From sauropsids to mammals and back: new approaches to comparative cortical development. J. Comp. Neurol. 524, 630–645.

Mooney, R., 1992. Synaptic basis for developmental plasticity in a birdsong nucleus. J. Neurosci. 12.

Mooney, R., 2014. Auditory–vocal mirroring in songbirds. Philos. Trans. R. Soc. B 369, 20130179.

Mooney, R., Rao, M., 1994. Waiting periods versus early innervation: the development of axonal connections in the zebra finch song system. J. Neurosci. 14, 6532–6543.

Moorman, S., Mello, C.V., Bolhuis, J.J., 2011. From songs to synapses: molecular mechanisms of birdsong memory. Molecular mechanisms of auditory learning in songbirds involve immediate early genes, including zenk and arc, the ERK/MAPK pathway and synapsins. Bioessays 33, 377–385.

Mora, C.V., Davison, M., Wild, J.M., Walker, M.M., 2004. Magnetoreception and its trigeminal mediation in the homing pigeon. Nature 432, 508–511.

Mouritsen, H., Janssen-Bienhold, U., Liedvogel, M., Feenders, G., Stalleicken, J., Dirks, P., Weiler, R., 2004. Cryptochromes and neuronal-activity markers colocalize in the retina of migratory birds during magnetic orientation. Proc. Natl. Acad. Sci. U.S.A. 101, 14294–14299.

Mouritsen, H., Feenders, G., Liedvogel, M., Wada, K., Jarvis, E.D., 2005. Night vision brain area in migratory songbirds. Proc. Natl. Acad. Sci. U.S.A. 102, 8339–8344.

Mouritsen, H., Heyers, D., Güntürkün, O., 2016. The neural basis of long-distance navigation in birds. Annu. Rev. Physiol. 78, 133–154.

Mpodozis, J., Letelier, J.-C., Concha, M.L., Maturana, H., 1995. Conduction velocity groups in the retino-tectal and retino-thalamic visual pathways of the pigeon (*Columba livia*). Int. J. Neurosci. 81, 123–136.

Müller, J.R., Metha, A.B., Krauskopf, J., Lennie, P., 1999. Rapid adaptation in visual cortex to the structure of images. Science 285, 1405–1408.

Mulligan, K.A., Ulinski, P.S., 1990. Organization of geniculocortical projections in turtles: isoazimuth lamellae in the visual cortex. J. Comp. Neurol. 296, 531–547.

Muñoz, A., Muñoz, M., González, A., ten Donkelaar, H.J., 1997. Spinal ascending pathways in amphibians: cells of origin and main targets. J. Comp. Neurol. 378, 205–228.

Naumann, R.K., Ondracek, J.M., Reiter, S., Shein-Idelson, M., Tosches, M.A., Yamawaki, T.M., Laurent, G., 2015. The reptilian brain. Curr. Biol. 25, R317–R321.

Necker, R., 1991. Cells of origin of avian postsynaptic dorsal column pathways. Neurosci. Lett. 126, 91–93.

Necker, R., 1999. Specializations in the lumbosacral spinal cord of birds: morphological and behavioural evidence for a sense of equilibrium. Eur. J. Morphol. 37, 211–214.

Necker, R., 2000. The avian ear and hearing. In: Whittow, G.C. (Ed.), Sturkie's Avian Physiology. Academic Press, Orlando, pp. 21–38.

Necker, R., 2000. The somatosensory system. In: Whittow, G.C. (Ed.), Sturkie's Avian Physiology. Academic Press, Orlando, pp. 57–69.

Necker, R., 2002. Mechanosensitivity of spinal accessory lobe neurons in the pigeon. Neurosci. Lett. 320, 53–56.

Necker, R., 2005. Are paragriseal cells in the avian lumbosacral spinal cord displaced ventral spinocerebellar neurons? Neurosci. Lett. 382, 56–60.

Necker, R., 2005. The structure and development of avian lumbosacral specializations of the vertebral canal and the spinal cord with special reference to a possible function as a sense organ of equilibrium. Anat. Embryol. 210, 59–74.

Necker, R., 2006. Specializations in the lumbosacral vertebral canal and spinal cord of birds: evidence of a function as a sense organ which is involved in the control of walking. J. Comp. Physiol. 192, 439–448.

Necker, R., Janßen, A., Beissenhirtz, T., 2000. Behavioral evidence of the role of lumbosacral anatomical specializations in pigeons in maintaining balance during terrestrial locomotion. J. Comp. Physiol. 186, 409–412.

Neuenschwander, S., Varela, F.J., 1993. Visually triggered neuronal oscillations in the pigeon: an autocorrelation of tectal activity. Eur. J. Neurosci. 5, 870–881.

Neves, K., Meireles Ferreira, F., Tovar-Moll, F., Gravett, N., Bennett, N.C., Kaswera, C., Herculano-Houzel, S., 2014. Cellular scaling rules for the brain of afrotherians. Front. Neuroanat. 8, 5.

Newman, E.A., Hartline, P.H., 1981. Integration of visual and infrared information in bimodal neurons in the rattlesnake optic tectum. Science 213, 789–791.

Ng, B.S.W., Grabska-Barwinska, A., Güntürkün, O., Jancke, D., 2010. Dominant vertical orientation processing without clustered maps: early visual brain dynamics imaged with voltage-sensitive dye in the pigeon visual wulst. J. Neurosci. 30, 6713–6725.

Ngo, T.D., Davies, D.C., Egedi, G.Y., Tömböl, T., 1994. A phaseolus lectin anterograde tracing study of the tectorotundal projections in the domestic chick. J. Anat. 184, 129–136.

Ngwenya, A., Patzke, N., Spocter, M.A., Kruger, J.L., Dell, L.A., Chawana, R., Mazengenya, P., Billings, B.K., Olaleye, O., Herculano-Houzel, S., Manger, P.R., 2013. The continuously growing central nervous system of the Nile crocodile (Crocodylus niloticus). Anat. Rec. Hob. 296, 1489–1500.

Ngwenya, A., Patzke, N., Manger, P.R., Herculano-Houzel, S., 2016. Continued growth of the central nervous system without mandatory addition of neurons in the nile crocodile (Crocodylus niloticus). Brain Behav. Evol. 87, 19–38.

Nieder, A., Wagner, H., 1999. Perception and neuronal coding of subjective contours in the owl. Nat. Neurosci. 2, 660–663.

Nieder, A., Wagner, H., 2001. Hierarchical processing of horizontal disparity information in the visual forebrain of behaving owls. J. Neurosci. 21, 4514–4522.

Nieuwenhuys, R., 1998. Morphogenesis and general structure. In: Nieuwenhuys, R., Ten Donkelaar, H.J., Nicholson, C. (Eds.), The Central Nervous System of Vertebrates. Springer, Berlin, pp. 158–228.

Nieuwenhuys, R., Ten Donkelaar, H.J., Nicholson, C. (Eds.), 1998. The Central Nervous System of Vertebrates. Springer, Berlin.

Nishizawa, K., Izawa, E.I., Watanabe, S., 2011. Neural-activity mapping of memory-based dominance in the crow: neural networks integrating individual discrimination and social behaviour control. Neuroscience 197, 307–319.

Nomura, T., Takahashi, M., Hara, Y., Osumi, N., 2008. Patterns of neurogenesis and amplitude of Reelin expression are essential for making a mammalian-type cortex. PLoS One 3, e1454.

Northcutt, R.G., 1981. Evolution of the telencephalon in nonmammals. Annu. Rev. Neurosci. 4, 301–350.

Northcutt, R.G., 2012. Variation in reptilian brains and cognition. Brain Behav. Evol. 82, 45–54.

Northcutt, R.G., 2013. Variation in reptilian brains and cognition. Brain Behav. Evol. 82, 45–54.

Nottebohm, F., Liu, W.-C., 2010. The origins of vocal learning: new sounds, new circuits, new cells. Brain Lang. 115, 3–17.

Nottebohm, F., Stokes, T.M., Leonard, C.M., 1976. Central control of song in the canary, Serinus canarius. J. Comp. Neurol. 165, 457–486.

Nowicki, S., Searcy, W.A., 2014. The evolution of vocal learning. Curr. Opin. Neurobiol. 28, 48–53.

Nowicki, S., Peters, S., Podos, J., 1998. Song learning, early nutrition and sexual selection in songbirds. Am. Zool. 38, 179–190.

Nowicki, S., Hasselquist, D., Bensch, S., Peters, S., 2000. Nestling growth and song repertoire size in great reed warblers: evidence for song learning as an indicator mechanism in mate choice. Proc. Biol. Sci. R. Soc. 267, 2419–2424.

Nowicki, S., Searcy, W.A., Peters, S., 2002. Brain development, song learning and mate choice in birds: a review and experimental test of the "nutritional stress hypothesis". J. Comp. Physiol. A Neuroethol. Sens. Neural. Behav. Physiol. 188, 1003–1014.

Olkowicz, S., Kocourek, M., Lučan, R.K., Porteš, M., Fitch, W.T., Herculano-Houzel, S., Němec, P., 2016. Birds have primate-like numbers of neurons in the forebrain. Proc. Natl. Acad. Sci. U. S. A 113 (26), 7255–7260.

Ölveczky, B.P., Andalman, A.S., Fee, M.S., Schultz, W., 2005. Vocal experimentation in the juvenile songbird requires a basal ganglia circuit. PLoS Biol. 3, e153.

Papi, F., Casini, G., 1990. Pigeons with ablated pyriform cortex home from familiar but not from unfamiliar sites. Proc. Natl. Acad. Sci. U.S.A. 87, 3783–3787.

Papi, F., Fiore, L., Fiaschi, V., Benvenuti, S., 1971. The influence of olfactory nerve section on the homing capacity of carrier pigeons. Monit. Zool. Ital. 5, 265–267.

Patzke, N., Manns, M., Güntürkün, O., Ioale, P., Gagliardo, A., 2010. Navigation induced ZENK expression in the olfactory system of pigeons (Columba livia). Eur. J. Neurosci. 31, 2062–2072.

Patzke, N., Manns, M., Güntürkün, O., 2011. Telencephalic organisation of the olfactory system in homing pigeons (Columba livia). Neuroscience 194, 53–61.

Peng, K., Feng, Y., Zhang, G., Liu, H., Song, H., 2010. Anatomical study of the brain of the African ostrich. Turkish J. Vet. Anim. Sci. 34, 235–241.

Penn, D.C., Povinelli, D.J., 2007. On the lack of evidence that nonhuman animals possess anything remotely resembling a 'theory of mind'. Philos. Trans. R. Soc. B 362, 731–744.

Pepperberg, I.M., 1999. The Alex Studies. Harvard University Press, Cambridge, MA.

Pepperberg, I.M., Koepke, A., Livingston, P., Girard, M., Hartsfield, L.A., 2013. Reasoning by inference: further studies on exclusion in grey parrots (Psittacus erithacus). J. Comp. Psychol. 127, 272–281.

Petkov, C.I., Jarvis, E.D., 2012. Birds, primates, and spoken language origins: behavioral phenotypes and neurobiological substrates. Front. Evol. Neurosci. 4, 12.

Pettigrew, J.D., 1979. Binocular visual processing in the owl's telencephalon. Proc. R. Soc. Lond. B Biol. Sci. 204, 435–454.

Pettigrew, J.D., Konishi, M., 1976. Neurons selective for orientation and binocular disparity in the visual Wulst of the barn owl (Tyto alba). Science 193, 675–678.

Pfenning, A.R., Hara, E., Whitney, O., Rivas, M.V., Wang, R., Roulhac, P.L., Howard, J.T., Wirthlin, M., Lovell, P.V., Ganapathy, G., Mouncastle, J., Moseley, M.A., Thompson, J.W., Soderblom, E.J., Iriki, A., Kato, M., Gilbert, M.T.P., Zhang, G., Bakken, T., Bongaarts, A., Bernard, A., Lein, E., Mello, C.V., Hartemink, A.J., Jarvis, E.D., 2014. Convergent transcriptional specializations in the brains of humans and song-learning birds. Sciience (New York, N.Y.) 346, 1256846.

Phan, M.L., Pytte, C.L., Vicario, D.S., 2006. Early auditory experience generates long-lasting memories that may subserve vocal learning in songbirds. Proc. Natl. Acad. Sci. U.S.A. 103, 1088–1093.

Pollok, B., Prior, H., Güntürkün, O., 2000. Development of object-permanence in the food-storing magpie (Pica pica). J. Comp. Psychol. 114, 148–157.

Ponzi, A., 2008. Dynamical model of salience gated working memory, action selection and reinforcement based on basal ganglia and dopamine feedback. Neural Netw. 21, 322–330.

Poole, J.H., Tyack, P.L., Stoeger-Horwath, A.S., Watwood, S., 2005. Animal behaviour: elephants are capable of vocal learning. Nature 434, 455–456.

Porciatti, V., Fontanesi, G., Rafaelli, A., Bagnoli, P., 1990. Binocularity in the little owl, *Athene noctua*. II. Properties of visually evoked potentials from the Wulst in response to monocular and binocular stimulation with sine wave gratings. Brain Behav. Evol. 35, 40–48.

Prather, J.F., 2013. Auditory signal processing in communication: perception and performance of vocal sounds. Hear. Res. 305, 144–155.

Prather, J.F., Peters, S., Nowicki, S., Mooney, R., 2010. Persistent representation of juvenile experience in the adult songbird brain. J. Neurosci. 30, 10586–10598.

Prior, H., Schwartz, A., Güntürkün, O., 2008. Mirror induced behaviour in the magpie *Pica pica*: evidence of self-recognition. PLoS Biol. 6, e202.

Pritz, M.B., 1974. Ascending connections of a midbrain auditory area in a crocodile *Caiman crocodilus*. J. Comp. Neurol. 153, 179–198.

Pritz, M.B., 1974. Ascending connections of a thalamic auditory area in a crocodile *Caiman crocodilus*. J. Comp. Neurol. 153, 199–214.

Pritz, M.B., 1975. Anatomical identification of a telencephalic visual area in crocodiles: ascending connections of nucleus rotundus in *Caiman crocodilus*. J. Comp. Neurol. 164, 323–338.

Pritz, M.B., 1980. Parallels in the organization of auditory and visual systems in crocodiles. In: Ebbesson, S. (Ed.), Comparative Neurology of the Telencephalon. Plenum Press, New York, pp. 331–342.

Pritz, M.B., 1997. Some morphological features of a visual thalamic nucleus in a reptile: observations on nucleus rotundus in *Caiman crocodilus*. Brain Behav. Evol. 49, 237–248.

Pritz, M.B., Northcutt, R.G., 1977. Succinate dehydrogenase activity in the telencephalon of crocodiles correlates with the projection areas of sensory thalamic nuclei. Brain Res. 124, 357–360.

Pritz, M.B., Northcutt, R.G., 1980. Anatomical evidence for an ascending somatosensory pathway to the telencephalon in crocodiles, *Caiman crocodilus*. Exp. Brain Res. 40, 342–345.

Pritz, M.B., Siadati, A., 1999. Calcium binding protein immunoreactivity in nucleus rotundus in a reptile, *Caiman crocodilus*. Brain Behav. Evol. 53, 277–287.

Pritz, M.B., Stritzel, M.E., 1989. Reptilian somatosensory midbrain: identification based on input from the spinal cord and dorsal column nucleus. Brain Behav. Evol. 33, 1–14.

Pritz, M.B., Stritzel, M.E., 1994. Anatomical identification of a telencephalic somatosensory area in a reptile, *Caiman crocodilus*. Brain Behav. Evol. 43, 107–127.

Pritz, M.B., Stritzel, M.E., 1994. Anatomical identification of a postsynaptic dorsal column system in a reptile, *Caiman crocodilus*. Brain Behav. Evol. 43, 233–243.

Prum, R.O., Berv, J.S., Dornburg, A., Field, D.J., Townsend, J.P., Lemmon, E.M., Lemmon, A.R., 2015. A comprehensive phylogeny of birds (Aves) using targeted next-generation DNA sequencing. Nature 526, 569–573.

Puelles, L., Kuwana, E., Puelles, E., Bulfone, A., Shimamura, K., Keleher, J., Smiga, S., Rubenstein, J.L.R., 2000. Pallial and subpallial derivatives in the embryonic chick and mouse telencephalon, traced by the expression of the genes Dlx-2, Emx-1, Nkx-2.1, Pax-6, and Tbr-1. J. Comp. Neurol. 424, 409–438.

Puelles, L., et al., 2016. Selective early expression of the orphan nuclear receptor Nr4a2 identifies the claustrum homologue in the avian mesopallium: impact on sauropsidian/mammalian pallium comparisons. J. Comp. Neurol. 524, 665–703.

Puig, M.V., Rose, J., Schmidt, R., Freund, N., 2014. Dopamine modulation of learning and memory in the prefrontal cortex: insights from studies in primates, rodents, and birds. Front. Neural Circuits 8, 93.

Raby, C.R., Alexis, D.M., Dickinson, A., Clayton, N.S., 2007. Planning for the future by Western scrub-jays. Nature 445, 919–921.

Rehkämper, G., Frahm, H.D., Zilles, K., 1991. Quantitative development of brain and brain structures in birds (galliformes and passeriformes) compared to that in mammals (insectivores and primates). Brain Behav. Evol. 37, 125–143.

Rehkämper, G., Schuchmann, K.L., Schleicher, A., Zilles, K., 1991. Encephalization in hummingbirds (Trochilidae). Brain Behav. Evol. 37, 85–91.

Reichmuth, C., Casey, C., 2014. Vocal learning in seals, sea lions, and walruses. Curr. Opin. Neurobiol. 28, 66–71.

Reiner, A., 1994. Laminar distribution of the cells of origin of ascending and descending tectofugal pathways in turtles: implications for the evolution of tectal lamination. Brain Behav. Evol. 43, 254–292.

Reiner, A., 2002. Functional circuitry of the avian basal ganglia: implications for basal ganglia organization in stem amniotes. Brain Res. Bull. 57, 513–528.

Reiner, A., Karten, H.J., 1985. Comparison of olfactory bulb projections in pigeons and turtles. Brain Behav. Evol. 27, 11–27.

Reiner, A., Northcutt, R.G., 2000. Succinic dehydrogenase histochemistry reveals the location of the putative primary visual and auditory areas within the dorsal ventricular ridge of *Sphenodon punctatus*. Brain Behav. Evol. 55, 26–36.

Reiner, A., Powers, A.S., 1980. The effects of extensive forebrain lesions on visual discriminative performance in turtles (*Chrysemys picta picta*). Brain Res. 192, 327–337.

Reiner, A., Brecha, N.C., Karten, H.J., 1982. Basal ganglia pathways to the tectum: the afferent and efferent connections of the lateral spiriform nucleus of pigeon. J. Comp. Neurol. 208, 16–36.

Reiner, A., Karten, H.J., Brecha, N.C., 1982. Enkephalin-mediated basal ganglia influences over the optic tectum: immunohistochemistry of the tectum and the lateral spiriform nucleus in pigeon. J. Comp. Neurol. 208, 37–53.

Reiner, A., Zhang, D., Eldred, W.D., 1996. Use of the sensitive anterograde tracer cholera toxin fragment B reveals new details of the central retinal projections in turtles. Brain Behav. Evol. 48, 307–337.

Reiner, A., Medina, L., Veenman, C., 1998. Structural and functional evolution of the basal ganglia in vertebrates. Brain Res. Rev. 28, 235–285.

Reiner, A., Bruce, L., Butler, A., Csillag, A., Kuenzel, W., Medina, L., Paxinos, G., Perkel, D., Powers, A., Shimizu, T., Striedter, G., Wild, M., Ball, G., Durand, S., Güntürkün, O., Lee, D., Mello, C., White, S., Hough, G., Kubikova, L., Smulders, T., Wada, K., Dugas-Ford, J., Husband, S., Yamamoto, K., Yu, J., Siang, C., Jarvis, E.D., 2004. Revised nomenclature for avian telencephalon and some related brainstem nuclei. J. Comp. Neurol. 473, 377–414.

Reiner, A., Perkel, D., Mello, C., Jarvis, J.D., 2004. Songbirds and the revised avian brain nomenclature. In: Zeigler, H.P., Marler, P.R. (Eds.), Behavioral Neurobiology of BirdsongAnnals of the New York Academy of Sciences, vol. 1016, pp. 77–108.

Reiner, A., Yamamoto, K., Karten, H.J., 2005. Organization and evolution of the avian forebrain. Anat. Rec. Part A Discov. Mol. Cell. Evol. Biol. 287, 1080–1102.

Remy, M., Güntürkün, O., 1991. Retinal afferents of the tectum opticum and the nucleus opticus principalis thalami in the pigeon. J. Comp. Neurol. 305, 57–70.

Repérant, J., Angaut, P., 1977. The retinotectal projections in the pigeon. An experimental optical and electron microscope study. Neuroscience 2, 119–140.

Repérant, J., Gallego, A., 1976. Fibres centrifuges dans la rétine humaine. Arch. Anat. Microsc. Morphol. Exp. 65, 103–120.

Repérant, J., Ward, R., Miceli, D., Rio, J.P., Médina, M., Kenigfest, N.B., Vesselkin, N.P., 2006. The centrifugal visual system of vertebrates: a comparative analysis of its functional anatomical organization. Brain Res. Rev. 52, 1–57.

2. The Brains of Fish, Amphibians, Reptiles and Birds

Repérant, J., Médina, M., Ward, R., Miceli, D., Kenigfest, N.B., Rio, J.P., Vesselkin, N.P., 2007. The evolution of the centrifugal visual system of vertebrates. A cladistic analysis and new hypotheses. Brain Res. Rev. 53, 161–197.

Ridgway, S.H., Hanson, A.C., 2014. Sperm whales and killer whales with the largest brains of all toothed whales show extreme differences in cerebellum. Brain Behav. Evol. 83, 266–274.

Ritz, T., Adem, S., Schulten, K., 2000. A model for photoreceptor-based magnetoreception in birds. Biophys. J. 78, 707–718.

Roberts, T.F., Mooney, R., 2013. Motor circuits help encode auditory memories of vocal models used to guide vocal learning. Hear. Res. 303, 48–57.

Roberts, T.F., Gobes, S.M.H., Murugan, M., Olveczky, B.P., Mooney, R., 2012. Motor circuits are required to encode a sensory model for imitative learning. Nat. Neurosci. 15, 1454–1459.

Rogers, J.J., Miles, F.A., 1972. Centrifugal control of the avian retina. V. Effects of lesions of the isthmo-optic nucleus on visual behaviour. Brain Res. 48, 147–156.

Romer, A.S., 1966. Vertebrate Paleontology, third ed. University of Chicago Press, Chicago.

Rose, M., 1923. Histologische Lokalisation des Vorderhirns der Reptilien. J. für Psychol. Neurologie 29, 219–272.

Roth, G., Dicke, U., 2005. Evolution of the brain and intelligence. Trends Cogn. Sci. 9, 250–257.

Russchen, F.T., Jonker, A.J., 1988. Efferent connections of the striatum and the nucleus accumbens in the lizard Gekko gecko. J. Comp. Neurol. 276, 61–80.

Ryan, J.M., Cushman, J., Jordan, B., Samuels, A., Frazer, H., Baier, C., 1998. Topographic position of forelimb motoneuron pools is conserved in vertebrate evolution. Brain Behav. Evol. 51, 90–99.

Saranathan, V., Hamilton, D., Powell, G.V.N., Kroodsma, D.E., Prum, R.O., 2007. Genetic evidence supports song learning in the three-wattled bellbird Procnias tricarunculata (Cotingidae). Mol. Ecol. 16, 3689–3702.

Sarko, D.K., Catania, K.C., Leitch, D.B., Kaas, J.H., Herculano-Houzel, S., 2009. Cellular scaling rules of insectivore brains. Front. Neuroanat. 3, 8.

Sayol, F., Lefebvre, L., Sol, D., 2016. Relative brain size and its relation with the associative pallium in birds. Brain Behav. Evol. 87, 69–77.

Scarf, D., Hayne, H., Colombo, M., 2011. Pigeons on par with primates in numerical competence. Science 334, 1664.

Schall, U., 1987. Vestibular, olfactory, and vibratory responses of nucleus basalis prosencephalis neurons in pigeons. Neurosci. Res. 4, 376–384.

Schall, U., Delius, J.D., 1986. Sensory inputs to the nucleus basalis prosencephali, a feeding- pecking centre in the pigeon. J. Comp. Physiol. 159, 33–41.

Schall, U., Güntürkün, O., Delius, J.D., 1986. Sensory projections to the nucleus basalis prosencephali of the pigeon. Cell Tissue Res. 245, 539–546.

Scharff, C., Nottebohm, F., 1991. A comparative study of the behavioral deficits following lesions of various parts of the zebra finch song system: implications for vocal learning. J. Neurosci. 11, 2896–2913.

Schnabel, R., Metzger, M., Jiang, S., Hemmings Jr., H.C., Greengard, P., Braun, K., 1997. Localization of dopamine D1 receptors and dopaminoceptive neurons in the chick fore-brain. J. Comp. Neurol. 388, 146–168.

Schroeder, D.M., Murray, R.G., 1987. Specializations within the lumbosacral spinal cord of the pigeon. J. Morphol. 194, 41–53.

Schubloom, H.E., Woolley, S.C., 2016. Variation in social relationships relates to song preferences and EGR1 expression in a female songbird. Dev. Neurobiol. 76 (9), 1029–1040.

Seamans, J.K., Yang, C.R., 2004. The principal features and mechanisms of dopamine modulation in the prefrontal cortex. Prog. Neurobiol. 74, 1–58.

Searcy, W.A., 1992. Song repertoire and mate choice in birds. Am. Zool. 32, 71–80.

Seed, A., Emery, N., Clayton, N.C., 2009. Intelligence in corvids and apes: a case of convergent evolution? Ethology 115, 401–420.

Shaffer, H.B., Minx, P., Warren, D.E., Shedlock, A.M., Thomson, R.C., Valenzuela, N., Abramyan, J., Amemiya, C.T., Badenhorst, D., Biggar, K.K., Borchert, G.M., Botka, C.W., Bowden, R.M., Braun, E.L., Bronikowski, A.M., Bruneau, B.G., Buck, L.T., Capel, B., Castoe, T.A., Czerwinski, M., Delehaunty, K.D., Edwards, S.V., Fronick, C.C., Fujita, M.K., Fulton, L., Graves, T.A., Green, R.E., Haerty, W., Hariharan, R., Hernandez, O., Hillier, L.W., Holloway, A.K., Janes, D., Janzen, F.J., Kandoth, C., Kong, L., de Koning, A.P., Li, Y., Literman, R., McGaugh, S.E., Mork, L., O'Laughlin, M., Paitz, R.T., Pollock, D.D., Ponting, C.P., Radhakrishnan, S., Raney, B.J., Richman, J.M., St John, J., Schwartz, T., Sethuraman, A., Spinks, P.Q., Storey, K.B., Thane, N., Vinar, T., Zimmerman, L.M., Warren, W.C., Mardis, E.R., Wilson, R.K., 2013. The western painted turtle genome, a model for the evolution of extreme physiological adaptations in a slowly evolving lineage. Genome Biol. 14 (3), R28.

Shanahan, M., 2012. The brain's connective core and its role in animal cognition. Philos. Trans. R. Soc. Lond. B 67, 2704–2714.

Shanahan, M., Bingman, V., Shimizu, T., Wild, M., Güntürkün, O., 2013. The large-scale network organization of the avian forebrain: a connectivity matrix and theoretical analysis. Front. Comput. Neurosci. 7, 89.

Shettleworth, S.J., 2010. Clever animals and killjoy explanations in comparative psychology. Trends Cogn. Sci. 14, 477–481.

Shimizu, T., Karten, H.J., 1990. Immunohistochemical analysis of the visual wulst of the pigeon (Columba livia). J. Comp. Neurol. 300, 346–369.

Shimizu, T., Karten, H.J., 1993. The avian visual system and the evolution of the neocortex. In: Zeigler, H.P., Bischof, H.-J. (Eds.), Vision, Brain and Behavior in Birds. MIT Press, pp. 25–46.

Shimizu, T., Cox, K., Karten, H.J., 1995. Intratelencephalic projections of the visual wulst in pigeons (Columba livia). J. Comp. Neurol. 359, 551–572.

Shine, R., 2005. Life-history evolution in reptiles. Annu. Rev. Ecol. Evol. 36, 23–46.

Shine, R., 2013. Reptiles. Curr. Biol. 23, R227–R231.

Shortess, G.K., Klose, E.F., 1977. Effects of lesions involving efferent fibers to the retina in pigeon (Columba livia). Physiol. Behav. 18, 409–414.

Shubin, N., Tabin, C., Carroll, S., 2009. Deep homology and the origins of evolutionary novelty. Nature 457, 818–823.

Simonyan, K., Horwitz, B., 2011. Laryngeal motor cortex and control of speech in humans. Neuroscience 17, 197–208.

Simonyan, K., Horwitz, B., Jarvis, E.D., 2012. Dopamine regulation of human speech and bird song: a critical review. Brain Lang. 122, 142–150.

Smith, L.M., Ebner, F.F., Colonnier, M., 1980. The thalamocortical projection in Pseudemys turtles: a quantitative electron microscopic study. J. Comp. Neurol. 461, 445–461.

Soma, M., Garamszegi, L.Z., 2011. Rethinking birdsong evolution: meta-analysis of the relationship between song complexity and reproductive success. Behav. Ecol. 22, 363–371.

Stacho, M., Ströckens, F., Xiao, Q., Güntürkün, O., 2016. Functional organization of telencephalic visual association fields in pigeons. Behav. Brain Res. 303, 93–102.

Stanford, L.R., Hartline, P., 1984. Spatial and temporal integration in primary trigeminal nucleus of rattlesnake infrared system. Brain Res. 185, 115–123.

Stanford, L.R., Schroeder, D.M., Hartline, P.H., 1981. The ascending projection of the nucleus of the lateral descending trigeminal tract: a nucleus in the infrared system of the rattlesnake (Crotalus viridis). J. Comp. Neurol. 201, 161–174.

Stapput, K., Güntürkün, O., Hoffmann, K.-P., Wiltschko, R., Wolfgang, W., 2010. Magnetoreception of directional information in birds requires non-degraded object vision. Curr. Biol. 20, 1259–1262.

Starck, D., 1979. Cranio-cerebral relations in recent reptiles. In: Gans, C., Northcutt, R.G., Ulinski, P. (Eds.), Biology of the Reptilia. Neurology, vol. 9. Academic Press, London, pp. 1–38.

Starosta, S., Güntürkün, O., Stüttgen, M., 2013. Stimulus-response-outcome coding in the pigeon nidopallium caudolaterale. PLoS One 8 (2), e57407.

Stewart, M., Kabai, P., Harrison, E., Steele, R., Kossut, M., Gierdalski, M., Csillag, A., 1996. The involvement of dopamine in the striatum in passive avoidance training in the chick. Neuroscience 70, 7–14.

Striedter, G.F., 2005. Principles of Brain Evolution. Sinauer, Sunderland.

Stripling, R., Kruse, A.A., Clayton, D.F., 2001. Development of song responses in the zebra finch caudomedial neostriatum: role of genomic and electrophysiological activities. J. Neurobiol. 48, 163–180.

Strutz, J., 1981. The origin of centrifugal fibers to the inner ear in Caiman crocodilus: a horseradish peroxidase study. Neurosci. Lett. 27, 95–100.

Sultan, F., 2002. Brain evolution (Communication arising): analysis of mammalian brain architecture. Nature 415, 133–134.

Sun, H., Frost, B.J., 1998. Computation of different optical variables of looming objects in pigeon nucleus rotundus neurons. Nat. Neurosci. 1, 296–303.

Tartaglia, E.M., Mongillo, G., Brunel, N., 2015. On the relationship between persistent delay activity, repetition enhancement and priming. Front. Psychol. 5, 1–11.

Taylor, A.E., Saint-Cyr, J.A., Lang, A.E., 1986. Frontal lobe dysfunction in Parkinson's disease. The cortical focus of neostriatal outflow. Brain 109, 845–883.

Taylor, A.H., Hunt, G.R., Holzhaider, J.C., Gray, R.D., 2007. Spontaneous metatool use by New Caledonian crows. Curr. Biol. 17, 1504–1507.

Taylor, A.H., Hunt, G.R., Medina, F.S., Gray, R.D., 2009. Do new caledonian crows solve physical problems through causal reasoning? Proc. R. Soc. Biol. Sci. 276, 247–254.

Timmermans, S., Lefebvre, L., Boire, D., Basu, P., 2000. Relative size of the hyperstriatum ventrale is the best predictor of feeding innovation rate in birds. Brain Behav. Evol. 56, 196–203.

Trinh, A.-T., Harvey-Girard, E., Teixeiri, F., Maler, L., 2016. Cryptic laminar and columnar organization in the dorsolateral pallium of a weakly electric fish. J. Comp. Neurol. 524, 408–428.

Troje, N.F., Frost, B.J., 1998. The physiological fine structure of motion sensitive neurons in the pigeon's tectum opticum. Soc. Neurosci. Abstr. 24, 642.9.

Uchiyama, H., 1999. The isthmo-optic nucleus: a possible neural substrate for visual competition. Neurocomputing 26–27, 565–571.

Uchiyama, H., Ito, H., 1993. Target cells for the isthmo-optic fibers in the retina of the Japanese quail. Neurosci. Lett. 154, 35–38.

Uchiyama, H., Ito, H., Tauchi, M., 1995. Retinal neurones specific for centrifugal modulation of vision. Neuroreport 6, 889–892.

Uchiyama, H., Nakamura, S., Imazono, T., 1998. Long-range competition among the neurons projecting centrifugally to the quail retina. Vis. Neurosci. 15, 417–423.

Uchiyama, H., Aoki, K., Yonezawa, S., Arimura, F., Ohno, H., 2004. Retinal target cells of the centrifugal projection from the isthmo-optic nucleus. J. Comp. Neurol. 476, 146–153.

Ulinski, P.S., 1978. Organization of anterior dorsla ventricular ridge in snakes. J. Comp. Neurol. 178, 411–449.

Ulinski, P.S., 1990. The cerebral cortex of reptiles. In: Jones, E.G., Peters, A. (Eds.), Cerebral Cortex. Comparative Structure and Evolution of Cerebral Cortex, Part I, vol. 8A. Plenum, New York, pp. 139–215.

Ulinski, P.S., Nautiyal, J., 1988. Organization of retinogeniculate projections in turtles of the genera Pseudemys and Chrysemys. J. Comp. Neurol. 276, 92–112.

Veenman, C.L., Wild, J.M., Reiner, A., 1995. Organization of the avian "corticostriatal" projection system: a retrograde and anterograde pathway tracing study in pigeons. J. Comp. Neurol. 354, 87–126.

Veenman, C.L., Medina, L., Reiner, A., 1997. Avian homologues of mammalian intralaminar, mediodorsal and midline thalamic nuclei: immunohistochemical and hodological evidence. Brain Behav. Evol. 49, 78–98.

Veit, L., Nieder, A., 2013. Abstract rule neurons in the endbrain support intelligent behaviour in corvid songbirds. Nat. Commun. 4, 2878.

Veit, L., Hartmann, K., Nieder, A., 2014. Neuronal correlates of visual working memory in the corvid endbrain. J. Neurosci. 34, 7778–7786.

Veit, L., Pidpruzhnykova, G., Nieder, A., 2015. Associative learning rapidly establishes neuronal representations of upcoming behavioral choices in crows. Proc. Natl. Acad. Sci. U.S.A. 112, 15208–15213.

Vergne, A.L., Pritz, M.B., Mathevon, N., 2009. Acoustic communication in crocodilians: from behaviour to brain. Biol. Rev. 84, 391–411.

Verhaal, J., Luksch, H., 2013. Mapping of the receptive fields in the optic tectum of chicken (Gallus gallus) using sparse noise. PLoS One 8 (4), e60782.

Verhaal, J., Luksch, H., 2016. Neuronal responses to motion and apparent motion in the optic tectum of chickens. Brain Res. 1635, 190–200.

Vicario, D.S., 1991. Organization of the zebra finch song control system: II. Functional organization of outputs from nucleus robustus archistriatalis. J. Comp. Neurol. 309, 486–494.

Vu, E.T., Mazurek, M.E., Kuo, Y.-C., 1994. Identification of a forebrain motor programming network for the learned song of zebra finches. J. Neurosci. 14, 6924–6934.

Wagner, H., Frost, B., 1993. Disparity-sensitive cells in the owl have a characteristic disparity. Nature 364, 796–798.

Wagner, H., Güntürkün, O., Nieder, B., 2003. Anatomical markers for the subdivisions of the barn owl's inferior-collicular complex and adjacent subventricular structures. J. Comp. Neurol. 465, 145–159.

Waldmann, C.M., Güntürkün, O., 1993. The dopaminergic innervation of the pigeon caudolateral forebrain: immunocytochemical evidence for a "prefrontal cortex" in birds? Brain Res. 600, 225–234.

Wallenberg, A., 1898. Das mediale opticus Bündel der Taube. Neurol. Zentralblatt 17, 532–537.

Wallraff, H.G., 1988. Olfactory deprivation in pigeons: examination of methods applied in homing experiments. Comp. Biochem. Physiol. A. Comp. Physiol. 89, 621–629.

Wallraff, H.G., 2005. Avian Navigation: Pigeon Homing as a Paradigm. Springer Verlag, Berlin.

Wang, S.R., 2003. The nucleus isthmi and dual modulation of the receptive field of tectal neurons in non-mammals. Brain Res. Rev. 41, 13–25.

Wang, Y., Frost, B.J., 1992. Time to collision is signalled by neurons in the nucleus rotundus of pigeons. Nature 356, 236–238.

Wang, Y., Jiang, S., Frost, B.J., 1993. Visual processing in pigeon nucleus rotundus: luminance, color, motion, and looming subdivisions. Vis. Neurosci. 10, 21–30.

Wang, Y., Luksch, H., Brecha, N.C., Karten, H.J., 2006. Columnar projections from the cholinergic nucleus isthmi to the optic tectum in chicks (Gallus gallus): a possible substrate for synchronizing tectal channels. J. Comp. Neurol. 1, 7–35.

Wang, Y., Brzozowska-Prechtl, A., Karten, H.J., 2010. Laminar and columnar auditory cortex in avian brain. Proc. Natl. Acad. Sci. U.S.A. 107, 12676–12681.

Wang, Z., Pascual-Anaya, J., Zadissa, A., Li, W., Niimura, Y., Huang, Z., Li, C., White, S., Xiong, Z., Fang, D., Wang, B., Ming, Y., Chen, Y., Zheng, Y., Kuraku, S., Pignatelli, M., Herrero, J., Beal, K., Nozawa, M., Li, Q., Wang, J., Zhang, H., Yu, L., Shigenobu, S., Wang, J., Liu, J., Flicek, P., Searle, S., Wang, J., Kuratani, S., Yin, Y., Aken, B., Zhang, G., Irie, N., 2013. The draft genomes of soft-shell turtle and green sea turtle yield insights into the development and evolution of the turtle-specific body plan. Nat. Genet. 45, 701–706.

Watanabe, M., Ito, H., Masai, H., 1983. Cytoarchitecture and visual receptive neurons in the Wulst of the Japanese quail (*Coturnix coturnix japonica*). J. Comp. Neurol. 213, 188−198.

Watts, D.J., Strogatz, S.H., 1998. Collective dynamics of 'small-world' networks. Nature 393, 440−442.

Weidner, C., Repérant, J., Desroches, A.M., Miceli, D., Vesselkin, N.P., 1987. Nuclear origin of the centrifugal visual pathway in birds of prey. Brain Res. 436, 153−160.

Weiler, R., 1985. Mesencephalic pathway to the retina exhibits enkephalin-like immunoreactivity. Neurosci. Lett. 55, 11−16.

Weisbach, W., Schwartzkopff, J., 1967. Nervöse Antworten auf Schallreiz im Grosshirn von Krokodilen. Naturwissenschaften 54, 650.

Wheatcroft, D., Qvarnström, A., 2015. A blueprint for vocal learning: auditory predispositions from brains to genomes. Biol. Lett. 11, 20150155.

Wilbrecht, L., Nottebohm, F., 2003. Vocal learning in birds and humans. Ment. Retard. Dev. Disabil. Res. Rev. 9, 135−148.

Wild, J.M., 1987. Nuclei of the lateral lemniscus project directly to the thalamic auditory nuclei in the pigeon. Brain Res. 408, 303−307.

Wild, J.M., 1989. Avian somatosensory system: II. Ascending projections of the dorsal column and external cuneate nuclei in the pigeon. J. Comp. Neurol. 287, 1−18.

Wild, J.M., 1995. Convergence of somatosensory and auditory projections in the avian torus semicircularis, including the central auditory nucleus. J. Comp. Neurol. 358, 465−486.

Wild, M.J., 1997. Neural pathways for control of birdsong production. Neural Pathw. Control Birdsong Prod 33, 653−670.

Wild, J.M., Williams, M.N., 2000. Rostral wulst in passerine birds. I. Origin, course, and terminations of an avian pyramidal tract. J. Comp. Neurol. 416, 429−450.

Wild, J.M., Zeigler, H.P., 1996. Central projections and somatotopic organisation of trigeminal primary afferents in pigeon (*Columba livia*). J. Comp. Neurol. 368, 136−152.

Wild, J.M., Arends, J.J.A., Zeigler, H.P., 1985. Telecephalic connections of the trigeminal system in the pigeon (*Columba livia*): a trigeminal sensorimotor circuit. J. Comp. Neurol. 234, 441−464.

Wild, J.M., Karten, H.J., Frost, B.J., 1993. Connections of the auditory forebrain in the pigeon (*Columba livia*). J. Comp. Neurol. 337, 32−62.

Wild, J.M., Reinke, H., Farabaugh, S.M., 1997. A non-thalamic pathway contributes to a whole body map in the brain of the budgerigar. Brain Res. 755, 137−141.

Wild, J.M., Kubke, M.F., Peña, J.L., 2008. A pathway for predation in the brain of the barn owl (*Tyto alba*): projections of the gracile nucleus to the "claw area" of the rostral wulst via the dorsal thalamus. J. Comp. Neurol. 509, 156−166.

Wild, J.M., Krützfeldt, N.O.E., Kubke, M.F., 2010. Connections of the auditory brainstem in a songbird, *Taeniopygia guttata*. III. Projections of the superior olive and lateral lemniscal nuclei. J. Comp. Neurol. 518, 2149−2167.

Wilkinson, A., Kuenstner, K., Mueller, J., Huber, L., 2010. Social learning in a non-social reptile (*Geochelone carbonaria*). Biol. Lett. 6, 614−616.

Wilson, M., Lindstrom, S.H., 2011. What the bird's brain tells the bird's eye: the function of descending input to the avian retina. Vis. Neurosci. 28, 337−350.

Wiltschko, W., Wiltschko, R., 1972. Magnetic compass of European robins. Science 176, 62−64.

Wiltschko, R., Wiltschko, W., 1995. Magnetic Orientation in Animals. Springer, Berlin, Germany.

Wiltschko, W., Traudt, J., Güntürkün, O., Prior, H., Wiltschko, R., 2002. Lateralisation of magnetic compass orientation in a migratory bird. Nature 419, 467−470.

Wiltschko, W., Traudt, J., Güntürkün, O., Prior, H., Wiltschko, R., 2011. Reply to "Robins have a magnetic compass in both eyes". Nature 471, E12−E13.

Winkowski, D.E., Knudsen, E.I., 2006. Top-down gain control of the auditory space map by gaze control circuitry in the barn owl. Nature 439, 336−339.

Witmer, L.M., Ridgely, R.C., 2009. New insights into the brain, braincase, and ear region of tyrannosaurs (Dinosauria, Theropoda), with implications for sensory organization and behavior. Anat. Rec. 292, 1266−1296.

Wohlgemuth, S., Adam, I., Scharff, C., 2014. FoxP2 in songbirds. Curr. Opin. Neurobiol. 28, 86−93.

Wolf-Oberhollenzer, F., 1987. A study of the centrifugal projections to the pigeon retina using two fluorescent markers. Neurosci. Lett. 73, 16−20.

Woodgate, J.L., Leitner, S., Catchpole, C.K., Berg, M.L., Bennett, A.T.D., Buchanan, K.L., 2011. Developmental stressors that impair song learning in males do not appear to affect female preferences for song complexity in the zebra finch. Behav. Ecol. 22, 566−573.

Woodgate, J.L., Mariette, M.M., Bennett, A.T., Griffith, S.C., Buchanan, K.L., 2012. Male song structure predicts reproductive success in a wild zebra finch population. Anim. Behav. 83, 773−781.

Woodson, W., Reiner, A., Anderson, K., Karten, H.J., 1991. Distribution, laminar location, and morphology of tectal neurons projecting to the isthmo-optic nucleus and the nucleus isthmi, pars parvocellularis in the pigeon (*Columba livia*) and chick (*Gallus domesticus*): a retrograde labelling study. J. Comp. Neurol. 305, 470−488.

Woodson, W., Shimizu, T., Wild, J.M., Schimke, J., Cox, K., Karten, H.J., 1995. Centrifugal projections upon the retina: an anterograde tracing study in the pigeon (*Columba livia*). J. Comp. Neurol. 362, 489−509.

Wu, L.Q., Niu, Y.Q., Yang, J., Wang, S.R., 2005. Tectal neurons signal impending collision of looming objects in the pigeon. Eur. J. Neurosci. 22, 2325−2331.

Wullimann, M.F., 2014. Ancestry of basal ganglia circuits: new evidence in teleosts. J. Comp. Neurol. 522, 2013−2018.

Wynne, B., Güntürkün, O., 1995. The dopaminergic innervation of the forebrain of the pigeon (*Columba livia*): a study with antibodies against tyrosine hydroxylase and dopamine. J. Comp. Neurol. 357, 446−464.

Xiao, Q., Li, D.P., Wang, S.R., 2006. Looming-sensitive responses and receptive field organization of telencephalic neurons in the pigeon. Brain Res. Bull. 68, 322−328.

Xu, X., Zhou, Z., Dudley, R., Mackem, S., Chuong, C.M., Erickson, G.M., Varricchio, D.J., 2014. An integrative approach to understand bird origins. Science 346, 1253293.

Yamagata, M., Sanes, J.R., 1995. Target-independent diversification and target-specific projection of chemically defined retinal ganglion cell subsets. Development 121, 3763−3776.

Yamazaki, Y., Aust, U., Huber, L., Güntürkün, O., 2007. Lateralized cognition: asymmetrical and complementary strategies of pigeons during discrimination of the "human" concept. Cognition 104, 315−344.

Yan, K., Wang, S.R., 1986. Visual responses of neurons in the avian nucleus isthmi. Neurosci. Lett. 3, 340−344.

Yin, H.H., 2016. The basal ganglia in action. Neuroscience pii: 1073858416654115. [Epub].

Young, S.R., Rubel, E.W., 1983. Frequency-sensitivity projections of individual neurons in chick brainstem auditory nuclei. J. Neurosci. 3, 1373−1378.

Yu, A.C., Margoliash, D., 1996. Temporal hierarchical control of singing in birds. Science 273, 1871−1875.

Yu, S., Huang, D., Singer, W., Nikolic, D., 2008. A small world of neuronal synchrony. Cereb. Cortex 18, 2891−2901.

Zapka, M., Heyers, D., Hein, C.M., Engels, S., Schneider, N.L., Hans, J., Weiler, S., Dreyer, D., Kishkinev, D., Wild, J.M., Mouritsen, H., 2009. Visual but not trigeminal mediation of magnetic compass information in a migratory bird. Nature 461, 1274−1277.

Zeier, H., Karten, H.J., 1971. The archistriatum of the pigeon: organization of afferent and efferent connections. Brain Res. 31, 313−326.

Function and Evolution of the Reptilian Cerebral Cortex

R.K. Naumann, G. Laurent

Max Planck Institute for Brain Research, Frankfurt, Hessen, Germany

Glossary

Analogy (or homoplasy) Similarity due to convergent or parallel evolution, but not common evolutionary origin.

Cell type A morphologically recognizable cell with sufficient similarities for identification across individuals of the same species and, in some cases, across species. In addition, molecular and physiological parameters are used for cell type classification.

Cerebral cortex A laminar telencephalic structure consisting of cells arranged in layers parallel to the ventricular and pial surfaces.

Dorsal cortex Dorsal region of the pallium in reptiles. The dorsal cortex is frequently compared to the dorsal pallium of mammals (neocortex). However, only part of the reptilian dorsal cortex may be a derivative of the dorsal pallium, and more comparative and developmental studies of this structure are needed.

Dorsal lateral geniculate nucleus A component of the thalamus that receives direct information from the retina.

DVR Dorsal ventricular ridge: a large region of the ventrolateral pallium of the telencephalon of birds and reptiles.

Homologous Having the same relative position (topological position), embryonic origin, and common ancestor.

Neocortex Derivative of the dorsal pallium in mammals that typically shows a six-layered organization. The name emphasizes its novelty, which is debated. A more neutral term is isocortex, emphasizing its homogeneous architecture.

Pallial thickening A lateral expansion of the dorsal cortex of reptiles. It is considered a lateral part of the dorsal cortex.

Pallium A major division of the telencephalon in all vertebrates, which in reptiles consists of the cortical regions, DVR, and part of the amygdalar regions. It is subdivided into medial, dorsal, lateral, and ventral pallium.

Subpallium Regions of the telencephalon below the pallium that include parts of the amygdala and the striatum.

elucidation of the fundamental plan of the organ of the mind.- Santiago Ramón y Cajal (1988).

The evolution and function of the cerebral cortex remains a mystery. Most textbooks of comparative neurobiology focus on reconstruction of structural brain evolution, whereas textbooks on brain function often ignore evolution altogether. The cerebral cortex is the part of the amniote vertebrate brain that changed most dramatically during recent evolution and the part considered to be responsible for the most important cognitive functions. Yet, the absence of the largest cortical axon tract, the corpus callosum with nearly 200 million fibers in humans (Paul et al., 2007), or even a dramatic reduction in cortical volume can have surprisingly mild effects on human behavior (Feuillet and Pelletier, 2007; Chiappedi et al., 2012). Thus, when trying to relate brain structure to function, we face a paradox: our techniques to study atomistic details of the brain genes, ion channels, cells—are becoming evermore sophisticated, but we still struggle to define an overarching functional understanding. Our situation is not entirely dissimilar to that of pre-Darwinian taxonomists comparing, naming, and cataloguing species around the globe. This is particularly true with the cerebral cortex. Concerning its function and evolutionary origin, researchers have acquired "a pile of sundry facts, some of them interesting or curious but making no meaningful picture as a whole" (Dobzhansky, 1973).

Neurobiological studies on nonmammalian vertebrates are often motivated by the view that these animals have simpler nervous systems. Such studies frequently use a single species chosen because of its availability, experimental convenience, and speculation regarding its supposed primitive attributes. As shown in Fig. 9.1, it is clear that the cortex has more complex patterns of lamination and afferent innervation in

9.1 Introduction

... the cerebral cortex of reptiles resembles [...] that of mammals, although simplified. This [...] gives to the study of the cerebrum of reptiles a capital importance for the

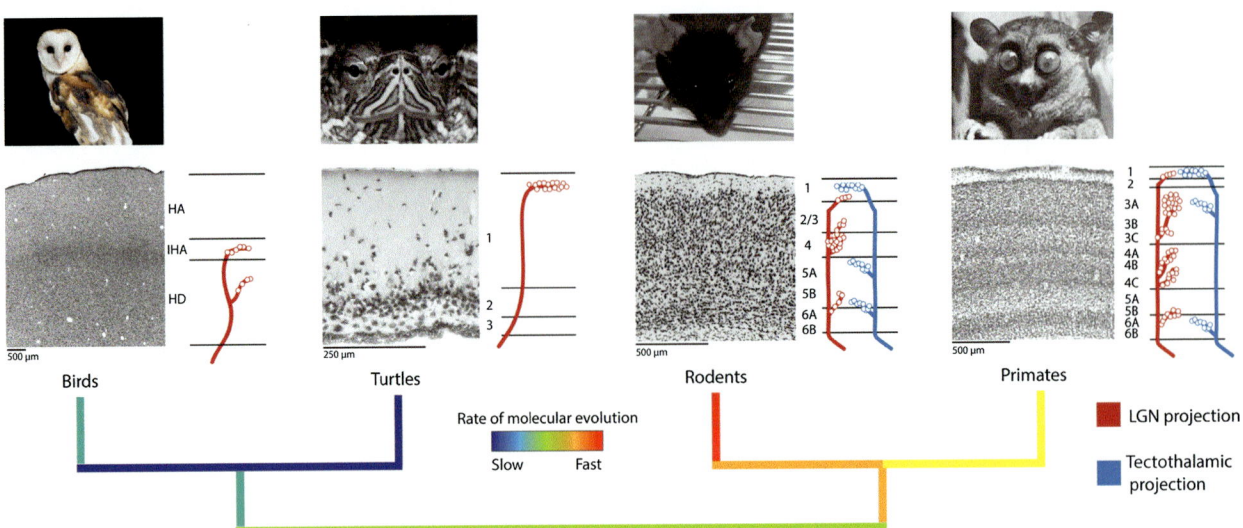

FIGURE 9.1 Evolution of the visual cortex in four species: barn owl, freshwater turtle, mouse, and tarsier (top row). The middle panels show the region of the forebrain that receives input from the lateral geniculate nucleus in these four species, often referred to as visual cortex. *Red axons* indicate direct thalamic input from the lateral geniculate nucleus (or lemnothalamic pathway); more complex laminar patterns are evident in mammals and especially in primates. *Blue axons* indicate input from the tectothalamic (or collothalamic) pathway: this pathway directly targets primary visual cortex in mammals but runs through a region called the dorsal ventricular ridge (DVR) in reptiles and birds. Visual cortices in turtles and mammals share a laminar structure and thalamic input into the superficial cell-poor layer, whereas the visual region in birds (wulst) is organized in stacked nuclei (HA, IHA, HD) without a superficial cell-poor domain. The lower panel shows a phylogenetic tree based on estimated molecular evolutionary rates inferred from DNA sequences. These rates indicate genome changes from a common a hypothetical common ancestor. Mammalian evolutionary rates are generally higher than in reptiles and birds. Parts of the figure simplified from: Green et al. (2014); Collins et al. (2005); Lund et al. (1994). *HA,* apical part of the hyperpallium; *IHA,* interstitial part of hyperpallium apicale; *HD,* densocellulare part of the hyperpallium. *Reproduced from Wang, Y., Shanbhag, S.J., Fischer, B.J., Peña, J.L., 2012. Population-wide bias of surround suppression in auditory spatial receptive fields of the owl's midbrain. J. Neurosci. 32 (31), 10470–10478; Brainmaps.org. Source Edward G. Jones; Collins, M., et al., 2005. Overview of the visual system of Tarsius. Anat. Rec. Part A Discov. Mol. Cell. Evol. Biol. 287 (1).*

mammals than in reptiles (for detailed reviews see Butler, 1994; Medina and Reiner, 2000; Bruce, 2006). But is reptilian cortex simply an "earlier version" of that in mammals? How should we compare the apparently more complex brains of birds with those of reptiles and mammals? Answering these questions requires an understanding of the evolution of the cortex in many different species. To understand principles of cortical evolution, should we look for homologous components of neural circuits or rather extract principles from the myriad details of neural hardware and compare computations? Should we try to reconstruct the phylogenetic history of the cortex or rather sample the variations across extant species and identify common patterns and their adaptive significance? Likewise, simplicity, even when evident at one level, such as the cytoarchitecture of reptilian cortex, still allows for complex synaptic connectivity. Thus, "simplicity" should not be assumed but, instead, quantified and tested for functional relevance. It is possible that we will discover that reptilian cortex is neither "simple" nor particularly illuminating about its mammalian analogue. We will take the opposing view that reptiles, arguably the first animals to evolve a cerebral cortex, offer a unique opportunity to look back to the origins of cortex and to understand

what it evolved for. We propose that understanding the "simple" cortex of reptiles can indeed help us to understand the avian and mammalian brains. Our review will be selective; for an in-depth survey of reptile forebrain structure, cortical cytoarchitecture, and thalamocortical connections, we refer the reader to excellent reviews (Bruce, 2006; Riss et al., 1969; Northcutt, 1981; Ulinski, 1990). The study of reptilian brains may shed light on many unresolved theories about cortical evolution, and recent issues in *Frontiers in Neuroscience* (Aboitiz and Montiel, 2015) and the *Journal of Comparative Neurology* (Martínez-Cerdeño and Noctor, 2016) cover many of the topics that we only touch upon.

9.1.1 Reptile Phylogeny

What are reptiles? To understand the differences among vertebrate brains (and how these differences evolved), it would be of great advantage to study the brains of dinosaurs and other earlier sauropsids. But brains do not fossilize well, so the best information we can get today from fossilized remains is about brain size and proportions (however, those measures are only approximate because the brain of living reptiles is generally significantly smaller than the cavity in which they lie). Unfortunately, we have no information about internal brain structures or physiology. Until we can recreate ancestral species in the laboratory, the best we can do is study extant reptiles. Reptiles, birds, and mammals are amniotes. Birds are nested within the reptiles; together, the birds and the nonavian reptiles form a group called the sauropsids. Here, we focus on nonavian reptiles, from this point on called "reptiles," as commonly understood. Turtles, crocodiles, and squamates (snakes and lizards) represent the major branches of the reptilian lineage. A fourth branch, the rhynchocephalians, is represented by a single genus, the tuatara, or *Sphenodon*. Among the main reptilian branches, species diversity is highest among squamates (around 10 000 species), while there are only about 300 turtle species and 24 crocodilian species (Pincheira-Donoso et al., 2013). Even though turtle species are few, their forebrains show dramatic structural differences (Riss et al., 1969; Striedter, 2016). Thus, one should be cautious when reading general statements about the brain of "lizards" or "turtles", as we will occasionally do in the text in the following section. Birds are most closely related to crocodiles and are the only surviving members of the now extinct dinosaurs. Crocodilians and birds are grouped together as archosaurs.

Understanding of reptile phylogeny was long clouded by the unique features of turtles: the enigmatic origin of the turtle shell and the turtle skull. The traditional view placed turtles at the base of the reptilian tree because their skull has no temporal opening (anapsid condition), whereas most other reptiles and mammals have temporal openings (parapsid or diapsid conditions). However, new fossil finds document that the closed skull of turtles evolved from a diapsid condition (Schoch and Sues, 2015, 2016; Bever et al., 2015), supporting the placement of turtles as sister group of archosaurs, as suggested by molecular phylogenetics (Hedges and Poling, 1999). Concerning the shell of turtles, fossils and developmental studies have shown that the upper (carapace) and lower (plastron) parts of the shell appeared at different times in evolution and have different developmental origin (Li et al., 2008; Lyson et al., 2013; Rice et al., 2016). Since the turtle shell evolved to its modern form in the Triassic, the general morphology of turtles has changed only little. Confirming this notion, molecular evolution studies reveal that turtles and crocodiles show the lowest rate of change among tetrapods, whereas birds, lizards, snakes, and mammals show significantly faster rates (Fig. 9.1; Green et al., 2014; Hugall et al., 2007). In summary, present-day turtles and crocodiles predate the rise of the dinosaurs and have outlived them with moderate changes over the last 100–200 million years. Despite their slow evolution, turtles and crocodilians are nested together with birds within the reptilian tree, whereas lizards and snakes form the earliest branch to diverge from the basal reptiles. This is also reflected in the high species diversity and rate of molecular evolution of squamates.

9.1.2 What Is the Cerebral Cortex?

While all the features to be here described are present in amphibia in a rudimentary form, it is only in the highly organized reptilian brain that they come out into prominence. Ludwig Edinger (1908)

9.1.2.1 Pallium Versus Cortex

Before examining cortical structure in detail, we provide a short reminder of important vertebrate brain terminology. The most anterior part of the brain is the forebrain (prosencephalon). It includes the eye, the hypothalamus, and the telencephalon. The telencephalon is divided into the pallium (mantle) and subpallium. The subpallium includes the basal ganglia, preoptic area, and the diagonal area that contains parts of the septum and amygdala. Finally, the pallium in mammals comprises the cortex, a number of cortical nuclei (that lie directly below the laminated cortex), and parts of the amygdala (Puelles et al., 2013). Thus, the pallium comprises both nuclear and layered structures. Among nonmammals, the pallium is composed more frequently of nuclear structures. Consequently, the term pallium appears more frequently in studies of fish and amphibian brains.

9.1.2.2 Pallial Subdivisions

Popular theory holds that the mammalian brain consists of three parts: the reptilian complex, the limbic system, and the neocortex (MacLean, 1990). While mammals, including humans, share the "reptilian" complex with lower animals, the limbic and cortical systems supposedly emerged to counteract those atavistic reptilian impulses. These ideas rest on the view that brains evolve toward more complex forms by addition of new parts (Edinger, 1908; Kappers, 1909), and many anatomical terms such as "paleopallium", "archipallium", and "neopallium" reflect their supposed order of appearance in evolution. A different view holds that brains evolved largely by modification of parts already present in a common ancestor (Johnston, 1923). Most modern studies support the latter view: structures once thought to be unique to some vertebrates are frequently found also in other species although occasionally dramatically modified (Butler and Hodos, 2005; Striedter, 2005).

The pallium is divided into three parts (medial, dorsal, and lateral), considered to correspond largely to the medial cortex (hippocampus, archicortex), dorsal cortex (isocortex, neocortex), and lateral cortex (olfactory cortex, paleocortex) (Striedter, 1997). The discovery of a fourth, ventral region of the pallium, characterized by the expression of the Tbr1 and the absence of the Emx1 gene (Puelles et al., 2000; Smith-Fernandez et al., 1998), has since complicated this picture (Fig. 9.2). The ventral pallium appears in early development, at the same time as the other regions of the pallium. It is located ventral to the lateral pallium and exists in all vertebrates (Brox et al., 2004; Northcutt, 2008; Nieuwenhuys, 2011; Mueller et al., 2011). (We will return later to the notions of homology, gene expression, and the evolutionary origin of the cortex but first try to define core elements of cortical structure.)

9.1.2.3 Some Essential Features of the Cerebral Cortex

Most of our knowledge about the cerebral cortex derives from studies of mammals. Here, we attempt to distil our knowledge about mammalian cortex to its essence and ask whether any aspect of it can be detected already in reptilian cortex. Some structural details of this comparison—eg, quantification of dendritic morphology and synaptic connectivity—are now also being clarified by large-scale electron microscopic reconstructions (Denk and Horstmann, 2004).

Dense internal reciprocal connectivity: The cortex is essentially a system of densely interconnected excitatory pyramidal cells (Braitenberg, 1978). Most of these short- and long-range connections contact other cortical cells, whereas inputs and outputs of cortex form only a minority of cortical connections (Braitenberg and Schüz, 1998;

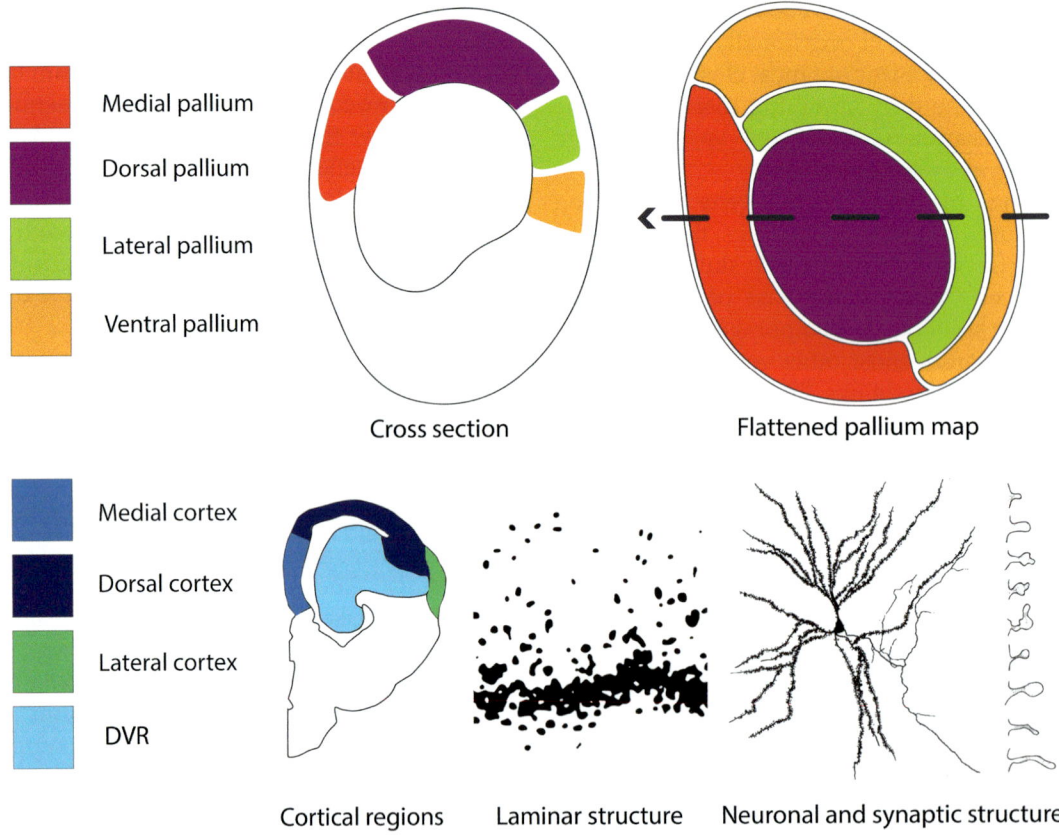

FIGURE 9.2   Two different approaches to subdividing the dorsal telencephalon and levels of analysis. The study of developmental gene expression patterns has led to a model of pallial organization into four parts (Puelles et al., 2000, 2013; Puelles, 2011). These parts are present in early development in all vertebrates, but their corresponding parts in the adult brain remain to be worked out in most groups of vertebrates, including reptiles. Cortex usually refers to laminar structures in the telencephalon, whereas the term pallium includes also nuclear regions and, thus, more generally applies to birds, reptiles, amphibians, and fishes where large parts of the dorsal telencephalon are not organized in layers. Therefore, the lower part of the figure based on the descriptive anatomy of adult reptiles uses different colors. Together with cortical development, cortical regions, layering/architectonics, neuronal, and synaptic structures are common levels of analysis (neuron from Poliakov, 1964; spines from Ulinski, 1977). *DVR*, dorsal ventricular ridge. *Reproduced from Puelles, L., 2011. Pallio-pallial tangential migrations and growth signaling: new scenario for cortical evolution? Brain Behav. Evol. 78 (1), 108—127; Poliakov, G.I., 1964. Development and complication of the cortical part of the coupling mechanism in the evolution of vertebrates. J. Hirnforsch. 7 (3), 253; Ulinski, P.S., 1977. Intrinsic organization of snake medial cortex: an electron microscopic and Golgi study. J. Morphol. 152 (2), 247—279.*

Costa and Martin, 2009). The number of cortical neurons is also larger than that of the incoming afferents; in the optic tectum of the frog, by contrast, the number of sensory afferents is about equal to the number of local neurons (Lázár and Székely, 1969; Kemali and Braitenberg, 1969). Intracortical connections are excitatory and made onto dendritic spines, structural specializations well suited for regulation (Yuste, 2011). Recurrent, excitatory, and modifiable connections with many other cortical neurons are arguably the key features that define the functional architecture of cortex (Douglas et al., 1995; Douglas and Martin, 2007) and distinguish cortex from other brain regions. For example, recurrent connections in the basal ganglia are mainly inhibitory; the granule cells of the cerebellum are excitatory, but not connected among themselves. The cortex is also typically a large system, containing at least 1 million neurons in all mammals investigated so far. Data for reptiles are scarce, but numbers range from an estimate of about 120 000 neurons in the dorsal cortex of turtles (Ulinski, 2006) to about 20 million in the entire telencephalon of a crocodile (Ngwenya et al., 2016). Whereas there are many more cerebellar granule cells than are cortical neurons, it is the recurrent connectivity of cortex that makes its architecture special and also particularly daunting. The overwhelming number of intracortical connections implies that wiring between individual cortical neurons cannot be encoded in the genome. Rather, genetic instructions are probably limited to the approximate distribution and probabilities of intracortical connections and to the targeting of specific neuronal types (Szentagothai, 1978).

Cell diversity and layering: Thanks to modern circuit reconstruction techniques, we can now examine circuit organization in exquisite detail (Denk et al., 2012). For example, the major class of cortical neurons—the pyramidal cells—are subdivided into many different types, based on patterns of projection and gene expression (Molyneaux et al., 2007). The second main neuronal class in the cortex—the interneurons—displays at least as much diversity (Freund and Buzsáki, 1996; Klausberger and Somogyi, 2008). In most cases, interneurons project only locally and inhibit their targets. We now know that interneurons are generated outside of cortex and invade cortical areas during development from a few specific regions of origin (Anderson et al., 1997; Marin and Rubenstein, 2003). Pyramidal cells are instead largely generated locally in the developing cortex, in an inside-out fashion. Via this radial migration from the progenitor cells near the ventricle, they create another key aspect of cortical structure: its laminar organization. In mammalian neocortex, we now recognize, by convention, six layers (Brodmann, 1909); every layer contains a specific combination of excitatory and inhibitory cell types. While ostensibly an idealization—one could

have easily picked 9 or 7 as Ramon y Cajal indeed did (Fig. 9.1)—this framework is still significant as one of the few commonly agreed upon characters for modeling cortical circuits (Douglas and Martin, 2004). Thus, despite the apparent differences in lamination of rodent and monkey cortex (Fig. 9.1), molding the many sublayers of the primate visual cortex into a common six-layer scheme (Balaram and Kaas, 2014) has been useful. However, given the essential nature of intracortical connections and circuits, the existence of six layers is by no means a necessary feature of cortex. Braitenberg's (1978) "skeleton" cortex simply consists of a superficial layer with afferent or intracortical axons and a pyramidal cell layer. This arrangement may suffice for the most elementary cortical computations (Larkum, 2013) and also happens to correspond to most reptilian cortices.

Modularity: Cortical thickness varies from about half a millimeter in reptiles and small mammals to several millimeters in larger mammals. Cortical surface area varies more dramatically across mammals, from <1 cm^2 in shrews to 7500 cm^2 in Orca whales (~2500 cm^2 in humans or about the size of a medium pizza per hemisphere once flattened). Cortical surface area did not increase dramatically during hominid evolution, nor is it dramatically smaller in cows or horses. Thus, tangential expansion by itself does not teach us much about changes in cortical function (but see Striedter, 2005). Yet, tangential expansion illustrates an important design feature: cortex is made up of modules that repeat, with local variations, across the cortical sheet. Lorente de Nó (1922, 1934) first suggested that repetitive cellular aggregations, or groups of neurons contacted by single thalamic axons, could form such a basic unit. Many have since tried to relate structural modularity to function without entirely convincing results. We observe a bewildering diversity of cortical modules (Rockland and Ichinohe, 2004; Rakic, 2008; Kaas, 2012; Molnár, 2013), without an entirely convincing, singular, and overarching elucidation of the cortical module's function (Horton and Adams, 2005; Costa and Martin, 2010). Inspired by the concept of a mammalian cortical column, Wang et al. (2010a) and others (Jarvis et al., 2013; Ahumada-Galleguillos et al., 2015; Trinh et al., 2016) attempted to define modular structures in non-cortical (though possibly homologous) regions of the pallium in birds and fish. Whereas most cells in a mammalian cortical column originate from a small part of the neuroepithelium, pallial columns in birds originate from broad regions of the neuroepithelium, and different columns originate from the same epithelial region (Jarvis et al., 2013; Montiel and Molnár, 2013). In the bird auditory system, tonotopic projections from the thalamus to the pallium, combined with local circuits that preserve tonotopy across nuclear slabs (Wang et al., 2010a), generate isofrequency columns (Scheich,

1983; Theurich et al., 1984). The benefits of this (possibly convergent) modular arrangement of neural elements are poorly understood, in particular for more complex phenomena, such as birdsong (Jarvis et al., 2013). This serves as a reminder that modular elements are not an exclusive feature of cortex (Rockland, 2010) nor are they exclusive to the brains of amniotes (Leise, 1990).

9.1.2.3.1 Reptilian Cortex

Three layers: The cortex of reptiles consists of three layers (Fig. 9.1). The outer and thickest layer (layer 1, plexiform or molecular layer) harbors scattered interneurons, whereas the intermediate layer (layer 2 or cellular layer) contains a large number of densely packed cell bodies, most of which belong to excitatory neurons. Layer 2, defined by the compact apposition of most excitatory neuron somata, is the most characteristic feature of reptilian cortex and is not subdivided into sublayers. The innermost layer (layer 3, deep plexiform or subcellular layer) is the thinnest layer (especially in medialmost areas of cortex) and contains a majority of interneurons, most of which are packed along the ependymal or ventricular surface. In discrete regions of the cortex in some reptiles, one observes an additional cellular layer, called cell plate, within the inner layer (Unger, 1906; Ulinski, 1974). Note that, as with mammalian cortex, the choice of three constitutive layers is a convention; some have subdivided the outer and innermost layers into further sublayers (Edinger, 1908; Ramón y Cajal, 1909). Layer 1, for instance, can be subdivided in a superficial zone (L1a) with distant incoming projections and a deeper "associational" L1b. In some areas, L1 contains a clear third intermediate zone identified with specific molecular markers. The best-studied example for this pattern is the medial cortex of certain lizards, where the outer third of L1 shows somatostatin (SST) immunoreactivity, the middle third acetylcholinesterase activity, and the inner third cytochrome oxidase activity (Regidor and Poch, 1988; Pérez-Clausell and Fredens, 1988). However, the acetylcholinesterase pattern in the medial cortex varies across lizard species and is more strongly correlated with the pattern of projections from the anterior thalamus (Hoogland et al., 1998).

Three subdivisions: Reptilian cortex is not, by any means, uniform over the mantle. Cortical regions are usually defined by converging evidence at three levels: normal anatomy or marker expression, intracortical or extracortical connections, and physiological responses. In all reptiles, we can recognize at least three major subdivisions: medial, dorsal, and lateral cortices (Ulinski, 1990; Fig. 9.2). In most species, these divisions form discrete cell plates, easily identified in Nissl preparations. In some species, the cell plates overlap at the medial and lateral edges of dorsal cortex, forming the medial and lateral superpositions (De Lange, 1911; Guirado and Davila, 2002;

Luzzati, 2015). Instead, in some reptiles, such as crocodiles and *Tuatara*, they form a continuous sheet (Hines, 1923; Ulinski, 1974). Lizards and turtles are the most studied reptiles, but the exact correspondence between lizard and turtle dorsal cortical areas, eg, is still debated (Ulinski, 1990; Hines, 1923). This lack of a common framework may have contributed to conflicting proposals regarding homology with mammalian areas (Striedter, 1997; Nieuwenhuys et al., 1998).

Sensory and thalamic inputs are a key feature used to define cortical regions. Aside from forming a separate cell plate in most reptiles, lateral cortex is defined by its direct input projections from the olfactory bulb (Reiner and Karten, 1985). Lateral cortex is frequently divided into two or more subregions (Riss et al., 1969; Skeen et al., 1984; Puelles et al., 2016).

The dorsal cortex consists of three zones called D1 (medial), D2 (intermediate), and D3 (lateral) (Beckers et al., 1972; Molowny et al., 1972; Platel et al., 1973); however, due to a lack of specific markers for these subdivisions, borders and names are not consistently defined (Ulinski, 1990; Nieuwenhuys et al., 1998). The most lateral zone of turtle dorsal cortex folds inward into a broad band of cells called the pallial thickening (Johnston, 1915). The pallial thickening receives visual thalamic input from the dorsal lateral geniculate nucleus in turtles and lizards (Hall and Ebner, 1970; Kenigfest et al., 1997), but only in turtles does this visual input extend further into the dorsal cortex (Desan, 1988). Desan (1988) divided dorsal cortex into two parts: D1 medially and D2 laterally, extending laterally to include the pallial thickening (Nieuwenhuys et al., 1998; Butler, 1980). This nomenclature has been used in studies of traveling waves in dorsal cortex (Prechtl, 1994; Prechtl et al., 1997, 2000), for example.

The medial cortex contains at least two (Ulinski, 1990), but according to some authors three, four, or more subregions with distinct cellular architecture (Striedter, 2016; Filimonoff, 1963). The major subdivisions are a medial part with small, densely packed cells and a dorsal region with larger, more loosely distributed cells. Most authors have considered these two regions as directly comparable across all reptiles (Ulinski, 1990; Nieuwenhuys et al., 1998), and some also consider the ontogenetic and functional equivalence of reptilian medial cortex to mammalian hippocampus as well established (Colombo and Broadbent, 2000). Therefore, most discussions on reptilian cortices and their potential mammalian homologues have centered instead on dorsal cortex. Striedter (2016), however, pointed out a number of potential flaws in the arguments used to support this equivalence and proposes a new scheme to align medial cortical regions of reptiles, birds, and mammals. More radically, he argues that the medial cortex extends well onto the dorsal surface in turtles, indicating that interareal borders may not be settled after all. In summary, cortical areas in reptiles have been variably defined, using cytoarchitecture either alone, or in combination with connectivity and some physiology. More modern techniques, including histochemical staining and gene expression patterns, will hopefully soon help clear up these issues.

Cortex in amphibians? Relying on (limited) gene expression patterns in developing and adult animals, it would appear as if the amphibian pallium contains four sectors similar to those in reptiles and mammals (Brox et al., 2004; Dugas-Ford, 2009; Amamoto et al., 2016). Like its reptilian counterpart, the amphibian pallium contains a roughly intermediate cellular zone and incoming afferent fibers near the pial surface. The principal neurons, however, are not grouped tightly into a clearly defined layer, and their radial density varies little between ventricular and pial boundaries. Because of this lack of lamination, eg, in the dorsal pallial sector, amphibians are often thought to lack cortical structures (Northcutt, 1981; but see Kemali and Braitenberg, 1969). However, identifying the key events that correspond to the transition between amphibian and reptilian pallial organizations would be of great value to understand the evolution of lamination and of cerebral cortex proper.

Cortex in birds? Birds are often thought to lack a laminated cortex. Indeed, the pallium of birds (and to some extent of the nonavian reptiles also) is dominated by a large nuclear structure called the DVR. The DVR has no obvious equivalent in amphibians or mammals, but it is of obvious functional importance to sauropsids, for it receives the largest share of sensory inputs from the thalamus. Consequently, understanding the relations between DVR and the other regions of the sauropsid pallium is key to understanding forebrain evolution (Northcutt and Kaas, 1995). The other components of the pallium in birds are the medial pallium, olfactory pallium, and the dorsal pallium (also called hyperpallium or wulst). In birds, medial and olfactory pallium are relatively small compared with the expanded DVR but both have, at least in part, a cell-free superficial layer above a condensed cellular layer and thus are genuine cortical structures in the pallium of birds (Craigie, 1935, 1936). Abellán et al. (2014) and Herold et al. (2014) have recently used developmental gene expression and the distribution of neurotransmitter receptors in adult brains to define corresponding parts between the mammalian hippocampus and the avian medial pallium. Thus far, the results indicate a mosaic of similarities and differences (Striedter, 2016).

9.1.3 Functional Architecture of Sensory Pathways to the Pallium in Reptiles

The sensory pathways to the pallium in reptiles are fundamentally similar to those in birds and mammals.

The most detailed studies on any reptile come from the turtle visual system, which we will use to illustrate how sensory projections influence pallial structure and function.

Retina and optic tract: The retina in turtle is among the best-studied retinas in vertebrates in terms of architecture and functional pathways (Ammermüller and Kolb, 1996). Seven types of photoreceptors capture light and produce electrical signals that are processed by more than fifty types of horizontal, bipolar, and amacrine cells. These signals are then transmitted by 24 types of ganglion cells to the rest of the brain. We will return later to the issue of cell types, but we note that the anatomy and physiology of retinal ganglion cells clearly covary in turtles (Jensen and Devoe, 1983; Granda and Sisson, 1992; Ammermüller and Kolb, 1995; Ammerüller et al., 1995; Dearworth and Granda, 2002). The optic tract in turtles contains about 400 000 axons (Geri et al., 1982) that form three groups based on conduction velocity and axon diameter (Woodbury and Ulinski, 1986). In other reptiles, such as dragon lizards, there are close to a million or more optic nerve axons (Wilhelm and Straznicky, 1992; Beazley et al., 1997). The optic tectum, the visual nuclei of the thalamus, and several regions of the pretectum and hypothalamus are the major targets of optic nerve fibers (Hergueta et al., 1991; Reiner et al., 1996). However, the large number of ganglion cell types suggests a finer organization of projections within these regions, ie, thus far unknown (Robles et al., 2014).

Tectum and thalamus: The optic tectum of reptiles is the homologue of the mammalian superior colliculus, but in reptiles, the tectal lamination is better defined and it is proportionally larger (relatively) when compared with the pallium. Much like in the cortex, the correspondence (if any) between tectal and collicular laminae is not clear (Foster and Hall, 1975). However, the reptilian tectum receives a major fraction of the axons from the retina and in turn projects to the largest sensory nucleus of the dorsal thalamus, the nucleus rotundus (or pulvinar in mammals) (Fig. 9.3). This pathway is called the tectothalamic, collothalamic, tectofugal, or indirect projection. The other, usually smaller, projection is the thalamofugal, lemnothalamic, or direct pathway, joining the retina to the dorsal lateral geniculate nucleus of the thalamus. Both visual pathways are present in all reptiles (Northcutt, 1981), but the projections from the dorsal lateral geniculate nucleus to the pallium vary substantially across species and appear to be most developed in turtles. The dorsal lateral geniculate nucleus is situated below and medial to the optic tract and contains two layers (Rainey and Ulinski, 1986). The cell plate layer contains densely packed somata, with dendrites extending into the neuropil layer, closer to the optic tract. There are two known types of retinal axon terminals in the turtle lateral

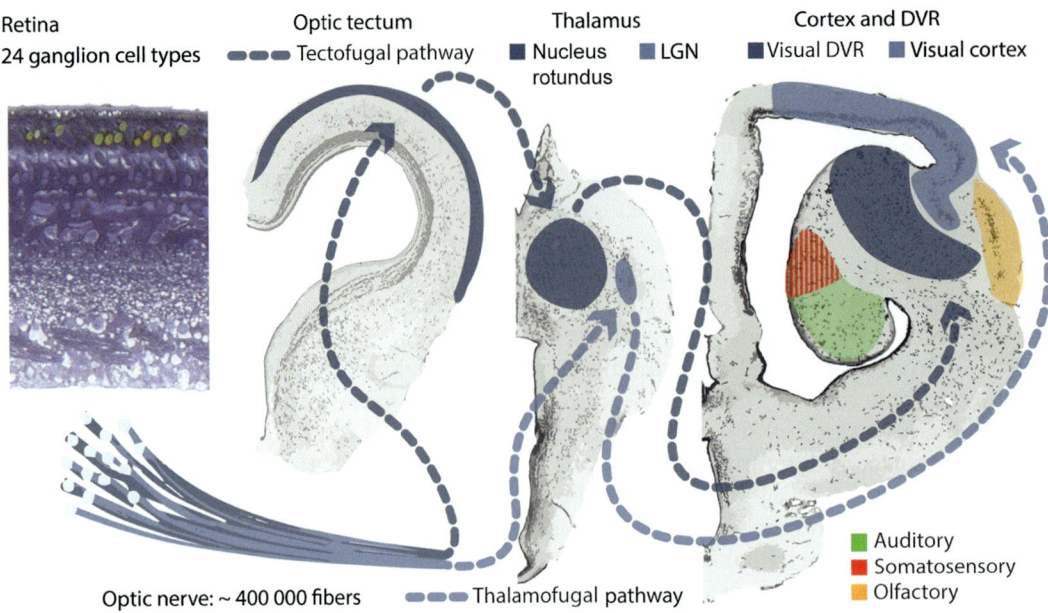

Retina
24 ganglion cell types

Optic tectum
Tectofugal pathway

Thalamus
Nucleus rotundus   LGN

Cortex and DVR
Visual DVR   Visual cortex

Auditory
Somatosensory
Olfactory

Optic nerve: ~ 400 000 fibers

Thalamofugal pathway

FIGURE 9.3 Visual pathways and sensory regions in the turtle pallium. The retina of turtles is among the most thoroughly studied retinas among vertebrates and contains 24 ganglion cell types (Ammermüller and Kolb, 1996) that transmit information to pallium via two main routes. The tectofugal pathway passes through the tectum and the thalamic nucleus rotundus to the visual part of the dorsal ventricular ridge, which also receives auditory and somatosensory thalamic inputs (Nieuwenhuys et al., 1998). A second optic path targets the lateral geniculate nucleus directly and from there projects to the dorsal cortex (Hall and Ebner, 1970; Desan, 1988). *N. Rot.*, nucleus rotundus; *LGN*, lateral geniculate nucleus; *DVR*, dorsal ventricular ridge.

geniculate nucleus: smaller fibers with small boutons that terminate in the outer part of the neuropil layer and larger fibers with larger boutons that terminate in the inner neuropil region (Sjöström and Ulinski, 1985; Ulinski, 1986). Retinal projections to the tectum and dorsal lateral geniculate nucleus are bilateral and contralateral projections by far the most extensive (Hergueta et al., 1991; Ulinski and Nautiyal, 1988). Note that, in addition to the main nuclei of the direct retinothalamic pathway, the anterior thalamic nuclei also receive direct retinal input (Hergueta et al., 1991; Reiner et al., 1996) and, in turn, project to the cortex (Zhu et al., 2005). Unfortunately, there are very few studies of physiological properties in any of the thalamic nuclei (Ulinski, 1990).

Thalamopallial projections: The largest visual thalamic projection originates in the nucleus rotundus and targets the visual region of the DVR in all reptiles (Pritz, 1975; Bruce and Butler, 1984a; Dacey and Ulinski, 1983). The DVR varies in size and structure across reptiles, and most of this variation is thought to result from a coevolution of visual input from the nucleus rotundus and DVR size in some orders of reptiles (Northcutt, 1978; Platel, 1979). Manger et al. (2002) used extracellular multiunit recordings to show that the largest part of the DVR in the iguana, a lizard with a well-developed visual system, responds preferentially to visual stimulation; only small regions respond to tactile or auditory stimulation. In turtles, the

somatosensory and auditory regions have been traced to adjacent regions of the DVR, but some authors also suggest that a somatosensory region is located in the dorsal cortex (Medina and Reiner, 2000; Ulinski, 1990). Retrograde tracing from the dorsal cortex (Desan, 1988; Balaban, 1978; Bruce and Butler, 1984b) and physiological evidence (Belekhova, 1979) show that the visual DVR sends projections to the dorsal cortex, providing an indirect route for the tectofugal pathway to dorsal cortex.

The discoveries of a direct thalamofugal pathway to turtle dorsal cortex (Hall and Ebner, 1970; Knapp and Kang, 1968) and to bird visual pallium (Karten et al., 1973) were important because they identified in non-mammals regions that correspond to mammalian primary visual cortex and set the stage for studies of visual functions in live animals (Pettigrew and Konishi, 1976) and in vitro preparations (Connors and Kriegstein, 1986). However, direct projections from LGN to the dorsal cortex (that define a "primary visual cortex") vary across reptiles. Kenigfest et al. (1997) showed in lizards that direct visual thalamic inputs reach the pallial thickening but not other parts of the dorsal cortex as they do in turtles. The significance of these differences is not yet understood, but a few details of axonal and synaptic morphology worked out in turtles are likely to hold in general for reptiles. In turtles, LGN axons reach the upper layer 1 (L1a) of visual cortex from the lateral edge of cortex, crossing the cell layer of the pallial thickening in thick bundles. Once in layer 1a, these axons run in slightly curved trajectories across the visual cortex along a lateromedial direction, bearing many en passant varicosities (Heller and Ulinski, 1987; Mulligan and Ulinski, 1990). Electron microscopic studies show that thalamocortical axons synapse upon dendritic spines, dendritic shafts, and somata of cortical neurons (Ebner and Colonnier, 1975, 1978; Smith et al., 1980). The synapses are glutamatergic and mediate fast excitatory postsynaptic potentials in cortical neurons (LoTurco et al., 1991; Blanton and Kriegstein, 1991, 1992; Larson-Prior et al., 1991).

In comparison, mammalian visual cortex receives visual input directly through both the tectofugal and thalamofugal pathways. The latter pathway (via the pulvinar) has received less attention because it is weak compared to the LGN input, which mainly targets layer 4 (Harris and Shepherd, 2015). LGN axons have more complex projections than usually acknowledged, with functional synapses in layer 1, layer 2/3, and deeper cortical layers (Cruz-Martín et al., 2014; Roth et al., 2016). In mammals, thalamic axons enter cortex from below and ascend radially, although some types of axons also spread tangentially in layer 1 (Rubio-Garrido et al., 2009). Large bundles of axons ascending to layer 1, as observed in turtles, are uncommon in mammals but have been observed in some small mammals (Valverde and Facal-Valverde, 1986; Flores, 1911) and in Dab1 knockout mice [Dab1 is a regulator of reelin signaling (Tomassy et al., 2014)]. At present, we do not know which aspect of mammalian thalamocortical projections is most comparable to the reptilian condition. We can conclude, however, that both cytoarchitecture and visual afferents in reptiles seem to represent a simpler condition.

Responses of dorsal cortex to visual stimulation: Current injections into turtle pyramidal cells in vitro elicit a regular firing pattern comparable to that seen in mammalian pyramidal cells. Somatic action potentials lead to actively backpropagating dendritic action potentials and an increase in intracellular calcium throughout the dendritic tree. Smaller action potentials, termed prepotentials or spikelets, may be more frequent in turtle than in mammalian pyramidal cells, but many of the electrophysiological properties measured in vitro are comparable (Connors and Kriegstein, 1986; Larkum et al., 2008). Single-unit studies in alert turtles suggest that neurons in this area have large receptive fields with little to no spatial selectivity (Mazurskaya, 1973). Neurons respond well to small moving objects all over the turtle's visual space and show strong habituation, in particular to repeated presentation of stimuli moving in the same direction (Gusel'nikov and Pivovarov, 1978). In contrast, there is no evidence for retinotopy and no indication for orientation selectivity. These features

suggest both spatial convergence and a possible involvement of cortex in motion computation (Ulinski, 1999, 2006). Ulinski et al. studied the convergence of visual projections in turtles and offered a simple model to explain the observed physiological responses (Ulinski, 1990; Ulinski and Nautiyal, 1988). However, these early tracing studies were conducted largely in vitro, and preliminary results from our laboratory indicate that both axonal topography and physiological response properties need to be reexamined with more modern tracing and recording methods (Fournier et al., 2015; Fournier et al., unpublished).

Brain stimulation and lesions: Although we have some knowledge of the visual projections to the reptilian pallium, we know very little about its functions. Removal of the dorsal cortex in turtles or lizards did not influence performance in visual, spatial, or pattern discrimination tasks (Hertzler and Hayes, 1967; Morlock, 1972; Bass et al., 1973; Peterson, 1980; Powers, 1990). Lesions of the visual DVR in turtles, however, impaired performance in both a simultaneous pattern discrimination and a simultaneous visual intensity discrimination (Reiner and Powers, 1980, 1983). The largest behavioral effects resulted from combined lesions of the thalamofugal and the tectofugal pathways (Hertzler and Hayes, 1967). In an interesting parallel to the strong habituation of dorsal cortex neuron responses to moving visual stimuli, lesions of the dorsal cortex impaired habituation to rapidly approaching objects (Killackey et al., 1972; Moran et al., 1997). The measure used for habituation was head withdrawal into the shell and lesioned animals withdrew their heads more often than controls. Thus, dorsal cortex may not relate directly to visual perception but rather make it sensitive to context. In addition, lesions of the dorsal cortex in lizards impair maze learning in lizards (Peterson, 1980).

While lesion studies can establish the involvement of brain structures in certain behaviors, other techniques such as electrical stimulation can be used to directly elicit behavioral reactions. Both lesion and electrical stimulation studies indicate that the cortex is probably not required for normal movements in reptiles (Peterson, 1980; Bremer et al., 1939; Schapiro and Goodman, 1969). Electrical microstimulation was used most extensively in *Iguana iguana* by Distel (1976, 1978a,b). Stimulation sites in the telencephalon evoked head movements or exploratory behavior, although the majority of stimulation sites in the DVR were ineffective. Stimulation of the medial cortex, septum, and nucleus accumbens often directly induced movements and exploration, possibly related to the recent finding that firing glutamatergic neurons in the mammalian septum initiates movements (Fuhrmann et al., 2015). Species-specific displays or aggressive behaviors were most reliably affected by stimulating or lesioning the striatum or the amygdaloid complex but not the cortex or DVR (Distel, 1978a; Keating et al., 1970; Greenberg, 1977; Greenberg et al., 1979, 1988; Sugerman and Demski, 1978). Since the advent of optogenetics and chemogenetics, it has become possible to introduce more controlled and reversible modulations of neural activity. We have screened a wide range of viral vectors that will facilitate the application of these tools to reptilian brains (Pammer et al., unpublished).

Open questions: Principal cells in reptile cortex are usually called "pyramidal" cells, but their morphology differs quite dramatically from prototypical, deep layer mammalian pyramidal cells (Larkum et al., 2008). How does morphology affect signal flow in dendrites, connectivity, and cellular function (Mainen and Sejnowski, 1996)? Do pyramidal cells in reptiles form a uniform group or do they form molecularly defined or functional subtypes (Luis de la Iglesia and Lopez-Garcia, 1997a)? Spine density in reptilian principal cells may be similar to those of mammalian pyramidal cells (Martínez-Guijarro et al., 1984), but we know little about spine numbers, morphology, or turnover. Some long-range efferent projections of reptilian dorsal cortex to other cortical regions, striatum, thalamus, and tectum have been described (see Nieuwenhuys et al., 1998), but evidence for a dorsal cortex projection to the brain stem is less compelling (Medina and Reiner, 2000). Locally, pyramidal cells give rise to axon collaterals that extend into layer 1. Axonal fields may be ovoid, extended along the mediolateral axis over an area of about one-third of diameter of dorsal cortex (Cosans and Ulinski, 1990), but the effect of local axons on other cortical neurons has not been investigated. In summary, while we have a blueprint for cortical circuits in reptiles (Connors and Kriegstein, 1986; Fournier et al., 2015; Shepherd, 2011a), we lack quantitative descriptions of these circuits (Braitenberg and Schüz, 1998; Binzegger et al., 2004; Oberlaender et al., 2012). We are still some distance away from understanding cortical computation (Hubel, 1982; DeFelipe, 2015), but we will try and sketch out the advantages of a comparative approach.

9.1.4 "Model Species" and the Need for Experimental Diversity

After all, the nervous system is the greatest reservoir of differences under the sun. Theodore Holmes Bullock (1984).

There are two major approaches in experimental studies of brain evolution. One emphasizes similarities in the construction of brains across vertebrates—and even across vertebrates and invertebrates (Strausfeld and Hirth, 2013; Hirth and Reichert, 1999). Species differences according to this school are what remains once all the similarities have been defined. The second approach emphasizes the differences. While both approaches are arguably two sides of the same coin, they have different strategic implications. The first accommodates itself to using only few model species; the second does not.

What is a model animal? In the early days of neurobiology, every problem was seemingly solved in a different species. The ionic basis of action potentials was discovered in the squid giant axon (Hodgkin and Huxley, 1952), chemical synaptic transmission was elucidated at the frog neuromuscular junction (Fatt and Katz, 1951), and motor cortex was discovered in dogs (Fritsch and Hitzig, 1870). This historical diversity of model systems is the expression of what is now known as Krogh's principle, which states that for any question in biology there exists a specialized or "champion" animal most appropriate for its elucidation (Krogh, 1929). Thus, because barn owls are supremely specialized in hunting using auditory cues, it was possible to identify a neural mechanism to calculate the location of prey with auditory delay lines (Carr and Konishi, 1988, 1990). With hindsight, ie, after investigating many other vertebrates, it seems unlikely that such a discovery would have been made were it not for the owl (and the scientists who identified it) (Ashida and Carr, 2011; Grothe et al., 2010). Therefore, certain animals are good models for understanding the implementation of neural coding principles precisely because of their unique specializations. Evolutionary neurobiologists also try to find unique animals that may have retained ancestral characters. Turtles and hedgehogs are frequently cited as slowly evolving species (Striedter, 2005), but by relying only on anatomical characters, it is not always easy to separate derived from ancestral or secondarily simplified features. Today, using information about evolutionary dynamics revealed by genome analysis, we may be in a better position to answer these questions.

Unfortunately, another trend, oriented toward translational medicine, is rapidly superceding these comparative approaches: scientific, economic, and political forces have led to a considerable loss of diversity of animal model systems used in biology, including neuroscience. "Genetically tractable" species (mostly mouse and fish among the vertebrates) constitute those with which most biological research is carried out today (see Fig. 9.4 for the evolution of research in the United Kingdom over the past 25 years) (Manger et al., 2008; Beery and Zucker, 2011).

There are more than 60 000 vertebrate species and over 10 000 species of reptiles (Pincheira-Donoso et al., 2013), but taxonomical sampling of animals in research laboratories shows that most research effort goes to no more than a dozen vertebrate species. It would be foolish to deny the importance and remarkable success of such a focus. It has led to new physiological and genetic tools to better standardized drug and vaccine tests (Koch and Reid, 2012). Yet, this lack of diversity has its

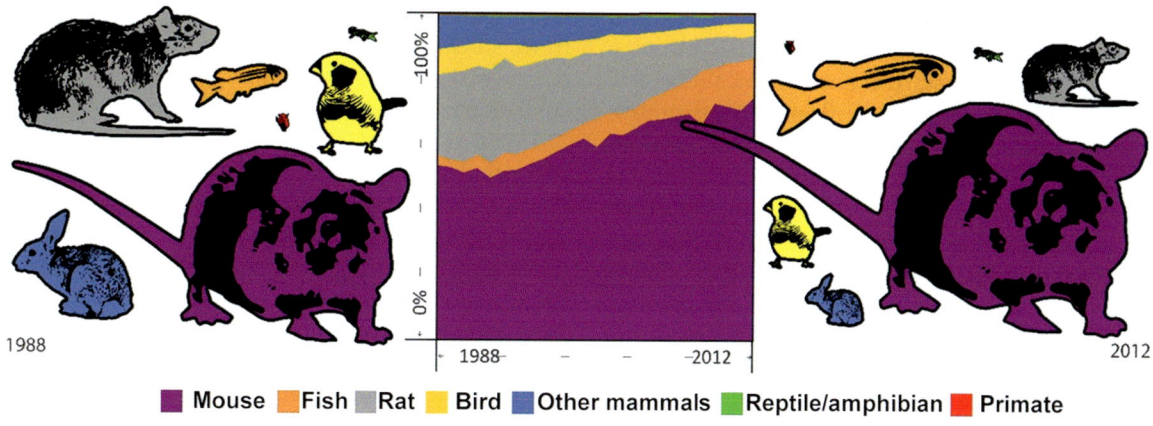

■ Mouse ■ Fish ■ Rat ■ Bird ■ Other mammals ■ Reptile/amphibian ■ Primate

FIGURE 9.4   Statistics on animal use for research in the UK between 1988 and 2012 (Annual Statistics). About 20 years ago, about half of the animals used for research were mice and fish; in 2012, they made up almost 80% of all animals used. Reptiles and amphibians are hardly studied at all. *Reproduced from Annual Statistics of Scientific Procedures on Living Animals Great Britain. https://www.gov.uk/government/collections/statistics-of-scientific-procedures-on-living-animals.*

risks and downsides too. First, animals and their brains differ across species, and we still know little about how to translate findings in mouse or fish to monkey or humans (Bullock, 1984; Preuss, 2000). For example, recent work demonstrated that scaling relations for total cortical neuron numbers exist, but that they differ across rodents and primates, possibly across all other groups of vertebrates (Herculano-Houzel et al., 2006, 2007). Beyond coarse measures of neural circuit complexity, such as volume and neuron numbers, general principles are still scarce.

Second, except for nonhuman primates, most current model species used in biological research today were selected for ease and speed of breeding and genetic tractability, not for ease of access to brain circuits or interesting behaviors. From the standpoint of evolutionary neurobiology, studying highly derived species such as *Drosophila melanogaster* may also have limited explanatory power for reconstructing the evolution of insect brains. Third, every species represents a mix of conserved and novel characters. Without studying species diversity and evolution, we cannot know which characters are innovations. Later, we will consider questions of cortical homology, which turn out to be both complex and disputed; arguments based on homology are important for translating findings from one species to another at a mechanistic level. Conversely, solutions to common problems may find a variety of forms, as revealed beautifully from biophysical and modeling work with small motor circuits (Prinz et al., 2004; Marder and Goaillard, 2006). Making sense of this diversity of solutions seems impossible without embracing the complexity of evolutionary history, and with it, the diversity of life forms and species.

Decades of data collection and analysis of mammalian cortex have not yet delivered a satisfactory explanation of cortical function, linking microcircuit operation to its emergent properties. Could a diversification of approaches help? We believe so. Aside from believing in the sheer value of discovery, we think that deep comparative approaches will play a critical role in modern neuroscience. First, we need to increase the overall diversity of species studied to understand general principles of neural circuit design and gene—structure—function relations (Striedter et al., 2014). In fact, this is the only way to test if any proposed circuit mechanisms can claim to be general. Second, to predict specific functions, we must study closely related species. Thus, research on nonhuman primates is essential if we wish to translate neurobiological findings to humans.

Third, while novel technologies for high-throughput circuit reconstruction and manipulation will continue to be developed and benchmarked in classical model animals, we should strengthen our efforts to translate these techniques to nonclassical organisms. Current techniques for recording and manipulating neural activity in nonclassical organisms are still unsatisfactory, but a string of recent molecular techniques such as cell type—specific viral vectors, transcriptome sequencing, CRISPR-Cas9, and the generation of transgenic lines have advanced to the stage where they can be applied to almost any species. Finally, understanding brain function can only be aided by the study of simpler animal model systems. But can we measure simplicity? We argued in the preceding chapter that reptilian cortical cytoarchitecture is simpler than in mammals. We now ask if that is, true also for cortical cell type diversity.

9.2 Cell Types in Reptilian Cortex

Central to an understanding of cortex is the identification of the role of specific components for computations.

A formidable challenge resides in the multiscale nature of cortical computation (eg, areas, layers, microcircuits, neurons, synapses). For historical and technical reasons, the cellular level has been dominant in systems neuroscience, most importantly because neurons have a well-defined output (action potentials) that we can try to correlate with behavioral performance. Neuronal cell types, unlike genes, do not directly replicate, yet retain their identity despite being generated anew within each generation (Wagner, 2014). Therefore, we suggest that neurons (neuronal cell types) constitute the lowest neuroanatomical level at which we can define unique characters that can be recognized across individuals. We differentiate cell types based on their molecular components, morphology, electrophysiological properties, firing patterns during behavior, and finally the computations they perform (Fig. 9.5). Influential theoretical work by David Marr on the visual system indicates that computational and implementation levels can be dealt with independently of one another (Marr, 1982; McCulloch and Pitts, 1943). However, we can rarely identify neuronal function at first sight, nor can we assess whether a computation is implemented at the synaptic, cellular, or circuit levels (Churchland and Sejnowski, 1992). One can recognize a hierarchy among properties, as depicted in Fig. 9.5, but even the full knowledge of the molecular and morphological makeup of a neuron will determine its physiological properties only to some degree. In turn, many different molecular, biophysical, morphological, and wiring properties can in principle enable the same computation.

How do we overcome these difficulties in practice? One approach is to identify cell types—an exercise in neuronal taxonomy—and later figure out their function (Fig. 9.6; Tyner, 1975; Masland, 2004). As we illustrate below for the retina, strong structure–function correlations can simplify this task, but more critical is the ability to access the same cell population repeatedly across many animals and brains. In the extreme, we can identify specific neurons because they are unique, such as the so-called LGMD, a looming-sensitive neuron in locusts (O'Shea and Williams, 1974; Hatsopoulos et al., 1995; Fotowat and Gabbiani, 2011). The concept of

identifiable neurons with distinct functions emerged from studies of arthropod brains (Bullock, 2000), but there exist a few comparable examples in vertebrates, such as the Mauthner cell (Zottoli and Faber, 2000).

Here, we illustrate the concept of computation in identified neurons with an example from the locust visual system and related studies in vertebrates. Detecting time-to-contact is needed to avoid obstacles, prepare for landing, capture prey, and evade predators. In locusts, looming-sensitive neurons respond to approaching objects according to a simple function, ie, proportional to the ratio of the angular velocity of the approaching object to an exponential function of its size. Accordingly, their firing rate consistently peaks a short time before the animal jumps to safety (Fotowat et al., 2011), with a peak time determined both by object size and approach velocity (Hatsopoulos et al., 1995; Gabbiani et al., 2002). Excitatory (velocity-dependent) and inhibitory (size-dependent) inputs are generated in different portions of the dendritic tree, sum near the axon initial segment, and are exponentiated to generate a spike output (Gabbiani et al., 2002). These findings show in detail how biophysical properties, firing patterns, and computation are interrelated, and they constitute one of the first pieces of evidence on how to implement a particular type of computation—a multiplication—in single neurons.

How do we translate such findings to vertebrate brains, where neurons are usually not unique and identifiable? Are similar computations carried out by single cells or distributed across cell populations? In mammals, looming-sensitive neurons are found in the retina (Münch et al., 2009; Yilmaz and Meister, 2013) and the optic tectum (Liu et al., 2011). Among vertebrates, looming-sensitive neurons are best characterized in the pigeon, where three distinct classes of neurons have been reported in the thalamic nucleus rotundus (Sun and Frost, 1998). One of these classes expresses firing profiles similar to those of the locust LGMD, suggesting a similar type of computation. Some of these responses are transmitted to the entopallium, the visual part of the DVR in birds (Xiao et al., 2006). In reptiles, the homologous regions have not been studied using

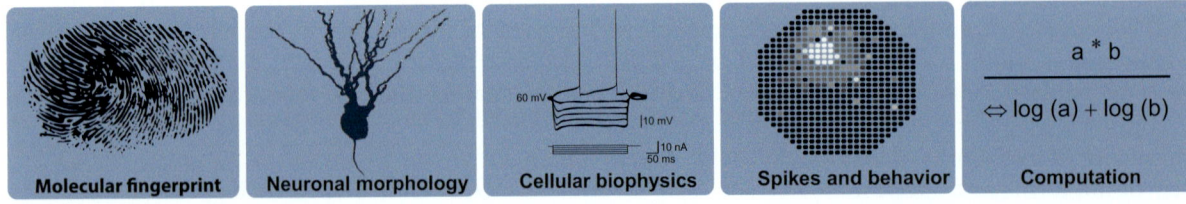

FIGURE 9.5 Hierarchy of cell type classification. The molecular fingerprint represents the constituent molecules of a neuron and their changes during development. Neuronal morphology reflects afferent and efferent connectivity. The biophysical properties of a cell are determined by morphology and molecular components.

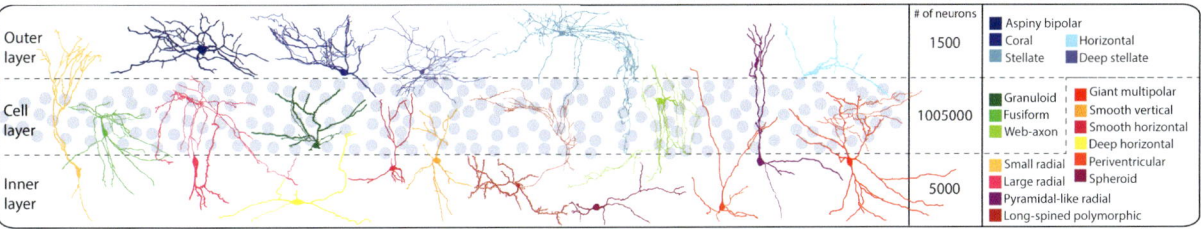

FIGURE 9.6 Morphology of interneuron types in the medial cortex of the lizard, based on axonal morphology, laminar position, dendritic morphology, and marker expression. Redrawn from Luis de la Iglesia, J., Lopez-Garcia, C., 1997. Neuronal circuitry in the medial cerebral cortex of lizards. In: International Work-Conference on Artificial Neural Networks. Springer, Berlin, Heidelberg, pp. 61–71; Lopez-Garcia, C., Tineo, P.L., Del Corral, J., 1983. Increase of the neuron number in some cerebral cortical areas of a lizard, *Podarcis hispanica*, (Steind., 1870), during postnatal periods of life. J. Hirnforsch. 25 (3), 255–259; Luis de la Iglesia, J.A., Lopez-Garcia, C., 1997. A Golgi study of the short-axon interneurons of the cell layer and inner plexiform layer of the medial cortex of the lizard *Podarcis hispanica*. J. Comp. Neurol. 385, 565–598; Luis de la Iglesia, J.A., Martinez-Guijarro, F.J., Lopez-Garcia, C., 1994. Neurons of the medial cortex outer plexiform layer of the lizard *Podarcis hispanica*: Golgi and immunocytochemical studies. J. Comp. Neurol. 341 (2), 184–203.

physiological methods, but turtles withdraw their heads into the shell when presented with looming stimuli (Hayes and Saiff, 1967). In the turtle dorsal cortex, which is reciprocally connected to the DVR, looming stimuli evoke persistent field potential oscillations (Prechtl et al., 2000) and ablating the dorsal cortex impairs behavioral responses to looming stimuli (Killackey et al., 1972; Moran et al., 1997). In conclusion, the most detailed studies of looming-computation come from nonmammals (Gabbiani et al., 2002; Sun and Frost, 1998), illustrating the value of model system diversity. These studies also indicate that it is, in the best of cases, possible to identify the essence of a neuronal computation directly from firing pattern analysis and to derive strong clues about physical implementation from morphological features.

9.2.1 Retinal Cell Types in Turtles, Ex Vivo Preparations of Nervous System in Reptiles

Although our focus is on reptilian cortex, we briefly discuss some aspects of cell-classification in the retina as an example of a highly organized, layered neural tissue with a large number of known cell types; the retina may, indeed, be the first part of the brain with a complete list of identified neurons. In the retina, photoreceptors transduce impinging light into electrical signals, which are processed by many different cell types in the inner cell layer and relayed to the brain via ganglion cells. Today, we recognize close to 100 cell types in the mouse retina (Sanes and Masland, 2015; Baden et al., 2016). Interestingly, a similar diversity of retinal cell types was described long before in turtles (Ammermüller and Kolb, 1995, 1996; Ammerüller et al., 1995; Kolb, 1982; Peterson, 1992). Early accounts dismissed the complexity of retinal cell types in reptiles as features of "lower" animals with "simple" brains. We now realize

that retinal cell diversity is common (if not the rule) across vertebrates. Research on the retina has been greatly helped by a close relation between structure and function. In many cases, molecular, morphological, and physiological characteristics define cell types uniquely. A second advantage of cell type identification in the retina was the development of reduced (in vitro) experimental preparations with complete control over the stimulus (light) and full access to the output (retinal ganglion cell activity). Because reptiles, and especially turtles, are resistant to anoxia, the turtle retina was long ago adopted for in vitro physiological recordings (Baylor and Fettiplace, 1976), with olfactory (Mori and Shepherd, 1979) and whole-brain ex vivo brain preparations (Kriegstein, 1987) following soon thereafter. In the cortex, only a few cell types were initially described using these preparations (Kriegstein, 1987), but in the retina, extensive studies led to the anatomical and physiological characterization of nearly 100 distinct cell types (Ammermüller and Kolb, 1995, 1996; Ammerüller et al., 1995).

One notable result was the discovery of a diverse set of retinal cell types related to color vision. Color vision in many reptiles is tetrachromatic (red, green, blue, and UV). In addition, colored oil droplets in the inner segment of the photoreceptors can act as color filters (Fig. 9.3). Together, the combination of oil droplets, multiple photoreceptor subtypes, and the direct connections between some red and green photoreceptors make the turtle retina a great preparation to study basic principles of color vision (Kolb and Jones, 1982; Ohtsuka, 1985a,b). The retina of many reptiles contains a region of higher photoreceptor density; some even share with primates and birds the existence of a specialized retinal region called the fovea. In primates and birds, the fovea is important for high-acuity vision and color discrimination due to its specialized arrangement and combination of cell types. The retina in reptiles also contains regions of high density of photoreceptors and ganglion cells, but the functional relevance of these features has not been explored in depth.

In summary, several lessons for cortex can be learned from studies of the retina. First, the high cell type diversity in the retina of reptiles is not a feature of "lower" vertebrates but could be a general characteristic of the retina in all vertebrates. Second, novel principles for cell type identification such as retinal tiling, the regular nonoverlapping spacing of distinct cell types, and innovative preparations with complete access to physiological parameters could be further explored by studying cortical cell type function in reptiles. Third, chromatic and high-acuity vision are well developed in primates, reptiles, and birds but poorly developed in most other mammalian species. This variability and diversity across orders is ideal for comparative insights.

9.2.2 Cell Types in the Cerebral Cortex, With a Focus on Interneurons

The main cell classes in reptilian cortex are, as in other vertebrates, glia, excitatory neurons, and inhibitory neurons.

Glia: In mammals, glial cells consist of microglia, oligodendrocytes, astrocytes, and radial glia. All of these types are also present in reptiles Ramón y Cajal, 1909; Kruger and Maxwell, 1967; Stensaas and Stensaas, 1968; Castellano et al., 1991) but with variations in morphology and gene or protein expression. For example, glial fibrillary acidic protein (GFAP) labels a group of astrocytes in mammals and birds, but only few such typically star-shaped cells are present in reptiles (Onteniente et al., 1983; Yanes et al., 1990; Kálmán et al., 1994, 1997; Kálmán and Pritz, 2001). The predominant pattern seen in GFAP-stained sections of reptilian cortex is radial fibers thought to belong to radial glia cells (Kriegstein et al., 1986; Yanes et al., 1990). Radial glia are proliferating cells, generating not only glial cells but most of the excitatory neurons of the cortex in vertebrates (Malatesta et al., 2000; Noctor et al., 2001; Weissman et al., 2003). Radial glia generate astrocytes and neurons in mammals but are present only in a few cortical regions in adult mammals. Radial glia in reptiles also generate neurons, but they do not give rise to stellate glia, and radial glia remain widely distributed in adult reptile cortex. In turtles, radial glia also generate cells that may correspond to other (not star shaped) types of mammalian astrocytes (Lazzari and Franceschini, 2006; Clinton et al., 2014). Interestingly, in the mammalian dentate gyrus, radial glia stay active in adult mammals and continue to generate neurons (Cameron et al., 1993). In reptiles, radial glia can also generate neurons well into adult age and even replace neurons from lesioned regions of cortex (López-García et al., 1992; Molowny et al., 1995; Font et al., 2002).

Pyramidal neurons: Estimates of cortical neuron diversity range from a handful to several hundred (Braitenberg, 1978; Stevens, 1998). The current tally in mammals suggests some 50 cortical neuron types (Zeisel et al., 2015; Tasic et al., 2016), comparable to prior estimates based on cellular morphology (Lorente de Nó, 1922). In reptiles, Kriegstein et al. (1986) discovered an antibody that labeled neurons in all regions of the cortex and the DVR, including a region later identified as composed of Tbr1-positive neurons (Moreno et al., 2010). Within this molecularly defined region, we distinguish two main classes of neurons: excitatory projection neurons and inhibitory (local) interneurons (Freund and Buzsáki, 1996). Only few studies have attempted to subdivide the excitatory neurons of the reptilian cortex into molecularly defined subtypes. Instead, classification has relied mainly on morphology as seen in Golgi preparations. The main excitatory neuron type in reptilian cortex is the pyramidal neuron, with several apical large-diameter spiny dendrites. Their axonal projections have not yet been studied comprehensively but are thought to extend both locally and into neighboring cortical regions. Note that reptilian cortical pyramidal neurons bear little morphological resemblance to typical mammalian pyramidal cells—yet, the term is widely used (Ulinski, 1990). Pyramidal cells in reptiles comprise a range of heterogeneous morphological subtypes (Ulinski, 1977; Ramón y Cajal, 1909; Guirado et al., 1984; Davila et al., 1985; Berbel et al., 1987). At one extreme are the so-called double pyramidal cells found in the medial cortex of lizards. These neurons possess two tufts of radially oriented dendrites, both densely covered with spines (Martínez-Guijarro et al., 1990). Based on dendritic pattern, spine coverage, and soma position, Luis de la Iglesia and Lopez-Garcia (1997a) distinguish eight types of projection neurons in the medial cortex of lizards but, so far, no study has linked these morphological types to specific functional or molecular cell classes.

Interneurons: Cortical interneurons in reptiles are found in all three layers. The majority of neurons above and below the main cellular layer (L2) are interneurons. The classification of interneurons is more advanced than that of principal cells. Interest in reptilian interneurons peaked in the early 1990s, when many reliable antibodies became available. Since then, research on interneurons in mammals has been revolutionized by the identification of the developmental origin of interneurons (Marin and Rubenstein, 2003), advances in morphological and molecular classification (DeFelipe et al., 2013), and functional studies (Klausberger and Somogyi, 2008; Kepecs and Fishell, 2014). With few exceptions (Tanaka et al., 2011), interneurons in reptiles have not been studied in the light of these new developments. However, we know that the subcortical origin and tangential migration of interneurons into the pallium predates the divergence of reptiles and mammals (Cobos et al., 2001; Métin et al., 2007; Tanaka and Nakajima, 2012).

In mammals, Kubota et al. (1994) and Gonchar and Burkhalter (1997) first classified cortical interneurons into three major groups encompassing the many different individual types. Two large groups of interneurons express either parvalbumin (PV) or somatostatin (SST). The third major group originates from a different zone during development and is more variably characterized by the expression of a type of serotonin receptor (5HT3A), vasoactive intestinal polypeptide (VIP), calretinin (CR), or cholecystokinin (CCK) (Kawaguchi and Kubota, 1997; Gonchar et al., 2008; Rudy et al., 2011; Xu et al., 2010). Details of these classifications are still being fine tuned, but given the power of recent techniques for morphological, physiological, and molecular categorization, a complete characterization of interneuron diversity is within reach (Zeisel et al., 2015; Tasic et al., 2016; Jiang et al., 2015; Fuzik et al., 2016; Cadwell et al., 2016).

The most detailed studies relating interneuron morphology, marker expression, and physiology in mammals derive from the hippocampal formation (Freund and Buzsáki, 1996; Klausberger and Somogyi, 2008). Similarly, in reptiles, the most extensive data set on morphology and molecular markers comes from the medial cortex of lizards (Fig. 9.6; Luis de la Iglesia and Lopez-Garcia, 1997b; Guirado and Davila, 1999). Most, if not all of the neurons, of the outer and inner layer are interneurons, while there are only few interneurons in the main cellular layer. Based on Lopez-Garcia et al. (1983) (Fig. 9.5), the fraction of interneurons can be estimated as 6–7% in lizard medial cortex (Luis de la Iglesia and Lopez-Garcia, 1997b,c), the remainder being excitatory neurons, that is, a proportion significantly lower than that in mammalian cortex (DeFelipe, 2015). In turtles, data on interneuron marker expression are extensive and reviewed in detail by Reiner (1991, 1992, 1993), but a major shortcoming is that there are few quantitative or morphological data on turtle interneurons. Table 9.1 summarizes the literature on lizard and turtle interneuron markers obtained largely by immunohistochemical techniques. In the absence of a hierarchical classification into interneuron subgroups for reptiles, we ordered the entries by marker types, ie, neurotransmitters, neuropeptides, and calcium-binding proteins.

At the top of the list is GABA, the defining transmitter for interneurons (Blanton et al., 1987; Schwerdtfeger and García, 1986). In addition, some interneurons can release other transmitters such as acetylcholine and nitric oxide, usually identified by the presence of their synthesizing enzymes. Nitric oxide synthase (NOS) is present in turtles and lizard cortex, but much more abundant in lizards (Brüning et al., 1994; Dávila et al., 1995). Choline acetyl transferase appears to be absent from turtle cortex but present in lizards (Reiner, 1993; Medina et al., 1993). Serotonergic nerve fibers innervate several types of interneurons in the cortex of lizards (Martínez-Guijarro et al., 1994), but there are no comparable data on serotonin receptors. In mammals, a specific type of serotonin receptor (5HTR3A) marks a major subgroup of interneurons (Tremblay et al., 2016). Preliminary observations from our laboratory indicate that GABAergic neurons expressing 5HTR3A are present in turtle cortex (Müller et al., unpublished).

Neuropeptides are a large (currently about 100), well-conserved group of small- to medium-sized signaling molecules often expressed in interneurons (Nässel and Homberg, 2006; Baraban and Tallent, 2004). As observed with neurotransmitters, the neuropeptides themselves, their synthesizing enzymes, or their receptors are often differentially expressed in neurons. Here, we focus on neuropeptides in reptile cortex that were shown to be present in interneurons by their laminar distribution, costaining with GABA, or identified as interneuron markers in mammals. For example, we exclude galanin and orexin neurons, involved in sleep regulation (Sakurai, 2007), because they are present only in subcortical regions in reptiles (Domínguez et al., 2010; Jiménez et al., 1994). In addition to neuropeptides, hormone receptors, eg, genes for estrogen and progesterone receptors are expressed in the dorsal cortex of lizards (Young et al., 1994), but it is not known whether they are restricted to interneurons. For simplicity, we also neglect differences between cortical regions and closely related species (eg, different lizard species).

Interestingly, some of the most commonly used neuropeptide interneuron markers for mammalian cortex (CCK, VIP) are not present in interneurons of reptile cortex (Reiner, 1993; Reiner and Beinfeld, 1985; Reiner et al., 1985; Petko and Ihionvien, 1988; Guirado et al., 1998). In contrast, both SST and NPY are present in both turtle and lizards, coexpressed in the same neurons (Weindl et al., 1984; Reiner and Oliver, 1987; Dávila et al., 1991). Whereas SST and NPY are expressed in many neurons, neuropeptide SP, FMRFamide, NPFF, Enkephalin, and CRF are restricted to very few neurons, often found only in the medial cortex. Reelin is a special case, for it is not a neuropeptide but rather a large protein and signaling molecule. Reelin was studied extensively for its role in cortical development (Bar et al., 2000). In adult mammals, most reelin-positive neurons are interneurons (Alcántara et al., 1998) and this appears to hold true for reptiles (Pérez-García et al., 2001). The last group of common interneuron markers is the calcium-binding proteins PV, calbindin (CB), and CR. CR-positive neurons in lizards preferentially synapse onto cell bodies

TABLE 9.1　Candidate interneuron markers in turtles and lizards

Interneuron marker	Turtle	Lizard	References
Neurotransmitters, their receptors or synthesizing enzymes			
Gamma aminobutyric acid (GABA)	Yes	Yes	Blanton et al. (1987) and Schwerdtfeger and García (1986)
Neuronal nitric oxide synthase (nNOS)	Yes	Yes	Regidor and Poch (1988), Brüning et al. (1994), and Dávila et al. (1995)
Choline acetyl transferase (ChAT)	No	Yes	Reiner (1993) and Medina et al. (1993)
Serotonin receptor 3A (5HTR3A)	Yes	Unknown	Müller et al. (unpublished)
Neuropeptides and reelin			
Vasoactive intestinal polypeptide (VIP)	No	No	Reiner (1992) and Petko and Ihionvien (1988)
Cholecystokinin (CCK)	No	No	Reiner and Beinfeld (1985), Reiner et al. (1985), and Guirado et al. (1998)
Somatostatin (SST)	Yes	Yes	Weindl et al. (1984), Reiner and Oliver (1987), and Dávila et al. (1991)
Neuropeptide Y (NPY)	Yes	Yes	Reiner and Oliver (1987) and Dávila et al. (1993)
Substance P (SP)	Yes	No	Reiner (1992), Reiner et al. (1985, 1984), Petko and Ihionvien (1988), and Bennis et al. (1994)
FMRFamide	Unknown	Yes	Reiner (1992) and Vallarino et al. (1994)
Neuropeptide FF (NPFF)	Yes	Yes	Smeets et al. (2006) and Muñoz et al. (2008)
Enkephalin/opioids	Yes	Yes	Reiner (1992) and Luis de la Iglesia et al. (1994)
Corticotrophin releasing factor (CRF)	Yes	Unknown	Fellmann et al. (1984)
Reelin	Yes	Yes	Pérez-García et al. (2001)
Calcium-binding proteins			
Parvalbumin (PV)	Unknown	Yes	Reiner (1993), Wang et al. (2011), and Martínez-Guijarro and Freund (1992)
Calbindin (CB)	Yes	Yes	Belekhova et al. (2003), Martínez-Guijarro and Freund (1992), and Dávila et al. (1993)
Calretinin (CR)	Yes	Yes	Davila et al. (1997), Belekhova et al. (2003), Wang et al. (2011), and Martínez-Guijarro and Freund (1992)

Most of the data were obtained by immunohistochemical staining methods.
No, not detected; unknown, not tested.

and proximal dendrites of principal cells (Guirado and Davila, 1999). In contrast to the medial cortex, some CR-positive interneurons in the dorsal cortex coexpress PV (Davila et al., 1997). In turtles, CR is found in some interneurons, albeit sparsely (Belekhova et al., 2003; Wang et al., 2011).

When we sort the interneuron markers according to the classical tripartite mammalian scheme (SST, PV, and VIP/5HTR3A), markers for the SST-positive group appear to be the most conserved ones in reptiles.

Variability among the other groups is larger; for example, VIP and CCK have not been detected in reptilian cortical interneurons. However, the number of known interneuron markers in reptiles is too small for a detailed comparison with mammalian interneurons. In general, single interneuron markers do not uniquely identify interneuron types. For example, reelin-positive interneurons in mammals can belong to all three major groups (Tasic et al., 2016). Thus, both in mammals and in reptiles, a combination of markers is necessary to assign interneuron identities. In turtles, very little is known about coexpression of interneuron markers (Reiner and Oliver, 1987), but work from our laboratory has quantified expression and overlap for all major interneuron markers in turtles (Müller et al., unpublished). Although studies of interneurons in lizards largely predate the widespread use of multicolor confocal fluorescence microscopy for colocalization, some estimates of overlap between interneuron markers are available. Martínez-Guijarro and Freund (1992) estimate that 20% of the GABAergic neurons in lizard cortex contain calcium-binding proteins. Only a small percentage of PV-positive interneurons express SST or neuropeptide Y, indicating a segregation of these major groups similar to that in mammals. In lizard cortex, the SST/NPY neurons comprise 20–30% of all GABAergic neurons, with smaller percentages of neurons staining only for SST or NPY (Dávila et al., 1993). CR and other interneuron markers are generally present in lower densities, but may show strong regional variations, eg, in the dorsomedial cortex in lizards (Davila et al., 1997). A rather large fraction of GABAergic neurons in lizard (about 40%) does not express SST, NPY, or PV (Dávila et al., 1993). As shown in Table 9.1, a wide range of other interneuron markers is expressed in lizard cortex, but their relation to the major interneuron markers is unknown.

The second major source of information on interneuron diversity in reptiles comes from morphological studies. More than 1000 Golgi-impregnated neurons of the lizard medial cortex were used to define 18 different morphological interneuron types in lizard medial cortex (Fig. 9.6; Luis de la Iglesia and Lopez-Garcia, 1997c; Luis de la Iglesia et al., 1994). These authors identified 5 types in the outer layer, 3 types in layer 2 and 10 in layer 3. In Fig. 9.6, we redraw these morphological types to illustrate dendritic diversity. Most of the interneuron types in the outer layer of lizard medial cortex are PV positive; a small fraction also contains opioids (Luis de la Iglesia et al., 1994). Substance P abundantly labels outer layer interneurons in turtle cortex, but has not been demonstrated in lizards (Reiner, 1993; Petko and Ihionvien, 1988). Overall, relatively few markers for outer layer interneurons are known, and because stellate and deep stellate neurons both express PV, it is not clear whether they constitute truly separate interneuron classes (Luis de la Iglesia et al., 1994). However, based on thalamic and corticocortical input, the outer layer can be subdivided into superficial and deep layers, and their differential inputs may confer different functions to stellate and deep stellate neurons.

Interneurons in the main cell layer of lizard cortex are scarce and in some cases express PV (Luis de la Iglesia and Lopez-Garcia, 1997c). In mammalian cortex, chandelier or axo-axonic cells are a unique type of interneuron. The axons of these neurons, first described in Golgi preparations (Szentagothai and Arbib, 1974; Somogyi, 1977), terminate in vertical cartridges that synapse on the axon initial segment of pyramidal neurons. Subsets of chandelier cells are either PV or CRF positive (Lewis and Lund, 1990; Taniguchi et al., 2013). Are there chandelier cells in reptiles? Martínez-Guijarro et al. (1990, 1993) report low numbers of PV-positive axoaxonic contacts onto principal cells but no full morphology of axoaxonic cells or chandelier cells. Based on axonal morphology, Luis de la Iglesia and Lopez-Garcia (1997c) conclude that the web-axon interneurons of the cell layer are the most likely candidate for reptilian axoaxonic cells. Turtles may lack PV interneurons (Reiner, 1993), but a few CRF-positive neurons are present in the medial cortex (Fellmann et al., 1984). In summary, axoaxonic cells are either absent from reptilian cortex or insufficiently differentiated from other interneuron types.

The largest and most differentiated interneurons in lizard medial cortex are located in the deep layer. PV-positive interneurons are more frequent superficially within L3, whereas SST/NPY-, CR-, and CB- neurons are more abundant closer to the ventricle. The long-spined polymorphic interneurons have probably the most distinctive morphology of all deep layer interneurons. They have two to four long, thick dendrites with a dense covering of 30–50 µm long "spines" or microdendrites. Their axons traverse the cell layer and ascend to the most superficial part of the outer layer where they ramify extensively. Long-spined polymorphic interneurons are NPY positive (Luis de la Iglesia and Lopez-Garcia, 1997c) and, correspondingly, SST/NPY-immunoreactive axons are particularly abundant in superficial L1. The synaptic targets of SST-positive boutons are distal dendrites of spiny, most likely excitatory neurons (Luis de la Iglesia and Lopez-Garcia, 1997c).

9.2.3 Some Limitations of Cell Classification

Both classic and recent anatomical, functional, developmental, and molecular studies directed at characterizing mammalian cortical neurons point to an increasingly large "zoo" of cell types (Lorente de Nó, 1922; Ramón y Cajal, 1909; Tasic et al., 2016; Jiang et al., 2015). What are we to make of this diversity? Are all the details important, and if not, which ones are critical? How do we derive "principles", if all we do is collect "properties"? We argued above that a sequential approach with large-scale molecular and anatomical screening of cell types is a good starting point, but there are practical limitations. For example, genetic or morphological cell type identification can be a prelude to selective study and manipulation of a specific cell type and uncovering variability within that class. However, it appears that single anatomical or molecular features are rarely enough to uniquely identify cell types. If we need many parameters to differentiate cell types, physiological methods that allow concurrent cell type identification are currently limited to studying single neurons (Fuzik et al., 2016; Cadwell et al., 2016; Tukker et al., 2007; Viney et al., 2013). The most complete data set on cortical interneuron diversity stems from decades of work on the CA1 region of the rodent hippocampus (Freund and Buzsáki, 1996; Klausberger and Somogyi, 2008), yet only recently did it become possible to construct quantitative models taking cell type diversity into account (Bezaire and Soltesz, 2013). Finally, even if we knew everything about a cell type within an interconnected network, it might not be possible to predict aggregate properties. For example, the effect of a neurotransmitter on a circuit depends on the postsynaptic receptors (of which there often are many subtypes), not on the transmitter itself. In addition, most of the aforementioned descriptions were static, but we know that functional properties can change dramatically with context.

9.3 Comparing Brain Areas and Cell Types Across Species

An idealized concept of novel brain parts considers them as structures for which there exists no trace in the ancestral phenotype. However, defining novelty can be even more challenging than defining cell types because brain structures are usually modifications of some prior elements and do not appear out of thin air (Striedter, 2005). Among examples for novel brain structures in vertebrates, the corpus callosum may be a relatively easy case (Wagner, 2014; Suárez et al., 2014). The corpus callosum, a massive fiber tract connecting the two pallial hemispheres, exists in placental mammals but not in marsupials. In lizards and snakes, commissural projections exist between several regions of the cortex (Voneida and Ebbesson, 1969; Butler, 1976; Martínez-García et al., 1986, 1990; Lohman and Hoogland, 1979), but evidence for commissural corticocortical connections in turtles is not conclusive (Ulinski, 1990). The commissural projections in reptiles arise mainly from pyramidal cells and run through the anterior pallial commissure. The best-studied commissural connections in reptiles originate in the dorsomedial and medial cortices from which they project to the contralateral septum and contralateral cortical regions (Lohman and Mentink, 1972; Ulinski, 1976; Hoogland and Vermeulen-VanderZee, 1989); therefore, they have been

compared to the hippocampal commissure in mammals. However, the main difference between the corpus callosum and other commissural fiber tracts, like the anterior commissure, is that, during development, corpus callosum fibers interrupt the integrity of the embryonic brain surface in order to travel from one forebrain hemisphere to the other. Since the corpus callosum has no morphological or developmental counterpart outside of placental mammals (Suárez et al., 2014), what can we learn here about reptilian brains?

In mammals, cortical principal neurons fall into several groups, such as brain stem—projecting neurons (PT), corticothalamic neurons (CT), and corpus callosum PT. For each of these projection neuron groups, specific markers have been identified that are, in some cases, involved in generating the projections. For example, Satb2 is expressed in callosal PT, and the absence of Satb2 prevents the formation of the corpus callosum. We can now test these markers (eg, CTIP2 for PT or TLE4 for CT) in reptiles and determine whether they define distinct sets of cortical projections. An interesting case is Satb2 which, despite the absence of a corpus callosum in reptiles, is widely expressed in dorsal cortex (Fig. 9.7; Nomura et al., 2013; Suzuki and Hirata, 2014). This suggests that Satb2 alone cannot be responsible for a callosal phenotype. Hence (and not surprisingly), a single gene is probably not enough to define a specific projection neuron type.

How many genes, then, are involved? As shown by Belgard et al. (2013), characterization of cell types and regions may in some cases involve thousands of genes. In their comparison of regional gene expression in the adult chicken and mouse forebrain, they found that selecting any single gene can support many different hypotheses, only when using a large pool of genes could they recover homologies of the hippocampal formation and the striatum. Belgard et al. (2013) restricted their analysis to the 5000 most highly expressed genes, but are highly expressed genes more informative than the rare ones? If expression of single genes is so variable, do the same genes even have the same function across species? Which layers and regions of the cortex should we really compare? Are any two adult animals comparable or do we need to compare also different developmental stages? For Satb2, a part of the answer may come from a study by García-Moreno and Molnár (2015). Using lineage tracing in developing mice and chickens, they followed a homologous subset of progenitor cells in both species. In mice, some of these cells become committed to generating callosal neurons and express Satb2. In chickens, homologous cells follow a different developmental trajectory and do not generate callosal projections; yet some of these cells express

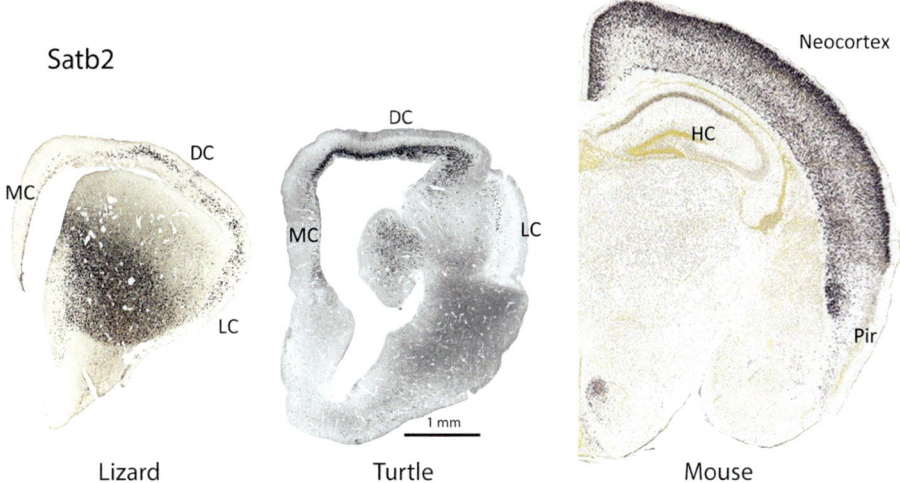

FIGURE 9.7 Interspecies comparisons of cortical regions based on single markers—here Satb2—(protein or gene expression). Satb2 is expressed mostly in the dorsal pallium and the dorsal ventricular ridge but very little in lateral or medial cortex. Despite dramatic differences in layering between mammalian neocortex and reptilian dorsal cortex, marker expression can be used to define comparable regions in the cortex of reptiles and mammals. Mouse brain section from Allen Brain Atlas (Thompson et al., 2014). *MC*, medial cortex; *DC*, dorsal cortex; *LC*, lateral cortex; *HC*, hippocampus; *pir*, piriform cortex. *Reproduced from Thompson, C.L., Ng, L., Menon, V., Martinez, S., Lee, C.K., Glattfelder, K., Garcia-Lopez, R., 2014. A high-resolution spatiotemporal atlas of gene expression of the developing mouse brain. Neuron 83 (2), 309–323.*

Satb2 (García-Moreno and Molnár, 2015; Suzuki et al., 2012), indicating that Satb2 expression and generation of callosal connections can occur independently.

9.3.1 Theories of Cortical Evolution and Their Predictions

Much of the discussion about the evolution of cortex revolves around the thorny and difficult issue of homology: namely, which elements of the cortex are homologous across reptiles, birds, and mammals? Evolutionary novelties, such as the corpus callosum, are convenient examples that may be helpful to define homologies; but in most cases, the situation is complicated. The definition of homology is itself a matter of ongoing debates (Wagner, 2014). In the study of brain evolution, some have stressed the importance of establishing homologies at different levels of analysis, such as cell types, connections, or larger brain regions (Striedter and Northcutt, 1991; Faunes et al., 2015). Others maintain that only one criterion can define homology: similarity in relative position in early development (Puelles and Medina, 2002; Nieuwenhuys and Puelles, 2016), with other criteria such as anatomical similarity, as helpful support.

Is the neocortex "new" after all? Even homologous brain regions can differ in size, connections, or function. These quantitative differences can be large enough to define a qualitative change, a novel brain structure. How do the neocortex and the dorsal ventricular ridge fit in? Early theories of cortical evolution held that

forebrain components without obvious counterparts in other species were novelties, added onto a common frame. Species characters are thus either primitive or derived in this scala naturae view of animal phylogeny. Thus, at some point in the evolution of mammals, a six-layered neocortex appeared. No other vertebrates possess a six-layered cortex, and there are no surviving intermediate forms with three-, four-, or five-layered neocortex (Puelles, 2011).

In the evolution of reptiles and birds, the DVR emerged as a large nuclear structure without apparent counterpart in, eg, fishes or amphibians. "Ladder" theories of evolution are long obsolete, and both the neocortex and the DVR must have had a precursor in evolution. But what was the nature of these precursors? Were they small regions, with little functional significance? Was the precursor of the neocortex and the dorsal ventricular ridge the same region? In the case of the DVR, there is a comparable region in amphibians based on developmental topology and a specific combination of markers (Brox et al., 2004). This provides strong evidence for the existence of a dorsal ventricular ridge precursor region in the common ancestor of reptiles and amphibians, although sensory connections and function of the DVR changed dramatically in reptiles. Finding the precursor of the neocortex is more complicated. So far, there is no marker combination that would uniquely identify the neocortex—or its precursor, in the absence of a six-layered structure. The genes used for establishing common developmental patterns of the cortex in different vertebrates, such as Emx1, are often expressed across several sectors of the pallium (Puelles et al., 2000; Smith-Fernandez et al., 1998). Most commonly, the dorsal cortex of reptiles and the neocortex of mammals are assumed to share a common precursor. Yet, the pathway that has played a key role in the debate about a reptilian homologue to mammalian neocortex, the visual projection from dorsal thalamus to dorsal cortex, is surprisingly variable across reptilian orders (Bruce and Butler, 1984b; Heller and Ulinski, 1987; Wang and Halpern, 1977; Lohman and van Woerden-Verkley, 1978). Instead, exploiting similarities in position, architecture, and efferent connections, several authors have compared the reptilian medial cortex and parts of the dorsal cortex to the hippocampal formation of mammals (Striedter, 2016; Unger, 1906; Filimonoff, 1963; Hoogland and Vermeulen-VanderZee, 1989; Kirsche, 1972; Schwerdtfeger and Germroth, 1990; Lohman and Smeets, 1991). In conclusion, this could mean that only a subregion of the reptilian dorsal cortex shares a common ancestry with the mammalian neocortex. Defining the extent, connections, and function of this region would help to decide if the neocortex is "new" after all.

Evolution of cortical microcircuits: The problem of neocortical evolution is often addressed using the neocortex as a starting point. Gordon Shepherd instead suggested starting with the simplest layered pallium, represented by the olfactory cortex (Rowe and Shepherd, 2016; Shepherd, 1974, 1988, 2011b). Arguably, the olfactory cortex is the region of the cortex least changed in vertebrate evolution and has provided a model for a basic cortical circuit that is reflected in circuit elements of the dorsal cortex of reptiles (Connors and Kriegstein, 1986; Smith et al., 1980) and probably also of the neocortex of mammals. The argument for similarity regarding the ancestral cortical circuits rests on the common features of tangential organization, with afferents running in the superficial-most layer and tangential associative networks beneath. Mammalian olfactory and reptilian olfactory and dorsal cortices all exhibit a similar laminar organization and no apparent topographic mapping of the input (visual for dorsal cortex) (Fournier et al., 2015; Rowe and Shepherd, 2016; Shepherd, 2011b), suggesting similar modes of associative processing. Thus, cerebral cortex probably evolved from a basic tangential associative circuit with little radial and modular structure (Aboitiz and Montiel, 2015). Note, however, that modular structures have been reported to appear transiently during development in the pallium of reptiles and birds (Davila et al., 1999; Kovjanic and Redies, 2003; Suárez et al., 2006).

Cortical projection neurons: Rowe and Shepherd (2016) suggest that the dominant projection neurons in the dorsal cortex of reptiles are intratelencephalic (Fig. 9.8). In mammals, these intratelencephalic projection neurons can be differentiated by laminar position,

morphology, inputs, and targets, such as the specialized thalamorecipient stellate cells of layer four of primary sensory cortices. They propose, eg, that brain stem PT and CT are novel, ie, found only in mammals. Indeed, evidence for brain stem PT in reptilian dorsal cortex is weak, but corticothalamic projections have been described (Ulinski, 1986). At this point, however, too little is known about reptilian cortical projections, and projection neuron molecular markers need to be identified.

Dugas-Ford et al. (2012), by contrast, propose that reptilian dorsal cortex contains both intratelencephalic and brain stem PT but that they are spatially segregated. Their work, consistent with the cell type—equivalence hypothesis of Karten (1969), uses two genes, Rorb and Eag2, that are preferentially expressed in the anterior dorsal cortex of turtles, and a third gene, Er81, that is expressed in the posterior dorsal cortex. Because Rorb and Eag2 are found in layer 4 and Er81 in layer 5 of mammalian neocortex, their data are interpreted as supporting the existence of different projection neuron types. Yet, Rorb is not exclusive to sensory neocortex but broadly expressed, in areas such as the motor cortex, the lateral amygdala and the medial pallium. Also, neither Rorb nor Eag2 is exclusive to layer 4—they are present also in layer 5 (Zeisel et al., 2015; Tasic et al., 2016). Finally, Er81, Rorb, and Eag2 are expressed relatively late in cortical development, and unlike Satb2, eg, they have no known association with generating specific projection neuron types. Thus, these genes may not be well suited to establishing cell type homologies (Medina et al., 2013). This work points to the importance of determining the genetic identity of subcortical projection neurons in reptilian cortex and their projections via pathway tracing studies.

Reptilian cortex versus mammalian deep layers: Reiner (1991, 1993) observed that interneuron types found only in the upper layers of mammalian cortex (VIP, CCK, ChaT interneurons) are absent in the turtle dorsal cortex; by contrast, cell types prominent in layers 1, 5, and 6 of mammalian neocortex (SST, NPY, SP) are found in turtle dorsal cortex. He used this evidence to support the idea that layers 2—4 are recent mammalian additions to a reptilian framework based on layers 1, 5, and 6 (Fig. 9.8). Yet, while some interneuron subtypes may be more abundant in certain layers of mammalian cortex, their distribution is often not exclusive. In mammals, VIP-, CCK-, SP-, and SST-positive interneurons can be found in all layers (Tasic et al., 2016; Peters et al., 1983; Kaneko and Mizuno, 1994). In addition, ChaT-positive neurons are found in lizard cortex, although they appear to be absent from all other reptiles investigated (Reiner, 1991; Mufson et al., 1984; Brauth et al., 1985; Hoogland and Vermeulen-VanderZee, 1990). Developmental mechanisms for interneuron generation and migration most likely preceded the

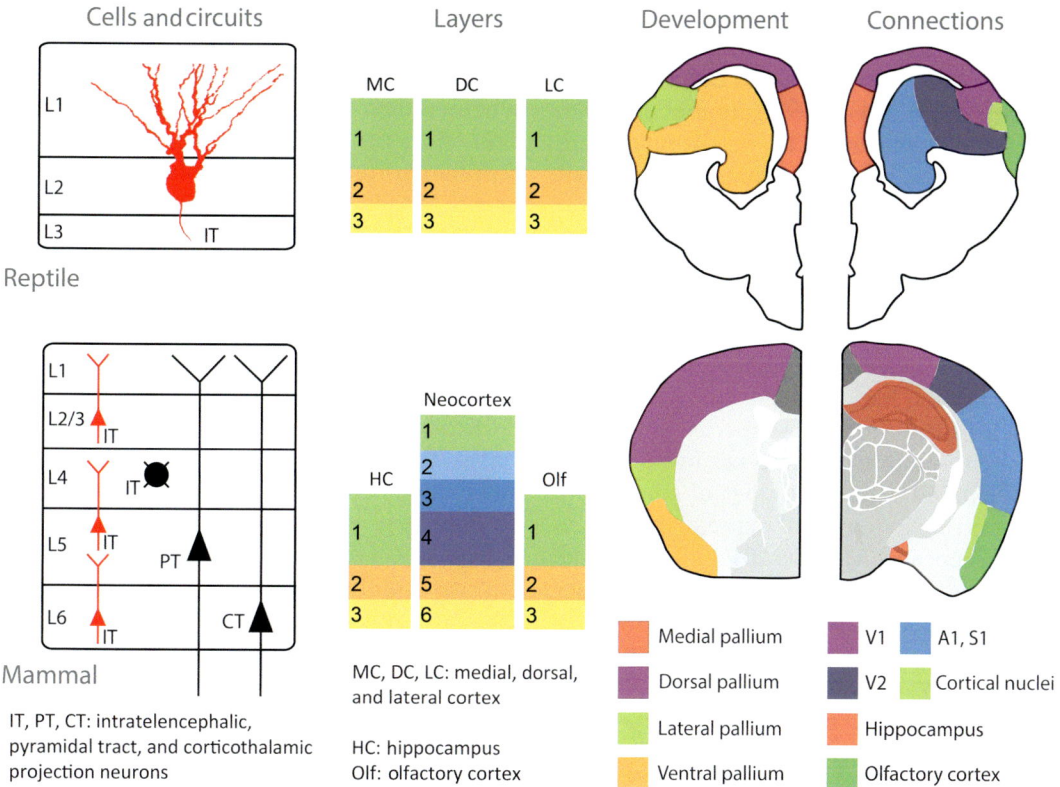

FIGURE 9.8   Corresponding cortical circuits, layers, and regions in mammals and reptiles. Simplified view of several theories based on projection cell types and circuits (Rowe and Shepherd, 2016), conserved cortical layers (Reiner, 1991, 1993), developmental topology (Puelles et al., 2016; Medina, 2006), and long-range functional connections (Reiner, 2000; Karten, 1969). Both the "development" and "connections" views consider the mammalian hippocampus and reptilian medial pallium as equivalent; therefore, we show the mammalian hippocampus only in the right panel. *IT*, intratelencephalic projection neuron; *PT*, pyramidal tract projection neuron; *CT*, corticothalamic projection neuron; *V1*, primary visual cortex; *V2*, secondary visual cortex; *A1*, primary auditory cortex; *S1*, primary somatosensory cortex. *Reproduced from Rowe, T.B., Shepherd, G.M., 2016. Role of ortho-retronasal olfaction in mammalian cortical evolution. J. Comp. Neurol. 524 (3), 471–495; Medina, L., 2006. Do birds and reptiles possess homologues of mammalian visual, somatosensory, and motor cortices? Evol. Nerv. Syst. 23; Puelles, L., et al., 2016. Selective early expression of the orphan nuclear receptor Nr4a2 identifies the claustrum homolog in the avian mesopallium: impact on sauropsidian/mammalian pallium comparisons. J. Comp. Neurol. 524 (3), 665–703; Reiner, A.J., 2000. A hypothesis as to the organization of cerebral cortex in the common amniote ancestor of modern reptiles and mammals. In: Novartis Foundation Symposium. John Wiley, 1999, Chichester, New York, pp. 83–113.*

emergence of neocortical structures. Also, interneurons are not generated locally in cortex but rather migrate from subpallial regions to all of the pallium, to extents comparable in birds (Cobos et al., 2001), frogs (Brox et al., 2004), and fishes (Carrera et al., 2008). In conclusion, current evidence for the laminar position of interneuron subtypes in mammalian cortex and developmental arguments make a suggestive, but not entirely tight, case for a layer equivalence scheme between reptilian and mammalian cortices.

Reptilian cortex versus mammalian subplate: In mammals, some of the first cortical neurons are generated in the subplate, a largely transient cell layer during development found below the cortical plate. The subplate contains a variety of cell types with respect to morphology, electrophysiological properties, and neurotransmitter content (Kanold and Luhmann, 2010; Judas et al., 2013; Hoerder-Suabedissen and Molnár, 2015).

Thalamic connections and migrating interneurons enter the subplate, but 50–80% of subplate neurons die before they are integrated into cortical layer 6B in adult rodents or survive as interstitial white matter neurons in primates (Kostovic and Rakic, 1980, 1990). There are two alternative arguments about the evolution of the subplate. The first states that, because layer 1 and the subplate appear very early in mammalian cortex development and because the subplate has a limited number of morphological cell types and lacks fully mature pyramidal types (Marín-Padilla, 1971, 1972; Hanganu et al., 2002; Marx et al., 2015), the mammalian subplate and reptilian dorsal pallium are equivalent structures (Marín-Padilla, 1971, 1998; Goffinet, 1983; Molnár and Cordery, 1999). If so, the cortical plate, which generates most of the cortical excitatory neurons in mammals, would have no equivalent in reptiles (Marín-Padilla, 1978). A broader interpretation states that both mammalian subplate and deep cortical layers correspond to the reptilian cortical cell layer, while the upper layers of mammalian neocortex could be recent additions (Reiner, 1991; Karten, 1997; Super et al., 1998; Aboitiz et al., 2005). However, we do not know if there is a dramatic neuron loss in reptiles comparable to that in the subplate.

The second hypothesis is that the subplate does not have a reptilian cortex equivalent, but is unique to mammals and increases in anatomical complexity as the neocortex increases in size in different lineages (Kostovic and Rakic, 1990; Super and Uylings, 2001). Identifying specific markers for subplate cell types and investigating their distribution across species can illuminate this issue. In mammals, screens for subplate markers have identified a large number of specifically activated genes and their temporal expression profiles (Hoerder-Suabedissen and Molnár, 2015). Wang et al. (2010b, 2011) showed that a number of mammalian subplate markers, such as Ctgf, Cplx3, and Moxd1, are present in the cortex of reptiles and birds. However, these neurons constitute only a subset of the neurons in layer 2, decreasing support for its subplate equivalence. Hence, both hypotheses may be partly true, implying that the mammalian subplate contains both new and ancestral cell populations. According to this scheme, a subplate was likely present in the common ancestor of reptiles and mammals, but additional populations of subplate cells evolved in mammals (Wang et al., 2011; Montiel et al., 2011).

Reptile cortex versus upper layers of mammalian cortex: There is little disagreement on the fact that the uppermost, cell-poor cortical layer 1, one can be compared across reptiles and mammals, both in terms of development and tangential connectivity (Marín-Padilla, 1971; Fig. 9.1). We reviewed previously some arguments based on interneuron subtypes suggesting the equivalence of reptilian layer 2 with mammalian deep layers. But elements of laminar position, neuronal connectivity, density, and projection neuron morphology could easily support an upper-layer equivalence. Indeed, the cellular morphology of turtle pyramidal cells bears little resemblance with mammalian deep layer pyramidal cells (Larkum et al., 2008). Turtle pyramidal cells instead resemble the atypical pyramidal cells of superficial layers in the mammalian dorsal and medial pallium (Sanides and Sanides, 1972). These excitatory neurons are frequently characterized by spiny dendrites growing directly into the cell-free layer, a paucity of basal dendrites and several larger apical branches. Gene expression studies of reptilian cortex have argued that mammalian upper layer markers are also expressed in the dorsal cortex of reptiles (Nomura et al., 2013; Suzuki and Hirata, 2014; Dugas-Ford et al., 2012). While the specificity of the chosen markers for

upper layers was not perfect, it is clear that reptile cortex layer 2 does not exclusively express mammalian deep layer markers. We should thus question more generally whether it makes sense to compare layer-specific expression of genes derived from a six-layer cortex with reptilian cortex, which counts only one bona fide cell layer. Indeed, a comparison of hundreds of layer-specific genes across different sectors of the pallium (olfactory cortex, neocortex, and entorhinal cortex) in a single mammalian species did not support any simple correspondence between cortical layers (Luzzati, 2015; Ramsden et al., 2015).

Conserved development versus conserved thalamo-cortical connections: In reptiles and even more so in birds, the DVR dwarfs all cortical structures. Using novel methods for tracing connections and enzyme histochemistry, Karten et al. (1973), Karten (1969), and Parent and Olivier (1969) demonstrated that the DVR is not subpallial but belongs to the pallium. These findings revived interest in comparative neurobiology, and explaining the origin and noncortical architecture of the DVR became central questions in brain evolution (Striedter, 2005). In mammals, the only pallial regions with a nuclear appearance are portions of the amygdala and the cortical nuclei located beneath the insular and olfactory cortices. These nuclei receive, at least in part, sensory input from the thalamus, but these inputs differ from those to DVR in terms of origin and/or neurochemical identity (Reiner, 2013). The DVR is the main recipient of sensory thalamic afferents that target three broadly segregated regions (visual, somatosensory, and auditory) (Pritz, 1975, 1974; Bruce and Butler, 1984a; Karten et al., 1973; Balaban and Ulinski, 1981). Whether integration exists between the three modality-specific regions is not known, especially because internal connections within DVR are predominantly perpendicular to the ventricle (Bruce and Butler, 1984a; Ulinski, 1978, 1983) and DVR outputs mainly target segregated zones of the striatum (Gonzalez et al., 1990). Finally, the DVR, while entirely composed of nuclei in birds, has a partially or fully laminar structure in some reptiles (Northcutt, 1978; Ulinski, 1983; Reiner and Northcutt, 2000).

Some investigators have proposed that sensory regions of the avian and reptile DVR and sensory regions of the mammalian neocortex are homologous in birds and mammals (Fig. 9.8; Reiner, 2000; Karten, 1969). To some extent, recent gene expression data support the equivalence of sensory regions in mammalian neocortex and DVR in reptiles and birds (Striedter et al., 2014; Dugas-Ford et al., 2012). However, while these data indicate the existence of common input—output connections, they do not necessarily imply homology, since connections may change more readily than developmental topology of neural centers. Karten (1969) proposed a specific mode of tangential migration of cortical neurons—one that would support homology of reptilian DVR and mammalian neocortex—but supporting evidence for this migration has so far proven elusive because migrating excitatory neurons of the pallium are largely transient populations (Teissier et al., 2010, 2012). Therefore, following the development of neuronal populations remains critical to establishing potential homologies. Such approaches suggest that the DVR derives from components of the ventral pallium (Fig. 9.8; Puelles et al., 2000, 2016; Smith-Fernandez et al., 1998; Medina, 2006; Striedter et al., 1998). Thus, the case of the DVR is interesting because evidence based on thalamocortical connections and evidence based on developmental gene expression appear to lead to different conclusions (Striedter, 2005; Reiner, 2000). Both scenarios are not mutually exclusive. For example, comparable (or homologous) thalamic connections could target pallial regions of different developmental origin in reptiles and mammals (Striedter, 2005; Montiel and Aboitiz, 2015). In conclusion, the issue of DVR-cortex equivalence has not yet been resolved.

9.3.2 Conclusions: Simplicity, Evolution, and Function of the Reptilian Cortex

The brains of reptiles should reveal, in a simplified form, the general pattern of the amniote forebrain, as well as the most basic subdivisions of the pallium. Reptilian cortical cytoarchitecture, cell type diversity, and thalamocortical connections are, at first glance, simpler than in mammals. However, it is critical to study a range of reptilian species to differentiate derived adaptations from ancestral features. Another key aspect to identifying "simplicity" in brain or circuit organization is to define the units appropriate for comparison. Defining homologous brain regions between reptiles and mammals is a largely unresolved issue (Striedter, 2016).

Homology, by definition a mark of common descent, fundamentally differs from analogy (which deceptively suggests common ancestry). But while defining homologous relationships is important to retrace the history of species, analogies may often be more useful to understanding general structure–function principles. For example, whether the DVR and parts of the neocortex are homologous is not yet resolved, but we can compare sensory responses and learn about general coding principles regardless of homology. Recently, work from our laboratory showed that sleep-related brain dynamics similar to mammalian sleep exist in the DVR of a lizard, *Pogona vitticeps* (Shein-Idelson et al., 2016). This suggests that the circuits underlying the electrophysiological signatures of sleep evolved in a common ancestor early in evolution, even though their homology relations remain unknown.

Indeed, convergent (ie, analogous) solutions emerge frequently when ecological/physical demands are similar. Hence, studying the closest analogue can be very enlightening by revealing replicated solutions (Gould, 1976): with similarities based on analogies, function can be studied as a natural experiment with replication. Thus, we can assess design principles and their implementation and by the same token reveal the functional constraints on brain structure (Sterling and Laughlin, 2015).

The relation of the reptilian cortex and its mammalian equivalent remain open questions (Abbie, 1938). To solve this and related problems about brain evolution, we need to integrate studies on neural development, comparative neurology, and brain function (Striedter, 1998). Combining the first two aspects is well underway, thanks to conceptual advances in integrating evolutionary and developmental neurobiology and remarkable technical advances in molecular biology. These combined approaches are now applicable to most species. Functional studies, however, lag behind in nontraditional experimental species, including the reptiles. The coming years should provide a wealth of new functional and molecular data to better understand the evolution of cortex and its original design.

References

Abbie, A.A., 1938. The relations of the fascia dentata, hippocampus and neocortex, and the nature of the subiculum. J. Comp. Neurol. 68 (3), 307–333.

Abellán, A., Desfilis, E., Medina, L., 2014. Combinatorial expression of Lef1, Lhx2, Lhx5, Lhx9, Lmo3, Lmo4, and Prox1 helps to identify comparable subdivisions in the developing hippocampal formation of mouse and chicken. Front. Neuroanat. 8.

Aboitiz, F., Montiel, J.F., 2015. Olfaction, navigation, and the origin of isocortex. Front. Neurosci. 9, 402.

Aboitiz, F., Montiel, J., García, R.R., 2005. Ancestry of the mammalian preplate and its derivatives: evolutionary relicts or embryonic adaptations? Rev. Neurosci. 16 (4), 359–376.

Ahumada-Galleguillos, P., Fernández, M., Marin, G.J., Letelier, J.C., Mpodozis, J., 2015. Anatomical organization of the visual dorsal ventricular ridge in the chick (*Gallus gallus*): layers and columns in the avian pallium. J. Comp. Neurol. 523 (17), 2618–2636.

Alcántara, S., Ruiz, M., D'Arcangelo, G., Ezan, F., de Lecea, L., Curran, T., Soriano, E., 1998. Regional and cellular patterns of reelin mRNA expression in the forebrain of the developing and adult mouse. J. Neurosci. 18 (19), 7779–7799.

Amamoto, R., Huerta, V.G.L., Takahashi, E., Dai, G., Grant, A.K., Fu, Z., Arlotta, P., 2016. Adult axolotls can regenerate original neuronal diversity in response to brain injury. eLife 5, e13998.

Ammermüller, J., Kolb, H., 1995. The organization of the turtle inner retina. I. ON-and OFF-center pathways. J. Comp. Neurol. 358 (1), 1–34.

Ammermüller, J., Kolb, H., 1996. Functional architecture of the turtle retina. Prog. Retin. Eye Res. 15 (2), 393–433.

Ammerüller, J., Muller, J.F., Kolb, H., 1995. The organization of the turtle inner retina. II. Analysis of color-coded and directionally selective cells. J. Comp. Neurol. 358 (1), 35–62.

Anderson, S.A., Eisenstat, D.D., Shi, L., Rubenstein, J.L.R., 1997. Interneuron migration from basal forebrain to neocortex: dependence on Dlx genes. Science 278 (5337), 474–476.

Annual Statistics of Scientific Procedures on Living Animals Great Britain. https://www.gov.uk/government/collections/statistics-of-scientific-procedures-on-living-animals.

Ashida, G., Carr, C.E., 2011. Sound localization: Jeffress and beyond. Curr. Opin. Neurobiol. 21 (5), 745–751.

Baden, T., Berens, P., Franke, K., Rosón, M.R., Bethge, M., Euler, T., 2016. The functional diversity of retinal ganglion cells in the mouse. Nature 529 (7586), 345–350.

Balaban, C.D., 1978. Structure of anterior dorsal ventricular ridge in a turtle (Pseudemys scripta elegans). J. Morphol. 158 (3), 291–322.

Balaban, C.D., Ulinski, P.S., 1981. Organization of thalamic afferents to anterior dorsal ventricular ridge in turtles. I. Projections of thalamic nuclei. J. Comp. Neurol. 200 (1), 95–129.

Balaram, P., Kaas, J.H., 2014. Towards a unified scheme of cortical lamination for primary visual cortex across primates: insights from NeuN and VGLUT2 immunoreactivity. Front. Neuroanat. 8.

Bar, I., de Rouvroit, C.L., Goffinet, A.M., 2000. The evolution of cortical development. An hypothesis based on the role of the Reelin signaling pathway. Trends Neurosci. 23 (12), 633–638.

Baraban, S.C., Tallent, M.K., 2004. Interneuron diversity series: interneuronal neuropeptides—endogenous regulators of neuronal excitability. Trends Neurosci. 27 (3), 135–142.

Bass, A.H., Pritz, M.B., Northcutt, R.G., 1973. Effects of telencephalic and tectal ablations on visual behavior in the side-necked turtle, Podocnemis unifilis. Brain Res. 55 (2), 455–460.

Baylor, D.A., Fettiplace, R., 1976. Transmission of signals from photoreceptors to ganglion cells in the eye of the turtle. In: Cold Spring Harbor Symposia on Quantitative Biology, vol. 40. Cold Spring Harbor Laboratory Press, pp. 529–536.

Beazley, L.D., Sheard, P.W., Tennant, M., Starac, D., Dunlop, S.A., 1997. Optic nerve regenerates but does not restore topographic projections in the lizard Ctenophorus ornatus. J. Comp. Neurol. 377 (1), 105–120.

Beckers, H.J.A., Platel, R., Nieuwenhuys, R., 1972. Les aires corticales de quelques reptiles squamates (Chamaeleo lateralis, Lacerta viridis, Monopeltis guentheri). Acta Morphol. Neerl. Scand. 9 (4), 337–364.

Beery, A.K., Zucker, I., 2011. Sex bias in neuroscience and biomedical research. Neurosci. Biobehav. Rev. 35 (3), 565–572.

Belekhova, M.G., 1979. Neurophysiology of the forebrain. Biol. Reptil. 10, 287–359.

Belekhova, M.G., Kenigfest, N.B., Minakova, M.N., Rio, J.P., Repérant, J., 2003. Calcium-binding proteins in the turtle thalamus. Analysis in the light of hypothesis of the "Core-Matrix" thalamic organization in relation to the problem of homology of thalamic nuclei among amniotes. J. Evol. Biochem. Physiol. 39 (6), 624–647.

Belgard, T.G., Montiel, J.F., Wang, W.Z., García-Moreno, F., Margulies, E.H., Ponting, C.P., Molnár, Z., 2013. Adult pallium transcriptomes surprise in not reflecting predicted homologies across diverse chicken and mouse pallial sectors. Proc. Natl. Acad. Sci. U.S.A. 110 (32), 13150–13155.

Bennis, M., Araneda, S., Calas, A., 1994. Distribution of substance P-like immunoreactivity in the chameleon brain. Brain Res. Bull. 34 (4), 349–357.

Berbel, P.J., Martínez-Guijarro, F.J., López-García, C., 1987. Intrinsic organization of the medial cerebral cortex of the lizard Lacerta pityusensis: a Golgi study. J. Morphol. 194 (3), 275–286.

Bever, G.S., Lyson, T.R., Field, D.J., Bhullar, B.A.S., 2015. Evolutionary origin of the turtle skull. Nature 525.

Bezaire, M.J., Soltesz, I., 2013. Quantitative assessment of CA1 local circuits: knowledge base for interneuron-pyramidal cell connectivity. Hippocampus 23 (9), 751–785.

Binzegger, T., Douglas, R.J., Martin, K.A., 2004. A quantitative map of the circuit of cat primary visual cortex. J. Neurosci. 24 (39), 8441–8453.

Blanton, M.G., Kriegstein, A.R., 1991. Spontaneous action potential activity and synaptic currents in the embryonic turtle cerebral cortex. J. Neurosci. 11 (12), 3907–3923.

Blanton, M.G., Kriegstein, A.R., 1992. Properties of amino acid neurotransmitter receptors of embryonic cortical neurons when activated by exogenous and endogenous agonists. J. Neurophysiol. 67 (5), 1185–1200.

Blanton, M.G., Shen, J.M., Kriegstein, A.R., 1987. Evidence for the inhibitory neurotransmitter γ-aminobutyric acid in aspiny and sparsely spiny nonpyramidal neurons of the turtle dorsal cortex. J. Comp. Neurol. 259 (2), 277–297.

Braitenberg, V., 1978. Cortical architectonics: general and areal. In: Brazier, M.A.B. (Ed.), Architectonics of the Cerebral Cortex. Raven Press, New York, USA, pp. 443–465.

Braitenberg, V., Schüz, A., 1998. COrtex: Statistics and Geometry of Neuronal Connectivity. .

Brauth, S.E., Kitt, C.A., Price, D.L., Wainer, B.H., 1985. Cholinergic neurons in the telencephalon of the reptile Caiman crocodilus. Neurosci. Lett. 58 (2), 235–240.

Bremer, F., Dow, R.S., Moruzzi, G., 1939. Physiological analysis of the general cortex in reptiles and birds. J. Neurophysiol. 2 (6), 473–487.

Brodmann, K., 1909. Vergleichende Lokalisationslehre der Grosshirnrinde in ihren Prinzipien dargestellt auf Grund des Zellenbaues. Barth.

Brox, A., Puelles, L., Ferreiro, B., Medina, L., 2004. Expression of the genes Emx1, Tbr1, and Eomes (Tbr2) in the telencephalon of Xenopus laevis confirms the existence of a ventral pallial division in all tetrapods. J. Comp. Neurol. 474 (4), 562–577.

Bruce, L.L., 2006. Evolution of the reptilian brain. Evol. Nerv. Syst. 2.

Bruce, L.L., Butler, A.B., 1984. Telencephalic connections in lizards. II. Projections to anterior dorsal ventricular ridge. J. Comp. Neurol. 229 (4), 602–615.

Bruce, L.L., Butler, A.B., 1984. Telencephalic connections in lizards. I. Projections to cortex. J. Comp. Neurol. 229 (4), 585–601.

Brüning, G., Wiese, S., Mayer, B., 1994. Nitric oxide synthase in the brain of the turtle Pseudemys scripta elegans. J. Comp. Neurol. 348 (2), 183–206.

Bullock, T.H., 1984. Comparative neuroscience holds promise for quiet revolutions. Science (New York, NY) 225 (4661), 473.

Bullock, T.H., 2000. Revisiting the concept of identifiable neurons. Brain Behav. Evol. 55 (5), 236–240.

Butler, A.B., 1976. Telencephalon of the lizard Gekko gecko (Linnaeus): some connections of the cortex and dorsal ventricular ridge. Brain Behav. Evol. 13 (5), 396–417.

Butler, A.B., 1980. Cytoarchitectonic and connectional organization of the lacertilian telencephalon with comments on vertebrate forebrain evolution. Comparative Neurology of the Telencephalon. Springer, US, pp. 297–329.

Butler, A.B., 1994. The evolution of the dorsal thalamus of jawed vertebrates, including mammals: cladistic analysis and a new hypothesis. Brain Res. Rev. 19 (1), 29–65.

Butler, A.B., Hodos, W., 2005. Comparative Vertebrate Neuroanatomy: Evolution and Adaptation. John Wiley & Sons.

Cadwell, C.R., Palasantza, A., Jiang, X., Berens, P., Deng, Q., Yilmaz, M., Sandberg, R., 2016. Electrophysiological, transcriptomic and morphologic profiling of single neurons using Patch-seq. Nat. Biotechnol. 34 (2), 199–203.

Cameron, H.A., Woolley, C.S., McEwen, B.S., Gould, E., 1993. Differentiation of newly born neurons and glia in the dentate gyrus of the adult rat. Neuroscience 56 (2), 337–344.

Carr, C.E., Konishi, M., 1988. Axonal delay lines for time measurement in the owl's brainstem. Proc. Natl. Acad. Sci. U.S.A. 85 (21), 8311–8315.

Carr, C.E., Konishi, M., 1990. A circuit for detection of interaural time differences in the brain stem of the barn owl. J. Neurosci. 10 (10), 3227–3246.

Carrera, I., Ferreiro-Galve, S., Sueiro, C., Anadón, R., Rodríguez-Moldes, I., 2008. Tangentially migrating GABAergic cells of subpallial origin invade massively the pallium in developing sharks. Brain Res. Bull. 75 (2), 405–409.

Castellano, B., Gonzalez, B., Dalmau, I., Vela, J.M., 1991. Identification and distribution of microglial cells in the cerebral cortex of the lizard: a histochemical study. J. Comp. Neurol. 311 (3), 434–444.

Chiappedi, M., Fresca, A., Baschenis, I.M.C., 2012. Complete corpus callosum agenesis: can it be mild? Case Rep. Pediatr 2012, 4.

Churchland, P.S., Sejnowski, T.J., 1992. The Computational Brain. MIT Press, Cambridge, MA.

Clinton, B.K., Cunningham, C.L., Kriegstein, A.R., Noctor, S.C., Martínez-Cerdeño, V., 2014. Radial glia in the proliferative ventricular zone of the embryonic and adult turtle, Trachemys scripta elegans. Neurogenesis 1 (1), e970905.

Cobos, I., Puelles, L., Martínez, S., 2001. The avian telencephalic subpallium originates inhibitory neurons that invade tangentially the pallium (dorsal ventricular ridge and cortical areas). Dev. Biol. 239 (1), 30–45.

Collins, M., et al., 2005. Overview of the visual system of Tarsius. Anat. Rec. Part A Discov. Mol. Cell. Evol. Biol. 287 (1).

Colombo, M., Broadbent, N., 2000. Is the avian hippocampus a functional homologue of the mammalian hippocampus? Neurosci. Biobehav. Rev. 24 (4), 465–484.

Connors, B.W., Kriegstein, A.R., 1986. Cellular physiology of the turtle visual cortex: distinctive properties of pyramidal and stellate neurons. J. Neurosci. 6 (1), 164–177.

Cosans, C.E., Ulinski, P.S., 1990. Spatial organization of axons in turtle visual cortex: intralamellar and interlamellar projections. J. Comp. Neurol. 296 (4), 548–558.

Da Costa, N.M., Martin, K.A., 2009. The proportion of synapses formed by the axons of the lateral geniculate nucleus in layer 4 of area 17 of the cat. J. Comp. Neurol. 516 (4), 264–276.

Da Costa, N.M., Martin, K.A., 2010. Whose cortical column would that be. Front. Neuroanat. 4.

Craigie, E.H., 1935. The hippocampal and parahippocampal cortex of the emu (Dromiceius). J. Comp. Neurol. 61 (3), 563–591.

Craigie, E.H., 1936. Notes on Cytoarchitectural Features of the Lateral Cortex and Related Parts of the Cerebral Hemisphere in a Series of Reptiles and Birds. Royal Society of Canada.

Cruz-Martín, A., El-Danaf, R.N., Osakada, F., Sriram, B., Dhande, O.S., Nguyen, P.L., Huberman, A.D., 2014. A dedicated circuit links direction-selective retinal ganglion cells to the primary visual cortex. Nature 507 (7492), 358–361.

Dacey, D.M., Ulinski, P.S., 1983. Nucleus rotundus in a snake, Thamnophis sirtalis: an analysis of a nonretinotopic projection. J. Comp. Neurol. 216 (2), 175–191.

Davila, J.C., Guirado, S., De la Calle, A., Marin-Giron, F., 1985. Electron microscopy of the medial cortex in the lizard Psammodromus algirus. J. Morphol. 185 (3), 327–338.

Dávila, J.C., de la Calle, A., Gutiérrez, A., Megías, M., Andreu, M.J., Guirado, S., 1991. Distribution of neuropeptide Y (NPY) in the cerebral cortex of the lizards Psammodromus algirus and Podarcis hispanica: Co-localization of NPY, somatostatin, and GABA. J. Comp. Neurol. 308 (3), 397–408.

Dávila, J.C., Megías, M., de la Calle, A., Guirado, S., 1993. Subpopulations of GABA neurons containing somatostatin, neuropeptide Y, and parvalbumin in the dorsomedial cortex of the lizard Psammodromus algirus. J. Comp. Neurol. 336 (2), 161–173.

Dávila, J.C., Megías, M., Andreu, M.J., Real, M.A., Guirado, S., 1995. NADPH diaphorase-positive neurons in the lizard hippocampus:

a distinct subpopulation of GABAergic interneurons. Hippocampus 5 (1), 60–70.

Davila, J.C., Padial, J., Andreu, M.J., Real, M.A., Guirado, S., 1997. Calretinin immunoreactivity in the cerebral cortex of the lizard Psammodromus algirus: a light and electron microscopic study. J. Comp. Neurol. 382 (3), 382–393.

Davila, J.C., Padial, J., Andreu, M.J., Guirado, S., 1999. Calbindin-D28k in cortical regions of the lizard Psammodromus algirus. J. Comp. Neurol. 405 (1), 61–74.

Dearworth, J.R., Granda, A.M., 2002. Multiplied functions unify shapes of ganglion-cell receptive fields in retina of turtle. J. Vis. 2 (3), 1.

DeFelipe, J., 2015. The anatomical problem posed by brain complexity and size: a potential solution. Front. Neuroanat. 9.

DeFelipe, J., López-Cruz, P.L., Benavides-Piccione, R., Bielza, C., Larrañaga, P., Anderson, S., Fishell, G., 2013. New insights into the classification and nomenclature of cortical GABAergic interneurons. Nat. Rev. Neurosci. 14 (3), 202–216.

Denk, W., Horstmann, H., 2004. Serial block-face scanning electron microscopy to reconstruct three-dimensional tissue nanostructure. PLoS Biol. 2 (11), e329.

Denk, W., Briggman, K.L., Helmstaedter, M., 2012. Structural neurobiology: missing link to a mechanistic understanding of neural computation. Nat. Rev. Neurosci. 13 (5), 351–358.

Desan, P.H., 1988. Organization of the cerebral cortex in turtle. The Forebrain of Reptiles. Karger Publishers, pp. 1–11.

Distel, H., 1976. Behavior and electrical brain stimulation in the green iguana, Iguana iguana L. Brain Behav. . Evol. 13 (6), 421–435.

Distel, H., 1978. Behavior and electrical brain stimulation in the green iguana, Iguana iguana L. II. Stimulation effects. Exp. Brain Res. 31 (3), 353–367.

Distel, H., 1978. Behavioral responses to the electrical stimulation of the brain in the green iguana. Behav. Neurol. Lizards 135–147.

Dobzhansky, T., 1973. Nothing in biology makes sense except in the light of evolution. Am. Biol. Teach. 35, 125–129.

Domínguez, L., Morona, R., Joven, A., González, A., López, J.M., 2010. Immunohistochemical localization of orexins (hypocretins) in the brain of reptiles and its relation to monoaminergic systems. J. Chem. Neuroanat. 39 (1), 20–34.

Douglas, R.J., Martin, K.A., 2004. Neuronal circuits of the neocortex. Annu. Rev. Neurosci. 27, 419–451.

Douglas, R.J., Martin, K.A., 2007. Recurrent neuronal circuits in the neocortex. Curr. Biol. 17 (13), R496–R500.

Douglas, R.J., Koch, C., Mahowald, M., Martin, K.A., Suarez, H.H., 1995. Recurrent excitation in neocortical circuits. Science 269 (5226), 981.

Dugas-Ford, J., 2009. A Comparative Molecular Study of the Amniote Dorsal Telencephalon. The University of Chicago.

Dugas-Ford, J., Rowell, J.J., Ragsdale, C.W., 2012. Cell-type homologies and the origins of the neocortex. Proc. Natl. Acad. Sci. U.S.A. 109 (42), 16974–16979.

Ebner, F.F., Colonnier, M., 1975. Synaptic patterns in the visual cortex of turtle: an electron microscopic study. J. Comp. Neurol. 160 (1), 51–79.

Ebner, F.F., Colonnier, M., 1978. A quantitative study of synaptic patterns in turtle visual cortex. J. Comp. Neurol. 179 (2), 263–276.

Edinger, L., 1908. Vorlesungen über den Bau der nervösen des Zentralorgane des Menschen und der Tiere: für Ärzte und Studierende. FCW Vogel.

Fatt, P., Katz, B., 1951. An analysis of the end-plate potential recorded with an intra-cellular electrode. J. Physiol. 115 (3), 320.

Faunes, M., Botelho, J.F., Galleguillos, P.A., Mpodozis, J., 2015. On the hodological criterion for homology. Front. Neurosci. 9.

Fellmann, D., Bugnon, C., Bresson, J.L., Gouget, A., Cardot, J., Clavequin, M.C., Hadjiyiassemis, M., 1984. The CRF neuron: immunocytochemical study. Peptides 5, 19–33.

Feuillet, L.H.D., Pelletier, J., 2007. Brain of a white-collar worker. Lancet (London, England) 370 (9583), 262.

Filimonoff, I.N., 1963. Homologies of the cerebral formations of mammals and reptiles. J. Hirnforsch. 7 (2), 229–251.

Flores, A., 1911. Die Myeloarchitektonik und die Myelogenie des Cortex cerebri beim Igel (*Erinaceus europaeus*). J. Psychol. Neurol. 17, 215–247.

Font, E., Desfilis, E., Perez-Canellas, M.M., Garcia-Verdugo, J.M., 2002. Neurogenesis and neuronal regeneration in the adult reptilian brain. Brain Behav. Evol. 58 (5), 276–295.

Foster, R.E., Hall, W.C., 1975. The connections and laminar organization of the optic tectum in a reptile (*Iguana iguana*). J. Comp. Neurol. 163 (4), 397–425.

Fotowat, H., Gabbiani, F., 2011. Collision detection as a model for sensory-motor integration. Annu. Rev. Neurosci. 34, 1–19.

Fotowat, H., Harrison, R.R., Gabbiani, F., 2011. Multiplexing of motor information in the discharge of a collision detecting neuron during escape behaviors. Neuron 69 (1), 147–158.

Fournier, J., Müller, C.M., Laurent, G., 2015. Unpublished observations Looking for the roots of cortical sensory computation in three-layered cortices. Curr. Opin. Neurobiol. 31, 119–126.

Freund, T.F., Buzsáki, G.Y., 1996. Interneurons of the hippocampus. Hippocampus 6 (4), 347–470.

Fritsch, G., Hitzig, E., 1870. Über die elektrische Erregbarkeit des Grosshirns. Arch. Anat. Physiol. Wiss. Med. 300.

Fuhrmann, F., Justus, D., Sosulina, L., Kaneko, H., Beutel, T., Friedrichs, D., Remy, S., 2015. Locomotion, theta oscillations, and the speed-correlated firing of hippocampal neurons are controlled by a medial septal glutamatergic circuit. Neuron 86 (5), 1253–1264.

Fuzik, J., Zeisel, A., Máté, Z., Calvigioni, D., Yanagawa, Y., Szabó, G., Harkany, T., 2016. Integration of electrophysiological recordings with single-cell RNA-seq data identifies neuronal subtypes. Nat. Biotechnol. 34 (2), 175–183.

Gabbiani, F., Krapp, H.G., Koch, C., Laurent, G., 2002. Multiplicative computation in a visual neuron sensitive to looming. Nature 420 (6913), 320–324.

García-Moreno, F., Molnár, Z., 2015. Subset of early radial glial progenitors that contribute to the development of callosal neurons is absent from avian brain. Proc. Natl. Acad. Sci. U.S.A. 112 (36), E5058–E5067.

Geri, G.A., Kimsey, R.A., Dvorak, C.A., 1982. Quantitative electron microscopic analysis of the optic nerve of the turtle, *Pseudemys*. J. Comp. Neurol. 207 (1), 99–103.

Goffinet, A.M., 1983. The embryonic development of the cortical plate in reptiles: a comparative study in *Emys orbicularis* and *Lacerta agilis*. J. Comp. Neurol. 215 (4), 437–452.

Gonchar, Y., Burkhalter, A., 1997. Three distinct families of GABAergic neurons in rat visual cortex. Cereb. Cortex 7 (4), 347–358.

Gonchar, Y., Wang, Q., Burkhalter, A.H., 2008. Multiple distinct subtypes of GABAergic neurons in mouse visual cortex identified by triple immunostaining. Front. Neuroanat. 2, 3.

Gonzalez, A., Russchen, F.T., Lohman, A.H.M., 1990. Afferent connections of the striatum and the nucleus accumbens in the lizard *Gekko gecko* (Part 1 of 2). Brain Behav. Evol. 36 (1), 39–48.

Gould, S.J., 1976. In defense of the analog: a commentary to N. Hotton. Evol. Brain Behav. Persistent Probl. 175–179.

Granda, A.M., Sisson, D.F., 1992. Retinal function in turtles. Sensorimot. Integr. 17, 136.

Green, R.E., et al., 2014. Three crocodilian genomes reveal ancestral patterns of evolution among archosaurs. Science 346 (6215).

Greenberg, N., 1977. A neuroethological study of display behavior in the lizard *Anolis carolinensis* (Reptilia, Lacertilia, Iguanidae). Am. Zool. 17 (1), 191–201.

Greenberg, N., MacLean, P.D., Ferguson, J.L., 1979. Role of the paleostriatum in species-typical display behavior of the lizard (*Anolis carolinensis*). Brain Res. 172 (2), 229–241.

Greenberg, N., Font, E., Switzer, R.C., 1988. The reptilian striatum revisited: studies on *Anolis* lizards. The Forebrain of Reptiles. Karger Publishers, pp. 162–177.

Grothe, B., Pecka, M., McAlpine, D., 2010. Mechanisms of sound localization in mammals. Physiol. Rev. 90 (3), 983–1012.

Guirado, S., Davila, J.C., 1999. GABAergic cell types in the lizard hippocampus. Eur. J. Morphol. 37 (2–3), 89–94.

Guirado, S., Davila, J.C., 2002. Thalamo-telencephalic connections: new insights on the cortical organization in reptiles. Brain Res. Bull. 57 (3), 451–454.

Guirado, S., De la Calle, A., Davila, J.C., Marin-Giron, F., 1984. Light microscopy of the medial wall of the cerebral cortex of the lizard *Psammodromus algirus*. J. Morphol. 181 (3), 319–331.

Guirado, S., Ángeles Real, M., Padial, J.U.S., Andreu, M.J., Carlos Dávila, J., 1998. Cholecystokinin innervation of the cerebral cortex in a reptile, the lizard *Psammodromus algirus*. Brain Behav. Evol. 51 (2), 100–112.

Gusel'nikov, V.I., Pivovarov, A.S., 1978. Postsynaptic mechanism of habituation of turtle cortical neurons to moving stimuli. Neurosci. Behav. Physiol. 9 (1), 1–7.

Hall, W.C., Ebner, F.F., 1970. Parallels in the visual afferent projections of the thalamus in the hedgehog (*Paraechinus hypomelas*) and the turtle (*Pseudemys scripta*). Brain Behav. Evol. 3 (1–4), 135–154.

Hanganu, I.L., Kilb, W., Luhmann, H.J., 2002. Functional synaptic projections onto subplate neurons in neonatal rat somatosensory cortex. J. Neurosci. 22 (16), 7165–7176.

Harris, K.D., Shepherd, G.M., 2015. The neocortical circuit: themes and variations. Nat. Neurosci. 18 (2), 170–181.

Hatsopoulos, N., Gabbiani, F., Laurent, G., 1995. Elementary computation of object approach by a wide-field visual neuron. Science 270 (5238), 1000.

Hayes, W.N., Saiff, E.I., 1967. Visual alarm reactions in turtles. Anim. Behav. 15 (1), 102–106.

Hedges, S.B., Poling, L.L., 1999. A molecular phylogeny of reptiles. Science 283 (5404), 998–1001.

Heller, S.B., Ulinski, P.S., 1987. Morphology of geniculocortical axons in turtles of the genera *Pseudemys* and *Chrysemys*. Anat. Embryol. 175 (4), 505–515.

Herculano-Houzel, S., Mota, B., Lent, R., 2006. Cellular scaling rules for rodent brains. Proc. Natl. Acad. Sci. U.S.A. 103 (32), 12138–12143.

Herculano-Houzel, S., Collins, C.E., Wong, P., Kaas, J.H., 2007. Cellular scaling rules for primate brains. Proc. Natl. Acad. Sci. U.S.A. 104 (9), 3562–3567.

Hergueta, S., Lemire, M., Ward, R., Rio, J.P., Repérant, J., 1991. A reconsideration of the primary visual system of the turtle *Emys orbicularis*. J. Hirnforsch. 33 (4–5), 515–544.

Herold, C., Bingman, V.P., Ströckens, F., Letzner, S., Sauvage, M., Palomero-Gallagher, N., Güntürkün, O., 2014. Distribution of neurotransmitter receptors and zinc in the pigeon (*Columba livia*) hippocampal formation: a basis for further comparison with the mammalian hippocampus. J. Comp. Neurol. 522 (11), 2553–2575.

Hertzler, D.R., Hayes, W.N., 1967. Cortical and tectal function in visually guided behavior of turtles. J. Comp. Physiol. Psychol. 63 (3), 444.

Hines, M., 1923. The development of the telencephalon in *Sphenodon punctatum*. J. Comp. Neurol. 35 (5), 483–537.

Hirth, F., Reichert, H., 1999. Conserved genetic programs in insect and mammalian brain development. Bioessays 21 (8), 677–684.

Hodgkin, A.L., Huxley, A.F., 1952. A quantitative description of membrane current and its application to conduction and excitation in nerve. J. Physiol. 117 (4), 500.

Hoerder-Suabedissen, A., Molnár, Z., 2015. Development, evolution and pathology of neocortical subplate neurons. Nat. Rev. Neurosci. 16 (3), 133–146.

Hoogland, P.V., Vermeulen-VanderZee, E., 1989. Efferent connections of the dorsal cortex of the lizard *Gekko gecko* studied with *Phaseolus vulgaris*-leucoagglutinin. J. Comp. Neurol. 285 (3), 289–303.

Hoogland, P.V., Vermeulen-VanderZee, E., 1990. Distribution of choline acetyltransferase immunoreactivity in the telencephalon of the lizard *Gekko gecko*. Brain Behav. Evol. 36 (6), 378–390.

Hoogland, P., Martinez-Garcia, F., Geneser, F., Vermeulen-VanderZee, E., 1998. Convergence of thalamic and cholinergic projections in the 'dentate area'of lizards. Brain Behav. Evol. 51 (2), 113–122.

Horton, J.C., Adams, D.L., 2005. The cortical column: a structure without a function. Philos. Trans. R. Soc. Lond. B Biol. Sci. 360 (1456), 837–862.

Hubel, D.H., 1982. Cortical neurobiology: a slanted historical perspective. Annu. Rev. Neurosci. 5 (1), 363–370.

Hugall, A.F., Foster, R., Lee, M.S., 2007. Calibration choice, rate smoothing, and the pattern of tetrapod diversification according to the long nuclear gene RAG-1. Syst. Biol. 56 (4), 543–563.

Jarvis, E.D., Yu, J., Rivas, M.V., Horita, H., Feenders, G., Whitney, O., Siang-Bakshi, C., 2013. Global view of the functional molecular organization of the avian cerebrum: mirror images and functional columns. J. Comp. Neurol. 521 (16), 3614–3665.

Jensen, R.J., Devoe, R.D., 1983. Comparisons of directionally selective with other ganglion cells of the turtle retina: intracellular recording and staining. J. Comp. Neurol. 217 (3), 271–287.

Jiang, X., Shen, S., Cadwell, C.R., Berens, P., Sinz, F., Ecker, A.S., Tolias, A.S., 2015. Principles of connectivity among morphologically defined cell types in adult neocortex. Science 350 (6264), aac9462.

Jiménez, A.J., Mancera, J.M., Pérez-Fígares, J.M., Fernández-Llebrez, P., 1994. Distribution of galanin-like immunoreactivity in the brain of the turtle *Mauremys caspica*. J. Comp. Neurol. 349 (1), 73–84.

Johnston, J.B., 1915. The cell masses in the forebrain of the turtle, *Cistudo carolina*. J. Comp. Neurol. 25 (5), 393–468.

Johnston, J.B., 1923. Further contributions to the study of the evolution of the forebrain. J. Comp. Neurol. 35 (5), 337–481.

Judas, M., Sedmak, G., Kostovic, I., 2013. The significance of the subplate for evolution and developmental plasticity of the human brain. Front. Hum. Neurosci. 7, 423.

Kaas, J.H., 2012. Evolution of columns, modules, and domains in the neocortex of primates. Proc. Natl. Acad. Sci. U.S.A. 109 (Suppl. 1), 10655–10660.

Kálmán, M., Pritz, M.B., 2001. Glial fibrillary acidic protein-immunopositive structures in the brain of a Crocodilian, *Caiman crocodilus*, and its bearing on the evolution of astroglia. J. Comp. Neurol. 431 (4), 460–480.

Kálmán, M., Kiss, A., Majorossy, K., 1994. Distribution of glial fibrillary acidic protein-immunopositive structures in the brain of the red-eared freshwater turtle (*Pseudemys scripta elegans*). Anat. Embryol. 189 (5), 421–434.

Kálmán, M., Martin-Partido, G., Hidalgo-Sanchez, M., Majorossy, K., 1997. Distribution of glial fibrillary acidic protein-immunopositive structures in the developing brain of the turtle *Mauremys leprosa*. Anat. Embryol. 196 (1), 47–65.

Kaneko, T., Mizuno, N., 1994. Glutamate-synthesizing enzymes in GABAergic neurons of the neocortex: a double immunofluorescence study in the rat. Neuroscience 61 (4), 839–849.

Kanold, P.O., Luhmann, H.J., 2010. The subplate and early cortical circuits. Annu. Rev. Neurosci. 33, 23–48.

Kappers, C.A., 1909. The Phylogenesis of the Palaeo-cortex and Archicortex compared with the evolution of the visual Neo-cortex. J. Truscott and Son.

Karten, H.J., 1969. The organization of the avian telencephalon and some speculations on the phylogeny of the amniote telencephalon. Ann. N.Y. Acad. Sci. 167 (1), 164–179.

Karten, H.J., 1997. Evolutionary developmental biology meets the brain: the origins of mammalian cortex. Proc. Natl. Acad. Sci. U.S.A. 94 (7), 2800–2804.

Karten, H.J., Hodos, W., Nauta, W.J., Revzin, A.M., 1973. Neural connections of the "visual wulst" of the avian telencephalon. Experimental studies in the pigeon (*Columba livia*) and owl (*Speotyto cunicularia*). J. Comp. Neurol. 150 (3), 253–277.

Kawaguchi, Y., Kubota, Y., 1997. GABAergic cell subtypes and their synaptic connections in rat frontal cortex. Cereb. Cortex 7 (6), 476–486.

Keating, E.G., Kormann, L.A., Horel, J.A., 1970. The behavioral effects of stimulating and ablating the reptilian amygdala (*Caiman sklerops*). Physiol. Behav. 5 (1), 55IN959–58IN10.

Kemali, M., Braitenberg, V., 1969. Atlas of the Frog's Brain.

Kenigfest, N., Martinez-Marcos, A., Belekhova, M., Font, C., Lanuza, E., Desfilis, E., Martinez-Garcia, F., 1997. A lacertilian dorsal retinorecipient thalamus: a re-investigation in the Old-World lizard *Podarcis hispanica* (Part 1 of 2). Brain Behav. Evol. 50 (6), 313–323.

Kepecs, A., Fishell, G., 2014. Interneuron cell types are fit to function. Nature 505 (7483), 318–326.

Killackey, H., Pellmar, T., Ebner, F.F., 1972. *The effects* of general cortex *ablation* on habituation in the *turtle*. Fed. Proc. 31, 819.

Kirsche, W., 1972. Die Entwicklung des Telencephalons der Reptilien und deren Beziehung zur Hirn-Bauplanlehre. Barth.

Klausberger, T., Somogyi, P., 2008. Neuronal diversity and temporal dynamics: the unity of hippocampal circuit operations. Science 321 (5885), 53–57.

Knapp, H., Kang, D.S., 1968. The visual pathways of the snapping turtle (*Chelydra serpentina*). Brain Behav. Evol. 1 (1), 19–42.

Koch, C., Reid, R.C., 2012. Neuroscience: observatories of the mind. Nature 483 (7390), 397–398.

Kolb, H., 1982. The morphology of the bipolar cells, amacrine cells and ganglion cells in the retina of the turtle *Pseudemys scripta elegans*. Philos. Trans. R. Soc. Biol. Sci. 298 (1092), 355–393.

Kolb, H., Jones, J., 1982. Light and electron microscopy of the photoreceptors in the retina of the red-eared slider, *Pseudemys scripta elegans*. J. Comp. Neurol. 209 (4), 331–338.

Kostovic, I., Rakic, P., 1980. Cytology and time of origin of interstitial neurons in the white matter in infant and adult human and monkey telencephalon. J. Neurocytol. 9 (2), 219–242.

Kostovic, I., Rakic, P., 1990. Developmental history of the transient subplate zone in the visual and somatosensory cortex of the macaque monkey and human brain. J. Comp. Neurol. 297 (3), 441–470.

Kovjanic, D., Redies, C., 2003. Small-scale pattern formation in a cortical area of the embryonic chicken telencephalon. J. Comp. Neurol. 456 (2), 95–104.

Kriegstein, A.R., 1987. Synaptic responses of cortical pyramidal neurons to light stimulation in the isolated turtle visual system. J. Neurosci. 7 (8), 2488–2492.

Kriegstein, A.R., Shen, J.M., Eshhar, N., 1986. Monoclonal antibodies to the turtle cortex reveal neuronal subsets, antigenic cross-reactivity with the mammalian neocortex, and forebrain structures sharing a pallial derivation. J. Comp. Neurol. 254 (3), 330–340.

Krogh, A., 1929. The progress of physiology. Am. J. Physiol. 90 (2), 243–251.

Kruger, L., Maxwell, D.S., 1967. Comparative fine structure of vertebrate neuroglia: teleosts and reptiles. J. Comp. Neurol. 129 (2), 115–141.

Kubota, Y., Hattori, R., Yui, Y., 1994. Three distinct subpopulations of GABAergic neurons in rat frontal agranular cortex. Brain Res. 649 (1), 159–173.

De Lange, S., 1911. Das Vorderhirn der Reptilien. Fol. Neurobiol. 5, 555–557.

Larkum, M., 2013. A cellular mechanism for cortical associations: an organizing principle for the cerebral cortex. Trends Neurosci. 36 (3), 141–151.

Larkum, M.E., Watanabe, S., Lasser-Ross, N., Rhodes, P., Ross, W.N., 2008. Dendritic properties of turtle pyramidal neurons. J. Neurophysiol. 99 (2), 683–694.

Larson-Prior, L.J., Ulinski, P.S., Slater, N.T., 1991. Excitatory amino acid receptor-mediated transmission in geniculocortical and intracortical pathways within visual cortex. J. Neurophysiol. 66 (1), 293–306.

Lázár, G., Székely, G., 1969. Distribution of optic terminals in the different optic centres of the frog. Brain Res. 16 (1), 1.

Lazzari, M., Franceschini, V., 2006. Glial cytoarchitecture in the central nervous system of the soft-shell turtle, *Trionyx sinensis*, revealed by intermediate filament immunohistochemistry. Anat. Embryol. 211 (5), 497–506.

Leise, E.M., 1990. Modular construction of nervous systems: a basic principle of design for invertebrates and vertebrates. Brain Res. Rev. 15 (1), 1–23.

Lewis, D.A., Lund, J.S., 1990. Heterogeneity of chandelier neurons in monkey neocortex: corticotropin-releasing factor-and parvalbumin-immunoreactive populations. J. Comp. Neurol. 293 (4), 599–615.

Li, C., Wu, X.C., Rieppel, O., Wang, L.T., Zhao, L.J., 2008. An ancestral turtle from the Late Triassic of southwestern China. Nature 456 (7221), 497–501.

Liu, Y.J., Wang, Q., Li, B., 2011. Neuronal responses to looming objects in the superior colliculus of the cat. Brain Behav. Evol. 77 (3), 193–205.

Lohman, A.H.M., Hoogland, P.V., 1979. Anatomy of cerebral commissures. Structure and Function of Cerebral Commissures. Macmillan Education UK, pp. 1–14.

Lohman, A.H.M., Mentink, G.M., 1972. Some cortical connections of the tegu lizard (*Tupinambis teguixin*). Brain Res. 45 (2), 325–344.

Lohman, A.H., Smeets, W.J., 1991. The dorsal ventricular ridge and cortex of reptiles in historical and phylogenetic perspective. The Neocortex. Springer, US, pp. 59–74.

Lohman, A.H.M., van Woerden-Verkley, I., 1978. Ascending connections to the forebrain in the tegu lizard. J. Comp. Neurol. 182 (3), 555–574.

Lopez-Garcia, C., Tineo, P.L., Del Corral, J., 1983. Increase of the neuron number in some cerebral cortical areas of a lizard, *Podarcis hispanica*, (Steind., 1870), during postnatal periods of life. J. Hirnforsch. 25 (3), 255–259.

López-García, C., Molowny, A., Martínez-Guijarro, F.J., Blasco-Ibáñez, J.M., Luis, D.L.I.J., Bernabeu, A., García-Verdugo, J.M., 1992. Lesion and regeneration in the medial cerebral cortex of lizards. Histol. Histopathol. 7 (4), 725–746.

Lorente de Nó, R., 1922. La corteza cerebral del ratón: (Primera contribución.-La corteza acústica). Tipogr. Artística 20.

Lorente de Nó, R., 1934. Studies on the structure of the cerebral cortex. I. J. Psychol. Neurol.

LoTurco, J.J., Blanton, M.G., Kriegstein, A.R., 1991. Initial expression and endogenous activation of NMDA channels in early neocortical development. J. Neurosci. 11 (3), 792–799.

Luis de la Iglesia, J.A., Lopez-Garcia, C., 1997. A Golgi study of the principal projection neurons of the medial cortex of the lizard *Podarcis hispanica*. J. Comp. Neurol. 385 (4), 528–564.

Luis de la Iglesia, J., Lopez-Garcia, C., 1997. Neuronal circuitry in the medial cerebral cortex of lizards. International Work-Conference on Artificial Neural Networks. Springer, Berlin, Heidelberg, pp. 61–71.

Luis de la Iglesia, J.A., Lopez-Garcia, C., 1997. A Golgi study of the short-axon interneurons of the cell layer and inner plexiform layer of the medial cortex of the lizard *Podarcis hispanica*. J. Comp. Neurol. 385, 565–598.

Luis de la Iglesia, J.A., Martinez-Guijarro, F.J., Lopez-Garcia, C., 1994. Neurons of the medial cortex outer plexiform layer of the lizard *Podarcis hispanica*: Golgi and immunocytochemical studies. J. Comp. Neurol. 341 (2), 184–203.

Lund, H.J., et al., 1994. Substrates for interlaminar connections in area V1 of macaque monkey cerebral cortex. Primary Visual Cortex in Primates. Springer, US.

Luzzati, F., 2015. A hypothesis for the evolution of the upper layers of the neocortex through co-option of the olfactory cortex developmental program. Front. Neurosci. 9, 162.

Lyson, T.R., Bever, G.S., Scheyer, T.M., Hsiang, A.Y., Gauthier, J.A., 2013. Evolutionary origin of the turtle shell. Curr. Biol. 23 (12), 1113–1119.

MacLean, P.D., 1990. The Triune Brain in Evolution: Role in Paleocerebral Functions. Springer Science & Business Media.

Mainen, Z.F., Sejnowski, T.J., 1996. Influence of dendritic structure on firing pattern in model neocortical neurons. Nature 382 (6589), 363–366.

Malatesta, P., Hartfuss, E., Gotz, M., 2000. Isolation of radial glial cells by fluorescent-activated cell sorting reveals a neuronal lineage. Development 127 (24), 5253–5263.

Manger, P.R., Slutsky, D.A., Molnár, Z., 2002. Visual subdivisions of the dorsal ventricular ridge of the iguana (*Iguana iguana*) as determined by electrophysiologic mapping. J. Comp. Neurol. 453 (3), 226–246.

Manger, P., Cort, J., Ebrahim, N., Goodman, A., Henning, J., Karolia, M., Strkalj, G., 2008. Is 21st century neuroscience too focussed on the rat/mouse model of brain function and dysfunction? Front. Neuroanat. 2, 5.

Marder, E., Goaillard, J.M., 2006. Variability, compensation and homeostasis in neuron and network function. Nat. Rev. Neurosci. 7 (7), 563–574.

Marin, O., Rubenstein, J.L., 2003. Cell migration in the forebrain. Annu. Rev. Neurosci. 26 (1), 441–483.

Marín-Padilla, M., 1971. Early prenatal ontogenesis of the cerebral cortex (neocortex) of the cat (*Felis domestica*). A Golgi study. Z. Anat. Entwicklungsgesch. 134 (2), 117–145.

Marín-Padilla, M., 1972. Prenatal ontogenetic history of the principal neurons of the neocortex of the cat (*Felis domestica*) a Golgi study. Z. Anat. Entwicklungsgesch. 136 (2), 125–142.

Marín-Padilla, M., 1978. Dual origin of the mammalian neocortex and evolution of the cortical plate. Anat. Embryol. 152 (2), 109–126.

Marín-Padilla, M., 1998. Cajal–Retzius cells and the development of the neocortex. Trends Neurosci. 21 (2), 64–71.

Marr, D., 1982. The philosophy and the approach. Vision: A Computational Investigation into the Human Representation and Processing of Visual Information.

Martíanez-García, F., Amiguet, M., Schwerdtfeger, W.K., Olucha, F.E., Lorente, M.J., 1990. Interhemispheric connections through the pallial commissures in the brain of *Podarcis hispanica* and *Gallotia stehlinii* (Reptilia, Lacertidae). J. Morphol. 205 (1), 17–31.

Martínez-Cerdeño, V., Noctor, S., 2016. Cortical evolution conference 2015. J. Comp. Neurol. 524 (3), 431–432.

Martínez-García, F., Amiguet, M., Olucha, F., Lopez-Garcia, C., 1986. Connections of the lateral cortex in the lizard *Podarcis hispanica*. Neurosci. Lett. 63 (1), 39–44.

Martínez-Guijarro, F.J., Freund, T.F., 1992. Distribution of GABAergic interneurons immunoreactive for calretinin, calbindin D28K, and parvalbumin in the cerebral cortex of the lizard *Podarcis hispanica*. J. Comp. Neurol. 322 (3), 449–460.

Martínez-Guijarro, F.J., Berbel, P.J., Molowny, A., García, C.L., 1984. Apical dendritic spines and axonic terminals in the bipyramidal neurons of the dorsomedial cortex of lizards (Lacerta). Anat. Embryol. 170 (3), 321–326.

Martínez-Guijarro, F.J., Desfilis, E., Lopez-Garcia, C., 1990. Organization of the dorsomedial cortex in the lizard *Podarcis hispanica*. The Forebrain in Nonmammals. New Aspects of Structure and Development, pp. 77–92.

Martínez-Guijarro, F.J., Soriano, E., Del Rio, J.A., Blasco-Ibáñez, J.M., López-Garcia, C., 1993. Parvalbumin-containing neurons in the cerebral cortex of the lizard *Podarcis hispanica*: morphology, ultrastructure, and coexistence with GABA, somatostatin, and neuropeptide Y. J. Comp. Neurol. 336 (3), 447–467.

Martínez-Guijarro, F.J., Blasco-Ibáñez, J.M., Freund, T.F., 1994. Serotoninergic innervation of nonprincipal cells in the cerebral cortex of the lizard *Podarcis hispanica*. J. Comp. Neurol. 343 (4), 542–553.

Marx, M., Qi, G., Hanganu-Opatz, I.L., Kilb, W., Luhmann, H.J., Feldmeyer, D., 2015. Neocortical layer 6B as a Remnant of the subplate-a morphological comparison. Cereb. Cortex bhv279.

Masland, R.H., 2004. Neuronal cell types. Curr. Biol. 14 (13), R497—R500.

Mazurskaya, P.Z., 1973. Organization of receptive fields in the forebrain of *Emys orbicularis*. Neurosci. Behav. Physiol. 6 (4), 311—318.

McCulloch, W.S., Pitts, W., 1943. A logical calculus of the ideas immanent in nervous activity. Bull. Math. Biophys. 5 (4), 115—133.

Medina, L., 2006. Do birds and reptiles possess homologues of mammalian visual, somatosensory, and motor cortices? Evol. Nerv. Syst. 23.

Medina, L., Reiner, A., 2000. Do birds possess homologues of mammalian primary visual, somatosensory and motor cortices? Trends Neurosci. 23 (1), 1—12.

Medina, L., Smeets, W.J., Hoogland, P.V., Puelles, L., 1993. Distribution of choline acetyltransferase immunoreactivity in the brain of the lizard *Gallotia galloti*. J. Comp. Neurol. 331 (2), 261—285.

Medina, L., Abellán, A., Desfilis, E., 2013. A never-ending search for the evolutionary origin of the neocortex: rethinking the homology concept. Brain Behav. Evol. 81 (3), 150—153.

Métin, C., Alvarez, C., Moudoux, D., Vitalis, T., Pieau, C., Molnár, Z., 2007. Conserved pattern of tangential neuronal migration during forebrain development. Development 134 (15), 2815—2827.

Molnár, Z., 2013. Cortical columns. In: Comprehensive Developmental Neuroscience: Neural Circuit Development and Function in the Brainvol, vol. 3, pp. 109—129.

Molnár, Z., Cordery, P., 1999. Connections between cells of the internal capsule, thalamus, and cerebral cortex in embryonic rat. J. Comp. Neurol. 413 (1), 1—25.

Molowny, A., Lopez, C., Martin, F., 1972. Estudio citoarquitectónico de la corteza cerebral de reptiles. Trab. Inst. Cajal Inv. Biol. 14, 125—152.

Molowny, A., Nacher, J., Lopez-Garcia, C., 1995. Reactive neurogenesis during regeneration of the lesioned medial cerebral cortex of lizards. Neuroscience 68 (3), 823—836.

Molyneaux, B.J., Arlotta, P., Menezes, J.R., Macklis, J.D., 2007. Neuronal subtype specification in the cerebral cortex. Nat. Rev. Neurosci. 8 (6), 427—437.

Montiel, J.F., Aboitiz, F., 2015. Pallial patterning and the origin of the isocortex. Front. Neurosci. 9.

Montiel, J.F., Molnár, Z., 2013. The impact of gene expression analysis on evolving views of avian brain organization. J. Comp. Neurol. 521 (16), 3604—3613.

Montiel, J.F., Wang, W.Z., Oeschger, F.M., Hoerder-Suabedissen, A., Tung, W.L., García-Moreno, F., Molnár, Z., 2011. Hypothesis on the Dual Origin of the Mammalian Subplate.

Moran, A., Wojcik, L., Cangiane, L., Schade Powers, A., 1997. Dorsal cortex lesions impair habituation in turtles (*Chrysemys picta*). Brain Behav. Evol. 51 (1), 40—47.

Moreno, N., Morona, R., López, J.M., Gonzalez, A., 2010. Subdivisions of the turtle *Pseudemys scripta* subpallium based on the expression of regulatory genes and neuronal markers. J. Comp. Neurol. 518 (24), 4877—4902.

Mori, K., Shepherd, G.M., 1979. Synaptic excitation and long-lasting inhibition of mitral cells in the in vitro turtle olfactory bulb. Brain Res. 172 (1), 155—159.

Morlock, H.C., 1972. Behavior following ablation of the dorsal cortex of turtles. Brain Behav. Evol. 5 (2—3), 256—263.

Mueller, T., Dong, Z., Berberoglu, M.A., Guo, S., 2011. The dorsal pallium in zebrafish, *Danio rerio* (Cyprinidae, Teleostei). Brain Res. 1381, 95—105.

Mufson, E.J., Desan, P.H., Mesulam, M.M., Wainer, B.H., Levey, A.I., 1984. Choline acetyltransferase-like immunoreactivity in the forebrain of the red-eared pond turtle (*Pseudemys scripta elegans*). Brain Res. 323 (1), 103—108.

Mulligan, K.A., Ulinski, P.S., 1990. Organization of geniculocortical projections in turtles: isoazimuth lamellae in the visual cortex. J. Comp. Neurol. 296 (4), 531—547.

Münch, T.A., da Silveira, R.A., Siegert, S., Viney, T.J., Awatramani, G.B., Roska, B., 2009. Approach sensitivity in the retina processed by a multifunctional neural circuit. Nat. Neurosci. 12 (10), 1308—1316.

Muñoz, M., Smeets, W.J.A.J., López, J.M., Moreno, N., Morona, R., Domínguez, L., González, A., 2008. Immunohistochemical localization of neuropeptide FF-like in the brain of the turtle: relation to catecholaminergic structures. Brain Res. Bull. 75 (2), 256—260.

Nässel, D.R., Homberg, U., 2006. Neuropeptides in interneurons of the insect brain. Cell Tissue Res. 326 (1), 1—24.

Ngwenya, A., Patzke, N., Manger, P.R., Herculano-Houzel, S., 2016. Continued growth of the central nervous system without mandatory addition of neurons in the Nile crocodile (*Crocodylus niloticus*). Brain Behav. Evol. 87 (1), 19—38.

Nieuwenhuys, R., 2011. The development and general morphology of the telencephalon of actinopterygian fishes: synopsis, documentation and commentary. Brain Struct. Funct. 215 (3—4), 141—157.

Nieuwenhuys, R., Puelles, L., 2016. Towards a New Neuromorphology. Springer International Publishing.

Nieuwenhuys, R., Donkelaar, H., Nicholson, C., 1998. The Central Nervous System of Vertebrates. Springer-Verlag.

Noctor, S.C., Flint, A.C., Weissman, T.A., Dammerman, R.S., Kriegstein, A.R., 2001. Neurons derived from radial glial cells establish radial units in neocortex. Nature 409 (6821), 714—720.

Nomura, T., Gotoh, H., Ono, K., 2013. Changes in the regulation of cortical neurogenesis contribute to encephalization during amniote brain evolution. Nat. Commun. 4.

Northcutt, R.G., 1978. Forebrain and midbrain organization in lizards and its phylogenetic significance. Behav. Neurol. Lizards 11—64.

Northcutt, R.G., 1981. Evolution of the telencephalon in nonmammals. Annu. Rev. Neurosci. 4 (1), 301—350.

Northcutt, R.G., 2008. Forebrain evolution in bony fishes. Brain Res. Bull. 75 (2), 191—205.

Northcutt, R.G., Kaas, J.H., 1995. The emergence and evolution of mammalian neocortex. Trends Neurosci. 18 (9), 373—379.

O'Shea, M., Williams, J.L.D., 1974. The anatomy and output connection of a locust visual interneurone; the lobular giant movement detector (LGMD) neurone. J. Comp. Physiol. A Neuroethol. Sens. Neural Behav. Physiol. 91 (3), 257—266.

Oberlaender, M., de Kock, C.P., Bruno, R.M., Ramirez, A., Meyer, H.S., Dercksen, V.J., Sakmann, B., 2012. Cell type—specific three-dimensional structure of thalamocortical circuits in a column of rat vibrissal cortex. Cereb. Cortex 22 (10), 2375—2391.

Ohtsuka, T., 1985. Spectral sensitivities of seven morphological types of photoreceptors in the retina of the turtle, *Geoclemys reevesii*. J. Comp. Neurol. 237 (2), 145—154.

Ohtsuka, T., 1985. Relation of spectral types to oil droplets in cones of turtle retina. Science 229 (4716), 874—877.

Onteniente, B., Kimura, H., Maeda, T., 1983. Comparative study of the glial fibrillary acidic protein in vertebrates by PAP immunohistochemistry. J. Comp. Neurol. 215 (4), 427—436.

Parent, A., Olivier, A., 1969. Comparative histochemical study of the corpus striatum. J. Hirnforsch. 12 (1), 73—81.

Paul, L.K., et al., 2007. Agenesis of the corpus callosum: genetic, developmental and functional aspects of connectivity. Nat. Rev. Neurosci. 8 (4), 287—299.

Pérez-Clausell, J., Fredens, K., 1988. Chemoarchitectonics in the telencephalon of the lizard *Podarcis hispanica*. The Forebrain of Reptiles. Karger Publishers, pp. 85—96.

Pérez-García, C.G., Gonzalez-Delgado, F.J., Suarez-Sola, M.L., Castro-Fuentes, R., Martin-Trujillo, J.M., Ferres-Torres, R., Meyer, G., 2001. Reelin-immunoreactive neurons in the adult vertebrate pallium. J. Chem. Neuroanat. 21 (1), 41—51.

Peters, A., Miller, M., Kimerer, L.M., 1983. Cholecystokinin-like immunoreactive neurons in rat cerebral cortex. Neuroscience 8 (3), 431—448.

Peterson, E., 1980. Behavioral studies of telencephalic function in reptiles. Comparative Neurology of the Telencephalon. Springer, US, pp. 343–388.

Peterson, E.H., 1992. Retinal structure. Biol. Reptil. 17, 1–135.

Petko, M., Ihionvien, M., 1988. Distribution of substance P, vasoactive intestinal polypeptide and serotonin immunoreactive structures in the central nervous system of the lizard, *Lacerta agilis*. J. Hirnforsch. 30 (4), 415–423.

Pettigrew, J.D., Konishi, M., 1976. Effect of monocular deprivation on binocular neurones in the owl's visual Wulst. Nature 264, 753–754.

Pincheira-Donoso, D., Bauer, A.M., Meiri, S., Uetz, P., 2013. Global taxonomic diversity of living reptiles. PLoS One 8 (3), e59741.

Platel, R., 1979. Comparative volumetric analysis of the principal subdivisions of the telencephalon in saurian reptiles. J. Hirnforsch. 21 (3), 271–291.

Platel, R., Beckers, H.J.A., Nieuwenhuys, R., 1973. Les champs corticaux chezTestudo hermanni (Reptile Chelonien) et chezCaiman crocodylus (Reptile Crocodylien). Acta Morphol. Neerl. Scand. 11, 121–150.

Poliakov, G.I., 1964. Development and complication of the cortical part of the coupling mechanism in the evolution of vertebrates. J. Hirnforsch. 7 (3), 253.

Powers, A.S., 1990. Brain mechanisms of learning in reptiles. Neurobiol. Comp. Cogn. 157–177.

Prechtl, J.C., 1994. Visual motion induces synchronous oscillations in turtle visual cortex. Proc. Natl. Acad. Sci. U.S.A. 91 (26), 12467–12471.

Prechtl, J.C., Cohen, L.B., Pesaran, B., Mitra, P.P., Kleinfeld, D., 1997. Visual stimuli induce waves of electrical activity in turtle cortex. Proc. Natl. Acad. Sci. U.S.A. 94 (14), 7621–7626.

Prechtl, J.C., Bullock, T.H., Kleinfeld, D., 2000. Direct evidence for local oscillatory current sources and intracortical phase gradients in turtle visual cortex. Proc. Natl. Acad. Sci. U.S.A. 97 (2), 877–882.

Preuss, T.M., 2000. Taking the measure of diversity: comparative alternatives to the model-animal paradigm in cortical neuroscience. Brain Behav. Evol. 55 (6), 287–299.

Prinz, A.A., Bucher, D., Marder, E., 2004. Similar network activity from disparate circuit parameters. Nat. Neurosci. 7 (12), 1345–1352.

Pritz, M.B., 1974. Ascending connections of a thalamic auditory area in a crocodile, *Caiman crocodilus*. J. Comp. Neurol. 153 (2), 199–213.

Pritz, M.B., 1975. Anatomical identification of a telencephalic visual area in crocodiles: ascending connections of nucleus rotundus in *Caiman crocodilus*. J. Comp. Neurol. 164 (3), 323–338.

Puelles, L., 2011. Pallio-pallial tangential migrations and growth signaling: new scenario for cortical evolution? Brain Behav. Evol. 78 (1), 108–127.

Puelles, L., Medina, L., 2002. Field homology as a way to reconcile genetic and developmental variability with adult homology. Brain Res. Bull. 57 (3), 243–255.

Puelles, L., Kuwana, E., Puelles, E., Bulfone, A., Shimamura, K., Keleher, J., Rubenstein, J.L., 2000. Pallial and subpallial derivatives in the embryonic chick and mouse telencephalon, traced by the expression of the genes Dlx-2, Emx-1, Nkx-2.1, Pax-6, and Tbr-1. J. Comp. Neurol. 424 (3), 409–438.

Puelles, L., Harrison, M., Paxinos, G., Watson, C., 2013. A developmental ontology for the mammalian brain based on the prosomeric model. Trends Neurosci. 36 (10), 570–578.

Puelles, L., et al., 2016. Selective early expression of the orphan nuclear receptor Nr4a2 identifies the claustrum homolog in the avian mesopallium: impact on sauropsidian/mammalian pallium comparisons. J. Comp. Neurol. 524 (3), 665–703.

Rainey, T.W., Ulinski, P.S., 1986. Morphology of neurons in the dorsal lateral geniculate complex in turtles of the genera *Pseudemys* and *Chrysemys*. J. Comp. Neurol. 253 (4), 440–465.

Rakic, P., 2008. Confusing cortical columns. Proc. Natl. Acad. Sci. U.S.A. 105 (34), 12099–12100.

Ramón y Cajal, S., 1909. Histologie du Système Nerveux de l'Homme et des Vertébrés.

Ramón y Cajal, S., De Felipe, J., Jones, E.G., 1988. Cajal on the Cerebral Cortex: An Annotated Translation on the Complete Writings. Oxford University Press.

Ramsden, H.L., Sürmeli, G., McDonagh, S.G., Nolan, M.F., 2015. Laminar and dorsoventral molecular organization of the medial entorhinal cortex revealed by large-scale anatomical analysis of gene expression. PLoS Comput. Biol. 11 (1), e1004032.

Regidor, J., Poch, L., 1988. Histochemical analysis of the lizard cortex: an acetylcholinesterase, cytochrome oxidase and NADPH-diaphorase study. The Forebrain of Reptiles. Karger Publishers, pp. 77–84.

Reiner, A., 1991. A comparison of neurotransmitter-specific and neuropeptide-specific neuronal cell types present in the dorsal cortex in turtles with those present in the isocortex in mammals: implications for the evolution of isocortex. Brain Behav. Evol. 38 (2–3), 73–82.

Reiner, A., 1992. *Neuropeptides* in the nervous system. In: Biology of the ReptiliaPhysiology E. Hormones, Brain, and Behavior, vol. 18.

Reiner, A., 1993. Neurotransmitter organization and connections of turtle cortex: implications for the evolution of mammalian isocortex. Comp. Biochem. Physiol. Part A Physiol. 104 (4), 735–748.

Reiner, A.J., 2000. A hypothesis as to the organization of cerebral cortex in the common amniote ancestor of modern reptiles and mammals. Novartis Foundation Symposium. John Wiley, Chichester, New York, pp. 83–113.

Reiner, A., 2013. You are who you talk with-a commentary on Dugas-Ford et al. PNAS, 2012. Brain Behav. Evol. 81 (3), 146–149.

Reiner, A., Beinfeld, M.C., 1985. The distribution of cholecystokinin-8 in the central nervous system of turtles: an immunohistochemical and biochemical study. Brain Res. Bull. 15 (2), 167–181.

Reiner, A., Karten, H.J., 1985. Comparison of olfactory bulb projections in pigeons and turtles. Brain Behav. Evol. 27 (1), 11–27.

Reiner, A., Northcutt, R.G., 2000. Succinic dehydrogenase histochemistry reveals the location of the putative primary visual and auditory areas within the dorsal ventricular ridge of *Sphenodon punctatus*. Brain Behav. Evol. 55 (1), 26–36.

Reiner, A., Oliver, J.R., 1987. Somatostatin and neuropeptide Y are almost exclusively found in the same neurons in the telencephalon of turtles. Brain Res. 426 (1), 149–156.

Reiner, A., Powers, A.S., 1980. The effects of extensive forebrain lesions on visual discriminative performance in turtles (*Chrysemys picta picta*). Brain Res. 192 (2), 327–337.

Reiner, A., Powers, A.S., 1983. The effects of lesions of telencephalic visual structures on visual discriminative performance in turtles (*Chrysemyspicta picta*). J. Comp. Neurol. 218 (1), 1–24.

Reiner, A., Krause, J.E., Keyser, K.T., Eldred, W.D., McKelvy, J.F., 1984. The distribution of substance P in turtle nervous system: a radioimmunoassay and immunohistochemical study. J. Comp. Neurol. 226 (1), 50–75.

Reiner, A., Eldred, W.D., Beinfeld, M.C., Krause, J.E., 1985. The co-occurrence of a substance P-like peptide and cholecystokinin-8 in a fiber system of turtle cortex. J. Neurosci. 5 (6), 1527–1544.

Reiner, A., Zhang, D., Eldred, W.D., 1996. Use of the sensitive anterograde tracer cholera toxin fragment B reveals new details of the central retinal projections in turtles. Brain Behav. Evol. 48 (6), 322–337.

Rice, R., Kallonen, A., Cebra-Thomas, J., Gilbert, S.F., 2016. Development of the turtle plastron, the order-defining skeletal structure. Proc. Natl. Acad. Sci. U.S.A. 201600958.

Riss, W., Halpern, M., Scalia, F., 1969. The quest for clues to forebrain evolution—the study of reptiles. Brain Behav. Evol. 2 (1), 26–50.

Robles, E., Laurell, E., Baier, H., 2014. The retinal projectome reveals brain-area-specific visual representations generated by ganglion cell diversity. Curr. Biol. 24 (18), 2085–2096.

Rockland, K.S., 2010. Five points on columns. Neocortical Column 6.

Rockland, K.S., Ichinohe, N., 2004. Some thoughts on cortical minicolumns. Exp. Brain Res. 158 (3), 265–277.

Rose, M., 1923. Histologische Lokalisation des Vorderhirns der Reptilien. J. Psychol. Neurol. 29, 219–272.

Roth, M.M., Dahmen, J.C., Muir, D.R., Imhof, F., Martini, F.J., Hofer, S.B., 2016. Thalamic nuclei convey diverse contextual information to layer 1 of visual cortex. Nat. Neurosci. 19 (2), 299–307.

Rowe, T.B., Shepherd, G.M., 2016. Role of ortho-retronasal olfaction in mammalian cortical evolution. J. Comp. Neurol. 524 (3), 471–495.

Rubio-Garrido, P., Pérez-de-Manzo, F., Porrero, C., Galazo, M.J., Clascá, F., 2009. Thalamic input to distal apical dendrites in neocortical layer 1 is massive and highly convergent. Cereb. Cortex 19 (10), 2380–2395.

Rudy, B., Fishell, G., Lee, S., Hjerling-Leffler, J., 2011. Three groups of interneurons account for nearly 100% of neocortical GABAergic neurons. Dev. Neurobiol. 71 (1), 45–61.

Sagan, C., 1977. The Dragons of Eden. Ballantine, New York, p. 4.

Sakurai, T., 2007. The neural circuit of orexin (hypocretin): maintaining sleep and wakefulness. Nat. Rev. Neurosci. 8 (3), 171–181.

Sanes, J.R., Masland, R.H., 2015. The types of retinal ganglion cells: current status and implications for neuronal classification. Annu. Rev. Neurosci. 38, 221–246.

Sanides, F., Sanides, D., 1972. The "extraverted neurons" of the mammalian cerebral cortex. Anat. Embryol. 136 (3), 272–293.

Schapiro, H., Goodman, D.C., 1969. Motor functions and their anatomical basis in the forebrain and tectum of the alligator. Exp. Neurol. 24 (2), 187–195.

Scheich, H., 1983. Two columnar systems in the auditory neostriatum of the chick: evidence from 2-deoxyglucose. Exp. Brain Res. 51 (2), 199–205.

Schoch, R.R., Sues, H.D.A., 2015. Middle Triassic stem-turtle and the evolution of the turtle body plan. Nature 523.

Schoch, R.R., Sues, H.D., 2016. The diapsid origin of turtles. Zoology 119.

Schwerdtfeger, W.K., García, C.L., 1986. GABAergic neurons in the cerebral cortex of the brain of a lizard (Podarcis hispanica). Neurosci. Lett. 68 (1), 117–121.

Schwerdtfeger, W.K., Germroth, P., 1990. Archicortical and periarchicortical areas in the vertebrate forebrain. The Forebrain in Nonmammals: New Aspects of Structure and Development. Springer, Berlin, pp. 197–212.

Shein-Idelson, M., Ondracek, J.M., Liaw, H.P., Reiter, S., Laurent, G., 2016. Slow waves, sharp waves, ripples, and REM in sleeping dragons. Science 352 (6285), 590–595.

Shepherd, G.M., 1974. The Synaptic Organization of the Brain. Oxford University Press.

Shepherd, G.M., 1988. A Basic Circuit for Cortical Organization. MIT-Press, pp. 93–134.

Shepherd, G.M., 2011. The microcircuit concept applied to cortical evolution: from three-layer to six-layer cortex. Front. Neuroanat. 5.

Shepherd, G.M., 2011. The microcircuit concept applied to cortical evolution: from three-layer to six-layer cortex. Front. Neuroanat. 5, 30.

Sjöström, A.M., Ulinski, P.S., 1985. Morphology of retinogeniculate terminals in the turtle, Pseudemys scripta elegans. J. Comp. Neurol. 238 (1), 107–120.

Skeen, L.C., Pindzola, R.R., Schofield, B.R., 1984. Tangential organization of olfactory, association, and commissural projections to olfactory cortex in a species of reptile (Trionyx spiniferus), bird (Aix sponsa), and mammal (Tupaia glis). Brain Behav. Evol. 25 (4), 206–216.

Smeets, W.J., López, J.M., González, A., 2006. Distribution of neuropeptide FF-like immunoreactivity in the brain of the lizard Gekko gecko and its relation to catecholaminergic structures. J. Comp. Neurol. 498 (1), 31–45.

Smith, L.M., Ebner, F.F., Colonnier, M., 1980. The thalamocortical projection in Pseudemys turtles: a quantitative electron microscopic study. J. Comp. Neurol. 190 (3), 445–461.

Smith-Fernandez, A., Pieau, C., Repérant, J., Boncinelli, E., Wassef, M., 1998. Expression of the Emx-1 and Dlx-1 homeobox genes define three molecularly distinct domains in the telencephalon of mouse, chick, turtle and frog embryos: implications for the evolution of telencephalic subdivisions in amniotes. Development 125 (11), 2099–2111.

Somogyi, P., 1977. A specific 'axo-axonal' interneuron in the visual cortex of the rat. Brain Res. 136 (2), 345–350.

Stensaas, L.J., Stensaas, S.S., 1968. Light microscopy of glial cells in turtles and birds. Z. Zellforsch. Mikrosk. Anat. 91 (3), 315–340.

Sterling, P., Laughlin, S., 2015. Principles of Neural Design. MIT Press.

Stevens, C.F., 1998. Neuronal diversity: too many cell types for comfort? Curr. Biol. 8 (20), R708–R710.

Strausfeld, N.J., Hirth, F., 2013. Deep homology of arthropod central complex and vertebrate basal ganglia. Science 340 (6129), 157–161.

Striedter, G.F., 1997. The telencephalon of tetrapods in evolution. Brain Behav. Evol. 49 (4), 179–194.

Striedter, G.F., 1998. Progress in the study of brain evolution: from speculative theories to testable hypotheses. Anat. Rec. 253 (4), 105–112.

Striedter, G.F., 2005. Principles of Brain Evolution.

Striedter, G.F., 2016. Evolution of the hippocampus in reptiles and birds. J. Comp. Neurol. 524.

Striedter, G.F., Northcutt, R.G., 1991. Biological hierarchies and the concept of homology. Brain Behav. Evol. 38 (4–5), 177–189.

Striedter, G.F., Marchant, T.A., Beydler, S., 1998. The "neostriatum" develops as part of the lateral pallium in birds. J. Neurosci. 18 (15), 5839–5849.

Striedter, G.F., Belgard, T.G., Chen, C.C., Davis, F.P., Finlay, B.L., Güntürkün, O., Hofmann, H.A., 2014. NSF workshop report: discovering general principles of nervous system organization by comparing brain maps across species. Brain Behav. Evol. 83 (1), 1–8.

Suárez, J., Dávila, J.C., Real, M.Á., Guirado, S., Medina, L., 2006. Calcium-binding proteins, neuronal nitric oxide synthase, and GABA help to distinguish different pallial areas in the developing and adult chicken. I. Hippocampal formation and hyperpallium. J. Comp. Neurol. 497 (5), 751–771.

Suárez, R., Gobius, I., Richards, L.J., 2014. Evolution and development of interhemispheric connections in the vertebrate forebrain. Front. Hum. Neurosci. 8, 497.

Sugerman, R.A., Demski, L.S., 1978. Agonistic behavior elicited by electrical stimulation of the brain in western collared lizards, Crotaphytus collaris. Brain Behav. Evol. 15 (5–6), 446–469.

Sun, H., Frost, B.J., 1998. Computation of different optical variables of looming objects in pigeon nucleus rotundus neurons. Nat. Neurosci. 1 (4), 296–303.

Super, H., Uylings, H.B.M., 2001. The early differentiation of the neocortex: a hypothesis on neocortical evolution. Cereb. Cortex 11 (12), 1101–1109.

Super, H., Soriano, E., Uylings, H.B.M., 1998. The functions of the preplate in development and evolution of the neocortex and hippocampus. Brain Res. Rev. 27 (1), 40–64.

Suzuki, I.K., Hirata, T., 2014. A common developmental plan for neocortical gene-expressing neurons in the pallium of the domestic chicken Gallus gallus domesticus and the Chinese softshell turtle Pelodiscus sinensis. Front. Neuroanat. 8, 20.

Suzuki, I.K., Kawasaki, T., Gojobori, T., Hirata, T., 2012. The temporal sequence of the mammalian neocortical neurogenetic program drives mediolateral pattern in the chick pallium. Dev. Cell 22 (4), 863–870.

Szentagothai, J., 1978. Specificity versus (Quasi-) Randomness in Cortical Connectivity. Raven Press, New York, pp. 77–97.

Szentagothai, J., Arbib, M.A., 1974. Conceptual models of neural organization. Neurosci. Res. Program Bull. 12.

Tanaka, D.H., Nakajima, K., 2012. Migratory pathways of GABAergic interneurons when they enter the neocortex. Eur. J. Neurosci. 35 (11), 1655–1660.

Tanaka, D.H., Oiwa, R., Sasaki, E., Nakajima, K., 2011. Changes in cortical interneuron migration contribute to the evolution of the neocortex. Proc. Natl. Acad. Sci. U.S.A. 108 (19), 8015–8020.

Taniguchi, H., Lu, J., Huang, Z.J., 2013. The spatial and temporal origin of chandelier cells in mouse neocortex. Science 339 (6115), 70–74.

Tasic, B., Menon, V., Nguyen, T.N., Kim, T.K., Jarsky, T., Yao, Z., Bertagnolli, D., 2016. Adult mouse cortical cell taxonomy revealed by single cell transcriptomics. Nat. Neurosci. 19 (2), 335–346.

Teissier, A., Griveau, A., Vigier, L., Piolot, T., Borello, U., Pierani, A., 2010. A novel transient glutamatergic population migrating from the pallial–subpallial boundary contributes to neocortical development. J. Neurosci. 30 (31), 10563–10574.

Teissier, A., Waclaw, R.R., Griveau, A., Campbell, K., Pierani, A., 2012. Tangentially migrating transient glutamatergic neurons control neurogenesis and maintenance of cerebral cortical progenitor pools. Cereb. Cortex 22 (2), 403–416.

Theurich, M., Müller, C.M., Scheich, H., 1984. 2-Deoxyglucose accumulation parallels extracellularly recorded spike activity in the avian auditory neostriatum. Brain Res. 322 (1), 157–161.

Thompson, C.L., Ng, L., Menon, V., Martinez, S., Lee, C.K., Glattfelder, K., Garcia-Lopez, R., 2014. A high-resolution spatiotemporal atlas of gene expression of the developing mouse brain. Neuron 83 (2), 309–323.

Tomassy, G.S., Berger, D.R., Chen, H.H., Kasthuri, N., Hayworth, K.J., Vercelli, A., Arlotta, P., 2014. Distinct profiles of myelin distribution along single axons of pyramidal neurons in the neocortex. Science 344 (6181), 319–324.

Tremblay, R., Lee, S., Rudy, B., 2016. GABAergic interneurons in the neocortex: from cellular properties to circuits. Neuron 91 (2), 260–292.

Trinh, A.T., Harvey-Girard, E., Teixeira, F., Maler, L., 2016. Cryptic laminar and columnar organization in the dorsolateral pallium of a weakly electric fish. J. Comp. Neurol. 524 (2), 408–428.

Tukker, J.J., Fuentealba, P., Hartwich, K., Somogyi, P., Klausberger, T., 2007. Cell type-specific tuning of hippocampal interneuron firing during gamma oscillations in vivo. J. Neurosci. 27 (31), 8184–8189.

Tyner, C.F., 1975. The naming of neurons: applications of taxonomic theory to the study of cellular populations. Brain Behav. Evol. 12 (1–2), 75–96.

Ulinski, P.S., 1974. Cytoarchitecture of cerebral cortex in snakes. J. Comp. Neurol. 158 (3), 243–266.

Ulinski, P.S., 1976. Intracortical connections in the snakes Natrix sipedon and Thamnophis sirtalis. J. Morphol. 150 (2 Pt. 2), 463–483.

Ulinski, P.S., 1977. Intrinsic organization of snake medial cortex: an electron microscopic and Golgi study. J. Morphol. 152 (2), 247–279.

Ulinski, P.S., 1978. Organization of anterior dorsal ventricular ridge in snakes. J. Comp. Neurol. 178 (3), 411–449.

Ulinski, P.S., 1983. Dorsal Ventricular Ridge: A Treatise on Forebrain Organization in Reptiles and Birds. Wiley-Interscience.

Ulinski, P.S., 1986. Organization of corticogeniculate projections in the turtle, Pseudemys scripta. J. Comp. Neurol. 254 (4), 529–542.

Ulinski, P.S., 1990. The Cerebral Cortex of Reptiles. Springer, US, pp. 139–215.

Ulinski, P.S., 1999. Neural mechanisms underlying the analysis of moving visual stimuli. Models of Cortical Circuits. Springer, US, pp. 283–399.

Ulinski, P.S., 2006. Visual cortex of turtles. Evol. Nerv. Syst. 2, 195–203.

Ulinski, P.S., Nautiyal, J., 1988. Organization of retinogeniculate projections in turtles of the genera Pseudemys and Chrysemys. J. Comp. Neurol. 276 (1), 92–112.

Unger, L., 1906. Untersuchungen über die Morphologie und Faserung des Reptiliengehirns. Anat. Embryol. 31 (2), 269–348.

Vallarino, M., Feuilloley, M., D'Aniello, B., Rastogi, R.K., Vaudry, H., 1994. Distribution of FMRFamide-like immunoreactivity in the brain of the lizard Podarcis sicula. Peptides 15 (6), 1057–1065.

Valverde, F., Facal-Valverde, M.V., 1986. Neocortical layers I and II of the hedgehog (Erinaceus europaeus). Anat. Embryol. 173 (3), 413–430.

Viney, T.J., Lasztoczi, B., Katona, L., Crump, M.G., Tukker, J.J., Klausberger, T., Somogyi, P., 2013. Network state-dependent inhibition of identified hippocampal CA3 axo-axonic cells in vivo. Nat. Neurosci. 16 (12), 1802–1811.

Voneida, T.J., Ebbesson, S.O.E., 1969. On the origin and distribution of axons in the pallial commissures in the tegu lizard (Tupinambis nigropunctatus). Brain Behav. Evol. 2 (5–6), 467–481.

Wagner, G.P., 2014. Homology, Genes, and Evolutionary Innovation. Princeton University Press.

Wang, R.T., Halpern, M., 1977. Afferent and efferent connections of thalamic nuclei of visual system of garter snakes. Anat. Rec. 187 (4), 741–742.

Wang, Y., Brzozowska-Prechtl, A., Karten, H.J., 2010. Laminar and columnar auditory cortex in avian brain. Proc. Natl. Acad. Sci. U.S.A. 107 (28), 12676–12681.

Wang, W.Z., Hoerder-Suabedissen, A., Oeschger, F.M., Bayatti, N., Ip, B.K., Lindsay, S., Clowry, G.J., 2010. Subplate in the developing cortex of mouse and human. J. Anat. 217 (4), 368–380.

Wang, W.Z., Oeschger, F.M., Montiel, J.F., García-Moreno, F., Hoerder-Suabedissen, A., Krubitzer, L., Molnár, Z., 2011. Comparative aspects of subplate zone studied with gene expression in sauropsids and mammals. Cereb. Cortex 21 (10), 2187–2203.

Weindl, A., Triepel, J., Kuchling, G., 1984. Somatostatin in the brain of the turtle Testudo hermanni Gmelin an immunohistochemical mapping study. Peptides 5, 91–100.

Weissman, T., Noctor, S.C., Clinton, B.K., Honig, L.S., Kriegstein, A.R., 2003. Neurogenic radial glial cells in reptile, rodent and human: from mitosis to migration. Cereb. Cortex 13 (6), 550–559.

Wilhelm, M., Straznicky, C., 1992. The topographic organization of the retinal ganglion cell layer of the lizard Ctenophorus nuchalis. Arch. Histol. Cytol. 55 (3), 251–259.

Woodbury, P.B., Ulinski, P.S., 1986. Conduction velocity, size and distribution of optic nerve axons in the turtle, Pseudemys scripta elegans. Anat. Embryol. 174 (2), 253–263.

Xiao, Q., Li, D.P., Wang, S.R., 2006. Looming-sensitive responses and receptive field organization of telencephalic neurons in the pigeon. Brain Res. Bull. 68 (5), 322–328.

Xu, X., Roby, K.D., Callaway, E.M., 2010. Immunochemical characterization of inhibitory mouse cortical neurons: three chemically distinct classes of inhibitory cells. J. Comp. Neurol. 518 (3), 389–404.

Yanes, C., Monzon-Mayor, M., Ghandour, M.S., De Barry, J., Gombos, G., 1990. Radial glia and astrocytes in developing and adult telencephalon of the lizard Gallotia galloti as revealed by immunohistochemistry with anti-GFAP and anti-vimentin antibodies. J. Comp. Neurol. 295 (4), 559–568.

Yilmaz, M., Meister, M., 2013. Rapid innate defensive responses of mice to looming visual stimuli. Curr. Biol. 23 (20), 2011–2015.

Young, L.J., Lopreato, G.F., Horan, K., Crews, D., 1994. Cloning and in situ hybridization analysis of estrogen receptor, progesterone and androgen receptor expression in the brain of whiptail lizards (Cnemidophorus uniparens and C. inornatus). J. Comp. Neurol. 347 (2), 288–300.

Yuste, R., 2011. Dendritic spines and distributed circuits. Neuron 71 (5), 772–781.

Zeisel, A., Muñoz-Manchado, A.B., Codeluppi, S., Lönnerberg, P., La Manno, G., Juréus, A., Rolny, C., 2015. Cell types in the mouse cortex and hippocampus revealed by single-cell RNA-seq. Science 347 (6226), 1138–1142.

Zhu, D., Lustig, K.H., Bifulco, K., Keifer, J., 2005. Thalamocortical connections in the pond turtle *Pseudemys scripta elegans*. Brain Behav. Evol. 65 (4), 278–292.

Zottoli, S.J., Faber, D.S., 2000. Review: the Mauthner cell: what has it taught us? The Neuroscientist 6 (1), 26–38.

10

The Cerebellum of Nonmammalian Vertebrates

Kara E. Yopak[1], J.M.P. Pakan[2], Douglas Wylie[3]

[1]University of North Carolina Wilmington, Department of Biology and Marine Biology, Wilmington, NC, United States; [2]University of Edinburgh, Edinburgh, United Kingdom; [3]Department of Biological Sciences, University of Alberta, Edmonton, AB, Canada

10.1 Introduction

As early as 1824, Flourens made the fundamental observation that movements are affected but not completely lost after cerebellar ablation in vertebrates, which led him to propose a critical role for the cerebellum in the coordination of movement (Flourens, 1824). While pioneering work such as this was extensive (for historical review, see Dow and Moruzzi, 1958), for more than a century the coordination of movement was the sole function attributed to the cerebellum. The functional role of the cerebellum expanded when, for example, it was proposed that the cerebellar cortex acts as a learning device (eg, Marr, 1969). At present, the role of the cerebellum has continued to expand well beyond its traditional role in motor behavior. The cerebellum is a major associative center for sensory input, and its role in sensory perception is just beginning to be elucidated (Baumann et al., 2015). Further, there has been an increasing amount of recent research that suggests the cerebellum plays a role in higher cognitive functions, such as emotion, speech, and memory (eg, Strata, 2015). This functional diversity highlights the importance of understanding the ontology, anatomy, and physiology of the cerebellum from an evolutionary perspective. In this chapter, we will discuss some comparative aspects of the cerebellum in nonmammalian vertebrates. In addition to the original research cited, we direct the reader to previous reviews by Larsell (1967), Llinas and Hillman (1969), Nieuwenhuys et al. (1998), Butler and Hodos (2005), and Glickstein et al. (2009).

10.2 Gross Morphology of the Cerebellum

The word cerebellum means "little brain," despite the fact that, in many species, the cerebellum accounts for more than 75% of the total number of neurons in the brain (Herculano-Houzel et al., 2015). The cerebellum always sits atop the fourth ventricle but, as shown in Fig. 10.1, there is a tremendous diversity with respect to the relative size and structure of cerebellum among vertebrates.

10.2.1 Agnathans

The presence of cerebellar tissue in the cyclostomes (lampreys and hagfish) has been an area of considerable debate and, until recently, it was a generally accepted view that the cerebellum appeared as an evolutionary innovation at the juncture between vertebrates and gnathostomes (jawed vertebrates; reviewed by Montgomery et al., 2012). A true cerebellum is not apparent in lampreys or hagfish, although lampreys do possess "cerebellar-like" structures: the dorsal and medial octavolateralis nuclei (DON and MON). The DON and MON are involved in the processing of electroreceptive and lateral line sensory information, respectively. The presence of these structures is questionable in hagfishes: they lack electroreceptors and have thus no DON, and have a variably developed lateral line system (for review, see Jørgensen et al., 1998). In other fishes, these cerebellar-like structures are adjacent to the "true" cerebellum (see below), and it has been proposed that the cerebellar-like structures may be the evolutionary antecedent of the true cerebellum (reviewed by Montgomery et al., 2012).

In addition, a recent genoarchitecture study has demonstrated that the precursor to the cerebellum is likely present in the brain of cyclostomes. In gnathostomes, the cerebellum develops from the "rhombic lip" region in the dorsal part of rhombomere 1 (Watson et al., 2011). *PAX6* and *Atoh1* are two genes expressed in this region that are important for the development of the cerebellum (Engelkamp et al., 1999; Wullimann

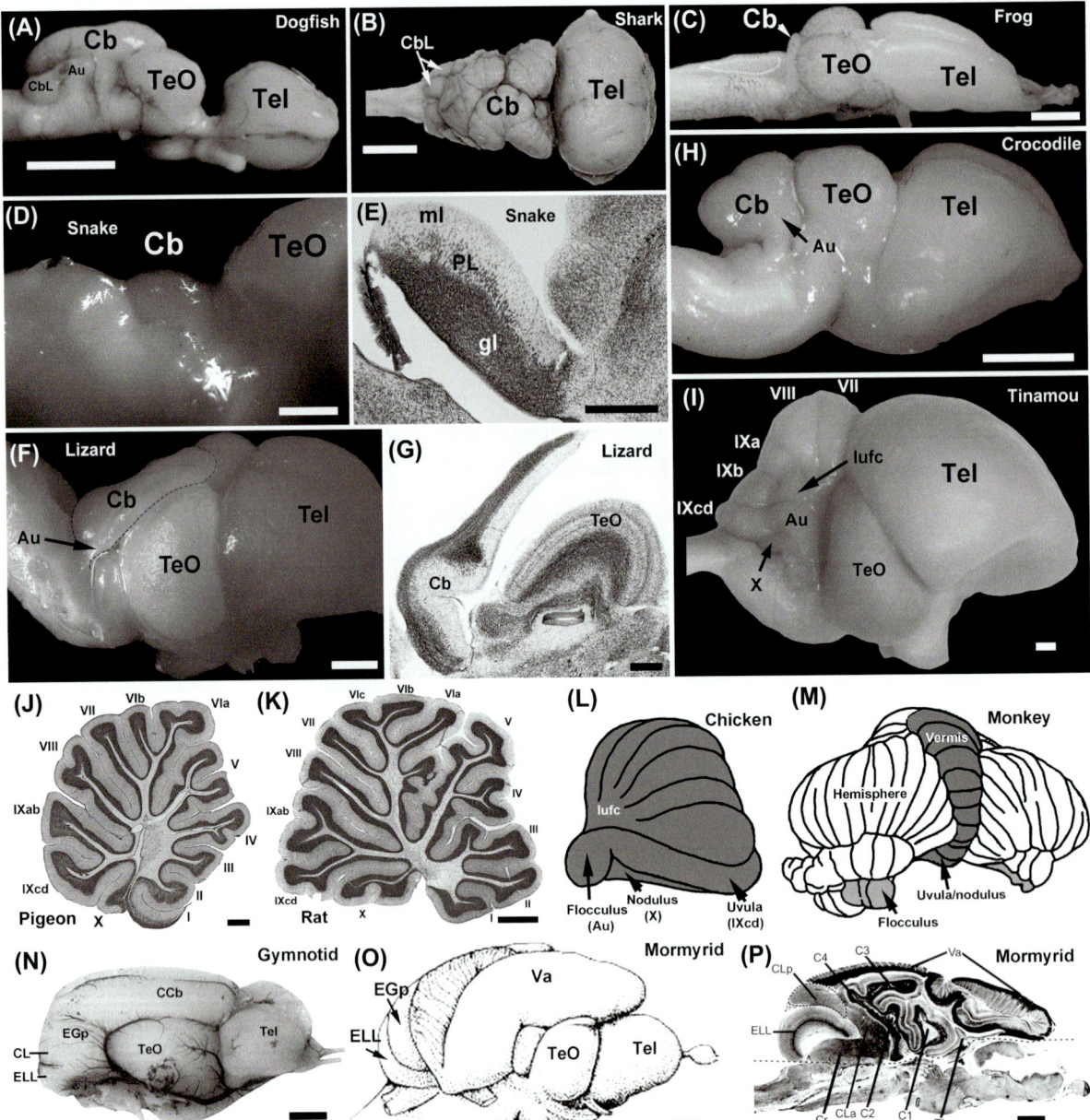

FIGURE 10.1 The cerebellum across vertebrates. Photomicrographs of lateral views of the brain in a northern spiny dogfish (*Squalus acanthias*) (A); a green tree frog (*Hyla cinerea*) (C, adapted from Wilczynski, W., 2009. Evolution, of the brain: in Amphibians. In: Binder, M.D., Hirokawa, N., Windhorst, U., Hirsch, M. (Eds.), Encyclopedia of Neuroscience. Springer, Berlin, Heidelberg, pp. 1301—1305; with permission.); a lizard species, the central netted dragon (*Ctenophorus nuchalis*) (F); a nile crocodile (*Crocodylus niloticus*) (H); a Chilean tinamou (*Nothoprocta perdicaria*) (I, from Corfield, J.R., Kolominsky, J., Marin, G.J., Craciun, I., Mulvany-Robbins, B.E., Iwaniuk, A.N., Wylie, D.R., 2015. Zebrin II expression in the cerebellum of a Paleognathous bird, the Chilean tinamou (*Nothoprocta perdicaria*). Brain Behav. Evol. 85, 94—106; with permission.), and a gymnotid (*Apteronotus leptorhynchus*) (N, adapted from Maler, L., Sas, E., Johnston, S., Ellis, W., 1991. An atlas of the brain of the electric fish *Apteronotus leptorhynchus*. J. Chem. Neuroanat. 4, 1—38; with permission.). A photomicrograph of the dorsal view of the brain of a smooth hammerhead shark (*Sphyrna zygaena*) is shown in (B) (B, from Yopak, K.E., 2012. Neuroecology of cartilaginous fishes: the functional implications of brain scaling. J. Fish Biol. 80, 1968—2023; with permission. (O) offers a drawing of the brain of a mormyrid (*Brienomyrus brachyistius*) (O, adapted from Carlson, B.A., Hopkins, C.D., 2004. Central control of electric signaling behavior in the mormyrid *Brienomyrus brachyistius*: segregation of behavior-specific inputs and the role of modifiable recurrent inhibition. J. Exp. Biol. 207, 1073—1084; with permission.). (P, Reproduced with permission from Nieuwenhuys, R., Pouwels, E. & Smulders-Kersten, E., 1974. The neuronal organization of cerebellar lobe C1 in the mormyrid fish gnathonemus petersii (teleostei). Z. Anat. Entwickl. Gesch. 144, 315—336). https://doi.org/10.1007/BF00522813). Drawings of the cerebellum of a chicken (*Gallus domesticus*) and a monkey are shown in (L) and (M), respectively (L and M, adapted from Voogd, J., Glickstein, M., 1998. The anatomy of the cerebellum. Trends Neurosci. 21, 370—375; with permission.). (D) shows a photomicrograph of the lateral view of the caudal part of the brain of a western diamondback rattlesnake (*Crotalus atrox*) (D, from Aspden, J.W., Armstrong, C.L., Gutierrez-Ibanez, C.I., Hawkes, R., Iwaniuk, A.N., Kohl, T., Graham, D.J., Wylie, D.R., 2015. Zebrin II/aldolase C expression in the cerebellum of the western diamondback rattlesnake (*Crotalus atrox*).

et al., 2011). As such, a region that expresses *Pax6* and *Atoh1* has been identified as the rhombic lip in hagfishes. Similarly, there is a rhombic-lip-like region in lampreys that expresses *Atoh1* and *Pax6B* (Sugahara et al., 2016). Thus, although the true cerebellum is an innovation found in the gnathostomes, its developmental precursor exists in the cyclostomes.

10.2.2 Cartilaginous Fishes

The first true cerebellum is documented in the cartilaginous fishes (comprised of sharks, batoids, and chimaerids). Compared to other jawed vertebrates, the cerebellum of cartilaginous fishes is highly developed (Fig. 10.1A and B). Two major divisions are seen, which have their homologues in other vertebrate classes: a vestibulocerebellum and a corpus cerebellum. The corpus cerebellum is unpaired, dorsally situated and, in many species of cartilaginous fishes, highly foliated (ie, convoluted). There is considerable variation across species with respect to the relative size, degree of foliation, and even symmetry of the foliation (see Fig. 10.3; reviewed in Yopak, 2012). In the cartilaginous fishes, the vestibulocerebellum is usually referred to as the vestibulateral cerebellum, which contains two subdivisions, namely a medial vestibulocerebellum and the lateral auricles (Smeets, 1998). The medial vestibulocerebellum is continuous with the anterior ends of the auricles. The posterior ends of the auricles are continuous with, and supply parallel fibers to, the cerebellar-like structures (MON and DON; Paul, 1982; Schmidt and Bodznick, 1987).

10.2.3 Amphibians and Nonavian Reptiles

In these groups, the cerebellum consists of a corpus cerebellum and a vestibulocerebellum (auricle), but the cerebellum is small and rudimentary (Larsell, 1967). Invariably, the corpus cerebellum is unfoliated. In amphibians, the cerebellum is very small, appearing as a single leaf overlying the fourth ventricle just posterior to the optic tectum (Fig. 10.1C). The situation is not appreciably different in snakes (Fig. 10.1D and E) and turtles. Although larger compared to snakes and turtles, the cerebellum in lizards also appears as a relatively simple single sheet. However, the lizard cerebellum is flipped anteriorly over the tectum, such that the granular layer is exposed dorsally (Fig. 10.1F and G). Among the nonavian reptiles, the cerebellum is largest among crocodilians (Fig. 10.1H). Their cerebellum is not foliated, but two shallow fissures divide it into three lobes (Larsell, 1967).

10.2.4 Birds

In birds, the cerebellum is rather large, averaging about 12% of total brain volume across species. It is highly developed, being divided into 10 major folia, indicated with Roman numerals I–X, from anterior to posterior (Fig. 10.1I, J, and L). Indeed, the foliation pattern in birds and mammals is strikingly similar when viewed in midsagittal sections (Fig. 10.1J and K). The laterally protruding auricle is prominent in all avian species, and it is generally referred to as the flocculus, as it is in mammals (and some other tetrapods). The flocculus is a site of visual–vestibular integration and is strikingly similar in mammals and birds with respect to physiological response properties and anatomical connections (Voogd and Wylie, 2004). The vestibulocerebellum in birds also includes the posterior parts of folium IX (IXcd) and folium X. Respectively, these are referred to as the uvula and nodulus, as they are in mammals (Fig. 10.1L and M).

The major difference between birds and mammals with respect to cerebellar morphology is that the neocerebellum (ie, the paired hemispheres) of the mammalian cerebellum is absent in birds. [It has been suggested that the lateral unfoliated cortex of the avian cerebellum (Fig. 10.1I and L) may represent a small rudimentary cerebellar hemisphere (Larsell, 1967; Pakan et al., 2007)]. As such, the traditional view is that the bulk of the cerebellum in birds is essentially equivalent to the vermis of the mammalian cerebellum (Fig. 10.1L and M). Some developmental studies have recently challenged this traditional view. In mouse the induction of the vermis is dependent on fibroblast growth factor 8 (Fgf8), whereas the hemispheres are Fgf8 independent. In the chick, the cerebellum is largely FGF independent; however, the development of the

PLoS One 10, e0117539.). Nissl-stained midsagittal sections through the cerebellum are shown for a western diamondback rattlesnake (E, from Aspden, J.W., Armstrong, C.L., Gutierrez-Ibanez, C.I., Hawkes, R., Iwaniuk, A.N., Kohl, T., Graham, D.J., Wylie, D.R., 2015. Zebrin II/aldolase C expression in the cerebellum of the western diamondback rattlesnake (*Crotalus atrox*). PLoS One 10, e0117539.), the central netted dragon (G), a pigeon (*Columba livia*) (J), a rat (*Rattus norvegicus*) (K), and a mormyrid (*Gnathonemus petersii*) (P, from Nieuwenhuys, R., Nicholson, C., 1969. A survey of the general morphology, the fiber connections and the possible functional significance of the gigantocerebellum of mormyrid fishes. In: Llinas, R. (Ed.), Neurobiology of Cerebellar Evolution and Development. American Medical Association, Chicago, pp. 107–134; with permission.). For all panels except (L) and (M), left is posterior, right is anterior. Abbreviations: *Au*, auricle; *Cb*, cerebellum; *CCb*, corpus cerebellum; *C1-4*, divisions of the CCb in mormyrids; *gl*, granular layer; *CL(a/p)*, caudal lobe (anterior/posterior); *CbL*, cerebellar-like structures; *Cr*, cerebellar crest; *EGp*, eminentia granularis, posterior; *ELL*, electrosensory lateral line lobe; *lufc*, lateral unfoliated cortex; *ml*, molecular layer; *Pl*, Purkinje layer; *Tel*, telencephalon; *TeO*, optic tectum; *TL*, lobus transitorius; *Va*, valvula. Scale bars: 1 mm in (C), (D), (F), (H–K), (N–P); 500 μm in (E), 400 μm in (G).

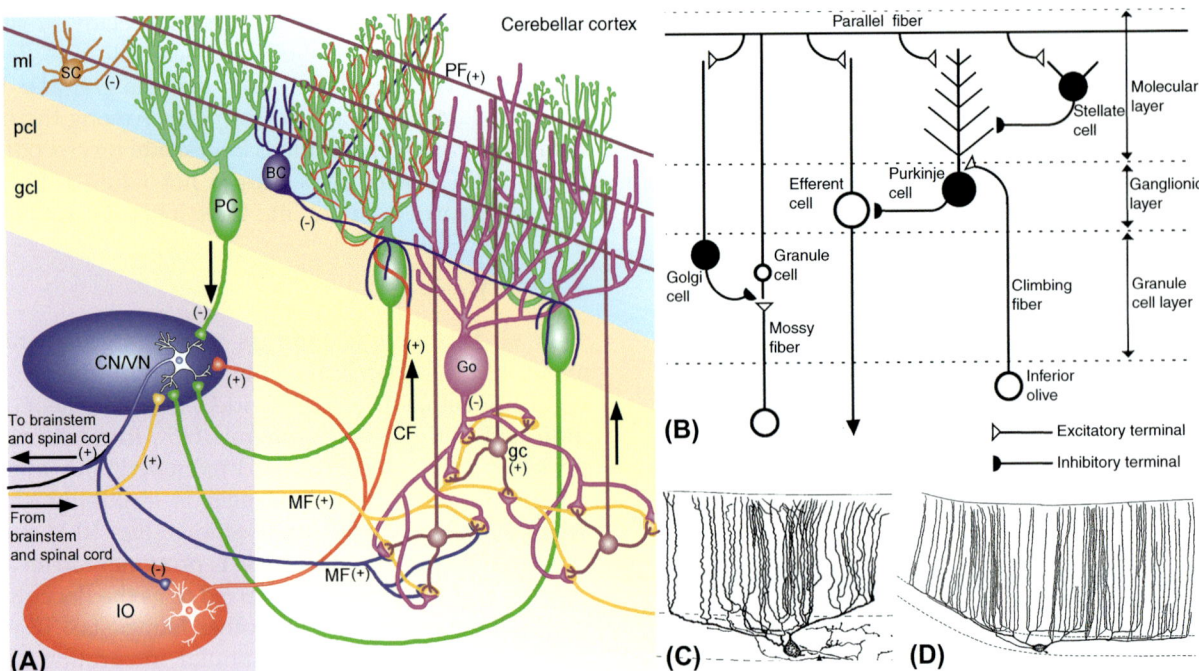

FIGURE 10.2   (A) An illustration of the connectivity and cytoarchitecture of neurons in the standard cerebellar circuit. (B) A schematic of the cerebellar circuitry in teleosts (From Ikenaga, T., 2013. Teleost fish. In: Manto, M., Gruol, D.L., Schmahmann, J., Koibuchi, N., Rossi, F. (Eds.), Handbook of the Cerebellum and Cerebellar Disorders, Springer, Netherlands, pp. 1463—1480; with permission.). (C) and (D) show drawings of a eurodendroid cell and a Purkinje cell, respectively, from a mormyrid fish (*Gnathonemus petersii*) (C and D, from Han, V.Z., Meek, J., Campbell, H.R., Bell, C.C., 2006. Cell morphology and circuitry in the central lobes of the mormyrid cerebellum. J. Comp. Neurol. 497, 309—325; with permission.). *BC*, basket cell; *CF*, climbing fiber; *CN*, cerebellar nuclei; *gc*, granule cell; *gcl*, granule cell layer; *Go*, Golgi cell; *IO*, inferior olive; *MF*, mossy fiber; *ml*, molecular layer; *PC*, Purkinje cell; *pcl*, Purkinje cell layer; *PF*, parallel fiber; *SC*, stellate cell; *VN*, vestibular nuclei.

isthmic region just rostral to the cerebellum is Fgf8 dependent. Moreover, in Fgf8-deficient mice, a cerebellar dysplasia results such that vermis is apparently absent. Thus, these data suggest that the vermis is absent in birds (and other nonmammalian vertebrates). Rather, the avian cerebellum is, by and large, equivalent to the hemispheres of the mammalian cerebellum. The vermis in mammals is the evolutionary novelty derived from tissue that leads to the development of the isthmic region in birds (Butts et al., 2014).

10.2.5 Lobed-Finned Fishes

Early sarcopterygians (the lobed-finned fishes) include the coelacanth (*Latimeria*) and the lungfishes. Like that of other fishes, the coelacanth cerebellum is comprised of pronounced, laterally positioned auricles and a distinct, dome-shaped corpus cerebellum. Although the coelacanth brain is relatively small, it has a well-developed cerebellum (17% of total brain mass) that is considerably larger than that of living amphibians, but smaller than that of most ray-finned fishes (Northcutt et al., 1978). In lungfishes, however, the cerebellum is relatively small and more similar in relative size to that of amphibians. It is comprised of a platelike

corpus cerebellum, which is almost fully obscured by the tectal hemispheres in some species, and paired auricles (Nieuwenhuys et al., 1998). As in cartilaginous fishes, the lower leafs of the auricles are continuous with the cerebellar-like structures in the octavolateralis regions of the hindbrain in lungfishes (eg, Northcutt, 2011).

10.2.6 Ray-Finned Fishes

The cerebellum of ray-finned fishes (Fig. 10.1N—P) is comprised of three main subdivisions: (1) a vestibulocerebellum, consisting of the caudal lobe with attached lateral auricles (usually called the eminentia granularis in teleosts); (2) an anteriorly positioned corpus cerebellum; and (3) the valvula cerebelli (Wullimann and Northcutt, 1989). Ray-finned fishes also possess cerebellar-like structures in the form of an MON. Electroreceptive teleosts (eg, Mormyriformes, Gymnotiformes, Siluriformes, and Xenomystinae) possess a cerebellar-like electrosensory lateral line lobe (ELL) as well (Bell, 2002).

The valvula is a shared derived feature in actinopterygian fishes and has no known homologue in other vertebrates (Meek et al., 2008). The valvula extends rostrally from the corpus into the midbrain ventricle,

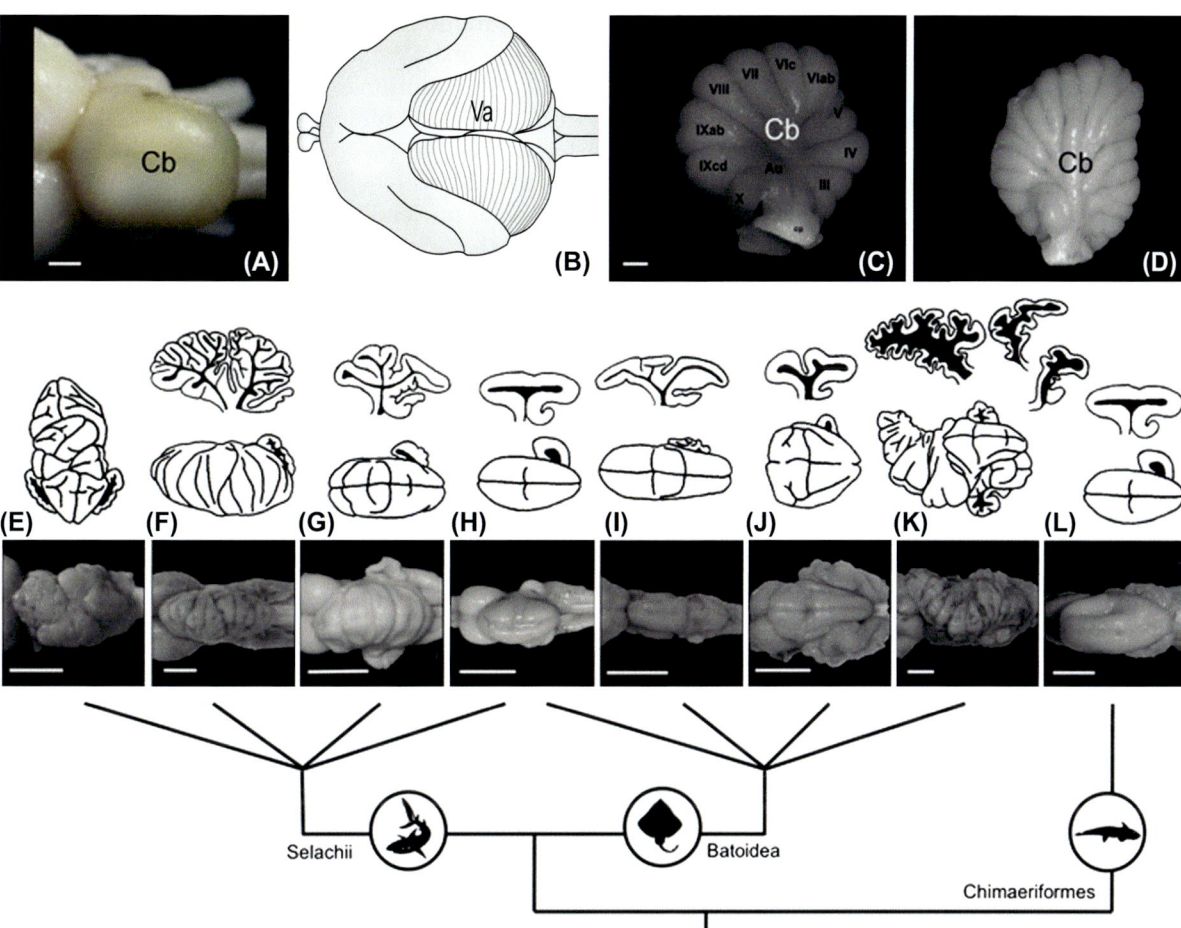

FIGURE 10.3 Variation in folding (foliation) of the corpus (Cb) and/or valvula cerebellum (Va) across taxa. (A–B) Representative bony fishes, such as the barramundi (*Lates calcarifer*) (A, photomicrograph after Ullmann, J.F., Cowin, G., Collin, S.P., 2010. Quantitative assessment of brain volumes in fish: comparison of methodologies. Brain Behav. Evol. 76, 261–270.) and a mormyrid (*Gnathonemus petersii*) (B, drawing adapted from Nieuwenhuys, R., Nicholson, C., 1969. A survey of the general morphology, the fiber connections and the possible functional significance of the gigantocerebellum of mormyrid fishes. In: Llinas, R. (Ed.), Neurobiology of Cerebellar Evolution and Development. American Medical Association, Chicago, pp. 107–134; with permission.). (C–D) Lateral views of the cerebellum from avian species: the pigeon (*Columba livia*) (C, from Pakan, J.M., Iwaniuk, A.N., Wylie, D.R., Hawkes, R., Marzban, H., 2007. Purkinje cell compartmentation as revealed by zebrin II expression in the cerebellar cortex of pigeons (*Columba livia*). J. Comp. Neurol. 501, 619–630; with permission.) and the Australian Pelican (*Pelecanus conspicillatus*) (D, from Iwaniuk, A.N., Hurd, P.L., Wylie, D.R., 2006. Comparative morphology of the avian cerebellum: I. Degree of foliation. Brain Behav. Evol. 68, 45–62; with permission.). (E–L) Photomicrographs of the dorsal view of the cerebellum across a range of cartilaginous fishes, including the bonnethead sharks (*Sphyrna tiburo*) (E), the shortfin mako (*Isurus oxyrinchus*) (F, from Yopak, K.E., Lisney, T.J., Collin, S.P., Montgomery, J.C., 2007. Variation in brain organization and cerebellar foliation in chondrichthyans: sharks and holocephalans. Brain Behav. Evol. 69, 280–300; with permission.), the gray smooth hound (*Mustelus californicus*) (G), northern spiny dogfish (*Squalus acanthias*) (H), the crossback stingaree (*Urolophus cruciatus*) (I, from Lisney, T.J., Yopak, K.E., Montgomery, J.C., Collin, S.P., 2008. Variation in brain organization and cerebellar foliation in chondrichthyans: batoids. Brain Behav. Evol. 72, 262–282; with permission.), the big skate (*Raja binoculata*) (J), the smooth stingray (*Dasyatis brevicaudata*) (K), and the giant chimaera (*Chimaera lignaria*) (L) (K–L, from Yopak, K.E., 2012. Neuroecology of cartilaginous fishes: the functional implications of brain scaling. J. Fish Biol. 80, 1968–2023; with permission.). The photomicrographs for cartilaginous fishes are accompanied by drawings in midsagittal (top, excluding E) and dorsal view (bottom), illustrating the wide degree of variation in corpus folding across this clade. *Au*, auricle; *Cb*, cerebellum; *Va*, valvula. Scale bars: 1 mm in (A), (C), (D); 2 mm in (B); 1 cm in (E–L). Adapted from Northcutt, R.G., 1989. Brain variation and phylogenetic trends in elasmobranch fishes. J. Exp. Zool. (Suppl. 2), 83–100; with permission.

ventral to the tectal lobes. Although the valvula is located in the mesencephalon, studies have shown that its cytoarchitecture and cellular physiology clearly designate it as part of the cerebellum (Shi et al., 2008; Zhang et al., 2011).

Despite being exclusive to actinopterygians, the valvula exhibits extensive variation in relative size and morphology among species. The most pronounced specialization of the valvula cerebelli is found in mormyrid electric fishes, which enlarge this structure beyond all allometric expectations (Fig. 10.1O and P; Bell and Szabo, 1986a; Meek, 1998). It extends beyond the tectal ventricle and is comprised of an enormous, 1 m long sheet of hyperfolded tissue that extends over the dorsal and dorsolateral aspects of the entire brain (Nieuwenhuys and Nicholson, 1969; Nieuwenhuys et al., 1998). In these species, the cerebellum in its entirety comprises nearly 80% of the brain's total mass, nearly 1% of the total body mass, with the valvula contributing the largest component. There are multiple direct and indirect pathways between electroreceptive nuclei and the valvula in mormyrid fishes that are absent in other actinopterygians (Bell and Szabo, 1986b). Thus, unlike other teleost fishes, the mormyrid valvula is highly specialized for the processing of electroreceptive stimuli. Based on electrophysiological and anatomical data, the mormyrid valvula is a major part of the electrosensory—electromotor system (for review, see Zhang et al., 2011).

10.3 Cellular Organization of the Cerebellum

The cellular organization of the cerebellar cortex has been described in numerous sources (eg, Butler and Hodos, 2005). The cerebellar cortex is divided into three layers in all jawed vertebrates, as evident in the Nissl-stained sections shown in Fig. 10.1E, G, J, and K. There is a superficial molecular layer, a Purkinje cell layer (referred to as the ganglionic cell layer in ray-finned fishes; see below), and the darkly stained granule cell layer. There are only a few main cerebellar cell types that comprise the cerebellar circuit, which has been known since Ramon Y Cajal (1909): granule cells, Purkinje cells, and inhibitory neurons (Golgi, stellate, and basket cells). Cerebellar granule cells are widely held to be the most abundant class of neurons in the brain. The circuitry is illustrated in Fig. 10.2A. Axons of the granule cells ascend into the molecular layer and bifurcate to form T-shaped branches called parallel fibers, because they run parallel to the long axis of the cerebellar folia that relay information by excitatory synapses onto the dendritic trees of Purkinje cells. Axons from a variety of brain stem pontine and spinal cord neurons project to the granular layer as mossy fibers—aptly named so because of the appearance of the synaptic terminals. Fine branches of the mossy fiber axons twist through the granular cell layer and slight enlargements, referred to as rosettes, give a knotted appearance indicating synaptic contacts. Purkinje cells present the most striking cytoarchitectural feature of the cerebellum. The Purkinje cell dendritic trees branch extensively in the molecular layer, but remain in a single plane that is at right angles to the trajectory of the parallel fibers (ie, the dendritic tree of the Purkinje cell lies in a parasagittal plane). Thus, each Purkinje cell is in a position to receive input from a large number of parallel fibers, and each parallel fiber contacts a large number of Purkinje cells. In this way, the Purkinje cells receive indirect inputs from the mossy fiber projections. Purkinje cells also receive input on their proximal dendritic shafts from climbing fibers, which arise from neurons in the inferior olive. Each Purkinje cell receives numerous contacts from a single climbing fiber. Purkinje cells project predominantly to the cerebellar nuclei, although those in the vestibulocerebellum project predominantly to the vestibular nuclei. Purkinje cells are thus the only output cells of the cerebellar cortex (except in ray-finned fishes, see below).

In most cartilaginous and lobe-finned fishes and amphibians, a single, bilaterally paired, cerebellar nucleus is present (ray-finned fishes lack cerebellar nuclei, see below). In reptiles, two cerebellar nuclei are recognized, the medial and lateral cerebellar nuclei, which are thought to be homologous with the mammalian fastigial and interposed nuclei, respectively (Larsell, 1967). Birds have a third paired cerebellar nucleus, the infracerebellar nucleus, which receives projections from Purkinje cells in the auricle, and is thought to be equivalent to the ventral dentate nucleus in mammals (Arends and Zeigler, 1991; Wylie et al., 2003).

As the Purkinje cells are GABAergic, the output of cerebellar cortex is wholly inhibitory. The cerebellar and vestibular nuclei also receive excitatory input from the collaterals of mossy and climbing fibers. Therefore, there is constant tonic excitation of neurons in the cerebellar and vestibular nuclei, and Purkinje cell inhibitory projections serve to modulate this level of excitation. Inputs from interneurons also modulate the inhibitory activity from Purkinje cells and occur on both dendritic shafts and the cell body. The most powerful of these local inputs are inhibitory complexes of synapses made around the Purkinje cell bodies by basket cells. Stellate cells receive input from the parallel fibers and produce an inhibitory input to the Purkinje cell dendrites. Golgi cell bodies are located in the granule cell layer, and their apical dendrites are located in the molecular layer. The Golgi cells receive input from the parallel fibers and provide an inhibitory feedback to the cells of origin of the parallel fibers (the granule cells). Therefore, there are many potential feedback loops within the circuitry of the cerebellar cortex itself, as well as within the afferent and efferent projection patterns between the deep cerebellar and vestibular nuclei and the inferior olive. This potential for feedback has been a driving force behind the theory that the cerebellum functions in the fine tuning of movements and motor coordination by providing error signals, and or precise timing information during sensorimotor behaviors (eg, Ito, 1984).

The cytoarchitecture and connections in the cerebellar circuit is highly conserved among vertebrates, although there are some differences (Butler and Hodos, 2005; Larsell, 1967; Voogd and Glickstein, 1998). The Purkinje cell layer is usually a monolayer, but in snakes the Purkinje layer is not as well defined, and Purkinje cell somata can be seen in clusters extending into the molecular layer (Fig. 10.1E). Similarly, in lungfishes and some cartilaginous fishes, Purkinje cells can be scattered in both the molecular and granular layers (Northcutt, 2011; Smeets et al., 1983). Whereas the granular layer is adjacent to the Purkinje layer in most vertebrates, in cartilaginous fishes, the granule cells form clusters that are separated from the Purkinje cells by white matter tracks. The climbing fibers contact mainly the proximal dendrites of Purkinje cells in fishes and amphibians and nonavian reptiles, but ascend higher up the dendritic tree in avian species (and higher still in mammals). Basket cells are not found in the cerebellum of fishes or amphibians.

Not drawn in the circuit shown in Fig. 10.2A are the unipolar brush cells (UBCs) and Lugaro cells. UBCs are found in the granular layer of the vermis and vestibulocerebellum, receive inputs from mossy fibers, and have extensive contacts with granule cells and other UBCs. They have been described in numerous mammalian species and two avian species (pigeons and chickens) (Mugnaini et al., 2011). Lugaro cells have also been reported in numerous mammalian species and are located in the granule layer, just below the Purkinje cells. The Lugaro cells receive serotonergic inputs from the raphe nuclei and provide inhibitory inputs to Golgi, basket, and stellate neurons (Laine and Axelrad, 2002). The only nonmammalian species in which they have been reported is a teleost, *Pholidapus dybowskii* (Pushchina and Varaksin, 2001) and a bird, the rock pigeon, Columba livia (Craciun et al., 2019).

The most striking difference in the structure of the cerebellar circuit among vertebrates is found in ray-finned fishes. Shown in Fig. 10.2B, the molecular and granular layers are divided by the ganglionic cell layer. It appears as a monolayer but consists of two cell types: Purkinje cells and eurydendroid cells (labeled "Efferent cell" in Fig. 10.2B). These two neuronal types are strikingly similar in their appearance, with vast dendritic trees spreading in the sagittal plane (Fig. 10.2C and D). These Purkinje cells do not project out of the cerebellar cortex, but rather their axons terminate on the eurydendroid cells within the ganglionic cell layer, whose axons project to the brain stem. Thus, in teleosts, the eurydendroid cells collectively seem to replace the deep cerebellar nuclei (Han et al., 2006; Ikenaga, 2013).

10.4 Variation in Relative Cerebellar Size and Cerebellar Foliation

Although the cytoarchitecture of the cerebellar circuitry is highly conserved, there is a high degree of variation in the relative size of the cerebellum. In addition, the cerebellum shows variability with respect to the

degree, complexity, and symmetry of foliation across a range of vertebrate clades (Figs. 10.1 and 10.3). Such variation can be seen in the valvula of mormyrid fishes and the corpus cerebellum of birds, sharks, and mammals (Iwaniuk et al., 2006a; Yopak, 2012). This variation in relative size and foliation almost certainly reflects variation in functional capability and/or in the relative importance of cerebellar-related functions.

Numerous studies have explored variation in relative cerebellar size in relation to cognitive and behavioral processes. For example, in bony and cartilaginous fishes, relative enlargement of the cerebellum has been linked to locomotor behaviors, habitat dimensionality, swimming speed/maneuverability, agile prey capture and manipulation, proprioception, and sensory acquisition (eg, Huber et al., 1997; Yopak, 2012). Conversely, a relatively small cerebellum has also been correlated with lower activity levels and a close association with the substrate in fishes (eg, Northcutt, 1989). Interestingly in cichlids, while cerebellum volume increases with increased depth and habitat complexity, there is a negative correlation between cerebellum size and female-only parental care, suggestive of an interplay between environmental parameters and sexual selection (Gonzalez-Voyer and Kolm, 2010). In amphibians, an increase in cerebellum size has been correlated with arboreality (Taylor et al., 1995). Finally, in birds, an enlarged cerebellum has been correlated with activity patterns, habitat complexity, and behavioral repertoire (Boire and Baron, 1994; Day et al., 2005; Iwaniuk et al., 2006b; Larsell, 1967; Sultan, 2005).

In additions to variations in size, documented variations in cerebellar surface complexity (ie, foliation) are noted across a range of vertebrate taxa. It has been proposed that an increase in foliation accommodates an increase in cerebellar surface area (Striedter, 2005), leading to an increase in Purkinje cell numbers, thereby increasing cerebellar processing capacity and facilitating the complexity of cerebellar-dependent functions and behaviors (Iwaniuk et al., 2007; Sudarov and Joyner, 2007; Sultan and Glickstein, 2007; Welker, 1990). Much of the variation in degree of foliation can be accounted for by allometric relationships with cerebellar size, brain size, and body size (Iwaniuk et al., 2006a; Pearson and Pearson, 1976; Senglaub, 1963; Yopak and Frank, 2009; Yopak et al., 2010); ie, larger animals have larger brains with larger, more foliated cerebella. Notwithstanding, differences in the degree of cerebellar foliation have been attributed to behavioral specializations in several vertebrate groups. Mormyrid fishes possess an exceptionally foliated cerebellum (particularly the cerebellar valvula; Fig. 10.1), attributed at least partly to specialized electroreceptive function. In birds, an increase in cerebellar foliation is significantly related to cognitive and/or behavioral differences (Iwaniuk et al., 2006a,b; Sultan and Glickstein, 2007) including nest complexity (Hall et al., 2013) and tool use (Iwaniuk et al., 2009).

The most dramatic variation in cerebellar foliation occurs in the cartilaginous fishes. The cerebellum ranges from a completely smooth corpus to having deeply branched convexities (Fig. 10.3; reviewed by Yopak, 2012). Moreover, foliation in the cerebellum of cartilaginous fishes is not symmetrical with respect to the midline (Fig. 10.1B) as it is in other vertebrates, and this asymmetry can vary even within a species (Ari, 2011; Puzdrowski and Gruber, 2009). Patterns of cerebellar foliation also vary ontogenetically, whereby the number and depth of folia appear to increase with age (Pose-Mendez et al., 2016; Yopak and Frank, 2009). High degrees of foliation appear to have arisen multiple times throughout the cartilaginous fishes, suggesting that this feature confers a functional advantage, where an increased cerebellar surface area may allow for more specific representations of sensory and motor components from the body surface. Generally, there is a clear trend toward higher levels of foliation occurring in large-bodied species that occupy spatially complex habitats, such as corals reefs or open oceanic habitats, which suggests a link between this characteristic and maneuverable predation strategies, proprioception, and sensory acquisition. Conversely, low levels of foliation are documented in small-bodied, less active benthic and deep-sea dwelling benthopelagic species, potentially related to more passive predation strategies and a close association with the substrate (reviewed in Yopak, 2012). In sum, the degree of folding in the shark cerebellum appears to be a morphological signature of relative brain size, but also is associated with greater behavioral complexity.

Because there are subdivisions within the cerebellum that serve different functional roles, it is possible that particular regions of the cerebellum show an increase in size and/or foliation in concert with an increase in the complexity of particular behaviors (following Jerison's "principle of proper mass"; Jerison, 1973). Although this has not been studied broadly or in much detail, the size of specific folia is reportedly associated with flight behavior (Iwaniuk et al., 2007) and cognitive ability (Sultan, 2005) in birds and with behavioral and motor adaptations in mammals (eg, Welker, 1990).

10.5 Sagittal Zones of the Cerebellum

Although the cerebellum is divided into transverse lobules in many species, it has become clear that the cerebellum is also organized into bands or "zones" that lie in the sagittal plane, cutting across the folia (for review,

see: Armstrong and Hawkes, 2000; Voogd, 2012). The origin of this concept began with the work of Jansen and Brodal (1940, 1942) and was championed by Jan Voogd and colleagues during the last five decades (for review, see Voogd, 2011; eg, Voogd and Bigaré, 1980). Fig. 10.4A shows a schematic of the parasagittal zones in the mammalian cerebellum. Two aspects of the parasagittal organization are shown in this figure. First, climbing fibers terminate in sagittal zones in the cerebellar cortex (for review, see Reeber et al., 2012). Second, the Purkinje cells in each sagittal zone project to different regions in the cerebellar and vestibular nuclei (eg, Andersson and Oscarsson, 1978). It has also been shown that mossy fiber afferents terminate in particular sagittal zones (eg, Voogd and Ruigrok, 1997). Electrophysiological evidence has shown that Purkinje cells in sagittal zones have similar response properties (eg, De Zeeuw et al., 1994) and tend to fire synchronously (eg, Llinas and Sasaki, 1989). Finally, it has been shown that numerous molecular markers are organized into parasagittal bands in the cerebellum (see below; eg, Herrup and Kuemerle, 1997).

Although the bulk of the research demonstrating the sagittal organization of the cerebellum has been done in mammals, comparative research has shown that this is a general feature among vertebrates. Fig. 10.4B shows that the climbing fiber projection from the inferior olive to the vermis in pigeons terminates as four sagittal zones, cutting across folia I through IXab (Arends and Voogd, 1989). Other studies in birds have shown that mossy fibers inputs, Purkinje cell response properties, and Purkinje cell projections also follow a sagittal zonal organization (Arends and Zeigler, 1991; Graham and Wylie, 2012; Pakan et al., 2010). Studies of species from other vertebrate classes are relatively scarce but reinforce the general idea that a sagittal organization of the cerebellum is highly conserved. For example, in snakes, turtles, and lizards, corticonuclear projections divide the cerebellum into four sagittal bands (ten Donkelaar and Bangma, 1992).

The sagittal zonal organization of the cerebellum is also revealed by the expression of numerous molecular markers (Herrup and Kuemerle, 1997), the most extensively studied of which is zebrin II (ZII, a.k.a. aldolase C). As first shown in several mammalian species (Brochu et al., 1990; Hawkes and Herrup, 1995) and subsequently in birds (Pakan et al., 2007), ZII is expressed heterogeneously, such that some Purkinje cells show a high level of ZII expression (ZII+) whereas others show little or no ZII expression (ZII−; Fig. 10.5B−D). The ZII+ and ZII− Purkinje cells tend to be organized into alternating sagittal stripes cutting across the folia (Fig. 10.5A). The pattern of the stripes is remarkably similar in birds and mammals (Fig. 5E−F) suggesting that this is either highly conserved, or an

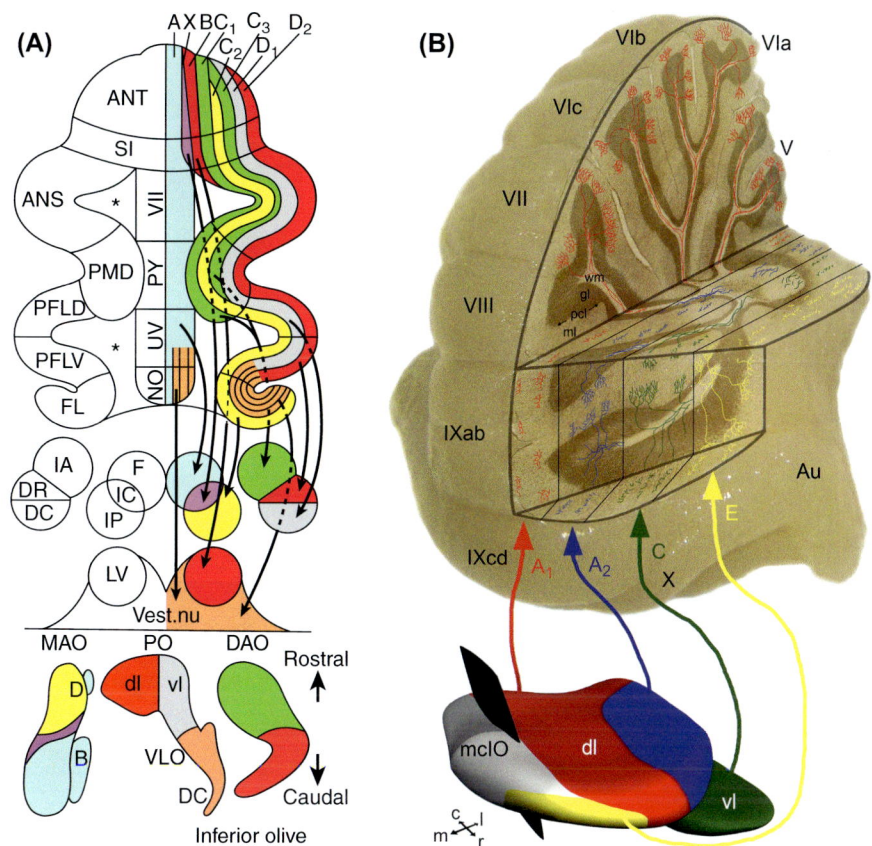

FIGURE 10.4 Sagittal organization of the cerebellum. (A) is the classic illustration of the sagittal zones in the mammalian cerebellar cortex (A, X, B, $C_{1,2,3}$, $D_{1,2}$) showing the corticonuclear projections to the cerebellar and vestibular nuclei (vest. nu.) and the climbing fiber afferents from subdivisions of the inferior olive (A, from Voogd, J., Glickstein, M., 1998. The anatomy of the cerebellum. Trends Neurosci. 21, 370–375; with permission.). (B) shows the olivocerebellar projection to the sagittal zones ($A_{1,2}$, C and E) in the pigeon cerebellum (Based on data from Arends, J., Voogd, J., 1989. Topographic aspects of the olivocerebellar system in the pigeon. Exp. Brain Res. Ser. 17, 52–57.). *ANS*, ansiform lobule; *ANT*, anterior lobe; *B*, beta subnucleus; *Au*, auricle; *c*, caudal; *D*, dorsomedial cell column; *DAO*, dorsal accessory olive; *DC/R*, dentate nucleus, caudolateral/rostromedial; *DC*, dorsal cap of inferior olive; *dl*, dorsal leaf in (A), dorsal lamella in (B); *F*, fastigial nucleus; *FL*, flocculus; *IA*, anterior interposed nucleus; *IC*, interstitial cell group; *IP*, posterior interposed nucleus; *l*, lateral; *LV*, lateral vestibular nucleus; *m*, medial; *MAO*, medial accessory olive; *mcIO*, medial column of the inferior olive; *NO*, nodulus; *PFLD/V*, ventral paraflocculus, dorsal/ventral; *PMD*, paramedian lobule; *PO*, principal olive; *Py*, pyramis; *r*, rostral; *SI*, lobulus simplex; *UV*, uvula; *vl*, ventral leaf in (A), ventral lamella in (B); *VLO*, ventrolateral outgrowth.

instance of remarkable convergent evolution (Marzban and Hawkes, 2010). Supporting the latter conclusion, all Purkinje cells are ZII+ in snakes (Fig. 5G–I) and turtles (eg, Aspden et al., 2015). However, a recent study showed that ZII is expressed in alternating ZII+ and ZII– stripes in a species of lizard (Fig. 10.5K; Wylie et al., 2015). Perhaps the striped expression of ZII represents the situation in stem reptiles, and this has been lost in snakes and turtles. Surprisingly, given their close relationship with birds, ZII is not expressed in sagittal stripes in alligators and crocodiles; in those species all Purkinje cells are ZII+ (Fig. 10.5J; Wylie et al., 2015).

In both cartilaginous and bony fishes, ZII is expressed in Purkinje cells (Fig. 10.5L). In stingrays, it has been shown that all Purkinje cells are ZII+ (Puzdrowski, 1997), but in teleosts, ZII is heterogeneously expressed (Meek et al., 1992). ZII– and ZII+ Purkinje cells are not organized in sagittal stripes, but rather there are large areas containing either ZII+ or ZII– cells. This is shown in Fig. 10.5M for a goldfish, a gymnotid and a mormyrid (from Meek et al., 1992). In all three species, the transitional area between the valvula and corpus cerebellum is ZII–, whereas the corpus is ZII+ (with the exception of the rostral portions of the corpus in the mormyrid). In all three species, a caudomedial portion of the valvula is ZII–, whereas the rostolateral region is ZII+. Parts of the caudal lobe in both the gymnotid and mormyrid are ZII– whereas the caudal lobe is entirely ZII+ in the goldfish.

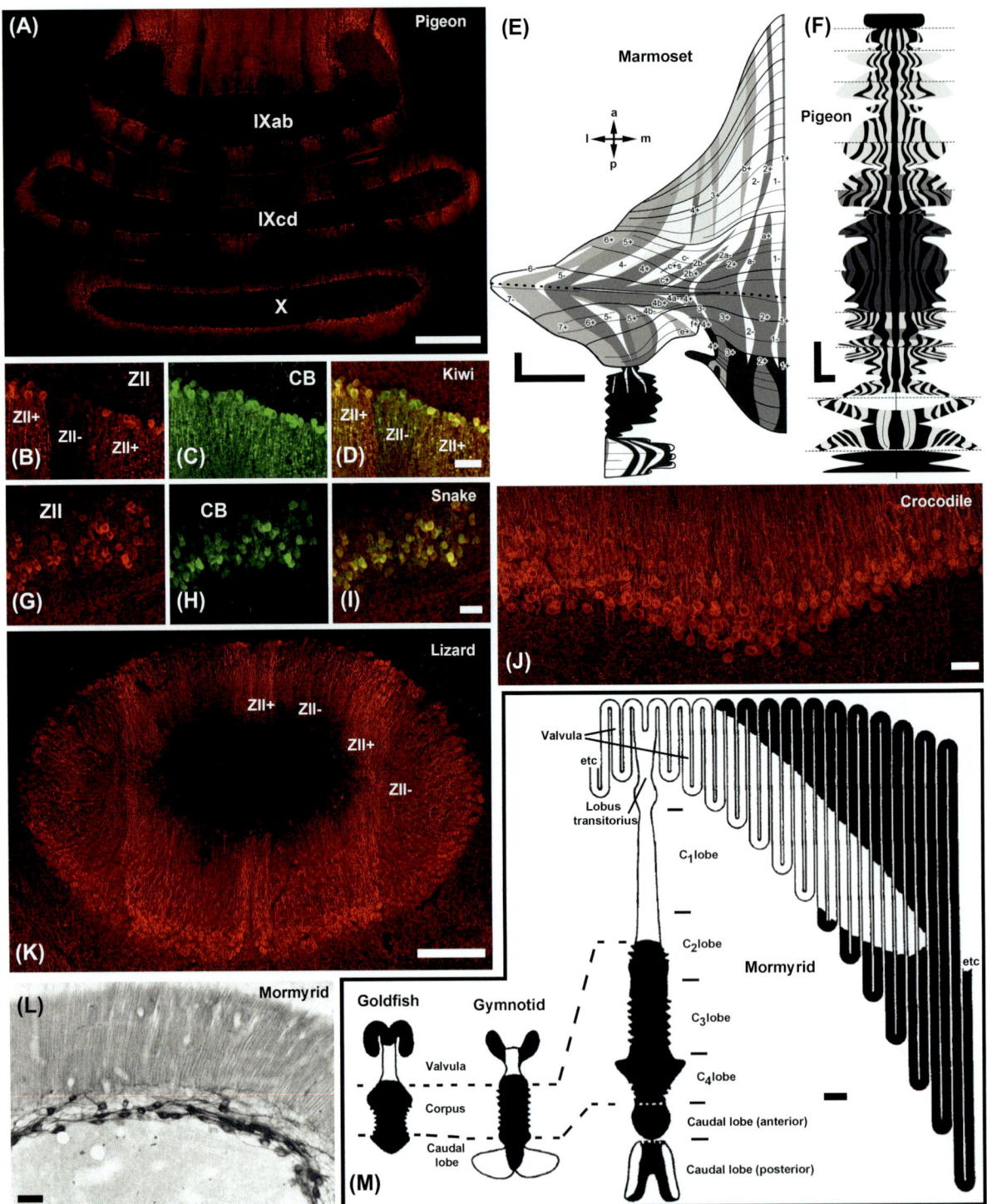

FIGURE 10.5 Zebrin II expression by Purkinje cells in the cerebellum of various vertebrates. (A) shows a coronal section through the posterior cerebellum of a pigeon (*Columba livia*) immunoprocessed for ZII expression. Sagittal stripes of high expression (ZII+) and low expression (ZII−) can be seen across folia IXab and IXcd (Pakan et al., 2007). (B–D) shows higher magnification ZII+/− stripes from the cerebellum of a kiwi (*Apteryx mantelli*). A single coronal section is shown as a triptych: ZII expression (*red*, B); calbindin expression (CB), which labels all Purkinje cells (*green*, C); and the overlay (D) (Corfield et al., 2016). (E) and (F) respectively illustrate the patterns of ZII+/− stripes (*black/gray* = Z+, *white* = Z−) on unfolded cerebella for a marmoset (*Callithrix jacchus*) (E, from Fujita, H., Oh-Nishi, A., Obayashi, S., Sugihara, I., 2010. Organization of the marmoset cerebellum in three-dimensional space: lobulation, aldolase C compartmentalization and axonal projection. J. Comp. Neurol. 518, 1764–1791; with permission.) and a pigeon (*Columba livia*) (F, adapted from Corfield, J.R., Kolominsky, J., Marin, G.J., Craciun, I., Mulvany-Robbins, B.E., Iwaniuk, A.N., Wylie, D.R., 2015. Zebrin II expression in the cerebellum of a Paleognathous bird, the Chilean tinamou (*Nothoprocta perdicaria*). Brain Behav. Evol. 85, 94–106; with permission.). (G–I) show ZII and CB expression in a coronal section through the cerebellum of a rattlesnake (*Crotalus atrox*). All Purkinje cells are ZII+. (J) and (K) respectively show ZII expression in coronal sections through the cerebellum

The significance of the heterogeneous expression of ZII is not well understood, but it emphasizes that the cerebellum should not be regarded as an assembly of uniform Purkinje cell microcircuits (Cerminara et al., 2015). Recently it has been suggested that ZII+ and ZII− Purkinje cells have differing roles with respect to motor learning, with ZII+ cells relying more on long-term potentiation (LTP) and ZII− cells relying more on long-term depression (LTD) as a mechanism for synaptic plasticity (eg, Wadiche and Jahr, 2005; Zhou et al., 2014). If one assumes that the ancestral condition is a uniformly ZII+ cerebellum, based on the above mentioned stingray study (Puzdrowski, 1997), then it could be argued that ZII heterogeneity evolved independently in teleosts and stem reptiles as a way to meet increased motor learning demands by creating cerebellar modules specialized for either LTD and LTP. Furthermore, this requirement for compartments relying on LTD versus LTP seems to have been lost in some species (turtles, snakes, crocodilians), which have only ZII+ Purkinje cells and, therefore, may be relying disproportionately on LTP to modify their synapses.

10.6 Conclusions and Future Directions

It has been argued that the cerebellum was an innovation to the central nervous system in the Devonian and has paved the way for some aspects of higher neural function and complexity in the evolution of vertebrate brains. Despite a remarkable conservation of architecture and circuitry throughout gnathostomes reviewed here, there are numerous aspects of cerebellar structure and function that remain unresolved. Suggested functional roles for this structure include control of certain classes of movement, motor learning, and other cognitive behaviors. However, a consensus on basic cerebellar functionality is required before we can ever hope to ascertain the drivers of cerebellum size and complexity across vertebrates.

Conserved patterns of cerebellar scaling across vertebrates have been given some attention in recent years and much more is needed to gain a basic understanding of the general principles governing relative cerebellar hypertrophy. Of particular interest is the potential for coevolution between the telencephalon and cerebellum, as the greatest proportional size increase has occurred in these two regions. Given the coordinated scaling between telencephalon and cerebellum across some taxa (Finlay et al., 2001), and the remarkable constancy of the ratio of the number of neurons in the cerebellum versus the neocortex in a range of mammals (Herculano-Houzel, 2010), there may in fact be some aspects of coordinated functionality. As the evolution of the circuitry between cerebellum and telencephalon is poorly defined (particularly in more basal vertebrate groups), a greater understanding of the interconnectivity between these two regions is a critical area of future cerebellar evolutionary research.

The cerebellum has also proven vital to our understanding of the organizational principles of the nervous system in general, and more widely to study the principles linking structure and function in brain evolution. Indeed, many functional and anatomical studies have already revealed a fundamental functional architecture of modular, parasagittal organization within cerebellar circuits. Although measures such as folia size vary among certain groups in somewhat predictable dimensions (emphasizing their utility), it is possible that the sagittal zones may prove to be more functionally important and, consequently, emphasis has shifted from lobules/folia to zones in recent years. While the functional significance of these zones still remains unclear, comparatively speaking, increasing the breadth or number of zones may somehow improve processing capacity. A similar case has been made for the addition of cortical areas in mammals (eg, Kaas, 2005). However, the cellular processes that guide cluster formation, the molecular mechanisms that trigger subsequent cluster wiring, as well as how the zones acquire their functional properties are all largely unknown. It is here that the discovery and investigation of patterned molecular markers and the correlation with anatomical and functional organizational patterns across various vertebrate species will continue to prove invaluable in our understanding of the function of the cerebellar cortex.

---

of a crocodile (*Crocodylus niloticus*) and a lizard (*Ctenophorus nuchalis*). Alternating sagittal ZII+ and ZII− sagittal stripes are present in the lizard, but all Purkinje cells in crocodile are ZII+ (Wylie et al., 2015; Wylie, unpublished observations) (G–I, from Aspden, J. W., Armstrong, C. L., Gutierrez-Ibanez, C. I., Hawkes, R., Iwaniuk, A. N., Kohl, T., Graham, D. J., Wylie, D. R., 2015. Zebrin II / aldolase C expression in the cerebellum of the western diamondback rattlesnake (*Crotalus atrox*). PLoS One 10, e0117539.). (L) shows ZII + Purkinje cells in cerebellum of a mormyrid (*Gnathonemus petersii*) (L, from Meek, J., Hafmans, T.G., Maler, L., Hawkes, R., 1992. Distribution of zebrin II in the gigantocerebellum of the mormyrid fish *Gnathonemus petersii* compared with other teleosts. J. Comp. Neurol. 316, 17–31; with permission.). (M) shows an illustration of the Z+ (*black*) and Z− (*white*) regions in the cerebellum for three different teleosts: a goldfish (*Carassius auratus*), a gymnotid (*Eigenmannia virescens*), and a mormyrid (*Gnathonemus petersii*) (M, adapted from Meek, J., Hafmans, T.G., Maler, L., Hawkes, R., 1992. Distribution of zebrin II in the gigantocerebellum of the mormyrid fish *Gnathonemus petersii* compared with other teleosts. J. Comp. Neurol. 316, 17–31; with permission.). Scale bars: 1 mm in (A) and (M); 50 μm in (B–D), (G–J) and (L); 250 μm in (K). In (E), vertical and horizontal scale bars are 10 and 2 mm, respectively. In (F), vertical and horizontal scale bars are 5 and 1 mm, respectively.

References

Andersson, G., Oscarsson, O., 1978. Projections to lateral vestibular nucleus from cerebellar climbing fiber zones. Exp. Brain Res. 32, 549–564.

Arends, J., Voogd, J., 1989. Topographic aspects of the olivocerebellar system in the pigeon. Exp. Brain Res. Ser. 17, 52–57.

Arends, J.J., Zeigler, H.P., 1991. Organization of the cerebellum in the pigeon (Columba livia): I. Corticonuclear and corticovestibular connections. J. Comp. Neurol. 306, 221–244.

Ari, C., 2011. Encephalization and brain organization of mobulid rays (Myliobatiformes, Elasmobranchii) with ecological perspectives. Open Anat. J. 3, 1–13.

Armstrong, C.L., Hawkes, R., 2000. Pattern formation in the cerebellar cortex. Biochem. Cell Biol. 78, 551–562.

Aspden, J.W., Armstrong, C.L., Gutierrez-Ibanez, C.I., Hawkes, R., Iwaniuk, A.N., Kohl, T., Graham, D.J., Wylie, D.R., 2015. Zebrin II/aldolase C expression in the cerebellum of the western diamondback rattlesnake (Crotalus atrox). PLoS One 10, e0117539.

Baumann, O., Borra, R.J., Bower, J.M., Cullen, K.E., Habas, C., Ivry, R.B., Leggio, M., Mattingley, J.B., Molinari, M., Moulton, E.A., Paulin, M.G., Pavlova, M.A., Schmahmann, J.D., Sokolov, A.A., 2015. Consensus paper: the role of the cerebellum in perceptual processes. Cerebellum 14, 197–220.

Bell, C.C., 2002. Evolution of cerebellum-like structures. Brain Behav. Evol. 59, 312–326.

Bell, C.C., Szabo, T., 1986. Electroreception in mormyrid fish: central anatomy. In: Bullock, T.H., Heiligenberg, W. (Eds.), Electroreception. John Wiley and Sons, New York.

Bell, C.C., Szabo, T., 1986. Electroreception: behavior anatomy and physiology. In: Bullock, T.H., Heiligenberg, W. (Eds.), Electroreception. John Wiley and Sons, New York, pp. 577–612.

Boire, D., Baron, G., 1994. Allometric comparison of brain and main brain subdivisions in birds. J. Hirnforsch. 35, 49–66.

Brochu, G., Maler, L., Hawkes, R., 1990. Zebrin II: a polypeptide antigen expressed selectively by Purkinje cells reveals compartments in rat and fish cerebellum. J. Comp. Neurol. 291, 538–552.

Butler, A.B., Hodos, W., 2005. Comparative Vertebrate Neuroanatomy: Evolution and Adaptation. Wiley-Liss, New York.

Butts, T., Green, M.J., Wingate, J.T., 2014. Development of the cerebellum: simple steps to make a 'little brain'. Development 141, 4031–4041.

Carlson, B.A., Hopkins, C.D., 2004. Central control of electric signaling behavior in the mormyrid Brienomyrus brachyistius: segregation of behavior-specific inputs and the role of modifiable recurrent inhibition. J. Exp. Biol. 207, 1073–1084.

Cerminara, N.L., Lang, E.J., Sillitoe, R.V., Apps, R., 2015. Redefining the cerebellar cortex as an assembly of non-uniform Purkinje cell microcircuits. Nat. Rev. Neurosci. 16, 79–93.

Corfield, J.R., Kolominsky, J., Craciun, I., Mulvany-Robbins, B.E., Wylie, D.R., 2016. Is cerebellar architecture shaped by sensory ecology in the New Zealand kiwi (Apteryx mantelli)? Brain Behav. Evol. 87, 88–104.

Corfield, J.R., Kolominsky, J., Marin, G.J., Craciun, I., Mulvany-Robbins, B.E., Iwaniuk, A.N., Wylie, D.R., 2015. Zebrin II expression in the cerebellum of a Paleognathous bird, the Chilean tinamou (Nothoprocta perdicaria). Brain Behav. Evol. 85, 94–106.

Craciun, I., Gutierrez-Ibanez, C., Chan, A.S.M., Luksch, H., Wylie, D.R., 2019. Secretagogin immunoreactivity reveals Lugaro cells in the pigeon cerebellum. Cerebellum 18, 544–555. https://doi.org/10.1007/s12311-019-01023-7.

Day, L.B., Westcott, D.A., Olster, D.H., 2005. Evolution of bower complexity and cerebellum size in bowerbirds. Brain Behav. Evol. 66, 62–72.

Dow, R.S., Moruzzi, G., 1958. The Physiology and Pathology of the Cerebellum. University of Minnesota Press, Minneapolis.

ten Donkelaar, H.J., Bangma, G.C., 1992. The cerebellum. In: Gans, C., Ulinski, P.S. (Eds.), Biology of the Reptilia. Neurology C. Sensorimotor Integration, vol. 17. University of Chicago Press, Chicago, pp. 496–586.

Engelkamp, D., Rashbass, P., Seawright, A., van Heyningen, V., 1999. Role of Pax6 in development of the cerebellar system. Development 126, 3585–3596.

Finlay, B.L., Darlington, B.R., Nicastro, N., 2001. Developmental structure in brain evolution. Behav. Brain Sci. 24, 263–278.

Flourens, P., 1824. Rechérches expérimentales sur les propriétés et les fonctions du système nerveux dans les animaux vertébrés. Crevot, Paris.

Fujita, H., Oh-Nishi, A., Obayashi, S., Sugihara, I., 2010. Organization of the marmoset cerebellum in three-dimensional space: lobulation, aldolase C compartmentalization and axonal projection. J. Comp. Neurol. 518, 1764–1791.

Glickstein, M., Strata, P., Voogd, J., 2009. Cerebellum: history. Neuroscience 162, 549–559.

Gonzalez-Voyer, A., Kolm, N., 2010. Sex, ecology and the brain: evolutionary correlates of brain structure volumes in Tanganyikan cichlids. PLoS One 5, e14355.

Graham, D.J., Wylie, D.R., 2012. Zebrin-immunopositive and -immunonegative stripe pairs represent functional units in the pigeon vestibulocerebellum. J. Neurosci. 32, 12769–12779.

Hall, Z.J., Street, S.E., Healy, S.D., 2013. The evolution of cerebellum structure correlates with nest complexity. Biol. Lett. 9, 20130687.

Han, V.Z., Meek, J., Campbell, H.R., Bell, C.C., 2006. Cell morphology and circuitry in the central lobes of the mormyrid cerebellum. J. Comp. Neurol. 497, 309–325.

Hawkes, R., Herrup, K., 1995. Aldolase C/zebrin II and the regionalization of the cerebellum. J. Mol. Neurosci. 6, 147–158.

Herculano-Houzel, S., 2010. Coordinated scaling of cortical and cerebellar numbers of neurons. Front. Neuroanat. 4, 12. https://doi.org/10.3389/fnana.2010.00012.

Herculano-Houzel, S., Catania, K., Manger, P.R., Kaas, J.H., 2015. Mammalian brains are made of these: a dataset of the numbers and densities of neuronal and nonneuronal cells in the brain of Glires, Primates, Scandentia, Eulipotyphlans, Afrotherians and Artiodactyls, and their relationship with body mass. Brain Behav. Evol. 86, 145–163.

Herrup, K., Kuemerle, B., 1997. The compartmentalization of the cerebellum. Annu. Rev. Neurosci. 20, 61–90.

Huber, R., Van Staaden, M.J., Kaufman, L.S., Liem, K.F., 1997. Microhabitat use, trophic patterns, and the evolution of brain structure in African cichlids. Brain Behav. Evol. 50, 167–182.

Ikenaga, T., 2013. Teleost fish. In: Manto, M., Gruol, D.L., Schmahmann, J., Koibuchi, N., Rossi, F. (Eds.), Handbook of the Cerebellum and Cerebellar Disorders. Springer, Netherlands, pp. 1463–1480.

Ito, M., 1984. The Cerebellum and Neural Control. Raven, New York.

Iwaniuk, A.N., Hurd, P.L., Wylie, D.R., 2006. Comparative morphology of the avian cerebellum: I. Degree of foliation. Brain Behav. Evol. 68, 45–62.

Iwaniuk, A.N., Hurd, P.L., Wylie, D.R., 2006. The comparative morphology of the cerebellum in caprimulgiform birds: evolutionary and functional implications. Brain Behav. Evol. 67, 53–68.

Iwaniuk, A.N., Hurd, P.L., Wylie, D.R., 2007. Comparative morphology of the avian cerebellum: II. Size of folia. Brain Behav. Evol. 69, 196–219.

Iwaniuk, A.N., Lefebvre, L., Wylie, D.R., 2009. The comparative approach and brain-behaviour relationships: a tool for understanding tool use. Can. J. Exp. Psychol. 63, 150–159.

Jansen, J., Brodal, A., 1940. Experimental studies on the intrinsic fibres of the cerebellum. II. The cortico-nuclear projection. J. Comp. Neurol. 73, 267–321.

Jansen, J., Brodal, A., 1942. Experimental studies on the intrinsic fibres of the cerebellum. III. The cortico-nuclear projection in the rabbit and the monkey (*Macacus rhesus*). Nor. Vid. Akad. Avh. I Math. Nat. Ki. 3, 1–50.

Jerison, H.J., 1973. Evolution of the Brain and Intelligence. Academic Press, New York.

Jørgensen, J.M., Lomholt, J.P., Weber, R.E., Malte, H., 1998. The Biology of Hogfishes. Chapman and Hall, London.

Kaas, J.H., 2005. From mice to men: the evolution of the large, complex human brain. J. Biosci. 30, 155–165.

Laine, J., Axelrad, H., 2002. Extending the cerebellar Lugaro cell class. Neuroscience 115, 363–374.

Larsell, O., 1967. The Cerebellum: From Myxinoids through Birds. The University of Minnesota Press, Minnesota.

Lisney, T.J., Yopak, K.E., Montgomery, J.C., Collin, S.P., 2008. Variation in brain organization and cerebellar foliation in chondrichthyans: batoids. Brain Behav. Evol. 72, 262–282.

Llinás, R., Hillman, D.E., 1969. Physiological and Morphological Organization of the Cerebellar Circuits in Various Vertebrates. AMA-ERF Institute for Biomedical Research, Chicago.

Llinas, R., Sasaki, K., 1989. The functional organization of the olivo-cerebellar system as examined by multiple Purkinje cell recordings. Eur. J. Neurosci. 1, 587–602.

Maler, L., Sas, E., Johnston, S., Ellis, W., 1991. An atlas of the brain of the electric fish *Apteronotus leptorhynchus*. J. Chem. Neuroanat. 4, 1–38.

Marr, D., 1969. A theory of cerebellar cortex. J. Physiol. 202, 437–470.

Marzban, H., Hawkes, R., 2010. On the architecture of the posterior zone of the cerebellum. Cerebellum 10, 422–434.

Meek, J., 1998. Holosteans and teleosts. In: Nieuwenhuys, R., Ten Donkelaar, H., Nicholson, C. (Eds.), The Central Nervous System of Vertebrates. Springer-Verlag, Berlin, pp. 759–937.

Meek, J., Hafmans, T.G., Maler, L., Hawkes, R., 1992. Distribution of zebrin II in the gigantocerebellum of the mormyrid fish *Gnathonemus petersii* compared with other teleosts. J. Comp. Neurol. 316, 17–31.

Meek, J., Yang, J.Y., Han, V.Z., Bell, C.C., 2008. Morphological analysis of the mormyrid cerebellum using immunohistochemistry, with emphasis on the unusual neuronal organization of the valvula. J. Comp. Neurol. 510, 396–421.

Montgomery, J.C., Bodznick, D., Yopak, K.E., 2012. The cerebellum and cerebellum-like structures of cartilaginous fishes. Brain Behav. Evol. 80, 152–165.

Mugnaini, E., Sekerkova, G., Martina, M., 2011. The unipolar brush cell: a remarkable neuron finally receiving deserved attention. Brain Res. Rev. 66, 220–245.

Nieuwenhuys, R., Nicholson, C., 1969. A survey of the general morphology, the fiber connections and the possible functional significance of the gigantocerebellum of mormyrid fishes. In: Llinas, R. (Ed.), Neurobiology of Cerebellar Evolution and Development. American Medical Association, Chicago, pp. 107–134.

Nieuwenhuys, R., Ten Donkelaar, H., Nicholson, C., 1998. The Central Nervous System of Vertebrate. Springer, Berlin.

Northcutt, R.G., 1989. Brain variation and phylogenetic trends in elasmobranch fishes. J. Exp. Zool. Suppl. 2, 83–100.

Northcutt, R.G., 2011. Paleontology. Evolving large and complex brains. Science 332, 926–927.

Northcutt, R.G., Neary, T.J., Senn, D.G., 1978. Observations on the brain of the coelacanth *Latimeria chalumnae*: external anatomy and quantitative analysis. J. Morphol. 155, 181–192.

Pakan, J.M., Graham, D.J., Wylie, D.R., 2010. Organization of visual mossy fiber projections and zebrin expression in the pigeon vestibulocerebellum. J. Comp. Neurol. 518, 175–198.

Pakan, J.M., Iwaniuk, A.N., Wylie, D.R., Hawkes, R., Marzban, H., 2007. Purkinje cell compartmentation as revealed by zebrin II expression in the cerebellar cortex of pigeons (*Columba livia*). J. Comp. Neurol. 501, 619–630.

Paul, D.H., 1982. The cerebellum of fishes: a comparative neurophysiological and neuroanatomical review. Adv. Comp. Physiol. Biochem. 8, 111–177.

Pearson, R., Pearson, L., 1976. The Vertebrate Brain. Academic Press, London.

Pose-Mendez, S., Candal, E., Mazan, S., Rodriguez-Moldes, I., 2016. Morphogenesis of the cerebellum and cerebellum-related structures in the shark *Scyliorhinus canicula*: insights on the ground pattern of the cerebellar ontogeny. Brain Struct. Funct. 221, 1691–1717.

Pushchina, E.V., Varaksin, A.A., 2001. Argyrophilic and nitroxydergic bipolar neurons (Lugaro cells) in the cerebellum of *Pholidapus dybowskii*. Zh. Evol. Biokhim. Fiziol. 37, 437–441.

Puzdrowski, R.L., 1997. Anti-Zebrin II immunopositivity in the cerebellum and octavolateral nuclei in two species of stingrays. Brain Behav. Evol. 50, 358–368.

Puzdrowski, R.L., Gruber, S., 2009. Morphologic features of the cerebellum of the Atlantic stingray, and their possible evolutionary significance. Integr. Zool. 4, 110–122.

Ramon Y Cajal, S., 1909. Histologie du systeme nerveux de l'homme et des vertebres. Maloine, Paris.

Reeber, S.L., White, J.J., George-Jones, N.A., Sillitoe, R.V., 2012. Architecture and development of olivocerebellar circuit topography. Front. Neural Circuits 6, 115.

Schmidt, A.W., Bodznick, D., 1987. Afferent and efferent connections of the vestibulolateral cerebellum of the little skate, *Raja erinacea*. Brain Behav. Evol. 30, 282–302.

Senglaub, K., 1963. Das Kleinhirn der Vögel in Beziehung zu phylogenetischer Stellung, Lebensweise and Körpergrösse. Z. Wiss. Zool. 169, 1–63.

Shi, Z., Zhang, Y., Meek, J., Qiao, J., Han, V.Z., 2008. The neuronal organization of a unique cerebellar specialization: the valvula cerebelli of a mormyrid fish. J. Comp. Neurol. 509, 449–473.

Smeets, W.J.A.J., 1998. Cartilaginous Fishes. Springer-Verlag, Berlin.

Smeets, W.J.A.J., Nieuwenhuys, R., Roberts, B.L., 1983. The Central Nervous System of Cartilaginous Fishes: Structural and Functional Correlations. Springer-Verlag, New York.

Strata, P., 2015. The emotional cerebellum. Cerebellum 14, 570–577.

Striedter, G., 2005. Principles of Brain Evolution. Sinauer Associates, Sunderland, Mass.

Sudarov, A., Joyner, A.L., 2007. Cerebellum morphogenesis: the foliation pattern is orchestrated by multi-cellular anchoring centers. Neural Dev. 2, 26.

Sugahara, F., Pascual-Anaya, J., Oisi, Y., Kuraku, S., Aota, S., Adachi, N., Takagi, W., Hirai, T., Sato, N., Murakami, Y., Kuratani, S., 2016. Evidence from cyclostomes for complex regionalization of the ancestral vertebrate brain. Nature 531, 97–100.

Sultan, F., 2005. Why some bird brains are larger than others. Curr. Biol. 15, R649–R650.

Sultan, F., Glickstein, M., 2007. The cerebellum: comparative and animal studies. Cerebellum 6, 168–176.

Taylor, G.M., Nol, E., Boire, D., 1995. Brain regions and encephalization in anurans: adaptation or stability? Brain Behav. Evol. 45, 96–109.

Ullmann, J.F., Cowin, G., Collin, S.P., 2010. Quantitative assessment of brain volumes in fish: comparison of methodologies. Brain Behav. Evol. 76, 261–270.

Voogd, J., 2011. Cerebellar zones: a personal history. Cerebellum 10, 334–350.

Voogd, J., 2012. A note on the definition and the development of cerebellar Purkinje cell zones. Cerebellum 11, 422–435.

Voogd, J., Bigaré, F., 1980. Topographical distribution of olivary and cortico nuclear fibers in the cerebellum: a review. In: Courville, J., de Montigny, C., Lamarre, Y. (Eds.), The Inferior Olivary Nucleus: Anatomy and Physiology. Raven Press, New York, pp. 207–234.

Voogd, J., Glickstein, M., 1998. The anatomy of the cerebellum. Trends Neurosci. 21, 370–375.

Voogd, J., Ruigrok, T.J., 1997. Transverse and longitudinal patterns in the mammalian cerebellum. Prog. Brain Res. 114, 21–37.

Voogd, J., Wylie, D.R., 2004. Functional and anatomical organization of floccular zones: a preserved feature in vertebrates. J. Comp. Neurol. 470, 107–212.

Wadiche, J.I., Jahr, C.E., 2005. Patterned expression of Purkinje cell glutamate transporters controls synaptic plasticity. Nat. Neurosci. 8, 1329–1334.

Watson, C., Paxinos, G., Puelles, L., 2011. The Mouse Nervous System. Academic Press, London.

Welker, W.I., 1990. The significance of foliation and fissuration of cerebellar cortex. The cerebellar folium as a fundamental unit of sensorimotor integration. Arch. Ital. Biol. 128, 87–109.

Wilczynski, W., 2009. Evolution, of the brain: in Amphibians. In: Binder, M.D., Hirokawa, N., Windhorst, U., Hirsch, M. (Eds.), Encyclopedia of Neuroscience. Springer, Berlin Heidelberg, pp. 1301–1305.

Wullimann, M.F., Mueller, T., Distel, M., Babaryka, A., Grothe, B., Köster, R.W., 2011. The long adventurous journey of rhombic lip cells in jawed vertebrates: a comparative developmental analysis. Front. Neuroanat. 5, 27.

Wullimann, M.F., Northcutt, R.G., 1989. Afferent connections of the valvula cerebelli in two teleosts, the common goldfish and the green sunfish. J. Comp. Neurol. 289, 554–567.

Wylie, D.R., Aspden, J., Gutierrez-Ibanez, C., Iwaniuk, A.N., Hoops, D., 2015. Evolution of the cerebellum in light of the expression of zebrin II (aldolase C) in mammals, birds, and non-avian reptiles. Brain Behav. Evol. 85, 292–293.

Wylie, D.R., Brown, M.R., Barkley, R.R., Winship, I.R., Crowder, N.A., Todd, K.G., 2003. Zonal organization of the vestibulocerebellum in pigeons (*Columba livia*): II. Projections of the rotation zones of the flocculus. J. Comp. Neurol. 456, 140–153.

Yopak, K.E., 2012. Neuroecology of cartilaginous fishes: the functional implications of brain scaling. J. Fish Biol. 80, 1968–2023.

Yopak, K.E., Frank, L.R., 2009. Brain size and brain organization of the whale shark, *Rhincodon typus*, using magnetic resonance imaging. Brain Behav. Evol. 74, 121–142.

Yopak, K.E., Lisney, T.J., Collin, S.P., Montgomery, J.C., 2007. Variation in brain organization and cerebellar foliation in chondrichthyans: sharks and holocephalans. Brain Behav. Evol. 69, 280–300.

Yopak, K.E., Lisney, T.J., Darlington, R.B., Collin, S.P., Montgomery, J.C., Finlay, B.L., 2010. A conserved pattern of brain scaling from sharks to primates. Proc. Natl. Acad. Sci. U.S.A. 107, 12946–12951.

de Zeeuw, C.I., Wylie, D.R., Digiorgi, P.L., Simpson, J.I., 1994. Projections of individual Purkinje cells of identified zones in the flocculus to the vestibular and cerebellar nuclei in the rabbit. J. Comp. Neurol. 349, 428–447.

Zhang, Y., Shi, Z., Magnus, G., Meek, J., Han, V.Z., Qiao, J.T., 2011. Functional circuitry of a unique cerebellar specialization: the valvula cerebelli of a mormyrid fish. Neuroscience 182, 11–31.

Zhou, H., Lin, Z., Voges, K., Ju, C., Gao, Z., Bosman, L.W., Ruigrok, T.J., Hoebeek, F.E., De Zeeuw, C.I., Schonewille, M., 2014. Cerebellar modules operate at different frequencies. Elife 3, e02536.

PART 3

Early Mammals and Subsequent Adaptations

11

The Emergence of Mammals

T.B. Rowe

The University of Texas at Austin, Austin, TX, United States

11.1 Introduction

Mammalia and Reptilia are the two living sister clades that comprise Amniota (Fig. 11.1). The fossil record indicates that mammals and reptiles diverged from their last common ancestor—the "ancestral amniote"—during the Carboniferous Period (Gauthier et al., 1988a,b). By ~310 million years ago, the mammalian lineage had diverged from the ancestral amniote onto its own unique evolutionary trajectory. This chapter on the subsequent emergence of mammals focuses on a reciprocal relationship between the skeleton and neurosensory systems. As a starting point, it first characterizes what features can be inferred in the ancestral amniote based on comparative anatomy and the fossil record. It then traces the sequential evolution of correlated skeletal and neurosensory traits that culminated in the origin of crown Mammalia, by the Early or Middle Jurassic, some ~170 million years ago. This historic sequence in the accumulation of morphological innovation took place over approximately 140 million years of geological time, and it established the bauplan from which subsequently proceeded diversification of the many living and extinct mammalian clades, from the Jurassic to the present (Fig. 11.2).

Paleontology affords a special signal on the emergence of mammalian neurosensory systems because many parts of that system require rigid armatures to function properly, and these armatures are provided by the skeleton and associated connective tissues. An epigenetic responsiveness of the skeleton during ontogeny is critical to the developing neurosensory system during its ontogenetic integration and the establishment of its proper mature functionality (Rowe, 1996a,b). Some measure of this profound degree of integrated morphogenesis can be seen by comparing the cartilaginous chondocranium of a 1-day-old opossum (Fig. 11.3) with a fully ossified mature skull (Fig. 11.4). In understanding the integral relationship between skeleton and neurosensory systems, new insights from experimental and ontogenetic studies in extant species are rapidly growing. Moreover, the rich fossil record of mammals and their extinct relatives is providing unique insights into the timing, patterns of evolutionary correlation, and sequences of events in the evolution of both skeletal and neurosensory systems that cannot be gleaned from the comparative anatomy and development of living species alone.

Amniote phylogeny has been the subject of intense study for more than a century. Thanks to a fairly dense fossil record, and to the many recent advances in molecular phylogenetics, the major features of amniote interrelationships are now known with considerable resolution (Gauthier et al., 1988a,b, 1989; Laurin and Reisz, 1995; Meyer and Zardoya, 2003). The position of turtles within Reptilia is the most significant problem facing the major living amniote clades (Lyson et al., 2010), and the relationships between certain fossil taxa are also problematic (discussed later). However, broad segments of amniote phylogeny are now highly resolved, with strong support from unprecedented volumes of morphological, developmental, and molecular data. These advances offer a rich, rigorous phylogenetic context in which to understand the evolutionary emergence of mammals in general, and the evolution of the mammalian neurosensory system in particular.

The rise of phylogenetic systematics in the last generation has led to a wholesale reconceptualization of the history of mammalian evolution. Phylogenetic systematics was driven by increasingly sophisticated computing algorithms and methodological rigor in phylogeny reconstruction. In its earliest days, molecular phylogenetics threatened to exterminate morphology (Patterson, 1987), thanks in part to the comparative simplicity and small data volumes of molecular sequences. But the importance of morphology and the special role of fossils in phylogeny reconstruction soon reasserted themselves forcefully (Gauthier et al.,

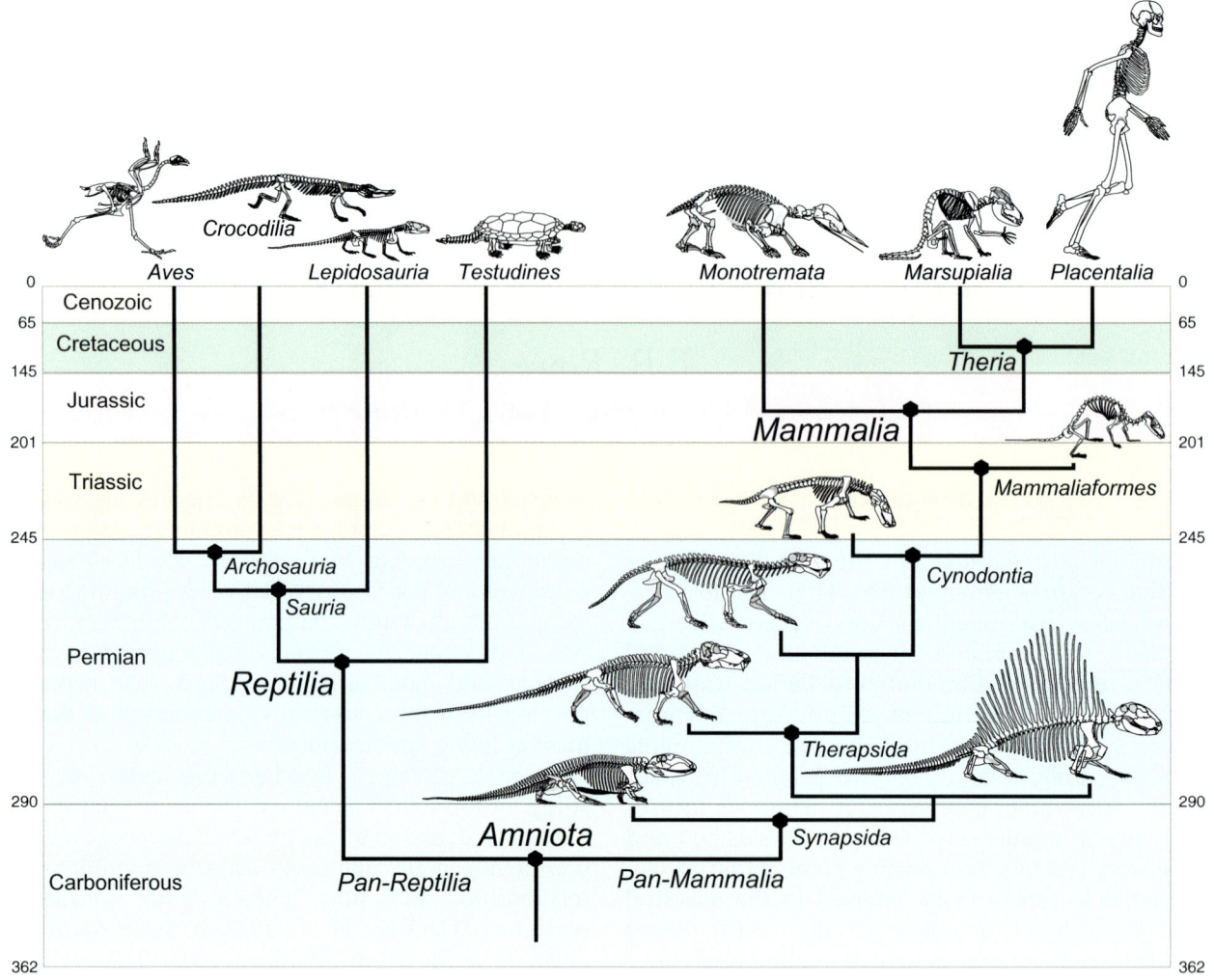

FIGURE 11.1   Phylogeny of the major clades of Amniota distributed over the geological timescale. *After Dingus, L., Rowe, T.B., 1998. The Mistaken Extinction — Dinosaur Evolution and the Origin of Birds. W. H. Freeman & Co, New York.*

1988a,b; Donoghue et al., 1989; Huelsenbeck, 1991). Shortly thereafter, a relentless increase in computing power catalyzed a renaissance in morphology through advances in digital imaging technologies and the ability to manipulate the huge data volumes they produce. X-ray computed tomography in particular has enabled the volumetric visualization and quantification of fossils and entire living organisms (Fig. 11.5) in unprecedented and ever-increasing resolution (Rowe et al., 1995, 1997, 1999; Carlson et al., 2003). Paleontologists can nondestructively "see" through external bony surfaces in fossils to visualize and explore internal volumes such as the endocranial cavity (Macrini et al., 2006, 2007a,b,c; Kirk et al., 2014), the inner ear (Berlin et al., 2013; Ekdale, 2013, 2016), organization of the olfactory capsule (Rowe et al., 2005; Green et al., 2012; Macrini, 2012, 2014) and to trace the passageways for nerves and blood vessels for the first time in large comparative samples of specimens.

The power of computed tomography is being augmented by new vital staining techniques that render different soft tissues in living taxa radiopaque, allowing their relationship to the skeleton to be visualized as never before in ontogenetic sequences and mature specimens (Metscher, 2009; Li and Clark, 2015). Magnetic resonance imaging (Fig. 11.6) has now gained sufficient resolution that soft tissues in even small laboratory mammals can be visualized with this technique (Rowe and Frank, 2011). The interpretation of fossil specimens has benefited immensely from the unprecedented levels of structural detail that can now be visualized in their living relatives.

Comparative and developmental anatomy of living mammals offers much insight into what the ancestral mammal must have been like, and such studies have postulated numerous evolutionary driving factors. These include increased brain size and emergence of

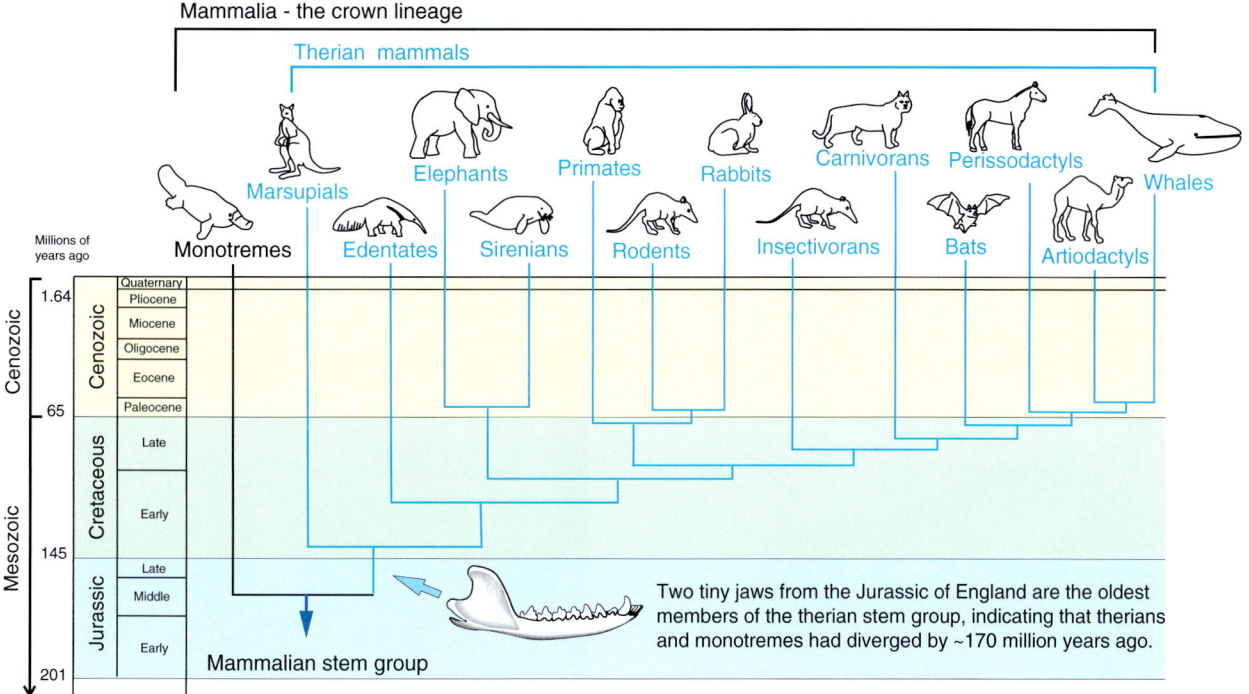

FIGURE 11.2 Phylogeny of the major clades of extant crown Mammalia distributed over the geological timescale. *After Rowe, T.B., 1999. At the roots of the mammalian tree. Nature 398, 283–284.*

the neocortex (Rowe, 1993, 1996a,b; Rowe et al., 2011; Krubitzer and Kaas, 2005; Kaas, 2009), innovations involving hearing (Allen, 1975; Kermack and Kermack, 1984), innovations in feeding and mastication (Crompton, 1989; Crompton and Parker, 1978), enhanced taste and olfaction (Rowe et al., 2011; Rowe and Shepherd, 2016; Aboitiz and Montiel, 2015), miniaturization (Eisenberg, 1981, 1990), parental care (Farmer, 2000, 2003), elevated metabolism and endothermy (Rubin, 1995; Kemp, 2006a; Hopson, 2012), nocturnality (Geisser et al., 2002; Kermack and Kermack, 1984), and others. Although deeply informative, living taxa offer only limited insight into the relative importance of the various proposed driving mechanisms at particular points in pre-mammalian history, or in regard to timing or sequences of historical events that culminated in the origin of mammals. As will be seen, each of these factors was influential at certain times, but some were influential prior to the origin of mammals and others only thereafter.

A detailed map of these events has grown from a succession of phylogenetic analyses conducted over the last three decades that explored the place of Mammalia within more inclusive clades containing its closest extinct relatives among amniotes (Rowe, 1988, 1993; Gauthier et al., 1988a,b; Rubidge and Sidor, 2001; Kielan-Jaworowska et al., 2004; Meng et al., 2006; Ji et al., 2006; Liu and Olsen, 2010). The more powerful analyses combed the entire skeleton of fossil and recent

taxa for phylogenetically informative characters without assuming a priori that any anatomical subsystem, for example, the dentition or the postcranium, is more or less susceptible to homoplasy. This approach, based on the most strongly supported trees, has illuminated new and unexpected correlations between morphological characters, as well as nuanced patterns in the historical sequence of morphological transitions in the skeleton, many of which are proxies for changes in the neurosensory system.

These patterns of transformation afford a test of proposed driving factors behind the emergence of mammals that extend into the genetic and epigenetic control of development. For example, we will see potential support in this history for the idea that peripheral sensory arrays influence central organization and that through epigenetic population matching cortical reorganization and relative increases in brain size may have been driven in part by connectional invasions from peripheral cell populations and sensory structures such as olfactory receptors, teeth, and hair (Katz and Lasek, 1978; Krubitzer and Kaas, 2005; Streidter, 2005). It is also informative to view this history in terms of Günter Wagner's (2014) suggestion that there are two basic types of morphological innovations or novelties in animal evolution. Type I novelties involve the origin of a novel "character identity," and as examples Wagner cites the vertebrate head and the insect wing. Type II innovations involve the origin of a novel "character-state" and

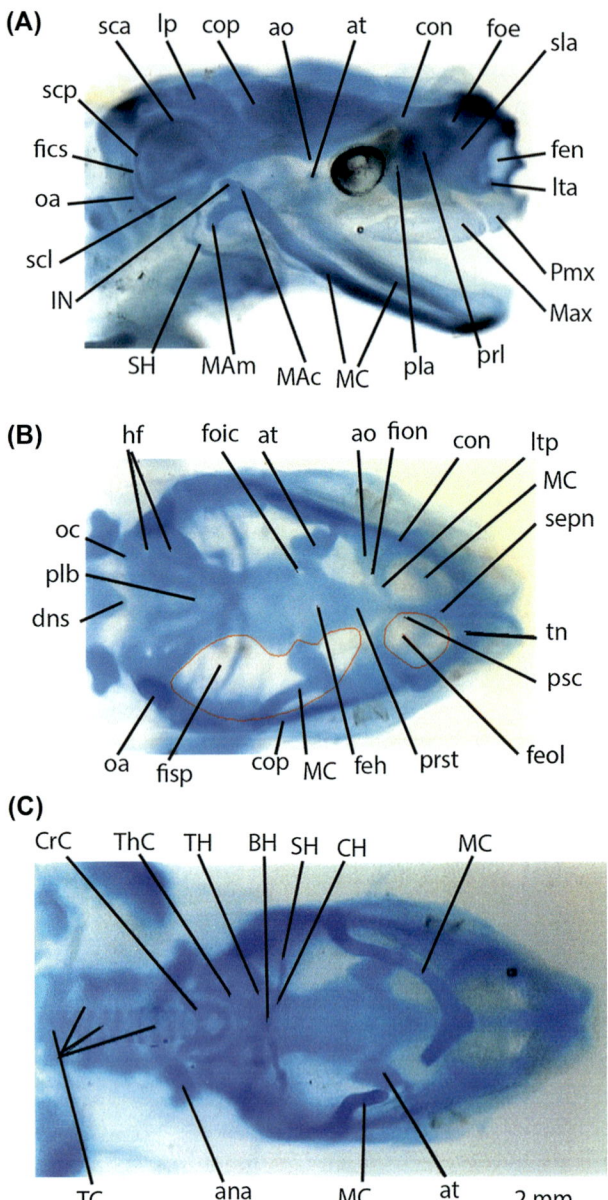

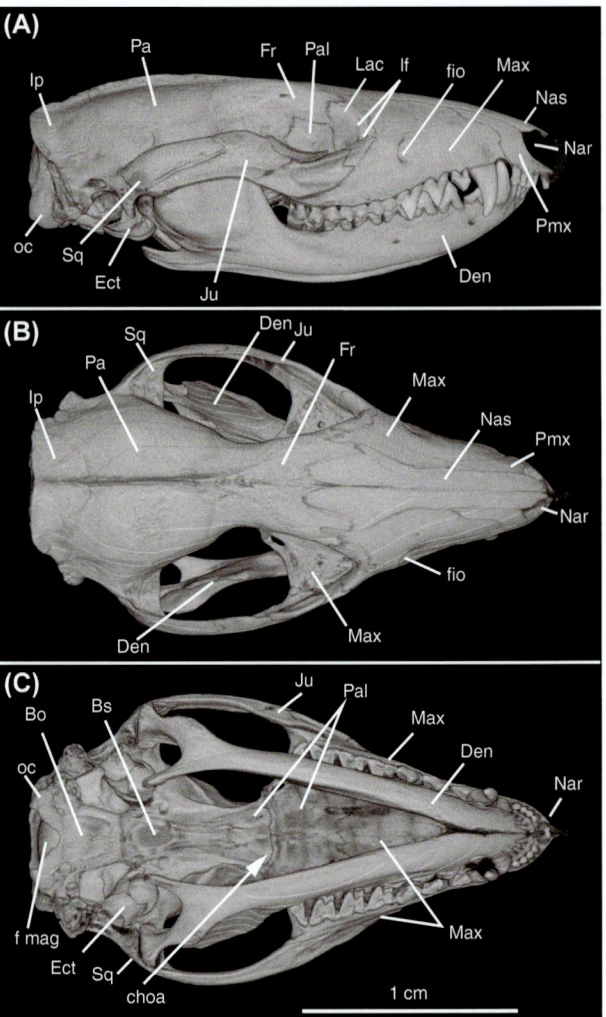

FIGURE 11.4 Skull of adult *Monodelphis domestica*, rendered in 3D from computed tomography data in (A) lateral, (B) dorsal, and (C) ventral views. See Table 11.1 for key to abbreviations. *After Rowe, T.B., Eiting, T.P., Macrini, T.E., Ketcham, R.A., 2005. Organization of the olfactory and respiratory skeleton in the nose of the gray short-tailed opossum* Monodelphis domestica. *J. Mammal. Evol. 12, 303–336.*

FIGURE 11.3 Head of the 1-day-old opossum *Monodelphis domestica*, cleared and stained in Alcian Blue to show the cartilaginous chondocranium in (A) lateral, (B) dorsal, and (C) ventral views. See Table 11.1 for key to abbreviations.

novel "variational modality" in systems of repeated structures, such as the origin of the tetrapod limb from a more primitive fin, and feathers from integumentary scales. The origins of Type I innovations are not predicted by conventional Darwinian natural selection, and instead Wagner recognizes a special role for gene duplication and for new gene regulatory networks in the emergence of innovation. The pattern of early mammalian history described later reveals effects on the brain and skeletal morphogenesis of gene duplications, particularly in the olfactory odorant receptor subgenome (Niimura, 2012), and in genes regulating the radial units of cortical organization (Rakic, 1988).

Throughout this history can also be seen the appearance of new variational modalities in systems of repeated elements that are distributed across the entire skeleton, including the dentition, vertebral column, and limbs, each with its own special relationship to the neurosensory system. These patterns of correlated evolutionary transformations raise the provocative question of whether the mammalian neocortex qualifies as a Type I innovation. The heuristic value of asking this question may be more important than arriving at a final answer in understanding the remarkable balance between individuation of new character identities, the appearance of new character states, and variational

(A)

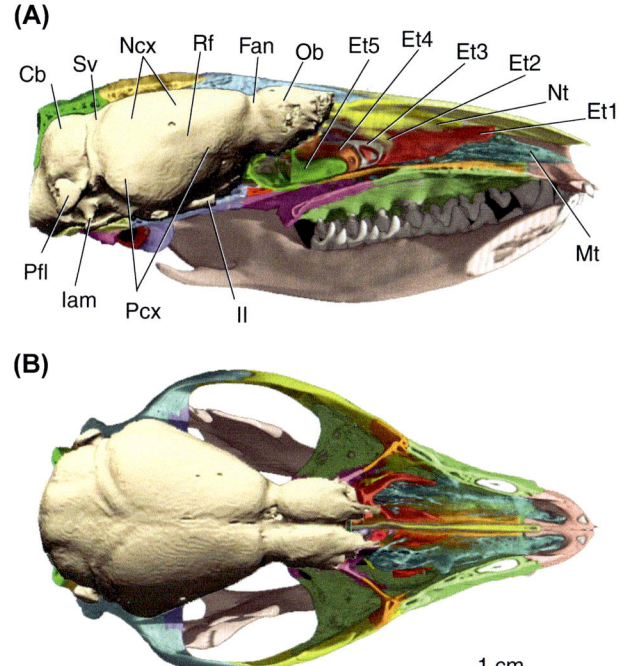

(B)

1 cm

FIGURE 11.5 Skull of mature *Monodelphis domestica*, reconstructed in 3D from computed tomography, in cutaway sagittal (A) and horizontal (B) views. The entire endocranial cavity was rendered solid beige to show the endocast of the brain in relation to the various bones of the skull, which were individually segmented and colored using VGStudio Max 2.0 software. See Table 11.1 for key to abbreviations. *After Rowe, T.B., Macrini, T.E., Luo, Z.-X., 2011. Fossil evidence on origin of the mammalian brain. Science 332, 955–957. https://doi.org/10.1126/science.1203117.*

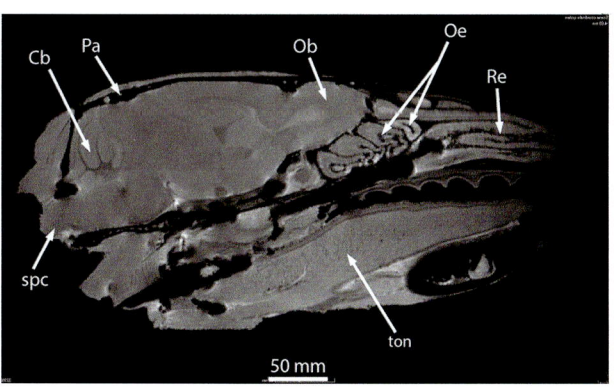

FIGURE 11.6 Magnetic resonance image of mature *Monodelphis domestica* taken along a parasagittal plane. See Table 11.1 for key to abbreviations.

modalities, with the functional integrating of these changes in individual organisms.

This chapter focuses on the first half of mammalian evolution, beginning with divergence of the mammalian total clade from the ancestral amniote ~310 million years ago and ending with the origin of crown Mammalia, ~170 million years ago. By way of introduction, a

brief historical review is presented of noteworthy events in the early in the history of zoology surrounding the initial scientific realization that Mammalia is the product of evolution, that it has an ancient fossil record, and how that evolutionary history was conceptualized at the start of the evolutionary era of scientific inquiry. There follows a summary of relevant points in the development of phylogenetic systematics that has led to a recent reconceptualization of the evolution of the mammalian total clade, with an explanation of new nomenclatural conventions that have developed from the application of modern phylogenetic methods. Next is an overview of the skeletal and neurosensory characters that can be inferred as present in the ancestral amniote. From this point of departure follows an outline of the sequential evolutionary changes in the skeleton of the mammalian stem group that have now been mapped from phylogenetic analysis of fossils, with insights from living species, and their implications for understanding some of the mechanisms underlying neurosensory innovations that culminated in the origin of Mammalia.

11.2 The Emergence of an Evolutionary View of Mammalia

In the wake of publication of *On the Origin of Species* (Darwin, 1859), Ernest Haeckel, Thomas Huxley, and Edward Drinker Cope (Fig. 11.7) were among the first to understand that paleontology had taken on an entirely new meaning as transitional fossils were discovered that connected living lineages in deep time. How to classify fossils in the context of preexisting Linnaean classifications of living taxa soon became problematic. As Huxley observed, "The root of the matter appears to me to be that the palaeontological facts which have come to light in the course of the last ten or fifteen years have completely broken down existing taxonomical conceptions, and that attempts to construct fresh classifications upon the old model are necessarily futile" (Huxley, 1880: p. 652).

Huxley went on to arrange living mammals into a "scala mammalium" as he called it, for which he created several names that reflected progressive stages in evolution. He divided living mammals into Prototheria (=Monotremata), Metatheria (=Marsupialia), and Eutheria (=Placentalia). He explained that "Our existing classifications have no place for the 'submammalian' stage of evolution… I propose to term the representatives of this stage Hypotheria, and I do not doubt that when we have a fuller knowledge of the terrestrial Vertebrata of the later Palaeozoic epochs, forms belonging to this stage will be found among them" (Huxley, 1880: p. 670).

Huxley's prediction soon proved true, and in the coming decades a bonanza of fossils recognized as belonging to the mammalian stem group were collected

FIGURE 11.7   Early evolutionists who pioneered the understanding of mammalian history. (A) Ernst Haeckel, (B) Thomas Huxley, (C) Edward Drinker Cope, (D) Robert Broom, (E) William King Gregory, and (F) William Diller Matthew.

in Russia, South Africa, India, and the United States. Although the name Hypotheria never caught on, Huxley's basic concept of the fossil record as a succession of typological stages dominated early paleontological and evolutionary thought. Similar views were expressed by Cope (1878) and Haeckel (1866, 1877) that the fossils documenting mammalian evolution could be classified into a succession of primitive, extinct groups that "gave rise" to Mammalia and its various living lineages.

Deeply entrenched during Huxley's time was the view that reptiles were the most generalized of living amniotes and that reptiles (or possibly amphibians) included the ancestors of living birds and mammals (Cope, 1878; Huxley, 1880; Broom, 1910; Romer, 1956, 1966). The fossils representing the "submammalian" stage of evolution soon presented taxonomists with a difficult choice necessitated by the Linnaean practice of taxonomic ranking and the century-old tradition of

placing Reptilia and Mammalia each in a class of its own (Linnaeus, 1758). Sir Richard Owen, the preeminent anatomist and outspoken anti-Darwinian, was the first to express an opinion. Since these fossils retained suites of "reptilian" features (meaning plesiomorphic features in modern terms), he classified them as newly discovered groups of extinct Reptilia (Owen, 1844, 1861, 1876). Owen's descriptions are remarkable in their observation of detailed similarities that these fossils shared with mammals, including histological minutiae of tooth structure and modes of dental growth. The very names that Owen coined for them, such as Cynodontia (doglike teeth), Dicynodontia (two doglike teeth, in reference to the canines), and Theriodontia (teeth like therians, like wild beasts), allude to similarities with living mammals. But "reptilian" features ruled their classification. Owen was deeply embroiled in the ongoing debate over evolution and aware of the apparent transitional nature of these fossils, but he could never reconcile their unique combinations of plesiomorphic ("reptilian") and derived (mammalian) characters. Late in his life, he summarized his view: "Certain it is that the lost reptilian structures specified in the present Catalogue [of 'reptiles' from the Permian and Triassic of South Africa] are now manifested by quadrupeds with a higher condition of cerebral, circulatory, respiratory, and tegumentary systems [i.e., Mammalia]—a condition the acquisition of which is unintelligible to the writer on either the Lamarckian or the Darwinian hypothesis" (Owen, 1876: p. 76).

Even after Cope and his successors recognized the evolutionary connections of these fossils to living mammals, the practice became widely accepted of naming extinct ancestral groups (ie, paraphyletic stem groups) that excluded living descendants (Fig. 11.8). Those Paleozoic and Mesozoic fossils recognized as possessing a combination of primitive (reptilian) and derived or mammal-like features came to be known collectively in the vernacular as the "mammal-like reptiles" (Broom, 1910, 1932).

The most primitive fossils in the mammalian stem group were also the oldest, and they came from Carboniferous and Early Permian sediments from North America and Europe. They had a few mammalian features,

first recognized in the shoulder girdle (Cope, 1878), and later in the single temporal fenestra and zygomatic arch (Broom, 1910), and they were grouped together as "Pelycosauria" (Cope, 1878). Owing to their preponderance of primitive traits, pelycosaurs were classified for the next century as an extinct group of Reptilia, while being widely recognized as involved in the distant ancestry of mammals.

The pelycosaurs were conceived as an ancestral group that became extinct just as it "gave rise" to Therapsida, ~270 million years ago, in the mid-Permian (Fig. 11.8). Therapsids were believed to have thrived during the Late Permian and Triassic, and they exhibited many more mammalian features than the pelycosaurs, notably a large mandibular adductor chamber and a more agile skeleton, and they were considered to occupy a new, more mammal-like adaptive level or stage of evolution (Kemp, 2006b). But like the pelycosaurs, they were still sufficiently plesiomorphic as to be classified as reptiles. Among the therapsids, Harry Seeley and Robert Broom soon recognized Cynodontia as containing the most proximal ancestors of mammals, thanks to the presence of a secondary palate and double-occipital condyle. Still, the taxonomic practice of balancing similarities to mammals against primitive "reptilian" differences was interpreted to mean that cynodonts too were assigned to class

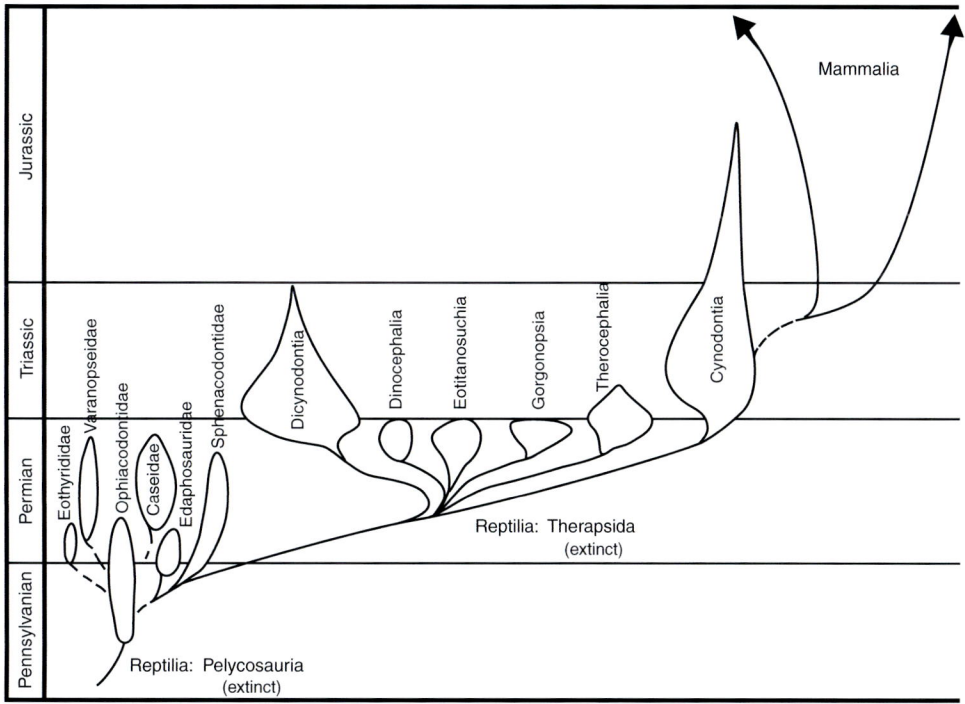

FIGURE 11.8 The archaic 19th- and 20th-century conceptualization of pan-mammalian history envisioned ancestral, paraphyletic "reptilian" taxa "giving rise" to more advanced paraphyletic groups that went extinct just as they gave rise to mammals. Thus the extinct pelycosaurian reptiles gave rise to therapsidan reptiles, who in turn gave rise to cynodont reptiles, and finally the cynodonts became extinct just as they gave rise to Mammalia, a taxon of questionable monophyly under this view. *Modified from Carroll, R.L., 1988. Vertebrate Paleontology and Evolution. W. H. Freeman & Co., New York.*

Reptilia. Several Triassic fossils that Owen and Broom had recognized as fossil mammals were later ejected from Mammalia because they retained a primitive jaw articulation or some other primitive feature deemed inconsistent with membership in class Mammalia (see Section 11.5.14). Cynodonts and Therapsida as a whole were portrayed as becoming extinct just as mammals appeared in the Late Triassic (Broom, 1932; Romer, 1956, 1966; Carroll, 1988).

But this was a start, and by the mid-20th century, a broadly correct general outline of mammalian evolution was established by Robert Broom (1910, 1932), William King Gregory (1910, 1916, 1922, 1953), and William Diller Matthew (1915) had published the first attempt at a synthesis of mammalian evolution in light of geological and climatic processes driving natural selection (Fig. 11.8).

Under this early evolutionary view, a fuzzy boundary separated mammals from reptiles, and for much of the 20th century generations of paleontologists battled over which fossil represented the oldest mammal, and which character or characters were the defining features that separated mammals from reptiles. This in turn led to disagreement on estimates of the evolutionary properties of Mammalia, such as its distribution in time and space. For example, today the oldest fossil that can be placed within the crown clade Mammalia is *Phascolotherium bucklandi* from the Middle Jurassic of England (Rowe, 1988, 1993). It is a member of the therian stem group, and it indicates that therians and monotremes had diverged from the last common ancestor of Mammalia by ~170 million years ago, following a major extinction event at the Triassic—Jurassic boundary. This is taken to be the minimum age for crown Mammalia by most contemporary paleontologists. However, others still include various fossil members of the mammalian stem group, with destabilizing consequences. For example, the Late Triassic *Morganucodon* is often considered to be the oldest "mammal" and as a consequence the temporal span represented by the name "Mammalia" extends back in time more than 30 million years, to the Rhaetian (~201—208 million years ago; Crompton and Jenkins, 1973; Kermack et al., 1973, 1981; Kemp, 1983, 1988a,b). Still another proposal would include the extinct stem group fossil *Adelobasileus* under the name "Mammalia" (Lucas and Luo, 1993; Benton, 2015), extending the origin of mammals still further into the Triassic, into the Carnian (~228—235 million years ago). Thus the meaning of the name "Mammalia" became fraught with equivocation. Depending on which fossil was considered the oldest mammal, estimates of the age of Mammalia differed by as much as 60 million years. Which particular fossil was deemed the oldest mammal was determined more by one's own academic line of descent than by the relationships among the fossils themselves.

To further characterize this world view, in 1802, Everard Home described the reproductive anatomy of living monotremes, which had been known for only a decade, and he raised the suspicion that monotremes laid eggs (not confirmed until 1884; see Burrell, 1927; Gould, 1985). This led to a debate that extended across the entire 19th century over whether monotremes should be classified as mammals. Home (1802a,b) maintained that monotremes were "an intermediate link between the classes Mammalia, Aves, and Amphibia" (Home, 1802b: p. 360). A number of 19th century zoological luminaries, including Lamarck (1809), advocated placing monotremes in a class by themselves that was separate from class Mammalia.

A central issue pervading 19th- and 20th-century zoology was the nature of taxonomic groups, for example, whether or not the class Mammalia was a grade of organization achieved independently multiple times. Did the egg-laying monotremes have a common ancestor with therians that would be called a mammal? Some authors suggested that monotremes may have evolved to a mammalian grade from the extinct dicynodont therapsids, while the viviparous therians independently evolved from extinct cynodont therapsids. Driving this issue was the debate over the quintessential defining characters of Mammalia. Even Meckel's discovery in 1826 of mammary glands in monotremes, "the outstanding characteristic of the mammalian class" (Burrell, 1927: p. 29) and the namesake character of Mammalia, did not solve the monotreme problem to general satisfaction. Nor did any other character-based definition resolve these problems.

11.3 The Phylogenetic System

Many of these questions became irrelevant as systematists began to explore the impact of Hennig's (1966) concept of monophyly, and as taxa came to be understood as clades—comprising an ancestor and all of its descendants. The idea of paraphyletic, extinct ancestral groups "giving rise" to other groups was replaced when systematists developed the methods to discover and map the internested monophyletic clades that are the products of evolution, and to define the meanings of their names in terms of ancestry, rather than characters. Having abandoned the Linnaean practice of ranking taxa, the presence of shared derived features, or synapomorphies, became the sole basis for phylogeny reconstruction. Systematists no longer used "defining" characters to circumscribe groups, but rather employed a new practice of diagnosing the existence of monophyletic clades based on the discovery of synapomorphies (or apomorphies) that were inferred to have arisen in some particular ancestor. Only after the distinction was made between definitions of taxonomic names based on ancestry, in contrast to names based on characters, did taxonomic names come to connote stable, unequivocal meanings in terms of their various evolutionary properties, such as their content and distribution in time and space (Rowe, 1987, 1988; Gauthier et al., 1988a; de Queiroz and Gauthier, 1990, 1992, 1994).

A major shift has been to move away from the practice of defining and naming successions of paraphyletic, extinct ancestral groups from which the living clades emerged. The idea that groups defined by the possession of primitive (plesiomorphic) characters could "give rise" to other, less-primitive groups and to living taxa has been replaced by a phylogenetic hierarchy consisting of internested monophyletic clades. In this view, living taxa are placed explicitly into the phylogenetic framework of their extinct relatives.

This shift was accompanied by the emergence of a phylogenetic nomenclature. In the phylogenetic system, the meanings of taxonomic names are defined in terms of ancestry, and their membership is diagnosed by characters (Rowe, 1987; de Queiroz and Gauthier, 1990, 1992, 1994; de Queiroz, 2007). This has led to a nomenclatural system that integrates living and extinct taxa in a uniform, rank-less system predicated on hypotheses of phylogeny, with practices set out in the International Code of Phylogenetic Nomenclature, or the Phylocode (Cantino and de Queiroz, 2000). To preserve historical continuity, the meanings of widely recognized names like Mammalia and Reptilia were redefined. Hence Mammalia came to be defined in reference to the last common ancestor shared by living monotremes and therians, and all of its descendants (Rowe, 1987, 1988), while Reptilia now refers to the clade arising from the last common ancestor shared by living turtles, lepidosaurs, crocodilians, and birds, and all its descendants (Gauthier et al., 1988a).

The redefinition of the names Mammalia and Reptilia reflects adoption of the "crown-clade" concept (Jeffries, 1979), that is, a clade specified by living taxa and their last common ancestor and all its descendants (Fig. 11.9). This convention cut the Gordian knot that had long bound widely used names like Mammalia and Reptilia into contentious debates over their precise meanings. Fossils that are more closely related to Reptilia, but that lie outside its crown clade, are now considered to be members of the reptilian stem lineage or the paraphyletic reptilian "stem group" while also belonging to the monophyletic "total clade" of Reptilia. Fossils that are more closely related to Mammalia, but that lie outside its crown clade, are now considered to be members of the mammalian stem or the paraphyletic mammalian "stem group" while also belonging to the monophyletic "total clade" of Mammalia. At onetime or another, the term Mammalia was placed on almost every well-supported node across the mammalian

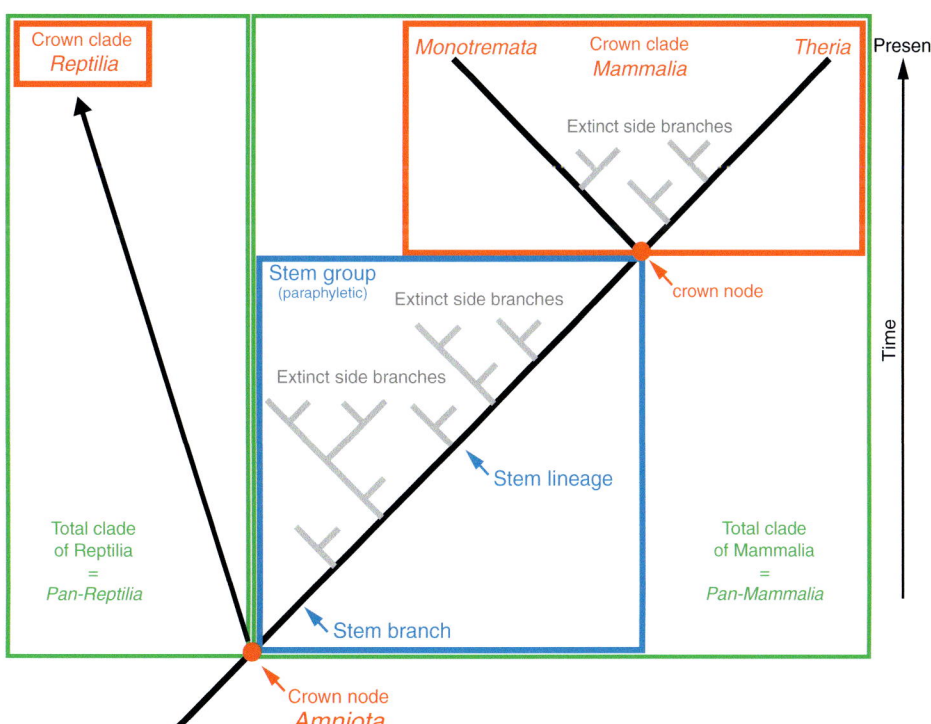

FIGURE 11.9   Categories of clades and groups employed under the phylogenetic system of taxonomic nomenclature. *Modified from de Queiroz, K., 2007. Toward an integrated system of clade names. Syst. Biol. 56, 956–974.*

stem group, generally in recognition of some feature deemed the essential mammalian character, yet no name was ever proposed as replacing that of the crown (Rowe and Gauthier, 1992). Growing in popularity is the "pan-clade" convention, in which the prefix Pan (for all) is added to the crown clade name to reflect its total clade (de Queiroz, 2007). Under this convention, Pan-Reptilia is the name designating the total clade of Reptilia, and Pan-Mammalia is the total clade of Mammalia. Together Pan-Reptilia and Pan-Mammalia comprise the crown clade Amniota.

To summarize, in the phylogenetic system, each taxonomic name represents a monophyletic clade, that is, an ancestor and all of its descendants, and clades are discovered and diagnosed by characters, whether morphological, molecular, or both. In many cases, large suites of correlated characters diagnose named clades. When these diagnoses are interpreted as an evolutionary sequence of correlated changes, entirely new patterns of morphological transformation help to illuminate the genetic and developmental mechanisms driving evolutionary history, and in ideal circumstances they can be tested in the laboratory.

New nomenclatural conventions may seem intrusive, but the rationale in this case is that an integrated nomenclatural system for monophyletic clades can more efficiently and unequivocally convey modern phylogenetic hypotheses, while also facilitating digital indexing and retrieval of information (de Queiroz, 2007). Perhaps the most dramatic example of how this modified nomenclature captures the reconceptualization of evolutionary history is the idea that only *some* of the dinosaurs are extinct (Gauthier, 1986; Dingus and Rowe, 1998). The ~10 000 species of living birds are now considered to be members of the successively more inclusive monophyletic clades Theropoda, Saurischia, Dinosauria, Archosauria, Diapsida, Reptilia, Amniota, Vertebrata, and so on. The idea that *birds are dinosaurs* captures the larger concept, that phylogenetic systematics catalyzed the reconceptualization of large swaths of evolutionary history.

The emergence of mammals has undergone a reconceptualization that may not seem quite so dramatic as our transformed understanding that birds are dinosaurs. But emerging from this newly mapped phylogenetic hierarchy is a nuanced sequence of evolutionary transformations that raise the question of whether the neocortex is a Type I novelty (Wagner, 2014), and new mechanisms are being revealed that may have driven the course of mammalian evolution. The discovery of sequences and correlations among transforming characters offer surprising new insights into the integrated evolution of systems that are unique to mammals among living taxa. This new perspective reveals a complex interplay between the evolution of development, the

emergence and integration of entirely new systems, and the evolution of genes, gene families, signaling pathways, and the epigenetic factors involved in the evolutionary emergence of mammals. The following narrative attempts to convey some flavor of the diversity achieved by the mammalian stem group during the first ~140 million years of evolution of the mammalian total clade, as a context for understanding the origin and subsequent diversification of the crown clade Mammalia.

The emergence of mammals is discussed later in terms of the diagnoses of 17 well-established internested nodes that include all the major extinct side branches of the mammalian stem group (Fig. 11.10). Uncertainty in phylogenetic relationships is reflected in polytomies of multiple taxa branching from one node. Some nodes are diagnosed by large numbers of characters whose importance is well understood. Outside of Mammalia, the more important of the successively more inclusive clades are Mammaliaformes, Mammaliamorpha, Cynodontia, and Therapsida. Other nodes are unnamed and are diagnosed by seemingly minor characters, but their inclusion helps to demonstrate another layer of complexity in this history involving transformations in the variational modalities (Wagner, 2014) of anatomical systems of repeated elements, as well as the emergence of entirely new features, behaviors, and systems that distinguish mammals from other living taxa.

11.4 The Ancestral Amniote

In contemporary terms, the fundamental dichotomy in amniote evolution was the divergence of the sister taxa Pan-Reptilia and Pan-Mammalia from the ancestral amniote. Today amniotes comprise approximately 5500 living mammal species and 17 000 living reptiles species (including birds), plus a fossil record that includes several thousand extinct amniote species and clades that date back in geological time to at least ~310 million years (Gauthier et al., 1988a,b). The ancestral amniote was a small predatory quadruped. Early amniote fossils are generally under a meter in length, half of which is accounted for by their tails. The Carboniferous *Limnoscelis paludis* is somewhat larger and far more complete than other early amniotes, and it will be used to illustrate the general features that can be inferred as present in the ancestral amniote (Fig. 11.11). Most accounts place *Limnoscelis* as one of the closest extinct sister taxa to crown Amniota (Gauthier et al., 1988a,b; 1989; Laurin and Reisz, 1995).

The early amniotes lived in the great Carboniferous forests along rivers and deltas that were later buried to become vast coal deposits that eventually fueled the human Industrial Revolution. From today's perspective, the terrestrial ecosystem occupied by early amniotes was a bizarre assemblage of mostly predatory tetrapods

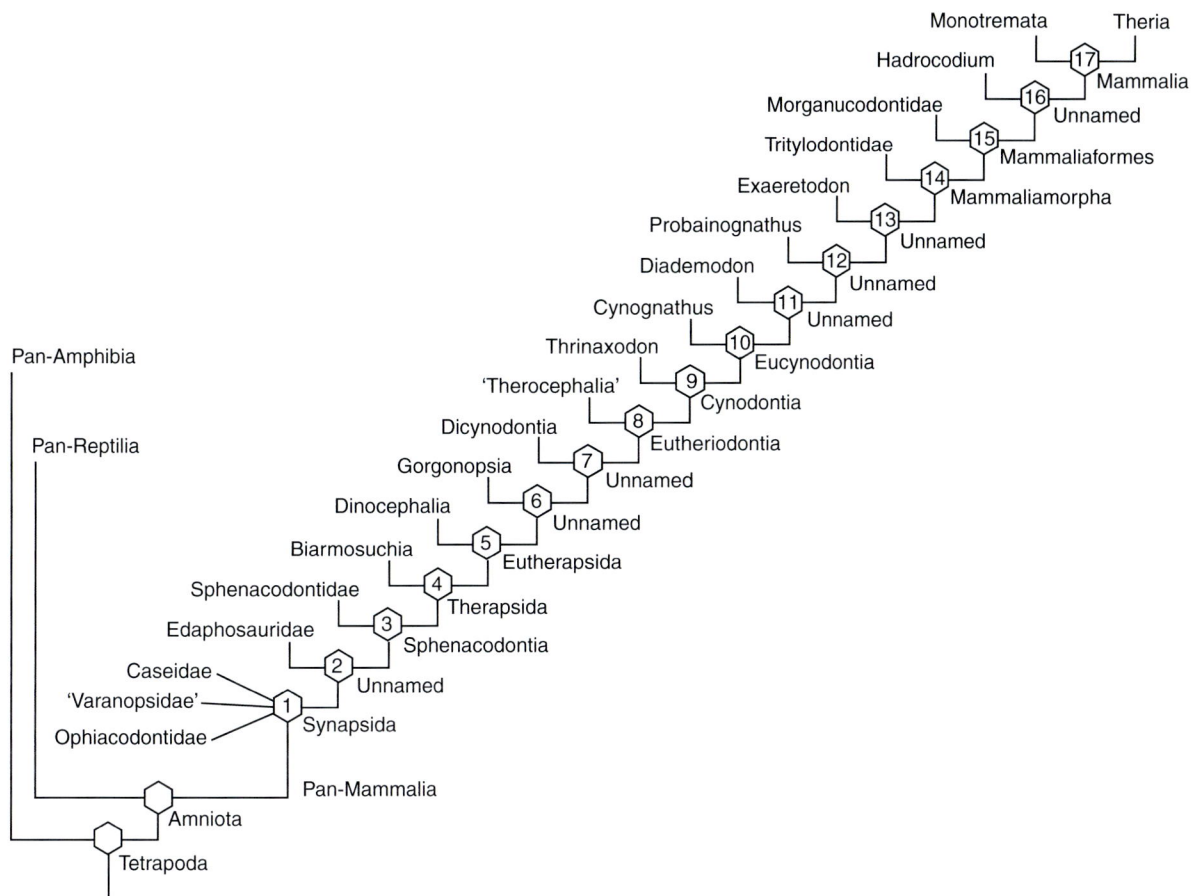

FIGURE 11.10   Basic cladogram of Pan-Mammalia, with major nodes on the mammalian stem lineage numbered and keyed to the text.

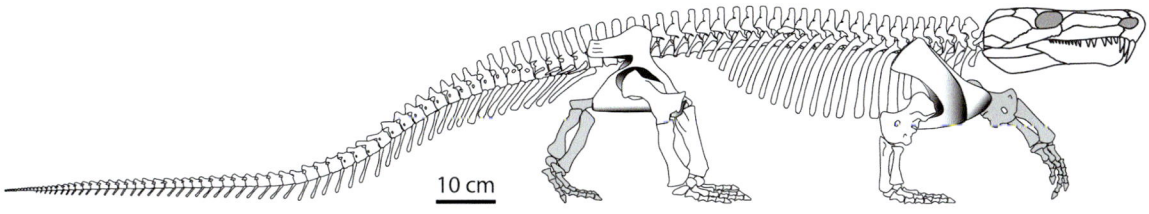

FIGURE 11.11   *Limnoscelis paludis*—an example of an early amniote. *After Romer, A.S., 1956. Osteology of the Reptilia. University of Chicago Press, Chicago.*

who preyed upon each other, and on the arthropods, mollusks, and other nonvertebrates that could digest cellulose and were the intermediates to the photosynthesizing base of the food pyramid (Benton, 2015).

Fossils of the earliest amniotes and pan-mammals are generally rare and incomplete. But some of the best known became trapped and fossilized under exceptional circumstances. The Carboniferous Joggins fauna of Nova Scotia preserves lycopod stumps whose rotted, cavitated cores became natural pit traps for early amniotes who fell in and could not escape. Rotten stumps from this vast forest were periodically buried in upright positions as inland lakes rose and inundated the area, carrying

sediment that entombed the small carcasses within. In 1852, geologists began to explore their contents and within the lycopod stumps they found a number of Carboniferous skeletons including *Archaeothyris*, now regarded as the oldest pan-mammal (Romer and Price, 1940; Romer, 1956, 1966; Carroll, 1988; Benton, 2015).

The namesake feature of Amniota is the amniotic egg, plus a complex suite of attendant equipment for internal fertilization that afforded reproductive independence from the water for the first time in vertebrate history. Incubation of the amniote embryo is a more protracted process than before, because a larval stage and metamorphosis are lost, and instead a fully formed precocial young emerges from the egg. Amniote eggs are larger than those of most nonamniotes, with larger volumes of yolk. As the embryo grows, its size produces special problems with respect to metabolic intensity, the exchange of respiratory gases, structural support, and the mobilization and transport of nutrients (Packard and Seymour, 1997; Stewart, 1997). These are mediated by an outer eggshell made of semipermeable collagen fibers and varying proportions of crystalline calcite, which permits respiration while preventing desiccation. The eggshell provides a calcium repository for the developing skeleton, and the embryo is also equipped with several novel extra-embryonic membranes. The amnion encloses a fluid-filled cavity in which the embryo develops. The allantois stores nitrogenous wastes, and the chorion is a respiratory membrane (Packard and Seymour, 1997; Stewart, 1997). A single penis with erectile tissue is also apomorphic of Amniota (Gauthier et al., 1988a; Rowe, 2004).

The amniote skeleton: Whereas aquatic vertebrates are effectively neutrally buoyant, those who successfully moved onto land faced the new effects of gravity, which increases exponentially with increase in body size such that kinetic energy scales to the fifth power of linear dimension (McMahon and Bonner, 1983). This may explain why the first amniotes were generally small. Responses to the effects of gravity are evident throughout the early amniote skeleton. Amniotes initiated the reversal of a mode of variation established in many aquatic clades that had led to increased complexity and numbers of individual bones. Instead, amniotes and pan-mammals carry forward a trend toward the simplification of the skeleton as a means of increasing its strength, by consolidating primitively compound structures into simpler elements. Strengthening the skeleton was key to feeding and locomotion diversification and increases in body size that occurred independently many times in amniote history. Skeletal simplification occurred through ontogenetic repatterning of regions of the skeleton in which some bones fail to differentiate altogether and a single element grows in their place, or where separate bones differentiate at an earlier time in ontogeny and quickly fuse.

Like their ancestors before them, the first amniotes and earliest pan-mammals were macropredators. But they no longer lived in a water column where the primitive behavior of gape-and-suck feeding could assist capturing and swallowing prey. Early amniotes, and especially early pan-mammals, had elongated faces, and their mouths became prehensile organs for seizing prey items. Their jaws were lined with powerful, conical marginal teeth. Present in the upper jaw were one or two enlarged caniniform teeth, and additional teeth were present on bones of the palate in the roof of the mouth (Fig. 11.12). Teeth were shed and replaced throughout

life. Large vertical flanges were formed by palatal bones alongside the inner surfaces of the jaws (pterygoid flanges). These served as a secondary, sliding articulation that prevented the mandible from twisting about its long axis when the mouth was open, and it stabilized the jaw and helped to direct the application of muscular forces to the teeth. The mandible was simplified by the loss of some of the numerous bones it once contained, and strengthened in the process. The primitive mode of ingesting food was inertial swallowing, lunging forward the head and mouth over a prey item, assisted by palatal teeth on the roof of the mouth. Inertial swallowing persists in many reptiles, and it persisted over a long segment of early pan-mammalian history. But this would profoundly change in mammals and their closest extinct relatives.

A consistent relationship between early ontogenetic segments of the brain and individual bones of the skull became deeply entrained in amniote and pan-mammal development patterns coincident with overall simplification of the numbers of bones forming the skull. For example, the frontal bone consistently covers the derivatives of the embryonic prosencephalon (viz., telencephalon and olfactory bulbs, diencephalon), and within pan-mammals later took on a constant relationship to the eyeballs and orbital rim, while the supraoccipital and basioccipital bones surround the derivatives of the rhombencephalon (viz., cerebellum and medulla). Dorsally, the parietal lies over the cerebellum and mesencephalon, and ventrally the basisphenoid lies beneath the hypophysis. Once established, these relations generally held across pan-mammals. However, we will see them disrupted from time to time, as new bones were added and others displaced or lost altogether in association with episodes of encephalization and new patterns of relative growth.

With loss of a larval stage and functional gills, ventilation in amniotes was achieved by coopting the former pharyngeal skeleton into a branchial pump, as the lungs became the primary or sole site of metabolic gas exchange. The series of former gill arches became variously modified, generally through reductions in the numbers of separate arches and elements per arch (Goodrich, 1930; Romer, 1956). In the ancestral amniote and for a long segment of pan-mammalian history, the pharyngeal skeleton supported musculature of the branchial pump that was the main driver of ventilation. These bones would later be converted to augment mobility of the tongue and unique swallowing behaviors, and in mammals they became the ossicles of a uniquely configured impedance-matching middle ear.

Another trend especially evident in the axial skeleton was regional specialization (Goodrich, 1930; Romer, 1956). The amniote skull articulated with the vertebral column via a specialized cranivertebral joint that gave

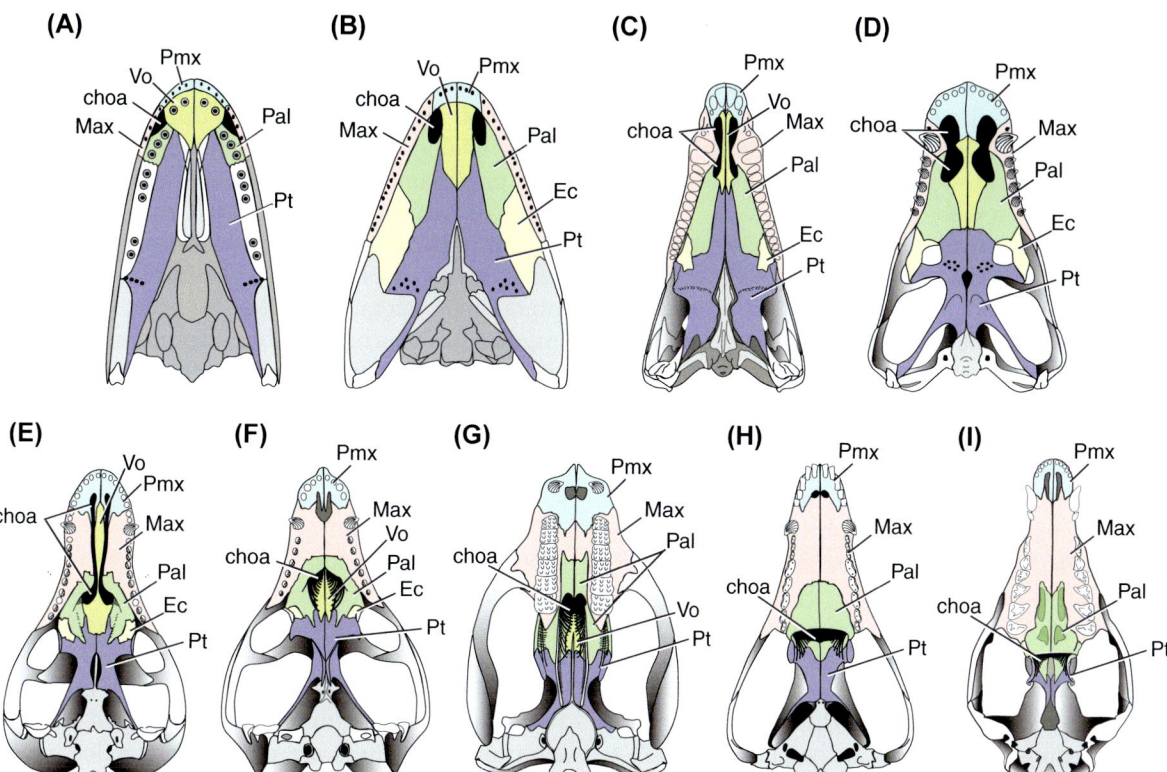

FIGURE 11.12 Stages in the evolution of mammalian secondary palate and the orthoretronasal olfaction duality. (A) *Eusthenopteron*, a stem tetrapod; (B) *Seymouria*, a stem amniote; (C) *Dimetrodon*, a basal synapsid; (D) *Syodon*, a more advanced noncynodontian synapsid; (E) *Procynosuchus*, the basal-most cynodont with an incipient secondary palate; (F) *Thrinaxodon*, an early cynodont with a complete secondary palate; (G) *Kayentatherium*, a basal mammaliamorph with a complex dentition; (H) *Morganucodon*, a basal mammaliaform, with secondary palate extending to back of tooth row; (I) *Didelphis*, with secondary extending behind tooth row. See Table 11.1 for key to abbreviations. *After Rowe, T.B., Shepherd, G.M., 2016. The role of ortho-retronasal olfaction in mammalian cortical evolution. J. Comp. Neurol. 524, 471–495. https://doi.org/10.1002/cne.23802.*

some measure of stable, increased mobility of the head. It involved a semispherical occipital condyle of the skull, built by the median basioccipital and the paired lateral exoccipital bones that fit into a shallow socket formed by specializations of the first two vertebrae, known together as the "atlas–axis complex." A primitive neck, which enabled the head to be raised, can be traced into early pan-tetrapods (Gauthier et al., 1989). But early amniotes carried elaboration of the craniovertebral joint and specialization of the cervical vertebrae a good deal further, facilitating prey capture, drinking, and inertial swallowing. Movement of the head also served to expand sensory horizons and refined directional perception of their sensory systems (discussed later). The early craniovertebral joint was positioned immediately below the foramen magnum, which affords passage of the spinal cord from the endocranial cavity. A combined design requirement of the occipital condyle and atlas–axis complex is to ensure that the spinal cord was not stretched or kinked by head movement. At many points in pan-mammalian history, the craniovertebral joint was subjected to subtle modifications that balanced the seemingly conflicting demands of increased head mobility and evolving posture

with increases in the diameter of the spinal cord that accompanied encephalization and appearance of the specialized corticospinal projections characteristic of living mammals.

The vertebral column also continued the pan-tetrapod trend toward regionalization of distinct thoracic and sacral regions. A single sacral vertebra appeared in stem tetrapods to attach the vertebral column to the pelvis and hindlimb, while the ancestral amniote had two sacral vertebrae, and more were added by early pan-mammals with size increases and diversified modes of locomotion. It is noteworthy that the spinal cord is itself a segmented structure at multiple levels of organization and that each segment forms a dorsal (afferent) and ventral (efferent) spinal nerves that correspond in the neck and trunk to the numbers of vertebral segments. Moreover, there are cervical and sacral enlargements associated with the paired limbs, and generally their size is correlated with the lengths of the corresponding extremities (Nieuwenhuys et al., 1998).

The early amniote and pan-mammalian vertebral column was robust compared to more primitive tetrapods, but it retained the plesiomorphic design in which each vertebral centrum was a composite of two elements (pleurocentrum, intercentrum) that were concave on either face (amphicoelous), and they retained a small central penetration for a persistent remnant of the notochord (notochordal vertebrae). Ribs extended along the neck to the pelvis, and most were double-headed with heads connected by a web of bone.

The limbs were relatively longer than in the first tetrapods, but by modern standards they were still very short and sprawled to the sides, and the mode of locomotion was altogether different from any living amniote. The forelimbs were exceedingly robust. The pectoral girdle no longer had direct attachment to the back of the head and was attached to the trunk via a muscular sling. The scapula was a broad plate of bone, and there were two coracoid ossifications in the shoulder girdle which collectively offered a broad attachment for the musculature of the forelimb. The humerus was a twisted, propeller-like bone that rotated at the shoulder. The hindlimb was generally shorter and less robustly developed than the forelimb. The hands and feet were angled outward with the fourth digit the longest. The hands and feet have finally arrived at a pattern of five digits each, with the phalangeal formula of 2-3-4-5-3 in the hand, and 2-3-4-5-4 in the foot (Romer, 1956; Gauthier et al., 1988a,b).

Axial and intercostal musculature provided the major force in locomotion, via sequential waves of undulating contraction passed along the trunk. Rotation of the humerus helped pull the body forward, while the hindlimbs served as props against the substrate for propulsion via axial undulation, and powerful muscles at the base of the tail and pelvis rotated and retracted the femur a short distance to push the body forward. Highly sculpted joint surfaces at the shoulder and hip indicate a combination of rotation and swing in a more or less horizontal plane, while the hands and feet pushed both to the side and backward while walking. This pull-squirm-push mode of locomotion in early amniotes differed from early tetrapods only in the greater degree of effectiveness of their slightly longer limbs, and by added stability from their longer and more robust tails. Still, the early amniotes and pan-mammals were sufficiently mobile to achieve a wide geographic range across Pangaea's equatorial belt by the Early Permian.

The ancestral amniote lacked bony scales in the skin and had ventral ribs (gastralia) imbedded in the skin of stomach to protect the viscera and to strengthen abdominal muscles that played the role of antagonist to the axial musculature during locomotion. By the standards of their later descendants, the first amniotes and early pan-mammals were limited in speed, agility, and in gait repertoire. They could walk and could probably still swim, but it is doubtful that they could run. From such an ancestor, running, galloping, jumping, hopping, climbing, gliding, diving, and flying would emerge over the subsequent ~310 million years of endurance and diversification played out by its descendants.

The amniote peripheral sensory system: The nature of information available on land is different in degree from that available in the water, and many aspects of amniote and pan-mammalian evolution can be explained in terms of increased or decreased sensitivity and range in their particular sensory modalities. For example, having moved out of the water, amniotes quickly lost the lateral line system, which had served to detect electrical impulses and turbulence in surrounding water. Neither signal is well conducted through air. It is remarkable, however, that some 200 million years later within Mammalia would evolve de novo compensatory systems. For example, an electroreceptive system of considerable sensitivity evolved in monotremes (Augee and Gooden, 1992; Pettigrew, 1999; Rowe et al., 2008), while the vibrissae and whisking system of some therian mammals became so sensitive that it can detect subtle air turbulence in their immediate environments (Catania and Catania, 2015).

Vision may have led the other senses in the first amniotes and pan-mammals. Compared to most aquatic environments, there are many more reflective surfaces on land and less light scatter or absorption in air. With more light energy in its new environment, the ancestral amniote crossed over a new threshold of information conveyed via vision, being diurnal with a pure cone retina (Walls, 1942). It may have traded light sensitivity for a marked increase in visual acuity and sharp resolving power, as predaceous vertebrates generally require sharp vision to pursue and capture prey. The distribution of opsin genes suggests that the ancestral amniote had trichromatic color vision (Davis et al., 2007; Wakefield et al., 2008), and that separate motor nuclei directed movement of the muscles that moved its eye (Wall, 1942). A notion deeply entrenched in the literature is that mammals evolved in a nocturnal milieu. However, the discovery of three opsin genes in the platypus and echidna suggests that the ancestral mammal, as well as the entire extinct mammalian stem group, was diurnal with trichromatic vision supported by the SWS1, SWS2, and M/LWS genes (Davies et al., 2007; Wakefield et al., 2008). Dichromatic nocturnality evolved later, following the divergence of monotremes and therians, with loss of the SWS2 gene in pantherian mammals. Remarkably, trichromatic vision returned in Old World primates thanks to a gene duplication (Wakefield et al., 2008).

Hearing in all vertebrates is based on transduction of environmental acoustic information by the sensory hair cell of the inner ear. The apical end of this cell projects into the lumen of the otolithic chamber and has a series of cilia that, when bent, produce a receptor potential in the hair cells that stimulates neurons of the auditory nerve (cranial nerve VIII) to auditory neurons in the brain stem (Roberts et al., 1988). In fishes, this function can be augmented by the lateral line system, and in teleosts by the swim bladder (derived from paired lungs) as well. A very wide range of acoustic adaptations have been described in fish (Coffin et al., 2014), but hearing in all vertebrates fundamentally involves frequency discrimination, enhancement of signal to noise ratio, and sound localization (Popper and Lu, 2000).

Amniotes never converted their lungs into a swim bladder and have lost the lateral line system, and the sensitivity and resolving power of hearing in the ancestral amniote and pan-mammal must have been profoundly diminished in the transition from waterborne sound to airborne acoustic information. Nevertheless, the ancestral amniote can be inferred to have conserved basic functions of hearing involving frequency discrimination, signal to noise ratio enhancement, and sound localization, as well as the plesiomorphic transmission pathway involving the transduction of acoustic information by sensory hair cells of the inner ear via the auditory nerve to brain stem auditory neurons (Carr and Soares, 2009; Carr and Christiansen-Dalsgaard, 2016).

Having left the water, hearing remained important in tetrapods and amniotes, and the fossil record leaves no doubt that a middle ear evolved independently in amphibians, reptiles, and mammals (Gauthier et al., 1988a,b, 1989). In each clade, the middle ear has its own distinct anatomical structure, and neural mechanisms for sound localization are also different (Carr and Soares, 2009). However, in each case, the middle ear was derived developmentally from elements of the first and/or second branchial arches, and each involved a tympanic membrane connected via a lever system of bone and/or cartilage to the fluid-filled inner ear that effectively coupled airborne sounds to the fluid of the inner ear (Grothe et al., 2005, 2010). In light of the differences in development, mature anatomy, and evolutionary pathways of these tympanic middle ears, we are left to speculate that terrestrial hearing in early amniotes was probably limited at first to receiving mechanical vibrations from the ground via jaws and branchial arch skeleton as these animals rested their heads on the ground. This may explain the independent derivation of an impedance-matching middle ear from components of the branchial arches.

As ventilation and metabolic gas exchange transformed from the medium of water to that of air, profound changes took place in the olfactory system. Amniotes inherited a dual olfactory system that can be traced back to the ancestral gnathostome. It consists of the main olfactory system, which was at first broadly sensitive to environmental odorant molecules in the water column, and became adapted to airborne odorant molecules with the transition onto land. Its axons make their first synapse on the rostral extremity of the telencephalon to induce differentiation of the main olfactory bulb, which in turn induces differentiation of the olfactory (piriform) cortex (Farbman, 1988, 1990; Schlosseer, 2010). The vomeronasal system was also inherited from the ancestral gnathostome. Its receptors were at first intermingled with those of the main olfactory system on the olfactory epithelium, but they are activated primarily by pheromones, and their axons form the "terminal nerve" (cranial nerve 0) and make their first synapse in the accessory olfactory bulb (Demski, 1993; Demski and Schwanzel-Fukuda, 1987). Odorant receptors of the main olfactory system and vomeronasal system are encoded by entirely separate subfamilies of odorant receptor genes. In the first stem tetrapods, the formerly diffuse vomeronasal receptors became organized and encapsulated within the vomeronasal organ, but their synaptic pathway remained the same.

In the main olfactory system of amniotes, genes that once coded for olfactory receptors activated by water-soluble molecules were mostly lost or were transformed into a new family of genes that encode odorant receptors activated by airborne odorant molecules. Current estimates suggest that approximately 100 odorant receptor genes were expressed in amniotes ancestrally and that in different amniote clades these founding genes underwent considerable diversification as olfactory ecology became a driving mechanism that influenced the histories of many different clades (Niimura and Nei, 2005, 2006; Niimura, 2009, 2012). Nowhere was olfactory diversification so extensive as it was in the history of pan-mammals, where the number of OR genes increased to an estimated 1200 by the time crown Mammalia originated, and further diversification occurred in subclades within the crown. With a few notable exceptions, the mammalian vomeronasal system was more conservative.

The ancestral amniote and early pan-mammals had small external nostrils that were directed laterally. Their internal nostrils (choanae) formed small openings near the front of the palate, and the space between nostril and choana afforded space for only a small nasal capsule and olfactory epithelium (Fig. 11.12). Olfaction consisted of only orthonasal olfaction, in which environmental scents were inhaled across the olfactory epithelium.

With the origin of Amniota, airflow through the adult derivatives of the nasal capsule became tied to two distinct functions. Each was supported by a primary "choncha" or an epithelial fold supported by a low ridge of cartilage protruding into the lumen from the lateral wall of the nasal capsule. The anterior choncha consists of mucociliary respiratory epithelium and represents the primordium of the mammalian respiratory turbinal, while the posterior concha comprises olfactory epithelium and represents the primordium of mammalian olfactory turbinals (Gauthier et al., 1988a). Delicate and highly elaborate skeletons arose in the last common ancestor of crown Mammalia to support hypertrophied respiratory and olfactory epithelia (discussed later).

Bony scales were lost from the skin in amniotes, and in their place are neural crest placode-induced epidermal structures that would eventually evolve into mammalian hair and reptilian scales and feathers. Amniote placodes share common spatial expression of placode molecular markers such as *Shh*, *Ctnnb1*, and *Edar*, as well as conserved localized signaling in the dermis underlying the placode by *Bmp4*, suggesting shared common ancestry (Di-Pöi and Milinkovitrch, 2016). With the appearance of placode-induced epidermal structures, so began an amazing diversification of integumentary specializations that would eventually evolve to prevent water loss, to protect the skin from solar radiation, to enhance sensory perception over the surface of the body and the space around it, to insulate the body, to assist locomotion, to provide camouflage, to attract mates, and to perform other functions. These placode-driven structures are innervated, and in mammals the tactile signals from developing hair follicles induce topotypic representation maps on the somatosensory cortex (Sengel, 1976; Zelená, 1994; Rowe et al., 2011). Remarkably, the fossil record preserves several lines of evidence allowing inferences regarding the timing of evolution of mammalian hair and that its full expression is correlated with a pulse of encephalization that probably indicates the emergence of neocortex (discussed later).

The amniote brain: The fossil record leaves no doubt that the ancestral amniote had a very small brain compared to living mammals (and birds). Only the floor and rear parts of its braincase were ossified, with a cerebellum that was wider than the forebrain. The olfactory bulbs were small and mounted on long stalks, a fairly large pineal eye was present, and the midbrain was exposed dorsally between the telencephalon and cerebellum (Romer, 1956; Jerison, 1973). One point that the fossil record of amniotes makes abundantly clear is that the evolutionary increases in encephalization that are so striking in birds and mammals occurred independently, from a small-brained ancestor. However, the picture typically painted by paleontologists (eg, Olson, 1944; Romer, 1956) of the brain in early amniotes and pan-mammals tends to understate the dimensions of neurosensory organization assembled from the comparative neurology of living amniotes.

Telencephalon: The telencephalon in the ancestral amniote consisted of four basic divisions that surrounded the ventricle. The piriform (olfactory) cortex was positioned laterally, the hippocampus formed the telencephalic medial wall, the telencephalic roof or pallium formed the dorsal cortex, and the basal ganglia differentiated in the telencephalic floor. Dorsal cortex, piriform cortex, and hippocampus each consisted of three layers. Peripheral afferent projections to the dorsal and piriform cortex coursed over the outer layer, while efferents project from the inner layer (Fig. 11.13). However, in mammals, in which the dorsal cortex is elaborated into neocortex, peripheral afferents may connect to the innermost layers of neocortex, in a fundamental reorganization of connectivity to and from dorsal cortex. In all amniotes, projections from the dorsal cortex

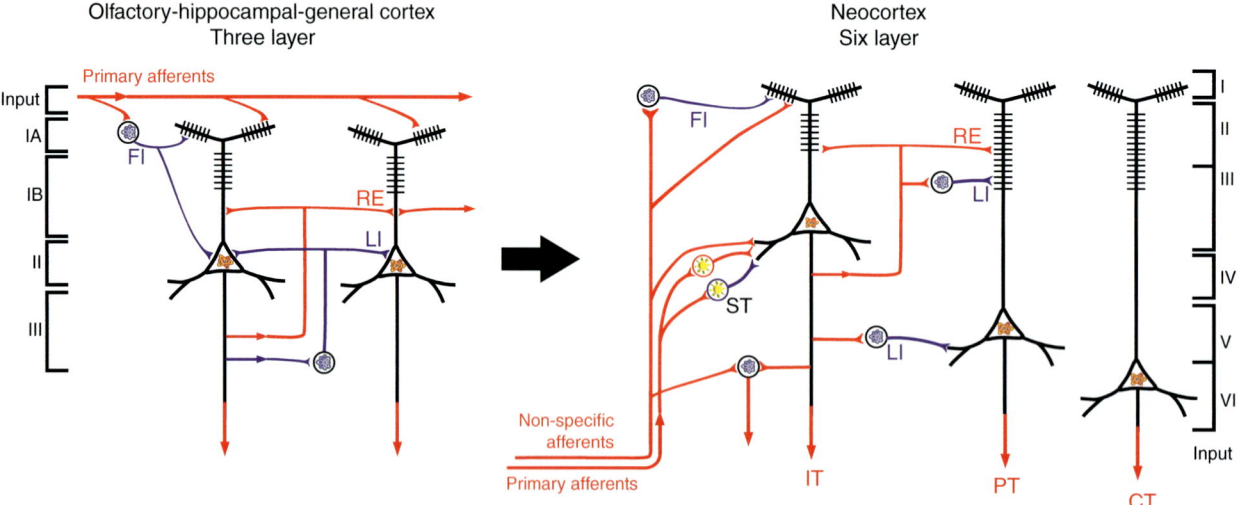

FIGURE 11.13 Circuit diagram showing three-layer cortex characteristic of Amniota ancestrally, and the six-layer neocortex characteristic of Mammalia. Pyramidal neurons are illustrated in *black* and interneurons in *blue*. *FI*, forward inhibition; *LI*, lateral inhibition; *RE*, recurrent excitation; *IT*, intratelencephalic projections; *PT*, pyramidal or corticospinal tract projections; *CT*, corticothalamic projections. *Modified after Shepherd, G.M., 2011. The microcircuit concept applied to cortical evolution: from three-layer to six-layer cortex. Front. Neuroanat. 5 (30), 1–15; Rowe, T.B., Shepherd, G.M., 2016. The role of ortho-retronasal olfaction in mammalian cortical evolution. J. Comp. Neurol. 524, 471–495. https://doi.org/10.1002/cne.23802.*

innervate the basal ganglia and brain stem, but in mammals, some project directly into the spinal cord as well. These corticospinal projections are associated with emergence of the neocortex and the mammalian pyramidal or corticospinal tract. One of the enduring questions of evolution is when and how the six-layered mammalian neocortex evolved from the three-layered cortex of the ancestral amniote (Fig. 11.13). As we shall see, the fossil record provides indications that the emergence of neocortex occurred late in the first half of pan-mammalian evolution and that it was correlated with elaboration of the olfactory system, an agile locomotor system, differentiation of integumentary hair, and to lesser degrees with other peripheral sensory systems.

The principal cells in the amniote forebrain are pyramidal cells. This cell type is present in amphibians but it lacks basal dendrites, whereas in amniotes the basal dendrites are not only present but also have become extensively branched and interconnected in a vast synaptic web (Striedter, 2005). In mammals, the pyramidal neuron populations are greatly expanded, and their cell bodies are densely packed and lie at several different layers in the six-layered neocortex. Pyramidal cells are present in the forebrains of all reptiles except birds, where they were secondarily lost (Striedter, 2005). The amniote dorsal cortex had a ventricular zone throughout its extent and a subventricular zone in its lateral regions, from which neurogenesis occurred in an inside-out pattern. Neurogenesis proceeded throughout much of ontogeny. In the ancestral amniote, there was neither a six-layered neocortex, nor a dorsoventricular ridge, but the basic neurogenerative pattern was established giving a degree of radial organization to the forebrain. This would be carried to its greatest extreme in mammals (Rakic, 1988; Butler, 1994).

In terms of its basic circuitry, the olfactory cortex has a similar neural organization in turtles and lizards (Ulinski, 1983; Bruce, 2007, 2009; Bruce and Braford, 2009) as in monotremes, marsupials, and placentals (Ashwell, 2010, 2013; Shepherd, 2011), supporting the inference that this organization was present in amniotes ancestrally. The olfactory cortex receives input from the olfactory bulb in the form of an odor image and transforms it into a central representation as an odor object (Shepherd, 1991; Wilson and Stevenson, 2006). Studies of opossum and rodent neurophysiology have shown that output fibers from the olfactory bulb course over the surface of the piriform (olfactory) cortex, emitting collaterals that terminate on the most distal dendrites of the pyramidal cells (Fig. 11.13). They excite both the spines of these cells and the smooth dendrites of interneurons which feed inhibition forward onto the pyramidal neurons. The activated pyramidal cells through their axon collaterals feed recurrent excitation back onto themselves and onto neighboring pyramidal neurons. They also excite

interneurons which feed inhibition back onto themselves and onto neighboring pyramidal neurons. Thus in the two critical operations of the circuit, processing of the input and processing of the output, excitation and inhibition are balanced (Shepherd, 2004; Rowe and Shepherd, 2016).

Anatomical and physiological studies in the hippocampus have shown that across amniotes the neurons and circuits are similar to those in the olfactory cortex, with similar long association fibers and interconnections for excitation and inhibition (Connors and Kriegstein, 1986; Haberly, 2001). In these regards, the intrinsic organization of olfactory cortex and hippocampus are similar to higher association cortical areas, for example, the face area of inferotemporal cortex (Haberly, 1985). Numerous studies show a close similarity between the intrinsic organization of the hippocampus and the basic organization of the olfactory cortex, in terms of layering of inputs on the apical dendrites and long association fibers (Neville and Haberly, 2004). Since the inputs to the hippocampus consist exclusively of central sites in the limbic regions, it is clear that the three-layered hippocampus was devoted to higher order processing such as learning and memory from the very start of amniote and pan-mammalian evolution (Rowe and Shepherd, 2016).

It is generally recognized that the three-layered dorsal cortex was the precursor of the six-layered neocortex in mammals (Fournier et al., 2014; Molnar et al., 2014; Montiel et al., 2016; Rowe and Shepherd, 2016). A pioneering anatomical study by Smith et al. (1980) on three-layered dorsal cortex of the turtle *Pseudemys scripta* reported connections through an interneuron that provide for feedforward and lateral inhibition. This was followed by an electrophysiological study which incorporated feedback and lateral excitation and inhibition into the local circuit (Kriegstein and Connors, 1986; Connors and Kriegstein, 1986). Close similarity across amniotes of this local circuit to the olfactory and hippocampal circuits pointed to a "basic circuit" or a "canonical circuit" common to all three forebrain regions (Kriegstein and Connors, 1986; Shepherd, 2011). This does not mean that each region does not have its own fine tuning for its particular types of input, but rather that there is a basic framework common to all three. Comparison with three-layer hippocampal and olfactory cortices suggests that, like those regions, three-layer dorsal cortex performs higher level association on visual input. In this view, the three-layer dorsal cortex from which the neocortex became elaborated was not a "simple" cortex for low-level processing, but rather had an organization that integrated high-level association functions analogous to those in olfactory cortex and hippocampus. Thus, evidence from mammals, lizards, and turtles indicates that these higher order cortical functions were present at the origin of Amniota (Rowe and Shepherd, 2016).

Thalamus: The thalamus switches circuits passing in both directions from the cortex to postcranial skeleton and muscles and from the body surface. Compared to other tetrapods, amniotes have an expanded and highly differentiated thalamus, consisting of the dorsal thalamus, ventral thalamus, hypothalamus, and epithalamus. This central relay center of the brain takes on a new level of complex organization in amniotes, one that is further elaborated during pan-mammalian history in association with the emergence of neocortex. Amniotes have an elaborated dorsal thalamus that is larger and contains many more individual cell masses or nuclei than anamniotes (Butler, 1994; Nieuwenhuys et al., 1998). Highly characteristic of amniotes is differentiation of the dorsal thalamus into two principle regions, the collothalamus and the lemnothalamus (Butler, 1994). In amniotes, both regions differentiate further into discrete nuclei that function as a complex of way stations interposed between the environmental sensory world and the telencephalon. In amniotes the collothalamic nuclei relay visual, auditory, and somatosensory inputs into the telencephalon.

A large series of nuclei differentiate in the amniote lemnothalamus, and this is carried to its extreme degree in mammals. The integral linkage that dorsal thalamus would eventually acquire in mammals is demonstrated in oblation experiments in which the neocortex was removed, with the result that all of the dorsal thalamic projection neurons died, leaving the dorsal thalamus severely degenerated (Kaas, 2009). In combination with the emergence of neocortex, thalamic differentiation endowed mammals with a remarkable capacity for detailed analysis of their environment (Nieuwenhuys et al., 1998).

The amniote hypothalamus differs from anamniotes in receiving input from the limbic system of the forebrain, from those regions with responsibility to memory and the resonance of experience (Butler, 1994; Butler and Hodos, 1996). Many functions of the hypothalamus are tied to light, to the daily cycle of light from dawn to dusk; the influence of light on the hypothalamus extends to seasonal variability, to the shorter winter days and longer summer days; but whether this was present ancestrally is difficult to infer since most fossils are known only from paleoequatorial regions of Pangea, with the least degree of seasonal variation (Parrish et al., 1986). The hypothalamus regulates water balance by directing kidney function—a crucial process of these terrestrial vertebrates. The hypothalamus also controls the production of hormones involved in reproductive physiology, involving the movement of ova in the oviduct, contractions of muscles of the reproductive organs, and many behaviors involved in courtship.

Adult neurogenesis: In amphibians and reptiles, neurogenesis occurs throughout life in many regions of the telencephalon (Chapouton et al., 2007). In reptiles, neurogenesis proceeds in all major subdivisions of the adult telencephalon and occurs to a much lesser extent within the cerebellum, while in frogs and salamanders, neurogenesis occurs in the telencephalon, preoptic nucleus, and tectum. It usually takes place in subdomains of the ventricular zones of several brain areas and can serve multiple adaptive functions. However, in mammals, adult neurogenesis is limited to the rostral migratory stream and dentate gyrus of the hippocampus (Ming and Song, 2005). This basic new pattern of neurogenesis in which extraordinarily rapid proliferation is almost completely restricted to early ontogeny and was tied to remarkable neuronal longevity that contributed to a unique sense of memory (Rakic, 2007, 2009) over the course of pan-mammalian history.

Spinal cord: The amniote spinal cord is thick and extends through the entire length of the vertebral column in most mammals. It has more different types of cells than nonamniotes, and many of these secondary neurons send axons across the midline to the contralateral side for left-right coordination of movement (Butler, 1994). A distinct lateral column of motor neurons provides innervation to the limbs; and there are now expanded cervical enlargements (segments 7–10) and lumbosacral enlargements (segments 19–22) that represent the initial integrating centers of the brachial and sacral plexi, which innervate muscle complexes during locomotion and control reflexive action in the limbs. In addition, the autonomic neuronal groups (ie, "fright and flight reflexes") of the brain stem and spinal cord were highly developed, indicating that the spinal cord is performing more internal decision-making processes that are independent of the brain (Streidter, 2005).

In summary, judging from living tetrapods, the ancestral amniote neurosensory system enjoyed an increase in numbers of genes, more neuronal types and more complex pyramidal cells with greater interconnectivity, faster rates of neuron proliferation that produced a larger forebrain, and elaboration in complexity and computing power of the brain compared to the first tetrapods. Overall, it controlled more highly coordinated body movements using a more complex muscular system. While it had abandoned the ancient lateral line system, it had begun to integrate peripheral information from more acute visual and olfactory systems, and its processing of sensory cues involved the higher level operations. This underscores that three-layer cortex of amniotes ancestrally was not a primitive "simple" cortex, but rather that it operated at the level of higher order associations underlying analysis, discrimination, learning, and memory (Rowe and Shepherd, 2016), and a remarkable capacity for detailed analysis of their environment (Nieuwenhuys et al., 1998). Amniotes were probably more reflective and introspective of experience, using a more highly developed sense of memory as a guide to action, including reflexive actions (Butler, 1994; Butler and Hodos, 1996).

Such was the general organization of the ancestral amniote skeletal and neurosensory systems. This account presents a considerably more sophisticated neurosensory system than traditionally depicted in accounts confined to evidence from the fossil record. From this point of departure we now turn to the fossil record of Pan-Mammalia and what it allows us to infer of neurosensory evolution culminating in the origin of crown clade Mammalia.

11.5 Pan-Mammalian History

11.5.1 Node 1: Synapsida

The namesake diagnostic feature of Synapsida is the presence of a single temporal fenestra, bounded below by the homologue of the mammalian zygomatic arch (Fig. 11.14A). At first the arch was built by the squamosal, jugal, and postorbital bones, but the latter element was reduced and lost prior to the origin of mammals. The single fenestra and arch comprise the "synapsid condition." This name was long used for the extinct, paraphyletic stem group (Romer, 1966; Carroll, 1988) and later converted to represent the monophyletic total clade of Mammalia (Gauthier et al., 1988a,b). In the following account, the name Synapsida is restricted to the node stemming from the last common ancestor that mammals share with Ophiacodontidae, Caseidae, and Varanopsidae (of doubtful monophyly), three clades whose relationships to one another are controversial. Synapsida, so defined, may or may not include a number of ancient fragmentary pan-mammals known only from isolated postcranial material that may lie basal to the synapsid node within Pan-Mammalia.

The synapsid temporal fenestra allowed the mandibular adductor musculature room to flex and expand as it snapped closed the jaws. At first the fenestra faced laterally, but it would eventually become open dorsally (see Section 11.5.4), in association with an increase in adductor muscle mass, its differentiation into subunits (temporalis, masseter, and the pterygoideus group), and increase in the complexity of mandibular function. At maturity, most of the early pan-mammals had longer faces than other early amniotes, with more than half of the skull lying in front of the orbits, and a jaw articulation displaced to a level behind the occiput to further widen the gape of the jaws. The mouth was lined with a long

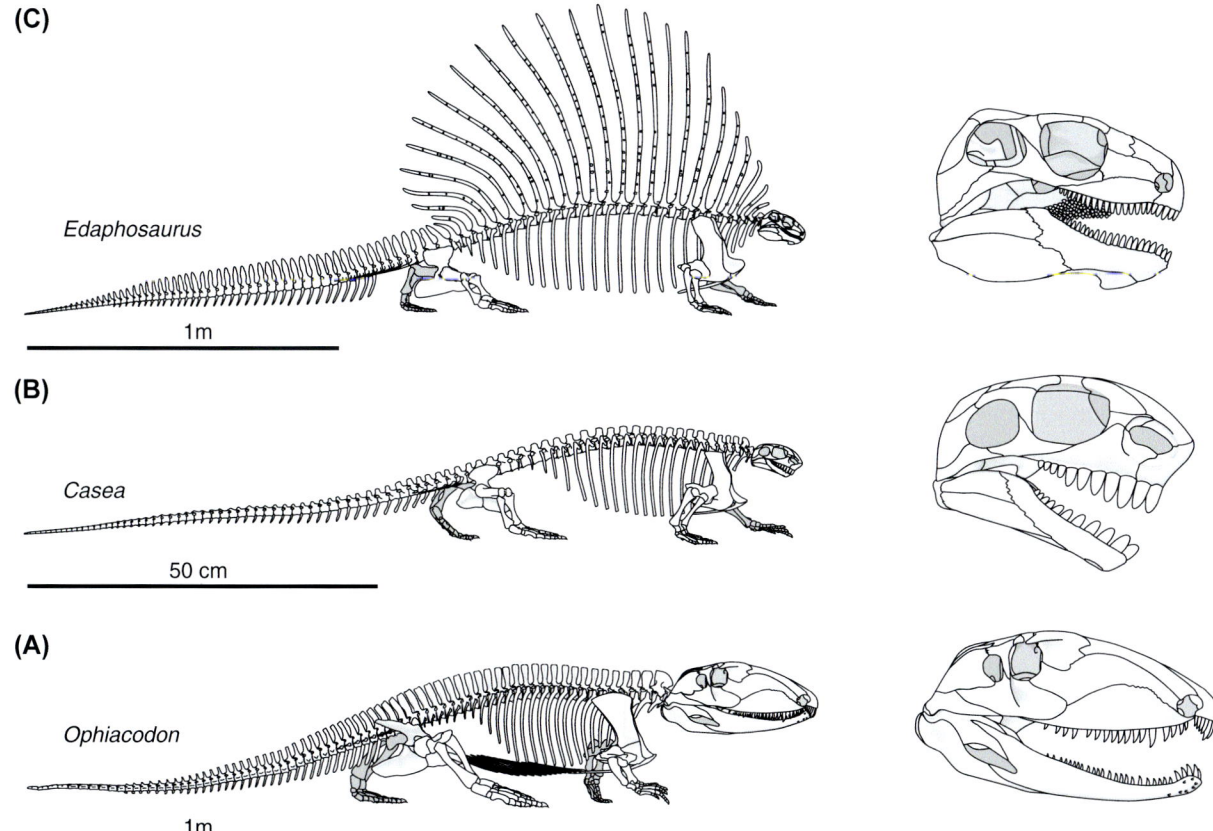

FIGURE 11.14 Skulls and skeletons of the early pan-mammals (A) *Ophiacodon*, (B) *Casea*, and (C) *Edaphosaurus*, drawn to the same lengths.

row of sharp, recurved teeth that were replaced continuously throughout life. Early pan-mammals had a faster and more powerful bite than the ancestral amniote.

From the start, the orbits were large and held large, mobile eyeballs, plus extrinsic musculature and adnexa. The latter consisted of bodies of fat and the lacrimal and Harderian glands that lubricate and moisten the eyeball. An important characteristic of living mammals involves their manner of eye movement. While its origin cannot be pinpointed in the fossil record, it is convenient to mention it here. As Gordon Wall (1942: pp. 310–311) described, "in the matter of eye movements, mammals are at once set off from all other vertebrates by the fact that whenever voluntary movements are possible at all, the two eyes are never independent but are always conjugated. This universal conjugation is associated with the fact that mammals (whales, rabbits, and some others excepted) examine things only binocularly — even the bats, small rodents, insectivores, and other nose- or ear-minded nocturnal forms whose eyes never move even reflexively. Where the eyes are placed laterally as in the rabbits, there usually is no area centralis, let alone a fovea, and there are no spontaneous movements at all. But even the rabbits have the gyroscopic reflex eye movement, including the optomotor reaction. These compensatory movements in mammals are always most extensive in the plane of greatest biological usefulness, which usually means horizontal…The voluntary eye movements of mammals are really best correlated with visual acuity, which, it so happens, does go pretty well with intelligence in this group of vertebrates."

Basal synapsids also had elongated choana (Fig. 11.12C), indicating a larger nasal capsule and olfactory epithelium. It was a subtle shift, but nevertheless marked the beginning of a long trend of skeletal modifications that hint at olfaction being a leading sense, behind that of vision. In addition, the angular bone of the mandible has become a tall, flat plate, with a deep keel. Only in retrospect can we see this as a first, almost imperceptible step in the evolution of the mammalian impedance-matching middle ear. However, some 20 million years or more would pass before a middle ear skeleton showing enhanced sensitivity evolved, and another 50 million before it gained sensitivity to high-frequency sounds (see Section 11.5.14). Thus, vision and olfaction led pan-mammalian senses for much of their early history.

Early pan-mammals showed increase in the power of the hindlimbs, with the two sacral ribs attaching to the ilium at a level above the acetabulum, lowering the hip joint beneath the vertebral column, and conveying greater stride and lunge capability. These were subtle changes, but enough for early pan-mammals to quickly become the largest and most commonly fossilized macropredators throughout the Permian.

Three taxa bear mention to convey the nature of early pan-mammalian diversification. Ophiacodontidae (Fig. 11.14A) includes *Archaeothyris*, the oldest pan-mammal recognized at present. Whether Ophiacodontidae or Caseidae (discussed later) is the sister taxon to all other synapsids is controversial. Ophiacodonts are distinct among early pan-mammals in having a hindlimb that is longer than the forelimb. In derived members of the clade, the wrists, ankles, hands, and feet are exceptionally broad, and some have flattened ungual phalanges that invited speculation that hands and feet were webbed and that ophiacodonts were piscivorous. Consistent with this is a deep incisure of the dorsal rim of the orbit directing the optic axis somewhat dorsally. Size increase characterized this clade, some reaching 3 m in length and approaching 200 kilos in weight. Ophiacodonts became extinct before the end of the Early Permain.

Caseidae (Fig. 11.14B) is a second basal pan-mammal clade that originated as small predatory creatures, but soon caseids grew in size and reached 4 m length and approached 500 kilos in body mass. The most derived caseids had proportionately tiny heads, with short faces supporting blunt marginal teeth suited for cropping vegetation, and voluminous abdominal cavities characteristic of living herbivores with large gastrointestinal fermentation tracts. Dense clusters of small palatal teeth and robust hyoid bones hint at an early attempt to masticate plants by grinding them with a muscular tongue against the roof of the mouth.

"Varanopsidae" is the third widely recognized basal group. Older accounts viewed varanopsids as a basal lineage of small predators. However, its representatives are rare and mostly based on fragmentary fossils from the Early Permian of the American southwest. *Varanops* is the best-known taxon, but all the material recovered is immature (Romer and Price, 1940: p. 270). It may be that the only character grouping "varanopsids" together is that the specimens so assigned were immature at time of death. More mature and complete fossils are needed to determine whether "varanopsidae" is monophyletic. By the Early Permian, basal pan-mammals were found all along equatorial Pangaea, from Utah and New Mexico to the Carpathian district of Russia. With geographic expansion came taxonomic diversification.

11.5.2 Node 2: Unnamed

The extinct side branch Edaphosauridae (Fig. 11.14C) is one step closer to the mammalian crown than those just mentioned. It is the first of two "sail-back" clades, whose members include independently developed elongated neural spines that project far above the trunk vertebrae. Speculation surrounding its function is described later. Features linking edaphosaurs closer to mammals than the groups just mentioned involve lateral placement of the orbit and establishment of the frontal bone in the orbital margin. This was the beginning of a constant relationship between the frontal bone and eyeball, one that maintained despite the eventual loss of other bones (prefrontal, postfrontal, and postorbital) that once participated in the orbital rim. The eyes now faced laterally and more distant horizons were coming into focus as the head was held a bit higher than before, although their binocular field was narrow if present at all. There were additional, seemingly minor modifications in relations between bones at the back of the mandible and at the craniomandibular joint that are difficult to assess functionally, except in the retrospective context of a long history of transformation that culminated in the mammalian middle ear. Emargination of the cheek had also begun, indicating that the masseter muscle had begun to differentiate and was making its way from the adductor chamber onto the face. Also present at Node 2 are three vertebrae in the sacrum of mature individuals, more strongly connecting the vertebral column to the pelvis and hindlimb.

Basal members of the edaphosaur clade had predaceous dentitions, but in more derived members the teeth were blunted and scissorslike. There also grew broad clusters of palatal teeth that were matched by broad shelves on the lower jaw that held opposing batteries of small teeth that could grind against the uppers. Edaphosaur ribs are curved to enclose a cylindrical gut cavity, but one less voluminous than in derived caseids.

11.5.3 Node 3: Sphenacodontia

Sphenacodontia is the clade stemming from the last common ancestor that mammals share with the extinct side branch Sphenacodontidae, the second "sail-backed" clade (Fig. 11.15A), which had arisen by ~300 million years ago. Its basal members have unspecialized neural spines, but the more derived sphenacodontines have elongate spines 1 m or more in length. Their skulls have a broad, powerful snout with a characteristic rounded profile that supported very powerful teeth with long canines and incisors. They were the largest and most charismatic macropredators of the Early Permian. A distinctive innovation arising in Sphenacodontia is a notch in the angular bone at the back of the jaw that freed a thin plate of bone which became reflected laterally, enclosing a narrow space between it and the jaw. This is the "reflected lamina" of the angular bone, and it is the distant, but recognizable homolog of the mammalian ectotympanic, the bone holding our tympanic membrane. Whether the notch above the reflected lamina held a functional tympanum is doubtful,

and the reflected lamina itself may have functioned as a very crude, insensitive tympanum. Once again its significance is clear only in retrospect, and apart from its anatomical relationships its overall size and form were unlike any auditory element in living mammals. Mechanically, the reflected lamina could only respond to loud, low frequency sound, and correspondingly the sacculus of the inner ear occupied only a shallow depression in the floor of the otic capsule. Sphenacodontids also possessed an enlarged coronoid eminence along the top of the mandible that incrementally increased leverage of the jaw adductors for bite forces exerted by the canines and incisors.

The edaphosaur and sphenacodontine clades independently evolved elongate neural spines along the neck and trunk. In sphenacodontines the spines are straight, whereas in edaphosaurs they are studded with short lateral tubercles. Doubtless, a web of skin connected the spines, and the name for these "sail-backs" arose in the distant past when it was thought that they literally functioned as sails to propel them across freshwater lakes. This strange apparatus surely hindered movement. Many specimens are known with neural spines that had broken and healed before death. Much speculation surrounds the sail-fin as an early experiment in endothermy. Any structure affecting surface area of the body, as the sail surely did, must affect thermal absorption and radiation. However, this was not the mechanism by which mammalian endothermy was achieved, and early pan-mammals lacking similar adornments gained wide geographic distribution. Sexual selection, difficult to test in this case, may be the best explanation.

11.5.4 Node 4: Therapsida

Therapsida is the clade stemming from the last common ancestor that mammals share with the mid-Permian Biarmosuchia and all its descendants. Most of the basal therapsid clades are represented in the fossil record by ~270 million years ago, in what is now the Old World. *Titanophoneus potens* is one of the best-known examples of a basal therapsid (Fig. 11.15B). Therapsida is diagnosed by characters of the skull related to enhanced sensory perception, further specializations for macropredation, and modifications in the postcranial skeleton that afforded more agile locomotion (Gauthier et al., 1988a; Rubidge and Sidor, 2001; Sidor, 2003; Kemp, 2005, 2006b).

The bones of the therapsid face were rearranged to suggest an increasingly forward or frontal axis of attention and activity and directional coordination of visual and olfactory fields (Fig. 11.15B). The nostrils are redirected from the lateral sides of the snout and now point anterolaterally, enhancing stereoscopic directional

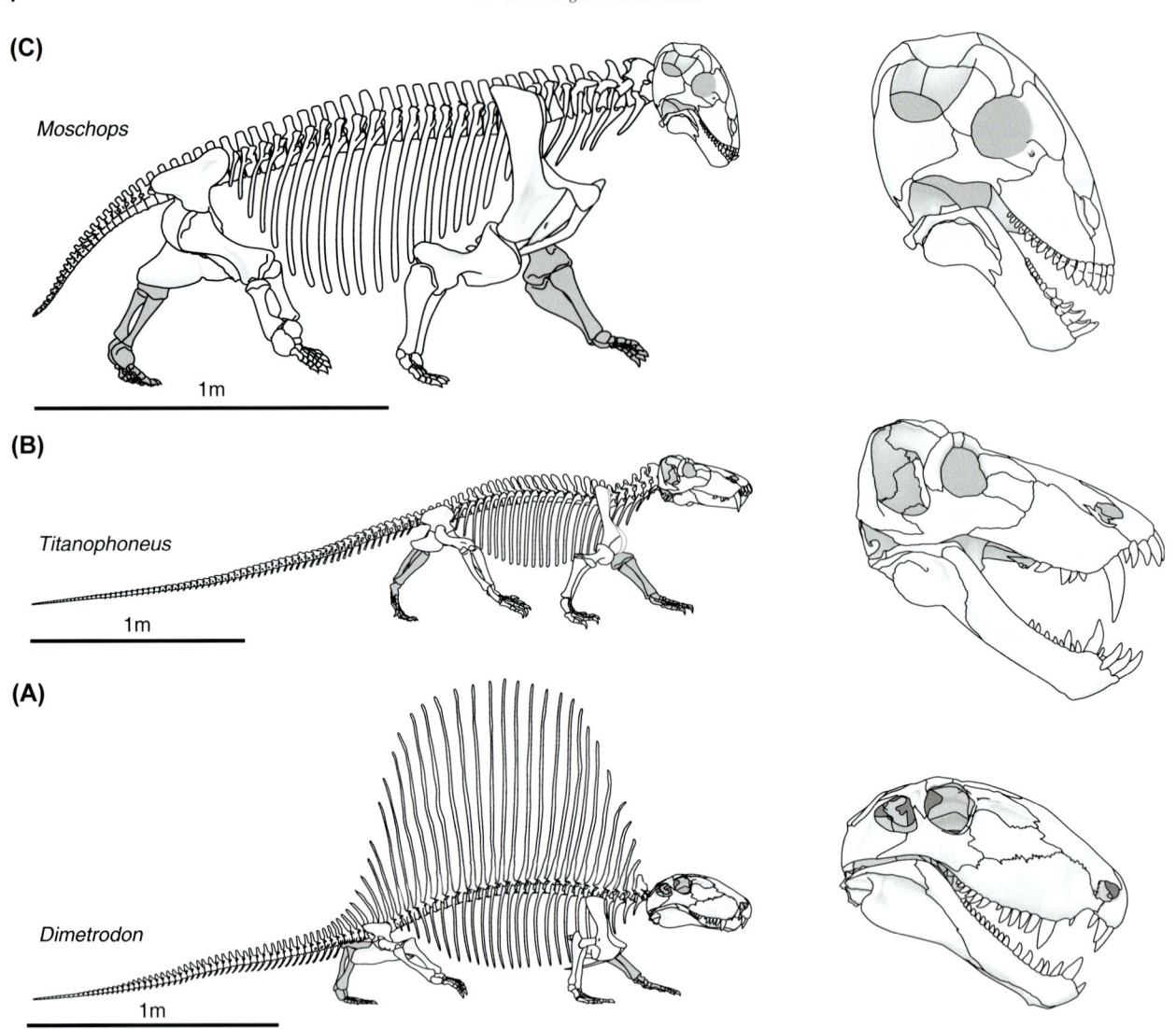

(C)

Moschops

1m

(B)

Titanophoneus

1m

(A)

Dimetrodon

1m

FIGURE 11.15   Skulls and skeletons of the early pan-mammals (A) *Dimetrodon*, (B) *Titanophoneus*, and (C) *Moschops*, drawn to the same lengths.

perception of olfactory cues, which is important to mammalian olfaction (Louis et al., 2008; Catania, 2013). The choanae are further elongated (Sidor, 2003) over the condition of the basal-most pan-mammals (Fig. 11.12D), indicating expansion of the nasal capsule and probably also the surface area of the olfactory epithelium. The trenchant upper canine is relatively longer than in earlier pan-mammals and separates specialized enlarged anterior incisors in front, from sharp, unicuspid, compressed, recurved postcanine teeth. Early therapsids were increasingly specialized in apprehending prey with a bite from their canines and incisors (Gauthier et al., 1988a; Kemp, 2005, 2006b). Correspondingly, the orbits are more frontal in their orientation, with an increased field of binocular stereoscopic vision focused in front of the nose and mouth, as is characteristic of predators (Walls, 1942).

In general, the therapsid skull is more solidly built and was subjected to greater masticatory forces than in early pan-mammals. The temporal arch is emarginated from below and opens dorsally in all but the most primitive therapsids. This marks not only a much larger adductor muscle mass but also differentiation and migration of a larger part of the masseter from within the temporal fossa onto the ventral edge of the zygomatic arch. The palate (Fig. 11.12) was strengthened by losing its mobile ball-in-socket joint with the braincase (the basipterygoid articulation), as the pterygoids became closely joined against the parabasisphenoid in a nonkinetic articulation. The transverse processes of the pterygoid shifted forward to a level beneath the orbit and are oriented vertically, the pterygoids meet on the midline in front of the transverse processes, and behind them the interpterygoidal vacuity shifted posteriorly and diminished in size or became closed altogether during ontogeny. The palate is now stronger and offers far more resistance to torsion of the mandible when the jaws were open (Gauthier et al., 1988a; Rubidge and Sidor, 2001). Much of this particular macropredatory equipment was eventually lost, but for the next ~50 million years the ancestral lineage of Mammalia made its living as a macropredator.

In the mandible, the reflected lamina of the angular is quite thin and is deeply incised along its dorsal margin. It may have functioned as a crude tympanum and a component of the middle ear. However, it remained attached to the mandible along with several other bones in the sound transduction pathway, and any transmitted vibrations had to pass across the craniomandibular joint. But the quadrate and quadratojugal were reduced in size, and they become loosely attached by a ligamentous sheet to a broad descending flange of the squamosal, and the therapsid stapes no longer contacted the paroccipital process and now lay directly between the fenestra vestibuli and the quadrate. Thus the elements of the middle ear chain closest to the sacculus had a new measure of freedom to vibrate (Allin, 1975). But there was no obvious change in the structure or size of the inner ear that might signal an improvement in hearing.

Postcranial innovations involved longer limbs and more powerful, agile locomotion (Fig. 11.15B). These changes also augmented the reach of the special senses by raising the head well above the ground, and by increasing mobility of the head on the neck, as the neck became longer and more flexible. Basal therapsids had six cervical vertebrae, and soon settled on the seven cervicals that are almost invariably present in mammals. It is noteworthy that the mammalian vestibular system helps direct muscles of the neck that are responsible for reflexive compensatory movements of the head and eyes that keep a stereo visual image stable and in focus as the head is otherwise jostled in walking and running (Walls, 1942). Maintenance of these reflexes may explain the invariance in number of cervical vertebrae in mammals, and we may speculate further that the conjugated eye movement observed in living mammals was established in early therapsid history.

During early development of the therapsid vertebral column, the intercentrum and pleurocentrum of the trunk vertebrae consolidate into a single robust centrum, and the primitive notochordal canal ossifies and closed in mature individuals. The neural arches and ribs are more robust, and there is a single, fused, ossified sternum to which the ends of the thoracic ribs attached via cartilage. The sternum would later become segmented at intercostal junctions into a series of sternebrae that facilitated parasagittal locomotion and diaphragmatic ventilation characteristic of many mammals (see Section 11.5.14). The entire therapsid vertebral column was now

both more robust and flexible, and it transmitted greater forces generated by body size increases and increased power of the longer limbs.

The shoulder girdle was comparatively slender and the shoulder socket (glenoid) was formed predominantly by the scapula and posterior coracoid while the contribution by the anterior coracoid was diminished and eventually disappeared altogether. The glenoid faced posteroventrally instead of laterally, and the head of the humerus was now dorsally inflected and smoothly convex, allowing greater excursion at the shoulder. At the elbow, the radius and ulna have separate, distinct articular surfaces with the humerus, also allowing greater excursion.

The hindlimb was also longer, and the pelvis was more robust. The iliac blade was expanded upward and forward and forms a supraacetabular crest above a circular acetabulum. The femur has a smoothly convex, rounded head that is inflected medially, and locomotor forces are directed more vertically into the acetabulum, allowing greater excursion at the hip. The knees turned forward, and modifications to the ankle included a small tuberosity ("heel") on the calcaneum that improved its leverage. The astragalus broadly overlaps the calcaneum, and the ankle was simplified as a single navicular bone ossifies in the place of separate medial and lateral centralia.

Collectively, changes to the limbs and girdles mark an early, incremental shift away from sprawling sigmoid vertebral propulsion, toward strident parasagittal gait in which the limbs played a much larger role. The shoulder and hip sockets allowed for greater ranges of motion, just as the humerus and femur lost the massive protuberances that once attached to massive muscles that drove rotation about their long axes—now the limbs could swing. The elbows turned backward a bit, while the knees and feet turned forward. The hands and feet are both simplified as their phalangeal formula is reduced to 2-3-4-4-3 or further (Rowe and van den Heever, 1986). The hands and feet are now more involved in forward propulsion and less in lateral stabilization of vertebral undulation. The tail was also reduced in massiveness, if not length, as the retractor musculature of the hindlimb began its long evolutionary migration onto the pelvis. Whether the earliest pan-mammals could run is questionable, but it seems certain that therapsids could.

11.5.5 Node 5: Eutherapsida

Eutherapsida is the clade stemming from the last common ancestor that mammals share with Dinocephalia (Fig. 11.15C). It is diagnosed by further expansion of the temporal fenestra, a more powerful articulation of the head and neck, and subtle alterations in the locomotor system. The temporal fenestra now opens dorsally, continuing the trend of expansion of the adductor muscles and increased bite force at front of the mouth. The mandible was simplified by the loss of the anterior coronoid bone. At the back of the skull, the lambdoidal crest appeared along the dorsal rim of the occiput, providing attachment for powerful cervical musculature. All of these changes are consistent with further escalation in macropredation.

In the neck, the atlas–axis complex takes one step closer to the condition in Mammalia as the atlas pleurocentrum differentiates and then fuses during late ontogeny to the axis pleurocentrum above the atlantal intercentrum, to create the odontoid process. This developmental repatterned joined together elements of the first and second vertebral segments, and increasingly the atlas–axis complex became individuated from the rest of the vertebral column, and adopted its own variational modality. The atlas–axis complex is perhaps the most variable region of the vertebral column across pan-mammalian history. In general, its variation effected a balance that enabled both greater flexibility of the head and stability at the craniovertebral joint, while also accommodating the diverse transformations in diet and behavior that influenced the profound diversification of the skull manifest across pan-mammalian history.

In the postcranial skeleton, the hand was simplified by losing phalanges as it achieved the formula of 2-3-3-3-3 that is highly characteristic of most mammals today (Rowe and van den Heever, 1986). It also became more symmetrical around the middle digit and pointed forward. In the hindlimb, the greater trochanter appeared on the femur, as the pelvis expanded in support of enlarged femoral retractor musculature, and the length of the tail was considerably shortened.

Dinocephalia was a short-lived extinct side branch that originated as a macropredator, and quickly adapted to herbivory, as their teeth became blunt and the adductor musculature was reduced. They were one of many clades to evolve larger body size, some approaching 3 m in length and 500 kilos in weight. Derived dinocephalians represented the first experiment in head-butting, a bizarre sexual selection behavior practiced by living mountain sheep and other mammals with horns and antlers that act both as visual signals and shock absorbers (Barghusen, 1975). But dinocephalians never enjoyed the luxury of horns or antlers and simply thickened their bony skull roofs to withstand the collision impacts. As body size increased, so too did relative thickness of the skull roof, and the occipital condyle rotated to a ventral position to direct the force of frontal blows in a straight vector via the condyle to the neck. Dinocephalia appeared in the mid-Permian by ~270 million years ago and became extinct before the end of the Permian.

11.5.6 Node 6: Unnamed

Node 6 is an unnamed clade stemming from the last common ancestor that mammals share with Gorgonopsia and all its descendants. It is diagnosed by simplification of the skull involving loss of the supratemporal bone. This clade also has somewhat greater frontality of the orbits and a long coronoid process on the tooth-bearing denary bone of the lower jaw (Fig. 11.16A). It also has an exceedingly long, deeply vaulted choana that marks further expansion of the nasal capsule and its contents. There was still a roof over the adductor chamber, but the temporal fenestra was enlarged, and the jaw adductor muscles had continued the trend toward increased relative size. Flexibility of the head at the craniovertebral joint was enhanced by reduction of the zygapophyses and enlargement of the intervertebral foramen between the atlas and axis, and loss of the atlantal epipophysis.

In the early part of the Late Permian gorgonopsians were the top predators of the terrestrial ecosystem. Their taxonomy has bewildered paleontologists. They were once divided across 19 families and believed to make up the bulk of taxa that became extinct at the end-Permian event, which was possibly the most severe extinction event in vertebrate history. More recent efforts to revise the group acknowledge that some named taxa were simply at different levels of maturity at time of death, and the clade is less diverse than once believed.

11.5.7 Node 7: Unnamed

Node 7 is an unnamed clade stemming from the last common ancestor that mammals share with Dicynodontia and all its descendants. It is diagnosed by transformations that affected many parts of the skeleton. Dicynodontia arose early in the Late Permian, and it was the first globally successful terrestrial adaptive radiation of herbivores, surviving until the very end of the Triassic.

In the skull, the face was shortened so that only about half the total skull length lies in front of the orbit. The skull roof between the temporal fenestrae was reduced in width and the temporal fenestrae correspondingly enlarged. Another step toward simplification of the skull occurred as the postfrontal bone was reduced to a thin splint or lost altogether.

Additional modifications affected the craniovertebral joint (Kemp, 1982; Gauthier et al., 1988a). Each exoccipital forms a distinct subspherical articular surface, presaging the double-occipital condyle of mammals and their extinct relatives among cynodonts (discussed later), and there is now a trefoil-shaped articular facet on the odontoid process that indicates a new measure of complexity in craniovertebral articulation geometry.

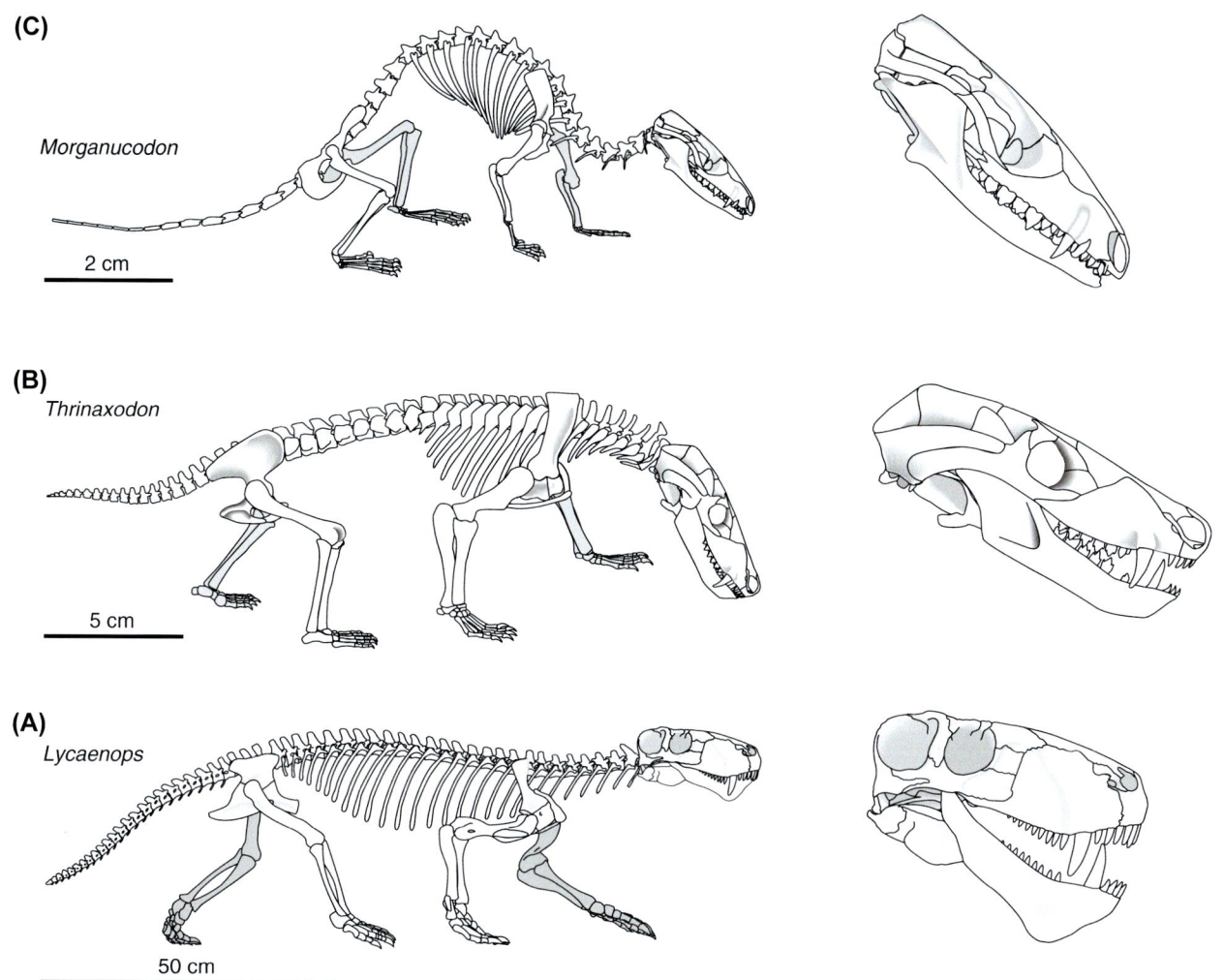

(C) *Morganucodon*

2 cm

(B) *Thrinaxodon*

5 cm

(A) *Lycaenops*

50 cm

FIGURE 11.16   Skulls and skeletons of the early therapsids (A) *Lycaenops*, (B) *Thrinaxodon*, and (C) *Morganucodon*, drawn to the same lengths.

Elsewhere in the vertebral column, the tail is shortened to 20 or fewer vertebrae, and the pelvis is now expanded with iliac "spines" present at either end. In effect, the anterior iliac spine has captured the femoral protractor musculature that once originated from beneath the vertebral transverse processes at the back of the trunk, while the posterior iliac spine now provided the origin of the femoral retractor muscles that primitively originated from the anterior caudal vertebrae. Thus the pelvis has come to support the bulk of the musculature that moved the hindlimb. The pedal phalangeal formula is now reduced to match that of the hand, to 2-3-3-3-3, and the foot faces forward.

Most of the known dicynodonts are highly derived herbivores who have cast off much of the macropredatory equipment of their ancestors. For this reason, their place in pan-mammal phylogeny was long held in doubt, and it was generally believed that they diverged very early in therapsid history (Hopson and Barghusen, 1986), or even evolved independently from herbivorous

pelycosaur grade pan-mammals (Olson, 1959). The more recent discovery of highly plesiomorphic dicynodonts has clarified the phylogenetic position (Gauthier et al., 1988a; Rubidge and Sidor, 2001). Dicynodonts are remarkable in reducing or completely losing their dentition, except for a pair of long canines—their namesake feature—and even these are lost in some. In place of the dentition was a turtle-like keratinous beak. They also developed a secondary palate that was lined with rugose keratinous pads, and broad shelves on the dentary that occluded against the palate. The coronoid process of the dentary was reduced and eventually lost, but it was compensated by enormous adductor muscles supported by a wide zygomatic arch that moved the mandible in fore—aft or propalinal jaw motion. Dicynodonts heralded an ecosystem of modern aspect in which predatory vertebrates were at the top of the food chain, and herbivorous vertebrates provided a link in the food pyramid between the top and the photosynthesizing base of the food chain. The smaller dicynodonts were also scratch diggers, and multiple specimens have been discovered preserved in burrows, in curled sleeping postures that may reflect that hibernation, a physiological trait that would persist to the present day, was established by Node 7.

11.5.8 Node 8: Eutheriodontia

Phylogenetic relationships among those therapsids most closely approaching Cynodontia (Node 9) is tangled. An extinct side branch known as Therocephalia was long recognized and treated as the sister taxon to Cynodontia (Hopson and Barghusen, 1986; Gauthier et al., 1988a; Rubidge and Sidor, 2001; Sigurdsen et al., 2012). However, it may be a paraphyletic group, with some members closer to Cynodontia than others, while one small clade (Bauriidae) may actually lie within Cynodontia. All of these taxa carried on the trend inherited from the ancestral amniote as macropredators. They share several additional apomorphic features with Mammalia, including the migration of the adductor musculature onto the skull roof to produce a sagittal crest, and the postcranial skeletons were simplified with loss of the cleithrum from the shoulder girdle. As their relationships become better known, some of the characters described later as diagnostic of Cynodontia may prove to have slightly wider distributions, and this will shed new light on the sequence of evolutionary assembly of the highly innovative cynodont skeleton.

11.5.9 Node 9: Cynodontia

Cynodontia arose in the Late Permian, and today it includes the ~5500 species of living mammals. Organizational changes in the skeleton of the earliest cynodonts had sweeping importance in the breathtaking diversification of mammalian clades and their distribution to all parts of the globe and into almost every imaginable habitat. The best known of the early cynodonts is *Thrinaxodon liorhinus* (Fig. 11.16B), which serves as a basal exemplar here. Reorganization of the cynodont skeleton partly reflects a shift from macropredation to insectivory and/or omnivory. However, major innovations in the mammalian neurosensory system can also be traced to the first cynodonts, as well as the first of several successive reductions in body size (Fig. 11.17) that effected shifts in ecology and life history strategy. The first cynodonts mark a first, modest pulse in encephalization, and this occurred in combination with the onset of new behaviors that would eventually come to be well represented in neocortex.

Secondary palate: The secondary palate is one of the first features noticed by 19th-century paleontologists Harry Seeley and Robert Broom, which indicated that some of the predatory therapsids were closer to mammals than others. The cynodont secondary palate created a new passageway, the nasopharyngeal passageway, which became separated from the oral cavity, and it displaced the choana from the front to the back of the mouth (Fig. 11.12). It forms as shelves of the maxillae and palatines grow toward the midline and fuse together to provide a bony floor beneath the nasal capsule and nasopharyngeal passageway, and a bony roof of the mouth. In the Late Permian cynodont *Procynosuchus*, the maxillae and palatines form a pair of shelves that extend two-thirds the length of the snout but they remain separated on the midline (Fig. 11.12E) and a narrow channel separates them from the vomer (Kemp, 1979). This condition resembles the human congenital deformity known as cleft-palate (Murry, 2002). In all other cynodonts, the maxillae and palatines meet on the midline beneath the vomer to form a complete secondary palate (reversed in many *Cetacea*). Intimately associated with the secondary palate is the emergence of an occlusal dentition (discussed later). We may infer that the tongue also took on a new role using the secondary palate as a substrate against which to move food within the oral cavity, but little of the pharyngeal skeleton is preserved in known fossils.

It is noteworthy that dicynodonts and possibly one other therapsid groups (Bauriidae) evolved a secondary palate. Dicynodonts (see Section 11.5.7) used their secondary palate with a masticatory system that involved abandonment of the teeth and their replacement by a keratinized beak in which keratin plates on the lower jaw occluded against opposing plates on the roof of the mouth. In dicynodonts, this innovation was tied to herbivory and contributed to their long temporal success, but it produced no obvious neurosensory innovations such as the increased encephalization that correlates with the secondary palate and occlusal dentition in cynodonts (discussed later). The secondary palate in the poorly known bauriids may be an indication that they are actually members of Cynodontia.

Occlusal dentition and mastication: In concert with the cynodont, secondary palate arose an occlusal dentition and the ability to masticate food items. This was a new behavior in cynodonts that produced a faster and enriched caloric return (Crompton, 1963, 1972, 1989; Rowe and Shepherd, 2016). Mastication occurs at the posterior (distal) part of the tooth row, where the mandibular adductor musculature had been reorganized to exert its greatest force. The cynodont mandible has a deep masseteric fossa on its lateral surface (Fig. 11.16B), indicating that the mammalian temporalis and masseter muscles were fully differentiated and that the largest mandibular forces had shifted from the front to the back of the tooth row (Gregory, 1953; Crompton, 1989; Kemp, 2005). This reflected a profound shift in

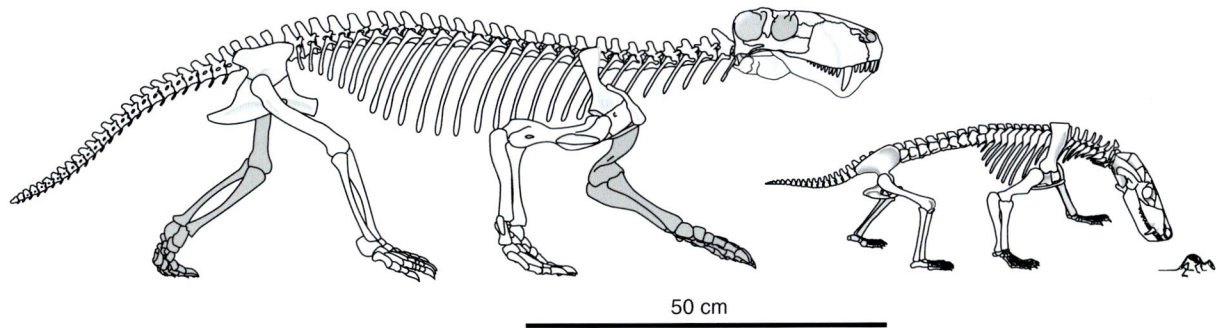

FIGURE 11.17 Skeletons drawn to scale of *Lycaenops* (a Permian gorgonopsian), *Thrinaxodon*, and *Morganucodon*. Note the differentiation of thoracic and lumbar vertebrae in *Thrinaxodon* and *Morganucodon* skeletons, indicating presence of the diaphragm. *From Rowe, T.B., Shepherd, G.M., 2016. The role of ortho-retronasal olfaction in mammalian cortical evolution. J. Comp. Neurol. 524, 471–495. https://doi.org/10.1002/cne.23802.*

feeding behavior. From the beginning of pan-mammalian history, macropredators had seized prey items with a bite delivered by the canines and incisors at front of the mouth and simply swallowed their prey whole or in large chunks. In cynodonts, mastication and oral breakdown of food items prior to swallowing not only expedited caloric return, but also enabled more thorough inspection and analysis of food items, and the ability to extract and process new kinds of information from food that led to a prolific diversification in dental morphology, and by inference in diet as well.

Over the course of cynodont evolution, as seen in living mammals, mastication eventually became linked to a complex array of orofacial muscles behaviors involving diverse orofacial motor skills including learned orofacial movements and swallowing. These were long attributed to brain stem regulatory mechanisms, but it is now apparent from anatomical, electrophysiological, imaging, and behavioral studies of the face sensorimotor cortex in humans and laboratory animals that the face primary motor cortex and the face primary somatosensory cortex make important contributions to the control of these elemental and learned movements (Avivi-Arber et al., 2011). Hence, the new function of mastication would take on a large and diverse neocortical presence.

Early cynodont teeth had "triconodont" crowns (Fig. 11.18) in which there are generally three principal cups aligned longitudinally, with the middle cusp the tallest, and with a row of smaller cuspules on a narrow shelf at the base of the inner surface (Crompton, 1963; Rowe et al., 1995). The outer (buccal) surfaces of lower molariform teeth occluded against the inner (lingual) surfaces of the upper molariforms and produced irregular wear facets that are evidence of crown-to-crown occlusion. This simple occlusal triconodont pattern set the stage for an unprecedented diversification of molariform tooth structure that eventually enabled cynodonts to pierce, slice, dice, shred, and grind their food in ever more complicated ways. In the process the cynodont

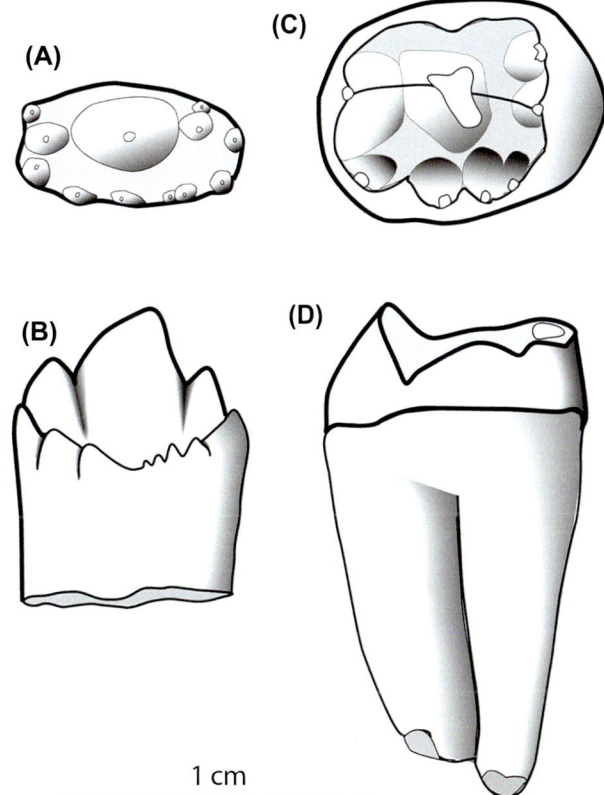

FIGURE 11.18 Molariform teeth of the Triassic cynodonts *Thrinaxodon* in (A) occlusal and (B) medial or lingual views, and *Diademodon* in (C) occlusal and (D) mesial or posterior views.

dentition became individuated into a new peripheral sensory array of considerable anatomical and neural complexity. The degree of structural diversity found in the dentition is perhaps best described by noting that nearly all species of Cynodontia, living and extinct, can be identified by the crowns of their molariform teeth alone (for reviews of mammalian crown diversity, see Gregory, 1910, 1916, 1922, 1953; Hillson, 2009; Ungar, 2010). This is not true for other clades of extinct pan-mammals, or for other amniotes.

To add a layer of complexity, the variational modality of the cynodont dentition also transformed, as it became increasingly individuated from the rest of the skull (Hlusko et al., 2011) and shifted to a pattern of rapid structural diversification that can be followed into nearly all of the living mammalian clades. Beginning with the first cynodonts, dental variation soon began to affect cusp numbers and shapes, crown sizes and geometry, enamel ultrastructure, occlusal relationships, growth and replacement rates, numbers and shapes of roots, and a corresponding increase in evolutionary rate (see discussion) of the dentition as a modular system unto itself.

Orthoretronasal olfaction: The evolution of a secondary palate and occlusal dentition had a cascading effect on the cynodont olfactory system that was only partly owing to the separation of the oral and nasopharyngeal passages. Mastication liberated an entirely new class of odors and scents from food as it was chewed and broken down, and from this new behavior a new duality was introduced into the main olfactory system, known as orthoretronasal olfaction (Rowe and Shepherd, 2016). It corresponded to the earliest measurable pulse in pan-mammalian encephalization, with enlargement of the piriform (olfactory) cortex, and it signals a first discernable transitional step toward differentiation of the mammalian neocortex from the three-layered dorsal cortex present in the ancestral amniote.

The behavior of inhaling external environmental odorant molecules through the naris is known as orthonasal olfaction (Fig. 11.19). This behavior was an innovation of Pan-Tetrapoda, and its history can be traced back into the fossil record to the earliest Devonian fossils belonging to the tetrapod stem (Jarvik, 1942). These transitional pan-tetrapods (formerly known as rhipidistian crossopterygians) were the first vertebrates in which the nasal capsule had an external opening, the naris (nostril), and a choana (internal naris) which opened through its floor into the roof of the mouth. They were the first vertebrates in which the naris conveyed airborne environmental molecules across the olfactory epithelium and into the mouth via the choana. The vomeronasal receptors had become segregated and encapsulated into the vomeronasal organ. Orthonasal smell is employed by nearly all tetrapod species, the exceptions being secondarily adapted to a committed aquatic lifestyle such as odontocete cetaceans (Colbert et al., 2005; Racicot and Rowe, 2014) and sirenians, in which the main olfactory system is largely or wholly abandoned, and possibly the terrestrial lungless plethodontid salamanders (Rowe and Shepherd, 2016).

The counterpart to orthonasal smell is "retronasal" smell, in which air exhaled from the lungs carries with it an entirely new information domain of odor molecules liberated in the mouth through the breakdown of food by chewing, saliva, and actions of the tongue (Fig. 11.19). These molecules pass forward from the caudal part of the mouth and via the choana they cross the main olfactory epithelium before being expelled through the nares (Fig. 11.20). In retronasal smell, *olfaction* combines with *taste* and other senses, including somatosensation, vision, and hearing, to generate our sensation of *flavor* (Shepherd, 2004, 2006, 2012). Orthonasal smell, retronasal smell, taste, and somatosensory signals from the lips, gums, cheeks, tongue, and teeth are passed to many different neural processing centers, but all converge on individual neurons in the neocortical area known as the orbitofrontal cortex (De Araujo et al., 2003Small et al., 2007; Rolls and Grabenhorst, 2008). The fact that flavor is a multisensory map in which distinct

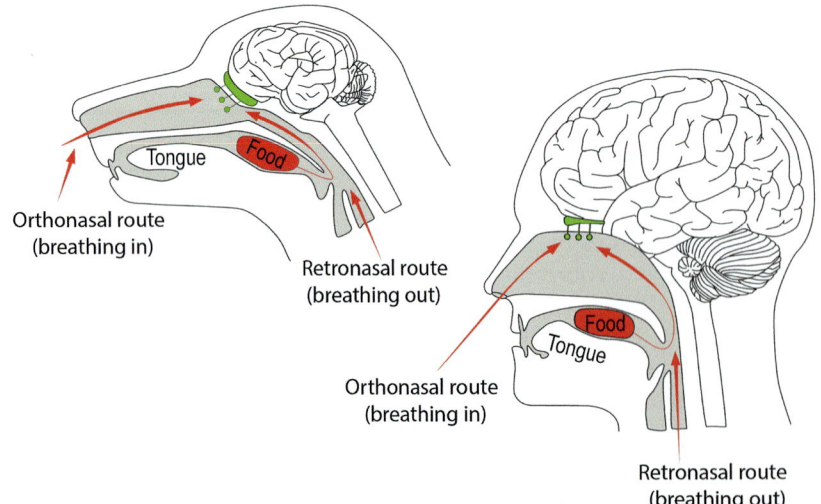

FIGURE 11.19 Diagrammatic representation of orthonasal and retronasal olfactory modes in a dog and human. *After Rowe, T.B., Shepherd, G.M., 2016. The role of ortho-retronasal olfaction in mammalian cortical evolution. J. Comp. Neurol. 524, 471–495. https://doi.org/10.1002/cne.23802.*

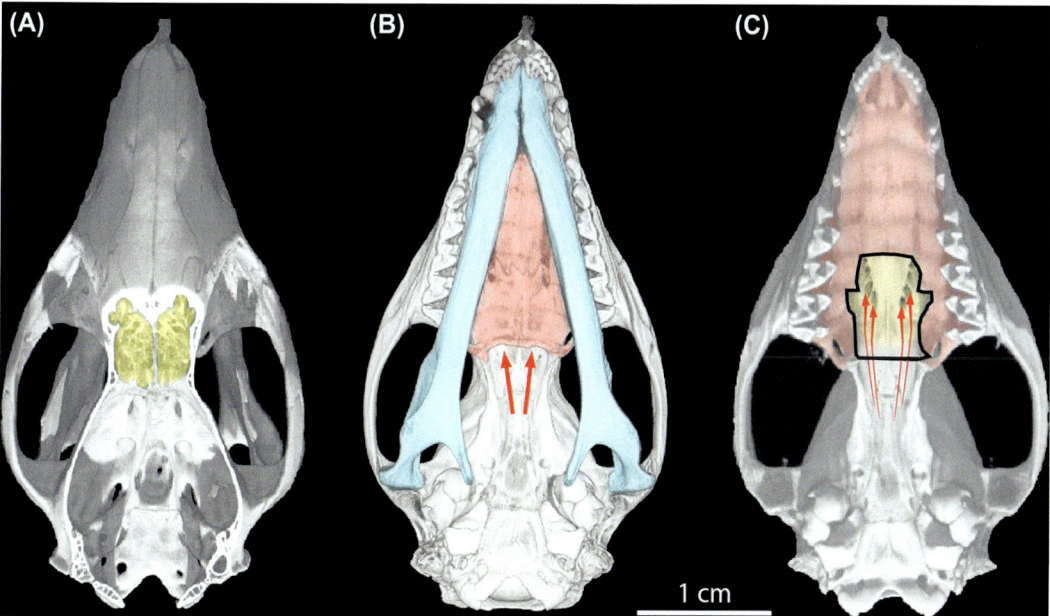

FIGURE 11.20 Mature skull of *Monodelphis* reconstructed from computed tomography data. (A) Dorsal view cutaway to show cribriform plate (*yellow*); (B) ventral view, with jaws (*blue*) and secondary palate (*red*) with *arrows* showing retronasal entrance to the nose via the choanae; (C) jaws and part of secondary palate removed with *arrows* showing the sphenethmoidal apertures in the ossified floor of the nasal capsule (*yellow*), which direct retronasal airflow across olfactory epithelium. After Rowe, T.B., Shepherd, G.M., 2016. The role of ortho-retronasal olfaction in mammalian cortical evolution. J. Comp. Neurol. 524, 471–495. https://doi.org/10.1002/cne.23802.

classes of information are integrated is evident in clinical data from patients who lost olfactory sensation following nasal infection or cranial trauma (Cullen and Leopold, 1999; Franselli et al., 2004; Bonfils et al., 2005) and from laboratory experiments (Heilmann and Humel, 2004; Sun and Halpern, 2005; Gautam and Verhagen, 2012). To emphasize the duality of mammalian olfaction, involving both external smells and internally generated volatiles, the term "orthoretronasal olfaction" was introduced (Rowe and Shepherd, 2016). This is yet another system that would eventually assume a large neocortical presence.

Different components of orthoretronasal olfaction are cognitively processed by many different cortical regions, and then integrated and referred to the mouth as our sense of flavor. Retronasal smell and the orthoretronasal duality in the construction of flavor are unique to mammals among living species, and the first indications of this duality can be traced back through the fossil record of pan-mammals to the origin of cynodonts. Many facets of orthoretronasal olfaction are dependent on the spatial organization and mechanical performance of the skull, dentition, and postcranial skeleton. As we will see, its influence in shaping the subsequent history of cynodonts is richly reflected in the fossil record.

Double-occipital condyle: Also apomorphic of *Cynodontia* is a specialized craniovertebral joint, in which the basioccipital recedes from the joint, and a "double-occipital condyle" is formed by the right and left exoccipitals that are positioned at the ventrolateral edges of the foramen magnum (Fig. 11.21). The new articulation expanded the degree of stable dorsoventral excursion of the head on the neck without impairing passage of the spinal cord through the foramen magnum and the cervical neural canal (Jenkins, 1969, 1971). The ventrolateral position of the condyles also suggests that the head was habitually held at a tilt with the nose toward the ground. Many mammals target their noses toward the ground and move their heads rapidly from side to side in scent-tracking and scent-guided navigation. More agile head movement potentially enabled cynodont olfaction to assert a new measure of importance in scent-tracking, navigation, and geographic memory (Rowe and Shepherd, 2016).

Ventilation: Augmenting this function is evidence of a shift from the plesiomorphic buccal pump ventilation system to diaphragmatic ventilation in the earliest cynodonts. In the axial skeleton, separate thoracic and lumbar regions become differentiated such that long ribs persist on the anterior thoracic vertebrae, while the posterior three to five ribs form attenuated processes that fuse to their respective neural arches (ie, lumbar ribs). Differentiation of separate thoracic and lumbar regions (Fig. 11.16B) marks the development of a muscular diaphragm, which separated the thoracic and abdominal cavities and initiated onset of the vacuum-chamber or

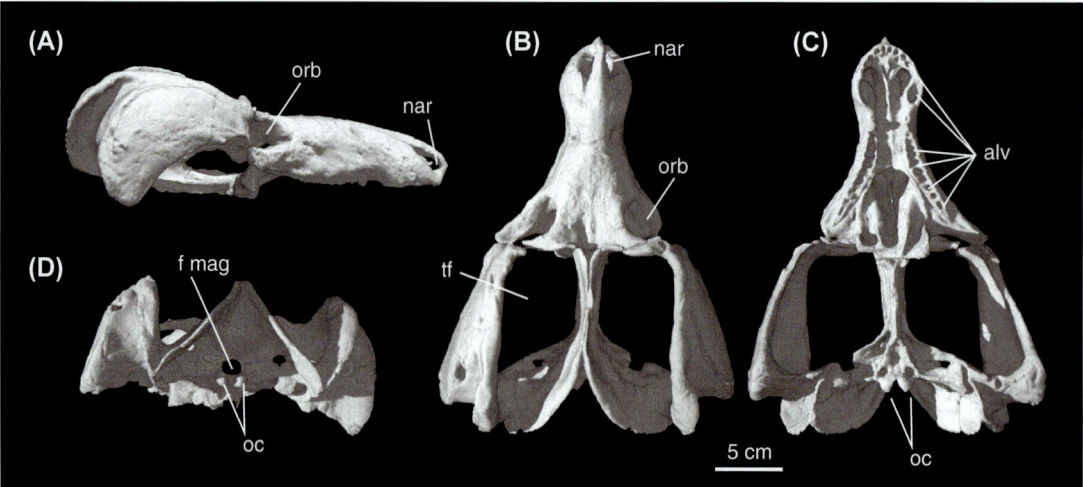

FIGURE 11.21   Skull of the Triassic cynodont *Diademodon*, rendered in 3D from computed tomography imagery, showing the double-occipital condyle, and the empty alveoli where the teeth fell out postmortem as the periodontal ligament decayed. See Table 11.1 for key to abbreviations. *From DigiMorph.org, with permission.*

bellows-like tidal diaphragmatic ventilation of extant *Mammalia* (Jenkins, 1971; Gauthier et al., 1988a; Hirasawa and Kuritani, 2013; Rowe and Shepherd, 2016).

While stationary or at rest, ventilation in living mammals is driven by the diaphragm (Bramble, 1989; Alexander, 2003). Chewing food is mostly a stationary action, thus orthoretronasal olfaction is driven by diaphragmatic ventilation. The rapid sniffing of scent tracking is a specialized form of ventilation that is highly characteristic of many mammals in exploring their olfactory environments (Stoddart, 1980; Shepherd, 2012). It is mostly done between steps or when moving slowly and is also driven by the diaphragm. In basal cynodonts, the proximal ends of the thoracic ribs are flattened and imbricate in a condition unknown in living mammals. Hence their modes of breathing and locomotion were not entirely modern, but the important new capacity of diaphragmatic ventilation was introduced. Early cynodonts were probably the first scent trackers in the ancestral lineage of mammals and could walk about with their noses to the ground, just as a living dogs and opossums. We can infer that olfactory navigation and memory began to assume a role in their lives that would eventually shape a wide range of mammalian behaviors.

Encephalization: Early cynodonts reflect the first measurable pulse in encephalization in pan-mammalian history up to this point. As we have seen, this pulse of encephalization was correlated with the new function of mastication and the added control requirements in jaw, tongue, and pharyngeal movements, as well as the integration of orthoretronasal olfaction, and a more agile head and body, and probably also the new behavior of scent tracking. The lateral walls of the braincase became more fully ossified as ventral sheets of bone from the frontal and parietal bones and a new

lamina of the prootic grew around the sides of the cortex and cerebellum. Most important was the newly formed alisphenoid bone, which arose as a compound element joining the endochondral ala temporalis of the epipterygoid with a new, membranous ossification in the sphenoobturator membrane (Gauthier et al., 1988a; Rowe, 1988). Collectively, these new ossifications enclosed a larger endocranial volume and indicate an increase in the relative size of the brain as measured using computed tomography. This was the first of several measurable pulses in encephalization that preceded the origin of crown Mammalia (Northcutt, 2011; Rowe et al., 2011).

In ontogeny of living mammals, ossification of the alisphenoid occurs in response to growth and inflation of the olfactory (piriform) cortex, and it may be an epigenetic response of connective tissues surrounding the brain to increased tension produced by rapid and expanded growth of the telencephalon (Rowe, 1996b). The onset of odorant receptor gene expression in the olfactory epithelium of the nose induces the development and controls the size of the olfactory bulb through the number of odorant receptor axons that reach the olfactory bulb. Projections from the olfactory bulb in turn determine the size and number of layers of the piriform cortex (Fig. 11.22), a division of the telencephalon that is invariably enclosed by the alisphenoid bone in living mammals (Rowe and Shepherd, 2016). The correlated appearance of an enlarged endocranial cavity and the ossified alisphenoid suggest that the origin of Cynodontia was tied to a pulse in encephalization that was driven specifically by enhanced olfaction, a sensory modality that would become developed to a higher degree in mammals than in any other vertebrates (Rowe et al., 2011; Rowe and Shepherd, 2016).

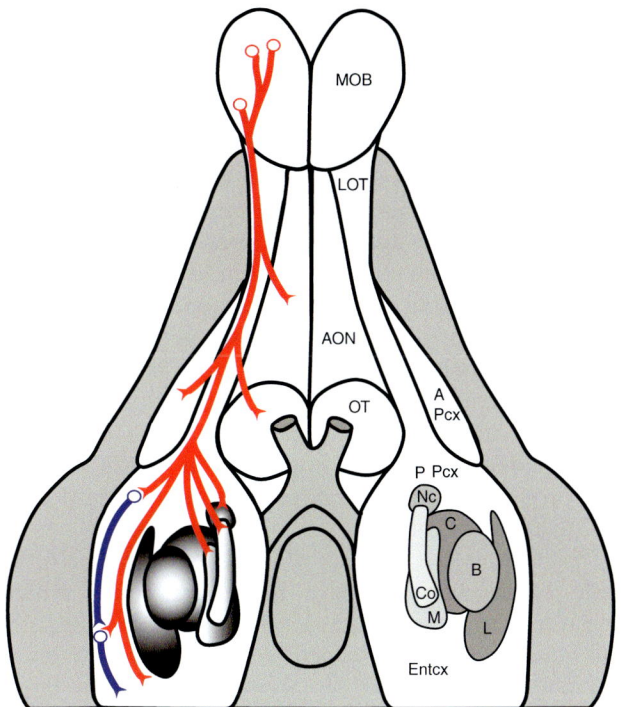

FIGURE 11.22 Projections from main olfactory bulb to piriform cortex and nuclei of the amygdala. Secondary projections of the main olfactory bulb are in *red*, while prepiriform projections are in *blue*. Nuclei of the amygdala include the basal (B), central (C), cortical (Co), lateral (L), medial (M), and nucleus of lateral olfactory tract (Nc). AOB, accessory olfactory bulb; MOB, main olfactory bulb. See Table 11.1 for key to other abbreviations. *After Nieuwenhuys, R., ten Donkelaar, H.J., Nicholson, C., 1998. The meaning of it all, pp. 2135–2196. In: Nieuwenhuys, R., ten Donkelaar, H.J., Nicholson, C. (Eds.), The Central Nervous System of Vertebrates. Springer, Berlin and Heidelberg.*

In summary, early cynodonts forged new linkages between biting, chewing, swallowing, breathing, orthonasal and retronasal olfaction, taste, flavor, and probably also scent-tracking behaviors. This increase in peripheral sensory information from olfaction, from an occlusal dental array and mastication, new swallowing behaviors, and a more agile head and postcranial skeleton commanded new levels of cortical sensory integration, and the demand for more highly coordinated motor skills that were all carried out by a relatively larger brain. Subtle variations on this theme, more so than any other, would characterize cynodont diversification prior to the origin of crown Mammalia, and it would influence the diversification of the major clades of mammals within the crown (Fig. 11.2).

11.5.10 Node 10: Eucynodontia

Eucynodontia refers to the last common ancestor that mammals share with *Cynognathus* and all its descendants (Rowe, 1993). *Cynognathus* was a large predatory cynodont from the early Triassic of South Africa and the first to manifest yet another key ingredient of the mammalian dentition. Up to this point, the teeth in therapsids and the earliest cynodonts had a short open root that was held to the jaws by a ring of bone at the base of the crown, and they were continuously replaced throughout life (Crompton, 1963). The roots occupied shallow sockets, and what loads the crowns could withstand were dependent on the bony collar that held the tooth in place. In contrast, in *Cynognathus* and other eucynodonts, the roots are longer and are held in place in deep alveoli in the premaxilla, maxilla, and dentary bones by the periodontal ligament (Figs. 11.18 and 11.21). This condition is known as thecodont implantation, and the ligament is sometimes referred to as the thecodont gomphosis (Rowe, 1993; Ten Cate, 1997).

It bears explanation that in the more basal pan-mammals, we often find jaws with teeth intact, or they are broken off at the jawline, where the collar of bone ankylosed them to the jaw. However, in the skulls and jaws of eucynodont fossils, the teeth have often slipped from their sockets as the periodontal ligament decayed, leaving an empty alveolus (Fig. 11.21D). If the jaws still hold their teeth, inevitably there is a thin sheet of matrix around the roots, ancient sediment filling the space once occupied by the ligament.

The periodontal ligament has mostly been studied in humans and primates where it is richly innervated. Recordings from single nerve fibers have shown that human periodontal receptors adapt slowly to maintained tooth loads. Most receptors are broadly tuned to the direction of force application, and about half respond to forces applied to adjacent teeth. Information about the magnitude of tooth loads is made available in the mean firing rate response of periodontal receptors, and they record precisely the intensity and the spatiotemporal aspects of the forces applied on the tooth. These mechanoreceptors are particularly important when biting and chewing because they efficiently encode tooth load during intraoral food manipulation and are involved in jaw motor control (Trullson, 2006; Trullson et al., 2010).

In living mammals, signals from tooth crown pulp, and periodontal mechanoreceptors, project to separate oral fields of the primary somatosensory cortex (Rempel et al., 2003; Kaas, 2006; Iyengar et al., 2007; Trulsson et al., 2010). Using fMRI, significant activations in the somatosensory cortex were found for all stimulated teeth, and the response in these cortical areas was dominant and robust. Periodontal receptors encode information about the teeth stimulated and the direction of forces applied to the individual teeth, and when a tooth is stimulated, the brain is most of the time able to recognize its location. These results demonstrate the dental pulp representation area in the primary somatosensory cortex and that it receives input from intradental A-beta neurons, providing a detailed organizational map of the orofacial area, and adds the representation of dental pulp to the classic neocortical sensory homunculus (Kubo, 2008). There is also strong evidence for bilateral representation of the teeth into the primary sensory cortex coming directly from the thalamus or via transcallosal projections (Kaas, 2006; Iyengar et al., 2007; Habre-Hallage, 2014). Projections from the somatosensory oral cavity integrate cutaneous stimuli and movements of the tongue and jaw that are important for mastication and for the ability to recognize and discriminate the form of objects by using intraoral or perioral sensors. The connections between the somatosensory representation of the teeth and tongue and adjoining motor and premotor representations of the oral cavity and jaw may help to coordinate motor control in chewing and swallowing (Iyengar et al., 2007). The tongue and teeth are also extremely important in vocalization. In the tongue, 80% or more of neurons are tactile, and 2–10% respond to taste (Iyengar et al., 2007).

These results strongly suggest that the pulp and periodontal ligament mechanoreceptors play a significant role in the specification of the forces used to hold, manipulate, and explore food between teeth. In these respects, the masticatory system appears analogous to fine finger-control mechanisms used during precision manipulation of small objects. Because fMRI reveals activations in posterior insular cortex, periodontal ligament mechanoreceptors and SA II-type receptors, in general, may be involved in one aspect of the feeling of body ownership (Iyengar et al., 2007).

During development, the tooth crowns erupt first (Ten Cate, 1969). As they emerge, they twist and shift and adjust to one another to establish occlusal relationships between upper and lower teeth. The roots then begin to grow into their bony sockets, anchoring the crowns in place as the occlusal relationship between postcanine crowns develops. It is this plasticity that dentists exploit in using braces to straighten and adjust the maturing teeth in humans. Implantation of the teeth via the periodontal ligament also enabled a stronger bite force applied to a broader occlusal surface (Kemp, 1982, 2005). At the same time, the molariform crowns began to increase in occlusal surface area and complexity (Figs. 11.18 and 11.23). The innovation of thecodont implantation was correlated with reduction in the rate and mode of tooth replacement (Cifelli et al., 1996; Luo et al., 2004), affording a greater measure of learning and memory in relation to food.

11.5.11 Node 11: Unnamed

Node 12 is the unnamed clade stemming from the last common ancestor that mammals share with *Diademodon*

(Fig. 11.21) and all its descendants. It is diagnosed by yet another modification of the dentition, in which the roots are much longer and each cheek tooth crown has an "incipiently divided" root (Fig. 11.18D). That is, there were two separate root canals, each conveying its own dental nerve to the pulp cavity, but a web of bone still connected them. This "incipient" division of the roots carried on in Early and Middle Triassic cynodonts and suggests that more information could be found in the differential loading of individual tooth crowns. This innovation is correlated with expansion of the temporal fossa, strengthening of the masticatory musculature, a heightened coronoid process of the dentary, more robust zygomatic arch, and a tall sagittal crest. One mechanical effect of deep thecodont implantation was a stronger bite force applied to a broader occlusal surface (Kemp, 1982, 2005; Rowe, 1993).

Diademodon is a member of a presumed herbivorous clade of modest diversity found in South Africa, Argentina, and the eastern United States, in which the first generation of cheek teeth were sectorial, for biting and shearing. But the replacement teeth had larger crowns that rotated 90 degrees as they erupted to form a tightly packed broad battery of teeth that occluded in a surface-to-surface occlusal relationship. The temporalis and masseter muscles were gigantic in these strange eucynodonts. Plant material can be highly abrasive to the teeth, whether woody, or from accumulated particulate dust, or from small siliceous crystals known as phytoliths that many plants secrete. It was once believed that a single great radiation of herbivorous cynodonts played out in the Triassic. It is now clear, however, that several different eucynodonts independently adapted to a herbivorous diet independently.

11.5.12 Node 12: Unnamed

This unnamed clade stems from the last common ancestor shared by the Mammalia and the mid-Triassic cynodont *Probainognathus* and all of its descendants (Fig. 11.24). The frontality of its orbits is remarkable, and it must have had a binocular visual field that approached 180 degrees and was amplified toward periscopy by the flexibility of its neck. The bones of the jaw lying behind the tooth-bearing dentary are considerably reduced, marking the onset of their negative allometric growth with respect to the skull and mandible (Rowe, 1996a,b), and their increasing individuation as components of the auditory chain of the middle ear and a trend toward higher-frequency sound sensitivity. These bones were still involved in the jaw articulation, but an elongate condylar process of the dentary now passed backward over the auditory chain and it would gradually take over the mechanical role of the jaw articulation,

TABLE 11.1 Key to Abbreviations used in the Figures

Ali	Alisphenoid
Alv	Alveoli for the dentition
ana	Atlantal neural arch
Ang	Angular
Ao	Ala orbitalis
Aon	Anterior olfactory nucleus
A pcx	Prepiriform cortex
Art	Articular
at	Ala temporalis (=processus ascendans)
BH	Basihyal
Bo	Basioccipital
Bs	Basisphenoid
c	Lower canine
C	Upper canine
Cb, cere	Cerebellum
CH	Ceratohyal
choa	Choana
Con	Commissura orbitonasalis
Cop	Commissura orbitoparietalis
CrC	Cricoid cartilage
cve	Cavum epipterycum
D cond	Condylar process of dentary
D cor	Coronoid process of dentary
D ctx	Dorsal cortex
Den	Dentary
dns	Dens
D ram	Dentary ramus
Ec	Ectopterygoid
Ect	Ectotympanic
Entcx	Entorhinal cortex
Eoc	Exoccipital
Et 1—5	Ethmoid turbinals 1—5
Fan	Annular fissure
feh	Fenestra hypophyseos

Continued

TABLE 11.1 Key to Abbreviations used in the Figures—cont'd

fen	Fenestra narina
feol	Fenestra olfactorium
fics	Fissura occipitocapsularis superior
fion	Fissura orbitonasalis
fisp	Fissura sphenoparietalis
f mag	Foramen magnum
foe	Foramen epiphaniale
foh	Hypoglossal foramina
foic	Internal carotid foramen
Fr	Frontal
Fv	Fenestra vestibuli
Hyp	Hypophysis
i 1—4	Lower incisors
I 1—3	Upper incisors
Iam	Internal auditory meatus
ifo	Infraorbital foramen
II	Cranial nerve II (optic)
IN	Incus
Ip	Interparietal
Ju	Jugal
Lac	Lacrimal
lf	Lacrimal foramina
LOT	Lateral olfactory tract
Lp	Lamina parietalis
lta	Lamina transversalis anterior
ltp	Lamina transversalis posterior
m 1—3	Lower molars
M 1—3	Upper molars
MAc	Caput mallei
MAm	Manubrium mallei
Max	Maxilla
MC	Meckel's cartilage
Me	Mesethmoid
Mt	Maxilloturbinal
Nar	Naris

Continued

3. Early Mammals and Subsequent Adaptations

TABLE 11.1   Key to Abbreviations used in the Figures—cont'd

Nas	Nasal
Ncx	Neocortex
Npp	Nasopharyngeal passageway
Nt	Nasoturbinal
oa	Occipital arch
ob	Olfactory bulb
oc	Occipital condyle
Ocx	Olfactory (piriform) cortex
Oe	Olfactory epithelium
Opl	Optic lobes
Orb	Orbit
osc	Occipital condyle
OT	Olfactory tubercle
p 1–5	Lower premolars
P 1–2	Upper premolars
Pa	Parietal
Pal	Palatine
Pcx	Piriform (olfactory) cortex
Pet	Petrosal
Pin	Pineal body
pfl	Paraflocculus
pla	Planum antorbitale
plb	Planum basale
Pmx	Premaxilla
prl	Prominentia lateralis
prom	Promontorium of petrosal
Pt	Pterygoid
prst	Pars trabecularis
psc	Paraseptal cartilage
Pt	Pterygoid
Qu	Quadrate
Re	Respiratory epithelium
Re lam	Reflected lamina of angular (=ectotympanic)
Rf	Rhinal fissure
sca	Anterior semicircular canal

TABLE 11.1   Key to Abbreviations used in the Figures—cont'd

scl	Lateral semicircular canal
scp	Posterior semicircular canal
sepn	Septum nasi
SH	Stylohyal
sla	Sulcus lateralis anterior
Smx	Septomaxilla
Soc	Supraoccipital
Sp	Secondary palate
spc	Spinal cord
Sq	Squamosal
sss	Superior sagittal sinus
Sv	Sinus venosus
TC	Trachial cartilages
Tf	Temporal fossa
ThC	Thyroid cartilage
TH	Thyrohyal
tn	Tectum nasi
ton	Tongue
V	Cranial nerve V (trigeminal)
Vo	Vomer

eventually freeing the "postdentary" bones increasingly for a singular role in hearing. In *Probainognathus* we see evidence that enhanced sound sensitivity was coming to join sharp binocular vision and olfaction.

11.5.13 Node 13: Unnamed

Still closer to Mammalia is *Exaeretodon*, a large herbivore from Argentina that was once lumped in a paraphyletic group with other presumed herbivores. It was among the largest therapsids of the Triassic, with a skull more than a half-meter long at full maturity. The brain was somewhat larger, in response to both sensory integration and motor coordination, as can be seen in the rearward displacement of the occipital plate and a downward bend in the floor of the braincase. One notable detail involves the trigeminal nerve (cranial nerve V). It has three major trunks, and collectively they innervate the jaw muscles, the teeth, and provide sensory innervation to the skin of the face and forehead. They converge at a large ganglion (the trigeminal or Gasserian) that integrates their individual signals before

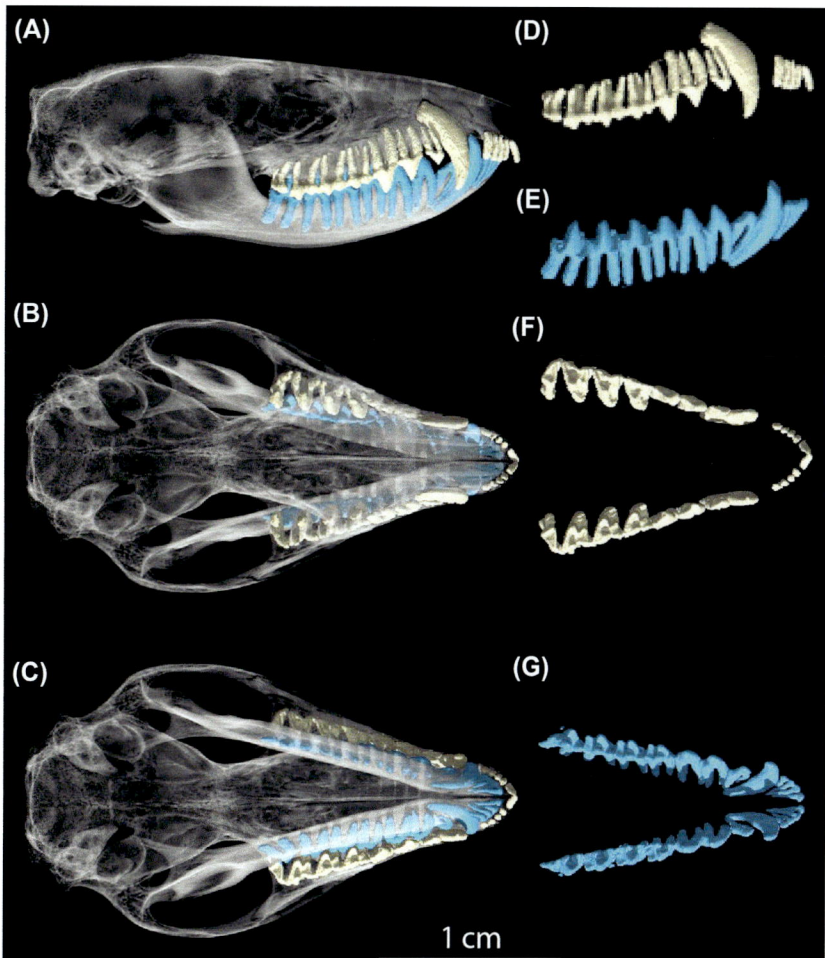

FIGURE 11.23   Mature skull of *Monodelphis* reconstructed from computed tomography data, with the bones of the skull rendered translucent, and the dentition opaque, to show the relationship of the dental array to the skull in lateral (A), ventral (B), and dorsal (C) views. Upper (D) and lower (E) dentitions in lateral view; upper dentition in occlusal view (F); and lower dentition in occlusal view (G).

either sending them on to the brain stem or evoking a reflexive movement of the jaws. The ganglion lies outside of the endocranial cavity, and the space it occupies is now partially enclosed by bone. This subtle feature reflects increasing competition for space between the brain, jaw muscles, and the tongue and pharynx, and its new partial floor suggests expansion of the ganglion that may indicate the tactile field of the trigeminal nerve the face was expanding (Huber, 1930a,b). The pineal foramen was lost, and the pineal stalk itself may have been covered by rear expansion of the telencephalon.

The thorax in *Exaeretodon* has at last lost the rhomboidal expansions of the ribs seen in basal cynodonts, and we can presume that another step toward the mature pattern of ventilation seen in mammals had taken place. Subtle modifications of the skeleton involve reduction of the interclavicle, and in the shoulder joint involved the scapula and only one of the two coracoid ossifications. A large process grew from the proximal end of the ulna at the elbow, increasing leverage in extending the forelimb. The pelvis had rotated down and backward relative to the sacrum in a way that gave the lumbar vertebrae more freedom in dorsoventral flexion and more propulsive power to the hindlimbs. The tail was further reduced in length.

Other possible members of Node 13 are small poorly known taxa which, owing to their small size, have all have been implicated in the proximal ancestry of mammals at onetime or another. These include the tritheledontids, prozostrodontids, and the brasilodontids, all groups of doubtful monophyly. Some of these may lie within Mammaliamorpha (Node 14), but until more complete skeletons are found, their phylogenetic relationships will remain ambiguous.

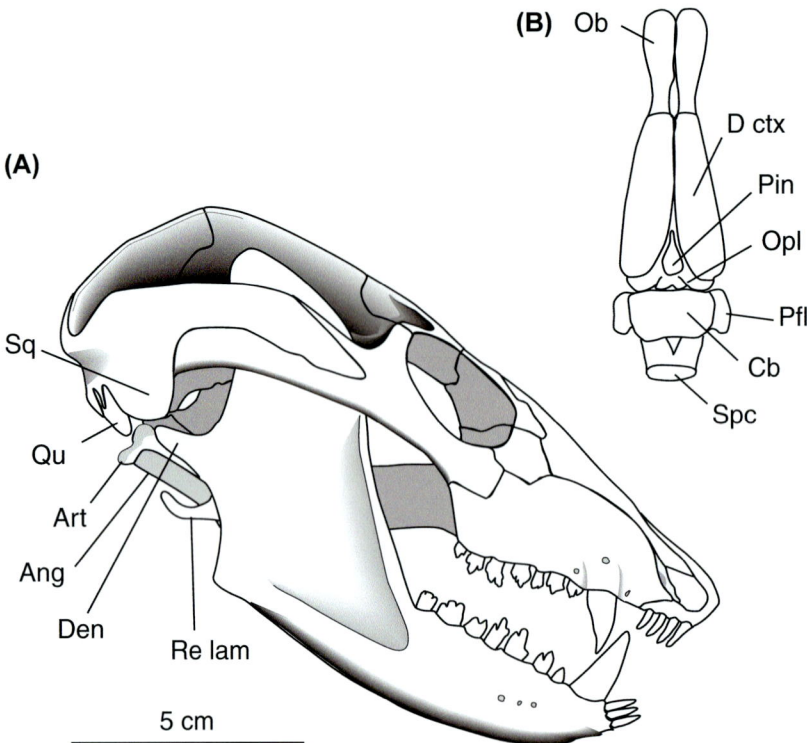

FIGURE 11.24   (A) Reconstructed skull of the Triassic cynodont *Probainognathus* in lateral view (After Romer, A.S., 1970. The Chanares (Argentina) Triassic reptile fauna: 6. A chiniquodontid cynodont with an incipient squamosal-dentary jaw articulation. Breviora 344, 1—18), Note that the dentary and squamosal are in very close approximation. (B) A hypothetical reconstruction of its brain. Modified after Quiroga, J.C., 1980. Sobre un molde endocraneano del cinodonte *Probainognathus jenseni* Romer, 1970 (Reptilia, Therapsida), de la Formación Ischichuca (Triasico Medio), La Rioja, Argentina. Ameghiniana, 17, 181—190. See Table 11.1 for key to abbreviations.

11.5.14 Node 14: Mammaliamorpha

The name Mammaliamorpha refers to the last common ancestor shared by Mammalia and the extinct side branch Tritylodontidae, a clade originating in the Late Triassic and surviving until the mid-Jurassic. When first discovered in the 19th century, tritylodontids were classified as mammals, but in the early 20th century a number of features were identified that place them outside of the crown clade. The tritylodont dentition took on a radical new organization that perplexed the mammalian paleontologists who interpreted the dentition a priori as key to understanding mammalian phylogeny at all taxonomic scales. Tritylodonts have long incisiform teeth, no canine, and their upper molariform teeth are the source of their name. Each upper molariform has a large crown with three longitudinal rows of backswept cusps. Between the three rows of cusps are two longitudinal troughs into which the two-rowed lower teeth slid. With fore—aft movement of the jaws, the teeth were exceptionally suited as shearing and grinding devices. It is generally presumed that trity-lodonts were herbivores, but they could eat virtually anything with such a dentition. In the mammaliamorph dentition, all of the postcanine teeth have two or more fully divided roots that are no longer connected by a web of dentine (Fig. 11.25), each with its own dental nerve canal, and the molariform crowns have complex occlusal patterns (Fig. 11.23). Mature molariform teeth were not replaced, and their permanence potentially enabled the subtle textural information from different kinds of food to be learned and remembered. The smell-taste-flavor system surged forward in the complexity of information it processed.

An overwhelming characteristic of the basal mammaliamorphs is that they underwent miniaturization (Fig. 11.17; Rowe, 1988, 1993; Rowe and Shepherd, 2016). While derived tritylodontids and numerous mammal clades would secondarily attain large and even huge sizes, the Mesozoic members of the mammalian stem lineage remained tiny for millions of years, and even today the greatest diversity of mammals is to be found in the small rodents and bats. With miniaturization, early mammaliamorphs encountered greater spatial and environmental heterogeneity. Entry into new microhabitats must have resulted in dietary diversification, where new food items such as seeds, grains, fungi, small fruiting bodies, and small invertebrates were available for the first time, and it altered activity patterns and life history strategies (Harvey et al., 1980; Mace et al., 1981; Hayden et al., 2010).

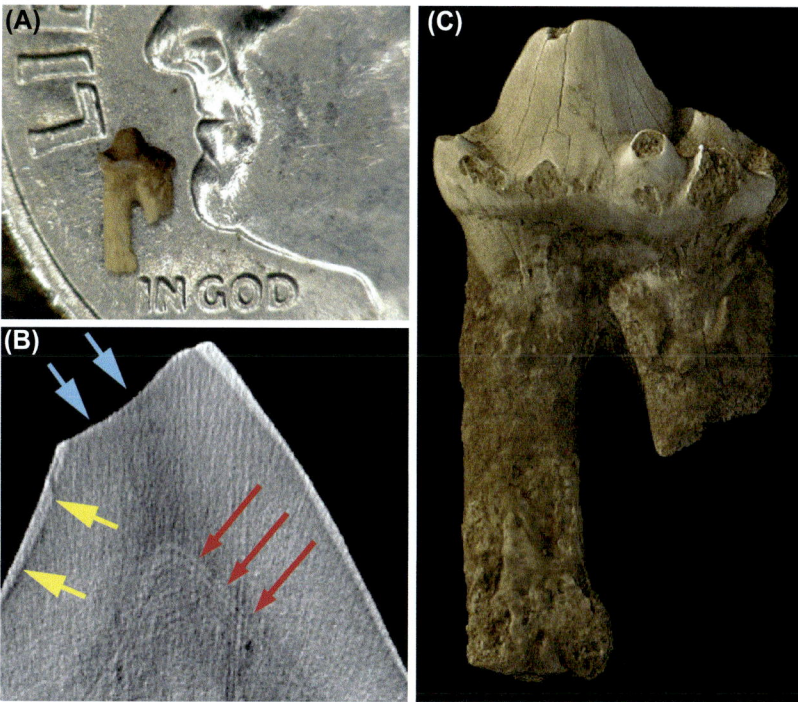

FIGURE 11.25 Molariform tooth of the Triassic mammaliamorph (and mammaliaform) *Morganucodon* showing fully divided molariform tooth roots. (A) Photomicrograph on US dime for scale. (B) Computed tomography slice showing occlusal wear facet (*blue arrows*), enamel dentine junction (*yellow arrows*), and what are probably daily growth lines (*red arrows*). (C) 3D reconstruction from computed tomography data showing one complete and one broken root. *After Rowe, T.B., Frank, L.R., 2011. The vanishing third dimension. Science 331, 712–714.*

Early mammaliamorphs attained miniaturization by accelerated maturation of the skeleton at smaller and smaller sizes. As a rule, small mammals reproduce very quickly, and more rapid generation times can equate with overall increases in speciation rate. Mammaliamorphs have added the epipubic bone, otherwise known as the marsupial bone to the pelvis. This bone supports the pouch in monotremes and marsupials (and persists in those species in which the pouch is secondarily lost). It seems likely that basal mammaliamorphs reared their young in pouches, a practice that would not change until the origin of placental mammals.

Miniaturization was accompanied by remodeling of the postcranial skeleton, in which the joint surfaces were more precisely sculpted, and the trochanteric attachments for limb muscles came to resemble those found in crown mammals. Locomotion at small size is metabolically far less expensive than at larger size, and climbing vertically costs little more than locomotion over flat surfaces (McMahon and Bonner, 1983). There is also a regular change in the mechanical advantage muscles have about the joints of the skeleton as a consequence of the stability of the joints under loads determined by inertia and gravity. Flexion angles at joints increase as size decreases. This implies that muscle spindles and joint proprioceptors were recording more information produced by the greater degrees of excursion by

all the joints than before and that a new level of agility had emerged. As miniature animals weighing only a few grams, mammaliamorphs were freed of some of the constraints of gravity, and highly characteristic remodeling of nearly all the joints speaks to enhanced agility. Scampering and climbing agility were now added to the locomotion repertoire of the mammalian stem group.

The entire mammaliamorph skeleton took on many detailed, if subtle, resemblances to the skeleton in Mammalia. Especially noteworthy are apomorphies that involve the nose, orbital wall, ear, and palate. In the skull, an ascending lateral process of the premaxilla forms the rear margin of naris. Computed tomography scans reveal what may be the first partial ossification of the rear parts of the nasal capsule (Kielan-Jaworowska et al., 2004). Around the orbit, the prefrontal and postorbital bones are absent, and the postorbital arch no longer separates the orbit from the temporal fenestra. Sheets of bone from the orbitosphenoid, frontal, and palatine join to form an extensively ossified medial orbital wall and a rigid anchor for the extrinsic ocular musculature (a complex feature absent in tritheledontids and brasilodontids). Ossification of the orbital wall may have also been a response to increased encephalization. However, an endocast is as yet unavailable for any tritylodontid, and the external shape of the skull leaves no doubt that the telencephalon beneath the frontals and parietals remained quite narrow.

The otic capsule and adjacent regions, including the craniomandibular joint, underwent multiple transformations that collectively indicate an ear that was increasingly sensitive to somewhat higher frequencies. The internal auditory meatus is walled medially with separate foramina for the vestibular and cochlear nerves, and the cochlea reflects a first pulse in elongation, curving over an arch of about 70 degrees (Kielan-Jaworowska et al., 2004). The middle ear is still bound to the mandible, but has continued its evolutionary decrease in relative size as the postdentary bones are reduced to a collective narrow rod that lies deeply set into a postdentary trough beneath an elongated posteriorly directed dentary condylar process. The articular has de novo dorsal and ventral processes, one being the homologue of the *manubrium mallei* in Mammalia. The surangular is no longer involved in the craniomandibular joint, which now lies primarily between the quadrate and articular.

Since the time of ancient stem tetrapods, the quadrate was the bone attached to the skull that met the articular bone of the lower jaw to form the craniomandibular joint. The quadrate had some measure of mobility with the squamosal bone of the skull throughout the history of pan-mammals recounted to this point. But in mammaliamorphs the quadrate gained a new measure of freedom. The paroccipital process is directed laterally (instead of ventrolaterally) and is bifurcated distally, with one distal process forming a separate condyle for a kinetic articulation with the quadrate, and the other apparently articulating with the hyoid. This carried early mammaliamorphs considerably closer to the impedance-matching middle ear of mammals.

The palate and basicranium present several apomorphic transformations that may have affected suspension of the musculature for swallowing, allowing speculation that this is where a liquid diet of milk was introduced to postnatal development. The parabasisphenoid and pterygoid no longer form a single continuous ventral sagittal ridge and instead form parallel parasagittal ridges separated by a shallow trough. The basicranium is broadly expanded to widely separate the pterygoid transverse processes. The parasphenoid wings (alae) are believed to form broad ventrolateral flanges (Sues, 1986); if so, they fuse indistinguishably to the basicranium at an early stage in ontogeny (the parasphenoid disappears as a separate ossification in *Mammalia*). All of these changes affected the roof of the mouth and pharynx behind the choana where orthonasal smell, taste, and retronasal smell all come together. At the same time, the hyoid bone of the tongue and its musculature took on new attachments. Collectively these changes suggest that new behaviors involving licking, sucking, and swallowing were introduced at this point in pan-mammalian history.

In the postcranium, the atlantal postzygapophyses are absent, the atlas centrum is flattened, and the dens is present on the anterior face of the axis centrum. Collectively these continue the trend of greater flexibility of the skull at the craniovertebral joint. The sternum became segmented at the costal joints to form a sequence of sternebrae, facilitating parasagittal flexion-extension of the trunk. This may represent a next step in mammalian ventilation by enhancing the overall flexibility of the thoracic rib cage. The iliac blade is low with a flat dorsal margin and divided into dorsal and ventral components by a longitudinal ridge, giving this bone a triangular shape in coronal cross section. The ischium, pubis, and acetabulum are rotated posteriorly to lie entirely behind the sacrum where they enclose an obturator foramen that exceeds the acetabulum in diameter. The femoral head is nearly spherical with a distinct fovea for attachment of the *ligamentum capitis femoris*. Finally, in the tail the caudal vertebrae are graded in length with elongated centra in the distalmost vertebrae. All of these changes signal greater excursion of the limbs and increased agility moving over complex three-dimensional habitats (Kemp, 1983, 1988b, 2005; Rowe, 1988, 1993; Rowe and Shepherd, 2016).

11.5.15 Node 15: Mammaliaformes

Mammaliaformes is the clade stemming from the last common ancestor that mammals share with *Morganucodon oehleri* (Fig. 11.26), and it arose by or before 235 million years ago. *Morganucodon* is the best-known member of the extinct clade Morganucodonta which diversified and became distributed across Pangea in the Late Triassic and Early Jurassic. The most striking feature of Mammaliaformes, as indicated by the bones of the cranium, is that their brains had almost doubled in relative size, and the endocast now looks very much like a mammalian brain with large olfactory bulbs and an expanded "pear-shaped" piriform (olfactory) cortex (Figs. 11.26 and 11.27). This is indicated in the parietals, basioccipital, and basisphenoid, which are expanded dorsally, posteriorly, and laterally. Except for the lack of a cribriform plate (and for neural and vascular penetrations), the endocranial cavity has become entirely enclosed by bone.

Brain volume measured relative to estimated body mass produces an encephalization quotient (EQ), based on an empirically derived equation that affords a quantitative comparison of relative brain size between different taxa (Jerison, 1973; Eisenberg, 1981, 1990). In this case, an EQ of 1 equals the average for living

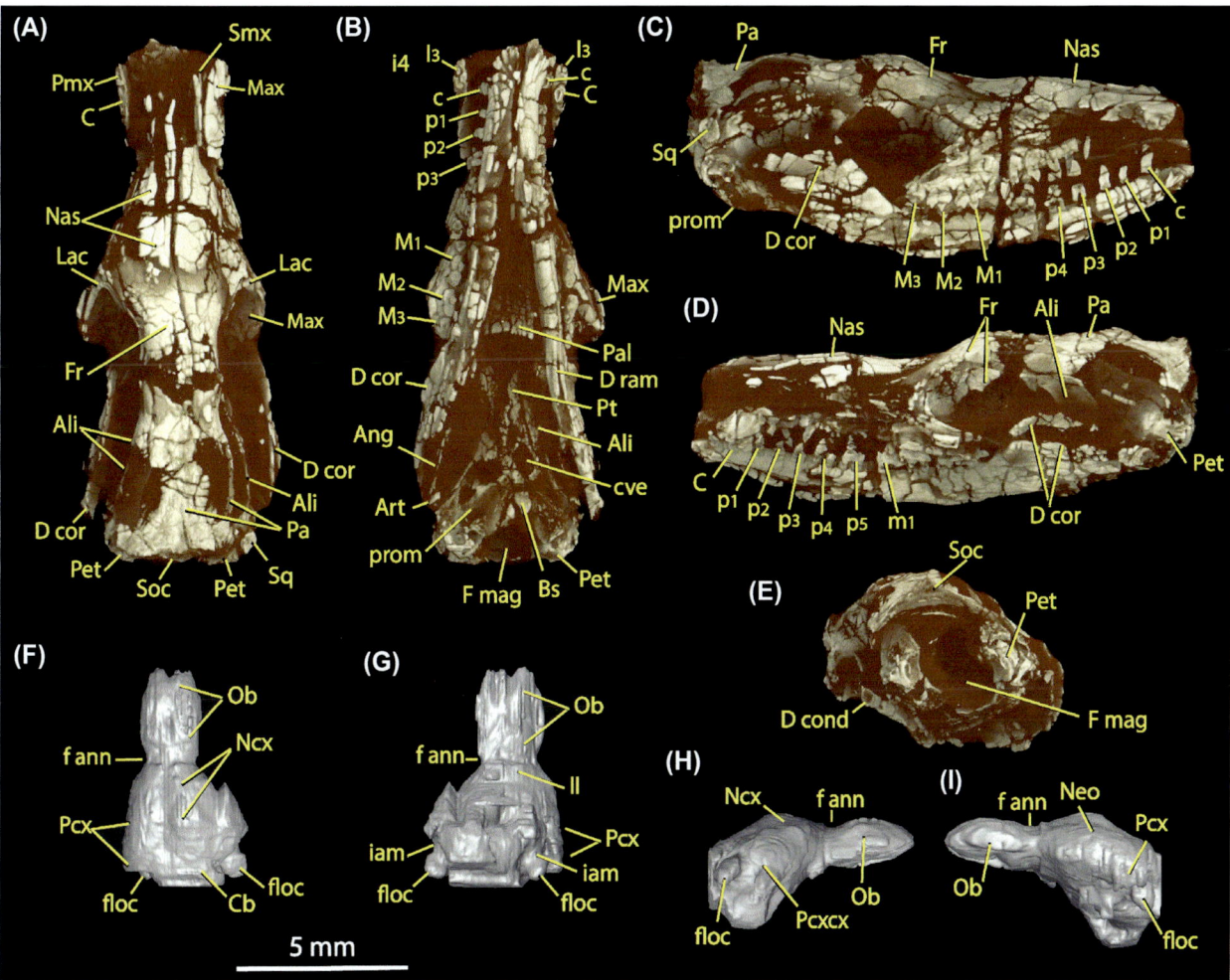

FIGURE 11.26 3D reconstructions of the skull and endocast of *Morganucodon*, based on computed tomography imagery (false colors). Skull in (A) dorsal, (B) ventral, (C) right lateral, (D) left lateral, and (E) occipital views. Endocast in (F) dorsal, (G) ventral, (H) right lateral, and (I) left lateral views. See Table 11.1 for abbreviations. *After Rowe, T.B., Macrini, T.E., Luo, Z.-X., 2011. Fossil evidence on origin of the mammalian brain. Science 332, 955–957. https://doi.org/10.1126/science.1203117.*

mammals. The brain in *Morganucodon* was nearly twice as large as the brain in early cynodonts such as *Thrinaxodon*. The EQ of early cynodonts ranges from ~0.16 to 0.23, whereas the EQ of *Morganucodon* is ~0.32, reflecting an increase of 30–50%. The olfactory bulb and olfactory cortex are by far the regions of greatest expansion (Figs. 11.26 and 11.27). A deep annular fissure encircles the olfactory tract, marking a distinctive external division of the brain between the inflated olfactory bulbs and the cortex. The cortex is inflated and wider than the cerebellum, covering the midbrain and the pineal stalk. The cerebellum is also enlarged, implying expansion of the basal nuclei, thalamus, and medulla, and the spinal cord is of greater relative diameter. In shape and proportions, the brain resembles the brain in a living mammal more than it resembles the brain in early cynodonts (Rowe et al., 2011).

In addition to the general shape of its endocast, there is indirect evidence that the origin of Mammaliaformes marks the emergence of the neocortex as a specialized region of the pallium, whether or not it had all six layers present in mammals. A pelt of modern aspect, with guard hairs and vellus underfur, was discovered in an early mammaliaform fossil from the Middle Jurassic (~165 million years ago) of China, *Castorocauda lutrasimilis* (Ji et al., 2006). Hair follicles have been described a "dynamic miniorgans" thanks to the remarkable complexity of each (Schneider et al., 2009). In all extant mammals, guard hairs are equipped with at least three different kinds of mechanoreceptors (Fig. 11.28) that induce the development of somatosensory maps on the outer layer of the neocortex (Rowe et al., 2011), and each is generally associated with its own musculature (the arrector pili muscles) and sebaceous glands. In

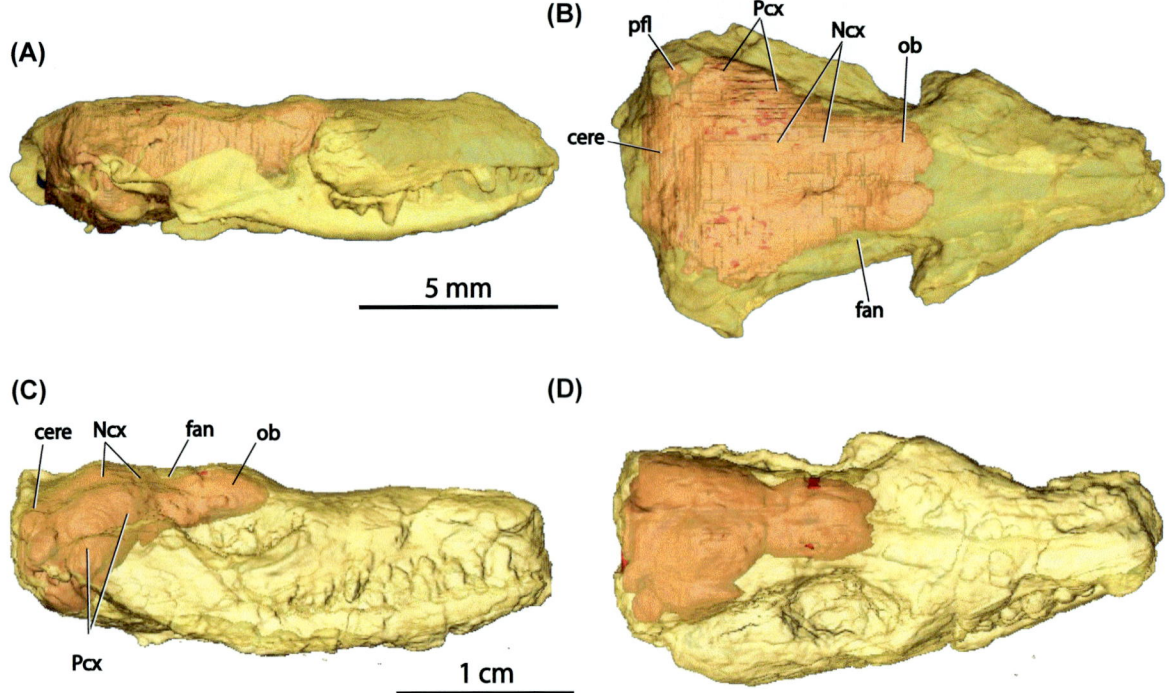

FIGURE 11.27 3D reconstruction of the skulls of *Hadrocodium* (A−B) and *Morganucodon* (C−D) from computed tomography data. Bones are rendered translucent to show the relationship of their endocasts to the surround skull. *After Rowe, T.B., Macrini, T.E., Luo, Z.-X., 2011. Fossil evidence on origin of the mammalian brain. Science 332, 955−957. https://doi.org/10.1126/science.1203117.*

living mammals with small brains (eg, *Monodelphis*, *Didelphis*), the small neocortex is dominated by a single primary somatosensory field that maps sensation from mechanoreceptors in the skin, hair follicles, muscle spindles, and joint receptors (Fig. 11.29). Its conscious component involves tactile exploration of the environment and body surface monitoring. Peripheral somatosensory input is mapped to the neocortex as an "animunculus" or a topographic representation. A parallel, underlying neocortical motor map is represented in pyramidal neurons whose axons form the pyramidal tract (corticospinal tract), which projects via the brain stem directly to the spinal column to program and execute skilled movements requiring precise control of distal musculature. Increased agility and coordination are correlative behaviors.

With the origin of Mammaliaformes, it appears likely that the characteristic architectonic and/or histochemical organization of the neocortex into a distinct primary somatosensory area (S1) a primary visual area (V1), and the primary auditory area (A1), had occurred. Despite the extreme morphological and behavioral specializations of many mammals, these fields are always present, even in the absence of apparent use. Their ubiquity, plus aspects of their corticocortical and thalamocortical connectivity, and their general geographic arrangement across mammals, and the fact that they cannot be eliminated under most experimental circumstances indicate

that they were present in the last common ancestor of living mammals (Krubitzer and Kaas, 2005), and the evidence summarized here suggests that their integration occurred at a deeper point in pan-mammalian history.

The discovery of fur in an ancient mammaliaform has additional implications for the emergence of mammalian features. During its ontogeny in living mammals, hair performs first as a tactile organ and only later does it insulate as underfur thickens and matures (Zelená, 1994). The body temperature of newborns is initially regulated by their mothers. This sequence in the ontogeny of the function of hair in Mammalia, in conjunction with the presence of a pelt in early mammaliforms, implies that parental care and endothermy were present in Mammaliaformes ancestrally (if not in Mammaliamorpha, Node 14). Endothermy may have been an evolutionary consequence of mammaliaform encephalization because a large brain operates properly only within narrow thermal tolerances, and it is metabolically the most expensive organ to maintain. Metabolism is under hormonal control that does not command large cerebral regions and thus did not itself drive encephalization (Rowe et al., 2011).

Whereas the auditory ossicles of the middle ear remained coupled to the mandible, their relative size and architecture suggests that they functioned much as the middle ear in mammals (Allin, 1975; Luo et al., 2001a). However, the cochlea in early mammaliforms

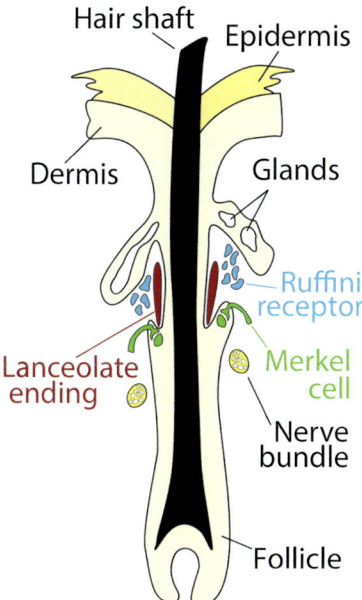

FIGURE 11.28 Diagram of a hair follicle and its innervation. *After Zelená, J., 1994. Nerves and Mechanoreceptors: The Role of Innervation in the Development and Maintenance of Mammalian Mechanoreceptors. Springer Science & Business Media; Rowe, T.B., Macrini, T.E., Luo, Z.-X., 2011. Fossil evidence on origin of the mammalian brain. Science 332, 955–957. https://doi.org/10.1126/science.1203117.*

differs from that of Mammalia in being much shorter in length and curving over only about 70 degrees, and they lack the bony lamina which supports the basilar membrane (Greybeal et al., 1989; Rosowski and Graybeal, 1991; Kielan-Jaworowska et al., 2004). Additional

diagnostic apomorphies of this region include expansion of the petrosal to provide a complete bony floor beneath the cavum epipterycum and presence of the petrosal promontorium (Gauthier et al., 1988a; Rowe, 1988, 1993).

The secondary palate extended to the caudal margin of the tooth row (Fig. 11.12H). Tooth replacement has a diphyodont pattern (Luo et al., 2004). The dentary has an expanded condyle that articulates with a well-developed glenoid fossa on the squamosal; the dentary and squamosal are now the primary load-bearing bones in the craniomandibular joint, although the quadrate and articular continued to participate. The dentition is differentiated into anterior bicuspid premolariform and posterior molariform teeth with three to five main cusps. Novel dental wear facets on the molariform teeth indicate that they occluded in a complex unilateral pattern in which rotational movement of the mandibles occurred during mastication (Kielan-Jaworowska et al., 2004; McKenna and Bell, 1997).

Early in mammaliaform history, the two minor cups of the molariform teeth become "circumducted" or rotated relative to the taller principal cusp, by moving to the outer side in the upper teeth, and to the inner side in the lowers, thus establishing cheek teeth that not only occluded, but also that interlocked to produce an elongated shearing edge for mastication. This condition is termed the "tritubercular" molar. Its fundamental importance in mammalian evolution was first recognized by Edward Drinker Cope (1883, 1888) who hypothesized that this was the basic form from which

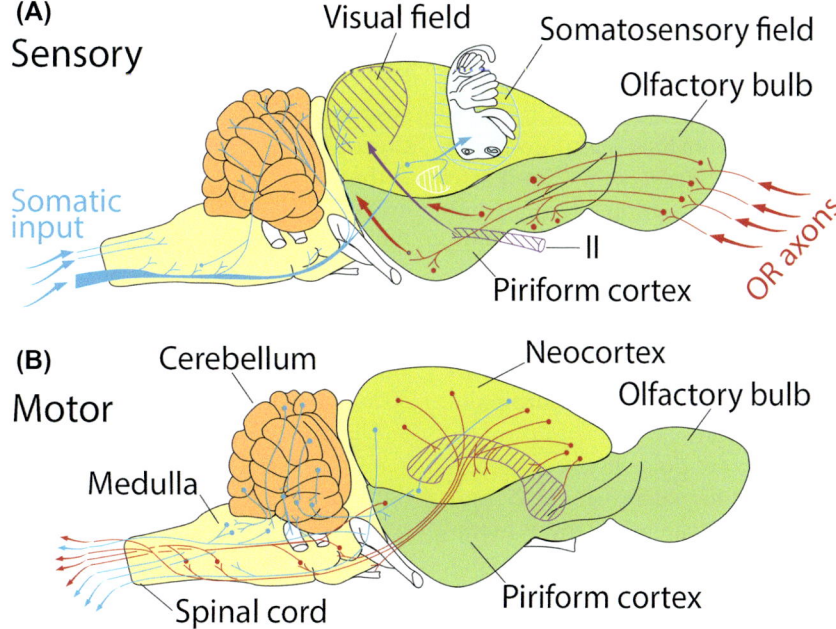

FIGURE 11.29 Circuitry schematic of brain of modern opossum (*Didelphis*) brain showing (A) sensory inputs and (B) motor outputs. *From Rowe, T.B., Macrini, T.E., Luo, Z.-X., 2011. Fossil evidence on origin of the mammalian brain. Science 332, 955–957. https://doi.org/10.1126/science.1203117.*

most or all mammalian teeth were derived, and it came to be known as Cope's tritubercular theory (Wortman, 1886, 1902; Gidley, 1906). Henry Fairfield Osborn (1888, 1907) traced the evolution of mammalian tooth crowns in considerable detail, and the theory came to be known as the Cope–Osborn theory (eg, Gregory, 1910, 1916, 1922). Although the detailed homologies of various cusps and cuspules in different mammalian subclades have been debated for decades, it is generally true that the tritubercular pattern represents the morphological primordium from which an immense diversity of crown morphologies ultimately evolved (Cope, 1883; Osborn, 1907; Gregory, 1922; Romer, 1966; Carroll, 1988; Benton, 2015; Luo et al., 2001b). A further step was taken later, as the tritubercular molariform tooth became elaborated into the dual-action "tribosphenic" molar along the stem of therian mammals (Osborn, 1907; Gregory, 1916, 1922, 1953). In tribosphenic dentitions, opposing upper and lower crowns have taken on the form of reversed-triangular interlocking molars, where the principal cusps support long crests that shear past one another, and where the principal upper molar cusp (protocone) bites into the lower talonid basin with a crushing action (Fig. 11.23).

In the mammaliaform postcranial skeleton, the lumbar vertebrae are markedly differentiated from the thoracic vertebrae (Fig. 11.16C), with vertical and anticlinal lumbar neural spines, and lumbar centra with articular faces that are inclined instead of vertical, such that the lumbar region is arched dorsally (Jenkins and Parrington, 1976). Otherwise, their postcranial skeletons were largely unchanged from those in basal mammaliamorphs.

11.5.16 Node 16: Unnamed

The Early Jurassic fossil *Hadrocodium wui* (Luo et al., 2001) is known from a single skull, from the Early Jurassic of China. Its systematic position is somewhat uncertain, with some analyses finding it to lie just within the mammalian crown (Rowe et al., 2008), and others finding it to be the sister taxon to Mammalia (Rowe et al., 2011; Luo et al., 2015). It is expedient to treat it as the sister taxon to Mammalia here, because it demonstrably lacks some of the features of crown Mammalia relevant to the present discussion.

In *Hadrocodium*, a third discrete pulse in encephalization is manifest, raising its EQ to ~0.5 (Figs. 11.27 and 11.30), a level that lies within the range of crown mammals (Rowe et al., 2011). This had occurred by about 190 million years ago. Most of this increase in relative size is in the olfactory bulbs and olfactory cortex. Its cerebellum has also expanded to such a degree that the occipital plate bulges backward, enclosing a relatively

large foramen magnum and spinal cord. Only a few features separate *Hadrocodium* from crown mammals. Most importantly, it lacks ossified turbinals, and large pterygoid processes indicate that its bilateral chewing and swallowing mechanics were transitional to those in crown Mammalia. Once the brain reached this level of relative size, further increases in encephalization occurred independently in many mammalian clades (Jerison, 1973; Rowe et al., 2011).

Once the brain reached the relative size found in *Hadrocodium*, and with the evidence that neocortex had by this time emerged, subsequent independent increases in encephalization came to characterize the diversification of the various mammalian clades. It is noteworthy that evolutionary decreases in encephalization are rare among mammals. One of the best documented cases is in the platypus lineage, where more highly encephalized Cenozoic fossils indicate that a ~10% reduction of both the olfactory bulb and overall EQ occurred as the living platypus *Ornithorhynchus* adapted to its semiaquatic habitus (Macrini et al., 2006). Decreases in encephalization are also associated with domestication in various mammalian species (Kruska, 2007).

An extraordinary morphogenic consequence of the expanded olfactory cortex in *Hadrocodium* and mammals is that the auditory chain was disrupted during ontogeny, and those ossicles directly involved in hearing were detached from their ancestral and embryonic position on the mandible, relocated a short distance behind the mandible, and suspended exclusively from beneath the braincase (Rowe, 1996a,b). The mammalian middle ear is a highly distinctive chain of three tiny ossicles that transmit vibrations received by the tympanic membrane to the fenestra vestibuli of the inner ear (Fig. 11.31). These bones become detached from (Fig. 11.32) the mandible as the brain grows, becoming suspended from a new position beneath the cranium during maturation (Fig. 11.33). As a result, the dentary is the only bone in the adult mandible, and the dentary articulates with the squamosal as the sole elements of the mature craniomandibular joint. Whether this is a unique autapomorphy of Mammalia, or Mammalia plus *Hadrocodium*, or if it represents convergent evolution among mammals is controversial (Rowe, 1988, 1996a,b; Bever et al., 2005; Luo et al., 2007; Ji et al., 2006; Meng et al., 2006).

Discovery of the Early Cretaceous basal mammal *Yanoconodon allini* (Luo, 2007; Luo et al., 2007) challenged the hypothesized ontogenetic relationship between telencephalic expansion and detachment of the middle ear ossicles (Rowe, 1996a,b), because it preserves both a large brain and an ossicular chain still connected to the jaw. However, numerous features of the only known skeleton attest to immaturity of this specimen at time of death and that it corresponds to a 3- to 4-week-old opossum in

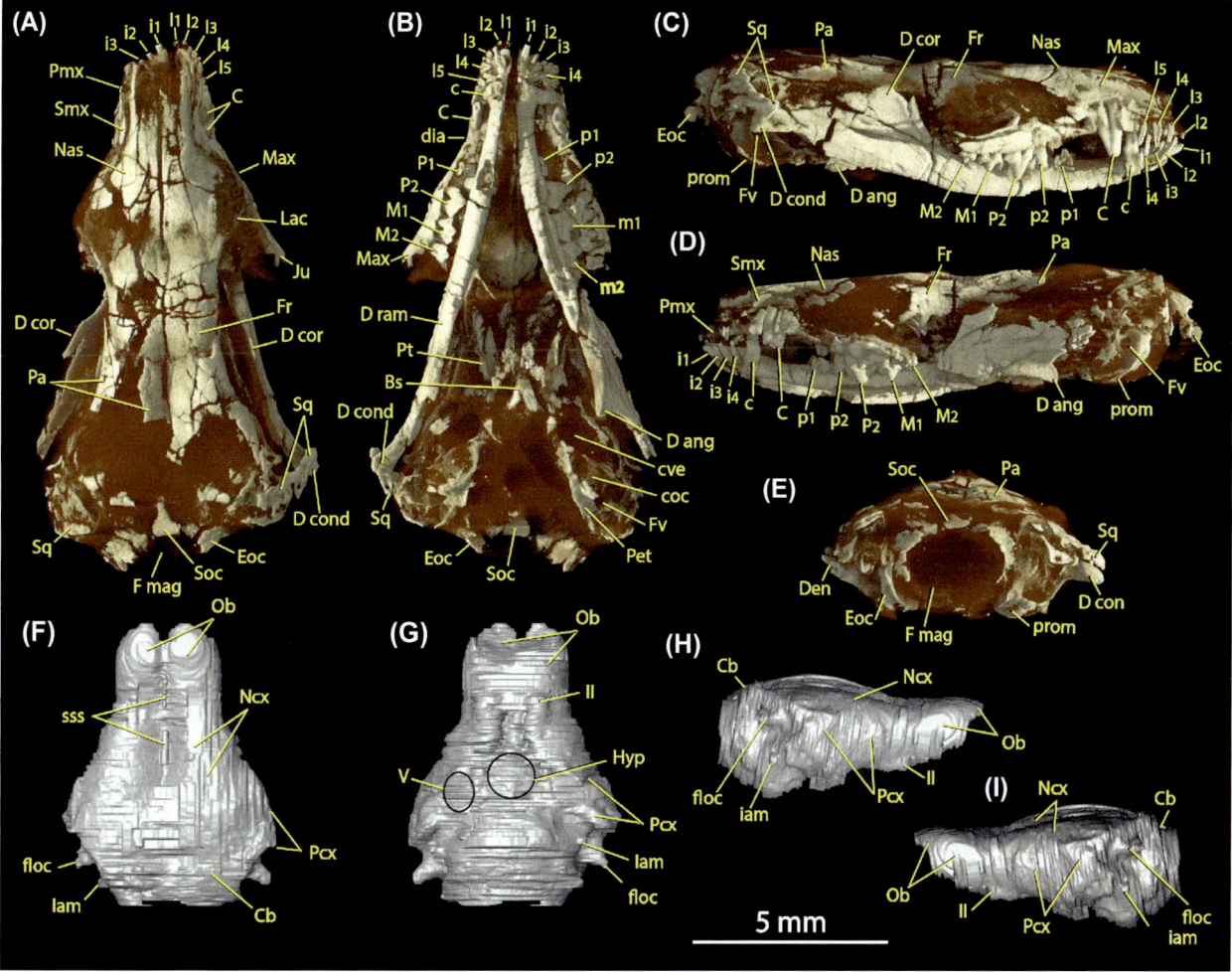

FIGURE 11.30 3D reconstructions of the skull and endocast of *Hadrocodium*, based on computed tomography imagery (false colors). Skull in (A) dorsal, (B) ventral, (C) right lateral, (D) left lateral, and (E) occipital views. Endocast in (F) dorsal, (G) ventral, (H) right lateral, and (I) left lateral views. See Table 11.1 for abbreviations. *After Rowe, T.B., Macrini, T.E., Luo, Z.-X., 2011. Fossil evidence on origin of the mammalian brain. Science 332, 955–957. https://doi.org/10.1126/science.1203117.*

which detachment of the ossicles has yet to occur. Lack of fusions in the atlas, axis, and along the rest of the vertebral column, and between pelvic elements indicates that *Yanoconodon* presents an ontogenetic transitional stage rather than a phylogenetic intermediate and that this individual died before the position of the ear ossicles had matured. Its dentition indicates that it is referable to the taxon *Jeholodens jenkinsi*, known from more mature specimens from the same formation, rather than being a separate taxon. An additional challenge to this interpretation of transformation of the mammalian ear ossicles came from the discovery of an ossified Meckel's cartilage in several euticonodont fossils from the Cretaceous of China (Wang et al., 2001; Luo et al., 2007). This feature is securely known only within euticonodonts, and it represents an autapomorphic condition of that clade rather than a transitional mechanism in evolution of the mammalian ear. Its functional implications are difficult to interpret in light of the fact that the auditory ossicles

making up the transmission chain otherwise resemble the middle ear chain in mammals and are suspended from beneath the cranium. The larger point is that cortical expansion had a remarkable morphogenic impact on cranial architecture during the evolution of Mammaliaformes and crown Mammalia and that it explains one of the historically most problematic transformations in early mammalian evolution.

11.5.17 Node 17: Crown Clade Mammalia

Mammalia comprises the last common ancestor of living therians and monotremes, and all its descendants. The fossil records indicate that monotremes and therians had diverged by the Middle Jurassic, by about 170 million years ago.

Compared to other living animals, Mammalia is distinguishable by features of virtually all anatomical systems, from genes and cells to organ systems, and

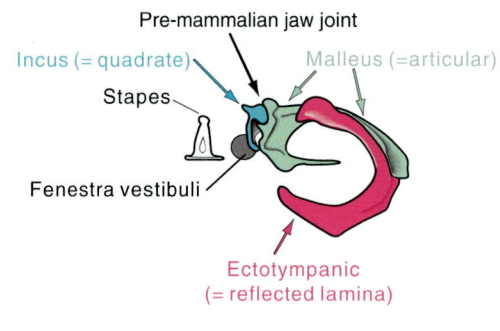

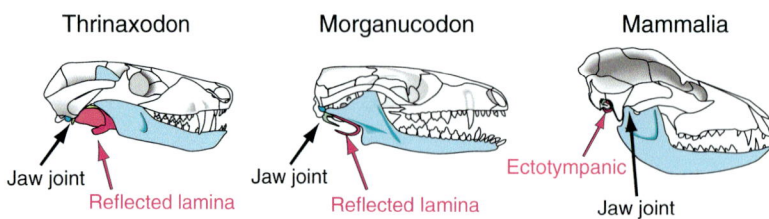

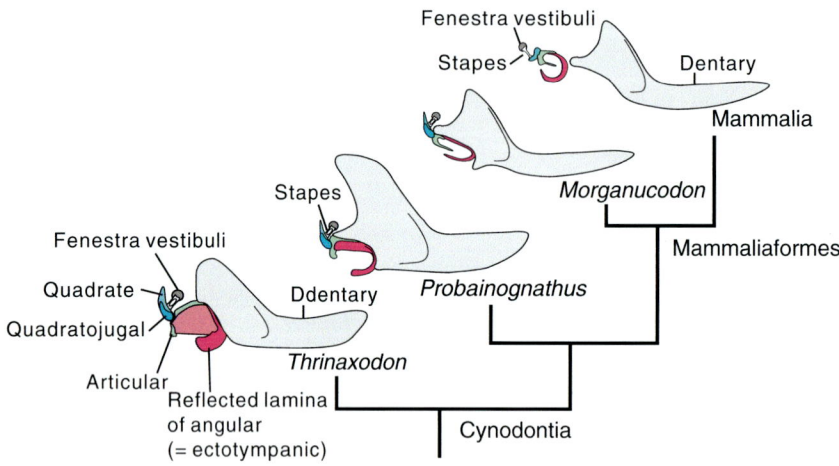

FIGURE 11.31   Phylogenetic transformations of middle ear ossicles from their ancestral position attached to the mandible to their detachment and displacement behind the craniomandibular joint in Mammalia. *After Rowe, T.B., 1996a. Coevolution of the mammalian middle ear and neocortex. Science 273, 651–654; Rowe, T.B., 1996b. Brain heterochrony and evolution of the mammalian middle ear, pp. 71–96. In: Ghiselin, M., Pinna, G. (Eds.), New Perspectives on the History of Life. California Academy of Sciences, Memoir 20.*

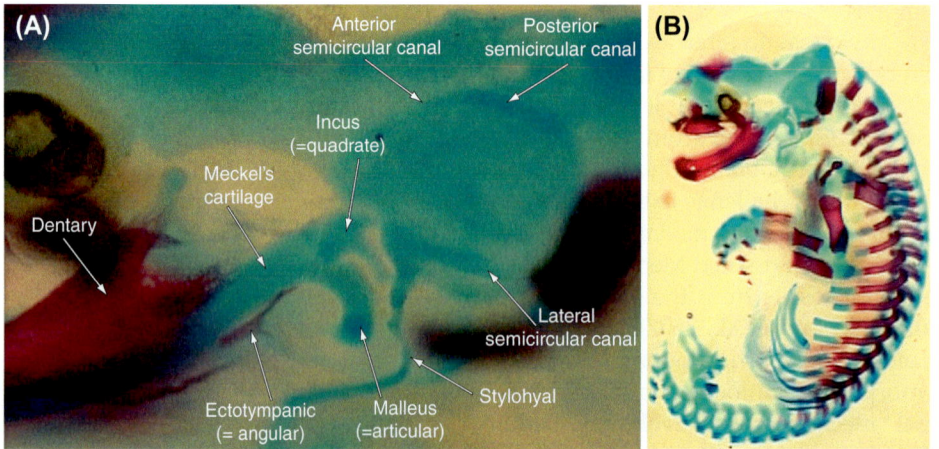

FIGURE 11.32   *Monodelphis* at day 3, cleared and stained to show developmental primordia of middle ear ossicles. (A) Close-up of temporal region showing developmental primordia of middle ear ossicles and (B) the entire skeleton.

3. Early Mammals and Subsequent Adaptations

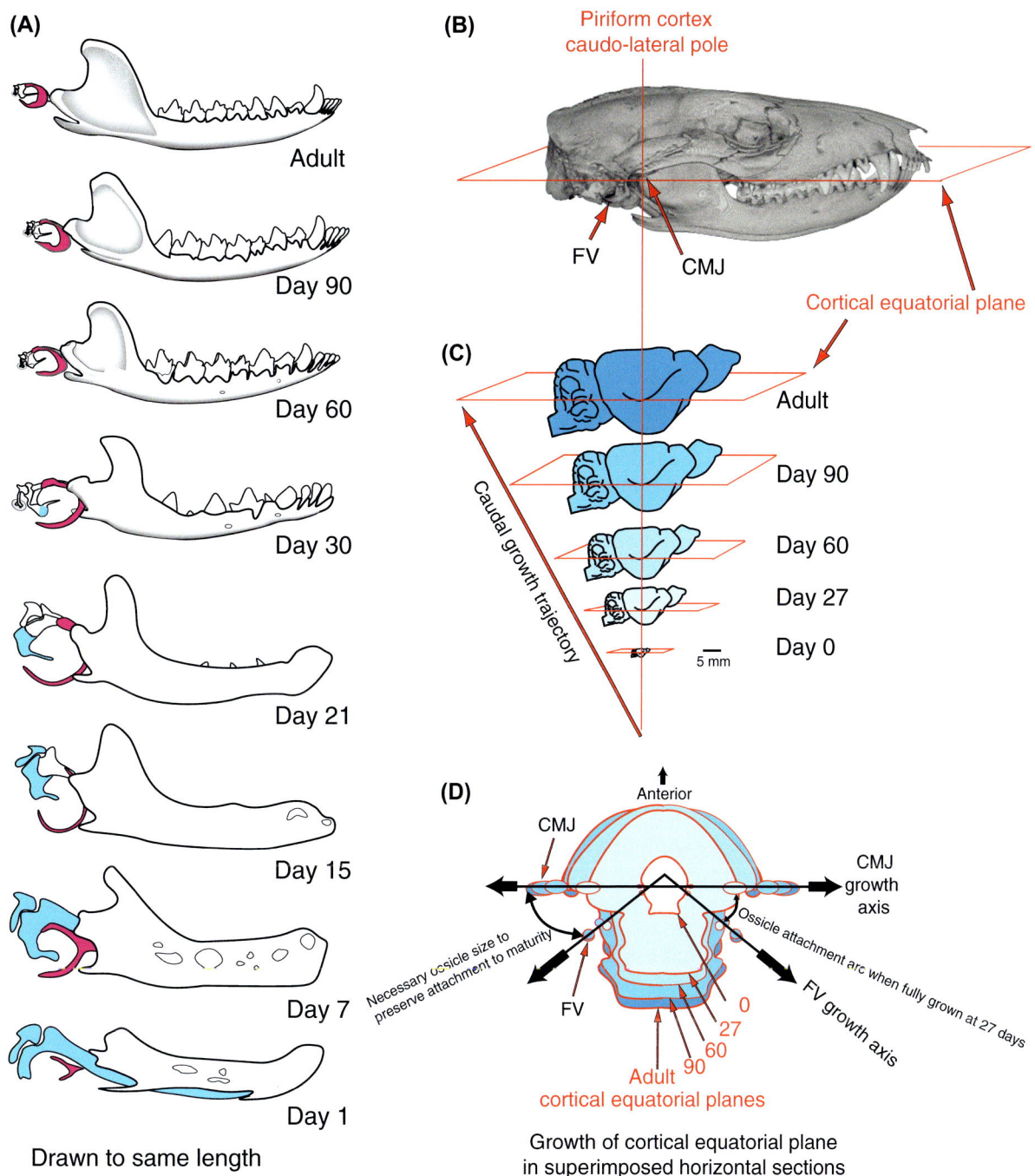

FIGURE 11.33 Ontogeny of the middle ear ossicles and brain in *Monodelphis domestica*, showing differential growth of the middle ear ossicles and brain. (A) Mandibles of a growth series drawn to same length, showing initial positive allometry of ear ossicles for first 3 weeks, followed thereafter by negative allometry. (B) Adult *Monodelphis* showing position of craniomandibular joint on plane of cortical equator. (C) Growth of the brain of *Monodelphis*, drawn to same scale, with plane of cortical equator in *red*. (D) Growth of the brain of *Monodelphis* depicted as horizontal slices taken at the cortical equator, showing relative repositioning of the craniomandibular joint (CMJ), the axis of migration of the fenestra vestibuli (FV), and arc of ossicular attachment *After Rowe, T.B., 1996a. Coevolution of the mammalian middle ear and neocortex. Science 273, 651–654; Rowe, T.B., 1996b. Brain heterochrony and evolution of the mammalian middle ear, pp. 71–96. In: Ghiselin, M., Pinna, G. (Eds.), New Perspectives on the History of Life. California Academy of Sciences, Memoir 20.*

from molecular sequences to behavior. Among its most distinctive features is a relatively large brain in which the central region of the telencephalic pallium is the six-layered neocortex (Figs. 11.5 and 11.29). It receives afferents from the cerebellum and basal ganglia that project via specialized regions of the thalamus to restricted neocortical regions. The neocortex is dominated by two or more layers of pyramidal cells whose association projections remain almost exclusively within the cortex and basal ganglia (ie, intratelencephalic). From the neocortex project specialized corticothalamic cells, and pyramidal tract motor output cells project throughout the brain stem and directly to the spinal cord (ie, the corticospinal tract). The medial telencephalic wall is a well-developed hippocampus in which dorsal and anterior commissures interconnect pallial structures of the two cerebral hemispheres. Well-developed specific motor nuclei are situated rostrally in the ventral half of the thalamus. Mammalian ependymal cells fail to reach the periphery of the brain in adults (Rowe and Shepherd, 2016), and adult neurogenesis is restricted to regions in the hippocampus (Chapouton et al., 2007; Ming and Song, 2005).

Whether this full measure of organization was achieved in early Mammaliaformes such as *Morganucodon* or *Hadrocodium* is uncertain. As we have now seen, however, a series of transitional fossils allow the inference of a deeper history for much of this organization. The increase in brain size diagnostic of Mammalia represents a continuation of the trend first established as diagnostic of early Cynodontia and further elaborated in Mammaliaformes in which successive increases in encephalization were driven largely, but not exclusively, by enhanced olfaction as the increased numbers of odorant receptor genes became the largest single gene subfamily in the mammalian genome. The fact that airborne olfaction was a driving factor in cynodont encephalization is further suggested by one of the rare instances in which reduction in encephalization is documented, in the platypus clade, whose main olfactory bulb and overall EQ were reduced with increasing commitment to a semiaquatic habitus (Macrini et al., 2006). Although some secondarily aquatic mammals have reduced olfactory capacities (cetaceans, platypus), their large olfactory genomes have converted to pseudogenes to varying degrees, reflecting their descent from hyperosmic ancestors (McGowen et al., 2008; Rowe et al., 2005).

The basic skeletal armature of a muscular diaphragm is diagnostic of Cynodontia, and in Mammalia its complete expression can be seen as a fundamental component of the mammalian system of ventilation. The diaphragm encloses the pleural cavities, marking the transition from thoracic to lumbar vertebrae, and it drives diaphragmatic breathing. The lungs are complex and divided into lobes, bronchioles, and alveoli; and they expand ventrally, surrounding the heart and almost meeting on the ventral midline, leaving only a ventral mediastinum connecting the pericardial sac with the ventral body wall.

Thanks to the remarkably preserved Chinese fossil *C. lutrasimilis* (Ji et al., 2006), the "mammalian" pelt is now known to diagnose the more inclusive clade Mammaliaformes (Node 15). Speculation abounds that vibrissae may have been present among members of the mammalian stem group (Benoit et al., 2016). However, there is no evidence of an integrated system of vibrissae with differentiated facial muscles, or a cortical motor-sensory region capable of dealing with such a system until therian mammals arose. Moreover, these specialized organs and associated cortical barrel columns occur in only a few therian clades (Catania and Catania, 2015). In mammals, the *musculus panniculus carnosus* forms a continuous sheath of muscle wrapping the trunk and neck and differentiates into groups of mobile facial musculature associated with the eye, ear (pina), and snout. Available evidence suggests that monotreme facial musculature was never highly differentiated and that more elaborate differentiation is a feature restricted to certain clades of therians (Huber, 1930a,b; Lightoller, 1942).

Mammalia can be also distinguished from its closest extinct relatives by many features of the skeleton. The internasal process of the premaxilla is absent, leaving the external nares confluent, and an intricate internal skeletal armature develops within the cartilaginous nasal capsule to form the ethmoid complex (Figs. 11.5 and 11.34). It forms an elaborate, delicate scaffolding of thin, interfolded bony sheets known as ethmoid turbinals (or turbinates). There has been much speculation that an elaborate cartilaginous armature may have been present in members of the mammalian stem group (Ruben, 1995), but the ethmoid turbinals of mammals are membranous ossifications, and no mammal is known to have an elaborate cartilaginous ethmoid precursor. Ossification of the ethmoid turbinals afforded a 10-fold increase in the surface area of olfactory epithelium that could be deployed inside the nasal cavity (Rowe et al., 2005). The ethmoid turbinals coalesce around the olfactory nerve fascicles to form the bony cribriform plate, a compound structure that separates the olfactory recess from the cavum cranii. The turbinals grow rostrally from the cribriform plate (Fig. 11.20A) as the olfactory epithelium matures, and their mature geometry is highly variable among mammals. Also ossifying in the nose is the maxillary turbinal (Fig. 11.5), which increases the epithelial surface area by nearly an order of magnitude and is involved in regulating respiratory moisture and heat exchange (Rowe et al., 2005; Green et al., 2012).

A double-occipital condyle arose in Cynodontia, and in Mammalia the condyles expand to surround the entire ventral half of the foramen magnum. Correspondingly, the mammalian atlas, or first vertebra, is highly distinctive in that it forms a bony ring via the ontogenetic fusion of three separate ossification centers that remained separate throughout life in all other members of Pan-Mammalia. The cervical ribs are also apomorphic

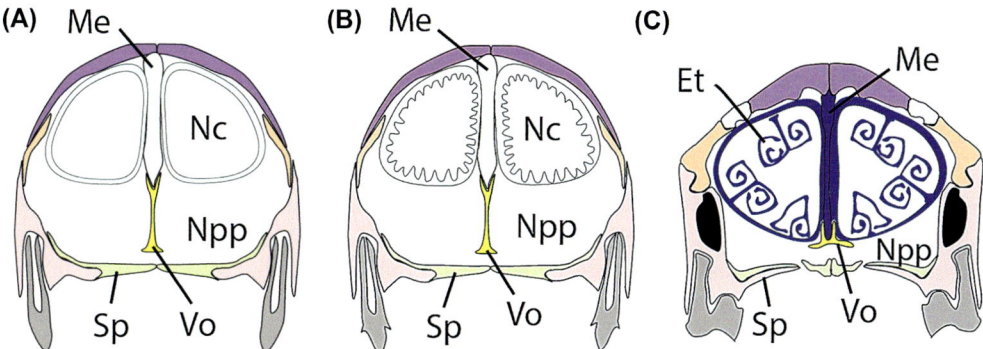

FIGURE 11.34 Schematic through nasal capsule in (A) hypothetical basal cynodont with smooth nasal capsule, (B) an early mammaliaform with expanded olfactory surface area, and (C) an extant mammal in which the ethmoid turbinals have ossified and invaded the lumen of the nasal capsule.

in fusing to their respective vertebrae in early ontogeny. The limbs and girdles develop secondary ossification centers, the most obvious of which are in the cartilaginous epiphyses of the long bones. Sesamoid bones form in tendons of the flexor muscles of the hands and feet, and in the hindlimb a single large sesamoid forms the patella (Rowe, 1988). These modifications correlate with increased thickness and regionalization of the spinal cord, owing in part to the advent of the corticospinal tract, and to increased agility to which the sesamoid bones may contribute.

The ancestral mammal was thus a tiny scansorial terrestrial creature that scurried and climbed over complex three-dimensional surfaces of its microhabitat, carrying its young in a pouch, and nurturing them with milk and warmth until they were self-sufficient in feeding and could regulate their own body temperatures. With a neocortex and corticospinal tract, it was exceedingly agile and quick and used olfaction in navigation, scent-tracking, and myriad social behaviors. It had color image—forming eyes, lived under the sunlight, and used its conjugated eyes to inspect the world around it in stereo, while it listened to the higher-frequency sounds of the environment. Above all, it was intensely sampling and processing a surrounding information sphere that was dominated to an unprecedented degree by smells, odors, and scents.

11.6 Discussion

There is a wealth of data in developing mammals which indicates that peripheral sensory arrays play an enormous role in the development of cortical organization. Peripheral sensory information can induce and determine the functionality and relative sizes of cortical fields, while also influencing their internal organization and connectivity (Zelená, 1994; Farbman, 1988, 1990; Krubitzer and Kaas, 2005), as well as the overall size of

the cortex and other parts of the brain (Rowe, 1996a,b; Rowe et al., 2011; Rowe and Shepherd, 2016). From this survey of the early history Pan-Mammalia, it also seems evident that evolutionary transformations in peripheral sensory arrays had long-term impacts on evolution of the mammalian brain and the emergence of neocortex. In both ontogenetic and phylogenetic transformations, interactions are evident between genetic and epigenetic factors that affected peripheral sensory information acquisition and processing, as well as in consequent morphogenesis of the supporting skeleton.

The single dominant pattern in the emergence of mammals is a cascade which followed from the order-of-magnitude increase in the olfactory odorant receptor subgenome through the mechanism of gene duplication. One cannot pinpoint the precise phylogenetic moments of these duplication events, because evidence from elsewhere in chordate phylogeny suggests that gene duplication events may occur well before the new genes find their initial expression, (Rowe, 2004). However, the onset of new gene expression is indicated by modifications of those parts of the skeleton involved in supporting expanded peripheral and central processing regions, owing to ontogenetic interdependencies regulating development of the mature system.

Odorant receptor gene expression determines the number of functional odorant receptors, while influencing olfactory epithelial surface area. This in turn influences the numbers of glomeruli that are induced as odorant receptor axons make their first synapse in the presumptive main olfactory bulb, where the numbers of arriving axons influence the size of the olfactory bulb and the numbers of olfactory glomeruli (Farbman, 1988, 1990; Chen and Shepherd, 2005; Schlosser, 2010). Glomeruli communicate to form odor "images" that project to the piriform cortex (Figs. 11.22 and 11.29) and then to the neocortex for higher level associative processing (Hildebrand and Shepherd, 1997; Bargman, 2006; Mori et al., 2006). Thus ontogenetic interdependencies exist between odorant receptor gene numbers, the numbers and types of expressed odorant receptors, olfactory epithelium surface areas and in mammals the olfactory turbinal surface area, the size of the main olfactory bulb, and size of the piriform cortex (Rowe and Shepherd, 2016). There is evidence that the numbers of odorant genes may even influence the degree of piriform cortical layering, which in the echidnas (monotremes) have the largest turbinal surface area yet measured in any mammal and a five-layer piriform cortex (Ashwell, 2013). Odorant receptors, populating as they do the olfactory epithelium of the nasal capsule, comprise a peripheral sensory array whose diverse evolutionary transformations profoundly influenced the organization the brain and supporting skeleton to a degree in mammals unique among vertebrates.

The remarkable hypertrophy in innervation of the dentition amounted to the advent of a virtually new peripheral array, which in crown mammals commands a considerable neocortical presence by inducing the individuation of dental sensory fields. The origins of the dental array trace back into earliest cynodont history (Node 9) beginning with the transformation of a simple prehensile dentition into an integrated system of repeated parts designed for complex occlusion. The addition of thecodont tooth implantation in eucynodonts (Node 10) and incipiently divided roots (Node 11) further magnified textural information while also integrating the dental array as a discrete module. As ever-replacing teeth gave way to "permanent" molariform teeth in Mammaliaformes (Node 15), new avenues to analysis, learning, and memory in regard to food evolved. Although it did not arise until some 50 million years of pan-mammalian history had passed, the cynodont dental array quickly took on its own modular identity, and a novel variational modality that would eventually affect dental morphology in virtually all mammalian clades, even to the level of marking the identity of most individual mammalian species.

To frame this transformation in dental variational modality in a simple quantitative perspective, consider the basis of a series of phylogenetic analyses that scored cynodont dental characters. Gauthier et al. (1988a) scored 207 characters across 29 taxa that characterized skeletal variation across the major amniote clades, including the entire stem groups of Mammalia and Reptilia. Only 8% (17) of these characters reflect dental variation. At this scale of analysis, most of the observed dental variation among pan-mammals presented itself as autapomorphy and was phylogenetically uninformative. But for data matrices designed to capture variation specifically among extinct cynodonts and early mammals, Meng et al. (2006) scored 435 total osteological characters for 58 taxa, and 25% (108 characters) reflect dental variation. Building on that matrix, Ji et al. (2006)

scored 445 characters for 103 taxa, of which 39% (173) describe dental variation. In a matrix of 3660 osteological characters for 86 fossil and living Mammaliaformes, O'Leary et al. (2013) found 40% of total skeletal variation (1450 characters) to reside in the dentition. Additionally, in a study of cynodont (including early mammalian) relationships, Rowe (1993) found in a matrix of 151 characters for 24 taxa that there was 30% more homoplasy in cynodont dental character states than in either the skull, the postcranium, or the combined skeleton. Whereas these numbers all beg systematic questions about character-state independence and weighting (Harjunmaa et al., 2014), they illustrate the general trend of accelerated rate in dental evolution and increased homoplasy that began with the first cynodonts and can be traced into many clades among Mammalia (Rowe and Shepherd, 2016).

In combination with the hypertrophied olfactory system, the new peripheral dental array was quickly subsumed into the larger integrated, centralized system of orthoretronasal olfaction in which information from orthonasal smell, retronasal smell, taste, and somatosensory signals from the lips, tongue, and teeth, not to mention sensations of expectation and anticipation, all converge on individual neurons in the neocortical area known as the orbitofrontal cortex (De Araujo et al., 2003; Small et al., 2007; Rolls and Grabenhorst, 2008; Shepherd, 2012). This higher multimodal system emerged during cynodont history as a multisensory map in which classes and combinations of information are integrated in ways unique to mammals (Rowe and Shepherd, 2016).

Another peripheral sensory array of cascading influence was the emergence of the mammalian pelt, derived from thousands of tiny placodes. Each matures into a highly complex hair follicle, a "dynamic miniorgan" (Schneider et al., 2009), that is innervated by three or more types of tactile receptors and has its own pair of arrector pili muscles and sebaceous gland (Fig. 11.28). Signals from developing hair follicles are known to induce sensory fields in the neocortex (Zelená, 1994). The presence in early Mammaliaformes (Node 15) of a pelt, and a brain approaching the size and shape of extant mammalian brains, offers the most compelling evidence that a small neocortex had differentiated by this time and that its primary somatosensory, visual, and auditory fields were probably also differentiated in these closest extinct relatives of Mammalia (Rowe et al., 2011). As explained earlier, one can infer that endothermy and parental care may also have been present in early Mammaliaformes.

In summary, this history of Pan-Mammalia supports the idea that transformations in peripheral sensory arrays profoundly affected the organization and evolution of the central nervous system as a dominant theme in the emergence of mammals. Whether through connectional invasions and epigenetic population matching (Katz and Lasek, 1978; Streidter, 2005), or some other developmental mechanism, hypertrophy in peripheral sensory arrays involving olfaction, dentition, and integument produced cascading influences on the general organization of the mammalian neurosensory system and indeed on general physiological and behavioral repertoires that are so distinctive of mammals today.

These transformations of the neurosensory system had profound morphogenetic effects on the skeleton. In numerous cases this is reflected in transformed variational modality in systems of repeated parts such as the vertebral column and limbs, for example, in the individuation of the atlas—axis complex (Node 1 or before), the establishment of seven cervical vertebrae (Node 4 or 5) and differentiation of thoracic and lumbar regions (Node 9). Another trend was simplification of the individual bones of the skull. However, this trend was disrupted with the induction of the compound alisphenoid bone, which formed as the caudal pole of the piriform cortex expanded posterolaterally in the encephalization event manifest in the earliest cynodonts (Node 9). A more profound transition occurred in Mammaliaformes (Node 15) with swelling of the endocranial cavity in response to a second pulse in encephalization, and in *Hadrocodium* (Node 16) plus crown Mammalia, as yet another pulse of encephalization carried its brain to within the size range of living mammals.

This latter pulse had the remarkable ontogenetic effect of detaching and relocating the developing middle ear ossicles (Fig. 11.33) from their embryonic (Fig. 11.32) and ancestral (Fig. 11.31) attachment to the mandible, and physically moving them to a new position behind the mandible, suspended solely from beneath the cranium. This transformation was driven by differential growth and famously recapitulates the phylogenetic transformation involving a shift of the middle ear ossicles to negative allometric growth (Nodes 12, 14, 15) and finally detachment from the mandible (Fig. 11.33; Node 16). A study of ontogeny of *Monodelphis domestica* revealed that the ear ossicles reach mature size by the end of the fourth week, when they stop growing, whereas the brain continues to grow for 90 days or more (Rowe, 1996a,b). The trajectory of brain growth, and specifically that of the piriform cortex precisely describes the timing of ossicle detachment, and the subsequent ontogenetic movements observed as the otic capsule and middle ear ossicles are swept backward with the expanding brain to achieve their mature positions (Fig. 11.34). Fossil-based objections have been raised and assertions of homoplasy notwithstanding (see Section 11.5.15), this mechanism of differential growth of the ossicles and brain in morphogenesis of the mature mammalian middle ear has yet to be falsified. Insofar as this morphogenic event indeed proves

ultimately traceable to the onset of expression of duplicated odorant receptor genes, as argued earlier, assertions of homoplasy take on an exceptional burden of evidence (Bever et al., 2005).

In his probing exploration of morphological innovation, Günter Wagner (2014) made the heuristically valuable distinction between Type I and Type II innovations or novelties. Type I novelties involve the origin of novel *character identity*, such as the vertebrate head and insect wing. Type II innovations involve establishment of a novel *character state*, such as the origin of tetrapod limb from paired fins and the modification of epidermal scales to form feathers. In Type I innovations, Wagner acknowledged a special role for gene duplications and new gene regulatory networks and that Type I innovations are generally associated with the far more common Type II innovations which involve new character states and new variational modalities in anatomical systems of repeated structures.

With development of the radial unit hypothesis of neocortical architecture (Rakic, 1988, 2000), it became generally appreciated that duplications in genes directing cell proliferation in the mammalian ventricular zone and subventricular zone may have been involved in the great numbers of radial units that produced a 1000-fold increase in cortical surface area compared to the plesiomorphic condition in amniotes (Rubenstein and Rakic, 1999; Rakic, 2009; Marín and Rubenstein, 2001; Ayoub et al., 2011). Less is known at present about the genetic mechanism that promoted the immense proliferation of epidermal placodes involved in producing the mammalian pelt (Rogers, 2003, Morris et al., 2004; Schneider et al., 2009). However, duplications in the olfactory subgenome are firmly established.

Over the course of pan-mammalian history, new variational modalities evolved in diverse anatomical regions, especially in systems of repeated parts. The brain and particularly the neocortex being a segmental structure at many levels of organization (Nieuwenhuys et al., 1998) can be counted among these, starting with the radial units that are fundamental to its organization (Rakic, 1988) and extending to its six-layer organization, to the immense proliferation of pyramidal neurons as its basic cell type, and the resulting proliferation of "canonical" neuronal circuits (Shepherd, 2011). For the first 50 million years of pan-mammalian history, little change in the relative size and organization of the brain can be discerned from the fossil record. However, with the origin of Cynodontia (Node 9), several discrete pulses in encephalization have been recognized that preceded the origin of Mammalia. Moreover, once the brain reached an EQ of ~0.5, and with evidence that neocortex had by this time differentiated (Nodes 15, 16), multiple independent increases in encephalization began to evolve (Fig. 11.35). Thus the variational modality of the brain itself shifted from virtual stasis to one of repeated independent episodes of encephalization that subsequently characterized many of the mammalian subclades. Independent increases in brain size, particularly neocortical surface area, would become a dominant pattern in the history of diversification of individual clades within the mammalian crown clade (Node 17) (Jerison, 1973; Rowe et al., 2011).

In closing, numerous mechanisms have been postulated as driving the emergence of mammals including relative enlargement of the brain (Nodes 9, 15, 16), differentiation of the neocortex, fur (Node 14), endothermy (Node 14), nocturnality (therian mammals), parental care of young (Node 14 or 15), miniaturization (Node 14), enhanced olfaction (Nodes 9, 15, 16) and hearing (Node 12, 14), and others. Whereas these features distinguish living mammals from other living clades, as we have seen conceptualizing Mammalia as internested within the more inclusive clades Mammaliaformes, Mammaliamorpha, Cynodontia, and Therapsida, and Pan-Mammalia brings a sharper focus and a nuanced phylogenetic context to this history. Current evidence suggests that most of these factors were influential at certain periods in pan-mammalian history, but all can now be

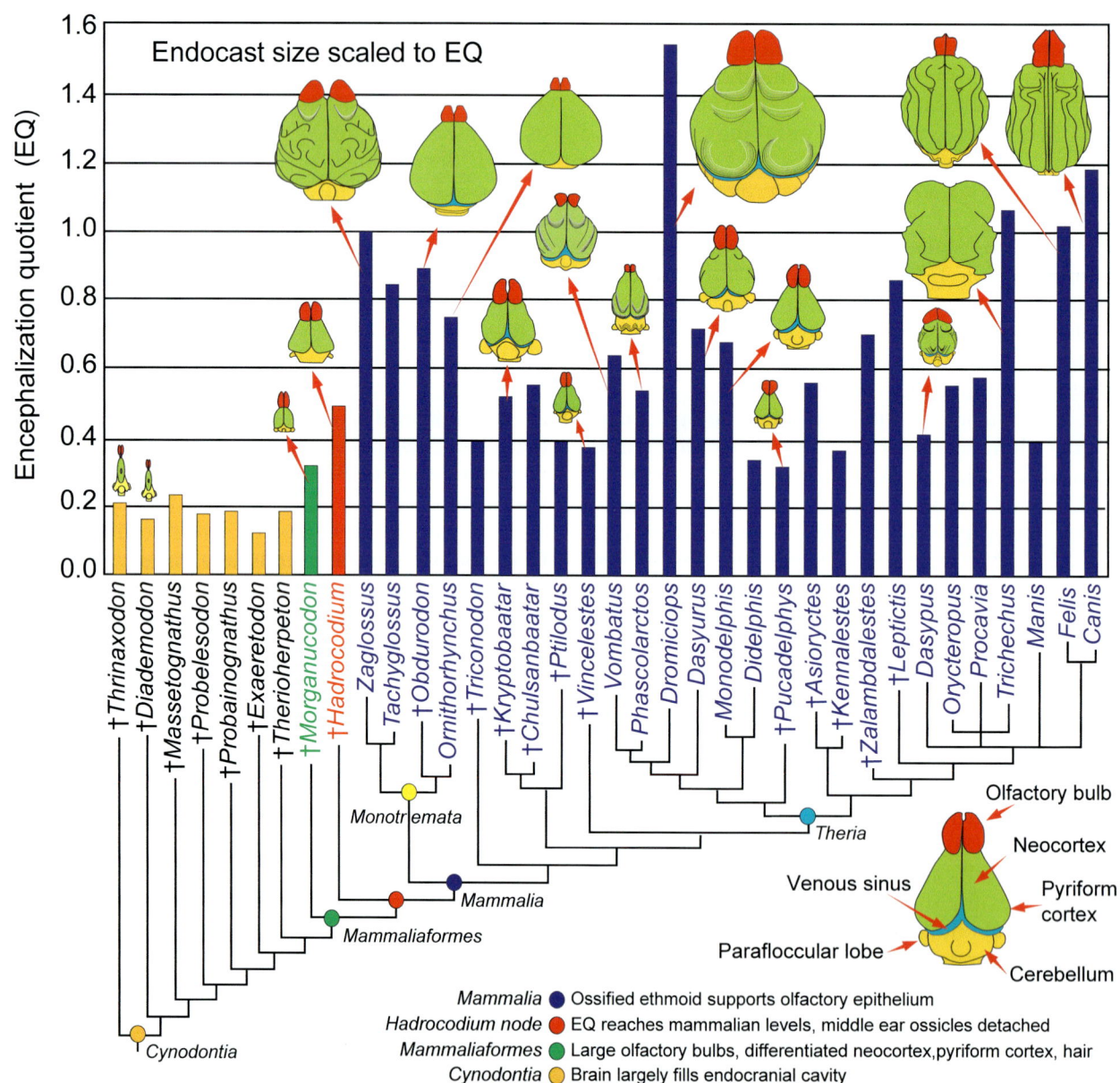

FIGURE 11.35   Patterns of brain evolution in phylogeny of basal Triassic cynodonts and selected crown Mammalia. Encephalization quotient (EQ) is shown as a histogram, and selected endocasts are scaled to EQ. *From Rowe, T.B., Macrini, T.E., Luo, Z.-X., 2011. Fossil evidence on origin of the mammalian brain. Science 332, 955–957. https://doi.org/10.1126/science.1203117.*

inferred as originating along the mammalian stem, prior to the origin of Mammalia, in the approximate historical sequence articulated above. Nocturnality and acute high-frequency hearing evolved only following the origin of Mammalia, in the therian stem group.

Viewing Mammalia as internested within its more inclusive clades among Pan-Mammalia and Amniota also helps to tease apart the many documented transformations in such a way as to identify episodes in which more or less separate evolutionary modules differentiated and to identify the appearance of new character states and new variational modalities. It also highlights the remarkable dichotomy and explains how, on the one hand, some of these systems became individuated from the others, while on the other hand they also became fully integrated with one another over the course of pan-mammalian evolution. In balancing the seemingly opposing forces of innovation and integration, the emergence of neocortex seems key to an explanation.

In broadest terms, the emergence of mammals appears to have proceeded as a flood of new peripheral information ascended to the mammalian brain. This was driven partly by duplication of the olfactory receptor genes and those encoding the radial units of the telencephalon and by the input from the dentition and tens of thousands of hair follicles. Additional new inputs arrived as the skeleton became more agile, from muscle spindles and joint receptors. High-frequency hearing also provided new input, but most of the auditory enhancements enjoyed by living mammals came long after the origin of the crown clade. And finally, mammals were immersed in a rich new source of environmental information in their new microhabitats, as the closest relatives of mammals miniaturized and came to occupy new niches dominated to unprecedented degrees by scents and odors.

Over the course of early pan-mammalian history, neocortex emerged to analyze and integrate these new kinds and levels of peripheral information, to process it in unique measures, and to respond in unique ways with unprecedented memory, introspection, and agility. Lying as it does at the integral core of mammalian organization and diversifying to a remarkable degree across the span of ~200 million of subsequent pan-mammalian history, neocortex would appear to meet all criteria as a Type I innovation. In any event, it seems beyond question that by the time cortical elaboration reached its unprecedented degree in humans, neocortex had unquestionably emerged as a Type I innovation. Simple reflection on the emergence of human technology, its effects on global population growth, and on human impact across Earth's climate and biota are unprecedented in the history of life. Surely this constitutes one of life's most singular innovations and the key to understanding the emergence of mammals.

References

Aboitiz, F., Montiel, J.F., 2015. Olfaction, navigation, and the origin of isocortex. Front. Neurosci. 9 (402), 1–12.

Alexander, R.M., 2003. Principles of Animal Locomotion. Princeton University Press, Princeton.

Allin, E.F., 1975. Evolution of the mammalian middle ear. J. Morphol. 147 (4), 403–437.

de Araujo, I.E., Rolls, E.T., Kringelbach, M.L., McGlone, F., Phillips, N., 2003. Taste-olfactory convergence, and the representation of the pleasantness of flavour, in the human brain. Eur. J. Neurosci. 18, 2059–2068.

Ashwell, K. (Ed.), 2010. The Neurobiology of Australian Marsupials. Cambridge University Press, Cambridge, UK.

Ashwell, K. (Ed.), 2013. Neurobiology of Monotremes: Brain Evolution in Our Distant Mammalian Cousins. CSIRO Publishing, Collingwood, Victoria.

Augee, M.L., Gooden, B.A., 1992. Evidence for electroreception from field studies of the echidna *Tachyglossus aculeatus*. In: Augee, M.L. (Ed.), Platypus and Echidna. Royal Zoological Society of New South Wales, Sydney, pp. 211–215.

Avivi-Arber, L., Martin, R., Lee, J.C., Sessle, B.J., 2011. Face sensorimotor cortex and its neuroplasticity related to orofacial sensorimotor functions. Arch. Oral Biol. 56 (12), 1440–1465.

Ayoub, A.E., Oh, S., Xie, Y., Leng, J., Cotney, J., Dominguez, M.H., Noonan, J.P., Rakic, P., 2011. Transcriptional programs in transient embryonic zones of the cerebral cortex defined by high-resolution mRNA sequencing. Proc. Natl. Acad. Sci. U.S.A. 108 (36), 14950–14955.

Barghusen, H.R., 1975. A review of fighting adaptations in dinocephalians (Reptilia, Therapsida). Paleobiology 1, 295–311.

Bargmann, C.I., 2006. Comparative chemosensation from receptors to ecology. Nature 444, 295–301.

Benoit, J., Manger, P.R., Rubidge, B.S., 2016. Palaeoneurological clues to the evolution of defining mammalian soft tissue traits. Sci. Rep. 6.

Benton, M., 2015. Vertebrate Palaeontology, fourth ed. Wiley Blackwell, West Sussex.

Berlin, J.C., Kirk, E.C., Rowe, T.B., 2013. Functional implications of ubiquitous semicircular canal non-orthogonality in mammals. PLoS One 8 (11), p. e79585. PONE-D-13–32641R1. https://doi.org/10.1371/journal.pone.0079585.

Bever, G.S., Rowe, T.B., Ekdale, E.G., Macrini, T.E., Colbert, M.W., Balanoff, A.M., 2005. Comment on "Independent origins of middle ear bones in monotremes and therians. Science 309, 1492a.

Bonfils, P., Avan, P., Faulcon, P., Malinvaud, D., 2005. Distorted odorant perception—analysis of a series of 56 patients with parosmia. Arch. Otolaryngol. Head Neck Surg. 131, 107–112.

Bramble, D.M., 1989. Axial-appendicular dynamics and the integration of breathing and gait in mammals. Am. Zool. 29, 171–186.

Broom, R., 1910. A comparison of the Permian reptiles of North America with those of South Africa. Bull. Am. Mus. Nat. Hist. 28 (20), 197–234.

Broom, R., 1932. The Mammal-like Reptiles of South Africa and the Origin of Mammals. H. F. & G. Witherby, London.

Bruce, L.L., 2007. Evolution of the nervous system in reptiles. In: Bullock, T.H., Rubenstein, L.R., Kaas, J.H. (Eds.), Evolution of Nervous Systems: A Comprehensive Referenc, the Evolution of Nervous Systems in Non-mammalian Vertebrates, vol. II. Elsevier Academic Press, Oxford, pp. 125–156.

Bruce, L.L., 2009. Evolution of the nervous system in reptiles. In: Kaas, J.H. (Ed.), Evolutionary Neuroscience. Academic Press, New York, pp. 233–265.

Bruce, L.L., Braford Jr., M.R., 2009. Evolution of the limbic system. In: Squire, L.R. (Ed.), New Encyclopedia of Neuroscience. Elsevier Academic Press, Oxford, pp. 43–55.

Burrell, H., 1927. The Platypus. Angus & Robertson Limited, Sydney.

Butler, A.B., 1994. The evolution of the dorsal pallium in the telencephalon of amniotes: cladistic analysis and a new hypothesis. Brain Res. Rev. 19 (1), 66–101.

Butler, A.B., Hodos, W., 1996. Comparative Vertebrate Neuroanatomy: Evolution and Adaptation. John Wiley & Sons, New York.

Cantino, P.D., de Queiroz, K., 2000. PhyloCode: a phylogenetic Code of biological nomenclature. Published at: http://www.ohiou.edu/phylocode/.

Carlson, W.D., Rowe, T.B., Ketcham, R.A., Colbert, M.W., 2003. Geological applications of high-resolution X-ray computed tomography in petrology, meteoritics and palaeontology. In: Mees, F., Swennen, R., Van Geet, M., Jacobs, P. (Eds.), Applications of X-ray Computed Tomography in the Geosciences, vol. 215. Geological Society, London, pp. 7–22.

Carr, E.C., Christiansen-Dalsgaard, J., 2016. Evolutionary trends in directional hearing. Curr. Opin. Neurobiol. 40, 111–117.

Carr, C.E., Soares, D., 2009. Shared and convergent features of the auditory system of vertebrates. In: Kaas, J.H. (Ed.), Evolutionary Neuroscience. Academic Press, New York, pp. 479–493.

Carroll, R.L., 1988. Vertebrate Paleontology and Evolution. W. H. Freeman & Co., New York.

Catania, K.C., 2013. Stereo and serial sniffing guide navigation to an odour source in a mammal. Nat. Commun. 4, 1441–1449.

Catania, K.C., Catania, E.H., 2015. Comparative studies of somatosensory systems and active sensing. In: Krieger, P., Groth, A. (Eds.), Sensorimotor Integration in the Whisker System. Springer, New York, pp. 7–30.

Ten Cate, A.R., 1997. The development of the periodontium—a largely ectomesenchymally derived unit. Periodontol 13 (1), 9–19, 2000.

Chapouton, P., Jagasia, R., Bally-Cuif, L., 2007. Adult neurogenesis in non-mammalian vertebrates. Bioessays 29 (8), 745–757.

Chen, W.R., Shepherd, G.M., 2005. The olfactory glomerulus: a cortical module with specific functions. J. Neurocytol. 34, 353–360.

Cifelli, R.L., Rowe, T.B., Luckett, W.P., Banta, J., Reyes, R., Howes, R.I., 1996. Fossil evidence for the origin of the marsupial pattern of tooth replacement. Nature 379, 715–718.

Coffin, A.B., Zeddies, D.G., Fay, R.R., Brown, A.D., Alderks, P.W., Bhandiwad, A.A., Mohr, R.A., Gray, M.D., Rogers, P.H., Sisneros, J.A., 2014. Use of the swim bladder and lateral line in near-field sound source localization by fish. J. Exp. Biol. 217 (12), 2078–2088.

Colbert, M.W., Racicot, R., Rowe, T.B., 2005. Anatomy of the cranial endocast of the bottlenose dolphin Tursiops truncatus, based on HRXCT. J. Mammal. Evol. 12, 197–209.

Connors, B.W., Kriegstein, A.R., 1986. Cellular physiology of the turtle visual cortex: distinctive properties of pyramidal and stellate neurons. J. Neurosci. 6, 164–177.

Cope, E.D., 1878. Descriptions of extinct Batrachia and Reptilia from the permian formation of Texas. Proc. Am. Philos. Soc. 17 (101), 505–530.

Cope, E.D., 1883. On the trituberculate type of molar tooth in the Mammalia. Proc. Am. Philos. Soc. 1883, 324–326.

Cope, E.D., 1888. On the tritubercular molar in human dentition. J. Morphol. 2, 7–23.

Crompton, A.W., 1963. Tooth replacement in the cynodont Thrinaxodon liorhinus. Ann. South Afr. Mus. 46, 479–521.

Crompton, A.W., 1972. Postcanine occlusion in cynodonts and tritylodontids. Bull. Br. Mus. Nat. Hist. Geol. 21, 27–71.

Crompton, A.W., 1989. The evolution of mammalian mastication. In: Wake, D.B., Roth, G. (Eds.), Complex Organismal Functions: Integration and Evolution in Vertebrates. John Wiley, New York, pp. 23–40.

Crompton, A.W., Jenkins Jr., F.A., 1973. Mammals from reptiles: a review of mammalian origins. Annu. Rev. Earth Planet Sci. 1, 131–155.

Crompton, A.W., Parker, P., 1978. Evolution of the Mammalian Masticatory Apparatus: the fossil record shows how mammals evolved both complex chewing mechanisms and an effective middle ear, two structures that distinguish them from reptiles. Am. Sci. 66 (2), 192–201.

Cullen, M.M., Leopold, D.A., 1999. Disorders of smell and taste. Med. Clin. N. Am. 83, 57–74.

Darwin, C., 1859. On the Origin of Species by Means of Natural Selection. Murray, London.

Davies, W.L., Carvalho, L.S., Cowing, J.A., Beazley, L.D., Hunt, D.M., Arrese, C.A., 2007. Visual pigments of the platypus: a novel route to mammalian colour vision. Curr. Biol. 17 (5), R161–R163.

The terminal nerve (nervus terminalis) structure, function and evolution. In: Demski, L.S., Schwanzel-Fukuda, M. (Eds.), 1987, 519. Ann. N.Y. Acad. Sci.

Demski, L.S., 1993. Terminal nerve complex. Cells Tissues Organs 148 (2–3), 81–95.

Di-Pöi, N., Milinkovitch, M.C., 2016. The anatomical placode in reptile scale morphogenesis indicates shared ancestry among skin appendages in amniotes. Sci. Adv. 2, e1600708.

Dingus, L., Rowe, T.B., 1998. The Mistaken Extinction – Dinosaur Evolution and the Origin of Birds. W. H. Freeman & Co, New York.

Donoghue, M.J., Doyle, J., Gauthier, J.A., Kluge, A.G., Rowe, T.B., 1989. Importance of fossils in phylogeny reconstruction. Annu. Rev. Ecol. Syst. 20, 431–460.

Edinger, T., 1974. Paleoneurology 1804–1966: an annotated bibliography. Adv. Anat. Embryol. Cell Biol. 49 (1–6), 3–258.

Eisenberg, J.F., 1981. The Mammalian Radiations: An Analysis of Trends in Evolution, Adaptation, and Behaviour. University of Chicago Press, Chicago.

Eisenberg, J.F., 1990. The behavioral/ecological significance of body size in the Mammalia. In: Damuth, J., MacFadden, B. (Eds.), Body Size in Mammalian Paleobiology: Estimation and Biological Implications. Cambridge University Press, Cambridge, UK, pp. 25–37.

Ekdale, E.G., 2013. Comparative anatomy of the bony labyrinth (inner ear) of placental mammals. PLoS One 8 (6), e66624.

Ekdale, E.G., 2016. Form and function of the mammalian inner ear. J. Anat. 228 (2), 324–337.

Farbman, A.I., 1988. Cellular interactions in the development of the vertebrate olfactory system. In: Margolis, F.L., Getchell, T.V. (Eds.), Molecular Neurobiology of the Olfactory System. Plenum Press, New York, pp. 319–332.

Farbman, A.I., 1990. Olfactory neurogenesis: genetic or environmental controls? Trends Neurosci. 13, 362–365.

Farmer, C.G., 2000. Parental care: the key to understanding endothermy and other convergent features in birds and mammals. Am. Nat. 155, 326–334.

Farmer, C.G., 2003. Reproduction: the adaptive significance of endothermy. Am. Nat. 162, 826–840.

Fournier, J., Muller, C.M., Laurent, G., 2014. Looking for the roots of cortical sensory computation in three-layered cortices. Curr. Opin. Neurobiol. 31, 119–128.

Franselli, J., Landis, B.N., Heilmann, S., Hauswald, B., Huttenbrink, K.B., Lacroix, J.S., Leopold, D.A., Hummel, T., 2004. Clinical presentation of qualitative olfactory dysfunction. Eur. Arch. Oto-Rhino-Laryngol. 261, 411–415.

Garrett, E.C., Steiper, M.E., 2014. Strong links between genomic and anatomical diversity in both mammalian olfactory chemosensory systems. Proc. R. Soc. Lond. B 281, 2013–2828.

Gautam, S.H., Verhagen, J.V., 2012. Retronasal odor representations in the dorsal olfactory bulb of rats. J. Neurosci. 32, 7949–7959.

Gauthier, J.A., 1986. Saurischian monophyly and the origin of birds. Memoirs Calif. Acad. Sci. 8, 1–55.

Gauthier, J.A., Kluge, A.G., Rowe, T.B., 1988. Amniote phylogeny and the importance of fossils. Cladistics 4, 105–209.

Gauthier, J.A., Kluge, A.G., Rowe, T., 1988. The early evolution of the Amniota. Systematics Association special volume No. 35a. In:

Benton, M. (Ed.), The Phylogeny and Classification of the Tetrapods, Amphibians, Reptiles and Birds, vol. 1. Clarendon Press, Oxford, pp. 103–155.

Gauthier, J.A., Cannatella, D., de Queiroz, K., Kluge, A.G., Rowe, T.B., 1989. Tetrapod phylogeny. In: Fernholm, B., Bremer, H., Jornvall, H. (Eds.), The Hierarchy of Life, Nobel Symposium, vol. 70. Excerpta Medica, Amsterdam, pp. 337–353.

Geiser, F., Goodship, N., Pavey, C.R., 2002. Was basking important in the evolution of mammalian endothermy? Naturwissenschaften 89 (9), 412–414.

Gidley, J.W., 1906. Evidence bearing on tooth-cusp development. Proc. Wash. Acad. Sci. 8, 91–110.

Goodrich, E.S., 1930. Studies on the Structure and Function of Vertebrates. Constable & Co, London.

Gould, S.J., 1985. To be a platypus. Nat. Hist. Mag. 94 (8), 10–15.

Graybeal, A., Rosowski, J.J., Ketten, D.R., Crompton, A.W., 1989. Inner-ear structure in Morganucodon, an early Jurassic mammal. Zool. J. Linn. Soc. 96 (2), 107–117.

Green, P.A., Valkenburgh, B., Pang, B., Bird, D., Rowe, T.B., Curtis, A., 2012. Respiratory and olfactory turbinal size in canid and arctoid carnivorans. J. Anat. 221 (6), 609–621.

Gregory, W.K., 1910. The orders of mammals. Bull. Am. Mus. Nat. Hist. 27, 1–524.

Gregory, W.K., 1916. Studies on the evolution of the primates. Part I. The Cope-Osborn "theory of trituberculy" and the ancestral molar patterns of the primates. Part II. Phylogeny of recent and extinct anthropoids, with special reference to the origin of man. Bull. Am. Mus. Nat. Hist. 35, 239–355.

Gregory, W.K., 1922. The origin and evolution of the human dentition. Journal of Dental Research, New York.

Gregory, W.K., 1953. Evolution Emerging. Macmillan, New York.

Grothe, B., Carr, C.E., Cassedy, J.H., Fritzsch, J.H., Koppl, C., 2005. The evolution of central pathways and their neural processing patterns. In: Manley, G.A., Popper, A.N., Fay, R.R. (Eds.), Evolution of the Vertebrate Auditory System. Springer, pp. 289–359.

Grothe, B., Pecka, M., McAlpine, D., 2010. Mechanisms of sound localization in mammals. Physiol. Rev. 90 (3), 983–1012.

Haberly, L.B., 1985. Neuronal circuitry in olfactory cortex: anatomy and functional implications. Chem. Senses 10, 219–238.

Haberly, L.B., 2001. Parallel-distributed processing in olfactory cortex: new insights from morphological and physiological analysis of neuronal circuitry. Chem. Senses 26, 551–576.

Haberly, L.B., Shepherd, G.M., 1973. Current-density analysis of summed evoked potentials in opossum prepyriform cortex. J. Neurophysiol. 36, 789–802.

Habre-Hallage, P., Dricot, L., Hermoye, L., Reychler, H., van Steenberghe, D., Jacobs, R., Grandin, C.B., 2014. Cortical activation resulting from the stimulation of periodontal mechanoreceptors measured by functional magnetic resonance imaging (fMRI). Clin. Oral Invest. 18 (8), 1949–1961.

Haeckel, E., 1866. Generelle Morphologie der Organismen. Allgemeine Grundzüge der organischen Formen-Wissenschaft, mechanisch begründet durch die von Charles Darwin reformirte Descendenz-Theorie. Georg Reimer, Berlin.

Haeckel, E., 1877. Anthropogenie oder Entwickelungsgeschichte des Menschen. Wilhelm Engelmann, Leipzig.

Harjunmaa, E., Seidel, K., Häkkinen, T., Renvoisé, E., Corfe, I.J., Kallonen, A., Zhang, Z.Q., Evans, A.R., Mikkola, M.L., Salazar-Ciudad, I., Klein, O.D., 2014. Replaying evolutionary transitions from the dental fossil record. Nature 512, 44–48.

Harvey, P.H., Clutton-Brock, T.H., Mace, G.M., 1980. Brain size and ecology in small mammals and primates. Proc. Natl. Acad. Sci. U.S.A. 77, 4387–4389.

Hayden, S., Bekaert, M., Crider, T.A., Mariani, S., Murphy, W.J., Teeling, E.C., 2010. Ecological adaptation determines functional mammalian olfactory subgenomes. Genome Res. 20, 1–9.

Heilmann, S., Hummel, T., 2004. A new method for comparing ortho-nasal and retronasal olfaction. Behav. Neurosci. 118, 412–419.

Hennig, W., 1966. Phylogenetic Systematics. In: Davis, D.D., Zangerl, R. (Eds.). University of Illinois Press.

Hildebrand, J.G., Shepherd, G.M., 1997. Mechanisms of olfactory discrimination: converging evidence for common principles across phyla. Ann. Rev. Neurosci. 20, 595–631.

Hillenius, W.J., 2000. Septomaxilla of nonmammalian synapsids: soft-tissue correlates and a new functional interpretation. J. Morphol. 245 (1), 29–50.

Hillson, S., 2009. In: Teeth, second ed. Cambridge University Press.

Hirasawa, T., Kuratani, S., 2013. A new scenario of the evolutionary derivation of the mammalian diaphragm from shoulder muscles. J. Anat. 222 (5), 504–517.

Hlusko, L.J., Sage, R.D., Mahaney, M.C., 2011. Modularity in the mammalian dentition: mice and monkeys share a common dental genetic architecture. J. Exp. Zool. Part B Mol. Dev. Evol. 316 (1), 21–49.

Home, E., 1802. A description of the anatomy of the Ornithorhynchus paradoxus. Philos. Trans. R. Soc. Lond. 92, 67–84.

Home, E., 1802. Description of the anatomy of Ornithorhynchus hystrix. Philos. Trans. R. Soc. Lond. 92, 348–364.

Hopson, J.A., 2012. The role of foraging mode in the origin of therapsids: implications for the origin of mammalian endothermy. Fieldiana Life Earth Sci 5, 126–148.

Hopson, J.A., Barghusen, H.R., 1986. An analysis of therapsid relationships. In: Hotton III, N., McLean, P.D., Roth, J.J., Roth, E.C. (Eds.), Ecology and Biology of Mammal-like Reptiles. Smithsonian Institution and National Institute of Mental Health, Washington, DC, pp. 83–106.

Huber, E., 1930. Evolution of facial musculature and cutaneous field of Trigeminus. Part I. Q. Rev. Biol. 5 (2), 133–188.

Huber, E., 1930. Evolution of facial musculature and cutaneous field of Trigeminus. Part II. Q. Rev. Biol. 5 (4), 389–437.

Huelsenbeck, J.P., 1991. When are fossils better than extant taxa in phylogenetic analysis? Syst. Biol. 40 (4), 458–469.

Huxley, T.H., 1880. On the application of the laws of evolution to the arrangement of the Vertebrata, and more particularly of the Mammalia. Proc. Zool. Soc. Lond. 1880, 649–662.

Iyengar, S., Qi, H.X., Jain, N., Kaas, J.H., 2007. Cortical and thalamic connections of the representations of the teeth and tongue in somatosensory cortex of new world monkeys. J. Comp. Neurol. 501 (1), 95–120.

Jarvik, E., 1942. On the structure of the snout of crossopterygians and lower gnathostomes in general. Zool. Bidr. Från Upps. 21, 235–675.

Jefferies, R.P.S., 1979. The origin of chordatesda methodological essay. In: House, M.R. (Ed.), The Origin of Major Invertebrate Groups. Academic Press, London, pp. 443–477.

Jenkins Jr., F.A., 1969. The evolution and development of the dens of the mammalian axis. Anat. Rec. 164, 173–184.

Jenkins, F.A., 1971. The postcranial skeleton of African cynodonts: problems in the early evolution of the mammalian postcranial skeleton. Bull. - Peabody Mus. Nat. Hist. 36, 1–216.

Jenkins Jr., F.A., Parrington, F.R., 1976. The postcranial skeleton of the triassic mammals Eozostrodon, Megazostrodon, and Erythrotherium. Philos. Trans. R. Soc. B 273, 387–431.

Jerison, H., 1973. Evolution of the Brain and Intelligence. Academic Press, New York.

Ji, Q., Luo, Z.-X., Yuan, C., Tabrum, A.R., 2006. A swimming mammaliaform from the Middle Jurassic and ecomorphological diversification of early mammals. Science 311, 1123–1127.

Kaas, J., 2009. The evolution of the dorsal thalamus in mammals. In: Kaas, J. (Ed.), Evolutionary Neuroscience. Elsevier, Oxford, pp. 569–586.

Kaas, J.H., Qi, H.X., Iyengar, S., 2006. Cortical network for representing the teeth and tongue in primates. Anat. Rec. Part A Discov. Mol. Cell. Evol. Biol. 288 (2), 182–190.

Katz, M.J., Lasek, R.J., 1978. Evolution of the nervous system: role of ontogenetic mechanisms in the evolution of matching populations. Proc. Natl. Acad. Sci. U.S.A. 75 (3), 1349–1352.

Kemp, T.S., 1979. The primitive cynodont *Procynosuchus*: functional anatomy of the skull and relationships. Philos. Trans. R. Soc. Lond. B 285, 73–122.

Kemp, T.S., 1982. Mammal-Like Reptiles and the Origin of Mammals. Academic Press, London.

Kemp, T.S., 1983. The relationships of mammals. Zool. J. Linn. Soc. 77, 353–384.

Kemp, T.S., 1988. Interrelationships of the Synapsida. Systematics association special volume, 35B. In: Benton, M.J. (Ed.), The Phylogeny and Classification of the Tetrapods. Mammals, vol. 2. Clarendon Press, Oxford, pp. 1–22.

Kemp, T.S., 1988. A note on the Mesozoic mammals, and the origin of therians. Systematics Association Special Volume, 35B. In: Benton, M.J. (Ed.), The Phylogeny and Classification of the Tetrapods. Mammals, vol. 2. Clarendon Press, Oxford, pp. 23–29.

Kemp, T.S., 2005. The Origin and Evolution of Mammals. Oxford University Press, Oxford.

Kemp, T.S., 2006. The origin of mammalian endothermy: a paradigm for the evolution of complex biological structure. Zool. J. Linn. Soc. 147 (4), 473–488.

Kemp, T.S., 2006. The origin and early radiation of the therapsid mammal-like reptiles: a palaeobiological hypothesis. J. Evol. Biol. 19 (4), 1231–1247.

Kermack, D.M., Kermack, K.E., 1984. The Evolution of Mammalian Characters. Springer Science & Business Media.

Kermack, K.A., Mussett, F., Rigney, H.W., 1973. The lower jaw of *Morganucodon*. Zool. J. Linn. Soc. 53, 87–175.

Kermack, K.A., Mussett, F., Rigney, H.W., 1981. The skull of *Morganucodon*. Zool. J. Linn. Soc. 71, 1–158.

Kielan-Jaworowska, Z., Cifelli, R.L., Luo, Z.-X., 2004. Mammals from the Age of Dinosaurs. Columbia University Press, New York.

Kirk, E.C., Daghighi, P., Macrini, T.E., Bhullar, B.A.S., Rowe, T.B., 2014. Cranial anatomy of the Duchesnean primate *Rooneyia viejaensis*: new insights from high resolution computed tomography. J. Hum. Evol. 74, 82–95.

Kriegstein, A.R., Connors, B.W., 1986. Cellular physiology of the turtle visual cortex: synaptic properties and intrinsic circuitry. J. Neurosci. 6, 178–191.

Krubitzer, L., Kaas, J., 2005. The evolution of the neocortex in mammals: how is phenotypic diversity generated? Curr. Opin. Neurobiol. 15 (4), 444–453.

Kruska, D.C.T., 2007. The effects of domestication on brain size. In: Bullock, T.H., Rubenstein, L.R., Kaas, J.H. (Eds.), Evolution of Nervous Systems: A Comprehensive Reference, the Evolution of Nervous Systems in Non-mammalian Vertebrates, vol. II. Elsevier Academic Press, Oxford, pp. 143–153.

Kubo, K., Shibukawa, Y., Shintani, M., Suzuki, T., Ichinohe, T., Kaneko, Y., 2008. Cortical representation area of human dental pulp. J. Dent. Res. 87 (4), 358–362.

Lamarck, J.B.P.A., 1809. Philosophie Zoologique, vol. 1. J. B. Baillière, Paris.

Laurin, M., Reisz, R.R., 1995. A reevaluation of early amniote phylogeny. Zool. J. Linn. Soc. 113, 165–223.

Li, Z., Clarke, J.A., 2015. The craniolingual morphology of waterfowl (Aves, anseriformes) and its relationship with feeding mode revealed through contrast-enhanced X-ray computed tomography and 2D morphometrics. Evol. Biol. 43 (1), 12–25.

Lightoller, G.S., 1942. Matrices of the facialis musculature: homologization of the musculature in monotremes with that of marsupials and placentals. J. Anat. 76 (3), 258–269.

Linnaeus, C., 1758. Systema Naturae Per Regna Tria Naturae, Secundum Classes, Ordines, Genera, Species, Cum Characteribus, Differentiis, Synonymis, Locis. Editio decima, reformata, Tomus I. Laurentius Salvius, Stockholm.

Liu, J., Olsen, P., 2010. The phylogenetic relationships of Eucynodontia (Amniota: Synapsida). J. Mammal. Evol. 17, 151–176.

Louis, M., Huber, T., Benton, R., Sakmar, T.P., Vosshall, L.B., 2008. Bilateral olfactory sensory input enhances chemotaxis behavior. Nat. Neurosci. 11 (2), 187–199.

Lucas, S.G., Luo, Z.-X., 1993. *Adelobasileus* from the Upper Triassic of west Texas: the oldest mammal. J. Vertebr. Paleontol. 13, 309–334.

Luo, Z.-X., 2007. Transformation and diversification in early mammal evolution. Nature 450, 1011–1019.

Luo, Z.-X., Crompton, A.W., Sun, A.-L., 2001. A new mammal from the Early Jurassic and evolution of mammalian characteristics. Science 292, 1535–1540.

Luo, Z.-X., Cifelli, R.L., Kielan-Jaworowska, Z., 2001. Dual origin of tribosphenic mammals. Nature 409 (6816), 53–57.

Luo, Z.-X., Kielan-Jaworowska, Z., Cifelli, R.L., 2004. Evolution of dental replacement in mammals. Bull. Carnegie Mus. Nat. Hist. 36, 159–175.

Luo, Z.-X., Quang, J., Yuan, C.-X., 2007. Convergent dental adaptations in pseudotribosphenic and tribosphenic mammals. Nature 450, 93–97.

Luo, Z.-X., Gatesy, S.M., Jenkins Jr., F.A., Amaral, W.W., Shubin, N.H., 2015. Mandibular and dental characteristics of Late Triassic mammaliaform *Haramiyavia* and their ramifications for basal mammal evolution. Proc. Natl. Acad. Sci. U.S.A. 112 (51), E7101–E7109. https://doi.org/10.1073/pnas.1519387112.

Lyson, T.R., Bever, G.S., Bhullar, B.A.S., Joyce, W.G., Gauthier, J.A., 2010. Transitional fossils and the origin of turtles. Biol. Lett. 6 (6), pp. 830–833. p.rsbl20100371.

Mace, G.M., Harvey, P.H., Clutton-Brock, T.H., 1981. Brain size and ecology in small mammals. J. Zool. 193, 333–354.

Macrini, T.E., 2012. Comparative morphology of the internal nasal skeleton of adult marsupials based on x-ray computed tomography. Bull. Am. Mus. Nat. Hist. 1–91.

Macrini, T.E., 2014. Development of the ethmoid in *Caluromys philander* (Didelphidae, Marsupialia) with a discussion on the homology of the turbinal elements in marsupials. Anat. Rec. 297 (11), 2007–2017.

Macrini, T.E., Rowe, T.B., Archer, M., 2006. Description of a cranial endocast from a fossil platypus, *Obdurodon dicksoni* (Monotremata, Ornithorhynchidae), and the relevance of endocranial characters to monotreme monophyly. J. Morphol. 267, 1000–1015.

Macrini, T.E., Rowe, T.B., VandeBerg, J., 2007. Cranial endocasts from a growth series of *Monodelphis domestica* (Didelphidae, Marsupialia): a study of individual and ontogenetic variation. J. Morphol. 268, 844–865.

Macrini, T.E., Rougier, G., Rowe, T.B., 2007. Description of a cranial endocast from the fossil mammal *Vincelestes neuquenianus* (Theriiformes) and its relevance to the evolution of endocranial characters in therians. Anat. Rec. 290, 875–892.

Macrini, T.E., de Muizon, C., Cifelli, R.L., Rowe, T., 2007. Digital endocast of *Pucadelphys andinus*, a Paleocene metatherian. J. Vertebr. Paleontol. 27, 99–107.

Marín, O., Rubenstein, J.L., 2001. A long, remarkable journey: tangential migration in the telencephalon. Nat. Rev. Neurosci. 2 (11), 780–790.

Matthew, W.D., 1915. Climate and evolution. Ann. N.Y. Acad. Sci. 24, 171–318.

McGowen, M.R., Clark, C., Gatesy, J., 2008. The vestigial olfactory receptor subgenome of odontocete whales: phylogenetic congruence between gene-tree reconciliation and supermatrix methods. Syst. Biol. 57 (4), 574–590.

McKenna, M.C., Bell, S.K., 1997. Classification of Mammals above the Species Level. Columbia University Press, New York.

McMahon, T.A., Bonner, J.T., 1983. On Size and Life. Scientific American Library, New York.

Meng, J., Hu, Y., Wang, Y., Wang, X., Li, C., 2006. A Mesozoic gliding mammal from northeastern China. Nature 444, 889–893.

Metscher, B.D., 2009. MicroCT for comparative morphology: simple staining methods allow high-contrast 3D imaging of diverse non-mineralized animal tissues. BMC Physiol. 9 (1), 1.

Meyer, A., Zardoya, R., 2003. Recent advances in the (molecular) phylogeny of vertebrates. Annu. Rev. Ecol. Evol. Syst. 311–338.

Ming, G.L., Song, H., 2005. Adult neurogenesis in the mammalian central nervous system. Annu. Rev. Neurosci. 28, 223–250.

Molnár, Z., Butler, A.B., 2002. Neuronal changes during forebrain evolution in amniotes: an evolutionary developmental perspective. Prog. Brain Res. 136, 21–38.

Molnár, Z., Kaas, J.H., de Carlos, J.A., Hevner, R.F., Lein, E., Němec, P., 2014. Evolution and development of the mammalian cerebral cortex. Brain Behav. Evol. 83, 126–139.

Montiel, J.F., Vasistha, N.A., Garcia-Moreno, F., Molnár, Z., 2016. From sauropsids to mammals and back: new approaches to comparative cortical development. J. Comp. Neurol. 524 (3), 630–645.

Mori, K., Takahashi, Y.K., Igarashi, K.M., Yamaguchi, M., 2006. Maps of odorant molecular features in the mammalian olfactory bulb. Physiol. Rev. 86, 409–433.

Morris, R.J., Liu, Y., Marles, L., Yang, Z., Trempus, C., Li, S., Lin, J.S., Sawicki, J.A., Cotsarelis, G., 2004. Capturing and profiling adult hair follicle stem cells. Nat. Biotechnol. 22 (4), 411–417.

Murray, J.C., 2002. Gene/environment causes of cleft lip and/or palate. Clin. Genet. 61 (4), 248–256.

Neville, K.R., Haberly, L.B., 2004. Olfactory cortex. In: Shepherd, G.M. (Ed.), The Synaptic Organization of the Brain. Oxford University Press, New York, pp. 415–454.

Nieuwenhuys, R., ten Donkelaar, H.J., Nicholson, C., 1998. The meaning of it all. In: Nieuwenhuys, R., ten Donkelaar, H.J., Nicholson, C. (Eds.), The Central Nervous System of Vertebrates. Springer, Berlin and Heidelberg, pp. 2135–2196.

Niimura, Y., 2009. On the origin and evolution of vertebrate olfactory receptor genes: comparative genome analysis among 23 chordate species. Genome Biol. Evol. 1, 34–44.

Niimura, Y., 2012. Olfactory receptor multigene family in vertebrates: from the viewpoint of evolutionary genomics. Curr. Genomics 13, 103–114.

Niimura, Y., Nei, M., 2005. Evolutionary dynamics of olfactory receptor genes in fishes and tetrapods. Proc. Natl. Acad. Sci. U.S.A. 102 (17), 6039–6044.

Niimura, Y., Nei, M., 2006. Evolutionary dynamics of olfactory and other chemosensory receptor genes in vertebrates. J. Hum. Genet. 51 (6), 505–517.

Northcutt, G.L., 2011. Evolving large and complex brains. Science 332, 926.

O'Leary, M., Bloch, J.I., Flynn, J.J., et al., 2013. The placental mammal ancestor and the post–K-Pg radiation of placentals. Science 339, 662–667.

Olson, E.C., 1944. Origin of mammals based upon cranial morphology of the therapsid suborders. Geol. Soc. Am. Special Pap. 55, 1–130.

Olson, E.C., 1959. The evolution of mammalian characters. Evolution 13, 344–353.

Osborn, H.F., 1888. The evolution of mammalian molars to and from the tritubercular type. Am. Nat. 22 (264), 1067–1079.

Osborn, H.F., 1907. In: Gregory, W.K. (Ed.), The Evolution of Mammalian Molar Teeth to and from the Triangular Type. Macmillan, New York.

Owen, R., 1844. Description of certain fossil crania, discovered by A. G. Bain, Esq., in sandstone rocks at the south-eastern extremity of Africa, referable to different species of an extinct Genus of Reptilia (Dicynodon), and indicative of a new tribe of suborder Sauria. Proc. Geol. Soc. Lond. 4, 500–504.

Owen, R., 1861. Paleontology or a Systematic Summary of Extinct Animals and Their Geological Relations, second ed. Adams and Charles Black, Edinburgh.

Owen, R., 1876. Descriptive and Illustrated Catalogue of the Fossil Reptilia of South Africa in the Collection of the British Museum. Trustees of the British Museum, London.

Packard, M.J., Seymour, R.S., 1997. Evolution of the amniote egg. In: Sumida, S.S., Martin, K.L.M. (Eds.), Amniote Origins. Academic Press, San Diego, pp. 265–290.

Parrish, J.M., Parrish, J.T., Ziegler, A.M., 1986. Permian-triassic paleogeography and paleoclimatology and implications for therapsid distribution. In: Hotton III, N., McLean, P.D., Roth, J.J., Roth, E.C. (Eds.), Ecology and Biology of Mammal-like Reptiles. Smithsonian Institution and National Institute of Mental Health, Washington, DC, pp. 109–132.

Patterson, C. (Ed.), 1987. Molecules and Morphology in Evolution: Conflict or Compromise?. Cambridge University Press, Cambridge.

Pettigrew, J.D., 1999. Electroreception in monotremes. J. Exp. Biol. 202, 1447–1454.

Pihlström, H., Fortelius, M., Hemilä, S., Forsman, R., Reuter, T., 2005. Scaling of mammalian ethmoid bones can predict olfactory organ size and performance. Proc. R. Soc. Lond. B 272, 957–962.

Popper, A.N., Lu, Z., 2000. Structure–function relationships in fish otolith organs. Fish. Res. 46 (1), 15–25.

de Queiroz, K., 2007. Toward an integrated system of clade names. Syst. Biol. 56, 956–974.

de Queiroz, K., Gauthier, J.A., 1990. Phylogeny as a central principle in taxonomy: phylogenetic definitions of taxon names. Syst. Biol. 39 (4), 307–322.

de Queiroz, K., Gauthier, J.A., 1992. Phylogenetic taxonomy. Annu. Rev. Ecol. Syst. 449–480, 1992.

de Queiroz, K., Gauthier, J.A., 1994. Toward a phylogenetic system of biological nomenclature. Trends Ecol. Evol. 9 (1), 27–31.

Quiroga, J.C., 1980. Sobre un molde endocraneano del cinodonte Probainognathus jenseni Romer, 1970 (Reptilia, Therapsida), de la Formación Ischichuca (Triasico Medio), La Rioja, Argentina. Ameghiniana 17, 181–190.

Racicot, R., Rowe, T.B., 2014. Endocranial anatomy of a new fossil porpoise (Odontoceti: Phocoenidae) from the Pliocene san Diego formation of California. J. Palaeontol. 88 (4), 652–663.

Rakic, P., 1988. Specification of cerebral cortical areas. Science 1988 (241), 170–176.

Rakic, P., 2000. Radial unit hypothesis of neocortical expansion. Evolutionary Developmental Biology of the Cerebral Cortex. Novartis Foundation Symposium. John Wiley, Chichester, New York, pp. 30–52, 1999.

Rakic, P., 2007. The radial edifice of cortical architecture: from neuronal silhouettes to genetic engineering. Brain Res. Rev. 55 (2), 204–219.

Rakic, P., 2009. Evolution of the neocortex: a perspective from developmental biology. Nat. Rev. Neurosci. 10, 724–735.

Remple, M.S., Henry, E.C., Catania, K.C., 2003. Organization of somatosensory cortex in the laboratory rat (Rattus norvegicus): evidence for two lateral areas joined at the representation of the teeth. J. Comp. Neurol. 467 (1), 105–118.

Roberts, W.M., Howard, J., Hudspeth, A.J., 1988. Hair cells: transduction, tuning and transmission in the inner ear. Annu. Rev. Cell Biol. 4, 63–92.

Rogers, G.E., 2003. Hair follicle differentiation and regulation. Int. J. Dev. Biol. 48 (2–3), 163–170.

Rolls, E.T., Grabenhorst, F., 2008. The orbitofrontal cortex and beyond: from affect to decision-making. Prog. Neurobiol. 86, 216–244.

Romer, A.S., 1956. Osteology of the Reptilia. University of Chicago Press, Chicago.

Romer, A.S., 1966. Vertebrate Paleontology. University of Chicago Press, Chicago.

Romer, A.S., 1970. The Chanares (Argentina) Triassic reptile fauna: 6. A chiniquodontid cynodont with an incipient squamosal-dentary jaw articulation. Breviora 344, 1–18.

Romer, A.S., Price, L.W., 1940. Review of the *Pelycosauria*. Geol. Soc. Am. Special Pap. 28, 1−534.

Rosowski, J.J., Graybeal, A., 1991. What did Morganucodon hear? Zool. J. Linn. Soc. 101 (2), 131−168.

Rowe, T.B., 1987. Definition and diagnosis in the phylogenetic system. Syst. Zool. 36 (2), 208−211.

Rowe, T.B., 1988. Definition, diagnosis and origin of *Mammalia*. J. Vertebr. Paleontol. 8 (3), 241−264.

Rowe, T.B., 1993. Phylogenetic systematics and the early history of mammals. In: Szalay, F.S., Novacek, M.J., McKenna, M.C. (Eds.), Mammalian Phylogeny. Springer-Verlag, New York, pp. 129−145.

Rowe, T.B., 1996. Coevolution of the mammalian middle ear and neocortex. Science 273, 651−654.

Rowe, T.B., 1996. Brain Heterochrony and Evolution of the Mammalian Middle Ear. In: Ghiselin, M., Pinna, G. (Eds.), New Perspectives on the History of Life. California Academy of Sciences. Memoir 20, pp. 71−96.

Rowe, T.B., 1999. At the roots of the mammalian tree. Nature 398, 283−284.

Rowe, T.B., 2004. Chordate phylogeny and development. In: Cracraft, J., Donoghue, M.J. (Eds.), Assembling the Tree of Life. Oxford University Press, Oxford and New York, pp. 384−409.

Rowe, T.B., Frank, L.R., 2011. The vanishing third dimension. Science 331, 712−714.

Rowe, T.B., Gauthier, J.A., 1992. Ancestry, paleontology, and definition of the name *Mammalia*. Syst. Biol. 41, 372−378.

Rowe, T.B., van den Heever, J., 1986. The hand of *Anteosaurus magnificus* (Therapsida, Dinocephalia) and its bearing on the origin of the mammalian manual phalangeal formula. South Afr. J. Sci. 82 (11), 641−645.

Rowe, T.B., Shepherd, G.M., 2016. The role of ortho-retronasal olfaction in mammalian cortical evolution. J. Comp. Neurol. 524, 471−495. https://doi.org/10.1002/cne.23802.

Rowe, T.B., Carlson, W., Bottorff, W., 1995. *Thrinaxodon*: Digital Atlas of the Skull. CD-ROM, second ed., for Windows and Macintosh Platforms. University of Texas Press. 547 megabytes.

Rowe, T.B., Kappelman, J., Carlson, W.D., Ketcham, R.A., Denison, C., 1997. High-resolution computed tomography: a breakthrough technology for Earth scientists. Geotimes 42 (9), 23−27.

Rowe, T.B., Brochu, C.A., Kishi, K., Colbert, M., Merck, J.W., 1999. Introduction to alligator: digital atlas of the skull. In: Rowe, T.B., Brochu, C.A., Kishi, K. (Eds.), Cranial Morphology of Alligator and Phylogeny of Alligatoroidae. Society of Vertebrate Paleontology Memoir 6, Journal of Vertebrate Paleontology, vol. 19, pp. 1−8 supplement to number 2.

Rowe, T.B., P Eiting, T., Macrini, T.E., Ketcham, R.A., 2005. Organization of the olfactory and respiratory skeleton in the nose of the gray short-tailed opossum *Monodelphis domestica*. J. Mammal. Evol. 12, 303−336.

Rowe, T.B., Rich, T.H., Vickers-Rich, P., Springer, M., Woodburne, M.O., 2008. The oldest platypus, and its bearing on divergence timing of the platypus and echidna clades. Proc. Natl. Acad. Sci. U.S.A. 105, 1238−1242.

Rowe, T.B., Macrini, T.E., Luo, Z.-X., 2011. Fossil evidence on origin of the mammalian brain. Science 332, 955−957. https://doi.org/10.1126/science.1203117.

Ruben, J., 1995. The evolution of endothermy in mammals and birds: from physiology to fossils. Annu. Rev. Physiol. 57 (1), 69−95.

Rubenstein, J.L., Rakic, P., 1999. Genetic control of cortical development. Cereb. Cortex 9 (6), 521−523.

Rubidge, B.S., Sidor, C.A., 2001. Evolutionary patterns among Permo-Triassic therapsids. Annu. Rev. Ecol. Syst. 32, 449−480.

Ruf, I., 2014. Comparative anatomy and systematic implications of the turbinal skeleton in Lagomorpha (Mammalia). Anat. Rec. 297 (11), 2031−2046.

Schlosser, G., 2010. Making senses: development of vertebrate cranial placodes. Int. Rev. Cell Mol. Biol. 283, 129−234.

Schneider, M.R., Schmidt-Ullrich, R., Paus, R., 2009. The hair follicle as a dynamic miniorgan. Curr. Biol. 19 (3), R132−R142.

Sengel, P., 1976. Morphogenesis of Skin. Cambridge University Press, Cambridge.

Shepherd, G.M., 1991. Computational structure of the olfactory system. In: Eichenbaum, H.M., Davis, J. (Eds.), Olfaction: A Model for Computational Neuroscience. MIT Press, Cambridge, Mass, pp. 3−42.

Shepherd, G.M., 2004. The human sense of smell: are we better than we think? PLoS Biol. 2, e146.

Shepherd, G.M., 2006. Smell images and the flavour system in the human brain. Nature 444, 316−321.

Shepherd, G.M., 2011. The microcircuit concept applied to cortical evolution: from three-layer to six-layer cortex. Front. Neuroanat. 5 (30), 1−15.

Shepherd, G.M., 2012. Neurogastronomy: How the Brain Creates Flavor and Why it Matters. Columbia University Press, New York.

Sidor, C.A., 2001. Simplification as a trend in synapsid cranial evolution. Evolution 55, 1419−1442.

Sidor, C.A., 2003. The naris and palate of *Lycaenodon longiceps* (*Therapsida: Biarmosuchia*), with comments on their early evolution in the Therapsida. J. Paleontol. 77, 977−984.

Sidor, C.A., Hopson, J.A., 1998. Ghost lineages and 'mammalness': assessing the temporal pattern of character acquisition in the Synapsida. Paleobiology 24, 254−273.

Sigurdsen, T., Huttenlocker, A.K., Modesto, S.D., Rowe, T.B., Damiani, R., 2012. Reassessment of the morphology and paleobiology of the therocephalian *Tetracynodon darti* (Therapsida), and the phylogenetic relationships of Baurioidea. J. Vertebr. Paleontol. 32 (5), 1113−1134.

Small, D.M., Bender, G., Veldhuizen, M.G., Rudenga, K., Nachtigal, D., Felsted, J., 2007. The role of the human orbitofrontal cortex in taste and flavor processing. Ann. N.Y. Acad. Sci. 1121, 136−151.

Smith, L.M., Ebner, F.F., Colonnier, M., 1980. The thalamocortical projection in *Pseudemys* turtles: a quantitative electron microscopic study. J. Comp. Neurol. 190, 445−461.

Stewart, J., 1997. Morphology and evolution of the egg of oviparous amniotes. In: Sumida, S.S., Martin, K.L.M. (Eds.), Amniote Origins. Academic Press, San Diego, pp. 291−326.

Stoddart, D.M., 1980. Olfaction in Mammals. Academic Press, London.

Streidter, G.F., 2005. Principles of Brain Evolution. Sinauer Associates, Inc., Sunderland, Massachusetts, USA.

Sues, H.-D., 1986. The skull and dentition of two tritylodont synapsids from the Lower Jurassic of western North America. Bull. Mus. Comp. Zool. 151, 217−268.

Sun, B.C., Halpern, B.P., 2005. Identification of air phase retronasal and orthonasal odorant pairs. Chem. Senses 30, 693−706.

Ten-Cate, A.R., 1969. The mechanism of tooth eruption. In: Melcher, A.H., Bowen, W.H. (Eds.), Biology of the Periodontium. Academic Press, New York, pp. 91−103.

Trulsson, M., 2006. Sensory-motor function of human periodontal mechanoreceptors. J. Oral Rehabil. 33 (4), 262−273.

Trulsson, M., Francis, S.T., Bowtell, R., McGlone, F., 2010. Brain activations in response to vibrotactile tooth stimulation: a psychophysical and fMRI study. J. neurophysiol. 104 (4), 2257−2265.

Ulinski, P.S., 1983. Dorsal Ventricular Ridge; a Treatise on Forebrain Organization in Reptiles and Birds. John Wiley & Sons, New York.

Ungar, P.S., 2010. Mammal Teeth: Origin, Evolution, and Diversity. Johns Hopkins University Press.

Wagner, G.P., 2014. Homology, Genes, and Evolutionary Innovation. Princeton University Press, Princeton, New Jersey.

Wakefield, M.J., Anderson, M., Chang, E., Wei, K.J., Kaul, R., Graves, J.A.M., Gruetzner, F., Deeb, S.S., 2008. Cone visual pigments of monotremes: filling the phylogenetic gap. Vis. Neurosci. 25 (3), 257−264.

Walls, G.L., 1942. The Vertebrate Eye and its Adaptive Radiation. Hafner Publishing Company, New York.

Wang, Y., Hu, Y., Meng, J., Li, C., 2001. An ossified Meckel's cartilage in two Cretaceous mammals and origin of the mammalian middle ear. Science 294 (5541), 357–361.

Wilson, D.A., Stevenson, R.J., 2006. Learning to Smell — Olfactory Perception from Neurobiology to Behavior. Johns Hopkins University Press, Baltimore, MD.

Wortman, J.L., 1886. Comparative anatomy of the teeth in Vertebrata. Am. Syst. Dent. 351–503.

Wortman, J.L., 1902. Origin of the tritubercular molar. Am. J. Sci. 8, 93–98.

Zelená, J., 1994. Nerves and Mechanoreceptors: The Role of Innervation in the Development and Maintenance of Mammalian Mechanoreceptors. Springer Science & Business Media.

12

Mammalian Evolution: The Phylogenetics Story

O.R.P. Bininda-Emonds, A.-M. Hartmann

Carl von Ossietzky University Oldenburg, Oldenburg, Germany

12.1 Introduction

Mammals as a group hold a natural interest for us as humans. They represent our closest relatives and so, in trying to better understand ourselves, our origins, and our (neuro)biology, other mammal species represent a natural reference point. This reference point, however, is an incredibly diverse group comprising 5825 species (www.catalogoflife.org as of April 28, 2016) with a truly worldwide distribution. Virtually all habitat types suitable for terrestrial vertebrates are occupied by mammals, and the group as a whole displays a remarkable difference in body size of nearly 80 million times from the smallest member, the bumblebee bat (*Craseonycteris thonglongyai*, <2 g), to the largest, the blue whale (*Balaenoptera musculus*, >150 tons; data from Jones et al., 2009).

To some extent, this phenotypic and ecological diversity has historically led to problems in satisfactorily unraveling mammalian evolutionary history more completely beyond the recognition of the numerous, obvious *Baupläne* within the group. This, in turn, confounds our efforts to better understand both the neurobiology of the group as a whole (and our neurobiology within this group) and the evolution of it. For instance, in performing any comparative study, it is important to distinguish between similarities that arise because of similar selection pressures (eg, in response to flight or the evolution of color vision) versus because of a shared evolutionary history (Felsenstein, 1985). Modern methods even go beyond this, recognizing that the question does not necessarily have an either-or answer, and try to more precisely quantify and correct for the extent to which a given variable is constrained by its evolutionary history (see Pagel, 1999).

This chapter has two goals. The first, more obvious goal is to provide an overview of mammalian evolutionary relationships, both from a historical and, more importantly, from a modern perspective. The development of thought along this time axis largely correlates with the molecular revolution and the rise and eventual domination of molecular phylogenetics over morphological analyses. This overview will also focus almost exclusively on the placental mammals, the group to which we as primates belong and for which the most research by far has been performed.

The second goal is to reinforce the importance of incorporating a phylogenetic perspective ("tree thinking") in biology. Although this is becoming increasingly a common practice in many other areas of biology, neurobiology has been slow to follow suit, perhaps in part because the data are comparatively more difficult to generate and so are available from relatively few (model) species, thereby hampering a more comprehensive, comparative perspective. In this chapter, we provide some examples where knowledge of mammalian phylogenetic history can or has influenced the questions asked by neurobiologists. As a consequence of this second goal, we will generally avoid using taxonomic categories (eg, order or family) because of their inherent subjectivity and the tendency of many to interpret them too literally despite this. Instead, we will refer more generically to "clades" that can be traced onto a phylogenetic tree.

12.2 The Evolutionary Tree of Mammals

12.2.1 The Historical Perspective

Even before the age of evolutionary biology and phylogenetics, mammals clearly presented a number of obvious subgroupings at several, hierarchical levels. Primary among these divisions is that at the most inclusive level, where mammals are split into the egg-laying monotremes (Prototheria), the pouched marsupials (Metatheria), and the placental mammals (Eutheria). Although of more or less equal rank taxonomically, the prefixes of the group names reveal the historical opinion

Evolutionary Neuroscience, Second Edition
https://doi.org/10.1016/B978-0-12-820584-6.00012-X

as to their relative degree of advancement, going from the first, "proto" mammals to an additional "meta"-level above that before finally reaching the true, "eu"-mammals. From a more phylogenetic perspective, except notably for many early molecular analyses (eg, Janke et al., 1996; Waddell et al., 1999; Zardoya and Meyer, 1998), it was always widely agreed that the monotremes formed the sister group to the remaining mammals (Theria) which together share some form of internal development (although obviously abridged in marsupials) in combination with viviparity.

[As an aside, there exists a widespread confusion over the proper names for the clades of marsupial versus placental mammals. When restricted to the crown groups, which are the smallest clades containing all extant species of either group, the proper names are Marsupialia and Placentalia, respectively. We will adhere to this convention, despite Eutheria (= all extinct and extant mammals that are more closely related to humans than to marsupials) being much more commonly, albeit wrongly used, for the latter group.]

Within these traditional subclasses, by far the most attention has been given to the placental mammals. Monotremes are species poor (five species; www.catalogoflife.org as of April 28, 2016) and, together with the 335 species of marsupials (www.catalogoflife.org as of April 28, 2016), are largely confined to relatively far-flung corners of the Earth from a western scientific perspective. Placentals, by contrast, constitute the bulk of extant mammal species (5485 species; www.catalogoflife.org as of April 28, 2016), are morphologically and geographically much more diverse, and hold innately more interest for us because of their greater familiarity, similarity, and phylogenetic relatedness to us. As such, placental mammals are also far better researched with respect to various aspects of their nervous systems, with representative model species in this context including the domestic cat (*Felis catus*), the domestic dog (*Canis lupus familiaris*), the house mouse (*Mus musculus*), the brown rat (*Rattus norvegicus*), and the rhesus monkey (*Macaca mulatta*), among others. Thus, this chapter will also largely confine itself to Placentalia, although the general message regarding the use of phylogenetic information is equally applicable to all mammals and beyond.

Historically, the major lineages within Placentalia were grouped into roughly 20 "orders" that represent the major *Baupläne* familiar to us (eg, rodents, primates, carnivores, even- and odd-toed ungulates, or whales). With only few exceptions, these lineages remain recognized today, revealing the relative accuracy of both early intuition and morphological analyses at this rough subdivision of placental mammals. One of these exceptions is Insectivora, which today has been divided into two lineages well separated in the placental tree

(Afrosoricida and Eulipotyphla; see later discussion). In part, Insectivora owed its existence to Victorian scientific views, which held them to be evolutionary holdovers of the earliest mammals (Huxley, 1880; Matthew, 1909): small, insectivorous creatures that scurried around at night, thereby surviving the Age of the Dinosaurs until the impact event at the K—Pg boundary *c*.65 million years ago turned climatic conditions to their favor. That Insectivora continued to exist into modern times represented in part the minority opinion that it might be a real taxon (when restricted to the "real" insectivores as Lipotyphla; eg, MacPhee and Novacek, 1993; Novacek, 1986), but more so its status as a necessary evil, a known wastebasket taxon for any small, brown mammal with sharp teeth that obviously was not a rodent, but could not be placed any more precisely than this.

The phylogenetic tree of Novacek (1992) (Fig. 12.1A) is perhaps typical of the premolecular view of mammalian phylogeny. It is conspicuous not so much for being very wrong, but more so for being relatively unresolved. Monotremes and marsupials are each represented by a single lineage only (again reflecting the general bias toward placental mammals), and the major placental lineages are grouped into only a handful of clades, with many of the lineages arising simultaneously. The placement of the lineages Artiodactyla, Hyracoidea, and Pholidota are also held to be uncertain. The lack of resolution present in the Novacek (1992) tree largely reflects the large amount of disagreement among morphological phylogeneticists with respect to placental interrelationships at this time rather than a lack of knowledge per se. Many other opinions were present, and another morphological summary tree appears in Springer et al. (2004). Although the tree, which is obtained from Shosani and McKenna (1998), is more resolved at first glance, many of the internal branches are dashed, representing areas of conflict with the Novacek (1992) tree (Fig. 12.2).

Important, however, is that Novacek (1992) does not have Insectivora originating at the base of the placentals despite being commonly held to hold a primitive status within the group and to strongly resemble Cretaceous mammals. As such, members of Insectivora represent a poor choice as exemplars for the primitive placental nervous system. More generally, the same is true for any lineage(s) placed at the base of the placental tree (eg, Xenarthra and possibly Pholidota in the case of Novacek, 1992) because it is frequently unappreciated that these "basal" lineages are as far removed from the placental common ancestor temporally as any other living species of placental mammal, even those from more "derived" lineages. As such, they will have evolved their own sets of special, derived adaptations like any other extant mammal species.

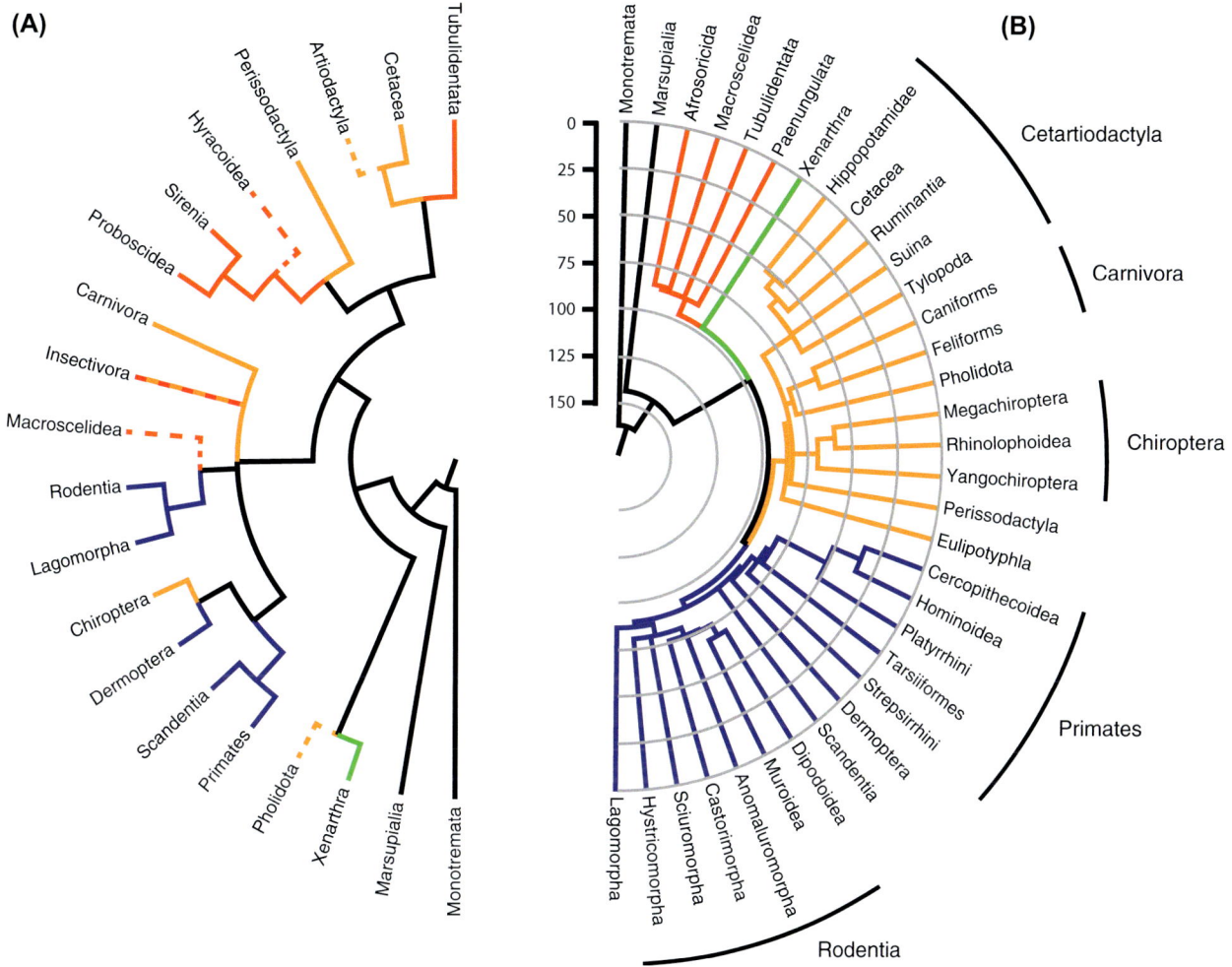

FIGURE 12.1 Two views on the evolutionary history of mammals. (A) The historical, largely morphological, perspective as exemplified by the study of Novacek (1992). *Dashed lines* indicate taxa of uncertain placement. (B) The current, largely molecular perspective as derived from a consensus of the studies of Bininda-Emonds et al. (2007, 2008), Meredith et al. (2011), and O'Leary et al. (2013). Divergence dates estimates in (B) follow Bininda-Emonds et al. (2008) and are in millions of years before the present. The major groups (superorders) of placental mammals are color-coded following Springer et al. (2004): *red* (Afrotheria or Afrosoricida), *orange* (Laurasiatheria), *green* (Xenarthra), and *blue* (Euarchontoglires).

12.2.2 The Mammal Tree Today

The molecular revolution, even in its earliest days, has done much to resolve the mammal family tree. Again, most of the major lineages escaped the revolution largely unscathed, but were often grouped in new constellations in part because of an unrealized degree of morphological convergence within placental mammals (for an overview, see Springer et al., 2004).

However, this is not to say that molecular analyses are infallible. One cogent example here comes from the many early molecular phylogenetic analyses indicating that Rodentia were not a monophyletic clade, with the guinea pig (*Cavia porcellus*) as the exemplar for all remaining hystricomorph rodents not clustering with the remaining rodents (eg, D'Erchia et al., 1996; Graur et al., 1991). Although the precise tree and sets of relationships differed between the various studies, this general result was obtained for analyses of both nuclear DNA markers and especially of complete mitochondrial genomes. Eventually, however, the result was deemed to be an artifact deriving from the limited species sampling (often <20) as well as the use of overly simplistic models of evolution, especially for the fast-evolvingmitochondrial data (Sullivan and Swofford, 1997), such that more comprehensive and sophisticated analyses quickly restored the otherwise long-undisputed Rodentia as a clade (Luckett and Hartenberger, 1993; Cao et al., 1994). For similar reasons, early molecular analyses often supported the grouping of Monotremata plus Marsupialia (= Marsupionta; see previous discussion) and were also slow to support Glires (= Rodentia + Lagomorpha), a clade that historically has been largely unquestioned on morphological grounds (eg, Novacek, 1992).

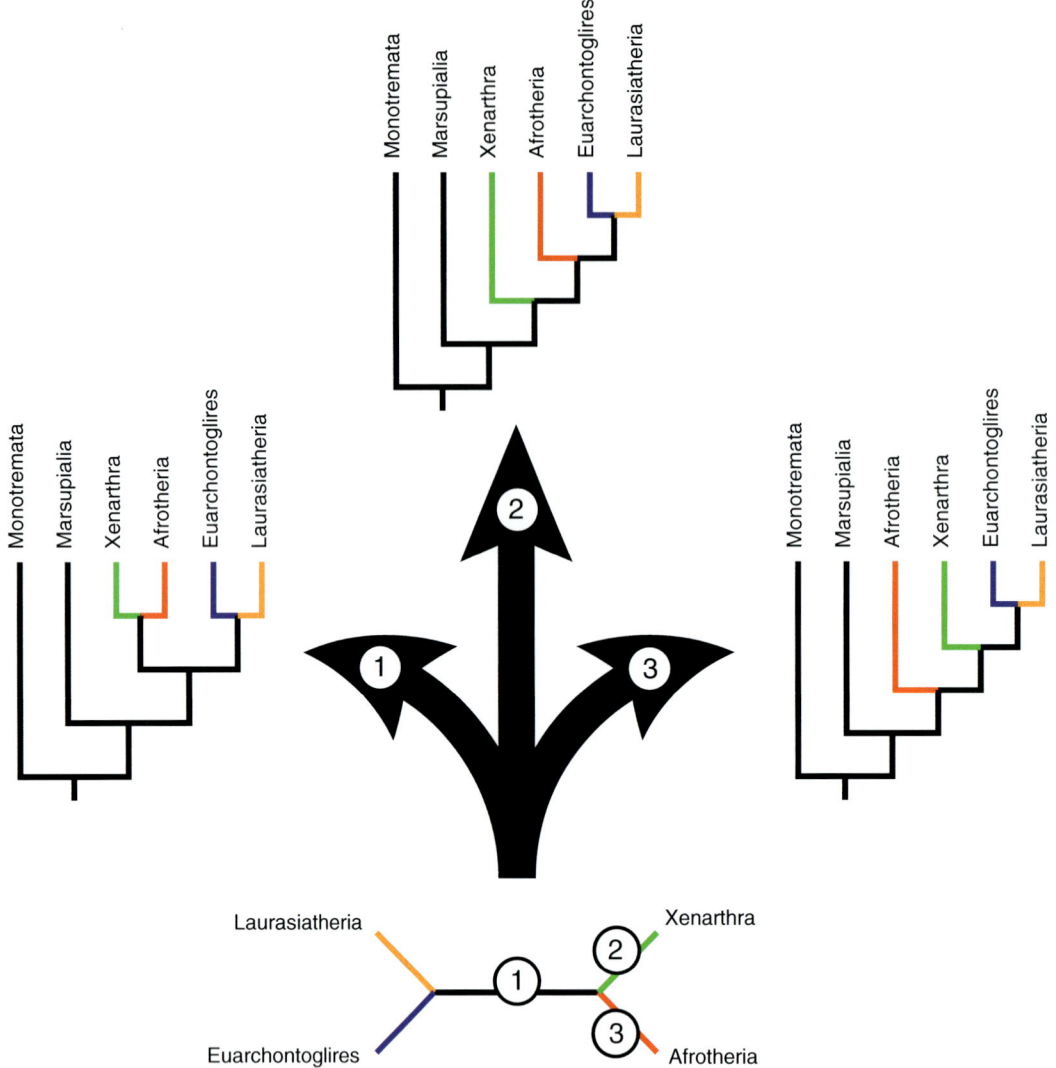

FIGURE 12.2 The three alternative positions for the root of the Placentalia given the consensus that Euarchontoglires and Laurasiatheria comprise the clade Boreoeutheria. From the unrooted tree of the Placentalia at the bottom of the figure, three different topologies are possible if the root is placed either at position (1) (as favored by Bininda-Emonds et al., 2007), (2) (as favored by paleontologists generally as well as O'Leary et al., 2013), or (3) (as favored by Meredith et al., 2011). For clarity, only the four superorders are shown for the placental mammals.

This chapter will summarize the findings of three major research groups using different analytical protocols: the purely molecular work of Mark Springer and colleagues (Meredith et al., 2011; *Murphy* et al., 2001b; Springer et al., 2004), the supertree work of Bininda-Emonds et al. (2007, 2008) that indirectly combines morphological and molecular data, and the direct, combined molecular and morphological approach of O'Leary et al. (2013).

All three research groups place the placental orders into four supergroups (commonly referred to as superorders): Afrosoricida, Euarchontoglires, Laurasiatheria, and Xenarthra, with Euarchontoglires and Laurasiatheria furthered clustered as Boreoeutheria. Although all three groups agree with respect to the membership of the superorders (Fig. 12.1B), the interrelationships among them, and thus the position of the placental root, remain an area of outstanding disagreement (see Teeling and Hedges, 2013). Indeed, the three research groups present all possible options given the consensus that Boreoeutheria is a clade. For instance, Bininda-Emonds et al. (2007) hold Afrotheria to be the sister group of the remaining placentals, whereas O'Leary et al. (2013) claim this position for Xenarthra, a placement favored generally by paleontologists (also Novacek, 1992). Meredith et al. (2011) instead place these two taxa in the clade Atlantogenata that itself is the sister taxon of Boreoeutheria in contrast to earlier studies by this group that favored an afrotherian root (Madsen et al., 2001; Murphy et al., 2001a,b). Even large, genomic-scale studies have been unable to resolve this question consistently (see Teeling and Hedges, 2013) because the extremely short, nearly simultaneous time frame in which the superorders apparently originated some 80+ million years ago (Tarver et al., 2016; Bininda-Emonds et al., 2008; Meredith et al., 2011; dos Reis et al., 2012; see Fig. 12.1B) precludes any easy solution to the problem. Similar adaptive radiations also appear to characterize the origins of the placental orders, albeit with a large difference of opinion between molecular phylogeneticists (c.80 million years ago; Bininda-Emonds et al., 2008; Meredith et al., 2011) and paleontologists (shortly after the K–Pg boundary; O'Leary et al., 2013) as to when this occurred. Bininda-Emonds et al. (2007, 2008) also hold that the basal divergences of many of the orders also occurred in the Cretaceous, whereas most other studies tend to place them in or much closer to the Paleogene following the extinction of the dinosaurs (Tarver et al., 2016; Meredith et al., 2011; O'Leary et al., 2013; dos Reis et al., 2012).

To a large extent, much of the redrafting of the placental family tree via molecular evidence occurred because of the recognition of Afrotheria as a clade. Previously, its members were distributed across all of Placentalia (see Fig. 12.1A). Although Afrotheria enjoys almost universal, strong support from diverse sources of molecular data (Stanhope et al., 1998; Nikaido et al., 2003; Nikolaev et al., 2007) and some subclades within it are largely undisputed morphologically (elephants, sea cows, and hyraxes = Paenungulata; Simpson, 1945), acceptance of the entire clade among morphological phylogeneticists was limited initially because of the early lack of any decisive morphological features shared among all its members. Indeed, apart from molecular information there appears to be little uniting species as diverse as aardvarks, tenrecs, golden moles, and the above-named groups except perhaps that all the groups are found at least in part on the African continent. This latter situation has ameliorated itself to some degree, even if the morphological synapomorphies that have been identified are arguably more obscure than those often defining other major mammalian clades: a high thoracolumbar vertebral number (Sánchez-Villagra et al., 2007), details regarding the formation of the placental membranes (Mess and Carter, 2006), the morphology of the ankle bones (Tabuce et al., 2007), and a comparatively late eruption of the permanent dentition (Asher and Lehmann, 2008).

A direct consequence of the erection of Afrotheria was the final dismantling of the problematic Insectivora (or, more accurately, its remains as Lipotyphla at this point) into two separate clades. Golden moles and tenrecs (Afrosoricida) left real shrews behind to cluster with elephant shrews within Afrotheria, with the remaining lipotyphlan taxa (= Erinaceidae, Soricidae, Talpidae, and Solenodontidae) now forming Eulipotyphla within Laurasiatheria. Indeed, "Insectivora" represents perhaps the most cogent example of convergent evolution between Afrotheria and Laurasiatheria (see Springer et al., 2004).

A second major alteration indicated by molecular data was the removal of the bats (Chiroptera) from Archonta (= Dermoptera, Primates, and Scandentia), thereby moving them from Euarchontoglires to Laurasiatheria (Pumo et al., 1998; Teeling et al., 2000; Murphy et al., 2001a,b). In so doing, the historical association between bats and colugos (Dermoptera) was dissolved, one that was largely based on species in both taxa being able to fly and therefore having similar morphological adaptations to accomplish this. In hindsight, however, it is clear that this situation is another case of convergent evolution and numerous differences between the two groups exist. For instance, although colugos are highly accomplished gliders, only bats achieve true powered flight among extant mammals. Moreover, whereas the main lifting surface area in colugos stretches between the forelimbs and hindlimbs, that in bats is present between the digits of the forelimb. Shifting bats out of Archonta also resolves problems with their general lack of typical archontan features (especially those relating to the ankle; Szalay and Drawhorn, 1980); however, their highly derived morphology relating to powered flight has also hindered the process of placing bats successfully within Laurasiatheria. Unfortunately, fossil evidence here is unhelpful, given that the oldest known bat fossils (eg, *Onychonycteris finneyi*, 52.5 million years old; Simmons et al., 2008) highly resemble extant species morphologically. Combined with bat-wing morphology having essentially remained unaltered for at least 50 million years (Sears et al., 2006), morphological evidence to definitively link bats with other laurasiatherian groups is thus lacking, and even molecular data have been unable to place bats within Laurasiatheria with any degree of consistency or confidence.

Molecular data also strongly reaffirmed the monophyly of Chiroptera (Mindell et al., 1991; Pumo et al., 1998; Van Den Bussche and Hoofer, 2004; Teeling et al., 2002), a historical given that had been called into question in the 1980s and 1990s, largely on the basis of novel neurological data (visual pathways) that indicated that the large, diurnal fruit bats (Megachiroptera) appeared to be more closely related to primates than they were to the remaining, nocturnal, echolocating bats (Microchiroptera), the "flying-primate" hypothesis (Smith and Madkour, 1980; Pettigrew, 1986; see later discussion). Unexpected in the reunion, however, was the dissolution of the historically uncontested megabat/microbat dichotomy with the finding that Megachiroptera clustered with rhinolophoid microbats (as Yinpterochiroptera) (Teeling et al., 2002; Meredith et al., 2011; Tsagkogeorga et al., 2013), thereby rendering Microchiroptera paraphyletic (but see O'Leary et al., 2013) and confusing the issue of how often laryngeal echolocation in bats arose (once for the entire group before being lost in megabats versus in parallel between the two clades of microbats; see later discussion and Springer et al., 2004).

Otherwise, the only other significant, high-level alteration brought about by molecular data was the placing of whales (Cetacea) as the sister group of hippos within the even-toed ungulates (Artiodactyla) to form Cetartiodactyla (Nikaido et al., 1999; Graur and Higgins, 1994; Shimamura et al., 1997; Montgelard et al., 1997; Gatesy et al., 1999). Although this example was sometimes championed as demonstrating the superiority of molecular phylogenetics, the change is arguably not as dramatic as it appears. The close affinities of Cetacea and Artiodactyla have been recognized since the late 1800s (eg, Flower, 1883), and especially since the 1960s, with the two taxa usually being placed as sister taxa among extant mammals (eg, Novacek, 1992). Indeed, suggestions of a placement of whales within the even-toed ungulates appeared even before the first molecular evidence for it (eg, Beddard, 1900). However, this new placement was again more of a case of unraveling a lack of resolution rather than correcting an outright error. Problematic here was that the defining characteristics of artiodactyls—their exclusive possession of a paraxonic (even-toed) limb and a double-pulleyed surface of the astragalus in the ankle joint (Rose, 1996)—could never be confirmed in the whales because the latter have greatly reduced their hindlimbs such that only remnants of the pelvis that no longer connect to the vertebral column remain. Confounding the problem further was other aspects of their highly specialized morphology, which often diverged greatly from that of terrestrial mammals. However, the description of the two archaeocete cetaceans *Pakicetus* and *Ichthyolestes* from Pakistan in 2001 confirmed both their partly terrestrial nature and their possession of a double-pulleyed astragalus (Gingerich et al., 2001; Thewissen et al., 2001), thereby placing cetaceans minimally as the immediate sister group of artiodactyls and not to the extinct, but closely related mesonychids (Van Valen, 1966). Nevertheless, support for a placement of whales inside of even-toed ungulates on morphological grounds alone remains difficult.

12.2.3 The Way Forward

As a result of this concerted effort and the inclusion of both morphological and molecular lines of evidence, higher-level relationships among placental mammals are relatively well agreed upon (*contra* Meredith et al., 2011) and resolved. In the admittedly subjective selection of major placental groups outlined in Fig. 12.1B, which is a composite of the studies of Bininda-Emonds et al. (2007, 2008), Meredith et al. (2011), and O'Leary et al.

(2013), only three outstanding areas of uncertainty remain. Two have already been mentioned: the ongoing debate regarding the position of the placental root and uncertainly regarding the basal laurasiatherian relationships, largely because of differing opinions regarding the placement of Chiroptera. The third regards the interrelationships of the three long-standing major morphological lineages recognized among a monophyletic Rodentia, the hystricomorph, sciuromorph, and myomorph rodents. Virtually all combinations of these three lineages have been proposed, even among the three major working groups highlighted in this chapter. Thus, although there appears to be little doubt regarding the monophyly of each lineage, a solid resolution of their interrelationships remains lacking.

The reason underlying these remaining problems areas might be the extremely short evolutionary timescales that are involved (<10 million years in some cases) combined with the events occurring some tens of millions of years ago (see Fig. 12.1B). The exact timing of the basal radiations within placentals as a group and also within the individual orders remains contentious (see earlier), although recent molecular studies would seem to have at least the latter happening closer to the K–Pg boundary than was thought a decade ago (Tarver et al., 2016; dos Reis et al., 2012). Nevertheless, it remains that we are trying to tease apart a series of apparently rapid evolutionary events from a long time in the past. Two problems are faced here in particular by molecular data.

First, the limited number of character states (ie, the 4 nucleotides and, to a lesser degree, the 21 amino acids) means that convergence and back mutations become increasingly common with divergence time, thereby progressively obfuscating any signal from before. Ideally, we need to find a molecular marker that is both informative in the desired time frame and, equally important, has retained this information. This seems unlikely for any pure sequence-based marker, with rare genomic changes (see Rokas and Holland, 2000) potentially holding more promises in such cases. Although the rarity of such changes means that few, if any, will occur at the proper point in time, this same property decreases the likelihood that the signal will get washed out with time through further homoplastic changes. For instance, unique families of short interspersed nuclear elements (SINEs, a class of retroposons that can move around the genome) characterize each of Afrotheria as well as Paenungulata within Afrotheria (Nikaido et al., 2003) as well as Cetartiodactyla and most major lineages within it, including the grouping of whales with hippos (Nikaido et al., 1999). Both examples include rapid radiation events many tens of millions of years ago that have otherwise proven difficult to resolve (see Fig. 12.1B).

Second, and more ominously, in working with sequence data of any sort (including SINEs), we need to be aware of the potential for conflicts between gene trees and species trees (see Maddison, 1997). Here, natural processes such as horizontal gene transfer (which is unlikely in mammals) or incomplete lineage sorting, whereby a polymorphism becomes fixed in only some of the descendants of a common ancestor (see Maddison, 1997), can cause the two topologies to differ, but yet still be accurate for the entities being tracked (ie, genes or species). Disturbingly, such problems do not necessarily disappear with time and are more likely to occur during a series of rapid speciation events, exactly as is presumed in the cases discussed earlier (Degnan and Rosenberg, 2006; Rosenberg, 2013). Thus, the conflicts above might be real and inherent and, in extreme cases, also might not be easily resolved by the stock solution of simply throwing more genes/markers at it because they themselves are providing conflicting, but accurate information (but see Maddison and Knowles, 2006).

12.3 Applying Tree-Thinking to Question in Neurobiology

An accompanying narrative in this chapter has been the relatively high degree of morphological convergence within placental mammals and how this convergence has confounded earlier views of mammalian phylogeny. An overview of this situation is clearly presented by Springer et al. (2004). However, recognizing convergence as such is dependent on the prevailing phylogenetic hypothesis. In this, molecular sequence data were instrumental as a more-or-less independent data source to help decide between competing, morphology-based hypotheses. Again, this is not to imply that morphological data are inherently inferior or untrustworthy (see also previous discussion), with recent studies having found much hidden support for "molecular hypotheses" within morphological data sets (eg, Lee and Camens, 2009) and the combined data analyses of the Bininda-Emonds and O'Leary working groups being able to include more of the entire phylogenetic data set than the molecular-only data set of Springer and colleagues. What is important, however, is recognizing convergence for what it is (using phylogenetic information) and then using it to direct the appropriate questions in both phylogeny reconstruction and the application of such data in a comparative biology framework.

An instructive example is provided by the rise of the "flying-primate" hypothesis in the 1980s mentioned earlier. Although the possible paraphyly of Chiroptera was first proposed in modern times on the basis of penile morphology (Smith, 1980; Smith and Madkour, 1980), perhaps the key support for this hypothesis was obtained from neurological data, particularly the different visual systems present in Mega- and Microchiroptera

(Pettigrew, 1986; Pettigrew et al., 1989). Pettigrew and colleagues (Pettigrew, 1986; Pettigrew et al., 2008) found that megabats have a retinotectal pathway with a vertical hemidecussation, something that is only known in primates and dermopterans, whereas the microbats possess the ancestral symplesiomorphic pattern of connections found in all remaining vertebrates. The crux of Pettigrew's (1986, 1991) arguments was to ask which one of the two necessary instances of convergence—powered flight and all the attendant morphological adaptations in the case of bat paraphyly versus the uniquely similar derived visual systems of megabats and primates in the case of bat monophyly—was more unlikely to have occurred. In so doing, he noted that similar morphological adaptations for flying were also present in Dermoptera (perhaps indicating general morphological constraints in getting a mammal to fly), that some slight, but notable, differences with respect to the flight apparatus also exist between the mega- and microbats (Pettigrew, 1991), and that virtually all morphological support for bat monophyly derived solely from the (constrained) flight apparatus. However, no apparent constraints could be found to explain the similar visual systems of Megachiroptera and Primates.

From a neurobiological perspective, the resolution of this debate is particularly interesting because of the implications of the current phylogenetic hypothesis on the degree of plasticity and convergence in neurobiological systems. For instance, assuming bat monophyly to be correct, the obvious biological question to ask is why the advanced visual system of primates + dermopterans (and numerous other neurological features; see Maseko and Manger, 2007; Maseko et al., 2007) also evolved in parallel in the completely unrelated megabats, but then nowhere else in any other mammal or vertebrate for that matter, a scenario which Pettigrew would understandably argue to be highly unlikely. Detailed investigations could then reveal if the two systems are really identical or merely "similar" (as is the case for bat and colugo wings) as well as any necessary preadaptations and/or selective forces needed for their convergent evolution. On the flip side, bat monophyly in association with probable microbat paraphyly (Tsagkogeorga et al., 2013) potentially raises a number of analogous questions regarding the evolution of laryngeal echolocation were it also to have arisen in parallel between the two major microbat clades (Yangochiroptera and Rhinolophoidea). Although it remains unclear at present how many times laryngeal echolocation evolved within Chiroptera, support for it having evolved once for the entire group comes from an examination of O. finneyi, which might have had the capacity for laryngeal echolocation because its stylohyal bones probably articulated with its tympanic bones in a fashion similar to those found in other laryngeal echolocating bats (Veselka et al., 2010; but see

Simmons et al., 2008). By contrast, Parker et al. (2013) detected strong, significant support for convergence at a genomic level between echolocating bats and dolphins, potentially indicating that convergence for this trait within the paraphyletic microbats might be more likely than previously thought.

A similar scenario for the pinnipeds—the semi-aquatic seals, sea lions, and walrus—exists within Carnivora. Although the placement of the pinnipeds within carnivores was long recognized (eg, Flower, 1869), their exact location has long been disputed and, in the 1960s, strong arguments against the monophyly of the group were raised. Similar to the flight apparatus of bats, it was argued that the aquatic adaptations of pinnipeds were too constrained functionally to serve as reliable phylogenetic characters, with many other more "neutral" characters indicating a separate origin of the seals from the sea lions and walrus (McLaren, 1960). This view persisted until the late 1980s when additional morphological analyses (eg, Bininda-Emonds and Russell, 1996; Wyss, 1987, 1988) as well as molecular analyses (eg, Arnason et al., 1995; Ledje and Arnason, 1996a,b) again unequivocally supported pinniped monophyly. Again, regardless of which hypothesis is correct, the key issue in this context is that the use of one hypothesis or the other can dramatically change the nature of the analyses and the questions that one asks as a neurobiologist. For instance, if the pinnipeds are indeed diphyletic, then their physiological adaptations to diving and sensory adaptations to the underwater environment must have evolved convergently, albeit from a very similar ground plan (bears vs weasels within arctoid carnivores), and one should expect slight, but possibly important differences in the adaptations. Under the monophyly hypothesis, by contrast, few, if any, differences are to be expected.

Implicit in this general discussion is that answering such evolutionary questions requires the inclusion of at least three species, one representing each of the two focal groups and a third, closely related outgroup species representing an independent reference point. Doing so automatically tempers notions of "primitive" versus "derived" or that one of the species might represent an earlier stage of evolution of the latter. An extreme example of the latter is provided by humans and chimps, with the common, but unsupported view being that humans evolved from a chimplike ancestor that was virtually identical to modern chimps (eg, Wrangham, 2001), when both species are really the end points of lineages, each of which has been diverging independently from their last common ancestor for anywhere between 4 and 13 million years (Arnason et al., 1998; Patterson et al., 2006).

In the same way, it is unjustified to take modern insectivore species as representative of the earliest mammals, although the detailed reconstruction of the mammalian common ancestor provided by O'Leary et al. (2013) does indicate that it was indeed very much insectivore-like. It remains that modern insectivore species are polyphyletic and thus have converged on the same body plan. Moreover, as individual species, each is also as far removed from the mammalian common ancestor as any other living species of mammals. That being said, it is justifiable to examine "insectivore" species within specific contexts that they happen to share with the earliest mammals, as was done by Catania (2007) in using shrews as an exemplar to understand the evolution of small brains and neocortices in mammals. Indeed, the polyphyly of insectivore-like placentals is actually beneficial here because it implies multiple, independent retentions and/or rederivations of a small brain, thereby increasing the statistical sample size when looking for neurological traits that characterize or are constrained by a small brain. By contrast, examining multiple species of shrews from the same genus as a means of increasing sample size potentially commits the statistical error of pseudoreplication because it is unclear how much of the brain morphology is determined because of the limited space versus simply being inherited from a common ancestor (Felsenstein, 1985). Instead, the better way forward statistically would be to examine additional exemplars from other, independent clades with species having small brain sizes (eg, from within rodents or bats).

Much of the previous discussion has concerned itself with the examination of neurobiological data within a phylogenetic framework. Although this is critical for any biological data, it is important to remember that neurobiological data, like any biological data, can also help to inform phylogenetic opinion. The potential for this was obvious in the "flying-primate" debate spurred on by Pettigrew (see earlier discussion). Although the visual systems data are now held by most to have evolved convergently, Pettigrew was absolutely correct in principle in challenging the notion that neural pathway data might not contain much (reliable) phylogenetic signal (see Wible and Novacek, 1988). Another interesting, but unappreciated example was provided by Hof et al. (1999) who examined the distribution of calcium-binding proteins in the mammalian neocortex. Although they framed their observations in an explicit phylogenetic context, they noted several discrepancies in their observations in the context of the Novacek-like (1992) framework that they were using. In addition to some shared similarities between megabats and primates (cf flying-primate hypothesis), they particularly made note of numerous similarities between carnivores, artiodactyls, cetaceans, and some insectivore species (eulipotyphlans). At least some of these discrepancies were attributed to retained, primitive features (Hof et al., 1999), but, instead, they are easier to reconcile under, and indeed provided early anatomical evidence for, the more molecular-influenced view of mammal phylogeny that is favored today given that all these groups are now held to belong to Laurasiatheria.

12.4 Conclusions

Placental mammals represent one of the best-studied groups from a phylogenetic perspective, both from early morphological studies to cutting-edge molecular analyses. Although there has been much made about molecular data rewriting the mammalian tree, the reality is more that, apart from the recognition of Afrotheria, such data have primarily refined it by resolving areas of previous uncertainty. Together the combination of morphological and molecular data have produced a mostly resolved and relatively stable evolutionary hypothesis of mammals, at least at the higher taxonomic levels such that remaining work with the group will focus more on filling in the numerous smaller branches leading to the individual species. As we have shown through a number of examples, it is important to take account of phylogenetic history in framing our (neuro)biological questions, but that the answers are often necessarily dependent on the tree being used. Fortunately, the apparent robustness of the mammalian phylogeny (again at the higher taxonomic levels) will remove much of this ambiguity and yield a solid framework upon which to base our analyses.

References

Arnason, U., Bodin, K., Gullberg, A., Ledje, C., Mouchaty, S., 1995. A molecular view of pinniped relationships with particular emphasis on the true seals. J. Mol. Evol. 40, 78−85.

Arnason, U., Gullberg, A., Janke, A., 1998. Molecular timing of primate divergences as estimated by two nonprimate calibration points. J. Mol. Evol. 47, 718−727.

Asher, R.J., Lehmann, T., 2008. Dental eruption in afrotherian mammals. BMC Biol. 6, 14.

Beddard, F.E., 1900. A Book of Whales. G.P. Putnam's Sons, New York.

Bininda-Emonds, O.R.P., Russell, A.P., 1996. A morphological perspective on the phylogenetic relationships of the extant phocid seals (Mammalia: Carnivora: Phocidae). Bonn. Zool. Monogr. 41, 1−256.

Bininda-Emonds, O.R.P., Cardillo, M., Jones, K.E., Macphee, R.D.E., Beck, R.M.D., Grenyer, R., Price, S.A., Vos, R.A., Gittleman, J.L., Purvis, A., 2007. The delayed rise of present-day mammals. Nature 446, 507−512.

Bininda-Emonds, O.R.P., Cardillo, M., Jones, K.E., Macphee, R.D.E., Beck, R.M.D., Grenyer, R., Price, S.A., Vos, R.A., Gittleman, J.L., Purvis, A., 2008. Corrigendum: the delayed rise of present-day mammals. Nature 456, 274.

Cao, Y., Adachi, J., Yano, T.A., Hasegawa, M., 1994. Phylogenetic place of Guinea pigs: no support of the rodent-polyphyly hypothesis from maximum-likelihood analyses of multiple protein sequences. Mol. Biol. Evol. 11, 593−604.

Catania, K.C., 2007. Organization of a miniature neocortex − what shrew brains suggest about mammalian evolution. In: Kaas, J.H., Krubitzer, L.A. (Eds.), Evolution of Nervous Systems. Elsevier Academic Press, Amsterdam.

D'Erchia, A.M., Gissi, C., Pesole, G., Saccone, C., Arnason, U., 1996. The Guinea-pig is not a rodent. Nature 381, 597−600.

Degnan, J.H., Rosenberg, N.A., 2006. Discordance of species trees with their most likely gene trees. PLoS Genet. 2, 762−768.

dos Reis, M., Inoue, J., Hasegawa, M., Asher, R.J., Donoghue, P.C.J., Yang, Z., 2012. Phylogenomic datasets provide both precision and accuracy in estimating the timescale of placental mammal phylogeny. Proc. R. Soc. Biol. Sci. 279, 3491−3500.

Felsenstein, J., 1985. Phylogenies and the comparative method. Am. Nat. 125, 1−15.

Flower, W.H., 1869. On the value of characters of the base of the cranium in the classification of the order Carnivora, and on the systematic position of Bassaris and other disputed forms. Proc. Zool. Soc. Lond. 1869, 4−37.

Flower, W.H., 1883. On the characters and divisions of the family Delphinidæ. Proc. Zool. Soc. Lond. 1883, 466−513.

Gatesy, J., Milinkovitch, M., Waddell, V., Stanhope, M., 1999. Stability of cladistic relationships between Cetacea and higher-level artiodactyl taxa. Syst. Biol. 48, 6−20.

Gingerich, P.D., Haq, M., Zalmout, I.S., Khan, I.H., Malkani, M.S., 2001. Origin of whales from early artiodactyls: hands and feet of Eocene Protocetidae from Pakistan. Science 293, 2239−2242.

Graur, D., Higgins, D.G., 1994. Molecular evidence for the inclusion of cetaceans within the order Artiodactyla. Mol. Biol. Evol. 11, 357−364.

Graur, D., Hide, W.A., Li, W.-H., 1991. Is the Guinea-pig a rodent? Nature 351, 649−652.

Hof, P.R., Glezer, I.I., Condé, F., Flagg, R.A., Rubin, M.B., Nimchinsky, E.A., Vogt Weisenhorn, D.M., 1999. Cellular distribution of the calcium-binding proteins parvalbumin, calbindin, and calretinin in the neocortex of mammals: phylogenetic and developmental patterns. J. Chem. Neuroanat. 16, 77−116.

Huxley, T.H., 1880. On the application of the laws of evolution to the arrangement of the Vertebrata, and more particularly of the Mammalia. Proc. R. Soc. Lond. 43, 649−662.

Janke, A., Feldmaier-Fuchs, G., Gemmell, N., Von Haeseler, A., Pääbo, S., 1996. The complete mitochondrial genome of the platypus (Ornithorhynchus anatinus). J. Mol. Evol. 42, 153−159.

Jones, K.E., Bielby, J., Cardillo, M., Fritz, S.A., O'dell, J., Orme, C.D.L., Safi, K., Sechrest, W., Boakes, E.H., Carbone, C., Christina Connolly, C., Cutts, M.J., Foster, J.K., Grenyer, R., Habib, M., Plaster, C.A., Price, S.A., Rigby, E.A., Rist, J., Teacher, A., Bininda-Emonds, O.R.P., Gittleman, J.L., Mace, G.M., Purvis, A., 2009. PanTHERIA: a species-level database of life history, ecology, and geography of extant and recently extinct mammals. Ecology 90, 2648.

Ledje, C., Arnason, U., 1996. Phylogenetic analyses of complete cytochrome b genes of the order Carnivora with particular emphasis on the Caniformia. J. Mol. Evol. 42, 135−144.

Ledje, C., Arnason, U., 1996. Phylogenetic relationships within caniform carnivores based on analyses of the mitochondrial 12s rRNA gene. J. Mol. Evol. 43, 641−649.

Lee, M.S., Camens, A.B., 2009. Strong morphological support for the molecular evolutionary tree of placental mammals. J. Evol. Biol. 22, 2243−2257.

Luckett, W.P., Hartenberger, J.-L., 1993. Monophyly or polyphyly of the order Rodentia: possible conflict between morphological and molecular interpretations. J. Mamm. Evol. 1, 127−147.

MacPhee, R.D.E., Novacek, M.J., 1993. Definition and relationships of Lipotyphla. In: Szalay, F.S., Novacek, M.J., Mckenna, M.C. (Eds.), Mammalian Phylogeny: Placentals. Springer-Verlag, New York.

Maddison, W.P., 1997. Gene trees in species trees. Syst. Biol. 46, 523−536.

Maddison, W.P., Knowles, L.L., 2006. Inferring phylogeny despite incomplete lineage sorting. Syst. Biol. 55, 21–30.

Madsen, O., Scally, M., Douady, C.J., Kao, D.J., Debry, R.W., Adkins, R., Amrine, H.M., Stanhope, M.J., De Jong, W.W., Springer, M.S., 2001. Parallel adaptive radiations in two major clades of placental mammals. Nature 409, 610–614.

Maseko, B.C., Manger, P.R., 2007. Distribution and morphology of cholinergic, catecholaminergic and serotonergic neurons in the brain of Schreiber's long-fingered bat, *Miniopterus schreibersii*. J. Chem. Neuroanat. 34, 80–94.

Maseko, B.C., Bourne, J.A., Manger, P.R., 2007. Distribution and morphology of cholinergic, putative catecholaminergic and serotonergic neurons in the brain of the Egyptian rousette flying fox, *Rousettus aegyptiacus*. J. Chem. Neuroanat. 34, 108–127.

Matthew, W.D., 1909. The Carnivora and insectivora of the Bridger basin. Mem. Am. Mus. Nat. Hist. 9, 291–567.

McLaren, I.A., 1960. Are the Pinnipedia biphyletic? Syst. Zool. 9, 18–28.

Meredith, R.W., Janecka, J.E., Gatesy, J., Ryder, O.A., Fisher, C.A., Teeling, E.C., Goodbla, A., Eizirik, E., Simao, T.L.L., Stadler, T., Rabosky, D.L., Honeycutt, R.L., Flynn, J.J., Ingram, C.M., Steiner, C., Williams, T.L., Robinson, T.J., Burk-Herrick, A., Westerman, M., Ayoub, N.A., Springer, M.S., Murphy, W.J., 2011. Impacts of the cretaceous terrestrial revolution and KPg extinction on mammal diversification. Science 334, 521–524.

Mess, A., Carter, A.M., 2006. Evolutionary transformations of fetal membrane characters in Eutheria with special reference to Afrotheria. J. Exp. Zool. B Mol. Dev. Evol. 306b, 140–163.

Mindell, D.P., Dick, C.W., Baker, R.J., 1991. Phylogenetic relationships among megabats, microbats, and primates. Proc. Natl. Acad. Sci. U.S.A. 88, 10322–10326.

Montgelard, C., Catzeflis, F.M., Douzery, E., 1997. Phylogenetic relationships of artiodactyls and cetaceans as deduced from the comparison of cytochrome b and 12s rRNA mitochondrial sequences. Mol. Biol. Evol. 14, 550–559.

Murphy, W.J., Eizirik, E., Johnson, W.E., Zhang, Y.P., Ryder, O.A., O'brien, S.J., 2001. Molecular phylogenetics and the origins of placental mammals. Nature 409, 614–618.

Murphy, W.J., Eizirik, E., O'brien, S.J., Madsen, O., Scally, M., Douady, C.J., Teeling, E., Ryder, O.A., Stanhope, M.J., De Jong, W.W., Springer, M.S., 2001. Resolution of the early placental mammal radiation using Bayesian phylogenetics. Science 294, 2348–2351.

Nikaido, M., Rooney, A.P., Okada, N., 1999. Phylogenetic relationships among cetartiodactyls based on insertions of short and long interspersed elements: hippopotamuses are the closest extant relatives of whales. Proc. Natl. Acad. Sci. U.S.A. 96, 10261–10266.

Nikaido, M., Nishihara, H., Hukumoto, Y., Okada, N., 2003. Ancient Sines from African endemic mammals. Mol. Biol. Evol. 20, 522–527.

Nikolaev, S., Montoya-Burgos, J.I., Margulies, E.H., Program, N.C.S., Rougemont, J., Nyffeler, B., Antonarakis, S.E., 2007. Early history of mammals is elucidated with the ENCODE multiple species sequencing data. PLoS Genet. 3, e2.

Novacek, M.J., 1986. The skull of leptictid insectivorans and the higher-level classification of eutherian mammals. Bull. Am. Mus. Nat. Hist. 183, 1–112.

Novacek, M.J., 1992. Mammalian phylogeny: shaking the tree. Nature 356, 121–125.

O'Leary, M.A., Bloch, J.I., Flynn, J.J., Gaudin, T.J., Giallombardo, A., Giannini, N.P., Goldberg, S.L., Kraatz, B.P., Luo, Z.X., Meng, J., Ni, X., Novacek, M.J., Perini, F.A., Randall, Z.S., Rougier, G.W., Sargis, E.J., Silcox, M.T., Simmons, N.B., Spaulding, M., Velazco, P.M., Weksler, M., Wible, J.R., Cirranello, A.L., 2013. The placental mammal ancestor and the post-K-Pg radiation of placentals. Science 339, 662–667.

Pagel, M., 1999. Inferring the historical patterns of biological evolution. Nature 401, 877–884.

Parker, J., Tsagkogeorga, G., Cotton, J.A., Liu, Y., Provero, P., Stupka, E., Rossiter, S.J., 2013. Genome-wide signatures of convergent evolution in echolocating mammals. Nature 502, 228–231.

Patterson, N., Richter, D.J., Gnerre, S., Lander, E.S., Reich, D., 2006. Genetic evidence for complex speciation of humans and chimpanzees. Nature 441, 1103–1108.

Pettigrew, J.D., 1986. Flying primates? Megabats have the advanced pathway from eye to midbrain. Science 231, 1304–1306.

Pettigrew, J.D., 1991. Wings or brain: convergent evolution in the origins of bats. Syst. Zool. 40, 199–216.

Pettigrew, J.D., Jamieson, B.G., Robson, S.K., Hall, L.S., Mcanally, K.I., Cooper, H.M., 1989. Phylogenetic relations between microbats, megabats and primates (Mammalia: Chiroptera and Primates). Philos. Trans. R. Soc. Lond. B Biol. Sci. 325, 489–559.

Pettigrew, J.D., Maseko, B.C., Manger, P.R., 2008. Primate-like retinotectal decussation in an echolocating megabat, *Rousettus aegyptiacus*. Neuroscience 153, 226–231.

Pumo, D., Finamore, P., Franek, W., Phillips, C., Tarzami, S., Balzarano, D., 1998. Complete mitochondrial genome of a neotropical fruit bat, *Artibeus jamaicensis*, and a new hypothesis of the relationships of bats to other eutherian mammals. J. Mol. Evol. 47, 709–717.

Rokas, A., Holland, P.W., 2000. Rare genomic changes as a tool for phylogenetics. Trends Ecol. Evol. 15, 454–459.

Rose, K.D., 1996. On the origin of the order Artiodactyla. Proc. Natl. Acad. Sci. U.S.A. 93, 1705–1709.

Rosenberg, N.A., 2013. Discordance of species trees with their most likely gene trees: a unifying principle. Mol. Biol. Evol. 30, 2709–2713.

Sánchez-Villagra, M.R., Narita, Y., Kuratani, S., 2007. Thoracolumbar vertebral number: the first skeletal synapomorphy for afrotherian mammals. Syst. Biodivers. 5, 1–7.

Sears, K.E., Behringer, R.R., Rasweiler, J.J., Niswander, L.A., 2006. Development of bat flight: morphologic and molecular evolution of bat wing digits. Proc. Natl. Acad. Sci. U.S.A. 103, 6581–6586.

Shimamura, M., Yasue, H., Ohshima, K., Abe, H., Kato, H., Kishiro, T., Goto, M., Munechika, I., Okada, N., 1997. Molecular evidence from retroposons that whales form a clade within even-toed ungulates. Nature 388, 666–670.

Shoshani, J., McKenna, M.C., 1998. Higher taxonomic relationships among extant mammals based on morphology, with selected comparisons of results from molecular data. Mol. Phylogenet. Evol. 9, 572–584.

Simmons, N.B., Seymour, K.L., Habersetzer, J., Gunnell, G.F., 2008. Primitive early Eocene bat from Wyoming and the evolution of flight and echolocation. Nature 451, 818–821.

Simpson, G.G., 1945. The principles of classification and a classification of mammals. Bull. Am. Mus. Nat. Hist. 85, 1–350.

Smith, J.D., 1980. Chiropteran phylogenetics: introduction. In: Wilson, D.E., Gardner, A.L. (Eds.), Proceedings Fifth International Bat Research Conference. Texas Tech Press, Lubbock, Texas.

Smith, J.D., Madkour, G., 1980. Penial morphology and the question of chiropteran phylogeny. In: Wilson, D.E., Gardner, A.L. (Eds.), Proceedings Fifth International Bat Research Conference. Texas Tech Press, Lubbock, Texas.

Springer, M.S., Stanhope, M.J., Madsen, O., De Jong, W.W., 2004. Molecules consolidate the placental mammal tree. Trends Ecol. Evol. 19, 430–438.

Stanhope, M.J., Waddell, V.G., Madsen, O., De Jong, W., Hedges, S.B., Cleven, G.C., Kao, D., Springer, M.S., 1998. Molecular evidence for multiple origins of insectivora and for a new order of endemic African insectivore mammals. Proc. Natl. Acad. Sci. U.S.A. 95, 9967–9972.

Sullivan, J., Swofford, D.L., 1997. Are Guinea pigs rodents? the importance of adequate models in molecular phylogenetics. J. Mamm. Evol. 4, 77−86.

Szalay, F.S., Drawhorn, G., 1980. Evolution and diversification of the Archonta in an arboreal milieu. In: Luckett, W.P. (Ed.), Comparative Biology and Evolutionary Relationships of Tree Shrews. Plenum Press, New York.

Tabuce, R., Marivaux, L., Adaci, M., Bensalah, M., Hartenberger, J.-L., Mahboubi, M., Mebrouk, F., Tafforeau, P., Jaeger, J.-J., 2007. Early Tertiary mammals from North Africa reinforce the molecular Afrotheria clade. Proc. R. Soc. Biol. Sci. 274, 1159−1166.

Tarver, J.E., Dos Reis, M., Mirarab, S., Moran, R.J., Parker, S., O'reilly, J.E., King, B.L., O'connell, M.J., Asher, R.J., Warnow, T., Peterson, K.J., Donoghue, P.C., Pisani, D., 2016. The interrelationships of placental mammals and the limits of phylogenetic inference. Genome Biol. Evol. 8, 330−344.

Teeling, E.C., Hedges, S.B., 2013. Making the impossible possible: rooting the tree of placental mammals. Mol. Biol. Evol. 30, 1999−2000.

Teeling, E.C., Scally, M., Kao, D.J., Romagnoli, M.L., Springer, M.S., Stanhope, M.J., 2000. Molecular evidence regarding the origin of echolocation and flight in bats. Nature 403, 188−192.

Teeling, E.C., Madsen, O., Van Den Bussche, R.A., De Jong, W.W., Stanhope, M.J., Springer, M.S., 2002. Microbat paraphyly and the convergent evolution of a key innovation in Old World rhinolophoid microbats. Proc. Natl. Acad. Sci. U.S.A. 99, 1431−1436.

Thewissen, J.G., Williams, E.M., Roe, L.J., Hussain, S.T., 2001. Skeletons of terrestrial cetaceans and the relationship of whales to artiodactyls. Nature 413, 277−281.

Tsagkogeorga, G., Parker, J., Stupka, E., Cotton, J.A., Rossiter, S.J., 2013. Phylogenomic analyses elucidate the evolutionary relationships of bats. Curr. Biol. 23, 2262−2267.

Van Den Bussche, R.A., Hoofer, S.R., 2004. Phylogenetic relationships among recent chiropteran families and the importance of choosing appropriate out-group taxa. J. Mammal. 85, 321−330.

Van Valen, L.M., 1966. Deltatheridia, a new order of mammals. Bull. Am. Mus. Nat. Hist. 132, 1−126.

Veselka, N., Mcerlain, D.D., Holdsworth, D.W., Eger, J.L., Chhem, R.K., Mason, M.J., Brain, K.L., Faure, P.A., Fenton, M.B., 2010. A bony connection signals laryngeal echolocation in bats. Nature 463, 939−942.

Waddell, P.J., Cao, Y., Hauf, J., Hasegawa, M., 1999. Using novel phylogenetic methods to evaluate mammalian mtDNA, including amino acid-invariant sites−Logdet plus site stripping, to detect internal conflicts in the data, with special reference to the positions of hedgehog, armadillo and elephant. Syst. Biol. 48, 31−53.

Wible, J.R., Novacek, M.J., 1988. Cranial evidence for the monophyletic origin of bats. Am. Mus. Novit. 2911, 1−19.

Wrangham, R.W., 2001. Out of the *Pan*, into the fire: from ape to human. In: De Waal, F.B.M. (Ed.), Tree of Origin: What Primate Behavior Can Tell Us about Human Social Evolution. Harvard University PRess, Cambridge, MA.

Wyss, A.R., 1987. The walrus auditory region and the monophyly of pinnipeds. Am. Mus. Novit. 2871, 1−31.

Wyss, A.R., 1988. Evidence from flipper structure for a single origin of pinnipeds. Nature 334, 427−428.

Zardoya, R., Meyer, A., 1998. Complete mitochondrial genome suggests diapsid affinities of turtles. Proc. Natl. Acad. Sci. U.S.A. 95, 14226−14231.

13

The Organization of Neocortex in Early Mammals

J.H. Kaas

Vanderbilt University, Nashville, TN, United States

13.1 Introduction

We are all a bit curious about how we came to be humans. Having large brains that are capable of finding reasons for almost everything, often to our great advantage. Various cultures have nearly universally created stories of how humans came into existence. But without an appreciation for the great timescales involved in our being and the power of evolution, these highly variable creation stories did not reach back into the prehuman part to hypothesize on how we evolved from earlier forms of life. Thanks to many investigators, we now have the outlines of a scientific understanding that has grown considerably since the time of Darwin (1859) as more and more relevant information has been acquired. In the simplest form, the story of human evolution is much like that explained by Richard Dawkins (2004) in *The Ancestors Tale* where he asks us to imagine a series of ancestors from our parents and grandparents to direct relatives evermore distant in the past, as a pace in time that we recognize the kinship between jumps across generations, while acknowledging that these individuals look less and less like ourselves. In helping us do this, Dawkins describes present-day species as stand-ins for actual ancestors. The logical end to this imaginary journey is at the beginning of life on earth. Here we would not do that, in part because no living species can adequately represent a distant ancestor, and in part because the story would both be too long and too incomplete. Instead, we consider one critical time point, 230 mya, when the first mammals emerged, with some of them being out distant ancestors and the ancestors of all 5825 or more species of mammals living today. More specifically, we focus on what their brains, and especially their neocortex was like so we can start to understand the many impressive differences in their brains

as early mammals continued to evolve, and specialize in so many ways.

The task is a difficult one, and perhaps we can only approximate an accurate understanding of what the ancient brains were like. The relevant information comes from several sources. Critically important information comes from the fossil record, and any theory of early mammalian brain organization must be consistent with the ever-growing record. Although brains themselves do not fossilize, fossil skulls do provide information about the sizes, shapes, and even parts of brains, including neocortex. As such fossils often can be somewhat accurately dated, they provide time stamps for the emergence of mammals and early and subsequent changes in brain sizes and shapes. Other aspects of these fossils provide information on sensory systems and behavioral adaptations related to brain organization and function. The time stamps of fossils also provide the necessary calibrations of molecular measures of genetic distance to provide estimate of rates of evolutionary change and time of divergences of different clades of extant mammals.

While the fossil record is extremely useful and important, it does not provide information about the internal structure of brains. However, such features can be reconstructed using a cladistic approach (eg, Hennig, 1966). This approach rests on the evidence that some features or traits of present-day brains have been retained from early ancestors while other are greatly changed or new. Thus, if most or all of the members of a clade, all descendants of a common ancestor, have a specific feature in common, then it is likely that the common ancestor had that trait, and it was passed on to the descendants. The less likely, but possible, alternative is that the trait evolved independently in the lines leading to some of most of the present member of the clade. Thus, the outcomes of a cladistic approach depend on the overall pattern of the distribution of traits across members of the clades, and outcomes can have different levels of confidence (Cunningham et al., 1998). In some instances, other types of information can guide interpretations of cladistic data. For example, an extensive and impressive cladistic analysis of 86 fossil and living species of placental mammals usefully reconstructed the common ancestor of all placental mammals (O'Leary et al., 2013). However, one proposed trait, that the neocortex of the common ancestor had fissures, seems unlikely in view of recent understandings of the mechanisms of cortical folding. Mota and Herculano-Houzel (2015) reported from an extensive data set that the amount of cortical folding is dependent on the surface area and thickness of the cortical sheet. Most importantly no present-day mammals of the small brain size of the postulated common ancestor have a neocortex with fissures. However, many independent lines of mammalian evolution have led to larger brains with fissures, so the common distribution of fissures in the larger brains incorrectly predicted that the common ancestor would have fissures in cortex, while the fossil record constrained the predicted size of the brain and cortex. As the cladistic approach depends on an understanding of the branching points of the mammalian evolution, this topic is considered next.

13.2 The Mammalian Family Tree

Comparative studies of morphological features of living and extinct mammals have long been used to deduce the evolutionary relations of different species of mammals. This approach worked reasonably well, but uncertainties existed because of the independent evolution of traits and the presence of shared traits that had been retained from a far distant relative. However, more recent studies of molecular similarities across species have added a great magnitude of convincing data on gene products for comparison, resulting in detailed geneologized trees of relationships of living and extinct mammals. Dated fossils have provided a timescale for mammalian evolution, with molecular studies suggesting younger and sometimes much younger dates for branching points than those based on fossils (as the oldest fossils of a type likely have not been found). While uncertainties remain, there is general agreement on the shape of the tree and the many relationships (Fig. 13.1). Here we use time points based on Murphy et al. (2004). Younger dates have been proposed by O'Leary et al. (2013), suggesting that the common ancestor of extant placental mammals existed 65 mya just after the mass extinction at the Cretaceous-Paleogene (K−Pa) boundary some 66 mya and that an "explosion" of placental mammal radiation occurred about 60 mya (also, see Bininda-Emonds et al., 2007). The Murphy et al. (2004) and related estimates depend more on the molecular evidence, and the O'Leary et al. (2013) estimates depend more on fossil evidence and influences from such evidence, raising the concern that branch lengths need to be studied more. (For further review, see Bininda-Emonds and Hartmann 2.03, Mammalian Evolution: The Phylogenetics Story this volume.)

According to more estimates, early amniotes, those vertebrates that had evolved egg-covering membranes that allowed eggs to survive on land, produced two

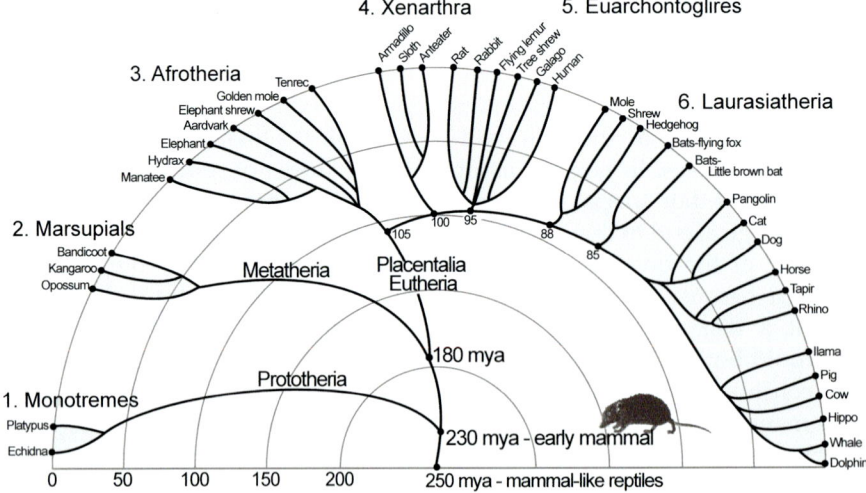

FIGURE 13.1 The phyletic radiation of mammals into the six major clades or superorders. The times of divergences have been variously estimated and remain somewhat uncertain, but the overall pattern is widely accepted. The times shown here are largely from Murphy et al. (2004). *From Kaas, J.H., Preuss, T.M., 2014. Published in the chapter Human Brain Evolution in Fundamental Neuroscience, fourth ed. In: Squire, L.R. (Ed.), Elsevier, pp. 901–918.*

surviving members some 315 mya, the synapsids and the sauropsids, distinguished by differences in skull morphology (Rose, 2006). The synapsids produced therapsids with some mammal-like features and then primitive mammals some 200–250 mya. A line of primitive mammals branched into a surviving prototherian line, monotremes, that led to present-day platypuses and echidnas, and a line of therian mammals that subsequently branched about 148 mya into prototherian (marsupial) and eutherian (placental) lines. Perhaps 100 mya or more, eutherian mammals first branched to form the beginning of the Xenarthra superorder, followed soon afterward by a branching into the Afrotherian, Euarchontoglire, and Laurasiatherian superorders. The sauropsids produced lines leading to present-day reptiles (lizards and snakes, tuatara, turtles, crocodiles, and birds) (Butler and Hodos, 2005). This tree of six superorders of mammals, each with many further branches and 5360 or more living species, offers many opportunities for comparing the brains of extant mammals to reveal shared and derived features. However, all nonmammalian branches of the synapsid line are extinct, including extinct branches of early mammals and therapsids (mammal-like "reptiles"). Thus, there is a long 150 mya period of synapsid evolution that is unrepresented, except for mammals, in living species for comparison. To compensate for this loss of representation, it has become common and productive to compare the brains of extant reptiles and birds with those of extant mammals. The assumption is that some basic features of the brains of present-day reptiles are likely to be similar to those of the direct ancestors of mammals. However, the 315 million years of independent evolution of the sauropsid line suggests caution in this assumption.

13.3 Dorsal Cortex of Reptiles and Neocortex

Neocortex is not a new part of the brain, as it is widely recognized as homologous to the dorsal cortex of reptiles and the *hyperpallium* or Wulst of birds (Butler and Hodos, 2005; Northcutt and Kaas, 1995). Thus, neocortex is sometimes called isocortex to avoid the impression that it is new with mammals. Instead, it is a part of the forebrain that existed in early amniotes and changed relatively little in reptiles. Dorsal cortex is a small part of the forebrain that consists of a single row of pyramidal cells and a scattering of a few (10–20%) stellate neurons that are largely inhibitory (Fig. 13.2). The activating inputs are largely from the dorsal lateral geniculate nucleus of the visual thalamus (Hall et al., 1977), but a small portion gets somatosensory inputs from the ventral posterior nucleus of the somatosensory thalamus, much as primary visual and somatosensory areas

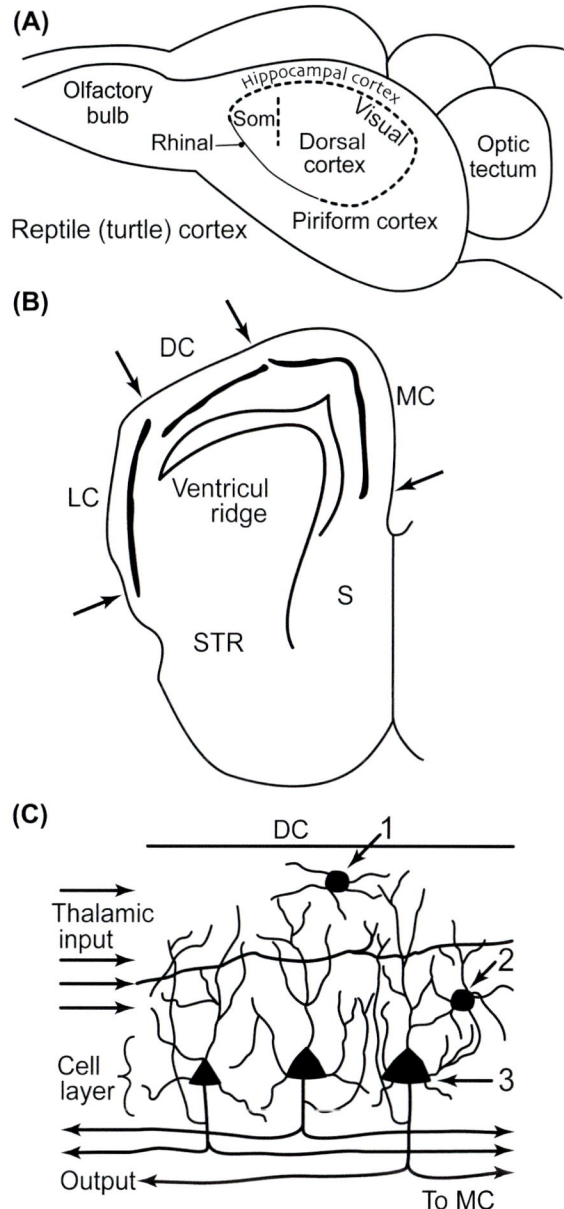

FIGURE 13.2 The dorsal cortex of reptiles, a homolog of mammalian neocortex. (A) Dorsolateral view of the brain of a turtle showing the olfactory bulb, piriform (olfactory) cortex, dorsal cortex, and hippocampal cortex. The optic tectum (superior colliculus of mammals) of the midbrain is exposed. Compare with early mammals (Fig. 13.3). (B) A frontal section through the forebrain showing the lateral (piriform) cortex, *LC*, the dorsal cortex, *DC*, and the medial cortex (hippocampus), *MC*, are all characterized by a single row of pyramidal cells (*dark lines*). The forebrain section also contains the septum, *S*, the striatum, *STR*, and the dorsal ventricular ridge. (C) The arrangement of neurons in dorsal cortex. A few subpial (1) and stellate (2) neurons are inhibitory on pyramidal (3) neurons after being activated by pyramidal neuron axons and input axons. The pyramidal neuron axons also branch to provide outputs to subcortical structures and to medial hippocampal cortex. *From Kaas, J.H., Preuss, T.M., 2004. Human brain evolution. In: Fundamental Neuroscience, Elsevier, Inc., pp. 1019–1037.*

do in mammals (Medina, 2007). There is no evidence of an auditory input, as the projections of the auditory thalamus remain subcortical. The individual axons of their visual and somatosensory inputs provide only a few synapses on the apical dendrites of each pyramidal neuron, while synapsing on many such neurons. Thus, pyramidal neuron activity depends on the coactivation of many visual or somatosensory thalamic inputs, so these inputs can be summed (Shepherd, 2011). Likely synaptic strengths can be rapidly changed, so responses to stimuli can be modified. The output axons from the pyramidal neurons branch to go medially to medial cortex, the hippocampus, and laterally to project to the sensory thalamus and midbrain. The medial hippocampal cortex and the lateral piriform olfactory cortex are similar in organization, having a single row of pyramidal cells that sum large numbers of inputs.

Note how different dorsal cortex is from neocortex. Most of neocortex consists of five layers of neurons, with each layer at least somewhat specialized in pyramidal neuron morphology and having different connections. Most importantly, a middle layer (layer 4 of Brodmann, 1909) that consists largely of small stellate or granular neurons (consider them modified pyramidal neurons) receives the bulk of the activating input from the dorsal thalamus, and each neuron is activated by only a few inputs. In contrast to dorsal cortex, these neurons preserve thalamic input information to distribute it to neurons in other layers of cortex, and their responses are not highly modifiable. Pyramidal neurons in other layers sum more thalamic inputs and receive more modulating inputs from subcortical structures, including terminations on their apical dendrites that extend into or near the layer 1 neuropil. Layer 2 neurons are most related to layer 1, layer 3 neurons provide outputs to other areas of neocortex, layer 5 projects to subcortical targets, and layer six provides feedback to the primary source of activating layer 4 inputs. Instead of two areas in dorsal cortex, there are at least 20 and as many as 200 areas in neocortex, depending on species (Kaas, 2017). Cortical areas vary in inputs, outputs, and function. Thus, neocortex is "new" in many ways, but it likely evolved from something that closely resembled the dorsal cortex of reptiles.

It is not clear how these many modifications of dorsal cortex occurred, or when. The fossil evidence on the long and narrow shape of the forebrain in therapsids suggests that most of these changes occurred with or just before the first mammals emerged. The use of the comparative method across extant vertebrates is limited in regard to understanding the evolution of neocortex because mammals with a well-developed neocortex are the only survivors of the synapsid radiation from early amniotes. Instead, more answers are likely to come from studies of the development of neocortex. It

is clear that a major change was in the generation time of neurons for neocortex, as by extending this time resulted in more neurons (Finlay and Darlington, 1995). The guiding role of the migration pathway of most of the developing neurons along radial glia was a critical factor as it increased cortical thickness instead of spreading it out (Rakic, 1995). The timing of the arrival of these cells in a radial unit appears to be an important factor in the formation of laminar specializations (Molnár, 2011), and genetic specification factors have a lot to do with the specification and location of cortical areas (O'Leary et al., 2007). Finally, activation patterns in developing cortex promote the functional segregation of connections in cortex so that the laminar specializations are enhanced and functional modules emerge (Kaas, 2012).

13.4 What the Fossil Record Tells About Brains and Behavior in Early Mammals

Fossils are important source of information about the sizes and shapes of the brains of extinct mammals. The interiors of the skulls of fossil mammals closely conform to the shape of the brain, so that under favorable conditions, the shape and size of the brain of early mammals can be determined (Kaas, 2016). The endocasts of fossil skulls of larger, more recent mammals may even reflect the locations of some of the larger fissures in the neocortex (eg, Radinsky, 1976), and such fissures may reflect aspects of the functional organization of cortex (Welker, 1990). However, even the rhinal fissure separating the piriform (olfactory) cortex from the more dorsal neocortex is often now apparent in the skull endocasts of early mammals (Kielan-Jaworowska et al., 2004; Rowe et al., 2011). This is because the rhinal fissure is no more than a dimple in small brains with little neocortex. However, in some cases, the rhinal dimple is apparent in the endocast of an early mammalian brain, and the border between neocortex and piriform cortex can be determined (Fig. 13.3). This allows close estimates of the sizes of olfactory and neocortex. Such endocasts clearly demonstrate that the forebrains of early mammals only had a small cap of neocortex and a proportionally large extent of olfactory cortex subserving a large olfactory bulb. Most notably, the neocortex fails to extend caudally to cover the midbrain and part of the cerebellum or extend laterally to hide the rhinal fissure, as in most extant mammals. The olfactory system was clearly important in early mammals, especially in comparison with neocortex. This early olfactory system included both the main olfactory system and the vomeronasal system, which is important in pheromone-mediated behaviors, such as mating and parenting (Grus and Zhang, 2006). The brains of even earlier

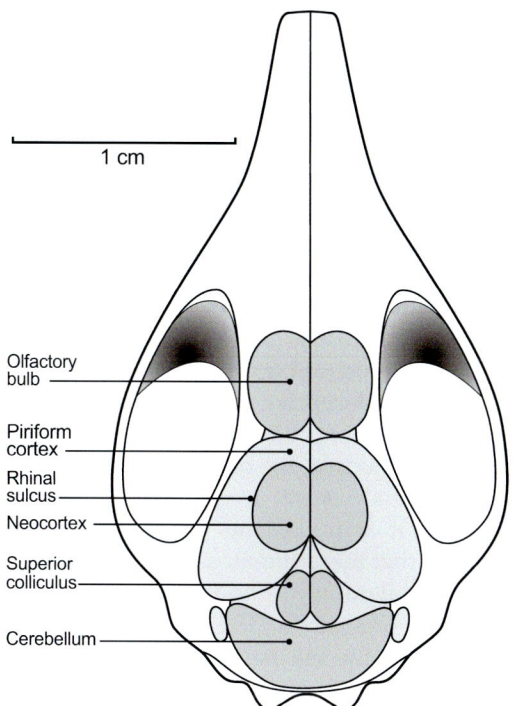

Olfactory
bulb

Piriform
cortex

Rhinal
sulcus

Neocortex

Superior
colliculus

Cerebellum

1 cm

FIGURE 13.3 A reconstruction of the skull and dorsal view of the brain from the skull endocast of an early Eutherian mammal of about 85 mya. Primitive features of the brain include a small cap of neocortex on the forebrain with large olfactory bulbs and piriform (olfactory) cortex. *Redrawn and modified from Kielan-Jaworowska, Z., Cifelli, R.L., Luo, Z.-K., 2004. Mammals From the Age of Dinosaurs, Columbia University Press.*

mammalian forms were even smaller and had smaller olfactory bulbs (Rowe et al., 2011). The cerebral hemispheres were narrow, and a dorsal pineal eye was exposed. The midbrain was also exposed. The eyes were small, and the middle ear bones were large, suggesting that high-frequency hearing had not developed. However, early mammal forebrains were larger, the cerebral hemispheres were wider, and the olfactory bulb and cortex were greatly expanded. The narrow cerebral hemispheres suggested that the thick layers of neocortex did not yet exist. The larger cerebellum and thicker spinal cord suggest that motor control had improved.

The small eyes of early mammals, revealed by the skulls, suggested that these mammals were nocturnal and not that dependent on vision. One possibility was that early mammals were less dependent on the superior colliculus of the midbrain for visual processing and more dependent on visual cortex (Hall et al., 2012; Diamond and Hall, 1969). The cochlea was longer and with more spiral turns, and the middle ear bones smaller, enabling more sensitive hearing extending into the higher sound frequencies, allowing parent—offspring communication beyond the hearing of reptilian predators (Coleman and Boyer, 2012; Luo et al., 2011; Allman, 1999). The teeth, which are more often preserved, were suited for eating

invertebrates and small vertebrates (Allman, 1999). In rare instances, signs of body hair and longer tactile hairs were preserved (Rowe et al., 2011; Rose, 2006), suggesting a somatosensory system modified for processing information from sensory vibrissae, as in most present-day mammals (Muchlinski, 2010). Body size was small, from shrew to rat size, but sometimes larger (Hu et al., 2005). However, for roughly the next 170 million years, most mammals remained small until the major Cretaceous—Paleogene extinction 66 mya allowed new opportunities for surviving mammals (Smith et al., 2010).

13.5 Which Brains Should Be Studied?

The brains of modern mammals typically contain a mixture or mosaic of primitive or old traits that have been retained from ancient ancestors and newer traits that have emerged in more recent ancestors. The cladistic method is used to deduce what traits or characters have been retained from the last common ancestor of all members of the clades (Hennig, 1966). The clade can be of any branch of mammals, some or all primates, or in this case, all extant mammals. There are complications to this approach, as characters can be lost of gained independently in the branches of evolution within the clade, so that decisions can be complicated (Cunningham et al., 1998), and conclusions are best made from observations on large numbers of species distributed across the clade. However, there are practical concerns as well. Because studies of brain organization can be difficult and costly, the brains of few mammals have been extensively studied. In addition, many of the larger brains, such as a human brain or an elephant brains, are very complex, and primitive traits may be difficult to identify in the array of changes (Kaas, 2002). As the brains of early mammals were small with little neocortex, it is useful to focus on the organization of brains of present-day mammals that are small, and, for cladistic reasons, distributed across the six superorders. We start with monotremes followed by marsupials, and then the four superorders of placental mammals.

13.6 Monotremes

Less than 1% of extant mammalian species are monotremes. These mammals represent the oldest surviving branch of the mammalian radiation (Phillips et al., 2009). They share the primitive feature of egg-laying with birds, reptiles, and cynodont ancestors. They have a reptile-like cloaca (hence the name, monotremes) and "sweat" milk for young from glands on the mother's stomach (Ashwell, 2014). Their sex determining chromosomes are different from other mammals and more like those of birds and reptiles (Grützner et al., 2004). They share many features with other present-day mammals including a covering of hair, providing milk to feed young, and a brain with a six-layered neocortex. The eyes and retinae contain a mixture of mammalian and sauropsid characteristics (Walls, 1942; Young and Vaney, 1990).

The monotreme taxon includes a single species of platypuses and three species of echidnas. Platypuses are semiaquatic, feeding on insects and other invertebrates in the waterways of Australia and Tasmania with their unique ducklike bills. Echidnas are terrestrial, feeding on termites, ants, and other invertebrates in Australia and New Guinea. Despite their specializations as an aquatic predator, platypuses are thought to be an older member of the order, with a fossil record going back over 100 mya. Echidnas branched off and reinvaded terrestrial ecosystems perhaps as recent as 20–32 mya (Phillips et al., 2009; Rowe et al., 2008). The platypus and echidna are the only mammals that have electroreception, and this system is well developed in platypuses where it is of great use in detecting prey in muddy water. Electroreception has been retained in echidnas, where it appears to be somewhat degenerate and of limited use.

The brains of platypuses and echidnas have been extensively studied, and a recent volume of chapters edited by Ashwell (2014) provides an excellent review. A lateral view of the two brains makes it clear that they both have much more neocortex than early mammals, as the neocortex in platypus and echidna brains extends laterally to cover the rhinal fissure, and caudally to cover the midbrain and rostral half of the cerebellum. As a clear difference, the platypus cortex is smooth, while the somewhat larger echidna brain has several prominent fissures. From a ventral view of the brain, it is obvious that the olfactory bulb and piriform cortex are well developed in echidnas but poorly developed in platypuses, as olfaction would not be used in hunting for prey in water. The echidna brain puzzled early investigators as it appeared to have an unusual amount of prefrontal cortex, which is important for higher-order cognition in humans, and there was little evidence for planning and cognition in echidnas. It now appears that the large region of frontal cortex in echidnas is largely orbitofrontal cortex, a region in other mammals that integrates olfactory information with taste and other sensory inputs, reflecting the large role that olfaction plays in echidna behavior (Ashwell, 2014). The renewed emphasis on olfaction is one of the marked specializations of the echidna brain. Both platypus and echidna brains have the main and the vomeronasal olfactory pathways, but these are poorly developedin platypuses and in the oldest fossil platypuses.

The areal organization of neocortex in the platypus reflects its extreme dependence on its bill for finding prey. A great portion of cortex represents the bill (Fig. 13.4), and there are at least three cortical representations of the bill, one of which is in primary somatosensory cortex, S1, which is subdivided in a pattern of alternating modules for touch or for electroreception (Krubitzer et al., 1995), reflecting the 10 000 receptors for electroreception on the bill and a similar number for receptors for touch (Pettigrew, 1999). The bill is represented again in a rostral area R, which may correspond to an area on the rostral border of S1 of other mammals that receives proprioceptive and tactile sensory information, and has an enhanced role in motor behavior. The caudal representation of the bill is part of an area identified as the parietal ventral area, PV, of other mammals. Alternatively, it may be the second somatosensory area, S2. Note that these somatosensory areas also represent the contralateral lower body, as in other mammals, but with much less cortex. The great dominance of the representation of the bill means that there is relatively little cortex devoted to auditory or visual inputs, but at least primary visual and auditory areas appear to be present, as are a small dorsal lateral geniculate nucleus for vision and medial lateral geniculate nucleus for hearing in the thalamus (Mikula et al., 2008; Ashwell, 2014). The thalamus is clearly distorted in organization by the large ventroposterior nucleus for somatosensation and electroreception. Yet, Mikula et al. (2008) conclude that the thalamus is mammalian rather than reptilian in overall organization. There are some uncertainties about the existence of motor cortex in monotremes. In most mammals, motor responses can be evoked by electrical stimulation from somatosensory cortex (see Lende,

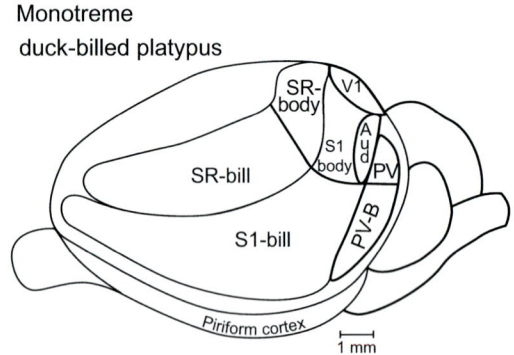

FIGURE 13.4 The forebrain of a platypus (a monotreme). Neocortex is expanded compared to piriform cortex, and it is largely devoted to representing the tactile receptors and electroreceptors of the bill, while less cortex is devoted to the contralateral body. *S1*, primary somatosensory cortex; *R*, the rostral proprioceptive area; *PV*, the parietal ventral somatosensory area. Little cortex is devoted to auditory (Aud) or visual (V1) inputs. *Based on Krubitzer, L., Manger, P., Pettigrew, J., Calford, M., 1995. Organization of somatosensory cortex in monotremes: in search of the prototypical plan. J. Comp. Neurol. 351, 261–306.*

1964 for monotremes), and the rostral cortical region denoted as R (Fig. 13.4) has been referred to as motor cortex, and a similar rostral "motor" region has been identified in echidna cortex. Alternatively, there are somatosensory areas of other mammals that also have motor functions, and primary motor cortex, M1, and premotor cortex did not emerge until the evolution of placental mammals (see Section 13.7). See Ashwell (2014) for an extensive review of brain organization in monotremes in comparison to other mammals.

13.7 Marsupials

These metatherian mammals have been called the 6% solution (Kirsch, 1977) as they represent close to 6% of the present-day species of mammals. The therian line of evolution branched into the prototherian and eutherian lines perhaps around 150 mya (Fig. 13.1). Although once more widespread (Luo et al., 2003), marsupials became restricted to what is now South America and they migrated via Gondwana to Australia and New Guinea. They presently consist of seven orders, with three orders remaining in the Americas and four orders in Australia and New Guinea as a result of a single migration (Nilsson et al., 2010). The American marsupials are all opossums or opossum-like, while the Australian radiation includes the opossum-like possums, as well as species adapted to a range of niches such as the Tasmanian devil, bandicoot, koala, kangaroo, and wallaby. The brains of opossums and possums would most likely reflect more of the characteristics of the brains of early mammals than the more derived marsupials.

Marsupials differ from eutherian (placental) mammals in many ways, but most notably in terms of the lack of a eutherian type of placenta, short gestation times (8–42 days), and the birth of very altricial young (Kirsch, 1977). As a result, pouch rearing is common (the marsupium), although not universal. Due to the early birth, some aspects of the nervous system need to be functional at a very early stage, such as those mediating the ability to climb into the pouch and find a nipple.

The brains of the smaller possums and opossums are highly relevant to reconstructions of the brains of the common ancestry as marsupials represent an early branching point, and opossums and possums appear to have changed little in body form and lifestyle. Remarkably, American opossums and Australian possums have very similar brains despite being separated for over 100 million years of independent evolution. As for monotremes, marsupials have no corpus callosum, a primitive trait. The smaller marsupials have small brains, with somewhat more neocortex than early mammals, but with no neocortical fissures. Ashwell (2010) provides an excellent source for studies of the brains of Australian marsupials.

A number of studies have focused on the organization of neocortex in the North American opossum, Didelphis (Fig. 13.5). While the olfactory bulb and the piriform cortex are large, the neocortex has expanded laterally and caudally so that more cortex is devoted to processing visual and somatosensory inputs. Recordings and anatomical studies have identified five somatosensory areas, all with apparent homologies in placental mammals. A large representation of the contralateral body surface is clearly the primary somatosensory area, S1, as identified by neurons responding to touch, a mediolateral somatotopy from tail to tongue, a well-developed layer 4 as in primary areas, and inputs from the ventroposterior thalamic nucleus (eg, Beck et al., 1996). A second somatosensory area, S2, and a parietal ventral area, PV, were identified lateral to S1. A narrow strip of somatosensory cortex, SR for somatosensory rostral, and SC for somatosensory caudal, were identified rostral and caudal to S1. SR and SC areas have been reported in most studied placental mammals, but names have varied. Similar results have been obtained from the South American opossum, Monodelphis (Frost et al., 2000; Catania et al., 2000) and in the Australian bush-tailed possum, Trichosurus (Elston and Manger, 1999). Thus, five somatosensory areas are present in opossums and possums, and likely all marsupials (Huffman et al., 1999).

These marsupials also reveal a lack of a primary motor M1 and premotor cortical areas. Early studies provided evidence for an overlap of somatosensory and motor cortex in opossums (Lende, 1963). However, the ability to

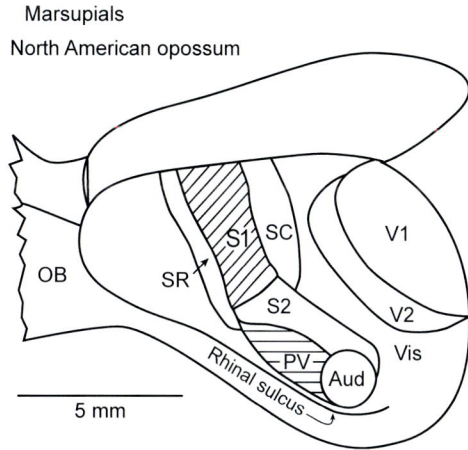

Marsupials
North American opossum

5 mm

FIGURE 13.5 The forebrain of a North American opossum (a marsupial). The slightly enlarged neocortex includes four somatosensory areas: *S1*, primary somatosensory area; *S2*, the second somatosensory area; *PV*, the parietal ventral area; and rostral and caudal somatosensory areas bordering S1. Primary and secondary visual areas, *V1* and *V2*, have been identified. Temporal cortex contains at least one visual area, *T*, and prostriata (not shown). *OB*, olfactory bulb. A primary motor area, and premotor areas have not been identified. *Based on Beck, P.D., Pospichal, M.W., Kaas, J.H., 1996. Topography, architecture, and connections of somatosensory cortex in opossums: evidence for five somatosensory areas. J. Comp. Neurol. 366, 109–133.*

evoke movements from somatosensory cortex with electrical stimulation is now well established, and this simply reflects the motor functions and connections of sensory cortex (Nudo and Masterton, 1990). Thus, the sensory mapping of S1 in marsupials is characteristic of S1 in other mammals, and not motor cortex. As for Beck et al. (1996), Frost et al. (2000) found "no evidence of a motor presentation rostral to S1." We conclude that along with the lack of a corpus callosum, marsupials and monotremes lack the primary motor area, M1, and the premotor areas of placental mammals. These features were thereby not part of the neocortex of early mammals.

The neocortex of the opossums and other marsupials is also characterized by a primary visual area, V1, a second visual area, V2, a medial visual area, prostriata (not shown), and one or more areas of temporal visual cortex (Fig. 13.6) (Beck et al., 1996; Martinich et al., 2000; Rosa and Krubitzer, 1999). Moreover, V2, as reported in several species of placental mammals, is not homogenous, but

consists of a number of modules with input from V1 separated by narrow regions with inputs from the other cerebral hemisphere via the anterior commissure in marsupials (Martinich et al., 2000) and corpus callosum in placental mammals (Kaas and Krubitzer, 1991).

In addition a lateral region of cortex responds to sounds and has the architectonic features of auditory cortex in opossums (Beck et al., 1996). The tonotopic organization of this auditory cortex has been studied in the *Virginia opossum*, brush-tailed possum, and northern quoll (see Aitkin, 1995 for review). Surprisingly, there appears to be only one representation of tone frequencies in these marsupials, with high-frequency tones represented centrally and low frequencies caudally in cortex with architectonic characteristics of A1. In placental mammals, the organization of auditory cortex is variable across species, but also more complex with typically two or more primary areas, and several secondary areas (Kaas, 2011c). More comparative evidence is needed, but the

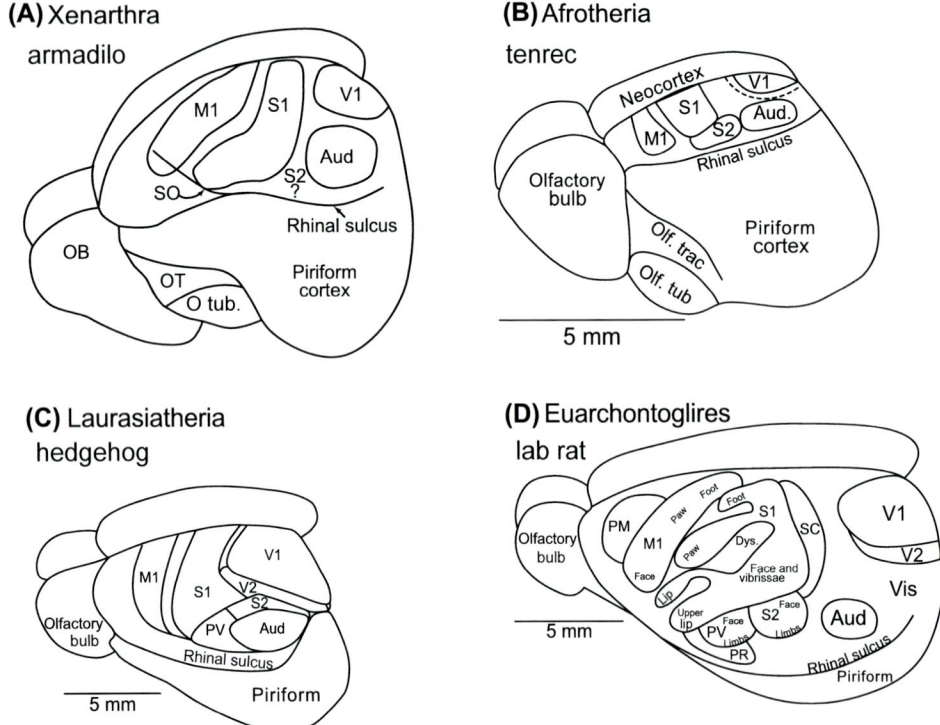

FIGURE 13.6   Dorsolateral views of the forebrains of small-brained members of the four major superorders of placental (eutherian) mammals. All have primary and secondary visual areas, V1 and V2, one or more primary auditory areas (Aud), a primary motor area (M1), and possibly one or more secondary (M2) premotor areas. *OB*, olfactory bulb; *OT*, olfactory tract; *O tub*, olfactory tubercle. (A) Representing the superorder Xenarthra, the brain of a North American armadillo (*Dasypus novemcinctus*). Retrosplenial (RS) areas have been defined in caudal cortex of the medial wall of the cerebral hemisphere. (B) Cortical organization in a member of the Afrotheria superorder, a tenrec (*Echinops telfari*). (C) Cortical organization in a member of the Laurasiatheria superorder, a hedgehog (*Atelerix albiventris*). (D) Cortical organization in a laboratory rat, a member of the Euarchontoglire superorder. *(A). (Based on Royce, G., Martin, G.F., Dom R.M., 1975. Functional localization and cortical architecture in the nine-banded armadilli (Dasypus novemcinctus mexicanus). J. Comp. Neurol. 164, 495–521.). (B). (Based on Krubitzer, L., Künzle, H., Kaas, J.H., 1997. Organization of sensory cortex in a Madagascan insectivore, the tenrec (Echinops telfairi). J. Comp. Neurol. 379, 399–414.). (C). (Based on Catania, K.C., Jain, N., Franca, J.G., Volchan, E., Kaas, J.H., 2000. The organization of somatosensory cortex in the short-tailed opossum (Monodelphis domestica). Somatosens. Mot. Res. 17, 39–51; Kaas, J.H., Hall, W.C., Diamond, I.T., 1970. Cortical visual areas I and II in the hedgehog: relation between evoked potential maps and architectonic subdivisions. J. Neurophysiol. 33, 595–615.). (D). Based on multiple sources.*

results from marsupials suggest that the neocortex of early mammals may have had only one auditory area.

Architectonic studies of neocortex in marsupials provide evidence for additional areas that likely correspond to those found in placental mammals (Wong and Kaas, 2009; see Ashwell, 2010 for review). On the medial wall of the cerebral hemisphere, these include dorsal, ventral, and rostral cingulate areas and dorsal (agranular) and ventral (granular) retrosplenial areas. Frontal cortex likely included an orbital-frontal region and a medioventral frontal region, as well as a dorsolateral "frontal myelinated" (FM) field (Karlen and Krubitzer, 2006) of uncertain identity. The perirhinal cortex does not have distinct subdivisions, but functionally distinct subdivisions, including taste cortex, may exist. Overall, in possums and opossums, there is evidence for roughly 16–20 cortical areas, but there may be more.

13.8 Placental Mammals

Eutherian or placental mammals emerged as the major branch of present-day mammals nearly 150 mya and then diverged around 100 mya with four major superorders: Xenarthra, Afrotheria, Laurasiatheria, and Euarchontoglires (Fig. 13.1). Each of these superorders produced a variety of surviving members, some resembling the reconstructed ancestral placental mammal, and others greatly modified and variously specialized. Here we focus on members of the four superorders that most closely resemble the proposed common ancestor of all placental mammals. The best estimate of what the features of the common ancestor of all placental mammals is probably the description of O'Leary et al. (2013) which is based on a character analysis of 86 fossils and living species representing the roughly 5825 living species of placental mammals. The placental mammal ancestor was small and insectivorous. The cortex was moderately expanded and had a corpus callosum. We counter the conclusion that the neocortex of this brain had fissures, as the results of a more recent study provided evidence that brains with little neocortex do not have fissures (Mota and Herculano-Houzel, 2015). The branch times of the superorder of mammalian evolution have been variously estimated. Here we use mainly those of Murphy et al. (2004).

Superorder Xenarthra, formerly known as Edentata, represents the oldest split of placentalia into Xenarthra and Epitheria (O'Leary et al., 2013). The order consists of three infraorders that are very different from each other and highly specialized: the sloths, anteaters, and armadillos (Eisenberg, 1981). While none of the living xenarthrans has incisors or canines, only anteaters have no teeth. These mammals are restricted to the New World from the southern United States to South America. The two-toed and three-toed sloths spend most of their lives hanging upside down in trees where they are largely motionless and feed on leaves. The four species of anteaters have powerful claws for opening ant and termite nests. The 20 species of armadillos are distinguished by an armorlike covering of horn and bone over the upper surface of the body. They eat insects and other invertebrates, plants, and carrion.

There have been few studies of brain organization, with most of those studies on the brains of armadillos. The armadillo brain has a proportionally large olfactory system, including the olfactory bulb and piriform cortex, probably for an aid to finding insect food. The neocortex is relatively small, but covers the midbrain and part of the cerebellum (Fig. 13.6A). Visual and auditory areas have been localized using electrode recordings and histological appearance. The somatotopy of S1 has been mapped, with the mouth and face represented most laterally in cortex, followed more medially by the forelimb and hindlimb as in other mammals (Royce et al., 1975). The motor cortex, as revealed by electrical stimulation, was just rostral to primary somatosensory cortex (Dom et al., 1971). As in other placental mammals, this primary motor cortex was characterized by a poorly developed layer 4 of granular cells and large pyramidal cells. Much of the lateral parts of motor and sensory cortex are devoted to the tongue.

Similar studies were conducted in the brains of the three-toed sloth (Saraiva and Magalhães-Castro, 1975; Meulders et al., 1966). The somewhat larger brain has proportionately more cortex with an expanded temporal region and two main fissures (not shown). A primary somatosensory area, S1, was identified by surface electrode recordings. In addition, a second somatosensory area just lateral to S1 was found. A region of auditory cortex was located just caudal to this secondary somatosensory area in the expected location for auditory cortex. A region of visually responsive cortex was indicated, but this was too lateral to be either V1 or V2. Motor cortex was described in the sloth as coextensive with S1. It is not surprising to evoke motor responses from S1, but it is uncertain why a separate motor area, M1, was not found, as it has been found in a wide range of placental mammals, including the armadillos. Clearly more research is needed in the organization of cortex in members of Xenarthra.

Afrotheria is a big superorder with a third of the mammalian orders, including tenrecs, golden moles, elephant shrews, aardvarks, hyraxes, sea cows, and elephants (Hedges, 2001). The tenrecs and golden moles were once included with shrews, moles, and hedgehogs in the now abandoned order of Insectivora based on shared primitive features. Molecular evidence reassigned tenrecs and golden moles to Afrotheria while shrews, moles, and hedgehogs are now in the superorder Laurasiatheria (O'Leary et al., 2013). Tenrecs are considered to be primitive and generally resembling early mammals in being small, having teeth suitable for feeding on insects, and having a small brain with a proportionally small neocortex (Stephen et al., 1991). They are nocturnal and are confined to the island of Madagascar. The tenrec species are somewhat varied in size and other features, and the lesser hedgehog tenrec resembles the European hedgehog, while being smaller. Both of these distantly related placental mammals have the protection of having evolved quills independently, and thus depend less on higher brain functions for avoiding predators. Also, the reduced competition from other placentals on the island of Madagascar has likely contributed to their survival (Eisenberg, 1981).

The organization of cortex in the hedgehog tenrec has been explored in several studies. In overall appearance, the tenrec brain closely resembles those of early mammals in having very little neocortex in comparison with the large olfactory bulb and piriform cortex (Fig. 13.6B). Much of what is known depends on the multielectrode mapping and architectonic studies of Krubitzer et al. (1997). The results provided clear evidence for primary visual, auditory, and somatosensory areas, as well as for a secondary field lateral to S1 termed PV/S2 that had characteristics of PV and S2, and for a rostral area R that seemed to respond best to proprioceptive stimuli. Other studies provided evidence from connections and architecture for a motor area rostral to R (Künzle, 1995, 2009). These studies also provided architectonic evidence for subdivisions of retrosplenial and cingulate cortex. As for other placental mammals, there is a corpus callosum.

Tenrecs are closely related to elephant shrews, and the neocortex of these shrews has also been mapped. Elephant shrews include 4 genera and 14 species that are confined to Africa. They range from mouse to small cat in size, and they have been considered "living fossils" for their primitive skeletal features (Rovero et al., 2008). Their common name comes from their long, trunklike noses that they use to probe the leaf litter of the forest floor for insects. The brains of elephant shrews have proportionately more neocortex than tenrec brains, as the cortex extends further laterally, but it does not extend caudally to cover the midbrain (Dengler-Crish et al., 2006). The somatosensory cortex was explored with microelectrodes, and the somatotopy of S1 was revealed in the expected location. Within S1, the snout and vibrissa representations dominated the lateral half of S1. Just lateral to S1, there was a small second representation that could be S2, PV, or both. Unfortunately, other subdivisions of neocortex were not determined.

The brains of other members of Afrotheria, the aardvarks, manatees, sea cows, dugongs, hyraxes, and elephants, have not been well studied, but all have larger brains and likely different specializations, while retaining the primary sensory and motor areas of placental mammals with smaller brains.

Laurasiatheria is another very large superorder of six orders that includes moles, shrews, hedgehogs, the very successful bat radiation, carnivores, ungulates, and whales. The order of Eulipotyphla has been newly formed and includes many of the former members of the insectivore clade: hedgehogs, shrews, and moles. Hedgehog brains have been most studied as their small brains have long been of interest to comparative neurobiologists on the assumption of primitive forebrain organization (eg, Ebner, 1969). Hedgehogs are small omnivorous nocturnal mammals that populate Eurasia and Africa. As their dorsal surface is covered with short spines, their primary defense is to curl up into a prickly ball. Thus, a lot of brain power is not needed for defense.

From a dorsolateral view of the brain (Fig. 13.6C), it is apparent that hedgehogs have relatively little neocortex, as the cortex does not have an expanded temporal region, and the cortex does not cover the midbrain. An early microelectrode mapping and architecture study established the existence of visual areas V1 and V2, as well as S1 and an auditory region (Kaas et al., 1970). The neocortex was poorly laminated, and yet laminar differences clearly distinguished V1 from other areas. V1 and S1 were densely myelinated as in other animals. Similar but more extensive results have been obtained in a more recent study (Catania et al., 2000). The results provide evidence for primary somatosensory, visual, and auditory fields, as well as two secondary somatosensory areas, PV and S2. The second visual area, V2, seems to occupy the small space between S2 and V1, and there is very little space between S1 and V2. A primary motor area has been identified rostral to S1 by thalamic connections and architecture (Dinopoulos, 1994), but the extent and borders of M1 have not been established.

Similar results have been obtained from other members of the Eulipotyphla clade. Shrews are small mammals that range from 3 g to more than 20 g. In one study, neocortex was mapped in five species of North American shrews (Catania et al., 1999). These small brains included a cap of neocortex on a larger mass of piriform cortex and subcortical structures. Their neocortex contained well-defined primary somatosensory, visual, and auditory areas, and a large S2. All these areas were tightly packed into the caudal half of the neocortex, leaving no or little room for other areas between them. For the smallest, the least shrew, it appeared that there was no room for secondary areas between V1 and S1 and S2. Comparable results have been obtained for the extremely small 2-g Etruscan shrew from Italy (Roth-Alpermann et al., 2010). The location of motor cortex and other areas of frontal cortex have not been identified.

Moles constitute another branch of the Eulipotyphla radiation. They are specialized for an underground life, and therefore have small eyes and small optic nerves. Their neocortex has been largely studied because of the great importance of somatosensory processing in these nearly blind mammals. The star-nosed mole has an especially modified nose for tactile exploration. The somatosensory cortex of the star-nosed mole includes two large representations of the contralateral 11 of the 22 fleshy appendages of the nose, as well as at least part of a third representation (Catania and Kaas, 1995). These representations may constitute part of S1, S2, and PV. A region of auditory cortex occupies the caudolateral extreme of the hemisphere, while primary visual cortex appears to be very small. In the less specialized Eastern mole, somatosensory cortex is again very lateral and caudal, with evidence for motor cortex more medial than rostral to somatosensory cortex (Catania and Kaas, 1997). A small auditory region just caudal to somatosensory cortex may mediate the sense of vibration. There was no evidence for a visual area.

Cortical organization in other members of the superorder Laurasiatheria has not been well studied except for cats and ferrets, which have more expanded brains and more cortical areas. These animals have primary and secondary visual and somatosensory areas, a primary motor area, and two or three primary auditory areas, as well as additional sensory and multisensory areas and as many as 53 estimated cortical areas (Scannell et al., 1995). Most or all of the cortical areas of early mammals are likely to have been retained in their greatly transformed brains.

Finally, we should consider the small brains of bats, while recognizing that these brains are specialized for echolocation and guiding flight (Covey, 2005). Bats belong to the order Chiroptera with the superorder Laurasiatheria. They include the larger fruit bats or megabats that do not echolocate and the smaller insectivorous bats or microbats that do echolocate. Flight has made bats very successful, as they have evolved with over 900 extant species, with some of them existing in huge numbers (Kunz and Racey, 1998). While microbat brains are small and have only a small expansion of neocortex, neocortex is more expanded in the larger megabats. In microbats neocortex is dominated by a large complex of auditory areas and visual cortex is reduced (see Kaas, 2011c for review). However, in megabats auditory cortex is not expanded, while primary visual cortex is large, and several secondary visual areas exist (Rosa and Krubitzer, 1999). In both clades of bats, somatosensory areas are large, with the representation of the wing dominating. Besides S1, there is evidence for PV and S2, as well as somatosensory representations rostral and caudal to S2, with a more rostral motor area (Calford et al., 1985; Krubitzer and Calford, 1992; Chadha et al., 2011).

The sixth superorder of mammals to consider, Euarchontoglires, contains rodents and rabbits, tree shrews and flying lemurs, and primates. The proportion of the forebrain that is neocortex in all these mammals is larger to much larger than in early mammals and some

of the small-brained mammals of the other clades already discussed. For example, all primates have at least 40–50 cortical areas, with as many as 200 in human brains (eg, Van Essen et al., 2012b), so these brains have obviously changed a lot. Even tree shrews and some rodents (eg, squirrels) have an expanded cortical visual system with enlarged and possibly new visual and multisensory areas. Perhaps the least modified brains in this superorder are those of rabbits and some of the rodents, such as rats and mice. Here we focus on the organization of neocortex in laboratory rats, and then briefly consider the neocortex of other Euarchontoglires.

The neocortex of rats is only moderately expanded over that of early mammals (Fig. 13.6D). The organization of sensory and motor areas in the neocortex of rats and other rodents has been extensively reviewed by Krubitzer et al. (2011). Many studies have been devoted to the somatosensory cortex, in part because of the anatomically distinct modules called barrels, with each "barrel" in an array representing a specific sensory whisker of the face in primary somatosensory cortex, S1 (see Ebner and Kaas, 2015 for review). The primary area, S1, of granular cortex is bordered rostrally by a region of dysgranular cortex, much as in other mammals, except that the dysgranular cortex protrudes into S1 to separate representations of the foot and paw, the paw and lower lip, and the lower and upper lips. These protrusions are somewhat unusual, but overall the dysgranular cortex has the location, architecture, and connections of the proprioceptive cortex that is along the rostral border of S1 in other mammals. In primates and cats, and sometimes in other mammals, this cortex is referred to as area 3a. Connections with S1 also reveal a narrow representation along the caudal border of S1, termed the posterior medial area, in rats and mice. Likewise, connection patterns and microelectrode maps have identified secondary representation of the contralateral body surface, S2 and PV (Remple et al., 2003). These two somatosensory areas, now commonly found in mammals, were first distinguished in squirrels (Krubitzer et al., 1986). Primary motor cortex, M1, is located just rostral and medial to the dysgranular cortex and a premotor area, M2, or is medial to M1.

The organization of visual cortex in rats and mice is currently a contentious issue, as some investigators have proposed the existence of a large number of secondary areas that seem too small to function as areas. The primary area, V1, has been clearly defined, and V1 is bordered medially by prostriata and laterally by V2 (Rosa and Krubitzer, 1999). Others have divided this V2 into several small visual areas, starting with the anatomical studies of Montero et al. (1973). However, there is general agreement that this territory in the more visual squirrels is V2, and V2 in squirrels is subdivided into a series of modules, each representing part of the contralateral visual hemifield in an overlaying sequence (Kaas et al., 1989). As it seems unlikely that rodent species differ in having a V2 or not, we surmise that modules of V2 in rats and mice have been identified as separate visual areas, as mapping evidence suggests (Garrett et al., 2014). Other visual areas exist in temporal–occipital cortex lateral to V2, but there is no good argument on these areas. A large, architectonically distinct temporal posterior visual area is obvious in temporal cortex of squirrels, and a small version of this area may exist in rats (Krubitzer et al., 2011).

Auditory cortex in rats consists of at least two primary fields, A1 and the anterior auditory field (Polley et al., 2007). Other proposed fields include a posterior auditory field and two ventral fields. The organization of auditory cortex has been variously interpreted and areas named in rodents (see Krubitzer et al., 2011; Kaas, 2011c), but areas similar to AAF and A1 in rats have been identified in mice, gerbils, and guinea pigs.

Other areas of neocortex in rats have been identified by architecture and patterns of connections. These include granular and agranular divisions of retrosplenial cortex, several divisions of cingulate cortex, at least medial and orbitofrontal divisions of prefrontal cortex, taste cortex, and other divisions of perirhinal cortex (see Burwell, 2001; Paxinos et al., 2009; Chen et al., 2011).

13.9 Summary and Conclusions

In this review, comparative evidence on the organization of neocortex in members of the six superorders of the mammalian radiation was used to reconstruct and infer the organization of neocortex in the common early mammal ancestor. This cladistic approach was guided by recent molecular studies of the branching evolution of mammals that reassigned some mammals, especially insectivores, to different clades. In addition, the fossil record of early mammals informed and constrained the process, in that such fossils indicated that early mammals were small and had small brains with relatively little neocortex. They were also nocturnal, with poor vision, but a more developed olfactory system, and sensitive hearing that extended with the range of high-frequency sounds. They had fur and sensory vibrissae that were important in sensing nearby objects. Their teeth indicated that they fed on small invertebrates and vertebrates. Thus, information on how neocortex is organized in the brains of present-day mammals that most closely resembles early mammals in relative neocortical size and lifestyle, was most fully considered in small-brained mammals across the six superorders.

The comparative evidence indicates that the brains of early mammals already had a thick cortex of six layers of neurons and other cells. Roughly 100 or so neurons would be included in a "minicolumn" across the depth of cortex (Herculano-Houzel, 2016), and each layer was specialized in neuron morphologies and types of inputs and outputs. This is in contrast to the dorsal cortex of present-day reptiles, which is composed of a single row of pyramidal cells with inputs from the dorsal thalamus and outputs to the medial cortex, the homolog of the hippocampus, and to subcortical centers. Reptiles and birds represent surviving members of the sauropsid branch of evolution from early stem amniotes some 340 mya, while mammals are the only surviving members of the many branches of the synapsid branch. Thus, we are limited in our comparisons of neocortex to the dorsal cortex of reptiles and the thicker hyperpallium or Wulst of birds. Dorsal cortex organization is relatively simple in reptiles such as turtles, with visual and perhaps somatosensory inputs, but not auditory inputs, and neocortex likely evolved from something like the dorsal cortex of reptiles. The elongated appearance of the forebrain in the mammalian-like synapsids that preceded mammals suggests that a mammalian type of cortex did not emerge until early mammals. As evolving neocortex would have been a big step, it is not clear how this happened. Obviously, the changes occurred in the generation of the precursors of cortical neurons, and their migration along radial glia to create the thickness of the cortex was a major advance (Rakic, 1995). The evolution of the specializations of the layers also needs to be understood, as well as the factors that led to the areal organization of cortex (Dehay et al., 2001; O'Leary et al., 2007). Given that comparative studies of extant synapsids are limited to mammals, studies of cortical development and gene expression are likely to be the most effective way of addressing questions about how cortex in mammals first emerged (Molnár and Butler, 2002; Super and Uylings, 2001).

It is now clear that the neocortex of early mammals contained a small number of cortical areas, perhaps around 20 (Fig. 13.7). This conclusion is based on the shared cortical areas that can be identified across mammals with little neocortex, one or more from each of the six superorders shown here. In addition, comparable evidence is available for a number of well-studied brains of other mammals with more neocortex that were not considered in detail here. It is not yet known how these early cortical areas or newer areas evolved. In part, new areas could reflect gene duplications (Allman and Kaas, 1974) and changes in gene expression patterns (O'Leary et al., 2007), and they could result from the formation of different sets of modules (columns) within an area, as we commonly see (Kaas, 2012), and a subsequent fusion and separation of the two sets (Kaas, 1989; Krubitzer et al., 1995). The formation of columns is likely to be driven by neuron activity patterns, but the process of

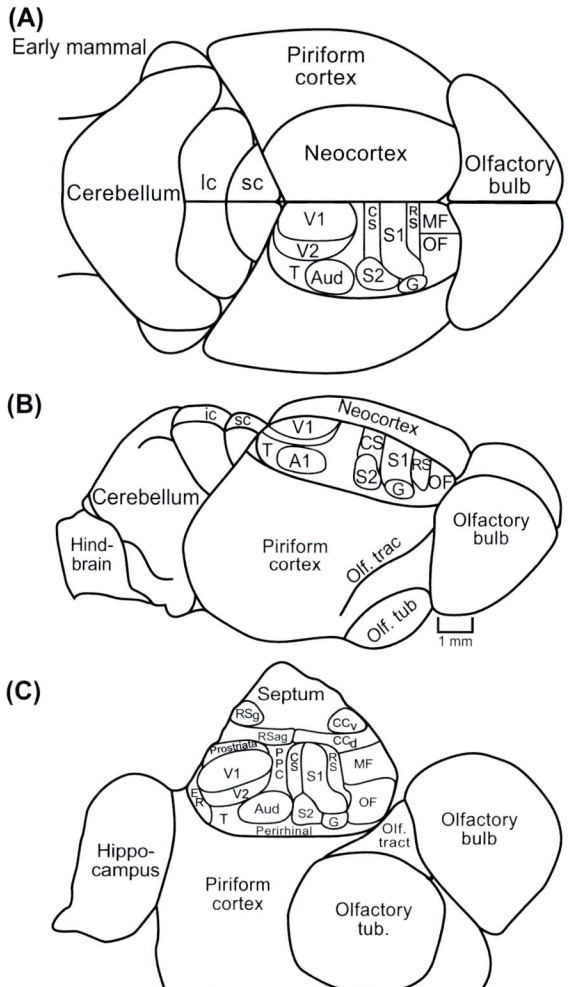

FIGURE 13.7 The inferred organization of neocortex of early mammals based on comparative evidence and the endocasts of the skulls of early mammals. Dorsal and dorsolateral views of the brain are in A and B, while C shows neocortex and other parts of the cerebral hemisphere as a flattened sheet. This summary uses the form of the tenrec brain (see Fig. 13.6B), a template for the relative sizes of parts of the forebrain, as this pattern is largely congruent with inferences from the fossil record. *CCd* and *CCv*, dorsal and ventral cingulate cortex; *RSag* and *RSg*, agranular and granular retrosplenial cortex. Other abbreviations as in previous figures. *Modified from Kaas, J.H., Preuss, T.M., 2014. Published in the chapter Human Brain Evolution in Fundamental Neuroscience, fourth ed. In: Squire, L.R. (Ed.), Elsevier, pp. 901–918.*

Baldwinian selection (Baldwin, 1896) would make the changes more certain and produce improvements (Krubitzer and Kaas, 2005). The evidence that the amount of neocortex increased greatly both absolutely and in proportion to the rest of the brain, many times, is clear (eg, Jerison, 1973). As cortical areas can be hard to define, how or if brains increase in numbers of areas with size is not obvious. However, using the more investigated primates as an example, the small amount of neocortex in the prosimian galago is divided into roughly 50 areas (Wong and Kaas, 2010), more than twice the number proposed here for early mammals, and cortex in marmosets

has a proposed 83 areas, macaques 144 areas (Van Essen et al., 2012a), and humans perhaps 200 areas (Van Essen et al., 2012b). These estimates may be somewhat off, but the evidence for a relationship of neocortical size to numbers of areas is convincing.

The cortical areas that were likely present in early mammals and are widely present in extant mammals tell us something about which cortical functions were important and remain important. All mammals have at least a primary somatosensory area, S1, and some contain secondary somatosensory areas. The somatosensory areas of early mammals likely included rostral and caudal somatosensory areas bordering S1, a rostral proprioceptive area known as area 3a in some mammals and a caudal secondary somatosensory area, variously named but called area 1 in primates. Lateral areas of somatosensory cortex included S2 and likely PV. The widespread presence of four to five somatosensory areas is impressive. In contrast, primary motor and premotor areas did not emerge until the evolution of placental mammals. This, in large part, is likely related to somatosensory areas also having motor functions. Yet, the partial separation of motor and sensory functions may provide advantages to placental mammals, as M1 and S1 have different types of internal organization. Visual cortex also included several areas in early mammals, the primary area V1, prostriata, V2, and at least one division of temporal cortex. This may reflect the very ancient, two visual systems that have been retained in reptiles and mammals: a retina to geniculate to V1, or dorsal cortex pathway, or a retina to tectum to pulvinar, or rotundus-to-temporal cortex pathway (Schneider, 1969; Medina, 2007). However, there are mammals that live underground and have little use for a cortical visual system, and it has been lost, or nearly lost (Cooper et al., 1993). Auditory processing would seem to be very important for early mammals, as nocturnal mammals likely depend more on hearing, and changes in the peripheral auditory system provided early mammals with more sensitive hearing and a sensitivity to higher-frequency sounds that could be used in communication without alerting reptilian predators (Allman, 1999). Yet, present evidence suggests that early mammals only had one cortical auditory area. Other cingulate, retrosplenial, prefrontal, and perirhinal areas that were present in early mammals have less obvious, but clearly important, functions that led to their retention.

References

Aitkin, L., 1995. The auditory neurobiology of marsupials: a review. Hear. Res. 82, 257–266.

Allman, J.M., Kaas, J.H., 1974. A crescent-shaped cortical visual area surrounding the middle temporal area (MT) in the Owl Monkey (*Aotus trivirgatus*). Brain Res. 81, 199–213.

Allman, J.M., 1999. Evolving Brains. H.W. Freeman and Company, New York, NY.

Ashwell, K.W.S., 2010. The Neurobiology of Australian Marsupials: Brain Evolution in the Other Mammalian Radiation. Cambridge Univ. Press, Cambridge, UK.

Ashwell, K., 2014. Neurobiology of Monotremes. CSIRO Publishing, Collingwood, Australia.

Baldwin, J.M., 1896. A new factor in evolution. Am. Nat. 30, 441–451.

Beck, P.D., Pospichal, M.W., Kaas, J.H., 1996. Topography, architecture, and connections of somatosensory cortex in opossums: evidence for five somatosensory areas. J. Comp. Neurol. 366, 109–133.

Bininda-Emonds, O.R.P., Cardillo, M., Jones, K.E., MacPhee Ross, D.E., Beck, R.M.D., Grenyer, R., Price, S.A., Vos, R.A., Gittleman, J.L., Purvis, A., 2007. The delayed rise of present-day mammals. Nature 446, 507–512.

Brodmann, K., 1909. Vergleichende Lokalisationslehre der Grosshirnrinde in ihren Prinzipien dargestellt auf Grund des Zellenbaues. Barth, Leipzig.

Burwell, R.D., 2001. Borders and cytoarchiture of the perirhinal and post-rhinal cortices in the rat. J. Comp. Neurol. 437, 17–41.

Butler, A.B., Hodos, W., 2005. Comparative Vertebrate Neuroanatomy, second ed. John Wiley and Sons, Hoboken.

Calford, M.B., Graydon, M.L., Huerta, M.F., Kaas, J.H., Pettigrew, J.D., 1985. A variant of the mammalian somatotopic map in a bat. Nature 313, 477–479.

Catania, K.C., Kaas, J.H., 1995. Organization of the somatosensory cortex of the star-nosed mole. J. Comp. Neurol. 351, 549–567.

Catania, K.C., Kaas, J.H., 1997. Organization of somatosensory cortex and distribution of corticospinal neurons in the eastern mole (*Scalopus aquaticus*). J. Comp. Neurol. 378, 337–353.

Catania, K.C., Jain, N., Franca, J.G., Volchan, E., Kaas, J.H., 2000. The organization of somatosensory cortex in the short-tailed opossum (*Monodelphis domestica*). Somatosens. Mot. Res. 17, 39–51.

Catania, K.C., Lyon, D.C., Mock, O.B., Kaas, J.H., 1999. Cortical organization in shrews: evidence from five species. J. Comp. Neurol. 410, 55–72.

Chadha, M., Moss, C.F., Sterbing-D'Angelo, S.J., 2011. Organization of the primary somatosensory cortex and wing representation in the Big Brown Bat (*Eptesicus fuscus*). J. Comp. Physiol. A. Neuroethol. Sens. Neural. Behav. Physiol. 197, 89–96.

Chakraborty, M., Jarvis, E.D., 2015. Brain evolution by brain pathway duplication. Philos. Trans. R. Soc. B 370, 20150056.

Chen, X., Gabitto, M., Peng, Y., Ryba, N.J., Zuker, C.S., 2011. A gustotopic map of taste qualities in the mammalian brain. Science 333, 1262–1266.

Coleman, M.N., Boyer, D.M., 2012. Inner ear evolution in primates through the Cenozoic: implications for the evolution of hearing. Anat. Rec. 295, 615–631.

Cooper, H.M., Herbin, M., Nevo, F., 1993. Visual system of a naturally microthalamic mammal: the blind mole rat, (*Spalax ehrenbergi*). J. Comp. Neurol. 328, 313–350.

Covey, E., 2005. Neurobiological specializations in echolocating bats. Anat. Rec. A Discov. Mol. Cell Evol. Biol. 287, 1103–1116.

Cunningham, C.W., Omland, K.E., Oakley, T.H., 1998. Reconstructing ancestral character states: a critical reappraisal. TREE 13, 361–366.

Darwin, C., 1859. The Origin of Species. Murray, London.

Dawkins, R., 2004. The Ancestor's Tale: A Pilgrimage to the Dawn of Evolution. Houghton Mifflin Co, Boston, USA.

Dehay, C., Savatier, P., Cortay, V., Kennedy, H., 2001. Cell-cycle kinetics of neocortical precursors are influenced by embryonic thalamic axons. J. Neurosci. 21, 201–214.

Dengler-Crish, C.M., Crish, S.D., O'Riain, M.J., Catania, K.C., 2006. Organization of the somatosensory cortex in elephant shrews (*E. edwardii*). Anat. Rec. A Discov. Mol. Cell Evol. Biol. 288, 859–866.

Diamond, I.T., Hall, W.C., 1969. Evolution of neocortex. Science 164, 251–262.

Dinopoulos, A., 1994. Reciprocal connections of motor neocortical areas with the contralateral thalamus in the hedgehog (*Erinaceus europaeus*) brain. Eur. J. Neurosci. 6, 374–380.

Dom, R., Martin, G.F., Fisher, B.L., Fisher, A.M., Harting, J.K., 1971. The motor cortex and corticospinal tract of the armadillos (*Dasypus novemcinctus*). J. Neurol. Sci. 14, 225–236.

Ebner, F.F., 1969. A comparison of primative forebrain organization in metatherian and eutherian mammals. Ann. N.Y. Acad. Sci. 167, 241–257.

Ebner, F.F., Kaas, J.H., 2015. Somatosensory system. In: Paxinos, G. (Ed.), The Rat Nervous System, fourth ed. Elsevier, London, pp. 673–699.

Elston, G.N., Manger, P.R., 1999. The organization and connections of somatosensory cortex in the brush-tailed possum (*Trichosurus vulpecula*): evidence for multiple, topographically organized and interconnected representations in an Australian marsupial. Somato Mot. Res. 16 (4), 312–337.

Eisenberg, J.F., 1981. The Mammalian Radiations. University of Chicago Press, Chicago.

Finlay, B.L., Darlington, R.B., 1995. Linked regularities in the development and evolution of mammalian brains. Science 268, 1578–1584.

Frost, S.B., Milliken, G.W., Plautz, E.J., Masterton, R.B., Nudo, R.J., 2000. Somatosensory and motor representations in cerebral cortex of a primitive mammal (*Monodelphis domestica*): a window into the early evolution of sensorimotor cortex. J. Comp. Neurol. 421, 29–51.

Grus, W.E., Zhang, J., 2006. Origin and evolution of the vertebrate vomeronasal system viewed through system-specific genes. Bioessays 28, 709–718.

Garrett, M.E., Nauhaus, I., Marshel, J.H., Callaway, E.M., 2014. Topography and areal organization of mouse visual cortex. J. Neurosci. 34, 12587–12600.

Grützner, F., Rens, W., Tsend-Ayush, E., El-Mogharbel, N., O'Brien, P.C.M., Jones, R.C., Ferguson-Smith, M.A., Graves, J.A.M., 2004. In the platypus a meiotic chain of ten sex chromosomes shares genes with the bird Z and mammal X chromosomes. Nature 432, 913–917.

Hall, J.A., Foster, R.E., Ebner, F.F., Hall, W.C., 1977. Visual cortex in a reptile, the turtle (*Pseudemys scripta*) and (*Chrysemys picta*). Brain Res. 130, 197–216.

Hall, M.I., Kamilar, J.M., Kirk, E.C., 2012. Eye shape and the nocturnal bottleneck of mammals. Proc. Biol. Sci. 279, 4962–4968.

Hedges, S.B., 2001. Afrotheria: plate tectonics meet genomics. Proc. Natl. Acad. Sci. U.S.A. 98, 1–2.

Hennig, W., 1966. Phylogenetic Systematics. University of Illinois Press, Urbana.

Herculano-Houzel, S., 2016. The Human Advantage. The MIT Press, Cambridge, Mass.

Hu, Y., Meng, J., Wang, Y., Li, C., 2005. Large Mesozoic mammals fed on young dinosaurs. Nature 433, 149–152.

Huffman, K.J., Nelson, J., Clarey, J., Krubitzer, L., 1999. Organization of somatosensory cortex in three species of marsupials, *Dasyurus hallucatus, Dactylopsila trivirgata*, and *Monodelphis domestica*: neural correlates of morphological specializations. J. Comp. Neurol. 403, 5–32.

Jerison, H.J., 1973. Evolution of the Brain and Intelligence. Academic Press, Cambridge, Mass.

Kaas, J.H., 1989. The evolution of complex sensory systems in mammals. J. Exp. Biol. 146, 165–176.

Kaas, J.H., 2002. Convergences in the modular and areal organization of the forebrain of mammals: implications for reconstruction of forebrain evolution. Brain Behav. Evol. 59, 262–272.

Kaas, J.H., 2011. Neocortex in early mammals and its subsequent variations. Ann. N.Y. Acad. Sci. 1225, 28–36.

Kaas, J.H., 2011. Reconstructing the areal organization of the neocortex of the first mammals. Brain Behav. Evol. 78, 7–21.

Kaas, J.H., 2011. The evolution of auditory cortex: the core areas. In: Winer, J.A., Schreiner, C.E. (Eds.), The Auditory Cortex. Springer, New York, pp. 407–427.

Kaas, J.H., 2012. Evolution of columns, modules, and domains in the neocortex of primates. Proc. Natl. Acad. Sci. U.S.A. 109 (Suppl. 1), 10655–10660.

Kaas, J.H., 2016. Approaches to the study of brain evolution. In: Shepherd, S.V. (Ed.), Handbook of Evolutionary Neuroscience. John Wiley and Sons, Chichester, UK (in press).

Kaas, J.H., 2017. The evolution of mammalian brains from early mammals to present-day primates. In: Watanabe, S., Hofman, M.A., Shimizu, T. (Eds.), Evolution of Brains, Cognition, and Emotion in Vertebrates. Springer, Japan, Tokyo (in press).

Kaas, J.H., Hall, W.C., Diamond, I.T., 1970. Cortical visual areas I and II in the hedgehog: relation between evoked potential maps and architectonic subdivisions. J. Neurophysiol. 33, 595–615.

Kaas, J.H., Krubitzer, L.A., 1991. The organization of extrastriate visual cortex. In: Leventhal, A.G. (Ed.), The Neural Basis of Visual Function. Macmillan Press, London, pp. 302–323.

Kaas, J.H., Krubitzer, L.A., Johanson, K.L., 1989. Cortical connections of area 17 (V-I) and 18 (V-II) of squirrels. J. Comp. Neurol. 281, 426–446.

Karlen, S.J., Krubitzer, L., 2006. Phenotypic diversity is the cornerstone of evolution: variation in cortical field size within short-tailed opossums. J. Comp. Neurol. 499, 990–999.

Kielan-Jaworowska, Z., Cifelli, R.L., Luo, Z.-X., 2004. Mammals from the Age of Dinosaurs. Columbia University Press, New York.

Kirsch, J.A.W., 1977. The six-percent solution: second thoughts on the adaptedness of the Marsupialia: features of their physiology and diversity suggest that marsupials represent an alternative but not inferior kind of mammal, valuable in understanding the course of mammalian evolution. Am. Sci. 65, 276–288.

Krubitzer, L.A., Sesma, M.A., Kaas, J.H., 1986. Microelectrode maps, myeloarchitecture, and cortical connections of three somatotopically organized representations of the body surface in the parietal cortex of squirrels. J. Comp. Neurol. 250, 403–430.

Krubitzer, L.A., Calford, M.B., 1992. Five topographically organized fields in the somatosensory cortex of the flying fox: microelectrode maps, myeloarchitecture, and cortical modules. J. Comp. Neurol. 317, 1–30.

Krubitzer, L., Kaas, J.H., 2005. The evolution of the neocortex in mammals: how is phenotypic diversity generated? Curr. Opin. Neurobiol. 15, 445–453.

Krubitzer, L., Künzle, H., Kaas, J.H., 1997. Organization of sensory cortex in a Madagascan insectivore, the tenrec (*Echinops telfairi*). J. Comp. Neurol. 379, 399–414.

Krubitzer, L., Campi, K.L., Cooke, D.F., 2011. All rodents are not the same, a modern synthesis of cortical organization. Brain Behav. Evol. 78, 51–93.

Krubitzer, L., Manger, P., Pettigrew, J., Calford, M., 1995. Organization of somatosensory cortex in monotremes: in search of the prototypical plan. J. Comp. Neurol. 351, 261–306.

Kunz, T.H., Racey, P.A., 1998. Bat Biology and Conservation. Smithsonian Institution Press, Washington, DC.

Künzle, H., 1995. Regional and laminar distribution of cortical neurons projecting to either the superior or inferior colliculus in the hedgehog tenrec. Cereb. Cortex 5, 338–352.

Künzle, H., 2009. Tracing the thalamo-cortical connections in tenrec. A further attempt to characterize poorly differentiated neocortical regions, particularly the motor cortex. Brain Res. 1253, 35–47.

Lende, R.A., 1963. Cerebral cortex: a sensorimotor amalgam in the Marsupialia. Science 141, 730–732.

Lende, R.A., 1964. Representation in the cerebral cortex of a primitive mammal: sensorimotor, visual, and auditory fields in the echidna (*Tachyglossus aculeatus*). J. Neurophysiol. 27, 37–48.

Luo, Z.X., Ji, Q., Wible, J.R., Yuan, C.X., 2003. An Early Cretaceous tribosphenic mammal and metatherian evolution. Science 302, 1934–1940.

Luo, Z.X., Ruf, I., Schultz, J.A., Martin, T., 2011. Fossil evidence on evolution of inner ear cochlea in Jurassic mammals. Proc. Biol. Sci. 278, 28–34.

Martinich, S., Pontes, M.N., Rocha-Miranda, C.E., 2000. Patterns of corticocortical, corticotectal, and commissural connections in the opossum visual cortex. J. Comp. Neurol. 416, 224–244.

Medina, L., 2007. Do birds and reptiles possess homologues of mammalian visual, somatosensory, and motor cortex?. In: Kaas, J.H., Bullock, T.H. (Eds.), Evolution of Nervous Systems. Non-mammalian Vertebrates, vol. 2. Elsevier, London.

Meulders, M., Gybels, J., Bergmans, J., Gerebtzoff, M.A., Goffart, M., 1966. Sensory projections of somatic, auditory and visual origin to the cerebral cortex of the sloth (*Choloepus hoffmanni* Peters). J. Comp. Neurol. 126, 535–546.

Mikula, S., Manger, P.R., Jones, E.G., 2008. The thalamus of the monotremes: cyto- and myelo-architecture and chemical neuroanatomy. Philos. Trans. R. Soc. B 363, 2415–2440.

Molnár, Z., 2011. Evolution of cerebral cortical development. Brain Behav. Evol. 78, 94–107.

Molnár, Z., Butler, A.B., 2002. Neuronal changes during forebrain evolution in amniotes: an evolutionary developmental perspective. Prog. Brain Res. 136, 21–38.

Montero, V.M., Bravo, H., Fernandez, V., 1973. Striate-peristriate cortico-cortical connections in the albino and gray rat. Brain Res. 53, 202–207.

Moran, A., Wojcik, L., Cangione, L., Powers, A.S., 1998. Dorsal cortex lesions impair habituation in turtles (*Chrysemys picta*). Brain Behav. Evol. 51, 40–47.

Mota, B., Herculano-Houzel, S., 2015. Brain Structure. Cortical folding scales universally with surface area and thickness, not number of neurons. Science 349, 74–77.

Muchlinski, M.N., 2010. A comparative analysis of vibrissa count, and infraorbital foramen area in primates and other mammals. J. Hum. Evol. 58, 447–473.

Murphy, W.J., Perzner, P.A., O'Brien, S.J., 2004. Mammalian phylogenomics comes of age. Trends Genet. 20, 631–639.

Nilsson, M.A., Churakov, G., Sommer, M., Van Tran, N., Zemann, A., Brosius, J., Schmitz, J., 2010. Tracking marsupial evolution using archaic genomic retroposon insertions. PLoS Biol. 8 (1–9), e1000436.

Northcutt, R.G., Kaas, J.H., 1995. The emergence and evolution of mammalian neocortex. TINS (Trends Neurosci.) 18, 373–379.

Nudo, R.J., Masterton, R.B., 1990. Descending pathways to the spinal cord. III: sites of origin of the corticospinal tract. J. Comp. Neurol. 296, 559–583.

O'Leary, D.D.M., Chou, S.-J., Sahara, S., 2007. Area patterning of the mammalian cortex. Neuron 56, 252–268.

O'Leary, M.A., Bloch, J.I., Flynn, J.J., Gaudin, T.J., Giallombardo, A., Giannini, N.P., Goldberg, S.L., Kraatz, B.P., Luo, Z.X., Meng, J., Ni, X., Novacek, M.J., Perini, F.A., Randall, Z.S., Rougier, G.W., Sargis, E.J., Silcox, M.T., Simmons, N.B., Spaulding, M., Velazco, P.M., Weksler, M., Wible, J.R., Cirranello, A.L., 2013. The placental mammal ancestor and the post-K-P radiation of placentals. Science 339, 662–667.

Paxinos, G., Watson, C.R.R., Carrive, P., Kirkcaldie, M.T.K., Ashwell, K.W.S., 2009. Chemoarchitectonic Atlas of the Rat Brain, second ed. Academic Press, San Diego.

Pettigrew, J.D., 1999. Electroreception in monotremes. J. Exp. Biol. 202, 1447–1454.

Phillips, M.J., Bennett, T.H., Lee, M.S., 2009. Molecules, morphology, and ecology indicate a recent, amphibious ancestry for echidnas. Proc. Natl. Acad. Sci. 106, 17089–17094.

Polley, D.B., Read, H.L., Storace, D.A., Merzenich, M.M., 2007. Multi-parametric auditory receptive field organization across five cortical fields in the albino rat. J. Neurophysiol. 97, 3621–3638.

Radinsky, L., 1976. Cerebral clues. Nat. Hist. 85, 54–59.

Rakic, P., 1995. A small step for the cell, a giant leap for mankind: a hypothesis of neocortical expansion during evolution. Trends Neurosci. 18, 383–388.

Remple, M.S., Henry, E.C., Catania, K.C., 2003. The organization of somatosensory cortex in the laboratory rat (*Rattus norvegicus*): evidence for two lateral areas joined at the representation of the teeth. J. Comp. Neurol. 467, 105–118.

Roth-Alpermann, C., Anjum, F., Naumann, R., Brecht, M., 2010. Cortical organization in the Etruscan shrew (*Suncus etruscus*). J. Neurophysiol. 104, 2389–2406.

Rosa, M.G., Krubitzer, L.A., 1999. The evolution of visual cortex: where is V2? Trends Neurosci. 22, 242–248.

Rose, K.D., 2006. The Beginning of the Age of Mammals. The Johns Hopkins University Press, Baltimore.

Rovero, F., Rathbun, G.B., Perkin, A., Jones, T., Ribble, D.O., Leonard, C., Mwakisoma, R.R., Doggart, N., 2008. A new species of giant sengi or elephant-shrew (genus *Rhynchocyon*) highlights the exceptional biodiversity of the Udzungwa Mountains of Tanzania. J. Zool. 274, 126–133.

Rowe, T., Rich, T.H., Vickers-Rich, P., Springer, M., Woodburne, M.O., 2008. The oldest platypus and its bearing on divergence timing of the platypus and echidna clades. Proc. Natl. Acad. Sci. 105, 1238–1242.

Rowe, T.B., Macrini, T.E., Luo, Z.X., 2011. Fossil evidence on origin of the mammalian brain. Science 332, 955–957.

Royce, G., Martin, G.F., Dom, R.M., 1975. Functional localization and cortical architecture in the nine-banded armadilli (*Dasypus novemcinctus mexicanus*). J. Comp. Neurol. 164, 495–521.

Saraiva, P.E., Magalhães-Castro, B., 1975. Sensory and motor representation in the cerebral cortex of the three-toed sloth (*Bradypus tridactylus*). Brain Res. 90, 181–193.

Scannell, J.W., Blakemore, C., Young, M.P., 1995. Analysis of connectivity in the cat cerebral cortex. J. Neurosci. 15, 1463–1483.

Schneider, G.E., 1969. Two visual systems. Science 163, 895–902.

Shepherd, G.M., 2011. The microcircuit concept applied to cortical evolution from three-layer to six-layer cortex. Front. Neuroanat. 5, 1–15.

Smith, F.A., Boyer, A.G., Brown, J.H., Costa, D.P., Dayan, T., Ernest, S.K., Evans, A.R., Fortelius, M., Gittleman, J.L., Hamilton, M.J., Harding, L.E., Lintulaakso, K., Lyons, S., McCain, C., Okie, J.G., Saarinen, J.J., Sibly, R.M., Stephens, P.R., Theodor, J., Uhen, M.D., 2010. The evolution of maximum body size of terrestrial mammals. Science 330, 1216–1219.

Stephen, H., Baron, G., Frahm, H.D., 1991. Comparative Brain Research in Mammals. Insectivora, vol. 1. Springer, New York.

Super, H., Uylings, H.B., 2001. The early differentiation of the neocortex: a hypothesis on neocortical evolution. Cereb. Cortex 11, 1101–1109.

Van Essen, D.C., Glasser, M.F., Dierker, D.L., Harwell, J., 2012. Cortical parcellations of the macaque monkey analyzed on surface-based atlases. Cereb. Cortex 22, 2227–2240.

Van Essen, D.C., Glasser, M.F., Dierker, D.L., Harwell, J., Coalson, T., 2012. Parcellations and hemispheric asymmetries of human cerebral cortex analyzed on surface-based atlases. Cereb Cortex 22, 2241–2262.

Walls, G.L., 1942. The Vertebrate Eye and its Adaptive Radiation. Hafner Publishing, New York.

Weil, A., 2005. Mammalian palaeobiology: living large in the Cretaceous. Nature 433, 116–117.

Welker, W.I., 1990. Why does cerebral cortex fissure and fold?. In: Jones, E.G., Peters, A. (Eds.), Cerebral Cortex. Comparative Structure and Evolution of Cerebral Cortex, vol. 8B. Plenum Press, New York, pp. 3–136.

Wong, P., Kaas, J.H., 2009. An architectonic study of the neocortex of the short-tailed opossum (*Monodelphis domestica*). Brain Behav. Evol. 73, 206–228.

Wong, P., Kaas, J.H., 2010. Architectonic subdivisions of neocortex in the Galago (*Otolemur garnetti*). Anat. Rec. 293, 1033–1069.

Young, H.M., Vaney, D.I., 1990. The retinae of prototherian mammals possess neuronal types that are characteristic of non-mammalian retinae. Vis. Neurosci. 5, 61–66.

14

What Modern Mammals Teach About the Cellular Composition of Early Brains and Mechanisms of Brain Evolution

S. Herculano-Houzel

Vanderbilt University, Nashville, TN, United States

14.1 Introduction

Mammalian brains all have the same gross morphology that makes them instantly recognizable, despite the enormous diversity in several aspects. Their mass varies from a fraction of a gram to several kilos, cortical surface area spans four orders of magnitude, cortical thickness varies by nearly 10-fold. The cerebral cortex can represent as little as 30% of the mass of the whole brain (as in the least shrew), or as much as 84% (as in humans; Hofman, 1988). How did such diversity come to be, in the 218 million years that have passed since the last common ancestor to all mammals lived (Meredith et al., 2011)? What makes brain size vary so much, and as it varies, what features in brain organization and its cellular composition are flexible, and have come to be characteristic of each mammalian clade, and what features are inflexible and universal, either due to biological constraints or basic physical properties? Moreover, what do those differences and similarities across extant mammals inform about what early mammalian brains were made of and looked like, and how they evolved over time, giving rise to the current diversity?

14.2 The Traditional View: All Brains Are Made of Same

It might seem surprising that neuroscience made so much progress during the 1900s without solid knowledge about the most basic facts about brains: what they are made of, what constraints and rules there are to how brains are built, and how that varies across

mammalian groups or related species with widely different brain sizes? At best, there were estimates of numbers of neurons in invertebrates: for instance, 302 in the hermaphrodite of the model nematode *Caenorhabditis elegans* (White et al., 1986), 851,000 neurons in the brain of the worker honeybee *Apis mellifera* (and 1.2 million neurons in the drone; Witthöft, 1967), counted by direct enumeration. Estimates for mammals were typically obtained from rough extrapolations from neuronal density in the cerebral cortex using brain volume. With no proper measurements, for instance, the human brain was widely held to consist of 100 billion neurons and 10–50 times more glial cells, numbers that even made it to one of the most popular textbooks in the field, Kandel, Jessel, and Schwartz's *Principles of Neural Science* (reviewed in Von Bartheld et al., 2016). But those (mistaken) numbers for the human brain had nothing to be compared to. At best, and probably from more extrapolations from neuronal density in the cerebral cortex and brain mass, there were estimates of "about 30 million for (…) small mammals such as shrews" (Campbell and Ryzen, 1953) and "well over 200 billion [neurons] for whales and elephants" (Williams and Herrup, 1988). Large mammalian brains were thus presumed to have three orders of magnitude more neurons than the smallest brains, and even more neurons than invertebrates.

Not that larger brains necessarily must have more neurons, though, nor, by the way, must larger animals necessarily come with larger brains. Given that mammalian bodies are fairly similar in their *bauplan*, operating and moving a large body should be exactly as complex as operating a small body, and therefore require just as many neurons [despite the expectation that larger bodies require more brain mass to operate them, behind Harry Jerison's (1973) concept of encephalization]. Thus, larger brains could in principle be made simply of larger neurons, not more neurons, with more glial cells providing myelination and support for the longer axons.

Another problem was that not enough data were available to compare what different mammalian brains were made of, or how that varied, much less to determine if that mattered. Until the 2000s, comparative studies of brain morphology and composition relied mostly on measurements of structure volume, neuronal density, and glia/neuron ratios in the cerebral cortex of assorted species (the parameters that were available for study), and typically analyzed mammalian species as an indiscriminate group (eg, Nissl, 1898; Tower and Elliott, 1952; Tower, 1954; Friede, 1954; Hawkins and Olszewski, 1957; Haug, 1987; Finlay and Darlington, 1995). Without any clade-specific analysis, those studies assumed that the same scaling rules applied as brains varied in size in mammalian evolution. Those rules were supposedly that as brains increased in volume, the cerebral cortex expanded homogeneously in surface area across species (Hofman, 1985; Finlay and Darlington, 1995), with decreasing neuronal densities (that is, larger neurons) and an increasing glia/neuron ratio (Tower and Elliott, 1952; Tower, 1954; Hawkins and Olszewski, 1957; Haug, 1987), but a constant number of neurons per surface area (Rockel et al., 1980). Not much was said about the cerebellum or the remaining brain structures; evolution was, anyway, all about cortical expansion (Hofman, 1985; Finlay and Darlington, 1995; Rakic, 1995). A few comparative studies did, however, point out that all brains were not made the same. For example, the expansion of the cerebral cortex does not happen similarly across all mammals with growing brain volume: in mammalian brains of a similar total volume, the cerebral cortex can occupy different proportions of the brain (Frahm et al., 1992; Clark et al., 2001).

Another problem was that brain size was for too long considered either a variable that depended on body size or, worse, an independent variable that controlled neuronal density, or numbers of neurons, or metabolic rate, when neither can be the case. Brain size in proportion to the body is too variable across individuals and species to be closely tied to body size (Riska and Atchley, 1985), and brain mass does not even scale universally with body mass across individuals of a same species and species of a same order (Armstrong, 1990). As to the second issue, how could the size of the brain control variables such as the number of cells that compose it or their size (and therefore density), when the size of a brain is the very result of the product of its number of cells and their size? According to what physiological rationale could brain size be directly, causally related to something global such as longevity? The overuse of

brain size as the key variable in comparative studies was brilliantly summarized in a review by Williams and Herrup (1988): "As has been repeatedly stressed and repeatedly ignored, brain mass is a compound variable, and little insight can be gained by reducing brain weight to a simple expression made up of one constant, one variable (body weight, surface area, metabolic rate, and even life span) and one coefficient" (Williams and Herrup, 1988: p. 424).

It is understandable, however, that brain volume (or mass) would necessarily end up overused, as for decades it was indeed the main variable available for comparative studies. With the understanding that the field needed proper data on what different brains were made of, so that brain size could finally become a dependent variable, as it should, and studied in its components, we developed in the early 2000s a new method to allow the rapid, direct estimation of numbers of neuronal and nonneuronal cells that compose any dissectable brain structure, the isotropic fractionator (Herculano-Houzel and Lent, 2005). The method initially met with some resistance; after all, it did, contrary to traditional histology methods, and in almost heretic fashion, consists of dissolving precious, fixed brain tissue in detergent and counting free-floating nuclei in isotropic (homogeneous) suspensions. Nevertheless, it has now been shown by two groups independently to yield results that are consistent with those obtained with stereology, the traditional approach, and in much less time (reviewed in Herculano-Houzel et al., 2015a). Our recent studies of brain composition using the isotropic fractionator have systematically addressed clade-specific relationships between brain mass and numbers and densities of neuronal and nonneuronal cells, and have uncovered both features of brain scaling that are indeed constant across mammalian species, and several others that are clade specific (Herculano-Houzel et al., 2014b). Those are reviewed in the following section.

14.3 The Many Ways of Putting a Brain Together

In our studies, which so far have included 56 mammalian species ranging in brain size from 0.2 g in the smoky shrew to 4619 g in the African elephant, and representing 8 mammalian clades (Fig. 14.1), we find that the total number of neurons in the brain ranges from 36.5 million to 257 billion in these same species (Herculano-Houzel et al., 2015b). The human brain, which is not the largest one analyzed, is also not the one with the most neurons as a whole—86 billion, by the way, not 100 billion (a difference that may seem small, but amounts to more than an entire baboon brain; Azevedo et al., 2009). The honor

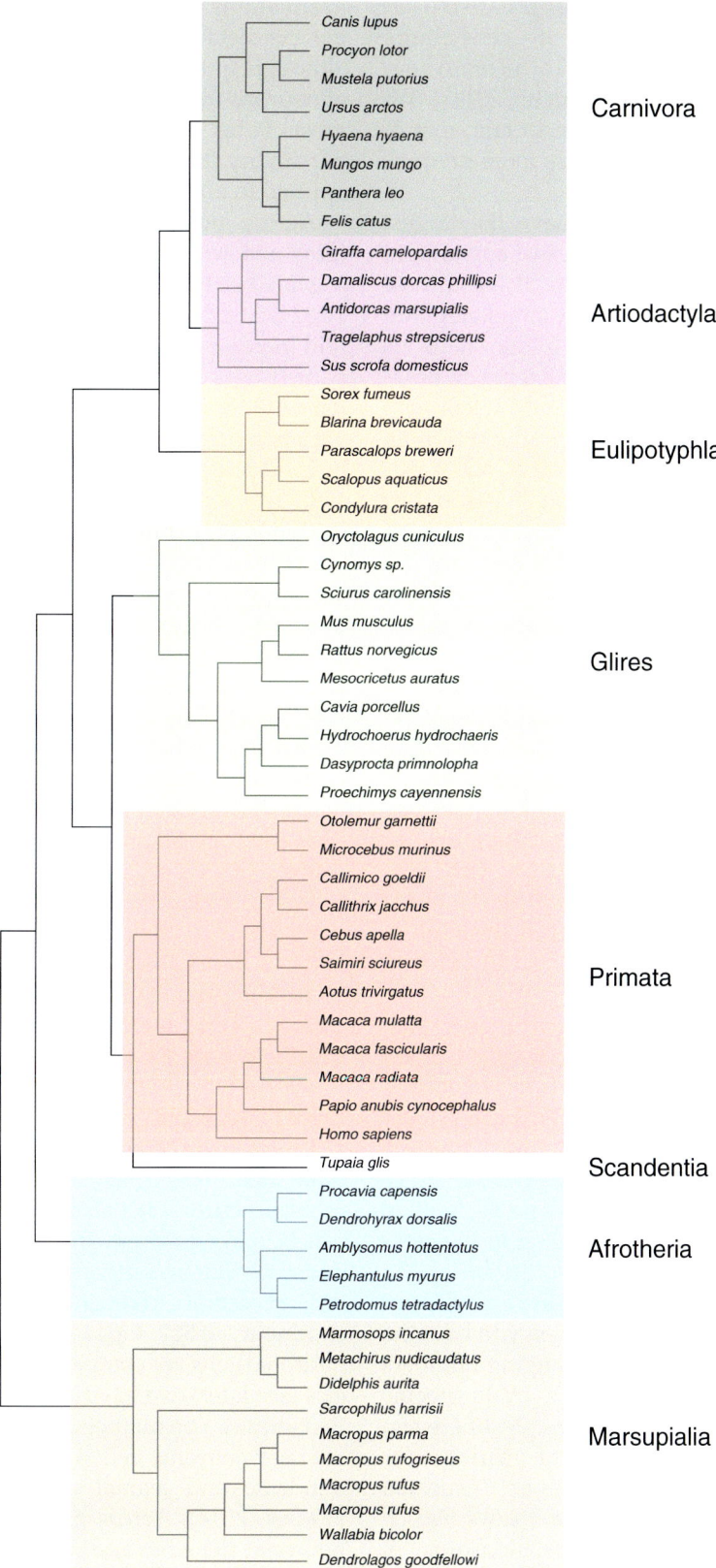

FIGURE 14.1 Species analyzed and their phylogenetic relationships. Relationships for carnivorans from Nyakatura and Bininda-Omonds (2012); primates from Purvis (1995); rodents from Blanga-Khanfi et al. (2009); eulipotyphlans from Douady et al. (2002) and Shinohara et al. (2003); afrotherians from Kuntner et al. (2011); marsupials from May-Collado et al. (2015). Relationships across artiodactyls and clades according to Murphy et al. (2004). Diagram does not include the naked mole-rat (*Heterocephalus glaber*), which we found to be an outlier among rodents and was therefore not included in the analyses shown in the figures in this chapter (Herculano-Houzel et al., 2011).

of the most brain neurons as a whole so far goes to the African elephant—but only because of its cerebellum, which alone concentrates an atypical 98% of all brain neurons in that species (Herculano-Houzel et al., 2014a). We have yet to determine whether the largest brain, that of the sperm whale, at nearly 8 kg, also has more neurons than the smaller elephant brain.

The human brain, however, does have the largest number of neurons in the cerebral cortex of any animal so far: at an average 16 billion neurons, it has almost three times as many neurons as the twice larger elephant cerebral cortex, which has only 5.6 billion neurons (Herculano-Houzel et al., 2014a). This discrepancy, however, does not apply to humans alone; primates as a whole have more neurons in the cerebral cortex than any other mammal with a similar cortical mass. For instance, the capuchin monkey cortex, at 39 g and 1.14 billion neurons, has almost four times as many neurons as the pig cortex, at a slightly larger 42 g but only 307 million neurons (Herculano-Houzel et al., 2015b). The larger number of neurons found in a primate brain structure than in the same structure of similar mass in other mammals also applies to the cerebellum and to the rest of brain. For example, the rhesus monkey cerebellum, at 7.7 g and 4.6 billion neurons, has more than three times as many neurons as the pig cerebellum, at a slightly larger 8.1 g and 1.8 billion neurons (Herculano-Houzel et al., 2015b). Similarly, the rest of the brain structures in the baboon (brain stem, diencephalon, and striatum), at 17.2 g, holds 278 million neurons, against only 108 million neurons found in the slightly larger rest of brain of the capybara, at 19.9 g (Herculano-Houzel et al., 2015b).

Our direct comparisons thus show that two mammalian brain structures of similar size are not necessarily made of similar numbers of neurons. Likewise, brain structures made of similar numbers of neurons can have widely different sizes. For example, there are 762 million neurons in the 213 g cerebral cortex of the greater kudu, an artiodactyl (Kazu et al., 2014), but a slightly larger 801 million neurons fit in a cerebral cortex of only 36 g in the crab-eating macaque (Herculano-Houzel et al., 2015b).

Importantly, these loose comparisons are examples of a trend, as there is mathematical consistency to how the mass of brain structures varies with changing numbers of neurons across species, with some clade-specific rules. The relationships depicted in Figs. 2—4 between brain structure mass and numbers of neurons (or neuronal density), which we refer to as "neuronal scaling rules", can be described as power laws, which means that these are scale-invariant relationships: in other words, a large cortex is not built in entirely new ways, but rather in ways that can be predicted from a small cortex (provided that one knows what rules to apply) —just like a large mammal is still recognizable

to anyone as a larger version of a small mammal, despite the predictably thicker bones and legs. Surprisingly, although, at this level of brain structure mass × number of neurons relationships, there is great uniformity across most clades—which comes in very handy when using parsimony to infer what scaling rules must have been applied to ancestral mammals, as will be explored later.

Across nonprimate mammals, the cerebral cortex (gray and white matter together) gains mass faster than it gains neurons: given the power function of exponent 1.582 that relates the two variables (Fig. 14.2A), any 10-fold increase in the number of cortical neurons is accompanied by a 38-fold increase in cortical mass (including the white matter). This function describes equally well variations across dozens of species of marsupials, glires, eulipotyphlans, afrotherians (including the African elephant), artiodactyls, and carnivorans. There are only three notable exceptions so far: the naked mole-rat, which has only half as many cortical neurons as predicted for its cortical mass, possibly due to its unusual living conditions of extreme hypoxia (Herculano-Houzel et al., 2011); the raccoon, which has almost twice as many cortical neurons as expected for a nonprimate cortex of its mass (Messeder et al., unpublished results); and the brown bear, which has only about one-fourth as many cortical neurons as expected for its cortical mass (which we have suggested that is due to metabolical limitations; Messeder et al., unpublished results). Primates, in contrast, have a more economical way of adding neurons to the cerebral cortex: with the power function of exponent 1.087 that relates cortical mass to number of cortical neurons in primates (Fig. 14.2A, red), close to linearity, a 10-fold increase in the number of cortical neurons in a primate is accompanied by only a 12-fold increase in cortical mass (compared to a 38-fold increase in other mammals).

Fitting more neurons in a primate cortex results from higher neuronal densities (ie, smaller neurons on average) in the cerebral cortex of primate than in nonprimate mammals (Fig. 14.2B). Across the latter, more cortical neurons are also larger neurons on average, as can be inferred from the rapidly decreasing neuronal densities that accompany larger numbers of neurons in the cerebral cortex (a power function of exponent −0.582; Fig. 14.2B). Because densities of non-neuronal cells are comparatively constant across all species (see later), we have shown that decreasing neuronal densities can be mathematically used to infer increasing average neuronal cell size (including the soma and all dendritic and axonal arbors; Mota and Herculano-Houzel, 2014). Across primate species, in comparison, neuronal densities are not only larger than in nonprimate species of similar cortical mass; they also do not decrease significantly as the number of neurons increases in primate cortices (Fig. 14.2B, red). This dissociation between increasing numbers of neurons and

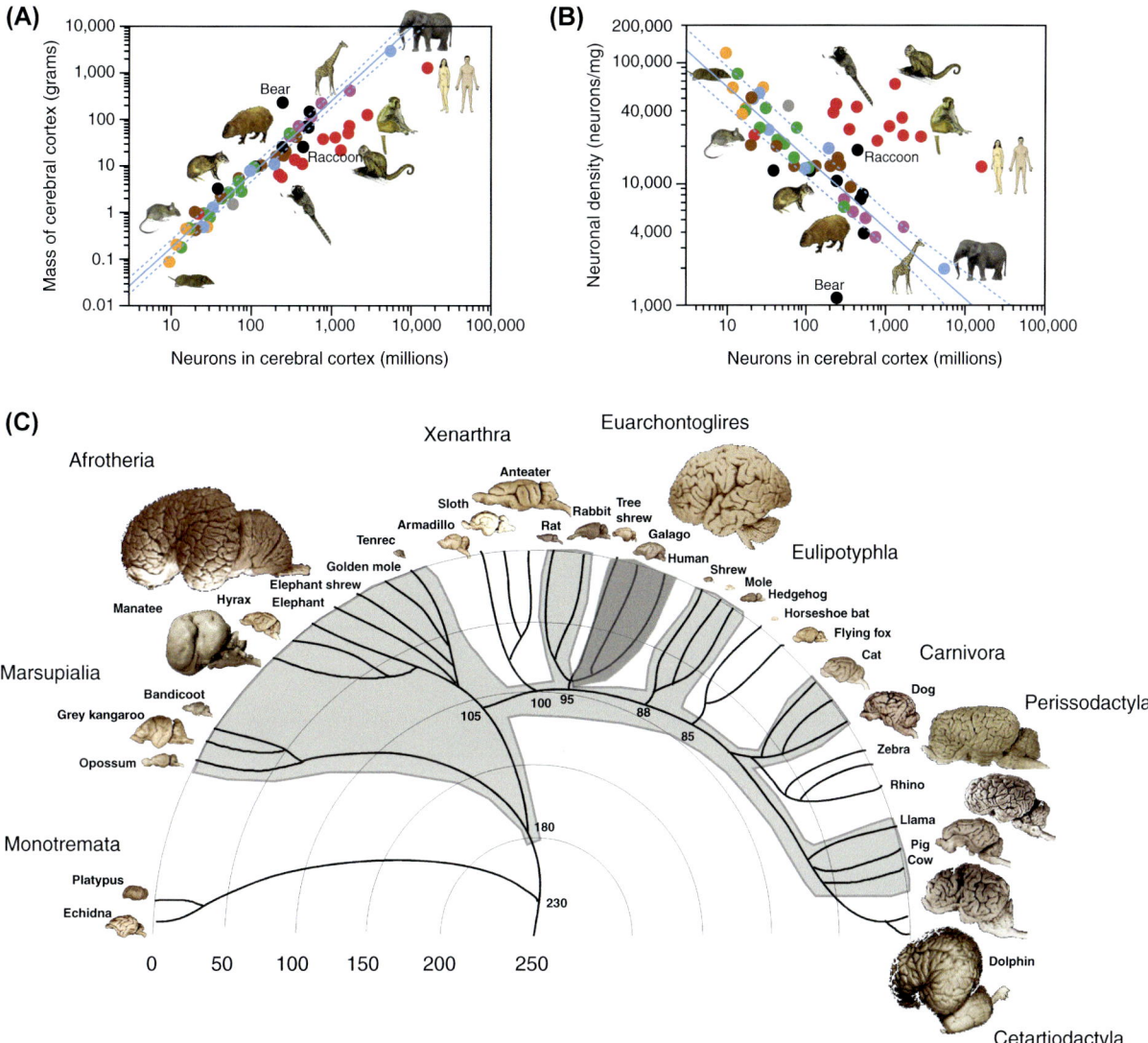

FIGURE 14.2 Neuronal scaling rules for the cerebral cortex are shared among extant nonprimate mammalian clades, and thus probably applied to early mammals. (A) The mass of the cerebral cortex (including the white matter) varies as a power function of the number of neurons in the structure that is shared across all nonprimate mammalian species examined so far (exponent: 1.582 ± 0.039, $r^2 = 0.978$, $p < 0.0001$; primates, exponent 1.087 ± 0.074, $r^2 = 0.956$, $p < 0.0001$). The raccoon and the brown bear are exceptions with more and fewer neurons than predicted for their cortical mass, respectively (Messeder et al., unpublished results). (B) The density of neurons in the cerebral cortex (expressed as neurons per milligram of tissue including the white matter) varies as a power function of the number of neurons in the structure that is shared across all nonprimate mammalian species examined so far (exponent: 0.582 ± 0.038, $r^2 = 0.858$, $p < 0.0001$). Again, the raccoon and the brown bear are exceptions, with higher and lower neuronal densities than predicted for their cortical mass, respectively. Average neuronal density does not decrease systematically across primate species with increasing numbers of cortical neurons (*red*; $r^2 = 0.123$, $p = 0.2644$). (C) Mammalian evolutionary tree indicating the nonprimate clades that share the same relationship between cerebral cortical mass and number of neurons, allowing the inference that the common ancestor also shared that relationship. Original data on 5 species of eulipotyphlans, 6 species of afrotherians, 10 species of glires, 12 species of primates, and 4 species of artiodactyls are available in Herculano-Houzel et al. (2014b). *Data on carnivorans from Messeder et al. (unpublished results). Data on marsupials from Dos Santos et al. (unpublished results).*

increasing average neuronal cell size is one of the key distinguishing features of primate cerebral cortices.

There is also consistency in how the mass of the cerebellum scales across nonprimate species—although for this structure, eulipotyphlans (shrews and moles of Europe and the Americas) also have their own neuronal scaling rule. Both primates and eulipotyphlans have cerebella that are systematically composed of more cells than other mammals with similar cerebellar mass (Fig. 14.3A). For instance, the cerebellum of the Eastern mole (a eulipotyphlan), at 0.153 g, is composed of 158 million neurons, while the slightly larger cerebellum of the elephant shrew (an afrotherian), at 0.168 g, is built with only 89 million neurons (Herculano-Houzel et al.,

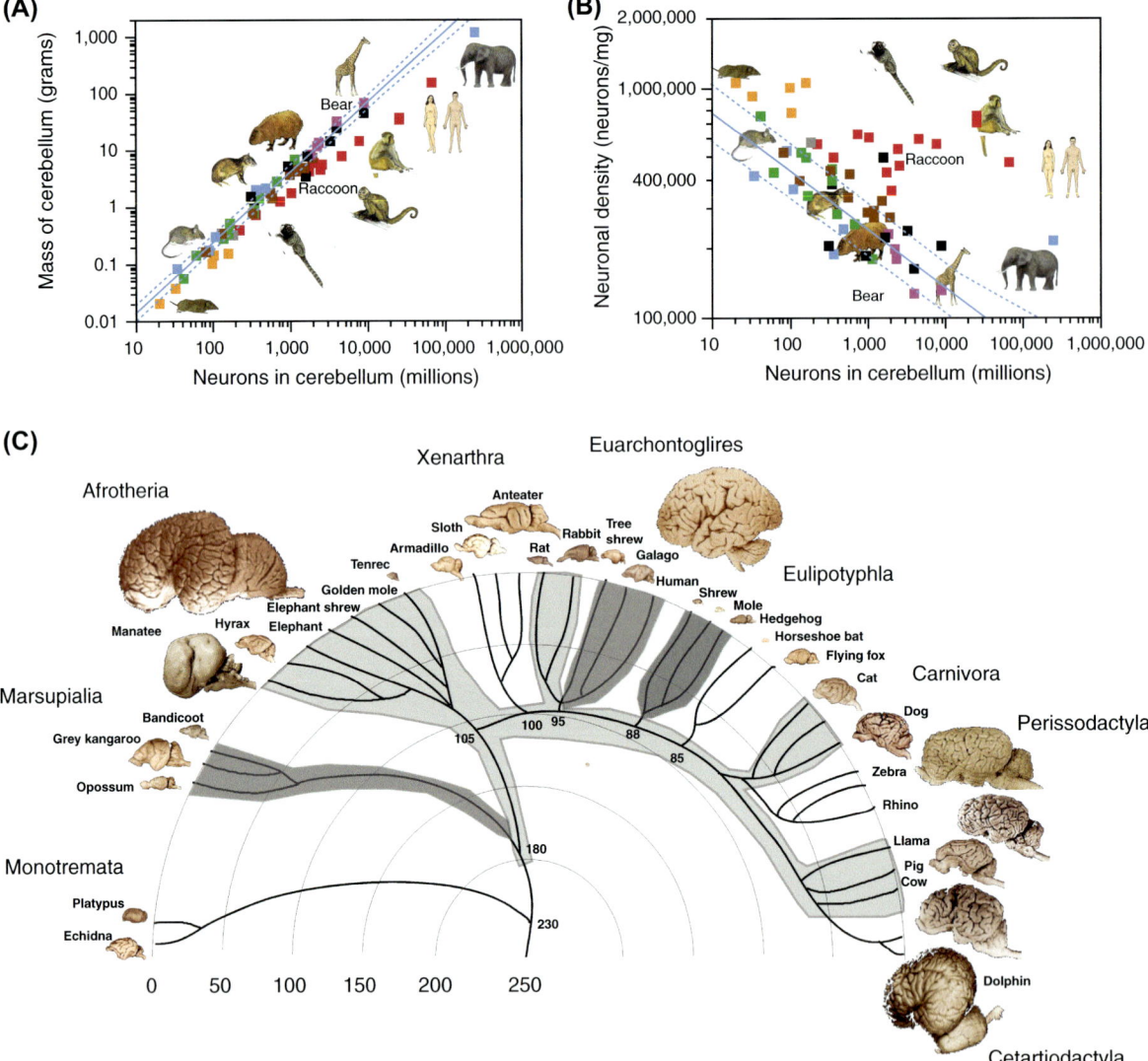

FIGURE 14.3   Neuronal scaling rules for the cerebellum are shared among several extant mammalian clades, and thus probably applied to early mammals. (A) The mass of the cerebellum (including the white matter and deep nuclei) varies as a power function of the number of neurons in the structure that is shared across afrotherians (with the exception of the elephant; Herculano-Houzel et al., 2014), glires, carnivorans, and artiodactyls (exponent: 1.267 ± 0.036, $r^2 = 0.982$, $p < 0.0001$). The raccoon is an exception with more neurons than predicted for its cerebellar mass (Messeder et al., unpublished results). Marsupials, primates, and eulipotyphlans deviate from this relationship, each in a different manner (marsupials, exponent 1.186 ± 0.037, $r^2 = 0.992$, $p < 0.0001$; primates, exponent 0.976 ± 0.036, $r^2 = 0.985$, $p < 0.0001$; eulipotyphla, exponent 1.028 ± 0.084, $r^2 = 0.980$, $p = 0.0012$) with more neurons than predicted for cerebellar mass in other clades. (B) The density of neurons in the cerebellum (expressed as neurons per mg of tissue including the white matter and deep nuclei) varies as a power function of the number of neurons in the structure that is shared across afrotherians (with the exception of the elephant; Herculano-Houzel et al., 2014), glires, carnivorans, and artiodactyls (exponent: -0.267 ± 0.036, $r^2 = 0.705$, $p < 0.0001$). Again the raccoon is an exception with higher neuronal densities than predicted for its cerebellar mass. Marsupials, primates, and eulipotyphlans deviate from this relationship, each in a different manner, with higher neuronal densities than predicted for cerebellar mass in other clades. There is no significant decrease in neuronal density in the cerebellum of primates with increasing numbers of neurons (*red*; $r^2 = 0.171$, $p = 0.1808$). (C) Mammalian evolutionary tree indicating the clades that share the same relationship between cerebellar mass and number of neurons, allowing the inference that the common ancestor also shared that relationship. Original data on 5 species of eulipotyphlans, 6 species of afrotherians, 10 species of glires, 12 species of primates, and 4 species of artiodactyls are available in Herculano-Houzel et al. (2014b). *Data on carnivorans from Messeder et al. (unpublished results). Data on marsupials from Dos Santos et al. (unpublished results).*

2015b). Likewise, while the cerebellum of the baboon (a primate), at 13.7 g, is built of 7.8 billion neurons, the cerebellum of the blesbok (an artiodactyl), at a similar 13.4 g, is composed of only 2.4 billion neurons (Herculano-Houzel et al., 2015b). Like the cerebral cortex, the mass of the cerebellum scales predictably with the number of neurons in the cerebellum across primates (with an exponent of 0.976), across eulipotyphlans (with an exponent of 1.028), and across all other mammalian species examined so far (with a larger exponent of 1.267;

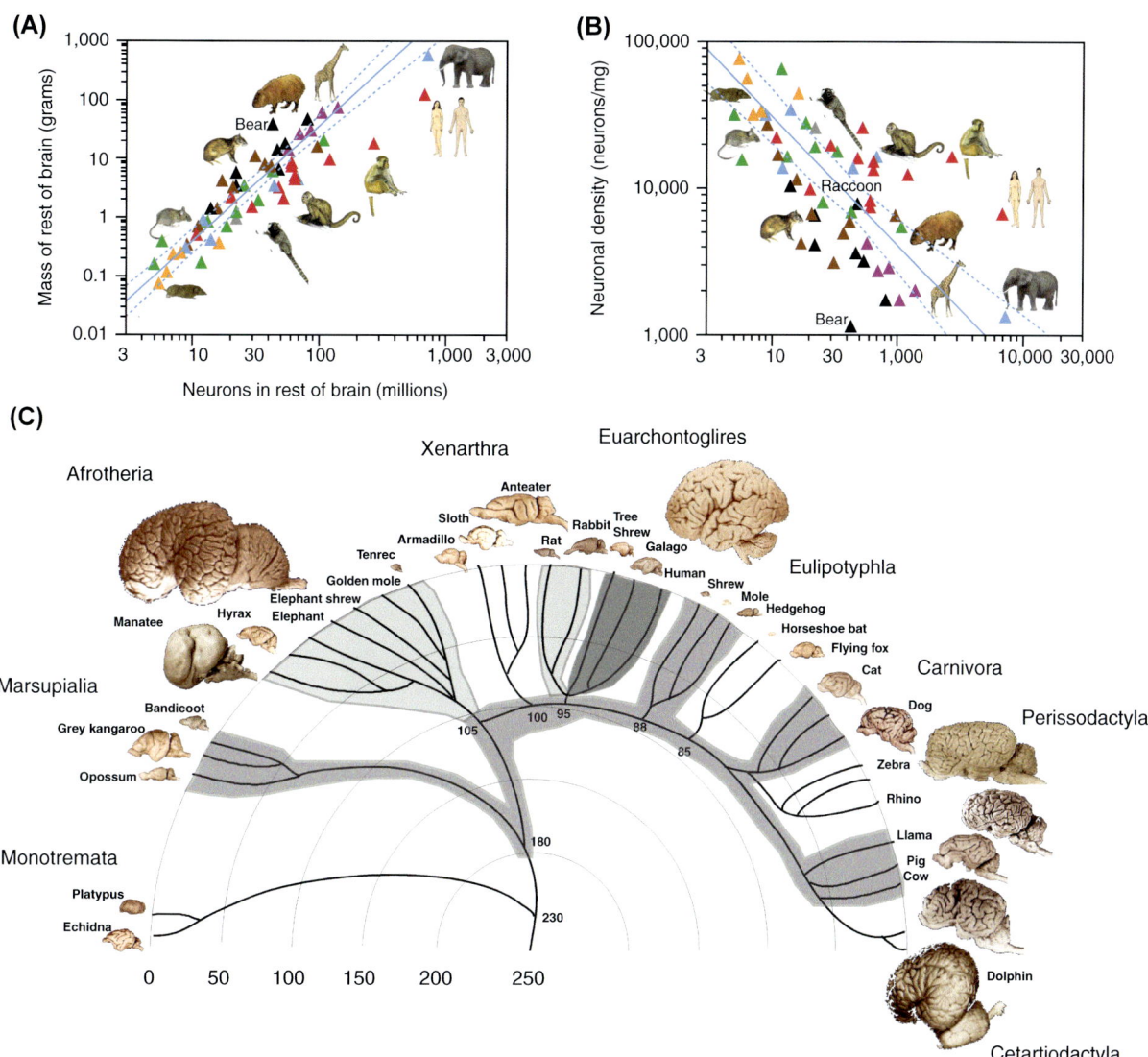

FIGURE 14.4 Neuronal scaling rules for the rest of brain are shared among several extant mammalian clades, and thus probably applied to early mammals. (A) The mass of the rest of brain (brain stem, diencephalon, and striatum) varies as a power function of the number of neurons in each structure across all nonprimate species (plotted exponent: 1.921 ± 0.120, $r^2 = 0.867$, $p < 0.0001$) and as a different function across primates (exponent: 1.198 ± 0.116, $r^2 = 0.915$, $p < 0.0001$). Different relationships (not plotted) apply to the ensemble of eulipotyphlans, marsupials, carnivorans, and artiodactyls (exponent: 2.057 ± 0.120, $r^2 = 0.925$, $p < 0.0001$) and to the ensemble of glires and afrotherians (exponent: 1.624 ± 0.120, $r^2 = 0.929$, $p < 0.0001$). (B) The density of neurons in the rest of brain (expressed as neurons per mg of tissue) varies as a power function of the number of neurons in the structure in nonprimates (exponent, -0.928 ± 0.130, $r^2 = 0.566$, $p < 0.0001$) whose 95% confidence interval (*dotted lines*) excludes most primates (*red*). The relationship for primates does not reach significance (exponent: -0.180 ± 0.113, $r^2 = 0.203$, $p = 0.1413$). Within nonprimates, species can be grouped in two power functions (not plotted): one that applies to glires and afrotheria (exponent, -0.621 ± 0.120, $r^2 = 0.658$, $p = 0.0001$), and one that applies to marsupials, eulipotyphla, carnivorans, and artiodactyla (exponent, -1.077 ± 0.136, $r^2 = 0.723$, $p < 0.0001$). (C) Mammalian evolutionary tree indicating the clades that share the same relationship between rest of brain mass and number of neurons (marsupials, eulipotyphlans, carnivorans, and artiodactyls), suggesting that the common ancestor also shared that relationship. Original data on 5 species of eulipotyphlans, 6 species of afrotherians, 10 species of glires, 12 species of primates, and 4 species of artiodactyls are available in Herculano-Houzel et al. (2014b). *Data on carnivorans from Messeder et al. (unpublished results). Data on marsupials from Dos Santos et al. (unpublished results).*

Fig. 14.3A). The difference in exponents means that although a primate or eulipotyphlan cerebellum made with 10× more neurons becomes only 10-fold larger, increasing the number of neurons in the cerebellum of other mammals by 10 times makes that cerebellum 18-fold larger.

As for the cerebral cortex, the different scaling rules for the cerebellum of primates, eulipotyphlans, and all other mammals result from how increasing numbers of neurons are accompanied by rapidly decreasing neuronal densities (and thus increasing average neuronal cell sizes) in the latter, but not in the former two clades (Fig. 14.3B). As a result of the dissociation between more neurons and larger neurons in eulipotyphlans and primates, neuronal densities in the cerebellum of these species are much larger than in similar-sized cerebellums of other mammals (Fig. 14.3B, compare yellow and red points with all others); in other words, the cerebella of the former hide a much larger number of neurons than could be suspected from their sheer mass.

The remaining brain structures (brain stem, diencephalon, and striatum, which we refer to as "rest of brain") also have more neurons in primate than in nonprimate species of similar rest of brain mass. For example, the rest of brain of the baboon, at 17.2 g, is composed of 278 million neurons, while the slightly larger rest of brain of the capybara, at 19.9 g, is made of only 108 million neurons (Herculano-Houzel et al., 2015b). The relationship between rest of brain mass and number of neurons is also well described by a power function of exponent 1.921 across all nonprimate species (Fig. 14.4A), although there is a larger spread across species. This larger variability in the relationship is, however, to be expected from the widely different structures that are thrown in the mix, from the neuron dense striatum to the neuron sparse brain stem, and their variable relative sizes across different species. As found in other structures, the primate rest of brain scales in mass with a different power function of its number of neurons, with a smaller exponent of 1.198 (Fig. 14.4A, red points). As a consequence, while the rest of brain has a fairly close 690 million neurons in the human and 741 million neurons in the African elephant, it is a much smaller structure in the primate, at 117.7 g versus 564.7 g in the afrotherian (Fig. 14.4A; Herculano-Houzel et al., 2015b). We initially reported that the rest of brain in glires shared its neuronal scaling rules with afrotherians, eulipotyphlans, and artiodactyls (Herculano-Houzel et al., 2014b), and marsupials, as well as primates, deviated from those scaling rules (Dos Santos et al., unpublished results). However, the recent addition of carnivorans to the analysis suggests that a similar set of rules apply to carnivorans, marsupials, and eulipotyphlans (and possibly artiodactyls); a second set to glires and afrotherians; and a third set to primates (Fig. 14.4B).

The difference between primates and nonprimate mammals again lies in the higher neuronal densities in the primate rest of brain compared to nonprimate species (Fig. 14.4B). As in the cerebral cortex and cerebellum, neuronal densities in the rest of brain decrease rapidly in nonprimate species as the structure gains neurons, as a power function of numbers of rest of brain neurons with exponent −0.928, but less rapidly in primate species, with an exponent of −0.180 that does not, however, reach significance (Fig. 14.4B; Herculano-

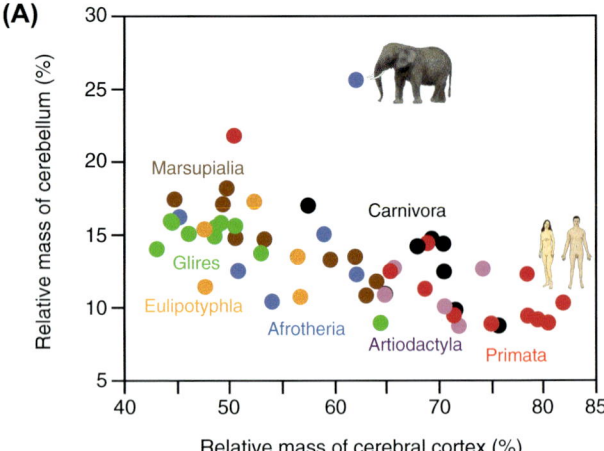

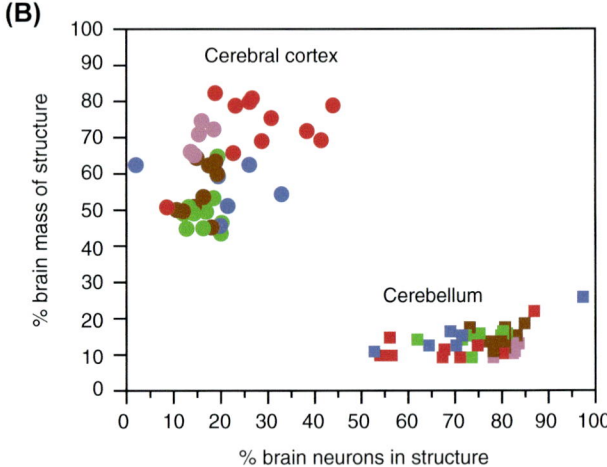

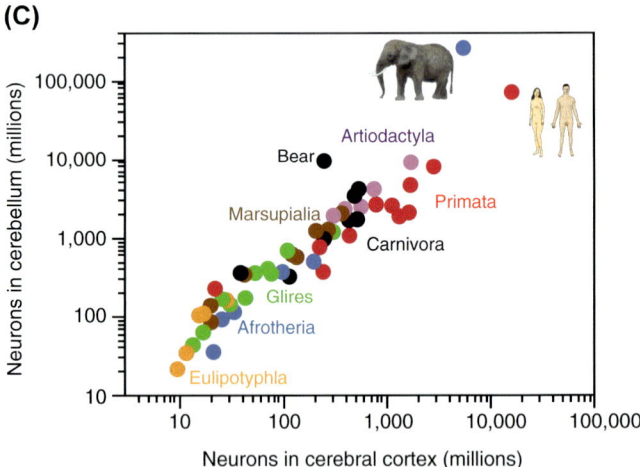

FIGURE 14.5 Faster increase in cerebral cortical mass over cerebellar mass does not reflect the proportional increase in numbers of neurons across these structures. (A) As the relative mass of the cerebral cortex (including the white matter and expressed as percentage of brain mass) increases in larger brains, the relative mass of the cerebellum decreases across mammalian species as a whole. (B) Although the cerebral cortex (*circles*) represents between 40% and 82% of brain mass in the species we have analyzed so far, it typically contains only between 15% and 25% of all brain neurons—and those animals that have relatively larger cortices do not have relatively more brain neurons in it. Likewise, although the cerebellum (*squares*) usually

Houzel et al., 2015b). Within nonprimates, neuronal densities in the rest of brain fall very fast across marsupials, eulipotyphlans, carnivorans, and artiodactyls as the structure gains neurons (with an exponent of −1.077), but less rapidly across glires and afrotherians (exponent, −0.621; Fig. 14.4B).

As seen by the exponents for the various structures, not only different neuronal scaling rules apply to primate and nonprimate clades, but also to the relationship between structure mass and number of neurons across structures. With much higher neuronal densities, the cerebellum always has more neurons than both a cerebral cortex of similar mass, and the larger cerebral cortex that forms the same brain with that cerebellum. One consequence of this difference is that the proportional mass of each structure in the brain does not reflect the proportional number of brain neurons that it contains.

In fact, examining numbers of neurons directly and their distribution across brain structures shows that the relative expansion of the cerebral cortical mass that is seen in larger brains, at the cost of a relatively smaller cerebellum (Fig. 14.5A) and rest of brain, does not reflect a relatively larger number of brain neurons in the cortex (Fig. 14.5B). In other words, the relatively larger cerebral cortex of some species does not hold relatively more brain neurons than in species with relatively smaller cortices. Take, for instance, two extremes: the human brain and the guinea pig brain. The human cerebral cortex represents 82% of the entire brain mass, and the guinea pig cerebral cortex represents a much smaller 53% of the brain—but both cortices contain a similar 19% of all brain neurons. In fact, across species, although the cerebral cortex represents between 50% and 80% of brain mass, it typically holds 15–25% of all brain neurons, while the cerebellum, which represents between 10% and 20% of brain mass, concentrates 70–85% of all brain neurons (Fig. 14.5B). Thus, in larger mammalian brains, the cerebral cortex expands in relative mass while retaining a fairly constant 15–25% of all brain neurons. This relative constancy is due to what we found to be a concerted addition of neurons to the cerebral cortex and cerebellum

represents only between 10% and 20% of brain mass, it holds typically between 70% and 80% of all brain neurons—but relative cerebellar mass and relative number of cerebellar neurons are not correlated. (C) With the sole exceptions of the African elephant and the brown bear, the number of neurons in the cerebellum and in the cerebral cortex increase proportionately, with a relatively steady average of four neurons in the cerebellum for every neuron in the cerebral cortex across species—that is, the cerebral cortex does not gain neurons disproportionately over the cerebellum, as would have been expected from its increase in relative mass. Original data on 5 species of eulipotyphlans, 6 species of afrotherians, 10 species of glires, 12 species of primates, and 4 species of artiodactyls are available in Herculano-Houzel et al. (2014b). *Data on carnivorans from Messeder et al. (unpublished results). Data on marsupials from Dos Santos et al. (unpublished results).*

(Fig. 14.5C). Across all mammalian species, primates included, the relationship between numbers of neurons in the cerebellum and the number of neurons in the cerebral cortex of a same species is linear across species, with a slope of roughly 4.0 (Fig. 14.5C). That is, for every neuron that is added to the cerebral cortex in evolution, about four neurons are added to the cerebellum (Herculano-Houzel, 2010; Herculano-Houzel et al., 2014b). To this coordinated, proportional addition of neurons to both structures follows a relatively faster expansion of the cerebral cortex over the cerebellum simply because of the difference in how average neuronal cell size (or neuronal density) increases faster in the cerebral cortex (Fig. 14.2A) than in the cerebellum (Fig. 14.3A) as neurons are added to these structures.

The ratio of neurons in cerebral cortex (and in the cerebellum) to the number of neurons in the rest of brain does, however, vary across mammalian clades. As shown in Fig. 14.6, eulipotyphlans, glires, small afrotherians, and American marsupials have a seemingly constant ratio of 2 neurons in the cerebral cortex to every neuron in the rest of brain, but australasian marsupials and artiodactyls have between 5 and 8 neurons in the cortex to every neuron in the rest of brain, carnivorans have between 5 and 10, and most primates have between 5 and 25 (Fig. 14.6A). These clade-specific ratios would not have been predicted from the expansion of the relative mass of the cerebral cortex as the rest of brain increases in mass (Fig. 14.6B).

14.4 The Many Ways of Putting a Brain in a Mammalian Body

The long tradition of considering body mass as an independent variable to which others, such as brain mass, could be normalized across widely diverse mammalian clades (eg, Jerison, 1973) disregarded the possibility of clade-specific relationships between brain mass and body mass. As shown in Fig. 14.7A, and for our data set alone, the scaling of brain mass to body mass differs across mammalian clades. For a similar body mass, primates have larger brains than nonprimates as a whole, and carnivorans show larger brains than rodents of similar body mass; for instance, both the prairie dog (a rodent) and the banded mongoose (a carnivoran) have body masses of about 1.5 kg, but the brain of the former weighs only 5.3 g, while the latter weighs 14.2 g (Messeder et al., unpublished results). Once again, there is order to the variation, and clade-specific power functions apply to each mammalian group (Fig. 14.7A). Thus, there is not a single way for nature to place a brain inside a mammalian body.

As seen earlier, there is also not a single way to build mammalian brains with neurons, and that diversity

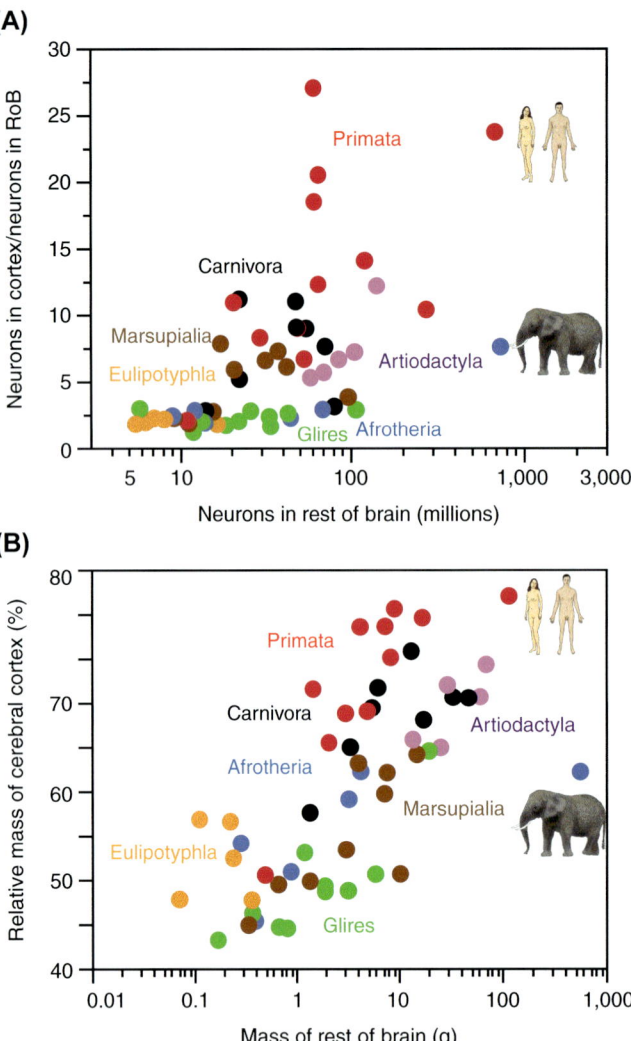

FIGURE 14.6   Increase in the relative mass of the cerebral cortex with increasing mass of the rest of brain does not reflect changes in the ratio between numbers of neurons in the two structures. (A) There is a clear distinction between Eulipotyphla, Glires, and Afrotheria on the one hand, and Primata, Carnivora, and Artiodactyla on the other in regard to the ratio between numbers of neurons in the cerebral cortex and rest of brain: while this ratio is of around two in the former, it is higher and more variable in the latter clades. (B) As the mass of the rest of brain (brain stem, diencephalon, and striatum) increases, the relative mass of the cerebral cortex increases across mammalian species as a whole—although not in some clades, such as Eulipotyphla. Original data on 5 species of eulipotyphlans, 6 species of afrotherians, 10 species of glires, 12 species of primates, and 4 species of artiodactyls are available in Herculano-Houzel et al. (2014b). *Data on carnivorans from Messeder et al. (unpublished results). Data on marsupials from Dos Santos et al. (unpublished results).*

appears in the relationship between the number of brain neurons and body mass across diverse mammalian species (Fig. 14.7B). Here, the main difference appears to be across primates (Fig. 14.7B, red) and nonprimate species (Fig. 14.7B, all other colors): for a similar body mass, primates have many more brain neurons than other mammalian species. However, the relationship between total number of brain neurons and body mass is skewed by the number of neurons in the cerebellum, typically 70–80% of all brain neurons. For example, it is due to the cerebellum that the elephant brain appears to hold many more neurons in the brain than expected for its body mass (Fig. 14.7B).

If only the cerebral cortex is considered, even more diversity is revealed across mammalian clades (Fig. 14.7C). Afrotherians (Fig. 14.7C, blue) have more cortical neurons than rodents (Fig. 14.7C, green) of similar body mass, and mid-sized carnivorans (Fig. 14.7C, black) have more cortical neurons than marsupials (Fig. 14.7C, brown) or glires of similar body mass. However, while larger mammals appear to have more cortical neurons within each clade, this correlation does not apply to carnivorans as a whole: larger felines do not have more cortical neurons than a large dog, and the brown bear actually has fewer cortical neurons than all these animals, and only as many as a house cat (Fig. 14.7C, black;

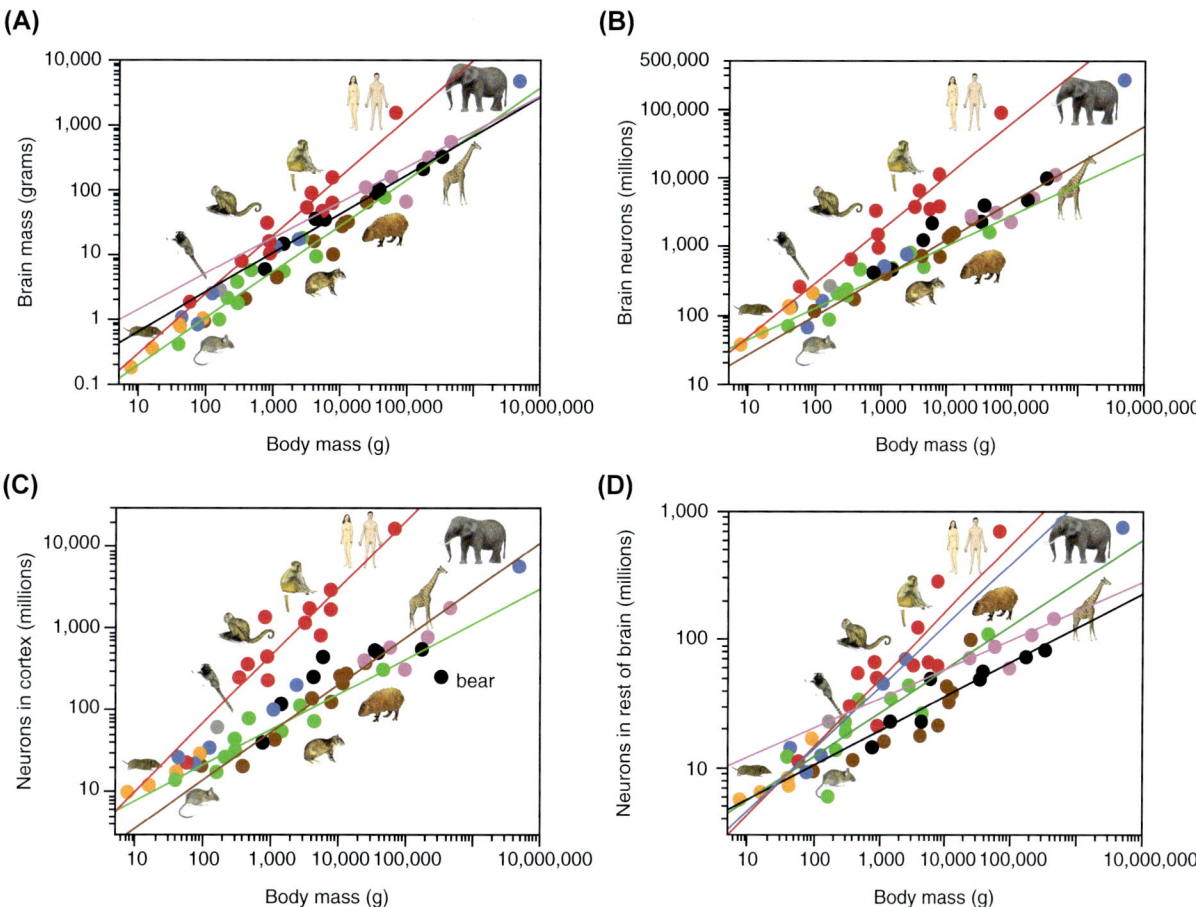

FIGURE 14.7 Clade-specific scaling of brain mass with body mass. (A) Brain mass scales faster with increasing body mass across primates (*red*; exponent, 0.903 ± 0.082, r^2 = 0.931, p < 0.0001) than across species of any other clade (plotted relationships: glires, *green*, exponent, 0.712 ± 0.071, r^2 = 0.927, p < 0.0001; carnivorans, *black*, exponent, 0.602 ± 0.041, r^2 = 0.973, p < 0.0001; artiodactyls, *pink*, exponent, 0.548 ± 0.038, r2 = 0.990, p = 0.0048). (B) The total number of neurons in the brain scales with increasing body mass much faster across primates (*red*, exponent: 0.777 ± 0.091, r^2 = 0.889, p < 0.0001) than across nonprimates as a whole (plotted relationships: glires, *green*, exponent 0.452 ± 0.071; marsupials, *brown*, exponent 0.554 ± 0.041, r2 = 0.964, p < 0.0001). (C) The number of neurons in the cerebral cortex scales much faster with increasing body mass across primates (*red*; exponent 0.825 ± 0.097, r^2 − 0.878, p < 0.0001) than across nonprimates (plotted relationships: glires, *green*, exponent 0.431 ± 0.056, r^2 = 0.881, p < 0.0001; marsupials, *brown*, exponent 0.582 ± 0.058, r^2 = 0.936, p < 0.0001). Notice that the carnivoran cerebral cortex does not gain neurons progressively as body mass increases; rather, lion, and hyena have similar numbers of cortical neurons as large dogs despite their larger body mass, and the brown bear has only as many cortical neurons as a house cat. (D) The number of neurons in the rest of brain, which contains structures that are directly connected to the body, varies with body mass in a clade-specific manner that indicates that the two variables are not causally related across mammals as a whole (exponents: primates, *red*, 0.525 ± 0.089, r^2 = 0.777, p = 0.0002; afrotherians, *blue*, exponent 0.480 ± 0.093, r^2 = 0.898, p = 0.0142; glires, *green*, exponent 0.338 ± 0.072, r^2 = 0.735, p = 0.0015; carnivorans, *black*, exponent 0.267 ± 0.0006, r^2 = 0.877, p = 0.0006; artiodactyls, *pink*, exponent 0.227 ± 0.027, r^2 = 0.973, p = 0.0136). Original data on 5 species of eulipotyphlans, 6 species of afrotherians, 10 species of glires, 12 species of primates, and 4 species of artiodactyls are available in Herculano-Houzel et al. (2014a,b). *Data on carnivorans from Messeder et al. (unpublished results). Data on marsupials from Dos Santos et al. (unpublished results).*

Messeder et al., unpublished results). Interestingly, this break away from the typical correlation between increasing body mass and increasing numbers of neurons only applies to the carnivoran cerebral cortex, not to the rest of brain (Fig. 14.7D, black). The inverted U-shape of the curve relating number of cortical neurons to body mass is indicative of a trade-off between the two, which we have suggested that is due to the high metabolic cost of cortical neurons when faced with energetic limitations, as is likely to be the case for large carnivorans (Messeder et al., unpublished results).

Importantly, the rest of brain contains those structures in the brain stem that are directly connected to the body, and therefore most directly responsible for operating it. Yet, it is for the rest of brain that scaling relationships to body mass are most clearly clade specific, with not only different exponents in the power relationship but also different constants. As a result, we find that for a given body mass, primates have the most neurons in the rest of brain (Fig. 14.7D, red), followed by rodents and artiodactyls (Fig. 14.7D, green and pink), then carnivorans (Fig. 14.7D, black), and marsupials

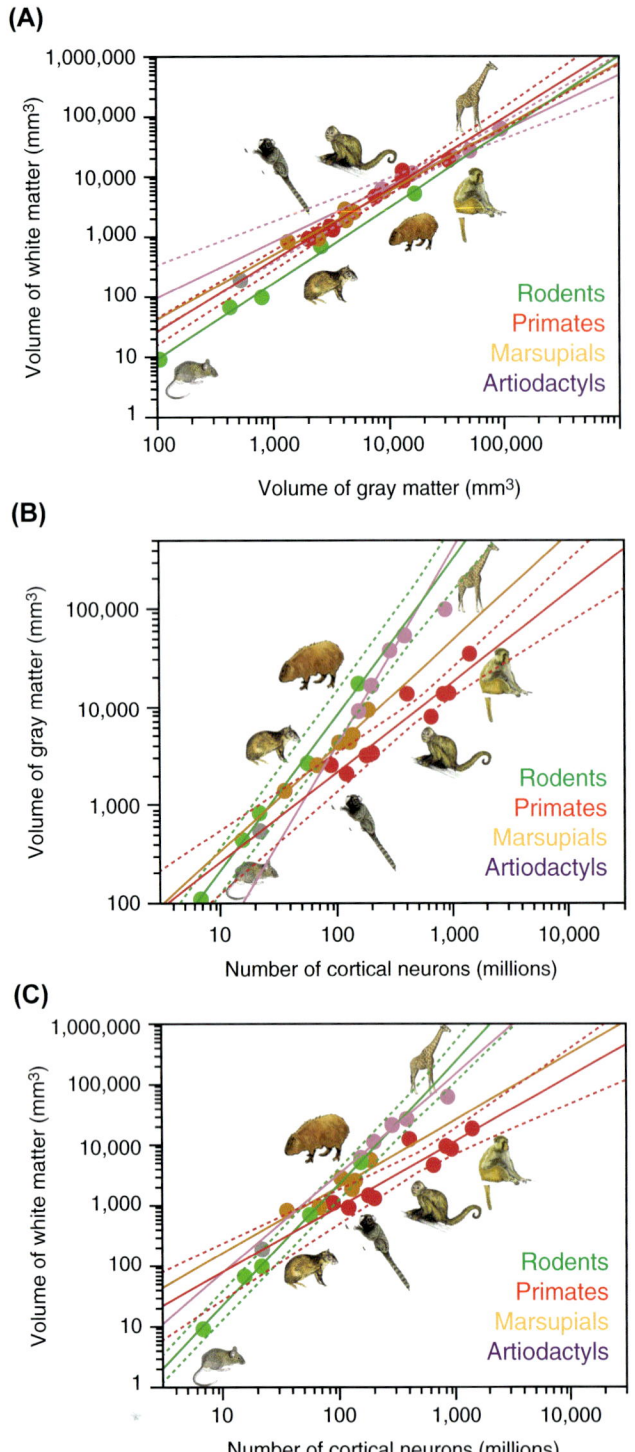

(A)

(B)

(C)

have the least neurons in the rest of brain (Fig. 14.7D, brown). The widely different numbers of neurons found in the rest of brain of different mammals of similar body mass indicate that the number of neurons in the brain is not dictated directly by the body. Further, the fact that all these animals are obviously capable of controlling their bodies well enough suggests that numbers of neurons in the rest of brain are not limiting, in the sense that there is not a certain number of neurons that is universally required to control a body of a given mass.

14.5 The Several Ways of Distributing the Cortical Volume Into Gray and White Matter

The cerebral cortical volume, or mass, analyzed previously consists of two components of different cellular makeup: the gray matter, containing neuronal cell bodies, and the white matter, with the axons leaving and entering the gray matter. The fact that the majority of those axons in the white matter belong to neuronal cell bodies in the gray matter (given that corticofugal projections outnumber corticopetal projections; Jones, 1985) justifies the analysis of the cortical mass as a whole, as estimates of average neuronal cell size of cortical neurons should include their axons. On the other hand, the separate analysis of the two components of the cerebral cortical tissue can inform on how the components that process information locally (within the gray matter) or across long distances (through the white matter) scale as cortices gain neurons.

It was once considered that gray and white matter volumes were universally related across all mammalian species alike (Zhang and Sejnowski, 2000), but we now know that is not the case. Similar volumes of gray matter are associated with larger volumes of white matter in primates, artiodactyls, and marsupials than in rodents (Fig. 14.8A; Ventura-Antunes et al., 2013; Herculano-Houzel et al., unpublished results). The difference in scaling is even more pronounced when numbers of neurons are taken into consideration, since, as shown

$r^2 = 0.964, p = 0.0005$; primates, 0.918 ± 0.083, $r^2 = 0.938, p < 0.0001$. (C) The volume of the white matter scales as a different function of the number of neurons in the cortex in each mammalian clade analyzed. Exponents: rodents, 2.009 ± 0.065, $r^2 = 0.997$, $p < 0.0001$; artiodactyls, 1.637 ± 0.225, $r^2 = 0.964$, $p = 0.0183$; marsupials, 1.101 ± 0.271, $r^2 = 0.805$, $p = 0.0005$; primates, 1.080 ± 0.120, $r^2 = 0.909$, $p < 0.0001$. *Data from Ventura-Antunes, L., Mota, B., Herculano-Houzel, S., 2013. Different scaling of white matter volume, cortical connectivity, and gyrification across rodent and primate brains. Front. Neuroanat. 7, 3 and Herculano-Houzel et al. (unpublished results). All values are for a single cortical hemisphere.*

FIGURE 14.8 Clade-specific scaling of cerebral cortical gray and white matter. (A) The volume of the white matter scales as different functions of the volume of the gray matter between rodents and other mammals. Exponents: rodents, 1.260 ± 0.062, $r^2 = 0.993$, $p = 0.0003$; primates, 1.184 ± 0.054, $r^2 = 0.983$, $p < 0.0001$; artiodactyls, 0.927 ± 0.017, $r^2 = 0.984$, $p = 0.0009$; marsupials, 1.066 ± 0.171, $r^2 = 0.907$, $p = 0.0034$. (B) The volume of the gray matter scales as a different function of the number of neurons in the cortex in each mammalian clade analyzed. Exponents: artiodactyls, 1.992 ± 0.180, $r^2 = 0.984$, $p = 0.0081$; rodents, 1.587 ± 0.064, $r^2 = 0.995$, $p = 0.0001$; marsupials, 1.076 ± 0.104,

previously, cortices of similar size are composed of much larger numbers of neurons in primates than in nonprimates. Indeed, the volumes of gray (Fig. 14.8B) and white matter (Fig. 14.8C) scale as very different functions of the number of cortical neurons across these four clades, and for similar numbers of neurons, primates have the smallest volume of white matter (Fig. 14.8C). The difference is very significant: the springbok (an artiodactyl) and the owl monkey (a primate) both have about 200 million neurons in a single cerebral cortical hemisphere, but the volume of the corresponding white matter in the owl monkey is only one-tenth of that in the springbok. There is not a single way to build gray and white matter, and obviously both the primate and artiodactyl ways work well enough that the animals can function successfully—but the primate-specific scaling rules result in cortices in which axons travel comparatively shorter distances for their numbers of neurons. The difference is associated not only with smaller neurons in primate gray matter compared to artiodactyl gray matter with similar numbers of neurons, but also a smaller fraction of cortical neurons connected through the white matter in primates (Ventura-Antunes et al., 2013; see also Chapter 3.02, "What Primate Brains Are Made of" in volume III of this series).

14.6 The Even More Numerous Ways of Distributing the Cortical Volume Into Surface Area and Thickness

As described earlier, we have found basically that two relationships apply between cortical mass and the number of cortical neurons across species: there is the scaling rule that applies to primates, and another scaling rule that applies to all other mammalian species examined so far, as a whole. As described in Section 14.5, the cortical volume is not divided uniformly across all species into gray and white matter volumes. However, clade-specific distinctions in the distribution of the cortical volume go even farther.

The volume of the cortical gray matter in any mammalian species is a result of the product of total cortical surface area and average gray matter thickness, which in principle are independent features of a cerebral cortex. It has been generally considered that cortical expansion is mostly due to two-dimensional expansion of the gray matter, with only minor changes in cortical thickness (Rakic, 1995; Zhang and Sejnowski, 2000). Indeed, average cortical thickness varies by less than one order of magnitude across mammalian species, while total cortical surface area spans over three orders of magnitude. However, the variation in cortical thickness is not only systematic

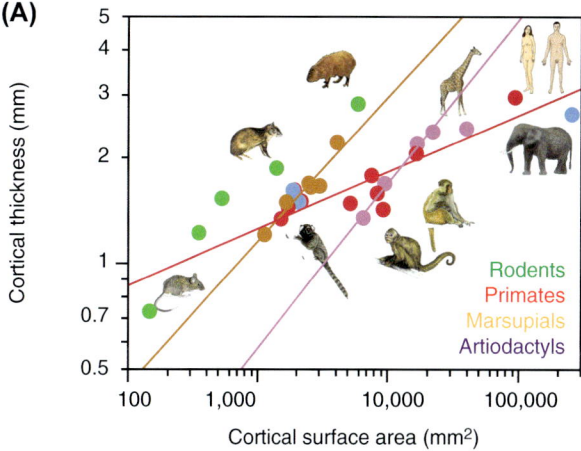

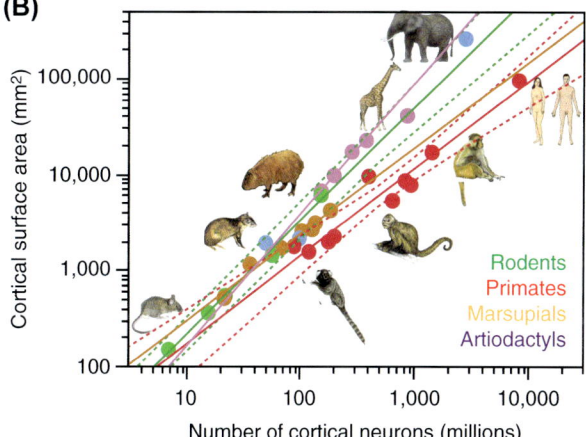

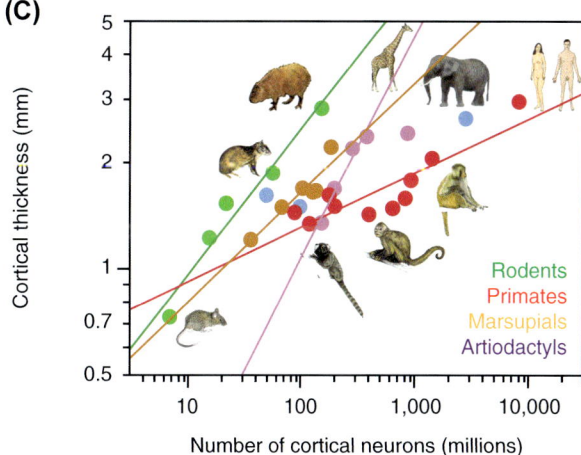

FIGURE 14.9   Clade-specific scaling of cortical thickness and surface area. (A) The average thickness of the cerebral cortical gray matter scales as different functions of the cortical surface area in each clade examined. Exponents: primates, 0.161 ± 0.032, $r^2 = 0.757$, $p = 0.0011$; artiodactyls, 0.466 ± 0.036, $r^2 = 0.988$, $p = 0.0059$; marsupials, 0.407 ± 0.063, $r^2 = 0.912$, $p = 0.0030$; rodents, not plotted since different functions apply to lissencephalic and gyrencephalic species. (B) The surface area of the cerebral cortex scales as a different function of the number of cortical neurons in each mammalian clade analyzed. Exponents: artiodactyls, 1.362 ± 0.094, $r^2 = 0.991$, $p = 0.0047$; rodents, 1.177 ± 0.049, $r^2 = 0.995$, $p = 0.0002$; marsupials, 0.897 ± 0.064, $r^2 = 0.975$, $p < 0.0001$; primates, 0.911 ± 0.083, $r^2 = 0.938$, $p < 0.0001$. (C) The average

but also clade specific. As shown in Fig. 14.9A, we have recently found that different mammalian clades show different scaling of average thickness as the cortex expands in surface area (Mota and Herculano-Houzel, 2015). This means that neurons are distributed in clade-specific combinations of cortical surface area and thickness, as shown in Fig. 14.9B and C. As the cerebral cortex gains neurons, it expands in surface area much faster in artiodactyls and in rodents than in marsupials of primates (Fig. 14.9B). At the same time, the cortex becomes more rapidly thicker in artiodactyls, followed by rodents, marsupials, and then primates—even though it is thinner in artiodactyls than in other nonprimate species (compare slopes in Fig. 14.9C). It is remarkable that even across nonprimates only, a given cortical volume, which consists of similar numbers of neurons across species, is distributed into clade-specific combinations of cortical surface area and thickness. For instance, a similar 100–200 million neurons in one cortical hemisphere are distributed into a larger but thinner cortex in artiodactyls than in rodents or marsupials (Fig. 14.9B and C). It seems, therefore, that even though a single relationship applies to building cortical volume as a function of increasing numbers of neurons that also become predictably larger across nonprimate species, there are multiple ways of distributing this volume sideways into different combinations of surface area and thickness.

One consequence of the clade-specific relationships between addition of neurons to the cortex, expansion of cortical surface area, and cortical thickening is that the number of neurons per unit area of cortical surface, N/A, is not constant across clades nor species, contrary to what used to be widely held (Rockel et al., 1980; Carlo and Stevens, 2013). For similar numbers of cortical neurons (Fig. 14.10A), cortical surface area (Fig. 14.10B), or cortical thickness (Fig. 14.10C), N/A is two to five times larger in primates than in other mammalian species. Moreover, while N/A does not vary systematically across primate or marsupial species, it decreases with increasing numbers of neurons, cortical surface area, and thickness in rodents and artiodactyls (Fig. 14.10).

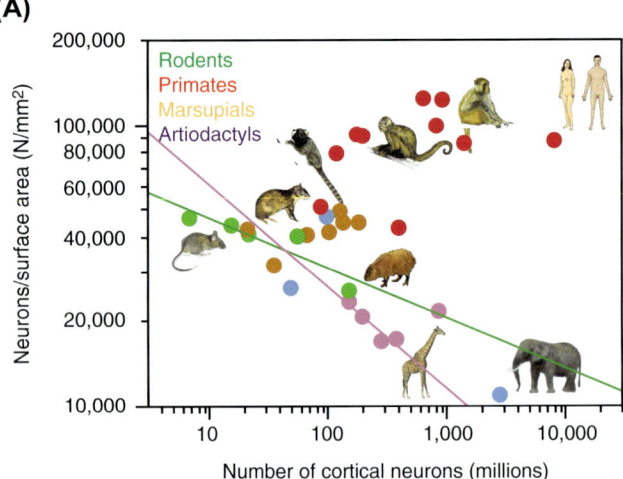

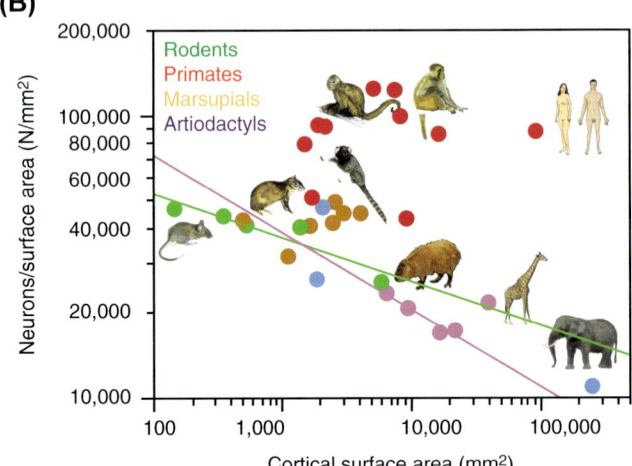

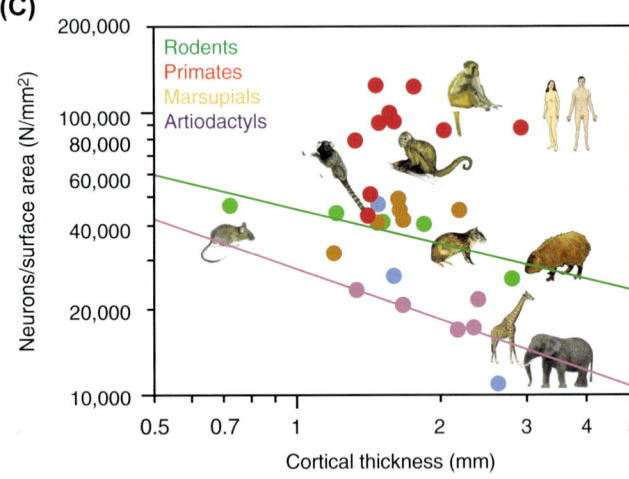

thickness of the cortical gray matter scales as a different function of the number of neurons in the cortex in each mammalian clade analyzed. Exponents: rodents, 0.410 ± 0.049, $r^2 = 0.958$, $p = 0.0037$; artiodactyls, 0.630 ± 0.089, $r^2 = 0.962$, $p = 0.0192$; marsupials, 0.309 ± 0.060, $r^2 = 0.869$, $p = 0.0067$; primates, 0.153 ± 0.029, $r^2 = 0.773$, $p = 0.0028$. *Data from Ventura-Antunes, L., Mota, B., Herculano-Houzel, S., 2013. Different scaling of white matter volume, cortical connectivity, and gyrification across rodent and primate brains. Front. Neuroanat. 7, 3 and Herculano-Houzel et al. (unpublished results). All values are for a single cortical hemisphere.*

FIGURE 14.10 Clade-specific scaling of neurons/surface area. (A) The average number of neurons per square millimeter of the cerebral cortical surface is much larger in primates than in marsupials, and larger in these than in rodent or artiodactyl cortices of similar numbers of neurons. The number of neurons per square millimeter does not vary significantly with increasing numbers of cortical neurons in primates ($r^2 = 0.127$, $p = 0.3117$) or in marsupials ($r^2 = 0.342$, $p = 0.1679$), but decreases significantly with increasing numbers of cortical neurons in rodents (exponent: -0.177 ± 0.049, $r^2 = 0.812$, $p = 0.0369$) and

14.7 What Does Not Change due to Biological Constraints

While numbers of neurons vary across mammalian structures of similar mass, we were surprised to find that the nonneuronal composition of brain structures (that is, the number of endothelial cells, microglial cells, astrocytes, and oligodendrocytes, defined as cells whose nuclei fail to express detectable levels of NeuN) is nearly invariant across brain structures of similar mass, regardless of the species (reviewed in Herculano-Houzel, 2014). Moreover, as shown in Fig. 14.11, the mass of all of the major brain structures (cerebral cortex, cerebellum, and the rest of brain) varies across all mammalian species analyzed as a single, shared power function of the number of nonneuronal cells in the structure. Although the number of nonneuronal cells is technically a ceiling for the number of glial cells in the tissue, the fact that the microvasculature of the brain represents only 1–6% of brain volume (Buchweitz and Weiss, 1986; Lauwers et al., 2008; Tsai et al., 2009) suggests that the vast majority of nonneuronal cells are in fact glial cells. The fact that this relationship is shared not only across species (including primates) but also across brain structures indicates that nonneuronal cells are added to brain tissue following nearly invariant, universal rules that have not been subject to change, that is, that have not evolved.

The power function that relates numbers of glial (nonneuronal) cells to brain structure mass across structures and species has an exponent of 1.020, close to linearity (Herculano-Houzel et al., 2014b), which translates mathematically into densities of nonneuronal cells in the various tissues that do not vary systematically across species or structures (Fig. 14.12A), in sharp contrast to the large and systematic variation of neuronal densities across species and structures (Fig. 14.12B). The lack of systematic variation in nonneuronal cell density

suggests that the average size of glial cells does not vary by much across brain structures and species, in contrast to the average size of neurons. Indeed, applying chi-square minimization to a simple mathematical model that used measured densities of neuronal and nonneuronal cells to obtain the most likely values for average neuronal and nonneuronal cell masses, we find that while the former (m_N) is highly variable across structures and species, the estimated average mass of nonneuronal cells (m_G) does not depart much from 3 to 4 pg across all structures and species examined (Mota and Herculano-Houzel, 2014; Fig. 14.13A; center panel). Estimated average neuronal cell mass varies between 5 and 144 pg in the cerebral cortex and rest of brain across all orders, a 29-fold variation that extends to over 500-fold when the cerebellum is included (Fig. 14.13A, center panel; Mota and Herculnao-Houzel, 2014). As predicted, the largest average neuronal cell masses in the cerebral cortex are found in the capybara and agouti, although these cortices are smaller than the macaque cortices investigated. A recent investigation of neuronal cell body and dendritic arbors confirms that layer III pyramidal neurons are indeed larger and more branched in rodents than in primates (Elston and Manger, 2014).

Interestingly, the fraction of brain tissue that is composed of neurons in each structure, f_N, varies little across structures, typically around 0.7; in other words, we estimate that around 70% of brain tissue is composed of neurons (Mota and Herculano-Houzel, 2014, Fig. 14.13A; top panel). There is, however, a narrow range of variation in f_N, which we find to depend not on the large variation in average neuronal cell mass m_N (Fig. 14.13B), but rather to vary closely with small variations in average nonneuronal cell mass m_G (Mota and Herculano-Houzel, 2014; Fig. 14.13C).

As a consequence of the large variation in average neuronal cell size, and thus in neuronal densities, in the absence of much systematic variation in glial cell densities, the glia/neuron ratio in mammalian brains is highly variable across brain structure and species (Fig. 14.14A)—but not as previously thought to be, as a simple function of brain mass (Hawkins and Olszewski, 1957; Marino, 2006). Rather, we find that the glia/neuron ratio varies uniformly across mammalian brain structures and species, including primates, as a function of neuronal density in the structure (Fig. 14.14B). In fact, the best predictor for the glia/neuron ratio in any brain structure is the estimated average neuronal cell mass (Fig. 14.15A).

The universality of the relationships between brain structure mass and number of glial cells, and between glia/neuron ratio and neuronal density in brain structures, indicates that the rules that govern how glial cells are added to brain tissue have not changed in

---

approaches significance in artiodactyls (exponent: -0.362 ± 0.094, $r^2 = 0.882$, $p = 0.0610$). (B) The number of neurons per square millimeter does not vary significantly with increasing cortical surface area in primates ($r^2 = 0.013$, $p = 0.7553$) or in marsupials ($r^2 = 0.202$, $p = 0.3113$), but decreases significantly with increasing numbers of cortical neurons in rodents (exponent: -0.155 ± 0.035, $r^2 = 0.865$, $p = 0.0220$) and in artiodactyls (exponent: -0.273 ± 0.050, $r^2 = 0.937$, $p = 0.0321$). (C) The number of neurons per square millimeter does not vary significantly with increasing cortical thickness in primates ($r^2 = 0.060$, $p = 0.4936$) or in marsupials ($r^2 = 0.555$, $p = 0.0894$), but decreases significantly with increasing numbers of cortical neurons in rodents (exponent: -0.408 ± 0.134, $r^2 = 0.754$, $p < 0.0001$) and in artiodactyls (exponent: -0.592 ± 0.074, $r^2 = 0.970$, $p = 0.0152$). *Data from Ventura-Antunes, L., Mota, B., Herculano-Houzel, S., 2013. Different scaling of white matter volume, cortical connectivity, and gyrification across rodent and primate brains. Front. Neuroanat. 7, 3 and Herculano-Houzel et al. (unpublished results). All values are for a single cortical hemisphere.*

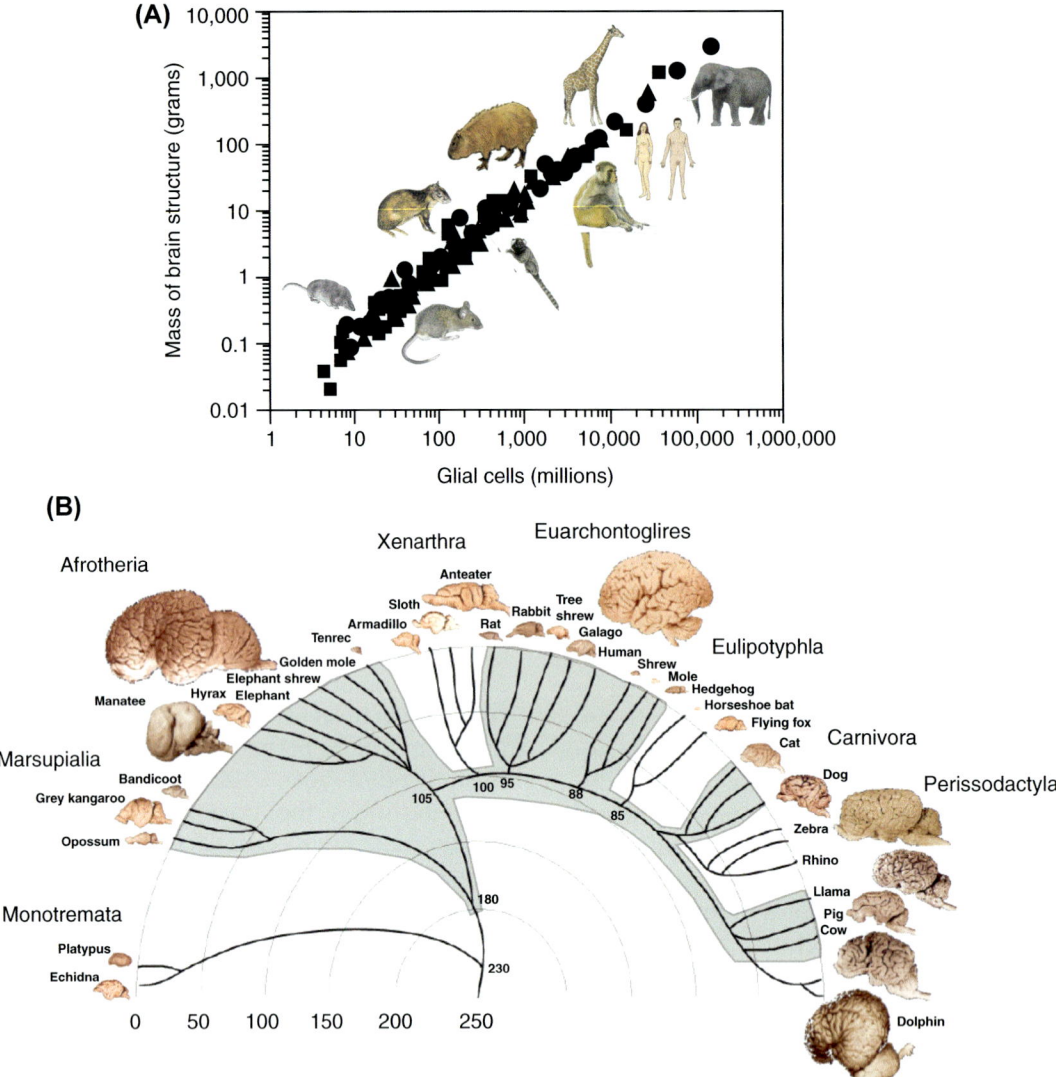

FIGURE 14.11   Glial scaling rules are shared among extant mammalian clades, and thus probably applied to early mammals. (A) The mass of major brain structures (cerebral cortex, *circles*; cerebellum, *squares*; rest of brain, *triangles*) varies as a power function of the number of nonneuronal cells in each structure that is shared across all mammalian species examined so far. (B) Mammalian evolutionary tree indicating the clades that share the same relationship between brain structure mass and number of glial cells, allowing the inference that the common ancestor also shared that relationship. Original data on 5 species of eulipotyphlans, 6 species of afrotherians, 10 species of glires, 12 species of primates, and 5 species of artiodactyls are available in Herculano-Houzel et al. (2014a,b). *Data on carnivorans from Messeder et al. (unpublished results). Data on marsupials from Dos Santos et al. (unpublished results).*

mammalian brain evolution (Herculano-Houzel, 2014). Absence of evolutionary change is indicative of either a biological constraint or a physical principle at play. In the case of glial cells, the little variation in their cell density is consistent with a mechanism whereby glial cells are added to developing brain tissue without significant systematic changes in average glial cell size (which is suggestive of a biological constraint to cell size variation), and in numbers that are physically constrained by the volume of the tissue, initially comprised almost exclusively of neurons. We have proposed that this happens as glial progenitors divide until

proliferation is inhibited by cell contact, as glial cells fill the tissue, whose initial volume is determined by the number of neurons and their average size (Mota and Herculano-Houzel, 2014). This proposal is supported by the finding that the number of glial cells in the tissue scales universally with the neuronal mass in the tissue, that is, with the product of number of neurons and estimated average neuronal cell mass (Fig. 14.15B; Mota and Herculano-Houzel, 2014). Based on these findings, we proposed further that there is a fundamental building block of brain tissue: the glial mass that accompanies a unit of neuronal mass (Fig. 14.15C). The fairly

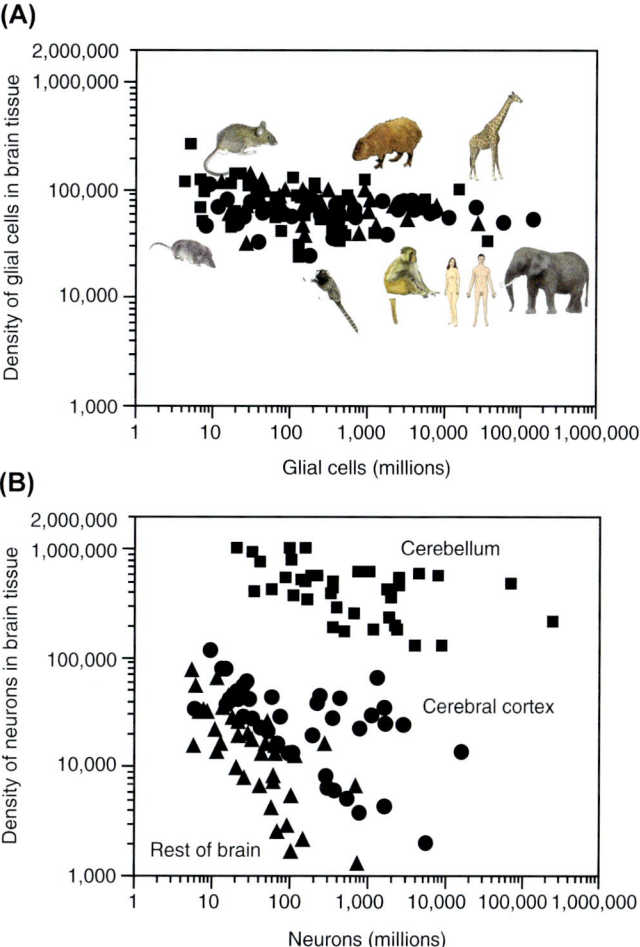

FIGURE 14.12 Neuronal densities are highly variable across brain structures and species, while glial cell densities are not. (A) Glial (non-neuronal) cell density in major brain structures (cerebral cortex, *circles*; cerebellum, *squares*; rest of brain, *triangles*) does not vary significantly or systematically with the number of nonneuronal cells in each structure across the mammalian species examined so far. (B) In the same scale as (A), neuronal density is seen to vary significantly and systematically with the number of neurons in each major brain structure across mammalian clades (though different rules apply for primates; see text). Original data on 5 species of eulipotyphlans, 6 species of afrotherians, 10 species of glires, 12 species of primates, and 5 species of artiodactyls are available in Herculano-Houzel et al. (2014b). *Data on carnivorans from Messeder et al. (unpublished results). Data on marsupials from Dos Santos et al. (unpublished results).*

stable composition of brain tissue, with c.70% neuronal mass and 30% glial mass, is a direct consequence of the constrained scaling of average glial cell mass and their addition to the tissue in numbers that depend on the total neuronal mass of the parenchyma (Mota and Herculano-Houzel, 2014).

It remains to be investigated whether the average size of particular glial cell types is indeed mostly invariant across unrelated species. However, given the role of astrocytes and oligodendrocytes in providing metabolic support to neurons and synapses (Pellerin and Magistretti, 1994; Lee et al., 2012), it is indeed likely that these cells are not as free to vary in size as neurons are (Mota and Herculano-Houzel, 2014). Importantly, whereas there seems to be biological evolutionary constraints that limit the variation of average glial cell size, no evolutionary constraint is required to account for the

universality of the dependency of the glia/neuron ratio on neuronal density (that is, on the average size of neurons in the tissue); a simple mechanism of contact inhibition that limits how many glial cells fill a given volume of tissue may suffice.

14.8 What Does Not Change due to Physical Properties

While all small mammalian cortices are smooth, all large mammalian cortices share one characteristic that jumps to the eye: although the cortex remains a single sheet, it is no longer smooth, but rather folded into sulci and gyri (Fig. 14.16). Because of this obvious distinction between small and large cortices, cortical expansion has long been associated with increased gyrification

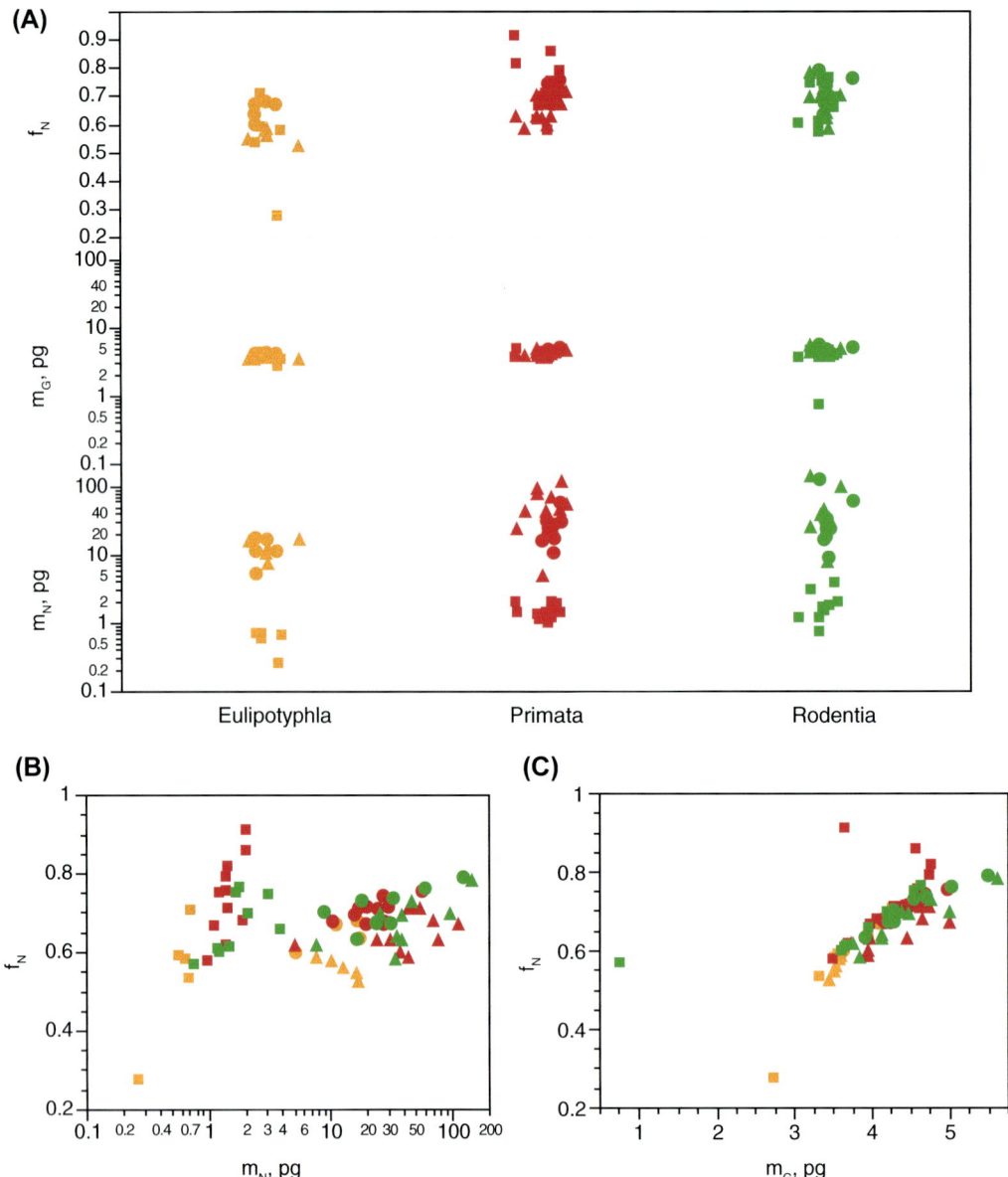

FIGURE 14.13 Estimated average mass of individual neuronal cells is highly variable across brain structures and species, but average mass of glial cells is not. (A) Estimated average cell mass of individual neuronal cells (m_N, in picograms) and of individual glial cells (m_G, also in picograms) are plotted for each of three clades, along with the neuronal fraction of the tissue (f_N), that is, the fraction of tissue mass that consists of neurons. Plots show m_N and m_G on a similar scale to evidence how the former is much more variable than the latter. Notice that f_N is typically between 0.6 and 0.7, that is, neurons represent 60–70% of the mass of different brain structures in different species. (B) Small variations in the neuronal fraction of brain tissue mass (f_N) do not correlate systematically with variations in the average neuronal cell mass (m_N). (C) Small variations in the neuronal fraction of brain tissue mass (f_N) do correlate systematically with variations in the average glial cell mass (m_G) across all structures and species analyzed. Cerebral cortex, *circles*; cerebellum, *squares*; rest of brain, *triangles*. *Data plotted from Mota, B., Herculano-Houzel, S., 2014. All brains are made of this: a fundamental building block of brain matter with matching neuronal and glial masses. Front. Neuroanat. 8, 127.*

(Jerison, 1973; Welker, 1990). It was intuitively considered that increases in the number of neurons in the cerebral cortex would lead directly to expanded cortical surfaces and thus to folding (Rakic, 1995), through mechanisms that were unclear but possibly related to the expansion of progenitor cell populations in the developing cortex (Reillo et al., 2010; Lui et al., 2011).

Against those expectations, we recently showed that the degree of folding of the cortical surface is not a simple function of increasing numbers of neurons across mammalian species, as animals with similar numbers of neurons in the cerebral cortex can have cortices that are much more or less folded, and animals with similar degrees of folding can have widely different numbers of

(A)

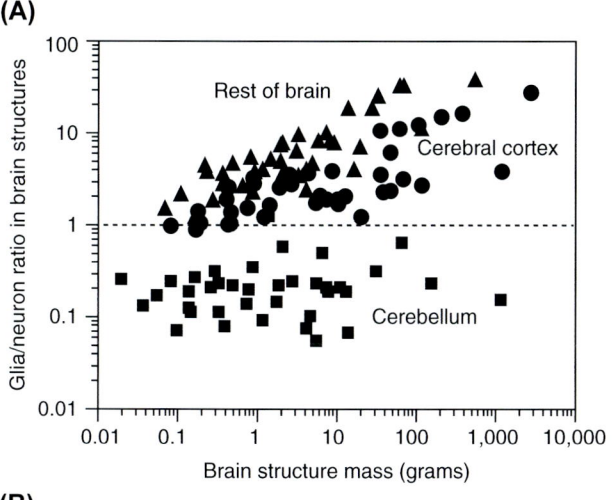

(B)

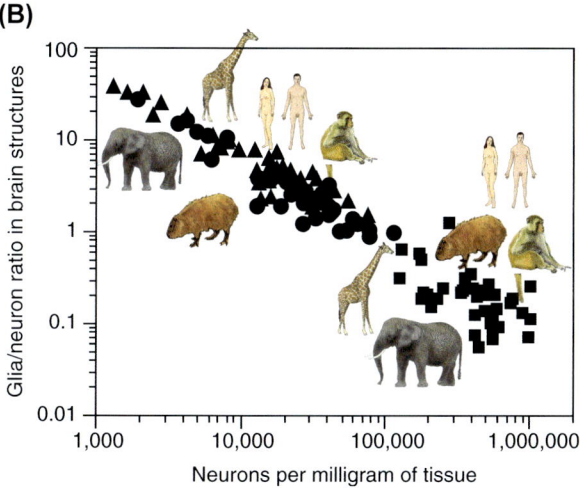

(A)

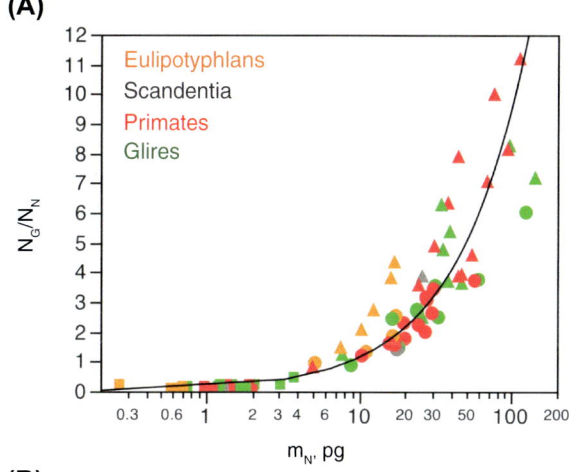

(B)

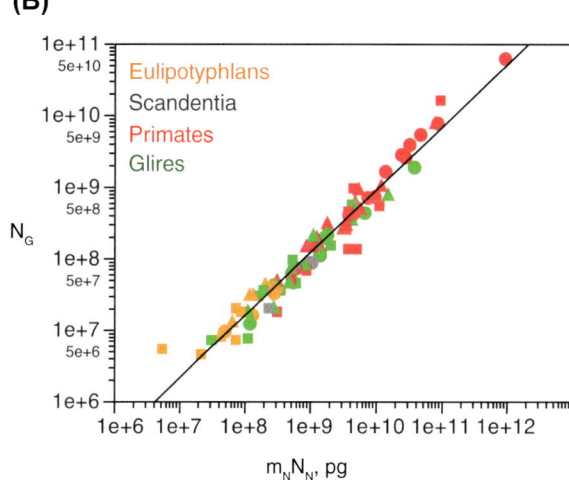

(C)

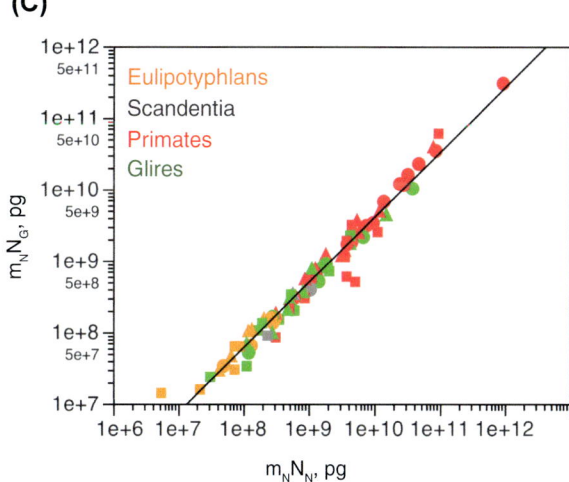

FIGURE 14.14 Glia/neuron ratio varies universally not with structure mass, but with neuronal density. (A) Glia/neuron (non-neuronal/neuronal) cell ratio in major brain structures (cerebral cortex, *circles*; cerebellum, *squares*; rest of brain, *triangles*) does not vary significantly or systematically with the mass of each major brain structure across the mammalian species examined so far. (B) However, glia/neuron ratio does vary universally across brain structures and species, increasing with decreasing neuronal density (which indicates larger average neuronal cell size). Original data on 5 species of eulipotyphlans, 6 species of afrotherians, 10 species of glires, 12 species of primates, and 5 species of artiodactyls are available in Herculano-Houzel et al. (2014a,b). *Data on carnivorans from Messeder et al. (unpublished results). Data on marsupials from Dos Santos et al. (unpublished results).*

cortical neurons (Fig. 14.17A; Mota and Herculano-Houzel, 2015). Instead, over 99% of the variation in the degree of folding of the cortex is accounted for by the product of total cortical surface area and average cortical thickness, which varied universally as a power function of the exposed surface area of the cortex across all species examined so far, and across lissencephalic and gyrencephalic species alike (Fig. 14.17B; Mota and Herculano-Houzel, 2015). This is the function predicted for the configuration of minimal effective free energy of

FIGURE 14.15 There is a universal relationship between neuronal mass and glial mass in brain structures across all species. (A) The glia/neuron ratio (N_G/N_N) (that is, the ratio between numbers of glial and neuronal cells) varies as a single power function of the estimated average cell mass of individual neuronal cells (m_N, in picograms) of exponent 0.922 ± 0.035 ($r^2 = 0.898$, $p < 0.0001$). (B) The number of glial cells in the tissue, N_G, varies across structure and species as a single power function of the neuronal mass in the tissue ($m_N N_N$), which is the product of average neuronal cell mass (m_N) and number of neurons in

Marsupialia

Rodentia

Primata

Carnivora

Artiodactyla

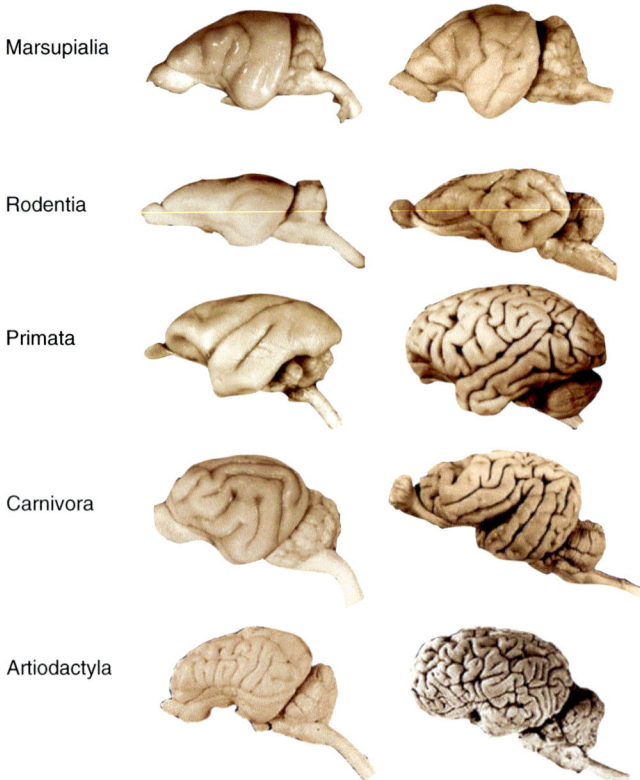

FIGURE 14.16 Clade-specific folding pattern of the cerebral cortex. Brains were scaled to appear with similar sizes for comparison and are thus not shown to scale. Species shown, from left to right: marsupials, *Macropus eugenii* and *Macropus fuliginosus*; rodents, *Cavia porcellus* and *Hydrochoerus hydrochoeris*; primates, *Aotus trivirgatus* and *Pan troglodytes*; carnivorans, *Felis catus* and *Ursus maritimus*; artiodactyls, *Pecari tajacu* and *Llama glama*. *Images from the Mammalian Brain Collection (www.brainmuseum.org).*

the cerebral cortex, and accordingly, we have proposed that cortical folding ensues as, during development, the expanding cortex, subject to uneven tension and pressure from various sources, folds and settles into the conformation that is most energetically favorable and thus most stable at each point in time, given its current thickness (Mota and Herculano-Houzel, 2015).

Cortical folding is thus a physical process that ensues universally across mammalian species. As such, we can now infer that the mechanism of folding itself did not evolve (given that it is a physical property). However, the degree to which a given cortex folds depends on the biological properties of each particular cortex, in particular the relationship between total surface area and cortical thickness—and, as shown earlier (Fig. 14.9A), this relationship is particular to each

mammalian order. Thus, what evolves is not folding per se, but how a cortex expands: how the addition of neurons to the cortex leads to different combinations of cortical surface and cortical thickness, even though the relationship between cortical volume (the product of surface area and thickness) and number of neurons is shared across most mammalian clades (see Fig. 14.2A).

Interestingly, there is a second variable in cortical folding that differs across clades: the exact placement of the folds, that is, the spatial pattern of folding. This pattern is characteristic enough of each clade that it allows the ready identification of the clade to which a gyrencephalic species belongs: carnivorans have fortune cookie-shaped cortices with concentric gyri; primates have a distinctive lateral sulcus and gyri at right angles to each other; artiodactyls have no major sulcus, but rather many gyri that run in parallel to each other and orthogonally to the pyriform cortex; rodents have folds concentrated in the occipital cortex, whereas marsupial cortices have a relatively smooth occipital cortex, with folds concentrated anteriorly (Fig. 14.16; Welker, 1990). We suggest that the clade-specific spatial pattern of cortical folding is related to the placement of first folds in early embryonic development of the cortex, which

the tissue (N_N). Exponent: 0.877 ± 0.022 ($r^2 = 0.952$, $p < 0.0001$). (C) Total glial mass in each structure ($m_G N_G$) varies as a universal power function of total neuronal mass in each structure ($m_N N_N$) of exponent 0.911 ± 0.019 ($r^2 = 0.968$, $p < 0.0001$) across all structures and species. Cerebral cortex, *circles*; cerebellum, *squares*; rest of brain, *triangles*. *Data plotted from Mota, B., Herculano-Houzel, S., 2014. All brains are made of this: a fundamental building block of brain matter with matching neuronal and glial masses. Front. Neuroanat. 8, 127.*

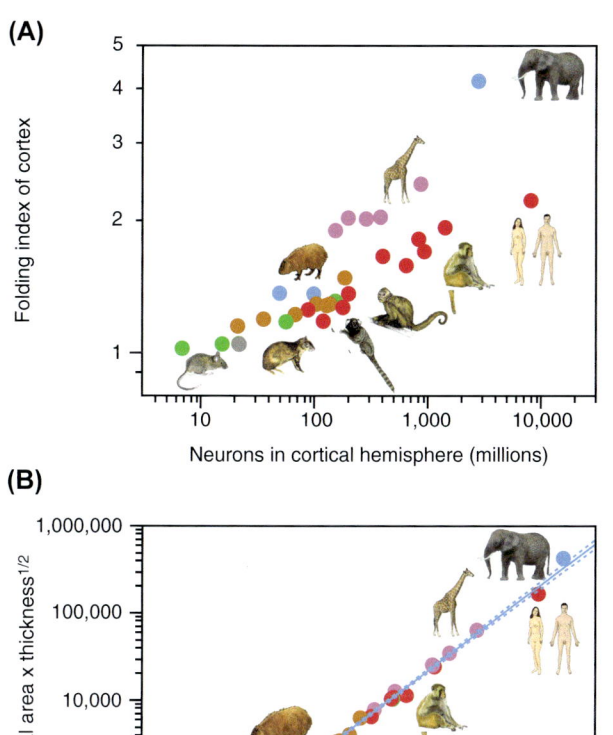

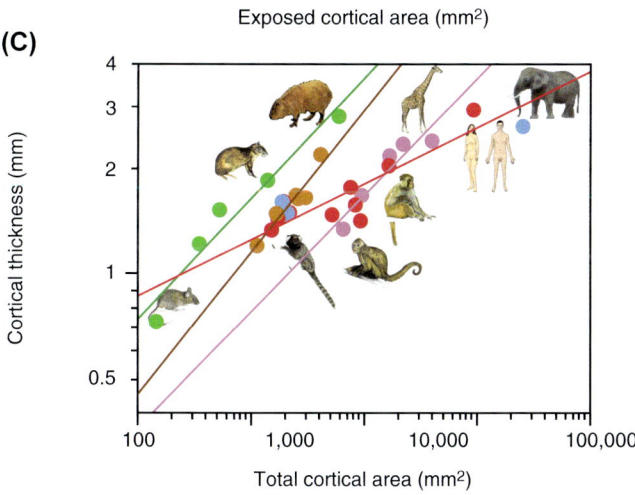

serve as seeds for the placement of first-order folds in cortical expansion (Tallinen et al., 2016).

14.9 Inferences About Early Mammalian Brains and Mechanisms of Brain Evolution

As reviewed earlier, it turns out that some features in mammalian brain organization and its cellular composition are indeed flexible across species, and clade specific, while others are universal, applying to all clades examined so far. At this point, there is reasonable consensus as to what are the main phylogenetic relationships across mammalian clades, although several issues persist, including the timing of mammalian origin and clade diversification (Murphy et al., 2004; Bininda-Emonds et al., 2007; Meredith et al., 2011). Mapping clade-specific scaling rules onto the mammalian phylogenetic tree, considering that the first mammals were very small (Rowe et al., 2011), and applying the principle of parsimony—that the most likely scenario in evolution is the simplest one, with the least changes—it is possible to use those differences and similarities across extant mammals to inform about what early mammalian brains were made of and looked like, and how they evolved over time, giving rise to the current diversity.

The simplest case is that of nonneuronal (glial) scaling rules: the relationship between brain structure mass and number of nonneuronal cells in the structure, as reviewed earlier, is shared not only across all mammalian species examined so far (and even birds; Olkowicz et al., 2016) but also across brain structures. The most complex explanation for the shared relationship across extant mammalian species would be that the precise same relationship between brain structure mass and number of glial cells appeared de novo in each brain structure, in each clade. The simplest scenario, on the other hand, is that the shared scaling rule that applies to modern mammalian species already applied to the ancestral mammal to all clades examined in our data set, which lived around 218 million years ago (Meredith et al., 2011), and has remained conserved since then. This is the scenario illustrated in the evolutionary tree depicted in Fig. 14.11B, one that we expect to extend further to Xenarthra, Chiroptera, and Perissodactyla as well, clades that we have not been able to examine yet.

FIGURE 14.17 Universal scaling of the degree of cortical folding depending on cortical surface area and thickness, not number of cortical neurons. (A) The folding index of the cortex of different species (the ratio between total surface area and exposed surface area of the cerebral cortex) does not vary universally with the number of neurons in the cortex; for a similar number of cortical neurons, artiodactyl cortices are much more folded than primate cortices, and many more neurons fit in primate cortices of similar folding as artiodactyl cortices. (B) The product of total cortical surface area and the square root of average cortical thickness scales universally with exposed cortical surface area across all brains examined. (C) Cortical thickness varies as different functions of cortical surface area across mammalian clades; the degree of folding of each cortex depends on the combination of thickness and surface area that hold the neurons that compose that cortex. However, for a given combination of cortical thickness and surface area, the degree of folding is predicted according to the relationship in (B). *Figures adapted from Mota, B., Herculano-Houzel, S., 2015. Cortical folding scales universally with surface area and thickness, not number of neurons. Science 349, 74–77.*

Interestingly, in the case that the modern glial scaling rules already applied 218 million years ago, we can infer how many glial cells composed the brain of the earliest mammalian species by simply applying the mathematical equation that relates brain (or brain structure) size to the number of glial cells in it that apply to modern mammals. Further, we can infer from the near-linearity of that equation, which stems from the lack of a systematic change in nonneuronal cell density across extant mammalian brain structures (Fig. 14.12A), that the average cell size of glial cells has varied little since the first mammals appeared (Fig. 14.13; Mota and Herculano-Houzel, 2014).

The next simplest scenario is that of the evolution of the neuronal composition of the cerebral cortex. As shown in Fig. 14.2, the same relationships between the mass of the cerebral cortex, neuronal density, and the number of neurons in the structure apply to all extant nonprimate mammalian species examined, whose phylogenetic relationship is shown in the tree in Fig. 14.2C. Given that the same neuronal scaling rules apply to the cerebral cortex of six widely diverse mammalian clades, only not to primates, the most parsimonious interpretation is that the shared rule already applied to the common ancestor to these six clades, around 180 million years ago, and has been conserved since then in all clades—except primates (Herculano-Houzel et al., 2014b). Thus, even though no fossil tissue of this common ancestor will most likely ever be found to be preserved enough for analysis, it can be inferred that the cortex of early mammalia forms such as *Hadrocordium wui* (Luo et al., 2001), with an estimated mass of only 0.020 g (Herculano-Houzel et al., 2014b), was composed of only 3.6 million neurons. Accordingly, we have proposed that as early mammals gained neurons in their cerebral cortex, these neurons also became larger on average, at a rate that was tied to the increase in numbers of cortical neurons, leading to the predictably lower neuronal cell densities shown in Fig. 14.2B. Primates later appeared with changes in the neuronal scaling rules that allowed numbers of neurons to increase without the same rapid increase in structure mass that is observed in other mammals (Herculano-Houzel et al., 2014b). Importantly, it is not the case that neurons became any smaller in primates as these animals diverged; rather, it appears that neuronal size simply no longer increased as rapidly in the cerebral cortex of the new animals as numbers of neurons increased. Recent evidence that dendritic size and branching increases faster across rodents than primates (Elston and Manger, 2014) supports this hypothesis.

A similar logic can be applied to the cerebellum (Fig. 14.3), in which case the neuronal scaling rules shared across modern afrotherians, glires, carnivorans, and artiodactyls can be inferred to have also applied to the eutherian ancestor (Fig. 14.3C), rather than to have appeared independently four times (Herculano-Houzel et al., 2014b). Resolving whether these rules also applied to the ancestor to all therians, from which marsupials then diverged, will require examining the cerebellum of monotremes. Again, we have proposed that as early mammals gained neurons in their cerebellum, these neurons also became larger on average, at a rate that was tied to the increase in numbers of cerebellar neurons, leading to the predictably lower neuronal cell densities shown in Fig. 14.3B. Primates, and also eulipotyphlans, later appeared with independent changes in the neuronal scaling rules that allowed increases in numbers of cerebellar neurons to become uncoupled from increasing average neuronal size (Herculano-Houzel et al., 2014b). Marsupials, on the other hand, still exhibit decreasing neuronal densities indicative of increasing average neuronal cell size as their cerebellum gains neurons (Dos Santos et al., unpublished results). However, these animals systematically display larger neuronal densities in the cerebellum compared to other nonprimate mammalian species, which suggests that when marsupials appeared, average neuronal cell size in the cerebellum became smaller—but, from that new "starting point," continued to scale predictably with increasing numbers of cerebellar neurons just as in the ancestral cerebellum.

While we have previously suggested that the neuronal scaling rules for the rest of brain were shared across afrotherians, glires, eulipotyphlans, and artiodactyls (Herculano-Houzel et al., 2015b), the recent addition of marsupials and carnivorans to the analysis suggests a different interpretation: that marsupials, eulipotyphlans, carnivorans, and artiodactyls share neuronal scaling rules for the rest of brain, and afrotherians and glires share a different set of rules (Fig. 14.4). Because of the conformity of the early branching marsupials to the scaling rules that apply to later-branching groups, it is thus possible that the neuronal scaling rules shared by these modern groups applied in early mammals (Fig. 14.4C).

Cortical expansion is a trend in most mammalian clades, with the exception of eulipotyphlans and chiropterans (Safi et al., 2005)—although it must be kept in mind that cortical expansion is not mandatory in mammalian evolution: very small animals with small brains and cortices are found in many mammalian clades, even if not all. Because ancestral mammals were small (Rowe et al., 2011), presumably with a very small cortex, cortical expansion has probably occurred independently in all mammalian lineages. We can infer that the first mammals had a ratio of twice as many neurons in the cerebral cortex as in the rest of brain (Fig. 14.6A), and four times as many neurons in the cerebellum as in the cerebral cortex (Fig. 14.5C). Australasian marsupials, carnivorans, primates, and artiodactyls seem to have diverged away from this ancestral distribution of neurons in the brain with increased ratios between numbers of neurons in the cerebral cortex and the rest of brain, while maintaining the same four neurons in the cerebellum for every neuron in the cerebral cortex. In other words, the expansion of the numbers of neurons in the cerebral cortex has been linked to a proportional expansion of numbers of neurons in the cerebellum in mammalian evolution (Herculano-Houzel, 2010; Herculano-Houzel et al., 2014b). In each clade, the volume of the cerebral cortex then expands over the volume of the cerebellum (and rest of brain) depending on the neuronal scaling rules that tie average neuronal cell size to numbers of neurons in each structure.

From the small size of the brain of early mammalian fossils (Rowe et al., 2011), we can also infer that early mammals were lissencephalic, and thus had small cortical surfaces that expanded with the square of their thickness, as in all modern lissencephalic mammals examined (Mota and Herculano-Houzel, 2015). We propose that gyrencephaly appeared in each mammalian clade not because of the evolution of new mechanisms that lead to folding, but because in each clade, the cerebral cortex starts expanding faster in surface area than with the square of average cortical thickness (the only condition that maintains lissencephaly; Mota and Herculano-Houzel, 2015). Clade-specific scaling of the distribution of the cortical volume into surface area and thickness in the face of a shared relationship between cortical volume and number of neurons can be reconciled in our model in which all morphological aspects of the brain covered here are accounted for by variations in only three parameters: (1) the number of symmetrical divisions of early progenitors (which expand the proliferating subventricular zone, and thus contribute directly to expanding the cortical surface), (2) the number of asymmetrical divisions that progenitors undergo (which expand the cortex radially and thus contribute directly to increasing the cortical thickness), and (3) the average size (volume or mass) of the adult neurons (which contributes to both cortical surface area and thickness; Mota and Herculano-Houzel, unpublished observations). According to our proposition, while non-primate mammalian cortical volumes scale similarly as they gain neurons, this volume is then distributed sideways, and radially, in different fashions across mammalian clades. The recent surge in interest in identifying genetic and cellular mechanisms that lead to cortical expansion in different mammals (eg, Martínez-Cerdeño et al., 2012; Stahl et al., 2013; Wong et al., 2015) should soon allow the investigation of how mechanisms of cortical expansion compare across species.

14.10 What Difference Does It Make?

If one considers that larger bodies of similar morphology have unchanging degrees of freedom of movement and the same relative disposition of sensory surfaces, then operating them should in principle not require larger brains with more neurons. Still, it is reasonable to expect growing sensory surfaces and numbers of muscle fibers to require more neurons to monitor or operate them—or perhaps larger sensory surfaces and more muscle fibers allow the survival of more neurons in development, thus adjusting neuronal populations to the size of their targets (Watson et al., 2012). However, it appears that the increase in number of motor neurons that accompanies larger bodies is very modest, described by a power function of exponent 0.184 across marsupial species (Watson et al., 2012) or an even smaller 0.127 in primates (Sherwood, 2005). With such small exponents, a 10-fold larger body, with presumably a 10-fold larger muscle mass to be innervated, would be expected to have at best 1.5-fold as many motor neurons. In contrast, we find that 10-fold larger bodies come with 2.5- to up to 6-fold more brain neurons (that is, the number of brain neurons scales with body mass raised to an exponent of 0.4—0.8, depending on the clade). Even in the spinal cord as a whole, directly connected to the body, we found that numbers of neurons scale with body mass raised to an exponent of only 0.360 across primate species (Burish et al., 2010); that is, a 10-fold larger body comes with only 2.3 times as many neurons. Thus, even if a larger body requires a larger number of neurons to operate it, the increase in number of brain neurons with increasing body mass cannot be accounted for by the simple demand for a larger number of neurons imposed by larger body mass.

The faster expansion of the total number of brain neurons compared to the expansion of the number of neurons that deal directly with bodily functions thus suggests that the majority of neurons added to larger brains are not strictly involved in operating the body. As a consequence, with increased numbers of brain neurons must come an increased capacity to process information beyond what would be strictly necessary to operate the body. This scenario is well illustrated by the scaling of brain over spinal cord: as the primate spinal cord gains neurons, the number of neurons in the rest of brain, which also contains structures directly associated with bodily organs, scales proportionately (that is, linearly), but the number of neurons in the cerebral cortex and cerebellum scales much faster, with the number of spinal cord neurons raised to 1.6—2.1 (Burish et al., 2010).

Interestingly, while all clades gain brain neurons faster than could be accounted for by the increase in body mass (that is, numbers of brain neurons scale with body mass raised to exponents significantly larger than 0.184; Watson et al., 2012), only primates and artiodactyls exhibit a significantly faster expansion in the number of neurons in the cerebral cortex over the number of neurons in the rest of brain. In particular, we have estimated that the number of neurons in the primate motor cortex (area M1) scales roughly with the number of neurons in the spinal cord squared. As a result, while there are fewer than two neurons in the motor cortex for every spinal cord neuron in small primates, the ratio reaches to 20:1 in human and chimpanzee (Herculano-Houzel et al., 2016). We have proposed that the increasing numerical preponderance of cortical motor neurons over the rest of brain and spinal neurons with increasing body mass in primates results in corticalization of motor control, with increasing dependence on the motor cortex for the control of fine movements in humans and chimpanzees (Herculano-Houzel et al., 2016). Similarly, we found that the numbers of neurons in visual and auditory cortices scale faster than the numbers of neurons in the respective thalamic and collicular nuclei (Collins et al., 2013; Wong et al., 2013). It seems therefore that, at least in primates, cortical expansion leads to increased ratios of cortical to subcortical neurons that could contribute to adding complexity and flexibility to information processing in larger brains.

But what drove cortical expansion? Why is there a trend not only for absolute enlargement of the cerebral cortex (ie, more neurons), but also its relative expansion over other brain structures, which occurs as more cortical neurons also become larger neurons? Larger average neuronal cell size used to be an expected consequence of adding more neurons because of the ensuing increase in cortical mass that comes with more neurons; with longer distances to cover, the average neuron would have to at least have longer axons of probably larger caliber as well. However, the fact that primate cortices gain neurons with no significant decrease in neuronal density argues against such a mandatory increase in average neuronal cell size. Additionally, the absolute mass of the cerebral cortex (or brain), not relative size, is the best correlate of cognitive abilities across species so far (Deaner et al., 2007; MacLean et al., 2014). Thus, it is in principle more cortical neurons, and not larger neurons, that provide a cognitive advantage.

A recent observation led to the suggestion that the addition of neurons to the cortex accompanied by larger average neuronal size (and therefore lower neuronal densities) has one advantage that would lead to brains of both increasing numbers of cortical neurons and relatively larger cortices: it is associated with decreasing daily sleep time (Herculano-Houzel, 2015). Across a variety of mammalian species, it appears that the best correlate of daily sleep time so far is the ratio between neuronal density and surface area, which should determine the rate at which sleep-inducing metabolites produced during waking accumulate in the parenchyma (Fig. 14.18A). The smaller this ratio, the more slowly metabolites should accumulate, and thus the longer the animal should be able to remain awake; daily sleep time,

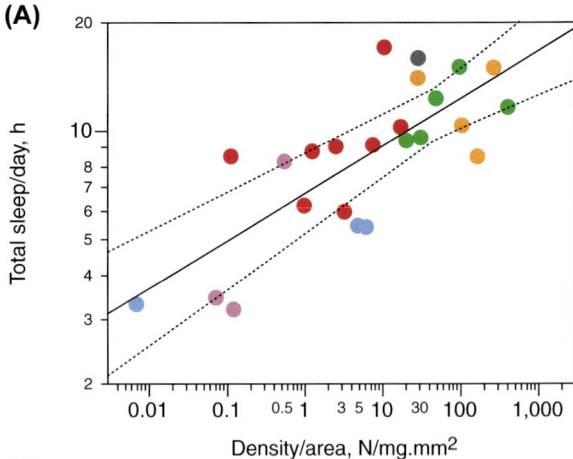

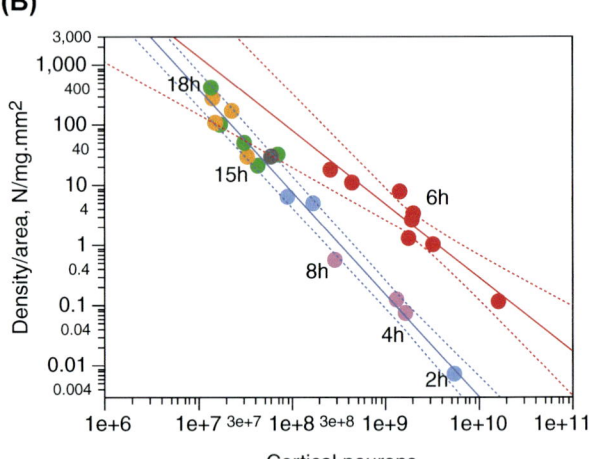

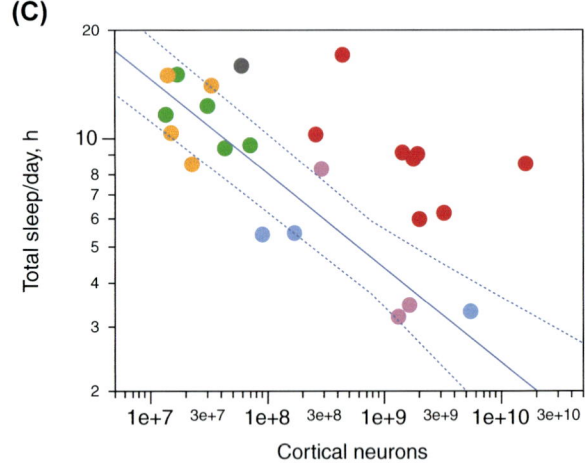

FIGURE 14.18   Increasing numbers of cortical neurons are associated with a decrease in total daily sleep across nonprimate species through a decrease in density per surface area (D/A) (neuronal density/mm^2). (A) Total daily sleep requirement scales as a power function of neuronal density/mm^2 with an exponent of 0.133 ± 0.023 across all 24 mammalian species examined ($r^2 = 0.601$, $p < 0.0001$). (B) Neuronal D/A decreases steeply with increasing number of cortical neurons as a power function of exponent -1.694 ± 0.080 across nonprimates ($r^2 = 0.971$, $p < 0.0001$) and less steeply across primates (exponent: -1.233 ± 0.144, $r^2 = 0.924$, $p < 0.0001$). Numbers indicate the number of daily sleep hours associated with different values of D/A. (C) Total daily sleep requirement decreases as a power function

therefore, should decrease with decreasing ratios of neuronal density per surface area (D/A). Importantly, across nonprimate species, the addition of neurons of increasing average cell size (that is, decreasing neuronal densities) leads to a steep decrease in D/A, and simultaneously to a highly significant decrease in daily sleep time (Fig. 14.18B and C; Herculano-Houzel, 2015). Across primate species, on the other hand, more neurons are not significantly larger neurons, and thus D/A does not fall steeply enough with increasing numbers of cortical neurons to lead to a significant drop in daily sleep time (see Chapter 2.06, The Evolution of Mammalian Sleep in this volume for a review).

Because early mammals can be inferred to have had small cerebral cortices made of small numbers of small neurons (that is, with very high neuronal densities), as reviewed earlier, they presumably also had very high D/A ratios, and therefore probably spent most hours of the day asleep. In that scenario, increased numbers of neurons that came with larger average neuronal cell size would have led to lower D/A ratios, and thus a slightly decreased total daily sleep time—with the accompanying advantage that more time would become available for the new species to feed, and thus afford both a larger body and a larger number of cortical neurons. Through the increase in waking hours and thus energy intake, adding larger neurons to the cerebral cortex would thus become self-reinforcing, possibly driving the tendency for brains and bodies to become larger in mammalian evolution.

14.11 Conclusions

The comparative analysis of the cellular composition of the brains of extant mammalian shows that there is not a single way to put brains together. Yet, there is remarkable consistency across enough clades that inferences can be drawn about how early mammalian brains were built. Such consistency implies that there are not many degrees of freedom to brain evolution. Indeed, the current range of diversity in cerebral cortical morphology may be accounted for by clade-specific changes in just three parameters that define the lateral expansion of the cortex, its radial expansion, and the average neuronal cell volume; an additional parameter then accounts for the relative distribution of neurons into the cerebral cortex and rest of brain, and another for the cerebellum. It will be interesting to see these predictions based solely on comparative neuroanatomy become complemented by direct investigations on the cellular developmental mechanisms that generate brain diversity.

of the number of cortical neurons across nonprimates (exponent: -0.266 ± 0.034, $r^2 = 0.809$, $p < 0.0001$), but not across primates ($p = 0.2597$). *Red*, primates; orange, eulipotyphlans; *green*, glires; *blue*, afrotherians; *pink*, artiodactyls; *gray*, scandentia.

References

Armstrong, E., 1990. Brains, bodies and metabolism. Brain Behav. Evol. 36, 166—176.

Azevedo, F.A.C., Carvalho, L.R.B., Grinberg, L.T., Farfel, J.M., Ferretti, R.E.L., Leite, R.E.P., Jacob Filho, W., Lent, R., Herculano-Houzel, S., 2009. Equal numbers of neuronal and nonneuronal cells make the human brain an isometrically scaled-up primate brain. J. Comp. Neurol. 513, 532—541.

Bininda-Emonds, O.R.P., Cardillo, M., Jones, K.E., MacPhee, R.D.E., Beck, R.M.D., Grenyer, R., Price, S.A., Vos, R.A., Gittleman, J.L., Purvis, A., 2007. The delayed rise of present-day mammals. Nature 446, 507—512.

Blanga-Khanfi, S., Miranda, H., Penn, O., Pupko, T., DeBry, R.W., Huchon, D., 2009. Rodent phylogeny revised: analysis of six nuclear genes from all major rodent clades. BMC Evol. Biol. 9, 71.

Buchweitz, E., Weiss, H.R., 1986. Alterations in perfused capillary morphometry in awake vs anesthetized brain. Brain Res. 377, 105—111.

Burish, M.J., Peebles, J.K., Baldwin, M.K., Tavares, L., Kaas, J.H., Herculano-Houzel, S., 2010. Cellular scaling rules for primate spinal cords. Brain Behav. Evol. 76, 45—59.

Campbell, B., Ryzen, M., 1953. The nuclear anatomy of the diencephalon of *Sorex cinereus*. J. Comp. Neurol. 99, 1—22.

Carlo, C.N., Stevens, C.F., 2013. Structural uniformity of neocortex, revisited. Proc. Natl. Acad. Sci. U.S.A. 110, 1488—1493.

Clark, D.A., Mitra, P.P., Wang, S.S., 2001. Scalable architecture in mammalian brains. Nature 411, 189—193.

Collins, C.E., Leitch, D.B., Wong, P., Kaas, J.H., Herculano-Houzel, S., 2013. Faster scaling of visual neurons in cortical areas relative to subcortical structures in non-human primate brains. Brain Struct. Funct. 218, 805—816.

Deaner, R.O., Isler, K., Burkart, J., van Schaik, C., 2007. Overall brain size, and not encephalization quotient, best predicts cognitive ability across non-human primates. Brain Behav. Evol. 70, 115—124.

Douady, C.J., Chatelier, C.I., Madsen, O., de Jong, W.W., Catzeflis, F., Springer, M.S., et al., 2002. Molecular phylogenetic evidence confirming the Eulipotyphla concept and in support of hegdehogs as the sister group to shrews. Mol. Phylogen. Evol. 25, 200—209.

Elston, G.N., Manger, P.R., 2014. Pyramidal cells in V1 of African rodents are bigger, more branched and more spiny than those in primates. Front. Neuroanat. 8, 4.

Finlay, B.L., Darlington, R.B., 1995. Linked regularities in the development and evolution of mammalian brains. Science 268, 1578—1584.

Frahm, H.D., Stephan, H., Stephan, M., 1992. Comparison of brain structure volumes in Insectivora and Primates. I. Neocortex. J. Hirnforsch. 23, 375—389.

Friede, R., 1954. Der quantitative Anteil der Glia an der Cortexentwicklung. Acta Anat. 20, 290—296.

Haug, H., 1987. Brain sizes, surfaces, and neuronal sizes of the cortex cerebri: a stereological investigation of man and his variability and a comparison with some mammals (primates, whales, marsupials, insectivores, and one elephant). Am. J. Anat. 180, 126—142.

Hawkins, A., Olszewski, J., 1957. Glia/nerve cell index for cortex of the whale. Science 126, 76—77.

Herculano-Houzel, S., 2010. Coordinated scaling of cortical and cerebellar numbers of neurons. Front. Neurosci. 4, 12.

Herculano-Houzel, S., 2014. The glia/neuron ratio: how it varies uniformly across brain structures and species and what that means for brain physiology and evolution. Glia 62, 1377—1391.

Herculano-Houzel, S., 2015. Decreasing sleep requirement with increasing numbers of neurons as a driver for bigger brains and bodies in mammalian evolution. Proc. R. Soc. B 282, 20151853.

Herculano-Houzel, S., Lent, R., 2005. Isotropic fractionator: a simple, rapid method for the quantification of total cell and neuron numbers in the brain. J. Neurosci. 25, 2518—2521.

Herculano-Houzel, S., Ribeiro, P., Campos, L., Valotta da Silva, A., Torres, L.B., Catania, K.C., Kaas, J.H., 2011. Updated neuronal scaling rules for the brains of Glires (rodents/lagomorphs). Brain Behav. Evol. 78, 302–314.

Herculano-Houzel, S., Avelino-de-Souza, K., Neves, K., Porfírio, J., Messeder, D., Feijó, L.M., Maldonado, J., Manger, P.R., 2014. The elephant brain in numbers. Front. Neuroanat. 8, 46.

Herculano-Houzel, S., Manger, P.R., Kaas, J.H., 2014. Brain scaling in mammalian evolution as a consequence of concerted and mosaic changes in numbers of neurons and average neuronal cell size. Front. Neuroanat. 8, 77.

Herculano-Houzel, S., Von Bartheld, C.S., Miller, D.J., Kaas, J.H., 2015. How to count cells: the advantages and disadvantages of the isotropic fractionator compared with stereology. Cell Tiss. Res. 360, 29–42.

Herculano-Houzel, S., Catania, K., Manger, P.R., Kaas, J.H., 2015. Mammalian brains are made of these: a dataset of the numbers and densities of neuronal and nonneuronal cells in the brain of glires, primates, scandentia, eulipotyphlans, afrotherians and artiodactyls, and their relationship with body mass. Brain Behav. Evol. 86, 145–163.

Herculano-Houzel, S., Kaas, J.H., Oliveira-Souza, R., 2016. Corticalization of motor control in humans is a consequence of brain scaling in primate evolution. J. Comp. Neurol. 524, 448–455.

Hofman, M.A., 1985. Size and shape of the cerebral cortex in mammals. I. The cortical surface. Brain Behav. Evol. 27, 28–40.

Hofman, M.A., 1988. Size and shape of the cerebral cortex in mammals. II. The cortical volume. Brain Behav. Evol. 32, 17–26.

Jerison, H.J., 1973. Evolution of the Brain and Intelligence. Academic Press, New York.

Jones, E.G., 1985. The Thalamus. Plenum Press, New York.

Kazu, R.S., Maldonado, J., Mota, B., Manger, P.R., Herculano-Houzel, S., 2014. Cellular scaling rules for the brain of Artiodactyla include a highly folded cortex with few neurons. Front. Neuroanat. 8, 128.

Kuntner, M., May-Collado, L.J., Agnarsson, I., 2011. Phylogeny and conservation priorities of afrotherian mammals (Afrotheria, Mammalia). Zool. Scr. 40, 1–15.

Lauwers, F., Cassot, F., Lauwers-Cances, V., Puwanarajah, P., Duvernoy, H., 2008. Morphometry of the human cerebral cortex microcirculation: general characteristics and space-related profiles. Neuroimage 39, 936–948.

Lee, Y., Morrisson, B.M., Li, Y., Lengacher, S., Farah, M.H., Hoffman, P.N., Liu, Y., Tsingalia, A., Jin, L., Zhang, P.W., Pellerin, L., Magistretti, P.J., Rothstein, J.D., 2012. Oligodendroglia metabolically support axons and contribute to neurodegeneration. Nature 487, 443–448.

Lui, J.H., Hansen, D.V., Kriegstein, A.R., 2011. Development and evolution of the human neocortex. Cell 146, 18–36.

Luo, Z.-X., Crompton, A.W., Sun, A.L., 2001. A new mammaliaform form the early Jurassic and evolution of mammalian characteristics. Science 292, 1535–1540.

MacLean, E., Hare, B., Nunn, C.L., Addessi, E., Amici, F., Anderson, R.C., et al., 2014. The evolution of self-control. Proc. Natl. Acad. Sci. 111, E2140–E2148.

Marino, L., 2006. Absolute brain size: did we throw the baby out with the bathwater? Proc. Natl. Acad. Sci. U.S.A. 103, 13563–13564.

Martínez-Cerdeño, V., Cunningham, C.L., Camacho, J., Antczak, J.L., Prakash, A.N., Cziep, M.E., Walker, A.I., Noctor, S.C., 2012. Comparative analysis of the subventricular zone in rat, ferret and macaque: evidence for an outer subventricular zone in rodents. PLoS One 7, e30178.

May-Collado, L.J., Kilpatrick, C.W., Agnarsson, I., 2015. Mammals from "down under": a multi-gene species-level phylogeny of marsupial mammals (Mammalia, Metatheria). PeerJ 3, e805.

Meredith, R.W., Janecka, J.E., Gatesy, J., Ryder, O.A., Fisher, C.A., Teeling, E.C., Goodbla, A., Eizirik, E., Simao, T.L.L., Stadler, T., Rabosky, D.L., Honeycutt, R.L., Flynn, J.J., Ingram, C.M., Steiner, C., Williams, T.L., Robinson, T.J., Burk-Herrick, A., Westerman, M., Ayoub, N.A., Springer, M.S., Murphy, W.J., 2011. Impacts of the cretaceous terrestrial revolution and KPg extinction on mammal diversification. Science 334, 521–524.

Mota, B., Herculano-Houzel, S., 2014. All brains are made of this: a fundamental building block of brain matter with matching neuronal and glial masses. Front. Neuroanat. 8, 127.

Mota, B., Herculano-Houzel, S., 2015. Cortical folding scales universally with surface area and thickness, not number of neurons. Science 349, 74–77.

Murphy, W.J., Pevzner, P.A., O'Brien, P.J., 2004. Mammalian phylogenomics comes of age. Trends Genet. 20, 631–639.

Nissl, F., 1898. Die Neuronenlehre und ihre Anhänger. Münch Med. Wsch 31.

Nyakatura, K., Bininda-Omonds, O.R.P., 2012. Updating the evolutionary history of Carnivora (Mammalia): a new species-level supertree complete with divergence time estimates. BMC Biol. 10, 12.

Olkowicz, S., Kocourek, M., Lucan, R.K., Portes, M., Fitch, W.T., Herculano-Houzel, S., Nemec, P., 2016. Birds have primate-like numbers of neurons in the forebrain. Proc. Natl. Acad. Sci. U.S.A. 113, 7255–7260.

Pellerin, L., Magistretti, P.L., 1994. Glutamate uptake into astrocytes stimulates aerobic glycolysis: a mechanism coupling neuronal activity to glucose utilization. Proc. Natl. Acad. Sci. U.S.A. 91, 10625–10629.

Purvis, A., 1995. A composite estimate of primate phylogeny. Philos. Trans. R. Soc. Lond. 348, 405–421.

Rakic, P., 1995. A small step for the cell, a giant leap for mankind: a hypothesis of neocortical expansion during evolution. Trends Neurosci. 18, 383–388.

Reillo, I., de Juan Romero, C., Garcia-Cabezas, M.A., Borrell, V., 2010. A role for intermediate radial glia in the tangential expansion of the mammalian cerebral cortex. Cereb. Cortex 21, 1674–1694.

Riska, B., Atchley, W.R., 1985. Genetics of growth predict patterns of brain-size evolution. Science 229, 668–671.

Rockel, A.J., Hiorns, R.W., Powell, T.P., 1980. The basic uniformity in structure of the neocortex. Brain 103, 221–244.

Rowe, T.B., Macrini, T.E., Luo, Z.X., 2011. Fossil evidence on origin of the mammalian brain. Science 332, 955–957.

Safi, K., Seid, M.A., Dechmann, D.K.N., 2005. Bigger is not always better: when brains get smaller. Biol. Lett. 1, 283–286.

Sherwood, C., 2005. Comparative anatomy of the facial nucleus in mammals, with an analysis of neuron numbers in primates. Anat. Rec. 287, 1067–1079.

Shinohara, A., Campbell, K.L., Suzuki, H., 2003. Molecular phylogenetic relationships of moles, shrew moles, and desmans from the new and old worlds. Mol. Phylogen. Evol. 27, 247–258.

Stahl, E., Walcher, T., Romero, C.D.J., Pilz, G.A., Cappello, S., Irmler, M., Sanz-Aquela, J.M., Beckers, J., Blum, R., BOrrell, V., Götz, M., 2013. Trnp1 regulates expansion and folding of the mammalian cerebral cortex by control of radial glial fate. Cell 153, 535–549.

Talllinen, T., Chung, J.Y., Rousseau, F., Girard, N., Lefèvre, J., Mahadevan, L., 2016. On the growth and form of cortical convolutions. Nat. Phys. 12, 588–593.

Tower, D.B., 1954. Structural and functional organization of mammalian cerebral cortex: the correlation of neurone density with brain size. J. Comp. Neurol. 101, 19–51.

Tower, D.B., Elliott, K.A.C., 1952. Activity of acetylcholine system in cerebral cortex of various unanesthetized mammals. Am. J. Physiol. 168, 747–759.

Tsai, P.S., Kaufhold, J.P., Blnder, P., Friedman, B., Frew, P.J., Karten, H.J., Lyden, P.D., Kleinfeld, D., 2009. Correlations of neuronal and microvascular densities in murine cortex revealed by direct counting and colocalization of nuclei and vessels. J. Neurosci. 29, 14553–14570.

Ventura-Antunes, L., Mota, B., Herculano-Houzel, S., 2013. Different scaling of white matter volume, cortical connectivity, and gyrification across rodent and primate brains. Front. Neuroanat. 7, 3.

Von Bartheld, C.S., Bahney, J., Herculano-Houzel, S., 2016. The search for true numbers of neurons and glial cells in the human brain: a review of 150 years of cell counting. J. Comp. Neurol. (in press).

Watson, C., Provis, J., Herculano-Houzel, S., 2012. What determines motor neuron number? Slow scaling of facial motor neuron numbers with body mass in marsupials and primates. Anat. Rec. 295, 1683–1691.

Welker, W., 1990. Why does cerebral cortex fissure and fold? In: Cortex, C., Jones, E.G., Peters, A. (Eds.), A Review of Determinants of Gyri and Sulci. Plenum Press, New York.

White, J.G., Southgate, E., Thomson, J.N., Brenner, S., 1986. The structure of the nervous system of the nematode Caenorhabditis elegans. Philos. Trans. R. Soc. Lond. 1165, 1–340.

Williams, R.W., Herrup, K., 1988. The control of neuron number. Annu. Rev. Neurosci. 11, 423–453.

Witthöft, W., 1967. Absolute Anzahl und Verteilung der Zellen inm Hirn der Honigbiene. Z Morph Tiere 61, 160–184.

Wong, P., Peebles, J.K., Asplund, C.L., Collins, C.E., Herculano-Houzel, S., Kaas, J.H., 2013. Faster scaling of auditory neurons in cortical areas relative to subcortical structures in primate brains. Brain Behav. Evol. 81, 209–218.

Wong, F.K., Fei, J.-F., Mora-Bermúdez, F., Taverna, E., Haffner, C., Fu, J., Anastassiadis, K., Stewart, A.F., Huttner, W.B., 2015. Sustained Pax6 expression generates primate-like basal radial glia in developing mouse neocortex. PLoS Biol. 13, e1002217.

Zhang, K., Sejnowski, T.J., 2000. A universal scaling law between gray matter and white matter of cerebral cortex. Proc. Natl. Acad. Sci. U.S.A. 97, 5621–5626.

15

Consistencies and Variances in the Anatomical Organization of Aspects of the Mammalian Brain stem

P.R. Manger

University of the Witwatersrand, Johannesburg, South Africa

15.1 Introduction

The mammalian brain stem, as referred to herein, is composed of three major distinct brain regions, the midbrain, pons, and medulla oblongata. The terminology used to demarcate these three regions of the brain varies among authors and are often used interchangeably. For example, the pons and medulla oblongata are often grouped and referred to as the hindbrain or rhombencephalon, while the midbrain, pons, and medulla oblongata can be referred to as the mesencephalon, metencephalon, and myelencephalon, respectively. In some cases, mostly in textbooks of human anatomy, the diencephalon, or thalamus, is combined with the midbrain, pons, and medulla oblongata and referred to as the brain stem. To avoid terminological difficulties, herein, the term brain stem is used to refer to the combination of the midbrain, pons, and medulla oblongata.

The brain stem is generally a solid cylinder of neural tissue, both white and gray matter, found between the diencephalon and spinal cord and appears to play a central role in the linkage of the spinal cord (and thereby the body) and nonneural cranium (via the cranial nerves) with the regions of the brain involved in higher order processing (diencephalon and telencephalon). This cylinder of neural tissue normally does not contain any major flexures; however, in the Hottentot golden mole (*Amblysomus hottentotus*), which is a small subterranean Afrotherian mammal, there is a distinct mesencephalic flexure of the brain stem (Fig. 15.1) that is not seen in other mammals (Calvey et al., 2013), even in closely related Afrotherians (Calvey et al., 2013; Peiters et al., 2010; Maseko et al., 2013). It appears this flexure is the result of rostrocaudal foreshortening of the cranium

associated with a subterranean habitat, forcing the brain into an unusual shape in this species, with a wide telencephalon, but foreshortened rostrocaudal length of the brain. This has some similarities to the telescoping of the odontocete cetacean skull and brain (Miller, 1923; Rommel et al., 2002), although the odontocete cetacean brain maintains the developmentally prominent hypothalamic flexure as opposed to the mesencephalic flexure observed in the golden mole (Fig. 15.1B).

Despite this variance, all other mammals studied to date display a similar superficial appearance of the brain stem. On the dorsal surface (or posterior surface in humans and some other primates), the colliculi demarcate the roof of the midbrain, with the middle cerebellar peduncle forming the lateral borders of the anterior portion of the fourth ventricle demarcating the pons, and the stria medullaris and the posterior portion of the fourth ventricle delimiting the medulla oblongata. On the ventral (or anterior) surface, the cerebral peduncles demarcate the midbrain, the ventral pontine nucleus the pons, and the pyramids, olives, and decussation of the pyramidal tracts marking the medulla oblongata (Fig. 15.2). This superficial appearance is very consistent across mammals, and the extensive descriptions of the surfaces of the brain stem provided in the numerous manuals of human neuroanatomy can be readily tra2nsferred to all mammals studied to date and do not need to be described in further detail herein.

Within the brain stem are numerous nuclei and fiber pathways, the anatomical organization of which can be quite confusing. To simplify this and provide a more accessible description, herein the brain stem is considered to be composed of five distinct basic components: the cranial nerve nuclei and associated nerves, the ascending and

(A)

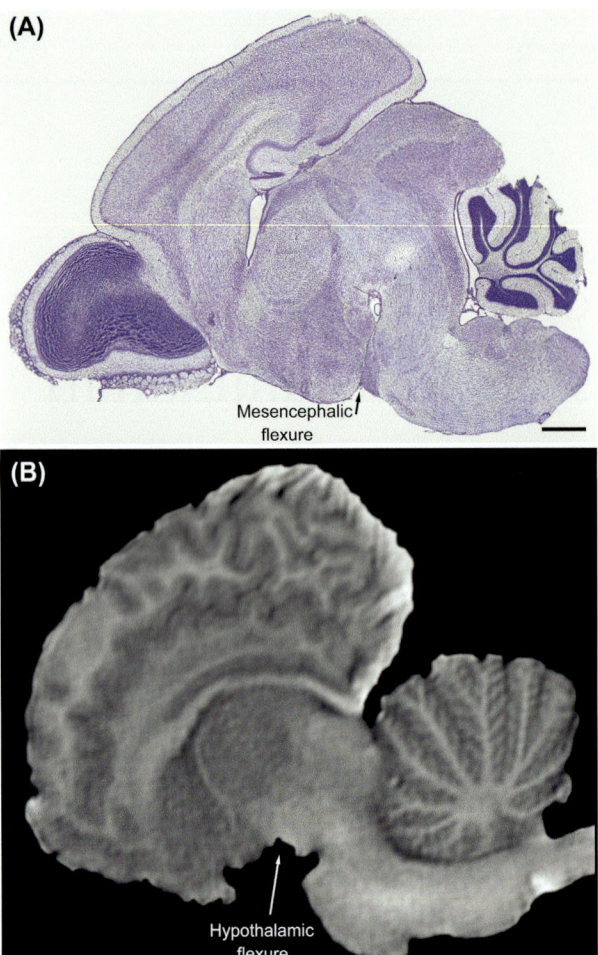

Mesencephalic flexure

(B)

Hypothalamic flexure

FIGURE 15.1   (A) Midsagittal Nissl-stained section through the brain of the Hottentot golden mole (*Amblysomus hottentotus*) showing the unusual mesencephalic flexure observed in this subterranean Afrotherian mammal (Calvey et al., 2013). Scale bar = 1 mm. (B) Midsagittal magnetic resonance image through the brain of the harbor porpoise (*Phocoena phocoena*) showing the hypothalamic flexure maintained in the adults of this species. Scale bar = 2 cm. Note that to date, the golden mole is the only species where a significant flexure of the brain stem has been observed in adult mammals; all other mammals have a mostly nonflexed brain stem.

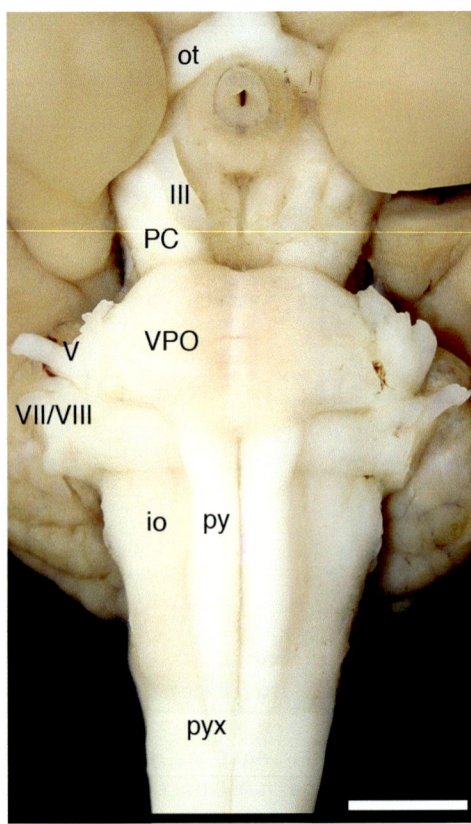

FIGURE 15.2   Ventral view of the brain stem of the African lion (*Panthera leo*) showing the typical subdivisions of the mammalian brain stem. Rostral is to the top of the image. Scale bar = 1 cm. III, oculomotor nerve; V, trigeminal nerve; VII/VIII, facial and vestibulocochlear nerves; *io*, inferior olive; *ot*, optic tract; *PC*, cerebral peduncle; *py*, pyramidal tract; *pyx*, decussation of the pyramidal tract; *VPO*, ventral pontine nucleus.

descending fiber pathways, the neuromodulatory nuclei (the cholinergic, catecholaminergic, and serotonergic nuclei that often have widespread projections throughout the central nervous system), the intrinsic nuclei (normally contained within one part of the brain stem), and the tegmental/reticular nuclei. In the following description, the structures making up each of these five components in the three different portions of the brain stem will be described separately, but it should be noted that strong interconnectivity between components, and between portions of the brain stem, are often present. In terms of the anatomical location of the various structures to be mentioned, templates of the mammalian brain stem, taken from atlases such as that of the rat brain (Paxinos et al.,

2009) or extensive analyses of a particular species (Maseko et al., 2013), are useful for the understanding of the anatomical location of the structures described within the brain stem and relative to each other. The current description is not an exhaustive analysis of the mammalian brain stem, as this would require a far more extensive review than provided here, but aims to provide a broad picture of this important neural structure and highlights how this structure stays consistent and varies across mammalian species that have been studied. Unfortunately, this means that an enormous amount of detail will be missing from this chapter, and the author apologizes in advance for this.

15.2   The Midbrain

15.2.1   The Cranial Nerve Nuclei of the Midbrain

Within the midbrain of most mammalian species, four distinct nuclei associated with specific cranial nerves are usually identified: the oculomotor nucleus, the Edinger-Westphal nucleus, the trochlear nucleus, and the trigeminal mesencephalic nucleus.

15.2.1.1 The Oculomotor Nucleus (III)

In all mammals studied to date, distinct oculomotor nuclei have been observed in the ventral medial aspect of the periaqueductal gray matter, located in a pararaphe position. The neurons forming this nucleus are generally large cholinergic motor neurons that innervate several of the extraocular muscles found within the orbit that are attached to the eye. The oculomotor nerve emerges from the ventral aspect of the nucleus, forming several distinct fasciculi that pass through the medial aspect of the midbrain tegmentum to exit at the medial aspect of the cerebral peduncle. The number of neurons forming this nucleus, and thus the size of the nucleus, varies across mammalian species, with those having large, highly mobile eyes, generally associated with higher visual acuity, having larger numbers of neurons and thus a large oculomotor nucleus. In some cases, such as the African Bathyergid mole rats, the oculomotor nucleus is substantially reduced in size (Da Silva et al., 2006; Bhagwandin et al., 2008), associated with the microphthalmic nature of these rodents, and the oculomotor and trochlear nuclei appear to merge (see as discussed in the following section).

15.2.1.2 The Preganglionic Component of the Edinger-Westphal Nucleus

Located on the rostral and medial aspect of the oculomotor nucleus is a cluster of small- to medium-sized motor neurons that form the preganglionic component of the Edinger-Westphal nucleus (Kozicz et al., 2011). This cluster of neurons is often contiguous across the midline (Kruger et al., 2010), but in some species can form two distinct pararaphe columns (Maseko et al., 2013). The neurons within this nucleus are often cholinergic, but in some species, these neurons are not immunopositive for antibodies to cholineacetyltransferase (Dell et al., 2010) and this lack of immunoreactivity appears to be related to phylogenetic history. The axons emerging from the neurons travel with the oculomotor nerve initially, but leave this nerve to terminate in the ciliary ganglion. The ciliary ganglion then provides parasympathetic innervation of the iris sphincter muscle and the ciliary muscles (Kozicz et al., 2011).

15.2.1.3 The Trochlear Nucleus (IV)

A second cluster of large cholinergic motor neurons found at the caudal level of the oculomotor nucleus but located just outside of the periventricular gray matter is the trochlear nucleus. This nucleus is generally smaller in size, with fewer neurons, than the oculomotor nucleus and serves to innervate a single extraocular muscle, the superior oblique muscle. The trochlear nerve is slender and difficult to identify in both histological sections and whole tissue, but has the distinction of being the only cranial nerve to exit the brain stem from the dorsal (or posterior) aspect of the brain stem, with the passage of this nerve marking the dorsal border between the midbrain and pons. In microphthalmic mammals, such as the Bathyergid mole rats, the trochlear nucleus is very small and difficult to distinguish histologically from the oculomotor nucleus (Da Silva et al., 2006; Bhagwandin et al., 2008).

15.2.1.4 The Midbrain Portion of the Trigeminal Mesencephalic Nucleus and Tract

In most mammals studied to date, a low-density cluster of large spherical ganglion cells are located on the lateral edge of the caudal aspect of the periventricular gray matter (Fig. 15.3A). These ganglion cells, the only true ganglion cells within the central nervous system, form the trigeminal mesencephalic nucleus and appear to be involved in the transmission of proprioceptive information from the muscles of the jaw (Weinberg, 1928). Surrounding and lateral to this nucleus is the trigeminal mesencephalic tract that arises from the trigeminal nerve within the pons. No recent major comparative studies of this nucleus across mammals are known to the author; however, a comparison across vertebrate species indicates a strongly conserved appearance, location, and number of cells bodies within this nucleus (Weinberg, 1928), although this nucleus does appear to possess more cells in the river hippopotamus than in other species examined by the author (Fig. 15.3B).

15.2.2 The Main Ascending and Descending Fiber Pathways of the Midbrain

Within the mammalian midbrain, several distinct ascending and descending fiber pathways are generally noted. Those most readily identified anatomically across species include: the descending medial longitudinal fasciculus; the descending corticobulbar and corticospinal tracts; and the ascending medial and lateral lemnisci. The medial longitudinal fasciculus is located in the dorsomedial aspect of the midbrain tegmentum, adjacent to the ventromedial border of the periventricular gray matter. The medial longitudinal fasciculus usually becomes evident at the level of the oculomotor nucleus and is seen to continue caudally to the pons. The fibers forming the medial longitudinal fasciculus carry information about the directions the eyes should move by facilitating vestibulo-ocular and optokinetic reflexes as well as for movement of the head (Gernandt, 1968). The corticobulbar, corticospinal, and corticopontine tracts are not readily distinguished on the basis of histological

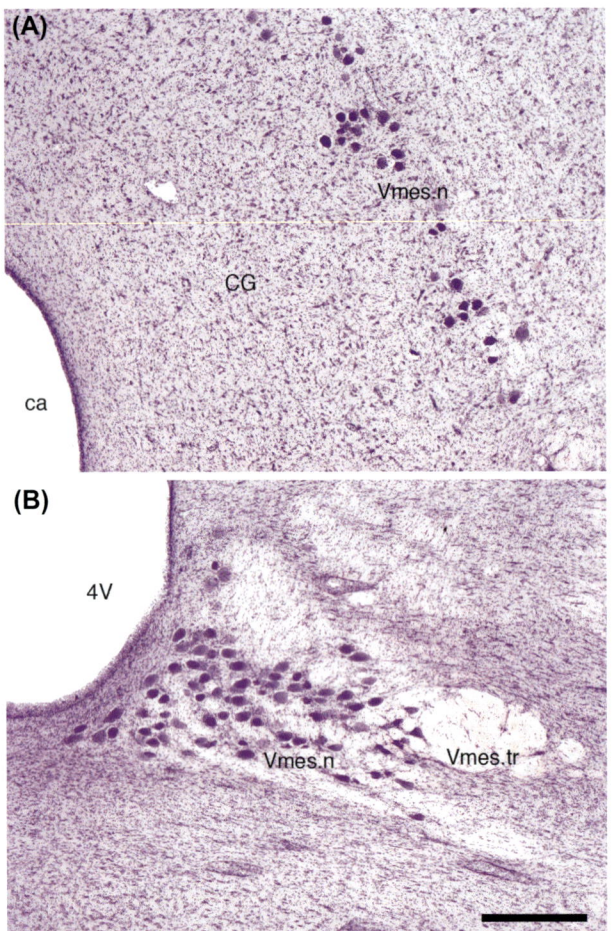

FIGURE 15.3   Photomicrographs of Nissl-stained coronal sections through the midbrain (A) and pons (B) of the river hippopotamus (*Hippopotamus amphibius*) showing the extent of the fifth mesencephalic nucleus (Vmes.n) and tract (Vmes.tr) in this species. While this nucleus is situated in a similar location in all mammalian species, the number of true ganglion cells comprising this nucleus varies between species, with the river hippopotamus being a species with comparatively many ganglion cells within this nucleus. These ganglion cells process proprioceptive information from the jaw, and given the massive gape of the river hippopotamus, it is not surprising that the nucleus is enlarged in this species. In both images, dorsal is to the top and medial to the left. Scale bar in (B) = 500 μm and applies to both images. *4V*, fourth ventricle; *ca*, cerebral aqueduct; *CG*, central or periaqueductal gray matter.

appearance, and thus, here, they are grouped together under the cerebral peduncle (also referred to as the cerebral crus or crus cerebri). The cerebral peduncle is a continuation of the internal capsule and forms a large arc defining the ventral (or anterior) aspect of the midbrain. The majority of the fibers forming the cerebral peduncle terminate in the ventral pontine nucleus, but some form the longitudinal fasciculus of the pons to continue as the pyramidal tract in the medulla oblongata.

The medial lemniscus is initially most readily identified as a coalescence of fibers around the longitudinal fasciculus of the pons and when traced rostrally forms

a distinct tract lateral to the midbrain tegmentum. As the medial lemniscus is traced rostrally, it is found in a more dorsal location on the lateral aspect of the midbrain, describing a course toward the somatic sensory nuclei of the dorsal thalamus. The lateral lemniscus is first observed as a distinct structure lateral to the longitudinal fasciculus of the pons and takes an almost dorsal course along the lateral caudal aspect of the midbrain to terminate in the inferior colliculus. Within the lateral lemniscus, occasional clusters of neurons have been described as specific nuclei of the lateral lemniscus, but the number of nuclei varies across species making a general description based on a few species misleading. These two ascending fiber pathways are usually of similar size, but in mammals where audition is a dominant sense, such as microchiropteran bats; these animals appear to have a relatively larger lateral lemniscus associated with the larger inferior colliculus (see as discussed in the following section).

15.2.3   The Neuromodulatory Nuclei of the Midbrain

15.2.3.1   The Catecholaminergic Nuclei of the Midbrain

Within the midbrain of all mammals studied to date (Calvey et al., 2013, 2015a, 2016; Dell et al., 2010), three distinct clusters of catecholaminergic, presumably dopaminergic, neurons are observed and best revealed with the use of tyrosine hydroxylase immunohistochemistry. These are the ventral tegmental area (A10), the substantia nigra complex (A9), and the retrorubral nucleus (A8). For the most part, the organization and appearance of these nuclei is consistent across mammals (Dahlström and Fuxe, 1964), but in some cases, there are distinct and unusual arrangements present (Maseko et al., 2013; Kruger et al., 2010; Manger et al., 2004).

15.2.3.1.1   The Ventral Tegmental Area Nuclei (A10 Complex)

Four distinct nuclei comprise the ventral tegmental area and include the ventral tegmental area (A10), the central ventral tegmental area (A10c), the dorsal ventral tegmental area (A10d), and the dorsal caudal ventral tegmental area (A10dc). The A10 nucleus is generally observed to be a high density of neurons found between the interpeduncular nucleus and the oculomotor nerve. The A10c nucleus is dense cluster of cells found dorsal to the interpeduncular nucleus, between it and the rostral border of the decussation of the superior cerebellar peduncle. Located dorsal to the A10c, between it and the oculomotor nucleus, is a triangular-shaped cluster of cells forming the A10d nucleus. Extending further dorsal to this nucleus into the periaqueductal gray matter and surrounding the ventrolateral edges of the cerebral aqueduct is a low-density cluster of neurons termed the A10dc nucleus, as these cells are found through to the caudal aspect of the cerebral aqueduct. The ventral tegmental area gives rise to the mesolimbic dopaminergic projection and is involved in the reward circuitry of the brain (Luo et al., 2011). This cluster of neurons appears to differ very little across mammalian species (Dell et al., 2010); however, three additional catecholaminergic nuclei, as identified with tyrosine hydroxylase immunohistochemistry, appear to be found within the midbrain of the bottlenose dolphin (Manger et al., 2004), although the functional significance of these extraneurons is unknown.

15.2.3.1.2 The Substantia Nigra Complex (A9 Complex)

Lying immediately above the cerebral peduncle, ventral to the midbrain tegmentum, is a large arc of cells generally assigned to the substantia nigra nuclear complex and associated with production of both dopamine and neuromelanin. This complex generally consists of four distinct nuclei, a pars medialis (A9m), a pars compacta (A9pc), a pars lateralis (A9l), and a pars ventralis (A9v). The A9m is usually found immediately lateral to the oculomotor nerve and consists of a high density of cells. Continuing laterally from the A9m is a dense arc of cells lying over the cerebral peduncle forms the A9pc. Lateral to the lateral edge of the A9pc and dorsal to the lateral edge of the cerebral peduncle are a moderate density of cells assigned to the A9l nucleus. Finally, ventral to the A9pc, intermingled with the fibers forming the cerebral peduncle, is the A9v nucleus. The substantia nigra gives rise to the nigrostriatal pathway, playing a major role in the modulation of movement (Nicola et al., 2000). The general appearance of these four nuclei is very similar across most mammalian species, but the A9pc is highly specialized and expanded in the African elephant (Maseko et al., 2013), the A9m is greatly expanded in the bottlenose dolphin (Manger et al., 2004), and the A9v is either incipient or absent in insectivores and microchiropteran bats (Kruger et al., 2010; Dell et al., 2010), reflecting a potential phylogenetic affinity (Calvey et al., 2016).

15.2.3.1.3 The Retrorubral Nucleus (A8)

The retrorubral nucleus is typically observed as a sparse collection of neurons located within the ventral half of the midbrain tegmentum in a position dorsal to the A9pc, but caudal to the red nucleus. This nucleus has been identified in all mammalian species studied to date (Calvey et al., 2013, a,b, 2016; Dell et al., 2010; Kitahama et al., 1994), but varies in its extent across species (Kitahama et al., 1994; Bux et al., 2010). This nucleus is thought to be involved in the modulation of oral motor behaviors (Arts et al., 1998).

15.2.3.2 The Serotonergic Nuclei of the Midbrain

The neurons producing serotonin within the midbrain of most mammals show a very similar organization across mammalian species (Calvey et al., 2013, 2015a, 2015b, 2016; Dell et al., 2010; Steinbusch, 1981). Within the midbrain, the serotonergic neurons form a series of distinct nuclei that are often grouped together as the rostral serotonergic nuclear cluster (Bjarkam et al., 1997). These serotonergic nuclei are found from the level of the decussation of the superior cerebellar peduncle to the most rostral level of the pons.

15.2.3.2.1 The Caudal Linear and Supralemniscal (B9) Nuclei

A small number of serotonergic neurons located around the ventral midline of the midbrain tegmentum immediately dorsal and rostral to the interpeduncular nucleus and rostral and ventral to the decussation of the superior cerebellar peduncle have been found in all mammals studied to date and form the caudal linear nucleus (Calvey et al., 2013, 2015a,b, 2016; Dell et al., 2010; Steinbusch, 1981). Serotonergic cells extending laterally from the caudal linear nucleus, lying dorsal to the lemnisci, are included in the supralemniscal, or B9, serotonergic nucleus. These two nuclei are, for the most part, consistent in appearance and location across mammalian species, although in some species, such as the highveld mole rat, the neurons of the B9 nucleus can extend substantially laterally over the cerebral peduncle (Bhagwandin et al., 2008).

15.2.3.2.2 The Median Raphe Nucleus

Across all mammals studied to date, the median raphe nucleus has been observed to consist of two distinct, but densely packed columns of serotonergic cells located in a pararaphe position dorsal and caudal to the superior cerebellar peduncle (Calvey et al., 2013, 2015a,b, 2016; Dell et al., 2010; Steinbusch, 1981; Patzke et al., 2014). The rostral border of this nucleus is coincident with the caudal level of the oculomotor nucleus, and this nucleus continues caudally to the border of the midbrain and pons, although it can invest a small way in the rostral pons in some species.

15.2.3.2.3 The Dorsal Raphe Nuclear Complex

The dorsal raphe complex is found in the ventral midline periaqueductal gray matter in all mammals studied to date. In most mammals, this nuclear complex continues into the periventricular gray matter of the pons; however, this caudal extension of serotonergic neurons, the caudal division of the dorsal raphe complex, is not observed in monotremes (Manger et al., 2002a). Within most mammals, six distinct nuclei form the dorsal raphe complex: interfascicular, ventral, dorsal, peripheral, lateral, and caudal. The interfascicular, ventral, and dorsal nuclei have a high density of similar ovoid-shaped bipolar cells, and while being contiguous across the midline and in a dorsoventral plane, the connectivity of these three nuclei differs, warranting their individual parcellation. Located laterally to the ventral nucleus in the ventrolateral periaqueductal gray is the peripheral nucleus, which usually has a moderate density of ovoid-shaped multipolar cells, some of which are found in the adjacent midbrain tegmentum. The number of these neurons in the midbrain tegmentum varies across species. Dorsal and lateral to the dorsal nucleus, in a position surrounding the ventrolateral borders of the cerebral aqueduct on either side, is a low-density cluster of large multipolar neurons that form the lateral nucleus. These neurons are found caudally in the brains of most mammals, and when the cerebral aqueduct opens into the fourth ventricle, merge across the ventricular border to form the caudal nucleus (which as mentioned previously is absent in monotremes). This appearance of the dorsal raphe complex is very consistent across mammalian species and only varies in the extent of each nucleus and the numbers of neurons present in each nucleus.

15.2.4 The Intrinsic Nuclei of the Midbrain

15.2.4.1 The Superior Colliculus

The superior colliculus is a clearly laminated neural structure that is found on the rostral aspect of the midbrain tectum. Generally seven or eight layers are identified within the superior colliculus, depending on the species examined, with each layer having a specific function. These layers, from dorsal to ventral, are: the optic tract layer, the zonal layer, the superficial gray matter layer, the optic nerve layer, the intermediate gray layer, the intermediate white layer, the deep gray layer, and the deep white matter layer. The first four layers, or the superficial layers, mainly receive input directly from the retina, but also from the visual cortex, the pretectum, and the cholinergic parabigeminal nucleus. The remaining, or deep, layers receive auditory and somatosensory input and are connected to a range of sensorimotor areas of the brain. Thus, the superior colliculus is involved in a range of visual functions as well as multisensory integration. The size of the superior colliculus varies significantly among mammals, with those mammalian species with vision as a dominant sense having large superior colliculi (Lane et al., 1971). In the microphthalmic Bathyergid mole rat, the superficial layers of the superior colliculus are substantially reduced in size compared to other mammals, although the deeper layers appear to maintain their normal appearance (Da Silva et al., 2006). In most mammals, the superior colliculus contains a complete

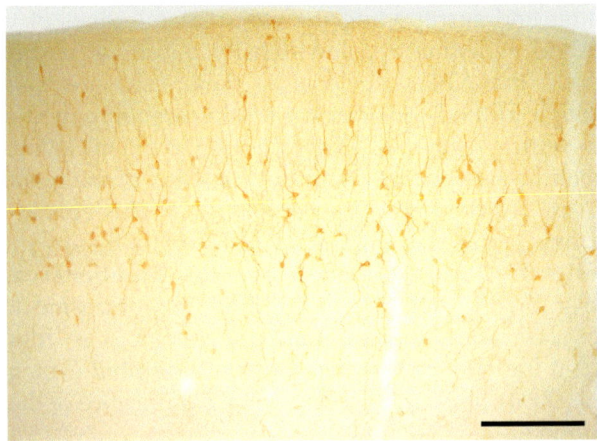

FIGURE 15.4 Photomicrograph showing cholinergic neurons within the superficial layers of the agouti (*Dasyprocta primnolopha*) superior colliculus. Despite having a distinct parabigeminal nucleus, the agouti has additional cholinergic neurons in the retinorecipient layers of the superior colliculus. Scale bar = 250 μm.

map of the contralateral retina (both hemifields), but in primates and megachiropterans, only the contralateral hemifield is represented in the superior colliculus (Pettigrew, 1986; Pettigrew et al., 2008). Cholinergic interneurons have been observed in the superior colliculus of some species, including laboratory rats and mice, the elephant shrews, the tree shrew (Calvey et al., 2013, 2015a; Dell et al., 2010), and the agouti (Fig. 15.4), although the functional significance of the presence of these neurons is unclear.

15.2.4.2 The Inferior Colliculus

In all mammals studied to date, the inferior colliculus occupies the caudal portion of the midbrain tectum, immediately posterior to the superior colliculus and medial to the terminal portion of the lateral lemniscus (which supplies the inferior colliculus with ascending auditory input from which a map of auditory space is formed). Generally, three distinct regions of the inferior colliculus are noted with different anatomical stains, including the dorsal cortex, the central nucleus, and the external cortex. While in most mammals these regions are rather homogeneous in architectonic appearance, the inferior colliculus of the bottlenose dolphin appears to have distinct anatomical segregations that have a potential functional significance (Glezer et al., 1998); however, this has not been observed in other species with specialized auditory systems (Maseko et al., 2013). While the input from the lateral lemniscus covers the caudal and lateral aspects of the inferior colliculus, fiber tracts emerging from the dorsal surface pass from lateral to medial and coalesce to form the commissure of the inferior colliculus, which again varies in size across species and their dependence on auditory processing.

15.2.4.3 The Red Nucleus

The red nucleus of mammals is usually composed of both a magnocellular and parvocellular division and is located in the rostral ventral medial half of the midbrain tegmentum, dorsal to the substantia nigra complex. While the magnocellular division of the red nucleus is readily identified across mammalian species, the more rostrally located parvocellular division is difficult to distinguish from the surrounding midbrain tegmentum due to the similarity in size and shape of the neurons. The size of the magnocellular division does vary across mammalian species, especially in species such as primates with a distinct corticospinal tract; however, it has been reported to be present in all species studied. The red nucleus forms part of the extrapyramidal motor system, receiving input from the cerebellum and the ipsilateral motor cortex and projects to the ipsilateral inferior olivary complex and the contralateral medullary tegmentum and spinal cord forming the rubrospinal tract.

15.2.4.4 The Periaqueductal Gray Matter

In mammals, the cerebral aqueduct is surrounded by a series of rostrocaudally oriented, longitudinal columns of gray matter that are collectively referred to as the periaqueductal, or central, gray matter. While classical architectural stains did not readily reveal the columnar organization of the periaqueductal gray matter, hodological and histochemical studies revealed these clearly (Bandler and Shipley, 1994; Behbehani, 1995). While six distinct columns can be observed in this region, two of these, the column made up of the visuomotor nuclei and that made up of the serotonergic dorsal raphe, are considered distinct regions and usually discussed separately (see aforementioned section). The remaining four columns include the dorsal, lateral dorsal, lateral ventral, and medial columns, although the terminology does vary (Bandler and Shipley, 1994; Behbehani, 1995). These four columns in particular appear to run the full rostrocaudal extent of the periaqueductal gray matter. These columns of the periaqueductal gray matter are involved in at least five distinct functions: pain processing/modulation (dorsal lateral and ventral lateral columns), vocalization (dorsal lateral and ventral lateral columns), autonomic regulation (dorsal lateral and ventral lateral columns), fear/anxiety/aversion/ defence (dorsal and dorsal lateral columns), and mammalian lordosis (female and male sexual presentation/positioning, dorsal and ventral lateral columns). Across mammals, there does not appear to be substantial variation in the appearance of the periaqueductal gray matter, although in elephants and both suborders of cetaceans, an enlargement of the ventral lateral column forms a distinct nucleus termed nucleus ellipticus (Maseko et al., 2013; Jansen, 1969). Given that both these

mammalian groups possess distinctive vocalization systems and that newborn dolphins vocalize prior to echolocating and the nucleus ellipticus is present in the newborn dolphin (Manger, 2006), it is likely that this expanded portion of the periaqueductal gray matter forms part of a specialized vocal pattern generator system in these species.

15.2.5 The Reticular/Tegmental Nuclei of the Midbrain

The reticular formation of the midbrain forms the core of this brain region across mammals. This region is bordered rostrally by the ascending medial lemniscus and caudally by the decussation of the superior cerebellar peduncle. Across mammals in general, the reticular formation is bordered dorsally by the tectum/ periaqueductal gray matter, medially by the median raphe nucleus, laterally by the dorsal thalamus, and ventrally by the substantia nigra complex. This region houses a range of both large and small cells forming a homogeneous mesh-like dendritic network around fascicles of axons. Within the midbrain, in well-studied mammals, parabrachial, cuneiform, and subcuneiform nuclei are often described; however, these are difficult to compare across mammalian species more generally due to a lack of distinctive anatomical compartmentalization of these reticular subdivisions. Housed within the midbrain reticular formation are the red nucleus and retrorubral catecholaminergic field (see aforementioned section).

15.3 Pons

15.3.1 The Cranial Nerve Nuclei of the Pons

Within the pons of all mammals studied to date, three distinct nuclear complexes/individual nuclei associated with specific cranial nerves have been identified. These include the trigeminal nerve and associated nuclei, the abducens nucleus, and the facial nerve and associated nuclei. The nuclei appear to maintain a similar position across species, although the size of these nuclei can vary substantially.

15.3.1.1 The Trigeminal Nerve and Associated Nuclei

Across mammals, the trigeminal nerve is typically one of, if not the, largest cranial nerve. The trigeminal nerve is observed to enter/exit the brain via the ventrolateral portion of the pons, investing (or divesting) itself into/out of the middle cerebellar peduncle, just lateral to the ventral pontine nucleus (see as discussed in the following section). Typically, the afferent (sensory) branch of the trigeminal nerve splits into three divisions, one that terminates within the subdivisions of the principal trigeminal nucleus, one that continues rostrally to form the trigeminal mesencephalic tract, and one that turns caudally to form the spinal trigeminal tract. The efferent fibers of the trigeminal motor nucleus join the afferent branches to exit the brain from the middle cerebellar peduncle.

15.3.1.2 The Principal Trigeminal Nucleus

The principal trigeminal nucleus is the main recipient of afferent inputs from the trigeminal nerve within the pons. While this nucleus can be subdivided into dorsomedial and ventrolateral divisions, such as seen in the laboratory rat (Paxinos et al., 2009), this nucleus is often considered one large single nucleus in comparative studies. The principal trigeminal nucleus is found in a location medial to the spinal trigeminal tract and lateral to the pontine parvocellular reticular column and the trigeminal motor nucleus. It is found in this position along the entire rostrocaudal extent of the pons, until the level of the descending arm of the facial nerve, where it transitions into the medullary portions of the trigeminal sensory nuclei (see as discussed in the following section).

15.3.1.3 The Pontine Portion of the Trigeminal Mesencephalic Nucleus and Tract

The ganglion cells that form this nucleus represent a caudal continuation of those present in the midbrain (see aforementioned section) and present as a low-density cluster of cells located at the lateralmost border of the periventricular gray matter (Fig. 15.3B). This nucleus is not found along the entire extent of the pons, but is restricted to the rostral half. From the trigeminal nerve, a cluster of axons passes rostral and dorsal to form the trigeminal mesencephalic tract which is observed as a distinct fiber pathway lateral to the location of the ganglion cells (Fig. 15.3B).

15.3.1.4 The Trigeminal Motor Nucleus

In all mammals studied to date, a moderately dense cluster of large, cholinergic motor neurons located in the dorsolateral part of the parvocellular reticular pontine region gives rise to the motor division of the trigeminal nerve and is classified as the trigeminal motor nucleus. This nucleus is located just rostral and medial to the descending arm of the facial nerve. The efferents of these neurons form a distinct cholinergic fasciculus that traverses dorsal to the dorsomedial division of the principal trigeminal nucleus to coalesce with the large afferent portion of the trigeminal nerve to exit the brain from the ventrolateral portion of the middle cerebellar peduncle. The number of neurons forming this nucleus does vary across mammals, likely related to the size of the animal and the volume of the masticatory muscles, but no correlation of this sort has yet been demonstrated.

15.3.1.5 The Abducens Nerve and Nucleus

A densely packed, but not numerous, cluster of cholinergic motor neurons located in the dorsomedial aspect of the pontine reticular formation, immediately adjacent to the periventricular gray matter forms the abducens nucleus in most mammals. This nucleus is usually nestled beneath the genu of the facial nerve. The efferent fiber bundle from this nucleus is not particularly large, as these efferents innervate a single extraocular muscle, the lateral rectus muscle, and descends from the nucleus to exit the brain at the pontomedullary junction. The appearance and location of this nucleus is similar across mammals, although the number of neurons within this nucleus appears to be greater in mammals that display a greater range of ocular movement.

15.3.1.6 The Facial Nerve and Facial Nuclear Complex

In all mammals studied to date, the facial nerve emerges from the dorsal aspect of the facial nuclear complex to travel to the roof of the caudal aspect of the pons and bends laterally over the abducens nucleus (sometimes forming the distinctive facial colliculus) to then extend ventrolaterally to exit the brain at the pontomedullary junction adjacent to the vestibulocochlear nerve. This large cranial nerve varies in size across species, being especially large in animals with extensive facial musculature, such as the elephant (Maseko et al., 2013). Typically, the facial nerve nuclear complex is comprised of three divisions, a large ventral division, a smaller dorsal division, and a parasympathetic component (the superior salivatory nucleus) (Dell et al., 2010). All these neurons are cholinergic. In some species, such as the laboratory rat (Paxinos et al., 2009) and the African elephant (Maseko et al., 2013), additional medially located subdivisions, the dorsomedial and ventromedial, of the facial nerve nuclear complex are observed, probably associated with complex movement of the vibrissae and trunk, respectively. Quantitative analyses of the numbers of neurons within the facial nerve nuclear complex in primates and marsupials have indicated that while there is an increase in number with associated body mass (and presumably mass of the facial musculature), this is not a rapid scaling increase in number (Sherwood, 2005; Watson et al., 2012).

15.3.2 The Ascending and Descending Fiber Pathways of the Pons

The mammalian pons typically houses several distinct large and small fiber pathways, some of which are clearly evident in basic anatomical preparation (Nissl and myelin stains) and some of which need either specialized immunostaining or hodological studies to reveal. Despite this, the general pattern across mammals is quite similar and varies mostly in accordance with specific specializations of the different species, be they sensory or motor specializations. The rostral border of the pons is demarcated by the decussation of the superior cerebellar peduncle (sometimes referred to as the brachium conjunctivum). From this decussation, the fibers coalesce to form a distinctive arc in a position dorsolateral to the pontine tegmentum. The superior cerebellar peduncle maintains this location within the pons to the level of the trigeminal motor nucleus, where it shifts dorsally to invest into the white matter of the cerebellum. In all mammals, the largest cerebellar peduncle is the middle cerebellar peduncle and the axons making up this structure arise from the ventral pontine nucleus. These invest into the cerebellar white matter from the lateral and ventral aspects of the pons.

Within the pons of all mammals, a distinct medial column of white matter lies between the raphe reticular column and the gigantocellular reticular nucleus (see as discussed in the following section). This column extends from the dorsal to the ventral aspect of the pons, being expanded in the ventral aspect and extends around the lateral aspect of the pons. This medial white matter column runs the entire rostrocaudal aspect of the pons and houses several distinctive pathways, such as the medial longitudinal fasciculus, the lateral lemniscus, and the longitudinal fasciculus of the pons, as well as several less distinct pathways such as the tectospinal tract. The mediodorsal aspect of this white matter column is typically occupied by the medial longitudinal fasciculus, and below this, several of the pathways aforementioned are found, but are difficult to distinguish in basic preparations. At the ventromedial aspect of the medial white matter column, the longitudinal fasciculus of the pons (the continuation of the cerebral peduncles) is normally found, immediately dorsal to the ventral pontine nucleus, and runs the entire length of the pons before continuing as the pyramidal tract in the medulla oblongata. The medial lemniscus is usually found dorsal to and around the longitudinal fasciculus of the pons and forms a consolidated mass dorsally and laterally around the pontine reticular formation before entering the midbrain (see aforementioned section). Lateral to the medial lemniscus is the lateral lemniscus, which invests into the inferior colliculus (see aforementioned section) after passing laterally around the pontine tegmentum. Occasionally small, reticulated nuclei are found within the pontine portion of the lateral lemniscus, but this varies across species.

15.3.3 The Neuromodulatory Nuclei of the Pons

15.3.3.1 The Cholinergic Nuclei of the Pons

Within the pons, three cholinergic nuclei are typically reported, these include the parabigeminal, pedunculopontine (PPT), and laterodorsal tegmental (LDT) nuclei. The parabigeminal nucleus is a small, but densely packed nucleus typically located on the lateral margin of the rostralmost pontine tegmentum in a location ventral to the brachium of the inferior colliculus. This nucleus is reciprocally connected to the superficial, or retinorecipient, layers of the superior colliculus and projects to the contralateral thalamic dorsal geniculate nucleus (Hashikawa et al., 1986), making it part of the visual system (Sherk, 1979). While present in most mammals studied (Dell et al., 2010), it is not clearly present in monotremes (Manger et al., 2002b), certain species of microchiropterans (Kruger et al., 2010; Maseko and Manger, 2007) or shrews (Calvey et al., 2016), which, in addition to other neural features, suggest a phylogenetic link between the microchiropterans and the shrews (Calvey et al., 2016).

The LDT and PPT nuclei have been reported in every mammal studied to date (Dell et al., 2010; Calvey et al., 2016; Patzke et al., 2014). The cholinergic neurons forming the LDT are generally located in the ventrolateral corner of the rostral periventricular gray matter of the pons and in some species extend further rostral into the midbrain. The cholinergic neurons of the PPT are found in the parvocellular pontine tegmentum adjacent to the LDT neurons, rostral to the trigeminal motor nucleus, and sometimes surrounding the superior cerebellar peduncle laterally. Typically, the neurons forming the LDT and PPT nuclei are similar in size and shape, but it does appear that the animals with larger brains have more neurons in each of these nuclei, but this has not been quantified across a range of species to determine whether neuronal numbers in these nuclei follow an allometric scaling or whether these numbers are species specific. Despite this general similarity across mammalian species, some variances in the LDT and PPT nuclei have been noted. In giraffe (*Giraffa camelopardalis*), the neurons of the LDT appear to be larger than those of the PPT (Bux et al., 2010), although the functional significance of this difference is unclear. In the rock hyrax (*Procavia capensis*), it was noted that surrounding both the PPT and LDT were clusters of smaller cholinergic neurons, with the appearance of interneurons (Gravett et al., 2009). Again, the functional significance of these novel cholinergic neurons is unclear, but along with differences in the orexinergic system of the rock hyrax (Gravett et al., 2011), they may play a role in the appearance of a novel sleep state that exhibits a mixture of features of both slow wave and REM sleep (Gravett et al., 2012). In the river hippopotamus (*Hippopotamus amphibius*), similar small cholinergic neurons are found intermingled with the larger LDT neurons, and this, along with other unusual features of the cholinergic system in the river hippopotamus, may lead to the generation of a novel sleep state in the river hippopotamus (Dell et al., 2016), but this has not yet been investigated. In the Tasmanian devil (*Sarcophilus harrisii*), the LDT nucleus appears to have both a ventral and dorsal subdivision of large cholinergic neurons, the ventral subdivision being homologous to that observed in all other mammals, and the dorsal subdivision being novel (Patzke et al., 2014), but again the functional significance of this novel nucleus is not understood, although it may be a feature of marsupials brains that relates to variances in sleep. Finally, the LDT nucleus has an unusual appearance in both microchiropteran bats and insectivores (Fig. 15.5), with a dorsomedial extension of the main cell group surrounding the posterodorsal

tegmental nucleus, indicating a phylogenetic affinity between the microchiropterans and insectivores (Calvey et al., 2016) as this feature is not observed in other mammals studied to date.

15.3.3.2 The Catecholaminergic Nuclei of the Pons: The Locus Coeruleus Complex

The locus coeruleus complex can be readily identified across mammalian species through the use of immunohistochemical techniques for tyrosine hydroxylase (Dell et al., 2010). The neurons forming the locus coeruleus complex are found throughout much of the rostrocaudal extent of the pons and are generally classified into five distinct nuclei, although the number and organization does differ across species (Dell et al., 2010). The four most commonly reported nuclei are the compact portion of the subcoeruleus (A7sc), the diffuse portion of the subcoeruleus (A7d), the diffuse portion of the locus coeruleus (A6d), and the fifth arcuate nucleus (A5). The A7sc is observed as a tightly packed cluster of tyrosine hydroxylase immunopositive neurons adjacent to

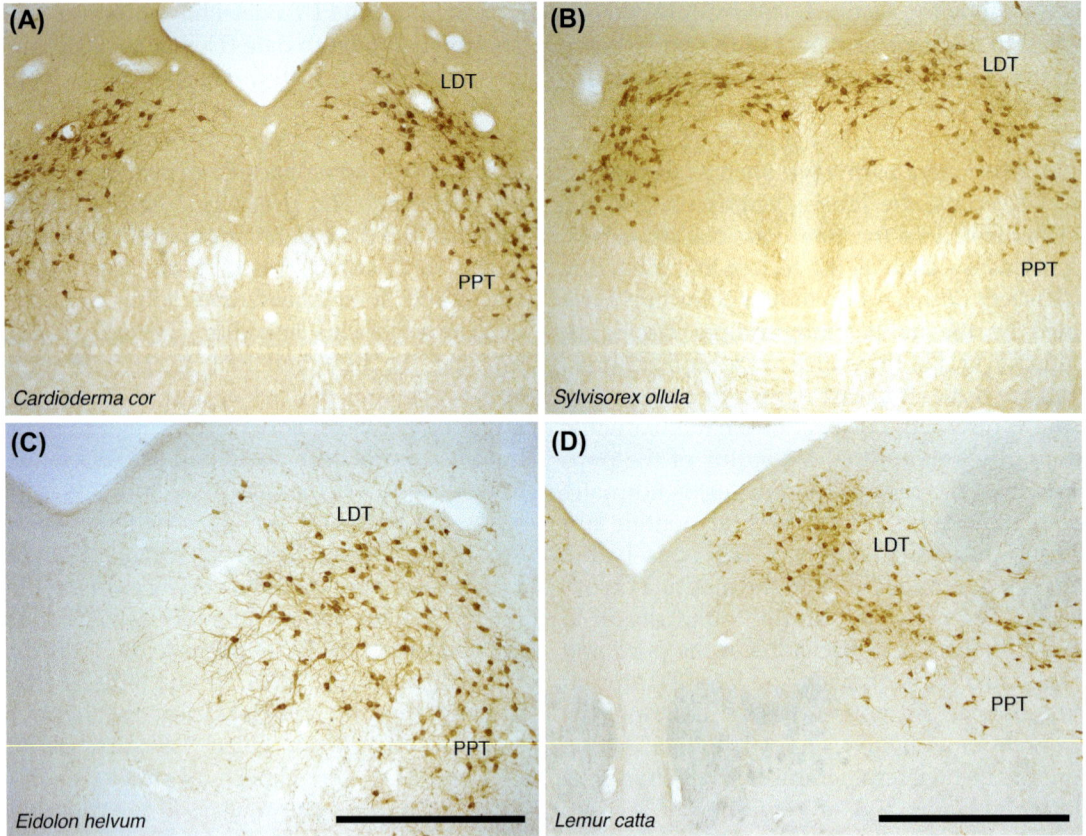

FIGURE 15.5 Photomicrographs of the laterodorsal tegmental (LDT) and pedunculopontine tegmental (PPT) cholinergic neurons, immunostained with cholineacetyltransferase, in the pons of four species of mammals: (A) the microchiropteran *Cardioderma cor*; (B) the shrew *Sylvisorex ollula*; (C) the megachiropteran *Eidolon helvum*; and (D) the prosimian primate *Lemur catta*. Note the dorsomedial extension of the LDT over the posterodorsal tegmental nucleus in the microchiropteran and shrew, a feature that indicates shared phylogeny in these species, compared to the more typical mammalian appearance of the LDT in the megachiropteran and primate which do not show this dorsomedial extension. Scale bar in (C) = 500 μm and applies to (A–C). Scale bar in (D) = 1000 μm and applies to (D) only.

the ventrolateral border of the periventricular gray matter within the dorsolateral portion of the parvocellular pontine tegmentum. This nucleus was previously termed the subcoeruleus (Dahlström and Fuxe, 1964). Ventral and lateral to this nucleus, a diffusely organized aggregation of tyrosine hydroxylase immunoreactive neurons forms the A7d nucleus. These neurons are found throughout the majority of the parvocellular pontine tegmentum rostral to the trigeminal motor nucleus and lateral to the superior cerebellar peduncle in the parabrachial region. Within the ventrolateral portion of the periventricular gray matter, often adjacent to or intermixed with the cholinergic LDT nucleus (Manger et al., 2002c), is what is often referred to as the locus coeruleus proper (Dahlström and Fuxe, 1964), but, with further comparative studies, is now termed the diffuse portion of the locus coeruleus complex (A6d). The last commonly reported nucleus of the locus coeruleus complex is the fifth arcuate nucleus (A5). This nucleus is found in the ventrolateral pontine tegmentum, from the level of the fifth motor nucleus to the caudal border of the pons, in a position within the divisions of the superior olivary complex and ventrolateral and caudal to the facial nerve nuclear complex. The neurons within the A5 form a mesh-like dendritic network around the fascicles located within the ventrolateral parvocellular pontine tegmentum.

Variations within the locus coeruleus complex have not been observed in the A7 or A5 nuclei, but several variations have been observed in the A6 region as well as the dorsolateral division of the locus coeruleus (A4). When present, the A4 nucleus is seen as a tightly packed, but small cluster of neurons in the dorsal lateral extremity of the periventricular gray matter adjacent to the floor of the fourth ventricle. This nucleus has been observed in primates, megachiropterans, lagomorphs, scandents, rodents, and some Afrotherian species (Calvey et al., 2013, 2015a; Dell et al., 2010). The A6 region is far more variable than other parts of the locus coeruleus complex, but as mentioned previously, the diffuse portion is what is most commonly observed across mammals (Dell et al., 2010). In the African elephant, an additional medial division of the A6 (A6m) was observed (Maseko et al., 2013), but this A6m division has not been seen in any other mammals studied. In primates and megachiropterans, the core of the A6d houses a very densely packed cluster of neurons termed the A6 compact (A6c) division of the locus coeruleus. This arrangement of the A6, with a densely packed core (A6c) and more loosely packed surround (A6d), is specific to primates and megachiropterans and has not been observed in other mammalian species (Fig. 15.6), indicating a phylogenetic affinity between the primates and megachiropterans (Dell et al., 2010; Calvey et al., 2015b). Another variation is specific to murid rodents,

but not other rodents (Kruger et al., 2012). In murid rodents, instead of having the typical A6d appearance of the locus coeruleus, a tightly packed, high-density cluster of tyrosine hydroxylase immunoreactive neurons forms an A6c, but no A6d is present. Moreover, in the murid rodents, these neurons stretch across the periventricular gray matter from the floor of the fourth ventricle to its border with the pontine tegmentum (Kruger et al., 2012) a feature not seen in other rodents or other mammals.

15.3.3.3 The Serotonergic Nuclei of the Pons

In the rostralmost portion of the pons in most mammals, two serotonergic nuclei are often found (Dell et al., 2010), these are the caudal division of the dorsal raphe and the median raphe nucleus. Both are caudal extensions of the midbrain nuclei. The caudal division of the dorsal raphe is an extension of the lateral division, and as mentioned previously, it is described as a distinct nucleus due to the lack of this nucleus in the monotremes (Manger et al., 2002a). Similarly, the median raphe columns can extend a short way into the midline of the pons. This occasional pontine extension of the median raphe nucleus is sometimes referred to as the pontine raphe nucleus (Paxinos et al., 2009; Dahlström and Fuxe, 1964), but it is evident that these neurons are a caudal extension of the midbrain nucleus and do not form a distinct pontine serotonergic nucleus.

15.3.4 The Intrinsic Nuclei of the Pons

15.3.4.1 The Periventricular Gray Matter

In all mammals studied to date, the periaqueductal gray matter continues as a distinct structure caudally from the midbrain, into the pons, where when the cerebral aqueduct opens up, the caudal extension of gray matter can be called the periventricular gray matter. The periventricular gray matter is located in the dorsal part of the pons, adjacent to the floor of the fourth ventricle, where it occupies this position for the full rostrocaudal extent of the pons. This gray matter is thicker rostrally and becomes gradually thinner caudally, before extending into the medulla oblongata to merge with the rostral portion of the gray matter of the tractus solitarius. In the rostral medial portion of the periventricular gray matter is a distinct nucleus termed the posterodorsal tegmental nucleus. The remainder of this gray matter is rather homogenous in appearance, but immunohistochemical stains reveal the presence of the cholinergic LDT nucleus, the catecholaminergic locus coeruleus, and the serotonergic caudal division of the dorsal raphe (see aforementioned section) within different portions of this gray matter. Additional topological subdivisions have been reported in the laboratory rat (Paxinos et al.,

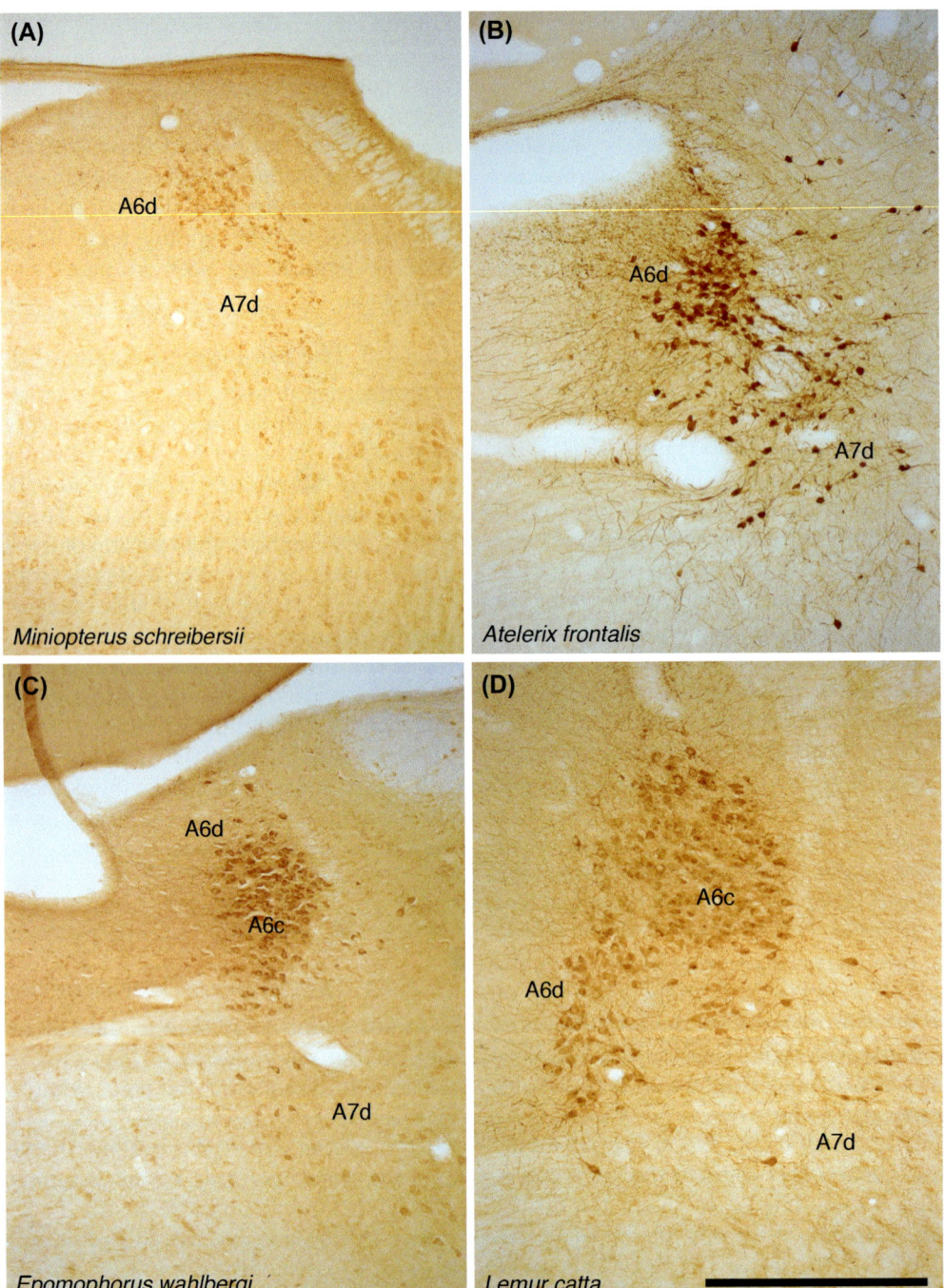

FIGURE 15.6 Photomicrographs of the locus coeruleus complex, immunostained with tyrosine hydroxylase, in four species of mammals: (A) the microchiropteran *Miniopterus schrebersii*; (B) the hedgehog *Atelerix frontalis*; (C) the megachiropteran *Epomophorus wahlbergi*; and (D) the prosimian primate *Lemur catta*. In the microchiropteran and hedgehog, the typically mammalian appearance of the diffuse division of the locus coeruleus complex (A6d) showing a cluster of cells that are not densely packed is observed within the periventricular gray matter adjacent to the diffuse division of the subcoeruleus (A7d). In contrast, in the megachiropteran and the primate, there is a densely packed compact division of the locus coeruleus (A6c) surrounded by the A6d. This arrangement of the locus coeruleus is only observed in megachiropterans and primates and is a feature that indicates shared ancestry in these species. Scale bar in (D) = 1000 μm and applies to all.

2009), but these are generally not reported in other mammals.

15.3.4.2 The Superior Olivary Complex and Trapezoid Body

The superior olivary nuclear complex is generally located ventral to the pontine parvocellular reticular column, in a position immediately rostral to the facial nerve nucleus. This complex usually consists of lateral superior olive, medial superior olive, medioventral periolivary, and superior paraolivary nuclei. This nuclear complex receives ascending input from the cochlear nuclei and is involved in detecting interaural intensity differences to assist with sound localization. In the African elephant, the lateral superior olive nucleus is particularly large (Maseko et al., 2013). Located medial and slightly caudal to the superior olivary complex is the trapezoid body, which is where some of the ascending fibers from the cochlear nucleus decussate. In the tree shrew (*Tupaia belangeri*), the nucleus of the trapezoid body exhibits cholinergic neurons, which potentially highlights the importance of awareness of sound localization in the tree shrews (Calvey et al., 2015a).

15.3.4.3 The Ventral Pontine Nucleus

At the base of the pons in all mammals studied to date, occupying the ventral aspect of the entire rostrocaudal extent of the pons is the large ventral pontine nucleus (Fig. 15.2). The neurons within this nucleus appear to be arranged in horizontally oriented sheets, between which longitudinal (corticobulbar, corticospinal, and corticopontine) and transverse (pontocerebellar) fiber bundles are found. From the lateral aspect of the ventral pontine nucleus, the large middle cerebellar peduncle can be seen to emerge and invest into the cerebellar white matter. There appears to be some variance in the size of this nucleus across species, with those animals having a relatively large cerebellum (Maseko et al., 2012) having a seemingly larger ventral pontine nucleus, but this has not been assessed in detail.

15.3.5 The Reticular/Tegmental Nuclei of the Pons

In mammals, the reticular formation of the pons forms the core of this segment of the brain stem. It is bordered anteriorly by the decussation of the superior cerebellar peduncle and caudally by the descending limb of the facial nerve. In mammals generally, the pontine reticular formation can be divided into three major rostrocaudally oriented cell columns: the raphe, the gigantocellular/magnocellular, and the parvocellular. The raphe column, which can be serotonergic at the rostralmost pole of the pons (see aforementioned

section), consists of two parasagittal, or pararaphe, cell columns, often only a cell or two wide, located along the rostrocaudal extent of the pons and separated by a thin sheet of white matter. The ventralmost portions of these columns can be seen to expand laterally, forming the distinct reticular tegmental nucleus of the pons along much of the rostrocaudal extent of the pons, although the size and the extent of this nucleus can vary across mammals. At most levels of the pons, the raphe columns and reticular tegmental nucleus are isolated from the gigantocellular column by a column of white matter that varies in thickness (see aforementioned section).

The gigantocellular column is found lateral to the medial white matter column and extends the full rostrocaudal extent of the pons. This column exhibits the typical reticular appearance, with large cells forming a homogeneous mesh-like dendritic network around descending or ascending fascicles. The parvocellular column is found lateral to the gigantocellular column throughout the pons and is bordered laterally by the superior cerebellar peduncle, the principal trigeminal nucleus, and the descending arm of the facial nerve. The marked reduction in size of the neurons provides a reliable delineation of this nucleus at the medial edge where it borders the gigantocellular column (Fig. 15.7). The typical reticular appearance is always present, and in the region of this nucleus anterior to the trigeminal motor nucleus, the cholinergic PPT nucleus and the catecholaminergic subcoeruleus are found. Caudal to the

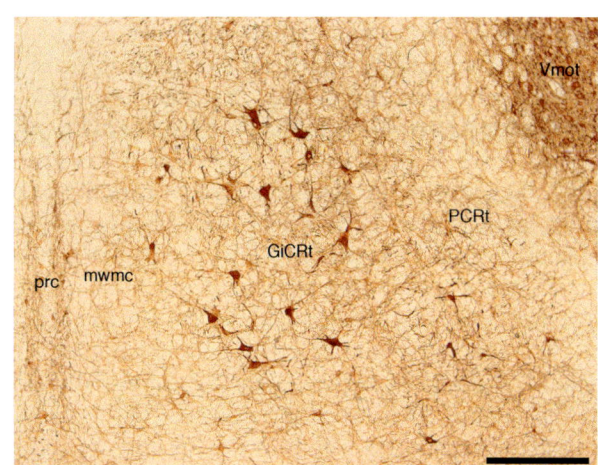

FIGURE 15.7 Photomicrograph showing the medial to lateral columnar organization of the pontine tegmentum in the banded mongoose (*Mungos mungo*) as immunostained with SMI 32 in a coronal section at the level of the trigeminal motor nucleus (Vmot). The SMI 32 antibody recognizes a nonphosphorylated epitope of neurofilament proteins and is especially useful in recognizing large neurons and their processes. The most medial columns, of two to three cells wide, are the pararaphe columns (prc). Immediately lateral to these columns are the medial white matter column (mwmc), followed by the gigantocellular reticular nucleus (GiCRt) and the parvocellular reticular nucleus (PCRt). Scale bar = 500 µm.

trigeminal motor nucleus, the catecholaminergic fifth arcuate nucleus is located in the ventral portion of the parvocellular column.

15.4 Medulla Oblongata

15.4.1 The Cranial Nerve Nuclei of the Medulla Oblongata

Within the medulla oblongata, six nuclear complexes/individual nuclei associated with specific cranial nerves, found in similar topological positions within the medulla oblongata, are consistently reported across mammals. This includes the spinal trigeminal tract and associated nuclei, the vestibulocochlear nerve and associated nuclei, the nucleus ambiguus, the inferior salivatory nucleus, the dorsal motor vagus nucleus, and the hypoglossal nucleus.

15.4.1.1 The Spinal Trigeminal Tract and Associated Nuclei

The spinal trigeminal tract and associated nuclei are found lateral to the medullary tegmentum along the entire rostrocaudal extent of the medulla oblongata and continue into the spinal cord to merge with the dorsal horn of the cervical spinal cord. In the rostral medulla, these structures are bordered laterally by the inferior cerebellar peduncle, but caudally they form the lateral border of the medulla oblongata. The spinal trigeminal tract arises in the pons, formed by a caudal branch of the main trigeminal afferent nerve. This tract is consistently located on the lateral aspects of the medullary trigeminal sensory nuclei. At the pontomedullary junction, the principal trigeminal nucleus of the pons is seen to transition into the oral subdivision of the spinal trigeminal nucleus. Progressing further caudal, at the caudal level of the facial nerve nuclear complex, the oral subdivision is seen to transition into the lateral subdivision of the spinal trigeminal nucleus. At the level of the rostral pole of the nucleus ambiguus, the dorsal division of the dorsomedial spinal trigeminal nucleus appears in a position dorsal to the lateral subdivision, and the ventral division of the dorsomedial spinal trigeminal nucleus appears ventrally to the lateral subdivision. There are differences in the exact arrangement of these nuclei across species, but mostly in relative location (Maseko et al., 2013; Paxinos et al., 2009).

At the level of the tractus solitarius, the lateral subdivision transitions into the caudal subdivision of the spinal trigeminal nucleus. While the lateral subdivision tapers away, the gelatinous layer of the caudal spinal trigeminal nucleus appears to become a solid nuclear mass at the level of the decussation of the pyramidal tracts. As the medulla transitions into the spinal cord, the gelatinous layer splits into two distinct sublamina, inner and outer, that eventually become the first and second layers of the dorsal horn of the cervical spinal cord. The caudal subdivision of the spinal trigeminal nucleus also tapers and flattens caudally to transition into the third layer of the dorsal horn of the cervical spinal cord. This complex arrangement of nuclei and transitions is difficult to follow with classical neuroanatomical stains, but immunostaining, especially for the calcium-binding proteins, assists in delineating these nuclei and their transitions in different species (Maseko et al., 2013; Paxinos et al., 2009).

15.4.1.2 The Vestibulocochlear Nerve and Cochlear and Vestibular Nuclei

In all mammals, the vestibulocochlear nerve enters the lateral aspect of the medulla at the pontomedullary junction. As this nerve enters the medulla, the nerve spreads to provide innervation to both the cochlear and vestibular nuclei. The cochlear portion of the nerve invests directly into the cochlear nuclei that are located on the lateral and laterodorsal aspects of the rostral medulla oblongata. The vestibular branch passes ventral to the inferior cerebellar peduncle and laterodorsal to the spinal trigeminal tract to supply the vestibular nuclear complex.

The cochlear nuclear complex is located in the dorsolateral portion of the rostral medulla oblongata, lateral to the inferior cerebellar peduncle. Generally, two distinct nuclei, a dorsal and ventral cochlear nucleus, are apparent in mammals. In some mammals, the dorsal cochlear nucleus is composed of two or more distinct lamina, with a granular layer forming a distinct bridge between the dorsal and ventral nuclei. The ventral nucleus is sometimes divided into anterior and posterior parts, but requires specific stains to reveal this division. The ventral cochlear nucleus is general located on the very lateral aspect of the medulla between the inferior and middle cerebellar peduncles. The specific parcellation and arrangement of the cochlear nuclear complex has not been investigated in great detail across species, although given its primary role in audition, it is likely to reveal variations across species.

The vestibular nuclear complex is generally located in the dorsal aspect of the rostral medulla oblongata in a position dorsal to the spinal trigeminal system and medial to the inferior cerebellar peduncle. In more intensely studied species, such as the laboratory rat, there appear to be several distinct nuclei within this complex, lateral, superior, vestibulocerebellar, and medial (Paxinos et al., 2009); however, across species, these subdivisions are not always readily apparent. The one nucleus that can usually be identified with confidence across species is the lateral vestibular nucleus due to the presence of large multipolar neurons that stain intensely with cresyl violet. The superior and vestibulocerebellar nuclei are typically found dorsal to the lateral nucleus, invested within the superior cerebellar peduncle. The medial nucleus is generally found medial to the lateral nucleus, extending close to the floor of the fourth ventricle. Again, cross-species comparisons of this nuclear complex are not common, although variants are likely to be present.

15.4.1.3 The Nucleus Ambiguus

The nucleus ambiguus is generally represented as a small cluster of large cholinergic motor neurons within the upper portion of the ventral half of the medullary tegmentum. This nucleus is generally observed from the level of the caudal pole of the facial nerve nuclear complex to the level of the area postrema (see as discussed in the following section). It is difficult to identify the efferents of this nucleus, although they are known to supply both the vagal and glossopharyngeal nerves with motor fibers.

15.4.1.4 The Preganglionic Motor Neurons of the Inferior Salivatory Nucleus (pIX)

A small cluster of medium-sized cholinergic neurons located in the very rostral aspect of the gray matter of the tractus solitarius, and rostral to the dorsal motor vagal nucleus, form the inferior salivatory nucleus. This nucleus has been observed in all mammals studied to date with immunohistochemistry (Dell et al., 2010; Calvey et al., 2015a,b, 2016).

15.4.1.5 The Dorsal Motor Vagal Nucleus (X)

The dorsal motor vagal nucleus is observed as a discreet and densely packed nucleus of medium-sized cholinergic neurons located in the lateral aspect of the periventricular gray matter of the dorsal medulla oblongata. This nucleus usually extends from the caudal level of the facial nerve nucleus through to the spinomedullary junction. The efferent axons emerging from this nucleus are difficult to identify, but generally, this nucleus supplies the vagal nerve with parasympathetic efferent fibers. In the African elephant, a small number of neurons extend caudally from this nucleus to spread along the borders of laminas V and VII of the gray matter of the spinal cord (Maseko et al., 2013), but this caudal expansion has not been observed in other mammals studied to date.

15.4.1.6 The Hypoglossal Nucleus (XII)

The hypoglossal nucleus has been reported in all mammals studied to date and is readily identified as a dense cluster of cholinergic neurons located in the periventricular gray matter of the dorsal medial aspect of the caudal medulla oblongata. The hypoglossal nerve is observed to emerge from the ventral aspect of this

nucleus, course lateral to the inferior olivary nuclear complex, and emerge from the ventral surface of the medulla oblongata in a position lateral to the bulge formed by the inferior olivary nuclear complex. No specific cross-species comparisons of this nucleus have been undertaken, but it might be predicted that those species with specialized glossal and subglossal musculature may have relatively larger nuclei with relatively more cells, but this awaits confirmation.

15.4.2 The Ascending and Descending Fiber Pathways of the Medulla Oblongata

Within the medulla oblongata of all mammals studied to date are two major regions of distinct white matter, a medial column and a lateral sheath. The medial column lies between the raphe reticular column and the gigantocellular reticular nucleus (see as discussed in the following section). The medial column extends from the dorsal to the ventral aspect of the medulla oblongata for the entire rostrocaudal extent of this segment of the brain stem, except for when the inferior olivary nuclear complex is present, and then the column is split above and below this nuclear complex. This medial column houses several distinctive pathways, such as the medial longitudinal fasciculus, the medial lemniscus, and the pyramidal tract, and several less distinct pathways such as the tectospinal tract. The lateral sheath of white matter courses around the ventral and lateral edges of the medulla oblongata, and within this sheath, the ventral, rostral, and dorsal spinocerebellar, cuneocerebellar, and olivocerebellar tracts are found. These tracts all ultimately merge to form the inferior cerebellar peduncle in the more dorsal and rostral portions of the medulla oblongata.

15.4.3 The Neuromodulatory Nuclei of the Medulla Oblongata

15.4.3.1 The Catecholaminergic Nuclei of the Medulla Oblongata

Within the medulla oblongata of most mammals studied to date, five distinct catecholaminergic nuclei are invariably reported, these being: the rostral ventrolateral (RVL) tegmental group (C1), the rostral dorsomedial group (C2), the caudal ventrolateral (CVL) tegmental group (A1), the caudal dorsomedial group (A2), and the area postrema (AP) (Dell et al., 2010). A low density of tyrosine hydroxylase immunoreactive neurons forms the C1 nucleus in the ventrolateral medullary tegmentum from the level of the superior olivary nuclear complex to the caudal pole of the nucleus ambiguus. These neurons are found medial to the facial nerve nucleus rostrally and the nucleus ambiguus caudally.

They normally extend as a column of one to two cells width from the ventral aspect of the medullary tegmentum to the midlevel of the tegmentum. This C1 nucleus is not continuous with the A5 nucleus (see aforementioned section), as there is a distinct break between these two very similar appearing catecholaminergic cell columns. Toward the caudal end of the nucleus ambiguus, a third ventrolateral catecholaminergic cell column appears, but this time on the lateral aspect of the nucleus ambiguus, the A1 nucleus. While again appearing very similar to both the A5 and C1 nuclei, the A1 nucleus is again topologically distinct from the C1 column. The neurons of the A1 nucleus continue through to the spinomedullary junction. This almost continuous, but topologically distinct, set of catecholaminergic nuclear columns, A5, C1, and A1, show a very similar appearance across all mammals studied to date (Dell et al., 2010; Calvey et al., 2015a, 2015b, 2016).

A cluster of tyrosine hydroxylase neurons located in the dorsalmost portion of the solitary nuclear mass (see as discussed in the following section) adjacent to the floor of the fourth ventricle is assigned to the C2 nucleus. Within the C2 nucleus, two subdivisions are usually reported, a rostral densely packed cluster and a caudal region in very close apposition to the floor of the fourth ventricle and often referred to as the dorsal strip of C2 (Maseko et al., 2013; Dahlström and Fuxe, 1964). The rostral portion of C2 is distinguished as a small, but densely packed, cluster of tyrosine hydroxylase immunoreactive neurons lying close to the floor of the fourth ventricle and that can extend into the nuclear mass of the tractus solitarius. Extending caudally from the dorsal strip, catecholaminergic neurons are seen to spread into the gray matter of the tractus solitarius between the dorsal motor vagus nucleus and the area postrema, forming the A2 nucleus. The number and packing density of the A2 neurons is substantially less than that observed in the dorsal strip of the C2 nucleus. The A2 nucleus continues in this position through to the spinomedullary junction. The last commonly observed nucleus is the area postrema, which is generally observed to be a large cluster of intensely stained catecholaminergic neurons overlying the solitary nuclear mass. While this nucleus is present in all species studied to date (Dell et al., 2010), this nucleus does vary in appearance, from a more often observed solid nuclear mass at the midline in smaller-brained mammals (Kruger et al., 2012), to two distinct, slightly laterally placed masses in larger-brained mammals (Maseko et al., 2013).

While these nuclei are very similar across mammalian species, one distinct variance has been reported regularly. Within the rodents and the cape hare (Calvey et al., 2015a), a distinct rostral dorsal midline nucleus (C3) is reported. This nucleus is constituted of a small number of catecholaminergic cells located within the dorsal medial medullary tegmentum, dorsal to the raphe obscurus (ROb) (see as discussed in the following section), and close to the floor of the fourth ventricle. Thus, this nucleus appears to be specific to rodents and probably the lagomorphs, but further lagomorphs need to be studied to determine whether it appears in all lagomorphs or just the cape hare (Calvey et al., 2015a). While it is possible that other small catecholaminergic nuclei are present in the medulla oblongata (Maseko et al., 2013), expansions of those nuclei mentioned here, rather than assignation of new nuclei, are a more likely scenario.

15.4.3.2 The Serotonergic Nuclei of the Medulla Oblongata

Within the medulla oblongata, up to five distinct serotonergic nuclei are regularly reported (Dell et al., 2010; Calvey et al., 2015a, 2015b, 2016), although there is some variance (Patzke et al., 2014; Manger et al., 2002a). These five commonly reported nuclei include the raphe magnus (RMg), the RVL and CVL columns, the raphe pallidus (RPa), and the ROb nuclei. Throughout the rostrocaudal extent of the medulla oblongata, running from the dorsal gray matter to the floor of the medulla, two pararaphe columns of cells, usually only a few cells in width, form the raphe nuclei. At certain levels in the medulla oblongata, these raphe neurons produce serotonin and are classified as serotonergic nuclei (the RMg and ROb). In the rostral medulla, rostral to the inferior olivary nuclear complex, bilateral columns of medium-to-large serotonergic neurons form the RMg nucleus. At the caudal end of the RMg nucleus, lateroventral expansions of serotonergic neurons extend over the medial lemniscus and the pyramidal tracts and consolidate as distinct columns lateral to the inferior olivary nuclear mass, forming the most rostral part of the bilateral RVL serotonergic columns. The RVL is found in the ventrolateral medullary tegmentum from the caudal level of the facial nerve nucleus through to the level of the nucleus ambiguus. At this level, the serotonergic neurons continue as a column caudally, but are referred to as the CVL serotonergic column, which extends to the spinomedullary junction. The number of neurons forming these columns steadily decreases from rostral to caudal. The reason that these columns are divided into a rostral and caudal component is that the caudal component, while present in all eutherian mammals studied to date (Dell et al., 2010; Calvey et al., 2015a, 2015b, 2016), is absent from monotremes (Manger et al., 2002a) and marsupials (Patzke et al., 2014; Crutcher and Humbertson, 1978).

The RPa nucleus generally consists of a small number of serotonergic neurons located in the ventral midline of the medulla oblongata, between the pyramidal tracts. This nucleus is normally reported to be located from the level of the caudal pole of the facial nerve nucleus through to the rostral pole of the nucleus ambiguus. Within the pararaphe columns, extending from the level of the nucleus ambiguus to the spinomedullary junction, are serotonergic neurons assigned to the ROb nucleus. While the majority of the serotonergic neurons forming this nucleus are found within the pararaphe columns, occasional neurons are found one to two cells away from the midline columns (Maseko et al., 2013; Manger et al., 2002a). The appearance of both the RPa and ROb is quite consistent across mammals.

15.4.4 The Intrinsic Nuclei of the Medulla Oblongata

15.4.4.1 *The Inferior Olivary Nuclear Complex*

Typically, in mammals, the inferior olivary nuclear complex is located in the ventromedial rostral half of the medulla oblongata. There are generally four distinct nuclei found within this complex, the principal, medial, dorsomedial, and dorsal nuclei, but this varies with terminology used and species investigated. In some species, such as the African elephant, this complex is quite large (Maseko et al., 2013), presumably in line with the enlarged and complex cerebellum of the elephant (Maseko et al., 2012; Jacobs et al., 2014; Herculano-Houzel et al., 2014). The principal nucleus is generally the largest of the nuclei, with the medial nucleus located between the principal nucleus and the pyramidal tract. The dorsomedial nucleus is generally located between the principal nucleus and the raphe, while the dorsal nucleus typically forms a thin sheet, or cap, surrounding the dorsal and lateral aspects of the principal nucleus. All the neurons forming these nuclei are generally strongly immunoreactive for calbindin. A distinct olivocerebellar tract emerges from the medial aspect of this nuclear complex to turn ventrally at the midline to decussate just dorsal to the pyramidal tract to form the olivocerebellar tracts in the ipsilateral, ventrolateral margin of the medulla oblongata. This tract continues around the lateral margin of the medulla to join the inferior cerebellar peduncle.

15.4.4.2 *The Nuclei of Tractus Solitarius*

The gray matter that forms the nuclear mass of the solitary tract appears as a rostral continuation of lamina X of the spinal cord in all mammals studied to date. This gray matter occupies a position adjacent to the floor of the fourth ventricle from the level of the caudal pole of the vestibular nuclear complex through to the spinomedullary junction, spreading in a mediolateral plane from the midline and being wider rostrally, but remaining quite thin in the dorsoventral plane. The tractus solitarius is usually evident as a distinct white matter tract

lateral to this gray matter column. While subdivisions of the nuclear mass are well known in certain species (Paxinos et al., 2009), in a comparative sense, it is generally difficult to discern all the subdivisions well studied in the laboratory rat. Despite this, some of the subdivisions are readily observed across species. The dorsal motor vagal nucleus and hypoglossal nucleus (see aforementioned section) are invested within the gray matter of the tractus solitarius. Caudally, the gray matter of the tractus solitarius is bordered laterally and superiorly by the gracile nucleus and tract and is bordered inferiorly by the medullary reticular nuclei. Cholinergic neurons within this gray matter form the preganglionic neurons of the inferior salivatory nucleus (see aforementioned section), and these nuclei likely indicate the location of the gustatory region of the solitary nuclear mass.

15.4.4.3 *The Dorsal Column Nuclei*

Three nuclei, the gracile, cuneate, and external cuneate, are those that typically form the dorsal column nuclei. The dorsal funiculus of the spinal cord houses the gracile and cuneate fasciculi, which, at the level of the spinomedullary junction, invest into the caudal aspect of the gracile and cuneate nuclei. These two nuclei are small caudally and expand rostrally, inverse to the fasciculi. The gracile nucleus occupies the medial dorsal aspect of the caudal medulla oblongata. Lateral to the gracile nucleus is the cuneate and external cuneate nuclei. The cuneate nucleus shows a very similar organization to that of the gracile nucleus, but it is smaller than the gracile nucleus, likely due to the presence of the external cuneate nucleus that also receives input from the cuneate fasciculus. The external cuneate nucleus is located dorsal and lateral to the cuneate nucleus and caudally appears lobulated, but rostrally appears as a solid nuclear mass. This pattern of nuclei within the dorsal column is quite similar across mammalian species, but size does vary in accordance with innervation densities in the periphery.

15.4.5 The Reticular/Tegmental Nuclei of the Medulla Oblongata

The core of the medulla oblongata is occupied by the reticular formation throughout its rostrocaudal length. As with the pons, the medullary reticular formation has three clear columns: the midline raphe, the medial gigantocellular/magnocellular, and the lateral parvocellular. In addition, at the caudolateral portion of the medulla oblongata, a distinct lateral reticular nucleus is observed across most mammals. An intermediate reticular nucleus has been reported for the laboratory rat (Paxinos et al., 2009), but this is not commonly observed in comparative studies. The raphe column consists of two parasagittal cell columns usually less than a few cells wide, and these columns are separated by a thin sheet of white matter. The neurons within these columns are often, but not always, serotonergic (see aforementioned section). At most levels of the medulla oblongata, these raphe columns are separated from the gigantocellular column by the medial white matter column (see aforementioned section).

The gigantocellular column occupies the medial half of the larger reticular formation and is readily distinguished by its large neurons. These large neurons form mesh-like dendritic networks around the medullary fascicles. The lateral margin of the gigantocellular column is marked by a cluster of catecholaminergic neurons (the RVL tegmental group, C1, and the CVL tegmental group, A1), while the medioventral border is marked by the presence of the serotonergic neurons of the rostroventrolateral and caudoventrolateral columns. Lying between the gigantocellular column and the spinal trigeminal system (see aforementioned section) is the parvocellular reticular nucleus/column. A marked reduction in size of the neurons distinguishes this column from the other medullary reticular nuclei. It should be noted here that the parvocellular column of the medulla oblongata is principally involved in the respiratory system and has been divided into four distinct nuclei from the caudal end of the facial nerve nucleus to the spinomedullary junction and include, rostrocaudally, the Bötzinger complex, the pre-Bötzinger complex, and the rostral and caudal ventral respiratory groups (Smith et al., 2009). The caudal ventral respiratory group is also often referred to as nucleus retroambiguus, which consists of premotor neurons that project to the nucleus ambiguus to control changes in eupneic breathing to produce vocalizations, vomiting, coughing, and sneezing among other actions (Subramanian and Holstege, 2009). While this region is well studied in rats and cats, this region has not been examined in detail in more exotic species. In the caudolateral aspect of the medulla oblongata, the lateral reticular nucleus becomes distinct. In the more caudal aspects of the medulla, the lateral reticular nucleus is separated from the parvocellular and gigantocellular columns by a distinct region of white matter. This nucleus typically has a high density of medium-sized cells, and this difference in cell size can be used to distinguish this nucleus from the other reticular nuclei at more rostral levels of the medulla oblongata.

15.5 Consistency and Variation in the Mammalian Brain stem

The mammalian brain stem is an anatomically complex region of the brain, with multiple regions subserving a range of different functions. Despite this, the anatomy of the brain stem, analyzed at the level of nuclei and pathways, is remarkably consistent across mammalian species, despite changes in brain size, phylogenetic history, life histories, and sensory and motor specializations. Atlases of the brain stem in commonly studied mammals provide an easy guide to understand similarities and differences across mammalian species. While it is conventionally thought that larger brains are more complexly organized, in terms of the number of identifiable subdivisions, than smaller brains, this line of reasoning does not appear to apply to the nuclei of the brain stem (Manger et al., 2003). In fact, phylogenetic history appears to be a more reliable predictor of nuclear complexity than brain size or life history (Dell et al., 2010; Manger, 2005). Thus, within a mammalian order, the organization of the brain stem nuclei and pathways appears to be very consistent (varying in size of nuclei, but not in number of homologous nuclei), yet the organization can change between mammalian orders (Dell et al., 2010). Despite this, the brain stem still retains a strong level of stasis within its organization across all mammals, suggesting that: (1) it is difficult to change the organization of the brain stem without producing lethal phenotypes; and (2) the system level of analysis described here is not sensitive enough to extract the information required to understand much of the precise phylogenetic variation of the brain stem. While this second conclusion does not negate the first, it does indicate that analysis at finer levels of organization within the brain stem, such as individual neurons, chemical anatomy, receptors and receptor complexes, molecular organization, and patterns of active gene expression, are potentially the levels at which changes that overtly affect the behavior of a specific species are to be found. Unfortunately, to date, this kind of analysis across species is yet to be undertaken, but is a fertile ground for future research into understanding how neural processing within a seemingly stable neural milieu may vary between individuals, species, and orders of mammals. Despite this, at the systems level of organization, there is a very robust general organization of the brain stem across mammalian species that is not readily amenable to evolutionary change, but the changes that do occur appear to be related to phylogeny at the level of the mammalian order (Manger, 2005).

References

Arts, P.M., Bemelmans, F.J., Cools, A.R., 1998. Role of the retrorubral nucleus in striatally elicited orofacial dyskinesia in cats: effects of muscimol and bicuculline. Psychopharmacology 140, 150–156.

Bandler, R., Shipley, M.T., 1994. Columnar organization in the midbrain periaqueductal gray: modules for emotional expression? Trends Neurosci. 17, 379–389.

Behbehani, M.M., 1995. Functional characteristics of the midbrain periaqueductal gray. Prog. Neurobiol. 46, 575–605.

Bhagwandin, A., Fuxe, K., Bennett, N.C., Manger, P.R., 2008. Nuclear organization and morphology of cholinergic, putative catecholaminergic and serotonergic neurons in the brains of two species of African mole-rat. J. Chem. Neuroanat. 35, 371–387.

Bjarkam, C.R., Sorensen, J.C., Genesser, F.A., 1997. Distribution and morphology of serotonin-immunoreactive neurons in the brainstem of the New Zealand white rabbit. J. Comp. Neurol. 380, 507–519.

Bux, F., Bhagwandin, A., Fuxe, K., Manger, P.R., 2010. Organization of cholinergic, putative catecholaminergic and serotonergic nuclei in the diencephalon, midbrain and pons of sub-adult male giraffes. J. Chem. Neuroanat. 39, 189–203.

Calvey, T., Patzke, N., Kaswera, C., Gilissen, E., Bennett, N.C., Manger, P.R., 2013. Nuclear organization of some immunohistochemically identifiable neural systems in three Afrotherian species — *Potamogale velox, Amblysomus hottentotus* and *Petrodromus tetradactylus*. J. Chem. Neuroanat. 50–51, 48–65.

Calvey, T., Alagaili, A.N., Bertelsen, M.F., Bhagwandin, A., Pettigrew, J.D., Manger, P.R., 2015. Nuclear organization of some immunohistochemically identifiable neural systems in two species of the Euarchontoglires: a Lagomorph, *Lepus capensis*, and a Scandentia, *Tupaia belangeri*. J. Chem. Neuroanat. 70, 1–19.

Calvey, T., Patzke, N., Kaswera-Kyamakya, C., Gilissen, E., Bertelsen, M.F., Pettigrew, J.D., Manger, P.R., 2015. Organization of cholinergic, catecholaminergic, serotonergic and orexinergic nuclei in three strepsirrhine primates: *Galago demidoff, Perodicticus potto* and *Lemur catta*. J. Chem. Neuroanat. 70, 42–57.

Calvey, T., Patzke, N., Bennett, N.C., Kaswera-Kyamakya, C., Gilissen, E., Alagaili, A.N., Mohammed, O.B., Pettigrew, J.D., Manger, P.R., 2016. Nuclear organisation of some immunohistochemically identifiable neural systems in five species of insectivore — *Crocidura cyanea, Crocidura olivieri, Sylvisorex ollula, Paraechinus aethiopicus* and *Atelerix frontalis*. J. Chem. Neuroanat. 72, 34–52.

Crutcher, K.A., Humbertson, A.O., 1978. The organization of monoamine neurons within the brainstem of the North American opossum (*Didelphis virginiana*). J. Comp. Neurol. 179, 195–222.

Da Silva, J.N., Fuxe, K., Manger, P.R., 2006. Nuclear parcellation of certain immunohistochemically identifiable neuronal systems in the midbrain and pons of the Hihgveld molerat (*Cryptomys hottentotus*). J. Chem. Neuroanat. 31, 37–50.

Dahlström, A., Fuxe, K., 1964. Evidence for the existence of monoamine-containing neurons in the central nervous system. 1. Demonstration of monoamine in the cell bodies of brainstem neurons. Acta Physiol. Scand. 62, 1–52.

Dell, L.A., Kruger, J.L., Bhagwandin, A., Jillani, N.E., Pettigrew, J.D., Manger, P.R., 2010. Nuclear organization of cholinergic, putative catecholaminergic and serotonergic systems in the brains of two megachiropteran species. J. Chem. Neuroanat. 40, 177–195.

Dell, L.A., Patzke, N., Spocter, M.A., Bertelsen, M.F., Siegel, J.M., Manger, P.R., 2016. Organization of the sleep-related neural systems in the brain of the river hippopotamus (*Hippopotamus amphibius*): a most unusual cetartiodactyl species. J. Comp. Neurol. 524 (10), 2036–2058.

Gernandt, B.E., 1968. Functional properties of the descending medial longitudinal fasciculus. Exp. Neurol. 22, 326–342.

Glezer, I.I., Hof, P.R., Morgane, P.J., 1998. Comparative analysis of calcium-binding protein-immunoreactive neuronal populations in the auditory and visual systems of the bottlenose dolphin (*Tursiops truncatus*) and the macaque monkey (*Macaca fascicularis*). J. Chem. Neuroanat. 15, 203–237.

Gravett, N., Bhagwandin, A., Fuxe, K., Manger, P.R., 2009. Nuclear organization and morphology of cholinergic, putative catecholaminergic and serotonergic neurons in the brain of the rock hyrax, *Procavia capensis*. J. Chem. Neuroanat. 38, 57–74.

Gravett, N., Bhagwandin, A., Fuxe, K., Manger, P.R., 2011. Distribution of orexin-A immunoreactive neurons and their terminal networks in the brain of the rock hyrax, *Procavia capensis*. J. Chem. Neuroanat. 41, 86–96.

Gravett, N., Bhagwandin, A., Lyamin, O.I., Siegel, J.M., Manger, P.R., 2012. Sleep in the rock hyrax, *Procavia capensis*. Brain Behav. Evol. 79, 155–169.

Hashikawa, T., van Lieshout, D., Harting, J.K., 1986. Projections from the parabigeminal nucleus to the dorsal lateral geniculate nucleus in the tree shrew *Tupaia glis*. J. Comp. Neurol. 246, 382–394.

Herculano-Houzel, S., Avelino-de-Souza, K., Neves, K., Porfirio, J., Messeder, D., Feijo, L.M., Maldonado, J., Manger, P.R., 2014. The elephant brain in numbers. Front. Neuroanat. 8, 46.

Jacobs, B., Johnson, N.L., Wahl, D., Schall, M., Maseko, B.C., Lewandowski, A., Raghanti, M.A., Wicinskim, B., Butti, C., Hopkins, W.D., Bertelsen, M.F., Walsh, T., Roberts, J.R., Reep, R.L., Hof, P.R., Sherwood, C.C., Manger, P.R., 2014. Comparative neuronal morphology of the cerebellar cortex in afrotherian, carnivores, cetartiodactyls, and primates. Front. Neuroanat. 8, 24.

Jansen, J., 1969. On cerebellar evolution and organization from the point of view of a morphologist. In: Llinas, R. (Ed.), Neurobiology of Cerebellar Evolution and Development. American Medical Association Education & Research Foundation, Chicago, pp. 881–893.

Kitahama, K., Nagatsum, L., Pearson, J., 1994. Catecholamine systems in mammalian midbrain and hindbrain: theme and variations. In: Smeets, W.J.A.J., Reiner, A. (Eds.), Phylogeny and Development of Catecholamine Systems in the CNS of Vertebrates. Cambridge University Press, Cambridge, pp. 183–205.

Kozicz, T., Bittencourt, J.C., May, P.J., Reiner, A., Gamlin, P.D., Palkovits, M., Horn, A.K., Toledo, C.A., Ryabinin, A.E., 2011. The Edinger-Westphal nucleus: a historical, structural, and functional perspective on a dichotomous terminology. J. Comp. Neurol. 519, 1413–1434.

Kruger, J.L., Dell, L.A., Bhagwandin, A., Jillani, N.E., Pettigrew, J.D., Manger, P.R., 2010. Nuclear organization of cholinergic, putative catecholaminergic and serotonergic systems in the brains of five microchiropteran species. J. Chem. Neuroanat. 40, 210–222.

Kruger, J.L., Patzke, N., Fuxe, K., Bennett, N.C., Manger, P.R., 2012. Nuclear organization of cholinergic, putative catecholaminergic, serotonergic and orexinergic systems in the brain of the African pygmy mouse (*Mus minutoides*): organizational complexity is preserved in small brains. J. Chem. Neuroanat. 44, 45–56.

Lane, R.H., Allman, J.M., Kaas, J.H., 1971. Representation of the visual field in the superior colliculus of the grey squirrel (*Sciurus carolinensis*) and the tree shrew (*Tupaia glis*). Brain Res. 26, 277–292.

Luo, A., Tahsili-Fahadan, P., Wise, R.A., Lupica, C.R., Aston-Jones, G., 2011. Linking context with reward: a functional circuit from hippocampal CA3 to ventral tegmental area. Science 333, 353–356.

Manger, P.R., 2005. Establishing order at the systems level in mammalian brain evolution. Brain Res. Bull. 68, 282–289.

Manger, P.R., 2006. An examination of cetacean brain structure with a novel hypothesis correlating thermogenesis to the evolution of a big brain. Biol. Rev. Camb. Philos. Soc. 81, 293–338.

Manger, P.R., Fahringer, H.M., Pettigrew, J.D., Siegel, J.M., 2002. Distribution and morphology of serotonergic neurons in the brain of the monotremes. Brain Behav. Evol. 60, 315–332.

Manger, P.R., Fahringer, H.M., Pettigrew, J.D., Siegel, J.M., 2002. Distribution and morphological characteristics of cholinergic cells in the brain of monotremes as revealed by ChAT immunohistochemistry. Brain Behav. Evol. 60, 275–297.

Manger, P.R., Fahringer, H.M., Pettigrew, J.D., Siegel, J.M., 2002. Distribution and morphological characteristics of catecholaminergic cells in the brain of monotremes as revealed by tyrosine hydroxylase immunohistochemistry. Brain Behav. Evol. 60, 298–314.

Manger, P.R., Ridgway, S.H., Siegel, J.M., 2003. The locus coeruleus complex of the bottlenose dolphin (*Tursiops truncatus*) as revealed by tyrosine hydroxylase immunohistochemistry. J. Sleep Res. 12, 149–155.

Manger, P.R., Fuxe, K., Ridgway, S.H., Siegel, J.M., 2004. The distribution and morphological characteristics of catecholaminergic cells in the diencephalon and midbrain of the bottlenose dolphin (*Tursiops truncatus*). Brain Behav. Evol. 64, 42–60.

Maseko, B.C., Manger, P.R., 2007. Distribution and morphology of cholinergic, catecholaminergic and serotonergic neurons in the brain of Schreiber's long-fingered bat, *Miniopterus schreibersii*. J. Chem. Neuroanat. 34, 80–94.

Maseko, B.C., Spocter, M.A., Haagensen, M., Manger, P.R., 2012. Elephants have relatively the largest cerebellum size of mammals. Anat. Rec. 295, 661–672.

Maseko, B.C., Patzke, N., Fuxe, K., Manger, P.R., 2013. Architectural organization of the African elephant diencephalon and brainstem. Brain Behav. Evol. 82, 83–128.

Miller, G.S., 1923. The telescoping of the cetacean skull. Smithson. Misc. Collect. 76, 1–71.

Nicola, S.M., Surmeier, D.J., Malenka, R.C., 2000. Dopaminergic modulation of neuronal excitability in the striatum and nucleus accumbens. Annu. Rev. Neurosci. 23, 185–215.

Patzke, N., Bertelsen, M.F., Fuxe, K., Manger, P.R., 2014. Nuclear organization of cholinergic, catecholaminergic, serotonergic and orexinergic systems in the brain of the Tasmanian devil (*Sarcophilus harrisii*). J. Chem. Neuroanat. 61–62, 94–106.

Paxinos, G., Watson, C., Carrive, P., Kirkcaldie, M.T.K., 2009. Chemoarchitectonic Atlas of the Rat Brain. Elsevier, New York.

Peiters, R.P., Gravett, N., Fuxe, K., Manger, P.R., 2010. Nuclear organization of cholinergic, putative catecholaminergic and serotonergic nuclei in the brain of the eastern rock elephant shrew, *Elephantulus myurus*. J. Chem. Neuroanat. 39, 175–188.

Pettigrew, J.D., 1986. Flying primates? Megabats have the advanced pathway from eye to midbrain. Science 231, 1304–1306.

Pettigrew, J.D., Maseko, B.C., Manger, P.R., 2008. Primate-like retinotectal decussation in an echolocating megabat, *Rousettus aegyptiacus*. Neuroscience 153, 226–231.

Rommel, S.A., Pabst, D.A., McLellan, W.A., 2002. Skull anatomy. In: Perrin, W.F., Wursig, B., Thewissen, J.G.M. (Eds.), Encyclopedia of Marine Mammals. Academic Press, Elsevier, Amsterdam, pp. 1033–1047.

Sherk, H., 1979. A comparison of visual-response properties in cat's parabigeminal nucleus and superior colliculus. J. Neurophysiol. 42, 1640–1655.

Sherwood, C.C., 2005. Comparative anatomy of the facial motor nucleus in mammals, with an analysis of neuron numbers in primates. Anat. Rec. 287, 1067–1079.

Smith, J.C., Abdala, A.P.L., Rybak, I.A., Paton, J.F.R., 2009. Structural and functional architecture of respiratory networks in the mammalian brainstem. Philos. Trans. R. Soc. Lond. B Biol. Sci. 364, 2577–2587.

Steinbusch, H., 1981. Distribution of serotonin-immunoreactivity in the central nervous system of the rat – cell bodies and terminals. Neuroscience 6, 557–618.

Subramanian, H.H., Holstege, G., 2009. The nucleus retroambiguus control of respiration. J. Neurosci. 29, 3824–3832.

Watson, C., Provis, J., Herculano-Houzel, S., 2012. What determines motor neurons number? Slow scaling of facial motor neuron numbers with body mass in marsupials and primates. Anat. Rec. 295, 1683–1691.

Weinberg, E., 1928. The mesencephalic root of the fifth nerve. A comparative anatomical study. J. Comp. Neurol. 46, 249–405.

16

Comparative Anatomy of Glial Cells in Mammals

A. Reichenbach, A. Bringmann
University of Leipzig Faculty of Medicine, Leipzig, Germany

16.1 Classification of Glial Cells

The term "glia" (Greek for "glue") was introduced in the 19th century by Rudolf Virchow (1846). Virchow searched for a connective tissue of the nervous system, and he discovered cells which were no nerve or vascular cells. Glial or glialike neuron-supporting cells are present in the central (CNS) and peripheral nervous systems (PNS). The term "glia" applies to all cells in the CNS that (1) are not neurons and (2) do not belong to mesenchymal structures such as blood vessels and meninges.

CNS glia can be divided into two "subfamilies," microglia and macroglia. Microglia are blood-derived macrophages which invade the brain during ontogenesis. The term "macroglia" summarizes a wide diversity of cell types arising from progenitor cells of the embryonic neuroepithelium together with neurons, including ependymoglia [radial glia including tanycytes and retinal Müller cells, ependymocytes, choroid plexus epithelial cells, and retinal pigment epithelial (RPE) cells], several subtypes of astrocytes, and oligodendrocytes (Fig. 16.1). In contrast to oligodendroglia, astroglia and ependymoglia are characterized by the fact that (at least some of) their processes bear endfeet, which contact a basal lamina around blood vessels, the pia mater (or the vitreous body of the eye), or both. Ependymoglia display a bipolar shape and additionally contact the ventricular surface (or the subretinal space) (Reichenbach and Robinson, 1995). Astrocytes may be radially oriented but never contact the ventricular system. Peripheral glia is specific to the PNS and can be classified as Schwann cells that form myelin, non–myelinforming Schwann cells, perisynaptic Schwann cells, perineuronal satellite cells of the dorsal root ganglia and the autonomic ganglia, enteric glia, and glialike cells

in sensory epithelia. A peculiar glial cell type is constituted by the olfactory ensheathing glial cells which accompany the olfactory nerve axons from PNS into the CNS.

Although glial cells are defined as being not neurons, it is often difficult to discriminate between both cell types because neurons and glial cells share a lot of functional characteristics, eg, expression of voltage-dependent ion channels, ligand receptors, transmembrane transporters, second messenger pathways, and release mechanisms for signaling molecules, and because macroglia can serve as source for neurons in the developing and adult CNS (Kimelberg, 2004; Wang and Bordey, 2008). However, a characteristic feature of glial cells is that these cells, in contrast to neurons, are not directly involved in information processing chains. Glia are neuron-supporting cells, providing homeostasis for neuronal networks. However, for many functions, it is difficult to differentiate between a supportive and an instructive role of glial cells.

16.2 General Principles of Glial Cell Phenotype and Distribution

16.2.1 Evolution of the Neuronal Support by Glial Cells

Glial cells are neuron-supporting cells which play crucial roles in the maintenance of the neuronal integrity, metabolism, and survival. One main task of glial cells is the creation of an optimal microenvironment for neuronal functioning. During the evolution of metazoa, the functional role of glial cells became more and more complex (Reichenbach and Pannicke, 2008). In small "primitive" animals such as polyps, single sensory

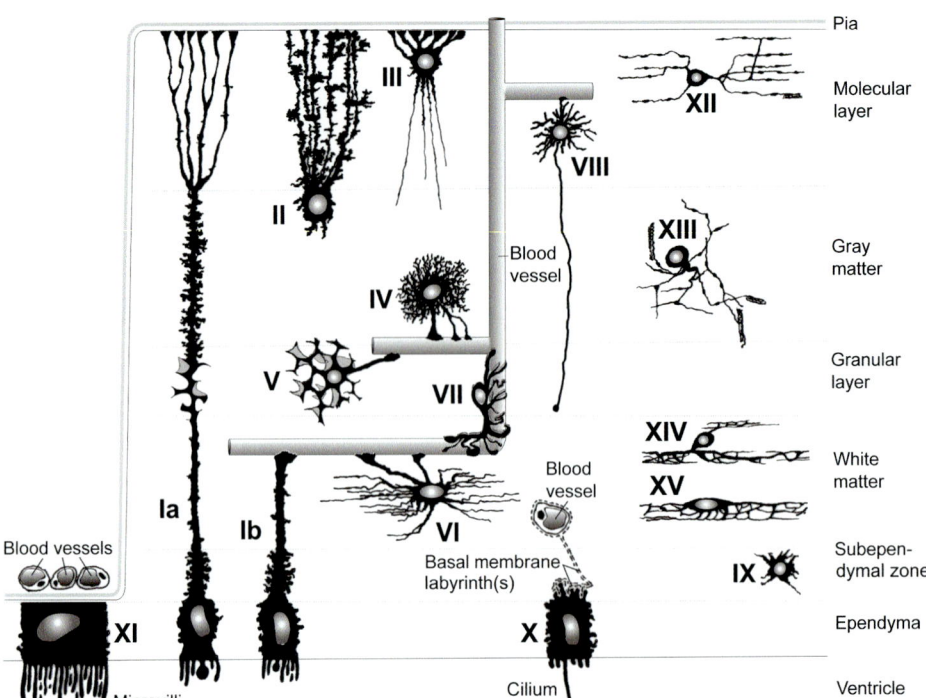

FIGURE 16.1   The main types of macroglial cells and the localization in different layers and specialized regions of the central nervous system. I, tanycyte (Ia, pial; Ib, vascular); II, radial astrocyte/Bergmann glial cell; III, marginal astrocyte; IV, protoplasmic astrocyte; V, velate astrocyte; VI, fibrous astrocyte; VII, perivascular astrocyte; VIII, interlaminar astrocyte; IX, immature glial cell/glioblast; X, ependymocyte; XI, choroid plexus cell; XII, type I oligodendrocyte; XIII, type II oligodendrocyte; XIV, type III oligodendrocyte; XV, type IV oligodendrocyte. *Modified from Fig. 2.1, Reichenbach, A., Wolburg, H., 2005. Astrocytes and ependymal glia. In: Kettenmann, H., Ransom, B.R. (Eds.), Neuroglia, second ed. Oxford University Press, Oxford, New York, pp. 19–35.*

and ganglion neurons are scattered throughout the tissues, without any associated specialized supporting cells (the single-cell stage of nervous system complexity; Fig. 16.2). The evolution of larger and more complex animals (eg, *Caenorhabditis elegans*) resulted in the development of specialized sensory organs and groups of loosely associated ganglion neurons (the oligocellular stage of nervous system complexity; Fig. 16.2). The sensory organs contain glialike cells, but ganglion neurons are, at best, touched by a glial cell process (Fig. 16.2). At this stage, glial cells are required to increase the sensitivity and specificity of sensory neurons (Bacaj et al., 2008), while the metabolism and survival of ganglion neurons do not need glial support. This situation can be also observed in oligocellular mechanoreceptive organs of the vertebrate skin (Reichenbach et al., 2004).

More complex animals develop large, sophisticated sensory organs and nervous centers (brains; multicellular level of nervous system complexity; Fig. 16.2). The sensory epithelia in these organs consist of neurons and glialike cells. As exemplified by the vertebrate retina, these glialike cells (Müller and RPE cells) support the receptive functions of neurons, eg, by providing light guidance toward photoreceptors and recycling of photopigments (Franze et al., 2007; Muniz et al., 2007). However, in such complex multicellular sensory organs,

glial cells delegate some of the sensory support functions to other structures outside the neural tissue such as cornea, iris, and lens in the eye, and nonneural structures of the ear. This may be a reason for a reduced number of glial cells per sensory neuron (in some mammalian retinas, one Müller cell is responsible for the support of up to 30 photoreceptors; Figs. 16.2 and 16.3B; Reichenbach, 1989a), while in oligocellular sensory organs, glialike cells may even outnumber neurons. In multicellular sensory organs, ganglia, and brains, glial cells become absolutely essential for the neuronal metabolism and survival; for example, induction of Müller cell death is followed by the degeneration of the entire neural retina (Dubois-Dauphin et al., 2000). In addition, the perisynaptic glia in ganglia or brains is essential for the synaptic transmission (Araque et al., 1999).

Generally, increasing brain size is associated with a rise in glial cell number per neuron (Fig. 16.2; Reichenbach, 1989a). In the mammalian brain, glial cells outnumber neurons by far. The total size of a mammal correlates with the size of the brain and with the size of neurons (Purves, 1988). These relations have been explained by the facts that in a big body, (1) more cells exist and need innervation (which requires increasing numbers of neurons and axon collaterals), and (2) the

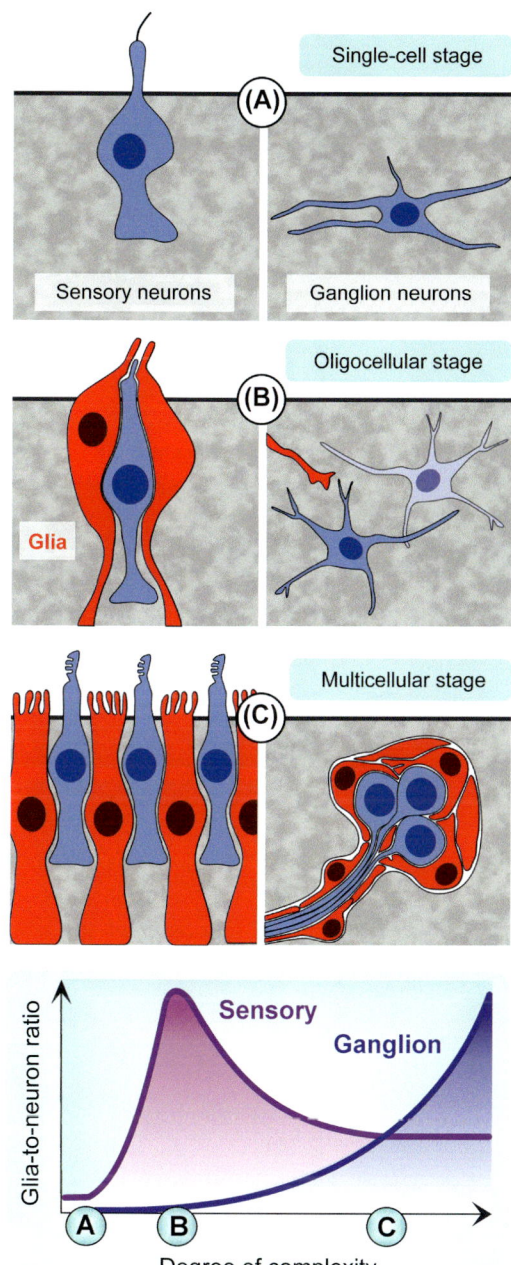

FIGURE 16.2 Evolution of increasing complexity of glial cell (*red*) support of sensory (*left*) and ganglion (*right*) neurons (*blue*). Below: The ratio between glial and neuronal cells is shown over three stages of increasing nervous system complexity. At stage A, single neurons do not require glial cell support to differentiate, function, or survive. At stage B, sensory neurons need glial support for selection, transformation, and transduction of adequate stimuli; the glial support improves the signal-to-noise ratio of stimulus perception. At stage C, neurons need glial support for the improvement of the signal-to-noise ratio of signal transmission, synaptic transmission, saltatory conduction, metabolism, clearance of waste products, extracellular homeostasis, and survival. *Modified from Fig. 16.1, Reichenbach, A., Pannicke, T., 2008. A new glance at glia. Science 322, 693–694.*

nerves have to bridge larger distances (which requires accelerated impulse propagation and, thus, increases of the axonal length and thickness, resulting in increased size of neuronal somata). In parallel, not only the absolute and relative number of glial cells (Reichenbach, 1989a) but also the size of glial cells increases (Fig. 16.4A).

The step from the single-neuron stage, which lacks glia, to the functionally associated neurons and glial cells of an oligocellular-stage sensory organ may have been triggered by the need for more sensitive and specific perception of environmental stimuli, forcing the development of accessory glialike cells as the most ancestral glia type. The primary glial cell function of improving the signal-to-noise ratio of perception appears to involve homeostasis of the extracellular milieu (Bacaj et al., 2008). Once established, these homeostatic glial functions may have been used to assist synaptic transmission (Araque et al., 1999), rapid axonal signal transport, and other neuronal functions. The step from the oligocellular stage to more complex neural tissues may have solved an additional quantitative problem: the dense crowding of neurons in large sensory epithelia or ganglia and brains. Both types of tissues are typically encapsulated against their nonneural environment, involving a blood–brain barrier to which glial cells contribute (Reichenbach and Wolburg, 2009). The insulated neurons depend on nutrient delivery and clearance of waste products. Thus, the need for extracellular homeostasis in an extended sense may have been the driving force for the multiplication of glia in complex animals (Reichenbach, 1989a). The homeostatic functions of glia may have been the basis for a more direct involvement in neuronal information processing, both by controlled modification of these functions (Oliet et al., 2001) and by further mechanisms including the release of "gliotransmitters" (Araque et al., 1999). After the glia-cellular system had been established in the phylogenesis, it became available for a variety of other functions including the guidance of neuron migration and axon pathfinding during embryogenesis, interaction with the immune system, and many others. Most of these functions rely upon the very special morphology of glial cells.

16.2.2 Glial Cell Phenotype: Cell Processes

The morphology of astro- and ependymoglia is very diverse (Fig. 16.1). Glial cells in the CNS differentiate relatively late in ontogenesis, at periods when most of the target structures (pia mater, vessels, neurons, and their axons) are already formed. Glial cells and processes must fit into this preexisting neuronal and vascular network. Thus, the shape and ultrastructure

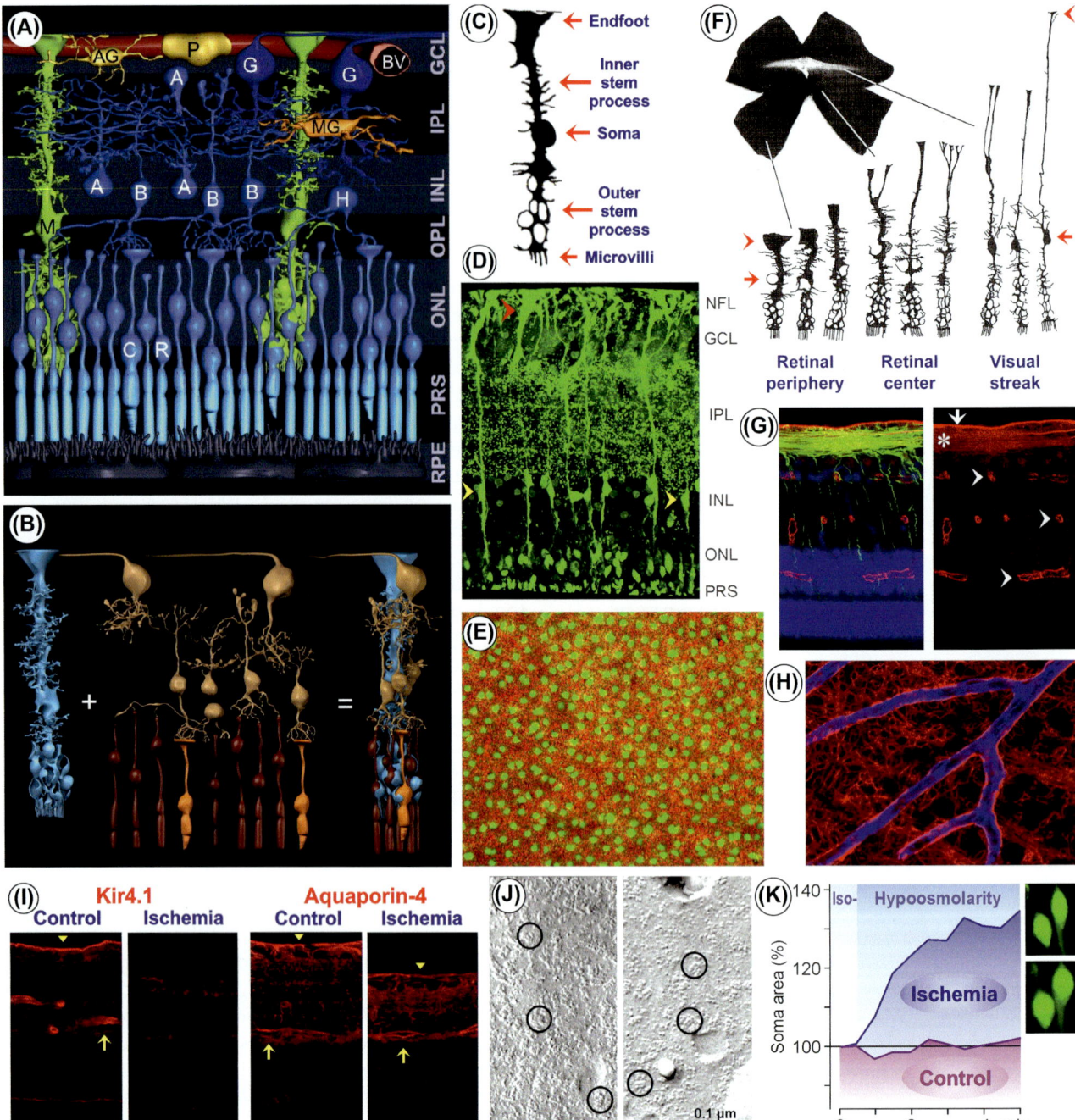

FIGURE 16.3 Müller cells as radial glia of the retina. (A) Schematic drawing of the cellular constituents of the mammalian retina. Müller cells (M) span the entire thickness of the neuroretina. The inner retina contacts the vitreal cavity (*top*) and contains the following retinal layers: nerve fiber layer, ganglion cell layer (GCL), inner plexiform layers (IPL), and inner nuclear layer (INL). The outer retina is directed to the outer surface of the eye and contains the following layers: outer plexiform layer (OPL), outer nuclear layer (ONL), the subretinal space containing the photoreceptor segments (PRS), and the retinal pigment epithelium (RPE). The inner retina is supplied with oxygen and nutrients by intraretinal blood vessels (BV). Astrocytes (AG) are localized in the nerve fiber/ganglion cell layers. Astrocytes and Müller cells contact the superficial blood vessels and the basal lamina at the inner surface of the retina. The perikarya of Müller cells are localized in the INL. From the perikaryon, two stem processes of Müller cells extend toward both surfaces of the neuroretina. The funnel-shaped endfeet of Müller cells form (in association with the basement membrane) the inner surface of the retina. In the IPL and OPL, side branches which form perisynaptic membrane sheets originate at the stem processes of Müller cells. In the ONL, the stem processes of Müller cells form membrane sheaths which envelop the perikarya of rods (R) and cones (C). Microvilli of Müller cells extend into the subretinal. Microglial (MG) cells are located in the inner retinal layers and the OPL. *A*, amacrine cell; *B*, bipolar cell; *G*, ganglion cell; *H*, horizontal cell; *P*, pericyte. (B) Müller cells constitute the cores of oligocellular domains of the retinal information processing. Every Müller cell is surrounded by a group of retinal neurons with which it interacts specifically during development, mature functioning, and retinal injury. *Blue*, Müller cell; *bright brown*, neurons; *orange*, cone; *dark brown*, rods. (C) A Golgi-stained human Müller

of glial processes are largely determined by the local environment, in particular by the type of the contacted element (cerebrospinal fluid, basal lamina, synapses, axons, neuronal somata) (Reichenbach and Wolburg, 2009). Generally, there are three types of contact elements toward which glial cell processes grow, resulting in three basic types of processes: (1) ventricle-contacting processes, (2) blood vessel- or pia-contacting processes, and (3) neuron-contacting processes (Fig. 16.5; Reichenbach, 1989b). Depending on the process morphology, glial cells can be classified as, eg, radial glia, astrocyte, or oligodendroglia (Fig. 16.5).

Epithelia or epitheloid aggregates are formed where macroglial cells constitute a "border sheath" against the ventricular space, pia, or blood vessels. This is observed in ependymocytes (Fig. 16.1, X), choroid plexus cells (Fig. 16.1, XI), and RPE cells (Fig. 16.6), but also, although less obvious, in marginal astrocytes (Fig. 16.1, III) and perivascular astrocytes (Fig. 16.1, VII). In the brain parenchyma, typical astrocytes are more or less star shaped (Fig. 16.1, IV and VIII), but this may be modified by adjacent neuronal cell bodies (Fig. 16.1, V) or axons (Fig. 16.1, VI), or by peculiar

relationships to the pial surface (Fig. 16.1, II). The term "radial glia" should be restricted to bipolar ependymoglial cells that extend long processes throughout (most of) the thickness of the tissue from one surface (the outer pial surface) to the other surface (the inner ventricular surface) of the brain.

The soma of astrocytes is usually rather poor in organelles. In most cases, bundles of intermediate filaments are present in astroglial somata (Fig. 16.7) which can be used as glial marker in electron microscopic sections. The somata of some types of ependymoglia contain melanin pigment granula (eg, choroid epithelial and RPE cells; Fig. 16.6). In most of these cells, the lateral membranes of the somata are interconnected by zonulae adherentes and tight junctions (Fig. 16.6), thus forming the blood–brain (blood–retinal) barrier (Reichenbach and Wolburg, 2005; Wolburg, 2006).

The soma of astro- and ependymoglia may give rise to one or several "primary" or "stem" processes, from which secondary branches may origin (Fig. 16.3C). Stem processes contain bundles of intermediate filaments ("glial fibrils"). High densities of intermediate filaments are found in the basal (endfoot-bearing)

cell. The main subcellular compartments are marked. (D) Slice of the pig retina. Müller cells (as well as PRS and some photoreceptor nuclei in the ONL) are *green* stained. *Yellow arrowheads,* Müller cell somata. *Red arrowhead,* Müller cell endfoot. Note the multitude of glial membrane sheaths that surround synapses in the IPL. (E) Supervision on the IPL of the guinea-pig retina, illustrating the regular pattern of profiles of Müller cell stem processes (*green*). Synaptic structures are *red* stained. In (D) and (E), Müller cells in freshly isolated retinal tissues were loaded with the vital dye MitoTracker Orange (*green*). (F) The phenotype of rabbit Müller cells varies with the retinal topography. Single Müller cells from the retinal periphery, retinal center, and the myelinated visual streak are shown. *Arrows,* cell somata; *arrowheads,* cell endfeet. (G–J) Basal laminae (at the inner retinal surface, *arrow*; and of the blood vessels, *arrowheads*) are contacted by glial endfeet membranes which contain ion and water channels. (G) A slice of the porcine retina immunostained against vimentin (*green*) and the glial potassium channel Kir4.1 (*red*). Cell nuclei are *blue* stained. Nerve fibers (*) are also surrounded by Kir4.1-containing glial membranes. (H) View onto the nerve fiber/ganglion cell layers of a rat retina. The superficial vessels of the retina are surrounded by perivascular glial membranes which contain aquaporin-4 water channels (*red*). Blood vessels are *blue* stained. (I) Transient retinal ischemia of 1 h induces a downregulation of Kir4.1 in glial endfeet membranes which abut the basal lamina at the inner retinal surface (*arrowheads*) and which surround the blood vessels (*arrows*); the endfeet expression of aquaporin-4 remains unaltered. Slices of a control (*left*) and a 3-day-postischemic retina (*right*) were immunostained. OLM, outer limiting membrane. (J) Freeze-fracture electron microscopy of replicas of vitreal glial endfeet. Orthogonal arrays of membrane particles (*encircled*) were found in control retinas (*left*) and 3 days after ischemia (*right*). (K) Müller cells in slices of 3-day-postischemic retinas swell rapidly during exposure to a hypoosmotic extracellular solution (60% osmolarity), while cells in control retinas do not alter the cell volume. The *images* show Müller cell somata before and after cell swelling. *Modified from (A) Fig. 16.1A, Reichenbach, A., Bringmann, A., 2013. New functions of Müller cells. Glia 61, 651–678; (B) Fig. 2.15, Reichenbach, A., Bringmann, A., 2010. Müller Cells in the Healthy and Diseased Retina. Springer, New York, Dordrecht, Heidelberg, London; (C) Fig. 16.2A, Reichenbach, A., Ziegert, M., Schnitzer, J., Pritz-Hohmeier, S., Schaaf, P., Schober, W., Schneider, H., 1994. Development of the rabbit retina. V. The question of 'columnar units'. Dev. Brain Res. 79, 72–84; (D) Fig. 16.2A, Wurm, A., Pannicke, T., Iandiev, I., Bühner, E., Pietsch, U.C., Reichenbach, A., Wiedemann, P., Uhlmann, S., Bringmann, A., 2006. Changes in membrane conductance play a pathogenic role in osmotic glial cell swelling in detached retinas. Am. J. Pathol. 169, 1990–1998; (E) Fig. 16.1C, Uckermann, O., Iandiev, I., Francke, M., Franze, K., Grosche, J., Wolf, S., Kohen, L., Wiedemann, P., Reichenbach, A., Bringmann, A., 2004. Selective staining by vital dyes of Müller glial cells in retinal wholemounts. Glia 45, 59–66; (F) Fig. 16.2, Reichenbach, A., Schneider, H., Leibnitz, L., Reichelt, W., Schaaf, P., Schümann, R., 1989. The structure of rabbit retinal Müller (glial) cells is adapted to the surrounding retinal layers. Anat. Embryol. (Berlin) 180, 71–79; from Fig. 16.4, Reichenbach, A., Eberhardt, W., Scheibe, R., Deich, C., Seifert, B., Reichelt, W., Dähnert, K., Rödenbeck, M., 1991. Development of the rabbit retina. IV. Tissue tensility and elasticity in dependence on topographic specializations. Exp. Eye Res. 53, 241–251; (G) Fig. 16.4A, Iandiev, I., Uckermann, O., Pannicke, T., Wurm, A., Pietsch, U.-C., Reichenbach, A., Wiedemann, P., Bringmann, A., Uhlmann, S., 2006. Glial cell reactivity in a porcine model of retinal detachment. Invest. Ophthalmol. Visual Sci. 47, 2161–2171; (H) Fig. 16.2A, Iandiev, I., Pannicke, T., Biedermann, B., Wiedemann, P., Reichenbach, A., Bringmann, A., 2006. Ischemia-reperfusion alters the immunolocalization of glial aquaporins in rat retina. Neurosci. Lett. 408, 108–112; (I) Fig. 16.2A, Pannicke, T., Iandiev, I., Uckermann, O., Biedermann, B., Kutzera, F., Wiedemann, P., Wolburg, H., Reichenbach, A., Bringmann, A., 2004. A potassium channel-linked mechanism of glial cell swelling in the postischemic retina. Mol. Cell. Neurosci. 26, 493–502; (J) Fig. 16.3A and C, Pannicke, T., Iandiev, I., Uckermann, O., Biedermann, B., Kutzera, F., Wiedemann, P., Wolburg, H., Reichenbach, A., Bringmann, A., 2004. A potassium channel-linked mechanism of glial cell swelling in the postischemic retina. Mol. Cell. Neurosci. 26, 493–502; (K) Fig. 16.1A and B, Pannicke, T., Iandiev, I., Uckermann, O., Biedermann, B., Kutzera, F., Wiedemann, P., Wolburg, H., Reichenbach, A., Bringmann, A., 2004. A potassium channel-linked mechanism of glial cell swelling in the postischemic retina. Mol. Cell. Neurosci. 26, 493–502.*

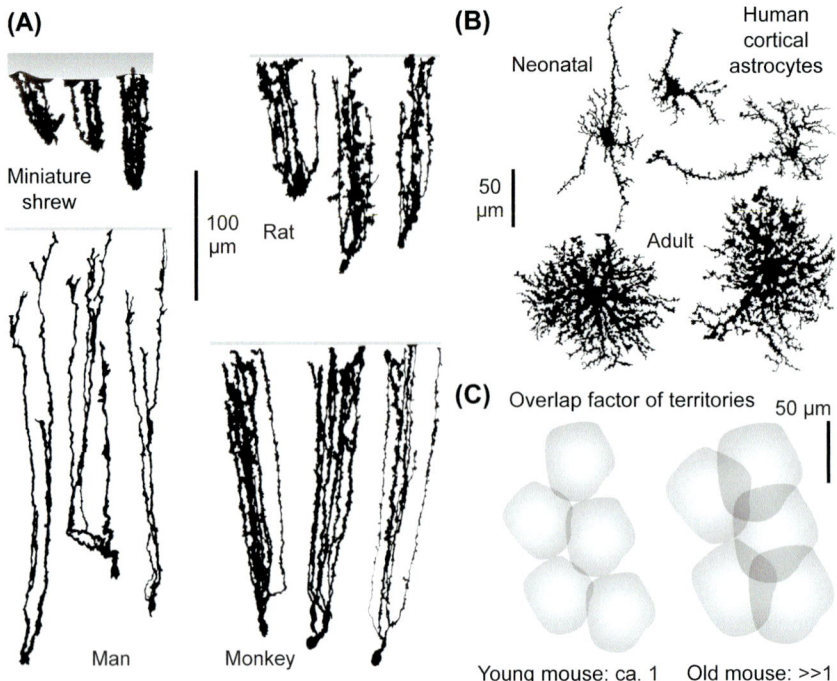

FIGURE 16.4 The size of astroglial cells is species dependent and varies with age. (A) As exemplified by cerebellar Bergmann glial cells, bigger mammalian species have not only bigger brains but also bigger glial cells. (B) As shown for human cortical astrocytes, there occurs a significant postnatal increase of the number, length, and complexity of cell processes. (C) The overlap factor of neuropilar territories occupied by the processes of astroglial cells in the mouse cortex varies with age. In young animals, the territories just touch each other (overlap factor ∼1), while in aged mice, astrocytic territories considerably overlap (overlap factor ∼2). *Modified from (A) Fig. 16.3, Siegel, A., Reichenbach, A., Hanke, S., Senitz, D., Brauer, K., Smith Jr., T.G., 1991. Comparative morphometry of Bergmann glial (Golgi epithelial) cells. A Golgi study. Anat. Embryol. (Berlin) 183, 605—612; (B) and (C) Figs. 16.2 and 3 Senitz, D., Reichenbach, A., Smith Jr., T.G., 1995. Surface complexity of human neocortical astrocytic cells: changes with development, aging, and dementia. J. Hirnforschung 36, 531—537.*

processes of tanycytes and retinal Müller cells (Fig. 16.7) and in processes of fibrous and radial astrocytes including Bergmann glia. Microtubuli are rarely found in astroglial processes; one of the few exceptions are the apical (microvilli-bearing) processes of Müller cells. Stem processes usually contain numerous mitochondria. An interesting exception are Müller cell processes in species with avascular retinas which contain mitochondria only at their apical pole (close to the choroid which is the only source of oxygen supply), whereas the stem processes are devoid of mitochondria. It has been suggested that due to the dominant glycolytic energy metabolism, Müller cells (and perhaps other astroglial cells) are free to move and place their mitochondria toward the sites of the highest pO_2 (Germer et al., 1998) rather than to sites of high energy demand as observed in neurons with a dominant aerobic metabolism.

Stem processes of astrocytes, tanycytes, and Müller cells do not show the rather regular dichotomic branching pattern characteristic of neuronal dendrites, but rather are the origins of specialized endings or side branches described in the following sections.

16.2.2.1 Apical Ventricle-Contacting Processes (Type I Processes)

Cell processes which contact the ventricular (or subretinal) space occur in ependymoglia such as tanycytes (Fig. 16.1, Ia and Ib) and retinal Müller cells (Fig. 16.3C), but never in astrocytes. The ventricle-contacting processes end with microvilli which extend into the fluid compartment (Figs. 16.3C and 16.5, I). Microvilli enlarge the surface area. The apical pole contains abundant mitochondria, a feature which indicates a high metabolic activity related to an active exchange of substances with the luminal fluid. The supporting cells of the inner ear (whose apical pole is confronted to endolymph resembling the intracellular milieu in many respects; Fig. 16.8A) seem to be an exception to this pattern, because apical microvilli and mitochondria are rare if not absent. The apical surface of some but not all ependymocytes (Fig. 16.1, X) is characterized by the presence of kinocilia.

Neighboring glial ventricular contact processes (and, when present, adjacent neuronal cell processes) are connected by various types of intercellular junctions which

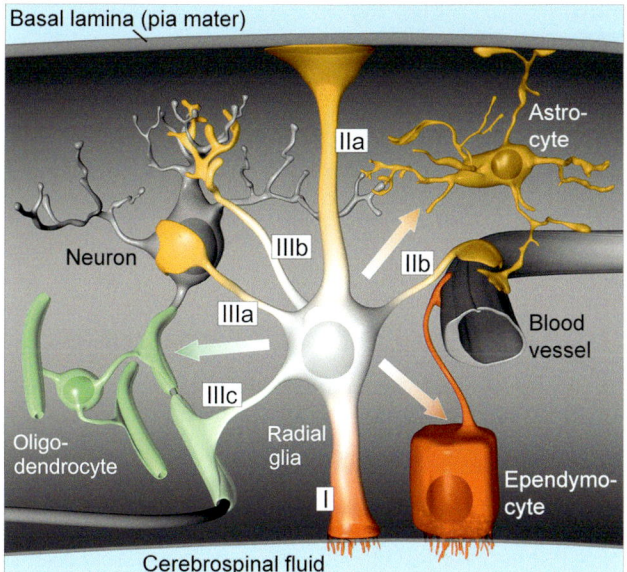

FIGURE 16.5 Hypothetical macroglial cell with all possible types of cell processes and microenvironmental contacts, and its real derivatives. The presence versus absence of the three basic types of cell processes defines the four basic macroglial cell types: radial glial cells (all three), ependymocytes (only two), astrocytes (only two), and oligodendrocytes (only one). I, ventricle-contacting process; II, pia- (IIa) or blood vessel- (IIb) contacting processes; III, neuron-contacting processes (IIIa, to neuronal somata; IIIb, to neuropil including synapses; IIIc, to axons). *Modified from Fig. 1.3, Reichenbach, A., Bringmann, A., 2010. Müller Cells in the Healthy and Diseased Retina. Springer, New York, Dordrecht, Heidelberg, London.*

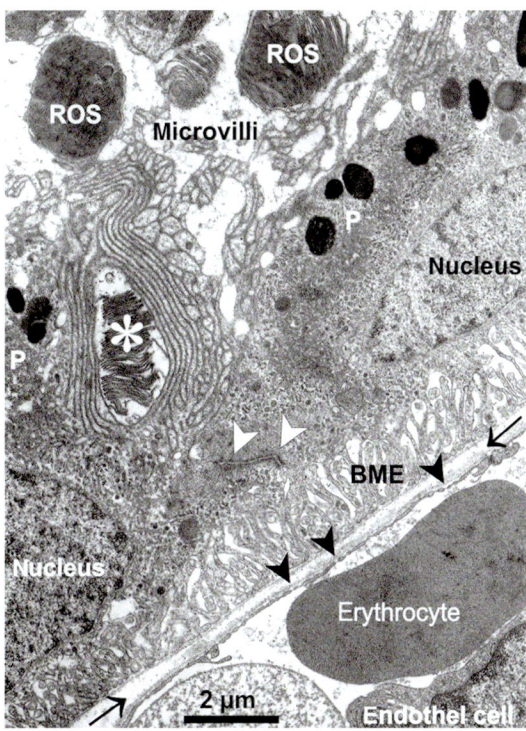

FIGURE 16.6 Retinal pigment epithelial (RPE) cells (example from rabbit) contain pigment granula (P). Apically (toward the outer segments of photoreceptor cells, ROS), RPE cells extend microvilli which may enclose the shed tips of ROS (*asterisk*) as the first step of phagocytosis. Basally, the cells face a basal lamina (Bruch's membrane, between the *arrows*) and display an enlarged surface area due to basal membrane enfoldings (BMEs). Because the capillaries of the underlying choroid possess a fenestrated endothelium (*black arrowheads*), RPE cells form the blood—retinal barrier by the expression of tight junctions (*white arrowheads*). *Modified from Fig. 2.5E, Reichenbach, A., Wolburg, H., 2005. Astrocytes and ependymal glia. In: Kettenmann, H., Ransom, B.R. (Eds.), Neuroglia, second ed. Oxford University Press, Oxford, New York, pp. 19—35.*

are important for the biomechanical tissue stability. These junctions (in particular, desmosomes) are a general feature of epithelial cells and occur very early in development (Revel and Brown, 1976). The nature of the intercellular junctions varies in dependence on the local microenvironment. In regions where no endothelial blood—brain barrier exists (in most circumventricular organs and RPE; Fig. 16.6), but not elsewhere, ependymoglial cells are connected by tight junctions which constitute the structural basis of the brain—cerebrospinal fluid barrier (Reichenbach and Wolburg, 2005). In addition to the type of intercellular junctions, the polarity of ependymoglial cells depends on the local microenvironment. When the basal pole of retinal Müller cells is exposed to the vitreous humor (after basal lamina defects in retinal wounds), the cells send microvilli into the vitreous cavity (Miller et al., 1986). Likewise, RPE cells proliferating under areas of retinal detachment may lose the basal lamina contact, and the whole cells face the subretinal fluid; in this case, the entire surface of the cells is covered by microvilli (Anderson et al., 1983). Thus, the microenvironment seems to determine not only the occurrence but also many of the particular features of distinct glial processes.

16.2.2.2 Basal Endfoot-Bearing Pia- and Vessel-Contacting Processes (Type II Processes)

While the contact with the apical surface of the neural epithelium is a primary feature of neuroectodermal cells that is retained even during mitotic cell proliferation (which occurs at the ventricular surface; Sauer, 1935), the basal pole of the neuroepithelium can be reached only by the growth of processes. In addition to the inherent tendency of ependymoglia toward bipolar growth (Hatten, 1984), the growth of basal processes may be stimulated by neighboring neurons (Hatten, 1985) and the chemoattractive influence of mesenchymal cells.

The basal processes of ependymoglia form endfeet onto the basal lamina of the mesenchymal layer underlying the nervous epithelium (pia mater; Figs. 16.1, Ia, II, III, and 16.5, IIa) or the basal lamina of blood vessels (Figs. 16.3G—16.I, 16.5, IIb, and 16.9B), or both (Fig. 16.9A). The latter occurs more frequently at later

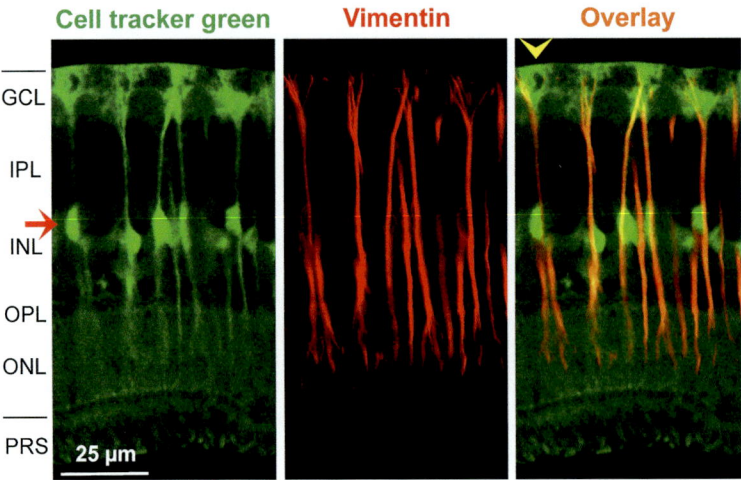

FIGURE 16.7 Vimentin bundles in the somata and stem processes of Müller glial cells. In a freshly isolated slice of the guinea pig retina, Müller cells and photoreceptor segments (PRSs) were loaded with the vital dye CellTracker Green; after fixation, the slice was immunostained against the glial intermediate filament vimentin which marks the axis of Müller cells. Note that the inner part of Müller cell endfeet (*arrowhead*) does not contain vimentin. *Arrow*, Müller cell soma. *GCL*, ganglion cell layer; *INL*, inner nuclear layer; *IPL*, inner plexiform layer; *ONL*, outer nuclear layer; *OPL*, outer plexiform layer. *Modified from Fig. 16.2A, Uckermann, O., Iandiev, I., Francke, M., Franze, K., Grosche, J., Wolf, S., Kohen, L., Wiedemann, P., Reichenbach, A., Bringmann, A., 2004. Selective staining by vital dyes of Müller glial cells in retinal wholemounts. Glia 45, 59–66.*

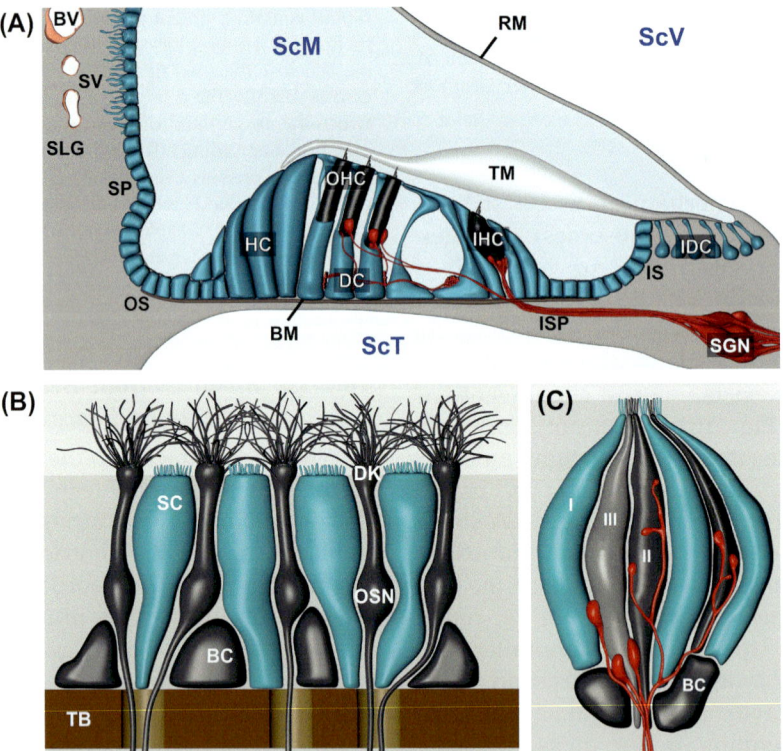

FIGURE 16.8 Cellular constituents of peripheral sensory epithelia and taste buds. (A) Organ of Corti. *BM*, basilar membrane; *BV*, blood vessel; *DC*, Deiters' cell; *HC*, Hensen's cell; *IDC*, interdental cells; *IHC*, inner hair cell; *IS*, inner sulcus; *ISP*, inner spiral plexus (type I SGN); *OHC*, outer hair cell; *OS*, outer sulcus; *RM*, Reissner's membrane; *ScM*, scala media; *ScT*, scala tympani; *ScV*, scala vestibuli; *SGN*, spiral ganglion neuron; *SLG*, spiral ligament; *SP*, spiral prominence; *SV*, stria vascularis; *TM*, tectorial membrane. (B) Olfactory epithelium. *BC*, basal cell; *DK*, dendritic knob; *OSN*, olfactory sensory neuron; *SC*, sustentacular (supporting) cell; *TB*, tubular bone. (C) Taste bud. I, supporting cell (*blue*); II, receptor cell; III, presynaptic cell; BC, (type IV) basal cell; *red*, chemosensory afferent fibers. *Modified from Figs. 16.2–16.4, Housley, G.D., Bringmann, A., Reichenbach, A., 2009. Purinergic signaling in special senses. Trends Neurosci. 32, 128–141.*

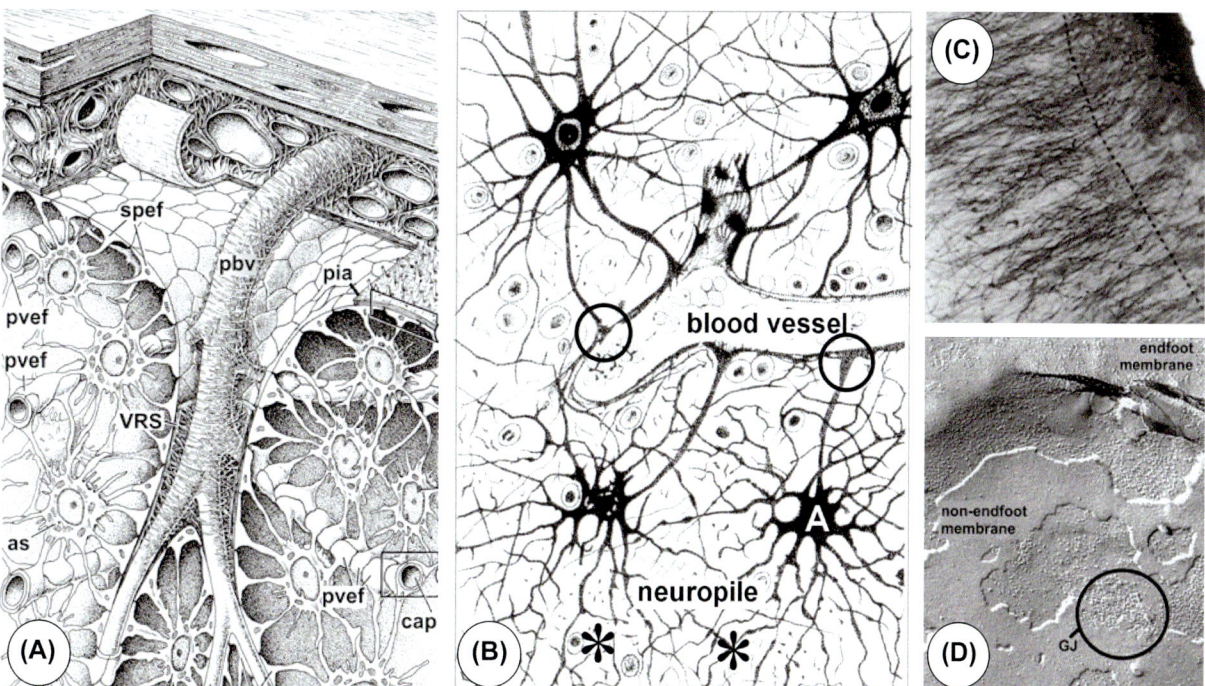

FIGURE 16.9 Astroglia of the brain. (A) Astrocytic endfeet around blood vessels and in the glia limitans of the Virchow-Robin spaces (VRSs). The layer of the subpial endfeet (spef) of astrocytes (as) is continuous with that of perivascular endfeet (pvef) because the VRSs extend from the pia mater. *pbv*, primary blood vessel; *cap*, capillary. (B). Astrocytes as bridges between neurons and blood vessels. Camera lucida drawing of silver-impregnated astrocytes (A). Astrocytes are polarized cells which display perivascular endfeet (*encircled*) on the one side, and neuron-abutting branches in the neuropil (*asterisks*) on the other side. (C) In the neocortex of higher primates, the radially aligned long processes of interlaminar astrocytes run in parallel to the columnar organization of neuronal elements. Astrocytes were labeled with antibodies against glial fibrillary acidic protein. The *dashed line* represents the border of layer I. (D) Electron microscopy of a freeze-fracture replica of an astrocytic process forming a perivascular endfoot (rat optic nerve). The endfoot membrane proper displays orthogonal arrays of membrane particles but no gap junctions. The membrane particle arrays of a gap junction (GJ) are found in the lateral nonendfoot membrane of the same process, probably constituting a coupling to another adjacent endfoot-bearing process. *Modified from (A) Plate 94, Krstic, R.V., 1991. Human Microscopic Anatomy. An Atlas for Students of Medicine and Biology. Springer, Berlin, Heidelberg, New York, London, Paris, Tokyo, Hongkong, Barcelona, Budapest; (B) Ramón y Cajál, S., 1911. Histologie du système nerveux de l'homme et des vertébrés. Maloine, Paris; Krstic, R.V., 1991. Human Microscopic Anatomy. An Atlas for Students of Medicine and Biology. Springer, Berlin, Heidelberg, New York, London, Paris, Tokyo, Hongkong, Barcelona, Budapest; (C) Fig. 16.1D, Colombo, J.A., Reisin, H.D., 2004. Interlaminar astroglia of the cerebral cortex: a marker of the primate brain. Brain Res. 1006, 126–131; (D) Fig. 10.11A, Reichenbach, A., Wolburg, H., 2009. Structural association of astrocytes with neurons and vasculature: defining territorial boundaries. In: Parpura, V., Haydon, P.G. (Eds.), Astrocytes in (Patho)physiology of the Nervous System. Springer, New York, pp. 251–286.*

stages of development when the tissue becomes thicker and vascularized. Endfoot (footplate) membranes are connected to the basal lamina by hemidesmosomes and form, together with the basal lamina and the mesenchymal cells (endothelial cells, pericytes, meningeal fibroblasts), the interface between the neural tissue and a fluid-filled extraneural compartment. Astrocytic endfeet almost completely cover all basal laminae in the CNS, along the vessels, the pia (Fig. 16.9A), and the vitreous body of the eye, here together with Müller cell endfeet (Fig. 16.7). The endfeet are often coupled to each other by gap junctions (and, sometimes, by zonulae adherentes) (Fig. 16.9D) which is the basis of dynamic interactions between astrocytes, neurons, and vascular structures (Giaume et al., 2010).

Astrocytes and radial glial (including retinal Müller) cells are positioned at a strategic position between neurons and blood vessels (Fig. 16.9B). The astroglial processes contacting neurons and blood vessels are the "opposite poles" of astrocytes as polarized cells. These glial cells constitute the anatomical and functional link between neurons and the compartments with which the neurons need to exchange molecules (blood vessels, vitreous chamber). Most nutrients, waste products, ions, water, and other molecules are transported through astrocytes between the vessels and neurons. In astrocytes, the exchange of molecules between neuronal compartments and blood vessels may occur within the processes of a single cell or via gap junctional coupling of entire populations of astrocytes (and between oligodendrocytes and astrocytes) (Robinson et al., 1993; Hampson and Robinson, 1995; Rash et al., 1997; Zahs, 1998). This coupling is spatially restricted and functionally regulated.

The glial envelope of larger blood vessels is rather similar to those on capillaries. Frequently, the glial envelope of larger vessels is formed by more than one layer of glial endfeet. Virchow-Robin spaces, localized between arteries, veins, and glial processes (Fig. 16.9A), represent perivascular extensions of the pia mater that accompany the arteries entering and the veins emerging from the cerebral cortex. Between the external surface of the blood vessels and the pial–glial peripheral lining, these spaces contain extracellular fluid "lakes" which may be very large. These perivascular spaces serve as transport routes for molecules exchanged between the circulation and the brain tissue, as they are interspaced between these compartments, may play an important role for the drainage of the cerebrospinal fluid, and represent reservoirs to buffer changes in extracellular ion concentrations and water. There may also be a communication between the perivascular spaces and lymphatic channels in the walls of major cerebral arteries. As a continuation of the glial envelope of the Virchow-Robin spaces, the parenchymal surface of the pia mater is covered by a so-called glia limitans. In higher mammals such as man, this compartment consists of a multilayered arrangement of astroglial endfeet. There are also specific "bordering astrocytes" which have no contacts other than to the pia and to neighboring astrocytes (Braak, 1975). Such cells may be involved in molecular exchange between the subarachnoid space and the brain parenchyma and may provide buffer capacity for ions and water.

The exchange of molecules between blood vessels and the brain (or retinal) parenchyma is limited by the blood–brain (blood–retinal) barrier. The barriers are constituted by vascular endothelial cells connected by tight junctions (Wolburg, 2006). Blood vessels in the subarachnoidal space express an endothelial blood–brain barrier like those of the brain parenchyma. The restrictive paracellular diffusion barrier established by tight junctions goes along with an extremely low rate of transcytosis and the expression of a high number of channels and transporters for such warranted molecules which cannot enter or leave the brain paracellularly. Perivascular glial endfeet contribute to the induction and maintenance of the blood–brain (blood–retinal) barrier (Wolburg, 2006; Reichenbach and Bringmann, 2015). However, there are exceptions constituted by the circumventricular organs and the outer retina which are supplied by fenestrated blood vessels. In these areas, the blood–brain (blood–retinal) barrier is generated by tight junctions among the tanycytes and RPE cells, respectively.

Endfoot-bearing processes are characterized by a cytoskeleton that is reinforced with abundant intermediate filaments. Bundles of intermediate filaments extend into the endfeet but fail to occupy the cytoplasm close to the basal lamina-contacting endfoot membrane (Fig. 16.7). The filaments often consist of vimentin when the endfoot is in contact with cerebrospinal fluid or vitreous humor and of glial fibrillary acidic protein (GFAP) when a blood vessel is contacted.

In tanycytes and retinal Müller cells, the endfeet are densely filled with smooth endoplasmic reticulum. With the exception of Müller cells in species with completely avascular retinas, the endfeet are rich in mitochondria. The occurrence of caveolae, coated pits, and vesicles concerned with endo-/exo- and pinocytosis indicates active material exchange between the endfeet and the fluid compartments behind the basal lamina (blood plasma, vitreous body, subarachnoidal fluid, or perilymph). A secretory function has been ascribed to the tanycytes of some circumventricular organs. Endfoot membranes contain ion channels (Fig. 16.3G and 16.I), transporters (Nagelhus et al., 1999), and orthogonal arrays of membrane particles (OAPs) (Fig. 16.3J; Wolburg, 2006). OAPs are composed (at least in part) of aquaporin-4 water channels (Fig. 16.3H and 16.I; Verbavatz et al., 1997). The expression of ion and water channels suggests that the compensation of ion and water gradients between the neural parenchyma and fluid-filled extraneural compartments occurs via glial endfeet. OAPs are accumulated in those parts of endfoot membranes which directly contact the perivascular or superficial basal lamina, while neuron-contacting membranes express only few OAPs if any (Fig. 16.9D; Wolburg, 2006; Reichenbach and Wolburg, 2009). This polarity develops concomitantly with the maturation of the blood–brain barrier and is lost or severely reduced under pathological conditions such as in tumors or inflammatory diseases and in cultured glial cells (Reichenbach and Wolburg, 2009).

It is an important question how the polarity of astrocytes is organized at the molecular level. Because the density of OAPs depends on the basal lamina contact, it is assumed that the extracellular matrix plays a critical role for the molecular arrangement of the endfoot membrane. Indeed, the OAP-related polarity of astrocytes seems to depend on the presence of agrin and of the dystrophin–dystroglycan complex (Noell et al., 2009, 2011). Agrin is a heparan sulfate proteoglycan of the extracellular matrix which is also present around brain microvessels (Barber and Lieth, 1997). Agrin binds to α-dystroglycan, a member of the dystrophin–dystroglycan complex which localizes at glial endfoot membranes in a similar distribution as aquaporin-4. This complex may be also responsible for the clustering of the glial potassium channel, Kir4.1 (Connors and Kofuji, 2002). Kir4.1 is normally restricted to the endfoot membrane in astrocytes and retinal Müller cells (Fig. 16.3G and 16.I; Kofuji and Newman, 2004). On the basis of the colocalization of aquaporin-4 and Kir4.1 in Müller cell endfeet (Fig. 16.316.I), and due to the fact that water fluxes are driven by ion currents, it was hypothesized that the glia-mediated water flux is coupled to the clearance of potassium (Nagelhus et al., 1999). Transglial water fluxes are also involved in mediating the volume regulation of glial cells. In the hypoxic retina, where Kir4.1 is downregulated in Müller cells (Fig. 16.316.I), the cellular water efflux is compromised, and the cells rapidly swell under hyposmotic conditions while normally the volume of the cells remains constant (Fig. 16.3K; Pannicke et al., 2004).

The occurrence and features of endfoot-bearing glial processes are determined by the local microenvironment. When, under pathological or experimental conditions, apical–lateral membranes of RPE cells, Müller cells, or ependymocytes come in contact with the mesenchymal compartment, these membranes develop an endfoot-like structure (Rosenstein and Brightman, 1979; Korte et al., 1986). This response suggests that the mesenchymal contact (ie, the extracellular matrix containing collagen, laminin, agrin, and other molecules) induces the production of basal laminae and the formation of endfeet with a membrane insertion of Kir4.1 potassium channels (Crawford, 1983; Sievers et al., 1986; Ishii et al., 1997). The formation of basal laminae by astro- and ependymoglial cells requires the presence of adjacent mesenchymal cells; mature Müller cells are not capable to restore the vitreal basal lamina after destruction by enzymatic digestion or microsurgical peeling (Miller et al., 1986; Halfter, 1998).

16.2.2.3 Lateral Neuron-Contacting Processes (Type III Processes)

Neuron-contacting glial processes, which are the main sites of glioneuronal interactions, ensheath neuronal somata (Fig. 16.5, IIIa), synapses, and processes (Fig. 16.5, IIIb), as well as axonal internodes (Fig. 16.5, IIIc) and nodal specializations of axons (Figs. 16.10A, B and 16.11A). According to the complex

shape of the ensheathed structures, this type of glial processes shows the most complex structure.

16.2.2.3.1 Processes of Protoplasmic Astrocytes

In the gray matter of the CNS which contains a high density of synapses, the dominant glial cells are protoplasmic astrocytes (Figs. 16.1, IV, 16.11E, and 16.12; Andriezen, 1893). Protoplasmic astrocytes have many irregularly shaped peripheral processes like lamellipodia and filopodia (Wolff, 1968; Derouiche and Frotscher, 2001; Chao et al., 2002; Reichenbach et al., 2004; Reichenbach and Wolburg, 2009). In contrast to the stem processes, peripheral processes contain only a minor portion of the glial cytoplasm and are virtually devoid of organelles. These processes contain ezrin and radixin (Derouiche and Frotscher, 2001), two actin-binding proteins which link the plasma membrane to the actin cytoskeleton and which are probably involved in the generation and stabilization of the complex process shape. Peripheral processes of protoplasmic astrocytes have a high surface-to-volume ratio of 12–18 μm^2 μm^{-3} (Fig. 16.13A; Rasmussen, 1975; Reichenbach et al., 1988). Despite the small size, these processes constitute about half of the astroglial volume and about 80% of the astrocytic surface area (Chao et al., 2002). The large membrane area gives space for the insertion of a wealth of membrane proteins such as ion channels, ligand receptors, and carrier molecules implicated in glioneuronal interactions.

Many synapses of the CNS are endowed with glial sheaths. One hippocampal astrocyte, for example, can contact up to 100 000 synapses (Bushong et al., 2002). Perisynaptic glial sheaths (Fig. 16.3D) are the prototypic

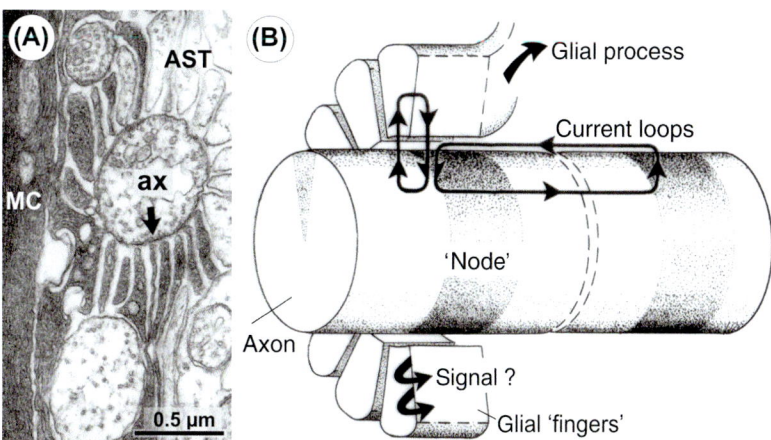

FIGURE 16.10 Periaxonal processes extending from retinal Müller cells and astrocytes. (A) Electron microphotograph obtained from the nerve fiber layer of a cat retina. The axon of a retinal ganglion cell (*ax*) displays local nodelike specializations (*arrow*) representing sites of high sodium channel density required for saltatory action potential propagation. Both an astrocyte (AST) and a Müller cell (MC) extend fingerlike processes toward this axolemmal area, constituting a corona of such processes. (B) Schematic view of such a corona of fingerlike glial processes at the "node" of a ganglion cell axon, together with the proposed flux of transmembrane ion currents. *Modified after (A) Fig. 16.7A, Holländer, H., Makarov, F., Dreher, Z., van Driel, D., Chan-Ling, T.L., Stone, J., 1991. Structure of the macroglia of the retina: sharing and division of labour between astrocytes and Müller cells. J. Comp. Neurol. 313, 587–603; (B) Fig. 16.7A, Chao, T.I., Skatchkov, S.N., Eberhardt, W., Reichenbach, A., 1994. Na^+ channels of Müller (glial) cells isolated from retinae of various mammalian species including man. Glia 10, 173–185.*

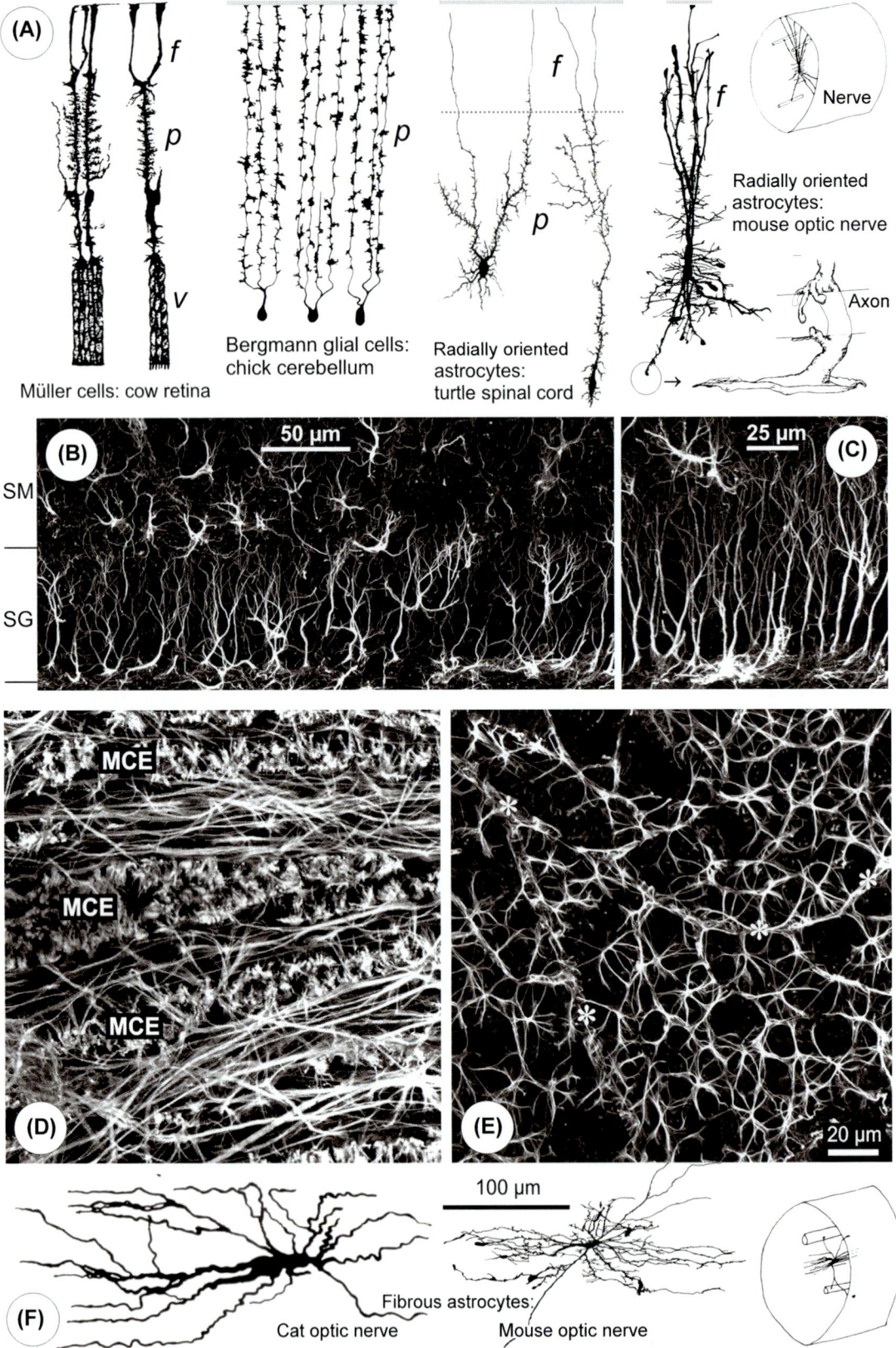

FIGURE 16.11　Radially oriented glial cells. (A) Camera lucida drawings of Golgi-stained or dye-injected cells. Müller cells represent radial glial cells ("tanycytes") of the retina while cerebellar Bergmann glial cells and radially oriented astrocytes in other CNS regions miss a contact to the ventricular surface and thus are specialized astrocytes. Note that the character of the processes is determined by the local microenvironment

sites of glioneuronal interactions and constitute the "third element" of the typical synapse, in addition to the pre- and postsynaptic terminals (Araque et al., 1999). Many synaptic clefts are "sealed" at the margins by astrocyte processes, constituting a physical barrier to diffusion, limiting spillover of neurotransmitters, and thus preventing extrasynaptic transmission. However, astrocytic processes are highly plastic. Like the dynamic dendritic spines that respond to changes in activity by structural alterations, astrocytic processes dynamically alter the coupling to neurons in response to environmental cues. The degree of synaptic ensheathing by glial lamellar processes varies considerably, even in the same brain area. In the rat neocortex, for example, about 56% of all synaptic perimeters are covered by glial processes (Chao et al., 2002). In the hippocampus, 57% of the synapses are in contact with astrocytic processes; astrocytic processes surround the synaptic cleft of 43% of these synapses (Ventura and Harris, 1999). The two types of excitatory inputs of cerebellar Purkinje cells, parallel and climbing fibers, exhibit different degrees of astrocytic ensheathment (67% vs. 94%) (Spacek, 1985; Xu-Friedman et al., 2001). In the supraoptic nucleus of the rat hypothalamus, astrocytes are capable of retracting from synapses in a reversible manner during maternal lactation (Oliet et al., 2001). On "synaptic glomeruli" or "complex synapses" (specialized subcortical structures where multiple synaptic junctions are enclosed in a common glial sheath), the glial coverage is very high (there are even multilamellar sheaths), but the glia does not penetrate the interior of the complexes and, thus, cannot seal the individual synaptic clefts. As an extreme, there are astroglia-free neuropil compartments, eg, in Rolando's substantia gelatinosa of the spinal cord and in the cochlear nucleus, where thin sensory axons terminate in "synaptic nests" that lack intrinsic glia. Incomplete ensheathment of synapses by glial

processes may favor neurotransmitter spillover between synapses and, thus, synchronized synaptic activity.

Glial uptake of neurotransmitters from the extracellular space is crucial for the maintenance of the spatial and temporal specificity of synaptic transmission (Oliet et al., 2001). The glutamate-neutralizing enzyme, glutamine synthetase, is present in the cytoplasm of perisynaptic astrocytic processes (Derouiche and Frotscher, 1991). Stimulation of glial neurotransmitter receptors, which are assembled in perisynaptic membranes, initiates glial metabolic responses, beneficial for the activated neuronal compartments (Araque et al., 1999). Glutamate, which modulates synaptic transmission, may be released from astrocytes through perisynaptic membranes (Bezzi et al., 2004). Perisynaptic glial processes are also involved in the maintenance and degradation of synapses and, thus, in synaptic plasticity (Chao et al., 2002; Reichenbach et al., 2004).

In Bergmann glial cells, a subtype of radial astrocytes, peripheral glial processes may form a "microdomain" which consists of a thin stalk and a small head from which lamellar processes arise which ensheath about five synapses (Fig. 16.13A, B, and D; Grosche et al., 1999, 2002). The heads of the microdomains may extend numerous "glial thimbles," forming complete caps on small neuronal protrusions which may represent dying or growing synapses. The microdomains occur as repetitive units on the stem processes of the cells, each interacting with a small group of synapses, or as appendages of another microdomain (Fig. 16.14). One Bergmann glial cell may have ~100 microdomains. A glial microdomain is thought to be capable of autonomous interactions with the ensheathed group of synapses, independent of other microdomains and the stem process. The energetic demands of each microdomain are supported by the mitochondria in the head (Fig. 16.13C). Multiple glial microdomains may overlap;

and may change from complex (protoplasmic, p) to smooth (fibrous, f) when the processes pass from gray to white matter; in nuclear layers with many small neuronal somata, the processes are velate (v). (B and C) Glial fibrillary acidic protein (GFAP) immunofluorescence of the dentate gyrus of the adult rat hippocampus; astroglial cell processes are radially aligned in the stratum granulare (SG) while typical "star-shaped" astrocytes are found in the stratum molecular (SM). (D—F) Nonradially oriented fibrous astrocytes in the mammalian retina (D and E) and optic nerve (F). (D and E) GFAP-labeled fibrous astrocytes in the murine retina. Close to the optic nerve (D), the processes of fibrous astrocytes run parallel to the bundles of ganglion cell axons and form (together with Müller cell endfeet, MCE) perivascular endfeet (mostly of the "en passant" type). In the retinal periphery (E), the density of nerve fibers is low; accordingly, the pattern of astrocytic processes is more or less star shaped. These processes form endfeet at blood vessels (*asterisk*). (F) Fibrous astrocytes in the cat and mouse optic nerves. The cell processes are mainly aligned longitudinally (parallel to the axons). The cells extend fingerlike perinodal processes; an artist's reconstruction of one of them is shown in (A) (*bottom right*). *Modified from (A) Ramón y Cajál, S., 1972. The Structure of the Retina. Thomas, Springfield, IL; from Figs. 16.2 and 9, Stensaas, L.J., Stensaas, S.S., 1968. Light microscopy of glial cells in turtles and birds. Z. Zellforsch. 91, 315—340; from Fig. 16.8A, B, and D, Butt, A.M., Duncan, A., Berry, M., 1994b. Astrocyte associations with nodes of Ranvier: ultrastructural analysis of HRP-filled astrocytes in the mouse optic nerve. J. Neurocytol. 23, 486—499; (B) and (C) Fig. 2.3H and I, Reichenbach, A., Wolburg, H., 2005. Astrocytes and ependymal glia. In: Kettenmann, H., Ransom, B.R. (Eds.), Neuroglia, second ed. Oxford University Press, Oxford, New York, pp. 19—35; (D) and (E) Fig. 2.4A and B, Reichenbach, A., Wolburg, H., 2005. Astrocytes and ependymal glia. In: Kettenmann, H., Ransom, B.R. (Eds.), Neuroglia, second ed. Oxford University Press, Oxford, New York, pp. 19—35; (F) Fig. 16.2E, Reichenbach, A., Siegel, A., Senitz, D., Smith Jr., T.G., 1992. A comparative fractal analysis of various mammalian astroglial cell types. Neuroimage 1, 69—77; from Fig. 16.4C, i, Butt, A.M., Colquhoun, K., Tutton, M., Berry, M., 1994. Three-dimensional morphology of astrocytes and oligodendrocytes in the intact mouse optic nerve. J. Neurocytol. 23, 469—485.*

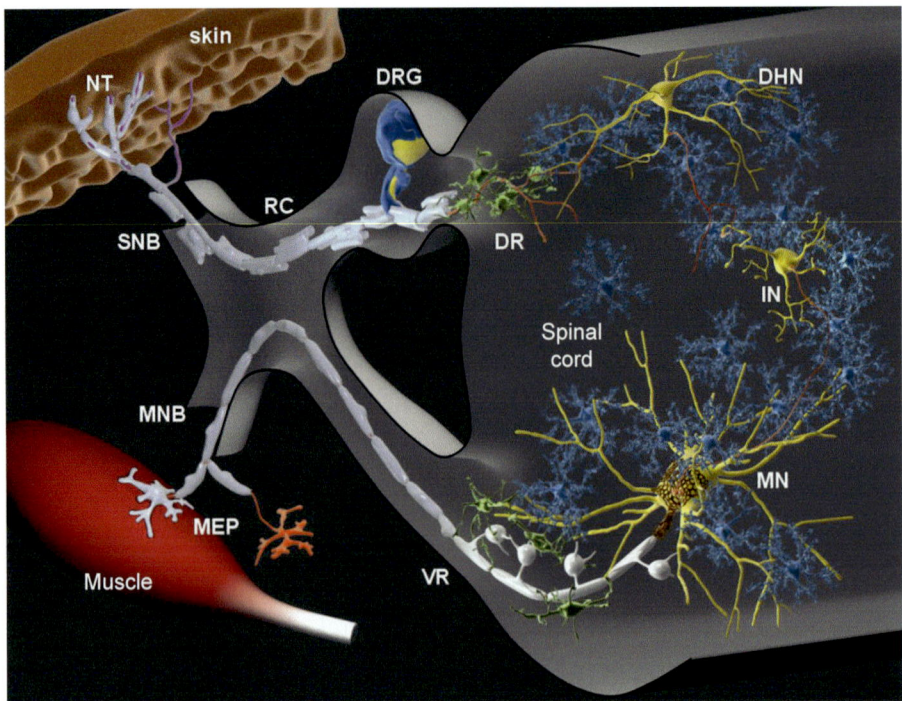

FIGURE 16.12   Glioneuronal associations along the signaling pathways from the receptor pole to the effector pole in the pain-induced withdrawal reflex arc. The neuronal circuit is simplified (eg, the reflex is actually multisynaptic and involves more interneurons in the spinal cord than shown; the peripheral pathways are unproportionally shortened). To visualize the ensheathed elements, parts of the covering of neuronal elements by glial sheaths (which actually is virtually complete, with the exception of the nonmyelinated axons in the central part of the dorsal root) have been omitted. The color codes are as follows: *yellow*, neuronal dendrites and somata; *red*, axons; *pink*, sensory processes; *white*, oligodendrocytes; *bluish white*, Schwann cells; *light blue*, satellite glial cells; *dark blue*, protoplasmic astrocytes; *green*, fibrous astrocytes; *brown*, perineuronal nets. *DHN*, dorsal horn neuron; *DR*, dorsal root; *DRG*, dorsal root ganglion; *IN*, (one of the) interneuron(s); *MEP*, motor endplate; *MN*, motoneuron; *MNB*, motor nerve bundle; *NT*, nociceptive terminal; *RC*, ramus communis; *SNB*, sensory nerve fiber bundle; *VR*, ventral root. *Original, courtesy of Jens Grosche, Leipzig.*

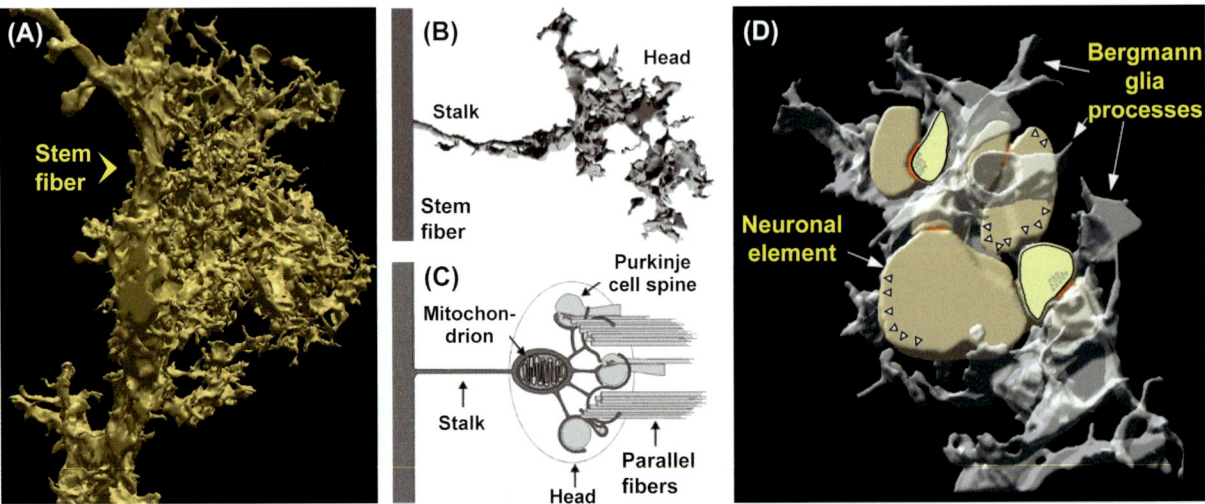

FIGURE 16.13   Bergmann glial cell processes and glial microdomains. (A) 3-D reconstruction of a part of a Bergmann glial cell process on the basis of electron microscopical images. (B) Glial microdomain as part of the 3-D reconstruction shown in (A). (C) Schematic drawing of such a microdomain and the relationships to the neuronal elements. (D) 3-D reconstruction of a group of neighboring cerebellar synapses (*bright brown*; synaptic clefts: *yellow*) together with the surrounding leaflets provided by a Bergmann glial cell (*blue-gray*). The *arrowheads* point to neuronal surfaces not covered by glial sheaths. *Modified after (A) Fig. 16.1B, Grosche, J., Matyash, V., Möller, T., Verkhratsky, A., Reichenbach, A., Kettenmann, H., 1999. Microdomains for neuron-glia interaction: parallel fiber signaling to Bergmann glial cells. Nat. Neurosci. 2, 139—143; (B) and (D) Figs. 16.6, 8A, and 10A, Grosche, J., Kettenmann, H., Reichenbach, A., 2002. Bergmann glial cells form distinct morphological structures to interact with cerebellar neurons. J. Neurosci. Res. 68, 138—149.*

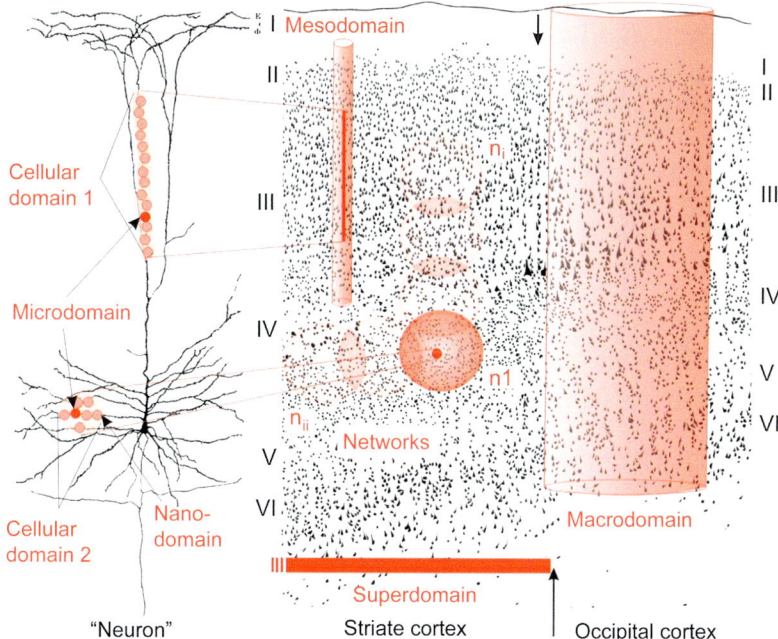

FIGURE 16.14 Schematic representation of the various types of glial domains showing the transition (*arrows*) between the striate and the occipital cortex in a human brain. At increasing levels of spatial organization, glial cells provide nanodomains up to superdomains for the interaction with neurons and blood vessels. With the possible exception of the nanodomains which probably interact with the neuronal nano-domains as long as these domains exist, the domains are not only determined by the (ultra)structural features of glial cells but also by the properties of the signal (or stimulus): (1) (a few) individual synapses are associated with the ensheathing glial microdomains but a higher level of stimulation of the related inputs may integrate (2) (oligo)cellular domains involving the whole glial cell (or a few of them) and the neuronal partners. It depends on the shape of the glial cells involved whether these domains are columnar ("type 1," eg, Bergmann glial cells; probably radial astrocytes in hippocampal stratum oriens, and hypothetically interlaminar astrocytes in the primate cortex) or rather spherical or ellypsoid ("type 2", star-shaped astrocytes). Strong, frequent, or synchronized stimulation may activate (via gap-junctional coupling) networks consisting of >50 astrocytes (the types n1 and n2 depend on the shape of the constituting glial cells). These networks are dynamic; because members of the coupled network at the margin are coupled to other cells, various networks overlap. Thus, when a neuronal stimulus will arrive later at such a cell, or when a calcium wave was triggered by the first "excited" astrocyte, (3) macrodomains will develop; this mechanism may proceed radially ($n1-n_{ii}$) or tangentially ($n1-n_i$). The size of the macrodomains may vary from small (macrodomain 1, corresponding, for example, to the orientation columns of Mountcastle, 1957) to large (such as ocular dominance columns or barrel fields; macrodomain 2). A further progression of integration will result in the generation of very large functional units, (4) superdomains, corresponding to entire cortical areas or gyri. Eventually, even whole hemispheres associated with huge astrocytic populations may transiently be involved, putatively mediating events such as spreading depression, seizures, and widespread neuronal degeneration. *Modified from Fig. 10.14, Reichenbach, A., Wolburg, H., 2009. Structural association of astrocytes with neurons and vasculature: defining territorial boundaries. In: Parpura, V., Haydon, P.G. (Eds.), Astrocytes in (Patho)physiology of the Nervous System. Springer, New York, pp. 251–286.*

in every given volume unit of the molecular layer, at least two microdomains, originating from different Bergmann glial cells, interdigitate (about eight Berg-mann glial cells are arranged around a Purkinje cell). This may fit with the observation that Purkinje cells ex-press two functionally distinct populations of synaptic spines and that individual spines are capable of inde-pendent activation (Denk et al., 1995); glial microdo-mains may be adjusted to meet this functional diversity of Purkinje cell synapses.

The axon hillock (initial segment) of neurons is indi-vidually covered by glial cell processes. Relatively large extracellular spaces, filled with a specialized extracel-lular matrix, are often interspaced between the initial segment and the glial sheath (Brückner et al., 1996). The nonsynaptic surface of neuronal dendrites may be covered by sheaths originating from several adjacent glial cells (Reichenbach and Wolburg, 2009). Many neu-rons in the brain (and motoneurons in the spinal cord; Fig. 16.12) are coated by perineuronal nets which consist of a particular extracellular matrix covered by netlike astrocyte processes (Brückner et al., 1996). Peridendritic glial sheaths and perineuronal nets may serve the buff-ering and clearance of ions and bioactive molecules in the adjacent extracellular clefts. The width of the extra-cellular cleft between neurons and astrocytic processes may vary considerably, depending on the develop-mental and functional state and on local specializations. This cleft contains extracellular matrix that also may vary in abundance and molecular composition and that contributes to the wealth of structural and func-tional glioneuronal interactions.

16.2.2.3.2 Processes of Fibrous Astrocytes

The white matter of the CNS is constituted by large numbers of axons or axon bundles. When the thickness of an axon exceeds a threshold of ~0.2 μm, the internodes (the sections between two consecutive nodes of Ranvier) are myelinated by the processes of oligodendrocytes which allows a saltatory conduction of action potentials (Waxman and Black, 1995). Bundles of thin, nonmyelinated axons are loosely ensheathed by cytoplasmic tongues arising from processes of fibrous astrocytes (Figs. 16.1, 16.VI, 16.11D, and 16.12; Andriezen, 1893). Because there are almost no synapses in the white matter, the processes of fibrous astrocytes are less complex than those of protoplasmic astrocytes, and the surface-to-volume ratio is significantly smaller (less than 4 $\mu m^2\ \mu m^{-3}$). The processes of fibrous astrocytes are preferentially aligned in parallel to the axon bundles; glial intermediate filaments are more prominent in white matter than gray matter astrocytes. The processes of fibrous astrocytes extend fingerlike tiny processes into the perinodal spaces (Fig. 16.11A; Raine, 1984; Black and Waxman, 1988). Such fingerlike processes originating from astrocytes and Müller cells also contact nodelike specializations of the unmyelinated axons in the retinal nerve fiber layer (Fig. 16.10A; Reichenbach et al., 1988). These "glial fingers" express the J-1 glycoprotein, an adhesion-modulating protein presumably involved in axon–glial interactions modulating the assembly and maintenance of the nodes (Ffrench-Constant et al., 1986). Where the "fingers" contact the axon, the extracellular cleft is extremely narrow (~6 nm) (Black and Waxman, 1988), whereas generally the perinodal spaces contain abundant extracellular matrix, comparable to that of perineuronal nets (Raine, 1984). The extracellular matrix may buffer large increases of the extracellular potassium level near the sites of action potential generation (Härtig et al., 1999). Perinodal glial processes may monitor neuronal activity and may trigger glial responses. Depolarization of glial membranes could be induced by ephaptic transmission, ie, a spillover of currents from the axons to the glial cells, which may result in metabolic responses of glial cells (Fig. 16.10B; Chao et al., 1994).

16.2.2.3.3 Processes of Velate Astrocytes

The nuclear layers of the CNS, such as the granule cell layer in the cerebellum, contain densely packed, small neuronal somata. Groups of such neuronal cell bodies are ensheathed by the honeycomb-like cytoplasmic tongues of velate astrocytes (Fig. 16.1, V; Chan-Palay and Palay, 1972). Velate processes have high surface-to-volume ratios of 20–25 $\mu m^2\ \mu m^{-3}$ (Rasmussen, 1975; Reichenbach et al., 1988, 1992). Depending on the size of a given neuronal soma, the glial sheath may be provided by one or more glial cells. Typically, large neurons

are ensheathed by velate terminal processes of several astrocytes, such as Purkinje cells in the cerebellum. By contrast, several densely packed small neuronal somata (such as the granule cells of the hippocampal fascia dentata and the cerebellar granule cells) are ensheathed by the velate processes of one or a few glial cells (Chan-Palay and Palay, 1972). Usually, glial sheaths cover the neuronal soma surface almost completely, with the exception of "holes" allowing for direct apposition of neuronal membranes at synaptic sites. Such holes are also visible in perineuronal nets that fill the interfaces between neuronal membranes and glial sheaths but leave space for synaptic contacts (Brückner et al., 1996). Some brain stem nuclei contain groups of neuronal somata without individual glial sheaths.

16.2.3 Glial Cell Phenotype in Development

The prime glia of the vertebrate CNS, the radial glia, traces back to the support cells of the starfish nervous system (Fig. 16.15). Radial glial cells are characterized by contacts to both surfaces of the neural epithelium (fluid environment and basal lamina) as well as to neurons. Radial glial cells are present everywhere in the embryonic vertebrate CNS (Fig. 16.16B) and constitute the dominant glia in the adult CNS of many lower vertebrates (Fig. 16.16A). When a further growth of the CNS increases the thickness of the neural tube wall to more than a few hundred micrometers, full-length radial glial cells appear to become inefficient (for example, for the transcellular

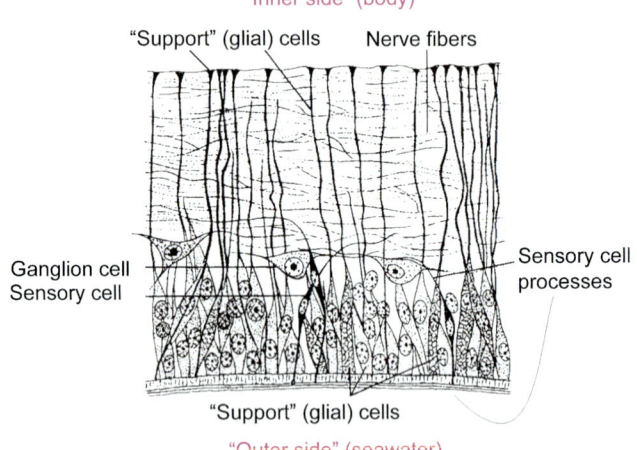

FIGURE 16.15   The starfish nervous system as a vertebrate "prototype central nervous system." Cross section through the radial nerve of *Asterias rubens*. The sensory and ganglion neurons are surrounded by supporting cells which send radial processes toward the basal lamina (which delimits the ectoderm from the mesoderm); conical endfeet of these processes contact the basal lamina (*above*). These cells may be considered as "ancestral radial glial cells." *Redrawn from Meyer, R., 1906. Untersuchungen über den feineren Bau des Nervensystems der Asteriden (Asterias rubens). Z. Wiss. Zool. 81, 96–144.*

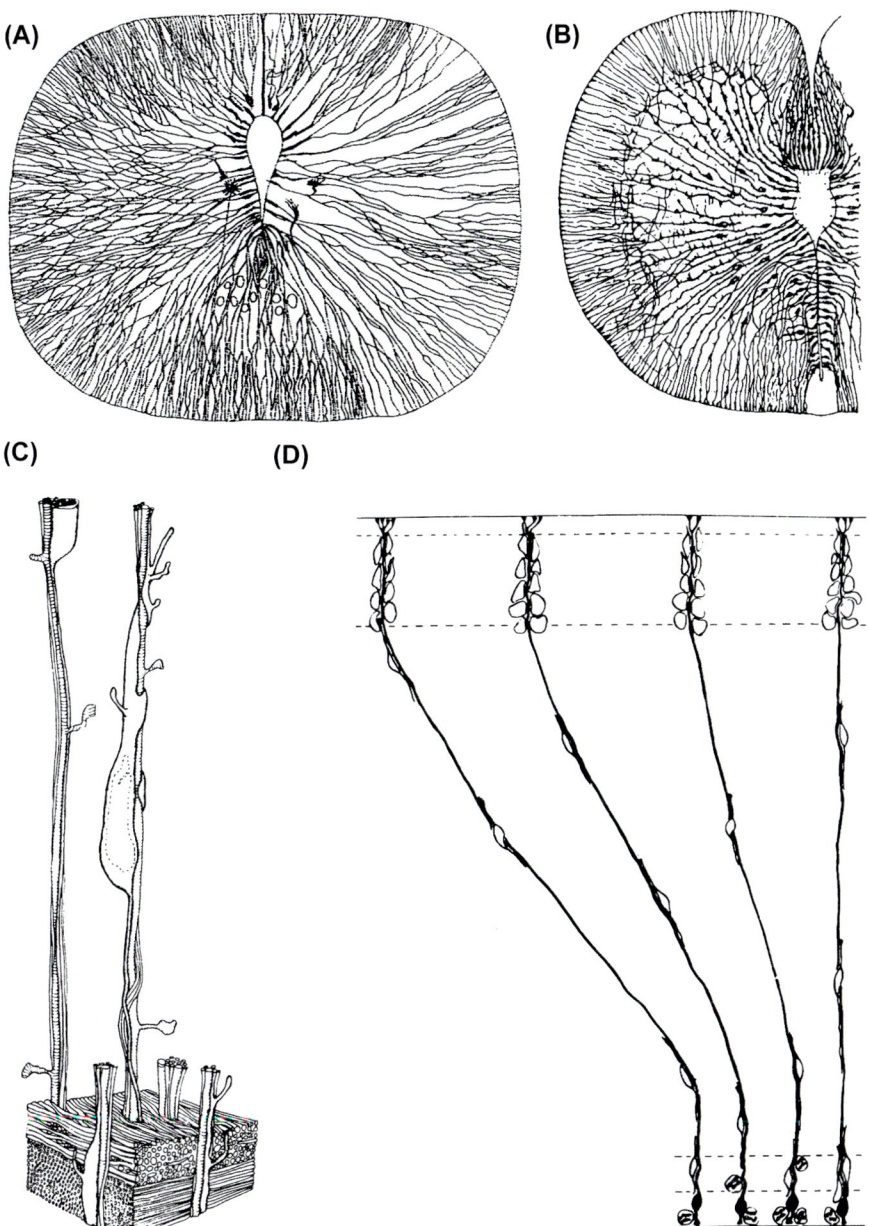

FIGURE 16.16 Radial glial cells. (A) Mature radial glial cells in the spinal cord of *Petromyzon* (lamprey, agnatha). (B) Fetal radial glial cells in the embryonic human spinal cord (44 mm long embryo; from Ramón y Cajál, S., 1911. Histologie du système nerveux de l'homme et des vertébrés. Maloine, Paris). (C and D) Fetal radial glial cells in the monkey brain cortex. The images show how newborn neurons migrate along (bundles of) radial glial cells from the place of birth (the ventricular zone) toward the site of destination, the future cortical plate. *Modified from (A) Fig. 16.15A, Retzius, G., 1893. Studien über Ependym und Neuroglia. Biologische Untersuchungen (Stockholm) Neue Folge 5, 9—26, Stockholm, Samson & Wallin; (B) Fig. 16.15B, Ramón y Cajál, S., 1911. Histologie du système nerveux de l'homme et des vertébrés. Maloine, Paris; (C) Fig. 16.14, Rakic, P., 1972. Mode of cell migration to the superficial layers of fetal monkey neocortex. J. Comp. Neurol. 145, 61—84; (D) Fig. 16.2, Rakic, P., 1981. Neuron-glial interaction during brain development. Trends Neurosci. 4, 184—187.*

potassium transport) or unable to survive as such; the cells then undergo mitotic division, and the daughter cells (and the later-generated glial cells) fail to establish one or more of the three principal contacts (Reichenbach, 1987, 1989b). This results in the emergence of multipolar astrocytes (basal lamina and neuron contact) and oligodendrocytes (neuron contact only) (Fig. 16.1).

Embryonic and fetal radial glial cells are neural progenitor cells in the developing brain and generate both neurons and glial cells after mitotic division (Noctor et al., 2001). Mature radial glial cells begin their morphological differentiation by sprouting lateral protrusions from the stem processes. Later, some of the cells begin to lose the contact to the ventricular system and, thus, are no longer members of the ependymoglia family. The number, length, and complexity of glial cell side branches grow rapidly during the early ontogenesis (Fig. 16.4B; Hanke and Reichenbach, 1987; Senitz et al., 1995; Grosche et al., 2002). As mentioned above, glial processes are shaped by the local microenvironment, because the processes had been grown into the available interspaces between established neuronal and vascular compartments. The growth of neuron-contacting glial processes may be triggered by the overall growth of the neural tissue, the ingrowth of blood vessels, and by specific neuron—glia interactions (for an overview of such interactions, see Giaume et al., 2010). In addition, differences of the viscoelastic properties between glial and neuronal cells (Lu et al., 2006) and the viscosity of the extracellular medium influence the growth of glial processes.

Specialized glioneuronal contacts cannot be elaborated until neuronal differentiation is completed. After the general shape of an astroglial cell has been established in ontogenesis (ie, after at least one endfoot-bearing stem process has grown), the neuron-contacting processes develop in mutual dependence on the developing neuronal cell processes and synapses (Reichenbach et al., 2010). The formation of glial processes may be stimulated by the onset of the neuronal activity, and the growth of glial processes may be attracted (or repelled) by signals from active neurons such as potassium ions, neurotransmitter molecules, and growth factors (Reichenbach and Reichelt, 1986; Reichelt et al., 1989; Cornell-Bell et al., 1990). Glial development may be modulated by the strength and pattern of the neuronal activity triggered by the sensory input and behavioral requirements. For instance, the complexity of glial processes differs significantly when animals are kept in enriched environments or in complete darkness (Stewart et al., 1986; Sirevaag and Greenough, 1991). Similar mechanisms are maintained in the mature CNS where changing needs of neurons modify the structure of glial processes even in the short-term range (eg, Hirrlinger et al., 2004). For a recent discussion of the activity-dependent astroglial plasticity in the mature brain, see, for example, Reichenbach et al. (2010).

The cell processes of each astrocyte occupy a neuropilar territory that constitutes the domain of the (autonomous) glial interactions with neuronal (and vascular) compartments. In several studies, the territories of neighboring astrocytes were found to abut each other with minimal overlap (Bushong et al., 2004; Wilhelmsson et al., 2006). Individual astrocytes specifically interact with the neuronal synapses within their territory (Henneberger et al., 2010). These individual domains are competitively organized during the early postnatal development and may be demarked by distinct molecules such as chondroitin sulfate (Horii-Hayashi et al., 2010). However, electron microscopical studies have revealed a substantial overlap of astrocytic domains in the murine

cerebellum and rat cortex (Wolff, 1968; Grosche et al., 1999). The presence of overlaps of astrocytic domains may depend on the species and brain compartment, as well as on age. For example, the domains of murine cortical astrocytes do not overlap in young (2-month-old) animals but show considerable overlap in 2-year-old animals (Fig. 16.4C; Reichenbach and Wolburg, 2005).

There exists an inverse relation between the surface complexity (surface-to-volume ratio) and the length of glial processes (Siegel et al., 1991; Reichenbach et al., 1992). This relationship may indicate that the absolute membrane surface area of a glial cell is limited (perhaps, by metabolic constraints). A similar relation exists in respect to the diameter of radial glia processes. In a given species and tissue, and at a given developmental stage, there is an inverse relation between the length and the diameter of a process; the absolute diameter ranges between several microns and (in embryonic radial glia) less than 400 μm (Ugrumov et al., 1979; Bruni et al., 1983; Reichenbach et al., 1987; Hanke and Reichenbach, 1987). When rat radial glial cells of very different lengths are compared at the same embryonic stage, the absolute cell volumes are almost the same (Reichenbach et al., 1987). It seems that ependymoglia require a certain period to produce a given cytoplasmic volume. Because radial glial cells are attached to both surfaces of the neuroepithelium, the cells may be stretched differently by local tissue growth rates, but the elongation occurs at the expense of process diameter until new material can be synthesized. Such "passive" growth (Jaeger, 1988) may be supported by stretch-activated signaling mechanisms (Christensen, 1987). Generally, older cells have larger volumes than younger cells (Reichenbach and Reichelt, 1986; Hanke and Reichenbach, 1987).

In addition to their role as neural progenitor cells, embryonic and fetal radial glia also serve as scaffolding for the migration of neurons from the ventricular surface of the neuroepithelium (where cell multiplication occurs) to the target sites (Fig. 16.16C and D), and for axon pathfinding; thus, fetal radial glia play a critical role in defining the cytoarchitecture of the CNS (Rakic, 1988). It has been suggested that the cohort of neurons migrating along the same radial glial cell later maintains a particular relationship or "symbiosis" with the glial cells which developed from this radial glial cell (Rakic, 1978; Reichenbach et al., 1993). This may be the ontogenetic basis of functional columnar units or domains in the mature CNS, exemplified by the orientation columns in the visual cortex (Mountcastle, 1957) and the microcolumns of retinal information processing (Fig. 16.3B; Reichenbach and Bringmann, 2010).

16.2.4 Glial Cells in Adult Neurogenesis

Specialized astrocytes in the adult neurogenic zones, the subventricular zone and the subgranular zone of the hippocampal dentate gyrus, are neural progenitors necessary for the genesis of neurons, astrocytes, and oligodendrocytes in the adult brain (Wang and Bordey, 2008). Progenitor cells in the subventricular zone are derived from radial glia (Merkle et al., 2004) and develop to neuroblasts that migrate to the olfactory bulb and piriform cortex and differentiate to interneurons. Progenitor cells in the subgranular zone generate granule neurons of the hippocampal dentate gyrus. The progenitor cells express GFAP and nestin, an intermediate filament of embryonic precursor cells (Hockfield and McKay, 1985) that is also expressed in the mature brain by reactive astrocytes (Clarke et al., 1994). Progenitor cells share morphological and functional properties with radial glia and astrocytes. For instance, these cells have long processes that envelop and contact blood vessels and neuroblasts, contain glycogen granules, and express astrocytic glutamate and γ-aminobutyric acid (GABA) transporters. However, these cells lack Kir channel—mediated inward potassium currents, a hallmark of astrocytic cell cycle exit and differentiation (Macfarlane and Sontheimer, 2000). Perhaps these cells are retained in a transitional stage between embryonic radial glia and mature astrocytes (Wang and Bordey, 2008).

16.2.5 Functional Astrocytic Syncytia

Astrocytes may be extensively coupled via gap junctions; the coupling constitutes a multicellular glial "syncytium" that is not coupled to neurons and that synchronizes the activities of neighboring astrocytes or (perivascular) astrocytic endfeet (Rash et al., 1997). Moreover, via gap junctional coupling, the astrocytic syncytium may extend to oligodendrocytes (Hampson and Robinson, 1995). Gap junction hemichannels are clustered in astrocytic membranes (Fig. 16.9D; Reichenbach and Wolburg, 2009). Open channels allow the flow of ionic currents (electrical coupling) and the intercellular exchange of larger molecules (up to a molecular mass of 1—1.4 kDa) like biocytin or fluorescent dyes; this "dye coupling" is often used to demonstrate the presence of functional gap junctions among cells. There is evidence for an active metabolic trafficking through astrocytic gap junctions which links glucose metabolism and proliferation (Tabernero et al., 2006).

Among the at least 11 connexins described to be present in astrocytes, connexin-43 is the most widely expressed connexin (Nagy et al., 2004). Astrocytes are coupled to oligodendrocytes via connexin-32 (Hampson and Robinson, 1995). The expression pattern of astrocytic connexins is highly heterogenous throughout the CNS (Reichenbach and Wolburg, 2009). Because connexins display distinct permeabilities and voltage dependencies (Giaume and Venance, 1995), different connexin expression may reflect different functional requirements of glial coupling. In addition to coupled astrocytes, there are astrocytes which are not coupled to neighboring cells (Reichenbach and Wolburg, 2009).

Astrocytic coupling may play an important role in the spatial potassium buffering (Wallraff et al., 2006). Neuronal activity causes local increases of the extracellular potassium level; excess potassium is redistributed by the uptake into astrocytic processes and the release into large extraneural, fluid-filled compartments ("sinks") such as capillaries, the brain surface, and the vitreous cavity of the eye. At the interfaces to these compartments, glial membranes display a particularly high potassium conductance mediated by glial potassium channels such as Kir4.1 (Fig. 16.3G and 16.I). When large areas of the neuropil are exposed to high extracellular potassium, the processes of individual astrocytes are not long enough to exit the overload area and to reach a sink; in these cases, spatial buffering is performed by an entire population of gap junction-coupled astrocytes. Via the coupling between oligodendrocytes and astrocytes, high extracellular potassium around the internodes and the nodes of Ranvier of myelinated axons may be buffered into brain capillaries (Zahs, 1998; Kamasawa et al., 2005).

Another function of astrocytic gap junctions is the propagation of calcium waves among the astrocytic population ("astrocytic excitation"). Glial calcium waves are a part of a nonsynaptic information processing in the CNS (Scemes and Giaume, 2006). Calcium waves may be propagated by different mechanisms: (1) a direct calcium flux through gap junction channels into neighboring cells; (2) a flux of inositol 1,4,5-triphosphate (IP$_3$) through gap junction channels; IP$_3$ induces calcium responses in neighboring cells; and (3) a gap junction—independent mechanism which involves a calcium-induced release of gliotransmitters [glutamate, adenosine 5'-triphosphate (ATP)] from one astrocyte that (via stimulation of ligand receptors at the plasma membrane) cause calcium responses in adjacent astrocytes (Stout et al., 2002; Peters et al., 2003). The latter mechanism also contributes to the propagation of calcium waves among noncoupled glial cells such as retinal Müller cells (Fig. 16.17B; Newman, 2001). The glial release of ATP is mediated by connexin or pannexin hemichannels (Stout et al., 2002; Voigt et al., 2015). The connexin- or pannexin-mediated release of ATP is also involved in the regulation of the glial cell volume (Brückner et al., 2012; Voigt et al., 2015).

The spatial extension of coupled astrocytic networks is limited; mostly some 50—100 astrocytes are coupled, and the diameter of a coupled network does not exceed ~200—300 μm. In the juvenile mouse cortex and hippocampus, for example, many of the coupled astrocytic

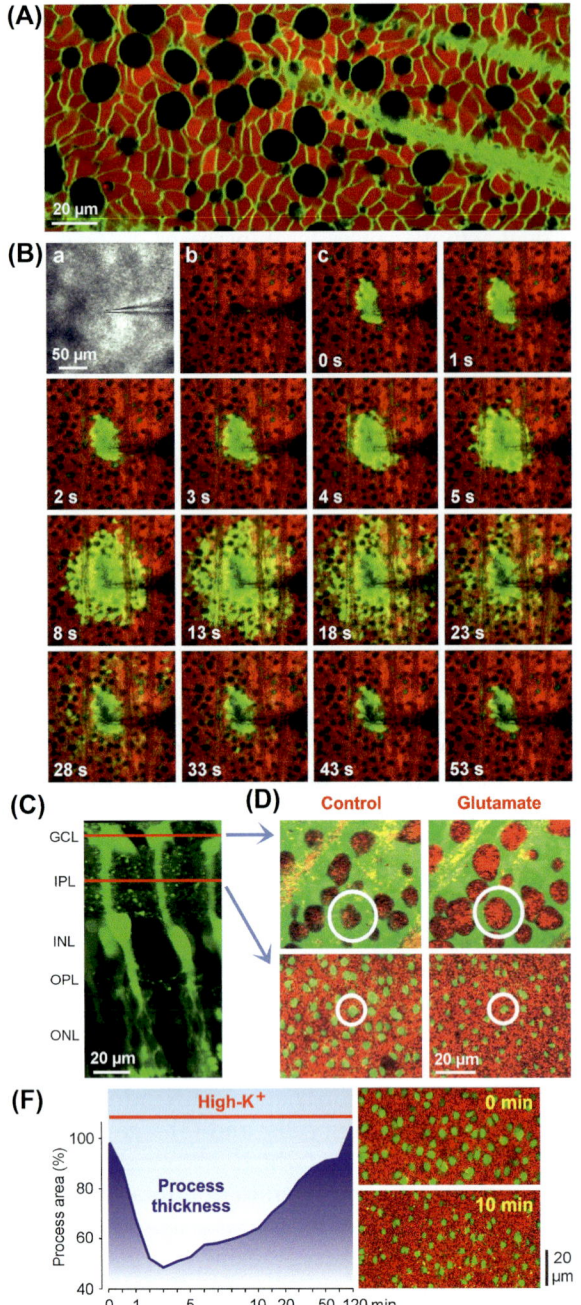

networks display a tendency to reside within a given layer (Houades et al., 2006). The actual degree of the coupling among adjacent astrocytes may vary considerably in dependence on physiological parameters such as pH, calcium, and the association of connexin molecules to the submembrane cytoskeleton (Butkevich et al., 2004). Astrocytic coupling and glial calcium waves may also play a role in the determination of functional territories in the brain (Reichenbach and Wolburg, 2009).

16.2.6 Glioneuronal Domains of Information Processing

The CNS is structurally and functionally compartmentalized into domains at various hierarchical levels (Figs. 16.14 and 16.18A—C; Reichenbach and Wolburg, 2009). One domain is constituted by neuronal and glial elements, can be structurally distinguished from other adjacent compartments, and may function

stimulation. Müller cell endfeet are *red* labeled, and neuronal somata are unlabeled (*black*). The elongated *dark* structures are nerve fiber bundles. (c) Time-dependent recording of the Fluo-4 fluorescence (*green*) in Müller cell endfeet which were mechanically stimulated by the pipette tip at 0 s. The Müller cell endfeet near the pipette tip displayed a calcium response (*green*) immediately after the mechanical stimulation. This response was long lasting and was observed until the end of the record. 3—13 s after mechanical stimulation, a calcium wave developed which propagated laterally through the Müller cell endfeet around the site of mechanical stimulation. This propagating calcium wave was transient and disappeared between 18 and 23 s after mechanical stimulation. (C—E) Glutamate-induced morphological alterations of Müller cells. (C) The retinal slice displays MitoTracker Orange—filled Müller cells (*green*) and the optical planes in the ganglion cell layer (GCL) and inner plexiform layer (IPL) which were used to record the morphological alterations of the cells. Neuronal structures are unstained (*black*). (D) A retinal wholemount was exposed to glutamate (1 mM) for 10 min. *Above*: Glutamate exposure resulted in a swelling of the neuronal somata in the GCL (*red*). The *elongated structures* are nerve fiber bundles. *Below*: Glutamate induced a reduction in the thickness of Müller cell stem processes that traverse the IPL (*green*) resulting from the swelling of the synapses between the Müller cell processes (*red*). (E) Time dependence of the effect of high (50 mM) potassium on the cross-sectional area of Müller cell processes in the IPL. When the high potassium solution was applied longer than 3 min, the thickness of the Müller cell processes returned slowly to the control value; a full recovery was observed after 120 min. High potassium induces a release of endogenous glutamate. *INL*, inner nuclear layer; *ONL*, outer nuclear layer; *OPL*, outer plexiform layer. *Modified from (A) Courtesy of Martina Prasse, Leipzig; (B) Courtesy of Silke Agte, Leipzig; (C) Fig. 16.1A, Uckermann, O., Vargová, L., Ulbricht, E., Klaus, C., Weick, M., Rillich, K., Wiedemann, P., Reichenbach, A., Syková, E., Bringmann, A., 2004. Glutamate-evoked alterations of glial and neuronal cell morphology in the guinea-pig retina. J. Neurosci. 24, 10149—10158; (D) Fig. 16.3A and B, Bringmann, A., Uckermann, O., Pannicke, T., Iandiev, I., Reichenbach, A., Wiedemann, P., 2005. Neuronal versus glial cell swelling in the ischaemic retina. Acta Ophthalmol. Scand. 83, 528—538; (E) Fig. 16.7A and B, Uckermann, O., Vargová, L., Ulbricht, E., Klaus, C., Weick, M., Rillich, K., Wiedemann, P., Reichenbach, A., Syková, E., Bringmann, A., 2004. Glutamate-evoked alterations of glial and neuronal cell morphology in the guinea-pig retina. J. Neurosci. 24, 10149—10158.*

FIGURE 16.17 Functional morphology of Müller glial cells in freshly isolated retinal wholemounts of the avascular guinea pig retina with vital dyes. (A) View on the nerve fiber/ganglion cell layers. The tissue was loaded with FM 1—43 (which labels plasma membranes, *green*) and MitoTracker Orange (*red*) which is selectively taken up by Müller cells. All available space, with the exceptions of that filled by unlabeled neuronal somata (*black*) and nerve fiber bundles (elongated *green* structures), is occupied by Müller cell endfeet (*red*). (B) Calcium imaging in a wholemount loaded with MitoTracker Orange (*red*) and the calcium fluorescence dye Fluo-4/AM (*green*). Calcium responses in Müller cell endfeet were induced by focal mechanical stimulation of the inner retinal surface for 10 ms with the tip of a micropipette (mechanical deformation by ~10 μm). (a) View on the nerve fiber/ganglion cell layers of the wholemount with the tip of the stimulation pipette. (b) The image was recorded 15 s before the mechanical

autonomously (ie, independent on hierarchically higher structures), at least under distinct conditions. The range of elements interacting within or across the limits of a hierarchical level is variable according to the present and previous activity of information processing and to the metabolic conditions of the tissue.

The smallest domains represent subregions of presynaptic terminals which contain specific sets of neurotransmitter receptors and uptake carriers (Dorostkar and Boehm, 2008); one such neuronal "nanodomain" may be faced to an adjacent glial nanodomain which specifically interacts with them (Figs. 16.14 and 16.18A–C). Such nanodomains are also the glial tongues which line the synaptic clefts (active in neurotransmitter uptake) and glial membranes near pre- or postsynaptic compartments (active in metabolic fueling of neuronal elements). Similarly, the fingerlike perinodal glial processes (Figs. 16.10 and 16.11A) may serve specific local glioneuronal interactions, such as molecule transfer and sensing of spike activity (Chao et al., 1994).

One hierarchical step upward, individual (or small groups of) synapses are the smallest units of neuronal information processing. These neuronal microdomains are accompanied by ensheathing glial microdomains (Figs. 16.13A–D and 16.14; Grosche et al., 1999). At the next level of integration, an individual neuron represents a cellular neuronal domain; it interacts with the surrounding glial cells which, thereby, constitute an (oligo) cellular glial domain (Figs. 16.3 and 16.18B). When a

syncytial network of >50 gap junction–coupled astrocytes is simultaneously activated, a mesodomain can be formed (Fig. 16.14); neuronal excitation within these functional units is accompanied by glial responses like cytosolic calcium rises (Aquado et al., 2002). Examples of such mesodomains may represent astrocytes and their processes, which envelope the glomeruli in the olfactory bulb (Chao et al., 1997), and interlaminar astrocytes in the adult cortex of higher primates, which display radially aligned processes in parallel to the columnar neuronal arrangement (Fig. 16.9C; Colombo and Reisin, 2004). Astrocytes, which envelope the glomeruli in the olfactory bulb and which have contact to small blood vessels, strongly express aquaporin-4 to drain water from the synaptic compartment; this suggests that the cellular domains of a glomerulum possess a distinct blood supply (Reichenbach and Wolburg, 2009).

At the next level of integration, columnar arrays of hundreds or thousands of neurons (Figs. 16.3E and 16.18C) may form functional units (macrodomains; Fig. 16.14); examples are the direction-sensitive and the ocular dominance columns in the visual cortex (Mountcastle, 1957; Müller and Best, 1989) and the barrel fields in the rodent somatosensory cortex (Rice and van der Loos, 1977). At the uppermost level, neuronal superdomains involve one entire cortical area or even several of them which are activated together during cognitive tasks (Horwitz, 2004), or even a whole hemisphere or the entire cortex during arousal/sleep or in pathological

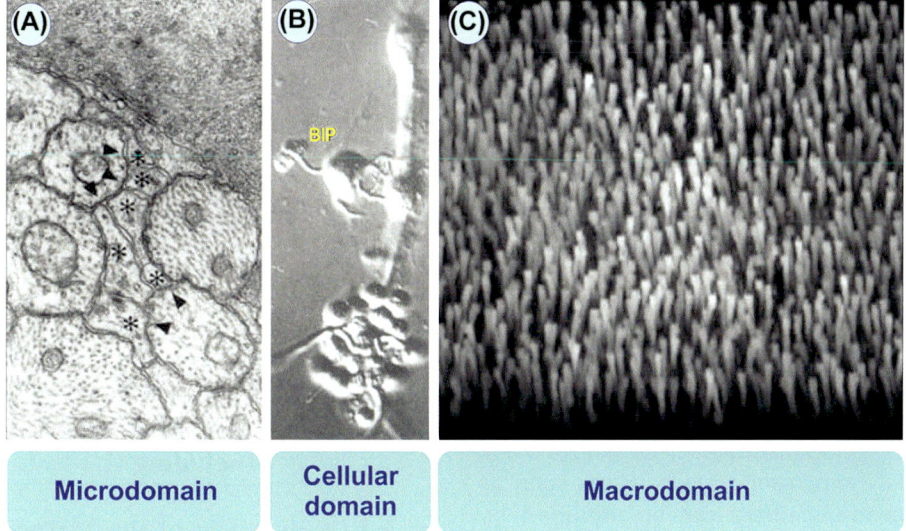

FIGURE 16.18 Ascending hierarchic levels of glial domains involved in glioneuronal interactions in the retina. (A) Example of a microdomain, consisting of a few fingerlike end branches which interact with the nodelike specialization of a ganglion cell axon. (B) Cellular domain, consisting of one Müller cell interacting with the neurons of "its" columnar unit. (C) A macrodomain, consisting of a large population of Müller cells, together interacting with the (light-stimulated) neurons and the intraretinal blood vessels of a given retinal area. (A) Transmission electron micrograph of a tree shrew retina. (B) Group of unstained cells enzymatically dissociated from a guinea pig retina. *BIP*, bipolar cell. (C) 3-D reconstruction of a series of confocal images of a guinea pig retina. Müller cells are visualized by vimentin immunohistochemistry. *Modified from (A) Fig. 16.5A, Reichenbach, A., Frömter, C., Engelmann, R., Wolburg, H., Kasper, M., Schnitzer, J., 1995. Müller glial cells of the tree shrew retina. J. Comp. Neurol. 360, 257–270; (B) and (C) Fig. 2.17B and C, Reichenbach, A., Bringmann, A., 2010. Müller cells in the healthy and diseased retina. Springer, New York, Dordrecht, Heidelberg, London.*

instances such as spreading depression, epileptiform activity, or migraine. This widespread neuronal activity is accompanied by glial responses in the same tissue compartments (Schipke and Kettenmann, 2004).

When a domain involves more than a few neuronal and glial cells (ie, when the size of the domain exceeds the maximum distance for the mere diffusion of oxygen and nutrients which is close to some 100 μm), the maintenance of the activity in the domain needs a third element, blood vessels. The vascular beds fit to the size and shape of the corresponding domains (Bär, 1980). Glial cells are crucially involved in angiogenesis and the formation of the blood—brain (blood—retinal) barrier (Wang and Bordey, 2008; Reichenbach and Bringmann, 2015). The neuronal activity within such a domain is continuously registered by glial cells (Schummers et al., 2008) which then, according to the metabolic needs, control the local blood flow by eliciting vasoconstriction or vasodilatation of the local arterioles (neurovascular coupling) (Gordon et al., 2008).

There are "gates" between neighboring domains of the same level which control whether a single domain functions in an autonomous manner or as a part of a larger domain. The long and thin stalk of glial microdomains (Fig. 16.13B and C) may represent such a gate. The cytoplasmic longitudinal resistance of the stalk constitutes a barrier against the spread of cytosolic calcium rises, triggered in the head of the microdomain by the synaptic activity, toward the glial stem process or adjacent microdomains. In addition, together with the shunt conductance of the stalk membrane, it prevents the electrotonic propagation of even large depolarizations of the head membrane (Grosche et al., 2002). Thus, individual glial microdomains may exclusively display calcium responses in response to low-intensity stimulation of single axons, while more extensive axonal stimulation may cause calcium rises in several neighboring microdomains or even in a whole Bergmann glial cell (Grosche et al., 2002). There are also active mechanisms of glial gate control. Glial networks are constituted by gap junctional coupling (Fig. 16.9D), the conductance of which is under the control of a variety of signals including intracellular second messengers (Rouach et al., 2000). In addition, there are extracellular gate control mechanisms, for instance, the propagation of glial calcium waves mediated by gliotransmitters (see Section 1.4). The activity of glial homeostatic mechanisms such as the uptake of neurotransmitters can be modified by these signals and by the metabolic state of the cells; this will modify the extracellular propagation of signal molecules released from neurons and glial cells ("volume transmission") (Syková and Chvátal, 2000). In addition to neuronal and glial mechanisms, a propagation of activity across the limiting "gates" of the hierarchical levels may be mediated by vascular elements which cross the borders between different domains.

16.2.7 Analysis of Glial Cell Morphology

In addition to the different cell phenotypes, astroglial and ependymoglial cell populations are not homogenous in respect to the expression of marker molecules. While some of these molecules may be expressed by all astrocytes (eg, GLT-1), others (eg, Kir4.1) are only expressed by a subset of astrocytes. While every cell in the brain which is immunopositive for GFAP is an astrocyte (or a tanycyte, eg, in the subventricular zone), there are many astrocytes which do not express GFAP. S-100β is only expressed by a subtype of astrocytes that ensheath blood vessels and by GFAP-immunonegative oligodendrocyte progenitor cells which express the proteoglycan nerve—glia antigen 2 (NG2). Some of the glial antigens (eg, intermediate filament proteins) allow mainly an immunocytochemical visualization of the larger stem processes (Fig. 16.7). Antibodies directed to cytoplasmic proteins (such as glutamine synthetase and S-100β) may stain fine side branches and may be used at the electron microscopical level to identify very thin cytoplasmic leaflets of glial sheaths. Antibodies against membrane proteins (eg, ion channels, ligand receptors, and transporter proteins) may label (parts of) the cell surface. Ependymo- and astroglial cells have the capability to accumulate exogenously applied fluorescent dyes (Figs. 16.3D, E, 16.7, and 16.18A and C). Such dyes can be used to monitor the levels of cytosolic free calcium (Fig. 16.17B) and glutathione, as well as the pH, oxidative stress level, and ion concentrations in living glial cells; simultaneously, the cell morphology can be studied (Figs. 16.3D, E, K and 16.17C—E). Microscopically controlled intracellular injections of fluorescent dyes can be used to visualize individual glial cells. In transgenic mice, astroglial cells can be induced to express fluorescent reporter proteins like green fluorescent protein by the coupling of the genes to a glia-specific promoter (eg, GFAP and glutamine synthetase). In any of these cases, fluorescent dyes are desirable because they can be used with high-resolution confocal microscopy.

16.3 Macroglia of the Central Nervous System Including the Retina

16.3.1 Radial Glia of the Mature Central Nervous System

In thin-walled brains or CNS regions, fetal radial glial cells may become postmitotic and differentiate into adult radial glia. Because these cells pass through different local tissue specializations which predominantly contain synapses, neuronal somata, or axons, radial glial cells of the mature CNS display all three principal types of glial processes and all three types of specialized neuron ensheathment.

16.3.1.1 Tanycytes

The term "tanycyte" ("stretched cell") was introduced by Horstmann (1954) to describe ependymoglial cells whose processes extended over large distances. This term is now commonly used for such ependymoglial cells of the adult brain and spinal cord. Tanycytes (Fig. 16.1, Ia and Ib) are the most common type of macroglia in the CNS of lower vertebrates (Figs. 16.15 and 16.16A). In adult mammals, tanycytes are restricted to certain brain regions where the tissue is rather thin. These regions are the wall of the diencephalic third ventricle (Millhouse, 1971; Akmayev et al., 1973; Seress, 1980), the dorsal and ventral walls of the mesencephalic aqueduct, the floor of the fourth ventricle (Lindemann and Leonhardt, 1973), and the ventral part of the spinal central canal. Tanycyte-like cells have also been reported in the lateral ventricles of adult rats (Hirano and Zimmerman, 1967) and in the velum medullare of adult primates (Reichenbach, 1990). The capillaries in the circumventricular organs (eg, subcommissural organ, subfornical organ, area postrema) are fenestrated; tanycytes (in addition to the choroid plexus epithelial cells) constitute the blood−cerebrospinal fluid barrier by expressing tight junctions.

The various types of tanycytes can be classified in different ways: (1) "ependymal tanycytes" (unipolar cells with somata in the ependymal layer) versus bipolar "extraependymal tanycytes" with somata elsewhere in the nervous tissue (Horstmann, 1954); (2) (short) "protoplasmic tanycytes" that have many side branches and protrusions in the gray matter versus (long) "fibrous tanycytes" with processes that are mainly smooth surfaced and that extend through the white matter; and (3) "pial tanycytes" that have processes extending to the pia mater versus short "vascular tanycytes" that form endfeet at blood vessels in the proximity of the ependymal layer. The processes of some cells may form "en passant" contacts to blood vessels on the way to farther distant targets. With the exceptions of tanycytes in the ventral margin of the diencephalic third ventricle and the velum medullare, all mammalian tanycytes are of the vascular variety. The cells may be born rather late in ontogenesis (Altman and Bayer, 1978), while pial tanycytes may directly arise by differentiation from embryonic radial glia.

Tanycytes have either one or no cilium and a varying number of microvilli and larger protrusions from their apical surface. Together with endo-/exocytotic membrane vesicles, these specializations indicate an active material exchange with the cerebrospinal fluid. Neighboring tanycytes are connected via intercellular junctions. In the hypothalamus and most circumventricular organs, adjacent tanycytes are connected by tight junctions that serve to separate the cerebrospinal fluid from the "blood milieu" surrounding the fenestrated capillaries (Brightman and Reese, 1969; Leonhardt, 1980). The basal processes of all tanycytes contain abundant microtubuli and intermediate filaments and extend lateral protrusions. These processes terminate in conical endfeet whose basal membranes contain OAPs (Hatton and Ellisman, 1982).

The hypothalamus contains several types of tanycytes (Akmayev et al., 1973): a1-tanycytes (mostly vascular tanycytes) face the ventromedial nucleus, and a2-tanycytes (long pial tanycytes) are found at the level of the arcuate nucleus. The endfeet of b-tanycytes abut the blood vessel system of the pars tuberalis. b1-Tanycytes are located in the lateral extensions of the infundibular recess, and b2-tanycytes line the floor of the ventricle (median eminence proper). b-Tanycytes contain abundant lipid inclusions and smooth endoplasmic reticulum, together with enzymes of lipid metabolism (Rodríguez et al., 1979). The basal processes of b2- (but not b1-) tanycytes receive up to 100 synapselike contacts that arise partly from fibers of the tuberohypophyseal tract (Wittkowski, 1967). These contacts consist of vesicle-containing presynaptic terminals and an unspecialized postsynaptic membrane (Knowles, 1967). The endfeet of infundibular tanycytes show varying organelle contents, probably related to their functional state. These endfeet may occupy modifiable surface areas on the fenestrated blood vessel plexus, and, thus, regulate the available contact areas of neurohemal axons in dependence on functional requirements (Wittkowski, 1973; Lichtensteiger et al., 1978). In rhesus monkeys, the structure of hypothalamic tanycytes differs between male and female animals; the size and number of the apical protrusions of tanycytes in females change in dependence on the estrus cycle (Knowles and Anand-Kumar, 1969).

In the fetal primate brain, tanycytes can be more than 20 mm long (Rakic, 1984). In adult rabbits, the floor of the fourth ventricle contains fibrous tanycytes with a length of 2 mm (Felten et al., 1981). Fibrous tanycytes of more than 500 mm length are present in the velum medullare and the hypothalamic region of adult monkeys (Knowles and Anand-Kumar, 1969; Reichenbach, 1990). In contrast, protoplasmic tanycytes are never longer than ∼250 mm (Millhouse, 1971; Lichtensteiger et al., 1978; Bruni et al., 1983).

The subcommissural organ is a circumventricular organ that lies at the border of the diencephalon and mesencephalon and surrounds the dorsal entry of the cerebral aqueduct. A subcommissural organ is present in early development in all vertebrates, but by adulthood it is often substantially reduced or missing (eg, in humans; Oksche, 1964). In fetal mammals, glial cells in the subcommissural organ have a tanycyte-like shape; the cells lose the long basal processes in adult animals (Oksche, 1961). The ultrastructure of these cells indicates a very active secretory function; abundant mitochondria, an elaborated endoplasmic reticulum, and secretory vacuoles are tightly packed in the cytoplasm (Herrlinger, 1970). The main secretory product is a glycoprotein which is released into the ventricular lumen where it aggregates to the Reissner's fiber (Reissner, 1860). The aggregation is supported and oriented by the kinocilia of the ependymocytes, by the flow of the cerebrospinal fluid, and by stretching forces of the already formed fiber. The fully formed fiber can be up to 50 μm thick and (because it extends down to the ventriculus terminalis of the spinal cord) more than 1 m long. A growth rate of several millimeters per day has been observed in most species studied (Ermisch et al., 1971). The function of Reissner's fiber is unknown, but there is some evidence that it may contribute to organize the axis of the developing body (Hauser, 1969).

16.3.1.2 Müller Cells

Müller cells, first described by Heinrich Müller (1851), are the radial glia of the retina (Fig. 16.3A and D). Because the retina is an embryonic outgrowth of the diencephalon, Müller cells belong to the diencephalic tanycyte group. Most mammals have vascularized retinas which contain astrocytes (in the two innermost retinal layers) and Müller cells (Fig. 16.3A). In mammals with avascular retinas (eg, echidna and horse) or avascular retinal areas (eg, rabbit, hare, guinea pig, cat), Müller cells are the only neuron-supporting macroglial cells of the retina.

Müller cells are bipolar cells. The somata of the cells lie in the inner nuclear layer (Fig. 16.3A and D); stem processes span the entire thickness of the neuroretina (Fig. 16.3A and D). Although there is a considerable variability in the shape of Müller cells from different species (Reichenbach and Robinson, 1995), some features are fairly universal. Müller cells contact virtually every neuronal and nonneuronal element of the retina with specialized branches of the stem processes (Fig. 16.3A−C and F). The inner stem process (Fig. 16.3C) ends with one or several funnel-shaped endfeet (Figs. 16.3F, G and 16.11A) which contact the basal lamina at the inner retinal surface. The vitreal endfoot membranes together with the basal lamina form the inner limiting membrane of the retina. Müller cells may express up to three different phenotypes as their processes pass through various retinal layers. In the nuclear layers of the retina (particularly, in the outer nuclear layer which contains photoreceptor cell somata; Fig. 16.3A), Müller cell processes assume the shape of velate astrocytes (Figs. 16.1, V, 16.3C, F, and 16.11A). In the plexiform layers (which mainly contain synapses),

Müller cell processes resemble those of protoplasmic astrocytes (Figs. 16.1, IV, 16.3B-D, F, and 16.11A). Generally, in rod-dominant retinas a Müller cell extends one stem process from the soma to the inner limiting membrane (Fig. 16.3A and C); this is the case in most mammals. In the cone-dominant retina of the tree shrew (as well as of many birds and reptiles), and in retinal areas with a very thick nerve fiber layer, the inner stem process of Müller cells is split into several branches; these branches are thin and smooth like those of fibrous astrocytes (Figs. 16.3F and 16.11A).

Müller cells send side branches into the two plexiform (synaptic) layers, where the processes form sheaths around most neuronal processes and synapses, particularly, around the photoreceptor pedicles in the outer plexiform layer (Reichenbach et al., 1988). In the nuclear layers, the lamellar processes of Müller cells form basketlike structures which envelop the neuronal cell bodies (Figs. 16.3C, F and 16.11A; Ramon y Cajál, 1972; Reichenbach et al., 1989). Like astrocytes around nodes of Ranvier, Müller cells send side branches that form fingerlike fine processes which abut the nodelike specializations of nerve fibers (Fig. 16.10A and B; Hildebrand and Waxmann, 1984; Reichenbach et al., 1988). Müller cell endfoot membranes which abut the basal lamina at the inner retinal surface (Fig. 16.3A, F, and G) express OAPs (Fig. 16.3J; Wolburg, 2006). In species with vascularized retinas, OAPs are also found in Müller cell membranes which form "en passant" endfeet onto retinal blood vessels.

At the level of the outer limiting membrane, Müller cells extend apical microvilli into the subretinal space, between the inner segments of photoreceptor cells (Fig. 16.3C). Here, Müller cells are laterally connected to neighboring Müller and photoreceptor cells by zonulae adherentes to form the outer limiting "membrane." In the inner retina, adherent junctions are also present among retinal astrocytes and Müller cells, and between adjacent astrocytes and Müller cells, but not between glial cells and neurons, or among neurons (Holländer et al., 1991). This suggests that the glial cell network constitutes the mechanical stability of the retina.

Mammalian Müller cells are normally not coupled by gap or tight junctions (Wolburg et al., 1990; Holländer et al., 1991; Robinson et al., 1993). In contrast, retinal astrocytes are coupled by gap junctions (Holländer et al., 1991). The gap junctional coupling of astrocytes creates a functional syncytium which allows the intercellular propagation of intracellular signals, the control of ionic and metabolic homeostasis of retinal ganglion cell somata and axons, and the neurovascular coupling (Reichenbach and Bringmann, 2015). Rat (but not cat and human) astrocytes are also coupled to Müller cells; one astrocyte is coupled to 13−88 astrocytes and to >100 Müller cells (Robinson et al., 1993; Zahs and

Newman, 1997). While the gap junctional coupling between astrocytes is symmetric, the coupling between astrocytes and Müller cells is asymmetric, allowing only a unidirectional transfer of small intracellular molecules from astrocytes to Müller cells but not vice versa (Robinson et al., 1993; Zahs and Newman, 1997). The diffusion of internal messengers, presumably IP_3, through gap junctions mediates the propagation of intercellular calcium waves between astrocytes. The calcium waves between astrocytes and Müller cells, and between Müller cells (Fig. 16.17B), propagate via paracrine extracellular ATP signaling (Newman, 2001).

Müller cells occupy a variable volume fraction of the retinal tissue, from 5–8% (most mammals, ie, those with vascularized retinas) to about 20% (mammals with avascular retinas, such as rabbit and guinea pig). The volume of individual Müller cells varies from 400 μm^3 (mouse) to >2000 μm^3 (rabbit, retinal periphery); the surface area is in the range of 6000–12 000 μm^2. Müller cells constitute a fairly uniform "lattice" in the retinal tissue (Figs. 16.3E and 16.18C); the density of Müller cells per square-millimeter retinal surface area ranges between 5000–12 000 (most mammals) and >25 000 (primate

fovea centralis). With the exception of Müller cells in the primate fovea centralis (Fig. 16.19), each Müller cell ensheathes and supports a columnar group of retinal neurons and photoreceptors (Fig. 16.3B); in mammals, the number of neurons per Müller cell ranges between 7 (tree shrew), about 16 (human nonfoveal retina, rabbit, and other herbivorous mammals), and up to 30 (rodents and carnivores with rod-dominant retinas). The total length of a Müller cell is determined by the local thickness of the retina (Fig. 16.3F). A special topographical adaptation is the Z-shaped phenotype of Müller cells in the primate perifovea (Fig. 16.19). Literature about Müller cells can be found in Reichenbach and Bringmann (2010).

16.3.2 Astrocytes

Radial astrocytes (Fig. 16.1, II) are common in the spinal cord, particularly, of nonmammalian species (Fig. 16.11A). As these cells cross white and gray matter, the properties of the cell processes may change from "protoplasmic" to "fibrous." Some radial astrocytes are also found in the optic nerve where these cells are

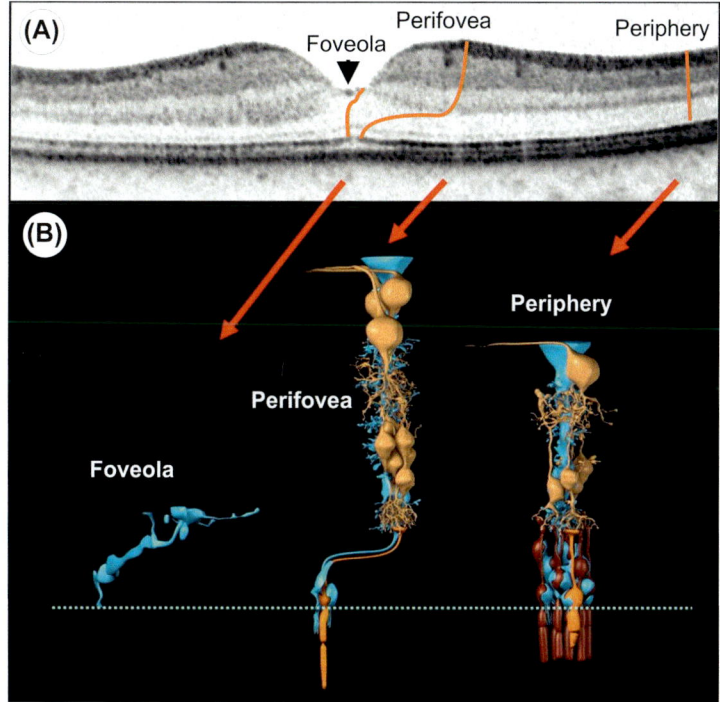

FIGURE 16.19   Morphological diversity of Müller cells in the human fovea. (A) Optical coherence tomographical image of the human fovea. (B) Müller cells and the neurons and photoreceptors of their columns in three parts of the retina. In the fovea centralis (foveola), where the inner retinal layers are shifted laterally and the light directly hits cone photoreceptors, short Müller cells extend from the outer limiting membrane (*dashed line*) to the vitreal (inner) surface of the retina. These cells do not form a column with retinal neurons and photoreceptors. In the perifovea, very long Müller cells display a Z-shaped morphology because (within the Henle fiber layer) Müller cell processes run centrifugally in association with cone axons. Each Müller cell of the perifovea interacts with one cone and many inner retinal neurons, while each Müller cell of the retinal periphery interacts with one cone, a varying number of rods, and inner retinal neurons. *Modified from (A) Fig. 2.29C, Reichenbach, A., Bringmann, A., 2010. Müller Cells in the Healthy and Diseased Retina. Springer, New York, Dordrecht, Heidelberg, London and original. Courtesy of Jens Grosche, Leipzig; (B) Courtesy of Jens Grosche, Leipzig.*

intermingled with the more abundant fibrous astrocytes (Fig. 16.11F). The hippocampus contains radially oriented astrocytes with processes which do not abut the pia; rather, these cells are confined to the stratum granulare of the dentate gyrus (Fig. 16.11B and C) and the stratum oriens of the CA1 region.

Bergmann glial (Golgi epithelial) cells are the radial astrocytes of the cerebellum (Figs. 16.4 and 16.11A). The cell bodies reside in the layer of the Purkinje cell somata, and the processes (usually three to six per cell) cross the molecular layer. (Fañanás cells are a subtype of short Bergmann glial cells; the somata are located in the molecular layer rather than in the layer of the Purkinje cell somata.) Because the processes do not reach the ventricular surface, Bergmann glial cells cannot be considered as radial glial cells, but rather belong to protoplasmic astrocytes. The many elaborate side branches (Fig. 16.13A) display complex morphological and functional interactions with the synapses on the dendrites of Purkinje cells (Fig. 16.13D); these side branches are characterized by a high surface-to-volume ratio of up to 25 $\mu m^2 \mu m^{-3}$ (Grosche et al., 1999). In the rodent cerebellum, there are about eight Bergmann glial cells per Purkinje cell ($\sim$8000 mm^{-2} cerebellar surface area). Each Bergmann glial cell ensheathes several thousands (2000–6000) of Purkinje cell synapses (Reichenbach et al., 2010). The size and shape of Bergmann glial cells are species dependent (Fig. 16.4A). The total length of the processes is determined by the thickness of the molecular layer. Generally, in small species (eg, shrews), Bergmann glial cell processes are short and densely covered with lateral appendages, while in large species (eg, man), the processes are much longer but show less dense lateral outgrowths (Fig. 16.4A).

Protoplasmic astrocytes (Figs. 16.1, IV, 16.4B, and 16.11E) are present in the gray matter. The numerous processes of these cells spread more or less radially from the soma, usually occupying a spheroid volume (Fig. 16.4B), and extend many fine and complex lamellar side branches. In rodent brain astrocytes, these surface extensions occupy $\sim$50% of the volume ($\sim$5500 μm^3) but as much as 80% of the surface area of an average cell ($\sim$80 000 μm^2) (Chao et al., 2002). Thus, although the volume fraction of astroglia in the cortical tissue amounts only to 10–20%, astrocytic processes and side branches contact much of the neuronal surfaces present in a given volume compartment. Human astrocytes are larger and more complex than astrocytes of laboratory mammals such as rodents (Oberheim et al., 2006). At least one of the cell processes bears one or several perivascular endfeet; the surfaces of the blood vessels in the CNS are virtually completely ensheathed by astroglial endfoot plates (Fig. 16.3G and H; Mathiisen et al., 2010). The density of astrocytes in the cerebral cortex is high, for example, 4000–10 000 mm^{-3} in layers III/IV of lissencephalic

cortices of insectivores and 12 000 to >30 000 mm^{-3} in the rat cortex (Stolzenburg et al., 1989; Distler et al., 1991). The cortical glia-to-neuron ratio (largely determined by protoplasmic astrocytes) increases with the thickness of the tissue, from $\sim$0.1 in shrew to $\sim$5 in whale, with intermediate values of $\sim$2 in man (for literature, see Reichenbach and Wolburg, 2009).

Fibrous astrocytes (Figs. 16.1, VI, and 16.11D, F) are present in white matter tracts, the optic nerve, and the retinal nerve fiber layer of mammals with vascularized retinas. A density of $\sim$200 000 fibrous astrocytes per cubic millimeter has been estimated for the anterior commissure of the mouse (Sturrock et al., 1977). The somata are often arranged in rows between the axon bundles; the processes are comparatively smooth and frequently oriented in parallel to the axons. In the mouse optic nerve, every astrocyte possesses several perivascular and/or subpial endfeet (Fig. 16.11F). Fibrous astrocytes in the retina may bear endfeet at both intraretinal vessels and the vitread surface (Fig. 16.3G and H). As a characteristic feature, the processes of fibrous astrocytes extend multiple fingerlike outgrowths into the perinodal space of adjacent axons (Figs. 16.10A, B and 16.11A). The processes of fibrous astrocytes are generally longer (up to 300 μm in mice) than those of protoplasmic astrocytes (<50 μm in mice), but the surface-to-volume ratio is significantly smaller (less than 4 versus 12–18 $\mu m^2 \mu m^{-3}$).

Velate astrocytes (Fig. 16.1, V) are present in the granule layer of the cerebellum where these cells surround several small neuronal granule cells with velate sheaths (for literature, see Reichenbach and Wolburg, 2009). Similar cells occur in the olfactory bulb where the cells ensheath several periglomerular neurons and dendritic segments. Obviously, this cell type develops in brain tissues which contain many small, densely packed neuronal somata.

Interlaminar astrocytes (Figs. 16.1, VIII, and 16.9C) are present in the supragranular layers of the cerebral cortex of higher primates including man (but not in other mammals studied so far). These cells are similar to protoplasmic astrocytes in the upper cortical layers (I–III) but are characterized by a long (up to 1.0 mm in humans) process arising from the cell soma usually located in lamina I and descending down to lamina IV where the process ends in a small bulb (Fig. 16.9C; Colombo and Reisin, 2004). Interlaminar astrocytes may optimize the modular (columnar) organization of the cortex (Colombo, 2001).

Close to the pia mater, specialized astrocytes are present (Fig. 16.1, III), which may form several layers of endfoot plates. Usually, these cells extend several long, smooth processes down into the neuropil. The main function of these marginal astrocytes is the formation of a glial "limiting zone" below the pia mater (glia limitans; Fig. 16.9A). Such surface-associated, gap junction—connected astrocytes in the posterior piriform cortex of the rat were suggested to contribute to neurovascular regulation, interstitial homeostasis, and neuromodulation (Feig and Haberly, 2011). In the retina and brain (eg, neurohypophysis), perivascular astrocytes (Fig. 16.1, VII), which are virtually devoid of neuron-contacting processes, form extensive endfoot contacts to blood vessels. In analogy to the marginal astrocytes, perivascular astrocytes constitute a glial "coating" of the vessels.

16.3.3 Diacytes

A recently described type of glial cells, diacytes (or nonastrocytic inner retinal glia), is scattered across the inner retinal layers of dogs and nonhuman primates (Fischer et al., 2010). Diacytes show features intermediate between those of astrocytes and oligodendrocytes. The function of diacytes in the retina is unclear. Diacytes accumulate at sites of retinal cell death; the survival of the cells seems to be dependent on the presence of microglial cells (Zelinka et al., 2012).

16.3.4 Ependymoglia, Choroid Plexus Cells, and Retinal Pigment Epithelial Cells

Ependymoglia, choroid plexus epithelial cells, and RPE cells are specialized glial cells which form "epithelia" lining the ventricle (or the subretinal space). The ventricular lining is mainly composed of cuboid "ciliated ependymocytes" (Fig. 16.1, X). Depending on the location, these cells may be columnar with a height of up to 15 mm or as flat as 0.1 mm. Ciliated ependymocytes tend to be taller over gray matter than white matter (Leonhardt, 1980). The apical (ventricular) surface of ependymocytes is characterized by the presence of a varying number of microvilli and protrusions, and of 12—60 kinocilia; the number of kinocilia varies with age and species (Brightman and Palay, 1963; Nakayama and Kohno, 1974). The cilia are 10—20 μm long and of the $9 \times 2 + 2$ type, and beat rhythmically at a frequency of about 200 min^{-1} (Singer and Goodman, 1966). The cilia movement appears to assist the rostrocaudal flow of the cerebrospinal fluid. At the basal pole, ependymocytes (Fig. 16.1, X) contact the basal lamina of a blood vessel or a "basement membrane labyrinth." Such labyrinths are the remnants of an embryonic vascular network in the subventricular zone (Booz et al., 1974). Fine basal processes of ciliated ependymocytes may extend over considerable distances to meet the basal lamina of subependymal blood vessels (Hirano and Zimmerman, 1967); in contrast to those of tanycytes, these basal processes are oriented transversely rather than orthogonally

to the ependymal surface (Inokuchi et al., 1988). Ependymocytes are connected by zonulae adherentes and gap junctions (Brightman and Reese, 1969). The lack of tight junctions between ciliated ependymocytes permits the free exchange between the extracellular space of the brain and the cerebrospinal fluid. Ependymocytes have a cytoplasm that is rich in intermediate filaments and large mitochondria.

The ependymal ceiling of the ventricular system is not uniform. Characteristic ependymal specializations, "circumventricular organs", are located at, or near, the median axis of the CNS. The distribution of these "organs" is a legacy of the brain development. Whereas the lateral plates of the neural anlage generate many neurons and differentiate into large tissue blocks, parts of the roof plate, floor plate, and lamina terminalis undergo very restricted neurogenesis. These regions remain "ependymal" and form, together with blood vessels, peculiar ependymal "organs." Because the vascular endothelia in most of these organs are fenestrated, the blood—cerebrospinal fluid barrier is formed by ependymocytes which are connected by tight junctions near the ventricular surface (Brightman and Reese, 1969). Each choroidal villus is composed of a single layer of cuboidal ependymocytes resting on a basal lamina, a layer of interposed connective tissue elements, and a blood vessel (mostly a thick capillary) beneath. The basal plasma membrane, which rests on the basal lamina, is elaborately infolded. Together with the apical microvilli, the basal infoldings characterize the choroid plexus as a transport epithelium. Choroid plexuses are responsible for most of the cerebrospinal fluid production by means of the apical Na^+/K^+ pumps that transport, into the lumen, an excess of sodium ions which drag along water and soluble molecules (Wright et al., 1977). This function is supported by numerous cristae mitochondria which are dispersed throughout the cytoplasm and are often seen in close approximation to components of the abundant rough endoplasmic reticulum; these organelles are subject of neuroendocrine regulatory mechanisms (Nilsson et al., 1992).

The RPE lines the subretinal space (a vestige of the embryonic optic ventricle) opposite to the neuroretina (Fig. 16.3A) and separates the subretinal space from the blood in the choriocapillaris. The RPE functions as epithelium, macrophage, and glia (Steinberg, 1985). The apical surface of RPE cells extends two types of microvilli: long (5—7 μm) thin microvilli which maximize the membrane area available for transepithelial transport, and shorter microvilli which ensheath the outer segments of rod and cone photoreceptor cells (Fig. 16.6). The basal cytoplasm is rich in mitochondria, and the basal plasma membrane displays numerous infoldings which increase the surface area (Fig. 16.6). Size and shape of RPE cells depend on the location in retina; in the human macula, the cells measure ~14 μm in diameter (12 μm height), while in the retinal periphery, the cells become wider (up to 60 μm diameter) and flatter. Like choroid plexus cells, RPE cells (1) are in close apposition to many blood vessels, (2) are specialized for transmembrane transport, and (3) form the blood—cerebrospinal (or, in this case, blood-retinal) fluid barrier by tight junctions (Fig. 16.6). A specific feature of RPE cells is the presence of melanin pigment granules (melanosomes) in the apical cytoplasm and villous processes (Fig. 16.6); melanosomes absorb light which passed the photoreceptor cells and, thus, avoid a back-scattering of light. The cytoplasm is densely packed with rough endoplasmic reticulum and Golgi complexes. There is continuous renewal of photoreceptor outer segments by disk shedding; the shed tips are taken up by RPE cells (Bok and Young, 1979). In the RPE cytoplasm, phagosomes are fused with lysosomes and become degraded.

Choroid plexus and RPE cells express aquaporin-1 water channels in their apical membranes. An important difference between choroid plexus and RPE cells is the direction of the transcellular water transport; while choroid plexus cells release water across the apical membranes into the ventricle, RPE cells absorb excess water from the subretinal space and transport the water to the choroidal vessels (when RPE cells would release water into the analogous space, ie, the subretinal space, the neuroretina would detach from the RPE; retinal detachment, however, is an important cause of retinal disease). The water transport through the RPE is mediated by an inwardly directed ion transport via the $Na^+/K^+/2Cl-$ cotransporter (NKCC1) and the Na^+/bicarbonate cotransporter in the apical membrane; the ion flux is associated with an aquaporin-1-mediated water flux (Strauss, 2005).

16.3.5 Oligodendroglia

Up to 50% of the volume of human brains is made up by white matter which contains myelinated nerve fibers. Oligodendroglia are the myelin-forming glia of the CNS (Baumann and Pham-Dinh, 2001; Bradl and Lassmann, 2010). These cells were first discovered by Robertson (1899) and named by Del Río Hortega (1921), a student of Ramón y Cajal. By allowing a saltatory conduction of action potentials, myelination increases the conduction velocity of action potentials (up to 200 m s^{-1}) without an increase of the axonal caliber (Tolhurst and Lewis, 1992). Axons with diameters of more than 0.2 μm may be ensheathed by myelin sheaths. Oligodendrocytes also secrete factors which induce, during axogenesis, clustering of sodium and potassium channels in the nodes of Ranvier (Kaplan et al., 1997). In the nodes, the sodium channels are clustered at a density

of some 120 000 μm^{-2}, the highest density in the nervous system. Oligodendrocytes also support the axonal stability, modify the rate of the axonal vesicular transport, and exert metabolic and trophic support for neurons.

Myelin, named by Virchow (1854), is formed by the spiral wrapping of oligodendrocyte plasma membrane extensions around an axon, followed by extrusion of the cytoplasm and compaction of the stacked membrane bilayers. Myelin sheaths are rich in lipids (70% of dry weight) and low in water (40% water in contrast to 80% in gray matter); this composition allows the electrical insulation of axons. Within the myelin sheaths, the intracellular compartments are very thin (30 Å).

Myelin-forming oligodendrocytes may have numerous processes linked with myelin segments which enwrap individual axonal internodes. The number of processes that form myelin sheaths from a single oligodendrocyte varies according to the CNS region and species, reaching from 1 in the spinal cord of large mammals to 40 in the rat optic nerve. On the same axon, adjacent myelin segments belong to different oligodendrocytes. Myelin segments have a length of 150—200 μm and end with paranodal loops that contain cytoplasm. Thicker axons have thicker myelin sheaths.

Del Río Hortega (1928) classified oligodendrocytes in four categories, in relation to the number of cell processes. Butt et al. (1995) also distinguished four types of myelinating oligodendrocytes, from small cells supporting the short, thin myelin sheaths of 15—30 small-diameter axons (type I) (Fig. 16.1, XII), through intermediate types (II and III) (Fig. 16.1, XIII, XIV), to the largest cells forming the long, thick myelin sheaths of 1—3 large-diameter axons (type IV) (Fig. 16.1, XV). At the electron microscopic level, oligodendrocytes are characterized by dark inclusions in the cytoplasm and by clumped chromatin. Aged oligodendrocytes contain lipofuscin. Mori and Leblond (1970) distinguished three types of oligodendrocytes: light, medium, and dark; the dark type displays the most dense cytoplasm. Light oligodendrocytes may represent the most actively dividing cells; as the cells mature, oligodendrocytes become progressively dark (Mori and Leblond, 1970). A number of features distinguish oligodendrocytes from astrocytes, in particular the smaller size, the greater densities of cyto- and nucleoplasm, the absence of intermediate filaments and glycogen, and the presence of a large number of microtubules in the processes.

Oligodendrocytes are most abundant in white matter (type III and IV) (Fig. 16.1, XIV, XV), but these cells also occur in gray matter, sometimes as satellites to neurons (type I and II) (Fig. 16.1, XII, XIII; Penfield, 1932). Perineuronal satellite oligodendrocytes normally do not form myelin sheaths but serve to regulate the microenvironment around neurons (Ludwin, 1997). However, in the gray matter, oligodendrocytes may also myelinate (parts of) very long dendrites, eg, of cerebellar Purkinje neurons and neocortical pyramidal cells (Pinching, 1971). The membranes of these long dendrites possess "hot spots" capable to generate calcium-mediated action potentials (Llinas et al., 1969).

Oligodendrocytes are coupled together, and with astrocytes, by gap junctions (Hampson and Robinson, 1995). The gap junctional coupling synchronizes the activities of astrocytes and oligodendrocytes, regulates the myelin formation in local oligodendrocyte populations (Baumann and Pham-Dinh, 2001), and may be implicated in spatial potassium buffering around myelinated axons (Zahs, 1998; Kamasawa et al., 2005).

Myelin sheaths are no static structures but serve the communication between axons and oligodendrocytes. Oligodendrocytes detect axonal activity by responding to axon-derived glutamate and potassium with cellular depolarization (Yamazaki et al., 2007). Oligodendrocytes also regulate the propagation of action potentials; an increase in the axonal action potential conduction velocity may be caused by osmotic swelling of the myelin sheath secondary to the transmembrane ion currents that cause cell depolarization (Yamazaki et al., 2007). This may promote neural synchrony among the multiple axons which are under the domain of an individual oligodendrocyte.

16.3.5.1 Oligodendroglia Development

During early stages of embryonic development, multipotent neural precursor cells generate oligodendrocyte precursor cells in multiple foci of the ventral ventricular/subventricular zone of the brain, in a developmental stage-dependent manner (Kessaris et al., 2006). Oligodendrocyte precursor cells, which have a bipolar morphology, populate the brain in three waves. A subpopulation of human oligodendrocytes has a dorsal origin, originating from cortical radial glial cells (Jakovcevski et al., 2009). The spinal cord is colonized by oligodendrocyte precursor cells mainly from the ventral ventricular zone of the embryonic spinal cord. However, the temporal and spatial distribution of oligodendrocyte differentiation differs between mammalian species (Jakovcevski et al., 2009).

The proliferation of oligodendrocyte precursor cells depends on the electrical activity of axons (Barres and Raff, 1993). The migration of oligodendrocyte precursors is controlled by their interaction with secreted signaling molecules such as semaphorins, sonic hedgehog, and growth factors, and with extracellular matrix and contact proteins such as tenascin, laminin, fibronectin, and integrins (Reichenbach and Bringmann, 2015). Sonic hedgehog stimulates the migration of oligodendrocyte precursors via enhanced secretion of growth factors from astrocytes (Dakubo et al., 2008).

Oligodendroglial cells are the last cell type which mature in the CNS. After migration, oligodendrocyte progenitors settle along the fiber tracts of the future white matter and transform into preoligodendrocytes, multiprocessed cells which keep the property of cell division but are less motile or even immobile. Preoligodendrocytes become immature oligodendrocytes which are ready for myelination. During development, oligodendrocytes are greatly overproduced; the cell number is adjusted by apoptosis of oligodendrocytes that failed to ensheath an axon. In the mouse, myelination starts at birth in the spinal cord; in the brain, myelination is achieved in almost all regions around 45–60 days postnatally. In humans, myelin formation starts during the second half of fetal life in the spinal cord; the peak of myelin formation occurs during the first year postnatally. Myelination continues up to 20–30 years of age in some axons, especially of associative cortical areas (Yakovlev and Lecours, 1966).

In addition to thyroid hormone and brain-derived neurotrophic factor, axon-derived soluble factors such as ATP, adenosine, sonic hedgehog, and neuregulin induce the differentiation and the myelinating competence of oligodendrocytes; such factors also regulate the thickness of the myelin sheath (Baumann and Pham-Dinh, 2001; Gao and Miller, 2006; Bradl and Lassmann, 2010). This regulation results in a ratio between the axon diameter and the outer diameter of the myelin sheath of 0.6–0.7. Axon-derived ATP does not directly act on oligodendrocytes, but triggers the release of leukemia inhibitory factor from astrocytes, which promotes myelination by mature oligodendrocytes (Ishibashi et al., 2006). Astrocytes also control the onset of myelination by promoting the adhesion of oligodendrocyte processes to axons (Meyer-Franke et al., 1999). The developmental vimentin-to-GFAP transition in the astrocyte cytoskeleton occurs at the time of myelination, suggesting a bidirectional developmental regulation between astrocytes and oligodendrocytes.

The optic nerve is colonized by oligodendrocyte progenitors from the brain. In the rat, myelination of the optic nerve starts at postnatal day 5 and proceeds the next 3 weeks. In humans, myelination of the optic nerve begins after 32 weeks of gestation and continues for several years into the early childhood. Because myelination of retinal ganglion cell axons would cause impaired image perception by the underlying photoreceptors, retinas of most mammals do not contain oligodendrocytes; exceptions are the myelinated retinal nerve fiber bundles of rabbits, hares, and dogs (Reichenbach and Bringmann, 2015). In most mammals, retinal ganglion cell axons are myelinated in the extraocular part of the optic nerve, but remain nonmyelinated in the retinal nerve fiber layer and the intraocular portion of the optic nerve (optic nerve head). The immigration of oligodendrocyte precursors from the optic nerve is inhibited by the lamina cribrosa or by fibrous astrocytes in the optic nerve head (Reichenbach and Bringmann, 2015). A similar glial barrier at the motor axon exit points prevents the migration of oligodendrocyte precursors from the spinal cord into peripheral nerves (Fig. 16.12; Kucenas et al., 2009).

Oligodendrocyte progenitor cells are also found in the adult CNS. These cells express NG2. NG2-expressing cells, which comprise 3–8% of all brain cells, are ubiquitously spread throughout the gray and white matter. NG2-expressing cells constitute the main cell population that proliferates in the adult brain (Dawson et al., 2003). Many of these cells are associated with axons and detect axonal activity in the nodes of Ranvier by glutamate receptors. Half of the NG2-expressing cells fire action potentials, receive synaptic inputs, form synapses with neurons, and regulate the synaptic transmission between neurons; this regulation is involved in learning and memory (Káradóttir et al., 2008; Sakry et al., 2014). NG2-expressing cells generate neurons, astrocytes, and oligodendrocytes; differentiation of the cells into oligodendrocytes, eg, in demyelinated lesions, is associated with a downregulation of NG2 and the loss of synapses.

16.3.5.2 Oligodendroglia in Axonal Injury

Neuronal axons are capable to regenerate after transection in the adult CNS of lower vertebrates (fish, amphibia, reptiles) but not of higher vertebrates (birds, mammals). In contrast, the PNS is, to varying degrees, capable of regeneration also in adult higher vertebrates. The failure of avian and mammalian CNS axons to regenerate is attributed to a programmed loss of the capability of neurons to elongate axons and to the presence of growth-inhibitory CNS myelin proteins such as tenascin and other glial molecules such as chondroitin sulfate proteoglycans (Reichenbach and Bringmann, 2015). The capacity for axon regeneration in the CNS of mammals declines sharply with the appearance of mature oligodendrocytes and myelin. Myelin-associated glycoprotein promotes the growth of axons in the developing tissue and inhibits axon growth in the mature tissue (Cai et al., 2001). After optic nerve injury, oligodendrocytes and reactive macroglia inhibit the regeneration of retinal ganglion cell axons by repulsive axon guidance molecules, such as ephrins and semaphorins, and by extracellular matrix constituents such as tenascin (Reichenbach and Bringmann, 2015). On the other hand, the regeneration of transected retinal ganglion cell axons is permitted in the optic nerve of the adult Browman-Wyse mutant rat in which oligodendrocytes and CNS myelin are absent and Schwann cells are present (Berry et al., 1992).

16.4 Microglia

Microglial cells are present in abundance in the CNS and constitute 10–20% of the total glial cell population in the adult brain (Banati, 2003). Microglia were originally identified by Franz Nissl (1899) as (activated) "rod cells." Del Río Hortega (1939) introduced the term "microglia." As this term suggests, these cells are the smallest members of the neuroglia family. Microglia are the resident cells of the innate immunity in the CNS and represent phagocytes and antigen-presenting cells. Microglia act as the first and main "responsible" of the neural tissue defense against invading microorganisms, by the use of phagocytic and cytotoxic mechanisms. Activated microglia initiate inflammatory processes and contribute to tissue repair.

The embryonic origin of microglia is distinct from the other types of neuroglia. Whereas macroglia is derived from the embryonic neuroectoderm, microglia descends from the extraembryonic mesoderm. Microglial precursors, originating from primitive macrophages in the yolk sac (Ginhoux et al., 2010), immigrate into the brain during development (from embryonic day 9.5 to postnatal day 10 in rodents) and differentiate to ramified parenchymal microglia. Another pool of microglia precursors may migrate into the brain along with vascular precursors and differentiate to perivascular microglia/macrophages.

In addition to the parenchymal and perivascular microglia, the brain hosts several other myeloid populations such as meningeal and choroid plexus macrophages. Though all of these macrophage populations share various myeloid- and macrophage-specific markers and exhibit similar immune regulatory functions (such as local immune surveillance and removal of debris), microglia and CNS macrophages represent two ontogenetically distinct populations, suggesting that these two myeloid populations have different functions (Prinz and Priller, 2014).

In the adult brain, parenchymal microglia are long lived and are not replaced by peripheral cells from the circulation; however, the cells are capable to proliferate in situ after tissue injury. Parenchymal microglia may be also repopulated from nestin-expressing CNS resident progenitor cells (Elmore et al., 2014). After elimination of ~99% of all microglial cells from the adult murine brain, the brain becomes fully repopulated from these progenitors with microglia within 7–14 days (Elmore et al., 2014). Microglia play critical roles in the histogenesis of the neural tissue during development through the induction of programmed neuronal cell death, removal of neuronal cell debris and inappropriate axons, and promotion of axonal growth (Hume et al., 1983; Frade and Barde, 1998). By phagocytosis of weak synapses, microglia is involved in the activity-dependent synaptic pruning and thus in the modeling of neuronal circuits (Paolicelli et al., 2011).

16.4.1 Resting Microglia

Resting (quiescent) microglia is found throughout the brain, spinal cord, and inner retina. There are more microglia in gray than in white matter, and in the neocortex and hippocampus than in the brain stem and cerebellum. In addition, species variations have been noted, as human white matter contains three times more microglia than rodent white matter.

Resting microglia are programmed for immunological tolerance, display an antiinflammatory phenotype, and have mainly homeostatic functions. Microglial cells display a dynamic cell phenotype in dependence on the location and the functional state. Resting microglia is characterized by a stellate, ramified phenotype with a small rod-shaped soma from which numerous elaborated thin processes extend in all directions (Fig. 16.20A and B). Similar to astrocytes, every microglial cell has its own territory, about 15−30 μm wide; there is little overlap between neighboring territories. Resting microglia exhibit pinocytotic activity and localized motility. While the soma does not translocate, the processes are highly motile and constantly survey the local microenvironment to remove metabolic products, toxic by-products, pathogens, and cell debris (Nimmerjahn et al., 2005). Microglial processes make transient contact with neuronal synapses and contribute to the modification and elimination of synaptic structures (Tremblay et al., 2010). Considering the velocity of process movement (2−3 μm min^{-1}), the brain parenchyma can be completely scanned by microglial processes every several hours (Nimmerjahn et al., 2005).

The process motility of resting microglia is not cell-autonomously regulated but modulated by endogenous neuro- and gliotransmission, for example, by extracellular ATP released from neurons and macroglial cells (Dibaj et al., 2010; Fontainhas et al., 2011; Damani et al., 2011; Wang and Wong, 2014). Extracellular ATP also induces membrane ruffling and filopodia extension from the cells (Haynes et al., 2006). The microglial process motility is increased by glutamatergic neurotransmission and decreased by GABAergic neurotransmission (Fontainhas et al., 2011). The neurotransmitter effects on microglia are not directly mediated but are induced by extracellular ATP, released in response to glutamatergic neurotransmission (Fontainhas et al., 2011).

The function of microglia in the adult healthy brain remains obscure. When the microglial cell population is nearly fully depleted during embryonic development, the animals die before adulthood (Ginhoux et al., 2010).

However, when microglia are depleted in adult mice, the animals are healthy and fully viable and have no impairments in behavior, cognition, or motor function, suggesting that microglia are not necessary for these functions in the adult brain (Elmore et al., 2014).

16.4.2 Activated Microglia

Once a pathogenic stimulus is detected, microglia become activated (polarized), proliferate (Fig. 16.20A and B), and migrate toward the site of injury (Fig. 16.20C) where the cells kill bacteria, release cytotoxic agents, and phagocytize dead cell debris. Focal neuronal damage induces rapid and concerted movements of many microglial processes toward the site of lesion which is completely surrounded by the processes within less than an hour. In acute lesions, the peak of microglial activation occurs 2−3 days post insult (Banati, 2003). Finally, microglia take on an antiinflammatory role and secrete antiinflammatory cytokines as part of the resolution and repair process, before returning to a surveillant state (Colton and Wilcock, 2010). If the size of the lesion exceeds a certain limit, microglial cells not only extend cell processes toward the lesion, but also migrate toward the lesion site from more distant locations. This migration is stimulated by nitric oxide (Dibaj et al., 2010).

Microglia activation is associated with a morphological change from a stellate, ramified phenotype to an amoeboid cell shape characteristic for differentiation toward migratory phagocytes (enlarged soma, retracted and thickened processes, vacuolation) (Fig. 16.20A). Activated microglia appear as "bushy" or "rods" depending on the activation level and contain numerous lysosomes and phagosomes. Amoeboid microglia exhibit a round cell body, possess pseudopodia and thin filopedia-like processes. Gitter cells (compound granular corpuscle), a characteristic of some brain lesions, are large microglial cells packed with granules which contain lipids from necrotic cells and degenerating myelin. In the retina, activated microglia increasingly express adhesion proteins which allows the cells to adhere to Müller cells; activated microglia migrate intraretinally in a radial direction using Müller cell processes as an adhesive scaffold (Fig. 16.20C; Wang et al., 2011). Activated microglia can be detected with antibodies raised against a number of macrophage-specific antigens such as Iba1, CD68, the OX-42 antibody, or lectins that recognize glycoproteins containing terminal α-D-galactose residues (Fig. 16.20C); however, these procedures does not distinguish microglia from macrophages (Fig. 16.20C).

Normally, perivascular microglia express high levels of MHC class II proteins used for the presentation of

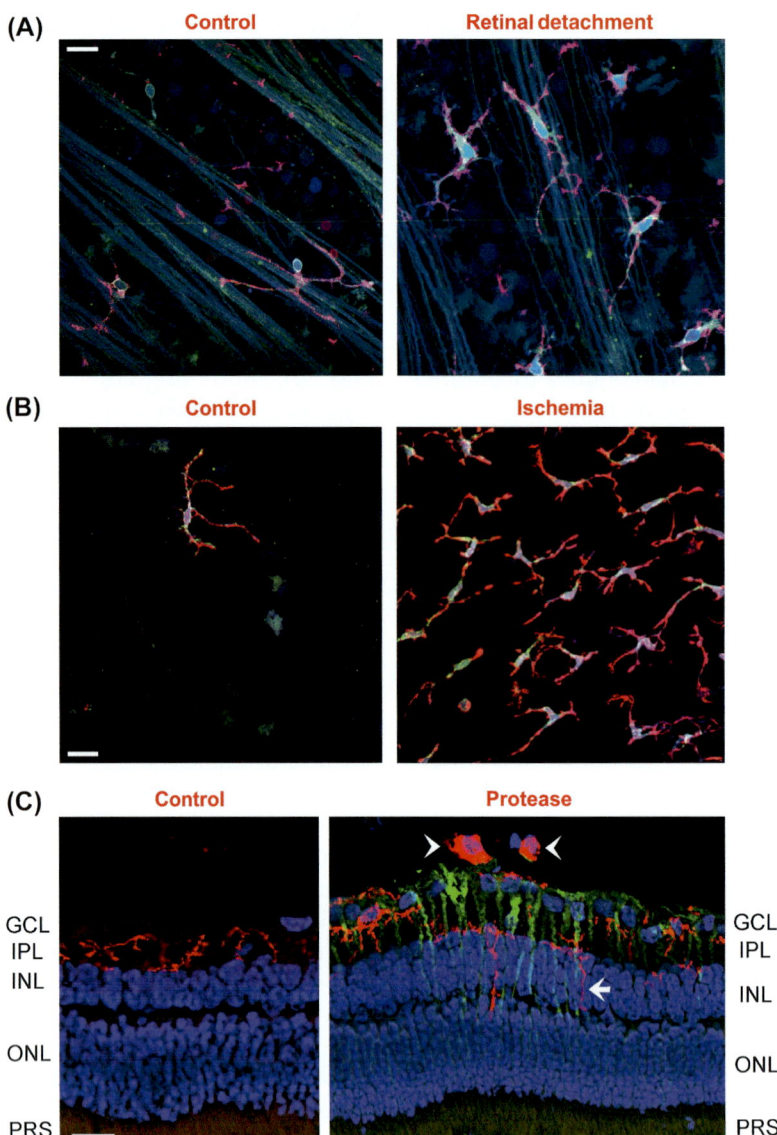

FIGURE 16.20 Microglia activation in the rabbit retina. (A) In avascular regions of the rabbit retina, microglial cells are located in the innermost retinal layers. Acutely isolated wholemounts of a control retina (*left*) and a retina obtained 2 days after experimental retinal detachment (*right*) are viewed at the vitreal surface (the nerve fiber/ganglion cell layers). Microglia are *red* stained; cell nuclei are *blue* stained. *Elongated structures* are light reflections on nerve fibers. (B) Microglia cells at the vitreal surface of a control retina (*left*) and a retina obtained 8 days after a 1-h transient retinal ischemia (*right*). (C) Interaction of blood-derived monocytes/macrophages and retinal glial cells in a rabbit model of protease-induced inner limiting membrane injury. The protease was injected into the vitreous body of the eye. Microglia and macrophages in retinal slices were labeled with *Griffonia simplicifolia* agglutinin isolectin (*red*). Activated Müller glial cells were labeled with an antibody against glial fibrillary acidic protein (GFAP) (*green*), and cell nuclei are *blue* stained. Injury of the inner limiting membrane induced retinal glial cell reactivity characterized by an upregulation of GFAP in Müller cells and activated microglia that begin to migrate toward the outer retina (*arrow*). Bloodborne monocytes/macrophages adhere to the vitreal surface of the site of injury (*arrowheads*), suggesting a relationship between the attachment of macrophages and retinal glial cell activation. *GCL*, ganglion cell layer; *INL*, inner nuclear layer; *ONL*, outer nuclear layer; *PRS*, photoreceptor segment. Scale bars, 20 μm. *Modified from (A) Fig. 16.6A and B, Uhlmann, S., Bringmann, A., Uckermann, O., Pannicke, T., Weick, M., Ulbricht, E., Goczalik, I., Reichenbach, A., Wiedemann, P., Francke, M., 2003 Early glial cell reactivity in experimental retinal detachment: effect of suramin. Invest. Ophthalmol. Visual Sci. 44, 4114–4122; (B) Fig. 16.6A, Uckermann, O., Uhlmann, S., Pannicke, T., Francke, M., Gamsalijew, R., Makarov, F., Ulbricht, E., Wiedemann, P., Reichenbach, A., Osborne, N.N., Bringmann, A., 2005. Ischemia-reperfusion causes exudative detachment of the rabbit retina. Invest. Ophthalmol. Visual Sci. 46, 2592–2600; (C) Fig. 16.2A and C, Francke, M., Uhlmann, S., Pannicke, T., Goczalik, I., Uckermann, O., Weick, M., Härtig, W., Wiedemann, P., Reichenbach, A., Bringmann, A., 2003. Experimental dispase-induced retinopathy causes up-regulation of P2Y receptor-mediated calcium responses in Müller glial cells. Ophthalmic Res. 35, 30–41.*

antigens to T lymphocytes (Hickey and Kimura, 1988). The parenchymal resting microglia lack or express MHC class II molecules at low level, but are capable to express the molecules after activation (Reichenbach and Bringmann, 2015). By releasing cytokines, microglia are important for recruiting leucocytes into the CNS.

Microglia activation has a Janus face nature. While activated microglia initially contribute to neuronal protection and tissue repair, persistent microglia activation has harmful effects on neuronal function and survival, thereby exacerbating disease processes. Phagocytotic cells release oxygen and nitrogen free radicals, and toxic cytokines, that induce neuronal apoptosis (Nakazawa et al., 2007). Alarm signals released from degenerating neurons (like ATP) trigger microglia activation which may also lead to attacks against healthy neurons by the release of cytotoxic cytokines and the initiation of immune responses against neuronal antigens (Langmann, 2007). In the retina, activated microglia containing phagocytosed photoreceptor debris may exit the tissue via retinal and choroidal vessels and may reach the spleen; here, microglia act as antigen-presenting cells and elicit systemic immune responses against retinal antigens (Raoul et al., 2008).

Activated microglia may also promote neuroprotection and regeneration via various mechanisms such as the production of neurotrophic factors (Reichenbach and Bringmann, 2015) and stimulation of the regrowth and remapping of damaged neural circuitry (Gehrmann et al., 1995). Perivascular microglia/macrophages are essential to repair of vascular walls and to clear amyloid-β from the brain, for example (Mildner et al., 2011). Until recently, it was unclear why microglia are sometimes damaging and other times protective. Nonneuronal macrophages are known to display at least two distinct states of activation: M1 (classical proinflammatory activation) and M2 polarization (alternative antiinflammatory activation) (Prinz and Priller, 2014). Despite the different ontogeny, microglia may also have the capacity to become polarized into M1-like and M2-like phenotypes (Prinz and Priller, 2014; Cherry et al., 2014). M1-polarized microglia exert cytotoxic effects on neurons and oligodendrocytes and inhibit axon extension, while M2-polarized microglia exhibit phagocytic capacity and promote neurite outgrowth and tissue repair (Kigerl et al., 2009) including remyelination by driving oligodendrocyte differentiation (Miron et al., 2013). Because long-term inflammation has harmful consequences in the CNS, the inflammatory response needs to be downregulated for proper tissue healing. Normally, a switch from an M1- to an M2-type microglial response completes tissue inflammation and initiates tissue repair (Cherry et al., 2014). Astrocytes and retinal Müller cells limit the magnitude of inflammatory responses by facilitating the quiescence of microglial cells (Wang and Wong, 2014).

With advancing age, microglia undergo changes in gene expression patterns that give rise to pathogenic phenotypes and dysregulation of phagocytosis (Ma et al., 2013). In the retina of old animals, for example, microglia, Müller cells, and astrocytes display increased signs of gliosis compared to retinas of young animals (Kim et al., 2004). Aged resting microglia have smaller and less branched dendritic arbors, and slower process motilities, which compromise the ability to survey and interact with the environment (Damani et al., 2011). While young microglia respond to extracellular ATP by increasing the process motility and becoming more ramified, aged microglia exhibit a contrary response, becoming less dynamic and ramified (Damani et al., 2011). Upon retinal injury, aged microglia show slower acute responses with lower rates of process motility and cell migration while the long-term response is more sustained compared with young microglia (Damani et al., 2011). Microglia are functionally impaired by amyloid plaque deposition in the brain, as indicated by the reduced process motility and phagocytic activity (Krabbe et al., 2013). In aged animals, microglia become less responsive to M2 induction signals, favoring the inflammatory microglial responses while the tissue repair function of activated microglia is disturbed (Cherry et al., 2014).

16.5 Glia of the Peripheral Nervous System

16.5.1 Schwann Cells

The Schwann cell is the most common cell type in the PNS. Peripheral nerve axons are ensheathed by nonmyelinating or myelinating Schwann cells (Fig. 16.12; Jessen and Mirsky, 2005). Unlike the vast majority of oligodendrocytes, each myelinating Schwann cell forms a single internode. The axonal diameter is directly related to the number of myelin lamellae which affects the saltatory conduction velocity; the more lamellae, the faster the conduction. Aδ nociceptor fibers are ensheathed by few compact myelin lamellae, while larger-diameter (>1 μm) sensory fibers are ensheathed by up to 100 spirals of compact myelin lamellae. The nodes of Ranvier are covered by specific, fingerlike Schwann cell processes (Baumann and Pham-Dinh, 2001), similar as formed by astrocytes and Müller cells in the CNS. Schwann cells (in contrast to oligodendrocytes) are surrounded by a basal lamina that is continuous with the adjacent internode. Myelinated axons in the PNS are separated by an extracellular compartment, the endoneurium. Although myelin produced by oligodendroglia and myelinating Schwann cells is similar at the ultrastructural level (with the exception that lamellae in the CNS myelin are ~30% thinner than in PNS myelin), there are differences in the protein composition (Baumann and Pham-Dinh, 2001).

The majority (~80%) of axons in peripheral nerves are unmyelinated. Nonmyelinating Schwann cells ensheath bundles of multiple thin axons within grooves of the plasma membrane (Remak fibers). Olfactory ensheathing cells are a peculiar type of nonmyelinating Schwann cells which associate with the primary olfactory axons throughout their way from the olfactory epithelium to the olfactory bulb. Thus, they are the only type of PNS glia which enters the CNS. Noteworthy, even there they maintain their capability to support axon regeneration (in contrast to oligodendrocytes) (Raisman, 2001). The synapses between motor nerve terminals and skeletal muscle fibers (neuromuscular junction) are covered by another specialized type of nonmyelinating Schwann cells, perisynaptic or terminal Schwann cells (Fig. 16.12). Perisynaptic Schwann cells regulate synapse formation, maturation, and stability, and synaptic transmission (Wu et al., 2010). Nonmyelinating Schwann cells of bone marrow sympathetic fibers regulate the hibernation and activation state of hematopoietic stem cells (Yamazaki and Nakauchi, 2014). Schwann cells provide trophic support required for axon survival (Riethmacher et al., 1997). Schwann cells are also facultative antigen-presenting cells and regulate, via release of ATP and adenosine, the excitability of axonal nociceptors (Armati and Mathey, 2013).

Many sensory nerve endings in the skin associate with terminal Schwann cells that form the inner parts of mechanoreceptors (Fig. 16.12). At the center of Pacinian corpuscles lies a nerve ending that is encapsulated by lamellar processes of Schwann cells which are coupled by gap junctions and desmosomes and which form the inner core of these mechanosensitive organs (Rico et al., 1996). Schwann cells regulate the axonal excitation by the release of GABA (Pawson et al., 2009). Schwann cells in mechanosensitive organs of the skin are trophically dependent on sensory axon terminals which release neuregulin (Kopp et al., 1997).

Schwann cells develop from migratory progenitors in the neural crest. Neural crest cells develop to Schwann cell precursors which, subsequently, become immature Schwann cells; finally, these cells diverge into myelinating or nonmyelinating Schwann cells (Jessen and Mirsky, 2005). Neuregulin-1 is implicated in the lineage specification of neural crest cells and is essential for the proliferation and survival of Schwann cell precursors (Dong et al., 1995). In the rat, Schwann cells start the myelination process around birth; PNS axons are fully myelinated at postnatal day 30.

Axons dictate which differentiation path Schwann cells follow. Active axons release ATP which inhibits the proliferation and differentiation of Schwann cells (Stevens and Fields, 2000). The type III isoform of neuregulin-1 on the axonal surface is required for the myelination by Schwann cells (Taveggia et al., 2005). The level of neuregulin-1 type III also defines the thickness of the myelin sheath and drives the dedifferentiation of myelinating cells in injured nerves. Neuregulin-1 type III may also regulate the differentiation of nonmyelinating Schwann cells and the formation of Remak bundles.

Mature Schwann cells retain a considerable phenotypic plasticity. Nerve roots have specialized transition zones that limit the migration of oligodendrocyte and Schwann cell precursor cells between the CNS and PNS (Fig. 16.12). However, at least in lower vertebrates, Schwann cells can dedifferentiate and migrate into the CNS after damage of peripheral nerves near the border, or after central nerves lose the myelin sheath; vice versa, oligodendrocyte precursor cells migrate from the spinal cord into peripheral nerves, and myelinate motor axons, when Schwann cells are lost (Kucenas et al., 2009). Peripheral nerves serve as a niche for glial multipotent neural crest—like cells. The peripheral nerve—associated glia can change the fate of, and produce tissues at remote destinations during development and regeneration. For example, peripheral glial cells generate multipotent mesenchymal stem cells that produce half of pulp cells and odontoblasts in the growing incisor (Kaukua et al., 2014).

16.5.2 Satellite Cells

Neuronal cell bodies in dorsal root sensory ganglia (Fig. 16.12), and in sympathetic and parasympathetic ganglia, are covered by flattened sheet—like glial cells known as satellite cells (Hanani, 2005). Satellite cells usually form envelopes around single neurons providing physical support and a protective barrier (Hanani, 2010). There are numerous fine invaginations between the neurons and satellite cell sheaths. Satellite cells share many properties with astrocytes of the CNS (eg, gap junctional coupling and ATP-induced calcium responses), support neurons with essential molecules for neural transmission, regulate the extracellular ion levels, and clear extracellular neurotransmitters (Hanani, 2005, 2010). Satellite cells participate in the information processing of sensory signals from afferent terminals to the spinal cord (Huang et al., 2013). These cells are highly sensitive to injury and inflammation, and regulate pathological states, such as chronic pain (Armati and Mathey, 2013). Satellite cells function as antigen-presenting cells involved in sensing the local environment and the control of local T cell responses to protect neuronal somata; thus, satellite cells also resemble CNS microglia (Van Velzen et al., 2009).

PNS glia represent multipotent progenitor cells that generate both glia and neurons (Kaukua et al., 2014). Peripheral glial cells arrive to late-developing tissues on the pioneer presynaptic nerve fibers. Subsequently, some glial cells change their fate, navigate for short distances, and then convert into neurons and satellite cells of parasympathetic ganglia (Dyachuk et al., 2014).

16.5.3 Enteric Glia

The enteric nervous system is a network of neurons and glia within the wall of the bowel that controls motility, blood flow, uptake of nutrients, secretion, and inflammatory processes in the gastrointestinal tract. Enteric glial cells, first described by Dogiel (1899), are more abundant (up to fourfold) than neurons (Bassotti et al., 2007). The enteric nervous system is subdivided into several ganglionated and nonganglionated plexuses. Most enteric glial cells are present in the ganglia. Enteric glial cells are also present in the interconnecting nerve strands of the ganglionated and in all nonganglionated plexuses (Bassotti et al., 2007).

In contrast to Schwann cells, enteric glial cells do not form basal laminae. Enteric glia are small cells, with several projecting processes of various length and shapes, which often confer them a starlike appearance (Gabella, 1981; Hanani and Reichenbach, 1994). In the ganglia, enteric glial cells resemble protoplasmic astrocytes; they are very tightly packed around neurons and extend several flat projections which incompletely insulate enteric neurons from extraganglionic cells. In the nerve strands, enteric glial cells resemble fibrous astrocytes of the CNS; their processes wrap up several axon bundles (Gabella, 1981; Hanani and Reichenbach, 1994). Enteric glial cells make numerous synaptoid contacts with vesicle-containing nerve varicosities, are coupled extensively by gap junctions, and anchored to ganglionic surfaces. Although these cells have a different embryological origin, enteric glia are remarkably similar to astrocytes in respect to both morphological and molecular properties. Glia in myenteric and submucous plexuses express astrocytic markers such as GFAP, glutamine synthetase, and vimentin (Jessen and Mirsky, 1983).

Developmentally, enteric glia, like Schwann cells, are derived from the neural crest (Gershon and Rothman, 1991). Precursor cells, which give rise to both enteric glia and neurons, migrate from vagal and sacral segments of the crest to the bowel. Enteric glia may also arise from Schwann cells that enter the gut with the extrinsic innervation, and may, vice versa, replace lost Schwann cells (Gershon and Rothman, 1991).

Enteric glia are active participants in almost every gut function including motility, mucosal secretion, and host defense (Bassotti et al., 2007; Gulbransen and Sharkey, 2012). Enteric glia support neuronal survival and proliferation, take up neurotransmitters, maintain the neuronal homeostasis, and present antigens (Bassotti et al., 2007). Glia in enteric ganglia are activated by synaptic stimulation (Gulbransen and Sharkey, 2012). Subepithelial glia have a trophic and supporting relationship with intestinal epithelial cells (Gulbransen and Sharkey, 2012). These glial cells contribute to the maintenance of the integrity of the intestinal epithelial barrier. Vagal nerve stimulation activates enteric glial cells through nicotinic cholinergic signaling; factors secreted from enteric glia increase the barrier function in the gut and induce a shift in epithelial cell phenotype toward increased cell adhesion and cell differentiation, eg, after intestinal mucosal injury (Yu and Li, 2014).

16.5.4 Glia in Peripheral Sensory Epithelia

Sensory epithelia are specialized for mechanoperception (statoacoustic systems with organs of Corti and vestibular organs; Fig. 16.8A) or olfaction (nasal olfactory epithelium; Fig. 16.8B). Neurons in sensory epithelia project their receptive organelles (generally a cilium) into a fluid- or mucous-filled lumen and are enclosed by the processes of support cells. The neuron-supporting cells closely resemble ependymoglia (Reichenbach and Robinson, 1995). In mechanoreceptive epithelia, the apical poles of supporting cells fuse with those of adjacent cells to form an apical "membrane" (Lindeman, 1969). Adjacent cells are joined by tight junctions. Most of these cells possess short microvilli that project into the lumen; numerous mitochondria and glycogen granules are found near the bases of the microvilli, indicating that these are sites of intense metabolic exchange. The ependymoglia-like cells span the entire thickness of the sensory structure; the basal endfeet abut the underlying basal lamina and surround the axons passing through the basal lamina.

The organ of Corti (Fig. 16.8A) contains many distinct and specialized types of supporting cells. Each type of support cell extends from the basilar membrane to the free surface of the organ of Corti where the processes unite to form an apical (reticular) "membrane." All of the spaces contacted by these cells are filled with endolymph. One type, the pillar cell, has a cementlike basal cone from which tonofilaments extend up to the apical end of the cell. Some of the tonofilaments are tubular (28 nm outer diameter) while others are microfilamentous (6 nm diameter); these filaments form a regular, tightly packed array (Angelborg and Engström, 1972). A second cell type, Deiters' cells (outer phalangeal cells) have similar tonofilaments and support the outer hair cells (Fig. 16.8A). A third type, phalangeal (border) cells are located at the medial side of the organ and support the inner hair cells. Laterally, Hensen's (Fig. 16.8A) and Claudius' cells form a columnar epithelium that decreases in height with increasing laterality. Vimentin is present in two types of supporting cells in the mammalian organ of Corti, Deiters' cells and inner pillar cells (Oesterle et al., 1990).

Hensen's cells and Deiters' cells of the organ of Corti exhibit pronounced calcium coupling which is induced by a release of calcium from internal stores, a connexin hemichannel–mediated release of ATP, and ATP-mediated activation of purinergic P2 receptors in surrounding support cells (Piazza et al., 2007). This calcium signaling mechanism can be induced by the death of a single hair cell which releases ATP (Gale et al., 2004) and may compromise the survival of further hair cells (Housley et al., 2009). This may be part of the injury response mechanism that is activated during acoustic overstimulation and ototoxic hair cell damage.

The nasal olfactory epithelium consists of olfactory receptor neurons, glialike sustentacular (supporting) cells, and basal cells which are progenitors for the continuous renewal of the epithelium (Fig. 16.8B; Graziadei, 1971). Sustentacular cells are located apically and form a barrier on the surface of the olfactory epithelium, while olfactory sensory neurons are located basal to these cells. Sustentacular cells possess many microvilli on the apical surface, which underlie the cilia layer of the olfactory epithelium, and extend processes across the width of the epithelium to the basement membrane. Sustentacular cells are thought to provide metabolic, structural, and trophic support for olfactory receptor neurons (Getchell and Getchell, 1992). Sustentacular cells resemble those of mechanosensory epithelia, except that these cells contain pigment granules and secretory vesicles, and large Golgi complexes in the apical cytoplasm, indicating a secretory function. Secretion from the apical surface contributes to the mucous covering of the epithelium. Sustentacular cells are functionally coupled via gap junctions (Vogalis et al., 2005).

Gustatory organs (taste buds), which are made up of 50–100 cells, contain a type of supporting cell ("light cell"; type I cell) with ependymoglial features (Fig. 16.8C; Pumplin et al., 1997). Supporting cells possess short microvilli on the apical surface and ensheath other taste bud cells (type II receptor cells, type III presynaptic cells) with lamellar processes. All taste bud cells are connected by tight junctions at the apical pole of the cells. The glialike support cells regulate the ionic milieu, may contribute to salt perception, and express glial glutamate transporters (Bartel et al., 2006). ATP-hydrolyzing exoenzymes at the surface of the glialike cells terminate the signals of the neurotransmitter ATP, which is released from receptor cells and which activates surrounding receptor and presynaptic cells, and the nerve endings of chemosensory afferent fibers (Bartel et al., 2006; Housley et al., 2009).

References

Akmayev, I.G., Fidelina, O.V., Kabolova, Z.A., Popov, A.P., Schitkova, T.A., 1973. Morphological aspects of the hypothalamic-hypophyseal system. IV. Medial basal hypothalamus. An experimental morphological study. Z. Zellforsch. 137, 493–512.

Altman, J., Bayer, S.A., 1978. Development of the diencephalon in the rat. III. Ontogeny of the specialized ventricular linings of the hypothalamic third ventricle. J. Comp. Neurol. 182, 995–1016.

Anderson, D.H., Stern, W.H., Fisher, S.K., Erickson, P.A., Borgula, G.A., 1983. Retinal detachment in the cat: the pigment epithelial-photoreceptor interface. Invest. Ophthalmol. Vis. Sci. 2, 906–926.

Andriezen, W.L., 1893. The neuroglia elements of the brain. Br. Med. J. Clin. Res. 2, 227–230.

Angelborg, C., Engström, H., 1972. Supporting elements in the organ of Corti. I. Fibrillar structures in the supporting cells of the organ of Corti of mammals. Acta Otolaryngol. Suppl. 301, 49–60.

Aquado, F., Espinosa-Parrilla, J.F., Carmona, M.A., Soriano, F., 2002. Neuronal activity regulates correlated network properties of spontaneous calcium transients in astrocytes in situ. J. Neurosci. 22, 9430–9444.

Araque, A., Parpura, V., Sanzgiri, R.P., Haydon, P.G., 1999. Tripartite synapses: glia, the unacknowledged partner. Trends Neurosci. 22, 208–215.

Armati, P.J., Mathey, E.K., 2013. An update on Schwann cell biology — immunomodulation, neural regulation and other surprises. J. Neurol. Sci. 333, 68–72.

Bacaj, T., Tevlin, M., Lu, Y., Shaham, S., 2008. Glia are essential for sensory organ function in C. elegans. Science 322, 744–747.

Banati, R., 2003. Neuropathological imaging: in vivo detection of glial activation as a measure of disease and adaptive change in the brain. Br. Med. Bull. 65, 121–131.

Bär, T., 1980. The vascular system of the cerebral cortex. Adv. Anat. Embryol. Cell Biol. 59, 1–62.

Barber, A.J., Lieth, E., 1997. Agrin accumulates in the brain microvascular basal lamina during development of the blood-brain barrier. Dev. Dyn. 208, 62–74.

Barres, B.A., Raff, M.C., 1993. Proliferation of oligodendrocyte precursor cells depends on electrical activity in axons. Nature 361, 258–260.

Bartel, D.L., Sullivan, S.L., Lavoie, E.G., Sevigny, J., Finger, T.E., 2006. Nucleoside triphosphate diphosphohydrolase-2 is the ecto-ATPase of type I cells in taste buds. J. Comp. Neurol. 497, 1–12.

Bassotti, G., Villanacci, V., Antonelli, E., Morelli, A., Salerni, B., 2007. Enteric glial cells: new players in gastrointestinal motility? Lab. Invest. 87, 628–632.

Baumann, N., Pham-Dinh, D., 2001. Biology of oligodendrocyte and myelin in the mammalian central nervous system. Physiol. Rev. 81, 871–927.

Berry, M., Hall, S., Rees, L., Carlile, J., Wyse, J.P., 1992. Regeneration of axons in the optic nerve of the adult Browman-Wyse (BW) mutant rat. J. Neurocytol. 21, 426–448.

Bezzi, P., Gundersen, V., Galbete, J.L., Seifert, G., Steinhäuser, C., Pilati, E., Volterra, A., 2004. Astrocytes contain a vesicular compartment that is competent for regulated exocytosis of glutamate. Nat. Neurosci. 7, 613–620.

Black, J.A., Waxman, S.G., 1988. The perinodal astrocyte. Glia 1, 169–183.

Bok, D., Young, R., 1979. Phagocytic properties of the retinal pigment epithelium. In: Zinn, K.M., Marmor, M.F. (Eds.), The Retinal

Pigment Epithelium. Harvard University Press, Cambridge, MA, pp. 148–174.

Booz, K.H., Desaga, U., Felsing, T., 1974. Über die Entstehung der Basalmembranlabyrinthe. Eine licht- und elektronenmikroskopische Untersuchung. Z. Anat. Entwicklungsgesch. 143, 185–203.

Braak, E., 1975. On the fine structure of the external glial layer in the isocortex of man. Cell Tissue Res. 157, 367–390.

Bradl, M., Lassmann, H., 2010. Oligodendrocytes: biology and pathology. Acta Neuropathol. 119, 37–53.

Brightman, M.W., Palay, S.L., 1963. The fine structure of ependyma in the brain of the rat. J. Cell Biol. 19, 415–439.

Brightman, M.W., Reese, T.S., 1969. Junctions between intimately apposed cell membranes in the vertebrate brain. J. Cell Biol. 40, 648–677.

Bringmann, A., Uckermann, O., Pannicke, T., Iandiev, I., Reichenbach, A., Wiedemann, P., 2005. Neuronal versus glial cell swelling in the ischaemic retina. Acta Ophthalmol. Scand. 83, 528–538.

Brückner, G., Härtig, W., Kacza, J., Seeger, J., Welt, K., Brauer, K., 1996. Extracellular matrix organization in various regions of rat brain gray matter. J. Neurocytol. 25, 333–346.

Brückner, E., Grosche, A., Pannicke, T., Wiedemann, P., Reichenbach, A., Bringmann, A., 2012. Mechanisms of VEGF- and glutamate-induced inhibition of osmotic swelling of murine retinal glial (Müller) cells: indications for the involvement of vesicular glutamate release and connexin-mediated ATP release. Neurochem. Res. 37, 268–278.

Bruni, J.E., Clattenburg, R.E., Millar, E., 1983. Tanycyte ependymal cells in the third ventricle of young and adult rats: a Golgi study. Anat. Anz. 153, 53–68.

Bushong, E.A., Martone, M.E., Jones, Y.Z., Ellisman, M.H., 2002. Protoplasmic astrocytes in CA1 stratum radiatum occupy separate anatomical domains. J. Neurosci. 22, 183–192.

Bushong, E.A., Martone, M.E., Ellisman, M.H., 2004. Maturation of astrocyte morphology and the establishment of astrocyte domains during postnatal hippocampal development. Int. J. Dev. Neurosci. 22, 73–86.

Butkevich, E., Hülsmann, S., Wenzel, D., Shirao, T., Duden, R., Majoul, I., 2004. Drebrin is a novel connexin-43 binding partner that links gap junctions to the submembrane cytoskeleton. Curr. Biol. 14, 650–658.

Butt, A.M., Colquhoun, K., Tutton, M., Berry, M., 1994. Three-dimensional morphology of astrocytes and oligodendrocytes in the intact mouse optic nerve. J. Neurocytol. 23, 469–485.

Butt, A.M., Duncan, A., Berry, M., 1994. Astrocyte associations with nodes of Ranvier: ultrastructural analysis of HRP-filled astrocytes in the mouse optic nerve. J. Neurocytol. 23, 486–499.

Butt, A.M., Ibrahim, M., Ruge, F.M., Berry, M., 1995. Biochemical subtypes of oligodendrocyte in the anterior medullary velum of the rat as revealed by the monoclonal antibody Rip. Glia 14, 185–197.

Cai, D., Qiu, J., Cao, Z., McAtee, M., Bregman, B.S., Filbin, M.T., 2001. Neuronal cyclic AMP controls the developmental loss in ability of axons to regenerate. J. Neurosci. 21, 4731–4739.

Chan-Palay, V., Palay, S.L., 1972. The form of velate astrocytes in the cerebellar cortex of monkey and rat: high-voltage electron microscopy of rapid-Golgi preparations. Z. Anat. Entwicklungsgesch. 138, 1–19.

Chao, T.I., Skatchkov, S.N., Eberhardt, W., Reichenbach, A., 1994. Na$^+$ channels of Müller (glial) cells isolated from retinae of various mammalian species including man. Glia 10, 173–185.

Chao, T.I., Kasa, P., Wolff, J.R., 1997. Distribution of astroglia in glomeruli of the rat main olfactory bulb: exclusion from the sensory subcompartment of neuropil. J. Comp. Neurol. 388, 191–210.

Chao, T.I., Rickmann, M., Wolff, J.R., 2002. The synapse-astrocyte boundary: anatomical basis for an integrative role of glia in synaptic transmission. In: Volterra, A., Magistretti, P., Haydon, P. (Eds.),

Tripartite Synapses: Synaptic Transmission with Glia. Oxford University Press, Oxford, pp. 3–23.

Cherry, J.D., Olschowka, J.A., O'Banion, M.K., 2014. Neuroinflammation and M2 microglia: the good, the bad, and the inflamed. J. Neuroinflammation 11, 98.

Christensen, O., 1987. Mediation of cell volume regulation by Ca^{2+} influx through stretch-activated channels. Nature 330, 66–68.

Clarke, S.R., Shetty, A.K., Bradley, J.L., Turner, D.A., 1994. Reactive astrocytes express the embryonic intermediate neurofilament nestin. Neuroreport 5, 1885–1888.

Colombo, J.A., 2001. A columnar-supporting mode of astroglial architecture in the cerebral cortex of adult primates? Neurobiology 9, 1–16.

Colombo, J.A., Reisin, H.D., 2004. Interlaminar astroglia of the cerebral cortex: a marker of the primate brain. Brain Res. 1006, 126–131.

Colton, C., Wilcock, D.M., 2010. Assessing activation states in microglia. CNS Neurol. Disord. - Drug Targets 9, 174–191.

Connors, N.C., Kofuji, P., 2002. Dystrophin Dp71 is critical for the clustered localization of potassium channels in retinal glial cells. J. Neurosci. 22, 4321–4327.

Cornell-Bell, A., Thomas, P.G., Smith, S.J., 1990. The excitatory neurotransmitter glutamate causes filopodia formation in cultured hippocampal astrocytes. Glia 3, 322–334.

Crawford, B.J., 1983. Some factors controlling cell polarity in chick retinal pigment epithelial cells in clonal culture. Tissue Cell 15, 993–1005.

Dakubo, G.D., Beug, S.T., Mazerolle, C.J., Thurig, S., Wang, Y., Wallace, V.A., 2008. Control of glial precursor cell development in the mouse optic nerve by sonic hedgehog from retinal ganglion cells. Brain Res. 1228, 27–42.

Damani, M.R., Zhao, L., Fontainhas, A.M., Amaral, J., Fariss, R.N., Wong, W.T., 2011. Age-related alterations in the dynamic behavior of microglia. Aging Cell 10, 263–276.

Dawson, M.R., Polito, A., Levine, J.M., Reynolds, R., 2003. NG2-expressing glial progenitor cells: an abundant and widespread population of cycling cells in the adult rat CNS. Mol. Cell. Neurosci. 24, 476–488.

Del Río Hortega, P., 1921. Histogenesis y evolucion normal; exodo y distribucion regional de la microglia. Memorias Real Soc. Espanola Hist. Nat. 11, 213–268.

Del Río Hortega, P., 1928. Tercera aportacion al conocimiento morfologico e interpretacion funcional de la oligodendroglia. Memorias Real Soc. Espanola Hist. Nat. 14, 5–122.

Del Río Hortega, P., 1939. The microglia. Lancet 233, 1023–1026.

Denk, W., Sugimori, M., Llinas, R., 1995. Two types of calcium response limited to single spines in cerebellar Purkinje cells. Proc. Natl. Acad. Sci. U.S.A. 92, 8279–8282.

Derouiche, A., Frotscher, M., 2001. Peripheral astrocyte processes: monitoring by selective immunostaining for the actin-binding ERM proteins. Glia 36, 330–341.

Dibaj, P., Nadrigny, F., Steffens, H., Scheller, A., Hirrlinger, J., Schomburg, E.D., Neusch, C., Kirchhoff, F., 2010. NO mediates microglial response to acute spinal cord injury under ATP control in vivo. Glia 58, 1133–1144.

Distler, C., Dreher, Z., Stone, J., 1991. Contact spacing among astrocytes in the central nervous system: an hypothesis of their structural role. Glia 4, 484–494.

Dogiel, A.S., 1899. Über den Bau der Ganglien in den Geflechten des Darmes und der Gallenblase des Menschen und der Saugethiere. Z. Naturforsch. B 5, 130–158.

Dong, Z., Brennan, A., Liu, N., Yarden, Y., Lefkowitz, G., Mirsky, R., Jessen, K.R., 1995. Neu differentiation factor is a neuron-glia signal and regulates survival, proliferation, and maturation of rat Schwann cell precursors. Neuron 15, 585–596.

Dorostkar, M.M., Boehm, S., 2008. Presynaptic ionotropic receptors. Handb. Exp. Pharmakol. 184, 479–527.

Dubois-Dauphin, M., Poitry-Yamate, C., de Bilbao, F., Julliard, A.K., Jourdan, F., Donati, G., 2000. Early postnatal Müller cell death leads to retinal but not optic nerve degeneration in NSE-Hu-Bcl-2 transgenic mice. Neuroscience 95, 9–21.

Dyachuk, V., Furlan, A., Shahidi, M.K., Giovenco, M., Kaukua, N., Konstantinidou, C., Pachnis, V., Memic, F., Marklund, U., Müller, T., Birchmeier, C., Fried, K., Ernfors, P., Adameyko, I., 2014. Neurodevelopment. Parasympathetic neurons originate from nerve-associated peripheral glial progenitors. Science 345, 82–87.

Elmore, M.R., Najafi, A.R., Koike, M.A., Dagher, N.N., Spangenberg, E.E., Rice, R.A., Kitazawa, M., Matusow, B., Nguyen, H., West, B.L., Green, K.N., 2014. Colony-stimulating factor 1 receptor signaling is necessary for microglia viability, unmasking a microglia progenitor cell in the adult brain. Neuron 82, 380–397.

Ermisch, A., Sterba, G., Mueller, A., Hess, J., 1971. Autoradiographische Untersuchungen am Subkommissuralorgan und am Reissnerschen Faden. I. Organsekretion und Parameter der Organleistung als Grundlagen zur Beurteilung der Organfunktion. Acta Zool 52, 1–21.

Feig, S.L., Haberly, L.B., 2011. Surface-associated astrocytes, not endfeet, form the glia limitans in posterior piriform cortex and have a spatially distributed, not a domain, organization. J. Comp. Neurol. 519, 1952–1969.

Felten, D.L., Harrigan, P., Burnett, B.T., Cummings, J.P., 1981. Fourth ventricular tanycytes: a possible relationship with monoaminergic nuclei. Brain Res. Bull. 6, 427–436.

Ffrench-Constant, C., Miller, R.H., Kruse, J., Schachner, M., Raff, M.C., 1986. Molecular specialization of astrocyte processes at nodes of Ranvier in rat optic nerve. J. Cell Biol. 102, 844–852.

Fischer, A.J., Zelinka, C., Scott, M.A., 2010. Heterogeneity of glia in the retina and optic nerve of birds and mammals. PLoS One 5, e10774.

Fontainhas, A.M., Wang, M., Liang, K.J., Chen, S., Mettu, P., Damani, M., Fariss, R.N., Li, W., Wong, W.T., 2011. Microglial morphology and dynamic behavior is regulated by ionotropic glutamatergic and GABAergic neurotransmission. PLoS One 6, e15973.

Frade, J.M., Barde, Y.A., 1998. Microglia-derived nerve growth factor causes cell death in the developing retina. Neuron 20, 35–41.

Francke, M., Uhlmann, S., Pannicke, T., Goczalik, I., Uckermann, O., Weick, M., Härtig, W., Wiedemann, P., Reichenbach, A., Bringmann, A., 2003. Experimental dispase-induced retinopathy causes up-regulation of P2Y receptor-mediated calcium responses in Müller glial cells. Ophthalmic Res. 35, 30–41.

Franze, K., Grosche, J., Skatchkov, S.N., Schinkinger, S., Foja, C., Schild, D., Uckermann, O., Travis, K., Reichenbach, A., Guck, J., 2007. Müller cells are living optical fibers in the vertebrate retina. Proc. Natl. Acad. Sci. U.S.A. 104, 8287–8292.

Gabella, G., 1981. Ultrastructure of the nerve plexuses of the mammalian intestine: the enteric glial cells. Neuroscience 6, 425–436.

Gale, J.E., Piazza, V., Ciubotaru, C.D., Mammano, F., 2004. A mechanism for sensing noise damage in the inner ear. Curr. Biol. 14, 526–529.

Gao, L., Miller, R.H., 2006. Specification of optic nerve oligodendrocyte precursors by retinal ganglion cell axons. J. Neurosci. 26, 7619–7628.

Gehrmann, J., Matsumoto, Y., Kreutzberg, G.W., 1995. Microglia: intrinsic immuneffector cell of the brain. Brain Res. Rev. 20, 269–287.

Germer, A., Biedermann, B., Schousboe, A., Wolburg, H., Mack, A., Reichenbach, A., 1998. Distribution of mitochondria within Müller cells. I. Correlation with retinal vascularization in different mammalian species. J. Neurocytol. 27, 329–345.

Gershon, M.D., Rothman, T.P., 1991. Enteric glia. Glia 4, 195–204.

Getchell, M.L., Getchell, T.V., 1992. Fine structural aspects of secretion and extrinsic innervation in the olfactory mucosa. Microsc. Res. Tech. 23, 111–127.

Giaume, C., Venance, L., 1995. Gap junctions in brain glial cells and development. Perspect. Dev. Neurobiol. 2, 335–345.

Giaume, C., Koulakoff, A., Roux, L., Holcman, D., Rouach, N., 2010. Astroglial networks: a step further in neuroglial and gliovascular interactions. Nat. Rev. Neurosci. 11, 87–99.

Ginhoux, F., Greter, M., Leboeuf, M., Nandi, S., See, P., Gokhan, S., Mehler, M.F., Conway, S.J., Ng, L.G., Stanley, E.R., Samokhvalov, I.M., Merad, M., 2010. Fate mapping analysis reveals that adult microglia derive from primitive macrophages. Science 330, 841–845.

Gordon, G.R.J., Choi, H.B., Rungta, R.L., Ellis-Davies, G.C.R., MacVicar, B.A., 2008. Brain metabolism dictates the polarity of astrocyte control over arterioles. Nature 456, 745–749.

Graziadei, P.P.C., 1971. The olfactory mucosa of vertebrates. In: Autrum, H., Jung, R., Loewenstein, W.R., MacKay, O.M., Teuber, H.L. (Eds.), Handbook of Sensory Physiology, vol. 4/1. Springer, Berlin, pp. 27–58.

Grosche, J., Matyash, V., Möller, T., Verkhratsky, A., Reichenbach, A., Kettenmann, H., 1999. Microdomains for neuron-glia interaction: parallel fiber signaling to Bergmann glial cells. Nat. Neurosci. 2, 139–143.

Grosche, J., Kettenmann, H., Reichenbach, A., 2002. Bergmann glial cells form distinct morphological structures to interact with cerebellar neurons. J. Neurosci. Res. 68, 138–149.

Gulbransen, B.D., Sharkey, K.A., 2012. Novel functional roles for enteric glia in the gastrointestinal tract. Nat. Rev. Gastroenterol. Hepatol. 9, 625–632.

Halfter, W., 1998. Disruption of the retinal basal lamina during early embryonic development leads to retraction of vitreal endfeet, an increased number of ganglion cells, and aberrant axonal growth. J. Comp. Neurol. 397, 89–104.

Hampson, E.C., Robinson, S.R., 1995. Heterogeneous morphology and tracer coupling patterns of retinal oligodendrocytes. Philos. Trans. R. Soc. Lond. Ser. B Biol. Sci. 349, 353–364.

Hanani, M., 2005. Satellite glial cells in sensory ganglia: from form to function. Brain Res. Rev. 48, 457–476.

Hanani, M., 2010. Satellite glial cells in sympathetic and parasympathetic ganglia: in search of function. Brain Res. Rev. 64, 304–327.

Hanani, M., Reichenbach, A., 1994. Morphology of horseradish peroxidase (HRP)-injected glial cells in the myenteric plexus of the Guinea-pig. Cell Tissue Res. 278, 153–160.

Hanke, S., Reichenbach, A., 1987. Quantitative-morphometric aspects of Bergmann glial (Golgi epithelial) cell development in rats. A Golgi study. Anat. Embryol. 177, 183–188.

Härtig, W., Derouiche, A., Welt, K., Brauer, K., Grosche, J., Mädfer, M., Reichenbach, A., Brückner, G., 1999. Cortical neurons immunoreactive for the potassium channel Kv3.1b subunit are predominantly surrounded by perineuronal nets presumed as a buffering system for cations. Brain Res. 842, 15–29.

Hatten, M.E., 1984. Embryonic cerebellar astroglia in vitro. Dev. Brain Res. 13, 309–313.

Hatten, M.E., 1985. Neuronal regulation of astroglial morphology and proliferation in vitro. J. Cell Biol. 100, 384–396.

Hatton, J.D., Ellisman, M.H., 1982. The distribution of orthogonal arrays in the freeze-fractured rat median eminence. J. Neurocytol. 11, 335–349.

Hauser, R., 1969. Abhängigkeit der normalen Schwanzregeneration bei Xenopus-Larven von einem diencephalen Faktor im Zentralkanal. Wilhelm Rouxs Arch. Dev. Biol. 163, 221–247.

Haynes, S.E., Hollopeter, G., Yang, G., Kurpius, D., Dailey, M.E., Gan, W.B., Julius, D., 2006. The $P2Y_{12}$ receptor regulates microglial activation by extracellular nucleotides. Nat. Neurosci. 9, 1512–1519.

Henneberger, C., Papouin, T., Oliet, S.H., Rusakov, D.A., 2010. Long-term potentiation depends on release of D-serine from astrocytes. Nature 463, 232–236.

Herrlinger, H., 1970. Licht- und elektronenmikroskopische Untersuchungen am Subkommissuralorgan der Maus. Ergeb. Anat. Entwicklungsgesch. 42, 1–73.

Hickey, W.F., Kimura, H., 1988. Perivascular microglial cells of the CNS are bone marrow-derived and present antigen in vivo. Science 239, 290–292.

Hildebrand, C., Waxman, S.G., 1984. Postnatal differentiation of rat optic nerve fibers: electron microscopic observations of nodes of Ranvier and axoglial relations. J. Comp. Neurol. 224, 25–37.

Hirano, A., Zimmerman, H.M., 1967. Some new cytological observations of the normal rat ependymal cell. Anat. Rec. 158, 293–302.

Hirrlinger, J., Hülsmann, S., Kirchhoff, F., 2004. Astroglial processes show spontaneous motility at active synaptic terminals in situ. Eur. J. Neurosci. 20, 2235–2239.

Hockfield, S., McKay, R.D., 1985. Identification of major cell classes in the developing mammalian nervous system. J. Neurosci. 5, 3310–3328.

Holländer, H., Makarov, F., Dreher, Z., van Driel, D., Chan-Ling, T.L., Stone, J., 1991. Structure of the macroglia of the retina: sharing and division of labour between astrocytes and Müller cells. J. Comp. Neurol. 313, 587–603.

Horii-Hayashi, N., Tatsumi, K., Matsusue, Y., Okuda, H., Okuda, A., Hayashi, M., Yano, H., Tsuboi, A., Nishi, M., Yoshikawa, M., Wanaka, A., 2010. Chondroitin sulfate demarcates astrocytic territories in the mammalian cerebral cortex. Neurosci. Lett. 483, 67–72.

Horstmann, E., 1954. Die Faserglia des Selachiergehirns. Z. Zellforsch. 39, 488–617.

Horwitz, B., 2004. Relating fMRI and PET signals to neural activity by means of large-scale neural models. Neuroinformatics 2, 251–266.

Houades, V., Rouach, N., Ezan, P., Kirchhoff, F., Koulakoff, A., Giaume, C., 2006. Shapes of astrocyte networks in the juvenile brain. Neuron Glia Biol. 2, 3–14.

Housley, G.D., Bringmann, A., Reichenbach, A., 2009. Purinergic signaling in special senses. Trends Neurosci. 32, 128–141.

Huang, L.Y., Gu, Y., Chen, Y., 2013. Communication between neuronal somata and satellite glial cells in sensory ganglia. Glia 61, 1571–1581.

Hume, D.A., Perry, V.H., Gordon, S., 1983. Immunohistochemical localization of a macrophage-specific antigen in developing mouse retina: phagocytosis of dying neurons and differentiation of microglial cells to form a regular array in the plexiform layers. J. Cell Biol. 97, 253–257.

Iandiev, I., Uckermann, O., Pannicke, T., Wurm, A., Pietsch, U.-C., Reichenbach, A., Wiedemann, P., Bringmann, A., Uhlmann, S., 2006. Glial cell reactivity in a porcine model of retinal detachment. Invest. Ophthalmol. Vis. Sci. 47, 2161–2171.

Iandiev, I., Pannicke, T., Biedermann, B., Wiedemann, P., Reichenbach, A., Bringmann, A., 2006. Ischemia-reperfusion alters the immunolocalization of glial aquaporins in rat retina. Neurosci. Lett. 408, 108–112.

Inokuchi, T., Satoh, H., Kawahara, S., Higashi, R., 1988. Scanning electron microscopic observations of the ependymal cell and the subependymal layer of the third brain ventricle in the rat. Physiol. Bohemia 37, 275–280.

Ishibashi, T., Dakin, K.A., Stevens, B., Lee, P.R., Kozlov, S.V., Stewart, C.L., Fields, R.D., 2006. Astrocytes promote myelination in response to electrical impulses. Neuron 49, 823–832.

Ishii, M., Horio, Y., Tada, Y., Hibino, H., Inanobe, A., Ito, M., Yamada, M., Gotow, T., Uchiyama, Y., Kurachi, Y., 1997. Expression and clustered distribution of an inwardly rectifying potassium channel, KAB-2/Kir4.1, on mammalian retinal Müller cell membrane: their regulation by insulin and laminin signals. J. Neurosci. 17, 7725–7735.

Jaeger, C.B., 1988. Plasticity of astroglia: evidence supporting process elongation by "stretch". Glia 1, 31–38.

Jakovcevski, I., Filipovic, R., Mo, Z., Rakic, S., Zecevic, N., 2009. Oligodendrocyte development and the onset of myelination in the human fetal brain. Front. Neuroanat. 3, 1–15.

Jessen, K.R., Mirsky, R., 1983. Astrocyte-like glia in the peripheral nervous system: an immunohistochemical study of enteric glia. J. Neurosci. 3, 2206–2218.

Jessen, K.R., Mirsky, R., 2005. The origin and development of glial cells in peripheral nerves. Nat. Rev. Neurosci. 6, 671–682.

Kamasawa, N., Sik, A., Morita, M., Yasumura, T., Davidson, K.G.V., Nagy, J.I., Rash, J.E., 2005. Connexin-47 and connexin-32 in gap junctions of oligodendrocyte somata, myelin sheaths, paranodal loops and Schmidt-Lanterman incisures: implications for ionic homeostasis and potassium siphoning. Neuroscience 136, 65–86.

Kaplan, M.R., Meyer-Franke, A., Lambert, S., Bennett, V., Duncan, I.D., Levinson, S.R., Barres, B.A., 1997. Induction of sodium channel clustering by oligodendrocytes. Nature 386, 724–728.

Káradóttir, R., Hamilton, N.B., Bakiri, Y., Attwell, D., 2008. Spiking and nonspiking classes of oligodendrocyte precursor glia in CNS white matter. Nat. Neurosci. 11, 450–456.

Kaukua, N., Shahidi, M.K., Konstantinidou, C., Dyachuk, V., Kaucka, M., Furlan, A., An, Z., Wang, L., Hultman, I., Ahrlund-Richter, L., Blom, H., Brismar, H., Lopes, N.A., Pachnis, V., Suter, U., Clevers, H., Thesleff, I., Sharpe, P., Ernfors, P., Fried, K., Adameyko, I., 2014. Glial origin of mesenchymal stem cells in a tooth model system. Nature 513, 551–554.

Kessaris, N., Fogarty, M., Iannarelli, P., Grist, M., Wegner, M., Richardson, W.D., 2006. Competing waves of oligodendrocytes in the forebrain and postnatal elimination of an embryonic lineage. Nat. Neurosci. 9, 173–179.

Kigerl, K.A., Gensel, J.C., Ankeny, D.P., Alexander, J.K., Donnelly, D.J., Popovich, P.G., 2009. Identification of two distinct macrophage subsets with divergent effects causing either neurotoxicity or regeneration in the injured mouse spinal cord. J. Neurosci. 29, 13435–13444.

Kim, K.Y., Ju, W.K., Neufeld, A.H., 2004. Neuronal susceptibility to damage: comparison of the retinas of young, old and old/caloric restricted rats before and after transient ischemia. Neurobiol. Aging 25, 491–500.

Kimelberg, H.K., 2004. The problem of astrocyte identity. Neurochem. Int. 45, 191–202.

Knowles, F., 1967. Neuronal properties of neurosecretory cells. In: Stutinsky, F. (Ed.), Neurosecretion. IV. International Symposium on Neurosecretion. Springer, Berlin, Heidelberg, New York, pp. 8–19.

Knowles, F., Anand-Kumar, T.C., 1969. Structural changes, related to reproduction, in the hypothalamus and in the pars tuberalis of the rhesus monkey. Philos. Trans. R. Soc. Lond. Ser. B Biol. Sci. 256, 357–375.

Kofuji, P., Newman, E.A., 2004. Potassium buffering in the central nervous system. Neuroscience 129, 1045–1056.

Kopp, D.M., Trachtenberg, J.T., Thompson, W.J., 1997. Glial growth factor rescues Schwann cells of mechanoreceptors from denervation-induced apoptosis. J. Neurosci. 17, 6697–6706.

Korte, G.E., Bellhorn, R.W., Burns, M.S., 1986. Remodelling of the retinal pigment epithelium in response to intraepithelial capillaries: evidence that capillaries influence the polarity of the epithelium. Cell Tissue Res. 245, 135–142.

Krabbe, G., Halle, A., Matyash, V., Rinnenthal, J.L., Eom, G.D., Bernhardt, U., Miller, K.R., Prokop, S., Kettenmann, H., Heppner, F.L., 2013. Functional impairment of microglia coincides with β-amyloid deposition in mice with Alzheimer-like pathology. PLoS One 8, e60921.

Krstic, R.V., 1991. Human Microscopic Anatomy. An Atlas for Students of Medicine and Biology. Springer, Berlin, Heidelberg, New York, London, Paris, Tokyo, Hongkong, Barcelona, Budapest.

Kucenas, S., Wang, W.D., Knapik, E.W., Appel, B., 2009. A selective glial barrier at motor axon exit points prevents oligodendrocyte migration from the spinal cord. J. Neurosci. 29, 15187−15194.

Langmann, T., 2007. Microglia activation in retinal degeneration. J. Leukoc. Biol. 81, 1345−1351.

Leonhardt, H., 1980. Ependym und circumventriculäre Organe. In: Oksche, A. (Ed.), Neuroglia I. Handbuch der mikroskopischen Anatomie des Menschen, Part 10, vol. 4. Springer, Berlin, Heidelberg, New York, pp. 177−666.

Lichtensteiger, W., Richards, J.G., Kopp, H.G., 1978. Changes in the distribution of non-neuronal elements in rat median eminence and in anterior pituitary hormone secretion after activation of tuberoinfundibular dopamine neurones by brain stimulation or nicotine. Brain Res. 157, 73−88.

Lindeman, H.H., 1969. Structure of the vestibular sensory epithelia. Adv. Anat. Embryol. Cell Biol. 42, 31−57.

Lindemann, B., Leonhardt, H., 1973. Supraependymale Neuriten, Gliazellen und Mitochondrienkolben im caudalen Abschnitt des Bodens der Rautengrube. Z. Zellforsch. 140, 401−412.

Llinas, R., Nicholson, C., Freeman, J.A., Hillman, D.E., 1969. Dendritic spikes versus cable properties. Science 163, 97.

Lu, Y.-B., Franze, K., Seifert, G., Steinhäuser, C., Kirchhoff, F., Wolburg, H., Guck, J., Janmey, P., Wei, E.-Q., Käs, J., Reichenbach, A., 2006. Viscoelastic properties of individual glial cells and neurons in the CNS. Proc. Natl. Acad. Sci. U.S.A. 103, 17759−17764.

Ludwin, S.K., 1997. The pathobiology of the oligodendrocyte. J. Neuropathol. Exp. Neurol. 56, 111−124.

Ma, W., Cojocaru, R., Gotoh, N., Gieser, L., Villasmil, R., Cogliati, T., Swaroop, A., Wong, W.T., 2013. Gene expression changes in aging retinal microglia: relationship to microglial support functions and regulation of activation. Neurobiol. Aging 34, 2310−2321.

Macfarlane, S.N., Sontheimer, H., 2000. Changes in ion channel expression accompany cell cycle progression of spinal cord astrocytes. Glia 30, 39−48.

Mathiisen, T.M., Lehre, K.P., Danbolt, N.C., Ottersen, O.P., 2010. The perivascular astroglial sheath provides a complete covering of the brain microvessels: an electron microscopic 3D reconstruction. Glia 58, 1094−1103.

Merkle, F.T., Tramontin, A.D., Garcia-Verdugo, J.M., Alvarez-Buylla, A., 2004. Radial glia give rise to adult neural stem cells in the subventricular zone. Proc. Natl. Acad. Sci. U.S.A. 101, 17528−17532.

Meyer, R., 1906. Untersuchungen über den feineren Bau des Nervensystems der Asteriden (Asterias rubens). Z. Wiss. Zool. 81, 96−144.

Meyer-Franke, A., Shen, S., Barres, B.A., 1999. Astrocytes induce oligodendrocyte processes to align with and adhere to axons. Mol. Cell. Neurosci. 14, 385−397.

Mildner, A., Schlevogt, B., Kierdorf, K., Böttcher, C., Erny, D., Kummer, M.P., Quinn, M., Brück, W., Bechmann, I., Heneka, M.T., Priller, J., Prinz, M., 2011. Distinct and non-redundant roles of microglia and myeloid subsets in mouse models of Alzheimer's disease. J. Neurosci. 31, 11159−11171.

Miller, B., Miller, H., Patterson, R., Ryan, S.J., 1986. Retinal wound healing. Cellular activity at the vitreoretinal interface. Arch. Ophthalmol. 104, 281−285.

Millhouse, O.E., 1971. A Golgi study of the third ventricle tanycytes in the adult rodent brain. Z. Zellforsch. 121, 1−13.

Miron, V.E., Boyd, A., Zhao, J.W., Yuen, T.J., Ruckh, J.M., Shadrach, J.L., van Wijngaarden, P., Wagers, A.J., Williams, A., Franklin, R.J., Ffrench-Constant, C., 2013. M2 microglia and macrophages drive oligodendrocyte differentiation during CNS remyelination. Nat. Neurosci. 16, 1211−1218.

Mori, S., Leblond, C.P., 1970. Electron microscopic identification of three classes of oligodendrocytes and a preliminary study of their proliferative activity in the corpus callosum of young rats. J. Comp. Neurol. 139, 1−30.

Mountcastle, V.B., 1957. Modality and topographic properties of single neurons of cat's somatic sensory cortex. J. Neurophysiol. 20, 408−434.

Müller, H., 1851. Zur Histologie der Netzhaut. Z. Wiss. Zool. 3, 234−237.

Müller, C.M., Best, J., 1989. Ocular dominance plasticity in adult cat visual cortex after transplantation of cultured astrocytes. Nature 342, 427−430.

Muniz, A., Villazana-Espinoza, E.T., Hatch, A.L., Trevino, S.G., Allen, D.M., Tsin, A.T., 2007. A novel cone visual cycle in the cone-dominated retina. Exp. Eye Res. 85, 175−184.

Nagelhus, E.A., Horio, Y., Inanobe, A., Fujita, A., Haug, F.M., Nielsen, S., Kurachi, Y., Ottersen, O.P., 1999. Immunogold evidence suggests that coupling of K^+ siphoning and water transport in rat retinal Müller cells is mediated by a coenrichment of Kir4.1 and AQP4 in specific membrane domains. Glia 26, 47−54.

Nagy, J.I., Dudek, F.E., Rash, J.E., 2004. Update on connexins and gap junctions in neurons and glia in the mammalian nervous system. Brain Res. Rev. 47, 191−215.

Nakayama, Y., Kohno, K., 1974. Number and polarity of the ependymal cilia in the central canal of some vertebrates. J. Neurocytol. 3, 449−458.

Nakazawa, T., Hisatomi, T., Nakazawa, C., Noda, K., Maruyama, K., She, H., Matsubara, A., Miyahara, S., Nakao, S., Yin, Y., Benowitz, L., Hafezi-Moghadam, A., Miller, J.W., 2007. Monocyte chemoattractant protein 1 mediates retinal detachment-induced photoreceptor apoptosis. Proc. Natl. Acad. Sci. U.S.A. 104, 2425−2430.

Newman, E.A., 2001. Propagation of intercellular calcium waves in retinal astrocytes and Müller cells. J. Neurosci. 21, 2215−2223.

Nilsson, C., Lindvall-Axelsson, M., Owman, C., 1992. Neuroendocrine regulatory mechanisms in the choroid plexus-cerebrospinal fluid system. Brain Res. Rev. 17, 109−138.

Nimmerjahn, A., Kirchhoff, F., Helmchen, F., 2005. Resting microglial cells are highly dynamic surveillants of brain parenchyma in vivo. Science 308, 1314−1318.

Nissl, F., 1899. Über einige Beziehungen zwischen Nervenzellenerkrankungen und gliösen Erscheinungen bei verschiedenen Psychosen. Arch. Psychiatr. 32, 1−21.

Noctor, S.C., Flint, A.C., Weissman, T.A., Dammerman, R.S., Kriegstein, A.R., 2001. Neurons derived from radial glial cells establish radial units in neocortex. Nature 409, 714−720.

Noell, S., Fallier-Becker, P., Deutsch, U., Mack, A.F., Wolburg, H., 2009. Agrin defines polarized distribution of orthogonal arrays of particles in astrocytes. Cell Tissue Res. 337, 185−195.

Noell, S., Wolburg-Buchholz, K., Mack, A.F., Beedle, A.M., Satz, J.S., Campbell, K.P., Wolburg, H., Fallier-Becker, P., 2011. Evidence for a role of dystroglycan regulating the membrane architecture of astroglial endfeet. Eur. J. Neurosci. 33, 2179−2186.

Oberheim, N.A., Wang, X., Goldman, S., Nedergaard, M., 2006. Astrocytic complexity distinguishes the human brain. Trends Neurosci. 29, 547−553.

Oesterle, E.C., Sarthy, P.V., Rubel, E.W., 1990. Intermediate filaments in the inner ear of normal and experimentally damaged Guinea pigs. Hear. Res. 47, 1−16.

Oksche, A., 1961. Vergleichende Untersuchungen über die sekretorische Aktivität des Subkommissuralorgans und den Gliacharakter seiner Zellen. Z. Zellforsch. 54, 549−612.

Oksche, A., 1964. Das Subcommissuralorgan des Menschen. Anat. Anz. 112 H, 373−383.

Oliet, S.H.R., Piet, R., Poulain, D.A., 2001. Control of glutamate clearance and synaptic efficacy by glial coverage of neurons. Science 292, 923–926.

Pannicke, T., Iandiev, I., Uckermann, O., Biedermann, B., Kutzera, F., Wiedemann, P., Wolburg, H., Reichenbach, A., Bringmann, A., 2004. A potassium channel-linked mechanism of glial cell swelling in the postischemic retina. Mol. Cell. Neurosci. 26, 493–502.

Paolicelli, R.C., Bolasco, G., Pagani, F., Maggi, L., Scianni, M., Panzanelli, P., Giustetto, M., Ferreira, T.A., Guiducci, E., Dumas, L., Ragozzino, D., Gross, C.T., 2011. Synaptic pruning by microglia is necessary for normal brain development. Science 333, 1456–1458.

Pawson, L., Prestia, L.T., Mahoney, G.K., Güçlü, B., Cox, P.J., Pack, A.K., 2009. GABAergic/glutamatergic-glial/neuronal interaction contributes to rapid adaptation in pacinian corpuscles. J. Neurosci. 29, 2695–2705.

Penfield, W., 1932. Neuroglia: normal and pathological. In: Penfield, W. (Ed.), Cytology and Cellular Pathology in the Nervous System, vol. 2. Hoeber, New York, pp. 421–479.

Peters, O., Schipke, C.G., Hashimoto, Y., Kettenmann, H., 2003. Different mechanisms promote astrocyte Ca^{2+} waves and spreading depression in the mouse neocortex. J. Neurosci. 23, 9888–9896.

Piazza, V., Ciubotaru, C.D., Gale, J.E., Mammano, F., 2007. Purinergic signalling and intercellular Ca^{2+} wave propagation in the organ of Corti. Cell Calcium 41, 77–86.

Pinching, A.J., 1971. Myelinated dendritic segments in the monkey olfactory bulb. Brain Res. 29, 133–138.

Prinz, M., Priller, J., 2014. Microglia and brain macrophages in the molecular age: from origin to neuropsychiatric disease. Nat. Rev. Neurosci. 15, 300–312.

Pumplin, D.W., Yu, C., Smith, D.V., 1997. Light and dark cells of rat vallate taste buds are morphologically distinct cell types. J. Comp. Neurol. 378, 389–410.

Purves, D., 1988. Body and Brain. A Trophical Theory of Neural Connections. Harvard University Press, Cambridge, MA.

Raine, C.S., 1984. On the association between perinodal astrocytic processes and the node of Ranvier in the CNS. J. Neurocytol. 13, 21–27.

Raisman, G., 2001. Olfactory ensheathing cells – another miracle cure for spinal cord injury? Nat. Rev. Neurosci. 2, 369–375.

Rakic, P., 1972. Mode of cell migration to the superficial layers of fetal monkey neocortex. J. Comp. Neurol. 145, 61–84.

Rakic, P., 1978. Neuronal migration and contact guidance in the primate telencephalon. Postgrad. Med. J. 54, 25–40.

Rakic, P., 1981. Neuron-glial interaction during brain development. Trends Neurosci. 4, 184–187.

Rakic, P., 1984. Organizing principles for development of primate cerebral cortex. In: Sharma, S.C. (Ed.), Organizing Principles of Neural Development. Plenum, New York, pp. 21–48.

Rakic, P., 1988. Specification of cerebral cortical areas. Science 241, 170–176.

Ramón y Cajál, S., 1911. Histologie du système nerveux de l'homme et des vertébrés. Maloine, Paris.

Ramón y Cajál, S., 1972. The Structure of the Retina. Thomas, Springfield, IL.

Raoul, W., Keller, N., Rodéro, M., Behar-Cohen, F., Sennlaub, F., Combadière, C., 2008. Role of the chemokine receptor CX3CR1 in the mobilization of phagocytic retinal microglial cells. J. Neuroimmunol. 198, 56–61.

Rash, J.E., Duffy, H.S., Dudek, F.E., Bilhartz, B.L., Whalen, L.R., Yasumura, T., 1997. Grid-mapped freeze-fractured analysis of gap junctions in gray and white matter of adult rat central nervous system, with evidence for a "panglial syncytium" that is not coupled to neurons. J. Comp. Neurology 388, 265–292.

Rasmussen, K.-E., 1975. A morphometric study of the Müller cell in rod and cone retinas with and without retinal vessels. Exp. Eye Res. 20, 151–166.

Reichelt, W., Dettmer, D., Brückner, G., Brust, P., Eberhardt, W., Reichenbach, A., 1989. Potassium as a signal for both proliferation and differentiation of rabbit retinal (Müller) glia growing in cell culture. Cell. Signal. 1, 187–194.

Reichenbach, A., 1987. Quantitative and qualitative morphology of rabbit retinal glia. A light microscopical study on cells both in situ and isolated by papaine. J. Hirnforschung 28, 213–220.

Reichenbach, A., 1989. Glia:neuron index: review and hypothesis to account for different values in various mammals. Glia 2, 71–77.

Reichenbach, A., 1989. Attempt to classify glial cells by means of their process specialization using the rabbit retinal Müller cell as an example of cytotopographic specialization of glial cells. Glia 2, 250–259.

Reichenbach, A., 1990. Radial glial cells are present in the velum medullare of adult monkeys. J. Hirnforschung 31, 269–271.

Reichenbach, A., Bringmann, A., 2010. Müller Cells in the Healthy and Diseased Retina. Springer, New York, Dordrecht, Heidelberg, London.

Reichenbach, A., Bringmann, A., 2013. New functions of Müller cells. Glia 61, 651–678.

Reichenbach, A., Bringmann, A., 2015. Retinal glia. In: Verkhratsky, A., Parpura, V. (Eds.), Colloquium Series on Neuroglia in Biology and Medicine: From Physiology to Disease. Morgan & Claypool Life Sciences, Philadelphia, PA.

Reichenbach, A., Pannicke, T., 2008. A new glance at glia. Science 322, 693–694.

Reichenbach, A., Reichelt, W., 1986. Postnatal development of radial glial (Müller) cells of the rabbit retina. Neurosci. Lett. 71, 125–130.

Reichenbach, A., Robinson, S.R., 1995. Ependymoglia and ependymoglia-like cells. In: Ransom, B., Kettenmann, H. (Eds.), Neuroglia Cells. Oxford University Press, Oxford, pp. 58–84.

Reichenbach, A., Wolburg, H., 2005. Astrocytes and ependymal glia. In: Kettenmann, H., Ransom, B.R. (Eds.), Neuroglia, second ed. Oxford University Press, Oxford, New York, pp. 19–35.

Reichenbach, A., Wolburg, H., 2009. Structural association of astrocytes with neurons and vasculature: defining territorial boundaries. In: Parpura, V., Haydon, P.G. (Eds.), Astrocytes in (Patho)physiology of the Nervous System. Springer, New York, pp. 251–286.

Reichenbach, A., Neumann, M., Brückner, G., 1987. Cell length to diameter relation of rat fetal radial glia – does impaired K^+ transport capacity of long thin cells cause their perinatal transformation into multipolar astrocytes? Neurosci. Lett. 73, 95–100.

Reichenbach, A., Schippel, K., Schümann, R., Hagen, E., 1988. Ultrastructure of rabbit nerve fibre layer – neuro-glial relationships, myelination, and nerve fiber spectrum. J. Hirnforschung 29, 481–491.

Reichenbach, A., Hagen, E., Schippel, K., Eberhardt, W., 1988. Quantitative electron microscopy of rabbit Müller (glial) cells in dependence on retinal topography. Z. Mikrosk.-Anat. Forsch. (Leipz.) 102, 721–755.

Reichenbach, A., Schneider, H., Leibnitz, L., Reichelt, W., Schaaf, P., Schümann, R., 1989. The structure of rabbit retinal Müller (glial) cells is adapted to the surrounding retinal layers. Anat. Embryol. 180, 71–79.

Reichenbach, A., Eberhardt, W., Scheibe, R., Deich, C., Seifert, B., Reichelt, W., Dähnert, K., Rödenbeck, M., 1991. Development of the rabbit retina. IV. Tissue tensility and elasticity in dependence on topographic specializations. Exp. Eye Res. 53, 241–251.

Reichenbach, A., Siegel, A., Senitz, D., Smith Jr., T.G., 1992. A comparative fractal analysis of various mammalian astroglial cell types. Neuroimage 1, 69–77.

Reichenbach, A., Stolzenburg, J.-U., Eberhardt, W., Chao, I.T., Dettmer, D., Hertz, L., 1993. What do retinal Müller (glial) cells do for their neuronal 'small siblings'? J. Chem. Neuroanat. 6, 201–213.

Reichenbach, A., Ziegert, M., Schnitzer, J., Pritz-Hohmeier, S., Schaaf, P., Schober, W., Schneider, H., 1994. Development of the rabbit retina. V. The question of 'columnar units'. Dev. Brain Res. 79, 72–84.

Reichenbach, A., Frömter, C., Engelmann, R., Wolburg, H., Kasper, M., Schnitzer, J., 1995. Müller glial cells of the tree shrew retina. J. Comp. Neurol. 360, 257–270.

Reichenbach, A., Derouiche, A., Grosche, J., Hanani, M., 2004. Structural association of glia with the various compartments of neurons. In: Hatton, G.I., Parpura, V. (Eds.), Glial Neuronal Signaling. Kluwer, Boston, MA, pp. 53–97.

Reichenbach, A., Derouiche, A., Kirchhoff, F., 2010. Morphology and dynamics of perisynaptic glia. Brain Res. Rev. 63, 11–25.

Reissner, E., 1860. Beiträge zur Kenntnis vom Bau des Rückenmarks von *Petromyzon fluviatilis* L. Arch. Anat. Physiol. 545–588.

Retzius, G., 1893. Studien über Ependym und Neuroglia. Biol. Untersuchungen Stockh. Neue Folge 5, 9–26.

Revel, J.-P., Brown, S., 1976. Cell junctions in development, with particular reference to the neural tube. Cold Spring Harb. Symp. Quant. Biol. 40, 443–455.

Rice, F.L., van der Loos, H., 1977. Development of the barrels and barrel field in the somatosensory cortex of the mouse. J. Comp. Neurol. 171, 545–560.

Rico, B., Solas, M.T., Clément, J., Suárez, I., Fernández, B., 1996. Ultrastructural study of the Pacinian corpuscles in the newborn and adult dog forefoot. Eur. J. Morphol. 34, 311–320.

Riethmacher, D., Sonnenberg-Riethmacher, E., Brinkmann, V., Yamaai, T., Lewin, G.R., Birchmeier, C., 1997. Severe neuropathies in mice with targeted mutations in the ErbB3 receptor. Nature 389, 725–730.

Robertson, J.D., 1899. On a new method of obtaining a black reaction in certain tissue elements of the central nervous system (platinum method). Scott. Med. Surg. J. 4, 23–30.

Robinson, S.R., Hampson, E.C.G.M., Munro, M.N., Vaney, D.I., 1993. Unidirectional coupling of gap junctions between neuroglia. Science 262, 1072–1074.

Rodríguez, E.M., González, C.B., Delannoy, L., 1979. Cellular organization of the lateral and postinfundibular regions of the median eminence in the rat. Cell Tissue Res. 201, 377–408.

Rosenstein, J.M., Brightman, M.W., 1979. Regeneration and myelination in autonomic ganglia transplanted to intact brain surfaces. J. Neurocytol. 8, 359–379.

Rouach, N., Glowinski, J., Giaume, C., 2000. Activity-dependent neuronal control of gap-junctional communication in astrocytes. J. Cell Biol. 149, 1513–1526.

Sakry, D., Neitz, A., Singh, J., Frischknecht, R., Marongiu, D., Binamé, F., Perera, S.S., Endres, K., Lutz, B., Radyushkin, K., Trotter, J., Mittmann, T., 2014. Oligodendrocyte precursor cells modulate the neuronal network by activity-dependent ectodomain cleavage of glial NG2. PLoS Biol. 12, e1001993.

Sauer, F.C., 1935. Mitosis in the neural tube. J. Comp. Neurol. 62, 377–397.

Scemes, E., Giaume, C., 2006. Astrocyte calcium waves: what they are and what they do. Glia 54, 716–725.

Schipke, C.G., Kettenmann, H., 2004. Astrocyte responses to neuronal activity. Glia 47, 226–232.

Schummers, J., Yu, H., Sur, M., 2008. Tuned responses of astrocytes and their influence on hemodynamic signals in the visual cortex. Science 320, 1638–1643.

Senitz, D., Reichenbach, A., Smith Jr., T.G., 1995. Surface complexity of human neocortical astrocytic cells: changes with development, aging, and dementia. J. Hirnforschung 36, 531–537.

Seress, L., 1980. Development and structure of the radial glia in the postnatal rat brain. Anat. Embryol. 160, 213–226.

Siegel, A., Reichenbach, A., Hanke, S., Senitz, D., Brauer, K., Smith Jr., T.G., 1991. Comparative morphometry of Bergmann glial (Golgi epithelial) cells. A Golgi study. Anat. Embryol. 183, 605–612.

Sievers, J., Pehlemann, F.-W., Berry, M., 1986. Influence of meningeal cells on brain development. Naturwissenschaften 73, 188–194.

Singer, I., Goodman, S.J., 1966. Mammalian ependyma: some physicochemical determinants of cilia activity. Exp. Cell Res. 43, 367–380.

Sirevaag, A.M., Greenough, W.T., 1991. Plasticity of GFAP-immunoreactive astrocyte size and number in visual cortex of rats reared in complex environments. Brain Res. 540, 273–278.

Spacek, J., 1985. Three-dimensional analysis of dendritic spines. III. Glial sheath. Anat. Embryol. 171, 245–252.

Steinberg, R.H., 1985. Interactions between the retinal pigment epithelium and the neural retina. Doc. Ophthalmol. 60, 327–346.

Stensaas, L.J., Stensaas, S.S., 1968. Light microscopy of glial cells in turtles and birds. Z. Zellforsch. 91, 315–340.

Stevens, B., Fields, R.D., 2000. Response of Schwann cells to action potentials in development. Science 287, 2267–2271.

Stewart, M.G., Bourne, R.C., Gabbott, P.L.A., 1986. Decreased levels of an astrocytic marker, glial fibrillary acidic protein, in the visual cortex of dark-reared rats: measurement by enzyme-linked immunosorbent assay. Neurosci. Lett. 63, 147–152.

Stolzenburg, J.-U., Reichenbach, A., Neumann, M., 1989. Size and density of glial and neuronal cells within the cerebral neocortex of various insectivorian species. Glia 2, 78–84.

Stout, C.E., Costantin, J.L., Naus, C.C.G., Charles, A.C., 2002. Intercellular calcium signaling in astrocytes via ATP release through connexin hemichannels. J. Biol. Chem. 277, 10482–10488.

Strauss, O., 2005. The retinal pigment epithelium in visual function. Physiol. Rev. 85, 845–881.

Sturrock, R.R., Smart, J.L., Dobbing, J., 1977. Effect of undernutrition during the suckling period on the indusium griseum and rostral part of the mouse anterior commissure. Neuropathol. Appl. Neurobiol. 3, 369–375.

Syková, E., Chvátal, A., 2000. Glial cells and volume transmission in the CNS. Neurochem. Int. 36, 397–409.

Tabernero, A., Medina, J.M., Giaume, C., 2006. Glucose metabolism and proliferation in glia: role of astrocytic gap junctions. J. Neurochem. 99, 1049–1061.

Taveggia, C., Zanazzi, G., Petrylak, A., Yano, H., Rosenbluth, J., Einheber, S., Xu, X., Esper, R.M., Loeb, J.A., Shrager, P., Chao, M.V., Falls, D.L., Role, L., Salzer, J.L., 2005. Neuregulin-1 type III determines the ensheathment fate of axons. Neuron 47, 681–694.

Tolhurst, D.J., Lewis, P.R., 1992. Effect of myelination on the conduction velocity of optic nerve fibres. Ophthalmic Physiol. Opt. 12, 241–243.

Tremblay, M.È., Lowery, R.L., Majewska, A.K., 2010. Microglial interactions with synapses are modulated by visual experience. PLoS Biol. 8, e1000527.

Uckermann, O., Vargová, L., Ulbricht, E., Klaus, C., Weick, M., Rillich, K., Wiedemann, P., Reichenbach, A., Syková, E., Bringmann, A., 2004. Glutamate-evoked alterations of glial and neuronal cell morphology in the Guinea-pig retina. J. Neurosci. 24, 10149–10158.

Uckermann, O., Iandiev, I., Francke, M., Franze, K., Grosche, J., Wolf, S., Kohen, L., Wiedemann, P., Reichenbach, A., Bringmann, A., 2004. Selective staining by vital dyes of Müller glial cells in retinal wholemounts. Glia 45, 59–66.

Uckermann, O., Uhlmann, S., Pannicke, T., Francke, M., Gamsalijew, R., Makarov, F., Ulbricht, E., Wiedemann, P., Reichenbach, A., Osborne, N.N., Bringmann, A., 2005. Ischemia-reperfusion causes exudative detachment of the rabbit retina. Invest. Ophthalmol. Vis. Sci. 46, 2592–2600.

Ugrumov, M.V., Chandrasekhar, K., Borisova, N.A., Mitskevich, M.S., 1979. Light and electron microscopical investigations on the tanycyte differentiation during the perinatal period in the rat. Cell Tissue Res. 201, 295–303.

Uhlmann, S., Bringmann, A., Uckermann, O., Pannicke, T., Weick, M., Ulbricht, E., Goczalik, I., Reichenbach, A., Wiedemann, P., Francke, M., 2003. Early glial cell reactivity in experimental retinal detachment: effect of suramin. Invest. Ophthalmol. Vis. Sci. 44, 4114–4122.

Van Velzen, M., Laman, J.D., Kleinjan, A., Poot, A., Osterhaus, A.D., Verjans, G.M., 2009. Neuron-interacting satellite glial cells in human trigeminal ganglia have an APC phenotype. J. Immunol. 183, 2456–2461.

Verbavatz, J.M., Ma, T., Gobin, R., Verkman, A., 1997. Absence of orthogonal arrays in kidney, brain and muscle from transgenic knockout mice lacking water channel aquaporin-4. J. Cell Sci. 110, 2855–2860.

Ventura, R., Harris, K.M., 1999. Three-dimensional relationships between hippocampal synapses and astrocytes. J. Neurosci. 19, 6897–6906.

Virchow, R., 1846. Ueber das granulirte Aussehen der Wandungen der Gehirnventrikel. Allg. Z. Psychiatr. 3, 242–250.

Virchow, R., 1854. Ueber das ausgebreitete Vorkommen einer dem Nervenmark analogen substanz in den tierischen Geweben. Virchows Arch. Pathol. Anat. 6, 562.

Vogalis, F., Hegg, C.C., Lucero, M.T., 2005. Electrical coupling in sustentacular cells of the mouse olfactory epithelium. J. Neurophysiol. 94, 1001–1012.

Voigt, J., Grosche, A., Vogler, S., Pannicke, T., Hollborn, M., Kohen, L., Wiedemann, P., Reichenbach, A., Bringmann, A., 2015. Nonvesicular release of ATP from rat retinal glial (Müller) cells is differentially mediated in response to osmotic stress and glutamate. Neurochem. Res. 40, 651–660.

Wallraff, A., Kohling, R., Heinemann, U., Theis, M., Willecke, K., Steinhäuser, C., 2006. The impact of astrocytic gap junctional coupling on potassium buffering in the hippocampus. J. Neurosci. 26, 5438–5447.

Wang, D.D., Bordey, A., 2008. The astrocyte odyssey. Prog. Neurobiol. 86, 342–367.

Wang, M., Wong, W.T., 2014. Microglia-Müller cell interactions in the retina. Adv. Exp. Med. Biol. 801, 333–338.

Wang, M., Ma, W., Zhao, L., Fariss, R.N., Wong, W.T., 2011. Adaptive Müller cell responses to microglial activation mediate neuroprotection and coordinate inflammation in the retina. J. Neuroinflammation 8, 173.

Waxman, S.G., Black, J.A., 1995. Axoglial interactions at the cellular and molecular levels in central nervous system myelinated fibers. In: Hettenmann, H., Ransom, B.R. (Eds.), Neuroglia. Oxford University Press, New York, pp. 587–612.

Wilhelmsson, U., Bushong, E.A., Price, D.L., Smarr, B.L., Phung, V., Terada, M., Ellisman, M.H., Pekny, M., 2006. Redefining the concept of reactive astrocytes as cells that remain within their unique domains upon reaction to injury. Proc. Natl. Acad. Sci. U.S.A. 103, 17513–17518.

Wittkowski, W., 1967. Synaptische Strukturen und Elementargranula in der Neurohypophyse des Meerschweinchens. Z. Zellforsch. 82, 434–458.

Wittkowski, W., 1973. Elektronenmikroskopische Untersuchungen zur funktionellen Morphologie des tubero-hypophysären Systems der Ratte. Z. Zellforsch. 139, 101–148.

Wolburg, H., 2006. The endothelial frontier. In: Dermietzel, R., Spray, D.C., Nedergaard, M. (Eds.), Blood-Brain Barriers. From Ontogeny to Artificial Interfaces. Wiley-VCH, Weinheim, pp. 77–107.

Wolburg, H., Reichelt, W., Stolzenburg, J.U., Richter, W., Reichenbach, A., 1990. Rabbit retinal Müller cells in cell culture show gap and tight junctions which they do not express in situ. Neurosci. Lett. 111, 58–63.

Wolff, J., 1968. The role of astroglia in the brain tissue. Acta Neuropathol. Suppl. IV 33–39.

Wright, E.M., Wiedner, G., Rumrich, G., 1977. Fluid secretion by the frog choroid plexus. Exp. Eye Res. Suppl. 149–155.

Wu, H., Xiong, W.C., Mei, L., 2010. To build a synapse: signaling pathways in neuromuscular junction assembly. Development 137, 1017–1033.

Wurm, A., Pannicke, T., Iandiev, I., Bühner, E., Pietsch, U.C., Reichenbach, A., Wiedemann, P., Uhlmann, S., Bringmann, A., 2006. Changes in membrane conductance play a pathogenic role in osmotic glial cell swelling in detached retinas. Am. J. Pathol. 169, 1990–1998.

Xu-Friedman, M.A., Harris, K.M., Regehr, W.G., 2001. Three-dimensional comparison of ultrastructural characteristics at depressing and facilitating synapses onto cerebellar Purkinje cells. J. Neurosci. 21, 6666–6672.

Yakovlev, P.I., Lecours, A.R., 1966. The myelinogenic cycles of regional maturation of the brain. In: Minkovski, A. (Ed.), Regional Development of the Brain in Early Life. Blackwell, Oxford, UK, pp. 3–70.

Yamazaki, S., Nakauchi, H., 2014. Bone marrow Schwann cells induce hematopoietic stem cell hibernation. Int. J. Hematol. 99, 695–698.

Yamazaki, Y., Hozumi, Y., Kaneko, K., Sugihara, T., Fujii, S., Goto, K., Kato, H., 2007. Modulatory effects of oligodendrocytes on the conduction velocity of action potentials along axons in the alveus of the rat hippocampal CA1 region. Neuron Glia Biol. 3, 325–334.

Yu, Y.B., Li, Y.Q., 2014. Enteric glial cells and their role in the intestinal epithelial barrier. World J. Gastroenterol. 20, 11273–11280.

Zahs, K.R., 1998. Heterotypic coupling between glial cells of the mammalian central nervous system. Glia 24, 85–96.

Zahs, K.R., Newman, E.A., 1997. Asymmetric gap junctional coupling between glial cells in the rat retina. Glia 20, 10–22.

Zelinka, C.P., Scott, M.A., Volkov, L., Fischer, A.J., 2012. The reactivity, distribution and abundance of Non-astrocytic Inner Retinal Glial (NIRG) cells are regulated by microglia, acute damage, and IGF1. PLoS One 7, e44477.

17

The Monotreme Nervous System

K.W.S. Ashwell

The University of New South Wales, Sydney, NSW, Australia

17.1 Introduction

Monotremes occupy a special and unique place in mammalian brain evolution. Nevertheless, they should neither be seen as primitive relics of a bygone era, nor as models of the archaic original mammalian brain. The living monotremes are highly specialized animals that occupy distinct ecological niches within Australia, New Guinea, and the associated islands. Those niches have shaped the unusual features of their neuroanatomy and neurophysiology to such an extent that they are rightly considered as advanced with respect to particular features, for example, electroreception and olfaction, just as they may show plesiomorphic neurosensory features (eg, partial coiling of the cochlear duct, simple visual pathways, or absence of a corpus callosum) (Ashwell, 2013b).

Modern monotremes belong to two families. The Ornithorhynchidae are represented by a single species, the platypus (*Ornithorhynchus anatinus*) that is confined to the freshwater streams of eastern Australia (excluding northern Queensland) and Tasmania (Nicol, 2013). The platypus is a semiaquatic insectivore/carnivore that feeds mainly on invertebrates (annelids, molluscs, insects, and decapod crustaceans). The Tachyglossidae include long- and short-beaked echidnas, all of which feed on ants, termites, and other soft-bodied invertebrates. Long-beaked echidnas (eg, *Zaglossus bruijnii*) are currently known only from New Guinea, where they occupy moist montane forests, although long-beaked forms lived in Pleistocene Australia and may even have survived in the Kimberley region of Western Australia into historical times (Helgen et al., 2012). The short-beaked echidna is without doubt the most successful monotreme, occupying diverse environments (from alpine forest to desert) across the entire Australian continent, nearby islands, and New Guinea. Indeed, the short-beaked echidna has a more widespread distribution than any other Australian mammal, including native and introduced eutherians.

17.2 Evolution and Fossil Record of Monotremes

17.2.1 The First Monotreme

How do the monotremes fit into the mammalian fossil record, and what did the first monotreme look like? Neither of these questions is easy to answer. Monotremes have their origins in the Mesozoic, but Mesozoic mammals or near-mammals are represented in the fossil record as only fragmentary remains (Musser, 2013). The first monotremes were probably neither platypus- nor echidna-like, and the form of the archetypal monotreme remains uncertain (Musser, 2013). Mesozoic monotremes are known almost exclusively from isolated lower jaws, some with dentition, and an isolated humerus (Musser, 2013). The oldest known monotremes come from the Early Cretaceous (115—108 million years ago, *Teinolophos trusleri*), and this species has been described as a gopherlike monotreme (Rich et al., 2001) or the oldest known platypus (Rowe et al., 2008). Given the fragmentary nature of the remains, such categorization is very uncertain.

17.2.2 The Monotreme Fossil Record

The platypus-like *Steropodon galmani* is slightly younger (108—103 mya) than *Teinolophos*, the oldest known platypus, and may have had ornithorhynchid dentition (Archer et al., 1985; Pascual et al., 1992). Nevertheless, *Steropodon* is more plesiomorphic than ornithorhynchids and has been described as a protoplatypus (Musser, 2013), that is, not an ornithorhynchid, but perhaps ancestral to later ornithorhynchids.

It is not until the Cenozoic, about 65 million years ago, that more anatomically complete monotreme fossils become available and that only for the ornithorhynchids. The most complete monotreme fossils from the middle Cenozoic are small to midsized platypuses from the Oligocene to Miocene (*Obdurodon insignis* and *Obdurodon dicksoni*, respectively) (Woodburne and Tedford, 1975; Archer et al., 1978, 1992; Musser and Archer, 1998). The skull of *O. dicksoni* is strikingly similar to the modern platypus, with an extended and flattened bill that suggests it may have lived much like the modern species. This has fueled speculation that the ornithorhynchid niche and body form preceded those of the tachyglossids (see the following section).

Echidnas have a much shorter fossil record than the ornithorhynchids. Definitively dated tachyglossid fossils are largely confined to the Pleistocene, although some remains may date to the Miocene (Musser, 2013). There is no dental fossil trail for tachyglossids because they lack teeth and modern tachyglossids live in environments that do not favor fossilization, so the absence of Early Cenozoic tachyglossid fossils may reflect lack of fossilization opportunity, rather than nonexistence or paucity. Long-beaked (*Megalibgwilia* and *Zaglossus*) are actually more prevalent that short-beaked echidnas in the fossil record and are known as part of the Pleistocene megafauna. Neurologically, the extinct long-beaked echidnas are very similar to the modern short- and long-beaked species (Ashwell et al., 2014), with striking similarities in sensory specializations and cortical gyri.

17.2.3 Which Monotreme Body Form Is the Oldest?

Despite the poor fossil record, the modern platypus and echidnas, as the only monotremes, must share an exclusive common ancestor between them. In this case, which body form and lifestyle came first in monotreme evolution: did that common ancestor have the body form and lifestyle of the modern platypus, that of the echidna, or something else altogether? Several authors have argued that the platypus body form preceded the tachyglossid form in monotreme evolution (Pettigrew, 1999; Phillips et al., 2009). In other words, they contend that the tachyglossids are derived from a platypus-like ancestor. Phillips et al. (2009) maintain on the basis of morphology and the fossil record that all extant monotremes are derived from a platypus-like ancestor that had aquadynamic streamlining, dorsally projecting hindlimbs acting as rudders, and locomotion based on a long-axis rotation of a hypertrophied humerus that provided an efficient swimming stroke. They contend that this ancestry is reflected in a number of anatomical features of the modern tachyglossids (Fig. 17.1). These include the dorsoventral flattening of the body, caudally pointing hind-foot posture, and forelimb-dominant locomotion reflected structurally in the rotation of the long axis of the humerus. They argued that each of those traits would be highly anomalous if they were derived directly from a more generalized terrestrial basal mammal morphotype. On the neurological side, Pettigrew (1999) also favored a semiaquatic platypus-like electrosensory-capable ancestor as the original monotreme body form. He interpreted the difference in the number of electroreceptors between the modern monotreme families as evidence of the regression of electrosensation among the tachyglossids as their distant ancestors moved into progressively drier environments where electrosensory ability would be less useful in prey detection. Molecular clock studies also support a relatively recent divergence of the ornithorhynchid and tachyglossid lineages (Phillips et al., 2009).

Nevertheless, the contention that echidnas are derived from the ornithorhynchid lineage is not universally accepted. Certainly the available fossil material would be consistent with the early emergence of a platypus-like body form, but we have already seen that there are good reasons why the Early Cenozoic fossil record would be deficient in tachyglossids. The platypus-first hypothesis is therefore based on an absence of evidence. The focus on electroreception in the literature has also ignored several highly derived features of tachyglossids (the large and gyrified olfactory bulb, the large and laminated olfactory cortex, the gyrified isocortex, and the expansion of the isocortex rostral to the primary motor cortex) all of which are inconsistent with tachyglossids being recent derivatives of an olfactorily deficient and lissencephalic ornithorhynchid ancestor (Ashwell, 2013b). Furthermore, the molecular clock data for monotremes do not take into account the possible slow rate of divergence that is likely for monotremes, and a much earlier date of divergence has been obtained when a relaxed molecular clock method is applied to DNA and amino acid databases (Rowe et al., 2008). Finally, as Musser (2003) has observed: "highly specialised animals seldom give rise to new taxa, particularly those that then develop novel but dissimilar specialisations of their own."

17.3 What is Different About Monotremes From Other Mammals?

There are some key points about monotreme physiology that have a bearing on nervous system function and development. Monotremes are mammals and

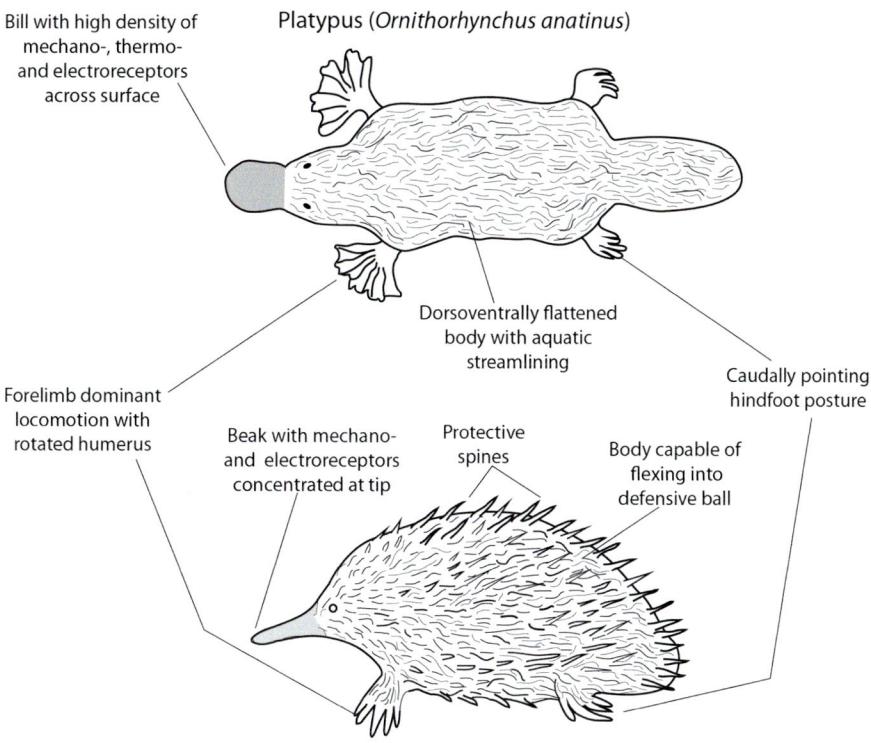

FIGURE 17.1 *Line diagrams* showing the key body-form features of the platypus and short-beaked echidna. The platypus is adapted for a semiaquatic lifestyle with a dense layer of fur for aquatic streamlining. The echidna is adapted for a fossorial lifestyle, breaking open termite mounds and rotten logs.

show similarity in hair and the ability to produce milk for their young with all other mammals, but they have many unusual features that clearly set them apart from other mammals.

17.3.1 Reproduction

Of course, the most striking feature of the monotremes is that they are egg-laying mammals. After an intrauterine developmental phase of about 11 days, an egg about 16 mm long is delivered from the cloaca and placed either into a depression on the maternal abdomen (echidnas) or a vegetation nest (platypus). Mother echidnas may leave the nest to forage with their egg firmly lodged in the abdominal pouch, whereas the semiaquatic platypus must leave its eggs and young in a secure nest within the riverbank while it searches for food. Hatching of the leathery-skinned egg occurs after an incubation phase of about 10 or 11 days (see review by Ashwell, 2013c). This requires that the young monotremes have the motor strength and neural control to break through the egg membranes and move around the nest or maternal abdomen to locate the milk source. Lactation is by secretion from nippleless milk patches on the maternal abdomen. The lactational phase extends for

about 130 days for the platypus (Holland and Jackson, 2002; Hawkins and Battaglia, 2009) and 200 days for the short-beaked echidna (Rismiller and McKelvey, 2003). Growth of the body is surprisingly rapid despite the lack of a secure attachment to the milk source, although cerebral metabolism and cortical growth may be slower in monotremes than in metatherians and eutherians (see the following section; Ashwell, 2015a,b).

17.3.2 Body Temperature and Metabolism

Like other mammals, monotremes maintain a body temperature above the ambient environment, although at a lower level (usually 30–32°C) than among eutherians and metatherians (Nicol, 2013). Body temperature in the case of the short-beaked echidna can fluctuate markedly across the year (Nicol, 2013) as the animal enters and leaves torpor. The platypus metabolic rate is low but similar to metatherians and eutherians of similar body weight, whereas the short-beaked echidna can undergo striking fluctuations in metabolic rate with season (Nicol, 2013). As we will see later on, the low body temperature and low and/or variable metabolic rate of the monotremes may have profound implications for brain development.

17.3.3 Monotreme Cognition

Brain size in the monotremes is comparable to many eutherians and greater than most metatherians. In fact, the brain of the short-beaked echidna (20–30 mL) is comparable in size to some New World primates (eg, squirrel monkey, *Saimiri sciureus* at 25 mL), and brain size of the long-beaked echidnas (averaging 50 mL, but up to 60 mL) exceeds the domestic cat (25–30 mL) (Ashwell, 2013d). The brain of the platypus is somewhat smaller (up to 15 mL) (Ashwell, 2013d). Studies of monotreme cognition are largely confined to short-beaked echidnas because they are much easier to keep in the laboratory (see review by Nicol, 2013). Laboratory tests of echidna behavior have shown an acquisition rate of choice behavior that is almost identical with laboratory rats at a similar level of training (Saunders et al., 1971). Even though echidnas have a rather small and simple visual system and lack a corpus callosum, they can match laboratory rodents in both visual discrimination and interocular transfer of discrimination (Gates, 1978). In studies using operant techniques, short-beaked echidnas have an overall performance that compares favorably with cats and rats (Buchmann and Rhodes, 1979). Of course, such laboratory-based studies do not adequately reflect the abilities of animals in a natural setting, and there is a great need for more studies of monotreme behavior in the wild (Nicol, 2013).

17.4 Electroreception and Mechanoreception

Monotremes are most famous for the ability to sense electrical fields, but their feats with mechanoreception are also impressive. Both of these depend on highly adapted bills and beaks and specialization of the trigeminal sensory pathway.

17.4.1 Overview of the Trigeminal System

The overall plan of the monotreme trigeminal sensory system is the same as in all other mammals, but there are some distinctive features peculiar to the group. All extant monotremes have electrosensory receptors on their bill (the platypus) or beaks (the long- and short-beaked echidnas), but these are much more numerous and spatially organized on the platypus bill (30 000 to 40 000 mucous sensory glands arranged in stripes parallel to the long axis of the bill) (Andres and von Düring, 1984; Manger and Pettigrew, 1996) than in the echidnas (only a few hundred glands forming a "hot spot" on the tip of the beak in the short-beaked echidna, with a few thousand in the long-beaked echidna species) (Andres et al., 1991; Manger and Hughes, 1992; Manger et al., 1997). The platypus appears to use its bill as an antenna for electrical teloreception (ie, sensation at a distance), whereas the echidnas use their beaks as push-probes for leaf-litter and soil. The beaks and bills are also exquisitely sensitive mechano- and thermoreceptors, which may be just as behaviorally important as electroreception, probably more so for the short-beaked echidna. The central trigeminal pathways of both platypus and echidna have the same nuclear groups seen in therians [trigeminal sensory nuclei in the brain stem, ventral posteromedial (VPM) nucleus or its homologue in the thalamus and S1, and other sensory fields in the isocortex], but the platypus has some very unusual features of these, as detailed later.

17.4.2 Peripheral Receptors

The platypus bill has three types of glands, two of which have sensory innervation (mucous and serous sensory glands) and a third which does not (mucous glands without sensory innervation). The mucous sensory glands are the putative electrosensory glands (Scheich et al., 1986) and consist of a coiled secretory tubule in the dermis with autonomic innervation, a papillary portion of 150 μm length surrounded by up to 30 myelinated nerve fibers forming a cuff around the base of the papilla and a 200- to 330-μm-long duct to the surface (Fig. 17.2A). Axons of the sensory cuff penetrate the papillary duct wall and presumably detect electrical impulses propagated down the duct lumen from the surface. Unlike electrosensitive fish, where there is a specialized sensory cell that detects the electrical field and then conveys this information to a sensory nerve terminal, the naked nerve terminals of the platypus mucous sensory gland are the actual sense organ (Manger and Pettigrew, 1996).

The echidnas also have (presumably electroreceptive) mucous sensory glands, but these are arranged in a spot at the end of the beak rather than in longitudinal stripes. As in the platypus, axons innervate a modified epidermal portion of the gland to form a sensory cuff, but the epidermal pore is quite simple in the echidnas, and there are only a few nerve terminals per sensory gland in the echidna (4–10), whereas as many as 30 may be found in each platypus sensory gland (Manger and Hughes, 1992).

Mechanosensory structures in the monotreme beaks and bills include vesicle chains, Merkel cells, and lamellated (paciniform) corpuscles, but these are grouped into distinctive and complex mechanosensory organs such as the push-rod mechanoreceptor. In both the platypus and short-beaked echidna, the central element of this sensory apparatus is a rod composed of flattened epithelial cells connected to each other by desmosomes (Manger and Pettigrew, 1996; Andres et al., 1991; Manger and

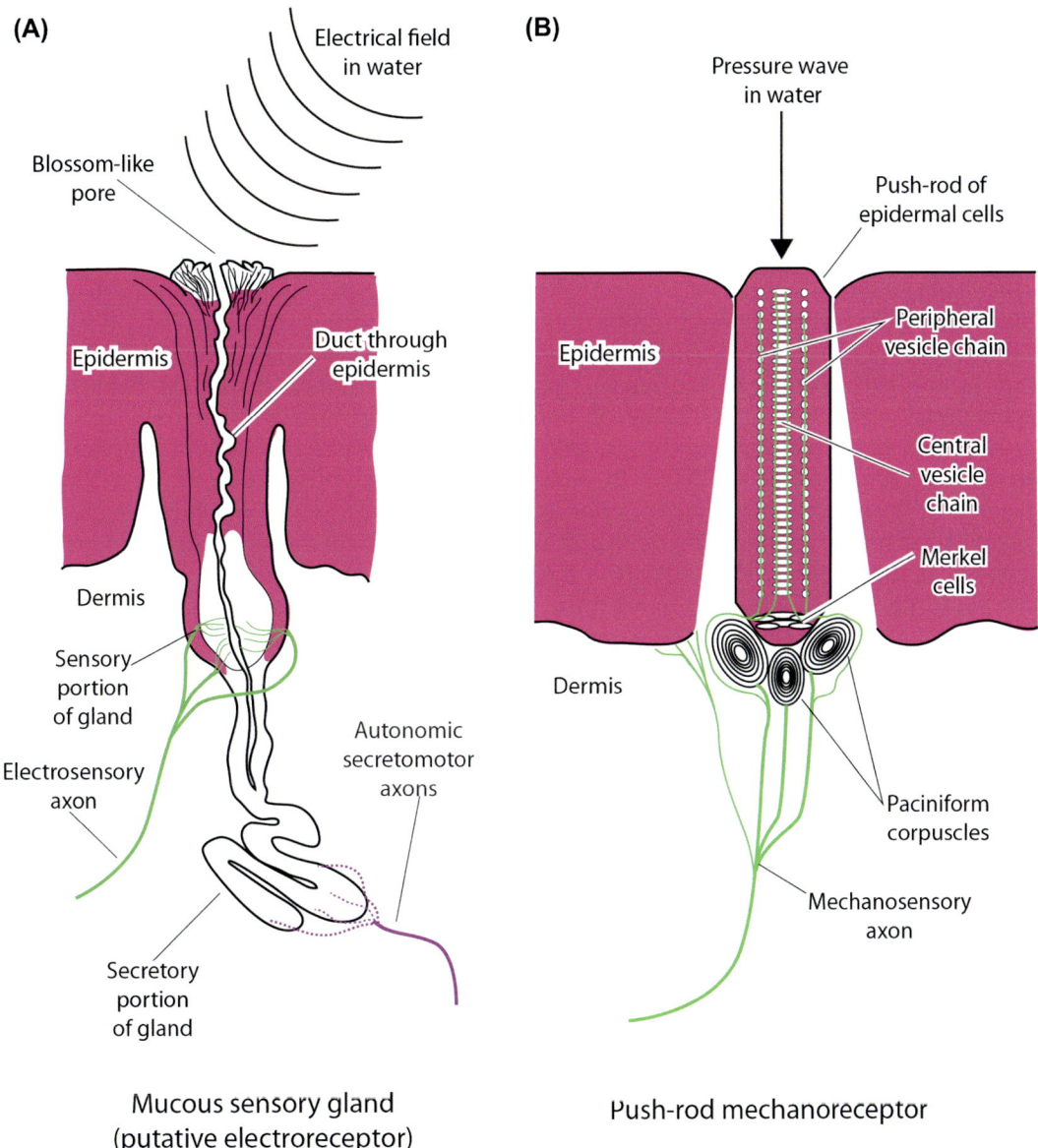

(A)

Electrical field
in water

Blossom-like
pore

Duct through
epidermis

Epidermis

Dermis

Sensory
portion
of gland

Electrosensory
axon

Autonomic
secretomotor
axons

Secretory
portion
of gland

Mucous sensory gland
(putative electroreceptor)

(B)

Pressure wave
in water

Push-rod of
epidermal cells

Peripheral
vesicle chain

Central
vesicle
chain

Merkel
cells

Epidermis

Dermis

Paciniform
corpuscles

Mechanosensory
axon

Push-rod mechanoreceptor

FIGURE 17.2   Schematic diagrams showing mucous sensory gland (putative electroreceptor) and the push-rod mechanoreceptor complex of the platypus bill. The mucous sensory gland (A) has a secretory portion under autonomic control and a sensory portion surrounded by a cuff of nerve fibers, which project naked nerve terminals into the duct. Electrical signals reaching the pore at the epidermal surface are transmitted down the duct filled with conductive secretions to the nerve terminals in the sensory portion. The push-rod mechanoreceptor complex (B) consists of a mobile epidermal rod about 400 μm long that is depressed below the epidermal surface by pressure waves in the surrounding water. Movement of the push-rod is detected by peripheral and central vesicle chains, and Merkel cells within the rod, and Paciniform corpuscles in the dermis immediately deep to the rod.

Hughes, 1992). In the platypus, this rod is about 400 μm long and can move relatively freely in a direction perpendicular to the skin surface. Three types of mechanosensory structures are located within the rod (central vesicle chain, peripheral vesicle chain, Merkel cells) and three to six paciniform corpuscles lie in the dermis deep to the rod (Fig. 17.2B). A key difference between the platypus and short-beaked echidna with respect to the push-rod organ is that the rod is linked to surrounding epithelial cells by desmosomes in the echidna and is

therefore limited in its mobility and sensitivity (Manger and Hughes, 1992). The echidna also has only two paciniform corpuscles at the base of the rod, instead of the three seen in the platypus.

17.4.3 Trigeminal Ganglion and Sensory Nuclei

Both bills and beaks are richly supplied with branches from trigeminal sensory ganglion cells, particularly for the maxillary and mandibular divisions in the case of the platypus, and estimates of the number of axons innervating the platypus have been as high as 1.34 million (Manger and Pettigrew, 1996). The size of the trigeminal sensory ganglion is very different between the platypus and short-beaked echidna, about 55 mm^3 for the platypus and only about 9 mm^3 for the larger-bodied short-beaked echidna, a difference that is reflected throughout the central parts of the trigeminal pathway (Ashwell et al., 2012).

The trigeminal sensory nuclei of the short-beaked echidna are rather like other mammals, being clearly divisible into mesencephalic, principal, and nucleus of trigeminal spinal tract components, but those of the platypus are hypertrophied, internally complex, and overlap rostrocaudally (Ashwell et al., 2006). In both platypus and echidna, the neurons of the mesencephalic nucleus are arranged in an arc spanning the dorsal rim of the periaqueductal gray (Hines, 1929), but it is the more caudal sensory nuclei that are most noteworthy. In the platypus there are distinct magno- and parvocellular components of the hypertrophied principal trigeminal nucleus (Ashwell et al., 2006; Watson et al., 1977), and neurons of this nucleus are intensely parvalbumin reactive. The oral and interpolar parts of the trigeminal spinal tract nucleus are also hypertrophied and strongly parvalbumin reactive in the platypus, with extensive rostrocaudal overlap in their distribution. The principal trigeminal sensory nucleus of the echidna has distinct dorsolateral and ventromedial components, but these are much like those seen in rodents (Ashwell et al., 2006).

17.4.4 Thalamus and Cortex

As in all other mammals, the ventral posterior (VP) thalamic nucleus is the dedicated somatosensory relay nucleus for monotremes. The VP nucleus is identifiable in both the platypus and short-beaked echidna, but in the platypus the structure is extraordinarily large as it serves the mechano- and electrosensory information from the large bill. Jones (2007) described three subdivisions of the monotreme VP on the basis of chemo- and cytoarchitecture, but did not feel that the therian division into VPM and ventral posterolateral (VPL) was obvious in either monotreme. On the other hand, Ashwell and Paxinos (2005) identified discrete VPL and VPM in the thalamus of the short-beaked echidna.

There are significant discrepancies between earlier and later studies of the platypus somatosensory cortex. Whereas Bohringer and Rowe identified in the 1970s only a single contralateral body representation in the isocortex, Krubitzer et al. found in 1995 as many as four topographically ordered contralateral body

representations (primary somatosensory cortex, S1; parietoventral somatosensory cortex, PV; manipulation field of somatosensory cortex, M; and rostral somatosensory cortex, R; Fig. 17.1). Krubitzer et al. also identified coextensive myelin and cytochrome oxidase (CO) bands within S1. Regions where myelin staining and CO reactivity are strongest (CO dark bands) contain neurons that respond only to cutaneous stimulation, whereas areas between the CO dark bands contain neurons that respond to both electrical and mechanical stimuli (but more to the former) (Manger et al., 1996). It has been suggested that the banded pattern within S1 allows for processing of temporal disparities between mechano- and electrosensory input (Pettigrew et al., 1998; Elston et al., 1999). Analysis of such a temporal disparity would allow the platypus to estimate the distance to crustacean prey, since the two forms of information are propagated at very different velocities through creek water.

There are also significant structural differences of pyramidal neurons between CO-dark and CO-light regions of S1 (Elston et al., 1999). Pyramidal neurons in the CO-light (mechano- and electroresponsive) regions have wider basal dendritic fields than in the CO-dark (purely mechanoresponsive) regions. Elston et al. suggested that this structural difference allows improved spatial and temporal discrimination of stimuli from the bill, but this has never been tested.

The somatosensory cortex of the short-beaked echidna is located in the caudal cortex between the α and ζ sulci (Lende, 1964; Allison and Goff, 1972; Rowe, 1990). Although shifted caudally (see Fig. 17.1), the somatosensory cortex of the short-beaked echidna has a very similar somatotopic organization to therians. Krubitzer et al. (1995) also identified four somatosensory fields in the short-beaked echidna (S1, PV, M, and R), but no interdigitating subdivisions that could allow comparison of mechano- and electrosensory information have been found within the echidna S1. Krubitzer reported a specialized forepaw region within the forelimb representation of PV that was not found in the platypus. This region appears to receive input from Pacinian corpuscles in the glabrous skin of the forelimb and may serve a role in fine motor control of the forelimb during digging.

17.4.5 Electroreception and Mechanoreception in the Natural Setting

Monotreme electroreception is purely passive, meaning that neither the platypus nor echidnas generate electrical fields of their own. There are many laboratory studies that demonstrate electrosensory function in both the platypus (Pettigrew, 1999; Manger and Pettigrew, 1996; Scheich et al., 1986) and short-beaked echidna (Gregory et al., 1989), but the question remains as to how useful electrosensation is in the natural environment. Behavioral studies with the platypus have reported electrical field strength thresholds of 50–200 μV cm^{-1} (Scheich et al., 1986) or 20 μV cm^{-1} (Pettigrew, 1999; Manger and Pettigrew, 1996), but thresholds for individual trigeminal afferent axons are much higher (around 4 mV cm^{-1}) (Gregory et al., 1988, 1989). This indicates that some extensive spatial and temporal summation would be required along the trigeminal pathway to achieve the observed behavioral threshold (Fjällbrant et al., 1998).

Laboratory studies have suggested that the platypus bill functions as a directional antenna. Electrical signals in the environment induce head saccades of free-swimming platypuses that allow the bill to be swept through the electrical fields. Pettigrew et al. (1998) have proposed that neural reconstruction of the isoelectric fields across the bill is achieved at the level of S1 cortex (Pettigrew et al., 1998) and provides the platypus with information on both range and direction of the prey. Directionality could also be detected by central analysis of the decay of electrical field strength across the bill (Fjällbrant et al., 1998).

It is when we move from the laboratory setting to the natural environment that questions of the behavioral value of electroreception arise. Most models of the use of mechano- and electroreception in prey detection focus on crustacean prey, because free-swimming crustaceans at about 15 cm produce mechanical and electrical signals that exceed the behavioral threshold for the platypus bill. However, the freshwater environment that the platypus inhabits is far more complex than a laboratory tank and is filled with many obstacles that distort and attenuate both mechanical and electrical signals. Furthermore, the platypus ingests many prey that produce only negligible electrical signals (Ashwell and Hardman, 2013). In fact, crustaceans make up only 12% of the platypus winter diet (Grant and Carrick, 1978). The bulk of the platypus diet is made up of immature mollusks, aquatic worms, and the larvae of many insects (Burrell, 1927), none of which produce a significant electrical signal. It may be that mechanosensation is the most important sense for detection of prey across the year and that electroreception has evolved to allow detection of electrically loud prey during the breeding and lactation season.

The role of electroreception in the natural environment is even more doubtful for the short-beaked echidna. Electroreceptors in the beak tip are poorly responsive to electrical signals in air (Gregory et al., 1989), so the snout tip must be kept moist to be effective as an electrical probe. This means that electroreception is unlikely to be an effective sense for the echidnas unless the beak tip has been inserted into soft, moist soil or leaf-litter. The behavioral threshold for the short-beaked echidna in the laboratory setting is 1.8 mV cm^{-1}, but this was for a field applied through a water trough. Field studies have suggested that trained short-beaked echidnas can detect buried 9-V batteries at field strengths of 15–20 mV cm^{-1} at distances of 12–14 cm (Augee and Gooden, 1992), but many of the echidna prey produce much lower electrical signals than buried batteries. The electrical signals of buried earthworms, for example, are only 3 μV cm^{-1} (Taylor et al., 1992). Although electroreception may be of considerable benefit for the long-beaked echidna, which inhabits moist environments in New Guinea, it is more likely that olfaction plays the major role in prey detection for the short-beaked echidna.

17.5 The Olfactory System in Monotremes

17.5.1 Overview of Monotreme Olfactory System Structure

The platypus and echidnas have very different olfactory systems. This system is small and anatomically unspecialized in the platypus, but large and complex in the short-beaked echidna. It is likely that olfaction is a much more behaviorally important sense than electroreception for the echidnas, contributing both to prey detection and mating behavior (Morrow et al., 2009; Morrow and Nicol, 2009; Harris et al., 2012).

The olfactory system of the platypus is quite small and simple. The nasal and ethmoidal turbinals are small (Griffiths, 1978), and the main olfactory pathway is tiny, but, despite the aquatic lifestyle, the vomeronasal organ and central accessory olfactory pathway are significant in size, and the platypus has a large array of vomeronasal chemoreceptor genes (see Section 17.5.2) (Shi and Zhang, 2006; Grus et al., 2007). In fact, the main and accessory olfactory bulbs of the platypus are similar in size (Ashwell, 2006), suggesting similar functional importance of the two systems. The olfactory tubercle is very small in the platypus (only 1.5 mm^2), and the anterior olfactory region and piriform cortex are similar in structure to therians, but small (Ashwell and Phillips, 2006).

By contrast, the olfactory systems of the short- and long-beaked echidnas are large and structurally complex. The tachyglossid nasal cavity has seven vertical endoturbinals suspended from the lamina cribrosa, many ectoturbinals, and sets of naso- and maxilloturbinals (Griffiths, 1978). The olfactory bulb of the short-beaked echidna is gyrified, involving all layers down to the granule cells, thereby increasing the surface area of the bulb by 21—39% (Ashwell, 2013e). The piriform cortex of the short-beaked echidna is complex, with regional variation of the laminar structure and chemoarchitecture around the circumference of the olfactory lobe. The size and complexity of the olfactory system is comparable to that seen in olfactorily proficient therians of similar body size. For example, the volume of the combined main and accessory olfactory bulbs of the short-beaked echidna averages 630 mm^3, but can reach 900 mm^3 in larger individuals, which is similar to the largest values reached among eulipotyphlids and prosimians (up to 800 mm^3 among eulipotyphlids and up to 900 mm^3 among prosimians) (Ashwell, 2013e).

17.5.2 The Olfactory Receptor Gene Repertoire of Monotremes

Information on this aspect is confined to the platypus, but the results are surprising. Both the main and accessory systems of vertebrates have distinct sets of chemoreceptors: in the main system there are olfactory receptors (ORs) and trace amine—associated receptors (TAARs), whereas in the vomeronasal or accessory olfactory system there are two types of vomeronasal receptors (V1 and V2) as well as formyl peptide receptor-like proteins. Grus et al. (2007) identified 270 intact genes and 579 pseudogenes in the platypus V1R family—83 more intact genes than the largest eutherian V1R family. Grus et al. also identified significant chromosomal clustering of V1R genes in the platypus, suggesting that the large repertoire has arisen by tandem gene duplication. In contrast, the numbers of OR and TAAR genes in the platypus are much less than in therians, consistent with reduced olfaction and an aquatic lifestyle. Nevertheless, the proportion of OR class I genes in the platypus is low (11.5%) and similar to the laboratory mouse (11.0%), whereas aquatic mammals usually show a preponderance of class I genes.

17.6 The Cortex in Monotremes

The somatosensory cortex has already been discussed briefly in the contact of electroreception, but there are many other interesting aspects of the isocortex in monotremes. One obvious difference between the platypus and echidnas is that the former has a smooth (or lissencephalic) cerebral cortex, whereas the latter have a folded (or gyrencephalic) cortex. In fact, short-beaked echidnas have a gyrification index comparable to many large-brained eutherians (eg, the

domestic cat or prosimians) (Hassiotis et al., 2003; Zilles et al., 1989).

17.6.1 Cortical Topography and Functional Areas

In the platypus, the somatosensory cortex occupies the bulk of the isocortical mantle. In maps based on the studies of Krubitzer et al. (1995) the four somatosensory fields (S1, M, R, PV) occupy almost 90% of the cortical area, and the auditory and visual areas are relegated to the caudomedial rim (Fig. 17.3). It should be noted that Bohringer and Rowe (1977) identified only one somatosensory field (S1) and found considerable overlap between motor and somatosensory cortex, so further studies are warranted. Nevertheless, both authors agree on the enormously expanded representation of the bill in the platypus somatosensory cortex. In the short-beaked echidna, the motor and somatosensory cortex is shifted caudally and laterally. The somatosensory cortex is located between the α and the ζ or η sulci with somatotopy arranged in a lateral to medial sequence, that is, snout and tongue laterally and tail medially. Krubitzer et al. also identified four fields in the somatosensory cortex of the short-beaked echidna, and there is some agreement between their physiologically based delineation and results from cyto- and chemoarchitectural studies (Hassiotis et al., 2004). The visual cortex is located dorsocaudally and auditory cortex laterocaudally. There are many sulcal similarities between short-beaked echidnas and both extant and extinct long-beaked echidnas that suggest the general cortical topography of all tachyglossids has been similar since the Pleistocene (Ashwell et al., 2014). Although the gyrification of the echidna cortex is reminiscent of that seen in primates, cetaceans, proboscideans, and carnivores, the cytoarchitecture of the echidna cortex exhibits less regional variation than is seen in large-brained eutherians, and the caudal displacement of the motor and sensory fields is quite unlike that seen in any other mammal (Hassiotis et al., 2004).

The function of the large area of cortex rostral to the motor cortex in the echidnas is an enduring mystery. Welker and Lende (1980) considered it to be an expanded prefrontal cortex, which would give echidnas the proportionally largest prefrontal cortex of all mammals (Divac et al., 1987a,b). The region is cyto- and chemoarchitecturally heterogeneous, with as many as three discrete fields (Fr1, Fr2, and Fr3) (Hassiotis et al., 2004). The region also engages in reciprocal topographic connections with the rostromediodorsal part of the thalamus (Divac et al., 1987a,b) and projects to the dorsomedial part of the striatum

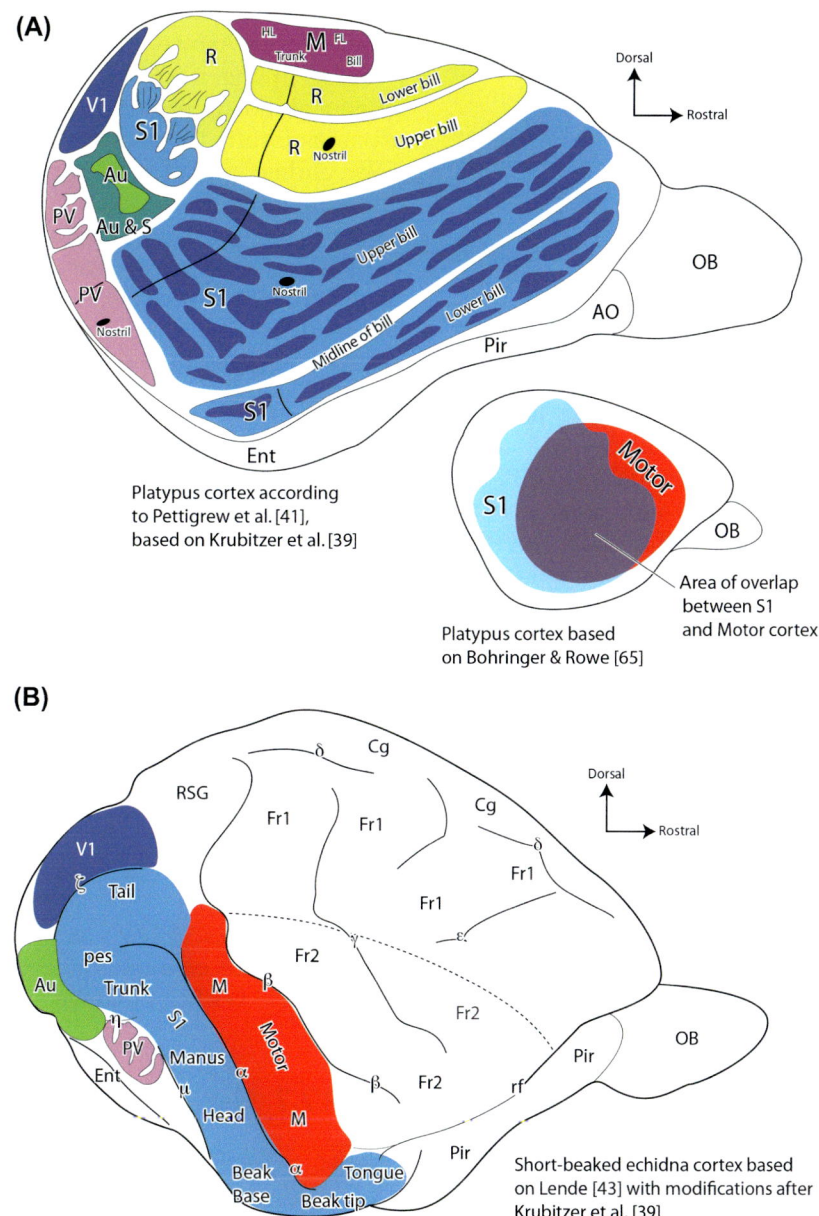

FIGURE 17.3 Diagrams showing the major cortical regions in the platypus (A) and the short-beaked echidna (B). The large diagram in (A) shows the sensory areas identified by Krubitzer et al. (1995) and is based on a broader map of the cortex illustrated in Pettigrew et al. (1998), whereas the smaller diagram shows the motor and somatosensory maps obtained by Bohringer and Rowe (1977). The darker bands within the platypus S1 indicate the cytochrome oxidase—dense zones. Note that the manipulation somatosensory cortex (M) overlaps with motor cortex in the maps by Krubitzer et al. (1995) and that the rostral somatosensory field (R) is within the depths of the α sulcus in the short-beaked echidna and therefore not visible externally. *Greek letters* indicate sulci of the echidna cortex. *AO*, anterior olfactory area; *Au*, auditory cortex; *Au and S*, combined auditory and somatosensory cortex; *Cg*, cingulate gyrus; *Ent*, entorhinal cortex; *FL*, forelimb region; *Fr1, 2*, frontal cortical fields; *HL*, hindlimb region; manus, forepaw; *OB*, olfactory bulb; *pes*, hindpaw; *Pir*, piriform cortex; *PV*, parietoventral area of somatosensory cortex; *rf*, rhinal fissure; *RSG*, retrosplenial gyrus; *S1*, primary somatosensory cortex; *V1*, primary visual cortex.

(Divac et al., 1987b), although these studies have only really examined cortical areas corresponding to Fr1 and Fr2. It may be that Fr1 is homologous to the prefrontal cortex of eutherians, but it might also be homologous to anterior cingulate, orbital, prelimbic, premotor, and/or rostral agranular insular cortex because those regions also receive input from the mediodorsal thalamus in eutherians (Ashwell, 2013f). The connections of Fr3 have never been studied, but on topographic grounds one could posit homology with the orbital frontal cortex of eutherians, which might provide an integrative center for chemosensory information and/or a limbic center for regulation of behavior (Ashwell, 2013f).

17.6.2 Thalamocortical Relationships

Anatomical connections between the monotreme thalamus and the cortex have only been studied in the short-beaked echidna (Welker and Lende, 1980; Divac et al., 1987a,b; Ulinski, 1984). The pattern of thalamocortical connections in the echidna is broadly similar to that in therians. The VP nucleus identified by Welker and Lende corresponds with that identified on chemoarchitectural grounds and occupies a similar position within the thalamus to the therian VP. The pattern of projection from the putative mediodorsal nucleus to frontal cortex is also consistent with the proposed homology of the large frontal cortex. The medial geniculate has been identified in the ventrolateral part of the most caudal thalamus, and the lateral geniculate nucleus (LGN) corresponds to the LGNb identified as a retinorecipient nucleus by Campbell and Hayhow (1971, 1972). Questions remain on the thalamocortical connections and homologies of midline and intralaminar nuclei in the monotreme thalamus (Ashwell, 2013a).

17.6.3 Is There a Monotreme Claustrum?

The claustrum is a pallial structure that lies deep to the cerebral cortex in all therians. There has been a long-running debate on this question in the scientific literature, and the importance of the question hinges around whether the mammalian claustrum is homologous to the anterior part of the dorsal ventricular ridge of turtles and diapsid reptiles, or to the avian neostriatum (Butler et al., 2002). Abbie (1940) was unable to identify a claustrum in either the platypus or echidna, and Divac et al. (1987a,b) also noted the apparent absence in their studies of the echidna. Indeed, the apparent absence has been listed as a primitive character of the monotreme nervous system (Johnson et al., 1982a,b, 1994). In more recent times, Butler et al. (2002) have concluded that the claustrum is missing, whereas Ashwell et al. (2004) argued for its presence in the short-beaked echidna, at least. Ashwell et al. identified a dorsal claustrum in the forebrain of the short-beaked echidna on the basis of cellular, fiber, and chemoarchitectural criteria. As in therians, the putative claustrum of the echidna engages in reciprocal connections with other parts of the pallium, but this needs further study (Ashwell, 2013f).

17.6.4 Cellular Composition and Neuronal Structure

There is perhaps 115 million years of divergence between the monotremes and the therians, and yet they have a superficially similar laminated isocortex. This naturally raises the question of how neuronal morphology differs between monotremes and therians. Hassiotis and Ashwell (2003) identified eight classes of neurons in the isocortex of the echidna (pyramidal, spinous bipolar, aspinous bipolar, spinous bitufted, aspinous bitufted, spinous multipolar, aspinous multipolar, and neurogliaform), all of which, except the pyramidal neurons, were similar to those types in therian isocortex. In particular, many pyramidal neurons in the echidna isocortex were atypical (30—42% depending on region) with inverted somata, short or branching apical dendrites, and/or few basal dendrites, unlike the classical pyramidal neurons familiar from therian isocortex. Hassiotis and Ashwell also found that dendritic spine density on the apical and basal dendrites of S1 and motor pyramidal neurons was lower than in rodents, although the functional implications of this remain to be explored. Nevertheless, the similarities suggest that both pyramidal neurons and many types of nonpyramidal neurons had already emerged very early in mammalian cortical evolution, that is, at the time of divergence of the therian and prototherian lineage.

These are also some intriguing reports concerning the glia in monotreme brains that warrant further study. Lambeth and Blunt (1975) found only a single morphological type of macroglia in the platypus brain, but found two types in the brain of the juvenile echidna. They reached the conclusion that monotremes have only a single type of macroglial cell, albeit with lighter and darker forms. If this is indeed true it would be a significant feature of monotreme neurocytology, but the report has never been followed up.

17.7 Nervous System Development in Monotremes

17.7.1 Overview of Monotreme Development

The neural tube forms during the intrauterine stage of development, and elaboration of forebrain vesicles occurs during incubation (Ashwell, 2013c). At the time of hatching (about 14 mm body length), the hindbrain regions of the platypus and echidna are similar to each other and to those of newborn metatherians. The pallium of the newly hatched monotreme is undifferentiated neuroepithelium with most of the wall taken up by the ventricular germinal zone. The most significant events in posthatching life are the differentiation of the forebrain and cerebellum, and it is at this stage that the pallial development diverges between the platypus and echidnas (see later discussion). The cortical plate emerges at the end of the first posthatching week, and afferent and efferent connections begin to form in the second posthatching week. The large neurons of the

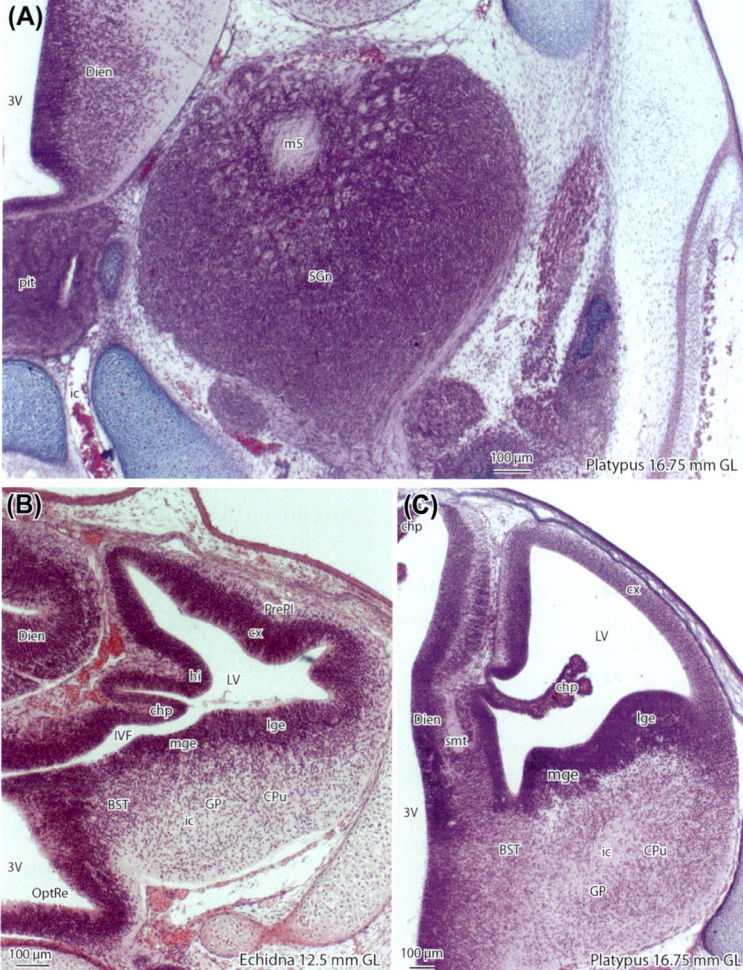

FIGURE 17.4 Photomicrographs of frontal sections through the developing brain of a platypus (A,C) and a short-beaked echidna (B) around the time of hatching [12.5 mm greatest length (GL) echidna and 16.75 mm GL platypus]. Note the striking size of the trigeminal ganglion (5Gn) in the newly hatched platypus (A). Even before hatching (B), the cerebral cortex (cx) of the short-beaked echidna has begin to show folding and a distinct preplate layer (PrePl) with postmitotic neurons, whereas in the newly hatched platypus (C) the cx is still smooth and consists of only a ventricular germinal zone. *3V*, third ventricle; *BST*, bed nuclei of the stria terminalis; *chp*, choroid plexus primordium; *CPu*, caudatoputamen; Dien, diencephalon; *GP*, globus pallidus; *hi*, hippocampal primordium; *ic*, internal capsule; *lge*, lateral ganglionic eminence; *LV*, lateral ventricle; *m5*, motor root of trigeminal nerve; *mge*, medial ganglionic eminence; *OptRe*, optic recess of third ventricle; *pit*, pituitary; *smt*, stria medullaris thalami.

cerebellum are generated late in incubation, but small neurons are not formed until the first few weeks after hatching (Ashwell, 2013c).

17.7.2 Trigeminal System Development

The platypus and echidnas have very different distributions of sensory glands and mechanoreceptors in their bills or beaks. The platypus has a distinctive linear pattern of mucous sensory glands on its bill, whereas the short-beaked echidna has a sensory gland hot spot on its beak tip. This naturally raises the question of when this distinct pattern of peripheral trigeminal receptors appears in the two species during

development. Ashwell et al. (2012) found that each monotreme showed a characteristic and different pattern of distribution of developing epidermal sensory gland specializations (electroreceptor primordia) from the time of hatching. These were linear in the platypus, but focal at the snout tip in the echidna. The cross-sectional areas of the trigeminal divisions and the volume of the trigeminal ganglion itself were also very different between the two species at embryonic ages; very large in the platypus (Fig. 17.4A), but comparable in size to therians in the echidna. This difference in size continues throughout posthatching development. Ashwell et al. interpreted these findings to indicate that the trigeminal pathways in the short-beaked echidna and the platypus follow very different developmental trajectories from the earliest embryonic appearance of epidermal specialization. They went on to argue that such an early developmental divergence is more consistent with the notion that the platypus and echidna have both diverged from an ancestor with rudimentary electroreception and/or trigeminal specialization, rather than the contention that the echidna is derived from a platypus-like ancestor.

17.7.3 Cortical Development

We have already seen that the platypus and echidna have very different adult structure, being lissencephalic and gyrencephalic, respectively. Studies of embryos in the Hill collection indicate that gyrencephaly begins to emerge in the echidna brain a few days before birth (at about 12.5 mm body length), whereas the cortex of the platypus remains lissencephalic throughout development (see Fig. 17.4B and C; Ashwell and Hardman, 2012). In both platypus and echidna the cortex is very immature at the time of hatching. During posthatching life (ie, from 14 mm body length), the cortices of both species have subventricular compartments within both the striatum and pallium. In particular, the platypus has a sheetlike extension of the subventricular zone that runs from the palliostriatal angle beneath the developing trigeminoreceptive part of the somatosensory cortex of the lateral cortex. This region beneath the trigeminal part of S1 has many mitotic figures and appears to be traversed by large numbers of thalamocortical axons as these grow toward the cortical plate. The association of putative thalamocortical fibers with this region suggests that it may also serve functions similar to the subplate zone of eutheria.

Quantitative analysis of somatosensory cortex development in monotremes compared to therians (Ashwell, 2015a) shows that growth of the cortex during posthatching development is at about half the pace of that in metatherians and rodents. This suggests a slow pace of mitosis in the developing cortex and may be due to the relatively low body temperature of monotremes compared to therians.

17.8 Conclusions. Not Primitive, Just Different!

17.8.1 Monotremes, Like All Other Mammals, Present a Mosaic of Primitive and Derived Features

In the 19th and 20th centuries, monotremes were often seen by the scientific community as relics of the Cretaceous, and the so-called primitive features of their anatomy were emphasized in descriptions (Lende, 1964). With the discovery of their ability to sense electrical fields and the more detailed analysis of monotreme neuroanatomy that occurred in the later 20th century, it has become clear that monotremes have many remarkable sensory specializations. These neural specializations of monotremes are just as significant as their plesiomorphic characters. Those who would search the monotreme nervous system for hallmarks of the primitive should remember that so-called primitive and advanced features can exist side by side in the human brain as much as the monotreme. No mammal is wholly primitive or wholly advanced, but rather a mosaic of both, and the nature of that mix need not necessarily predict success in the natural world. Although the long-term survival of the platypus is at risk due to degradation of its natural environment, the short-beaked echidna is one of the most successful and widespread native animals on the Australian continent.

17.8.2 What Can Monotremes Tell Us About Mammalian Brain Evolution?

Although monotremes should not simply be seen as living basal mammals, such that every feature their brains show can be posited as a character of the archaic mammalian brain, careful analysis and comparison of the monotreme nervous system with that in therians can inform thinking about how the mammalian nervous system evolved. There are two areas where this approach has potential: cortical cytoarchitecture in the adult brain and developmental mechanisms in the cortex.

The monotremes and therians have followed separate evolutionary paths for as much as 115 million years (Pettigrew, 1999). In both therians and monotremes the cortex has undergone expansion and specialization, producing a large gyrified cortex in the case of the tachyglossids and the remarkable somatosensory field in the case of ornithorhynchids. Nevertheless, the cellular, chemical, and ultrastructural features of the monotreme cerebral cortex are remarkably similar to those in eutherian isocortex. This suggests that the key precursors of these similarities (eg, neuronal morpho- and chemotypes) were present in the cortex of the common ancestor for both lineages, since the details of cortical cytoarchitecture are so alike that convergent evolution seems unlikely. On the other hand, there are some very distinctive features of cytological organization that may have emerged in the isocortex during the long period of separate evolution of the prototherian and therian lineages. These include the large proportion of atypical pyramidal neurons in monotreme isocortex and the poor development of bundling of pyramidal neuron apical dendrites (Hassiotis and Ashwell, 2003). So, the commonalities point to archaic features of the cortex of the putative stem mammal, but idiosyncrasies may indicate unique features peculiar to monotremes.

Isocortical development in eutherians involves a two-zone pattern of neurogenesis (ventricular germinal zone and subventricular zone) with the further evolution of a transient cortical subplate to support incoming thalamo-cortical afferents. The substantial size and regional complexity of the isocortex in the living monotremes (thick and lissencephalic in the platypus, but thinner and gyrencephalic in the echidnas) suggests that sophisticated developmental mechanism(s) for building the isocortex are at play in both groups of monotremes. Mitotic figures are present some distance away from the ventricular lining in developing monotreme cortex, suggesting the presence of a pallial subventricular zone. This has been seen as a critically important event in mammalian brain evolution, because it permitted the prolonged and amplified neurogenesis of micro-neurons for a thicker and functionally more effective iso-cortex (Charvet and Striedter, 2011; Charvet et al., 2009; Cheung et al., 2007; Kriegstein et al., 2006). Findings in the developing platypus cortex suggest that there may be a region serving the role of a subplate, at least beneath the electrosensory area of the somatosensory cortex. If this were the case, then molecular analysis of events in monotreme cortical development would provide important clues as to how complex cortical development is controlled across mammals and would indicate that the key developmental mechanisms for building a complex mammalian cortex were a very ancient acquisition during mammalian evolution.

Acknowledgment

I would like to thank Dr. Peter Giere of the Museum für Naturkunde, Berlin, for access to the specimens that were photographed for Figure 17.2.

References

Abbie, A.A., 1940. Cortical lamination in the monotremata. J. Comp. Neurol. 72, 429–467.

Allison, T., Goff, W.R., 1972. Electrophysiological studies of echidna, *Tachyglossus aculeatus* 3. Sensory and interhemispheric evoked responses. Arch. Ital. Biol. 110, 195–216.

Andres, K.H., von Düring, M., 1984. The platypus bill. A structural and functional model of pattern-like arrangement of different cutaneous sensory receptors. In: Hamann, W., Iggo, A. (Eds.), Sensory Receptor Mechanisms. World Scientific Publishing Company, Singapore, pp. 81–89.

Andres, K.H., von Düring, M., Iggo, A., Proske, U., 1991. The anatomy and fine–structure of the echidna *Tachyglossus aculeatus* snout with respect to its different trigeminal sensory receptors including the electroreceptors. Anat. Embryol. 184, 371–393.

Archer, M., Plane, M.D., Pledge, N.S., 1978. Additional evidence for interpreting the Miocene *Obdurodon insignis* Woodburne and Tedford, 1975, to be a fossil platypus (Ornithorhynchidae: monotremata) and a reconsideration of the status of *Ornithorhynchus agilis* De Vis, 1885. Aust. Zool. 20, 9–27.

Archer, M., Flannery, T.F., Ritchie, A., Molnar, R.E., 1985. First Mesozoic mammal from Australia — an early Cretaceous monotreme. Nature 318, 363–366.

Archer, M., Jenkins Jr., F.A., Hand, S.J., Murray, P., Godthelp, H., 1992. Description of the skull and non-vestigial dentition of a Miocene platypus (*Obdurodon dicksoni* n. sp.) from Riversleigh, Australia, and the problem of monotreme origins. In: Augee, M.L. (Ed.), Platypus and Echidnas. Royal Zoological Society of New South Wales, Sydney, pp. 15–27.

Ashwell, K.W.S., Hardman, C.D., 2012. Distinct development of the cerebral cortex in platypus and echidna. Brain Behav. Evol. 79, 57–72.

Ashwell, K.W.S., Hardman, C.D., 2013. Somatosensory and electrosensory systems. In: Ashwell, K.W.S. (Ed.), Neurobiology of Monotremes: Brain Evolution in Our Distant Mammalian Cousins. CSIRO, Collingwood, pp. 179–217.

Ashwell, K.W.S., Paxinos, G., 2005. Cyto– and chemoarchitecture of the dorsal thalamus of the monotreme *Tachyglossus aculeatus*, the short beaked echidna. J. Chem. Neuroanat. 30, 161–183.

Ashwell, K.W.S., Phillips, J.M., 2006. The anterior olfactory nucleus and piriform cortex of the echidna and platypus. Brain Behav. Evol. 67, 203–227.

Ashwell, K.W.S., Hardman, C.D., Paxinos, G., 2004. The claustrum is not missing from all monotreme brains. Brain Behav. Evol. 64, 223–241.

Ashwell, K.W.S., Hardman, C.D., Paxinos, G., 2006. Cyto- and chemo-architecture of the sensory trigeminal nuclei of the echidna, platypus and rat. J. Chem. Neuroanat. 31, 81–107.

Ashwell, K.W.S., Hardman, C.D., Giere, P., 2012. Distinct development of peripheral trigeminal pathways in the platypus (*Ornithorhynchus anatinus*) and short-beaked echidna (*Tachyglossus aculeatus*). Brain Behav. Evol. 79, 113–127.

Ashwell, K.W., Hardman, C.D., Musser, A., 2014. Brain and behaviour of living and extinct echidnas. Zoology 117, 349–361.

Ashwell, K.W.S., 2006. Chemoarchitecture of the monotreme olfactory bulb. Brain Behav. Evol. 67, 69–84.

Ashwell, K.W.S., 2013. Diencephalon and deep telencephalic structures. In: Ashwell, K.W.S. (Ed.), Neurobiology of Monotremes: Brain Evolution in Our Distant Mammalian Cousins. CSIRO, Collingwood, pp. 107–130.

Ashwell, K.W.S., 2013. Reflections: monotreme neurobiology in context. In: Ashwell, K.W.S. (Ed.), Neurobiology of Monotremes: Brain Evolution in Our Distant Mammalian Cousins. CSIRO, Collingwood, pp. 285–298.

Ashwell, K.W.S., 2013. Embryology and post-hatching development of the monotremes. In: Ashwell, K.W.S. (Ed.), Neurobiology of Monotremes: Brain Evolution in Our Distant Mammalian Cousins. CSIRO, Collingwood, pp. 31–46.

Ashwell, K.W.S., 2013. Overview of monotreme nervous system structure and evolution. In: Ashwell, K.W.S. (Ed.), Neurobiology of Monotremes: Brain Evolution in Our Distant Mammalian Cousins. CSIRO, Collingwood, pp. 47–68.

Ashwell, K.W.S., 2013. Chemical senses: olfactory and gustatory systems. In: Ashwell, K.W.S. (Ed.), Neurobiology of Monotremes: Brain Evolution in Our Distant Mammalian Cousins. CSIRO, Collingwood, pp. 235–250.

Ashwell, K.W.S., 2013. Cerebral cortex and claustrum/endopiriform complex. In: Ashwell, K.W.S. (Ed.), Neurobiology of Monotremes: Brain Evolution in Our Distant Mammalian Cousins. CSIRO, Collingwood, pp. 131–159.

Ashwell, K.W.S., 2015. Quantitative analysis of somatosensory development in metatherians and monotremes, with comparison to the laboratory rat. Somatosens. Mot. Res. 32, 87–98.

Ashwell, K.W.S., 2015. Quantitative analysis of somatosensory development in eutherians, with a comparison with metatherians and monotremes. Somatosens. Mot. Res. 32, 137–152.

Augee, M.L., Gooden, B.A., 1992. Evidence for electroreception from field studies of the echidna Tachyglossus aculeatus. In: Augee, M.L. (Ed.), Platypus and Echidnas. Royal Zoological Society of New South Wales, Sydney, pp. 211–215, 1992.

Bohringer, R.C., Rowe, M.J., 1977. The organization of the sensory and motor areas of cerebral cortex in the platypus (Ornithorhynchus anatinus). J. Comp. Neurol. 174, 1–14.

Buchmann, O.L.K., Rhodes, J., 1979. Instrumental discrimination: reversal learning in the monotreme Tachyglossus aculeatus setosus. Anim. Behav. 27, 1048–1053.

Burrell, H., 1927. The Platypus. Angus and Robertson, Sydney.

Butler, A.B., Molnar, Z., Manger, P.R., 2002. Apparent absence of claustrum in monotremes: implications for forebrain evolution in amniotes. Brain Behav. Evol. 60, 230–240.

Campbell, C.B.G., Hayhow, W.R., 1971. Primary optic pathways in echidna, Tachyglossus aculeatus – experimental degeneration study. J. Comp. Neurol. 143, 119–136.

Campbell, C.B.G., Hayhow, W.R., 1972. Primary optic pathways in the duckbill platypus, Ornithorhynchus anatinus: an experimental degeneration study. J. Comp. Neurol. 145, 195–208.

Charvet, C.J., Striedter, G.F., 2011. Causes and consequences of expanded subventricular zones. Eur. J. Neurosci. 34, 988–993.

Charvet, C.J., Owerkowicz, T., Striedter, G.F., 2009. Phylogeny of the telencephalic subventricular zone in sauropsids: evidence for the sequential evolution of pallial and subpallial subventricular zones. Brain Behav. Evol. 73, 285–294.

Cheung, A.F.P., Pollen, A.A., Tavare, A., DeProto, J., Molnár, Z., 2007. Comparative aspects of cortical neurogenesis in vertebrates. J. Anat. 211, 164–176.

Divac, I., Holst, M.–C., Nelson, J., McKenzie, J.S., 1987. Afferents of the frontal cortex in the echidna (Tachyglossus aculeatus). Indication of an outstandingly large prefrontal cortex. Brain Behav. Evol. 30, 303–320.

Divac, I., Pettigrew, J.D., Holst, M.-C., McKenzie, J.S., 1987. Efferent connections of the prefrontal cortex of echidna (Tachyglossus aculeatus). Brain Behav. Evol. 30, 321–327.

Elston, G.N., Manger, P.R., Pettigrew, J.D., 1999. Morphology of pyramidal neurons in cytochrome oxidase modules of the S-I bill representation of the platypus. Brain Behav. Evol. 53, 87–101.

Fjällbrant, T.T., Manger, P.R., Pettigrew, J.D., 1998. Some related aspects of platypus electroreception: temporal integration behaviour, electroreceptive thresholds and directionality of the bill acting as an antenna. Philos. Trans. R. Soc. Lond. Ser. B. Biol. Sci. 353, 1211–1219.

Gates, G.R., 1978. Vision in the monotreme echidna. In: Augee, M.L. (Ed.), Monotreme Biology: The Australian Zoologist Special SymposiumThe Australian Zoologist, vol. 20, pp. 147–169.

Grant, T.R., Carrick, F.N., 1978. Some aspects of the ecology of the platypus, Ornithorhynchus anatinus, in the upper Shoalhaven river, New South Wales. Aust. Zool. 20, 181–199.

Gregory, J.E., Iggo, A., McIntyre, A.K., Proske, U., 1988. Receptors in the bill of the platypus. J. Physiol. Lond. 400, 349–366.

Gregory, J.E., Iggo, A., McIntyre, A.K., Proske, U., 1989. Responses of electroreceptors in the snout of the echidna. J. Physiol. Lond. 414, 521–538.

Griffiths, M., 1978. The Biology of the Monotremes. Academic, New York, p. 17.

Grus, W.E., Shi, P., Zhang, J., 2007. Largest vertebrate vomeronasal type 1 receptor gene repertoire in the semiaquatic platypus. Mol. Biol. Evol. 24, 2153–2157.

Harris, R.L., Davies, N.W., Nicol, S.C., 2012. Chemical composition of odorous secretions in the Tasmanian short-beaked echidna (Tachyglossus aculeatus setosus). Chem. Senses 37, 819–836.

Hassiotis, M., Ashwell, K.W.S., 2003. Neuronal classes in the isocortex of a monotreme, the Australian echidna (Tachyglossus aculeatus). Brain Behav. Evol. 61, 6–27.

Hassiotis, M., Paxinos, G., Ashwell, K.W.S., 2003. The anatomy of the cerebral cortex of the echidna (Tachyglossus aculeatus). Comp. Biochem. Physiol. Part A. Mol. Integr. Physiol. 136, 827–850.

Hassiotis, M., Paxinos, G., Ashwell, K.W.S., 2004. Cyto– and chemoarchitecture of the cerebral cortex of the Australian echidna (Tachyglossus aculeatus). I. Areal organization. J. Comp. Neurol. 475, 493–517.

Hawkins, M., Battaglia, A., 2009. Breeding behaviour of the platypus (Ornithorhynchus anatinus) in captivity. Aust. J. Zool. 57, 283–293.

Helgen, K.M., Portela Miguez, R., Kohen, J., Helgen, L., 2012. Twentieth century occurrence of the long-beaked echidna Zaglossus bruijnii in the Kimberley region of Australia. ZooKeys 255, 103–132.

Hines, M., 1929. The brain of Ornithorhynchus anatinus. Philos. Trans. R. Soc. Ser. B. Biol. Sci. 217, 155–288.

Holland, N., Jackson, S.M., 2002. Reproductive behaviour and food consumption associated with the captive breeding of platypus (Ornithorhynchus anatinus). J. Zool. Lond. 256, 279–288.

Johnson, J.I., Kirsch, J.A.W., Switzer III, R.C., 1982. Phylogeny through brain traits: fifteen characters which adumbrate mammalian genealogy. Brain Behav. Evol. 20, 72–83.

Johnson, J.I., Switzer III, R.C., Kirsch, J.A.W., 1982. Phylogeny through brain traits: the distribution of categorizing characters in contemporary mammals. Brain Behav. Evol. 20, 97–117.

Johnson, J.I., Kirsch, J.A.W., Reep, R.L., Switzer III, R.C., 1994. Phylogeny through brain traits: more characters for the analysis of mammalian evolution. Brain Behav. Evol. 43, 319–347.

Jones, E.G., 2007. The Thalamus, second ed. Cambridge, Cambridge, p. 1385.

Kriegstein, A., Noctor, S., Martinez-Cerdeno, V., 2006. Patterns of neural stem and progenitor cell division may underlie evolutionary cortical expansion. Nat. Rev. Neurosci. 7, 883–890.

Krubitzer, L., Manger, P., Pettigrew, J., Calford, M., 1995. Organization of somatosensory cortex in monotremes. In search of the prototypical plan. J. Comp. Neurol. 351, 261–306.

Lambeth, L., Blunt, M.J., 1975. Electron–microscopic study of monotreme neuroglia. Acta Anat. Basel 93, 115–125.

Lende, R.A., 1964. Representation in cerebral cortex of primitive mammal. Sensorimotor, visual and auditory fields in echidna (Tachyglossus aculeatus). J. Neurophysiol. 27, 37–48.

Manger, P.R., Hughes, R.L., 1992. Ultrastructure and distribution of epidermal sensory receptors in the beak of the echidna, Tachyglossus aculeatus. Brain Behav. Evol. 40, 287–296.

Manger, P.R., Pettigrew, J.D., 1996. Ultrastructure, number, distribution and innervation of electroreceptors and mechanoreceptors in the bill skin of the platypus, Ornithorhynchus anatinus. Brain Behav. Evol. 48, 27–54.

Manger, P.R., Calford, M.B., Pettigrew, J.D., 1996. Properties of electrosensory neurons in the cortex of the platypus (Ornithorhynchus anatinus): implications for processing of electrosensory stimuli. Proc. R. Soc. Lond. Ser. B Biol. Sci. 263, 611–617.

Manger, P.R., Collins, R., Pettigrew, J.D., 1997. Histological observations on presumed electroreceptors and mechanoreceptors in the beak skin of the long–beaked echidna. Zaglossus bruijnii. Proc. R. Soc. Lond. Ser. B. Biol. Sci. 264, 165–172.

Morrow, G., Nicol, S.C., 2009. Cool sex? Hibernation and reproduction overlap in the echidna. PLoS One 4, e6070.

Morrow, G., Andersen, N.A., Nicol, S.C., 2009. Reproductive strategies of the short–beaked echidna – a review with new data from a long–term study on the Tasmanian subspecies (Tachyglossus aculeatus setosus). Aust. J. Zool. 57, 275–282.

Musser, A.M., Archer, A., 1998. New information about the skull and dentary of the Miocene platypus Obdurodon dicksoni, and a discussion of ornithorhynchid relationships. Philos. Trans. R. Soc. Lond. Ser. B. Biol. Sci. 353, 1063−1079.

Musser, A.M., 2003. Review of the monotreme fossil record and comparison of paleontological and molecular data. Comp. Biochem. Physiol. Part A. Mol. Integr. Physiol. 136, 927−942.

Musser, A.M., 2013. Classification and evolution of the monotremes. In: Ashwell, K.W.S. (Ed.), Neurobiology of Monotremes: Brain Evolution in Our Distant Mammalian Cousins. CSIRO Publishing, Collingwood, pp. 1−16.

Nicol, S.C., 2013. Behaviour and ecology of monotremes. In: Ashwell, K.W.S. (Ed.), Neurobiology of Monotremes: Brain Evolution in Our Distant Mammalian Cousins. CSIRO, Collingwood, pp. 17−30.

Pascual, R., Archer, M., Ortiz Jaureguizar, E., Prado, J.L., Godthelp, H., Hand, S.J., 1992. The first non-Australian monotreme: an early Paleocene South American platypus (Monotremata: Ornithorhynchidae). In: Augee, M.L. (Ed.), Platypus and Echidnas. Royal Zoological Society of New South Wales, Sydney, pp. 1−14.

Pettigrew, J.D., Manger, P.R., Fine, S.L.B., 1998. The sensory world of the platypus. Philos. Trans. R. Soc. Ser. B. Biol. Sci. 353, 1199−1210.

Pettigrew, J.D., 1999. Electroreception in monotremes. J. Exp. Biol. 202, 1447−1454.

Phillips, M.J., Bennett, T.H., Lee, M.S.Y., 2009. Molecules, morphology, and ecology indicate a recent, amphibious ancestry for echidnas. Proc. Natl. Acad. Sci. U.S.A. 106, 17089−17094.

Rich, T.H., Vickers-Rich, P., Trusler, P., Flannery, T.F., Cifelli, R.L., Constantine, A., 2001. Monotreme nature of the Australian early Cretaceous mammal *Teinolophos trusleri*. Acta Paleontol. Pol. 46, 113−118.

Rismiller, P.D., McKelvey, M.W., 2003. Body mass, age and sexual maturity in short−beaked echidnas, *Tachyglossus aculeatus*. Comp. Biochem. Physiol. Part A. Mol. Integr. Physiol. 136, 851−865.

Rowe, T., Rich, T.H., Vickers-Rich, P., Springer, M., Woodburne, M.O., 2008. The oldest platypus and its bearing on divergence timing of the platypus and echidna clades. Proc. Natl. Acad. Sci. U.S.A. 105, 1238−1242.

Rowe, M.J., 1990. Organization of the cerebral cortex in monotremes and marsupials. In: Jones, E.G., Peters, A. (Eds.), Cerebral Cortex, vol. 8B. Plenum, New York, pp. 263−334.

Saunders, J.C., Teague, J., Slonim, D., Pridmore, P.A., 1971. A position habit in the monotreme *Tachyglossus aculeatus* (the spiny anteater). Aust. J. Psychol. 23, 47−51.

Scheich, H., Langner, G., Tidemann, C., Coles, R.B., Guppy, A., 1986. Electroreception and electrolocation in platypus. Nature 319, 401−402.

Shi, P., Zhang, J., 2006. Contrasting modes of evolution between vertebrate sweet/umami receptor genes and bitter receptor genes. Mol. Biol. Evol. 23, 292−300.

Taylor, N.G., Manger, P.R., Pettigrew, J.D., Hall, L.S., 1992. Electromyographic potentials of a variety of platypus prey items: an amplitude and frequency analysis. In: Augee, M.L. (Ed.), Platypus and Echidnas. Royal Zoological Society of New South Wales, Sydney, pp. 216−224.

Ulinski, P.S., 1984. Thalamic projections to the somatosensory cortex of the echidna, *Tachyglossus aculeatus*. J. Comp. Neurol. 229, 153−170.

Watson, C.R.R., Provis, J.M., Bohringer, R., 1977. The subdivisions of the trigeminal nucleus of the platypus, *Ornithorhynchus anatinus*: a comparative study. J. Anat. 124, 533. Note to Publisher: this is a single page abstract.

Welker, W., Lende, R.A., 1980. Thalamocortical relationships in echidna (*Tachyglossus aculeatus*). In: Ebbesson, S.O.E. (Ed.), Comparative Neurology of the Telencephalon. Plenum, New York, pp. 449−481.

Woodburne, M.O., Tedford, R.H., 1975. The first Tertiary monotreme from Australia. Am. Mus. Novitates 2588, 1−11.

Zilles, K., Armstrong, E., Moser, K.H., Schleicher, A., Stephan, H., 1989. Gyrification in the cerebral cortex of primates. Brain Behav. Evol. 34, 143−150.

Evolution of Flight and Echolocation in Bats

S.J. Sterbing-D'Angelo[1,2], C.F. Moss[2]

[1]University of Maryland, College Park, MD, United States; [2]Johns Hopkins University, Baltimore, MD, United States

18.1 Introduction

Evolutionary pressures result in different sensory specializations that are best suited for an animal to explore its natural environment and exploit its particular ecological niche. In the modality of tactile sensing, for example, rodents have exquisitely sensitive whiskers (Hartmann et al., 2003), subterranean star-nosed moles have tactile fingerlike protrusions on the nose (Catania, 2011), and primates explore and manipulate objects with large hands capable of high tactile acuity (Phillips and Johnson, 1981). Bats (Chiroptera) are one of the largest mammalian groups. With over 1300 species, they account for about 20% of extant mammals (Fenton and Simmons, 2015). Bats are the only mammals with true powered flight. Their remarkable flight capabilities are made possible by an evolutionary modification of the forelimbs and greatly elongated digits to support a thin, flexible flight membrane. This specialized membrane is poised to serve as a tactile organ to detect features of boundary layer airflow and wing stretch during flight. The bat's flexible wing skin is equipped with an array of microscopic hairs that have been shown to modulate flight control (Sterbing-D'Angelo et al., 2011). Bats forage under low-light conditions, mostly from dusk until dawn, and many bat species have developed a sophisticated echolocation system to facilitate orientation and navigation at night. However, not all species of bats echolocate, and those that do use different echolocation signals and strategies.

Also, not all bat species fly. The flightless New Zealand lesser short-tailed bat, *Mystacina tuberculata*, is one of only two bat species to use their arms and legs for a true walking gait while moving on the ground. The other species capable of walking, and even galloping, is the neotropical common vampire bat, *Desmodus rotundus*, which is also capable of nimble flight (Riskin et al., 2006). *M. tuberculata* is the last surviving member of the family Mystacinidae, the only mammalian family endemic to New Zealand, and a member of the Gondwanan superfamily of bats, the Noctilionoidea. In both bat species, the quadrupedal terrestrial locomotion is a secondary, derived condition, as suggested by skeletal and muscular specializations absent in other extant bats (Hand et al., 2009). In the case of *Mystacina*, the lack of mammalian predators in New Zealand is most likely the cause for this adaptation. *Desmodus*, which bites terrestrial mammals' lower extremities to feed on blood, approaches and departs from its prey by walking on the carpal joints and the feet, the latter being necessary because the bats are too heavy to fly after a blood meal up to 60% of their body mass (Breidenstein, 1982).

The fossil record of bats extends over 50 million years, starting in the early Eocene, but remains sparse to date. In particular, the early evolution of the group is poorly understood (Springer et al., 2001; Eick et al., 2005; Teeling et al., 2005; Simmons and Geisler, 1998). Recent genetic studies strongly support the notion that bats are monophyletic (Simmons, 1994; Murphy et al., 2001; Teeling et al., 2000). Also, the evolutionary relationships between extant lineages have been clarified (Springer et al., 2001; Eick et al., 2005; Teeling et al., 2005; Van Den Bussche and Hoofer, 2004). Nevertheless, the bat fossil record is estimated to be only 12% complete at the genus level. Particularly for the large noncholocating Pteropodidae, most of the fossil history is missing. It has been estimated that the vast majority of bats either never left a fossil record or that such record has yet to be discovered (Eiting and Gunnell, 2009).

18.2 Evolution of Bat Flight

Bats evolved later than birds (Jepsen, 1970) and are the only mammals with true powered flight. Whether the earliest bats were gliders or could actively fly remains a mystery, given the absence of transition forms. If protobats were gliders, avian predation, coupled with competition for insect resources, may have driven these protobats to a nocturnal lifestyle, which they share with all extant gliding marsupial and placental mammals, such as gliding squirrels and opossums. One hypothesis proposes that the bat ancestor's webbed forelimbs were first used for gliding to capture nocturnal insects (Speakman, 2001). Most likely this bat precursor was a diurnal frugivorous animal that used gliding as an efficient means of transport from tree to tree. This animal may have begun supplementing its diet with insects encountered while gliding and would have been forced into the nocturnal realm by both competition for insect prey and the predation forces of birds (Speakman, 2001).

Developmental processes suggest morphological transitions in the evolution of flight. In particular, developmentally regulated programmed cell death sculpts the limbs in all vertebrates. Bone morphogenetic proteins (*Bmp*) have been shown to cause apoptosis of the interdigital webbing, leading to isolated digits in both nonaquatic birds such as chickens and the mouse. In waterfowl, such as ducks, *Bmp* antagonists inhibit apoptosis, leading to webbed feet (Weatherbee et al., 2006). The same mechanism is responsible for the development of the bat's wing membranes that are found between digits 2, 3, 4, and 5. Forelimb and hindlimb interdigital tissues both express *Bmp* signaling components, but only hindlimbs undergo interdigital apoptosis in bats. Additional expression patterns of cell-fate genes, eg, sonic hedgehog (*Shh*), fibroblast growth factor 8 (*Fgf8*), and the *Bmp* inhibitor *Gremlin*, results in the retention of interdigital webbing in the bat's forelimb. This expression pattern of the bat seems to be unique across amniotes and is one of the key factors in the evolution of the bat wing (Weatherbee et al., 2006; Nissim et al., 2006; Bandyopadhyay et al., 2006). The processes during embryological development of the dactylopatagium, which derives from embryonic interdigital webbing that is not subject to apoptosis in bats, also promote hair growth (Weatherbee et al., 2006). This finding has important implications, namely that there is no glabrous skin on the bat's hand-wing and that both the dorsal and ventral wing surfaces are covered with fine, microscopical, airflow-sensing hairs that are involved in sensorimotor flight control (Sterbing-D'Angelo et al., 2011; Marshall et al., 2015).

During the early stages of limb development the bat's forelimb buds are very similar to those of equivalently staged mouse embryos (Cretekos et al., 2005). Later, however, the shape and size of the bat forelimb bud expands in posterior-distal direction (Chen et al., 2005;

Giannini et al., 2006; Tokita, 2006). In addition, the digital rays of both the hand- and foot-plates are evident during these early stages in the bat, while only appearing later in other mammals (Tokita, 2006; Lawrence, 1991). By stage 20 of development the radius and ulna, as well as the carpals and digits, are significantly longer than the humerus in the developing bat forelimb (Cretekos et al., 2005). This increase in length corresponds to an increase in size of the hypertrophic zone, the most mature area of the developing bone, in the growing metacarpals (Sears et al., 2006). The size of this mature zone relative to the length of the metacarpals is twice as large in the bat when compared to an equivalently staged mouse limb (Sears et al., 2006). Apparently, the bat forelimb elongates via an increase in the number of cartilage cells in comparison to the hindlimb. It should be noted that the relative length of bat forelimb digits has not changed in 50 million years. *Icaronycteris index*, the oldest known bat fossil, looks remarkably similar to modern bats (Sears et al., 2006).

The bat wing does not exclusively consist of interdigital webbing. Indeed, the largest partition of the wing, the plagiopatagium, connects the flank of the bat with digit 5, the caudal upper arm (humerus), forearm (ulna), and hindlimb (Fig. 18.1). Furthermore, bats with a tail also have an additional flight membrane, the uropatagium, which connects the tail with the caudal hindlimbs. During embryogenesis, the plagiopatagium originates from the flank of the bat (Cretekos et al., 2005). Consistently, the target of sensory innervation of the plagiopatagium is in the thoracic spinal dorsal root ganglia (DRGs), and not in cervical DRGs (Marshall et al., 2015). The unique embryology of the bat, the development of flight membranes through several processes on molecular level, and the resulting neuroanatomical findings of wing membrane innervation have to be regarded as the result of the evolutionary processes for flight development.

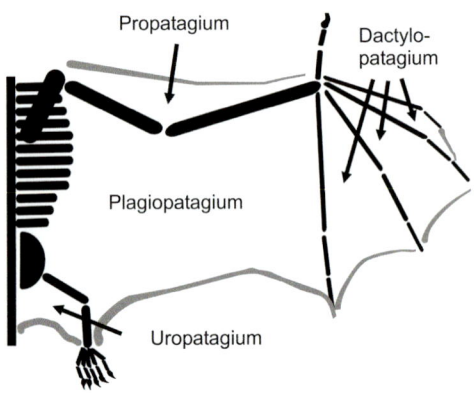

FIGURE 18.1 Schematic of bat wing.

Comparative data can guide insights into the evolution of flight. Besides bats, retained interdigital webbing that is used for gliding can be found in extant reptiles, the "flying" geckos, eg, *Ptychozoon kuhli*. These geckos also grow cutaneous flaps from the flanks of their trunks. The flaps play a functional role for gliding by reducing gliding velocity and angle. Among extant reptiles, varying degrees of aerial locomotion have been described (Russell, 1979; Losos et al., 1989). Best known for gliding are the lizards, eg, the agamid *Draco* (Colbert, 1967), and the geckonid *Ptychozoon* (Marcellini and Keefer, 1976). The wing of *Draco* is an extension of the flank of the animal (Fig. 18.2A). It is well supported by elongated ribs and associated skeletal musculature, and wing extension and folding is under voluntary control (Colbert, 1967). In contrast, the multiple "wings" of *Ptychozoon*, ie, the cutaneous flaps and the interdigital webbing, have neither skeletal elements nor striated muscle (Fig. 18.2B): these wings are not under voluntary

control but passively respond to air resistance (Russell, 1979; Russell et al., 2001).

Gliding mammals, eg, the marsupial sugar glider (*Petaurus breviceps*) and the many gliding squirrels (15 genera of the tribe Pteromyini; Thorington Jr et al., 2005), neither have interdigital webbing, nor elongated digits, and their gliding membrane spans from the flank to the carpal joints of the limb (Fig. 18.2C).

Most similar to the bat wing is the wing of a recently described fossil dinosaur of the Scansoriopterygidae group, which is of small size, and possesses a longer third finger than other theropods. This newly described species has a long, rodlike (styliform) bone in each wrist, not seen in any other dinosaur, but with similarities to other flying and gliding animals, including bats, flying squirrels, and pterosaurs. Its wings appear to be featherless membranes, similar to the bat (Fig. 18.2D). Whether this animal could actively fly or was a glider cannot be determined for sure, because the fossil is incomplete (Xu et al., 2015).

18.3 Evolution of Bat Echolocation

Cochlear size is correlated with echolocation function in bats: nonecholocating bats tend to have shorter basilar membranes and fewer cochlear turns than bats that use laryngeal echolocation and also the stylohyal and tympanic bones differ (Simmons and Geisler, 1998; Veselka et al., 2010; Novacek, 1985; Davies et al., 2013). While there is agreement that many Eocene bats were capable of echolocation, the transitional form, *Onychonycteris finneyi*, dated at 52.5 Mya, had a cochlea comparable in size to the cochlea of nonecholocating fruit bats and smaller than that found in most echolocating bats. Veselka et al. (2010), however, found that *O. finneyi* may have used laryngeal echolocation because its stylohyal bones may have articulated with its tympanic bones like in modern echolocating bats. Its flight apparatus indicated its ability for powered flight (Simmons et al., 2008). Hence, the current fossil record still leaves the question open whether early Eocene bats developed flight earlier than echolocation.

Not all extant bats echolocate. Bats in the family Pteropodidae (Old World fruit bats, eg, flying foxes) do not use laryngeal echolocation, and only pteropodid bats of one genus, *Rousettus*, echolocate by tongue clicking. In contrast, all echolocating nonpteropodid bats produce sonar calls with their larynx. Laryngeal echolocation calls vary significantly between different bat families that apparently use different echolocation strategies as they forage under different conditions (Denzinger and Schnitzler, 2013).

The question has been raised whether echolocation evolved only once or several times independently. The

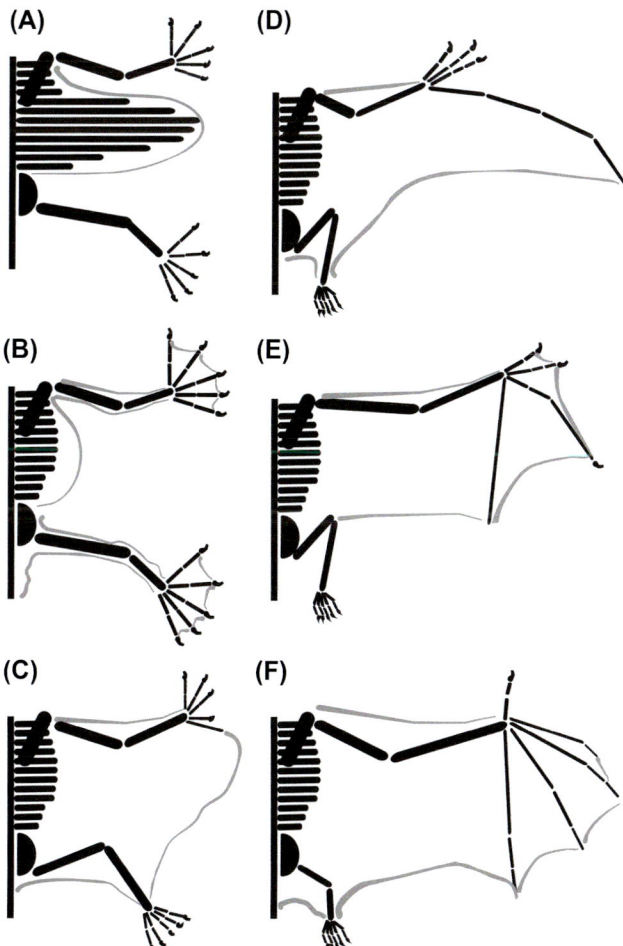

FIGURE 18.2 Comparison of forelimb and flight membrane anatomy of gliding and flying vertebrates. (A) "Flying" lizard (*Draco*), (B) "flying" gecko (*Ptychozoon*), (C) gliding squirrel, (D) pterosaur, (E) *Yi qi*, (F) bat.

phylogenetic tree of bats has been subject of controversial discussion. Historically, two suborders of bats (Chiroptera) were considered, the Megachiroptera, which were comprised of all Pteropodidae including the genus *Rousettus*, and the Microchiroptera, which included all laryngeal echolocators. Based on molecular data, Teeling et al. (2005) proposed a revised set of two suborders: grouping the nonecholocating Pteropodidae with the laryngeal echolocating superfamily Rhinolophoidea (now including the family of Rhinopomatidae which were historically part of the superfamily Emballonuroidea) in the suborder Yinpterochiroptera, and leaving all other echolocating lineages together in the monophyletic suborder Yangochiroptera. This grouping suggests that either laryngeal echolocation evolved once in the ancestor of bats and later had been lost in the Pteropodids, with echolocation evolving secondarily by tongue clicking in cave roosting bats in the genus *Rousettus* (Mohres and Kulzer, 1956; Yovel et al., 2011), or that echolocation evolved twice or even more times independently in the laryngeal echolocating bats.

The laryngeal echolocating Yinpterochiroptera, eg, the horseshoe bats (Rhinolophidae) and Old World leaf-nosed bats (Hipposideridae), have a particularly specialized echolocation system involving the emission of constant frequency (CF, Fig. 18.3E) calls enabling the bats to classify of fluttering insect prey based on the spectrotemporal features of "acoustic glints" in the echoes (von der Emde and Schnitzler, 1990), combined with broadband sweeps for localizing targets and the ability to adjust the frequency of emitted calls to compensate for Doppler shifts induced by the bat's own flight velocity (Schnitzler, 1968; Trappe and Schnitzler, 1982; Hiryu et al., 2005). Some CF/FM bats have a specialized area on the cochlear partition in which the frequency range of the CF component is overrepresented by the amount of sensory hair cells (acoustic fovea). If the bat experiences Doppler shifts, it adjusts the frequency of the emitted echolocation call so that the returning echo falls on this acoustic fovea. The other 15 families of bats that use laryngeal echolocation, including the recently proposed Miniopteridae (Miller-Butterworth et al., 2007) and Cistugidae (Lack et al., 2010), were pooled in the suborder Yangochiroptera (Teeling et al., 2000, 2005; Jones and Teeling, 2006).

Recent studies revealed, however, that there are conflicting findings between molecular and morphological data sets, regarding the issue of the evolution of echolocation. The recent and surprising finding that combined large-scale phenomic and gene data sets recover a monophyletic group of echolocating bats (O'Leary et al., 2013) warrants further study. Teeling et al. (2012) point out that ultimately, a single origin of echolocation followed by secondary loss in pteropodids would be better supported if fossilized ancestral pteropodids with anatomical characteristics of echolocation were found, or if pseudogenization of genes known to be specific for echolocation could be identified in nonecholocating taxa. This is challenging given that there is a poor fossil record for pteropodids, anatomical features may become damaged during fossilization and also for the reason that genes associated with echolocation are likely to be variants of genes fundamental to vocalization and hearing in more general contexts. However, it is only through the integration of information gathered in these different fields that the evolution of echolocation in bats will be elucidated.

Acknowledgments

We wish to acknowledge the following grants, which supported research in the lab and the preparation of this article: Human Frontiers Science Program, RGP0040; ONR, N00014-12-1-0339; AFOSR FA9550-14-1-0398, and NSF Collaborative Research in Computational Neuroscience, IOS1460149.

References

Bandyopadhyay, A., Tsuji, K., Cox, K., Harfe, B.D., Rosen, V., Tabin, C.J., 2006. Genetic analysis of the roles of BMP2, BMP4, and BMP7 in limb patterning and skeletogenesis. PLoS Genet. 2 (12), e216.

Breidenstein, C.P., 1982. Digestion and assimilation of bovine blood by a vampire bat (*Desmodus rotundus*). J. Mammal. 482—484.

Catania, K.C., 2011. The sense of touch in the star-nosed mole: from mechanoreceptors to the brain. Philos. Trans. R. Soc. Biol. Sci. 366 (1581), 3016—3025.

Chen, C.H., Cretekos, C.J., Rasweiler, J.J., Behringer, R.R., 2005. Hoxd13 expression in the developing limbs of the short-tailed fruit bat, *Carollia perspicillata*. Evol. Dev. 7 (2), 130—141.

Colbert, E.H., 1967. Adaptations for gliding in the lizard *Draco*. Am. Mus. Novit. 2283, 1—20.

Cretekos, C.J., Weatherbee, S.D., Chen, C.H., Badwaik, N.K., Niswander, L., Behringer, R.R., Rasweiler, J.J., 2005. Embryonic staging system for the short-tailed fruit bat, *Carollia perspicillata*, a model organism for the mammalian order Chiroptera, based upon timed pregnancies in captive-bred animals. Dev. Dyn. 233 (3), 721—738.

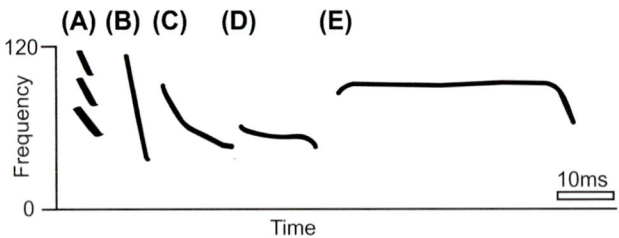

FIGURE 18.3 Echolocation calls of extant bat species. (A) Multiharmonic frequency-modulated (FM) sweep typical for Phyllostomidae, (B) short FM sweep (Vespertilionidae), (C) "Doppler-resistant" longer FM sweep (Vespertilionidae), (D) variable constant frequency (CF)/FM (Mormoopidae), (E) long CF/FM (Rhinolophoidae). It should be noted that echolocation calls can be quite variable in each individual, depending on the environment and foraging situation.

Davies, K.T.J., Maryanto, I., Rossiter, S.J., 2013. Evolutionary origins of ultrasonic hearing and laryngeal echolocation in bats inferred from morphological analyses of the inner ear. Front. Zool. 10, 2–15.

Denzinger, A., Schnitzler, H.U., 2013. Bat guilds, a concept to classify the highly diverse foraging and echolocation behaviors of microchiropteran bats. Front. Physiol. 4 (164), 1–15.

Eick, G.N., Jacobs, D.S., Matthee, C.A., 2005. A nuclear DNA phylogenetic perspective on the evolution of echolocation and historical biogeography of extant bats (Chiroptera). Mol. Biol. Evol. 22, 1869–1886.

Eiting, T.P., Gunnell, G.F., 2009. Global completeness of the bat fossil record. J. Mamm. Evol. 16 (3), 151–173.

Fenton, M.B., Simmons, N.B., 2015. Bats: A World of Science and Mystery. University of Chicago Press.

Giannini, N., Goswami, A., Sánchez-Villagra, M.R., 2006. Development of integumentary structures in *Rousettus amplexicaudatus* (Mammalia: Chiroptera: Pteropodidae) during late-embryonic and fetal stages. J. Mammal. 87 (5), 993–1001.

Hand, S.J., Weisbecker, V., Beck, R.M., Archer, M., Godthelp, H., Tennyson, A.J., Worthy, T.H., 2009. Bats that walk: a new evolutionary hypothesis for the terrestrial behaviour of New Zealand's endemic mystacinids. BMC Evol. Biol. 9 (1), 169.

Hartmann, M.J., Johnson, N.J., Towal, R.B., Assad, C., 2003. Mechanical characteristics of rat vibrissae: resonant frequencies and damping in isolated whiskers and in the awake behaving animal. J. Neurosci. 23 (16), 6510–6519.

Hiryu, S., Katsura, K., Lin, L.K., Riquimaroux, H., Watanabe, Y., 2005. Doppler-shift compensation in the Taiwanese leaf-nosed bat (*Hipposideros terasensis*) recorded with a telemetry microphone system during flight. J. Acoust. Soc. Am. 118 (6), 3927–3933.

Jepsen, G.L., 1970. Bat origins and evolution. In: Wimsatt, W.A. (Ed.), Biology of Bats, vol. 1. Academic Press, London, pp. 1–64.

Jones, G., Teeling, E.C., 2006. The evolution of echolocation in bats. Trends Ecol. Evol. 21 (3), 149–156.

Lack, J.B., Wilkinson, J.E., Bussche, R.A.V.D., 2010. Range-wide population genetic structure of the pallid bat (*Antrozous pallidus*)-incongruent results from nuclear and mitochondrial DNA. Acta Chiropterol. 12 (2), 401–413.

Lawrence, M.A., 1991. Biological observations on a collection of new Guinea *Syconycteris australis* (Chiroptera, Pteropodidae) in the American Museum of natural history. Am. Mus. Novit. 3024, 1–27.

Losos, J.B., Papenfuss, T.J., Macay, J.R., 1989. Correlates of sprinting, jumping, and parachuting performance in the butterfly lizard, *Leiolepis belliani*. J. Zool. Lond. 217, 559–568.

Marcellini, D.L., Keefer, T.E., 1976. Analysis of gliding behavior in *Ptychozoon lionatum* (Reptilia: Gekkonidae). Herpetologica 32, 362–366.

Marshall, K.L., Chadha, M., Sterbing-D'Angelo, S.J., Moss, C.F., Lumpkin, E.A., 2015. Somatosensory substrates of flight control in bats. Cell Rep. 11 (6), 851–858.

Miller-Butterworth, C.M., Murphy, W.J., O'Brien, S.J., Jacobs, D.S., Springer, M.S., Teeling, E.C., 2007. A family matter: conclusive resolution of the taxonomic position of the long-fingered bats. *Miniopterus*. Mol. Biol. Evol. 24 (7), 1553–1561.

Mohres, F.P., Kulzer, E., 1956. Ueber die Orientierung der Flughunde (Chiroptera Pteropodidae). Zeitschr. Vergl. Physiol. 38, 1–29.

Murphy, W.J., Eizirik, E., Johnson, W.E., Zhang, Y.P., Ryder, O.A., O'Brien, S.J., 2001. Molecular phylogenetics and the origin of placental mammals. Nature 409, 614–618.

Nissim, S., Hasso, S.M., Fallon, J.F., Tabin, C.J., 2006. Regulation of Gremlin expression in the posterior limb bud. Dev. Biol. 299 (1), 12–21.

Novacek, M.J., 1985. Evidence for echolocation in the oldest known bats. Nature 315, 140–141.

O'Leary, M.A., Bloch, J.I., Flynn, J.J., Gaudin, T.J., Giallombardo, A., Giannini, N.P., Goldberg, S.L., Kraatz, B.P., Luo, Z.X., Meng, J.,

Ni, X., 2013. The placental mammal ancestor and the post−K-Pg radiation of placentals. Science 339 (6120), 662–667.

Phillips, J.R., Johnson, K.O., 1981. Tactile spatial resolution. II. Neural representation of bars, edges, and gratings in monkey primary afferents. J. Neurophysiol. 46 (6), 1192–1203.

Riskin, D.K., Parsons, S., Schutt, W.A., Carter, G.G., Hermanson, J.W., 2006. Terrestrial locomotion of the New Zealand short-tailed bat *Mystacina tuberculata* and the common vampire bat *Desmodus rotundus*. J. Exp. Biol. 209 (9), 1725–1736.

Russell, A.P., 1979. The origin of parachuting locomotion in gekkonid lizards (Reptilia: Gekkonidae). Zool. J. Linn. Soc. 65 (3), 233–249.

Russell, A.P., Dijkstra, L.D., Powell, G.L., 2001. Structural characteristics of the patagium of *Ptychozoon kuhli* (Reptilia:Gekkonidae) in relation to parachuting locomotion. J. Morphol. 247, 252–263.

Schnitzler, H.U., 1968. Die Ultraschall-Ortungslaute der Hufeisen-Fledermäuse (Chiroptera-Rhinolophidae) in verschiedenen Orientierungssituationen. Z. Vgl. Physiol. 57 (4), 376–408.

Sears, K.E., Behringer, R.R., Rasweiler, J.J., Niswander, L.A., 2006. Development of bat flight: morphologic and molecular evolution of bat wing digits. Proc. Natl. Acad. Sci. U.S.A. 103 (17), 6581–6586.

Simmons, N.B., 1994. The case for chiropteran monophyly. Am. Mus. Novitat. 3103, 1–54.

Simmons, N.B., Geisler, J.H., 1998. Phylogenetic relationships of *Icaronycteris*, *Archaeonycteris*, *Hassianycteris*, and *Palaeochiropteryx* to extant bat lineages, with comments on the evolution of echolocation and foraging strategies in Microchiroptera. Bull. Am. Mus. Nat. Hist. 235, 1–182.

Simmons, N.B., Seymour, K.L., Habersetzer, J., Gunnell, G.F., 2008. Primitive Early Eocene bat from Wyoming and the evolution of flight and echolocation. Nature 451 (7180), 818–821.

Speakman, J.R., 2001. The evolution of flight and echolocation in bats: another leap in the dark. Mammal. Rev. 31 (2), 111–130.

Springer, M.S., Teeling, E.C., Madsen, O., Stanhope, M.J., de Jong, W.W., 2001. Integrated fossil and molecular data reconstruct bat echolocation. Proc. Natl. Acad. Sci. U.S.A. 98, 6241–6246.

Sterbing-D'Angelo, S., Chadha, M., Chiu, C., Falk, B., Xian, W., Barcelo, J., Zook, J.M., Moss, C.F., 2011. Bat wing sensors support flight control. Proc. Natl. Acad. Sci. U.S.A. 108 (27), 11291–11296.

Teeling, E.C., Scally, M., Kao, D.J., Romagnoli, M.L., Springer, M.S., Stanhope, M.J., 2000. Molecular evidence regarding the origin of echolocation and flight in bats. Nature 403, 188–192.

Teeling, E.C., Springer, M.S., Madsen, O., Bates, P., O'brien, S.J., Murphy, W.J., 2005. A molecular phylogeny for bats illuminates biogeography and the fossil record. Science 307, 580–584.

Teeling, E.C., Dool, S., Springer, M.S., 2012. Phylogenies, fossils and functional genes: the evolution of echolocation in bats. In: Gunnell, G.F., Simmons, N.B. (Eds.), Evolutionary History of Bats: Fossils, Molecules and Morphology. Cambridge University Press, Cambridge, pp. 1–22.

Thorington Jr., R.W., Schennum, C.E., Pappas, L.A., Pitassy, D., 2005. The difficulties of identifying flying squirrels (Sciuridae: Pteromyini) in the fossil record. J. Vert. Paleontol. 25 (4), 950–961.

Tokita, M., 2006. Normal embryonic development of the Japanese pipistrelle, *Pipistrellus abramus*. Zoology 109 (2), 137–147.

Trappe, M., Schnitzler, H.U., 1982. Doppler-shift compensation in insect-catching horseshoe bats. Naturwissenschaften 69 (4), 193–194.

Van Den Bussche, R.A., Hoofer, S.R., 2004. Phylogenetic relationships among recent chiropteran families and the importance of choosing appropriate out-group taxa. J. Mamm. 85, 321–330.

Veselka, N., McErlain, D.D., Holdsworth, D.W., Eger, J.L., Chhem, R.K., Mason, M.J., Mason, M.J., Brain, K.L., Faure, P.A., Fenton, M.B., 2010. A bony connection signals laryngeal echolocation in bats. Nature 463 (7283), 939–942.

von der Emde, G., Schnitzler, H.U., 1990. Classification of insects by echolocating greater horseshoe bats. J. Comp. Physiol. 167 (3), 423–430.

Weatherbee, S.D., Behringer, R.R., Rasweiler, J.J., Niswander, L.A., 2006. Interdigital webbing retention in bat wings illustrates genetic changes underlying amniote limb diversification. Proc. Natl. Acad. Sci. U.S.A. 103, 15103–15107.

Xu, X., Zheng, X., Sullivan, C., Wang, X., Xing, L., Wang, Y., Zhang, X., O'Connor, J.K., Zhang, F., Pan, Y., 2015. A bizarre Jurassic maniraptoran theropod with preserved evidence of membranous wings. Nature 521 (7550), 70–73.

Yovel, Y., Geva-Sagiv, M., Ulanovsky, N., 2011. Click-based echolocation in bats: not so primitive after all. J. Comp. Physiol. 197 (5), 515–530.

19

Carnivoran Brains: Effects of Sociality on Inter- and Intraspecific Comparisons of Regional Brain Volumes

S.T. Sakai[1], B.M. Arsznov[2]

[1]Michigan State University, East Lansing, MI, United States; [2]Minnesota State University Mankato, Mankato, MN, United States

19.1 Introduction

Much of what we know about brain evolution is based on studies in primate species. Relatively few studies have examined the relationship between the brain and behavioral traits in the other mammalian orders. The order Carnivora offers a unique opportunity to examine how brain structure varies as a function of behavioral traits, as it consists of over 270 species exhibiting a range of lifestyles and diverse ecological niches. The social brain hypothesis (Dunbar, 1992, 1998) proposes that the neural processing of social information selectively drives the evolution of large brains, particularly the frontal cortex. In the following, we will examine this hypothesis by comparing brain differences at both the family and species level in carnivorans using the virtual endocast method.

19.2 Factors Related to Brain Size Variation

19.2.1 Principle of Proper Mass

Jerison (1973) first proposed the principle of proper mass, the idea that specialized functions correspond to relatively enlarged brain regions devoted to the information-processing demands of that function. Comparative studies of sensory and motor cortex provide substantial support for this principle. For example, Welker and Seidenstein (1959) first demonstrated that the raccoon, a species noted for its dexterous use of the forepaws, possesses a correspondingly enlarged hand representation in the primary somatosensory cortex. This finding provided the impetus for many other studies demonstrating that specialized functions, such as whisking behavior in rodents (mouse somatosensory cortical barrels) (Woolsey and Van der Loos, 1970) and the fine manipulative skills in monkeys (macaque primary motor cortex) (Plautz et al., 2000), are accompanied by an increase in the cortical volume devoted to the representation of that behavior. This idea has been extended beyond sensory and motor processing to include a broad range of complex cognitive skills correlated with increasing relative brain volumes. More recent studies have inquired whether other factors including environmental and behavioral complexity, including sociality, positively correlate with brain size (Dunbar, 1998), and also whether these factors selectively drive the evolution of large brains (Iwaniuk et al., 1999; Finarelli and Flynn, 2009).

19.2.2 Social Brain Hypothesis

One proposal to explain the evolution of large brains is the "social brain" hypothesis. This hypothesis asserts that the relatively large brain size found in primates is due to factors associated with living in large complex social environments (Jolly, 1966; Humphrey, 1976; Byrne and Whiten, 1988; Dunbar, 2003). Variables such as social group size (Sawaguchi and Kudo, 1990; Dunbar, 1992, 1998; Barton, 1996; Barton and Dunbar, 1997), grooming clique size (Kudo and Dunbar, 2001), the extent to which social skills are used in male mating strategies (Pawlowski et al., 1998), the frequency of tactical deception (Byrne, 1995), and the frequency of social play (Lewis, 2001) have been used as measures of social complexity and are found to positively correlate with increases in relative brain size. The term relative brain size has been used in various ways and requires further explanation. Because absolute brain volume scales allometrically with body size, one way for controlling the influence of body size on brain size is to use a measure of relative brain size. Relative brain size has been measured as (1) a fractional volume such as total brain volume divided by body weight, body length, or a skull measure (Swanson et al., 2012; Radinsky, 1967) or as (2) a residual deviation from a regression of brain size plotted against body size, sometimes adjusted for specific genera or family (Clutton-Brock and Harvey, 1980; Gittleman, 1986) or as (3) the deviation of observed brain volume as a function of expected brain volume for a given body weight (encephalization) (Jerison, 1973; Finarelli and Flynn, 2009). Which measure of relative brain size is most appropriate in predicting cognitive and behavioral abilities or is the most useful in comparative studies is arguable (Jerison, 1973; Deacon, 1990; Dunbar, 1998; Deaner et al., 2000, 2007; Clark et al., 2001; Roth and Dicke, 2005; Smaers et al., 2012).

An alternative to the use of the relative brain size measure is to use ratio or proportional analysis of regional brain volume relative to total brain volume. Ratio analysis can avoid some of the problems associated with using body mass as a baseline measure (Dunbar, 1998; Clark et al., 2001). In a comparison of multiple measures of brain size including residuals of brain volume and body mass and absolute brain volume, the strongest predictor of social group size among primates was found to be the ratio of neocortex to the rest of brain volume (Dunbar, 1998). In primates, the greatest enlargement in brain size is due to an increase in the amount of neocortex relative to the brain as a whole (Finlay and Darlington, 1995; Barton and Harvey, 2000; Clark et al., 2001). In particular, a strong association between sociality and frontal lobe volume has been observed (Dunbar, 2011). The frontal cortex is known to mediate complex social behaviors including selection of socially appropriate responses in humans and other mammals (for review: Adolphs, 2001; Amodio and Frith, 2006). These findings suggest that selection pressures related to social information processing may act on functionally discrete areas including the frontal cortex resulting in changes in the relative size of the brain regions (Barton and Harvey, 2000).

19.2.3 Comparative Studies in Carnivorans

The majority of studies examining factors that influence mammalian brain size have focused on primate species (Gould, 1975; Clutton-Brock and Harvey, 1980;

Dunbar and Shultz, 2007), while comparatively few studies have sought to investigate these factors in other mammals (Gittleman, 1986; Marino, 1996; Dunbar and Bever, 1998; Bush and Allman, 2004; Perez-Barberia et al., 2007; Finarelli, 2006; Finarelli and Flynn, 2009; Swanson et al., 2012). Nonprimate species that face the same selection pressures as those seen in primate species should also exhibit corresponding increases in either relative brain size or regional brain volumes. Carnivorans offer an excellent alternative mammalian model in which to further investigate the influence of social complexity on variations in brain size. The order Carnivora includes 13 extant families and over 270 species (Agnarsson et al., 2010) that display variations in brain size as well as variation in social complexity. A broad analysis of three mammalian orders, primates, ungulates, and carnivorans, showed a correlation between relative brain size and sociality (Perez-Barberia et al., 2007), but supporting data have been equivocal. Gittleman (1986) failed to find a relationship between sociality and relative brain size (brain volume relative to body size) in his large study of 7 carnivoran families including 153 species. An analysis of fossil and extant carnivoran species found that the relationship between sociality and encephalization is solely driven by the highly social Canidae family (the doglike family including foxes, wolves, jackals, and others) and that no relationship was found in the families Felidae, the cat family, or Hyaenidae, the hyaena family (Finarelli and Flynn, 2009). A more recent large-scale analysis reported that increased encephalization found in carnivores likely evolved independently, notably in families Canidae, Ursidae (bear family), and Mustelidea (a diverse family including badgers, ferrets, otters, and weasels) (Boddy et al., 2012). Many member species of Ursidae and Mustelidae are solitary. Taken together, these studies suggest that family-level analyses are important to our understanding of processes underlying the evolution of the brain. Study of brain variations in closely related species that differ along a behavioral dimension will help identify common principles underlying the brain and behavior relationship (Striedter et al., 2014). Yet studies examining brain variations at the family level in carnivorans are rare. Most studies in neuroscience are conducted in very few species, and it is difficult to study species outside of the laboratory. Also, when interspecies brain comparisons are made, the results are often based on a small sample pool of individuals. These limitations are due in large part to the lack of available specimens for study. However, an alternative technique for studying brains that capitalizes on the diversity of skeletal material already available in museum collections is the endocast method. Most recently, this method has been successfully applied to the study of a wide variety of species using computed tomography (CT) to create a three-dimensional virtual brain (Swanson et al., 2012). In the following, we describe how this method can be applied to studies of the carnivoran brain.

19.3 The Virtual Endocast

To determine how brain size varies in relation to behavior and to identify the factors that influence the course of brain evolution, accurate and reliable assessment of brain volume is necessary in a broad range of species including large sample sizes of each species. However, comparative brain studies are relatively rare. It is noteworthy that over 75% of neuroscientific research is based on findings from only three species: the rat, mouse, and human (Manger et al., 2008). Moreover, a single data set generated by Stephan and colleagues (Stephan et al., 1981; Baron et al., 1996) is the basis for some of the most influential theories of brain evolution including mosaic evolution of brain structures (Barton and Harvey, 2000; de Winter and Oxnard, 2001), developmental constraints and brain size (late equals large) theory of Finlay and Darlington (1995), the social brain hypothesis (Dunbar, 1998; Shultz and Dunbar, 2007), and the correlation of overall brain size with cognitive ability (Deaner et al., 2007). This comprehensive data set consists of total brain and nuclear volume measurements in 131 species (Stephan et al., 1981; Baron et al., 1996). Reep et al. (2007) expanded this data set to include 29 additional species. The brain measurements were based on analysis of histological sections. However, shrinkage rates due to fixation and histology were estimated to range from 35% to 58% and were noted to vary inconsistently across species and brain structures (Stephan et al., 1981). Moreover, species data were frequently based on very small sample sizes within species (Stephan et al., 1981; Bush and Allman, 2004; Reep et al., 2007). To explore the similarities and differences in brain organization across species, it is critical to supplement and expand the existing brain volume data in a variety of species.

One drawback to comparative brain studies is that intact brain tissue is often not available for the majority of mammalian species. Brain tissue is fragile, requiring fixation prior to histological processing and anatomical evaluation. However, external brain features and volumes have been studied in a variety of species using the endocast method. This method utilizes the impression of the brain made on the internal surface of skull, the endocranium, to create a mold bearing this impression, an endocast. The endocast replicates the surface features of the brain including the gyral and sulcal pattern. This method has been widely used to study brain evolution in extinct and extant species (Radinsky,

1969, 1975; Conroy et al., 1990; Falk et al., 2000; Holloway et al., 2004). Older studies utilized latex molds to create the endocast (Radinsky, 1969; Jerison, 1973), but current studies use CT to create a virtual endocast (Conroy et al., 1990; Falk et al., 2000). Advantages of the CT technique are numerous. The surface impressions left on the endocranial surface are captured in high resolution and are digitally compiled into a three-dimensional reconstruction of the endocast. The endocast can be imaged, sectioned, and volumetrically measured using current CT software leaving the original specimen undamaged. The virtual endocast methodology has advanced our understanding of hominid evolution (Falk et al., 2009; Bienvenu et al., 2011; Neubauer, 2014). Moreover, these methods can be effectively used to study a wide range of species including a large sample size of a single species using specimens already available in natural history museum collections.

19.3.1 Computed Tomography Analysis

The CT technique has been previously used to create virtual endocasts in a variety of species where preservation of the brain is difficult or impossible, but skull specimens are readily available (Arsznov et al., 2010; Sakai et al., 2011a,b; Arsznov and Sakai, 2012). A series of coronal CT images through the skull is analyzed using MIMICS (Materialise, Inc., Ann Arbor, MI, USA) image-processing software. Using thresholds to separate matter of different densities, the bony internal skull surface is isolated from the endocranial air cavity in each coronal section starting rostrally where the cribriform plate forms the floor of the intracranial cavity, and continuing caudally through the foramen magnum. The slices are aligned and stacked to create the three-dimensional reconstruction (see Relevant Website). Finally, algorithms are applied to enhance the image and eliminate uneven surfaces. The surface features of the brain are identifiable on the virtual endocast and compare favorably to whole brain photographs (Fig. 19.1). The gyral and sulcal features including the cruciate sulcus, lateral sulcus, ansate sulcus, and suprasylvian sulcus are clearly identifiable in both the brain photograph and virtual endocast in the raccoon (Fig. 19.1) and lion (Fig. 19.2).

Total endocranial volume can be reliably measured using this method. In humans, endocranial volume exceeds fixed brain volume by 10% (Holloway et al., 2004), but studies in nonhuman primates estimate the endocranial volume exceeds brain mass by 4% (Isler et al., 2008). A large study of 82 bird species reported no significant differences in endocranial volume and brain volume (Iwaniuk and Nelson, 2002). Our own comparison found that endocranial volume exceeded

Raccoon
Procyon lotor

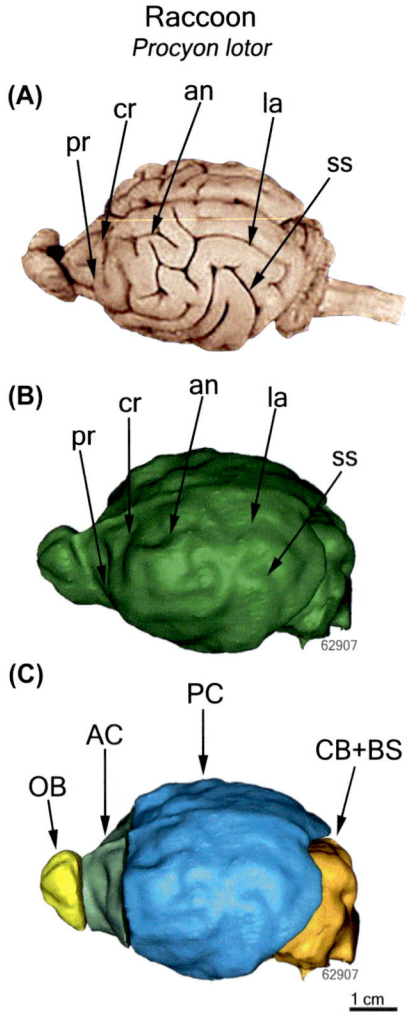

FIGURE 19.1 (A) Dorsolateral views of the whole brain of a raccoon from the Comparative Mammalian Brain Collection (#57-88), (B) virtual endocast reconstructed from a raccoon skull (Michigan State University Museum specimen #62907), and (C) the regional brain subdivisions. The major sulci are noted in both the brain and virtual endocast. *an*, ansate sulcus; *cr*, cruciate sulcus; *la*, lateral sulcus; *pr*, proreal sulcus; *ss*, suprasylvian sulcus. The subdivided endocast in (C) shows the anterior cerebrum (AC—*green*) just caudal to the olfactory bulbs (OB—*yellow*) and rostral to the posterior cerebrum (PC—*blue*). The cerebellum and brain stem (CB + BS—*gold*) is the region just caudal to the cerebellar tentorium and rostral to the foramen magnum.

brain volume as measured from magnetic resonance imaging (MRI) by 3.65% in the lion (Arsznov and Sakai, 2012; Fig. 19.2C and D). Since overall this difference in brain volume is relatively small, the endocast method provides a reasonable measure of brain volume.

19.3.2 Regional Brain Volumes

The virtual endocast method can also be used to obtain regional brain volumetric measurements by using landmarks including bony features and sulcal

patterns to subdivide the model into three regions of interest: anterior cerebrum, posterior cerebrum, and cerebellum/brain stem (Fig. 19.1C).

A reliable measure of frontal cortex in carnivores is necessary to assess the relationship between sociality and frontal cortex volume. Although frontal cortex in primate species is defined as cortex anterior to the central sulcus, a landmark that separates that motor and somatosensory cortex, the central sulcus is absent in carnivores. Instead, in some carnivores, the postcruciate sulcus or dimple, the boundary between motor and somatosensory cortex, varies in both size and shape (Radinsky, 1969). However, the cruciate sulcus is a reliable sulcus in almost all carnivore species (exception: some species in families Viverridae and Nandiniidae) (Radinsky, 1975) and is found to functionally correspond to rostral motor cortex in the cat (Nieoullon and Rispal-Padel, 1976), dog (Gorska, 1974), ferret (McLaughlin et al., 1998), and raccoon (Hardin et al., 1968). Anatomically, this cortical area contains giant and large layer V pyramidal cells characteristic of motor cortex in these species as well as in the spotted hyena (Sakai et al., 2011a, 2011) and lion (Arsznov and Sakai, 2012). For these reasons, the cruciate sulcus was chosen as a reliable boundary separating the cerebrum into the anterior cerebrum including frontal cortex and posterior cerebrum. The anterior cerebrum is defined as the area rostral to the cruciate sulcus and midline and caudal to the olfactory bulbs (Fig. 19.1C). Anterior cerebrum volume includes frontal cortex, rostral motor cortex, and portions of subcortical structures (rostral striatum, ventral pallidum, olfactory tubercle, prepiriform cortex). This measure proves to be reliable and consistent and as used here includes frontal cortex volume (Sakai et al., 2011a; Arsznov and Sakai, 2012).

The cruciate sulcus also serves as the boundary for the posterior cerebrum, the region extending from the cruciate sulcus caudally to the bony tentorium (Fig. 19.3D). The posterior cerebrum region consists of cortex posterior to the cruciate sulcus, the diencephalon, hippocampus, and the rostral mesencephalon. The third region of interest is the cerebellum and brain stem and includes the area posterior to the tentorium and anterior to the foramen magnum. In addition to including the cerebellum and brain stem, this regional volume includes the pons and medulla. Since these three regions of interest (anterior cerebrum, posterior cerebrum, cerebellum, and brain stem) can be reliably measured from the virtual endocast in a variety of carnivore species, they provide important and consistent measurements that can be used to assess brain variations enabling both within species as well as between species comparisons.

Here, the relative proportion of the brain region of interest as a function of total endocranial volume is used in our analysis. This proportional approach is

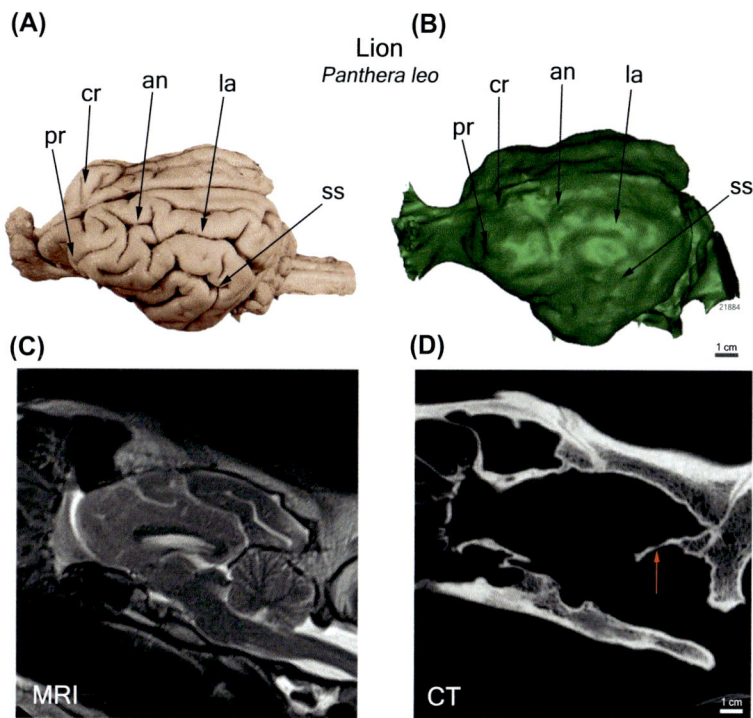

FIGURE 19.2 (A) Dorsolateral views of the whole brain of a lion from the Comparative Mammalian Brain Collection (#61-193) and (B) virtual endocast reconstructed from a lion skull (Michigan State University Museum specimen #21884). Abbreviations of the major sulci listed in Fig. 19.1(C) and (D) are sagittal views through a lion brain of a magnetic resonance image (MRI) (courtesy of Dr. A. Pease, MSU) and computed tomography (CT) image, respectively. *Arrow* denotes the cerebellar tentorium used to separate the posterior cerebrum region (PC) and cerebellum and brain stem region (CB + BS). *an*, ansate sulcus; *cr*, cruciate sulcus; *la*, lateral sulcus; *pr*, proreal sulcus; *ss*, suprasylvian sulcus.

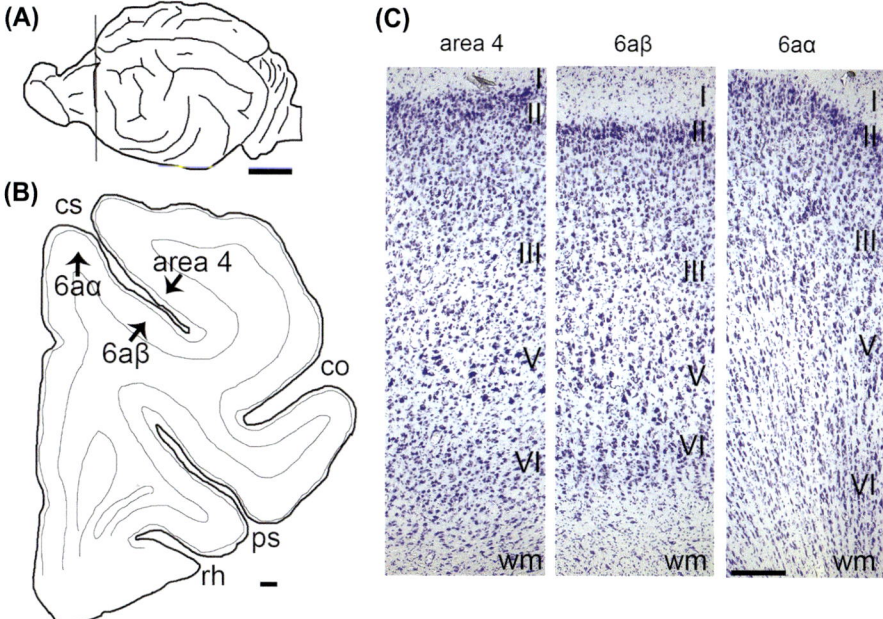

FIGURE 19.3 Cruciate sulcal region in the raccoon. (A) Line drawing of a raccoon brain. *Vertical line* indicates the level of the coronal section through the cruciate sulcus shown in (B). Scale bar = 1 cm. (B) Line drawing of a coronal section through the cruciate sulcus (cs) showing the relative locations of cytoarchitectonic areas 4, 6aα, and 6aβ. Scale bar = 1 mm. (C) Photomicrographs of the Nissl-stained section. Area 4 contains giant and large layer V pyramidal cells, whereas cytoarchitectonic features of area 6aβ are large layer V pyramidal cells and a compact layer VI. Area 6aα contains a wide layer III, medium-sized pyramidal cells in layer V, and a palisade arrangement of small cells in layer VI. Scale bar = 25 μm.

preferable (for discussion: see Dunbar, 1998) and allows for direct comparisons within and between species, particularly at the family level, without the uncertainty introduced by differing measures of relative brain size as discussed earlier. Relative brain size is dependent on body size, typically, body weight. But individual body weight is known to fluctuate dramatically by season in wild animals and is not always known for individuals. An alternative measure such as skull basal length has also been used as an indicator of body size in calculating relative brain size. However, the relationship between behavioral or ecological variables and brain size/body size varies depending on what measure of body size is used (Radinsky, 1975; Swanson et al., 2012). For these reasons, proportional analysis of regional brain volume relative to total brain volume is preferred over analyses of relative brain size measures (Dunbar, 1998; Clark et al., 2001; de Winter and Oxnard, 2001; Willemet, 2012), particularly since the allometric relationship between brain and body size differs across taxa (Deacon, 1990).

19.4 Interspecies Comparisons

The possibility that the same factors influencing relative brain size, total brain volume as function of body size or proportional regional brain size such as the frontal cortex relative to total brain volume, might also influence variations at the family level in a comparative (interspecific) analysis have been largely overlooked. As noted previously, the influence of social complexity on brain size is unclear based on large interspecific analyses (Finarelli and Flynn, 2009; Boddy et al., 2012; Swanson et al., 2012), but a family-level comparative analysis between closely related extant species exhibiting divergent behaviors can identify potential brain structures involved in the mediation of species-specific behaviors. If large brain size evolved independently in different carnivoran lines, then family-level analysis may reveal selection factors that act on the brain by examining closely related species with distinctive and different behavioral traits. Here we have examined brain variations in Hyaenidae and Procyonidae species that differ in social complexity and other behavioral traits.

19.4.1 Family Hyaenidae

The four extant species in Hyaenidae vary in sociality ranging from the most social spotted hyena (*Crocuta crocuta*), to the brown hyena (*Hyaena brunnea*) noted for living in small groups of up to 14 individuals, to the sometimes solitary to small group living striped hyena (*Hyaena hyaena*), and lastly to the aardwolf (*Proteles*

cristata) that pair-bonds during breeding season but is otherwise solitary. An interspecific analysis of the influence of social complexity on relative and regional brain size within the carnivore family Hyaenidae revealed that among the four extant species, the highly gregarious spotted hyena (*C. crocuta*) possesses the largest brain volume relative to body size and the greatest ratio of frontal cortex to total brain volume compared to the less social hyena species (Sakai et al., 2011a). Both relative brain size and relative frontal cortex volume increased with group size in the hyaenid species (Fig. 19.4).

These data in the Hyaenidae are consistent with the social brain hypothesis (Dunbar, 1998). It is notable that spotted hyenas share many similarities in social organization with cercopithecine primates (Holekamp et al., 2007). Spotted hyenas live in fission–fusion groups consisting of upward of 90 individuals. These groups, called clans, are both matrilineally and hierarchically organized. Members may form groups or remain solitary depending on resources and activities. In addition, individuals show the ability to recognize third-party relationships and discriminate between individual group members during coalition formation

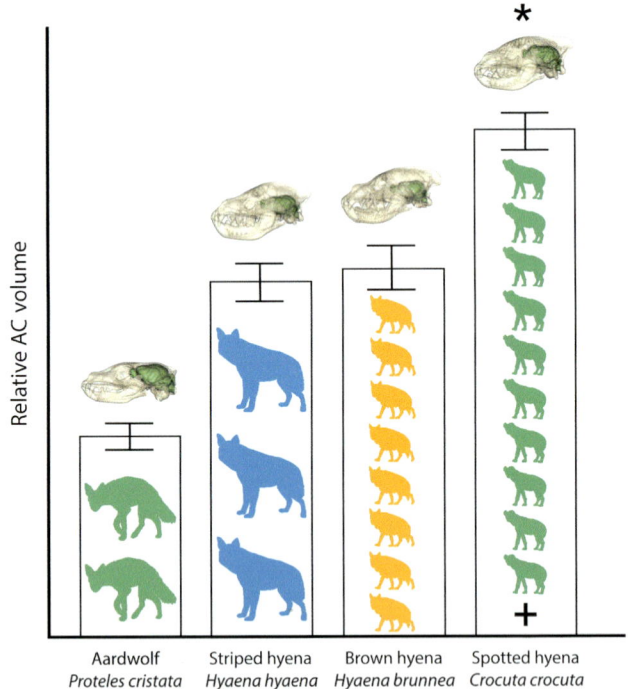

FIGURE 19.4 Relative frontal cortex (AC) volume (to total endocranial volume) in four hyaenid species differing in sociality: aardwolf (pair-bonds), striped hyena (sometimes solitary or lives in small groups), brown hyena (group size of up to 14 individuals), and spotted hyena (clans of up to 90 individuals). Error bars indicate ±1 SEM. Relative AC volume is significantly greater in spotted hyenas compared to the other hyena species ($p = .001$). A lateral view of the endocast and skull for each species is shown above each bar.

(Holekamp et al., 2007). The high degree of social complexity in the spotted hyena may impose greater cognitive demands and be associated with the relatively larger brain and expansion of frontal cortex. Although there are no studies on frontal cortex functions in the spotted hyena, frontal cortex has been implicated in behavioral inhibition in the dog (Brutkowski, 1965). Inhibiting a conditioned response by exerting impulse control is functionally linked to frontal cortex. Inhibitory control is especially important in negotiating a complex hierarchical society as that of the spotted hyena. Since access to food and mates is determined by rank, an individual may dominate over lower-ranking individuals but must also refrain from attacking a higher-ranking individual. Rank within the hierarchy is an important social information, and spotted hyenas retain this information throughout their life (Holekamp et al., 2007).

At the same time, these findings in Hyaenidae may also be explained by differences in cognitive complexity other than social information processing. Foraging strategies vary between hyena species in a manner consistent with these data. Spotted hyenas obtain most of their food by hunting medium to large size antelope (Kruuk, 1972). Striped and brown hyenas mainly forage for carrion and aardwolves forage for termites (Holekamp and Kolowski, 2009). The cognitive demands associated with predation on ungulates are expected to be greater than those related to either scavenging on carrion or foraging for termites. Whether increased brain size including frontal cortex volume is due to differences in sociality or cognitive complexity cannot be determined based on studies of this small carnivore family. Additional family-level comparisons will be helpful in determining the relationship of social complexity and brain variations.

19.4.2 Family Procyonidae

Variations in regional brain size were investigated in three extant species in the small family Procyonidae (Arsznov and Sakai, 2013). These three procyonid species were selected to study the relationship between sociality and brain structure since they display a continuum of behaviors related to social complexity. First, the highly social coatimundi (*Nasua nasua*, *Nasua narica*) lives in large female-bonded social groups, called bands, consisting of adult females, juveniles, and dependent young of both sexes (Kaufmann, 1982; Gompper, 1996; Romero and Aureli, 2008). These bands may consist of up to 40 individuals and are organized in linear dominance hierarchies (Gompper, 1995; Gompper and Decker, 1998; Hirsch, 2011). Coatimundis also exhibit a fission—fusion organization that is similar to that observed in spotted hyenas, where subgroups

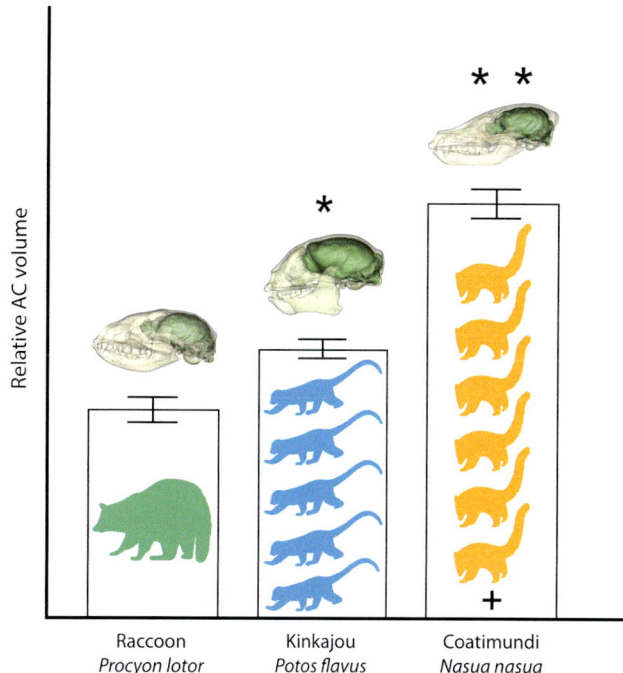

FIGURE 19.5 Relative frontal cortex (AC) volume (to total endocranial volume) in three procyonid species differing in sociality: raccoons (solitary), kinkajous (small family group), coatimundi (bands of up to 20 individuals). Error bars indicate ±1 SEM. Relative AC volume is significantly greater in coatimundis than both kinkajous and raccoons ($p = .001$) and is significantly greater in kinkajous than raccoons ($p = .004$). A lateral view of the endocast and skull for each species is shown above each bar.

have been observed to forage independent of the larger social band (Gompper, 1995). The other two species are less social; the kinkajou (*Potos flavus*) interacts within small polyandrous family groups of five (Kays and Gittleman, 2001), and the raccoon (*Procyon lotor*) is mostly solitary (Kaufmann, 1982). The interspecies virtual endocast analysis revealed that the relative amount of frontal cortex was greater in social than solitary species, and the highly social coatimundi possessed the greatest frontal cortex volume compared to other two species. Relative frontal cortex volume positively correlated with social group size among the three procyonid species (Fig. 19.5).

The previous study of the hyena family raised the possibility that differences in foraging strategies may account for the species differences in relative frontal cortex volume. However, this variable is unlikely to contribute to the difference in frontal cortex observed among the procyonid species studied here since all are omnivorous (Gittleman, 1986) subsisting on a diet primarily of fruit, invertebrates, and carrion when available. Nonetheless, these procyonid species do exhibit lifestyle differences with corresponding brain variations. The raccoon displays the greatest amount of forepaw use during tactile exploration compared to the other two species.

Welker and Campos (1963) demonstrated that the dexterous forepaw of the raccoon is accompanied by an enhanced representation of the forepaw in the somatosensory cortex compared to closely related procyonids, including coatimundis and kinkajous. Moreover, raccoons possess an expansion in the neural pathways associated with the representation of the forepaw in somatosensory cortex (Welker and Campos, 1963; Welker and Seidenstein, 1959), thalamus (Welker and Johnson, 1965), and dorsal column (Pubols et al., 1965). The kinkajou is also noted for specialized grasping behavior during forepaw use (McClearn, 1992); however, this superior grasping ability is not associated with a concomitant enhancement in the forepaw representation in somatosensory cortex (Welker and Campos, 1963). The kinkajou's grasping ability may reflect its exclusively arboreal lifestyle. While both coatimundis and raccoons are able to climb, the kinkajou displays locomotor flexibility and stability in a complex arboreal environment (ie, handling food while being suspended by the tail and hind feet) (McClearn, 1992). Interspecific comparisons of regional brain variations revealed that the raccoon possessed the greatest amount of relative posterior cerebrum, a region that includes somatosensory cortex and thalamus, compared to the coatimundi and kinkajou. In addition, the arboreal kinkajou had the largest relative cerebellum and brain stem volume compared to the other two species. The expansion in posterior cerebrum volume in the raccoon is hypothesized to be associated with the species-specific forepaw use in tactile exploration of the physical environment. The relatively large cerebellum and brain stem volume in the kinkajou in comparison to the other procyonid species is hypothesized to be related to motor coordination necessary to navigate a complex arboreal environment. In summary, these analyses within Procyonidae support the idea that specialized behaviors are associated with concomitant differences in regional brain size including an expansion of frontal cortex associated with social complexity.

19.5 Intraspecies Comparisons

In addition to interspecific family-level comparisons, intraspecific comparisons are helpful in revealing patterns of brain variation related to divergent social life history patterns. Intraspecific or intrapopulation comparison is an approach largely overlooked in comparative brain studies (Willemet, 2012; Smaers et al., 2012; Gonda et al., 2013). Differences in behavioral traits corresponding to brain variations within a species suggest adaptive changes that may be evolutionarily and functionally significant. Intraspecific comparisons can help identify the role of specific behaviors in brain

organization by eliminating confounding variables. Diet, foraging strategies, ecology, activity patterns, and cognitive complexity are all hypothesized to influence brain size and regional brain variations (Gittleman, 1986; Sol et al., 2008). The influence of any one of these variables on brain organization is difficult to determine from comparative studies since closely related species are often similar with respect to these characteristics. However, individuals of the same species that do exhibit behavioral variations provide a unique opportunity to evaluate the brain structure/function relationship. For example, reproductive behaviors differ between males and females, and sexually dimorphic brain structures mediating these behaviors are well documented (Cooke et al., 1998). However, the role of sexual dimorphism in social cognition is less clear. In humans, females are thought to possess better social cognitive skills than males with corresponding size differences in the brain regions mediating this information. Females possess proportionately greater orbital frontal cortex (Gur, 2002) and ventral frontal cortex (Wood et al., 2008) than males. But other studies conclude that the ratio of frontal lobe volume to total intracranial volume in humans does not differ between the sexes (Allen et al., 2002; DeCarli et al., 2005; Ellis et al., 2008). A sex-specific effect of frontal cortex and social behavior has also been found in a study of 10 primate species (Smaers et al., 2012). Based on these data, it is suggested that sexual dimorphism influences the evolution of the frontal cortex. But it is not known if a sex-specific effect of sociality and frontal cortex is limited to primates. Studies of intraspecific brain variations in carnivore species that exhibit differing social life histories will provide important information on the generality of the primate findings.

19.5.1 Family Hyaenidae: Sex Differences in the Spotted Hyena

Of all extant hyaenidae species, the spotted hyena is the most gregarious, living in clans of as many as 90 members (Kruuk, 1972; Fig. 19.6). However, the social life history experienced by adult male and female spotted hyenas differs. Female spotted hyenas are socially dominant to males and display higher frequencies of aggression compared to males both in captivity (Glickman et al., 1993) and in the wild (Szykman et al., 2003). Female spotted hyenas are usually philopatric and remain with their natal clan throughout life (Henschel and Skinner, 1987; Mills, 1990; Smale et al., 1997; Boydston et al., 2005), whereas most males voluntarily emigrate from their natal clans and join neighboring clans after puberty. This time of emigration presents postpubertal males with a novel set of cognitive

FIGURE 19.6 Both the spotted hyena (left) and lion (right) are highly social carnivores, but the social life history differs between the sexes along with differing volumes of frontal cortex. *Photo credits: Barbara Lundrigan*

demands related to navigation of the social hierarchy of the new clan. When immigrant males join a new clan, they enter at the bottom of a male dominance queue (East and Hofer, 2001) and acquire higher social ranks only upon the dispersal or death of natal and nonnatal resident males. Immigrant males follow the higher-ranking females in the dominance hierarchy for weeks to establish relationships to ensure survival and reproductive opportunities within the new clan (Kruuk, 1972; East and Hofer, 2001; Szykman et al., 2001). Natal males maintain the social rank status of their mothers within the hierarchy and display acts of aggression at higher frequencies and intensities compared to immigrant males. Shifting social status, access to resources and potential mates imposed on emigrating postpubertal males is expected to be more cognitively demanding than what is required of females who retain maternal rank and remain with their natal clan. An emigrating male spotted hyena must learn a new social hierarchy and assess the risks and benefits of approaching specific individuals. In addition, male spotted hyenas may require enhanced inhibitory control over inappropriate behaviors, such as aggression, to successfully navigate the social system in the new clan.

Increased inhibitory control and enhanced social information processing are functions often associated with frontal cortex. In view of the sexually dimorphic social life histories in the spotted hyena, we analyzed regional brain volumes in male and female spotted hyenas and found that males possess a greater relative amount of frontal cortex than females (Fig. 19.7; Arsznov et al., 2010). Interestingly, brain volume relative to body size did not differ between the sexes, but the ratio of anterior and posterior cerebrum volume relative to total brain volume did differ. Males possess relatively more frontal cortex than females, but females possessed relatively more posterior cerebrum than males. The larger frontal cortex found in male compared to female spotted

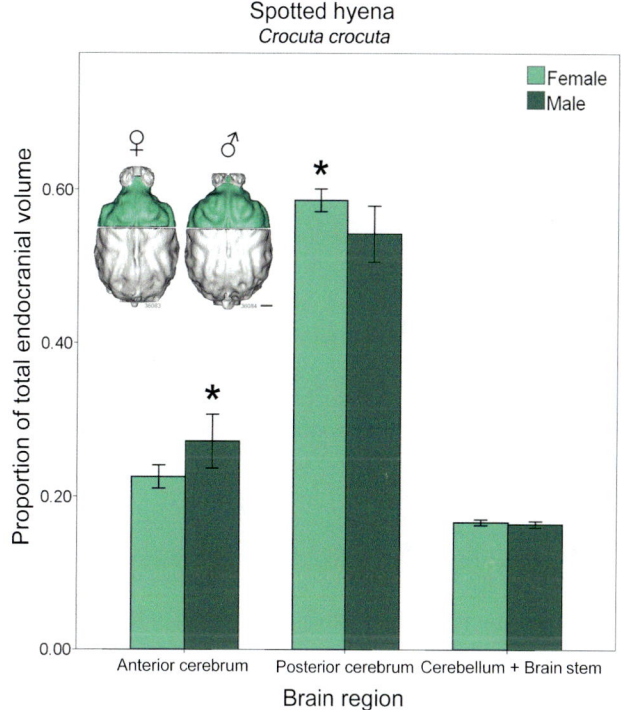

FIGURE 19.7 Proportional regional brain volumes (AC, PC, and CB + BS, all relative to total endocranial volume) in the spotted hyena (males: *dark green* and females: *light green*). Error bars indicate ±1 SEM. AC volume is significantly larger in male than female spotted hyenas ($p = .006$), while PC volume is significantly larger in female than male spotted hyenas ($p = .008$). Inset shows frontal cortex (AC) highlighted in *green* in a dorsal view of the female and male spotted hyena endocasts.

hyenas may reflect cognitive processing demands related to social life history, including the need for greater inhibitory control during social interactions with the highly aggressive and dominant females. Conversely, female spotted hyenas possess a relatively greater amount of posterior cerebrum than males. The posterior cerebrum measure includes all cortex posterior to the cruciate sulcus and subcortical regions including the diencephalon, hippocampus, amygdala, midbrain, and much of the basal ganglia. These regions subserve such a broad range of functions that it is difficult to offer reasonable explanations for the larger posterior cerebrum volume found in the females compared to males in our sample. Therefore, while this finding may be a consequence of the smaller relative amount of frontal cortex in females compared to males, many additional factors may also influence the sex differences in cortex posterior to cruciate sulcus.

19.5.2 Family Procyonidae: Sex Differences in the Coatimundi

As noted earlier the Procyonid family is a model family for examining the relationship of sociality and regional brain variations. A positive relationship between social group size and the frontal cortex volume was noted. Among the procyonid species, the only truly social species is the coatimundi living in bands of up to 40 members (Gompper, 1996). In addition, the social life histories of male and female coatimundis diverge. The coatimundi social organization is female bonded consisting of philopatric adult females along with juvenile males and females (Kaufmann, 1982). In this social structure, adult females show a high level of tolerance for juvenile aggressive behaviors with juvenile males displaying higher levels of aggression than females (Hirsch et al., 2012). One unusual feature of the coatimundi social organization is a dominance hierarchy in which juveniles rank higher than adult females or subadults (Hirsch, 2007). In contrast to the social life history of females, adult males emigrate from their natal band and become solitary (Kaufmann, 1982). Furthermore, adult males are known to behave aggressively toward bands and in some instances prey on juvenile coatis (Russell, 1981). Given the divergent social life histories experienced by male and female coatimundis, we predicted that female coatimundis should possess larger frontal cortical volumes than male coatimundis related to differing social information—processing demands experienced by sex. Indeed, we found that females possess significantly larger frontal cortical volumes relative to total brain volume than male coatimundis (Arsznov and Sakai, 2013; Fig. 19.8). At the same time, the ratio of posterior cerebrum volume to total brain volume is slightly but insignificantly larger in males than in female coatimundis. Interestingly, sexual dimorphic differences were also found in cerebellum and brain stem volume. Male coatimundis possess a greater proportion of cerebellum and brain stem volume to total brain volume than females. It is possible that males require

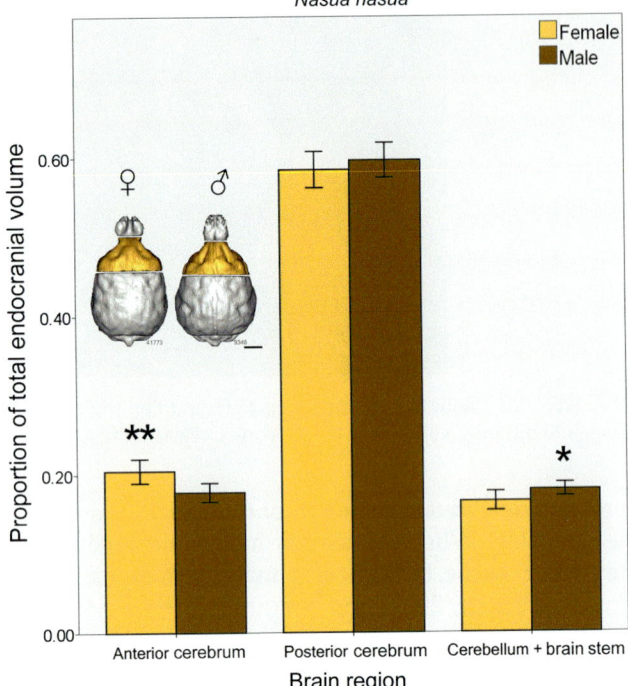

FIGURE 19.8 Proportional regional brain volumes (AC, PC, and CB + BS, all relative to total endocranial volume) in coatimundi (males: *brown* and females: *gold*). Error bars indicate ±1 SEM. AC volume is significantly larger in female than male coatimundis ($p = .005$), while PC volume did not significantly differ between the sexes. CB + BS volume was significantly larger in male than female coatimundis ($p = .04$). Inset shows frontal cortex (AC) highlighted in *gold* in a dorsal view of the female and male coatimundi endocast.

greater motor coordination than females since male coatimundis tend to be larger than females (Allen, 1987). However, these discrepant results suggest that further investigation is warranted.

In contrast to the sexual dimorphic social life histories of the coatimundi, social life history does not significantly differ in either the solitary raccoon or the small family dwelling kinkajou. No sex differences were found in regional brain volumes in either raccoons or kinkajous. These findings further support the idea that differing social life histories can contribute to differences in relative frontal cortex volume.

19.5.3 Family Felidae: Sex Differences in Lion and Cougar

An analysis of the brain volume variations in two Felidae species, the lion (*Panthera leo*) and cougar (*Puma concolor*), further evaluated the role of divergent social life histories on relative frontal cortex volume (Arsznov and Sakai, 2012). The family Felidae, the cat family, consists largely of solitary species, and solitary life typically does not differ between the sexes except for maternal care of dependent offspring. However, the lion is one notable exception as the only consistently gregarious species among felids (Packer and Pusey, 1987; Fig. 19.6). Lions live in complex fission—fusion social groups called prides that consist of up to 21 lions (Packer and Pusey, 1987). Among lions, sexually divergent social life histories exist in both dispersal and dominance behavior. Female lions are philopatric and typically remain in the maternal pride (Packer et al., 2001). Female lions are egalitarian and lack a formal dominance hierarchy (Packer et al., 2001). Within the lion pride, females form symmetrical relationships providing protection for their young through a communal cub rearing system with multiple reproducing females, which provides protection for juvenile lions against attacks from intruding males (Packer et al., 1990; Packer et al., 2001). In contrast to females, postpubertal males emigrate from the natal pride and enter a nomadic phase where they live in solitude or form a coalition with other male kin mates (Packer and Pusey, 1987). Male lions are dominant toward females and have been observed using lethal aggression toward adult females (Mosser and Packer, 2009). When a pride's resident male coalition is

defeated by an immigrant coalition, juvenile males and prepubertal females will disperse or be killed (Hanby and Bygott, 1987). An intraspecific comparison of relative regional brain volumes among lions was conducted to determine if the differing social life histories may correspond to the relative size of the frontal cortex. We found that female lions do possess a significantly larger ratio of frontal cortex volume to total brain volume than male lions (Fig. 19.9).

These results are similar to the previous findings of intraspecific differences in the spotted hyena and coatimundi. However, unlike the social life history of either the male spotted hyena or female coatimundi, the female lion resides in an egalitarian group. Female members of the lion pride share similar status, all subordinate to the dominant male. Consequently, the sex difference in frontal cortex volume in lions is not likely to be due to social information processing demands related to navigation of a social hierarchy. One explanation for the finding of larger frontal cortex in the female than male lion may be related to greater demand for inhibitory control in mediating appropriate social behavior in the presence of a dominant male aggressor. While the precise role of the frontal cortex

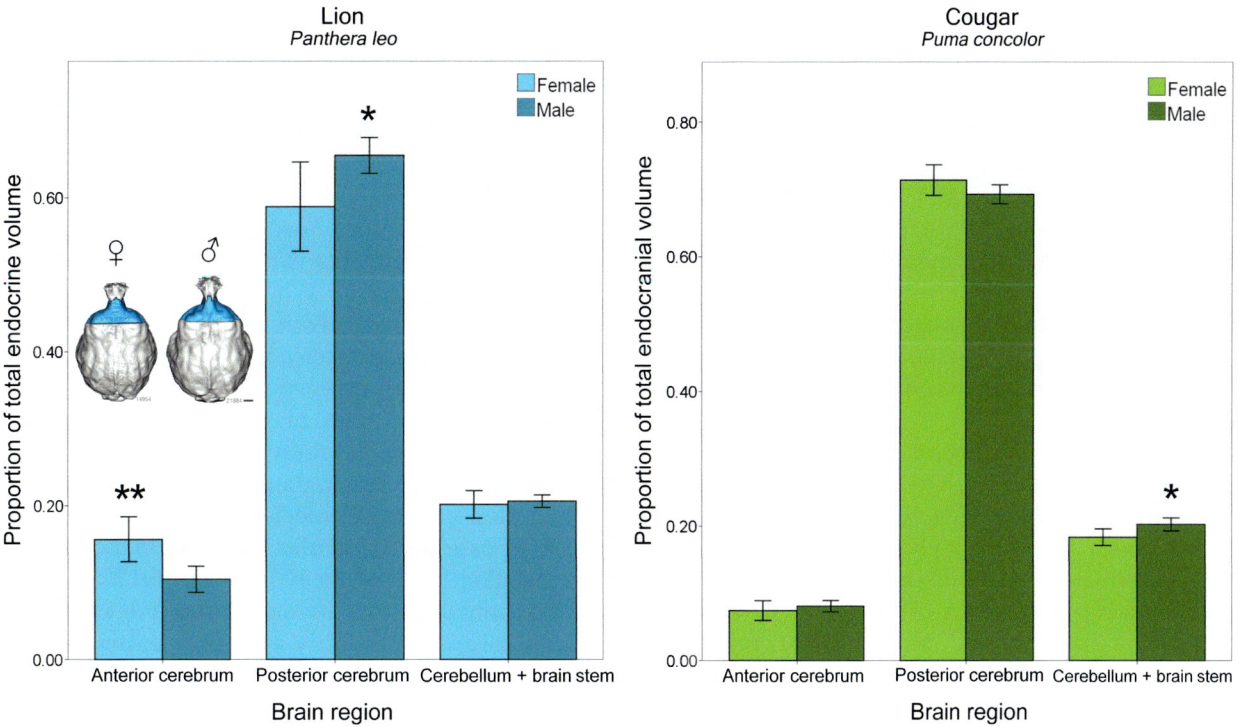

FIGURE 19.9   Proportional regional brain volumes (AC, PC, and CB + BS, all relative to total endocranial volume) for lions (left) (males: *dark blue* and females: *light blue*) and cougar (right) (males: *dark green* and females: *light green*). Error bars indicate ±1 SEM. AC volume is significantly larger in female than male lions ($p = .002$), while PC volume is significantly larger in male than female lions ($p = .012$). Inset shows frontal cortex (AC) highlighted in *blue* in a dorsal view of the female and male lion endocast. Both AC and PC volumes did not differ in male and female cougars, but CB + BS volume was larger in male than female cougars ($p = .03$).

in lion behavior is, of course, unknown, behavioral studies report previously learned inhibitory responses are disinhibited following lesions of the prefrontal cortex in dogs (Brutkowski and Dabrowski, 1963; Brutkowski, 1965) and cats (Warren et al., 1969).

Finally, the analysis of total brain volume relative to body size did not reveal any differences between the sexes. It is possible the proportional increase in frontal cortex to total brain volume observed in female lions is associated with a corresponding proportional decrease in posterior cortex to total brain volume, whereas male lions possess proportionately larger posterior cerebrum with correspondingly smaller frontal cortex.

In contrast to the social behavior of the lion, most other felid species including cougars (*P. concolor*) (Sunquist and Sunquist, 2002) are primarily solitary, only briefly interacting with conspecifics during the breeding season when they form temporary breeding pairs (Nowak and Paradiso, 1983). Intraspecific analysis in cougars yielded no sex differences in the relative amount of frontal cortex or posterior cerebrum (Fig. 19.9). The lack of sex differences in these regions may be related to the absence of divergent social life histories in male and female cougars. Thus, these findings suggest that the cognitive demands also do not differ between the sexes. However, male cougars possess a relatively greater amount of cerebellum and brain stem than females. This measurement, consisting primarily of cerebellum, may reflect greater demands related to voluntary movement, gait, posture, and motor functions (Ghez and Fahn, 1985). Male cougars have 1.4 times the body mass and occupy territories up to 3 times than that of females (Logan and Sweanor, 2001). The differences in physique and muscular build of males may require a greater degree of motor coordination than in females, particularly in negotiating mountainous terrain. Thus, male cougars might require greater motor coordination and agility while navigating larger and more difficult physical environments than their smaller female counterparts.

The lion provides an interesting felid model where the cognitive demands related to social complexity differ between the sexes. The significant sex difference in frontal cortex found in the lion lends support to our previous finding of sexual dimorphism of frontal cortex volume in the spotted hyena (Arsznov et al., 2010) and coatimundi (Arsznov and Sakai, 2013) related to differing social life histories.

Taken together, the intraspecific comparisons show that in three highly social carnivore species exhibiting divergent social life histories, the social sex possesses relatively more frontal cortex than its solitary conspecific counterpart. Moreover, this effect of enhanced frontal cortex volume is not determined by sex since it is the male spotted hyena that possessed more frontal cortex

volume than the female while both female lion and coatimundis possessed greater relative frontal cortex than their male counterparts. One interpretation of the sex difference associated with the enlarged frontal cortex in these species is that cognitive demands are greater when navigating a social environment dominated by the presence of an aggressor and that this increased information processing is accompanied by an increase in frontal cortex volume. Frontal cortex volumes do not differ between the sexes in the carnivore species studied when the social life histories do not differ.

19.6 Limitations

Our interspecific and intraspecific studies were based on analysis of regional brain volumes obtained from measuring virtual endocasts reconstructed from CT-scanned skulls. While this method is invaluable for expanding the number and variety of species studied, it has some drawbacks. The measurements of regional brain volumes are based on areas delineated by either bony landmarks or endocranial sulcal impressions. Nuclear details, gray/white matter distinctions, cortical lamination pattern, and subcortical features are not detectable with this method. Only histological examination of brain sections or MRI could provide this information, and species in the wild are not typically available for such studies. Another limitation of these studies lies in the definition of frontal cortex volume in carnivores. Here, anterior cerebrum volume as a proxy for frontal cortex volume is based on using the cruciate sulcus as a landmark. This sulcus corresponds to motor areas in the few carnivores species studied (cat, dog, raccoon, ferret), and this is presumed to be the case in other carnivore species. Whether the cortex anterior to the motor cortex in carnivore species resembles prefrontal cortex in primate species is not clear. Most investigators agree that the presence of a distinct granular layer IV in the frontal cortex of primates is unique and is linked to executive functions including working memory (Preuss, 2007; Passingham and Wise, 2012). Among the limited carnivore species studied, arguably, only the dog possesses a granular layer within the proreal gyrus (Tanaka, 1987 but for Rajkowska and Kosmal, 1988) of the frontal cortex. In most nonprimate mammals, frontal cortex is primarily agranular either possessing an incipient layer IV or lacking this layer entirely (for review: see Passingham and Wise, 2012). Agranular cortex predominantly occupies medial and orbital frontal cortex in primates and is typically characteristic of frontal cortex in nonprimate mammalian species (for review: see Ongur and Price, 2000; Fuster, 2015). This part of frontal cortex is functionally associated with emotional regulation, response selection, and impulse control in primate species (for review, see Fuster, 2015; Ongur and Price, 2000; Passingham and Wise, 2012). Whether these are functions of the carnivore frontal cortex remains unclear, but lesion studies in dogs (Dabrowska, 1971) and cats (Warren et al., 1972) suggest that the canine and feline frontal cortex plays a role in the inhibition of prepotent responses. Clearly, additional behavioral and anatomical studies addressing the role of the frontal cortex in carnivores are needed.

The correlative nature of these studies also limits the interpretation of these data. Although relative frontal cortex volume correlates with sociality in these interspecific and intraspecific analyses, the factors that drive frontal cortical expansion are not known. Since many carnivore species are solitary and some, particularly in the families Ursidae and Mustelidae, possess large brains relative to body size and possibly by extension, relatively large frontal cortex, it is likely that frontal cortex in carnivores plays a broader role in behavior beyond social information processing. But direct experimental evidence of the behavioral changes resulting from frontal cortex manipulation is rare. Finally, whether the sex difference in frontal cortex volume observed in the spotted hyena, lion, and coatimundi is due to differing life experience or other factors cannot be known from these data. However, the accumulated evidence based on frontal cortex volumes in these three species and differing social life histories is suggestive.

Finally, based on these comparisons, it is not clear if enhanced social information processing leads to recruitment of more frontal cortical neurons that then lead to the relative increase in regional brain volume. The data suggest that this interpretation is one possibility but further studies are warranted. Large relative brain volume may reflect increased neuron density, larger and more neurons, more glial cells, or increased neuropil or white matter. Neuron density is known to vary across a variety of mammals (Herculano-Houzel et al., 2008) with an increase in brain size associated with a decrease in neuron density in a study of 28 mammalian species (Herculano-Houzel, 2011). It is not known if neuron density, number, and cell size in frontal cortex differs among carnivore species. Future studies analyzing the cytoarchitectural characteristics and functional attributes of the frontal cortex and its subdivisions in the carnivores are needed.

19.7 Summary and Conclusions

Identifying the factors that influence variations in total and regional brain size is an essential aspect of comparative neuroscience. A number of hypotheses have been proposed to explain such brain variations, and the majority of these hypotheses are based on

studies in primates. Relatively few studies have examined brain variations in carnivore species (Healy and Rowe, 2007), and these studies primarily compared differences across broad taxonomic groups (Gittleman, 1994; Iwaniuk et al., 1999; Pérez-Barbería et al., 2007; Swanson et al., 2012). While these broad comparisons are important to our understanding of brain variations between taxa, variations within families or between sexes have been greatly overlooked. Analysis of closely related species from a single family that exhibit brain variations corresponding to behavioral or ecological differences provides important information about the relationship between structure and function. Total brain volume, regional brain volume, and morphological variations are expected to reflect differences in an individual's phenotype. Phenotypic selection for a specific behavioral attribute may influence heritability of a brain trait. For example, artificial selection of high wheel-running mice led to changes in midbrain volume compared to controls within 52 generations (Kolb et al., 2013). This study suggests the regional brain variation associated with running behavior is heritable over a relatively short time span and may be important in evolution as a factor contributing ultimately to species differentiation. Analysis of inter- and intraspecific differences in total and regional brain size will provide insight into the factors that influence brain evolution in carnivores.

The social brain hypothesis proposes that large brain size including the expansion of the frontal cortex is a result of the increased neural processing required in navigating socially complex environments (Dunbar, 1992, 2003), and there are many primate studies that support this hypothesis. Data presented here on selected social carnivore species and regional brain variations also lend support to the social brain hypothesis. The interspecific evidence in select species from the carnivore families Hyaenidae (Sakai et al., 2011a,b) and Procyonidae (Arsznov and Sakai, 2013) suggest that an increase in the relative amount of frontal cortex, a brain region related to enhanced social cognition and social information processing, is associated with social group size. Additionally, intraspecific comparisons among select species from Hyaenidae (Arsznov et al., 2010), Procyonidae (Arsznov and Sakai, 2013), and Felidae (Arsznov and Sakai, 2012) suggest that when species display sexually divergent social life histories, sexual dimorphism in the relative amount of frontal cortex also exists. Here, the results of these analyses suggest that social life history factors including living in social groups marked by the presence of a dominant aggressor may be associated with an increase in relative size of frontal cortex. We hypothesize that the expanded frontal cortex present in social carnivores plays a role in the inhibition of prepotent responses, possibly including inhibition of socially inappropriate behavior. No sex differences in the relative amount of frontal cortex were found in species that do not exhibit differences in social life history between the sexes such as cougars, raccoons, and kinkajous (Arsznov and Sakai, 2012, 2013). Still, until the influence of differing social life history on variations in regional brain size is investigated in additional carnivore species, our interpretation of these findings is tentative.

The inter- and intraspecific comparative analyses reviewed throughout this chapter highlight the use of measures obtained from three-dimensional virtual endocasts created from computed tomographic scans from skull specimens to investigate questions related to variations in total and regional brain size in carnivores. While natural fossilized and latex endocranial casts have long been utilized in the field of comparative neurology (Radinsky, 1969; Jerison, 1973, 2007), more recently the application of imaging technologies such as CT to create three-dimensional virtual models of the endocranial cavity has proven to be an exciting method to further investigate comparative questions (Sakai et al., 2011a,b). Taken all together, these results support the principle of proper mass suggesting that behavioral specializations are associated with increases in regional size of brain regions principally mediating those specialized functions.

Acknowledgment

This work was supported by NSF grant IOS 1146614.

References

Adolphs, R., 2001. The neurobiology of social cognition. Curr. Opin. Neurobiol. 11, 231–239.

Agnarsson, I., Kuntner, M., May-Collado, L., 2010. Dogs, cats, and kin: a molecular species-level phylogeny of Carnivora. Mol. Phylogenet. Evol. 54, 726–745.

Allen, T.B., 1987. Wild Animals of North America. The National Geographic Society, Washington DC, USA.

Allen, J.S., Damasio, H., Grabowski, T.J., 2002. Normal neuroanatomical variation in the human brain: an MRI-volumetric study. Am. J. Phys. Anthropol. 118, 341–358.

Amodio, D.M., Frith, C.D., 2006. Meetings of minds: the medial frontal cortex and social cognition. Nat. Rev. Neurosci. 7, 268–277.

Arsznov, B.M., Lundrigan, B.L., Holekamp, K.E., Sakai, S.T., 2010. Sex and the frontal cortex: a developmental CT study in the spotted hyena. Brain Behav. Evol. 76, 185–197.

Arsznov, B.M., Sakai, S.T., 2012. Pride diaries: sex, brain size and sociality in the African lion (Panthero leo) and cougar (Puma concolor). Brain Behav. Evol. 79, 275–289.

Arsznov, B.M., Sakai, S.T., 2013. The procyonid social club: comparison of brain volumes in the coatimundi (Nasua nasua, N. narica), kinkajou (Potos flavus), and raccoon (Procyon lotor). Brain Behav. Evol. 82, 129–145.

Baron, G., Stephan, H., Frahm, H.D., 1996. Comparative neurobiology. In: Chiroptera, vol. 3. Birkhäuser Verlag, Basel, Switzerland.

Barton, R.A., 1996. Neocortex size and behavioural ecology in primates. Proc. R. Soc. Lond. Ser. B Biol. Sci. 263, 173–177.

Barton, R.A., Dunbar, R.I.M., 1997. Evolution of the social brain. In: Byrne, R., Whiten, A. (Eds.), Machiavellian Intelligence II: Extensions and Evaluations. Cambridge University Press, Cambridge, UK, pp. 240–263.

Barton, R.A., Harvey, P., 2000. Mosaic evolution of brain structure in mammals. Nature 405, 1055–1058.

Bienvenu, T., Guy, F., Coudyzer, W., Gilissen, E., Roualdes, G., Vignaud, P., Brunet, M., 2011. Assessing endocranial variations in great apes and humans using 3D data from virtual endocasts. Am. J. Phys. Anthropol. 145, 231–246.

Boddy, A.M., McGowen, M.R., Sherwood, C.C., Grossman, L.I., Goodman, M., Wildman, D.E., 2012. Comparative analysis of encephalization in mammals reveals relaxed constraints on anthropoid primate and cetacean brain scaling. J. Evol. Biol. 25, 981–994.

Boydston, E.E., Kapheim, K.M., Van Horn, R.C., Smale, L., Holekamp, K.E., 2005. Sexually dimorphic patterns of space use throughout ontogeny in the spotted hyena (Crocuta crocuta). J. Zool. 267, 271–281.

Bush, E.C., Allman, J.M., 2004. The scaling of frontal cortex in primates and carnivores. Proc. Natl. Acad. Sci. U.S.A. 101, 3962–3966.

Brutkowski, S., Dabrowski, J., 1963. Disinhibition after prefrontal lesions as a function of duration of intertrial intervals. Science 139, 505–506.

Brutkowski, S., 1965. Functions of prefrontal cortex in animals. Physiol. Rev. 45, 721–746.

Byrne, R.W., 1995. The Thinking Ape. Oxford University Press, Oxford, UK.

Byrne, R., Whiten, A., 1988. Machiavellian Intelligence: Social Expertise and the Evolution of Intellect in Monkeys, Apes, and Humans. Oxford University Press, Oxford, UK.

Clark, D.A., Mitra, P.P., Wang, S.S.-H., 2001. Scalable architecture in mammalian brains. Nature 411, 189–193.

Clutton-Brock, T., Harvey, P., 1980. Primates, brains and behavior. J. Zool. Lond. 190, 309–323.

Conroy, G.C., Vannier, M.W., Tobias, P.V., 1990. Endocranial features of Australopithecus africanus revealed by 2 and 3-D computed tomography. Science 247, 838–841.

Cooke, B., Hegstrom, C.D., Villeneuve, L.S., Breedlove, S.M., 1998. Sexual differentiation of the vertebrate brain: principles and mechanisms. Front. Neuroendocrinol. 19, 323–362.

Dabrowska, J., 1971. Dissociation of impairment after lateral and medial prefrontal lesions in dogs. Science 171, 1037–1038.

DeCarli, C., Massaro, J., Harvey, D., Hald, J., Tullberg, M., Au, R., Beiser, A., D'Agostino, R., Wolf, P.A., 2005. Measures of brain morphology and infarction in the framingham heart study: establishing what is normal. Neurobiol. Aging 26, 491–510.

Deacon, T.W., 1990. Fallacies of progression in theories of brain-size evolution. Int. J. Primatol. 11, 193–236.

Deaner, R.O., Nunn, C.L., van Schaik, C.P., 2000. Comparative tests of primate cognition: different scaling methods produce different results. Brain Behav. Evol. 55, 44–52.

Deaner, R.O., Isler, K., Burkhart, J., van Schaik, C., 2007. Overall brain size, and not encephalization quotient, best predicts cognitive ability across non-human primates. Brain Behav. Evol. 70, 115–124.

Dunbar, R.I.M., Bever, J., 1998. Neocortex size predicts group size in carnivores and some insectivores. Ethology 104, 695–708.

Dunbar, R.I.M., Shultz, S., 2007. Evolution in the social brain. Science 317, 1344–1347.

Dunbar, R.I.M., 1992. Neocortex size as a constraint on group size in primates. J. Hum. Evol. 20, 469–493.

Dunbar, R.I.M., 1998. The social brain hypothesis. Evol. Anthropol. 6, 178–190.

Dunbar, R.I.M., 2003. The social brain: mind, language and society in evolutionary perspective. Annu. Rev. Anthropol. 325, 163–181.

Dunbar, R.I.M., 2011. Evolutionary basis of the social brain. In: Decety, J., Cacioppo, J. (Eds.), Oxford Handbook of Social Neuroscience. Oxford University Press, Oxford, UK, pp. 28–38.

East, M., Hofer, H., 2001. Male spotted hyenas (Crocuta crocuta) queue for status in social groups dominated by females. Behav. Ecol. 12, 558–568.

Ellis, L., Herschberger, S., Field, E., Wersinger, S., Pellis, S., Geary, D., Palmer, C., Hoyenga, K., Hetsroni, A., Karadi, K., 2008. In: Ellis, L. (Ed.), Sex Differences: Summarizing More than a Century of Scientific Research. Psychology Press, New York.

Falk, D., Redmond Jr., J.C., Guyer, J., Conroy, C., Recheis, W., Weber, G.W., Seidler, H., 2000. Early hominid brain evolution: a new look at old endocasts. J. Hum. Evol. 38, 695–717.

Falk, D., Hildebolt, C., Smith, K., Morwood, M.J., Sutikna, T., Jatmiko, E., Saptomo, W., Prior, F., 2009. LB1's virtual endocast, microcephaly, and hominin brain evolution. J. Hum. Evol. 57, 597–607.

Finarelli, J.A., Flynn, J.J., 2009. Brain size evolution and sociality in Carnivora. Proc. Natl. Acad. Sci. U.S.A. 106, 9345–9349.

Finarelli, J.A., 2006. Estimation of endocranial volume through the use of external skull measures in the Carnivora (Mammalia). J. Mammal. 87, 1027–1036.

Finlay, B.L., Darlington, R.B., 1995. Linked regularities in the development and evolution of mammalian brains. Science 268, 1578–1584.

Fuster, J.M., 2015. The Prefrontal Cortex, fifth ed. Elsevier, Amsterdam.

Ghez, C., Fahn, S., 1985. The cerebellum. In: Kandel, E., Schwartz, J. (Eds.), Principles of Neural Science, second ed. Elsevier, New York, pp. 502–522.

Gittleman, J.L., 1986. Carnivore brain size, behavioral ecology, and phylogeny. J. Mammal. 67, 23–36.

Gittleman, J.L., 1994. Female brain size and parental care in carnivores. Proc. Natl. Acad. Sci. U.S.A. 91, 5495–5497.

Glickman, S.E., Frank, L.G., Holekamp, K.E., Smale, L., Licht, P., 1993. Costs and benefits of 'androgenization' in the female spotted hyena: the natural selection of physiological mechanisms. Perspect. Ethol. 10, 87–117.

Gompper, M.E., Decker, D., 1998. Nasua nasua. Mamm. Species 580, 1–9.

Gompper, M.E., 1995. Nasua narica. Mamm. Species 487, 1–10.

Gompper, M.E., 1996. Foraging costs and benefits of coati (Nasua narica) sociality and asociality. Behav. Ecol. 7, 254–263.

Gonda, A., Herczeg, G., Merilä, J., 2013. Evolutionary ecology of intraspecific brain size variation: a review. Ecol. Evol. 3, 2751–2764.

Gorska, T., 1974. Functional organization of cortical motor areas in adult dogs and puppies. Acta Neurobiol. Exp. 34, 171–203.

Gould, S.J., 1975. Allometry in primates, with emphasis on scaling and the evolution of the brain. Contrib. Primatol. 5, 244–292.

Gur, R.C., Gunning-Dixon, F., Bilker, W.B., Gur, R.E., 2002. Sex differences in temporo-limbic and frontal brain volumes of healthy adults. Cereb. Cortex 12, 998–1003.

Hanby, J., Bygott, J., 1987. Emigration of subadult lions. Anim. Behav. 35, 161–169.

Hardin Jr., W.B., Arumugasamy, N., Jameson, H.D., 1968. Pattern of localization in 'precentral' motor cortex of raccoon. Brain Res. 11, 611–627.

Healy, S.D., Rowe, C., 2007. A critique of comparative studies of brain size. Proc. R. Soc. B 274, 453–464.

Henschel, J., Skinner, J., 1987. Social relationships and dispersal patterns in a clan of spotted hyaenas (Crocuta crocuta) in the Kruger National Park. South Afr. J. Zool. 22, 18–24.

Herculano-Houzel, S., Collins, C.E., Wong, P., Kaas, J.H., Lent, R., 2008. The basic non-uniformity of the cerebral cortex. Proc. Natl. Acad. Sci. U.S.A. 105, 12593–12598.

Herculano-Houzel, S., 2011. Not all brains are made the same: new views on brain scaling in evolution. Brain Behav. Evol. 78 (1), 22–36.

Hirsch, B.T., Stanton, M.A., Maldonado, J.E., 2012. Kinship shapes affiliative social networks but not aggression in ring-tailed coatis. PLoS One 7, e37301.

Hirsch, H.T., 2007. Spoiled brats: is extreme juvenile agonism in ring-tailed coatis (Nasua nasua) dominance or tolerated aggression? Ethology 113, 446–456.

Hirsch, B.T., 2011. Within-group spatial position in ring-tailed coatis: balancing predation, feeding competition, and social competition. Behav. Ecol. Sociobiol. 65, 391–399.

Holekamp, K.E., Kolowski, J.M., 2009. Hyaenidae. In: Wilson, D., Mittermeier, R., Fonseca, G. (Eds.), Handbook of Mammals of the World. Lynx Edicions, Madrid, Spain, pp. 234–260.

Holekamp, K.E., Sakai, S.T., Lundrigan, B.L., 2007. Social intelligence in the spotted hyena (Crocuta crocuta). Philos. Trans. R. Soc. Lond. Ser. B Biol. Sci. 362, 523–538.

Holloway, R.L., Broadfield, D.C., Yuan, M.S., 2004. The Human Fossil Record. Brain Endocasts the Paleoneurological Evidence, vol. 3. John Wiley and Sons, New York.

Humphrey, N.K., 1976. The social formation of intellect. In: Bateson, P., Hinde, R. (Eds.), Growing Points in Ethology. Cambridge University Press, Cambridge, pp. 303–317.

Isler, K., Christopher Kirk, E., Miller, J.M.A., Albrecht, G.A., Gelvin, B.R., Martin, R.D., 2008. Endocranial volumes of primate species: scaling analyses using a comprehensive and reliable data set. J. Hum. Evol. 55, 967–978.

Iwaniuk, A.N., Nelson, J.E., 2002. Can endocranial volume be used as an estimate of brain size in birds? Can. J. Zool. 80, 16–23.

Iwaniuk, A.N., Pellis, S.M., Whishaw, I.Q., 1999. Brain size is not correlated with forelimb dexterity in fissiped carnivores (Carnivora): a comparative test of the principle of proper mass. Brain Behav. Evol. 54, 167–180.

Jerison, H., 1973. Evolution of the Brain and Intelligence. Academic Press, London.

Jerison, H., 2007. What fossils tell us about the evolution of the neocortex. In: Kaas, J., Krubitzer, L. (Eds.), Evolution of Nervous System. Elsevier, New York, pp. 1–12.

Jolly, A., 1966. Lemur social behavior and primate intelligence. Science 153, 501–506.

Kaufmann, J.H., 1982. Raccoon and allies. In: Chapman, J.A., Feldhamer, G.A. (Eds.), Wild Mammals of North America: Biology, Management, and Economics. The Johns Hopkins Univ. Press, Baltimore, pp. 567–585.

Kays, R., Gittleman, J., 2001. The social organization of the kinkajou Potos flavus (Procyonidae). J. Zool. Lond. 253, 491–504.

Kolb, E.M., Rezende, E.L., Holness, L., Radtke, A., Lee, S.K., Obenaus, A., Garland Jr., T., 2013. Mice selectively bred for high voluntary wheel running have larger midbrains: support for the mosaic model of brain evolution. J. Exp. Biol. 216, 515–553.

Kruuk, H., 1972. The Spotted Hyena: A Study of Predation and Social Behavior. University of Chicago Press, Chicago.

Kudo, H., Dunbar, R.I.M., 2001. Neocortex size and social network size in primates. Anim. Behav. 62, 711–722.

Lewis, K., 2001. A comparative study of primate play behaviour: implications for the study of cognition. Folia Primatol. 71, 417–421.

Logan, K., Sweanor, L., 2001. Desert Puma: Evolutionary Ecology and Conservation of an Enduring Carnivore. Island Press, Covelo, California.

Manger, P.R., Cort, J., Ebrahim, N., Goodman, A., Henning, J., Karolia, M., Rodrigues, S.-L., Štrkal, G., 2008. Is 21st century neuroscience too focussed on the rat/mouse model of brain function and dysfunction? Front. Neuroanat. 2, 1–7.

Marino, L., 1996. What can dolphins tell us about primate evolution? Evol. Anthropol. 5, 81–86.

McClearn, D., 1992. Locomotion, posture, and feeding behavior of kinkajous, coatis, and raccoons. J. Mammal. 73, 245–261.

McLaughlin, D.F., Sonty, R.V., Juliano, S.L., 1998. Organization of the forepaw representation in ferret somatosensory cortex. Somatosens. Mot. Res. 15, 253–268.

Mills, M., 1990. Kalahari Hyenas: The Behavioural Ecology of Two Species. Unwin Hyman, London.

Mosser, A., Packer, C., 2009. Group territoriality and the benefits of sociality in the African lion, Panthera leo. Anim. Behav. 78, 359–370.

Neubauer, S., 2014. Endocasts: possibilities and limitations for the interpretation of human brain evolution. Brain Behav. Evol. 84, 117–134.

Nieoullon, A., Rispal-Padel, L., 1976. Somatotopic localization in cat motor cortex. Brain Res. 105, 405–422.

Nowak, R., Paradiso, J., 1983. Walker's Mammals of the World. Johns Hopkins University Press, Baltimore.

Öngür, D., Price, J.L., 2000. The organization of networks within the orbital and medial prefrontal cortex of rats, monkeys, and humans. Cereb. Cortex 10, 206–219.

Packer, C., Pusey, A., 1987. Intrasexual cooperation and the sex ratio in African lions. Am. Nat. 130, 636–642.

Packer, C., Scheel, D., Pusey, A.E., 1990. Why lions form groups: food is not enough. Am. Nat. 136, 1–19.

Packer, C., Pusey, A.E., Eberly, L.E., 2001. Egalitarianism in female African lions. Science 293, 690–693.

Passingham, R.E., Wise, S.P., 2012. The Neurobiology of the Prefrontal Cortex: Anatomy, Evolution, and the Origin of Insight. Oxford University Press, Oxford, UK.

Pawlowski, B.P., Lowen, C.B., Dunbar, R.I.M., 1998. Neocortex size, social skills and mating success in primates. Behaviour 135, 357–368.

Perez-Barberia, F.J., Shultz, S., Dunbar, R.I.M., 2007. Evidence for coevolution of sociality and relative brain size in three orders of mammals. Evolution 61, 2811–2821.

Plautz, E.J., Milliken, G.W., Nudo, R.J., 2000. Effects of repetitive motor training on movement representations in adult squirrel monkeys: role of use versus learning. Neurobiol. Learn. Mem. 74, 27–55.

Preuss, T.M., 2007. Primate brain evolution in phylogenetic context. In: Kaas, J.H., Preuss, T.M. (Eds.), Evolution of Nervous Systems, The Evolution of Primate Nervous Systems, vol. 4. Elsevier, Oxford, pp. 3–34.

Pubols, B.H., Welker, J.I., Johnson, J.R., 1965. Somatic sensory representation of forelimb in dorsal root fibers of raccoon, coatimundi, and cat. J. Neurophysiol. 28, 312–341.

Radinsky, L., 1967. Relative brain size: a new measure. Science 155, 836–838.

Radinsky, L., 1969. Outlines of canid and felid brain evolution. Ann. N.Y. Acad. Sci. 167, 277–288.

Radinsky, L., 1975. Viverrid neuroanatomy: phylogenetic and behavioral implications. J. Mammal. 56, 130–150.

Rajkowska, G., Kosmal, A., 1988. Intrinsic connections and cytoarchitectonic data of the frontal association cortex in the dog. Acta Neurobiol. Exp. 48, 169–192.

Reep, R.L., Finlay, B.L., Darlington, R.B., 2007. The limbic system in mammalian brain evolution. Brain Behav. Evol. 70, 57–70.

Romero, T., Aureli, F., 2008. Reciprocity of support in coatis (Nasua nasua). J. Comp. Psych. 112, 19–25.

Roth, G., Dicke, U., 2005. Evolution of the brain and intelligence. Trends Cogn. Sci. 9, 250–257.

Russell, J.K., 1981. Exclusion of adult male coatis from social groups: protection from predation. J. Mammal. 62, 206–208.

Sakai, S.T., Arsznov, B.M., Lundrigan, B.L., Holekamp, K.E., 2011. Brain size and social complexity: a computed tomography study in hyaenidae. Brain Behav. Evol. 77, 91–104.

Sakai, S.T., Arsznov, B.M., Lundrigan, B.L., Holekamp, K.E., 2011. Virtual endocasts: an application of computed tomography in the study of brain variation among hyenas. Ann. N.Y. Acad. Sci. 1225 (Suppl. 1), E160–E170.

Sawaguchi, T., Kudo, H., 1990. Neocortical development and social-structure in primates. Primates 31, 283–289.

Shultz, S., Dunbar, R.I.M., 2007. The evolution of the social brain: anthropoid primates contrast with other vertebrates. Proc. R. Soc. Biol. Sci. 274, 2429–2436.

Smaers, J.B., Dechmann, D.K.N., Goswami, A., Soligo, C., Safi, K., 2012. Comparative analyses of evolutionary rates reveal different pathways to encephalization in bats, carnivorans, and primates. Proc. Natl. Acad. Sci. U.S.A. 109, 8006–18011.

Smale, L., Nunes, S., Holekamp, K.E., 1997. Sexually dimorphic dispersal in mammals: patterns, causes, and consequences. In: Slater, P., Rosenblatt, J., Snowden, C., Milinski, M. (Eds.), Advances in the Study of Behavior, vol. 26. Academic Press, San Diego, pp. 181–250.

Sol, D., Bacher, S., Reader, S.M., Lefebvre, L., 2008. Brain size predicts the success of mammal species introduced into novel environments. Am. Nat. 172, S63–S71.

Stephan, H., Frahm, H., Baron, G., 1981. New and revised data on volumes of brain structures in insectivores and primates. Folia Primatol. 35, 1–29.

Striedter, G.F., Belgard, T.G., Chen, C.C., Davis, F.P., Finlay, B.L., Güntürkün, O., Hale, M.E., Harris, J.A., Hecht, E.E., Hof, P.R., Hofmann, H.A., Holland, L.Z., Iwaniuk, A.N., Jarvis, E.D., Karten, H.J., Katz, P.S., Kristan, W.B., Macagno, E.R., Mitra, P.P., Moroz, L.L., Preuss, T.M., Ragsdale, C.W., Sherwood, C.C., Stevens, C.F., Stüttgen, M.C., Tsumoto, T., Wilczynski, W., 2014. NSF workshop report: discovering general principles of nervous system organization by comparing brain maps across species. Brain Behav. Evol. 83, 1–8.

Sunquist, M., Sunquist, F., 2002. Wild Cats of the World. University of Chicago Press, Chicago.

Swanson, E.M., Holekamp, K.E., Lundrigan, B.L., Arsznov, B.M., Sakai, S.T., 2012. Multiple determinants of whole and regional brain volume among terrestrial carnivorans. PLoS One 7, e38447.

Szykman, M., Engh, A.L., Van Horn, R.C., Funk, S.M., Scribner, K.T., Holekamp, K.E., 2001. Association patterns among male and female spotted hyenas (Crocuta crocuta) reflect male mate choice. Behav. Ecol. Sociobiol. 50, 231–238.

Szykman, M., Engh, A.L., Van Horn, R.C., Boydston, E.E., Scribner, K.T., Holekamp, K.E., 2003. Rare male aggression directed toward females in a female-dominated society: baiting behavior in the spotted hyenas. Aggress. Behav. 29, 457–474.

Tanaka, D., 1987. Differential laminar distribution of corticostriatal neurons in the prefrontal and pericruciate gyri of the dog. J. Neurosci. 7, 4095–4106.

Warren, J.M., Coutant, L.W., Cornwall, P.R., 1969. Cortical lesions and response inhibition in cats. Neuropsychologia 7, 245–257.

Warren, J.M., Warren, H.B., Akert, K., 1972. The behavior of chronic cats with lesions in the frontal association cortex. Acta Neurobiol. Exp. 32, 361–392.

de Winter, W., Oxnard, C.E., 2001. Evolutionary radiations and convergences in the structural organization of mammalian brains. Nature 409, 710–714.

Welker, W.I., Campos, G.B., 1963. Physiological significance of sulci in somatic sensory cerebral cortex in mammals of the family Procyonidae. J. Comp. Neurol. 120, 19–36.

Welker, W.I., Johnson, J.I., 1965. Correlation between nuclear morphology and somatotopic organization in ventrobasal complex of the raccoon's thalamus. J. Anat. 99, 761–790.

Welker, W.I., Seidenstein, S., 1959. Somatic sensory representation in the cerebral cortex of the raccoon (*Procyon lotor*). J. Comp. Neurol. 111, 469–501.

Willemet, R., 2012. Understanding the evolution of mammalian brain structures; the need for a (new) cerebrotype approach. Brain Sci. 2, 203–224.

Wood, J.L., Heitmiller, D., Andreasen, N.C., Nopoulos, P., 2008. Morphology of the ventral frontal cortex: relationship to femininity and social cognition. Cereb. Cortex 18, 534–540.

Woosley, T.A., Van der Loos, H., 1970. The structural organization of layer IV in the somatosensory region (SI) of mouse cerebral cortex. Brain Res. 17, 205–242.

Relevant Website

Vimeo (last accessed on 07.09.16.).

PART 4

Primates

The Phylogeny of Primates

J.G. Fleagle[1], E.R. Seiffert[2]

[1]Stony Brook University, Stony Brook, NY, United States; [2]University of Southern California, Los Angeles, CA, United States

20.1 Introduction

Primates are one of the most diverse orders of mammals alive today. The most recent systematic revisions recognize almost 80 genera and over 500 species of living primates (Appendix 1; Mittermeier et al., 2013; Rylands and Mittermeier, 2014). This makes our order the third most speciose order of living mammals, following bats (Chiroptera) and rodents (Rodentia). Moreover, living primates are just a small representation of the primate species and genera that have inhabited the earth over the past 65 million years or so (eg, Fleagle, 2013; Fig. 20.1).

The number of recognized primate species has increased dramatically in recent years (Rylands and Mittermeier, 2014). This is due to many factors, including more extensive field surveys in previously inaccessible parts of the developing world, increased sampling of genetic diversity among wild populations of primates, and changes in criteria by which species are identified (see Fleagle, 2014 and papers therein). This increase in species diversity is particularly important for laboratory scientists because it means that the names (and identity) of the primates involved in many studies from earlier decades may be different from what the same taxa are called today. More significantly, many studies that were reported to have been based on a single species may well have actually involved different species with different characteristics and different genetic identities.

While our understanding of primate diversity has changed dramatically in recent decades and is being updated on an almost weekly basis, our understanding of how different primate genera are related to one another has become relatively stable (Fig. 20.2). After a half century of molecular systematics, previously debated phylogenetic relationships among living members of the order Primates are largely resolved in the

views of most authorities. However, the relationships among primates, and primate relatives known from the fossil record, are more contentious because of the serendipitous sampling of the fossil record, the incomplete nature of most fossils, and the inaccessibility of DNA for all but the youngest fossil species.

In this chapter, we describe the phylogenetic relationships among major clades of extant primates as well as a brief summary of the fossil record of the major clades. For each clade of extant primates, from the major subdivisions of the order, down to individual families and subfamilies, we discuss the biodiversity (number of genera and species) as well as aspects of the behavior and ecology of these living primates that may be relevant to understanding their neurobiology.

20.2 Primate Origins

It is unclear when the order Primates diverged from other orders of placental mammals. The fossil record suggests that this event occurred sometime near the beginning of the Cenozoic Era, the so-called "Age of Mammals." However, molecular estimates regularly suggest that Primates arose much earlier, in the Late Cretaceous 90 to 70 million years ago (eg, Springer et al., 2012; Steiper and Young, 2008). It has been suggested that the discrepancy between the ancient molecular divergence estimates, and the more recent origin suggested by the fossil record, is due to a convergent slowdown in molecular "clock" rate among multiple primate groups that has led to artificially ancient DNA-based estimates (Steiper and Seiffert, 2012). Among extant mammalian orders, Primates are most closely related to two small groups of Asian mammals— Dermoptera, the flying lemurs or colugos, and Scandentia, the tree shrews. The other group of mammals

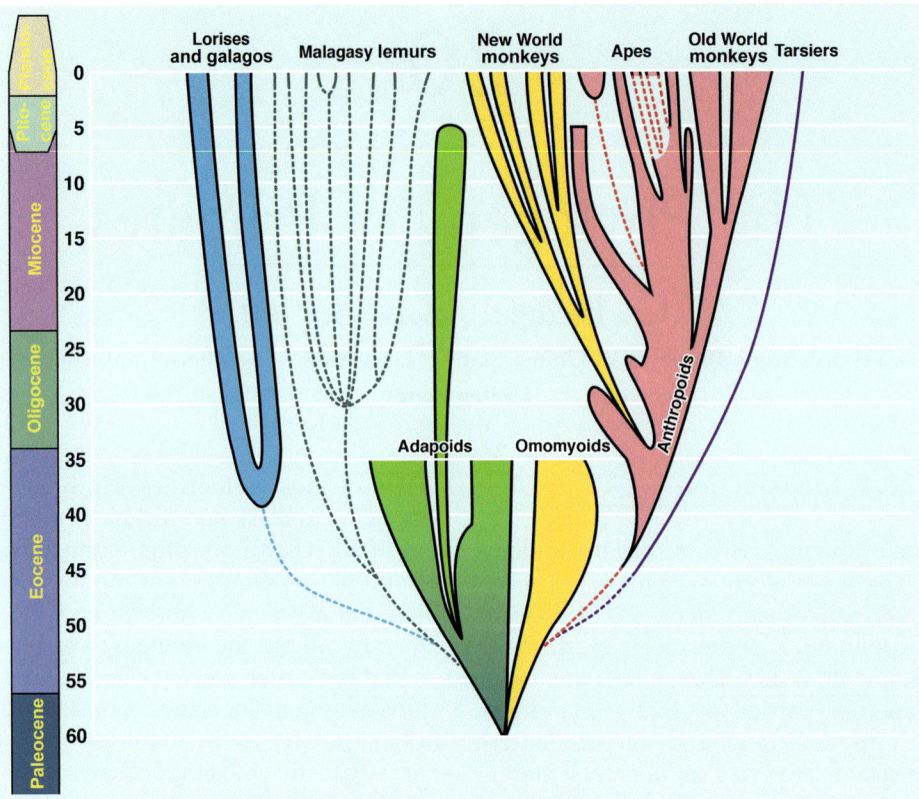

FIGURE 20.1   A phylogenetic tree showing temporal span, relative diversity, and relationships among major groups of living and fossil primates.

commonly identified as the closest relatives, or sister group, of Primates are the extinct Plesiadapiformes, known from an extensive fossil record in the Paleocene and Eocene of North America and Europe, and a more limited record in Asia (eg, Bloch et al., 2007). Some authors include plesiadapiforms within the order Primates. However, the interrelationships among Plesiadapiformes, Dermoptera, and Primates are far from resolved.

20.3 Order Primates

Extant primates, and fossil primates known from the past 56 million years or so, differ from other mammals in many morphological features that reflect aspects of their behavior (eg, Fleagle, 2013). Primates have grasping hands and feet, with nails rather than claws on the tips of their fingers, which are used for grasping branches and manipulating objects with more dexterity than other mammals. Primates also have forward-facing eyes that give primates overlapping visual fields with stereoscopic vision, and each eye is surrounded by a protective

bony ring (a "postorbital bar"). There is debate over whether the earliest primates were nocturnal, with relatively large eyes, and presumably no color vision, or diurnal with some degree of color vision (eg, Heesy and Ross, 2004). The fossil record does not offer any clear answers. In general, primates have relatively larger brains for their body size than other mammals (dolphins and elephants are notable exceptions). However, in Primates, as in other groups of mammals, brain sizes seem to have increased independently in multiple lineages over the past epochs, presumably due at least in part to competition and evolutionary arms races between predators and prey (Simons et al., 2007; Steiper and Seiffert, 2012; Gonzales et al., 2015).

The major division among living (and fossil) Primates (Fig. 20.2) is into two semiorders: Strepsirrhini, the lemurs and lorises (and their fossil relatives), and Haplorhini, the anthropoids and tarsiers (and their fossil relatives). The earliest members of both groups first appear in the fossil record at the beginning of the Eocene Epoch about 56 million years ago, but the earliest and most primitive members (stem taxa) of these clades do not have all of the specialized

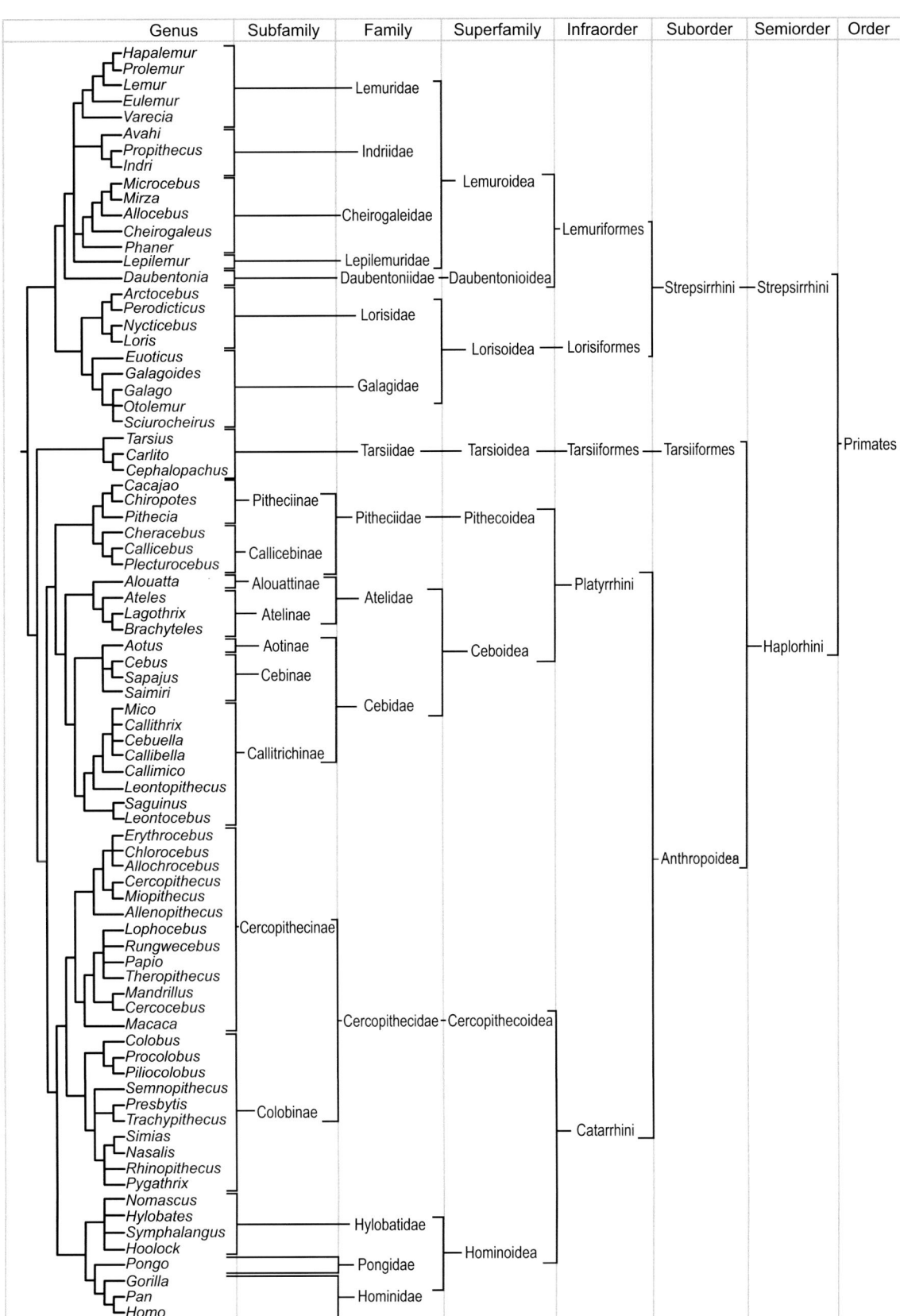

FIGURE 20.2   A cladogram showing phyletic relationships and taxonomy of extant primate genera. *Updated from Fleagle, J.G., 2013. Primate Adaptation and Evolution, third ed. Academic Press, New York.*

morphological features shared by the living members (the crown group).

20.4 Semiorder Strepsirrhini

The living strepsirrhines are divided into two groups: the African and Asian Lorisiformes and the Lemuriformes of Madagascar, both of which share the unusual feature of having lower front teeth (incisors and canines) that are elongated and incorporated into a comblike structure (called a toothcomb) that is variously used for grooming and feeding. Because they have a "wet nose" like most other mammals, the nasolacrimal (tear) duct, which transfers tears from the orbit to the nasal cavity, extends anteriorly to help moisten the nose (Rossie and Smith, 2007). As a group, lemuriforms have smaller brains for their body size than haplorhines (Figs. 20.3 and 20.4). Molecular studies typically place their last common ancestor at well over 50 million years ago, but the earliest tooth-combed primates in the fossil record are about 37 million years old, from sites in North Africa (Seiffert et al., 2003). Stem strepsirrhines, placed in the superfamily Adapoidea, were widespread and often diverse in the Eocene of North America, Europe, Africa, and Asia. They largely went extinct in North America, Europe, and Africa by the end of the Eocene, but one group, the sivaladapids, persisted in Asia until the end of the Miocene (Fig. 20.1).

The Lorisiformes have a fossil record in Africa going back at least 37 million years (Seiffert et al., 2003; Pozzi et al., 2014). Living lorisoids are all small (<2 kg) and nocturnal, but the two families have very different appearances and modes of locomotion. The galagos (Galagidae) of Africa are primarily leapers with very long hindlimbs and feet, long tails, and big ears. The five genera of galagos (*Galagoides*, *Galago*, *Sciurocheirus*, *Euoticus*, and *Otolemur*) show a range of diets with some species specializing on fruits, some on insects, and others on gum found on tree trunks. The living lorisids consist of two separate radiations—one in Africa and one in Asia. In contrast with the galagids, all lorisids are (usually) slow-moving quadrupeds with similar sized forelimbs and hindlimbs, a short stubby tail, and smaller ears that are often hidden in their fur. In both the African and Asian radiations, there is a slender genus with sticklike limbs (Asian *Loris* and African *Arctocebus*) and a heavier, thickly furred genus (Asian *Nycticebus* and African *Perodicticus*), but these similarities are evidently due to convergent evolution (eg, Springer et al., 2012; Fig. 20.5). The slender ones seem to be carnivores that specialize on invertebrates or small vertebrates, while the larger, stockier species eat more fruit and gums. Stem members of the galagid family are first documented in 37-million-year-old (late Eocene) rocks of Egypt, but crown members of Galagidae are not known from fossils until the Miocene epoch (Seiffert et al., 2003; Seiffert, 2007). Molecular studies suggest

FIGURE 20.3 Plots of relative brain size (ECV) in genera of extant primates with Tarsius, Daubentonia, Miopithecus, and Macaca sylvanus labeled. *From Isler, K., Kirk, E.C., Miller, J.M.A., Albrecht, G.A., Gelvin, B.R., Martin, R.D., 2008. Endocranial volumes of primate species: scaling analyses using a comprehensive and reliable data set. J. Hum. Evol. 55, 967–978. Courtesy of Karin Isler.*

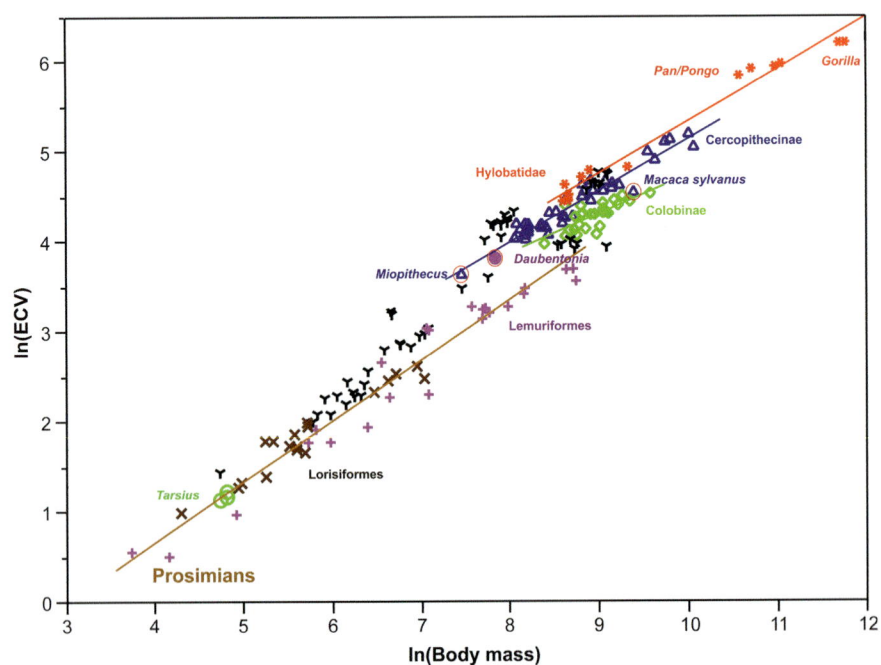

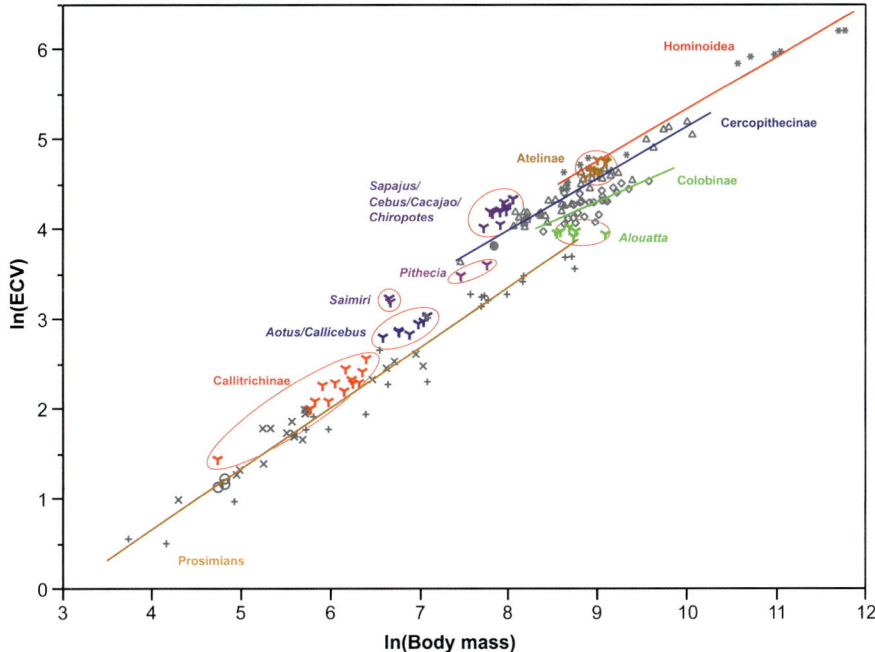

FIGURE 20.4 Plots of relative brain size (ECV) in genera of extant primates with New World Monkeys labeled. *From Isler, K., Kirk, E.C., Miller, J.M.A., Albrecht, G.A., Gelvin, B.R., Martin, R.D., 2008. Endocranial volumes of primate species: scaling analyses using a comprehensive and reliable data set. J. Hum. Evol. 55, 967 −978. Courtesy of Karin Isler.*

FIGURE 20.5 Two Asian lorises, the slender loris, *Loris tardigradus*, and the slow loris, *Nycticebus coucang*. *Illustration by Stephen Nash. From Fleagle, J.G., 2013. Primate Adaptation and Evolution, third ed. Academic Press, New York.*

that the Asian and African lorises diverged in the Eocene or Oligocene (Springer et al., 2012; Pozzi et al., 2014).

The most diverse and well-known radiation of strepsirrhines is Lemuriformes, found exclusively on Madagascar and a few nearby islands, with 15 genera and over 100 species, at last count. The extant species and genera are just a part of an extraordinary radiation of lemurs that evolved on Madagascar. Even though that island has a very poor (effectively nonexistent) Cenozoic fossil record, subfossils from the last 20 000 years or so document many recently extinct large-bodied subfossil lemurs with adaptations that are very different from those found in the relatively small extant taxa (Fig. 20.6; Godfrey and Jungers, 2003; Goodman and Jungers, 2014). Many of the extant Malagasy lemuriforms are nocturnal, and it seems likely that this was the primitive condition for the group. The extant taxa range in size from some of the smallest living primates with a body mass of 50 g or less to species up to 5 kg. The extinct species were all much larger with one estimated to be as large as a gorilla.

Living lemuriforms are normally divided into five families, but many of the relationships among the families are unresolved. The bizarre aye-aye (*Daubentonia*) is placed in its own superfamily Daubentonioidea and is the sister taxon to the other lemuriforms. This strange creature (Fig. 20.7) differs from other lemurs in having ever-growing incisors (like a rodent) rather than a toothcomb, and claws rather than nails on all of its fingers and toes except the hallux. It has a long, slender middle finger that it uses to probe for and retrieve insect larvae. Aye-ayes have enormous ears and very sensitive hearing that extends into the ultrasound range. *Daubentonia* also has an exceptionally large brain for a strepsirrhine of its size (Fig. 20.3).

Lemurids (Lemuridae; Fig. 20.8) include the well-known ring-tailed lemur, *Lemur catta*, as well as numerous species of the genus *Eulemur*, the bamboo

FIGURE 20.6   A reconstruction of some of the extinct subfossil lemurs from Madagascar. *Drawing by Stephen Nash. From Fleagle, J.G., 2013. Primate Adaptation and Evolution, third ed. Academic Press, New York.*

FIGURE 20.8   Three lemurids; the brown lemur *Eulemur fulvus* (above), the ruffed lemur, *Varecia variegata* (middle), and the ring-tailed lemur, *Lemur catta* (below). *Drawing by Stephen Nash. From Fleagle, J.G., 2013. Primate Adaptation and Evolution, third ed. Academic Press, New York.*

FIGURE 20.7   The aye-aye, Daubentonia madagascariensis. *Drawing by Stephen Nash. From Fleagle, J.G., 2013. Primate Adaptation and Evolution, third ed. Academic Press, New York.*

lemurs (*Hapalemur* and *Prolemur*), and ruffed lemurs (*Varecia*). They are all medium-sized species, 1–4 kg in mass, with long tails. Most are arboreal quadrupeds

although *L. catta* often comes to the ground and some species are quadrupedal leapers. Their diets include varying amounts of fruit and leaves, with some species specializing on nectar, and the bamboo lemurs specializing on various parts and species of bamboo. Some lemurids live in monogamous social groups, while others live in larger social groups. Most are diurnal, but some are cathemeral; they are active throughout the 24-h cycle.

Indriids are a relatively small family with only three living genera (*Avahi*, *Propithecus*, and *Indri*). They are all specialized vertical clingers and leapers with very long hindlimbs. They are generally more folivorous than other lemuriforms, but some taxa eat large amounts of fruit seasonally. *Propithecus* and *Indri* are diurnal, while the smaller *Avahi* is nocturnal. The palaeopropithecids (the sloth lemurs) and archaeolemurids (the monkey lemurs) are two extinct families that were closely related to the indriids (Kistler et al., 2015), but much larger and quite different in their habits.

FIGURE 20.10   The sportive lemur, *Lepilemur mustelinus* in a dry forest.   *Drawing by Stephen Nash. From Fleagle, J.G., 2013. Primate Adaptation and Evolution, third ed. Academic Press, New York.*

FIGURE 20.9   Five genera of cheirogaleids. Counterclockwise from upper left: the mouse lemur *Microcebus murinus*; the hairy-eared dwarf lemur, *Allocebus trichotis*; a greater dwarf lemur, *Cheirogaleus major*; a fork-marked lemur, *Phaner furcifer* licking gum from a tree trunk; and above, Coquerel's dwarf lemur, *Mirza coquereli*.   *Drawing by Stephen Nash. From Fleagle, J.G., 2013. Primate Adaptation and Evolution, third ed. Academic Press, New York.*

Cheirogaleids (Fig. 20.9) are the smallest (30–500 g) of the lemuriforms and are a very speciose family, with over 20 species recognized in the genus *Microcebus* alone. They are all nocturnal, mostly quadrupedal animals with long tails that eat mostly fruits and invertebrates. They exhibit a variety of social patterns, including communal nesting. The fork-marked lemurs, genus *Phaner*, are gum specialists with clawlike nails for clinging to tree trunks. They have traditionally been grouped with cheirogaleids, but some molecular phylogenies find that they are more closely related to the sportive or weasel lemurs of the genus *Lepilemur*. Either way, *Phaner* is a more ancient and distinct branch in the lemur tree than had previously been appreciated.

Lepilemurids (Lepilemuridae), the sportive or weasel lemurs, are a family of lemuriforms comprised of the single widespread genus *Lepilemur* (Fig. 20.10), with over 25 recognized species. Lepilemurs are smallish (most less than 1 kg), nocturnal, leaf-eating specialists

with no upper incisor teeth and unusual intestinal features for digesting leaves. They are all leapers with long legs and a long tail. The place of lepilemurs in lemuriform phylogeny has long been difficult to resolve. In the past decades, they have been linked to lemurids, indriids, cheirogaleids, as well as the large, extinct "koala lemur" *Megaladapis*; and some studies suggest that they are especially close to the genus *Phaner*. Recent molecular studies tend to place lepilemurids closest to cheirogaleids (Springer et al., 2012; Herrera and Dávalos, 2016).

20.5 Semiorder Haplorhini

Living haplorhines consist of two very distinct groups, tarsiers and anthropoids, usually separated at the subordinal level. As a group, haplorhines share a suite of derived features, mostly in the cranium, that distinguish them from the more primitive strepsirrhines. All extant haplorhines have a retinal fovea on the posterior surface of their eyeballs, and their orbits are partitioned off from the temporal fossa by a bony plate, the postorbital septum. Haplorhines have relatively smaller olfactory bulbs with a reduced number of nasal turbinates from the condition found in strepsirrhines and most other mammals (Cave, 1973; Smith et al., 2007). They also have a reduced vomeronasal (Jacobson's) organ and lack the moist rhinarium (wet nose) characteristic of strepsirrhines and most other mammals. In conjunction with their "dry nose," haplorhines have a vertical nasolacrimal (tear) duct linking the orbit with the nasal cavity (Rossie and Smith, 2007).

The oldest fossil haplorhines are known as omomyi-forms or omomyoids, and first appear at the beginning of the Eocene epoch about 56 million years ago (Fig. 20.1; Smith et al., 2006). However, the phyletic relationships among anthropoids, tarsiers, and omomyiforms are a source of considerable debate. Some authors argue that anthropoids and tarsiers are sister taxa, to the exclusion of omomyiforms. Others argue that anthropoids and tarsiers are derived from separate clades within omomyiforms. Still others argue that all known omomyiforms are specifically related to tarsiers. In any case, it is clear that tarsiers and anthropoids have been distinct for tens of millions of years. Molecular analyses place the divergence time at over 60 million years. Both tarsier-like fossils and stem anthropoids appear in the fossil record in the middle Eocene of China (Beard et al., 1994, 1998).

Living tarsiids (family Tarsiidae), currently restricted to islands of Southeast Asia (Philippines, Malaysia, Indonesia), are among the strangest of living primates (Fig. 20.11). Once grouped in the genus *Tarsius*, there are now three recognized genera (Groves and Shekelle, 2010)—*Tarsius*, *Cephalopachus*, and *Carlito*—whose common ancestor likely traces back to the early Miocene or

FIGURE 20.11 The spectral or Selayar tarsier, *Tarsius tarsier*. *Drawing by Stephen Nash. From Fleagle, J.G., 2013. Primate Adaptation and Evolution, third ed. Academic Press, New York.*

late Oligocene (Springer et al., 2012). They are all small (<150 g) with extremely long legs, long ankle bones, a long slender tail, enormous hands, large ears, and enormous eyes. All tarsiers are nocturnal, but they share with anthropoids a very specialized retina with an all-cone fovea centralis. Fossil tarsiids have been described on the basis of teeth from the Eocene of China (Beard et al., 1994, 1998) and the Miocene of Thailand and Pakistan (eg, Chaimanee et al., 2010; Zijlstra et al., 2013).

20.5.1 Anthropoids

Anthropoids (monkeys, apes, and humans) are the most widespread and morphologically diverse (Fleagle et al., 2010) of living primates. Nonhuman primate species are most abundant in tropical areas of North and South America, Africa, and Asia. Anthropoids are generally larger than extant strepsirrhines or tarsiids, have relatively larger brains (Fig. 20.4), and, with the exception of the New World night monkey *Aotus*, are all diurnal. The earliest fossil anthropoids are from the middle Eocene of Asia (Beard et al., 1994). They had a large adaptive radiation in the late Eocene and Oligocene of Africa (Seiffert et al., 2010; Seiffert, 2012), and most major clades probably originated in Africa and subsequently spread to other continents. There are two very diverse, and (with the exception of humans) geographically distinct clades—the platyrrhines of South and Central America, and the catarrhines of Africa and Asia, which were formerly widespread in Europe as well.

20.5.2 Platyrrhines (Infraorder Platyrrhini)

There are over 20 genera and 170 species of platyrrhines, or New World monkeys, in tropical areas of Central and South America. The group is almost certainly derived from early anthropoids that rafted from Africa to the Neotropics sometime late in the Eocene epoch. The earliest fossil monkeys from South America are strikingly like fossils of the same age from North Africa (Bond et al., 2015). There is an extensive record of fossil platyrrhines from many parts of South America and the Caribbean, some of which were much larger than any extant New World monkeys (Kay et al., 2012; Kay, 2015).

Living platyrrhines are small to medium in size (50 g–10 kg). All but one (*Aotus*) are diurnal, and all have polymorphic trichromatic color vision, except for *Alouatta* which has routine trichromacy and *Aotus* which lacks color vision. Like strepsirrhines and tarsiers, most platyrrhines seem to have a functioning vomeronasal (Jacobson's) organ. As a group, platyrrhines are all arboreal and mostly quadrupedal, with leaping common in many smaller species and suspensory behavior more common in the largest. Living platyrrhines are divided into three families.

20.5.3 Pitheciids

The pitheciids (family Pitheciidae) are medium-sized monkeys (1–4 kg). They are all arboreal quadrupeds and leapers, and many specialize as seed predators. There are six genera placed in two subfamilies. The subfamily Callicebinae consists of the titi monkeys now placed in three genera *Callicebus*, *Cheracebus*, and *Plecturocebus*, with 34 recognized species (Byrne et al., 2016). The subfamily Pitheciinae has three genera, the saki monkeys of the genus *Pithecia*; the bearded sakis of the genus *Chiropotes*; and the uakaries of the genus *Cacajao*. These are all characterized by enlarged front teeth that they use to open the hard shells of seeds, and the latter two genera have relatively large brains for their size (Fig. 20.4). Most phylogenies place pitheciids as the sister group of all other living platyrrhines.

20.5.4 Cebids

Cebidae is a very heterogeneous family of small- to medium-sized monkeys placed in three distinct subfamilies (eg, Kiesling et al., 2015). Cebines include squirrel monkeys (*Saimiri*) and capuchins (*Cebus* and *Sapajus*). All are primarily arboreal quadrupeds who also leap. They eat fruits and invertebrates and live in large social groups. All cebines have relatively large brains for their body size (Fig. 20.4), and many capuchin species use tools in their foraging behavior (Fig.20.12; Ottoni and Izar, 2008; Visalberghi and Fragaszy, 2013). According to all molecular studies, the nocturnal owl monkeys (*Aotus*) are closely related to cebines. As noted previously, *Aotus* (Fig. 20.13) is the only nocturnal anthropoid. They are small, monogamous, fruit- and insect-eating monkeys with very large eyes.

The other subfamily in the Cebidae is the Callitrichinae—marmosets (*Callithrix*, *Cebuella*, and *Mico*) (Schneider et al., 2012), tamarins (*Leontopithecus* and *Saguinus*), and Goeldi's monkey (*Callimico*). There are all small monkeys (<700 g), with long tails and claws rather than nails on all of their digits except the big toe. They eat fruits, invertebrates, and gums, and all except *Callimico* have two rather than three molar teeth in each quadrant. In addition, they all have twin births, and a cooperative breeding system, except *Callimico*, which has single births.

20.5.5 Atelids

Atelids are the largest platyrrhines (7–10 kg), and all are suspensory to some degree and have a prehensile

FIGURE 20.12   A tufted capuchin (*Sapajus libidinosus*) using a stone to break open nuts. *Photo courtesy of Barth Wright.*

FIGURE 20.13   The night monkey, *Aotus trivirgatus*. *Drawing by Stephen Nash. From Fleagle, J.G., 2013. Primate Adaptation and Evolution, third ed. Academic Press, New York.*

tail. There are four genera. The howler monkeys (*Alouatta*) are the most folivorous platyrrhines and have the smallest brains (relative to body mass) of any anthropoid primate (Fig. 20.4). They are usually placed in their own subfamily, the Alouattinae. The other three genera, placed in a separate family, the Atelinae, have relatively larger brains. The spider monkeys, *Ateles*, are the most frugivorous and live in a fluid, fission–fusion system similar to that of chimpanzees. The phyletic relationship of *Brachyteles*, the muriqui, to other atelids has been debated for many decades, but it now seems resolved that *Brachyteles* is most closely related to the woolly monkeys, *Lagothrix* (Springer et al., 2012;

Kiesling et al., 2015), even though it shares some specializations of its limbs, including lack of a thumb, with spider monkeys and shares dental features (probably related to folivory) with howler monkeys.

20.5.6 Catarrhines (Infraorder Catarrhini)

The sister taxon to the New World platyrrhines are the catarrhines, which, except for humans, are found only in Africa and Asia. During the Miocene epoch, there was also a diverse group of catarrhines living in Europe. Catarrhines diverged from primitive anthropoids and platyrrhines in the middle or late Eocene of Africa (Seiffert et al., 2010; Seiffert, 2012). Compared with platyrrhines, catarrhines have two rather than three premolar teeth in each quadrant; they all have routine trichromacy and a reduced vomeronasal organ. Relative brain sizes of catarrhines overlap with those of platyrrhines (Fig. 20.4), but it is now clear that relative brain size increased independently, not only in catarrhines and platyrrhines (eg, Simons et al., 2007), but also within different groups of catarrhines (Fig. 20.3; Gonzales et al., 2015).

There are two living superfamilies of catarrhines, the Cercopithecoidea, or Old World monkeys, and the Hominoidea, apes and humans. Most molecular studies place the divergence between cercopithecoids and hominoids at 25—30 million years ago, which agrees with the earliest fossil appearance of the two in the late Oligocene (Stevens et al., 2013). Both groups underwent widespread adaptive radiations during the Miocene with that of hominoids occurring earlier than that of cercopithecoids. Although stem catarrhines and stem hominoids were more diverse and abundant than early cercopithecoids were in the Miocene, living Old World monkeys are now more diverse, more widespread, and more abundant than living apes, except humans.

20.5.7 Old World Monkeys (Cercopithecoidea)

Living Old World monkeys are distinct from living apes in the bilophodont morphology of their teeth, in having arms that are generally equal in length to, or shorter than, their legs, and in that all have tails (although they may be very short). Cercopithecoids are divided into two clades, usually recognized at either the family or subfamily level, the leaf-eating colobines (or colobids) and the more fruit-eating cercopithecines (or cercopithecids).

Colobines are medium to large monkeys (4—20 kg) and differ from cercopithecines in having higher cusps on their molar teeth and in having a sacculated stomach (like cows), both adaptations for a diet of leaves. Colobines also have thumbs that are reduced or absent.

They tend to be leapers and have relatively longer legs than cercopithecines. Although the group originated in Africa in the middle Miocene (eg, Nakatsukasa et al., 2010), they are now more diverse in Asia (7 genera and over 50 species; Fig. 20.14) than in Africa (3 genera and about 25 species; Fig. 20.15).

Cercopithecines are small to large monkeys (1—30 kg). They have lower, more distinct cusps on their molar teeth than do colobines, and have cheek pouches for storing food. They are predominantly arboreal quadrupeds, but many are also terrestrial. Cercopithecines have a moderate fossil record from the late Miocene onward. There is roughly the same number of extant species of cercopithecines (76) as colobines (80), but cercopithecines are divided into many more genera (13) than colobines (10). They are overwhelmingly African, with only one very speciose genus (*Macaca* with 23 species) found in Asia (Fig. 20.16). Extant cercopithecines are divided into two major groups, usually recognized as tribes. The larger papionins [macaques,

FIGURE 20.14 Two species of colobine monkeys in Malaysia, the spectacled langur, *Trachypithecus obscurus* (above) and the white-thighed surili, *Presbytis siamensis* (below). *Drawing by Stephen Nash. From Fleagle, J.G., 2013. Primate Adaptation and Evolution, third ed. Academic Press, New York.*

FIGURE 20.15   Two species of African colobine monkeys, the red colobus, *Piliocolobus badius* (above and below), and the black-and-white colobus, *Colobus guereza* (center). *Drawing by Stephen Nash. From Fleagle, J.G., 2013. Primate Adaptation and Evolution, third ed. Academic Press, New York.*

FIGURE 20.16   Two species of Asian macaques, the crab-eating or long-tailed macaque, *Macaca fascicularis*, and the larger, more terrestrial pigtailed macaque *Macaca nemestrina*. *Drawing by Stephen Nash. From Fleagle, J.G., 2013. Primate Adaptation and Evolution, third ed. Academic Press, New York.*

mangabeys, baboons (Fig. 20.17), geladas, and mandrills] are the largest cercopithecoids and have a rich fossil record in the Pliocene and Pleistocene. Within the group, macaques form one clade; baboons (*Papio*), kipungis (*Rungwecebus*), geladas (*Theropithecus*), and *Lophocebus* mangabeys form another clade; and mandrills and drills (*Mandrillus*) and *Cercocebus* mangabeys form another clade. However, phyletic relationships within and among those clades are unresolved.

20.5.8 Hominoids

Hominoids, apes and humans, are one of the least diverse clades of living primates, but have a very extensive fossil record that begins in the late Oligocene (Stevens et al., 2013) and that includes many extinct genera in Africa, Europe, and Asia (Fleagle, 2013). There are eight extant hominoid genera and 26 species (most of which are gibbons). Living hominoids are commonly divided into three families: Hylobatidae (gibbons), Pongidae (orangutans), and Hominidae (gorillas, chimpanzees, and humans). Although the superfamily

FIGURE 20.17   A group of olive baboons *Papio anubis*, on the savannah of East Africa. *Drawing by Stephen Nash. From Fleagle, J.G., 2013. Primate Adaptation and Evolution, third ed. Academic Press, New York.*

originated in Africa, their evolution took place throughout Africa and Eurasia. Only humans have expanded to other continents. There are numerous genera of stem hominoids and stem hominids in the fossil record from the late Oligocene through Miocene.

There are Asian fossils from the Miocene and later that are related to orangutans (*Pongo*), but there are few, if any, identifiable fossil gibbons not attributable to extant taxa (Jablonski and Chapman, 2009), or fossil relatives of the African ape genera, *Pan* and *Gorilla*. In contrast, there are numerous genera and species of extinct humans (subfamily Homininae).

Compared to other primates, living hominoids are medium to very large in size (5–175 kg). Gibbons are totally arboreal and suspensory, but other hominoids, while more suspensory than most other primates, show varying amounts of terrestrial locomotion. Living hominoids include predominantly frugivorous as well as predominantly folivorous taxa; humans are the most faunivorous/carnivorous hominoids, and our dietary behavior is greatly augmented by cooking and other cultural habits. Great apes (but not gibbons) show varying amounts of tool use and other cultural abilities. All hominoids have relatively larger brains than other primates, and human brains are among the largest, relative to body mass, in the animal kingdom (Fig. 20.4). As with other catarrhines, hominoids have routine trichromatic vision and a very reduced or absent vomeronasal organ.

There are 19 species of gibbons (family Hylobatidae) placed in four genera (*Hylobates*, *Symphalangus*, *Hoolock*, and *Nomascus*), all from tropical parts of Asia. They are medium-sized primates (5–12 kg), totally arboreal and suspensory, with diets that include varying proportions of fruits (especially figs), leaves, and invertebrates. Their rapid brachiating locomotor abilities would seem to require exceptional agility and coordination (Fig. 20.18). There are no descriptions of tool use or other cultural behaviors by gibbons.

The two species of Orangutans (Pongo) in the family Pongidae are from the islands of Borneo and Sumatra, but there are remains of fossil pongids from many parts of eastern and southeastern Asia.

Hominids are an African radiation. There are two species of gorillas (Fig. 20.19) and two species of chimpanzees with numerous subspecies of Pan troglodytes. Chimpanzees and humans share many anatomical and behavioral features, including tool use and cultural traditions (Fig. 20.20; Matsuzawa et al., 2011; Whiten et al., 1999).

20.6 Broad-Scale Trends in Primate Brain and Sensory Evolution

As noted earlier, primates as a group can be distinguished from other mammals by their relatively large brains, and it has been commonly assumed that the last common ancestor of primates was similarly large brained when compared to members of other placental mammalian orders that were diversifying in the early

FIGURE 20.18   Two hylobatids from Malaysia, the white-handed gibbon *Hylobates lar* (upper left), and the siamang, *Symphalangus syndactylus* (below). *Drawing by Stephen Nash. From Fleagle, J.G., 2013. Primate Adaptation and Evolution, third ed. Academic Press, New York.*

part of the Cenozoic. Recent fossil discoveries have challenged this view, however, and instead support a scenario in which relative brain size has probably increased independently in multiple primate lineages. Ancestral reconstructions of body size, endocranial volume, and relative endocranial volume that incorporate the evidence from early fossil haplorhines (omomyiforms) and stem strepsirrhines (adapiforms) indicate that the ancestor of all living primates was probably tiny (about 55 g) and had a smaller endocranial volume and smaller relative endocranial volume than any living primate (Steiper and Seiffert, 2012). This pattern is driven in part by the relatively small endocranial volumes of Eocene primates (eg, Silcox et al., 2010). Evidence from the stem anthropoid *Parapithecus* (Bush et al., 2004), the stem catarrhine *Aegyptopithecus* (Simons et al., 2007), the stem platyrrhines *Chilecebus* and *Homunculus* (Sears et al., 2008; Kay et al., 2012), and the stem cercopithecoid *Victoriapithecus* (Gonzales et al., 2015) suggests that relative brain size has increased independently within Platyrrhini, Cercopithecoidea, and Hominoidea (Fig. 20.21). Estimates of relative endocranial volume provide only a rough proxy for "intelligence," however, particularly given the problems associated with estimating body mass in fossil primates.

FIGURE 20.19 A group of mountain gorillas *Gorilla beringei*, in the Virunga volcanoes of Rwanda. *Drawing by Stephen Nash. From Fleagle, J.G., 2013. Primate Adaptation and Evolution, third ed. Academic Press, New York.*

FIGURE 20.20 A young chimpanzee watches its mother use a stone tool. *Photo provided by Tetsuro Matsuzawa, Kyoto University.*

The early evolution of activity patterns, color vision, and high-acuity vision within Primates are also topics of great interest at present. An analysis that took into account both fossil and genetic evidence concluded that the common ancestor of living primates was nocturnal and lacked color vision (Heesy and Ross, 2004), and

the hypothesis that nocturnality characterized the ancestral primate has more recently been supported by various phylogenetically informed comparative analyses of extant taxa (Griffin et al., 2012; Santini et al., 2015). The activity pattern of the ancestral haplorhine has been, and continues to be, much more controversial. Some have argued that nocturnal haplorhines (tarsiers and *Aotus*) have greatly enlarged eyeballs because a reflective tapetum lucidum was lost, and retinal fovea acquired, along the haplorhine stem lineage during a diurnal phase (eg, Cartmill, 1980; Ross, 1996); reversions to nocturnality consequently required evolution of exceptionally large eyeballs to capture adequate light for nighttime activity (eg, Kay and Kirk, 2000). However, Heesy and Ross (2001) reconstructed the last common ancestor of extant haplorhines as having been nocturnal, and the same result was found by Santini et al. (2015), while the ancestral reconstruction of Griffin et al. (2012) was equivocal. And while most authorities have assumed that anthropoids were ancestrally diurnal [supported by the analyses of Ross (1996), Heesy and Ross (2001), and Griffin et al. (2012)], the ancestral reconstruction of Santini et al. (2015) for this node was equivocal. Fossil evidence may come to play an interesting role in helping to resolve this issue, as one of the oldest fossil anthropoids from Africa (*Biretia*) exhibits some morphological features of the orbital floor that are consistent with ocular hypertrophy (Seiffert et al., 2005); nevertheless, all other fossil anthropoids appear to have been diurnal, and all extant diurnal anthropoids are characterized by high-acuity vision and decreased retinal summation, consistent with them having had a diurnal common ancestor. While it is most parsimonious to interpret this high-acuity vision as having been acquired along the anthropoid stem lineage, the situation could be more complex if tarsiers are secondarily nocturnal (Kay and Kirk, 2000). Among anthropoids, the only clades that have evolved routine trichromatic color vision are catarrhines and the platyrrhine *Alouatta*.

Olfaction is important for primates in both feeding and social interaction, but measurement of olfactory abilities is difficult and subject to much debate (Nevo and Heymann, 2015). However, Primates appear to show considerable variation in the relative emphasis on visual versus olfactory exploration of their environments, with the evolution of high-acuity vision clearly having been the primary adaptive strategy, at the expense of the olfactory system, in anthropoids. Primitively, primates likely had a complex system of four ethmoturbinal bones in an ethmoturbinal recess, but this number has been reduced to two in haplorhines (Smith et al., 2007). Correlated with this reduction in the ethmoturbinals in haplorhines is a reduction of olfactory epithelium. The main olfactory system and vomeronasal organ appear to have been functional in most primate

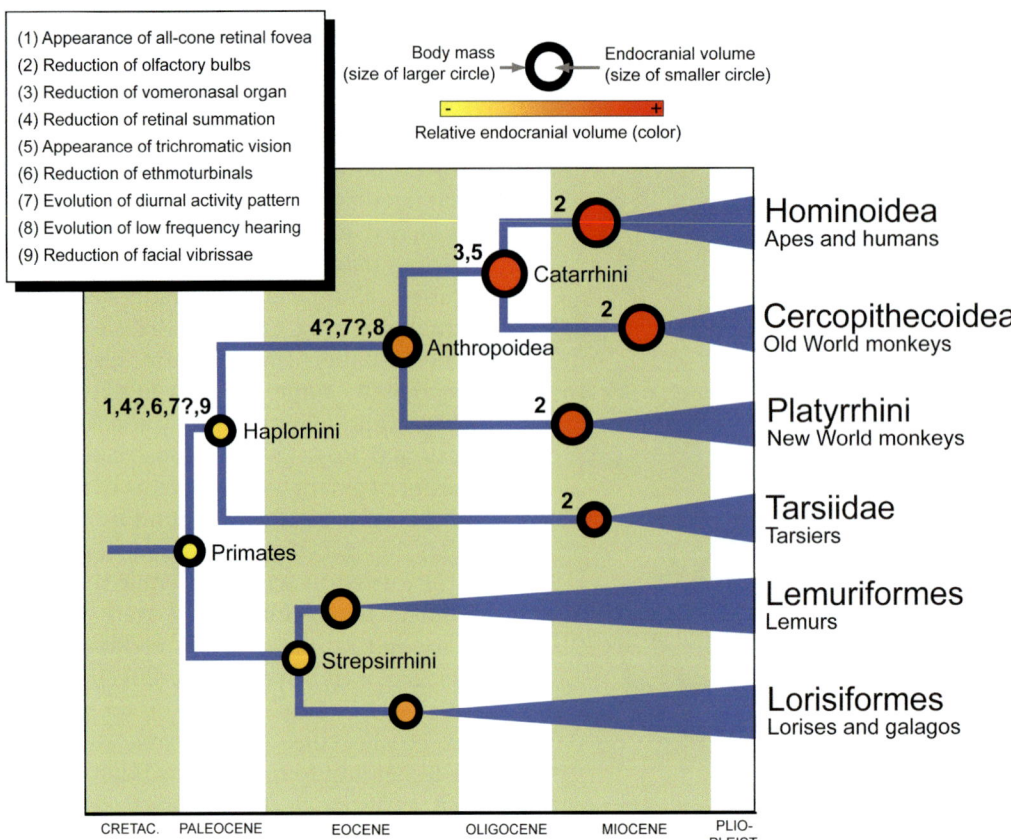

FIGURE 20.21 A plot showing evolutionary changes and ancestral states in relative brain size and numerous aspects of sensory biology during primate evolution. *Modified in part from Steiper, M.E., Seiffert, E.R., 2012. Evidence for a convergent slowdown in primate molecular rates and its implications for the timing of early primate evolution. Proc. Natl. Acad. Sci. U.S.A. 109, 6006–6011.*

clades throughout the evolutionary history of the order, but the vomeronasal organ was greatly reduced or lost along the catarrhine stem lineage and appears to be nonfunctional in members of that group (Smith et al., 2001; Gonzales et al., 2015). Interestingly, however, Gonzales et al. (2015) have recently argued on the basis of olfactory bulb size in the middle Miocene stem cercopithecoid *Victoriapithecus* that the relative size of the olfactory bulbs decreased independently within Platyrrhini, Cercopithecoidea, and Hominoidea.

Facial vibrissae (or "whiskers") also provide primates with sensory signals as they move through their environments, and it has long been suggested that primates have fewer facial vibrissae than nonprimates and that haplorhines have fewer vibrissae than strepsirrhines, reflecting a reduced emphasis on this aspect of the trigeminal system (eg, Le Gros Clark, 1959). However, Muchlinksi (2010) found that primates do not differ significantly from nonprimates, and haplorhines do not differ significantly from strepsirrhines, in vibrissae counts. Nevertheless, she found that primates do have

relatively small infraorbital foramina, which form the bony conduit for the infraorbital nerve that innervates midfacial vibrissae, when compared with nonprimates other than dermopterans.

Finally, the hearing capacities of primates are not particularly notable when compared with other mammals, but variation within the order does exist along phylogenetic lines. In particular, strepsirrhines appear to be particularly sensitive to high-frequency sounds, while anthropoids, and particularly catarrhines, are sensitive to low-frequency sounds; also, anthropoids typically have two peaks of sensitivity, while strepsirrhines have only one (eg, Coleman and Colbert, 2010). Haplorhines also have longer cochleae than strepsirrhines (Coleman and Boyer, 2012). Coleman and Boyer (2012) predicted hearing sensitivity in various fossil primates and found that omomyiforms and adapiforms were likely most similar to strepsirrhines in their hearing capacities, while the characteristic anthropoid pattern was present by the Oligocene, as shown by the stem catarrhine *Aegyptopithecus*.

APPENDIX 1 Currently recognized living Primate taxa (species and subspecies)

Cheirogaleidae	Gray Mouse Lemur	*Microcebus murinus*
	Gray-brown Mouse Lemur	*Microcebus griseorufus*
	Madame Berthe's Mouse Lemur	*Microcebus berthae*
	Peters' Mouse Lemur	*Microcebus myoxinus*
	Golden-brown Mouse Lemur	*Microcebus ravelobensis*
	Bongolava Mouse Lemur	*Microcebus bongolavensis*
	Ambarijeby Mouse Lemur	*Microcebus danfossi*
	Margot Marsh's Mouse Lemur	*Microcebus margotmarshae*
	Sambirano Mouse Lemur	*Microcebus sambiranensis*
	Nosy Be Mouse Lemur	*Microcebus mamiratra*
	Tavaratra Mouse Lemur	*Microcebus tavaratra*
	Montagne d'Ambre Mouse Lemur	*Microcebus arnholdi*
	Rufous Mouse Lemur	*Microcebus rufus*
	Jolly's Mouse Lemur	*Microcebus jollyae*
	Gerp's Mouse Lemur	*Microcebus gerpi*
	Goodman's Mouse Lemur	*Microcebus lehilahytsara*
	Simmons' Mouse Lemur	*Microcebus simmonsi*
	Anjiahely Mouse Lemur	*Microcebus macarthurii*
	Mittermeier's Mouse Lemur	*Microcebus mittermeieri*
	Marohita Mouse Lemur	*Microcebus marohita*
	Anosi Mouse Lemur	*Microcebus tanosi*
	Bemanasy Mouse Lemur	*Microcebus manitatra*
	Ganzhorn's Mouse Lemur	*Microcbeus ganzhorni*
	Nosy Boraha Mouse Lemur	*Microcebus boraha*
	Coquerel's Giant Mouse Lemur	*Mirza coquereli*
	Northern Giant Mouse Lemur	*Mirza zaza*
	Hairy-eared Dwarf Lemur	*Allocebus trichotis*
	Fat-tailed Dwarf Lemur	*Cheirogaleus medius*
	Sibree's Dwarf Lemur	*Cheirogaleus sibreei*
	Greater Dwarf Lemur	*Cheirogaleus major*
	Crossley's Dwarf Lemur	*Cheirogaleus crossleyi*
	Lavasoa Dwarf Lemur	*Cheirogaleus lavasoensis*
	Thomas' Dwarf Lemur	*Cheirogaleus thomasi*
	Montagne d'Ambre Dwarf Lemur	*Cheirogaleus andysabini*
	Ankarana Dwarf Lemur	*Cheirogaleus shethi*
	Groves' Dwarf Lemur	*Cheirogaleus grovesi*
	Masoala Fork-marked Lemur	*Phaner furcifer*
	Pale Fork-marked Lemur	*Phaner pallescens*
	Sambirano Fork-marked Lemur	*Phaner parienti*
	Montagne d'Ambre Fork-marked Lemur	*Phaner electromontis*

Continued

Lepilemuridae	Weasel Sportive Lemur	*Lepilemur mustelinus*
	Betsileo Sportive Lemur	*Lepilemur betsileo*
	Small-toothed Sportive Lemur	*Lepilemur microdon*
	Manombo Sportive Lemur	*Lepilemur jamesorum*
	Wright's Sportive Lemur	*Lepilemur wrightae*
	Andohahela Sportive Lemur	*Lepilemur fleuretae*
	Mananara-Nord Sportive Lemur	*Lepilemur hollandorum*
	Seal's Sportive Lemur	*Lepilemur seali*
	Masoala Sportive Lemur	*Lepilemur scottorum*
	Daraina Sportive Lemur	*Lepilemur milanoii*
	Ankarana Sportive Lemur	*Lepilemur ankaranensis*
	Sahafary Sportive Lemur	*Lepilemur septentrionalis*
	Gray's Sportive Lemur	*Lepilemur dorsalis*
	Nosy Be Sportive Lemur	*Lepilemur tymerlachsoni*
	Mittermeier's Sportive Lemur	*Lepilemur mittermeieri*
	Sahamalaza Sportive Lemur	*Lepilemur sahamalaza*
	Anjiamangirana Sportive Lemur	*Lepilemur grewcockorum*
	Ambodimahabibo Sportive Lemur	*Lepilemur otto*
	Milne-Edwards' Sportive Lemur	*Lepilemur edwardsi*
	Antafia Sportive Lemur	*Lepilemur aeeclis*
	Tsiombikibo Sportive Lemur	*Lepilemur ahmansonorum*
	Bemaraha Sportive Lemur	*Lepilemur randrianasoloi*
	Red-tailed Sportive Lemur	*Lepilemur ruficaudatus*
	Zombitse Sportive Lemur	*Lepilemur hubbardorum*
	Petter's Sportive Lemur	*Lepilemur petteri*
	White-footed Sportive Lemur	*Lepilemur leucopus*
Lemuridae	Gray Bamboo Lemur	*Hapalemur griseus*
	Gray Bamboo Lemur	*Hapalemur griseus griseus*
	Beanamalao Bamboo Lemur	*Hapalemur griseus gilberti*
	Ranomafana Bamboo Lemur	*Hapalemur griseus ranomafanensis*
	Southern Bamboo Lemur	*Hapalemur meridionalis*
	Northern Bamboo Lemur	*Hapalemur occidentalis*
	Lac Alaotra Bamboo Lemur	*Hapalemur alaotrensis*
	Golden Bamboo Lemur	*Hapalemur aureus*
	Greater Bamboo Lemur	*Prolemur simus*
	Ring-tailed Lemur	*Lemur catta*
	Brown Lemur	*Eulemur fulvus*
	Rufous Brown Lemur	*Eulemur rufus*
	Red-fronted Brown Lemur	*Eulemur rufifrons*
	White-fronted Brown Lemur	*Eulemur albifrons*

APPENDIX 1 Currently recognized living Primate taxa (species and subspecies)—cont'd

	Sanford's Brown Lemur	*Eulemur sanfordi*
	White-collared Brown Lemur	*Eulemur cinereiceps*
	Red-collared Brown Lemur	*Eulemur collaris*
	Black Lemur	*Eulemur macaco*
	Blue-eyed Black Lemur	*Eulemur flavifrons*
	Crowned Lemur	*Eulemur coronatus*
	Red-bellied Lemur	*Eulemur rubriventer*
	Mongoose Lemur	*Eulemur mongoz*
	Black-and-White Ruffed Lemur	*Varecia variegata*
	Variegated Ruffed Lemur	*Varecia variegata variegata*
	Southern Black-and-White Ruffed Lemur	*Varecia variegata editorum*
	Northern Black-and-White Ruffed Lemur	*Varecia variegata subcincta*
	Red Ruffed Lemur	*Varecia rubra*
Indriidae	Eastern Woolly Lemur	*Avahi laniger*
	Masoala Woolly Lemur	*Avahi mooreorum*
	Peyriéras' Woolly Lemur	*Avahi peyrierasi*
	Betsileo Woolly Lemur	*Avahi betsileo*
	Manombo Woolly Lemur	*Avahi ramanantsoavani*
	Southern Woolly Lemur	*Avahi meridionalis*
	Western Woolly Lemur	*Avahi occidentalis*
	Bemaraha Woolly Lemur	*Avahi cleesei*
	Sambirano Woolly Lemur	*Avahi unicolor*
	Verreaux's Sifaka	*Propithecus verreauxi*
	Decken's Sifaka	*Propithecus deckenii*
	Crowned Sifaka	*Propithecus coronatus*
	Coquerel's Sifaka	*Propithecus coquereli*
	Tattersall's Sifaka	*Propithecus tattersalli*
	Diademed Sifaka	*Propithecus diadema*
	Milne-Edwards' Sifaka	*Propithecus edwardsi*
	Silky Sifaka	*Propithecus candidus*
	Perrier's Sifaka	*Propithecus perrieri*
	Indri	*Indri indri*
Daubentoniidae	Aye-aye	*Daubentonia madagascariensis*
Galagidae	Demidoff's Dwarf Galago	*Galagoides demidoff*
	Demidoff's Dwarf Galago	*Galagoides demidoff demidoff*
	Bioko Dwarf Galago	*Galagoides demidoff poensis*
	Thomas's Dwarf Galago	*Galagoides thomasi*
	Angolan Dwarf Galago	*Galagoides kumbirensis*
	Mountain Dwarf Galago	*Paragalago orinus*
	Rondo Dwarf Galago	*Paragalago rondoensis*

Continued

4. Primates

	Mozambique Dwarf Galago	*Paragalago granti*
	Kenya Coast Dwarf Galago	*Paragalago cocos*
	Tanzania Coast Dwarf Galago	*Paragalago zanzibaricus*
	Zanzibar Dwarf Galago	*Paragalago zanzibaricus zanzibaricus*
	Udzungwa Dwarf Galago	*Paragalago zanzibaricus udzungwensis*
	Northern Lesser Galago	*Galago senegalensis*
	Senegal Lesser Galago	*Galago senegalensis senegalensis*
	Kenya Lesser Galago	*Galago senegalensis. braccatus*
	Ethiopia Lesser Galago	*Galago senegalensis dunni*
	Uganda Lesser Galago	*Galago senegalensis sotikae*
	Somali Lesser Galago	*Galago gallarum*
	Southern Lesser Galago	*Galago moholi*
	Spectacled Lesser Galago	*Galago matschiei*
	Allen's Squirrel Galago	*Sciurocheirus alleni*
	Bioko Squirrel Galago	*Sciurocheirus alleni alleni*
	Cross River Squirrel Galago	*Sciurocheirus alleni cameronensis*
	Gabon Squirrel Galago	*Sciurocheirus gabonensis*
	Makandé Squirrel Galago	*Sciurocheirus makandensis*
	Southern Needle-clawed Galago	*Euoticus elegantulus*
	Northern Needle-clawed Galago	*Euoticus pallidus*
	Bioko Needle-clawed Galago	*Euoticus pallidus pallidus*
	Nigeria Needle-clawed Galago	*Euoticus pallidus talboti*
	Thick-tailed Greater Galago	*Otolemur crassicaudatus*
	South African Large-eared Galago	*Otolemur crassicaudatus crassicaudatus*
	Northern Silver Galago	*Otolemur crassicaudatus argentatus*
	Tanganyika Large-eared Galago	*Otolemur crassicaudatus kirkii*
	Miombo Silver Galago	*Otolemur crassicaudatus monteiri*
	Garnett's Greater Galago	*Otolemur garnettii*
	Zanzibar Small-eared Galago	*Otolemur garnettii garnettii*
	Kikuyu Small-eared Galago	*Otolemur garnettii kikuyuensis*
	White-tailed Small-eared Galago	*Otolemur garnettii lasiotis*
	Pangani Small-eared Galago	*Otolemur garnettii panganiensis*
Lorisidae	Calabar Angwantibo	*Arctocebus calabarensis*
	Golden Angwantibo	*Arctocebus aureus*
	West African Potto	*Perodicticus potto*
	Bosman's Potto	*Perodicticus potto potto*
	Benin Potto	*Perodicticus potto juju*
	East African Potto	*Perodicticus ibeanus*

APPENDIX 1   Currently recognized living Primate taxa (species and subspecies)—cont'd

	Eastern Potto	*Perodicticus ibeanus ibeanus*
	Mount Kenya Potto	*Perodicticus ibeanus stockleyi*
	Milne-Edwards' Potto	*Perodicticus edwardsi*
	Gray Slender Loris	*Loris lydekkerianus*
	Mysore Slender Loris	*Loris lydekkerianus lydekkerianus*
	Malabar Slender Loris	*Loris lydekkerianus malabaricus*
	Highland Slender Loris	*Loris lydekkerianus grandis*
	Northern Ceylon Slender Loris	*Loris lydekkerianus nordicus*
	Red Slender Loris	*Loris tardigradus*
	Red Slender Loris	*Loris tardigradus tardigradus*
	Horton Plains Slender Loris	*Loris tardigradus nycticeboides*
	Bengal Slow Loris	*Nycticebus bengalensis*
	Sunda Slow Loris	*Nycticebus coucang*
	Javan Slow Loris	*Nycticebus javanicus*
	Philippine Slow Loris	*Nycticebus menagensis*
	Philippine Slow Loris	*Nycticebus menagensis menagensis*
	Natuna Islands Slow Loris	*Nycticebus menagensis natunae*
	Pygmy Slow Loris	*Nycticebus pygmaeus*
	Bangka Slow Loris	*Nycticebus bancanus*
	Bornean Slow Loris	*Nycticebus borneanus*
	Hiller's Slow Loris	*Nycticebus hilleri*
	Kayan River Slow Loris	*Nycticebus kayan*
Tarsiidae	Western Tarsier	*Cephalopachus bancanus*
	Horsfield's Tarsier	*Cephalopachus bancanus bancanus*
	Bornean Tarsier	*Cephalopachus bancanus borneanus*
	Natuna Islands Tarsier	*Cephalopachus bancanus natunensis*
	Belitung Tarsier	*Cephalopachus bancanus saltator*
	Philippine Tarsier	*Carlito syrichta*
	Philippine Tarsier	*Carlito syrichta syrichta*
	Mindanao Tarsier	*Carlito syrichta carbonarius*
	Bohol Tarsier	*Carlito syrichta fraterculus*
	Selayar Tarsier	*Tarsius tarsier*
	Makassar Tarsier	*Tarsius fuscus*
	Dian's Tarsier	*Tarsius dentatus*
	Peleng Tarsier	*Tarsius pelengensis*
	Great Sangihe Tarsier	*Tarsius sangirensis*
	Siau Island Tarsier	*Tarsius tumpara*
	Sulawesi Mountain Tarsier	*Tarsius pumilus*
	Lariang Tarsier	*Tarsius lariang*

Continued

APPENDIX 1　Currently recognized living Primate taxa (species and subspecies)—cont'd

	Wallace's Tarsier	*Tarsius wallacei*
	Bumbulan Tarsier	*Tarsius supriatnai*
	Manado Spectral Tarsier	*Tarsius spectrumgurskyae*
	Togean Tarsier	*Tarsius niemitzi*
Callitrichidae	Western Pygmy Marmoset	*Cebuella pygmaea*
	Eastern Pygmy Marmoset	*Cebuella niveiventris*
	Black-crowned Dwarf Marmoset	*Mico humilis*
	Silvery Marmoset	*Mico argentatus*
	Golden-white Bare-ear Marmoset	*Mico leucippe*
	Snethlage's Marmoset	*Mico emiliae*
	Black-tailed Marmoset	*Mico melanurus*
	Rio Aripuanã Marmoset	*Mico intermedius*
	Rondon's Marmoset	*Mico rondoni*
	Black-headed Marmoset	*Mico nigriceps*
	Marca's Marmoset	*Mico marcai*
	Rio Acarí Marmoset	*Mico acariensis*
	Maués Marmoset	*Mico mauesi*
	Golden-white Tassel-ear Marmoset	*Mico chrysoleucos*
	Sateré Marmoset	*Mico saterei*
	Santarém Marmoset	*Mico humeralifer*
	Munduruku marmoset	*Mico munduruku*
	Goeldi's Monkey	*Callimico goeldii*
	Buffy-tufted-ear Marmoset	*Callithrix aurita*
	Buffy-headed Marmoset	*Callithrix flaviceps*
	Geoffroy's Tufted-ear Marmoset	*Callithrix geoffroyi*
	Wied's Black-tufted-ear Marmoset	*Callithrix kuhlii*
	Black-tufted-ear Marmoset	*Callithrix penicillata*
	Common Marmoset	*Callithrix jacchus*
	Black-mantled Tamarin	*Leontocebus nigricollis*
	Spix's Black-mantled Tamarin	*Leontocebus nigricollis nigricollis*
	Graell's Black-mantled Tamarin	*Leontocebus nigricollis graellsi*
	Hernández-Camacho's Black-mantled Tamarin	*Leontocebus nigricollis hernandezi*
	Lesson's Saddle-back Tamarin	*Leontocebus fuscus*
	Golden-mantled Saddle-back Tamarin	*Leontocebus tripartitus*
	Red-mantled Saddle-back Tamarin	*Leontocebus lagonotus*
	Andean Saddle-back Tamarin	*Leontocebus leucogenys*
	Illiger's Saddle-back Tamarin	*Leontocebus illigeri*
	Geoffroy's Saddle-back Tamarin	*Leontocebus nigrifrons*
	Cruz Lima's Saddle-back Tamarin	*Leontocebus cruzlimai*
	Spix's Saddle-back Tamarin	*Leontocebus fuscicollis*

APPENDIX 1   Currently recognized living Primate taxa (species and subspecies)—cont'd

	Lako's Saddle-back Tamarin	*Leontocebus fuscicollis fuscicollis*
	Ávila Pires' Saddle-back Tamarin	*Leontocebus fuscicollis avilapiresi*
	Gray-fronted Saddle-back Tamarin	*Leontocebus fuscicollis mura*
	Hershkovitz's Saddle-back Tamarin	*Leontocebus fuscicollis primitivus*
	Weddell's Saddle-back Tamarin	*Leontocebus weddelli*
	Weddell's Saddle-back Tamarin	*Leontocebus weddelli weddelli*
	White Saddle-back Tamarin	*Leontocebus weddelli melanoleucus*
	Crandall's Saddle-back Tamarin	*Leontocebus weddelli crandalli*
	Mustached Tamarin	*Saguinus mystax*
	Spix's Mustached Tamarin	*Saguinus mystax mystax*
	Red-cap Mustached Tamarin	*Saguinus mystax pileatus*
	White-rump Mustached Tamarin	*Saguinus mystax pluto*
	Red-bellied Tamarin	*Saguinus labiatus*
	Geoffroy's Red-bellied Tamarin	*Saguinus labiatus labiatus*
	Gray's Red-bellied Tamarin	*Saguinus labiatus rufiventer*
	Thomas's Red-bellied Tamarin	*Saguinus labiatus thomasi*
	Emperor Tamarin	*Saguinus imperator*
	Black-chinned Emperor Tamarin	*Saguinus imperator imperator*
	Bearded Emperor Tamarin	*Saguinus imperator subgrisescens*
	Midas Tamarin	*Saguinus midas*
	Western Black-handed Tamarin	*Saguinus niger*
	Eastern Black-handed Tamarin	*Saguinus ursulus*
	Pied Tamarin	*Saguinus bicolor*
	Martins' Bare-faced Tamarin	*Saguinus martinsi*
	Martins' Bare-faced Tamarin	*Saguinus martinsi martinsi*
	Ochraceous Bare-faced Tamarin	*Saguinus martinsi ochraceus*
	Mottled-face Tamarin	*Saguinus inustus*
	White-footed Tamarin	*Saguinus leucopus*
	Cotton-top Tamarin	*Saguinus oedipus*
	Geoffroy's Tamarin	*Saguinus geoffroyi*
	Golden Lion Tamarin	*Leontopithecus rosalia*
	Golden-headed Lion Tamarin	*Leontopithecus chrysomelas*
	Black Lion Tamarin	*Leontopithecus chrysopygus*
	Black-faced Lion Tamarin	*Leontopithecus caissara*
Cebidae	Central American Squirrel Monkey	*Saimiri oerstedii*
	Black-crowned Central American Squirrel Monkey	*Saimiri oerstedii oerstedii*
	Grey-crowned Central American Squirrel Monkey	*Saimiri oerstedii citrinellus*
	Humboldt's Squirrel Monkey	*Saimiri cassiquiarensis*

Continued

4. Primates

APPENDIX 1   Currently recognized living Primate taxa (species and subspecies)—cont'd

Colombian Squirrel Monkey	*Saimiri cassiquiarensis cassiquarensis*
Colombian Squirrel Monkey	*Saimiri cassiquiarensis albigena*
Ecuadorian Squirrel Monkey	*Saimiri cassiquiarensis macrodon*
Golden-backed Squirrel Monkey	*Saimiri ustus*
Guianan Squirrel Monkey	*Saimiri sciureus*
Collins' Squirrel Monkey	*Saimiri collinsi*
Black-capped Squirrel Monkey	*Saimiri boliviensis*
Bolivian Squirrel Monkey	*Saimiri boliviensis boliviensis*
Peruvian Sqiurrel Monkey	*Saimiri boliviensis peruviensis*
Black-headed Squirrel Monkey	*Saimiri vanzolinii*
Black-horned Capuchin	*Sapajus nigritus*
Northern Black-horned Capuchin	*Sapajus nigritus nigritus*
Southern Black-horned Capuchin	*Sapajus nigritus cucullatus*
Hooded Capuchin	*Sapajus cay*
Crested Capuchin	*Sapajus robustus*
Bearded Capuchin	*Sapajus libidinosus*
Yellow-breasted Capuchin	*Sapajus xanthosternos*
Blond Capuchin	*Sapajus flavius*
Guianan Brown Capuchin	*Sapajus apella*
Guianan Brown Capuchin	*Sapajus apella apella*
Margarita Island Capuchin	*Sapajus apella margaritae*
Peruvian White-fronted Capuchin	*Cebus yuracus*
Shock-headed Capuchin	*Cebus cuscinus*
Spix's White-fronted Capuchin	*Cebus unicolor*
Humboldt's White-fronted Capuchin	*Cebus albifrons*
Guianan Weeper Capuchin	*Cebus olivaceus*
Chestnut Weeper Capuchin	*Cebus castaneus*
Ka'apor Capuchin	*Cebus kaapori*
Brown Capuchin	*Cebus brunneus*
Sierra de Perijá White-fronted Capuchin	*Cebus leucocephalus*
Varied White-fronted Capuchin	*Cebus versicolor*
Varied White-fronted Capuchin	*Cebus versicolor versicolor*
Río Cesar White-fronted Capuchin	*Cebus versicolor cesarae*
Santa Marta White-fronted Capuchin	*Cebus malitiosus*
Trinidad White-fronted Capuchin	*Cebus trinitatis*
Ecuadorian White-fronted Capuchin	*Cebus aequatorialis*
Colombian White-faced Capuchin	*Cebus capucinus*
Colombian White-faced Capuchin	*Cebus capucinus capucinus*
Gorgona White-faced Capuchin	*Cebus capucinus curtus*
Panamanian White-faced Capuchin	*Cebus imitator*

APPENDIX 1   Currently recognized living Primate taxa (species and subspecies)—cont'd

Aotidae	Lemurine Night Monkey	*Aotus lemurinus*
	Gray-legged Night Monkey	*Aotus griseimembra*
	Panamanian Night Monkey	*Aotus zonalis*
	Brumback's Night Monkey	*Aotus brumbacki*
	Humboldt's Night Monkey	*Aotus trivirgatus*
	Spix's Night Monkey	*Aotus vociferans*
	Hernández-Camacho's Night Monkey	*Aotus jorgehernandezi*
	Andean Night Monkey	*Aotus miconax*
	Ma"s Night Monkey	*Aotus nancymaae*
	Black-headed Night Monkey	*Aotus nigriceps*
	Azara's Night Monkey	*Aotus azarae*
	Azara's Night Money	*Aotus azarae azarae*
	Bolivian Night Monkey	*Aotus azarae boliviensis*
	Feline Night Monkey	*Aotus azarae infulatus*
Pitheciidae	Rio Beni Titi	*Plecturocebus modestus*
	White-eared Titi	*Plecturocebus donacophilus*
	White-coated Titi	*Plecturocebus pallescens*
	Olalla Brother's Titi	*Plecturocebus olallae*
	Rio Mayo Titi	*Plecturocebus oenanthe*
	Coppery Titi	*Plecturocebus cupreus*
	Red-crowned Titi	*Plecturocebus discolor*
	Ornate Titi	*Plecturocebus ornatus*
	Chestnut-bellied Titi	*Plecturocebus caligatus*
	Doubtful Titi	*Plecturocebus dubius*
	Stephen Nash's Titi	*Plecturocebus stephennashi*
	Caquetá titi	*Plecturocebus caquetenesis*
	Ashy Titi	*Plecturocebus cinerascens*
	Hoffmann's Titi	*Plecturocebus hoffmannsi*
	Lake Baptista Titi	*Plecturocebus baptista*
	Red-bellied Titi	*Plecturocebus moloch*
	Vieira's Titi	*Plecturocebus vieirai*
	Brown Titi	*Plecturocebus brunneus*
	Prince Bernhard's Titi	*Plecturocebus bernhardi*
	Madidi Titi	*Plecturocebus aureipalatii*
	Milton's Titi	*Plecturocebus miltoni*
	Toppin's Titi	*Plecturocebus toppini*
	Urubamba Titi	*Plecturocebus urubambensis*
	Groves' Titi	*Plecturocebus grovesi*
	Parecis Titi	*Plecturocebus parecis*

Continued

4. Primates

APPENDIX 1   Currently recognized living Primate taxa (species and subspecies)—cont'd

Medem's Titi	*Cheracebus medemi*
White-chested Titi	*Cheracebus lugens*
Yellow-handed Titi	*Cheracebus lucifer*
Rio Purus Titi	*Cheracebus purinus*
Juruá Collared Titi	*Cheracebus regulus*
Black-fronted Titi	*Callicebus nigrifrons*
Masked Titi	*Callicebus personatus*
Southern Bahian Titi	*Callicebus melanochir*
Blond Titi	*Callicebus barbarabrownae*
Coimbra-Filho's Titi	*Callicebus coimbrai*
White-faced Saki	*Pithecia pithecia*
Golden-faced Saki	*Pithecia chrysocephala*
Hairy Saki	*Pithecia hirsuta*
Miller's Saki	*Pithecia milleri*
Geoffroy's Monk Saki	*Pithecia monachus*
Burnished Saki	*Pithecia inusta*
Cazuza's Saki	*Pithecia cazuzai*
Equatorial Saki	*Pithecia aequatorialis*
Napo Saki	*Pithecia napensis*
Isabel's Saki	*Pithecia isabela*
Buffy Saki	*Pithecia albicans*
Gray's Bald-faced Saki	*Pithecia irrorata*
Vanzolini's Bald-faced Saki	*Pithecia vanzolinii*
Mittermeier's Tapajós Saki	*Pithecia mittermeieri*
Rylands' Bald-faced Saki	*Pithecia rylandsi*
Pissinatti's Bald-faced Saki	*Pithecia pissinattii*
White-nosed Bearded Saki	*Chiropotes albinasus*
Black Saki	*Chiropotes satanas*
Rio Negro Bearded Saki	*Chiropotes chiropotes*
Uta Hick's Bearded Saki	*Chiropotes utahickae*
Guianan Bearded Saki	*Chiropotes sagulatus*
Rio Negro Bearded Saki	*Chiropotes israelita*
Bald Uakari	*Cacajao calvus*
White Uakari	*Cacajao calvus calvus*
Novaes's Bald Uakari	*Cacajao calvus novaesi*
Red Uakari	*Cacajao calvus rubicundus*
Ucayali Bald Uakari	*Cacajao calvus ucayalii*
Golden-Brown Uakari	*Cacajao melanocephalus*
Neblona Uakari	*Cacajao hosomi*
Acarí Uakari	*Cacajao ayresi*

APPENDIX 1 Currently recognized living Primate taxa (species and subspecies)—cont'd

Atelidae	Colombian Red Howler	*Alouatta seniculus*
	Juruá Red Howler	*Alouatta juara*
	Purus Red Howler	*Alouatta puruensis*
	Ursine Red Howler	*Alouatta arctoidea*
	Guianan Red Howler	*Alouatta macconnelli*
	Bolivian Red Howler	*Alouatta sara*
	Amazon Black Howler	*Alouatta nigerrima*
	Red-Handed Howler	*Alouatta belzebul*
	Spix's Howler	*Alouatta discolor*
	Maranhão Red-handed Howler	*Alouatta ululata*
	Brown Howler Monkey	*Alouatta guariba*
	Northern Brown Howler	*Alouatta guariba guariba*
	Southern Brown Howler	*Alouatta guariba clamitans*
	Paraguayan Howler	*Alouatta caraya*
	Mantled Howler	*Alouatta palliata*
	Golden-mantled Howler	*Alouatta palliata palliata*
	Azuero Peninsula Howler	*Alouatta palliata trabeata*
	Coiba Island Howler	*Alouatta palliata coibensis*
	Ecuadorian Mantled Howler	*Alouatta palliata aequatorialis*
	Mexican Howler	*Alouatta palliata mexicana*
	Central American Black Howler	*Alouatta pigra*
	Central American Spider Monkey	*Ateles geoffroyi*
	Geoffroy's Spider Monkey	*Ateles geoffroyi geoffroyi*
	Azuero Spider Monkey	*Ateles geoffroyi azuerensis*
	Black-browed Spider Monkey	*Ateles geoffroyi frontatus*
	Hooded Spider Monkey	*Ateles geoffroyi grisescens*
	Ornate Spider Monkey	*Ateles geoffroyi ornatus*
	Mexican Spider Monkey	*Ateles geoffroyi vellerosus*
	Brown-headed Spider Monkey	*Ateles fusciceps*
	Brown-headed Spider Monkey	*Ateles fusciceps fusciceps*
	Colombian Black Spider Monkey	*Ateles fusciceps rufiventris*
	Black Spider Monkey	*Ateles chamek*
	Red-faced Black Spider Monkey	*Ateles paniscus*
	White-whiskered Spider Monkey	*Ateles marginatus*
	White-bellied Spider Monkey	*Ateles belzebuth*
	Variegated Spider Monkey	*Ateles hybridus*
	Humboldt's Woolly Monkey	*Lagothrix lagothricha*
	Humboldt's Woolly Monkey	*Lagothrix lagothricha lagothricha*
	Brown Woolly Monkey	*Lagothrix lagothricha lugens*

Continued

4. Primates

APPENDIX 1   Currently recognized living Primate taxa (species and subspecies)—cont'd

	Geoffroy's Woolly Monkey	*Lagothrix lagothricha cana*
	Peruvian Woolly Monkey	*Lagothrix lagothricha tschudii*
	Poeppig's Woolly Monkey	*Lagothrix lagothricha poeppigii*
	Peruvian Yellow-tailed Woolly Monkey	*Lagothrix flavicauda*
	Southern Muriqui	*Brachyteles arachnoides*
	Northern Muriqui	*Brachyteles hypoxanthus*
Cercopithecidae	Barbary Macaque	*Macaca sylvanus*
	Lion-tailed Macaque	*Macaca silenus*
	Crested Macaque	*Macaca nigra*
	Gorontalo Macaque	*Macaca nigrescens*
	Heck's Macaque	*Macaca hecki*
	Tonkean Macaque	*Macaca tonkeana*
	Moor Macaque	*Macaca maura*
	Booted Macaque	*Macaca ochreata*
	Botted Macaque	*Macaca ochreata ochreata*
	Buton Macaque	*Macaca ochreata brunnescens*
	Pagai Macaque	*Macaca pagensis*
	Siberut Macaque	*Macaca siberu*
	Sunda Pig-tailed Macaque	*Macaca nemestrina*
	Northern Pig-tailed Macaque	*Macaca leonina*
	Toque Macaque	*Macaca sinica*
	Common Toque Macaque	*Macaca sinica sinica*
	Pale-fronted Toque Macaque	*Macaca sinica aurifrons*
	Hill Zone Toque Macaque	*Macaca sinica opisthomelas*
	Bonnet Macaque	*Macaca radiata*
	Dark-bellied Bonnet Macaque	*Macaca radiata radiata*
	Pale-bellied Bonnet Macaque	*Macaca radiata diluta*
	Assamese Macaque	*Macaca assamensis*
	Eastern Assamese Macaque	*Macaca assamensis assamensis*
	Western Assamese Macaque	*Macaca assamensis pelops*
	Arunachal Macaque	*Macaca munzala*
	White-cheeked Macaque	*Macaca leucogenys*
	Tibetan Macaque	*Macaca thibetana*
	Sichuan Tibetan Macaque	*Macaca thibetana thibetana*
	Fujian Tibetan Macaque	*Macaca thibetana pullus*
	Stump-tailed Macaque	*Macaca arctoides*
	Long-tailed Macaque	*Macaca fascicularis*
	Common Long-tailed Macaque	*Macaca fascicularis fascicularis*
	Dark-crowned Long-tailed Macaque	*Macaca fascicularis atriceps*
	Myanmar Long-tailed Macaque	*Macaca fascicularis aureus*

APPENDIX 1   Currently recognized living Primate taxa (species and subspecies)—cont'd

Con Son Long-tailed Macaque	*Macaca fascicularis condorensis*
Simeulue Long-tailed Macaque	*Macaca fascicularis fuscus*
Lasia Long-tailed Macaque	*Macaca fascicularis lasiae*
Karimunjawa Long-tailed Macaque	*Macaca fascicularis karimondjawae*
Maratua Long-tailed Macaque	*Macaca fascicularis tua*
Nicobar Long-tailed Macaque	*Macaca fascicularis umbrosus*
Rhesus Macaque	*Macaca mulatta*
Japanese Macaque	*Macaca fuscata*
Japanese Macaque	*Macaca fuscata fuscata*
Yaku Macaque	*Macaca fuscata yakui*
Taiwanese Macaque	*Macaca cyclopis*
Tana River Mangabey	*Cercocebus galeritus*
Agile Mangabey	*Cercocebus agilis*
Golden-bellied Mangabey	*Cercocebus chrysogaster*
Sanje River Mangabey	*Cercocebus sanjei*
Sooty Mangabey	*Cercocebus atys*
White-naped Mangabey	*Cercocebus lunulatus*
Red-capped Mangabey	*Cercocebus torquatus*
Mandrill	*Mandrillus sphinx*
Drill	*Mandrillus leucophaeus*
Mainland Drill	*Mandrillus leucophaeus leucophaeus*
Bioko Drill	*Mandrillus leucophaeus poensis*
Gray-cheeked Mangabey	*Lophocebus albigena*
Western Gray-cheeked Mangabey	*Lophocebus albigena albigena*
Rusty-mantled Mangabey	*Lophocebus albigena osmani*
Johnston's Mangabey	*Lophocebus albigena johnstoni*
Ugandan Crested Mangabey	*Lophocebus albigena ugandae*
Black Mangabey	*Lophocebus aterrimus*
Northern Black Crested Mangabey	*Lophocebus aterrimus aterrimus*
Southern Black Crested Mangabey	*Lophocebus aterrimus opdenboschi*
Kipunji	*Rungwecebus kipunji*
Yellow Baboon	*Papio cynocephalus*
Central Yellow Baboon	*Papio cynocephalus cynocephalus*
Northern Yellow Baboon	*Papio cynocephalus ibeanus*
Kinda Baboon	*Papio kindae*
Chacma Baboon	*Papio ursinus*
Southern Chacma Baboon	*Papio ursinus ursinus*
Grey-footed Chacma Baboon	*Papio ursinus griseipes*
Namibian Chacma Baboon	*Papio ursinus ruacana*
Olive Baboon	*Papio anubis*

Continued

4. Primates

APPENDIX 1 Currently recognized living Primate taxa (species and subspecies)—cont'd

Guinea Baboon	*Papio papio*
Hamadryas Baboon	*Papio hamadryas*
Gelada	*Theropithecus gelada*
Northern Gelada	*Theropithecus gelada gelada*
Southern Gelada	*Theropithecus gelada obscurus*
Allen's Swamp Monkey	*Allenopithecus nigroviridis*
Southern Talapoin Monkey	*Miopithecus talapoin*
Northern Talapoin Monkey	*Miopithecus ogouensis*
Patas Monkey	*Erythrocebus patas*
Western Patas Monkey	*Erythrocebus patas patas*
Eastern Patas Monkey	*Erythrocebus patas pyrrhonotus*
Air Massif Pata Monkey	*Erythrocebus patas villiersi*
Southern Patas Monkey	*Erythrocebus baumstarki*
Heuglin's Patas Monkey	*Erythrocebus poliophaeus*
Grivet Monkey	*Chlorocebus aethiops*
Tantalus Monkey	*Chlorocebus tantalus*
Tantalus Monkey	*Chlorocebus tantalus tantalus*
Budgett's Tantalus	*Chlorocebus tantalus budgetti*
Jebbel Marra Tantalus	*Chlorocebus tantalus marrensis*
Green Monkey	*Chlorocebus sabaeus*
Malbrouck Monkey	*Chlorocebus cynosurus*
Vervet Monkey	*Chlorocebus pygerythrus*
Black-chinned Vervet Monkey	*Chlorocebus pygerythrus pygerythrus*
Hilgert's Vervet Monkey	*Chlorocebus pygerythrus hilgerti*
Pemba Vervet Monkey	*Chlorocebus pygerythrus nesiotes*
Reddish-Green Vervet Monkey	*Chlorocebus pygerythrus rufoviridis*
de Beaux's Grivet	*Chlorocebus pygerythrus zavattarii*
Bale Monkey	*Chlorocebus djamdjamensis*
Bale Monky	*Chlorocebus djamdjamensis djamdjamensis*
Harenna Bale Monkey	*Chlorocebus djamdjamensis harennaensis*
L'Hoest's Monkey	*Allochrocebus lhoesti*
Preuss's Monkey	*Allochrocebus preussi*
Cameroon Preuss's Monkey	*Allochrocebus preussi preussi*
Bioko Preuss's Monkey	*Allochrocebus preussi insularis*
Sun-tailed Monkey	*Allochrocebus solatus*
Dryad Monkey	*Cercopithecus dryas*
Diana Monkey	*Cercopithecus diana*
Roloway Monkey	*Cercopithecus roloway*

APPENDIX 1 Currently recognized living Primate taxa (species and subspecies)—cont'd

De Brazza's Monkey	*Cercopithecus neglectus*
Mona Monkey	*Cercopithecus mona*
Campbell's Monkey	*Cercopithecus campbelli*
Lowe's Monkey	*Cercopithecus lowei*
Crowned Monkey	*Cercopithecus pogonias*
Golden-bellied Crowned Monkey	*Cercopithecus pogonias pogonias*
Gray's Crowned Monkey	*Cercopithecus pogonias grayi*
Black-footed Crowned Monkey	*Cercopithecus pogonias nigripes*
Wolf's Monkey	*Cercopithecus wolfi*
Congo Basin Wolf's Monkey	*Cercopithecus wolfi wolfi*
Fire-bellied Wolf's Monkey	*Cercopithecus wolfi pyrogaster*
Lomami River Wolf's Monkey	*Cercopithecus wolfi elegans*
Dent's Monkey	*Cercopithecus denti*
Owl-faced Monkey	*Cercopithecus hamlyni*
Lesula	*Cercopithecus lomamiensis*
Spot-nosed Monkey	*Cercopithecus petaurista*
Eastern Lesser Spot-nosed Monkey	*Cercopithecus petaurista petaurista*
Western Lesser Spot-nosed Monkey	*Cercopithecus petaurista buettikoferi*
Red-bellied Monkey	*Cercopithecus erythrogaster*
Red-bellied Monkey	*Cercopithecus erythrogaster erythrogaster*
Nigeria White-throated Monkey	*Cercopithecus erythrogaster pococki*
Sclater's Monkey	*Cercopithecus sclateri*
Red-eared Monkey	*Cercopithecus erythrotis*
Bioko Red-eared Monkey	*Cercopithecus erythrotis erythrotis*
Cameroon Red-eared Monkey	*Cercopithecus erythrotis camerunensis*
Mustached Monkey	*Cercopithecus cephus*
Red-tailed Mustached Monkey	*Cercopithecus cephus cephus*
Gray-tailed Mustached Monkey	*Cercopithecus cephus cephodes*
White-nosed Mustached Monkey	*Cercopithecus cephus ngottoensis*
Red-tailed Monkey	*Cercopithecus ascanius*
Black-cheeked Red-tailed Monkey	*Cercopithecus ascanius ascanius*
Black-nosed Red-tailed Monkey	*Cercopithecus ascanius atrinasus*
Katanga Red-tailed Monkey	*Cercopithecus ascanius katangae*
Schmidt's Red-tailed Monkey	*Cercopithecus ascanius schmidti*
Yellow-nosed Red-tailed Monkey	*Cercopithecus ascanius whitesidei*
Putty-nosed Monkey	*Cercopithecus nictitans*
Eastern Putty-nosed Monkey	*Cercopithecus nictitans nictitans*
Nigerian Putty-nosed Monkey	*Cercopithecus nictitans insolitus*
Red-rumped Putty-nosed Monkey	*Cercopithecus nictitans ludio*

Continued

4. Primates

	Bioko Putty-nosed Monkey	*Cercopithecus nictitans martini*
	Stampfli's Putty-nosed Monkey	*Cercopithecus nictitans stampflii*
	Blue Monkey	*Cercopithecus mitis*
	Pluto Monkey	*Cercopithecus mitis mitis*
	Lomami River Blue Monkey	*Cercopithecus mitis heymansi*
	Stuhlmann's Blue Monkey	*Cercopithecus mitis stuhlmanni*
	Boutourlini's Blue Monkey	*Cercopithecus mitis boutourlinii*
	Rump-spotted Blue Monkey	*Cercopithecus mitis opisthostictus*
	Silver Monkey	*Cercopithecus mitis doggetti*
	Virungas Golden Monkey	*Cercopithecus mitis kandti*
	Zanzibar Sykes's Monkey	*Cercopithecus mitis albogularis*
	Pousargues's White-collared Monkey	*Cercopithecus mitis albotorquatus*
	Stairs's White-collared Monkey	*Cercopithecus mitis erythrarchus*
	Kolb's White-collared Monkey	*Cercopithecus mitis kolbi*
	Samango Monkey	*Cercopithecus mitis labiatus*
	Moloney's White-collared Monkey	*Cercopithecus mitis moloneyi*
	Tanzania Sykes's Monkey	*Cercopithecus mitis monoides*
	Somalia White-collared Monkey	*Cercopithecus mitis zammaranoi*
Colobinae	Angolan Colobus	*Colobus angolensis*
	Sclater's Angolan Colobus	*Colobus angolensis angolensis*
	Cordier's Angolan Colobus	*Colobus angolensis cordieri*
	Powell-Cotton's Angolan Colobus	*Colobus angolensis cottoni*
	Peters' Angolan Colobus	*Colobus angolensis palliatus*
	Prigogine's Angolan Colobus	*Colobus angolensis prigoginei*
	Adolf Friedrichs's Angolan Colobus	*Colobus angolensis ruwenzorii*
	Sharpe's Angolan Colobus	*Colobus angolensis sharpei*
	Mahale Mountains Angolan Colobus	*Colobus angolensis ssp.*
	Guereza	*Colobus guereza*
	Omo River Guereza	*Colobus guereza guereza*
	Dodinga Hills Guereza	*Colobus guereza dodingae*
	Djaffa Mountains Guereza	*Colobus guereza gallarum*
	Mt. Kenya Guereza	*Colobus guereza kikuyuensis*
	Mau Forest Guereza	*Colobus guereza matschiei*
	Western Guereza	*Colobus guereza occidentalis*
	Mt. Uaraguess Guereza	*Colobusa guereza percivali*
	Mt. Kilimanjaro Guereza	*Colobus caudatus*
	King Colobus	*Colobus polykomos*
	Black Colobus	*Colobus satanas*
	Bioko Black Colobus	*Colobus satanas satanas*
	Gabon Black Colobus	*Colobus satanas anthracinus*

APPENDIX 1 Currently recognized living Primate taxa (species and subspecies)—cont'd

White-thighed Colobus	*Colobus vellerosus*
West African Red Colobus	*Piliocolobus badius*
Upper Guinea Red Colobus	*Piliocolobus badius badius*
Temminck's Red Colobus	*Piliocolobus badius temminckii*
Miss Waldron's Red Colobus	*Piliocolobus waldroni*
Zanzibar Red Colobus	*Piliocolobus kirkii*
Uzungwa Red Colobus	*Piliocolobus gordonorum*
Pennant's Red Colobus	*Piliocolobus pennantii*
Bouvier's Red Colobus	*Piliocolobus bouvieri*
Niger Delta Red Colobus	*Piliocolobus epieni*
Preuss's Red Colobus	*Piliocolobus preussi*
Tana River Red Colobus	*Piliocolobus rufomitratus*
Semliki Red Colobus	*Piliocolobus semlikiensis*
Foa's Red Colobus	*Piliocolobus foai*
Lulindi River Red Colobus	*Piliocolobus lulindicus*
Lang's Red Colobus	*Piliocolobus langi*
Oustalet's Red Colobus	*Piliocolobus oustaleti*
Lomami Red Colobus	*Piliocolobus parmentieri*
Ashy Red Colobus	*Piliocolobus tephrosceles*
Tshuapa Red Colobus	*Piliocolobus tholloni*
Olive Colobus	*Procolobus verus*
Robinson's Banded Langur	*Presbytis robinsoni*
Raffles' Banded Langur	*Presbytis femoralis*
Pale-thighed Langur	*Presbytis siamensis*
Malayan Pale-thighed Langur	*Presbytis siamensis siamensis*
Riau Pale-thighed Langur	*Presbytis siamensis cana*
Mantled Pale-thighed Langur	*Presbytis siamensis paenulata*
Bintan Pale-thighed Langur	*Presbytis siamensis rhionis*
Thomas' Langur	*Presbytis thomasi*
Black Sumatran Langur	*Presbytis sumatrana*
East Sumatran Banded Langur	*Presbytis percura*
Black-and-white Langur	*Presbytis bicolor*
Black-crested Sumatran Langur	*Presbytis melalophos*
Mitered Langur	*Presbytis mitrata*
Siberut Langur	*Presbytis siberu*
Mentawai Langur	*Presbytis potenziani*
Javan Langur	*Presbytis comata*
Javan Grizzled Langur	*Presbytis comata comata*
Javan Fuscous Langur	*Presbytis comata fredericae*
Natuna Islands Langur	*Presbytis natunae*

Continued

4. Primates

APPENDIX 1 Currently recognized living Primate taxa (species and subspecies)—cont'd

Maroon Langur	*Presbytis rubicunda*
Maroon Red Langur	*Presbytis rubicunda rubicunda*
Red-naped Red Langur	*Presbytis rubicunda carimatae*
Orange-backed red Langur	*Presbytis rubicunda chrysea*
Orange-naped Red Langur	*Presbytis rubicunda ignita*
Maroon Langur	*Presbytis rubicunda rubida*
Cross-marked Langur	*Presbytis chrysomelas*
Bornean Banded Langur	*Presbytis chrysomelas chrysomelas*
Tricolored Langur (cruciger)	*Presbytis chrysomelas cruciger*
Hose's Langur	*Presbytis hosei*
Sabah Grizzled Langur	*Presbytis sabana*
Miller's Langur	*Presbytis canicrus*
White-fronted Langur	*Presbytis frontata*
Red-shanked Douc	*Pygathrix nemaeus*
Gray-shanked Douc	*Pygathrix cinerea*
Black-shanked Douc	*Pygathrix nigripes*
Tonkin Snub-nosed Monkey	*Rhinopithecus avunculus*
Yunnan Snub-nosed Monkey	*Rhinopithecus bieti*
Guizhou Snub-nosed Monkey	*Rhinopithecus brelichi*
Golden Snub-nosed Monkey	*Rhinopithecus roxellana*
Moupin Golden Snub-nosed Monkey	*Rhinopithecus roxellana roxellana*
Hubei Golden Snub-nosed Monkey	*Rhinopithecus roxellana hubeiensis*
Qinling Golden Snub-nosed Monkey	*Rhinopithecus roxellana qinlingensis*
Myanmar Snub-nosed Monkey	*Rhinopithecus strykeri*
Pig-tailed Langur	*Simias concolor*
Pagai Pig-tailed Langur	*Simias concolor concolor*
Siberut Pig-tailed Langur	*Simias concolor siberu*
Proboscis Monkey	*Nasalis larvatus*
Bengal Sacred Langur	*Semnopithecus entellus*
Chamba Sacred Langur	*Semnopithecus ajax*
Tarai Sacred Langur	*Semnopithecus hector*
Nepal Sacred Langur	*Semnopithecus schistaceus*
Malabar Sacred Langur	*Semnopithecus hypoleucos*
Travancore Langur	*Semnopithecus hypoleucos hypoleucos*
Northern Malabar Langur	*Semnopithecus hypoleucos achates*
Black-legged Langur	*Semnopithecus hypoleucos iulus*
Tufted Gray Langur	*Semnopithecus priam*
Madras Gray Langur	*Semnopithecus priam priam*
Central Indian Gray Langur	*Semnopithecus priam anchises*

APPENDIX 1   Currently recognized living Primate taxa (species and subspecies)—cont'd

Sri Lankan Gray Langur	*Semnopithecus priam thersites*
Purple-faced Langur	*Semnopithecus vetulus*
Southern Purple-faced Langur	*Semnopithecus vetulus vetulus*
Highland Purple-faced Langur	*Semnopithecus vetulus monticola*
Western Purple-faced Langur	*Semnopithecus vetulus nestor*
Northern Purple-faced Langur	*Semnopithecus vetulus philbricki*
Nilgiri Langur	*Semnopithecus johnii*
Golden Langur	*Trachypithecus geei*
Capped Langur	*Trachypithecus pileatus*
Blond-bellied Capped Langur	*Trachypithecus pileatus pileatus*
Buff-bellied Capped Langur	*Trachypithecus pileatus brahma*
Orange-bellied Capped Langur	*Trachypithecus pileatus durga*
Tenebrous Capped Langur	*Trachypithecus pileatus tenebricus*
Shortridge's Langur	*Trachypithecus shortridgei*
East Javan Langur	*Trachypithecus auratus*
West Javan Langur	*Trachypithecus mauritius*
Silvered Langur	*Trachypithecus cristatus*
Silvered Langur	*Trachypithecus cristatus cristatus*
Miller's Silvered Langur	*Trachypithecus cristatus vigilans*
Selangor Silvery Langur	*Trachypithecus selangorensis*
Germain's Langur	*Trachypithecus germaini*
Annamese Langur	*Trachypithecus margarita*
Tenasserim Langur	*Trachypithecus barbei*
Dusky Langur	*Trachypithecus obscurus*
Spectacled Langur	*Trachypithecus obscurus obscurus*
Tarulao Langur	*Trachypithecus obscurus carbo*
Blond-tailed Langur	*Trachypithecus obscurus flavicauda*
Cantor's Langur	*Trachypithecus obscurus halonifer*
Zadetkyi Kyun Langur	*Trachypithecus obscurus sanctorum*
Phangan Langur	*Trachypithecus obscurus seimundi*
Perhentian Langur	*Trachypithecus obscurus styx*
Phayre's Langur	*Trachypithecus phayrei*
Phayre's Langur	*Trachypithecus phayrei phayrei*
Shan Langur	*Trachypithecus phayrei shanicus*
Indochinese Gray Langur	*Trachypithecus crepusculus*
Cat Ba Langur	*Trachypithecus poliocephalus*
White-headed Langur	*Trachypithecus leucocephalus*
Delacour's Langur	*Trachypithecus delacouri*
Francois' Langur	*Trachypithecus francoisi*

Continued

4. Primates

APPENDIX 1　Currently recognized living Primate taxa (species and subspecies)—cont'd

	Black Langur	*Trachypithecus ebenus*
	Laos Langur	*Trachypithecus laotum*
	Hatinh Langur	*Trachypithecus hatinhensis*
Hylobatidae	Western Hoolock Gibbon	*Hoolock hoolock*
	Western Hoolock Gibbon	*Hoolock hoolock hoolock*
	Mishmi Hills Hoolock Gibbon	*Hoolock hoolock mishmiensis*
	Skywalker Hoolock Gibbon	*Hoolock tianxing*
	Eastern Hoolock Gibbon	*Hoolock leuconedys*
	Agile Gibbon	*Hylobates agilis*
	Bornean White-bearded Gibbon	*Hylobates albibarbis*
	Kloss's Gibbon	*Hylobates klossii*
	Lar Gibbon	*Hylobates lar*
	Malayan Lar Gibbon	*Hylobates lar lar*
	Carpenter's Lar gibbon	*Hylobates lar carpenteri*
	Central Lar Gibbon	*Hylobates lar entelloides*
	Sumatran Lar Gibbon	*Hylobates lar vestitus*
	Yunnan Lar Gibbon	*Hylobates lar yunnanensis*
	Javan Silvery or Moloch Gibbon	*Hylobates moloch*
	Müller's Gibbon	*Hylobates muelleri*
	Abbott's Gray Gibbon	*Hylobates abbotti*
	East Bornean Gray Gibbon	*Hylobates funereus*
	Pileated Gibbon	*Hylobates pileatus*
	Western Black Crested Gibbon	*Nomascus concolor*
	Tonkin black crested gibbon	*Nomascus concolor concolor*
	Laotian black crested gibbon	*Nomascus concolor lu*
	Central Yunnan Black Crested Gibbon	*Nomascus concolor jingdongensis*
	Eastern Black Crested Gibbon	*Nomascus nasutus*
	Hainan Crested Gibbon	*Nomascus hainanus*
	Northern Yellow-cheeked Crested Gibbon	*Nomascus annamensis*
	Northern White-cheeked Crested Gibbon	*Nomascus leucogenys*
	Southern White-cheeked Crested Gibbon	*Nomascus siki*
	Southern Yellow-cheeked Crested Gibbon	*Nomascus gabriellae*
	Siamang	*Symphalangus syndactylus*
Hominidae	Bornean Orangutan	*Pongo pygmaeus*
	Northwest Bornean Orangutan	*Pongo pygmaeus pygmaeus*
	Northeast Bornean Orangutan	*Pongo pygmaeus morio*
	Southwest Bornean Orangutan	*Pongo pygmaeus wurmbii*
	Sumatran Orangutan	*Pongo abelii*
	Tapanuli Orangutan	*Pongo tapanuliensis*
	Western Gorilla	*Gorilla gorilla*

APPENDIX 1   Currently recognized living Primate taxa (species and subspecies)—cont'd

Western Lowland Gorilla	*Gorilla gorilla gorilla*
Cross River Gorilla	*Gorilla gorilla diehli*
Eastern Gorilla	*Gorilla beringei*
Mountain Gorilla	*Gorilla beringei beringei*
Grauer's Gorilla	*Gorilla beringei graueri*
Chimpanzee	*Pan troglodytes*
Central Chimpanzee	*Pan troglodytes troglodytes*
Nigeria-Cameroon Chimpanzee	*Pan troglodytes ellioti*
Eastern Chimpanzee	*Pan troglodytes schweinfurthii*
Western Chimpanzee	*Pan troglodytes verus*
Bonobo	*Pan paniscus*

Currently recognized living Primate taxa (species and subspecies). Courtesy of A.B., Rylands, R.A., Mittermeier, 2020. IUCN SSC Primate Specialist Group and Global Wildlife Conservation, Austin, Texas, USA.

References

Beard, K.C., Qi, T., Dawson, M.R., Wang, B., LI, C., 1994. A diverse new primate fauna from middle Eocene fissure-fillings in southeastern China. Nature 368, 604–609.

Beard, K.C., 1998. A new genus of Tarsiidae (Mammalia: primates) from the middle Eocene of Shanxi Province, China, with notes on the historical biogeography of tarsiers. Bull. Carnegie Mus. Nat. Hist. 34, 260–277.

Bloch, J.I., Silcox, M.T., Boyer, D.M., Sargis, E.J., 2007. New Paleocene skeletons and the relationship of plesiadapiforms to crown-clade primates. Proc. Natl. Acad. Sci. U.S.A. 104, 1159–1164.

Bond, M., Tejedor, M.F., Campbell, K.E., Chornogubsky, L., Novo, N., Francisco, G., 2015. Eocene primates of South America and the African origins of new world monkeys. Nature 520, 538–542.

Bush, E.C., Simons, E.L., Dubowitz, D.J., Allman, J.M., 2004. Endocranial volume and optic foramen size in *Parapithecus grangeri*. In: Ross, C.F., Kay, R.F. (Eds.), Anthropoid Origins: New Visions. Kluwer/Plenum, New York, pp. 603–614.

Byrne, H., Rylands, A.B., Carneiro, J.C., Alfaro, J.W.L., Bertuol, F., et al., 2016. Phylogenetic relationships of the New World titi monkeys (*Callicebus*): first appraisal of taxonomy based on molecular evidence. Front. Zool. 13 (10).

Cartmill, M., 1980. Morphology, function and evolution of the anthropoid postorbital septum. In: Ciochon, R.L., Chiarelli, A.B. (Eds.), Evolutionary Biology of the New World Monkeys and Continental Drift. Plenum, New York, pp. 243–274.

Cave, A.J.E., 1973. The primate nasal fossa. J. Linn. Soc. 5, 377–387.

Chaimanee, Y., Lebrun, R., Yamee, C., Jaeger, J.-J., 2010. A new Middle Miocene tarsier from Thailand and the reconstruction of its orbital morphology using a geometric-morphometric method. Proc. R. Soc. Biol. Sci. 278, 1956–1963.

Coleman, M.N., Colbert, M.W., 2010. Correlations between auditory structures and hearing sensitivity in non-human primates. J. Morphol. 271, 511–532.

Coleman, M.N., Boyer, D.M., 2012. Inner ear evolution in Primates through the Cenozoic: implications for the evolution of hearing. Anat. Rec. 295, 615–631.

Fleagle, J.G., Gilbert, G.C., Baden, A.L., 2010. Primate cranial diversity. Am. J. Phys. Anthropol. 142, 565–578.

Fleagle, J.G., 2013. Primate Adaptation and Evolution, third ed. Academic Press, New York.

Godfrey, L.R., Jungers, W.L., 2003. The extinct sloth lemurs of Madagascar. Evol. Anthropol. 12, 252–263.

Gonzales, L.A., Benefit, B.R., McCrossin, M.L., Spoor, F., 2015. Cerebral complexity preceded enlarged brain size and reduced olfactory bulbs in Old World monkeys. Nat. Commun. 6. https://doi.org/10.1038/ncomms8580.

Goodman, S.M., Jungers, W.L., 2014. Extinct Madagascar. University of Chicago Press, Chicago.

Griffin, R.H., Matthews, L.J., Nunn, C.L., 2012. Evolutionary disequilibrium and activity period in primates: a Bayesian phylogenetic approach. Am. J. Phys. Anthropol. 147, 409–416.

Groves, C., Shekelle, M., 2010. The genera and species of Tarsiidae. Int. J. Primatol. 31 (6), 1071–1082.

Heesy, C.P., Ross, C.F., 2001. Evolution of activity patterns and chromatic vision in primates: morphometrics, genetics, and cladistics. J. Hum. Evol. 40, 111–149.

Heesy, C.P., Ross, C.F., 2004. Mosaic evolution of activity pattern, diet, and color vision in haplorhine primates. In: Ross, C.F., Kay, R.F. (Eds.), Anthropoid Origins: New Visions. Kluwer/Plenum, New York.

Herrera, J.P., Dávalos, L.M., 2016. Phylogeny and divergence times of lemurs inferred with recent and ancient fossils in the tree. Syst. Biol. 65 (5), 772–791.

Identifying primate species. In: Fleagle, J.G. (Ed.), Evol. Anthropol. 23, 1–40.

Isler, K., Kirk, E.C., Miller, J.M.A., Albrecht, G.A., Gelvin, B.R., Martin, R.D., 2008. Endocranial volumes of primate species: scaling analyses using a comprehensive and reliable data set. J. Hum. Evol. 55, 967–978.

Jablonski, N.G., Chapman, G., 2009. The fossil record of gibbons. In: Lappan, S., Whittaker, D.J. (Eds.), The Gibbons: New Perspectives on Small Ape Socioecology and Population Biology. Springer, New York.

Kay, R.F., Kirk, E.C., 2000. Osteological evidence for the evolution of activity pattern and visual acuity in primates. Am. J. Phys. Anthropol. 113, 235–262.

Kay, R.F., Perry, J.M.G., Malinzak, M.D., Allen, K.L., Kirk, E.C., Plavcan, J.M., Fleagle, J.G., 2012. The paleobiology of Santacrucian primates. In: Vizcaíno, S.F., Kay, R.F., Bargo, M.S. (Eds.), Early

Miocene Paleobiology in Patagonia: High-Latitude Paleocommunities of the Santa Cruz Formation. Cambridge University Press, Cambridge, pp. 306–330.

Kay, R.F., 2015. Biogeography in deep time – what do phylogenetics, geology, and paleoclimate tell us about early platyrrhine evolution? Mol. Phylogenet. Evol. 82, 358–374.

Kiesling, N.M.J., Yi, S.V., Xu, K., Sperone, F.G., Wildman, D.E., 2015. The tempo and mode of New World monkey evolution and biogeography in the context of phylogenomic analysis. Mol. Phylogenet. Evol. 82, 386–399.

Kistler, L., Ratan, A., Godfrey, L.R., Crowley, B.E., Hughes, C.E., et al., 2015. Comparative and population mitogenomic analyses of Madagascar's extinct, giant 'subfossil' lemurs. J. Hum. Evol. 79, 45–54.

Le Gros Clark, W.E., 1959. The Antecedents of Man. An Introduction to the Evolution of the Primates. The University Press, Edinburgh.

Matsuzawa, T., Humle, T., Sugiyama, Y. (Eds.), 2011. The Chimpanzees of Bossou and Nimba. Springer, New York.

Mittermeier, R.A., Rylands, A.B., Wilson, D.E. (Eds.), 2013. Handbook of the Mammals of the World: 3. Primates. Lynx Edicions, Barcelona.

Muchlinksi, M., 2010. A comparative analysis of vibrissa count and infraorbital foramen area in primates and other mammals. J. Hum. Evol. 58, 447–473.

Nakatsukasa, M., Mbua, E., Sawada, Y., Sakai, T., Nakaya, H., et al., 2010. Earliest colobine skeletons from Nakali, Kenya. Am. J. Phys. Anthropol. 143, 365–382.

Nevo, O., Heymann, E.W., 2015. Led by the nose: olfaction in primate feeding ecology. Evol. Anthropol. 24, 137–148.

Ottoni, E.B., Izar, P., 2008. Capuchin monkey tool use: overview and implications. Evol. Anthropol. 17, 171–178.

Pozzi, L., Disotell, T.R., Masters, J.C., 2014. A multilocus phylogeny reveals deep lineages within African galagids (Primates: Galagidae). BMC Evol. Biol. 14, 72.

Ross, C.F., 1996. An adaptive explanation for the origin of the Anthropoidea (Primates). Am. J. Primatol. 40, 205–230.

Rossie, J.B., Smith, T.D., 2007. Ontogeny of the nasolacrimal duct in primates: functional and phylogenetic implications. J. Anat. 210, 195–208.

Rylands, A.B., Mittermeier, R.A., 2014. Primate taxonomy: species and conservation. Evol. Anthropol. 23, 8–10.

Santini, L., Rojas, D., Donati, G., 2015. Evolving through day and night: origin and diversification of activity pattern in modern primates. Behav. Ecol. 26, 789–796.

Schneider, H., Bernardi, J.A.R., da Cunha, D.B., Tagliaro, C.H., Vallinoto, M., Ferrari, S.F., Sampaio, I., 2012. A molecular analysis of the evolutionary relationships in the Callitrichinae, with emphasis on the position of the dwarf marmoset. Zool. Scr. 41, 1–10.

Sears, K.E., Finarelli, J.A., Flynn, J.J., Wyss, A.R., 2008. Estimating body mass in new world "monkeys" (Platyrrhini, primates), with a consideration of the Miocene platyrrhine, Chilecebus carrascoensis. Am. Mus. Novit. 3617, 1–29.

Seiffert, E.R., Simons, E.L., Attia, Y., 2003. Fossil evidence for an ancient divergence of lorises and galagos. Nature 422, 421–424.

Seiffert, E.R., Simons, E.L., Clyde, W.C., Rossie, J.B., Attia, Y., et al., 2005. Basal anthropoids from Egypt and the antiquity of Africa's higher primate radiation. Science 310, 300–304.

Seiffert, E.R., 2007. Early evolution and biogeography of lorisiform strepsirrhines. Am. J. Primatol. 69, 27–35.

Seiffert, E.R., Simons, E.L., Fleagle, J.G., Godinot, M., 2010. In: Werdelin, L., Sanders, W.J. (Eds.), Paleogene Anthropoids. Cenozoic Mammals of Africa. University of California Press, Berkeley, pp. 369–391.

Seiffert, E.R., 2012. Early primate evolution in Afro-Arabia. Evol. Anthropol. 21, 239–253.

Silcox, M.T., Benham, A.E., Bloch, J.I., 2010. Endocasts of Microsyops (Microsyopidae, Primates) and the evolution of the brain in primitive primates. J. Hum. Evol. 58, 505–521.

Simons, E.L., Seiffert, E.R., Ryan, T.M., Attia, Y., 2007. A remarkable female cranium of the Oligocene anthropoid Aegyptopithecus zeuxis (Catarrhini, Propliopithecidae). Proc. Natl. Acad. Sci. U.S.A. 104, 8731–8736.

Smith, T.D., Siegel, M.I., Bhatnagar, K.P., 2001. Reappraisal of the vomeronasal system of catarrhine primates: ontogeny, morphology, functionality, and persisting questions. Anat. Rec. 265, 176–192.

Smith, T.D., Rossie, J.B., Bhatnagar, K.P., 2007. Evolution of the nose and nasal skeleton in primates. Evol. Anthropol. 16, 132–146.

Smith, T., Rose, K.D., Gingerich, P.D., 2006. Rapid Asia-Europe-North America geographic dispersal of earliest Eocene primate Teilhardina during the Paleocene-Eocene thermal maximum. Proc. Natl. Acad. Sci. U.S.A. 103 (30), 11223–11227.

Springer, M.S., Meredith, R.W., Gatesy, J., Emerling, C.A., Park, J., et al., 2012. Macroevolutionary dynamics and historical biogeography of primate diversification inferred from a species supermatrix. PLoS One 7, e49521.

Steiper, M.E., Seiffert, E.R., 2012. Evidence for a convergent slowdown in primate molecular rates and its implications for the timing of early primate evolution. Proc. Natl. Acad. Sci. U.S.A. 109, 6006–6011.

Steiper, M.E., Young, N.M., 2008. Timing primate evolution: lessons from the discordance between molecular and paleontological estimates. Evol. Anthropol. 17, 179–188.

Stevens, N.J., Seiffert, E.R., O'Connor, P.M., Roberts, E.M., Schmitz, M.D., Krause, C., Gorscak, E., Ngasala, S., Hieronymus, T.L., Temu, J., 2013. Palaeontological evidence for an Oligocene divergence between Old World monkeys and apes. Nature 497, 611–614.

Visalberghi, E., Fragaszy, D., 2013. The EthoCebus project. Stone tool use by wild capuchin monkeys. In: Sanz, C., Call, J., Boesch, C. (Eds.), Multidisciplinary Perspectives on the Cognition and Ecology of Tool Using Behaviors. Cambridge University Press, Cambridge, pp. 203–222.

Whiten, A., Goodall, J., McGrew, W.C., Nishida, T., Reynolds, V., Sugiyama, Y., Tutin, C.E.G., Wrangham, R.W., Boesch, C., 1999. Cultures in chimpanzees. Nature 399, 682–685.

Zijlstra, J.S., Flynn, L.J., Wessels, W., 2013. The westernmost tarsier: a new genus and species from the Miocene of Pakistan. J. Hum. Evol. 65, 544–550.

21

The Expansion of the Cortical Sheet in Primates

S. Mayer, A.R. Kriegstein

University of California, San Francisco, San Francisco, CA, United States

Nomenclature

IPC Intermediate progenitor cell
ISVZ Inner subventricular zone
oRG Outer radial glia cell
OSVZ Outer subventricular zone
RG Radial glia cell
vRG Ventricular radial glia cell
VZ Ventricular zone

21.1 Cortical Sheet Formation During Development

Comparative developmental and genomic studies can often provide insights into evolutionary processes that are inaccessible to studies of the fossil record, particularly those involving soft tissues such as the brain whose fine structure escapes fossilization (Karten, 2015). One of the key questions in developmental biology is: How does a large organism that is composed of many different dynamically interacting cell and tissue types emerge from a single cell, the fertilized oocyte? Framing this question in the context of cortical sheet development will ultimately allow us to appreciate how the diverse areas and cell types of the neocortex coevolve with the highly sophisticated cognitive behaviors that define primates. Understanding how the complex anatomical layout of the cerebral cortex is generated in development is particularly important because function depends on structure. Further, phenotypic diversity is generated by alterations in developmental mechanisms that give rise to some aspect of the phenotype, such as the size and structure of the neocortex. Thus, understanding how different developmental processes have been modified in different lineages and generate an expanded cortical sheet will provide important insights into how the primate brain evolved.

In all chordates, the nervous system develops from one of the three primary germ layers, the ectoderm, through an interaction with the mesoderm. During neural induction, the mesoderm sends signals to a sheet of the overlying ectoderm causing it to invaginate to form the neural tube. The neural tube is subsequently patterned along its anterior—posterior axis to form the different segmental regions of the nervous system. As anterior areas proliferate more than posterior regions, brain vesicles are formed anteriorly. The cerebral cortex develops from the most anterior vesicle, the telencephalon. Neural stem cells in the telencephalon replicate and give rise to all the cells that constitute the mature cortex and also seed the limited cell reservoirs that continue to produce newborn neurons in the adult cerebral cortex. During evolution, a laminar cortex appears to have developed in an ancestral reptile and likely consisted of only three layers (Cheung et al., 2007). The cerebral cortex of mammals, consisting mostly of the neocortex, is characterized by six distinct layers of cells (Fig. 21.1A). This distinguishes it from the hippocampus, a more ancient part of the cerebral cortex that resembles the trilaminar arrangement of the archicortex in reptiles, and differs from the nuclear arrangement of the forebrain in birds. Neurons in each layer have distinct morphologies and connections with each other, as well as with neurons in other layers or neocortical areas (corticocortical projections) or other brain regions (subcortical projections).

The cells of the primate neocortex originate from subtypes of neural stem and progenitor cells that are located in the germinal zones of the telencephalon. Neural stem cells can be characterized by their self-replicative potential as well as their ability to generate daughter cells of multiple neuronal and glial lineages. Progenitor cells are more lineage-restricted. These features can be determined by analyzing the patterns of neural progenitor cell divisions. Are the divisions symmetric in the sense that they generate progeny that share the same fate? Or do they divide asymmetrically to produce cells

519

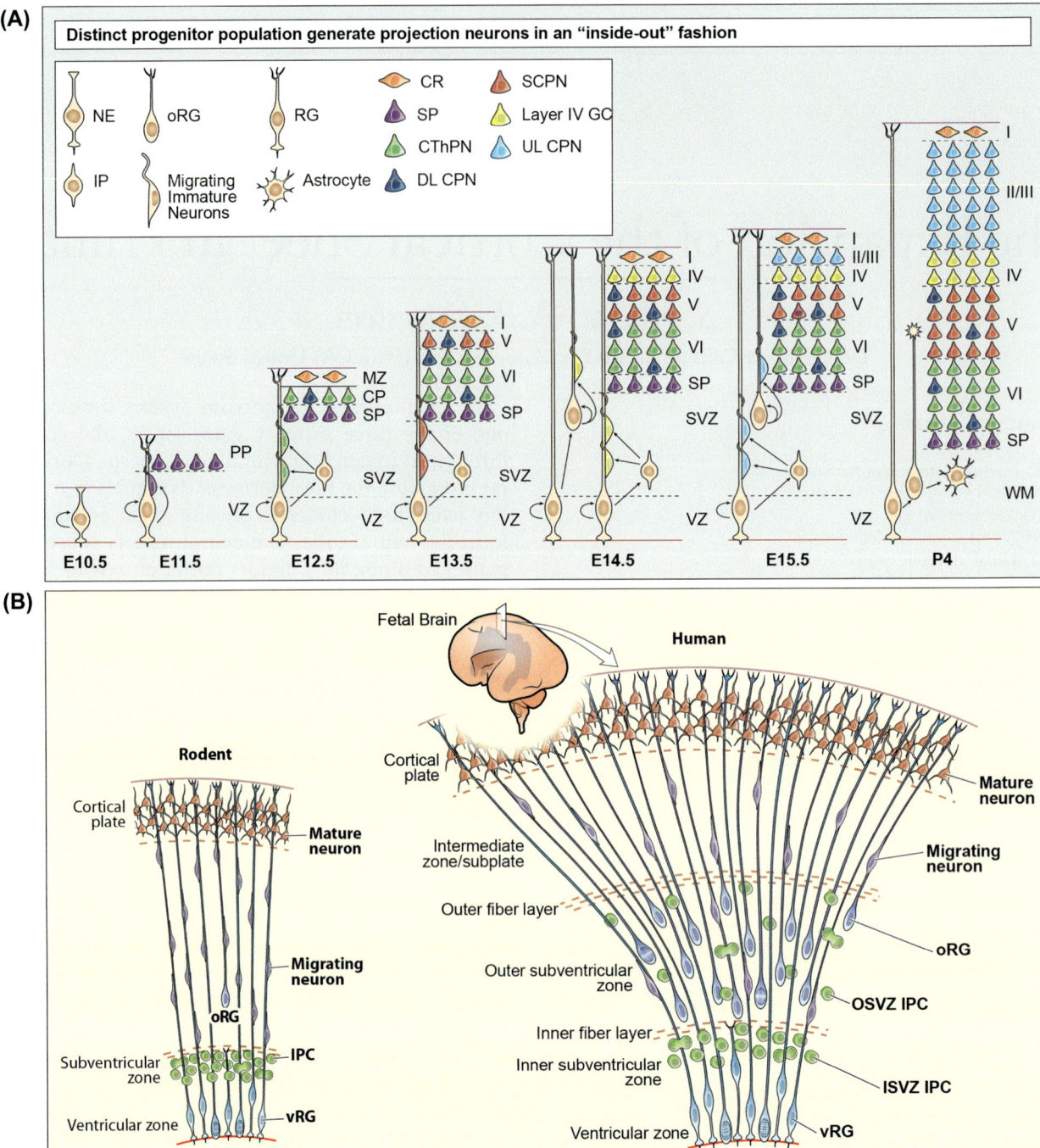

FIGURE 21.1 Neurogenesis in the mammalian cerebral cortex. (A) Neurogenesis by distinct types of progenitors in the mouse neocortex. Progenitor cells (Neuroepithelial cells, *NE*; outer radial glia cells, *oRG*; (ventricular) radial glia cell, *(v)RG*; intermediate progenitor cell, *IPC*) in the ventricular zone (VZ) and later in development in the subventricular zone (SVZ) divide generating different types of neurons in a time-dependent manner. Newborn immature neurons migrate radially into the overlying cortical plate to give rise to the 6-layered neocortex in an inside-out fashion. Subsequently radial glia cells give rise to astrocytes. Preplate, *PP*; Subplate (neurons), *SP*; Cortical Plate, *CP*; Marginal Zone, *MZ*; layers of the cortex, I-VI; White Matter, *WM*; Cajal-Retzius Cell, *CR*; Corticothalamic Projection Neuron, *CThPN*; Deep Layer Callosal Projection Neuron, *DL CPN*; Subcerebral Projection Neuron, *SCPN*; Granule Cell, *GC*; Upper Layer Callosal Projection Neuron, *UL CPN*. Days of embryonic (E) development in the mouse are indicated. (B) Comparison of neocortical development between mice and humans. Note the enlargement of the subventricular zone due to the presence of large numbers of oRGs, which leads to the tangential expansion of the cortical sheet. Inner subventricular zone, *ISVZ*; outer subventricular zone, *OSVZ*. *Modified from Woodworth, M.B., Custo Greig, L., Kriegstein, A.R., Macklis, J.D., 2012. SnapShot: cortical development. Cell 151 (4), 918–918.e1; Lui, J.H., Hansen, D.V., Kriegstein, A.R., 2011. Development and evolution of the human neocortex. Cell 146 (1), 18–36; Ostrem, B.E.L., Lui, J.H., Gertz, C.C., Kriegstein, A.R., 2014. Control of outer radial glial stem cell mitosis in the human brain. Cell Rep. 8 (3), 656–664. Reprinted with Permission from Elsevier Inc., 2011, 2012 and 2014.*

with different fates? Does their mode of division change during development? In this chapter we will describe the properties of the different stem and progenitor cell types currently known to be present in the primate neocortex and outline how they emerge developmentally. Since the lineage restriction for the presumed stem and progenitor cell types has not been unequivocally shown so far (Taverna et al., 2014), we will refer to them as neural progenitor cells in this text. We will discuss findings made by studying cortical development in the mouse, because features of most progenitor cell types are shared between all mammals. Subsequently, we will discuss the presumed emergence of each of these cell types in evolutionary time, and how defects in progenitor cell biology can cause disease. By understanding how different types of progenitors contribute to cortical development in different species, including mammals, birds, and reptiles, we can start to unravel how the evolutionary expansion of the neocortical sheet may have occurred in primates.

In early cortical development, a single layer of neuroepithelial cells spans the entire width of the neural tube from the ventricle to the pial surface [Fig. 21.1A, embryonic day (E) 10.5 in the mouse]. These cells display hallmark features of epithelial cells: they adhere tightly to each other through adherens junctions at their apical side and are polarized along their apical–basal axis resulting in the asymmetric distribution of proteins (Aaku-Saraste et al., 1996; Manabe et al., 2002). For example, Prominin-1, a membrane protein expressed in different types of stem cells, is localized at the apical surface at the interface with the ventricle (Götz and Huttner, 2005). In contrast, the integrin cell adhesion proteins are localized at the basal fiber contact with basal lamina extracellular matrix components, just underneath the pia (Götz and Huttner, 2005). The early neuroepithelium is pseudostratified, since neuroepithelial cells undergo interkinetic nuclear migration (Sauer, 1935). During this process the soma is always located at the ventricular surface during mitosis, while during S-phase it is located at a distance from the ventricle, at the most basal part of the neuroepithelium. During the G1 and G2 stages of the cell cycle the soma moves between these locations. Neuroepithelial cells generally divide symmetrically to self-renew and increase their pool size (Götz and Huttner, 2005), a process that is controlled, in part, by β-catenin signaling (Chenn and Walsh, 2002).

As neocortical development progresses, FGF and Notch signaling as well as microRNA-mediated signaling promote the transformation of neuroepithelial cells to radial glia (Fig. 21.1A; E11.5 in the mouse) (Sahara and O'Leary, 2009; Hatakeyama et al., 2004; Gaiano et al., 2000; Nowakowski et al., 2011). Radial glia predominantly divide asymmetrically to self-renew and generate neurons or other types of progenitor cells (Fig. 21.1A; E12.5–15.5) (Noctor et al., 2001; Miyata et al., 2001; Malatesta et al., 2000). Radial glia share many hallmark features with neuroepithelial cells, such as adherens junctions, the asymmetrical distribution of proteins along the apical–basal axis, and interkinetic nuclear migration within the ventricular zone. Additionally, they display characteristics of astroglial cells such as glycogen granules, and they express glia-specific intermediate filaments such as vimentin and glial fibrillary acidic protein (Götz and Huttner, 2005). Radial glia maintain contact with the ventricle, which may be important for trophic support (Lehtinen et al., 2011). Thus, to distinguish them from other types of neural progenitor cells as well as from radial glia-like cells that do not contact the ventricle, they have also been referred to as apical progenitors or ventricular radial glia (vRGs). vRGs generally span the entire cortical sheet, and their radial processes serve as a scaffold for the migration of newly born neurons into the cortical plate (Noctor et al., 2001; Rakic, 1972).

Abventricular progenitors, which we define as progenitor cells that do not contact the ventricle, have expanded in number and type during evolution, as we will discuss in detail below. Intermediate progenitor cells (IPCs) and outer radial glia cells (oRGs) are two types of abventricular progenitors that derive from vRGs. At early stages of neurogenesis, IPCs generated from vRGs intermingle with these in the ventricular zone, but IPCs produced at later stages migrate away to form a secondary progenitor zone, the subventricular zone (SVZ), adjacent to the ventricular zone (Noctor et al., 2004; Haubensak et al., 2004; Fig. 21.1). oRGs are generated from vRGs at later stages of development and also migrate to the SVZ (LaMonica et al., 2013; Fig. 21.1B). The loss of direct contact with the ventricle may lead to a limited capacity for self-renewal or the requirement of other mechanisms that allow the maintenance of stem cell characteristics in these abventricular progenitor cells.

As the development of the cortical sheet progresses, the different progenitor cell types undergo neurogenic divisions, where they divide to give rise to one or two postmitotic neurons, if the division is asymmetric or symmetric, respectively. The newborn neurons migrate radially along the processes of radial glia cells toward the pia until they reach their target position in the cortical plate. Most often the daughter cells of radial glia cells migrate along their parental fibers (Noctor et al., 2001), at least in rodents, while at later developmental stages, a more complex migratory pattern emerges in gyrencephalic mammals (Gertz and Kriegstein, 2015). In the cortical plate newborn neurons form the six layers of the mature neocortex in an inside-out manner—deeper layers are formed before upper layers (Fig. 21.1A; E12.5–15.5 in the mouse) (Angevine and Sidman, 1961). Sequentially born neurons form minicolumns, or radial units, in the neocortex, presumably as a result of migrating along the same or neighboring radial glial fibers (Rakic, 2009). Later in development, glial cells are generated, partly from radial glia that transform into astrocytes (Fig. 21.1A; P4 in the mouse).

Neurons generated in the cortical ventricular zone are destined to become excitatory neurons. The inhibitory interneurons of the cortex are generated from progenitor cells in the ventricular zone of the ventral forebrain, a region known as the ganglionic eminence, from which they migrate tangentially to reach the cortical plate of the developing neocortex. Additionally, interneurons may be generated directly in the germinal zone of the dorsal cortex in primates (Radonjić et al., 2014; Clowry, 2014). We will not discuss the developmental origins of interneurons in detail here; the interested reader is referred to recent reviews on this topic (Anastasiades and Butt, 2011; Marín, 2013).

21.2 Cortical Sheet Expansion Is Possible in Two Dimensions

The neocortex has undergone a rapid expansion throughout mammalian evolution, increasing in size and complexity. While humans do not have the largest cortex among mammals—whales and elephants, for example, have larger and more highly folded cortices— the human neocortex has the largest number of neurons, comprising 16 billion neurons, which constitute two-thirds of the total brain mass (Azevedo et al., 2009). The expansion of the cortical sheet has occurred in two dimensions: it has increased in surface area (tangential expansion) and thickness (radial expansion) (Fig. 21.2). Proportionally, the cortical sheet has expanded primarily tangentially generating a larger surface area, with only a small relative increase in its thickness (Rakic, 2009).

The evolutionary expansion of the neocortex has been accompanied by the addition of distinct functional areas. Over 20 years ago, Pasko Rakic proposed the radial unit hypothesis to explain how tangential cortical expansion could occur through the addition of radial units, groups of cells that arise over time from a single

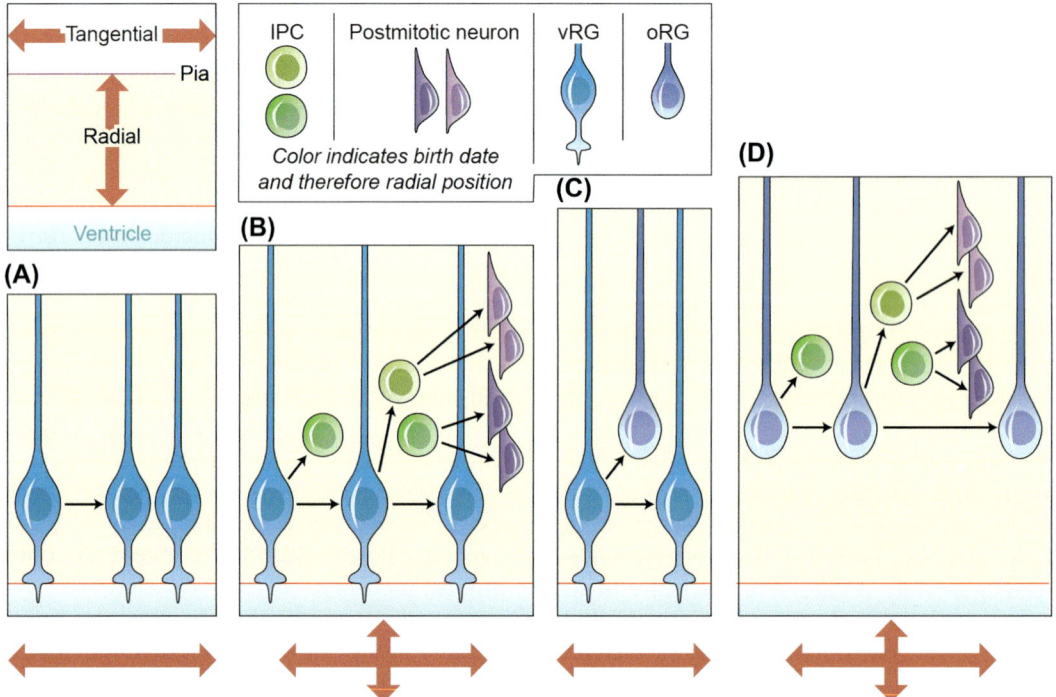

FIGURE 21.2 Different progenitor types drive the radial and tangential expansion of the cortical sheet. (A) When ventricular radial glia cells (vRGs) undergo self-renewing symmetric divisions, this leads to an increase in the number of radial units and hence the tangential expansion of the cortical sheet. (B) When vRGs divide asymmetrically to generate an intermediate progenitor cell (IPC) and a vRG this leads to the amplification of the same radial unit with newborn neurons migrating along the same radial fiber and hence radial expansion. Subsequently IPCs can divide to generate two neurons, which are born simultaneously and will thus assume the same radial position leading to tangential expansion. (C) When vRGs divide asymmetrically to give rise to an outer radial glia cell (oRG) and a vRG this increases the number of radial units and hence causes the tangential expansion of the cortical sheet. (D) When oRGs divide asymmetrically to generate an IPC and an oRG this leads to the amplification of the same radial unit with newborn neurons migrating along the same radial fiber and hence radial expansion. Subsequently IPCs can divide to generate two neurons, which are born simultaneously and will thus assume the same radial position leading to tangential expansion.

progenitor cell and then migrate along the same or neighboring radial glia to form a column of neurons (Rakic, 1988). At that time it was not suspected that radial glia were neural progenitor cells, but the radial scaffold of glial fibers suggested that cells in a radial column could have a lineage relationship, giving rise to the concept of an ontogenetic column (Rakic, 1995). Tangential expansion of the cortex, both across developmental time and across species during evolution, could then be explained by the addition of more radial units, which in turn can be driven by an increase in the number of radial glia, through increased proliferation, reduced apoptosis, or both (Fig. 21.2A).

The question then arises how different areas of the cortex are produced—are neurons in different regions fated to be different from each other at their time of origin, or are they specified by local signals (Rakic et al., 2009)? The protomap hypothesis proposed by Rakic suggests that a primordial map of the different functional areas of the cortex already exists at the level of the progenitor cells in the germinal zone (Rakic, 1988). As neurons migrate to the overlying cortical plate along radial glia fibers, the areal specification is maintained and projected onto the cortical mantle. Thus, a crude cortical map is established before environmental signals such as synaptic input from other cortical and subcortical areas can have an effect. While subcortical axons might directly influence regional identity, early-arriving neurons may also help to attract the appropriate axons to establish synaptic input from specific thalamic nuclei (Rakic et al., 2009).

Different lines of research have provided support for the protomap model. Interestingly, histological differences between cortical areas are more apparent in the expanded primate fetal cortex than in the rodent embryonic cortex, reflecting the evolutionary importance of arealization concurrent with cortical sheet expansion (Rakic et al., 2009). The kinetics of proliferation and the ensuing changes in the thickness of the germinal zones have been shown to vary between cortical areas in primates (Miyama et al., 1997; Smart et al., 2002; Zecevic et al., 2005; Lukaszewicz et al., 2005). The molecular mechanisms that underlie the areal specification of progenitor cells in the germinal zones are largely unexplored. Different morphogenetic gradients are established by patterning centers in the proliferative zones in the mammalian neocortex, and these control transcription factor expression in progenitor cells (Rakic et al., 2009). How these gradients might influence progenitor cell biology and the fate of newly born neurons is not yet clear. Genes regulating the cell cycle have been shown to be differentially expressed in progenitor cells in the germinal zones of macaques thereby potentially accounting for differences in neuron number of different cortical areas (Lukaszewicz et al., 2005).

Differential gene expression in different regions of the developing neocortex has been reported in many studies (eg, Abraham et al., 2007; Kang et al., 2011). Kang and colleagues found for example that one transcript of the ANKRD32 gene is expressed in a frontal-occipital gradient in the human brain, but not in the mouse brain during the peak of neurogenesis (Kang et al., 2011). This finding hints at regional patterns of gene expression during early development that may play a role in the development of different functional areas in the expanded cortical sheet of primates. Besides shedding light on the formation of different functional areas of the cortex during development, analyzing regional patterns of progenitor abundance and gene expression in the germinal zones may also help us understand the mechanisms that produce cortical folding in gyrencephalic mammals (Reillo et al., 2011; de Juan Romero et al., 2015).

In addition to the tangential expansion of the cortical sheet, the cortex has also expanded radially. The evolution of the six-layered mammalian neocortex from the three-layered archicortex is one way in which radial expansion has contributed to cortical development. Furthermore, during mammalian evolution, the upper layers of the neocortex have dramatically expanded radially in certain cortical regions, particularly in primates (DeFelipe et al., 2002). Neurons in the upper layers form intracortical connections, which may be particularly important for higher cognitive functions. Changes in the thickness of the cortical sheet may not only impact the functionality but also the topography of the cortex as indicated in a recent study by Mota and Herculano-Houzel (2015), who used mathematical modeling to conclude that the ratio of cortical thickness to cortical area directly correlates with the degree of folding.

The radial unit hypothesis suggests that prolonging the period of neurogenesis would lead to radial cortical expansion (Rakic, 1988). The addition of cells to cortical columns can be achieved through an increase in the number of asymmetric radial glia divisions, based on the observation that cells with different birthdates will generally migrate to different radial positions or cortical layers (Desai and McConnell, 2000). Changes in cell cycle length regulate a switch of divisions from ones that expand the progenitor pool to ones that produce neurons (Lukaszewicz et al., 2005; Nowakowski et al., 2002; Calegari, 2003). Moreover, more neurons are generated when the neurogenic period is lengthened, as has happened during primate evolution (Dehay et al., 1993; Takahashi et al., 1996; Noctor et al., 1997; Finlay et al., 1998; Kornack and Rakic, 1998).

The addition of more cells to each layer of the neocortex could also be achieved by the introduction of IPCs, an additional progenitor cell type that originates from radial glia cells (Kriegstein et al., 2006). These

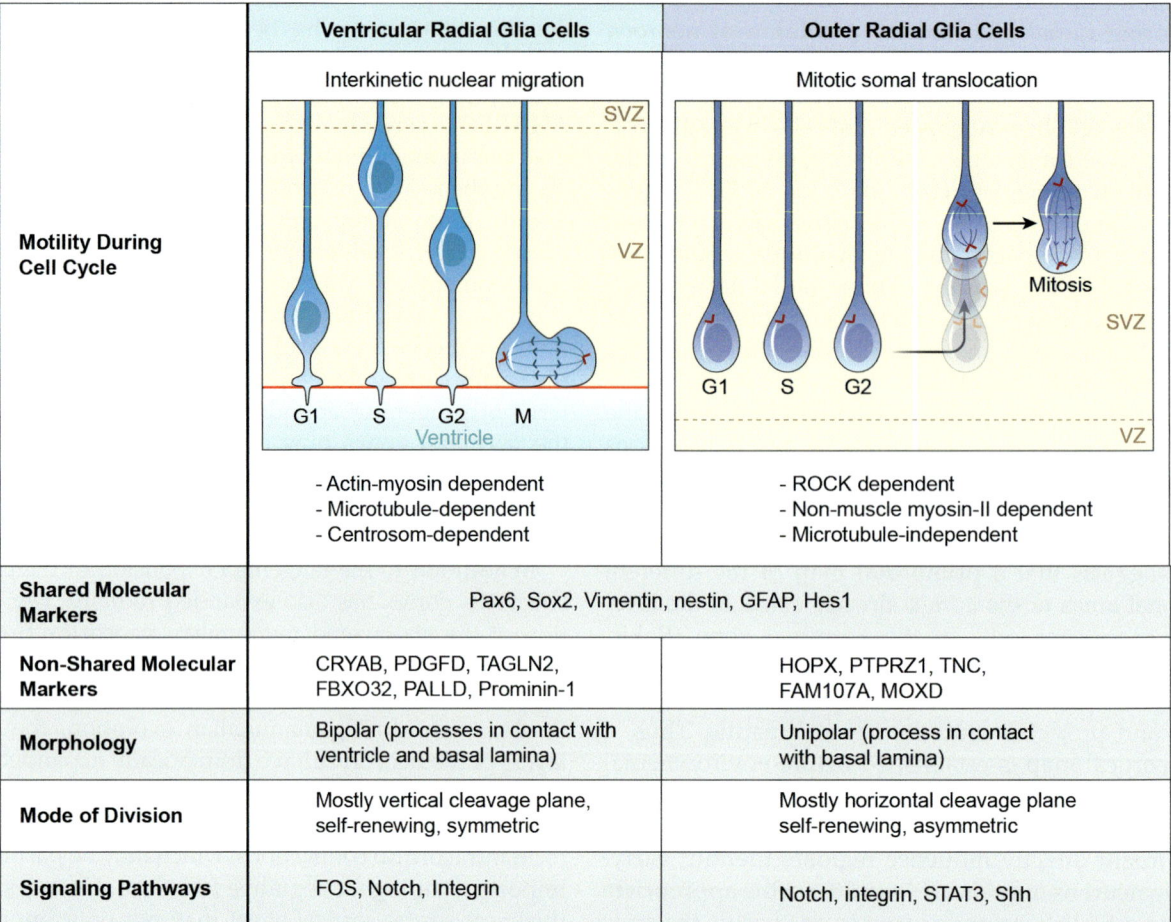

	Ventricular Radial Glia Cells	Outer Radial Glia Cells
Motility During Cell Cycle	Interkinetic nuclear migration - Actin-myosin dependent - Microtubule-dependent - Centrosom-dependent	Mitotic somal translocation - ROCK dependent - Non-muscle myosin-II dependent - Microtubule-independent
Shared Molecular Markers	Pax6, Sox2, Vimentin, nestin, GFAP, Hes1	
Non-Shared Molecular Markers	CRYAB, PDGFD, TAGLN2, FBXO32, PALLD, Prominin-1	HOPX, PTPRZ1, TNC, FAM107A, MOXD
Morphology	Bipolar (processes in contact with ventricle and basal lamina)	Unipolar (process in contact with basal lamina)
Mode of Division	Mostly vertical cleavage plane, self-renewing, symmetric	Mostly horizontal cleavage plane self-renewing, asymmetric
Signaling Pathways	FOS, Notch, Integrin	Notch, integrin, STAT3, Shh

FIGURE 21.3 Comparison of cell biological characteristics of ventricular radial glia cells (vRGs) and outer radial glia cells (oRGs).

transit amplifying-type cells undergo symmetric division to produce pairs of neurons, or to increase the pool of progenitors (Noctor et al., 2004, 2008). IPCs had not been appreciated as a neurogenic population at the time when the radial unit hypothesis was formulated. As discussed in detail below, IPCs are present in large numbers in the ventral forebrain in reptiles, but only few IPCs are present in developing reptile cortex. As the number of these progenitors increases throughout development in species such as rodents, and presumably also as their numbers increased during mammalian evolution, they came to reside in an important secondary proliferative region, at a distance form the ventricle, the SVZ. The Intermediate Cell hypothesis suggests that through the symmetric division of IPCs, a larger number of neurons can be generated in a spatially constricted area derived from a relatively small population of founding neuroepithelial stem cells (Kriegstein et al., 2006). As outlined below, IPCs may have contributed both to the radial expansion of upper cortical layers as well as to tangential cortical expansion (Fig. 21.2B).

21.3 Emergence of Transit Amplifying Cells in Vertebrate Evolution

As the brain expanded during evolution, the space at the ventricular surface became limited (Smart, 1965). This restriction can be overcome by cells detaching from the adhesion belt and migrating away. This happens when vRGs undergo asymmetric divisions, where one daughter remains attached to the ventricular surface as a vRG and the other becomes an abventricular progenitor, delaminating from the ventricular epithelium.

IPCs are one type of abventricular progenitor cell. The earliest evolutionary evidence for IPCs in the developing forebrain comes from lizards (Fig. 21.4), where there are cells scattered in the ventricular zone that express the marker protein for IPCs, Tbr2 (also referred to by its gene name EOMES) (Martínez-Cerdeño et al., 2015). Turtles, which are evolutionarily closer to birds than to lizards (Crawford et al., 2015), also have small numbers of proliferating Tbr2-positive IPCs scattered in the ventricular zone of the dorsal cortex. Unlike

FIGURE 21.4 Evolutionary emergence of abventricular progenitors and distinct proliferative zones. Phylogenetic tree (not to scale) showing the emergence of intermediate progenitor cells (IPCs) and outer radial glia cells (oRGs) and the formation of distinct proliferative zones during vertebrate evolution as the cortical sheet expanded. As examples, the forebrain germinal zones of fish, reptiles, birds, ferrets, mice, and primates are shown. Ventricular Zone, *VZ*; Subventricular Zone, *SVZ*; Intermediate Zone, *IZ*; Cortical Plate, *CP*; Inner Subventricular Zone, *ISVZ*; Outer Subventricular Zone, *OSVZ*.

lizards, they additionally have much larger numbers of IPCs in a rudimentary SVZ in the ventral forebrain, in a structure called the dorsal ventricular ridge (Cheung et al., 2007; Martínez-Cerdeño et al., 2015, 2006). Similarly, alligators possess an SVZ in some parts of their forebrain (Charvet et al., 2009). It is unclear whether there is an SVZ in birds (Montiel et al., 2016). Chickens and doves have a distinct proliferative zone within the developing telencephalon containing Tbr2-positive cells (Martínez-Cerdeño et al., 2015). This organization is also present in the six-layered neocortex in mammals, with mouse, opossum, and wallaby sharing the same organization of Tbr2-positive IPCs in a distinct proliferative layer (Englund, 2005; Cheung et al., 2010). The similarity between birds and mammals with regard to the presence and organization of IPCs in the germinal zones of the forebrain parallels similarities in neuronal subtypes and circuits in the mature brain, and may relate to similarities in some aspects of their cognitive functions (Scharff et al., 2013).

In mammals, and especially primates, the SVZ has become enlarged and can be subdivided into two parts, the inner SVZ (ISVZ) directly adjacent to the VZ, and the outer SVZ (OSVZ) between the ISVZ and the cortical plate (Fig. 21.1B). Initially these two zones were defined in the macaque neocortex based on cytoarchitectural features: while cells in the ISVZ are randomly organized like cells in the SVZ in rodents, cells in the OSVZ are radially oriented (Smart et al., 2002). Additionally, an axon fiber bundle, the inner fiber layer, was found to run between the ISVZ and the OSVZ at the occipital pole at certain developmental time points in macaques and humans (Smart et al., 2002; Zecevic et al., 2005). Recently, the definitions of the ISVZ and OSVZ have been suggested to be dependent on the expression of marker genes such as Tbr2 (Martínez-Cerdeño et al., 2012). According to this definition, Tbr2-positive cells, that is, IPCs, are found in both layers of the SVZ, as a dense band in the ISVZ and a diffuse band in the OSVZ (Martínez-Cerdeño et al., 2012). Recent evidence indicates that a distinction between the ISVZ and the OSVZ may be present in all mammals, rather than being unique to primates or gyrencephalic mammals (Martínez-Cerdeño et al., 2012; Garcia-Moreno et al., 2012). The abundance of IPCs increases in later developmental stages in all the different species analyzed, with the highest proportion of abventricular mitoses occurring in late development in animals with a folded (gyrencephalic) cortex such as primates and ferrets (Cheung et al., 2007; Martinez-Cerdeno et al., 2006).

In mammals, IPCs are multipolar, and they usually arrest migration and retract their processes when they undergo mitosis (Noctor et al., 2004). They appear to divide symmetrically producing two daughter cells with similar, if not identical fates, either neurons or IPCs (Noctor et al., 2004). In mammals, IPCs rather than radial glia have been suggested to generate most neurons (Kowalczyk et al., 2009). Several studies suggest that IPCs are particularly responsible for the production of upper layer neurons in the mouse (Tarabykin et al., 2001; Wu et al., 2005). Thus, their increased abundance in mammals compared to birds and reptiles can be correlated to the addition of further layers on top of the three-layered cortex to generate the six-layered neocortex of mammals.

21.4 Emergence of a Further Progenitor Subtype in Mammalian Evolution

As first described in the macaque, the SVZ has been enlarged during mammalian evolution (Smart et al., 2002). This was concomitant with an increase in the length of neurogenesis allowing more cell divisions that amplify the progenitor pool and ultimately more neurogenic cell divisions to occur. Interestingly, the major wave of neurogenesis in the primate cortex coincides with the peak of cell divisions in the OSVZ (Lukaszewicz et al., 2005; Rakic, 1974). Thus, the majority of supragranular neurons, which reside in layers I—III of the mature cortex, originate from progenitors located in the OSVZ (Lukaszewicz et al., 2005). Since these layers are particularly pronounced in primates compared to rodents, it is likely that the OSVZ plays a significant role in primate evolution. Recently, a new progenitor cell type that resides primarily in the OSVZ was identified in the developing human cortex (Hansen et al., 2010). Termed an oRG, it was soon also identified in the developing cortex of ferrets and nonhuman primates (Reillo et al., 2011; Fietz et al., 2010). While first only described in gyrencephalic mammals, oRGs were subsequently identified in nongyrencephalic species, such as the mouse (Reillo et al., 2011; Hansen et al., 2010; Fietz et al., 2010; Wang et al., 2011). However, the abundance of oRGs has increased dramatically in mammalian evolution, reminiscent of the evolutionary increase in the abundance IPCs (Lui et al., 2011; Fig. 21.4).

As their name implies, oRGs are similar to radial glia and express many of the same marker genes, such as PAX6 and SOX2 (Fig. 21.3). Their morphology is classically characterized by a single basal process, which often contacts the pial surface. oRGs lack a process contacting the ventricular surface. Hence, they also lack the typical apicobasal polarity of epithelial cells and do not express related markers such as Prominin-1(Fietz et al., 2010). Recent studies indicate that oRGs may adopt various morphologies; they may, for example, have a multipolar morphology (Betizeau et al., 2013; Gertz et al., 2014). oRGs are generated from vRGs through horizontal divisions (LaMonica et al., 2013). Instead of performing interkinetic nuclear migration, oRGs undergo a characteristic mode of division called mitotic somal translocation, in which their soma "jumps" several dozens of micrometers prior to mitosis (Hansen et al., 2010). However, not all oRG mitoses are characterized by mitotic somal translocation, some oRGs are also stationary when they divide (Gertz et al., 2014). Their cleavage plane forms at right angles to their basal process, so that one daughter inherits the basal fiber (LaMonica et al., 2013; Gertz et al., 2014). In the human, oRGs mostly undergo self-renewing asymmetric divisions where the basal daughter inherits the basal fiber and maintains oRG features while the apical daughter attains IPC characteristics (Hansen et al., 2010; Fietz et al., 2010; Ostrem et al., 2014; Pollen et al., 2015). In the ferret, however, most divisions are symmetric, generating two oRG daughter cells and thereby expanding the progenitor pool (Gertz et al., 2014). The identification of a characteristic mitotic behavior has allowed the elucidation of the molecular pathways underlying this unusual movement. Interestingly, unlike in the G2 to M phase transition in vRGs undergoing interkinetic nuclear migration, the mitotic somal translocation of the oRG does not rely on the microtubule cytoskeleton. Instead myosin-II motor proteins moving along actin fibers are responsible for the fast movement of the soma along the basal process during M phase (Ostrem et al., 2014). Recently, molecular markers of oRGs in the human brain were identified through single cell sequencing, making this cell type more amenable for cell biological studies (Pollen et al., 2015; Thomsen et al., 2016). Among the newly identified oRG marker proteins are TNC, PTPRZ1, FAM107A, and HOPX. Beyond serving as markers, these proteins may help to demonstrate how oRGs undergo epithelial–mesenchymal transition to delaminate from the apical surface and provide their own niche through the cell autonomous provision of extracellular matrix components and trophic factors (Pollen et al., 2015). These features allow them to maintain a proliferative stem cell identity in the absence of contact with the cerebrospinal fluid at the ventricle (Lehtinen et al., 2011; Pollen et al., 2015). The radial fibers of oRGs have been suggested to be the main substrate for the migration of newborn neurons to the cortical plate at later stages of human cortical development, since at that stage vRG radial fibers no longer make contact with the pial surface (Nowakowski et al., 2016). With antibodies that specifically label oRGs now at hand, it will be possible to further explore the role of these cells in cortical neurogenesis and neuronal migration in the developing primate cortex.

21.5 Principles Underlying the Increase in Progenitor Cell Types and Numbers

During the last two decades, careful analysis of cortical development resulted in the identification of several distinct neural progenitor cell types, including vRGs, IPCs, and oRGs. IPCs were originally described based on time-lapse lineage tracing experiments in rodents that revealed a novel form of neural progenitor cell derived from vRGs (Noctor et al., 2004). This same study demonstrated that most IPCs underwent a single symmetric division to produce pairs of neurons. In this way IPCs are more properly termed progenitor cells,

while radial glia are often referred to as neural stem cells. More recently, similar time-lapse imaging of human cortical slice cultures has revealed that primate oRG cells derive from vRGs as well as by self-renewing divisions of oRGs (LaMonica et al., 2013) and also that oRG cells produce IPCs that appear to undergo more than one division before producing neurons (Hansen et al., 2010; Betizeau et al., 2013). In this way the IPCs that are produced from oRG cells may be different from those derived from vRGs. In fact there is evidence that oRGs have neural stem cell characteristics since they have been shown to give rise to cells of both the neuronal and glial lineage in the ferret and human (Reillo et al., 2011; Pollen et al., 2015).

Comparative studies have unraveled the emergence and accumulation of the two types of abventricular progenitors in vertebrates that have evolved at different times (Fig. 21.4). To date, we cannot ascertain if these two progenitor types arose in the last common ancestor of the various species analyzed, or whether they have evolved independently in the different lineages. The most parsimonious explanation, however, is that they were inherited from the last common ancestor. In both cases, the evolutionary appearance of a new cell type was followed by a gradual increase in its number in certain species, presumably due to a conferred adaptive advantage. For example, IPCs (defined by the expression of Tbr2) are present, albeit in small numbers, in the ventricular zone of reptiles such as lizards. They are present in larger numbers in certain areas of the brain in birds and turtles and are sufficiently increased in number to form a discrete layer, the SVZ, in the rodent forebrain. The emergence of oRGs is characterized by a very similar pattern, which presumably occurred at a later evolutionary time point. oRGs are found sparsely in the SVZ of the mouse brain. As their numbers increase in gyrencephalic species, most of them coalesce in a new layer, the OSVZ. Recent evidence indicates that progenitors displaying features of radial glia, such as long processes and expression of the markers Pax6 and Sox2, may also be present at abventricular locations in birds (Martínez-Cerdeño et al., 2015; Nomura et al., 2016). Whether these are related to oRGs in mammals remains to be shown.

Based on these observations, we hypothesize that the expansion of the cortical sheet has been achieved through the expansion of different types of stem and progenitor cells that generate the large output of cells making up the mature neocortex. It has been proposed previously that cell types and circuits that emerged during evolution of the brain of different nonmammalian vertebrates have subsequently been reorganized in mammals (Karten, 2015). It is therefore interesting to note that this notion already applies to different progenitor cell populations and thus may provide a developmental rational for the similarities in cell types in the mature brain. Additionally, other cell types (for example, Cajal—Retzius cells) and brain structures (for example, the subplate) essential for the development of the mammalian neocortex are present, although sparsely, in nonmammalian vertebrates (Aboitiz and Zamorano, 2013).

The different classes of stem and progenitor cells may contribute to cortical expansion in distinct ways. A comparison of cortical size and progenitor cell type and number across mammalian species suggests the interesting, though simplified notion, that IPCs generate neurons that contribute to the radial expansion of the cortex, while vRGs and oRGs may both provide additional radial units that contribute to the tangential expansion of the cortex (Fig. 21.2). For example, reptiles have a three-layered cortex derived mainly from radial glia cells, with very few IPCs (Fig. 21.4). Rodents, however, have significant numbers of IPCs, particularly from midcorticogenesis onward, when a distinct SVZ appears. The rodent cortex is expanded when compared to the reptilian cortex, particularly in the radial dimension, in which rodents have six layers. The addition of three cortical layers has been attributed, in part, to additional neurogenesis provided by IPCs (Cheung et al., 2007; Kriegstein et al., 2006). oRGs are particularly numerous in the developing primate cortex and may have played a significant role in the large tangential expansion of primate cortex (Fig. 21.2D). The concept that IPCs contribute to radial cortical expansion while radial glia—like cells contribute to tangential expansion is no doubt an oversimplification, but may serve as a model for closer analysis of the relative contributions of different progenitor cell types to major features of cortical architecture.

Upon the generation of large numbers of specific abventricular progenitor types, such as IPCs or oRGs, new progenitor zones have emerged. Is this anatomical organization functionally relevant? One possible answer is that the grouping of cells with identical function and fate facilitates their environmental regulation. Cell fate is not only controlled by genetically determined cell intrinsic mechanisms, but also by environmental signals. Paracrine signaling between progenitor cells in the germinal zones as well as signals from newborn neurons migrating tangentially or radially through the germinal zones can regulate the cell biology of SVZ progenitor cells (Cheung et al., 2007). Indeed, the different classes of neuronal progenitors may require different kinds of signaling to maintain their identity. A recent transcriptomic study identified genes differentially expressed between proliferative zones within the same species and between human and mouse to be mostly related to different means of cell—cell communication, for example, through secretory pathways, extracellular matrix, or cell adhesion (Fietz et al., 2012). Indeed, several studies have found that oRGs, with their large proliferative potential and lack of apical contact to the ventricle, require other types

of trophic support for maintenance and self-renewal. This is achieved through Notch signaling from more differentiated cells in the SVZ, integrin signaling at the basal lamina, and by creating their own niche through LIF-STAT3 signaling (Hansen et al., 2010; Fietz et al., 2010; Pollen et al., 2015). Additionally, a recent study also showed a role for Sonic hedgehog (Shh) signaling in the generation of oRGs and IPCs (Wang et al., 2016).

The use of single cell transcriptomics provides a promising approach for understanding how the cortical sheet became expanded based on changes in progenitor cell biology. First, identifying transcripts uniquely expressed in the different types of progenitors will allow new markers to be determined that will enable studies on the neuronal and glial lineages produced by different types of progenitors. Second, further progenitor types and subtypes may be identified based on differences in gene expression. An integrated approach characterizing cell types by their transcriptomes as well as by morphology might help to deconvolve the report of cortical progenitor subtypes with various morphologies that have been described in the macaque (Betizeau et al., 2013). Single cell transcriptomics also has the potential of revealing mechanisms that drive neurogenesis in the developing brain (Pollen et al., 2015). Recent studies have highlighted that alternative splicing can contribute substantially to neuronal differentiation (Irimia et al., 2014; Quesnel-vallières et al., 2015; Li et al., 2015). Alternative splicing has also been shown to be of evolutionary importance (Gueroussov et al., 2015), so it will be interesting to determine if posttranscriptional control of gene expression may also have driven the expansion of the cortical sheet in primates.

Moving forward, a better understanding of progenitor subtypes may also be achieved when different dimensions that define cell identity are combined (Shapiro et al., 2013). Recent technological advances have made it possible to combine multiple readouts of cell identity such as morphology, physiology, chromatin state, and location (Fuzik et al., 2016; Baden et al., 2016; Li et al., 2015; Angermueller et al., 2015). New challenges, however, are likely to arise when parsing genomic signatures into novel cell types (Shapiro et al., 2013). How different do the transcriptomes and the physiological characteristics of two cells need to be in order for them to be considered two different cell types? What biological mechanisms can be used to clarify the distinction between different cell types and differentiation states?

21.6 Insights Into the Possible Adaptive Benefit of Cortical Expansion in Evolution

Mechanisms underlying the expansion of the cortical sheet may be inferred from studies of patients with microcephaly. These patients have genetic mutations that prevent the generation of the large number of neurons that characterize the human neocortex. Often, these patients have intellectual disability (Gilmore and Walsh, 2013). Microcephaly mutations have been mapped to a handful of different genes, most of which are involved in centrosomal function (Gilmore and Walsh, 2013). Centrosomes are multifunctional cellular organelles that play a key role in the organization of microtubules, especially during cell division. Additionally, they are involved in regulating cell migration and the formation of cilia. Considering that microcephaly is most likely caused by a defect in neurogenesis, it is tempting to speculate that the convergence of microcephaly-associated mutations in genes encoding centrosomal proteins is related to their involvement in the control of the spindle orientation of progenitor cells and therefore their mode of cell division (symmetric or asymmetric). Experimentally, mouse models, in which microcephaly genes such as CDK5RAP2 or MCPH1 have been knocked out, show evidence of an increase in apoptosis (Lizarraga et al., 2010; Gruber et al., 2011). The most parsimonious explanation for microcephaly lies in the disruption of cell division in neuroepithelial or radial glia cells leading to a reduction in the founder population and a consequent reduction in the number of neurons produced. The finding that mouse models of human microcephaly often show very minimal reductions in brain size may implicate defects specifically in primate-specific progenitor cell types, possibly oRGs or their daughter IPCs, in the development of microcephaly in humans.

In addition to genetic causes, microcephaly can also be a result of a viral infection of the developing brain. Cytomegalovirus and rubella have been known to cause microcephaly for a long time (Yazigi et al., 2016; Naing et al., 2016). In 2015, the Zika virus outbreak in Brazil has resulted in a large increase in microcephaly cases, and current research aims at unraveling the pathogenic mechanism (Beckham et al., 2016). While genetic causes of microcephaly predominantly prevent the generation of large numbers of cells from progenitors, viral infections often (additionally) cause cell death of postmitotic neurons and thereby lead to more severe malformations (Mlakar et al., 2016; Tardieu et al., 2013).

Lissencephaly is a neurodevelopmental disorder that leads to the absence of cortical folding resulting in a smooth brain. It is also associated with a defect in cortical layering and reduced thickness of the cortical sheet. While lissencephaly is primarily thought to be a neuronal migration disorder, there may be some overlap with microcephaly as defects in neurogenesis may also contribute to the disease phenotype (Moon and Wynshaw-Boris).

Not surprisingly, many of the genes implicated in these neurodevelopmental diseases have been positively selected during evolution in the primate lineage (Gilmore and Walsh, 2013). Recently, the genomes of several early human species, the Neanderthals and the Denisovans, have been sequenced, making a comparison with the human genome possible (Prüfer et al., 2014). Interestingly, of the 87 proteins in modern humans whose amino acid sequence has changed compared to our early human ancestors, an unexpectedly large number (5) are expressed in the proliferative zones of the developing neocortex, underscoring the evolutionary importance of neurogenesis (Prüfer et al., 2014). Akin to the observations of the genes mutated in microcephaly, three of the five genes whose amino acid sequence has changed in human evolution and that are expressed in the ventricular zone are associated with microtubule function during cell division (Prüfer et al., 2014). Mutations in CASC5, one of these three recently mutated genes, have been shown to cause microcephaly in humans (Genin et al., 2012; Morris-Rosendahl and Kaindl, 2015). This highlights that research in many different areas, including human paleontology, neurology, evolutionary biology, and developmental biology, can together contribute to furthering our understanding of how the cortical sheet in primates expanded during evolution.

Intriguingly, different early hominids have diverse head sizes—while Neanderthals and Denisovans had larger cranial volumes than modern humans, the crania of other species of the genus *Homo* are characterized by smaller sizes (Somel et al., 2013). An example of a species with a small head circumference, but the large body and hand sizes characteristic of modern humans, is the newly discovered species *Homo naledi* (Berger et al., 2015). This fossil record shows that human evolution has not been a linear process and that different aspects of human physiology may have evolved independently from each other. Since skull development is correlated with brain development, we can make inferences about the brain size of different early hominids. Future efforts to sequence more genomes of our ancestors, together with a better understanding of their cognitive skills and their anatomy will undoubtedly bring us closer to understanding our evolutionary origins and the genetic changes that produced the morphological and functional changes embodied in the human brain.

References

Aaku-Saraste, E., Hellwig, A., Huttner, W.B., 1996. Loss of occludin and functional tight junctions, but not ZO-1, during neural tube closure—remodeling of the neuroepithelium prior to neurogenesis. Dev. Biol. 180 (2), 664–679.

Abrahams, B.S., Tentler, D., Perederiy, J.V., Oldham, M.C., Coppola, G., Geschwind, D.H., November 6, 2007. Genome-wide analyses of human perisylvian cerebral cortical patterning. Proc. Natl. Acad. Sci. U. S. A. 104 (45), 17849–17854 Epub 2007 Oct 31.

Aboitiz, F., Zamorano, F., 2013. Neural progenitors, patterning and ecology in neocortical origins. Front. Neuroanat. 7, 38.

Anastasiades, P.G., Butt, S.J.B., 2011. Decoding the transcriptional basis for GABAergic interneuron diversity in the mouse neocortex. Eur. J. Neurosci. 34 (10), 1542–1552.

Angermueller, C., Clark, S.J., Lee, H.J., Macaulay, I.C., Teng, M.J., Hu, T.X., Krueger, F., Smallwood, S.A., Ponting, C.P., Voet, T., et al., 2015. Parallel single-cell sequencing links transcriptional and epigenetic heterogeneity. Nat. Methods 13.

Angevine, J.B., Sidman, R.L., 1961. Autoradiographic study of cell migration during histogenesis of cerebral cortex in the mouse. Nature 192 (4804), 766–768.

Azevedo, F.A.C., Carvalho, L.R.B., Grinberg, L.T., Farfel, J.M., Ferretti, R.E.L., Leite, R.E.P., Filho, W.J., Lent, R., Herculano-Houzel, S., 2009. Equal numbers of neuronal and nonneuronal cells make the human brain an isometrically scaled-up primate brain. J. Comp. Neurol. 513 (5), 532–541.

Baden, T., Berens, P., Franke, K., Roman-Roson, M., Bethge, M., Euler, T., 2016. The functional diversity of mouse retinal ganglion cells. Nature 1–21.

Beckham, J.D., Pastula, D.M., Massey, A., Tyler, K.L., 2016. Zika virus as an emerging global pathogen. JAMA Neurol 1, 1–5.

Berger, L.R., Hawks, J., de Ruiter, D.J., Churchill, S.E., Schmid, P., Delezene, L.K., Kivell, T.L., Garvin, H.M., Williams, S.A., DeSilva, J.M., et al., 2015. *Homo naledi*, a new species of the genus *Homo* from the Dinaledi Chamber, South Africa. Elife 4, e09560.

Betizeau, M., Cortay, V., Patti, D., Pfister, S., Gautier, E., Bellemin-Ménard, A., Afanassieff, M., Huissoud, C., Douglas, R.J., Kennedy, H., et al., 2013. Precursor diversity and complexity of lineage relationships in the outer subventricular zone of the primate. Neuron 80 (2), 442–457.

Calegari, F., 2003. An inhibition of cyclin-dependent kinases that lengthens, but does not arrest, neuroepithelial cell cycle induces premature neurogenesis. J. Cell Sci. 116 (24), 4947–4955.

Charvet, C.J., Owerkowicz, T., Striedter, G.F., 2009. Phylogeny of the telencephalic subventricular zone in sauropsids: evidence for the sequential evolution of pallial and subpallial subventricular zones. Brain Behav. Evol. 73, 285–294.

Chenn, A., Walsh, C.A., 2002. Regulation of cerebral cortical size by control of cell cycle exit in neural precursors. Science 297 (5580), 365–369.

Cheung, A.F.P., Pollen, A.A., Tavare, A., Deproto, J., Molnár, Z., 2007. Comparative aspects of cortical neurogenesis in vertebrates. J. Anat. 211 (2), 164–176.

Cheung, A.F., Kondo, S., Abdel-Mannan, O., Chodroff, R.A., Sirey, T.M., Bluy, L.E., Webber, N., DeProto, J., Karlen, S.J., Krubitzer, L., et al., 2010. The subventricular zone is the developmental milestone of a 6-Layered neocortex: comparisons in Metatherian and Eutherian mammals. Cereb. Cortex 20 (5), 1071–1081.

Clowry, G.J., 2014. An enhanced role and expanded developmental origins for gamma-aminobutyric acidergic interneurons in the human cerebral cortex. J. Anat. 1–10.

Crawford, N.G., Parham, J.F., Sellas, A.B., Faircloth, B.C., Glenn, T.C., Papenfuss, T.J., Henderson, J.B., Hansen, M.H., Simison, W.B., 2015. A phylogenomic analysis of turtles. Mol. Phylogenet. Evol. 83, 250–257.

DeFelipe, J., Alonso-Nanclares, L., Arellano, J.I., 2002. Microstructure of the neocortex: comparative aspects. J. Neurocytol. 31 (3–5), 299–316.

Dehay, C., Giroud, P., Berland, M., Smart, I., Kennedy, H., 1993. Modulation of the cell cycle contributes to the parcellation of the primate visual cortex. Nature 366 (6454), 464–466.

Desai, A.R., McConnell, S.K., 2000. Progressive restriction in fate potential by neural progenitors during cerebral cortical development. Development 127 (13), 2863–2872.

Englund, C., 2005. Pax6, Tbr2, and Tbr1 are expressed sequentially by radial glia, intermediate progenitor cells, and postmitotic neurons in developing neocortex. J. Neurosci. 25 (1), 247–251.

Fietz, S.A., Kelava, I., Vogt, J., Wilsch-Bräuninger, M., Stenzel, D., Fish, J.L., Corbeil, D., Riehn, A., Distler, W., Nitsch, R., et al., 2010. Osvz progenitors of human and ferret neocortex are epithelial-like and expand by integrin signaling. Nat. Neurosci. 13 (6), 690–699.

Fietz, S.A., Lachmann, R., Brandl, H., Kircher, M., Samusik, N., Schroder, R., Lakshmanaperumal, N., Henry, I., Vogt, J., Riehn, A., et al., 2012. Transcriptomes of germinal zones of human and mouse fetal neocortex suggest a role of extracellular matrix in progenitor self-renewal. Proc. Natl. Acad. Sci. U.S.A. 109 (29), 11836–11841.

Finlay, B.L., Hersman, M.N., Darlington, R.B., 1998. Patterns of vertebrate neurogenesis and the paths of vertebrate evolution. Brain Behav. Evol. 52 (4–5), 232–242.

Fuzik, J., Zeisel, A., Máté, Z., Calvigioni, D., Yanagawa, Y., Szabó, G., Linnarsson, S., Harkany, T., 2016. Integration of electrophysiological recordings with single-cell RNA-Seq data identifies neuronal subtypes. Nat. Biotechnol. 34 (2), 175–183.

Gaiano, N., Nye, J.S., Fishell, G., 2000. Radial glial identity is promoted by Notch1 signaling in the Murine forebrain. Neuron 26 (2), 395–404.

Garcia-Moreno, F., Vasistha, N.A., Trevia, N., Bourne, J.A., Molnar, Z., 2012. Compartmentalization of cerebral cortical germinal zones in a lissencephalic primate and gyrencephalic rodent. Cereb. Cortex 22 (2), 482–492.

Genin, A., Desir, J., Lambert, N., Biervliet, M., Van der Aa, N., Pierquin, G., Killian, A., Tosi, M., Urbina, M., Lefort, A., et al., 2012. Kinetochore KMN network gene CASC5 mutated in primary microcephaly. Hum. Mol. Genet. 21 (24), 5306–5317.

Gertz, C.C., Kriegstein, A.R., 2015. Neuronal migration dynamics in the developing ferret cortex. J. Neurosci. 35 (42), 14307–14315.

Gertz, C.C., Lui, J.H., LaMonica, B.E., Wang, X., Kriegstein, A.R., 2014. Diverse behaviors of outer radial glia in developing ferret and human cortex. J. Neurosci. 34 (7), 2559–2570.

Gilmore, E.C., Walsh, C.A., 2013. Genetic causes of microcephaly and lessons for neuronal development. Wiley Interdisip. Rev. Dev. Biol. 2 (4), 461–478.

Götz, M., Huttner, W.B., 2005. The cell biology of neurogenesis. Nat. Rev. Mol. Cell Biol. 6 (10), 777–788.

Gruber, R., Zhou, Z., Sukchev, M., Joerss, T., Frappart, P.-O., Wang, Z.-Q., 2011. MCPH1 regulates the neuroprogenitor division mode by coupling the centrosomal cycle with mitotic entry through the Chk1–Cdc25 pathway. Nat. Cell Biol. 13 (11), 1325–1334.

Gueroussov, S., Gonatopoulos-Pournatzis, T., Irimia, M., Raj, B., Lin, Z.-Y., Gingras, A.-C., Blencowe, B.J., 2015. An alternative splicing event amplifies evolutionary differences between vertebrates. Science 349 (6250), 868–873.

Hansen, D.V., Lui, J.H., Parker, P.R.L., Kriegstein, A.R., 2010. Neurogenic radial glia in the outer subventricular zone of human neocortex. Nature 464 (7288), 554–561.

Hatakeyama, J., Bessho, Y., Katoh, K., Ookawara, S., Fujioka, M., Guillemot, F., Kageyama, R., 2004. Hes genes regulate size, shape and histogenesis of the nervous system by control of the timing of neural stem cell differentiation. Development 131 (22), 5539–5550.

Haubensak, W., Attardo, A., Denk, W., Huttner, W.B., 2004. From the cover: neurons arise in the basal neuroepithelium of the early mammalian telencephalon: a major site of neurogenesis. Proc. Natl. Acad. Sci. U.S.A. 101 (9), 3196–3201.

Irimia, M., Weatheritt, R.J., Ellis, J.D., Parikshak, N.N., Gonatopoulos-pournatzis, T., Tapial, J., Raj, B., Hanlon, D.O., Barrios-rodiles, M., Babor, M., et al., 2014. A highly conserved program of neuronal microexons is misregulated in autistic brains. Cell 1511–1523.

de Juan Romero, C., Bruder, C., Tomasello, U., Sanz-Anquela, J.M., Borrell, V., Romero, C.D.J., Bruder, C., Tomasello, U., Sanz-Anquela, J.M., 2015. Discrete domains of gene expression in germinal layers distinguish the development of gyrencephaly. EMBO J. 1–16.

Kang, H.J., Kawasawa, Y.I., Cheng, F., Zhu, Y., Xu, X., Li, M., Sousa, A.M.M., Pletikos, M., Meyer, K.A., Sedmak, G., et al., 2011. Spatio-temporal transcriptome of the human brain. Nature 478 (7370), 483–489.

Karten, H.J., 2015. Vertebrate brains and evolutionary connectomics: on the origins of the mammalian " neocortex". Philos. Trans. R. Soc. Lond. B Biol. Sci. 370 (1684).

Kornack, D.R., Rakic, P., 1998. Changes in cell-cycle kinetics during the development and evolution of primate neocortex. Proc. Natl. Acad. Sci. U.S.A. 95 (3), 1242–1246.

Kowalczyk, T., Pontious, A., Englund, C., Daza, R.A.M., Bedogni, F., Hodge, R.D., Attardo, A., Bell, C., Huttner, W.B., Hevner, R.F., 2009. Intermediate neuronal progenitors (basal progenitors) produce pyramidal–projection neurons for all layers of cerebral cortex. Cereb. Cortex 19 (10), 2439.

Kriegstein, A., Noctor, S., Martínez-cerdeño, V., November, 2006. Patterns of neural stem and progenitor cell division may underlie evolutionary cortical expansion. Neuroscience 7, 883–890.

LaMonica, B.E., Lui, J.H., Hansen, D.V., Kriegstein, A.R., 2013. Mitotic spindle orientation predicts outer radial glial cell generation in human neocortex. Nat. Commun. 4, 1665.

Lehtinen, M.K., Zappaterra, M.W., Chen, X., Yang, Y.J., Hill, A.D., Lun, M., Maynard, T., Gonzalez, D., Kim, S., Ye, P., et al., 2011. The cerebrospinal fluid provides a proliferative niche for neural progenitor cells. Neuron 69 (5), 893–905.

Li, C.-L., Li, K.-C., Wu, D., Chen, Y., Luo, H., Zhao, J.-R., Wang, S.-S., Sun, M.-M., Lu, Y.-J., Zhong, Y.-Q., et al., 2015. Somatosensory neuron types identified by high-coverage single-cell RNA-sequencing and functional heterogeneity. Cell Res. 1–20.

Li, Y.I., Sanchez-pulido, L., Haerty, W., Ponting, C.P., 2015. RBFOX and PTBP1 proteins regulate the alternative splicing of micro-exons in human brain transcripts. Genome Res. 25, 1–13.

Lizarraga, S.B., Margossian, S.P., Harris, M.H., Campagna, D.R., Han, A.P., Blevins, S., Mudbhary, R., Barker, J.E., Walsh, C.A., Fleming, M.D., 2010. Cdk5rap2 regulates centrosome function and chromosome segregation in neuronal progenitors. Development 137 (11), 1907–1917.

Lui, J.H., Hansen, D.V., Kriegstein, A.R., 2011. Development and evolution of the human neocortex. Cell 146 (1), 18–36.

Lukaszewicz, A., Savatier, P., Cortay, V., Giroud, P., Huissoud, C., Berland, M., Kennedy, H., Dehay, C., 2005. G1 phase regulation, area-specific cell cycle control, and cytoarchitectonics in the primate cortex. Neuron 47 (3), 353–364.

Malatesta, P., Hartfuss, E., Götz, M., 2000. Isolation of radial glial cells by fluorescent-activated cell sorting reveals a neuronal lineage. Development 127 (24), 5253–5263.

Manabe, N., Hirai, S.I., Imai, F., Nakanishi, H., Takai, Y., Ohno, S., 2002. Association of ASIP/mPAR-3 with adherens junctions of mouse neuroepithelial cells. Dev. Dyn. 225 (1), 61–69.

Marin, O., 2013. Cellular and molecular mechanisms controlling the migration of neocortical interneurons. Eur. J. Neurosci. 38 (1), 2019–2029.

Martinez-Cerdeño, V., Noctor, S.C., Kriegstein, A.R., 2006. The role of intermediate progenitor cells in the evolutionary expansion of the cerebral cortex. Cereb. Cortex 16 (Suppl. 1), i152–i161.

Martínez-Cerdeño, V., Cunningham, C.L., Camacho, J., Antczak, J.L., Prakash, A.N., Cziep, M.E., Walker, A.I., Noctor, S.C., 2012. Comparative analysis of the subventricular zone in rat, ferret and macaque: evidence for an outer subventricular zone in rodents. PLoS One 7 (1), e30178.

Martínez-Cerdeño, V., Cunningham, C.L., Camacho, J., Keiter, J.A., Ariza, J., Lovern, M., Noctor, S.C., 2015. Evolutionary origin of Tbr2-expressing precursor cells and the subventricular zone in the developing cortex. J. Comp. Neurol. 524.

Miyama, S., Takahashi, T., Nowakowski, R.S., Caviness, V.S., 1997. A gradient in the duration of the G1 phase in the Murine neocortical proliferative epithelium. Cereb. Cortex 7 (7), 678–689.

Miyata, T., Kawaguchi, A., Okano, H., Ogawa, M., 2001. Asymmetric inheritance of radial glial fibers by cortical neurons. Neuron 31 (5), 727–741.

Mlakar, J., Korva, M., Tul, N., Popović, M., Poljšak-Prijatelj, M., Mraz, J., Kolenc, M., Resman Rus, K., Vesnaver Vipotnik, T., Fabjan Vodušek, V., et al., 2016. Zika virus associated with microcephaly. N. Engl. J. Med. 374 (10), 951–958.

Montiel, J.F., Vasistha, N.A., Garcia-Moreno, F., Molnár, Z., 2016. From Sauropsids to mammals and back: new approaches to comparative cortical development. J. Comp. Neurol. 524 (3), 630–645.

Moon, H.M., Wynshaw-Boris, A., 2013. Cytoskeleton in action: lissencephaly, a neuronal migration disorder. Wiley Interdiscip. Rev. Dev. Biol. 2 (2), 229–245.

Morris-Rosendahl, D.J., Kaindl, A., 2015. What next-generation sequencing (NGS) technology has enabled us to learn about primary autosomal recessive microcephaly (MCPH). Mol. Cell. Probes 29.

Mota, B., Herculano-Houzel, S., 2015. Cortical folding scales universally with surface area and thickness, not number of neurons. Science 349 (6243), 74–77.

Naing, Z.W., Scott, G.M., Shand, A., Hamilton, S.T., van Zuylen, W.J., Basha, J., Hall, B., Craig, M.E., Rawlinson, W.D., 2016. Congenital cytomegalovirus infection in pregnancy: a review of prevalence, clinical features, diagnosis and prevention. Aust. N. Z. J. Obstet. Gynaecol. 56 (1), 9–18.

Noctor, S.C., Scholnicoff, N.J., Juliano, S.L., 1997. Histogenesis of ferret somatosensory cortex. J. Comp. Neurol. 387 (2), 179–193.

Noctor, S.C., Flint, A.C., Weissman, T.A., Dammerman, R.S., Kriegstein, A.R., 2001. Neurons derived from radial glial cells establish radial units in neocortex. Nature 409 (6821), 714–720.

Noctor, S.C., Martínez-Cerdeño, V., Ivic, L., Kriegstein, A.R., 2004. Cortical neurons arise in symmetric and asymmetric division zones and migrate through specific phases. Nat. Neurosci. 7 (2), 136–144.

Noctor, S.C., Martínez-Cerdeño, V., Kriegstein, A.R., 2008. Distinct behaviors of neural stem and progenitor cells underlie cortical neurogenesis. J. Comp. Neurol. 508 (1), 28–44.

Nomura, T., Ohtaka-Maruyama, C., Yamashita, W., Wakamatsu, Y., Murakami, Y., Calegari, F., Suzuki, K., Gotoh, H., Ono, K., 2016. The evolution of basal progenitors in the developing nonmammalian brain. Development 143 (1), 66–74.

Nowakowski, T.J., Pollen, A.A., Sandoval-Espinosa, C., Kriegstein, A.R., September 21, 2016. Transformation of the radial glia scaffold Demarcates two stages of human cerebral cortex development. Neuron 91 (6), 1219–1227.

Nowakowski, R.S., Caviness, V.S., Takahashi, T., Hayes, N.L., 2002. Population dynamics during cell proliferation and neuronogenesis in the developing Murine neocortex. Results Probl. Cell Differ. 39, 1–25.

Nowakowski, T.J., Mysiak, K.S., Pratt, T., Price, D.J., 2011. Functional dicer is necessary for appropriate specification of radial glia during early development of mouse telencephalon. PLoS One 6 (8), e23013.

Ostrem, B.E.L., Lui, J.H., Gertz, C.C., Kriegstein, A.R., 2014. Control of outer radial glial stem cell mitosis in the human brain. Cell Rep. 8 (3), 656–664.

Pollen, A.A., Nowakowski, T.J., Chen, J., Retallack, H., Sandoval-Espinosa, C., Nicholas, C.R., Shuga, J., Liu, S.J., Oldham, M.C., Diaz, A., et al., 2015. Molecular identity of human outer radial glia during cortical development. Cell 163 (1), 55–67.

Prüfer, K., Racimo, F., Patterson, N., Jay, F., Sankararaman, S., Sawyer, S., Heinze, A., Renaud, G., Sudmant, P.H., de Filippo, C., et al., 2014. The complete genome sequence of a Neanderthal from the Altai Mountains. Nature 505 (7481), 43–49.

Quesnel-vallières, M., Irimia, M., Cordes, S.P., Blencowe, B.J., 2015. Essential roles for the splicing regulator nSR100/SRRM4 during nervous system development. Genes Dev. 746–759.

Radonjić, N.V., Ayoub, A.E., Memi, F., Yu, X., Maroof, A., Jakovcevski, I., Anderson, S.A., Rakic, P., Zecevic, N., 2014. Diversity of cortical interneurons in primates: the role of the dorsal proliferative niche. Cell Rep. 9 (6), 2139–2151.

Rakic, P., Ayoub, A.E., Breunig, J.J., Dominguez, M.H., 2009. Decision by division: making cortical maps. Trends Neurosci. 32 (5), 291–301.

Rakic, P., 1972. Mode of cell migration to the superficial layers of fetal monkey neocortex. J. Comp. Neurol. 145 (1), 61–83.

Rakic, P., 1974. Neurons in rhesus monkey visual cortex: systematic relation between time of origin and eventual disposition. Science 183 (4123), 425–427.

Rakic, P., 1988. Specification of cerebral cortical areas. Science 241 (4862), 170–176.

Rakic, P., 1995. A small step for the cell, a giant leap for mankind: a hypothesis of neocortical expansion during evolution. Trends Neurosci. 18 (9), 383–388.

Rakic, P., 2009. Evolution of the neocortex: perspective from developmental biology. Nat. Rev. Neurosci. 10 (10), 724–735.

Reillo, I., De Juan Romero, C., García-Cabezas, M.Á., Borrell, V., 2011. A role for intermediate radial glia in the tangential expansion of the mammalian cerebral cortex. Cereb. Cortex 21 (7), 1674–1694.

Sahara, S., O'Leary, D.D.M., 2009. Fgf10 regulates transition period of cortical stem cell differentiation to radial glia controlling generation of neurons and basal progenitors. Neuron 63 (1), 48–62.

Sauer, F.C., 1935. Mitosis in the neural tube. J. Comp. Neurol. 62 (2), 377–405.

Scharff, C., Friederici, A.D., Petrides, M., 2013. Neurobiology of human language and its evolution: primate and non-primate perspectives. Front. Evol. Neurosci. 5, 1.

Shapiro, E., Biezuner, T., Linnarsson, S., 2013. Single-cell sequencing-based technologies will revolutionize whole-organism science. Nat. Rev. Genet. 14 (9), 618–630.

Smart, I.H.M., Dehay, C., Giroud, P., Berland, M., Kennedy, H., 2002. Unique morphological features of the proliferative zones and postmitotic compartments of the neural epithelium giving rise to striate and extrastriate cortex in the monkey. Cereb. Cortex 12 (1), 37–53.

Smart, I., 1965. The operation of ependymal "choke" in neurogenesis. J. Anat. 99, 941–943.

Somel, M., Liu, X., Khaitovich, P., 2013. Human brain evolution: transcripts, metabolites and their regulators. Nat. Rev. Neurosci. 14 (2), 112–127.

Takahashi, T., Nowakowski, R.S., Caviness, V.S., 1996. The leaving or Q Fraction of the Murine cerebral proliferative epithelium: a general model of neocortical neuronogenesis. J. Neurosci. 16 (19), 6183–6196.

Tarabykin, V., Stoykova, A., Usman, N., Gruss, P., 2001. Cortical upper layer neurons derive from the subventricular zone as indicated by Svet1 gene expression. Development 128 (11), 1983–1993.

Tardieu, M., Tejiokem, M., Nguefack, S., 2013. Virus-induced Lesions and the Fetal Brain: Examples of the Transmission of HIV-1 and CMV from Mother to Offspring, first ed., vol. 112. Elsevier B.V.

Taverna, E., Götz, M., Huttner, W.B., 2014. The cell biology of neurogenesis: toward an understanding of the development and evolution of the neocortex. Annu. Rev. Cell Dev. Biol. 30 (1), 465–502.

Thomsen, E.R., Mich, J.K., Yao, Z., Hodge, R.D., Doyle, A.M., Jang, S., Shehata, S.I., Nelson, A.M., Shapovalova, N.V., Levi, B.P., et al., 2016. Fixed single-cell transcriptomic characterization of human radial glial diversity. Nat. Methods 13 (1), 87–93.

Wang, L., Hou, S., Han, Y.G., July, 2016. Hedgehog signaling promotes basal progenitor expansion and the growth and folding of the neocortex. Nat. Neurosci. 19 (7), 888–896.

Wang, X., Tsai, J.-W., LaMonica, B., Kriegstein, A.R., 2011. A new subtype of progenitor cell in the mouse embryonic neocortex. Nat. Neurosci. 14 (5), 555–561.

Wu, S.-X., Goebbels, S., Nakamura, K., Nakamura, K., Kometani, K., Minato, N., Kaneko, T., Nave, K.-A., Tamamaki, N., 2005. Pyramidal neurons of upper cortical layers generated by NEX-positive progenitor cells in the subventricular zone. Proc. Natl. Acad. Sci. U.S.A. 102 (47), 17172–17177.

Yazigi, A., De Pecoulas, A.E., Vauloup-Fellous, C., Grangeot-Keros, L., Ayoubi, J.-M., Picone, O., 2016. Fetal and neonatal abnormalities due to congenital rubella syndrome: a review of literature. J. Matern. Fetal Neonatal Med. 1–5.

Zecevic, N., Chen, Y., Filipovic, R., 2005. Contributions of cortical subventricular zone to the development of the human cerebral cortex. J. Comp. Neurol. 491 (2), 109–122.

22

Scaling Up the Simian Primate Cortex: A Conserved Pattern of Expansion Across Brain Sizes

T.A. Chaplin, M.G.P. Rosa, H.-H. Yu

Monash University, Melbourne, VIC, Australia

Glossary

ACC: Anterior cingulate cortex.

Allometry: The study of the relationship between brain size and body size, or the size of brain structures and overall brain size. The relationship is usually nonlinear.

Association cortex: Cortical regions that are not unisensory or motor related, eg, multisensory areas, areas related to cognition, etc.

Concerted evolution (in brain evolution): Enlargement of certain brain structures as a by-product of overall brain size increases.

Cortical area: A subdivision of the cerebral cortex based on cellular or functional criteria.

Isocortex: The largest component of the cerebral cortex, consisting of six layers. Also known as the neocortex.

Isometrically scaled: Scaled uniformly in all directions.

Late equals large: A principle of brain development stating that late-developing structures tend to be large in species with larger brains.

LPFC: Lateral prefrontal cortex.

Mirror neurons: Neurons that are activated both when an animal observes and performs an action.

Mosaic evolution (in brain evolution): Evolutionary changes in one brain component without simultaneous changes in another brain component.

Phyletic dwarfism: The reduction in size of animals over time.

Registration: The computational procedure of aligning or matching models of two brains.

Simian: The group of primates including New World monkeys, Old World monkeys, apes, and humans (which excludes tarsiers, as well as prosimians: lemurs, galagos, pottos, and lorises).

Spandrel: A characteristic that is a by-product of development, rather the result of selection pressures.

TPJ: Temporal parietal junction.

22.1 Variations in Brain Size Among Simian Primates

Simian primates consist of New World monkeys, Old World monkeys, apes, and humans. The lineages that led to New World monkeys diverged from a common ancestor ~40 million years ago (Fleagle, 1998; Chatterjee et al., 2009), and the hominoids (apes and humans) diverged from Old World monkeys ~25 million years ago (Fleagle, 1998). Reflecting their diverse ecological niches of the ~200 living species, simian primates have a wide range of body sizes, which are strongly correlated with differences in brain size (Stephan et al., 1981; Striedter, 2005).

This variability is dramatically illustrated by comparing the human brain to that of the common marmoset (*Callithrix jacchus*) (Fig. 22.1), a species of New World monkey. Weighting only ~350 g, the common marmoset is one of the smallest simian primates (the smallest being the pygmy marmoset *Cebuella pygmaea*, which, however, is not commonly available as a laboratory animal). In general, the miniature body size of *Callitrichid* monkeys, such as the common and pygmy marmoset, is considered the result of a process of phyletic dwarfism (Ford, 1980; Montgomery and Mundy, 2013). The human brain is ~180 times larger in volume than the common marmoset brain (Stephan et al., 1981), and yet the two brains share a common general plan of organization in terms of the relative location of cortical areas and thalamic nuclei linked to analogous functions.

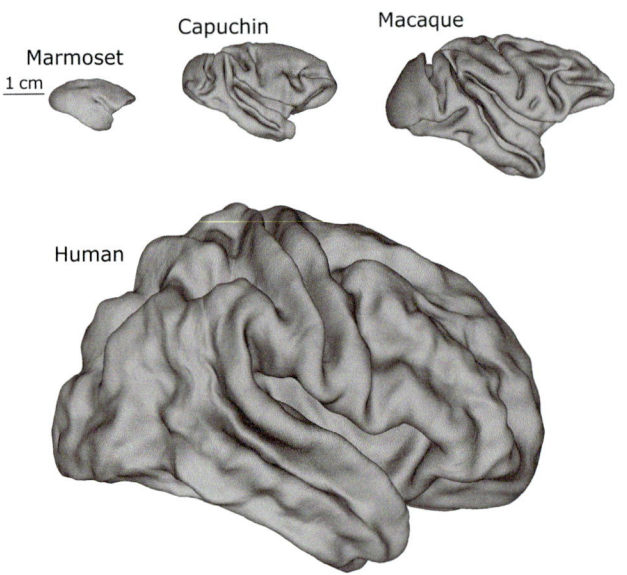

FIGURE 22.1 Surface models of the cerebral cortex of several primate species, scale is 1 cm, showing the dramatic differences in size across species.

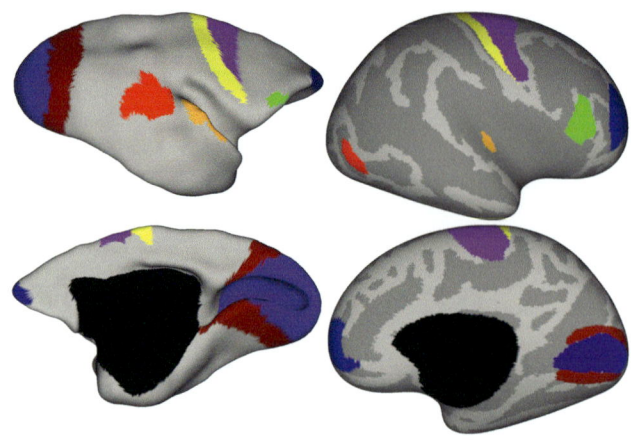

FIGURE 22.2 The position of several cortical areas with well-established homology in the marmoset (left) and human brain (right, not to scale), top row is the lateral view, bottom row is the medial view. Note that even though the areas are positioned similarly, there are clearly some enlarged regions in the human cerebral cortex, meaning that human brain is not a linearly scaled-up monkey brain. *Blue*, primary visual cortex; *maroon*, secondary visual area; *red*, MT+ complex; *orange*, auditory cortex; *yellow*, primary somatosensory cortex; *purple*, primary motor cortex; *green*, frontal eye field. *From Chaplin, T.A., Yu, H.H., Soares, J.G., Gattass, R., Rosa, M.G., 2013. A conserved pattern of differential expansion of cortical areas in simian primates. J. Neurosci. 33 (38), 15120–15125.*

Of particular interest here is the organization of the cerebral cortex, the largest structure in primate brains. A similar layout of homologous sensory, motor, and association areas has been recognized in both species (Rosa and Tweedale, 2005; Paxinos et al., 2009, 2012; Chaplin et al., 2013). However, it is not the case that the human cortex is simply a larger (isometrically scaled) version of the cortex of the marmoset, or indeed the cortex of any nonhuman primate species. Several previous studies have made clear that some regions of the cerebral cortex have become disproportionately larger in the human brain, relative to their homologs in monkeys and apes (Deacon, 1990; Rilling and Insel, 1999; Semendeferi et al., 2001; Orban et al., 2004; Rosa and Tweedale, 2005; Rakic, 2009; Preuss, 2011; Chaplin et al., 2013; Kaas, 2013). This can be appreciated in Fig. 22.2, which compares the locations of several homologous cortical areas in the common marmoset and human brain. For example, in the marmoset brain, the middle temporal visual area (MT, labeled in red) and the auditory core area (orange) are in close proximity, but in the human brain, they are widely separated by an expanded temporal lobe. Similarly, the spatial separation between the primary motor cortex (area 4; highlighted in purple) and the prefrontal area 45 (in green) is clearly significantly wider in the human brain relative to the marmoset brain. Regional differences such as these have long fascinated scientists, because they may help understand differences in cognition that distinguish us from other primates.

In this chapter, we will compare the patterns of cortical organization in species of simian primates that have been mapped in some detail, to examine evolutionary and developmental factors that might contribute nonuniform expansion of the cortex across brains of different sizes, and to try to understand the significance of this differential expansion for the physiology and pathology of cognitive processes.

22.2 Mosaic Versus Concerted Evolution

The principal drivers of the expansion of brain structures remain a topic of debate (for review, see Chapter 5 of Striedter, 2005). The theory of mosaic evolution contends that evolution can alter the size of one brain structure independently of other structures, according to the adaptive advantages afforded by such modifications (Barton and Harvey, 2000; Krubitzer and Seelke, 2012; Smaers and Soligo, 2013). However, it has been noted that, in many conditions, the developmental programs that underlie brain expansion impose significant constraints on independent regional changes, causing structures to covary with one another in size. Under the theory of *concerted evolution*, the enlargement of a brain structure in a larger brain is not necessarily due to increases in adaptive advantages, but can be the by-product (or, a "spandrel" according to Gould and Lewontin, 1979) of making a larger brain (Finlay and Darlington, 1995; Finlay et al., 2001). Finlay and Darlington

(1995), for example, observed that although the proportional sizes of mammalian brain components (isocortex, cerebellum, striatum, hippocampus, etc.) vary, most of the variance could be predicted by the overall size of the brain, suggesting that the pattern was due to a common cause, rather than to the unique evolutionary history of the individual species. It is also important to note that, at least within a range of closely related taxa, brain and body sizes are highly correlated (Stephan et al., 1981; Striedter, 2005), raising the possibility that brain growth may, itself, be partially driven to evolutionary pressures applied to other body systems. As a result, constraints related to mechanisms of brain development could naturally give rise to variations around a common pattern. We will return to this topic later in Section 22.6.

22.3 Measuring Expansion by Measuring the Size of Areas

The cerebral cortex can be subdivided into many regions (or "areas"), based on anatomical and functional criteria. To determine which regions of the isocortex have increased in size, a useful approach is to measure the size (volume, or surface area) of each cortical area and compare it to the homologous area in a different species. The concerted evolution hypothesis would be supported if a careful comparison of simian brains of different sizes reveals that the preferential expansion occurs in the same set of areas, irrespective of the species used in this analysis. If, on the other hand, this analysis shows unique combinations of enlarged areas, then the mosaic evolution would be favored (particularly, if this can be linked to species-specific adaptations).

Several studies have employed this approach. For example, the volume of cortical area 10 (Semendeferi et al., 2001), area 13 (Semendeferi et al., 1998), the insular cortex (Bauernfeind et al., 2013), and other areas (Sherwood et al., 2012) have been measured in humans and compared to homologs in other primates. However, until recently, this methodology had not been applied to the entire cerebral cortex, partially due to the difficulty in defining cortical areas and establishing homology between species. Primary sensory areas have well-known anatomical and physiological characteristics that permit a high degree of confidence in identification and homology, but other cortical areas, usually high-level association areas, are often less well defined. Even in some sensory areas, there is sometimes considerable disagreement on layout and homology (eg, Lyon and Connolly, 2012; Angelucci and Rosa, 2015). Therefore, in practice, the comparison needs to be preferentially anchored on those areas that can be reliably identified on histological and functional criteria.

22.4 Measuring Expansion by Surface Registration

To avoid the issues and complexities associated with identifying homology in cortical areas, computational methods have been utilized to measure expansion continuously across the whole cerebral cortex. The essence of this approach is to "morph" or stretch a 3D model of cerebral cortex of one species into another and measure the resulting regional expansion. This method, known as *surface-based registration*, has been adapted from human MRI studies in which the brains of individuals need to be registered to a common template brain.

The registration of human brains has been a challenging problem for algorithms based on volumetric image data, because human brains typically have a large degree of variability in the locations of cortical foldings, making perfect alignments of individual sulci and gyri computationally difficult. An alternative approach is surface-based algorithms. By exploiting the sheet-like structure of the isocortex, surface-based methods temporarily ignore the thickness of the cortex and treat the entire cortex as an infinitely thin surface. This level of abstraction allows registration algorithms to concentrate on aligning the topological relationships among cortical landmarks, and thus improving the precision of registration (Dale et al., 1999; Fischl et al., 1999; Van Essen et al., 2001; Van Essen, 2004; Ghosh et al., 2010). These advantages are particularly important for cross-species comparisons, given the large variations in brain morphology among primates (Fig. 22.1). Surface registration methods require only a small number of homologous areas to anchor the registration procedure (Fig. 22.3A). This is a useful feature, given the putative nature of our current understanding of brain homology among primates.

Surface-based methods begin by modeling the isocortex as a wireframe mesh consisting of many points connected as triangles (Fig. 22.3). The goal of registration is to spatially match each triangle in one brain to exactly one corresponding triangle in the second brain, ie, to match each part of the source brain to the target brain. There are several algorithms for performing surface registration, but they typically proceed as follows. Registration with a surface begins by first unfolding and inflating the cortical mesh models to create perfect spheres (Fig. 22.3B). The advantage of a spherical representation is twofold. Firstly, it normalizes brains of any size and shape from different species to standard shape and size. Secondly, the sphere has a regular geometry in which it is easy to calculate distances, angles, and areas, which is either computationally difficult or undefined on arbitrary shape like the cortex. Next, the source brain sphere is warped to match a target brain sphere by matching a set of landmarks (Fig. 22.3B). The anchoring to landmarks pulls and stretches the rest of

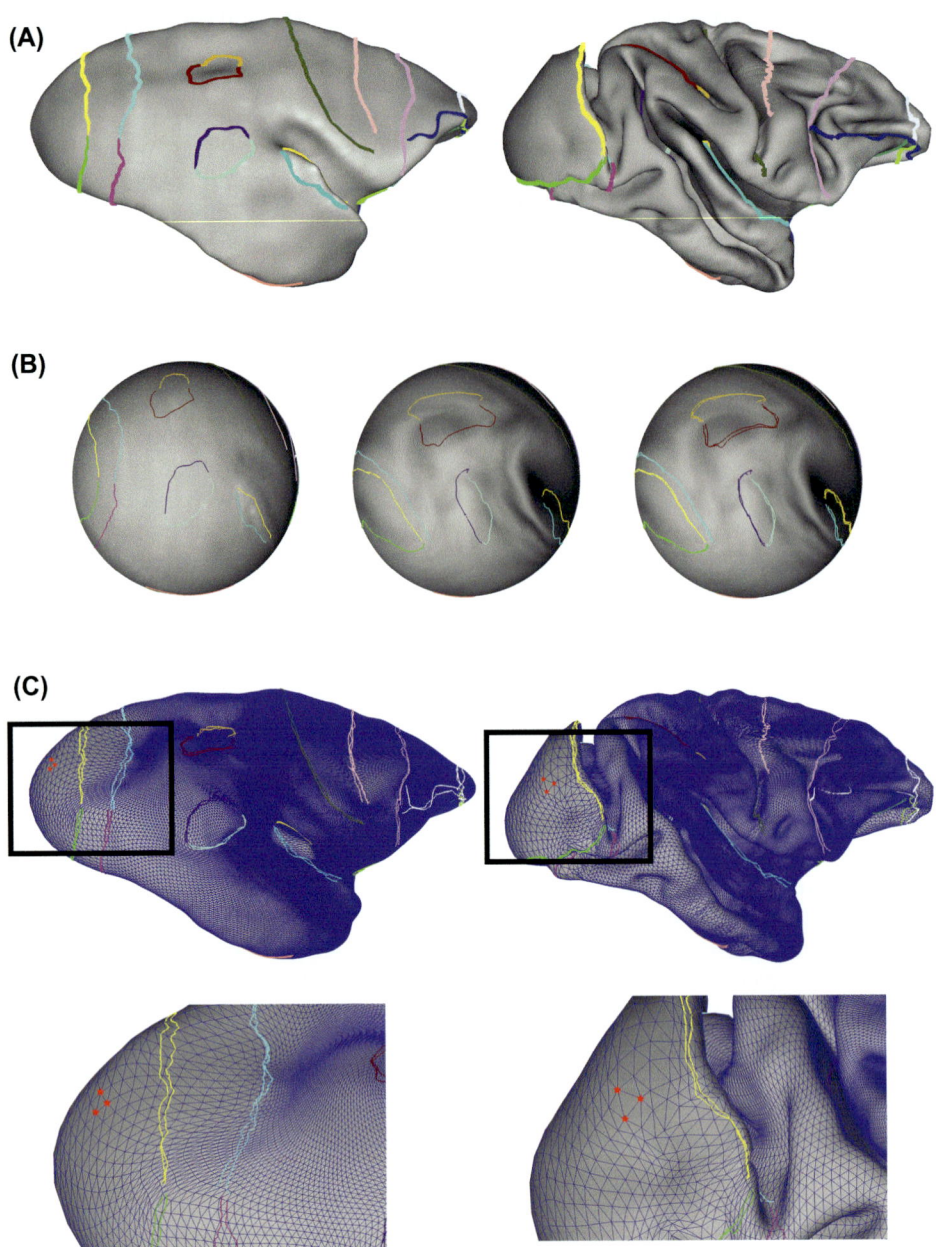

FIGURE 22.3 The surface registration method. (A) Models of the marmoset and macaque cerebral cortex with borders of homologous areas drawn on as colored lines. (B) Spherical models of the cortex. The marmoset and macaque models are expanded to spheres, shown in the left and middle, with the anchoring landmarks from (A). The surface registration procedure stretches the marmoset sphere to match the macaque sphere so that the landmarks align, as shown on the right. (C) Now each triangle in the model of the marmoset corresponds to a triangle in the macaque. One example triangle is denoted by *red stars* and shown in close-up on both models. Expansion is measured by simply measure the area of each corresponding pair of triangles between two models.

the sphere into place. Once complete, each triangle in the original brain mesh is mapped to a triangle in the target brain mesh (Fig. 22.3C) in both the spherical and original models. It is now possible to measure the size of each corresponding triangle on the source and target brains (Fig. 22.3C, see inset) in physical units (eg, mm^2). The ratio of the areas of a particular pair of triangles is used as the measure of increase in cortical surface area at this region of the cortex between two brains.

22.5 The Spatial Pattern of Expansion in the Simian Cerebral Cortex

The first application of surface-based registration to interspecies comparisons was by Van Essen and colleagues (Orban et al., 2004; Van Essen and Dierker, 2007; Hill et al., 2010). They compared the human cortex to that of the macaque monkey (*Macaca mulatta*), constrained by several well-known homologous landmarks, mostly primary sensory areas and functional MRI data. The result was the first map of cortical expansion from a smaller-brained primate to humans (Fig. 22.4A). This provided a quantitative description of the nonuniform changes of cortical surface area across the entire cortex between these two species. It was abundantly clear from these data that there were a few hot spots that had expanded disproportionately—the lateral prefrontal cortex (LPFC), the temporal parietal junction (TPJ), and the anterior cingulate cortex (ACC) (Fig. 22.4A). In contrast, the sensory, motor, retrosplenial, and orbital prefrontal cortex show low to moderate increases in surface area. Interestingly, the areas that are disproportionately expanded are high-level association areas involved in complex human behaviors such as language. This supports the idea that it may not simply be the increase in brain size that caused the development of advanced human behaviors, but the increase in these particular regions of the cortex. However, it was not apparent from these data alone if these changes in certain areas of the

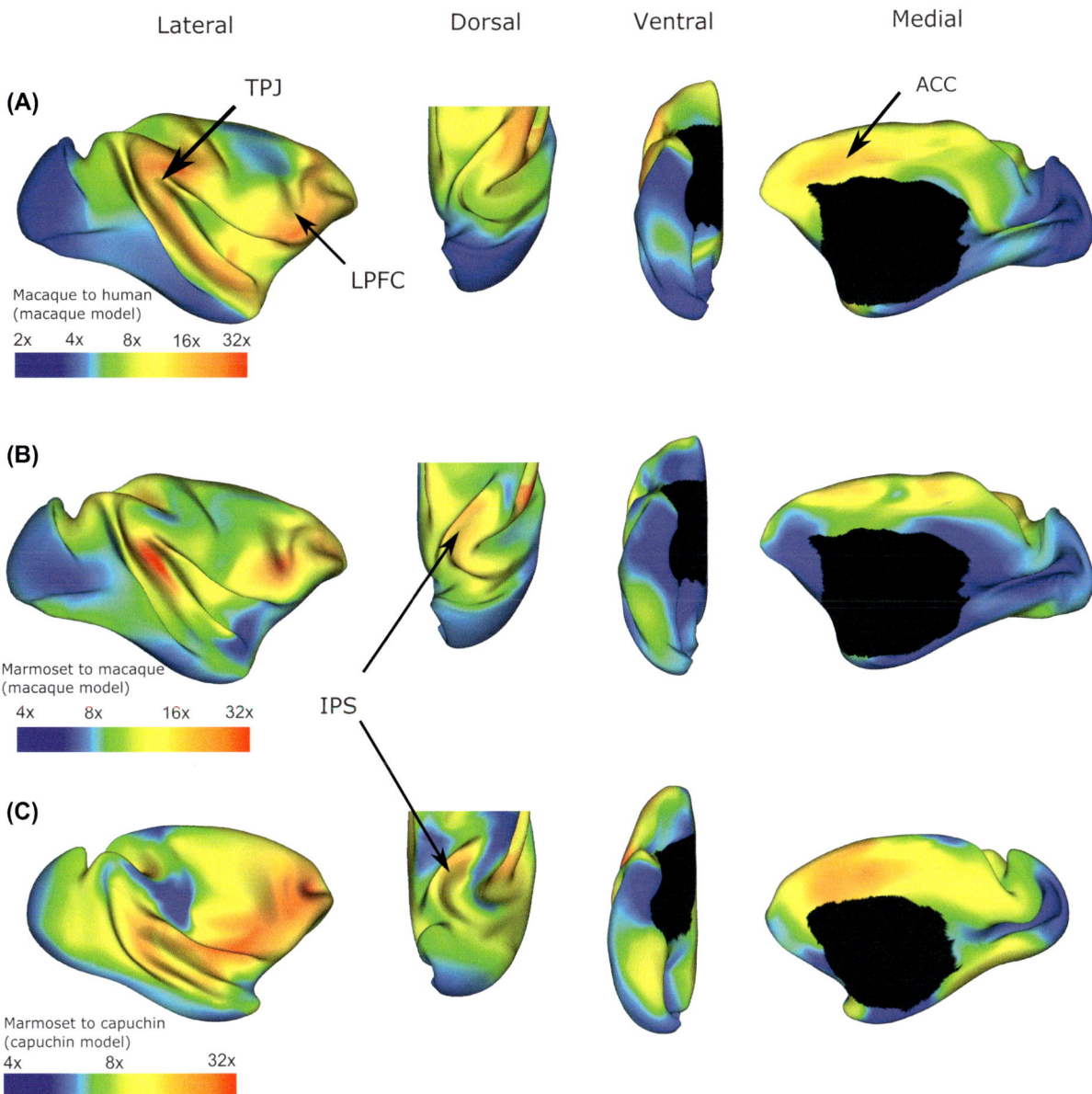

FIGURE 22.4   The expansion of the cortex in different species of primate. (A) The expansion from macaque to human (Van Essen and Dierker, 2007), showing different views of an inflated macaque cortex colored for expansion. Note the key hot spots of expansion at the lateral prefrontal cortex (LPFC), the temporal parietal junction (TPJ), and the anterior cingulate cortex (ACC). Black region is noncortical medial wall (ie, the location of noncortical structures). (B and C) The expansion from marmoset to capuchin and marmoset to macaque, respectively. Note that as in (A) the areas of high expansion are the LPFC, the TPJ, and the ACC, although there is some expansion in the intraparietal sulcus (IPS), indicating a slight deviation to the macaque to human patter. *From Chaplin, T.A., Yu, H.H., Soares, J.G., Gattass, R., Rosa, M.G., 2013. A conserved pattern of differential expansion of cortical areas in simian primates. J. Neurosci. 33 (38), 15120–15125.*

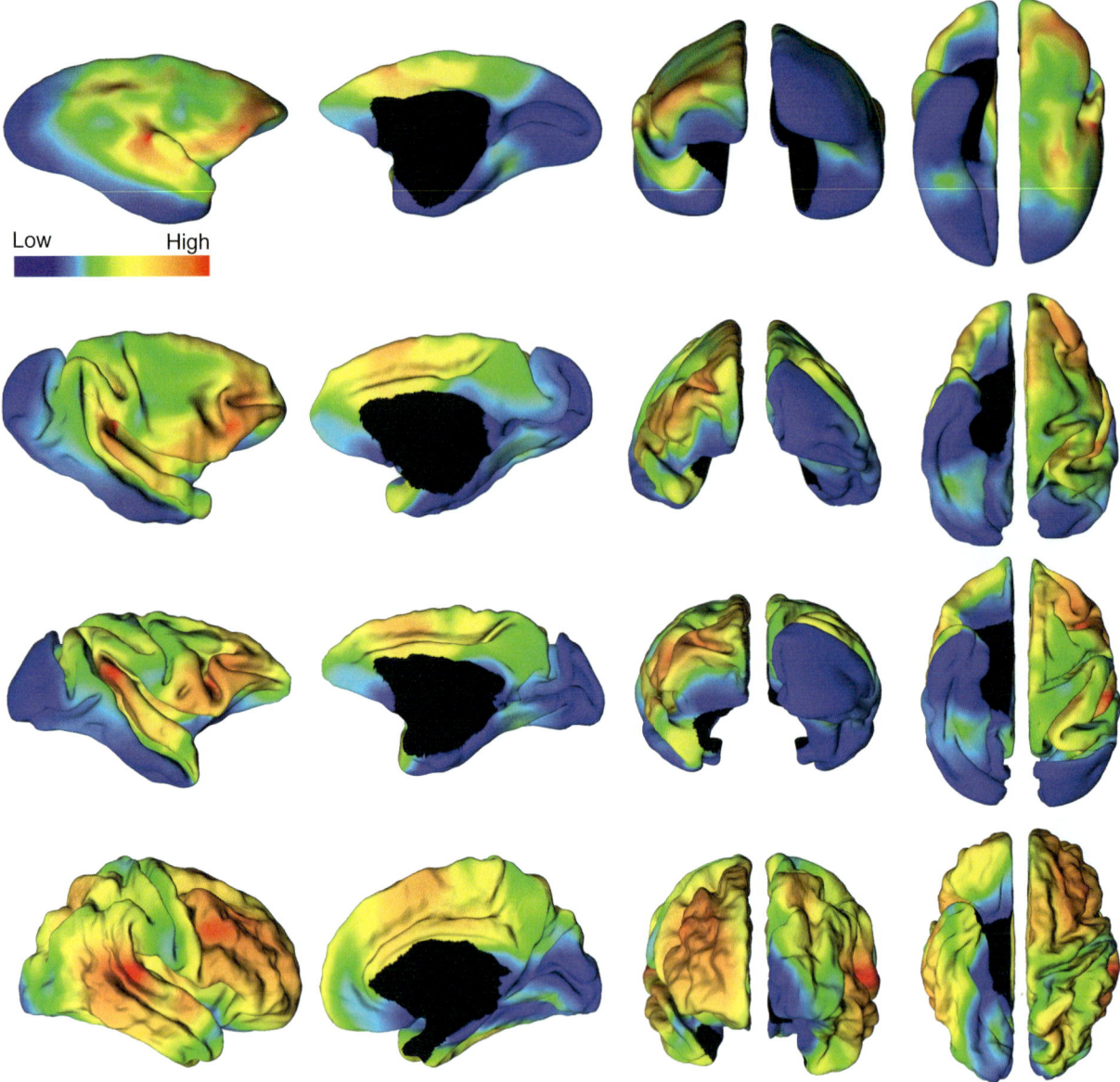

FIGURE 22.5 The averaged map of primate cortical expansion shown on four species. Each row shows a different species, from top to bottom: marmoset, capuchin, macaque, and human. Each column shows a different view of the cortex, from left to right: lateral, medial, anterior, posterior, ventral, and dorsal. The surface model is colored for expansion where *red* is the highest and *blue* is the lowest. *Black* is the noncortical medial wall (ie, the location of noncortical structures). An animated version can be viewed at https://www.youtube.com/watch?v=7wEjOfi9DiU.

cortex are human-specific adaptations (mosaic evolution, perhaps driven by evolutionary pressure to develop cognitive abilities) or instead are a consequence of the developmental mechanisms that generate a larger cerebral cortex (concerted evolution).

This issue was addressed by Chaplin et al. (2013), who compared the cerebral cortex of marmosets (*C. jacchus*) to both macaques (*M. mulatta*) and capuchin monkeys (*Cebus apella*). The cortical surface areas of macaques and capuchins are approximately 12 and 7 times larger than than those of the marmosets, respectively. The aim was to determine if the pattern of cortical expansion between the diminutive marmoset and the larger monkeys matched that of the macaque to human. If so, it would suggest there is a conserved pattern of expansion in the cerebral cortex of primates as the cortex increases in size. Indeed that was the case: as in the macaque-to-human comparisons, the LPFC, TPJ, and ACC are disproportionately larger in all comparisons between macaques, capuchins, and marmosets (Fig. 22.4). Thus the pattern of expansion in the cortex is conserved across primates, regardless of brain size. This supports the model that concerted evolution is the main factor in the expansion of the primate cerebral cortex, that is, the disproportionate increase in certain regions is a by-product of scaling up the overall size of the brain. By averaging the maps of expansion across all species, it was possible to obtain a general map of this conserved pattern of primate cortical expansion (Fig. 22.5).

There were, however, some species differences—the posterior parietal cortex is noticeably expanded in macaques and capuchins compared to marmosets (Fig. 22.4B and C) but not in the macaque-to-human registrations. The areas of the inferior parietal lobule show greater expansion in the human brain compared to the macaque brain (Fig. 22.4A), but this was not present in the monkey-to-monkey registrations. Given the current state of knowledge, this may be due to imperfect data on the arrangement of parietal areas, particularly in the marmoset and capuchin monkeys, or might represent genuine differences in the cortical expansion between species, reflecting a degree of mosaic evolution.

It should be noted that the smaller simian species used in this analysis cannot be regarded as ancestors of the larger species; rather, all species of living primates have evolved from extinct common ancestors. In fact, as noted above, there is strong evidence to suggest that marmosets, as a group, have evolved from larger animals, through a process of phyletic dwarfism (Ford, 1980; Montgomery and Mundy, 2013; Kelava et al., 2013).

22.6 The Late Equals Large Principle

Why does the primate cerebral cortex follow a particular pattern of expansion when scaling up in size? It is likely that the answer is related to the way the cortex develops. The "late equals large" principle states that late-developing regions of the brain tend to grow larger in larger-brained species. However, at different stages of development the mechanisms behind this principle may differ.

During prenatal development in primates, neural stem cells in the ventricular zone can symmetrically divide before neurogenesis (ie, before they become neurons), which increase the total pool of neural precursor cells. This results in an increase in the number of cortical columns and therefore surface area, but without increases in cortical thickness (Rakic, 2000, 2009). Thus protracted development times may allow a larger pool of neural stem cells to form in some regions, particularly those maturing after more cycles of cell division. This would result in an increase in cortical surface area in these late-developing regions (Finlay and Darlington, 1995; Striedter, 2005; Rakic, 2009).

Postnatally, neurogenesis and migration is complete, yet the human cerebral cortex continues to develop rapidly in a highly heterogeneous fashion (Shaw et al.,

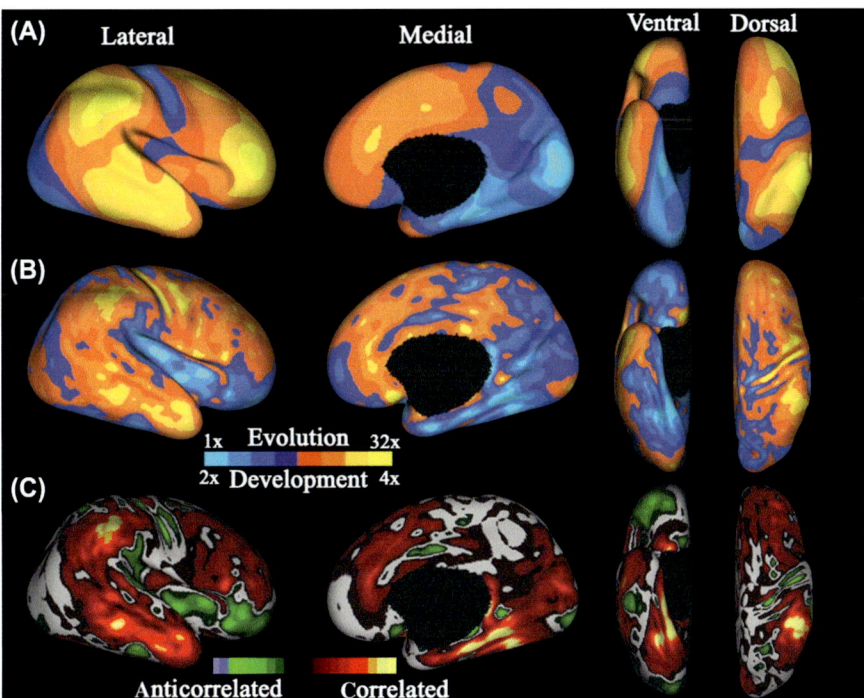

FIGURE 22.6 Comparison of macaque to human expansion with postnatal surface area changes in humans. (A) The macaque to human expansion map. (B) The postnatal expansion map. (C) Correlation of the two maps, showing high level of agreement (positive correlation in *red*). *From Hill, J., Inder, T., Neil, J., Dierker, D., Harwell, J., Van Essen, D., 2010. Similar patterns of cortical expansion during human development and evolution. Proc. Natl. Acad. Sci. U.S.A. 107, 13135—13140.*

2008; Hill et al., 2010), with large, nonuniform increases in surface area without corresponding increases in cortical thickness (in fact, cortical thickness decreases after birth). The pattern is reminiscent of the expansion of the primate cerebral cortex in evolution. Indeed, comparisons of the primate cortical expansion map and cortical expansion during human development have shown a large degree of overlap (Fig. 22.6). The key areas of expansion in the primate cortex, the LPFC, the TPJ, and the ACC, expand greatly after birth in humans (Hill et al., 2010; Amlien et al., 2016). Conversely, the cortical thickness of these areas actually decreases, showing a strong negative correlation with primate cortical expansion (Amlien et al., 2016). Critically, these correlations exist for macaque-to-human expansion (ie, human-specific expansion) as well as marmoset to macaque/capuchin expansion (ie, monkey only expansion) (Amlien et al., 2016). This suggests that changes in surface area and cortical thickness during human development are not human-specific adaptations, but instead are related to general scaling laws of the primate cerebral cortex.

It has been hypothesized that cellular and functional differences between different regions of the cortex may explain the postnatal nonuniform expansion of the cortex (Hill et al., 2010). Higher-expanding regions are less mature at birth (in terms of synaptic density and metabolic activity), have greater cellular complexity (dendritic basal field area, arbor complexity, and spine number), and take longer to reach maturity (Elston and Rosa, 1998; Elston et al., 1999; Bourne and Rosa, 2006; Hill et al., 2010). Thus, it may be necessary to delay maturation and growth of these high-level cortical areas to prioritize the development of areas critical for early survival, eg, sensory and motor areas, or that high-level cortical areas require postnatal experience to develop (Weiner et al., 2008; Hill et al., 2010). Whatever the case, the protracted development timelines of larger species of primate may allow certain regions of the cortex to grow disproportionately even after the process of prenatal neurogenesis is completed.

22.7 Characteristics of the Expanded Regions

Sensory and motor areas have been subject to the least amount of expansion in primate evolution. The most expanded regions tend to be association areas that mediate high-level cognitive functions. For example, the human TPJ cortex has been attributed with functions such as spatial awareness (Karnath, 2001), theory of mind (the ability to attribute mental states to others; Saxe and Kanwisher, 2003), and moral judgments (Young et al., 2010). The LPFC is involved in cognitive control, working memory, and planning (Miller and Cohen,

2001; Curtis and D'Esposito, 2003; Tanji and Hoshi, 2008), and the ACC is implicated in motor control, emotion, error detection, and conflict monitoring (Paus, 2001; Kuwabara et al., 2014; Mansouri et al., 2015). In addition, Broca's area in LPFC and Wernicke's area in TPJ in the dominant hemisphere have received tremendous amount of attention, because they are the two most well-known "language areas" in the human brain (although linguistic functions involves a larger network of areas in the prefrontal and the temporal lobe).

Given that these cognitive functions appear to be the basis of human intelligence, it is natural to expect that the associated cortical regions are enlarged in the human brain. Indeed, the size of the prefrontal cortex has long been considered the most salient anatomical feature that sets the human brain apart from the brains of monkeys and apes (Brodmann, 1909; Passingham, 1973; Deacon, 1990). The proportion of the prefrontal cortex relative to the isocortex is estimated to be ~30% in human and ~10% in macaque (Brodmann, 1909; Passingham, 1973), indicating a considerable expansion of the prefrontal cortex in human evolution. In addition to the prefrontal cortex, the temporal lobe is another highly relevant region, due to the presence of Wernicke's area in the human left temporal lobe. However, the temporal lobe does not appear to be proportionally larger in human. Based on volumetric MRI data, Rilling and Seligman (2002) reported that the temporal lobe occupies 16.6% of the human cortex, but in macaque it occupies 23.0% of the cortex. Finally, the human ACC has also been reported to be enlarged disproportionally in comparison to macaque ACC (Vogt et al., 2013).

Complementing and extending these data, registration analyses (Avants et al., 2006; Van Essen and Dierker, 2007; Hill et al., 2010; Chaplin et al., 2013) have revealed a consistent pattern of expansion that primarily involves the frontal lobe, the temporal lobe, and medial wall of the cortex. Expansions within these regions are nonuniform, with "hot spots" centering in the LPFC, TPJ, and the ACC. It is worth noting that some measurements of the prefrontal cortex have emphasized an expansion of the temporal pole area (or area 10; Semendeferi et al., 2001; Holloway, 2002), but surface registration studies have consistently located the region of highest expansion at the lateral prefrontal cortex (approximately area 44 and 45).

A more difficult, and more controversial, issue is to go beyond the observation that the human prefrontal cortex is proportionally larger than that of the macaque, and further inquire if the size of the human prefrontal cortex is predictable, given the trend observed in monkeys or apes. In other words, by establishing a potentially nonlinear relationship (allometry) between the size of the prefrontal cortex and the entire cortex, it is possible to evaluate if the size of the human prefrontal cortex is comparable to that of a human-sized monkey or ape. Semendeferi et al. (2002) argued that the human prefrontal cortex is the expected size of a human-sized ape, therefore casting doubt on the uniqueness of the human prefrontal cortex. Barton and Venditti (2013) further claimed that the human prefrontal cortex is the expected size of a human-sized monkey (see also Passingham, 2002; Sherwood and Smaers, 2013). With respect to the temporal lobe, Rilling and Seligman (2002) argued that although the human temporal lobe is not proportionally larger than that of the apes, its size is actually larger than the expected given the allometry of the apes.

Although surface registration analysis does not address the issue of allometry, allometric analyses show that the expansions of cortical areas are under powerful constraints during evolution (see Section 22.6), and thus provide additional evidence for concerted evolution.

22.8 The Reorganization of the Cortex in Primate Evolution

Broca's area in the human prefrontal cortex and Wernicke's area in the human temporal lobe are the two most well-known cortical areas involved in the production and comprehension of speech. That homologs of these two areas could be identified in apes, monkeys, and prosimians with cytoarchitectonic criteria testifies the high degree of conservation that is commonly observed in brain evolution. The discovery of "mirror neurons" (a class of neuron that is important for action comprehension) in area F5, part of the macaque homolog of Broca's area (Petrides et al., 2005; Rilling, 2014; see also Suzuki et al., 2015 for mirror neurons in the marmoset), suggests that Broca's area evolved from an existing area in the primate brain, whose functions are considered precursors to semantics. Similarly, auditory neurons in the macaque superior temporal gyrus, particularly those responsible for discriminating species-specific vocalizations, have been proposed to be the precursors to the speech-related functions of Wernicke's area (Rilling, 2014). Also noteworthy is the proposal that the human TPJ is homologous with the middle section of the macaque superior temporal cortex (Mars et al., 2013). According to this view, face-selective neurons in the macaque superior temporal cortex are the precursors of the theory of mind functions in the human TPJ.

That these human areas appear to be derived from existing areas does not eliminate the possibility that new cortical areas could emerge in evolution. The current estimates of the number of cortical areas in humans (150–200; Kaas, 2008; Van Essen et al., 2012b) are substantially higher than in the macaque and marmoset (∼140 and ∼110, respectively; Van Essen et al., 2012a; Paxinos et al., 2012), which reflects the general trend toward an increase in the number of areas in larger brains (Changizi and Shimojo, 2005). Where are those "new" areas? While this is a controversial issue, and much is unknown about the process through which new cortical fields are formed (Striedter, 2005; Krubitzer and Huffman, 2000), it is noteworthy that some of the human areas that have been proposed to be without homologs in the macaque brain are in the regions that underwent preferential expansion in the human brain. For example, the frontal cortex went through reorganization in the evolution of primates (Preuss and Goldman-Rakic, 1991a). In particular, it has been proposed that the lateral part of the frontal pole cortex (area 10) in the human brain may have no correspondence in the macaque prefrontal cortex (Koechlin, 2011; Sallet et al., 2013; also see Burman et al., 2011 for evidence of subdivisions in marmosets). Likewise in the ACC, areas d32 and 32′ appear to have no homologs in monkeys (Cole et al., 2009; Vogt et al., 2013). Finally, in the inferior parietal lobule—a highly expanded region in the macaque-to-human registration (although the expansion is not as obvious in the marmoset to macaque registration), areas 39 and 40 are thought to be unique in the human brain (Preuss and Goldman-Rakic, 1991b; De La Mothe et al., 2012; Karnath, 2001). Given these examples, a plausible scenario is that expansion of the cortex leads to reorganization of the connectivity within the region, which eventually causes the formation of new cortical areas.

The expansion of the isocortex not only allows new areas to form, it also creates opportunities for new cortical networks to emerge. Resting-state MRI experiments, for example, have uncovered frontal-parietal networks in the human brain that seemingly cannot be found in the macaque brain (Mantini et al., 2013). Since both the prefrontal and the parietal cortex are expanded in the human brain relative to the macaque brain, the result is consistent with a scenario in which cortical expansions cause the reorganization of the cortex. The idea is elaborated by Buckner and Krienen (2013), who proposed that the expansion of the cortex can "untether" the hierarchical networks that are more commonly observed in sensory and motor cortices, causing densely connected, nonhierarchical networks to form in the expanded association areas.

22.9 Implications of Primate Cortical Expansion

Understanding and quantifying the pattern of expansion of the primate cerebral cortex has wider implications in brain mapping. Modern brain atlases contain maps of many different types of data or "modalities," for example, traditional cytoarchitectonic maps, connectivity maps, functional network maps, myelin maps, and cortical thickness maps to name a few. Combining these maps has already led to a deeper understanding of brain architecture and function, and incorporating maps of primate cortical expansion has also added to our understanding of the cerebral cortex.

The map of cortical expansion shows interesting correlations with other data sets. Regions of the human cortex that are correlated with intellectual abilities are also correlated with regions of high expansion (Fjell et al., 2015a; Fig. 22.6A). The cortical expansion map also aligns well the individual variability in functional connectivity (Mueller et al., 2013; Fig. 22.6B). There are also several interesting negative correlations. Areas of high expansion are associated with low cortical myelination (Glasser et al., 2014; Fig. 22.7C) and, according to some criteria, lower potential for cortical plasticity (Walhovd et al., 2015).

Cortical expansion has also been useful for understanding the origin of human neurological diseases. Alzheimer's disease is seemingly a human-only disease since it is rarely found in monkeys, so one hypothesis of Alzheimer's disease was that it affected the more recently evolved regions of the cortex that had rapidly expanded in human evolution. Fjell et al. (2015b) examined the overlap of maps of Alzheimer's disease and primate cortical expansion and found that there was actually no clear relationship between them (Fig. 22.7D). On the one hand, there appears to be some correlation—the highly expanded TPJ is susceptible to Alzheimer's disease, and likewise, the low-

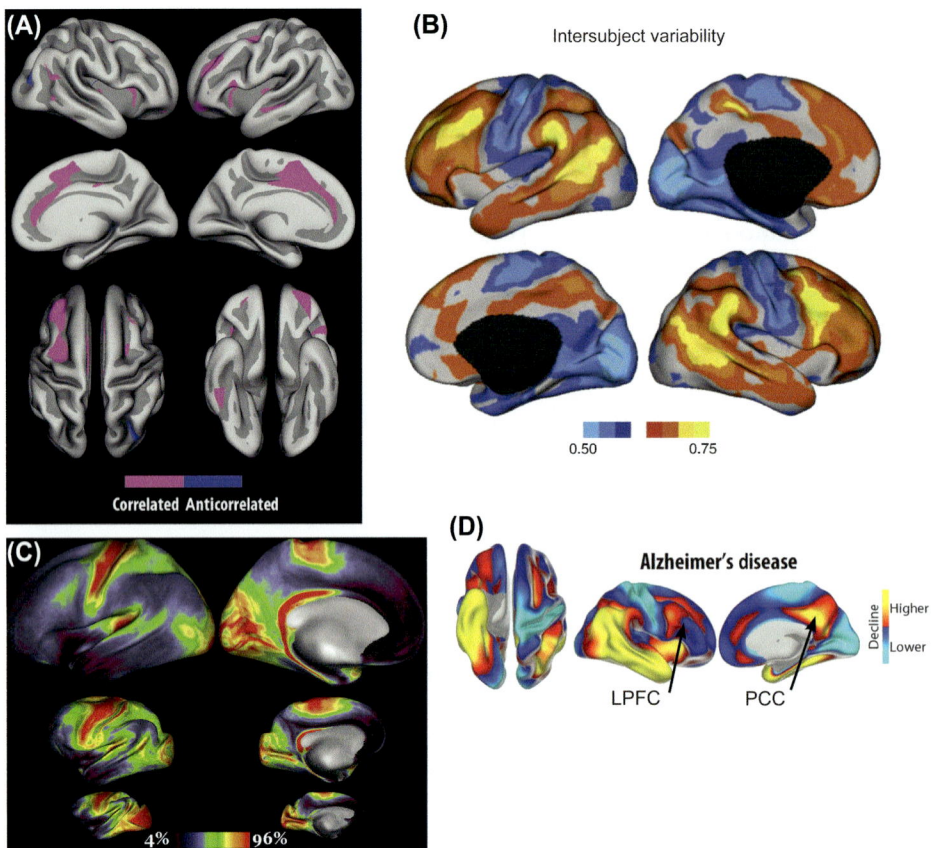

FIGURE 22.7   Applications of the primate cortical expansion map in human brain mapping studies. (A) Regions of cortex that are associated with higher intellectual abilities show correlation with the map of primate cortical expansion. (B) Map of individual variability in functional connectivity, which shows many similarities to the maps of primate cortical expansion. (C) Myelin maps of the human, chimpanzee, and macaque cortex show an inverse correlation with maps of primate cortical expansion, high-expanded regions being most lightly myelinated. (D) Maps of Alzheimer's disease—affected regions of the cortex do not correlate well with the primate cortical expansion map, despite Alzheimer's disease being a human-specific disease. Although there is partial agreement, there are some critical differences. For example the lateral prefrontal cortex (LPFC) is highly expanded but not susceptible to Alzheimer's disease, and conversely the posterior cingulate cortex (PCC) is often strongly affected by Alzheimer's disease despite being highly conserved. *From (A) Fjell, A.M., Westlye, L.T., Amlien, I., Tamnes, C.K., Grydeland, H., Engvig, A., Espeseth, T., Reinvang, I., Lundervold, A.J., Lundervold, A., Walhovd, K.B., 2015. High-expanding cortical regions in human development and evolution are related to higher intellectual abilities. Cereb Cortex 25 (1), 26–34; (B) Mueller, S., Wang, D., Fox, M.D., Yeo, B.T., Sepulcre, J., Sabuncu, M.R., Shafee, R., Lu, J., Liu, H., 2013. Individual variability in functional connectivity architecture of the human brain. Neuron 6, 77 (3), 586–595; (C) Glasser, M.F., Goyal, M.S., Preuss, T.M., Raichle, M.E., Van Essen, D.C., 2014. Trends and properties of human cerebral cortex: correlations with cortical myelin content. Neuroimage 93 (Pt 2), 165–175; (D) Fjell, A.M., Amlien, I.K., Sneve, M.H., Grydeland, H., Tamnes, C.K., Chaplin, T.A., Rosa, M.G., Walhovd, K.B., 2015. The roots of Alzheimer's disease: are high-expanding cortical areas preferentially targeted? Cereb. Cortex 25 (9), 2556–2565.*

expanding visual cortex is not. However, other highly expanded areas such as the LPFC are not heavily affected by Alzheimer's disease, and the highly conserved cingulate cortex is actually very susceptible (Fig. 22.6D). Instead, the localization of Alzheimer's disease was better explained by susceptibility to aging-related effects and association with the default mode network (Fjell et al., 2015b). These comparisons could only have been made by incorporating primate cortical expansion, Alzheimer's disease, aging, and default mode network data into a single, computationally based representation of primate brains.

22.10 Summary

Simian primates vary greatly in brain size. Yet it is clear that larger-brained species are not linearly scaled-up versions of smaller species—several key regions of the isocortex are disproportionately large in larger species. Computational methods have demonstrated that the spatial pattern of expansion is conserved across species—the changes between marmosets and macaques are similar to the changes between macaques and humans. This lends supports to the theory that concerted evolution largely drives evolution of the primate isocortex, since it is likely there are developmental constraints when scaling up overall brain size, and these changes may be responsible for the advanced cognitive abilities of humans. However, there still may be smaller but significant species-specific adaptations that arise from mosaic evolution.

References

Amlien, I.K., Fjell, A.M., Tamnes, C.K., Grydeland, H., Krogsrud, S.K., Chaplin, T.A., Rosa, M.G., Walhovd, K.B., 2016. Organizing principles of human cortical development-thickness and area from 4 to 30 years: insights from comparative primate neuroanatomy. Cereb. Cortex 26 (1), 257–267.

Angelucci, A., Rosa, M.G., 2015. Resolving the organization of the third tier visual cortex in primates: a hypothesis-based approach. Vis. Neurosci. 32, E010.

Avants, B.B., et al., 2006. Lagrangian frame diffeomorphic image registration: morphometric comparison of human and chimpanzee cortex. Med. Image Anal. 10 (3), 397–412.

Barton, R.A., Harvey, P.H., 2000. Mosaic evolution of brain structure in mammals. Nature 405 (6790), 1055–1058.

Barton, R.A., Venditti, C., 2013. Human frontal lobes are not relatively large. Proc. Natl. Acad. Sci. U.S.A. 110 (22), 9001–9006.

Bauernfeind, A.L., de Sousa, A.A., Avasthi, T., Dobson, S.D., Raghanti, M.A., Lewandowski, A.H., Zilles, K., Semendeferi, K., Allman, J.M., Craig, A.D., Hof, P.R., Sherwood, C.C., 2013. A volumetric comparison of the insular cortex and its subregions in primates. J. Hum. Evol. 64 (4), 263–279.

Bourne, J.A., Rosa, M.G.P., 2006. Hierarchical development of the primate visual cortex, as revealed by neurofilament immunoreactivity: early maturation of the middle temporal area (MT). Cereb. Cortex 16, 405–414.

Brodmann, K., 1909. VergleichendeLokalisationslehre der Grosshirnrinde in ihrenPrinzipiendargestellt auf Grund des Zellenbaues. JA Barth, Leipzig (L.J. Garey Trans. 1994. As localisation in the cerebral cortex. Smith-Gordon, London).

Buckner, R.L., Krienen, F.M., 2013. The evolution of distributed association networks in the human brain. Trends Cogn. Sci. 7 (12), 648–665.

Burman, K.J., Reser, D.H., Yu, H.-H., Rosa, M.G.P., 2011. Cortical input to the frontal pole of the marmoset monkey. Cereb. Cortex 21, 1712–1737.

Changizi, M.A., Shimojo, S., 2005. Parcellation and area-area connectivity as a function of neocortex size. Brain Behav. Evol. 66, 88–98.

Chaplin, T.A., Yu, H.H., Soares, J.G., Gattass, R., Rosa, M.G., 2013. A conserved pattern of differential expansion of cortical areas in simian primates. J. Neurosci. 33 (38), 15120–15125.

Chatterjee, H.J., Ho, S.Y., Barnes, I., Groves, C., 2009. Estimating the phylogeny and divergence times of primates using a supermatrix approach. BMC Evol. Biol. 9.

Cole, M.W., Yeung, N., Freiwald, W.A., Botvinick, M., 2009. Cingulate cortex: diverging data from humans and monkeys. Trends Neurosci. 32 (11), 566–574.

Curtis, C.E., D'Esposito, M., 2003. Persistent activity in the prefrontal cortex during working memory. Trends Cogn. Sci. 7 (9), 415–423.

Dale, A.M., Fischl, B., Sereno, M.I., 1999. Cortical surface-based analysis. I. Segmentation and surface reconstruction. Neuroimage 9 (2), 179–194.

De La Mothe, L.A., Blumell, S., Kajikawa, Y., Hackett, T.A., 2012. Cortical connections of auditory cortex in marmoset monkeys: lateral belt and parabelt regions. Anat. Rec. 295 (5), 800–821.

Deacon, T.W., 1990. Problems of ontogeny and phylogeny in brain-size evolution. Int. J. Primatol. 11, 237–282.

Elston, G.N., Rosa, M.G.P., 1998. The occipitoparietal pathway of the macaque monkey: comparison of pyramidal cell morphology in layer III of functionally related cortical visual areas. Cereb. Cortex 7 (5), 432–452.

Elston, G.N., Tweedale, R., Rosa, M.G., 1999. Cortical integration in the visual system of the macaque monkey: large-scale morphological differences in the pyramidal neurons in the occipital, parietal and temporal lobes. Proc. Biol. Sci. 266 (1426), 1367–1374.

Finlay, B.L., Darlington, R.B., 1995. Linked regularities in the development and evolution of mammalian brains. Science 268, 1578–1584.

Finlay, B.L., Darlington, R.B., Nicastro, N., 2001. Developmental structure in brain evolution. Behav. Brain Sci. 24, 263–278.

Fischl, B., Sereno, M.I., Dale, A.M., 1999. Cortical surface-based analysis. II: Inflation, flattening, and a surface-based coordinate system. Neuroimage 9 (2), 195–207.

Fleagle, J.G., 1998. Primate Adaptation and Evolution. Academic Press.

Ford, S.M., 1980. Callitrichids as phyletic dwarfs, and the place of the Callitrichidae in Platyrrhini. Primates 21, 31–34.

Ghosh, S.S., Kakunoori, S., Augustinack, J., Nieto-Castanon, A., Kovelman, I., Gaab, N., Christodoulou, J.A., Triantafyllou, C., Gabrieli, J.D., Fischl, B., 2010. Evaluating the validity of volume-based and surface-based brain image registration for developmental cognitive neuroscience studies in children 4 to 11 years of age. Neuroimage 53 (1), 85–93.

Gould, S.J., Lewontin, R.C., 1979. The spandrels of San Marco and the Panglossian paradigm: a critique of the adaptationist programme. Proc. R. Soc. Lond. B Biol. Sci. 205 (1161), 581–598.

Hill, J., Inder, T., Neil, J., Dierker, D., Harwell, J., Van Essen, D., 2010. Similar patterns of cortical expansion during human development and evolution. Proc. Natl. Acad. Sci. U.S.A. 107, 13135–13140.

Holloway, R.L., 2002. How much larger is the relative volume of area 10 of the prefrontal cortex in humans? Am. J. Phys. Anthropol. 118 (4), 399–401.

Kaas, J.H., 2008. The evolution of the complex sensory and motor systems of the human brain. Brain Res. Bull. 75, 384–390.

Kaas, J.H., 2013. The evolution of brains from early mammals to humans. Wiley Interdisc. Rev. Cogn. Sci. 4, 33–45.

Karnath, H.O., 2001. New insights into the functions of the superior temporal cortex. Nat. Rev. Neurosci. 2 (8), 568–576.

Kelava, I., Lewitus, E., Huttner, W.B., 2013. The secondary loss of gyrencephaly as an example of evolutionary phenotypical reversal. Front. Neuroanat. 7, 16.

Koechlin, E., 2011. Frontal pole function: what is specifically human? Trends Cogn. Sci. 15 (6), 241.

Krubitzer, L.A., Huffman, K.J., 2000. Arealization of the neocortex in mammals: genetic and epigenetic contributions to the phenotype. Brain Behav. Evol. 55 (6), 322–335.

Krubitzer, L.A., Seelke, A.M.H., 2012. Cortical evolution in mammals: the bane and the beauty of phenotypic variability. Proc. Natl. Acad. Sci. U.S.A. 109, 10647–10654.

Kuwabara, M., Mansouri, F.A., Buckley, M.J., Tanaka, K., 2014. Cognitive control functions of anterior cingulate cortex in macaque monkeys performing a Wisconsin Card Sorting Test Analog. J. Neurosci. 34 (22), 7531–7547.

Lyon, D.C., Connolly, J.D., 2012. The case for primate V3. Proc. Biol. Sci. 279 (1729), 625–633.

Mansouri, F.A., Rosa, M.G., Atapour, N., 2015. Working memory in the service of executive control functions. Front. Syst. Neurosci. 9, 166.

Mantini, D., Corbetta, M., Romani, G.L., Orban, G.A., Vanduffel, W., 2013. Evolutionarily novel functional networks in the human brain? J. Neurosci. 33 (8), 3259–3275.

Mars, R.B., Sallet, J., Neubert, F.X., Rushworth, M.F., 2013. Connectivity profiles reveal the relationship between brain areas for social cognition in human and monkey temporoparietal cortex. Proc. Natl. Acad. Sci. U.S.A. 110 (26), 10806–10811.

Miller, E.K., Cohen, J.D., 2001. An integrative theory of prefrontal cortex function. Annu. Rev. Neurosci. 2001 (24), 167–202.

Montgomery, S.H., Mundy, N.I., 2013. Parallel episodes of phyletic dwarfism in callitrichid and cheirogaleid primates. J. Evol. Biol. 26, 810–819.

Orban, G.A., Van Essen, D., Vanduffel, W., 2004. Comparative mapping of higher visual areas in monkeys and humans. Trends Cogn. Sci. 8, 315–324.

Passingham, R.E., 1973. Anatomical differences between neocortex of man and other primates. Brain Behav. Evol. l7 (5), 337–359.

Passingham, R.E., 2002. The frontal cortex: does size matter? Nat. Neurosci. 5 (3), 190–192.

Paus, T., 2001. Primate anterior cingulate cortex: where motor control, drive, and cognitive interface. Nat. Rev. Neurosci. 2 (6), 417–424.

Paxinos, G., Huang, X.F., Petrides, M., Toga, A.W., 2009. The Rhesus Monkey Brain in Stereotaxic Coordinates. Academic Press.

Paxinos, G., Watson, C., Petrides, M., Rosa, M.G.P., Tokuno, H., 2012. The Marmoset Brain in Stereotaxic Coordinates. Academic Press.

Petrides, M., Cadoret, G., Mackey, S., 2005. Orofacial somatomotor responses in the macaque monkey homologue of Broca's area. Nature 435 (7046), 1235–1238.

Preuss, T.M., Goldman-Rakic, P.S., 1991. Myelo- and cytoarchitecture of the granular frontal cortex and surrounding regions in the strepsirhine primate Galago and the anthropoid primate Macaca. J. Comp. Neurol. 310 (4), 429–474.

Preuss, T.M., Goldman-Rakic, P.S., 1991. Architectonics of the parietal and temporal association cortex in the strepsirhine primate Galago compared to the anthropoid primate Macaca. J. Comp. Neurol. 310 (4), 475–506.

Preuss, T.M., 2011. The human brain: rewired and running hot. Ann. N.Y. Acad. Sci. 1225, 182–191.

Rakic, P., 2000. Radial unit hypothesis of neocortical expansion. Novartis Found. Symp. 228, 30–42.

Rakic, P., 2009. Evolution of the neocortex: a perspective from developmental biology. Nat. Rev. Neurosci. 10, 724–735.

Reser, D.H., Burman, K.J., Yu, H.-H., Chaplin, T.A., Richardson, K.E., Worthy, K.H., Rosa, M.G.P., 2013. Contrasting patterns of cortical input to architectural subdivisions of area 8: a retrograde tracing study in marmoset monkeys. Cereb. Cortex 23, 1901–1922.

Rilling, J.K., Insel, T.R., 1999. The primate neocortex in comparative perspective using magnetic resonance imaging. J. Hum. Evol. 37, 191–223.

Rilling, J.K., Seligman, R.A., 2002. A quantitative morphometric comparative analysis of the primate temporal lobe. J. Hum. Evol. 42 (5), 505–533.

Rilling, J.K., 2014. Comparative primate neuroimaging: insights into human brain evolution. Trends Cogn. Sci. 18 (1), 46–55.

Rosa, M.G.P., Tweedale, R., 2005. Brain maps, great and small: lessons from comparative studies of primate visual cortical organization. Proc. R. Soc. B 360, 665–691.

Sallet, J., Mars, R.B., Noonan, M.P., Neubert, F.X., Jbabdi, S., O'Reilly, J.X., Filippini, N., Thomas, A.G., Rushworth, M.F., 2013. The organization of dorsal frontal cortex in humans and macaques. J. Neurosci. 33 (30), 12255–12274.

Saxe, R., Kanwisher, N., 2003. People thinking about thinking people. The role of the temporo-parietal junction in "theory of mind". Neuroimage 19, 1835–1842.

Semendeferi, K., Armstrong, E., Schleicher, A., Zilles, K., Van Hoesen, G.W., 1998. Limbic frontal cortex in hominoids: a comparative study of area 13. Am. J. Phys. Anthropol. 106 (2), 129–155.

Semendeferi, K., Armstrong, E., Schleicher, A., Zilles, K., Van Hoesen, G.W., 2001. Prefrontal cortex in humans and apes: a comparative study of area 10. Am. J. Phys. Anthropol. 114, 224–241.

Semendeferi, K., Lu, A., Schenker, N., Damasio, H., 2002. Humans and great apes share a large frontal cortex. Nat. Neurosci. 5 (3), 272–276.

Shaw, P., Kabani, N.J., Lerch, J.P., Eckstrand, K., Lenroot, R., Gogtay, N., Greenstein, D., Clasen, L., Evans, A., Rapoport, J.L., Giedd, J.N., Wise, S.P., 2008. Neurodevelopmental trajectories of the human cerebral cortex. J. Neurosci. 28 (14), 3586–3594.

Sherwood, C.C., Smaers, J.B., 2013. What's the fuss over human frontal lobe evolution? Trends Cogn. Sci. 17 (9), 432–433.

Sherwood, C.C., Bauernfeind, A.L., Bianchi, S., Raghanti, M.A., Hof, P.R., 2012. Human brain evolution writ large and small. In: Hofman, M.A., Falk, D. (Eds.), Progress in Brain Research, vol. 195.

Smaers, J.B., Soligo, C., 2013. Brain reorganization, not relative brain size, primarily characterizes anthropoid brain evolution. Proc. R. Soc. B 280.

Stephan, H., Frahm, H., Baron, G., 1981. New and revised data on volumes of brain structures in insectivores and primates. Folia Primatol. 35, 1–29.

Striedter, G.F., 2005. Principles of Brain Evolution. Sinauer Associates, Sunderland, MA.

Suzuki, W., Banno, T., Miyakawa, N., Abe, H., Goda, N., Ichinohe, N., 2015. Mirror neurons in a new world monkey, common marmoset. Front. Neurosci. 10 (9), 459.

Tanji, J., Hoshi, E., 2008. Role of the lateral prefrontal cortex in executive behavior control. Physiol. Rev. 88 (1), 37–57.

Van Essen, D.C., 2004. Surface-based approaches to spatial localization and registration in primate cerebral cortex. Neuroimage 23, S97–S107.

Van Essen, D.C., Dierker, D.L., 2007. Surface-based and probabilistic atlases of primate cerebral cortex. Neuron 56, 209–225.

Van Essen, D.C., Drury, H.A., Dickson, J., Harwell, J., Hanlon, D., Anderson, C.H., 2001. An integrated software suite for surface-based analyses of cerebral cortex. J. Am. Med. Inf. Assoc. 8, 443–459.

Van Essen, D.C., Glasser, M.F., Dierker, D.L., Harwell, J., 2012. Cortical parcellations of the macaque monkey analyzed on surface-based atlases. Cereb. Cortex 22, 2227–2240.

Van Essen, D.C., Glasser, M.F., Dierker, D.L., Harwell, J., Coalson, T., 2012. Parcellations and hemispheric asymmetries of human cerebral cortex analyzed on surface-based atlases. Cereb. Cortex 22, 2241–2262.

Vogt, B.A., Hof, P.R., Zilles, K., Vogt, L.J., Herold, C., Palomero-Gallagher, N., 2013. Cingulate area 32 homologies in mouse, rat, macaque and human: cytoarchitecture and receptor architecture. J. Comp. Neurol. 521 (18), 4189–4204.

Walhovd, K.B., Westerhausen, R., de Lange, A.G., Bråthen, A.C., Grydeland, H., Engvig, A., Fjell, A.M., 2015. Premises of plasticity

– and the loneliness of the medial temporal lobe. Neuroimage. https://doi.org/10.1016/j.neuroimage.2015.10.060.

Weiner, S., Monge, J., Mann, A., 2008. Bipedalism and parturition: an evolutionary imperative for cesarean delivery? Clin. Perinatol. 35, 469–478.

Young, L., Camprodon, J.A., Hauser, M., Pascual-Leone, A., Saxe, R., 2010. Disruption of the right temporoparietal junction with transcranial magnetic stimulation reduces the role of beliefs in moral judgments. Proc. Natl. Acad. Sci. U.S.A. 107, 6753–6758.

23

Evolution of Visual Cortex in Primates

J.H. Kaas

Vanderbilt University, Nashville, TN, United States

23.1 Introduction

Our understanding of how visual cortex was organized in early primates and evolved in the various branches of the primate radiation depends greatly on inferences made from the results of studies on visual system organizations of present-day primates and their closest relatives (Kaas, 2016). The reason for this is that the fossil record of skull endocasts provides useful information about brain sizes and shapes, and sometimes fissure patterns, but nothing about the internal organization of the brain. Thus, comparative studies of the visual systems of primates are very important. The comparative or cladistic approach to studies of brain evolution rest on the assumption that brain characteristics widely shared across members of a related group of species are more likely to have been retained from their common ancestors than to have been acquired independently (Hennig, 1966). Obviously, conclusions are more strongly supported as observations are made on more members of the group (clade), something that can be difficult to do in the studies of brains, as relevant research can be costly, difficult, and even impossible. Some observations are easily made, such as noting whether a corpus callosum is present or not, and thereby providing clear support for the conclusion that only placental mammals have this forebrain structure. Other observations depend on experimental approaches and sometimes on observations that seem to support conflicting conclusions, the presence or not of a V3, for example (see Kaas and Lyon, 2001). It is also important to remember that some traits do evolve independently a number of times (Kaas, 2002). For example, larger brains with cortical fissures have evolved independently many times in mammals. The common appearance of brains with cortical fissures recently lead the authors of a massive cladistic analysis of placental mammals to conclude that the ancestral placental mammals had a neocortex with fissures (see O'Leary et al., 2013). We now know that brains of the proposed size of the placental ancestor would not have fissures in neocortex (Mota and Herculano-Houzel, 2015). Thus, comparative evidence can lead to incorrect conclusions, and especially limited evidence needs to be interpreted with caution.

Here we consider current theories on how visual cortex is organized in primates and attempt to identify those features and cortical areas that have been retained in primates from nonprimate ancestors and those features and areas that likely emerged with early primates, and can be considered to be primate innovations. In addition, we consider the ways that the organization of visual cortex may have changed as visual cortex evolved differently in the major branches of the primate radiation. Because visual cortex depends on the eye, the retina, and the retinal projection to subcortical structures, we briefly discuss the evolution of those parts of the visual system as well.

23.2 The Primate Radiation and Other Members of Euarchontoglires

The overall structure of the primate radiation and its relation to those of other mammals is now well known as the recent use of comparative molecular data has produced compelling evidence and widespread agreement on phylogenetic relationships (eg, Murphy et al., 2004; O'Leary et al., 2013; Bininda-Emonds and Hartmann, 2016), but less agreement about when times of divergence occurred (Bininda-Emonds and Hartmann, 2016). Evolutionary changes can occur at different rates, and the evolutionary relationships appear to be the most important for comparative studies. The values of Murphy et al. (2004) are largely used here.

In brief, primates are placental mammals of the superorder Euarchontoglires that emerged 80–100 mya. The closest relatives of primates are colugos or gliding lemurs (Dermoptera), which have not been adequately studied, and their closest relatives, the tree shrews (Scandentia). It is not clear if colugos or tree shrews are more closely related to primates, but relevant comparative data are available only for tree shrews. More distantly related members of the superorder include rodents and lagomorphs, where many studies of the visual system have focused on rats and mice, and, to a more limited extent, on the more visual squirrels. The ancestors of present-day primates emerged from a group of archaic primates. All extant primates evolved from a common euprimate ancestor perhaps 70–80 mya. Early euprimates resembled some of the current strepsirrhine primates in body and brain size, as well as nocturnal lifestyle (Martin, 1990; Bloch and Boyer, 2002; Silcox et al., 2010). Early strepsirrhine primates radiated into present-day lorises lemurs and galagos that mostly remained nocturnal. Anthropoids or haplorhines emerged around 60 mya as a diurnal branch of primate evolution that gave rise to the now nocturnal tarsiers and the now almost exclusively diurnal simians (Steiper and Seiffert, 2012). These early simians largely occupied Africa, where some few, around 35–40 mya, successfully rafted to South America to form the platyrrhine radiation of New World monkeys (Schrago, 2007; Kay, 2015), with owl monkeys becoming the only nocturnal monkey (Tyler, 1991). The surviving African monkeys radiated from a common ancestor to form the catarrhine radiation of Old World monkeys, with one branch of about 25–30 mya leading to modern apes and humans (Stewart and Disotell, 1998). Humans, as we now know, shared a common ancestor with chimpanzees as recently as 6 mya.

23.3 The Eye, Retina, and Retinal Projections

The fossil record indicates that early stem primates were small, nocturnal, and fed in the fine branches of tropical trees and bushes (eg, Block and Boyer, 2002). Early primates likely fed on insects, other small invertebrates and vertebrates, fruits, buds, and leaves (Milton, 2006). As an adaptation for this environment, early primates had forward facing eyes that enhanced binocular vision and provided advantages for reaching and grasping food items (Melmoth and Grant, 2006). A reflecting tapetum behind the retina was a likely way of aiding vision in dim light. Retinal ganglion cells were of the three major types of present-day primates and a wide range of other mammals, referred to as the P, M, and K classes in primates and the X, Y, and W classes in other mammals (eg, Yamada et al., 1998; Casagrande, 1994; Perry and Cowey, 1981; Sherman et al., 1976). In present-day primates, P(X) cells have the smaller receptive fields important for detailed vision,

and they are concentrated in the central retina to form an "area centralis" of high ganglion cell density, thereby promoting a small 5–10° circle of high resolution vision. Judging from present-day nocturnal primates, there was no foveal pit, which would generate even higher visual acuity for the central one-degree of vision. The retina would have been rod dominated, with the fewer cones being of two types, a less frequent "blue" cone sensitive to short wavelengths, the S cone, and a cone for longer red–green wavelengths, the M–L cone. The P ganglion cells carried color and light information to the dorsal lateral geniculate nucleus (LGN) to provide information about the details of the visual image. These P ganglion cells project exclusively to the LGN in monkeys, and likely all primates (Weller and Kaas, 1989), where they constitute about 80% of retinal ganglion cells. This proportion may have been less in early primates, but it was likely way more than the much smaller proportion of X (p) cells in nonprimate mammals. Thus, early primates already had important adaptations for detailed binocular, dichromatic form vision in dim light. The M cell classes of ganglion cells important for detecting visual motion and change, projected to both the LGN and superior colliculus, as did the less well-studied K (W) cell class of ganglion cells.

Patterns of the terminations of retinal ganglion cell inputs to the LGN of mammals are highly variable, with laminar segregations related to eye of origin being nearly universal. Other laminae segregations in the LGN involve ganglion cell classes in different ways, including sometimes segregations of ganglion cell inputs with light "on" or "off" receptive field center responses. This variability complicates an understanding of the pattern of LGN lamination in the nonprimate ancestors of primates, but it likely involved a simple pattern of two layers with mixed X (P) and Y (M) inputs, one for each eye, and a third layer for W (K) inputs (Kaas, 2014). In contrast, all primates have LGNs with two dorsal P layers and two ventral M layers. In strepsirrhine primates, thought to most closely resemble early primates, the most dorsal P layer and the most ventral M layer get inputs from the contralateral eye, with the internal P and M layers getting inputs from the ipsilateral eye, as in all anthropoid primates (Kaas et al., 1978) except tarsiers (Rosa et al., 1996). The P layers subdivide to varying extents in anthropoid monkeys, apes, and humans as more and more P ganglion cells from central vision project to the LGN, but this is not the case in nocturnal strepsirrhine primates, tarsiers, and owl monkeys, and was thereby unlikely in early primates. In strepsirrhine primates, the K ganglion cells project mainly to small LGN K cells between the two parvocellular layers. These K cells are divided into a dorsal K layer for the contralateral eye, and a ventral K layer for the ipsilateral eye. K cell inputs also project in smaller numbers to small cells between other layers. In diurnal primates, the K cell inputs to the LGN are reduced to the few small cells between layers. However, both the nocturnal tarsiers and owl monkeys have a large K cell region in the LGN between the P and M layers. Thus, K cell retinal inputs appear to be more important in nocturnal primates, while P cell inputs are less dominant. The LGN of early primates most likely resembled those of present-day strepsirrhine primates, but the locations of the K cell inputs may have differed.

The superior colliculus is another important target of the retinal projections in primates. In contrast to other studied mammals, all primates have a superior colliculus that represents only the contralateral visual hemifield (eg, Lane et al., 1973), but via both eyes, as in the LGN and primary visual cortex of primates and other mammals. In nonprimates the complete retina of the contralateral eye projects to the superior colliculus, and the inputs from the ipsilateral eye are typically sparse (eg, Kaas et al., 1974; Lane et al., 1974). Early primates likely had the pattern of inputs from the two eyes, of current primates. Thus, binocular, frontal vision was emphasized in the superior colliculus, as well as in visual cortex, of early primates.

Tarsiers, as the earliest surviving branch of the haplorhine radiation, are especially interesting as they evolved from diurnal ancestors and soon thereafter reverted to a nocturnal lifestyle. The ancestral haplorhines had become diurnal, resulting in a greater emphasis on vision for detecting predators and food at a distance. Visual acuity was greatly increased by the evolution of a fovea, and an increased packing of P ganglion cells with small receptive fields around the fovea. The reflecting tapetum of the back of the retina, which degraded the retinal image and was not needed in daylight, was lost. The ancestors of tarsiers, in becoming nocturnal again, had to readapt to dim light, as tarsiers depended on vision in dim light to detect their insect and small vertebrate prey that became their only food. Tarsiers did not reinvent the eye of the nocturnal early primates, as tarsiers do not have a reflecting tapetum and they have retained the diurnal adaptation of a fovea (Collins et al., 2005). The longer wavelength cones are concentrated in the central retina as expected for their diurnal anthropoid ancestors. The main adaptation to dim light was to evolve huge eyes, as their eye size relative to body size is greater than any other mammal (Polyak, 1957). The ganglion cell densities are lower in the central retina than in diurnal monkeys, suggesting more convergence of receptors and bipolar cells on ganglion cells, allowing for more sensitivity in dim light, but reduced visual acuity as in nocturnal owl monkeys (Tetreault et al., 2004). Another adaptation to dim light in the retinae of tarsiers is having fewer P ganglion cells and more K ganglion cells, judging from retinal projections to the LGN. Thus, the two P cell LGN layers are

smaller in tarsiers and undivided compared to diurnal monkeys. However, the K cells largely terminate on small LGN cells between the M and P layers, as in nocturnal owl monkeys, rather than in K cell layers between the P layers as in strepsirrhine primates. Thus, the LGN of tarsiers resembles that of other anthropoid primates in laminar organization, except for a strange inversion of the M layers so that the M layer with input from the ipsilateral eye is most ventral (Rosa et al., 1996). Likewise as an anthropoid (haplorhine) primate, the visual pulvinar has an organization more similar to that of monkeys than prosimian galagos and lemurs (Collins et al., 2005).

In a similar manner, owl monkeys adapted to feeding and moving about the canopy in dim light (Wright, 1994) by evolving large eyes, more convergence of receptors on to bipolar cells and then ganglion cells, and replacing the fovea with area centralis (Webb and Kaas, 1976; Silveira et al., 1993). The highly nocturnal retina of owl monkeys has fewer ganglion cells, fewer cones, and more rod receptors, and more rod bipolar cells (Dyer et al., 2009). These specializations of the eye for dim light are thought to have evolved within the last 15 million years.

Other innovations of the retina in primates involved color vision (Jacobs, 1996). Early haplorhines, such as most present-day New World monkeys, were dichromats having an S cone class in the short wavelength range and an M—L cone in the middle-to-long wavelength range. This limited type of color vision was retained in the early anthropoids that somehow rafted from Africa to South America some 30—40 mya, but was subsequently modified in some New World monkeys in two different ways. Some New World monkeys have two versions of the gene for M—L cones on the X chromosome. As each version of the gene codes for a photopigment with slightly different wavelength sensitivity, and females have two X chromosomes, some females have different M—L cone genes on each X chromosome, while others have the same version on each X chromosome. Males, having only one X chromosome, have only one variant. Those females with two versions have at least a weak type of trichromatic color vision, which would presumably provide a great advantage in the selection of ripe fruit or new leaves and buds for food. Another line of New World monkey evolution emerged with a new gene for a third cone pigment, thereby obtaining sensitive trichromatic color vision in males as well as females. The common ancestor of all Old World monkeys, as well as apes and humans, independently acquired a third gene for color vision, so that all these anthropoid primates have trichromatic vision. As this added benefit is carried by the P ganglion cell pathway to primary visual cortex via the P geniculate layers, this system has expanded to include 80% of the ganglion cells in catarrhine primates, and elaborations

4. Primates

of the P cell layer of the LGN to four or more (six to eight) sublayers in most of these primates (Kaas et al., 1978). The cortical visual system likely adapted rapidly to the addition of a new cone type.

In summary, we have reviewed evidence that the eyes and retinal projections of primates differ from those of nonprimates in several ways. (1) A clear laminar segregation of P, M, K ganglion cell projections to the LGN exists, the P, M, and K layers. (2) The P ganglion cells constitute as much as 80%, of the total, and these ganglion cells project only to the LGN P layers. As the P ganglion cells are specialized for detailed vision, those changes demonstrate the importance of detailed vision for early primates, and especially for anthropoid primates, where the greater mass of P cell LGN layers has subdivided. (3) The eyes of early primates, as in present-day primates, were more frontal, allowing for more binocular overlap and the advantages that it brings to cortical processing. (4) This emphasis on frontal vision is also apparent in the superior colliculus of all primates, where only projections from the nasal hemiretina of the contralateral eye project to the superior colliculus, and the inputs from the temporal retina of the ipsilateral eye have greatly increased. This is in contrast to rodents and tree shrews, where the complete retina of the contralateral eye projects to the superior colliculus, and there is little input from the ipsilateral eye.

Other changes involved those that emerged in various branches of the primate radiation. (1) The ancestors of anthropoid primates became diurnal and evolved a fovea, which allowed central vision to be represented in even more detail. (2) Some New World monkeys evolved a type of trichromatic color vision that was restricted to some females as it was based on having two versions of a retinal pigment gene on the two X chromosomes. (3) Another line of primates evolved a new gene for medium wavelength pigment so that all individuals acquired trichromatic color vision. (4) The common ancestor of all Old World monkeys, apes, and humans, independently acquired trichromatic color vision by evolving a new gene for a cone pigment. (5) Two separate lines of evolution in anthropoid primates became nocturnal, returned to foraging in dim light, and adjusted by evolving larger eyes, having retinal ganglion cells with larger receptive fields, and reducing the proportion of P ganglion cells while increasing the proportion of K cells. The fovea was lost in owl monkeys, but not tarsiers.

23.4 Primary Visual Cortex V1 or Area 17

All mammals have a primary visual area, V1, although this area can regress in size and become almost nonfunctional, as in the blind mole rat, where only a narrow sliver of histological area 17 remains (Cooper et al., 1993). Thus,

V1 is not new with primates, although it has been variously differentiated in the branches of the primate radiation, but has been retained from nonprimate ancestors going back to early mammals and even early amniotes (eg, Kaas, 2017). Here we use the comparative method to reconstruct the likely anatomical and functional organization of V1 in early primates; and describe how this V1 differed from V1 of the nonprimate ancestors who gave rise to tree shrews, rodents, lagomorphs, and primates.

In all mammals, V1 represents the contralateral half of the visual field, with more input coming from the contralateral eye which mediates monocular peripheral vision, and less input from the ipsilateral eye to form the binocular field. The early ancestors of the Euarchontoglire superorder had laterally placed eyes that produced a large 300 degrees or so of visual field, but only a narrow 60% or so of binocular field. These mammals were likely nocturnal, without a major emphasis on vision. This started to change with the evolution of the ancestors of early primates as the eyes became more frontal; and the binocular field increased. Early primates already had a large binocular field approaching 180 degrees, and only small 10–20 degrees monocular crescents, judging from the fossil evidence for eye position (Ravosa and Savakova, 2004), and the visual fields of nocturnal prosimian primates. In prosimian galagos, in which the visual system is not likely to be much different from those of early primates, the visual hemifield is close to 90 degrees along the horizontal meridian, with perhaps 80 degrees of the representation of the contralateral hemifield being binocular (Rosa et al., 1997a). The distinctiveness of cortical layers in area 17 of galagos is less pronounced than in simian primates, including nocturnal owl monkeys (Casagrande and Kaas, 1994; Wong and Kaas, 2010; Balaram and Kaas, 2014; Balaram et al., 2011). As in other primates and other mammals, the representation of the contralateral visual hemifield in V1 of galagos is retinotopic, with more of V1 representing central vision where retinal ganglion cells are concentrated (De Bruyn et al., 1993), than for other parts of the retina. However, galagos devote only 20% of V1 to the central 10% of the visual field, while this value is 50–60% in diurnal simians and 30% in nocturnal owl monkeys (Rosa et al., 1997). In galagos, more of the outer boundary of V1 represents the monocular visual field, and less of the outer boundary represents the vertical meridian through gaze than in simian primates. As we would expect in early, nocturnal primates, V1 in galagos is histologically less differentiated, smaller relative to the rest of the brain, and less devoted to central vision than V1 in anthropoid primates, while being closer in these ways to V1 of non-primate ancestors of primates.

Our analysis of the laminar organization of V1 in primates is complicated by the common use of Brodmann's designations for layers in primates (Brodmann, 1909).

We believe Brodmann became confused by a specialization of a sublayer of layer 3 in anthropoid primates, as part of the middle of layer 3, layer 3B, has densely packed small neurons, as does layer 4. Thus, this sublayer of 3 became layer 4A of Brodmann, with a deeper part of layer 3 becoming layer 4B. Only layer 4C of Brodmann (1909) is homologous to layer 4 of other mammals, including tree shrews and rodents. This error has been recognized by Hässler (1967) and others (for reviews see Balaram and Kaas, 2014; Henry, 1991; Casagrande and Kaas, 1994). Yet the well-established labels for layers of V1 of Brodmann continue to dominate the field, and are used even by those who recognize the error. As we need to make valid comparisons of laminar organization of V1 in both primates and nonprimates, we use the laminar terms of Hässler (1967) here.

Histologically, V1 of galagos looks like V1 of nonprimate mammals that have only a moderately developed visual system. Thus, in Nissl preparations, only two layers are densely packed with neurons, layers 4 and 6. The layer 4 neurons are small but not as small as the granular cells of simian primates, the dust-like koniocells of layer 4 (Wong and Kaas, 2010). Galagos and other strepsirrhine primates lack the specialization of having a granular sublayer of layer 3, a feature emphasized by Hässler (1967) in his comparative study of primate visual systems. The vesicular glutamate transporter 2 (VGLUT2) is densely expressed in layer 4 of galagos, mouse lemurs, and other primates (Balaram and Kaas, 2014), as this transporter reflects the location of synapses of geniculate axon terminations in cortex. This concentration of terminations is an identifying feature of layer 4 and is present in layer 4 of V1 in tree shrews, squirrels and other rodents, and other mammals (Wong and Kaas, 2010; Balaram and Kaas, 2014; Balaram et al., 2014). Less dense expressions of VGLUT2 occur in layers 6 and 3B, reflecting less dense inputs from the LGN. Reactions for the metabolic enzyme, cytochrome oxidase (CO), label these layers and sublayers in a similar manner, as the geniculate terminations and synapses are locations of high metabolic activity. As in other cortical areas, layer 4 of galagos also can be identified by a low expression of VGLUT1 mRNA (Balaram et al., 2011). Thus, layer 4 can be clearly marked with several procedures, and the Brodmann scheme for lamination in V1 adds considerable confusion and misdirection in comparative studies.

The way that inputs from the LGN terminate in layer 4 of area 17 varies across mammalian taxa. In all primates that have been studied, inputs from the parvocellular (P) geniculate layers terminate in the inner half of layer 4 (4B) while those from the magnocellular (M) layers terminate in the outer half of layer 4 (4A) (see Casagrande and Kaas, 1994 for review). This input pattern differs from those of most mammals, including rodents and tree shrews (Kretz et al., 1986; Heimel et al., 2005) where the X (P) and Y (M) pathways are mixed in LGN layers projecting to layer 4. Tree shrews differ from most mammals in having separate LGN layers for light-on center receptive fields or light-off center receptive fields, with light-on inputs terminating in the upper half of layer 4 (4a) and light-off inputs in the lower half (4b) (see Takahata and Kaas, 2017 for review). The isolation of M and P cell inputs in upper and lower halves of layer 4 occurs in all primates, and thus likely was present in early primates. How this sublaminar, layer 4 pattern emerged is uncertain, but it likely depended on the segregation of M and P pathways in the LGN.

Mammals also vary in the ways the inputs relayed from the ipsilateral eye segregate from inputs relayed from the contralateral eye. Many of the anthropoid primates, macaque monkeys, for example, have highly segregated inputs to layer 4 from each eye that form band-like ocular dominance "columns" or bands that alternate for right and left eyes, and branch to form more bands (eg, Florence and Kaas, 1992). However, the alternating ocular dominance bands are weak and nearly absent in some anthropoids, such as nocturnal owl monkeys, where they can be demonstrated by more sensitive methods of activity-dependent gene expression as a dot and surround pattern, or a variable banding pattern (Takahata et al., 2014). Among strepsirrhine primates, only galagos have been studied for ocular dominance columns, where there is evidence for only a weak pattern of alternation (see Casagrande and Kaas, 1994). Perhaps, more sensitive methods could reveal more. In tree shrews, there is no evidence for ocular dominance banding, but there is some sublaminar isolation of contralateral and ipsilateral eye activations (Takahata and Kaas, 2017). In rodents, there is little evidence for ocular dominance segregation in layer 4, but a strip of cortex in the binocular field of V1 in rats has been shown to be mainly activated by the contralateral eye, while having patches of mainly ipsilateral eye activation (Laing et al., 2015). Thus, patterns of activation via each eye in the binocular part of V1 vary in the extent of the isolation of LGN inputs for each eye in layer 4, but some form of isolation is common. The pattern of results suggests that a weak pattern of isolation existed in the small brains of early primates, and this pattern was more like that in galagos than in rodents and tree shrews. The potential for ocular dominance columns of various types to emerge in evolution seems high, so the potential was there in early primates, and similar patterns of expression emerged independently in some New World monkeys with larger brains, and in all Old World anthropoid primates (Kaas, 2012).

Another widespread feature of V1 is the presence of neurons with receptive fields that are sensitive to stimulus orientation. Thus, a bar in the receptive field more effectively activates the cortical neuron if the bar has a particular orientation, and bars that are progressively different from that orientation are progressively less effective stimuli. The receptive field location in the visual field, of course, reflects the location of the activated neurons in the retina, and locations of the neurons in the cortical map of the retina. Thus, in many mammals, the neurons with different orientation selectivity are randomly placed in V1 according to their contribution to the retinotopy of V1. However, all primates differ from this organization by grouping V1 neurons with similar orientation selectivity and similar retinotopic receptive fields to form local hypercolumns of neurons that are spatially organized to form orderly progressions of orientation preferences around a hypercolumn center, thus forming "pinwheels" of orientation-selective neurons relative to the cortical surface. Within a pinwheel, all neurons have receptive fields devoted to roughly the same restricted part of visual field space. The pinwheels repeat themselves many times across the extent of V1 to represent the complete visual hemifield (see Kaas, 2012 for review). Tree shrews, the closest studied relative of primates, also have this orderly organization of orientation-selective neurons into pinwheels of orientation-selective columns (Fitzpatrick, 1996), but rodents do not (Van Hooser et al., 2005; Van Hooser, 2007). These results suggest that the common ancestor of all extant primates also had pinwheels of orientation-selective neurons in V1. Perhaps the common ancestor of primates and tree shrews did as well, but independent evolution of this trait is a possibility. However, it is likely that such an organization of orientation-selective neurons in V1 evolved in archontans after the split from glires (rodents and lagomorphs).

Another distinctive feature of V1 in all primates is the presence of cytochrome oxidase "blobs" in layer 3. These "blobs" (this is the most commonly used term) are scattered throughout V1 of all primates (see Casagrande and Kaas, 1994; Preuss and Kaas, 1996 for review). These blobs appear to be centered in ocular dominance columns in those anthropoid primates that have band-like ocular dominance columns. The blobs reflect foci of dense terminations of geniculate K cell axons, which also terminate more diffusely in layer 1. Such blobs are not found in tree shrews, rodents, or lagomorphs, or in most other mammals, and thus they are a primate innovation that likely emerged in early primates or their immediate ancestors. In addition, a similar blob-like pattern of CO-dense patches in layer 3 of V1 may have evolved independently in carnivores (Boyd and Matsubara, 1994; however, see Rockland, 1985). The functional roles of the blobs are not clear, but the K cell terminations in blobs carry information from the short S or "blue" cones of the retina (Shostak et al., 2002). Other specializations of layer 3 occurred in V1 of humans and apes and are not yet well understood (Preuss et al., 1999).

A further specialization of V1 in primates is the almost total lack of callosal connection in some species. The general impression that many investigators have is that V1 is completely devoid of callosal connection in all primates, but this is far from the case (Innocenti, 1986). Early studies of callosal connections were in macaque monkeys and they showed restrictions of callosal connection to the region of the V1/V2 border with terminations and projecting neurons restricted almost completely to V2 rather than V1 (Zeki, 1970; Pandya et al., 1971; Van Essen et al., 1982; Kennedy et al., 1986). Similarly, the callosal connections in squirrel monkeys are "virtually confined to area 18 and avoid area 17" (Gould et al., 1987). However, in owl monkeys and marmosets, callosal projections and projecting neurons extend 1 mm or more into V1, and in strepsirrhine galagos the callosal connections extend even further into V1 for 2—3 mm, most densely in locations with cytochrome oxidase blobs (Cusick et al., 1984). In tree shrews, callosal connections are concentrated along both sides of the V1/V2 border, while extending well toward the center of V1 as they gradually become less dense (Cusick et al., 1985). In rodents, including squirrels (Gould, 1984), and rats (Cusick and Lund, 1981; Olavarria and Van Sluyters, 1983), callosal connections extend past the V1/V2 border into much of lateral V1. Across nonprimate mammals, the general observation is that callosal connections extend into V1 (area 17) while gradually decreasing in density. As neurons within V1 can have widespread interactions with other neurons within that V1, callosal connections can contribute to these interactions for neurons near the V1 border. For uncertain reasons, some anthropoid primates, but not all, and not strepsirrhine primates, have eliminated this form of neural interactions in V1 as another primate innovation.

In summary, V1 of primates has several innovative features that likely emerged in early primate ancestors or somewhat earlier. These include the separation of upper and lower parts of layer 4 to represent M or P ganglion cell inputs, respectively, the blobs in layer 3 with K cell inputs, the pinwheels of orientation selective neurons, and at least a weak form of ocular dominance segregation. Within the primate radiation, cortical circuits for trichromatic vision emerged independently in some New World monkeys and Old World anthropoids. Pronounced ocular dominance bands emerged independently in some New World monkeys and Old World anthropoids. Some diurnal anthropoid primates lost all V1 callosal connections away from the V1/V2 border. A greater representation of ventral stream and fovea vision emerged in anthropoid primates. These V1 specializations for detailed vision were correlated with an almost total lack of callosal connections within V1.

23.5 The Second Visual Area, V2, and Prostriata

The two visual areas that border V1 in primates and other mammals are V2 and prostriata. V2 borders the portion of V1 that represents the zero vertical meridian through the center of gaze. Prostriata occupies the much smaller one-tenth of the border that represents the monocular periphery of the contralateral visual field. V2 is also called area 18, although area "18" of Brodmann (1909) variably relates to V2, and does not even closely correspond to V2, as presently defined in Old World monkeys, apes, or humans. However, Brodmann's area 18 in marmosets is reasonably close to the now known extent of V2.

V2 in primates is a second-order representation of the contralateral visual hemifield (Allman and Kaas, 1974a,b). This means that the representation is split along most of the representation of the zero horizontal meridians such that the outer boundary of V2 is formed by the representation of the horizontal meridian, while the inner boundary with V1 is retinotopically matched along a joint representation of the vertical meridian. Similar microelectrode maps of V2 have been produced for a number of primate species.

Current use of the term "area 18" is for the architectonic field that is coextensive with the physiologically defined V2. In anthropoid primates, V2 is histologically characterized by a repeating sequence of four band-like modular subdivisions, the cytochrome oxidase (CO) dense "thick" and "thin" stripes that cross the width of V2, separated by two functionally different CO pale stripes (Tootell et al., 1983). Because the stripes differ in architecture, as the CO dense stripes also are more myelinated, it can be difficult to precisely determine the rostral border V2 in traditional coronal or sagittal cut brain sections, but the border can be very clear from the stripe pattern in sections from flattened cortex cut parallel to the surface. In macaque monkeys, for example, V2 in flattened cortex has 28 sets of the four stripes (Olavarria and Van Essen, 1997). This number varies across primate species, as about half that number of sets exists in the smaller V2 of marmosets (Rosa et al., 1997). Thus, V2 can be defined by its retinotopic organization, and even more precisely by appropriate architecture. Additionally, connection patterns with V1 can be used to identify V2. The thick stripes are further distinguished by being activated by the M cell pathway relayed from V1. The thick stripes project to the middle temporal visual area, MT. Thin stripes and interstripes get different inputs from V1, and project to the dorsolateral visual area, DL (Allman and Kaas, 1974a), commonly known as V4 (Shipp and Zeki, 1985; DeYoe and Van Essen, 1985). Neurons in each of the four types of stripes differ in how they respond to stimulus color, brightness, orientation, motion, and binocular disparity (see Felleman et al., 2015 for review). While most of the research has been done on macaque monkeys, the stripe or band-like modular organization of V2 has also been found in New World cebus monkeys (Nascimento-Silva et al., 2014), squirrel monkeys (Malach et al., 1994), marmosets (Rosa et al., 1997), and owl monkeys (Kaskan et al., 2009). Thus, this feature of V2 is likely present in all anthropoid primates. In contrast, this band-like modular feature of V2 does not appear to be well developed in strepsirrhine primates. Neurons in V2 of galagos are selective for stimulus orientation, but there appears to be little or no stripe pattern, either functionally or histologically (Xu et al., 2005; Fan et al., 2012). Thus, it appears to be unlikely that early primates had a V2 with anthropoid-like modules, as they are weakly developed or absent in strepsirrhine primates and do not exist in rodents or tree shrews, or, as far as is known, in other mammals. However, another type of modular V2 of organization does exist in other mammals.

An extensive amount of evidence suggests that nearly all mammals have a V2 (see Rosa and Krubitzer, 1999; Kaas, 2017), although there is some confusion on this for mice and rats. Partly this follows Brodmann's (1909) misidentification of the medial part of V1 as "area 18" in squirrels, and a subsequent retention of this error in rats (eg, Krieg, 1946). Currently, it appears that a modular organization of V2 in mice is being mistaken for a series of visual areas, as each of these visual areas represent only part of the visual hemifield (Garrett et al., 2014). Perhaps V2 is degenerated or has been otherwise modified in those rodents with poor vision and small brains, such as rats or mice, as V2 has been clearly recognized by position, retinotopy, connection pattern with V1, and histological characteristics in more visual rodents, such as squirrels (Kaas et al., 1989; Wong and Kaas, 2008; Gould, 1984; Hall et al., 1971; Sereno et al., 1991; Paolini and Sereno et al., 1998), and the agouti (Dias et al., 2014; Picanco-Diniz et al., 1991), a large diurnal rodent with a well-developed visual system from the Amazon region of South America. Yet, squirrels, and perhaps other less studied diurnal rodents, do have modular subdivisions of V2. In myelin stains of flattened visual cortex, V2 is densely myelinated, but has a series of five or more myelin-poor "holes" with the "holes" receiving V1 projections (Kaas et al., 1989). The holes are relatively devoid of callosal connections, but they are dense in the myelinated surround (Gould, 1984). As for the region of the presumptive V2 of rats and mice, V2 of squirrels consists of a sequence of partial representations of the visual hemifield, from lower to higher parts of the visual hemifield, in a rostrocaudal sequence in "holes" separated by callosal inputs. Likewise in tree shrews, a closer relative of primates, V2 has been well defined by microelectrode mapping, connections with V1, and architecture (Kaas et al., 1972; Sesma et al., 1984; Lyon et al., 1998; Jain et al., 1994; Wong and Kaas, 2009). Again, the V2 organization appears to be modular, as injection of tracers at restricted sites in V1 label several patches of terminations in V2, with rostral V1 injections labeling more rostral V2 patches, and caudal injections labeling more caudal patches, and callosal termination label the "surrounds" of seven or more callosal-free "holes" along the length of V2 (Cusick et al., 1985). Likely the V1 projections are focused on the holes. The functional roles of V2 modules with extensive callosal input or little callosal input in rodents and tree shrews is yet unknown, but the callosal connection could provide the receptive fields of neurons in the callosal regions with a widespread surrounds, while neurons in the holes would have more restricted surrounds. The long intrinsic connections in V2 of monkeys are thought to contribute to the receptive field structure (Levitt et al., 1994), and one of the functions of callosal connections is to act as intrinsic connections between hemifield representations in the two hemispheres. The longer intrinsic connections, and by implication, the callosal connections, would likely contribute to the wide, extra-classical receptive field surround (Angelucci et al., 2006). The ancestors of primates likely had the callosal and noncallosal modules in V2, and there are suggestions of that pattern in the callosal connections of V2 in monkeys, as band-like extensions of callosal connections extend from the V1 border into V2 at regular intervals (Cusick et al., 1984; Gould et al., 1987). Possibly, this type of modular organization based on callosal connections was replaced by a more complex organization of four types of modules in anthropoid primates, but not fully in early primates, now represented by strepsirrhine galagos.

Prostriata is a small visual area located along the border of V1 that represents the periphery of the contralateral monocular visual hemifield (Allman and Kaas, 1971a; Rosa et al., 1997; Xu et al., 2012). Neurons in prostriata respond to visual stimuli, although not as reliably as neurons in V2 or V1, and their large receptive fields do not correspond retinotopically to those in adjacent parts of V1 and V2. Architectonically, prostriata has less pronounced lamination and is less myelinated than V1 and V2. The connections of prostriata have not been well studied, but they are thought to be widespread with visual, auditory, and cingulate areas (see Xu et al., 2012). In macaques projections to cingulate motor cortex have been demonstrated (Morecraft et al., 2000). Prostriata appears to exist in other mammals, where it is sometimes called the splenial or limbic visual area (see Rosa et al., 1997 for review). The evidence suggests that prostriata is a very old visual area that was present in early mammals.

23.6 The Third Visual Area, V3

A third visual area, V3, has long been a proposed part of the primate visual system. The early evidence was based on patterns of connections of extrastriate cortex with V1 in macaque monkeys, which revealed evidence for both V2 and V3 (Cragg, 1969; Zeki, 1969). Subsequently, more extensive and convincing evidence for V2 and V3 in macaques came from microelectrode mapping studies (Gattass et al., 1988). The V3 concept was possibly favored because Brodmann (1909) had described two ringlike architectonic areas around most of V1 (area 17), area 18, and area 19. Thus, the idea of a ringlike third area, much like V2, was attractive. Additionally, Hubel and Wiesel (1965) had already provided evidence for two visual areas, V2 and V3, successively lateral to V1 (area 17) in cats, and these two areas reasonably conformed to the locations of areas 18 and 19 in carnivores as defined by Brodmann (1909). Subsequently, studies provided further support for areas V2 and V3 in cats (eg, Tusa et al., 1979). Thus, results from primates that were possibly open to other interpretations were likely to be related to the V3 and area 19 model. Zeki (1969), for example, used the terms area 18 and area 19 for the two targets of V1 projections he later called V2 and V3, although the proposed locations of V2 and V3 were well within the cortex that Brodmann (1909) defined as area 18 in Old World monkeys. The recognition of this lack of a correspondence of V3 and V2 with Brodmann areas 18 and 19 likely accounts for a general reluctance by researchers to refer to area V3 as area 19, although V2 is often termed as area 18 (a better correspondence of at least area 18 with V2 exists for Brodmann's (1909) depiction of area 18 in a New World marmoset).

Despite continuing support for the V3 model of visual cortex organization in monkeys and other primates, there have been serious concerns about the existence of V3 for several reasons. An early concern was over the common belief that only the territory of dorsal V3, representing the lower visual quadrant, receives projections from V1, while the territory of ventral V3, representing the upper visual quadrant, does not. This conclusion reflects the fact that nearly all evidence for V1 connections was based on injections of tracers or lesions in the more accessible dorsal V1. Thus, ventral V3 was redefined as another visual area, the ventroposterior (VP) area, while dorsal V3, with well-established V1 connections, remained V3d. Overall, many summaries of visual cortex organization in primates contained two "improbable visual areas" (Kaas, 1993), with VP representing only the upper visual quadrant and V3 or V3d representing only the lower visual quadrant. Although such an organization would seem difficult to explain in terms of visual function, other supporting evidence accumulated, including the claim that VP and V3d have neurons that differ substantially in how they respond to visual stimuli. However, Lyon and Kaas demonstrated that the VP region does have connections with V1, just as dorsal V3 (V3d) does, completing the overall pattern of V1 connections expected for V3. The studies included New World marmoset monkeys (Lyon and Kaas, 2001), strepsirrhine (prosimian) galagos (Lyon and Kaas, 2002a), macaque monkeys (Lyon and Kaas, 2002b), and three other species of New World monkeys, owl, squirrel, and titi monkeys (Lyon and Kaas, 2002c; see Sousa et al., 1991 for Cebus monkeys). Given the evidence across primate species, it should be clear now that VP does not differ from V3d in V1 connections, and the evidence for other proposed differences should be reevaluated. For example, proposed differences between VP and V3d in neuron response properties could be due to the use of different procedures and contaminations of results by recordings from adjacent visual areas.

The other concern about the V3 concept is ongoing. While there is considerable evidence from microelectrode mapping in macaque monkeys (Gattass et al., 1988) and optical imaging maps in owl monkeys (Lyon et al., 2002a) for the retinotopic organization expected for V3d, consistent with the evidence from connection patterns, other evidence suggests that V3d is not complete, and may have one or more breaks or disjunctions that are occupied by part of another visual area (see Angelucci and Rosa, 2015 for an extensive review). Most notably, early studies of visual cortex organization in owl monkeys presented evidence that parts of dorsal V2 are bordered by parts of the representation of the upper visual quadrant, and this part or parts could not be V3d, which represents only the lower visual quadrant. More specifically, parts of three proposed visual areas were thought to border dorsal V2, leaving little room for V3d: a dorsomedial visual area, DM (Allman and Kaas, 1975); a medial visual area, M (Allman and Kaas, 1976); and a less well-described dorso-intermediate area, DI (Allman and Kaas, 1975). These results and subsequent findings led to recent proposals that rejected the V3 model and proposed that the lateral part of V3d is an extension of VP, renamed VLP (see Angelucci and Rosa, 2015). This created another "improbable" visual area, as dorsal VLP would represent only a central part of the lower visual quadrant. Thus, the proposed VLP is an incomplete area.

Another model that seems to be compatible with much of the data is that dorsal V3 exists, and it includes all of the lower visual quadrant in V3d as expected, but it has a gap in which another visual area protrudes to separate medial and lateral parts of V3d with a partial representation of the upper visual quadrant (Kaas et al., 2015). Allman and Kaas (1975) proposed that

DM borders V2, and others have come to this same conclusion, but this does not seem to be the case. If we recognize DM as a visual area that receives direct input from V1, and projects to MT and other "dorsal stream" targets, then projections from V1 can be used to localize DM relative to V2 and the proposed territory of V3d. In the studies of V1 connections in six species of primates by Lyon and Kaas (see above), injections in all locations in V1 labeled retinotopically expected locations in V2, V3, DM, and MT, providing evidence that all these areas are complete representations, or nearly so. All of the labeled sites that were judged to be in DM, whether from dorsal V1 (lower quadrant) or ventral V1 (upper quadrant) labeled expected retinotopic locations in DM that were rostral to V3d. Others may define DM in other ways, but clearly, the visual area in dorsal extrastriate cortex that receives a retinotopic pattern of projection from V1 is along the rostral border of V3, and it does not intrude into V3. V1 has sparse connections with other cortical areas, but the strongest and most consistent targets are V2, V3, DM, and MT. Other areas may intrude into V3d, DI. for example, but this remains uncertain. More medially, area M appears to border V2, suggesting that V3d is not quite as long as V2d, and this remains a possibility. As much of recent results on the areas along V2d have been in marmosets, one suggestion is that marmosets, as very small monkeys that evolved from larger ancestors (Leutenegger, 1980), have an altered organization of parts of extrastriate cortex, and the proposed break or breaks in V3d do not occur in monkeys with larger brains (Gattass et al., 2015). However, Angelucci and Rosa (2015) suggest instead that the VLP model holds for all primates. These differences in opinion and interpretation have not been resolved. The model of V3 with a medial area M and DM rostral to V3 is favored here (Fig. 23.1), with the added possibility that DI or some other visual area protrudes into V3d and places part of a representation of upper visual quadrant along V2, in at least some primates. The existence of a V3 with dorsal and ventral halves, and thereby forming a complete representation of the contralateral visual hemifield is well supported, and new names for V3 seem unnecessary and confusing. How or if V3 varies across primate taxa is not yet understood, but retinotopic maps of V3 in humans are consistent with the concept of a V3 with representations of both the dorsal and the ventral visual quadrants (eg, Wandell and Winawer, 2011; Arcaro and Kastner, 2015).

A remaining question is if V3 evolved in early primates or earlier in nonprimate ancestors. At first it would seem that V3 must be an area common to many mammals, as V3 was defined in cats, and it likely exists in other carnivores. But this idea does not seem to be the case. Most notably, a V3 does not appear to exist in tree shrews, the closest studied relative of primates, or in

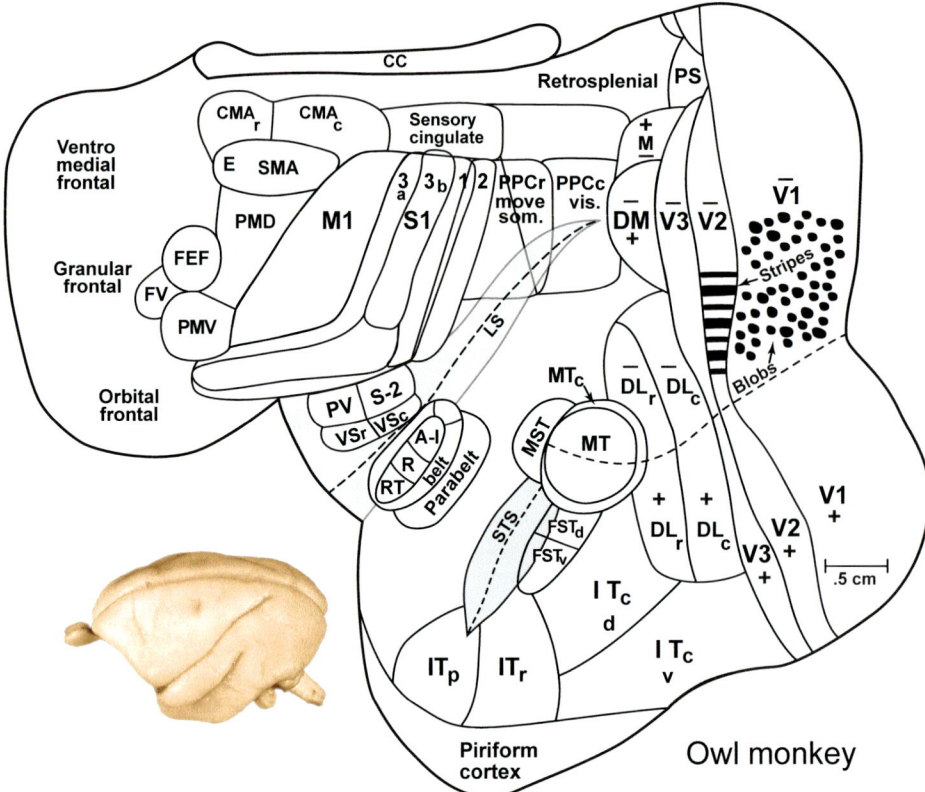

FIGURE 23.1 Visual areas and other areas of neocortex of an owl monkey. To view cortex hidden in fissures or on the medial wall or ventral surface, cortex has been separated from the rest of the brain and flattened. The cortical areas are outlined. The intact brain is shown on the lower left. The visual areas discussed in the text include the primary area V1 or area 17, which has modular subdivisions of blobs and nonblob regions throughout V1. V1 represents the contralateral visual hemifield with the lower visual quadrant (−) represented dorsal to that of the lower visual quadrant (+). The − and + code is also used for other visual areas. The dashed line through V1 and other visual areas marks the representation of the zero horizontal meridian through the center of gaze. To flatten cortex V1 was partially split along the representation of the horizontal meridian. V1 is bordered by the second visual area, V2, which has a band-like modular organization of thick and thin myelin dense stripes separated by myelin-light interstripes (shown for only part of V2). The part of V1 representing monocular peripheral vision is bordered by a small visual area, prostriata (PS). Most of V2 is bordered by the third visual area, V3, which also has a band-like modular organization (not shown). The thin central part of V3 is bordered by DLc (caudal part of the dorsolateral area). DLr is the rostral part. The dorsal part of V3 is bordered by the dorsomedial visual area, DM, and the medial visual area, M. Some proposals also include a dorsointermediate area, DI, between DM and DL, and suggest that DM, M, and sometimes DI directly border dorsal V2 (see text). The MT complex is just rostral to DL. The middle temporal visual area, MT, is nearly surrounded by a narrow visual area, the MT crescent (MTc) while the most rostral part of MT is adjoined by the medial superior temporal area, MST. Ventral to MT and MTc, the fundal area of the superior temporal sulcus, FST, named by the location in macaques, has dorsal FSTd, and ventral FSTv, subdivisions. Other visual areas have been proposed, including dorsal and ventral subdivisions of caudal inferior temporal cortex, ITc, and rostral and polar divisions of IT cortex. These and other proposed visual areas of temporal and posterior parietal cortex are not discussed here. Other subdivisions of cortex are indicated for reference. Posterior parietal cortex has been divided into a caudal portion, PPCc, dominated by visual inputs (vis), and a movement related rostral portion, PPCr, with more somatosensory inputs. A medial portion, PPCm has other functions. Anterior parietal cortex has the four fields of Brodmann, areas 3a, 3b, 1, and 2, with area 3b corresponding to S1 of other mammals. The parietal ventral, PV, second somatosensory, S2, and rostral and caudal divisions of the ventral somatosensory area, VSr and VSc, are indicated. Motor areas include the primary motor area, M1, dorsal and ventral premotor areas, PMD and PMV, the frontal eye field, FEF, for eye movements, and the frontal visual area, FV, for visual fixation. Rostral and caudal cingulate motor areas, CMAr and CMAc are shown. Auditory areas include three primary fields, A1, a rostral area, R, and a rostrotemporal area, RT, as well as a belt and parabelt that have subdivisions (not shown). *CC*, corpus callosum; *STS*, superior temporal sulcus; *LS*, lateral sulcus.

rodents, as somewhat more distant relatives. In tree shrews (Lyon et al., 1998) and gray squirrels (Kaas et al., 1989), patterns of cortical connections support the view that several visual areas border V2, not a single area with the retinotopy of V3. Likewise, no clear evidence has been obtained for a V3 in the large South American rodent, the agouti, although one of two proposed areas along the outer border of V2 was called

(unfortunately) V3 (Elston et al., 2006; Dias et al., 2014). Possibly the two areas, AL and "V3" together could form a V3 with representations of the lower quadrant in AL and upper quadrant in "V3." A very narrow V3 along the outer border of V2 has been proposed for ground squirrels (see Paolini and Sereno, 1998), but no evidence has been presented, and this proposed V3 is so narrow that confirming connectional or microelectrode mapping evidence would be difficult to obtain. Overall, present evidence suggests that V3 evolved in early primates or their immediate ancestors. Yet, these are reasons to be cautious about this conclusion, given the history of interpretations of visual cortex organization in New World monkeys that omitted a V3, or proposed a very short V3. As cats and other carnivores are members of the Laurasiatheria superorder, and the nonprimate members of the Euarchontoglire superorder do not appear to have a V3, carnivores likely have evolved a V3-like area independently.

23.7 The Dorsomedial Visual Area, DM

The dorsomedial visual area, DM, was first described by Allman and Kaas (1975) in owl monkeys as a complete representation of the contralateral visual hemifield at the caudal tip of the extended lateral sulcus, and extending from dorsomedial cortex onto the cortex of the medial wall of the hemisphere along the dorsal border of V2. DM was soon found by degeneration methods to project to the middle temporal visual area, MT, and to posterior parietal cortex (Wagor et al., 1975), but connections with V1 were not revealed until later with more sensitive tracers (Krubitzer and Kaas, 1993). An early microelectrode mapping study in prosimian galagos also placed DM along the rostral border of V2 (Allman et al., 1979). On the basis of earlier mapping data, published only recently, Sereno et al. (2015) again located DM along the dorsal border of V2 in owl monkeys. The issue of a complete V3 with a ventral half, V3v, occupying the territory of VP of some proposals, and a dorsal half, V3d, with or without a "hole" is addressed in Section 23.6. On the basis of V1 connections, we provided evidence for a retinotopically complete DM rostral to V3d (Kaas et al., 2015). Others continue to place part of DM, but not all, along the V2 border in marmosets as a model for all primates (eg, Jeffs et al., 2015; Angelucci and Rosa, 2015). The DM of these authors includes most or all the DM of Fig. 23.1, but also parts of cortex included here in V3, and possibly the hypothesized V3 "hole," which could be occupied by part of DI, an area that has not been consistently included in proposals of visual cortex organization (see Sereno et al., 2015).

In macaques, other areas have been commonly portrayed in the expected location of DM, and the proposed territories of these areas could be part or all of DM. These areas include V3A (Van Essen and Zeki, 1978), PO (Colby et al., 1988) and V6 (Zeki, 1986). These areas have been redefined and subdivided in various ways (eg, V6, V6 Av, and V6 Ad; V7; Pitzalis et al., 2013). Additional visual areas adjoin V3A in posterior parietal cortex (eg, Arcaro et al., 2011). There have been few attempts to compare these areas with those in New World monkeys. More efforts are needed to determine how New and Old World monkeys are similar and different; and more comparisons with strepsirrhine (prosimian) primates are needed. V6 and PO are in the expected location of DM, and PO (Colby et al., 1988) and V6 (Galletti et al., 2001) get inputs from V1, but the proposed retinotopies appear to differ. Yet, DM, PO, and V6 are likely different names for the same area, but with poorly defined boundaries that may include parts of other areas, such as area M (Allman and Kaas, 1976).

In summary, there have been several versions of DM, as parts of other visual areas have probably been included in DM. We define DM as an area along the outer boundary of V3d that is one of the 4 visual areas with significant inputs from V1 (V2, V3, DM, MT). This DM has both upper and lower visual quadrant representations, and does not intrude into V3. As V3d may not extend as far as the terminal border of V2 with prostriata, an area M medial to DM may border V2 (Allman and Kaas, 1976). V3 may have a gap in some individuals or species, or in all primates, where part of the upper visual field representation intrudes toward the border of V2. Possibly this area is the poorly studied DI. As DM appears to exist in all primates, DM likely evolved with early primates. Yet DM could have a homolog with an area along the outer border of V2 of other mammals. Less is known about the existence of M, and other proposed areas, other than in the primate species where they have been studied.

23.8 Area DL or V4

The dorsolateral visual area was identified by microelectrode mapping of retinotopy in owl monkeys as a region of cortex between V2 and MT that represented the lower visual quadrant dorsally and the upper visual quadrant ventrally (Allman and Kaas, 1974a). The area has subsequently been modified by recognizing that the narrow strips of DM bordering MT on ventral and dorsal sides belong to another visual area, MTc (see Section 23.9). There have also been efforts to use connection patterns with V2 to determine the dorsal and ventral extents of DL more precisely (Stepniewska et al., 2005; Gattass et al., 1997). In addition, DL may have rostral and caudal divisions, DLr and DLc, or have a modular organization that suggests these divisions. DL is largely coextensive with area V4, as identified by V2 and V3 projections in macaques (Zeki, 1971). Gattass et al. (1988) provided the first extensive retinotopic map of V4 in macaque monkeys. A similar detailed map of retinotopy in V4 was later provided for New World cebus monkeys (Pinon et al., 1998). Connections with V2 became one of the defining features of V4 in macaque monkeys (Gattass et al.,

1997). Zeki (1971) initially divided V4 into two areas, V4 and a somewhat more rostral V4a, but this convention has not been continued. However, Cusick and Kaas (1988) subsequently divided DL into rostral and caudal halves, DLr and DLc. The DLc was dominated by connections with V2 (and also V3, see Kaas et al., 2015), while DLr had major connections with DM and MT (also see Beck and Kaas, 1999). Thus, Cusick and Kaas (1988) considered DLr to be part of the dorsal stream of visual processing for action and DLc part of the ventral stream for object identification (Ungerleider and Haxby, 1994; Goodale and Milner, 1992). Different subcortical connections of DLr and DLc with the pulvinar complex provided further evidence for this interpretation (Kaas and Lyon et al., 2007). In macaques, Baizer et al. (1991) also provided evidence that a part of V4 has connections that relate to the dorsal stream. Sereno and Allman (1991) and Sereno et al. (2015) have divided the original territory of DL differently into posterior (DLp), intermediate (DLi), and anterior (DLa) areas, with much of DLa likely corresponding to MTc (see below) and parts of DLp and their VP corresponding to much of V3, but possibly also part of DLc of Cusick and Kaas (1988). DLr of Cusick and Kaas also included the caudal part of MTc. Thus, evidence from mapping and connection patterns for rostral and caudal divisions of DL could have been contaminated by involvements with V3 or MTc, so the question of divisions of DL/V4 remains open. Another possibility rests on the evidence for domains or large modules in V4 with quite different functions. One class of domains contains neurons sensitive to color while the other class of domains has orientation-selective neurons (Tanigawa et al., 2010). As domains were as large as 500 μm in width, they could be confused with rostral and caudal divisions of V4. Probably the safe conclusion is that the DL or V4 territory is not structurally or functionally homogeneous. For now, it is uncertain if DL/V4 is divided into two or more areas, or large modules, and if this differs across primate taxa. Other differences in DL/V4 are expected between nocturnal and diurnal primates, and in primates with or without trichromatic color vision. While there are cortical areas lateral to V2 in tree shrews and squirrels that could be homologous to DL/V4, clear criteria for identifying them as homologs have not been specified.

23.9 The MT Complex (MT, MTc, MST, FSTd, and FSTv)

The middle temporal (MT) visual area was first described in owl monkeys as a first-order retinotopic representation of the contralateral visual hemifield by Allman and Kaas (1971). The representation was coextensive with a densely myelinated region of cortex in upper temporal lobe. Surprisingly, this feature of MT was not recognized in earlier architectonic studies. Neurons in MT responded to both flashes of a spot of light in the receptive field, but more strongly to a bar of light moving through the receptive field. Area 17 (V1) was already known to project to the region of MT in squirrel monkeys (Spatz et al., 1970) and in macaque monkeys (Kuypers et al., 1965; Cragg, 1969; Zeki, 1969). Zeki (1971) originally referred to the MT region as the projection cortex of the posterior bank of the superior temporal sulcus, and then the motion area (Zeki, 1980), but subsequently introduced the term V5, while noting a lack of wavelength-selective neurons (Zeki, 1983). However, Zeki (1980) used the term MT in studying neuron response properties of MT in owl monkeys, while arguing that MT in owl monkeys is not homologous to V5 in macaque monkeys. In contrast, other investigators have consistently recognized MT and V5 as different terms for as the same area across primate taxa. Microelectrode recordings have now demonstrated the similarities of MT in retinotopy, location relative to other visual areas, and inputs from V1 and V2 in prosimian galagos (eg, Allman et al., 1973; Symonds and Kaas, 1978), marmosets (eg, Rosa and Elston, 1998), cebus monkeys (eg, Fiorani et al., 1989; Rosa et al., 1993), squirrel monkeys (Krubitzer and Kaas, 1990), and macaque monkeys (Maunsell and Van Essen, 1987; Weller and Kaas, 1983; Gattass and Gross, 1981). In addition, MT has been identified by architecture in a number of other primates (eg, Preuss et al., 1993), and by fMRI (eg, Huk et al., 2002; Tootell et al., 1995) and myelination (Tootell and Taylor, 1995; Annese et al., 2005) in humans. Thus, there should be no question about the existence and homology of MT across all extant primates. The question of the origin and evolution of MT is, however, unsettled, as no region of cortex in nonprimate mammals has enough of the features of MT be identified as MT with confidence.

Although there is no obvious cortical area in temporal cortex of the tree shrews and rodents that could suggest that MT emerged from an area in nonprimate ancestors, a possibility is suggested by the finding in marmosets that part of the visual pulvinar gets a direct projection from the retina, and that this projection is most dense soon after birth and then decreases substantially as the marmosets mature (Warner et al., 2012). In addition, this part of the visual pulvinar projects to MT. Normally, MT is highly dependent on inputs from V1 for activation (eg, Kaas and Krubitzer, 1992; Collins et al., 2003). Furthermore, MT and other members of the MT complex receive inputs from nuclei of the inferior pulvinar, and the only nucleus projecting densely to MT is the medial nucleus of the inferior pulvinar, IPm, the nucleus without inputs from the superior colliculus (Kaas and Lyon, 2007). This suggests that MT emerged out of the temporal cortex that receives visual information from the superior colliculus that is relayed to temporal cortex via the visual pulvinar in most mammals (see Lyon et al., 2003; Baldwin and Bourne, 3.10, The Evolution of Subcortical Pathways to the Extrastriate Cortex, this volume). Possibly, direct retinal inputs to part of the pulvinar in the immediate ancestors of early primates displaced the superior colliculus inputs from part of its territory in the pulvinar, providing a subdivision of temporal cortex with a direct retina-to-pulvinar-to-cortex pathway of activation. This direct visual pathway may have caused primary-like histological features of MT to evolve. Somehow, this emerging MT acquired direct inputs from V1, so that V1 started to outcompete the retina-to-pulvinar-to-cortex pathways in early primates. Thus, early MT became dependent on V1 inputs, while cortex around MT remained dependent on a superior colliculus to cortex pathway. According to this model, MT is a transformed part of the temporal cortex that is responsive to superior colliculus-to-pulvinar-to-cortex activation in other mammals, but less so or not at all in primates. The other members of the MT complex with inputs relayed via the pulvinar from the superior colliculus also have been derived from the projection zone of the inferior pulvinar in temporal cortex. A temporal cortex projection zone from the superior colliculus via the pulvinar has been retained in tree shrews, rodents, and many other mammals (Lyon et al., 2003), and it appears to be a very old pathway going back to early amniotes (Major et al., 2003).

The other members of the MT complex (or cluster) include the medial superior temporal area, MST; dorsal and ventral subdivision, FSTd, and FST of the fundal area of the superior temporal sulcus, and the MT crescent, MTc. MST was first defined in monkeys by Maunsell and Van Essen (1983), and FST by Desimone and Ungerleider (1986) in macaque monkeys. The territory of FST was divided into a dorsal half with connections with MT and a ventral half with connections with MTc in owl monkeys by Kaas and Morel (1993). They also defined MTc architectonically as a narrow band around most of MT consisting of a series of bead-like patches that stain densely for myelin and cytochrome oxidase. Only these patches receive dense inputs from FSTv. The territory of MTc was included within the dorsolateral visual area, DL, by Allman and Kaas (1974a). This part of DL was later distinguished as DLa (Sereno et al., 2015). MTc or DLa represents the contralateral visual hemifield with the lower quadrant laterally, with congruent or matched retinotopic borders with MT (Allman and Kaas, 1974a; Rosa and Elston, 1998; Sereno et al., 2015). A part of MTc that represents the lower visual quadrant was termed V4t by Desimone and Ungerleider (1986) as cortex "transitional" between V4 and MT. MT connections have provided evidence for MST

in owl monkeys (Weller et al., 1984) and in other New World monkeys (Krubitzer and Kaas, 1990; Rosa et al., 1993). The patchy cytochrome oxidase staining of MTc has been demonstrated in cortex around MT in green monkeys, macaque monkeys, and humans (Tootell et al., 2015). Sereno et al. (2015) have provided evidence in owl monkeys that FSTd, FSTv, and MST have separate representations of the contralateral visual hemifield, Less is known about prosimian primates, but MT connections include the locations expected for MST and FSTd, and cytochrome oxidase stains provided clear evidence for a CO dark MST that is distinct from the CO dark MT, and for the CO dark patches of MTc (Kaskan and Kaas, 2007). There was also evidence for MTc connections with FSTv. In humans, these are visual areas around MT (see Orban, 4.14, Evolution of Human Visual System Specializations), but the relationship of these areas to MST, FSTv, and MTc is not clear.

In summary, the evidence suggests that MT, MTc, MST, FSTd, and FSTv are visual areas shared by most or all primates. They likely emerged out of the zone of temporal cortex that is commonly activated via the superior colliculus-pulvinar path in other mammals, but homologous areas in nonprimate mammals have not been identified.

23.10 Epilogue

There are other regions of cortex that are predominantly involved in processing visual stimuli that are less well understood in terms of comparative anatomy and functional organization, including the greatly expanded temporal lobe of primates, which contain several functional subdivisions, and has areas or domains with neurons responsive to objects or faces as part of the ventral stream of visual processing. In addition, especially areas MT, MST, and DM, but also V3 and V2, provide information to a greatly expanded posterior parietal cortex, where all or most primates have a series of action-specific domains with outputs to frontal motor areas (see Kaas et al., 3.15, Evolution of Parietal-Frontal Networks in Primates, this volume). The areas of these expanded regions of cortex are likely to be primate innovations in that they likely emerged with early primates from smaller, less differentiated regions of cortex, and they likely were modified further in various lines of primate evolution. Our conclusions about the evolution of visual cortex in primates are constrained by the sparseness of the comparative data, but also by the varying interpretations of these data that sometimes seem conflicting. Nevertheless, primates clearly have a large number of visual areas that are shared, and some further differentiation of visual areas may have occurred in the evolution of Old World monkeys, apes, and humans, with some parallel changes in the organization of visual cortex of the larger New World monkeys. The focus here has been on similarities in cortical organization across primate taxa, as evidence for similarities can often be more convincing than evidence for differences. Yet, our history of using microelectrode mapping methods, optical and fMRI imaging, connection patterns, and a full range of histological procedures has not been that long, and tremendous progress has been made. Thus, we can be optimistic about future progress.

References

Allman, J.M., Kaas, J.H., 1971. Representation of the visual field in striate and adjoining cortex of the owl monkey (*Aotus trivirgatus*). Brain Res. 35, 89–106.

Allman, J.M., Kaas, J.H., 1971. A representation of the visual field in the caudal third of the middle tempral gyrus of the owl monkey (*Aotus trivirgatus*). Brain Res. 31, 85–105.

Allman, J.M., Kaas, J.H., 1974. A crescent-shaped cortical visual area surrounding the middle temporal area (MT) in the owl monkey (*Aotus trivirgatus*). Brain Res. 81, 199–213.

Allman, J.M., Kaas, J.H., 1974. The organization of the second visual area (V II) in the owl monkey: a second order transformation of the visual hemifield. Brain Res. 76, 247–265.

Allman, J.M., Kaas, J.H., 1975. The dorsomedial cortical visual area: a third tier area in the occipital lobe of the owl monkey (*Aotus trivirgatus*). Brain Res. 100, 473–487.

Allman, J.M., Kaas, J.H., 1976. Representation of the visual field on the medial wall of the occipital-parietal cortex in the owl monkey. Science 191, 572–575.

Allman, J.M., Kaas, J.H., Lane, R.H., 1973. The middle temporal visual area (MT)in the bushbaby, *Galago senegalensis*. Brain Res. 57, 197–202.

Allman, J., Campbell, C.B., McGuinness, E., 1979. The dorsal third tier area in *Galago senegalensis*. Brain Res. 179, 355–361.

Angelucci, A., Bressloff, P.C., 2006. Contribution of feedforward, lateral and feedback connections to the classical receptive field center and extra-classical receptive field surround of primate V1 neurons. Prog. Brain Res. 154, 93–120.

Angelucci, A., Rosa, M.G., 2015. Resolving the organization of the third tier visual cortex in primates: a hypothesis-based approach. Vis. Neurosci. 32, E010.

Annese, J., Gazzaniga, M.S., Toga, A.W., 2005. Localization of the human cortical visual area MT based on computer aided histological analysis. Cereb. Cortex 15, 1044–1053.

Arcaro, M.J., Kastner, S., 2015. Topographic organization of areas V3 and V4 and its relation to supra-areal organization of the primate visual system. Vis. Neurosci. 32, E014.

Arcaro, M.J., Pinsk, M.A., Li, X., Kastner, S., 2011. Visuotopic organization of macaque posterior parietal cortex: a functional magnetic resonance imaging study. J. Neurosci. 31, 2064–2078.

Baizer, J.S., Ungerleider, L.G., Desimone, R., 1991. Organization of visual inputs to the inferior temporal and posterior parietal cortex in macaques. J. Neurosci. 11, 168–190.

Balaram, P., Hackett, T.A., Kaas, J.H., 2011. VGLUT1 mRNA and protein expression in the visual system of prosimian galagos (*Otolemur garnetti*). Eye Brain 2011, 81–98.

Balaram, P., Kaas, J.H., 2014. Towards a unified scheme of cortical lamination for primary visual cortex across primates: insights from NeuN and VGLUT2 immunoreactivity. Front. Neuroanat. 8, 81.

Balaram, P., Young, N.A., Kaas, J.H., 2014. Histological features of layers and sublayers in cortical visual areas V1 and V2 of chimpanzees, macaque monkeys, and humans. Eye Brain 2014, 5–18.

Beck, P.D., Kaas, J.H., 1999. Cortical connections of the dorsomedial visual area in old world macaque monkeys. J. Comp. Neurol. 406, 487–502.

Bininda-Emonds, O.R.P., Hortman, A.M., 2016. Mammalian Evolution: The Phylogenetics story. Chapter 031 of this series.

Bloch, J.I., Boyer, D.M., 2002. Grasping primate origins. Science 298, 1606–1610.

Brodmann, K., 1909. Vergleichende Lokalisationslehre der Grosshirnrinde in ihren Prinzipien dargestellt auf Grund des Zellenbaues. Barth, Leipzig.

Boyd, J., Matsubara, J., 1994. Tangential organization of callosal connectivity in the cat's visual cortex. J. Comp. Neurol. 347, 197–210.

Casagrande, V.A., Kaas, J.H., 1994. The afferent, intrinsic, and efferent connections of primary visual cortex in primates. In: Peters, A., Rockland, K.S. (Eds.), Cerebral Cortex, vol. 10. Plenum Press, New York, pp. 201–259.

Colby, C.L., Gattass, R., Olson, C.R., Gross, C.G., 1988. Topographical organization of cortical afferents to extrastriate visual area PO in the macaque: a dual tracer study. J. Comp. Neurol. 269, 392–413.

Collins, C.E., Lyon, D.C., Kaas, J.H., 2003. Responses of neurons in the middle temporal visual area after long-standing lesions of the primary visual cortex in adult new world monkeys. J. Neurosci. 23, 2251–2264.

Collins, C.E., Hendrickson, A., Kaas, J.H., 2005. Overview of the visual system of Tarsius. Anat. Rec. A Discov. Mol. Cell Evol. Biol. 287, 1013–1025.

Cooper, H.M., Herbin, M., Nevo, E., 1993. Visual system of a naturally microphthalmic mammal: the blind mole rat. *Spalax ehrenbergi*. J. Comp. Neurol. 328, 313–350.

Cragg, B.G., 1969. The topography of the afferent projections in the circumstriate visual cortex of the monkey studied by the Nauta method. Vis. Res. 9, 733–747.

Cusick, C.G., Gould, H.J., Kaas, J.H., 1984. Interhemispheric connections of visual cortex of owl monkeys (*Aotus trivirgatus*), marmosets (*Callithrix jacchus*), and galagos (*Galago crassicaudatus*). J. Comp. Neurol. 230, 311–336.

Cusick, C.G., Kaas, J.H., 1988. Cortical connections of area 18 and dorsolateral visual cortex in squirrel monkeys. Vis. Neurosci. 1, 211–237.

Cusick, C.G., Lund, R.D., 1981. The distribution of the callosal projection to the occipital visual cortex in rats and mice. Brain Res. 214, 239–259.

Cusick, C.G., MacAvoy, M.G., Kaas, J.H., 1985. Interhemispheric connections of cortical sensory areas in tree shrews. J. Comp. Neurol. 235, 111–128.

DeBruyn, E.J., Casagrande, V.A., Beck, P.D., Bonds, A.B., 1993. Visual resolution and sensitivity of single cells in the primary visual cortex (V1) of a nocturnal primate (bush baby): correlations with cortical layers and cytochrome oxidase patterns. J. Neurophysiol. 69, 3–18.

Desimone, R., Ungerleider, L.G., 1986. Multiple visual areas in the caudal superior temporal sulcus of the macaque. J. Comp. Neurol. 248, 164–189.

DeYoe, E.A., Van Essen, D.C., 1985. Segregation of efferent connections and receptive field properties in visual area V2 of the macaque. Nature 317, 58–61.

Dias, L., Silveira, L.C., Franca, J.G., Houzel, J.C., Lent, R., Mayer, A.O., Santiago, L.F., Silveira, L.C., Picanco-Diniz, C.W., Pereira, A., 2014. Topography and architecture of visual and somatosensory areas of the agouti. J. Comp. Neurol. 522, 2576–2593.

Dyer, M.A., Martins, R., daSilva Filho, M., Muniz, M., Silveira, L.C., Cepko, C.L., Finlay, B.L., 2009. Developmental sources of conversation and variation in the evolution of the primate eye. Proc. Natl. Acad. Sci. U.S.A. 106, 8963–8968.

Elston, G.N., Elston, A., Aurelio-Freire, M., Gomes Leal, W., Dias, I.A., Pereira Jr., A., Silveira, L.C., Picanco Diniz, C.W., 2006. Specialization of pyramidal cell structure in the visual areas V1, V2 and V3 of the South American rodent, *Dasyprocta primnolopha*. Brain Res. 1106, 99–110.

Fan, R.H., Baldwin, M.K., Jermakowicz, W.J., Casagrande, V.A., Kaas, J.H., Roe, A.W., 2012. Intrinsic signal optical imaging evidence for dorsal V3 in the prosimian galago (*Otolemur garnettii*). J. Comp. Neurol. 520, 4254–4274.

Felleman, D.J., Lim, H., Xiao, Y., Wang, Y., Eriksson, A., Parajuli, A., 2015. The representation of orientation in macaque V2: four stripes not three. Cereb. Cortex 25, 2354–2369.

Fiorani Jr., M., Gattass, R., Rosa, M.G., Sousa, A.P., 1989. Visual area MT in the Cebus monkey: location, visuotopic organization, and variability. J. Comp. Neurol. 287, 98–118.

Fitzpatrick, D., 1996. The functional organization of local circuits in visual cortex: insights from the study of tree shrew striate cortex. Cereb. Cortex 6, 329–341.

Florence, S.L., Kaas, J.H., 1992. Ocular dominance columns in area 17 of Old World macaque and talapoin monkeys: complete reconstructions and quantitative analyses. Vis. Neurosci. 8, 449–462.

Galletti, C., Gamberini, M., Kutz, D.F., Fattori, P., Luppino, G., Matelli, M., 2001. The cortical connections of area V6: an occipito-parietal network processing visual information. Eur. J. Neurosci. 13, 1572–1588.

Garrett, M.E., Nauhaus, I., Marshel, J.H., Callaway, E.M., 2014. Topography and areal organization of mouse visual cortex. J. Neurosci. 34, 12587–12600.

Gattass, R., Gross, C.G., 1981. Visual topography of striate projection zone (MT) in posterior superior temporal sulcus of the macaque. J. Neurophysiol. 46, 621–638.

Gattass, R., Sousa, A.P., Gross, C.G., 1988. Visuotopic organization and extent of V3 and V4 of the macaque. J. Neurosci. 8, 1831–1845.

Gattass, R., Sousa, A.P., Mishkin, M., Ungerleider, L.G., 1997. Cortical projections of area V2 in the macaque. Cereb. Cortex 7, 110–129.

Gattass, R., Lima, B., Soares, J.G., Ungerleider, L.G., 2015. Controversies about the visual areas located at the anterior border of area V2 in primates. Vis. Neurosci. 32, E019.

Goodale, M.A., Milner, A.D., 1992. Separate visual pathways for perception and action. Trends Neurosci. 15, 20–25.

Gould 3rd, H.J., 1984. Interhemispheric connections of the visual cortex in the grey squirrel (*Sciurus carolinensis*). J. Comp. Neurol. 223, 259–301.

Gould 3rd, H.J., Weber, J.T., Rieck, R.W., 1987. Interhemispheric connections in the visual cortex of the squirrel monkey (*Saimiri sciureus*). J. Comp. Neurol. 256, 14–28.

Hall, W.C., Kaas, J.H., Killackey, H., Diamond, I.T., 1971. Cortical visual areas in the grey squirrel (*Sciurus carolinesis*): a correlation between cortical evoked potential maps and architectonic subdivisions. J. Neurophysiol. 34, 437–452.

Hässler, R., 1967. Comparative anatomy of the central visual systems in day-and nigh-active primates. In: Hässler, R., Stevens, S. (Eds.), Evolution of the Forebrain. Stuttgart Thieme, pp. 419–434.

Hennig, W., 1966. Phylogenetic Systematics. Univ. Ill. Press, Urbana, USA.

Heimel, J.A., Van Hooser, S.D., Nelson, S.B., 2005. Laminar organization of response properties in primary visual cortex of the gray squirrel (*Sciurus carolinensis*). J. Neurophysiol. 94, 3538–3554.

Hendrickson, A., Djajadi, H.R., Nakamura, L., Possin, D.E., Sajuthi, D., 2000. Nocturnal tarsier retina has both short and long/medium-wavelength cones in an unusual topography. J. Comp. Neurol. 424, 718–730.

Henry, G.H., 1991. Afferent inputs, receptive field properties and morphological cell types in different laminae of the striate cortex. In: Leventhal, A.G. (Ed.), Vision and Visual Dysfunction, vol. 4. Macmillan press, London, pp. 223–245.

Hubel, D.H., Wiesel, T.N., 1965. Receptive fields and functional architecture in two nonstriate visual areas (18 and 19) of the cat. J. Neurophysiol. 28, 229–289.

Huk, A.C., Dougherty, R.F., Heeger, D.J., 2002. Retinotopy and functional subdivision of human areas MT and MST. J. Neurosci. 22, 7195–7205.

Innocenti, G., 1986. General organization of callosal connections i the cerebral cortex. In: Jones, E.G., Peters, A. (Eds.), Cerebral Cortex, vol. 5. Plenum Press, New York, pp. 291–353.

Jacobs, G.H., 1996. Primate photopigments and primate color vision. Proc. Natl. Acad. Sci. U.S.A. 93, 577–581.

Jain, N., Preuss, T.M., Kaas, J.H., 1994. Subdivisions of the visual system labeled with Cat-301 antibody in tree shrews. Vis. Neurosci. 11, 731–741.

Jeffs, J., Federer, F., Angelucci, A., 2015. Corticocortical connection patterns reveal two distinct visual cortical areas bordering dorsal V2 in marmoset monkey. Vis. Neurosci. 32, E012.

Kaas, J.H., Huerta, M.F., Weber, J.T., Harting, J.K., 1978. Patterns of retinal terminations and laminar organization of the lateral geniculate nucleus of primates. J. Comp. Neurol. 182, 517–553.

Kaas, J.H., 1993. Evolution of multiple areas and modules within neocortex. Perspect. Dev. Neurobiol. 1, 101–107.

Kaas, J., 2002. Convergences in the modular and areal organization of the forebrain of mammals: implications for the reconstruction of forebrain evolution. Brain Behav. Evol. 59, 262–272.

Kaas, J.H., 2014. The evolution of the visual system in primates. In: Werner, J.S., Chalupa, L.M. (Eds.), The New Visual Neurosciences. MIT Press, Cambridge, Mass, pp. 1233–1246.

Kaas, J.H., 2016. Approaches to the study of brain evolution. In: Shepherd, S.V. (Ed.), Handbook of Evolutionary Neuroscience. John Wiley and Sons, Chichester, UK.

Kaas, J.H., 2012. Evolution of columns, modules, and domains in the neocortex of primates. Proc. Natl. Acad. Sci. U.S.A. 109 (Suppl. 1), 10655–10660.

Kaas, J.H., 2017. The organization of neocortex in early mammals. In: Herculano-Houzel, S. (Ed.), Evolution of Nervous Systems, second ed., vol. 2. Elsevier, London.

Kaas, J.H., Krubitzer, L.A., 1992. Area 17 lesions deactivate area MT in owl monkeys. Vis. Neurosci. 9, 399–407.

Kaas, J.H., Guillery, R.W., Allman, J.M., 1972. Some principles of organization in the dorsal lateral geniculate nucleus. Brain Behav. Evol. 6, 253–299.

Kaas, J.H., Hall, W.C., Killackey, H., Diamond, I.T., 1972. Visual cortex of the tree shrew (*Tupaia glis*): architectonic subdivisions and representations of the visual field. Brain Res. 42, 491–496.

Kaas, J.H., Harting, J.K., Guillery, R.W., 1974. Representation of the complete retina in the contralateral superior colliculus of some mammals. Brain Res. 65, 343–346.

Kaas, J.H., Krubitzer, L.A., Johanson, K.L., 1989. Cortical connections of areas 17 (VI) and 18 (VII) of squirrels. J. Comp. Neurol. 281, 426–446.

Kaas, J.H., Lyon, D.C., 2007. Pulvinar contributions to the dorsal and ventral streams of visual processing in primates. Brain Res. Rev. 55, 285–296.

Kaas, J.H., Lyon, D.C., 2001. Visual cortex organization in primates: theories of V3 and adjoining visual areas. Prog. Brain Res. 134, 285–295.

Kaas, J.H., Morel, A., 1993. Connections of visual areas of the upper temporal lobe of owl monkeys: the MT crescent and dorsal and ventral subdivisions of FST. J. Neurosci. 13, 534–546.

Kaas, J.H., Roe, A.W., Baldwin, M.K., Lyon, D.C., 2015. Resolving the organization of the territory of the third visual area: a new proposal. Vis. Neurosci. 32, E016.

Kaskan, P.M., Kaas, J.H., 2007. Cortical connections of the middle temporal and middle temporal crescent visual areas in prosimian galagos (*Otolemur garnetti*). Anat. Rec. 290, 349–366.

Kaskan, P.M., Dillenburger, B.C., Lu, H.D., Kaas, J.H., Roe, A.W., 2009. The organization of orientation-selective, luminance-change and binocular-preference domains in the second (V2) and third (V3) visual areas of New World of owl monkeys as revealed by intrinsic-signal optical imaging. Cereb. Cortex 19, 1394–1407.

Kay, R.F., 2015. Anthropology. New World monkey origins. Science 347, 1068–1069.

Kennedy, H., Dehay, C., Bullier, J., 1986. Organization of the callosal connections of visual areas V1 and V2 in the macaque monkey. J. Comp. Neurol. 247, 398–415.

Kretz, R., Rager, G., Norton, T.T., 1986. Laminar organization of ON and OFF regions and ocular dominance in the striate cortex of the tree shrew (*Tupaia belangeri*). J. Comp. Neurol. 251, 135–145.

Krieg, W.J., 1946. Connections of the cerebral cortex; the albino rat; topography of the cortical areas. J. Comp. Neurol. 84, 221–275.

Krubitzer, L.A., Kaas, J.H., 1990. Cortical connections of MT in four species of primates: areal, modular, and retinotopic patterns. Vis. Neurosci. 5, 165–204.

Krubitzer, L.A., Kaas, J.H., 1993. The dorsomedial visual area of owl monkeys: connection, myeloarchitecture, and homologies in other primates. J. Comp. Neurol. 334, 497–528.

Kuypers, H.G., Szwarcbart, M.K., Mishkin, M., Rosvold, H.E., 1965. Occipitotemporal corticocortical connections in the rhesus monkey. Exp. Neurol. 11, 245–262.

Laing, R.J., Turecek, J., Takahata, T., Olavarria, J.F., 2015. Identification of eye-specific domains and their relation to callosal connections in primary visual cortex of long Evans rats. Cereb. Cortex 25, 3314–3329.

Lane, R.H., Allman, J.M., Kaas, J.H., Miezin, F.M., 1973. The visuotopic organization of the superior colliculus of the owl monkey (*Aotus trivirgatus*) and the bushbaby (*Galago senegalensis*). Brain Res. 60, 335–349.

Lane, A.H., Kaas, J.H., Allman, J.M., 1974. Visuotopic organization of the superior colliculus in normal and Siamese cats. Brain Res. 70, 413–430.

Leutenegger, W., 1980. Monogamy in Callitrichide: a consequence of phyletic dwarfism? Int. J. Primatol. 1, 95–98.

Levitt, J.B., Yoshioka, T., Lund, J.S., 1994. Intrinsic cortical connections in macaque visual area V2: evidence for interaction between different functional streams. J. Comp. Neurol. 342, 551–570.

Lyon, D.C., Jain, N., Kaas, J.H., 1998. Cortical connections of striate and extrastriate visual areas in tree shrews. J. Comp. Neurol. 401, 109–128.

Lyon, D.C., Kaas, J.H., 2001. Connectional and architectonic evidence for dorsal and ventral V3, and dorsomedial area in marmoset monkeys. J. Neurosci. 21, 249–261.

Lyon, D.C., Jain, N., Kaas, J.H., 2003. The visual pulvinar in tree shrews II. Projections of four nuclei to areas of visual cortex. J. Comp. Neurol. 467, 607–627.

Lyon, D.C., Kaas, J.H., 2002. Evidence for a modified V3 with dorsal and ventral halves in macaque monkeys. Neuron 33, 453–461.

Lyon, D.C., Kaas, J.H., 2002. Evidence from V1 connections for both dorsal and ventral subdivisions of V3 in three species of New World monkeys. J. Comp. Neurol. 449, 281–297.

Lyon, D.C., Kaas, J.H., 2002. Connectional evidence for dorsal and ventral V3, and other extrastriate areas in the prosimian primate, *Galago garnetti*. Brain Behav. Evol. 59 (3), 114–129.

Lyon, D.C., Xu, X., Casagrande, V.A., Stefansic, J.D., Shima, D., Kaas, J.H., 2002. Optical imaging reveals retinotopic organization of dorsal V3 in New World owl monkeys. Proc. Natl. Acad. Sci. U. S. A 99, 15735–15742.

Major, D.E., Rodman, H.R., Libedinsky, C., Kartin, H.J., 2003. Pattern of retinal projections in the California ground squirrel (*Spermophilus beecheyi*): anterograde tracing study using cholera toxin. J. Comp. Neurol. 403, 317–334.

Malach, R., Tootell, R.B., Malonek, D., 1994. Relationship between orientation domains, cytochrome oxidase stripes, and intrinsic horizontal connections in squirrel monkey area V2. Cereb. Cortex 4, 151–165.

Martin, R.D., 1990. Primate Origin and Evolution. Princeton University Press, Princeton, New Jersey.

Maunsell, J.H., Van Essen, D.C., 1983. The connections of the middle temporal visual area (MT) and their relationship to a cortical hierarchy in the macaque monkey. J. Neurosci. 3, 2563–2586.

Maunsell, J.H., Van Essen, D.C., 1987. Topographic organization of the middle temporal visual area in the macaque monkey: representational biases and the relationship to callosal connections and myeloarchitectonic boundaries. J. Comp. Neurol. 266, 535–555.

Melmoth, D.R., Grant, S., 2006. Advantages of binocular vision for the control of reaching and grasping. Exp. Brain Res. 171, 371–388.

Milton, K., 2006. Diet and primate evolution. Sci. Am. 16, 22–29.

Morecraft, R.J., Rockland, K.S., Van Hoesen, G.W., 2000. Localization of area prostriata and its projection to the cingulate motor cortex in the rhesus monkey. Cereb. Cortex 10, 192–203.

Mota, B., Herculano-Houzel, S., 2015. Cortical folding scales universally with surface area and thickness, not number of neurons. Science 349, 74–77.

Murphy, W.J., Pevzner, P.A., O'Brien, S.J., 2004. Mammalian phylogenomics comes of age. Trends Genet. 20, 631–639.

Nascimento-Silva, S., Pinon, C., Soares, J.G., Gattass, R., 2014. Feedforward and feedback connections and their relation to the cytox modules of V2 in Cebus monkeys. J. Comp. Neurol. 522, 3091–3105.

Olavarria, J., Van Sluyters, R.C., 1983. Widespread callosal connections in infragranular visual cortex of the rat. Brain Res. 279, 233–237.

Olavarria, J.F., Van Essen, D.C., 1997. The global pattern of cytochrome oxidase stripes in visual area V2 of the macaque monkey. Cereb. Cortex 7, 395–404.

O'Leary, M.A., Bloch, J.I., Flynn, J.J., Gaudin, T.J., Giallombardo, A., et al., 2013. The placental mammal ancestor and the post-K-Pg radiation of placentals. Science 339, 662–667.

Pandya, D.N., Karol, E.A., Heilbronn, D., 1971. The topographical distribution of interhemispheric projections in the corpus callosum of the rhesus monkey. Brain Res. 32, 31–43.

Paolini, M., Serano, M.I., 1998. Direction selectivity in the middle lateral and lateral (ML and L) visual areas in the California ground squirrel. Cereb. Cortex 8, 362–371.

Perry, V.H., Cowey, A., 1981. The morphological correlates of X- and Y-like retinal ganglion cells in the retina of monkeys. Exp. Brain Res. 43, 226–228.

Picanco-Diniz, C.W., Silveira, L.C., de Carvalho, M.S., Oswaldo-Cruz, E., 1991. Contralateral visual field representation in area 17 of the cerebral cortex of the agouti: a comparison between the cortical magnification factor and retinal ganglion cell distribution. Neuroscience 44, 325–333.

Pinon, M.C., Gattass, R., Sousa, A.P., 1998. Area V4 in Cebus monkey: extent and visuotopic organization. Cereb. Cortex 8, 685–701.

Pitzalis, S., Sereno, M.I., Committeri, G., Fattori, P., Galati, G., Tosoni, A., Galletti, C., 2013. The human homologue of macaque area V6A. Neuroimage 82, 517–530.

Polyak, S.T., 1957. The Vertebrate Visual System. University of Chicago Press, Chicago.

Preuss, T.M., Beck, P.D., Kaas, J.H., 1993. Areal, modular, and connectional organization of visual cortex in a prosimian primate, the slow loris (*Nycticebus coucang*). Brain Behav. Evol. 42, 321–335.

Preuss, T., Goldman-Rakic, P.S., 1991. Myelo- and cytoarchitecture of the granular frontal cortex and surrounding regions in the strepsirrhine primate Galago and the anthropoid primate Macaca. J. Comp. Neurol. 310, 429–474.

Preuss, T.M., Kaas, J.H., 1996. Cytochrome oxidase 'blobs' and other characteristics of primary visual cortex in a lemuroid primate, *Cheirogaleus medius*. Brain Behav. Evol. 47, 103–112.

Preuss, T.M., Qi, H., Kaas, J.H., 1999. Distinctive compartmental organization of human primary visual cortex. Proc. Natl. Acad. Sci. U.S.A. 96, 11601–11606.

Ravosa, M.J., Savakova, D.G., 2004. Euprimate origins: the eyes have it. J. Hum. Evol. 46, 357–364.

Rockland, K.S., 1985. Anatomical organization of primary visual cortex (area 17) in the ferret. J. Comp. Neurol. 241, 225–236.

Rosa, M.G., Elston, G.N., 1998. Visuotopic organization and neuronal response selectivity for direction of motion in visual areas of the caudal temporal lobe of the marmoset monkey (*Callithrix jacchus*): middle temporal area, middle temporal crescent, and surrounding cortex. J. Comp. Neurol. 393, 505–527.

Rosa, M.G., Soares, J.G., Fiorani Jr., M., Gattass, R., 1993. Cortical afferents of visual area MT in the Cebus monkey: possible homologies between New and Old World monkeys. Vis. Neurosci. 10, 827–855.

Rosa, M.G., Pettigrew, J.D., Cooper, H.M., 1996. Unusual pattern of retinogeniculate projections in the controversial primate Tarsius. Brain Behav. Evol. 48, 121–129.

Rosa, M.G., Casagrande, V.A., Preuss, T., Kaas, J.H., 1997. Visual field representation in striate and prestriate cortices of a prosimian primate (*Galago garnetti*). J. Neurophysiol. 77, 3193–3217.

Rosa, M.G., Fritsches, K.A., Elston, G.N., 1997. The second visual area in the marmoset monkey: visuotopic organization, magnification factors, architectonical boundaries, and modularity. J. Comp. Neurol. 387, 547–567.

Rosa, M.G., Krubitzer, L.A., 1999. The evolution of visual cortex: where is V2? Trends Neurosci. 22, 242–248.

Schrago, C.G., 2007. On the time scale of New World primate diversification. Am. J. Phys. Anthropol. 132, 344–354.

Sereno, M.I., Rodman, H.R., Karten, H.J., 1991. Organization of visual cortex in the California ground squirrel. Soc. Neurosci. Abstr. 17, 844.

Sereno, M.I., McDonald, C.T., Allman, J.M., 2015. Retinotopic organization of extrastriate cortex in the owl monkey—dorsal and lateral areas. Vis. Neurosci. 32, E021.

Sesma, M.A., Casagrande, V.A., Kaas, J.H., 1984. Cortical connections of area 17 in tree shrews. J. Comp. Neurol. 230, 337–351.

Sherman, S.M., Wilson, J.R., Kaas, J.H., Webb, S.V., 1976. X- and Y-cells in the dorsal lateral geniculate nucleus of the owl monkey (*Aotus trivirgatus*). Science 192, 475–477.

Shipp, S., Zeki, S., 1985. Segregation of pathways leading from area V2 to areas V4 and V5 of macaque monkey visual cortex. Nature 315, 322–325.

Shostak, Y., Ding, Y., Mavity-Hudson, J., Casagrande, V.A., 2002. Cortical synaptic arrangements of the third visual pathway in three primate species: *Macaca mulatta*, *Saimiri sciureus*, and *Aotus trivirgatus*. J. Neurosci. 22, 2885–2893.

Silcox, M.T., Benham, A.E., Bloch, J.I., 2010. Endocasts of Microsyops (Microsyopidae, Primates) and the evolution of the brain in primitive primates. J. Hum. Evol. 58, 505–521.

Silveira, L.C., Perry, V.H., Yamada, E.S., 1993. The retinal ganglion cell distribution and the representation of the visual field in area 17 of the owl monkey. *Aotus trivirgatus*. Vis. Neurosci. 10, 887–897.

Sousa, A.P., Pinon, M.C., Gattass, R., Rosa, M.G., 1991. Topographic organization of cortical input to striate cortex in the Cebus monkey: a fluorescent tracer study. J. Comp. Neurol. 308, 665–682.

Spatz, W.B., Tigges, J., Tigges, M., 1970. Subcortical projections, cortical associations, and some intrinsic interlaminar connections of the striate cortex in the squirrel monkey (Saimiri). J. Comp. Neurol. 140, 155–174.

Stepniewska, I., Collins, C.E., Kaas, J.H., 2005. Reappraisal of DL/V4 boundaries based on connectivity patterns of dorsolateral visual cortex in macaques. Cereb. Cortex 15, 809–822.

Stewart, C.B., Disotell, T.R., 1998. Primate evolution - in and out of Africa. Curr. Biol. 8, R582–R588.

Steiper, M.E., Seiffert, E.R., 2012. Evidence for a convergent slowdown in primate molecular rates and its implications for the timing of early primate evolution. Proc. Natl. Acad. Sci. U. S. A 109, 6006–6011.

Symonds, L.L., Kaas, J.H., 1978. Connections of striate cortex in the prosimian, *Galago senegalensis*. J. Comp. Neurol. 181, 477–512.

Takahata, T., Kaas, J.H., 2017. c-FOS expression in the visual system of tree shrews after monocular inactivation. J. Comp. Neurol. 525, 151–165. https://doi.org/10.1002/cne.24053.

Takahata, T., Miyashita, M., Tanaka, S., Kaas, J.H., 2014. Identification of ocular dominance domains in New World owl monkeys by immediate-early gene expression. Proc. Natl. Acad. Sci. U.S.A. 111, 4297–4302.

Tanigawa, H., Lu, H.D., Roe, A.W., 2010. Functional organization for color and orientation in macaque V4. Nat. Neurosci. 13, 1542–1548.

Tetreault, N., Hakeem, A., Allman, J.M., 2004. The distribution and size of retinal ganglion cells in *Micvocebus murinus, Cheirogaleus medius*, and *Tavsius syrichta*: implications for the evolution of sensory systems in primates. In: Ross, C.F., Kay, R.F. (Eds.), Anthropoid Origins: New Visions. Plenum Publisher, New York, pp. 463–475.

Tootell, R.B., Reppas, J.B., Kwong, K.K., Malach, R., Born, R.T., Brady, T.J., Rosen, B.R., Belliveau, J.W., 1995. Functional analysis of human MT and related visual cortical areas using magnetic resonance imaging. J. Neurosci. 15, 3215–3230.

Tootell, R.B.H., Silverman, M.S., DeValois, R.L., Jacobs, G.H., 1983. Functional organization of the second cortical visual area in primates. Science 220, 237–739.

Tootell, R.B., Taylor, J.B., 1995. Anatomical evidence for MT and additional cortical visual areas in humans. Cereb. Cortex 5, 39–55.

Tootell, R.B., Echavarria, C., Nasr, S., 2015. A problem of overlap. Vis. Neurosci. 32, E001.

Tusa, R.J., Rosenquist, A.C., Palmer, L.A., 1979. Retinotopic organization of areas 18 and 19 in the cat. J. Comp. Neurol. 185, 657–678.

Tyler, D.E., 1991. The evolutionary relationship of Aotus. Folia Primatol. 56, 50–52.

Ungerleider, L.G., Haxby, J.V., 1994. What" and "where" in the human brain. Curr. Opin. Neurobiol. 4, 157–165.

Van Essen, D.C., Zeki, S.M., 1978. The topographic organization of rhesus monkey prestriate cortex. J. Physiol. 277, 193–226.

Van Essen, D.C., Newsome, W.T., Bixby, J.L., 1982. The pattern of interhemispheric connections and its relationship to extrastriate visual areas in the macaque monkey. J. Neurosci. 2, 265–283.

Van Hooser, S.D., 2007. Similarity and diversity in visual cortex: is there a unifying theory of cortical computation? The Neuroscientist 1, 639–656.

Van Hooser, S.D., Heimel, J.A., Chung, S., Nelson, S.B., Toth, L.J., 2005. Orientation selectivity without orientation maps in visual cortex of a highly visual mammal. J. Neurosci. 25, 19–28.

Van Hooser, S.D., Heimel, J.A., Nelson, S.B., 2003. Receptive field properties and laminar organization of lateral geniculate nucleus in the gray squirrel (*Sciurus carolinensis*). J. Neurophysiol. 90, 3398–3418.

Wagor, E., Lin, C.S., Kaas, J.H., 1975. Some cortical projections of the dorsomedial visual area (DM) of association cortex in the owl monkey, *Aotus trivirgatus*. J. Comp. Neurol. 163, 227–250.

Wandell, B.A., Winawer, J., 2011. Imaging retinotopic maps in the human brain. Vis. Res. 51, 718–737.

Warner, C.E., Kwan, W.C., Bourne, J.A., 2012. The early maturation of visual cortical area MT is dependent on input from the retinorecipient medial portion of the inferior pulvinar. J. Neurosci. 32, 17073–17085.

Webb, S.B., Kaas, J.H., 1976. The sizes and distribution of ganglion cells in the retina of the owl monkey. *Aotus trivirgatus*. Vis. Res. 16, 1247–1254.

Weller, R.E., Kaas, J.H., 1983. Retinotopic patterns of connections of area 17 with visual areas V-II and MT in macaque monkeys. J. Comp. Neurol. 220, 253–279.

Weller, R.E., Kaas, J.H., 1989. Parameters affecting the loss of ganglion cells of the retina following ablations of striate cortex in primates. Vis. Neurosci. 3, 327–349.

Weller, R.E., Wall, J.T., Kaas, J.H., 1984. Cortical connections of the middle temporal visual area (MT) and the superior temporal cortex in owl monkeys. J. Comp. Neurol. 228, 81–104.

Wong, P., Kaas, J.H., 2008. Architectonic subdivisions of neocortex in the gray squirrel (*Sciurus carolinensis*). Anat. Rec. Hob. 291, 1301–1333.

Wong, P., Kaas, J.H., 2009. Architectonic subdivisions of neocortex in the tree shrew (*Tupaia belangeri*). Anat. Rec. Hob. 292, 994–1027.

Wong, P., Kaas, J.H., 2010. Architectonic subdivisions of neocortex in the Galago (*Otolemur garnetti*). Anat. Rec. Hob. 293, 1033–1069.

Wright, P.C., 1994. Night watch on the Amazon. Nat. Hist. 5, 45–50.

Xu, H.H., Chaplin, T.A., Davies, A.J., Verma, R., Rosa, M.G., 2012. A specialized area in limbic cortex for fast analysis of peripheral vision. Curr. Biol. 22, 1351–1357.

Xu, X., Bosking, W.H., White, L.E., Fitzpatrick, D., Casagrande, V.A., 2005. Functional organization of visual cortex in the prosimian bush baby revealed by optical imaging of intrinsic signals. J. Neurophysiol. 94, 2748–2762.

Yamada, E.S., Marshak, D.W., Silveira, L.C., Casagrande, V.A., 1998. Morphology of P and M retinal ganglion cells of the bushbaby. Vis. Res. 38, 3345–3352.

Zeki, S.M., 1969. Representation of central visual fields in prestriate cortex of monkey. Brain Res. 14, 271–291.

Zeki, S.M., 1970. Interhemispheric connections of prestriate cortex in monkey. Brain Res. 19, 63–75.

Zeki, S.M., 1971. Cortical projections from two prestriate areas in the monkey. Brain Res. 34, 19–35.

Zeki, S., 1980. The response properties of cells in the middle temporal area (area MT) of own monkey visual cortex. Proc. R. Soc. Lond. B Biol. Sci. 207, 239–248.

Zeki, S., 1983. The distribution of wavelength and orientation selective cells in different areas of monkey visual cortex. Proc. R. Soc. Lond. B Biol. Sci. 217, 449–470.

Zeki, S., 1986. The anatomy and physiology of area V6 of macaque monkey visual cortex. J. Phys. Lond. 381, 62.

24

The Evolution of Subcortical Pathways to the Extrastriate Cortex

M.K.L. Baldwin, J.A. Bourne

University of California, Davis, Davis, CA, United States; Monash University, Clayton, VIC, Australia

24.1 Introduction

There are multiple routes through which visual information can reach the neocortex from the retina. The most well-studied are the parallel pathways involving projections of magnocellular (M), parvocellular (P), and koniocellular (K) retinal ganglion cells (RGCs) that project to the lateral geniculate nucleus (LGN) of the thalamus, which relays predominantly to primary visual cortex (V1). From V1, visual information is propagated in a hierarchical manner to extrastriate visual areas (visual cortical areas outside of area 17/V1). These pathways constitute the retinogeniculate visual pathway/system, which is present in all studied mammals and has been well documented in the literature (Casagrande and Ichida, 2002; Casagrande et al., 2007; Kahn and Krubitzer, 2002; Van Hooser and Nelson, 2006). There are, however, a few less well-studied pathways that bypass V1 and directly target extrastriate visual cortex. These include pathways that route through the K layers of the LGN or the pulvinar, either by direct retinal input or indirect retinal input by way of the superior colliculus (SC).

Although the relay from the LGN to V1 is considered to provide the necessary input for conscious visual perception (Polonsky et al., 2000; Tong and Engel, 2001; but see Maier et al., 2008; Leopold, 2012), many visual abilities, such as visually guided movements and the ability to discriminate some visual stimuli, are retained in the absence of V1 (mice: Prusky and Douglas, 2014; rats: Mize et al., 1971; squirrels: Levey et al., 1973; Wagor, 1978; tree shrews: Snyder et al., 1969; Ward and Masterton, 1970; Killackey et al., 1971; Casagrande and Diamond, 1974; Ware et al., 1974; galagos: Marcotte and Ward, 1980; macaques and humans: Perenin, 1991; Barbur et al., 1993; Weiskrantz et al., 1995; ffytche et al., 1996; Bridge et al., 2008; Tamietto et al., 2010).

However, retention of these visually mediated behaviors in the absence of V1 is highly variable in different species. This variability could, in part, be due to the nature of the stimuli tested or the required physical response to the visual stimuli (see Schneider, 1969). Alternatively, such variability in visual performance after V1 lesions could reflect fundamental differences in the relative importance of the alternative visual pathways leading to striate or extrastriate cortical areas. Thus, although this chapter focuses on the primate visual system, it is informative to compare the anatomical and functional similarities and differences of these pathways across other extant mammals, in addition to primates, to gain insights into both the evolution of the visual system and how it has been altered to generate adaptive visually mediated behavior in different lineages.

In anthropoid primates, while the majority of RGCs project to the LGN, ~10% of retinal projections terminate in the SC (Perry and Cowey, 1984). In contrast, nearly all RGCs have projections to the SC in many of the mammals that share a close phylogenetic relationship with primates, such as rabbits (Vaney et al., 1981) and rodents (rat: Linden and Perry, 1983; mouse: Hofbauer and Dräger, 1985; hamster: Chalupa and Thompson, 1980), indicating a significant difference in the routing of visual information from the retina to neocortex between primates and their close mammalian relatives.

Certainly, the high percentage of RGC projections to structures outside the geniculostriate pathway suggests that these pathways were critically important to visual processing in early mammals, perhaps most important for their nocturnal lifestyle and the avoidance of large, fast-moving predators. Furthermore, the continued presence of such visual pathways to the extrastriate cortex in extant primates, including humans, suggests that these pathways still serve an important role in visual

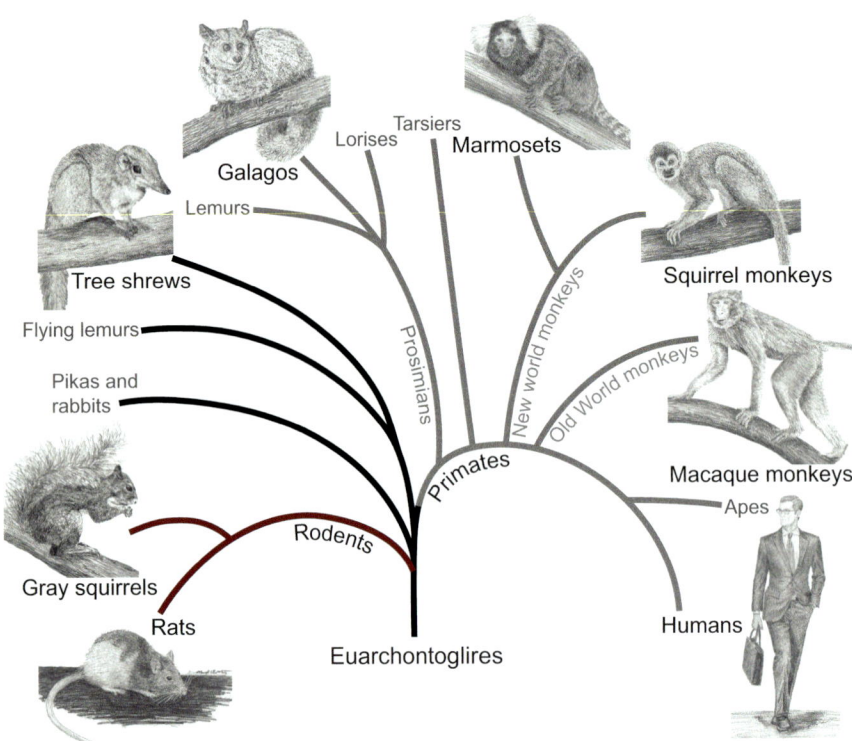

FIGURE 24.1   Phylogenetic tree representing the relationship of extant mammals within the Euarchontoglires clade, which includes rodents, pikas and rabbits, tree shrews, and primates, including humans.

function; however, the function of these alternative pathways and their contribution to visual processing in primates is not completely understood.

The current chapter will focus on these alternate visual pathways that ultimately project to the extrastriate cortex in primates and their close relatives, with a focus on those mammals within the Euarchontoglires clade (Fig. 24.1). This will provide possible insights into the evolutionary changes these pathways have undergone and the potential role they play in our own visually mediated behavior, in addition to the extent to which they contribute to visual function following visuocortical injury.

24.2   Subcortical Structures Associated With Extrastriate Pathways

The retina sends projections to numerous brain stem and thalamic structures, and these structures, in turn, send this visual information to telencephalic structures including the neocortex. The retina also projects to the brain stem and thalamic structures involved in orienting the eyes and body toward objects of interest. Here we will focus on three structures, which receive retinal input and participate in relaying information to the neocortex: the SC in the midbrain, the pulvinar/lateral posterior (LP) nucleus, and the LGN of the dorsal

thalamus. Each of these three structures is present in all studied mammals, including primates and possibly all vertebrates (Butler and Hodos, 2005); however, further research is needed to corroborate the evidence for homology across vertebrates.

24.2.1   Thalamus

The vertebrate thalamus is a complex nuclear structure subdivided into four major parts: the hypothalamus, epithalamus, prethalamus, and dorsal thalamus. Arising from the diencephalon, of which it is the largest structure, it has evolved to be an important relay of information. And while 250 million years older than the neocortex (Butler, 2008, 2011), it is the dorsal thalamus, most commonly referred to as the thalamus, that serves as the connectional hub for information to the neocortex and between cortical areas. In mammalian evolution, especially for the primates, the dorsal thalamus has also seen the greatest expansion.

In many ways, the dorsal thalamus has adapted to the needs of a particular species, with many of the associated thalamic nuclei demonstrating an elaborate structure, especially the LGN and pulvinar of primates, which rely heavily on vision. While the evolutionary complexity of the LGN can be fully described in species that are more visual, doing this for the pulvinar nucleus has been more difficult. Although the pulvinar nucleus is the largest of the dorsal thalamic nuclei in primates, it is the least well understood and has been surrounded by significant controversy regarding its role and relationship to vision and higher cognitive functions observed in the primates.

24.2.1.1 Lateral Geniculate Nucleus

The LGN is located dorsolaterally within the thalamus and is a well-defined laminar structure in primates (see Fig. 24.2). The LGN is the most studied thalamic nucleus and has been extensively studied across a great variety of extant mammals. Therefore, much is known about the organization and function of this structure. The anatomical borders and laminar organization of the LGN are well defined in all studied primates and many other mammals (Casagrande and Ichida, 2002; Casagrande et al., 2007; Kahn and Krubitzer, 2002; Van Hooser and Nelson, 2006). However, the laminar pattern of the LGN is not always apparent in some of the close relatives to primates, such as nocturnal rats and mice. Furthermore, the functional organization of the various cell types within the LGN is markedly different across taxonomic orders, and even the variability within closely related mammals is substantial. For example, in primates the layers of the LGN are segregated based on the morphology and functional characteristics of the M, P, and K pathways (Casagrande et al., 2007). That is, cells across layers of the LGN are heterogeneous in their response properties based on M, P, or K inputs from the retina. However, cells within a layer are homogeneous, such that cells within M layers have similar response properties and are quite different from the response properties of cells within the P layers (galagos: Norton and Casagrande, 1982; Irvin et al., 1986, 1993, Norton et al., 1988; marmosets: Kremers et al., 1997; White et al., 1998; owl monkeys: Usrey and Reid, 2000; Xu et al., 2001; macaque monkeys: Kaplan and Shapley, 1982; Derrington and Lennie, 1984; Hubel and Livingstone, 1990; Reid and Shapley, 1992). Although the layers of the LGN are segregated into M, P, and K layers, the organization and number of specific layers are not consistent across all primates (see Fig. 24.2). In tree shrews, one of the closest extant relatives to primates, cells within the discrete layers of the LGN do not have many differences in their response properties. In fact, the only main difference across cells in different layers of the LGN is in their response to luminance. That is, layers are segregated based on receiving input from RGCs that either have "ON-center" or "OFF-center" receptive fields (Van Hooser et al., 2013). Finally, in both the rat and mouse, which are becoming increasingly popular in vision studies, the LGN does not have discrete layers but is instead divided into a core and a shell. Interestingly, when analyzing patterns of retinal projections in these species, a laminar structure emerges in the medial portion, which receives binocular input (rat: Reese and Jeffery, 1983; mouse: Godement et al., 1984; Coleman et al., 2009). The lamination follows a contra—ipsi—contralateral pattern, reminiscent of the laminar pattern of other species (Sherman and Guillery, 2006).

In summary, although the LGN is easily identified across mammalian species and is superficially well characterized by direct retinal input and connections with striate cortex, the details of the anatomical and functional organization of this structure are vastly different across extant species. This diversity suggests that there has been a high degree of divergence in the evolution of this structure, and therefore the geniculostriate pathway across closely related mammals. Thus, elucidating the principles that underlie the evolution of the LGN and the associated subpathways that traverse it is extremely challenging.

24.2.1.2 Pulvinar Complex

The pulvinar complex, or lateral posterior (LP) nucleus, of the thalamus is located medial and dorsal to the LGN within the most caudal aspect of the thalamus in most mammals. Much less is known about the function of the pulvinar in comparison to other thalamic structures such as the LGN. Part of the reason why the function of the pulvinar has been such a mystery is likely due to its complex anatomical substructure. The pulvinar is comprised of multiple subnuclei that are not arranged in an easily discernable order, unlike the laminar organization of the LGN or SC. However, within the last two decades, considerable progress has been made in elucidating the anatomical organization of the visual pulvinar (consisting of the inferior and lateral pulvinar) in anthropoid primates (Fig. 24.3) due to more advanced functional and anatomical techniques.

Unlike the LGN, which shares most of its cortical connections with a single visual area, namely V1, the pulvinar shares connections with multiple extrastriate cortical areas (squirrels: Robson and Hall, 1977; Wong et al., 2008; tree shrews: Lyon et al., 2003b; galagos: Wong et al., 2009; New and Old World monkeys: Adams et al., 2000; Shipp, 2001; Soares et al., 2001; Kaas and Lyon, 2007). Additionally, the number of pulvinar subdivisions found in any given species seems to correlate with the complexity and number of extrastriate cortical areas that a given species has. For instance, in gray squirrels, only four pulvinar divisions have been reported (Baldwin et al., 2011). On the other hand, in squirrel monkeys, there are four divisions within the inferior pulvinar, and the inferior pulvinar itself is only one of four main subdivisions of the entire pulvinar complex (Gutierrez et al., 1995; Stepniewska and Kaas, 1997; Stepniewska et al., 1999, 2000; Gray et al., 1999; Adams et al.,

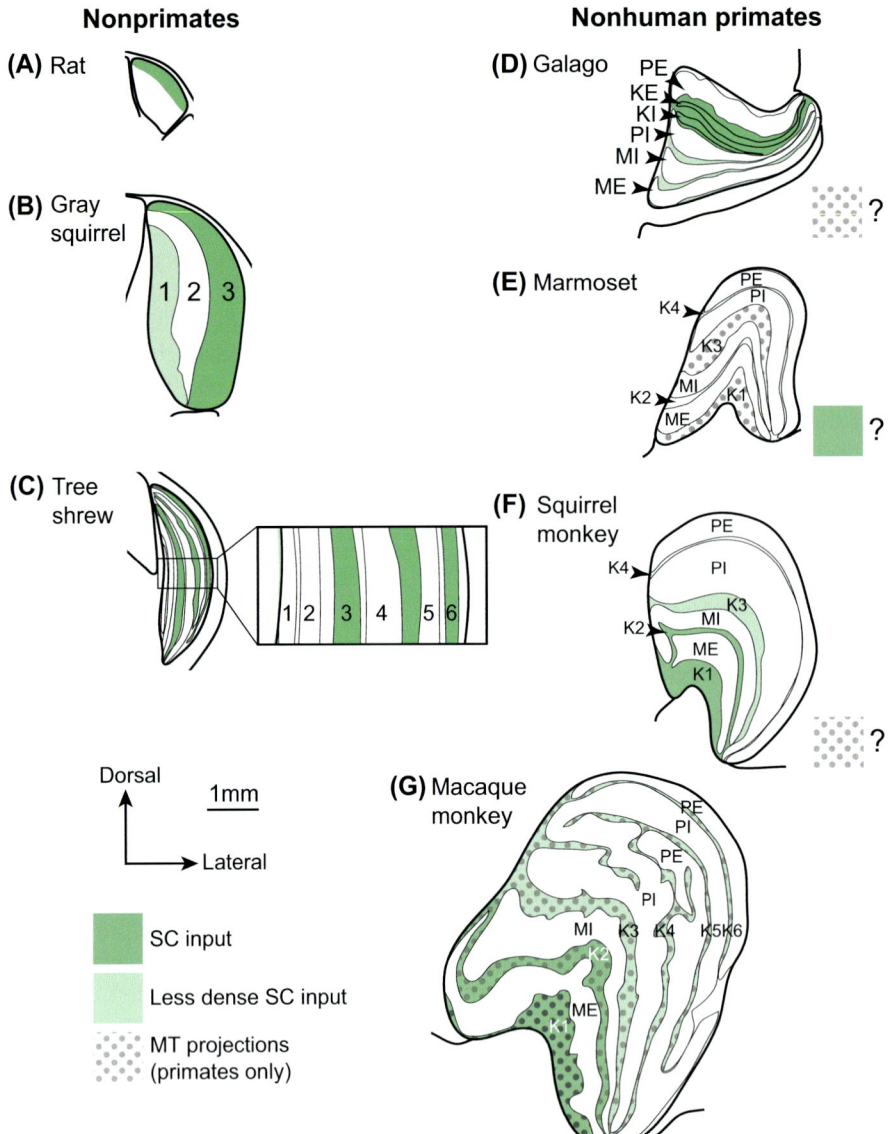

FIGURE 24.2 Organization of layers within the lateral geniculate nucleus (LGN) of various mammals within the Euarchontoglires clade including the rat (A), gray squirrel (B), tree shrew (C), galago (D), marmoset (E), squirrel monkey (F), and macaque monkey (G). Layers receiving superior colliculus (SC) input are highlighted in green, while layers that project to the middle temporal area (MT) are highlighted with *polka dots*. *Shaded boxes with question marks* to the right side of the LGN illustration for a given species indicate that, to the best of our knowledge at time of publication, the location of the given projection within the LGN is unknown. The layers of the LGN for squirrel and tree shrew are indicated by numbers. In primate species MI and ME stand for magnocellular internal and external layers, respectively. PI and PE stand for parvocellular internal and external layers, respectively. Koniocellular layers are indicated by a K1-6, K1 being the most ventral layer.9 *Sources of information: (A) Harting, J.K., Huerta, M.F., Hashikawa, T., Van Lieshout, D.P., 1991. Projections of the mammalian superior colliculus upon the dorsal lateral geniculate nucleus: organization of tectogeniculate pathways in nineteen species. J. Comp. Neurol. 304: 275–306; (B) Harting, J.K., Huerta, M.F., Van Lieshout, D.P., 1991. Projections of the mammalian superior colliculus upon the dorsal lateral geniculate nucleus: organization of tectogeniculate pathways in nineteen species. J. Comp. Neurol. 304: 275–306; Baldwin, M.K., Wong, P., Reed, J.L., Kaas, J.H., 2011. Superior colliculus connections with visual thalamus in gray squirrels (Sciurus carolinensis): evidence for four subdivisions within the pulvinar complex. J. Comp. Neurol. 519 (6), 1071–1094; (C) Fitzpatrick, D.R., Carey, G., Diamond, I.T., 1980. The projection of the superior colliculus upon the lateral geniculate body in Tupaia glis and Galago senegalensis. Brain Res. 194 (2), 494–499; Luppino, G.M., Matelli, C.R., Fitzpatrick, G.D., Diamond, I.T., 1988. New view of the organization of the pulvinar nucleus in Tupaia as revealed by tectopulvinar and pulvinar-cortical projections. J. Comp. Neurol. 273, 67–86; Diamond, I.T., Conley, M., Fitzpatrick, D., Raczkowski, D., 1991. Evidence for separate pathways within the tecto-geniculate projection in the tree shrew. Proc. Natl. Acad. Sci. U.S.A. 88 (4), 1315–1319; (D) Fitzpatrick, D., Carey, R.G., Diamond, I.T., 1980. The projection of the superior colliculus upon the lateral geniculate body in Tupaia glis and Galago senegalensis. Brain Res. 194 (2), 494–499; Baldwin et al. (2013); (E) Warner, C.E., Goldshmit, Y., Bourne, J.A., 2010. Retinal afferents synapse with relay cells targeting the middle temporal area in the pulvinar and lateral geniculate nuclei. Front. Neuroanat. 4, 8; (F) Stepniewska, I., Hui-Xin, Q., Kaas, J.H., 2000. Projections of the superior colliculus to subdivisions of the inferior pulivnar in New World and Old World monkeys. Vis. Neurosci. 17, 529–549; (G) Harting, J.K., Huerta, M.F., Hashikawa, T., Van Lieshout, D.P., 1991. Projections of the mammalian superior colliculus upon the dorsal lateral geniculate nucleus: organization of tecto-geniculate pathways in nineteen species. J. Comp. Neurol. 304, 275–306; Stepniewska, I., Hui-Xin, Q., Kaas, J.H., 2000. Projections of the superior colliculus to subdivisions of the inferior pulivnar in New World and Old World monkeys. Vis. Neurosci. 17, 529–549; Sincich, L.C., Park, K.F., Wohlgemuth, M.J., Horton, J.C., 2004. Bypassing V1: a direct geniculate input to area MT. Nat. Neurosci. 7 (10), 1123–1128. All drawings are to scale.*

2000; Jones, 2007). The disparity in the number of pulvinar divisions between gray squirrels and squirrel monkeys is not surprising, considering the greater number of extrastriate cortical areas and the internal modularity within individual extrastriate areas (eg, V2 stripe types) found in squirrel monkeys (see Fig. 24.4).

Because the pulvinar, unlike the LGN, is composed of subdivisions that are not arranged in a manner that is easily discernable on a 2D plane and encompasses a large area, considerable inconsistencies in how different research groups have subdivided and named the pulvinar nuclei are quite varied, even within the same species (Benevento and Standage, 1983; Gutierrez et al., 1995; Stepniewska and Kaas, 1997; Gray et al., 1999; Adams et al., 2000; Jones, 2007; Lyon et al., 2010). However, even though these inconsistencies exist, there are some clear and reliable anatomical characteristics that are shared across those primates that have been extensively studied.

For example, the divisions of the medial aspect of the inferior pulvinar in both New and Old World monkeys contain two divisions that stain darkly for calbindin (the posterior inferior pulvinar, PIp, and the central medial inferior pulvinar, PIcm) and are separated by a "calbindin hole" (the medial inferior pulvinar, PIm) (Gutierrez et al., 1995; Stepniewska and Kaas, 1997; Gray et al., 1999; Stepniewska et al., 2000; Adam's et al., 2000; Jones, 2007; Warner et al., 2010). Inferior posterior pulvinar (PIp) and PIcm also stain weakly for parvalbumin and darkly for VGLUT2, while PIcm has the opposite staining patterns for these proteins (Stepniewska et al., 2000; Rovo et al., 2012; Balaram et al., 2013). What separates PIp from PIcm is their respective AChE and CO staining characteristics, in that PIp stains darkly for AChE and CO, while PIcm stains much more weakly (see Fig. 24.5). The AChE staining characteristics of the inferior pulvinar of chimps and humans are also similar to those observed in New and Old World monkeys (Cola et al., 1999, 2005). Although the staining characteristics of the pulvinar of prosimian primates have not been fully reported, some of the anatomical features are much like those observed in New and Old World monkeys (Baldwin et al., 2013a). Thus, the organization of at least the inferior pulvinar of primates seems to be relatively conserved. Similar anatomical features of some subdivisions of the primate pulvinar are also apparent in other members of the Euarchontoglires clade (see Fig. 24.3: Lyon et al., 2003a; Baldwin et al., 2011, 2013), where the homologue of the pulvinar is often referred to as the LP nucleus. From here on out, we will only use the term pulvinar, including in those instances where the term lateral posterior nucleus is more commonly used.

Unfortunately, unlike the LGN where the physiological characteristics have been well described and related to discrete layers, there are few studies that have described the physiological properties of neurons in anatomically defined subdivisions of the pulvinar, especially those within the inferior pulvinar that is most well defined anatomically. Those studies that have described the response properties of the inferior pulvinar did so before our current understanding of the internal organization and extent of this division (Chalupa, 1977; Gattass et al., 1978; Bender, 1982; Felsten et al., 1983; Petersen et al., 1985). Thus, it does not seem clear how cytoarchitecturally defined divisions correlate with the reported electrophysiological responses, which in themselves were seemingly weak, with variable reports of direction selectivity and broad orientation selectivity (Bender, 1982; Felsten et al., 1983; Petersen et al., 1985).

24.2.2 Superior Colliculus

The superior colliculus (SC) (also known as the optic tectum) is considered to be one of the most conserved brain structures in vertebrates with considerable similarities in its efferent and afferent connectivity, topographic and laminar organization, and multiple cell types (Butler and Hodos, 2005; May 2006). However, there are considerable differences in the proportional size, lamination, and cortical input across mammals, including between members of the primate order and their close relatives (Fig. 24.1).

The SC is primarily involved in producing orientating movements toward sensory stimuli of various modalities (visual, auditory, and somatosensory) and receives sensory input from the neocortex, other brain stem structures, and the retina. It also provides input to various brain stem structures involved in moving the eyes, head, and body toward objects of interest. Therefore, the variation in the morphology and the density of particular sensory inputs to the SC likely reflects the importance of a particular sensory system to the species studied (May 2006). In primates, the SC receives a substantial input from almost all visual cortical areas (Collins et al., 2005; Baldwin et al., 2013a; Cerkevich et al., 2014), suggesting that it is heavily involved in the visual behavior of primates. Additionally, differences in the complexity of cell types and overall proportional size of specific layers may also reveal the degree of complexity/importance of general functions of those layers within the SC.

24.3 Extrastriate Cortex

Studies of the posterior cortex of eutherian mammals suggest that most mammals have at least one additional visual area beyond V1, yet the organization and the total

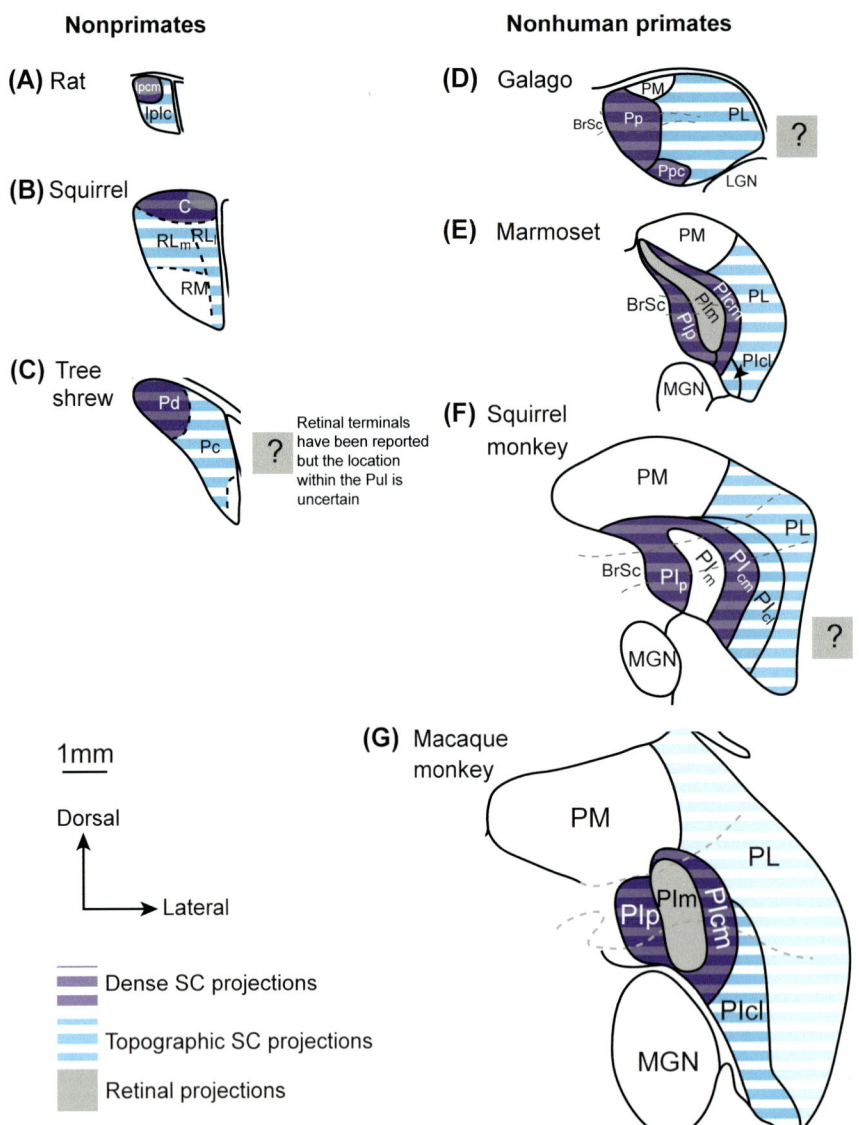

FIGURE 24.3 Organization of the pulvinar and its subdivisions across different Euarchontoglires mammals including rats (A), squirrels (B), tree shrews (C), galagos (D), marmosets (E), squirrel monkeys (F), and macaque monkeys (G). Divisions of the pulvinar that receive direct retinal input are highlighted with gray shading, while divisions receiving dense superior colliculus (SC) input or topographic SC input are shaded with *dark blue* or *light blue horizontal stripes*, respectively. Those projections that are unknown for any given species are indicated by the *respective colored boxes accompanied with question marks* to the right of the associated pulvinar illustration. (Source data for the rat illustration were taken from Takahashi, T., 1985. The organization of the lateral thalamus of the hooded rat. J. Comp. Neurol. 231, 281–309.); for lateralis posterior pars caudomedialis nucleus (lpcm) and lateralis posterior pars lateralis nucleus (lplc) (From Perry, V.H., Cowey, A., 1982. A sensitive period for ganglion cell degeneration and the formation of aberrant retino-fugal connections following tectal lesions in rats. Neuroscience 7 (3), 583–594.). (Data for the squirrel illustration are taken from both data in *gray squirrels* and *ground squirrels*; Major, D.E., Rodman, H.R., Libedinsky, C., Karten, H.J., 2003. Pattern of retinal projections in the California ground squirrel (*Spermophilus beecheyi*): anterograde tracing study using cholera toxin. J. Comp. Neurol. 463 (3), 317–340; Baldwin, M.K., P. Wong, J.L. Reed and J.H. Kaas (2011). "Superior colliculus connections with visual thalamus in gray squirrels (*Sciurus carolinensis*): evidence for four subdivisions within the pulvinar complex. J. Comp. Neurol. 519 (6), 1071–1094; caudal pulvinar) (C), rostral lateral medial pulvinar (PLm), rostral lateral lateral pulvinar (PLl), rostral medial pulvinar (RM). (Data for the tree shrew drawing were taken from Luppino, G., M. Matelli, Carey, R.G., Fitzpatrick, D., Diamond, I.T., (1988). New view of the organization of the pulvinar nucleus in *Tupaia* as revealed by tectopulvinar and pulvinar-cortical projections. J. Comp. Neurol. 273, 67–86.), and although many reports have retinal terminals within the pulvinar, none to our knowledge have shown the location of the terminal label with respect to architectonic boundaries; dorsal pulvinar (Pd), central pulvinar (Pc). (Data for galagos were taken from Baldwin, M.K., Balaram, P., Kaas, J.H., 2013. Projections of the superior colliculus to the pulvinar in prosimian galagos (Otolemur garnettii) and VGLUT2 staining of the visual pulvinar. J. Comp. Neurol. 521 (7), 1664–1682.); posterior pulvinar (Pp), posterior central pulvinar (Ppc), lateral pulvinar (PL) and medial pulvinar (PM). (Data for the marmoset were taken from Stepniewska, I., Hui-Xin, Q. Kaas, J.H., 2000. Projections of the superior colliculus to subdivisions of the inferior pulivnar in New World and Old World monkeys. Vis. Neurosci. 17, 529–549; Warner, C.E., Goldshmit, Y., Bourne, J.A., 2010. Retinal afferents synapse with relay cells targeting the middle temporal area in the pulvinar and lateral geniculate nuclei. Front. Neuroanat. 4, 8.). Abbreviations in

number of extrastriate visual areas in the neocortex is unclear. Most data on extrastriate cortex has been obtained from Euarchontoglires mammals, and of these, only a few species of nonprimate mammals have been systematically analyzed. These include studies of tree shrews, rabbits, and a few rodents. Unfortunately, the results of these studies are not consistent. For instance, the organization and number of extrastriate visual areas in some rodents are quite variable from study to study, even within the same species. For example, it is contentious whether there is a single area (V2) or multiple visual areas bordering the lateral and medial aspects of V1 (Olavarria and Montero, 1984, 1998; 1990; Bravo et al., 1990a,b; Espinoza et al., 1992; Montero, 1993; Coogan and Burkhalter, 1993; Rosa and Krubitzer, 1999; Kalatsky and Stryker, 2003; Van der Gucht et al., 2007; Van Brussel et al., 2009). This ambiguity makes assigning homology to extrastriate visual areas across species exceptionally difficult. In primates, including humans, a significant number of extrastriate cortical areas have emerged during evolution, and their demarcation has been established from physiological, connectional, and anatomical properties. There is nevertheless some dispute over the topographic organization of extrastriate visual areas. Thus, regarding the evolutionary principles for additional extrastriate areas, much has to be clarified. A general understanding of changes to the corticothalamic networks of additional extrastriate areas, and the associated genetic and epigenetic influences across species, could be used to better explore the origins of "new" cortical fields.

Furthermore, the developmental profile of the extrastriate cortex in an individual species can divulge a lot about the origin and functional importance of a particular area. The traditional view on the development of the extrastriate cortex proposes a hierarchical arrangement in which the primary visual area (V1) develops first, followed by the higher-order areas in a wave from V1 (Felleman and Van Essen, 1991; Kaas and Collins, 2001; Zhang et al., 2005; Guillery, 2005; Bourne, 2010). More recent evidence from monkeys, however, shows that this is not the case and that the middle temporal (MT) extrastriate area matures early in parallel to V1 (Condé et al., 1996; Bourne and Rosa, 2006; Warner et al., 2012). Furthermore, in the marmoset, it has been observed that the developmental maturation of the

extrastriate cortex occurs in a nonlinear manner, with significant differences between the dorsal and ventral streams, which is likely a consequence of teleological requirements in primates (Mundinano et al., 2015).

One of the main extrastriate cortical areas that we will be focusing on in the current chapter is MT, which is only observed in primates and is the recipient of input from the pulvinar and LGN (see further discussion). MT is one of the more prominent cortical visual areas outside of V1 and V2 and has many features similar to V1, such as heavy myelination, a clear topographic organization (first order), and early maturation (Bourne and Rosa, 2006 Bourne et al., 2007: Mundinano et al., 2015).

It is unclear when MT emerged in eutherian mammals, and if there are homologues of MT in nonprimate mammals. In tree shrews, some have suggested that an area bordering V2, the temporal dorsal area (TD), is a possible homologue to MT, as it contains a complete representation of the contralateral visual hemifield and has similar anatomical characteristics as those observed in MT in primates (Jain et al., 1994; Wong et al., 2009), including moderate myelination and strong connections with V1 (Sesma et al., 1984; Lyon et al., 1998). In gray squirrels, V1 has strong projections to an area termed the occipital-temporal area (OT), which directly borders the rostral aspect of V2, as does TD in tree shrews. OT may be the homologue of MT (Kaas, 1989), but this area has a patchy and relatively light myelination pattern. Alternatively, the temporal posterior area (TP) of tree shrews or squirrels seems quite similar to MT, as these areas in their respective species are more heavily myelinated; however, projections from V1 to TP are relatively weak (Lyon et al., 1998; Sesma et al., 1984). Outside the Euarchontoglires clade, area OT of flying foxes (Megachiroptera) is similar to the MT of primates in that it lies rostral to V3 and has connections with the pulvinar (Rosa, 1999; Manger and Rosa, 2005). In carnivores, such as cats and ferrets, strong evidence suggests that the posteromedial lateral suprasylvian area (PMLS) is analogous, but unlikely homologous, to MT and middle superior temporal area (Dreher et al., 1996; Homman-Ludiye et al., 2010). Finally, in short-tailed opossums, an area called the caudotemporal area could also be a candidate for an MT homologue, as this area shares connections with V1 and is also heavily myelinated. However, although similarities based on

the marmoset are the same as those for the squirrel monkey and macaque monkey. PM, medial pulvinar; PL, lateral pulvinar; Pip, posterior inferior pulvinar; PIm, medial inferior pulvinar; PIcm, central medial inferior pulvinar; PIcl, central lateral inferior pulvinar. (Data for the squirrel monkey illustration are based on Stepniewska, I., Hui-Xin, Q., Kaas, J.H., 2000. Projections of the superior colliculus to subdivisions of the inferior pulivnar in New World and Old World monkeys. Vis. Neurosci. 17, 529–549.). (Data for the macaque monkey are based on Stepniewska, I., Hui-Xin, Q., Kaas, J.H., 2000. Projections of the superior colliculus to subdivisions of the inferior pulivnar in New World and Old World monkeys. Vis. Neurosci. 17, 529–549; Benevento, L.A., Standage, G.P., 1983. The organization of projections of the retinorecipient and nonretinorecipient nuclei of the pretectal complex and layers of the superior colliculus and the lateral pulvinar and medial pulvinar in the macaque monkey. J. Comp. Neurol. 217, 307–336.). The projection to the lateral pulvinar in macaque monkeys has had variable reports (see Benevento and Standage, 1983; Stepniewska et al., 2000) and therefore, this division is shaded with blue stripes. All drawings are to scale.

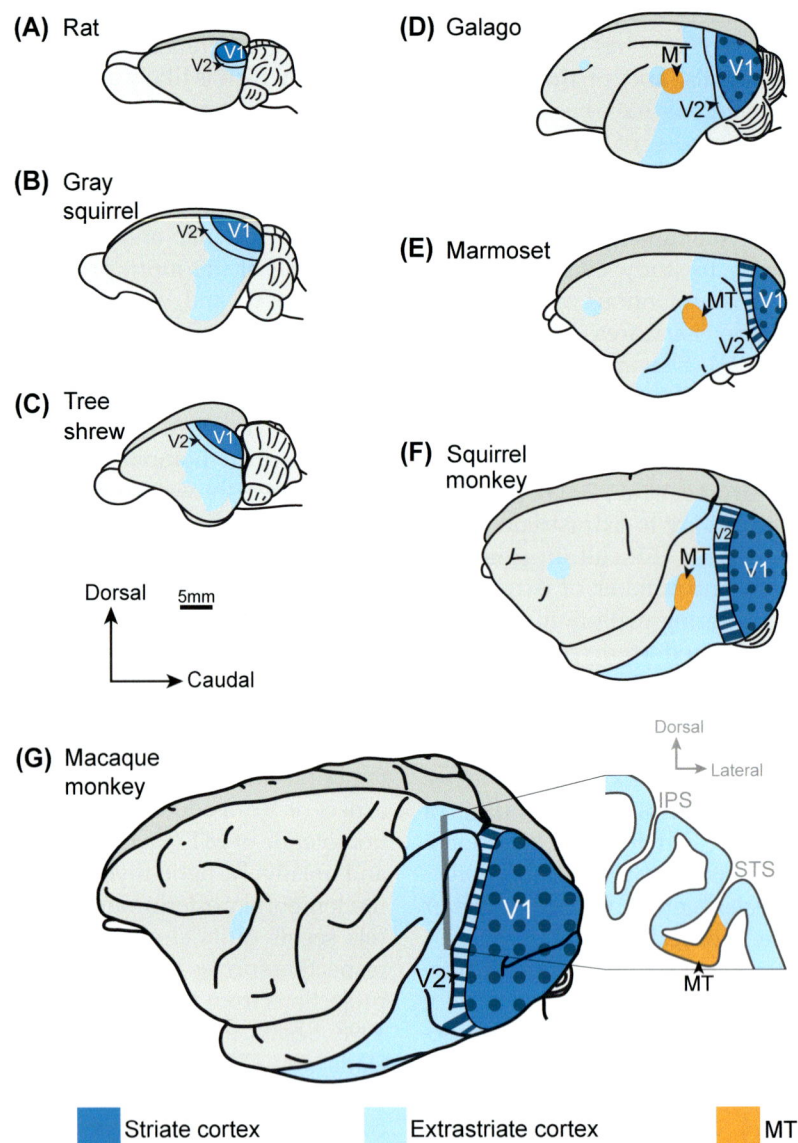

FIGURE 24.4 Extent of both striate (V1; *dark blue*) and extrastriate cortical areas (*light blue*), in the rat (A), squirrel (B), tree shrew (C), galago (D), marmoset (E), squirrel monkey (F), and macaque monkey (G). The middle temporal area (MT: *orange*) for the primates is also depicted and is buried within the superior temporal sulcus in macaque monkeys and is therefore, illustrated in a cross section to the right of the image of the whole brain—the cross section is at the level of the *gray line*. *Dark blue dots* and *stripes* within V1 and V2 of primates represent the internal modularity of 'stripes' and 'blobs' described in those species. The separate region rostral to the main extent of the extrastriate cortex represents the frontal eye fields.

connection patterns with V1 and myelination suggest that areas in other mammals are akin to MT in primates, the data are too sparse to draw any reliable conclusions regarding homology. More detailed patterns of connections and physiological properties of these fields need to be demonstrated first. Further, the presence of MT-like cortical areas needs to be demonstrated for all related species sharing a common ancestor for homology to be assigned. For instance, similarities in the thalamocortical connections within the extrastriate visual pathways of nonprimate mammals could provide some key insights into the origins of the MT of primates (see further discussion).

24.4 Extrageniculostriate Pathways

The next section of this chapter will cover the details by which the subcortical structures outlined in the previous section connect with the extrastriate cortex to relay

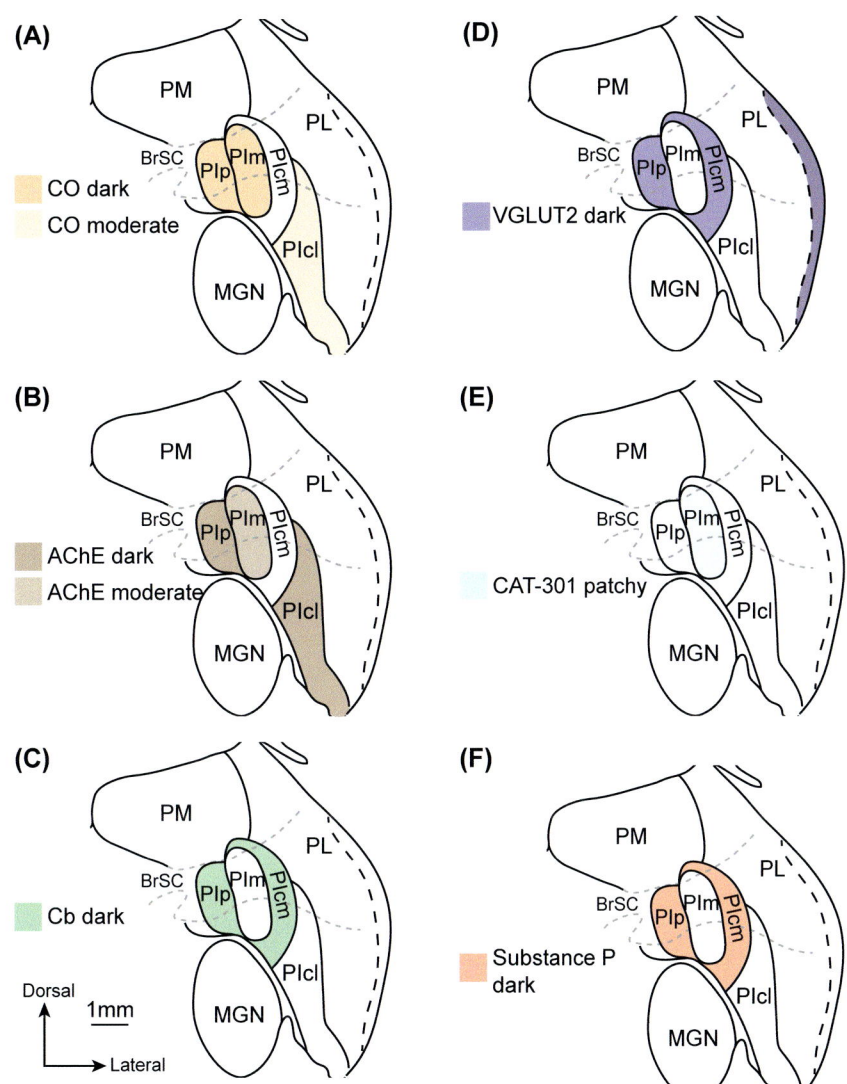

FIGURE 24.5 Cytoarchitectonic characteristics of the inferior pulvinar have been well described for macaque monkeys. (A) Is a representation of cytochrome oxidase (CO) staining patterns. (B) Shows the staining pattern for acetylcholinesterase (AChE). (C) Depicts the staining pattern for calbindin (Cb). (D) Shows the representation of vesicular glutamate transporter type 2 staining patterns (VGLUT2). (E) and (F) Depict the staining pattern for CAT-301 and substance P, respectively. *Source: Gutierrez, C., Yaun, A., Cusick, C.G., 1995. Neurochemical subdivisions of the inferior pulvinar in macaque monkeys. J. Comp. Neurol. 363, 545–562.; Stepniewska, I., Kaas, J.H., 1997. Architectonic subdivisions of the inferior pulvinar in New World and Old World monkeys. Vis. Neurosci. 14 (6), 1043–1060; Stepniewska, I., Hui-Xin, Q., Kaas, J.H., 1999. Do superior colliculus projection zones in the inferior pulvinar project to MT in primates? Eur. J. Neurosci. 11, 460–480; Stepniewska, I., Hui-Xin, Q. Kaas, J.H., 2000. Projections of the superior colliculus to subdivisions of the inferior pulvinar in New World and Old World monkeys. Vis. Neurosci. 17, 529–549; Adams, M.M., Hof, P.R., Gattass, R., Webster, M.J., Ungerleider, L.G., 2000. Visual cortical projections and chemoarchitecture of macaque monkey pulvinar. J. Comp. Neurol. 419, 377–393; Balaram, P., Hackett, T.A., Kaas, J.H., 2013. Differential expression of vesicular glutamate transporters 1 and 2 may identify distinct modes of glutamatergic transmission in the macaque visual system. J. Chem. Neuroanat. 50–51, 21–38; Rovo, Z., Ulbert, I., Acsady, L., 2012. Drivers of the primate thalamus. J. Neurosci. 32 (49), 17894–17908. Similar staining patterns are also present in marmoset (Warner, C.E., Goldshmit, Y., Bourne, J.A., 2010. Retinal afferents synapse with relay cells targeting the middle temporal area in the pulvinar and lateral geniculate nuclei. Front. Neuroanat. 4, 8, and unpublished data). Divisions and abbreviations of pulvinar subdivisions are based on Stepniewska, I., Kaas, J.H., 1997. Architectonic subdivisions of the inferior pulvinar in New World and Old World monkeys. Vis. Neurosci. 14 (6), 1043–1060 and are the same as those for anthropoid primates in Fig. 3.*

information from the periphery. While there are many decades of work that underpin the pathways that relay information through subcortical areas to the extrastriate cortex, there is still much conjecture about their actual existence and role.

24.4.1 Disynaptic Projection From Retina to Extrastriate Cortex via the Lateral Geniculate Nucleus

In both New and Old World monkeys, there have been many reports of projections to the extrastriate cortex via the LGN. The most consistently reported projection arises from koniocellular cells, present mostly within the K or interlaminar layers of the LGN, which projects to area MT (Stepniewska et al., 2000; Sincich et al., 2004; Dick et al., 1991; Warner et al., 2010). There are, however, additional reports of connections between the LGN and other extrastriate visual areas, such as the second visual area (V2; Benevento and Yoshida, 1981; Fries, 1981; Bullier and Kennedy, 1983), fourth visual area (V4; Lysakowski et al., 1988; Tanaka et al., 1990; Cowey and Stoerig, 1989; Lyon and Rabideau, 2012), and the inferior temporal area (Yukie and Iwai, 1981) in macaque monkeys, as well as area 19 in marmosets (Dick et al., 1991) or the dorsal medial area in prosimian galagos and owl monkeys (Beck and Kaas, 1998). Connections between the LGN and these additional extrastriate areas have not been well replicated, and there have been contradictory reports (Benevento and Yoshida, 1981; Benevento and Standage, 1982; Sorenson and Rodman, 1996; Lyon and Rabideau, 2012). Thus, we will focus primarily on the projection from the LGN to MT in the next section. However, it is important to note that such projections to MT from the LGN may be lacking in prosimian primates (Wall et al., 1982), suggesting that this pathway may be a specialization of anthropoid primates. Regarding the normal function of this relay, the majority of what we know has been determined from studies of connections and the molecular aspects of the circuitry.

While most work on the geniculate relays to extrastriate cortex has predominately focused on primates, projections from the LGN to extrastriate cortex have been reported in rats (Coleman and Clerici, 1980), sheep (Karamanlidis et al., 1979), and cats (Garey and Powell, 1967; Glickstein et al., 1967), to name a few. However, it is unclear if such projections are homologous to those described in primates, especially regarding their function. Understanding the extrastriate cortical projections of the LGN across mammals could shed much needed light on their possible function, as well as the evolution of the purported geniculo-extrastriate pathways, including those that possibly terminate outside of MT. As stated before, studying such pathways could even provide insights into the evolutionary origins of area MT.

24.4.1.1 Diversity of K Cells Across Primates

The majority of LGN input to extrastriate areas, such as MT, arises from K cells (Sincich et al., 2004). In primates, the K cells are typically organized into K or "interlaminar" layers and receive direct and indirect (via SC) retinal input (see earlier discussion) (Fig. 24.2). Because the connections between the LGN and extrastriate cortex have yet to be examined in prosimians or tarsiers, it is unclear whether this pathway is a specialization of anthropoid primates, or if it was present in the common ancestor to all primates. However, projections from the retina and the SC to the K layers of the LGN have been reported in all studied primates.

The organization of the K layers within the LGN varies greatly among primate species, especially between prosimian and anthropoid primates (see Fig. 24.2) (Casagrande et al., 2007). This suggests that there has been a considerable divergence in the K pathway in primate evolution, making it difficult to infer the organization of K cells/layers in the common ancestor of primates or their close relatives. Furthermore, there is significant heterogeneity in the anatomical characteristics of cells within the K layers. Although K cells can be characterized by the expression of the calcium-binding protein, calbindin, in prosimians (Diamond et al., 1993; Johnson and Casagrande, 1995; Beck and Kaas, 1998), New (Goodchild and Martin, 1998) and Old World monkeys (Jones and Hendry, 1989; Yan et al., 1996; Rodman et al., 2001), apes, and humans (Leuba and Saini, 1996), the chemical composition of the projections to K cells may not be consistent. For example, VGLUT2-immunopositive terminals are present within the K layers of the LGN in prosimian galagos (Balaram et al., 2011), but not in the interlaminar layers of macaques (Balaram et al., 2013), suggesting a possible functional difference in those afferent projections in galagos versus macaques.

K cells are also diverse in their physiological response properties (Irvin et al., 1986; Xu et al., 2001; Martin et al., 1997; White et al., 2001; Solomon et al., 2002). For instance, in New World monkeys, cells within K1 and K2 layers (Fig. 24.2) are highly sensitive to contrast, respond well to high temporal frequencies, and have large diameter receptive fields (marmoset: White et al., 2001; Solomon et al., 2002; owl monkey: Xu et al., 2001). However, more dorsal K layers, such as K3 and K4, contain cells that are responsive to blue light (marmoset only: Martin et al., 1997).

Although we have been able to gain some insights into the anatomical and functional characteristics of K cells within visual relays, the heterogeneity of these cells, and the inconsistencies across primates, suggests a complex and exceptionally flexible evolution. Thus, further investigation is needed to fully characterize this pathway's role in vision across a variety of primate species.

24.4.1.2 Retinal Input to Koniocellular/ Interlaminar Layers of Lateral Geniculate Nucleus

Direct evidence for disynaptic retinal input to MT has been well studied in the New World marmoset monkey, where retinal projections have been found to synapse directly onto cells in layers K1 and K3 that project to MT (see Fig. 24.2), with the densest synapses found within K1 (Warner et al., 2010). These retinal inputs synapse directly onto the proximal dendrites of relay cells to MT (Warner et al., 2010), suggesting a modulatory influence.

The majority of studies looking at the retinal origins of input to the K layers have focused on input from S-cone sources, or melanopsin input (Martin et al., 1997; Szmajda et al., 2008), and such sources have also been found to provide input to the LGN of rodents (Hattar et al., 2006), suggesting that such a pathway from the retina to the LGN is homologous in primates and rodents. However, it seems incongruous that the type of information relayed by S-cones would be relayed to extrastriate areas such as MT given the response properties of MT neurons, but such a possibility remains unexplored. As stated earlier, there is a great degree of variability in the response properties of cells within the different K layers, which suggests that there are likely differences in the types of retinal cells projecting to those layers. Recent evidence from marmosets does suggest that input to K1 cells may originate from a different retinal source than those that project to K3 and could be the source giving rise to the high contrast and temporal frequency sensitivity characteristics of cells within K1, and ultimately the source of input to MT (Percival et al., 2014). However, this hypothesis requires further studies.

24.4.1.3 Superior Colliculus Input to the Lateral Geniculate Nucleus

Projections from the SC to the LGN have been found to be present in all studied mammals (Harting et al., 1991) (Fig. 24.1). The cells projecting to the LGN originate within the upper stratum griseum superficiale (SGS: see Fig. 24.6) of the SC in hamsters (Mooney et al., 1984), tree shrews (Graham and Casagrande, 1980; Diamond et al., 1991), and primates (Harting et al., 1991).

Projections from the SC to the LGN of marmosets have not been studied, although it is likely that such projections are present considering that such projections have been reported in prosimian primates, Old World Monkeys, and other New World Monkeys. Furthermore, the cytoarchitecture of the SC is very similar to other primates (Bourne and Rosa, 2003). For most primates, dense SC input to the LGN is found within the most ventral layer (K1) of the LGN, adjacent to the optic tract (Fig. 24.2). In prosimian galagos, however, dense SC input terminates within the most dorsal layers of the LGN, between the parvocellular layers (Fig. 24.2). This region stains uniformly for calbindin (Beck and Kaas, 1998); however, two layers are distinguishable within this region of the LGN using VGLUT2 immunohistological staining (Balararm et al., 2011). It is not entirely clear whether SC projections terminate within the K layers, or within the interlaminar region surrounding them (Baldwin et al., 2013a). More sparse SC projections are found in the layer of the LGN that is adjacent to the optic tract, as well as within the interlaminar zones in more ventral layers of the LGN (Harting et al., 1991; Baldwin et al., 2013a) (Fig. 24.2). Thus, the SC input to the LGN seems to be quite different between prosimian primates and their anthropoid relatives (see Fig. 24.2), suggesting a possible functional change in this connectional network.

24.4.2 Direct Retinal Projections to the Pulvinar With Efferent Targets in Extrastriate Visual Cortex

An alternate route for retinal information to enter the extrastriate cortex, bypassing the geniculostriate system, is through the pulvinar nucleus. This monosynaptic pathway originates within the retina, which projects directly onto MT relay cells within the inferior pulvinar (PIm). While knowledge of this pathway is still rather limited, it likely serves an important role in the establishment and development of the visual cortex in primates (Warner et al., 2015).

Direct retinal projections to the pulvinar have been reported in many species, including, but not limited to short-tailed opossums (Kahn and Krubitzer, 2002), cats (Itoh et al., 1983; Matteau et al., 2003), a number of rodent species (mice: Sun et al., 1996; Real et al., 2008; Morin and Studholme, 2014; Allen et al., 2016; rats: Perry and Cowey, 1982; hamsters: Crain and Hall, 1980; Ling et al., 1997; Nile grass rats: Gaillard et al., 2013; Japanese voles: Uchiumi et al., 1995; and squirrels: Major et al., 2003), tree shrews (Ohno et al., 1975; Somogyi et al., 1981; Casagrade and Diamond, 1974), New World monkeys (Warner et al., 2010, 2012; 2015), Old World monkeys (Campos-Ortega, 1970; Mizuno et al., 1982; Itaya and Van Hoesen, 1983; Cowey et al., 1994; O'Brien et al., 2001), and baboons (Campos-Ortega et al., 1970), suggesting that characteristics of the division of the pulvinar receiving retinal input may be present in all mammals. However, in primates, the retinorecipient division of the pulvinar (PIm), which possesses strong connectivity with MT, has no clear homologue in nonprimate species.

The retinal projection to PIm observed in primates is seemingly separate from the SC projections to the pulvinar, which likely project to the flanking areas PIp and PIcm (Stepniewska et al., 1999, 2000 but see Berman and Wurtz, 2011). The retinopulvinar projection comprises primarily of the P-gamma subtype of RGC, which also projects to the SC, in macaques (Cowey et al., 1994). Although retinal terminations within the pulvinar nucleus have been reported for some other mammals, it is unclear whether they overlap with projections from

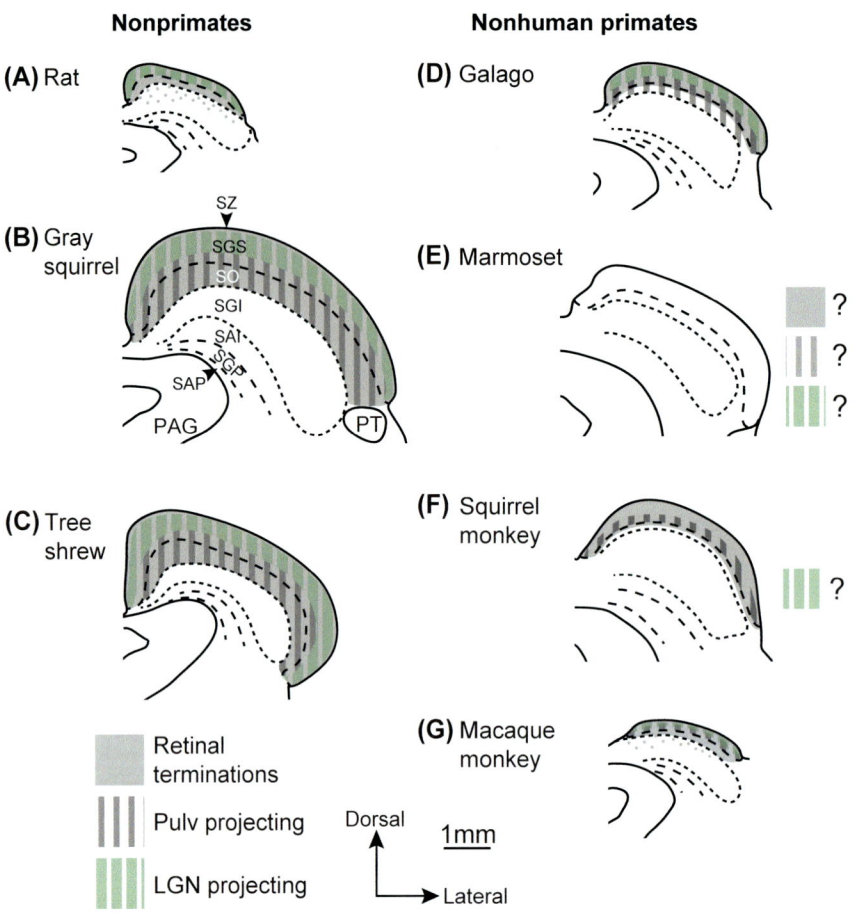

Nonprimates   **Nonhuman primates**

(A) Rat   **(D)** Galago

(B) Gray squirrel   **(E)** Marmoset

(C) Tree shrew   **(F)** Squirrel monkey

(G) Macaque monkey

Retinal terminations

Pulv projecting

LGN projecting

Dorsal
Lateral
1mm

FIGURE 24.6   Organization of layers within the superior colliculus (SC; optic tectum) of the rat (A), gray squirrel (B), tree shrew (C), galago (D), marmoset (E), squirrel monkey (F), and macaque monkey (G). Seven main layers are present across all Euarchontoglires species. These layers are listed in the depiction of the gray squirrel superior colliculus and include the stratum zonal (SZ), stratum griseum superficiale (SGS), stratum opticum (SO), stratum griseum intermediate (SGI), stratum album intermedium (SAI), stratum griseum profundum (SGP). The location of retinal terminals is indicated with gray shading, while layers known to project to the lateral geniculate nucleus (LGN) and pulvinar are indicated with *green or gray vertical stripes. Where information is not known, we have placed a box with a question mark to the right of the illustration. Sources: (A) Beckstead, R.M., Frankfurter, A., 1983. A direct projection from the retina to the intermediate gray layer of the superior colliculus demonstrated by anterograde transport of horseradish peroxidase in monkey, cat and rat. Exp. Brain Res. 52, 261–268; Harting, J.K., Huerta, M.F., Hashikawa, T., Van Lieshout, D.P., 1991. Projections of the mammalian superior colliculus upon the dorsal lateral geniculate nucleus: organization of tectogeniculate pathways in nineteen species. J. Comp. Neurol. 304: 275–306; (B) Fredes, F., Vega-Zuniga, T., Karten, H., Mpodozis, J., 2012. Bilateral and ipsilateral ascending tectopulvinar pathways in mammals: a study in the squirrel (Spermophilus beecheyi). J. Comp. Neurol. 520 (8), 1800–1818; Harting, J.K., Huerta, M.F., Hashikawa, T., Van Lieshout, D.P., 1991. Projections of the mammalian superior colliculus upon the dorsal lateral geniculate nucleus: organization of tectogeniculate pathways in nineteen species. J. Comp. Neurol. 304: 275–306; (C) Graham, J., Casagrande, V.A., 1980. A light microscopic and electron microscopic study of the superficial layers of the superior colliculus of the tree shrew (Tupaia glis). J. Comp. Neurol. 191, 133–151; Chomsung, R.D., Petry, H.M., Bickford, M.E., 2008. Ultrastructural examination of diffuse and specific tectopulvinar projections in the tree shrew. J. Comp. Neurol. 510 (1), 24–46; Harting, J.K., Huerta, M.F., Hashikawa, T., Van Lieshout, D.P., 1991. Projections of the mammalian superior colliculus upon the dorsal lateral geniculate nucleus: organization of tectogeniculate pathways in nineteen species. J. Comp. Neurol. 304: 275–306; Diamond, I.T., Conley, M., Fitzpatrick, D., Raczkowski, D., 1991. Evidence for separate pathways within the tecto-geniculate projection in the tree shrew. Proc. Natl. Acad. Sci. U.S.A. 88 (4), 1315–1319; (D) Feig, S., Van Lieshout, D.P., Harting, J.K., 1992. Ultra-structural studies of retinal, visual cortical (area 17), and parabigeminal terminals within the superior colliculus of Galago crassicaudatus. J. Comp. Neurol. 319 (1), 85–99; Harting, J.K., Huerta, M.F., Hashikawa, T., Van Lieshout, D.P., 1991. Projections of the mammalian superior colliculus upon the dorsal lateral geniculate nucleus: organization of tectogeniculate pathways in nineteen species. J. Comp. Neurol. 304: 275–306; Raczkowski, D., Diamond, I.T., 1981. Projections from the superior colliculus and the neocortex to the pulvinar nucleus in galago. J. Comp. Neurol. 200, 231–254 (F) Tigges, J., Tigges, M., 1981. Distribution of retinofugal and corticofugal axon terminals in the superior colliculus of squirrel monkey. Invest. Ophthalmol. Vis. Sci. 149–158; Huerta, M.F., Harting, J.K., 1983. Sublamination within the superficial gray layer of the squirrel monkey: an analysis of the tectopulvinar projection using anterograde and retrograde transport methods. Brain Res. 261 (1), 119–126. (G) Beckstead, R.M., Frankfurter, A., 1983. A direct projection from the retina to the intermediate gray layer of the superior colliculus demonstrated by anterograde transport of horseradish peroxidase in monkey, cat and rat. Exp. Brain Res. 52, 261–268; Harting, J.K., Huerta, M.F., Hashikawa, T., Van Lieshout, D.P., 1991. Projections of the mammalian superior colliculus upon the dorsal lateral geniculate nucleus: organization of tectogeniculate pathways in nineteen species. J. Comp. Neurol. 304: 275–306; Benevento, L.A., Standage, G.P., 1982. Demonstration of lack of dorsal lateral geniculate nucleus input to extrastriate areas MT and visual 2 in the macaque monkey. Brain Res. 252 (1), 161–166. All drawings are to scale.

the SC (Fig. 24.3). This would be an important avenue for research as the results of such a study could help elucidate the possible evolutionary origins of PIm and the possible changes that have occurred between this division of the pulvinar in primates and their close relatives. Superficially, the inferior pulvinar is the main architectonic division that receives both retinal and SC input (see Fig. 24.3: Major et al., 2003; Perry and Cowey, 1982). Furthermore, developmental studies in both rats and hamsters suggest that the SC and retinal input compete for space within the pulvinar, and if the SC is left intact throughout development, the retinal projections disappear, or at least become quite sparse (Crain and Hall, 1980; Perry and Cowey, 1982). However, the retinal projections to the pulvinar are intact in adult squirrels (Major et al., 2003). Descriptions of the retinal projections to the pulvinar in most species have not been well documented on current known architectonically defined divisions.

There are data from marmosets that suggest that the input to MT via the pulvinar (PIm) is more influential than the projection to MT via the K layers of the LGN due to multiple factors. First, there is a greater density of connections between PIm and MT versus the LGN and MT. Second, there are nearly 10 times more synapses from RGCs onto MT relay cells within PIm than onto MT relay cells within the LGN (Warner et al., 2010). Finally, the retinal projections to PIm directly synapse onto the cell bodies of MT relay cells, while retinal projections synapse onto proximal dendrites in the LGN.

Recent studies have suggested that this monosynaptic relay through PIm to MT plays an important role in the early maturation of area MT and the dorsal visual stream (Warner et al., 2012, 2015). Furthermore, it was demonstrated that this pathway is plastic in the postnatal period and is pruned when the direct input from V1 to MT is fully established (Warner et al., 2015). This pruning process is inhibited if V1 is removed in early life (Warner et al., 2015). Unfortunately, there are few comparisons of the development of the visual system in different primate species, especially on changes in the retinal and cortical projections associated with this pathway.

Although, the presence of a homologue of MT in nonprimate species is unclear, understanding the cortical connections of the retinal projection zone within the pulvinar of other mammals could provide valuable insights into the evolution of MT.

24.4.3 The Extrageniculate Pathway Through the Superior Colliculus and Pulvinar

As stated before, the SC receives direct retinal projections. The SC, in turn, sends projections conveying visual information to two main nuclei within the dorsal thalamus, the LGN (described in Section 24.4.1) and

the pulvinar, which is the focus of this section. The progression of retinal information through the SC and pulvinar destined for extrastriate cortical areas is commonly referred to as the extrageniculate pathway and has been described for all studied mammals (Harting et al., 1973; Diamond, 1973; Kaas and Huerta, 1988). Similar pathways from the tectum to the telencephalon have been described in both anamniotes (fish and amphibians) and amniotes (birds and reptiles), with the suggestion that the tectorotundal projection is homologous to the SC—pulvinar projection (Karten et al., 1997; Butler and Hodos, 2005); however, direct evolutionary relationships between the tectofugal pathways across all vertebrates need to be further explored.

The extrageniculate pathway has received a great deal of attention for many decades, yet the exact role this pathway plays in vision is still uncertain. Recent evidence from rodent and primate studies suggests that the SC/pulvinar pathway is composed of at least two subpathways, which may have been present in the common ancestor of all mammals (Baldwin et al., 2013b) and have different functional roles. The details of these pathways will be discussed later. However, it is clear that these pathways have changed considerably across primates and their close relatives, yet many of the basic features do seem to be conserved (Lyon et al., 2003a,b; Baldwin et al., 2011).

24.4.3.1 Superior Colliculus Projections to the Pulvinar

24.4.3.1.1 Superior Collicular Terminals within the Pulvinar

There are two main projection patterns from the SC to the pulvinar that have been described across different species. One projection results in topographically organized SC terminals, while the second pattern is seemingly denser and either does not have a clear topographic projection pattern, or the topography of the SC terminals is course, at best (tree shrews: Luppino et al., 1988; Chomsung et al., 2008, 2010; gray squirrels: Baldwin et al., 2011; galagos: Baldwin et al., 2013a; anthropoid monkeys: Lin and Kaas, 1979; Stepniewska et al., 2000). The topographically organized SC projection terminates within more rostral and lateral locations of the pulvinar complex, while the second type of projection terminates in more caudomedial locations (see Baldwin et al., 2013a for review).

The pulvinar nucleus of primates looks architecturally different from other mammals and has far more subdivisions than the pulvinar nucleus of nonprimate species (see Fig. 24.3). Therefore, it can sometimes be quite difficult to determine what divisions of the primate pulvinar are homologous or similar to those in other mammals. However, converging evidence based on SC projection patterns, architectonic features, and cortical connections are providing significant insights into what divisions of the pulvinar are likely homologous across species.

In anthropoid primates, the SC projects to different divisions of the inferior and lateral pulvinar. Specifically, in New World and Old World monkeys, the PIp and the PIcm receive dense, crudely topographic projections from the SC, while the inferior central lateral pulvinar (PIcl, although it is important to note that this division of the pulvinar has varying names across studies, in this chapter we have adopted the nomenclature from Stepniewska and Kaas, 1997) and portions of the lateral pulvinar (PL) receive less dense, more punctate, topographic projections (Benevento and Standage, 1983; Stepniewska et al., 2000). PIp and PIcm share similar staining characteristics for calbindin, VGLUT2, parvalbumin, and substance P (Fig. 24.5), but have different staining properties for CO and AChE (Gutierrez et al., 1995; Stepniewska and Kaas, 1997; Gray et al., 1999; Stepniewska et al., 2000; Adams et al., 2000; Jones, 2007). In many instances, it looks as though PIp and PIcm merge either dorsally or ventrally depending on the plane of section; however, the presence of different staining patterns for CO and AChE among the two divisions (Fig. 24.5) suggests that they have separate functions.

There are claims from physiological studies, that the SC projects to PIm (Berman and Wurtz, 2008, 2010, 2011), although many anatomical studies suggest that such projections are minimal to nonexistent (Stepniewska et al., 1999, 2000; Lyon et al., 2010). Alternatively, the physiological data supporting SC influences on MT could be due to connections through a division of the pulvinar at the most lateral extent (Lyon et al., 2010), which has been given various names by different investigators, such as the inferior lateral pulvinar (Lyon et al., 2010; Benevento and Standage, 1983) or the pulvinar "shell" (Gutierrez et al., 1995; Gray et al., 1999). Interestingly, this division of the pulvinar stains darkly for VGLUT2 protein (Rovo et al., 2012), as well as moderately for AChE and calbindin and patchy for Cat-301 (Gray et al., 1999), but the overlap of these staining features needs to be more properly assessed with respect to SC terminal label in this region of the pulvinar.

Subdivisions within the pulvinar of prosimian galagos are less well studied, and therefore, the exact divisions receiving SC input are not as clear. However, a recent study suggests that there are two divisions of the pulvinar such as PIp and PIcm in monkeys, and these divisions are located within the caudal aspect of the pulvinar and receive dense SC projections that are crudely topographic (Baldwin et al., 2013a). Further, these divisions also stain darkly for VGLUT2 (Baldwin

et al., 2013a), similar to PIp and PIcm in macaque monkeys (Rovo et al., 2012; Balaram et al., 2013). There are also two projections from the SC that terminate within more rostral and lateral divisions of the galago pulvinar that are less dense but focused, and topographically organized, similar to projections observed in PIlm and PL in New and Old World monkeys (Glendenning et al., 1975; Baldwin et al., 2013a). Where exactly these projections terminate on homologous divisions of PIlm or PL is not clear, but likely one of the topographic projections is within PIlm and the second within PL.

Finally, although connectional studies are limited in humans and other great apes, histological staining patterns within the pulvinar nucleus are consistent with reports on the number and divisions of the pulvinar in New and Old World monkeys (Cola et al., 1999, 2005; Arcaro et al., 2015).

Regarding nonprimate species, the SC projections to the pulvinar of tree shrews have been studied extensively. Although the pulvinar of tree shrews is not as large or complex as that observed in primates (Fig. 24.3), many of the features of tectal projections and histological staining patterns are similar to primate species (Luppino et al., 1988; Lyon et al., Chomsung et al., 2008). These include a dense and coarsely topographic projection to a caudal division of the pulvinar, known as the dorsal pulvinar (Pd). This division also stains darkly for CO, AChE, substance P, and VGLUT2, similar to PIp in anthropoid primates. More punctate and topographic projections to more rostral lateral regions of the pulvinar, which is known as the central pulvinar (Pc), are also observed in tree shrews (Luppino et al., 1988; Chomsung et al., 2008, 2010). Pc has similar staining patterns to those observed within the PL in anthropoids; however, VGLUT2 staining within Pc is seemingly much stronger than that found in the PL of primates. Although similarities in the connection and staining patterns of these divisions are present, further analysis is required before homologies can be fully assigned.

In squirrels, a popular diurnal rodent model for studying vision (Van Hooser and Nelson, 2006), dense projections from the SC terminate within caudal aspects of the pulvinar, while more focused and topographic terminations are found in more rostrolateral locations (Robson and Hall, 1977; Baldwin et al., 2011). Similar findings have also been reported in other rodents, including hamsters (Crain and Hall, 1980; Ling et al., 1997) and rats (Takahashi, 1985). Furthermore, the staining properties of the different divisions of the pulvinar complex of squirrels are identical to those observed in tree shrews with the caudal division of the pulvinar (C), staining darkly for VGLUT2, CO, and AChE similar to Pd of tree shrews (see Baldwin et al., 2011 for review).

Additionally, rostral lateral divisions also stain moderately for VGLUT2, CO, and AChE similar to observations in tree shrews.

Therefore, it seems likely that the most caudal aspect of the pulvinar across rodents, tree shrews, and primates are homologous given the similarities in the SC projection and staining patterns. Further, the two basic types of projections (one coarse, one more focused and topographic) were likely present in the common ancestor of Euarchontoglires. However, the increase in the number of projections found in primates (two dense yet coarse projections to two caudal divisions of the pulvinar) and multiple additional retinotopic projections to more rostral lateral locations (in PIcm and PL) may be due to the increase in extrastriate cortical areas.

24.4.3.1.2 Origins of Superior Collicular Projections

The cells that project to the pulvinar are located within the superficial layers of the SC. Specifically, cells within the lower SGS and the stratum opticum (SO) have been found to project to the pulvinar across most studied species (macaques: Benevento and Standage, 1983; squirrel monkeys: Huerta and Harting, 1983; galagos: Raczkowski and Diamond, 1981; tree shrews: Graham and Casagrande, 1980; Harting et al., 1991; squirrels: Fredes et al., 2012).

Recent studies in squirrels indicate that cells within the SGS and SO have different targets within the pulvinar. For instance, cells within the SGS project to the caudal pulvinar, while cells within the SO project to the rostrolateral (RL) divisions. Although similar studies have not been conducted in other mammals, based on correlating information on connection patterns and architectonic staining properties, it seems these pathways likely exist. For instance, the region of the caudal pulvinar in squirrels stains darkly for VGLUT2. As stated before. This protein is involved in repackaging glutamate in the synaptic terminals of cells. Additionally, VGLUT2 is abundant in caudal aspects of the pulvinar for many primate and nonprimate mammals and is associated with coarse or nontopographic SC projections. Cells that contain VGLUT2 in their terminals are most likely to be transcribing VGLUT2 mRNA in their cell bodies. In galagos, the majority of cells within the lower SGS express VGLUT2 mRNA (Balaram et al., 2011, 2015). Thus, it seems likely that these cells are also projecting to the caudalmost aspect of the pulvinar in galagos, the portion that stains darkly for VGLUT2 protein. There are only a few cells within the SO of galagos that express VGLUT2 mRNA; therefore, it seems likely that few, if any, projections of cells within the SO to the pulvinar have terminals that contain VGLUT2 protein. Similar patterns of VGLUT2 mRNA are also found in the SC of the macaque monkey (Balaram et al., 2013). Furthermore, tree shrews also share similarities in VGLUT2 mRNA expression in the SC, with the majority of cells within the lower SGS expressing VGLUT2 mRNA, but only a few cells within the SO expressing VGLUT2 mRNA. Although it is important to note that terminals within Pc of tree shrews (the division that receives punctate and topographic SC projections) do contain VGLUT2 protein (Chomsung et al., 2008), the VGLUT2 staining pattern within Pc is much stronger than divisions of the pulvinar in primates that receive topographic SC projections (PIcl and the PL). Such a difference in VGLUT2 staining could reflect a difference in the "driving" abilities of this type of SC projection between primates and tree shrews (Rovo et al., 2012). Regardless, it seems likely that the two pathways present within squirrels, with one originating within the lower SGS and the other originating within the SO, may be present in other mammals such as primates and tree shrews.

24.4.3.2 Pulvinar Connections With Extrastriate Cortex

The pulvinar shares connections with almost all of the visual cortex; however, the divisions of the pulvinar that also receive connections from the SC share connections with visual cortical areas outside of V1. Therefore, any possible visual information flowing through the SC and the pulvinar terminates within extrastriate cortex.

Some researchers have suggested that different divisions of the pulvinar are connected with different cortical areas associated with the dorsal or ventral streams of visual processing (Kaas and Lyon, 2007). Interestingly, those divisions in primates that receive dense SC projections that are only coarsely topographic tend to share connections with visual areas associated with dorsal visual stream cortical areas. Divisions of the pulvinar that receive the more refined topographic projections from the SC share connections with areas primarily related to the ventral visual stream, although these divisions also share connections with dorsal stream cortical areas such as MT (Lin and Kaas, 1979; Wong et al., 2009 Adams et al., 2000). Therefore, it may be that these two projection pathways through the SC to the pulvinar could be relaying different information associated with their specific cortical targets (one associated with maintaining topographic information, with the second more focused on motion) in much the same way as the LGN has multiple segregated pathways associated with dorsal and ventral visual processing.

As the LGN relays evolved to become the more prominent pathways for visual information flow to the extrastriate cortex in primates, especially concerning the ventral stream, the pathways through the SC became less pronounced. Such a theory could explain why there may be a lack or a significant reduction of SC projections to the PL in Old World Monkeys (Stepniewska et al., 2000; but see Benevento and Standage, 1983; Lyon et al., 2010).

24.5 Role of Extrageniculostriate Pathways Following Ablation of V1

Much of the discussion of the functional role of the extrageniculostriate pathways has emerged following experimental and accidental lesions of V1, in both humans and monkeys. Traditionally, geniculostriate lesions were considered to result in complete and permanent visual loss in the topographically related area of the visual field (Huber, 1992). However, a distinctive phenomenon, referred to as "blindsight," is often observed in patients with geniculostriate pathway damage, whereby subjects can respond to visual stimuli that they do not consciously see (Pöppel et al., 1973; Weiskrantz, 2004; Stoerig and Cowey, 2007). For example, these patients often demonstrate visually guided behavior within the blind field (scotoma), such as pre-shaping their hand to an object in front of them when asked to pick it up, although they cannot report the size of the object (Whitwell et al., 2011; Vision Research). By comparing the MRI/tracer and behavioral results of monkey studies with those of patient G.Y., all whom had complete bilateral lesions of V1, it has been concluded that there was much similarity between the two species and that analogous pathways were likely invoking the preserved visual capacity (eg, Cowey, 2010b). To this end, both the geniculo-extrastriate and pulvino-extrastriate pathways have been suggested to be responsible for providing the neural substrate underpinning this phenomenon (Cowey, 2010a; Schmid and Maier, 2015; Bertini et al., 2016). However, while much conjecture still exists about the exact pathway, or whether it is a combination of these pathways, experimental evidence has provided significant insights into the notion that these pathways do play an important role in visual processing at some level.

Many monkey studies have demonstrated residual activity of cortical area MT neurons following a lesion of V1 (eg, Rodman et al., 1989; Girard et al., 1992; Rosa et al., 2000; Azzopardi et al., 2003; Yu et al., 2013), even though the responses of MT neurons become substantially weakened in the absence of V1 inputs and may even be abolished under some protocols (Girard et al., 1992; Kaas and Krubitzer, 1992; Collins et al., 2003). There are multiple schools of thought on the neural substrate responsible for this residual activity in MT observed in primates, each with specific caveats that must be applied when making conclusions from the results. The first focuses on the geniculo-MT pathway (Sincich et al., 2004; Bridge et al., 2008; Schmid et al., 2010; Schmid and Maier, 2015). Opponents to this pathway being the neural substrate highlight that following a lesion of V1 there is significant transneuronal degeneration of the disynaptic pathway originating in the retina, resulting in significant loss of LGN neurons and RGCs (Cowey et al., 2011; Hendrickson et al., 2015; Warner et al., 2015; Sorenson and Rodman, 1999), although some cells remain intact in the K layers of the marmoset LGN (Yu et al., 2013) and in that of macaque monkeys (Rodman et al., 2001). The second hypothesis suggests that it is the pulvino-MT that relay drives the residual visual activity (Barleben et al., 2015; Warner et al., 2015; Bridge et al., 2016).

Another interesting aspect of this story is that monkeys and humans that receive damage to V1 very early in life possess a higher level of residual vision (Moore et al., 1996; Sorenson and Rodman, 1999; Gross et al., 2004). For example, humans who had experienced perinatal infarctions to V1, when tested as adults, were much better in their visual performance than those who had acquired comparable damage during adolescence (Tinelli et al., 2013; Kiper et al., 2002). Also, monkeys that received a V1 lesion during the second month of life had greater residual vision than those having received adult lesions (Moore et al., 1996). Furthermore, the recent finding in the marmoset monkey of a more extensive pulvino-MT connection in early life (Warner et al., 2015), and while typically transient reducing by adulthood, remains intact following a lesion of V1. This would suggest a role of this pulvinar pathway in the preserved visual capacity. Furthermore, the degeneration of the retinogeniculostriate pathway is more apparent and occurs more rapidly if the lesion to V1 is in early life (Hendrickson et al., 2015). Therefore, it could be that both the LGN and pulvinar relays to MT play a significant role in the absence of V1, but at different stages of life.

In terms of the role of the SC, studies in adult macaques suggest that it may play a role in sustaining responsiveness and direction selectivity in MT in the absence of V1, even though visual information passing through the SC does not normally contribute significantly to MT responses when V1 is intact (Rodman et al., 1990). Furthermore, studies in humans also suggest that the SC may play a prominent role in residual visual functions in blindsight patients (Leh et al., 2006; Tamietto et al., 2010). However, it is not entirely clear whether the SC pathway through the LGN or the pulvinar is maintaining responsiveness of MT in these and other studies. But because the projections from the SC to the pulvinar, at least in macaque monkeys, are denser than those from the SC to the LGN (Lyon et al., 2010), the SC—pulvinar projection may be more influential in transferring visual signals to extrastriate cortex than the SC—LGN pathway.

While research has principally focused on primates, including humans, other mammals have been shown to produce varying visual deficits after V1 lesions (rodents: Dean, 1981; Exp brain res; Prusky and Douglas, 2014; Vis Res; squirrels: Levey et al., 1973; Wagor, 1978; Diamond, 1973 tree shrews: Snyder et al., 1969; Ward and Masterton, 1970; Killackey et al., 1971; Casagrande and Diamond, 1974; Ware et al., 1974; galagos: Marcotte and Ward, 1980 but see Atencio et al., 1975). Interestingly, unlike primates, many of the visual capabilities remain intact in these species when V1 is damaged, even if that damage is sustained in adulthood. For instance, in ground squirrels pattern vision remains intact after V1 lesions, and it is only when degeneration is observed within the pulvinar (Levey et al., 1973) or the lesion includes both V1 and much of extrastriate cortex (Wagor 1978) that animals show significant visual deficits. Tree shrews maintain both pattern and color vision discrimination capabilities after V1 lesions (Diamond and Hall, 1969; Day-Brown et al., 2010). However, if the SC is damaged, tree shrews act as if they are blind even toward threatening gestures (Casagrande et al., 1972; Casagrande and Diamond, 1974). Therefore, the significance of the extrageniculate pathways in visual perception/performance in tree shrews and squirrels is significantly different than that observed in primates. It is likely that an understanding of how the geniculate and extrageniculate pathways are different across these closely related species may actually shed further light on the role the extrageniculate pathways in normal vision in primates, including humans, and could provide insights into therapeutic strategies when damage occurs in the geniculate pathways of primates.

While it is likely that the debate will continue for many years as to the neural substrate underpinning "blindsight" phenomenon, much information has been garnered as to the function of the extrageniculostriate pathways and their evolution. Future studies must, however, involve multiple species and of different ages, as it is apparent that these pathways have the potential to be modulated during development.

24.6 Summary

Since the previous edition of this book, there have been seminal advances in our understanding of the anatomy, connectivity, and physiology of the subcortical projections to the extrastriate cortex of mammals. We have highlighted three main visual pathways that bypass V1 and terminate directly in extrastriate cortical visual areas. Some of these pathways may be a derived feature found only in primates, including humans, such as the LGN projections to MT, or pulvinar to MT projection. Other pathways may be present in all mammals, and possibly all vertebrates, such as the extrageniculate

pathway through the SC and pulvinar. However, the features of these pathways are variable across mammals and within the primate order, and although the final terminal location in primates of some of these pathways is MT, the presence of homologous pathways in other mammals should not be ignored or discounted. The presence of these extrastriate pathways suggests that they serve some role in visual processing, although at present their particular function is unknown. Furthermore, the functional role these pathways play for any given mammal will likely reflect the needs of species-specific visually mediated behavior, the proportion of retinal input to these alternative pathways, the elaboration and complexity of the cortical targets, and also changes within the geniculate pathway to V1, which has elaborated considerably within primates.

24.7 Future Perspectives

While significant advances have been made, there are still many remaining questions with implications for helping us better understand the evolution of the multiple pathways and the purpose they serve in the developing and adult brain of mammals, with implications for higher-order cognition. A majority of studies have focused on the anatomy and connectivity of the pathways, while much is still to be learned about their physiology. Furthermore, considerable knowledge can be gained from understanding the genetics of the circuitry. RNA sequencing (RNA-seq) is a powerful technique that can be used to analyze the continually changing cellular transcriptome. Mapping and comparing the molecular architecture of the pathways in different species will likely shed significant light on their evolution and role in various species. This research, of course, is underpinned by genome sequencing, with many mammals, including some primates, already having their genome sequenced (eg, Rogers and Gibbs, 2014; www.ensembl.org).

Regarding pathway function, tools developed in the rodent are now successfully translating into other species to assist in probing specific elements of neural pathways. These include genetically based technologies to activate/inhibit specific circuits such as the Designer Receptors Exclusively Activated by Designer Drugs (DREADDs) and optogenetics, which have already advanced our knowledge of circuits in nonhuman primates (Eldridge et al., 2016; Klein et al., 2016). Therefore, it is likely that in the next decade we will be able to transcend many of the questions that remain unanswered to provide a more fruitful conclusion to the function of these pathways in various species, and how they evolved.

References

Adams, M.M., Hof, P.R., Gattass, R., Webster, M.J., Ungerleider, L.G., 2000. Visual cortical projections and chemoarchitecture of macaque monkey pulvinar. J. Comp. Neurol. 419, 377–393.

Allen, A.E., Procyk, C.A., Howarth, M., Walmsley, L., Brown, T.M., 2016. Visual input to the mouse lateral posterior and posterior thalamic nuclei: photoreceptive origins and retinotopic order. J. Physiol. 594 (7), 1911–1929.

Arcaro, M.J., Pinsk, M.A., Kastner, S., 2015. The anatomical and functional organization of the human visual pulvinar. J. Neurosci. 35 (27), 9848–9871.

Atencio, F.W., Diamond, I.T., Ward, J.P., 1975. Behavioral study of the visual cortex of Galago senegalensis. J Comp. Pysiol. Psychol. 89 (10), 1109–1135.

Azzopardi, P., Fallah, M., Gross, C.G., Rodman, H.R., 2003. Response latencies of neurons in visual areas MT and MST of monkeys with striate cortex lesions. Neurospsychologia 41 (13), 1738–1756.

Balaram, P., Hackett, T.A., Kaas, J.H., 2013. Differential expression of vesicular glutamate transporters 1 and 2 may identify distinct modes of glutamatergic transmission in the macaque visual system. J. Chem. Neuroanat. 50–51, 21–38.

Balaram, P., Isaamullah, M., Petry, H.M., Bickford, M.E., Kaas, J.H., 2015. Distributions of vesicular glutamate transporters 1 and 2 in the visual system of tree shrews (Tupaia belangeri). J. Comp. Neurol. 523 (12), 1792–1808.

Balaram, P., Takahata, T., Kaas, J.H., 2011. VGLUT2 mRNA and protein expression in the visual thalamus and midbrain of prosimian galagos (Otolemur garnettii). Eye Brain 2011 (3), 5–15.

Baldwin, M.K., Balaram, P., Kaas, J.H., 2013. Projections of the superior colliculus to the pulvinar in prosimian galagos (Otolemur garnettii) and VGLUT2 staining of the visual pulvinar. J. Comp. Neurol. 521 (7), 1664–1682.

Baldwin, M.K., Nguyen, H., Sekizaki, D., Krubitzer, L.A., 2013. Subcortical Connections of the Superior Colliculus and VGLUT2 Staining in Short-Tailed Opossums (Monodelphis Domestica). SfN Abstr, San Diego.

Baldwin, M.K., Wong, P., Reed, J.L., Kaas, J.H., 2011. Superior colliculus connections with visual thalamus in gray squirrels (Sciurus carolinensis): evidence for four subdivisions within the pulvinar complex. J. Comp. Neurol. 519 (6), 1071–1094.

Barbur, J.L., Watson, J.D.G., Frackowiak, R.S.J., Zeki, S., 1993. Conscious visual perception with V1. Brain 116, 1293–1302.

Barleben, M., Stoppel, C.M., Kaufmann, J., Merkel, C., Wecke, T., Goertler, M., Heinze, H.J., Hopf, J.M., Schoenfeld, M.A., 2015. Neural correlates of visual motion processing without awareness in patients with striate cortex and pulvinar lesions. Hum. Brain Mapp. 36 (4), 1585–1594.

Beck, P.D., Kaas, J.H., 1998. Thalamic connections of the dorsomedial visual area in primates. J. Comp. Neurol. 396, 381–398.

Beckstead, R.M., Frankfurter, A., 1983. A direct projection from the retina to the intermediate gray layer of the superior colliculus demonstrated by anterograde transport of horseradish peroxidase in monkey, cat and rat. Exp. Brain Res. 52, 261–268.

Bender, D.B., 1982. Receptive-field properties of neurons in the macaque inferior pulvinar. J. Neurophysiol. 48 (1), 1–17.

Benevento, L.A., Standage, G.P., 1982. Demonstration of lack of dorsal lateral geniculate nucleus input to extrastriate areas MT and visual 2 in the macaque monkey. Brain Res. 252 (1), 161–166.

Benevento, L.A., Standage, G.P., 1983. The organization of projections of the retinorecipient and nonretinorecipient nuclei of the pretectal complex and layers of the superior colliculus and the lateral pulvinar and medial pulvinar in the macaque monkey. J. Comp. Neurol. 217, 307–336.

Benevento, L.A., Yoshida, K., 1981. The afferent and efferent organization of the lateral geniculo-prestriate pathways in the macaque monkey. J. Comp. Neurol. 203 (3), 455–474.

Berman, R.A., Wurtz, R.H., 2008. Exploring the pulvinar path to visual cortex. Prog. Brain Res. 171, 467–473.

Berman, R.A., Wurtz, R.H., 2010. Functional identification of a pulvinar path from superior colliculus to cortical area MT. J. Neurosci. 30 (18), 6342–6354.

Berman, R.A., Wurtz, R.H., 2011. Signals conveyed in the pulvinar pathway from superior colliculus to cortical area MT. J. Neurosci. 31 (2), 373–384.

Bertini, C., Grasso, P.A., Làdavas, E., May, 11, 2016. The role of the retino-colliculo-extrastriate pathway in visual awareness and visual field recovery. Neuropsyhologia 90, 72–79.

Bickford, M.E., Zhou, N., Krahe, T.E., Govindaiah, G., Guido, W., 2015. Retinal and tectal "Driver-Like" inputs converge in the shell of the mouse dorsal lateral geniculate nucleus. J. Neurosci. 35 (29), 10523–10534.

Bourne, J.A., 2010. Unravelling the development of the visual cortex: implications for plasticity and repair. J. Anat. 217 (4), 449–468.

Bourne, J.A., Rosa, M.G., 2003. Laminar expression of neurofilament protein in the superior colliculus of the marmoset monkey (Callithrix jacchus). Brain Res. 973 (1), 142–145.

Bourne, J.A., Rosa, M.G., 2006. Hierarchical development of the primate visual cortex, as revealed by neurofilament immunoreactivity: early maturation of the middle temporal area (MT). Cereb. Cortex 16 (3), 405–414.

Bourne, J.A., Warner, C.E., Upton, J.D., Rosa, M.G., 2007. Chemoarchitecture of the middle temporal visual area in the marmoset monkey (Callithrix jacchus): laminar distribution of calcium-binding proteins (calbindin, parvalbumin) and nonphosphorylated neurofilament. J. Comp. Neurol. 500 (50), 832–849.

Bravo, H., Olavarria, J., Martinich, S., 1990. Patterns of interhemispheric and striate-peristriate connections in visual cortex of the South American marsupial Marmosa elegans (mouse opossum). Anat. Embryol. Berl. 182 (6), 583–589.

Bravo, H., Olavarria, J., Torrealba, F., 1990. Comparative study of visual inter and intrahemispheric cortico-cortical connections in five native Chilean rodents. Anat. Embryol. Berl. 181 (1), 67–73.

Bridge, H., Leopold, D.A., Bourne, J.A., 2016. Adaptive pulvinar circuitry supports visual cognition. Trends Cogn. Sci. 20 (2), 146–157.

Bridge, H., Thomas, O., Jbabdi, S., Cowey, A., 2008. Changes in connectivity after visual cortical brain damage underlie altered visual function. Brain 131, 1433–1444.

Bullier, J., Kennedy, H., 1983. Projection of the lateral geniculate nucleus onto cortical area V2 in the macaque monkey. Exp. Brain Res. 53 (1), 168–172.

Butler, A.B., 2008. Evolution of brains, cognition, and consciousness. Brain Res. Bull. 75 (2–4), 442–449.

Butler, A.B., 2011. Evolution of the amniote pallium and the origins of mammalian neocortex. Ann. N.Y. Acad. Sci. 1225, 14–27.

Butler, A.B., Hodos, W., 2005. Comparative Vertebrate Neuroanatomy: Evolution and Adaptation. John Wiley & Sons Inc.

Campos-Ortega, J.A., Hayhow, W.R., Cluver, P.F., 1970. A note on the problem of retinal projections to the inferior pulvinar nucleus of primates. Brain Res. 22 (1), 126–130.

Casagrande, V.A., Diamond, I.T., 1974. Ablation study of the superior colliculus in the tree shrew (Tupaia glis). J. Comp. Neurol. 156, 207–238.

Casagrande, V.A., Harting, J.K., Hall, J.A., Diamond, I.T., Martin, G.F., 1972. Superior colliculus of the tree shrew: a structural and functional subdivision into superficial and deep layers. Science 177 (4047), 444–447.

Casagrande, V.A., Ichida, J.M., 2002. The lateral geniculate nucleus. Physiology of the Eye. Mosby Co., St. Louis, pp. 655–668.

Casagrande, V.A., Ichida, J.M., Boyd, J., 2007. The evolution of parallel pathways in the brains of primates. In: Kaas, J.H. (Ed.), Evolutionary Neuroscience. Academic Press, NY, pp. 871–892.

Casagrande, V.A., Kaas, J.H., 1994. The afferent, intrinsic, and efferent connections of primary visual cortex in primates. In: Peters, A., Rockland, K. (Eds.), Primary Viusal Cortex of Primates, vol. 10. Plenum Press, N.Y, pp. 201–259.

Cerkevich, C.M., Lyon, D.C., Balaram, P., Kaas, J.H., 2014. Distribution of cortical neurons projecting to the superior colliculus in macaque monkeys. Eye Brain 6, 121–137.

Chalupa, L.M., 1977. A review of cat and monkey studies implicating the pulvinar in visual function. Behav. Biol. 20 (2), 149–167.

Chalupa, L.M., Thompson, I., 1980. Retinal ganglion cell projections to the superior colliculus of the hamster demonstrated by the horseradish peroxidase technique. Neurosci. Lett. 19 (1), 13–19.

Chomsung, R.D., Petry, H.M., Bickford, M.E., 2008. Ultrastructural examination of diffuse and specific tectopulvinar projections in the tree shrew. J. Comp. Neurol. 510 (1), 24–46.

Chomsung, R.D., Wei, H., Day-Brown, J.D., Petry, H.M., Bickford, M.E., 2010. Synaptic organization of connections between the temporal cortex and pulvinar nucleus of the tree shrew. Cereb. Cortex 20 (4), 997–1011.

Chow, K.L., 1950. A retrograde cell degeneration study of the cortical projection field of the pulivnar in the monkey. J. Comp. Neurol. 93 (3), 313–340.

Cola, M.G., Gray, D.N., Seltzer, B., Cusick, C.G., 1999. Human Thalamus: neurochemical mapping of inferior pulvinar complex. Neuroreport 10 (18), 3733–3738.

Cola, M.G., Seltzer, B., Preuss, T.M., Cusick, C.G., 2005. Neurochemical organization of chimpanzee inferior pulvinar complex. J. Comp. Neurol. 484 (3), 299–312.

Coleman, J., Clerici, W.J., 1980. Extrastriate projections from thalamus to posterior occipital-temporal cortex in rat. Brain Res. 194 (1), 205–209.

Coleman, J.E., Law, K., Bear, M.F., 2009. Anatomical origins of ocular dominance in mouse primary visual cortex. Neuroscience 161 (2), 561–571.

Collins, C.E., Lyon, D.C., Kaas, J.H., 2003. Responses of neurons in the middle temporal visual area after long-standing lesions of the primary visual cortex in adult new world monkeys. J. Neurosci. 23 (6), 2251–2264.

Collins, C.E., Lyon, D.C., Kaas, J.H., 2005. Distribution across cortical areas of neurons projecting to the superior colliculus in new world monkeys. Anat. Rec. A Discov. Mol. Cell Evol. Biol. 285 (1), 619–627.

Condé, F., Lund, J.S., Lewis, D.A., 1996. The hierarchical development of monkey visual cortical regions as revealed by the maturation of parvalbumin-immunoreactive neurons. Brain Res. Dev. Brain Res. 96 (1–2), 261–276.

Coogan, T.A., Burkhalter, A., 1993. Hierarchical organization of areas in rat visual cortex. J. Neurosci. 13 (9), 3749–3772.

Cowey, A., Stoerig, P., 1989. Projection patterns of surviving neurons in the dorsal lateral geniculate nucleus following discrete lesions of striate cortex: implications for residual vision. Exp. Brain Res. 75 (3), 631–638.

Cowey, A., 2010. Visual system: how does blindsight arise? Curr. Biol. 20 (17), 702–704.

Cowey, A., 2010. The blindsight saga. Exp. Brain Res. 200 (1), 3–24. https://doi.org/10.1007/s00221-009-1914-2.

Cowey, A., Alexander, I., Stoerig, P., 2011. Transneuronal retrograde degeneration of retinal ganglion cells and optic tract in hemianopic monkeys and humans. Brain 143 (7), 2149–2157.

Cowey, A., Stoerig, P., Bannister, M., 1994. Retinal ganglion cells labelled from the pulvinar nucleus in macaque monkeys. Neuroscience 61 (3), 691–705.

Crain, B.J., Hall, W.C., 1980. The organization of afferents to the lateral posterior nucleus of the golden hamster after different combinations of neonatal lesions. J. Comp. Neurol. 193, 403–412.

Cusick, C.G., Scripter, J.L., Darensbourg, J.G., Weber, J.T., 1993. Chemoarchitectonic subdivisions of the visual pulvinar in monkeys and their connectional relations with the middle temporal and rostral dorsolateral viusal areas, MT and DLr. J. Comp. Neurol. 336, 1–30.

Day-Brown, J.D., Wei, H., Chomsung, R.D., Petry, H.M., Bickford, M.E., 2010. Pulvinar projections to the striatum and amygdala in the tree shrew. Front. Neuroanat. 15 (4), 143. https://doi.org/10.3389/fnana.2010.00143.

Derrington, A.M., Lennie, P., 1984. Spatial and temporal contrast sensitivities of neurons in lateral geniculate nucleus of macaque. J. Physiol. 357, 219–240.

Diamond, I.T., 1973. The evolution of the tectal-pulvinar system in mammals: structural and behavioral studies of the visual system. Symp. Zool. Soc. Lond. 33, 205–233.

Diamond, I.T., Conley, M., Fitzpatrick, D., Raczkowski, D., 1991. Evidence for separate pathways within the tecto-geniculate projection in the tree shrew. Proc. Natl. Acad. Sci. U.S.A. 88 (4), 1315–1319.

Diamond, I.T., Fitzpatrick, D., Schmechel, D., 1993. Calcium binding proteins distinguish large and small cells in the ventral posterior and lateral geniculate nucleus of prosimian galago and the tree shrew (Tupaia belangeri). Proc. Natl. Acad. Sci. U.S.A. 90, 1425–1429.

Diamond, I.T., Hall, W.C., April, 18 1969. Evolution of neocortex. Science 164 (3877), 251–262.

Dick, A., Kaske, A., Creutzfeldt, O.D., 1991. Topographical and topological organization of thalamocortical projection to the striate and prestriate cortex in the marmoset (Callithrix jacchus). Exp. Brain Res. 84, 233–253.

Dreher, B., Wang, C., Burke, W., 1996. Limits of parallel processing: excitatory convergence of different information channels on single neurons in striate and extrastriate visual cortices. Clin. Exp. Pharmacol. Physiol. 23 (10–11), 913–925.

Eldridge, M.A., Lerchner, W., Saunders, R.C., Kaneko, H., Krausz, K.W., Gonzalez, F.J., Ji, B., Higuchi, M., Minamimoto, T., Richmound, B.J., 2016. Chemogenetic disconnection of monkey orbitofrontal and rhinal cortex reversibly disrupts reward value. Nat. Neurosci. 19 (1), 37–39.

Espinoza, S.G., Subiabre, J.E., Thomas, H.C., 1992. Retinotopic organization of striate and extrastriate visual cortex in the golden hamster (Mesocricetus auratus). Biol. Res. 25 (2), 101–107.

Feig, S., Van Lieshout, D.P., Harting, J.K., 1992. Ultrastructural studies of retinal, visual cortical (area 17), and parabigeminal terminals within the superior colliculus of Galago crassicaudatus. J. Comp. Neurol. 319 (1), 85–99.

Felleman, D.J., Van Essen, D.C., 1991. Distributed hierarchical processing in the primate cerebral cortex. Cereb. Cortex 1 (1), 1–47.

Felsten, G., Benevento, L.A., Burman, D., 1983. Opponent-color responses in macaque extrageniculate visual pathways: the lateral pulvinar. Brain Res. 288, 363–367.

ffytche, D.H., Guy, C.N., Zeki, S., 1996. Motion specific responses from a blind hemifield. Brain 119, 1971–1982.

Fitzpatrick, D., Carey, R.G., Diamond, I.T., 1980. The projection of the superior colliculus upon the lateral geniculate body in Tupaia glis and Galago senegalensis. Brain Res. 194 (2), 494–499.

Fredes, F., Vega-Zuniga, T., Karten, H., Mpodozis, J., 2012. Bilateral and ipsilateral ascending tectopulvinar pathways in mammals: a study in the squirrel (Spermophilus beecheyi). J. Comp. Neurol. 520 (8), 1800–1818.

Fries, W., 1981. The projection from the lateral geniculate nucleus to the prestriate cortex of the macaque monkey. Proc. R. Soc. Lond. B Biol. Sci. 213 (1190), 73−86.

Gaillard, F., Karten, H.J., Sauvé, Y., 2013. Retinorecipient areas in the diurnal murine rodent Arvicanthis niloticus: a disproportionally large superior colliculus. J. Comp. Neurol. 521 (8), 1699−1726.

Garey, L.J., Powell, T.P.S., 1967. The projection of the lateral geniculate nucleus upon the cortex in the cat. Proc. R. Soc. Biol. Sci. 169 (1014).

Gattass, R., Sousa, A.P., Oswaldo-Cruz, E., 1978. Single unit response types in the pulvinar of the Cebus monkey to multisensory stimulation. Brain Res. 158 (1), 75−87.

Girard, P., Salin, P.A., Bullier, J., 1992. Response selectivity of neurons in area MT of the macaque monkey during reversible inactivation of area V1. J. Neurophysiol. 67 (6), 1437−1446.

Glendenning, K.K., Hall, J.A., Diamond, I.T., Hall, W.C., 1975. The pulvinar nucleus of *Galago senegalensis*. J. Comp. Neurol. 161, 419−458.

Glickstein, M., King, R.A., Miller, J., Berkley, M., 1967. Cortical projections from the dorsal lateral geniculate nucleus of cats. J. Comp. Neurol. 130 (1), 55−75.

Godement, P., Salaün, J., Imbert, M., 1984. Prenatal and postnatal development of retinogeniculate and retinocollicular projections in the mouse. J. Comp. Neurol. 230 (4), 552−575.

Goodchild, A.K., Martin, P.R., 1998. The distribution of calcium-binding proteins in the lateral geniculate nucleus and visual cortex of a New World monkey, the marmoset. *Callithrix jacchus*. Vis. Neurosci. 15 (4), 625−642.

Graham, J., Casagrande, V.A., 1980. A light microscopic and electron microscopic study of the superficial layers of the superior colliculus of the tree shrew (*Tupaia glis*). J. Comp. Neurol. 191, 133−151.

Gray, D., Gutierrez, C., Cusick, C.G., 1999. Neurochemical organization of inferior pulvinar complex in squirrel monkeys and macaque monkeys revealed by acetylcholinesterase histochemistry, calbindin and Cat-301 immunostaining and Wisteria floribunda agglutinin binding. J. Comp. Neurol. 409 (3), 452−468.

Gross, C.G., Moore, T., Rodman, H.R., 2004. Visually guided behavior after V1 lesions in young and adult monkeys and its relationship to blindsight in humans. Prog. Brain Res. 144, 279−294.

Guillery, R.W., 2005. Is postnatal neocortical maturation hierarchical? Trends Neurosci. 28 (10), 512−517.

Gutierrez, C., Yaun, A., Cusick, C.G., 1995. Neurochemical subdivisions of the inferior pulvinar in macaque monkeys. J. Comp. Neurol. 363, 545−562.

Harting, J.K., Glendenning, K.K., Diamond, I.T., Hall, J.A., 1973. Evolution of the primate visual system: anterograde degeneration studies of the tecto-pulvinar system. Am. J. Phys. Anthropol. 38 (2), 383−392.

Harting, J.K., Huerta, M.F., Hashikawa, T., Van Lieshout, D.P., 1991. Projections of the mammalian superior colliculus upon the dorsal lateral geniculate nucleus: organization of tectogeniculate pathways in nineteen species. J. Comp. Neurol. 304, 275−306.

Hattar, S., Kumar, M., Park, A., Tong, P., Tung, J., Yau, K.W., Berson, D.M., 2006. Central projections of melanopsin-expressing retinal ganglion cells in the mouse. J. Comp. Neurol. 497 (3), 326−349.

Hendrickson, A., Warner, C.E., Possin, D., Huang, J., Kwan, W.C., Bourne, J.A., 2015. Retrograde degeneration in the retina and lateral geniculate nucleus of the V1-lesioned marmoset monkey. Brain Struct. Funct. 220 (1), 351−360.

Hofbauer, A., Dräger, U.C., 1985. Depth segregation of retinal ganglion cells projecting to mouse superior colliculus. J. Comp. Neurol. 234 (4), 465−474.

Homman-Ludiye, J., Manger, P.R., Bourne, J.A., 2010. Immunohistochemical parcellation of the ferret (*Mustela putorius*) visual cortex reveals substantial homology with the cat (*Felis catus*). J. Comp. Neurol. 518 (21), 4439−4462.

Hubel, D.H., Livingstone, M.S., 1990. Color and contrast sensitivity in the lateral geniculate body and primary visual cortex of the macaque monkey. J. Neurosci. 10, 2223−2237.

Huber, A., 1992. Homonymous hemianopia. Neuro. Ophtalmol. 12, 351−366. https://doi.org/10.3109/01658109209036989.

Huerta, M.F., Harting, J.K., 1983. Sublamination within the superficial gray layer of the squirrel monkey: an analysis of the tectopulvinar projection using anterograde and retrograde transport methods. Brain Res. 261 (1), 119−126.

Irvin, G.E., Casagrande, V.A., Norton, T.T., 1993. Center/surround relationships of magnocellular, parvocellular, and koniocellular relay cells in primate lateral geniculate nucleus. Vis. Neurosci. 10, 254−270.

Irvin, G.E., Norton, T.T., Sesma, M.A., Casagrande, V.A., 1986. W-like response properties of interlaminar zone cells in the lateral geniculate nucleus of a primate (*Galago crassicaudatus*). Brain Res. 363−373.

Itaya, S.K., Van Hoesen, G.W., 1983. Retinal projections to the inferior and medial pulvinar nuclei in the old-world monkey. Brain Res. 269 (2), 223−230.

Itoh, K., Mizuno, N., Kudo, M., 1983. Direct retinal projections to the lateroposterior and pulvinar nuclear complex (LP-Pul) in the cat, as revealed by the anterograde HRP method. Brain Res. 276 (2), 325−328.

Jain, N., Preuss, T.M., Kaas, J.H., 1994. Subdivisions of the visual system labeled with the Cat-301 antibody in tree shrews. Vis. Neurosci. 11 (4), 731−741.

Johnson, J.K., Casagrande, V.A., 1995. Distribution of calcium-binding proteins within the parallel visual pathways of a primate (*Galago crassicaudatus*). J. Comp. Neurol. 356, 238−320.

Jones, E.G., 2007. The Thalamus. Cambridge University Press, NY.

Jones, E.G., Hendry, S.H., 1989. Differential calcium binding protein immunoreactivity distinguishes classes of relay neurons in monkey thalamic nuclei. Eur. J. Neurosci. 1 (3), 222−246.

Kaas, J.H., 1982. The segregation of function in the nervous system: why do sensory systems have so many subdivisions?. In: Neff, W.P. (Ed.), Contributions of Sensory Physiology, vol. 7. Academic Press, New York, pp. 201−240.

Kaas, J.H., 1989. The evolution of complex sensory systems in mammals. J. Exp. Biol. 146, 165−176.

Kaas, J.H., Collins, C.E., 2001. The organization of sensory cortex. Curr. Opin. Neurobiol. 11 (4), 498−504.

Kaas, J.H., Huerta, M.F., 1988. The subcortical system of primates. In: Steklis, H.D. (Ed.), Comparative Priamte Biology. Liss. Wiley, New York.

Kaas, J.H., Krubitzer, L., 1992. Area 17 lesions deactivate area MT in owl monkeys. Vis. Neurosci. 9 (3−4), 399−407.

Kaas, J.H., Lyon, D.C., 2007. Pulvinar contributions to the dorsal and ventral streams of visual processing in primates. Brain Res. Rev. 55 (2), 285−296.

Kahn, D.M., Krubitzer, L., 2002. Retinofugal projections in the short-tailed opossum (*Monodelphis domestica*). J. Comp. Neurol. 447, 114−127.

Kalatsky, V.A., Stryker, M.P., 2003. New paradigm for optical imaging: temporally encoded maps of intrinsic signal. Neuron 38 (4), 529−545.

Kaplan, E., Shapley, R.M., 1982. X and Y cells in the lateral geniculate nucleus of macaque monkey. J. Physiol. 330, 125−143.

Karamanlidis, A.N., Saigal, R.P., Giolli, R.A., Mangana, O., Michaloudi, H., 1979. Visual thalamocortical connections in sheep studied by means of the retrograde transport of horseradish-peroxidase. J. Comp. Neurol. 187 (2), 245−260.

Karten, H.J., Cox, K., Mpodozis, J., 1997. Two distinct populations of tectal neurons have unique connections within the retinotectorotundal pathway of the pigeon (*Columba livia*). J. Comp. Neurol. 387 (3), 449−465.

Killackey, H., Snyder, M., Diamond, I.T., 1971. Function of striate and temporal cortex in the tree shrew. J. Comp. Physiol. Psychol. 74 (1), 1–29.

Kiper, D.C., Zesiger, P., Maeder, P., Deonna, T., Innocenti, G.M., 2002. Vision after early-onset lesions of the occipital cortex: I. Neuropsychological and psychophysical studies. Neural Plast. 9 (1), 1–25.

Klein, C., Evrard, H.C., Shapcott, K.A., Havekamp, S., Logothetis, N.K., Schmid, M.C., 2016. Cell-targeted optogenetics and electrical microstimulation reveal the primate koniocellular projection to supragranular visual cortex. Neuron 90 (1), 143–151.

Kremers, J., Weiss, S., Zrenner, E., 1997. Temporal properties of marmoset lateral geniculate cells. Vis. Res. 37, 2649–2660.

Leh, S.E., Mullen, K.T., Ptito, M., 2006. Absence of S-cone input in human blindsight following hemispherectomy. Eur. J. Neurosci. 24 (10), 2954–2960.

Leopold, D.A., 2012. Primary visual cortex: awareness and blindsight. Ann. Rev. Neurosci. 35, 91–109.

Leuba, G., Saini, K., 1996. Calcium-binding proteins immunoreactivity in the human subcortical and cortical visual structures. Vis. Neurosci. 13 (6), 997–1009.

Levey, N.H., Harish, J., Jane, J.A., 1973. Effects of visual cortical ablation on pattern discrimination in ground squirrel (Citellus tridecemlineatus). Exp. Neurol. 38, 270–276.

Lin, C.S., Kaas, J.H., 1979. The inferior pulvinar complex in owl monkeys: architectonic subdivisions and patterns of input from the superior colliculus and subdivisions of visual cortex. J. Comp. Neurol. 187, 655–678.

Linden, V.H., Perry, V.H., 1983. Massive retinotectal projection in rats. Brain Res. 272 (1), 145–149.

Ling, C., Schneider, G.E., Northmore, D., Jhaveri, S., 1997. Afferents from the colliculus, cortex, and retina have distinct terminal morphologies in the lateral posterior thalamic nucleus. J. Comp. Neurol. 388 (3), 467–483.

Luppino, G., Matelli, M., Carey, R.G., Fitzpatrick, D., Diamond, I.T., 1988. New view of the organization of the pulvinar nucleus in Tupaia as revealed by tectopulvinar and pulvinar-cortical projections. J. Comp. Neurol. 273, 67–86.

Lyon, D.C., Jain, N., Kaas, J.H., 1998. Cortical connections of striate and extrastriate visual areas in tree shrews. J. Comp. Neurol. 401 (1), 109–128.

Lyon, D.C., Jain, N., Kaas, J.H., 2003. The visual pulvinar in tree shrews I. Multiple subdivisions revealed through acetylcholinesterase and Cat-301 chemoarchitecture. J. Comp. Neurol. 467 (4), 593–606.

Lyon, D.C., Jain, N., Kaas, J.H., 2003. The visual pulvinar in tree shrews II. Projections of four nuclei to areas of visual cortex. J. Comp. Neurol. 467 (4), 607–627.

Lyon, D.C., Nassi, J.J., Callaway, E.M., 2010. A disynaptic relay from superior colliculus to dorsal stream visual cortex in macaque monkey. Neuron 65, 270–279.

Lyon, D.C., Rabideau, C., 2012. Lack of robust LGN label following transneuronal rabies virus injections into macaque area V4. J. Comp. Neurol. 520 (11), 2500–2511.

Lysakowski, A., Standage, G.P., Benevento, L.A., 1988. An investigation of collateral projections of the dorsal lateral geniculate nucleus and other subcortical structures to cortical areas V1 and V4 in the macaque monkey: a double label retrograde tracer study. Exp. Brain Res. 69, 651–661.

Maier, A., Wilke, M., Aura, C., Zhu, C., Ye, F.Q., Leopold, D.A., 2008. Divergence of fMRI and neural signals in V1 during perceptual suppression in the awake monkey. Nat. Neurosci. 10, 1193–1200.

Major, D.E., Rodman, H.R., Libedinsky, C., Karten, H.J., 2003. Pattern of retinal projections in the California ground squirrel (Spermophilus beecheyi): anterograde tracing study using cholera toxin. J. Comp. Neurol. 463 (3), 317–340.

Manger, P.R., Rosa, M.G., 2005. Visual thalamocortical projections in the flying fox: parallel pathways to striate and extrastriate areas. Neuroscience 130 (2), 497–511.

Marcotte, R.R., Ward, J.P., 1980. Preoperative overtraining protects against form learning deficits after lateral occipital lesions in Galago senegalensis. J. Comp. Neurol. 94 (2), 305–312.

Martin, P.R., White, A.J., Goodchild, A.K., Wilder, H.D., Sefton, A.J., 1997. Evidence that blue-on cells are part of the third geniculocortical pathway in primates. Eur. J. Neurosci. 9 (7), 1536–1541.

Matteau, I., Boire, D., Ptito, M., 2003. Retinal projections in the cat: a cholera toxin B subunit study. Vis. Neurosci. 20 (5), 481–493.

May, P.J., 2006. The mammalian superior colliculus: laminar structure and connections, 151, 321–378.

Mize, R.R., Wetzel, A.B., Thompson, V.E., 1971. Contour discrimination in the rat following removal of posterior neocortex. Physiol. Behav. 6, 241–246.

Mizuno, N., Itoh, K., Uchida, K., Uemura-Sumi, M., Matsushima, R., 1982. A retino-pulvinar projection in the macaque monkey as visualized by the use of anterograde transport of horseradish peroxidase. Neurosci. Lett. 30 (3), 199–203.

Montero, V.M., 1993. Retinotopy of cortical connections between the striate cortex and extrastriate visual areas in the rat. Exp. Brain Res. 94 (1), 1–15.

Mooney, R.D., Fish, S.E., Rhodes, R.W., 1984. Anatomical and functional organization of pathway from superior colliculus to lateral posterior nucleus in hamster. J. Neurophysiol. 51 (3), 407–431.

Moore, T., Rodman, H.R., Reep, A.B., Gross, C.G., Mezrich, R.S., 1996. Greater residual vision in monkey after striate cortex damage in infancy. J. Neurophysiol. 76 (6), 3928–3933.

Morin, L.P., Studholme, K.M., 2014. Retinofugal projections in the mouse. J. Comp. Neurol. 522 (16), 3733–3753.

Mundinano, I.C., Kwan, W.C., Bourne, J.A., October 20, 2015. Mapping the mosaic sequence of primate visual cortical development. Front. Neuroanat. 9, 132.

Nassi, J.J., Lyon, D.C., Callaway, E.M., 2006. The parvocellular LGN provides a robust disynaptic input to the visual motion area MT. Neuron 50 (2), 319–327.

Norton, T.T., Casagrande, V.A., 1982. Laminar organization of receptive-field properties in lateral geniculate nucleus of bush baby (Galago crassicaudatus). J. Neurophysiol. 47, 715–741.

Norton, T.T., Casagrande, V.A., Irvin, G.E., Sesma, M.A., Petry, H.M., 1988. Contrast-sensitivity functions of W-, X-, and Y-like relay cells in the lateral geniculate nucleus of bush baby, Galago crassicaudatus. J. Neurophysiol. 59, 1639–1656.

O'Brien, B.J., Abel, P.L., Olavarria, J.F., 2001. The retinal input to calbindin-D28k-defined subdivisions in macaque inferior pulvinar. Neurosci. Lett. 312 (3), 145–148.

Ohno, T., Misgeld, U., Kitai, S.T., Wagner, A., 1975. Organization of the visual afferents into the LGd and the pulvinar of the tree shrew Tupaia glis. Brain Res. 90 (1), 153–158.

Olavarria, J., Montero, V., 1990. Elaborate organization of visual cortex in the hamster. Neurosci. Res. 8 (1), 40–47.

Olavarria, J., Montero, V.M., 1984. Relation of callosal and striate-extrastriate cortical connections in the rat: morphological definition of extrastriate visual areas. Exp. Brain Res. 54 (2), 240–252.

Olavarria, J., Montero, V.M., 1998. Organization of visual cortex in the mouse revealed by correlating callosal and striate-extrastriate connections. Vis. Neurosci. 3 (1), 59–69.

Percival, K.A., Koizumi, A., Masri, R.A., Buzas, P., Martin, P.R., Grunert, U., 2014. Identification of a pathway from the retina to koniocellular layer K1 in the lateral geniculate nucleus of marmoset. J. Neurosci. 34 (11), 3821–3825.

Perenin, M.T., 1991. Discrimination of motion direction in perimetrically blind fields. Neuroreport 2, 397–400.

Perry, V.H., Cowey, A., 1982. A sensitive period for ganglion cell degeneration and the formation of aberrant retino-fugal connections following tectal lesions in rats. Neuroscience 7 (3), 583–594.

Perry, V.H., Cowey, A., 1984. Retinal ganglion cells that project to the superior colliculus and pretectum in the macaque monkey. Neuroscience 12 (4), 1125–1137.

Petersen, S.E., Robinson, D.E., Keys, W., 1985. Pulvinar nucleus of the behaving rhesus monkey: visual responses and their modulation. J. Neurophysiol. 54 (4), 867–886.

Polonsky, A., Blake, R., Braun, J., Heeger, D.J., 2000. Neural activity in human primary visual cortex correlates with perception during binocular rivalry. Nat. Neurosci. 11, 1153–1159.

Pöppel, E., Held, R., Frost, D., 1973. Residual visual function after brain wounds involving the central visual pathways in man. Nature 243, 295–296.

Prusky, G.T., Douglas, R.M., 2014. Characterization of mouse cortical spatial vision. Vis. Res. 44 (28), 3411–3418.

Raczkowski, D., Diamond, I.T., 1980. Cortical connections of the pulvinar nucleus in galago. J. Comp. Neurol. 193, 1–40.

Raczkowski, D., Diamond, I.T., 1981. Projections from the superior colliculus and the neocortex to the pulvinar nucleus in galago. J. Comp. Neurol. 200, 231–254.

Real, M.A., Heredia, R., Dávila, J.C., Guirado, S., 2008. Efferent retinal projections visualized by immunohistochemical detection of the estrogen-related receptor beta in the postnatal and adult mouse brain. Neurosci. Lett. 438 (1), 48–53.

Reese, B.E., Jeffery, G., 1983. Crossed and uncrossed visual topography in dorsal lateral geniculate nucleus of pigmented rat. J. Neurophysiol. 49, 877–885.

Reid, R.C., Shapley, R.M., 1992. Spatial structure of cone inputs to receptive fields in primate lateral geniculate nucleus. Nature 356, 716–718.

Robson, J.A., Hall, W.C., 1977. The organization of the pulvinar in the grey squirrel (Sciurus carolinensis). J. Comp. Neurol. 173, 355–388.

Rodman, H.R., Gross, C.G., Albright, T.D., 1989. Afferent basis of visual response properties in area MT of the macaque. I. Effects of striate cortex removal. J. Neurosci. 9 (6), 2033–2050.

Rodman, H.R., Gross, C.G., D, A.T., 1990. Afferent basis of visual response properties in area MT of the macaque. II. Effects of superior colliculus removal. J. Neurosci. 10 (4), 1154–1164.

Rodman, H.R., Sorenson, K.M., Shim, A.J., Hexter, D.P., 2001. Calbindin immunoreactivity in the geniculo-extrastriate system of the macaque: implications for heterogeneity in the koniocellular pathway and recovery from cortical damage. J. Comp. Neurol. 431 (2), 168–181.

Rogers, J., Gibbs, R.A., 2014. Comparative primate genomics: emerging patterns of genome content and dynamics. Nat. Rev. Genet. 15 (5), 347–359.

Rosa, M.G., 1999. Topographic organisation of extrastriate areas in the flying fox: implications for the evolution of mammalian visual cortex. J. Comp. Neurol. 411 (3), 503–523.

Rosa, M.G., Krubitzer, L.A., 1999. The evolution of visual cortex: where is V2? Trends Neurosci. 22 (6), 242–248.

Rosa, M.G., Tweedale, R., Elston, G.N., 2000. Visual responses of neurons in the middle temporal area of new world monkeys after lesions of striate cortex. J. Neurosci. 20 (14), 5552–5563.

Rovo, Z., Ulbert, I., Acsady, L., 2012. Drivers of the primate thalamus. J. Neurosci. 32 (49), 17894–17908.

Schmid, M.C., Maier, A., 2015. To see or not to see—thalamo-cortical networks during blindsight and perceptual suppression. Prog. Neurobiol. 126 (36–48).

Schmid, M.C., Mrowka, S.W., Turchi, J., Saunders, R.C., Wilke, M., Peters, A., Ye, F.Q., Leopold, D.A., 2010. Blindsight depends on the lateral geniculate nucleus. Nature 466 (7304), 373–377.

Schneider, G.E., 1969. Two visual systems. Science 163, 895–902.

Sesma, M.A., Casagrande, V.A., Kaas, J.H., 1984. Cortical connections of area 17 in tree shrews. J. Comp. Neurol. 230 (3), 337–351.

Sherman, S.M., Guillery, R.W., 2002. The role of the thalamus in the flow of information to the cortex. Philos. Trans. R. Soc. Lond. B Biol. Sci. 357 (1428), 1695–1708.

Sherman, S.M., Guillery, R.W., 2006. Exploring the Thalamus and its Role in Cortical Function. MIT Press, MA.

Shipp, S., 2001. Corticopulvinar connections of areas V5, V4, and V3 in macaque monkey: a dual model of retinal and cortical topographies. J. Comp. Neurol. 439, 469–490.

Sincich, L.C., Park, K.F., Wohlgemuth, M.J., Horton, J.C., 2004. Bypassing V1: a direct geniculate input to area MT. Nat. Neurosci. 7 (10), 1123–1128.

Snyder, M., Killackey, H., Diamond, I.T., 1969. Color vision in the tree shrew after removal of posterior neocortex. J. Neurophysiol. 32, 554–563.

Soares, J.G.M., Gattass, R., Souza, A.P.B., Rosa, M.G., Fiorani, M., Brandao, B.L., 2001. Connectional and neurochemical subdivisions of the pulvinar in Cebus monkey. Vis. Neurosci. 18, 25–41.

Solomon, S.G., White, A.J., Martin, P.R., 2002. Extraclassical receptive field properties of parvocellular, magnocellular, and koniocellular cells in the primate lateral geniculate nucleus. J. Neurosci. 22 (1), 338–349.

Somogyi, G., Hajdu, F., Hassler, R., Wanger, A., 1981. An experimental electron microscopical study of a direct retino-pulvinar pathway in the tree shrew. Exp. Brain Res. 43 (3–4), 447–450.

Sorenson, K.M., Rodman, H.R., 1996. The lateral geniculate nucleus does not project to area TE in infant or adult macaques. Neurosci. Lett. 217 (1), 5–8.

Sorenson, K.M., Rodman, H.R., 1999. A transient geniculo-extrastriate pathway in macaques? Implications for 'blindsight'. Neuroreport 10 (16), 3295–3299.

Stepniewska, I., Hui-Xin, Q., Kaas, J.H., 1999. Do superior colliculus projection zones in the inferior pulvinar project to MT in primates? Eur. J. Neurosci. 11, 460–480.

Stepniewska, I., Hui-Xin, Q., Kaas, J.H., 2000. Projections of the superior colliculus to subdivisions of the inferior pulivnar in New World and Old World monkeys. Vis. Neurosci. 17, 529–549.

Stepniewska, I., Kaas, J.H., 1997. Architectonic subdivisions of the inferior pulvinar in new world and old world monkeys. Vis. Neurosci. 14 (6), 1043–1060.

Stoerig, P., Cowey, A., 1997. Blindsight in man and monkey. Brain 120, 535–559.

Stoerig, P., Cowey, A., 2007. Blindsight. Curr. Biol. 17 (19), 822–842.

Sun, N., Cassell, M.D., Perlman, S., 1996. Anterograde, transneuronal transport of herpes simplex virus type 1 strain H129 in the murine visual system. J. Virol. 70 (8), 5405–5413.

Szmajda, B.A., Grunert, U., Martin, P.R., 2008. Retinal ganglion cell inputs to the koniocellular pathway. J. Comp. Neurol. 510 (3), 251–268.

Takahashi, T., 1985. The organization of the lateral thalamus of the hooded rat. J. Comp. Neurol. 231, 281–309.

Tamietto, M., Cauda, F., Corazzini, L.L., Savazzi, S., Marzi, C.A., Gobel, R., Weiskrantz, L., de Gelder, B., 2010. Collicular vision guides nonconscious behavior. J. Cogn. Neurosci. 22 (5), 888–902.

Tanaka, M., Lindsley, E., Lausmann, S., Creutzfeldt, O.D., 1990. Afferent connections of the prelunate visual association cortex (areas V4 and DP). Anat. Embryol. Berl. 181 (1), 19–30.

Tigges, J., Tigges, M., 1981. Distribution of retinofugal and corticofugal axon terminals in the superior colliculus of squirrel monkey. Invest. Ophthalmol. Vis. Sci. 149–158.

Tinelli, F., Cicchini, G.M., Arrighi, R., Tosetti, M., Cioni, G., Morrone, M.C., 2013. Blindsight in children with congenital and acquired cerebral lesions. Cortex 49 (6), 1636–1647.

Tohmi, M., Meguro, R., Tsukano, H., Hishida, R., Shibuki, K., 2014. The extrageniculate visual pathway generates distinct response properties in the higher visual areas of mice. Curr. Biol. 24 (6), 587–597.

Tong, F., Engel, S.A., 2001. Interocular rivalry revealed in the human cortical blind spot representation. Nature 411 (6834), 195–199.

Uchiumi, O., Sugita, S., Fukuta, K., 1995. Retinal projections to the subcortical nuclei in the Japanese field vole (Microtus montebelli). Exp. Anim. 44 (3), 193–203.

Usrey, W.M., Reid, R.C., 2000. Visual physiology of the lateral geniculate nucleus in two species of New World monkey: Saimiri sciureus and Aotus trivirgatis. J. Physiol. 523, 755–769.

Van Brussel, L., Gerits, A., Arckens, L., 2009. Identification and localization of functional subdivisions in the visual cortex of the adult mouse. J. Comp. Neurol. 514 (1), 107–116.

Van der Gucht, E., Hof, P.R., van Brussel, L., Burnat, K., Arkens, L., 2007. Neurofilament protein and neuronal activity markers define regional architectonic parcellation in the mouse visual cortex. Anat. Rec. Hob. 290 (1), 1–31.

Van Hooser, S.D., Heimel, J.A., Nelson, S.B., 2003. Receptive field properties and laminar organization of lateral geniculate nucleus in the gray squirrel (Sciurus carolinensis). J. Neurophysiol. 90 (5), 3398–3418.

Van Hooser, S.D., Nelson, S.B., 2006. The squirrel as a rodent model of the human visual system. Vis. Neurosci. 23, 765–778.

Van Hooser, S.D., Roy, A., Rhodes, H.J., Culp, J.H., Fitzpatrick, D., 2013. Transformation of receptive field properties from lateral geniculate nucleus to superficial V1 in the tree shrew. J. Neurosci. 33 (28), 11494–11505.

Vaney, D.I., Peichi, L., Wassle, H., B, I.R., 1981. Almost all ganglion cells in the rabbit retina project to the superior colliculus. Brain Res. 212 (2), 447–453.

Wagor, E., 1978. Pattern vision in the gray squirrel after visual cortex ablation. Behav. Biol. 22, 1–22.

Wall, J.T., Symonds, L.L., Kaas, J.H., 1982. Cortical and subcortical projections of the middle temporal area (MT) and adjacent cortex in galagos. J. Comp. Neurol. 211, 193–214.

Wang, Q., Sporns, O., Burkhalter, A., 2012. Network analysis of corticocortical connections reveals ventral and dorsal processing streams in mouse visual cortex. J. Neurosci. 32 (13), 4386–4399.

Ward, J.P., Masterton, B., 1970. Encephalization and visual cortex in the tree shrew (Tupaia glis). Brain Behav. Evol. 3, 421–469.

Ware, C.B., Diamond, I.T., Casagrande, V.A., 1974. Effects of ablating the striate cortex on a successive pattern discrimination: further study of the visual system in the tree shrew (Tupaia glis). Brain Behav. Evol. 9, 264–279.

Warner, C.E., Goldshmit, Y., Bourne, J.A., 2010. Retinal afferents synapse with relay cells targeting the middle temporal area in the pulvinar and lateral geniculate nuclei. Front. Neuroanat. 4, 8.

Warner, C.E., Kwan, W.C., Bourne, J.A., 2012. The early maturation of visual cortical area MT is dependent on input from the retinorecipient medial portion of the inferior pulvinar. J. Neurosci. 32 (48), 17073–17085.

Warner, C.E., Kwan, W.C., Wright, D., Johnston, L.A., Egan, G.F., Bourne, J.A., 2015. Preservation of vision by the pulvinar following early-life primary visual cortex lesions. Curr. Biol. 25 (4), 424–434.

Weiskrantz, L., 2004. Roots of blindsight. Prog. Brain Res. 144, 229–241.

Weiskrantz, L., Barbur, J.L., Sahraie, A., 1995. Parameters affecting conscious versus unconscious visual discrimination with damage to visual cortex (V1). Proc. Natl. Acad. Sci. U.S.A. 92, 6122–6126.

Weller, R.E., Kaas, J.H., 1981. Preservation of retinal ganglion cells and normal patterns of retinogeniculate projections in prosimian primates with long-term ablations of striate cortex. Invest. Ophthalmol. Vis. Sci. 20 (2), 139–147.

White, A.J., Solomon, S.G., Martin, P.R., 2001. Spatial properties of koniocellular cells in the lateral geniculate nucleus of the marmoset Callithrix jacchus. J. Physiol. 533 (2), 519–535.

White, A.J., Wilder, H.D., Goodchild, A.K., Sefton, A.J., Martin, P.R., 1998. Segregation of receptive field properties in the lateral geniculate nucleus of a New World monkey, the marmoset Callithrix jacchus. J. Neurophysiol. 80, 2063–2076.

Whitwell, R.L., Striemer, C.L., Nicolle, D.A., Goodale, M.A., 2011. Grasping the non-conscious: preserved grip scaling to unseen objects for immediate but not delayed grasping following a unilateral lesion to primary visual cortex. Vis. Res. 51 (8), 908–924. https://doi.org/10.1016/j.visres.2011.02.005.

Wong, P., Collins, C.E., Baldwin, M.K., Kaas, J.H., 2009. Cortical connections of the visual pulvinar complex in prosimian galagos (Otolemur garnetti). J. Comp. Neurol. 517 (4), 493–511.

Wong, P., Gharbawie, O.A., Luethke, L.E., Kaas, J.H., 2008. Thalamic connections of architectonic subdivisions of temporal cortex in grey squirrels (Sciurus carolinensis). J. Comp. Neurol. 510 (4), 440–461.

Xu, X., Ichida, J.M., Allison, J.D., Boyd, J.D., Bonds, A.B., Casagrande, V.A., 2001. A comparison of koniocellular, magnocellular and parvocellular receptive field properties in the lateral geniculate nucleus of the owl monkey (Aotus trivirgatus). J. Physiol. 531 (1), 203–218.

Yan, Y.H., Winarto, A., Mansjoer, I., Hendrickson, A., 1996. Parvalbumin, calbindin, and calretinin mark distinct pathways during development of monkey dorsal lateral geniculate nucleus. J. Neurophysiol. 31 (2), 189–209.

Yu, H.H., Chaplin, T.A., Egan, G.W., Reser, D.H., Worthy, K.H., Rosa, M.G., 2013. Visually evoked responses in extrastriate area MT after lesions of striate cortex in early life. J. Neurosci. 33 (30), 12479–12489.

Yukie, M., Iwai, E., 1981. Direct projection from the dorsal lateral geniculate nucleus to prestriate cortex in macaque monkeys. J. Comp. Neurol. 201, 81–97.

Zhang, B., Zheng, J., Watanabe, I., Maruko, I., Bi, H., Smith 3rd., E.L., Chino, Y., 2005. Delayed maturation of receptive field center/surround mechanisms in V2. Proc. Natl. Acad. Sci. U.S.A. 102 (16), 5862–5867.

25

Evolved Mechanisms of High-Level Visual Perception in Primates

D.A. Leopold, J.F. Mitchell, W.A. Freiwald

National Institutes of Health, Bethesda, MD, United States; University of Rochester, Rochester, NY, United States;
The Rockefeller University, New York, NY, United States

25.1 Introduction

Among the sensory modalities, it may be vision that places an animal at the greatest advantage over its environment. This is because seeing, unlike hearing, smelling, or feeling, offers a detailed map of the surrounding world from a comfortable distance. Vision allows an animal to quickly and quietly make judgments about potential threats, food sources, territories, and, in some species, social relationships. It can also guide the planning and execution of detailed motor actions, such as reaching for a food item or pursuing another animal as prey. The broad utility of visual perception, along with its automatic and effortless nature, tends to obscure the fact that vision poses immensely challenging computational problems. In fact, the mechanisms by which the brain reads, interprets, and interacts with image information on our retina remain largely mysterious despite decades of scientific inquiry. The present chapter examines this problem from an ecological and comparative perspective. We focus on how high-level visual specialization in the primate brain may have evolved.

Primates are mammals, as well as vertebrates. As there is much conservation in the visual system, the primate brain can be considered and studied in multiple evolutionary contexts. Here we largely restrict the comparisons to other mammals, mentioning only briefly connections to more distantly related clades, and acknowledging throughout that certain visual operations and corresponding neural circuitry may have evolved long before the first mammal. The visual brain draws upon multiple maps in the retina, which are processed and transmitted to different target locations by parallel visual pathways composed of different cell types (Leopold, 2012; Nassi and Callaway, 2006). This

parallelism is present in all mammals and is thought to reflect the multiple ways in which the brain can use information in the retinal images. At the same time, the primate eye and brain have certain distinctive features that set primate vision apart from other mammals (Kaas, 2012; Kaas and Stepniewska, 2016; Mitchell and Leopold, 2015; Preuss, 2007). Some of these features support complex social interaction, with an obvious example being the broad networks in the cerebral cortex specialized for faces (Tsao et al., 2003, 2008a) and bodies (Popivanov et al., 2012). Other features relate more to particular behaviors, such as visual guidance for reaching and grasping (Janssen and Scherberger, 2015), or interpreting others' goal-directed actions (Nelissen et al., 2011; Rizzolatti and Sinigaglia, 2010). The cortex also controls socially important actions, such as outward expression of emotional expression (Gothard, 2014; Müri, 2016) and internal regulation of autonomic arousal systems (Barbas et al., 2011). Thus a hallmark of the primate brain is a strong investment of the cerebral cortex in comprehending social information and executing precise actions based on complex social context. These important aspects of a primate's life are critically orchestrated by a command of vision.

At the same time, not much is known about high-level visual processing and the cortical control of visual behaviors in other mammals. Are the aforementioned functions of the primate brain de novo evolutionary inventions, or are they in fact a refinement of a preexisting mammalian repertoire? Drawing a distinction between derived ("new") and ancestral ("old") traits is at the core of comparative neuroscience and is of great importance when attempting to understand the nature of complex and composite systems in the brain. In high-level visual perception, the parallax afforded by investigating multiple species may ultimately be critical for attaining a deeper understanding of how adaptive specializations during different epochs of primate evolution have come to support some of the most complex human behaviors (Leopold and Rhodes, 2010).

In this chapter, we adopt this perspective to consider the evolutionary layering of the primate brain with regards to high-level vision. We begin by defining terms and breaking down "high-level" visual problems into workable components. We next provide a broad view of mammalian vision, sampling the landscape of visual behaviors and reviewing a few known features of high-level vision in nonprimate mammals. We then focus on primates, reviewing recent discoveries pertaining to neural mechanisms of high-level visual specializations. Throughout the chapter we discuss overlap and divergence in the visual capacities of primates and other mammals and tentatively speculate on the evolution of the supporting mechanisms in the brain.

25.2 Natural Vision

25.2.1 What Is the Problem?

Vision starts at the eye and affords sensation from a distance. The extreme selective advantage of vision is evident in the multiple, independent inventions of image-forming eyes during evolution (Fernald, 2006). One of these inventions, which was inherited by all extant vertebrates, is the powerful "camera" eye, which is shared with species as distinct as the agnathan lamprey and the human (Lamb et al., 2007). The camera eye, its transmission lines into the brain, and the tight control of gaze direction through muscles of the eye and head are basic features of the vertebrate nervous system. These visual components, together with yet poorly specified perceptual mechanisms in the brain, support navigation, foraging, mate selection, territory defense, predation, and many other actions across vertebrates. The visual systems of invertebrates vary widely in their composition, but need to solve many of the same problems.

Thus monkeys, minnows, monarchs, and mollusks all use visual signals to interact with their environment and with one another. In which ways is their perception like ours? Visual scientists have long pondered the essence of seeing. Helmholtz (1867) conceived of visual perception as an active, inferential process, whereas Gibson (1986) believed perception involved the direct analysis of the rich information available in a natural image. Marr (1982) influenced a generation of visual scientists by conceptually separating computational and algorithmic aspects of vision from their neural implementation. These theoretical starting points place emphasis on

different aspects of the visual problem, although it may be said that none is particularly in tune with the capricious nature of the biological world.

An interesting perspective is offered by Ramachandran (1985), who suggests that the brain's strategy for understanding the visual environment is well conceived as a "bag of tricks." According to this view, an animal's brain need not always take a generalized or elegant approach when interpreting retinal information, as if it were designed to solve all possible visual problems with equal efficiency. Instead, "the visual system often *cheats*, ie, uses rules of thumb, shortcuts, and clever sleight-of-hand tricks that were acquired by trial and error through millions of years of natural selection." Likewise, any particular stimulus attribute, such as a specific pattern of motion, color, or shape, can be directed to a shortcut visual channel that links to a highly specific behavioral response. This evolutionary perspective, Ramachandran argues, resonates with the biologist who understands the inherent opportunism of evolution, but often absent from discussion among vision scientists.

Vision scientists are accustomed to thinking of vision as a fundamentally serial problem, though with two streams and subject to modulation from higher cognitive centers. The starting point is the extraction of local image features, first in the retina and then among neurons in retinotopically organized brain areas. In the visual cortex, neurons selective for simple features converge their projections onto neurons selective for more complex patterns. The resulting progression of feature complexity is thought to culminate in specialized neurons (eg, "face cells") that support visual recognition and action. From this perspective, it makes sense to speak of low-level vision as pertaining to the coding of local image features and high-level vision as pertaining to the representation of complex objects. The sensory responses of these neurons are further subject to top-down modulation by attention, memory, object value, and other cognitive operations. The multiple stages of serial processing are captured in the concept of the cortical visual hierarchy, which has become the centerpiece of visual neuroscience (Felleman and Van Essen, 1991).

The above view of vision represents a synthesis that has greatly advanced knowledge of brain function and thus been highly influential. At the same time, this textbook framework is strongly tied to a few dominant experimental paradigms, most notably the brief flashing of stimuli onto the retina of an immobilized animal, and may thus fail to capture important aspects of vision. Here we approach vision from a rather different theoretical perspective centered on evolution and comparative biology. We advance a framework that considers the universality of certain visual operations and distinguishes them from adaptive specializations that are more recent and clade specific (see Fig. 25.1). Our hope is that this somewhat different starting point will shed new light on some of the more difficult and poorly understood problems in visual neuroscience.

25.2.2 Toward a Conceptual Hierarchy for Natural Vision

Vision can be conceived as an evolved composite of operations, each of which is supported by particular neural systems. Thus an evolutionary and comparative perspective can be useful, and may be essential, to create a framework for understanding its multifarious nature. This is different and complementary to the more conventional framing of the visual problem, which is often centered on a set of serial processing stages through which local features are synthesized into complex representations (Felleman and Van Essen, 1991). Whether an evolutionary theoretical perspective offers fresh insights into the local-to-global and simple-to-complex accounts of vision is yet unknown, as there has not been much exploration in this area.

At the foundation of this framework are those operations that are essential for virtually all sighted creatures under natural conditions. Any animal that uses vision must have mechanisms to contend with at least three of its fundamental challenges. We term these capabilities *core vision*. The first core vision feature is the extraction of the third dimension of space from the 2D projections on the retina (Cavanagh, 1987). Understanding depth is critical for moving through the environment, interacting with other animals, and perceiving object structure. It is achieved through the implicit understanding of a range of depth cues, including texture gradients, occlusion, shadows, color gradients, binocular disparity, perspective, motion parallax, and many others. Different species draw upon different combinations of these and other cues, which often reinforce one another. The second core vision feature is the discounting of self-generated movements that stimulate the retina. This so-called visual reafference is a potentially grave disturbance to

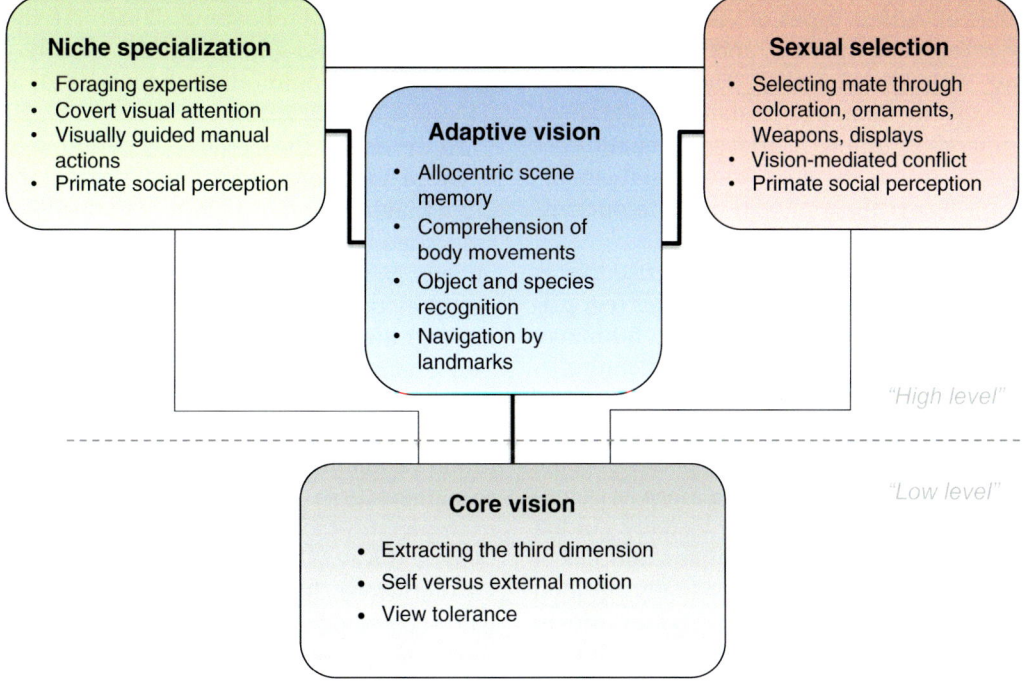

FIGURE 25.1 A proposed hierarchy of natural vision based on the evolution of visual operations and the extent to which they are shared among animals. The *core* visual operations are the foundational elements of vision, which are thought to be present to some degree in all visually oriented vertebrate species. At a higher level, certain *adaptive* elements of vision afford individuals improved cognitive capacities and are critical for a range of basic visual actions, such as scene recognition and predation. At a somewhat higher level are examples of clade-specific visual expertise that tailor a given species' needs for given *niche*, or for a given intraspecific contract for *sexual selection*. Within this hierarchy, the social perception and visually guided manual actions associated with primates represent clade-specific specializations. At the same time, these visual faculties draw upon cognitive building blocks, such as the comprehension of body movements, which are more broadly shared among mammals. Each of these visual operations engages mechanism of core vision, including the perception of three-dimensional depth and spatial relationships, the accounting of reafferent self-motion, and some capacity for tolerance to transformation of the retinal images cast by objects, animals, or scenes. Although the core vision operations are thought to engage multiple stages of visual processing, and are poorly understood in terms of mechanism, they are conceived here as "low level" because they are foundational for virtually all sighted creatures.

visual analysis in every moving animal (Holst and Mittelstaedt, 1950). Evidence suggests an early evolution of active mechanisms to contend with this challenge to vision (Crapse and Sommer, 2008). The third core feature is tolerance for natural image transformations, where the brain is able to "see past" the precise pattern of light on the retina, and ever changing projections based on spatial distance, viewing angle, or illumination, to robustly recognize and interact with objects, scenes, and other animals. These core visual problems are seldom put forward as "low-level" problems in vision. However, within the present framework, they are the most fundamental and ubiquitous and thus at the foundation. They are also among the most vexing and poorly understood problems in vision.

Built upon core visual operations is an intermediate tier, which we term *adaptive vision*, referring to a set of visual operations that selectively draw upon the fundamental elements of vision, but are nonetheless widely shared across taxa. Adaptive vision includes operations such as allocentric scene memory (eg, for understanding territories and food sources), a basic comprehension of quadruped bodily movements (eg, for predation and predator avoidance), certain types of visual recognition (eg, for certain objects and species), and spatial navigation (eg, using visual landmarks). These operations contribute critically to foraging, predation, migration, and territory defense within terrestrial and aquatic ecosystems. Adaptive vision employs core visual operations to solve problems that are critical for a species' survival.

One level higher, we identify visual expertise that is more specialized to meet the needs of an order (eg, primates), family (eg, hominids), or species (eg, humans). We term this level of vision *clade vision* and distinguish two aspects. *Niche specialization* refers to those behaviors that adapt vision in a unique way to the needs of a specific ecosystem. This could be certain species-specific expertise in foraging, such as identifying areas as potentially rich in food, caching behaviors, or visually guided predation strategies. For primates, familiar examples of niche specialization might include visually guided reaching and grasping, covert attention, and an unusually detailed perception of others' actions and intentions. In contrast, *sexual selection* relates to the visual assessment of individuals for the purposes of mating or competition. Here clade-specific expertise pertains to the visual perception of features, which can be the vividness of coloration, the quality of displays, or the size of muscles or antlers. In primates, sexual attraction is often mediated through visual assessment of dominance or estrous phase through features of the body and face.

This sketch of the multiple tiers of specialization in natural vision is simply a theoretical starting point,

without reference to the brain. In the next section, we investigate the visual repertoire of mammals, first considering how different species' behaviors map onto this taxonomy, and then how known specializations in the brain might support the associated visual operations.

25.3 The Ancestry of Primate Visual Abilities

Given the excellent vision of primates, it can be surprising to learn that the taxonomic class of mammals, of which primates are members, is renown for its poor vision. The low visual acuity and poor color perception of most mammals are products of an extended period of evolution in which mammals were nocturnal (Heesy and Hall, 2010). During this nocturnal period, the eye of early mammals diverged from that of their diurnal ancestors, altering its shape to collect more light (Hall et al., 2012) and shifting its complement of photoreceptors toward rods (Heesy and Hall, 2010). Cone vision largely degenerated through a decrease in photoreceptor density and abandonment of all but two photopigment genes (Yokoyama, 2000). At the same time, other senses strengthened, most notably olfaction and hearing, to accommodate the needs of nocturnal life (Heesy and Hall, 2010).

At the end of the Cretaceous period, mammals began to adopt a higher proportion of diurnal niches, displacing nonavian dinosaurs whose extinction is commonly associated with the impact of a massive asteroid approximately 66 million years ago (Renne et al., 2013). While recent evidence suggests that this shift may have been more gradual than immediate (Bininda-Emonds et al., 2007), mammals eventually grew in size and diversified in their sensory specializations, including vision. Nonetheless, those nocturnal adaptations that shaped the evolution of mammals for many millions of years are still recognizable in the mammalian eye and brain. For example, most extant mammals have low spatial acuity (Kirk and Kay, 2004) and dichromatic or monochromatic vision (Jacobs, 2013), and an eye shape that is characteristic of a nocturnal vertebrate (Hall et al., 2012). Furthermore, mammals are more dependent than most other amniotes on olfaction and chemical signaling, which they use for foraging and a range of social interactions including sexual and maternal behaviors.

Primates constitute about 5% of the approximately 5400 mammalian species. Although primates diverged from other orders before the Cretaceous period was brought to an abrupt end, the first recognizable primate fossils are from well after this event, approximately 55 million years ago (Martin, 2012; Tavaré et al., 2002). Primates are atypical mammals, as they are strongly committed to vision in aspects of life related to feeding, mate selection, and manual actions. This shift is reflected in a relative increase in the proportion of visual areas of the brain, along with a relative decrease in olfactory areas (Barton et al., 1995). It is also reflected in an expansion of circuits used to support visually guided grasping and locomotion through an arboreal environment (Martin, 2012). It is these specializations that invite the question of whether primates' unique use of vision is best seen as a modification of the mammalian ancestral repertoire. Before attempting to address this thorny question directly, we survey how a sample of nonprimate mammals uses vision to solve difficult problems.

25.3.1 Mammalian Visual Behaviors

A short survey of literature on the natural ethology of mammals makes it clear that most studied species use vision to solve complex problems needed for survival. Vision is often used in combination with other senses. The specific examples in this section were selected to illustrate one or more from each of four major branches of mammals and are thus arranged according to their mammalian superorder (see Fig. 25.2A).

25.3.1.1 "High-Level" Vision in Nonprimate Mammals

In light of the taxonomy of natural vision proposed earlier, we use the term "high-level" here to indicate aspects of vision that extends beyond the core operations identified in Fig. 25.1, and we focus largely on examples of intraspecific signaling.

Superorder Euarchontoglires. This group of placental mammals includes primates and their closest relatives, including the approximately 1500 species of rodents. Recent work has characterized aspects of spatial and object vision of laboratory rats and mice (Vinken et al., 2016; Zoccolan, 2015). Diurnal squirrels, whose most recent common ancestor with the rat and mouse lived >80 MYA (Bininda-Emonds et al., 2007), may be better suited to illustrate the range of rodent visual capabilities. The squirrel retina, similar to that of a primate, is dominated by cones, although like most mammals squirrels have only two photopigment types (Blakeslee et al., 1988; Van Hooser and Nelson, 2006). Squirrels' large eyes afford them visual acuity that, at 4 cyc/deg, is good compared to other rodents, though still an order of magnitude lower than most monkeys (Kirk and Kay, 2004). Squirrels depend on their vision for their impressive climbing and leaping through trees (Orkin and Pontzer, 2011) as well as their renown caching and retrieval of food (Macdonald, 1997; Samson and Manser, 2016). While squirrels, like most mammals, depend on chemical signals (Betts, 1976; Steiner, 2011), they also use their vision for foraging (Duncan and Jenkins,

2011), nest construction (Owings et al., 1977; Quanstrom, 1971), predator detection (Robinson, 1980), and social communication (Putman and Clark, 2015). Visual communication is central to their sexual behaviors, aggression, and play (Steiner, 1971). Their intraspecific visual signaling, while not well understood, consists of bodily postures and tail flicks (Betts, 1976; Owings et al., 1977; Quanstrom, 1971).

Superorder Laurasiatheria. This superorder contains a large number of diverse and widely dispersed species, including carnivores, ungulates, bats, shrews, and cetaceans. In the past decades, much has been learned about the visual cognitive abilities in several species in this group, and in particular dogs and sheep. Here we focus on sheep, which are ungulates with large eyes, dichromatic vision, and an acuity estimated to be 20 cyc/deg, which rivals that of a small primate (Shinozaki et al., 2010). In familiar terms, sheep vision is considered equivalent to that of a colorblind human with 20:60 vision (Kendrick, 2008); however, as their retina contains a high photoreceptor density streak rather than a foveal pit, their visual experience of the world must be somewhat different. The retinal streak enables sheep and related ungulates to monitor the horizon with high acuity, with vision allowing them to detect predator threats up to 1 km away, out of range for olfactory detection (Kendrick, 2008). While sheep use olfactory cues for certain aspects of social recognition, such as a ewe recognizing a newborn lamb (Baldwin and Shillito, 1974), vision is the primary sense by which adults recognize one another (Kendrick, 2008). Systematic testing in the laboratory has demonstrated that sheep recognize photographs of individual sheep and will preferentially approach photographs of familiar faces over unfamiliar ones and over faces of other species (Kendrick et al., 1995). Using photographs of sheep faces, individual sheep can visually discriminate at least 50 individuals, which is better than their discrimination of human faces (Kendrick, 2008). They have been shown to retain the visual memory of a familiar face for periods exceeding 1 year (Kendrick et al., 2001). Sheep also recognize and respond selectively to the emotional content of their conspecifics' faces, such as the opening of the eyes and positioning of the ears (Tate et al., 2006). As in humans, sheep face perception requires experience during early development, with lambs first able to visually recognize their mothers only after 2—4 weeks (Kendrick, 1994). As discussed later, this expertise for faces is prominently reflected the selective responses of neurons within the sheep cerebral cortex and constitutes one of the very few examples in which concrete information is available about high-level visual specialization in the brains of nonprimates.

Superorder Afrotheria. This small radiation of mammals includes aardvarks, elephants, and manatees. The

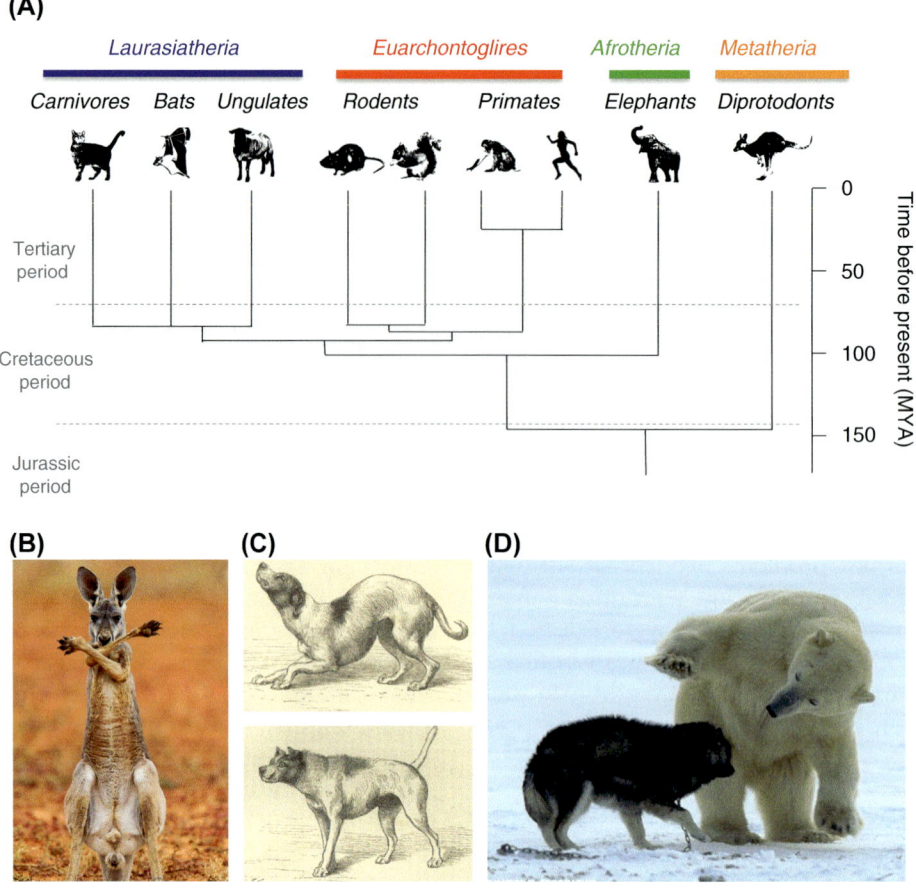

FIGURE 25.2 Complex mammalian visual behaviors considered from an ancestral vantage point. (A) Cladogram of select taxa in which species have been shown to solve complex visual problems, as described in the main text. The taxa are arranged according to their mammalian superorder, with time since common ancestry shown in the vertical dimension (Bininda-Emonds et al., 2007). (B–D) Examples of visual body comprehension within and between species. (B) Upright threat display of in red kangaroo (Russell, 1984) (Photograph from http://www. jamitarris.photoshelter.com/image/I00008sAMYYaJaf8.). (C) Darwin's illustration of the principle of antithesis in mammalian emotional expression of the body, contrasting body language that is "humble and affectionate" with that having "hostile" intentions (Darwin, 1872). (D) Bodily interactions between a leashed dog and a polar bear during their "friendship" established over the course of multiple visits (Rosing and Stirling, 2010).

best-studied family is elephants, which are large, diurnal mammals with brains four times larger than a human. They are dichromats with surprisingly low visual acuity (4.1 cyc/deg) given their large size (Kirk and Kay, 2004). Elephants live in large, complex groups of interacting families, each centered around a matriarch (Wittemyer and Getz, 2007). Like most mammals, they use their vision for navigation and rely on a combination of chemical, tactile, and visual signals for foraging and communication (Langbauer, 2000). Acoustic playback experiments have shown that elephants readily determine individual identity through infrasonic vocalizations and can distinguish between contact calls from at least 100 individuals from different families (McComb et al., 2000). They are also able to recognize and remember individual identity based on the chemosensory cues in urine, even if decades have passed since the last meeting (Hart et al., 2008). In the domain of

vision, elephants perform poorly on explicit tests of visual discrimination learning and insightful tool use (Hart et al., 2008; Nissani et al., 2005; Savage et al., 1994). However, several observations indicate that they are able to use their vision to recognize individual identity and bodily actions. They have recently been reported to locate hidden food based on human pointing (Smet and Byrne, 2013) and to more generally recognize the focus of attention from a human face and body (Smet and Byrne, 2014). They are also one of the few animal species to pass the mirror self-recognition test (Plotnik et al., 2010) and have been reported to use vision for social learning (Greco et al., 2013). While their use of visual signaling under natural conditions is a topic of study and is not completely understood, elephants exhibit a range of gestures and displays using their body, ears, tail, and trunk that are thought to advertise social information, such as hormonal state, emotion, and possibly the communication of information about the environment (Langbauer, 2000).

Superorder Metatheria. This large group of mammals commonly known as marsupials has diverse visual abilities, although most species are nocturnal and depend less on vision than on olfaction and audition for survival (Russell, 1984). Nonetheless, vision is important for some marsupials, and there is considerable diversity among visual systems in this superorder. For example, recent work has shown that at least four diurnal species have trichromatic vision, whereas other closely related species do not (Arrese et al., 2002; Ebeling et al., 2010). Among species with visual acuity exceeding that of a rat, such as the dunnart (2.3 cyc/deg) (Bonney and Wynne, 2002), numbat (6.3 cyc/deg) (Arrese et al., 2000), and tammar wallaby (6 cyc/deg) (Hemmi and Mark, 1998), vision appears important for predator detection (Arrese et al., 2000; Blumstein et al., 2000; Lippolis et al., 2010), some types of social signaling (Russell, 1984), and foraging and predation (Haythornthwaite and Dickman, 2000). Studies of visual social signaling in the whiptail wallaby revealed aggressive signals in threat postures, grass pulling, and stiff-legged walking, whereas courtship was accompanied by head bobbing and tail-lashing (Russell, 1984). Likewise, the red kangaroo exhibits various visual displays, including grass pulling and an upright threat display with crossed forelimbs (Fig. 25.2B; Russell, 1984).

This brief survey serves only to illustrate common themes that are distributed broadly in mammalian vision. Countless other examples exist in which mammals use their vision in a specialized way to solve important problems related to sexual selection, territorial displays, nest construction, navigation, species recognition, predation, and many other purposes.

25.3.1.2 A Shared Understanding of the Body

One common denominator among mammals appears to be their shared understanding of the bodily actions of other animals, including the way they orient, run, turn, crouch, leap, and open their mouths (Andrew, 1963; Bekoff, 1977; Henry and Herrero, 1974; Johnson, 2006). Experiments using point light walkers (Johansson, 1973) have shown that even birds and fish perceive biological movements and prefer them to nonbiological ones (Rosa-Salva et al., 2014; Troje and Aust, 2013; Vallortigara et al., 2005). Among mammals, cats (Blake, 1993), dogs (Kovács et al., 2016), and rats (MacKinnon et al., 2010) have been shown to distinguish between biological motion from other forms of motion. This understanding of the quadruped body plan and its movements may have been stabilized through evolution not only by predation and species-specific gestures, but also aggression, play, and filial attachment (Johnson, 2006).

A shared comprehension of bodily actions was noted by Darwin, in his "The Expression of Emotions in Man and Animals," who suggested that opponent body language ("the principle of antithesis") evolved to indicate opposite valence emotions in a broad range of species (Fig. 25.2C). That different species share a common visual language is evident not only through predation, but also through examples of interspecific attachment and "friendships." While there is minimal rigorous scientific research studying the sensory basis for such interactions, they often capture public interest and are well documented in photographic essays (Smuts, 2001), newspaper articles (Goode, 2015), and monographs (Buckley, 2009; Rosing and Stirling, 2010). These interspecific interactions, which are often affiliative, nurturing, or playful, are at least suggestive of a broader mammalian comprehension of biological actions mediated through vision (Fig. 25.2D). Examples of interspecific attachment that have been studied center around humans, with some systematic efforts to document effective visual communication between humans and dogs (Miklósi et al., 2005) and between humans and horses (McGreevy et al., 2009).

The perception of gestures and actions, which appears largely shared among mammals, suggests a basal foundation for action interpretation upon which primate visual social perception may be built. In the taxonomy above, this was referred to as adaptive vision. While it is always important to proceed with caution when considering evolutionary continua and interspecies homology, it would be equally perilous to ignore these similarities and assume on principle that the specializations for social vision in the primate brain are recent adaptations.

25.3.1.3 The Role of Early Visual Experience

One challenge in studying the brain, which becomes particularly poignant in the consideration of visual expertise, is that it is designed to configure *itself* based on its environmental input as well as the experiences it seeks. These factors come strongly into play early in life, when the specializations within the brain are forming. While a newborn's face perception is not good by adult standards, observations in humans and monkeys have shown that their brains are innately wired to look at upright faces, and particularly that of the mother (Morton and Johnson, 1991; Pascalis et al., 1995; Sugita, 2009). Newborns are similarly drawn to gaze at upright biological movements (Simion et al., 2008). An influential theory holds that brain regions responsible for the initial detection and selective orientation toward such stimuli are in the midbrain, and thus distinct from those better studied cortical areas (for a recent review, see Johnson et al., 2015). These findings illustrate the complexity in understanding the brain's specialization for such stimuli and also indicate the complex interaction of drive and experience during development, where one brain area may be enlisted to train the sensory selectivity of another. The important role of development in shaping visual expertise is likely to have evolved early in vertebrate evolution, since similar learning effects are common in certain precocial birds, with hatchlings having a preference for biological stimuli (Regolin et al., 2000; Salva et al., 2011), and subsequent visual imprinting on the stimulus (ie, "mother") to whom they initially attach (Bateson, 1990; Bolhuis and Honey, 1998).

Less is known about the importance of visual development in other mammals, although for some precocial ungulates, certain visually guided neonatal behaviors appear similar to that of birds. As many ungulates live in open, unprotected areas, they are particularly vulnerable to predation in the first weeks of life. There are two basic strategies, "following" and "hiding," that are used by different species to avoid early predation (Roberts and Rubenstein, 2014; Rutberg, 1987). Among "followers," such as wildebeest, American buffalo, moose, and equids, the newborn imprints on the mother, possibly through visual cues, and begins to follow her less an hour after birth (Roberts, 2012). While olfactory signals likely play a role, vision clearly mediates the tracking and following of the mother's movements. Thus this visual analysis must be either innate or learned very quickly. By contrast, the "hiders" take the strategy of visually concealing themselves from predators by lying in tall vegetation for a few days to a few weeks.

Experience and interaction with the mother and siblings is also important to refine visual expertise, skills, and even sexual preferences. One cross-fostering experiment in which goats were reared by sheep mothers, and vice versa, is particularly illuminating in this regard (Kendrick et al., 1998). These experiments revealed that certain social actions, such as juvenile play and grooming behavior, resembled the traits of the foster mother and siblings, whereas other species-specific behaviors related to aggression, climbing, feeding, and vocalization appeared to be genetically determined. Importantly, the maternal influence had long-term social consequences. Male lambs raised by goat mothers preferred to affiliate and mate with female goats, whereas male kids raised by sheep mothers preferred to affiliate and mate with female sheep. This effect appeared irreversible and persisted even after kids and lambs had 3 years of reintroduction to living with their genetic species. The cross-fostering preferences among females were weaker and generally reversible.

The examples of mammalian visual behavior reviewed in this section were by necessity drawn together from a range of different research areas.

Nonetheless, the collection of ostensibly complex visual behaviors exhibited in diverse mammalian taxa suggests that there may be some widespread components of the primate visual repertoire that was already present early in mammalian evolution. If so, it is natural to ask whether specializations in the brain that support the corresponding perception might also be shared. In the next section, we approach this challenging question, first by outlining the types of general principles that might affect the capacity for high-level visual specialization in a given clade's brain, and then by reviewing three examples of nonprimate mammalian species in which experiments have shed light on particular types of visual behaviors.

25.3.2 Cortical Visual Specializations in the Mammalian Brain

It is clear that some mammalian species such as primates have evolved to be highly dependent on the use of vision, whereas others are much less dependent on vision. At the same time, certain "high-level" visual operations, such as the comprehension of bodily movements and the recognition of territories, appear to be widely shared among mammals, even though such operations might normally be associated with evolved visual specializations. The real question is whether early-evolved homologous pathways support equivalent complex visual behaviors in distant species, and to what extent primates have inherited important features of their visual repertoire from their nonprimate ancestors living in the Cretaceous period. Answering this question with any precision is out of reach at this point, since relatively little information is available about the neural mechanisms of high-level perception in nonprimate mammals, and even in primates the connection between electrophysiological features and natural visual behaviors is poorly understood. Nonetheless, it is possible to approach this question indirectly by considering comparative electrophysiological studies that have shed light on the basic organization of the mammalian brain, and, in a few cases, specific mechanisms of high-level visual perception. In this section we draw upon a large number of comparative studies to first identify the evolutionary and developmental factors that influence visual specialization in the brain and second show that neural mechanisms in the few mammals that have been tested bear surprising resemblance to observations in primates. We focus on the cerebral cortex and mention only in passing that subcortical structures such as superior colliculus and related structures are also important points of comparison (Arendes, 1994; Isbell, 2006; Nguyen et al., 2013, 2014), and particularly with regard to their role in early development (Johnson, 2005).

25.3.2.1 Principles of Cortical Visual Specialization

By what path did the cerebral cortex of primates come to support a range of high-level visual operations? In considering how the primate brain distinguishes itself from that of other mammals, comparative neuroscientists have placed much focus on charting the cortical territories supporting different types of sensory input. These studies have demonstrated that the relative positions of the olfactory, visual, auditory, and somatosensory cortex are conserved among mammals; however, the proportions of their respective cortical territories differ greatly, as does the span of "association" cortex between them (Fig. 25.3). Through these studies, it has been suggested that the cortex of early mammals was likely dominated by olfaction, with extant insectivores taken as a modern approximation of such a "protomammal" (Catania, 2005; Kaas, 2013; Krubitzer et al., 1997). The visual capabilities of this hypothetical ancestor are, of course, unknown, but may have included at least some of the features commonly considered to be "high-level" vision, such as the visual comprehension of predator movements or the recognition of navigational landmarks. To the extent that primates evolved from this protomammal, it is interesting to consider the factors that shape the brain's adoption of high-level visual specializations.

The evolution of the brain is inseparable from that of the sensory periphery. The prominence of different sensory organs shapes the spatial extents of the primary and secondary cortical sensory areas across mammalian species (Fig. 25.3). In primates, olfaction is greatly diminished compared to most other mammals. Even within the domain of vision, the retinal adaptations within a clade strongly affect the manner in which the cortex is able to specialize for features in the environment. For example, the brain of an owl monkey (a monochromat) would not be expected to specialize for color-specific features in the same way as that of the closely related howler monkey (a trichromat). For high-level vision, the size of the eye, degree of ocular convergence, number of chromatic cone types, and receptor density and distribution all strongly shape visual capacities and the corresponding cortical specialization (Mitchell and Leopold, 2015). As selective pressures expand the sensory capabilities of one sense relative to another, the proportion of representation in the cortex changes accordingly. On timescales of evolution, these shifts can proceed relatively rapidly.

A second and often overlooked factor in cortical evolution is the surprisingly strict developmental regulation of the mammalian brain, at least with respect to the basic wiring and layout of brain components. A careful analysis of the growth of the brain, and in particular the relative timing of development events, reveals striking regularities in onset of genetically controlled ontological events (Finlay and Darlington, 1995; Finlay and Uchiyama, 2015; Workman et al., 2013) (see also Finlay Chapter X; Halley and Deacon Chapter X). This conserved program largely restricts the evolutionary degrees of freedom of the mammalian brain. The extent of developmental determinism, and its implication for mosaicism, and the specialness of the primate brain are important topics that are vigorously debated (Barton and Harvey, 2000; Finlay and Uchiyama, 2015; Herculano-Houzel, 2012). Fig. 25.4A depicts the notion of a conserved "event scale," taken from Finlay's work (Workman et al., 2013), and illustrates the shared progression of developmental events across several mammals, including those with vastly different brain sizes and phyletic relations. Importantly, the strong central constraints on brain evolution can be contrasted with the rapid and flexible evolution of the sensory periphery, illustrated in Fig. 25.4B, which shows that the photoreceptor density for two different species of domestic dog diverges markedly, reflecting rapid specialization of the retina within a single species (McGreevy et al., 2003). Considerable evidence suggests that such peripheral changes within a given sensory domain, or a change in the proportion of different senses, initiate and drive important aspects of evolutionary change (Kaskan et al., 2005; Krubitzer, 2009). Thus the flexible evolution of the sensory epithelia, serving as a counterpoint to the slow and steady evolution of the brain's structure, accounts for much of the observed species differences in the cortical allotment to the different senses.

A third factor often exploited by evolution is the overall size of the brain. Increases and decreases in mammalian brain size are thought to be driven by selection pressures to accommodate specific niches and to some extent the complexity and control of behavior. Absolute brain size covaries with the length of the gestational period and the need for experience-dependent learning after birth. Comparing large versus small brains, it is obvious that larger brains have more neurons and more overall territory and should thus be able to accommodate a higher level of sensory specialization in associative regions between the primary sensory areas (see Fig. 25.3). Less obvious, however, is the fact that larger brains have a different composition and internal connectivity than smaller brains simply because of their size. This latter principle stems from the allometric, rather than proportional, scaling of different brain areas, which in turn derives from the strong developmental regulation mentioned above. Simply put, if the brain from species B is double the size of the brain from species A, it does not follow that each of the components in B's brain

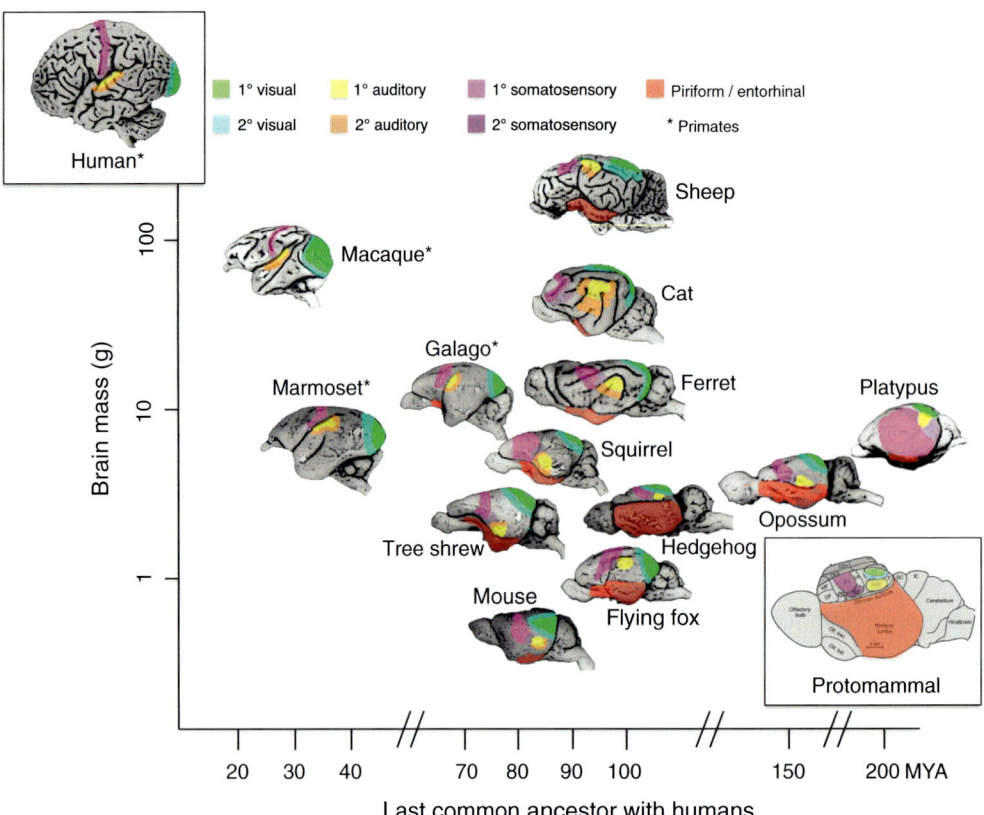

FIGURE 25.3 Sensory specialization mapped in the brains of mammalian species. Mammals vary greatly in the extent of cortical area that is not occupied by primary and secondary sensory areas. Two principal factors are considered in the axes of this graph. Position in the y dimension corresponds to brain size, and it is evident that larger brains have higher proportion of cortical tissue that is not strictly linked to sensory input. Position in the x dimension approximates the time of each species' common ancestry with humans. For the present purposes, this dimension is taken as a proxy for establishment of primate sensory specializations, including those high-level visual specializations that are the focus of this chapter. *The data depicted in this figure are approximations of coverage drawn from the following references: sheep (Broad, K.D., Mimmack, M.L., Kendrick, K.M., 2000. Is right hemisphere specialization for face discrimination specific to humans? Eur. J. Neurosci. 12, 731–741; Clarke, P.G., Whitteridge, D., 1976. The cortical visual areas of the sheep. J. Physiol. 256, 497–508; Ebinger, P., 1975. Quantitative investigations of visual brain structures in wild and domestic sheep. Anat. Embryol. 146, 313–323; Johnson, J.I., Rubel, E.W., Hatton, G.I., 1974. Mechanosensory projections to cerebral cortex of sheep. J. Comp. Neurol. 158, 81–107. https://doi.org/10.1002/cne.901580106; Kendrick, K.M., 1991. How the sheep's brain controls the visual recognition of animals and humans. J. Anim. Sci. 69, 5008–5016; Pettigrew, J.D., Ramachandran, V.S., Bravo, H., 1984. Some neural connections subserving binocular vision in ungulates. Brain Behav. Evol. 24, 65–93; Woolsey, C.N., Fairman, D., 1946. Contralateral, ipsilateral, and bilateral representation of cutaneous receptors in somatic areas I and II of the cerebral cortex of pig, sheep, and other mammals. Surgery 19, 684–702. https://doi.org/10.5555/uri:pii:003960604690061X), cat (Clemo, H.R., Sharma, G.K., Allman, B.L., Meredith, M.A., 2008. Auditory projections to extrastriate visual cortex: connectional basis for multisensory processing in "unimodal" visual neurons. Exp. Brain Res. 191, 37–47. https://doi.org/10.1007/s00221-008-1493-7; Grant, S., Hilgetag, C.C., 2005. Graded classes of cortical connections: quantitative analyses of laminar projections to motion areas of cat extrastriate cortex. 22, 681–696. Eur. J. Neurosci. https://doi.org/10.1111/j.1460-9568.2005.04232.x; Palmer, L.A., Rosenquist, A.C., Tusa, R.J., 1978. The retinotopic organization of lateral suprasylvian visual areas in the cat. J. Comp. Neurol. 177, 237–256. https://doi.org/10.1002/cne.901770205; Payne, B., 1993. Evidence for visual cortical area homologs in cat and macaque monkey. Cereb. Cortex; Sprague, J.M., Levy, J., DiBerardino, A., Berlucchi, G., 1977. Visual cortical areas mediating form discrimination in the cat. J. Comp. Neurol. 172, 441–488. https://doi.org/10.1002/cne.901720305), mouse (Krubitzer, L., 2009. In search of a unifying theory of complex brain evolution. Ann. N.Y. Acad. Sci.; Wang, Q., Burkhalter, A., 2007. Area map of mouse visual cortex. J. Comp. Neurol. 502, 339–357. https://doi.org/10.1002/cne.21286), hedgehog (Catania, K.C., Collins, C.E., Kaas, J.H., 2000. Organization of sensory cortex in the East African hedgehog (Atelerix albiventris). J. Comp. Neurol. 421, 256–274; Kaas, J., Hall, W.C., Diamond, I.T., 1970. Cortical visual areas I and II in the hedgehog: relation between evoked potential maps and architectonic subdivisions. J. Neurophysiol. 33, 595–615), ferret (Bizley, J.K., Nodal, F.R., Nelken, I., King, A.J., 2005. Functional organization of ferret auditory cortex. Cereb. Cortex 15, 1637–1653. https://doi.org/10.1093/cercor/bhi042; Cantone, G., Xiao, J., McFarlane, N., Levitt, J.B., 2005. Feedback connections to ferret striate cortex: direct evidence for visuotopic convergence of feedback inputs. J. Comp. Neurol. 487, 312–331. https://doi.org/10.1002/cne.20570; Manger, P.R., Engler, G., Moll, C.K.E., Engel, A.K., 2008. Location, architecture, and retinotopy of the anteromedial lateral suprasylvian visual area (AMLS) of the ferret (Mustela putorius). Vis. Neurosci. 25, 27–37. https://doi.org/10.1017/S0952523808080036), tree shrew (Kaas, J.H., Gharbawie, O.A., Stepniewska, I., 2011. The organization and evolution of dorsal stream multisensory motor pathways in primates. Front. Neuroanat. 5, 34. https://doi.org/10.3389/fnana.2011.00034; Wong, P., Kaas, J.H., 2009. Architectonic subdivisions of neocortex in the tree shrew (Tupaia belangeri). Anat. Rec. (Hoboken) 292, 994–1027. https://doi.org/10.1002/ar.20916), macaque (', E., Litinas, E., Recanzone, G.H., Padberg, J., Krubitzer, L., 2003. Cortical connections of the second somatosensory area and the parietal ventral area in macaque monkeys. J. Comp. Neurol. 462, 382–399. https://doi.org/10.1002/cne.10731; Fellemann, D.J., Van Essen, D.C., 1991. Distributed hierarchical processing in the primate cerebral cortex. Cereb. Cortex 1, 1–47; Kaas, J.H., Hackett, T.A., 1998. Subdivisions of auditory cortex and levels of processing in primates. Audiol. Neurootol. 3, 73–85), squirrel (Kaas, J.H., Hall, W.C., Diamond, I.T., 1972. Visual cortex of the grey squirrel (Sciurus carolinensis):*

will be twice as large as the corresponding components in A's brain. Instead, each brain area scales with a specific allometry, an exponential factor that varies with the rate of neurogenesis among other things (Finlay et al., 2001; Herculano-Houzel et al., 2007; Kaas, 2000). Allometric scaling has significant consequences for understanding evolution. For example, larger brains have proportionately larger frontal regions of the cortex simply because their overall size is larger. A related point is that larger brains have a higher proportion of cortical neurons that are in a position to exert direct control over primary motor neurons in the spinal cord and brain stem via the corticospinal and corticobulbar tracts (Herculano-Houzel et al., 2015), although this remains to be experimentally tested. Thus, in the absence of specific selective adaptations, larger brains have more overall cortical territory, a higher proportion in the prefrontal cortex (PFC), and proportionally stronger descending projections, all of which can significantly influence an animal's behavioral capacities.

In light of the large range of mammalian brain sizes, and the calculable differences in proportion brought about by allometric scaling, the preservation of a shared developmental program is that much more remarkable. The brain of a masked shrew (0.094 g) has the same basic developmental trajectory and organization as that of the sperm whale (8000 g), despite a mass ratio of 85 000 between the two. This robustness in overall brain organization with scale reflects the fact that growth rate and size of neural progenitor pools are among the few degrees of developmental modulation that evolution tolerates. Brains can change greatly in size (and internal proportion) as long as the basic developmental recipe is followed (Fig. 25.4A).

Finally, as indicated earlier, the role of experience during development, and particularly the relationship of an offspring with its mother, is a critical factor shaping brain evolution. In the case of mammalian vision, the examples of the importance of early visual experience, initially for survival and later for the development of skills in hunting, foraging, social perception, and motherhood itself, are largely circumstantial and based on behavioral observations in the field. Recent work in primates suggests that for certain important stimuli such as faces, experience critically shapes visual specialization in the brain, even if artificially delayed relative to birth. Macaques reared with no exposure to faces for up to 2 years then showed an immediate preference for viewing faces, followed by a selective and lasting expertise for either human or macaque faces, depending on which they then had exposure to (Sugita, 2008). When infants of the same species were trained extensively on symbolic stimulus categories, they exhibited an unusual fluency in discriminating them (Srihasam et al., 2012), and, unlike control animals, their brains were marked by selective responses to the trained stimuli (Srihasam et al., 2014). Thus early visual experience, in addition to developmental programs, overall brain size, and peripheral adaptation, appears to critically influence the establishment of functional specialization within the brain. Recent studies have begun to uncover the epigenetic mechanisms responsible for experience-driven alterations to brain organization (Stolzenberg Chapter X).

25.3.2.2 Examples of High-Level Visual Specialization in the Mammalian Brain

The mammalian capacities for understanding certain visual categories of objects, spatial representations, actions of others, and one's own visually guided actions are thought to depend upon more than one visual pathway. The strong projection from the retina to the lateral geniculate nucleus (LGN) is a main entry point of visual information to the cortical hierarchy feeding

architectonic subdivisions and connections from the visual thalamus. J. Comp. Neurol. 145, 273−305. https://doi.org/10.1002/cne.901450303; Sur, M., Nelson, R.J., Kaas, J.H., 1978. The representation of the body surface in somatosensory area I of the grey squirrel. J. Comp. Neurol. 179, 425−449. https://doi.org/10.1002/cne.901790211; Wong, P., Kaas, J.H., 2008. Architectonic subdivisions of neocortex in the gray squirrel (*Sciurus carolinensis*). Anat. Rec (Hoboken) 291, 1301−1333. https://doi.org/10.1002/ar.20758), flying fox (Calford, M.B., Graydon, M.L., Huerta, M.F., Kaas, J.H., Pettigrew, J.D., 1985. A variant of the mammalian somatotopic map in a bat. Nature 313, 477−479; Rosa, M.G., 1999. Topographic organisation of extrastriate areas in the flying fox: implications for the evolution of mammalian visual cortex. J. Comp. Neurol. 411, 503−523), human (Toga, A.W., Thompson, P.M., Mori, S., Amunts, K., Zilles, K., 2006. Towards multimodal atlases of the human brain. Nat. Rev. Neurosci. 7, 952−966. https://doi.org/10.1038/nrn2012), galago (Wong, P., Kaas, J.H., 2010. Architectonic subdivisions of neocortex in the Galago (*Otolemur garnetti*). Anat. Rec. (Hoboken) 293, 1033−1069. https://doi.org/10.1002/ar.21109), marmoset (Bendor, D., Wang, X., 2008. Neural response properties of primary, rostral, and rostrotemporal core fields in the auditory cortex of marmoset monkeys. J. Neurophysiol. 100, 888−906. https://doi.org/10.1152/jn.00884.2007; Krubitzer, L.A., Kaas, J.H., 1990. The organization and connections of somatosensory cortex in marmosets. J. Neurosci. 10, 952−974; Solomon, S.G., Rosa, M.G.P., 2014. A simpler primate brain: the visual system of the marmoset monkey. Front. Neural Circuits 8, 96. https://doi.org/10.3389/fncir.2014.00096), opossum (Dooley, J.C., Franca, J.G., Seelke, A.M.H., Cooke, D.F., Krubitzer, L.A., 2014. Evolution of mammalian sensorimotor cortex: thalamic projections to parietal cortical areas in *Monodelphis domestica*. Front. Neuroanat. 8, 163. https://doi.org/10.3389/fnana.2014.00163; Wong, P., Kaas, J.H., 2009b. An architectonic study of the neocortex of the short-tailed opossum (Monodelphis domestica). Brain Behav. Evol. 73, 206−228. https://doi.org/10.1159/000225381), platypus (Krubitzer, L., Manger, P., Pettigrew, J., Calford, M., 1995. Organization of somatosensory cortex in monotremes: in search of the prototypical plan. J. Comp. Neurol. 351, 261−306. https://doi.org/10.1002/cne.903510206).

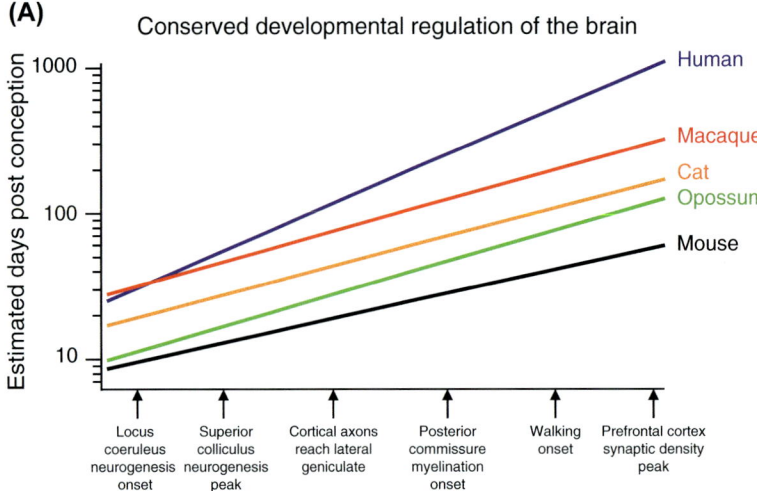

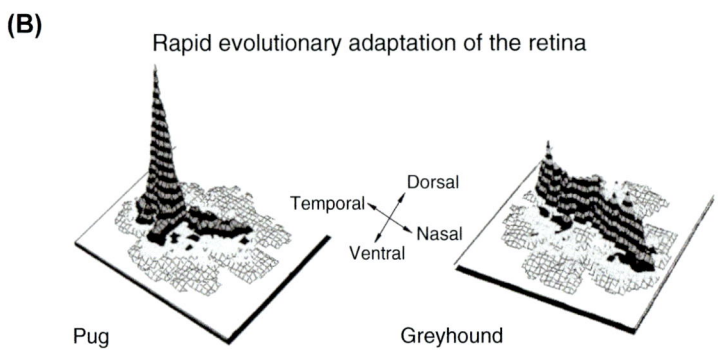

FIGURE 25.4 Contrasting slow evolutionary change of the brain with rapid adaptation of the periphery. (A) Development of the mammalian brain is tightly constrained by linked regularities in shared developmental events in brain. For example, the temporal ordering of events is very similar across different mammalian taxa. Accordingly, differences in the overall size and composition of the brain are most strongly affected by two free parameters, one relates to the initial buildup of progenitor pools preceding neurogenesis (y-intercept) and the second relates to the rate of brain development (slope) (Adapted from Workman, A.D., Charvet, C.J., Clancy, B., Darlington, R.B., Finlay, B.L., 2013. Modeling transformations of neurodevelopmental sequences across mammalian species. J. Neurosci. 33, 7368–7383. https://doi.org/10.1523/JNEUROSCI.5746-12.2013.). Strict developmental constraints limit the capacity for selective pressures to fundamentally alter the composition of the brain in the form of mosaic evolution, where one area expands based on selection pressures. (B) In contrast, peripheral sensory adaptations are linked to more rapidly evolving features and thus differ widely in closely related species, and even the same species. For example, retinal specializations in the domestic dog vary from a focal area centralis in dogs with flat faces to a visual streak in dogs with long snouts and lateralized eyes. Adapted from McGreevy, P., Grassi, T.D., Harman, A.M., 2003. A strong correlation exists between the distribution of retinal ganglion cells and nose length in the dog. Brain Behav. Evol. 63, 13–22. https://doi.org/10.1159/000073756.

dorsal and ventral visual streams (Felleman and Van Essen, 1991). Other visual pathways, such as those involving the superior colliculus, which projects to both the LGN (Harting et al., 1991) and pulvinar (Baldwin et al., 2011; Day-Brown et al., 2010) are less well understood (see Baldwin and Bourne, Chapter X). Recent work also highlights the direct retinal innervation of the pulvinar (Boire et al., 2004; Cowey et al., 1994; O'Brien et al., 2001; Warner et al., 2010), which has been hypothesized to play a critical role in the development of the cortical hierarchy (Bridge et al., 2015; Warner et al., 2015).

Historically, visual neuroscience has employed a range of animal models to address questions of neural wiring and function. Over time, studies have concentrated primarily on a small number of species, notably the mouse and the macaque, with a decreasing emphasis on explicitly comparative work. A few other species, such as the cat, ferret, and tree shrew, are still used to study functional architecture of the primary visual cortex (Kremkow et al., 2016; Lee et al., 2016). Due to a gradual decline in the diversity of animal models, relatively little is known about neural mechanisms supporting high-level vision in mammals other than primates. Here we review three exceptions to this rule, where good traction has been made in certain aspects of visual processing: visual attention in the cat, face processing in the sheep, and visual navigation in the rat.

The first example is that the role of the parietal cortex in controlling visual attention in the cat. Ablating or cooling a region of the cat cortex in the vicinity of the temporooccipitoparietal junction leads to deficits in visual attention of the contralateral hemifield that resemble human hemineglect (Lynch and McLaren, 1989; Payne et al., 1996; Fig. 25.5A). As in humans and macaques, these effects of parietal disruption can be dissociated from the inactivation or lesioning of ventral temporal cortex, which in cats leads to deficits in the learning and recognition of objects (Lomber et al., 1996). Mapping the metabolic effects of cat parietal cortex inactivation using 2-deoxyglucose revealed a reduction in activity of specific ipsilateral cortical and midbrain regions, with an increase in activity of the contralateral superior colliculus (Rushmore et al., 2006). Similar behaviors have been reported in rats following parietal disruption (Reep and Corwin, 2009). Together, these observations indicate that at least certain

aspects of visual attention are directed by the parietal cortex in nonprimate mammals. As such, they lend support to an early mammalian origin of the primate dorsal visual pathway (Kravitz et al., 2011; Payne, 1993).

The second example is that of face-selective responses in the temporal cortex of sheep. Using a range of methods, Kendrick and colleagues discovered selective neural responses to faces in the sheep temporal cortex, amygdala, and medial PFC (da Costa et al., 2004; Kendrick et al., 2001; Kendrick and Baldwin, 1987; Fig. 25.5B). Single-unit recordings, primarily in the temporal cortex, revealed neurons responsive to face images of sheep, humans, goats, and dogs, as well as neurons sensitive to face viewing angle, specific postures, schematic faces, and dynamic bodily movements (Kendrick, 1991; Kendrick and Baldwin, 1989). More neurons were selective for familiar than novel faces (Kendrick et al.,

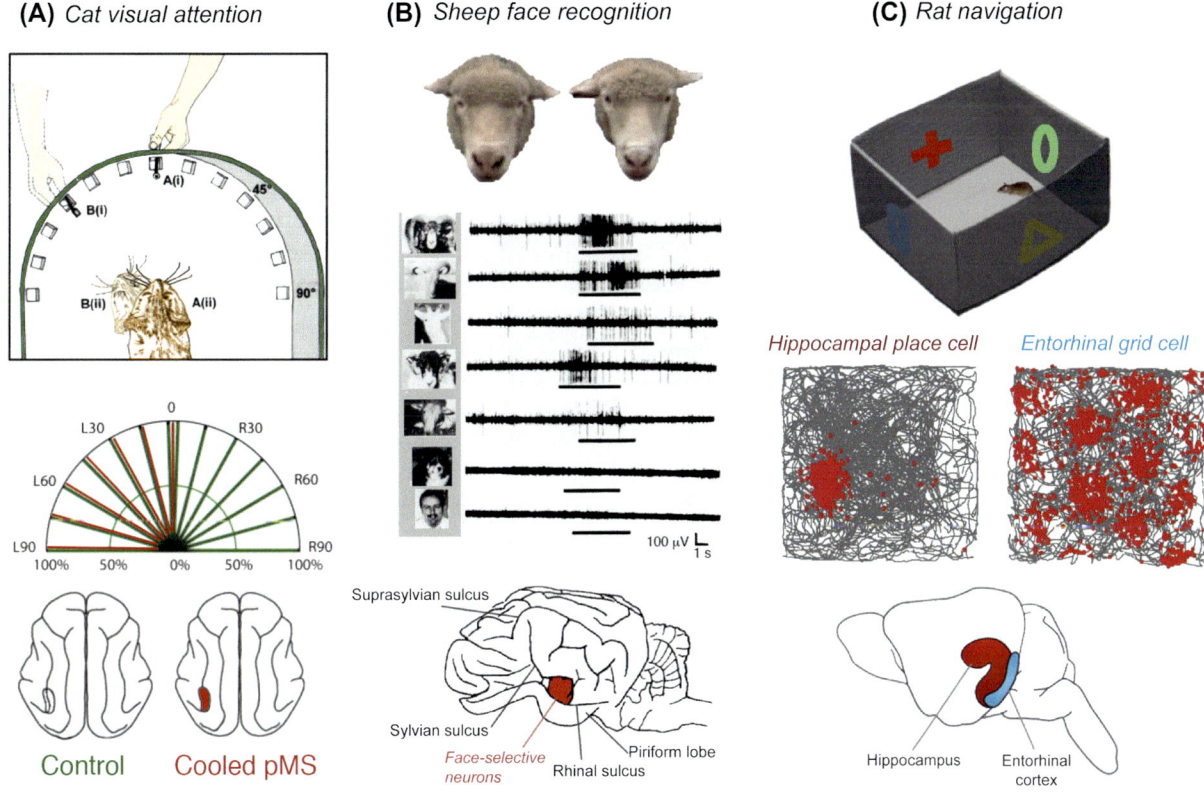

FIGURE 25.5 Three examples of visual cognitive abilities whose neural basis has been studied in nonprimate mammals. (A) Lesion and cooling studies in the domestic cat have demonstrated a parietal cortex contribution to spatial orienting that resembles that observed in humans and monkeys. Prior to cooling of the parietal area posterior middle suprasylvian (pMS), cats are able to orient to targets across the visual azimuth. However, after cooling, animals ceased to orient to targets in the contralateral hemisphere (Adapted from Rushmore, R.J., Valero-Cabre, A., Lomber, S.G., Hilgetag, C.C., Payne, B.R., 2006. Functional circuitry underlying visual neglect. Brain 129, 1803–1821. https://doi.org/10.1093/brain/awl140.). (B) Sheep readily recognize individuals visually, including through images of their face, and can learn and remember a large number of individual faces. Electrophysiological studies have revealed neurons in the sheep temporal cortex that are selective for faces, and particularly for faces of familiar individuals (Reviewed in Tate, A., Fischer, H., Leigh, A., Kendrick, K., 2006. Behavioural and neurophysiological evidence for face identity and face emotion processing in animals. Philos. Trans. R. Soc. B Biol. Sci. 361, 2155.). (C) Visual cues figure prominently into the navigation strategies of rats and play a role in shaping and reinforcing the firing patterns of well-described place and grid cells in the temporal lobe. In the two examples shown, *gray traces* are the exploration trajectories of a rat, and the *red dots* correspond to spatial locations in which the example cells were active. *Adapted from Derdikman, D., Moser, E.I., 2010. A manifold of spatial maps in the brain. Trends Cogn. Sci. Regul. Ed. 14, 561–569. https://doi.org/10.1016/j.tics.2010.09.004.*

2001), and neurons in the lateral amygdala, but not other areas, responded more to faces of sheep than to those of goats (da Costa et al., 2004; Kendrick, 2008). In sheep, an interesting feature of the responses to faces is the graded responses to horn size (Kendrick, 1994; Kendrick and Baldwin, 1987; see Fig. 25.8B). As horns are an important feature of sheep identity that are thought to be used in mate selection (Coltman et al., 2002), the analogue and continuous tuning of neurons to this feature may be an example of a visual specialization evolved for sexual selection (see Fig. 25.1). These findings in the sheep, together with the effect of temporal cortex cooling in the cat and a few scattered observations, are consistent with an early mammalian origin of the primate ventral visual pathway serving important aspects of social vision (Kravitz et al., 2013; Payne, 1993).

The third example relates to mechanisms underlying navigation in the rat and specifically to the visual contribution to the well-studied "place cells" and "grid cells" (Moser et al., 2008). As mentioned earlier, one widespread use of vision among mammals is allocentric memory for scenes. Visual cues are sufficient to stimulate place and grid cells (Aronov and Tank, 2014), with the most important visual information present in distal cues corresponding to natural landmarks in a scene (Scaplen et al., 2014). When such landmarks are unreliable, place cells no longer anchor their responses to those visual elements (Jeffery and O'Keefe, 1999). While the correspondence of these mechanisms to neural mechanisms of spatial navigation in primates is currently a topic of investigation, one set of results suggests that neurons in the macaque entorhinal cortex responds in a gridlike manner according not to physical position within a scene, but the direction of eye gaze upon a complex scene (Killian et al., 2012). Hippocampal and entorhinal cortex activity in primates and rodents is strongly shaped through direct and indirect connections by activity in the parietal and visual areas (Koganezawa et al., 2015; Kravitz et al., 2011). Again comparative work is informative, in this case demonstrating that place and grid cells are not a clade-specific adaptation, since similar neural responses were found in fruit bats (Yartsev et al., 2011), whose common ancestor with rodents lived in the Cretaceous period (see Fig. 25.2A). That such responses are expressed in three dimensions in that species (Yartsev and Ulanovsky, 2013) and are contributed by different senses than in rodents (Geva-Sagiv et al., 2015), are important considerations when attempting to understand the functioning of this neural system.

25.3.3 Summary

In this section, we focused on visual behaviors and corresponding brain regions in nonprimate mammals.

The brains of extant mammals have been shaped by a complex mixture of ancestral, shared traits, highly conserved principles of brain development, rapidly adapting peripheral organs, and the demands natural and sexual selection within highly specified niches. Data from example mammalian species indicate a recognizable similarity to the nonhuman primate in the neural basis of complex visual behaviors as diverse as selective attention, face perception, and visual navigation. While this observation is consistent with the notion that much of primate vision draws upon mechanisms present before the divergence of placental mammals, it is also clear that primate vision is unique in many ways. The next section provides a more detailed account of visual specializations in primate vision, centered necessarily on the well-studied macaque, with consideration throughout the way in which such specializations might have evolved from the visual capacities of mammalian ancestors.

25.4 Clade-Specific Visual Specializations in Primates

Primate vision stands out from that of other mammals in fundamental ways, which are often emphasized. While the stem adaptations of mammals emphasized olfaction and hearing more than vision (Heesy and Hall, 2010), primates came to depend on vision for gathering social information, such as dominance relationships (Deaner et al., 2005; Haude et al., 1976; Keverne et al., 1978), suitability of a potential mate (Gerald et al., 2007; Gerald et al., 2009; Waitt and Little, 2006; Waitt et al., 2003), and the focus of a conspecific's attention (Emery and Clayton, 2009; Shepherd et al., 2006; Tomasello et al., 1998). Vision also figures centrally into their unique capacity for reaching, grasping, and fine manual manipulation, including a precision grip, which together are thought to have evolved for an arboreal lifestyle (Bloch and Boyer, 2002; Cartmill, 1992).

The perceptual side of these derived traits originates largely from specialization of the eyes, including the emergence of the fovea, binocular convergence, and trichromatic complement of cone photopigment types (Mitchell and Leopold, 2015). Visual acuity in large primates exceeds that of most vertebrates, notably in the center of gaze, where the ability to resolve detail is surpassed by only a handful of predatory birds (Kirk and Kay, 2004). Binocular convergence facilitates stereoscopic information for active vision (Parker, 2007) and redundant sampling of a cluttered arboreal environment (Changizi and Shimojo, 2008). Trichromatic color vision is speculated to have played a role in primates' commitment to visual cues for foraging and sexual selection, which is reinforced by the observation that trichromatic primates have a deterioration of olfactory receptor genes (Bradley and Mundy, 2008). These adaptations to the eyes and retina are matched by characteristic specialization in the brain, where neural activity supports face and body perception (Perrett et al., 1992; Tsao and Livingstone, 2008), reaching and grasping movements (Cohen and Andersen, 2002; Kaas and Stepniewska, 2016), and social information such as interpreting goal-directed actions of others (Rizzolatti and Craighero, 2004).

If the unique aspects of primate vision are to be summarized in two words, they may be *coordination* and *manipulation*, both of which equally apply to manual behaviors and social interactions. Manual behaviors involve the visual coordination of movements with the environment, which is critical for arboreal life, as in the reaching and grasping for branches for locomotion through trees, and the use of hands in conflict and defensive postures. Manual manipulation is central to other actions, such as foraging, grooming, and tool use in certain species (van Schaik et al., 1999) (see Fragazy Chapter X). In the social domain, coordination and manipulation refer to the use of social vision to interact in a large group, judge dominance relationships, and climb the social ladder (Maestripieri, 2008; Miller et al., 2016). We next take a closer look at the unique uses of vision in primates that go beyond those found in nonprimate mammals and are thus likely to have evolved to serve the needs of primates.

25.4.1 Adaptations for Social Vision

Unique among mammals, every species of the primate order is highly social (Sussman, 1999). Primate sociality may in fact have been the key factor driving the evolution of primate intelligence (Humphrey, 1976). Primate social behaviors frequently involve communication through face and body that can include tactile signals (eg, kissing Goodall, 1990) or scents secreted from glands on the head (Zeller, 1987), but is primarily mediated through auditory and visual channels and thus suited for long-range communication (Dobson, 2009a). Much of primate social information is mediated through visual perception of faces, bodies, and their actions. The primate face conveys a particularly rich set of information about qualities such as species, identity, age, gender, mood, attention, attractiveness, and even trustworthiness (Kanwisher and Barton, 2011; Willis and Todorov, 2006), which can be gleaned from the face in just a fraction of a second. Rich nonverbal social information is also gathered through the capacity to perceive and judge body shape, gait, actions, emotional expressions, and goal-directed interactions between individuals (Weisbuch and Ambady, 2011). The majority of research into primate social perception has focused on

faces, and the neural systems supporting face recognition are now sufficiently well understood in two primate species to allow for a cross-species comparison (Tsao et al., 2008a; Yovel and Freiwald, 2013). Parallel studies provide a theoretical, behavioral, and neural basis for understanding perceptual specializations for bodies and actions (Nelissen et al., 2011; Popivanov et al., 2014; Rizzolatti and Sinigaglia, 2010; Singer and Sheinberg, 2010). Here we place most focus on how faces are perceived, mentioning the less studied processing of bodies and behaviors as space permits. We first provide an overview of visual social perception in monkeys, apes, and humans, which is followed by a review of the neural structures thought to support primates' high-level visual social perception.

25.4.1.1 How Primates Look at Faces

As faces carry diverse types of information, face perception entails multiple different operations. As described earlier, some of these operations, such as the detection of a face or the understanding of a coarse facial posture (eg, opening of the mouth, turning of the head), are likely to be shared widely among vertebrates, and particularly among mammals (Leopold and Rhodes, 2010; Parr, 2011). The analysis of bodies is similarly subject to multiple levels of perceptual operation, for example, determining bodies' direction of orientation, posture, and dynamic actions. Little is known about whether each of these perceptual operations for bodies and faces map onto unique and identifiable circuits, let alone whether similar operations in different species draw upon homologous or analogous neural circuits. Where primates excel is in the capacity to see beyond the basic structural and mechanistic aspects of faces, and recognize more implicit information. The word "recognition" as applied to faces itself holds a range of meanings with respect to the attributes being recognized (discussed now, see Fig. 25.6).

For species recognition, primates use vision to assess head shape, the details and configuration of internal facial features, and specific colorations, markings, tufts, swellings, or regions of bare skin. These adaptations are important for both New and Old World monkey species, and particularly those which are closely related and occupy the same territory (Allen et al., 2014; Bradley and Mundy, 2008; Santana et al., 2013, 2012; Fig. 25.6A). One role of species recognition pertains to sexual selection, where the recognition of conspecific facial features protects against hybrid breeding and the production of low-fitness offspring of related species (Allen et al., 2014). Experimentally, multiple species of primates have been shown to preferentially direct gaze toward images of their own species (Fujita, 1987; Neiworth et al., 2007; Tanaka, 2003). Species preferences are driven in part by early visual experience, such that

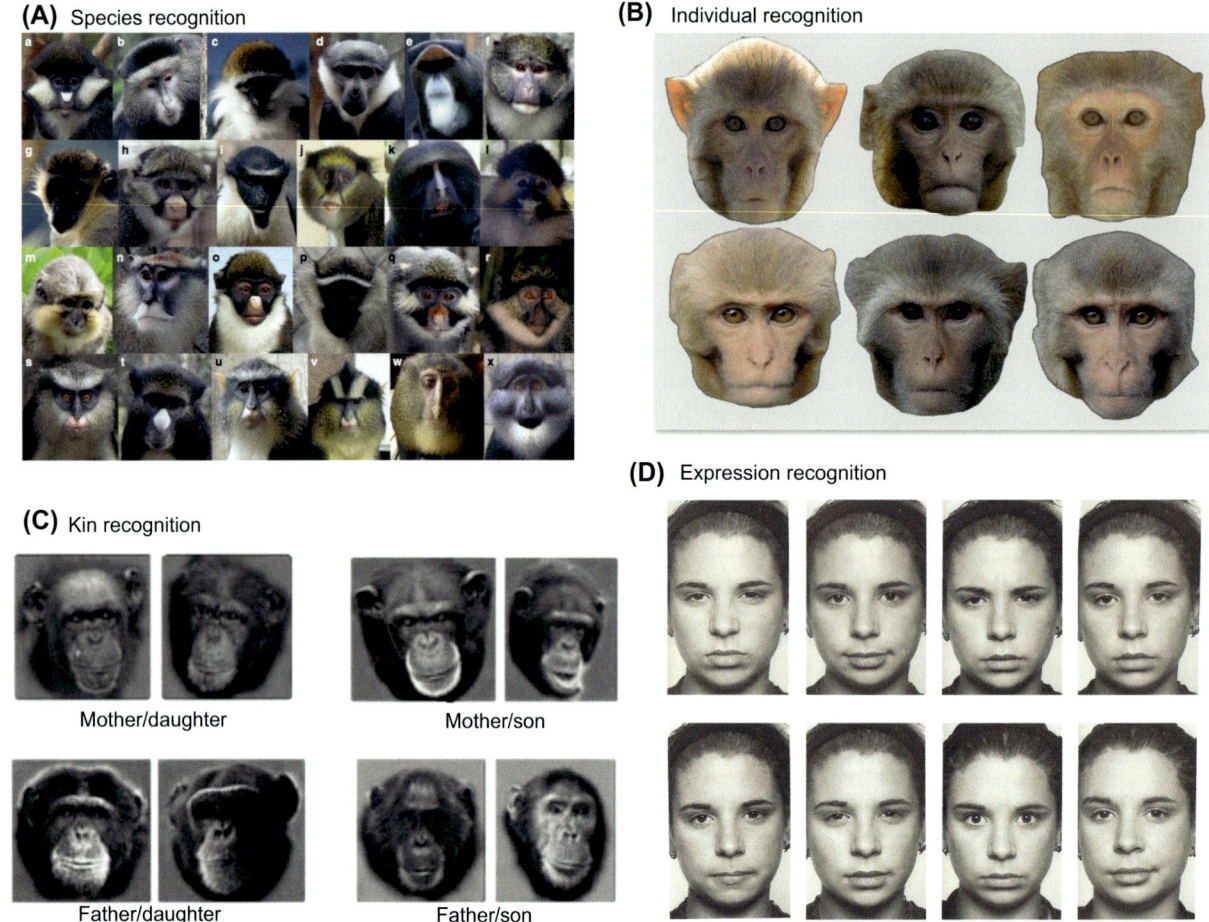

(A) Species recognition

(B) Individual recognition

(C) Kin recognition

Mother/daughter　　Mother/son

Father/daughter　　Father/son

(D) Expression recognition

FIGURE 25.6　Examples of the multiple dimensions of visual face recognition among primates. (A) Species recognition. At a basic level, primate face coloration and patterning serves as an important signal to identify conspecifics and limit breeding with heterospecifics, particularly for related species that forage and travel together, such as the guenon species depicted here (Adapted from Allen, W.L., Stevens, M., Higham, J.P., 2014. Character displacement of Cercopithecini primate visual signals. Nat. Commun. 5, 4266. https://doi.org/10.1038/ncomms5266.). (B) Individual recognition. Multiple primate species have been shown to recognize one another based on the distinctive characteristics of their faces, with typical facial variation among rhesus macaques shown here. (C) Chimps are able to determine the kin resemblance through photographs of unfamiliar individuals, which is thought to play an important role in avoiding inbreeding. (Adapted from Parr, L.A., Heintz, M., Lonsdorf, E., Wroblewski, E., 2010. Visual kin recognition in nonhuman primates: (*Pan troglodytes* and *Macaca mulatta*): inbreeding avoidance or male distinctiveness? J. Comp. Psychol. 124, 343–350. https://doi.org/10.1037/a0020545.) (D) Humans in particular are experts at recognizing very subtle facial expressions, which can guide their perceptions and actions in a complex social situation. The large repertoire of human facial expressions is enabled by a flat and bare face, together with exquisite control over a large number of facial muscles. Adapted from Ekman, P., 2003. Emotions Revealed. Henry Holt and Company. http://us.macmillan.com/books/9780805083392.

primates raised in a human environment can show a preference and expertise for human faces rather than those of their conspecifics (Martin-Malivel and Okada, 2007; Sugita, 2008; Tanaka, 2007).

Individual recognition has received the greatest attention in the study of human and nonhuman primates (Fig. 25.6B). While most mammals rely on other senses, notably olfaction and hearing, for individual recognition (with some exceptions such as sheep (Kendrick, 2008), mentioned earlier), primates recognize individuals by visual cues, including the face (Dufour et al., 2006; Parr et al., 2000; Rosenfeld and Van Hoesen, 1979). Some evidence suggests that chimps, and to some extent rhesus

monkeys, are even able to determine kinship from photographs of unfamiliar individuals (Parr et al., 2010a; Fig. 25.6C). In humans, much work has been done to pinpoint the expertise underlying individual recognition. One critical factor appears to be the sensitivity to the second-order configuration of facial parts, such as the precise geometrical layout of the eyes, nose, mouth, brow cheeks, and chin (Carey, 1992). This configurational aspect, together with the shape and coloration of individual features, serves as the basis for much of human individual recognition. Psychophysical evidence suggests that the brain approaches individual identity by structurally comparing faces to an internal reference, or norm face (Rhodes et al., 1987; Rhodes and Leopold, 2011; Valentine, 1991). Face norms are thought to form naturally over a range of timescales based on the brain's accumulated estimation of average facial structure. The extent to which human and nonhuman primates use norm-based mechanisms is debated (Kahn and Aguirre, 2012; Leopold and Bondar, 2005; Leopold et al., 2006; Parr et al., 2012; Storrs and Arnold, 2015), and we discuss this issue further in the next section.

In addition to species and individual identity, primates are able to read a wealth of other types of social information from faces. Most anthropoid primate species have trichromatic vision, allowing for the inspection of facial color. Facial coloration in primates involves the pelage, as well as the coloration of bare skin related to the distribution of pigment and the perfusion of blood (Bradley and Mundy, 2008). The latter of these factors is particularly important, as it relates to factors such as dominance, emotion, pregnancy, and possibly the phase of the estrous cycle visible in bare-faced primates, including humans (Changizi et al., 2006; Gerald et al., 2007; Waitt et al., 2003, 2006). Some primates are also able to determine conspecific's orientation of gaze using the head and eyes. Several species, including chimps, macaques and marmosets, Diana monkeys, and lemurs, have some capacity to follow gaze (Burkart and Heschl, 2006; Goossens et al., 2008; Okamoto-Barth et al., 2007; Overduin-de Vries et al., 2014; Scerif et al., 2004; Shepherd and Platt, 2008; Tomasello et al., 1998), particularly when orienting movements are coupled with certain facial expressions (Overduin-de Vries et al., 2016). This ability has the potential to direct them to objects of attention or feed into cognitive operations such as joint attention or visual perspective taking. While there are clear demonstrations of gaze following in individual species, the particular abilities and uses of gaze following appear to vary substantially among primates (for a recent review, see Rosati and Hare, 2009) (see Rosati Chapter X).

Primate species exhibiting a wide range of facial expressions also have particularly large eyes (Dobson, 2009b; van Hooff, 1962), which affords them a higher visual acuity. While this relationship may derive naturally from the allometric scaling of eyes and facial musculature in larger primates, it has the consequence that species with the most complex facial musculature and larger repertoire of expressions also have the most powerful vision. Facial expressions are controlled by a complex set of central pathways (Gothard, 2014; Müri, 2016) that innervate facial muscles with independent attachment sites that show a high degree of conservation across primates (Burrows, 2008; Burrows et al., 2009; Huber, 1931). Some facial behavior is used in similar ways (Parr et al., 2010b; Waller et al., 2008), consistent

with Darwin's original ideas about the evolutionary continuum of emotional expression (Darwin, 1872). Darwin also proposed that facial expressions evolved as adaptations of sensory regulation (Darwin, 1872; Schmidt and Cohn, 2002), for which there is now experimental evidence: during fearful expressions, human subjects had a subjectively larger visual field, faster eye movements during target localization, and an increase in nasal volume and air velocity during inspiration, while the opposite was the case during expressions of disgust (Susskind et al., 2008). These functional aspects may explain certain conserved features of facial expression among mammals. At the same time, facial behavior does vary significantly across primate species, and the number of facial muscles does not necessarily predict the level of facial expressions for a given species, as other factors come into play. For example, highly expressive faces tend to be those that are devoid of complex patterning (Santana et al., 2014), possibly because facial coloration and markings can reduce the perceptual salience of facial expression. Nonetheless, in humans, facial mobility and visual acuity are both at the extreme among primates, and thus humans are unparalleled in their capacity produce and read subtle facial expressions (Fig. 25.6D).

In summary, primates are able to extract diverse types of information from the face that supports their complex social interaction. Analogous operations for perceiving the body likely also provide important information about species, identity, estrous, direction of attention, and emotional state, although this has been studied less and is not reviewed here. In the next section, we address neural machinery underlying the perception of social stimuli, again placing primary focus on faces.

25.4.1.2 Neural Specializations

Comparative retinotopic mapping and histochemical studies in a host of primate species have painted a clear picture of visual field representations and areal boundaries in the primate visual cortex (Kaas, 2013; Krubitzer and Kaas, 1990a; Preuss, 2007; Rosa and Tweedale, 2005) (see Kaas Chapter X). This work has shown that primates have a distinct constellation of visual cortical areas whose layout and connections distinguish their brains from those of nonprimates. Many of these primate visual specializations are present in prosimians such as the galago, whose common ancestor with simian primates lived in the Cretaceous period, whereas others appear more restricted to simian primates, including humans. Thus in evaluating functionally defined visual specializations, such as those for faces, bodies, and other types of social information, there exists a backdrop of conserved cortical visual areas.

25.4.1.2.1 The fMRI Layout of Social Processing Systems in the Primate Brain

Initial comparative explorations regarding the processing of faces tentatively point to conserved face-processing systems in simian primates. These comparisons derive primarily from fMRI studies of face-selective hemodynamic responses in humans and trained monkeys, which so far has been restricted to just four species, namely the rhesus and crab-eating macaque, human, and marmoset (Fig. 25.7A). Face processing in the brains of other primate species, such as the chimpanzee (Parr et al., 2009) and vervet monkey (Zangenehpour and Chaudhuri, 2005), has been explored using other methods and is broadly consistent with the findings in macaques and humans.

In human and macaque, multiple labs have asked the straightforward question of which brain regions respond more strongly to images of faces than to images of other stimulus categories, which typically include bodies, common objects, abstract patterns, and scenes. In both species, this type of experiment leads to the reliable identification of several circumscribed regions in the temporal and frontal cortex that appear dedicated to the processing of faces over other visual stimuli (Duchaine and Yovel, 2015; Hoffman and Haxby, 2000; Ishai et al., 2005; Kanwisher et al., 1997; Logothetis et al., 1999; Pinsk et al., 2005; Tsao et al., 2003). These functionally defined areas, sometimes known as "patches," not only outline possible nodes in an important network for social perception (Grimaldi et al., 2016; Moeller et al., 2008; Schwiedrzik et al., 2015), but have also come to serve as important landmarks or reference points for a broad range of studies of the ventral visual pathway.

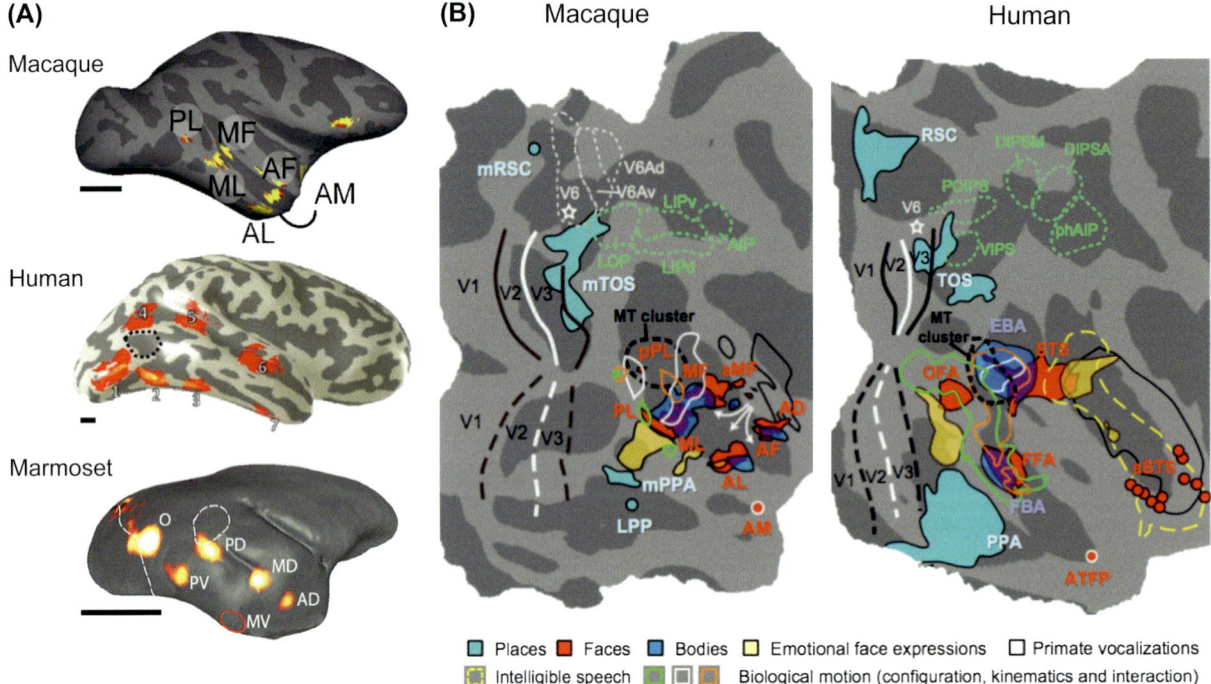

FIGURE 25.7 Functional layout of areas related to high-level vision based on fMRI experiments in primates. (A) Comparison of the face-selective visual areas in three primate species that differ markedly in size (bar represents 1 cm). Macaque: *PL*, posterior lateral; *MF*, middle fundus; *ML*, middle lateral; *AF*, anterior fundus; *AL*, anterior lateral; *AM*, anterior medial. Marmoset: *O*, occipital; *PD*, posterior dorsal; *PV*, posterior ventral; *MD*, middle dorsal; *MV*, middle ventral; *AD*, anterior dorsal. (Adapted from Dubois, J., de Berker, A.O., Tsao, D.Y., 2015. Single-unit recordings in the macaque face patch system reveal limitations of fMRI MVPA. J. Neurosci. 35, 2791–2802. https://doi.org/10.1523/JNEUROSCI.4037-14.2015; Hung, C.-C., Yen, C.C., Ciuchta, J.L., Papoti, D., Bock, N.A., Leopold, D.A., Silva, A.C., 2015. Functional mapping of face-selective regions in the extrastriate visual cortex of the marmoset. J. Neurosci. 35, 1160–1172. https://doi.org/10.1523/JNEUROSCI.2659-14.2015; Weiner, K.S., Grill-Spector, K., 2015. The evolution of face processing networks. Trends Cogn. Sci. Regul. Ed. 19, 240–241. https://doi.org/10.1016/j.tics.2015.03.010.) Correspondence of individual face patches between species is an active area of investigation. (B) Face patches (*red*) in the macaque and human in the context of body patches and other regions selectively responsive to social and biological information. Macaque: *mRSC*, macaque retrosplenial cortex; *LIP*, lateral intraparietal area; *AIP*, anterior intraparietal area; *LOP*, lateral occipital parietal area; *mTOS*, macaque transverse occipital sulcus; *MT*, middle temporal area; *mPPA*, macaque parahippocampal place area; *LPP*, lateral place patch. Human: *RSC*, retrosplenial cortex; *DIPSM*, dorsal intraparietal sulcus medial; *DIPSA*, dorsal intraparietal sulcus anterior; *POIPS*, parieto-occipital intraparietal sulcus; *phAIP*, putative homolog of the anterior intraparietal area; *TOS*, transverse occipital sulcus; *MT*, middle temporal area; *EBA*, extrastriate body area; *STS*, superior temporal sulcus; *ATFP*, anterior temopral face patch; *OFA*, occipital face area; *FBA*, fusiform body area; *PPA*, parahippocampal place area. *Adapted from Vanduffel, W., Zhu, Q., Orban, G.A., 2014. Monkey cortex through fMRI glasses. Neuron 83, 533–550. https://doi.org/10.1016/j.neuron.2014.07.015.*

The correspondence across species is not yet established, and there is no guarantee that any strict homology will be demonstrable. However, the superficial appearance of functional responses is conspicuously similar, with a similar number and spatial arrangement of the face patches. In both macaques and humans, there appear to be two series of patches running in parallel along the occipitotemporal pathway, including one in the ventral temporal cortex and the other more dorsally in the superior temporal sulcus (STS) (for a review, see Yovel and Freiwald, 2013). The marmoset is less studied, but a dorsal and ventral set of patches in its temporal cortex is consistent with this view (Hung et al., 2015). It bears mentioning that the primate species tested in this way serve as outstanding points of reference for a comparative study, since, at least for simian primates, they represent a very broad range of brain sizes. The marmoset, macaque, and human brains differ in mass by steps of approximately one order of magnitude, with the marmoset brain being less than 10 g, the macaque brain approximately 100 g, and the human brain exceeding 1000 g. That these functionally defined networks take a similar form in the largest and one of the smallest primate brains suggests that a multinode face-processing network evolved early in primate evolution.

Using fMRI to compare macaque and human functional specialization more broadly, face-selective patches can be visualized in the context of other types of responses (Fig. 25.7B) (for a recent review, see Vanduffel et al., 2014). In both humans and macaques, face and body patches sit in close proximity and, depending on how they are defined, show some degree of overlap (Fisher and Freiwald, 2015a; Peelen and Downing, 2005; Popivanov et al., 2014; Schwarzlose et al., 2005). The layout of body patches in the two species shows enough topological similarity to allow for cautious speculation about correspondence in macaques and humans between the posterior patches (mid-STS body area and extrastriate body area, EBA, respectively) and between the anterior patches (ant-STS body area, and fusiform body area, FBA, respectively). These patches overlap with regions sensitive to different types of biological motion (Fig. 25.7B). The correspondence between *individual* face patches is not yet clear, although plausible mappings exist (Tsao et al., 2008a; Yovel and Freiwald, 2013). Interestingly, face patches are closely associated with other functionally defined regions, such as color-selective areas in both species (Lafer-Sousa and Conway, 2013; Lafer-Sousa et al., 2016). This association could potentially be related to important aspects of social perception, given the aforementioned hypothesized role of primate trichromacy in face perception and mate selection, mentioned earlier.

The similarity in the layout of patches extends beyond the temporal cortex. Face-selective neurons and areas have been found in the macaque in three major parts of PFC (O Scalaidhe et al., 1997; Rolls et al., 2006; Tsao et al., 2008b), with fMRI work confirmed the presence of similarly face-selective areas in the human PFC (Axelrod and Yovel, 2013; Chan and Downing, 2011; Pitcher et al., 2011).

A major question in considering this layout is whether functionally selective regions such as face patches should be considered as nodes within a processing hierarchy. There is significant evidence from fMRI experiments that supports this notion. For example, the selectivity for faces compared to other objects increases in more rostral face patches (Bell et al., 2009), position dependence decreases (Hemond et al., 2007), mirror-symmetric confusion of facial profile views emerges (Axelrod and Yovel, 2012; Kietzmann et al., 2012), and facial identity selectivity grows stronger (Yang et al., 2016). Similarly, in both macaque and human face areas, the response to faces is augmented in the presence of an anatomically correctly placed body, and this augmentation grows stronger from posterior to anterior face areas (Bernstein et al., 2014; Fisher and Freiwald, 2015a; Song et al., 2013). Furthermore selectivity for facial motion is highly pronounced in dorsal areas, but not in ventral ones (Fisher and Freiwald, 2015b; Fox et al., 2009; Furl et al., 2012; Pitcher et al., 2011; Polosecki et al., 2013). Similar to the macaque, the ventral face areas appear to be interconnected based on diffusion tractography, whereas connectivity between the dorsal and ventral streams appears more limited (Gschwind et al., 2012) and is still the subject of active research (Yeatman et al., 2014). Thus considerable data point to a hierarchical organization of the face system, possibly with parallel streams that differ in their processing of dynamic aspects of facial behavior (Bernstein and Yovel, 2015; Duchaine and Yovel, 2015; Fisher and Freiwald, 2015b; Pitcher et al., 2014). In the next section, we connect these observations to electrophysiological and anatomical studies in the macaque.

25.4.1.2.2 Understanding the Activity of Macaque "Face Cells"

Neurons in the temporal cortex of the macaque that respond selectively to faces have been described for nearly half a century (Gross, 1994; Gross et al., 1972; Perrett et al., 1982), including their approximate spatial distribution (Perrett et al., 1992). Face-selective neurons in the PFC were described more recently (O Scalaidhe et al., 1997; Rolls et al., 2006). Following the systematic fMRI mapping of face patches in the macaque, multiple labs have targeted individual face patches to investigate how nodes in the face network may differ.

It is clear that fMRI-defined face patches contain high proportions of face-selective neurons (Freiwald and Tsao, 2010; Issa et al., 2013; McMahon et al., 2014b;

Tsao et al., 2006). Whatever the computational reason for local grouping of face cells, they constitute, at some level of description, domain-specific specialized hardware (Kanwisher, 2010). Microstimulation/fMRI (Moeller et al., 2008), resting state connectivity (Schwiedrzik et al., 2015), and anatomical tracing studies (Grimaldi et al., 2016) reveal that the face patches are selectively interconnected with one another. Interestingly, these anatomical data have not provided strong clues about the hierarchical organization of face processing, aside from the fact that the most posterior patch (PL) receives its projections broadly from early visual areas and thus may serve as the input to the face-processing system (Grimaldi et al., 2016). Nonetheless, electrophysiological scrutiny of individual patches using controlled images reinforces the idea of a hierarchical progression in face processing (Freiwald and Tsao, 2010). First, face areas are organized along a posterior–anterior axis, along which neural response latencies systematically increase. Moreover, facial information appears to be systematically transformed from early view-specific representations into late identity-specific representations (Freiwald and Tsao, 2010; Meyers et al., 2015). In the most posterior area, PL, selectivity for a single feature, the eye, seems to dictate neural responses (Issa and DiCarlo, 2012). At the next level, in areas MF and ML, a wide range of features are represented, and this selectivity is modulated by the embedding of features into the facial whole (Freiwald et al., 2009). While face representations in MF and ML are view selective, this dependence is reduced at the next level, AL, through construction of a mirror-symmetric invariance to head orientation, and reduced even further in AM, where representations for facial identity are robust against variation in head orientation (Dubois et al., 2015; Freiwald and Tsao, 2010; Meyers et al., 2015). The response tolerance for stimulus position and stimulus size also appears to increase in parallel with tolerance for view and selectivity for identity.

Orthogonal to this hierarchical principle of information processing, a second functional distinction between face areas along the ventral–dorsal axis exists. Neurons in the face areas located in the fundus of the STS (dorsal) respond with a much shorter latency and more strongly to facial motion than do neurons in ventral areas (Furl et al., 2012; Polosecki et al., 2013). In fact, neurons in these dorsal areas exhibit a pronounced preference for natural facial motion that neurons in face areas located laterally in the lower bank (ventral) lack (Fisher and Freiwald, 2015b). These results suggest different kinds of facial information are processed through two parallel streams. However, given the complex connectivity between face areas (Grimaldi et al., 2016; Moeller et al., 2008) and the fact that information processed in either stream is likely informative for computations in the

other, this parallelism does not appear to result from strict functional isolation between streams. Several other lines of research have been directed to address particular questions with respect to the firing of face-selective neurons, and we review a few of these now.

Some studies have attempted to understand how single-unit responses might pertain to known psychophysical parameters of face perception. As mentioned earlier, a long-standing theory of face recognition holds that a prototype or "norm" face serves as an internal reference against which facial structure is compared (Rhodes and Leopold, 2011). While not universally accepted, there is much evidence in favor of this model, including the special role of the average face in defining psychological dimensions for identity and caricaturization (Rhodes et al., 1998), as well as the existence face-identity aftereffect (Leopold et al., 2001). How the brain might implement an internal reference for face perception was previously unknown. Single-unit recordings in face patches MF/ML (Freiwald et al., 2009) and in face patch AM (Leopold et al., 2006) provide some answers to this question, with a tentative link to the observed perceptual effects (Fig. 25.8A). In MF/ML, neurons were found to be tuned in a ramplike manner to the parametric variation of facial features in cartoon faces (Freiwald et al., 2009). Tuning most often yielded a maximal response at one extreme end of the spectrum (eg, large pupils, eyes very close together) and a minimal response at the other end (eg, small pupils, eyes very far apart). Many feature dimensions were reflected across the population, often with multiple dimensions represented by individual cells. Different cells showed opposite polarity in their feature modulation. Neurons in MF/ML responded most strongly to upright faces, and their feature responses were highly dependent on the context of the entire face (Freiwald et al., 2009). In a different study, neurons in area ML were shown to lose their sensitivity to details of internal facial features once faces were inverted (Taubert et al., 2015). More anteriorly and ventrally, neurons in AM were tested with morphed realistic faces that varied in their level of caricaturization (Leopold et al., 2006). In that study, individual features were manipulated independently, but covaried naturally through the morphing. Like the MF/ML neurons, the AM neurons showed broad tuning curves to and responded most strongly to extreme values of the faces. However, unlike with the cartoon faces in ML/MF, the neural responses in AM responded least for the average face, or "norm." Recent results in face patch AF revealed a very similar pattern to that observed in AM, again with the average face eliciting the lowest responses to morphed faces (Jones et al., 2014). Whether the observed differences between these areas reflect steps in the brain's processing of identity, or are more tightly linked to differences in stimuli

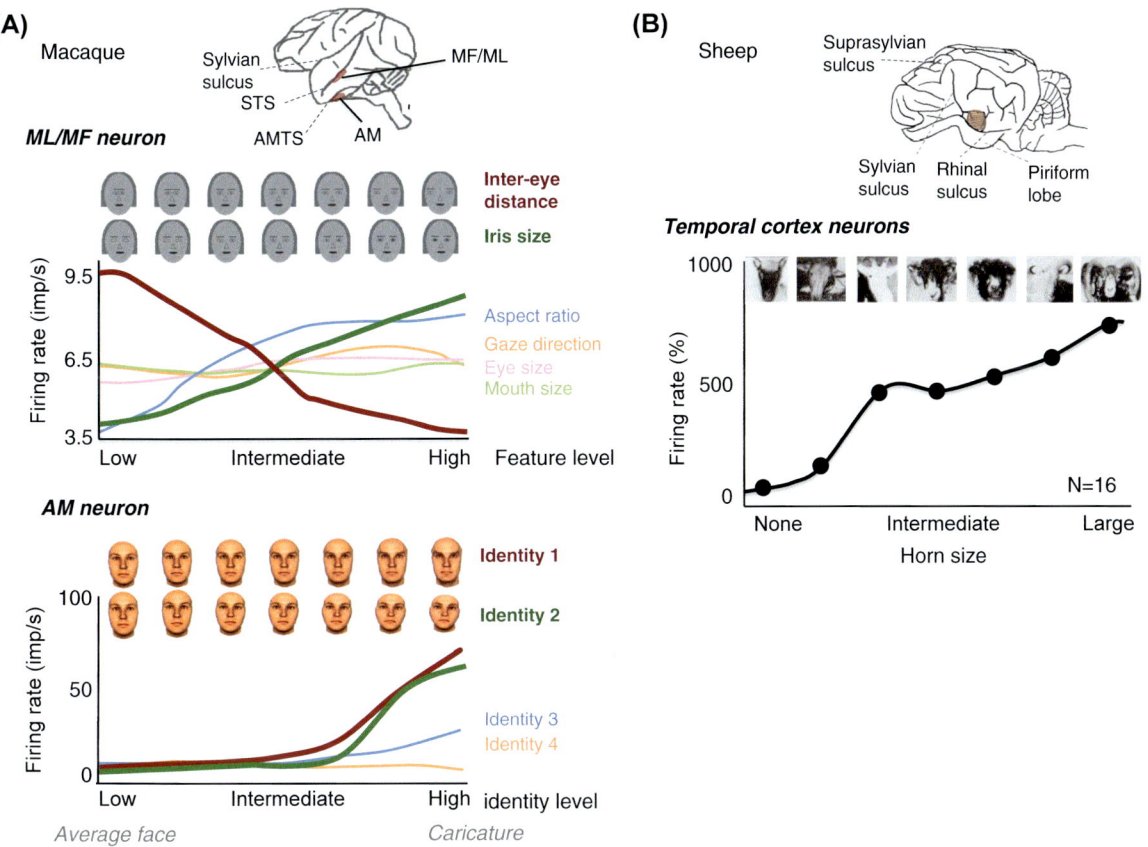

FIGURE 25.8 Neural tuning patterns to facial identity features that provide mechanistic insights into aspects of visual face recognition. (A) Macaque tuning related to facial identity. A study of ML/MF neurons found that the level of individual features in cartoon faces, such as the size of the irises or distance between the eyes, strongly influenced the firing of neurons, with extreme responses corresponding to extreme feature levels (Freiwald et al., 2009). In face patch AM, a study investigating morphed identity of realistic human faces found that the average face tended to elicit lowest response, while the high identity and caricaturized faces elicited the highest responses (Leopold et al., 2006). Both results suggest that neurons show extreme responses for the most distinctive facial features. (B) An electrophysiology study showing photographs of sheep faces to awake sheep found that many neurons in the temporal cortex showed graded responses as a function of the size of the horns in the picture. STS, superior temporal sulcus; AMTS, anterior medial temporal sulcus; AM, anterior medial face patch; MF/ML, middle fundus/middle lateral face patch. *Adapted from Kendrick, K.M., 1991. How the sheep's brain controls the visual recognition of animals and humans. J. Anim. Sci. 69, 5008–5016.*

between the studies, remains to be determined. Importantly, none of the above studies found neurons that were sharply tuned for intermediate values of features or caricaturization level. Thus these physiological results fit well with principles of coding identity that stem from opponent, and possibly norm-based, mechanisms. They also bear resemblance to at least one feature of the single-unit responses observed in sheep, the only nonprimate species in which similar studies have been conducted (Fig. 25.8B).

Other electrophysiological studies have asked whether neural responses are fine-tuned for other types of social information, such as the inferred direction of social attention or emotional expression. Many face cells have tuning properties that suggest an involvement in some of these social inferences. Some face cells are tuned

to head orientation or gaze (De Souza et al., 2005; Desimone et al., 1984; Perrett et al., 1985; Freiwald and Tsao, 2010), and similar selectivity has been reported in the STS of the human brain (Carlin et al., 2011). Such selectivity could support an analysis of the direction of social attention in the dorsal face pathways (Shepherd et al., 2009). Similar head-view selectivity was also found in the parallel, ventral face pathway, but there it may reflect an initial view-dependent step preceding a view-invariant face representation. Other face-selective cells appear to be tuned to facial expressions (Gothard et al., 2007; Hasselmo et al., 1989; Sugase et al., 1999). It is important to point out that the areas involved in processing different attributes of the face are unlikely to be completely segregated. Thus while the large-scale fMRI mapping approach has been enormously useful for determining the large-scale organization of category selective regions (Kanwisher, 2010), it is inherently limited in its capacity to ask deeper mechanistic questions, particularly within populations of neurons in which different types of information are integrated in the same region of space (Dubois et al., 2015).

This latter point is particularly salient in light of recent findings about how face cells respond under more natural viewing conditions. Given that most of our knowledge about the physiology of face-selective neurons derives from their responses to briefly presented stimuli, a recent set of studies asked the question how face-selective neurons respond under the conditions in which monkeys viewed natural videos depicting macaque social behavior. An fMRI study found that activation in face patches could be readily extracted when the time course of voxels throughout the brain was correlated with the evolving face content of the videos (Russ and Leopold, 2015). However, when examining the activity of a tight population of face-selective neurons within 1 mm^3 of face patch AF, most cells were not obviously linked to the appearance and disappearance of faces in the videos, but instead responded to a wide range of features (McMahon et al., 2015). Moreover, response time courses of neighboring neurons were often uncorrelated with one another, despite all residing within the space of a single fMRI voxel. More research is needed here to understand whether individual face patches should be considered functional units, or whether the building blocks of face processing are spatially distributed over multiple patches.

Finally, recent research has demonstrated that, while it is likely certain face-selective neurons in the brain must show plasticity, many retain their precise selectivity at least over periods of weeks and months. Recent advances in chronic microwire electrodes have made it possible to track individual neurons longitudinally across sessions and over extended time periods (Bondar et al., 2009; McMahon et al., 2014a). In face patch AF, the selectivity profile of individual neurons to a few dozen stimuli was used as a fingerprint that could readily differentiate it from its neighbors, including those neurons recorded simultaneously from the same electrode. Remarkably, the selectivity fingerprint of face-selective neurons was retained precisely, over the course of months, with no detectable changes despite the normal diet of visual experience, including monkey and human faces (McMahon et al., 2014b). Nor was the tuning of isolated AF neurons significantly altered by weeks of intensive perceptual learning with morphed faces (Leopold laboratory, unpublished observations). Future research will investigate the extent to which neurons in different face patches are plastic in their tuning responses, as the capacity to learn may be an important feature that

distinguishes different areas specialized for face processing (Freiwald and Tsao, 2014).

25.4.1.2.3 Mechanisms of Action Perception and Their Importance

The focus of this section has been on faces. A number of studies have begun to investigate body-selective responses in the macaque at the single-unit level (Popivanov et al., 2014; Puce and Perrett, 2003; Vangeneugden et al., 2011, 2009). The visual analysis of bodily actions may be extremely important when attempting to understand the evolution of high-level visual specialization, for at least two reasons. First, as described earlier in the chapter, there is likely to be a stronger link between species, and deeper evolutionary roots, in the perception of bodies. Most mammals have a visual understanding of bodily postures and actions that facilitates their conspecific interactions and to some extent heterospecific interactions, such as predation and predator avoidance. Second, for primates, one potentially important implementation of visual body and action perception involves one's *own* actions, for example, the visually guided manipulation of objects with the hand (Janssen and Scherberger, 2015). The ramifications of this connection are profound and may have shaped important features that are specific to the primate brain, such as the interplay between the dorsal and ventral streams (Van Dromme et al., 2016), the emergence of so-called "mirror neurons" (Rizzolatti and Sinigaglia, 2010), the transformation of reference frames (Cohen and Andersen, 2002), and the capacity to use tools (Iriki, 2006; Orban and Caruana, 2014). The next section does not explore each of these areas in depth, but instead touches on a few of the better studied areas in which primate vision has evolved in novel ways to monitor and steer complex actions.

25.4.2 Adaptations for Visually Guided Actions

Primates depend on vision extensively for navigation and manipulation of objects. The frontoparietal networks of the primate brain comprise a highly sophisticated system for representing space near the body and for planning visually guided actions. All simian primates have a pronounced foveal specialization with the two eyes positioned to be highly convergent, thus enabling highly accurate depth perception. When planning a reach or jump to a specific target, primates typically verge their eyes on its location, aligning its position on the two foveae to give fine estimates of depth that, when combined with the position of the two eyes, provides the target's position in space. Primates use eye movements during their self-motion to track objects in space and orient their bodies in relation to them, and do so with speed and fluidity. This is evident by the dexterity they display moving through their arboreal environments and their use of reaching movements. In this section we discuss the unique brain specializations for high-level spatial vision that make these complex movements possible. Ultimately the increasingly sophisticated processing of these frontoparietal networks gives rise to abilities that far exceed visually directed reaching or reach-to-grasp movements, including the use of tools and more complicated aspects of cognitive planning.

25.4.2.1 How Primates Steer Their Hands

Primates are unique among mammals in their capacity to use vision to guide reaching and grasping, which is critical for such diverse actions as movement through an arboreal environment, manipulation of food and other objects, and social interactions such as grooming. In terms of fine control of the hands, humans have the most advanced skills, which serve as the basis for the production and use of tools, and the many other features of modern society. The visual control over precise and flexible actions is linked in part to a refinement of mammalian dorsal visual stream, with an expansion of posterior parietal cortex (PPC) and elaboration in the primary motor cortex that affords direct cortical control over the hands and fingers (Bortoff and Strick, 1993; Herculano-Houzel et al., 2015).

From an evolutionary perspective, reaching movements likely originated in locomotor stepping movements in quadrupedal mammals, which increasingly came to be guided and refined by visual or tactile cues. This independence in forelimb control is thought to have initiated a range of adaptive behaviors for nonlocomotor movements such as pushing, swatting, or digging (Beloozerova and Sirota, 1993). Across mammals, the extent to which different species use vision to guide such movements is highly variable. For example, rats do not rely on vision during stepping movements partly because their eyes are upward facing, and instead use tactile cues to adjust paw position after landing (Hermer-Vazquez et al., 2007). Other mammals, such as cats, regularly employ vision during stepping and other movements related to reaching (Marigold and Drew, 2011). For example, cats adjust their steps on a treadmill to avoid obstacles (Drew et al., 2008), a behavior that is severely impaired by lesions to area 5 of parietal cortex (Fabre and Buser, 1981; Lajoie and Drew, 2007). The overlap between stepping movements and reaching movements is further supported by evidence from microstimulation studies where it is found that stimulation of the same points in parietal cortex can evoke either type of movement (Stepniewska et al., 2009).

Grasping is a rather different operation that in most mammals appears to rely on tactile cues, with primates being a notable exception (Karl and Whishaw, 2013). Rodents are unable to preshape their forepaw when retrieving a distal object, even following extensive training (Metz and Whishaw, 2000). However, if information about the size and shape of a graspable object is available from another modality, for example, from tactile cues for a food item held in the mouth, rodents are then able to preshape their forepaw in a way that resembles primates (Whishaw and Coles, 1996; Whishaw et al., 1992). Thus while rodents have no intrinsic motor limitation, they appear unable to adjust their grasp based on a visual assessment of object shape.

Primate reaching and grasping has a level of precision, dexterity, and visual coordination that makes it qualitatively different from that of other mammals. The evolutionary origin of these capacities is debated, but is thought to stem initially from the use of hands for either the predation of insects or the extraction of small fruits from the terminal branches of angiosperms (Bloch and Boyer, 2002; Cartmill, 1992; Kirk et al., 2003; Sussman et al., 2013). The fluid "reach-to-grasp" actions of primates are composed of the transport of the hand to the location of an object through control of the proximal shoulder and arm musculature together with metric movements of the hand and digits. Normally, reaching and grasping are coordinated, as the hand's grasp aperture opening and orientation is preshaped to match the object during the transport phase of the reach movement. Interestingly, this coordination depends on vision, and the two movements are dissociated when a human is blindfolded (Karl and Whishaw, 2013). These distinct phases of movement are consistent with the decomposition of coarse followed by fine movement in the dual visuomotor theory (Woodworth, 1899). Both the reaching and grasping components rely critically on the high-resolution fovea, however, in rather different ways.

Primate reaching movements are closely associated with the direction of gaze, which suggests an important role for visual perception. However, considerable evidence points to nearly the opposite conclusion: in many cases visually guided actions can be dissociated from perceptual experience (Goodale and Milner, 1992). To orient gaze to an object, the brain must process and compute metric information about the location of the target to precisely adjust the head and eye position, including details such as vergence angle. Considerable evidence suggests this mode of visual analysis is distinct from that supporting perception. The parietal cortex is thought to be central to the transformations involved, acting upon retinotopically organized maps to compute coordinate systems that facilitate action, including eye, head, and body reference frames (Cohen and Andersen,

2002). However, a popular hypothesis holds that the metrics of visually guided reaching movements are computed not based on the perceptual contents of vision per se, but instead on the internal signals associated with the positioning of the head and eyes (Bosco et al., 2016; Hadjidimitrakis et al., 2012; Horstmann and Hoffmann, 2005). According to this view, accurate reaching to a visual target depends on proprioceptive or corollary discharge information involved in directing the foveae. In humans, the direction of gaze affects reach accuracy even when visual information is removed, as when a subject is blindfolded, which illustrates the inherent coordination of orienting actions and manual reaching, even in the absence of visual perception (Jackson et al., 2005).

The close link between gaze and reaching is further underscored by certain features of optic ataxia, which refers to a diverse range of orienting and reaching impairments following parietal lesions (Bálint, 1909; Rizzo and Vecera, 2002). Visual control appears to be at the center of the reaching deficits, since reaching is not affected by targets that are defined by auditory or somatosensory cues (Jackson et al., 2009). Impairments are strongest when subjects are not allowed to foveate the target of their reach, called "nonfoveal" reaching (Buxbaum and Coslett, 1997; Jackson et al., 2005), an effect that has also been demonstrated in the macaque (Hwang et al., 2012). Thus optic ataxia is specific to the problem of reconciling the eye-based coordinate frame of a peripheral target with body-centered coordinates (Jackson et al., 2009). The dependence of reaching on foveation points to the primacy of directing gaze, including vergence, where internal motor and proprioceptive signals carry information about target position that can then be translated into goal-directed reaches.

Distinct from reaching, grasping in primates relies on an entirely different use of the fovea for its guidance shaping hand posture. Grasping must draw information about object shape from pathways involved in object analysis. It is thought that the dorsal stream receives such input from ventral visual pathways and integrates it to plan precise reach-to-grasp manual actions (Nelissen et al., 2011). In humans there is a developmental progression of the reach and grasp movements that parallels the development of higher level vision and may recapitulate aspects of the evolution of visually guided actions. In the first months of life, reach and grasp movements appear somewhat independent and do not obviously incorporate visual cues. Infants use haptic cues to shape their hands around objects. By 4 months of age, they are able to use haptic cues to preshape their hands for certain targets, such as parts of their own body (Newell et al., 1989). By 5 months, reaching movements begin to show rudimentary visual guidance to distal targets, and at 7–9 months, there is a small

degree of coordination of the hand for matching target orientation (Sacrey et al., 2012). However, the visual coordination of reaching and grasping does not begin until much later. Manual actions continue to conform to the two phases of movement, reach followed by grasp after tactile exploration, much like blindfolded human subjects, for an extended period over the first 2 years of life (Karl and Whishaw, 2013). The incorporation of vision for uniting these actions begins relatively late, emphasizing the primacy of somatosensory inputs in establishing these movement systems independently, and perhaps the computational difficulty in using vision to coordinate their action.

25.4.2.2 Neural Specializations

The advanced facility for visually guided reaching and grasping in primates is thought to have drawn, at least in part, upon the expansion and specialization of the PPC, together with the development of motor and premotor cortex (Kaas and Stepniewska, 2016). The PPC lies along the dorsal visual stream, which gains much of its input through projections from area MT. Area MT, in turn, is heavily innervated not only through the geniculostriate visual pathway, but also a pathway through the inferior pulvinar. Details of this pathway are a matter of debate (Berman and Wurtz, 2008; Bridge et al., 2015; Lyon et al., 2010; Stepniewska et al., 1999), and its function is largely unknown, although it has long been conceived as a potential parallel route for action information from the superior colliculus to reach the cerebral cortex (Schneider, 1969). Recent work in marmosets suggests that, in addition, a direct retinal projection to the inferior pulvinar is a significant but transient input to area MT during the first months of life (Warner et al., 2015, 2012). This pathway could be important both for the visual guidance of action in the postnatal period, but also for setting up the initial maturation of the dorsal visual pathway, which is known to develop before its ventral stream counterpart. Throughout life, area MT is involved in the analysis of motion and stereoscopic depth information (Albright et al., 1984; DeAngelis and Newsome, 1999; Maunsell and Van Essen, 1983). Major projections from MT to subregions of the PPC are critical for guiding visually directed actions to mediate specific behaviors and spatial judgments, whereas further projections from the PPC carry visual information further to retrosplenial and parahippocampal regions and are thought to be important for navigation (Kaas and Stepniewska, 2016; Kravitz et al., 2011). While elements of this pathway are clearly more developed in primates than in other mammals, the uniqueness of MT and its projections are not yet clear (Homman-Ludiye et al., 2010; Krubitzer, 2009; Payne, 1993; Preuss, 2007). As MT is a critical input stage in the primate dorsal pathway, gaining a deeper understanding of its uniqueness through comparatives studies may provide important insights into the evolution and composition of the human brain.

Comparative studies of the PPC in a few species indicate that its contribution to spatial perception, attentional orienting (see Fig. 25.5A), and landmark-based navigation is present in multiple mammalian orders (Kolb et al., 1994; Kolb and Walkey, 1987; Lomber et al., 2006). Such deficits are consistent with the role of this region in integrating visual inputs into body-centered (egocentric) and world-centered (allocentric) coordinate frames to guide actions (Buneo et al., 2002; Wilber et al., 2014). In rodents and other mammals the parietal cortex further plays a role in stepping movements during locomotion, which may be an evolutionary antecedent to reaching movements in primates (Karl and Whishaw, 2013). The same phases of movement described earlier for reaching are also recognizable in stepping movements (Whishaw et al., 2010), and although such movements are typically yoked to locomotion, most animals have some degree of independent control of the forelimbs during stepping. Thus the parietal cortex appears to contribute shared functions related to forelimb control among mammals. Nonetheless, the elaboration of the PPC in primates compared to other studied mammals is considerable, which likely reflects the arboreal niche, mode of locomotion, and upright postures that shaped the evolution of early primates (for a review, see Kaas, 2008). This elaboration likely relates, at least in part, to the development of specific circuits that exploit foveal vision, control the focus of attention, and coordinate complex manual movements—features that are greatly enhanced to primates.

The organization of the primate parietal cortex breaks down into distinct subnetworks that mediate different visually guided behaviors that include reaching, defensive, and grasping movements (Kaas and Stepniewska, 2016). These areas have been studied in detail in the macaque, most notably the parietal reach region (PRR) for reaching movements (Batista et al., 1999), the lateral intraparietal area for eye movements (Colby et al., 1996), the anterior intraparietal area (AIP) for grasping movements (Sakata et al., 1995), and the ventral intraparietal area for defensive movements of the head and arm (Cooke et al., 2003). In particular, there is a dissociation between those areas involved in reach movements (PRR and V6A) and grasp movements (AIP) along different pathways through parietal cortex. Specifically, area AIP lies along a dorsolateral route through parietal cortex with connections to ventral premotor cortex and regions of motor cortex controlling movements of the hands and digits in grasp (Davare et al., 2011; Jeannerod et al., 1995). Areas PRR and V6A lie along a dorsomedial parietal route with connections to dorsal premotor

cortex and regions of motor cortex controlling upper shoulder and arm movements (Grafton, 2010; Rizzolatti and Luppino, 2001). In humans there is a double-disassociation of reach and grasp functions, with lesions of dorsolateral pathway (AIP) impairing grasping performance (Binkofski et al., 1998; Jeannerod et al., 1994), and conversely lesions of the dorsomedial pathway (PRR and V6A) impairing reach performance (Cavina-Pratesi et al., 2010). Areas with similar functional segregation have been found across primate species including nocturnal prosimians such as the galago (Stepniewska et al., 2009, 2005) as well as New World monkeys. Among all these primates, each subregion of the parietal cortex projects to corresponding subregions of premotor and motor cortex that specialize in similar movements, and thus form distinct networks for particular categories of actions. The extent to which these pathways and functions appear across primate species suggests that they were established relatively early in a primate ancestor, perhaps to facilitate the particular conditions of arboreal living.

The primate network for grasping is partially distinct from that for reaching, as it is interconnected directly with object processing areas in the ventral visual pathway. For example, parietal area AIP, involved in grasping movements, receives direct inputs from object vision pathways in the inferior temporal cortex in addition to somatosensory inputs (Borra et al., 2008) and also receives inputs from area CIP in parietal cortex where neurons have been found selective to object shape (Tsutsui et al., 2001). A recent study investigating the mirror system may cast some light on how the parietal cortex draws upon object representations (Nelissen et al., 2011). That study first identified regions showing selective fMRI responses to videos of manual actions. They then correlated the pattern of activation with the cortico-cortical connections established in previous studies (see Fig. 25.9). The analysis found evidence for two largely segregated pathways carrying ventral object information into the parietal cortex and then onto the frontal cortex. These two pathways were interpreted as carrying different types of visual information related to interpreting others' actions. The first pathway, from the upper bank of STS to area PFG to area F5c, was thought to carry information about the observed agent and their intentions. The second pathway, from the lower bank of the STS to area AIP to area F5a/p, was thought to be more involved in understanding the motor acts themselves. It is possible that this latter pathway, which includes area AIP known to contain neurons sensitive to grasping, is jointly specialized in the visual guidance of one's own grasping, but also exhibits "mirror" properties such that it also responds to the observations of others' actions (Rizzolatti and Sinigaglia, 2010). These pathways may be uniquely evolved in primates,

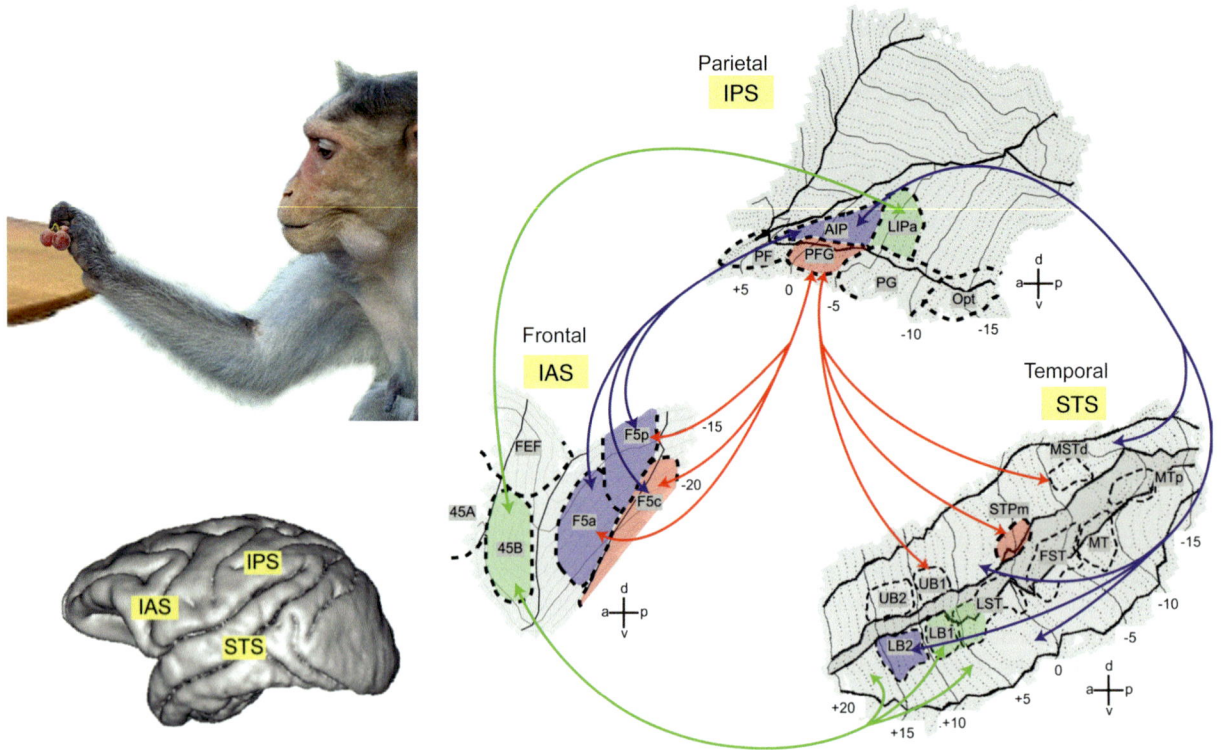

FIGURE 25.9 Complex network of temporal, parietal, and frontal network associated of observed manual actions in primates. Grasping areas such as anterior intraparietal are draw upon visual object information from the superior temporal sulcus, in addition to interacting with frontal areas thought to be involved in the initiation of action. Such circuitry may serve the dual role of interpreting others' goal-directed actions and visually guiding one's own behavior. This interplay represents an important meeting of visual signals in the dorsal and ventral visual pathways, which have abundant but selective anatomical interconnections. *IAS*, inferior arcuate sulcus; *IPS*, intraparietal sulcus; *STS*, superior temporal sulcus; *FEF*, frontal eye fields; *AIP*, anterior intraparietal area; *LIP*, lateral intraparietal area; *UB*, upper bank; *LB*, lower bank; *STP*, superior temporal polysensory area; *FST*, fundus of the superior temporal sulcus; *MT*, middle temporal area; *MST*, medial superior temporal area. *Adapted from Nelissen, K., Borra, E., Gerbella, M., Rozzi, S., Luppino, G., Vanduffel, W., Rizzolatti, G., Orban, G.A., 2011. Action observation circuits in the macaque monkey cortex. J. NEUROSCI. 31, 3743–3756. https://doi.org/10.1523/JNEUROSCI.4803-10.2011.*

although at present there is little evidence to support or refute this hypothesis.

Finally, the specialization of the parietal cortex is particularly important for understanding the human brain. Human bipedalism has fundamentally changed the manner in which extrapersonal space is encoded and has further led to exceedingly complex visually guided actions, and associated brain specializations, that facilitate tool use (Orban and Caruana, 2014). Humans have an expansion of the PPC that includes areas that appear to be absent in other primates (Chaplin et al., 2013). This expansion of PPC is thought to give rise to a more sophisticated repertoire of motor behaviors including the fine manipulation of tools and other objects (Orban et al., 2006; Peeters et al., 2009). One hypothesis for the expansion of parietal cortex and its role in tool use is that it occurred due to a duplication event in the left hemisphere of area AIP (Orban and Caruana, 2014). This duplication event would have occurred some 2 million years ago between the development of bipedalism, which freed the hands for greater tool use, and

the advent of language more recently (Orban and Caruana, 2014). The duplicated area would correspond to area aSMG identified in the left hemisphere in human fMRI studies, which is found to be selectively active during observation of tool actions (Peeters et al., 2009). Like area AIP, area aSMG receives inputs from inferior temporal cortex. In the case of AIP this inferior temporal input is thought to carry information about object shape to guide grasp, but in the case of aSMG might also carry semantic information about the object function. Although the origin of these additions to PPC remains speculative, their addition does appear to contribute to the more advanced tool use among humans, as lesions to the left hemisphere produce impairments that are more specific to tool use (Goldenberg and Hagmann, 1998).

25.4.3 Summary

In this section, we described the behaviors and neural mechanisms of two of the most conspicuous primate adaptations, both of which contribute to primates' unusual capacities for coordination and manipulation. In its steering of manual actions, vision allows for the fluid coordination of reaching and grasping for the purpose of locomotion, foraging, or in the case of humans and a few other species, tool use. Social vision allows primates to read, interpret, and respond to each others' behaviors in the context of a large group. It coordinates complex social dynamics and alliances. At the same time, the inspection of others, and the prospect of being inspected, enables aggression and deceit in the struggle to advance in the social hierarchy. Both sets of behaviors rely on brain systems that have clear mammalian antecedents, but appear to have become much more specialized in primates. Vision is at the center of these specializations, and particularly the high acuity fovea and its targeting on objects by eye movements.

25.5 Conclusions

In this chapter, we have placed the most prominent visual specializations in primates into a mammalian context. We first investigated the fabric of vision and advanced a new evolutionary framework for understanding the perceptual components of vision. With this in hand, we reviewed a broad range of comparative literature that informed the visual capabilities, and to some extent visual mechanisms, supporting high-level behaviors in mammals. Then turning to primates, we focused on two areas in which primate vision surpasses that of other mammals, and reviewed recent experimental work outlining mechanisms of vision for social cognition and action. Together, this large amount of material points to the multiple evolved layers of visual analysis that shape the cognition of primates, including humans. Our perceptions are shaped by these layers, where measurements in neural circuits have in some cases offered a semblance of mechanistic understanding, and in others allow only for speculation. Ultimately, the elegance of vision lies in its biology, whose invisible hand has guided a miraculous genetic and developmental program that gives form to our eyes and brain, our perceptual world, and all its trappings.

Acknowledgments

D.A.L was supported by the Intramural Research Program of the National Institute of Mental Health (ZIA MH002898). J.F.M. was supported by grants from the National Institute for Mental Health (R21 MH104756-01) and the National Institute of Health (U01-NS094330). W.A.F. was supported by grants from the National Eye Institute (5 R01 EY021594-03, NIH/NEI), the National Institute for Mental Health (5 R01 MH105397-02, NIH/NIMH), the Human Frontier Science Program (RGP0015/2013), the National Science Foundation funded Center for Brains, Minds, and Machines (CBMM), and the New York Stem Cell Foundation (NYSCF-R-NI23). W.A.F. is an NYSCF-Robertson Investigator.

References

Albright, T.D., Desimone, R., Gross, C.G., 1984. Columnar organization of directionally selective cells in visual area MT of the macaque. J. Neurophysiol. 51, 16–31.

Allen, W.L., Stevens, M., Higham, J.P., 2014. Character displacement of Cercopithecini primate visual signals. Nat. Commun. 5, 4266. https://doi.org/10.1038/ncomms5266.

Andrew, R., 1963. Evolution of facial expression. Science 142, 1034–1041.

Arendes, L., 1994. Superior colliculus activity related to attention and to connotative stimulus meaning. Brain Res. Cogn. Brain Res. 2, 65–69.

Aronov, D., Tank, D.W., 2014. Engagement of neural circuits underlying 2D spatial navigation in a rodent virtual reality system. Neuron 84, 442–456. https://doi.org/10.1016/j.neuron.2014.08.042.

Arrese, C., Archer, M., Runham, P., Dunlop, S.A., Beazley, L.D., 2000. Visual system in a diurnal marsupial, the numbat (Myrmecobius fasciatus): retinal organization, visual acuity and visual fields. Brain Behav. Evol. 55, 163–175. https://doi.org/10.1159/000006650.

Arrese, C.A., Hart, N.S., Thomas, N., Beazley, L.D., Shand, J., 2002. Trichromacy in Australian marsupials. Curr. Biol. 12, 657–660. https://doi.org/10.1016/S0960-9822(02)00772-8.

Axelrod, V., Yovel, G., 2012. Hierarchical processing of face viewpoint in human visual cortex. J. Neurosci. 32, 2442–2452. https://doi.org/10.1523/JNEUROSCI.4770-11.2012.

Axelrod, V., Yovel, G., 2013. The challenge of localizing the anterior temporal face area: a possible solution. Neuroimage 81, 371–380. https://doi.org/10.1016/j.neuroimage.2013.05.015.

Baldwin, B.A., Shillito, E.E., 1974. The effects of ablation of the olfactory bulbs on parturition and maternal behaviour in soay sheep. Anim. Behav. 22, 220–223. https://doi.org/10.1016/S0003-3472(74)80072-2.

Baldwin, M.K.L., Wong, P., Reed, J.L., Kaas, J.H., 2011. Superior colliculus connections with visual thalamus in gray squirrels (Sciurus carolinensis): evidence for four subdivisions within the pulvinar complex. J. Comp. Neurol. 519, 1071–1094. https://doi.org/10.1002/cne.22552.

Bálint, D.R., 1909. Seelenlähmung des "Schauens," optische Ataxie, räumliche Störung der Aufmerksamkeit. Eur. Neurol. 25, 51–66. https://doi.org/10.1159/000210464.

Barbas, H., Zikopoulos, B., Timbie, C., 2011. Sensory pathways and emotional context for action in primate prefrontal cortex. Biol. Psychiatry 69, 1133–1139. https://doi.org/10.1016/j.biopsych.2010.08.008.

Barton, R.A., Harvey, P.H., 2000. Mosaic evolution of brain structure in mammals. Nature 405, 1055–1058. https://doi.org/10.1038/35016580.

Barton, R.A., Purvis, A., Harvey, P.H., 1995. Evolutionary radiation of visual and olfactory brain systems in primates, bats and insectivores. Philos. Trans. R. Soc. Lond. B Biol. Sci. 348, 381–392. https://doi.org/10.1098/rstb.1995.0076.

Bateson, P., 1990. Is imprinting such a special case? Philos. Trans. Biol. Sci. 329, 125–131.

Batista, A.P., Buneo, C.A., Snyder, L.H., Andersen, R.A., 1999. Reach plans in eye-centered coordinates. Science 285, 257–260.

Bekoff, M., 1977. Social communication in canids: evidence for the evolution of a stereotyped mammalian display. Science 197, 1097.

Bell, A.H., Hadj-Bouziane, F., Frihauf, J.B., Tootell, R.B.H., Ungerleider, L.G., 2009. Object representations in the temporal cortex of monkeys and humans as revealed by functional magnetic resonance imaging. J. Neurophysiol. 101, 688–700. https://doi.org/10.1152/jn.90657.2008.

Beloozerova, I.N., Sirota, M.G., 1993. The role of the motor cortex in the control of accuracy of locomotor movements in the cat. J. Physiol. 461, 1–25.

Bendor, D., Wang, X., 2008. Neural response properties of primary, rostral, and rostrotemporal core fields in the auditory cortex of marmoset monkeys. J. Neurophysiol. 100, 888–906. https://doi.org/10.1152/jn.00884.2007.

Berman, R.A., Wurtz, R.H., 2008. Exploring the pulvinar path to visual cortex. Prog. Brain Res. 171, 467–473. https://doi.org/10.1016/S0079-6123(08)00668-7.

Bernstein, M., Yovel, G., 2015. Two neural pathways of face processing: a critical evaluation of current models. Neurosci. Biobehav. Rev. 55, 536–546. https://doi.org/10.1016/j.neubiorev.2015.06.010.

Bernstein, M., Oron, J., Sadeh, B., Yovel, G., 2014. An integrated face-body representation in the fusiform gyrus but not the lateral occipital cortex. J. Cogn. Neurosci. 26, 2469–2478. https://doi.org/10.1162/jocn_a_00639.

Betts, B.J., 1976. Behaviour in a population of Columbian ground squirrels, Spermophilus columbianus columbianus. Anim. Behav. 24, 652–680. https://doi.org/10.1016/S0003-3472(76)80079-6.

Bininda-Emonds, O.R.P., Cardillo, M., Jones, K.E., MacPhee, R.D.E., Beck, R.M.D., Grenyer, R., Price, S.A., Vos, R.A., Gittleman, J.L., Purvis, A., 2007. The delayed rise of present-day mammals. Nature 446, 507–512. https://doi.org/10.1038/nature05634.

Binkofski, F., Dohle, C., Posse, S., Stephan, K.M., Hefter, H., Seitz, R.J., Freund, H.J., 1998. Human anterior intraparietal area subserves prehension: a combined lesion and functional MRI activation study. Neurology 50, 1253–1259.

Bizley, J.K., Nodal, F.R., Nelken, I., King, A.J., 2005. Functional organization of ferret auditory cortex. Cereb. Cortex 15, 1637–1653. https://doi.org/10.1093/cercor/bhi042.

Blake, R., 1993. Cats perceive biological motion. Psychol. Sci. 4, 54–57. https://doi.org/10.1111/j.1467-9280.1993.tb00557.x.

Blakeslee, B., Jacobs, G.H., Neitz, J., 1988. Spectral mechanisms in the tree squirrel retina. J. Comp. Physiol. A Neuroethol. Sens. Neural Behav. Physiol. 162, 773–780. https://doi.org/10.1007/BF00610966.

Bloch, J.I., Boyer, D.M., 2002. Grasping primate origins. Science 298, 1606–1610. https://doi.org/10.1126/science.1078249.

Blumstein, D.T., Daniel, J.C., Griffin, A.S., Evans, C.S., 2000. Insular tammar wallabies (Macropus eugenii) respond to visual but not acoustic cues from predators. Behav. Ecol. 11, 528–535. https://doi.org/10.1093/beheco/11.5.528.

Boire, D., Matteau, I., Casanova, C., Ptito, M., 2004. Retinal projections to the lateral posterior-pulvinar complex in intact and early visual cortex lesioned cats. Exp. Brain Res. 159, 185–196. https://doi.org/10.1007/s00221-004-1946-6.

Bolhuis, J.J., Honey, R.C., 1998. Imprinting, learning and development: from behaviour to brain and back. Trends Neurosci. 21, 306–311.

Bondar, I.V., Leopold, D.A., Richmond, B.J., Victor, J.D., Logothetis, N.K., 2009. Long-term stability of visual pattern selective responses of monkey temporal lobe neurons. PLoS One 4, e8222. https://doi.org/10.1371/journal.pone.0008222.

Bonney, K.R., Wynne, C.D.L., 2002. Visual discrimination learning and strategy behavior in the fat-tailed dunnart (Sminthopsis crassicaudata). J. Comp. Psychol. 116, 55–62.

Borra, E., Belmalih, A., Calzavara, R., Gerbella, M., Murata, A., Rozzi, S., Luppino, G., 2008. Cortical connections of the macaque anterior intraparietal (AIP) area. Cereb. Cortex 18, 1094–1111. https://doi.org/10.1093/cercor/bhm146.

Bortoff, G.A., Strick, P.L., 1993. Corticospinal terminations in two new-world primates: further evidence that corticomotoneuronal connections provide part of the neural substrate for manual dexterity. J. Neurosci. 13, 5105–5118.

Bosco, A., Breveglieri, R., Hadjidimitrakis, K., Galletti, C., Fattori, P., 2016. Reference frames for reaching when decoupling eye and target position in depth and direction. Sci. Rep. 6, 21646. https://doi.org/10.1038/srep21646.

Bradley, B.J., Mundy, N.I., 2008. The primate palette: the evolution of primate coloration. Evol. Anthropol. 17, 97–111. https://doi.org/10.1002/evan.20164.

Bridge, H., Leopold, D.A., Bourne, J.A., 2015. Adaptive pulvinar circuitry supports visual cognition. Trends Cogn. Sci. Regul. https://doi.org/10.1016/j.tics.2015.10.003.

Broad, K.D., Mimmack, M.L., Kendrick, K.M., 2000. Is right hemisphere specialization for face discrimination specific to humans? Eur. J. Neurosci. 12, 731–741.

Buckley, C., 2009. Tarra & Bella. Penguin.

Buneo, C.A., Jarvis, M.R., Batista, A.P., Andersen, R.A., 2002. Direct visuomotor transformations for reaching. Nature 416, 632–636. https://doi.org/10.1038/416632a.

Burkart, J., Heschl, A., 2006. Geometrical gaze following in common marmosets (Callithrix jacchus). J. Comp. Psychol. 120, 120–130. https://doi.org/10.1037/0735-7036.120.2.120.

Burrows, A., 2008. The facial expression musculature in primates and its evolutionary significance. Bioessays 30.

Burrows, A.M., Waller, B.M., Parr, L.A., 2009. Facial musculature in the rhesus macaque (Macaca mulatta): evolutionary and functional contexts with comparisons to chimpanzees and humans. J. Anat. 215, 320–334. https://doi.org/10.1111/j.1469-7580.2009.01113.x.

Buxbaum, L.J., Coslett, H.B., 1997. Subtypes of optic ataxia: reframing the disconnection account. Neurocase 3, 159–166. https://doi.org/10.1080/13554799708404050.

Calford, M.B., Graydon, M.L., Huerta, M.F., Kaas, J.H., Pettigrew, J.D., 1985. A variant of the mammalian somatotopic map in a bat. Nature 313, 477–479.

Cantone, G., Xiao, J., McFarlane, N., Levitt, J.B., 2005. Feedback connections to ferret striate cortex: direct evidence for visuotopic convergence of feedback inputs. J. Comp. Neurol. 487, 312–331. https://doi.org/10.1002/cne.20570.

Carey, S., 1992. Becoming a face expert. Philos. Trans. R. Soc. Lond. B Biol. Sci. 335, 95–102. https://doi.org/10.1098/rstb.1992.0012 discussion 102-3.

Carlin, J.D., Calder, A.J., Kriegeskorte, N., Nili, H., Rowe, J.B., 2011. A head view-invariant representation of gaze direction in anterior superior temporal sulcus. Curr. Biol. 21, 1817–1821. https://doi.org/10.1016/j.cub.2011.09.025.

Cartmill, M., 1992. New views on primate origins. Evol. Anthropol. 1, 105–111. https://doi.org/10.1002/evan.1360010308.

Catania, K.C., 2005. Evolution of sensory specializations in insectivores. Anat. Rec. 287, 1038–1050. https://doi.org/10.1002/ar.a.20265.

Catania, K.C., Collins, C.E., Kaas, J.H., 2000. Organization of sensory cortex in the East African hedgehog (Atelerix albiventris). J. Comp. Neurol. 421, 256–274.

Cavanagh, P., 1987. Reconstructing the third dimension: interactions between color, texture, motion, binocular disparity, and shape. Comput. Vis. Graph Image Process 37, 171–195. https://doi.org/10.1016/S0734-189X(87)80001-4.

Cavina-Pratesi, C., Monaco, S., Fattori, P., Galletti, C., McAdam, T.D., Quinlan, D.J., Goodale, M.A., Culham, J.C., 2010. Functional magnetic resonance imaging reveals the neural substrates of arm transport and grip formation in reach-to-grasp actions in humans. J. Neurosci. 30, 10306–10323. https://doi.org/10.1523/JNEUROSCI.2023-10.2010.

Chan, A.W.-Y., Downing, P.E., 2011. Faces and eyes in human lateral prefrontal cortex. Front. Hum. Neurosci. 5, 51. https://doi.org/10.3389/fnhum.2011.00051.

Changizi, M.A., Shimojo, S., 2008. "X-ray vision" and the evolution of forward-facing eyes. J. Theor. Biol. 254, 756–767. https://doi.org/10.1016/j.jtbi.2008.07.011.

Changizi, M., Zhang, Q., Shimojo, S., 2006. Bare skin, blood and the evolution of primate colour vision. Biol. Lett. 2, 217.

Chaplin, T.A., Yu, H.-H., Soares, J.G.M., Gattass, R., Rosa, M.G.P., 2013. A conserved pattern of differential expansion of cortical areas in simian primates. J. Neurosci. 33, 15120−15125. https://doi.org/10.1523/JNEUROSCI.2909-13.2013.

Clarke, P.G., Whitteridge, D., 1976. The cortical visual areas of the sheep. J. Physiol. 256, 497−508.

Clemo, H.R., Sharma, G.K., Allman, B.L., Meredith, M.A., 2008. Auditory projections to extrastriate visual cortex: connectional basis for multisensory processing in "unimodal" visual neurons. Exp. Brain Res. 191, 37−47. https://doi.org/10.1007/s00221-008-1493-7.

Cohen, Y.E., Andersen, R.A., 2002. A common reference frame for movement plans in the posterior parietal cortex. Nat. Rev. Neurosci. 3, 553−562. https://doi.org/10.1038/nrn873.

Colby, C.L., Duhamel, J.R., Goldberg, M.E., 1996. Visual, presaccadic, and cognitive activation of single neurons in monkey lateral intraparietal area. J. Neurophysiol. 76, 2841−2852.

Coltman, D.W., Festa-Bianchet, M., Jorgenson, J.T., Strobeck, C., 2002. Age-dependent sexual selection in bighorn rams. Proc. Biol. Sci. 269, 165−172. https://doi.org/10.1098/rspb.2001.1851.

Cooke, D.F., Taylor, C.S.R., Moore, T., Graziano, M.S.A., 2003. Complex movements evoked by microstimulation of the ventral intraparietal area. Proc. Natl. Acad. Sci. U.S.A. 100, 6163−6168. https://doi.org/10.1073/pnas.1031751100.

Cowey, A., Stoerig, P., Bannister, M., 1994. Retinal ganglion cells labelled from the pulvinar nucleus in macaque monkeys. Neuroscience 61, 691−705.

Crapse, T.B., Sommer, M.A., 2008. Corollary discharge across the animal kingdom. Nat. Rev. Neurosci. 9, 587−600. https://doi.org/10.1038/nrn2457.

da Costa, A., Leigh, A., Man, M., Kendrick, K., 2004. Face pictures reduce behavioural, autonomic, endocrine and neural indices of stress and fear in sheep. Proc. R. Soc. Biol. Sci. 271, 2077.

Darwin, C., 1872. The Expression of the Emotions in Man and Animals.

Davare, M., Kraskov, A., Rothwell, J.C., Lemon, R.N., 2011. Interactions between areas of the cortical grasping network. Curr. Opin. Neurobiol. 21, 565−570. https://doi.org/10.1016/j.conb.2011.05.021.

Day-Brown, J.D., Wei, H., Chomsung, R.D., Petry, H.M., Bickford, M.E., 2010. Pulvinar projections to the striatum and amygdala in the tree shrew. Front. Neuroanat. 4, 143. https://doi.org/10.3389/fnana.2010.00143.

De Souza, W.C., Eifuku, S., Tamura, R., Nishijo, H., Ono, T., 2005. Differential characteristics of face neuron responses within the anterior superior temporal sulcus of macaques. J. Neurophysiol. 94, 1252−1266. https://doi.org/10.1152/jn.00949.2004.

Deaner, R.O., Khera, A.V., Platt, M.L., 2005. Monkeys pay per view: adaptive valuation of social images by rhesus macaques. Curr. Biol. 15, 543−548. https://doi.org/10.1016/j.cub.2005.01.044.

DeAngelis, G.C., Newsome, W.T., 1999. Organization of disparity-selective neurons in macaque area MT. J. Neurosci. 19, 1398−1415.

Derdikman, D., Moser, E.I., 2010. A manifold of spatial maps in the brain. Trends Cogn. Sci. Regul. Ed. 14, 561−569. https://doi.org/10.1016/j.tics.2010.09.004.

Desimone, R., Albright, T., Gross, C., Bruce, C., 1984. Stimulus-selective properties of inferior temporal neurons in the macaque. J. Neurosci. 4.

Disbrow, E., Litinas, E., Recanzone, G.H., Padberg, J., Krubitzer, L., 2003. Cortical connections of the second somatosensory area and the parietal ventral area in macaque monkeys. J. Comp. Neurol. 462, 382−399. https://doi.org/10.1002/cne.10731.

Dobson, S., 2009. Socioecological correlates of facial mobility in nonhuman anthropoids. Am. J. Phys. Anthropol. 139.

Dobson, S., 2009. Allometry of facial mobility in anthropoid primates: implications for the evolution of facial expression. Am. J. Phys. Anthropol. 138, 138.

Dooley, J.C., Franca, J.G., Seelke, A.M.H., Cooke, D.F., Krubitzer, L.A., 2014. Evolution of mammalian sensorimotor cortex: thalamic projections to parietal cortical areas in Monodelphis domestica. Front. Neuroanat. 8, 163. https://doi.org/10.3389/fnana.2014.00163.

Drew, T., Andujar, J.-E., Lajoie, K., Yakovenko, S., 2008. Cortical mechanisms involved in visuomotor coordination during precision walking. Brain Res. Rev. 57, 199−211. https://doi.org/10.1016/j.brainresrev.2007.07.017.

Dubois, J., de Berker, A.O., Tsao, D.Y., 2015. Single-unit recordings in the macaque face patch system reveal limitations of fMRI MVPA. J. Neurosci. 35, 2791−2802. https://doi.org/10.1523/JNEUROSCI.4037-14.2015.

Duchaine, B., Yovel, G., 2015. A Revised Neural Framework for Face Processing. https://doi.org/10.1146/annurev-vision-082114-035518.

Dufour, V., Pascalis, O., Petit, O., 2006. Face processing limitation to own species in primates: a comparative study in brown capuchins, Tonkean macaques and humans. Behav. Process. 73, 107−113.

Duncan, R.D., Jenkins, S.H., 2011. Use of visual cues in foraging by a diurnal herbivore, Belding's ground squirrel. Can. J. Zool. https://doi.org/10.1139/z98-119.

Ebeling, W., Natoli, R.C., Hemmi, J.M., 2010. Diversity of color vision: not all Australian marsupials are trichromatic. PLoS One 5, e14231. https://doi.org/10.1371/journal.pone.0014231.

Ebinger, P., 1975. Quantitative investigations of visual brain structures in wild and domestic sheep. Anat. Embryol. 146, 313−323.

Ekman, P., 2003. Emotions Revealed. Henry Holt and Company.

Emery, N.J., Clayton, N.S., 2009. Comparative social cognition. Annu. Rev. Psychol. 60, 87−113. https://doi.org/10.1146/annurev.psych.60.110707.163526.

Fabre, M., Buser, P., 1981. Effects of lesioning the anterior suprasylvian cortex on visuo-motor guidance performance in the cat. Exp. Brain Res. 41, 81−88.

Felleman, D.J., Van Essen, D.C., 1991. Distributed hierarchical processing in the primate cerebral cortex. Cereb. Cortex 1, 1−47.

Fernald, R.D., 2006. Casting a genetic light on the evolution of eyes. Science 313, 1914−1918. https://doi.org/10.1126/science.1127889.

Finlay, B.L., Darlington, R.B., 1995. Linked regularities in the development and evolution of mammalian brains. Science 268, 1578−1584.

Finlay, B.L., Uchiyama, R., 2015. Developmental mechanisms channeling cortical evolution. Trends Neurosci. 38, 69−76. https://doi.org/10.1016/j.tins.2014.11.004.

Finlay, B.L., Darlington, R.B., Nicastro, N., 2001. Developmental structure in brain evolution. Behav. Brain Sci. 24, 263−278. https://doi.org/10.1017/S0140525X01333952. discussion 278−308.

Fisher, C., Freiwald, W.A., 2015. Whole-agent selectivity within the macaque face-processing system. Proc. Natl. Acad. Sci. U.S.A. 112, 14717−14722. https://doi.org/10.1073/pnas.1512378112.

Fisher, C., Freiwald, W.A., 2015. Contrasting specializations for facial motion within the macaque face-processing system. Curr. Biol. 25, 261−266. https://doi.org/10.1016/j.cub.2014.11.038.

Fox, C.J., Moon, S.Y., Iaria, G., Barton, J.J.S., 2009. The correlates of subjective perception of identity and expression in the face network: an fMRI adaptation study. Neuroimage 44, 569−580. https://doi.org/10.1016/j.neuroimage.2008.09.011.

Freiwald, W.A., Tsao, D.Y., 2010. Functional compartmentalization and viewpoint generalization within the macaque face-processing system. Science 330, 845−851. https://doi.org/10.1126/science.1194908.

Freiwald, W.A., Tsao, D.Y., 2014. Neurons that keep a straight face. Proc. Natl. Acad. Sci. U.S.A. 111, 7894−7895. https://doi.org/10.1073/pnas.1406865111.

Freiwald, W.A., Tsao, D.Y., Livingstone, M.S., 2009. A face feature space in the macaque temporal lobe. Nat. Neurosci. 12, 1187−1196. https://doi.org/10.1038/nn.2363.

Fujita, K., 1987. Species recognition by five macaque monkeys. Primates 28, 353−366.

Furl, N., Hadj-Bouziane, F., Liu, N., Averbeck, B.B., Ungerleider, L.G., 2012. Dynamic and static facial expressions decoded from motion-sensitive areas in the macaque monkey. J. Neurosci. 32, 15952–15962. https://doi.org/10.1523/JNEUROSCI.1992-12.2012.

Gerald, M., Waitt, C., Little, A., Kraiselburd, E., 2007. Females pay attention to female secondary sexual color: an experimental study in *Macaca mulatta*. Int. J. Primatol. 28, 1–7.

Gerald, M.S., Waitt, C., Little, A.C., 2009. Pregnancy coloration in macaques may act as a warning signal to reduce antagonism by conspecifics. Behav. Process. 80, 7–11. In: https://doi.org/10.1016/j.beproc.2008.08.001.

Geva-Sagiv, M., Las, L., Yovel, Y., Ulanovsky, N., 2015. Spatial cognition in bats and rats: from sensory acquisition to multiscale maps and navigation. Nat. Rev. Neurosci. 16, 94–108. https://doi.org/10.1038/nrn3888.

Gibson, J.J., 1986. The Ecological Approach to Visual Perception. Psychology Press.

Goldenberg, G., Hagmann, S., 1998. Tool use and mechanical problem solving in apraxia. Neuropsychologia 36, 581–589.

Goodale, M.A., Milner, A.D., 1992. Separate visual pathways for perception and action. Trends Neurosci. 15, 20–25.

Goodall, J., 1990. Through a Window. Houghton Mifflin Harcourt.

Goode, E., 2015. Learning from Animal Friendships. N.Y. Times.

Goossens, B., Dekleva, M., Reader, S., Sterck, E., Bolhuis, J., 2008. Gaze following in monkeys is modulated by observed facial expressions. Anim. Behav. 75, 1673–1681.

Gothard, K.M., 2014. The amygdalo-motor pathways and the control of facial expressions. Front. Neurosci. 8, 43. https://doi.org/10.3389/fnins.2014.00043.

Gothard, K., Battaglia, F., Erickson, C., Spitler, K., Amaral, D., 2007. Neural responses to facial expression and face identity in the monkey amygdala. J. Neurophysiol. 97, 1671.

Grafton, S.T., 2010. The cognitive neuroscience of prehension: recent developments. Exp. Brain Res. 204, 475–491. https://doi.org/10.1007/s00221-010-2315-2.

Grant, S., Hilgetag, C.C., 2005. Graded classes of cortical connections: quantitative analyses of laminar projections to motion areas of cat extrastriate cortex. Eur. J. Neurosci. 22, 681–696. https://doi.org/10.1111/j.1460-9568.2005.04232.x.

Greco, B.J., Brown, T.K., Andrews, J.R.M., Swaisgood, R.R., Caine, N.G., 2013. Social learning in captive African elephants (*Loxodonta africana africana*). Anim. Cogn. 16, 459–469. https://doi.org/10.1007/s10071-012-0586-7.

Grimaldi, P., Saleem, K.S., Tsao, D., 2016. Anatomical connections of the functionally defined "face patches" in the macaque monkey. Neuron. https://doi.org/10.1016/j.neuron.2016.05.009.

Gross, C., 1994. How inferior temporal cortex became a visual area. Cereb. Cortex 4, 455.

Gross, C.G., Rocha-Miranda, C.E., Bender, D.B., 1972. Visual properties of neurons in inferotemporal cortex of the Macaque. J. Neurophysiol. 35, 96–111.

Gschwind, M., Pourtois, G., Schwartz, S., Van De Ville, D., Vuilleumier, P., 2012. White-matter connectivity between face-responsive regions in the human brain. Cereb. Cortex 22, 1564–1576. https://doi.org/10.1093/cercor/bhr226.

Hadjidimitrakis, K., Breveglieri, R., Bosco, A., Fattori, P., 2012. Three-dimensional eye position signals shape both peripersonal space and arm movement activity in the medial posterior parietal cortex. Front. Integr. Neurosci. 6, 37. https://doi.org/10.3389/fnint.2012.00037.

Hall, M.I., Kamilar, J.M., Kirk, E.C., 2012. Eye shape and the nocturnal bottleneck of mammals. Proc. Biol. Sci. https://doi.org/10.1098/rspb.2012.2258 rspb20122258.

Hart, B.L., Hart, L.A., Pinter-Wollman, N., 2008. Large brains and cognition: where do elephants fit in? Neurosci. Biobehav. Rev. 32, 86–98. https://doi.org/10.1016/j.neubiorev.2007.05.012.

Harting, J.K., Huerta, M.F., Hashikawa, T., Van Lieshout, D.P., 1991. Projection of the mammalian superior colliculus upon the dorsal lateral geniculate nucleus: organization of tectogeniculate pathways in nineteen species. J. Comp. Neurol. 304, 275–306. https://doi.org/10.1002/cne.903040210.

Hasselmo, M.E., Rolls, E.T., Baylis, G.C., 1989. The role of expression and identity in the face-selective responses of neurons in the temporal visual cortex of the monkey. Behav. Brain Res. 32, 203–218.

Haude, R.H., Graber, J.G., Farres, A.G., 1976. Visual observing by rhesus monkeys: some relationships with social dominance rank. Anim. Learn. Behav. 4, 163–166. https://doi.org/10.3758/BF03214028.

Haythornthwaite, A.S., Dickman, C.R., 2000. Foraging strategies of an insectivorous marsupial, *Sminthopsis youngsoni* (Marsupialia: Dasyuridae), in Australian sandridge desert. Austral Ecol. 25, 193–198. https://doi.org/10.1046/j.1442-9993.2000.01037.x.

Heesy, C.P., Hall, M.I., 2010. The nocturnal bottleneck and the evolution of mammalian vision. Brain Behav. Evol. 75, 195–203. https://doi.org/10.1159/000314278.

Helmholtz, H. von, 1867. Treatise on Physiological Optics, vol. III.

Hemmi, J.M., Mark, R.F., 1998. Visual acuity, contrast sensitivity and retinal magnification in a marsupial, the tammar wallaby (*Macropus eugenii*). J. Comp. Physiol. 183, 379–387.

Hemond, C.C., Kanwisher, N.G., Op De Beeck, H.P., 2007. A preference for contralateral stimuli in human object- and face-selective cortex. PLoS One 2, e574. https://doi.org/10.1371/journal.pone.0000574.

Henry, J., Herrero, S., 1974. Social play in the American black bear: its similarity to canid social play and an examination of its identifying characteristics. Integr. Comp. Biol. 14, 371.

Herculano-Houzel, S., 2012. Neuronal scaling rules for primate brains: the primate advantage. Prog. Brain Res. 195, 325–340. https://doi.org/10.1016/B978-0-444-53860-4.00015-5.

Herculano-Houzel, S., Collins, C.E., Wong, P., Kaas, J.H., 2007. Cellular scaling rules for primate brains. Proc. Natl. Acad. Sci. U.S.A. 104, 3562–3567. https://doi.org/10.1073/pnas.0611396104.

Herculano-Houzel, S., Kaas, J.H., de Oliveira-Souza, R., 2015. Corticalization of motor control in humans is a consequence of brain scaling in primate evolution. J. Comp. Neurol. https://doi.org/10.1002/cne.23792.

Hermer-Vazquez, L., Hermer-Vazquez, R., Chapin, J.K., 2007. The reach-to-grasp-food task for rats: a rare case of modularity in animal behavior? Behav. Brain Res. 177, 322–328. https://doi.org/10.1016/j.bbr.2006.11.029.

Hoffman, E.A., Haxby, J.V., 2000. Distinct representations of eye gaze and identity in the distributed human neural system for face perception. Nat. Neurosci. 3, 80–84. https://doi.org/10.1038/71152.

Holst, E., Mittelstaedt, H., 1950. Das Reafferenzprinzip. Naturwissenschaften 37, 464–476. https://doi.org/10.1007/BF00622503.

Homman-Ludiye, J., Manger, P.R., Bourne, J.A., 2010. Immunohistochemical parcellation of the ferret (*Mustela putorius*) visual cortex reveals substantial homology with the cat (*Felis catus*). J. Comp. Neurol. 518, 4439–4462. https://doi.org/10.1002/cne.22465.

Horstmann, A., Hoffmann, K.-P., 2005. Target selection in eye-hand coordination: do we reach to where we look or do we look to where we reach? Exp. Brain Res. 167, 187–195. https://doi.org/10.1007/s00221-005-0038-6.

Huber, E., 1931. Evolution of Facial Musculature and Facial Expression.

Humphrey, N., 1976. The social function of intellect. Grow. Points Ethol.

Hung, C.-C., Yen, C.C., Ciuchta, J.L., Papoti, D., Bock, N.A., Leopold, D.A., Silva, A.C., 2015. Functional mapping of face-selective regions in the extrastriate visual cortex of the marmoset. J. Neurosci. 35, 1160–1172. https://doi.org/10.1523/JNEUROSCI.2659-14.2015.

Hwang, E.J., Hauschild, M., Wilke, M., Andersen, R.A., 2012. Inactivation of the parietal reach region causes optic ataxia, impairing reaches but not saccades. Neuron 76, 1021—1029. https://doi.org/10.1016/j.neuron.2012.10.030.

Iriki, A., 2006. The neural origins and implications of imitation, mirror neurons and tool use. Curr. Opin. Neurobiol. 16, 660—667. https://doi.org/10.1016/j.conb.2006.10.008.

Isbell, L.A., 2006. Snakes as agents of evolutionary change in primate brains. J. Hum. Evol. 51, 1—35. https://doi.org/10.1016/j.jhevol.2005.12.012.

Ishai, A., Schmidt, C.F., Boesiger, P., 2005. Face perception is mediated by a distributed cortical network. Brain Res. Bull. 67, 87—93. https://doi.org/10.1016/j.brainresbull.2005.05.027.

Issa, E.B., DiCarlo, J.J., 2012. Precedence of the eye region in neural processing of faces. J. Neurosci. 32, 16666—16682. https://doi.org/10.1523/JNEUROSCI.2391-12.2012.

Issa, E.B., Papanastassiou, A.M., DiCarlo, J.J., 2013. Large-scale, high-resolution neurophysiological maps underlying fMRI of macaque temporal lobe. J. Neurosci. 33, 15207—15219. https://doi.org/10.1523/JNEUROSCI.1248-13.2013.

Jackson, S.R., Newport, R., Mort, D., Husain, M., 2005. Where the eye looks, the hand follows; limb-dependent magnetic misreaching in optic ataxia. Curr. Biol. 15, 42—46. https://doi.org/10.1016/j.cub.2004.12.063.

Jackson, S.R., Newport, R., Husain, M., Fowlie, J.E., O'Donoghue, M., Bajaj, N., 2009. There may be more to reaching than meets the eye: re-thinking optic ataxia. Neuropsychologia 47, 1397—1408. https://doi.org/10.1016/j.neuropsychologia.2009.01.035.

Jacobs, G.H., 2013. Losses of functional opsin genes, short-wavelength cone photopigments, and color vision — a significant trend in the evolution of mammalian vision. Vis. Neurosci. 30, 39—53. https://doi.org/10.1017/S0952523812000429.

Janssen, P., Scherberger, H., 2015. Visual guidance in control of grasping. Annu. Rev. Neurosci. 38, 69—86. https://doi.org/10.1146/annurev-neuro-071714-034028.

Jeannerod, M., Decety, J., Michel, F., 1994. Impairment of grasping movements following a bilateral posterior parietal lesion. Neuropsychologia 32, 369—380.

Jeannerod, M., Arbib, M.A., Rizzolatti, G., Sakata, H., 1995. Grasping objects: the cortical mechanisms of visuomotor transformation. Trends Neurosci. 18, 314—320.

Jeffery, K.J., O'Keefe, J.M., 1999. Learned interaction of visual and idiothetic cues in the control of place field orientation. Exp. Brain Res. 127, 151—161. https://doi.org/10.1007/s002210050785.

Johansson, G., 1973. Visual perception of biological motion and a model for its analysis. Percept. Psychophys 14, 201—211. https://doi.org/10.3758/BF03212378.

Johnson, M., 2005. Subcortical face processing. Nat. Rev. Neurosci.

Johnson, M.H., 2006. Biological motion: a perceptual life detector? Curr. Biol. 16, R376—R377. https://doi.org/10.1016/j.cub.2006.04.008.

Johnson, J.I., Rubel, E.W., Hatton, G.I., 1974. Mechanosensory projections to cerebral cortex of sheep. J. Comp. Neurol. 158, 81—107. https://doi.org/10.1002/cne.901580106.

Johnson, M.H., Senju, A., Tomalski, P., 2015. The two-process theory of face processing: modifications based on two decades of data from infants and adults. Neurosci. Biobehav. Rev. 50, 169—179. https://doi.org/10.1016/j.neubiorev.2014.10.009.

Jones, A.P., McMahon, D.B.T., Leopold, D.A., 2014. Norm-based responses to identity in the macaque face patch AF. Soc. Neurosci. Abstr.

Kaas, J.H., 2000. Why is brain size so important: design problems and solutions as neocortex gets bigger or smaller. Brain Mind 1, 7—23. https://doi.org/10.1023/A:1010028405318.

Kaas, J.H., 2008. The evolution of the complex sensory and motor systems of the human brain. Brain Res. Bull. 75, 384—390. https://doi.org/10.1016/j.brainresbull.2007.10.009.

Kaas, J.H., 2012. Evolution of columns, modules, and domains in the neocortex of primates. Proc. Natl. Acad. Sci. U.S.A. 109 (Suppl. 1), 10655—10660. https://doi.org/10.1073/pnas.1201892109.

Kaas, J.H., 2013. The evolution of brains from early mammals to humans. Wiley Interdisc. Rev. Cogn. Sci. 4, 33—45. https://doi.org/10.1002/wcs.1206.

Kaas, J.H., Hackett, T.A., 1998. Subdivisions of auditory cortex and levels of processing in primates. Audiol. Neuro. Otol. 3, 73—85.

Kaas, J.H., Stepniewska, I., 2016. Evolution of posterior parietal cortex and parietal-frontal networks for specific actions in primates. J. Comp. Neurol. 524, 595—608. https://doi.org/10.1002/cne.23838.

Kaas, J., Hall, W.C., Diamond, I.T., 1970. Cortical visual areas I and II in the hedgehog: relation between evoked potential maps and architectonic subdivisions. J. Neurophysiol. 33, 595—615.

Kaas, J.H., Hall, W.C., Diamond, I.T., 1972. Visual cortex of the grey squirrel (Sciurus carolinensis): architectonic subdivisions and connections from the visual thalamus. J. Comp. Neurol. 145, 273—305. https://doi.org/10.1002/cne.901450303.

Kaas, J.H., Gharbawie, O.A., Stepniewska, I., 2011. The organization and evolution of dorsal stream multisensory motor pathways in primates. Front. Neuroanat. 5, 34. https://doi.org/10.3389/fnana.2011.00034.

Kahn, D.A., Aguirre, G.K., 2012. Confounding of norm-based and adaptation effects in brain responses. Neuroimage 60, 2294—2299. https://doi.org/10.1016/j.neuroimage.2012.02.051.

Kanwisher, N., 2010. Functional specificity in the human brain: a window into the functional architecture of the mind. Proc. Natl. Acad. Sci. U.S.A. 107, 11163—11170. https://doi.org/10.1073/pnas.1005062107.

Kanwisher, N.G., Barton, J.J.S., 2011. The functional architecture of the face system: integrating evidence from fMRI and patient studies. In: Calder, A.J., Rhodes, G., Johnson, M.H., Haxby, J.V. (Eds.), Oxford Handbook of Face Perception. The Oxford Handbook of Face.

Kanwisher, N., McDermott, J., Chun, M., 1997. Norm-based responses to identity in the macaque face patch AF. J. Neurosci.

Karl, J.M., Whishaw, I.Q., 2013. Different evolutionary origins for the reach and the grasp: an explanation for dual visuomotor channels in primate parietofrontal cortex. Front. Neurol. 4, 208. https://doi.org/10.3389/fneur.2013.00208.

Kaskan, P.M., Franco, E.C.S., Yamada, E.S., Silveira, L.C., Darlington, R.B., Finlay, B.L., 2005. Peripheral variability and central constancy in mammalian visual system evolution. Proc. Biol. Sci. 272, 91—100. https://doi.org/10.1098/rspb.2004.2925.

Kendrick, K.M., 1991. How the sheep's brain controls the visual recognition of animals and humans. J. Anim. Sci. 69, 5008—5016.

Kendrick, K.M., 1994. Neurobiological correlates of visual and olfactory recognition in sheep. Behav. Process. 33, 89—111. https://doi.org/10.1016/0376-6357(94)90061-2.

Kendrick, K.M., 2008. Sheep senses, social cognition and capacity for consciousness. In: The Welfare of Sheep, Animal Welfare. Springer, Netherlands, Dordrecht, pp. 135—157. https://doi.org/10.1007/978-1-4020-8553-6_4.

Kendrick, K.M., Baldwin, B.A., 1987. Cells in temporal cortex of conscious sheep can respond preferentially to the sight of faces. Science 236, 448—450.

Kendrick, K.M., Baldwin, B.A., 1989. Visual responses of sheep temporal cortex cells to moving and stationary human images. Neurosci. Lett. 100, 193—197.

Kendrick, K., Atkins, K., Hinton, M., Broad, K., Fabre-Nys, C., Keverne, B., 1995. Facial and vocal discrimination in sheep. Anim. Behav. 49, 1665—1676.

Kendrick, K.M., Hinton, M.R., Atkins, K., Haupt, M.A., Skinner, J.D., 1998. Mothers determine sexual preferences. Nature 395, 229—230. https://doi.org/10.1038/26129.

Kendrick, K.M., da Costa, A.P., Leigh, A.E., Hinton, M.R., Peirce, J.W., 2001. Sheep don't forget a face. Nature 414, 165–166. https://doi.org/10.1038/35102669.

Keverne, E.B., Leonard, R.A., Scruton, D.M., Young, S.K., 1978. Visual monitoring in social groups of talapoin monkeys (*Miopithecus talapoin*). Anim. Behav. 26, 933–944. https://doi.org/10.1016/0003-3472(78)90157-4.

Kietzmann, T.C., Swisher, J.D., Konig, P., Tong, F., 2012. Prevalence of selectivity for mirror-symmetric views of faces in the ventral and dorsal visual pathways. J. Neurosci. 32, 11763–11772. https://doi.org/10.1523/JNEUROSCI.0126-12.2012.

Killian, N.J., Jutras, M.J., Buffalo, E.A., 2012. A map of visual space in the primate entorhinal cortex. Nature 491, 761–764. https://doi.org/10.1038/nature11587.

Kirk, E., Kay, R., 2004. The evolution of high visual acuity in the Anthropoidea. Anthr. Orig. New Visions 539–602.

Kirk, E.C., Cartmill, M., Kay, R.F., Lemelin, P., 2003. Comment on "Grasping primate origins". Science 300. https://doi.org/10.1126/science, 1081587 741–author reply 741.

Koganezawa, N., Gisetstad, R., Husby, E., Doan, T.P., Witter, M.P., 2015. Excitatory postrhinal projections to principal cells in the medial entorhinal cortex. J. Neurosci. 35, 15860–15874. https://doi.org/10.1523/JNEUROSCI.0653-15.2015.

Kolb, B., Walkey, J., 1987. Behavioural and anatomical studies of the posterior parietal cortex in the rat. Behav. Brain Res. 23, 127–145.

Kolb, B., Buhrmann, K., McDonald, R., Sutherland, R.J., 1994. Dissociation of the medial prefrontal, posterior parietal, and posterior temporal cortex for spatial navigation and recognition memory in the rat. Cereb. Cortex 4, 664–680.

Kovács, K., Kis, A., Kanizsár, O., Hernádi, A., Gácsi, M., Topál, J., 2016. The effect of oxytocin on biological motion perception in dogs (*Canis familiaris*). Anim. Cogn. 19, 513–522. https://doi.org/10.1007/s10071-015-0951-4.

Kravitz, D.J., Saleem, K.S., Baker, C.I., Mishkin, M., 2011. A new neural framework for visuospatial processing. Nat. Rev. Neurosci. 12, 217–230. https://doi.org/10.1038/nrn3008.

Kravitz, D.J., Saleem, K.S., Baker, C.I., Ungerleider, L.G., Mishkin, M., 2013. The ventral visual pathway: an expanded neural framework for the processing of object quality. Trends Cogn. Sci. Regul. Ed. 17, 26–49. https://doi.org/10.1016/j.tics.2012.10.011.

Kremkow, J., Jin, J., Wang, Y., Alonso, J.M., 2016. Principles underlying sensory map topography in primary visual cortex. Nature 533, 52–57. https://doi.org/10.1038/nature17936.

Krubitzer, L., 2009. In search of a unifying theory of complex brain evolution. Ann. N.Y. Acad. Sci.

Krubitzer, L., Kaas, J., 1990. Cortical connections of MT in four species of primates: areal, modular, and retinotopic patterns. Vis. Neurosci. 5, 165–204.

Krubitzer, L.A., Kaas, J.H., 1990. The organization and connections of somatosensory cortex in marmosets. J. Neurosci. 10, 952–974.

Krubitzer, L., Manger, P., Pettigrew, J., Calford, M., 1995. Organization of somatosensory cortex in monotremes: in search of the prototypical plan. J. Comp. Neurol. 351, 261–306. https://doi.org/10.1002/cne.903510206.

Krubitzer, L., Künzle, H., Kaas, J., 1997. Organization of sensory cortex in a Madagascan insectivore, the tenrec (*Echinops telfairi*). J. Comp. Neurol. 379, 399–414 https://doi.org/10.1002/(SICI)1096-9861(19970317)379:3<399::AID-CNE6>3.0.CO;2-Z.

Lafer-Sousa, R., Conway, B.R., 2013. Parallel, multi-stage processing of colors, faces and shapes in macaque inferior temporal cortex. Nat. Neurosci. 16, 1870–1878. https://doi.org/10.1038/nn.3555.

Lafer-Sousa, R., Conway, B.R., Kanwisher, N.G., 2016. Color-biased regions of the ventral visual pathway lie between face- and place-selective regions in humans, as in macaques. J. Neurosci. 36, 1682–1697. https://doi.org/10.1523/JNEUROSCI.3164-15.2016.

Lajoie, K., Drew, T., 2007. Lesions of area 5 of the posterior parietal cortex in the cat produce errors in the accuracy of paw placement during visually guided locomotion. J. Neurophysiol. 97, 2339–2354. https://doi.org/10.1152/jn.01196.2006.

Lamb, T.D., Collin, S.P., Pugh, E.N., 2007. Evolution of the vertebrate eye: opsins, photoreceptors, retina and eye cup. Nat. Rev. Neurosci. 8, 960–976. https://doi.org/10.1038/nrn2283.

Langbauer, W., 2000. Elephant communication. Zoo Biol. 19, 425–445.

Lee, K.-S., Huang, X., Fitzpatrick, D., 2016. Topology of ON and OFF inputs in visual cortex enables an invariant columnar architecture. Nature 533, 90–94. https://doi.org/10.1038/nature17941.

Leopold, D.A., 2012. Primary visual cortex: awareness and blindsight. Annu. Rev. Neurosci. 35, 91–109. https://doi.org/10.1146/annurev-neuro-062111-150356.

Leopold, D.A., Bondar, I., 2005. Adaptation to complex visual patterns in humans and monkeys. In: Clifford, C.W.G., Rhodes, G. (Eds.), Fitting the Mind to the World Adaptation and After-Effects in High-Level Vision. Oxford University Press. https://doi.org/10.1093/acprof:oso/9780198529699.001.0001.

Leopold, D.A., Rhodes, G., 2010. A comparative view of face perception. J. Comp. Psychol. 124, 233–251. https://doi.org/10.1037/a0019460.

Leopold, D.A., O'Toole, A.J., Vetter, T., Blanz, V., 2001. Prototype-referenced shape encoding revealed by high-level aftereffects. Nat. Neurosci. 4, 89–94. https://doi.org/10.1038/82947.

Leopold, D.A., Bondar, I.V., Giese, M.A., 2006. Norm-based face encoding by single neurons in the monkey inferotemporal cortex. Nature 442, 572–575. https://doi.org/10.1038/nature04951.

Lippolis, G., Westman, W., McAllan, B., Rogers, L., 2010. Lateralisation of escape responses in the stripe-faced dunnart, *Sminthopsis macroura* (Dasyuridae: Marsupialia). Laterality Asymmetries Body Brain Cogn 10, 457–470. https://doi.org/10.1080/13576500442000210.

Logothetis, N.K., Guggenberger, H., Peled, S., Pauls, J., 1999. Functional imaging of the monkey brain. Nat. Neurosci. 2, 555–562. https://doi.org/10.1038/9210.

Lomber, S.G., Payne, B.R., Cornwell, P., Long, K.D., 1996. Perceptual and cognitive visual functions of parietal and temporal cortices in the cat. Cereb. Cortex 6, 673–695.

Lomber, S.G., Yi, S.K., Woller, E.M., 2006. Relocation of specific visual functions following damage of mature posterior parietal cortex. Prog. Brain Res. 157, 157–172. https://doi.org/10.1016/S0079-6123(06)57010-4.

Lynch, J.C., McLaren, J.W., 1989. Deficits of visual attention and saccadic eye movements after lesions of parietooccipital cortex in monkeys. J. Neurophysiol. 61, 74–90.

Lyon, D.C., Nassi, J.J., Callaway, E.M., 2010. A disynaptic relay from superior colliculus to dorsal stream visual cortex in macaque monkey. Neuron 65, 270–279. https://doi.org/10.1016/j.neuron.2010.01.003.

Macdonald, I., 1997. Field experiments on duration and precision of grey and red squirrel spatial memory. Anim. Behav. 54, 879–891.

MacKinnon, L.M., Troje, N.F., Dringenberg, H.C., 2010. Do rats (*Rattus norvegicus*) perceive biological motion? Exp. Brain Res. 205, 571–576. https://doi.org/10.1007/s00221-010-2378-0.

Maestripieri, D., 2008. Macachiavellian Intelligence. University of Chicago Press.

Manger, P.R., Engler, G., Moll, C.K.E., Engel, A.K., 2008. Location, architecture, and retinotopy of the anteromedial lateral suprasylvian visual area (AMLS) of the ferret (*Mustela putorius*). Vis. Neurosci. 25, 27–37. https://doi.org/10.1017/S0952523808080036.

Marigold, D.S., Drew, T., 2011. Contribution of cells in the posterior parietal cortex to the planning of visually guided locomotion in the cat: effects of temporary visual interruption. J. Neurophysiol. 105, 2457–2470. https://doi.org/10.1152/jn.00992.2010.

Marr, D., 1982. Vision. Mit Press.

Martin, R.D., 2012. Primates. Curr. Biol. 22, R785–R790. https://doi.org/10.1016/j.cub.2012.07.015.

Martin-Malivel, J., Okada, K., 2007. Human and chimpanzee face recognition in chimpanzees (*Pan troglodytes*): role of exposure and impact on categorical perception. Behav. Neurosci. 121, 1145–1155. https://doi.org/10.1037/0735-7044.121.6.1145.

Maunsell, J.H., Van Essen, D.C., 1983. Functional properties of neurons in middle temporal visual area of the macaque monkey. I. Selectivity for stimulus direction, speed, and orientation. J. Neurophysiol. 49, 1127–1147.

McComb, K., Moss, C., Sayialel, S., Baker, L., 2000. Unusually extensive networks of vocal recognition in African elephants. Anim. Behav. 59, 1103–1109. https://doi.org/10.1006/anbe.2000.1406.

McGreevy, P., Grassi, T.D., Harman, A.M., 2003. A strong correlation exists between the distribution of retinal ganglion cells and nose length in the dog. Brain Behav. Evol. 63, 13–22. https://doi.org/10.1159/000073756.

McGreevy, P.D., Oddie, C., Burton, F.L., McLean, A.N., 2009. The horse-human dyad: can we align horse training and handling activities with the equid social ethogram? Vet. J. 181, 12–18. https://doi.org/10.1016/j.tvjl.2009.03.005.

McMahon, D.B.T., Bondar, I.V., Afuwape, O.A.T., Ide, D.C., Leopold, D.A., 2014. One month in the life of a neuron: longitudinal single unit electrophysiology in the monkey visual system. J. Neurophysiol. 112, 1748–1762. https://doi.org/10.1152/jn.00052.2014.

McMahon, D.B.T., Jones, A.P., Bondar, I.V., Leopold, D.A., 2014. Face-selective neurons maintain consistent visual responses across months. Proc. Natl. Acad. Sci. U.S.A. 111, 8251–8256. https://doi.org/10.1073/pnas.1318331111.

McMahon, D.B.T., Russ, B.E., Elnaiem, H.D., Kurnikova, A.I., Leopold, D.A., 2015. Single-unit activity during natural vision: diversity, consistency, and spatial sensitivity among AF face patch neurons. J. Neurosci. 35, 5537–5548. https://doi.org/10.1523/JNEUROSCI.3825-14.2015.

Metz, G.A., Whishaw, I.Q., 2000. Skilled reaching an action pattern: stability in rat (*Rattus norvegicus*) grasping movements as a function of changing food pellet size. Behav. Brain Res. 116, 111–122.

Meyers, E.M., Borzello, M., Freiwald, W.A., Tsao, D., 2015. Intelligent information loss: the coding of facial identity, head pose, and non-face information in the macaque face patch system. J. Neurosci. 35, 7069–7081. https://doi.org/10.1523/JNEUROSCI.3086-14.2015.

Miklósi, A., Pongrácz, P., Lakatos, G., Topál, J., Csányi, V., 2005. A comparative study of the use of visual communicative signals in interactions between dogs (*Canis familiaris*) and humans and cats (*Felis catus*) and humans. J. Comp. Psychol. 119, 179–186. https://doi.org/10.1037/0735-7036.119.2.179.

Miller, C.T., Freiwald, W.A., Leopold, D.A., Mitchell, J.F., Silva, A.C., Wang, X., 2016. Marmosets: a neuroscientific model of human social behavior. Neuron 90, 219–233. https://doi.org/10.1016/j.neuron.2016.03.018.

Mitchell, J.F., Leopold, D.A., 2015. The marmoset monkey as a model for visual neuroscience. Neurosci. Res. 93, 20–46. https://doi.org/10.1016/j.neures.2015.01.008.

Moeller, S., Freiwald, W., Tsao, D., 2008. Patches with links: a unified system for processing faces in the macaque temporal lobe. Science 320, 1355.

Morton, J., Johnson, M.H., 1991. CONSPEC and CONLERN: a two-process theory of infant face recognition. Psychol. Rev. 98, 164–181.

Moser, E.I., Kropff, E., Moser, M.-B., 2008. Place cells, grid cells, and the brain's spatial representation system. Annu. Rev. Neurosci. 31, 69–89. https://doi.org/10.1146/annurev.neuro.31.061307.090723.

Müri, R.M., 2016. Cortical control of facial expression. J. Comp. Neurol. 524, 1578–1585. https://doi.org/10.1002/cne.23908.

Nassi, J.J., Callaway, E.M., 2006. Multiple circuits relaying primate parallel visual pathways to the middle temporal area. J. Neurosci. 26, 12789–12798. https://doi.org/10.1523/JNEUROSCI.4044-06.2006.

Neiworth, J.J., Hassett, J.M., Sylvester, C.J., 2007. Face processing in humans and new world monkeys: the influence of experiential and ecological factors. Anim. Cogn. 10, 125–134. https://doi.org/10.1007/s10071-006-0045-4.

Nelissen, K., Borra, E., Gerbella, M., Rozzi, S., Luppino, G., Vanduffel, W., Rizzolatti, G., Orban, G.A., 2011. Action observation circuits in the macaque monkey cortex. J. Neurosci. 31, 3743–3756. https://doi.org/10.1523/JNEUROSCI.4803-10.2011.

Newell, K.M., Scully, D.M., McDonald, P.V., Baillargeon, R., 1989. Task constraints and infant grip configurations. Dev. Psychobiol. 22, 817–831. https://doi.org/10.1002/dev.420220806.

Nguyen, M.N., Hori, E., Matsumoto, J., Tran, A.H., Ono, T., Nishijo, H., 2013. Neuronal responses to face-like stimuli in the monkey pulvinar. Eur. J. Neurosci. 37, 35–51. https://doi.org/10.1111/ejn.12020.

Nguyen, M.N., Matsumoto, J., Hori, E., Maior, R.S., 2014. Neuronal responses to face-like and facial stimuli in the monkey superior colliculus. Front. Behav.

Nissani, M., Hoefler Nissani, D., Lay, U.T., Htun, U.W., 2005. Simultaneous visual discrimination in Asian elephants. J. Exp. Anal. Behav. 83, 15–29. https://doi.org/10.1901/jeab.2005.34-04.

O Scalaidhe, S.P., Wilson, F.A., Goldman-Rakic, P.S., 1997. Areal segregation of face-processing neurons in prefrontal cortex. Science 278, 1135–1138.

O'Brien, B.J., Abel, P.L., Olavarria, J.F., 2001. The retinal input to calbindin-D28k-defined subdivisions in macaque inferior pulvinar. Neurosci. Lett. 312, 145–148.

Okamoto-Barth, S., Call, J., Tomasello, M., 2007. Great apes" understanding of other individuals" line of sight. Psychol. Sci. 18, 462–468.

Orban, G.A., Caruana, F., 2014. The neural basis of human tool use. Front. Psychol. 5, 310. https://doi.org/10.3389/fpsyg.2014.00310.

Orban, G.A., Claeys, K., Nelissen, K., Smans, R., Sunaert, S., Todd, J.T., Wardak, C., Durand, J.-B., Vanduffel, W., 2006. Mapping the parietal cortex of human and non-human primates. Neuropsychologia 44, 2647–2667. https://doi.org/10.1016/j.neuropsychologia.2005.11.001.

Orkin, J.D., Pontzer, H., 2011. The narrow niche hypothesis: gray squirrels shed new light on primate origins. — Orkin — 2011-American Journal of Physical Anthropology — ary Am. J. Phys. https://doi.org/10.1002/ajpa.21450/pdf.

Overduin-de Vries, A.M., Spruijt, B.M., Sterck, E.H.M., 2014. Long-tailed macaques (*Macaca fascicularis*) understand what conspecifics can see in a competitive situation. Anim. Cogn. 17, 77–84. https://doi.org/10.1007/s10071-013-0639-6.

Overduin-de Vries, A.M., Bakker, F.A.A., Spruijt, B.M., Sterck, E.H.M., 2016. Male long-tailed macaques (*Macaca fascicularis*) understand the target of facial threat. Am. J. Primatol. 78, 720–730. https://doi.org/10.1002/ajp.22536.

Owings, D.H., Borchert, M., Virginia, R., 1977. The behaviour of California ground squirrels. Anim. Behav. 25, 221–230. https://doi.org/10.1016/0003-3472(77)90085-9.

Palmer, L.A., Rosenquist, A.C., Tusa, R.J., 1978. The retinotopic organization of lateral suprasylvian visual areas in the cat. J. Comp. Neurol. 177, 237–256. https://doi.org/10.1002/cne.901770205.

Parker, A.J., 2007. Binocular depth perception and the cerebral cortex. Nat. Rev. Neurosci. 8, 379–391. https://doi.org/10.1038/nrn2131.

Parr, L.A., 2011. The evolution of face processing in primates. Philos. Trans. R. Soc. Biol. Sci. 366, 1764—1777. https://doi.org/10.1098/rstb.2010.0358.

Parr, L., Winslow, J., Hopkins, W., de, F., 2000. Recognizing facial cues: individual discrimination by chimpanzees (*Pan troglodytes*) and rhesus monkeys (*Macaca mulatta*). J. Comp. Psychol. Wash.

Parr, L.A., Hecht, E., Barks, S.K., Preuss, T.M., Votaw, J.R., 2009. Face processing in the chimpanzee brain. Curr. Biol. 19, 50—53. https://doi.org/10.1016/j.cub.2008.11.048.

Parr, L.A., Heintz, M., Lonsdorf, E., Wroblewski, E., 2010. Visual kin recognition in nonhuman primates: (*Pan troglodytes* and *Macaca mulatta*): inbreeding avoidance or male distinctiveness? J. Comp. Psychol. 124, 343—350. https://doi.org/10.1037/a0020545.

Parr, L.A., Waller, B.M., Burrows, A.M., Gothard, K.M., Vick, S.J., 2010. Brief communication: MaqFACS: a muscle-based facial movement coding system for the rhesus macaque. Am. J. Phys. Anthropol. 143, 625—630. https://doi.org/10.1002/ajpa.21401.

Parr, L.A., Taubert, J., Little, A.C., Hancock, P.J.B., 2012. The organization of conspecific face space in nonhuman primates. Q. J. Exp. Psychol. 65, 2411—2434. https://doi.org/10.1080/17470218.2012.693110.

Pascalis, O., de Schonen, S., Morton, J., Deruelle, C., Fabre-Grenet, M., 1995. Mother's face recognition by neonates: a replication and an extension. Infant Behav. Dev. 18, 79—85.

Payne, B., 1993. Evidence for visual cortical area homologs in cat and macaque monkey. Cereb. Cortex.

Payne, B.R., Lomber, S.G., Geeraerts, S., Van Der Gucht, E., Vandenbussche, E., 1996. Reversible visual hemineglect. Proc. Natl. Acad. Sci. U.S.A. 93, 290—294.

Peelen, M.V., Downing, P.E., 2005. Selectivity for the human body in the fusiform gyrus. J. Neurophysiol. 93, 603—608. https://doi.org/10.1152/jn.00513.2004.

Peeters, R., Simone, L., Nelissen, K., Fabbri-Destro, M., Vanduffel, W., Rizzolatti, G., Orban, G.A., 2009. The representation of tool use in humans and monkeys: common and uniquely human features. J. Neurosci. 29, 11523—11539. https://doi.org/10.1523/JNEUROSCI.2040-09.2009.

Perrett, D.I., Rolls, E.T., Caan, W., 1982. Visual neurones responsive to faces in the monkey temporal cortex. Exp. Brain Res. 47, 329—342.

Perrett, D.I., Smith, P.A., Mistlin, A.J., Chitty, A.J., Head, A.S., Potter, D.D., Broennimann, R., Milner, A.D., Jeeves, M.A., 1985. Visual analysis of body movements by neurones in the temporal cortex of the macaque monkey: a preliminary report. Behav. Brain Res. 16, 153—170.

Perrett, D.I., Hietanen, J.K., Oram, M.W., Benson, P.J., 1992. Organization and functions of cells responsive to faces in the temporal cortex. Philos. Trans. R. Soc. Lond. B Biol. Sci. 335, 23—30. https://doi.org/10.1098/rstb.1992.0003.

Pettigrew, J.D., Ramachandran, V.S., Bravo, H., 1984. Some neural connections subserving binocular vision in ungulates. Brain Behav. Evol. 24, 65—93.

Pinsk, M.A., DeSimone, K., Moore, T., Gross, C.G., Kastner, S., 2005. Representations of faces and body parts in macaque temporal cortex: a functional MRI study. Proc. Natl. Acad. Sci. U.S.A. 102, 6996—7001. https://doi.org/10.1073/pnas.0502605102.

Pitcher, D., Dilks, D.D., Saxe, R.R., Triantafyllou, C., Kanwisher, N., 2011. Differential selectivity for dynamic versus static information in face-selective cortical regions. Neuroimage 56, 2356—2363. https://doi.org/10.1016/j.neuroimage.2011.03.067.

Pitcher, D., Duchaine, B., Walsh, V., 2014. Combined TMS and fMRI reveal dissociable cortical pathways for dynamic and static face perception. Curr. Biol. 24, 2066—2070. https://doi.org/10.1016/j.cub.2014.07.060.

Plotnik, J.M., de Waal, F.B.M., Moore, D., Reiss, D., 2010. Self-recognition in the Asian elephant and future directions for cognitive research with elephants in zoological settings. Zoo Biol. 29, 179—191. https://doi.org/10.1002/zoo.20257.

Polosecki, P., Moeller, S., Schweers, N., Romanski, L.M., Tsao, D.Y., Freiwald, W.A., 2013. Faces in motion: selectivity of macaque and human face processing areas for dynamic stimuli. J. Neurosci. 33, 11768—11773. https://doi.org/10.1523/JNEUROSCI.5402-11.2013.

Popivanov, I.D., Jastorff, J., Vanduffel, W., Vogels, R., 2012. Stimulus representations in body-selective regions of the macaque cortex assessed with event-related fMRI. Neuroimage 63, 723—741. https://doi.org/10.1016/j.neuroimage.2012.07.013.

Popivanov, I.D., Jastorff, J., Vanduffel, W., Vogels, R., 2014. Heterogeneous single-unit selectivity in an fMRI-defined body-selective patch. J. Neurosci. 34, 95—111. https://doi.org/10.1523/JNEUROSCI.2748-13.2014.

Preuss, T., 2007. Evolutionary Specializations of Primate Brain Systems. Primate Origins: Developments in Primatology: Progress and Prospects. Springer US, New York, NY, pp. 625—675.

Puce, A., Perrett, D., 2003. Electrophysiology and brain imaging of biological motion. Philos. Trans. R. Soc. Lond. B Biol. Sci. 358, 435—445. https://doi.org/10.1098/rstb.2002.1221.

Putman, B.J., Clark, R.W., 2015. The fear of unseen predators: ground squirrel tail flagging in the absence of snakes signals vigilance. Behav. Ecol. 26, 185—193. https://doi.org/10.1093/beheco/aru176.

Quanstrom, W.R., 1971. Behaviour of richardson's ground squirrel *Spermophilus richardsonii richardsonii*. Anim. Behav. 19, 646—652. https://doi.org/10.1016/S0003-3472(71)80166-5.

Ramachandran, V.S., 1985. The neurobiology of perception. Perception 14, 97—103.

Reep, R.L., Corwin, J.V., 2009. Posterior parietal cortex as part of a neural network for directed attention in rats. Neurobiol. Learn. Mem. 91, 104—113. https://doi.org/10.1016/j.nlm.2008.08.010.

Regolin, L., Tommasi, L., Vallortigara, G., 2000. Visual perception of biological motion in newly hatched chicks as revealed by an imprinting procedure. Anim. Cogn. 3, 53—60. https://doi.org/10.1007/s100710050050.

Renne, P.R., Deino, A.L., Hilgen, F.J., Kuiper, K.F., Mark, D.F., Mitchell, W.S., Morgan, L.E., Mundil, R., Smit, J., 2013. Time scales of critical events around the Cretaceous-Paleogene boundary. Science 339, 684—687. https://doi.org/10.1126/science.1230492.

Rhodes, G., Leopold, D.A., 2011. Adaptive norm-based coding of face identity. In: Handbook of Face Perception. Oxford University Press, pp. 263—286.

Rhodes, G., Brennan, S., Carey, S., 1987. Identification and ratings of caricatures: implications for mental representations of faces. Cogn. Psychol. 19, 473—497.

Rhodes, G., Carey, S., Byatt, G., Proffitt, F., 1998. Coding spatial variations in faces and simple shapes: a test of two models. Vis. Res. 38, 2307—2321.

Rizzo, M., Vecera, S.P., 2002. Psychoanatomical substrates of Bálint's syndrome. J. Neurol. Neurosurg. Psychiatr. 72, 162—178. https://doi.org/10.1136/jnnp.72.2.162.

Rizzolatti, G., Craighero, L., 2004. The mirror-neuron system. Annu. Rev. Neurosci. 27, 169—192. https://doi.org/10.1146/annurev.neuro.27.070203.144230.

Rizzolatti, G., Luppino, G., 2001. The cortical motor system. Neuron 31, 889—901.

Rizzolatti, G., Sinigaglia, C., 2010. The functional role of the parieto-frontal mirror circuit: interpretations and misinterpretations. Nat. Rev. Neurosci. 11, 264—274. https://doi.org/10.1038/nrn2805.

Roberts, B.A., 2012. Perinatal behavior of a wild Grevy's zebra (*Equus grevyi*) mare and foal. J. Ethol. 30, 205—209. https://doi.org/10.1007/s10164-011-0308-7.

Roberts, B.A., Rubenstein, D.I., 2014. Maternal tactics for mitigating neonate predation risk during the postpartum period in Thomson's gazelle. Behaviour 151, 1229—1248. https://doi.org/10.1163/1568539X-00003181.

Robinson, S.R., 1980. Antipredator behaviour and predator recognition in Belding's ground squirrels. Anim. Behav. 28, 840–852. https://doi.org/10.1016/S0003-3472(80)80144-8.

Rolls, E.T., Critchley, H.D., Browning, A.S., Inoue, K., 2006. Face-selective and auditory neurons in the primate orbitofrontal cortex. Exp. Brain Res. 170, 74–87. https://doi.org/10.1007/s00221-005-0191-y.

Rosa, M.G., 1999. Topographic organisation of extrastriate areas in the flying fox: implications for the evolution of mammalian visual cortex. J. Comp. Neurol. 411, 503–523.

Rosa, M.G.P., Tweedale, R., 2005. Brain maps, great and small: lessons from comparative studies of primate visual cortical organization. Philos. Trans. R. Soc. Lond. B Biol. Sci. 360, 665–691. https://doi.org/10.1098/rstb.2005.1626.

Rosa-Salva, O., Sovrano, V.A., Vallortigara, G., 2014. What can fish brains tell us about visual perception? Front. Neural Circuits 8, 119. https://doi.org/10.3389/fncir.2014.00119.

Rosati, A.G., Hare, B., 2009. Looking past the model species: diversity in gaze-following skills across primates. Curr. Opin. Neurobiol. 19, 45–51. https://doi.org/10.1016/j.conb.2009.03.002.

Rosenfeld, S., Van Hoesen, G., 1979. Face recognition in the rhesus monkey. Neuropsychologia 17, 503–509.

Rosing, N., Stirling, I., 2010. The World of the Polar Bear. Firefly Books Limited.

Rushmore, R.J., Valero-Cabre, A., Lomber, S.G., Hilgetag, C.C., Payne, B.R., 2006. Functional circuitry underlying visual neglect. Brain 129, 1803–1821. https://doi.org/10.1093/brain/awl140.

Russ, B.E., Leopold, D.A., 2015. Functional MRI mapping of dynamic visual features during natural viewing in the macaque. Neuroimage 109, 84–94. https://doi.org/10.1016/j.neuroimage.2015.01.012.

Russell, E.M., 1984. Social behaviour and social organization of marsupials. Mammal. Rev. 14, 101–154. https://doi.org/10.1111/j.1365-2907.1984.tb00343.x.

Rutberg, A.T., 1987. Adaptive hypotheses of birth synchrony in ruminants: an interspecific test on JSTOR. Am. Nat. https://doi.org/10.2307/2461714.

Sacrey, L.-A.R., Karl, J.M., Whishaw, I.Q., 2012. Development of rotational movements, hand shaping, and accuracy in advance and withdrawal for the reach-to-eat movement in human infants aged 6-12 months. Infant Behav. Dev. 35, 543–560. https://doi.org/10.1016/j.infbeh.2012.05.006.

Sakata, H., Taira, M., Murata, A., Mine, S., 1995. Neural mechanisms of visual guidance of hand action in the parietal cortex of the monkey. Cereb. Cortex 5, 429–438.

Salva, O.R., Farroni, T., Regolin, L., Vallortigara, G., Johnson, M.H., 2011. The evolution of social orienting: evidence from chicks (Gallus gallus) and human newborns. PLoS One 6, e18802. https://doi.org/10.1371/journal.pone.0018802.

Samson, J., Manser, M.B., 2016. Caching in the presence of competitors: are Cape ground squirrels (Xerus inauris) sensitive to audience attentiveness? Anim. Cogn. 19, 31–38. https://doi.org/10.1007/s10071-015-0910-0.

Santana, S.E., Lynch Alfaro, J., Alfaro, M.E., 2012. Adaptive evolution of facial colour patterns in Neotropical primates. Proc. R. Soc. Biol. Sci. 279, 2204–2211. https://doi.org/10.1098/rspb.2011.2326.

Santana, S.E., Alfaro, J.L., Noonan, A., Alfaro, M.E., 2013. Adaptive response to sociality and ecology drives the diversification of facial colour patterns in catarrhines. Nat. Commun. 4, 2765. https://doi.org/10.1038/ncomms3765.

Santana, S.E., Dobson, S.D., Diogo, R., 2014. Plain faces are more expressive: comparative study of facial colour, mobility and musculature in primates. Biol. Lett. 10, 20140275. https://doi.org/10.1098/rsbl.2014.0275.

Savage, A., Rice, J.M., Brangan, J.M., Martini, D.P., Pugh, J.A., Miller, C.D., 1994. Performance of African elephants (Loxodonta africana) and California sea lions (Zalophus californianus) on a two-choice object discrimination task. Zoo Biol. 13, 69–75. https://doi.org/10.1002/zoo.1430130109.

Scaplen, K.M., Gulati, A.A., Heimer-McGinn, V.L., Burwell, R.D., 2014. Objects and landmarks: hippocampal place cells respond differently to manipulations of visual cues depending on size, perspective, and experience. Hippocampus 24, 1287–1299. https://doi.org/10.1002/hipo.22331.

Scerif, G., Gomez, J.-C., Byrne, R.W., 2004. What do Diana monkeys know about the focus of attention of a conspecific? Anim. Behav. 68, 1239–1247. https://doi.org/10.1016/j.anbehav.2004.01.011.

Schmidt, K.L., Cohn, J.F., 2002. Human facial expressions as adaptations: evolutionary questions in facial expression research. Am. J. Phys. Anthropol. 116, 3–24. https://doi.org/10.1002/ajpa.20001.

Schneider, G.E., 1969. Two visual systems. Science 163, 895–902.

Schwarzlose, R.F., Baker, C.I., Kanwisher, N., 2005. Separate face and body selectivity on the fusiform gyrus. J. Neurosci. 25, 11055–11059. https://doi.org/10.1523/JNEUROSCI.2621-05.2005.

Schwiedrzik, C.M., Zarco, W., Everling, S., Freiwald, W.A., 2015. Face patch resting state networks link face processing to social cognition. PLoS Biol. 13, e1002245. https://doi.org/10.1371/journal.pbio.1002245.

Shepherd, S.V., Platt, M.L., 2008. Spontaneous social orienting and gaze following in ringtailed lemurs (Lemur catta). Anim. Cogn. 11, 13–20. https://doi.org/10.1007/s10071-007-0083-6.

Shepherd, S.V., Deaner, R.O., Platt, M.L., 2006. Social status gates social attention in monkeys. Curr. Biol. 16, R119–R120. https://doi.org/10.1016/j.cub.2006.02.013.

Shepherd, S.V., Klein, J.T., Deaner, R.O., Platt, M.L., 2009. Mirroring of attention by neurons in macaque parietal cortex. Proc. Natl. Acad. Sci. 106, 9489–9494.

Shinozaki, A., Hosaka, Y., Imagawa, T., Uehara, M., 2010. Topography of ganglion cells and photoreceptors in the sheep retina. J. Comp. Neurol. 518, 2305–2315. https://doi.org/10.1002/cne.22333.

Simion, F., Regolin, L., Bulf, H., 2008. A predisposition for biological motion in the newborn baby. Proc. Natl. Acad. Sci. U.S.A. 105, 809–813. https://doi.org/10.1073/pnas.0707021105.

Singer, J.M., Sheinberg, D.L., 2010. Temporal cortex neurons encode articulated actions as slow sequences of integrated poses. J. Neurosci. 30, 3133–3145. https://doi.org/10.1523/JNEUROSCI.3211-09.2010.

Smet, A.F., Byrne, R.W., 2013. African elephants can use human pointing cues to find hidden food. Curr. Biol. 23, 2033–2037. https://doi.org/10.1016/j.cub.2013.08.037.

Smet, A.F., Byrne, R.W., 2014. African elephants (Loxodonta africana) recognize visual attention from face and body orientation. Biol. Lett. 10, 20140428. https://doi.org/10.1098/rsbl.2014.0428.

Smuts, B., 2001. Encounters with Animal Minds. J. Conscious. Stud.

Solomon, S.G., Rosa, M.G.P., 2014. A simpler primate brain: the visual system of the marmoset monkey. Front. Neural Circuits 8, 96. https://doi.org/10.3389/fncir.2014.00096.

Song, Y., Luo, Y.L.L., Li, X., Xu, M., Liu, J., 2013. Representation of contextually related multiple objects in the human ventral visual pathway. J. Cogn. Neurosci. 25, 1261–1269. https://doi.org/10.1162/jocn_a_00406.

Sprague, J.M., Levy, J., DiBerardino, A., Berlucchi, G., 1977. Visual cortical areas mediating form discrimination in the cat. J. Comp. Neurol. 172, 441–488. https://doi.org/10.1002/cne.901720305.

Srihasam, K., Mandeville, J.B., Morocz, I.A., Sullivan, K.J., Livingstone, M.S., 2012. Behavioral and anatomical consequences of early versus late symbol training in macaques. Neuron 73, 608–619. https://doi.org/10.1016/j.neuron.2011.12.022.

Srihasam, K., Vincent, J.L., Livingstone, M.S., 2014. Novel domain formation reveals proto-architecture in inferotemporal cortex. Nat. Neurosci. 17, 1776–1783. https://doi.org/10.1038/nn.3855.

Steiner, A.L., 1971. Play activity of Columbian ground squirrels1. Z. Tierpsychol. 28, 247–261. https://doi.org/10.1111/j.1439-0310.1971.tb00817.x.

Steiner, A.L., 2011. Body-rubbing, marking, and other scent-related behavior in some ground squirrels (Sciuridae), a descriptive study. Can. J. Zool. 52, 889–906. https://doi.org/10.1139/z74-120.

Stepniewska, I., Qi, H.X., Kaas, J.H., 1999. Do superior colliculus projection zones in the inferior pulvinar project to MT in primates? Eur. J. Neurosci. 11, 469–480.

Stepniewska, I., Fang, P.-C., Kaas, J.H., 2005. Microstimulation reveals specialized subregions for different complex movements in posterior parietal cortex of prosimian galagos. Proc. Natl. Acad. Sci. U.S.A. 102, 4878–4883. https://doi.org/10.1073/pnas.0501048102.

Stepniewska, I., Fang, P.-C.Y., Kaas, J.H., 2009. Organization of the posterior parietal cortex in galagos: I. Functional zones identified by microstimulation. J. Comp. Neurol. 517, 765–782. https://doi.org/10.1002/cne.22181.

Storrs, K.R., Arnold, D.H., 2015. Face aftereffects involve local repulsion, not renormalization. J. Vis. 15, 1. https://doi.org/10.1167/15.8.1.

Sugase, Y., Yamane, S., Ueno, S., Kawano, K., 1999. Global and fine information coded by single neurons in the temporal visual cortex. Nature 400, 869–873. https://doi.org/10.1038/23703.

Sugita, Y., 2008. Face perception in monkeys reared with no exposure to faces. Proc. Natl. Acad. Sci. U.S.A. 105.

Sugita, Y., 2009. Innate face processing. Curr. Opin. Neurobiol. 19, 39–44. https://doi.org/10.1016/j.conb.2009.03.001.

Sur, M., Nelson, R.J., Kaas, J.H., 1978. The representation of the body surface in somatosensory area I of the grey squirrel. J. Comp. Neurol. 179, 425–449. https://doi.org/10.1002/cne.901790211.

Susskind, J.M., Lee, D.H., Cusi, A., Feiman, R., Grabski, W., Anderson, A.K., 2008. Expressing fear enhances sensory acquisition. Nat. Neurosci. 11, 843–850. https://doi.org/10.1038/nn.2138.

Sussman, R.W., 1999. Primate Ecology and Social Structure. Pearson Custom Pub.

Sussman, R.W., Tab Rasmussen, D., Raven, P.H., 2013. Rethinking primate origins again. Am. J. Primatol. 75, 95–106. https://doi.org/10.1002/ajp.22096.

Tanaka, M., 2003. Visual preference by chimpanzees (*Pan troglodytes*). Primates 44, 157–165. https://doi.org/10.1007/s10329-002-0022-8.

Tanaka, M., 2007. Development of the visual preference of chimpanzees (*Pan troglodytes*) for photographs of primates: effect of social experience. Primates 48, 303–309. https://doi.org/10.1007/s10329-007-0044-3.

Tate, A., Fischer, H., Leigh, A., Kendrick, K., 2006. Behavioural and neurophysiological evidence for face identity and face emotion processing in animals. Philos. Trans. R. Soc. Biol. Sci. 361, 2155.

Taubert, J., Van Belle, G., Vanduffel, W., Rossion, B., Vogels, R., 2015. Neural correlate of the thatcher face illusion in a monkey face-selective patch. J. Neurosci. 35, 9872–9878. https://doi.org/10.1523/JNEUROSCI.0446-15.2015.

Tavaré, S., Marshall, C.R., Will, O., Soligo, C., Martin, R.D., 2002. Using the fossil record to estimate the age of the last common ancestor of extant primates. Nature 416, 726–729. https://doi.org/10.1038/416726a.

Toga, A.W., Thompson, P.M., Mori, S., Amunts, K., Zilles, K., 2006. Towards multimodal atlases of the human brain. Nat. Rev. Neurosci. 7, 952–966. https://doi.org/10.1038/nrn2012.

Tomasello, M., Call, J., Hare, B., 1998. Five primate species follow the visual gaze of conspecifics. Anim. Behav. 55, 1063–1069.

Troje, N.F., Aust, U., 2013. What do you mean with "direction?" Local and global cues to biological motion perception in pigeons. Vis. Res. 79, 47–55. https://doi.org/10.1016/j.visres.2013.01.002.

Tsao, D., Livingstone, M., 2008. Mechanisms of face perception. Annu. Rev.

Tsao, D.Y., Freiwald, W.A., Knutsen, T.A., Mandeville, J.B., Tootell, R.B.H., 2003. Faces and objects in macaque cerebral cortex. Nat. Neurosci. 6, 989–995. https://doi.org/10.1038/nn1111.

Tsao, D.Y., Freiwald, W.A., Tootell, R.B.H., Livingstone, M.S., 2006. A cortical region consisting entirely of face-selective cells. Science 311, 670–674. https://doi.org/10.1126/science.1119983.

Tsao, D.Y., Moeller, S., Freiwald, W.A., 2008. Comparing face patch systems in macaques and humans. Proc. Natl. Acad. Sci. U.S.A. 105, 19514–19519. https://doi.org/10.1073/pnas.0809662105.

Tsao, D.Y., Schweers, N., Moeller, S., Freiwald, W.A., 2008. Patches of face-selective cortex in the macaque frontal lobe. Nat. Neurosci. 11, 877–879. https://doi.org/10.1038/nn.2158.

Tsutsui, K., Jiang, M., Yara, K., Sakata, H., Taira, M., 2001. Integration of perspective and disparity cues in surface-orientation-selective neurons of area CIP. J. Neurophysiol. 86, 2856–2867.

Valentine, T., 1991. A unified account of the effects of distinctiveness, inversion, and race in face recognition. Q. J. Exp. Psychol. 43, 161–204.

Vallortigara, G., Regolin, L., Marconato, F., 2005. Visually inexperienced chicks exhibit spontaneous preference for biological motion patterns. PLoS Biol. 3, e208. https://doi.org/10.1371/journal.pbio.0030208.

Van Dromme, I.C., Premereur, E., Verhoef, B.-E., Vanduffel, W., Janssen, P., 2016. Posterior parietal cortex drives inferotemporal activations during three-dimensional object vision. PLoS Biol. 14, e1002445. https://doi.org/10.1371/journal.pbio.1002445.

van Hooff, J.A., 1962. Facial expression in higher primates. Symp. Zool. Soc. Lond. 8, 97–125.

van Hooff, J.A., 1962. Facial expression in higher primates. Symp. Zool. Soc. Lond. 8, 97–125.

Van Hooser, S.D., Nelson, S.B., 2006. The squirrel as a rodent model of the human visual system. Vis. Neurosci. 23, 765–778. https://doi.org/10.1017/S0952523806230098.

van Schaik, C.P., Deaner, R.O., Merrill, M.Y., 1999. The conditions for tool use in primates: implications for the evolution of material culture. J. Hum. Evol. 36, 719–741. https://doi.org/10.1006/jhev.1999.0304.

van Schaik, C.P., Deaner, R.O., Merrill, M.Y., 1999. The conditions for tool use in primates: implications for the evolution of material culture. J. Hum. Evol. 36, 719–741. https://doi.org/10.1006/jhev.1999.0304.

Vanduffel, W., Zhu, Q., Orban, G.A., 2014. Monkey cortex through fMRI glasses. Neuron 83, 533–550. https://doi.org/10.1016/j.neuron.2014.07.015.

Vangeneugden, J., Pollick, F., Vogels, R., 2009. Functional differentiation of macaque visual temporal cortical neurons using a parametric action space. Cereb. Cortex 19, 593–611. https://doi.org/10.1093/cercor/bhn109.

Vangeneugden, J., De Mazière, P.A., Van Hulle, M.M., Jaeggli, T., Van Gool, L., Vogels, R., 2011. Distinct mechanisms for coding of visual actions in macaque temporal cortex. J. Neurosci. 31, 385–401. https://doi.org/10.1523/JNEUROSCI.2703-10.2011.

Vinken, K., Van den Bergh, G., Vermaercke, B., Op De Beeck, H.P., 2016. Neural representations of natural and scrambled movies progressively change from rat striate to temporal cortex. bhw 111 Cereb. Cortex. https://doi.org/10.1093/cercor/bhw111.

Waitt, C., Little, A.C., 2006. Preferences for symmetry in conspecific facial shape among *Macaca mulatta*. Int. J. Primatol. 27, 133–145. https://doi.org/10.1007/s10764-005-9015-y.

Waitt, C., Little, A.C., Wolfensohn, S., Honess, P., Brown, A.P., Buchanan-Smith, H.M., Perrett, D.I., 2003. Evidence from rhesus macaques suggests that male coloration plays a role in female primate mate choice. Proc. Biol. Sci. 270 (Suppl. 2), S144–S146. https://doi.org/10.1098/rsbl.2003.0065.

Waitt, C., Gerald, M., Little, A., Kraiselburd, E., 2006. Selective attention toward female secondary sexual color in male rhesus macaques. Am. J. Primatol. 68, 738.

Waller, B., Parr, L., Gothard, K., Burrows, A., Fuglevand, A., 2008. Mapping the contribution of single muscles to facial movements in the rhesus macaque. Physiol. Behav. 95, 93–100.

Wang, Q., Burkhalter, A., 2007. Area map of mouse visual cortex. J. Comp. Neurol. 502, 339–357. https://doi.org/10.1002/cne.21286.

Warner, C.E., Goldshmit, Y., Bourne, J.A., 2010. Retinal afferents synapse with relay cells targeting the middle temporal area in the pulvinar and lateral geniculate nuclei. Front. Neuroanat. 4, 8. https://doi.org/10.3389/neuro.05.008.2010.

Warner, C.E., Kwan, W.C., Bourne, J.A., 2012. The early maturation of visual cortical area MT is dependent on input from the retinorecipient medial portion of the inferior pulvinar. J. Neurosci. 32, 17073–17085. https://doi.org/10.1523/JNEUROSCI.3269-12.2012.

Warner, C.E., Kwan, W.C., Wright, D., Johnston, L.A., Egan, G.F., Bourne, J.A., 2015. Preservation of vision by the pulvinar following early-life primary visual cortex lesions. Curr. Biol. 25, 424–434. https://doi.org/10.1016/j.cub.2014.12.028.

Weiner, K.S., Grill-Spector, K., 2015. The evolution of face processing networks. Trends Cogn. Sci. Regul. Ed. 19, 240–241. https://doi.org/10.1016/j.tics.2015.03.010.

Weisbuch, M., Ambady, N., 2011. Thin-slice vision. In: Adams, R.B., Ambady, N., Nakayama, K., Shimojo, S. (Eds.), The Science of Social Vision. The Science of Social Vision.

Whishaw, I.Q., Coles, B.L., 1996. Varieties of paw and digit movement during spontaneous food handling in rats: postures, bimanual coordination, preferences, and the effect of forelimb cortex lesions. Behav. Brain Res. 77, 135–148.

Whishaw, I.Q., Dringenberg, H.C., Pellis, S.M., 1992. Spontaneous forelimb grasping in free feeding by rats: motor cortex aids limb and digit positioning. Behav. Brain Res. 48, 113–125.

Whishaw, I.Q., Travis, S.G., Koppe, S.W., Sacrey, L.-A., Gholamrezaei, G., Gorny, B., 2010. Hand shaping in the rat: conserved release and collection vs. flexible manipulation in overground walking, ladder rung walking, cylinder exploration, and skilled reaching. Behav. Brain Res. 206, 21–31. https://doi.org/10.1016/j.bbr.2009.08.030.

Wilber, A.A., Clark, B.J., Forster, T.C., Tatsuno, M., McNaughton, B.L., 2014. Interaction of egocentric and world-centered reference frames in the rat posterior parietal cortex. J. Neurosci. 34, 5431–5446. https://doi.org/10.1523/JNEUROSCI.0511-14.2014.

Willis, J., Todorov, A., 2006. First impressions making up your mind after a 100-Ms exposure to a face. Psychol. Sci. J. Am. Psychol. Soc. APS 17, 592–598. https://doi.org/10.1111/j.1467-9280.2006.01750.x.

Wittemyer, G., Getz, W.M., 2007. Hierarchical dominance structure and social organization in African elephants, Loxodonta africana. Anim. Behav. 73, 671–681. https://doi.org/10.1016/j.anbehav.2006.10.008.

Wong, P., Kaas, J.H., 2008. Architectonic subdivisions of neocortex in the gray squirrel (Sciurus carolinensis). Anat. Rec. 291, 1301–1333. https://doi.org/10.1002/ar.20758.

Wong, P., Kaas, J.H., 2009. Architectonic subdivisions of neocortex in the tree shrew (Tupaia belangeri). Anat. Rec. 292, 994–1027. https://doi.org/10.1002/ar.20916.

Wong, P., Kaas, J.H., 2009. An architectonic study of the neocortex of the short-tailed opossum (Monodelphis domestica). Brain Behav. Evol. 73, 206–228. https://doi.org/10.1159/000225381.

Wong, P., Kaas, J.H., 2010. Architectonic subdivisions of neocortex in the Galago (Otolemur garnetti). Anat. Rec. 293, 1033–1069. https://doi.org/10.1002/ar.21109.

Woodworth, R.S., 1899. Accuracy of voluntary movement. Psychol. Rev. Monogr. Suppl. 3, i. https://doi.org/10.1037/h0092992.

Woolsey, C.N., Fairman, D., 1946. Contralateral, ipsilateral, and bilateral representation of cutaneous receptors in somatic areas I and II of the cerebral cortex of pig, sheep, and other mammals. Surgery 19, 684–702. https://doi.org/10.5555/uri pii:003960604690061X.

Workman, A.D., Charvet, C.J., Clancy, B., Darlington, R.B., Finlay, B.L., 2013. Modeling transformations of neurodevelopmental sequences across mammalian species. J. Neurosci. 33, 7368–7383. https://doi.org/10.1523/JNEUROSCI.5746-12.2013.

Yang, H., Susilo, T., Duchaine, B., 2016. The anterior temporal face area contains invariant representations of face identity that can persist despite the loss of right FFA and OFA. Cereb. Cortex 26, 1096–1107. https://doi.org/10.1093/cercor/bhu289.

Yartsev, M.M., Ulanovsky, N., 2013. Representation of three-dimensional space in the hippocampus of flying bats. Science 340, 367–372. https://doi.org/10.1126/science.1235338.

Yartsev, M.M., Witter, M.P., Ulanovsky, N., 2011. Grid cells without theta oscillations in the entorhinal cortex of bats. Nature 479, 103–107. https://doi.org/10.1038/nature10583.

Yeatman, J.D., Weiner, K.S., Pestilli, F., Rokem, A., Mezer, A., Wandell, B.A., 2014. The vertical occipital fasciculus: a century of controversy resolved by in vivo measurements. Proc. Natl. Acad. Sci. U.S.A. 111, E5214–E5223. https://doi.org/10.1073/pnas.1418503111.

Yokoyama, S., 2000. Molecular evolution of vertebrate visual pigments. Prog. Retin. Eye Res. 19, 385–419. https://doi.org/10.1016/S1350-9462(00)00002-1.

Yovel, G., Freiwald, W.A., 2013. Face recognition systems in monkey and human: are they the same thing? F1000Prime. For. Rep. 5, 10. https://doi.org/10.12703/P5-10.

Zangenehpour, S., Chaudhuri, A., 2005. Patchy organization and asymmetric distribution of the neural correlates of face processing in monkey inferotemporal cortex. Curr. Biol. 15, 993–1005.

Zeller, A.C., 1987. Communication by sight and smell. In: Smuts, B.B., Cheney, D.L., Seyfarth, R.M., Wrangham, R.W., wrang (Eds.), Primate Societies. University of Chicago Press, pp. 433–439.

Zoccolan, D., 2015. Invariant visual object recognition and shape processing in rats. Behav. Brain Res. 285, 10–33. https://doi.org/10.1016/j.bbr.2014.12.053.

26

Evolution of Parietal Cortex in Mammals: From Manipulation to Tool Use

A.B. Goldring, L.A. Krubitzer

University of California, Davis Center for Neuroscience, Davis, CA, United States

26.1 Introduction

The extraordinary degree to which humans can interact with and physically change the world around them can be attributed, in large part, to the ability to use our hands in a sophisticated and coordinated fashion. This was achieved by the coevolution of the hand and motor and posterior parietal cortical areas that program and control reaching, grasping, and object exploration and acquisition (Fig. 26.1). Primate posterior parietal cortex (PPC) is comprised of a constellation of areas that differ in their function and connectivity, as well as the effectors, such as the eyes and the hands, which they help control. However, these areas are not strictly involved in motor control, but rather sit at the interface of perception and action, combining sensory information from several modalities with effector kinematics to compute various movements tailored to specific objects and contexts (see Andersen and Cui, 2009; Gottlieb and Snyder, 2010 for review). Together with motor cortices, several areas within the primate PPC form the frontoparietal reaching and grasping network that allows primates to so successfully interact with and manipulate their environment.

This ability to physically interact with our environment requires the PPC to construct an internal model of the physical self: our body's configuration, the boundary between our body and external objects, and the temporary expansion of that self as we wield a tool that extends our reach and manual capabilities. Yet comprehension of where the self is and even the ability to manipulate objects and use them as tools did not evolve de novo in humans, but emerged from simpler networks that were present in early primates (60 million years ago) and possibly even early mammals (over 200 million years ago). PPC networks in different mammals often

evolved to solve very similar problems (eg, grasping food and bringing it to the mouth), but it is not clear if the solution to executing these fundamental manual behaviors is the same across species, especially when animals differ in body and forelimb morphology, use different major effectors to explore their environments, and have been independently evolving for millions of years (Figs. 26.1 and 26.2). The solution, computed in PPC, is the integration of sensory inputs providing information on the size, shape, texture, and location of objects in space (eg, food) with the internal representation of the body's current posture, dimensions, and physical capabilities. The ultimate behavioral outcome (eg, feeding) is similar, but it is not known how PPC networks in different mammals produce this behavior. One major problem when considering the evolution of frontoparietal networks in mammals, in addition to the species differences in morphology and effector use noted above, is determining the best criteria to define cortical areas in PPC in a wide range of mammals so that direct comparisons can be made, accurate inferences regarding the ancestral state can be articulated, and an appreciation of the evolution of this region of cortex can be fully understood.

Traditionally, sensory cortical areas, such as the primary and secondary somatosensory, visual areas, and auditory areas (Table 26.1) have been precisely defined using multiple criteria, including architectonic appearance, functional map organization, and neuroanatomical connections. For example, the primary somatosensory area (S1) in all species examined has a distinct myeloarchitectonic and cytoarchitectonic appearance (Fig. 26.3). This architectonically defined field is coextensive with a complete map of cutaneous receptors of the contralateral body (Fig. 26.4), as well as a distinct set of thalamocortical, corticocortical, and

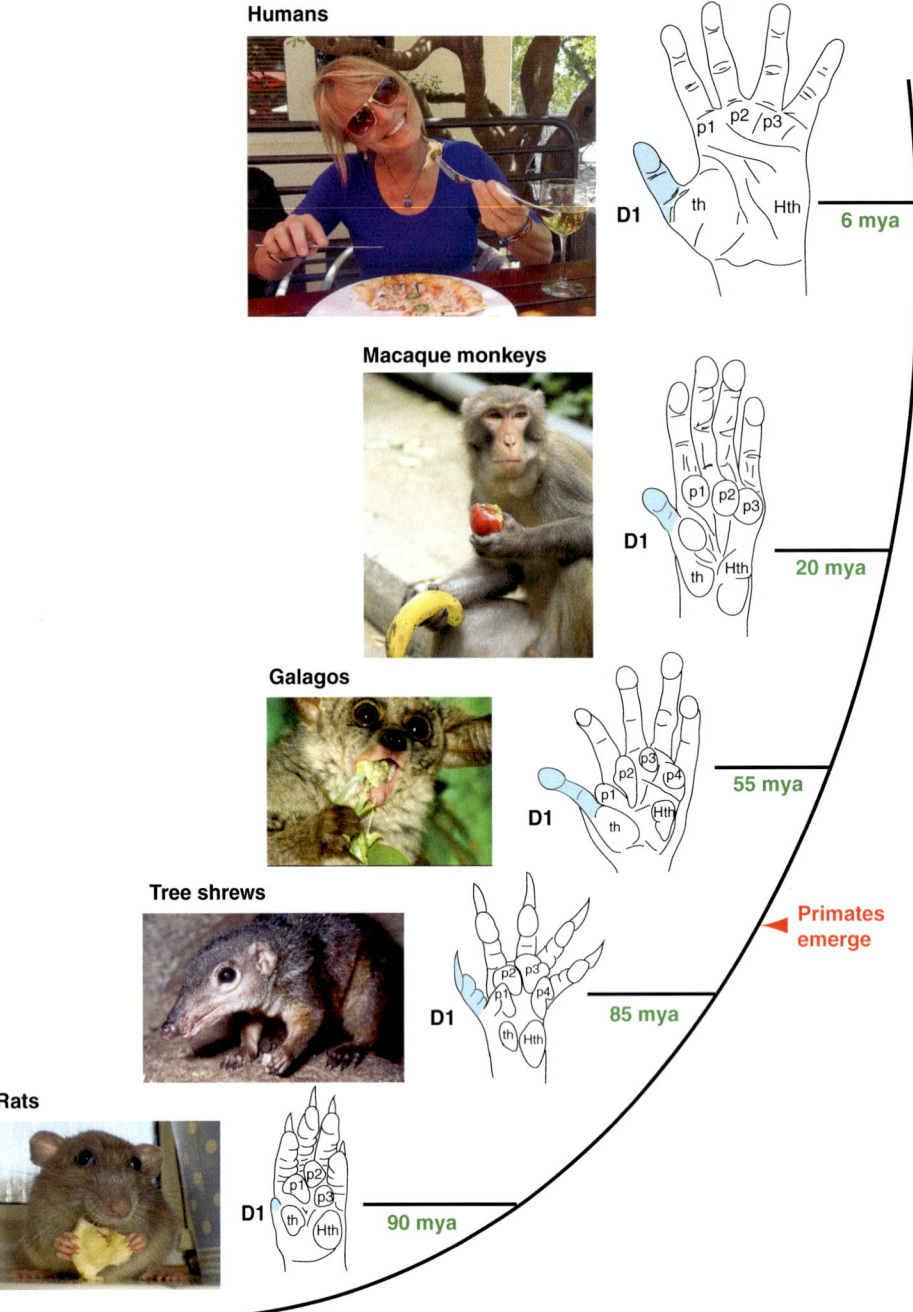

FIGURE 26.1   Five different mammals shown grasping a food item, and corresponding hand morphology. Note the well-developed hand of the human and macaque monkey compared to the paw of the rat with a recessed digit 1 (D1). *p*, pads; *th*, thenar; *Hth*, hypothenar. Hands are not drawn to scale.

interhemispheric connections. While these features clearly define S1 in all mammals tested, there are also derivations in S1 that are species specific. One such specialization is the cortical magnification of the representation of behaviorally relevant body parts such as the vibrissae of many rodents, the nose of star-nosed moles, the bill of platypuses, and the hand of primates (eg, Dooley et al., 2014; Fig. 26.5). Despite these specializations, S1 can still be defined across species and is considered homologous (inherited from a common ancestor).

Unfortunately, these same criteria are difficult to utilize when defining cortical fields in PPC. Cortical fields in PPC are not always distinct using traditional cytoarchitectonic methods. In addition, most often there is not a complete topographic representation of the sensory receptor array and a cortical magnification of behaviorally relevant body parts.

Expansion of PPC in primates

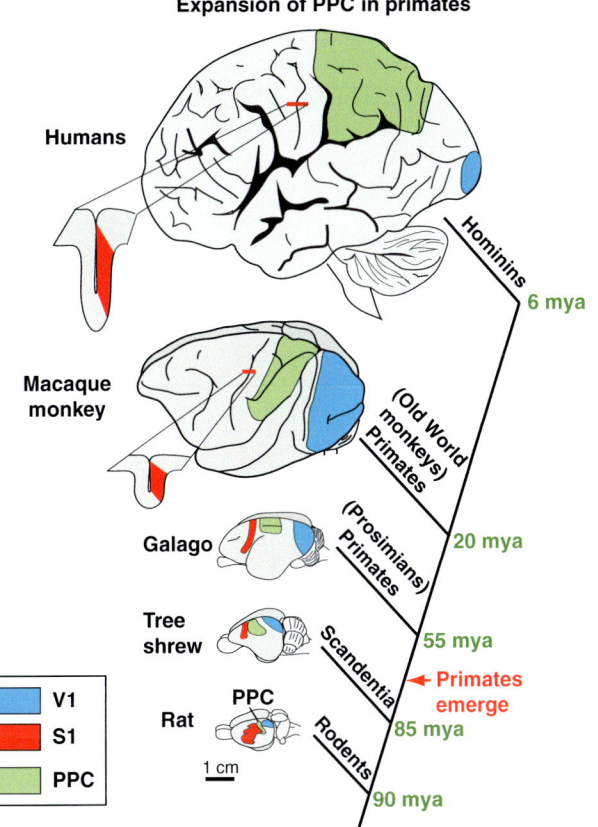

FIGURE 26.2 Cladogram showing the phylogenetic relationship of five different mammals with each branching point indicating the time of the last common ancestor. Brains are drawn to scale except the human brain, which is greatly expanded compared to the other mammals depicted. The location of the primary somatosensory area, S1 (*red*), the primary visual area, V1 (*blue*), and the posterior parietal cortex (PPC; *green*) are depicted. Note that the relative size of the PPC is greatly expanded in primates, particularly in humans.

TABLE 26.1   Common abbreviations of cortical fields, thalamic nuclei, body parts, and anatomical directions in different mammals

Cortical fields	
1	Anterior parietal area 1
1+2	Anterior parietal areas 1 and 2
2	Anterior parietal area 2
3a	Anterior parietal area 3a
3b	Anterior parietal area 3b; primary somatosensory area
5	Brodmann area 5
5D	Area 5 (dorsal division)
5V	Area 5 (ventral division)
7	Brodmann area 7
A1	Primary auditory area
A	Auditory cortex
AIP	Anterior intraparietal area
DG	Dysgranular zone
IPd	Intraparietal depth area (in depth of IPS, adjacent to POa and and PEa)
IPS	Intraparietal sulcus
LIPd	Lateral intraparietal area (dorsal division)
LIPv	Lateral intraparietal area (ventral division)
M1	Primary motor area
M1/PM	Primary motor area/premotor area
OPT	Area OPT, overlaps caudomedial 7a
OTr	Occipital temporal area (rostral division)
PE	Superior parietal lobule
PEa	Superior parietal lobule (anterior division)
PEc	Superior parietal lobule (caudal division)
PF	Rostral inferior parietal lobule area
PFG	Rostral inferior parietal lobule area (transition area between PF and PG)
PG	Rostral inferior parietal lobule area
PGm	Rostral inferior parietal lobule area (medial division)
PM	Parietal medial area
PO	Parietal occipital area (V6 + V6a)
POa	Parietal occipital area (anterior division)
PPC	Posterior parietal cortex
PPCc	Posterior parietal cortex (caudal division)
PPCr	Posterior parietal cortex (rostral division)
PV	Parietal ventral area

Continued

Other factors also make it difficult to clearly identify PPC. Neurons in PPC often do not respond in anesthetized animals. While a number of investigators have examined specific areas in PPC using single unit techniques in awake, behaving macaque monkeys (see later discussion), there is a paucity of similar data in other mammals. Further, even studies where such data do exist in other mammals (eg, Marigold and Drew, 2011; Harvey et al., 2012; Whitlock et al., 2012; Whitlock, 2014), it is unclear if investigators are recording in homologous cortical areas. There are 5500 species of mammals distributed across 29 orders (Roskov, 2015). The order Rodentia alone is composed of 29 families. Despite the extraordinary diversity of extant mammals, most of what we know about PPC is from studies in primates, specifically macaque monkeys, and more recently, rats. Because of this, it is difficult to appreciate if there are homologous cortical areas in PPC across mammals and if there is a basic frontoparietal network that has evolved

TABLE 26.1 Common abbreviations of cortical fields, thalamic nuclei, body parts, and anatomical directions in different mammals—cont'd

Cortical fields

S1	Primary somatosensory area
S2	Second somatosensory area
S2/PV	Second somatosensory area/parietal ventral area
S3	Third somatosensory area
TA	Temporal anterior area
TD	Temporal dorsal area
V1	Primary visual area
V2	Second visual area
VIPl	Ventral intraparietal area (lateral division)
VIPm	Ventral intraparietal area (medial division)

Sulci

CS	Central sulcus
LS	Lateral sulcus
IPS	Intraparietal sulcus
PCS	Precentral sulcus

Thalamic nuclei

LD	Lateral dorsal nucleus
LP	Lateral posterior nucleus

Body parts

cn	Chin
D1	Digit 1
dig	Digits
fa	Forearm
fl	Forelimb
gen	Genitals
h to m	Hand to mouth
hl	Hindlimb
Hth	Hypothenar pad
ll	Lower lip
nk	Neck
occ	Occiput
p1	Pad 1
p2	Pad 2
p3	Pad 3
p4	Pad 4
sh	Shoulder
sn/j	Snout/jaw

TABLE 26.1 Common abbreviations of cortical fields, thalamic nuclei, body parts, and anatomical directions in different mammals—cont'd

Cortical fields

T1	Toe 1
T1−2	Toes 1−2
T5−2	Toes 5−2
th	Thenar pad
tr	Trunk
ul	Upper lip

Antaomical directions

M	Medial
R	Rostral

Other

mya	Millions of years ago

and been modified in different lineages. Importantly, because of the limited number of species examined, it is difficult, if not impossible, to directly relate data collected in rodents with data collected in primates.

To understand how PPC and frontoparietal networks emerged in mammals, became expanded in primates, and ultimately became one of the hallmarks of human brain evolution, a comparative approach drawing on data from a variety of mammals is crucial. Making direct comparisons across a number of species, and ultimately linking these data to the disproportionally large data set in primates, will require the equal application of multiple techniques that can be used in animals that are less amenable to awake, behaving preparations.

In this chapter we will begin by outlining techniques that can be used effectively to explore PPC in primate and nonprimate mammals to help us define homologies across species. We will then briefly discuss what and where PPC is and examine the incomplete data sets for the existence of PPC in rodents and other nonprimate mammals. This chapter is not meant to be exhaustive, but aims to provide an understanding of the difficulty in determining homologies in PPC, and to explore the limited data that allow us to begin to construct a story on the evolution of PPC in mammals in general and primates in particular. Following this we will review both the classic and contemporary studies in primates that have allowed researchers to distinguish PPC from surrounding cortex based on anatomical and physiological criteria complemented by loss of function studies such as lesions and reversible deactivation. We focus mainly on areas in and adjacent to the intraparietal sulcus (IPS) in macaque monkeys, particularly in the context of manual behavior and tool use.

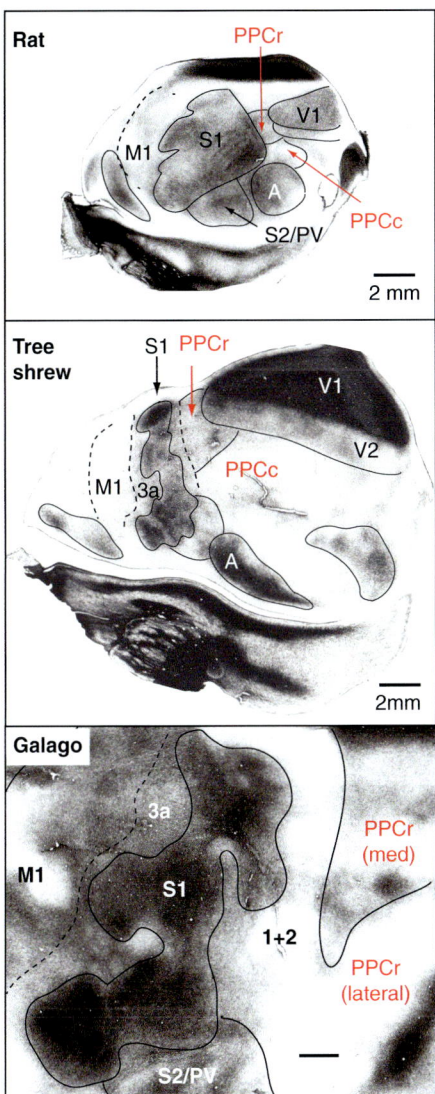

FIGURE 26.3 Myeloarchitecture of parietal and motor cortex in rats, tree shrews, and galagos. In all three species, S1 is darkly stained, and its rostral and caudal boundaries are easily identified. In all three species, PPC has at least one medial moderately staining field and, in rats and tree shrews, lateral lightly staining field. We have tentatively termed these fields "PPCm" and "PPCl" in rats and tree shrews and have added the term "medial" to the PPCr subdivision in galagos. In galagos, PPCc is caudal to the divisions of PPCr and is not shown in this image. To determine homology of these fields across species, multiple criteria should be used including connections with motor cortex as well as the presence of motor maps generated using intracortical microstimulation. Abbreviations are in Table 26.1.

26.2 The Use of Long-Train Intracortical Microstimulation to Define Movement Representations in Motor Cortex and Posterior Parietal Cortex in Mammals

Over a decade ago it became clear that traditional views of motor cortex organization in macaque monkeys did not capture the complexity in organization that

actually exists. Rather than a relatively simple map of movements of a selected portion of the body evoked with short-train intracortical microstimulation (ST-ICMS), the use of long-train intracortical microstimulation (LT-ICMS) revealed that in macaque monkeys there were domains of ethologically relevant movements overlaid upon this simple, grossly topographic motor map of the body (Graziano et al., 2005; Gharbawie et al., 2011b; Cooke and Graziano, 2004). Recently, using LT-ICMS, such domains have been demonstrated in squirrel monkeys, owl monkeys, and galagos, indicating that this type of organization of M1 is a general primate phenomenon (eg, Gharbawie et al., 2011a; Stepniewska et al., 2005, 2014). Similarly, this technique has been used recently to explore motor cortex in rats and tree shrews (Brown and Teskey, 2014; Baldwin et al., 2017; Cooke, 2016). As in primates, movement domains have been discovered, although they are fewer in number. Despite these differences, these data in nonprimate mammals suggest that this type of organization of motor cortex evolved prior to the emergence of primates.

Important for our discussion, LT-ICMS has been used to subdivide divisions of PPC in New World monkeys (see Kaas et al., 2011, 2013 for review), galagos (Stepniewska et al., 2005, 2011; Cooke et al., 2015), and recently macaque monkeys (Baldwin, 2016). Recent work demonstrates that LT-ICMS can be used effectively in nonprimate mammals such as tree shrews and rats to appreciate the functional organization of PPC (Baldwin et al., 2017; Cooke, 2016). Thus, the use of LT-ICMS across cortical areas, especially within PPC, can provide insights into similar features of organization as well as the possible changes and specializations that have emerged across different cortical fields and mammalian orders. Indeed, with a lack of robust anatomical markers, and with the difficulty of obtaining and comparing neural responses to sensory stimulation within PPC of various species, LT-ICMS may be the best tool at this point in time for revealing homologies of this cortical region. However, we do not suggest that this is the only technique that can be used to help identify homologies in PPC. Rather, this technique promises to be an important first step in defining regions of PPC across primate and nonprimate mammals, and thus can guide future studies, using other techniques (eg, single unit recordings in awake animals), as to the location and size of possible homologous fields in PPC across mammals.

26.3 Where and What Is Posterior Parietal Cortex in Nonprimate Mammals?

To establish how the primate PPC may have evolved from the very small neocortex dominated by sensory

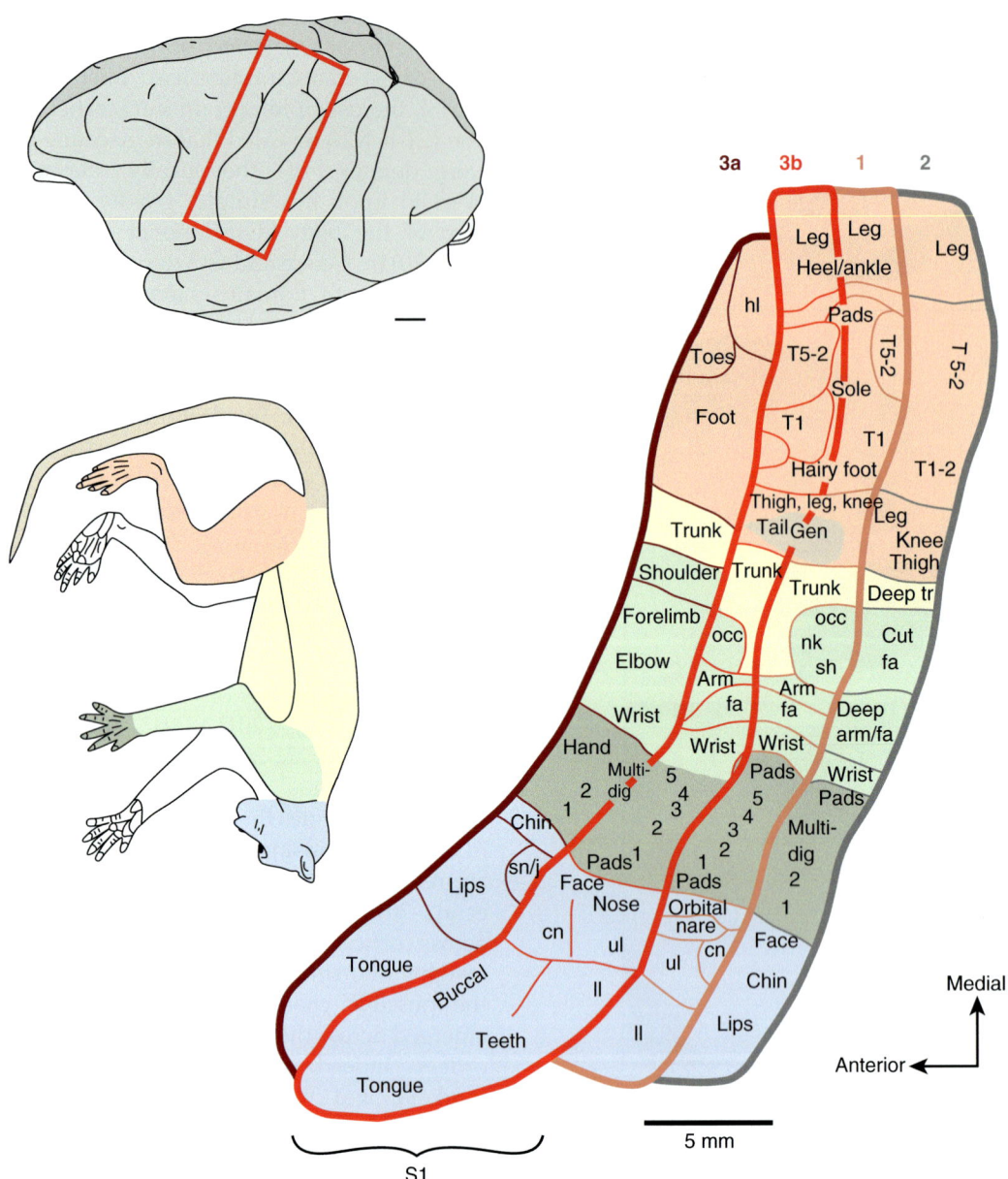

FIGURE 26.4   Topographic maps of the body in four anterior parietal fields in macaque monkeys including areas 3a, 3b, 1, and 2. These maps were generated by recording from multiple sites in anterior parietal cortex (see *red box* in brain drawing at the top) and determining receptive fields on the body for neurons at each site. The body parts represented in the different maps are color coded and correspond to the body image to the left. Areas 3b and 1 contain maps of cutaneous receptors of the contralateral body, while areas 3a and 2 contain maps of deep receptors, muscle spindles, and Golgi tendon organs. Only area 3b corresponds to S1 in other mammals. Note that all body parts are represented, but more space is devoted to representing specific body parts (such as the hand and forelimb) compared to others (such as the trunk). Abbreviations are in Table 26.1. *Combined maps have been adapted from Seelke, A.M., Padberg, J.J., Disbrow, E., Purnell, S.M., Recanzone, G., Krubitzer, L., 2012. Topographic Maps within Brodmann's Area 5 of macaque monkeys. Cereb. Cortex 22, 1834–1850.*

areas in early mammals, one must have a definition of PPC that can span multiple species. As will be discussed later in the chapter, PPC in macaque monkeys is traditionally comprised of Brodmann architectonically defined areas 5 and 7. In nonprimate mammals, however, no easily identifiable homologues of these regions, distinguished using traditional architectonic methods, exist. While a number of studies of "PPC" have been conducted on rats, criteria for defining rodent PPC are not stated, are insufficient, or are omitted entirely. Indeed, making direct extrapolations from rats to monkeys is difficult if not impossible without examining the status of PPC in other mammals.

In this portion of the chapter we review data from a number of nonprimate mammals in which multiple criteria have been used, either in isolation or in

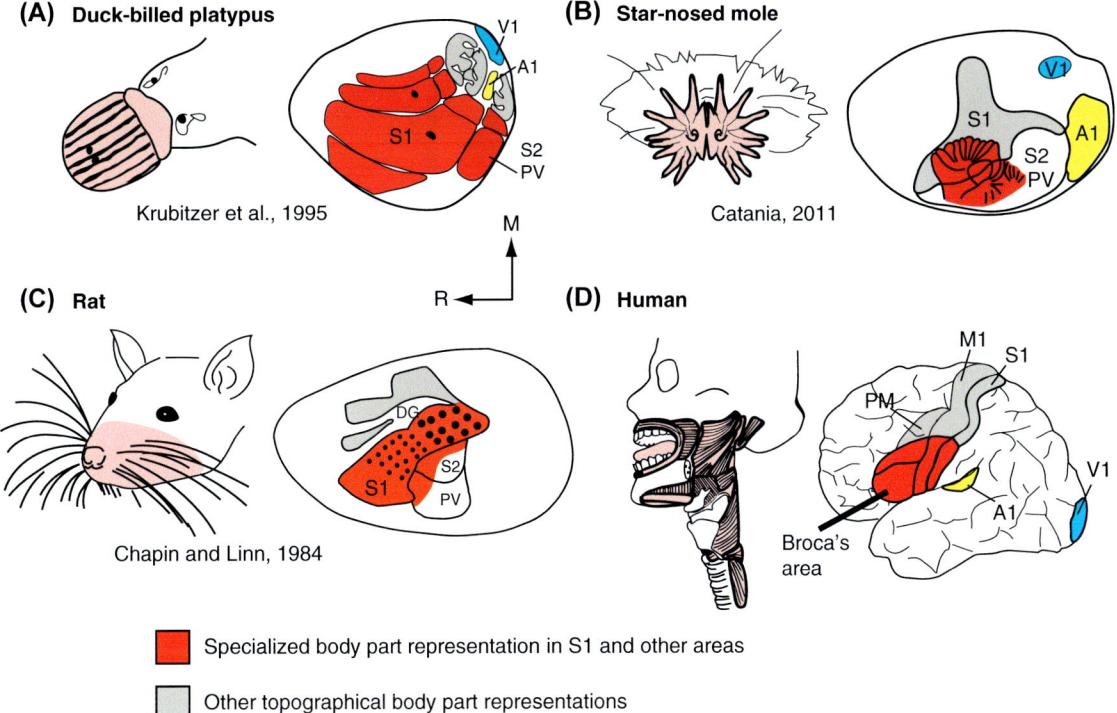

FIGURE 26.5 Examples of magnification of behaviorally relevant body parts in sensory and motor cortex of different mammals. The duck-billed platypus has an enormous representation of the bill in S1 and two adjacent regions of somatosensory cortex (A). The representation of the nose of the star-nosed mole is also magnified in S1 and surrounding sensory cortex (B) as is the whisker representation in the rat (C). Similar magnifications of behaviorally relevant body parts such as the superlaryngeal tract and oral structures are also observed in S1 of humans (D). Magnification of these body parts are also observed for movement representations in motor (M1) and premotor (PM) cortex, and together this region is often referred to as Broca's area. However, it should be noted that while Broca's area is a specialization of humans, the principle of cortical magnification is common to all mammals investigated. Abbreviations are in Table 26.1.

combination, to define PPC. Probably the simplest definition of PPC is based on location. PPC can be defined as the region of cortex located between visual (V2) and somatosensory cortex (S1) (Fig. 26.6). However, this definition presents a problem in large-brained mammals in which there is a huge expanse of cortex between V2 and S1, and additional somatosensory fields (eg, areas 1 and 2) are present caudal to S1 proper (area 3b), and additional extrastriate fields are present rostral to V2. These additional fields are usually not considered part of PPC, but defining the exact boundaries between them and PPC can be difficult. Thus, even if a mammal (be it large or small brained) has an architectonically distinct region between S1 and V2, one must first eliminate the possibility that it is an additional somatosensory or extrastriate visual field. PPC can also be defined by its connections, particularly direct and dense connections between motor cortex and rostral portions of PPC (PPCr), and direct inputs from visual cortex to caudal portions of PPC (PPCc). While some divisions of PPC are architectonically distinct in nonprimate mammals, there are little comparative data in which similar stains and techniques have been used across species and directly compared. However, our own

laboratory has begun an architectonic analysis combined with LT-ICMS in several species of mammals including primates in an effort to make direct comparisons and establish homologous areas within PPC across species (eg, Fig. 26.3). Electrophysiological recordings have also been made in PPC in nonprimate mammals, but most of this work has been done in anesthetized preparations in which neurons in PPC often respond poorly (eg, squirrels, tree shrews, ferrets). As noted earlier, modern studies using LT-ICMS indicate that this technique in combination with others (eg, neuroanatomy, cortical architecture) may well be the best combination of current methods for use in a variety of primate and nonprimate mammals to determine homologous cortical areas within PPC as well as species-specific derivations.

26.3.1 Rodents

In rats, Krieg (1947) and later Kolb and Walkey (1987) identified a thin strip of cortex between "S1" and areas 18a and b as presumptive PPC, which corresponds to Krieg area 7. Several lines of evidence have been put forth to support the assertion that this region is homologous to

Posterior parietal cortex in nonprimate mammals

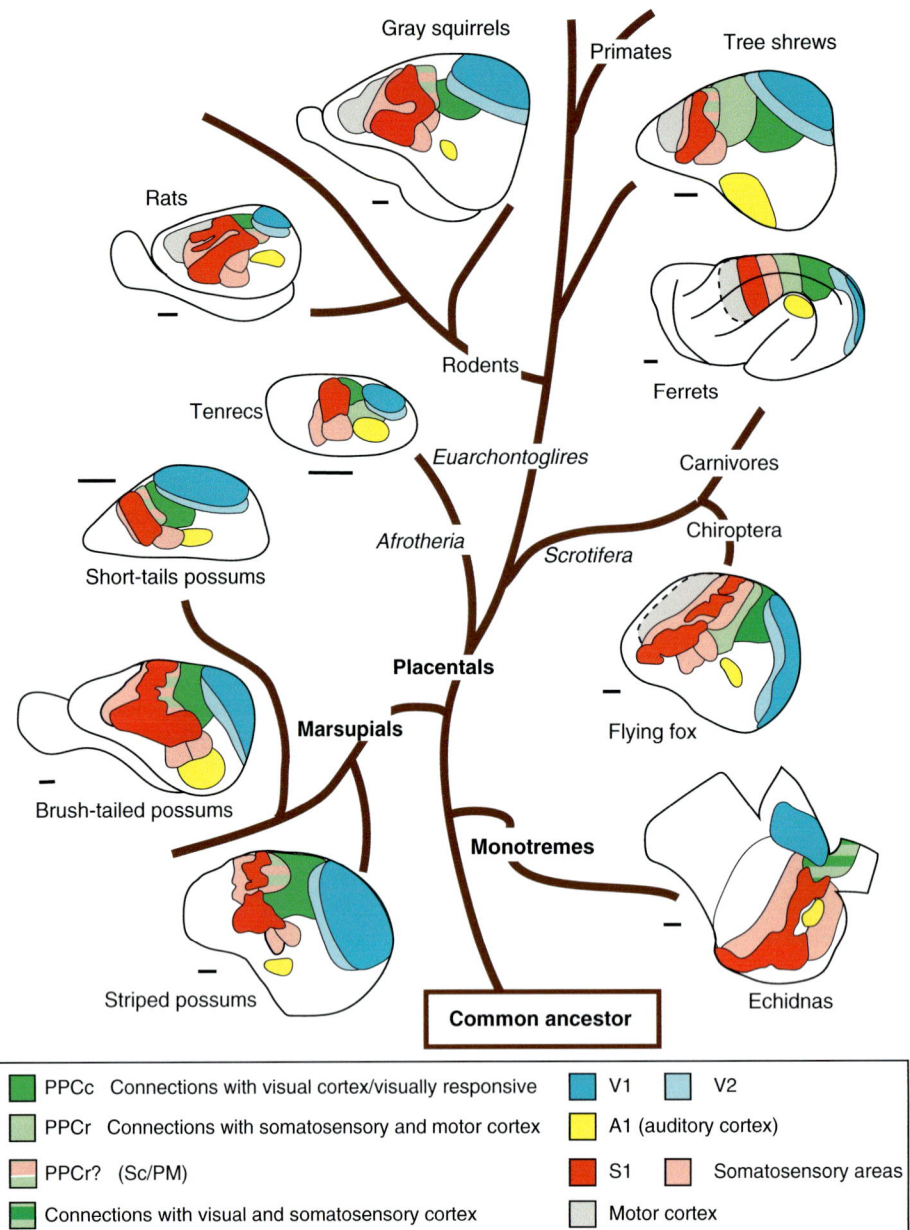

FIGURE 26.6   Posterior parietal cortex in nonprimate mammals. Based on a number of different criteria including location relative to S1 and V2, connections with somatosensory and motor cortex versus visual cortex, and neural response properties (see text), this figure illustrates our hypothesis regarding the status of PPC in nonprimate mammals. Although the data are limited and not all criteria are used conjointly in the same animals, there is evidence for a rostral division (PPCr) and a caudal division (PPCc) in eutherian mammals and marsupials. There are limited data in monotremes, but a very small region of cortex in which neurons receive inputs from both visual and somatosensory cortex and in which neurons respond to visual and somatosensory stimulation is present. While these similarities suggest homologies with primates, they do not imply analogy (similar function), since primate PPC is greatly expanded and contains multiple divisions associated with the contextual use of major effectors (eg, hands and eyes). Rostral is left and medial is to the top. These figures are not drawn to scale. The scale bar = 2 mm. Cortical fields are color coded (see key). Abbreviations are in Table 26.1.

primate PPC. This region in rats, called PM (parietal medial area) or Sc (somatosensory caudal area) by others (see Krubitzer et al., 2011; Ebner, 2015 for review), has been shown to have a pattern of thalamocortical connections that distinguish it from surrounding visual and

somatosensory cortices (Chandler et al., 1992; Reep et al., 1994). Specifically, rat PM/PPC receives most of its input from thalamic nuclei LD and LP while receiving no input from the visual (dorsal lateral geniculate) or somatosensory [ventrobasal (VB) complex] nuclei that characterize V2 and S1, respectively. While this is indeed a compelling and distinctive feature of this region of rat cortex, the thalamocortical connections with PPC in other species have not been consistently documented and thalamocortical connections of PPC in primates have significant differences. Although both primate areas 5 and 7 receive a large amount of input from LD and LP as in rat PPC, they also receive input from several divisions of the VB complex (Padberg et al., 2009; Schmahmann and Pandya, 1990).

Perhaps the most compelling neuroanatomical evidence for rat PM/PPC being homologous to portions of primate PPC is the convergence of afferents from multiple sensory modalities. Miller and Vogt (1984) demonstrated that cortex immediately caudal to rat S1 (PM/Sc/PPC) receives dense projections from areas 18a/b as well as some sparse projections from area 17. Similarly, this region receives inputs from S1 (Fabri and Burton, 1991), particularly the dysgranular zone (Kim and Lee, 2013), and M1 (Mohammed and Jain, 2016) as well as from orbital and agranular frontal cortices (Kolb and Walkey, 1987; Reep et al., 1994). Unfortunately, since the rostrocaudal extent of PM/PPC is relatively thin (Fig. 26.6), it is quite difficult to inject neuroanatomical tracers into this area without contaminating adjacent somatosensory and visual cortices. This makes interpreting the results of neuroanatomical tracing studies of this area challenging. Similarly, although lesioning PM/PPC can evoke a suite of deficits that resemble those of primate PPC lesions, it is difficult to do so without encroaching on adjacent sensory cortices (Kolb and Walkey, 1987; Pinto-Hamuy et al., 1987).

Given the multimodal inputs to rat PM/PPC, one would expect to find an abundance of neurons responsive to multiple sensory modalities. Wallace et al. (2004) explored this by recording the proportion of neurons responsive to multimodal stimulation throughout sensory cortices in the anesthetized rat. While some of these neurons were found within PM/PPC, almost all were recorded at the border between this area and visual cortices. This phenomenon was also observed at the border between visual and auditory cortices as well as somatosensory and auditory cortices, something that has been observed by other groups exploring these border regions (Di et al., 1994). This suggests that these multimodal responses may represent a feature of cortical border regions rather than PM/PPC representing a site of multimodal convergence. Additionally, a much larger proportion of neurons that respond to multimodal stimulation was observed in anterior parietal (somatosensory) cortex compared with PM/PPC, a finding supported by connections between areas 18a/b and S1 (Miller and Vogt, 1984). Thus, neurons responding to multisensory stimulation are at the very least not a

unique feature of PM/PPC when compared with adjacent cortices in rats. This raises the question of whether multisensory neurons are a defining feature of PPC in general, and if so, in what proportions must they be present. Interestingly, preliminary data from our laboratory indicate that movements can be evoked in rat PPC when LT-ICMS is used (Fig. 26.7). We find that PPC is dominated by representations of vibrissae movements, the major effector in the rat.

Data from other rodents such as squirrels indicate that cortex caudal to S1 and rostral to V2 has similar features as those described in the rat. Cortex just caudal to S1 has also been termed PM/PPC in squirrels (Krubitzer et al., 1995; Slutsky et al., 2000; see Krubitzer et al., 2011 for review), and neurons in this region are responsive to stimulation of deep receptors of the skin and joints. As in rats, squirrel PM/PPC has dense interconnections with M1 and area 3a (probably homologous to rat dysgranular zone in S1) and moderate connections with S1 (Krubitzer et al., 1986; Cooke et al., 2012), but these connections are with the more rostral portion of PPC. The caudolateral portion of this PPC region receives inputs from V1 and V2, although it was called OTr in an earlier study (Kaas et al., 1989). For consistency with the rat and other species, we refer to the rostromedial, somatomotor-receiving portion of PPC in rodents as PPCr and the caudolateral, visual portion of PPC as PPCc (Fig. 26.6). Obviously, more data on thalamocortical and corticocortical connections to these regions combined with architectonic analysis and LT-ICMS are needed in squirrels and other rodents to come to any firm conclusions about homology of PPC in different rodents and between rodents and primates.

Taken together, the data from rodents such as squirrels and rats indicate that there is a region of cortex between S1 and V2 that shares some of the features of some of the anterior portions of PPC in primates (PPCr). These include dense connections with motor and somatosensory cortex, some aspects of architecture, and the ability to evoke movements of major effectors when using LT-ICMS. While this region in some studies in rodents and other mammals was originally considered a somatosensory area (termed PM/Sc), we believe that the data indicate that this might be part of PPC. Further, a caudal portion of PPC (PPCc) appears to be more associated with the visual system, as evidenced by connections from V1 and V2. However, as noted later, PPC in primates, even those with a relatively small neocortex, is expansive and contains multiple cortical fields. Thus, without comparisons with other mammals, it is unclear if either part of PPC in rodents is homologous to any of the multiple PPC fields identified in primates.

The earlier section is a brief overview of the relative location and divisions of PPC in rodents, some of the neuroanatomical connections of PPCr and PPCc, and to

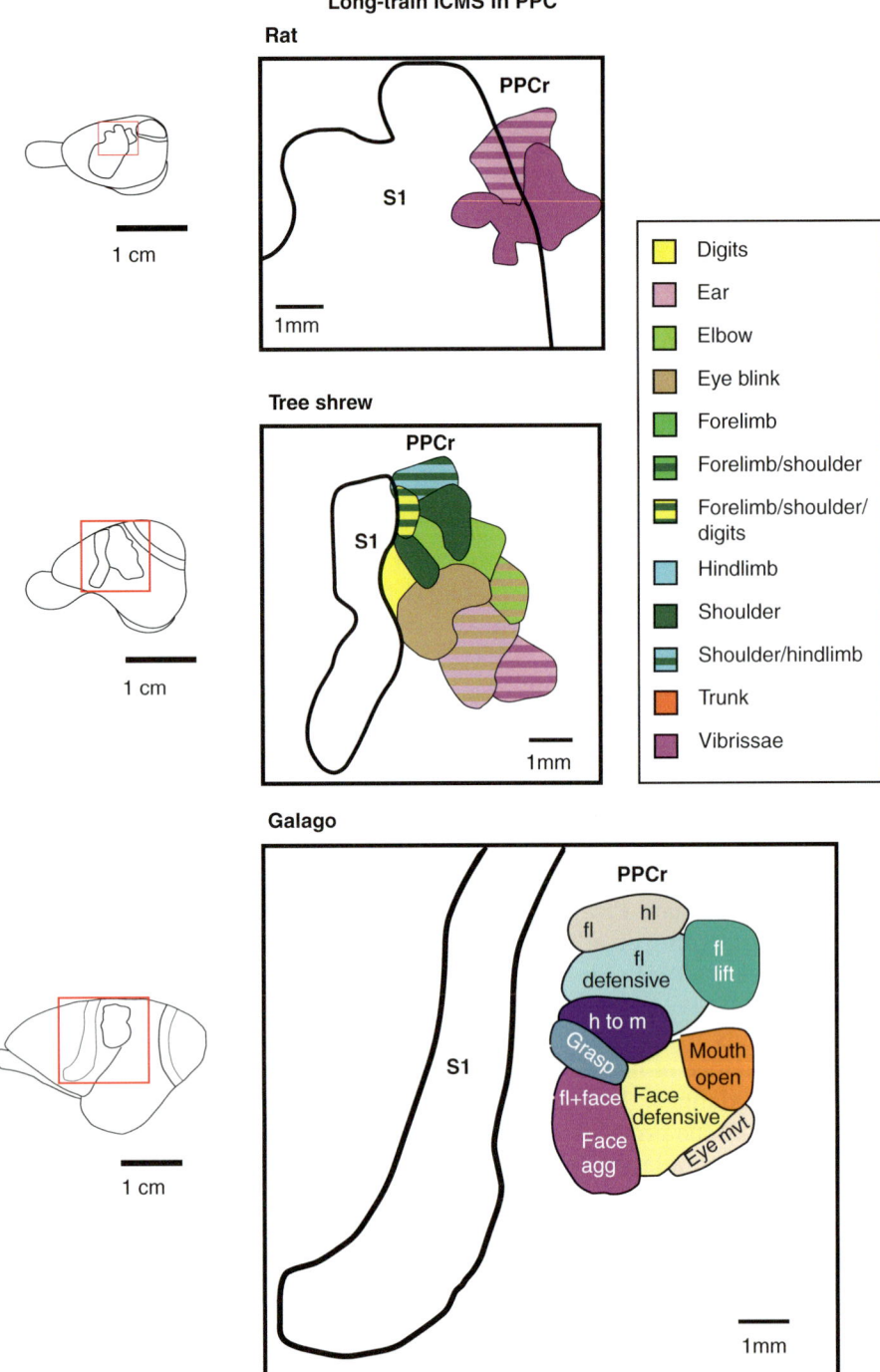

FIGURE 26.7   Motor maps of frontal and parietal cortex based on movements evoked by intracortical microstimulation (ICMS) at multiple sites in rats, tree shrews, and galagos. Different body part representations are color coded (see key at right); multiple body part movements are indicated by multicolor stripes. Note that rat PPC is dominated by vibrissae domain, the galago PPC is composed of multiple domains associated with movements of the hand and forelimb and represents ethologically relevant behaviors. Tree shrew PPC contains domains associated with the vibrissae and face, but also has domains associated with movements of the forelimb and hindlimb. See Table 26.1 for abbreviations.

a more limited extent, neural responses of the PPCr in anesthetized animals. There have been electrophysiological recording studies in awake rats and mice, and these studies indicate that cortex between S1 and V2 (although the actual location of recording is often times not demonstrated) is involved in a variety of functions, such as generating frames of reference for navigating the environment, spatial attention, perceptual decision-making (see Harvey et al., 2012; Krubitzer et al., 2011 for review; Wilber et al., 2014) as well as visual attention, working memory, and movement planning (see Whitlock, 2014 for review). Thus, PPC in rats computes navigational movement plans to generate meaningful interactions (eg, finding, acquiring, and eating food) within a spatial environment in an analogous manner to reaching and grasping movements that are computed in PPC in monkeys for acquiring target objects (Whitlock, 2014). Our recent preliminary data (Fig. 26.7) indicate that movements can be elicited in parietal and portions of PPC in rats and many of these evoked movements involve the vibrissae. These data suggest that comparisons between rats and monkeys may now be tractable if (1) standard, parallel techniques are utilized to define fields in PPC, and (2) species with intermediate types of organization and appropriate phylogenetic relationships are used as a link when making these comparisons.

26.3.2 Carnivores

Studies in both cats and ferrets in which neuroanatomical connections and neural response properties have been described indicate that cortex caudal to somatosensory cortex and rostral to visual cortex has some of the properties of PPC described in rodents, tree shrews (see later discussion), and primates. Immediately caudal to S1 in cats and ferrets is a field termed S3 (Garraghty et al., 1987; Foxworthy and Meredith, 2011). In cats a complete representation of the contralateral body has been described (somewhat similar to PM in squirrels), while in ferrets only representations of the face have been described in this same region. Just caudal to S3, cortex has been termed area 5 or 5a in cats (eg, Babb et al., 1984; Lajoie et al., 2010) and PPr in ferrets (Manger et al., 2002; Foxworthy et al., 2013a). Because we hypothesize that PPr in ferrets is homologous to PPCr in other species, we will term this field PPCr in ferrets for consistency and clarity. PPCr in ferrets contains neurons that respond to multimodal stimulation (Foxworthy et al., 2013b), and this region of cortex receives dense inputs from somatosensory cortex with moderate-to-small inputs from visual areas. Manger et al. (2002) report a gross retinotopy and somatotopy in PPCr, and this field appears to be dominated by the representation of the forelimb. Just caudal to PPCr in ferrets is a field termed PPc by Manger et al. (2002). Again, because of presumed homology with other species, we will term this field PPCc, which has also been explored by these same investigators and found to contain a gross retinotopic organization. Thus, PPCc appears to be more closely aligned with the visual system. This region of cortex in ferrets and cats receives inputs from V1 (Rockland, 1985; Han et al., 2008) and from extrastriate cortex in cats (eg, Olson and Lawler, 1987). While the overall organization has not been described for these fields in cats, lesion and single unit recording studies in the rostral portion of PPC (cat area 5) suggest that this region is involved in visually guided locomotion and computing the visuomotor transformations necessary for gait modification during locomotion (Lajoie and Drew, 2007; Lajoie et al., 2010; Marigold and Drew, 2011). This rostral region of PPC in cats, as in ferrets, has interconnections with motor cortex (eg, Babb et al., 1984; Andujar and Drew, 2007). Taken together, the data in ferrets and cats indicate that PPC can be divided into two gross divisions, one rostral that has neurons with multimodal responses and interconnections with motor cortex, and a more caudal region, more closely aligned to the visual system. These may correspond to similar regions in rodents although the data are too sparse to support any firm conclusion. The status of S3 when compared to regions in rodent cortex is unclear. While it may resemble PM in squirrels and rats, it may also be a region unique to cats, since it appears to be much more like a somatosensory field than a posterior parietal field.

26.3.3 Tree Shrews

Tree shrews are the closest living relative to primates and therefore represent an important animal model that can help link data from nonprimate mammals with that from primates. Early electrophysiological recording studies reported that immediately caudal to S1 lies a very narrow strip of cortex (less than 1 mm wide) that contains neurons responsive to somatic stimulation (Sur et al., 1980); these investigators termed this field Sc. However, no data were actually shown to support the existence of this field so its presence in tree shrews is questionable. Recent work in tree shrews in which neuroanatomical data were combined with ST-ICMS indicates that cortex caudomedial to S1 has dense interconnections with M1 (Remple et al., 2006). Based on location, this region includes Remple et al.'s (2006) Sc, as well as their areas PPd and TA (Remple et al., 2007). While the use of ST-ICMS failed to elicit movements in cortex caudal to S1 in tree shrews, recently our laboratory utilized LT-ICMS and found that movements can be evoked in a large swath of cortex in this region caudomedial to S1 which we term here PPCr (Fig. 26.7). Unlike movements evoked from a similar location in rat cortex, the suite of movements which can be evoked by LT-ICMS in tree shrew PPCr is not dominated by movements of the vibrissae, but has a more complete representation that includes not only portions of the face, but the forelimb and hindlimb as well. When compared with galagos, a prosimian primate, the organization of tree shrew PPC has a number of similarities (Fig. 26.7; see later discussion). However, additional work examining the connections of this field with other cortical areas as well as thalamic nuclei will allow us to more accurately establish homologous cortical fields between tree shrews and primates (if they are present).

As in rodents and carnivores, tree shrew PPC can be grossly divided into two regions based on inputs from different regions of cortex. The rostromedial region noted earlier (PPCr) contains movement domains identified using ICMS and has dense projections from M1 (and moderate projections from S1). A caudolateral region, referred to in previous studies as TD (Sesma et al., 1984) but here referred to as PPCc, receives input from both V1 and V2. Although sparse, these data appear similar to that described in rodents and carnivores.

26.3.4 Marsupials and Monotremes

Marsupials appear to have a PPC, or at least a small region of cortex between S1 and V2 that has several features of organization that are similar to rodents, carnivores, and tree shrews. Cortex immediately caudal to S1 has been demonstrated to contain neurons responsive to stimulation of deep receptors of the contralateral body in several marsupials (see Huffman et al., 1999; Elston and Manger, 1999). While receptive fields for neurons in this location have been defined for some of these species (eg, native cat, striped possum, brush-tailed opossum), in some species this region of cortex is extremely narrow in width and thus is difficult to record from. It has been termed C or Sc (Beck et al., 1996; Huffman et al., 1999), but like the data reviewed earlier, we believe that an alternate interpretation is that this cortex is part of the rostral division of PPC (termed PPCr). As in the species discussed earlier, in the marsupials that have been investigated PPCr receives input from S1 (Beck et al., 1996; Elston and Manger, 1999; Dooley et al., 2013) and from cortex just rostral to S1 [although it is not known if this frontal region is homologous with motor cortex in other mammals (Beck et al., 1996; Karlen and Krubitzer, 2007)]. Somewhat caudal to this, neurons are unresponsive or respond to multimodal stimulation including some combination of visual + somatosensory + auditory stimulation (Huffman et al., 1999). Further, the few studies of connections that have been done indicate that this caudal region (termed differently by different investigators) receives inputs from visual cortex (Crewther et al., 1984; Elston and Manger, 1999; Dooley et al., 2013; Martinich et al., 2000), much like PPCc in other species.

There is very little information available regarding PPC in monotremes. Extensive electrophysiological recording in these animals indicates that location is not a reliable criterion for defining PPC because the layout of the usual neighboring cortical fields in the two species that have been studied (duck-billed platypus and echidna) is remarkably different than in other mammals.

Specifically, the monotreme primary visual cortex is located medial, rather than caudal to S1, and auditory cortex is embedded between S1 and S2/PV (Krubitzer et al., 1995). However, there is a small strip of cortex between V1 and S1 in which neurons respond to both somatosensory and visual stimulation (Fig. 26.6). There are no data on connections of this region in platypus and only limited data on connections of S1 and V1 in echidnas indicating that it receives inputs from both S1 and V1 (Krubitzer, 1998). Thus, we tentatively call this area PPC.

26.3.5 Conclusion

Taken together, the data reviewed here from rodents, tree shrews, carnivores, marsupials, and monotremes, as well as data from other nonprimate mammals such as flying foxes indicate that cortex that resides caudal to S1 and rostral to V2 has some of the features of PPC described in primates. We tentatively term this region PPC and suggest that in all nonprimate mammals tested there is evidence that this region has at least two subdivisions. In species in which LT-ICMS has been used to delineate PPC, movements can be evoked in the rostromedial portion of PPC, in PPCr (eg, tree shrews). The comparative data in marsupials indicate that they appear to have two divisions of PPC with general characteristics of those of their eutherian counterparts, but the relative proportion of this cortex is much smaller. Thus, the presumptive PPCr and PPCc may have been present in the last common ancestor of marsupials and eutherians some 180 million years ago. The presence of a PPC-like region in monotremes seems unlikely since so little cortex exists between known sensory areas.

It is unclear if the rostral portion of PPC should be considered a caudal somatosensory area (terms PM, Sc, and analogous to primate area 1/2) that is situated between PPC and S1 in most of the species examined, or if this region of cortex should be included as part of PPC, specifically the rostral division of PPC termed PPCr, analogous to parts of primate area 5. This region of cortex is often characterized by neurons responsive to somatic stimulation (and in ferrets multimodal stimulation) and is interconnected with somatosensory cortex in all of the marsupials and eutherian mammals examined, and in eutherians with motor cortex as well. In the caudal portion of nonprimate PPC shares connections with visual cortex.

While this brief overview of the general organization of PPC in nonprimate mammals provides some insights into its evolution, it does not discuss how this cortex has coevolved with the unique behavioral repertoires that different animals possess. Importantly, we do not wish to imply that only two subdivisions of PPC exist in nonprimate mammals, but rather we tried to evaluate data in a way that would allow us to extract similarities between species. The methodological challenges we outlined earlier make defining homology in PPC incredibly difficult. Doing so will require multiple criteria and converging data from a variety of species. Even then, appreciating homologous divisions between species as disparate as mice and macaque monkeys may prove impossible. Perhaps the most critical element in this research program will be electrophysiological data in awake behaving animals from a few well-chosen species, which may reveal common features of the PPC cortex (if they exist), and how they evolved and expanded in primates.

26.4 Primates

The vast majority of what we know of PPC has come from studies in primates. Throughout the course of primate evolution, alterations in the morphology of the hand have allowed various species to exploit specific ecological niches and execute a variety of foraging and predation strategies (see Boyer et al., 2013; Almecija/Sherwood chapter 3.16, Hands, Brains, and Precision Grips: Origins of Tool Use Behaviors for review). Among these alterations, the evolution of an opposable thumb capable of performing a precision grip distinguishes catarrhines (humans, apes, and Old World monkeys) from most platyrrhines (New World monkeys), who primarily employ whole-hand power grasps to manipulate objects (see Napier and Tuttle, 1993). The exception to this is the New World tufted capuchin (also called cebus) monkey (*Sapajus apella*), which not only possesses an opposable thumb capable of performing a precision grip, but also uses these features to create and employ tools to perform various tasks (Christel and Fragaszy, 2000; Fragaszy et al., 2004; Spinozzi et al., 2004; Fig. 26.8). Perhaps unsurprisingly, these monkeys appear to have evolved a frontoparietal network that bears striking resemblance to that of Old World monkeys (Padberg et al., 2007). This remarkable example of homoplasy highlights the fact that the evolution of the hand and the brain areas supporting its movement are constrained, perhaps by the contingent nature of the genetic cascades involved in cortical development. This represents a compromise between contextual and experientially driven cortical plasticity and the homologous features of mammalian cortical organization that dictate the ways in which the cortical phenotype can be changed to solve a given problem (see Krubitzer, 2007 for review). In this case, this combination of form and function represents a common solution to a common evolutionary puzzle: how to analyze object affordances and compute movements that allow for interacting with that object in different ways depending on the current behavioral circumstances.

Direct measures and manipulations of the PPC have historically been performed in various species of macaque, which represent ideal animal models given the similarity in hand morphology to humans as well as the perceptual abilities critical to its use (ie, vision and somatosensation). Though it has been extensively subdivided in more contemporary investigations, the PPC is usually characterized as being comprised of Brodmann areas 5 and 7 (Fig. 26.9). Traditionally termed "association" cortex, most early investigations of these regions involved electrical stimulation in humans and other primates, with stimulation of area 7 generally eliciting arm, hand, and face movements laterally and eye movements medially. While stimulation of lateral portions of area 5 also elicited arm and face movements, more medial stimulation evoked trunk and hindlimb movements (eg, Fleming and Crosby, 1955). Early lesion and ablation experiments converged on several themes with regard to removal of various areas of the parietal lobe: unlike motor cortical lesions, parietal lesions did not cause paralysis of the contralateral body. Instead, subjects would display a reluctance to use the affected limb, hypotonia of that effector's muscle groups, weak or incomplete grasping, general ataxia, and erroneous limb trajectories which could be somewhat ameliorated by direct visual attention (see Peele, 1944 for review). Peele (1944) performed one of the first systematic, albeit qualitative, investigations of the effects of lesioning different areas of the parietal lobe alone or in combination. These results were generally congruent with electrical stimulation studies: lesions of areas 5 or 7 produced deficits which were less severe than ablation of anterior parietal cortex (APC; areas 3, 1, and 2) resulting in awkward movements and an ataxia that was ameliorated by visual attention or a heightened emotional state. Lesions to area 5 produced more severe deficits in movements of the hindlimbs while lesioning area 7 produced greater forelimb deficits. In addition, lesions to either area resulted in poor detection of various somatosensory stimuli and resulted in an inability to discriminate items by feel (such as food) without visual guidance.

Though these early studies generated important hypotheses about the function of these "associative" cortical areas, elucidating the response properties of individual neurons remained difficult due to the lack of stimulus-driven activity that could be evoked under general anesthesia. However, with the advent of awake (Duffy and Burchfiel, 1971) and awake-behaving monkey preparations (eg, Sakata et al., 1973; Hyvarinen and Poranen, 1974; Mountcastle et al., 1975), researchers began to appreciate the complexity of the neural

Phylogeny of cortical areas and manual abilities in primates

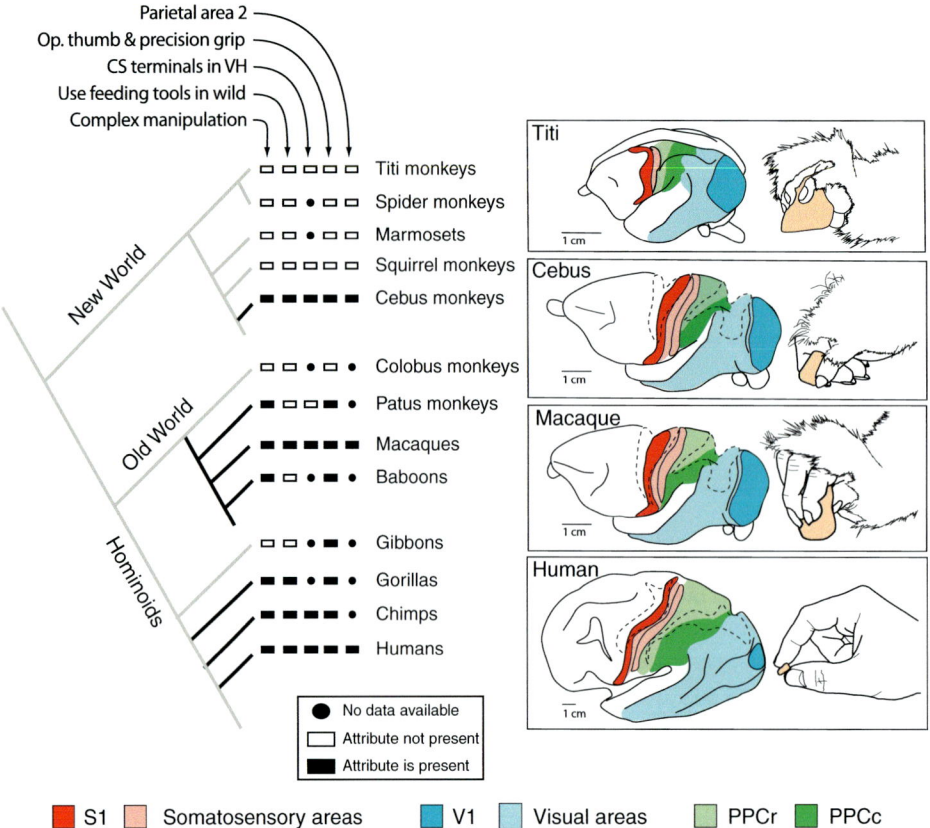

FIGURE 26.8 A cladogram showing different primate taxa and characteristics of hand and hand use (eg, opposable thumb, complex manipulation, tool use) and cortical organization (eg, presence of an area 2, corticospinal terminals in the ventral horn of the spinal cord) that each species possess. Note that the only New World monkeys to possess features of the hand, uses of the hand, and cortical organization similar to that of many (but not all) Old World monkeys and humans are the cebus monkeys. Cebus monkeys possess additional somatosensory areas caudal to S1 (areas 1 and 2) while these areas are not apparent in other New World monkeys. Further, like anthropoid primates, cebus monkeys have a large expansion of posterior parietal cortex. Abbreviations are in Table 26.1. *The cladogram was adapted from van Schaik, C.P., Deaner, R.O., Merrill, M.Y., 2003. The conditions for tool use in primates: implications for the evolution of material culture. J. Hum. Evol. 36, 719—741 and the illustration of the brain and hands is adapted from Padberg, J., Franca, J.G., Cooke, D.F., Soares, J.G., Rosa, M.G., Fiorani Jr., M., Gattass, R., Krubitzer, L., 2007. Parallel evolution of cortical areas involved in skilled hand use. J. Neurosci. 27, 10106—10115.*

networks that comprise the PPC. This wealth of physiological data produced by awake preparations coincided with ever-increasing refinements in both connection tracing and architectonics, which allowed researchers to subdivide the two large fields in PPC initially described by Brodmann into multiple cortical areas (eg, Seltzer and Pandya, 1986, 1980; Pandya and Seltzer, 1982; Preuss and Goldman-Rakic, 1991; Lewis and Van Essen, 2000a,b; see Fig. 26.9).

In the following sections we describe both early and more contemporary investigations of the neuronal response properties of primate areas 5 and 7, focusing mainly on the rostrolateral portions that have been studied in the context of complex manual dexterity. Compared to other mammals, we devote more space to what is known about these areas in primates (specifically rhesus macaques). The reason for this is that these

studies represent the largest and most detailed set information about the functional properties of PPC neurons in any mammal. This is largely due to the fact that, given the current state of technology, macaques represent the animal model most amenable to the types of awake-behaving electrophysiology necessary to reveal the unique sensorimotor responses of PPC neurons. As such, almost any investigation of PPC in other mammals must be considered in the context of monkey data to determine not only what might constitute PPC, but also whether a given PPC region is homologous.

26.4.1 Brodmann Area 5: Early Studies

Brodmann area 5 in monkeys occupies a large section of cortex caudal to area 2 on the postcentral gyrus and

Different schemes for organization of PPC in macaque monkeys

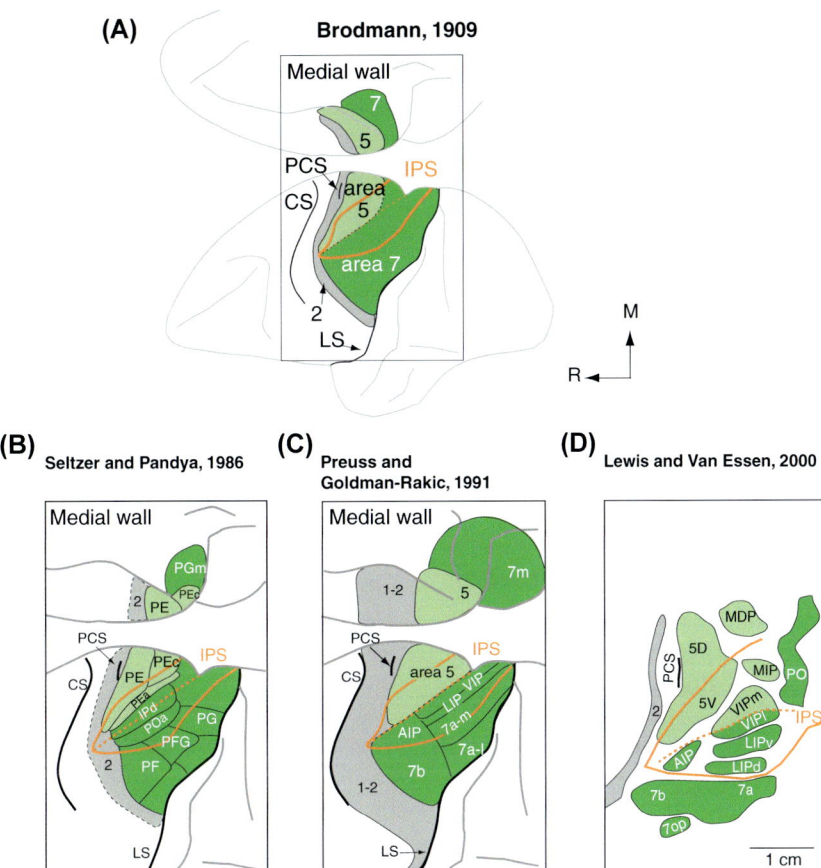

FIGURE 26.9 Architectonic maps of posterior parietal cortex described in different studies. One of the first architectonic studies of this region of cortex was by Brodmann (1909). In this early study (A), two large divisions of cortex were described, area 5 (*light green*) and area 7 (*dark green*). Subsequently, different schemes of how this cortex was organized in macaque monkeys were proposed by different investigators (B—D). While different investigators have proposed different schemes of organization, all agree that multiple cortical fields exist within this large expanse of cortex. Abbreviations are in Table 26.1. *This figure was adapted from Seelke, A.M., Padberg, J.J., Disbrow, E., Purnell, S.M., Recanzone, G., Krubitzer, L., 2012. Topographic Maps within Brodmann's Area 5 of macaque monkeys. Cereb. Cortex 22, 1834—1850.*

extending into the IPS where it abuts a portion of area 7. Its share of the postcentral gyrus is widest medially where at the midline it wraps onto the medial wall where it extends just beyond the cingulate sulcus (Brodmann, 1909). Subsequent to these early descriptions of area 5, modern anatomical techniques have parcellated this rather large cortical field into several overlapping subfields with differing nomenclature (Fig. 26.9; Seltzer and Pandya, 1986; Pandya and Seltzer, 1982; Lewis and Van Essen, 2000b).

The earliest studies of area 5 in awake macaques noted that, although neurons within this field responded primarily to tactile and proprioceptive stimuli, their receptive fields were larger and more complex than neurons in APC and often responded to the movement of multiple joints (Duffy and Burchfiel, 1971). Sakata et al. (1973) noted that neurons in this area sometimes possessed ipsilateral or bilateral receptive fields and were often highly selective to certain stimulus

parameters. Arguably the most seminal electrophysiological study of this area was performed by Mountcastle et al. (1975) in which the monkeys experienced both active and passive stimulation of their extremities in various behavioral contexts. Mountcastle and his colleagues found that in comparison with APC neurons, neurons in area 5 were less active when the monkey was in a resting state, and were driven less directly by sensory input. For instance, when the monkey was falling asleep it was difficult to evoke responses from these neurons. While many of the neurons which responded to manipulations of the joints had receptive fields which resembled those of area 2 neurons, they were often more responsive during active movements. Conversely, neurons that responded to cutaneous stimulation displayed receptive fields that were much larger than those found in APC neurons and often spanned the entire palm or volar surface of the arm. Many of these neurons responded more vigorously to moving stimuli and were often directionally selective. Perhaps most significantly, a group of neurons dubbed "projection" and "hand manipulation" neurons were strongly modulated by the behavioral context of a given movement. These neurons tended to be most active when reaching for food or something that would dispense a reward and were silent when the same movements were made in a nonrewarding context.

From these early studies it became clear that, to appreciate the unique response properties of area 5 (and other PPC) neurons, awake-behaving electrophysiology in carefully controlled conditions is essential. Advances in technology have allowed for more refined behavioral measures that have provided more detailed and precise accounts of these response properties. However, as described in the next section, the precise function and organization of area 5 and the subfields within it is still contentious.

26.4.1.1 Brodmann Area 5: Contemporary Studies

Given the size of Brodmann area 5 and the complex stimuli that seem to drive neurons within it, it is not surprising that subsequent investigations attempting to characterize its function have produced a myriad of different results. One reason for this is that the portion of area 5 from which recordings were made varied between laboratories (see Figure 26.2 of Seelke et al., 2012). This, combined with the variability in the behavioral tasks employed by different investigators, has implicated Brodmann area 5 in such diverse processes as coding of reach intention (Snyder et al., 1997; Debowy et al., 2001; Calton et al., 2002), reach and grasp kinematics (Kalaska, 1996; Wise et al., 1997), online monitoring of different reach styles during object approach (Gardner et al., 2007a,b; Chen et al., 2009), and the coordinate transformation of reach targets into body- and shoulder-centered coordinates (Lacquaniti et al., 1995; Ferraina and Bianchi, 1994) or eye-centered coordinates modulated by limb position (Pesaran et al., 2006). Complicating matters further, some investigations of area 5 describe subfields in purely functional terms without reference to architectonics of the cortical tissue. The most well-studied example of this is the parietal reach region (PRR), originally proposed by Snyder et al. (1997). This is defined as the region within the superior parietal lobule that predominantly contains neurons that are more responsive to reaches than to saccades (Batista and Andersen, 2001). While numerous subsequent studies have implicated this region in effector-specific movement intention (as opposed to attention, see Andersen and Cui, 2009 for review), it appears to overlap several architectonically distinct regions such as the medial intraparietal area (MIP), the medial dorsal parietal area (MDP), and V6a (Snyder et al., 2000). In light of the diverse stimuli and effector

movements to which neurons in Brodmann area 5 respond, it is very likely composed of several distinct cortical areas. Recent efforts by our laboratory have characterized one such subfield, area 5L, which has a distinct electrophysiological, architectonic (Seelke et al., 2012), and connectional (Cooke, 2013) profile, setting it apart from adjacent fields such as area 2 and MIP. Unlike anterior parietal fields, the functional topography of this region is fractured and contains an incomplete representation of the body notable for its extreme magnification of the hand and forelimb (Seelke et al., 2012). Similarly, MIP is characterized by a unique set of architectonic (Seelke et al., 2012; Lewis and Van Essen, 2000b), functional (Colby and Duhamel, 1991; Eskandar and Assad, 2002), and connectional (Lewis and Van Essen, 2000a) characteristics that, when considered together, make a compelling case for it being another distinct region within area 5.

Recently, PPC has been explored using long-train intracortical stimulation, and ethologically relevant movement domains have been demonstrated in squirrel monkeys, owl monkeys (see Kaas et al., 2011, 2013 for review), and galagos (Fig. 26.7; Stepniewska et al., 2005, 2011; Cooke et al., 2015; see Kaas Chapters 2.04, The Early Mammalian Brain, 3.11, Evolution of Visual Cortex in Primates, and 3.15, Evolution of Parietal-Frontal Networks in Primates for review). These studies demonstrate that the rostral portion of PPC (galago PPCr/monkey area 5) contains domains of ethologically relevant movements such as reach, grasp, and hand-to-mouth. Importantly, this rostral portion of PPC has dense interconnections with motor cortex. In macaques the rostral portion of area 5, particularly area 5L, is interconnected with motor, premotor, and supplementary motor cortex (Jones et al., 1978; Deacon, 1992; Darian-Smith et al., 1993; Luppino et al., 1993; Ghosh and Gattera, 1995; Tanne-Gariepy et al., 2002; Gharbawie et al., 2011b; Cooke, 2013). While the rostral portion of PPC is connected to both motor, premotor, and to a lesser extent supplementary motor cortices in squirrel monkeys (Gharbawie et al., 2011a), galagos (Fang et al., 2005; Stepniewska et al., 2009), and marmosets (Burman et al., 2014a,b); this portion of PPC seems to have few connections to primary motor cortex in owl monkeys (Stepniewska et al., 1993), although premotor connections are still present (Gharbawie et al., 2011a). While movement domains in macaque monkeys are found in multiple noncontiguous areas of parietal cortex [area 2, ventral (VIP) and lateral (LIP) intraparietal areas; Thier and Andersen, 1996, 1998; Cooke et al., 2003; Gharbawie et al., 2011b], their PPC has not been thoroughly explored using LT-ICMS, as in smaller-brained primates. Preliminary data from our own laboratory indicate that a larger amount of cortex on the postcentral gyrus, within the IPS and on the marginal gyrus (caudal to IPS), contains these movement domains, underscoring the need for a systematic study of movement domains in macaque PPC (Baldwin, 2016). Thus, even in primates, it is unclear which areas of PPC are homologous in Old and New World monkeys and prosimian galagos. This is further complicated by the fact that connections with motor cortices are a common but not unique feature of rostral PPC, as all of these species show varying degrees of connectivity between APC and motor cortical regions (see earlier discussion; Burton and Fabri, 1995). As noted in the previous section, the use of LT-ICMS combined with architectonics and studies of connections may be the best way in which to make comparisons across PPC in primates and other mammals.

26.4.2 Brodmann Area 7: Early Studies

As is the case with area 5, monkey area 7 occupies a large section of cortex within and caudal to the IPS. By Brodmann's original definitions, it occupies the entirety of the inferior parietal lobule, the caudal bank of the IPS, the upper bank of the lateral sulcus, and portions of the medial wall (Fig. 26.9; Brodmann, 1909). In 1919, Vogt and Vogt used additional architectonic criteria to divide this region into areas 7a and 7b. Much like area 5, more contemporary anatomical investigations have divided these areas further into overlapping regions with various naming schemes (Seltzer and Pandya, 1980, 1986; Pandya and Seltzer, 1982; Preuss and Goldman-Rakic, 1991; Lewis and Van Essen, 2000a,b; Gregoriou et al., 2006). The earliest electrophysiological investigations of these areas demonstrated that, while neurons within 7a primarily responded to movements of the eyes and visual fixation, neurons within 7b primarily responded to somatosensory stimulation as well as passive movements of the arms and hand (Hyvarinen and Poranen, 1974; Mountcastle et al., 1975; Leinonen and Nyman, 1979; Leinonen et al., 1979). Like area 5, cells in both 7a and 7b were most active in awake animals when the monkey reached, grasped, and manipulated various visually targeted objects, although it was noted that area 7 contained more neurons that responded to stimulation of the ipsilateral and bilateral body compared to area 5 (Hyvarinen and Poranen, 1974; Mountcastle et al., 1975). The receptive fields of area 7 neurons were generally large, with complex response properties. For instance, cells responding to cutaneous stimulation were often directionally selective. Additionally, some cells responded to both visual and somatosensory stimuli focused on the same region of the body (Fig. 26.10; ie, a cell with a cutaneous receptive field on the arm also responded to objects approaching that arm).

26.4.2.1 Brodmann Area 7: Contemporary Studies

Contemporary architectonic investigations of the inferior parietal lobule have subdivided monkey 7a and 7b into four distinct zones, PF, PFG, PG, and OPT (Fig. 26.9; Pandya and Seltzer, 1982; Gregoriou et al., 2006). In addition to architectonic differences, investigators have recently found that each zone possesses distinct connection patterns and that these regions form a functional gradient of neurons responsive to certain sensory stimuli and movements as one proceeds from caudal to rostral. Area 7a is comprised of areas OPT and PG, fields dominated by neurons responsive to visual stimuli, eye movements, and arm movements related to reaching and grasping. Area 7b is comprised of areas PFG and PF and is dominated by neurons responsive to somatosensory stimulation of the arm, hand, and mouth. Within PFG/PF, neurons which respond to motor actions are predominantly related to hand use, orofacial movements, and hand—mouth coordination (Yokochi et al., 2003; Rozzi et al., 2008; Fogassi et al., 2005; Bonini et al., 2011). Recently, long-train microstimulation of 7b and the rostrolateral portion of 7a has revealed that, much like areas 1 and 2, movement representations are dominated by flexion and extension of the wrist and digits. Interestingly, stimulation of these regions reveals a greater representation of digit 1 and 2 (d1 and d2) movements compared with APC. Specifically, within these domains, stimulation elicits a d1-d2

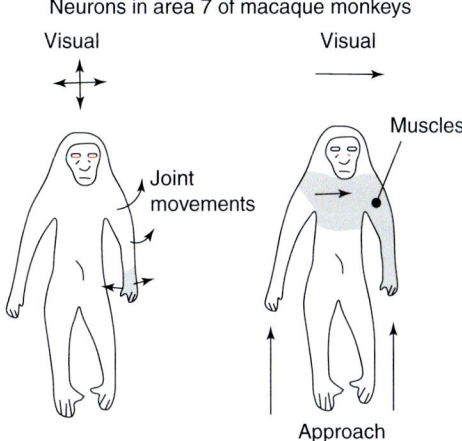

FIGURE 26.10 Examples of some of the characteristics of stimuli that proved to be effective for driving neurons in area 7 of macaque monkeys. All neurons responded to visual stimuli that were moving in a given direction (see *arrows* above monkey illustrations). Some cells responded to stimuli approaching the hands (right figure). All neurons had receptive fields in which neurons responded to cutaneous stimulation (*light gray* shading), and all neurons were activated by palpation of the muscles (see *circle* in right figure) or rotation of the joints (*curved arrows* in left figure). *This figure was redrawn from Leinonen, L., Hyvarinen, J., Nyman, G., Linnankoski, I., 1979. I. Functional properties of neurons in lateral part of associative area 7 in awake monkeys. Exp. Brain Res. 34, 299—320.*

precision grip rather than a whole-hand power grasp (Baldwin et al., 2017).

Much like area 5, several regions within monkey area 7 are located in the IPS, including the anterior intraparietal area (AIP) and LIP (Lewis and Van Essen, 2000b). Of particular interest to prehension, neurons within AIP are active during grasping, with different neurons tuned to certain objects that require a particular grasp type. A subpopulation of AIP neurons tuned to a certain grasp type will respond during passive viewing of an object that requires use of that same grasp type (Taira et al., 1990; Sakata et al., 1995; Murata et al., 2000). The visual receptive fields of many AIP neurons have been found to be nonuniform with multiple maxima and minima in both visual hemifields and the highest firing rate usually found foveally or parafoveally in a retinotopic reference frame (Romero and Janssen, 2016). These receptive fields can vary dramatically in size and are significantly modulated by stimulus type, with preferred and nonpreferred stimuli evoking differently shaped receptive fields. Although visual stimulation modulates the response of most AIP neurons, many respond equally well to memory-guided object manipulations made in darkness (Murata et al., 1996). Recently, it was found that this subpopulation of AIP neurons was also active when the monkey passively viewed a video of the same grasping action, even when the object itself was occluded (Pani et al., 2014). Other groups have also found mirrorlike neurons in AIP, though not as frequently as in the subdivisions of area 7b (Maeda et al., 2015). Neurons in AIP also appear capable of representing grasp types independent of object features, a property that would allow for the manipulation of the same object in different ways (Baumann et al., 2009). While it is still an open question whether AIP neurons primarily represent grasps or objects that require a grasp, it is clear that at some point the visual properties of an object must be extracted to determine what type of grasp is most appropriate. How AIP neurons accomplish this is an active area of research.

While many AIP neurons respond robustly to two-dimensional visual representations of small objects (Durand et al., 2007; Romero et al., 2012), a large portion also appear capable of extracting three-dimensional features of objects, although in a coarser and faster manner than neurons located in inferotemporal (IT) cortex (Srivastava et al., 2009). For many AIP neurons, this responsiveness to 3D objects seems to arise from a selectivity for particular binocular disparities (Verhoef et al., 2010; Romero et al., 2012, 2013). When examined in the context of decision-making (ie, discriminating a concave vs. convex surface), AIP neuronal activity correlates with a perceptual choice more slowly than IT neurons unless the monkey is required to make the decision quickly, in

which case this correlation arises much earlier (Verhoef et al., 2010, 2015). This suggests that the sensitivity to curvature and depth found in these neurons serves to make rapid perceptual decisions based on an object's properties rather than categorizing that object. Presumably, these neurons serve to match the 3D features of an object to specific grasp postures, depending on the object's affordances. Indeed, when examined during actual acts of prehension, most AIP neurons that show disparity-driven 3D object selectivity are active during object grasping (Theys et al., 2013). It should be noted, however, that while many AIP neurons exhibit this disparity-driven selectivity, many of those same neurons retain their object selectivity when 3D stimuli are reduced to two-dimensional silhouettes and outlines (Romero et al., 2012, 2013) and even retain their selectivity when image outlines are broken into elementary fragments (Romero et al., 2014). These responses to elementary spatial features combined with the fact that object selectivity is highly dependent on the stimulus' position in space (Romero et al., 2012, 2014) further highlights the differences in how neurons in AIP and IT represent the shape of objects.

Studies comparing the activity of monkey AIP neurons to other parietal and motor neurons during reaching and grasping behavior have found that AIP neurons respond maximally during the early stages of prehension, increasing their firing prior to contacting an object (Gardner et al., 2007a), consistent with AIP's role in preshaping the hand prior to a grasp (Gallese et al., 1994; Debowy et al., 2001). More recent studies have found that at a population level, object orientation (Townsend et al., 2011), grasp types, and movement kinematics can be reliably decoded from groups of AIP neurons with the most reliable decoding occurring during the movement planning epoch (Lehmann and Scherberger, 2013; Menz et al., 2015; Schaffelhofer et al., 2015). When considered at the level of local field potentials (LFPs), however, grasp type can be decoded across all task epochs (not just during movement planning) in all frequency bands analyzed (Lehmann and Scherberger, 2015). Interestingly, whether decoded from multiunits or LFPs, both gaze direction and target position can be extracted from the activity of AIP neurons. Like the AIP receptive fields themselves, these position signals seem to be coded in retinotopic coordinates. What role these signals play in the planning of reaching and grasping movements has yet to be elucidated. However, given that chemical inactivation of AIP produces deficits in grasping but not reaching (Gallese et al., 1994; Fig. 26.11), this spatial information may serve to provide further context for the selection of appropriate grasp types.

Though not directly related to the act of grasping, in one view LIP neurons are thought to create priority

Normal animal   **Muscimol in area AIP**

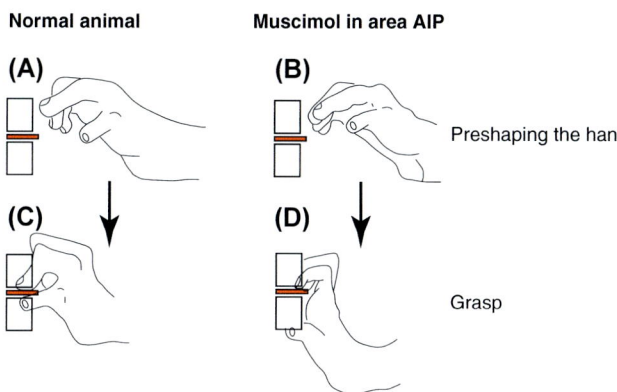

Preshaping the hand

Grasp

FIGURE 26.11 Drawings of video frames of an animal preshaping the hand prior to the use of a precision grip under normal conditions (A) and following an injection of Muscimol into anterior intraparietal area (AIP) of macaque monkeys (B). A precision grip used to grasp a small target (*red*) between two plates in a normal macaque monkey (C) and in a monkey following injections of Muscimol into AIP (D). Note that not only is the actual grasp abnormal, but also the animal is unable to match hand grasping posture with the target prior to the grasp. *Redrawn from Gallese, V., Murata, A., Kaseda, M., Niki, N., Sakata, H., 1994. Deficit of hand preshaping after muscimol injection in monkey parietal cortex. NeuroReport 5, 1525—1529.*

maps of space, directing visual attention and therefore saccades to important objects and locations that the animal can subsequently evaluate and/or act upon (see Bisley and Goldberg, 2003; Bisley and Goldberg, 2010 for review). However, Andersen and colleagues have emphasized the intentional component of neuronal responses in LIP and the PPC in general (see Snyder et al., 2000; Andersen and Cui, 2009 for review). For instance, Quian Quiroga et al. (2006) have shown that in a reach versus saccade task, trial by trial decoding predicts effector choice (ie, reach vs. saccade) far better than target location. However, given that intention and attention are inextricably linked during normal behavior, the distinction is only useful insofar as it demonstrates that elements of each can be found depending on what one is looking for.

Another well-studied area within the IPS is the monkey VIP, which straddles the area 5/7. Originally defined based on projections from the medial temporal area (Maunsell and van Essen, 1983; Colby et al., 1993), investigators distinguished this region from surrounding cortex based both on myelination patterns and the response properties of neurons within these borders. Neurons within this region were noted for having large visual receptive fields sensitive to moving stimuli, with a large proportion displaying directional selectivity. Some of these neurons were highly selective for the distance at which the visual stimuli were presented, especially for locations very close to the monkey's head. That same study found that many neurons fired in response to somatosensory stimuli with many responding to bimodal visual and somatosensory stimulation

with an overlapping receptive field. Interestingly, some bimodal neurons tuned to objects moving toward the monkey had receptive fields that corresponded to a point of impact on the monkey's face rather than some sort of retinal vector. This study and later investigations examining eye movement during microstimulation of VIP suggested that VIP neurons code visual stimuli in a head-centered reference frame for the monitoring of objects in immediate peripersonal space (Thier and Andersen, 1996, 1998), a perspective supported by data showing that motion detection in these neurons is modulated by attention (Cook and Maunsell, 2002).

Since then numerous studies have been conducted to try and elucidate the function of this region, especially with regards to the vestibular and multimodal stimuli. The use of three-dimensional motion platforms has greatly expanded the types of simulated movements that researchers can use to stimulate the vestibular system beyond isolated translations and rotations (Chen et al., 2011a,b, 2013a,c). These studies have demonstrated that VIP occupies an intermediate level in the processing hierarchy for vestibular stimuli, with neurons exhibiting peak directional tuning more quickly than the dorsal aspect of the medial superior temporal area (MSTd) but more slowly than parietoinsular vestibular cortex (Chen et al., 2011a). The tuning curves of VIP neurons are invariant to both head and eye movements, indicating a body (or world)-centered reference frame for vestibular stimuli (Chen et al., 2013c). Neurons tuned to different types of vestibular motion (ie, rotation and translation) can be found throughout VIP in addition to neurons that are optimally tuned to a combination of these movements (ie, curvilinear motion; Chen et al., 2016).

Many monkey VIP neurons are tuned to optic flow stimuli, have a preference for expanding flow patterns, and process optic flow stimuli in discrete clusters tuned to similar headings (Bremmer et al., 2002a; Zhang et al., 2004; Chen et al., 2011b). The heading tuning curves of most VIP neurons have been shown to remain largely unaltered during eye movements (Zhang et al., 2004; Kaminiarz et al., 2014) but generally shift with eye position, indicating an eye-centered reference frame for optic flow stimuli (Chen et al., 2013b). While previous studies using moving bar stimuli have reported head-centered reference frames for VIP neurons (Duhamel et al., 1997; Avillac et al., 2005), those using large-field stimuli generally support an eye-centered reference frame (Chen et al., 2013b, 2014). This highlights the fact that the coordinate system a PPC neuron employs can be dependent on both the sensory modality (ie, vestibular vs. visual) as well as the quality of the stimulus itself (small shapes vs. full-field stimuli). The invariance of heading tuning curves in the face of eye movements suggests that reafferent signals of eye movement may play a role in compensating for this distortion of the visual field (Kaminiarz et al., 2014). Indeed, neurons have been found within VIP that carry a fast and accurate eye position signal that when modeled at the population level can mimic the spatiotemporal dynamics of the eye during saccades (Morris et al., 2012, 2013, 2016). However, recent studies have found that VIP neurons can achieve invariant heading tuning using purely visual cues, such as motion parallax and global perspective cues, and have identified a population of neurons capable of jointly representing visual translation and rotation (Sunkara et al., 2015, 2016). Neurons that respond to combined visual-vestibular information have been well documented in VIP (Bremmer et al., 2002b; Chen et al., 2011b, 2013a; Yang et al., 2011), lending weight to the idea that VIP may represent a site of multimodal convergence for the purpose of spatial navigation (Bremmer et al., 2002a,b). Cells that respond to the same direction of visual and vestibular stimuli (ie, congruent stimuli) have been found to have lower heading discrimination thresholds compared to those neurons that have incongruent visual and vestibular tuning (Chen et al., 2013a). While it has been shown that the response properties of incongruent cells are not an artifact of head-fixation and viewing geometry (Yang et al., 2011), their exact function remains unclear. Despite the fact that these neurons are relatively rare within populations tuned for behaviorally relevant headings (Chen et al., 2013a), the presence of incongruent visual-vestibular direction preferences in this region suggests that it may also function to distinguish self-motion from object motion (Schlack et al., 2002). This possibility is bolstered by the fact that bilateral deactivation of VIP seems to have no effect on heading discrimination ability (Chen et al., 2016), although this may be an artifact of fixation as previous work by Zhang and Britten (2011) demonstrated that short-train microstimulation of VIP neuronal clusters can bias heading judgments during smooth-pursuit eye movements.

Interestingly, studies employing high-amplitude long-train microstimulation have provided a slightly different perspective. Cooke et al. (2003) found that stimulation of different sites in VIP evoked a relatively small set of movements seemingly related to defensive behavior. These movements included blinking, folding the pinnae against the head, contraction of facial muscles, shrugging of the shoulders, and a lateral movement of the arm which resembled a blocking posture. These movements often occurred in combination and were nearly identical to a set of movements evoked by puffing air at the monkey's face (Cooke et al., 2003). This has led to the idea that this region may serve, at least in part, to maintain a margin of safety around the animal's head and face in response to violation of peripersonal space or during natural navigation behaviors as the animal

passes by obstacles in its environment (Graziano and Cooke, 2006). Likely VIP serves multiple functions related to the monitoring of peripersonal space and relating the movement of the head, eyes, and arms to the movement of nearby objects in terms of a range of reference frames (Duhamel et al., 1997, 1998; Avillac et al., 2005; Chen et al., 2013b,c) and sensory modalities (Duhamel et al., 1998; Schlack et al., 2002). Indeed, whether measured using optic flow (Yang et al., 2011) or two-dimensional motion stimuli (Bremmer et al., 2013), the majority of VIP neurons prefer binocular disparities which correspond to near stimuli. Interestingly, this monitoring of peripersonal space may extend to the monitoring of other individuals as mirror neurons have recently been found in this region (Ishida et al., 2010).

Much like AIP, VIP is reciprocally connected with both subdivisions of 7b, another region in which mirror neurons can be found (Rozzi et al., 2006). In addition to defense and spatial navigation, it may be the case that neurons in VIP are tuned to other instances of self versus object motion such as hand—mouth coordination and feeding behavior. However, motion perception may not be the only function subserved by the region. Neurons have been found in VIP that appear to be tuned to different elements of numerosity, a quality also observed in different portions of area 5 (see Nieder, 2013 for review) and a subject which will be discussed in further detail in Canlon chapter.

26.4.3 Somatosensory Input to the Posterior Parietal Cortex in Primates

The preponderance of PPC neurons sensitive to tactile and proprioceptive stimulation likely reflects their role in monitoring the position of the limbs during reaching as well as the haptics (ie, active sensory exploration) of object manipulation. While neurons within these regions receive their own input from somatosensory regions of the thalamus (eg, Padberg et al., 2009; Schmahmann and Pandya, 1990), the majority of the somatosensory input to these regions ascends from the thalamus and through areas within APC before converging on these higher-order neurons (Burton et al., 1995; Burton and Fabri, 1995; Cusick et al., 1985; Darian-Smith et al., 1993; Gharbawie et al., 2010,b; Lewis and Van Essen, 2000a; Pons and Kaas, 1986; Rozzi et al., 2006). Classical models of somatosensory processing liken these pathways to the visual system, where information is processed in a serial, hierarchical manner (see Iwamura, 1998; Dijkerman and de Haan, 2007 for review). Support for this model comes not just from the increasing size and complexity of somatosensory receptive fields as one proceeds from rostral to caudal in the parietal lobe (Hyvarinen and Poranen, 1978a,b; Costanzo and Gardner, 1980; Iwamura et al., 1985a,b, 1980, 1993, 1994;

Nelson et al., 1980; Sur, 1980; Iwamura, 1983a,b; Gardner, 1988; Pei et al., 2010; Papadelis et al., 2011; Wacker et al., 2011; Sanchez-Panchuelo et al., 2012; Yau et al., 2013; Ashaber et al., 2014) but also from the fact that ablation of "lower-order" regions in this hierarchy can abolish neural responses in "higher-order" areas (Garraghty et al., 1990). Similarly, the degree of qualitative behavioral deficits observed after lesioning APC are not exacerbated by later lesioning PPC (Peele, 1944), suggesting that input from APC is critical for PPC function as it pertains to movement and prehension. Indeed, even within APC it seems as though areas become more "PPC-like" the further caudal one goes, with receptive fields not only becoming larger, but also being more strongly modulated by attention (Hyvarinen et al., 1980; Hsiao et al., 1993; Burton et al., 1997, 1999; Burton and Sinclair, 2000; Meftah el et al., 2002; Spingath et al., 2011, 2013; Wang et al., 2012).

Despite this apparent hierarchy, the connections between APC and PPC are reciprocal, with portions of both areas 5 and 7 sending feedback to APC either directly through corticocortical connections (Pons and Kaas, 1986; Cavada and Goldman-Rakic, 1989a,b; Felleman and Van Essen, 1991; Burton and Fabri, 1995; Rozzi et al., 2006) or indirectly through the thalamus (Weber and Yin, 1984; Yeterian and Pandya, 1985; Padberg et al., 2009). While the exact function of these feedback connections is still being investigated, our laboratory has demonstrated that the shape and size of receptive fields of anterior parietal neurons can be altered by reversibly deactivating portions of area 5L and 7b (Fig. 26.12; Goldring et al., 2014; Cooke et al., 2014). Thus neurons within the PPC have the potential to gate and refine incoming somatosensory information from APC neurons, possibly via selective attention in a manner analogous to the visual system.

26.5 Posterior Parietal Cortex and Tool Use

One of the most complex and cognitively advanced examples of manual dexterity is the temporary extension and specialization of peripheral morphology via the use of tools. Tool use as defined by Shumaker et al. (2011) is "the external employment of an unattached or manipulable attached environmental object to alter more efficiently the form, position, or condition of another object, another organism, or the user itself, when the user holds and directly manipulates the tool during or prior to use and is responsible for the proper and effective orientation of the tool." Although this category of manual ability was originally thought to be an exclusive and defining feature of human behavior, researchers now appreciate that New World monkeys, Old World monkeys, and great apes also use tools in

Receptive field changes following deactivation

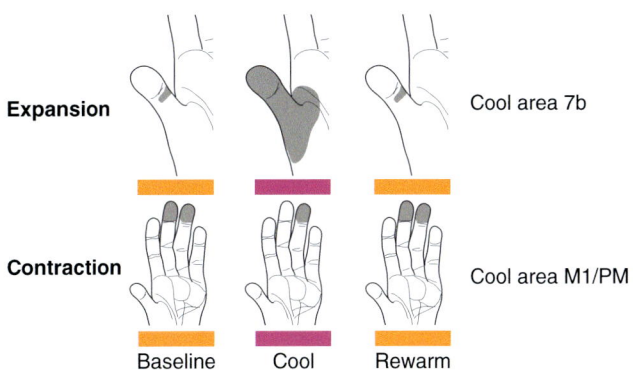

FIGURE 26.12   Different types of receptive field changes following deactivation of areas in posterior parietal cortex (area 7b) and motor cortex (M1 and PM) in macaque monkeys. Reversible deactivation was accomplished via cooling, and neurons were recorded in areas 1 and 2 during cooling. Both expansions and contractions of receptive fields for neurons in areas 1 and 2 were seen following deactivation of area 7b (and M1/PM), but the most common type of alterations to receptive fields when area 7b was cooled were expansions. *Adapted from Cooke, D.F., Goldring, A.B., Baldwin, M.K., Recanzone, G.H., Chen, A., Pan, T., Simon, S.I., Krubitzer, L., 2014. Reversible deactivation of higher-order posterior parietal areas. I. Alterations of receptive field characteristics in early stages of neocortical processing. J. Neurophysiol. 112, 2529–2544.*

the wild to varying degrees, a subject explored in Dorothy Fragazy chapter 3.17, Tool Use in Nonhuman Primates: Natural History, Ontogenetic Development and Social Supports for Learning. Given the multisensory integration and contextual deployment that tool use requires, it is perhaps not surprising that these behaviors are likely correlated with an expansion and specialization of posterior parietal cortical areas. As discussed in Almecija/Sherwood chapter 3.16, Hands, Brains, and Precision Grips: Origins of Tool Use Behaviors, humans have evolved specialized portions of the PPC not seen in macaques that have been implicated in complex tool use that goes beyond simple object grasping and manipulation. Namely, activity in the anterior portion of Brodmann area 40 on the supramarginal gyrus of the left hemisphere is correlated with the execution and observation of tool use behavior, while damage to this general region is associated with apraxias characterized by deficits in tool use ability (see Goldenberg and Spatt, 2009; Orban and Caruana, 2014 for review).

Of the species amendable to direct electrophysiological recording and experimental manipulation of function, cebus monkeys (*Sapajus* sp.) represent ideal model organisms for studying the neurological basis of tool use behavior. Unlike rhesus macaques, which only use tools under experimental conditions (eg, Iriki et al., 1996; Quallo et al., 2012; Umilta et al., 2008), these monkeys use tools in the wild and appear capable of extracting the affordances of different tools based on their physical properties and select the appropriate tool for the appropriate action or action sequence (see Fragazy chapter 3.17, Tool Use in Nonhuman Primates: Natural History, Ontogenetic Development and Social Supports for Learning). Like humans, they possess an opposable thumb capable of executing a thumb and forefinger precision grip that allows for the kind of fine object manipulation that just is not possible using a power grasp. Interestingly, they seemed to have independently evolved a proprioceptive cortical area 2 and an architectonically distinct area 5 that contains neurons which exhibit a similar response profile to those of macaque monkeys when measured in an anesthetized preparation (Padberg et al., 2007; Seelke et al., 2012). Recently, Mayer et al. (2016) conducted a detailed architectonic survey of the cebus monkey parietal lobe using a combination of stains (SMI-32, nissl, and myelin) that have been used successfully to anatomically parcellate PPC in rhesus macaques. Using these techniques, they identified a suite of anterior and posterior parietal subdivisions of the postcentral gyrus and anterior IPS that seem to mirror the architectonic schema of macaques. This is striking because all other New World monkeys tested lack most if not all of these divisions and therefore they appear to have arisen independently of their analogues in Old World monkeys. Future studies paralleling those in macaques (ie, anatomical connections, LT-ICMS, and awake-behaving electrophysiology) will be necessary to determine which aspects of PPC organization and function have independently arisen in these species. In addition, these types of studies will allow researchers to determine if nodes specialized for tool use exist in the cebus frontoparietal reaching and grasping network. These data would not only reveal potential mechanisms for how human tool use regions evolved but also shed light on the basic organizational principles of how the cortex can change to accommodate this kind of complex sensorimotor behavior.

26.6 Posterior Parietal Cortex in Humans

Some of the earliest knowledge regarding the function of the PPC arose from case studies of humans who sustained damage to different parts of the parietal lobe (see Fleming and Crosby, 1955 for review). Advances in functional imaging and the advent of transcranial magnetic stimulation have allowed researchers to characterize the organization of the human frontoparietal reaching and grasping network. Establishing homologies between human and monkey PPC areas, however, must be undertaken with caution as a number of factors differ in the methodologies of each type of investigation including: temporal and spatial resolution of neuroimaging versus neurophysiology, the dynamics of excitation versus inhibition in the BOLD (blood oxygenation level dependent) response underlying fMRI, and the criteria used to define what constitutes an area (Culham et al., 2006). Even with the aid of architectonics, individual differences in the location of cortical regions complicate matters when mapping onto a normalized brain (eg, Caspers et al., 2006; Krubitzer et al., 2004). That is, if an fMRI signal is mapped onto a standard "reference" brain, this does not take into account that the exact location and extent of a brain area in one subject may not match that of another subject in the study. Nevertheless some general patterns have emerged that allow one to draw parallels between functionally similar areas in monkeys and humans.

One method of parcellating the human parietal lobe is by examining topographic maps (ie, representations of adjacent regions of sensory space in adjacent parts of cortex) during the presentation of different visual stimuli or the performance of saccades to different regions of space. Borders can then be drawn based on the locations of reversals in the orientation of the visual field (Silver and Kastner, 2009). Using this methodology, at least seven different regions can be identified along the human IPS and superior parietal lobule in which grasping-related activation transitions to reaching-related as one proceeds anterolaterally to posteromedially through these regions (Konen et al., 2013). Similar trends can be seen when one relates the loci of activation for a given task to the location of sulci and gyri located in approximately the same areas in humans and macaques: as one proceeds from the anterior end of the IPS in a posteromedial direction, activation foci respond to grasps, then reaches, and then saccades (eg, Hinkley et al., 2009). The anterior end of the human IPS is often equated with macaque AIP, and like monkey AIP has been shown to be active during both passive viewing and active nonvisual manipulation of different objects (eg, Grefkes et al., 2002). A human area dubbed the "posterior eye field" (see Grefkes and Fink, 2005) has been shown to exhibit similar saccade-related activity to monkey LIP when comparing both species in an identical fMRI saccade paradigm (Koyama et al., 2004). Interestingly, in humans this posterior eye field is actually located medial to the IPS, on the superior parietal lobule, rather than within it. This highlights the fact that homologous areas may be located in different regions in different species. A region near the medial portion of the IPS in humans, roughly analogous in location to monkey MIP and PRR, has been shown to display reaching-related BOLD responses (Prado et al., 2005) and is purported to perform coordinate transformations (Grefkes et al., 2004) when employing a joystick paradigm similar to that employed by Eskandar and Assad (2002) to study such phenomena in macaque MIP.

Similarly, a region in the ventral portion of the human IPS has been shown to respond to combined visual, tactile, and auditory stimuli in a manner analogous to monkey VIP (Bremmer et al., 2001). In addition, like language, there are myriad behavioral and cognitive functions that have been specialized and lateralized to one hemisphere or the other in human PPC in ways not seen in other primates. However, a full discussion of this is beyond the scope of this chapter (see Rosati chapter 3.23, The Evolution of Primate Executive Function: From Response Control to Strategic Decision-Making).

26.7 Conclusion

When considering the organization of the neocortex, establishing homologies across species can be difficult. This is particularly true for PPC. One reason for this difficulty is that we currently do not have a species-agnostic definition of PPC nor do we use a robust set of criteria for identifying PPC across primate and nonprimate mammals. In addition, species differ in body morphology, sensory effector specialization, and ethologically relevant behavior, and these differences are reflected in sensory cortex, such as somatosensory cortex, and amplified in PPC. However, recent studies utilizing LT-ICMS combined with studies of connections from motor cortex and architectonic analysis have allowed us to begin to establish PPC homologues between primates. These techniques have also allowed us to examine cortex in the relative location of PPC in nonprimate mammals and make comparisons of this region with primates. However, data of this sort have only been gathered from a few species so that establishing homologies across groups is possible, but more data are needed to determine if this cortex is in any way analogous to PPC regions in primates. Further, while these techniques have proved useful for establishing a basic framework for PPC organization in all mammals, they do not speak to how PPC generates behavior.

While rats have proven to be a tractable nonprimate model in the context of awake-behaving electrophysiology, contemporary studies of rat "PPC" indicate that it is involved in navigation, spatial attention, and decision-making: all complex behaviors that make parsing the specific functional organization of the presumptive rodent PPC difficult, especially since a species-agnostic definition of PPC has largely been overlooked.

By far most work on PPC has been done in nonhuman primates, particularly the macaque monkey. The macaque PPC is comprised of subdivisions whose function and connectivity distinguish them from adjacent visual and somatosensory cortical areas. While those adjacent "early" sensory processing areas are characterized by features such as a complete topographic representation of the sensory epithelium and mirror reversals at areal borders (eg, Kaas et al., 1979), regions within the PPC contain fractured and sparse representations of somatosensory (eg, area 5L; Seelke et al., 2012) and visual (eg, LIP; Gottlieb et al., 1998) space, as well as magnified representations of ethologically relevant effectors (Seelke et al., 2012). PPC is the ultimate destination of what Mishkin and Ungerleider (1982) described as the "where" pathway, noting that the destruction of this dorsal stream of visual processing severely impaired a monkey's ability to localize objects in space (as opposed to the ventral, "what" stream responsible for object identification). Later work by Goodale and Milner (1992) modified this dichotomy by suggesting that the "what" and "where" streams would be better thought of as the ventral "perception" and the dorsal "action" streams. Indeed, since then the view of the dorsal stream as an "action" pathway has dominated theories regarding the function of PPC, which is now recognized as the cortical hub of attention, intention, and reference-frame transformation. Recently, this schema has been further elaborated into an "action affordances" model (Cisek, 2007), which seeks to supplant the information processing perspective that segregates perception, cognition, and action into separate domains that operate in a serial manner to analyze sensory data and make a single decision. Instead, as reviewed by Cisek and Kalaska (2010), the experimental data argue for a more distributed set of processes which operate *throughout* sensorimotor cortex and an organizational schema that is optimized for a continuous and dynamic interaction with the environment. From this perspective, a number of different actions compete for execution as sensory data accumulate, and are biased by contextual signals from prefrontal cortices and the basal ganglia.

Finally there is a recent plethora of studies of PPC in humans. While some homologies have been established between humans and monkeys (eg, AIP and VIP), the overall number of cortical fields, their connectivity, and their integrated function are not well understood. What is clear is that this region of the neocortex has been greatly enlarged in primates, especially humans, and has coevolved with the elaboration of the hand and sophisticated manual behaviors that define the human condition. To answer ultimate questions about how PPC has evolved, parallel studies employing the same methodology (ie, anatomy, LT-ICMS, reversible inactivation) in multiple mammalian species are essential. Similarly, to answer proximate questions about the treatment of PPC damage or the use of PPC in brain–machine interfaces in humans, an understanding of PPC organization in intermediate species will allow researchers to more easily extrapolate results from rodent models in which more precise methods of neuronal manipulation (ie, optogenetics) are readily available. As we enter a new era where our ability to process information and interact with our surroundings is inextricably linked with technology dependent on skilled movements of the hands and eyes, an understanding of the brain networks guiding those actions is imperative to understanding ourselves and how will we adapt to the future that we construct.

References

Andersen, R.A., Cui, H., 2009. Intention, action planning, and decision making in parietal-frontal circuits. Neuron 63, 568–583.

Andujar, J.E., Drew, T., 2007. Organization of the projections from the posterior parietal cortex to the rostral and caudal regions of the motor cortex of the cat. J. Comp. Neurol. 504, 17–41.

Ashaber, M., Palfi, E., Friedman, R.M., Palmer, C., Jakli, B., Chen, L.M., Kantor, O., Roe, A.W., Negyessy, L., 2014. Connectivity of somatosensory cortical area 1 forms an anatomical substrate for the emergence of multifinger receptive fields and complex feature selectivity in the squirrel monkey (*Saimiri sciureus*). J. Comp. Neurol. 522, 1769–1785.

Avillac, M., Deneve, S., Olivier, E., Pouget, A., Duhamel, J.R., 2005. Reference frames for representing visual and tactile locations in parietal cortex. Nat. Neurosci. 8, 941–949.

Babb, R.S., Waters, R.S., Asanuma, H., 1984. Corticocortical connections to the motor cortex from the posterior parietal lobe (areas 5a, 5b, 7) in the cat demonstrated by the retrograde axonal transport of horseradish peroxidase. Exp. Brain Res. 54, 476–484.

Baldwin, M.K., Cooke, D.F., Krubitzer, L., 2017. Intracortical microstimulation maps of motor, somatosensory, and posterior parietal cortex in tree shrews (*Tupaia belangeri*) reveal complex movement representations. Cereb. Cortex 27 (2), 1439–1456.

Baldwin, M.K., Cooke, D.F., Goldring, A., Krubitzer, L., 2016. Intracortical microstimulation maps of motor, somatosensory, and posterior parietal cortex in macaque monkeys. Neurosci. Abstr.

Batista, A.P., Andersen, R.A., 2001. The parietal reach region codes the next planned movement in a sequential reach task. J. Neurophysiol. 85, 539–544.

Baumann, M.A., Fluet, M.C., Scherberger, H., 2009. Context-specific grasp movement representation in the macaque anterior intraparietal area. J. Neurosci. 29, 6436–6448.

Beck, P.D., Pospichal, M.W., Kaas, J.H., 1996. Topography, architecture, and connections of somatosensory cortex in opossums: evidence for five somatosensory areas. J. Comp. Neurol. 366, 109–133.

Bisley, J.W., Goldberg, M.E., 2003. Neuronal activity in the lateral intraparietal area and spatial attention. Science 299, 81–86.

Bisley, J.W., Goldberg, M.E., 2010. Attention, intention, and priority in the parietal lobe. Annu. Rev. Neurosci. 33, 1–21.

Bonini, L., Serventi, F.U., Simone, L., Rozzi, S., Ferrari, P.F., Fogassi, L., 2011. Grasping neurons of monkey parietal and premotor cortices encode action goals at distinct levels of abstraction during complex action sequences. J. Neurosci. 31, 5876–5886.

Boyer, D.M., Yapuncich, G.S., Chester, S.G., Bloch, J.I., Godinot, M., 2013. Hands of early primates. Am. J. Phys. Anthropol. 152 (Suppl. 57), 33–78.

Bremmer, F., Duhamel, J.R., Ben Hamed, S., Graf, W., 2002. Heading encoding in the macaque ventral intraparietal area (VIP). Eur. J. Neurosci. 16, 1554–1568.

Bremmer, F., Klam, F., Duhamel, J.R., Ben Hamed, S., Graf, W., 2002. Visual-vestibular interactive responses in the macaque ventral intraparietal area (VIP). Eur. J. Neurosci. 16, 1569–1586.

Bremmer, F., Schlack, A., Kaminiarz, A., Hoffmann, K.P., 2013. Encoding of movement in near extrapersonal space in primate area VIP. Front. Behav. Neurosci. 7, 8.

Bremmer, F., Schlack, A., Shah, N.J., Zafiris, O., Kubischik, M., Hoffmann, K., Zilles, K., Fink, G.R., 2001. Polymodal motion processing in posterior parietal and premotor cortex: a human fMRI study strongly implies equivalencies between humans and monkeys. Neuron 29, 287–296.

Brodmann, K., 1909. Vergleichende Lokalisationslehre der Grobhirnrinde. JA Barth, Leipzeig (Germany).

Brown, A.R., Teskey, G.C., 2014. Motor cortex is functionally organized as a set of spatially distinct representations for complex movements. J. Neurosci. 34, 13574–13585.

Burman, K.J., Bakola, S., Richardson, K.E., Reser, D.H., Rosa, M.G., 2014. Patterns of afferent input to the caudal and rostral areas of the dorsal premotor cortex (6DC and 6DR) in the marmoset monkey. J. Comp. Neurol. 522, 3683–3716.

Burman, K.J., Bakola, S., Richardson, K.E., Reser, D.H., Rosa, M.G., 2014. Patterns of cortical input to the primary motor area in the marmoset monkey. J. Comp. Neurol. 522, 811–843.

Burton, H., Abend, N.S., Macleod, A.M., Sinclair, R.J., Snyder, A.Z., Raichle, M.E., 1999. Tactile attention tasks enhance activation in somatosensory regions of parietal cortex: a positron emission tomography study. Cereb. Cortex 9, 662–674.

Burton, H., Fabri, M., 1995. Ipsilateral intracortical connections of physiologically defined cutaneous representations in areas 3b and 1 of macaque monkeys: projections in the vicinity of the central sulcus. J. Comp. Neurol. 355, 508–538.

Burton, H., Fabri, M., Alloway, K., 1995. Cortical areas within the lateral sulcus connected to cutaneous representations in areas 3b and 1: a revised interpretation of the second somatosensory area in macaque monkeys. J. Comp. Neurol. 355, 539–562.

Burton, H., Sinclair, R.J., 2000. Attending to and remembering tactile stimuli: a review of brain imaging data and single-neuron responses. J. Clin. Neurophysiol. 17, 575–591.

Burton, H., Sinclair, R.J., Hong, S.Y., Pruett Jr., J.R., Whang, K.C., 1997. Tactile-spatial and cross-modal attention effects in the second somatosensory and 7b cortical areas of rhesus monkeys. Somatosens. Mot. Res. 14, 237–267.

Calton, J.L., Dickinson, A.R., Snyder, L.H., 2002. Non-spatial, motor-specific activation in posterior parietal cortex. Nat. Neurosci. 5, 580–588.

Caspers, S., Geyer, S., Schleicher, A., Mohlberg, H., Amunts, K., Zilles, K., 2006. The human inferior parietal cortex: cytoarchitectonic parcellation and interindividual variability. Neuroimage 33, 430–448.

Catania, K.C., 2011. The sense of touch in the star-nosed mole: from mechanoreceptors to the brain. Philos. Trans. R. Soc. Lond. B Biol. Sci. 366, 3016–3025.

Chapin, J.K., Lin, C.S., 1984. Mapping the body representation in the SI cortex of anesthetized and awake rats. J. Comp. Neurol. 229, 199–213.

Cavada, C., Goldman-Rakic, P.S., 1989. Posterior parietal cortex in rhesus monkey: I. Parcellation of areas based on distinctive limbic and sensory corticocortical connections. J. Comp. Neurol. 287, 393–421.

Cavada, C., Goldman-Rakic, P.S., 1989. Posterior parietal cortex in rhesus monkey: II. Evidence for segregated corticocortical networks linking sensory and limbic areas with the frontal lobe. J. Comp. Neurol. 287, 422–445.

Chandler, H.C., King, V., Corwin, J.V., Reep, R.L., 1992. Thalamocortical connections of rat posterior parietal cortex. Neurosci. Lett. 143, 237–242.

Chen, A., Deangelis, G.C., Angelaki, D.E., 2011. A comparison of vestibular spatiotemporal tuning in macaque parietoinsular vestibular cortex, ventral intraparietal area, and medial superior temporal area. J. Neurosci. 31, 3082–3094.

Chen, A., Deangelis, G.C., Angelaki, D.E., 2011. Representation of vestibular and visual cues to self-motion in ventral intraparietal cortex. J. Neurosci. 31, 12036–12052.

Chen, A., Deangelis, G.C., Angelaki, D.E., 2013. Functional specializations of the ventral intraparietal area for multisensory heading discrimination. J. Neurosci. 33, 3567–3581.

Chen, X., Deangelis, G.C., Angelaki, D.E., 2013. Diverse spatial reference frames of vestibular signals in parietal cortex. Neuron 80, 1310–1321.

Chen, X., Deangelis, G.C., Angelaki, D.E., 2013. Eye-centered representation of optic flow tuning in the ventral intraparietal area. J. Neurosci. 33, 18574–18582.

Chen, A., Gu, Y., Liu, S., Deangelis, G.C., Angelaki, D.E., 2016. Evidence for a causal contribution of macaque vestibular, but not intraparietal, cortex to heading perception. J. Neurosci. 36, 3789–3798.

Chen, J., Reitzen, S.D., Kohlenstein, J.B., Gardner, E.P., 2009. Neural representation of hand kinematics during prehension in posterior parietal cortex of the macaque monkey. J. Neurophysiol. 102, 3310–3328.

Chen, X., Deangelis, G.C., Angelaki, D.E., 2014. Eye-centered visual receptive fields in the ventral intraparietal area. J. Neurophysiol. 112, 353–361.

Christel, M.I., Fragaszy, D., 2000. Manual function in *Cebus apella*. Digital mobility, preshaping, and endurance in repetitive grasping. Int. J. Primatol. 21, 697–719.

Cisek, P., 2007. Cortical mechanisms of action selection: the affordance competition hypothesis. Philos. Trans. R. Soc. Lond. B Biol. Sci. 362, 1585–1599.

Cisek, P., Kalaska, J.F., 2010. Neural mechanisms for interacting with a world full of action choices. Annu. Rev. Neurosci. 33, 269–298.

Colby, C.L., Duhamel, J.R., 1991. Heterogeneity of extrastriate visual areas and multiple parietal areas in the macaque monkey. Neuropsychologia 29, 517–537.

Colby, C.L., Duhamel, J.R., Goldberg, M.E., 1993. Ventral intraparietal area of the macaque: anatomic location and visual response properties. J. Neurophysiol. 69, 902–914.

Cook, E.P., Maunsell, J.H., 2002. Attentional modulation of behavioral performance and neuronal responses in middle temporal and ventral intraparietal areas of macaque monkey. J. Neurosci. 22, 1994–2004.

Cooke, D.F., Baldwin, M.K.L., Donaldson, M., Helton, J., Stolzenberg, D., Krubitzer, L., 2016. Rats gone wild: how seminatural rearing of laboratory animals shapes behavioral development and alters somatosensory and motor cortex organization. Neurosci. Abstr.

Cooke, D.F., Goldring, A.B., Baldwin, M.K., Recanzone, G.H., Chen, A., Pan, T., Simon, S.I., Krubitzer, L., 2014. Reversible deactivation of higher-order posterior parietal areas. I. Alterations of receptive field characteristics in early stages of neocortical processing. J. Neurophysiol. 112, 2529–2544.

Cooke, D.F., Graziano, M.S., 2004. Sensorimotor integration in the precentral gyrus: polysensory neurons and defensive movements. J. Neurophysiol. 91, 1648–1660.

Cooke, D.F., Padberg, J., Zahner, T., Krubitzer, L., 2012. The functional organization and cortical connections of motor cortex in squirrels. Cereb. Cortex 22, 1959–1978.

Cooke Df, P.J., Cerkevich, C., Kaas, J., Krubitzer, L., 2013. Corticocortical connections of area 5 in macaque monkeys support the existence of functionally distinct medial and lateral regions. Neurosci. Abstr. 551, 09.

Cooke, D.F., Stepniewska, I., Miller, D.J., Kaas, J.H., Krubitzer, L., 2015. Reversible deactivation of motor cortex reveals functional connectivity with posterior parietal cortex in the prosimian Galago (*Otolemur garnettii*). J. Neurosci. 35, 14406–14422.

Cooke, D.F., Taylor, C.S., Moore, T., Graziano, M.S., 2003. Complex movements evoked by microstimulation of the ventral intraparietal area. Proc. Natl. Acad. Sci. U.S.A. 100, 6163–6168.

Costanzo, R.M., Gardner, E.P., 1980. A quantitative analysis of responses of direction-sensitive neurons in somatosensory cortex of awake monkeys. J. Neurophysiol. 43, 1319–1341.

Crewther, D.P., Crewther, S.G., Sanderson, K.J., 1984. Primary visual cortex in the brushtailed possum: receptive field properties and corticocortical connections. Brain Behav. Evol. 24, 184–197.

Culham, J.C., Cavina-Pratesi, C., Singhal, A., 2006. The role of parietal cortex in visuomotor control: what have we learned from neuroimaging? Neuropsychologia 44, 2668–2684.

Cusick, C.G., Steindler, D.A., Kaas, J.H., 1985. Corticocortical and collateral thalamocortical connections of postcentral somatosensory cortical areas in squirrel monkeys: a double-labeling study with radiolabeled wheatgerm agglutinin and wheatgerm agglutinin conjugated to horseradish peroxidase. Somatosens. Res. 3, 1–31.

Darian-Smith, C., Darian-Smith, I., Burman, K., Ratcliffe, N., 1993. Ipsilateral cortical projections to areas 3a, 3b, and 4 in the macaque monkey. J. Comp. Neurol. 335, 200–213.

Deacon, T.W., 1992. Cortical connections of the inferior arcuate sulcus cortex in the macaque brain. Brain Res. 573, 8–26.

Debowy, D.J., Ghosh, S., Ro, J.Y., Gardner, E.P., 2001. Comparison of neuronal firing rates in somatosensory and posterior parietal cortex during prehension. Exp. Brain Res. 137, 269–291.

Di, S., Brett, B., Barth, D.S., 1994. Polysensory evoked potentials in rat parietotemporal cortex: combined auditory and somatosensory responses. Brain Res. 642, 267–280.

Dijkerman, H.C., De Haan, E.H., 2007. Somatosensory processes subserving perception and action. Behav. Brain Sci. 30, 189–201 discussion 201–39.

Dooley, J.C., Franca, J.G., Seelke, A.M., Cooke, D.F., Krubitzer, L.A., 2013. A connection to the past: *Monodelphis domestica* provides insight into the organization and connectivity of the brains of early mammals. J. Comp. Neurol. 521, 3877–3897.

Dooley, J.C., Franca, J.G., Seelke, A.M., Cooke, D.F., Krubitzer, L.A., 2014. Evolution of mammalian sensorimotor cortex: thalamic projections to parietal cortical areas in *Monodelphis domestica*. Front. Neuroanat. 8, 163.

Duffy, F.H., Burchfiel, J.L., 1971. Somatosensory system: organizational hierarchy from single units in monkey area 5. Science 172, 273–275.

Duhamel, J.R., Bremmer, F., Ben Hamed, S., Graf, W., 1997. Spatial invariance of visual receptive fields in parietal cortex neurons. Nature 389, 845–848.

Duhamel, J.R., Colby, C.L., Goldberg, M.E., 1998. Ventral intraparietal area of the macaque: congruent visual and somatic response properties. J. Neurophysiol. 79, 126–136.

Durand, J.B., Nelissen, K., Joly, O., Wardak, C., Todd, J.T., Norman, J.F., Janssen, P., Vanduffel, W., Orban, G.A., 2007. Anterior regions of monkey parietal cortex process visual 3d shape. Neuron 55, 493–505.

Ebner Ff, K.J., 2015. Somatosensory system. In: Paxinos, G. (Ed.), The Rat Nervous System. Elsevier Sciences, Academic Press, Oxford.

Elston, G.N., Manger, P.R., 1999. The organization and connections of somatosensory cortex in the brush-tailed possum (*Trichosurus vulpecula*): evidence for multiple, topographically organized and interconnected representations in an Australian marsupial. Somatosens. Mot. Res. 16, 312–337.

Eskandar, E.N., Assad, J.A., 2002. Distinct nature of directional signals among parietal cortical areas during visual guidance. J. Neurophysiol. 88, 1777–1790.

Fabri, M., Burton, H., 1991. Ipsilateral cortical connections of primary somatic sensory cortex in rats. J. Comp. Neurol. 311, 405–424.

Fang, P.C., Stepniewska, I., Kaas, J.H., 2005. Ipsilateral cortical connections of motor, premotor, frontal eye, and posterior parietal fields in a prosimian primate, *Otolemur garnetti*. J. Comp. Neurol. 490, 305–333.

Felleman, D.J., Van Essen, D.C., 1991. Distributed hierarchical processing in the primate cerebral cortex. Cereb. Cortex 1, 1–47.

Ferraina, S., Bianchi, L., 1994. Posterior parietal cortex: functional properties of neurons in area 5 during an instructed-delay reaching task within different parts of space. Exp. Brain Res. 99, 175—178.

Fleming, J.F., Crosby, E.C., 1955. The parietal lobe as an additional motor area; the motor effects of electrical stimulation and ablation of cortical areas 5 and 7 in monkeys. J. Comp. Neurol. 103, 485—512.

Fogassi, L., Ferrari, P.F., Gesierich, B., Rozzi, S., Chersi, F., Rizzolatti, G., 2005. Parietal lobe: from action organization to intention understanding. Science 308, 662—667.

Foxworthy, W.A., Allman, B.L., Keniston, L.P., Meredith, M.A., 2013. Multisensory and unisensory neurons in ferret parietal cortex exhibit distinct functional properties. Eur. J. Neurosci. 37, 910—923.

Foxworthy, W.A., Clemo, H.R., Meredith, M.A., 2013. Laminar and connectional organization of a multisensory cortex. J. Comp. Neurol. 521, 1867—1890.

Foxworthy, W.A., Meredith, M.A., 2011. An examination of somatosensory area SIII in ferret cortex. Somatosens. Mot. Res. 28, 1—10.

Fragaszy, D., Izar, P., Visalberghi, E., Ottoni, E.B., De Oliveira, M.G., 2004. Wild capuchin monkeys (Cebus libidinosus) use anvils and stone pounding tools. Am. J. Primatol. 64, 359—366.

Gallese, V., Murata, A., Kaseda, M., Niki, N., Sakata, H., 1994. Deficit of hand preshaping after muscimol injection in monkey parietal cortex. Neuroreport 5, 1525—1529.

Gardner, E.P., 1988. Somatosensory cortical mechanisms of feature detection in tactile and kinesthetic discrimination. Can. J. Physiol. Pharmacol. 66, 439—454.

Gardner, E.P., Babu, K.S., Ghosh, S., Sherwood, A., Chen, J., 2007. Neurophysiology of prehension. III. Representation of object features in posterior parietal cortex of the macaque monkey. J. Neurophysiol. 98, 3708—3730.

Gardner, E.P., Babu, K.S., Reitzen, S.D., Ghosh, S., Brown, A.S., Chen, J., Hall, A.L., Herzlinger, M.D., Kohlenstein, J.B., Ro, J.Y., 2007. Neurophysiology of prehension. I. Posterior parietal cortex and object-oriented hand behaviors. J. Neurophysiol. 97, 387—406.

Garraghty, P.E., Florence, S.L., Kaas, J.H., 1990. Ablations of areas 3a and 3b of monkey somatosensory cortex abolish cutaneous responsivity in area 1. Brain Res. 528, 165—169.

Garraghty, P.E., Pons, T.P., Huerta, M.F., Kaas, J.H., 1987. Somatotopic organization of the third somatosensory area (SIII) in cats. Somatosens. Res. 4, 333—357.

Gharbawie, O.A., Stepniewska, I., Burish, M.J., Kaas, J.H., 2010. Thalamocortical connections of functional zones in posterior parietal cortex and frontal cortex motor regions in New World monkeys. Cereb. Cortex 20, 2391—2410.

Gharbawie, O.A., Stepniewska, I., Kaas, J.H., 2011. Cortical connections of functional zones in posterior parietal cortex and frontal cortex motor regions in New World monkeys. Cereb. Cortex 21, 1981—2002.

Gharbawie, O.A., Stepniewska, I., Qi, H., Kaas, J.H., 2011. Multiple parietal-frontal pathways mediate grasping in macaque monkeys. J. Neurosci. 31, 11660—11677.

Ghosh, S., Gattera, R., 1995. A comparison of the ipsilateral cortical projections to the dorsal and ventral subdivisions of the macaque premotor cortex. Somatosens. Mot. Res. 12, 359—378.

Goldenberg, G., Spatt, J., 2009. The neural basis of tool use. Brain 132, 1645—1655.

Goldring, A.B., Cooke, D.F., Baldwin, M.K., Recanzone, G.H., Gordon, A.G., Pan, T., Simon, S.I., Krubitzer, L., 2014. Reversible deactivation of higher-order posterior parietal areas. II. Alterations in response properties of neurons in areas 1 and 2. J. Neurophysiol. 112, 2545—2560.

Goodale, M.A., Milner, A.D., 1992. Separate visual pathways for perception and action. Trends Neurosci. 15, 20—25.

Gottlieb, J., Snyder, L.H., 2010. Spatial and non-spatial functions of the parietal cortex. Curr. Opin. Neurobiol. 20, 731—740.

Gottlieb, J.P., Kusunoki, M., Goldberg, M.E., 1998. The representation of visual salience in monkey parietal cortex. Nature 391, 481—484.

Graziano, M.S., Aflalo, T.N., Cooke, D.F., 2005. Arm movements evoked by electrical stimulation in the motor cortex of monkeys. J. Neurophysiol. 94, 4209—4223.

Graziano, M.S., Cooke, D.F., 2006. Parieto-frontal interactions, personal space, and defensive behavior. Neuropsychologia 44, 2621—2635.

Grefkes, C., Fink, G.R., 2005. The functional organization of the intraparietal sulcus in humans and monkeys. J. Anat. 207, 3—17.

Grefkes, C., Ritzl, A., Zilles, K., Fink, G.R., 2004. Human medial intraparietal cortex subserves visuomotor coordinate transformation. Neuroimage 23, 1494—1506.

Grefkes, C., Weiss, P.H., Zilles, K., Fink, G.R., 2002. Crossmodal processing of object features in human anterior intraparietal cortex: an fMRI study implies equivalencies between humans and monkeys. Neuron 35, 173—184.

Gregoriou, G.G., Borra, E., Matelli, M., Luppino, G., 2006. Architectonic organization of the inferior parietal convexity of the macaque monkey. J. Comp. Neurol. 496, 422—451.

Han, Y., Yang, X., Chen, Y., Shou, T., 2008. Evidence for corticocortical connections between areas 7 and 17 in cerebral cortex of the cat. Neurosci. Lett. 430, 70—74.

Harvey, C.D., Coen, P., Tank, D.W., 2012. Choice-specific sequences in parietal cortex during a virtual-navigation decision task. Nature 484, 62—68.

Hinkley, L.B., Krubitzer, L.A., Padberg, J., Disbrow, E.A., 2009. Visualmanual exploration and posterior parietal cortex in humans. J. Neurophysiol. 102, 3433—3446.

Hsiao, S.S., O'shaughnessy, D.M., Johnson, K.O., 1993. Effects of selective attention on spatial form processing in monkey primary and secondary somatosensory cortex. J. Neurophysiol. 70, 444—447.

Huffman, K.J., Nelson, J., Clarey, J., Krubitzer, L., 1999. Organization of somatosensory cortex in three species of marsupials, Dasyurus hallucatus, Dactylopsila trivirgata, and Monodelphis domestica: neural correlates of morphological specializations. J. Comp. Neurol. 403, 5—32.

Hyvarinen, J., Poranen, A., 1974. Function of the parietal associative area 7 as revealed from cellular discharges in alert monkeys. Brain 97, 673—692.

Hyvarinen, J., Poranen, A., 1978. Movement-sensitive and direction and orientation-selective cutaneous receptive fields in the hand area of the post-central gyrus in monkeys. J. Physiol. 283, 523—537.

Hyvarinen, J., Poranen, A., 1978. Receptive field integration and submodality convergence in the hand area of the post-central gyrus of the alert monkey. J. Physiol. 283, 539—556.

Hyvarinen, J., Poranen, A., Jokinen, Y., 1980. Influence of attentive behavior on neuronal responses to vibration in primary somatosensory cortex of the monkey. J. Neurophysiol. 43, 870—882.

Iriki, A., Tanaka, M., Iwamura, Y., 1996. Coding of modified body schema during tool use by macaque postcentral neurones. Neuroreport 7, 2325—2330.

Ishida, H., Nakajima, K., Inase, M., Murata, A., 2010. Shared mapping of own and others' bodies in visuotactile bimodal area of monkey parietal cortex. J. Cogn. Neurosci. 22, 83—96.

Iwamura, Y., 1998. Hierarchical somatosensory processing. Curr. Opin. Neurobiol. 8, 522—528.

Iwamura, Y., Iriki, A., Tanaka, M., 1994. Bilateral hand representation in the postcentral somatosensory cortex. Nature 369, 554—556.

Iwamura, Y., Tanaka, M., Hikosaka, O., 1980. Overlapping representation of fingers in the somatosensory cortex (area 2) of the conscious monkey. Brain Res. 197, 516—520.

Iwamura, Y., Tanaka, M., Sakamoto, M., Hikosaka, O., 1985. Diversity in receptive field properties of vertical neuronal arrays in the crown of the postcentral gyrus of the conscious monkey. Exp. Brain Res. 58, 400—411.

Iwamura, Y., Tanaka, M., Sakamoto, M., Hikosaka, O., 1985. Vertical neuronal arrays in the postcentral gyrus signaling active touch: a receptive field study in the conscious monkey. Exp. Brain Res. 58, 412–420.

Iwamura, Y., Tanaka, M., Sakamoto, M., Hikosaka, O., 1993. Rostrocaudal gradients in the neuronal receptive field complexity in the finger region of the alert monkey's postcentral gyrus. Exp. Brain Res. 92, 360–368.

Iwamura, Y., T.M., Sakamoto, M., Hikosaka, O., 1983. Converging patterns of finger representation and complex response properties of neurons in area 1 of the first somatosensory cortex of the conscious monkey. Exp. Brain Res. 51, 327–337.

Iwamura, Y., T.M., Sakamoto, M., Hikosaka, O., 1983. Functional subdivisions representing different finger regions in area 3 of the first somatosensory cortex of the conscious monkey. Exp. Brain Res. 51, 315–326.

Jones, E.G., Coulter, J.D., Hendry, S.H., 1978. Intracortical connectivity of architectonic fields in the somatic sensory, motor and parietal cortex of monkeys. J. Comp. Neurol. 181, 291–347.

Kaas, J.H., Gharbawie, O.A., Stepniewska, I., 2011. The organization and evolution of dorsal stream multisensory motor pathways in primates. Front. Neuroanat. 5, 34.

Kaas, J.H., Gharbawie, O.A., Stepniewska, I., 2013. Cortical networks for ethologically relevant behaviors in primates. Am. J. Primatol. 75, 407–414.

Kaas, J.H., Krubitzer, L.A., Johanson, K.L., 1989. Cortical connections of areas 17 (V-I) and 18 (V-II) of squirrels. J. Comp. Neurol. 281, 426–446.

Kaas, J.H., Nelson, R.J., Sur, M., Lin, C.S., Merzenich, M.M., 1979. Multiple representations of the body within the primary somatosensory cortex of primates. Science 204, 521–523.

Kalaska, J.F., 1996. Parietal cortex area 5 and visuomotor behavior. Can. J. Physiol. Pharmacol. 74, 483–498.

Kaminiarz, A., Schlack, A., Hoffmann, K.P., Lappe, M., Bremmer, F., 2014. Visual selectivity for heading in the macaque ventral intraparietal area. J. Neurophysiol. 112, 2470–2480.

Karlen, S.J., Krubitzer, L., 2007. The functional and anatomical organization of marsupial neocortex: evidence for parallel evolution across mammals. Prog. Neurobiol. 82, 122–141.

Kim, U., Lee, T., 2013. Intra-areal and corticocortical circuits arising in the dysgranular zone of rat primary somatosensory cortex that processes deep somatic input. J. Comp. Neurol. 521, 2585–2601.

Kolb, B., Walkey, J., 1987. Behavioural and anatomical studies of the posterior parietal cortex in the rat. Behav. Brain Res. 23, 127–145.

Konen, C.S., Mruczek, R.E., Montoya, J.L., Kastner, S., 2013. Functional organization of human posterior parietal cortex: grasping- and reaching-related activations relative to topographically organized cortex. J. Neurophysiol. 109, 2897–2908.

Koyama, M., Hasegawa, I., Osada, T., Adachi, Y., Nakahara, K., Miyashita, Y., 2004. Functional magnetic resonance imaging of macaque monkeys performing visually guided saccade tasks: comparison of cortical eye fields with humans. Neuron 41, 795–807.

Krieg, W.J., 1947. General conclusions from an experimental study of the cerebral cortex of the albino rat. Anat. Rec. 97, 350.

Krubitzer, L., 1998. What can monotremes tell us about brain evolution? Philos. Trans. R. Soc. Lond. B Biol. Sci. 353, 1127–1146.

Krubitzer, L., 2007. The magnificent compromise: cortical field evolution in mammals. Neuron 56, 201–208.

Krubitzer, L., Campi, K.L., Cooke, D.F., 2011. All rodents are not the same: a modern synthesis of cortical organization. Brain Behav. Evol. 78, 51–93.

Krubitzer, L., Huffman, K.J., Disbrow, E., Recanzone, G., 2004. Organization of area 3a in macaque monkeys: contributions to the cortical phenotype. J. Comp. Neurol. 471, 97–111.

Krubitzer, L., Manger, P., Pettigrew, J., Calford, M., 1995. Organization of somatosensory cortex in monotremes: in search of the prototypical plan. J. Comp. Neurol. 351, 261–306.

Krubitzer, L.A., Sesma, M.A., Kaas, J.H., 1986. Microelectrode maps, myeloarchitecture, and cortical connections of three somatotopically organized representations of the body surface in the parietal cortex of squirrels. J. Comp. Neurol. 250, 403–430.

Lacquaniti, F., Guigon, E., Bianchi, L., Ferraina, S., Caminiti, R., 1995. Representing spatial information for limb movement: role of area 5 in the monkey. Cereb. Cortex 5, 391–409.

Lajoie, K., Andujar, J.E., Pearson, K., Drew, T., 2010. Neurons in area 5 of the posterior parietal cortex in the cat contribute to interlimb coordination during visually guided locomotion: a role in working memory. J. Neurophysiol. 103, 2234–2254.

Lajoie, K., Drew, T., 2007. Lesions of area 5 of the posterior parietal cortex in the cat produce errors in the accuracy of paw placement during visually guided locomotion. J. Neurophysiol. 97, 2339–2354.

Lehmann, S.J., Scherberger, H., 2013. Reach and gaze representations in macaque parietal and premotor grasp areas. J. Neurosci. 33, 7038–7049.

Lehmann, S.J., Scherberger, H., 2015. Spatial representations in local field potential activity of primate anterior intraparietal cortex (AIP). PLoS One 10, e0142679.

Leinonen, L., Hyvarinen, J., Nyman, G., Linnankoski, I., 1979. I. Functional properties of neurons in lateral part of associative area 7 in awake monkeys. Exp. Brain Res. 34, 299–320.

Leinonen, L., Nyman, G., 1979. II. Functional properties of cells in anterolateral part of area 7 associative face area of awake monkeys. Exp. Brain Res. 34, 321–333.

Lewis, J.W., Van Essen, D.C., 2000. Corticocortical connections of visual, sensorimotor, and multimodal processing areas in the parietal lobe of the macaque monkey. J. Comp. Neurol. 428, 112–137.

Lewis, J.W., Van Essen, D.C., 2000. Mapping of architectonic subdivisions in the macaque monkey, with emphasis on parieto-occipital cortex. J. Comp. Neurol. 428, 79–111.

Luppino, G., Matelli, M., Camarda, R., Rizzolatti, G., 1993. Corticocortical connections of area F3 (SMA-proper) and area F6 (pre-SMA) in the macaque monkey. J. Comp. Neurol. 338, 114–140.

Maeda, K., Ishida, H., Nakajima, K., Inase, M., Murata, A., 2015. Functional properties of parietal hand manipulation-related neurons and mirror neurons responding to vision of own hand action. J. Cogn. Neurosci. 27, 560–572.

Manger, P.R., Masiello, I., Innocenti, G.M., 2002. Areal organization of the posterior parietal cortex of the ferret (Mustela putorius). Cereb. Cortex 12, 1280–1297.

Marigold, D.S., Drew, T., 2011. Contribution of cells in the posterior parietal cortex to the planning of visually guided locomotion in the cat: effects of temporary visual interruption. J. Neurophysiol. 105, 2457–2470.

Martinich, S., Pontes, M.N., Rocha-Miranda, C.E., 2000. Patterns of corticocortical, corticotectal, and commissural connections in the opossum visual cortex. J. Comp. Neurol. 416, 224–244.

Maunsell, J.H., Van Essen, D.C., 1983. The connections of the middle temporal visual area (MT) and their relationship to a cortical hierarchy in the macaque monkey. J. Neurosci. 3, 2563–2586.

Mayer, A., Nascimento-Silva, M.L., Keher, N.B., Bittencourt-Navarrete, R.E., Gattass, R., Franca, J.G., 2016. Architectonic mapping of somatosensory areas involved in skilled forelimb movements and tool use. J. Comp. Neurol. 524, 1399–1423.

Meftah el, M., Shenasa, J., Chapman, C.E., 2002. Effects of a cross-modal manipulation of attention on somatosensory cortical neuronal responses to tactile stimuli in the monkey. J. Neurophysiol. 88, 3133–3149.

Menz, V.K., Schaffelhofer, S., Scherberger, H., 2015. Representation of continuous hand and arm movements in macaque areas M1, F5, and AIP: a comparative decoding study. J. Neural Eng. 12, 056016.

Miller, M.W., Vogt, B.A., 1984. Direct connections of rat visual cortex with sensory, motor, and association cortices. J. Comp. Neurol. 226, 184–202.

Mishkin, M., Ungerleider, L.G., 1982. Contribution of striate inputs to the visuospatial functions of parieto-preoccipital cortex in monkeys. Behav. Brain Res. 6, 57–77.

Mohammed, H., Jain, N., 2016. Ipsilateral cortical inputs to the rostral and caudal motor areas in rats. J. Comp. Neurol.

Morris, A.P., Bremmer, F., Krekelberg, B., 2013. Eye-position signals in the dorsal visual system are accurate and precise on short timescales. J. Neurosci. 33, 12395–12406.

Morris, A.P., Bremmer, F., Krekelberg, B., 2016. The dorsal visual system predicts future and remembers past eye position. Front. Syst. Neurosci. 10, 9.

Morris, A.P., Kubischik, M., Hoffmann, K.P., Krekelberg, B., Bremmer, F., 2012. Dynamics of eye-position signals in the dorsal visual system. Curr. Biol. 22, 173–179.

Mountcastle, V.B., Lynch, J.C., Georgopoulos, A., Sakata, H., Acuna, C., 1975. Posterior parietal association cortex of the monkey: command functions for operations within extrapersonal space. J. Neurophysiol. 38, 871–908.

Murata, A., Gallese, V., Kaseda, M., Sakata, H., 1996. Parietal neurons related to memory-guided hand manipulation. J. Neurophysiol. 75, 2180–2186.

Murata, A., Gallese, V., Luppino, G., Kaseda, M., Sakata, H., 2000. Selectivity for the shape, size, and orientation of objects for grasping in neurons of monkey parietal area AIP. J. Neurophysiol. 83, 2580–2601.

Napier, J.R., Tuttle, R., 1993. Hands. Princeton University Press.

Nelson, R.J., Sur, M., Felleman, D.J., Kaas, J.H., 1980. Representations of the body surface in postcentral parietal cortex of *Macaca fascicularis*. J. Comp. Neurol. 192, 611–643.

Nieder, A., 2013. Coding of abstract quantity by 'number neurons' of the primate brain. J. Comp. Physiol. A Neuroethol. Sens. Neural Behav. Physiol. 199, 1–16.

Olson, C.R., Lawler, K., 1987. Cortical and subcortical afferent connections of a posterior division of feline area 7 (area 7p). J. Comp. Neurol. 259, 13–30.

Orban, G.A., Caruana, F., 2014. The neural basis of human tool use. Front. Psychol. 5, 310.

Padberg, J., Cerkevich, C., Engle, J., Rajan, A.T., Recanzone, G., Kaas, J., Krubitzer, L., 2009. Thalamocortical connections of parietal somatosensory cortical fields in macaque monkeys are highly divergent and convergent. Cereb. Cortex 19, 2038–2064.

Padberg, J., Franca, J.G., Cooke, D.F., Soares, J.G., Rosa, M.G., Fiorani Jr., M., Gattass, R., Krubitzer, L., 2007. Parallel evolution of cortical areas involved in skilled hand use. J. Neurosci. 27, 10106–10115.

Pandya, D.N., Seltzer, B., 1982. Intrinsic connections and architectonics of posterior parietal cortex in the rhesus monkey. J. Comp. Neurol. 204, 196–210.

Pani, P., Theys, T., Romero, M.C., Janssen, P., 2014. Grasping execution and grasping observation activity of single neurons in the macaque anterior intraparietal area. J. Cogn. Neurosci. 26, 2342–2355.

Papadelis, C., Eickhoff, S.B., Zilles, K., Ioannides, A.A., 2011. BA3b and BA1 activate in a serial fashion after median nerve stimulation: direct evidence from combining source analysis of evoked fields and cytoarchitectonic probabilistic maps. Neuroimage 54, 60–73.

Peele, T.L., 1944. Acute and chronic parietal lobe ablations in monkeys. J. Neurophysiol. 7, 269–286.

Pei, Y.C., Hsiao, S.S., Craig, J.C., Bensmaia, S.J., 2010. Shape invariant coding of motion direction in somatosensory cortex. PLoS Biol. 8, e1000305.

Pesaran, B., Nelson, M.J., Andersen, R.A., 2006. Dorsal premotor neurons encode the relative position of the hand, eye, and goal during reach planning. Neuron 51, 125–134.

Pinto-Hamuy, T., Olavarria, J., Guic-Robles, E., Morgues, M., Nassal, O., Petit, D., 1987. Rats with lesions in anteromedial extrastriate cortex fail to learn a visuosomatic conditional response. Behav. Brain Res. 25, 221–231.

Pons, T.P., Kaas, J.H., 1986. Corticocortical connections of area 2 of somatosensory cortex in macaque monkeys: a correlative anatomical and electrophysiological study. J. Comp. Neurol. 248, 313–335.

Prado, J., Clavagnier, S., Otzenberger, H., Scheiber, C., Kennedy, H., Perenin, M.T., 2005. Two cortical systems for reaching in central and peripheral vision. Neuron 48, 849–858.

Preuss, T.M., Goldman-Rakic, P.S., 1991. Architectonics of the parietal and temporal association cortex in the strepsirhine primate Galago compared to the anthropoid primate Macaca. J. Comp. Neurol. 310, 475–506.

Quallo, M.M., Kraskov, A., Lemon, R.N., 2012. The activity of primary motor cortex corticospinal neurons during tool use by macaque monkeys. J. Neurosci. 32, 17351–17364.

Quian Quiroga, R., Snyder, L.H., Batista, A.P., Cui, H., Andersen, R.A., 2006. Movement intention is better predicted than attention in the posterior parietal cortex. J. Neurosci. 26, 3615–3620.

Reep, R.L., Chandler, H.C., King, V., Corwin, J.V., 1994. Rat posterior parietal cortex: topography of corticocortical and thalamic connections. Exp. Brain Res. 100, 67–84.

Remple, M.S., Reed, J.L., Stepniewska, I., Kaas, J.H., 2006. Organization of frontoparietal cortex in the tree shrew (*Tupaia belangeri*). I. Architecture, microelectrode maps, and corticospinal connections. J. Comp. Neurol. 497, 133–154.

Remple, M.S., Reed, J.L., Stepniewska, I., Lyon, D.C., Kaas, J.H., 2007. The organization of frontoparietal cortex in the tree shrew (*Tupaia belangeri*): II. Connectional evidence for a frontal-posterior parietal network. J. Comp. Neurol. 501, 121–149.

Rockland, K.S., 1985. Anatomical organization of primary visual cortex (area 17) in the ferret. J. Comp. Neurol. 241, 225–236.

Romero, M.C., Janssen, P., 2016. Receptive field properties of neurons in the macaque anterior intraparietal area. J. Neurophysiol. 115, 1542–1555.

Romero, M.C., Pani, P., Janssen, P., 2014. Coding of shape features in the macaque anterior intraparietal area. J. Neurosci. 34, 4006–4021.

Romero, M.C., Van Dromme, I., Janssen, P., 2012. Responses to two-dimensional shapes in the macaque anterior intraparietal area. Eur. J. Neurosci. 36, 2324–2334.

Romero, M.C., Van Dromme, I.C., Janssen, P., 2013. The role of binocular disparity in stereoscopic images of objects in the macaque anterior intraparietal area. PLoS One 8, e55340.

Roskov, Y., A.L., Orrell, T., Nicolson, D., Kunze, T., Culham, A., Bailly, N., Kirk, P., Bourgoin, T., Dewalt, R., Decock, W., De Wever, A., 2015. Species 2000 & ITIS Catalogue of Life, 2015 Annual Checklist. www.catalogueoflife.org/annual-checklist/2015.

Rozzi, S., Calzavara, R., Belmalih, A., Borra, E., Gregoriou, G.G., Matelli, M., Luppino, G., 2006. Cortical connections of the inferior parietal cortical convexity of the macaque monkey. Cereb. Cortex 16, 1389–1417.

Rozzi, S., Ferrari, P.F., Bonini, L., Rizzolatti, G., Fogassi, L., 2008. Functional organization of inferior parietal lobule convexity in the macaque monkey: electrophysiological characterization of motor, sensory and mirror responses and their correlation with cytoarchitectonic areas. Eur. J. Neurosci. 28, 1569–1588.

Sakata, H., Taira, M., Murata, A., Mine, S., 1995. Neural mechanisms of visual guidance of hand action in the parietal cortex of the monkey. Cereb. Cortex 5, 429–438.

Sakata, H., Takaoka, Y., Kawarasaki, A., Shibutani, H., 1973. Somatosensory properties of neurons in the superior parietal cortex (area 5) of the rhesus monkey. Brain Res. 64, 85–102.

Sanchez-Panchuelo, R.M., Besle, J., Beckett, A., Bowtell, R., Schluppeck, D., Francis, S., 2012. Within-digit functional parcellation of Brodmann areas of the human primary somatosensory cortex using functional magnetic resonance imaging at 7 tesla. J. Neurosci. 32, 15815–15822.

Schaffelhofer, S., Agudelo-Toro, A., Scherberger, H., 2015. Decoding a wide range of hand configurations from macaque motor, premotor, and parietal cortices. J. Neurosci. 35, 1068–1081.

Schlack, A., Hoffmann, K.P., Bremmer, F., 2002. Interaction of linear vestibular and visual stimulation in the macaque ventral intraparietal area (VIP). Eur. J. Neurosci. 16, 1877–1886.

Schmahmann, J.D., Pandya, D.N., 1990. Anatomical investigation of projections from thalamus to posterior parietal cortex in the rhesus monkey: a WGA-HRP and fluorescent tracer study. J. Comp. Neurol. 295, 299–326.

Seelke, A.M., Padberg, J.J., Disbrow, E., Purnell, S.M., Recanzone, G., Krubitzer, L., 2012. Topographic maps within Brodmann's area 5 of macaque monkeys. Cereb. Cortex 22, 1834–1850.

Seltzer, B., Pandya, D.N., 1980. Converging visual and somatic sensory cortical input to the intraparietal sulcus of the rhesus monkey. Brain Res. 192, 339–351.

Seltzer, B., Pandya, D.N., 1986. Posterior parietal projections to the intraparietal sulcus of the rhesus monkey. Exp. Brain Res. 62, 459–469.

Sesma, M.A., Casagrande, V.A., Kaas, J.H., 1984. Cortical connections of area 17 in tree shrews. J. Comp. Neurol. 230, 337–351.

Shumaker, R., W.,K., Beck, B., 2011. Animal Tool Behavior: The Use and Manufacture of Tools by Animals. John Hopkins University Press, Baltimore, MD.

Silver, M.A., Kastner, S., 2009. Topographic maps in human frontal and parietal cortex. Trends Cogn. Sci. 13, 488–495.

Slutsky, D.A., Manger, P.R., Krubitzer, L., 2000. Multiple somatosensory areas in the anterior parietal cortex of the California ground squirrel (Spermophilus beecheyii). J. Comp. Neurol. 416, 521–539.

Snyder, L.H., Batista, A.P., Andersen, R.A., 1997. Coding of intention in the posterior parietal cortex. Nature 386, 167–170.

Snyder, L.H., Batista, A.P., Andersen, R.A., 2000. Intention-related activity in the posterior parietal cortex: a review. Vis. Res. 40, 1433–1441.

Spingath, E., Kang, H.S., Blake, D.T., 2013. Task-dependent modulation of SI physiological responses to targets and distractors. J. Neurophysiol. 109, 1036–1044.

Spingath, E.Y., Kang, H.S., Plummer, T., Blake, D.T., 2011. Different neuroplasticity for task targets and distractors. PLoS One 6, e15342.

Spinozzi, G., Truppa, V., Lagana, T., 2004. Grasping behavior in tufted capuchin monkeys (Cebus apella): grip types and manual laterality for picking up a small food item. Am. J. Phys. Anthropol. 125, 30–41.

Srivastava, S., Orban, G.A., De Maziere, P.A., Janssen, P., 2009. A distinct representation of three-dimensional shape in macaque anterior intraparietal area: fast, metric, and coarse. J. Neurosci. 29, 10613–10626.

Stepniewska, I., Fang, P.C., Kaas, J.H., 2005. Microstimulation reveals specialized subregions for different complex movements in posterior parietal cortex of prosimian galagos. Proc. Natl. Acad. Sci. U.S.A. 102, 4878–4883.

Stepniewska, I., Fang, P.C., Kaas, J.H., 2009. Organization of the posterior parietal cortex in galagos: I. Functional zones identified by microstimulation. J. Comp. Neurol. 517, 765–782.

Stepniewska, I., Friedman, R.M., Gharbawie, O.A., Cerkevich, C.M., Roe, A.W., Kaas, J.H., 2011. Optical imaging in galagos reveals parietal-frontal circuits underlying motor behavior. Proc. Natl. Acad. Sci. U.S.A. 108, E725–E732.

Stepniewska, I., Gharbawie, O.A., Burish, M.J., Kaas, J.H., 2014. Effects of muscimol inactivations of functional domains in motor, premotor, and posterior parietal cortex on complex movements evoked by electrical stimulation. J. Neurophysiol. 111, 1100–1119.

Stepniewska, I., Preuss, T.M., Kaas, J.H., 1993. Architectonics, somatotopic organization, and ipsilateral cortical connections of the primary motor area (M1) of owl monkeys. J. Comp. Neurol. 330, 238–271.

Sunkara, A., Deangelis, G.C., Angelaki, D.E., 2015. Role of visual and non-visual cues in constructing a rotation-invariant representation of heading in parietal cortex. Elife 4.

Sunkara, A., Deangelis, G.C., Angelaki, D.E., 2016. Joint representation of translational and rotational components of optic flow in parietal cortex. Proc. Natl. Acad. Sci. U.S.A. 113, 5077–5082.

Sur, M., 1980. Receptive fields of neurons in areas 3b and 1 of somatosensory cortex in monkeys. Brain Res. 198, 465–471.

Sur, M., Weller, R.E., Kaas, J.H., 1980. Representation of the body surface in somatosensory area I of tree shrews, Tupaia glis. J. Comp. Neurol. 194, 71–95.

van Schaik, C.P., Deaner, R.O., Merrill, M.Y., 2003. The conditions for tool use in primates: implications for the evolution of material culture. J. Hum. Evol. 36, 719–741.

Taira, M., Mine, S., Georgopoulos, A.P., Murata, A., Sakata, H., 1990. Parietal cortex neurons of the monkey related to the visual guidance of hand movement. Exp. Brain Res. 83, 29–36.

Tanne-Gariepy, J., Rouiller, E.M., Boussaoud, D., 2002. Parietal inputs to dorsal versus ventral premotor areas in the macaque monkey: evidence for largely segregated visuomotor pathways. Exp. Brain Res. 145, 91–103.

Theys, T., Pani, P., Van Loon, J., Goffin, J., Janssen, P., 2013. Three-dimensional shape coding in grasping circuits: a comparison between the anterior intraparietal area and ventral premotor area F5a. J. Cogn. Neurosci. 25, 352–364.

Thier, P., Andersen, R.A., 1996. Electrical microstimulation suggests two different forms of representation of head-centered space in the intraparietal sulcus of rhesus monkeys. Proc. Natl. Acad. Sci. U.S.A. 93, 4962–4967.

Thier, P., Andersen, R.A., 1998. Electrical microstimulation distinguishes distinct saccade-related areas in the posterior parietal cortex. J. Neurophysiol. 80, 1713–1735.

Townsend, B.R., Subasi, E., Scherberger, H., 2011. Grasp movement decoding from premotor and parietal cortex. J. Neurosci. 31, 14386–14398.

Umilta, M.A., Escola, L., Intskirveli, I., Grammont, F., Rochat, M., Caruana, F., Jezzini, A., Gallese, V., Rizzolatti, G., 2008. When pliers become fingers in the monkey motor system. Proc. Natl. Acad. Sci. U.S.A. 105, 2209–2213.

Verhoef, B.E., Vogels, R., Janssen, P., 2010. Contribution of inferior temporal and posterior parietal activity to three-dimensional shape perception. Curr. Biol. 20, 909–913.

Verhoef, B.E., Michelet, P., Vogels, R., Janssen, P., 2015. Choice-related activity in the anterior intraparietal area during 3-D structure categorization. J. Cogn. Neurosci. 27, 1104–1115.

Vogt, C., Vogt, O., 1919. Allgemeinere Ergebnisse unserer Hirnforschung: 1.-4. Mitt. J. Psychol. Neurol. 25, 275–461.

Wacker, E., Spitzer, B., Lutzkendorf, R., Bernarding, J., Blankenburg, F., 2011. Tactile motion and pattern processing assessed with high-field fMRI. PLoS One 6, e24860.

Wallace, M.T., Ramachandran, R., Stein, B.E., 2004. A revised view of sensory cortical parcellation. Proc. Natl. Acad. Sci. U.S.A. 101, 2167–2172.

Wang, L., Li, X., Hsiao, S.S., Bodner, M., Lenz, F., Zhou, Y.D., 2012. Behavioral choice-related neuronal activity in monkey primary somatosensory cortex in a haptic delay task. J. Cogn. Neurosci. 24, 1634–1644.

Weber, J.T., Yin, T.C., 1984. Subcortical projections of the inferior parietal cortex (area 7) in the stump-tailed monkey. J. Comp. Neurol. 224, 206–230.

Whitlock, J.R., 2014. Navigating actions through the rodent parietal cortex. Front. Hum. Neurosci. 8, 293.

Whitlock, J.R., Pfuhl, G., Dagslott, N., Moser, M.B., Moser, E.I., 2012. Functional split between parietal and entorhinal cortices in the rat. Neuron 73, 789–802.

Wilber, A.A., Clark, B.J., Forster, T.C., Tatsuno, M., Mcnaughton, B.L., 2014. Interaction of egocentric and world-centered reference frames in the rat posterior parietal cortex. J. Neurosci. 34, 5431–5446.

Wise, S.P., Boussaoud, D., Johnson, P.B., Caminiti, R., 1997. Premotor and parietal cortex: corticocortical connectivity and combinatorial computations. Annu. Rev. Neurosci. 20, 25–42.

Yang, Y., Liu, S., Chowdhury, S.A., Deangelis, G.C., Angelaki, D.E., 2011. Binocular disparity tuning and visual-vestibular congruency of multisensory neurons in macaque parietal cortex. J. Neurosci. 31, 17905–17916.

Yau, J.M., Connor, C.E., Hsiao, S.S., 2013. Representation of tactile curvature in macaque somatosensory area 2. J. Neurophysiol. 109, 2999–3012.

Yeterian, E.H., Pandya, D.N., 1985. Corticothalamic connections of the posterior parietal cortex in the rhesus monkey. J. Comp. Neurol. 237, 408–426.

Yokochi, H., Tanaka, M., Kumashiro, M., Iriki, A., 2003. Inferior parietal somatosensory neurons coding face-hand coordination in Japanese macaques. Somatosens. Mot. Res. 20, 115–125.

Zhang, T., Britten, K.H., 2011. Parietal area VIP causally influences heading perception during pursuit eye movements. J. Neurosci. 31, 2569–2575.

Zhang, T., Heuer, H.W., Britten, K.H., 2004. Parietal area VIP neuronal responses to heading stimuli are encoded in head-centered coordinates. Neuron 42, 993–1001.

27

Evolution of Parietal-Frontal Networks in Primates

J.H. Kaas, H.-X. Qi, I. Stepniewska
Vanderbilt University, Nashville, TN, United States

27.1 Introduction

A prominent feature of the neocortex of monkeys and humans is the collection of parallel parietofrontal sensorimotor networks that originate in a portion of posterior parietal cortex (PPC) and terminate in premotor cortex (PMC) and motor cortex (M1). Most of the research on these networks has been on macaque monkeys where the region of the intraparietal sulcus has been the most explored (for review, see Andersen et al., 1997; Battaglia-Mayer et al., 2006; Andersen and Buneo, 2002; Caminiti et al., 2015; Rizzolatti et al., 1997; Sereno and Huang, 2014). In addition, important comparable efforts have been made to study the intraparietal region of humans (eg, Filimon, 2010; Konen et al., 2013; Peeters et al., 2009; Ptak and Muri, 2013). Especially in macaque monkeys, an extensive amount of research has established that PPC contains a number of distinct regions or areas that are highly involved in different, specific classes of motor behavior (Fig. 27.1A). Thus, a proposed lateral intraparietal area (LIP) of the lateral bank of the intraparietal sulcus is thought to be involved in the production of saccadic eye movements. The most compelling evidence for this conclusion comes from the observation that electrical stimulation of the LIP region with microelectrodes evokes saccadic eye movements into the contralateral visual hemifield (Thier and Andersen, 1996; Shihutani et al., 1984; Kurylo and Skavenski, 1991). Other compelling evidence in support of this proposed functional role comes from anatomical studies showing that LIP projects to the frontal eye field (FEF) and to the superior colliculus (Andersen et al., 1985; Huerta et al., 1987; Lynch et al., 1985; Blatt et al., 1990), which are both parts of the brain where eye movements can be evoked by electrical stimulation (eg, Wurtz and Albano, 1980; Huerta et al., 1986). In addition, LIP receives inputs from higher level visual areas, such as the middle temporal (MT) visual area (eg, Blatt et al., 1990) that provide target information for the eye movements. Neurons in LIP respond to visual stimuli just before and during eye movements (eg, Duhamel et al., 1992) and represent the contralateral visual hemifield as a place map for eye movements (Savaki et al., 2010). Thus, LIP has long been considered to be a parietal eye field with sensory and motor functions related to producing saccades to visual targets (Andersen et al., 1992).

In a similar manner, other functionally distinct regions of PPC have been identified, including a parietal reach region (PRR) which appears to be analogous to LIP in function, but related to preparing the hand for reaching rather than eye movements (Cohen and Andersen, 2002; Andersen and Buneo, 2002). Another part of PPC, the ventral intraparietal area (VIP) was originally defined as a projection target of visual area MT (Mausell and Van Essen, 1983) and then redefined as a region with neurons responsive to visual motion (Colby et al., 1993) and touch on the face (Duhamel et al., 1992). More recently, electrical stimulation of the VIP region with long trains of current has been found to produce defensive movements of the arm and face (Cooke et al., 2003). In cortex more rostral in PPC on the posterior (lateral) bank of the intraparietal sulcus, investigators have found neurons that are activated during grasping and manipulation of objects by the hand (eg, Sakata et al., 1995; Baumann et al., 2009). This anterior intraparietal area (AIP) is thought to be a region where neurons use visual information to shape and guide appropriate hand movements (Murata et al., 2000). In a more recent study, electrical stimulation was used to evoke grasping movements from cortex just medial to AIP on the medial or rostral bank of the intraparietal sulcus (Gharbawie

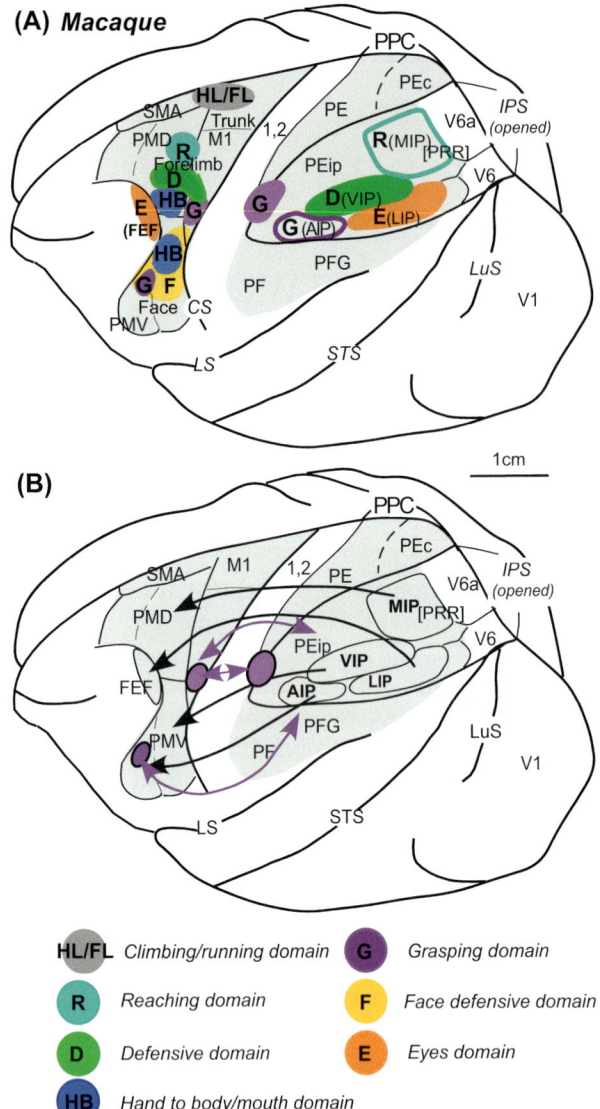

(A) Macaque

1cm

(B)

HL/FL *Climbing/running domain*   **G** *Grasping domain*

R *Reaching domain*   **F** *Face defensive domain*

D *Defensive domain*   **E** *Eyes domain*

HB *Hand to body/mouth domain*

FIGURE 27.1 Dorsolateral view of the macaque brain (left hemisphere). (A) Movement domains physiologically identified in primary motor cortex (M1), dorsal and ventral premotor cortex (PMD and PMV), and rostral half of posterior parietal cortex (PPCr) are marked with color. The intraparietal sulcus, IPS, has been opened to show buried areas (domains): PRR—parietal reach region (involving MIP and part of V6a), VIP—ventral intraparietal area, AIP—anterior intraparietal area, and LIP—lateral intraparietal area. VIP and LIP are potential domains since their stimulation evoked defensive and eye movements, respectively. AIP and MIP were identified, respectively, as grasp and reach areas on the basis of recordings from their neurons during grasping and reaching performance. These areas are outlined but not filled with color. The grasp domain identified by our long-train ICMS study in area 2/5 is also shown. (B) Major cortical connections between intraparietal areas and frontal areas are marked with *black arrows*. Some connections of the grasp domain in area 2/5 with matching domains in M1 and PMV are marked by *violet arrows*. Some connections of other movement related subdivisions of PPC are indicated by *black arrows*. Evidence for connections between matching grasp zones identified by microstimulation is from Gharbawie, O.A., Stepniewska, I., Qi, H., Kaas, J.H., 2011. Multiple parietal-frontal pathways mediate grasping in macaque monkeys. J. Neurosci. 31,

et al., 2011b). This region of cortex where grasping movements can be evoked may extend to include AIP, or the two regions may be functionally related. Nevertheless, it is clear that different parts of PPC have roles in producing different types of complex movements, such as looking, reaching, body defense, and grasping. The most direct evidence for this conclusion is that these biologically relevant movements can be evoked from specific regions of PPC by electrical stimulation.

These motor functions of PPC movement zones appear to depend, to a large part, on projections from these PPC zones to different parts of premotor and M1 (Fig. 27.1B). Thus, the AIP region projects to the ventral premotor area, the PRR region projects to the dorsal premotor area, and LIP projects to the FEF (Luppino et al., 1999; Borra et al., 2008; Andersen et al., 1985; Huerta et al., 1987; Caminiti et al., 1996). These three regions of PPC have provided most of the evidence for the existence of different parietal-frontal motor behaviors in primates. Functional imaging and other procedures have provided additional results that suggest that these networks also exist in humans, although they might be more elaborate (eg, Patel et al., 2015), perhaps, for example, mediating a more skilled use of hands (eg, Leone et al., 2015). Here, we review evidence that all primates have such parietal-frontal networks, for as many as eight or nine different types of behavior. The evidence for these networks naturally raises the questions of how these networks emerged in mammalian evolution and how do they vary across the different branches of primate evolution.

A second reason for being interested in the evolution of parietal-frontal sensorimotor networks relates to our changing views on the functional organization of M1 and PMC in primates. Classical concepts of primary motor cortex, M1, portrayed as a single somatotopic representation of the movements of parts of the contralateral half of the body, from toes (or tails) medially to mouth, and tongue laterally (eg, Penfield and Boldrey, 1937). Thus, M1 was seen as a general purpose motor area, with different regions of M1 involved in the production of voluntary movements of leg, trunk, arm, digits, face, and mouth in a variable pattern for whatever task was necessary. Premotor areas were thought to be functionally organized in much the same way. This view of M1 was compatible with what was known about the somatotopic organization of M1 as long as only the crude overall pattern of somatotopy was considered. When this somatotopy was explored in more detail with

11660–11677. *FEF*, frontal eye field; *LS*, lateral sulcus; *SMA*, supplementary motor area; *PE*, parietal area E; *PF*, parietal area F; *PFG*, parietal area FG; *STS*, superior temporal sulcus.

stimulating microelectrodes and near threshold levels of current, it became apparent that cortical locations for evoking specific movements of digits, wrist, elbow, or shoulder were both repeated and mixed in what appeared to be a "mosaic" of functionally different sites in M1 (Gould et al., 1986; Donoghue et al., 1992; Huntley and Jones, 1991; see Schieber, 2001 for review). This "mosaic" of movement sites, something like the "fractured" somatotopy of the sensorimotor map in the cerebellum (Shambes et al., 1978), is quite unlike the fine-grained somatotopy of maps in somatosensory cortex. Such motor mosaics suggest the possibility of functionally distinct subregions within M1 and within premotor areas. Clear evidence for this came from the studies of Graziano (2009). These investigators used half-second trains of electrical pulses to explore M1 with microelectrodes in awake macaque monkeys. These longer than commonly used (40–60 ms) trains of electrical pulses allowed a complex behavior to emerge and be completed during the time of the stimulation. The results indicated that primary M1 and the adjoining dorsal and ventral premotor areas were subdivided into at least seven "action zones" where electrical stimulation evoked chewing and licking, hand-to-mouth movements, body defensive movements of the arm and head, manipulation movements of the hand, reaching out as if to grasp, placing the hand in lower space, and climbing or leaping movements of the limbs. The organization of M1, with multiple sites evoking each movement, is better explained by the model of multiple "action zones" than the all-purpose model of M1. The results also suggest the existence of several parietal-frontal parallel networks that function to produce a number of distinct, complex behaviors. These results raise questions about how motor and premotor "action zones" relate to the sensorimotor zones of PPC and how did these "action zones" evolved.

27.2 Parietal-Frontal Networks in Other Primates

Primates are part of the euarchontoglire superorder that includes glires (rodents and rabbits) and archontans (gliding lemurs, tree shrews, and primates). Early primates emerged about 100 million years ago (mya), although the fossil evidence for primates goes back only about 55 mya (Steiper and Seiffert, 2012). Modern primates diverged from a common ancestor something like 80 or more mya, and major changes in brain size and shape started to emerge about 50 mya (Silcox et al., 2010). Primates diverged into the suborder Strepsirrhini of lemurs, lorises, and galagos and suborder Haplorhini of tarsiers, monkeys, and apes. The tarsiers are the only family of the Tarsiiforme infraorder, while

the Simiiform infraorder includes New World (Platyrrhine) monkeys, and the Catarrhine primates include Old World monkeys, apes, and humans. As part of our interest in the evolution of brain differences in primates, we have extensively studied the brains of African galagos, as they resemble early primates in brain size and body form (eg, Silcox et al., 2010).

Our early efforts to study parietal-frontal networks revealed that galagos have a primary motor area, M1, dorsal and ventral premotor areas (PMD and PMV), a FEF, and supplementary motor area (SMA) and cingulate motor area, with motor maps similar to those in macaque monkeys (Wu et al., 2000; also see Fogassi et al., 1994). M1, PMD, and PMV have connections with the rostral half of PPC (Fang et al., 2005) that receives inputs from somatosensory areas, and visual areas project to the caudal half of PPC (eg, Lyon and Kaas, 2002; also see Stepniewska et al., 2016). Thus, galagos appear to have a large region of posterior parietal cortex that resembles that of monkeys by having visual and somatosensory inputs and having projections to M1, PMC, and the FEF. In view of the evidence for "action zones" in M1 of macaques, we started experiments using long trains of microstimulation pulses in motor, premotor, and posterior parietal cortex of galagos.

Our first studies were on galago's PPC. Following the stimulation methods of Graziano et al. (2002), we stimulated different parts of PPC with long (500 msec) trains of electrical pulses. In our initial experiments, we were able to identify movement zones, we now called movement domains, for face defensive movements, face aggressive movements, grasping with the hand, hand-to-mouth movements, forelimb defensive movements, reaching, and forelimb and hindlimb movements as if running or climbing (Fig. 27.2A, Stepniewska et al., 2005). We found an eye movement domain next to the face defensive domain, as shown in Fig. 27.2. We subsequently identified a face aggressive domain with mouth opening and teeth showing just rostral to the face defense domain. Overall, we had evidence for nine different movement domains in the region of the shallow intraparietal sulcus. These domains roughly reflect a somatotopic motor pattern from hindlimb movements, most medially to face and eye movements most laterally, with different hand and forelimb domains in between (Fig. 27.2A).

In our early study of PPC domains in galagos, we also injected tracers to reveal the connections of domains (Stepniewska et al., 2005). These injections demonstrated that each domain (except eye movement domain) had connections with different parts of M1 and PMC (Fig. 27.2B). The eye movement domain (homolog of LIP of macaques) is expected to project to the FEF, but this has not been established yet. These early and subsequent results (Stepniewska et al., 2009a,b) suggested that

(A) Galago

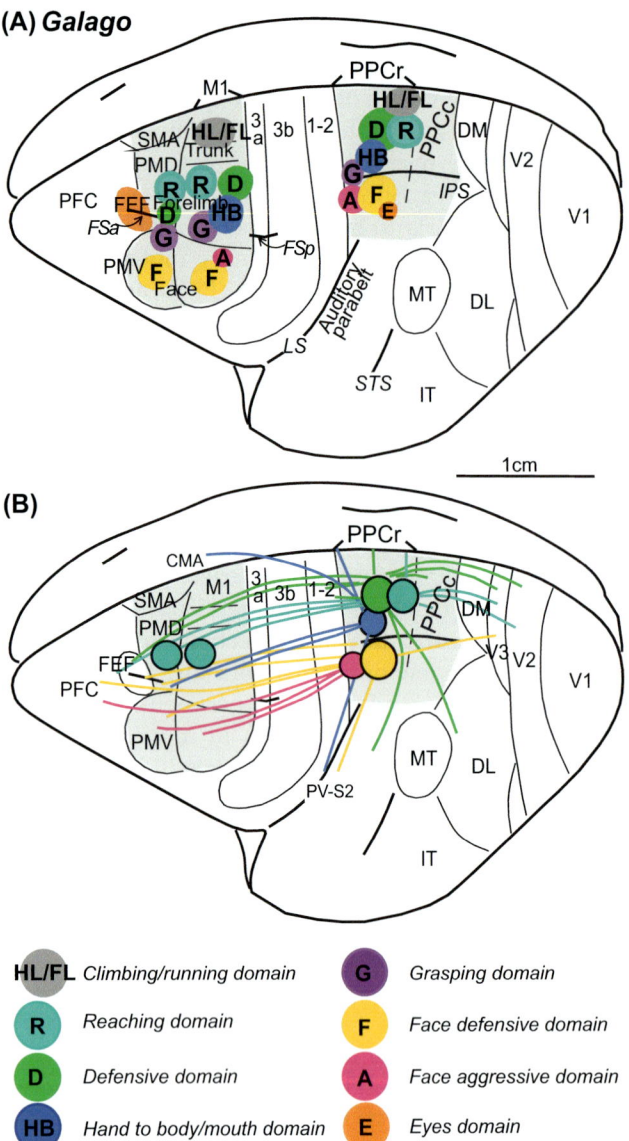

HL/FL *Climbing/running domain*

R *Reaching domain*

D *Defensive domain*

HB *Hand to body/mouth domain*

G *Grasping domain*

F *Face defensive domain*

A *Face aggressive domain*

E *Eyes domain*

FIGURE 27.2 Dorsolateral view of galago brain (left hemisphere). (A) Movement domains identified by long-train microstimulation are shown in areas M1, PMD, PMV, and PPC. Corresponding domains in three areas are marked with the same color. (B) Major cortical connections of PPC domains are marked by colored lines. Note connections between reach domains in PPC, PMD, and M1. *DM*, the dorsomedial visual area; *DL*, the dorsolateral visual area; *FEF*, frontal eye field; *IT*, inferiortemporal cortex; *MT*, middle temporal; *PMD*, dorsal premotor; *PMV*, ventral premotor; *PPC*, posterior parietal cortex; *SMA*, supplementary motor area; *M1*, motor cortex. *Based on Stepniewska, I., Fang, P.C., Kaas, J.H., 2005. Microstimulation reveals specialized subregions for different complex movements in posterior parietal cortex of prosimian galagos. Proc. Natl. Acad. Sci. U.S.A. 102, 4878–4883; Stepniewska, I., Fang, P.C., Kaas, J.H., 2009a. Organization of the posterior parietal cortex in galagos: I. Functional zones identified by microstimulation. J. Comp. Neurol. 517, 765–782; Stepniewska, I., Cerkevich, C.M., Fang, P.C., Kaas, J.H., 2009. Organization of the posterior parietal cortex in galagos: II. Ipsilateral cortical connections of physiologically identified zones within anterior sensorimotor region. J. Comp. Neurol. 517, 783–807 and unpublished results. PV-S2 areas projecting to some PPC domains are partly hidden in lateral sulcus, LS.*

movement domains exist in M1 and PMC, much as they do in macaques. This proved to be the case, as we used long-train microstimulation to identify action domains in M1 and PMC that matched those in PPC. Using optical imaging, we found that electrically stimulating PPC domains selectively activated matching M1 and PMC domains (Stepniewska et al., 2011). Furthermore, inactivating M1 domains with GABA receptors agonist muscimol (Stepniewska et al., 2014) or cooling (Cooke et al., 2015) abolished or greatly reduced the movements normally evoked by stimulating a functionally matched domain in PPC, while stimulation of nonmatched PPC domains usually remained effective.

These results from galagos clearly demonstrated the existence of 8–9 action-specific domains in PPC that function through connections with matching PMC and M1 domains (Fig. 27.2). We propose that these domains reflect three levels of decision making, where outcomes depend on the activation patterns across domains in PPC, PMC, or M1 (see Kaas and Stepniewska, 2016). In PPC, the outcome largely depends on the patterns of higher order sensory inputs that activate one domain more strongly and sooner than others. In addition, the most active domain most effectively suppresses others and provides the dominant output to matching PMC and M1 domains. While the dominant PPC output excites matching domains in frontal cortex, the output of PMC can be altered by other activating inputs, mainly those from prefrontal cortex but also thalamic inputs related to the basal ganglia (Graybiel and Grafton, 2015). Activity patterns in M1 would be altered by inputs from PMC, the SMA, and cingulate motor areas, but also thalamic inputs related to the cerebellum. We propose that these three levels of decision making, based on different sources of information, reflect a major function of the parietal-frontal parallel networks. It may be that domains in PPC, PMC, and M1 determine the goal of a movement pattern, while the specific sequences of movements are dependent on subcortical circuits (Adams et al., 2013; Esposito et al., 2014; Tovote et al., 2016).

We next address the question, do such parietal-frontal networks exist in other primates? We already know from a great bulk of research that such circuits exist in Old World macaque monkeys. However, we do not yet know how closely these circuits resemble those of galagos or what any differences might mean. This is because the macaque studies have largely focused on only three domains, the reaching, looking, and grasping domains, and much of the evidence depends on recording from neurons during motor activity, rather than in evoking movement with electrical stimulation. There is evidence that reaching and grasping behaviors activate neurons that are widely distributed in PPC (eg, Kalaska et al., 1997), and given that sensory inputs to PPC are widely distributed, this is not too surprising.

Therefore, we suggest that the regions of PPC where complex movements can be evoked by electrical stimulation, the domains, are the core centers of nodes for specific behaviors, while surrounding regions are less directly involved in the specific behaviors and may even be related to other domains. Given that the movements produced by electrical stimulation provide the most comparable data across taxa, there is evidence that the VIP region of PPC in macaques contains a domain where electrical stimulation evokes defensive movements, and another region, LIP, is where electrical stimulation evokes eye movements. As in galagos, LIP of macaques is connected preferentially to the FEF and VIP projects to PMC and M1. Thus, we have good evidence for at least two homologous networks. In addition, grasping movements of the hand can be evoked by stimulating cortex on the medial bank of the intraparietal sulcus in macaques (Gharbawie et al., 2011b). This suggests that a grasping domain exists in macaques and that it is close, but medial to the AIP grasp region defined by recordings.

For humans, direct stimulation evidence for action domains in PPC, PMC, and M1 is limited. However, Desmurget et al. (2014) evoked hand-to-mouth movements in humans by stimulation of M1 with bipolar electrode placed on the cortical surface. In addition, indirect evidence, largely from fMRI studies, suggests that grasping, looking, and reaching domains exist in PPC of humans (eg, Sereno and Huang, 2014). There is also evidence for further elaborations of functional representations in PPC of humans (eg, Hartwigsen et al., 2016; Peeters et al., 2009; Jerde et al., 2012). For now, we suspect that a basic set of domains has been retained in humans, but they have been elaborated and subdivided to mediate human specific abilities.

Our other primate comparisons are with New World owl and squirrel monkeys (Fig. 27.4). These are small monkeys with few cortical fissures and more of PPC and frontal M1 exposed on the brain surface, a clear advantage for our research. While our studies are still ongoing, we found clear microstimulation evidence for face defense, grasp, body defense, hand to mouth, reach, and running domains in PPC of squirrel and owl monkeys (Fig. 27.3A; Stepniewska et al., 2014; Gharbawie et al., 2011a). More recently, we have identified a domain in caudolateral PPC for eye movements in these primates (unpublished data). Matching domains have been identified in PMC and in M1. The eye movement domain in PPC was matched in the FEF, but not in M1. The running, hindlimb movement domain in PPC was matched in M1, but a match has not been found in PMC. Likewise, we have not found an eye movement (looking) domain in M1 and a running domain in PMC in galagos. It remains possible that a running domain will be found in part of dorsal PMC or in the

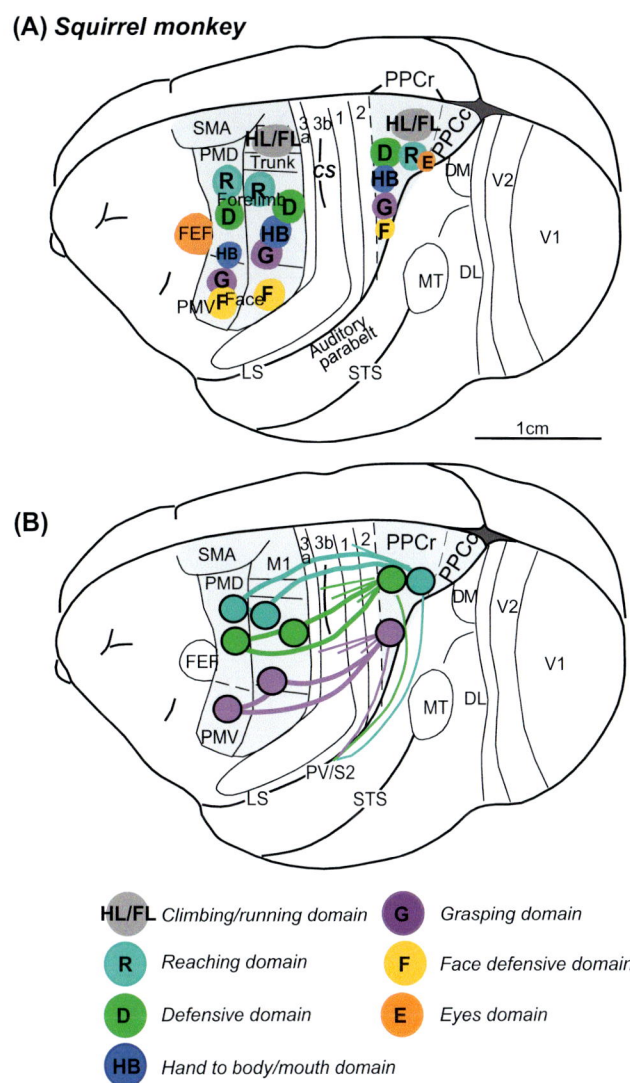

(A) Squirrel monkey

(B)

HL/FL — Climbing/running domain
R — Reaching domain
D — Defensive domain
HB — Hand to body/mouth domain
G — Grasping domain
F — Face defensive domain
E — Eyes domain

FIGURE 27.3 Dorsolateral view of New World squirrel monkey brain (left hemisphere). (A) Movement domains identified by long-train microstimulation are shown in areas M1, PMD, PMV, and PPC. An eye movement domain in PPC is mostly hidden in the upper bank of lateral sulcus, LS. (B) Major cortical connections of three PPC domains, reaching, defensive, and grasping. Note the strong connections between matching domains in PPC, PMC, and M1. *FEF*, frontal eye field; *MT*, middle temporal; *PMD*, dorsal premotor; *PMV*, ventral premotor; *PPC*, posterior parietal cortex; *SMA*, supplementary motor area; *M1*, motor cortex. *Based on Gharbawie, O.A., Stepniewska, I., Kaas, J.H., 2011. Cortical connections of functional zones in posterior parietal cortex and frontal cortex motor regions in new world monkeys. Cereb. Cortex 21, 1981–2002 and unpublished results.*

SMA. As in galagos, PPC domains in owl and squirrel monkeys are preferentially interconnected with functionally match domains in PMC and M1 (Fig. 27.3B; Gharbawie et al., 2011a). As in galagos, selective inactivations of functional domains in M1 of monkeys abolish or greatly reduce the movements evoked by electrically stimulating matching PPC domains (Stepniewska et al., 2014). With slight differences, the mediolateral

(A)

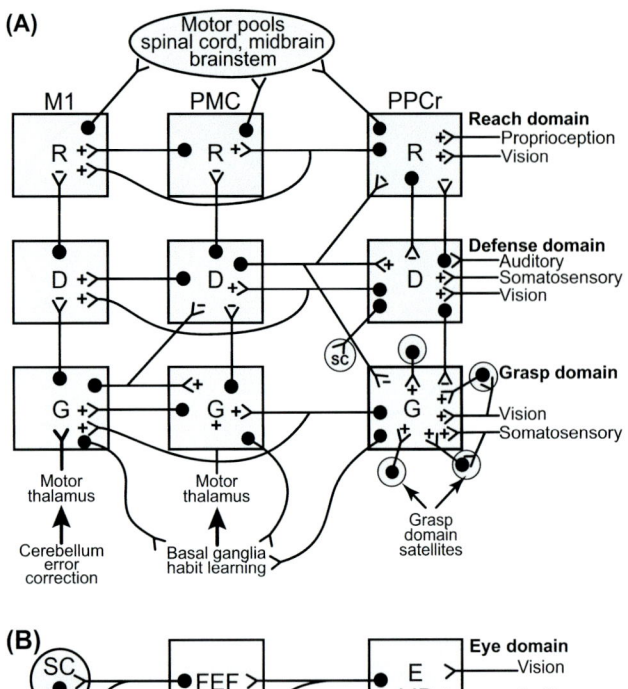

(B)

FIGURE 27.4   A model of domains interaction in primates. (A) The reaching (R), defense (D), and grasping (G) domains have been identified in galagos, New World owl, and squirrel monkeys and in macaque monkey (see Figs. 1–3). Each domain in rostral PPC (PPCr) receives multisensory information from higher order sensory areas and interacts via widespread PPC interconnections that may selectively activate local inhibitory neurons. We propose that domains most activated by sensory inputs use these interconnections to suppress other domains, resulting in excitatory outputs to premotor cortex (PMC) and primary motor cortex (M1) that promote the action of the most activated PPCr domain. Thus, a major function of PPCr is to "decide" on an appropriate action out of an array of approximately eight alternatives. Cortical neurons in clusters around domains provide additional excitatory inputs that promote domain activity (presumably for all domains, but shown for only the G domain). All domains project to subcortical targets, including the basal ganglia, but projections from the eye movement or looking domain (E or LIP) and the defense domain (D or VIP) to the superior colliculus may be especially important in promoting eye or defensive movements. Domains in PMC may interact in a similar competitive mode, based on PPCr domains activating inputs, motor thalamus inputs, but especially inputs from prefrontal cortex (not shown). M1 domains ultimately promote a final decision in the motor action based on PPCr and PMC inputs, as well as other cortical inputs and those from the motor thalamus. (B) The eye movement domain (E or LIP) in PPCr differs from other PPCr domains by having only a premotor target in the frontal eye field (FEF) and not in M1. For both LIP and FEF, a critical output is to the superior colliculus (SC) as well as motor pools in the pons. *LIP*, lateral intraparietal area; *PPC*, posterior parietal cortex; *VIP*, ventral intraparietal area.

arrangement of 7–9 domains in PPC, PMC, and M1 is similar in all three primates. If we consider the more limited evidence for such domains in macaque monkeys, their PPC domains are tilted so that medial domains are situated more caudal than medial. A similar rotation

seems to occur in PPC of humans. The rotation of medial PPC domains caudally could reflect the expansion of rostromedial PPC in macaques and humans. Most importantly, the combined evidence suggests that all primates share at least three or four action-specific PPC domains for different complex behaviors, and these behaviors are mediated via connections to matching domains in M1 and/or PMC.

27.3 The Functions of PPC Domains and Parallel Parietal-Frontal Networks

The primary functions of the domains region of PPC, as well as M1 and PMC, have been variously described. The evidence that PPC played a direct role in motor behavior led Mountcastle et al. (1975) to propose that this cortex contains "a command apparatus for the operation of the limbs, hands, and eyes." Another proposal, in consideration of other observations on the clinical syndrome of hemispatial neglect, is that PPC plays a critical role in directing attention to a subset of the incoming sensory information (Behrmann et al., 2004; Caminiti et al., 2010). Others have emphasized the role of PPC in decision making (Gold and Shadlen, 2007; Kiani and Shadlen, 2009; Katz et al., 2016). Most commonly, investigators have proposed a leading role for PPC areas in the planning of sensory-guided movements based on transformed terms of different spatial frames of reference (eg, Andersen and Buneo, 2002; Battaglia-Mayer et al., 2003; Kuang et al., 2016). For us, the existence of action-specific domains in PPC that are densely interconnected with each other suggests that a primary role for the domain region of PPC is to select one type of action over others based on sensory information (Fig. 27.4). Given expected differences in activation patterns of the 8 or 9 domains by the types and magnitudes of highly processed sensory inputs, one domain is likely to be activated more than others, depending on circumstances. During a short period of differences in activation, the interconnected domains could be inhibiting or facilitating each other in ways that produce a dominant output as sensory evidence accumulates, as in the "race model" of decision making of Logan et al. (1984). The decision-making process, in our view, is repeated in PMC and again in M1, as different factors come into play. For example, a decision to look at a visual target that is relayed to the FEF from LIP could be suppressed by inputs from adjoining frontal cortex that provide information about instructions held in short-term memory. There is already considerable evidence that signals for where to look or not to look are processed in the FEF in facilitation-suppression race-like model (Schall, 2005; Boucher et al., 2007). Thus, PMC and M1 are regions of frontal cortex where the action selection process can continue based on additional information. PPC provides a major source of information to PMC that already has gone through or is going through the PPC selection process. Other major influences on PMC domains come from PMC and the basal ganglia via the motor thalamus. M1 domains combine PPC and PMC influences, while adding inputs from cingulate motor areas, the SMA, and cerebellar outputs relayed from the motor thalamus. The widely distributed feedback connections from PMC to PPC or from M1 to PMC may largely excite inhibitory neurons, thus allowing M1 to widely suppress ongoing processing in PMC and PMC to widely suppress PPC domains (Falkner et al., 2010). Other possibilities further complicate this model, including the postulated existence of neurons around domains that assist in the processing (domain satellites), and domains providing sensory information to other domains (such as LIP providing visual information to AIP; eg, Nakamura et al., 2001). We suggest that the basic components of the behaviors evoked by electrical stimulation of domains are largely mediated by subcortical structures in the midbrain, brainstem, and spinal cord, as complex behaviors can be evoked by electrical stimulation from these structures (Mussa-Ivaldi and Bizzi, 2000; Flash and Hochner, 2005; Bizzi et al., 2008).

Our model of parietal-frontal domains and their interactions is a specific version of the "distributed consensus" model described elsewhere (Cisek, 2012). The distributed consensus model holds that decisions about actions are made through competitions taking place at multiple levels within the sensorimotor system. Usually, this model is concerned with the outcomes of similar actions with slightly different targets, such as where to look or where to reach, in which different parts of the FEF or the PRR compete with each other for dominance (see Schall, 2005). We hold that such competition within domains is important for decision about targets for action and versions of a class of action, but also for competition between different classes of actions. We add here the significance of reconsidering the action decision at each level in a sequence of regions involved in the process, that is PPC, PMC, and M1. The reason for decision making at multiple levels is clear: at each level, different factors come into play. Thus, sensory and perceptual information is dominant in PPC, but it also comes into play in frontal cortex. PMC is highly influenced by prefrontal cortex, which provides information about the values of choices, as well as information about recent instructions and learning from working and long-term memory. The presence of a greatly expanded PPC in primates, compared to most other mammals (eg, Fig. 27.5), allows primates the advantage of allowing higher order visual areas to more fully influence action choices. The less processed visual and other sensory information more directly influences action outcomes in M1 of rodents and tree shrews, allowing fast, but less

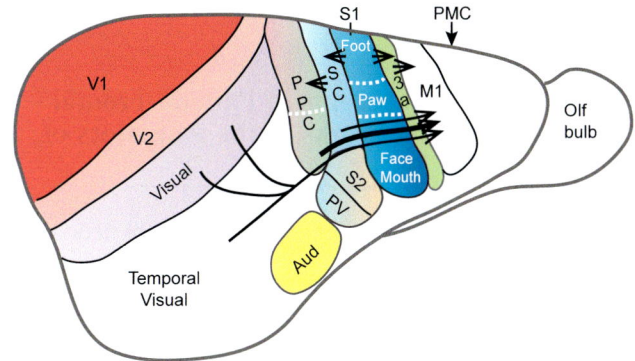

(A) Tree shrew

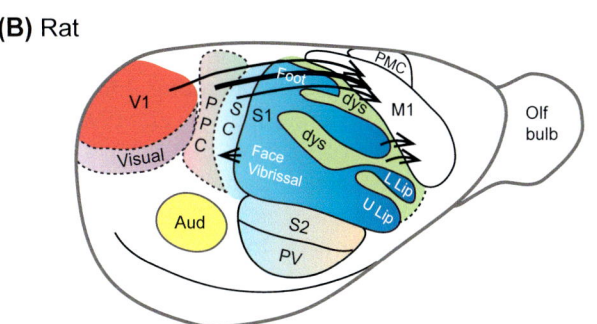

(B) Rat

FIGURE 27.5 Parietal-frontal pathways in relatives of primates. (A) Tree shrews are the closest living relatives of primates that are available for research. Visual cortex is expansive and includes primary (V1) and secondary (V2) visual areas, as well as a third level of cortical processing (visual) of several areas. Temporal visual cortex has additional subdivisions. Somatosensory areas include the primary area S1, a proprioceptive area 3a, just rostral to S1, a caudal somatosensory area, SC, just caudal to S1, a second somatosensory area S2, and the parietal ventral area PV. The region that appears to be posterior parietal cortex, PPC, is small with somatosensory and visual inputs and projections to primary motor cortex, M1. Movements can be evoked from PPC, but they tend to be simple. Direct inputs of visual and somatosensory areas to M1 likely play a more important role in motor behavior than those from PPC. M1 has suggestions of domain-like subregions. A premotor or cingulate motor area, PMC, on the medial wall of the frontal cortex has connections with M1. (B) Rodents are members of the superorder euarchontoglires with primates, but less closely related to primates than tree shrews. Cortical organization has been extensively studied in rats where a small PPC has been identified as in primates and tree shrews. PPC provides inputs to M1, while somatosensory and visual area provide direct sensory inputs to M1 and to PPC. There is a premotor area, PMC. The dysgranular cortex (dys) is homologue to area 3a of the shrews and primates. Electrical stimulation has provided some evidence for action-specific domains in M1 (see text). SC, superior colliculus. *(A). (Based on Remple, M.S., Reed, J.L., Stepniewska, I., Lyon, D.C., Kaas, J.H., 2007. The organization of frontoparietal cortex in the tree shrew (Tupaia belangeri): II. Connectional evidence for a frontal-posterior parietal network. J. Comp. Neurol. 501, 121–149; Lyon, D.C., Jain, N., Kaas, J.H., 1998. Cortical connections of striate and extrastriate visual areas in tree shrews. J. Comp. Neurol. 401, 109–128; Baldwin, M.K., Cooke, D.F., Krubitzer, L., 2017. Intracortical microstimulation maps of motor, somatosensory, and posterior parietal cortex in tree shrews (Tupaia belangeri) reveal complex movement representations. Cereb. Cortex 27(2), 1439–1456)*

27.4 The Antecedents of Parietal-Frontal Domains and Networks in the Ancestors of Primates

Comparative studies of the organization of neocortex of present-day mammals can provide some understanding of how this cortex was organized in the ancestors of primates. The comparative method allows ancestral traits or characters to be reconstructed on the premise that traits that are widely shared by extant members of a phylogenetically related group, a clade, are more likely to have been retained from a common ancestor than having been evolved independently (Henning, 1966; Brooks and McLennon, 1991; Cunningham et al., 1998). Such reasoning supports the conclusion that early primates had most or all of the action domains now found in PPC, PMC, and M1 of galagos and monkeys. To go back further in time, we need to consider present-day members of a large clade with a more distinct common ancestor, the last common ancestor of the superorder euarchontoglire, which emerged perhaps 100 mya (eg, Murphy et al., 2004). This clade includes primates, tree shrews, flying (gliding) lemurs, lagomorphs, and rodents. Tree shrews constitute the closet relatives of primates that have been studied (Fig. 27.5A). The cortical organizations of more distinctly related squirrels, rats (Fig. 27.5B), and mice have also been studied. These mammals have a primary motor area, M1, and a less well-studied secondary or premotor area. There is no clear evidence for separate dorsal and ventral premotor areas, a SMA, or a FEF in these mammals. Overall, M1 of rodents and tree shrews has a mediolateral somatotopic organization much as in M1 of other mammals (Baldwin et al., 2017; Young et al., 2012; Cooke et al., 2012; Remple et al., 2006). Maps of M1 with traditional methods of using brief periods of electrical microstimulation at near threshold levels have revealed a pattern of "fractured" or "mosaic" representation with separate, repeated representations of movements of body parts as seen more clearly in the bigger brains of squirrels (Cooke et al., 2012) and tree shrews (Remple et al., 2006). In this way, the near threshold maps of M1 resemble those of M1 in primates, although in less detail and with less emphasis on forepaw (hand) movements. Yet, this fractured somatotopy provides strong evidence for the emergence of an organization in M1 that is compatible with the existence of domains in the nonprimate ancestors of primates. It is also possible that this fractured somatotopy emerged with the evolution of M1 in the common ancestor of placental mammals (see Beck et al., 1996). Furthermore, long-train, suprathreshold

electrical stimulation of some sites in M1 in tree shrews produced complex movements such as defensive postures, feeding, or locomotion. Thus, there is some evidence for a domain type of organization in M1 of tree shrews that is superimposed on the overall crude somatotopy (Baldwin et al., 2017). In a similar manner, long trains of above threshold stimulation of sites in M1 of rats (Ramanathan et al., 2006; Bonazzi et al., 2013; Brown and Tesky, 2014) and mice (Hira et al., 2015) produced complex movements suggestive of domains. These findings suggest that some type of domain-like organization might have been a feature of the organization of M1 in early euarchontoglires.

There is less evidence for domains in PPC of rodents and tree shrews. Long-train stimulation has been applied to PPC only in tree shrews, and these mammals, in contrast to primates, have very little cortex that could be called PPC. Yet, tree shrews have a small PPC region, defined by location and visual and somatosensory inputs, and this cortex projects to M1 (eg, Remple et al., 2007). While movements were evoked from PPC in tree shrews, these movements were difficult to evoke, were mainly simple, and often consisted of an eye squint or ear movement. Thus, the limited evidence from stimulation, and the small size of PPC, suggests that the PPC of early euarchontoglire mammals had only a minor role in producing movements, and a domain-like organization did not exist.

We propose that the fractured somatotopy of M1 in early placental mammals set the stage for the formation of a small number of M1 domains and for a like organization for emerging PMC. This in turn influenced the evolution in early primates, of a much larger PPC, with a portion of PPC having domains and an enhanced role in the selection of complex actions (motor primitives). PPC domains differentiated further in the recent ancestors of humans to facilitate our abilities in tool use and manufacture and aspects of language. In early primates, the expansion and magnified motor role of PPC was accompanied by a great expansion of visual cortex and addition of visual areas, so that higher order processing of visual inputs became a dominant source of information for action selection in PPC (Kaas, 2017).

References

Adams, R.A., Shipp, S., Friston, K.J., 2013. Predictions not commands: active inference in the motor system. Brain Struct. Funct. 218, 611–643.

Andersen, R.A., Asanuma, C., Cowan, W.M., 1985. Callosal and prefrontal associational projecting cell populations in area 7A of the macaque monkey: a study using retrogradely transported fluorescent dyes. J. Comp. Neurol. 232, 443–455.

Andersen, R.A., Brotchie, P.R., Mazzoni, P., 1992. Evidence for the lateral intraparietal area as the parietal eye field. Curr. Opin. Neurobiol. 2, 840–846.

Andersen, R.A., Buneo, C.A., 2002. Intentional maps in posterior parietal cortex. Annu. Rev. Neurosci. 25, 189–220.

Andersen, R.A., Snyder, L.H., Bradley, D.C., Xing, J., 1997. Multimodal representation of space in the posterior parietal cortex and its use in planning movements. Annu. Rev. Neurosci. 20, 303–330.

Baldwin, M.K., Cooke, D.F., Krubitzer, L., 2017. Intracortical microstimulation maps of motor, somatosensory, and posterior parietal cortex in tree shrews (*Tupaia belangeri*) reveal complex movement representations. Cereb. Cortex 27 (2), 1439–1456.

Battaglia-Mayer, A., Archambault, P.S., Caminiti, R., 2006. The cortical network for eye-hand coordination and its relevance to understanding motor disorders of parietal patients. Neuropsychologia 44, 2607–2620.

Battaglia-Mayer, A., Caminiti, R., Lacquaniti, F., Zago, M., 2003. Multiple levels of representation of reaching in the parieto-frontal network. Cereb. Cortex 13, 1009–1022.

Baumann, M.A., Fluet, M.C., Scherberger, H., 2009. Context-specific grasp movement representation in the macaque anterior intraparietal area. J. Neurosci. 29, 6436–6448.

Beck, P.D., Pospichal, M.W., Kaas, J.H., 1996. Topography, architecture, and connections of somatosensory cortex in opossums: evidence for five somatosensory areas. J. Comp. Neurol. 366, 109–133.

Behrmann, M., Geng, J.J., Shomstein, S., 2004. Parietal cortex and attention. Curr. Opin. Neurobiol. 14, 212–217.

Bizzi, E., Cheung, V.C., d'Avella, A., Saltiel, P., Tresch, M., 2008. Combining modules for movement. Brain Res. Rev. 57, 125–133.

Blatt, G.J., Andersen, R.A., Stoner, G.R., 1990. Visual receptive field organization and cortico-cortical connections of the lateral intraparietal area (area LIP) in the macaque. J. Comp. Neurol. 299, 421–445.

Bonazzi, L., Viaro, R., Lodi, E., Canto, R., Bonifazzi, C., Franchi, G., 2013. Complex movement topography and extrinsic space representation in the rat forelimb motor cortex as defined by long-duration intracortical microstimulation. J. Neurosci. 33, 2097–2107.

Borra, E., Belmalih, A., Calzavara, R., Gerbella, M., Murata, A., et al., 2008. Cortical connections of the macaque anterior intraparietal (AIP) area. Cereb. Cortex 18, 1094–1111.

Boucher, L., Palmeri, T.J., Logan, G.D., Schall, J.D., 2007. Inhibitory control in mind and brain: an interactive race model of countermanding saccades. Psychol. Rev. 114, 376–397.

Brooks, D.R., McLennon, D.A., 1991. Phylogeny, Ecology, and Behavior. University of Chicago press, Chicago, IL.

Brown, A.R., Teskey, G.C., 2014. Motor cortex is functionally organized as a set of spatially distinct representations for complex movements. J. Neurosci. 34, 13574–13585.

Caminiti, R., Chafee, M.V., Battaglia-Mayer, A., Averbeck, B.B., Crowe, D.A., Georgopoulos, A.P., 2010. Understanding the parietal lobe syndrome from a neurophysiological and evolutionary perspective. Eur. J. Neurosci. 31, 2320–2340.

Caminiti, R., Ferraina, S., Johnson, P.B., 1996. The sources of visual information to the primate frontal lobe: a novel role for the superior parietal lobule. Cereb. Cortex 6, 319–328.

Caminiti, R., Innocenti, G.M., Battaglia-Mayer, A., 2015. Organization and evolution of parieto-frontal processing streams in macaque monkeys and humans. Neurosci. Biobehav. Rev. 56, 73–96.

Cisek, P., 2012. Making decisions through a distributed consensus. Curr. Opin. Neurobiol. 22, 927–936.

Cohen, Y.E., Andersen, R.A., 2002. A common reference frame for movement plans in the posterior parietal cortex. Nat. Rev. Neurosci. 3, 553–562.

Colby, C.L., Duhamel, J.R., Goldberg, M.E., 1993. Ventral intraparietal area of the macaque: anatomic location and visual response properties. J. Neurophysiol. 69, 902–914.

Cooke, D.F., Padberg, J., Zahner, T., Krubitzer, L., 2012. The functional organization and cortical connections of motor cortex in squirrels. Cereb. Cortex 22, 1959–1978.

Cooke, D.F., Stepniewska, I., Miller, D.J., Kaas, J.H., Krubitzer, L., 2015. Reversible deactivation of motor cortex reveals functional connectivity with posterior parietal cortex in the prosimian Galago (*Otolemur garnettii*). J. Neurosci. 35, 14406–14422.

Cooke, D.F., Taylor, C.S., Moore, T., Graziano, M.S., 2003. Complex movements evoked by microstimulation of the ventral intraparietal area. Proc. Natl. Acad. Sci. U.S.A. 100, 6163–6168.

Cunningham, C.W., Omland, K.E., Oakley, T.H., 1998. Reconstructing ancestral character states: a critical reappraisal. Trends Ecol. Evol. 13, 361–366.

Desmurget, M., Richard, N., Harquel, S., Baraduc, P., Szathmari, A., et al., 2014. Neural representations of ethologically relevant hand/mouth synergies in the human precentral gyrus. Proc. Natl. Acad. Sci. U. S. A 111, 5718–5722.

Donoghue, J.P., Leibovic, S., Sanes, J.N., 1992. Organization of the forelimb area in squirrel monkey motor cortex: representation of digit, wrist, and elbow muscles. Exp. Brain Res. 89, 1–19.

Duhamel, J.R., Colby, C.L., Goldberg, M.E., 1992. The updating of the representation of visual space in parietal cortex by intended eye movements. Science 255, 90–92.

Duhamel, J.R., Colby, C.L., Goldberg, M.E., 1998. Ventral intraparietal area of the macaque: congruent visual and somatic response properties. J. Neurophysiol. 79, 126–136.

Esposito, M.S., Capelli, P., Arber, S., 2014. Brainstem nucleus MdV mediates skilled forelimb motor tasks. Nature 508, 351–356.

Falkner, A.L., Krishna, B.S., Goldberg, M.E., 2010. Surround suppression sharpens the priority map in the lateral intraparietal area. J. Neurosci. 30, 12787–12797.

Fang, P.C., Stepniewska, I., Kaas, J.H., 2005. Ipsilateral cortical connections of motor, premotor, frontal eye, and posterior parietal fields in a prosimian primate, Otolemur garnetti. J. Comp. Neurol. 490, 305–333.

Filimon, F., 2010. Human cortical control of hand movements: parieto-frontal networks for reaching, grasping, and pointing. The Neuroscientist 16, 388–407.

Flash, T., Hochner, B., 2005. Motor primitives in vertebrates and invertebrates. Curr. Opin. Neurobiol. 15, 660–666.

Fogassi, L., Gallese, V., Gentilucci, M., Luppino, G., Matelli, M., Rizzolatti, G., 1994. The fronto-parietal cortex of the prosimian Galago: patterns of cytochrome oxidase activity and motor maps. Behav. Brain Res. 60, 91–113.

Gharbawie, O.A., Stepniewska, I., Kaas, J.H., 2011. Cortical connections of functional zones in posterior parietal cortex and frontal cortex motor regions in new world monkeys. Cereb. Cortex 21, 1981–2002.

Gharbawie, O.A., Stepniewska, I., Qi, H., Kaas, J.H., 2011. Multiple parietal-frontal pathways mediate grasping in macaque monkeys. J. Neurosci. 31, 11660–11677.

Gold, J.I., Shadlen, M.N., 2007. The neural basis of decision making. Annu. Rev. Neurosci. 30, 535–574.

Gould 3rd, H.J., Cusick, C.G., Pons, T.P., Kaas, J.H., 1986. The relationship of corpus callosum connections to electrical stimulation maps of motor, supplementary motor, and the frontal eye fields in owl monkeys. J. Comp. Neurol. 247, 297–325.

Graybiel, A.M., Grafton, S.T., 2015. The striatum: where skills and habits meet. Cold Spring Harb. Perspect. Biol. 7, a021691.

Graziano, M.S., 2009. The Intelligent Movement Machine. Oxford University Press, New York.

Graziano, M.S.A., Taylor, C.S.R., Moore, T., 2002. Complex movements evoked by microstimulation of precentral cortex. Neuron 34, 841–851.

Hartwigsen, G., Weigel, A., Schuschan, P., Siebner, H.R., Weise, D., et al., 2016. Dissociating parieto-frontal networks for phonological and semantic word decisions: a condition-and-perturb TMS study. Cereb. Cortex 26, 2590–2601.

Hennig, W., 1966. Phylogenetic Systematics. Univ. of Ill, Urbana, USA.

Hira, R., Terada, S., Kondo, M., Matsuzaki, M., 2015. Distinct functional modules for discrete and rhythmic forelimb movements in the mouse motor cortex. J. Neurosci. 35, 13311–13322.

Huerta, M.F., Krubitzer, L.A., Kaas, J.H., 1986. Frontal eye field as defined by intracortical microstimulation in squirrel monkeys, owl monkeys, and macaque monkeys: I. Subcortical connections. J. Comp. Neurol. 253, 415–439.

Huerta, M.F., Krubitzer, L.A., Kaas, J.H., 1987. Frontal eye field as defined by intracortical microstimulation in squirrel monkeys, owl monkeys, and macaque monkeys. II. Cortical connections. J. Comp. Neurol. 265, 332–361.

Huntley, G.W., Jones, E.G., 1991. Relationship of intrinsic connections to forelimb movement representations in monkey motor cortex: a correlative anatomic and physiological study. J. Neurophysiol. 66, 390–413.

Jerde, T.A., Merriam, E.P., Riggall, A.C., Hedges, J.H., Curtis, C.E., 2012. Prioritized maps of space in human frontoparietal cortex. J. Neurosci. 32, 17382–17390.

Kaas, J.H., 2017. Evolution of visual cortex in primates. In: Kaas, J.H., Krubitzer, L. (Eds.), Evolution of Nervous Systems, Primates, second ed., vol. 2. Elsevier, Oxford, UK (in press).

Kaas, J.H., Stepniewska, I., 2016. Evolution of posterior parietal cortex and parietal-frontal networks for specific actions in primates. J. Comp. Neurol. 524, 595–608.

Kalaska, J.F., Scott, S.H., Cisek, P., Sergio, L.E., 1997. Cortical control of reaching movements. Curr. Opin. Neurobiol. 7, 849–859.

Katz, L.N., Yates, J.L., Pillow, J.W., Huk, A.C., 2016. Dissociated functional significance of decision-related activity in the primate dorsal stream. Nature 535, 285–288.

Kiani, R., Shadlen, M.N., 2009. Representation of confidence associated with a decision by neurons in the parietal cortex. Science 324, 759–764.

Konen, C.S., Mruczek, R.E., Montoya, J.L., Kastner, S., 2013. Functional organization of human posterior parietal cortex: grasping- and reaching-related activations relative to topographically organized cortex. J. Neurophysiol. 109, 2897–2908.

Kuang, S., Morel, P., Gail, A., 2016. Planning movements in visual and physical space in monkey posterior parietal cortex. Cereb. Cortex 26, 731–747.

Kurylo, D.D., Skavenski, A.A., 1991. Eye movements elicited by electrical stimulation of area PG in the monkey. J. Neurophysiol. 65, 1243–1253.

Leone, F.T., Monaco, S., Henriques, D.Y., Toni, I., Medendorp, W.P., 2015. Flexible reference frames for grasp planning in human parietofrontal cortex(1,2,3). eNeuro 2.

Logan, G.D., Cowan, W.B., Davis, K.A., 1984. On the ability to inhibit simple and choice reaction time responses: a model and a method. J. Exp. Psychol. Hum. Percept. Perform. 10, 276–291.

Luppino, G., Murata, A., Govoni, P., Matelli, M., 1999. Largely segregated parietofrontal connections linking rostral intraparietal cortex (areas AIP and VIP) and the ventral premotor cortex (areas F5 and F4). Exp. Brain Res. 128, 181–187.

Lynch, J.C., Graybiel, A.M., Lobeck, L.J., 1985. The differential projection of two cytoarchitectonic subregions of the inferior parietal lobule of macaque upon the deep layers of the superior colliculus. J. Comp. Neurol. 235, 241–254.

Lyon, D.C., Kaas, J.H., 2002. Connectional evidence for dorsal and ventral V3, and other extrastriate areas in the prosimian primate, Galago garnetti. Brain Behav. Evol. 59, 114–129.

Lyon, D.C., Jain, N., Kaas, J.H., 1998. Cortical connections of striate and extrastriate visual areas in tree shrews. J. Comp. Neurol. 401, 109–128.

Maunsell, J.H., van Essen, D.C., 1983. The connections of the middle temporal visual area (MT) and their relationship to a cortical hierarchy in the macaque monkey. J. Neurosci. 3, 2563–2586.

Mountcastle, V.B., Lynch, J.C., Georgopoulos, A., Sakata, H., Acuna, C., 1975. Posterior parietal association cortex of the monkey: command functions for operations within extrapersonal space. J. Neurophysiol. 38, 871–908.

Murata, A., Gallese, V., Luppino, G., Kaseda, M., Sakata, H., 2000. Selectivity for the shape, size, and orientation of objects for grasping in neurons of monkey parietal area AIP. J. Neurophysiol. 83, 2580–2601.

Murphy, W.J., Pevzner, P.A., O'Brien, S.J., 2004. Mammalian phylogenomics comes of age. Trends Genet. 20, 631–639.

Mussa-Ivaldi, F.A., Bizzi, E., 2000. Motor learning through the combination of primitives. Philos. Trans. R. Soc. Lond. B Biol. Sci. 355, 1755–1769.

Nakamura, H., Kuroda, T., Wakita, M., Kusunoki, M., Kato, A., et al., 2001. From three-dimensional space vision to prehensile hand movements: the lateral intraparietal area links the area V3A and the anterior intraparietal area in macaques. J. Neurosci. 21, 8174–8187.

Patel, G.H., Yang, D., Jamerson, E.C., Snyder, L.H., Corbetta, M., Ferrera, V.P., 2015. Functional evolution of new and expanded attention networks in humans. Proc. Natl. Acad. Sci. U.S.A. 112, 9454–9459.

Peeters, R., Simone, L., Nelissen, K., Fabbri-Destro, M., Vanduffel, W., et al., 2009. The representation of tool use in humans and monkeys: common and uniquely human features. J. Neurosci. 29, 11523–11539.

Penfield, W., Boldrey, E., 1937. Somatic motor and sensory representation in the cerebral cortex of man as studied by electrical stimulation. Brain 37, 389–443.

Ptak, R., Muri, R.M., 2013. The parietal cortex and saccade planning: lessons from human lesion studies. Front. Hum. Neurosci. 7, 254.

Ramanathan, D., Conner, J.M., Tuszynski, M.H., 2006. A form of motor cortical plasticity that correlates with recovery of function after brain injury. Proc. Natl. Acad. Sci. U.S.A. 103, 11370–11375.

Remple, M.S., Reed, J.L., Stepniewska, I., Kaas, J.H., 2006. Organization of frontoparietal cortex in the tree shrew (Tupaia belangeri). I. Architecture, microelectrode maps, and corticospinal connections. J. Comp. Neurol. 497, 133–154.

Remple, M.S., Reed, J.L., Stepniewska, I., Lyon, D.C., Kaas, J.H., 2007. The organization of frontoparietal cortex in the tree shrew (Tupaia belangeri): II. Connectional evidence for a frontal-posterior parietal network. J. Comp. Neurol. 501, 121–149.

Rizzolatti, G., Fogassi, L., Gallese, V., 1997. Parietal cortex: from sight to action. Curr. Opin. Neurobiol. 7, 562–567.

Sakata, H., Taira, M., Murata, A., Mine, S., 1995. Neural mechanisms of visual guidance of hand action in the parietal cortex of the monkey. Cereb. Cortex 5, 429–438.

Savaki, H.E., Gregoriou, G.G., Bakola, S., Raos, V., Moschovakis, A.K., 2010. The place code of saccade metrics in the lateral bank of the intraparietal sulcus. J. Neurosci. 30, 1118–1127.

Schall, J.D., 2005. Decision making. Curr. Biol. 15, R9–R11.

Schieber, M.H., 2001. Constraints on somatotopic organization in the primary motor cortex. J. Neurophysiol. 86, 2125–2143.

Sereno, M.I., Huang, R.S., 2014. Multisensory maps in parietal cortex. Curr. Opin. Neurobiol. 24, 39–46.

Shambes, G.M., Gibson, J.M., Welker, W., 1978. Fractured somatotopy in granule cell tactile areas of rat cerebellar hemispheres revealed by micromapping. Brain Behav. Evol. 15, 94–140.

Shibutani, H., Sakata, H., Hyvarinen, J., 1984. Saccade and blinking evoked by microstimulation of the posterior parietal association cortex of the monkey. Exp. Brain Res. 55, 1–8.

Silcox, M.T., Benham, A.E., Bloch, J.I., 2010. Endocasts of Microsyops (Microsyopidae, Primates) and the evolution of the brain in primitive primates. J. Hum. Evol. 58, 505–521.

Steiper, M.E., Seiffert, E.R., 2012. Evidence for a convergent slowdown in primate molecular rates and its implications for the timing of early primate evolution. Proc. Natl. Acad. Sci. U.S.A. 109, 6006–6011.

Stepniewska, I., Fang, P.C., Kaas, J.H., 2009. Organization of the posterior parietal cortex in galagos: I. Functional zones identified by microstimulation. J. Comp. Neurol. 517, 765—782.

Stepniewska, I., Cerkevich, C.M., Fang, P.C., Kaas, J.H., 2009. Organization of the posterior parietal cortex in galagos: II. Ipsilateral cortical connections of physiologically identified zones within anterior sensorimotor region. J. Comp. Neurol. 517, 783—807.

Stepniewska, I., Cerkevich, C.M., Kaas, J.H., 2016. Cortical connections of the caudal portion of posterior parietal cortex in prosimian galagos. Cereb. Cortex 26, 2753—2777.

Stepniewska, I., Fang, P.C., Kaas, J.H., 2005. Microstimulation reveals specialized subregions for different complex movements in posterior parietal cortex of prosimian galagos. Proc. Natl. Acad. Sci. U.S.A. 102, 4878—4883.

Stepniewska, I., Friedman, R.M., Gharbawie, O.A., Cerkevich, C.M., Roe, A.W., Kaas, J.H., 2011. Optical imaging in galagos reveals parietal-frontal circuits underlying motor behavior. Proc. Natl. Acad. Sci. U.S.A. 108, E725—E732.

Stepniewska, I., Gharbawie, O.A., Burish, M.J., Kaas, J.H., 2014. Effects of muscimol inactivations of functional domains in motor, premotor, and posterior parietal cortex on complex movements evoked by electrical stimulation. J. Neurophysiol. 111, 1100—1119.

Thier, P., Andersen, R.A., 1996. Electrical microstimulation suggests two different forms of representation of head-centered space in the intraparietal sulcus of rhesus monkeys. Proc. Natl. Acad. Sci. U.S.A. 93, 4962—4967.

Tovote, P., Esposito, M.S., Botta, P., Chaudun, F., Fadok, J.P., Markovic, M., Wolff, S.B., Ramakrishnan, C., Fenno, L., Deisseroth, K., Herry, C., Arber, S., Luthi, A., 2016. Midbrain circuits for defensive behaviour. Nature 534, 206—212.

Wu, C.W., Bichot, N.P., Kaas, J.H., 2000. Converging evidence from microstimulation, architecture, and connections for multiple motor areas in the frontal and cingulate cortex of prosimian primates. J. Comp. Neurol. 423, 140—177.

Wurtz, R.H., Albano, J.E., 1980. Visual-motor function of the primate superior colliculus. Annu. Rev. Neurosci. 3, 189—226.

Young, N., Stepniewska, I., Kaas, J.H., 2012. Motor cortex. In: Watson, C., Paxinos, G., Puelles, L. (Eds.), The Mouse Nervous System. Elsevier, London, pp. 527—537.

28

The Evolution of the Prefrontal Cortex in Early Primates and Anthropoids

S.P. Wise

Olschefskie Institute for the Neurobiology of Knowledge, Potomac, MD, United States

28.1 Introduction

28.1.1 Lunar Primates

"Like the progress a monkey makes toward the moon by climbing a tree": so goes a simian simile that sneers at the formative stages of a seemingly quixotic quest. People who employ this expression might do well to consider that, from an evolutionary perspective, the journey that took primates to the moon actually began with climbing in trees—along with feeding, breeding, and living there—and a series of adaptations set in train by that way of life.

In fact, it did not take all that long—in the context of geologic time—for primates to reach the moon. Animals first evolved approximately 700–750 million years ago (Ma), and the first vertebrates appeared during the Cambrian explosion, ~530 Ma (Erwin and Valentine, 2013). Mammals emerged mainly in the Jurassic, ~170–180 Ma (Luo et al., 2015), and the first primates evolved in the Paleocene, 55–65 Ma (Silcox et al., 2015). Hominins and panins diverged in the late Miocene, approximately 6–7 Ma, with anatomically and behaviorally modern humans appearing some 50–150 thousand years ago (ka) (Klein, 2009). These people hunted and gathered for a while, then settled down to grow grassy grains ~10 ka. So, in the chronicle of life on earth, it took only ~11% of vertebrate history and ~8% of animal history for primates to emerge, diversify, and launch a lunar lander.

Shortly after the first two primates landed on the moon, a blurry, black-and-white image showed one of them descending a ladder, feet first. As he leapt off the ladder and landed in the lunar dust, he announced that he was taking "one small step for a man, one giant leap for mankind." A little while later, as he walked around the Sea of Tranquility on his hindlimbs, he directed his attention to minute distinctions among the colors, shapes, and visual textures of some rocks, using his foveal vision. After evaluating and choosing a particular rock, he used a hand tool to grasp it, bringing a gripper precisely to the rock's location under visual guidance.

The 500 million crown primates who witnessed these activities on television heard a great deal about the technological advances that enabled the first lunar landing; advances in rocketry, guidance systems, and computers all received their due adulation. Yet the audience heard little about the evolutionary developments that made equally necessary contributions. The lunar primate backed down the ladder and walked on his legs in part because early primates developed a hindlimb-dominated form of locomotion; he searched for moon rocks using cortical areas that evolved in primates; and he picked up moon rocks by drawing on an early-primate mechanism for visually guided movement and the foveal, trichromatic vision that evolved later.

28.1.2 Steps and Leaps

A quarter century ago, the scientific understanding of the primate prefrontal cortex took both a "small step" and a "giant leap" when Preuss and Goldman-Rakic (1991a,b,c) published a series of landmark articles. The small step involved the application of standard neuroanatomical methods to the brain of a strepsirrhine primate, variously called the bushbaby, *Galago*, or *Otolemur*. Preuss and Goldman-Rakic described some of the connections among its cortical areas, along with the cytoarchitecture and myeloarchitecture of its frontal cortex. The giant leap came from a comparison of the bushbaby frontal cortex with that of other mammals and other primates, especially rodents and macaque monkeys. Preuss and Goldman-Rakic advanced three key conclusions about the prefrontal cortex:

- one set of areas, shared among mammals, has an agranular cytoarchitecture;
- a second set, characterized by a homotypical cytoarchitecture and a moderate degree of myelination, evolved de novo in early primates; and
- a third set, with homotypical architecture but less myelination than the second set, emerged during anthropoid evolution.

28.1.3 Terms

One word in the previous section, strepsirrhine, is relatively uncommon in the neuroscience literature; and two anatomical expressions are unusual in writings on evolution: cytoarchitecture and myeloarchitecture.

As for the anatomical terms, neocortical areas are traditionally divided into three basic types of cytoarchitecture: agranular, homotypical, and granular, along with various intermediate forms. Homotypical areas, also known as eulaminate cortex, have a pattern of cell-body layering that dominates the neocortex of

humans, with a reasonable representation of the six main layers; agranular areas lack the fourth layer, also called the internal granular layer; and granular areas, also known as koniocortex, have an unusually thick layer 4. Among the intermediate forms, some cortical areas have a cytoarchitecture between agranular and homotypical and are called dysgranular. For historical reasons, and somewhat confusingly, homotypical and dysgranular areas are often called "granular" in the primate frontal lobe. Myeloarchitecture takes many forms, but cortical areas are often classified as lightly, moderately, or heavily myelinated.

Primate taxonomy can be as arduous as neuroanatomy. This chapter is aimed at neuroscientists and therefore uses common names as often as possible. Even so, readers unfamiliar with terms such as strepsirrhine and plesiadapiform might mark this paragraph and Fig. 28.1 for reference. Modern primates belong to one of two groups: strepsirrhines or haplorhines. The many kinds of modern strepsirrhines include lemurs, lorises, and bushbabies. Because of their relatively small size and nocturnal habit, one group of haplorhines, tarsiers, are commonly grouped with the strepsirrhines in a paraphyletic group called prosimians. The remaining

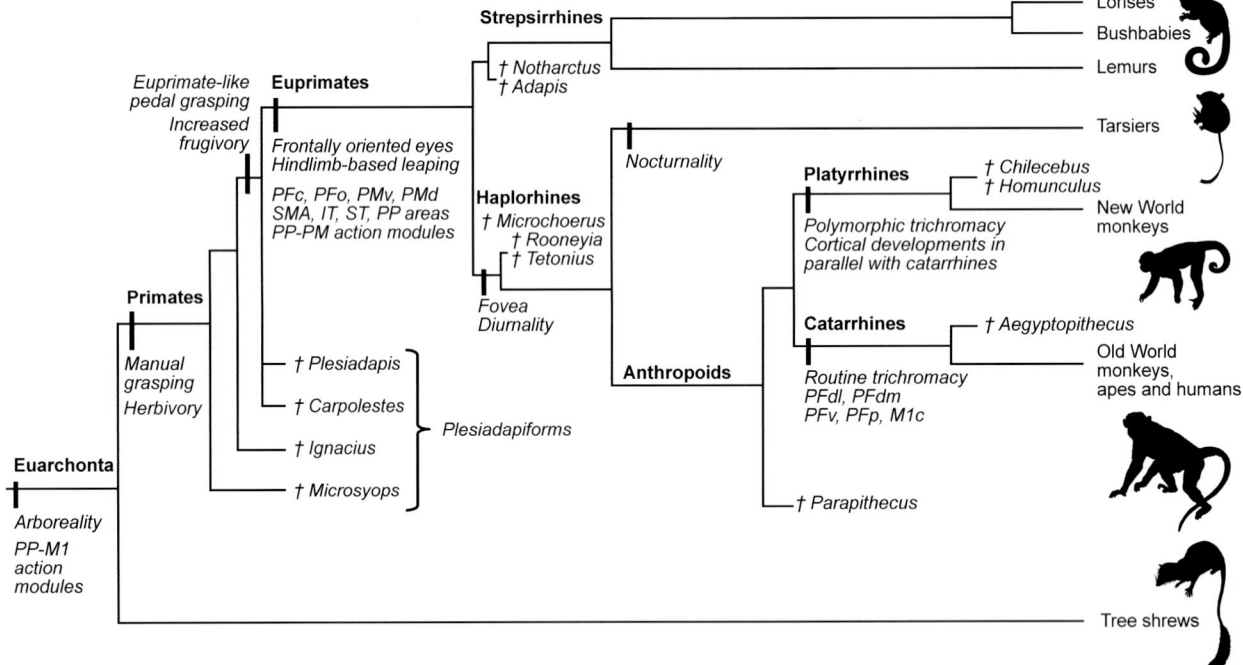

FIGURE 28.1   Cladogram of modern and extinct primates mentioned in this chapter, along with tree shews. *Black vertical bars* show the lineage associated with evolutionary innovations indicated beneath them. The *dagger* (†) indicates an extinct species. The placement of *Rooneyia* as a stem haplorhine follows Ni et al. (2013) and Ramdarshan and Orliac (2015). Alternatively, it could be a very primitive stem euprimate (Ross et al., 1998) or a very primitive stem strepsirrhine (Kay et al., 2004). The trifurcation for *Carpolestes, Plesiadapis*, and euprimates follows Bloch and Silcox (2006). Synapomorphies in early primates come from Silcox et al. (2015). Abbreviations: *IT*, inferior temporal cortex; *PFc*, caudal prefrontal cortex, which includes the frontal eye fields; *PFdl*, dorsolateral prefrontal cortex; *PFdm*, dorsomedial prefrontal cortex; *PFo*, granular orbitofrontal cortex; *PFp*, polar prefrontal cortex; *PFv*, ventral prefrontal cortex, *PM*, premotor cortex; *PMd*, dorsal premotor cortex; *PMv*, ventral premotor cortex; *M1c*, caudal primary motor cortex; *PP*, posterior parietal cortex; *SMA*, supplementary motor area; *ST*, superior temporal cortex. Silhouettes from open source http://phylopic.org/.

haplorhines, collectively called anthropoids, include all monkeys, apes, and humans. Anthropoids eventually split into two groups, platyrrhines (New World monkeys) and catarrhines (Old World monkeys, humans, and apes). Fig. 28.1 also shows a sister group of primates, tree shrews (*Scandentia*). [It omits another outgroup, flying lemurs (*Dermoptera*).] Collectively, primates, flying lemurs, and tree shrews compose a clade called "euarchonta." More basally among mammals, a clade that includes euarchontans, rabbits, and rodents is called "euarchontoglires."

This chapter also mentions a number of fossil primates, in three groups: plesiadapiforms, stem euprimates, and anthropoids. Plesiadapiforms are allied with stem primates (Bloch and Silcox, 2006; Bloch et al., 2007) and are probably a paraphyletic group. This chapter mentions four plesiadapiforms: *Carpolestes*, *Plesiadapis*, *Microsyops*, and *Ignacius*. Fossil euprimates fall into two major groups: omomyoids (such as *Tetonius*, *Microchoerus*, and *Rooneyia*), which are either stem euprimates or stem haplorhines, and adapoids (such as *Adapis* and *Notharctus*), which are either stem euprimates or stem strepsirrhines. All modern primates are "euprimates." This chapter also refers to four extinct anthropoid species: *Aegyptopithecus*, *Parapithecus*, *Homunculus*, and *Chilecebus*. Fig. 28.1 illustrates one view of the evolutionary relationships among these fossils and modern primates.

28.1.4 Advances

After describing the bushbaby prefrontal cortex in his papers with Goldman-Rakic, Preuss (1993, 1995) published review articles placing their findings in the context of then-current ideas about the prefrontal cortex. Much has changed in the interim. Most importantly, recent work has challenged two theories about the prefrontal cortex that dominated the field at that time. In the 1990s most experts believed that the primate orbitofrontal cortex functions mainly in behavioral inhibition. Some still do, but a great deal of experimental evidence contradicts this idea (Section **28.6.2.3.2**). Similarly, it was widely believed that lateral parts of the prefrontal cortex function in working memory—and little else. Experimental results have overturned this idea as well (Section **28.6.2.3.1**). Considerable uncertainty remains about the functions of the prefrontal cortex, but recent advances provide some insight into the advantages that these areas afforded to ancestral primates in their time and place.

Previous publications have dealt with many of the topics in this chapter, in various combinations and in association with several coauthors (Wise, 2006, 2007, 2008, 2014; Murray et al., 2011; Genovesio et al., 2014; Murray et al., 2016), but the largest debt is owed to the book *The Neurobiology of the Prefrontal Cortex* by Passingham and Wise (2012).

28.2 Primate Adaptations

28.2.1 Early Primates

According to most paleontologists, stem euprimates occupied a nocturnal foraging niche, living a life confined to the fine branches of trees and shrubs (Fleagle, 1999; Rose, 2006). Their shared derived traits included a leaping–grasping form of locomotion, finger nails instead of claws for grasping, and forward-facing eyes.

Fossil evidence indicates that the grasping specializations developed first and probably preceded the emergence of euprimates (Silcox et al., 2015). The plesiadapiform *Carpolestes*, which dates to the early Eocene ($\sim$55 Ma), lacked frontally directed eyes and either the hindlimb or pelvic specializations associated with leaping. However, *Carpolestes* had limbs specialized for grasping and the finger nails characteristic of primates (Bloch and Boyer, 2002).

Fig. 28.1 illustrates one idea about the relationship between euprimates and plesiadapiforms (Bloch and Silcox, 2006). Along with forward-facing eyes, the features that developed in euprimates and differentiated them from other primates included adaptations of hand morphology for two related functions: a leaping–grasping form of locomotion and clinging to branches with prehensile hands (Boyer et al., 2013b). Developments in foot structure also promoted leaping (Boyer et al., 2013a). Accordingly, frontally directed vision, hindlimb-dominated leaping, and nail-assisted grasping seem to have come together in stem euprimates. Grasping hands and feet provide obvious advantages for navigating through the fine-branch niche (Sussman, 1991) and for foraging (Cartmill, 1992; Boyer et al., 2013b).

Like other mammals, early primates lacked both a fovea and trichromatic vision, but their frontally directed eyes enhanced stereopsis and depth perception (Barton, 1998, 2004; Allman, 2000; Preuss, 2007a,b; Kaas, 2012). Although stereopsis contributes somewhat to distance vision, in space near the body it provides information crucial for the accurate guidance of reaching and grasping movements (Hadjidimitrakis et al., 2015). Furthermore, frontally oriented eyes sometimes permit at least one eye to have a clear line of sight to items in a cluttered environment (Changizi, 2009). For example, if a leaf blocks the path from one forward-facing eye to a food item, it might not block the other. Forward-facing orbits might also have enhanced the ability to see items in front of and beneath the snouts of early primates, a region of space in which they manipulated food items and nutrient-bearing branches (Barton, 2004).

Euprimates also evolved a derived form of locomotion (Larson, 1998). By generating the most force (Schmitt, 2010) and strongest grip (Patel et al., 2015) with their hindlimbs, and using less force overall than their ancestors did, these species probably decreased both the visible oscillations and rustling noises that can attract predators (Schmitt, 2010). Given the visual clutter around them, the mitigation of noise and limb shaking would have limited their vulnerability to detection. Importantly, hindlimb-dominated locomotion enabled the forelimbs to specialize in other functions, such as steering the body and manipulating food items (Schmitt, 2010). Early euprimates might also have used a feeding habit that involved bringing food to their mouth with one hand while stabilizing their body with the other hand, as some modern strepsirrhines do (MacNeilage et al., 1987). This technique not only involves grasping food items directly, but also bending thin branches to gnaw off their nutrients.

Tooth wear and other dental evidence suggest that stem euprimates adapted to an omnivorous diet with a preference for fruits or leaves (Long et al., 2015) but including tree gums and insects as well. Some controversy remains about early primate diets, with some experts favoring insectivory (Cartmill, 1974, 1992). However, according to Silcox et al. (2015) the preponderance of dental evidence indicates that stem euprimates had a predominantly herbivorous diet.

Regardless of their dietary preferences, the visual innovations of early primates and their shift to hindlimb-dominated locomotion empowered the kind of movements that predominate in modern primates: visually guided reaching and grasping with the forelimb and hand. In an especially important development, early primates evolved a new mechanism for guiding these movements, using a stereoscopic reference frame based on retinal coordinates. Section **28.6.1.1** summarizes this idea and some of the evidence for it.

In addition to the cranial and limb adaptations mentioned thus far, the emergence of a suite of new cortical areas accompanied adaptation to the fine-branch niche (Preuss, 1993, 1995, 2007a,b). Section **28.2.1** identifies these areas, and Section **28.2.1** considers their functions. Together, the forward-facing eyes, hindlimb-dominated locomotion, grasping adaptations, and cortical innovations of euprimates supported both visually guided locomotion and visually mediated foraging (Cartmill, 1972, 1974, 1992; Martin, 1990; Schmitt, 2010).

28.2.2 Anthropoids

Most modern strepsirrhines have conserved the nocturnal foraging habit inferred for stem euprimates (Heesy and Ross, 2004). In contrast, fossil evidence in the form of small orbits indicates that early haplorhines foraged diurnally (Fleagle, 1999), and most modern anthropoids have retained this trait. Fig. 28.1 reflects the consensus view that tarsiers reverted to nocturnal foraging.

The haplorhine shift to diurnal foraging had several consequences. It promoted the detection of resources at a distance, but at a cost: an increased vulnerability to predation because of greater visibility. These animals could see more during foraging excursions, but they also could be seen better. Partly as a result, daylight foraging also placed a premium on vision-based predator detection and social signaling.

In a particularly significant development, the primate fovea emerged in stem haplorhines (Ross, 2004). This retinal specialization established the distinction between central and ambient vision, with exquisitely acute vision at central retinal coordinates and lower acuity at peripheral locations. Although several vertebrate lineages evolved a fovea convergently (Ross, 2004), the primate fovea is a haplorhine innovation, one that all anthropoids inherited.

In conjunction with foveal vision, early anthropoids enhanced the forward eye orientation that characterized stem euprimates, a development that promoted the high-acuity stereopsis of central vision. According to Barton (1998), the degree of binocular vision correlates positively with the volume of visual cortex and the size of the brain in primates. Binocularity also correlates with the extent of retinothalamic projections, especially for the parvocellular (high-acuity) pathways (Barton, 2004).

In addition to the fovea, haplorhines developed enhancements in color vision, some of which evolved independently in tarsiers, platyrrhines, and catarrhines. Most mammals, along with most strepsirrhines, share the ancestral trait of dichromacy, which depends on two photosensitive opsin proteins in foveal receptors. Dichromacy results in color vision with two high-sensitivity peaks along the color spectrum. Recent work has suggested that stem tarsiers developed opsin polymorphisms that increased the number of high-sensitivity peaks from two to three (Melin et al., 2013), and several platyrrhines have similar adaptations (Williams et al., 2010). In this type of trichromacy, two versions of an existing gene underlie an enhancement in color vision. However, early in catarrhine evolution a different form of trichromacy developed, one that depended on the emergence of a novel gene, which codes for a third opsin pigment. Because it is common to Old World monkeys, apes, and humans, this way of generating a third high-sensitivity spectral peak has come to be called the "routine" form. In addition to stem catarrhines, New World howler monkeys (*Alouatta*) developed the routine form of trichromacy convergently.

During the diversification of anthropoids, both their brains and bodies enlarged. According to fossil evidence, early anthropoids weighed 100—300 g (Fleagle, 1999; Rose, 2006), about the size range encompassing pygmy marmosets and common marmosets today.

Significant body expansion occurred mainly after the platyrrhine—catarrhine split in the early Oligocene (~34 Ma) (Williams et al., 2010), and body sizes in the 1—5 kg range became typical of anthropoids. Section **28.5.2** takes up the increase in brain size that occurred as anthropoids became larger animals.

Locomotion also changed as anthropoids departed the fine-branch niche. Initially, anthropoids shifted from a leaping—grasping form of movement to become arboreal quadrupeds (Fleagle, 1999; Kay et al., 2004). Many modern anthropoids have retained this form of locomotion (Schmitt, 2010), although others have moved on, so to speak. Notable examples include macaques and baboons, which became terrestrial quadrupeds. Some modern modes of primate locomotion facilitated reentry into the fine-branch niche, including the development of prehensile tails in spider and howler monkeys and brachiation in gibbons. Notwithstanding later developments, the initial shift to quadrupedal locomotion had important consequences. Arboreal quadrupeds require high expenditures of energy, especially because of the changes in elevation that commonly occur while traversing large branches (Janson, 1988). In association with the high energy needs of large arboreal quadrupeds, the home ranges of anthropoids increased during their diversification (Martin, 1981). This development presumably permitted these animals to gain access to more nutrients.

Dietary preferences are difficult to reconstruct from tooth-wear patterns alone, but the first large anthropoids probably had a diet much like modern species: fruits, tender leaves, insects, flowers, bark, tree gums, nuts, and seeds (Williams et al., 2010). Among their favored resources, fruits probably played a particularly prominent role. One factor favoring frugivory is the fact that much of the remaining plant material in the tropics—where anthropoids evolved and most modern species still live—is completely indigestible. Some digestible resources, such as insects, seem prevalent but have relatively low biomass. They serve as a source of important amino acids and some vitamins but cannot satisfy the needs of most anthropoids (Oates, 1987). On the other hand, fruits provide high levels of energy and can be available in large quantities. Crown anthropoids have developed a variety of dietary specializations but an early preference for fruit seems likely.

Despite the abundance of resources in the tropics, long-range diurnal foraging posed several challenges for Oligocene anthropoids, as it does for modern ones. First, although daylight lasts a long time in the tropics, most of the day is associated with dangerous levels of heat stress (Oates, 1987), which limits foraging to periods near dawn and dusk (crepuscular foraging). Second, as mentioned earlier, foraging in daylight conditions exposes individuals to a serious risk of predation—even in dim light. Third, their foraging niche brings anthropoids into competition with frugivorous birds, which have the advantage of flying from tree to tree. Anthropoids face additional competition from other primate species, other mammals, and members of their own species, including their own social group. Fourth, and of particular importance, their reliance on nutrients produced by angiosperm plants makes anthropoids highly vulnerable to seasonal and weather-related fluctuations in resource availability.

Tree resources fluctuate dramatically in the tropics. Each species of angiosperm has a characteristic fruiting time and ripening rate, which varies among trees of that species, often in a way that reflects highly localized environments. Resource fluctuations—and shortfalls—differ from year to year, place to place, and season to season, with important longer-term variations as well (Janmaat et al., 2006). A given tree species might produce fruit in all of its trees in a given area, yielding an abundance of nutritional resources. On the other hand, the same species might not produce fruit every year or might do so only for a brief period. A given tree species might produce fruit in half its trees in one part of an anthropoid's home range, only 5% in other parts, and elsewhere no fruit at all (Chapman et al., 2004). Although an anthropoid's home range will include thousands of trees, field studies have found that fewer than 5% of them will have fruit during any given foraging excursion (Janmaat et al., 2006; Zuberbühler and Janmaat, 2010).

The increased risk of predation caused by diurnal foraging was partially offset by the protection that social groups afforded. However, this relative safety came at the expense of increased competition for resources. Many modern strepsirrhines live dispersed lives with limited social interactions (Kappeler, 2012), and, although there are dramatic exceptions in modern species, this trait probably represents the social systems in stem euprimates as well (Müller and Thalmann, 2000). Accordingly, anthropoid societies probably evolved after the haplorhine—strepsirrhine split. Modern anthropoids have especially complex social systems, with a high degree of diversity among species (Dunbar and Shultz, 2007; Dunbar, 2009; Shultz et al., 2011). In their analysis, Dunbar and Shultz (2007) emphasized the size of social cliques: the number of individuals that contact each other more frequently than expected by chance. Most strepsirrhine species have clique sizes no larger than two individuals; catarrhine monkeys, in contrast, typically have clique sizes that range from two to five members and often more.

In summary, a dependence on foods produced by angiosperm plants rendered anthropoids vulnerable to shortfalls in necessary resources, and, although diurnal foraging over a large range enabled these sizeable animals to reach many resources, doing so required a large expenditure of energy and entailed a high risk of predation.

28.3 New Prefrontal Areas

28.3.1 Early Primates

28.3.1.1 *Granular Prefrontal Cortex*

The work of Preuss and Goldman-Rakic (1991a,b) established two key points about primate brain evolution (Section **28.1.2**): strepsirrhines have prefrontal areas that nonprimates lack (Fig. 28.2C, *dark blue shading*) and anthropoids have prefrontal areas that strepsirrhines lack (Fig. 28.2A, *light blue shading*). Fig. 28.2 illustrates the prefrontal areas shared among mammals with green shading, and other agranular areas are shaded in gold.

In their analysis, Preuss and Goldman-Rakic described a part of the bushbaby frontal cortex with homotypical cytoarchitecture, which they called the posterior granular

cortex. This area lies immediately rostral to area 6 (an agranular, premotor area) and is characterized by both a conspicuous internal granular layer and moderate myelination. (As noted earlier, it is conventional to call homotypical and dysgranular areas "granular" in the frontal cortex, although this usage differs from the usual meaning of granular cortex, which refers to primary sensory areas). On the basis of its location relative to the premotor cortex, its architectonics, and aspects of its connectivity, Preuss and Goldman-Rakic concluded that the posterior granular area in bushbabies is the homologue of area 8 in anthropoids. Wong and Kaas (2010) confirmed this conclusion and extended it by employing additional architectonic methods.

In anthropoids, as typified by the well-studied rhesus macaque, area 8 includes the frontal eye field (FEF).

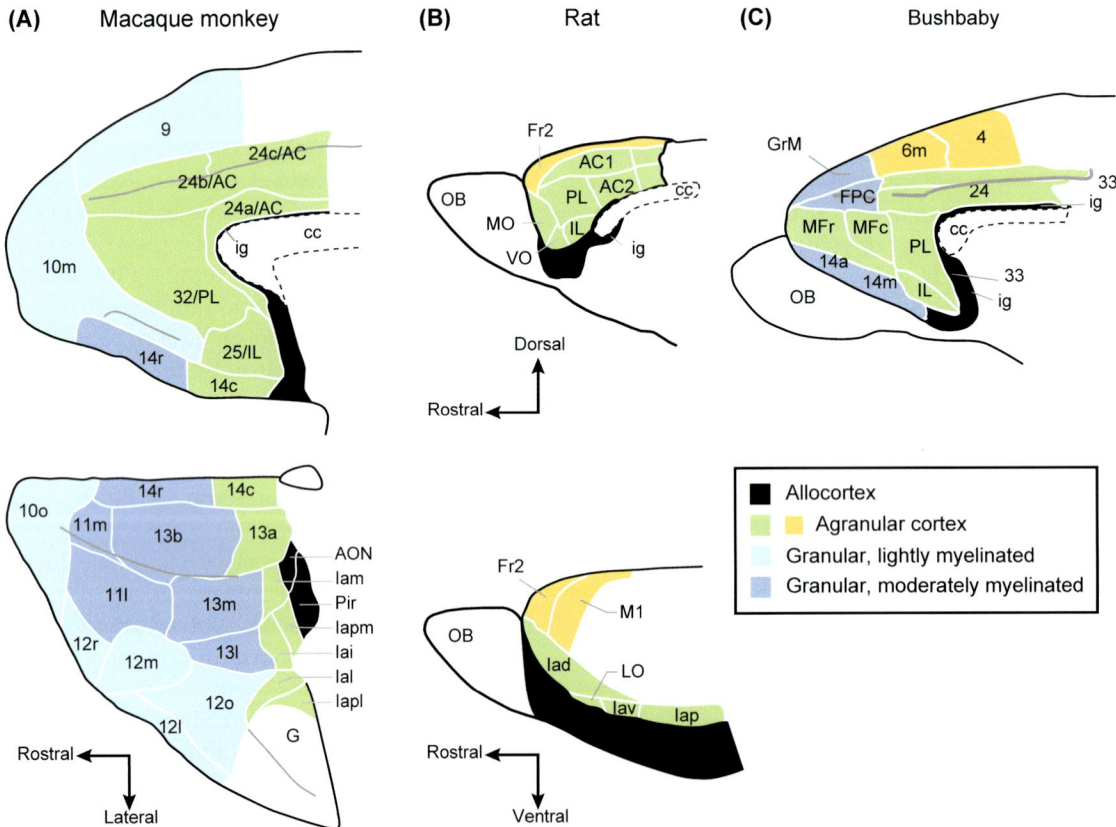

FIGURE 28.2 Topological relationship among allocortex, agranular frontal cortex, and granular frontal cortex in macaque monkeys (A), rats (B), and bushbabies (C). The top row shows a medial view of the cerebral hemisphere for all three species. In the bottom row, (A) the macaque orbital–insular cortex is best viewed from a ventral perspective and (B) the rat orbital–insular cortex is best seen from a lateral viewpoint. The two colors used for the agranular cortex correspond to prefrontal (*green*) and motor (*gold*) areas, respectively. *AC*, anterior cingulate cortex; *AON*, anterior olfactory nucleus; *c*, caudal; *cc*, corpus callosum; *FPC*, frontal pole cortex; *Fr2*, medial agranular motor area; *G*, gustatory cortex; *i*, inferior; *Ia*, agranular insular cortex; *ig*, indusium griseum; *l*, lateral; *IL*, infralimbic cortex (also known as area 25); *LO*, lateral orbitofrontal cortex; *m*, medial; *o*, orbital; *p*, posterior; *r*, rostral; *M1*, primary motor cortex (also known as area 4); *MFc* and *MFr*, caudal and rostral medial frontal cortex; *MO*, medial orbitofrontal cortex; *OB*, olfactory bulb; *Pir*, piriform cortex; *PL*, prelimbic cortex (also known as area 32); *tt*, tenia tecta; *VO*, ventral orbitofrontal cortex. (A) Redrawn from Carmichael, S.T., Price, J.L., 1994. Architectonic subdivision of the orbital and medial prefrontal cortex in the macaque monkey. J. Comp. Neurol. 346, 366–402. (B) Palomero-Gallagher, N., Zilles, K., 2004. Isocortex. In: Paxinos, G. (Ed.), The Rat Nervous System, third ed. Elsevier Academic Press, San Diego CA, pp. 729–757. (C) Redrawn from Preuss, T.M., Goldman-Rakic, P.S., 1991. Myelo- and cytoarchitecture of the granular frontal cortex and surrounding regions in the strepsirhine primate *Galago* and the anthropoid primate *Macaca*. J. Comp. Neurol. 310, 429–474.

Numerous experiments have demonstrated that intracortical microstimulation of the FEF evokes saccadic eye movements with low amplitudes of electrical current (eg, Bruce, 1985). A comparable level of microstimulation also evokes saccades from the posterior granular area of bushbabies, which supports the idea that this strepsirrhine area includes a homologue of the FEF (Wu et al., 2000). Preuss and Goldman-Rakic (1991a) described several additional granular prefrontal areas medial and lateral to the posterior granular area, which probably correspond to other parts of area 8 in macaques and other anthropoids.

In addition to these prefrontal areas, the orbitofrontal cortex also has homologous, granular areas in both bushbabies and macaques. An area that Preuss and Goldman-Rakic (1991a) called the ventral granular cortex is, according to their analysis, homologous with area 11, which is the rostral component of the granular orbitofrontal cortex in anthropoids. Fig. 28.2A (bottom) divides area 11 into medial and lateral parts: 11-m and 11-l, respectively. Wong and Kaas (2010) identified additional granular fields in the orbitofrontal cortex of bushbabies and considered them to be homologues of areas 13 and 14, in general agreement with Preuss and Goldman-Rakic.

Although the conclusions of Preuss and Goldman-Rakic are often expressed in a way that provokes controversy, their work does not imply that rodents (or other nonprimate mammals) "lack a prefrontal cortex," as is sometimes said. Nonprimate mammals have areas that can be called prefrontal, if for no other reason than the absence of a ruling authority such as the *Terminologia Anatomica* to dictate cortical nomenclature. The problem arises when researchers assume fallacious homologies between nonprimate and primate areas. The literature contains many examples of authors attempting to bolster an argument or interpretation about some frontal area in rodents by invoking findings from nonhomologous areas in primates—and vice versa. Similar problems pertain to posterior parietal areas and other parts of the cortex, but for some reason this problem has been particularly pervasive for prefrontal areas. Such arguments are often given more weight than they deserve, and they have contributed to several persistent misconceptions about the function of primate prefrontal cortex, including the behavioral inhibition and working memory theories. Section **28.6.2.3** returns to these topics.

28.3.1.2 Terminology

The terminology used in discussions of the prefrontal cortex can cause some confusion, especially when applied across species. These terms sometimes imply a degree of anatomical precision that, while laudable in intent and authoritatively expressed, is not reproducible. To mitigate this difficulty, this chapter uses general terms that can be applied across primates (Passingham and Wise, 2012).

Most pertinent to this section, the term "caudal prefrontal cortex" applies to areas homologous with area 8 of the macaque cortex, including the FEF. The term "orbitofrontal cortex" encompasses area 11 and the rostral, granular parts of areas 13 and 14 (*shaded blue* in Fig. 28.2A, bottom) along with the caudal, agranular parts of areas 13 and 14 and the agranular insular areas (*shaded green*). Collectively, the agranular areas on the orbital surface of the hemisphere can be called the "agranular orbital—insular cortex." Although the traditional anatomical literature often treats the orbitofrontal and insular cortex as distinct entities, a detailed analysis of their architectonics, connections, and topology revealed that the agranular insular areas are integral parts of an "orbital prefrontal network" (Carmichael and Price, 1996).

28.3.1.3 Diversity Denial: General Considerations

This chapter emphasizes the diversity among mammals, recognizing that different lineages have characteristic sets of prefrontal areas. The countervailing view denies this diversity and usually takes one of two forms: (1) nonprimate mammals have amalgams of the anthropoid prefrontal areas squeezed into a few small areas or (2) rodents and other nonprimate mammals have a replica in miniature of the anthropoid prefrontal cortex. Two versions of these ideas remain popular among rodent researchers:

- The medial frontal cortex of rodents (and other nonprimate mammals) is said to be homologous with the dorsolateral prefrontal cortex (area 46) in anthropoids (Uylings et al., 2003; Kolb, 2007; Seamans et al., 2008). [A variant holds that the homology might be with the ventral prefrontal cortex (mostly area 12), instead, or some amalgamated combination of the two (Birrell and Brown, 2000; Brown and Bowman, 2002).] The definition of the rodent medial prefrontal cortex differs among commentators, but it usually includes the infralimbic and prelimbic cortex (abbreviated as IL and PL in Fig. 28.2B, top), and sometimes a small amount of more dorsally situated cortex as well. (In other circumstances, the anterior cingulate cortex can be included among the medial prefrontal areas, but for this purpose it is usually excluded.) Taken together, the infralimbic, prelimbic, and dorsally adjacent areas are characterized by an agranular cytoarchitecture, unlike either the dorsolateral or ventral prefrontal cortex of anthropoids, which have homotypical lamination.
- The agranular orbital—insular cortex of rodents, also called the orbitofrontal cortex or the lateral prefrontal cortex, is said to be homologous to the whole orbitofrontal cortex of primates, including its rostral, granular components (Kolb, 2007; Schoenbaum et al., 2009).

Some of these proposals are based on a misconception about how homologies are established, one that places excessive emphasis on similarities. For example, proponents of the amalgam and replica in miniature theories have assembled lists of similarities between rodent and primate prefrontal areas. These lists usually include neurophysiological properties, the behavioral effects of cortical lesions, and features of connectivity, among other traits. Such lists can never provide a convincing case for homologies, regardless of their length and breadth, because the concept of homology concerns ancestry, not similarity. Similarities arise from many sources in addition to homology, as demonstrated by the many common features of wings that evolved in insects, bats, birds, and pterosaurs. Instead of depending on a list of similarities, the identification of homologies requires the documentation of specific diagnostic traits, which homologues share but otherwise similar structures lack. Put another way, the recognition of homologies depends in large measure on differences in traits—in addition to similarities. In fact, homologues sometimes have very few similarities. In a textbook example, the inner ear ossicles of mammals are homologous with jawbones in nonmammalian amniotes, but they have little in common with them, either in terms of structure or function. In another example, the vertebrate jaw derives from—but shares very few traits with—the most anterior brachial arches of agnathan vertebrates (Smith and Johanson, 2015).

Another common misconception is that the cortex of laboratory rats and mice replicates that of the last common ancestor of rodents, which gave rise to more than 2000 modern species. The last common ancestor of rodents lived approximately 70 Ma, and its descendant lineages have changed a great deal in the interim—in both body and brain. Likewise, both rodents and primates descended from the euarchontoglires common ancestor, but it is a mistake to assume that modern rodent brains are replicas of that ancestor's brain. Even for homologous areas, 70 Ma allows ample time for these areas, their associated networks, and their neurons to develop different properties, as they almost certainly did in response to the selective pressures faced by each independent lineage.

In broad perspective, the amalgam and replica in miniature theories of prefrontal cortex resemble long-discredited concepts in biology. In centuries past, biologists debated whether animals began with all their parts somehow miniaturized or mixed in a nascent form, a theory called preformationism. Proponents of the amalgam and replica in miniature theories essentially adopt a preformationist stance, either for the cortex as a whole or for some part of it.

Despite the persistence of such diversity denial, few contemporary neuroanatomists dispute the architectonic findings of Preuss and Goldman-Rakic. Some commentators argue that architectonic features are of little (or unknown) functional significance, in themselves, but this line of attack places unwarranted weight on the function of diagnostic traits. To cite just one example, the dentate gyrus can be identified across species by its uniquely high concentration of heavy metals, including zinc, as revealed by the Timms stain. It does not matter for identifying these homologues whether anyone understands the function of heavy metals in neural tissue. Furthermore, comparative neuroanatomy provides considerable additional support for the conclusions of Preuss and Goldman-Rakic (Section **28.3.1.5**), beyond the single diagnostic trait of granular cytoarchitecture. For example, the frontal pole cortex of rodents not only lacks the granular cytoarchitecture of the anthropoid frontal pole, but it has direct connections with the spinal cord, primary somatosensory cortex, and primary motor cortex, identifying it as a part of the somatic sensorimotor cortex (Wise and Donoghue, 1986; Cooke et al., 2011).

An underlying (but usually unstated) assumption of both the amalgam and replica in miniature theories is that although an abundance of new traits arose in the bodies of primates (Section28.2), for some reason their prefrontal cortex remained more or less the same, except for its size and perhaps the "granularization" of previously agranular cortex. These theories also assume that although other parts of the primate cerebral cortex developed a large number of new areas—such as the well-established proliferation of visual, somatosensory, and auditory areas (Krubitzer and Seelke, 2012; Kaas, 2013, 2014)—somehow nothing similar occurred in the prefrontal cortex.

These assumptions have little plausibility, and their weaknesses become obvious once the discussion shifts from the primate—rodent comparison to a less controversial one. For example, an auditory area in echolocating bats is specialized for Doppler-shift analysis (Fitzpatrick et al., 1998). As they pursue flying insects, bats can use this information to calculate whether a target is approaching or receding. Other auditory areas specialize in calculating the target's distance based on echo delays. No one thinks that modern rodents have homologues of these areas mixed up in some sort of cortical amalgam. Because rodents do not hunt mobile prey with echolocation, it would be odd if they did. Likewise, the last common ancestor of rodents and bats did not use echolocation for foraging and would have gained no benefit from representing Doppler shifts or sonar signals in an auditory amalgam. So it is clear that echolocating bats evolved new auditory areas, which have neurons with innovative representations such as tuning for Doppler shifts and echo delays. Rodents lack these specializations because they have a different and independent evolutionary history. It is incongruous to assume that what happened in the auditory cortex of echolocating bats could not have occurred in the expansive prefrontal cortex of primates.

Sections **28**.3.1.4 and **28**.3.1.5 discuss arguments against and in favor of new prefrontal areas in primates, respectively. To set the stage for this discussion, it is worth considering how new areas might arise during evolution and what the word "new" means in this context. One common source of evolutionary innovation outside the brain involves the replication of ancestral structures and their subsequent differentiation. When this occurs in the neocortex (Krubitzer and Kaas, 2005), the replicated areas become inserted among preexisting ones. Like most neighboring cortical areas, the inserted and preexisting areas tend to have connections with a similar set of brain structures and with each other. Partly as a result, they share many properties. After the insertions occur, some areas will change their properties less than others, and so it is sensible to say that they are "old" and homologous with the ancestral areas. Others will change their properties more, in part due to selective pressures that favor new kinds of neural representations, and so it is reasonable to designate them as "new." Like the examples presented previously for inner ear ossicles and jaws, the word "new" does not imply that a structure emerged from nothing, only that its properties differ so much from the ancestral condition that it should be considered an evolutionary innovation.

28.3.1.4 Diversity Denial: Specific Issues

In addition to a list of similarities, proponents of the amalgam and replica in miniature theories typically advance some combination of four specific arguments: (1) thalamic connections support homologies between rodent and primate prefrontal cortex; (2) inputs from dopaminergic neurons in the midbrain identify the prefrontal cortex across species; (3) the behavioral effect of lesions are said to be "the same" in rodents and macaque monkeys; and (4) nonprimates have a granular prefrontal cortex, as primates do. None of these arguments can survive critical analysis (Preuss, 1995, 2007b; Wise, 2008; Passingham and Wise, 2012).

28.3.1.4.1 Thalamic Inputs

The claim that thalamic connections serve as a defining feature of the prefrontal cortex dates to an early era of connectional neuroanatomy (Rose and Woolsey, 1948; Akert, 1964) and persists in the literature (Uylings et al., 2003). The contention is that connections with the mediodorsal nucleus of the thalamus (MD) serve to identify the prefrontal cortex. However, fiber-tracing studies in macaque monkeys have demonstrated that MD has direct connections with nearly the entire frontal lobe, including primary motor and premotor areas (Jones, 1985; Matelli et al., 1989). As a result, projections

to or from MD do not serve as a diagnostic feature of the prefrontal cortex.

28.3.1.4.2 Dopaminergic Inputs

Some early descriptions of dopaminergic inputs to the rodent cortex suggested that these fibers mainly target the prefrontal cortex (Divac et al., 1978; Porrino and Goldman-Rakic, 1982). However, such projections have a much wider distribution in primates. For example, in macaque monkeys dopaminergic neurons project more strongly to posterior parietal and premotor areas than to the prefrontal cortex (Gaspar et al., 1992; Williams and Goldman-Rakic, 1998). Therefore, like connections with MD, dopaminergic inputs cannot identify the prefrontal cortex.

28.3.1.4.3 Spatial Memory Impairments

A persistent claim in the rodent literature is that lesions of the medial prefrontal cortex in rats cause spatial memory impairments like those seen after prefrontal cortex lesions in macaques. The pertinent findings are generally accepted. Statistically significant spatial memory impairments follow medial prefrontal cortex lesions in rats (Kolb et al., 1974), and lesions of the dorsolateral prefrontal cortex produce spatial memory impairments in macaque monkeys, among other anthropoids (Jacobsen, 1936; Mishkin, 1957; Goldman et al., 1971). Some neuroscientists take these findings as support for the idea that the rodent medial prefrontal cortex is homologous with the primate dorsolateral prefrontal cortex. Two points undermine this proposed homology, however.

First, other cortical areas also contribute to spatial memory. For example, spatial memory impairments follow lesions of the anterior cingulate cortex in macaque monkeys (Meunier et al., 1997; Rushworth et al., 2003). In rodents the hippocampus plays a major role in the performance of spatial memory tasks, especially when they include a navigational component (eg, Markowska et al., 1989). Accordingly, the finding of a spatial memory impairment after a cortical lesion does not indicate a homology with the dorsolateral prefrontal cortex.

Second, the standard spatial memory tasks are poorly constrained, which means that they pose problems that animals can solve with different strategies and a variable set of brain structures. As a result, the lesion effects in rodents and anthropoids resemble each other only superficially. Two tests of spatial memory have been used extensively in macaques: the spatial delayed response task and the spatial delayed alternation task. In the delayed response task, a spatial cue on each trial indicates the location that should serve as the goal of the next movement, after a delay period has elapsed. In the delayed alternation task, a delay period separates a choice between two spatial goals, which should alternate from trial to trial.

Intact macaque monkeys learn these tasks quickly, but monkeys with preexisting lesions of the dorsolateral prefrontal cortex do not. In a typical experiment of this kind, monkeys with lesions of the dorsolateral prefrontal cortex continued to perform the delayed response task at chance levels for hundreds of trials—long after intact (control) monkeys had mastered the task (Battig et al., 1960). In a similar experiment, lesioned monkeys performed the delayed alternation task at chance level after thousands of trials (Goldman and Galkin, 1978).

In an alternative experimental design, monkeys learn the tasks before the surgical removal of their dorsolateral prefrontal cortex. This approach emphasizes task performance rather than a combination of learning and performance. Monkeys that could perform the delayed response task proficiently with a 5-s memory period before surgery reverted to chance level for at least 1000 trials after surgery, after which testing stopped (Mishkin, 1957; Goldman et al., 1971). In another experiment, monkeys that mastered the task with a 4-s delay period before surgery reverted to chance-level performance after surgery, even when the memory period was relaxed to 2 s (Kojima et al., 1982).

In contrast to these severe and permanent impairments in monkeys, lesions of the medial prefrontal cortex in rats cause only mild and transient effects on spatial memory. In one task, rats with such lesions reached 90% correct performance at a median memory period of 2 s, and they could do so in just 140 trials (Kolb et al., 1994). In another spatial memory task, lesioned rats relearned to control levels of performance in just 60 trials after surgery (Kolb et al., 1974).

In summary, these experiments fail to provide convincing evidence for either homologies or analogies between the medial frontal cortex of rodents and the granular prefrontal areas of primates. Passingham and Wise (2012) discussed this issue at length, based in part on a specific proposal about the strategies used by macaque monkeys to perform the delayed response and delayed alternation tasks. According to this account, monkeys with lesions of the dorsolateral prefrontal cortex have severe and permanent impairments on these tasks because they cannot use the order (or recency) of previous events to generate their next goal according to an abstract rule—"choose the last (or most recent) cued location"—and because they lack a fallback strategy. As a result, they succumb to trial-to-trial interference because every possible goal has been cued in the recent past. Rodents with medial prefrontal cortex lesions have spatial memory impairments for an entirely different reason. They lose the ability to integrate the spatial-navigational functions of the hippocampus with behavioral contexts represented in the

medial prefrontal cortex (Barker and Warburton, 2011; Navawongse and Eichenbaum, 2013; Spellman et al., 2015). Rats therefore have mild and transient impairments on spatial memory tasks until the hippocampus adapts to functioning without these frontal areas.

28.3.1.4.4 Granular Prefrontal Areas in Nonprimates

There have been reports of granular areas in the frontal cortex of various nonprimates, including dogs (Markow-Rajkowska and Kosmal, 1987), cats (Rose and Woolsey, 1948), squirrels (Wong and Kaas, 2008), and tree shrews (Wong and Kaas, 2009). The tree shrew area, which Wong and Kaas (2009) called area DFC, might be homologous to one of the granular prefrontal areas in primates. However, given the lack of granular frontal cortex in most rodents, it seems likely that the other reports resulted from either convergent evolution or a mischaracterization of the cytoarchitectonics. For example, the area called prefrontal cortex in cats lacks any granular areas that this neuroanatomist could discern (Wise et al., 1978), notwithstanding claims to the contrary.

28.3.1.5 Supportive Evidence

The idea that primates developed new granular prefrontal areas—but inherited the agranular prefrontal areas—finds additional support in four lines of evidence: topology, autonomic outputs, corticostriatal projections, and sensory inputs. The next four sections take up these topics, in turn.

28.3.1.5.1 Topology

When the cerebral cortex is viewed as a two-dimensional sheet, allocortical areas surround the neocortex, and much of the agranular prefrontal cortex lies adjacent to the allocortex. The agranular insular cortex, for example, abuts the piriform cortex. The anterior cingulate, prelimbic, and infralimbic cortex lie next to small allocortical areas called the indusium griseum and the tenia tecta (Carmichael and Price, 1994; Palomero-Gallagher and Zilles, 2004). Parts of the agranular orbital–insular cortex adjoin the tenia tecta as well. Fig. 28.2 illustrates the topological relationship between the allocortex and the agranular prefrontal areas in a representative anthropoid, rodent, and strepsirrhine; it shades the allocortex in black and the agranular areas in green.

In contrast to these agranular prefrontal areas, granular prefrontal areas (*shaded blue* in Fig. 28.2A and C) do not border allocortex. Instead, they lie adjacent either to agranular prefrontal areas or to each other. Accordingly, the granular prefrontal areas have a trait that distinguishes them from the agranular prefrontal areas: topological separation from the allocortex.

28.3.1.5.2 Autonomic Outputs

Electrical stimulation of the agranular prefrontal cortex evokes autonomic outputs, but stimulation of the granular prefrontal cortex does not or does so only at much higher levels of current (Bailey and Sweet, 1940; Smith, 1945; Delgado and Livingston, 1948; Kaada et al., 1949; Wall and Davis, 1951). Stimulation effects in these classic physiological experiments have included piloerection; pupillary dilation; and changes in blood pressure, respiratory rate, and heart rate. For the primate orbitofrontal cortex, stimulation evoked autonomic outputs only from its caudal, agranular part and not from its rostral, granular part. For the medial prefrontal cortex, stimulation of the anterior cingulate, prelimbic, and infralimbic cortex evoked autonomic responses but stimulation of the medial granular cortex did not. In macaques, lesions of a medial agranular area, the infralimbic cortex, decreased the pupillary dilatation seen in trace conditioning (Rudebeck et al., 2014).

In nonprimate mammals, autonomic outputs have been evoked from the entire agranular frontal cortex of rodents, rabbits, cats, and dogs. Stimulation of the medial agranular cortex caused a slowing of heart rate in rabbits (Powell and Ginsberg, 2005) and rats (Scopinho et al., 2009), and lesions of these areas diminished or blocked conditioned autonomic responses in both species (Frysztak and Neafsey, 1994; Powell and Ginsberg, 2005). Like the results from macaque monkeys, lesions of the infralimbic cortex in rats blocked autonomic trace conditioning (Frysztak and Neafsey, 1994).

Agranular prefrontal areas therefore have a particular diagnostic trait that distinguishes them from the granular prefrontal cortex: relatively direct motor outputs via the autonomic nervous system.

28.3.1.5.3 Corticostriatal Projections

Corticostriatal projections from the rat infralimbic and prelimbic cortex terminate in the shell of the nucleus accumbens, for the most part (Brog et al., 1993; Reynolds and Zahm, 2005), and homologous areas in macaque monkeys—areas 25 and 32, respectively—have similar projections (Haber et al., 1995; Ferry et al., 2000; Haber et al., 2006). The rat orbitofrontal cortex projects to the nucleus accumbens and adjacent striatal territory (Berendse et al., 1992), and the agranular orbital–insular areas in macaques do so, as well (Ferry et al., 2000; Haber et al., 1995, 2006).

In contrast to these projections, which tend to target the ventral striatum or nearby regions, the granular prefrontal areas of primates project mainly to the head of the caudate nucleus, a part of the dorsal striatum (Haber et al., 1995, 2006; Selemon and Goldman-Rakic, 1985; Ferry et al., 2000; Öngur and Price, 2000). The existence of new prefrontal areas in primates, along with their new striatal territories (specifically, the head of the caudate nucleus), has two consequences. First, it creates the impression of a ventral shift in the corticostriatal projections that are homologous in primates and rodents (Schilman et al., 2008). Second, it means that the dorsomedial part of the rostral striatum is not homologous in rodents and primates, as some neuroscientists have assumed (Balleine and O'Doherty, 2010).

These aspects of comparative neuroanatomy indicate that a projection to and near the ventral striatum characterizes the agranular prefrontal areas and distinguishes them from the granular prefrontal areas, which project to dorsal striatal regions far from the ventral striatum.

28.3.1.5.4 Sensory Inputs

Neuroanatomical studies show that the agranular orbital–insular cortex receives convergent gustatory, visceral, and olfactory inputs in both rodents and macaque monkeys (Ray and Price, 1993), which supports the homology of these cortical areas. The granular prefrontal cortex, in contrast, receives such information less directly. The shared sensory inputs of these agranular prefrontal areas therefore serve as an additional diagnostic trait that supports their homology and differentiates them from the granular prefrontal areas.

28.3.1.6 Summary

Early primates adapted to an arboreal life, developed forward-facing eyes, adopted a hindlimb-dominated, leaping–grasping mode of locomotion, and evolved a new mechanism for guiding reaching and grasping movements in an extrinsic, stereoscopic frame of reference. It seems likely that many of these traits emerged at various times during the evolution of stem euprimates, along with their new granular prefrontal areas. Comparative neuroanatomy indicates that two granular prefrontal areas emerged in early primates: the caudal prefrontal cortex, which includes the FEF, and the granular orbitofrontal cortex (Fig. 28.3A and B, *dark gray*). Rodents and other nonprimate mammals lack homologues of these areas, although they share a number of agranular prefrontal areas with each other and with primates (*shaded green* in Fig. 28.2).

Along with the emergence of new prefrontal areas, the brains of euprimates enlarged relative to body mass (encephalization) and increased in their percentage of neocortex (corticalization) (Martin, 1990; Long et al., 2015). In light of the visual developments of primates, most discussions of brain or cortical expansion have focused on visual areas (Long et al., 2015), with considerable justification. However, visual specializations have consequences for many other parts of the neocortex as well, including posterior parietal, premotor, and prefrontal areas. To consider brain expansion in the context of this suite of cortical adaptations (Section

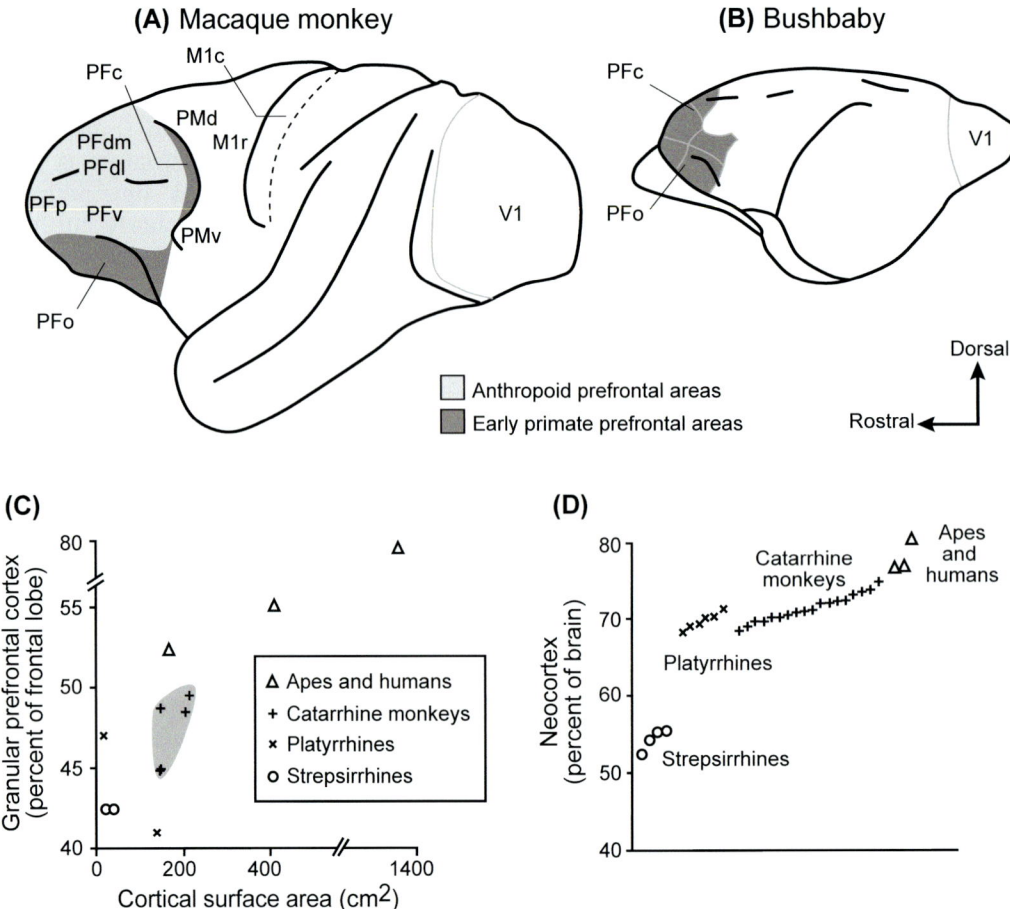

FIGURE 28.3 (A) One hemisphere of a representative anthropoid species, a macaque monkey. The *dashed line* near the central sulcus represents the fundus of the sulcus, and the cortex between the *dashed and solid lines* corresponds to the rostral bank of the central sulcus. *Dark shading* indicates the granular prefrontal areas that first emerged in early primates; *light shading* indicates granular prefrontal areas that first appeared in anthropoid primates. (B) One hemisphere of a representative strepsirrhine, a bushbaby, in the format of (A). (C) The percentage of the total frontal lobe consisting of granular prefrontal cortex. The *shading* marks the catarrhine range. Each data point comes from a different species. (D) The percentage of the brain consisting of neocortex. The x-axis in (D) is arbitrary; each data point comes from a different species, ranked from left to right according to the degree of corticalization within each group. Abbreviations: *PFc*, caudal prefrontal cortex; *PFdl*, dorsolateral prefrontal cortex; *PFdm*, dorsomedial prefrontal cortex; *PFo*, granular orbitofrontal cortex; *PFp*, polar prefrontal cortex; *PFv*, ventral prefrontal cortex, *PMd*, dorsal premotor cortex; *PMv*, ventral premotor cortex; *M1c*, caudal primary motor cortex; *M1r*, rostral primary motor cortex; *V1*, primary visual (striate) cortex. (C) Redrawn from Elston, G.N., Benavides-Piccione, R., Elston, A., Zietsch, B., DeFelipe, J., Manger, P., Casagrande, V., Kaas, J.H., 2006. Specializations of the granular prefrontal cortex of primates: implications for cognitive processing. Anat. Rec. A Discov. Mol. Cell. Evol. Biol. 288, 26–35. (D) Data from Dunbar, R., Shultz, S., 2007. Understanding primate brain evolution. Philos. Trans. R. Soc. Lond. Part B Biol. Sci. 362, 649–658.

28.4.1), as well as related developments in anthropoids (Section **28**.4.2), further consideration of brain size is deferred to Section **28**.5. To preview that discussion, it seems likely that new posterior parietal, premotor, and prefrontal areas developed during the transition to euprimates, and they could account for much of the encephalization and corticalization that occurred.

28.3.2 Anthropoids

Once one accepts the idea that new granular prefrontal areas evolved in stem euprimates, the additional conclusions of Preuss and Goldman-Rakic (1991a,b,c) seem

less controversial. As Section **28**.3.1.1 explained, they described certain strepsirrhine prefrontal areas as having a homotypical cytoarchitecture—"granular" in a frontal context—along with a moderate degree of myelination. Areas caudal to the prefrontal cortex can have much heavier concentrations of myelin, sensory areas in particular, so the degree of myelination in the caudal and orbital prefrontal cortex seems moderate in comparison. However, the rostral part of the anthropoid prefrontal cortex has very light myelination to go along with its homotypical cytoarchitecture. Fig. 28.3A depicts these areas in light gray for the lateral cortex, and Fig. 28.2A shades them in light blue for the medial (top) and orbital (bottom) cortex. The new anthropoid prefrontal areas correspond to the dorsolateral (area 46), dorsomedial (area 9), ventral (areas 12 and 45), and polar (area 10) prefrontal cortex.

One finding, in particular, supports the idea that granular prefrontal areas became more prominent in anthropoids, compared to their ancestral condition. Fig. 28.3C shows that, with the exception of capuchin monkeys (a platyrrhine species), a greater percentage of the anthropoid frontal lobe consists of granular cortex compared to two representative strepsirrhine species (black lemurs and dwarf lemurs). Obviously, this finding contradicts the idea that nothing happened during the evolution of the prefrontal cortex except various forms of allometric scaling (Charvet and Finlay, 2012; Barton and Venditti, 2013). New prefrontal areas appeared. Furthermore, in view of the subtle selective advantage that Section **28**.6.2.2 attributes to the granular prefrontal cortex, it should not be surprising that such an expensive tissue would regress in some lineages, as seems to have happened in capuchin monkeys and other cebids (Allen and Kay, 2012a). Section **28**.5.2 returns to this topic.

28.4 Other Neocortical Areas

Although this chapter emphasizes granular prefrontal areas by choice, they evolved as components of a complex suite of cortical adaptations.

28.4.1 Early Primates

Along with their new granular prefrontal areas, early primates also developed several premotor, posterior parietal, superior temporal, and inferior temporal areas. Expressed in this way, it might seem as though all of these cortical innovations of early primates arose simultaneously at the base of the euprimate stem. However, this scenario is unlikely for several reasons. In major evolutionary transitions, several combinations of synapomorphies typically develop, partially in parallel, as a lineage diversifies. The basal members of our order probably had different sets of new areas, with one combination ultimately prevailing.

The evolution of premotor and posterior parietal areas has received less attention than developments in the visual and auditory cortex. This emphasis is understandable, especially given the characteristic shape of the euprimate brain. The lateral fissure separates a large temporal lobe from more dorsally situated cortex, a feature that reflects the expansion of the visual and auditory cortex (Fig. 4A–D). However, in view of the adaptations of euprimates for reaching and grasping

movements, the premotor and posterior parietal areas deserve attention as well. Section **28**.6.1.1 explains why.

Nonprimate mammals have one or two small cortical fields that might be homologous with the primate posterior parietal (Wilber et al., 2014) and premotor areas, but not nearly as many as observed in primates (Wu et al., 2000; Remple et al., 2007; Krubitzer, 2009; Krubitzer and Padberg, 2009; Cooke et al., 2014). A popular brain atlas labels an area in the rodent frontal cortex "M2," which is a standard abbreviation for the supplementary motor area. This premotor area lies on the medial surface of the hemisphere in anthropoids, but there is virtually no comparative evidence for a homologue in rodents. Simply put, placing the label "M2" on an area in an atlas does not make it so. Instead, the comparative evidence indicates that the supplementary motor area is a primate innovation (Wise, 1996), along with several other premotor areas. Modern primates have more posterior parietal and premotor areas than the euarchontoglires (rabbit–rodent–tree shrew–primate) common ancestor, and this conclusion applies to the euarchontan (tree shrew–primate) common ancestor as well. Although tree shews have some of these areas (Baldwin et al., 2016), they do not have the large number of premotor and posterior parietal areas seen in modern primates (Remple et al., 2007). As Fang et al. (2005: p. 331) put it, based on a study of corticocortical connectivity:

the evidence suggests that the rather complex framework for sensorimotor processing that has been revealed in... [bushbabies] is similar to that seen in anthropoid primates. Thus, this framework likely emerged with or before early primates, but not earlier with the common ancestor of tree shrews... and primates, insofar as tree shrews have a much simpler cortical system.

28.4.2 Anthropoids

In addition to their new granular prefrontal areas (Fig. 28.3A), other areas also emerged in the visual, auditory, somatosensory, motor, insular, and posterior parietal areas of anthropoids, especially catarrhines (Wu et al., 2000; Remple et al., 2007; Krubitzer, 2009; Krubitzer and Padberg, 2009; Kaas, 2012; Bauernfeind et al., 2013; Cooke et al., 2014; Wilber et al., 2014). The visual cortex has received the most attention, in part because its areas are the easiest to define and identify. Kaas (2014: p. 1243) summarized the comparative data as follows:

In Old World monkeys, over 30 visual areas have been proposed [Felleman and Van Essen (1991)]..., and these primates have greatly expanded regions of visual cortex in the temporal and parietal lobes compared to most New World monkeys and all prosimians.

Developments in the superior temporal cortex, which processes auditory and combined auditory–visual

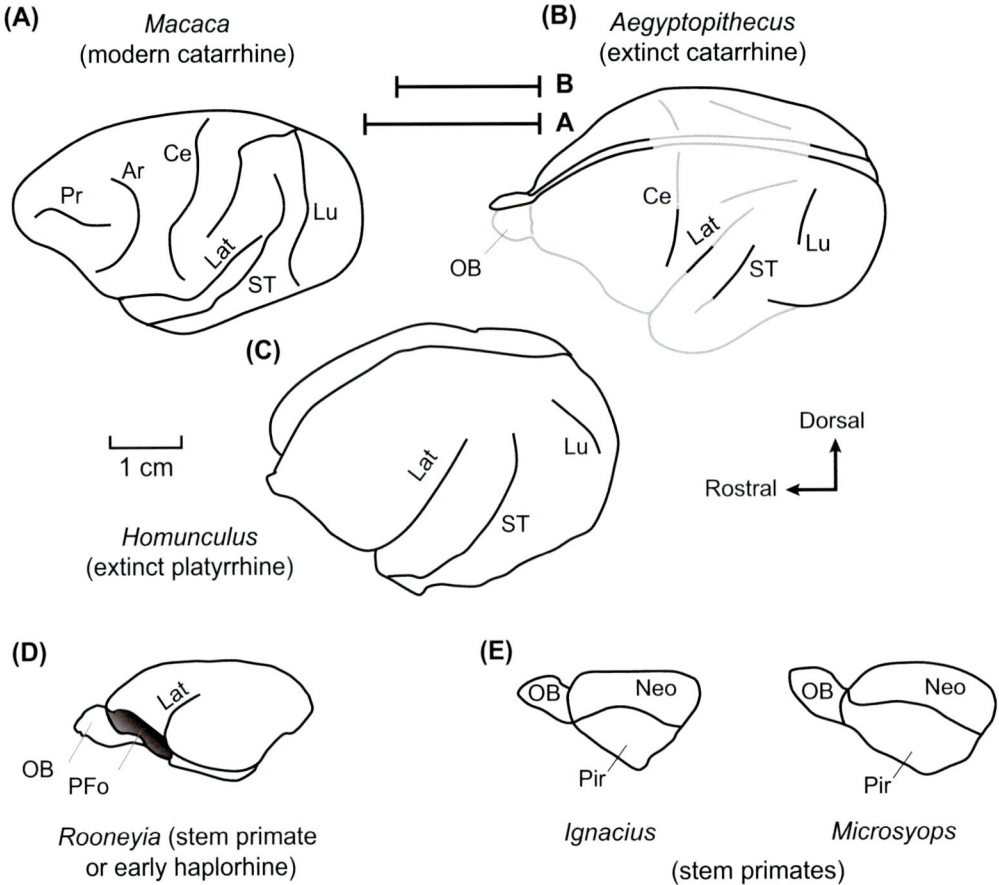

FIGURE 28.4 Cerebral surfaces in a modern anthropoid and five extinct primate species. (A) A modern catarrhine, as represented by a macaque brain. (B) An extinct catarrhine species, *Aegyptopithecus*. The *bar-ended lines* above and between (A) and (B) show the distance from the central sulcus to the frontal pole in the corresponding drawings. (C) An extinct platyrrhine, *Homunculus*. (D) An extinct primate, *Rooneyia*. According to Kirk et al. (2014: p. 94), *Rooneyia* is either "a very primitive stem haplorhine…, a very primitive stem strepsirrhine…, or an advanced stem primate." (E) Two extinct stem primates (plesiadapis), *Microsyops* and *Ignacius*. All brains in this figure are drawn to the same scale. Abbreviations: *Ar*, arcuate sulcus; *Ce*, central sulcus; *Lat*, lateral fissure; *Lu*, lunate sulcus; *Neo*, neocortex; *OB*, olfactory bulb; *Pr*, principal sulcus; *ST*, superior temporal sulcus. (A) and (B) Redrawn from Radinsky, L., 1979. The fossil record of primate brain evolution. In: 49th lecture James Arthur Lecture on the Evolution of the Human Brain. American Museum of Natural History, New York, pp 1–27. (C) Redrawn from Kay, R.F., Perry, J.M.G., Malinzak, M., Allen, K.L., Kirk, E.C., Plavcan, J.M., Fleagle, J.G., 2013. Paleobiology of Santacrucian primates. In: Vizcaíno, S.F., Kay, R.F., Bargo, M.S. (Eds.), Early Miocene Paleobiology in Patagonia: High Latitude Paleocommunities of the Santa Cruz Formation. Cambridge University Press, Cambridge UK, pp. 306–330. (D) Redrawn from Kirk, E.C., Daghighi, P., Macrini, T.E., Bhullar, B.A., Rowe, T.B., 2014. Cranial anatomy of the Duchesnean primate *Rooneyia viejaensis*: new insights from high resolution computed tomography. J Human Evol. 74, 82–95. (E) Redrawn from Long, A., Bloch, J.I., Silcox, M.T., 2015. Quantification of neocortical ratios in stem primates. Am. J. Phys. Anthropol. 157, 363–373.

information, occurred along with those in the visual inferior temporal areas. These distance senses played an important role in the life of Oligocene anthropoids, and their specialized representations in modern macaques provide some hints about their initial adaptive advantages.

Passingham and Wise (2012) proposed that anthropoid specializations represent the "signs" of distant resources. As anthropoids became larger and foraged over extensive home ranges, they came to rely more on vision and audition, and less on olfaction, for their foraging choices. Ancestral visual and auditory areas, mostly inherited from stem mammals (Krubitzer and Seelke, 2012; Kaas, 2013), represent combinations of sensory inputs at particular levels of complexity, called

feature conjunctions. Low-order areas, such as primary visual, primary auditory, extrastriate, and belt auditory areas, represent simple feature conjunctions or elemental sensory features. High-order areas, such as the perirhinal cortex, represent more complex feature conjunctions, at the level of objects. Since the advent of stem mammals, these low- and high-order representations have served to disambiguate sensory stimuli from each other (Saksida and Bussey, 2010), which enables mammals to recognize and differentiate not only elemental sights and sounds, but also a wide array of multidimensional objects.

These low- and high-order areas composed the ancestral condition of primates and were almost certainly present in the plesiadapiforms illustrated in Fig. 28.4E. During the diversification of primates, additional visual and auditory areas emerged, which represented feature conjunctions at new levels of complexity. It is common to assume that when new levels of representational complexity arise, they do so at levels higher than previous representations. However, in this case the new levels were midlevel conjunctions, more complex than the low-order conjunctions of striate and extrastriate areas but simpler than the object-level representations of the perirhinal cortex. New midlevel feature conjunctions arose in both the superior and inferior temporal cortex, which improved the ability of far-ranging anthropoids to discriminate and identify the acoustic and visual signs of resources, as exemplified by the sounds made by frugivorous birds and the pattern of colors and shapes in distant scenes.

Once new midlevel representations evolved, they could serve as exaptations for face recognition and other social traits. Passingham and Wise (2012) presented a detailed case for foraging-related factors as the principal selective force driving the elaboration of inferior temporal areas. This idea contrasts with an alternative proposal, which maintains that the mechanisms for face recognition developed in the inferior temporal cortex of stem mammals, if not earlier—long before the emergence of primates. According to Leopold and Rhodes (2010), sheep have a homologue of the anthropoid inferior temporal cortex, which requires an origin no later than the primate–ferungulate common ancestor (~ 85 Ma, during the Cretaceous period). Three arguments contradict this idea. First, the social systems of many strepsirrhines (Kappeler, 2012), and by inference those of stem euprimates (Müller and Thalmann, 2000), are relatively simple compared to those of anthropoids (Section **28.2.2**). Face recognition was probably of minor importance to the social life of early primates, which were nocturnal and dispersed. Accordingly, the most characteristic features of anthropoid societies, including face recognition, probably evolved long after the emergence of primates. Second, the brain of the primate–ferungulate common ancestor had only a few visual areas, in contrast to the two or three dozen visual areas in modern anthropoids (Krubitzer and Seelke, 2012; Kaas, 2013). Again, reference to the glires-like endocasts in Fig. 28.4E indicates how unlikely it is that stem primates had the complex set of visual areas seen in modern anthropoids. Clearly, the last common ancestor of macaque monkeys and sheep did not have the elaborate inferior temporal cortex of either. Third, although neurons encoding midlevel conjunctions respond selectively to face stimuli, as they do to objects, the high-level conjunctions of the perirhinal cortex are more appropriate to face discrimination, as they are for objects. The fact that midlevel representations contribute modestly to face or object discrimination means little from an evolutionary perspective. It is much more likely that face recognition evolved convergently in several crown mammals, as independently evolving (and predominantly diurnal) lineages increased in both brain size and body mass and developed their own social systems.

In addition to new visual areas, two new somatosensory areas—area 2 and an opercular area called area PR—also emerged during anthropoid evolution (Krubitzer, 2009), as did a new, caudal part of the primary motor cortex. Unlike the rostral part of the primary motor cortex, the caudal part has direct, monosynaptic projections to α-motor neurons in the spinal cord of monkeys (Rathelot and Strick, 2009). Although such projections receive a great deal of attention in medical textbooks, monosynaptic corticomotoneuronal connections are rare among mammals (Wise and Donoghue, 1986). The rostral part of the primary motor cortex in macaques resembles the entire primary motor cortex of other mammals in this regard (Rathelot and Strick, 2009). In prosimians and nonprimate mammals, there is little evidence of a distinction between rostral and caudal components of the primary motor cortex. Accordingly, it seems likely that the caudal primary motor area—including its monosynaptic projection to spinal motor neurons—is an anthropoid innovation (Kaas, 2004). Humans also have distinguishable rostral and caudal components of the primary motor cortex (Geyer et al., 1996).

The caudal part of the primary motor cortex is characterized by the predominance of inputs from cutaneous mechanoreceptors (Tanji and Wise, 1981; Strick and Preston, 1982), with a dramatic magnification factor for inputs from the digits of the hand. The glabrous surface of the fingers has several specializations for sensing small amounts of skin deformation and stretching. Meissner corpuscles are concentrated in the ridges of anthropoid fingertips in prodigious numbers, and they probably play a significant role in guiding the fine finger movements controlled by the caudal primary motor cortex. These movements, in turn, are important for the precise manipulation of fruits and other objects. Meissner corpuscles are also important for judging the ripeness of fruits by their mechanical compliance, a capacity of special importance for fruits that do not change color as they ripen. A significant positive correlation between the degree of frugivory in anthropoid species and the density of Meissner corpuscles on their fingertips highlights the importance of somatic sensation for their foraging choices (Hoffmann et al., 2004). For this reason, some experts draw an analogy between the fovea and anthropoid fingertips—a so-called "tactile fovea."

28.5 Size and Shape

Most discussions of cortical evolution rely heavily on allometry, which captures the influence of genetics and ontogenetic mechanisms on the overall amount of cortex or some part of the cortex. Many factors constrain the volume of neocortex that a species can support, but organizational variation within that limit is subject to selection. A simple example from outside the nervous system illustrates this point. No mammals have more than four limbs, but within that limit mammals have developed a variety of claws, paws, wings, flippers, hands, and feet—in various proportions. Limbs have also regressed in some lineages, such as cetaceans and pinniped carnivores. Likewise, the overall extent of neocortex is limited by body mass, brain size, basal metabolic rate, dietary specializations, foraging strategies, reproductive rate, longevity, and other factors.

Within these constraints new areas have emerged, others have expanded and elaborated, while others have regressed.

28.5.1 Brain Enlargement in Stem Euprimates

Fig. 28.5C illustrates the upward grade shift in brain size that occurred in euprimates. A "grade shift," in this sense, takes into account both brain size and body mass, and it corresponds to a displacement in brain–body regressions up or down on plots like those in Fig. 5A–C.

Although some uncertainty remains about its precise affinities, *Rooneyia* was most likely a stem euprimate, a primitive stem haplorhine (Ross et al., 1998; Ni et al., 2013), or a

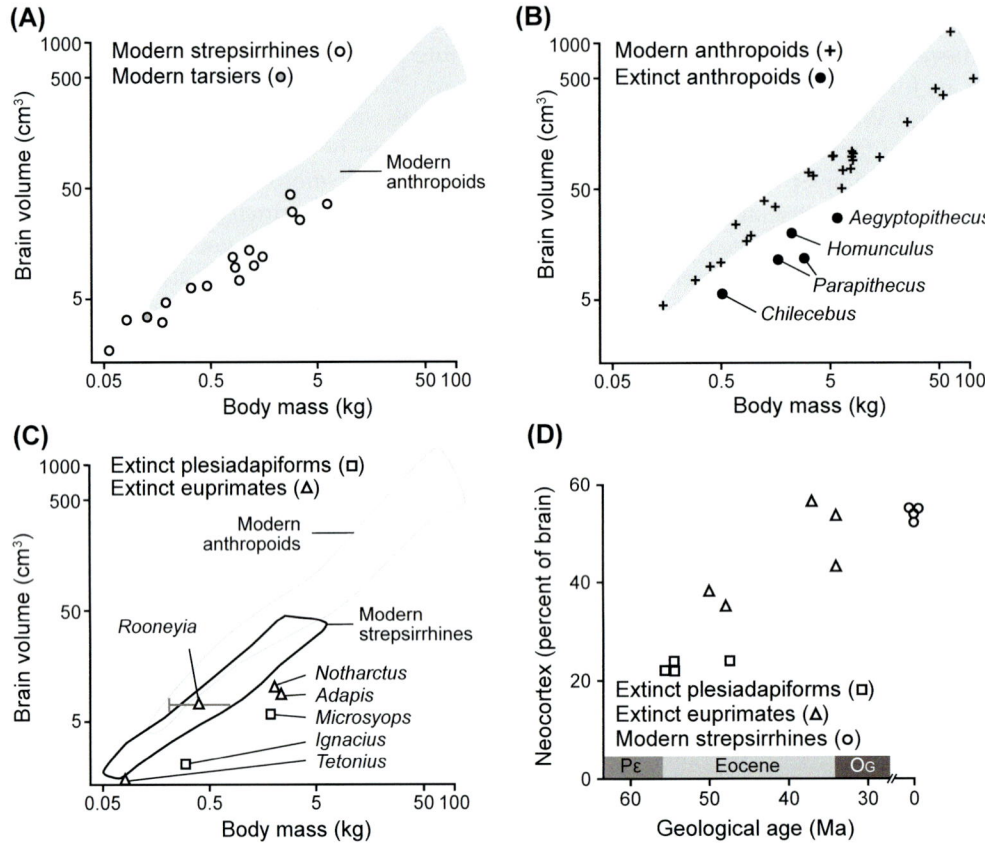

FIGURE 28.5 Upward grade shifts in brain size, relative to body size. (A) Modern strepsirrhines and tarsiers. The *gray shading* bounds the data points for modern anthropoids in both (A) and (B), plotted in the same coordinates. (B) Modern anthropoids contrasted with four extinct anthropoid species, with brain size measured as endocranial volume in fossil endocasts. *Parapithecus* is represented by two points due to different estimates of body size. (C) Modern anthropoids and strepsirrhines contrasted with several fossil basal euprimates, including *Rooneyia*, and fossil plesiadapiforms. The *error bar* for *Rooneyia* corresponds to 95% confidence bounds for body size estimates based on cranial length. (D) The percent of brain surface consisting of neocortex, as a function of the geologic age of fossil specimens. Abbreviations: *OG*, Oligocene; *Pɛ*, Paleocene. (A) and (B) Most data from Bush, E.C., Simons, E.L., Allman, J.M., 2004. High-resolution computed tomography study of the cranium of a fossil anthropoid primate, *Parapithecus grangeri*: new insights into the evolutionary history of primate sensory systems. Anat. Rec. A Discov. Mol. Cell. Evol. Biol. 281, 1083−1087; One data point, for *Homunculus*, comes from Kay, R.,F., Perry, J.M.G., Malinzak, M., Allen, K.L., Kirk, E.C., Plavcan, J.M., Fleagle, J.G., 2013. Paleobiology of Santacrucian primates. In: Vizcaíno, S.F., Kay, R.F., Bargo, M.S. (Eds.), Early Miocene Paleobiology in Patagonia: High Latitude Paleocommunities of the Santa Cruz Formation. Cambridge University Press, Cambridge UK, pp. 306−330. (C) Data from Kirk, E.C., Daghighi, P., Macrini, T.E., Bhullar, B.A., Rowe, T.B., 2014. Cranial anatomy of the Duchesnean primate *Rooneyia viejaensis*: new insights from high resolution computed tomography. J. Human Evol. 74, 82−95. (D) Redrawn from Long, A., Bloch, J.I., Silcox, M.T., 2015. Quantification of neocortical ratios in stem primates. Am. J. Phys. Anthropol. 157, 363−373; Modern strepsirrhine values are from Dunbar, R., Shultz, S., 2007. Understanding primate brain evolution. Philos. Trans. R. Soc. Lond. Part B Biol. Sci. 362, 649−658.

primitive stem strepsirrhine (Kay et al., 2004). Fig. 28.1 depicts it as a stem haplorhine, and Fig. 28.4D presents a drawing of its brain. This fossil specimen had the large, forward-oriented orbits characteristic of euprimates (Kirk et al., 2014). Fig. 28.5C shows that its endocranial volume, in relation to body mass, falls within the modern strepsirrhine range (Kirk et al., 2014), unlike the brains of plesiadapiforms (*Ignacius* and *Microsyops*) and extinct adapoids with strepsirrhine affinities (*Adapis* or *Notharctus*) (Ramdarshan and Orliac, 2015). Likewise, according to Orliac et al. (2014: p. 1), *Plesiadapis*, the index plesiadapiform, had "a very small and simple brain … [with] midbrain exposure, and minimal encephalization and neocorticalization, making it comparable with that of stem rodents and lagomorphs." Fig. 28.1 illustrates one possibility for relations among these species. The strepsirrhine-size brain of *Rooneyia* provides evidence for a grade shift in brain size during the evolution of euprimates.

In addition to brain–body allometry, the amount of neocortex—as a function of total brain size—increased in stem euprimates (Long et al., 2015). Fig. 28.5D plots the neocortex as a percentage of the brain in plesiadapiforms (*squares*), stem euprimates (*triangles*), and modern strepsirrhines (*circles*), as a function of geologic time. Omitted from Fig. 28.5D, another recently described fossil euprimate, *Microchoerus*, also falls into the euprimate range for both neocortical percentage (41%) and geologic age (late middle to late Eocene) (Ramdarshan and Orliac, 2015). (It is excluded from the plot because of uncertainty about its precise value on the abscissa, within a broad range from the middle to late Eocene.) In general, mammalian brains tend to increase in size over geologic time (Jerison, 2012), but the upward grade shift in euprimate brains seems to have been particularly significant.

Allometry aside, it is difficult for this neuroanatomist to look at Fig. 28.4 without noticing that the euprimate *Rooneyia* (Fig. 28.4D) had the basic form of a primate brain, whereas the plesiadapiforms *Ignacius* and *Microsyops* (Silcox et al., 2009; Silcox et al., 2010) had brains with a very different shape (Fig. 28.4E). Something fundamental seems to have occurred in the transition to euprimates; "primates of modern aspect" have primate-shaped brains. Future work might be directed toward comprehensive morphometry, with as much emphasis on the shape of the neocortex as on its size. An analysis contrasting the shapes of strepsirrhine and anthropoid brains exemplifies this approach (Allen and Kay, 2012b).

In summary, euprimate evolution seems to have been associated with an increase in both encephalization and corticalization. In overall shape and size, the brains of stem primates (Fig. 28.4E) seem more similar to rabbit and rodent brains than to euprimate brains (Fig. 4A–D). The conclusion that some plesiadapiform species, such as *Microsyops*, had entered the arboreal omnivorous niche characteristic of basal euprimates implies that life in the

fine-branch niche and omnivory did not lead directly to an upward grade shift in brain size (Silcox et al., 2010). Instead, it seems more likely that visual innovations in basal euprimates provided the main impetus for cortical enlargement (Silcox et al., 2010; Long et al., 2015). Discussions on this point usually emphasize the visual cortex and its role in the identification of stimuli. However, as Sections **28**.2.1 and **28**.3.1.6 mentioned—and Section **28**.6.1.1 elaborates—vision also guides reaching and grasping movements, which depend on a network of posterior parietal and premotor areas.

28.5.2 Brain Enlargement in Anthropoids

Allometric analysis, based on modern primates, indicates that an upward grade shift in brain size also occurred in crown anthropoids (Allman, 2000; Striedter, 2005; Fig. 28.5A and B). However, this observation says little about which areas contributed to the shift.

Evidence from fossil endocasts indicates that an expansion of the visual cortex preceded expansion of the frontal lobe (Radinsky, 1975, 1979). *Aegyptopithecus*, an early Oligocene catarrhine dated to ~33 Ma (Kay et al., 2004), had less cortex rostral to its central sulcus than modern catarrhine monkeys do (Fig. 28.4A and B). Radinsky concluded that by ~33 Ma, shortly after the platyrrhine–catarrhine split, the visual cortex had reached or approached its modern size. Nevertheless, the overall size of the brain remained beneath the anthropoid range in *Aegyptopithecus*, as it did in other fossil anthropoids (Fig. 28.5B). *Parapithecus* (Bush et al., 2004) probably lived prior to the divergence of catarrhines and platyrrhines and therefore represents a basal anthropoid, and *Chilecebus* (Sears et al., 2008) was probably an early platyrrhine. Both had endocranial volumes beneath the range of modern anthropoids (Fig. 28.5B), as did another platyrrhine fossil, *Homunculus*. As Kay et al. (2013: p. 315) summarized the latter specimen:

It… appears that living platyrrhines evolved larger brains relative to body size than stem platyrrhines. A parallel increase in encephalization also occurred in catarrhine primates… The small size of the brain in Homunculus is partly an effect of the small size of the frontal lobe….

Accordingly, it appears that the expansion of the brain into the modern anthropoid range occurred after the platyrrhine–catarrhine split. Allen and Kay (2012a) concluded that brain expansion occurred four or more times during platyrrhine evolution, alone: in stem platyrrhines and in three other lineages. In one of these lineages, cebid monkeys, brain reduction occurred later, which could account for the aberrantly low data point for capuchin monkeys in Fig. 28.3C. If the visual cortex had reached or approached its modern size prior to the platyrrhine–catarrhine divergence, then it probably did not contribute much to the upward grade shift in anthropoid brain size. Expansion of the granular prefrontal cortex, among other areas, remains a candidate for producing this allometric shift.

28.5.3 Social Factors

As mentioned in Sections **28**.2.2 and **28**.4.2, most modern strepsirrhines lack the large, physically interacting social groups that developed later in primate evolution (Kappeler, 2012), and it seems likely that early primates had a similar social system (Müller and Thalmann, 2000). However, sociality might be underestimated in these species because it depends to a significant extent on acoustic signaling among dispersed nocturnal animals. Nevertheless, the contrast between most strepsirrhine and most anthropoid social systems is considerable. Anthropoids are usually characterized by groups of socially interacting individuals (Dunbar and Shultz, 2007; Dunbar, 2009; Shultz et al., 2011), with a significant diversity among these species. As noted in Section **28**.4.2, this pattern suggests that the diverse social systems in anthropoids evolved after the strepsirrhine—haplorhine split and possibly substantially afterward.

Given the significance of sociality to the life of anthropoids, it comes as little surprise that group (or clique) size correlates positively with both the volume of neocortex (Dunbar and Shultz, 2007; Dunbar, 2009, 2014) and the size of particular cortical areas (Lewis et al., 2011; Von Der et al., 2014), although there are contradictions in the details. In a study by Dunbar and Shultz (2007), three factors accounted for about two-thirds of the variance in the total size of the neocortex: the complexity of social groups, total brain size, and longevity.

These analyses are somewhat controversial because many factors affect the size of the neocortex, which likely differ among lineages. Some authors dispute the idea that social factors provided a selective pressure toward an increase in the size of the neocortex or the brain (Charvet and Finlay, 2012), but the existence of additional selective factors does not negate the importance of social ones (Dunbar and Shultz, 2007; Pasquaretta et al., 2014). Although there is an understandable tendency to find "the" selective factor for increased brain size, the reality is invariably more complex.

28.5.4 Summary

Allometric analysis tends to dominate discussions of cortical evolution, but the comparative studies of Preuss and Goldman-Rakic (1991a,b,c) highlighted the importance of reorganization within the cortical mantle. In addition to size analysis, allometric studies have also pointed to factors such as an increase in the number of neurons (Gabi et al., 2014), interconnections (Sherwood et al., 2005; Elston, 2007; Semendeferi et al., 2010), dendritic

branching, spine density, and spine number (Elston et al., 2006; Elston, 2007) as characteristics of the primate prefrontal cortex. The functional significance of these traits remains unknown, although it is commonly assumed that they reflect an enhanced integrative capacity.

Allometry is and will remain an important analytical approach, but it has several drawbacks, especially regarding the prefrontal cortex. First, interpretations of allometry often gainsay the importance of absolute expansion. Yet, as the properties of artificial neural networks demonstrate, their properties depend to a considerable extent on absolute size. For example, emergent representational properties arise in computational networks after they reach a particular number of simulated neurons and interconnections (Plaut et al., 1996). Second, estimates of the size of the prefrontal cortex often depend on how an anatomical analysis distinguishes prefrontal from premotor areas. These delineations are often inconsistent and sometimes little more than guesswork, especially when applied across species. When this problem is avoided by focusing on the volume of the frontal lobe as a whole, the results might have little to do with the prefrontal cortex per se. Third and most importantly, allometric analysis rarely deals with organizational changes such as the emergence of new granular prefrontal areas. Obviously, the idea that the cortex does nothing but scale during evolution contradicts the conclusion that new granular prefrontal areas arose in euprimates and in anthropoids, perhaps at the expense of other cortical areas.

For these reasons, among others, it is difficult to draw conclusions about the relative expansion of a region as specific as the granular prefrontal cortex, let alone its components. In general, the anthropoid frontal lobe has a higher proportion of granular prefrontal cortex than in other primates (Fig. 28.3C), but the contribution of these new areas to overall neocortical extent remains uncertain. The endocast data indicate that an expansion of the brain, driven mainly by the neocortex, has occurred at least three times in primate evolution: during the evolution of basal euprimates (Fig. 28.5C) and in parallel for both platyrrhines and catarrhines after their split (Fig. 28.5B). Later, the brain expanded in several anthropoid lineages, often independently (Allen and Kay, 2012a).

28.6 Cortical Functions and Specializations

28.6.1 Early Primates

Preuss (1993) first proposed that a suite of new and elaborated cortical areas were among the primate adaptations to life in the fine-branch niche. As Section **28**.3.1 explained, two parts of the granular prefrontal cortex emerged in early primates: the caudal prefrontal cortex and the granular orbitofrontal cortex. Since his reviews, however, ideas about the function of the prefrontal cortex have changed (Passingham and Wise, 2012) and a new understanding of primate visuomotor mechanisms has crystallized (Shadmehr and Wise, 2005).

To preview the discussion in this section, Section **28.6.1.1** describes new premotor and posterior parietal mechanisms that underlie visually guided movements in an extrinsic frame of reference; Section **28.6.1.2** explains how the granular orbitofrontal cortex provides advantages in updating the valuation of expected outcomes—and especially visual their visual features—in accord with an animal's current nutritional and physiological state; Section **28.6.1.3** discusses the contribution of the caudal prefrontal cortex to the search for and maintenance of attention on items of value; and Section **28.6.1.4** briefly mentions developments in sensory areas of cortex.

28.6.1.1 Parietal–Premotor Networks

Most vertebrates use vision to guide movement, as do a wide variety of protostomes, so visually guided movements are hardly unique to primates. Nevertheless, early primates seem to have evolved a new cortical mechanism for the control of reaching and grasping (Wise, 2007).

Critics of this conclusion point to the fact that many nonprimates make reaching and grasping movements, including several well-studied rodent and carnivore species. However, these observations have little bearing on the cortical mechanisms that evolved in primates because of the prevalence of both convergent and parallel evolution. The lineages leading to modern rodents and primates diverged from the euarchontoglires common ancestor as early as the Cretaceous (according to molecular evidence) and no later than the earliest Paleocene (according to fossil evidence). A divergence time of ~ 70 Ma seems like a reasonable estimate, although some experts prefer more remote dates. The ferungulate–euarchontoglires lineages diverged much earlier, long before the Cretaceous period ended and probably more than ~ 85 Ma. Both of these ancestral species had small brains with relatively few cortical areas compared to crown primates, crown carnivores, and crown ungulates, which makes parallel and convergent evolution especially likely as brains and bodies enlarged in separate lineages. Along with brain enlargement came new areas, specialized neural representations, and innovative cortical functions.

Section **28.2.1** listed new premotor and posterior parietal areas among primate innovations, but simply identifying and naming areas does little to elucidate their functions. The term "premotor" might seem to imply a motor function, but when this name was coined in the 1930s it referred mainly to a location: "before" (ie, rostral to) the primary motor cortex (Wise, 1984). According to current theories of motor control, premotor and posterior parietal areas compute visuomotor transforms that convert the location of motor targets from retinal coordinates into the kinematics (joint-angle changes) and dynamics (forces) required to reach to a target. Parietal–premotor networks function in conjunction with the cerebellum and the basal ganglia to adjust movements based on errors made during previous and ongoing movements.

Shadmehr and Wise (2005) presented a detailed analysis of the various coordinate frames and effector systems involved in primate motor control, while Wise and Willingham (2009) offered a much briefer, more accessible summary. A few examples should suffice to convey a general impression of how these parietal–premotor networks function.

In the most general sense, posterior parietal and premotor areas interact to compute a variety of visuomotor transforms for the control of reaching, grasping, head, trunk, hindlimb, and eye movements. For example, the lateral intraparietal area represents visual locations in a body-centered coordinate frame (Colby and Duhamel, 1996; Colby et al., 1996; Snyder et al., 1998; Mullette-Gillman et al., 2005) and controls eye movements, in part via projections to the superior colliculus (Paré and Wurtz, 2001). The medial intraparietal area represents visual locations in a retinal coordinate frame (Buneo et al., 2002; Chang et al., 2009) and controls reaching movements (Kalaska and Crammond, 1995; Pesaran et al., 2006, 2008; Cui and Andersen, 2007, 2011), mainly via connections with the dorsal premotor cortex (Johnson et al., 1996; Wise et al., 1997). The anterior intraparietal area represents locations in an object-centered coordinate frame (Nakamura et al., 2001; Baumann et al., 2009) and controls the grasping and manipulation of objects, mainly via connections with the ventral premotor cortex (Fluet et al., 2010). In a key experiment that tested the latter idea, inactivation of either the anterior intraparietal or ventral premotor areas caused an impairment in the fine finger control needed for grasping and manipulating objects (Gallese et al., 1994; Kermadi et al., 1997; Fogassi et al., 2001). These and similar coordinate frames also control movements of other body parts—such as the head, eyes, hindlimbs, and trunk—via various premotor areas.

The prevailing theory of visually guided movement holds that parietal–premotor networks compute movements in terms of vectors represented in a coordinate frame external to the body (extrinsic coordinates) rather than one based on the body (intrinsic coordinates) (Bullock and Grossberg, 1988; Kermadi et al., 1997; Cisek et al., 1998; Burnod et al., 1999; Andersen and Buneo, 2002). One of these vectors, which depends mainly on the posterior parietal cortex, represents the location of movement targets in a retinal coordinate frame. This "target vector" has its tail at the fixation point and its head at the target. A second vector, which uses the same reference frame, represents the location of the hand or some other "end effector," such as a tool held in the hand or some other part of the body. This "end-effector vector" also depends on the posterior parietal cortex (Rushworth et al., 1997a,b). It has its tail at the fixation point and its head at the location of the hand or some other end-effector. Motor-systems specialists often refer to a retinal coordinate frame as "eye centered," but the origin of this coordinate frame is not in or near the eyes. Instead, the posterior parietal cortex represents locations in a three-dimensional (stereoscopic) coordinate system with its origin at some distance the animal's eyes and its axes extending into visual space. In anthropoids, the foveal fixation point serves as this origin, but any retinal coordinate can do so in principle.

In addition to estimating the location of the hand in visual coordinates, the primate motor system has access to proprioceptive signals that can provide an estimate based on joint angles and muscle lengths. Through a form of motor learning, these neural networks bring the intrinsic, proprioception-based estimate of hand position into alignment with the extrinsic, vision-based one, with vision predominating. This mechanism enables hand position to be represented in a visual coordinate frame even without vision of the hand. In addition to guiding movements in the dark, proprioceptive inputs enable visual attention to be focused on the target of a reaching movement. As a consequence of this proprioception–vision alignment mechanism, when refracting prisms have been used to displace visual inputs on the retina, both macaque monkeys and humans reached inaccurately for several consecutive attempts until parietal–premotor networks adjusted their computations to once again bring these two estimates of hand position into alignment (see Shadmehr and Wise, 2005, for a discussion of this literature).

By representing both hand position and target location in an extrinsic, retina-based coordinate frame, parietal–premotor networks can compute a motor plan in that reference frame, as well. A simple vector-subtraction computation yields the instantaneous distance and direction from the hand to the target, a neural representation called the "difference vector." The conclusion that this computation occurs in an extrinsic, visual frame of reference is somewhat counterintuitive because the difference vector has its tail (origin) at the hand and its head at the target. Accordingly, it is often assumed that this computation occurs in a hand- or body-centered (intrinsic) coordinate frame. This might be the case for nonprimate species, but in primates ample experimental evidence shows that the difference vector is—like target location and hand position—represented in extrinsic, vision-based coordinates. In these experiments:

- People made reaching movements that were straight in visual space, even when they needed to make slightly curved movements to generate a visually straight trajectory, as judged by experimentally manipulated visual feedback (Wolpert et al., 1995).

- Small, systematic errors (overshoots) occurred when people reached to targets in peripheral vision, and they made the same errors when reaching to acoustic targets (Pouget et al., 2002). This could only happen if the auditory signals that represent target locations are transformed into a vision-based coordinate frame.
- Congenitally blind people made straighter "visually guided" hand movements than sighted subjects did (Sergio and Scott, 1998). The sighted subjects, of course, had seen their own movements for a lifetime, and their slightly curved hand movements resulted from small optical distortions that induced minor deviations from straight-line trajectories. Congenitally blind people did not make these small errors because they had never experienced such distortions.
- Neurons in the posterior parietal cortex have been observed to recompute the difference vector and target vector either just before or just after every saccadic eye movement (Duhamel et al., 1992; Heiser and Colby, 2006). Because neither the hand nor the target moved near the time of these saccades, the only reason for recalculating these vectors is that they are represented in extrinsic, vision-based coordinates, which changed whenever the eyes moved.

Taken together, these and other (Shadmehr and Wise, 2005) experimental results have demonstrated that primate parietal–premotor networks represent motor plans (the difference vector), hand position (the end-effector vector), and target locations (the target vector) in extrinsic, visual coordinates.

To execute a movement plan, the difference vector needs to be transformed into muscle forces (dynamics) and changes in joint angles (kinematics). These transformations change over time, as the body grows, weakens or strengthens, or overcomes injuries and fatigue. They also differ for various kinds of targets, such as food items with different strengths of attachment to a tree branch. Parietal–premotor networks, functioning in conjunction with the cerebellum and the basal ganglia, learn these transforms based on feedback from previous and ongoing movements (see Wise and Willingham, 2009, for a brief summary of this literature).

Within the parietal–premotor network, localized areas have a modular pattern of organization, in which small regions of cortex control coordinated actions. Sometimes called "action modules," examples include reaching, grasping, manipulation, feeding, locomotion, and defensive movements. Evidence for this pattern of organization comes from the motor effects of sustained electrical stimulation of the primary motor, premotor, and posterior parietal cortex (Graziano et al., 2002a,b; Cooke and Graziano, 2004; Stepniewska et al., 2005; Graziano, 2006; Gharbawie et al., 2011). Such stimulation evokes movements that resemble those made in foraging, locomotion, or defensive behaviors (Stepniewska et al., 2005, 2009a,b, 2014, 2011; Gharbawie et al., 2011, 2011). Directly interconnected areas influence the same actions, which ultimately depend on outputs via the primary motor cortex (Stepniewska et al., 2014).

The ventral premotor cortex contains one of these action modules, which appears to control feeding movements (Graziano, 2006; Wise, 2007). In their comparative study of corticospinal projections, Nudo and Masterton (1988, 1990) called the ventral premotor cortex "Region C," as distinguished from the main sources of corticospinal projections, Regions A and B (mainly premotor, motor, somatosensory, and posterior parietal areas). They identified a homologue of both Regions A and B in a wide variety of mammals, but only primates had Region-C corticospinal projections. Furthermore, the extent of Region C correlated positively with whether a species lives in an arboreal niche, as opposed to other potential correlates such as its degree of hand—eye coordination or manual dexterity. Studies of cortico-cortical connections among the ventral premotor, posterior parietal, and primary motor areas in strepsirrhines, platyrrhines, and catarrhines have provided further support for the idea that the ventral premotor cortex emerged early in primate evolution (Lu et al., 1994; Preuss et al., 1996; Kaas, 2004; Dum and Strick, 2005a; Burish et al., 2008; Gharbawie et al., 2010).

The details of its corticospinal projection provide some insight into the functional specializations of the ventral premotor cortex. In contrast to the corticospinal projections from motor, somatosensory, and posterior parietal areas, those from the ventral premotor cortex rarely extend caudal to the cervical spinal cord (Dum and Strick, 2005b). Instead, these axons terminate predominantly in rostral cervical segments, which have motor neurons controlling muscles in the head, neck, and shoulder, along with intercostal muscles and the diaphragm, both of which regulate respiratory movements. Other projections from the ventral premotor cortex terminate in the facial nucleus, which has motor neurons controlling the jaw, lips, and the lower part of the face (Morecraft et al., 2001, 2004). The ventral premotor cortex also plays a role in manipulating objects, as indicated by the finding that inactivation of this area impaired the ability of macaque monkeys to adjust their hand posture to a target object's size (Fogassi et al., 2001). In aggregate, outputs from the ventral premotor cortex can control and coordinate movements of the hand, head, and mouth with each other as well as with breathing. Stem euprimates might have required this kind of coordinated movement for a hand-to-mouth feeding technique (MacNeilage et al., 1987).

Tree shrews (*Tupaia belangeri*), a sister group of primates (Fig. 28.1), have similar action modules. Sustained cortical stimulation in these animals evoked coordinated movements from both the posterior parietal and primary motor cortex (Baldwin et al., 2016). Although it is possible that tree shrews and primates developed action modules convergently (Silcox et al., 2010), it seems most likely that primates elaborated a euarchontan mode of cortical organization that included this trait, in part by developing several new parietal and premotor areas. Fig. 28.1 incorporates this suggestion.

In summary, among the many synapomorphies of euprimates, a new visuomotor network emerged in their neocortex. In addition to a few ancestral areas, this network included new posterior parietal and premotor areas, which probably contributed to both encephalization and corticalization (Section **28.5.1**, Fig. 28.5C and D). Likewise, the complex pattern of parietal—premotor connections seen in modern primates seems to be an innovation of early primates, possibly emerging during the transition to euprimates (Section **28.4.1**).

Early primates also developed two new parts of the prefrontal cortex, the granular orbitofrontal and caudal prefrontal cortex (Section **28.3.1**), as well as new and elaborated temporal areas (Section **28.4.1**). The next three sections take up these innovations.

28.6.1.2 Granular Orbitofrontal Cortex

A large literature—stemming mainly from research on monkeys but probably applying to strepsirrhines and early primates as well—has explored the specialized functions of the granular orbitofrontal cortex. In the past, it was thought that this research implicated these areas in four general functions: (1) behavioral inhibition; (2) emotional reactivity; and two kinds of value updating, (3) one based on external contingencies (eg, food availability) and (4) the other based on internal states (eg, satiety). The first kind of value updating has been called "credit assignment," and it reflects an estimation of whether a particular outcome will occur based on some sensory event or some action. In a typical experiment, this outcome consists of a food or fluid becoming available for consumption. The second kind of value updating reflects what such outcomes are worth at any given time, based on current needs. Different tasks have been used to assess these four functions:

1. Traditionally, the object-reversal task has been assumed to measure behavioral inhibition. In a typical version of this task, subjects first learn to choose one of two objects to receive a reward. In subsequent testing, they need to suppress this choice in favor of the alternative, and such reversals can occur repeatedly. Similar experiments involve reversals between patterns on a video monitor or a particular sensory dimension of such stimuli. Section **28.6.2.3.2** discusses several lines of evidence that cast

doubt on the assumption that these tasks measure behavioral inhibition, but results from experiments employing them are commonly interpreted in this context nevertheless.

2. A task called the "snake test" has been used to assess emotional reactivity in macaque monkeys. It measures a delay in reaching over an artificial snake to obtain food, compared to reaching over neutral objects.

3. Credit assignment has been assessed with the "three-arm bandit task." In this task, monkeys choose repeatedly among three visual images as the probability of reward associated with each image changes.

4. Value updating based on internal states has been assessed with the devaluation task, which is described later.

Behavioral impairments caused by aspiration lesions of the granular orbitofrontal cortex have suggested that it plays a necessary role in all four functions, and the results obtained from correlational methods, such as single-cell neurophysiology and functional brain imaging, have been interpreted as providing support for this conclusion. However, recent experiments have overturned this idea. They made use of a more selective lesion method, which spared fibers of passage, and they led to an entirely different conclusion about the function of the granular orbitofrontal cortex. This work ruled out the first three functions listed earlier—behavioral inhibition, emotional reactivity, and credit assignment—at least for the tasks used. For the sake of discussion, the remainder of this section assumes that these results apply to all primates, although nearly all of this work involves macaque monkeys.

Two points from Section 28.3 are especially pertinent to this discussion: (1) the granular orbitofrontal cortex (traditionally defined as areas 11, 13, and 14) was an early primate innovation and (2) the ventral prefrontal cortex (areas 12 and 45) was an anthropoid innovation. A recent report has suggested redefining the granular orbitofrontal cortex to include a part of area 12, area 12o (Chau et al., 2015). However, recall from Section 28.3.1.1 that Wong and Kaas (2010) did not identify any homologue of area 12 among the orbitofrontal areas in the bushbaby cortex. Instead, they considered the granular orbitofrontal areas in bushbabies to be homologous with areas 11, 13, and 14 of macaques. Accordingly, the present discussion treats all of area 12 as belonging to the ventral prefrontal cortex (also known as the ventrolateral prefrontal cortex) and not to the granular orbitofrontal cortex.

Until recently, lesion experiments have used the surgical aspiration technique to remove the granular orbitofrontal cortex in macaque monkeys. Unfortunately, this procedure caused inadvertent damage to fiber pathways running near or through the lesioned area. These axons connect various brain structures with the ventral prefrontal cortex, among other prefrontal areas (Lehman et al., 2011; Jbabdi et al., 2013). As a result, the lesions made in these experiments were much less localized to the granular orbitofrontal cortex than the investigators thought at the time.

To avoid damaging these pathways, recent experiments have used excitotoxic lesions to remove the gray matter of the granular orbitofrontal cortex selectively. To the surprise of many experts, these lesions had no effect on the performance of either the object-reversal task (Rudebeck et al., 2013), the snake test (Rudebeck et al., 2013), or the three-arm bandit task (Rudebeck and Murray, 2014). Lesions that severed subcortical fibers, but left most of the orbitofrontal gray matter intact, caused an impairment on the first two tasks (Rudebeck et al., 2013). The third task has not yet been tested in this way. Nevertheless, the functions spared by complete, excitotoxic lesions of the granular orbitofrontal cortex indicate that—for all three tasks—the impairments seen after aspiration lesions resulted from inadvertent damage to fiber pathways. Accordingly, these findings demonstrate that the granular orbitofrontal cortex does not make a necessary contribution to behavioral inhibition, emotional reactivity, or credit assignment, at least as assessed by the tasks used to date. Section 28.6.2.3.2 expands on this point for behavioral inhibition.

Given all of these findings, what function remains for the granular orbitofrontal cortex? Although fiber-sparing, excitotoxic lesions of granular orbitofrontal cortex had no effect on performance of the object-reversal task, the snake test, or the three-arm bandit task, they did cause an impairment on the devaluation task. In this task, macaque monkeys choose between two objects (Baxter et al., 2000; Rudebeck et al., 2013) or two actions (Rhodes and Murray, 2013) based on updated valuations of the predicted outcomes. An "outcome," in this sense, corresponds both to foods or fluids that become available as the result of a choice, as well as the specific sensory features of those resources, such as their sight, smell, and taste.

In one object-based version of the task, macaque monkeys first learned associations between two sets of objects and two foods. One set of objects was associated with a food of one type, peanuts, for example, while another set of objects was associated with a different kind of food, such as raisins. Later, the monkeys faced a test in which they had to choose, for the first time, between two objects drawn from the different sets. One object was always associated with the first food, the other with the second. There was no wrong choice because the monkeys obtained some food either way. The monkeys in these experiments usually chose the object associated with the food that they preferred. The devaluation task then entered its final phase. The monkeys consumed one of the two foods to satiety, which lowered the current value of that food in terms of desirability and palatability, and they later faced another choice between two objects. Like the previous tests, their choice determined which of the two foods they received. Control monkeys usually shifted their choice to the non-devalued food, but monkeys with selective lesions of the granular orbitofrontal cortex did so poorly, if at all (Rudebeck et al., 2013). Complete excitotoxic lesions of the granular orbitofrontal cortex had a similar effect on a choice between two actions that had different food outcomes (Rhodes and Murray, 2013).

These findings clarify the specific function of the granular orbitofrontal cortex, a set of areas that evolved in euprimates (Section 28.3.1.1). These areas update the valuation of predicted outcomes (or, more likely, the sensory features of predicted outcomes) based on current biological needs, and they confer an ability to use these updated valuations to choose among objects and actions. However, this statement needs at least three qualifications. First, the lesioned monkeys could shift their choices to a non-devalued food item when they chose directly between two foods. So they had normal satiety and motivation mechanisms. Their impairment specifically involved choices among objects or actions that predicted a particular outcome. Second, because the monkeys picked an object without having previously chosen between it and the alternative, stimulus–response conditioning could not explain their behavior. They instead made their choices based on predictions about the kind of outcomes that should occur and their current subjective value. Third, the fact that the same result was obtained for the object- and action-based versions of the task shows that the valuations were linked to the features of outcomes and not solely to the features of chosen objects.

This latter qualification is particularly relevant for understanding the advantage provided by the granular orbitofrontal cortex to early primates. In light of their visual innovations (Section 28.2.1), it seems likely that these animals made foraging choices based on vision, as modern primates do. Contributing to their vision were new, specialized sensory representations in the inferior temporal cortex. As explained in Section 28.4, these areas represent conjunctions of visual features at new levels: more complex than the elemental and low-level conjunctions of the striate and extrastriate areas, but less complex than the object-level representations of the perirhinal cortex. Both the perirhinal and inferior temporal cortex have reciprocal connections with the granular orbitofrontal cortex in primates (Kondo et al., 2005), and they almost certainly covey information about the visual features of objects, including the special kinds of objects—food items—that served as outcomes in the devaluation experiments discussed here. According to the present account, primates with lesions of the granular orbitofrontal cortex had impairments on both

the object- and action-based versions of the devaluation task because, although they could update their valuation of a food item per se, they could not use their neural representations of actions or nonfood objects to access that information and use it to choose future goals. An example that might apply to foraging in the fine-branch niche involves the prediction that pulling on a branch should reveal a nutritious flower or fruit, although neither the pulling movement nor the branch has any inherent nutritional value.

To sum up this line of research, selective lesions of the granular orbitofrontal cortex cause an impairment on the devaluation task, which requires updating the valuations of predicted outcomes based on current biological states and needs. In contrast, lesions of the ventral prefrontal cortex cause an impairment on the three-arm bandit task, which requires updating the estimate of reward probabilities—contingencies in the jargon of the field (Rudebeck et al., 2015). Thus the granular orbitofrontal and ventral prefrontal cortex play complementary roles in updating valuations and—by extension—in making foraging choices. The granular orbitofrontal cortex guides choices based on the current desirability or palatability of a predicted outcome (or the sensory features of that outcome); the ventral prefrontal cortex guides choices based on the likelihood of a desired outcome, a variable linked to resource availability. Put another way, the valuations mediated by the granular orbitofrontal cortex depend on dynamic internal states; those mediated by the ventral prefrontal cortex depend on dynamic external contingencies. Section 28.6.2.2.1 returns to this topic in its consideration of anthropoid-specific areas, which include the ventral prefrontal cortex.

In macaques, the granular orbitofrontal cortex influences the choice of future goals mainly through indirect connections with premotor areas, often via other parts of the granular prefrontal cortex. According to a recent finding, the caudal part of the granular orbitofrontal cortex, area 13, contributes mostly to updating the valuations, whereas the rostral part, area 11, specializes in transforming updated valuations into choices and actions (Murray et al., 2015).

Neurophysiological research sheds some light on how this valuation-to-action transform occurs. The granular orbitofrontal cortex appears to influence foraging choices in two ways: one described by the goods model (Padoa-Schioppa, 2011), the other by the affordance-competition model (Cisek, 2012). From an evolutionary perspective, these two models correspond to different primate adaptations, both of which involve the ability to choose foraging goals based on updated valuations. Foraging goals come in two forms: metrically concrete and metrically abstract. To say that a goal is metrically concrete means that the distance, direction, and other parameters of movement can be calculated, such as a choice between two visuospatial stimuli. To say that a goal is metrically abstract means that the monkey must choose between two or more goals without knowing when or where they might later become available. The affordance-competition model deals exclusively with metrically concrete goals; the goods model emphasizes metrically abstract goals. The difference lies in the timing of a valuation computation. In the goods model, metrically abstract options are evaluated prior to a choice among them; in the affordance-competition model, potential actions are planned first (Platt and Glimcher, 1997; Basso and Wurtz, 1998; Schall and Bichot, 1998; Powell and Goldberg, 2000; Cisek and Kalaska, 2005, 2010; Scherberger and Andersen, 2007; Baumann et al., 2009), then evaluated based on predicted outcomes (Pastor-Bernier and Cisek, 2011). Eventually, the most valuable target prevails and progresses to execution (Cisek, 2012). According to both models, the decision variables that affect foraging choices include factors such as effort and delay costs, reward probability, and an updated estimation of an outcome's value. Neurons in the posterior parietal and premotor areas encode all of these variables, among others (Platt and Glimcher, 1999; Gold and Shadlen, 2000; Dorris and Glimcher, 2004; Kim and Lee, 2011; Leathers and Olson, 2012).

In summary, the granular orbitofrontal cortex has a long history of association with behavioral inhibition and emotional reactivity, along with a more recent linkage to the representation of outcome probabilities (credit assignment). Correlational methods, such as functional brain imaging and single-unit neurophysiology, suggest some contribution to these functions—but not necessary ones. In terms of necessary contributions, the function of the granular orbitofrontal cortex is more specific. This primate-specific set of areas appears to specialize in linking the representations of nonfood items and actions to updated valuations of predicted outcomes, especially their visual features, based on current biological needs. Its rostral area, area 11, then generates a bias toward either metrically concrete or metrically abstract foraging goals based on these linkages. The advantages of this specialization to early primates probably involved the exploitation of their visual adaptations in combination with the visuomotor mechanisms discussed in Section 28.6.1.1.

28.6.1.3 Caudal Prefrontal Cortex

The other new granular prefrontal region of early primates, the caudal prefrontal cortex (including the FEF), makes a different contribution to primate fitness: an enhanced ability to search for and maintain attention on items of value in a cluttered environment.

Like the orbitofrontal cortex, ideas about the caudal prefrontal cortex have changed recently. The prevailing view of the FEF has long been that it functions as a premotor area for the selection and control of eye movements. At the same time, its granular cytoarchitecture in macaque monkeys has both linked it to the prefrontal cortex and distinguished it from the premotor areas. Passingham and Wise (2012) proposed that its functions are fundamentally attentional rather than motor. The reason for the latter impression is that eye movements correspond to a shift in overt attention. But in addition to overt attention, the FEF also functions in covert attention, which does not involve eye movements. So, viewed generally, the functions of the caudal prefrontal cortex involve search and attention rather than the control of movements per se. A simple example illustrates the ecological significance of this point. In the natural habitat of primates, simply looking at some item produces nothing of value, unlike the movements of the hand, head, and mouth, which can contribute directly to the acquisition and consumption of resources. The same goes for covert attention; simply attending to an item contributes nothing directly to fulfilling an animal's needs.

The search and attention functions of the caudal prefrontal cortex operate in two ways: top-down biases that favor the dominance of particular sensory modalities or features at the expense of others (Armstrong and Moore, 2007), and a bottom-up perceptual phenomenon called "'pop-out," in which relatively rare but salient stimulus features come to dominate sensory information processing (Sato and Schall, 2003). The caudal prefrontal cortex contributes to both aspects of attentional control.

It might seem of little adaptive value to track, covertly, the locations of valuable items in retinal coordinates, as primates do. After all, only one item can be the most valuable one, and it presumably is the item an animal will choose. Indeed, in laboratory experiments, the rejected options usually disappear at the end of a behavioral trial. However, things are different in the fine-branch niche. Foraging targets that lose an initial competition remain available as subsequent goals. As a result, sequences of foraging choices can be planned and updated even as other reaching and eye movements take place.

28.6.1.4 Temporal Cortex

Other chapters in these volumes consider the visual and auditory cortex in detail, so this chapter mentions them only briefly (Section 28.4.1). New and specialized visual representations encode the color, shape, and visual texture of stimuli and provide this information to the granular orbitofrontal cortex for its valuation updating function. Visual areas also provide inputs to the caudal prefrontal cortex for its search and attention functions, and auditory areas make a similar contribution.

28.6.1.5 Summary

A suite of cortical innovations emerged in stem euprimates, including granular orbitofrontal, caudal prefrontal, posterior parietal, premotor, and temporal areas.

These new cortical areas accompanied musculoskeletal adaptations to the fine-branch niche, and they supported an improved ability to identify, search for, attend to, evaluate, grasp, manipulate, and consume valuable items while foraging on a mechanically unstable substrate.

28.6.2 Anthropoids

Along with the appearance of several new granular prefrontal areas (Section 28.3.2), the posterior parietal and temporal cortex enlarged and elaborated in anthropoids (Section 28.4.2). These developments produced the sensory processing pathways known as the dorsal and ventral visual streams, along with analogous auditory specializations.

28.6.2.1 Brain Changes and Foraging

At the beginning of the Oligocene, ~34 Ma, the global climate cooled abruptly by ~2 °C, which followed a longer, gradual cooling trend that had decreased global temperature by ~7 °C since the early Eocene (~50 Ma). Given that the platyrrhine–catarrhine split also occurred ~34 Ma, the relative brain expansion that occurred in both lineages (Section 28.5.2; Fig. 28.5B) might have reflected a response to this environmental challenge. Shortfalls in resources probably accompanied global cooling, so a diversification of foraging strategies and an increased reliance on fallback resources would have been likely. For example, Dominy (2004) has suggested that a shift to folivory occurred in some lineages, which might have provided the principal selective pressure for trichromatic vision. According to this idea, the new opsin developed by catarrhines had a spectral sensitivity that promoted the detection of relatively nutritious leaves, which tend to be the younger, more tender, and redder ones.

28.6.2.2 Using Events to Reduce Foraging Errors

Given the comparative evidence indicating both an upward grade shift in brain size and the development of new granular prefrontal areas in crown anthropoids (Sections 28.3.2 and 3.22.5.2), Passingham and Wise (2012) advanced the idea that these areas provided advantages in the face of periodic shortfalls in resources and high predation risks. According to their proposal, these new areas built on the prefrontal areas that evolved in early primates: (1) anthropoid prefrontal areas extended the function of the granular orbitofrontal cortex by linking its valuation of predicted outcomes to contexts, goals, and actions and (2) they extended the function of the caudal prefrontal cortex by mediating the top-down attentive control of sensory processing.

In broad perspective, these anthropoid innovations augmented the ancestral reinforcement learning mechanisms, which update action–outcome and stimulus–outcome associations based on a weighted average of several events. According to Passingham and Wise (2012), anthropoid prefrontal areas provide an advantage over this phylogenetically ancient mechanism by storing representations of single goal-related events. These representations are composed of conjunctions among contexts, goals, actions, and outcomes, especially outcome representations that rely on the visual specializations of primates. (The term "goal," as used here, refers to an object that serves as the target of a reaching or grasping movement.) When the context of a successful goal choice or action recurs, the memories stored in the anthropoid prefrontal cortex can generate an appropriate goal, sometimes based on just one or two prior experiences. The consequent reduction in foraging errors mitigates the risk of predation, among other cost factors.

In support of these ideas, Passingham and Wise (2012) reviewed findings from neuropsychology, neuroimaging, neurophysiology, and neuroanatomy—at book length. The most telling support comes from the behavioral effects of experimental lesions in macaque monkeys, and so the remainder of this section focuses on these findings.

28.6.2.2.1 Credit Assignment

Section 28.6.1.2 discussed the credit-assignment functions of the prefrontal cortex, which have been subject to a recent reinterpretation. Originally attributed to the granular orbitofrontal cortex (Walton et al., 2010), recent experiments have shown that the ventral prefrontal cortex is necessary for these functions instead (Rudebeck and Murray, 2014; Rudebeck et al., 2015).

Section 28.6.1.2 also mentioned that a leading research group has extended their concept of the orbitofrontal cortex beyond the areas traditionally included (areas 11, 13, and 14) to encompass area 12o (Chau et al., 2015). At one level, this idea seems like a minor redefinition of the granular orbitofrontal cortex, but it changes something important from an evolutionary perspective. According to Preuss and Goldman-Rakic (1991a,b,c), the traditional granular orbitofrontal areas evolved in early primates and the ventral prefrontal cortex evolved much later—in anthropoids (Section 28.3).

Despite this change in emphasis, the principal ideas developed for credit assignment (Walton et al., 2010) can be transferred to the ventral prefrontal cortex. In these experiments, efficient credit assignment resulted from establishing choice–outcome or stimulus–outcome associations based on specific events. In essence, intact (control) monkeys accurately assigned reward feedback to the choice that led to a particular outcome. Macaque monkeys with a functionally compromised ventral prefrontal cortex did not assign credit as effectively. Instead, the lesioned monkeys made goal choices based on broader averages, which took into account several events. As explained earlier, the ancestral reinforcement learning mechanisms depend on this kind of cumulative, average-based learning, which adjusts associations gradually. One consequence of such averaging is that feedback from one event becomes partially attributed to other goals chosen in the same general time frame.

Passingham and Wise (2012) ascribed efficient credit assignment to the memory of single events, which augmented the ancestral, reinforcement learning mechanisms. More generally, this adaptation can be described as a change in the time frame used for assigning feedback to previous choices (Bernacchia et al., 2011; Murray et al., 2014). Any shortening of that time frame, up to the limit of a single event, would have benefitted Oligocene anthropoids by reducing the frequency of foraging errors, each of which enhanced the risk of predation. Lengthening the time frame would also have provided advantages, and later evolutionary developments in anthropoids exploited this possibility as well.

28.6.2.2.2 Discrimination and Reversal Learning Set

The term "learning set" refers to an ability to learn faster with experience on a particular type of problem. As a practical matter, learning-set experiments have nearly always concentrated on an animal's performance on the first few trials of the serial discrimination learning task.

In one such experiment, the serial version of the visual discrimination task was contrasted with a different version, called "concurrent discrimination learning." For serial discrimination learning, the investigators began a block of trials by selecting two novel objects and designating one of them as correct. The monkeys then saw these two visual stimuli on a series of consecutive trials. To obtain a reward, they needed to choose the correct object, and they learned by trial and error to do so. The investigators then substituted two new stimuli and another block of trials ensued. This cycle continued for a large number of stimulus pairs until the monkeys began choosing the correct stimulus earlier in a block of trials.

For concurrent discrimination learning, the experimenters interleaved several pairs of stimuli, so the monkeys faced a given choice—between a given pair of stimuli—only after several intervening trials involving different stimuli. Macaque monkeys developed a strong learning set for the serial discrimination task (Fig. 28.6A, *gray circles*), as evidenced by the relatively low percentage of errors (less than 20% on the third trial). However, for the concurrent discrimination task (Fig. 28.6A, *gray triangles*), they did not (Murray and Gaffan, 2006).

In interpreting these results, Murray and Gaffan (2006) emphasized the concept of prospective coding, which refers to the maintenance of a goal in short-term memory. According to this idea, the monkeys performed the serial discrimination task by making their choice for the next trial based on events that occurred during an ongoing trial, and they prospectively coded that choice until the two stimuli appeared again. A simple strategy, called win–stay, lose–shift, yielded the correct choice. When their choice on the first trial led to a reward, the monkeys could simply repeat that choice on the subsequent trial. Alternatively, for choices that failed to produce a reward, the monkeys could shift their choice to the alternative. This strategy works well for the serial discrimination task, but not for the concurrent version of the task because of the large number of intervening stimuli and choices.

Lesions that disconnected the granular prefrontal cortex from the inferior temporal cortex blocked the development of a strong learning set (Browning et al., 2007). This lesion, called a crossed disconnection, involved the removal of the prefrontal cortex in one hemisphere and of the inferior temporal cortex in the other. This surgical procedure left the monkey with both a functioning prefrontal cortex and a functioning inferior temporal cortex but degraded cortical function by blocking any interaction between the two. As a consequence, monkeys with such disconnections made significantly more errors (Fig. 28.6A, *black circles*) than monkeys that had a fully functional granular prefrontal cortex (Fig. 28.6A, *gray circles*). Importantly, this impairment occurred only on the serial (single-pair) version of the task. Put another way, these results showed that prefrontal–inferior temporal (PF × IT) disconnections eliminated the advantage provided by the granular prefrontal cortex (Fig. 28.6A, *downward-pointing arrow*), which brought the monkey's learning rate into line with that on the concurrent version of the task (Fig. 28.6A, *triangles*). In this experiment, either of two manipulations blocked the ability of monkeys to reduce errors in one or a few trials: (1) introducing intervening trials with different stimuli, as occurred in the concurrent version of the task, or (2) disconnecting the granular prefrontal from the inferior temporal cortex, as occurred in the lesion experiment. The same lesions caused a similar impairment on the object-reversal task (Wilson and Gaffan, 2008).

Taken together, these experiments demonstrated that prefrontal–inferior temporal interactions underlie the ability of macaque monkeys to solve novel problems rapidly, based on experience with such problems. In similar experiments, rodents and carnivores have also decreased their errors with experience, but not to the extent that macaque monkeys have achieved in these and other experiments. Rhesus monkeys have come close to choosing the correct stimulus 90% of the time based on a single prior trial. Some corvids appear to have developed this capacity convergently (Bond et al., 2007), and perhaps some nonprimate mammals have done so, as well. The comparative data remain controversial, in part because of claims for a macaque-like learning set in rodents. However, as Passingham and Wise (2012) explained in detail, these experiments employed simpler

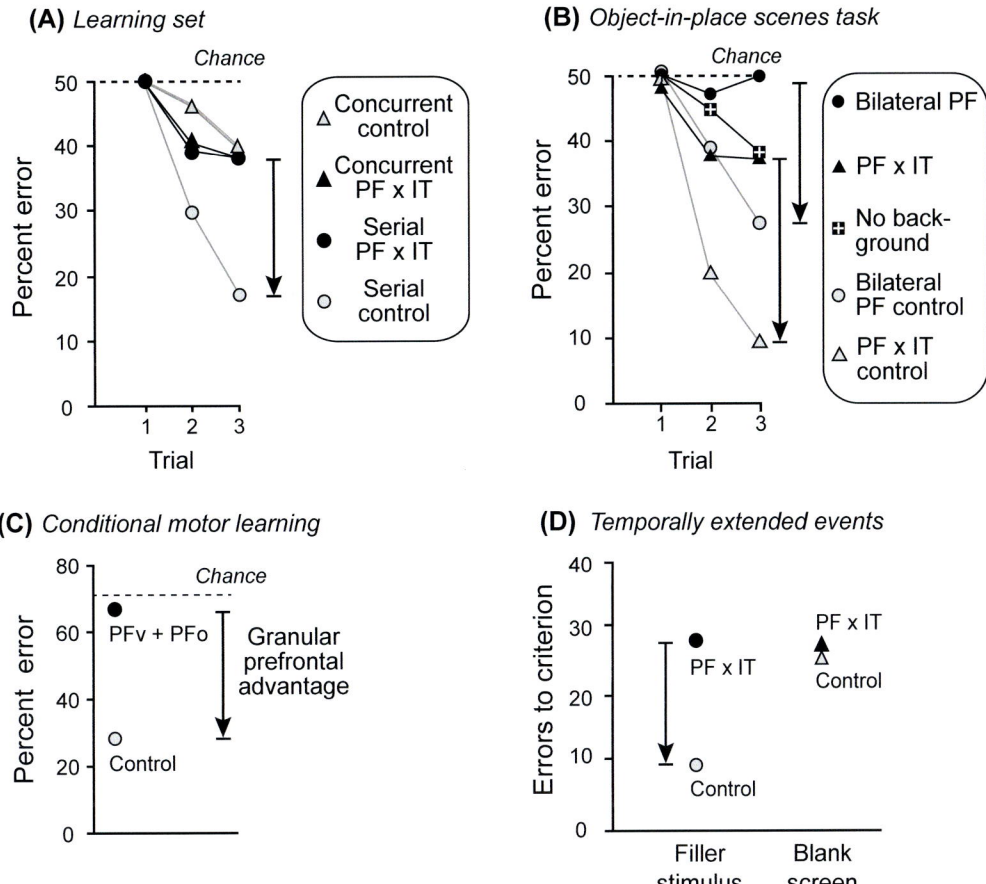

FIGURE 28.6 Effects of prefrontal cortex lesions in macaque monkeys. (A) Learning set over the first three trials. Disconnection lesions that block granular prefrontal–inferior temporal interactions (PF × IT) eliminated the advantage that experienced macaque monkeys had on learning new serial discriminations (*gray circles*) and brought the error rate into the range for concurrent discriminations (*triangles*). (B) Object-in-place scenes task, performance over the first three trials with a given scene and choice stimuli. Both bilateral prefrontal cortex lesions and prefrontal–inferior temporal disconnections (PF × IT) eliminated the advantage gained on the basis of one or two trials with the background scenes. (C) Conditional motor learning, mean over 48 trials. Combined lesions of the ventral and granular orbitofrontal cortex eliminated the advantage gained from one or a few experiences with a novel stimulus–goal associations. (D). Temporally extended events. Prefrontal–inferior temporal disconnections eliminated the performance advantage conferred by the filler stimulus. Abbreviations: *IT,* inferior temporal cortex; *PF,* prefrontal cortex; *pre-op,* preoperative; *post-op,* postoperative. (A) Data from Browning, P.G.F., Easton, A., Gaffan, D., 2007. Frontal-temporal disconnection abolishes object discrimination learning set in macaque monkeys. Cerebral Cortex 17, 859–864. (B) Data from Browning, P.G.F., Easton, A., Buckley, M.J., Gaffan, D., 2005. The role of prefrontal cortex in object-in-place learning in monkeys. Eur. J. Neurosci. 22, 3281–3291. (C) Data from Bussey, T.J., Wise, S.P., Murray, E.A., 2001. The role of ventral and orbital prefrontal cortex in conditional visuomotor learning and strategy use in rhesus monkeys. Behav. Neurosci. 115, 971–982. (D) Adapted from Wilson, C.R., Gaffan, D., Browning, P.G.F., Baxter, M.G., 2010. Functional localization within the prefrontal cortex: missing the forest for the trees? Trends Neurosci. 33, 533–540.

and easier tasks than the learning-set experiments on macaques, which undermines such claims.

28.6.2.2.3 Object-in-Place Scenes Task

A variant of concurrent discrimination learning is called the object-in-place scenes task. In the experiment described here, macaque monkeys saw a video monitor that had two object-like stimuli embedded in a complex background scene. The subjects needed to choose between them, as in other discrimination learning tasks, and each pair of stimuli had its own unique background scene. Macaque monkeys learned much faster with the

background scenes (Fig. 28.6B, *gray symbols*) than without them (*black square with plus signs*). Without the background scenes it took macaque monkeys ~10 trials to match the low error rate they typically achieved after just one trial with the backgrounds.

Both bilateral lesions of the granular prefrontal cortex and prefrontal–inferior temporal disconnections caused a severe impairment on this task (Browning et al., 2005). Fig. 28.6B illustrates the first three trials in two groups of monkeys. Before surgery (Fig. 28.6B, *gray symbols*), the two groups performed at different rates, so it is important to compare pre- and postoperative performance group by group. For bilateral prefrontal cortex lesions, the monkeys reverted to chance levels of performance; for prefrontal–inferior temporal disconnections (PF × IT), the impairment was of similar magnitude (*downward-pointing arrows*). After either method for compromising the function of the prefrontal cortex, the monkeys reverted to the same slow learning rates, as other monkeys could achieve without the background scenes (Fig. 28.6B, *black square with plus signs*).

This experiment showed that macaque monkeys could learn in one or a few trials to choose an object-like stimulus, but only when it appeared in a particular place within a large visual scene. It also demonstrated that a fully functioning granular prefrontal cortex reduced the number of errors that the animals made as they learned. It seems likely that this task measured a cognitive capacity that is especially important to the life of anthropoids, which typically forage over long distances based on choices that depend on complex visual scenes.

28.6.2.2.4 Conditional Motor Learning

In another experiment on one-trial learning, macaque monkeys learned arbitrary associations between an object-like stimulus and a spatial goal or action—a task called conditional motor learning (Bussey et al., 2001). Monkeys with little experience on this task learned new stimulus–goal associations relatively slowly, as rodents have done when facing similar problems (Murray et al., 2000).

In an extensive series of experiments, monkeys with experience on this task developed the ability to learn novel stimulus–goal associations in one or a few trials. Quite often, and with a little luck, the monkeys could learn a new stimulus–goal mapping without ever having made an error.

A combined lesion of the ventral prefrontal and granular orbitofrontal cortex blocked the fast learning of these associations (Fig. 28.6C; Bussey et al., 2001). Macaque monkeys with these lesions performed little better than naïve monkeys (Murray et al., 2000) or rodents (Bussey et al., 1996), although they eventually learned the stimulus–goal associations over several days of testing (Bussey et al., 2001). In another experiment, inactivation of the ventral prefrontal cortex, alone, caused a similar impairment (Wang et al., 2000).

These findings provided further support for the idea that the granular prefrontal cortex is necessary for fast learning—in one or a few trials—and therefore reduces foraging errors.

28.6.2.2.5 Temporally Extended Events

An experiment on the memory of temporallly extended events also provided support for this idea. In this experiment, macaque monkeys performed a standard serial discrimination task with two variants. In both versions, the chosen and unchosen stimuli disappeared during a delay period between the choice and the delivery of a reward, conditional upon a correct choice. The two variants differed in that in one version the screen was blank during the delay period, in the other, the experimenters replaced the chosen stimulus with a "filler" stimulus during the delay period.

The linkage provided by the filler stimulus reduced the number of errors to 90% correct performance (Fig. 28.6D, *gray circle*), and prefrontal–inferior temporal disconnections eliminated this advantage (Fig. 28.6D, *black circle*) (Browning and Gaffan, 2008). Put another way, even when a filler stimulus appeared, monkeys with a functionally compromised granular prefrontal cortex made roughly the same number of errors as when there was no filler stimulus (Fig. 28.6D, *triangles*). The filler stimulus seemed to promote the representation of a temporally extended event, which in this case consisted of a stimulus–choice–filler–reward sequence. Representations of this event, supported by prefrontal–inferior temporal interactions, enabled macaque monkeys to reduce their error rate.

28.6.2.2.6 Rules and Strategies

Section 28.6.2.2.2 discussed the win–stay and lose–shift strategies in the context of learning set. Macaque monkeys learn and apply similar strategies while performing other tasks, as well.

In an experiment on conditional motor learning (Section 28.6.2.2.4), monkeys learned three stimulus–goal associations concurrently, as a set. Under these conditions, many subjects made use of two abstract strategies, called repeat–stay and change–shift. Both strategies reduced errors, an advantage that was most conspicuous prior to learning the novel stimulus–goal associations. In one experiment, a series of "correction trials" ensured that the monkeys made a correct choice by the end of an extended trial (Bussey et al., 2001), but the experimenters scored only the first part of each trial to measure learning. Using the repeat–stay strategy, the subjects stayed with their most recent goal choice when the stimulus had repeated from the previous trial (called repeat trials). Using the change–shift strategy, the monkeys consistently rejected their most recent choice in favor of an alternative whenever the stimulus had changed from the previous trial (called change trials). For a task with three novel stimuli and three potential goals, the change–shift strategy can reduce the error rate from 67% to 50% errors on change trials; the repeat–stay strategy can eliminate errors entirely on repeat trials. A combined lesion of the ventral prefrontal and granular orbitofrontal cortex eliminated the benefits of these strategies, as well as those of a lose–shift strategy that the monkeys used on correction trials (Bussey et al., 2001).

28.6.2.2.7 Summary

A broad range of experiments on macaque monkeys have demonstrated that a fully functioning granular prefrontal cortex reduces foraging errors. Several capacities contribute to this advantage: the generation of goals from the memory of single events and from abstract strategies; the representation of goals associated with temporally extended events; the representation of both metrically concrete or metrically abstract goals; and the maintenance of goals in prospective memory.

The ability to represent single, goal-related events is especially important. The ancestral reinforcement learning mechanisms do not store these individuated memories and instead update the state of associative linkages, strengthening or weakening them based on feedback. Prefrontal areas in macaque monkeys retain these memories. Accordingly, when a context recurs, these memories elicit the retrieval of goals, actions, and outcomes that have occurred in that context, even if this has only happened a few times previously—and sometimes only once. These memories not only generate a bias among competing goals and actions (Cisek, 2012), but also among competing sensory representations (top-down, selective attention) (Desimone and Duncan, 1995).

The new anthropoid prefrontal areas are uniquely situated to perform their error-reduction function because they, alone among cortical areas, have relatively direct access to the orbitofrontal, premotor, posterior parietal, and temporal cortex (Passingham and Wise, 2012), all of which are dominated by the cortical innovations of primates. Only the anthropoid prefrontal areas have a combination of (1) direct information about the updated valuations of predicted outcomes, based on the animal's current state, which are represented in the granular orbitofrontal cortex (Section **28**.6.1.2); (2) inputs from the specialized context representations of the expanded visual and auditory cortex (inferior and superior temporal cortex); and (3) a relatively direct route to bias the planning and execution of actions via connections with the premotor and posterior parietal cortex. Along with the prominent intrinsic connections of the prefrontal cortex, their unique anatomical position enables the anthropoid prefrontal areas to represent conjunctions of contexts, goals, actions, and outcomes, which characterize goal-related events (Wilson et al., 2010; Passingham and Wise, 2012). The prefrontal cortex also empowers anthropoids to use these representations to influence foraging choices, thereby augmenting ancestral reinforcement learning mechanisms.

28.6.2.3 *Working Memory and Behavioral Inhibition*

Section **28**.1.4 mentioned two prevalent ideas about the function of the primate prefrontal cortex: working memory and behavioral inhibition. Rejection of these ideas enables an improved understanding of the adaptive advantages that the prefrontal cortex provides to primates. Accordingly, this section summarizes the experimental findings that argue against these long-standing doctrines.

28.6.2.3.1 Against Working-Memory Theory

The working-memory theory of the prefrontal cortex can be captured by some combination of the following four statements:

- Prefrontal areas have some working-memory functions.
- Only prefrontal areas have working-memory functions.
- All prefrontal areas have working-memory functions.
- Prefrontal areas have no functions other than working memory.

Except for the first of these four statements, the evidence conclusively refutes these ideas. Passingham and Wise (2012) presented the detailed evidence and arguments, but simply by unpacking the working-memory theory in this way its weaknesses become clear. Many areas of cortex function in working memory, the prefrontal cortex has many functions beyond working memory, and some prefrontal areas appear to lack working-memory functions.

Macaque monkeys with lesions of the ventral prefrontal cortex have impairments that have nothing to do with working memory. Such lesions have caused impairments on (1) the "go—no-go" object discrimination task (Iversen and Mishkin, 1970), (2) a matching-to-sample task modified to eliminate working-memory requirements (Rushworth et al., 1997), and (3) conditional motor learning tasks that also required little in the way of working memory (Bussey et al., 2001). Although these tasks had no memory delay period, they all depended on the integrity of the ventral prefrontal cortex in macaque monkeys. Proponents of the working-memory theory have claimed that the ventral prefrontal cortex performs working-memory functions (Wilson et al., 1993), as indeed it does. However, this idea does not characterize its functional specializations very effectively because the fundamental function of the ventral prefrontal cortex involves the generation of goals. The idea that the entire prefrontal cortex encodes working memories was also contradicted by the finding that few, if any, neurons in the polar prefrontal cortex encoded task-relevant events during a memory delay period (Tsujimoto et al., 2010, 2011, 2012).

The work of Petrides (2000) extended this conclusion to the dorsolateral prefrontal cortex, the area that originally gave rise to the working-memory theory. As Section **28**.3.1.4.3 explained, lesions of that prefrontal area in macaque monkeys have consistently impaired performance on the delayed response and delayed alternation tasks. Such lesions have also caused a severe impairment on a task that required macaque monkeys to remember a list of object-like stimuli. However, the lesioned monkeys performed like intact (control) subjects as a memory delay period increased in duration, thus taxing working memory. These findings showed that something other than the maintenance of information in working memory characterizes the function of the dorsolateral prefrontal cortex. In contrast, the same experiment showed that lesions of the anterior inferior temporal cortex caused an impairment that worsened as the memory period lengthened, consistent with a working-memory function. This finding should not be surprising because the specialized representations that underlie visual perception also underlie visual memory (Murray et al., 2007), but it contradicts some currently popular ideas about the segregation of perception and memory functions to different parts of the cortex (Buffalo et al., 1999, 2000).

Early neuroimaging studies of the human prefrontal cortex also emphasized working memory, but these interpretations eventually gave way to a broader perspective (Postle, Druzgal and D'Esposito, 2003). Experiments contrasting attention and working memory have indicated a predominance for the former (Rowe et al., 2000), which agrees with the idea that this area monitors representations in active memory (Petrides et al., 2002). Other human neuroimaging results have pointed to the same conclusion (Rowe and Passingham, 2001; Petrides et al., 2002; Manly et al., 2003; Lau et al., 2004), as did a neurophysiological result from macaque monkeys (Lebedev et al., 2004). In the latter experiment, monkeys had to remember one place while directing their attention somewhere else. Attention-related activity occurred much more frequently in the dorsolateral prefrontal cortex than did memory-related activity. Furthermore, the dorsolateral prefrontal cortex encoded attended locations much better than remembered locations (Lebedev et al., 2004).

28.6.2.3.2 Against Behavioral Inhibition Theory

As for behavioral inhibition, also known as inhibitory control and response inhibition, it is a small part of what the granular prefrontal cortex does, but this theory neglects the affirmative functions of the prefrontal cortex. Behavioral inhibition theory predicts that lesions of this area should always cause an abnormally high persistence of an established behavior (perseveration), but such lesions have consistently caused abnormally low persistence as well. Among the pertinent findings:

- Patients with lesions of the granular prefrontal cortex failed to stick with the correct rule on the Wisconsin card sorting task (random errors) as often as they perseverated (Barcelo and Knight, 2002).

- In experiments that employed action-reversal and object-reversal tasks, macaques (Kennerley et al., 2006; Rudebeck and Murray, 2008) and humans (Camille et al., 2011) with granular orbitofrontal cortex lesions used positive feedback inefficiently (Rudebeck and Murray, 2008). In contrast to this result, behavioral inhibition theory predicts that lesioned subjects should use negative feedback inefficiently and therefore perseverate on unrewarded choices.

- On the conditional motor learning task, macaques with granular prefrontal cortex lesions had impairments on both stay and shift strategies (Bussey et al., 2001), not mainly on shift strategies as behavioral inhibition theory predicts.

- The behavioral inhibition theory predicts that granular prefrontal cortex lesions should spare habitual, overlearned stimulus−response associations. The reason is that such behavior should benefit from a lack of inhibitory control over automatic responses. In contrast to this prediction, large combined lesions of the granular orbitofrontal and ventral prefrontal cortex caused a severe impairment in performing overlearned stimulus−response associations (Bussey et al., 2001).

- On the three-arm bandit task (Section **28**.6.1.2), macaques with granular prefrontal cortex lesions monkeys should have perseverated, but they did not (Walton et al., 2010). Instead, their impairment was better described as an insufficient persistence of behavior instead of the excessive persistence that behavioral inhibition theory predicts.

- According to the behavioral inhibition theory, lesions of the granular orbitofrontal cortex should have impaired the ability to suppress prepotent responses, such as choosing more food over less food. However, complete removal of this area had no such effect (Chudasama et al., 2007).

- On some "go−no-go" tasks, lesions of the ventral prefrontal cortex have caused a tendency to "go" incorrectly on "no-go" trials. Iversen and Mishkin (1970) interpreted these "failures of commission" as support for the behavioral inhibition theory, but their experimental design had a crucial flaw. In their variant of the task, only correct "go" responses were rewarded, which established "go" responses as the default choice. When, in a different experiment, "no-go" responses were made the default choice, monkeys with similar lesions tended to wrongly withhold movement on "go" trials ("failures of omission") (McEnaney and Butter, 1969). A balanced experimental design thus revealed that monkeys with lesions of the ventral prefrontal cortex were slow to change their behavior, but not in the way that behavioral inhibition theory predicts. Instead, all of these results reflected slower learning after prefrontal cortex lesions.

- Combined lesions of the granular orbitofrontal and ventral prefrontal cortex caused persistent, below-chance performance on a reversal task, which Jones and Mishkin (1972) attributed to behavioral inhibition. The results of Bussey et al. (2001) on the conditional motor learning task pointed to a different conclusion, however. In their experiment, the same lesion caused both slower learning and the loss of a lose—shift strategy, which accounts for the results of Jones and Mishkin without relying on the doctrine of behavioral inhibition.
- Lesions of the ventral prefrontal cortex in marmosets caused impairments on an extradimensional-shift task that Dias et al. (1996) interpreted in terms of behavioral inhibition. However, because the lesioned marmosets performed well on a second extradimensional shift (Dias et al., 1997), the initial impairment appears to have reflected slower learning rather than a failure of rule suppression.
- Impairments in extinction learning after lesions of the granular orbitofrontal cortex have also been interpreted in terms of behavioral inhibition (Butter, 1969), but slower learning accounts for these findings as well.

The generation of a new goal often involves canceling others, including previously planned or relatively automatic ones. This facet of goal generation has created the mistaken impression that the principal role of the prefrontal cortex is to negate unwanted behaviors. Instead of the negative function of behavioral inhibition, the prefrontal cortex performs the affirmative function of generating goals in a variety of circumstances, in part based on fast learning (Passingham and Wise, 2012).

28.7 Summary

28.7.1 Early Primates

Early primates adapted to life in the fine branches of trees and shrubs, with stem euprimates adopting a leaping—grasping form of locomotion dominated by hindlimb-generated forces. This way of moving not only reduced predator-attracting oscillations, but it also enabled forelimb specializations for grasping and manipulating both branches and food items. These animals lacked the fovea and trichromacy that would evolve in later primates, but they developed forward-facing eyes that increased the scope of stereoscopic vision, among other visual specializations.

The first granular prefrontal areas evolved in these animals, specifically the granular orbitofrontal cortex and the caudal prefrontal cortex, including the FEF.

These new areas accompanied a suite of new and elaborated premotor, posterior parietal, and temporal areas, which collectively provided early primates with several adaptive advantages:

- The caudal prefrontal cortex improved the ability to search for and maintain attention upon valuable items, using both overt and covert attention and both top-down and bottom-up mechanisms.
- The granular orbitofrontal cortex updated the valuations of predicted outcomes—based largely on their visual features—taking into account current states and needs. Vision-based valuations depended on inputs from the new and elaborated temporal areas of primates, such as the inferior temporal cortex, as well as information from older areas, such as the perirhinal cortex.
- Parietal—premotor networks implemented an innovative mechanism for reaching toward and grasping food items and food-bearing branches, as well as for manipulating food items and bringing them to the mouth. These cortical networks used extrinsic, stereoscopic coordinates to plan and guide reaching and grasping movements. The feeding technique used by these animals required the coordination of arm and hand movements with breathing and movements of the head, lips, and jaw. Similar parietal—premotor networks supported defensive movements and actions involving the forelimbs, hindlimbs, and body axis.

This suite of cortical adaptations contributed to the upward grade shift in brain size that occurred in euprimates (Fig. 28.5C) as well as to an increase in the proportion of brain consisting of neocortex (Fig. 28.5D).

28.7.2 Anthropoids

Anthropoids began as small, diurnal animals. As they diversified, these species increased in size and abandoned both the fine-branch niche and leaping—grasping locomotion. They became arboreal quadrupeds, at first, guided by high-acuity stereoscopic vision and a fovea inherited from their haplorhine ancestors. Anthropoid visual adaptations included an improved capacity for discriminating shape, translucence, glossiness, visual texture, and color (trichromacy). With their large size and high energy requirements, these animals needed to forage over an extended home range, based in part on distant visual and auditory signs. A decreased reliance on olfaction accompanied this way of life.

Like euprimate brains (Fig. 28.5C), anthropoid brains underwent another upward grade shift in size (Fig. 28.5A and B), which occurred mostly after the divergence of platyrrhines and catarrhines (Fig. 28.5B).

The main contributors to this development remain subject to debate, but it does not depend on simple scaling of preexisting areas. Instead, new cortical areas emerged during anthropoid evolution, including new granular prefrontal areas (Fig. 28.3A). The comparative evidence also points to three additional evolutionary developments in anthropoids:

- more neocortex as a proportion of the brain (Fig. 28.3D);
- a relative expansion of the frontal lobe (Fig. 28.4A and B), which occurred independently in platyrrhines and catarrhines (Kay et al., 2013); and
- more granular prefrontal cortex as a proportion of the frontal lobe (Fig. 28.3C).

The emergence of new granular prefrontal areas accompanied a broad suite of cortical innovations in anthropoids, as it did in euprimates. New somatosensory areas appeared, as did a new, caudal part of the primary motor cortex. These areas probably contributed to manipulating and evaluating food items, especially fruits, based on tactile cues. Expansion and elaboration also occurred in the inferior temporal, superior temporal, and posterior parietal cortex, among other areas. Of special importance to the function of the prefrontal cortex, the posterior parietal cortex came to represent a broad variety of metrics, such as relative number, duration, distance, and order (Genovesio et al., 2014), along with the spatial reference frames that guide movement via parietal–premotor networks (Section 28.6.1.1). Posterior parietal areas provided the dorsolateral and dorsomedial prefrontal cortex with the metric information that it used as contexts for generating foraging goals. In parallel, the inferior and superior temporal cortex provided the ventral prefrontal cortex with information about the signs of distant resources—visual and acoustic, respectively—that also led to the generation of foraging goals (Passingham and Wise, 2012).

Foraging errors can be costly under any circumstances, but especially so for far-ranging, diurnally foraging anthropoids, subject as they are to predation. Several interacting factors posed challenges for Oligocene anthropoids, as they do for many modern species: long-distance foraging that exposed them to predators; the need for high energy consumption to fuel arboreal quadrupedal locomotion; heat stress that limited foraging duration and enabled predators to concentrate on periods near dusk and dawn; and competition from other diurnal species, including members of their own social groups. Furthermore, the risks of long-range, daylight foraging must have escalated during shortfalls in necessary resources. These factors made it advantageous to limit foraging errors, even by a small amount.

According to a proposal by Passingham and Wise (2012), such selective pressures contributed to the emergence of the new granular prefrontal areas of anthropoids—specifically the dorsolateral, dorsomedial, ventral, and polar prefrontal areas (Fig. 28.3A). As Section 28.6.2.2 explained, these areas reduce foraging errors through the memory of single goal-related events, the learning and application of abstract rules and strategies, the representation of metrically abstract goals (independent of the means of achieving them), maintaining goals in prospective memory until the opportunity arises for achieving them, and the representation of temporally extended events. The consequent reduction in foraging errors increases fitness mainly by mitigating the risk of predation.

Other primates and other mammals are highly successful without the prefrontal areas that evolved in anthropoids, so the fitness advantage that they afford must be both subtle and significant. A reduction in foraging errors meets both criteria. It is subtle because it does not require some unprecedented cognitive capacity. Error reduction merely provides an advantage over ancestral mechanisms, one that has little impact on routine problems. It is significant because—for anthropoids facing a shortfall in necessary resources—the cost of poor foraging choices can often enough be the ultimate one.

28.7.3 From Trees to Tranquility

This chapter ends where many begin. Early anthropoids eventually gave rise to ~260 species of monkeys (Mitani et al., 2012), along with apes and the grain-growing, lunar-landing primates that were mentioned in Section 28.1.1. The trip from the trees to the Sea of Tranquility took time, of course, but much else besides. Before our species could undertake a voyage to the moon, our primate ancestors had to solve the problems that they faced in their time and place. Among many adaptations of brain and body, their prefrontal cortex made a singular contribution to their success.

References

Akert, K., 1964. Comparative Anatomy of Frontal Cortex and Thalamofrontal Connections. In: Warren, J.M., Akert, K. (Eds.), The Frontal Granular Cortex. McGraw-Hill, New York, pp. 372–396.

Allen, K.L., Kay, R.F., 2012. Dietary quality and encephalization in platyrrhine primates. Proc. R. Soc. Biol. Sci. 279, 715–721.

Allen, K.L., Kay, R.F., 2012. Endocast shape and brain proportions in primates. J. Vert. Paleontol 32, 55.

Allman, J., 2000. Evolving Brains. Freeman, New York.

Andersen, R.A., Buneo, C.A., 2002. Intentional maps in posterior parietal cortex. Annu. Rev. Neurosci. 25, 189–220.

Armstrong, K.M., Moore, T., 2007. Rapid enhancement of visual cortical response discriminability by microstimulation of the frontal eye field. Proc. Natl. Acad. Sci. U.S.A. 104, 9499–9504.

Bailey, P., Sweet, W.H., 1940. Effects on respiration, blood pressure and gastric motility of stimulation of orbital surface of frontal lobe. J. Neurophysiol 3, 276–281.

Baldwin, M.K., Cooke, D.F., Krubitzer, L., 2016. Intracortical microstimulation maps of motor, somatosensory, and posterior parietal cortex in tree shrews (Tupaia belangeri) reveal complex movement representations. Cereb. Cortex pii:bhv329. [Epub ahead of print].

Balleine, B.W., O'Doherty, J.P., 2010. Human and rodent homologies in action control: corticostriatal determinants of goal-directed and habitual action. Neuropsychopharmacology 35, 48–69.

Barcelo, F., Knight, R.T., 2002. Both random and perseverative errors underlie WCST deficits in prefrontal patients. Neuropsychologia 40, 349–356.

Barker, G.R., Warburton, E.C., 2011. When is the hippocampus involved in recognition memory? J. Neurosci. 31, 10721–10731.

Barton, R.A., 1998. Visual specialization and brain evolution in primates. Proc. R. Soc. Biol. Sci. 265, 1933–1937.

Barton, R.A., 2004. Binocularity and brain evolution in primates. Proc. Natl. Acad. Sci. U.S.A. 101, 10113–10115.

Barton, R.A., Venditti, C., 2013. Reply to Smaers: getting human frontal lobes in proportion. Proc. Natl. Acad. Sci. U. S. A 110, E3683–E3684.

Basso, M.A., Wurtz, R.H., 1998. Modulation of neuronal activity in superior colliculus by changes in target probability. J. Neurosci. 18, 7519–7534.

Battig, K., Rosvold, H.E., Mishkin, M., 1960. Comparison of the effects of frontal and caudate lesions on delayed response and alternation in monkeys. J. Comp. Physiol. Psychol. 53, 400–404.

Bauernfeind, A.L., de Sousa, A.A., Avasthi, T., Dobson, S.D., Raghanti, M.A., Lewandowski, A.H., Zilles, K., Semendeferi, K., Allman, J.M., Craig, A.D., Hof, P.R., Sherwood, C.C., 2013. A volumetric comparison of the insular cortex and its subregions in primates. J. Hum. Evol. 64, 263–279.

Baumann, M.A., Fluet, M.C., Scherberger, H., 2009. Context-specific grasp movement representation in the macaque anterior intraparietal area. J. Neurosci. 29, 6436–6448.

Baxter, M.G., Parker, A., Lindner, C.C., Izquierdo, A.D., Murray, E.A., 2000. Control of response selection by reinforcer value requires interaction of amygdala and orbital prefrontal cortex. J. Neurosci. 20, 4311–4319.

Berendse, H.W., Galis-de Graaf, Y., Groenewegen, H.J., 1992. Topographical organization and relationship with ventral striatal compartments of prefrontal corticostriatal projections in the rat. J. Comp. Neurol. 316, 314–347.

Bernacchia, A., Seo, H., Lee, D., Wang, X.J., 2011. A reservoir of time constants for memory traces in cortical neurons. Nat. Neurosci 14, 366–372.

Birrell, J.M., Brown, V.J., 2000. Medial frontal cortex mediates perceptual attentional set shifting in the rat. J. Neurosci. 20, 4320–4324.

Bloch, J.I., Boyer, D.M., 2002. Grasping primate origins. Science 298, 1606–1610.

Bloch, J.I., Silcox, M.T., 2006. Cranial anatomy of the Paleocene plesiadapiform Carpolestes simpsoni (Mammalia, Primates) using ultra high-resolution X-ray computed tomography, and the relationships of plesiadapiforms to euprimates. J. Hum. Evol. 50, 1–35.

Bloch, J.I., Silcox, M.T., Boyer, D.M., Sargis, E.J., 2007. New Paleocene skeletons and the relationship of plesiadapiforms to crown-clade primates. Proc. Natl. Acad. Sci. U.S.A. 104, 1159–1164.

Bond, A.B., Kamil, A.C., Balda, R.P., 2007. Serial reversal learning and the evolution of behavioral flexibility in three species of North American corvids (Gymnorhinus cyanocephalus, Nucifraga columbiana, Aphelocoma californica). J. Comp. Physiol. Psychol. 121, 372–379.

Boyer, D.M., Seiffert, E.R., Gladman, J.T., Bloch, J.I., 2013. Evolution and allometry of calcaneal elongation in living and extinct primates. Public Libr. Sci. One 8, e67792.

Boyer, D.M., Yapuncich, G.S., Chester, S.G., Bloch, J.I., Godinot, M., 2013. Hands of early primates. Am. J. Phys. Anthropol. 152 (Suppl. 57), 33–78.

Brog, J.S., Salyapongse, A., Deutch, A.Y., Zahm, D.S., 1993. The patterns of afferent innervation of the core and shell in the "accumbens" part of the rat ventral striatum: immunhistochemical detection of retrogradely transported fluoro-gold. J. Comp. Neurol. 338, 255–278.

Brown, V.J., Bowman, E.M., 2002. Rodent models of prefrontal cortical function. Trends Neurosci. 25, 340–343.

Browning, P.G., Gaffan, D., 2008. Prefrontal cortex function in the representation of temporally complex events. J. Neurosci. 28, 3934–3940.

Browning, P.G.F., Easton, A., Buckley, M.J., Gaffan, D., 2005. The role of prefrontal cortex in object-in-place learning in monkeys. Eur. J. Neurosci. 22, 3281–3291.

Browning, P.G.F., Easton, A., Gaffan, D., 2007. Frontal-temporal disconnection abolishes object discrimination learning set in macaque monkeys. Cereb. Cortex 17, 859–864.

Bruce, C.J., 1985. Primate frontal eye fields. II. Physiological and anatomical correlates of electrically evoked eye movements. J. Neurophysiol 54, 714–734.

Buffalo, E.A., Ramus, S.J., Clark, R.E., Teng, E., Squire, L.R., Zola, S.M., 1999. Dissociation between the effects of damage to perirhinal cortex and area TE. Learn. Mem. 6, 572–599.

Buffalo, E.A., Ramus, S.J., Squire, L.R., Zola, S.M., 2000. Perception and recognition memory in monkeys following lesions of area TE and perirhinal cortex. Learn. Mem. 7, 375–382.

Bullock, D., Grossberg, S., 1988. The Vite Model: A Neural Command Circuit for Generating Arm and Articulator Trajectories. In: Kelso, J.A.S., Mandell, A.J., Shlesinger, M.F. (Eds.), Dynamic Patterns in Complex Systems. World Scientific Publishers, Singapore.

Buneo, C.A., Jarvis, M.R., Batista, A.P., Andersen, R.A., 2002. Direct visuomotor transformations for reaching. Nature 416, 632–636.

Burish, M.J., Stepniewska, I., Kaas, J.H., 2008. Microstimulation and architectonics of frontoparietal cortex in common marmosets (Callithrix jacchus). J. Comp. Neurol. 507, 1151–1168.

Burnod, Y., Baraduc, P., BattagliaMayer, A., Guigon, E., Koechlin, E., Ferraina, S., Lacquaniti, F., Caminiti, R., 1999. Parieto-frontal coding of reaching: an integrated framework. Exp. Brain Res. 129, 325–346.

Bush, E.C., Simons, E.L., Allman, J.M., 2004. High-resolution computed tomography study of the cranium of a fossil anthropoid primate, Parapithecus grangeri: new insights into the evolutionary history of primate sensory systems. Anat. Rec. A Discov. Mol. Cell. Evol. Biol. 281, 1083–1087.

Bussey, T.J., Muir, J.L., Everitt, B.J., Robbins, T.W., 1996. Dissociable effects of anterior and posterior cingulate cortex lesions on the acquisition of a conditional visual discrimination: facilitation of early learning vs. impairment of late learning. Behav. Brain Res. 82, 45–56.

Bussey, T.J., Wise, S.P., Murray, E.A., 2001. The role of ventral and orbital prefrontal cortex in conditional visuomotor learning and strategy use in rhesus monkeys. Behav. Neurosci 115, 971–982.

Butter, C.M., 1969. Perseveration in extinction and in discrimination reversal tasks following selective frontal ablations in Macaca mulatta. Physiol. Behav. 4, 163–171.

Camille, N., Tsuchida, A., Fellows, L.K., 2011. Double dissociation of stimulus-value and action-value learning in humans with orbitofrontal or anterior cingulate cortex damage. J. Neurosci. 31, 15048–15052.

Carmichael, S.T., Price, J.L., 1994. Architectonic subdivision of the orbital and medial prefrontal cortex in the macaque monkey. J. Comp. Neurol. 346, 366–402.

Carmichael, S.T., Price, J.L., 1996. Connectional networks within the orbital and medial prefrontal cortex of macaque monkeys. J. Comp. Neurol. 371, 179–207.

Cartmill, M., 1972. Arboreal Adaptations and the Origin of the Order Primates. In: Tuttle, R. (Ed.), Functional and Evolutionary Biology of Primates. Aldine-Atherton, Chicago, IL, pp. 97−122.

Cartmill, M., 1974. Rethinking primate origins. Science 184, 436−443.

Cartmill, M., 1992. New views on primate origins. Evol. Anthropol. 1, 105−111.

Chang, S.W., Papadimitriou, C., Snyder, L.H., 2009. Using a compound gain field to compute a reach plan. Neuron 64, 744−755.

Changizi, M.A., 2009. The Visual Revolution. Benbella, Dallas, TX.

Chapman, C.A., Chapman, L.J., Struhsaker, T.T., Zanne, A.E., Clark, C.J., Pouslen, J.R., 2004. A long-term evaluation of fruit phenology: importance of climate change. J. Trop. Ecol. 21, 1−14.

Charvet, C.J., Finlay, B.L., 2012. Embracing covariation in brain evolution: large brains, extended development, and flexible primate social systems. Prog. Brain Res. 195, 71−87.

Chau, B.K., Sallet, J., Papageorgiou, G.K., Noonan, M.P., Bell, A.H., Walton, M.E., Rushworth, M.F., 2015. Contrasting roles for orbitofrontal cortex and amygdala in credit assignment and learning in macaques. Neuron 87, 1106−1118.

Chudasama, Y., Kralik, J.D., Murray, E.A., 2007. Rhesus monkeys with orbital prefrontal cortex lesions can learn to inhibit prepotent responses in the reversed reward contingency task. Cereb. Cortex 17, 1154−1159.

Cisek, P., 2012. Making decisions through a distributed consensus. Curr. Opin. Neurobiol. 22, 927−936.

Cisek, P., Grossberg, S., Bullock, D., 1998. A cortico-spinal model of reaching and proprioception under multiple task constraints. J. Cogn. Neurosci. 10, 425−444.

Cisek, P., Kalaska, J.F., 2005. Neural correlates of reaching decisions in dorsal premotor cortex: specification of multiple direction choices and final selection of action. Neuron 45, 801−814.

Cisek, P., Kalaska, J.F., 2010. Neural mechanisms for interacting with a world full of action choices. Annu. Rev. Neurosci. 33, 269−298.

Colby, C.L., Duhamel, J.R., 1996. Spatial representations for action in parietal cortex. Brain Res. Cogn. Brain Res. 5, 105−115.

Colby, C.L., Duhamel, J.R., Goldberg, M.E., 1996. Visual, presaccadic, and cognitive activation of single neurons in monkey lateral intraparietal area. J. Neurophysiol 76, 2841−2852.

Cooke, D.F., Goldring, A., Recanzone, G.H., Krubitzer, L.A., 2014. The Evolution of Parietal Areas Associated with Visuomanual Behavior: From Grasping to Tool Use. In: Chalupa, L.M., Werner, J. (Eds.), The Visual Neurosciences. MIT Press, Cambridge, MA, pp. 1049−1064.

Cooke, D.F., Graziano, M.S.A., 2004. Sensorimotor integration in the precentral gyrus: polysensory neurons and defensive movements. J. Neurophysiol 91, 1648−1660.

Cooke, D.F., Padberg, J., Zahner, T., Krubitzer, L., 2011. The functional organization and cortical connections of motor cortex in squirrels. Cereb. Cortex 22, 1959−1978.

Cui, H., Andersen, R.A., 2007. Posterior parietal cortex encodes autonomously selected motor plans. Neuron 56, 552−559.

Cui, H., Andersen, R.A., 2011. Different representations of potential and selected motor plans by distinct parietal areas. J. Neurosci. 31, 18130−18136.

Delgado, J.R., Livingston, R.B., 1948. Some respiratory, vascular and thermal responses to stimulation of the orbital surface of the frontal lobe. J. Neurophysiol 11, 39−55.

Desimone, R., Duncan, J., 1995. Neural mechanisms of selective visual attention. Annu. Rev. Neurosci. 18, 193−222.

Dias, R., Robbins, T.W., Roberts, A.C., 1996. Primate analogue of the Wisconsin card sorting test: effects of excitotoxic lesions of the prefrontal cortex in the marmoset. Behav. Neurosci. 110, 872−886.

Dias, R., Robbins, T.W., Roberts, A.C., 1997. Dissociable forms of inhibitory control within prefrontal cortex with an analog of the Wisconsin card sort test: restriction to novel situations and independence from 'on-line' processing. J. Neurosci. 17, 9285−9297.

Divac, I., Bjorklund, L.O., Passingham, R.E., 1978. Converging projections from the mediodorsal thalamic nucleus and mesencephalic dopaminergic neurons to the neocortex in three species. J. Comp. Neurol. 180, 59−72.

Dominy, N.J., 2004. Fruits, fingers, and fermentation: the sensory cues available to foraging primates. Integr. Comp. Biol. 44, 295−303.

Dorris, M.C., Glimcher, P.W., 2004. Activity in posterior parietal cortex is correlated with the relative subjective desirability of action. Neuron 44, 365−378.

Duhamel, J.-D., Colby, C.L., Goldberg, M.E., 1992. The updating of the representation of visual space in parietal cortex by intended eye movements. Science 255, 90−92.

Dum, R.P., Strick, P.L., 2005. Frontal lobe inputs to the digit representations of the motor areas on the lateral surface of the hemisphere. J. Neurosci. 25, 1375−1386.

Dum, R.P., Strick, P.L., 2005. Motor Areas of the Frontal Lobe: The Anatomical Substrate for the central Control of Movement. In: Riehle, A., Vaadia, E. (Eds.), Motor Cortex in Voluntary Movements. CRC Press, Boca Raton, FL, pp. 3−47.

Dunbar, R., 2014. Human Evolution. Pelican-Penguin, London.

Dunbar, R., 2009. The social brain hypothesis and its implications for social evolution. Ann. Hum. Biol. 36, 562−572.

Dunbar, R., Shultz, S., 2007. Understanding primate brain evolution. Philos. Trans. R. Soc. Biol. Sci. 362, 649−658.

Elston, G.N., 2007. Specialization of the Neocortical Pyramidal Cell during Primate Evolution. In: Preuss, T.M., Kaas, J.H. (Eds.), The Evolution of Nervous Systems. Elsevier, New York, pp. 191−242.

Elston, G.N., Benavides-Piccione, R., Elston, A., Zietsch, B., DeFelipe, J., Manger, P., Casagrande, V., Kaas, J.H., 2006. Specializations of the granular prefrontal cortex of primates: implications for cognitive processing. Anat. Rec. A Discov. Mol. Cell. Evol. Biol. 288, 26−35.

Erwin, D.H., Valentine, J.W., 2013. The Cambrian explosion: the Construction of animal diversity. Roberts, Greenwood Village, CO.

Fang, P.C., Stepniewska, I., Kaas, J.H., 2005. Ipsilateral cortical connections of motor, premotor, frontal eye, and posterior parietal fields in a prosimian primate, Otolemur garnetti. J. Comp. Neurol. 490, 305−333.

Felleman, D.J., Van Essen, D.C., 1991. Distributed hierarchical processing in the primate cerebral cortex. Cereb. Cortex 1, 1−47.

Ferry, A.T., Öngur, D., An, X.H., Price, J.L., 2000. Prefrontal cortical projections to the striatum in macaque monkeys: evidence for an organization related to prefrontal networks. J. Comp. Neurol. 425, 447−470.

Fitzpatrick, D.C., Suga, N., Olsen, J.F., 1998. Distribution of response types across entire hemispheres of the mustached bat's auditory cortex. J. Comp. Neurol. 391, 353−365.

Fleagle, J.G., 1999. Primate Adaptation and Evolution. Academic Press, San Diego, CA.

Fluet, M.C., Baumann, M.A., Scherberger, H., 2010. Context-specific grasp movement representation in macaque ventral premotor cortex. J. Neurosci. 30, 15175−15184.

Fogassi, L., Gallese, V., Buccino, G., Craighero, L., Fadiga, L., Rizzolatti, G., 2001. Cortical mechanism for the visual guidance of hand grasping movements in the monkey: a reversible inactivation study. Brain 124, 571−586.

Frysztak, R.J., Neafsey, E.J., 1994. The effect of medial frontal cortex lesions on cardiovascular conditioned emotional responses in the rat. Brain Res. 643, 181−193.

Gabi, M., Neves, K., Masseron, C., Ribeiro, P., Ventura-Antunes, L., Kaas, J.H., Herculano-Houzel, S., 2014. Human and non-human primates have similar distributions of neurons along the cerebral cortex, including the prefrontal cortex. In: Society for Neuroscience, Annual Meeting Planner. Online 2014 Program number 446.05.

Gallese, V., Murata, A., Kaseda, M., Niki, N., Sakata, H., 1994. Deficit of hand preshaping after muscimol injection in monkey parietal cortex. Neuroreport 5, 1525−1529.

Gaspar, P., Stepniewska, I., Kaas, J.H., 1992. Topography and collateralization of the dopaminergic projections to motor and lateral prefrontal cortex in owl monkeys. J. Comp. Neurol. 325, 1–21.

Genovesio, A., Wise, S.P., Passingham, R.E., 2014. Prefrontal–parietal function: from foraging to foresight. Trends Cogn. Sci. 18, 72–81.

Geyer, S., Ledberg, A., Schleicher, A., Kinomura, S., Schormann, T., Bürgel, U., Klingberg, T., Larsson, J., Zilles, K., Roland, P.E., 1996. Two different areas within the primary motor cortex of man. Nature 382, 805–807.

Gharbawie, O.A., Stepniewska, I., Burish, M.J., Kaas, J.H., 2010. Thalamocortical connections of functional zones in posterior parietal cortex and frontal cortex motor regions in New World monkeys. Cereb. Cortex 20, 2391–2410.

Gharbawie, O.A., Stepniewska, I., Qi, H., Kaas, J.H., 2011. Multiple parietal-frontal pathways mediate grasping in macaque monkeys. J. Neurosci. 31, 11660–11677.

Gold, J.I., Shadlen, M.N., 2000. Representation of a perceptual decision in developing oculomotor commands. Nature 404, 390–394.

Goldman, P.S., Galkin, T.W., 1978. Prenatal removal of frontal association cortex in the fetal rhesus monkey: anatomical and functional consequences in postnatal life. Brain Res. 152, 451–485.

Goldman, P.S., Rosvold, H.E., Vest, B., Galkin, T.W., 1971. Analysis of the delayed-alternation deficit produced by dorsolateral prefrontal lesions in the rhesus monkey. J. Comp. Physiol. Psychol. 77, 212–220.

Graziano, M., 2006. The organization of behavioral repertoire in motor cortex. Annu. Rev. Neurosci. 29, 105–134.

Graziano, M.S., Taylor, C.S., Moore, T., 2002. Complex movements evoked by microstimulation of precentral cortex. Neuron 34, 841–851.

Graziano, M.S., Taylor, C.S., Moore, T., Cooke, D.F., 2002. The cortical control of movement revisited. Neuron 36, 349–362.

Haber, S.N., Kim, K.S., Mailly, P., Calzavara, R., 2006. Reward-related cortical inputs define a large striatal region in primates that interface with associative cortical connections, providing a substrate for incentive-based learning. J. Neurosci. 26, 8368–8376.

Haber, S.N., Kunishio, K., Mizobuchi, M., Lynd-Balta, E., 1995. The orbital and medial prefrontal circuit through the primate basal ganglia. J. Neurosci. 15, 4851–4867.

Hadjidimitrakis, K., Dal, B.G., Breveglieri, R., Galletti, C., Fattori, P., 2015. Overlapping representations for reach depth and direction in caudal superior parietal lobule of macaques. J. Neurophysiol 114, 2310–2352.

Heesy, C.P., Ross, C.F., 2004. Mosaic Evolution of Activity Patterns, Diet, and Color Vision in Haplorhine Primates. In: Ross, C.F., Kay, R.F. (Eds.), Anthropoid Origins: New Visions. Academic/Plenum, New York, pp. 665–698.

Heiser, L.M., Colby, C.L., 2006. Spatial updating in area LIP is independent of saccade direction. J. Neurophysiol 95, 2751–2767.

Hoffmann, J.N., Montag, A.G., Dominy, N.J., 2004. Meissner corpuscles and somatosensory acuity: the prehensile appendages of primates and elephants. Anat. Rec. A Discov. Mol. Cell. Evol. Biol. 281, 1138–1147.

Iversen, S.D., Mishkin, M., 1970. Perseverative interference in monkeys following selective lesions of the inferior prefrontal convexity. Exp. Brain Res. 11, 376–386.

Jacobsen, C.F., 1936. Studies of cerebral function in primates. I. The functions of the frontal associations areas in monkeys. Comp. Psychol. Monogr 13, 3–60.

Janmaat, K.R.L., Byrne, R.W., Zuberbühler, K., 2006. Evidence for a spatial memory of fruiting states of rainforest trees in wild mangabeys. Anim. Behav. 72, 797–807.

Janson, C.H., 1988. Food competition in brown capuchin monkeys (Cebus apella): quantitative effects of group size and tree productivity. Behaviour 105, 53–76.

Jbabdi, S., Lehman, J.F., Haber, S.N., Behrens, T.E., 2013. Human and monkey ventral prefrontal fibers use the same organizational principles to reach their targets: tracing versus tractography. J. Neurosci. 33, 3190–3201.

Jerison, H.J., 2012. Digitized fossil brains: neocorticalization. Biolinguistics 6, 383–392.

Johnson, P.B., Ferraina, S., Bianchi, L., Caminiti, R., 1996. Cortical networks for visual reaching. II. Gradients of functional properties and extrinsic cortical connections of frontal and parietal lobe arm regions. Cereb. Cortex 6, 102–119.

Jones, B., Mishkin, M., 1972. Limbic lesions and the problem of stimulus-reinforcement associations. Exp. Neurol. 36, 362–377.

Jones, E.G., 1985. The Thalamus. Plenum Press, New York.

Kaada, B.R., Pribram, K.H., Epstein, J.A., 1949. Respiratory and vascular responses in monkeys from temporal pole, insula, orbital surface and cingulate gyrus. J. Neurophysiol 12, 347–356.

Kaas, J.H., 2004. Evolution of somatosensory and motor cortex in primates. Anat. Rec. A Discov. Mol. Cell. Evol. Biol. 281, 1148–1156.

Kaas, J.H., 2012. The evolution of neocortex in primates. Prog. Brain Res. 195, 91–102.

Kaas, J.H., 2013. The evolution of brains from early mammals to humans. Wiley Interdiscip. Rev. Cogn. Sci. 4, 33–45.

Kaas, J.H., 2014. The Evolution of the Visual System in Primates. In: Warner, J., Chalupa, L. (Eds.), The New Visual Neurosciences. MIT Press, Cambridge, MA, pp. 1233–1246.

Kalaska, J.F., Crammond, D.J., 1995. Deciding not to go: neuronal correlates of response selection in go/no go task in primate premotor and parietal cortex. Cereb. Cortex 5, 410–428.

Kappeler, P.M., 2012. The Biological Ecology of Strepsirrhines and Tarsiers. In: Mitani, J.C., Call, J., Kappeler, P.M., Palombit, R.A., Silk, J.B. (Eds.), The Evolution of Primate Societies. University of Chicago Press, Chicago, pp. 17–41.

Kay, R.F., Perry, J.M.G., Malinzak, M., Allen, K.L., Kirk, E.C., Plavcan, J.M., Fleagle, J.G., 2013. Paleobiology of Santacrucian Primates. In: Vizcaíno, S.F., Kay, R.F., Bargo, M.S. (Eds.), Early Miocene Paleobiology in Patagonia: High Latitude Paleocommunities of the Santa Cruz Formation. Cambridge University Press, Cambridge UK, pp. 306–330.

Kay, R.F., Williams, B.A., Ross, C.F., Takai, M., Shigehara, N., 2004. Anthropoid origins: a phylogenetic analysis. In: Ross, C.F., Kay, R.F. (Eds.), Anthropoid Origins: New Visions. Academic/Plenum, New York, pp. 91–135.

Kennerley, S.W., Walton, M.E., Behrens, T.E., Buckley, M.J., Rushworth, M.F., 2006. Optimal decision making and the anterior cingulate cortex. Nat. Neurosci. 9, 940–947.

Kermadi, I., Liu, Y., Tempini, A., Rouiller, E.M., 1997. Effects of reversible inactivation of the supplementary motor area (SMA) on unimanual grasp and bimanual pull and grasp performance in monkeys. Somatosens. Mot. Res. 14, 268–280.

Kim, S., Lee, D., 2011. Prefrontal cortex and impulsive decision making. Biol. Psychiatry 69, 1140–1146.

Kirk, E.C., Daghighi, P., Macrini, T.E., Bhullar, B.A., Rowe, T.B., 2014. Cranial anatomy of the Duchesnean primate Rooneyia viejaensis: new insights from high resolution computed tomography. J. Hum. Evol. 74, 82–95.

Klein, R.G., 2009. The human Career: human biological and Cultural origins. University of Chicago Press, Chicago.

Kojima, S., Kojima, M., Goldman-Rakic, P.S., 1982. Operant behavioral analysis of memory loss in monkeys and prefrontal lesions. Brain Res. 248, 51–59.

Kolb, B., 2007. Do All Mammals Have a Prefrontal Cortex? In: Kaas, J.H., Krubitzer, L. (Eds.), The Evolution of Primate Nervous Systems. Elsevier, Amsterdam, pp. 443–450.

Kolb, B., Buhrmann, K., McDonald, R., Sutherland, R.J., 1994. Dissociation of the medial prefrontal, posterior parietal, and posterior temporal cortex for spatial navigation and recognition memory in the rat. Cereb. Cortex 4, 664–680.

Kolb, B., Nonneman, A.J., Singh, R.K., 1974. Double dissociation of spatial impairments and perseveration following selective prefrontal lesions in rats. J. Comp. Physiol. Psychol. 87, 772–780.

Kondo, H., Saleem, K.S., Price, J.L., 2005. Differential connections of the perirhinal and parahippocampal cortex with the orbital and medial prefrontal networks in macaque monkeys. J. Comp. Neurol. 493, 479–509.

Krubitzer, L., 2009. In search of a unifying theory of complex brain evolution. Ann.. N.Y. Acad. Sci 1156, 44–67.

Krubitzer, L., Kaas, J., 2005. The evolution of the neocortex in mammals: how is phenotypic diversity generated? Curr. Opin. Neurobiol. 15, 444–453.

Krubitzer, L., Padberg, J., 2009. Evolution of Association Pallial Areas: Parietal association areas in mammals. In: Butler, A.B. (Ed.), Encyclopedic Reference of Neuroscience. Springer-Verlag, Berlin, pp. 1225–1231.

Krubitzer, L., Seelke, A.M., 2012. Cortical evolution in mammals: the bane and beauty of phenotypic variability. Proc. Natl. Acad. Sci. U.S.A. 109 (Suppl. 1), 10647–10654.

Larson, S.G., 1998. Unique Aspects of Quadrupedal Locomotion in Nonhuman Primates. In: Strasser, E., Fleagle, J.G., Rosenberger, A.L., McHenry, H.M. (Eds.), Primate Locomotion: Recent Advances. Plenum, New York, pp. 157–173.

Lau, H.C., Rogers, R.D., Ramnani, N., Passingham, R.E., 2004. Willed action and attention to the selection of action. Neuroimage 21, 1407–1415.

Leathers, M.L., Olson, C.R., 2012. In monkeys making value-based decisions, LIP neurons encode cue salience and not action value. Science 338, 132–135.

Lebedev, M.A., Messinger, A., Kralik, J.D., Wise, S.P., 2004. Representation of attended versus remembered locations in prefrontal cortex. Public Libr. Sci. Biol. 2, 1919–1935.

Lehman, J.F., Greenberg, B.D., McIntyre, C.C., Rasmussen, S.A., Haber, S.N., 2011. Rules ventral prefrontal cortical axons use to reach their targets: implications for diffusion tensor imaging tractography and deep brain stimulation for psychiatric illness. J. Neurosci. 31, 10392–10402.

Leopold, D.A., Rhodes, G., 2010. A comparative view of face perception. J. Comp. Psychol. 124, 233–251.

Lewis, P.A., Rezaie, R., Brown, R., Roberts, N., Dunbar, R., 2011. Ventromedial prefrontal volume predicts understanding of others and social network size. Neuroimage 57, 1624–1629.

Long, A., Bloch, J.I., Silcox, M.T., 2015. Quantification of neocortical ratios in stem primates. Am. J. Phys. Anthropol. 157, 363–373.

Lu, M.-T., Preston, J.B., Strick, P.L., 1994. Interconnections between the prefrontal cortex and the premotor areas in the frontal lobe. J. Comp. Neurol. 341, 375–392.

Luo, Z.-X., Gatesy, S.M., Jenkins, F.A., Amaral, W.W., Shubin, N.H., 2015. Mandibular and dental characteristics of Late Triassic mammaliaform *Haramiyavia* and their ramifications for basal mammal evolution. Proc. Natl. Acad. Sci. U. S. A 12, E7101–E7109.

MacNeilage, P.F., Studdert-Kennedy, M.G., Lindblom, B., 1987. Primate handedness reconsidered. Behav. Brain Sci. 10, 247–303.

Manly, T., Owen, A.M., McAvinue, L., Datta, A., Lewis, G.H., Scott, S.K., Rorden, C., Pickard, J., Robertson, I.H., 2003. Enhancing the sensitivity of a sustained attention task to frontal damage: convergent clinical and functional imaging evidence. Neurocase 9, 340–349.

Markow-Rajkowska, G., Kosmal, A., 1987. Organization of cortical afferents to the frontal association cortex in dogs. Acta Neurobiol. Exp. Wars 47, 137–161.

Markowska, A.L., Olton, D.S., Murray, E.A., Gaffan, D., 1989. A comparative analysis of the role of fornix and cingulate cortex in memory: rats. Exp. Brain Res. 74, 187–201.

Martin, R.D., 1981. Relative brain size and basal metabolic rate in terrestrial vertebrates. Nature 293, 60.

Martin, R.D., 1990. Primate Origins and Evolution: A Phylogenetic Reconstruction. Princeton University Press, Princeton, NJ.

Matelli, M., Luppino, G., Rogassi, L., Rizzolatti, G., 1989. Thalamic input to inferior area 6 and area 4 in the macaque monkey. J. Comp. Neurol. 280, 468–488.

McEnaney, K.W., Butter, C.M., 1969. Perseveration of responding and nonresponding in monkeys with orbital frontal ablations. J. Comp. Physiol. Psychol. 68, 558–561.

Melin, A.D., Matsushita, Y., Moritz, G.L., Dominy, N.J., Kawamura, S., 2013. Inferred L/M cone opsin polymorphism of ancestral tarsiers sheds dim light on the origin of anthropoid primates. Proc. R. Soc. Biol. Sci. 280, 20130189.

Meunier, M., Bachevalier, J., Mishkin, M., 1997. Effects of orbital frontal and anterior cingulate lesions on object and spatial memory in rhesus monkeys. Neuropsychologia 35, 999–1015.

Mishkin, M., 1957. Effects of small frontal lesions on delayed alternation in monkeys. J. Neurophysiol 20, 615–622.

Mitani, J.C., Call, J., Kappeler, P.M., Palombit, R.A., Silk, J.B., 2012. Primate Behavioral Diversity. In: Mitani, J.C., Call, J., Kappeler, P.M., Palombit, R.A., Silk, J.B. (Eds.), The Evolution of Primate Societies. University of Chicago Press, Chicago, pp. 5–16.

Morecraft, R.J., Cipolloni, P.B., Stilwell-Morecraft, K.S., Gedney, M.T., Pandya, D.N., 2004. Cytoarchitecture and cortical connections of the posterior cingulate and adjacent somatosensory fields in the rhesus monkey. J. Comp. Neurol. 469, 37–69.

Morecraft, R.J., Louie, J.L., Herrick, J.L., Stilwell-Morecraft, K.S., 2001. Cortical innervation of the facial nucleus in the non-human primate: a new interpretation of the effects of stroke and related subtotal brain trauma on the muscles of facial expression. Brain 124, 176–208.

Müller, A.E., Thalmann, U., 2000. Origin and evolution of primate social organization: a reconstruction. Biol. Rev. Camb. Philos. Soc. 75, 405–435.

Mullette-Gillman, O.A., Cohen, Y.E., Groh, J.M., 2005. Eye-centered, head-centered, and complex coding of visual and auditory targets in the intraparietal sulcus. J. Neurophysiol 94, 2331–2352.

Murray, E.A., Bussey, T.J., Saksida, L.M., 2007. Visual perception and memory: a new view of medial temporal lobe function in primates and rodents. Annu. Rev. Neurosci. 30, 99–122.

Murray, E.A., Bussey, T.J., Wise, S.P., 2000. Role of prefrontal cortex in a network for arbitrary visuomotor mapping. Exp. Brain Res. 133, 114–129.

Murray, E.A., Gaffan, D., 2006. Prospective memory in the formation of learning sets by rhesus monkeys (*Macaca mulatta*). J. Exp. Psychol. Anim. Behav. Process. 32, 87–90.

Murray, E.A., Moylan, E.J., Saleem, K.S., Basile, B.M., Turchi, J., 2015. Specialized areas for value updating and goal selection in the primate orbitofrontal cortex. eLife 4, e11695. https://doi.org/10.7554/eLife.11695.

Murray, E.A., Wise, S.P., Graham, K.S., 2016. The Evolution of Memory Systems: Ancestors, Anatomy, and Adaptations. Oxford University Press, Oxford.

Murray, E.A., Wise, S.P., Rhodes, S.E.V., 2011. What Can Different Brains Do with Reward? In: Gottfried, J.A. (Ed.), The Neurobiology of Sensation and Reward. Taylor & Francis, Boca Raton, FL, pp. 57–92.

Murray, J.D., Bernacchia, A., Freedman, D.J., Romo, R., Wallis, J.D., Cai, X., Padoa-Schioppa, C., Pasternak, T., Seo, H., Lee, D., Wang, X.J., 2014. A hierarchy of intrinsic timescales across primate cortex. Nat. Neurosci 17, 1661–1663.

Nakamura, H., Kuroda, T., Wakita, M., Kusunoki, M., Kato, A., Mikami, A., Sakata, H., Itoh, K., 2001. From three-dimensional space vision to prehensile hand movements: the lateral intraparietal area links the area V3A and the anterior intraparietal area in macaques. J. Neurosci. 21, 8174–8187.

Navawongse, R., Eichenbaum, H., 2013. Distinct pathways for rule-based retrieval and spatial mapping of memory representations in hippocampal neurons. J. Neurosci. 33, 1002–1013.

Ni, X., Gebo, D.L., Dagosto, M., Meng, J., Tafforeau, P., Flynn, J.J., Beard, K.C., 2013. The oldest known primate skeleton and early haplorhine evolution. Nature 498, 60–64.

Nudo, R.J., Masterton, R.B., 1988. Descending pathways to the spinal cord: a comparative study of 22 mammals. J. Comp. Neurol. 277, 53–79.

Nudo, R.J., Masterton, R.B., 1990. Descending pathways to the spinal cord, III. Sites of origin of the corticospinal tract. J. Comp. Neurol. 296, 559–583.

Oates, J.F., 1987. Food Distribution and Foraging Behavior. In: Smuts, B., Cheney, D.L., Seyfarth, R.M., Wrangham, R.W., Struhsaker, T.T. (Eds.), Primate Societies. University of Chicago Press, Chicago, IL, pp. 197–209.

Öngur, D., Price, J.L., 2000. The organization of networks within the orbital and medial prefrontal cortex of rats, monkeys and humans. Cereb. Cortex 10, 206–219.

Orliac, M.J., Ladeveze, S., Gingerich, P.D., Lebrun, R., Smith, T., 2014. Endocranial morphology of Palaeocene *Plesiadapis tricuspidens* and evolution of the early primate brain. Proc. R. Soc. Biol. Sci. 281, 20132792.

Padoa-Schioppa, C., 2011. Neurobiology of economic choice: a good-based model. Annu. Rev. Neurosci. 34, 333–359.

Palomero-Gallagher, N., Zilles, K., 2004. Isocortex. In: Paxinos, G. (Ed.), The Rat Nervous System, third ed. Elsevier Academic Press, San Diego CA, pp. 729–757.

Paré, M., Wurtz, R.H., 2001. Progression in neuronal processing for saccadic eye movements from parietal cortex area lip to superior colliculus. J. Neurophysiol 85, 2545–2562.

Pasquaretta, C., Leve, M., Claidiere, N., van de Waal, E., Whiten, A., MacIntosh, A.J., Pele, M., Bergstrom, M.L., Borgeaud, C., Brosnan, S.F., Crofoot, M.C., Fedigan, L.M., Fichtel, C., Hopper, L.M., Mareno, M.C., Petit, O., Schnoell, A.V., di Sorrentino, E.P., Thierry, B., Tiddi, B., Sueur, C., 2014. Social networks in primates: smart and tolerant species have more efficient networks. Sci. Rep. 4, 7600.

Passingham, R.E., Wise, S.P., 2012. The Neurobiology of the Prefrontal Cortex: Anatomy, Evolution, and the Origin of Insight. Oxford University Press, Oxford.

Pastor-Bernier, A., Cisek, P., 2011. Neural correlates of biased competition in premotor cortex. J. Neurosci 31, 7083–7088.

Patel, B.A., Wallace, I.J., Boyer, D.M., Granatosky, M.C., Larson, S.G., Stern, J.T., 2015. Distinct functional roles of primate grasping hands and feet during arboreal quadrupedal locomotion. J. Hum. Evol. 88, 79–84.

Pesaran, B., Nelson, M.J., Andersen, R.A., 2006. Dorsal premotor neurons encode the relative position of the hand, eye, and goal during reach planning. Neuron 51, 125–134.

Pesaran, B., Nelson, M.J., Andersen, R.A., 2008. Free choice activates a decision circuit between frontal and parietal cortex. Nature 453, 406–409.

Petrides, M., 2000. Dissociable roles of mid-dorsolateral prefrontal and anterior inferotemporal cortex in visual working memory. J. Neurosci. 20, 7496–7503.

Petrides, M., Alivisatos, B., Frey, S., 2002. Differential activation of the human orbital, midventrolateral, and mid-dorsolateral prefrontal cortex during the processing of visual stimuli. Proc. Natl. Acad. Sci. U. S. A 99, 5649–5654.

Platt, M.L., Glimcher, P.W., 1997. Responses of intraparietal neurons to saccadic targets and visual distractors. J. Neurophysiol 78, 1574–1589.

Platt, M.L., Glimcher, P.W., 1999. Neural correlates of decision variables in parietal cortex. Nature 400, 233–238.

Plaut, D.C., McClelland, J.L., Seidenberg, M.S., Patterson, K., 1996. Understanding normal and impaired word reading: computational principles in quasi-regular domains. Psychol. Rev. 103, 56–115.

Porrino, L.J., Goldman-Rakic, P.S., 1982. Brainstem innervation of prefrontal and anterior cingulate cortex in the rhesus monkey revealed by retrograde transport of HRP. J. Comp. Neurol. 205, 63–76.

Postle, B.R., Druzgal, T.J., D'Esposito, M., 2003. Seeking the neural substrates of visual working memory storage. Cortex 39, 927–946.

Pouget, A., Ducom, J.C., Torri, J., Bavelier, D., 2002. Multisensory spatial representations in eye-centered coordinates for reaching. Cognition 83, B1–B11.

Powell, D.A., Ginsberg, J.P., 2005. Single unit activity in the medial prefrontal cortex during pavlovian heart rate conditioning: effects of peripheral autonomic blockade. Neurobiol. Learn. Mem. 84, 200–213.

Powell, K.D., Goldberg, M.E., 2000. Response of neurons in the lateral intraparietal area to a distractor flashed during the delay period of a memory-guided saccade. J. Neurophysiol 84, 301–310.

Preuss, T.M., 1993. The role of the neurosciences in primate evolutionary biology: historical commentary and prospectus. In: MacPhee, R.D.E. (Ed.), Primates and Their Relatives in Phylogenetic Perspective. Plenum Press, New York, pp. 333–362.

Preuss, T.M., 1995. Do rats have prefrontal cortex? the Rose-Woolsey-Akert program reconsidered. J. Cogn. Neurosci. 7, 1–24.

Preuss, T.M., 2007. Evolutionary Specializations of Primate Brain Systems. In: Ravosa, M.J., Dagasto, M. (Eds.), Primate Origins: Adaptations and Evolution. Springer-Verlag, New York, pp. 625–675.

Preuss, T.M., 2007. Primate Brain Evolution in Phylogenetic Context. In: Kaas, J.H., Preuss, T.M. (Eds.), Evolution of Nervous Systems. Elsevier, Amsterdam, pp. 2–34.

Preuss, T.M., Goldman-Rakic, P.S., 1991. Myelo- and cytoarchiteture of the granular frontal cortex and surrounding regions in the strepsirhine primate *Galago* and the anthropoid primate Macaca. J. Comp. Neurol. 310, 429–474.

Preuss, T.M., Goldman-Rakic, P.S., 1991. Ipsilateral cortical connections of granular frontal cortex in the strepsirhine primate *Galago*, with comparative comments on anthropoid primates. J. Comp. Neurol. 310, 507–549.

Preuss, T.M., Goldman-Rakic, P.S., 1991. Architectonics of the parietal and temporal association cortex in the strepsirhine primate *Galago* compared to the anthropoid primate Macaca. J. Comp. Neurol. 310, 475–506.

Preuss, T.M., Stepniewska, I., Kaas, J.H., 1996. Movement representation in the dorsal and ventral premotor areas of owl monkeys: a microstimulation study. J. Comp. Neurol. 371, 649–675.

Radinsky, L., 1975. Primate brain evolution. Am. Sci. 63, 656–663.

Radinsky, L., 1979. The fossil record of primate brain evolution. In: 49th James Arthur lecture on the evolution of the human brain. American Museum of Natural History, New York, pp. 1–27.

Ramdarshan, A., Orliac, M.J., 2015. Endocranial morphology of *Microchoerus erinaceus* (*Euprimates*, Tarsiiformes) and early evolution of the euprimates brain. Am. J. Phys. Anthropol. 159, 5–16.

Rathelot, J.A., Strick, P.L., 2009. Subdivisions of primary motor cortex based on cortico-motoneuronal cells. Proc. Natl. Acad. Sci. U.S.A. 106, 918–923.

Ray, J.P., Price, J.L., 1993. The organization of projections from the mediodorsal nucleus of the thalamus to orbital and medial prefrontal cortex in macaque monkeys. J. Comp. Neurol. 337, 1–31.

Remple, M.S., Reed, J.L., Stepniewska, I., Lyon, D.C., Kaas, J.H., 2007. The organization of frontoparietal cortex in the tree shrew (*Tupaia belangeri*): II. Connectional evidence for a frontal-posterior parietal network. J. Comp. Neurol. 501, 121–149.

Reynolds, S.M., Zahm, D.S., 2005. Specificity in the projections of prefrontal and insular cortex to ventral striatopallidum and the extended amygdala. J. Neurosci. 25, 11757–11767.

Rhodes, S.E., Murray, E.A., 2013. Differential effects of amygdala, orbital prefrontal cortex, and prelimbic cortex lesions on goal-directed behavior in rhesus macaques. J. Neurosci. 33, 3380–3389.

Rose, J.E., Woolsey, C.N., 1948. The orbitofrontal cortex and its connections with the mediodorsal nucleus in rabbit, sheep and cat. Res. Publ. Assoc. Res. Nerv. Ment. Dis. 27, 210–232.

Rose, K.D., 2006. The Beginnings of the Age of Mammals. Johns Hopkins Press, Baltimore, MD.

Ross, C., Williams, B., Kay, R.F., 1998. Phylogenetic analysis of anthropoid relationships. J. Hum. Evol. 35, 221–306.

Ross, C.F., 2004. The tarsier fovea: functionless vestige or nocturnal adaptation? In: Ross, C.F., Kay, R.F. (Eds.), Anthropoid Origins: New Visions. Academic/Plenum, New York, pp. 477–537.

Rowe, J.B., Passingham, R.E., 2001. Working memory for location and time: activity in prefrontal area 46 relates to selection rather than maintenance in memory. Neuroimage 14, 77–86.

Rowe, J.B., Toni, I., Josephs, O., Frackowiak, R.S., Passingham, R.E., 2000. The prefrontal cortex: response selection or maintenance within working memory? Science 288, 1656–1660.

Rudebeck, P.H., Murray, E.A., 2008. Amygdala and orbitofrontal cortex lesions differentially influence choices during object reversal learning. J. Neurosci. 28, 8338–8343.

Rudebeck, P.H., Murray, E.A., 2014. The role of the macaque orbital prefrontal cortex in learning and reversing probabilistic reward associations.. In: Society for Neuroscience, Annual Meeting Planner. Online 2014 Program number 206.10.

Rudebeck, P.H., Putnam, P.T., Daniels, T.E., Yang, T., Mitz, A.R., Rhodes, S.E., Murray, E.A., 2014. A role for primate subgenual cingulate cortex in sustaining autonomic arousal. Proc. Natl. Acad. Sci. U.S.A. 111, 5391–5396.

Rudebeck, P.H., Saunders, R.C., Lundgren, D., Murray, E.A., 2015. Dissociating the functions of the macaque ventrolateral and orbitofrontal cortex. In: Society for neuroscience, Annual meeting planner online 2015 program number 176.09.

Rudebeck, P.H., Saunders, R.C., Prescott, A.T., Chau, L.S., Murray, E.A., 2013. Prefrontal mechanisms of behavioral flexibility, emotion regulation and value updating. Nat. Neurosci. 16, 1140–1145.

Rushworth, M.F., Nixon, P.D., Eacott, M.J., Passingham, R.E., 1997. Ventral prefrontal cortex is not essential for working memory. J. Neurosci. 17, 4829–4838.

Rushworth, M.F.S., Hadland, K.A., Gaffan, D., Passingham, R.E., 2003. The effect of cingulate cortex lesions on task switching and working memory. J. Cogn. Neurosci. 15, 338–353.

Rushworth, M.F.S., Nixon, P.D., Passingham, R.E., 1997. Parietal cortex and movement. I. Movement selection and reaching. Exp. Brain Res. 117, 292–310.

Rushworth, M.F.S., Nixon, P.D., Passingham, R.E., 1997. Parietal cortex and movement. II. Spatial representation. Exp. Brain Res. 117, 311–323.

Saksida, L.M., Bussey, T.J., 2010. The representational-hierarchical view of amnesia: translation from animal to human. Neuropsychologia 48, 2370–2384.

Sato, T.R., Schall, J.D., 2003. Effects of stimulus-response compatibility on neural selection in frontal eye field. Neuron 38, 637–648.

Schall, J.D., Bichot, N.P., 1998. Neural correlates of visual and motor decision processes. Curr. Opin. Neurobiol. 8, 211–217.

Scherberger, H., Andersen, R.A., 2007. Target selection signals for arm reaching in the posterior parietal cortex. J. Neurosci. 27, 2001–2012.

Schilman, E.A., Uylings, H.B.M., Graaf, Y.G., Joel, D., Groenewegen, H.J., 2008. The orbital cortex in rats topographically projects to central parts of the caudate-putamen complex. Neurosci. Lett. 432, 40–45.

Schmitt, D., 2010. Primate Locomotor Evolution: Biomechanical Studies of Primate Locomotion and Their Implications for Understanding Primate Neuroethology. In: Platt, M.L., Ghazanfar, A.A. (Eds.), Primate Neuroethology. Oxford University Press, New York, pp. 31–63.

Schoenbaum, G., Roesch, M.R., Stalnaker, T.A., Takahashi, Y.K., 2009. A new perspective on the role of the orbitofrontal cortex in adaptive behaviour. Nat. Rev. Neurosci. 10, 885–892.

Scopinho, A.A., Tavares, R.F., Correa, F.M., 2009. The medial forebrain bundle mediates cardiovascular responses to electrical stimulation of the medial prefrontal cortex. Aut. Neurosci 147, 38–47.

Seamans, J.K., Lapish, C.C., Durstewitz, D., 2008. Comparing the prefrontal cortex of rats and primates: insights from electrophysiology. Neurotoxicol. Res. 14, 249–262.

Sears, K.E., Finarelli, J.A., Flynn, J.J., Wyss, A., 2008. Estimating body mass in new world "monkeys" (Platyrrhini, primates), with consideration of the Miocene platyrrhine, Chilecebus carrascoensis. Am. Mus. Novitates 3617, 1–29.

Selemon, L.D., Goldman-Rakic, P.S., 1985. Longitudinal topography and interdigitation of corticostriatal projections in the rhesus monkey. J. Neurosci. 5, 776–794.

Semendeferi, K., Teffer, K., Buxhoeveden, D.P., Park, M.S., Bludau, S., Amunts, K., Travis, K., Buckwalter, J., 2010. Spatial organization of neurons in the frontal pole sets humans apart from great apes. Cereb. Cortex 21, 1485–1497.

Sergio, L.E., Scott, S.H., 1998. Hand and joint paths during reaching movements with and without vision. Exp. Brain Res. 122, 157–164.

Shadmehr, R., Wise, S.P., 2005. The Computational Neurobiology of Reaching and Pointing: A Foundation for Motor Learning. The MIT Press, Cambridge MA.

Sherwood, C.C., Holloway, R.L., Semendeferi, K., Hof, P.R., 2005. Is prefrontal white matter enlargement a human evolutionary specialization? Nat. Neurosci. 8, 537–538.

Shultz, S., Opie, C., Atkinson, Q.D., 2011. Stepwise evolution of stable sociality in primates. Nature 479, 219–222.

Silcox, M.T., Benham, A.E., Bloch, J.I., 2010. Endocasts of Microsyops (Microsyopidae, Primates) and the evolution of the brain in primitive primates. J. Hum. Evol. 58, 505–521.

Silcox, M.T., Dalmyn, C.K., Bloch, J.I., 2009. Virtual endocast of Ignacius graybullianus (Paromomyidae, Primates) and brain evolution in early primates. Proc. Natl. Acad. Sci. U.S.A. 106, 10987–10992.

Silcox, M.T., Sargis, E.J., Bloch, J.I., Boyer, D.M., 2015. Primate origins and supraordinal relationships: morphological evidence. In: Henke, K., Tattersall, I. (Eds.), Handbook of Paleoanthropology. Springer-Verlag, Berlin, pp. 1053–1081.

Smith, M.M., Johanson, Z., 2015. Origin of the Vertebrate Dentition: Teeth Transform Jaws into Biting Force. In: Dial, K.P., Shubin, N., Brainerd, E.L. (Eds.), Great Transformations in Vertebrate Evolution. University of Chicago Press, Chicago, pp. 9–30.

Smith, W.K., 1945. The functional significance of the rostral cingular cortex as revealed by its responses to electrical excitation. J. Neurophysiol 8, 242–255.

Snyder, L.H., Batista, A.P., Andersen, R.A., 1998. Change in motor plan, without a change in the spatial locus of attention, modulates activity in posterior parietal cortex. J. Neurophysiol 79, 2814–2819.

Spellman, T., Rigotti, M., Ahmari, S.E., Fusi, S., Gogos, J.A., Gordon, J.A., 2015. Hippocampal-prefrontal input supports spatial encoding in working memory. Nature 522, 309–314.

Stepniewska, I., Cerkevich, C.M., Fang, P.C., Kaas, J.H., 2009. Organization of the posterior parietal cortex in galagos: II. Ipsilateral cortical connections of physiologically identified zones within anterior sensorimotor region. J. Comp. Neurol. 517, 783–807.

Stepniewska, I., Fang, P.C., Kaas, J.H., 2009. Organization of the posterior parietal cortex in galagos: I. Functional zones identified by microstimulation. J. Comp. Neurol. 517, 765–782.

Stepniewska, I., Fang, P.C., Kaas, J.H., 2005. Microstimulation reveals specialized subregions for different complex movements in posterior parietal cortex of prosimian galagos. Proc. Natl. Acad. Sci. U.S.A. 102, 4878–4883.

Stepniewska, I., Friedman, R.M., Gharbawie, O.A., Cerkevich, C.M., Roe, A.W., Kaas, J.H., 2011. Optical imaging in galagos reveals parietal-frontal circuits underlying motor behavior. Proc. Natl. Acad. Sci. U.S.A. 108, E725—E732.

Stepniewska, I., Gharbawie, O.A., Burish, M.J., Kaas, J.H., 2014. Effects of muscimol inactivations of functional domains in motor, premotor, and posterior parietal cortex on complex movements evoked by electrical stimulation. J. Neurophysiol 111, 1100—1119.

Strick, P.L., Preston, J.B., 1982. Two representations of the hand in area 4 of a primate. II. Somatosensory input organization. J. Neurophysiol 48, 150—159.

Striedter, G.F., 2005. Principles of Brain Evolution. Sinauer, Sunderland, MA.

Sussman, R.W., 1991. Primates origins and the evolution of angiosperms. Am. J. Primatol. 23, 209—233.

Tanji, J., Wise, S.P., 1981. Submodality distribution in sensorimotor cortex of the unanesthetized monkey. J. Neurophysiol 45, 467—481.

Tsujimoto, S., Genovesio, A., Wise, S.P., 2010. Evaluating self-generated decisions in frontal pole cortex of monkeys. Nat. Neurosci. 13, 120—126.

Tsujimoto, S., Genovesio, A., Wise, S.P., 2011. Frontal pole cortex: encoding ends at the end of the endbrain. Trends Cogn. Sci. 15, 169—176.

Tsujimoto, S., Genovesio, A., Wise, S.P., 2012. Neuronal activity during a cued strategy task: comparison of dorsolateral, orbital, and polar prefrontal cortex. J. Neurosci. 32, 11017—11031.

Uylings, H.B.M., Groenewegen, H.J., Kolb, B., 2003. Do rats have a prefrontal cortex? Behav. Brain Res. 146, 3—17.

Von Der, H.R., Vyas, G., Olson, I.R., 2014. The social network-network: size is predicted by brain structure and function in the amygdala and paralimbic regions. Soc. Cogn. Affect. Neurosci. 9, 1962—1972.

Wall, P.D., Davis, G.D., 1951. Three cerebral cortical systems affecting autonomic function. J. Neurophysiol 14, 507—517.

Walton, M.E., Behrens, T.E., Buckley, M.J., Rudebeck, P.H., Rushworth, M.F., 2010. Separable learning systems in the macaque brain and the role of orbitofrontal cortex in contingent learning. Neuron 65, 927—939.

Wang, M., Zhang, H., Li, B.-M., 2000. Deficit in conditional visuomotor learning by local infusion of bicuculline into the ventral prefrontal cortex in monkeys. Eur. J. Neurosci. 12, 3787—3789.

Wilber, A.A., Clark, B.J., Forster, T.C., Tatsuno, M., McNaughton, B.L., 2014. Interaction of egocentric and world-centered reference frames in the rat posterior parietal cortex. J. Neurosci. 34, 5431—5446.

Williams, B.A., Kay, R.F., Kirk, E.C., 2010. New perspectives on anthropoid origins. Proc. Natl. Acad. Sci. U.S.A. 107, 4797—4804.

Williams, S.M., Goldman-Rakic, P.S., 1998. Widespread origin of the primate mesofrontal dopamine system. Cereb. Cortex 8, 321—345.

Wilson, C.R., Gaffan, D., 2008. Prefrontal-inferotemporal interaction is not always necessary for reversal learning. J. Neurosci. 28, 5529—5538.

Wilson, C.R., Gaffan, D., Browning, P.G., Baxter, M.G., 2010. Functional localization within the prefrontal cortex: missing the forest for the trees? Trends Neurosci. 33, 533—540.

Wilson, F.A., O'Scalaidhe, P., Goldman-Rakic, P.S., 1993. Dissociation of object and spatial processing domains in primate prefrontal cortex. Science 260, 1955—1958.

Wise, S.P., 1984. Nonprimary Motor Cortex and its Role in the Cerebral Control of Movement. In: Edelman, G., Gall, W.E., Cowan, W.M. (Eds.), Dynamic Aspects of Neocortical Function. John Wiley and Sons, New York, pp. 525—555.

Wise, S.P., 1996. Evolutionary and Comparative Neurobiology of the Supplementary Sensorimotor Area. In: Lüders, H. (Ed.), The Supplementary Sensorimotor Area. Raven Press, New York, pp. 71—83.

Wise, S.P., 2006. The ventral premotor cortex, corticospinal region C, and the origin of primates. Cortex 42, 521—524.

Wise, S.P., 2007. The Evolution of Ventral Premotor Cortex and the Primate Way of Reaching. In: Preuss, T.M., Kaas, J.H. (Eds.), The Evolution of Nervous Systems. Elsevier, Amsterdam, pp. 157—166.

Wise, S.P., 2008. Forward frontal fields: phylogeny and fundamental function. Trends Neurosci. 31, 599—608.

Wise, S.P., 2014. The Prefrontal Cortex in Comparative Perspective. In: Gazzaniga, M.S., Mangun, G.R. (Eds.), The Cognitive Neurosciences, fifth ed. MIT Press, Cambridge, MA, pp. 51—58.

Wise, S.P., Boussaoud, D., Johnson, P.B., Caminiti, R., 1997. Premotor and parietal cortex: corticocortical connectivity and combinatorial computations. Annu. Rev. Neurosci. 20, 25—42.

Wise, S.P., Donoghue, J.P., 1986. The Motor Cortex of Rodents. In: Peters, A., Jones, E.G. (Eds.), Cerebral Cortex. Plenum, New York, pp. 243—270.

Wise, S.P., Hendry, S.H.C., Jones, E.G., 1978. Prenatal development of sensory-motor cortical projections in cats. Brain Res. 138, 538—544.

Wise, S.P., Willingham, D.T., 2009. Motor Skill Learning. In: Squire, L.R. (Ed.), Encyclopedia of Neuroscience, vol. 5. Academic Press, Oxford, pp. 1057—1066.

Wolpert, D.M., Ghahramani, Z., Jordan, M.I., 1995. Are arm trajectories planned in kinematic or dynamic coordinates? an adaptation study. Exp. Brain Res. 103, 460—470.

Wong, P., Kaas, J.H., 2008. Architectonic subdivisions of neocortex in the gray squirrel (Sciurus carolinensis). Anat. Rec. 291, 1301—1333.

Wong, P., Kaas, J.H., 2009. Architectonic subdivisions of neocortex in the tree shrew (Tupaia belangeri). Anat. Rec. 292, 994—1027.

Wong, P., Kaas, J.H., 2010. Architectonic subdivisions of neocortex in the galago (Otolemur garnetti). Anat. Rec. 293, 1033—1069.

Wu, C.W.H., Bichot, N.P., Kaas, J.H., 2000. Converging evidence from microstimulation, architecture, and connections for multiple motor areas in the frontal and cingulate cortex of prosimian primates. J. Comp. Neurol. 423, 140—177.

Zuberbühler, K., Janmaat, K.R.L., 2010. Foraging Cognition in Nonhuman Primates. In: Platt, M.L., Ghazanfar, A.A. (Eds.), Primate Neuroethology. Oxford University Press, Oxford, pp. 64—83.

PART 5

Evolution of Human Brains

29

An Introduction to Human Brain Evolutionary Studies

T.M. Preuss

Emory University, Atlanta, GA, United States

29.1 Introduction

How we understand human brain evolution is central to our understanding of human nature. Yet the study of human brain evolution is not a unitary scientific discipline: as the contributions to this volume illustrate, scientists who address human brain evolution come from a remarkable range of specialties within the life sciences, including paleoanthropology, genetics, life history theory, and multiple branches of psychology and neuroscience. Moreover, the study of human brain evolution lies at the intersection of evolutionary biology and the neurosciences, and these are fields that have not always played well together, representing as they do two quite distinct approaches to biology. While evolutionary biology does have an experimental component, evolutionists are mainly engaged in comparative studies of living and extinct species, with the goal of reconstructing the history of life and unraveling the processes that govern changes in populations and species through time. Neuroscientists, with some important exceptions, are principally engaged in experimental studies of single species rather than comparative studies of multiple species. Indeed, until quite recently, most neuroscientific research involved one of a few "model animal" species (eg, Logan, 2002; Manger et al., 2008; Preuss and Robert, 2014; Striedter et al., 2014), although with the development of noninvasive imaging technologies, more and more research is being carried out in humans.

While evolutionary biologists and other nonneuroscientists are surely aware of the growing body of knowledge obtained from direct studies of the human brain, they probably are not aware of the growing body of research comparing humans to other species, obtained with both neuroimaging and traditional experimental approaches over the past several decades, and how that is transforming our understanding of brain evolution (eg, Preuss, 2001, 2007, 2010, 2011; Rilling, 2014; Rilling and Stout, 2014; Sherwood et al., 2012). For their part, neuroscientists are probably not aware that the core concepts and methods of evolutionary biology have undergone transformations no less profound than those in the neurosciences: evolution today differs in many ways from the evolution of Darwin's day.

An important difference between the two traditions is that evolutionary biologists tend not to have strongly held views or compelling intuitions about the brain, whereas neuroscientists—as well as nearly everybody else, scientist or layperson—has compelling intuitions about evolution. We cannot help it: ideas about human nature and our relationship to other species are embedded in our culture. This creates a problem: ideas about human brain evolution put forth by neuroscientists are sometimes more strongly rooted in popular understandings of evolution—folk evolutionism, as it were—than in scientific evolutionism.

The material in this chapter will, I hope, promote more sophisticated evolutionary approaches to human neuroscience and, more generally, serve to build bridges between the evolutionary and neuroscientific research traditions, providing some information about the history of ideas and methods in evolutionary biology and human evolutionary studies that will be useful to neuroscientists who read this volume, as well as information about the history of concepts and methods of neuroscience useful for non-neuroscientists. I conclude the chapter by offering some views on the current state of knowledge in this field and describing opportunities and challenges for future research.

Evolutionary Neuroscience, Second Edition
https://doi.org/10.1016/B978-0-12-820584-6.00029-5

29.2 Evolutionary Background

29.2.1 From the Phylogenetic Scale to the Tree of Life

If the publication of Charles Darwin's (1859) *On the Origin of Species* marked the beginning of evolutionary biology, formal scientific discussion of human evolution began with T.H. Huxley's *Evidence as to Man's Place in Nature* (1863) and Ernst Haeckel's *The Evolution of Man* (1874, 1879). There is an interesting contrast between Darwin's graphical depiction of evolutionary history in the *Origin*—a schematic drawing of a hypothetical branching tree of life—with the later, more human-centered accounts, all of which characterize humans as being the most advanced or perfect products of the evolutionary process, perched atop the phylogenetic scale. Haeckel's striking and frequently reproduced figure (see, eg, Preuss, 2009) is particularly intriguing, because although it is a tree, it is an extremely narrow tree with humans at the top, surmounting a crown of great apes, with successively lower life forms occupying lower branches. What Haeckel accomplished with this was to reconcile Darwin's notion of evolution as a diversifying process with an older, pre-Darwinian idea about the nature of life, specifically, that of the scale of being (or *scala naturae*), in which life forms can be ranked by their degree of perfection relative to humans (Richards, 1987, 1992). That idea ultimately has its roots in Medieval and Classical philosophical and theological traditions (Lovejoy, 1964) and remains a powerful idea in Western culture, congenial with the idea of progress. For most members of our culture, including many scientists, the phylogenetic scale is a given.

Although Darwin has sometimes been cast as a champion of tree-thinking over scale-thinking (eg, by Gould, 1989), the writings of Darwin and the early evolutionists are filled with references to the biological superiority of humans, and it is evident that they did not view these two views of life as being in tension: humans were simply superior, as a result of a particularly grueling history of natural selection. Thus, Darwin and his contemporaries did not so much replace the scale of being as propose a naturalistic explanation for it.

The tension between scale-thinking and tree-thinking grew in the first half of the 20th century as Darwin's theory of natural selection and adaptation was elaborated in the so-called "Modern Synthesis" of genetics and population biology (Mayr and Provine, 1980). At the heart of the Modern Synthesis is the idea that evolutionary processes act locally: natural selection preserves those random genetic mutations that adapt phenotypes to the specific circumstances governing survival and reproduction within an interbreeding population. Over time, each species tracks its changing circumstances, accumulating a unique history of adaptation. The forces of evolution thus tend to pull species apart, except in cases where similar circumstances act on different species to produce similar adaptations (convergent evolution). In this view, there is no force—internal or external—driving evolution globally in a particular direction, a view much more aptly captured by the tree metaphor than the scale metaphor.

With the success of the Modern Synthesis, tree-thinking began to supplant scale-thinking. Scientists who studied evolution on a large scale began to think more concretely about the information represented by a tree and, more importantly, about how to reconstruct trees that accurately reflect the history of genealogical relationships between different species or lineages. A set of methods was developed to group lineages in nested clusters, each cluster being defined by traits its members share but that other relatives lack. As another term for "lineage" is "clade," this methodology became known as "cladism." First developed in the 1950s by the entomologist Willi Hennig (1950, 1966), these methods came to the fore when the availability of computers made it possible to sort large numbers of species and large numbers of traits, which were usually anatomical in early work, but now often include gene sequences. Methods for identifying the best reconstructions continue to be refined (Baum and Smith, 2013).

These developments have led to a much better understanding of who is related to whom among living species. Notably, we now have—for the first time—a reliable and detailed phylogenetic tree of the various mammalian orders (Murphy et al., 2001a,b, 2004; Fig. 29.1). This tree has a number of important lessons. For one thing, it clarifies the relationship of rodents to other mammals, a matter long debated: rodents belong to a larger group (Euarchontoglires) composed of lagomorphs (rabbits and pikas) and the primate—tree shrew group. For another, it makes it apparent that large brains evolved independently in multiple mammalian lineages (Striedter, 2005).

Perhaps the most general lesson from the Modern Synthesis and from tree-thinking is that evolution is not human directed—modern humans are just one tip on a branch of a very large tree. This does not imply that humans are not unique or that it is inappropriate to talk about human specializations; rather, the lesson is that every branch of the phylogenetic tree is unique in some ways (Preuss, 2012a).

29.2.2 Primate and Human Evolution

The new trees and the new thinking they have spurred have many other important implications for primate and human evolution. As humans were the

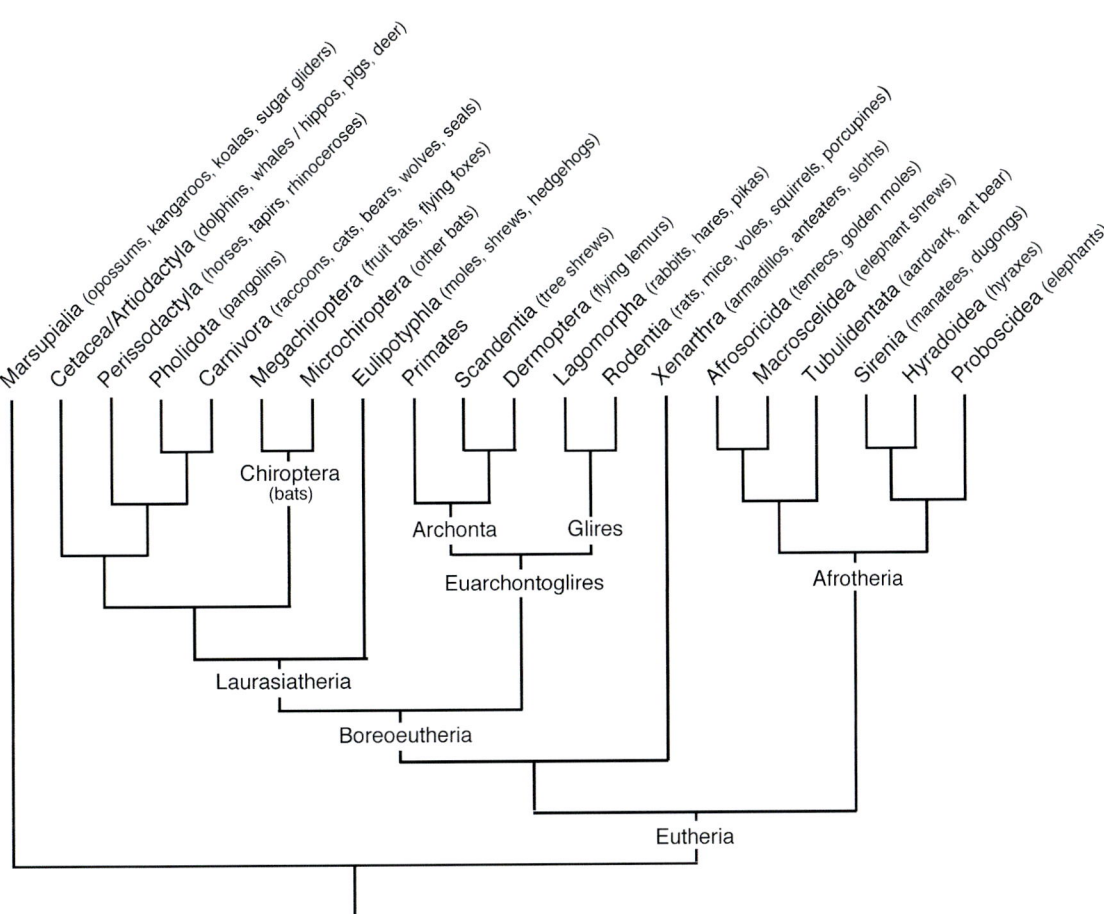

FIGURE 29.1 Evolutionary relationships among the living orders of therian (live-bearing) mammals. The terms arrayed along the tips of the branches as the modern orders of mammals, which are clustered into higher-order taxa. Primates belong to the higher-order group Euarchontoglires, which consists of primates, the tree shrew—flying lemur group, and the rodent—lagomorph group. *This figure is based on the work of Murphy, W.J., Eizirik, E., Johnson, W.E., Zhang, Y.P., Ryder, O.A., O'Brien, S.J., 2001. Molecular phylogenetics and the origins of placental mammals. Nature 409, 614−618; Murphy, W.J., Pevzner, P.A., O'Brien, S.J., 2004. Mammalian phylogenomics comes of age. Trends Genet. 20, 631−639 and is reproduced from Preuss, T.M., 2007. Primate Brain Evolution in Phylogenetic Context. Evolution of Nervous Systems. In: Primates, vol. 4. Elsevier, pp. 1−34.*

centerpiece of the phylogenetic scale, it is not surprising that early accounts of human evolution tended to paint a picture of irresistible progress from brutish ancestry to polished modernity—"up from the ape," as one early anthropologist put it (Hooton, 1931).

In Darwin's account, the path to humanity started with bipedalism, which freed the hands, which were modified for handling tools and weapons, which led to the reduction of the canines (used as weapons by apes); all of which was accompanied by the gradual enlargement of the brain and increase in mental powers (Darwin, 1859). Later writers followed a similar template. One very influential view of human evolution was articulated by the eminent anatomist, W.E. Le Gros Clark, in his widely read text, *The Antecedents of Man* (1959). In the *Antecedents*, human evolution history was presented as a steady ascent up a scale of primates, each step represented by living forms, beginning with insectivores, then tree shrews, followed by prosimians,

tarsiers, monkeys, apes, and finally humans. The unifying theme in human evolution was the steady increase in reliance on vision and brain power over olfaction and bodily specializations and the triumph of general abilities over specialized anatomy and behaviors. Other writers emphasized the acquisition of culture (with its material, cognitive, and linguistic components) as the primary adaptation or niche of humans, with culture again substituting for organismic specializations. In the middle years of the 20th century, this had the effect of reinforcing the view of human phylogeny as a linear progression, with little room for diversity, for if one understands culture to be the human ecological niche, and one applies the competitive exclusion principle (two species cannot long occupy the same niche), there could only be one human species at any point in time (Mayr, 1950; Wolpoff, 1971). Culture has remained an important theme in human evolutionary studies, but mainly as a factor shaping human biology, and to the extent that specific hominin cultures created specific niches, they were not necessarily so broad as to exclude other hominin species (eg, Boyd and Richerson, 1985; Henrich, 2015; Holloway, 1992; Richerson and Boyd, 2005; Tomasello, 1999, 2014).

The progressivist views exemplified by Le Gros Clark were rendered moot within a few years, as primatologists increasingly came to appreciate the unique specializations of the different groups that made up his supposed ascending primate scale. The distinction is readily apparent by contrasting Le Gros Clark (1959) with a widely used modern textbook of primatology (Fleagle, 2013). For example, in their social organization, locomotor behavior, and postcranial anatomy, the apes—chimpanzees, bonobos, gorillas, orangutans, and gibbons—could scarcely be more different. There can be no ape stage of primate evolution. Similarly, the idea of a monkey stage of primate evolution is belied by the incredible number and diversity of monkey species, not to mention the fact that the English term "monkey" covers two distinct groups, the New World monkeys and the Old World monkeys, the latter being more closely related to apes and humans than they are to New World monkeys (Fig. 29.2).

This change of perspective was reinforced by the results of formal phylogenetic analysis using cladistic methodology, increasingly informed by comparative molecular and genetic information. Up until the 1970s, anatomists had tended to lump the great apes (chimpanzees, bonobos, gorillas, and orangutans) into a single family (Pongidae) distinct from the human family (Hominidae). Comparative molecular studies indicated, however, that humans are part of the African great ape group, with the Asiatic orangutans being their sister group. Today, most specialists see humans (*Homo*) as being most closely related to the chimpanzee—bonobo clade (genus *Pan*), with gorillas as a closely related sister group. In the taxonomy adopted in this volume by Boyle and Wood (2016), the human clade is classified as tribe Hominini, and the *Homo—Pan* clade is classified as subfamily Homininae (Fig. 29.2). In this view, the family Hominidae ("hominids," in the vernacular) consists of humans, African great apes, and orangutans. Thus, taxonomy has been modified to reflect our understanding of relationships. Not everyone would adopt this classification, but almost all specialists accept the evidence—genetic, anatomical, and paleontological—that humans are part of the African great ape radiation and that the closest relatives of humans are members of the genus *Pan*.

The change in perspective from scale-thinking to tree-thinking is also dramatically apparent when one considers the extinct members of the human tribe. The documenting of the rich fossil record of human

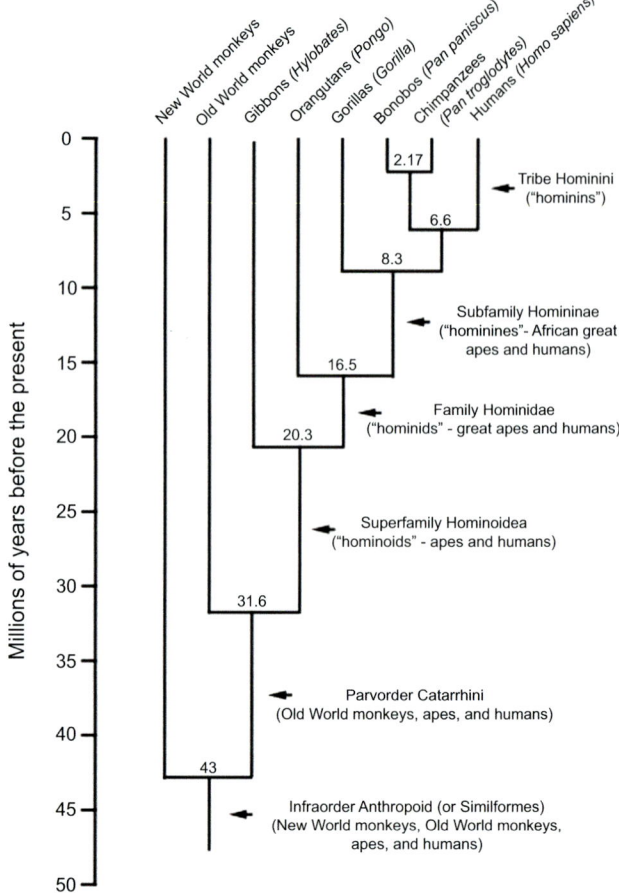

FIGURE 29.2 Evolutionary relationships among the living groups of anthropoid primates (ie, monkeys, apes, and humans), showing formal taxonomic nomenclature and estimates of divergence. *The hominoid taxonomy is based on de Sousa, A., Wood, B., 2007. The hominin fossil record and the emergence of the modern human central nervous system. In: Kaas, J.H., Preuss, T.M. (Eds.), Evolution of Nervous Systems, Primates, vol. 4. Elsevier, Oxford, pp. 291—336; and is consistent with Boyle, E.K., Wood, B., 2016. Human evolutionary history. In: Preuss, T.M. (Ed.), Evolution of Nervous Systems, Human Brain Evolution, second ed., vol. 4. Elsevier (in press); Divergence dates are from Perelman, P., Johnson, W.E., Roos, C., Seuánez, H.N., Horvath, J.E., Moreira, M.A.M., Kessing, B., Pontius, J., Roelke, M., Rumpler, Y., Schneider, M.P.C., Silva, A., O'Brien, S.J., Pecon-Slattery, J., 2011. A molecular phylogeny of living primates. PLoS Genet. 7, e1001342; The figure is reproduced from Preuss, T.M., 2016. The human brain: evolution and distinctive features. In: Tibayrenc, M., Ayala, F.J. (Eds.), On Human Nature. Biology, Psychology, Ethics, Politics, and Religion. Elsevier (in press).*

evolution must surely be counted as one of the most significant accomplishments of modern science. Beginning with the unearthing of the Neanderthals and "Java Man" in the 19th century, and the identification of *Australopithecus* in South Africa in the early 20th century, the pace of discovery has steadily accelerated. When relatively few forms were known, and dating was uncertain, it was possible to believe in a simple progression from australopithecines to early *Homo* to archaic *Homo sapiens* to modern *H. sapiens*, with one form transforming gradually into another. What we perceive instead today is an astonishing adaptive radiation of hominins, with multiple species present at many points in time over the last several million years (Anton et al., 2014; Stringer, 2016; Boyle and Wood, 2016; this volume). Perhaps most remarkable is the persistence of some ancestral phenotypes, represented specifically by the *Homo floresiensis* (the "hobbits") from Indonesia (Sutikna et al., 2016), and perhaps also *Homo naledi*, from South Africa (Dembo et al., 2016), at times when animals much more like modern humans in brain size and skeletal anatomy also walked the Earth. Moreover, it is likely that brain size increased convergently in different hominin lineages, so that, for example, *H. sapiens* and *Homo neanderthalensis* achieved comparably sized brains independently (Bruner, 2016, in this volume). Diversity is so extensive within the hominin clade, and convergent evolution so common, that while the broad outline of human evolutionary history can be reconstructed—with the genus *Homo* evolving from a species of australopithecine, and *H. sapiens* evolving from an earlier species of *Homo*—it has been difficult to achieve a robust, detailed reconstruction of the relationships of the different hominin species currently recognized (see Boyle and Wood, 2016; this volume).

Whereas some early scholars believed that encephalization (brain enlargement, adjusted for changes in body size) was a hallmark of early hominin evolution, with skeletal changes coming later, current research indicates that the earliest hominins had brains perhaps just slightly larger than those of African apes. However, in agreement with Darwin, the skeletal evidence suggests that bipedalism evolved very early. Evidence for stone-tool technologies comes somewhat later. Traditionally, the appearance of the stone tools has been associated with the appearance of the genus *Homo*, at around 2 million years ago (Mya). However, we now know that chimpanzees can make and use simple stone tools, which suggests that the capacity to do so was present in the earliest hominins (Lewis and Harmand, 2016). Currently, the oldest stone tools attributed to hominins in the archeological record date to ~3.3 Mya and are only little different than from those produced by chimpanzees (Harmand et al., 2015). The oldest distinctively hominin tools, assigned to the Oldowan tradition, date to ~2.6 Mya and consist of carefully chosen and deliberately transported stones with a few flakes removed to produce edges for defleshing bone (Lewis and Harmand, 2016). Conventionally, Oldowan tools have been attributed to members of the genus *Homo*, the earliest anatomical evidence of which currently dates to ~2.8 Mya (Villmoore et al., 2015), although it is possible that Oldowan tools were manufactured by australopithecines. Oldowan implements were replaced by tools of the Acheulean tradition after ~1.8 Mya and are

associated with *Homo erectus* and later hominins. Prominent in the Acheulean tool kit are the distinctive, almost stereotyped handaxes, which require more elaborate planning and more separate actions to produce—as well as additional brain areas—than does the making of Oldowan tools (eg, Hecht et al., 2015a,b). The last major stone-tool tradition recognized by Paleolithic archeologists is the Levallois technology, which appeared about 0.25 Mya, associated with the Neanderthals and early *H. sapiens* and is marked by the production of fine, elongated flakes (blades, in essence) from prepared cores.

While dependence on stone tools for subsistence is a hallmark of the genus *Homo*, human evolution produced additional, remarkable innovations, including important changes in social behavior and in modes of representing and conveying meaning, and these left their imprint on the human brain (reviewed by Rilling and Stout, 2014; by Boeckx, 2016; Rilling, 2016; in this volume). Reconstructing the evolutionary history of language and aspects of social behavior that leave no definite traces in the archeological or paleontological record represents an enormous challenge to students of human evolution. Even for some behaviors that should be datable—the control of fire and deliberate burial of the deceased, for example—the dates remain highly uncertain. Yet, there is much about the behavior of extinct hominins that can be reconstructed from physical evidence, including locomotor behavior, diet, and resource transport and utilization (Anton et al., 2014), and the later archeological record, contemporaneous with the appearance our species, *H. sapiens*, includes items of personal adornment and symbolic artifacts, suggestive of a new behavioral complex (McBrearty and Brooks, 2000). Although it has been argued that this complex emerged very rapidly— a "human revolution"—recent work in archeology and paleontology point to emergence of our species in East Africa in a period beginning perhaps as early as 0.2 Mya, with a successive accretion of new behavioral characteristics (McBrearty and Brooks, 2000). Nevertheless, exactly what distinguishes the cognitive psychology of *H. sapiens* from that of *H. neanderthalensis* and other late hominin species remains a vexing question.

29.3 Human Brain Evolution: Classical Views

29.3.1 Brain Size

Bigger animals tend to have bigger brains than smaller animals, and it has long been appreciated that humans have unusually large brains for a primate of our body size. The human brain is about three to four times the volume of that of chimpanzees and other great apes, animals with an adult body-size range that overlaps that of humans (Barks et al., 2015a; de Sousa and Wood, 2007; Schoenemann, 2006). What this signifies about the structure and function of the human brain is one of the key questions in human brain evolutionary studies, and how one answers that question depends on one's understanding of how brains are organized and what kinds of variations occur across species. How one conceptualizes brain organization and interspecific variation depends in turn on the methods available for studying brain organization. So, to track the development of our understanding of human brain evolution, it is important to consider the development of concepts and methods in comparative neuroscience.

Brain size is an attractive variable for scientists, as it can be measured in both living species and in fossil taxa. Because brain size varies with body size, however, it has seemed useful to have a measure of brain size that factors out body size—that is, a measure of relative brain size, or encephalization—and assessing encephalization rigorously is a more difficult task than one might suppose. This is mainly because brain size does not scale to body size in a one-to-one (isometric) manner: brain size increases with body size, but not to the same extent. As a result, larger-bodied species tend to have brains that are smaller relative to total body mass than smaller-bodied species, a pattern characterized as "negative allometry." Quantitative assessment of relative brain size depends, therefore, on having a baseline to provide an expected value for the brain size of an animal with a given body size. The classical approach to this problem is to take brain and body weights for a large set of species, perform log transformations on the weights (which changes the curvilinear brain:body relationship to something more nearly linear), and then perform a regression analysis. Relative brain size is measured as the ratio of brain size observed for a species to that expected for a mammal of that species' body size. This ratio was dubbed the "encephalization quotient" (EQ) by Jerison (1973). The value of EQ depends on the slope and intercept of the regression line, which provides the expected brain size as a function of body size. Jerison (1955) fit a line with a two-thirds slope to a set of mammalian brain—body data and adopted that slope in his subsequent work. Stephan and colleagues took a somewhat different strategy (eg, Stephan, 1972; Stephan and Andy, 1969): rather than fitting a line to a broad sample of mammals, they derived a baseline from log-transformed brain and body data for a set of "basal insectivores," a group of living mammals they considered to represent the state of the ancestral eutherians from which primates are derived. The slope of their regression line (0.65) was very similar to that of Jerison, however. They then calculated a Progression Index (PI) as the ratio of observed brain size to expected brain

size for a basal insectivore of equal body size. Stephan and colleagues also calculated PIs for different components of the brain to identify structures that were differentially enlarged or diminished in evolution. For Stephan and colleagues, PIs were simply indicators of relative size differences across species. For Jerison, however, EQ had a deeper significance, being a measure of what he termed "biological intelligence"—essentially, cognitive capacity. Positing a direct relationship between relative brain size and cognitive capacity entails certain assumptions about brain organization and cognitive organization, assumptions that will be explored in the following section.

Jerison's allometric methodology has been enormously influential in the study of human brain evolution, even by those who do not accept his assumptions about neural or cognitive organization, if only because it is a convenient means of characterizing the relative brain sizes of different species. In recent years, however, it is being supplanted by the more sophisticated techniques of "phylogenetic regression," which take account of the fact that samples are composed of clusters of related species, and thus the elements are not statistically independent (Garland et al., 1992; Harvey and Pagel, 1991). In addition, there is less emphasis today on determining "the" baseline for mammals, perhaps because the diversity of mammalian brain organization—and also body composition, the other part of the equation (Schoenemann, 2004)—is increasingly being appreciated. It seems unlikely that comparisons of EQ between very distantly related groups, with cortices built along different lines, such as primates and cetaceans, mean very much.

29.3.2 Brain Organization

A straightforward relationship between relative brain size and functional capacity makes sense if the brain (or at least the cortex, which makes up the largest part of the brain in primates) is relatively homogeneous in its organization. In fact, Jerison took his cue from the work of Karl Lashley (1931, 1949), who famously argued that the cerebral cortex is largely uncompartmentalized. According to Lashley, the cortex consists of a few sensory and motor areas, plus a region of general cortex, corresponding to the classical association cortex that is relatively undifferentiated. Lashley argued that this cortex is equipotential in function and operates *en masse*. What matters for a species' cognitive abilities is how much of this general cortex it possesses. Jerison's achievement was to take this idea and give it quantitative expression [although he later acknowledged that the cortex is more compartmentalized than indicated by Lashley (Jerison, 1977, 2001)].

Lashley's theory is an example of a broader class of ideas about brain structure, referred to as globalism, that stands in contrast to localizationism—the idea that the brain (or the cortex) consists of multiple, anatomically separate units that make different contributions to function. The original localizationists were the phrenologists, who believed the cortex consisted of a set of structures contributing to certain dispositional traits (eg, secretiveness, perfectionism, impulsiveness) that varied in their degree of structural development between individuals and that the differential enlargement of particular regions left bulges in the cranium that could be felt through the scalp. A more scientifically sound localizationism had its origins in the middle part of the 19th century, when Broca and Wernicke presented clinical cases suggesting damage to specific cortical regions of the left hemisphere contributed to specific deficits of speech and language (Broca, 1865; Wernicke, 1874). Subsequent studies by clinical neurologists identified other regions of the cortex with a wide variety of other cognitive and behavioral functions, such that in the period after World War I, the brain maps of clinical neurology rivaled in detail those of the phrenologists (eg, Kleist, 1934).

Studies of humans with brain damage, and early electrophysiological studies, indicated that the cortex consisted of a set of core or "primary" sensory areas (for vision, hearing, and touch, at least), a primary motor area, bands of secondary sensory and motor areas, and (in humans, especially) large regions of cortex without clear commitment to a particular sensory or motor function—higher-order association cortex. These functional differences must have an anatomical basis. If one looks at unstained cerebral cortex, however, it is hard to imagine it as any regional differentiation: it is pretty much all a thin, featureless band of gray, lying atop a thick mass of white. When tissue stains came into common use in the late 1800s, however, the thin gray band began to reveal its structure. Certain dyes stained the cell bodies of neurons, while other treatments stained the fatty, myelinated sheaths of axons, which extend into gray matter from white matter. Because the sizes, shapes, and spatial arrangements of neurons vary across the cortex, as does the density and spatial distribution of myelinated fibers, it became possible to parcel the cortex into small structural territories ("areas") and map the distribution of those areas across the cortex. These two approaches are referred to as "cytoarchitectonics" and "myeloarchitectonics," respectively; subsequently, it was found that the distribution of a number of molecules vary across the cortical mantle, as revealed with specific histochemical procedures.

A number of scientists adopted this approach (Zilles and Amunts, 2010), the best known today being Brodmann (1909), whose maps were constructed using the cytoarchitectonic method. Brodmann's work is notable for several reasons. For one, he produced a monograph with a well-developed theory of different cortical types and their developmental histories, as well as excellent photomicrographs. For another, he took a broad comparative approach, studying a remarkable array of mammalian species. In addition, he produced beautiful schematic cortical maps of those species, with the different areas numbered; the numbering system was as consistent across species as he could make it, so as to indicate homologous areas in different species. Those maps have so much eye appeal that they continue to be reproduced in textbooks today, and Brodmann's numbers are still widely used to designate cortical areas, especially in humans and other primates—area 17 for primary visual cortex, area 4 for primary motor cortex, and so forth—notwithstanding the fact that we have long known his cortical parcellations have a number of deficiencies.

Much of Brodmann's theoretical framework, and some of his important conclusions, also remains with us today. For example, the idea that cortex consists of two broad types, allocortex and isocortex (also known as neocortex), the latter consisting of six cell layers that are homologous across species and across the different areas that make up the isocortex. Virtually all the major subdivisions of sensory, motor, and association cortex fall within the isocortex; allocortex includes the hippocampus and other limbic territories and comprises a variety of layering schemes. Among Brodmann's important conclusions is that the number of cortical areas varies between species and that larger-brained species tend to have more cortical areas. According to Brodmann, humans possess areas that other primates lack. For example, in human inferior frontal cortex, he recognized areas 44 and 45, which correspond to Broca's area, and in the association cortex of the inferior parietal lobe, areas 39 and 40, but he did not recognize homologues of those areas in monkeys. (Regrettably, he never published a detailed map of chimpanzee cortex.) The implication is that human-distinctive higher-order cognitive functions of humans are supported by human-unique cortical areas.

The idea of human-unique areas was challenging for neuroscientists steeped in the tradition of Darwin and Huxley, both of whom emphasized phylogenetic continuity between humans and other living species, in the mind (Darwin, 1871) and brain (Huxley, 1863) particularly. In Darwin's words, "the difference in mind between man and the higher animals, great as it is, certainly is one of degree and not of kind" (Darwin, 1871). In fact, while Darwin and Huxley invoked continuity when discussing human evolution, neither adhered to it consistently as a general principle of evolution (Preuss, 1993). For example, Darwin, in his big species book—the expanded version of the *Origin* that went unpublished during his lifetime—produced a plausible scenario to explain how flying mammals could have evolved from nonflying ancestors (Darwin and Stauffer, 1975), and Huxley (1863) was untroubled by the fact that a conspicuous fiber tract such as the corpus callosum is unique to placental mammals. Today, evolutionists construe "continuity" historically, accepting that an unbroken genealogical chain that connects species, but rejecting the idea that every feature of humans (or any other species, for that matter) must be present in incipient form in other species. Yet, although the Darwin–Huxley construction of continuity was a problem for early neuroscientists, microscopic inspection of cortical tissue in different species clearly reveals they are not identical and that there are more regional differences in appearance between some species (especially large-brained primates) than in others. How can one understand this within a traditional Darwinian framework? One way is to suppose that all the cell types and circuitries found in humans are present in even the smallest-brained mammals and that when the cortex enlarges in evolution, the components become increasingly segregated, so that the distinctions between cortical zone are more apparent—although at the level of cells and circuitry, there really is not anything new (Preuss, 1993). This is the interpretation of cortical evolution adopted by Elliott Smith (1924), an early student of comparative neuroanatomy and primatology, and by Le Gros Clark (1959), and later by Lende (1969) and Ebbesson (1980). Basically, this amounts to an evolutionary trend from less organized to more organized—a process of "differentiation," comparable to the differentiation of tissues in embryology. This contrasts with Brodmann, who regarded novel areas as new cerebral organs. Brodmann's views are, to my mind, more congruent with most modern thinking about cortical areas, to be discussed in the following paragraphs. Note, however, that people who think that the cortex evolved by the addition of new cortical areas, with new structural and functional properties, often characterize that as "differentiation." Clearly, that term means different things to different people.

As noted, Brodmann was not the only person to produce cortical maps, and cytoarchitecture was not the only technique used, myeloarchitectonics and electrophysiological mapping also contributed. By the 1940s, at least, it had become clear that there were important limitations to the architectonic technique: different researchers studying the same species often came up with quite different cortical parcellations. Lashley, in collaboration with George Clark, made this point dramatically, by showing their independently drawn, and quite divergent, cytoarchitectonic parcellations of

a single spider monkey brain (Lashley and Clark, 1946). They concluded that the cytoarchitectonic method is unreliable and that evidence from other sources did not support the existence of a large number of cortical areas—certainly not as many as proposed by Brodmann—nor did they support the idea that cortical evolution entailed the proliferation of cortical areas.

The Lashley–Clark critique did not bring architectonic studies to an end, nor did it result in the general demise of localizationism. In neurology, its influence was limited, if only because neurologists had to deal with brain-damaged individuals, and they knew from long practice that it is often possible to localize a lesion with considerable accuracy based on the behavioral deficits a patient displayed. In the Soviet Union, the Moscow Brain Institute continued to refine Brodmann's human brain map (Sarkissov et al., 1955). Among neurologists, the idea that there are areas of the human brain that are uniquely human or, at least, highly modified in humans, remained strong. In addition to the classical language areas, dorsolateral prefrontal, posterior parietal, and middle temporal cortex have been highlighted as regions of likely human specialization (eg, Benson, 1994; Luria, 1973; Geschwind, 1965).

In other circles, however, the Lashley–Clark critique had a profound effect. The next generation of comparative primate architectonic maps, produced by Bonin and Bailey (1947, 1961), recognized many fewer areas than did Brodmann and denoted areal borders with shading rather than sharp lines (Bailey and Bonin, 1951; Bailey et al., 1950). Importantly, they also identified the same set of cortical areas in humans and nonhuman primates. The implication was that the enlargement of cortex in human evolution was *not* accompanied by the emergence of new areas, but rather by the expansion of established areas.

Another consequence of the Lashley–Clark critique was an emphasis on alternative techniques. Clinton Woolsey and his students carried out extensive comparative studies of cortical organization using electrophysiological recording and stimulation techniques (eg, Woolsey, 1958). In addition, Woolsey, in collaboration with Jerzy Rose, advocated using patterns of connectivity (especially connectivity between the thalamus and the cortex) as a potentially more reliable means of parceling the cortex and identifying homologous areas than architectonics (Rose and Woolsey, 1948, 1949). The results of both these projects remain fixtures of modern neuroscience textbooks, particularly in the form of eye-catching textbook figures showing the somatotopically organized "homunculi" revealed by electrophysiology in sensorimotor cortex and figures relating projections from different thalamic nuclei to specific cortical regions. However, the electrophysiological methods and anatomical techniques for tracking connections available at the time were very imprecise by modern standards. As a result, the cortical maps derived using them contain a number of important inaccuracies (which, regrettably, continue to be reproduced in textbooks), and these approaches did not ultimately yield much progress in our understanding of brain evolution. Nevertheless, the idea of using criteria in addition to architectonics for parceling cortical areas was sound and did contribute to progress when improved neuroanatomical methods became available, as discussed later.

29.3.3 The State of the Art Prior to the Current Era

Given the lack of good data about the brain organization of different species, it was possible for neuroscientists at mid-century to maintain radically different views about the nature of human brain evolution. On the one hand, one could follow Jerison in seeing it as a matter of the enlargement of undifferentiated association cortex. On the other, one could follow Brodmann and the neurologists in maintaining that enlargement was accompanied by the addition of new, human-specific areas to support new functions. As a third alternative, Holloway (1966a,b, 1968) proposed that humans possessed a cortex with a similar complement of areas as nonhuman primates, but that the association areas were differentially (but not necessarily uniformly) enlarged compared to sensory areas, and that those areas were reorganized internally and in their systems of interaction with each other.

Despite these controversies, one should not lose sight of the fact that a number of points were widely, if not universally, agreed upon. First, few questioned that the brain underwent enormous enlargement during human evolution, mainly due to the expansion of association cortex, and that this constitutes the principal neural correlate of specialized human cognitive abilities. Second, probably most workers accepted that the brain underwent some degree of increasing differentiation, marked by an increase in the number of areas, as well as an increase in the degree of hemispheric specialization related to the evolution of language and right-hand dominance. In addition, it seemed obvious to some that human-specific functions, such as language, implied that Broca's and Wernicke's language areas are human specific (eg, Crick and Jones, 1993). Each of these ideas, it should be noted, has been challenged— although not necessarily refuted—based on recent evidence. Thus, the idea that human brain evolution involved differential expansion of association cortex has been questioned in MRI studies of brain morphometry (Semendeferi and Damasio, 2000; Semendeferi et al., 1997), as has the idea that humans possess new,

human-specific cortical areas in association regions (Petrides et al., 2012). Moreover, while humans are unquestionably unusual among primates in the extent to which we are *functionally* lateralized, related to handedness and language, these functional specializations are not clearly related to any specialized human anatomical asymmetries (see Skiba and Taglialatela, 2016; in this volume).

29.4 Developments in Modern Neuroscience

29.4.1 The Neuroanatomical Revolution of the 1970s

Some major innovations in neuroscientific methodology were introduced in the 1960s and 1970s that made it possible to probe brain organization and connectivity in much greater detail and with much greater precision than in earlier decades. These included intracortical microelectrode techniques, as used, for example, by Hubel and Wiesel (1959, 1968, 1977), with which one can record from or stimulate very small regions of cortex (on the order of tens of microns across), in contrast to the much coarser resolution (on the order of millimeters or more) achievable with the large surface electrodes, such as those employed in the well-known work of Woolsey (1958). In addition, for the first time, sensitive and reliable techniques became available for tracing connections between cortical areas and between the cortex and subcortical structures, using chemicals that are taken up by neurons and actively transported anterogradely and retrogradely along their axons.

The result, beginning in the 1970s and continuing today, has been nothing short of a scientific revolution. Using microelectrodes, Zeki and colleagues (Dubner and Zeki, 1971; Zeki, 1969), in the United Kingdom, and Allman and Kaas (1971a,b), in the United States, began to remap the visual areas of Old World macaque monkeys and New World owl monkeys, respectively and discovered a number of new sensory areas, including multiple areas within cortex designated as "area 19" by Brodmann. Areas were recognizable because they have orderly representations of the visual field, which can be reconstructed if one records from enough sites. Importantly, however, the status of these territories as distinct areas could, in many cases, be validated by differences in connectivity (using recording sites to localize tracer injections) and by differences in cyto- and myeloarchitecture (Allman, 1982; Kaas, 1987a). Similar methods were applied to auditory, somatosensory, and motor cortex, and the number of cortical areas grew ever larger. Significantly, the techniques were extended to study prosimian primates, which have much smaller brains than New World and

Old World monkeys, and to nonprimate mammalian species, including many small taxa such as rodents, bats, and opossums. Results of these comparative studies suggested that larger-brained mammals such as primates and carnivores tend to have more cortical areas than smaller-brained mammals, much as Brodmann had argued (eg, Allman, 1990; Kaas, 1987b).

Several additional, important results came from these lines of research. In the 1970s and early 1980s, it was popular to argue that cerebral cortex has a very uniform organization, with a common cellular architecture and local circuitry shared across areas, and even across species—the doctrine of "basic uniformity" (Rockel et al., 1980; see also Szentagothai, 1975; Creutzfeldt, 1977; Mountcastle, 1978; Rockel et al., 1980; Shepherd, 1988). In this view, the internal information-processing architecture of all cortical areas was held to be basically similar, any differences in function being attributable to differences in inputs and outputs. Comparative studies subsequently rendered this view untenable—cortical areas can differ dramatically in their intrinsic organization (for reviews, see Herculano-Houzel et al., 2008; Hof and Sherwood, 2005, 2007; Preuss, 1995a, 2001, 2010; Sherwood et al., 2012, 2016). What is more, the accumulation of information about areas and their connections made it possible to build real wiring diagrams, representing the network organization of the cortex (eg, Casagrande and Kaas, 1994; Felleman and Van Essen, 1991; Young, 1992; more recently, Markov et al., 2012; van den Heuvel et al., 2015). Today, we understand cortical areas as nodes in sets of networks, each network having specific functional characteristics. One limitation of this research program, however, is that it has focused largely on sensory and motor cortex. Indeed, one result of these studies has been to expand the amount of cortex ascribed to sensory and motor regions and reduce the territory of the classical association regions of the parietal, temporal, and frontal lobes. Neurons in these latter regions do not typically respond to simple sensory stimulation. Nevertheless, progress has been made in parceling the classical association regions, based mainly on studies of connectivity and histology, but also on the effects of lesions on behavior, and the response characteristics of neurons during behavior. The association regions have remained less well understood than the sensory and motor regions, although their secrets are beginning to yield to new technology.

29.4.2 The Neuroimaging Revolution

The techniques that dominated the neuroanatomical revolution described previously—chemical tract-tracing and microelectrode recording and stimulation—require invasive and terminal procedures and therefore generally can only be used in experimental animals.

Thus, humans, being off-limits for most invasive neuroscientific research, were largely bypassed by the neuroanatomical revolution, as were the great apes, which have been under similar restrictions. As a result, while the neuroanatomical revolution provided new insights into how brains might change in evolution, it provided us with little insight into what actually happened in human brain evolution, which requires detailed structural comparisons of humans with our closest-living relatives (especially chimpanzees and the other African great apes) to determine what those species possess in common, and which presumably existed in the last common ancestor of humans and chimpanzees, and what changes occurred in its descendent lineages. Almost the only technique available for such comparisons involved examination of postmortem tissue. Recently, the potential of such comparisons has increased greatly with advances in genomics and molecular biology that have made it possible to compare the distribution of specific molecules in the brain, using techniques such as immunohistochemistry, in situ hybridization, and mass spectrometry (Preuss, 2010; Sherwood et al., 2012, 2016). Nevertheless, the biggest breakthrough in comparative methods for studying human brain evolution has been the introduction of neuroimaging technologies that can be used to study humans and nonhuman primates noninvasively (Mars et al., 2012; Preuss, 2010; Rilling, 2014; Rilling and Stout, 2014).

There are two broad classes of relevant noninvasive neuroimaging technologies (see especially Passingham and Rowe, 2016; for an introduction). One involves subjects ingesting radiolabeled, positron-emitting chemical ligands (usually a glucose analogue or oxygen isotope), followed after an appropriate interval by a positron-emission tomography (PET) brain scan to map the location of the radiolabeled ligand. PET scans can be used as anatomical tools, for example, to map the distribution of neurotransmitter receptors or other molecules with relatively stable distributions, or they can be used to map functional activity: scans of glucose utilization, for example, identify regions of differential metabolic activity. One can also collect functional PET scans that compare brain activity under different task conditions. One of the advantages of using PET with a glucose analogue is that one can avoid the difficulty of putting nonhuman primates in delicate and expensive scanners: animals can be dosed with the glucose analogue, tested in their usual testing environment, and then scanned for the accumulated radiolabeled ligand while sedated. So far, this has been the only practical method for doing functional neuroimaging in the large and very powerful great apes (Barks et al., 2015b; Parr et al., 2009; Rilling et al., 2007).

The second class of techniques is magnetic resonance imaging (MRI), which also consists of structural and functional variants. MRI uses electromagnetic pulses in the presence of a strong magnetic field to induce synchronized spins in protons (in most applications) and then maps spin desynchronization (and consequent photon emission) over time.

Because different molecules and tissues desynchronize at different rates, MRI can be used to distinguish tissue types—gray and white matter, for example. This is the basis for standard structural MRI scans (including so-called T1- and T2-weighted scans) that are used to reconstruct the gross morphology of the brain. Notably, structural MRI can also be used to map the myelin content of gray matter across the cerebral cortex, yielding myeloarchitectonic maps very similar to those obtained with classical histological techniques (eg, Glasser et al., 2014; Glasser and Van Essen, 2011). These studies also highlight some of the largely unappreciated advantages of neuroimaging compared to classical neuroanatomical methods, including the ability to collect data from large numbers of individuals from different species, overlay data obtained from different imaging modalities, and display results in a variety of different anatomical orientations, including flattened or expanded views of the cortical surface, advantages that have been exploited by the Human Connectome Project (Van Essen et al., 2013).

MRI can also be tuned to detect the direction and magnitude of water diffusion, using so-called diffusion-weighted imaging (DWI; sometimes referred to as diffusion-tensor imaging, or DTI). DWI can provide information about the connectivity of the brain: since water tends to diffuse along, rather than across, axons, owing to their fatty, hydrophobic sheaths, the distribution of axons can be reconstructed. DWI-based tractography is not as accurate as the chemical tract-tracing techniques employed in experimental species—there are both false positives and negatives—but they do yield some substantial agreement (Donahue et al., 2016; Gutman et al., 2012; Li et al., 2011, 2013; van den Heuvel et al., 2015), and DWI remains the most direct source of information we have about anatomical connectivity that can be obtained with MRI. Significantly, DWI (and other structural MRI modalities) can be used in both in vivo and ex vivo (postmortem) preparations, and it can be used in both humans and nonhuman primates, including great apes, which can be scanned under anesthesia (eg, Hecht et al., 2013, 2015a, Li et al., 2011, 2013, Rilling et al., 2008, 2012).

The use of MRI to localize functional activity (fMRI) rests on its ability to distinguish oxygenated from deoxygenated hemoglobin. When neural activity increases in brain regions, they become saturated with oxygenated blood, so that the ratio of oxygenated to deoxygenated hemoglobin increases (notwithstanding the increased oxygen utilization in the neural tissue). MRI can detect such changes with high temporal resolution (a few seconds) and high spatial resolution (a few millimeters, typically). These properties make fMRI ideal for studying functional localization in the brain, especially because in task-based paradigms, one can run blocks of different tasks in the same subject during the course of a single experiment and compare patterns of activation under different conditions. Functional MRI has, of course, been widely used in humans, but it can also be used (with special equipment and training) in smaller nonhuman primates, such as macaques.

Surprisingly, some of the most interesting results from functional neuroimaging have come from studies that involve no structured task at all—that is, with subjects just resting inside the scanner. These so-called "resting state" experiments have revealed a number of cortical regions that are highly active when subjects are not engaged in most externally cued tasks; collectively, these are referred to as the "default-mode network" or DMN (Raichle et al., 2001). Activation of the DMN correlates with human subject reports of mental simulations of possible actions, recalling events, and other types of daydreaming (Buckner and Carroll, 2007). In addition to being activated by "default," however, the DMN areas can be activated by certain tasks, specifically those involving social cognition (Mars et al., 2012). Current evidence indicates that the DMN is present in chimpanzees (Barks et al., 2015b) and macaques (Mantini et al., 2011; Vincent et al., 2007) as well as in humans.

In addition to providing information about brain function, fMRI scans also provide information about connectivity. Consider that the activity of a brain region will be affected by the activity of brain regions with which it is connected. By examining patterns of correlated activity (usually in resting-state scans), it is possible to extract networks of "functional connectivity." In part, because structural connectivity analyses are known to yield false-positives and false-negatives, fcMRI has become an important, and perhaps preferred, tool for inferring connections in humans as well as for comparing humans and macaques (Mantini et al., 2013; Neubert et al., 2014; Orban, 2016; Sallet et al., 2013). (At present, we lack fcMRI studies of great apes.) While it is clear that there is a correspondence between functional connectivity and structural connectivity, however, it seems unlikely that fcMRI yields a complete, picture of the monosynaptic connections between brain regions.

29.4.3 Parcellation

At the core of any attempt to reconstruct the evolution of the human brain (or any other biological system), is the need to decompose the system into its constituent parts and to identify corresponding parts across different individuals and species. In the study of human brain evolution, the relevant parts are the anatomical subdivisions of the brain, and in particular, given the enormous expansion and modification of the cerebral cortex in human evolution, the different subdivisions—usually called "areas"—of the cerebral cortex. How to divide up the cortex into areas has been a central problem in neuroscience from the beginning, and how this has been dealt with has changed as new methods have been brought to bear on the problem.

As discussed previously, a variety of methods have been employed to try to parcel the cortex into its constituent areas based on local differences in histology (architectonics), connectivity, and function, including the activity patterns of neurons. In addition, in sensory and motor areas, which typically map the body surface or sensory receptor organs (eg, the retina, the basilar membrane of the cochlea) in an orderly fashion, adjoining areas often exhibit mirror-reversed maps, and the locations of the reversals have been identified as area borders. In an influential review, Kaas (1987b) considered all these approaches and argued, first, that areal parcellations are most secure when they are based on the conjunction of multiple criteria; second, that sensory and motor areas should be complete, continuous representations of a sensory or body surface; and third, that the borders between areas should be sharply demarcated.

There remain important difficulties in parceling areas. For one thing, there are now examples of areas of incomplete sensory or motor representations. For example, the ventral premotor area of primates represents the face, forelimb, and possibly the upper trunk, but has little if any representation of lower limb or tail (Preuss et al., 1996). For another, even in sensory areas, the borders between areas are not always clear. In auditory cortex, for example, Morel and Kaas (1992) found that in some individual owl monkeys there was a clear reversal of frequency mapping ("tonotopy") between the primary (A1) and rostral (R) auditory areas, whereas in others, a single tonotopic representation spanned A1 and R. Moreover, A1 and R are sometimes difficult to distinguish architectonically. Results such as these suggest that areas—even lower-order sensory areas—need not always be static or sharply demarcated. The difficulties of parcellation are even greater for divisions of association cortex, which typically do not have well-ordered sensory or motor maps. They might map more abstract or derived dimensions of function in an orderly fashion, but until we understand what functions those areas represent, and have the means to map them in fine detail, we are limited in our ability to use internal functional organization to parcel association areas. Moreover, the architectonic differences between regions

of association cortex are often very modest. It is possible that association areas are in the main less strongly "canalized" in development than sensory and motor areas, and thus more variable in their normal adult organization both within and between species (Krubitzer and Dooley, 2013; Padberg et al., 2007; Buckner and Krienen, 2016; this volume).

Recently, there has been a very notable attempt to improve the parcellation of human cerebral cortex using neuroimaging that employs multiple criteria, identifying borders based on changes in myeloarchitecture, function, functional connectivity, and representational topography (Glasser et al., 2016). The authors take advantage of the large number of subjects that neuroimaging makes practical to study ($n = 210$, in this case) and the ability to accurately align results from different imaging modalities in a single individual, as well as across all the subjects in the data set. Parcellations were derived by visual inspection of borders by the authors and validated independently using a machine-learning classifier. The study identified 180 cortical areas, of which only 83 had been identified in previous studies.

Clearly, the work of Glasser and colleagues represents a major advance in human cortical cartography. Yet it raises a central question: which, if any, of the subdivisions they recognize are unique to humans? With such a large number of areas, it is tempting to assume that at least some of them must be unique to humans. Yet it is possible that some of the areas recognized in humans by Glasser and colleagues are actually subdivisions of areas that are present in other primates, but less anatomically or functionally distinct. At present, we cannot resolve these issues because we currently lack comparable neuroimaging data for nonhuman primates. This is a remarkable situation: for several classes of important data, we actually have better data for humans than for nonhuman primates.

29.4.4 Homology and Comparative Analysis

As students of brain evolution, we want answers to some of these questions: Did new areas evolve in the human lineage? Did certain areas enlarge differentially? Was the internal organization of certain areas modified? Did evolution alter the extrinsic connectivity of an area, and with it its position in cortical processing networks? Answering any of these questions definitively requires accurate brain maps not only for humans, but also for nonhuman primates. Ideally, we would like to have maps of comparable quality and detail. Although we do not have data for nonhuman primates comparable in quality to the best data currently available for humans, we do have useful comparative data, so the question is not hypothetical. In fact, the correspondence between cortical areas of different species is a matter raised repeatedly in this volume (Mars et al., 2016; Orban, 2016; Rauschecker, 2016; Schall et al., 2016).

When we talk about the correspondences of structures across species, we are invoking the concept of homology. Homology is the cornerstone upon which generalization in biology rests, the conceptual basis for comparing organisms, no two of which are ever identical to each another. As the concept of homology is frequently misunderstood and misapplied, some discussion is useful.

"Homology" is sometimes employed as a label for "similarity." But homology is not similarity, although similarities can serve as evidence of homology. Homology is fundamentally about genealogy: homologies are features of organisms that are derived from a common ancestor (see Mayr, 1982, for discussion of the concept and its history). You, the reader, and I have hands with five digits, as do most apes and monkeys, because we are descended from a common ancestor that had hands similarly organized. The hands of all primates are thus homologous. This does not imply that all primate hands are similar in all respects: some species—apes and spider monkeys, for example—have long fingers for hanging from branches, while others—patas monkeys, in particular—have short, stubby fingers, less vulnerable to injury when running on the ground. Most primates have broad nails on the tips of their digits, useful for grasping fine branches, but some, such as marmosets, have transformed the nails into claws, better suited for clinging to vertical supports. Despite these dissimilarities, the hands of all primates, as well as their nails, are homologous. Furthermore, understanding that the hands of all primates are homologous is essential for understanding the functional and structural changes that evolution produced in the hands of different primate groups. This is a recurring theme in evolutionary anatomy: evolutionary change most commonly involves functional and structural modifications of existing structures rather than the addition of new structures.

How, then, do we identify homologous structures? This is an old problem in biology, predating even the theory of evolution (Mayr, 1982), and over time, some practical guidelines have been developed. One of the most reliable guides is location in the body: that is, homologous structures are likely to be located in similar locations relative to other structures in related species. Presumably this heuristic works because the locations of structures are strongly canalized developmentally; shifts in structure location (heterotopies) can occur, but they appear to be rare. We can also make use of other information, for example, about the neuronal connections, histology, and internal organization of a structure—the same kinds of information used to delineate brain structures in the first place. To combine these principles with a brain example, consider the primary visual cortex (area V1). In almost all mammals that have been studied, it is located at or close to the posterior pole of the cerebrum, and in most mammals it has a better developed internal granular layer than neighboring cortex, has stronger connections with the lateral geniculate nucleus, and supports brisker responses to visual stimulation. This is not to say that area V1 is identical in all mammals: in fact, even among primates, there are numerous structural and functional differences across species (Preuss, 2004). What is more, no individual criterion for assessing whether an area is a V1 homologue is essential: the subterranean blind mole rat, for example, with vestigial eyes, possesses an area that responds strongly to auditory stimulation and minimally, if at all, to visual stimulation, but is unquestionably V1 based on other features (eg, projections from the lateral geniculate) and the possession of a structurally similar area located at the posterior pole of the cerebral in other mammals (Bronchti et al., 1989; Cooper et al., 1993; Sadka and Wollberg, 2004). In judgments of homology, then, dissimilarities are not necessarily disqualifying: rather, one must consider the location and properties of a structure in the context of other structures in related species. This principle is important for our understanding of human brain evolution, as it is becoming clear that many areas, including divisions of posterior parietal and prefrontal cortex, differ in their physiological and functional properties between humans and macaques, although on grounds of location, architectonics, and connectivity, they are almost certainly homologous (see especially Mars et al., 2016; Orban, 2016; in this volume).

Over time, evolution can change the organization of structures, and the context in which structures are located, to such an extent as to render homology determination problematic. Mammalian cerebral cortex presents particular challenges, owing to the large species differences in the size of the cortical mantle, the numbers of areas present within it, the diverse behavioral adaptations of mammals, and important differences in how cortex develops in different mammalian groups (eg, Letinic and Rakic, 2001; Letinic et al., 2002). In primates, the cortex of the frontal lobe anterior to the primary motor cortex (M1) consists of multiple premotor areas, multiple lateral prefrontal areas, orbital frontal areas, medial frontal areas, and cingulate areas. Rodent frontal cortex possesses fewer frontal areas, but what are they? The problem is constrained by the fact that M1 and orbital cortex can be recognized in both taxa, but the status of the other areas has been a matter of dispute. In rodents, almost all this cortex is connected with the mediodorsal nucleus, similar to the prefrontal cortex of primates, which led to the interpretation that rodents possess homologues of the prefrontal cortex of primates, including its lateral divisions, which are important components of higher-order cortical association networks (Goldman-Rakic, 1988). More detailed analyses of the connectivity, neurochemistry, function, and the topographic context of areas, however, strongly suggest that while rodents and primates share a number of divisions of premotor, cingulate, and orbital cortex, the lateral prefrontal areas are present only in primates (Preuss, 1995b; Passingham et al., 2016; this volume). Nevertheless, exactly which divisions of primate premotor and orbital cortex are homologous to specific divisions of rodent cortex remains to be resolved. Given the accumulation of differences produced by evolutionary processes, the assessment of homologies is generally easier when comparing more closely related species, such as different species of primates (Preuss, 2007).

The choice of species to compare matters is another, vital respect when considering how human evolution modified the brain structures we share with other species and whether humans possess structures that other species lack. Human brain specializations are, by definition, features of the human brain that evolved in our lineage after it separated from the lineage leading to our closest relatives, the chimpanzee–bonobo group (Fig. 29.2). What we would like to have, then—and what is essential for any detailed understanding of human brain evolution—is a reconstruction of the brain organization of the last common ancestor of the human and chimpanzee lineages. This requires comparisons of humans and chimpanzees to determine what they have in common and how they differ. In addition, it requires examination of related primate species ("outgroups"), to determine whether any particular difference found between humans and chimpanzees evolved in the human lineage or in the chimpanzee lineage: if a feature is shared by chimpanzees and by an outgroup (macaques, for example), but differs between humans and chimpanzees, the difference is likely a human specialization. If humans resemble the outgroup but chimpanzees do not, the difference is likely a chimpanzee specialization. This is a straightforward application of the principles of comparative analysis, as practiced in other branches of biology (Baum and Smith, 2013; Salemi et al., 2009). Currently, we have rather little information about great ape brain organization. We do, however, have a rapidly growing body of information about the similarities and differences between humans and macaque monkeys, some of which has been obtained with comparable methods (eg, Hubel and Wiesel, 1959; Kaas, 1987a; Mantini et al., 2013; Neubert et al., 2014; Sallet et al., 2013; Mars et al., 2016; Orban, 2016; Rauschecker, 2016; Schall et al., 2016; this volume). This is useful, of course, but comparisons of humans and macaques alone do not permit strong evolutionary inferences, since any differences could have arisen in the macaque lineage rather than the human lineage in the 25 million years or so since humans and macaques last shared a common ancestor. Of course, given the greater size of the human brain, any additional structures found in humans that are lacking in macaques would likely have evolved along the human line—although not necessarily, since structures are commonly lost or simplified in evolution. What is more, even if a feature did evolve in the human line, it might well be a feature that evolved prior to the split between humans and apes, who after all share much more recent common ancestry than humans and macaques. Such a feature would be a hominoid (ape–human) specialization, but not a hominin (human) specialization. The point is, information about great apes, and in particular chimpanzees, is critical for understanding human brain evolution.

29.5 Conclusions and Challenges

29.5.1 Some Lessons Learned

One cannot help but be impressed by the extent of progress the field of human brain evolution has made in the past few years. Consider that this field has historically been rife with speculation, but hamstrung by the lack of basic data about the brain organization of humans and our close relatives, as well as by outmoded ideas about evolution. Today, we have the technical means and theoretical orientation required to probe human brain evolution at levels of organization that are likely to yield deep insights into human nature, and with these have come a flood of new and useful data. Speculation is still welcome, but increasingly, our speculations can be translated into testable hypotheses. Based on these new findings, many of which are reviewed in the chapters of the present volume, I would like to offer some lessons learned and highlight issues for additional research (for a more comprehensive review, see Preuss, 2016).

Enlargement of association cortex. Although the classical claim of association cortex expansion in human evolution has recently been questioned, based on the finding that the relative sizes of the frontal, parietal, and temporal lobes are similar in humans and great apes, it now seems firmly established that the frontal, parietal, and temporal association cortices of humans underwent dramatic evolutionary enlargement, both in absolute size as well as relative to sensory and motors areas (Avants et al., 2006; Glasser et al., 2014; Passingham and Smaers, 2014; Passingham et al., 2016). The finding that the major cerebral lobes of humans and apes are similar in relative size is not necessarily incompatible with the finding of association cortex expansion, because the major lobes include primary cortical areas, which underwent little size change in evolution, and secondary and higher-order association areas, which did undergo expansion.

New areas. Despite the widespread conviction that the enlargement of human association cortex was accompanied by, or even driven by, the addition of new areas, we currently lack clear evidence for the existence of human-specific cortical areas. We do have evidence that humans possess areas in addition to those that macaque monkeys possess, specifically in prefrontal cortex [(Neubert et al., 2014; Sallet et al., 2013; Mars et al., 2016), this volume] and in cingulate cortex (Vogt et al., 2013). There are few such areas, however, and they would not seem to begin to account for the differences in the relative extent of association cortex in the two species. On current evidence, therefore, it seems likely that most of the enlargement of human cerebral cortex was driven by the enlargement of its constituent areas, especially in the parietal and frontal lobes. In the temporal cortex, however, there are large parts of the middle temporal region (including middle and anterior temporal regions involved in semantic representation and language) for which there are no clear counterparts in macaques (as discussed by Orban, 2016; in this volume), and to my mind, claims of homology between the fusiform gyrus areas of humans and the inferior temporal cortex of macaques are not currently strongly supported (Preuss, 2016). This discussion ignores the status of chimpanzees, for which we currently lack a cortical map with anything like the level of detail now available for humans and macaques, and that is crucial, because if humans possess areas that macaques lack, we cannot conclude they are human-specific areas unless we can show them to be absent in chimpanzees.

One conclusion we *can* draw with confidence about human areal organization is that the evolution of new functions does not require the evolution of new areas: there is abundant evidence that homologues of Broca's and Wernicke's areas exist in great apes and in other nonhuman primates (reviewed by Skiba and Taglialatela, 2016, in this volume). That is not to say that the evolution of language may not have been accompanied by the evolution of some new cortical territories (the middle and anterior temporal regions are likely candidates), but rather emphasizes the familiar lesson that the evolution of new functions commonly involves the modification of existing structures. It also raises the question of exactly how Broca's and Wernicke's areas were modified to support language.

Cerebral asymmetries. One way the cortex might have been modified to support language, or other lateralized, human-specific abilities, involves hemispheric specialization. It is reasonable to expect, for example, that humans possess differences between the left and right hemispheres in size or organization of Broca's and/or Wernicke's areas that are lacking on other primates. Surprisingly, although there are such left–right differences in humans, similar asymmetries have been reported in great apes (Skiba and Taglialatela, 2016). In fact, Wernicke's area is even more asymmetrical (left > right) in chimpanzees than in humans (Gannon et al., 1998; Hopkins et al., 1998), as is the arcuate fasciculus, the fiber bundle that connects Wernicke's area with the frontal lobe (Rilling et al., 2012). Currently, there is no compelling evidence for major differences in anatomical hemispheric asymmetry between humans and our closest nonhuman primate relatives, although given the likelihood of lateralized functional specializations in humans; we should ultimately expect to find anatomical differences, although perhaps not at the level of gross morphology.

Specializations of connectivity and microstructure. If evolution can change the functions of areas, those changes should be manifested anatomically, by changes in their connections and in their internal organization. With the advent of comparative neuroimaging studies, we now have evidence for changes in extrinsic connectivity, both from structural and functional connectivity studies (see Mars et al., 2016, in this volume). Moreover, because chimpanzees have been studied with DWI, we have data that are likely to relate to human (as opposed to hominoid) specializations (Hecht et al., 2013, 2015a; Rilling et al., 2008, 2012). Despite the strong conviction among neuroscientists of a "basic uniformity" in the intrinsic organization of the cortex, we also have a growing body of evidence documenting human and hominoid specializations at microstructural levels of organization (reviewed by Sherwood et al., 2016, in this volume).

29.5.2 Opportunities and Challenges for the Future

This chapter not only bears witness to the progress our field has made, but also highlights matters that deserved closer attention in the future. Much of the chapter concerns changes in association cortex, and reasonably so, since the areas of association cortex are critical elements of neural systems that instantiate higher-order cognitive processes (Passingham, 2008), and there clearly were important changes in these regions. We still have much to learn about the nature of these changes, however. Without cortical maps for chimpanzees of comparable detail to those available for humans, however, we cannot resolve the matter of new areas, and we are limited in our ability to determine which association areas were differentially enlarged, and to what extent, although we have some data for the frontal lobe (Semendeferi et al., 2001). What is more, given the evidence that evolution modified the functional properties of areas that are shared by humans and nonhuman primates, we should expect that the internal anatomical and physiological organization of those areas were modified as well. We have evidence that modifications took place [eg, (Elston, 2007; Schenker et al., 2005, 2008; Sherwood et al., 2016), in this volume], but a great deal more could be done.

It is clear, furthermore, that evolutionary changes were not restricted to association cortex: the human brain is not simply a monkey brain with a new top floor. When other brain regions have been examined in detail, evidence for human or hominoid specializations—often quantitative, and sometimes qualitative—is commonly found. Histological studies of visual cortex, including area V1, have revealed a number of human and hominoid specializations (Bryant et al., 2012; Preuss, 2004; Sherwood et al., 2007), the functional significance of which have not been explored in any depth. The histology of other sensory regions has not been compared at similar levels of detail. Some quantitative specializations of limbic structures are discussed by Lew and Semendeferi (2016) in this volume, and as noted, Vogt et al. (2013) have described territories of human cingulate cortex not present in monkeys or rodents. Subcortically, there is evidence for modifications involving the basal ganglia and cerebellum (eg, Balsters et al., 2010; MacLeod et al., 2003; Raghanti et al., 2016; Rilling and Insel, 1998). The profound differences in the embryology of the thalamus reported between humans and macaques suggest this structure was also an important nexus of evolutionary change (Letinic and Rakic, 2001).

The matters raised earlier can probably be addressed productively with our current methods and theoretical perspectives. Other important issues are not so easily addressed. A particular challenge for future research concerns how to integrate results from comparative neuroscience and paleontology and the related matter of specifying a chronology of changes in human brain evolution. Comparative neuroscience can, at best, provide us with answers to what happened in the 5–8 million years since humans last shared a common ancestor with chimpanzees, but by itself, it can say little about the sequence of changes or assign dates to them. Paleoneurology could help, based on the dating of fossil endocasts, if we could relate changes in endocranial morphology to changes in internal organization. Recent work suggests this may be possible in some cases (see Gunz, 2016; Bruner, 2016, in this volume, and also Bruner et al., 2017), but the extent to which this is possible is still unclear. Likewise, we need to find new ways to link behavioral and cognitive changes to neurobiological changes. The archeological record is indispensible, of course, in helping contribute to the chronology of behavioral changes, but human cognition is often only

very indirectly reflected in material artifacts or other physical evidence. Significantly, despite the growing body of comparative data on the cognition of humans, great apes, and other primates, we also lack significant consensus on exactly what the similarities and differences in cognition are between these species, and thus it is difficult to relate human neurobiological specializations to human cognitive and behavioral specializations (reviewed in this volume by Boeckx, 2016; Barrett, 2016), much less understand the chronology of specialization.

One might suppose that comparative genomics and molecular biology would provide important clues to understanding the evolutionary transformation of the human brain, given the number of human specializations of gene expression and gene sequences that have now been identified (reviewed in this volume by Vallender, 2016) and the fact that some of these changes can be dated. Unfortunately, it has proven very difficult to relate confidence specific genetic changes to specific changes in anatomical and behavioral phenotypes. Even for *FOXP2*, a gene that underwent selection in the human lineage and is associated with speech disorder in individuals with a loss-of-function mutation, we cannot say for certain that the changes it underwent in human evolution are related to the evolution of language (Preuss, 2012b). Relating genetic and molecular changes to phenotypic changes is a major, unsolved challenge, largely because our best tools for doing so currently involve transfection of human genes into cell lines, which are not convincing proxies for the organs we are most interested in, and "humanized" transgenic mouse models, in which human genes are expressed in a mouse—rather than human—biochemical, cellular, and histological context. So far, the potential of comparative genetic and molecular studies has yet to be fully realized. One way comparative molecular biology can contribute at present, however, is to help elucidate changes in cell physiological and biochemical phenotypes, given their close relationship of gene and protein expression (Preuss, 2012b; Varki et al., 2008).

Despite these caveats, there is reason for great optimism about the future of human brain evolutionary biology, mainly because of the efflorescence of comparative studies of the brain, cognition, and the genome. Yet there is also reason for pessimism, because of declining access to one of the essential resources required for understanding human brain evolution, and human nature more generally: the great apes. It cannot be stated too strongly that our prospects for understanding human brain evolution depend critically on our ability to compare humans to the great apes, and especially to chimpanzees, our closest relatives. We now have tools to allow us to study great apes with the same noninvasive methods we use to study humans—tools that provide us the means to answer fundamentally questions about the biology of humans and apes. Both captive and wild populations of all the great apes are in decline, and without them, our understanding of our place in the world will be forever impoverished (Povinelli and Preuss, 2012).

References

Allman, J.M., 1982. Reconstructing the evolution of the brain in primates through the use of comparative neurophysiological and anatomical data. In: Armstrong, E., Falk, D. (Eds.), Primate Brain Evolution. Methods and Concepts. Plenum, New York, pp. 13–28.

Allman, J.M., 1990. Evolution of neocortex. In: Jones, E.G., Peters, A. (Eds.), Cerebral Cortex. Comparative Structure and Evolution of Cerebral Cortex, Part II, vol. B. Plenum, pp. 269–283.

Allman, J.M., Kaas, J.H., 1971. Representation of the visual field in striate and adjoining cortex of the owl monkey (*Aotus trivirgatus*). Brain Res. 35, 89–106.

Allman, J.M., Kaas, J.H., 1971. A representation of the visual field in the caudal third of the middle temporal gyrus of the owl monkey (*Aotus trivirgatus*). Brain Res. 31, 85–105.

Anton, S.C., Potts, R., Aiello, L.C., 2014. Human evolution. Evolution of early *Homo*: an integrated biological perspective. Science 345, 1236828.

Avants, B.B., Avants, B.B., Schoenemann, P.T., Schoenemann, P.T., Gee, J.C., Gee, J.C., 2006. Lagrangian frame diffeomorphic image registration: morphometric comparison of human and chimpanzee cortex. Med. Image Anal. 10, 397–412.

Bailey, P., Bonin, G., 1951. The Isocortex of Man. Univ. of Illinois Press, Urbana.

Bailey, P., Bonin, G., Mcculloch, W.S., 1950. The Isocortex of the Chimpanzee. Univ. of Illinois Press, Urbana, IL.

Balsters, J.H., Cussans, E., Diedrichsen, J., Phillips, K.A., Preuss, T.M., Rilling, J.K., Ramnani, N., 2010. Evolution of the cerebellar cortex: the selective expansion of prefrontal-projecting cerebellar lobules. Neuroimage 49, 2045–2052.

Barks, S.K., Calhoun, M.E., Hopkins, W.D., Cranfield, M.R., Mudakikwa, A., Stoinski, T.S., Patterson, F.G., Erwin, J.M., Hecht, E.E., Hof, P.R., Sherwood, C.C., 2015. Brain organization of gorillas reflects species differences in ecology. Am. J. Phys. Anthropol. 156, 252–262.

Barks, S.K., Parr, L.A., Rilling, J.K., 2015. The default mode network in chimpanzees (*Pan troglodytes*) is similar to that of humans. Cereb. Cortex 25, 538–544.

Barrett, H.C., 2016. The search for human cognitive specializations. In: Preuss, T.M. (Ed.), Evolution of Nervous Systems, Human Brain Evolution, second ed., vol. 4. Elsevier (in press).

Baum, D.A., Smith, S.D., 2013. Tree Thinking: An Introduction to Phylogenetic Biology. Roberts and Company Publishers, Greenwood Village, Colo.

Benson, D.F., 1994. The Neurology of Thinking. Oxford University Press, New York.

Boeckx, C., 2016. Language evolution. In: Preuss, T.M. (Ed.), Evolution of Nervous Systems, Human Brain Evolution, second ed., vol. 4. Elsevier (in press).

Bonin, G., Bailey, P., 1947. The Neocortex of *Macaca mulatta*. University of Illinois Press, Urbana.

Bonin, G., Bailey, P., 1961. Patterns of the cerebral isocortex. In: Hofer, H., Schultz, A.H., Starck, D. (Eds.). Primatologia. S. Karger, Basel, pp. 1–42.

Boyd, R., Richerson, P.J., 1985. Culture and the Evolutionary Process. University of Chicago Press, Chicago.

Boyle, E.K., Wood, B., 2016. Human evolutionary history. In: Preuss, T.M. (Ed.), Evolution of Nervous Systems, Human Brain Evolution, second ed., vol. 4. Elsevier (in press).

Broca, P., 1865. Sur le siège de la faculté du langage articulé. Bull. la Société d'anthropologie Paris 6, 337–393.

Brodmann, K., 1909. In: Garey, L.J. (Ed.), Vergleichende Lokalisationslehre der Grosshirnrhinde. Barth, Leipzig (reprinted as Brodmann's 'Localisation in the Cerebral Cortex. Smith-Gordon, London, 1994.

Bronchti, G., Heil, P., Scheich, H., Wollberg, Z., 1989. Auditory pathway and auditory activation of primary visual targets in the blind mole rat (*Spalax ehrenbergi*): I. 2-deoxyglucose study of subcortical centers. J. Comp. Neurol. 284, 253–274.

Bruner, E., 2016. The fossil evidence of human brain evolution. In: Preuss, T.M. (Ed.), Evolution of Nervous Systems, Human Brain Evolution, second ed., vol. 4. Elsevier (in press).

Bruner, E., Preuss, T.M., Chen, X., Rilling, J.K., 2017. Evidence for expansion of the precuneus in human evolution. Brain Struct. Funct. 222 (2), 1053–1060.

Bryant, K.L., Suwyn, C., Reding, K.M., Smiley, J.F., Hackett, T.A., Preuss, T.M., 2012. Evidence for ape and human specializations in geniculostriate projections from VGLUT2 immunohistochemistry. Brain Behav. Evol. 80, 210–221.

Buckner, R.L., Carroll, D.C., 2007. Self-projection and the brain. Trends Cogn. Sci. 11, 49–57.

Buckner, R.L., Krienen, F.M., 2016. Human association cortex: expanded, untethered, neotenous, and plastic. In: Preuss, T.M. (Ed.), Evolution of Nervous Systems, Human Brain Evolution, second ed., vol. 4. Elsevier (in press).

Casagrande, V.A., Kaas, J.H., 1994. The afferent, intrinsic, and efferent connections of primary visual cortex in primates. In: Peters, A., Rockland, K. (Eds.), Cerebral Cortex. Primary Visual Cortex in Primates, vol. 10. Plenum, New York, pp. 201–259.

Cooper, H.M., Herbin, M., Nevo, E., 1993. Visual system of a naturally microphthalmic mammal: the blind mole rat, *Spalax ehrenbergi*. J. Comp. Neurol. 328, 313–350.

Creutzfeldt, O.D., 1977. Generality of the functional structure of the neocortex. Die Naturwiss 64, 507–517.

Crick, F., Jones, E., 1993. Backwardness of human neuroanatomy. Nature 361, 109–110.

Darwin, C., 1859. On the Origin of Species. John Murray, London [Facsimile of first edition: Cambridge, MA: Harvard University Press, 1984].

Darwin, C., 1871. The Descent of Man, and Selection in Relation to Sex. John Murray, London [Facsimile edition: Princeton, NJ: Princeton University Press, 1981].

Darwin, C., Stauffer, R.C., 1975. Charles Darwin's Natural Selection: Being the Second Part of His Big Species Book Written from 1856 to 1858. Cambridge University Press, London, New York.

de Sousa, A., Wood, B., 2007. The hominin fossil record and the emergence of the modern human central nervous system. In: Kaas, J.H., Preuss, T.M. (Eds.), Evolution of Nervous Systems. Primates, vol. 4. Elsevier, Oxford, pp. 291–336.

Dembo, M., Radovcic, D., Garvin, H.M., Laird, M.F., Schroeder, L., Scott, J.E., Brophy, J., Ackermann, R.R., Musiba, C.M., De Ruiter, D.J., Mooers, A.Ø., Collard, M., 2016. The evolutionary relationships and age of *Homo naledi*: an assessment using dated Bayesian phylogenetic methods. J. Hum. Evol. 97, 17–26.

Donahue, C.J., Sotiropoulos, S.N., Jbabdi, S., Hernandez-Fernandez, M., Behrens, T.E., Dyrby, T.B., Coalson, T., Kennedy, H., Knoblauch, K., Van Essen, D.C., Glasser, M.F., 2016. Using diffusion tractography to predict cortical connection strength and distance: a quantitative comparison with tracers in the monkey. J. Neurosci. 36, 6758–6770.

Dubner, R., Zeki, S.M., 1971. Response properties and receptive fields of cells in an anatomically defined region of the superior temporal sulcus in the monkey. Brain Res. 35, 528–532.

Ebbesson, S.O., 1980. The parcellation theory and its relation to interspecific variability in brain organization, evolutionary and ontogenetic development, and neuronal plasticity. Cell Tissue Res. 213, 179–212.

Elliot Smith, G., 1924. The Evolution of Man: Essays. Oxford University Press, London.

Elston, G.N., 2007. Specialization of the neocortical pyramidal cell during primate evolution. In: Kaas, J.H., Preuss, T.M. (Eds.), Evolution of Nervous Systems. Primates, vol. 4. Elsevier, Oxford, pp. 191–242.

Felleman, D.J., Van Essen, D.C., 1991. Distributed hierarchical processing in the primate cerebral cortex. Cereb. Cortex 1, 1–47.

Fleagle, J.G., 2013. Primate Adaptation and Evolution. Academic, Oxford.

Gannon, P.J., Holloway, R.L., Broadfield, D.C., Braun, A.R., 1998. Asymmetry of chimpanzee planum temporale: humanlike pattern of Wernicke's brain language area homolog. Science 279, 220–222.

Garland, T., Harvey, P., Ives, A., 1992. Procedures for the Analysis of Comparative Data Using Phylogenetically Independent Contrasts, vol. 41, pp. 18–32.

Geschwind, N., 1965. Disconnexion syndromes in animals and man. I. Brain 88, 237–294.

Glasser, M.F., Van Essen, D.C., 2011. Mapping human cortical areas in vivo based on myelin content as revealed by T1- and T2-weighted MRI. J. Neurosci. 31, 11597–11616.

Glasser, M.F., Goyal, M.S., Preuss, T.M., Raichle, M.E., Van Essen, D.C., 2014. Trends and properties of human cerebral cortex: correlations with cortical myelin content. Neuroimage 93 (Pt 2), 165–175.

Glasser, M.F., Coalson, T.S., Robinson, E.C., Hacker, C.D., Harwell, J., Yacoub, E., Ugurbil, K., Andersson, J., Beckmann, C.F., Jenkinson, M., Smith, S.M., Van Essen, D.C., 2016. A multimodal parcellation of human cerebral cortex. Nature 536, 171–178.

Goldman-Rakic, P.S., 1988. Topography of cognition: parallel distributed networks in primate association cortex. Annu. Rev. Neurosci. 11, 137–156.

Gould, S.J., 1989. Wonderful Life: The Burgess Shale and Nature of History. W.W. Norton, New York.

Gunz, P., 2016. Differences in brain organization between Neanderthals and modern humans. In: Preuss, T.M. (Ed.), Evolution of Nervous Systems, Human Brain Evolution, second ed., vol. 4. Elsevier.

Gutman, A., Keifer, O.P., Magnuson, M.E., Choi, D.C., Majeed, W., Keilholz, S., Ressler, K.J., 2012. A DTI tractography analysis of infralimbic and prelimbic connectivity in the mouse using high-throughput MRI. Neuroimage 63, 800–811.

Haeckel, E.H.P.A., 1874. Anthropogenie, oder, Entwickelungsgeschichte des Menschen. W. Engelmann, Leipzig.

Haeckel, E.H.P.A., 1879. The Evolution of Man: A Popular Exposition of the Principal Points of Human Ontogeny and Phylogeny. Appleton, New York.

Harmand, S., Lewis, J.E., Feibel, C.S., Lepre, C.J., Prat, S., Lenoble, A., Boes, X., Quinn, R.L., Brenet, M., Arroyo, A., Taylor, N., Clement, S., Daver, G., Brugal, J.P., Leakey, L., Mortlock, R.A., Wright, J.D., Lokorodi, S., Kirwa, C., Kent, D.V., Roche, H., 2015. 3.3-million-year-old stone tools from Lomekwi 3, west Turkana, Kenya. Nature 521, 310–315.

Harvey, P.H., Pagel, M.D., 1991. The Comparative Method in Evolutionary Biology. Oxford University Press, Oxford.

Hecht, E.E., Gutman, D.A., Preuss, T.M., Sanchez, M.M., Parr, L.A., Rilling, J.K., 2013. Process versus product in social learning: comparative diffusion tensor imaging of neural systems for action

execution-observation matching in macaques, chimpanzees, and humans. Cereb. Cortex 23, 1014–1024.

Hecht, E.E., Gutman, D.A., Bradley, B.A., Preuss, T.M., Stout, D., 2015. Virtual dissection and comparative connectivity of the superior longitudinal fasciculus in chimpanzees and humans. Neuroimage 108, 124–137.

Hecht, E.E., Gutman, D.A., Khreisheh, N., Taylor, S.V., Kilner, J., Faisal, A.A., Bradley, B.A., Chaminade, T., Stout, D., 2015. Acquisition of Paleolithic toolmaking abilities involves structural remodeling to inferior frontoparietal regions. Brain Struct. Funct. 220, 2315–2331.

Hennig, W., 1950. Grundzüge einer Theorie der Phylogenetischen Systematik. Deutscher Zentralverlag, Berlin.

Hennig, W., 1966. Phylogenetic Systematics. University of Illinois Press, Urbana.

Henrich, J., 2015. The Secret of Our Success: How Culture Is Driving Human Evolution, Domesticating Our Species, and Making Us Smarter. Princeton University Press, Princeton, NJ.

Herculano-Houzel, S., Collins, C.E., Wong, P., Kaas, J.H., Lent, R., 2008. The basic nonuniformity of the cerebral cortex. Proc. Natl. Acad. Sci. U.S.A. 105, 12593–12598.

Hof, P.R., Sherwood, C.C., 2005. Morphomolecular neuronal phenotypes in the neocortex reflect phylogenetic relationships among certain mammalian orders. Anat. Rec. Part A Discov. Mol. Cell. Evol. Biol. 287, 1153–1163.

Hof, P.R., Sherwood, C.C., 2007. The evolution of neuron classes in the neocortex of mammals. In: Kaas, J.H., Krubitzer, L.A. (Eds.), Evolution of Nervous Systems. Mammals, vol. 3. Elsevier, Oxford, pp. 113–124.

Holloway Jr., R.L., 1966. Cranial capacity and neuron number: a critique and proposal. Am. J. Phys. Anthropol. 25, 305–314.

Holloway Jr., R.L., 1966. Cranial capacity, neural reorganization, and hominid evolution: a search for more suitable parameters. Am. Anthropol. 68, 103–121.

Holloway Jr., R.L., 1968. The evolution of the primate brain: some aspects of quantitative relations. Brain Res. 7, 121–172.

Holloway Jr., R.L., 1992. Culture: a human domain. Curr. Anthropol. 47–64.

Hooton, E.A., 1931. Up from the Ape. The Macmillan Company, New York.

Hopkins, W.D., Marino, L., Rilling, J.K., Macgregor, L.A., 1998. Planum temporale asymmetries in great apes as revealed by magnetic resonance imaging (MRI). Neuroreport 9, 2913–2918.

Hubel, D.H., Wiesel, T.N., 1959. Receptive fields of single neurones in the cat's striate cortex. J. Physiol. 148, 574–591.

Hubel, D.H., Wiesel, T.N., 1968. Receptive fields and functional architecture of monkey striate cortex. J. Physiol. 195, 215–243.

Hubel, D.H., Wiesel, T.N., 1977. Ferrier lecture. Functional architecture of macaque monkey visual cortex. Proc. R. Soc. Lond. Ser. B Biol. Sci. 198, 1–59.

Huxley, T.H., 1863. Evidence as to Man's Place in Nature. Williams and Norgate, London [1959, Ann Arbor, University of Michigan].

Jerison, H.J., 1955. Brain to body ratios and the evolution of intelligence. Science 121, 447–449.

Jerison, H.J., 1973. Evolution of the Brain and Intelligence. Academic Press, New York.

Jerison, H.J., 1977. Should phrenology be rediscovered? Curr. Anthropol. 18, 744.

Jerison, H.J., 2001. The study of primate brain evolution: where do we go from here. In: Falk, D., Gibson, K. (Eds.), Evolutionary Anatomy of the Primate Cerebral Cortex. Cambridge University Press, Cambridge, pp. 305–337.

Kaas, J.H., 1987. The organization and evolution of neocortex. In: Wise, S.P. (Ed.), Higher Brain Function: Recent Explorations of the Brain's Emergent Properties. John Wiley, New York, pp. 347–378.

Kaas, J.H., 1987. The organization of neocortex in mammals: implications for theories of brain function. Annu. Rev. Psychol. 38, 129–151.

Kleist, K., 1934. Gehirnpathologie. Barth, Leipzig.

Krubitzer, L., Dooley, J.C., 2013. Cortical plasticity within and across lifetimes: how can development inform us about phenotypic transformations? Front. Hum. Neurosci. 7, 620.

Lashley, K.S., 1931. Mass action in cerebral function. Science 73, 245–254.

Lashley, K.S., 1949. Persistent problems in the evolution of mind. Q. Rev. Biol. 24, 28–42.

Lashley, K.S., Clark, G., 1946. The cytoarchitecture of the cerebral cortex of Ateles: a critical examination of architectonic studies. J. Comp. Neurol. 85, 223–306.

Le Gros Clark, W.E., 1959. The Antecedents of Man. Edinburgh University Press, Edinburgh.

Lende, R.A., 1969. A comparative approach to the neocortex: localization in monotremes, marsupials and insectivores. Ann. N.Y. Acad. Sci. 167, 262–276.

Letinic, K., Rakic, P., 2001. Telencephalic origin of human thalamic GABAergic neurons. Nat. Neurosci. 4, 931–936.

Letinic, K., Zoncu, R., Rakic, P., 2002. Origin of GABAergic neurons in the human neocortex. Nature 417, 645–649.

Lew, C.H., Semendeferi, K., 2016. Evolutionary specializations of the human limbic system. In: Preuss, T.M. (Ed.), Evolution of Nervous Systems, Human Brain Evolution, second ed., vol. 4. Elsevier.

Lewis, J.E., Harmand, S., 2016. An earlier origin for stone tool making: implications for cognitive evolution and the transition to *Homo*. Philos. Trans. R. Soc. Lond. Ser. B Biol. Sci. 371.

Li, L., Rilling, J.K., Preuss, T.M., Glasser, M.F., Hu, X., 2011. The effects of connection reconstruction method on the interregional connectivity of brain networks via diffusion tractography. Hum. Brain Mapp. 33, 1894–1913.

Li, L., Hu, X., Preuss, T.M., Glasser, M.F., Damen, F.W., Qiu, Y., Rilling, J.K., 2013. Mapping putative hubs in human, chimpanzee and rhesus macaque connectomes via diffusion tractography. Neuroimage 80, 462–474.

Logan, C.A., 2002. Before there were standards: the role of test animals in the production of empirical generality in physiology. J. Hist. Biol. 35, 329–363.

Lovejoy, A.O., 1964. The Great Chain of Being: A Study of the History of an Idea. Harvard University Press, Cambridge.

Luria, A.R., 1973. The Working Brain: An Introduction to Neuropsychology. Basic Books, New York.

Macleod, C.E., Zilles, K., Schleicher, A., Rilling, J.K., Gibson, K.R., 2003. Expansion of the neocerebellum in Hominoidea. J. Hum. Evol. 44, 401–429.

Manger, P.R., Cort, J., Ebrahim, N., Goodman, A., Henning, J., Karolia, M., Rodrigues, S.L., Strkalj, G., 2008. Is 21st century neuroscience too focussed on the rat/mouse model of brain function and dysfunction? Front. Neuroanat. 2, 5.

Mantini, D., Gerits, A., Nelissen, K., Durand, J.B., Joly, O., Simone, L., Sawamura, H., Wardak, C., Orban, G.A., Buckner, R.L., Vanduffel, W., 2011. Default mode of brain function in monkeys. J. Neurosci. 31, 12954–12962.

Mantini, D., Corbetta, M., Romani, G.L., Orban, G.A., Vanduffel, W., 2013. Evolutionarily novel functional networks in the human brain? J. Neurosci. 33, 3259–3275.

Markov, N.T., Ercsey-Ravasz, M.M., Ribeiro Gomes, A.R., Lamy, C., Magrou, L., Vezoli, J., Misery, P., Falchier, A., Quilodran, R., Gariel, M.A., Sallet, J., Gamanut, R., Huissoud, C., Clavagnier, S., Giroud, P., Sappey-Marinier, D., Barone, P., Dehay, C., Toroczkai, Z., Knoblauch, K., Van Essen, D.C., Kennedy, H., 2012. A weighted and directed interareal connectivity matrix for macaque cerebral cortex. Cereb. Cortex 24, 17–36.

Mars, R.B., Neubert, F.-X., Noonan, M.P., Sallet, J., Toni, I., Rushworth, M.F.S., 2012. On the relationship between the "default mode network" and the "social brain". Front. Hum. Neurosci. 6, 189.

Mars, R.B., Passingham, R.E., Neubert, F.X., Verhagen, L., Sallet, J., 2016. Evolutionary specializations of human association cortex. In: Preuss, T.M. (Ed.), Evolution of Nervous Systems, Human Brain Evolution, second ed., vol. 4. Elsevier.

Mayr, E., 1950. Taxonomic categories in fossil hominids. Cold Spring Harb. Symp. Quant. Biol. 15, 109–118.

Mayr, E., 1982. The Growth of Biological Thought: Diversity, Evolution, and Inheritance. Belknap Press, Cambridge, Mass.

Mayr, E., Provine, W.B., 1980. The Evolutionary Synthesis: Perspectives on the Unification of Biology. Harvard University Press, Cambridge, Mass.

Mcbrearty, S., Brooks, A.S., 2000. The revolution that wasn't: a new interpretation of the origin of modern human behavior. J. Hum. Evol. 39, 453–563.

Morel, A., Kaas, J.H., 1992. Subdivisions and connections of auditory cortex in owl monkeys. J. Comp. Neurol. 318, 27–63.

Mountcastle, V.B., 1978. An organizing principle for cerebral function: the unit module and the distributed system. In: Edelman, G.M. (Ed.), The Mindful Brain. The Mindful Brain, Cambridge, MA, pp. 7–50.

Murphy, W.J., Eizirik, E., Johnson, W.E., Zhang, Y.P., Ryder, O.A., O'Brien, S.J., 2001. Molecular phylogenetics and the origins of placental mammals. Nature 409, 614–618.

Murphy, W.J., Eizirik, E., O'Brien, S.J., Madsen, O., Scally, M., Douady, C.J., Teeling, E., Ryder, O.A., Stanhope, M.J., De Jong, W.W., Springer, M.S., 2001. Resolution of the early placental mammal radiation using Bayesian phylogenetics. Science 294, 2348–2351.

Murphy, W.J., Pevzner, P.A., O'Brien, S.J., 2004. Mammalian phylogenomics comes of age. Trends Genet. 20, 631–639.

Neubert, F.-X., Mars, R.B., Thomas, A.G., Sallet, J., Rushworth, M.F.S., 2014. Comparison of human ventral frontal cortex: areas for cognitive control and language with areas in monkey frontal cortex. Neuron 81, 700–713.

Orban, G.A., 2016. Evolution of human visual-system specializations. In: Preuss, T.M. (Ed.), Evolution of Nervous Systems, Human Brain Evolution, second ed., vol. 4. Elsevier.

Padberg, J., Franca, J.G., Cooke, D.F., Soares, J.G., Rosa, M.G., Fiorani Jr., M., Gattass, R., Krubitzer, L., 2007. Parallel evolution of cortical areas involved in skilled hand use. J. Neurosci. 27, 10106–10115.

Parr, L.A., Hecht, E., Barks, S.K., Preuss, T.M., Votaw, J.R., 2009. Face processing in the chimpanzee brain. Curr. Biol. 19, 50–53.

Passingham, R.E., 2008. What Is Special about the Human Brain? Oxford University Press, Oxford.

Passingham, R.E., Rowe, J.B., 2016. A Short Guide to Brain Imaging: The Neuroscience of Human Cognition. Oxford University Press, Oxford.

Passingham, R.E., Smaers, J.B., 2014. Is the prefrontal cortex especially enlarged in the human brain allometric relations and remapping factors. Brain Behav. Evol. 84, 156–166.

Passingham, R.E., Smaers, J.B., Sherwood, C.C., 2016. Evolutionary specializations of the human prefrontal cortex. In: Preuss, T.M. (Ed.), Evolution of Nervous Systems, Human Brain Evolution, second ed., vol. 4. Elsevier.

Perelman, P., Johnson, W.E., Roos, C., Seuánez, H.N., Horvath, J.E., Moreira, M.A.M., Kessing, B., Pontius, J., Roelke, M., Rumpler, Y., Schneider, M.P.C., Silva, A., O'Brien, S.J., Pecon-Slattery, J., 2011. A molecular phylogeny of living primates. PLoS Genet. 7, e1001342.

Petrides, M., Tomaiuolo, F., Yeterian, E.H., Pandya, D.N., 2012. The prefrontal cortex: comparative architectonic organization in the human and the macaque monkey brains. Cortex 48, 46–57.

Povinelli, D.J., Preuss, T.M., 2012. Afterword: world without apes? In: Povinelli, D.J. (Ed.), World without Weight: Perspectives on an Alien Mind. Oxford University Press, Oxford, pp. 330–332.

Preuss, T.M., 1993. The role of the neurosciences in primate evolutionary biology: historical commentary and prospectus. In: Macphee, R.D.E. (Ed.), Primates and Their Relatives in Phylogenetic Perspective. Plenum Press, New York, pp. 333–362.

Preuss, T.M., 1995. The argument from animals to humans in cognitive neuroscience. In: Gazzaniga, M.S. (Ed.), The Cognitive Neurosciences. MIT Press, Cambridge, MA, pp. 1227–1241.

Preuss, T.M., 1995. Do rats have prefrontal cortex? The Rose-Woolsey-Akert program reconsidered. J. Cogn. Neurosci. 7, 1–24.

Preuss, T.M., 2001. The discovery of cerebral diversity: an unwelcome scientific revolution. In: Falk, D., Gibson, K. (Eds.), Evolutionary Anatomy of the Primate Cerebral Cortex. Cambridge University Press, Cambridge, pp. 138–164.

Preuss, T.M., 2004. Specializations of the human visual system: the monkey model meets human reality. In: Kaas, J.H., Collins, C.E. (Eds.), The Primate Visual System. CRC Press, Boca Raton, FL, pp. 231–259.

Preuss, T.M., 2007. Primate brain evolution in phylogenetic context. In: Evolution of Nervous Systems. Primates, vol. 4. Elsevier, pp. 1–34.

Preuss, T.M., 2009. The cognitive neuroscience of human uniqueness. In: Gazzaniga, M.S. (Ed.), The Cognitive Neurosciences, fourth ed. MIT Press, Cambridge, MA, pp. 49–64.

Preuss, T.M., 2010. Reinventing primate neuroscience for the twenty-first century. In: Platt, M.L., Ghazanfar, A.A. (Eds.), Primate Neuroethology. Oxford University Press, Oxford, pp. 422–454.

Preuss, T.M., 2011. The human brain: rewired and running hot. Ann. N.Y. Acad. Sci. 1225 (Suppl. 1), E182–E191.

Preuss, T.M., 2012. Alien brains: a defense of exceptionalism. In: Povinelli, D.J. (Ed.), World without Weight: Perspectives on an Alien Mind. Oxford University Press, Oxford, pp. 30–31.

Preuss, T.M., 2012. Human brain evolution: from gene discovery to phenotype discovery. Proc. Natl. Acad. Sci. U.S.A. 109 (Suppl. 1), 10709–10716.

Preuss, T.M., 2016. The human brain: evolution and distinctive features. In: Tibayrenc, M., Ayala, F.J. (Eds.), On Human Nature. Biology, Psychology, Ethics, Politics, and Religion. Elsevier (in press).

Preuss, T.M., Robert, J.S., 2014. Animal models of the human brain: repairing the paradigm. In: Gazzaniga, M.S., Mangun, G. (Eds.), The Cognitive Neurosciences. MIT Press, pp. 59–66.

Preuss, T.M., Stepniewska, I., Kaas, J.H., 1996. Movement representation in the dorsal and ventral premotor areas of owl monkeys: a microstimulation study. J. Comp. Neurol. 371, 649–676.

Raghanti, M.A., Edler, M.K., Stephenson, A.R., Wilson, L.J., Hopkins, W.D., Ely, J.J., Erwin, J.M., Jacobs, B., Hof, P.R., Sherwood, C.C., 2016. Human-specific increase of dopaminergic innervation in a striatal region associated with speech and language: a comparative analysis of the primate basal ganglia. J. Comp. Neurol. 524, 2117–2129.

Raichle, M.E., Macleod, A.M., Snyder, A.Z., Powers, W.J., Gusnard, D.A., Shulman, G.L., 2001. A default mode of brain function. Proc. Natl. Acad. Sci. U.S.A. 98, 676–682.

Rauschecker, J.P., 2016. The evolution of auditory cortex in humans. In: Preuss, T.M. (Ed.), Evolution of Nervous Systems, Human Brain Evolution, second ed., vol. 4. Elsevier.

Richards, R.J., 1987. Darwin and the Emergence of Evolutionary Theories of Mind and Behavior. University of Chicago Press, Chicago.

Richards, R.J., 1992. The Meaning of Evolution: The Morphological Construction and Ideological Reconstruction of Darwin's Theory. University of Chicago Press, Chicago.

Richerson, P.J., Boyd, R., 2005. Not by Genes Alone: How Culture Transformed Human Evolution. University of Chicago Press, Chicago.

Rilling, J.K., 2014. Comparative primate neuroimaging: insights into human brain evolution. Trends Cogn. Sci. 18, 46–55.

Rilling, J.K., 2016. Evolution of human language circuits revealed with comparative diffusion tensor imaging. In: Preuss, T.M. (Ed.), Evolution of Nervous Systems, Human Brain Evolution, second ed., vol. 4. Elsevier.

Rilling, J.K., Insel, T.R., 1998. Evolution of the cerebellum in primates: differences in relative volume among monkeys, apes and humans. Brain Behav. Evol. 52, 308–314.

Rilling, J.K., Stout, D., 2014. Evolution of the neural bases of higher cognitive function in humans. In: Gazzaniga, M.S., Mangun, G. (Eds.), The Cognitive Neurosciences. MIT Press, Cambridge, MA, pp. 41–49.

Rilling, J.K., Barks, S.K., Parr, L.A., Preuss, T.M., Faber, T.L., Pagnoni, G., Bremner, J.D., Votaw, J.R., 2007. A comparison of resting-state brain activity in humans and chimpanzees. Proc. Natl. Acad. Sci. U.S.A. 104, 17146–17151.

Rilling, J.K., Glasser, M.F., Preuss, T.M., Ma, X., Zhao, T., Hu, X., Behrens, T.E.J., 2008. The evolution of the arcuate fasciculus revealed with comparative DTI. Nat. Neurosci. 11, 426–428.

Rilling, J.K., Glasser, M.F., Jbabdi, S., Andersson, J., Preuss, T.M., 2012. Continuity, divergence and the evolution of brain language pathways. Front. Evol. Neurosci. 3.

Rockel, A.J., Hiorns, R.W., Powell, T.P.S., 1980. The basic uniformity in structure of the neocortex. Brain 103, 221–244.

Rose, J.E., Woolsey, C.N., 1948. The orbitofrontal cortex and its connections with the mediodorsal nucleus in rabbit, sheep and cat. Res. Publ. Assoc. Res. Nerv. Ment. Dis. 27, 210–232.

Rose, J.E., Woolsey, C.N., 1949. Organization of the mammalian thalamus and its relationships to the cerebral cortex. Electroencephalogr. Clin. Neurophysiol. 1, 391–403.

Sadka, R.S., Wollberg, Z., 2004. Response properties of auditory activated cells in the occipital cortex of the blind mole rat: an electrophysiological study. J. Comp. Physiol. A Neuroethol. Sens. Neural Behav. Physiol. 190, 403–413.

Salemi, M., Vandamme, A.-M., Lemey, P., 2009. The Phylogenetic Handbook: A Practical Approach to Phylogenetic Analysis and Hypothesis Testing. Cambridge University Press, Cambridge, UK, New York.

Sallet, J., Mars, R.B., Noonan, M.P., Neubert, F.-X., Jbabdi, S., O'reilly, J.X., Filippini, N., Thomas, A.G., Rushworth, M.F., 2013. The organization of dorsal frontal cortex in humans and macaques. J. Neurosci. 33, 12255–12274.

Sarkissov, S., Filiminoff, I., Kononowa, E., Preobraschenskaya, S., Kukuew, L., 1955. Atlas of the Cytoarchitecture of the Human Cerebral Cortex. Medgiz, Moscow.

Schall, J.D., Zinke, W., Cosman, J.D., Schall, M.S., Paré, M., Pouget, P., 2016. On the evolution of the frontal eye field: comparisons of monkeys, apes and humans. In: Preuss, T.M. (Ed.), Evolution of Nervous Systems, Human Brain Evolution, second ed., vol. 4. Elsevier.

Schenker, N.M., Desgouttes, A.M., Semendeferi, K., 2005. Neural connectivity and cortical substrates of cognition in hominoids. J. Hum. Evol. 49, 547–569.

Schenker, N.M., Buxhoeveden, D.P., Blackmon, W.L., Amunts, K., Zilles, K., Semendeferi, K., 2008. A comparative quantitative analysis of cytoarchitecture and minicolumnar organization in Broca's area in humans and great apes. J. Comp. Neurol. 510, 117–128.

Schoenemann, P.T., 2004. Brain size scaling and body composition in mammals. Brain Behav. Evol. 63, 47–60.

Schoenemann, P.T., 2006. Evolution of the size and functional areas of the human brain. Annu. Rev. Anthropol. 35, 379–406.

Semendeferi, K., Damasio, H., 2000. The brain and its main anatomical subdivisions in living hominoids using magnetic resonance imaging. J. Hum. Evol. 38, 317–332.

Semendeferi, K., Damasio, H., Frank, R., Van Hoesen, G.W., 1997. The evolution of the frontal lobes: a volumetric analysis based on three-dimensional reconstructions of magnetic resonance scans of human and ape brains. J. Hum. Evol. 32, 375—388.

Semendeferi, K., Armstrong, E., Schleicher, A., Zilles, K., Van Hoesen, G.W., 2001. Prefrontal cortex in humans and apes: a comparative study of area 10. Am. J. Phys. Anthropol. 114, 224—241.

Shepherd, G.M., 1988. A basic circuit of cortical organization. In: Gazzaniga, M.S. (Ed.), Perspectives on Memory Research. MIT Press, Cambridge, MA, pp. 93—134.

Sherwood, C.C., Raghanti, M.A., Stimpson, C.D., Bonar, C.J., De Sousa, A.A., Preuss, T.M., Hof, P.R., 2007. Scaling of inhibitory interneurons in areas v1 and v2 of anthropoid primates as revealed by calcium-binding protein immunohistochemistry. Brain Behav. Evol. 69, 176—195.

Sherwood, C.C., Bauernfeind, A.L., Bianchi, S., Raghanti, M.A., Hof, P.R., 2012. Human brain evolution writ large and small. In: Hofman, M.A., Falk, D. (Eds.), Evolution of the Primate Brain, first ed. Elsevier, Amsterdam, Boston, pp. 237—254.

Sherwood, C.C., Bauernfeind, A.L., Verendeev, A., Raghanti, M.A., Hof, P.R., 2016. Evolutionary specializations of human brain microstructure. In: Preuss, T.M. (Ed.), Evolution of Nervous Systems, Human Brain Evolution, second ed., vol. 4. Elsevier (in press).

Skiba, S.A., Taglialatela, J.P., 2016. Evolution of laterality and language in primates. In: Preuss, T.M. (Ed.), Evolution of Nervous Systems, Human Brain Evolution, second ed., vol. 4. Elsevier.

Stephan, H., 1972. In: Tuttle, R. (Ed.), Evolution of Primate Brains: A Comparative Anatomical Investigation. Aldine, Chicago, pp. 155—174.

Stephan, H., Andy, O.J., 1969. Quantitative Comparative Neuroanatomy of Primates: An Attempt at a Phylogenetic Interpretation, vol. 167, pp. 370—387.

Stout, D., Toth, N., Schick, K., Chaminade, T., 2008. Neural correlates of Early Stone Age toolmaking: technology, language and cognition in human evolution. Philos. Trans. R. Soc. Lond. Ser. B Biol. Sci. 363, 1939—1949.

Stout, D., Hecht, E.E., Khreisheh, N., Bradley, B., Chaminade, T., 2015. Cognitive demands of lower paleolithic toolmaking. PLoS One 10, e0121804.

Striedter, G.F., 2005. Principles of Brain Evolution. Sinauer Associates, Sunderland, Mass.

Striedter, G.F., Belgard, T.G., Chen, C.C., Davis, F.P., Finlay, B.L., Gunturkun, O., Hale, M.E., Harris, J.A., Hecht, E.E., Hof, P.R., Hofmann, H.A., Holland, L.Z., Iwaniuk, A.N., Jarvis, E.D., Karten, H.J., Katz, P.S., Kristan, W.B., Macagno, E.R., Mitra, P.P., Moroz, L.L., Preuss, T.M., Ragsdale, C.W., Sherwood, C.C., Stevens, C.F., Stuttgen, M.C., Tsumoto, T., Wilczynski, W., 2014. NSF workshop report: discovering general principles of nervous system organization by comparing brain maps across species. Brain Behav. Evol. 83, 1—8 (also published in The Journal of Comparative Neurology: 522, 1445—1453, 2014).

Stringer, C., 2016. The origin and evolution of *Homo sapiens*. Philos. Trans. R. Soc. Lond. Ser. B Biol. Sci. 371.

Sutikna, T., Tocheri, M.W., Morwood, M.J., Saptomo, E.W., Jatmiko, A.R.D., Wasisto, S., Westaway, K.E., Aubert, M., Li, B., Zhao, J.-X., Storey, M., Alloway, B.V., Morley, M.W., Meijer, H.J.M., Van Den Bergh, G.D., Gruen, R., Dosseto, A., Brumm, A., Jungers, W.L., Roberts, R.G., 2016. Revised stratigraphy and chronology for *Homo floresiensis* at Liang Bua in Indonesia. Nature 532, 366.

Szentagothai, J., 1975. The 'module-concept' in cerebral cortex architecture. Brain Res. 95, 475—496.

Tomasello, M., 1999. The Cultural Origins of Human Cognition. Harvard University Press, Cambridge, Mass, London.

Tomasello, M., 2014. A Natural History of Human Thinking. Harvard University Press, Cambridge, Massachusetts.

Vallender, E.J., 2016. Molecular evolution and phenotypic change. In: Preuss, T.M. (Ed.), Evolution of Nervous Systems, Human Brain Evolution, second ed., vol. 4. Elsevier (in press).

van den Heuvel, M.P., De Reus, M.A., Feldman Barrett, L., Scholtens, L.H., Coopmans, F.M.T., Schmidt, R., Preuss, T.M., Rilling, J.K., Li, L., 2015. Comparison of diffusion tractography and tract-tracing measures of connectivity strength in rhesus macaque connectome. Hum. Brain Mapp. 36, 3064—3075.

Van Essen, D.C., Smith, S.M., Barch, D.M., Behrens, T.E., Yacoub, E., Ugurbil, K., Consortium, W.U.-M.H., 2013. The Wu-Minn human connectome project: an overview. Neuroimage 80, 62—79.

Varki, A., Geschwind, D.H., Eichler, E.E., 2008. Explaining human uniqueness: genome interactions with environment, behaviour and culture. Nat. Rev. Genet. 9, 749—763.

Villmoare, B., Kimbel, W.H., Seyoum, C., Campisano, C.J., Dimaggio, E.N., Rowan, J., Braun, D.R., Arrowsmith, J.R., Reed, K.E., 2015. Paleoanthropology. Early *Homo* at 2.8 Ma from Ledi-Geraru, Afar, Ethiopia. Science 347, 1352—1355.

Vincent, J.L., Patel, G.H., Fox, M.D., Snyder, A.Z., Baker, J.T., Van Essen, D.C., Zempel, J.M., Snyder, L.H., Corbetta, M., Raichle, M.E., 2007. Intrinsic functional architecture in the anaesthetized monkey brain. Nature 447, 83—86.

Vogt, B.A., Hof, P.R., Zilles, K., Vogt, L.J., Herold, C., Palomero-Gallagher, N., 2013. Cingulate area 32 homologies in mouse, rat, macaque and human: cytoarchitecture and receptor architecture. J. Comp. Neurol. 521, 4189—4204.

Wernicke, C., 1874. Der aphasische Symptomen-Complex: Eine Psychologische Studie auf anatomischer Basis. Cohn and Weigert, Breslau.

Wolpoff, M.H., 1971. Competitive exclusion among lower Pleistocene hominids: the single species hypothesis. Man 6, 601.

Woolsey, C.N., 1958. Organization of somatic sensory and motor areas of the monkey. In: Harlow, H.F., Woolsey, C.N. (Eds.), Biological and Biochemical Basis of Behavior. University of Wisconsin Press, Madison, WI, pp. 63—81.

Young, M.P., 1992. Objective analysis of the topological organization of the primate cortical visual system. Nature 358, 152—155.

Zeki, S.M., 1969. Representation of central visual fields in prestriate cortex of monkey. Brain Res. 14, 271—291.

Zilles, K., Amunts, K., 2010. Centenary of Brodmann's map—conception and fate. Nat. Rev. Neurosci. 11, 139—145.

30

Human Evolutionary History

E.K. Boyle, B. Wood

George Washington University, Washington, DC, United States

30.1 Introduction

All living species are at the end of twigs that reach the surface of the Tree of Life (TOL), and all species that have ever lived in the past are situated somewhere on the branches and twigs within the tree. The extinct species on the branch within the TOL that connect modern humans directly to the base of the tree are our ancestors. The species on branches that are closer to ours than to any other living taxon, but which do not reach the surface of the TOL, are our extinct close relatives.

Not that long ago it was assumed that the extant nonhuman great apes—chimpanzees, bonobos, gorillas, and orangutans—were more closely related to each other than they were to modern humans. This was reflected in a taxonomy that put modern humans in their own family, the Hominidae, and the great apes in a separate family, the Pongidae. To reflect this family-level division, the informal term "hominid" was used to refer to modern humans and any extinct species more closely related to modern humans than to any of the great apes. The equivalent informal term for the nonhuman great apes was "pongid."

30.2 Comparative Context

That traditional, premolecular, understanding of how modern humans are related to the other great apes has been superseded by a hypothesis that was initially driven by molecular evidence (Zuckerkandl et al., 1960; Goodman, 1962), but which is now supported by copious genetic, molecular, and morphological evidence. Draft sequences of the nuclear genomes of the chimpanzee (Consortium, 2005), orangutan (Locke et al., 2011), gorilla (Scally, 2012), and bonobo (Prüfer et al., 2012); better-quality annotated sequences (Gordon et al., 2016); and morphological evidence (Shoshani et al., 1996; Gibbs et al., 2002;

Lockwood et al., 2004; Diogo and Wood, 2011) are all consistent with the hypothesis that chimpanzees/bonobos are more closely related to modern humans than they are to gorillas. So the options for what a review of human evolutionary history should cover range from a long version, for example, one that starts c.400 million years ago (Ma) at the base of the major branch of the TOL tree that contains vertebrates with four limbs, to a much shorter version that would begin with the first appearance of modern humans c.200 thousand years ago (ka). In between those extremes are versions that would start at the base of the branches that contain, respectively, all mammals, all primates, all monkeys and apes, all apes, and all of the great apes including modern humans. For the purposes of this review we interpret human evolutionary history as what happened within the branch, also known as a clade or monophyletic group, of the TOL that has the most recent common ancestor of modern humans and chimpanzees/bonobos at its base and modern humans at its tip.

The contemporary understanding of the relationships among the great apes that is supported by genetic, molecular, and comparative anatomical evidence is reflected in taxonomies that demote modern humans from their own family and instead assign them to their own tribe, or even subtribe (Andrews and Harrison, 2005). In this review we use the tribe Hominini for modern humans and their ancestors and close relatives after the split between modern humans and chimpanzees/bonobos, and we use the informal equivalent of Hominini, "hominin," to refer extinct species, or individuals that belong to an extinct species that is more closely related to modern humans than to any of the other living great apes. The equivalent term for an extinct species that is more closely related to chimpanzees/bonobos is "panin." The term "hominid" now refers to modern humans, the extant great apes, and any extinct species that is more closely related to modern humans and the great apes than to any other living taxon.

If the observed genetic differences are calibrated using paleontological evidence for the split between the apes and the Old World monkeys, and if we make the assumption that most DNA differences are neutral, this suggests that the hypothetical ancestor of modern humans and chimpanzees/bonobos lived between about 8 and 5 Ma and probably closer to 5 than to 8 Ma (Bradley, 2008). Other estimates that lean more heavily on estimates of generation times (Langergraber et al., 2012) and mutation rates (Venn et al., 2014) suggest that the date may be closer to 8 and 13 Ma, respectively, but a recent analysis based on a large data set (Prado-Martinez et al., 2013) lends support to a date of between 8 and 5 Ma. However, recalibration of the molecular clock to take account of *Rukwapithecus fleaglei* (Stevens et al., 2013), an Oligocene catarrhine that affects our understanding of the timing of the split between the apes and the Old World monkeys and attempts to better reflect mutation rates and generation times in the models, could further influence these estimates.

30.3 Fossil Evidence

The temporal context of the species and site collections we will consider in this review are set out in Fig. 30.1. This interpretation of the hominin fossil record deliberately recognizes a larger rather than a smaller number of species, but species we consider to be idiosyncratic (eg, *Homo antiquus*; Ferguson, 1984) were excluded. The vertical axis is geological time, and morphology is used to distribute the taxa across the horizontal axis, with larger-brained taxa on the left and those with large chewing teeth on the right. The widths of the columns for species or site collections are all the same despite the fact that the fossil records range widely in quantity and quality. Researchers use the term first appearance datum, or FAD, to refer to the date of a taxon's first appearance in the fossil record, and last appearance datum, or LAD, to refer to the date of that taxon's last occurrence in the fossil record, and the location of the tops and bottoms of each column reflects our current understanding of the FAD and LAD (Appendix 1). However, the observed FAD of a taxon is almost certainly later than the time the taxon actually originated

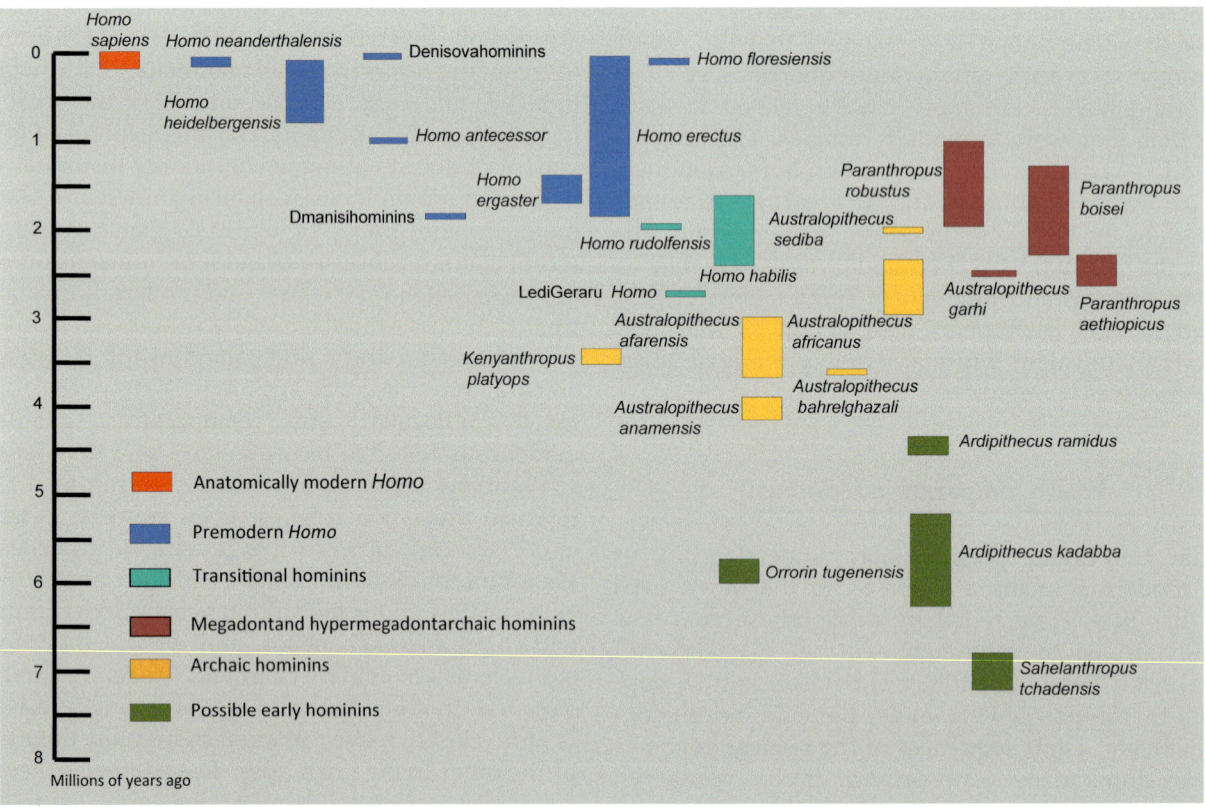

FIGURE 30.1 The temporal context of the formal taxa and site samples recognized in a deliberately speciose interpretation of the hominin fossil record. The vertical axis is geological time in Ma. Morphology is used to distribute the taxa across the horizontal axis, with larger-brained taxa on the left and those with large chewing teeth on the right. The widths of the columns are the same despite large differences in the quantity and quality of the fossil record from one column to another. The grade categories are described in the text.

in, or migrated into, that region, just as the equivalent observed LAD of a taxon is almost certainly earlier than the time the taxon actually became extinct, or emigrated from that region. How much earlier than the observed FAD the actual origination or migration occurred, and just how much later than the observed LAD of a taxon the actual extinction, or emigration occurred, is determined by many factors. These include technical considerations such as dating errors, but a much more potent factor is the nature of the relevant mammalian fossil record before and after the current observed FAD and LAD. The problem is the old adage "absence of evidence is not evidence of absence." It is yet another sampling problem, because at most sites early hominins are such a rare component of the mammalian faunal record (c.1—2%) that researchers need to find a substantial number of nonhominin mammalian fossils (at least several hundred) without finding *any* evidence of a particular hominin species, before it can be reasonably assumed that hominin was not part of the faunal assemblage being sampled at that location.

30.4 Hominin Taxonomy

The definition of taxonomic categories is a vexed issue. With respect to species, Smith (2009) usefully divides contemporary species concepts into those that emphasize the processes involved in the generation and maintenance of species and those that use patterns of variation to recognize species in the fossil record. The three main definitions in the process category are the biological species concept (BSC), the evolutionary species concept (ESC), and the recognition species concept (RSC). The BSC stresses reproductive isolation, so it is not a "stand-alone" species definition. The ESC is an attempt by Simpson (1961) to add a temporal dimension to the BSC, thus he suggested that under the ESC, a species is "an ancestral-descendant sequence of populations evolving separately from others and with its own evolutionary role and tendencies." The third concept in the process-related category, the RSC, shuns reproductive isolation and instead focuses on the processes that promote interbreeding. Paterson (1985) refers to these processes as the "specific mate recognition system" (or SMRS), and as long as a species' SMRS signal fossilizes the RSC can potentially be applied to the fossil record. The two main pattern-based species concepts are the phenetic species concept (PeSC) and the phylogenetic species concept (PySC). The PeSC gives equal weight to all aspects of the phenotype by assembling a matrix of characters and then uses multivariate analysis to detect clusters of individual specimens that share similar phenotypes. The PySC differs from the PeSC

by emphasizing only the diagnostic aspects of the phenotype. According to Nixon and Wheeler (1990), a species defined under the PySC is "the smallest aggregation of populations diagnosable by a unique combination of character states." In practice, most human evolution researchers use a version of the PySC in the sense that they search for the smallest cluster of individual organisms that is "diagnosable" on the basis of the morphology preserved in the fossil record.

One of the many factors paleoanthropologists must take into account when trying to identify species in the paleontological record is that the hominin fossil record is predominantly confined to the remains of hard tissues (ie, bones and teeth). We know from living animals that many uncontested species are difficult to distinguish using bones and teeth (eg, Old World monkey species of the genus *Cercopithecus*—see Manaster, 1979), so there are sound, logical, reasons to suspect that a fossil record that is confined to hard tissues, which because of differential preservation focuses on cranial and dental evidence, is always likely to underestimate the number of species.

30.5 The Case for Grades Within the Hominin Clade

A genus is even more difficult than a species to identify in the fossil record, but our view is that a genus should be both a clade (see previous discussion) *and* a grade (see later discussion). One of us is skeptical about our ability to recognize subclades (aka monophyletic groups) within the hominin clade, so our genus-level distinctions are based more on evidence about grades than on hypotheses about the details of the branching structure of the hominin clade.

Taxa are put in the same grade if their morphology suggests they share important behavioral adaptations, such as diet, posture, and locomotion. No store is set on how they came by those behaviors. It does not matter whether they were inherited from a recent common ancestor or were the result of convergent evolution. The judgment about how different two diets, or two locomotor strategies, have to be before the taxa being scrutinized are considered to belong to different grades is a subjective one, but until we can be sure we have access to ways of generating reliable hypotheses about the presence of subclades within the hominin clade, the grade concept helps sort taxa into broad functional categories.

In this review we use five grades, "anatomically modern *Homo*" (the grade that includes modern humans), "premodern *Homo*," "transitional hominins," "archaic hominins," and "megadont and hypermegadont archaic hominins" (Fig. 30.1), for taxa we are confident should be included within the hominin clade. In the following sections we describe each grade in terms of its characteristic regional morphology, and within each grade we describe the species in the historical order the taxa were recognized, not according to their estimated first appearance datum. In Appendix 1 we summarize currently available estimates for endocranial volume and body mass, two variables particularly relevant for a publication that focuses of the evolution of the nervous system.

30.6 Criteria for Including Taxa Within the Hominin Clade

One of the major developments in human evolution in the past two decades has been the discovery of fossil evidence that dates between 7 and 4 Ma. As we will see below all of that evidence has been interpreted by its discoverers as being a potential common ancestor of all later hominins. But if you subscribe to a "bushy" interpretation of hominin evolution and are prepared to contemplate the possibility of homoplasy, which is when similar morphology can evolve more than once, so shared morphology does not always mean shared recent evolutionary history, then if the fossil evidence for a taxon is relatively fragmentary, it may not always be possible to label a taxon as anything other than a possible or probable early hominin. We also need to entertain the possibility that such a taxon is neither a hominin nor a panin. Remember that most clades in the TOL have no living representatives, so a taxon between 7 and 4 Ma *could* belong to an extinct clade that is more closely related to modern humans and chimpanzees/bonobos than to gorillas, or to an extinct clade that is more closely related to modern humans than to chimpanzees/bonobos. Labeling *any* taxon as the earliest hominin, especially a taxon with a fossil record that samples only a few individuals, is always going to be a risky enterprise.

Four species, namely *Sahelanthropus tchadensis*, *Orrorin tugenensis*, *Ardipithecus kadabba*, and *Ardipithecus ramidus*, have been put forward as primitive members of the hominin clade. The reasons for including the *c*.7 Ma remains assigned to *S. tchadensis* (Brunet et al., 2002; Guy et al., 2005), the *c*.6 Ma remains assigned to *O. tugenensis* (Senut et al., 2001), the *c*.5.8–5.2 Ma remains assigned to *Ar. kadabba* (Haile-Selassie, 2001, 2004), and the *c*.4.5–4.4 Ma remains assigned to *Ar. ramidus* (White et al., 1994, 2009; White, 2010) in the hominin clade differ according to which anatomical regions are represented in their fossil records.

Claims that these taxa should be included in the hominin clade rest on three common lines of evidence. The first involves a reduction in size and a change in morphology of the canines, which is linked with the partial or complete loss of upper canine/lower first premolar honing, and reduced canine sexual dimorphism. The second involves the location and orientation of the foramen magnum and inferences about posture and gait. The third involves features of the pelvis and other preserved postcranial elements that imply a dependence on bipedalism.

The canine morphology that *Ar. ramidus* and *S. tchadensis* share with later hominins is the most convincing evidence in support of their hominin status. But it is important to recognize that during the late Miocene a number of Eurasian hominids (eg, *Oreopithecus*, *Ouranopithecus*, and *Gigantopithecus*) also developed smaller canines and a reduction in the extent to which the upper canine and the lower first premolar were against each other (aka honing). The anterior location and horizontal orientation of the foramen magnum in modern humans and later hominins compared to the extant great apes has been assumed to relate to the upright posture and bipedal locomotion of the former. However, comparisons with other primates suggest these features may also be linked with differences in head carriage and relative brain size rather than uniquely with having an upright trunk (Strait, 2001). Also the differences in the position and orientation of the foramen magnum seen in bonobos and chimpanzees, and the overlap in those variables between bonobos, *Sahelanthropus*, and *Ardipithecus*, suggest that we should exercise caution before assuming that a relatively anteriorly positioned and more horizontal foramen magnum is linked exclusively with the adoption of habitual bipedalism. The postcranial evidence for bipedalism in *Ar. kadabba* mainly involves the morphology of a toe bone—a proximal pedal phalanx presumed to belong to *Ar. kadabba*, but from an older geological horizon and with no associated craniodental remains—whereas in *O. tugenensis* the evidence mainly involves the morphology of the proximal femur. The case for the femur being that of a committed biped is stronger than the case for the phalanx. The claim that *Ar. ramidus* was a biped is mainly based on highly speculative inferences about the presence of curved lower back (aka lumbar lordosis) and a few features of the pelvis and foot, but they are based on questionable reconstructions and/or they involve characters whose link to habitual bipedalism has yet to be convincingly demonstrated. Researchers who support *S. tchadensis*, *O. tugenensis*, *Ar. kadabba*, and *Ar. ramidus* being given hominin status do so on the assumption that within the great apes, canine honing and bipedalism are *confined* to the hominin clade. But even if all hominins are bipedal and lack canine honing, the converse proposition, that among the extinct and extant members of the great ape clade bipedalism and the loss of canine honing are confined to the hominin clade, is not a logical corollary.

So, how strong are the cases for each of the four taxa being a hominin? The argument for including *Ar. kadabba* in the hominin clade at the present time is weak because it has a sparse fossil record, its teeth are apelike, and there is not enough evidence to be sure it is a committed biped. As for *O. tugenensis*, although the external morphology of the proximal femur is consistent with it being bipedal, the evidence from the internal morphology of the femoral neck is equivocal. The morphological evidence that *S. tchadensis* and *Ar. ramidus* should be included in the hominin clade is stronger but not compelling. Their age could also be against them being hominins. In the case of *Ar. ramidus*, if both it and the *c*.4.2 Ma *Australopithecus anamensis* (see later discussion) are lineal ancestors of later hominins, as its discoverers claim, then there is little time for the cranial and post-cranial morphology of the former to evolve into the latter. Moreover, if the Burtele foot (see later discussion) belongs to *Ar. ramidus* then that species would have lived alongside *Au. anamensis*. Thus, for all of these and other reasons, one of us has referred to *S. tchadensis*, *O. tugenensis*, *Ar. kadabba*, and *Ar. ramidus* as "possible hominins" (eg, Wood, 2010), and this is how we treat them in this review.

30.6.1 Archaic Hominins

This grade includes all the unambiguously hominin taxa not included in *Homo* and *Paranthropus*. All archaic hominins, no matter what their absolute size, have relatively larger chewing teeth (premolars and molars) and a more primitive postcranial skeleton than premodern *Homo*. They were all likely to be predominantly bipedal, but unlike premodern *Homo* the anatomy of their upper limb suggests that they were still effective and regular climbers. What is known of the life history of archaic hominins suggests that it is more like that of the extant apes than modern humans.

The first species to be recognized in this grade was *Australopithecus africanus* Dart 1925. The type specimen, Taung 1, a juvenile skull with a partial natural endocast, was recovered in 1924 from the Limeworks at Taung (formerly Taungs) now in South Africa. Most of the other fossil evidence for *Au. africanus* comes from two caves, Sterkfontein and Makapansgat, with other evidence coming from Gladysvale cave. The temporal range of *Au. africanus* is *c*.3–2.4 Ma. The cranium, mandible, and the dentition are well sampled, but the postcranial skeleton, and particularly the vertebral column, are less well represented in the fossil record, and many of the fossils that do exist have been crushed and deformed by rocks falling on the bones before

they were fully fossilized. The picture that has emerged from morphological and functional analyses suggests that although *Au. africanus* was capable of walking bipedally, it was probably more arboreally adapted than other archaic hominin taxa such as *Australopithecus afarensis* (see later discussion). It had relatively large chewing teeth and apart from the reduced canines, the skull is relatively apelike. Its mean endocranial volume is *c*.460 cm^3. The Sterkfontein evidence suggests that males and females of *Au. africanus* differed substantially in body size, but probably not to the degree they did in *Au. afarensis*.

The taxon *Au. afarensis* Johanson et al. 1978 is only known from East Africa, unless *Australopithecus bahrelghazali* from Chad proves to be belong to *Au. afarensis* (ie, it is conspecific). The type specimen is an adult mandible, LH 4, recovered in 1974 from Laetoli, Tanzania, but the largest contribution to the *Au. afarensis* fossil record (ie, its hypodigm) comes from Hadar in Ethiopia and from other sites in Ethiopia (Belohdelie, Brown Sands, Dikika, Fejej, Maka, White Sands, and Woranso-Mille) and from sites in Kenya (Allia Bay, Koobi Fora, Tabarin, and West Turkana). The temporal range of *Au. afarensis* is *c*.3.8–3 Ma (*c*.4–3 Ma if the presence of *Au. afarensis* is confirmed at Belohdelie and Fejej). The *Au. afarensis* hypodigm includes a well-preserved skull, other skulls, partial and fragmented crania, many lower jaws, sufficient limb bones to be able to estimate stature and body mass, and a specimen A.L.-288 ("Lucy") that preserves just less than half of the skeleton of a small adult female. Although most individual body mass estimates range from *c*.30–45 kg, some individuals are larger (*c*.60 kg) and some have endocranial volumes (*c*.550 cm^3) that come close to those of *Homo habilis*. The incisors of *Au. afarensis* are smaller than those of extant chimps/bonobos, but the premolars and molars are relatively larger. Comparative evidence suggests that the forelimbs of A.L.-288 are substantially longer than those of a modern human of similar stature. The discovery at Laetoli of several trails of fossil footprints provided very graphic direct evidence that at least one contemporary hominin, presumably *Au. afarensis*, but possibly *Kenyanthropus platyops*, was capable of bipedal locomotion, but the Laetoli prints are less modern human–like than *c*.1.5 Ma footprints from Koobi Fora presumed to be made by a premodern *Homo*. The upper limb of *Au. afarensis*, especially the hand and the shoulder girdle, retains morphology that most likely reflects a significant element of climbing. Although a recent study argues that sex differences in the bones and teeth (hard tissues) in this taxon are relatively poorly developed, most researchers interpret it as showing a substantial level of sexual dimorphism.

The taxon *Au. anamensis* Leakey et al. 1995 is presently restricted to East Africa. The type specimen, KNM-KP

29281, was recovered in 1994 from Kanapoi, Kenya. Other sites contributing to the hypodigm are Allia Bay, also in Kenya, and the Middle Awash study area, Ethiopia. The temporal range of *Au. anamensis* is *c*.4.2—3.9 Ma. The fossil evidence consists of jaws, teeth, and postcranial elements from the upper and lower limbs. Most of the differences between *Au. anamensis* and *Au. afarensis* relate to details of the teeth and jaws. In some respects the teeth of *Au. anamensis* are more primitive than those of *Au. afarensis* (eg, the asymmetry of the premolar crowns and the relatively simple crowns of the deciduous first mandibular molars), but in others (eg, the low cross-sectional profiles and bulging sides of the molar crowns) they show some similarities to *Paranthropus*. The upper limb remains are similar to those of *Au. afarensis*, and a tibia attributed to *Au. anamensis* has features associated with bipedality.

Australopithecus bahrelghazali Brunet et al. 1996 is most likely a regional variant of *Au. afarensis*. Nonetheless, the Chad discovery is significant because it substantially extended the geographical range of early probable hominins beyond sites in and around the Rift Valley in East Africa. It reminds us that important events in human evolution (eg, speciation, extinction) may have been taking place well away from the very small (relative to the size of the African continent) region sampled by the existing early hominin sites.

The penultimate archaic hominin taxon to be recognized is *K. platyops* Leakey et al. 2001. The type specimen, KNM-WT 40000, a *c*.3.5—3.3 Ma relatively complete but distorted cranium, was found in 1999 at Lomekwi, West Turkana, Kenya. The main reasons Meave Leakey and her colleagues did not assign this material to *Au. afarensis* are that its lower face is less prominent, the origin of the cheek-bone is further forward and its surface is flat and vertical, the molars are relatively small-crowned but with thick enamel, and the first upper molar crown is unusually small compared with the size of the crowns of the adjacent premolar and the third molar. Despite this unique combination of facial and dental morphology, some suggest that the new taxon is not justified because they claim that KNM-WT 40000 is a distorted *Au. afarensis* cranium, but this explanation is not consistent with the shape of the face of KNM-WT 40000, nor does it account for the overall small size of the postcanine teeth (ie, premolars and molars).

The most recent archaic hominin taxon to be recognized is *Australopithecus sediba* Berger et al. 2010, which was recovered from Malapa cave in the Blauuwbank Valley in southern Africa. The initial discoveries consisted of two associated skeletons; MH1, a juvenile, was made the type specimen. Although the lower limb of *Au. sediba* is like those of other archaic hominins, Berger et al. (2010) claim that aspects of its cranial (eg,

more globular neurocranium, gracile face), mandibular (eg, more vertical chin profile), dental (eg, simple canine crown, small anterior and postcanine tooth crowns), and pelvic morphology (eg, a broad, buttressed, iliac blade, and short ischium) are only shared with early and later *Homo* taxa. But the immaturity of MH1, plus the many overall similarities to *Au. africanus*, suggests the Malapa hominins may sample *Au. africanus* at a later stage of its evolution than is the case for the existing fossil evidence from Sterkfontein and Makapansgat. The demonstration that the Malapa hominins differ from the Sterkfontein and Makapansgat samples of *Au. africanus* does not exclude the possibility that the three samples were drawn from the same fossil taxon, but the finding that the diet of *Au. sediba* differed from that of *Au. africanus* (Henry et al., 2012) supports the hypothesis that the hominins found at Malapa sample a taxon other than *Au. africanus*.

30.6.2 Megadont and Hypermegadont Archaic Hominins

This grade includes hominin taxa conventionally included in the genus *Paranthropus*, plus *Australopithecus garhi*. As the term megadont suggests, the criterion for inclusion in this grade are postcanine tooth crowns that are either large, both absolutely and relatively (eg, in relation to the anterior dentition and to estimates of body mass), or excessively large (ie, hypermegadont). The genus *Paranthropus*, into which *Zinjanthropus* and *Paraustralopithecus* are subsumed, was reintroduced when cladistic analyses suggested that the first three species discussed in this section most likely formed a separate clade. The postcanine teeth of *Paranthropus robustus* (see later discussion) are not much larger than those of *Au. africanus*, but those of the East African taxa are substantially larger, hence they are referred to as hypermegadont. The enamel covering the crowns of the teeth of all of the taxa in this grade is thick, with the enamel of the two *Paranthropus* taxa from East Africa, *Paranthropus boisei* and *Paranthropus aethiopicus*, being exceptionally thick.

The taxon *P. robustus* Broom 1938 was established to accommodate an associated skeleton, the type specimen TM 1517, which was recovered in 1938 from the southern African site of Kromdraai B. The other sites contributing to the evidence for *P. robustus*, Swartkrans, Gondolin, Drimolen, and Cooper's caves, are all in the Blauuwbank Valley near Johannesburg, South Africa. The dentition is well represented, but many of the cranial remains are crushed or distorted and there are few associated postcranial remains. The temporal range of the taxon is *c*.2.0—1.5 Ma. The brain, face, and chewing teeth of *P. robustus* are on average larger than those of

Au. africanus, yet the incisor teeth are smaller. The morphology of the pelvis and the hip joint is much like that of *Au. africanus*; *Paranthropus robustus* was most likely capable of bipedal walking, but it was probably not an obligate biped. It has been suggested that the thumb of *P. robustus* would have been capable of the type of grip necessary for the manufacture of simple stone tools, but this claim has been disputed by some researchers.

In 1959 Louis Leakey suggested that a new genus and species, *Zinjanthropus boisei* Leakey, 1959, was needed to accommodate OH 5, a cranium belonging to a young adult male, recovered in 1959 from Bed I, Olduvai Gorge, Tanzania. A year later John Robinson suggested that *Z. boisei* be subsumed into the genus *Paranthropus* as *P. boisei*, and in 1967 Phillip Tobias suggested it should be subsumed into *Australopithecus* as *Australopithecus boisei*. In this review we refer to it as *P. boisei* (Leakey, 1959) Robinson 1960. Additional fossils from Olduvai Gorge were subsequently added to the new species, as was fossil evidence from Peninj, Omo Shungura, Konso, Koobi Fora, Chesowanja, and West Turkana, all of which are sites in East Africa. *Paranthropus boisei*, which has a temporal range of *c*.2.3—1.4 Ma, has a comprehensive craniodental fossil record, comprising several skulls and well-preserved crania, and many mandibles and isolated teeth. The size range suggests a substantial degree of body size sexual dimorphism, despite the evidence for modest canine sexual dimorphism. *Paranthropus boisei* is the only hominin to combine a wide, flat face, massive premolars and molars, small anterior teeth, and a modest endocranial volume (Appendix 1). The body of the mandibles of *P. boisei* is larger and wider, and the tooth crowns grow at a faster rate, than in any other early hominin. For a long time, there was no postcranial evidence that could, with certainty, be attributed to *P. boisei*, but a fragmentary associated with upper limb skeleton from Olduvai Gorge (Domínguez-Rodrigo et al., 2013) and a better preserved associated with upper limb skeleton from Koobi Fora (Richmond et al., 2011) almost certainly belong to that taxon. Some of the postcranial fossils from Bed I at Olduvai Gorge currently attributed to *H. habilis* (see later discussion) may also belong to *P. boisei*.

The taxon *Paranthropus aethiopicus* (Arambourg and Coppens, 1968) was introduced as *Paraustralopithecus aethiopicus* by Chamberlain and Wood (1985) to accommodate Omo 18.18 (or 18.1967.18) a toothless adult mandible recovered in 1967 from Omo Shungura in Ethiopia. The hypodigm is small, but it includes a well-preserved adult cranium (KNM-WT 17000) and mandibles (eg, KNM-WT 16005) from West Turkana together with isolated teeth from Omo Shungura—some also assign the Omo 338y-6 cranium to this taxon. No published postcranial fossils have been assigned to

P. aethiopicus, but the proximal end of a leg bone, a tibia, from Laetoli may belong to it. The temporal range of *P. aethiopicus* is *c*.2.5—2.3 Ma. *Paranthropus aethiopicus* is similar to *P. boisei* except that the face is more projecting, the cranial base is flatter, the anterior teeth are larger, and the postcanine tooth crowns are not so large or so morphologically specialized as those of *P. boisei*.

The most recent addition to the hypermegadont archaic hominin hypodigm is *Au. garhi* Asfaw et al. 1999, which was introduced to accommodate specimens recovered in 1997 from Aramis in the Middle Awash study area, Ethiopia. The evidence for *Au. garhi* is presently restricted to fossils recovered from the Hata Member in the Middle Awash study area, Ethiopia. The type specimen, the *c*.2.5 Ma BOU-VP-12/130, combines a primitive-looking cranium with large-crowned postcanine teeth with, unusually, a lower first premolar crown that is larger than that of the lower second premolar. However, unlike in *P. boisei*, the incisors and canines are large and the enamel apparently lacks the extreme thickness seen in the latter taxon. A partial skeleton with a long femur and forearm was found nearby, but is not associated with the type cranium and it has not been formerly assigned to *Au. garhi*. If further evidence suggests that the type specimen of *P. aethiopicus*, Omo 18.18, cannot be distinguished from mandibles that match the *Au. garhi* cranium, then *P. aethiopicus* would have priority instead of *Au. garhi*.

30.6.3 Transitional Hominins

For the purposes of this review *H. habilis* and *Homo rudolfensis* are retained within *Homo*, but they are treated separately from the premodern *Homo* grade. This is because the fossils assigned to these taxa show a mix of morphology some of which is seen in premodern *Homo* and some in archaic hominins.

The taxon *H. habilis* Leakey, Tobias, and Napier 1964 was introduced for fossils recovered from Olduvai Gorge, Tanzania. The rest of *H. habilis*, mostly cranial and dental evidence, consists of other fossils found at Olduvai Gorge and of fossils from Ethiopia (Omo Shungura and Hadar) and Kenya (Koobi Fora and perhaps Chemeron), and some researchers claim there is evidence of *H. habilis* in southern Africa at Sterkfontein, Swartkrans, and Drimolen. The endocranial volume of *H. habilis* ranges from *c*.500 cm^3 to *c*.700 cm^3—but a reassessment of the endocranial volume of OH 7 suggests that it may be closer to 800 cm^3. All *H. habilis* crania are wider at the base than across the vault, but the face is broadest in its upper part. The only limb fossils that can be assigned to *H. habilis* with confidence are the postcranial bones associated with the type specimen, OH 7, and the associated skeleton, OH 62: isolated postcranial bones from Olduvai Gorge assigned to *H. habilis* (eg, OH 10) could also belong to *P. boisei*. If OH 62 is representative of *H. habilis* the skeletal evidence suggests that its limb proportions and locomotion would have allowed it to walk bipedally (ie, it was a facultative and not an obligate biped). Its wrist bones were archaic hominin-like, and the curvature and well-developed muscle markings on the finger bones of OH 7 indicate that *H. habilis* was capable of powerful grasping. The size of the mandible and postcanine teeth suggest that the diet of *H. habilis* was as mechanically demanding as that of archaic hominins.

Some researchers suggest the transitional hominin grade contains a second taxon, *H. rudolfensis* (Alexeev, 1986) sensu Wood 1992, but not all researchers are convinced the scale and nature of the variation within early *Homo* justifies the recognition of two taxa. The temporal range of *H. rudolfensis* would be *c*.2.0–1.8 Ma if its fossil record is limited to the KNM-ER 1470 cranium from Koobi Fora, plus other Koobi Fora fossils (eg, KNM-ER 1482, 1801, 1590, 3732, 60000, 62000, 62003). Compared to *H. habilis*, the absolute size of the brain case in *H. rudolfensis* is a little greater, and its face is widest in its midpart, whereas the face of *H. habilis* is widest superiorly, and the dental arcades are differently shaped. Despite the mean absolute size of the *H. rudolfensis* brain (*c*.725 cm^3), when it is related to estimates of body mass based on orbit size, the brain is not so much larger than those of the archaic hominins. At present, no limb remains can be reliably linked with *H. rudolfensis*. As with *H. habilis*, the size of the mandible and postcanine teeth suggest that its diet made similar mechanical demands as those of the archaic hominins.

Currently the earliest evidence for *Homo* is LD 350-1, the left side of adult hominin mandible, found in 2.8–2.75 Ma sediments in the Lee Adoyta region of the Ledi-Geraru research area in the Afar region of Ethiopia (Villmoare et al., 2014). Although its sloping chin region and thick buttress inside the chin (the inferior transverse torus) resemble the mandibles of archaic hominins (see later discussion), the consistent height and reduced thickness (robusticity) of the mandibular body are reminiscent of *Homo* mandibles.

30.6.4 Premodern *Homo*

This very broad grade grouping includes *Homo* species that lack the derived and distinctive size and shape of the modern human cranium and the gracility of the modern human postcranial skeleton, but all of the included taxa have limb proportions that resemble those of modern humans, and postcranial morphology that is consistent with obligate bipedalism. The teeth are generally a little larger and the jaws are somewhat more robust than those of anatomically modern *Homo*. One of the reasons why this is such a broad grade grouping is the wide range of observed endocranial volume (*c*.600 cm^3 to *c*.>1300 cm^3).

The first fossil taxon to be recognized in the premodern *Homo* grade is *Homo neanderthalensis* King 1864, whose temporal range is *c*.200–28 ka (but if the fossils from Sima de los Huesos are included then it is *c*.>400–28 ka). The type specimen, the Neanderthal 1 skeleton, was found in 1856 at the Kleine Feldhofer Grotte in Elberfield, Germany, and fossil evidence for *H. neanderthalensis* has since been found in Europe as well as in the Near East, the Levant and Western and Central Asia. The distinctive features of the *H. neanderthalensis* cranium include thick, double-arched brow ridges, a face that projects anteriorly in the midline, a large nose, laterally projecting and rounded parietal bones, and a rounded, posteriorly-projecting, occipital bone. Mandibular and dental features include a space between the back of the third molar and the anterior border of the ramus of the mandible, distinctively high incidences of some nonmetrical mandibular and dental traits and thinner tooth enamel than in modern humans. The average endocranial volume of *H. neanderthalensis* is larger than that of living modern humans. Postcranially, *H. neanderthalensis* individuals were stout with a broad rib cage, a long clavicle, a wide pelvis, and limb bones that are generally robust, with large joint surface areas. The distal extremities tend to be short compared to most modern *Homo sapiens*, and the generally well-marked muscle attachments and robust long bone shafts point to a strenuous lifestyle. Some researchers restrict *H. neanderthalensis* to fossils from Europe and the Near East that used to be referred to as "Classic" Neanderthals, but others interpret the taxon more inclusively and include fossil evidence that is generally older and less distinctive (eg, Steinheim, Swanscombe, and from the Sima de los Huesos). The first DNA recovered from a fossil hominin was from the type specimen of *H. neanderthalensis* (Krings et al., 1997), and more recently researchers have been able to reconstruct the nuclear genome to high levels of precision and accuracy (Green et al., 2010; Prüfer et al., 2014).

The next fossil hominin taxon in the premodern *Homo* grade to be discovered was *Homo erectus* (Dubois, 1893) Weidenreich 1940. Its temporal range is conservatively *c*.1.8 Ma to *c*.140 ka. The initial discovery at Kedung Brubus on Java was made in 1890, but the type specimen was recovered in 1891 from Trinil, also in Java. *Homo erectus* is also known from other sites in Java (eg, Sambungmachan, Sangiran), and from sites in China (eg, Lantian, Zhoukoudian), Africa (eg, Melka Kunturé, Olduvai Gorge), and possibly the Caucasus (Dmanisi) (see later discussion). The fossil record of *H. erectus* is dominated by cranial remains, and while there is some postcranial evidence (mainly femora) the hand and foot is poorly represented in the fossil record.

Homo erectus crania have a low vault, a continuous supraorbital torus, a sharply angulated occipital region, and relatively thick inner and outer tables of the cranial vault. The body of the mandible is more robust than that of *H. sapiens*, it lacks a chin, and the mandibular tooth crowns are generally larger and the roots of the premolars more complex than those of modern humans. The limb proportions of *H. erectus* are similar to those of modern humans, but the shafts of the long bones of the lower limb are flatter (the femur from front to back and the tibia from side to side) relative to those of modern humans. Overall, the cranial vault, and the spongy layer in particular, is thicker in *H. erectus* than it is in modern humans. All of the dental and cranial evidence points to a modern humanlike diet for *H. erectus*, and the postcranial elements are consistent with an upright posture and obligate bipedalism. Those who support *Homo ergaster* Groves and Mazák 1975 as a species distinct from *H. erectus* point to features that are more primitive (eg, mandibular premolar root and crown morphology) and those that are less derived (eg, vault and cranial base morphology) than *H. erectus*. However, most researchers are not convinced there are sufficient consistent differences between the hypodigms of *H. ergaster* and *H. erectus* to justify the former being a separate species.

The hominins recovered from Dmanisi in Georgia were initially allocated to their own taxon, *Homo georgicus* (Gabounia et al., 2002), but overall the fossils from Dmanisi are most similar to *H. erectus*, and where there are differences they involve morphology inferred to be more primitive than that seen in *H. erectus*. Several of the authors of the original publication (eg, Lordkipanidze et al., 2013) no longer support a separate taxon for this material, but we treat the Dmanisi site sample separately in Appendix 1.

After *H. erectus*, the next taxon recognized within the premodern *Homo* grade was *Homo heidelbergensis* (Schoetensack 1908). Although the type specimen, Mauer 1, was an adult mandible found in 1907 in a sand quarry near Heidelberg, Germany, researchers have suggested that *c*.600–100 ka fossils from sites in Europe (eg, Petralona), the Near East (eg, Zuttiyeh), Africa (eg, Bodo, Kabwe), China (eg, Dali, Jinniushan, Xujiayao, Yunxian), and possibly India (Hathnora) have been included in *H. heidelbergensis*. What sets this material apart from *H. sapiens* and *H. neanderthalensis* is its cranial morphology and robusticity of the postcranial skeleton. Some *H. heidelbergensis* specimens have endocranial volumes as large as those of modern humans (Appendix 1), but the crania are always more robustly built than those of modern humans with a thickened occipital region and a projecting face and with large separate ridges above

the orbits. Researchers who see the African part of this hypodigm as distinctive refer it to a separate species, *Homo rhodesiensis*. Researchers who interpret the European component of the *H. heidelbergensis* hypodigm (eg, Sima de los Huesos) as already showing signs of the distinctive features of *H. neanderthalensis* would subsume *H. heidelbergensis* into *H. neanderthalensis*.

The taxon *Homo antecessor* Bermúdez de Castro et al. 1997 was introduced for hominins recovered from the Gran Dolina site at Atapuerca, Spain. The researchers who found the remains claim the combination of a modern humanlike facial morphology with large and relatively primitive tooth crowns and roots is not seen in *H. heidelbergensis*, and they see *H. antecessor* and not *H. heidelbergensis* as the likely recent common ancestor of *H. neanderthalensis* and *H. sapiens*.

The most recent taxon to be added to premodern *Homo* is *Homo floresiensis* Brown et al. 2004. It is currently only known from Liang Bua, a cave in Flores. Its published temporal range is *c*.74–17 ka, but it is more likely between 50 and 100 ka (Sutikna et al., 2016). The initial discovery and type specimen is LB1, an associated partial adult skeleton, but a second associated skeleton and close to 100 separate fossils representing up to 10 individuals have subsequently been recovered. *Homo floresiensis* displays a unique combination of early *Homo*-like cranial and dental morphology, a hitherto unknown suite of pelvic and femoral features, a small brain (*c*.417 cm^3) and a small body (25–30 kg) with an estimated stature of 1 m. When it was first described, researchers interpreted it as either a population of *H. erectus*, or of a *H. erectus*–like species, that had undergone endemic dwarfing, but others have suggested it could be a dwarfed *H. habilis*–like transitional hominin.

30.6.5 Anatomically Modern *Homo*

This grade includes hominin fossil evidence that is not significantly different from the morphology found in at least one regional population of modern humans. Presently, the earliest evidence of anatomically modern human morphology in the fossil record comes from Omo Kibish in Ethiopia (McDougall et al., 2005). Elsewhere in Africa there is evidence of crania (eg, Jebel Irhoud from North Africa, Laetoli 18 from East Africa, and Florisbad and the Cave of Hearths from southern Africa) that are generally more robust and archaic-looking than those of anatomically modern humans, yet they are not archaic enough to justify being allocated to *H. heidelbergensis* or to *H. neanderthalensis*. The gradual nature of the morphological change between *H. heidelbergensis* and anatomically modern humans makes setting the boundary between these two taxa challenging, but variation in the later *Homo* fossil record is too great to be accommodated in a single taxon (Mounier et al., 2009). Researchers who want to make a formal distinction between subrecent and living modern humans and fossils such as Florisbad and Laetoli 18 do so by referring the latter specimens to a separate species, *Homo helmei* Dreyer, 1935. Informally these specimens are sometimes referred to as "archaic *H. sapiens*."

30.7 Different Taxonomic Interpretations

We have deliberately presented a speciose interpretation of the hominin fossil record. This is in part because that is how we prefer to interpret the fossil evidence and also because combining exclusively defined taxa into a smaller number of more inclusive taxa is generally easier than subdividing inclusively defined taxa. However, our speciose interpretation looks restrained compared to interpretations (eg, Schwartz and Tattersall, 2003) that see evidence of new hominin species where even we see intraspecific variation.

Interpretations that recognize fewer species do so by either failing to recognize new species that are proposed, or by suggesting that new discoveries blur previously accepted distinctions between established species. The rejection of *K. platyops* by White (2003) is an example of the former, and the proposal of Lordkipanidze et al. (2013) to sink *H. habilis* into *H. erectus* is an example of the latter. The latter researchers claim that the amount of variation seen in the lateral profiles of crania from Dmanisi, which they interpret as a temporally and geographically constrained sample of a single hominin species, was much the same as the variation among crania assigned to *H. habilis* and *H. erectus*. Lordkipanidze et al. (2013) argued that this undermines the case for a species-level distinction between the hypodigms of *H. habilis* and *H. erectus*, and they proposed that the former taxon be sunk into the latter.

However, Lordkipanidze et al.'s argument assumes that (1) the distinction between *H. habilis* and *H. erectus* rested solely on the shape of the lateral cranial profile of the two taxa and (2) that the method they used to capture the shape of the lateral cranial profile was capable of discriminating among other early hominin taxa. In fact the method is so lacking in discrimination that it failed to distinguish the lateral profiles of crania in different hominin grades. For example, it was unable to distinguish the Dmanisi 4 cranium from a Neanderthal cranium, and it was unable to distinguish the lateral cranial profiles of crania belonging to *Au. africanus* (Sts 71 and Sts 5) from crania assigned to early African *H. erectus* (ie, KNM-ER 3733 and KNM-WT 15000). There may be good reasons to recognize fewer formal taxa or site samples than are

set out in Fig. 30.1 (Wood and Boyle, 2016), but the reasoning used by Lordkipanidze et al. (2013) to sink *H. habilis* into *H. erectus* does not pass muster.

30.8 Tempo and Mode

The study of macroevolution within the hominin clade is complicated by a number of factors unrelated to taxonomy. First, hominin remains are extremely rare in the fossil record, and most of the fossils that are found are frustratingly incomplete, so the same morphological regions are not always well enough represented in the fossil record to allow meaningful comparisons to be made among taxa (Wood and Boyle, 2016). Second, most hominin taxa, particularly early hominins, have no obvious ancestors, and in most cases ancestor–descendant sequences (fossil time-series) cannot be reliably constructed—two possible exceptions are mentioned later. Third, when error from many sources (eg, measurement, reconstruction, sampling, dating) is combined with the patchiness of the hominin fossil record, it can result in spurious conclusions about evolutionary patterns within the hominin clade. Finally, differences in scale can lead to differences in interpretation of the tempo and mode of macroevolutionary change. Gradual steady-rate evolutionary changes can appear to be punctuated, and punctuated changes can appear to be gradual. None of these complications are unique to the hominin clade, but many of these issues are exacerbated in paleoanthropology given the intense and detailed scrutiny our own lineage receives.

At one time, or another, most of the early hominins we have discussed have been presented as "the" ancestor of later hominins, but in our opinion only two pairs of taxa, *Au. anamensis* and *Au. afarensis* (Kimbel et al., 2006), and *P. aethiopicus* and *P. boisei* (Wood and Schroer, 2013), are plausible examples of an ancestor/descendant relationship (ie, they are examples of anagenesis). In the case of *Au. anamensis* and *Au. afarensis*, they are most likely time successive taxa within a single lineage with the Laetoli hypodigm of the former taxon intermediate between *Au. anamensis* and the Hadar hypodigm of *Au. afarensis*. This hypothesis has been given support by the discovery of fossil evidence from Woronso-Mille in Ethiopia that is both temporally and morphologically intermediate between *Au. anamensis* and *Au. afarensis* (Haile-Selassie et al., 2010). As for *P. aethiopicus* and *P. boisei*, although there are differences between the taxa (Suwa, 1988; Wood et al., 1994), they are consistent with the older, less-derived taxon being the ancestor of the younger more derived taxon. Indeed, some researchers have taken the view that the hypodigms of the two taxa are so similar they should both be included in *P. boisei* (Walker and Leakey, 1988). There is an obvious caveat attached to both of these suggested examples of anagenesis. This relates to the fact that because we know so little about the postcranial skeletons of *Au. anamensis*, *P. aethiopicus*, and *P. boisei*, we cannot eliminate the possibility that postcranial evidence could refute the hypothesis of an anagenetic relationship between these taxon pairs.

To investigate the tempo and mode of evolution within an early hominin species, (1) the taxon needs to be distinctive; (2) it must have a good, well-dated fossil record, and (3) the fossil sample needs to span enough time (at least several hundred thousand years) for any temporal trends to manifest. One, or more, of these criteria rules out most early hominin taxa, especially those only found in the southern African cave sites, and the only early hominin taxa that comply with these criteria are *Au. afarensis* and *P. boisei*. In both cases, when researchers tracked morphology that is well represented in the fossil record, essentially craniodental morphology, stasis was the predominant signal across approximately a million years of evolutionary history (Wood et al., 1994; Lockwood et al., 2000). In the case of a third taxon, *H. erectus*, researchers have reached different conclusions about the likelihood of stasis (Tobias, 1985; Wood et al., 1994; Ruff et al., 1997; Lockwood et al., 2000). Rightmire (1981) suggested that there was *no* consistent evidence of directional change in skull and tooth dimensions, but when Wolpoff (1984) analyzed a more narrowly defined sample of *H. erectus* mandibular, cranial, and dental features, he suggested there *was* evidence for evolutionary change within that taxon.

30.9 Temporal Trends in Hominin Brain Size

Only a few traits are known from a wide enough range of fossil hominins to allow for quantitative, rather than qualitative, comparisons to be made across long periods of time. One of them, endocranial volume, which is of particularly relevance to researchers interested in the evolution of the nervous system within the hominin clade, has been regularly pointed out as a classic example of a macroevolutionary trend (Haldane, 1949; Tobias, 1971; Jerison, 1973). Although current evidence on hominin brain size evolution points to a general increase over time (Holloway et al., 2004) and numerous hypotheses have been suggested as to the factors that caused this increase (eg, Clutton-Brock and Harvey, 1980; Martin, 1996; Dunbar, 1998), there is debate as to the tempo of brain size increase during hominin evolution. Some have argued for a gradual increase in hominin brain size over time (Lestrel and Read, 1973; Lestrel, 1976; but see Godfrey and Jacobs, 1981; Lee and Wolpoff, 2003), others for an increasing rate of change (Tobias, 1971; Holloway et al., 2004), and yet

another sees evidence of stasis followed by more rapid change (Ruff et al., 1997). A consistent theme of these interpretations is that there was a grade shift in endocranial volume *c.*1.8 Ma.

One problem with past studies is that they present endocranial volume values as if they had neither dating nor measurement error. When Du et al. (2016) analyzed a comprehensive data set on hominin brain size spanning the period from 3.5 to 0.5 Ma that takes account of dating and measurement error, they found that a gradual model of brain size received by far the strongest support.

30.10 Challenges to Conventional Wisdom

Popular perceptions of human evolution research oscillate between a sense that researchers "know what is going on," and one in which each new discovery sends researchers "back to the drawing board." The former perception is rooted in the worldview that more than 150 years of accumulating fossil evidence has resulted in a broad consensus into which most individual fossil discoveries, and most newly recognized hominin taxa, can be accommodated. The latter perception is fed by suggestions, mostly made to the media when new discoveries are announced, that this or that each new discovery challenges our basic understanding of hominin evolution because it "mixes and matches" morphology both within (eg, the foot) and among (eg, brain size and hand morphology) regions of the body in ways that are both novel and unpredictable.

The reality lies somewhere between these two extremes. It is true that discoveries such as those recently made in the Dinaledi chamber of the Rising Star system in southern Africa (Berger et al., 2015) probably represent a new species, *Homo naledi*, with a novel combination of regional morphology, but none of that morphology comes "out of left field." Moreover, the material is currently undated (Dirks et al., 2015), and until we know how old it is we cannot assess what impact it will have on our conventional wisdom. If the Dinaledi fossils turn out to be several millions of years old, then its particular mix of derived and primitive morphology will mean one thing, but if the remains are much more recent, *H. naledi* would be consistent with what might expect of a locally endemic hominin isolated from evolutionary trends that are being sampled in other regions of Africa. We must also remember that our knowledge of the early stages of human evolutionary history is based on fossil evidence recovered from sites that sample *c.*3% of the African continent. There is much we do not know, so we should not expect to have a comprehensive understanding of human evolutionary history. We should expect challenges to our conventional wisdom, but those challenges should be predictable within our current paradigm.

What *is* evident, however, even from our incomplete data set, is that over the last five million years, or so, of evolutionary history there is evidence of substantial taxic and functional diversity within the hominin clade. For one reason or another that diversity is mainly manifest in the fossil evidence of systems other than the nervous system. We do not have the CNS equivalent of information about bone and tooth microstructure, and although we all know intuitively that there must be a link between cognition and the evolution of hominin behaviors such as toolmaking, we are still ignorant about the details of any link. We are also woefully ignorant about how cognition relates to absolute brain size. Thankfully, much has been discovered but, also thankfully, much remains to be discovered about the evolution of the hominin nervous system.

Appendix 1

The formal and informal taxa recognized in a speciose interpretation of human evolution are listed in alphabetical order. The main sites for each are also listed in alphabetical order. The estimates of brain size and body mass are rounded up or down to the nearest whole number. The second age given for the first appearance datum (FAD) and last appearance datum (LAD) takes into account dating error. For more information about each taxon see Wood and Boyle (2016).

Ardipithecus kadabba Haile-Selassie, 2001

Key publications: Haile-Selassie, 2001; Haile-Selassie et al., 2004.

Main sites: Localities in Ethiopia. In the Middle Awash study area: Saitune Dora, Alayla, Asa Koma, and Digiba Dora in the Western Margin, and Amba East in the Central Awash Complex. Localities in the Gona Paleoanthropological Research Project study area are in the Western Ethiopian Escarpment and Asbole Dora.

First appearance date: 6.3/6.7 Ma. *Last appearance date*: 5.2/5.11 Ma.

Endocranial volume: No data available.

Body mass: No data available.

Ardipithecus ramidus White et al., 1995

Key publications: White et al., 1994, 1995, 2009; Semaw et al., 2005; Suwa et al., 2009.

Main sites: Localities in Ethiopia, and possibly also in Kenya (see later). In the Central Awash Complex in the Middle Awash study area: Aramis, Kuseralee Dora, Sagantole. Also, As Duma in the Gona Western Margin.

First appearance date: 4.51/4.6 Ma. *Last appearance date*: 4.3/4.262 Ma.

Endocranial volume: 300–350 cm^3 based on the digital reconstruction of a single cranium (Suwa et al., 2009).

Body mass: 32 kg. Univariate estimate based on the acetabular diameter of one individual (Grabowski et al., 2015).

Australopithecus afarensis Johanson et al., 1978

Key publications: Leakey, 1976; Johanson et al., 1978, 1982a,b; White et al., 1981; Clark et al., 1984; Kimbel et al., 1984, 1988, 1994, 2004; Suwa, 1990; Alemseged et al., 2005, 2006; Kimbel and Delezene, 2009; Haile-Selassie et al., 2010.

Main sites: Localities in Ethiopia in the Middle Awash study area [Belohdelie (tentative), Dikika, Hadar, Maka, and Omo (Usno)], and the Korsi Dora in the Woronso-Mille study area. Also, localities in Koobi Fora, Kenya, and Laetoli, Tanzania.

First appearance date: 3.7/3.89 Ma. *Last appearance date*: 3.0/2.9 Ma.

Endocranial volume: Mean = 433 cm^3 (N = 6). Range = 385–550 cm^3. Estimates based on endocast reconstruction, direct water displacement, volume formulas (Holloway et al., 2004), and filling a reconstructed cranium with mustard seed (Kimbel and Rak, 2010).

Body mass: Mean = 39 kg. Multivariate estimate based on 12 individuals. Range = 26–63 kg (Grabowski et al., 2015).

Australopithecus africanus Dart, 1925, 1948

Key publications: Dart, 1925, 1948; Berger, 1992; Clarke and Tobias, 1995; Lockwood and Tobias, 1999; Falk et al., 2000; Partridge et al., 2003; Toussaint et al., 2003; Moggi-Cecchi et al., 2006.

Main sites: All are in South Africa. The type locality, Taung, is in the North West Province. Gladysvale, Makapansgat (Members 3 and 4), Sterkfontein (Member 4, possibly Member 2), and possibly Jacovec cavern are in Gauteng Province.

First appearance date: 3.0/4.02 Ma. *Last appearance date*: 2.4/1.9 Ma.

Endocranial volume: Mean = 454 cm^3 (N = 9). Range = 391–568 cm^3. Estimates based on natural endocasts, endocast reconstruction, direct water displacement, and volume formulas in Holloway et al. (2004), and CT-generated virtual reconstructions of endocasts are in Neubauer et al. (2012).

Body mass: 31 kg. Multivariate estimate based on six individuals. Range = 23–43 kg (Grabowski et al., 2015).

Australopithecus anamensis Leakey et al., 1995

Key publications: Fleagle et al., 1991; Coffing et al., 1994; Leakey et al., 1995, 1998; Ward et al., 1999, 2001; White, 2002; White et al., 2006.

Main sites: Localities in the Omo-Turkana Basin in Kenya (Allia Bay and Kanapoi) and Ethiopia (Fejej), in the Middle Awash Study Area (Aramis and Asa Issie) and in the Woranso-Mille study area of Ethiopia.

First appearance date: 4.2/4.37 Ma. *Last appearance date*: 3.9/3.82 Ma.

Endocranial volume: No data available.

Body mass: 46 kg. Multivariate estimate based on a single individual (Grabowski et al., 2015).

Australopithecus bahrelghazali Brunet et al., 1996

Key publications: Brunet et al., 1995, 1996.

Main site: Bahr el Ghazal, Koro Toro, Chad.

First appearance date: 3.58/3.85 Ma. *Last appearance date*: 3.58/3.31 Ma.

Endocranial volume: No data available.

Body mass: No data available.

Australopithecus garhi Asfaw et al., 1999

Key publication: Asfaw et al., 1999.

Main site: Bouri, Ethiopia in the Middle Awash study area.

First appearance date: 2.5 Ma. *Last appearance date:* 2.45/2.488 Ma.

Endocranial volume: 450 cm^3. Estimate based on endocast reconstruction and direct water displacement of a single individual (Holloway et al., 2004).

Body mass: No data available.

Australopithecus sediba Berger et al., 2010

Key publication: Berger et al., 2010; Carlson et al., 2011.

Main site: Malapa, Gauteng, South Africa.

First appearance date: 1.98/2.05 Ma. *Last appearance date*: 1.98/1.91 Ma.

Endocranial volume: 420 cm^3. Minimum estimate based on CT-generated virtual reconstruction of the only available endocast (Carlson et al., 2011).

Body mass: Mean = 26 kg. Multivariate estimate based on three individuals. Range = 23–30 kg (Grabowski et al., 2015).

Denisova hominins

Key publication: Krause et al., 2010.

Main site: Denisova Cave, Siberia, Russia

First appearance date: 48.65/50.63 ka. *Last appearance date*: 29.2/28.84 ka.

Endocranial volume: No data available.

Body mass: No data available.

Dmanisi hominins

Key publications: Gabounia et al., 2002; Vekua et al., 2002; Lordkipanidze et al., 2007, 2013.

Main site: Dmanisi, Georgia.

First appearance date: 1.85 Ma. *Last appearance date*: 1.77 Ma

Endocranial volume: Mean = 645 cm^3 (N = 5). Range = 546–790 cm^3. Estimates based on endocast reconstruction, direct water displacement, volume formulas, and CT-data of crania (Grimaud-Hervé and Lordkipanidze, 2010; Lordkipanidze et al., 2013)

Body mass: 41 kg. Multivariate estimate based on a single individual (Grabowski et al., 2015).

Homo antecessor Bermúdez de Castro et al., 1997

Key publications: Bermúdez de Castro et al., 1997; Carbonell et al., 2005; Bermúdez de Castro et al., 2008.

Main site: Gran Dolina, Sierra de Atapuerca, Spain

First appearance date: 1.0/1.2 Ma. *Last appearance date*: 0.936 Ma.

Endocranial volume: The fragmentary nature of the cranial remains of *H. antecessor* precludes a reliable estimate of endocranial volume. However, Carbonell et al. (1995) suggest a cranial capacity of >1000 cm^3 based on the breadth of the frontal bone from one individual.

Body mass: No data available.

Homo erectus (Dubois, 1893) Weidenreich, 1940

Key publications: Dubois, 1893; Black, 1927; von Koenigswald, 1936; Weidenreich, 1936a,b, 1937, 1944, 1951; Leakey, 1961; von Koenigswald, 1968, 1975; Jacob, 1973; Rightmire, 1979; Santa Luca, 1980; Widianto and Grimaud-Hervé, 1993; Antón, 1999; Kaifu et al., 2008; Zaim et al., 2011.

Main sites: Localities in Asia including Java, Indonesia (Kedung Brubus, Mojokerto, Ngandong, Ngawi, Sambungmacan, Sangiran, and Trinil), and Zhoukoudian, China. Also Bouri from the Middle Awash study area in Ethiopia and Olduvai Gorge, Tanzania.

First appearance date: 1.81/1.85 Ma. *Last appearance date*: 27 ka.

Endocranial volume: Mean = 981 cm^3 (N = 36). Range = 656–1300 cm^3. Estimates based on endocast reconstruction, direct water displacement, volume formulas, filling a reconstructed cranium with teff seed, and CT-generated virtual reconstructions of crania and endocasts (Antón, 1997; Asfaw et al., 2002; Grimaud-Hervé and Lordkipanidze, 2010; Grimaud-Hervé et al., 2012; Holloway, 1983; Holloway et al., 2004; Indriati and Antón, 2010; Kaifu et al., 2011; Li and Etler, 1992; Macchiarelli et al., 2002; Spoor et al., 2007; Tyler, 1996; Weidenreich, 1943; Wu and Dong, 1982; Wu et al., 2011)

Body mass: Mean = 53 kg. Multivariate estimate based on eight individuals. Range = 49–62 kg (Grabowski et al., 2015).

Homo ergaster Groves and Mazák, 1975

Key publications: Leakey, 1974; Groves and Mazák, 1975; Leakey and Walker, 1985; Wood, 1994.

Main sites: Localities in Kenya (Koobi Fora and West Turkana), and perhaps also Swartkrans, Gauteng Province, South Africa.

First appearance date: 1.7/2.27 Ma. *Last appearance date*: 1.4/0.87 Ma.

Endocranial volume: Mean = 796 cm^3 (N = 4). Range = 715–909 cm^3. Estimates based on natural endocasts, endocast reconstruction, direct water displacement, and volume formulas (Begun and Walker, 1993;

Grimaud-Hervé and Lordkipanidze, 2010; Holloway et al., 2004).

Body mass: Mean = 51 kg. Combination of univariate and multivariate estimates based on three individuals. Range = 39–64 kg (Grabowski et al., 2015).

Homo floresiensis Brown et al., 2004

Key publications: Brown et al., 2004; Morwood et al., 2005; Sutikna et al., 2016.

Main site: Liang Bua, Flores, Indonesia

First appearance date: c.100 ka. *Last appearance date*: c.60 ka

Endocranial volume: 417 cm^3. Estimate based on CT-generated virtual reconstruction of the only endocast (Falk et al., 2005).

Body mass: 28 kg. Multivariate estimate based on a single individual (Grabowski et al., 2015).

Homo habilis sensu stricto Leakey et al., 1964

Key publications: Leakey et al., 1964, 1971, 1989; Leakey, 1974; White et al., 1981; Leakey and Walker, 1985; Johanson et al., 1987; Tobias, 1991; Wood, 1991; Grine et al., 1993; Schrenk et al., 1993; Kimbel et al., 1996; Grine, 2001; Blumenschine et al., 2003; Spoor et al., 2015.

Main sites: Localities in Kenya (Koobi Fora and West Turkana), Ethiopia [Hadar (tentative) and Omo Shungura], Tanzania (Olduvai), Gauteng Province, South Africa (Sterkfontein and Swartkrans), and Malawi (Uraha).

First appearance date: 2.35/2.6 Ma. *Last appearance date*: 1.65 Ma.

Endocranial volume: Mean = 611 cm^3 (N = 6). Range = 478–796 cm^3. Estimates based on near-complete and partial endocast reconstruction, direct water displacement, and volume formulas (Grimaud-Hervé and Lordkipanidze, 2010; Holloway et al., 2004; Tobias, 1991), as well as CT-generated virtual reconstructions (Benazzi et al., 2014; Spoor et al., 2015).

Body mass: Mean = 33 kg. Multivariate estimate based on two individuals. Range = 27–38 kg (Grabowski et al., 2015).

Homo heidelbergensis Schoetensack, 1908

Key publications: Schoetensack, 1908; Kokkoros and Kanellis 1960; de Lumley and de Lumley 1973; Rightmire, 1995; Mounier et al., 2009.

Main sites: Several localities in Europe: Mauer, Germany, Caune de l'Arago, France, Petralona, Greece, Montmaurin, France, plus Tighennif, Algeria.

First appearance date: 700 ka. *Last appearance date*: 100 ka.

Endocranial volume: Mean = 1158 cm^3 (N = 4). Range = 1057–1250 cm^3. Estimates based on several methods including CT-generated virtual reconstructions of crania (Conroy et al., 2000), filling a reconstructed cranium with millet seed (Ascenzi et al., 2000), regression based on internal biasterionic breadth (Rightmire, 1983), and volume formulas (Olivier and Tissier, 1975).

Body mass: Mean = 67 kg. Range = 59–81 kg. Carretero et al. (2004) derive these estimates by applying formulas from Ruff et al. (1997) to the femoral head diameters of five individuals.

Homo neanderthalensis King, 1864

Key publications: King, 1864; Schmitz et al., 2002; White et al., 2014.

Main sites: Many sites in Europe, Asia, and the Near East.

First appearance date: 130/197 ka. *Last appearance date*: 40/39 ka.

Endocranial volume: Mean = 1420 cm^3 (N = 29). Range = 1172–1740 cm^3. Estimates based on endocast reconstruction, direct water displacement, and volume formulas (Ruff et al., 1997; Holloway et al., 2004).

Endocranial shape: *H. neanderthalensis*, which is the only fossil hominin taxon to provide reliable evidence about endocranial shape and brain organization, shares relatively wide frontal lobes and laterally expanded parietals with *Homo sapiens* (Bruner, 2004; Bruner and Holloway, 2010). It has been suggested that these changes, which are not predicted by known allometric relationships, may be linked with cognitive function (Bruner and Holloway, 2010). The endocranial morphology of juvenile Neanderthal crania enables studies of the ontogeny of brain shape. Gunz et al. (2010) found that the different growth trajectories in Neanderthals and modern human brains result in differences in adult endocranial shape. The different growth trajectories apparently allow Neanderthals to achieve brain sizes as large, or larger, than those of modern humans without their brains becoming globularized like those of modern humans (Gunz et al., 2010).

Body mass: Mean = 76 kg. Range = 67–85 kg. Estimate based on femoral head diameter and bi-iliac breadth for 17 individuals (Ruff et al., 1997—supplementary information).

Homo rudolfensis (Alexeev, 1986) *sensu* Wood, 1992

Key publications: Alexeev, 1986; Groves, 1989; Wood, 1992; Leakey et al., 2012; Spoor et al., 2015.

Main sites: Koobi Fora, Kenya.

First appearance date: 2.0/2.09 Ma. *Last appearance date*: 1.95/1.78 Ma.

Endocranial volume: Mean = 793 cm^3 (N = 2). Range = 760–825 cm^3. Estimates based on endocast reconstruction, direct water displacement, and volume formulas (Grimaud-Hervé and Lordkipanidze, 2010; Holloway et al., 2004).

Body mass: No data available.

Homo sapiens

Key publications: Keith, 1912; Day, 1967; White et al., 2003; Schwartz and Tattersall, 2010.

Main sites: Many sites in the Old World and some in the New World.

First appearance date: 195/200 ka. *Last appearance date*: Present.

Endocranial volume: Mean = 1457 cm³ ($N = 56$). Range = 1090–1775 cm³. Estimates based on fossil *H. sapiens* taken by Ruff et al. (1997) and other sources, as reported in Holloway et al. (2004).

Body mass: Based on femoral head diameter and bi-iliac breadth, Ruff et al. (1997) provide a mean body mass estimate of 68 kg for 10 early Late Pleistocene *Homo* individuals and a mean estimate of 67 kg for 33 *H. sapiens* individuals dated to the early Upper Paleolithic.

Kenyanthropus platyops Leakey et al., 2001
Key publications: Leakey et al., 2001.
Main site: Lomekwi, West Turkana, Kenya.
First appearance date: 3.54/3.65 Ma. *Last appearance date*: 3.35 Ma.

Endocranial volume: Distortion of the KNM-WT 40000 cranium prevents a reliable numerical estimate, but Leakey et al. (2001, p. 436) suggest the cranial capacity of *K. platyops* is "within the range of *Australopithecus* or *Paranthropus*" based on glabella–opisthion arc length.

Body mass: No data available.

Orrorin tugenensis Senut et al., 2001
Key publications: Senut et al., 2001.
Main sites: Localities in the Lukeino Formation at Aragai, Cheboit, Kapcheberek, and Kapsomin, Tugen Hills, Baringo District, Kenya.
First appearance date: 6.0/6.14 Ma. *Last appearance date*: 5.7/5.52 Ma.

Endocranial volume: No data available.

Body mass: Mean = 36 kg. Multivariate estimate based on two individuals. Range = 30–43 kg (Grabowski et al., 2015).

Paranthropus aethiopicus (Arambourg and Coppens, 1968) Chamberlain and Wood, 1985
Key publications: Arambourg and Coppens, 1968; Walker et al., 1986; Wood and Chamberlain, 1987; Suwa, 1988.
Main sites: Omo, Ethiopia, West Turkana, Kenya, and Laetoli, Tanzania.
First appearance date: 2.66/2.73 Ma. *Last appearance date*: 2.3/2.23 Ma.

Endocranial volume: Mean = 443 cm³ ($N = 3$). Range = 410–491 cm³. Estimates based on endocast reconstruction, direct water displacement, and volume formulas (Brown et al., 1993; Holloway et al., 2004; Walker et al., 1986).

Body mass: A potential *P. aethiopicus* specimen has an estimated body mass of 32 kg, based on measurements of the proximal and distal ends of the femur (Grabowski et al., 2015).

Paranthropus boisei (Leakey, 1959) Robinson, 1960
Key publications: Leakey, 1959; Robinson, 1960; Leakey and Leakey, 1964; Tobias, 1967; Day, 1969; Carney et al.,

1971; Leakey and Walker, 1988; Suwa, 1988; Suwa et al., 1997; Kullmer et al., 1999; Wood and Constantino, 2007.

Main sites: Localities in Ethiopia (Konso and Omo), Kenya (Chesowanja, Koobi Fora, and West Turkana), Malawi (Malema), and Tanzania (Olduvai Gorge and Peninj).
First appearance date: 2.3/2.5. *Last appearance date*: 1.3/1.15 Ma.

Endocranial volume: Mean = 493 cm³ ($N = 8$). Range = 400–545 cm³. Estimates based on endocast reconstruction, direct water displacement, and volume formulas (Brown et al., 1993; Holloway et al., 2004).

Body mass: 46 kg. Univariate estimate based on the mediolateral width of the femur below the lesser trochanter from one individual (Grabowski et al., 2015).

Paranthropus robustus Broom, 1938, 1949
Key publications: Broom, 1938; Grine, 1989; Berger et al., 1995; Menter et al., 1999; Keyser, 2000; Keyser et al., 2000; Steininger and Berger, 2000; Thackeray et al., 2001; de Ruiter et al., 2009.

Main sites: Localities in Gauteng, South Africa: Cooper's, Drimolen, Kromdraai (Member 3); Sterkfontein (Member 5b); and Swartkrans (Members 1, 2, and 3). Gondolin in the North West Province, South Africa.
First appearance date: 2.0/2.27 Ma. *Last appearance date*: 1.0/0.87 Ma.

Endocranial volume: Mean = 536 cm³ ($N = 4$). Range = 465–650 cm³. Estimates based on direct water displacement of endocasts and volume formulas (Holloway et al., 2004; Schepers, 1946).

Body mass: Mean = 30 kg. Multivariate estimate based on four individuals. Range = 24–32 kg (Grabowski et al., 2015).

Sahelanthropus tchadensis Brunet et al., 2002
Key publications: Brunet et al., 2002; Zollikofer et al., 2005.
Main site: TM 266 locality, Toros-Menalla, Chad.
First appearance date: 7.2/7.43 Ma. *Last appearance date*: 6.8/6.38 Ma.

Endocranial volume: 365 cm³. Estimate based on a CT-generated virtual reconstruction of the type cranium (Zollikofer et al., 2005).

Body mass: No data available.

References

Alemseged, Z., Wynn, J.G., Kimbel, W.H., et al., 2005. A new hominin from the basal member of the Hadar formation, Dikika, and its geological context. J. Hum. Evol. 49, 499–514.

Alemseged, Z., Spoor, F., Kimbel, W.H., et al., 2006. A juvenile early hominin skeleton from Dikika, Ethiopia. Nature 443, 296–301.

Alexeev, V., 1986. The Origin of the Human Race. Progress Publishers, Moscow.

Andrews, P., Harrison, T., 2005. The last common ancestor of apes and humans. In: Lieberman, D.E., Smith, R.J., Kelly, J. (Eds.), Interpreting the Past: Essays on Human, Primate, and Mammal Evolution. Brill Academic Publishers, Boston, pp. 103–121.

Antón, S.C., 1997. Developmental age and taxonomic affinity of the Mojokerto child, Java, Indonesia. Am. J. Phys. Anthropol. 102, 497–514.

Antón, S., 1999. Cranial growth in Homo erectus: how credible are the Ngandong juveniles? Am. J. Phys. Anthropol. 108, 223–236.

Arambourg, C., Coppens, Y., 1968. Decouverte d'un australopithecien nouveau dans les Gisements de L'Omo (Ethiopie). S. Afr. J. Sci. 64, 58–59.

Ascenzi, A., Mallegni, F., Manzi, G., et al., 2000. A re-appraisal of Ceprano calvaria affinities with Homo erectus, after the new reconstruction. J. Hum. Evol. 39, 443–450.

Asfaw, B., White, T., Lovejoy, O., et al., 1999. Australopithecus garhi: a new species of early hominid from Ethiopia. Science 284, 629–635.

Asfaw, B., Gilbert, W.H., Beyene, Y., et al., 2002. Remains of Homo erectus from Bouri, middle Awash, Ethiopia. Nature 416, 317–320.

Begun, D., Walker, A., 1993. The endocast. In: Leakey, R., Walker, A. (Eds.), The Nariokotome Homo Erectus Skeleton. Harvard University Press, Massachusetts, pp. 326–358.

Benazzi, S., Gruppioni, G., Strait, D.S., Hublin, J.-J., 2014. Technical note: virtual reconstruction of KNM-ER 1813 Homo habilis cranium. Am. J. Phys. Anthropol. 153, 154–160.

Berger, L.R., 1992. Early hominid fossils discovered at Gladysvale cave, South Africa. South Afr. J. Sci. 88, 362.

Berger, L.R., Pickford, M., Thackeray, F., 1995. A Plio-Pleistocene hominid upper central incisor from the Cooper's site, South Africa. South Afr. J. Sci. 91, 541–542.

Berger, L.R., de Ruiter, D.J., Churchill, S.E., et al., 2010. Australopithecus sediba: a new species of Homo-like australopith from South Africa. Science 328, 195–204.

Berger, L.R., Hawks, J., de Ruiter, D.J., et al., 2015. Homo naledi, a new species of the genus Homo from the Dinaledi Chamber, South Africa. eLIFE 4, e09560.

Bermúdez de Castro, J.M., Arsuaga, J.L., Carbonell, E., et al., 1997. A hominid from the Lower Pleistocene of Atapuerca, Spain: possible ancestor to Neandertals and modern humans. Science 276, 1392–1395.

Bermúdez de Castro, J.M., Pérez-González, A., Martinón-Torres, M., et al., 2008. A new early Pleistocene hominin mandible from Atapuerca-TD6, Spain. J. Hum. Evol. 55, 729–735.

Black, D., 1927. On a lower molar hominid tooth from Chou kou tien deposit. Palaeontol. Sin. 7, 1–28.

Blumenschine, R.J., Peters, C.R., Masao, F.T., et al., 2003. Late Pliocene Homo and hominid land use from Western Olduvai Groge, Tanzania. Science 299, 1217–1221.

Bradley, B.J., 2008. Reconstructing phylogenies and phenotypes: a molecular view of human evolution. J. Anat. 212, 337–353.

Broom, R., 1938. The Pleistocene anthropoid apes of South Africa. Nature 142, 377–379.

Broom, R., 1949. Another new type of fossil ape-man. Nature 163, 57.

Brown, B., Walker, A., Ward, C.V., Leakey, R.E., 1993. New Australopithecus boisei calvaria from East lake Turkana, Kenya. Am. J. Phys. Anthropol. 91, 137–159.

Brown, P., Sutikna, T., Morwood, M.J., et al., 2004. A new small-bodied hominin from the Late Pleistocene of Flores, Indonesia. Nature 431, 1055–1061.

Bruner, E., 2004. Geometric morphometrics and paleoneurology: brain shape evolution in the genus Homo. J. Hum. Evol. 47, 279–303.

Bruner, E., Holloway, R.L., 2010. A bivariate approach to the widening of the frontal lobes in the genus Homo. J. Hum. Evol. 58, 138–146.

Brunet, M., Beauvilain, A., Coppens, Y., et al., 1995. The first australopithecine 2,500 kilometers west of the Rift Valley (Chad). Nature 376, 273–275.

Brunet, M., Beauvilain, A., Coppens, Y., et al., 1996. Australopithecus bahrelghazali, une nouvelle espece d'Hominide ancien de la region de Koro Toro (Tchad). Comptes Rendus Acad. Sci. 322, 907–913.

Brunet, M., Guy, F., Pilbeam, D., et al., 2002. A new hominid from the upper Miocene of Chad, central Africa. Nature 418, 145–151.

Carbonell, E., Bermúdez de Castro, J.M., Arsuaga, J.L., et al., 1995. Lower Pleistocene hominids and artefacts from Atapuerca–TD6 (Spain). Science 269, 826–832.

Carbonell, E., Bermúdez de Castro, J.M., Arsuaga, J.L., et al., 2005. An early Pleistocene hominin mandible from Atapuerca-TD6, Spain. Proc. Natl. Acad. Sci. U.S.A. 102, 5674–5678.

Carlson, K.J., Stout, D., Jashervili, T., et al., 2011. The endocast of MH1, Australopithecus sediba. Science 333, 1402–1407.

Carney, J., Hill, A., Miller, J.A., Walker, A.C., 1971. Late australopithecine from Baringo District, Kenya. Nature 230, 509–514.

Carretero, J.M., Arsuaga, J.L., Martínez, I., et al., 2004. Los humanos de la Sima de los Huesos (Sierra de Atapuerca) y la evolución del cuerpo en el género Homo. In: Baquedano, E., Rubio Jara, S. (Eds.), Miscelánea en Homenaje a Emiliano Aguirre III: Paleoantropología, vol. 3. Museo Arqueológico Regional, Alcalá de Henares, pp. 121–135.

Chamberlain, A.T., Wood, B.A., 1985. A reappraisal of variation in hominid mandibular corpus dimensions. Am. J. Phys. Anthropol. 66, 399–405.

Clark, J.D., Asfaw, B., Assefa, G., et al., 1984. Paleontological discoveries in the middle Awash valley, Ethiopia. Nature 307, 423–428.

Clarke, R.J., Tobias, P.V., 1995. Sterkfontein Member 2 foot bones of the oldest South African hominid. Science 269, 521–524.

Clutton-Brock, T.H., Harvey, P.H., 1980. Primates, brains and ecology. J. Zool. 190, 309–323.

Coffing, K., Feibel, C., Leakey, M., Walker, A., 1994. Four million-year-old hominids from East lake Turkana, Kenya. Am. J. Phys. Anthropol. 93, 55–65.

Conroy, G.C., Weber, G.W., Seidler, H., et al., 2000. Endocranial capacity of the Bodo cranium determined from three-dimensional computed tomography. Am. J. Phys. Anthropol. 113, 111–118.

Consortium, T.C.S., Consortium, A., 2005. Initial sequence of the chimpanzee genome and comparison with the human genome. Nature 437, 69–87.

Dart, R., 1925. Australopithecus africanus: the man-ape from South Africa. Nature 115, 195–199.

Dart, R.A., 1948. The Makapansgat proto-human Australopithecus prometheus. Am. J. Phys. Anthropol. 6, 259–284.

Day, M.H., 1967. Who were the first hominids? Eugen. Rev. 59, 150–152.

Day, M.H., 1969. Femoral fragment of a robust australopithecine from the Olduvai Gorge, Tanzania. Nature 221, 230–233.

Diogo, R., Wood, B., 2011. Soft-tissue anatomy of the primates: phylogenetic analyses based on the muscles of the head, neck, pectoral region and upper limb, with notes on the evolution of these muscles. J. Anat. 219, 273–359.

Dirks, P.H.G.M., Berger, L.R., Roberts, E.M., et al., 2015. Geological and taphonomic context for the new hominin species Homo naledi from the Dinaledi Chamber, South Africa. eLIFE 4, e09561.

Domínguez-Rodrigo, M., Pickering, T.R., Baquedano, E., et al., 2013. First partial skeleton of a 1.34-million-year-old Paranthropus boisei from bed II, Olduvai Gorge, Tanzania. PLoS One 8, e80347.

Dreyer, T.F., 1935. A human skull from Florisbad, Orange Free State, with a note on the endocranial cast, by C.U. Ariëns Kappers. Proc. Acad. Sci. Amsterdam 38, 119–128.

Du, A., Zipkin, A.M., Hatala, K.G., et al., 2016. Alternating Micro- and Macroevolutionary Processes Led to the Gradual Brain Size Increase in the Hominin Clade (in press).

Dubois, E., 1893. Palaeontologische onderzoekingen op Java. Extra bijvoegsel der Javasche Courant Verslag van het Mijnwezen over het 3e kwartaal 1892, 10–14.

Dunbar, R.I.M., 1998. The social brain hypothesis. Evol. Anthropol. 6, 178–190.

Falk, D., Hildebolt, C., Smith, K., et al., 2005. The brain of *Homo floresiensis*. Science 308, 242–245.

Falk, D., Redmond Jr., J.C., Guyer, J., et al., 2000. Early hominid brain evolution: a new look at old endocasts. J. Hum. Evol. 38, 695–717.

Fleagle, J.G., Rasmussen, D.T., Yirga, S., Brown, T.M., Grine, F.E., 1991. New hominid fossils from Fejej, southern Ethiopia. J. Hum. Evol. 21, 145–152.

Ferguson, W.W., 1984. Revision of fossil hominid jaws from the Plio/Pleistocene of Hadar, in Ethiopia including a new species of the genus *Homo* (Hominoidea: Homininae). Primates 25, 519–529.

Gabounia, L., de Lumley, M.-A., Vekua, A., Lordkipanidze, D., de Lumley, H., 2002. Découverte d'un nouvel hominidé à Dmanissi (Transcaucasie, Géorgie). Comptes Rendus Palevol 1, 243–253.

Gibbs, S., Collard, M., Wood, B.A., 2002. Soft-tissue anatomy of the extant hominoids: a review and phylogenetic analysis. J. Anat. 200, 3–49.

Godfrey, L., Jacobs, K.H., 1981. Gradual, autocatalytic and punctuational models of hominid brain evolution: a cautionary tale. J. Hum. Evol. 10, 255–272.

Goodman, M., 1962. Immunochemistry of the primates and primate evolution. Ann. N.Y. Acad. Sci. 102, 219–234.

Gordon, D., Huddleston, J., Chaisson, M.J.P., et al., 2016. Long-read sequence assembly of the gorilla genome. Science 352, aae0344.

Grabowski, M., Hatala, K.G., Jungers, W.L., Richmond, B.G., 2015. Body mass estimates of hominin fossils and the evolution of human body size. J. Hum. Evol. 85, 75–93.

Green, R., Krause, J., Briggs, A.W., et al., 2010. A draft sequence of the Neandertal genome. Science 328, 710–722.

Grimaud-Hervé, D., Lordkipanidze, D., 2010. The fossil hominids' brain of Dmanisi: D 2280 and D 2282. In: Broadfield, D.C., Yuan, M.S., Schick, K., Toth, N. (Eds.), The Human Brain Evolving: A Symposium in Honor of Professor Ralph L. Holloway. No. 4. Indiana University, Bloomington, pp. 60–82.

Grimaud-Hervé, D., Widianto, H., Detroit, F., Semah, F., 2012. Comparative morphological and morphometric description of the hominin calvaria from Bukuran (Sangiran, Central Java, Indonesia). J. Hum. Evol. 63, 637–652.

Grine, F.E., 1989. New hominid fossils from the Swartkrans formation (1979–1986 excavations): craniodental specimens. Am. J. Phys. Anthropol. 79, 409–449.

Grine, F.E., 2001. Implications of morphological diversity in early *Homo* crania from eastern and southern Africa. In: Tobias, P.V., Raath, M.A., Moggi-Cecchi, J., Doyle, G.A. (Eds.), Humanity from the African Naissance to the Coming Millennia. Florence University Press, Florence, pp. 107–115.

Grine, F.E., Demes, B., Jungers, W.L., Cole, T.M., 1993. Taxonomic affinity of early *Homo* cranium from Swartkrans, South Africa. Am. J. Phys. Anthropol. 92, 411–426.

Groves, C.P., 1989. A Theory of Human and Primate Evolution. Clarendon Press, Oxford.

Groves, C.P., Mazák, V., 1975. An approach to the taxonomy of the Hominidae: gracile Villafranchian hominids of Africa. Časopis pro Mineral. a Geol. 20, 225–247.

Gunz, P., Neubauer, S., Maureille, B., Hublin, J.-J., 2010. Brain development after birth differs between Neanderthals and modern humans. Curr. Biol. 20, R921–R922.

Guy, F., Lieberman, D.E., Pilbeam, D., et al., 2005. Morphological affinities of the *Sahelanthropus tchadensis* (Late Miocene hominid from Chad) cranium. Proc. Natl. Acad. Sci. U.S.A. 102, 18836–18841.

Haile-Selassie, Y., 2001. Late Miocene hominids from the middle Awash, Ethiopia. Nature 412, 178–181.

Haile-Selassie, Y., 2004. Late Miocene teeth from Middle Awash, Ethiopia, and early hominid dental evolution. Science 303, 1503–1505.

Haile-Selassie, Y., Saylor, B., Deino, A., Alene, M., Latimer, B., 2010. New hominid fossils from Woranso-Mille (Central Afar, Ethiopia) and taxonomy of early *Australopithecus*. Am. J. Phys. Anthropol. 141, 406–417.

Haile-Selassie, Y., Suwa, G., White, T.D., 2004. Late Miocene teeth from Middle Awash, Ethiopia, and early hominid dental evolution. Science 303, 1503–1505.

Haldane, J.B.S., 1949. Suggestions as to quantitative measurement of rates of evolution. Evolution 3, 51–56.

Henry, A.G., Ungar, P.S., Passey, B.H., et al., 2012. The diet of *Australopithecus sediba*. Nature 487, 90–93.

Holloway, R.L., 1983. Human paleontological evidence relevant to language behavior. Hum. Neurobiol. 2, 105–114.

Holloway, R.L., Broadfield, D.C., Yuan, M.S., 2004. Endocasts of early hominids. In: Schwartz, J.H., Tattersall, I. (Eds.), The Human Fossil Record: Brain Endocasts–The Paleoneurological Evidence, vol. 3. John Wiley & Sons Publishers, New York, pp. 39–107.

Indriati, E., Antón, S.C., 2010. The calvaria of Sangiran 38, Sendangbusik, Sangiran Dome, Java. HOMO J. Comp. Hum. Biol. 61, 225–243.

Jacob, T., 1973. Palaeoanthropological discoveries in Indonesia with special reference to the finds of the last two decades. J. Hum. Evol. 2, 473–485.

Jerison, H.J., 1973. Evolution of the Brain and Intelligence. Academic Press, New York.

Johanson, D.C., Masao, F.T., Eck, G.G., et al., 1987. New partial skeleton of *Homo habilis* from Olduvai Gorge, Tanzania. Nature 327, 205–209.

Johanson, D.C., Taieb, M., Coppens, Y., 1982. Pliocene hominids from the Hadar Formation, Ethiopia (1973–1977): stratigraphic, chronologic, and paleoenvironmental contexts, with notes on hominid morphology and systematics. Am. J. Phys. Anthropol. 57, 373–402.

Johanson, D.C., White, T.D., Coppens, Y., 1978. A new species of the genus *Australopithecus* (primates: Hominidae) from the Pliocene of East Africa. Kirtlandia 28, 1–14.

Johanson, D.C., White, T.D., Coppens, Y., 1982. Dental remains from the Hadar formation, Ethiopia: 1974–1977 collections. Am. J. Phys. Anthropol. 57, 545–603.

Kaifu, Y., Aziz, F., Indriati, E., et al., 2008. Cranial morphology of Javanese *Homo erectus*: new evidence for continuous evolution, specialization, and terminal extinction. J. Hum. Evol. 55, 551–580.

Kaifu, Y., Zaim, Y., Baba, H., et al., 2011. New reconstruction and morphological description of a *Homo erectus* cranium: skull IX (Tjg–1993.05) from Sangiran, central Java. J. Hum. Evol. 61, 270–294.

Keith, A., 1912. Ancient Types of Man. Harper & Brothers, London.

Keyser, A.W., 2000. The Drimolen skull: the most complete australopithecine cranium and mandible to date. South Afr. J. Sci. 96, 189–197.

Keyser, A.W., Menter, C.G., Moggi-Cecchi, J., Pickering, T.R., Berger, L.R., 2000. Drimolen: a new hominid-bearing site in Gauteng, South Africa. South Afr. J. Sci. 96, 193–197.

Kimbel, W.H., 1988. Identification of a partial cranium of *Australopithecus afarensis*, from the Koobi Fora formation, Kenya. J. Hum. Evol. 17, 647–656.

Kimbel, W.H., Delezene, L.K., 2009. "Lucy" redux: a review of research on *Australopithecus afarensis*. Am. J. Phys. Anthropol. 140, 2–48.

Kimbel, W.H., Rak, Y., 2010. The cranial base of *Australopithecus afarensis*: new insights from the female skull. Philosophical Trans. R. Soc. B Biol. Sci. 365, 3365–3376.

Kimbel, W.H., Johanson, D.C., Rak, Y., 1994. The first skull and other new discoveries of *Australopithecus afarensis* at Hadar, Ethiopia. Nature 368, 449–451.

Kimbel, W.H., Lockwood, C.A., Ward, C.V., et al., 2006. Was *Australopithecus anamensis* ancestral to *A. afarensis*? A case of anagenesis in the hominin fossil record. J. Hum. Evol. 51, 134−152.

Kimbel, W., Rak, Y., Johanson, D.C., 2004. The Skull of Australopithecus Afarensis. Oxford University Press, New York.

Kimbel, W.H., Walter, R.C., Johanson, D.C., et al., 1996. Late Pliocene *Homo* and Oldowan tools from the Hadar formation (kadar hadar member), Ethiopia. J. Hum. Evol. 31, 549−561.

Kimbel, W.H., White, T.D., Johanson, D.C., 1984. Cranial morphology of *Australopithecus afarensis*: a comparative study based on a composite reconstruction of the adult skull. Am. J. Phys. Anthropol. 64, 337−388.

King, W., 1864. The reputed fossil man of the Neanderthal. Q. J. Sci. 1, 88−97.

von Koenigswald, G.H.R., 1936. Ein fossiler hominide aus dem Altpleistocän Ostjavas. De. Ing. Nederl. 8, 149−158.

von Koenigswald, G.H.R., 1968. Observation upon two *Pithecanthropus* mandibles from Sangiran, central Java. Proc. K. Ned. Akad. Wet. 71, 99−107.

von Koenigswald, G.H.R., 1975. Early man in Java: catalogue and problems. In: Tuttle, R.H. (Ed.), Paleoanthropology, Morphol. Paleoecology. Mouton Publishers, Den Haag, pp. 303−309.

Kokkoros, P., Kanellis, A., 1960. Découverte d'un crâne d'homme paléolithique dans la péninsule Chalchidique. L'Anthropologie 64, 438−446.

Krause, J., Fu, Q., Good, J.M., et al., 2010. The complete mitochondrial DNA genome of an unknown hominin from southern Siberia. Nature 464, 894−897.

Krings, M., Stone, A., Schmitz, R.W., et al., 1997. Neandertal DNA sequences and the origin of modern humans. Cell 90, 19−30.

Kullmer, O., Sandrock, O., Abel, R., et al., 1999. The first *Paranthropus* from the Malawi Rift. J. Hum. Evol. 37, 121−127.

Langergraber, K.E., Prüfer, K., Rowney, C., et al., 2012. Generation times in wild chimpanzees and gorillas suggest earlier divergence times in great ape and human evolution. Proc. Natl. Acad. Sci. U.S.A. 109, 15716−15721.

Leakey, L.S.B., 1959. A new fossil skull from Olduvai. Nature 184, 491−493.

Leakey, L.S.B., 1961. The Progress of Man in Africa. Oxford University Press, London.

Leakey, L.S.B., Leakey, M.D., 1964. Recent discoveries of fossil hominids in Tanganyika, at Olduvai and near lake Natron. Nature 202, 5−7.

Leakey, L.S.B., Tobias, P.V., Napier, J., 1964. A new species of the genus *Homo* from Olduvai Gorge. Nature 202, 7−9.

Leakey, M.D., Clarke, R.J., Leakey, L.S.B., 1971. New hominid skull from bed I, Olduvai Gorge, Tanzania. Nature 232, 308−312.

Leakey, M.G., Feibel, C.S., McDougall, I., Walker, A., 1995. New four-million year old hominid species from Kanapoi and Allia Bay, Kenya. Nature 376, 565−571.

Leakey, M.G., Feibel, C.S., McDougall, I., Ward, C., Walker, A., 1998. New specimens and confirmation of an early age for *Australopithecus anamensis*. Nature 393, 62−66.

Leakey, M.G., Spoor, F., Brown, F.H., et al., 2001. New hominin genus from eastern Africa shows diverse middle Pliocene lineages. Nature 410, 433−440.

Leakey, M.G., Spoor, F., Dean, M.C., et al., 2012. New fossils from Koobi Fora in northern Kenya confirm taxonomic diversity in early *Homo*. Nature 488, 201−204.

Leakey, R.E.F., 1974. Further evidence of lower Pleistocene hominids from East Rudolf, North Kenya 1973. Nature 248, 653−656.

Leakey, R.E.F., 1976. New hominid fossils from the Koobi Fora formation in North Kenya. Nature 261, 574−576.

Leakey, R.E.F., Walker, A.C., 1985. Further hominids from the Plio-Pleistocene of Koobi Fora, Kenya. Am. J. Phys. Anthropol. 67, 135−163.

Leakey, R.E.F., Walker, A.C., 1988. New *Australopithecus boisei* specimens from East and West lake Turkana, Kenya. Am. J. Phys. Anthropol. 76, 1−24.

Leakey, R.E.F., Walker, A., Ward, C.V., Grausz, H.M., 1989. A partial skeleton of a gracile hominid from the upper Burgi member of the Koobi Fora formation, east lake Turkana, Kenya. In: Giacobini, G. (Ed.), Hominidae: Proceedings of the 2nd International Congress of Human Paleontology, Turin, September 28-October 3, 1987. Jaca Book, Milan, pp. 167−174.

Lee, S.H., Wolpoff, M.H., 2003. The pattern of evolution in Pleistocene human brain size. Paleobiology 29, 186−196.

Lestrel, P.E., 1976. Hominid brain size versus time: revised regression estimates. J. Hum. Evol. 5, 207−212.

Lestrel, P.E., Read, D.W., 1973. Hominid cranial capacity versus time: a regression approach. J. Hum. Evol. 2, 405−411.

Li, T., Etler, D.A., 1992. New middle Pleistocene hominid crania from Yunxian in China. Nature 357, 404−407.

Locke, D.P., Hillier, L.W., Warren, W.C., et al., 2011. Comparative and demographic analysis of orangutan genomes. Nature 469, 529−533.

Lockwood, C.A., Kimbel, W.H., Johanson, D.C., 2000. Temporal trends and metric variation in the mandibles and teeth of *Australopithecus afarensis*. J. Hum. Evol. 39, 23−55.

Lockwood, C.A., Kimbel, W.H., Lynch, J.M., 2004. Morphometrics and hominioid phylogeny: support for a chimpanzee-human clade and differentiation among great ape subspecies. Proc. Natl. Acad. Sci. U.S.A. 101, 4356−4360.

Lockwood, C.A., Tobias, P.V., 1999. A large male hominin cranium from Sterkfontein, South Africa and the status of *Australopithecus africanus*. J. Hum. Evol. 36, 637−685.

Lockwood, C.A., Tobias, P.V., 2002. Morphology and affinities of new hominin cranial remains from member 4 of the Sterkfontein formation, Gauteng Province, South Africa. J. Hum. Evol. 42, 389−450.

Lordkipanidze, D., Jashashvili, T., Vekua, A., et al., 2007. Postcranial evidence from early *Homo* from Dmanisi, Georgia. Nature 449, 305−310.

Lordkipanidze, D., Ponce de León, M.S., Margvelashvili, A., et al., 2013. A complete skull from Dmanisi, Georgia, and the evolutionary biology of early *Homo*. Science 342, 326−331.

de Lumley, H., de Lumley, M., 1973. Preneandertal human remains from Arago cave in southeastern France. Yearb. Phys. Anthropol. 17, 162−168.

Macchiarelli, R., Bondioli, L., Coppa, A., et al., 2002. The one-million-year-old human remains from the Danakil (Afar) Depression of Eritrea. Am. J. Phys. Anthropol. 117, 104.

Manaster, B.J., 1979. Locomotor adaptations within the *Cercopithecus* genus: a multivariate approach. Am. J. Phys. Anthropol. 50, 169−182.

Martin, R.D., 1996. Scaling of the mammalian brain: the maternal energy hypothesis. News Physiol. Sci. 11, 149−156.

McDougall, I., Brown, F.H., Fleagle, J.G., 2005. Stratigraphic placement and age of modern humans from Kibish, Ethiopia. Nature 433, 733−736.

Menter, C.G., Kuykendall, K.L., Keyser, A.W., Conroy, G.C., 1999. First record of hominid teeth from the Plio-Pleistocene site of Gondolin, South Africa. J. Hum. Evol. 37, 299−307.

Moggi-Cecchi, J., Grine, F.E., Tobias, P.V., 2006. Early hominid dental remains from Members 4 and 5 of the Sterkfontein Formation (1966−1996 excavations): catalogue, individual associations, morphological descriptions, and initial metrical analysis. J. Hum. Evol. 50, 239−328.

Morwood, M.J., Brown, P., Jatmiko, et al., 2005. Further evidence for small-bodied hominins from the late Pleistocene of Flores, Indonesia. Nature 437, 1012−1017.

Mounier, A., Marchal, F., Condemi, S., 2009. Is *Homo heidelbergensis* a distinct species? New insight on the Mauer mandible. J. Hum. Evol. 56, 219−246.

Neubauer, S., Gunz, P., Weber, G.W., Hublin, J.-J., 2012. Endocranial volume of *Australopithecus africanus*: new CT-based estimates and the effects of missing data and small sample size. J. Hum. Evol. 62, 498–510.

Nixon, K.C., Wheeler, Q.D., 1990. An amplification of the phylogenetic species concept. Cladistics 6, 211–223.

Olivier, G., Tissier, H., 1975. Determination of cranial capacity in fossil men. Am. J. Phys. Anthropol. 43, 353–362.

Partridge, T.C., Granger, D.E., Caffee, D.E., Clarke, R.J., 2003. Lower Pliocene hominid remains from Sterkfontein. Science 300, 607–612.

Paterson, H.E.H., 1985. The recognition concept of species. In: Vrba, E.S. (Ed.), Species and Speciation Transvaal Museum Monograph, vol. 4. Transvaal Museum, Pretoria, pp. 21–29.

Prado-Martinez, J., Sudmant, P.H., Kidd, J.M., et al., 2013. Great ape genetic diversity and population history. Nature 499, 471–475.

Prüfer, K., Munch, K., Hellmann, I., et al., 2012. The bonobo genome compared with the chimpanzee and human genomes. Nature 486, 527–531.

Prüfer, K., Racimo, F., Patterson, N., et al., 2014. The complete genome sequence of a Neanderthal from the Altai Mountains. Nature 505, 43–49.

Richmond, B.G., Green, D.J., Braun, D.R., et al., 2011. New fossils from Ileret, Kenya, and the evolution of hominin hand function. Proceedings of American Association of Physical Anthropology Meetings. Wiley-Blackwell Commerce Place, 350 Main Street, Malden 02148, MA, USA, p. 253.

Rightmire, G.P., 1979. Cranial remains of *Homo erectus* from Beds II and IV, Olduvai Gorge, Tanzania. Am. J. Phys. Anthropol. 51, 99–116.

Rightmire, G.P., 1981. Patterns in the evolution of *Homo erectus*. Paleobiology 7, 241–246.

Rightmire, G.P., 1983. The lake Ndutu cranium and early *Homo sapiens* in Africa. Am. J. Phys. Anthropol. 61, 245–254.

Rightmire, G.P., 1995. Geography, time and speciation in Pleistocene *Homo*. South Afr. J. Sci. 91, 450–454.

Robinson, J.T., 1960. The affinities of the new Olduvai australopithecine. Nature 186, 456–458.

Ruff, C.B., Trinkaus, E., Holliday, T.W., 1997. Body mass and encephalization in Pleistocene *Homo*. Nature 387, 173–176.

de Ruiter, D.J., Pickering, R., Steininger, C.M., et al., 2009. New *Australopithecus robustus* fossils and associated U-Pb dates from Cooper's cave (Gauteng, South Africa). J. Hum. Evol. 56, 497–513.

Santa Luca, A.P., 1980. The Ngandong fossil hominids: a comparative study of a Far Eastern *Homo erectus* group. Yale Univ. Publ. Anthropol. 78, 1–175.

Scally, A., Dutheil, J.Y., Hillier, L.W., et al., 2012. Insights into hominid evolution from the gorilla genome sequence. Nature 483, 169–175.

Schepers, G.W.H., 1946. The endocranial casts of South African ape men. In: Broom, R., Schepers, G.W.H. (Eds.), The South African Fossil Ape-Men: The Australopithecinae, vol. 2. Transvaal Museum, Pretoria, pp. 167–272.

Schmitz, R.W., Serre, D., Bonani, G., et al., 2002. The Neandertal type site revisited: interdisciplinary investigations of skeletal remains from the Neander Valley, Germany. Proc. Natl. Acad. Sci. U.S.A. 99, 13342–13347.

Schoetensack, O., 1908. Der Unterkiefer des *Homo heidelbergensis* aus den Sanden von Mauer bei Heidelberg. W. Engelmann, Leipzig.

Schrenk, F., Bromage, T.G., Betzler, C.G., Ring, U., Juwayei, Y.M., 1993. Oldest *Homo* and Pliocene biogeography of the Malawi Rift. Nature 365, 833–836.

The human fossil record. In: Schwartz, J.H., Tattersall, I. (Eds.), 2003. Craniodental Morphology of Genus *Homo* (Africa and Asia), vol. 2. Wiley–Liss, New York.

Schwartz, J.H., Tattersall, I., 2010. Fossil evidence for the origin of *Homo sapiens*. Am. J. Phys. Anthropol. 143, 94–121.

Semaw, S., Simpson, S.W., Quade, J., et al., 2005. Early Pliocene hominids from Gona, Ethiopia. Nature 433, 301–305.

Senut, B., Pickford, M., Gommery, D., et al., 2001. First hominid from the Miocene (Lukeino Formation, Kenya). Comptes rendus l'Academie Sci. 332, 137–144.

Shoshani, J., Groves, C.P., Simons, E.L., Gunnell, G.F., 1996. Primate phylogeny: morphological vs. molecular results. Mol. Phylogenet. Evol. 5, 101–153.

Simpson, G.G., 1961. Principles of Animal Taxonomy. Columbia University Press, New York.

Smith, A.B., 2009. Systematics and the Fossil Record: Documenting Evolutionary Patterns. Wiley, New York.

Spoor, F., Leakey, M.G., Gathogo, P.N., et al., 2007. Implications of new early *Homo* fossils from Ileret, east of Lake Turkana, Kenya. Nature 448, 688–691.

Spoor, F., Gunz, P., Neubauer, S., et al., 2015. Reconstructed *Homo habilis* type OH 7 suggests deep-rooted species diversity in early *Homo*. Nature 519, 83–86.

Steininger, C.M., Berger, L.R., 2000. Taxonomic affinity of a new specimen from Cooper's, South Africa. Am. J. Phys. Anthropol. Suppl. 30, 291.

Stevens, N.J., Seiffert, E.R., O'Connor, P.M., et al., 2013. Palaeontological evidence for an Oligocene divergence between Old World monkeys and apes. Nature 497, 611–614.

Strait, D.S., 2001. Integration, phylogeny, and the hominid cranial base. Am. J. Phys. Anthropol. 114, 273–297.

Sutikna, T., Tocheri, M.W., Morwood, M.J., et al., 2016. Revised stratigraphy and chronology for *Homo floresiensis* at Liang Bua in Indonesia. Nature. https://doi.org/10.1038/nature17179.

Suwa, G., 1988. Evolution of the "robust" australopithecines in the Omo succession: evidence from mandibular premolar morphology. In: Grine, F.E. (Ed.), Evolutionary History of the "Robust" Australopithecines. Aldine de Gruyter, New York, pp. 199–222.

Suwa, G., 1990. A Comparative Analysis of Hominid Dental Remains from the Shungura and Usno Formations, Omo Valley, Ethiopia. Ph.D. Dissertation, University of California, Berkeley.

Suwa, G., Asfaw, B., Beyene, Y., et al., 1997. The first skull of *Australopithecus boisei*. Nature 389, 489–492.

Suwa, G., Kono, R.T., Simpson, S.W., et al., 2009. Paleobiological implications of the *Ardipithecus ramidus* dentition. Science 326, 69–99.

Thackeray, J.F., de Ruiter, D.J., Berger, L.R., van der Merwe, N.J., 2001. Hominid fossils from Kromdraai: a revised list of specimens discovered since 1938. Ann. Transvaal Mus. 38, 43–56.

Tobias, P.V., 1967. The Cranium and Maxillary Dentition of *Australopithecus* (*Zinjanthropus*) *Boisei*. Cambridge University Press, Cambridge.

Tobias, P.V., 1971. The Brain in Hominid Evolution. Columbia University Press, New York.

Tobias, P.V., 1985. Single characters and total morphological pattern redefined the sorting effected by a selection of morphological features of the early hominids: the Hard Evidence. In: Delson, E. (Ed.), Ancestors: The Hard Evidence. Alan R. Liss, New York, pp. 94–101.

Tobias, P.V., 1991. Olduvai Gorge: The Skulls, Endocasts and Teeth of Homo Habilis. Cambridge University Press, Cambridge.

Toussaint, M., Macho, G.A., Tobias, P.V., Partridge, T.C., Hughes, A.R., 2003. The third partial skeleton of a late Pliocene hominin (Stw 431) from Sterkfontein, South Africa. South Afr. J. Sci. 99, 215–223.

Tyler, D.E., 1996. The taxonomic status of the "*Meganthropus*" cranium Sangiran 31, and the "*Meganthropus*" occipital fragment III. Bull. Indo Pacific Prehistory Assoc. 15, 235–241.

Vekua, A., Lordkipanidze, D., Rightmire, G.P., et al., 2002. A new skull of early *Homo* from Dmanisi, Georgia. Science 297, 85–89.

Venn, O., Turner, I., Mathieson, I., et al., 2014. Strong male bias drives germline mutation in chimpanzees. Science 344, 1272–1275.

Villmoare, B., Kimbel, W.H., Seyoum, C., et al., 2014. Early *Homo* at 2.8 Ma from Ledi-Geraru, Afar, Ethiopia. Science 347, 1352–1355.

Walker, A., Leakey, R.E., 1988. The evolution of *Australopithecus boisei*. In: Grine, F.E. (Ed.), Evolutionary History of the "Robust" Australopithecines. Transaction Publishers, New Jersey, pp. 247–258.

Walker, A.C., Leakey, R.E.F., Harris, J.M., Brown, F.H., 1986. 2.5 Myr *Australopithecus boisei* from West of lake Turkana, Kenya. Nature 322, 517–522.

Ward, C., Leakey, M., Walker, A., 1999. The new hominid species *Australopithecus anamensis*. Evol. Anthropol. 7, 197–205.

Ward, C.V., Leakey, M.G., Walker, A., 2001. Morphology of *Australopithecus anamensis* from Kanapoi and Allia Bay, Kenya. J. Hum. Evol. 41, 255–368.

Weidenreich, F., 1936. The mandibles of *Sinanthropus pekinensis*: a comparative study. Palaeontol. Sin. New Ser. D. 72, 1–162.

Weidenreich, F., 1936. The new discoveries of *Sinanthropus pekinensis* and their bearing on the *Sinanthropus* and *Pithecanthropus* problems. Bull. Geol. Soceity China 16, 439–466.

Weidenreich, F., 1937. The dentition of *Sinanthropus pekinensis*: a comparative odontography of the hominids. Palaeontol. Sin. 1, 1–180.

Weidenreich, F., 1940. Some problems dealing with ancient man. Am. Anthropol. 42, 375–383.

Weidenreich, F., 1943. The skull of *Sinanthropus pekinensis*: a comparative study on a primitive hominid skull. Paleontol. Sin. 127.

Weidenreich, F., 1944. Giant early man from Java and South China. Science 99, 479–482.

Weidenreich, F., 1951. Morphology of Solo man. Anthropol. Pap. Am. Mus. Nat. Hist. 43, 205–290.

Widianto, H., Grimaud-Hervé, D., 1993. Le crâne de Ngawi. Les. Dossiers Archéologie 184, 36.

White, S., Gowlett, J.A.J., Grove, M., 2014. The place of the Neanderthals in hominin phylogeny. J. Anthropol. Archaeol. 35, 32–50.

White, T.D., 2002. Earliest hominids. In: Hartwig, W. (Ed.), The Primate Fossil Record. Cambridge University Press, Cambridge, pp. 407–417.

White, T.D., 2003. Early hominids - diversity or distortion? Science 299, 1994–1997.

White, T.D., 2010. Human origins and evolution: Cold Spring Harbor, deja vu. Cold Spring Harb. Symp. Quant. Biol. 74, 335–344.

White, T.D., Asfaw, B., Beyene, Y., et al., 2009. *Ardipithecus ramidus* and the paleobiology of early hominids. Science 326, 64–86.

White, T.D., Asfaw, B., DeGusta, D., et al., 2003. Pleistoecne *Homo sapiens* from middle Awash, Ethiopia. Nature 423, 742–747.

White, T.D., Johanson, D.C., Kimbel, W.H., 1981. *Australopithecus africanus*: its phyletic position reconsidered. South Afr. J. Sci. 77, 445–470.

White, T.D., Suwa, G., Asfaw, B., 1994. *Australopithecus ramidus*, a new species of early hominid from Aramis, Ethiopia. Nature 371, 306–312.

White, T.D., Suwa, G., Asfaw, B., 1995. *Australopithecus ramidus*, a new species of early hominid from Aramis, Ethiopia - a corrigendum. Nature 375, 88.

White, T.D., WoldeGabriel, G., Asfaw, B., et al., 2006. Asa Issie, Aramis and the origin of *Australopithecus*. Science 440, 883–889.

Wolpoff, M.H., 1984. Evolution in *Homo erectus*: the question of stasis. Paleobiology 10, 389–406.

Wood, B.A., 1991. Koobi Fora research project. In: Hominid Cranial Remains, vol. 4. Clarendon Press, Oxford.

Wood, B.A., 1992. Origin and evolution of the genus *Homo*. Nature 355, 783–790.

Wood, B.A., 1994. Taxonomy and evolutionary relationships of *Homo erectus*. Cour. Forsch Senckenb. 171, 159–165.

Wood, B.A., 2010. Systematics, taxonomy, and phylogenetics: ordering life, past and present. In: Larsen, C.S. (Ed.), A Companion to Biological Anthropology. Wiley-Blackwell, Oxford, pp. 56–73.

Wood, B., Boyle, E.K., 2016. Hominin taxic diversity: fact or fantasy? Am. J. Phys. Anthropol. 159, 37–78.

Wood, B.A., Chamberlain, A.T., 1987. The nature and affinities of the "robust" australopithecines: a review. J. Hum. Evol. 6, 625–641.

Wood, B.A., Constantino, P., 2007. *Paranthropus boisei*: fifty years of evidence and analysis. Yearb. Phys. Anthropol. 50, 106–132.

Wood, B., Schroer, K., 2013. Paranthropus. In: Begun, D. (Ed.), A Companion to Paleoanthropology. Wiley–Blackwell, Malden, pp. 457–478.

Wood, B., Wood, C., Konigsberg, L., 1994. *Paranthropus boisei*: an example of evolutionary stasis? Am. J. Phys. Anthropol. 95, 117–136.

Wu, R.K., Dong, X.R., 1982. Preliminary study of *Homo erectus* remains from Hexian, Anhui. Acta Anthropol. Sin. 1, 2–13.

Wu, X., Holloway, R.L., Schepartz, L.A., Xing, S., 2011. A new brain endocast of *Homo erectus* from Hulu Cave, Nanjing, China. Am. J. Phys. Anthropol. 145, 452–460.

Zaim, Y., Ciochon, R.L., Polanski, J.M., et al., 2011. New 1.5 million-year-old *Homo erectus* maxilla from Sangiran (central Java, Indonesia). J. Hum. Evol. 61, 363–376.

Zollikofer, C.P., Ponce de León, M.S., Lieberman, D.E., et al., 2005. Virtual cranial reconstruction of *Sahelanthropus tchadensis*. Nature 434, 755–759.

Zuckerkandl, E., Jones, R.T., Pauling, L., 1960. A comparison of animal hemoglobins by tryptic peptide pattern analysis. Proc. Natl. Acad. Sci. U.S.A. 46, 1349–1360.

31

Evolution of Human Life History

B. Bogin[1], C. Varea[2]

[1]Loughborough University, Loughborough, United Kingdom; [2]Universidad Autónoma de Madrid, Madrid, Spain

31.1 Introduction

Just as the human body, behaviors, and emotions have evolved, so has human development. Indeed, it has long been recognized that changing patterns of growth and development underlie biological evolution and speciation (Thompson, 1917; Bonner, 1965). "Growth" and "development" refer to changes in the size, structure, and function of various body parts, including the brain, that occur during the lifetime of an organism. Shown in Fig. 31.1 is the typical pattern of human growth after birth in height. The "Height distance" curves (part A) indicate how tall girls and boys are at various ages. The "Height velocity" curves indicate how fast height is accumulating each year. We explain later in this chapter the human evolutionary changes in both the total amount of height growth and the changing rates of growth velocity.

Development also means the changes that lead to sexual maturity and reproduction and when these occur. Shown in Fig. 31.2 are growth curves for different body tissues. In this figure the curves are based on the percentage of gain in amount or maturity between birth and 20 years of age. The "Body" curve represents growth in stature or total body weight and is another way of representing the distance curve in Fig. 31.1. The "Brain" curve represents the total weight of the brain, which reaches more than 95% of final adult weight by 7 years of age. At that age the brain has not completed its development, which continues through the teenage years and into the fifth decade (Sowell et al., 2003). The "Dentition" curve is the median maturity score for girls of the permanent teeth. Dental maturity is based on the formation of the tooth crown (the enamel-covered upper portion of each tooth) and the tooth root. The amount of crown and root formation is assessed by a panoramic X-ray of the lower jaw and based on the first seven teeth on the left side (the central and lateral incisors, the canine, the two premolars, and

the first and second molar, but not the third molar). Finally, the "Reproductive" curve represents the weight of the gonads and primary reproductive organs, which is a strong correlate of sexual maturity.

The four curves of Fig. 31.2 show that different body parts or systems grow and mature at different rates. These curves are for the human species. Other species have different patterns of growth and maturity for the same systems. Growth curves for the rat are shown in Fig. 31.3. Note that the completion of reproductive maturity takes precedence over the growth of both the brain and the body. The rat pattern is typical of most species of mammals.

In this chapter, We present a view of the evolution of human development that is based on three branches of research. The first is anthropological studies of living humans, especially the ways in which biological and social factors interact to influence human development and health. The second is auxology, that is, the scientific study of the physical growth and maturation of human beings and closely related species such as monkeys and apes. The third area is life history evolution, which includes both theory and empirical studies of biological development in living and fossil species. Weaving together these three strands of research in human biology, auxology, and life history is a "biocultural perspective" of human development and evolution (Bogin and Smith, 2012). The biocultural perspective is complementary to approaches to human development from neuroscience and psychology, but it is also distinct in the use of some words and phrases. For this reason, prior to discussing the evolution of human life history, a few definitions are needed (all adapted from Bogin, 1999).

31.1.1 Some Definitions

"Life history theory" is a field of biology concerned with the strategy an organism uses to allocate its energy

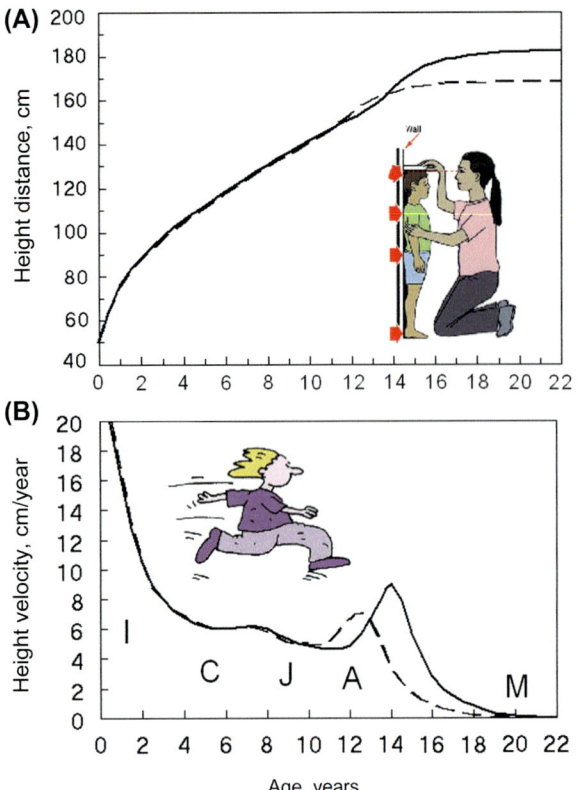

FIGURE 31.1 Average distance (A) and velocity (B) curves of growth in height for healthy girls (*dashed lines*) and boys (*solid lines*). Distance is the amount of height achieved at a given age. In part A, the image shows a child's height being measured. Velocity is the rate of growth at a given time, in this case shown as centimeters per year. In part B the running figure represents "velocity." The velocity curves show the postnatal periods of the pattern of human growth. Note the spurts in growth rate at midchildhood and adolescence for both girls and boys. The postnatal periods: *I*, infancy; *C*, childhood; *J*, juvenile; *A*, adolescence; *M*, mature adult.

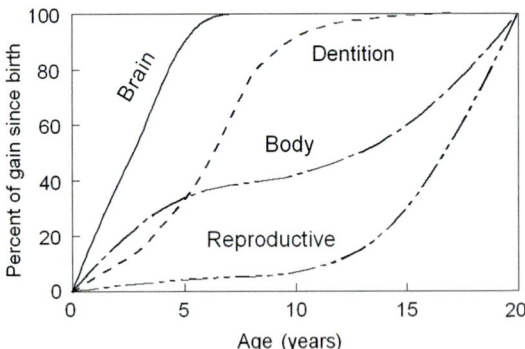

FIGURE 31.2 Growth curves for different body tissues. The brain curve is for total weight of the brain. The dentition curve is the median maturity score for girls based on the seven left mandibular teeth (the central and lateral incisors, the canine, the two premolars, and the first and second molar. The body curve represents growth in stature or total body weight and the reproductive curve represents the weight of the gonads and primary reproductive organs. *The figure is based on the work of Richard Scammon (1930) with additional data as described in Bogin (1999: p. 73).*

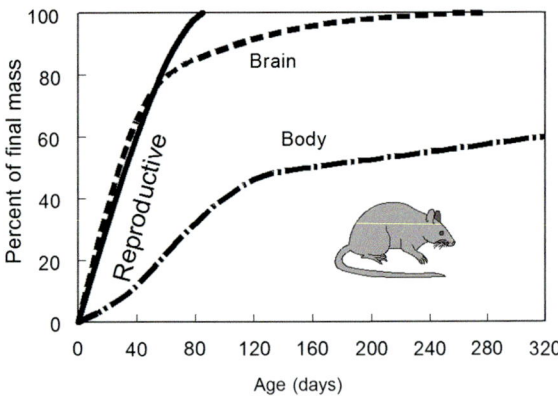

FIGURE 31.3 Growth of different types of tissue in the rat (Bogin, 1999: p. 131).

toward growth, body maintenance, defense against infection, reproduction, raising offspring to independence, and avoiding death (Varea and Bernis, 2013). For a mammal, it is the strategy of when to be born, when to be weaned, how many and what type of prereproductive stages of development to pass through, when to reproduce, and when to die. A central principle of life history theory is the concept of biological "trade-offs." These are life history strategies used when competition between two biological or behavioral traits requires a partial allocation of energy or materials to each trait. An example is the trade-off between investments of time or energy in one's own physical body versus investments in the physical needs of one's offspring. A list of some life history principles and trade-off is given in Table 31.1.

Human biologists define the word "growth" as a quantitative increase in size or mass. Measurements such as height in centimeters or total body mass (weight) in kilograms indicate how much growth has taken place in an infant, child, juvenile, or adolescent. "Development" is defined as a series of changes, either quantitative or qualitative, from a relative undifferentiated or immature state to a highly organized and specialized state. An example is the development of the female or the male reproductive systems, which are derived from essentially identical tissues in the human embryo. Genetic and hormonal information provided from the inheritance of pairs of XX or XY sex chromosomes promotes the development of the embryonic tissues toward one sex or the other, with Y chromosomes carrying the sex-determining region SRY genetic sequence that promotes maleness. The degree of development toward femaleness or maleness is influenced by a variety of epigenetic and environmental interactions. The common belief that there are only two sexes is proved false by studies of intersex people, who have physical and behavioral characteristics of both females and males (Fausto-Sterling, 2015). Estimates of the prevalence of

TABLE 31.1 Life history traits and trade-offs. This is a partial list of the most important traits. The list is based on the discussion in Cole (1954) and Stearns (1992), who provided additional traits

Traits	Trade-offs
1. Size at birth	1. Current reproduction vs. future reproduction.
2. Brain size	2. Current reproduction vs. survival
3. Growth patterns ○ Number of life cycle stages ○ Duration of each stage 4. Age at eruption of first permanent molar 5. Rate of maturation ○ Age at first reproduction ○ Age of last reproduction 6. Size at maturity 7. Number and sex ratio of offspring 8. Reproductive investment in each offspring 9. Length of life ○ Rate of aging/senescence ○ Age at death	3. Number vs. size offspring 4. Parental reproduction vs. growth 5. Brain size vs. body size 6. Parental health vs. offspring growth 7. Parental vs. offspring reproduction

intersex phenotypes at birth in the human population range from 0.018% to 1.7% (Sax, 2002). Surgery and social convention may force people to choose femaleness or maleness. The important point is that the study of human development requires an appreciation of biological and behavioral variability.

31.2 Human Life History Stages

Humans are no exception to biological evolution. In humans, part of what evolved during evolution were new life history stages of growth and development that allowed for greater flexibility in reproductive strategies than those of other species, living or extinct. No living species of mammal, including the nonhuman primates, has all the characteristics of human growth, development, and maturation. In common with all mammals, humans share the "infancy" stage of life history, the period of feeding by lactation, and the "adulthood" stage, the period of reproduction. In common with the social mammals, such as most primates, elephants, and some carnivores (wolves, lions, hyenas), humans share the "juvenile" stage of development, characterized by feeding independence but reproductive immaturity. In contrast to all other living species, humans have two unusual periods of development, childhood and adolescence (Bogin, 1999).

"Childhood" is defined as the period of time after infancy, from about the age of 3.0 to 6.9 years, when infants are "weaned," that is, infant feeding by lactation has ceased, but the youngster is still dependent on older people for care and food provisioning. In other species of mammals, including monkeys and apes, after individuals are weaned they have to get their own food. Children also have a moderate growth rate of about 6 cm per year. The childhood stage ends with the eruption of the first permanent molar and incisor teeth and the virtual completion of brain growth in total weight.

The evolution of childhood may explain, in part, the human-specific traits of relatively early weaning, prolonged growth and development between birth and adulthood, complex kinship systems, language, marriage, intergenerational transfers of wealth and political power, relatively slow aging, as well as senescence and greater longevity. Each of these characteristics of modern humans evolved, mostly likely, at different times since the split between ape and human ancestors occurred about 6–7 million years ago. This is the concept of "mosaic evolution" which was originally proposed in 1954 by Gavin de Beer in his analysis of the birdlike dinosaur *Archaeopteryx* (De Beer, 1954) and later applied to the evolution of *Australopithecus* (De Beer, 1975). We describe the mosaic nature of human evolution in more detail later in this chapter. Here we state that childhood is, in essence, the first piece of the mosaic of human life history evolution, including the evolution of human culture.

Human "adolescence" is the stage of life when social and sexual maturation takes place. Adolescence begins with puberty, or more technically with gonadarche, which is initiation of the adult pattern of activation of the gonadotropin-releasing hormone (GnRH) pulse generator of the hypothalamus. The transition from juvenile to adolescent stages requires not only the renewed production of GnRH but also its secretion from the hypothalamus in a specific frequency and amplitude of pulses (Ramaswamy et al., 2013). The adolescent stage also includes development of secondary sexual characteristics, such as development of the external genitalia, sexual dimorphism in body size and composition, and the onset of greater interest and practice of adult patterns of sociosexual and economic behavior. These physical and behavioral changes occur with puberty in many species of social mammals. What makes human adolescence unusual among the primates are two important differences. The first is the length of time between age at puberty and age at first birth. Most other mammals initiate reproduction within weeks or months of puberty. Chimpanzees, our closest genetic relatives, initiate female reproduction 1–3 years after puberty. Humans take, on average, at least 10 years for this transition (Bogin, 1999, 2001; Walker et al., 2006). Part of the reason for this is "adolescent infertility"—a time of about 3 years after menarche (first menstruation) with absent or infrequent ovulations. Equally important is that, cross-culturally, human births usually take place within marriage, which is most often a late-adolescent event (Bogin, 2001). The biocultural interaction between adolescent sterility and late-adolescent marriage results in an average age at first birth of 19 years for women in traditional premodern societies, such as foragers, pastoralists, and horticulturalists. The role of marriage and kinship in human reproduction are discussed later in this chapter.

The second human difference is that during the adolescent life stage both boys and girls experience a rapid acceleration in the growth velocity of almost all skeletal tissue—the adolescent growth spurt (see Fig. 31.1, velocity curve). Most primate species have rapid growth in length and body weight during infancy and then a declining rate of growth from weaning to adulthood. Some primate species may show a rapid acceleration in soft tissue growth at puberty, especially of muscle mass in male monkeys and apes. Sexually maturing nonhuman primates may have skeletal spurts in the face, for example, due to the eruption of large canine teeth in male baboons (Bogin, 1999; Leigh, 2001). However, unlike humans, other primate species either have no adolescent acceleration in total skeletal growth or a very small increase in growth rate (Hamada and Udono, 2002). The velocity of human long bone growth may be five times as rapid as that of apes. The human skeletal growth spurt is unequaled by other species, and when viewed graphically, the duration and intensity of the growth spurt defines human adolescence (Fig. 31.1)—it is a species-specific characteristic. More detailed explanation of the evolution of human

adolescence may be found elsewhere (Bogin, 2009; Bogin and Smith, 2012).

A summary of human life history stages is given in Table 31.2 and is depicted in terms of growth curves in Fig. 31.1.

31.3 The Primate Roots of Human Life History

Human life history was derived from processes of growth, development, and maturation that first evolved in ancestral Primates, the taxonomic order that includes prosimians, tarsiers, monkeys, apes, and humans. The earliest known primate fossils date to about 55 MYA, but it is likely that primatelike mammals are even older (Tavaré et al., 2002). An outline of evolutionary history of living apes and people is shown in Fig. 31.4. Primate evolutionary history provides considerable time for a variety of developmental patterns to evolve in different primate species. Even so, all primates share a number of biological and behavioral characteristics.

The British anatomist Wilfred Le Gros Clark (1895–1971) defined the Order Primates by nine categories of traits (Le Gros Clark, 1960). The first seven trends are associated with the history of primates as tree-living mammals, with specializations for rapid movement in trees and small object feeding, such as eating insects, flowers, fruits, and seeds. These traits are as following:

1. a generalized limb structure, such as five fingers and toes and retention of the clavicle (collar bone) that are often absent in other mammals;
2. flexible digits, especially the thumb and large toe;
3. flattened nails and tactile pads that occur at the end of digits;
4. less emphasis on the sense of smell;
5. reduced muzzle size;
6. more dependence on sense of vision (ie, binocular, stereoscopic, and color vision);
7. retention of the four kinds of mammalian teeth (incisor, canine, premolar, and molar) and a simple cusp pattern of molars;
   (Traits 8 and 9 are of more direct relevance to life history evolution.)
8. longer gestation and increased vascularization of placental membranes; and
9. expanded and elaborated brain, especially the cerebral cortex.

Longer gestation, along with a more efficient placenta, allows for primates to have more brain growth and development before birth. As explained later in this chapter, the energy need of a larger brain is a major factor driving life history evolution. A survey of the living primates (DeSilva and Lesnik, 2008) reports that humans are at

TABLE 31.2 Stages in the human life cycle

Stage	Growth events/duration (approximate or average)
Prenatal Life	
Fertilization	
First trimester	Fertilization to 12th week: embryogenesis
Second trimester	Fourth through sixth lunar month: rapid growth in length
Third trimester	Seventh lunar month to birth: rapid growth in weight and organ maturation
Birth	
Postnatal life	
Neonatal period	Birth to 28 days: extrauterine adaptation, most rapid rate of postnatal growth and maturation
Infancy	Second month to end of lactation, usually by 36 months: rapid growth velocity, but with steep deceleration in growth rate, feeding by lactation, deciduous tooth eruption, many developmental milestones in physiology, behavior, and cognition
Childhood	Years 3–7: moderate growth rate, dependency on older people for care and feeding, midgrowth spurt, eruption of the first permanent molar and incisor, cessation of brain growth by end of stage
Juvenile	Years 7–10 for girls, 7–12 for boys: slower growth rate, capable of self-feeding, cognitive transition leading to learning of economic and social skills
Puberty	Occurs at the end of juvenile stage and is an event of short duration (days or a few weeks): reactivation of central nervous system of sexual development, dramatic increase in secretion of sex hormones
Adolescence	The stage of development that lasts for 5–10 years after the onset of puberty: growth spurt in height and weight; permanent tooth eruption almost complete; development of secondary sexual characteristics; sociosexual maturation; intensification of interest in and practice of adult social, economic, and sexual activities
Adulthood	
Prime and transition	From 20 years old to end of child-bearing years: homeostasis in physiology, behavior, and cognition; menopause for women by age 50
Old age and senescence	From the end of child-bearing years to death: decline in the function of many body tissues or systems

the extreme upper end of the range of variation for primate brain size at birth. Human brain size at birth averages 375 g with a range from 255.0 to 540.0 g. Chimpanzees average only about 150 g with a range of 109–181 g. The human advantage in brain size increases rapidly with age so that adult chimpanzee brains average 382 g in contrast to 1350 g for adult humans (Fig. 31.5).

The gestation length of chimpanzees is 230–250 days and human gestation length ranges from 259 to 280 days. The average difference in gestation length is 29 days, which by itself cannot account for the more than doubling of human brain size at birth. DeSilva and Lesnik (2008) note that human newborn body size relative to maternal body size also lies at the extreme upper end of the primate range of values. Human babies are bigger and much fatter than the newborns of chimpanzees and other primates (Bogin and Smith, 2012). The greater fatness of the newborn, which is comprised of both white and brown adipose tissue, in turn, provides energy to support the larger brain at birth. Human mothers are able to provide

more energy to their gestating young to develop a larger newborn body and brain. Human mothers can supply this energy because, compared with chimpanzees, they have greater fatness (a reserve of energy) before pregnancy, as well as a greater intake of food, and often less physical labor, during pregnancy due to provisioning and care by other members of their social group (Bogin et al., 2014). More detail on the human style of social group care is given later in this chapter.

The overall adaptive benefit of the larger human brain is likely for more complex social organization (Dunbar, 2009). Compared with non-primate mammalian species, a larger brain allows for more neurons, neuronal connections, and glia, which support brain function (Herculano-Houzel et al., 2007). Together, these allow for greater learning, more information storage, such as the location of foods or social allies (both of which may change from day-to-day and season-to-season), more rapid processing of information, and greater flexibility in behavior.

5. Evolution of Human Brains

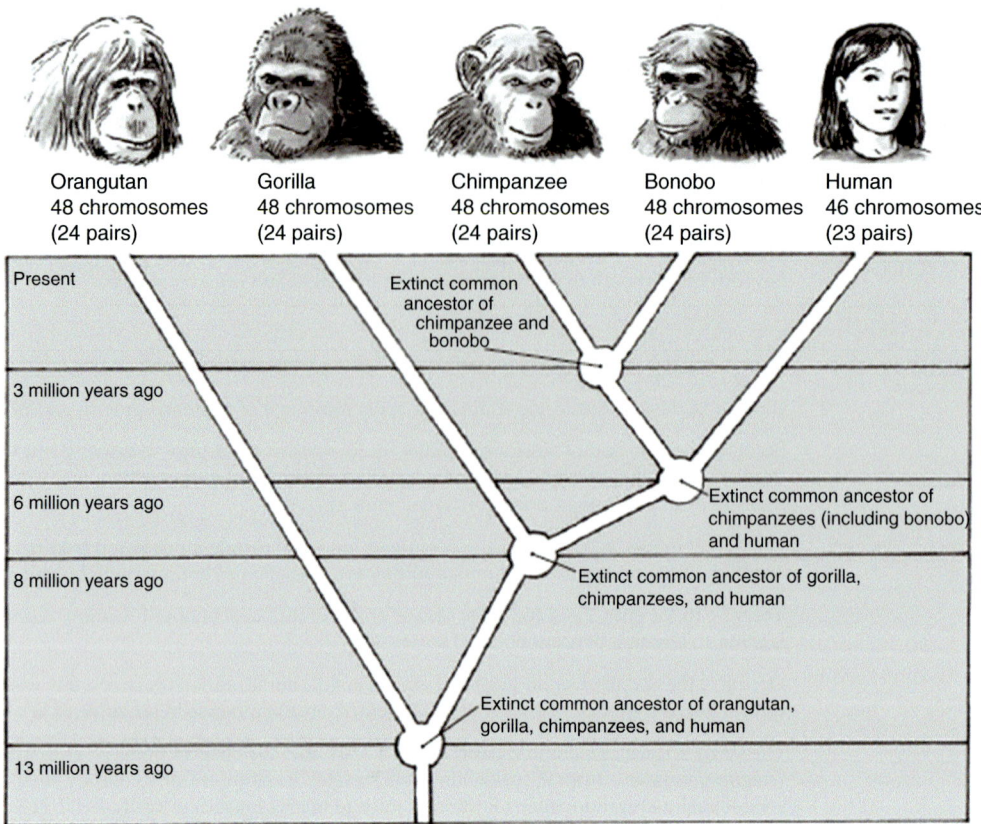

Orangutan
48 chromosomes
(24 pairs)

Gorilla
48 chromosomes
(24 pairs)

Chimpanzee
48 chromosomes
(24 pairs)

Bonobo
48 chromosomes
(24 pairs)

Human
46 chromosomes
(23 pairs)

Present

3 million years ago

6 million years ago

8 million years ago

13 million years ago

Extinct common
ancestor of
chimpanzee and
bonobo

Extinct common ancestor of
chimpanzees (including bonobo)
and human

Extinct common ancestor of gorilla,
chimpanzees, and human

Extinct common ancestor of orangutan,
gorilla, chimpanzees, and human

FIGURE 31.4 Phylogenetic history of the Hominidae—living apes and people. Dates are based on genomic analysis and are approximate. *Reproduced with permission from NOVA Teachers, Judgment Day: Intelligent Design on Trial, https://www-tc.pbs.org/wgbh/nova/teachers/activities/pdf/ 3416_id_02.pdf, © 2007–2020 WGBH Educational Foundation.*

31.4 Unique and Unusual Features of Humans Life History

In addition to a large and complex brain, anthropologists usually add several other physical and behavioral traits to the list of human-evolved specialties. These traits include bipedal locomotion, reduced anterior dentition with molar dominance, dependence on material culture for survival, and use of symbolic language. The evolution of these traits is well treated in many textbooks (Muehlenbein, 2015). Described above were unusual features of human physical growth and sexual maturation, including alternating cycles of rapid and slow growth velocity, prolonged childhood and adolescent periods of development, delayed sexual maturation along with several years of adolescent infertility following menarche.

A species-specific neuroendocrine physiology underlies all of these traits (Bogin, 1999; Hwa et al., 2013). The human pattern of endocrine physiology results in the menstrual cycle of women, no obvious signals of ovulation, the continuous sexual receptivity of both sexes, the development of secondary sexual characteristics (eg, patches of hair in the groin and armpits), and menopause in women. The living apes have some but not all of these traits. Compared with humans, apes have more and longer hair on most of their body, including axial and pubic hair. In humans, however, the size of the aggregation of axial apocrine glands is larger than in chimpanzee or gorilla (Stoddart, 1992: p. 51). Ape females have estrous cycles, with no menstruation, and sexual activity tends to peak with the visible anogenital swellings of estrus (the time of ovulation) and wane when females are not swollen.

31.5 Reproductive Strategies

Another unusual feature of human reproduction is that women almost always receive a great deal of help from other people. Most human reproduction takes place within the cultural institution of marriage, which establishes rules of behavior for the marriage partners and their families. A network of kinfolk, related to the mother via biological and social ties, provide care, feeding, protection, education, and more to the mother and her offspring (Hrdy, 1999, 2009). Especially

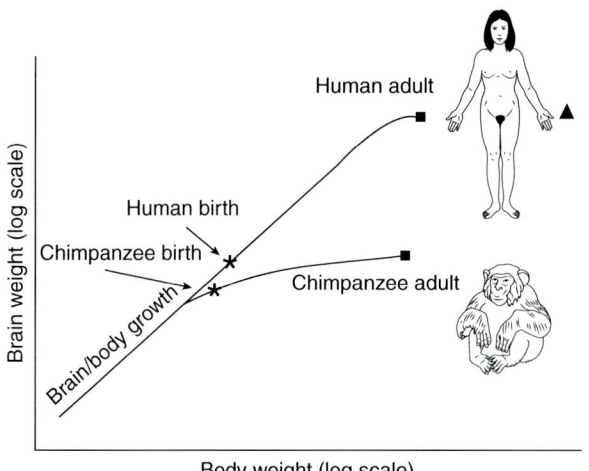

FIGURE 31.5 Growth curve for human brain and body compared with the chimpanzee. The length of the human fetal phase, in which brain and body grow at the same rate for both species, is extended for humans. Brain growth of chimpanzee begins to slow during late gestation and continues to slow after birth, but humans maintain the high rate of brain growth during infancy and childhood (as per Fig. 31.2). In contrast, the rate of human body growth slows after birth. If human brain/body growth rate were equal to the chimpanzee rate, then adult humans would weigh 454 kg and stand nearly 3.1 m tall, indicated by the "▲" symbol. *From Bogin, B., 1999. Patterns of Human Growth, second ed., Cambridge University Press, Cambridge; based on the data and concepts of Martin, R.D., 1983. Human brain evolution in an ecological context. Fifty-Second James Arthur Lecture. American Museum of Natural History, New York.*

important in many human societies are postmenopausal women who contribute to the reproductive success of their daughters and other younger women (Hawkes and Coxworth, 2013; Bogin, 2009; Hawkes, 2003). In contrast, no other species of animal practices marriage.

The only nonhuman primate species that provide the type of assistance seen in human families are the South American marmosets and tamarins and a few species of Madagascar lemurs. These species are cooperative and communal breeders and provide "allomaternal care," that is, feeding and protection of infants provided by other than the biological parents. Allomothering is rare in mammals, with only 1.8% all living species practicing any type of "cooperative breeding" (Lukas and Clutton-Brock, 2012). Cooperative breeding is defined by a reproductive strategy in which most of the females of a social group do not breed regularly and instead provide alloparental care to the offspring of breeding females. In these species the dominant female is responsible for at least 90% of all breeding attempts (Lukas and Clutton-Brock, 2012). The only nonhuman primates practicing this form of cooperative breeding are the South American marmosets and tamarins (Burkart et al., 2014). Many species of tamarins, such as *Saguinus imperator and Saguinus oedipus*, have a polyandrous mating system, with two or more males mating

with one female (Zahed et al., 2010). The female ovulates from both left and right ovaries simultaneously, so it is possible for different males to sire her litter of offspring, which usually numbers two or three babies. The multiple breeding males, or "fathers," support the breeding female by taking care of the twins or triplets in order that the mother can feed herself. Some species of lemurs practice "communal breeding," where most adult females breed regularly, but also share care such as "allonursing," which is the nursing of infants not their own (Lukas and Clutton-Brock, 2012; Vasey, 2007).

31.6 The Evolution of Hominin Communal Breeding

No living ape species practice any form of cooperative or communal breeding. Both the common chimpanzee (*Pan troglodytes*) and bonobos (*Pan paniscus*) practice social breeding, meaning that the females live in groups and virtually all breed, but rarely, if ever provide allomaternal care to others (Lukas and Clutton-Brock, 2012). There is no evidence that any fossil apes, or indeed any fossil or living Old World monkeys, practiced cooperative or communal breeding. In contrast, living humans provide cooperative and communal assistance, support, and mutual aid in relation to reproduction and care of pregnant women and their offspring. Indeed, humans have been called "hypercooperators" in these regards (Burkart et al., 2014).

It seems that some form of cooperative breeding, or more likely communal breeding, was independently evolved in some hominin species after the ape–hominin evolutionary split at about 6 million years ago (Fig. 31.4). Hominins are defined as modern humans and extinct species of the genera *Homo, Australopithecus, Paranthropus, Kenyanthropus, Ardipithecus, Orrorin,* and *Sahelanthropus*. All of these species were capable of some degree of bipedal walking. Which hominin species is the first to evolve communal or cooperative breeding is, at present, unknown. The most likely candidates are *Homo ergaster* or *Homo erectus* due to the estimate that females of these species were 50% heavier than australopithecine and earlier *Homo* females. Their larger body size would have necessitated greater energy consumption, which may have been supplied via cooperation in division of labor and the provisioning of food (Aiello and Key, 2002).

31.6.1 From Cooperative/Communal Breeding to Human Biocultural Reproduction

Humans not only show hyper-cooperation and communal assistance in reproduction, but are also superflexible in the ways in which communal support and cooperation are provided. In the wide spectrum of living and historic human societies virtually all possibilities of support are, or have been, shown. Alloparental care may be provided by fathers, older siblings, aunts and uncles, grandmothers and grandfathers, other kinsfolk of varying genetic relationship, friends, servants, slaves, paid professional caregivers, teachers, and religious or secular state institutions. In all of these societies, the tendency is for all fertile women to reproduce, which is more like the communal breeding of the lemurs than the cooperative breeding of the marmosets and tamarins. In the living apes, all fertile females reproduce. While speculative, it seems possible, and likely, that human superassistance in reproduction and offspring care evolved from an ancestor, such as *H. erectus*, with a more traditional type of communal breeding system, rather than a cooperative breeding system with reproduction confined mainly to dominant females.

The intensity and complexity of human assistance in the production and care of offspring is both quantitatively and qualitatively distinct from that of other mammalian species. Moreover, if human reproductive behavior evolved from a communal-breeding hominin, then it is not appropriate to label people as cooperative breeders, as is often done. The human reproductive strategy seems to require a different terminology to capture the novel type of alloparental care, marriage, and kinship relationships that we practice. One proposal is to use the phrase "biocultural reproduction" (Bogin et al., 2014).

Biocultural reproduction describes a suite of biological and sociocultural adaptations that include:

1. early weaning of infants by mothers, leading to an increased rate of reproduction compared with apes;
2. a life-cycle period of childhood, characterized by the absence of nursing but considerable ongoing nutritional dependence, creating extended opportunities and needs for the provision of care by individuals other than the parents;
3. cognitive capacities that allow a shift from genetically based prosocial behavior to hyper-cooperation based on nongenetic marriage and kinship relationships;
4. reproductive strategies that provide for culturally universal alloparental care of offspring that is demographically and ecologically flexible to local environments; and
5. decreased lifetime reproductive effort for women compared with ape females.

Each of these is discussed briefly below (see Bogin et al., 2014 for more detail).

31.7 Early Weaning and the Childhood Stage of Life History

Chimpanzee, bonobo, and orangutan mothers nurse their infants for 5–7 years. If the mother dies before this time, then the infant usually dies as well. No other female of the social group will adopt and nurse an orphan, as these other females are either pregnant or have their own infants. In addition, the infant ape is not able to easily shift its biopsychological attachment from its deceased mother to another caregiver (Hrdy, 2009). Only after weaning, do ape females have another successful pregnancy. The energetic demands of nursing a current infant either prevent ovulation and a subsequent pregnancy, or if a pregnancy occurs, then the result is loss of the embryo, fetus, or newborn. The time between successful births, coupled with the loss of 65% of liveborn infants to accidents, disease, and predation, means that chimpanzee females have, on average, only two offspring who survive to adulthood. This barely maintains chimpanzee populations at replacement levels.

In contrast, human women in traditional premodern societies nurse their infants for an average of only 3 years (Bogin, 1999; Humphrey, 2010). Weaning at this relatively early age allows the mother to become pregnant sooner than chimpanzee mothers. Human women may stack their offspring, having a current infant and one or more children, juveniles, and adolescents. In this sense, early weaning is an evolutionary adaptation to increase the mother's reproductive fitness. Early weaning entails a life history trade-off. The 3-year-old child potentially suffers because she is denied a ready source of food, as well as the emotional comfort and contact with her mother. Three-year-old humans are coevolved to avoid these dangers as they transition from the infant to the childhood stage, slow their rate of growth and, thus, their energy demands, and extend their social attachments beyond their mother.

The human capacity for a wide range of social attachments usually is well developed in infancy and both infants and children will accept feeding and care from a variety of older people. Moreover, these older people have an evolutionarily based desire to care for children (Bogin, 1999), which may be based on selection for prosocial behaviors unique to the genus *Homo* (Cowell and Decety, 2015). Walter Goldschmidt (2006) emphasized the importance of this human desire to be cared for and to care for others. Goldschmidt called this desire *affect hunger*, which he defined as the urge to get expressions of affection from others, often from many others. Chimpanzee infants are much less capable of redirecting emotional attachments to others, which is a reason why they usually die if they lose their mother.

A possible biological mechanism for human affect hunger is the ontogeny of brain neocortical myelination which follows a distinctively delayed and prolonged pattern of development in humans compared with chimpanzees (Miller et al., 2012). The delay in humans is especially noticeable postpuberty and may be associated with unique cognitive–emotional capacities of adolescents and young adults. Human adolescents seek out and form strong emotional bonds with other members of their social groups of younger, older, and similar age. In this regard, it is interesting to note that some of the genomic loci that have shown an intense selection in modern humans compared with other hominins, such as Denisovans (Neanderthal-like hominins from East Eurasia), favor both pre- and postpubertal nervous system development including neuronal growth, synaptic transmission, greater synaptic plasticity, and language (Meyer et al., 2012).

In summary, the evolution of early weaning followed by a protracted period of childhood dependence in hominin evolution greatly expanded opportunities and needs for care to be provided by other people in addition to the mother.

31.8 Cognitive Capacities for Nongenetically Based Marriage and Kinship

Cooperative/communal breeding by hominin ancestors may have facilitated and necessitated the evolution of larger brains and cognitive development due to the social complexities required to effectively carry this out (Dunbar, 2009). One of the evolved cognitive capacities that allowed for alloparental care is a complex mental skill that permits one individual to impute or attribute mental states to self and to others. This skill is sometimes called "theory of mind" or "shared intentionality" (Povinelli and Preuss, 1995; Povinelli and Vonk, 2003; Tomasello et al., 2005). Tomasello et al. define shared intentionality as, "…the ability to participate with others in collaborative activities with shared goals and intentions…" (p. 675). One essential collaborative activity in human social affairs is the creation of shared fictional realities based on common intentions. This is referred to as "collaborative pretense" (Bretherton, 1989) and is a basis for much of human social life.

In many ways, the rules for marriage, kinship, and alloparental care of offspring are a collaborative pretense of the genetic relatedness that drives cooperative/communal breeding in other species. In most human societies, marriage unites two families of people, not only the bride and groom, into larger kinship groups with contractual obligations and privileges (Bogin et al., 2014). Marriage contracts extend formal kinship systems with the use of language to specify kinship relations, such as words for "father/father-in-law," "mother/mother-in-law," "sister/sister-in-law," "cousin," etc. In the 19th century, the anthropologist Lewis Henry Morgan (1818–81) identified six basic types of kinship systems practiced by human societies (https://www.umanitoba.ca/faculties/arts/anthropology/tutor/kinterms/termsys.html). Each system is named for a human culture in which it was first described. People descended from western European culture generally practice the Eskimo kinship system.

Another type is called Hawaiian kinship. The four other kinship classification systems are designated Crow, Iroquois, Omaha, and Sudanese. Each of these systems establishes a formal network of named kinship relationships that forms the social basis for hyper-cooperative behavior. The degree of genetic relatedness between members of the network has little or no bearing on their obligations or privileges toward each other.

Human social kinship and the language of kinship terminologies reveal much about human shared intentionality. Povinelli and colleagues proposed that humans have a deeper type of shared intentionality than that found in any ape (Povinelli and Preuss, 1995; Povinelli and Vonk, 2003). In both apes and humans, the skills of shared intentionality develop during the first 24 months of life. Both ape and human infants come to understand others as animate, goal-directed, and intentional agents. Human infants develop further and come to possess a species-specific motivation to share emotions, experience, and activities with other persons. Tomasello and other evolutionary psychologists (Tomasello et al., 2005) demonstrate this human emotional difference using experimental studies of humans and apes. Goldschmidt (2006) called the human difference "affect hunger," which he described from anthropological field studies with dozens of human societies. This human emotional motivation is likely to be one basis of our species' evolution from a hominin ancestor practicing genetically based communal or cooperative breeding to the socially based biocultural reproduction that humans now practice.

Melvin Konner (2010) discussed examples of the importance of infant affection with many people, in addition to the mother, based on his research on the healthy development of human children in forager societies. Especially noteworthy are Konner's descriptions of the affection between children with members of the social group who serve as alloparents and as teachers of essential survival skills. Membership in these African forager groups is fluid and people living in any group may be quite distantly related to each other, and to each other's infants and children, in any hereditary sense (Hill et al., 2011). Even so, older individuals willingly provide alloparental care to youngsters. The essential message from these forager studies is that in the human species, relationships defined by marriage and social kinship rules often supersede genetic closeness.

31.9 Decreased Lifetime Reproductive Effort

Early weaning, the childhood period of development, human shared intentionality, marriage, and kinship based on collaborative pretense all work together to enhance "human lifetime reproductive effort" (LRE).

LRE may be defined as the energy a woman devotes to reproduction during her lifetime versus all other energy costs she expends to maintain and operate her body. Eric Charnov et al. (2007) calculated LRE as the product of adult reproductive lifespan in years E, annual birth rate b, and offspring mass at feeding independence m_i, normalized to adult size, m_a:

$$\text{LRE} = E \times b \times m_i/m_a$$

Charnov et al. found that for 54 lizard species and 40 mammal species LRE averaged about 1.4. This means that for those species analyzed, a female requires an energy investment equal to 1.4 times her own body mass for an average amount of reproduction. No primate species and no cooperatively breeding species were included in the analysis. LRE has not been estimated for any nonhuman primate.

Bogin et al. (2014) applied the LRE formula to human women from 17 small-scale, traditional societies, without modern contraception or medical care, including the South American Ache, Hiwi, Yanomamo and Tsimane, the Philippine Agta, African pygmy Baka, Gambian villagers, East African Turkana, the southern African Ju/'hoansi, and Central American Maya. Estimates of energy inputs from both the mother and from alloparents were included to calculate the terms of the equation above, and it was found that human LRE averages only about 1.1. Compared to the value of 1.4 reported by Charnov and colleagues for nonprimate mammals, human LRE represents an energy savings of about 21%. It is not that human offspring are less expensive than offspring of other species, rather it is that people other than the mother contribute the food and labor required to successfully produce and rear infants. Each human mother invests less than the total amount of energy required, and this saving allows each mother to produce either more total offspring, larger offspring, or both. Each offspring stands a good chance of survival due to the allocare provided by other family and social group members. The net result is greater reproductive success, which is one of the primary drivers of evolutionary biology. Along with early weaning and the stacking of offspring, human population size has expanded over time, while ape population sizes remain, at best, constant.

31.10 When Did Modern Human Life History Evolve?

It is not known when modern human life history evolved. It is likely that its evolution was mosaic, meaning that different aspects appeared at different times in hominin evolutionary history. Given the more humanlike features of *Homo habilis* and *Homo rudolfensis*, including their 610–776 cc brain size, it is possible that pieces of the human life history mosaic appear at about 2 MYA. The first bit of the human mosaic may have been a reduction of the infancy (nursing) stage and the beginnings of the childhood stage of life history. The selective advantage of earlier weaning would have been more a rapid rate of reproduction for adult females (Bogin and Smith, 2012).

Some incipient form of communal/cooperative breeding behavior may have coevolved soon after earlier weaning, as a strategy to provide alternate foods care for the newly weaned infants. As described earlier in this entry, a recognizable form of communal/cooperative breeding behavior may have evolved with the appearance of *H. erectus*, perhaps as early as 1.7 million years ago. This species was more humanlike in body size and shape than earlier hominins, had brains averaging 865 cc, and had improved use of technology, including the controlled use of fire. *Homo erectus* dental development was apelike, with eruption of all their permanent teeth by 12 years of age (Dean and Liversidge, 2015). Living humans complete dental eruption at about 18 years of age, and the additional 6 years allows for a great deal more of physical, cognitive, and emotional maturation (Bogin, 1999). Body growth in height was humanlike, with both males and females achieving an average adult stature within the range of height of modern humans (Dean and Smith, 2010). Genomic surveys have identified human-specific alleles at the *LIN28B* and *SIX6* loci that are associated with taller adult height and delayed puberty (Perry et al., 2015). Perhaps these loci, and others, were under selection to extend the childhood stage of *H. erectus* life history and allow for larger body size.

If *H. erectus* evolved an extended childhood, then more of the mosaic of human life history would have been in place. This included larger body size, large brain size, greater cognitive capacities as evidenced by more sophisticated stone tool technology and possible symbolic engravings (Joordens et al., 2015), and quite possibly communal or cooperative breeding strategies. Communal breeding would have helped to overcome the trade-off between the energy requirements for body and brain maintenance of adult females and the food provisioning and caregiving to their offspring. Initially, a communal breeding strategy would have been based on close genetic relationships, as is the case for similar breeding strategies in other mammals. Over time, positive feedback between cooperation in division of labor and the provisioning of food to mothers and their offspring, along with further increases of brain size, improved technology, cognition, and social–emotional complexity in later hominin species, would have provided the impetus to evolve from communal breeding toward biocultural reproduction. The final pieces of the mosaic of the modern human pattern of growth, development, and life history, as well as biocultural reproduction, likely evolved after the origins of the Neanderthals and perhaps only with the appearance of anatomically and behaviorally modern *Homo sapiens*, our own species, ~160 000 years ago (Mcbrearty and Brooks, 2000; White et al., 2003).

31.11 What We Know and What We Need to Know

It is generally accepted that human life history, with the stages of childhood, adolescence, and a postreproductive life, is distinct from all other species. Less clear is how these stages evolved and how they contributed to the evolution of biocultural reproduction. It is argued here that a decoupling from genetic relatedness of alloparental care of human infants and children is, more than anything, that which sets humans apart from other communal/cooperative breeding species (see Bogin et al., 2014 for more details). To be sure, not all researchers accept the proposal for biocultural reproduction advocated here. Additional research support is needed to link the evolution of human life history stages with human developmental biology, cognition, and behavior that underlie biocultural reproduction.

Future research should be directed toward a better understanding of human lifetime reproductive effort and other unusual traits such as menopause. A lower LRE for humans may have contributed the metabolic "fuel" necessary to slow the pace of ageing and extend lifespan, thereby expanding the pool of alloparents to older individuals and offering the possibilities for intergenerational transfers of knowledge and material wealth. These transfers from older generations are critical to human biocultural reproduction.

The evolution of human menopause and postmenopausal longevity are often explained as a consequence of selection for vigorous grandmothers who provide care toward their daughters and grand-offspring (Hawkes and Coxworth, 2013). Captive chimpanzees females who live to age 50 years also experience menopause, but are so frail and decrepit that they have very short postmenopausal lives (Herndon et al., 2012). Wild chimpanzees almost never live to age 40 years (Hill et al., 2001). A human life history model that includes biocultural reproduction may parsimoniously account for human postmenopausal longevity as the consequence of better-cared-for infants and children who, with the assistance of a network of kinship-based allocarers, develop into healthier and longer-lived adults. In our evolutionary past, healthier women who outlived their biological allotment of ova, and experienced menopause at about age 50 years, may have turned their energies toward culturally defined behaviors for allocare of their grand-offspring and the offspring of others in their social groups (Bogin and Smith, 2012).

A critical gap in knowledge about the evolution of human life history is the genomic correlates of the stages of development. The human pattern of growth shown in Fig. 31.1 is universally found in all healthy people. Given the neuroendocrine regulation for the transition in growth velocity from infancy to childhood and then to later stages of development a genomic basis is assumed (Hochberg, 2011), but the exact genomic regions for this control are unknown. For more than century the tried-and-true approach to understand normal growth and development was to study pathology and find its underlying cause. This has revealed much about the physiology and genetics of growth, but only shows what happens when things go wrong, not the healthy range of developmental variation. More recently, genome-wide association studies, or GWAS, have complimented and, perhaps, supplanted the study of pathology. GWAS depends on a combination of methods from genomics and bioinformatics—GWAS processes mega amounts of data—and GWAS benefits from a "hypothesis-free" approach to the identification of genomic regions as small as single-nucleotide polymorphisms (SNPs) that associate with variation in phenotype development (Sainani, 2015).

This strength is also one of the limitations of GWAS, as it can only identify correlations and not causal mechanisms. The associations of *LIN28B* and *SIX6* loci with taller adult height and delayed puberty mentioned above, for example, do not explain how these loci influence growth and development. Moreover, the influence of polymorphisms at these two loci and most other loci is quite small. The polymorphisms of the *LIN28B* locus are estimated to change the age at puberty by about 6 weeks (Perry et al., 2015). Another issue is that other loci identified by GWAS, such as *CENPW*, are associated with an increased age at menarche but reduced adult height. In these GWAS projects, menarche for girls or voice-breaking in boys are used as a proxies for age at puberty, but both of these events occur at or after the peak of the adolescent growth spurt. This is at least 2 years after the neuroendocrine changes of puberty take place in the hypothalamic–pituitary–gonadal axis, so these loci do not really identify the fundamental mechanisms for the onset of puberty. Even so, GWAS identifies fruitful lines for new research and at the time of this writing ~175 common genomic polymorphisms and variants have been identified with some association with the timing of puberty in girls and boys or menopause (Perry et al., 2015; Day et al., 2015). Nearly 700 common variants have been identified for variation in normal human height (Chan et al., 2015). Most of these are low-frequency or rare variants, with uncertainty regarding their mechanisms of action and difficulty in devising experimental methods to detect their probably small molecular effects.

Despite these limitations the genetic architecture and genomic regulation for some of human life history are being revealed. To be sure, there are likely to be important environmental and epigenetic factors involved as well that regulate human life history biology within each stage and the transitions between stages.

One environmental factor, with epigenetic influences, is diet. Energy availability is a major driver of biological development and evolution. One basis for human biocultural reproduction is large brain size and complexity. Compared with chimpanzees and other primates the human brain requires a higher input of energy and other nutrients at all life history stages, and especially prior to adulthood (Leonard et al., 2007). How dietary requirements interact with the genetic basis for the pattern of human brain growth and development is an active area of research. The National Center for Biotechnology Information (http://www.ncbi.nlm.nih.gov/gene) lists nearly 2500 genes related to the human brain. GWAS finds that more than 100 of these genomic elements are associated with brain expansion and elaboration of subcortical and cortical regions (Hibar et al., 2015; Florio et al., 2015; Geschwind and Konopka, 2012). After birth, brain growth and development is most rapid during infancy and childhood and energy requirements peak during childhood, measured as resting metabolic rate or as daily energy requirement expressed in grams of glucose per day (Kuzawa et al., 2014). Body growth rate decreases throughout infancy and childhood in a strong, inverse manner with glucose demand for the brain.

This seems an obvious life history trade-off between brain and soma, driven by a genetic architecture for brain growth and nutrient sensing by somatic cells and tissues. According to the data presented by Kuzawa et al. (2014; Fig. 31.1), the life history transition from infancy to childhood takes place when brain glucose uptake exceeds $100-110$ g day^{-1}. This association is only an observation and may be further studied using existing data from epidemiological research of poorly nourished infants and children in low-income regions, for example, the COHORTS collaboration (Richter et al., 2012) and the Healthy Birth, Growth, Development Knowledge Integration initiative (Varela-Silva et al., 2016). Some of these existing studies have data on energy intake and expenditure, as well as growth velocity of the body and the brain, that may be analyzed to see how nutrient availability influences the infant–childhood transition and later life transitions between stages.

Genomic influences on human longevity and the importance of postreproductive adults to human biocultural reproduction are being revealed by GWAS. A

unique human variant at the CD33 locus, an immunoregulatory receptor, seems to be protective against late onset Alzheimer dementia (Schwarz et al., 2016). Only modern humans seem to have this CD33 variant, a substitution of adenine for cytosine in allele rs3865444, as all samples of gorillas (Gorilla gorilla), chimpanzees (P. troglodytes and P. paniscus), Neanderthals, and Denisovans so far sequenced have only the cytosine allele. As mentioned above, maintaining physical and cognitive health after reproductive age, especially after age 40 years, is especially difficult for most primates, yet humans regularly do so in all known societies. The human CD33 variant may be one of many loci that may contribute to human health after age 40 years—Schwarz et al. (2016) list 11 uniquely human alleles with disease-protective associations that maintain cognitive function at postreproductive ages. CD33 and these other alleles may have only evolved in the last 50 000–100 000 years of human evolution, and modern human longevity may be no older than this (Caspari and Lee, 2004).

In living humans, health at age 40 years and beyond depends on the many factors operating across an individual's lifespan, especially during gestation, infancy, and childhood (Bogin, 2009). The CD33 allele may promote healthy growth and development for infants and children as it codes for an immune function protein with antiinflammatory, cell adhesion, and endocytosis functions (Walker et al., 2015), all of which are important throughout life.

There may be a role for many of the proposed "healthy aging" alleles, throughout the human life cycle and across generations. A case in point is the human variation of the APOE locus. Chimpanzees and other primates have only the ε4 isoform, but humans have three isoforms, ε2, ε3, and ε4. The APOE locus regulates lipid metabolism, and the ε2 and ε3 isoforms may have been selected for as a defense against the harmful consequences of meat eating. Chimpanzees and people on high animal food diets and with the ε4 allele have increased risks for cardiovascular disease, stroke, vascular dementia, and late onset Alzheimer dementia (Finch and Stanford, 2004). People with the ε2 and ε3 isoforms have lower risks for these problems. The fact that the ε4 allele remains in human populations may be evidence of a life history trade-off, in this case an example of "antagonistic pleiotropy." When a biological characteristic is beneficial during development or during the reproductive years of life, but harmful at older ages, that trait is said to exhibit antagonistic pleiotropy.

The APOε4 allele, which seems a detriment for older age—onset pathology, may be beneficial at younger ages. Experimental studies find that young adults with the ε4 allele perform better on measures of attention, even showing greater brain activation as detected by fMRI (Rusted et al., 2013). Research with Brazilian shantytown children investigated the associations between *APOE* genotype, infection, nutrition, physical growth, and cognitive performance. In a series of studies (Oriá et al., 2007), children suffering from intestinal infection and severe diarrhea were shown to have better long-term outcomes for physical growth and school performance if they carried the *APOE ε4* allele. In another intervention study, shantytown infants and children, aged 1.4–6.6 years, were given a variety of combinations of retinol, zinc, and glutamine supplementation over a 1-year period (Mitter et al., 2012). Participants with the *APOE ε4* allele showed significant, positive correlations with height, weight, and weight-for-age z-scores; verbal learning; and nonverbal learning in response to all supplementation regimes. In contrast, participants with the ε2 and ε3 isoforms showed declines in both verbal and nonverbal learning, regardless of the type of intervention. The authors of these studies propose that *APOE ε4* is a "thrifty allele" and that during development it may function interactively with gut physiology and essential nutrients to benefit both immediate nutritional and growth status, as well as long-term cognitive outcomes.

31.12 Conclusion

Human life history from birth to death includes six stages: infancy, childhood, juvenile, adolescence, adulthood, and the postreproductive stage. The last is especially important for woman as they become infertile around the time of menopause. Hominins prior to the genus *Homo* probably had life histories more similar to living apes, growing, developing, reproducing, and dying at a pace nearly twice as fast as modern humans. The evidence from fossil remains of the hominins suggests that the elements of human life history evolved as a mosaic over more than 1 million years. It is increasingly evident that the complete "package" of modern human life history took shape with the evolution of modern *H. sapiens*, and not before.

It is hypothesized that the evolution of the new childhood life stage represented a feeding and reproductive adaptation of the genus *Homo*. The evolution of human childhood may have been the "kick-off" needed to evolve communal or cooperative breeding strategies for *H. erectus*. Further selective pressures for increased social cooperation, brain size, and brain complexity seem to have resulted in lower lifetime reproductive effort for human women and the transition to a new form of hyper-cooperation called biocultural reproduction. Selection for improved care and feeding of infants and children lead to healthier growth and development prior to adulthood and a greater reserve of physical, social, and cognitive capacities into older ages. Healthier life after the reproductive years may help explain the creation of novel social, economic, and political roles for postreproductive women and a new stage of life history that might be called the "valuable grandmother." It is also likely that lifespan extension and novel roles for older men accrued at about the same time as they did for women. There is nothing like the total package of human life history in the development, biology, and behavior of any other species.

References

Aiello, L.C., Key, C., 2002. Energetic consequences of being a *Homo erectus* female. Am. J. Hum. Biol. 14 (5), 551–565.

Bogin, B., 2009. Childhood, adolescence, and longevity: a multilevel model of the evolution of reserve capacity in human life history. Am. J. Hum. Biol. 21 (4), 567–577.

Bogin, B., 1999. Patterns of Human Growth, second ed. Cambridge University Press, Cambridge.

Bogin, B., 2001. The Growth of Humanity. John Wiley & Sons, New York.

Bogin, B., Bragg, J., Kuzawa, C., 2014. Humans are not cooperative breeders but practice biocultural reproduction. Ann. Hum. Biol. 41 (4), 368–380.

Bogin, B., Smith, B.H., 2012. Evolution of the human life cycle. In: Stinson, S., Bogin, B., O'Rourke, D. (Eds.), Human Biology: An Evolutionary and Biocultural Perspective. Wiley, New York, pp. 515–586.

Bonner, J.T., 1965. Size and Cycle. Princeton University Press, Princeton, New Jersey.

Bretherton, I., 1989. Pretense: the form and function of make-believe play. Dev. Rev. 9 (4), 383–401.

Burkart, J.M., et al., 2014. The evolutionary origin of human hyper-cooperation. Nat. Commun. 5, 4747. Available at: http://www.nature.com/doifinder/10.1038/ncomms5747.

Caspari, R., Lee, S.-H., 2004. Older age becomes common late in human evolution. Proc. Natl. Acad. Sci. U.S.A. 101 (30), 10895–10900.

Chan, Y., et al., 2015. Genome-wide analysis of body proportion classifies height-associated variants by mechanism of action and implicates genes important for skeletal development. Am. J. Hum. Genet. 96 (5), 695–708. Available at: http://www.ncbi.nlm.nih.gov/pubmed/25865494.

Charnov, E.L., Warne, R., Moses, M., 2007. Lifetime reproductive effort. Am. Nat. 170 (6), E129–E142.

Cole, L.C., 1954. The population consequences of life history phenomena. Q. Rev. Biol. 29 (2), 103–137. Available at: http://www.ncbi.nlm.nih.gov/pubmed/13177850.

Cowell, J.M., Decety, J., 2015. Precursors to morality in development as a complex interplay between neural, socioenvironmental, and behavioral facets. Proc. Natl. Acad. Sci. U.S.A. 112 (41), 12657–12662. Available at: http://www.ncbi.nlm.nih.gov/pubmed/26324885.

De Beer, G.R., 1954. Archaeopteryx and evolution. Adv. Sci. 11, 160–170.

De Beer, G.R., 1975. Mosaic evolution. In: Hulse, F.S. (Ed.), Man and Nature. Random House, New York, pp. 36–54.

Day, F.R., et al., 2015. Shared genetic aetiology of puberty timing between sexes and with health-related outcomes. Nat. Commun. 6, 8842. Available at: http://www.nature.com/doifinder/10.1038/ncomms9842.

Dean, M.C., Liversidge, H.M., 2015. Age estimation in fossil hominins: comparing dental development in early Homo with modern humans. Ann. Hum. Biol. 42 (4), 415–429. Available at: http://www.tandfonline.com/doi/full/10.3109/03014460.2015.1046488.

Dean, M.C., Smith, B.H., 2010. Growth and development of the Nariokotome youth, KNM-WT 15000. In: Grine, F.E., Fleagle, J.G., Leakey, R.E. (Eds.), The First Humans — Origin and Early Evolution of the Genus Homo. Springer, New York, pp. 101—120.

DeSilva, J.M., Lesnik, J.J., 2008. Brain size at birth throughout human evolution: a new method for estimating neonatal brain size in hominins. J. Hum. Evol. 55 (6), 1064—1074. Available at: http://linkinghub.elsevier.com/retrieve/pii/S0047248408001401.

Dunbar, R.I.M., 2009. The social brain hypothesis and its implications for social evolution. Ann. Hum. Biol. 36 (5), 562—572.

Fausto-Sterling, A., 2015. Intersex: concept of multiple sexes is not new. Nature 519 (7543), 291. Available at: http://www.nature.com/doifinder/10.1038/519291e.

Finch, C.E., Stanford, C.B., 2004. Meat-adaptive genes and the evolution of slower aging in humans. Q. Rev. Biol. 79 (1), 3—50.

Florio, M., et al., 2015. Human-specific gene ARHGAP11B promotes basal progenitor amplification and neocortex expansion. Science 347 (6229), 1465—1470. Available at: http://www.ncbi.nlm.nih.gov/pubmed/25721503.

Geschwind, D.H., Konopka, G., 2012. Genes and human brain evolution. Nature 486, 481—482.

Goldschmidt, W., 2006. The Bridge to Humanity: How Affect Hunger Trumps the Selfish Gene. Oxford University Press, Oxford.

Hamada, Y., Udono, T., 2002. Longitudinal analysis of length growth in the chimpanzee (Pan troglodytes). Am. J. Phys. Anthropol. 118 (3), 268—284. Available at: http://www.ncbi.nlm.nih.gov/pubmed/12115283.

Hawkes, K., 2003. Grandmothers and the evolution of human longevity. Am. J. Hum. Biol. 15 (3), 380—400. Available at: http://www.ncbi.nlm.nih.gov/pubmed/12704714.

Hawkes, K., Coxworth, J.E., 2013. Grandmothers and the evolution of human longevity: a review of findings and future directions. Evol. Anthropol. 22 (6), 294—302. Available at: http://doi.wiley.com/10.1002/evan.21382.

Herndon, J.G., et al., 2012. Menopause occurs late in life in the captive chimpanzee (Pan troglodytes). Age (Dordr. Neth.) 34 (5), 1145—1156. Available at: http://www.ncbi.nlm.nih.gov/pubmed/22189910.

Herculano-Houzel, S., et al., 2007. Cellular scaling rules for primate brains. Proc. Natl. Acad. Sci. U.S.A. 104, 3562—3567.

Hibar, D.P., et al., 2015. Common genetic variants influence human subcortical brain structures. Nature 520 (7546), 224—229. Available at: http://www.nature.com/doifinder/10.1038/nature14101.

Hill, K., et al., 2001. Mortality rates among wild chimpanzees. J. Hum. Evol. 40 (5), 437—450. Available at: http://www.ncbi.nlm.nih.gov/pubmed/11322804.

Hill, K.R., et al., 2011. Co-residence patterns in hunter-gatherer societies show unique human social structure. Sci. (New York N.Y.) vol. 331 (6022), 1286—1289. Available at: http://www.ncbi.nlm.nih.gov/pubmed/21393537.

Hochberg, Z., 2011. Evolutionary perspective in child growth. Rambam Maimonides Med. J. 2 (3), e0057. Available at: http://www.ncbi.nlm.nih.gov/pubmed/23908815.

Hrdy, S.B., 1999. Mother Nature: A History of Mothers, Infants, and Natural Selection. Pantheon, Random House, New York.

Hrdy, S.B., 2009. Mothers and Others: The Evolutionary Origins of Mutual Understanding. The Belknap Press of Harvard University Press, Cambridge.

Humphrey, L.T., 2010. Weaning behaviour in human evolution. Semin. Cell Dev. Biol. 21 (4), 453—461. Available at: http://linkinghub.elsevier.com/retrieve/pii/S1084952109002286.

Hwa, V., et al., 2013. IGF-I in human growth: lessons from defects in the GH-IGF-I axis. Nestle Nutr. Inst. Workshop Ser. 71, 43—55.

Joordens, J.C.A., et al., 2015. Homo erectus at Trinil on Java used shells for tool production and engraving. Nature 518 (7538), 228—231. Available at: http://www.ncbi.nlm.nih.gov/pubmed/25470048.

Konner, M., 2010. The Evolution of Childhood: Relationships, Emotion, Mind. Harvard University Press, Cambridge, MA.

Kuzawa, C.W., et al., 2014. Metabolic costs and evolutionary implications of human brain development. Proc. Natl. Acad. Sci. U.S.A. 13010—13015. Available at: http://www.pnas.org/content/early/2014/08/21/1323099111/. http://www.ncbi.nlm.nih.gov/pubmed/25157149/. http://www.pnas.org/content/early/2014/08/21/1323099111.full.pdf.

Le Gros Clark, W.E., 1960. The Antecedents of Man, third ed. Edinburgh University Press, Edinburgh.

Leigh, S.R., 2001. Evolution of human growth. Evol. Anthropol. 10, 223—236.

Leonard, W.R., Snodgrass, J.J., Robertson, M.L., April, 2007. Effects of brain evolution on human nutrition and metabolism. Annu. Rev. Nutr. 27, 311—327. Available at: http://www.ncbi.nlm.nih.gov/pubmed/17439362.

Lukas, D., Clutton-Brock, T., 2012. Cooperative breeding and monogamy in mammalian societies. Proc. R. Soc. B 279 (1736), 2151—2156.

Mcbrearty, S., Brooks, A.S., 2000. The revolution that wasn't: a new interpretation of the origin of modern human behavior. J. Hum. Evol. 39 (5), 453—563. Available at: http://www.ncbi.nlm.nih.gov/pubmed/11102266.

Meyer, M., et al., October 2012. A high coverage genome sequence from an archaic denisovan individual. Science 222—226.

Miller, D.J., et al., 2012. Prolonged myelination in human neocortical evolution. Proc. Natl. Acad. Sci. U.S.A. 109 (41), 16480—16485.

Mitter, S.S., et al., 2012. Apolipoprotein E4 influences growth and cognitive responses to micronutrient supplementation in shantytown children from northeast Brazil. Clin. (São Paulo Brazil) 67 (1), 11—18. Available at: http://onlinelibrary.wiley.com/o/cochrane/clcentral/articles/299/CN-00860299/frame.html.

Muehlenbein, M.P., 2015. Basics in Human Evolution. Academic Press, Amsterdam.

Oriá, R.B., et al., 2007. Role of apolipoprotein E4 in protecting children against early childhood diarrhea outcomes and implications for later development. Med. Hypotheses 68 (5), 1099—1107. Available at: http://www.sciencedirect.com/science/article/pii/S0306987706006980.

Perry, J.R.B., et al., 2015. Molecular insights into the aetiology of female reproductive ageing. Nat. Rev. Endocrinol. 11 (12), 725—734. Available at: http://www.ncbi.nlm.nih.gov/pubmed/26460341.

Povinelli, D.J., Preuss, T.M., 1995. Theory of mind: evolutionary history of a cognitive specialization. Trends Neurosci. 18 (9), 418—424. Available at: http://www.ncbi.nlm.nih.gov/pubmed/7482808.

Povinelli, D.J., Vonk, J., 2003. Chimpanzee minds: suspiciously human? Trends Cogn. Sci. 7 (4), 157—160. Available at: http://www.ncbi.nlm.nih.gov/pubmed/12691763.

Ramaswamy, S., et al., 2013. The decline in pulsatile GnRH release, as reflected by circulating LH concentrations, during the infant-juvenile transition in the agonadal male rhesus monkey (Macaca mulatta) is associated with a reduction in kisspeptin content of KNDy neurons of the arc. Endocrinology 154 (5), 1845—1853. Available at: http://www.ncbi.nlm.nih.gov/pubmed/23525220.

Richter, L.M., et al., 2012. Cohort profile: the consortium of health-orientated research in transitioning societies. Int. J. Epidemiol. 41 (3), 621—626.

Rusted, J.M., et al., 2013. APOE e4 polymorphism in young adults is associated with improved attention and indexed by distinct neural signatures. Neuroimage 65, 364—373. Available at: https://doi.org/10.1016/j.neuroimage.2012.10.010.

Sainani, K., 2015. Unlocking the genetics of complex diseases: GWAS and beyond. Biomed. Comput. Rev. 11 (1). Available at: http://biomedicalcomputationreview.org/content/unlocking-genetics-complex-diseases-gwas-and-beyond.

Scammon, R.E., 1930. The measurement of the body in childhood. In: Harris, J.A., et al. (Eds.), The Measurement of Man. University of Minnesota Press, Minneapolis, pp. 173–215.

Sax, L., 2002. How common is intersex? A response to Anne Fausto-Sterling. J. Sex. Res. 39 (3), 174–178. Available at: http://www.ncbi.nlm.nih.gov/pubmed/12476264.

Schwarz, F., et al., 2016. Human-specific derived alleles of *CD33* and other genes protect against postreproductive cognitive decline. Proc. Natl. Acad. Sci. U.S.A. 113 (1), 74–79. Available at: http://www.pnas.org/lookup/doi/10.1073/pnas.1517951112.

Sowell, E.R., et al., 2003. Mapping cortical change across the human life span. Nat. Neurosci. 6 (3), 309–315. Available at: http://www.ncbi.nlm.nih.gov/pubmed/12548289.

Stearns, S.C., 1992. The Evolution of Life Histories. Oxford University Press, Oxford.

Stoddart, D.M., 1992. The Scented Ape: The Biology and Culture of Human Odour. Cambridge University Press, Cambridge.

Tavaré, S., et al., 2002. Using the fossil record to estimate the age of the last common ancestor of extant primates. Nature 416 (6882), 726–729. Available at: http://www.nature.com/doifinder/10.1038/416726a.

Thompson, D.W., 1917. On Growth and Form. Cambridge University Press, Cambridge.

Tomasello, M., et al., 2005. Understanding and sharing intentions: the origins of cultural cognition. Behav. Brain Sci. 28 (5), 675–691 discussion 691–735.

Varea, C., Bernis, C., 2013. Encephalization, reproduction and life history. Hum. Evol. 28 (1–2), 1–16.

Varela-Silva, M.I., et al., 2016. Deep data science to prevent and treat growth faltering in Maya children. Eur. J. Clin. Nutr. 70 (6), 679–680.

Vasey, N., 2007. The breeding system of wild red ruffed lemurs (*Varecia rubra*): a preliminary report. Primates 48 (1), 41–54. Available at: http://www.ncbi.nlm.nih.gov/pubmed/17024514.

Walker, D.G., et al., 2015. Association of CD33 polymorphism rs3865444 with Alzheimer's disease pathology and CD33 expression in human cerebral cortex. Neurobiol. Aging 36 (2), 571–582. Available at: http://www.ncbi.nlm.nih.gov/pubmed/25448602.

Walker, R., et al., 2006. Growth rates and life histories in twenty-two small-scale societies. Am. J. Hum. Biol. 18 (3), 295–311. Available at: http://www.ncbi.nlm.nih.gov/pubmed/16634027.

White, T.D., et al., 2003. Pleistocene *Homo sapiens* from Middle Awash, Ethiopia. Nature 423 (6941), 742–747. Available at: http://www.ncbi.nlm.nih.gov/pubmed/12802332.

Zahed, S.R., Kurian, A.V., Snowdon, C.T., 2010. Social dynamics and individual plasticity of infant care behavior in cooperatively breeding cotton-top tamarins. Am. J. Primatol. 72 (4), 296–306. Available at: http://www.ncbi.nlm.nih.gov/pubmed/20014011.

CHAPTER

32

The Fossil Evidence of Human Brain Evolution

E. Bruner

Centro Nacional de Investigación sobre la Evolución Humana, Burgos, Spain

32.1 Human Paleoneurology

32.1.1 Brains and Fossils

Brain evolution has always represented a major issue in paleoanthropology, largely because it deals with our cognitive capacities. A patent diversity in cerebral morphology and organization among primates can be easily recognized at first glance, suggesting that the brain has been an important target of selective processes. Nonetheless, despite its importance, we can say that in the last couple of centuries there has been little progress in the study of brain evolution in fossil humans (ie, the fossil species of the human genus), when compared with other evolutionary fields and when taking into account the attention devoted to this topic. We can hypothesize that such difficulties can be due to at least four main reasons. First, the fossil record is scarce and fragmented, and it can barely provide an exhaustive perspective of the anatomical variations, in terms of biological representativeness and statistical reliability. Second, this topic is often investigated as a general issue in paleoanthropology, and not necessarily interpreted as a field requiring specific competence or expertise. This generalized approach is probably also due to the appeal these issues have on mass media and dissemination. The fact that evolutionary hypotheses on brain evolution are very attractive and at the same time "harmless" (in the sense that they generally do not directly affect any applicative aspects of science) has probably generated an excess of speculations in this discipline. Third, in recent decades, morphology and gross anatomy as fields have been perceived as less "sexy" than other methodological approaches, like those based on molecular or functional techniques. Consequently, neurobiologists have probably paid less attention to paleoanthropology and other fields which largely rely on macroscopic anatomy and morphometrics. Finally, the fourth reason is intrinsically related to the object of the study: the smooth and blurred geometry of the brain,

its complex and complicated mechanisms, and its noticeable individual variability make any research associated with its morphology very entangled and difficult to develop within fixed methodological approaches. This situation is further complicated by the fact that we still ignore basic information even regarding the brain morphology of living humans, introducing a large speculative component in each hypothesis and debate.

Whatever the reason, after almost two centuries of hypotheses and opinions, our knowledge on brain evolution in fossil human species is still scanty and probably still full of expectations more than of certainties. It is interesting to note that, despite all the limits mentioned above, the field has been characterized by stringent statements, firm theories, and sharp debates, which clearly contrast with the scarce information that is really available on traits and mechanisms associated with the actual evolutionary processes involved. Apart from providing new data and analyses, it hence seems mandatory to reconsider the many prejudices and stereotypes that still influence many expectations in this field.

The study of brain evolution in fossil specimens was named *paleoneurology* (Kochetkova, 1978). The term is a bit out of date, because at present neurology is generally interpreted as a clinical field. *Paleoneurobiology* could be a more precise alternative. In any case, paleoneurology deals with the morphology of the brain as deduced by the information available from the cranial fossil record.

32.1.2 Brains and Endocasts

Paleoneurology is founded on a basic physical principle: the anatomy of the soft tissues of the brain can be inferred from the morphology of the hard tissues of the skull (Holloway, 1978, 2008; Falk, 1987; Bruner, 2015). It is not a matter of "phrenology," but just of recognizing that brain and braincase have a direct and reciprocal morphological influence because of structural and

Evolutionary Neuroscience, Second Edition
https://doi.org/10.1016/B978-0-12-820584-6.00032-5

functional relationships during growth and development (Moss and Young, 1960). *Structural relationships* are the result of physical contacts, growth forces, and biomechanical factors. *Functional relationships* are the result of shared physiological and biochemical mechanisms. The elements of the endocranial cavity (brain cortex, meninges, and vessels) mold the braincase and shape its surface (Kobayashi et al., 2014a). As a consequence, the form of the braincase matches the form of the brain, and its surface displays the imprints of its superficial structures (gyri, sulci, arteries, and veins). The "form" correspondence is due to a morphogenetic fact: during growth and development, the bones of the braincase adapt their morphology to the growing brain, by active modeling of its outer, lateral, and inner surfaces (Enlow, 1990). It has been hypothesized that neurocranial growth (variation in *size*) is due to brain pressure, and neurocranial development (variation in *shape*) is due to biomechanical redistribution of the growth forces through meningeal tensors like the *falx cerebri* and the *tentorium cerebelli* (Moss and Young, 1960). The "surface" correspondence is also due to a biomechanical induction of histological activation: pressure induces activation of osteoclasts (cells removing bone), and tension induces activation of osteoblasts (cells depositing bone). Accordingly, the pressure of cortical convolutions and vessels will induce the remodeling of the endocranial surface, leaving traces and imprints of their passage on the bone. The morphological correspondence between soft tissues and imprints is not complete, because it depends on many factors like the thickness of the meninges, the presence of cerebrospinal fluid, and the brain pressure. Accordingly, sometimes vessels and gyri can leave only faint traces, or no imprints at all. However, if the absence of a trace cannot be interpreted as the absence of the character, its presence is the actual witness of its existence. Surgical practice suggests that the correspondence between vessels and their traces is nonetheless pretty good.

Dealing with sulcal patterns, there are more visible imprints on the orbital and temporal areas, in which there is a spatial constraint with the underlying cranial counterparts (orbits and mandible, respectively) and, as a consequence, the bone is forced between two different shaping environments. The sulcal patterns are more visible on smaller and younger crania. Size-related effects possibly associated with meningeal thickness or endocranial pressure may be the reasons why larger skulls display more blurred cortical imprints. Similarly, the spatial relationship between brain and braincase is tighter during brain growth, and at older ages a successive shrinking of the brain mass may decrease the degree of expression of endocranial traces. However, the mechanisms behind this inverse relationship between size of the brain and degree of expression of the anatomical traces are not clear.

The morphogenetic relationships between cerebral cortex, meninges, vessels, and bones generate the *endocranial cavity*, a mold of the brain formed by the braincase that supplies information on brain size, brain shape, brain proportions, brain position, and brain vascularization. It is clear that, from brain to endocranial cavity (endocranium), there is an important loss of biological information. The endocranium can only provide data on the morphology of some superficial elements. Hence, the analysis is generally limited to some gross anatomical features. At the same time, it represents the only direct information available on the morphology of the brain in extinct species, deserving attention.

Interestingly, since the very beginning of the field, paleoneurologists preferred to work on molds of the cranial cavity, termed *endocasts* (Holloway et al., 2004; Zollikofer and de León, 2013; Neubauer, 2014; Fig. 32.1). In some cases, the endocranial cavity is not fully accessible, and endocasts are a way to "take out" the information. However, there is also an important sensorial component: as primates, we largely "think" with our eyes and our

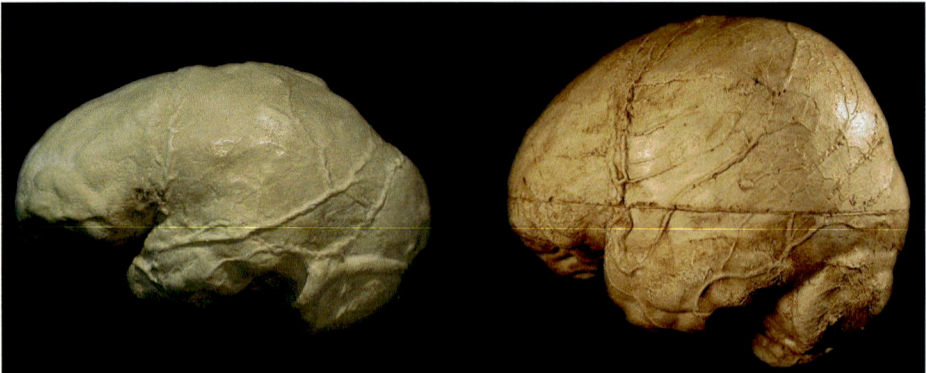

FIGURE 32.1 Traditionally, endocasts were made with physical materials, ranging from plaster to epoxy resins. Endocasts can be used to estimate cranial capacity, brain proportions, sulcal patterns, and craniovascular traits. In this image: (left) an endocast from the Asian *Homo erectus* Zhoukoudian III (Istituto Italiano di Paleontologia Umana, Rome), and (right) an endocast from a recent modern human (Institut de Paléontologie Humaine, Paris).

hands, and both things are difficult to do with a cavity, which is but an empty space. We definitely better perceive a positive form than its negative mold. The final chain is a little fuzzy: the brain is the original object, molding its negative casts (endocranium) which is used to mold a positive one (endocast). Of course, the endocast has the same information of the endocranium, and it does not add any additional data to the anatomical analysis.

Because the endocranium is used as a morphological proxy for the brain, and because the endocasts are used as proxy for the endocranium, an effort must be made to remember that we are working with partial information: an endocast is not a brain, and the endocranium is part of a larger integrated system called the skull (Fig. 32.2).

32.1.3 Reading Endocasts

The large variability and complex geometry of the brain surface has frustrated most attempts to investigate and analyze patterns and schemes of the sulcal organization. We have conventionally divided the cortical areas into lobules, gyri, and sulci which, to some extent, can be recognized in any living individual. However, at the same time, we know that individual differences are so strong that the recognition of macroanatomical areas and boundaries may be largely a matter of personal experience, with an important subjective component. Whatever the functional meaning of this "brain geography," it is clear that recognizing these same cortical areas on an endocast is definitely more complicated, adding

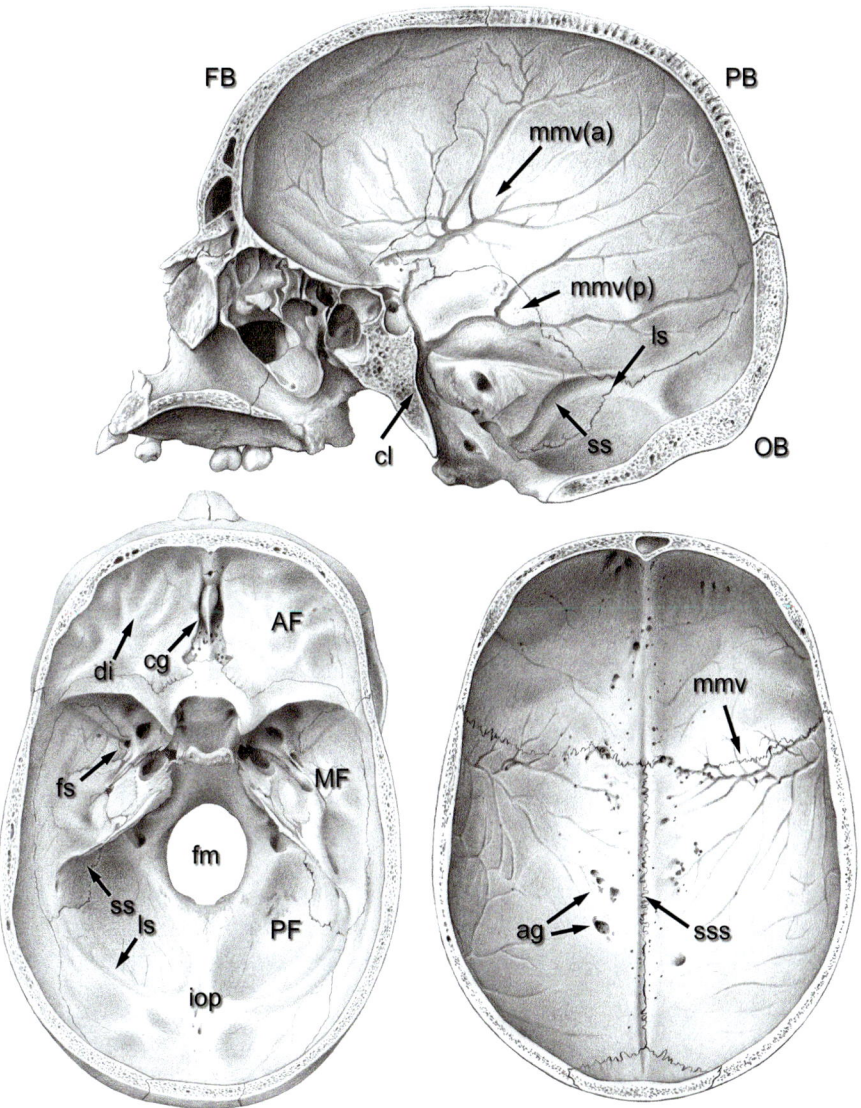

FIGURE 32.2 Basic endocranial traits, the three endocranial fossae (*AF*, anterior fossa; *MF*, middle fossa; *PF*, posterior fossa), and the vault bones (*FB*, frontal bone; *PB*, parietal bone; *OB*, occipital bone): *ag*, arachnoid granulations; *cg*, crista galli; *cl*, clivus; *di*, digital impressions; *fm*, foramen magnum; *fs*, foramen spinosum; *iop*, internal occipital protuberance; *ls*, lateral (transverse) sinus; *mmv*, traces of middle meningeal vessels (*a*, anterior branch; *p*, posterior branch); *ss*, sigmoid sinus; *sss*, trace of the superior sagittal sinus.

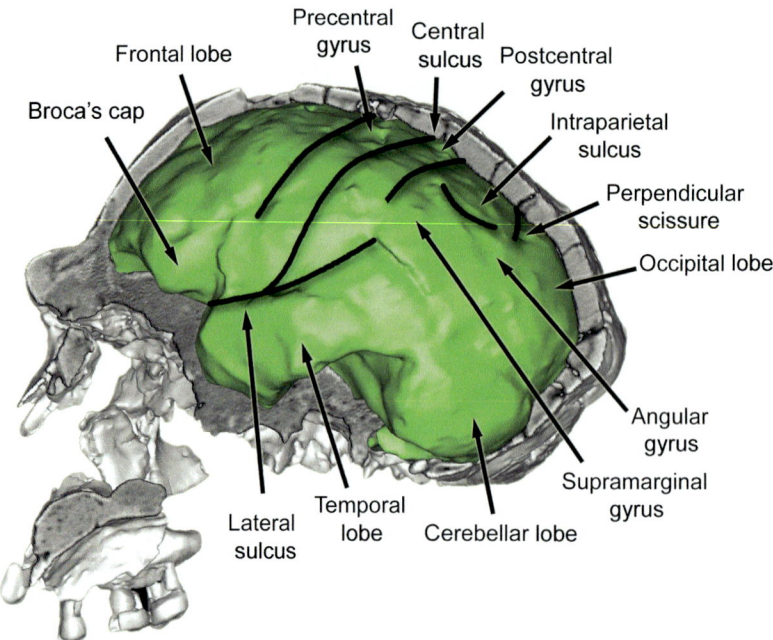

FIGURE 32.3   Endocranial cortical features that can be tentatively recognized on an endocranial cast. Most traces are but faint bosses or depressions. Therefore, apart from a specific experience on brain anatomical variation, crucial information on their position can be obtained from their relative spatial arrangement. That is, identification of these areas should be based not only on the expression of the area itself, but also on the spatial information available from the surrounding elements.

further challenges and crucial operational decisions (Fig. 32.3). This is probably why most paleoneurological analyses have only dealt with brain size, a variable that is easier to rely on, because it concerns the whole structure with no inference regarding its subdivisions. Chords and arcs have been largely used to quantify some general morphological aspects of the endocast, but more specific metrics, associated with localized cortical areas, are rarely found in the literature. Many students suspect that specific cortical elements cannot be recognized on the endocast surface. Skepticism in this case is, indeed, reasonable. Nonetheless, the morphology of the endocast has the same difficulties that one can find in other fields with similar conventional anatomical situations, ranging from cerebral to vascular anatomy. Uncertainty due to variation is constant in anatomy, and a proper method is needed to deal with it adopting adequate operational choices. As in any sector of anatomy, experience is a major issue. The variability of brain patterns requires specific training, in which physical and digital techniques are used to explore the morphological variations and their combinations. The cortical elements on the endocast surface will rarely be localized as points, but rather as "probable areas." This limit, however, does not hamper the descriptive and quantitative approaches used to study their evolutionary changes.

To understand the possibility and limitations of "reading endocasts," it must be taken into account that the estimations of the position and extension of the cortical areas are not simply based on a specific morphological trait, but rather on the overall distribution of all the features. More specifically, the position of an element is estimated by using the information coming from the whole endocranial surface, at least from its available portions. Therefore, a specific gyrus or sulcus can be localized by virtue of its own expression, but also thanks to the localization of the surrounding areas. In some cases, the element may not be visible as such, but the surrounding areas limit its position to a given spatial range. Hence, the concept of "probable area" is not intrinsic to that area, but it is a property of the whole endocranial system.

The possible meaning of these morphological brain differences has been, for centuries, a matter of debate, with contrasting positions. Phrenologists opted for a strict reductionist approach to sulcal morphology, interpreting individual sulcal variations in terms of cognitive capacities. On the other hand, many fields reject the possibility that brain anatomy can supply any functional information. Between these two extremes, there are different intermediate perspectives, which may merit attention. Nonetheless, strictly speaking, paleoneurology deals with brain anatomy, and inferences on functions are not the main target of the field, because they require additional information that is beyond what can be obtained from the fossil record. In paleoneurology the sulcal organization provides spatial and topological indications useful to evaluate geometry and proportions (for example, position or dimensions of specific brain areas). In contrast, inferences on specific sulcal traits (like the presence or position of particular gyri and sulci) have not provided to date any consistent result, because of the uncertainty associated with their possible functional meaning and because of the noticeable individual variability. We should therefore focus at least on two targets when dealing with specific cortical areas and endocasts. First, localizing specific structures of the cortical surface is necessary to supply anatomical references for comparative morphometrics. The localization of homologous areas is necessary to provide morphological variables suitable for statistical modeling or to evaluate hypotheses. Second, elements of the cortical surface are necessary to define boundaries or larger areas. Such boundaries are needed to estimate brain proportions. "Lobes" of the brain, such as gyri and sulci, are largely a matter of conventional labeling. Nevertheless, they represent larger units for which changes and variations can be easier to identify at evolutionary levels. Therefore, cortical elements, beyond their intrinsic significance, may be relevant as boundaries for identifying larger anatomical districts. Because of the limits already mentioned, it is possible that subtle anatomical differences cannot be detected on a macroscopic scale. However, larger changes can be localized and quantified.

It is clear that such anatomical considerations are also essential when dealing with possible functional (cognitive) consequences. But, in this case, the paleoneurological evidence can only provide additional complementary information, within a larger and more comprehensive perspective in evolutionary neuroanatomy.

32.1.4 Computing Paleoneurology

The geological matrix can enter the endocranial cavity of a fossil skull and create a "natural mold" of its endocranial surface and volume. In some special cases (like in the Taung child, the earliest *Australopithecus* ever described, discovered back in 1924), the braincase can even be lost, and the natural mold can represent the only remnant evidence of the neurocranial form. Such occurrences are, however, rare, and traditionally endocasts have been prepared artificially with molding materials, such as plaster or high-quality epoxy resins (Holloway, 2008, 2016). The endocast maker had a very specialized methodological expertise, being able to handle different materials and the chemical processes necessary to carry out the molding procedure. Needless to say, these procedures were somewhat invasive for the fossils. The chemical products came into contact with the fossil surfaces, with possible contamination or damage of the original fossil matrix, and the separation of the mold could even affect the integrity of the specimen. Furthermore, the passages from the extraction of the cast to its fixation may suffer

some minor deformation. First-generation casts were used to perform successive generations of casts, losing information at each successive molding step. Taking all these problems into consideration, the people involved in casting fossils did an amazing job to develop the collections, which, during more than one century, have made it possible to develop paleoneurological studies.

At the end of the past century, two new disciplines supplied techniques and methods for a comprehensive revolution in all fields based on morphology, including paleontology and, most of all, paleoneurology: digital anatomy and computed morphometrics (Spoor et al., 2000; Zollikofer and Ponce de León, 2005; Gunz et al., 2009; Gunz, 2015).

Digital anatomy deals with biomedical imaging, engineering, and graphics (Fig. 32.4). Although many different techniques have been developed in recent decades, in osteology and paleontology the most effective has been computed tomography (CT). Tomographic scans are planar X-rays sections that generate a map of the density of that specific section. For this reason, this technique is particularly useful when studying bone and bony elements. Differences in the density values can be used to separate structures and to localize different bone components. Multiple contiguous sections can be then assembled to form 3D virtual objects (digital replicas). Traditional CT machines supply a resolution of tenths of millimeters, and microCT can easily achieve much finer anatomical levels. In the last decade, many commercial and free programs have been developed to allow a versatile handling of tomographic data, and also the number of free, commercial, and private databases has increased consistently. In human anatomy, these methods make it possible to analyze the anatomical components in vivo and in large samples. In paleontology, they can be used to analyze the inner volumes of the fossil remains, to work with a completely noninvasive approach, to reconstruct the fossil specimens according to computed and standardized tools, and to exchange or share data within the scientific community. This advance was significant, indeed. In paleoneurology, it was even more essential, allowing complete access to the endocranial cavity and supplying new tools for endocast reconstructions, with extreme resolution and remarkable reliability (Schoenemann et al., 2007). The experience of the anatomist is still an issue when dealing with these methods, and the final reconstruction of a digital endocast often requires an integration of automatic, semiautomatic, and manual steps. Automatic filtering and selections of the pixels according to densitometric scales are based on algorithms and numerical principles, while manual segmentation requires subjective choices aimed at delimitating the extensions of different densitometric components that cannot be clearly distinguished by fixed or iterative thresholds. Beyond the anatomical

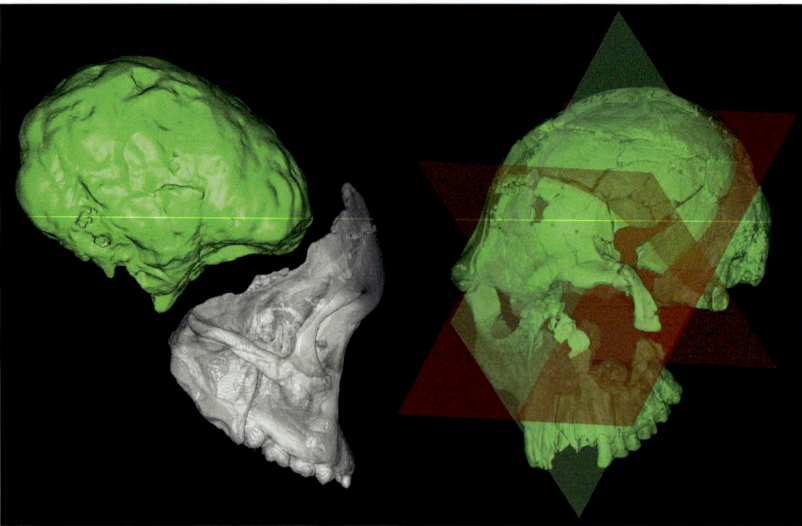

FIGURE 32.4 The Taung child (on the left) was the first *Australopithecus* described, discovered back in 1924. The vault was lost, but the geological matrix had filled the endocranial cavity, producing a natural endocast. Digital anatomy allows the virtual reconstruction of the fossil remains, the analysis of the inner surfaces, and a completely noninvasive handling of the specimens. On the right: a digital reconstruction of *Homo habilis* KNM-ER1813.

expertise, knowledge in informatics is essential to deal with these kinds of procedures.

The different degree of bone fossilization and the presence of the geological matrix within the endocranial cavity are the main factors influencing the possibility of properly delimitating the endocranial surface (Fig. 32.5). Endocasts can be reconstructed by multiple approaches or by different teams, by using traditional anatomical expertise or numerical criteria based on interpolation of the fossil fragments and extrapolation of the fossil surfaces. Apart from these advantages, these methods can be also used to provide a range of reconstruction of the fossil specimen, able to give a quantitative perspective of its reliability (Neubauer et al., 2012).

The second technical advance was represented by *computed morphometrics*, which was mainly supported by two related fields: multivariate statistics and shape analysis. Multivariate statistics was already available for many decades, but it experienced a fundamental improvement when computers became widely available. Shape analysis was promoted after the introduction of landmark-based techniques (Bookstein, 1991; Rohlf and Marcus, 1993; Slice, 2004; Zelditch et al., 2004). In this case, coordinates are used to analyze two- or three-dimensional models of anatomical systems, instead of chords, arcs, and other traditional variables. Coordinates are normalized according to a given criterion (a common baseline, or iterative processes minimizing the differences among the specimens of the sample), and their residuals after normalization are analyzed through multivariate tool kits. These approaches have three main advantages. First, they supply very powerful statistical frameworks, increasing the efficiency of the

morphometric analysis. Second, they allow the localization and quantification of the correlation patterns hidden behind the observed phenotypic variation. These correlation patterns are the results of those "biological rules" associating traits and processes in ontogeny and evolution. Third, these methods are based on geometry, and hence they allow the visualization of the results directly in terms of spatial variations and differences. This improvement was essential in every anatomical field and soon acknowledged in craniology and paleoanthropology (Lieberman et al., 2002; Zollikofer and Ponce De León, 2002; Bookstein et al., 2003).

Digital anatomy and computed morphometrics represented the way to improve the objectivity of paleontological studies, stepping into a more quantitative and experimental context. All these methods and techniques have generated new standards for professional training, in which morphology and evolution are associated with competence in statistics and informatics. Needless to say, the fascinating analytical power of these tools cannot replace a proper expertise in biology, which is still strictly necessary to work with the many variations of cranial and cerebral geography, with the design of an adequate methodology and with the interpretation of the numerical outputs. Experience is still, as always, an issue.

32.1.5 Statistics and the Fossil Record

An important limitation in paleoneurology (and in paleontology in general) is the sample size and composition. Bones can supply much biological information, but there are many relevant factors that cannot be

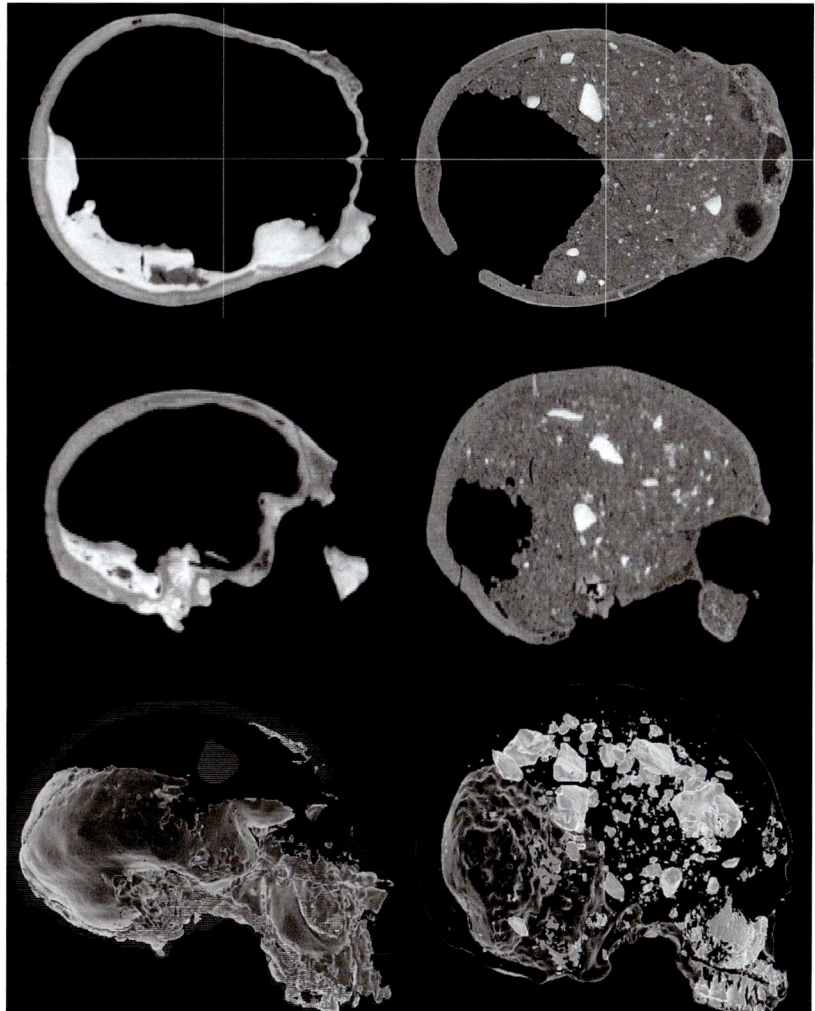

FIGURE 32.5 Computed tomography allows the digital analysis of the fossil remains and of the geological matrix filling the endocranial cavity. Fossils can be digitally cleaned, removing sediments and incrustations. The degree of separation between fossilized bone and geological filling matrix can be variable, depending upon the local fossilization context. On the left, sections and endocranial matrix reconstruction of Saccopastore 1, a Neanderthal dated to more than 120 ka, in which the density of the endocranial matrix is easily distinguished from the density of the bone. On the right, sections and endocranial matrix reconstruction of Fonterossi 1, a Mesolithic specimen in which the separation between bone and endocranial sediment is difficult to localize. *Modified from Bruner, E., Manzi, G., 2006. Digital tools for the preservation of the human fossil heritage: Ceprano, Saccopastore, and other case studies. Hum. Evol. 21, 33—44.*

investigated with bones alone. Furthermore, the fossil record generally includes few individuals, which may not be biologically representative of their species or population. But the paucity of the fossils also hampers proper statistical approaches. Inferential statistics needs tens of specimens per group to reach sufficient statistical power, defined as the possibility to reveal actual differences between groups. The more subtle the differences, the more specimens are necessary to reach statistical significance. As far as we know, endocranial morphology in humans is often characterized by large individual variations but minor (or null) species-specific differences. To deal with most such differences statistically, probably we would need hundreds of specimens, which is, for extinct human species, not feasible. Most parametric

tests require assumptions (like normality) which cannot even be tested in fossil samples. Nonparametric tests can be more useful in this context, but still suffer from limited statistical power. These limits are even more serious for multivariate inferential methods (such as canonical variates or discriminant analysis). As alternatives, heuristic descriptive approaches (such as principal components or cluster analysis) can be more useful, being based more on the observed variation. Nonetheless, multivariate analysis is also sensitive to the patterns of correlation among variables: the smaller the sample size, the more the noise introduced in the variation structure, with consequent instability of the resulting multivariate vectors. In these cases, a proper analysis of the stability and reliability of the outputs is required. It is worth noting that intraspecific and interspecific covariation patterns are generally structured on different rules (Martin and Barbour, 1989). Intraspecific variation is often based on normal distribution and random factors, while interspecific variation is nonnormal and based on adaptive factors. Therefore, multivariate approaches based on covariation structure should not mix these two sources of variability. Otherwise, the different rules of variation may clash, generating hybrid multivariate vectors.

Although the main criticism of fossil evidence concerns the small sample size, it must be noted that this is a methodological limit often shared with neurobiology. Also, genetic and functional studies are commonly based on a very small number of specimens, because of economical, logistic, and ethical limits when working with humans and other primates. Nonetheless, small sample sizes seem to be more accepted in neuroscience and in other biological fields than in paleontology. This limit is nowadays less stringent in anthropology and neuroanatomy, because of the availability of large digital samples.

Two more points merit attention, specifically regarding the information associated with endocasts: the preservation of the endocranial areas and the nature of the endocranial traits. The preservation of the endocranial areas is not homogeneous. The endocranial base is formed by a thin and fragile bone layer and is hence more rarely preserved in fossils. Consequently, all those cortical areas housed in the endocranial floor (orbital cortex, temporal lobes, cerebellar lobes) are less well represented than the frontal, parietal, and occipital areas, which are associated with thick vault bones.

A second limit concerns the type of characters usually investigated in paleoneurology. Beyond brain form and proportions, endocranial anatomy supplies information on craniovascular features (arteries, veins, and foramina) and osteological epigenetic traits associated with bone deposition (excess and defect of ossification). These characters are called nonmetric or discrete traits, because they are generally scored by their presence/absence or degree of expression. Statistics for these features is definitely more complicated and less reliable than for metric variables. The use of these traits requires operational choices when scoring the features (like thresholds, conventional scales of expression, or conventional grouping patterns) and when computing their variation (to transform nominal and ordinal distributions into continuous grade or groups, by using indexes, distances, or transformations). Therefore, the uneven distribution and incompleteness of data and the difficulties in representing variation according to proper metrics requires additional prudence when interpreting quantitative outputs. Caution is, more than in other fields, recommended.

Taking into account the large individual brain variation, the subtle species-specific differences, the incompleteness of the fossil record, small sample sizes, and the limits associated with the endocranial characters, it is apparent that paleoneurological analyses should be generally interpreted within the context of an explicit phylogenetic hypothesis, based on multiple and independent evidence. Although the endocranial variation can hide a phylogenetic signal, in general it is not a reliable source of taxonomic or evolutionary inference. On the other hand, the paleoneurological information is essential when the phylogenetic context has been previously evaluated or tested according to other kinds of information. It is important to remember that for most endocranial traits (sulcal patterns, craniovascular traits, brain—braincase dynamics, etc.), we still lack basic information on our species. Such information must therefore be investigated in living humans before it can be properly analyzed in fossil species.

32.2 Functional Craniology

32.2.1 Morphogenesis

In 1960, Moss and Young proposed their model of neurocranial morphogenesis, in which size variation (*growth*) is mainly due to brain enlargement, and shape variation (*development*) is mainly due to the mechanical redistribution of the growth forces along endocranial tensors formed by the main connective elements (the meninges and their interhemispheric folds, namely the *falx cerebri* and the *tentorium cerebelli*; Adeeb et al., 2012). The overall hypothesis is nevertheless difficult to test because of experimental limitations. A similar perspective based on biomechanical forces exerted by the different anatomical components was proposed at the histological level to explain patterns of cortical folding, assuming that neurons have physical properties (resistance to tension, torsion, and pressure) that can influence the morphogenetic processes, shaping the cortical surface (Van Essen, 1997; Hilgetag and Barbas, 2005, 2006; Toro and Burnod, 2005). If all these factors are effective in influencing brain morphology, macroscopic changes can be the secondary consequences of variations of the mechanical properties of the anatomical elements (Tallinen et al., 2016). In this case, specific genetic programs can influence the general morphogenetic schedule (like time and rate of growth), but the physical properties of the histological components have a crucial role in shaping the final phenotype.

In recent decades, computed morphometrics led to a further development of concepts and principles in morphological integration and modularity (Cheverud, 1982; Mitteroecker and Bookstein, 2008; Klingenberg,

2010, 2013), involving the relationships and reciprocal influences among different anatomical elements. This has represented a major advance in anatomy, revealing mechanisms of variation based on underlying phenotypic rules that can channel or influence evolutionary changes. Cranial evolution has represented, since the beginning, a good case study to test all these principles (Bookstein et al., 2003).

Despite the attention devoted to this topic in evolutionary anthropology, to date there is, however, scant evidence of a marked overall integration within the cranial and endocranial system. This is particularly evident at the endocranial base, which is a very complex morphogenetic structure influenced by many different factors and by many local relationships (Lieberman et al., 2000b; McCarthy, 2001; Bruner and Ripani, 2008; Bastir and Rosas, 2009; Fig. 32.6). The cranial base influences the overall cranial architecture (Lieberman et al., 2000a) and is integrated with the posterior face through factors ranging from maxillary position to respiration and bioenergetics (Bastir et al., 2004; Bastir and Rosas, 2006, 2016). The three endocranial fossae must rely on some pattern of reciprocal integration to adjust their respective spatial arrangement. However, they are involved in very different morphogenetic and evolutionary processes and are therefore influenced by

different local dynamics. The anterior fossa houses the prefrontal cortex and is mechanically constrained by the upper face and orbital area. The middle fossa houses the temporal lobes and is mechanically constrained by the mandibular structures. The posterior fossa houses the cerebellar lobes and is mechanically constrained by factors associated with posture and locomotion. The midsagittal area (clivus) supports the brain stem and is constrained by the ethmomaxillary complex and by the mechanisms associated with cranial base flexion (Bruner and Jeffery, 2007). This heterogeneous situation generates little integration within the whole endocranial system, and the morphology of each area is largely molded by local and independent factors. That is, although all the parts must be properly integrated according to a general scheme, the relationships are not linear, and each district has an important modular independence. A similar situation probably holds for the brain itself, formed as it is by areas that are subject to different functional and evolutionary influences, and that may vary according to independent patterns (Bruner et al., 2010; Gómez-Robles et al., 2013).

The vault and the upper brain regions show more integrated schemes of variation, because of minor biomechanical constraints between bones and cortical surface (Enlow, 1990). The position of the bones is less

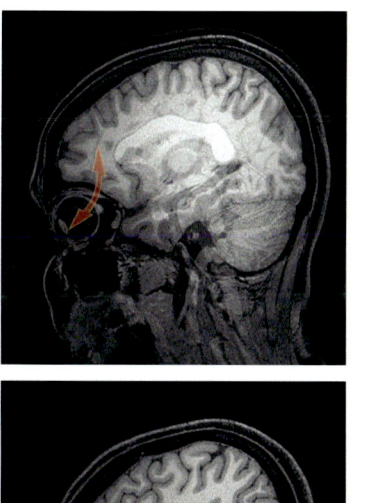
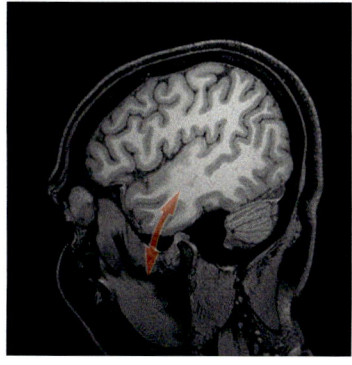
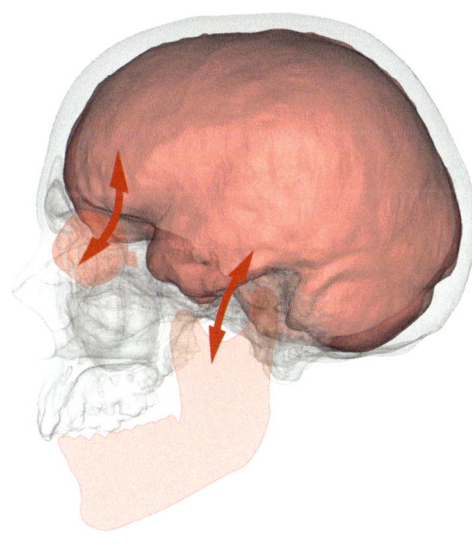

FIGURE 32.6   The morphology of upper brain regions is less influenced by cranial constraints. Conversely, in the lower areas of the endocranial cavity, the spatial relationships between skull and brain are generally more complex. The endocranial base is involved in many structural and functional processes. The frontal cortex is spatially influenced by the arrangement of the upper facial block, in particular by orbital and eye morphology, especially in modern humans. The temporal lobes are in direct contact with the mandibular system.

constrained by spatial limits, and the relationship between bones and lobes is more direct and simple. The frontoparietal cortex generally represents an integrated unit in evolutionary and functional anatomy (Caminiti et al., 2015), and the parietooccipital bones display a clear pattern of morphological integration (Gunz and Harvati, 2007). Therefore, in this region, more integration is to be expected when compared with the complex situation of the cranial base. Nonetheless, even in this case, similar morphological effects can be associated with different mechanisms, which may be difficult to analyze in terms of causes and consequences. For example, the simple and recurrent separation between brachycephalic (relatively short and wide) and dolichocephalic (relatively long and narrow) head phenotypes, a grade variation frequently found in interspecific and intraspecific samples, can be the result of different morphogenetic processes (Zollikofer and Ponce De León, 2002).

In any case, the growth and development of brain and braincase must coevolve, changing as a balanced morphogenetic system. Beyond specific growth programs influencing the time and rate of tissue production, osteoblasts are activated by tensions, and osteoclasts are activated by pressure. Therefore, genetic and environmental factors act together to generate species-specific patterns of osteoblast and osteoclast distribution, which finally mold the cranial system (Duterloo and Enlow, 1970; Enlow, 1990; Lieberman et al., 2000b; Martínez-Maza et al., 2006). Accordingly, the mechanical forces exerted by the soft tissues (brain and connectives) on the hard tissues (bones) must be adequate to support balanced morphogenetic changes. Excesses of ossification (*hyperostosis*) and defects of ossification (*hypostosis*) can appear as minor nonpathological or subpathological conditions, in which the morphogenetic relationship between brain and braincase is not properly equilibrated (Manzi et al., 1996). Hypostotic and hyperostotic traits can therefore be used to investigate the morphogenetic processes in extinct species, so as to indicate whether or not size and shape changes between brain and braincase were adequately balanced during ontogeny (Manzi et al., 2000).

32.2.2 Brain and Braincase

The anatomical relationships between brain and braincase are an essential topic in paleoneurology as well as in medicine (Ribas et al., 2006; Bruner et al., 2014a). In many pathological conditions (from craniosynostoses to Down syndrome), such relationships are patently impaired (Richtsmeier et al., 2006). Imbalances can generate many functional problems, and at the same time they can supply important hints on the underlying mechanisms associated with the normal and pathological schemes of variation. Taking also into account that both anthropology and medicine currently adopt similar digital approaches, collaborations between evolutionary neuroanatomists and medical doctors can be extremely fruitful, in both directions.

The anterior cranial fossa houses the frontal lobes, in particular the prefrontal cortex. The posterior limit of the frontal bone (bregma) is much more anterior than the posterior limit of the frontal lobe (central sulcus), which instead lies under the parietal bone. In macaques, the position of the coronal suture, separating the frontal and parietal bones, displays a constant position relative to the frontal cortex (Kobayashi et al., 2014b), but in humans such information on the relative position between cranial sutures and frontal lobes is still vague. The pteric area, within the temporal fossa, is a hinge between the neurocranial and facial axes, and major changes are then to be expected within and among different human lineages (Shea, 1985; Lieberman, 2000). In living apes and most extinct human species, the frontal lobes are positioned posteriorly to the facial plane, and there is only minor (or null) overlapping between the frontal lobes and the orbital roof (Bookstein et al., 1999; Bruner and Manzi, 2005). In contrast, in modern humans and Neanderthals, the frontal lobes are positioned almost entirely on the orbital roof (Bruner and Manzi, 2008). The orbital roof is formed by a very thin bone layer, in which the marked sulcal details reveal a spatial conflict between the cortical brain surface and the eye/orbit system. Furthermore, orbits are proportional to facial size, and eyes are proportional to brain size. Therefore, in large-brained hominoids (most of all in our species), small orbits, larger eyes, and frontal lobes overlapping with the orbits may generate spatial constraints leading to minor deformation of the eye axis, possibly with vision impairments (Masters, 2012). In fact, increased overlap between frontal lobes and eyes (as in women, in Asian populations, or in brachycephalic phenotypes) is associated with increasing levels of myopia.

The middle cranial fossa houses the temporal lobes, in particular their anterior and lateral portions. The temporal gyri are generally well visible on the bone surface. The middle cranial fossa is biomechanically contiguous with the underlying mandibular system, generating covariance patterns between temporal and mandibular morphology (Bastir and Rosas, 2005, 2006). Interestingly, there is a little integration between the endocranial sagittal and parasagittal variations. Midsagittaly, the cranial base flexion influences the spatial organization of the subcortical brain areas (Bruner and Jeffery, 2007).

The posterior cranial fossa houses the cerebellar lobes, and there is a good correspondence between cranial and cerebellar metrics (Kubo et al., 2014). In primates, this area is strongly influenced by posture and locomotion. This raises some relevant questions on the further necessity to integrate the cranial and cerebral system with the whole body organization (Bastir, 2008).

The parietal bone covers the parietal lobes, the posterior part of the frontal lobes, and the anterior part of the occipital lobes (Bruner et al., 2015; Fig. 32.7). There is a correlation between parietal lobe and bone size in adult humans, although the relative spatial position may vary. In particular, larger parietal lobes approach anteriorly the boundary between the frontal and parietal bones. Therefore, it seems that the spatial relationships between parietal bones and lobes depend upon some allometric effect, due to differential growth of the two areas or else to the effect of different morphogenetic stages, repositioning the anatomical elements by virtue of specific growth sequences. Endocranial morphogenesis is in fact characterized by independent steps (Neubauer et al., 2009), and the final morphotype is the result of the contribution of those independent but sequential phases. The parietal areas grow and develop very early, and their relative position is hence sensitive to later displacements of other components. There is a pattern of integration between the parietal and occipital bones: the more one is bulging, the more the other is flattened (Gunz and Harvati, 2007). Interestingly, such inverse correlation has not been found with parietal and occipital cortical volumes. In fact, at least in adult humans the relative volume of the parietal lobe is inversely proportional to the relative volume of the frontal and temporal lobes, but not to the occipital lobes (Allen et al., 2002).

According to these results, we must conclude that the integration of the parietooccipital block is a matter of shape (curvature) but not of size (cortical volume).

This functional system formed by bones and brain must be considered when interpreting paleoneurological differences. One of the main objectives in paleoneurology is to distinguish whether bones or brain are the primary cause of a given evolutionary change. In some regions (like the endocranial base), it is to be expected that cranial anatomy can constrain brain anatomy, while in some others (like the vault) the reverse may be more likely. Nonetheless, the overall situation is probably more complex than this. Structures with a clear mechanical role (like the mandible) may exert more constraints than structures with less rigid limitations (like the brain). At the same time, the sequence of maturation establishes a chain of causes and consequences which channels any possible variation. The braincase matures earlier during ontogeny, followed by the endocranial base and then by the facial block (Bastir et al., 2006). Given this, it is plausible that elements maturing earlier influence the shape and position of the elements maturing later, constraining the phenotypic expression along a chain of progressive steps.

In sum, some changes of the brain form are associated with *intrinsic* factors (biology and proportions of the brain areas), while others are secondary consequences as responses to *extrinsic* influences (constraints of the cranial system). In paleoneurology, a proper analysis of

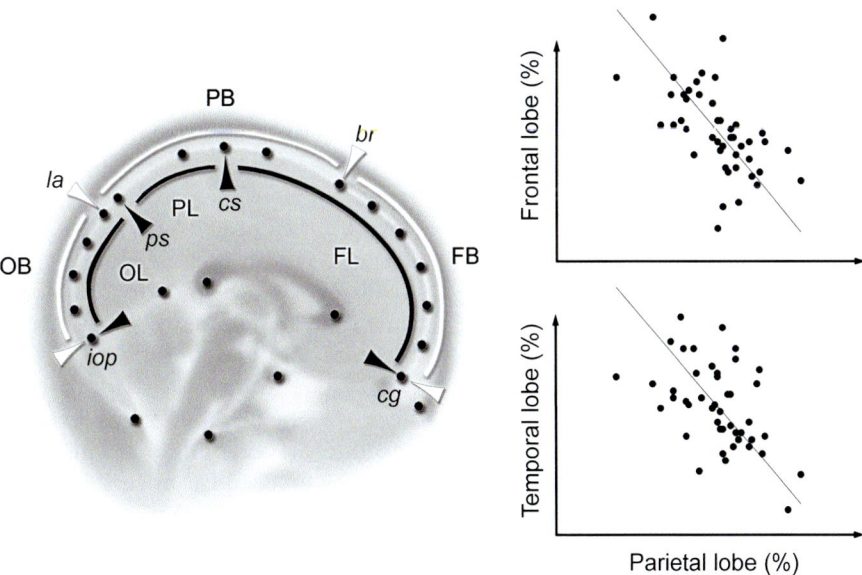

FIGURE 32.7   Relative position of the frontal, parietal, and occipital bones (FB, PB, OB) and lobes (FL, PL, OL) in adult humans, separated by bregma (br), lambda (la), central sulcus (cs), and perpendicular sulcus (ps), between crista galli (cg) and internal occipital protuberance (iop) (After Bruner, E., Amano, H., de la Cuétara, J.M., Ogihara, N., 2015. The brain and the braincase: a spatial analysis on the midsagittal profile in adult humans. J. Anat. 227, 268–276.). In adult humans, relative parietal lobe volume is inversely correlated with relative frontal and temporal lobe volume. *Original data from Allen, J.S., Damasio, H., Grabowski, T.J., 2002. Normal neuroanatomical variation in the human brain: an MRI-Volumetric Study. Am. J. Phys. Anthropol. 118, 341–358.*

these different mechanisms is mandatory when an endocranial morphological difference is detected. Moreover, although brain and bone are the main elements involved in the endocranial evolutionary and morphogenetic variations, the role of other components (like connective tissues and vessels) must also be taken into account.

32.3 Brain Size

32.3.1 Brain Size and Human Fossils

Brain size represents a dominant issue in paleoneurology, suggesting important changes in our species and lineage associated with the ontogenetic regulation of the brain and braincase morphogenetic schemes (Leigh, 2012; Zollikofer, 2012; Hublin et al., 2015; Neubauer, 2015). Cognitive issues aside, brain size is also directly involved in a complex network of allometric effects that influence folding architecture and neural wiring, energetics and metabolism, ecology and social structure (Hofman, 2014; Isler and Van Schaik, 2014). Modern humans have a brain size more or less three times the figure expected for a primate of our same body size, and this fact has not passed unnoticed, making brain size a major topic in human evolution. After cranial/endocast reconstruction, cranial capacity is relatively easy to calculate, traditionally by water displacement (from the mold) or filling the endocranial cavity with seeds (from the skull). Furthermore, the analysis of similarities and differences can be easily computed by standard univariate and bivariate techniques. So brain size is at the same time a very relevant topic, it is easy to calculate, and it is easy to analyze. This is probably why this issue has been so dominant during an entire century of paleoneurology. Despite this, however, few agreements have been achieved on the dynamics associated with brain size evolution, and the topic has generated more debates than solutions.

Following a traditional perspective (unfortunately still dominant in many professional and dissemination contexts), evolution has long been interpreted as linear, gradual, and progressive. The *scala naturae* required a progressive enhancement from incomplete to successful creatures, with living humans on the top. With this scheme in mind, we paleoanthropologists began recovering skulls with different brain volumes scattered in the fossil record and, driven by a predetermined scheme, we aligned all these figures on a same linear pathway, culminating with *Homo sapiens*. Then, we tried to apply different mathematical models to explain the numerical rule behind this change, but the results were inconclusive. A main limit was statistical: taking into consideration the few fossils available, derived from three continents and 5 million years, many different kinds of

curve can fit reasonably well to explain that distribution. However, the major limitation was conceptual, assuming that all those "points" (ie, species) were lying along the same curve. Since the late 1970s, however, many paleoanthropologists have suspected that things went in a different way. It is likely that different species underwent independent processes of brain size increase, and not always through gradual or continuous steps. This perspective requires two main corrections to our approach: we have to analyze separate processes, and those processes need not necessarily to rely on the same mechanisms.

In terms of methods, there is a relevant difference between brain size and endocranial volume. The endocranial cavity houses the brain, but also its vascular system, its connective protection (the meninges), and the cerebrospinal fluid. As a rule of thumb, approximately 10% of the endocranial cavity can be occupied by noncerebral tissues. So some corrections need to be applied when converting endocranial volume into brain size. Necessarily, these corrections are computed according to data available from modern humans, and hence may be biased by possible species-specific differences. A further correction must be applied if a conversion from brain volume to brain weight is necessary. Nevertheless, the problem of defining "brain volume" is intrinsic of the anatomical nature of the brain itself. What we call brain has actually no specific form because, as soft tissue, its geometry and spatial properties depend upon supporting forces and elements. The brain volume itself, intended as the space occupied by the brain, is a result of the hydrostatic pressure exerted by blood and cerebrospinal fluid. Therefore, what we call brain volume is necessarily an "index of spatial occupation," and not a real biological property of the neural mass. The endocranial volume is a proxy of cerebral volume, which in turn is a proxy for neural tissue volume, which in turn is assumed to be the real interesting figure we are trying to quantify. Nonetheless, a brain is probably more than a bunch of neurons, and such insistence to approximate a volumetric estimate may be unproductive. A distinction between endocranial volume and brain volume may be helpful, but only when taking into account this limit of the definition.

Current multivariate approaches supply tools that improve the estimation of complete and incomplete specimens, generating a range of values compatible with the anatomical uncertainty associated with fossils (Zhang et al., 2016). Fossil specimens are generally incomplete, and their reconstruction may be influenced by individual decision. Traditionally, cranial capacity was estimated following a specific reconstruction and a final volumetric result. Digital tools allow multiple reconstructions based on alternative models or iterative procedures, giving a range of values instead of one single figure (Neubauer et al., 2012). This statistical approach is definitely more adequate, bracketing the value according to the confidence of the available anatomical information. Such a quantitative perspective to the anatomical uncertainty in paleontology has probably represented one of the major advances of this field.

Techniques like principal component analysis can provide allometric vectors able to isolate size-related signals from the individual random variation of a reference sample (Wu and Bruner, 2016). In general, the first multivariate component of a morphometric data set is a size vector or a size-related vector. Therefore, it is able to supply a multivariate combination of the available variables which optimizes the size information, excluding other sources of variability. Also in this case, the observed residuals associated with the reference sample can supply an estimation of the uncertainty, giving a range of values and an index of the quality of the model.

With this in mind, brain volume can be estimated in fossils, and statistical parameters compared within different human groups. Of course, such estimations concern the whole brain volume, but they provide no information on what elements of the volume are involved in the observed evolutionary changes. Brain proportions (lobes and circumvolutions) apart, evolutionary changes in brain size are generally attributed to the number of neurons. Certainly, there is no reason to assume this is always true. Volume changes can also be due to the connections (axons and dendrites), to the vascular or connective elements, or to the supporting cells (glia).

Many efforts have been made to calculate, with the best approximation, the cranial capacity of many complete and incomplete fossils specimens. Thanks to this permanent commitment, we currently have stable and reliable estimates for the endocranial volume of many human specimens and taxa (Grimaud-Hervé, 1997; Holloway et al., 2004). Australopithecines had an average cranial capacity comparable with living apes, between 300 and 500 cc. The genus *Paranthropus* displays larger values than the genus *Australopithecus*, which may suggest an encephalization process. Nevertheless, the sample available is small, and any conclusion must be interpreted as provisional. The earliest groups possibly belonging to the human genus (*Homo habilis*) averaged 600 cc, and the earliest specimens with full human characters (*Homo ergaster*) approach 800 cc. Asian and African *Homo erectus* had a mean endocranial volume of approximately 1000 cc, and *Homo heidelbergensis* displayed a larger value of around 1200 cc. Modern humans and Neanderthals show an average cranial capacity of around 1400—1500 cc, a figure which is probably a bit larger for the latter group. Modern humans reached their maximum endocranial volume soon after their phylogenetic origin, approximately 100—150 thousand years ago (ka). In contrast, in the same period, early Neanderthals had a smaller cranial capacity, comparable

with *H. heidelbergensis*, and reached larger values much later, only around 60—50 ka (Bruner and Manzi, 2008; Bruner, 2014). This sequence makes the traditional iconographic view of linear brain size increase concluding in modern humans simply not in accordance with the fossil evidence: Neanderthals could have evolved a large cranial capacity after modern humans did.

Species-specific means and range estimations may vary depending upon the fossil samples used for their computation. Many groups are represented by only a few specimens, and statistical results are therefore sensitive to inclusion/exclusion of few single individuals. The brain size increase in modern humans, Neanderthals, *H. ergaster* and *H. heidelbergensis*, has been interpreted as a real encephalization process, namely an increase in brain size which was not dependent upon the increase of body size (Holloway, 1995; Tobias, 1995; Rightmire, 2004). Conversely, the brain size increase in *H. erectus* is probably a secondary consequence of a generalized body size increase.

Group-wise statistics are valuable and necessary to develop and test hypotheses regarding brain evolutionary changes. In contrast, individual values associated with specific specimens are less informative, because of the large variation of this trait. Taking into consideration the ratio between range and mean value (De Sousa and Cunha, 2012), *H. sapiens* and *H. erectus* show the largest variation in brain size (54%), followed by *H. heidelbergensis* (46%) and Neanderthals (40%) (Fig. 32.8). Such estimations may be biased by limited sample sizes and by decisions concerning the assignment and inclusion of groups and specimens. Nonetheless, they evidence a notable variability in all taxa. In our species, for which there is no taxonomic or statistical uncertainty, differences in cranial capacity can vary by more than 1000 cc, and the normal variation spans between 900 and 2000 cc (Holloway et al., 2004). Such large intraspecific ranges suggest caution when discussing cranial capacity in single fossil specimens, or in small fossil samples.

32.3.2 Absolute and Relative Brain Size

A major longstanding debate on brain size evolution concerns whether absolute or relative increase of the brain volume is the essential factor, especially when dealing with the possible cognitive changes associated with such variations (Alba, 2010; Isler and Van Schaik, 2014). After many decades of discussions, we still lack general agreement on this point, although brain size increase is traditionally interpreted as suggesting enhanced cognitive capacity.

Absolute increases deal with the rough volumetric enlargement, while relative increases are proportional

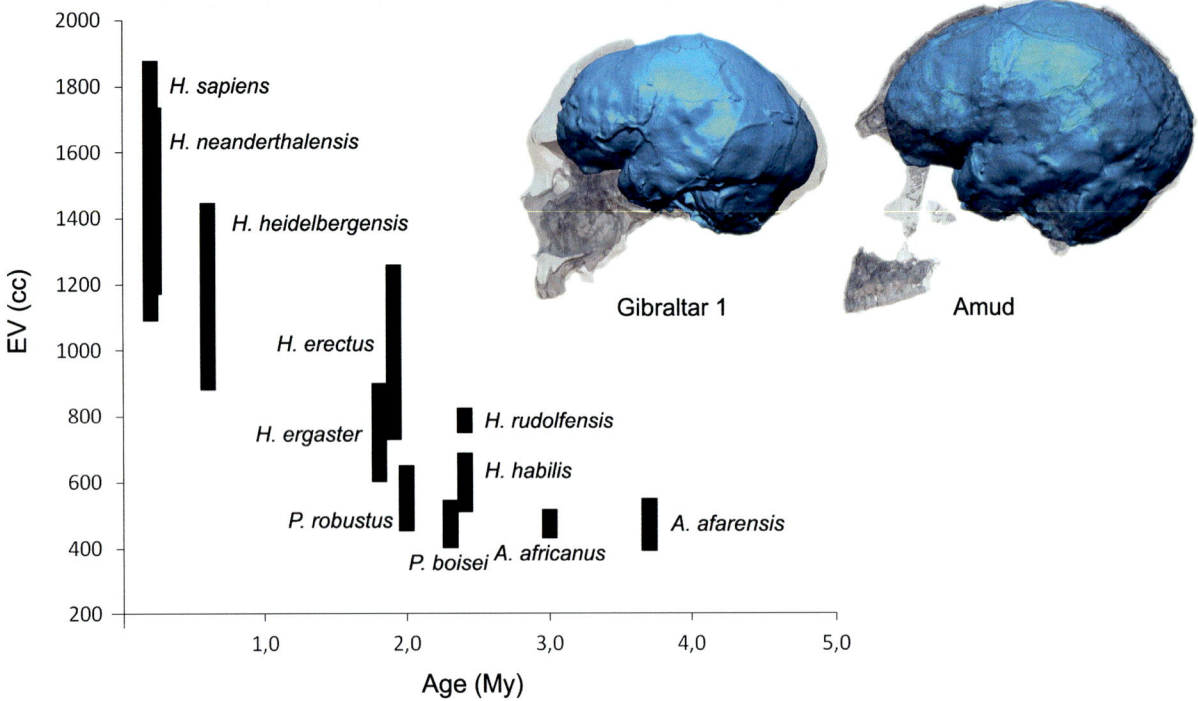

FIGURE 32.8 Species are reported according to their first appearance in the fossil record (My: million years) and to their range of endocranial volume (EV) (Original data from De Sousa, A., Cunha, E., 2012. Hominins and the emergence of the modern brain. Prog. Brain Res. 195, 293–322.). Neanderthals include specimens such as Gibraltar 1 (EV = 1270 cc) and Amud (EV = 1740 cc) (digital replicas after Amano, H., Kikuchi, T., Morita, Y., Kondo, O., Suzuki, H., Ponce de León, M.S., Zollikofer, C.P.E., Bastir, M., Stringer, C., Ogihara, N., 2015. Virtual reconstruction of the Neanderthal Amud 1 cranium. Am. J. Phys. Anthropol. 158, 185–197; Bruner, E., Amano, H., Pereira-Pedro, S., Ogihara, N., 2016. The evolution of the parietal lobe in the genus *Homo*. In: Bruner, E., Ogihara, N., Tanabe, H. (Eds.), Digital Endocasts. Springer, Tokyo (in press).). *H, Homo; P, Paranthropus; A, Australopithecus.*

to other biological factors. Absolute size is gaining popularity since recent studies have found that many biological parameters in primates are strongly correlated with the total number of neurons (Herculano-Houzel, 2012). Nonetheless, encephalization, strictly speaking, refers to a relative increase in brain size (Ruff et al., 1997). In general, proportional size changes are computed relatively to body size or metabolism. Needless to say, both variables are unknown in fossils and require further estimations and approximations based on the information only available for modern humans and other living primates. The traditional approach uses regression analysis to compute a linear prediction, so as to calculate deviations from the expectation (the so-called *encephalization indexes*). Brain size is regressed onto another variable (eg, body size or metabolism), a regression rule is computed for a specific taxon, and the residuals of each individual value (species or specimens) are calculated to evidence whether they have a larger or smaller value than expected. Such an approach is essential within a comparative framework. Nevertheless, it is worth noting that the resulting values are not real figures of the species/specimens, but instead specific results of that specific sample and analysis. The

regression approach requires the election of a contrasting biological variable (body size or other factors), of a comparative group (species, genera, families, etc.), and of a specific regression approach (for example, symmetrical or asymmetrical regression techniques). Hence, the results will depend upon all these operational choices and will be valid only within that specific analytical setting. Accordingly, decisions on groups or specimens to include, and on the procedure to be used, may seriously influence the final results. It is also worth noting that encephalization, as a relative increase of brain volume, can be achieved not only by increasing brain size but also by decreasing body size. In both cases, more neurons are available to a given body mass.

A special note concerns the current (living) variation of our phylogenetic group, namely the hominoids. Apart from our species, and excluding gibbons because of their much smaller body and brain size, any comparative and statistical analyses can be computed with only three "biological models," or genera: chimpanzees (*Pan* spp.), gorillas (*Gorillas* spp.), and orangutans (*Pongo* spp.). These three taxa have similar or at least largely overlapping cranial capacities, although their brain form is different because of specialized brain

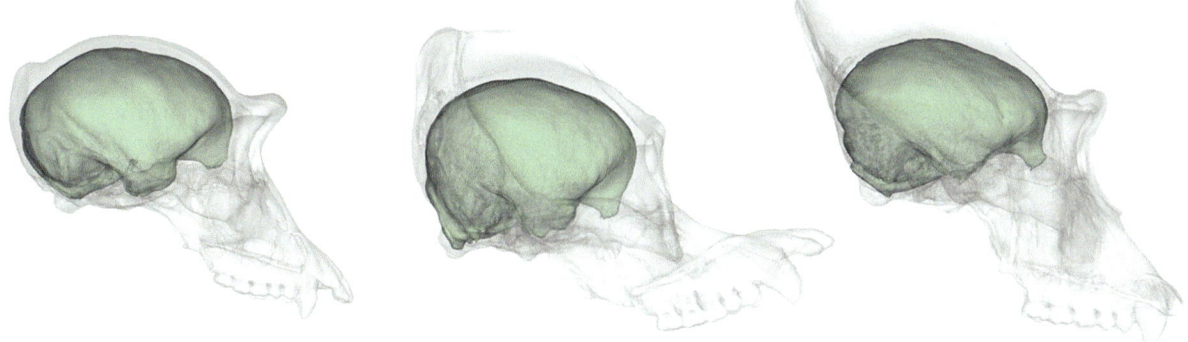

FIGURE 32.9   Digital skulls and endocasts of adult male chimp, orang, and gorilla. Australopithecines had a comparable brain size with the range of living apes. Specific morphological differences between *Australopithecus* and living apes (possibly associated with larger frontal and parietal cortical areas in the former) remain, to date, unclear.

proportions (Aldridge, 2011) and different cranial and neurocranial architecture (Bienvenu et al., 2011; Fig. 32.9). Gorillas and orangs are larger than chimpanzees, but their cranial structure is totally different, with a marked dolichocephaly and prognathism in the former and extreme brachycephaly and airorynchy (dorsal rotation of the facial axis) in the latter (Shea, 1985). Therefore, comparative regression approaches within our strict zoological group are limited to few units (N = 3) with similar size but different organization. Accordingly, any allometric analysis will be particularly sensitive to minor random factors (because of the few statistical units) and to phylogenetic bias (because of the derived characters of these genera). Furthermore, the noticeable separations between the volumetric ranges for humans, apes, and monkeys generate morphological and numerical gaps between these three groups, hampering robust quantitative approaches based on linear fits. It is therefore difficult to supply numerical models able to detect grade shifts or, conversely, to reveal allometric shared trajectories. Because of the outlier position of our species in the regression models, any minimal fluctuation of the database can involve major changes of the *H. sapiens* residuals from the model. This limitation, due to the scarce species diversity in living apes, is inherent to the current comparative framework and must be properly acknowledged when dealing with regression models in evolutionary neuroanatomy.

Through these regression procedures, we know that we modern humans have a brain roughly three times bigger than expected for a primate of our brain size. Cognition apart, this large brain size has also metabolic (Martin, 1981; Armstrong, 1983) and ecological (Leonard et al., 2007) consequences. Relatively larger brains have also been described for cetaceans and pinnipeds, but our metabolism is the one expected for our body size, while cetaceans and pinnipeds have a more expensive

energetic turnover, probably because they live in water and often in cold climates. Therefore, after proper regressions and adjustments to take into account all these factors, we can establish one of our unique features: we are the only mammals that evolved larger brains without increasing overall energetic costs. Taking into consideration that the brain is a very expensive organ, this can be achieved only in one way, which is through the redistribution of the energy budget and an energy restriction of other body parts, probably the digestive system (Aiello and Wheeler, 1995). In primates, most life—history parameters show a strong, positive correlation with brain size and metabolism: larger brains are associated with slower growth and development, delay and prolongation of all the life stages, increased longevity, and consequential adjustments of the ecological and social organization (Harvey and Clutton-Brock, 1985; Dunbar, 1992, 1998; Holly-Smith, 1992; Hammer and Foley, 1996). It is therefore apparent that brain size is not only a matter of cognition.

32.4 Brain Morphology

32.4.1 Sulcal Pattern and Brain Proportions

Changes associated with absolute and relative brain size are only part of the scenario. Volumetric increase or decrease of specific areas can generate changes in cortical proportions, by means of allometric scaling or by the addition of novel evolutionary characters (Rilling, 2006; Sherwood et al., 2012). Such differential changes are likely to be associated with variation of the neural organization, and possibly of neural functions (De Sousa and Proulx, 2014).

External brain morphology can supply information about the scheme of cortical gyri and sulci, and on the relative dimensions of specific brain areas. We are still far from understanding the exact relationships between cortical traits and cerebral functions, and the correspondence between macroscopic features and cytoarchitectonic areas is not necessarily constant (Amunts et al., 1999). Nonetheless, we know that different districts are involved in different functional tasks, and some specific cortical areas can be crucial for specific cognitive processes. Therefore, sulcal patterns can be useful as references for analyzing the relative size of specific cortical areas, and changes in the internal brain proportions can supply information on undergoing functional differences. It is worth noting that the brain is extremely plastic and sensitive to environmental and behavioral influences. In macaques, morphological brain changes can be observed after a short period of task-based training (Quallo et al., 2009). In humans, such plasticity is probably even larger than in nonhuman primates (Gómez-Robles et al., 2015), suggesting that plasticity per se may have been targeted by selection, more than specific cortical characters. Therefore, although paleoneurological changes are usually interpreted in terms of traditional evolutionary mechanisms (ie, genes and selection), environmental or epigenetic factors influencing macroscopic brain traits should not be excluded. This is particularly true when dealing with humans and with possible feedbacks between biological and cultural adaptations. Thus, some changes in brain proportions can be the result of a selective evolutionary process, but also a consequence of a specific behavior. A change in cortical proportions can influence a specific cognitive capability or, conversely, be the result of a persistent behavioral training. Currently, we ignore the degree and ranges of expression of these two mechanisms.

A second kind of cortical analysis concerns the morphology of the sulci, namely their geometrical and spatial organization. To date, there is no clear evidence of major differences of this sort among the species of the human genus. Regardless of brain volume, all extinct humans display a general sulcal pattern similar to the modern human basic scheme of circumvolutions. Among living humans there is notable individual variation, and fossil species do not depart patently from a comparable general organization of gyri and sulci. Some changes in the prefrontal and orbital cortex have been proposed for australopithecines, although they are based on few or single specimens, with different degrees of preservation (Carlson et al., 2011; Falk, 2014). However, recent evidence strongly supports the hypothesis that sulcal patterns may largely be a passive mechanical consequence of force redistribution during brain growth (Tallinen et al., 2016). In this model, the basic form of the brain, under the effects of pressures and tensions associated with growth and with the biomechanical properties of the brain tissues, passively folds to keep the pace with volume-to-surface

relationships, generating the characteristic human folding patterns, as well as random individual variations. This would explain the large intraspecific variability and the lack of strong correspondence with the cytoarchitectonic areas, warning against using these characters for phylogenetic or functional inferences.

Another frequently investigated morphological trait in paleoneurology is the expression of cerebral asymmetries, because of their relevance in human cognition (Balzeau et al., 2012). To date, however, there is no clear evidence of different specializations of cerebral asymmetry within the human lineage. Modern humans display a consistent asymmetric pattern, in which the right frontal lobes and left occipital lobes are larger than their respective counterparts. This pattern has been frequently associated with handedness or language. Nevertheless, all the human fossil species display this same scheme (Holloway, 1995). Living apes show a less frequent combination of right−left asymmetries (Holloway and De La Coste-Lareymondie, 1982), but allometric effects on the grade and expression of such frequencies cannot be excluded (Gómez-Robles et al., 2013). It is worth noting that the structural causes of these asymmetries are not yet totally clear. The skull and the connectives respond to the pressure of the growing cortical mass as an integrated biomechanical system and, in this case, we ignore the mechanisms of forces behind the asymmetrical distribution of the neural mass, as well as its correlation with macroscopic morphological appearance.

Taking all these considerations into account, it seems that the main information supplied by the sulcal pattern does not concern local features, but rather deals with relative cortical proportions within the entire brain volume. Changes of brain proportions have been described and hypothesized for different human taxa (Fig. 32.10) and, in a paleontological context, they can be detected following three different analytical approaches. First, dimensions of specific areas can be tentatively measured, through linear or volumetric variables. Sulcal patterns can be used to delimitate specific areas with a given approximation and uncertainty, so as to measure and compare chords, arcs, surfaces, or volumes. A second approach applies shape analysis to reveal differential size variations within the endocranial system. In this case, specific changes in the volume of some cortical areas will induce variation in the overall spatial arrangement of the endocranial morphology, and this geometrical difference can be detected through spatial analysis. A third method is indirect and relies on correlation and covariation patterns between cranial and cerebral features, using the former to reveal the latter. When a correlation is sufficiently robust to interpolate or extrapolate a dependent (brain) variable from an independent (bone) variable, soft tissue morphology can be inferred from cranial remains. In

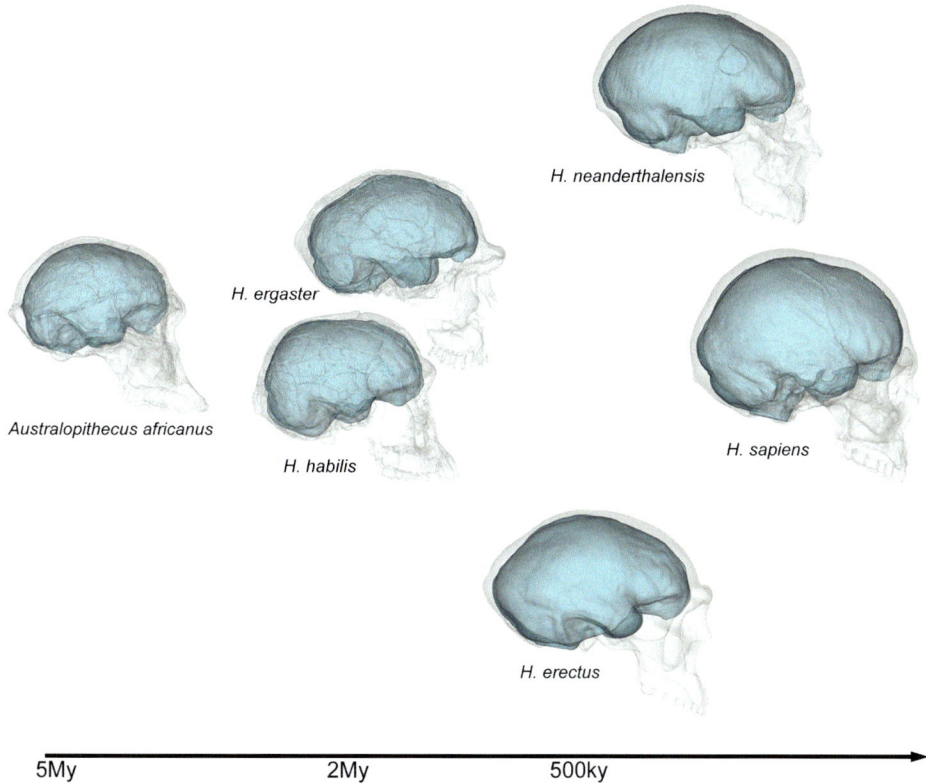

FIGURE 32.10 Digital skull and endocasts of some principal fossil species, with an approximate chronology (*My*, million years; *ky*, thousand years). For other fossil taxa associated with the human evolutionary radiation, the paleoneurological information is scarce or absent. Digital replicas are not to scale. *H.*, *Homo*.

this indirect method, proper knowledge of the causal reason of the correlation is not decisive. Because of its heuristic value, this method can reveal aspects of the soft tissues that cannot be investigated using more traditional and direct anatomical approaches.

32.4.2 Frontal Lobes

Despite the many claims stating that the evolution of the frontal cortex was patently apparent in the fossil record, paleoneurological evidence is far from being decisive. Some authors recognize sulcal differences in australopithecines that may suggest "incipient" human traits of the prefrontal cortex (Carlson et al., 2011; Falk et al., 2012; Falk, 2012, 2014). Nonetheless, such inference is associated with individual cases or incomplete specimens and at present cannot be probably generalized to a wider taxonomic context. The limited paleoneurological record and the uncertainty associated with sulcal details do not allow, at present, a consistent recognition of possible brain differences between australopiths and living apes, or of possible sulcal differences in the frontal cortex among human species. It is not known whether sulcal differences between apes and humans are due to size enlargement or else to the development of specific

cortical areas. The sulcal pattern associated with the lower frontal gyrus, including a bulging boss at Broca's area, can be roughly recognized in every specimen that has been assigned to the human genus, dating back to 2 million years ago (Tobias, 1987, 1995).

Apart from sulcal patterns, most studies on the frontal morphology deal with the form and proportions of the anterior cranial fossa. Modern humans are characterized by bulging of the frontal squama, when compared with fossil human species, and this was claimed to be a result of frontal lobe enlargement (Lieberman et al., 2002). However, the importance of this trait in paleoneurology has been largely revised and partially devalued. Human frontal lobe volume is to be expected for an ape with our brain size, and this fact apparently does not confirm noticeable evolutionary changes strictly associated with the frontal areas, at least in terms of size (Semendeferi et al., 1997, 2002; Rilling, 2006). There is a longstanding debate on whether or not minor volumetric human–ape differences in the frontal cortex can be significant or not (Barton and Venditti, 2013; Smaers, 2013). Of course, there may be many differences that cannot be demonstrated statistically, and also minor anatomical changes can have important functional effects. In any case, we must assume that if modern humans have evolved some specific morphological differences in their frontal lobes, these are not prominent, at least when dealing with gross size and when compared with living apes.

The degree of frontal curvature is accentuated in modern humans, but this is not exclusive to our species, because its variation overlaps with nonmodern human species (Bruner et al., 2013). Shape analysis suggests that the morphology of the frontal profile has undergone only minor changes in the human genus, and most of the variation is generally associated with the outer superstructures (the browridge) and with the relationship between neurocranial (the braincase) and splanchnocranial (the face) blocks (Bookstein et al., 1999). Therefore, taking into account that major volumetric changes in the frontal lobes have not been documented for modern human beyond a general brain size increase, and that in contrast the spatial relationships between face and braincase underwent notable adjustments, it is likely that the curvature of the frontal squama (a proxy for the frontal lobe bulging) increased in modern humans because of structural variations associated with the face versus brain spatial organization, and not because of a specific increase of the frontal cortex.

A change in frontal proportions has nevertheless been demonstrated in Neanderthals and modern humans, two species that display wider prefrontal regions and anterior cranial fossae (Bruner and Holloway, 2010). In earlier humans, the frontal lobes are definitely narrower than the maximum endocranial (parietotemporal) region, while in Neanderthals and modern humans, the anterior and posterior widths are more similar, a fact that is easily recognizable in upper view. A bivariate analysis showed that this is a real grade shift and change of allometric proportions, and not a size effect. Taking into account that there is no patent evidence of frontal lobe expansion beyond a common brain enlargement, the frontal widening described in modern humans and Neanderthals is likely to be a redistribution of the cortical volume, more than an actual size increase of the frontal cortex. As already mentioned, Neanderthals and modern humans are the only human species with their frontal lobes largely lying on the orbital roof (Bookstein et al., 1999; Bruner and Manzi, 2005, 2008). Therefore, the widening of the frontal lobes can be a structural secondary consequence of this vertical constraint, reorienting the cortical mass laterally. The fact that this widening involves Broca's area, however, leaves open the possibility of functional changes, in terms of relative brain mass or in terms of connections and spatial adjustments.

In archaic humans, the narrow frontal lobes are matched ectocranially by a large temporal fossae due to a marked postorbital constriction. This space, which is extremely delicate in terms of bone resistance, is occupied by the large temporal muscle whose size is, in primates, proportional to the muzzle size (Cachel, 1978). In modern humans, we observe the widening of the frontal lobes, the reduction of the facial block, and at the same time an extreme reduction of the temporal muscle. This reduction of the muscle is also associated with a change in its protein structure, which has led to the hypothesis that the reduction of mechanical forces exerted by the temporal muscle could have had a role in form changes associated with the human brain (Stedman et al., 2004). This hypothesis is, however, not properly validated (McCollum et al., 2006). The widening of the anterior fossa in modern humans also involved widening of the olfactory bulbs and frontoorbital cortex, which prompted suggestion that this region underwent functional specialization (Bastir et al., 2011). Although neural changes associated with these increased lateral diameters cannot be ruled out, the existence of complex structural constraints between the frontal cortex and the upper facial block suggest caution when making cognitive inferences from the gross proportions of the anterior cranial fossa.

Internal morphological changes of the prefrontal cortex that could be associated with modern humans, such as an increase in the relative white matter proportions (Schoenemann et al., 2005), cannot be at present evaluated in the paleoneurological record.

32.4.3 Parietal Lobes

The parietal lobes have attracted the attention of the paleoanthropologist since the beginning of the past century. Raymond Dart (1925) hypothesized parietal changes when observing the Taung child, the first *Australopithecus* ever described. Franz Weidenreich (1936, 1941) noted the marked parietal differences between our species and *H. erectus*. Half a century later, Ralph Holloway (1981a) published the first spatial analysis on endocasts, describing the noticeable variation of these areas. He was also the student most involved in supporting the hypothesis of Dart of a parietal expansion in australopiths, by virtue of a purported posterior displacement of the lunate sulcus, when compared with the position described in living apes (Holloway et al., 2003). After decades of debate (Falk, 2014), this controversial issue has yet to be definitely resolved, with contrasting views concerning the position or the actual possibility to recognize reliably the lunate sulcus in all species or specimens (Allen et al., 2006; Aldridge, 2011).

Despite their relevance for human brain functions, the lower parietal areas have not supplied any particular information in paleoneurology. The supramarginal and angular gyri are involved in specialized human cognitive functions, but their gross appearance or sulcal pattern does not show any specific change among human species, and their proportions seem to be largely influenced by brain size. In contrast, the upper parietal areas are associated with more patent changes within the human genus, both in terms of bones (Bruner et al., 2004, 2011) and lobes (Bruner et al., 2003; Bruner, 2004; Fig. 32.11). All archaic humans display a depressed upper parietal surface, with a distinctive parasagittal flattening. This leads to the midline longitudinal bulging (keel) and to a classic "tentlike" profile of the endocast in rear view. Neanderthals have a distinct morphology, and the upper parasagittal surface is laterally bulging, with a consequent "en bombe" profile in rear view. The morphogenesis of these upper vault areas is relatively simple, and the morphology of the bone is largely shaped by the pressure exerted by the brain mass. Therefore such localized upper parietal bulging is likely the result of expansion of the upper parietal cortical surface.

In modern humans, the changes are even more apparent. The rounded shape characteristic of the modern human head is largely due to a noticeable enlargement of the parietal bone. Such species-specific morphology of the bone is associated with a similar enlargement of the parietal lobe (Bruner et al., 2014a). In modern humans, this bulging of the parietal surface takes place early during ontogeny (Neubauer et al., 2009), probably in a postnatal stage absent in Neanderthals and living apes (Neubauer et al., 2010; Gunz et al., 2010; Scott et al., 2014).

Shape variation associated with the modern human parietal bulging is very similar to the variation associated with enlargement of the precuneus, which is a main factor in brain variability among living adult humans (Bruner et al., 2014b, 2015). The size of the precuneus also represents a major difference in the midsagittal brain geometry between humans and chimpanzees, being much larger relatively and absolutely in our species (Bruner et al., 2017). The correspondence between phylogenetic, evolutionary, and intraspecific patterns suggests that the precuneus may be involved in recent neuroanatomical changes associated with our lineage. It is worth noting that a supposed early modern human fossil like Jebel Irhoud, a Moroccan specimen dated approximately to 150—220 ka, did not show such typical modern brain morphology (Holloway, 1981b; Bruner and Pearson,

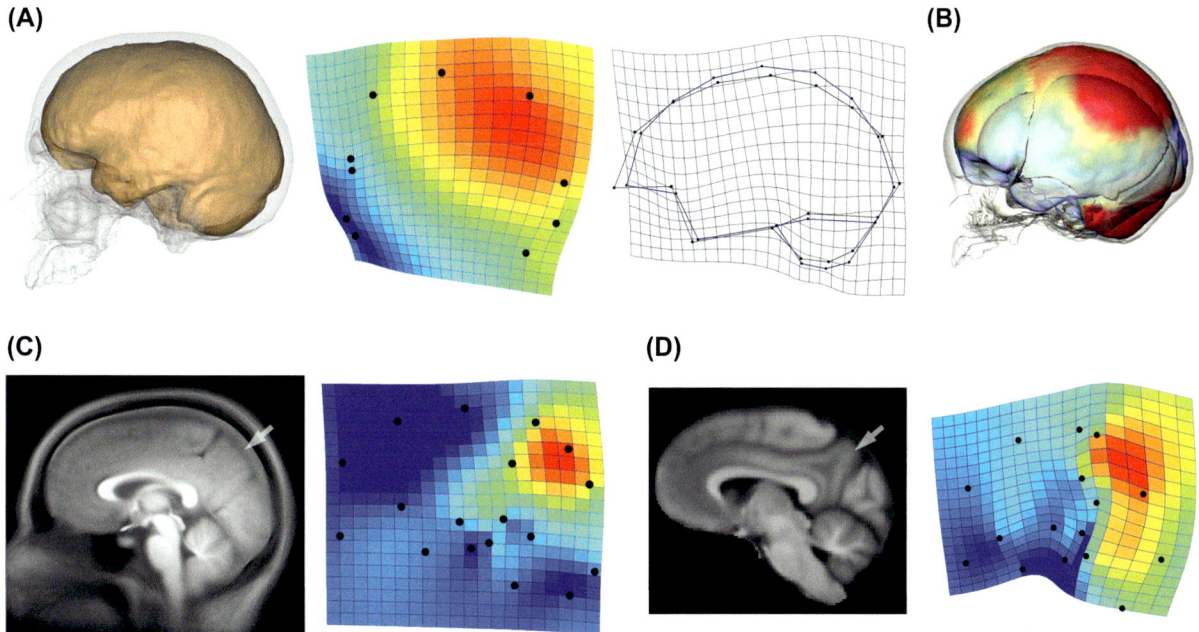

FIGURE 32.11   Parietal changes in *Homo sapiens*: (A) modern humans are characterized by bulging of the parietal bone (chromatic map; *red*: expansion) and lobe (wireframes); (B) modern humans also display a unique species-specific stage of parietal bulging after birth (in *red*: surface expansion); (C) in midsagittal adult brain, the main source of variation is the precuneus dimension; (D) comparing chimp with human midsagittal brain, the main difference is a larger precuneus in our species. The *arrows* show the precuneus in humans and chimps. *Images after Bruner, E., 2004. Geometric morphometrics and paleoneurology: brain shape evolution in the genus* Homo. *J. Hum. Evol. 47, 279—303; Gunz, P., Neubauer, S., Golovanova, L., Doronichev, V., Maureille, B., Hublin, J.J., 2012. A uniquely modern human pattern of endocranial development. Insights from a new cranial reconstruction of the Neandertal newborn from Mezmaiskaya. J. Hum. Evol. 62, 300—313; Bruner, E., de la Cuétara, J.M., Masters, M., Amano, H. Ogihara, N., 2014. Functional craniology and brain evolution: from paleontology to biomedicine. Front. Neuroanat. 8, 19; Bruner, E., Rangel de Lázaro, G., de la Cuétara, J.M., Martín-Loeches, M., Colom, R., Jacobs H.I.L., 2014. Midsagittal brain variation and MRI shape analysis of the precuneus in adult individuals. J. Anat. 224, 367—376; Bruner, E., Preuss, T., Chen, X., Rilling, J., 2017. Evidence for expansion of the precuneus in human evolution. Brain Struct. Funct. 222(2), 1053—1060.*

2013). Later modern human populations like those from Qafzeh and Skhul from the Near East, dated approximately to 100 ka, show a somewhat intermediate phenotype (Bruner et al., 2016d). These observations are associated with single specimens and so must be interpreted with caution. Nonetheless, these mosaic or intermediate phenotypes raise the question on whether the modern brain morphology was achieved by discrete or gradual changes and whether it has occurred due to genetic or environmental influences. In any case, they suggest that the modern brain morphology may not have originated together with the modern human lineage, but instead it may represent a later change.

Owing to their location, lateral parietal bulging has been tentatively associated with a relative enlargement (and outfolding) of the intraparietal sulcus, while the longitudinal additional enlargement described in modern humans has been tentatively associated with a cortical surface increase of the precuneus (Bruner, 2010; Pereira-Pedro and Bruner, 2016). The human intraparietal sulcus is expanded when compared with nonhuman primates, and it shows specific functions in our species (Vanduffel et al., 2002; Grefkes and Fink, 2005; Orban et al., 2006). This region is essential for hand—eye coordination, particularly relevant for primates, being directly involved in tool use and probably also in language. The precuneus is essential in the Default Mode Network (Utevsky et al., 2014), being a structural and functional hub in the large-scale network of cortical connectivity (Cavanna and Trimble, 2006; Hagmann et al., 2008; Zhang and Li, 2012). The precuneus is also critical for a proper coordination of spatial, temporal, and social relationships (Hills et al., 2015; Maister et al., 2015; Peer et al., 2015), linking visual representation, body experience, and self-awareness (Fletcher et al., 1995; Freton et al., 2014; Margulies et al., 2009; Land, 2014). All the superior parietal areas are deeply involved in visuospatial integration, coordinating the spatial relationships between brain, body, and environment (Bruner and Iriki, 2016). Such functions, apart from their mechanical roles, are crucial to integrate the body and the external object in cognitive processing, consistent with theories and hypotheses of embodiment and cognitive extension (Malafouris, 2013). It is likely, therefore, that some of these functions have undergone some profound changes in our lineage, by genetic selection, enhanced plasticity, or behavioral training and cultural feedback.

The parietal cortex is particularly integrated with the prefrontal areas, generating a complex frontoparietal system (Jung and Haier, 2007; Caminiti et al., 2015). It is thus relevant to note that the two human species with larger brain size, namely Neanderthals and modern humans, both display changes in frontal and parietal proportions.

32.4.4 Temporal Lobes

Temporal lobes support many cognitive functions, and modern humans have a relatively larger temporal cortex when compared with living apes (Rilling and Seligman, 2002; Rilling, 2006). Despite such evidence, the paleoneurological information on their evolutionary changes is not definitive. Temporal lobes are less well represented in the fossil record because of their fragility, so the number of samples available which can provide statistical evaluations is small. Modern humans have a longer middle endocranial fossa, when compared with other human species, and this change may be the consequence of an expansion of the temporal lobes (Bastir et al., 2008). However, the structural relationships with the endocranial base make any morphological changes difficult to explain only in terms of cerebral changes. When observing spatial variations in this area, the influence of basicranial and facial constraints cannot be easily excluded (Bastir and Rosas, 2016). The middle cranial fossa is also directly in biomechanical contact with the mandible, adding further extrinsic factors to the overall form variations. As with the orbital region, because of the tight relationships between bone and brain elements and a possible spatial conflict between soft and hard tissues, the temporal sulcal pattern can be easily recognized on the lower and lateral bone surface of the endocranial temporal areas. Nonetheless, at least modern humans and Neanderthals do not display any patent sulcal differences, apart from a larger subarcuate fossa in the former (Rosas et al., 2008). It is worth noting that the temporal lobes have important medial areas whose variation can have no direct influences on the endocranial morphology. Taking into account the complexity of endocranial base morphogenesis, more information on the structural relationships between temporal lobes and middle cranial fossa is necessary to evaluate properly any possible paleoneurological changes.

32.4.5 Occipital Lobes

Apparently, the occipital lobes did not show any specific or prominent change within the human genus. Many fossil groups (especially Asian *H. erectus*) show flattened temporoparietal regions and a marked bulging of the occipital lobe (Grimaud-Hervé, 1997; Balzeau et al., 2012). Nonetheless, it is not possible to evaluate whether this is a matter of phylogeny or local variation (Antón, 2003; Baab, 2008; Bruner et al., 2015c). The Chinese specimens are the most representative and homogeneous in this sense, but they all came from one single locality, Zhoukoudian (Wu et al., 2010), although similar bulging occipital areas can be found in earlier African fossils (Bruner et al., 2016c). The projection of the occipital lobes can be a secondary consequence of platycephaly (low and flattened vault) and parietal flattening, due to a morphological integration between parietal and occipital blocks (Gunz and Harvati, 2007). Modern humans have been hypothesized to show relatively smaller occipital lobes when compared with living apes (De Sousa et al., 2010), but isometric parietooccipital volume (Semendeferi and Damasio, 2000). These two conclusions implicitly suggest larger parietal volume in our species and plesiomorphic proportions for the occipital areas.

In primates, there is a robust correlation between orbit size and occipital lobe size which, taking into consideration the large orbital volume in Neanderthals, led some workers to hypothesize larger occipital lobes in this species, when compared with *H. sapiens* (Pearce et al., 2013). If confirmed, it remains to be determined which of the two groups display a derived condition. Taking into account the parietal enlargement in modern humans, the Neanderthal condition (large occipital lobes, small parietal lobes) is likely to be the plesiomorphic one.

32.4.6 Cerebellar Lobes

The cerebellar lobes have been little investigated in paleoneurology. This region is one of the less preserved in the fossil record, with clear limits for any statistical approach. Similar to the temporal region, this district is influenced not only by neural factors (cerebellar morphology) but also by structural factors associated with cranial base flexion, posture, and locomotion. All these factors have hampered, to date, a robust approach to its morphological variations within and between species. There is nonetheless a good correspondence between cerebellar dimensions and posterior fossa dimensions (Kubo et al., 2014). The position of the

cerebellar lobes is different among human species, probably depending upon the structural influence of the endocranial base. In modern humans, the cerebellar lobes are positioned under the temporal lobes, while in Neanderthals they are positioned under the temporal and occipital lobes, and in many archaic hominins they are almost entirely positioned below the occipital lobes (Grimaud-Hervé, 1997). Although this may be a secondary consequence of cranial base architecture, this variation in the spatial position involves a change in the relative distances of the cerebrocerebellar connections. Some metric changes in cerebellar proportions have been described for recent modern humans (Weaver, 2005). However, also taking into account the scarce information available on the cerebellar functions, any paleoneurological evaluation is, at the moment, preliminary.

32.4.7 The Paleoneurological Variation of the Human Genus

In summary, we can conclude that, despite the remarkable phylogenetic variation described for extinct humans and australopiths, the paleoneurological record suggests only a few consistent changes, at least when dealing with brain form as extrapolated from endocranial morphology (Fig. 32.12).

The genus *Australopithecus* had a brain size comparable to that of living apes. Larger parietal proportions and more complex sulcal pattern in the frontal lobes have been suggested, but without any conclusive evidence. Therefore, to date, we are unable to identify specific paleoneurological traits for this taxon, when compared with other apes. The same can be stated for the genus *Paranthropus* (robust australopithecines), whose study is further hampered by the limited sample available. The

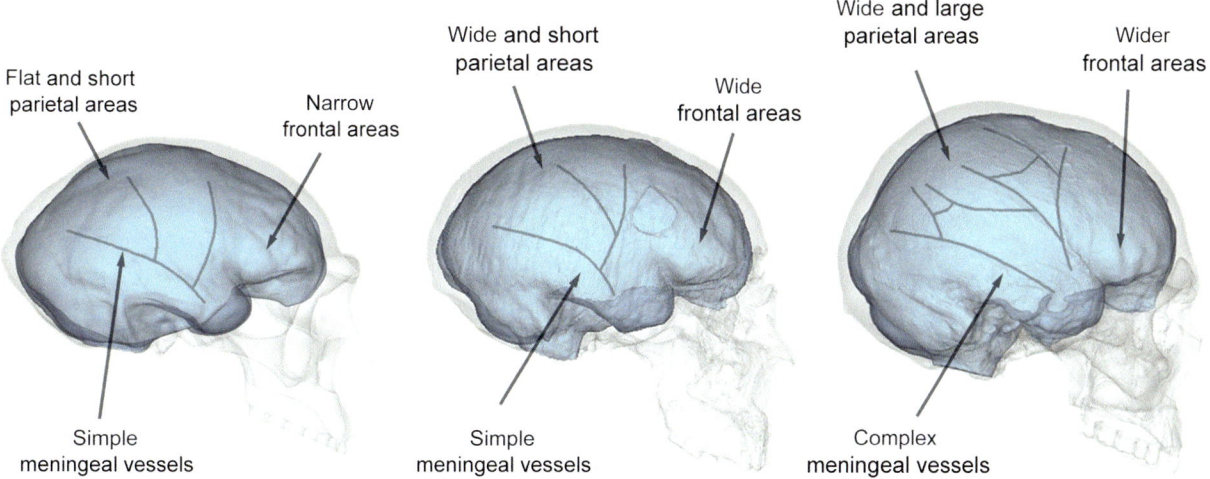

FIGURE 32.12 Main traits associated with *Homo erectus* (left), Neanderthal (center), and modern human (right) endocasts.

phylogenetic position of taxa such as *H. habilis*, *Homo rudolfensis*, and *Kenyanthropus platyops* is still debated (Lieberman, 2001), and there is insufficient information to interpret their endocranial morphology beyond individual variations, although all these species display larger brain size when compared with living apes.

Among undisputed members of the human genus, we can distinguish three separate endocranial morphotypes. The archaic one is formed by small- to medium-brained human species, including *H. ergaster*, *H. erectus*, and *H. heidelbergensis*. *Homo ergaster* and *H. heidelbergensis* display a relatively larger brain size compared with their ancestors. The taxonomic distinction between *H. ergaster* and *H. erectus* is cause for debate but, in any case, the latter shows a larger brain size only in terms of absolute value, not in terms of encephalization. Differences in brain size apart, all these taxa display similar endocranial traits and proportions, apparently all being simply scaled versions of the same morphological model. There is a noticeable variation between individuals, but no general trends or group-wise patterns. Therefore, at least according to the paleoneurological evidence, we must assume no specific neuroanatomical differences among these species, beyond brain size variations. Of course, we cannot discard the existence of subtle differences that cannot be detected because of the small samples, or nonmorphological differences that cannot be identified at the macroscopic anatomical level.

The two derived endocranial morphotypes are Neanderthals and modern humans. In both lineages, there is a relative increase of brain size, a relative widening of the frontal cortex, and a relative enlargement of the upper parietal areas. Nevertheless, in Neanderthals, the upper parietal lobules undergo only a lateral widening, while in modern humans there is also an overall longitudinal enlargement, associated with an increased complexity of the vascular networks.

It is important to note that here we are dealing strictly with the fossil evidence. There are many important brain features and processes for human evolution that cannot be considered in paleoneurology. We must be aware that many changes do not involve macroanatomical differences and that the scarce information available from the fossil record can hamper a reliable and robust evaluation of the morphological variations. As in every paleontological field, individual differences based on single traits or single specimens should not be used to put forward generalized and comprehensive evolutionary hypotheses. Descriptive results and speculative conclusions are necessary to influence perspectives and future research, but evolutionary hypotheses should rely on quantitative approaches (models and patterns), comparative analyses (samples), and converging multidisciplinary evidence (different complementary fields).

32.5 Craniovascular Traits and Brain Thermoregulation

32.5.1 Brain Morphology and Vascular Biology

The study of the vascular system in fossil species can be termed *paleoangiology*. There are three main reasons for investigating the vascular networks in paleoneurology. First, there are many vascular imprints on the endocranial surface and within the bone thickness (Scremin, 2004; Patel, 2009). Such traces reveal much morphological information on the vascular anatomy (like size and distribution of the vessels or complexity of the branching patterns), and there is considerable variation among fossil species. Second, morphogenesis is an integrated process between soft and hard tissues, and the vascular system is a crucial part of this functional and developmental neurocranial network (Percival and Richtsmeier, 2013). Third, these imprints are the remnants of physiological processes associated with blood circulation and represent a rare opportunity to investigate dynamic metabolic processes in extinct taxa, in this case specifically associated with brain evolution (Píšová et al., 2017).

As far as the cortical anatomy is concerned, the traces of the vessels can supply only limited information on the original soft tissues, but they also offer the only direct evidence we can obtain on the vascular system (Fig. 32.13). Therefore, the information may be little, but valuable. Beyond the importance of blood circulation per se, in the case of the brain there is a matter of special interest: metabolism and thermoregulation (Wang et al., 2016). The brain is the most expensive organ of the body, and its intrinsic metabolism represents a significant source of heat production. At the same time, the brain is extremely sensitive to temperature, and heat accumulation can be particularly dangerous for its cells and functions. Despite the high metabolic load and the sensitivity to thermal stress, to date no specific cooling system has been detected. The two contrasting hypotheses on this topic suggest the absence of such a specific system, or else an unfruitful search for a well-hidden mechanism (Brengelman, 1993; Cabanac, 1993; Zenker and Kubik, 1996). In both cases, the vascular system is thought to play a major role in heating or cooling the endocranial cavity. Taking into account that humans have a brain that is three times larger than expected for their body size, integrated coevolution of the brain and its vascular network is likely.

The endocranial vascular traces are the result of a dynamic interaction between soft and hard tissues: the pressure exerted by the vessels induces the activation of osteoclasts, remodeling the endocranial wall. Not every superficial vessel leaves a trace on the bone, and the size of the imprint is not necessarily proportional to the diameter of the vessel. The final shape of the

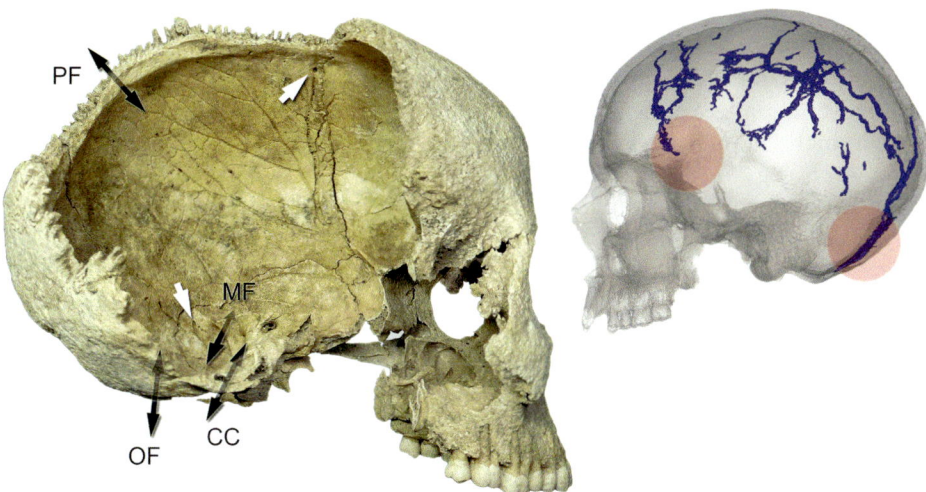

FIGURE 32.13   The imprints of the middle meningeal artery are easily recognizable on the endocranial wall in bioarchaeological and fossil cranial samples (left). The channels of the diploic vessels can be reconstructed using computed tomography (right). The *black arrows* indicates the main foramina (*CC*, condylar canal; *MF*, mastoid foramen; *OF*, occipital foramen; *PF*, parietal foramen). *White arrows* show in this specimen some connections between meningeal arteries and diploic channels. The *red circles* show areas of frequent connection among middle meningeal artery, diploic channels, and venous sinuses.   *Images after Píšová, H., Rangel de Lázaro, G., Velemínský, P., Bruner, E., 2017. Craniovascular traits in anthropology and evolution: from bones to vessels. J. Anthropol. Sci. (in press); Rangel de Lázaro, G., de la Cuatara, J.M., Píšová, H., Lorenzo, C., Bruner, E., 2016. Diploic vessels and computed tomography: segmentation and comparison in modern humans and fossil hominids. Am. J. Phys. Anthropol. 159, 313–324.*

imprint will depend upon the dimensions of the vessels, but also upon other factors like the thickness of the meninges or endocranial pressure, due to the brain volume or to the cerebrospinal fluid. Despite a lack of complete reciprocity between vessels and traces, common surgical practice suggests a good correspondence, at least with respect to gross anatomy.

The morphogenetic relationship between bones and vessels implies an active bone remodeling process: a trace will be necessarily associated with an effective pressure and a functional vessel. If the vessel ceases its activity and disappears, the bone surface will soon be remodeled and the trace will disappear too. This theoretical perspective on blood and bone dynamics unfortunately has a major limit: current knowledge on functions and variations of the endocranial vascular system is extremely scanty, even in our own species.

The braincase is supplied by four vascular systems: the *cerebral* vessels within the brain, the *meningeal* vessels between brain and bone, the *diploic* vessels within the bone, and the *pericranial* vessels on the external surface of the braincase (Saban, 1995). These distinctions are largely conventional: these systems do not originate through independent developmental processes and are all connected to each other in a single functional network. Hence, this terminology merely reflects the position and localization relative to brain/skull volumes and surfaces. The mechanisms behind the growth of these vessels are not fully understood. Both molecular and mechanical signaling are probably involved in the

degree of development and in the direction of growth of the vessels (Eichman et al., 2005).

32.5.2 Middle Meningeal Artery

The largest vascular trace in the lateral wall of the endocranial cavity is due to the middle meningeal artery (Bruner and Sherkat, 2008). This artery is accompanied by parameningeal veins, so the general term *middle meningeal vessels* would be more appropriate. Usually, the artery stems from the external carotid artery, and it enters the endocranial cavity through the *foramen spinosum*, a small passage on the endocranial floor in the middle cranial fossa (Ma et al., 2012). The homology between humans and apes for these vessels is uncertain, because of a contribution of the ophthalmic artery coming from the anterior endocranial districts (Diamond, 1991; Falk and Nicholls, 1992; Falk, 1993). There is a notable difference in the degree of complexity of these vessels among modern and nonmodern humans. All extinct human species have generally simpler networks with few branches and, generally, no anastomoses between the main branches. In contrast, modern humans generally display more complex reticulation patterns, especially at the parietal surface (Saban, 1995; Grimaud-Hervé, 1997; Bruner et al., 2005). Nonetheless, the variation of modern and nonmodern humans partially overlaps, and the character cannot be used to discriminate specimens or to support individual taxonomic interpretations. In many archaic human taxa (especially in Asian *H. erectus*), the posterior branches are equally or more developed than the anterior ones. This is likely a secondary effect of surface distribution, given that the same specimens also display flatter parietal areas and large and projected occipital areas. Among adult modern humans, cranial shape has no patent influence on the morphology of the vascular traces (Bruner et al., 2009), although some modifications have been described in artificially deformed crania (O'Laughlin, 1996). To date, there is no evidence of asymmetries or sex differences.

Despite their conspicuous development in modern humans, the exact function of these vessels is not clear (Bruner et al., 2011). During neurosurgery, the middle meningeal artery is often removed, being a common source of epidural hematomas and limiting access to the brain. There are no data on possible long-term effects of such elimination. In adults at rest, the artery has little or no blood flow. This suggests that it may be more active during exercise, pathological conditions, or emergency situations, when high metabolic activity may require an increase of blood exchange. Because these large vessels run within the meninges with no direct contact with the brain tissue, an oxygenation function is unlikely. Given account the complexity of their branching patterns, a role in thermal management has been hypothesized.

32.5.3 Venous Sinuses

The venous sinuses (also called sinuses of the dura mater) are enlarged vessels, able to relax, and act as blood reservoirs (Saban, 1995; Patel, 2009). Generally, they run within the *falx cerebri* and *tentorium cerebelli*, along the anchoring sites in which these connective sheets are attached to the braincase (Fig. 32.14). Few of them run within the brain mass and leave no traces on the bone walls. The largest imprint belongs to the superior sagittal sinus, which runs midsagittally with increasing diameter from the *foramen caecum* (a tiny midsagittal passage in the most anterior part of the anterior fossa) to the internal occipital protuberance. The internal occipital protuberance is the main meeting point of the largest sinuses, and this vascular area is called the confluence of sinuses, or *torcular herophili* (Fukusumi et al., 2010). On each side, a large transverse sinus runs between the cerebral and cerebellar hemispheres, continuing as the sigmoid sinus into the jugular fossa, draining the blood outside the endocranial cavity. Often, the superior sagittal sinus continues into the large right transverse/sigmoid/jugular system, while the internal cerebral sinuses (inferior sagittal sinus and *sinus rectus*) continue into the smaller left transverse/sigmoid/jugular system (Singh et al., 2004). However, the morphology of the confluence of the sinuses is very variable, the right and left flows generally mix together, and the asymmetry can be minimal or even reversed. The fact that usually the largest lateral sinus is on the right side and that the occipital lobes tend to be larger on the left side may suggest that different pressures of the brain mass can have a role in shaping the vascular pathways, with the larger vessels running in association with the smaller cortical mass. In Neanderthals, the vessels of the left and right side are more frequently separated, possibly because they had relatively wider brains (Peña-Melián et al., 2011). This separation probably decreased the possibility of admixture between the right and left side, although the functional significance of this variation is not known. Neanderthals also display a high prevalence of the sphenoparietal sinus, a large vessel running behind the coronal suture, parallel to the anterior branch of the middle meningeal artery (Grimaud-Hervé, 1997). However, there is disagreement regarding the anatomical interpretation of this sinus even in modern humans, given that it is possibly formed by multiple vascular elements (San Millán Ruíz et al., 2004).

As an alternative to the transverse-sigmoid system, at the confluence of sinuses the blood may flow along a complementary route: the occipital sinus between the

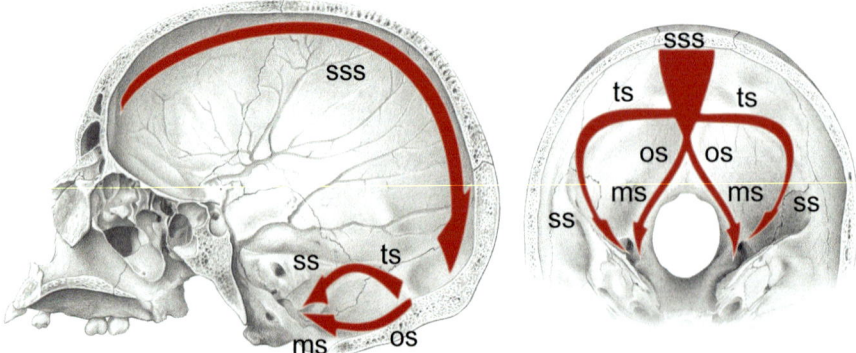

FIGURE 32.14 Main venous sinuses detectable as traces on the endocranial walls: marginal sinus (ms), occipital sinus (os), sigmoid sinus (ss), superior sagittal sinus (sss), transverse sinus (ts). The meeting area at the internal occipital protuberance is called confluence of sinuses. The blood leaves the endocranial cavity from the jugular fossae. The right vessels are usually larger than the left ones.

cerebellar hemispheres and then the marginal sinuses lateral to the foramen magnum. The transverse-sigmoid system and the occipitomarginal system both develop early during ontogeny. Later, the occipitomarginal system generally undergoes reduction or disappears, although it can be retained in some individuals. No functional significance is currently known for these two alternative routes, although they are thought to be complementary: the larger the one is, the smaller is the other. In *Paranthropus* and in *Australopithecus afarensis*, there is an unusually high prevalence of occipitomarginal vessels, but the variation in their prevalence among populations makes their phylogenetic significance difficult to interpret (Kimbel, 1984). Encephalization and bipedal posture may have influenced the expression of these features, considering that the blood flows mainly within the transverse-sigmoid system in the supine position and within the vertebral plexus in the upright position.

Some depressions of the endocranial bone table called *arachnoid foveae* can be found laterally to the main sinuses. These are the imprints of the arachnoid granulations (also called Pacchionian bodies), which are prolapses of the dura mater under the pressure of the arachnoid (Mack et al., 2009; Patel and Kirmi, 2009). These granulations are active in filtering the cerebrospinal fluid into the vascular system, and their expression may increase with age. Larger and smoother depressions of the bone surface can be due to lateral expansions of the sinuses and are called *venous lacunae*.

32.5.4 Diploic Vessels and Emissary Foramina

The diploic channels run within the bone thickness, generating complex morphological patterns influenced by morphogenetic factors that are to date only partially known (Jivraj et al., 2009; Toriumi et al., 2011; Tsutsumi et al., 2013). The study of these vessels is largely hampered by their secluded position. However, for the same reason, the channels are protected from the outer environment and can supply important paleontological, archaeological, and forensic information (Hershkovitz et al., 1999; Rangel de Lázaro et al., 2016). The veins running into the channels have no valves, so blood flow can be inverted. The vessels are connected with the meningeal artery (often at the pteric areas) and with the venous sinuses (often at the occipital bone). Their growth and dimensions have no apparent relationship to the thickness and size of the bones (Eisová et al., 2015). The diploic channels are more developed within the parietal bones, followed by the frontal and occipital bones. There is no evidence of asymmetric distribution or sex differences. As with the meningeal vessels, modern humans display a more complex vascular

network when compared with extinct human species or with nonhuman primates.

It is worth noting that the diploic channels reveal the passages of large diploic veins, but the network is not limited to these large vessels. The trabecular space between the inner and outer cranial tables hosts a complex vascular system, formed by microvessels that are smaller than the intertrabecular distance. This extended network of microvessels bridges the endocranial cavity through a large number of microscopic pores on the endocranial surface. The extension of this network suggests that these microscopic vascular elements may have a determining role in the overall metabolic and thermal balance. Nevertheless, information on this topic is still largely missing.

The cranial foramina are visible vascular passages on the endocranial and ectocranial surface (Berge and Bergman, 2001). The most common and important in humans are the parietal foramina in the upper vault, and the mastoid, occipital, and condylar foramina in the nuchal area (Louis et al., 2009). Their presence and dimensions may vary, with individual and population differences. Although the associated vascular elements are called "emissary veins," the vessels have no valves and the blood flow can occur in both directions. These passages are particularly active in draining the endocranial cavity in situations of heat stress. Otherwise, the blood flow is reduced or absent. Accordingly, the emissary foramina have been directly implicated in hypotheses concerning the evolution of brain thermal management (Falk, 1986; Braga and Boesch, 1997; Falk and Gage, 1997).

32.5.5 Endocranial Thermal Maps

At rest, the human brain can use 20% of total body energy, representing a highly metabolically active organ, being a major heat source while at the same time being sensitive to heat damage. Because of the large brain size changes in human evolution, it has been hypothesized that the vascular system may have evolved in response to demands for thermal management (Falk, 1990). Heat production and thermal regulation largely depend upon physiological factors, probably with a main role of the vascular system (Bertolizio et al., 2011; Wang et al., 2016). Despite the importance of thermal balance for brain functioning, we still lack much basic information on the mechanisms involved (Caputa, 2004; Rango et al., 2012). Brain thermal mechanisms are generally investigated through animal models, computer models, or numerical simulations (Karbowski, 2009; Zhu et al., 2006). In paleoneurology, the vascular system can supply many details regarding the possible cooling system, but the information is partial and the conclusions are largely speculative, especially in the absence of comparative experimental data. Additional information can be obtained by analyzing brain form. In fact, heat production and dissipation also depends in part on the geometry of the object. Apart from the role of the vascular system, the brain volume is responsible for heat production, and the cortical surface is responsible for heat dissipation. Accordingly, it is possible to compute a numerical model based on the heat equation in which each point of the volume releases heat, which is distributed all through the object which, in this case, is an endocast (Bruner, 2012; Fig. 32.15). The heat at each point will hence depend upon its position within the endocranial space, as a function of its distance from all the other points and from the surface. Although this is a simplified model, it can show how brain form variation can influence brain heat distribution. Heat accumulation can be calculated in terms of absolute differences between areas or in terms of percentage of the total heat produced. The basal areas are the ones with largest heat accumulation and, not surprisingly, these areas are also the most vascularized

(basilar artery and the circle of Willis). The "coldest" areas are the frontal, occipital, and cerebellar poles.

If we apply such an approach to modern humans, nonmodern humans, and apes, as expected brain size turns out to be the main factor involved in heat balance, most of all if we consider the overall heat accumulation (Bruner et al., 2014a). However, minor shape changes can influence heat distribution on specific cortical areas. It is interesting to compare Neanderthals and modern humans, because they have roughly the same brain size but different brain shapes. The cortical distribution in these two species is different, and the main difference is associated with the upper vault. The flat upper parietal surface in Neanderthals (as in other extinct hominins and living apes) is a "hot area": by virtue of the proximity with the geometrical core of the brain, the dorsal parietal surface shows high thermal load. Therefore, this region may be effective in exchanging heat with the meningeal vascular system and with the skull bones. The parietal enlargement characteristic of modern humans changes this cortical distribution, because it

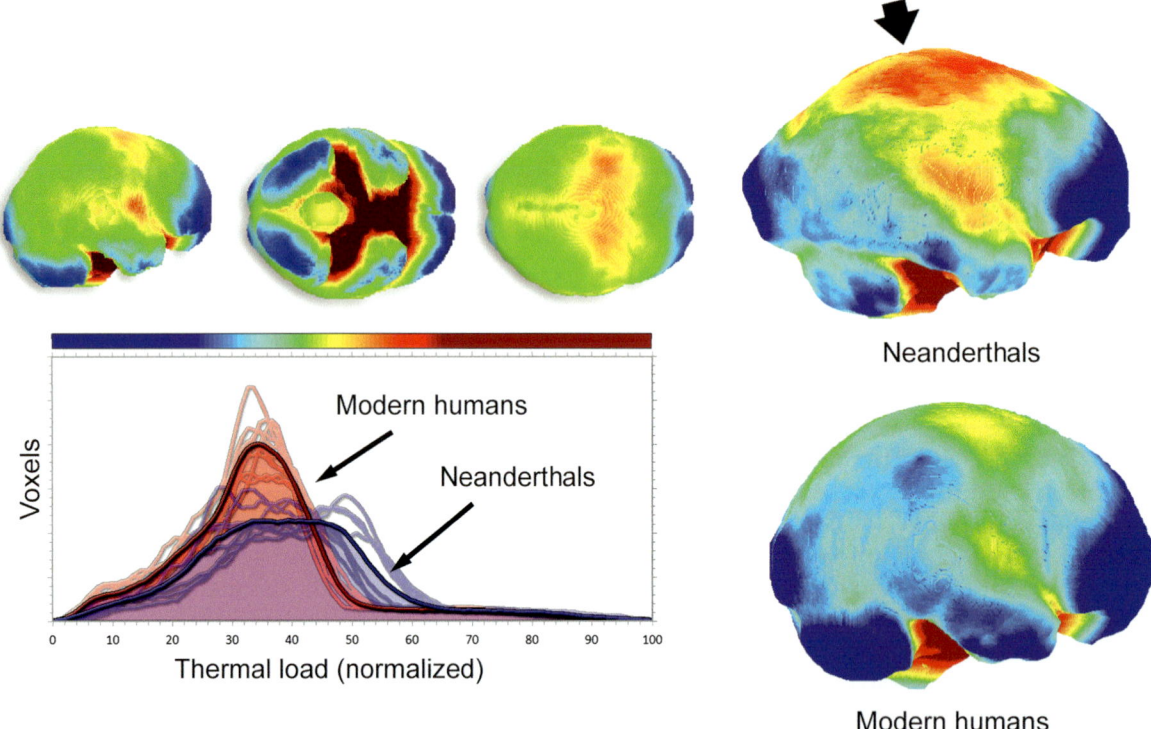

FIGURE 32.15 The distribution of heat produced by the brain mass can be modeled according to the shape of the endocranial cavity. The heat load of each cortical area will depend upon the distance from the thermal core of the brain volume and the distance from the surface. The areas with higher thermal load according to their position (*red*) are the lower brain areas, at the endocranial base (above). Neanderthals and modern humans display different curves of heat distribution, although they have similar brain mass and hence produce similar overall heat load. In Neanderthals, the upper vault has high heat loads, representing an important surface for dissipating heat (*arrow*). In modern humans, this region is positioned in a deeper location because of parietal bulging and is no longer available to exchange heat with the surface. *Images after Bruner, E., De la Cuétara, M., Musso, F., 2012. Quantifying patterns of endocranial heat distribution: brain geometry and thermoregulation. Am. J. Human Biol. 24, 753–762; Bruner, E., de la Cuétara, J.M., Masters, M., Amano, H., Ogihara, N., 2014. Functional craniology and brain evolution: from paleontology to biomedicine. Front. Neuroanat. 8, 19.*

displaces the vault surface far from the brain thermal core, decreasing its thermal load and the capacity to exchange heat. As a consequence, the deep parietal elements (like the precuneus) are more exposed to thermal stress, because of their position nearer the thermal core and further from the surface. Actually, the metabolic and thermal figures of these parietal areas show very high values (Cavanna and Trimble, 2006; Sotero and Iturria-Medina, 2011). Metabolic impairments in the deep parietal cortex have been described in the early stages of Alzheimer's disease, a pathology associated only with our species (Jacobs et al., 2012). We need to consider whether the parietal changes associated with our lineage added to thermal, vascular, or oxidative stresses, increasing the vulnerability or the sensitivity of these areas to neurodegenerative processes (Bruner and Jacobs, 2013).

32.6 Cognition, Fossils, and Future Steps

32.6.1 Cognitive Archaeology and Neuroarchaeology

Paleoneurology deals only with endocranial and cranial anatomy. However, brain evolution is clearly a multidisciplinary issue, and any paleoneurological evidence must be properly integrated within a reasonable and wider framework in evolutionary neuroanatomy. A real functional change in brain evolution is supposed to be associated with corresponding changes in the behavioral or ecological repertoire. The neural, behavioral, and cognitive niches exert reciprocal influences, integrating cognition, ecology, and social structure (Iriki and Sakura, 2008; Iriki and Taoka, 2012). Behavioral or cognitive changes are not necessarily associated with visible neuroanatomical variations, but the reverse is less likely: a real neuroanatomical change associated with functional variations should match some behavioral difference. As the archaeological record is incomplete, detecting such behavioral changes may be difficult. Nonetheless, it is clear that we must integrate paleontological and archaeological information to develop reliable evolutionary hypotheses, which should be properly tested within the experimental and quantitative perspectives of neuroscience.

Cognitive archaeology deals with neuropsychological interpretations of the archaeological record (Wynn and Coolidge, 2003; Coolidge and Wynn, 2005; Coolidge et al., 2015). *Neuroarchaeology* deals with the neurobiological evidence associated with archaeologically detectable behaviors (Stout and Chaminade, 2007, 2012; Stout and Hecht, 2015). There have been important developments in both fields in the last decade. Clear associations between anatomical and behavioral changes can be hard

to verify statistically, so these fields must rely on direct evidence (archaeology and paleontology) as well as on indirect information (living species), trying to focus perspectives and hypotheses through distinct approaches and looking for convergent results. It is interesting to note that the two more advanced tool modes, the Mousterian and the Aurignacian, are both associated with changes in the frontoparietal proportions, in Neanderthals and modern humans, respectively. In the case of modern humans, early populations like those from the Near East of Skhul and Qafzeh display only moderate parietal bulging and are associated with Mousterian-like tools. At the same time, these same populations also display some specific modern traits in their hand anatomy (Niewoehner, 2001). Upper Pleistocene modern humans, with large parietal regions, display a derived Aurignacian culture, which includes complex tools, ornaments, and cave paintings. So, at first glance, it looks like parietal morphology, hand anatomy, and cultural complexity display some parallel changes. Correlations, however, do not mean causality. If we detect a neuroanatomical variation associated with a behavioral change, we tend to think that the former has caused the latter. Even if this may be likely, the inverse may be also true (Quallo et al., 2009).

Nevertheless, the fossil and archeological evidence is very scattered in time and space, which makes it difficult to rigorously correlate between anatomy and behavior. Furthermore, the available "units" (eg, biological species and technical modes) are too few to support any robust correlation study between paleoneurological and cultural variation. Given these restrictions, longstanding debates on the processes behind "modernity" and cultural evolution are still unresolved, and contrasting views can all be defended in the absence of any definitive evidence (Klein, 2008; Mc Brearty and Brooks, 2000; Stringer, 2002; Tattersall, 2004).

An interesting case study of integration between paleoneurological and behavioral data concerns a behavioral trait described in nonmodern humans, especially in Neanderthals: the use of teeth for manipulation, which may evidence a different visuospatial perception and embodiment capacity (Bruner and Lozano, 2014, 2015; Bruner et al., 2017). The labial side of the incisors in many archaic humans shows scratches due to the use of the mouth as a "third hand." In fact, there are invasive marks on the dental surface left by a cutting edge. This behavior, measured in terms of number of scratches per tooth and percentage of individuals with scratches, is extremely high in Neanderthals. Most modern human populations do not use their teeth for handling, even in presence of complex cultural patterns. Some modern populations partially rely on the mouth for handling, but leave no scratches, or very few, on their teeth. Neanderthals therefore had at the same time a complex culture but small parietal proportions and had to rely on their mouth to handle their material culture. Taking into account that the parietal areas are strongly involved in visuospatial integration, body—object relationships, and eye—hand management, such an association may not be coincidental. The mouth, because of its central ecological role, looks like an inappropriate and risky choice for handling tools. Of course, according to the somatosensory map (the cortical *homunculus*), the mouth is the second largest territory after the hands. It might be expected that, if the hands are not sufficient to manage the body—object relationship, the mouth is the first alternative and additional complement. The fact that modern humans display a larger parietal cortex, more complex and visual culture, and do not need (or need to a very minor extent) the use of mouth for handling suggests the possibility of a visuospatial specialization and enhanced body—object integration. The hypothesis of an association between these factors can be currently tested by psychometric approaches or ethnological records and by improving our knowledge on the ontogenetic sensorial transition from a mouth-based exploration to a hand-based exploration. Different spatial cognition in Neanderthals and modern humans has been also hypothesized based on evidence associated with land use and ecology (Burke, 2012).

Such differences could be qualitative (brand-new functions) or quantitative (a matter of degree). At present there is no evidence of different functions for the parietal areas between human and nonhuman primates (Rilling et al., 2007; Barks et al., 2015; Caminiti et al., 2015). Also, for Neanderthals, some authors have suggested possible evidence of ornaments, burials, or even graphic marks. Agreement on the meaning of these cultural associations is still lacking, but even if this evidence is accepted, the level of cultural complexity in Neanderthals is not closely comparable with the level of complexity displayed by modern humans and their cultural resources (Wynn et al., 2016), from hunter-gatherers to modern industrial societies. Minor cognitive change can generate important behavioral changes, especially when the feedback between biology and culture is taken into account. The change of the parietal proportions is the most apparent macroscopic difference between modern human and Neanderthal brains. The fact that this difference is associated with changes in toolmaking and tool use, as well as with other visuospatial behavioral and social changes, cannot be neglected.

A final remark concerns the adaptive and selective significance behind these differences. Neanderthal and modern human evolutionary histories diverged for more than 100 ka, and it is likely that these two taxa underwent some divergent cognitive changes (Langbroek, 2012). For example, if a larger occipital cortex is confirmed for Neanderthals (Pearce et al., 2013), we should consider whether this may have been associated with specific cognitive capacities. The existence of distinct and divergent cognitive trajectories may offer a further interesting alternative to a traditional perspective in which cognitive evolution follows a gradual, linear, and progressive pathway. But it also suggests caution when trying to associate such cognitive differences with specific evolutionary fates. A plesiomorphic cognitive scheme or even a mismatch between brain and culture should not be necessarily interpreted in terms of decreased reproductive success or fitness, not to mention extinction. Unless a clear reproductive deficit is proven, or direct competition for resources is revealed, cognitive differences alone should not be used to make inferences on the survival capacity of a species. After all, *H. erectus* displayed far more primitive traits when compared with both modern humans and Neanderthals but, to date, it has represented the most successful human species in terms of phylogenetic endurance.

32.6.2 Future Steps in Paleoneurology

Paleoneurological evidence is the only direct evidence on brain evolution. Comparing living species can be extremely useful, but in that case we analyze independent products of evolution, not the process itself. There is no doubt that we must be aware of the limits of the paleoneurological record. Fossils can only provide information on the macroscopic organization of the brain, and the sample sizes are generally not sufficient to support reliable quantitative analyses. We can consider at least four main directions for current studies in this field. The first one is the most obvious: more fossils. Our record is very scattered and spotty, and any new specimen can change the evolutionary perspective in its general or specific components. Stability can be reached only by increasing the fossil record and the sample sizes. The second direction does not involve fossils, but living humans. We still lack much information on the biology and variation of most endocranial characters, and it is therefore mandatory, before investigating these traits in fossils, to support comprehensive research projects in living populations. A third advance could be technical and deals with the improvement of the analytical tools. For example, more detailed shape analyses can reveal specific areas of expansion/reduction on the cortical surface of different human species. The fourth enhancement regards statistics and aims at revealing associations and correlations among anatomical characters, behaviors, or cognitive processes. For example, correlations, when based on robust and reliable approaches, can be used to quantify brain proportions from cranial features. It is worth noting that these indirect approaches can be fruitful regardless of the causal explications behind them, being able to provide estimates of elusive features and variables (with their associated uncertainty) that are independent of their biological interpretations.

It is clear that these future perspectives must be accompanied by a proper evaluation of the results already obtained in the field. For example, dealing with the parietal differences described for modern humans, histological and cytological comparative analyses are necessary to investigate the anatomical mechanism behind the observed morphological changes.

Recent theories on extended cognition (Clark, 2007, 2008; Malafouris, 2010, 2013) are revealing an intimate relationship between brain, body, and material culture. The integration among these components is probably more crucial than we had previously assumed. We have been long stated that environment can influence brain biology, but now there is even the suspicion that the body and the environment are part of the cognitive process itself. In this case, the potential effect of the interaction between brain, culture, and social structure has yet to be properly evaluated. If these theories of extended cognition are further supported, we should be ready to change much of our past perspectives on evolutionary neuroanatomy.

Acknowledgments

I am grateful to the many friends and colleagues involved in the studies presented in this chapter, including Ralph Holloway, Atsushi Iriki, Jim Rilling, Heidi Jacobs, Giorgio Manzi, Roberto Colom, Manuel Martín-Loeches, Michael Masters, Marina Lozano, Naomichi Ogihara, Xiujie Wu, Sheela Athreya, Philipp Gunz, Simon Neubauer, Aida Gómez-Robles, Fred Coolidge, Leee Overmann, Enza Spinapolice, Ariane Burke, Dietrich Stout, Duilio Garofoli, and Markus Bastir. Many of the figures and images of this chapter are included thanks to the precious collaboration of my students, namely José Manuel de la Cuétara, Sofia Pedro, Hana Píšová, Gizéh Rangel, Hideki Amano, Annapaola Fedato, and Stana Eisová. Todd Preuss supplied many relevant comments and detailed suggestions on the early version of this chapter. Research at the Laboratory of Paleoneurology of the National Research Center for Human Evolution in Burgos is funded by the Spanish Government, by the Italian Institute of Anthropology, by the Atapuerca Foundation and by the Wenner-Gren Foundation.

References

Adeeb, N., Mortazavi, M., Tubbs, R., Cohen-Gadol, A., 2012. The cranial dura mater: a review of its history, embryology, and anatomy. Childs Nerv. Syst 28, 827–837.

Aiello, L., Wheeler, P., 1995. The expensive-tissue hypothesis: the brain and the digestive-system in human and primate evolution. Curr. Anthropol. 36, 199–221.

Alba, D.M., 2010. Cognitive inferences in fossil apes (Primates, Hominoidea): does encephalization reflect intelligence? J. Anthropol. Sci. 88, 11–48.

Aldridge, K., 2011. Patterns of differences in brain morphology in humans as compared to extant apes. J. Hum. Evol. 60, 94–105.

Allen, J.S., Damasio, H., Grabowski, T.J., 2002. Normal neuroanatomical variation in the human brain: an MRI-Volumetric Study. Am. J. Phys. Anthropol. 118, 341–358.

Allen, J.S., Bruss, J., Damasio, H., 2006. Looking for the lunate sulcus: a magnetic resonance imaging study in modern humans. Anat. Rec. 288A, 867–876.

Amano, H., Kikuchi, T., Morita, Y., Kondo, O., Suzuki, H., Ponce de León, M.S., Zollikofer, C.P.E., Bastir, M., Stringer, C., Ogihara, N., 2015. Virtual reconstruction of the Neanderthal Amud 1 cranium. Am. J. Phys. Anthropol. 158, 185–197.

Amunts, K., Schleicher, A., Burgel, U., Mohlberg, H., Uylings, H.B.M., Zilles, K., 1999. Broca's region revisited: cytoarchitecture and inter-subject variability. J. Comp. Neurol. 412, 319–341.

Antón, S.C., 2003. Natural history of Homo erectus. Yearb. Phys. Anthropol. 46, 126–170.

Armstrong, E., 1983. Relative brain size and metabolism in mammals. Science 220, 1302–1304.

Baab, K.L., 2008. The taxonomic implications of cranial shape variation in Homo erectus. J. Hum. Evol. 54, 827–847.

Balzeau, A., Holloway, R.L., Grimaud-Hervé, D., 2012. Variations and asymmetries in regional brain surface in the genus Homo. J. Hum. Evol. 62, 696–706.

Barks, S.K., Parr, L.A., Rilling, J.L., 2015. The default mode network in chimpanzees (Pan troglodytes) is similar to that of humans. Cereb. Cortex 25, 538–544.

Barton, R.A., Venditti, C., 2013. Human frontal lobes are not relatively large. Proc. Natl. Acad. Sci. U.S.A. 110, 9001–9006.

Bastir, M., 2008. A systems-model for the morphological analysis of integration and modularity in human craniofacial evolution. J. Anthropol. Sci. 86, 37–58.

Bastir, M., Rosas, A., 2005. Hierarchical nature of morphological integration and modularity in the human posterior face. Am. J. Phys. Anthropol. 128, 26–34.

Bastir, M., Rosas, A., 2006. Correlated variation between the lateral basicranium and the face: a geometric morphometric study in different human groups. Arch. Oral Biol. 51, 814–824.

Bastir, M., Rosas, A., 2009. Mosaic evolution of the basicranium in Homo and its relation to modular development. Evol. Biol. 36, 57–70.

Bastir, M., Rosas, A., 2016. Cranial base topology and basic trends in the facial evolution of Homo. J. Hum. Evol. 91, 26–35.

Bastir, M., Rosas, A., Kuroe, K., 2004. Petrosal orientation and mandibular ramus breadth: evidence for an integrated petroso-mandibular developmental unit. Am. J. Phys. Anthropol. 123, 340–350.

Bastir, M., Rosas, A., O'Higgins, P., 2006. Craniofacial levels and the morphological maturation of the human skull. J. Anat. 209, 637–654.

Bastir, M., Rosas, A., Lieberman, D.E., O'Higgins, P., 2008. Middle craneal fossa and the origin of modern humans. Anat. Rec. 291, 130–140.

Bastir, M., Rosas, A., Gunz, P., Peña-Melian, A., Manzi, G., Harvati, K., Kruszynski, R., Stringer, C., Hublin, J.J., 2011. Evolution of the base of the brain in highly encephalized human species. Nat. Commun. 2, 588.

Berge, J.K., Bergman, R.A., 2001. Variations in size and in symmetry of foramina of the human skull. Clin. Anat. 14, 406–413.

Bertolizio, G., Mason, L., Bissonnette, B., 2011. Brain temperature: heat production, elimination and clinical relevance. Paediatr. Anesth 21, 347–358.

Bienvenu, T., Guy, F., Coudyzer, W., Gilissen, E., Roualdès, G., Vignaud, P., Brunet, M., 2011. Assessing endocranial variations in great apes and humans using 3D data from virtual endocasts. Am. J. Phys. Anthropol. 145, 231–246.

Bookstein, F.L., 1991. Morphometric Tools for Landmark Data. Cambridge University Press, Cambridge.

Bookstein, F., Schafer, K., Prossinger, H., Seidler, H., Fieder, M., Stringer, C., Weber, G.W., Arsuaga, J.L., Slice, D.E., Rohlf, F.J.,

Recheis, W., Mariam, A.J., Marcus, L.F., 1999. Comparing frontal cranial profiles in archaic and modern *Homo* by morphometric analysis. Anat. Rec. 257, 217–224.

Bookstein, F.L., Gunz, P., Mitteroecker, P., Prossinger, H., Schaefer, K., Seidler, H., 2003. Cranial integration in *Homo*: singular warps analysis of the midsagittal plane in ontogeny and evolution. J. Hum. Evol. 44, 167–187.

Braga, J., Boesch, C., 1997. Further data about venous channels in South African Plio-Pleistocene hominids. J. Hum. Evol. 33, 423–447.

Brengelmann, G.L., 1993. Specialized brain cooling in humans? FASEB J. 7, 1148–1153.

Bruner, E., 2004. Geometric morphometrics and paleoneurology: brain shape evolution in the genus *Homo*. J. Hum. Evol. 47, 279–303.

Bruner, E., 2010. Morphological differences in the parietal lobes within the human genus: a neurofunctional perspective. Curr. Anthropol. 51, S77–S88.

Bruner, E., 2014. Functional craniology, human evolution, and anatomical constraints in the Neanderthal braincase. In: Akazawa, T., Ogihara, N., Tanabe, H.C., Terashima, H. (Eds.), Dynamics of Learning in Neanderthals and Modern Humans, vol. 2. Springer, Tokyo, pp. 121–129.

Bruner, E., 2015. Functional Craniology and Brain Evolution. In: Bruner, E. (Ed.), Human Paleoneurology. Springer, Cham, pp. 57–94.

Bruner, E., Manzi, G., 2005. CT-based description and phyletic evaluation of the archaic human calvarium from Ceprano. Italy. Anat. Rec 285A, 643–658.

Bruner, E., Manzi, G., 2008. Paleoneurology of an "early" Neandertal: endocranial size, shape, and features of Saccopastore 1. J. Hum. Evol. 54, 729–742.

Bruner, E., Jeffery, N., 2007. Extracting functional, phylogenetic and structural data from the subcortical brain: an exploratory geometric morphometric survey. J. Anthropol. Sci. 85, 125–138.

Bruner, E., Ripani, M., 2008. A quantitative and descriptive approach to morphological variation of the endocranial base in modern humans. Am. J. Phys. Anthropol. 137, 30–40.

Bruner, E., Sherkat, S., 2008. The middle meningeal artery: from clinics to fossils. Childs Nerv. Syst 24, 1289–1298.

Bruner, E., Holloway, R., 2010. A bivariate approach to the widening of the frontal lobes in the genus *Homo*. J. Hum. Evol. 58, 138–146.

Bruner, E., Jacobs, H.I.L., 2013. Alzheimer's Disease: the downside of a highly evolved parietal lobe? J. Alzheimer's Dis. 35, 227–240.

Bruner, E., Pearson, O., 2013. Neurocranial evolution in modern humans: the case of Jebel Irhoud 1. Anthropol. Sci. 121, 31–41.

Bruner, E., Lozano, M., 2014. Extended mind and visuospatial integration: three hands for the Neandertal lineage. J. Anthropol. Sci. 92, 273–280.

Bruner, E., Lozano, M., 2015. Three hands: one year later. J. Anthropol. Sci. 93, 191–195.

Bruner, E., Iriki, A., 2016. Extending mind, visuospatial integration, and the evolution of the parietal lobes in the human genus. Quat. Int. 405, 98–110.

Bruner, E., Manzi, G., Arsuaga, J.L., 2003. Encephalization and allometric trajectories in the genus *Homo*: evidence from the Neanderthal and modern lineages. Proc. Natl. Acad. Sci. U.S.A. 100, 15335–15340.

Bruner, E., Mantini, S., Ripani, M., 2009. Landmark-based analysis of the morphological relationship between endocranial shape and traces of the middle meningeal vessels. Anat. Rec. 292, 518–527.

Bruner, E., Martin-Loeches, M., Colom, R., 2010. Human midsagittal brain shape variation: patterns, allometry and integration. J. Anat. 216, 589–599.

Bruner, E., de la Cuétara, J.M., Holloway, R., 2011. A bivariate approach to the variation of the parietal curvature in the genus *Homo*. Anat. Rec. 294, 1548–1556.

Bruner, E., Mantini, S., Musso, F., de la Cuétara, J.M., Ripani, M., Sherkat, S., 2011. The evolution of the meningeal vascular system in the human genus: from brain shape to thermoregulation. Am. J. Human Biol 23, 35–43.

Bruner, E., Saracino, B., Passarello, P., Ricci, F., Tafuri, M., Manzi, G., 2004. Midsagittal cranial shape variation in the genus *Homo* by geometric morphometrics. Coll. Antropol. 28, 99–112.

Bruner, E., Preuss, T., Chen, X., Rilling, J., 2017. Evidence for expansion of the precuneus in human evolution. Brain Struct 222 (2), 1053–1060.

Bruner, E., Lozano, M., Lorenzo, C., 2016. Visuospatial integration and human evolution: the fossil evidence. J. Anthropol. Sci. 94, 81–97.

Bruner, E., Bondioli, L., Coppa, A., Frayer, D.W., Holloway, R.L., Libsekal, Y., Medin, T., Rook, L., Macchiarelli, R., 2016. The endocast of the one-million-year-old human cranium from Buia (UA 31), Danakil Eritrea. Am. J. Phys. Anthropol. 160, 458–468.

Bruner, E., Amano, H., Pereira-Pedro, S., Ogihara, N., 2016. The Evolution of the Parietal Lobe in the Genus *Homo*. In: Bruner, E., Ogihara, N., Tanabe, H. (Eds.), Digital Endocasts. Springer, Tokyo (in press).

Bruner, E., De la Cuétara, M., Musso, F., 2012. Quantifying patterns of endocranial heat distribution: brain geometry and thermoregulation. Am. J. Human Biol. 24, 753–762.

Bruner, E., Mantini, S., Perna, A., Maffei, C., Manzi, G., 2005. Fractal dimension of the middle meningeal vessels: variation and evolution in *Homo erectus*, Neanderthals, and modern humans. Eur. J. Morphol. 42, 217–224.

Bruner, E., Athreya, S., De la Cuétara, J.M., Marks, T., 2013. Geometric variation of the frontal squama in the genus *Homo*: frontal bulging and the origin of modern human morphology. Am. J. Phys. Anthropol. 150, 313–323.

Bruner, E., de la Cuétara, J.M., Masters, M., Amano, H., Ogihara, N., 2014. Functional craniology and brain evolution: from paleontology to biomedicine. Front. Neuroanat. 8, 19.

Bruner, E., Rangel de Lázaro, G., de la Cuétara, J.M., Martín-Loeches, M., Colom, R., Jacobs, H.I.L., 2014. Midsagittal brain variation and MRI shape analysis of the precuneus in adult individuals. J. Anat. 224, 367–376.

Bruner, E., Amano, H., de la Cuétara, J.M., Ogihara, N., 2015. The brain and the braincase: a spatial analysis on the midsagittal profile in adult humans. J. Anat. 227, 268–276.

Bruner, E., Román, F.J., de la Cuétara, J.M., Martin-Loeches, M., Colom, R., 2015. Cortical surface area and cortical thickness in the precuneus of adult humans. Neuroscience 286, 345–352.

Bruner, E., Grimaud-Hervé, D., Wu, X., de la Cuétara, J.M.A., Holloway, R., 2015. A paleoneurological survey of *Homo erectus* endocranial metrics. Quat. Int 368, 80–87.

Burke, A., 2012. Spatial abilities, cognition and the pattern of Neanderthal and modern human dispersal. Quat. Int. 247, 230–235.

Cabanac, M., 1993. Selective brain cooling in humans: 'fancy' or fact? FASEB J. 7, 1143–1146.

Cachel, S.M., 1978. A functional analysis of the primate masticatory system and the origin of the anthropoid postorbital septum. Am. J. Phys. Anthropol. 50, 1–17.

Caminiti, R., Innocenti, G.M., Battaglia-Mayer, A., 2015. Organization and evolution of parieto-frontal processing streams in macaque monkeys and humans. Neurosci. Biobehav. Rev. 56, 73–96.

Caputa, M., 2004. Selective brain cooling: a multiple regulatory mechanism. J. Therm. Biol. 29, 691–702.

Carlson, K.J., Stout, D., Jashashvili, T., de Ruiter, D.J., Tafforeau, P., Carlson, K., Berger, L.R., 2011. The endocast of MH1, *Australopithecus sediba*. Science 333, 1402–1407.

Cavanna, A.E., Trimble, M.R., 2006. The precuneus: a review of its functional anatomy and behavioural correlates. Brain 129, 564–583.

Cheverud, J.M., 1982. Phenotypic, genetic, and environmental morphological integration in the cranium. Evolution 36, 499–516.

Clark, A., 2007. Re-inventing ourselves: the plasticity of embodiment, sensing, and mind. J. Med. Philos. 32, 263–282.

Clark, A., 2008. Supersizing the mind. Embodiment, Action, and Cognitive Extension. Oxford University Press, Oxford.

Coolidge, F., Wynn, T., 2005. Working memory, its executive functions, and the emergence of modern thinking. Camb. Archaeol. J. 15, 5–26.

Coolidge, F.L., Wynn, T., Overmann, K.A., Hicks, J., 2015. Cognitive Archaeology and the Cognitive Sciences. In: Bruner, E. (Ed.), Human Paleoneurology. Springer, Cham, pp. 177–208.

Dart, R.A., 1925. *Australopithecus africanus*: the man-ape of South Africa. Nature 2884, 195–199.

De Sousa, A., Cunha, E., 2012. Hominins and the emergence of the modern brain. Prog. Brain Res. 195, 293–322.

De Sousa, A.A., Proulx, M.J., 2014. What can volumes reveal about human brain evolution? A framework for bridging behavioural, histometric, and volumetric perspectives. Front. Neuroanat. 8, 51.

De Sousa, A.A., Sherwood, C.C., Mohlberg, H., Amunts, K., Schleicher, A., MacLeod, C.E., Hof, P.R., Frahm, H., Zilles, K., 2010. Hominoid visual brain structure volumes and the position of the lunate sulcus. J. Hum. Evol. 58, 281–292.

Diamond, M.K., 1991. Homologies of the meningeal-orbital arteries of humans: a reappraisal. J. Anat. 178, 223–241.

Dunbar, R.I.M., 1992. Neocortex size as a constraint on group size in primates. J. Hum. Evol. 20, 469–493.

Dunbar, R.I.M., 1998. The social brain hypothesis. Evol. Anthropol. 6, 178–190.

Duterloo, H.S., Enlow, D.H., 1970. A comparative study of cranial growth in *Homo and Macaca*. Am. J. Anat. 127, 357–368.

Eichmann, A., Le Noble, F., Autiero, M., Carmeliet, P., 2005. Guidance of vascular and neural network formation. Curr. Opin. Neurobiol. 15, 108–115.

Eisová, S., Rangel de Lázaro, G., Písová, H., Pereira-Pedro, S., Bruner, E., 2015. Parietal bone thickness and vascular diameters in adult modern humans: a survey on cranial remains. Anat. Rec. 299, 888–896.

Enlow, D.H., 1990. Facial Growth. WB Saunders Company, Philadelphia.

Falk, D., 1986. Evolution of cranial blood drainage in hominids: enlarged occipital/marginal sinuses and emissary foramina. Am. J. Phys. Anthropol. 70, 311–324.

Falk, D., 1987. Hominid paleoneurology. Annu. Rev. Anthropol. 16, 13–30.

Falk, D., 1990. Brain evolution in *Homo*: the "radiator" theory. Behav. Brain Sci. 13, 333–381.

Falk, D., 1993. Meningeal arterial patterns in great apes: implications for hominid vascular evolution. Am. J. Phys. Anthropol. 92, 81–97.

Falk, D., 2012. Hominin paleoneurology: where are we now? Prog. Brain Res. 195, 255–272.

Falk, D., 2014. Interpreting sulci on hominin endocasts: old hypotheses and new findings. Front. Human Neurosci 8, 134.

Falk, D., Nicholls, P., 1992. Meningeal arteries in rhesus macaques (*Macaca mulatta*): implications for vascular evolution in anthropoids. Am. J. Phys. Anthropol. 89, 299–308.

Falk, D., Gage, T., 1997. Flushing the radiator? A reply to Braga Boesch. J. Hum. Evol. 33, 495–502.

Falk, D., Zollikofer, C.P., Morimoto, N., Ponce de León, M.S., 2012. Metopic suture of Taung (*Australopithecus africanus*) and its implications for hominin brain evolution. Proc. Natl. Acad. Sci. U.S.A. 109, 8467–8470.

Fletcher, P.C., Frith, C.D., Baker, S.C., Shallice, T., Frackowiak, R.S.J., Dolan, R.J., 1995. The mind's eye – precuneus activation in memory-related imagery. Neuroimage 2, 195–200.

Freton, M., Lemogne, C., Bergouignan, L., Delaveau, P., Lehéricy, S., Fossati, P., 2014. The eye of the self: precuneus volume and visual perspective during autobiographical memory retrieval. Brain Struct. Funct. 219, 959–968.

Fukusumi, A., Okudera, T., Takahashi, S., Sakamoto, T., Nakagawa, H., Takayama, K., Iwasaki, K., 2010. Anatomical evaluation of the dural sinuses in the torcular herophili using three dimensions venography. Acad. Radiol. 17, 1103–1111.

Gómez-Robles, A., Hopkins, W.D., Sherwood, C.C., 2013. Increased morphological asymmetry, evolvability and plasticity in human brain evolution. Proc. R. Soc. B 280, 20130575.

Gómez-Robles, A., Hopkins, W.D., Schapiro, S.J., Sherwood, C.C., 2015. Relaxed genetic control of cortical organization in human brains compared with chimpanzees. Proc. Natl. Acad. Sci. U.S.A. https://doi.org/10.1073/pnas.1512646112.

Grimaud-Hervé, D., 1997. L'évolution de l'encéphale chez l'Homo erectus et l'Homo sapiens. CNRS, Paris.

Grefkes, C., Fink, G.R., 2005. The functional organization of the intraparietal sulcus in humans and monkeys. J. Anat. 207, 3–17.

Gunz, P., 2015. Computed tools for paleoneurology. In: Bruner, E. (Ed.), Human Paleoneurology. Cham: Springer, pp. 39–55.

Gunz, P., Harvati, K., 2007. The Neanderthal "chignon": variation, integration, and homology. J. Hum. Evol. 52, 262–274.

Gunz, P., Mitteroecker, P., Neubauer, S., Weber, G.W., Bookstein, F.L., 2009. Principles for the virtual reconstruction of hominin crania. J. Hum. Evol. 57, 48–62.

Gunz, P., Neubauer, S., Maureille, B., Hublin, J.J., 2010. Brain development after birth differs between Neanderthals and modern humans. Curr. Biol. 20, R921–R922.

Gunz, P., Neubauer, S., Golovanova, L., Doronichev, V., Maureille, B., Hublin, J.J., 2012. A uniquely modern human pattern of endocranial development. Insights from a new cranial reconstruction of the Neandertal newborn from Mezmaiskaya. J. Hum. Evol. 62, 300–313.

Hagmann, P., Cammoun, L., Gigandet, X., Meuli, R., Honey, C.J., Wedeen, V.J., Sporns, O., 2008. Mapping the structural core of human cerebral cortex. PLoS Biol. 6, e159.

Hammer, M.L.A., Foley, R.A., 1996. Longevity and life history in hominid evolution. Hum. Evol. 11, 61–66.

Harvey, P.H., Clutton-Brock, T.H., 1985. Life history variation in primates. Evolution 39, 559–581.

Herculano-Houzel, S., 2012. Neuronal scaling rules for primate brains: the primate advantage. Prog. Brain Res. 195, 325–340.

Hershkovitz, I., Greenwald, C., Rothschild, B.M., Latimer, B., Dutour, O., Jellema, L.M., Wish-Baratz, S., Pap, I., Leonetti, G., 1999. The elusive diploic veins: anthropological and anatomical perspective. Am. J. Phys. Anthropol. 108, 345–358.

Hilgetag, C.C., Barbas, H., 2005. Developmental mechanics of the primate cerebral cortex. Anat. Embryol. 210, 411–417.

Hilgetag, C.C., Barbas, H., 2006. Role of mechanical factors in the morphology of the primate cerebral cortex. PLoS Comput. Biol. 2, e22.

Hills, T.T., Todd, P.M., Lazer, D., Redish, A.D., Couzin, I.D., Cognitive Search Research Group, 2015. Exploration versus exploitation in space, mind, and society. Trends Cogn. Sci. 19, 46–54.

Hofman, M.A., 2014. Evolution of the human brain: when bigger is better. Front. Neuroanat. 8, 15.

Holloway, R.L., 1978. The Relevance of the Endocasts for Studying Primate Brain Evolution. In: Noback, C.R. (Ed.), Sensory Systems in Primates. Academic Press, New York, pp. 181–200.

Holloway, R.L., 1981. Exploring the dorsal surface of hominoid brain endocasts by stereoplotter and discriminant analysis. Philos. Trans. R. Soc. Lond. B 292, 155–166.

Holloway, R.L., 1981. Volumetric and asymmetry determinations on recent hominid endocasts: Spy I and II, Djebel IhroudI, and the Salé *Homo erectus* specimens, with some notes on Neandertal brain size. Am. J. Phys. Anthropol. 55, 385–393.

Holloway, R.L., 1995. Toward a Synthetic Theory of Human Brain Evolution. In: Changeaux, J.P., Chavaillon, J. (Eds.), Origins of the Human Brain. Clarendon Press, Oxford, pp. 42–54.

Holloway, R.L., 2008. The human brain evolving: a personal retrospective. Annu. Rev. Anthropol. 37, 1–19.

Holloway, R.L., 2016. On the Making of Endocasts: The New and the Old in Paleoneurology. In: Bruner, E., Ogihara, N., Tanabe, H. (Eds.), Digital Endocasts. Springer, Tokyo (in press).

Holloway, R.L., De La Coste-Lareymondie, M.C., 1982. Brain endocast asymmetry in pongids and hominids: some preliminary findings on the paleontology of cerebral dominance. Am. J. Phys. Anthropol. 58, 101–110.

Holloway, R.L., Broadfield, D.C., Yuan, M.S., 2003. Morphology and histology of chimpanzee primary visual striate cortex indicate that brain reorganization predated brain expansion in early hominid evolution. Anat. Rec. 273A, 594–602.

Holloway, R.L., Broadfield, D.C., Yuan, M.S., 2004. Brain endocasts – the paleoneurological evidence. The Human Fossil Record, Volume Three. Alan R. Liss, New York.

Holly Smith, B., 1992. Life history and the evolution of human maturation. Evol. Anthropol. 1, 134–142.

Hublin, J.J., Neubauer, S., Gunz, P., 2015. Brain ontogeny and life history in Pleistocene hominins. Philos. Trans. R. Soc. B 370, 20140062.

Iriki, A., Sakura, O., 2008. The neuroscience of primate intellectual evolution: natural selection and passive and intentional niche construction. Philos. Trans. R. Soc. Lond. B 363, 2229–2241.

Iriki, A., Taoka, M., 2012. Triadic (ecological, neural, cognitive) niche construction: a scenario of human brain evolution extrapolating tool use and language from the control of reaching actions. Philos. Trans. R. Soc. Lond. B 367, 10–23.

Isler, K., Van Schaik, C.P., 2014. How humans evolved large brains: comparative evidence. Evol. Anthropol. 23, 65–75.

Jacobs, H.I.L., Van Boxtel, M.P.J., Jolles, J., Verhey, F.R.J., Uylings, H.B.M., 2012. Parietal cortex matters in Alzheimer's disease: an overview of structural, functional and metabolic findings. Neurosci. Biobehav. Rev. 36, 297–309.

Jivraj, K., Bhargava, R., Aronyk, K., Quateen, A., Walji, A., 2009. Diploic venous anatomy studied in-vivo by MRI. Clin. Anat. 22, 296–301.

Jung, R.E., Haier, R.J., 2007. The Parieto-Frontal Integration Theory (P-FIT) of intelligence: converging neuroimaging evidence. Behav. Brain Sci. 30, 135–187.

Karbowski, J., 2009. Thermodynamic constraints on neural dimensions, firing rates, brain temperature and size. J. Comput. Neurosci. 27, 415–436.

Kimbel, W.H., 1984. Variation in the pattern of cranial venous sinuses and hominid phylogeny. Am. J. Phys. Anthropol. 63, 243–263.

Klein, R.G., 2008. Out of Africa and the evolution of human behavior. Evol. Anthropol. 17, 267–281.

Klingenberg, C.P., 2010. Evolution and development of shape: integrating quantitative approaches. Nat. Rev. 11, 623–635.

Klingenberg, C.P., 2013. Cranial integration and modularity: insights into evolution and development from morphometric data. Hystrix 24, 43–58.

Kobayashi, Y., Matsui, T., Haizuka, Y., Ogihara, N., Hirai, N., Matsumura, G., 2014. Cerebral sulci and gyri observerd on macaque endocasts. In: Akazawa, T., Ogihara, N., Tanabe, H.C., Terashima, H. (Eds.), Dynamics of Learning in Neanderthals and Modern Humans, vol. 2. Springer, Tokyo, pp. 131–137.

Kobayashi, Y., Matsui, T., Haizuka, Y., Ogihara, N., Hirai, N., Matsumura, G., 2014. The coronal suture as an indicator of the caudal border of the macaque monkey prefrontal cortex. In: Akazawa, T., Ogihara, N., Tanabe, H.C., Terashima, H. (Eds.), Dynamics of Learning in Neanderthals and Modern Humans, vol. 2. Springer, Tokyo, pp. 139–143.

Kochetkova, V.I., 1978. Paleoneurology. Winston/Wiley, Washington.

Kubo, D., Tanabe, H.C., Kondo, O., Ogihara, N., Yogi, A., Murayama, S., Ishida, H., 2014. Cerebellar size estimation from endocranial measurements: an evaluation based on MRI data. In: Akazawa, T., Ogihara, N., Tanabe, H.C., Terashima, H. (Eds.), Dynamics of Learning in Neanderthals and Modern Humans, vol. 2. Springer, Tokyo, pp. 209–215.

Land, M.F., 2014. Do we have an internal model of the outside world? Philos. Trans. R. Soc. B 369, 20130045.

Langbroek, M., 2012. Trees and ladders: a critique of the theory of human cognitive and behavioural evolution in Palaeolithic archaeology. Quat. Int. 270, 4–14.

Leigh, S.R., 2012. Brain size growth and life history in human evolution. Evol. Biol. 39, 587–599.

Leonard, W., Snodgrass, J., Robertson, M., 2007. Effects of brain evolution on human nutrition and metabolism. Annu. Rev. Nutr. 27, 311–327.

Lieberman, D.E., 2000. Ontogeny, Homology, and Phylogeny in the Hominid Craniofacial Skeleton: The Problem of the Browridge. In: O'Higgins, P., Cohn, M. (Eds.), Development, Growth and Evolution. Academic Press, London, pp. 85–122.

Lieberman, D.E., 2001. Another face in our family tree. Nature 410, 419–420.

Lieberman, D.E., Pearson, O.M., Mowbray, K.M., 2000. Basicranial influence on overall cranial shape. J. Hum. Evol. 38, 291–315.

Lieberman, D.E., Ross, C., Ravosa, M., 2000. The primate cranial base: ontogeny function and integration. Yearb. Phys. Anthropol. 43, 117–169.

Lieberman, D.E., McBratney, B.M., Krovitz, G., 2002. The evolution and development of cranial form in Homo sapiens. Proc. Natl. Acad. Sci. U.S.A. 99, 1134–1139.

Louis Jr., R.G., Loukas, M., Wartmann, C.T., Tubbs, R.S., Apaydin, N., Gupta, A.A., Spentzouris, G., Ysique, J.R., 2009. Clinical anatomy of the mastoid and occipital emissary veins in a large series. Surg. Radiol. Anat. 31, 139–144.

Ma, S., Baillie, J.M., Stringer, M.D., 2012. Reappraising the surface anatomy of the pterion and its relationship to the middle meningeal artery. Clin. Anat. 25, 330–339.

Mack, J., Squier, W., Eastman, J.T., 2009. Anatomy and development of the meninges: implications for subdural collections and CSF circulation. Pediatr. Radiol. 39, 200–210.

Maister, L., Slater, M., Sanchez-Vives, M.V., Tsakiris, M., 2015. Changing bodies changes minds: owning another body affects social cognition. Trends Cogn. Sci. 19, 6–12.

Malafouris, L., 2010. The brain-artefact Interface (Bai): a challenge for archaeology and cultural neuroscience. Soc. Cogn. Affect. Neurosci. 5, 264–273.

Malafouris, L., 2013. How Things Shape the Mind: A Theory of Material Engagement. MIT Press, Cambridge.

Manzi, G., Vienna, A., Hauser, G., 1996. Developmental stress and cranial hypostosis by epigenetic trait occurrence and distribution: an exploratory study on the Italian Neandertals. J. Hum. Evol. 30, 511–527.

Manzi, G., Gracia, A., Arsuaga, J.L., 2000. Cranial discrete traits in the Middle Pleistocene humans from Sima de los Huesos (Sierra de Atapuerca, Spain). Does hypostosis represent any increase in "ontogenetic stress" along the Neandertal lineage? J. Hum. Evol. 38, 425–446.

Margulies, D.S., Vincent, J.L., Kelly, C., Lohmann, G., Uddin, L.Q., Biswal, B.B., Villringer, A., Castellanos, F.X., Milham, M.P., Petrides, M., 2009. Precuneus shares intrinsic functional architecture in humans and monkeys. Proc. Natl. Acad. Sci. U.S.A. 106, 20069–20074.

Martin, R., 1981. Relative brain size and basal metabolic rate in terrestrial vertebrates. Nature 293, 57–60.

Martin, R., Barbour, A.D., 1989. Aspects of line-fitting in bivariate allometric analyses. Folia Primatol. 53, 65–81.

Martínez-Maza, C., Rosas, A., García-Vargas, S., 2006. Bone paleohistology and human evolution. J. Anthropol. Sci. 84, 33–52.

Masters, M.P., 2012. Relative size of the eye and orbit: an evolutionary and craniofacial constraint model for examining the etiology and disparate incidence of juvenile-onset myopia in humans. Med. Hypotheses 78, 649–656.

Mc Brearty, S., Brooks, A., 2000. The revolution that wasn't: a new interpretation of the origin of modern human behavior. J. Hum. Evol. 39, 453–563.

McCarthy, R.C., 2001. Anthropoid cranial base architecture and scaling relationships. J. Hum. Evol. 40, 41–66.

McCollum, M.A., Sherwood, C.C., Vinyard, C.J., Lovejoy, C.O., Schachat, F., 2006. Of muscle-bound crania and human brain evolution: the story behind the MYH16 headlines. J. Hum. Evol. 50, 232–236.

Mitteroecker, P., Bookstein, F., 2008. The evolutionary role of modularity and integration in the hominoid cranium. Evolution 62, 943–958.

Moss, M.L., Young, R.W., 1960. A functional approach to craniology. Am. J. Phys. Anthropol. 18, 281–292.

Neubauer, S., 2014. Endocasts: possibilities and limitations for the interpretation of human brain evolution. Brain Behav. Evol. 84, 117–134.

Neubauer, S., 2015. Human brain evolution: ontogeny and phylogeny. In: Bruner, E. (Ed.), Human Paleoneurology. Cham: Springer, pp. 95–120.

Neubauer, S., Gunz, P., Hublin, J.J., 2009. The pattern of endocranial ontogenetic shape changes in humans. J. Anat. 215, 240–255.

Neubauer, S., Gunz, P., Hublin, J.J., 2010. Endocranial shape changes during growth in chimpanzees and humans: a morphometric analysis of unique and shared aspects. J. Hum. Evol. 59, 555–566.

Neubauer, S., Gunz, P., Weber, G.W., Hublin, J.J., 2012. Endocranial volume of *Australopithecus africanus*: new CT-based estimates and the effects of missing data and small sample size. J. Hum. Evol. 62, 498–510.

Niewoehner, W.A., 2001. Behavioral inferences from the Skhul/Qafzeh early modern human hand remains. Proc. Natl. Acad. Sci. U.S.A. 98, 2979–2984.

O'Laughlin, V.D., 1996. Comparative endocranial vascular changes due to craniosynostosis and artificial cranial deformation. Am. J. Phys. Anthropol. 101, 369–385.

Orban, G.A., Claeys, K., Nelissen, K., Smans, R., Sunaert, S., Todd, J.T., Wardak, C., Durand, J.B., Vanduffel, W., 2006. Mapping the parietal cortex of human and non-human primates. Neuropsychologia 44, 2647–2667.

Patel, N., 2009. Venous anatomy and imaging of the first centimeter. Semin. Ultrasound CT MRI 30, 513–524.

Patel, N., Kirmi, O., 2009. Anatomy and imaging of the normal meninges. Semin. Ultrasound CT MRI 30, 559–564.

Pearce, E., Stringer, C., Dunbar, R.I.M., 2013. New insights into differences in brain organization between Neanderthals and anatomically modern humans. Proc. R. Soc. B 280, 20130168.

Peer, M., Salomon, R., Goldberg, I., Blanke, O., Arzy, S., 2015. Brain system for mental orientation in space, time, and person. Proc. Natl. Acad. Sci. U.S.A. 112, 11072–11077.

Peña-Melián, A., Rosas, A., García-Tabernero, A., Bastir, M., De La Rasilla, M., 2011. Paleoneurology of two new Neandertal occipitals from El Sidrón (Asturias, Spain) in the context of *Homo* endocranial evolution. Anat. Rec. 294, 1370–1381.

Percival, C.J., Richtsmeier, J.T., 2013. Angiogenesis and intramembranous osteogenesis. Dev. Dyn 242, 909–922.

Pereira-Pedro, S., Bruner, E., 2016. Sulcal pattern, extension, and morphology of the precuneus in adult humans. Ann. Anat. https://doi.org/10.1016/j.aanat.2016.05.001.

Písová, H., Rangel de Lázaro, G., Velemínský, P., Bruner, E., 2017. Craniovascular traits in anthropology and evolution: from bones to vessels. J. Anthropol. Sci. (in press).

Quallo, M.M., Price, C.J., Ueno, K., Asamizuya, T., Cheng, K., Lemon, R.N., Iriki, A., 2009. Gray and white matter changes associated with tool-use learning in macaque monkeys. Proc. Natl. Acad. Sci. U.S.A. 106, 18379–18384.

Rangel de Lázaro, G., de la Cuatara, J.M., Písová, H., Lorenzo, C., Bruner, E., 2016. Diploic vessels and computed tomography: segmentation and comparison in modern humans and fossil hominids. Am. J. Phys. Anthropol. 159, 313–324.

Rango, M., Arighi, A., Bresolin, N., 2012. Brain temperature: what do we know? Neuroreport 23, 483–487.

Ribas, G.C., Yasuda, A., Ribas, E.C., Nishikuni, K., Rodrigues, A.J., 2006. Surgical anatomy of microneurosurgical sulcal key-points. Neurosurgery 59, S177–S208.

Richtsmeier, J.T., Aldridge, K., de Leon, V.B., Panchal, J., Kane, A.A., Marsh, J.L., Yan, P., Cole, T.M., 2006. Phenotypic integration of neurocranium and brain. J. Exp. Zool. 306B, 360–378.

Rightmire, G.P., 2004. Brain size and encephalization in early to mid-Pleistocene *Homo*. Am. J. Phys. Anthropol. 124, 109–123.

Rilling, J.K., 2006. Human and non-human primate brains: are they allometrically scaled versions of the same design? Evol. Anthropol. 15, 65–67.

Rilling, J.K., Seligman, R.A., 2002. A quantitative morphometrics comparative analysis of the primate temporal lobe. J. Hum. Evol. 42, 505–533.

Rilling, J.K., Barks, S.K., Parr, L.A., Preuss, T.M., Faber, T.L., Pagnoni, G., Bremner, J.D., Votaw, J.R., 2007. A comparison of resting-state brain activity in humans and chimpanzees. Proc. Natl. Acad. Sci. U.S.A. 104, 17146–17151.

Rohlf, F.J., Marcus, L.F., 1993. A revolution in morphometrics. Trends Ecol. Evol. 8, 129–132.

Rosas, A., Peña-Melián, A., García-Tabernero, A., Bastir, M., De La Rasilla, M., Fortea, J., 2008. Endocranial occipito-temporal anatomy of SD-1219 from the Neandertal El Sidron site (Asturias, Spain). Anat. Rec. 291, 502–512.

Ruff, C.B., Trinkaus, E., Holliday, T.W., 1997. Body mass and encephalization in Pleistocene *Homo*. Nature 387, 173–176.

Saban, R., 1995. Image of the Human Fossil Brain: Endocranial Casts and Meningeal Vessels in Young and Adult Subjects. In: Changeaux, P., Chavaillon, J. (Eds.), Origins of the Human Brain. Clarendon Press, Oxford, pp. 11–38.

San Millán Ruíz, D., Fasel, J.H., Rufenacht, D.A., Gailloud, P., 2004. The sphenoparietal sinus of breschet: does it exist? An anatomic study. Am. J. Neuroradiol. 25, 112–120.

Schoenemann, P.T., Gee, J., Avants, B., Holloway, R.L., Monge, J., Lewis, J., 2007. Validation of plaster endocast morphology through 3D CT image analysis. Am. J. Phys. Anthropol. 132, 183–192.

Schoenemann, P.T., Sheehan, M.J., Glotzer, L.D., 2005. Prefrontal white matter volume is disproportionately larger in humans than in other primates. Nat. Neurosci. 8, 242–252.

Scott, N., Neubauer, S., Hublin, J.J., Gunz, P., 2014. A shared pattern of postnatal endocranial development in extant hominoids. Evol. Biol. 41, 572–594.

Scremin, O.U., 2004. Cerebral Vascular System. In: Paxinos, G., Mai, J.K. (Eds.), The Human Nervous System. Elsevier Academic Press, San Diego, pp. 1325–1348.

Semendeferi, K., Damasio, H., 2000. The brain and its main anatomical subdivision in living hominoids using magnetic resonance imaging. J. Hum. Evol. 38, 317–332.

Semendeferi, K., Lu, A., Schenker, N., Damasio, H., 2002. Humans and great apes share a large frontal cortex. Nat. Neurosci. 5, 272–276.

Semendeferi, K., Damasio, H., Frank, R., Van Hoesen, G.W., 1997. The evolution of the frontal lobes: a volumetric analysis based on three-dimensional reconstructions of magnetic resonance scans of human and ape brains. J. Hum. Evol. 32, 375–388.

Shea, B.T., 1985. On aspects of skull form in African apes and orangutans, with implications for hominoid evolution. Am. J. Phys. Anthropol. 68, 329–342.

Sherwood, C.C., Bauernfeind, A.L., Bianchi, S., Raghanti, M.A., Hof, P.R., 2012. Human brain evolution writ large and small. Prog. Brain Res. 195, 237–254.

Singh, M., Nagashima, M., Inoue, Y., 2004. Anatomical variations of occipital bone impressions for dural venous sinuses around the torcular herophili, with special reference to the consideration of clinical significance. Surg. Radiol. Anat. 26, 480–487.

Slice, D.E., 2004. Modern Morphometrics in Physical Anthropology. Kluwer Academic – Plenum Publishers, New York.

Smaers, J.B., 2013. How humans stand out in frontal lobe scaling. Proc. Natl. Acad. Sci. U.S.A. 110, E3682.

Sotero, R.C., Iturria-Medina, Y., 2011. From blood oxygenation level dependent (BOLD) signals to brain temperature maps. Bull. Math. Biol. 73, 2731–2747.

Spoor, F., Jeffery, N., Zonneveld, F., 2000. Using diagnostic radiology in human evolutionary studies. J. Anat. 197, 61–76.

Stedman, H.H., Kozyak, B.W., Nelson, A., Thesier, D.M., Su, L.T., Low, D.W., Bridges, C.R., Shrager, J.B., Minugh-Purvis, N., Mitchell, M.A., 2004. Myosin gene mutation correlates with anatomical changes in the human lineage. Nature 428, 415–418.

Stout, D., Chaminade, T., 2007. The evolutionary neuroscience of tool making. Neuropsychologia 45, 1091–1100.

Stout, D., Chaminade, T., 2012. Stone tools, language and the brain in human evolution. Philos. Trans. R. Soc. B 367, 75–87.

Stout, D., Hecht, H., 2015. Neuroarchaeology. In: Bruner, E. (Ed.), Human Paleoneurology. Cham: Springer, pp. 145–175.

Stringer, C., 2002. Modern human origins: progress and prospects. Philos. Trans. R. Soc. Lond. B 357, 563–579.

Tallinen, T., Chung, J.Y., Rosseau, F., Girard, N., Lefèvre, J., Mahadevan, L., 2016. On growth and form of cortical convolutions. Nat. Phys. https://doi.org/10.1038/NPHYS3632.

Tattersall, I., 2004. What happened in the origin of human consciousness? Anat. Rec. 276B, 19–26.

Tobias, P.V., 1987. The brain of Homo habilis: a new level of organization in cerebral evolution. J. Hum. Evol. 16, 741–761.

Tobias, P.V., 1995. The Brain of the First Hominids. In: Changeaux, J.P., Chavaillon, J. (Eds.), Origins of the Human Brain. Clarendon Press, Oxford, pp. 61–83.

Toriumi, H., Shimizu, T., Shibata, M., Unekawa, M., Tomita, Y., Tomita, M., Suzuki, N., 2011. Developmental and circulatory profile of the diploic veins. Microvasc. Res. 81, 97–102.

Toro, R., Burnod, Y., 2005. A morphogenetic model for the development of cortical convolutions. Cereb. Cortex 15, 1900–1913.

Tsutsumi, S., Nakamura, M., Tabuchi, T., Yasumoto, Y., Ito, M., 2013. Calvarial diploic venous channels: an anatomic study using high-resolution magnetic resonance imaging. Surg. Radiol. Anat. 35, 935–941.

Utevsky, A.V., Smith, D.V., Huettel, S.A., 2014. Precuneus is a functional core of the default-mode network. J. Neurosci. 34, 932–940.

Van Essen, D.C., 1997. A tension-based theory of morphogenesis and compact wiring in the central nervous system. Nature 385, 313–318.

Vanduffel, W., Fize, D., Peuskens, H., Denys, K., Sunaert, A., Todd, J.T., Orban, G.A., 2002. Extracting 3D from motion: differences in human and monkey intraparietal cortex. Science 298, 413–415.

Wang, H., Kim, M., Normoyle, K.P., Llano, D., 2016. Thermal regulation of the brain – an anatomical and physiological review for clinical neuroscientists. Front. Neurosci. 9, 528.

Weaver, A.H., 2005. Reciprocal evolution of the cerebellum and neocortex in fossil humans. Proc. Natl. Acad. Sci. U.S.A. 102, 3576–3580.

Weidenreich, F., 1936. Observations on the form and proportions of the endocranial casts of Sinanthropus pekinensis, other hominids and the great apes: a comparative study of brain size. Paleontol. Sin D7, 4.

Weidenreich, F., 1941. The brain and its role in the phylogenetic transformation of the human skull. Trans. Am. Philos. Soc. XXXI 321–442.

Wu, X., Bruner, E., 2016. The endocranial anatomy of Maba 1. Am. J. Phys. Anthropol. 160, 633–643.

Wu, X.J., Schepartz, L., Norton, C.J., 2010. Morphological and morphometric analysis of variation in the Zhoukoudian Homo erectus brain endocasts. Quat. Int. 211, 4–13.

Wynn, T., Coolidge, F., 2003. The role of working memory in the evolution of managed foraging. Before Farming 2, 1–16.

Wynn, T., Overmann, K.A., Coolidge, F.L., 2016. The false dichotomy: a refutation of the Neandertal indistinguishability claim. J. Anthropol. Sci. 94. https://doi.org/10.4436/jass.94022.

Zelditch, M.L., Swidersky, D.L., Sheets, H.D., Fink, W.L., 2004. Geometric Morphometrics for Biologists. Elsevier, San Diego.

Zenker, W., Kubik, S., 1996. Brain cooling in humans—anatomical considerations. Anat. Embryol. 193, 1–13.

Zhang, S., Li, C.S.R., 2012. Functional connectivity mapping of the human precuneus by resting state fMRI. Neuroimage 59, 3548–3562.

Zhang, Y., Wu, X., Schepartz, L., 2016. Comparing methods for estimating cranial capacity in incomplete human fossils using the Jingchuan 1 partial cranium as an example. Quat. Int. https://doi.org/10.1016/j.quaint.2015.12.008.

Zollikofer, C.P., Ponce De León, M.S., 2002. Visualizing patterns of craniofacial shape variation in Homo sapiens. Proc. R. Soc. Lond. B 269, 801–807.

Zollikofer, C.P.E., Ponce de León, M.S., 2005. Virtual Reconstruction: A Primer in Computer-Assisted Paleontology and Biomedicine. Wiley-Liss, New York.

Zollikofer, C.P., de León, M.S., 2013. Pandora's growing box: inferring the evolution and development of hominin brains from endocasts. Evol. Anthropol. 22, 20–33.

Zollikofer, C.P., 2012. Evolution of hominin cranial ontogeny. Prog. Brain Res. 195, 273–292.

Zhu, M., Ackerman, J.J.H., Sukstanskii, A.L., Yablonskiy, D.A., 2006. How the body controls brain temperature: the temperature shielding effect of cerebral blood flow. J. Appl. Physiol. 101, 1481–1488.

33

Remarkable, But Not Special: What Human Brains Are Made of

S. Herculano-Houzel

Vanderbilt University, Nashville, TN, United States

33.1 The Former View: The Human Brain Is Special

Humans are formidable creatures, capable of transforming their food to refined "molecular" gastronomy, building skyscrapers, modifying their own bodies, and inquiring about themselves. Once the brain became recognized as the seat of mental functions, in the 19th century, it became clear that all those abilities must come from the brain. Yet the human brain appears nowhere as formidable as its feats: at best, it ranks 14th in size, after 2 species of elephants and 11 cetaceans, whose brains can weigh as much as 9 kg, six times larger than the human brain.

Cortical folds were once considered an indication of expansion of the cortical sheet and its number of neurons, and therefore the degree of gyrification was presumably an indication of intelligence (Jerison, 1973). Yet, the human cerebral cortex is not the most folded; larger cortices, such as those of the elephant and various cetaceans, have much larger folding indices (Hofman, 1985). The human cerebral cortex, although not the largest (the honor, again, goes to cetaceans), is the largest relative to the brain as a whole, in terms of the percentage of brain volume that it occupies—but, at 76–84% (Stephan et al., 1981; Hofman, 1988), it ranks first by very little, followed closely by the cerebral cortex of the chimpanzee, horse, and short-finned pilot whale, at 73–74% of brain volume (Hofman, 1988). It is unlikely that such a small difference accounts for the distance that we like to put between our cognitive feats and those of all other animals. The cerebral brain does cost a seemingly extraordinary 25% of all the energy that runs the body, when its mass represents but 2% of the body (Kety, 1957), and it has often been proposed that the extraordinary cost of the human brain reflects its

extraordinary prowess and would have resulted from human-specific changes in genes related to metabolism (Cáceres et al., 2003; Uddin et al., 2004; Somel et al., 2013). But that is at odds with the very low specific cost of the human brain, when expressed as glucose use per gram of tissue: gram per gram, the human brain costs less than the mouse brain (Karbowski, 2007).

How could the human brain be so remarkable when it did not rank an obvious first in any category that could reasonably explain its outstanding cognition? The paradox seemed to be resolved when Harry Jerison introduced the use of the encephalization quotient: a unitless measure of how much larger or smaller a brain is compared to how large it is expected to be, given the relationship between brain mass and body mass that applies across mammals as a whole (Jerison, 1973). The human brain finally ranked first: in comparison to all other mammals, it was over seven times larger than it "should" be. The outstanding encephalization quotient, which finally matched the outstanding cognitive powers of humans, is often illustrated by comparing humans and the largest great apes. Gorillas can weigh up to 200 kg, and because larger species tend to have larger brains, gorillas, who are obviously larger than humans, should also have larger brains to go with those bodies—and yet, their brains weigh about only one-third as much as human brains (Marino, 1998). That large encephalization quotient supposedly meant that humans had over seven times as many neurons as necessary to operate the body—an excess that would then be available for cognitive functions, thus accounting for the formidable capabilities of human brains (Jerison, 1973). Moreover, human encephalization quotients stood a distant first from quotients for the species ranked next: at best, they reached values of around 3, against 7 for humans (Jerison, 1973; Marino, 1998).

For almost 40 years, the view that the human brain is special, an outlier in its relative size compared to the body, remained the standard view, disseminated not only in the scientific literature but also in popular books for the lay public (for example, Gazzaniga, 2008). At the very least, the rules that relate brain and body size in all other mammals did not seem to apply to humans. Round numbers accompanied that tag: the human brain contained 100 billion neurons and 10 times as many glial cells. Along the same vein, a search for genes that set the human brain apart from all other primates soon ensued, and sure enough, there appear to be human-specific genes related to brain size (Evans et al., 2004; Dumas et al., 2012), synapse formation (Dennis et al., 2012; Charrier et al., 2012), speech and language development (Enard et al., 2009), cell metabolism (Somel et al., 2013), and the shape of the human wrist and thumb (Prabhakar et al., 2008).

The reign of the highly encephalized human species lasted about four decades. Recently, novel data on the numbers of neurons that compose different brains and how they relate to body mass and the energetic cost of the brain forced a reexamination of the view of the human brain as an outlier, the sole brain that escapes the scaling rules that apply to bodies in evolution. As it turns out, humans are no rebels or mavericks: the same rules of evolution that apply to every other species also apply to them. Our species does seem, however, to have come up with a neat trick that led them to escape one of the natural limitations that curb the number of brain neurons that other species can afford: the sole species to modify its food before it is ingested, humans can afford the largest number of neurons in any cerebral cortex. The findings that led to this new view of what makes the human brain remarkable were reviewed in depth recently (Herculano-Houzel, 2016) and are summarized in this chapter.

33.2 The Human Brain as a Scaled-Up Primate Brain

Prompted by the realization that the often cited "100 billion neurons and 10 times as many glial cells" in the human brain was at best an order of magnitude estimate (reviewed in von Bartheld et al., 2016), rather than an estimate based on actual measurements, and armed with a new method as reliable as stereology that allowed the rapid and precise estimation of numbers of cells in whole brains (Herculano-Houzel and Lent, 2005; Herculano-Houzel et al., 2015a), we recently determined the numbers of cells that compose human brains (Azevedo et al., 2009). We found that the (male) human brain has on average 86 billion neurons, 16 billion of which located in the cerebral cortex and 69 billion in

the cerebellum, leaving fewer than 1 billion neurons in the "rest of brain": the ensemble of brain stem, diencephalon, and striatum (Fig. 33.1; Azevedo et al., 2009). While this number falls in the 10^{11} order of magnitude, it is more than a baboon brain's worth short of the mythical 100 billion neurons so widespread in the literature (von Bartheld et al., 2016). Not a single human brain examined so far came close to that round mark; individual brains varied around 80 and 90 billion neurons—and interestingly, the largest number was found in the oldest brain examined, of 71 years of age (Azevedo et al., 2009).

The most important aspect of finally knowing the number of neurons that compose the human brain and their distribution is that we are now able to compare it to other primates and determine whether the human brain is out of the ordinary—an exception to the rules that apply to other primates—or just a scaled-up primate brain. That comparison is possible because we have by now examined the cellular composition of the brains of enough primate species to be able to determine the mathematical pattern that describes the relationship between the number of neurons that compose each structure, their density (the inverse of average cell size; Mota and Herculano-Houzel, 2014), and the resulting mass of the brain structure (reviewed in Herculano-Houzel et al., 2014a, and Chapter 2.07, What Modern Mammals Teach About the Cellular Composition of Early Brains and Mechanisms of Brain Evolution). These mathematical patterns are power laws, in which the mass of a brain structure (or its neuronal density) varies with the number of neurons in the structure raised to a certain exponent. These exponents are significantly larger than 1 for nonprimate species, indicating that brain structure mass scales up rapidly as numbers of neurons increase across species, but close to 1 for primates, indicating that brain structure mass in these animals scales close to linearly as numbers of neurons increase. As a result, primate brain structures are composed of many more neurons than nonprimate structures of similar mass—and the larger the structure, the larger the difference in numbers of neurons favoring primates (Herculano-Houzel et al., 2014a; see Chapter 3.02, "What Primate Brains Are Made of" in Volume II of this series for a more recent review).

Given these relationships between number of neurons, neuronal density, and brain structure mass, to which we refer collectively as the "neuronal scaling rules" that apply to a part of the brain across primates and nonprimates, we could determine how the human brain compared to other species. Did the different brain structures in the human brain have the mass predicted for a primate, given their number of neurons? If they had a significantly larger or smaller mass than predicted that would indicate that the human brain was indeed extraordinary, unique in its cellular makeup—special,

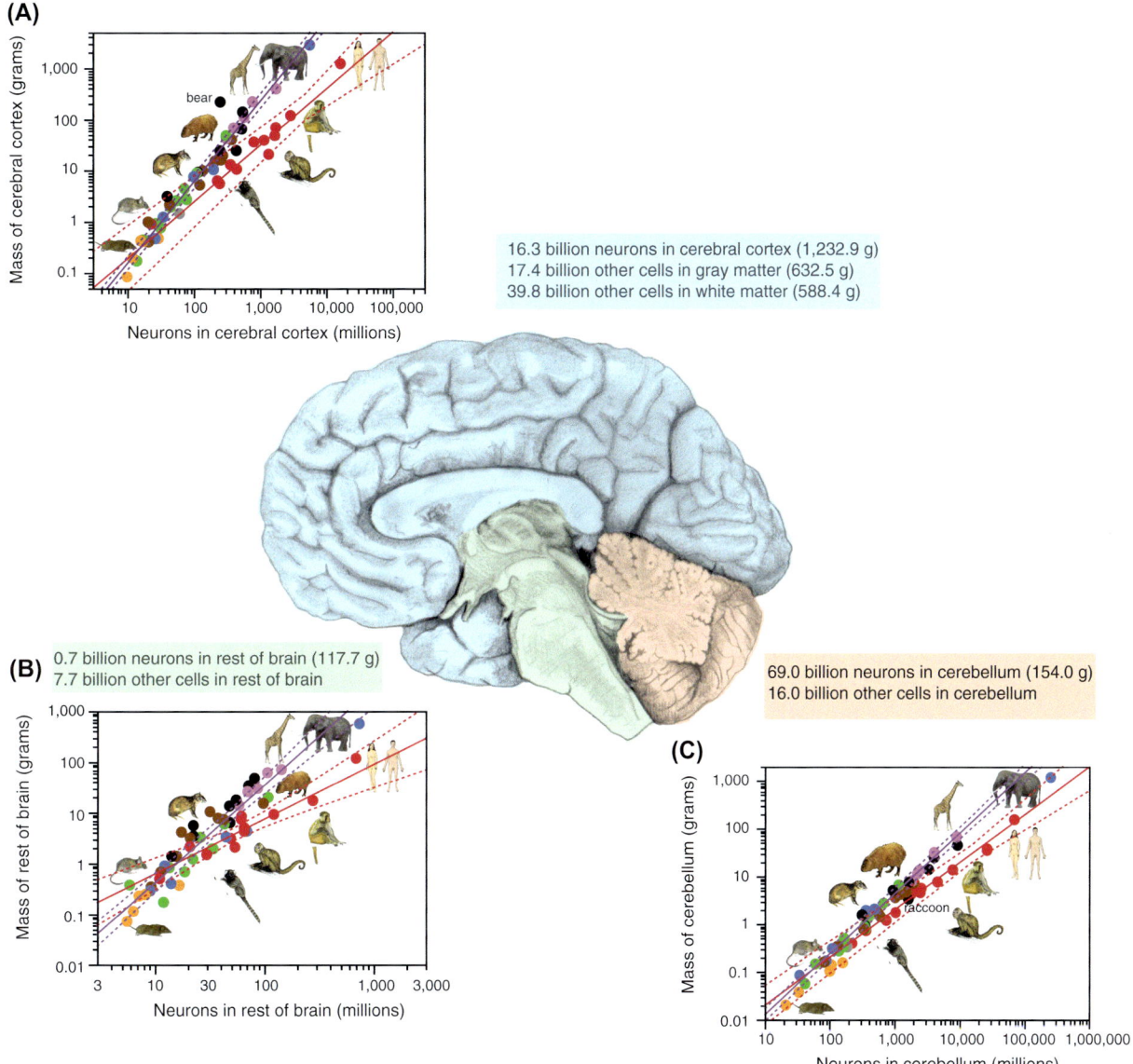

(A)

16.3 billion neurons in cerebral cortex (1,232.9 g)
17.4 billion other cells in gray matter (632.5 g)
39.8 billion other cells in white matter (588.4 g)

(B) 0.7 billion neurons in rest of brain (117.7 g)
7.7 billion other cells in rest of brain

69.0 billion neurons in cerebellum (154.0 g)
16.0 billion other cells in cerebellum

(C)

FIGURE 33.1 Numbers of neurons in the human brain make it a generic scaled-up primate brain. Numbers of neurons that compose each major brain structure are shown in colored boxes for the cerebral cortex (*blue*), cerebellum (*red*), and rest of brain (the ensemble of brain stem, diencephalon, and striatum, not including the olfactory bulbs; *red*). (A) The mass of the cerebral cortex (including the white matter) varies as a power function of the number of neurons in the structure that is shared across all nonprimate mammalian species examined so far (*purple*; exponent, 1.582 ± 0.039, $r^2 = 0.978$, $p < .0001$; primates, *red*; exponent, 1.087 ± 0.074, $r^2 = 0.956$, $p < .0001$). The human cerebral cortex has a total mass that falls within the 95% confidence interval of the mass predicted for its number of neurons. (B) The mass of the rest of brain varies as a power function of the number of neurons in each structure across all nonprimate species (plotted in *purple*; exponent, 1.921 ± 0.120, $r^2 = 0.867$, $p < .0001$) and as a different function across primates (plotted in *red*; exponent, 1.198 ± 0.116, $r^2 = 0.915$, $p < .0001$). The human rest of brain conforms to the scaling relationship that applies to other primates. (C) The mass of the cerebellum (including the white matter and deep nuclei) varies as a power function of the number of neurons in the structure that is shared across afrotherians (with the exception of the elephant; Herculano-Houzel et al., 2014b), glires, carnivorans, and artiodactyls (plotted in *purple*; exponent: 1.267 ± 0.036, $r^2 = 0.982$, $p < .0001$). Marsupials, primates, and eulipotyphlans deviate from this relationship, each in a different manner (marsupials, exponent 1.186 ± 0.037, $r^2 = 0.992$, $p < .0001$; primates, plotted in *red*, exponent 0.976 ± 0.036, $r^2 = 0.985$, $p < .0001$; eulipotyphla, exponent 1.028 ± 0.084, $r^2 = 0.980$, $p = .0012$), with more neurons than predicted for cerebellar mass in other clades. The human cerebellum conforms to the scaling relationship that applies to primates. *Illustrations by Lorena Kaz.*

indeed. But if their brain structures had the mass predicted for their number of neurons, that would be evidence that the human brain was built according to the same rules that apply to other primates.

We found that the latter was the case: the mass of each major brain structure in the human brain is very close to the predicted value for a generic primate with that many neurons, falling well within the 95% confidence interval

(Fig. 33.1). In contrast, when compared to a nonprimate, the human brain and its structures would indeed appear to be special, with 7.4, 9.1, 6, and 3.1 times more neurons in the whole brain, cerebral cortex (gray and white matter combined), cerebellum, and rest of brain than predicted for a nonprimate of similar mass in each structure. When the cortical gray matter and white matter are considered separately, the human brain still obeys the same relationship between structure volume and number of neurons that applies to other primates (Fig. 33.2). That is, the volume of the gray matter of the human cerebral cortex is not significantly larger than expected for a primate with its number of cortical neurons, and neither is the volume of the human cortical white matter. Altogether, these findings indicate that the human brain is just a scaled-up primate brain in its number of neurons.

The 86 billion neurons in the human brain are accompanied not by 10 times more glial cells, as legend had it (reviewed in von Bartheld et al., 2016), but rather by only as many nonneuronal cells (84.6, to be exact; Azevedo et al., 2009). Of the 84.6 billion nonneuronal cells in the human brain, 60.8 billion are found in the cerebral cortex, 16.0 billion are located in the cerebellum, and 7.7 billion in the rest of brain (Fig. 33.1; Azevedo et al., 2009). Importantly, these numbers match the universal relationship between numbers of nonneuronal cells and the mass of each of the major brain structures that we observed for any mammal, primate or not (Fig. 33.3A; Herculano-Houzel, 2014). Thus, the human brain has the number of nonneuronal cells in each major brain structure expected for the mass of those structures.

Because only a small minority of nonneuronal cells can be expected to be endothelial cells (about 1—6%; Buchweitz and Weiss, 1986; Lauwers et al., 2008; Tsai et al., 2009), most of these nonneuronal cells must be glial cells. That means that at best, the glia/neuron ratio for the human brain as a whole is 1.0. That number, however, hides the fact that the glia/neuron ratio is widely different across brain structures. Within the human brain, the average glia/neuron ratio (that is, the nonneuronal/neuronal cell ratio) is 1.48 in the cortical gray matter, a much larger 11.35 in the rest of brain (the ensemble of brain stem, diencephalon, and striatum), but only 0.23 in the cerebellum (Azevedo et al., 2009). As different as they are, these ratios are still just the expected for the neuronal density in each structure of the human brain, given the universal relationship that we have found to apply between the glia/neuron ratio and neuronal density across brain structures and mammalian species alike (Fig. 33.3B; Herculano-Houzel, 2014).

There is variation in the G/N ratio within the human cerebral cortex, as reported by other authors (Sherwood et al., 2006), but again we found that local G/N ratios

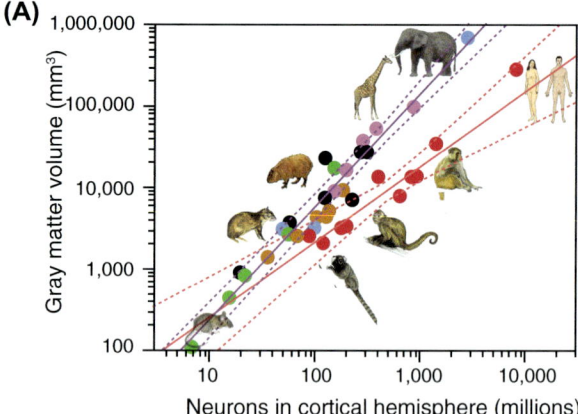

(A)

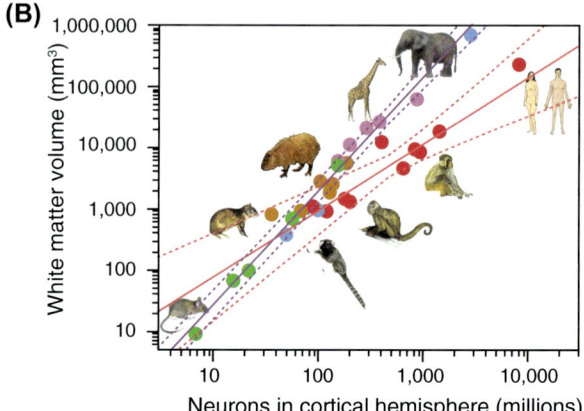

(B)

FIGURE 33.2 The human cerebral cortex conforms to the relationships between numbers of neurons and volumes of the gray and white matter. The volume of the gray matter scales as a different function of the number of neurons in the cortex in each mammalian clade analyzed, in such a way that for similar numbers of neurons, primates have the smallest volume of gray (A) and white matter (B). Importantly, humans do not depart significantly from the expected gray and white matter volumes given the number of cortical neurons in a primate. Exponents in (A): artiodactyls, 1.992 ± 0.180, $r^2 = 0.984$, $p = .0081$; rodents, 1.587 ± 0.064, $r^2 = 0.995$, $p = .0001$; marsupials, 1.076 ± 0.104, $r^2 = 0.964$, $p = .0005$; primates, 0.918 ± 0.083, $r^2 = 0.938$, $p < .0001$. Exponents in (B): rodents, 2.009 ± 0.065, $r^2 = 0.997$, $p < .0001$; artiodactyls, 1.637 ± 0.225, $r^2 = 0.964$, $p = .0183$; marsupials, 1.101 ± 0.271, $r^2 = 0.805$, $p = .0005$; primates, 1.080 ± 0.120, $r^2 = 0.909$, $p < .0001$. *Data from Ventura-Antunes, L., Mota, B., Herculano-Houzel, S., 2013. Different scaling of white matter volume, cortical connectivity, and gyrification across rodent and primate brains. Front. Neuroanat. 7, 3 and Herculano-Houzel et al. (unpublished observations). All values are for a single cortical hemisphere.*

within the human cerebral cortical gray matter are predicted by local variations in neuronal density (Ribeiro et al., 2013). Glia/neuron ratios, which once were considered an indicator of "progressive cognitive development" (Friede, 1954), are now known to simply reflect the average size of neuronal cells in the tissue (Herculano-Houzel, 2014; Mota and Herculano-Houzel, 2014).

(A)

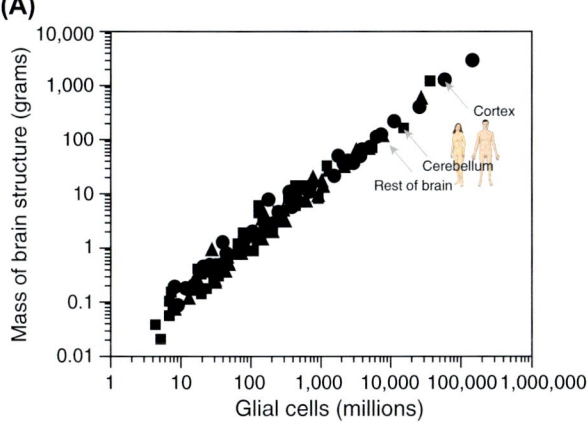

(B)

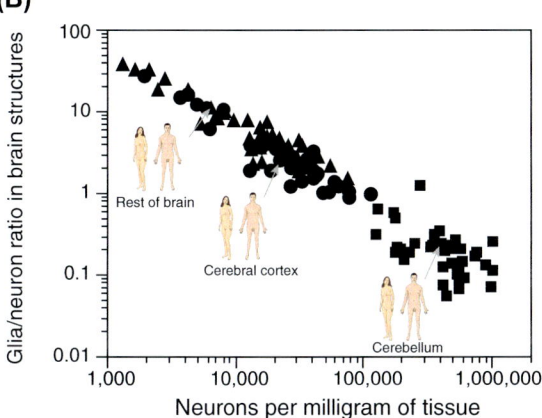

FIGURE 33.3 Human brains have expected numbers of glial cells and glia/neuron ratios. (A) The mass of major brain structures (cerebral cortex, *circles*; cerebellum, *squares*; rest of brain, *triangles*) varies as a power function of the number of nonneuronal cells in each structure that is shared across all mammalian species examined so far. Numbers of nonneuronal cells in the human cerebral cortex, cerebellum, and rest of brain (*arrows*) conform to the predicted for a generic mammalian species. (B) The glia/neuron ratio in each major brain structure (symbols as in A) varies universally across brain structures and species, increasing with decreasing neuronal density (which indicates larger average neuronal cell size). The glia/neuron ratios in human cerebral cortex, cerebellum, and rest of brain conform to the scaling relationship that applies across mammalian species as a whole. Original data on 5 species of eulipotyphlans, 6 species of afrotherians, 10 species of glires, 12 species of primates, and 5 species of artiodactyls are available in Herculano-Houzel, S., Catania, K., Manger, P.R., Kaas, J.H., 2015b. Mammalian brains are made of these: a dataset of the numbers and densities of neuronal and nonneuronal cells in the brain of glires, primates, scandentia, eulipotyphlans, afrotherians and artiodactyls, and their relationship with body mass. Brain Behav. Evol. 86, 145–163.

33.3 The Energetic Cost of the Human Brain

Well before our group showed that the glia/neuron ratio was inversely proportional to neuronal density across structures and species, and thus directly proportional to the average size of neurons in the tissue (Herculano-Houzel, 2014; Mota and Herculano-Houzel, 2014), it was widely held that this was the

case due to the expected larger metabolic cost of larger neurons (Hawkins and Olzewski, 1957). However, once data on the numbers of neurons that compose the brains of enough species that also had their metabolic cost measured directly, we found that the average metabolic cost per neuron did not vary significantly with neuronal density in the cerebral cortex or cerebellum of different species (Herculano-Houzel, 2011). In the human cerebral cortex, the average cost per neuron was 1.32×10^{-8} μmol of glucose per minute, as much as the 1.39×10^{-8} μmol of glucose per minute per neuron in the mouse cerebral cortex (Herculano-Houzel, 2011). Per gram of tissue, the mouse cerebral cortex costs about three times more glucose than the human cerebral cortex because the density of neurons is about three times higher in mouse than in human cerebral cortex.

Without a significant variation in average glucose cost per neuron across species, the total glucose use of a cerebral cortex, or of the brain as a whole, turned out to be a simple linear function of the number of neurons in the cortex or whole brain across rodents and primates alike (Herculano-Houzel, 2011)—species that we then knew to have different relationships between numbers of neurons and brain mass (Herculano-Houzel et al., 2006, 2007). It thus became evident that the previously reported hypometric power function describing the scaling of the metabolic cost of brains with increasing brain mass (Karbowski, 2007) was a result of the particular combination of neuronal densities in the data set. Most importantly, the data showed that, given an average cost of 6 kcal per billion neurons per day, the human brain cost just as much energy as expected for a rodent or primate with its number of neurons: about 516 kcal day^{-1} (Herculano-Houzel, 2011).

The reason why the human brain is so expensive in proportion to the body then turned out to be simply that a different relationship between body mass and number of brain neurons applies to primates: compared to other mammals of similar body mass, primates of increasing body mass have larger and larger numbers of brain neurons (reviewed in Herculano-Houzel et al., 2015b). The human brain does not cost a relatively enormous amount of energy because it is "special," but simply because it is the largest primate brain, with the largest number of neurons among primates (Herculano-Houzel, 2016).

It became apparent then that the question regarding the human encephalization quotient could be reframed: given that the energetic cost of a brain depending on its number of neurons was so large, what if instead of humans having larger brains than expected for their bodies, it was great apes who could not afford the large brains that would be expected for their very large bodies? Once we learned that also the brains of great apes conformed to the relationship between number of neurons and brain mass that applied to primates as a whole (Herculano-Houzel and Kaas, 2011) and that humans conformed to the relationship between body mass and number of brain neurons that applied to non-great ape primates (Azevedo et al., 2009), it appeared that great apes, not humans, were the outliers. Actually, it was the inclusion of great apes to the brain versus body comparison that made humans appear as outliers (Herculano-Houzel, 2016).

The mathematical consideration of the increasing energetic costs of bodies of increasing mass and of brains of increasing numbers of neurons, which must be balanced by energetic intake that depends on body mass and number of hours of feeding per day, showed that great apes indeed could not afford a brain any larger than it is (Fonseca-Azevedo and Herculano-Houzel, 2012). Because of the energetic limitation imposed by the primate diet, there is a trade-off between body mass and number of brain neurons: past the optimal combination, where numbers of neurons are maximal, larger bodies can only be sustained at the expense of decreasing numbers of brain neurons. According to our calculations, for gorillas to have the number of brain neurons that would be expected for their body masses, in a brain that corresponded to 2% of body mass as in humans and other primates, they would need to afford an additional 122 billion neurons—and that would require ingesting an extra 733 kcal, which would take over two more hours of feeding per day (Fonseca-Azevedo and Herculano-Houzel, 2012). For a primate that already averages almost 8 h of eating per day, that seems to be unviable. Orangutans, which similarly feed for about 8 h day^{-1}, do not seem capable of eating for longer periods during periods of low food availability and lose body mass (Knott, 1998). Feeding 8 h day^{-1} seems to be the practical limit for primates.

According to the same calculations, and considering the raw diet of extant nonhuman primates, *Homo habilis* would have required daily feeding times close to 7.5 h day^{-1}, close to the 8 h day^{-1} limit, and late *Homo erectus* individuals, such as modern *Homo sapiens*, would be required to feed consistently more than 9 h day^{-1} to afford their estimated body mass and number of neurons (Fonseca-Azevedo and Herculano-Houzel, 2012). Modern humans, therefore, would not be viable on the raw diet typical of other primates. Given that both body and brain in humans cost the expected amount of energy for their mass and number of neurons, respectively, and eating longer hours is not an option, there remains but one alternative to explain how human brains attained their modern numbers of neurons: with a change in diet that significantly increased caloric intake, preferably while decreasing feeding time.

Cooking does just that. The controlled use of fire to break down foods prior to ingestion increases the caloric yield of foodstuffs to 100% of their theoretical nutritional value, while also decreasing the time required to consume them (reviewed in Wrangham, 2009). Evidence of habitual use of fire to cook exists about 1–1.5 million years ago (Gwolett et al., 1981; Berna et al., 2012), a time that coincides with the steep increase in brain size in human ancestors—but, as Wrangham's critics like to point, might be too late to explain the earlier beginning of the ascent, about 2 million years ago.

In a looser definition, however, "cooking" has been around for much longer than 2 million years. Cutting, bashing, crushing, and otherwise tenderizing foods prior to chewing are forms of predigestion that also increase caloric yield significantly (Zink and Lieberman, 2016) and were available to early *Homo* and their hunter-gatherer ancestors at least 3.3 million years ago (Harmand et al., 2015). Between cold and hot versions of cooking, early *H. habilis*, with modernlike hands capable of precision grips by 2 million years ago (Susman, 1998; Alba et al., 2003), and certainly *H. erectus* of 1 million years ago must have enjoyed a luxury that no modern primate, or mammal, for that matter, seems to enjoy: having enough calories to feed an increased number of neurons as well as the free time to do more interesting things with them than looking for food and eating. Such "optional" activities, such as socializing and organizing hunts, which rely on memory, planning, reasoning, self-control, and awareness of the mental state of others, probably exerted selective pressure for even more brain neurons. Once over the energetic constrain that a raw diet imposes to other animals, the brain of *Homo* was then free to keep scaling as primate brains do. At this moment, this is the best narrative that we know to account for how *H. sapiens* came to have the largest primate brain, with the largest number of neurons in any cerebral cortex, without ever deviating from the biological rules of evolution. We never stopped being primates (Herculano-Houzel, 2016).

33.4 The Expanded Human Cerebral Cortex Does Not Have Relatively More Neurons

The expansion of the cerebral cortex relative to the remaining brain structures is probably the feature that is most equated with "brain evolution" in mammals, and in primates in particular (Stephan and Andy, 1969; Jerison, 1973; Frahm et al., 1982). A cerebral cortex that expands in relative size within the brain, coming to represent over 80% of brain volume in humans, would be expected to also expand in relative number of brain neurons. That is, the relatively larger cortex of humans would be expected to contain a relatively larger proportion of all brain neurons than in other primate species.

Contrary to that expectation, however, we found that the human cerebral cortex, which represents 82% of brain mass, has only 19% of all neurons in the human brain—a proportion that is similar to that found in the guinea pig, although the cerebral cortex of the latter represents a much smaller 53% of the brain (Azevedo et al., 2009; Herculano-Houzel et al., 2006). This similarity is not particular to these two species: across mammalian species, we find that although the cerebral cortex represents between 50% and 80% of brain mass, it typically holds 15–25% of brain neurons, while the cerebellum, at only 10–20% of brain mass, holds 70–85% of all brain neurons (Herculano-Houzel, 2010; Herculano-Houzel et al., 2014a). As illustrated in Fig. 33.4A, relatively larger cerebral cortices do not hold significantly larger percentages of all brain neurons. The expansion of the human cerebral cortex in evolution thus represented a gain in absolute numbers of neurons without a gain in relative numbers of neurons within the brain.

The reason is that the cerebellum, which alone holds the majority of all brain neurons, gains neurons together with the cerebral cortex, keeping a fairly steady average proportion of four neurons in the cerebellum to every neuron in the cerebral cortex across extant species (Fig. 33.4B; Herculano-Houzel, 2010; Herculano-Houzel et al., 2014a). This joint scaling of the two structures agrees with the modern view that they work in tandem (Leiner et al., 1989; Ramnani, 2006; Ito, 2008) and is in line with the concerted increase of the volume of connected cerebral cortical and cerebellar regions in primates (Ramnani et al., 2006; Balsters et al., 2010). Together, however, they do gain neurons significantly faster than the rest of brain across primate species, which means that larger primate brains have increasing ratios of numbers of cortical (and cerebellar) neurons over numbers of neurons in the rest of brain (Herculano-Houzel et al., 2014a). Across primate species, for example, the ratio between numbers of cortical motor neurons and neurons in the spinal cord increases from less than 1 in the mouse lemur to 4 in the macaque and 20 in the chimpanzee and human (Herculano-Houzel et al., 2016). We have proposed that this faster scaling of numbers of cortical versus brain stem and also spinal cord neurons in primates explains the corticalization of motor control in humans (Herculano-Houzel et al., 2016), that is, the dependence of motor control on the corticospinal tract in humans.

We find a similar expansion of numbers of cortical neurons over numbers of thalamic neurons across primate species. Although the primary visual cortex represents a relatively constant 36% of all cortical neurons across nonhuman primate species, the ratio between numbers of neurons in the primary visual cortex and

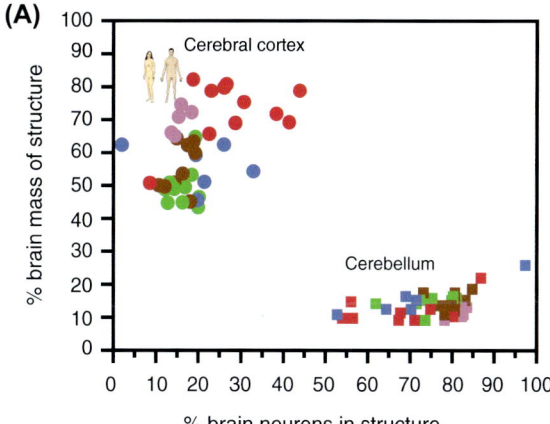

(A)

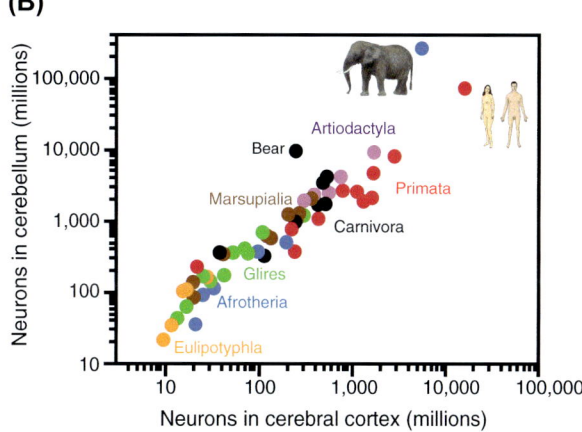

(B)

FIGURE 33.4 Relatively larger cerebral cortices do not have relatively larger proportions of brain neurons. (A) Although the cerebral cortex (*circles*) represents between 40% and 82% of brain mass in the species we have analyzed so far, it typically contains only between 15% and 25% of all brain neurons—and those animals that have relatively larger cortices do not have relatively more brain neurons in it (Herculano-Houzel, 2010; Herculano-Houzel et al., 2014a). The human cerebral cortex, in particular, represents 82% of brain mass, but holds only 19% of all brain neurons (Azevedo et al., 2009). Likewise, although the cerebellum (*squares*) usually represents only between 10% and 20% of brain mass, it holds typically between 70% and 80% of all brain neurons—but relative cerebellar mass and relative number of cerebellar neurons are not correlated. The human cerebellum, in particular, represents only 10% of brain mass, but holds 69% of all brain neurons (Azevedo et al., 2009). (B) The cerebral cortex and cerebellum gain neurons coordinately across mammalian species. With the sole exceptions of the African elephant and the brown bear, there is a relatively steady average of four neurons in the cerebellum for every neuron in the cerebral cortex across species, regardless of the relative size of the cerebral cortex as a percentage of brain mass. The human brain, in particular, conforms to the same proportionality that applies to other primates and to mammals as a whole. Original data on 5 species of eulipotyphlans, 6 species of afrotherians, 10 species of glires, 12 species of primates, and 4 species of artiodactyls are available in Herculano-Houzel, S., Manger, P.R., Kaas, J.H., 2014. Brain scaling in mammalian evolution as a consequence of concerted and mosaic changes in numbers of neurons and average neuronal cell size. Front. Neuroanat. 8, 77. *Data on carnivorans from Messeder et al. (unpublished observations). Data on marsupials from Dos Santos et al. (unpublished observations).*

in the lateral geniculate nucleus (LGN) (the visual nucleus of the thalamus) increases together with brain size, from 50 times more neurons in the visual cortex compared to the LGN in the marmoset to 133 in the baboon (Collins et al., 2013). Thus, it appears that larger primates have more cortical motor neurons to control each downstream effector neuron that operates the body, and also more cortical sensory neurons to elaborate on sensory input received, adding complexity and flexibility to behavior. In that vein, the human brain seems to benefit from being the most extreme case of faster addition of neurons to the cerebral cortex over the rest of brain across primates (Herculano-Houzel et al., 2016).

Interestingly, while the expansion of the cerebral cortex over other brain structures has been considered evidence of increased cortical functioning, it should not be taken as evidenced of diminished processing abilities by other structures. A case in point is the olfactory bulb. The residual volume of the primate olfactory bulb has long been known to decrease with increasing brain size across primate species, in contrast to the increasing residual volume of the primate cerebral cortex (Stephan and Andy, 1969)—meaning that given the relationship between structure volume and body mass, the primate cerebral cortex becomes larger and larger than predicted, while the olfactory bulb becomes smaller and smaller. It was based on these findings, confirmed by later repeated analyses (Baron et al., 1983; Finlay et al., 2001), that primates—and humans in particular—came to be considered microsmatic.

However, once absolute numbers of neurons in the olfactory bulb were determined directly in a number of mammalian species, it became clear that the supposedly microsmatic primates had olfactory bulbs composed of as many neurons as species highly reliant on olfaction, such as shrews: both groups have a similar range of number of neurons in the olfactory bulb, between 2 and 30 million (Ribeiro et al., 2014). Humans, in particular, have an estimated 15–16 million neurons in the olfactory bulb, which is more than the 10 million found in the highly olfactory star-nosed mole, an animal that relies on olfaction to catch its prey. The information processing capabilities of a sensory system should depend on the number of neurons available for that system, regardless of how many neurons are available to other sensory modalities. Thus, decreased reliance on one sensory modality because another came to be favored, such as vision in primates, by no means is synonymous of decreased abilities of the first modality; given the new information on numbers of olfactory bulb neurons, it seems that the increased reliance of primates on vision over olfaction did not occur at the expense of olfaction, but in addition to it. Indeed, recent behavioral studies showed that primate olfaction is far better than

presumed, both in squirrel monkeys (Laska et al., 2000) and humans (Porter et al., 2006): When blindfolded humans were invited to go down on all fours and track a chocolate scent on the grass using only their noses, most were perfectly capable of doing so. Given the similar numbers of neurons in the olfactory bulbs of highly visual humans and highly olfactory shrews, humans, like other primates, should no longer be considered microsmatic. Perhaps the difference is simply that, contrary to small mammals, we live with our noses over 5 feet above the ground.

33.5 The Expanded Human Cerebral Cortex Does Not Have Relatively More Neurons in the Prefrontal Region

As reviewed previously, our quantitative analyses showed that while cortical expansion in primate evolution did not amount to significantly increased percentages of total brain neurons, larger cortices do gain neurons significantly faster than the rest of brain (ie, brain stem, diencephalon, and striatum; Herculano-Houzel et al., 2014a). In the literature on brain evolution, a pervasive notion has been that the expansion of the primate cerebral cortex has been accompanied by a relative expansion of the prefrontal region within it— associative cortex required for so-called higher cognitive functions such as planning and logical reasoning. This view dates back to German neurologist Korbinian Brodmann, who estimated that the frontal cortex (prefrontal plus motor cortex) occupied 29% of the cerebral cortex in humans, but only 17% in the chimpanzee and 11% in macaque monkey (Brodmann, 1912).

However, using modern imaging techniques, Katerina Semendeferi and colleagues showed in 2002 that the frontal cortex of humans, bonobos, chimpanzees, gorillas, and orangutans occupies the same 35–37% of the cortical volume—percentages that, while higher than the 30% in macaques, did not single out humans. The studies that followed alternately argued that the human prefrontal white matter (although not the prefrontal gray matter) was larger in humans than expected (Schoenemann et al., 2005; Smaers et al., 2011), or simply that both human prefrontal gray and white matter had the volumes expected for a primate of human brain mass (Barton and Venditti, 2013).

Our own recent study comparing numbers of cells rather than volumes support Barton and Venditti's view that the prefrontal region of the cerebral cortex is not disproportionately expanded in humans (Gabi et al., 2016). Actually, we found no expansion at all of the percentage of cortical neurons located in prefrontal regions, anterior to the callosum (the same criterion used by Schoenemann et al., in 2005): almost all primate species analyzed, including humans, had a similar 8% of all cortical neurons located anterior to the callosum. Humans also fell well within the expected values of numbers of other cells in the prefrontal white matter compared to numbers of neurons in prefrontal gray matter or nonprefrontal white matter, indicating that the prefrontal white matter is also not preferentially expanded in the human brain (Gabi et al., 2016).

What does distinguish the human brain from others seems to be simply its absolute number of neurons in the cerebral cortex, the largest found so far (Fig. 33.5). The human cerebral cortex has about three times as many neurons as the twice larger cerebral cortex of the African elephant (Herculano-Houzel et al., 2014b). This "human advantage" lies simply in the fact that the human brain is built in the image of other primate brains, with more neurons per mass compared to other brains of similar brain mass (Herculano-Houzel, 2012). Across primates, even though humans share with other species the same 8% of all cortical neurons in prefrontal, associative regions, that percentage translates into a much larger absolute number of prefrontal neurons in the human cerebral cortex than in other primates: 1.3 billion prefrontal neurons in the human cortex, but only 230

million in the baboon, 137 million in the macaque, and a meager 20 million in the marmoset (Gabi et al., 2016). It is not yet clear what percentage of all cortical neurons have associative, prefrontal-like functions in elephant and whale brains, larger than human brains, but it seems that prefrontal areas are but a sliver of the cortical volume of those large brains (Morgane et al., 1980). However, because of the larger number of neurons in the human cerebral cortex even when compared to larger cortices, we can safely expect that our 1.3 billion neurons dedicated to associative, prefrontal functions will be matched by none. Just as more cortical neurons in the human cerebral cortex dedicated to sensory and motor functions can provide the basis for more complex and flexible cognition and motor control (Collins et al., 2013; Herculano-Houzel et al., 2016), more prefrontal neurons may provide the basis for refined integrative functions, planning, and modulation of behavior in the human brain compared to others, especially given that the general pattern of connectivity has been found to be similar across humans and other mammalian species (Shanahan et al., 2013). Although human-specific genes may well exist, their contribution to human cognition is probably in addition to the benefit of the human cerebral cortex having very large numbers of neurons to refine the processing of information relayed by subcortical structures. No exceptionality in regard to evolutionary rules is required, nor does it seem to exist, to account for the remarkable capabilities of the human brain.

33.6 Biological Capabilities × Developed Abilities

Having the largest number of neurons in the cerebral cortex certainly makes humans stand out from all other animals, while no longer singling us out as outliers. However, it is important to keep in mind that while having enough cortical neurons seems to be a necessary condition for complex and flexible behaviors, it is not a sufficient condition for such behaviors. There is a crucial distinction to be made here between the cognitive *capabilities* made possible by a large number of cortical neurons and the cognitive *abilities* that an individual cortex may develop. All the popular writing that hails the human brain as a wonder tends to omit the fact that it never starts life like so—although it certainly holds great promise. Even for the cortex with the most neurons, turning biological cognitive capabilities into actual abilities takes a lifetime of learning, practicing, and honing skills—if not generations.

Indeed, a key component of human evolution has been the development of technologies and their cultural transmission to the next generations (Herculano-Houzel,

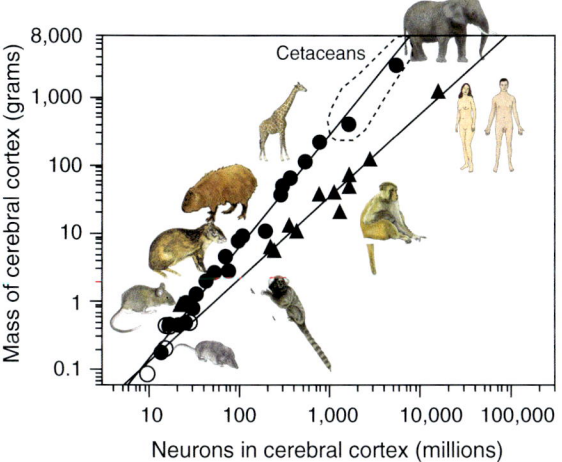

FIGURE 33.5 Because it conforms to the primate neuronal scaling rules, the human cerebral cortex has more neurons than any other mammalian cortex. The graph illustrates how the mass of the cerebral cortex varies across primate and nonprimate species, including our prediction for numbers of neurons in cetacean cortices (*dashed area*), around 1—5 billion neurons, given the neuronal scaling rules that apply to artiodactyls, with which cetaceans share the order Cetartiodactyla (Kazu et al., 2014). Although stereological studies have estimated that the cortex of the minke whale and the harbor porpoise are composed of much larger numbers of 13 and 15 billion neurons (Eriksen and Pakkenberg, 2007; Walloe et al., 2010), those studies grossly undersampled the cortices examined (Kazu et al., 2014). Our recent study using the isotropic fractionator estimated a total of fewer than 3 billion neurons in the cerebral cortex of the minke whale (Avelino-de-Souza et al., unpublished observations), in alignment with the prediction made from artiodactyls.

2016), thus leading to a cumulative increase in abilities not only of individuals, but also populations as a whole (Henrich, 2015). Perhaps the most striking evidence of the divide between cognitive capacities afforded by biology and the abilities that can be fostered with cultural transmission of technologies is the chasm that separates modern humans from their relatives of 100 000–200 000 years ago: judging from brain size, both had similar numbers of neurons in the brain, and in the cerebral cortex in particular—but the feats of the Ice Age version of our species pale in comparison to our modern achievements. Ever since then, humanity has known a history of identifying problems, inventing technologies to deal with them, and passing the knowledge on to the next generations, who can then put their biological capabilities to work on evermore complicating problems and technologies (Herculano-Houzel, 2016). Endowed with a brain that has enough cortical neurons to learn from others, create new knowledge and pass it on, our species stands on the shoulders of all those who came before us. At this point, the achievements of the human species far surpass those of any one individual: humankind has long transcended the individual human brain. That is why science (the knowledge) and engineering (the crafts) must be carefully cultivated, documented, and passed on to the next generations.

References

Alba, D., Moyà-Solà, S., Köhler, S., 2003. Morphological affinities of the *Australopithecus afarensis* hand on the basis of manual proportions and relative thumb length. J. Hum. Evol. 44, 225–254.

Azevedo, F.A.C., Carvalho, L.R.B., Grinberg, L.T., Farfel, J.M., Ferretti, R.E.L., Leite, R.E.P., Jacob Filho, W., Lent, R., Herculano-Houzel, S., 2009. Equal numbers of neuronal and nonneuronal cells make the human brain an isometrically scaled-up primate brain. J. Comp. Neurol. 513, 532–541.

Balsters, J.H., Cussans, E., Diedrichsen, J., Phillips, K.A., Preuss, T.M., Rilling, J.K., Ramnani, N., 2010. Evolution of the cerebellar cortex: the selective expansion of prefrontal-projecting cerebellar lobules. Neuroimage 49, 2045–2052.

Baron, G., Frahm, H.D., Bhatnagar, K.P., Stephan, H., 1983. Comparison of brain structure volumes in insectivora and primates. III. Main olfactory bulb (MOB). J. Hirnforsch. 24, 551–568.

Barton, R.A., Venditti, C., 2013. Human frontal lobes not relatively large. Proc. Natl. Acad. Sci. U.S.A. 110, 9001–9006.

Berna, F., Goldberg, P., Horwitz, L.K., Brink, J., Holt, S., Bamford, M., Chazan, M., 2012. Microstratigraphic evidence of in situ fire in the Acheulean strata of Wonderwerk Cave, Northern Cape province, South Africa. Proc. Natl. Acad. Sci. U.S.A. 109, E1215–E1220.

Brodmann, K., 1912. Localisation in the Cerebral Cortex. Springer, Berlin.

Buchweitz, E., Weiss, H.R., 1986. Alterations in perfused capillary morphometry in awake vs anesthetized brain. Brain Res. 377, 105–111.

von Bartheld, C.S., Bahney, J., Herculano-Houzel, S., 2016. The search for true numbers of neurons and glial cells in the human brain: a review of 150 years of cell counting. J. Comp. Neurol. 524 (18), 3865–3895.

Cáceres, M., Lachuer, J., Zapala, M.A., Redmond, J.C., Kudo, L., Geschwind, D.H., et al., 2003. Elevated gene expression levels distinguish human from non-human primate brains. Proc. Natl. Acad. Sci. U.S.A. 100, 13030–13035.

Charrier, C., Joshi, K., Coutinho-Budd, J., Kim, J.E., Lambert, N., de Marchena, J., et al., 2012. Inhibition of SRGAP2 function by its human-specific paralogs induces neoteny during spine maturation. Cell 149, 923–935.

Collins, C.E., Leitch, D.B., Wong, P., Kaas, J.H., Herculano-Houzel, S., 2013. Faster scaling of visual neurons in cortical areas relative to subcortical structures in non-human primate brains. Brain Struct. Funct. 218, 805–816.

Dennis, M.Y., Nuttle, X., Sudmant, P.H., Antonacci, F., Graves, T.A., Nefedoc, M., et al., 2012. Evolution of human-specific neural SRGAP2 genes by incomplete segmental duplication. Cell 149, 912–922.

Dumas, L., O'Bleness, M.S., Davis, J.M., Dickens, C.M., Anderson, N., Keeney, J.G., et al., 2012. DUR1220-domain copy number implicated in human brain-size pathology and evolution. Am. J. Hum. Genet. 91, 444–454.

Enard, W., Gehre, S., Hammerschmidt, K., Hölter, S.M., Blass, T., Somel, M., et al., 2009. A humanized version of Foxp2 affects cortico-basal ganglia circuits in mice. Cell 137, 961–971.

Eriksen, N., Pakkenberg, B., 2007. Total neocortical cell number in the mysticete brain. Anat. Rec. 290, 83–95.

Evans, P.D., Anderson, J.R., Vallender, E.J., Choi, S.S., Lahn, B.T., 2004. Reconstructing the evolutionary history of microcephalin, a gene controlling brain size. Hum. Mol. Genet. 13, 1139–1145.

Finlay, B.L., Darlington, R.B., Nicastro, N., 2001. Developmental structure in brain evolution. Behav. Brain Sci. 24, 263–308.

Fonseca-Azevedo, K., Herculano-Houzel, S., 2012. Metabolic constraint imposes trade-off between body size and number of brain neurons in human evolution. Proc. Natl. Acad. Sci. U.S.A. 109, 18571–18576.

Frahm, H.D., Stephan, H., Stephan, M., 1982. Comparison of brain structure volumes in Insectivora and Primates. I. Neocortex. J. Hirnforsch. 23, 375–389.

Friede, R., 1954. Der quantitative Anteil der Glia an der Cortexentwicklung. Acta Anat. 20, 290–296.

Gabi, M., Neves, K., Masseron, C., Ventura-Antunes, L., Ribeiro, P., Torres, L., Kaas, J.H., Herculano-Houzel, S., 2016. No expansion in numbers of prefrontal neurons in primate and human evolution. Proc. Natl. Acad. Sci. U.S.A. 113, 9617–9622.

Gazzaniga, M.S., 2008. Human: The Science behind what Makes Us Unique. Harper Collins, New York.

Gowlett, J.A.J., Harris, J.W.K., Walton, D., Wood, B.A., 1981. Early archaeological sites, hominid remains, and traces of fire from Chesowanja, Kenya. Nature 294, 125–129.

Harmand, S., et al., 2015. Stone tools. Nature 521, 310–315.

Hawkins, A., Olszewski, J., 1957. Glia/nerve cell index for cortex of the whale. Science 126, 76–77.

Henrich, J., 2015. The Secret of our Success. How Culture Is Driving Human Evolution, Domesticating Our Species, and Making Us Smarter. Princeton University Press, Princeton.

Herculano-Houzel, S., Kaas, J.H., 2011. Gorilla and orangutan brains conform to the primate scaling rules: implications for hominin evolution. Brain Behav. Evol. 77, 33–44.

Herculano-Houzel, S., Lent, R., 2005. Isotropic fractionator: a simple, rapid method for the quantification of total cell and neuron numbers in the brain. J. Neurosci. 25, 2518–2521.

Herculano-Houzel, S., Mota, B., Lent, R., 2006. Cellular scaling rules for rodent brains. Proc. Natl. Acad. Sci. U.S.A. 103, 12138–12143.

Herculano-Houzel, S., Collins, C., Wong, P., Kaas, J.H., 2007. Cellular scaling rules for primate brains. Proc. Natl. Acad. Sci. U.S.A. 104, 3562–3567.

Herculano-Houzel, S., Manger, P.R., Kaas, J.H., 2014. Brain scaling in mammalian evolution as a consequence of concerted and mosaic changes in numbers of neurons and average neuronal cell size. Front. Neuroanat. 8, 77.

Herculano-Houzel, S., Avelino-de-Souza, K., Neves, K., Porfírio, J., Messeder, D., Feijó, L.M., Maldonado, J., Manger, P.R., 2014. The elephant brain in numbers. Front. Neuroanat. 8, 46.

Herculano-Houzel, S., von Bartheld, C.S., Miller, D.J., Kaas, J.H., 2015. How to count cells: the advantages and disadvantages of the isotropic fractionator compared with stereology. Cell Tiss. Res. 360, 29–42.

Herculano-Houzel, S., Catania, K., Manger, P.R., Kaas, J.H., 2015. Mammalian brains are made of these: a dataset of the numbers and densities of neuronal and nonneuronal cells in the brain of glires, primates, scandentia, eulipotyphlans, afrotherians and artiodactyls, and their relationship with body mass. Brain Behav. Evol. 86, 145–163.

Herculano-Houzel, S., 2010. Coordinated scaling of cortical and cerebellar numbers of neurons. Front. Neurosci. 4, 12.

Herculano-Houzel, S., 2011. Scaling of brain metabolism with a fixed energy budget per neuron: implications for neuronal activity, plasticity and evolution. PLoS One 6, e17514.

Herculano-Houzel, S., 2012. The remarkable, yet not extraordinary human brain as a scaled-up primate brain and its associated costs and advantages. Proc. Natl. Acad. Sci. U.S.A. 109, 10661–10668.

Herculano-Houzel, S., 2014. The glia/neuron ratio: how it varies uniformly across brain structures and species and what that means for brain physiology and evolution. Glia 62, 1377–1391.

Herculano-Houzel, S., 2016. The Human Advantage: A New Understanding of How Our Brains Became Remarkable. MIT Press, Cambridge.

Herculano-Houzel, S., Kaas, J.H., de Oliveira-Souza, R., 2016. Corticalization of motor control in primate evolution. J. Comp. Neurol. 524, 448–455.

Hofman, M.A., 1985. Size and shape of the cerebral cortex in mammals. I. The cortical surface. Brain Behav. Evol. 27, 28–40.

Hofman, M.A., 1988. Size and shape of the cerebral cortex in mammals. II. The cortical volume. Brain Behav. Evol. 32, 17–26.

Ito, M., 2008. Control of mental activities by internal models in the cerebellum. Nat. Rev. Neurosci. 9, 304–313.

Jerison, H.J., 1973. Evolution of the Brain and Intelligence. Academic Press, New York.

Karbowski, J., 2007. Global and regional brain metabolic scaling and its functional consequences. BMC Biol. 5, 18.

Kazu, R.S., Maldonado, J., Mota, B., Manger, P.R., Herculano-Houzel, S., 2014. Cellular scaling rules for the brains of Artiodactyla. Front. Neuroanat. 8, 128.

Kety, S.S., 1957. The general metabolism of the brain in vivo. In: Richter, D. (Ed.), Metabolism of the Nervous System. Pergamon Press, London.

Knott, C., 1998. Changes in orangutan caloric intake, energy balance, and ketones in response to fluctuating fruit availability. Int. J. Prim. 19, 1061–1079.

Laska, M., Seibt, A., Weber, A., 2000. 'Microsmatic' primates revisited: olfactory sensitivity in the squirrel monkey. Chem. Senses 25, 47–53.

Lauwers, F., Cassot, F., Lauwers-Cances, V., Puwanarajah, P., Duvernoy, H., 2008. Morphometry of the human cerebral cortex microcirculation: general characteristics and space-related profiles. Neuroimage 39, 936–948.

Leiner, H.C., Leiner, A.L., Dow, R.S., 1989. Reappraising the cerebellum: what does the hindbrain contribute to the forebrain? Behav. Neurosci. 103, 998–1008.

Marino, L., 1998. A comparison of encephalization between odontocete cetaceans and anthropoid primates. Brain Behav. Evol. 51, 230–238.

Morgane, P.J., Jacobs, M.S., MacFarland, W.L., 1980. The anatomy of the brain of the bottlenose dolphin (*Tursiops truncatus*). Surface configurations of the telencephalon of the bottlenose dolphin with comparative anatomical observations in four other cetacean species. Brain Res. Bull. 5 (Suppl. l), 1–107.

Mota, B., Herculano-Houzel, S., 2014. All brains are made of this: a fundamental building block of brain matter with matching neuronal and glial masses. Front. Neuroanat. 8, 127.

Porter, J., Craven, B., Khan, R.M., Chang, S.-J., Kang, I., Judkewitz, B., Volpe, J., Settles, G., Sobel, N., 2006. Mechanisms of scent-tracking in humans. Nat. Neurosci. 10, 27–29.

Prabhakar, S., Visell, A., Akiyama, J.A., Shoukry, M.P., Lewis, K.D., Holt, A., et al., 2008. Human-specific gain of function in a developmental enhancer. Science 321, 1346–1350.

Ramnani, N., Behrens, T.E., Johansen-Berg, H., Richter, M.C., Pinsk, M.A., Andersson, J.L., Rudebeck, P., Ciccarelli, O., Richter, W., Thompson, A.J., Gross, C.G., Robson, M.D., Kastner, S., Matthews, P.M., 2006. The evolution of prefrontal inputs to the cortico-pontine system: diffusion imaging evidence from macaque monkeys and humans. Cereb. Cortex 16, 811–818.

Ramnani, N., 2006. The primate cortico-cerebellar system: anatomy and function. Nat. Rev. Neurosci. 7, 511–522.

Ribeiro, P.F.M., Manger, P.R., Catania, K., Kaas, J.H., Herculano-Houzel, S., 2014. Greater addition of neurons to the olfactory bulb than to the cerebral cortex of insectivores but not rodents or primates. Front. Neuroanat. 8, 23.

Ribeiro, P.F.M., Ventura-Antunes, L., Gabi, M., Mota, B., Grinberg, L.T., Farfel, J.M., Ferretti, R.E.L., Leite, R.E.P., Jacob Filho, W., Herculano-Houzel, S., 2013. The human cerebral cortex is neither one nor many: neuronal distribution reveals two quantitatively different zones in the grey matter, three in the white matter, and explains local variations in cortical folding. Front. Neuroanat. 7, 28.

Schoenemann, P.F., Sheehan, M.J., Glotzer, L.D., 2005. Prefrontal white matter volume is disproportionately larger in humans than in other primates. Nat. Neurosci. 8, 242–252.

Semendeferi, K., Lu, A., Schenker, N., Damasio, H., 2002. Humans and great apes share a large frontal cortex. Nat. Neurosci. 5, 272–276.

Shanahan, M., Bingman, V.P., Shimizu, T., Güntürkün, O., 2013. Large-scale network organization in the avian forebrain: a connectivity matrix and theoretical analysis. Front. Comput. Neurosci. 7, 89.

Sherwood, C., Stimpson, C.D., Raghanti, M.A., Wildman, D.E., Uddin, M., Grossman, L.I., Goodman, M., Redmond, J.C., Bonar, C.J., Erwin, J.M., Hof, P.R., 2006. Evolution of increased glia-neuron ratios in the human frontal cortex. Proc. Natl. Acad. Sci. U.S.A. 103, 13606–13611.

Smaers, J.B., Steele, J., Case, C.R., Cowper, A., Amunts, K., Zilles, K., 2011. Primate prefrontal cortex evolution: human brains are the extreme of a lateralized ape trend. Brain Behav. Evol. 77, 67–78.

Somel, M., Xiling, L., Khaitovich, P., 2013. Human brain evolution: transcripts, metabolites and their regulators. Nat. Rev. Neurosci. 14, 112–127.

Stephan, H., Andy, O., 1969. Quantitative comparative neuroanatomy of primates: an attempt at a phylogenetic interpretation. Ann. N.Y. Acad. Sci. 167, 370–387.

Stephan, H., Frahm, H., Baron, G., 1981. New and revised data on volumes of brain structures in insectivores and primates. Folia Primatol. 35, 1–29.

Susman, R.L., 1998. Hand function and tool behavior in early hominids. J. Hum. Evol. 35, 23–46.

Tsai, P.S., Kaufhold, J.P., Blnder, P., Friedman, B., Frew, P.J., Karten, H.J., Lyden, P.D., Kleinfeld, D., 2009. Correlations of neuronal and microvascular densities in murine cortex revealed by direct counting and colocalization of nuclei and vessels. J. Neurosci. 29, 14553–14570.

Uddin, M., Wildman, D.E., Liu, G., Grossman, L.I., Goodman, M., 2004. Sister grouping of chimpanzees and humans as revealed by genome-wide phylogenetic analysis of brain gene expression profiles. Proc. Natl. Acad. Sci. U.S.A. 101, 2957–2962.

Walloe, S., Eriksen, N., Torben, D., Pakkenberg, B., 2010. A neurological comparative study of the harp seal (*Pagophilus groenlandicus*) and harbor porpoise (*Phocoena phocoena*) brain. Anat. Rec. 293, 2129–2135.

Wrangham, R.W., 2009. Catching Fire: How Cooking Made Us Humans. Basic Books, New York.

Zink, K.D., Lieberman, D.E., 2016. Impact of meat and lower Paleolithic food processing techniques on chewing in humans. Nature 531, 500–503.

The Timing of Brain Maturation, Early Experience, and the Human Social Niche

B.L. Finlay, R. Uchiyama

Cornell University, Ithaca, NY, United States

34.1 Introduction

Descriptions of maturation of brain and behavior in human infants and children have been multiplying in spectacular fashion along with our ability to acquire a massive corpora of data, but our mechanistic understanding of the relationship of emergence of behavior wit neural maturation remains comparatively meager. Though modern imaging techniques generate increasingly more accurate renditions of the state of the brain, the translation of particular changes in brain measurements to the emergence or stepwise maturation of specific computational or behavioral capacities remains largely unrealized. Concentrating on humans and restricting coverage arbitrarily to approximately the last 5 years, we find documented the changing volume of every structural component, from brain subdivisions like cortex and brainstem (Choe et al., 2013; Ge et al., 2015; Gilmore et al., 2012; Kuklisova-Murgasova et al., 2011; Makropoulos et al., 2016; Oishi et al., 2013; Raznahan et al., 2014; Shiraishi et al., 2015; Tate et al., 2015); changing morphology, as in gyral and sulcal development (Habas et al., 2012; Rajagopalan et al., 2011; Wierenga et al., 2013; Zilles et al., 2013); the volume and arrangement of multiple cellular constituents, such as myelin, "gray" and "white" matter (Li et al., 2015; Miller et al., 2012; Raznahan et al., 2011, 2012; Simmonds et al., 2013); growth of tracts (Takahashi et al., 2012, 2014; Walker et al., 2016); and metabolism (Bluml et al., 2013); "functional networks" (Damaraju et al., 2014; Gao et al., 2015) as well as an increasing library of gene and gene product expression (Huang et al., 2013; Kang et al., 2011; Miller et al., 2014; Silbereis et al., 2016). In parallel, remarkable inventories of early infant behavior have been gathered, including instruments such as the MacArthur—Bates Communicative Development Inventory (CDI) as well as studies, particularly in the language domain (Bornstein et al., 2015; Lieven and Stoll, 2013). Some aspects of the environment of infants can now be gathered and to a degree, catalogued (Roy et al., 2015). Yet, with all these inventories, even basic questions such as "What are the features of brain maturational state necessary and sufficient to allow language learning?" remain unanswered. This is no insult to the researchers in this field, as the possibilities for experimentation in most primates, and particularly humans, is limited, but there are multiple ways to progress toward causal explanation without direct experimentation (Gu et al., 2015; Hedden et al., 2016). In the present overview, we will review the possibilities and recurring pitfalls of comparative developmental work as a solution to this central developmental question. Next, we will describe work we have performed looking at the comparative timing of brain development, focusing first on comparison of the mechanisms involved in basic neural development and brain growth. Finally, we will extend the close analysis of neural maturation to contextualize early life history events such as weaning, early locomotion, and adolescence, placing primates in the context of general mammalian development.

The central problem, universally acknowledged, is covariation of gradual change in most of the physical parameters of brain maturation, matched against mixtures of stepwise and gradual changes in behavioral capacities. Close attention to any inflections, accelerations, bumps or dips that can be detected in the rate of progress in any physical measure (synaptogenesis, myelination, gene expression, and so forth) as potential indicators of organizational changes in the physical substrate, is a first-stage approach to the covariation problem (Amunts et al., 2003; Goldman-Rakic, 1987; Scheibel, 1993). For example, making the assumption that "Broca's area" and "Wernicke's area" are the critical regions for language competence, researchers looked for stepwise changes in volume or bursts of synaptogenesis or myelination in those regions associated with language acquisition milestones (Brauer et al., 2011; Leroy et al., 2011; Simonds and Scheibel, 1989) or development of tracts between language-associated regions (Brauer et al., 2011), though whether language development was illuminated by such explanations was debated (for example, McMurray, 2007). Although the integration of all the comprehensive catalogues described at the outset has hardly been attempted, the search for stepwise indicators in brains or cognitive stages has not proved very fruitful. Various measures of gene expression, which might have been imagined to reveal stepwise changes underlying levels of functional maturation, show essentially uninterpretable complexity (eg, London et al., 2009). Much as the specification of a punctate location or unique structure in the body and brain typically appears to be specified by the relative expression of tens to thousands of genes (Kiecker and Lumsden, 2012; Lumsden and Krumlauf, 1996), so far, the specification of a unique maturational time point seems likely to have much the same description, an otherwise unremarkable intersection of multiple gene profiles each changing gradually over time.

Comparative approaches can unravel this knot. The possibilities for comparison are large, but three are central: individual variation, the variation produced by disease or developmental disorder, or variation between species. We will concentrate on brain evolution, the differences between species, as a key way of illuminating what differences in the timing of brain construction influence eventual brain structure and eventual behavioral capabilities. For many years, we have been amassing a database of the timing of neurodevelopmental and behavioral events in mammals, beginning in early neurogenesis, extending to as many morphological and functional developmental events as can be measured, and presently extending to perinatal and postnatal behavior, and early life history (see Relevant Websites) (Clancy et al., 2000, 2001, 2007; Darlington et al., 1999; Finlay et al., 2005; Finlay and Uchiyama, 2015; Workman et al., 2013). This research has enabled us to begin to address just how developmental events should be compared, a prerequisite to identifying which aspects of neural and behavioral development are species-general, versus species-unique.

34.1.1 "Allometrically Expected"

A traditional problem in brain allometry illustrates the problems of comparing species "fairly". If the volume of neocortex (isocortex henceforth, as the most appropriate nomenclature) is visually compared across a selection of mammalian species (Fig. 34.1, top), the fact that the human brain has a disproportionately large cortex is obvious. With this kind of comparison, cortex volume would appear to be an object of special evolutionary selection in primates, and in humans particularly. However, if either the volume or the number of neurons in the cortex is represented on a logarithmic scale, it is clear that the human isocortex is exactly the size it "should" be, following its allometric position (Fig. 34.1, bottom; from Reep et al., 2007). The entire human brain is large compared to other primates, with respect to body size, but given this large brain size, each part falls onto its "expected" position. The slope "a" representing the rate of increase of each brain component "y" with respect to whole brain volume "x" (or any index of brain volume or neuron number) is different, as given by the logarithmic transformation of the basic allometric equation ($\log y = a \log x + \log k$). The vertebrate brain itself has "negative allometry" with respect to body mass, so as the body enlarges, brains become a progressively smaller component of whole body mass (Jerison, 1973). With respect to brain mass, the isocortex has "positive allometry," so larger mammalian brains become progressively more composed of cortex, ranging from under 20% in relative volume in small shrews and rodents to over 80% in humans (Finlay and Darlington, 1995; Hofman, 1989). This point is driven home by the visual comparison of the proportion of cortex in humans versus the one-humped camel (*Camelus dromedarius*), which is roughly equivalent in mass to the human brain (Fig. 34.1, right) though camel bodies are much larger. The relative proportion of isocortex in those several brains that are absolutely larger in mass than the human brain, including some cetaceans and ungulates, continues the allometric equation, so that they have both absolutely and relatively more cortex.

Much energy has gone into a debate about whether a specific region of cortex, the prefrontal cortex, is "allometrically unexpected" in humans, which should hopefully illuminate why getting allometry right matters. Every brain region (in this case, a collection of cortical areas) has its own exponent (slope in the log-transformed equation) for its change in relative volume compared to overall isocortex volume. In mammals, the prefrontal and parietal cortex regions have an exponent that is larger than the cortex's overall exponent, or the exponent of primary sensory regions, "positive allometry" (Jerison, 1997). The question under debate is whether frontal cortex in humans is even larger still than would be expected from its already high positive allometry (Barton and Venditti, 2013; Chaplin et al., 2013; Passingham and Smaers, 2014; Semendeferi et al., 2002). But, why should anyone care about this issue? Although issues of measurement, correct statistical

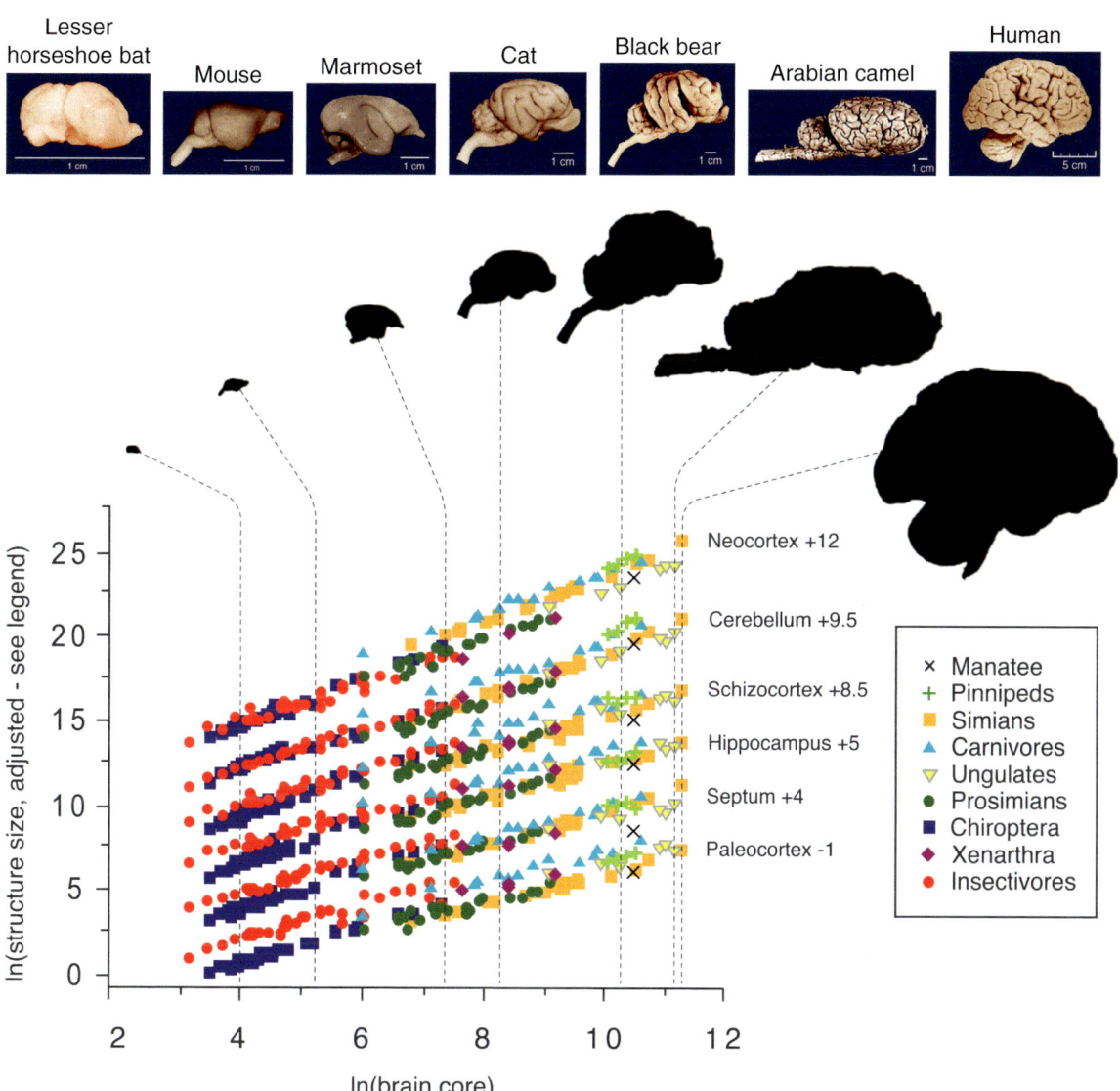

FIGURE 34.1   Comparisons of brain volume representations. Top: Brain images, normalized to show varying sizes of components. Species: Lesser horseshoe bat, (*Rhinolophus hipposideros*); mouse (*Mus musculus*); common marmoset (*Callithrix jacchus*); domestic cat (*Felis catusI*); American black bear (*Ursus americanus*); one-humped camel (*Camelus dromedaries*); human (*Homo sapiens*) Images courtesy of University of Wisconsin and Michigan State Comparative Mammalian Brain Collections (Neurosciencelibrary website (see Relevant Websites) — This site supported by the National Institutes of Health and the National Science foundation). Middle: Same images, vignetted midsagittal view above, to show absolute size. Image sources as above. Bottom: Brain component sizes from 160 mammalian species including the individual species shown above, both scales natural log scales, to demonstrate predictable scaling of brain components with respect to a brain volume index. The sizes of the 6 brain structures in the 160 species from the 9 keyed taxonomic groups are plotted relative to "brain core volume" (medulla, mesencephalon, diencephalon, and striatum). The six individual structure volumes have been adjusted by the indicated arbitrary constants to the right of each structure's name to separate the six scatterplots visually. *Data replotted from Reep, R., Darlington, R.B., Finlay, B.L., 2007. The limbic system in mammalian brain evolution. Brain Behav. Evol. 70, 57—70.*

procedure, and interpretation are all intertwined in such controversies, if researchers claim a region's volume is "allometrically unexpected" in a species of interest, they are claiming that it must have been the target of special selection, typically because of special importance of the function ascribed to that brain region in that species. Structures that change their volume according to

their allometric rules, even if they seem disproportionately large on a linear scale, by contrast, require no special explanation. If the entire brain has been under special selection for larger size in any species, all of the changes in the proportionality of its parts come "for free," presumably because of the lawful extension of the developmental mechanisms that produce neuron numbers and their associated volumes applied to every part (eg, Cahalane et al., 2014; Charvet and Finlay, 2014; Finlay and Darlington, 1995). So, the argument over allometry in the case of human prefrontal cortex is whether its developmental rules have been altered by natural selection to enhance the cognitive features associated with frontal cortex, such as cognitive control, or planning for the future, even more than they may have been enhanced by the already positive allometry of frontal cortex in mammals.

The idea of "allometrically expected" also applies to translations of developmental time from one species to another. The corpus of data is smaller, and some questions of how to compare time are rather less well worked out than the more traditional questions of energy metabolism or mass of structures, but basic allometric considerations apply. First, the appropriate coordinate system to represent time translations is a logarithmic, not a linear scale. Despite our everyday use of the 7x linear transform to represent "What is that in dog years?" such a transformation actually works poorly, overestimating the dog's relative age compared to humans in early development, and underestimating it in old age. Moreover, just as each part of the brain has its own relative rate of enlargement with respect to total brain size, each brain part has its own relative duration or rate of development with respect to the overall duration of that species' maturation (Workman et al., 2013).

To compare developmental schedules among animals, enough data must be collected to generate these allometric relationships from a number of relevant species. For example, if you wish to show that Broca's area is the subject of special selection in humans compared to a rhesus monkey, you cannot compare the relative size of a "control structure" such as primary visual cortex in the two species to "normalize" the comparison, see that the ratio of relative sizes of Broca's area is greater than that of the two primary visual cortices, and conclude that the size of Broca's area has been specially selected in humans. If Broca's area has "positive allometry" in primates, every contrast of a large and small brain will give this result. Rather, it is necessary to show that the size of Broca's area in humans exceeds its expected allometric position compared to Broca's area in other primates (Schoenemann, 2006).

Comparing relative developmental durations with an inappropriate "norming" procedure will produce the identical errors seen in comparing brain volumes inappropriately: you cannot, for example, compare the time from birth to adolescence in chimpanzees versus humans, see that the duration is longer in humans and conclude that human have been specially selected for a longer childhood, as the duration may be entirely predictable from body or brain size, or the position of birth with respect to brain and body maturation. We will

discuss in more detail various issues about how to represent or normalize developmental schedules with respect to each other. In addition to the rate and slope considerations discussed so far, an additional issue in such comparisons is what date in development represents "zero," the intercept or constant "k" in the allometric equation. Although the date of birth is often chosen as a natural zero, we will argue that the choice of birth as zero is often quite misleading, given the wide range of brain maturational states at birth in mammals and primates.

34.1.2 Human Exceptionalism

Much research into comparative brain organization and comparative development is organized to answer the question "What makes humans unique?" A perusal of the literature on this subject will soon return the answer "everything." A few such claims and counterclaims have included (but are in no way limited to) the presence of Von Economo cells in the cortex as related to human sociality (Allman et al., 2010; but see Niewenhuys, 2012), the depth of layer IV in the visual cortex (Bryant et al., 2012), the size of the prefrontal cortex (cited previously), the descended larynx in relation to language (Lieberman, 1984; but see Fitch and Reby, 2001), the duration of childhood (Locke and Bogin, 2006), and so forth. On closer examination, many of these claims prove to be simply hopeful, without the appropriate comparative database. In much of this literature, a human feature of some kind is compared to that of another species, chimpanzee, rhesus, or rodent. Any disproportionate difference is accorded human uniqueness, with causal significance in human evolution usually implied. For example, the rate of cerebral growth in humans compared to chimpanzees and macaques (Sakai et al., 2013; Fig. 34.2) was described this way:

> ...the rapid increase in cerebral total volume and proportional dynamic change in the cerebral tissue in humans during early infancy, when white matter volume increases dramatically, did not occur in chimpanzees. A dynamic reorganization of cerebral tissues of the brain during early infancy, driven mainly by enhancement of neuronal connectivity, is likely to have emerged in the human lineage after the split between humans and chimpanzees and to have promoted the increase in brain volume in humans.

Replotting these data in a common framework showed near perfect prediction of the human rate of brain growth from parameters determined from other primates and mammals generally, directly related to the time required to generate brains of particular mass (Finlay and Workman, 2013). The apparently exceptional "rapid increase" in early human brain volume is illusory, due to the combined effects of the linear scale and the relative placement of human birth with respect

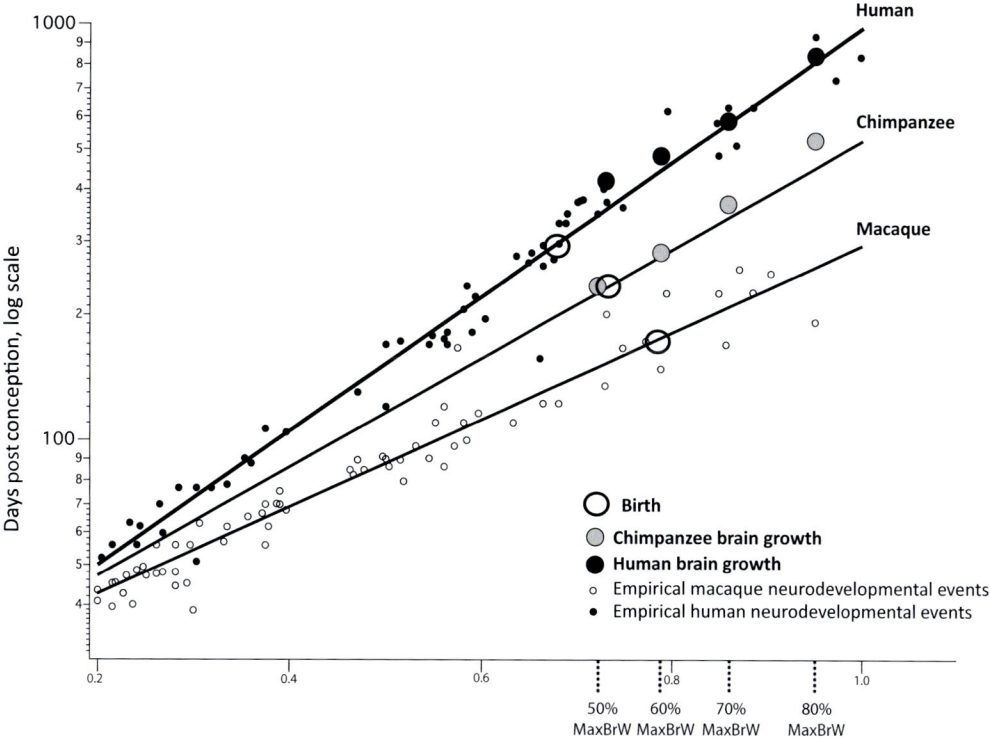

Event Scale: best ordering of 271 neurodevelopmental events across 18 species

FIGURE 34.2   Comparison of postnatal and prenatal brain growth in the rhesus macaque, *Macaca mulatta*; chimpanzee (*Pan troglodytes*), and human (*Homo sapiens*). *Data replotted from Sakai, T., Matsui, M., Mikami, A., Malkova, L., Hamada, Y., Tomonaga, M., et al., 2013. Developmental patterns of chimpanzee cerebral tissues provide important clues for understanding the remarkable enlargement of the human brain. Proc. R. Soc. B Biol. Sci. 280, 753–757 to show postnatal volumetric growth of the monkey, chimpanzee and human brains on linear scales of volume (y-axis) and years (x-axis). Volume change and birth date position (open circle) plotted against a model overall of neural maturation in the same species from Workman, A.D., Charvet, C.J., Clancy, B., Darlington, R.B., Finlay, B.L., 2013. Modeling transformations of neurodevelopmental sequences across mammalian species. J. Neurosci. 33, 7368–7383. The "Event Scale," the x-axis, is a set of 271 normalized neural maturational events drawn from 18 species, including primates and consists of such items as neurogenesis in multiple structures, myelination, and brain growth. The Y-axis, days postconception, allows the day each species reaches each neural maturational milestone to be represented.*

to the common progress of neural maturation, a feature of human development we will return to.

There is absolutely no doubt that humans are unique in many respects, but to understand human evolution, the impulse to label every apparent deviation as singular should be resisted vigorously. Systematic determination of which human features represent the normal unfolding of a vertebrate, mammalian or primate plan, and which are actually unexpected, is the only way to understand our basic biology.

34.1.3 Life History

The comparative organization of life history has been the subject of much investigation, both for comparative physiology and anatomy, and in primatology (Stearns, 1992; Smith and Tompkins, 1995). Study of life history, as its name suggests, looks at the relative timing and organization of major life events, beginning from birth, and including such events as weaning or other parental provision of food, adolescence and reproductive

maturity, dispersal, menopause, and death. Though generally allometrically predictable, life history events are much more variable than the early neurodevelopmental events, with demonstrated species differences, niche, and effects of environmental variability pervasive. In this chapter, however, we will begin to bring these two literature and types of modeling closer together, to see if any aspects of life history event prediction are better informed or predicted by using the general time representation, data, and methodology of the "translating time" project in neural development. We would like to address issues about the expectedness or unpredictability of the pattern of childbirth and weaning in humans in this framework, as well as examine claims about how neural maturation actually intersects behavioral and cognitive development in humans, and whether there is evidence of deviation of any "behavioral modules," critical periods, and the like. We will not settle these issues in this review but will begin the attempt to bring together primate life history with its neurodevelopment.

34.2 Comparative Approaches to Translating Time

34.2.1 Brief Review of Translating Time Methodology

Over the past 20 years, we initiated and then progressively expanded a database and methodology to compare the progress of neural development across species, which we call "translating time." The fundamental point of this work is to describe a mammalian "Bauplan" for neural development, and in doing so, identify deviations from this plan that might mark taxon- or species-specific alterations corresponding to evolutionary adaptations, a version of SJ Gould's heterochrony (Gould, 1977). Alternatively, the same information can be used to identify deviations related to individual differences or developmental disorders. We will briefly review the historical development of the model to highlight the changing aspects of development it revealed as taxonomic coverage expanded and the number of developmental events increased.

In the first analysis, only seven commonly employed laboratory species (including four rodents, a marsupial, cat and monkey) and peak day of neurogenesis were gathered for as many neural structures and cell groups that we could locate, from spinal cord to telencephalon (Finlay and Darlington, 1995). Quite surprisingly, the fit of a minimal two-factor model (the factors were "species" and "structure") to these empirical data was very high, and the addition of an additional term to slow the rate of marsupial development brought the fit of data to model to 0.98. We also identified systematic deviations in neurogenesis onset in limbic and olfactory structures in rodents versus the remaining species but did not yet add these terms to the model. These first observations of this set of taxonomically scattered species suggested that variations in the order of neurogenesis were relatively rare in mammals but that wholesale alterations in the rate of maturation were possible between taxonomic groups.

In further work, we added many more species, including humans and many more types of neuroembryological events. The statistical models employed continued to evolve as well. Expanding the set of noneutherian (marsupial) mammals, we confirmed the relative slowing of maturational rate in six further marsupial species and quantified deviations between them (Darlington et al., 1999). One possibility to limit the scope of the initial analysis was the hypothesis that the basic generation of neurons was predictable and clock like but that variations associated with species-specific adaptations might be abundant in subsequent organizational events, such as the initiation and elaboration of axonal and dendritic connectivity. Surprisingly again, however,

addition of neuroembryological events such as axon extension, synaptogenesis, and early aspects of myelination, which extended the time frame of the model into the early neonatal period in humans, did not make the model noisier but in fact improved the fit of data to model (Clancy et al., 2001). The predictability of neurogenesis was no different from the timing of the elaboration of connectivity (the events measured occur just up to the onset of "real-world" function). The addition of new species now allowed us to add a "limbic factor" and a "cortical factor," which for primates alone, moved the predicted days of olfactory-limbic neurogenesis forward (correlated with relatively smaller size in adulthood) but protractedcortical development (the longer period of neurogenesis making the cortex relatively larger) (Clancy et al., 2007). Thus, the unusual conservation of neurodevelopmental schedules observed at first was not confined to neurogenesis but to every measured event in neurodevelopment reaching until about the time of birth in humans. One important case of "heterochrony" of neurodevelopmental events was confirmed for the limbic/isocortex factor, though these kinds of changes seemed rare.

In the most recent analysis (Workman et al., 2013), the database was considerably extended, to 18 species, and new classes of data were added, including introduction of continuous processes like changes in brain volume and myelination (by digitization), and by adding many more postnatal events, extending to approximately 3 years postnatal in humans. The increased data volume allowed us to fit a single "event scale" to all the data, the best order, and interval relationship of our now 271 distinct neurodevelopmental events (x-axis, Fig. 34.3), and describe the speed of progress of each individual species through these events as a regression equation, in days (y-axis), identifying sources of variation from both taxonomic groups and individual species. The differences in each slope may be thought of as differences in maturational rate, with steeper slopes associated with slower progress through maturational stages in absolute time: the mouse takes only about 30 days to execute the 271 events, whereas the human takes 1000 days, the human generating greater numbers of neurons and volumes of connectivity per event. The fit of model results to empirically measured results is quite close, 0.9929, including only two interaction terms, a delay in corticogenesis in primates, marsupials, and carnivores associated with a larger isocortex in these species, and a delay in neurogenesis in the retina of the nocturnal cat, associated with greater numbers of rods and rod-associated neurons. Interestingly, the timing of birth appears to be quite decoupled from neural development (Fig. 34.4). For example, cortical and cerebellar neurogenesis is ongoing at birth in some rodents but completely concluded in primates.

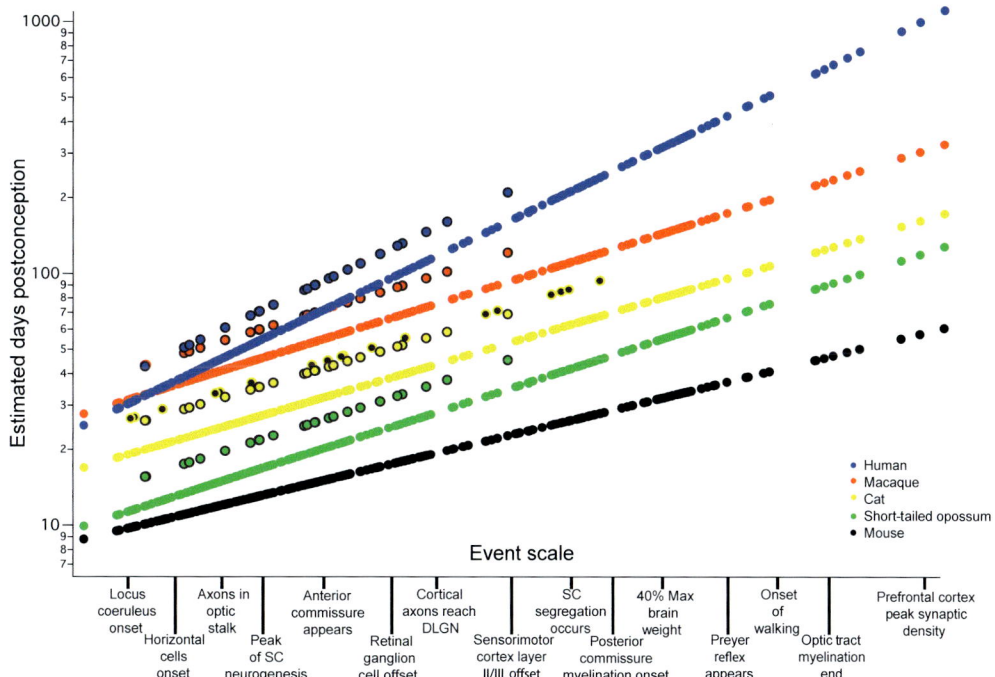

FIGURE 34.3 Translating time model. Predicted developmental schedules for human (*blue*), macaque (*red*), cat (*yellow*), short-tailed opossum (*green*), and mouse (*black*), selected from the modeled 18 species to illustrate the full range of developmental durations. The x-axis, the "Event Scale," is a common ordering of developmental events across all species and shows a subset of the 271 observed events. This scale ranges from 0 to 1, but in this case, event scale numerical values are replaced by these example events. The y-axis, log scale, is the estimated date of occurrence of each event in each species, measured from conception. Also represented on this graph are interaction terms for corticogenesis and retinogenesis, with interaction terms always associated with individual species. The *parallel lines* for a subset of events in four of the species (*black bordered circles* for human, macaque, cat, and possum) represent delays in cortical neurogenesis with respect to their time of occurrence in the rodent and rabbit. In the cat, a second *parallel line* can be seen representing the delay of retinal neurogenesis, (*yellow circle* with a *black dot*). *SC*, superior colliculus; *DLGN*, dorsal lateral geniculate nucleus. *Reproduced with permission, Figure 34.4 in Workman, A.D., Charvet, C.J., Clancy, B., Darlington, R.B., Finlay, B.L., 2013. Modeling transformations of neurodevelopmental sequences across mammalian species. J. Neurosci. 33, 7368–7383.*

Empirical support for this claim of highly conserved and predictable, but nonlinear, neurodevelopmental schedule can be found in multiple independent sources. Passingham inferred that the curves of change in brain volume across eutherian mammalian species when measured from conception were essentially superimposable, offset of growth being the only distinction (1985). Within the visual system, the timing of various developmental events could be predicted from an "anchor event," in this case, eye opening. Eye opening, which behaves as if it is a "neural event," happens at a time specified by neurodevelopmental state, whether that time occurs in utero, in an underground burrow, or even occasionally, synchronized with birth (Dreher and Robinson, 1988). Halley (2016), in a much larger and more closely measured data set of initial brain volumes, confirmed the same. More surprising still, in a study of multiple mammals ranging from laboratory rodents to antelopes, the time of the first unsupported step was similarly predictable from adult brain mass, with an additional factor for plantigrade standing position (Fig. 34.5; Garwicz et al., 2009). Therefore, precocial ungulates such as sheep and elk, who must be ready to run by birth, have accomplished this (in evolutionary time) by extending gestation and delaying birth to match conserved parameters of brain development, not by selectively advancing the general rate of brain maturation or the specific ability to walk.

The significance of a fixed neurodevelopmental program for understanding primate life history and its potential adaptations has multiple aspects. It is important to recall that in this model, only events in brain and some early behavioral capacities are included—no measures of body or organ maturation or volume, or interactional, life history events such as birth or weaning are part of the data set. In the present paper, we are outlining the first steps toward integration of somatic development and life histories with these observations.

First, recalling the idea of "allometrically expected," nothing as yet appears as yet to be unexpected about the duration, rate, or deviations from linearity in brain and behavioral development for primates in general or for humans in particular, as predicted from the brain sizes of this group. Humans have the duration and rate of neural development appropriate to produce a brain of typical human size. Therefore, though it is

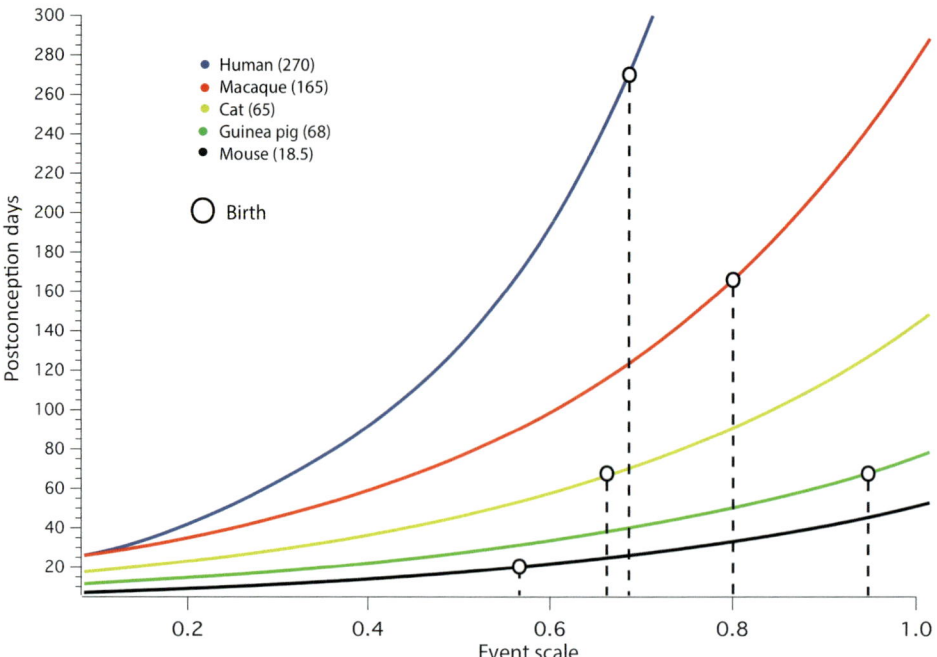

FIGURE 34.4   Variability in position of birth with respect to neural maturation. The position of birth (*open circle*, listed after each species name) for six placental mammals relative to the event scale (x-axis); the age of each mammal in PC days can be read for birth (or any event scale value) on the y-axis. The five placental mammals are chosen to represent close to the full range of the data set and include one highly precocial mammal, the guinea pig. For an example, in the mouse at birth cortical neurogenesis is still underway and synaptogenesis in the forebrain is only beginning, whereas in the guinea pig at birth, cortical neurogenesis, cortical cell migration, and basic axonogenesis are entirely complete, and the point of peak synaptic density has passed. Human, *blue*; macaque, *red*; cat, *yellow*; guinea pig, *green*; and mouse, *black*. *Reproduced with permission from Figure 34.9 in Workman, A.D., Charvet, C.J., Clancy, B., Darlington, R.B., Finlay, B.L., 2013. Modeling transformations of neurodevelopmental sequences across mammalian species. J. Neurosci. 33, 7368–7383.*

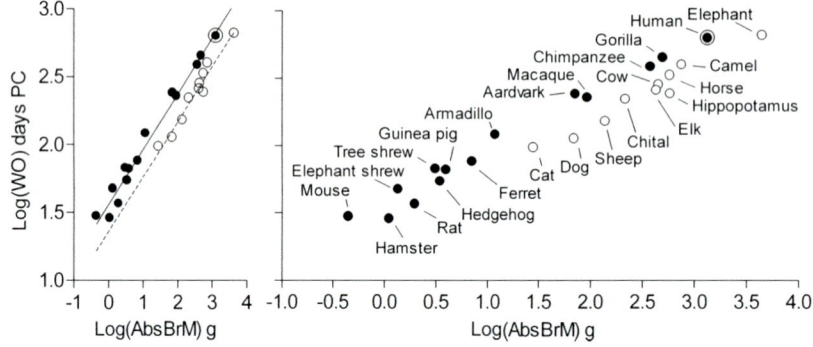

FIGURE 34.5   Absolute adult brain mass and hindlimb standing position as predictors of time to walking onset. Left, time to walking onset, log WO, as a function of absolute adult brain mass log(AbsBrM) and hindlimb standing position. *Solid symbols* and *solid regression line* represent species that can assume a plantigrade standing position and reach this milestone relatively earlier (most ungulates), *open symbols* represent those which cannot. Relative brain mass is an excellent predictor of overall maturational duration (Fig. 34.2). Right, same graph, but with x-axis expanded to allow species identification. *Reprinted with permission from Garwicz, M., Christensson, M., Psouni, E., 2009. A unifying model for timing of walking onset in humans and other mammals. Proc. Natl. Acad. Sci. 106, 21889–21893.*

entirely accurate to say that humans have the longest period of brain development of primates, as this parameter is virtually perfectly correlated with brain size (Workman et al., 2013), the claim that humans have been specially selected for a long developmental duration is unjustified—it comes "free" for selection for a large brain, or any large brain part (Barrickman et al., 2008; Charvet and Finlay, 2012). Theoretically, one could

argue that a large brain is a necessary by-product of selection for extended development.

Second, the timing of birth is quite uncorrelated with neural maturation (Fig. 34.4). Some rodents (mice and rats) are born at a stage of maturation equivalent to a human of 4–5 months gestation, whereas others like the guinea pig correspond to a human of approximately 3 years postnatal. Primates, in general, are born at a middle stage of neural maturation, not so mature as the precocial ungulates nor as immature as a number of rodents. Primates also show a fairly wide range of (neural) maturational states at birth, rhesus macaques relatively mature at birth, chimpanzees intermediate, and humans least mature (Fig. 34.2). There are a wide number of factors entering into the allocation of resources to brain versus somatic tissue, and for allowing early maturation and its nutritional source to proceed mostly in utero or mostly outside, via lactation, which we cannot discuss fully here but will list a few to illustrate the range of phenomena to consider. Depending on eventual brain size, somatic development may be delayed to prioritize energy allocation to brain (Halley, 2016); the animal may have adopted the strategy of producing a large number of embryos with short gestation, or a single one requiring a larger investment (Charnov and Downhower, 1995); the possibility of alloparenting or biparental care appears also to be a factor, allowing greater investment in brain volume (Isler and van Schaik, 2009, 2012). Finally, as has been discussed extensively for primates, the physical consideration of the size of the pelvis versus size of the cranium will be material in the timing of birth (reviewed in Trevathan, 2010). The information we have gathered on the stability of the timing of neural maturation adds a useful fixed point to this multivariate problem, as neural mass appears to be produced at a very stable rate in eutherian mammals. Any particular duration from conception to birth (if adult brain size is known) is associated with a predictable neural maturational state.

After reviewing one final aspect of the interaction of brain size and duration of development on cortical maturation, to place the state of the maturation of the cortex at birth in humans in a more general context, we will ask what is known about the significance of the particular state of the brain at birth for the types of behavioral capacities available to the animal, and particularly for learning specializations. One particular problem we will underline, and return to later, is that in most studies which compare life history events across species, time "zero" is quite naturally taken to be birth, but for many purposes, it is a very poor zero to use, as it is associated with a wide range of maturational states whose significance is sometimes assumed to be known, but often is not. Time from conception, appropriately modeled, can be a much more stable and informative way to compare neural maturational stages and aspects of life history across species.

34.2.2 Crossing Gradients in the Cortex and Their Phylogenetic Significance

We need to know what the brain is like around birth, in humans and in other primates, and in other species employed as "animal models" for research, to determine the significance of placement of birth and other early events like weaning, or establishment of early communication. For primates in general, and humans particularly, the maturational state of the isocortex has always been given particular attention, though we will argue later that this excess attention to the cortex may be misplaced for understanding species-specific adaptations in evolution. The following description of cortical development is intended to give an overview of the maturational state of the cortex around birth in terms of features of organization already formed and yet to form and mechanisms available for further maturation of the cortex. We assume the reader is familiar with the basic mechanism of the inside-out generation of the cortical layers (Rakic, 1974) within the cortical column and will focus on temporal features of this organizational feature of the cortex as it varies between species. For reasons of space, we must unfortunately neglect the rest of the brain, but the reader is directed to Workman et al. (2013) for a comprehensive catalogue of the maturational state of other brain components.

The isocortex is formed in the confluence of at least two maturational gradients, one associated with neurogenesis of the cortex itself, and the second deriving from thalamic maturation and the innervation of the cortex by the thalamus (Fig. 34.6) (reviewed in Cahalane et al., 2014; Charvet and Finlay, 2014; Finlay and Uchiyama, 2015). In all species yet described, including marsupials, rodents, carnivores, and primates, generation of neural stem cells proceeds uniformly over the cortical primordium, extending itself in surface area first, then depth, and finally in the addition of a second rank of undifferentiated cells, the subventricular zone, in the brains which will eventually be the largest (Dehay et al., 2015). The production of the first neurons (terminal neurogenesis) begins first in the rostrolateral cortex, marked by cells leaving the ventricular zone and migrating increasing distances to form the cortical plate, formed in the well-described "inside-out" relationship of neuronal "birthdate" to cortical laminar position (Rakic, 1974). Neurogenesis ceases altogether in the rostrolateral cortex first, with cessation progressing caudomedially. In the smallest brains, although the gradient can be observed, its morphological and organizational consequences are minor (Caviness et al., 1995; Charvet et al., 2014). In large brains, however, the gradient amplifies the relative time over which the anterior versus the posterior cortex is formed, giving the posterior cortex a greatly extended period of stem cell genesis and neurogenesis compared to anterior cortex (Fig. 34.7). For a within-species comparison, cortical neurogenesis begins at approximately embryonic day 38–40 throughout the isocortex in rhesus monkey (Rakic, 2002). However, neurogenesis in the anterior cortex (ie, motor or frontal

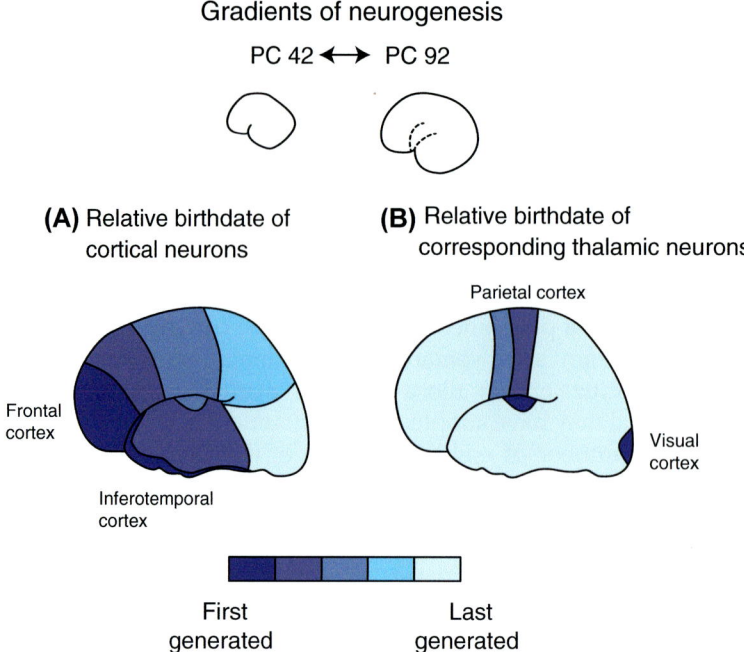

FIGURE 34.6   Contrasting gradients of neurogenesis of cortical neurogenesis of the isocortex versus thalamic innervation. Maturational gradients in early postnatal development superimposed on a lateral view of the human cortex; the relative size of the cortex during this period is represented above both diagrams. (A) Neurogenesis of cortical neurons begins at the rostrolateral margin of the cortex and proceeds posteriorly through parietal cortex to primary visual cortex, framing a period of genesis of about 50 days (PC day 42 to PC 92). (B) Neurogenesis of corresponding thalamic neurons begins with the medial geniculate body (auditory cortex), the lateral geniculate body (visual cortex), and the ventrobasal complex (somatosensory cortex), followed by neurons that innervate motor cortex. The last thalamic neurons to be produced are located in the nuclei that innervate the frontal, parietal, and inferotemporal cortex.

cortex) ends before embryonic day 80, whereas posteriorly, neurogenesis continues to embryonic day 100 in the somatosensory cortex and to embryonic day 102 in area V1 at the posterior pole of the cortex (Rakic, 2002). Note that this gradient is the opposite from the gradient usually presumed to be the case for capacities dependent on the cortex, where sensory abilities are thought to arise first in posterior occipital and parietal cortex, while frontal organizational and decision-making abilities appear later. We will resolve this apparent contradiction shortly.

This gradient becomes progressively more pronounced in the largest cortices and has large consequences (Fig. 34.8): there are many fewer neurons "per column" in frontal cortex, though the convergence of other cortical areas on frontal cortex expands the neuropil and renders the cortex approximately the same depth per column across the cortical surface(Cahalane et al., 2012). Posteriorly, cortex is neuron rich, particularly in primary sensory areas, in which additional developmental processes beyond neurogenesis allow the survival of excess neurons in layer IV, the principal thalamorecipient region (Finlay and Slattery, 1983). The gradient of neurogenesis may predispose the frontal cortex and the anterior parts of the parietal cortex to their "association" role. Neurons begin to spin out axonal processes even while migrating,

and axodendritic development commences as soon as neurons are in place (Schwartz et al., 1991). As thalamic and cortical neurons have been shown to compete for sites on the same cortical neurons, the establishment of principally intracortical connectivity in frontal cortex before any thalamic input is able to arrive may predispose it to its "association" role (Windrem et al., 1988; Finlay, 1991). By contrast, the arrival of thalamic input (from the lateral geniculate nucleus) in primary visual cortex actually precedes the arrival of migrating thalamorecipient neurons in the cortical plate of the primary visual cortex, which may give thalamic axons some privilege in the competition for terminal space (Allendoerfer and Shatz, 1994).

The second gradient (Fig. 34.6, right), produced by the thalamic input to the cortex, is organized quite differently. Of all the thalamic nuclei, the primary sensory nuclei, the lateral geniculate for visual input, the medial geniculate for auditory input, and the ventrobasal (also termed ventroposteriolateral, VPL) are generated earliest, migrate to their thalamic positions, and begin to form axonal processes immediately (Bayer and Altman, 1991; Rakic, 1977). In small-brained animals whose cortices are formed over a period of days, this difference, and its interaction with the cortical gradient is not particularly striking. In larger-brained

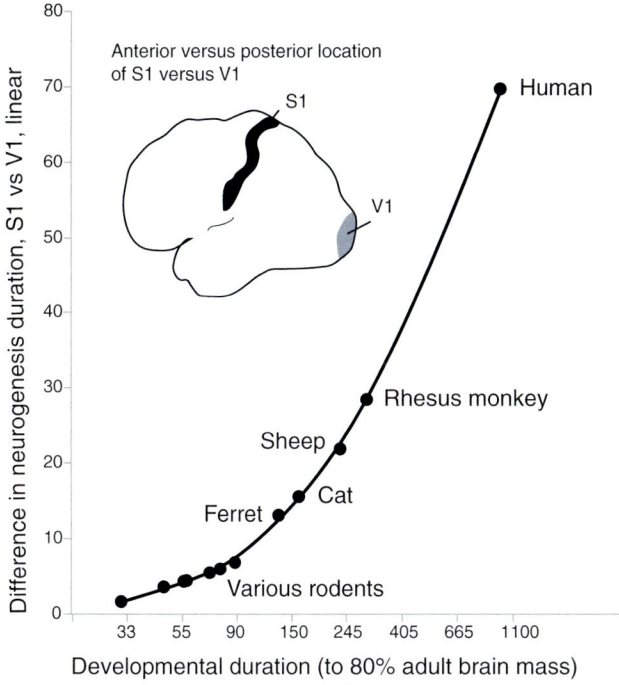

FIGURE 34.7 Difference in duration of neurogenesis of neurons contributing to anterior versus posterior regions of cortex across species. The difference in terminal neurogenesis between the anterior (ie, primary somatosensory cortex, S1) and posterior cortex (ie, primary visual cortex, V1) is plotted against the natural-logged values (ln) of developmental duration in 11 mammalian species (human, macaque, sheep, cat, ferret, rat, gerbil, hamster, mouse). This extended duration corresponds to an increasing relative number of neurons per cortical column in posterior cortex. *Replotted from Charvet, C.J., Cahalane, D.J., Finlay, B.L., 2014. Systematic, cross-cortex variation in neuron numbers in rodents and primates. Cereb. Cortex 25, 147—160.*

animals, whose cortices are formed over weeks, the thalamic input can reach the cortical plate before the migrating neuroblasts destined to become the stellate cells of layer IV arrive, and the thalamic axons remain under the cortical plate for a "waiting period" until they are signaled to enter—their target cells destined to become layer IV actually migrate through their eventual thalamic innervating axons on the way to the cortical plate. These thalamic axons exert a trophic, or perhaps a specifying, effect on these migrating neuroblasts (Oishi et al., 2016): the number of neurons in layer IV "per cortical column" is higher in primary sensory cortical areas. Across cortical areas, the differentiation of layer IV roughly matches the volume of thalamic input (Finlay, 1992). If the thalamic input is removed, by early damage to a particular thalamic nucleus, or by cytotoxic agents, the amount of early neuron death in the appropriate cortex is increased, and the relative number of neurons in layer IV is reduced (Woo and Finlay, 1996; Woo et al., 1996). Other thalamic nuclei, such as the multiple nuclei of the pulvinar, or the anterodorsal

group are generated much later, and innervate their parietal and frontal targets later (Altman and Bayer, 1988a,b,c; Chalfin et al., 2007). These nuclei, however, have been studied in much less detail than the primary visual and somatosensory nuclei.

The further emergence of morphological differentiation and sensorimotor capacities follows the thalamic gradient of innervation (primary sensory areas out) rather than the original cortical gradient of cell generation (anterior to posterior) (Bates et al., 2002; Finlay and Uchiyama, 2015). It is important to realize, however, that "sensorimotor capacities" reflecting real-world information and action, versus the electrical activity (as realized in action potentials in axons, field potentials, and somewhat later, active connections arising from neuron-to-neuron junctions and synapses) are not the same. Self-organizational activity, as has been described systematically in the case of "retinal waves," which segregate projections by eye-of-origin in the lateral geniculate body, and perhaps in the cortex as well, begins before photoreceptors are generated or the eyes are open (Wong, 1999). Similar self-organization begins in the somatosensory system, for example, as the number and topology of vibrissae are established in their central representations (Senft and Woolsey, 1991; Brown et al., 1995). Regular, periodic bouts of motor activity begin soon after motoneuron synaptic connections are made, one of the earlier events in central nervous system development (Robertson et al., 1982). This self-organization, subcortical and cortical, generally begins as soon as neurons have differentiated and in time, merges with externally generated organizational activity. For example, the noncorresponding activity in the two eyes produced independent by "retinal waves" in the two retinas, which initially serves to segregate the projections from the two eyes in the thalamus and cortex, falls away just before the time when the eyes first regard the visual world and are replaced by real-world correlations which reintegrate information from places in the two retinas representing corresponding places in the external world (Wong, 1999). However, all of the events in neurogenesis and establishment of initial connectivity that have been described so far are early embryonic events, and in humans occur well before establishment of ex utero viability, graded over 3—5 months postconception. In rats, mice and hamsters, however, birth, the end of corticogenesis and the beginning of thalamocortical innervation occur at approximately the same time.

By the time of 6 months postconception, many basic somatomotor functions are established (though we make no assumption such functions would depend on the cortex), such as responsiveness to somatosensory stimulation, basic motor programs such as alternating limb movements and suckling. Interestingly, multiple

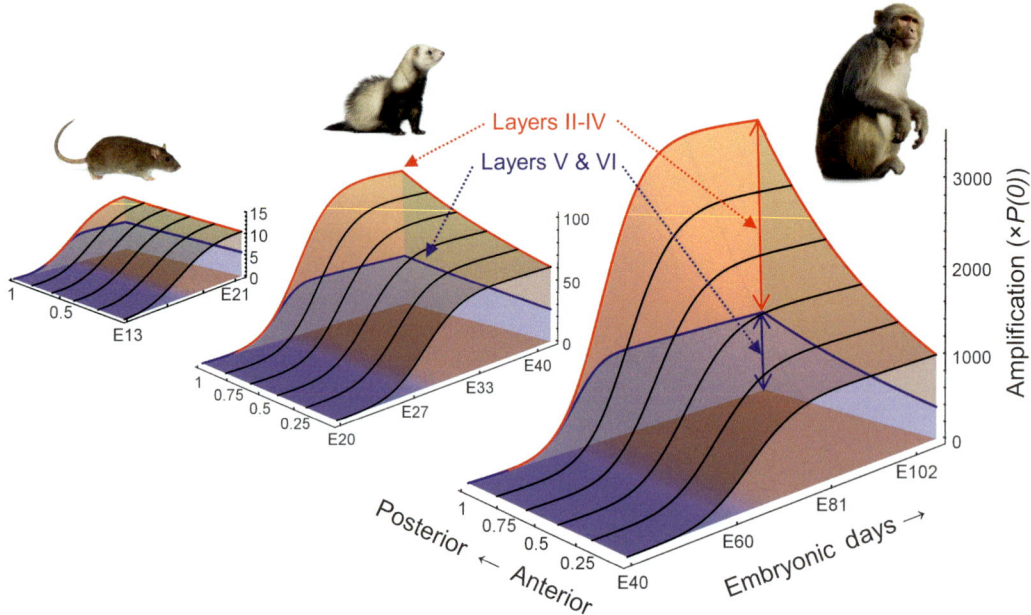

FIGURE 34.8   Model-predicted interspecies and intracortex differences in the timing, extent, and layer assignment of cortical neuron output. Shown are the predicted amounts of neuronal output (in terms of amplification of a unit precursor pool) across the anterior–posterior (spatial) axis of the cortex over the course of embryonic neurogenesis (time axis) for three different cortex adult volume values (1.0, similar to a rat; 1.75, similar to a ferret; 2.5, similar o a macaque monkey). The larger cortices have a longer developmental interval, produce orders of magnitude more neurons in total, and, in particular, have a greater complement of upper layer neurons. The anterior–posterior gradient in neuron number becomes more pronounced in larger cortices, and it is the upper layers that accommodate the greater proportion of the increasing quantities of neurons. *Rat and ferret images courtesy of iStockphoto/GlobalP; Macaque image courtesy of iStockphoto/JackF; From unpublished doctoral dissertation Cahalane, D.J., Charvet, C.J., Finlay, B.L., 2014. Modeling local and cross-species neuron number variations in the cerebral cortex as arising from a common mechanism. Proc. Natl. Acad. Sci. 111, 17642–17647.*

kinds of learning have been shown to occur in the prenatal period in humans, other primates and laboratory rodents, including habituation, associative learning, and reinforcement learning, including preferences for particular chemosensory and language prosody patterns (Alberts and Ronca, 1993; Beauchemin et al., 2011). Most interesting for the present discussion is that preference for mother's voice, and for the prosody of the infant's native language can be established in this prenatal period, which if not represented in the isocortex directly must surely interact with the later cortical representation of these features (Jusczyk et al., 1993; Mahmoudzadeh et al., 2013; Werker and Tees, 1992).

The immediate postnatal (but eyes closed, in burrow) period in rodents, up until the time of eye opening and first exit from the nest has been extensively studied and often used as a "model system" for the effects of early stress on attachment and later learning ((Sanchez et al., 2001; reviewed in Moriceau et al., 2010; Gunnar and Quevedo, 2006)). It is instructive to contrast just how different this maturational state is from the human immediate postnatal period (Fig. 34.4). Birth in rodents is roughly coincident with the end of cortical neurogenesis, which happens at about 70 days, a little over 2 months postconception in humans. Rodents at this age are capable of suckling, and rather uncoordinated

motion to reach the teat, which improves progressively up to eye opening, which happens two to two and a half weeks after birth (about a month postconception), somewhat before coordinated, unsupported walking (Fig. 34.5). As mentioned earlier, from birth until eye opening, several kinds of learning have been demonstrated including learning of olfactory preferences in rodents and the establishment of some chemosensory preferences. From birth to unsupported walking in rat pups ranges from PC day 15–20 to PC 30 (about 2 weeks postnatal). The corresponding period for humans from birth to unsupported walking begins 6 months prior to birth and extends to about 1 year postnatal. The same neural maturational state can bracket wildly different birth-related states. These rodent pups use learning to apply environmental information for immediate functional purposes (the "olfactory tether") at neural maturational states far earlier than humans do, though they are quite helpless sensorily and motorically for the most part (Blass, 1987). The capacity for in utero learning has been demonstrated in humans during roughly similar maturational states, and though no immediate functional purpose is attached to it, it has been implicated in channeling the future course of attachment and language learning (Mampe et al., 2009; Moon et al., 2013).

We have laid out the progress of morphological and functional development, focusing on the cortex, to get an idea of what stage of neural development the human infant presents to the world at birth and how that might relate to critical periods and perinatal behavior and learning in other species. The ultimate goal is to see if there is evidence of special junctures or states in the progress of neural maturation that might inform us about the significance of the peri- and postnatal period for learning in general and social learning in particular. So, to review, by about 3 months postconception, the human infant has generated almost all of its cortical plate as described in Fig. 34.8, including its richly detailed anterior to posterior organization, and thalamic innervation in primary sensory and motor areas is established. As we have argued elsewhere, those two features carry the seeds of hierarchically organized information extraction and prediction, progressively more powerful in larger brains (Finlay and Uchiyama, 2015). Intracortical connectivity is also being established, well along in frontal cortex at 3 months, but just beginning in occipital cortex.

In the second trimester, self-organizational activity is much in evidence, with the onset of retinal waves coincident both with establishment of retinal and thalamocortical connectivity, phasic motor activity, as well as locally correlated activity within structures (Wong, 1999). In addition to cell-to-cell genetic recognition mechanisms, this endogenous activity serves to separate axonal and dendritic projections with uncorrelated activity and helps refine topographic maps in projections between brain regions, based on correlated activity.

A first phase of "exuberance" of numbers of neurons, connections, and neurotransmitter and receptor expression peaks during this period, related to the first establishment of excitatory connectivity between regions, both synaptic and nonsynaptic (Innocenti and Price, 2005; O'Leary and Koester, 1993). A corresponding regressive event, normally occurring neuron loss is at its peak (Janowsky and Finlay, 1983, 1986). Neurotransmitters and neuromodulators are overexpressed in neurons at this time; that is, neurons of the type that will express a single neurotransmitter at adulthood may express several (reviewed in Finlay et al., 1991). One particularly interesting feature of this organizational period is the method by which inhibitory synapses are incorporated into circuitry: inhibitory neurons typically express excitatory transmitters first, which allows them to be integrated by excitatory Hebbian mechanisms into local circuitry. After they are thus inserted into topographically appropriate circuitry, their excitatory synaptic machinery is lost, and they begin to express GABA-ergic mechanisms or other neuromodulators (Ben-Ari, 2002; Cancedda et al., 2007). The first synaptic specializations appear, but in much reduced numbers compared to the eventual perinatal and adult states, to

be discussed at length in the next section. During this period of first organization of connectivity, altricial rodents and carnivores such as the ferret have already been born; cats are born somewhat later, just before eye opening, corresponding to the end of the second trimester in humans. In cats and ferrets at this point, the self-organizational feature of retinal waves drops out, to be replaced by real-world, visually driven activity (Wong et al., 1995). Primates remain securely in utero, though their eyes have opened, and the status of retinal waves is largely unknown.

Thus, by the end of the period corresponding to the end of the second trimester in humans, all neurons in the isocortex have been generated and deployed, and normal neuronal death has been completed in this population. The fundamental pattern of neuron density and laminar organization seen in adults is established. Gross patterns of axonal connectivity, including aspects of nearest-neighbor topology and self-segregation of neurons with uncorrelated activity is laid down both by cell-to-cell recognition and activity-driven processes. Basic distinctions of excitatory and inhibitory connections are expressed. At this point, however, differences in the position of birth and in the consequent "real-world" environmental stimulation cause divergence in maturation: animals born at early-to-middle stages of neural maturation will begin to embark on activity-driven critical periods such as the segregation of ocular dominance columns, whereas animals that are largely precocial, such as primates and ungulates, apparently remain "on hold."

34.2.3 The Case of the Synaptic Surge

One component of neural maturation dissociates itself from the rest of the highly integrated schedule of neural maturation and associates with birth, a surge of synaptogenesis raising synaptic density in the cortical neuropil nearly two orders of magnitude. The number of species forming the basis for this claim is limited, involving necessarily partial data from humans (Huttenlocher and Dabholkar, 1997; Huttenlocher et al., 1982), extensive data from the rhesus macaque (Bourgeois et al., 1994, 1989; Bourgeois and Rakic, 1993; Rakic et al., 1994, 1986; Zecevic et al., 1989; Zecevic and Rakic, 1991) and complete data sets from marmosets (Missler et al., 1993) and rodents (Blue and Parnevelas, 1983). Within and across species, the patterns are consistent.

Although the corpus of data collected by Huttenlocher (1997, 1982) on the development of human synaptic density is typically cited to show a sequential series of changes in synaptic density across cortical areas, contrasting with the "synchronous" macaque, the overall picture is quite similar to the macaque and marmoset when they are placed in corresponding time frames

(Fig. 34.9, top and bottom right; Bates et al., 2002). As the claims of sequential maturation of cortical areas are often supported by only one or two data points per area in the human work, with a methodology not employing important stereological corrections to capture relative synaptic and neuron densities (see extended discussion in Bates et al., 2002), the data collected from experimental animals are empirically better sources for this claim. All cited studies, including those from human material, show a burst of synaptogenesis at birth, of the "Type 1" morphology associated with excitatory neurotransmission. In the macaque, all isocortical areas show a burst of synaptogenesis around the time of birth, in every area simultaneously (Rakic et al., 1986), as well as in various noncortical targets, including the basal ganglia and cerebellum (Brand and Rakic, 1979; Eckenhoff and Rakic, 1991). Interestingly, the rat shows a similar surge, though its surge is not timed to the animal's birth, but rather to eye opening, the time when rat pups begin to first exit the nest (Blue and Parnevelas, 1983).

A major point of interest is the cause of the synaptic surge, whether it might be initiated by the onslaught of experience surrounding birth or entrance into the light directly, or whether it is initiated endogenously in anticipation of birth. While it is apparent an upswing

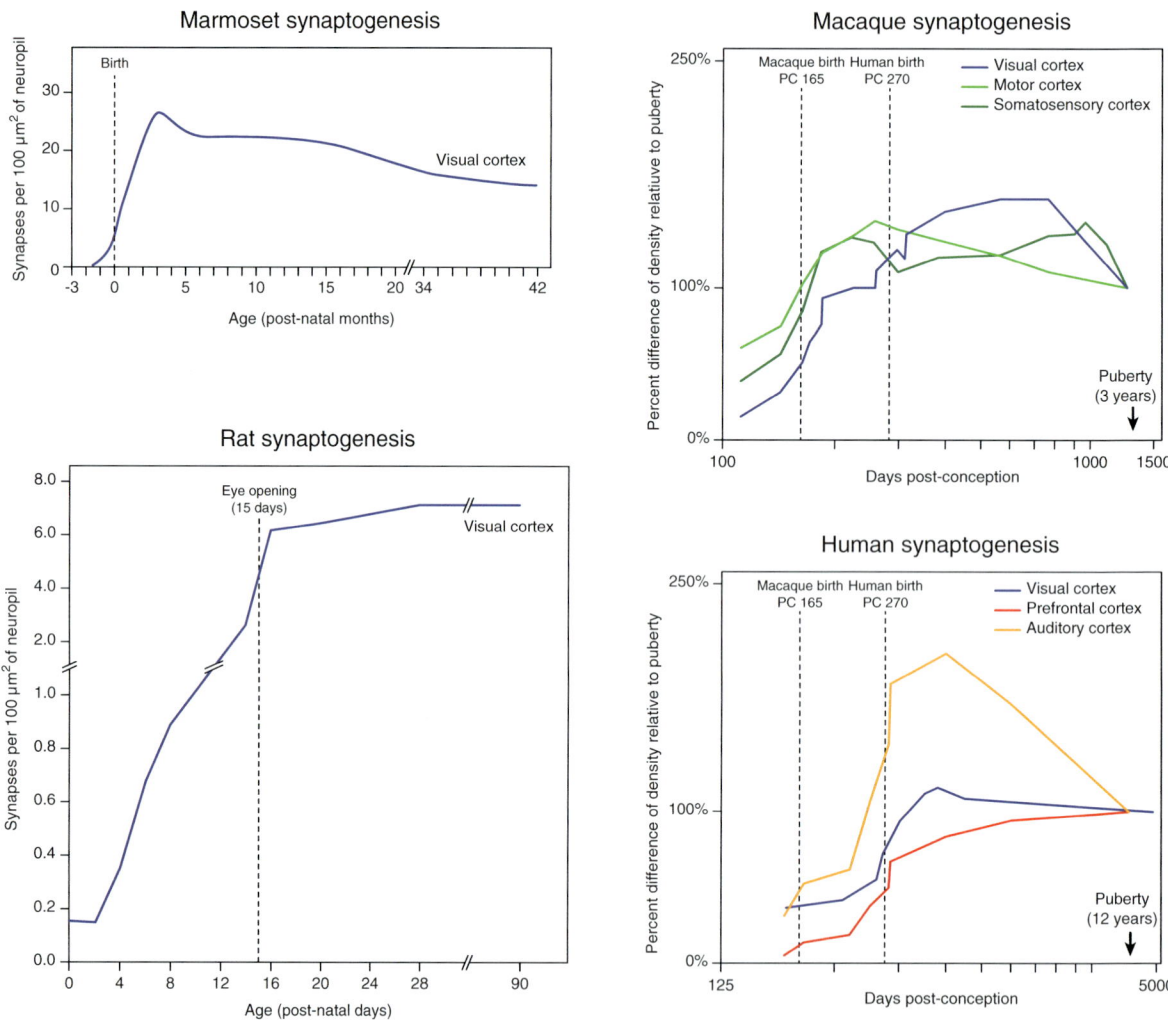

FIGURE 34.9   A surge in excitatory synaptic numbers in the isocortex occurs at birth or at exit from nest. Top left: changing density of synapses in the primary visual cortex of the marmoset; birth indicated by the dotted line. Redrawn from data of Figure 34.4 in Missler, M., Eins, S., Merker, H.J., Rothe, H., Wolff, J.R., 1993. Prenatal and postnatal development of the primary visual cortex of the common marmoset. 1. a changing space for synaptogenesis. J. Comp. Neurol. 333, 41–52. Bottom left: Densities of axodendritic synapses for the total number (type 1 plus type 2) in the visual cortex of rats of various postnatal ages. Rat eye opening and nest exit occurs between postnatal days 16–20. Redrawn from data of Figure 34.1, Blue, M.E., Parnevelas, J.G., 1983. The formation and maturation of synapses in the visual cortex of the rat: quantitatve analysis. J. Neurocytol. 12, 697–712. Top and bottom right: Synaptogenesis in macaque and human showing coordinated perinatal increases in three cortical regions. The corresponding neural maturational state for both species is shown on both graphs, to show that the surge in synaptic density relates to birth, not a particular state of neural maturation determined by the translating time models. Top right: macaques Bottom right) Human (model version, Finlay and Darlington, 1995).

begins before birth, the question was addressed directly in macaques by delivering the macaque infants several weeks prematurely, prior to the onset of the synaptic surge, to determine whether the onset of visual (and other) stimulation would initiate synaptogenesis (Bourgeois et al., 1989). While small changes in the proportion of excitatory and inhibitory synapses were observed immediately following this Caesarian delivery, the burst of synaptogenesis still occurred when the animal would have been born, not when it was born artificially. As it is the case that the fetus initiates birth (Nathanielscz, 1998), it is quite plausible that a maturational signal originating in the fetus prior to birth also initiates the coordinated synaptic surge across structures, though the nature of the signal is not known. A likely reason for the sudden addition of a massive number of synapses, whose state of activity is unknown, is that synaptic sorting and stabilization can then involve the addition, stabilization, and the subtraction of synapses rather than synapse addition alone, employing various empirically demonstrated additive and subtractive mechanisms (Bear, 2003; Bi and Poo, 2001).

A second important feature is that the surge of synaptogenesis is dissociated from the lockstep of neural maturation so dominant up until this point. As can be seen in the comparison of human and macaque synaptogenesis with their births (Parallel vertical lines; right top and bottom graphs for macaque versus human; Fig. 34.9), the surge occurs at quite different maturational points. Both of these points occur well after the organizational events occurring in the human second trimester we have just described but are still quite separated in maturational time. This observation, and several others we will discuss, all point at a long plateau beginning around birth and continuing well through infancy to early childhood, where events termed "critical periods" or "cortical plasticity" may be both initiated or delayed, and where the neural substrate of the cortex appears to undergo rather few known self-initiated organizational changes. This plateau stands in marked contrast to the rapid, time-locked sequence of organizational changes seen in the second trimester.

Finally, as described earlier, learning has been shown to occur in utero in humans, particularly in the language domain, for both descriptive studies demonstrating prenatal or immediate postnatal preference for mother's voice and accent, and experimental studies demonstrating evidence of postnatal recognition for text or music repeatedly presented before birth (Jusczyk et al., 1993; Mahmoudzadeh et al., 2013; Werker and Tees, 1992). It is interesting to speculate how such learning survives the perinatal synaptic onslaught; perhaps there are "tunable" subcortical structures not affected by the synaptic burst, which later serve to channel isocortex-based language learning, in the manner described in

models both for avian imprinting and for face recognition (Johnson, 2005). Alternatively, in several systems, "reference" topographic maps have been described which remain stable in the case of induced sensory conflict, such as between visual and auditory topographic maps in the owl, where the auditory map is accommodated to the visual map, independent of their relative volume, or which one has been altered from its normal state (Konishi, 2003).

34.2.4 Early and Late Processes in Myelination

The developmental study of myelination has two aspects. The first is the initial production of myelin sheaths, beginning not long after axonogenesis in long pathways, and time-locked in the way typical of most other early neural events (reviewed in Workman et al., 2013). The proximate stimulus for myelogenesis has been studied most extensively in the early development of motor neurons, and the optic nerve (Sherman and Brophy, 2005), and is initiated when axon diameter reaches a critical value (Foster et al., 1982). Therefore, in typical development, the highly predictable nature of this event is not initiated by an independent genetic signal directly derived in some fashion from fetal age, but by local signaling of a relevant maturational state. This mechanism, however, allows large deviations to occur in myelination should a developmental challenge of some kind occur. For example, after a bout of perinatal hypoxia, if the corpus callosum is inspected for gross morphology, its appearance in gross morphology and size is unchanged, but the number of axons comprising it is much larger than normal, with thinner axon diameters and less myelin sheathing (Miller et al., 1993). In macaques, the immediate perinatal period is the peak of callosal (and probably intracortical) axon loss (LaMantia and Rakic, 1990). As the reduction in number of callosal axons is unaccompanied by neuron loss in the relevant populations, neuron survival is most likely allowed by maintenance of branches in either long-range or short-range connections (Ivy and Killackey, 1981).

Initial interest in myelination initiation or surges in early development as markers of sensitive periods in sensory or cognitive capacities (eg, Pulvermüller and Schumann, 1994) has not generated much follow-up work. Complementing difficult developmental fMRI studies of the development of functional connectivity (Geerlings et al., 2015), experience-dependent axon organization or myelination, or some correlate of it, however, has become a staple of neuroimaging experiments in development and adulthood, principally using diffusion tensor imaging (Greene et al., 2014; Ingalhalikar et al., 2015; May, 2011; Pandit et al., 2014). This technique examines the vibration or diffusion of water molecules resulting from the presence of extended, "anisotropic" membranes characteristic of both axons and myelin sheathing to infer axon presence. Large tracts can be detected with this method, though branching and crossing tracts cause interpretative problems when it is used to infer details of connectivity. Precise comparisons of connection details of this technique with results obtained by traditional tract-tracing in both monkeys and macaques have produced rather disappointing results (Azadbakht et al., 2015; Jbabdi et al., 2015; Donahue et al., 2016). Nevertheless, plausible features of gross aspects of normal growth can be demonstrated, such as the first appearance of long tracts, growth, and changing within-tract organization. In addition to volume differences, a decrease in "fractional anisotropy" of identified tracts has been shown repeatedly, both in normal development and consequent to particular learning regimes (such as juggling or piano playing) from early childhood to middle adulthood (Dick et al., 2011; May, 2011; Zatorre et al., 2012). In these cases, the signal from identified tracts increases, as if axon bundles are becoming more orderly and aligned as the result of either growth or organizing experience (Fields, 2015). Just what feature of normal axon topology in tracts would become better "aligned" by piano playing or reading is not obvious, though one unintuitive feature of axon growth might be considered. For axons, the process of "growth" should be distinguished from addition of "length" as very little of total axon length is produced by growth by extension at axon terminal regions, as might reasonably be assumed (Pfister et al., 2004). Rather, neuron cell bodies act as if they are essentially fixed, as are their terminal zones, and as the brain otherwise grows through addition of mass of all kinds (other neural processes, glia; vasculature), tension thus exerted on axons causes new membrane to be interpolated into existing axon length. This traction has other consequences for brain growth and function and has been shown to warp the dimensions of receptive fields in the retina conforming to the directions of force and expansion (Schall and Leventhal, 1987; Wong, 1990) and has been proposed as one cause of the development of gyri and sulci in the cortex (Van Essen, 1997). Normal traction, interacting with neural activity in some as yet unknown manner, might contribute to the decreasing anisotropy of long tracts, as the order in a collection of overlaid textile fibers might emerge as any two points of the mass are put under tension.

The extensive use of neuroimaging to describe human development (the extent of this literature prevents a completely comprehensive review) has led to measurement of multiple morphological features whose physiological or functional interpretation is unclear but which are reliably associated with maturation. The relative volume of "grey matter volume," "cortical thickness," or "gray matter density" in the cortex is such a

phenomenon, where grey matter corresponds to the cell-dense layers of the cerebral cortex, contrasted with "white matter", the myelinated tracts located under the cortex or interposed between cortical areas in gyri. Typically, gray matter is measured to increase in total volume or, more typically, relative depth throughout early development and then regresses by a small amount, depending on cortical region (a selection of such surveys, in several species, emphasizing cortex (Aggarwal et al., 2015; Conrad et al., 2012; Croteau-Chonka et al., 2016; Giedd et al., 1999, 1997, 1996; Giedd and Rapoport, 2010; Gogtay et al., 2004; Hammelrath et al., 2016; Li et al., 2015; Makropoulos et al., 2016; Raznahan et al., 2012; Shaw et al., 2006; Shiraishi et al., 2015; Silbereis et al., 2016). The contributions to growth and loss of volume include any change in the volume (but not number) of neuronal cell bodies and their processes, including axonal and dendritic arbors, and the numbers and associated volumes of glia and vasculature. Excluding the perinatal "synaptic surge" discussed earlier, synapse number and density in cortical areas roughly follows the same pattern of small early increase, then decrease. In addition, simple segregation of cell and axonal processes, causing changes in delineation of borders, and the geometrical consequence of the expanding "sphere" of the cortex on the measured depth of its outer shell will also impact relative sizes of these zones. Additive and regressive changes should be assumed to cooccur, resulting in the repeatedly described small net increase in total volume in early development, and a small net negative near the end. The increase and then regression of gray matter depth changes progressively throughout childhood and occurs earlier in primary sensory areas than in posterior parietal and frontal cortex (Gogtay et al., 2004). Precocity as measured by IQ in childhood changes this measure: the peak grey matter depth in parietal and frontal cortex and its following regression occurs later in high-IQ children (Shaw et al., 2006).

Changes in gray matter thickness or total volume parallel some preconceptions about the nature of brain plasticity in some ways and contradict others. The early stabilization of primary sensory areas and the general back-to-front progression of the rate of maturation is in accord with well-known earlier neurodevelopmental features discussed earlier, but well more than half of the cortex is essentially asymptotic in depth and volume change well before adolescence, frontal cortex the exception in its tardiness (Petanjek et al., 2011). Consider also that the very definition of high IQ in childhood is precocity, while cortical measures show delay. Such a "paradoxical" effect is not hard to interpret as an extended period of plasticity, but it also indicated that a particular level of ability (in this case a precocious one) can be decoupled from gray matter thickness: the same brain thickness could be thus associated with quite different vocabulary attainment depending upon the overall trajectory of an individual's growth. Overall, an overarching structural and mechanistic similarity of the process of cortical maturation from the early postnatal period to adolescence emerges from this literature, given that the morphological changes described are produced by multiple components. Gradients across the cortical surface are often the focus of these papers, while mosaic, area-by-area distinctions are seldom, if ever, noted (eg, statements to the effect that "many investigators note the precocious emergence of adult white matter thickness specific to frontal area xyz"). Nor is there any particular pattern of genetic expression, nor description of a special subclass of synaptic plasticity, for example, associated only with early, middle, or late phases of cortical maturation. As always, "absence of evidence is not evidence of absence," but the vast amount of descriptive material now available to us begins to require at least an attempt at appreciating its message. Few signals of part-by-part maturation of the isocortex either idiosyncratic by species or general over species have emerged.

34.2.5 What Are "Critical Periods" in Morphological Terms?

In adults, the computational structures supporting "learning" are quite diverse, varying in organization in different brain locations. Associative or "statistical" learning is distributed widely throughout the nervous system, with a slow form in the isocortex contrasted with much more rapid association in the hippocampus (McClelland et al., 1995). Reinforcement and habit learning are best linked to the basal ganglia and associated structures (Doya, 2001; Graybiel, 2008) and are known to be gated by social and other specific reward systems by neuromodulators (Young and Wang, 2004; Goodson and Thompson, 2010). For the cerebellum, numerous models of circuitry optimizing error correction and prediction exist (Bastian, 2006; Doya, 2001). By contrast, for the mechanisms of early learning in humans, and in initial "animal model" work, researchers have focused almost exclusively on associative, Hebbian and anti-Hebbian learning (Berardi et al., 2000). No reason exists, however, to believe that associative learning is in any sense primary, or that reinforcement and error-driven learning do not participate in the early organization of behavior. The initial research focus on synapse stabilization at the neuromuscular junction (Sanes and Lichtman, 1999) and ocular dominance columns (Bi and Poo, 2001; Katz and Crowley, 2002) was reinforced by the early prominence of connectionist modeling (Elman et al., 1996) and demonstrations of statistical learning (Saffran et al., 1996), later resolving into "deep learning" of statistical and contingent

structure in complex input (Schmidhuber, 2015). All of these research threads highlight mechanisms that discover by fundamentally associative mechanisms correlated structure in input, and represent it efficiently, compared to, for example, selective stabilization of input according to behavioral or motivational importance, or efficiency in executing behavioral goals. Since the following brief review will compound the error of this idiosyncratic focus on a single mechanism, we point out at the outset the formal similarities of "habit learning" in the reinforcement literature to "critical period" and "sensitive period" effects, even though the latter studied almost entirely in the context of associative mechanisms.

34.2.5.1 Initial Parameter Setting: One-To-One Connections, Topographic Maps, and Brain—Body Alignment

The first work on learning in the mammalian nervous system began with two model systems: the formation of the "ocular dominance columns" of the primary visual cortex, integrating and segregating like and unlike input from the two eyes (Katz and Crowley, 2002; Wiesel, 1999) and the establishment of one-to-one neuron-to-muscle fiber connections at the neuromuscular junction (Sanes and Lichtman, 1999). Research in several other domains followed from these. The study of ocular dominance columns became naturally associated with the more general issue of topographic map formation in the visual system (Udin and Fawcett, 1988). Multimodal map registration, particularly between visuomotor and auditory maps, concentrated on midbrain representation (Stein and Stanford, 2008). Work in somatosensory system, particularly the rodent vibrissal system, addressed the question of how to match the number and arrangement of peripheral elements, the whiskers, onto a central cortical representation (Van der Loos and Welker, 1985). Every one of these domains depends on mechanisms involving Hebbian "fire-together, wire-together" mechanisms that bring like-responding elements together, increases efficiency and the sparseness of representations, and generate (in combination with other mechanisms) the ubiquitous mapping of nearest-neighbor to nearest-neighbor topological arrays in the nervous system (Kaas, 1997). Numerous interesting generalities emerged from this work. First, they all show continuity between endogenous and exogenous ordered activity. Like early neuron death (Oppenheim, 1991), the initiation of organization in these networks often depends on some increase in synaptic activity and is delayed or might not begin in the absence of activity, self-initiated and self-terminating. Thus the onset is not time-locked to an endogenous maturational signal but rather to meaningful stimulation (one exception being changes associated with metamorphosis (Hoskins,

1990)). The organizational processes are competitive, and imbalance in inputs can reduce the unfavored input, such as a closed eye in competition with an open eye for cortical synaptic space (Wiesel and Hubel, 1965). These processes are "critical periods" or "sensitive periods," because while they are somewhat plastic in timing, they are not infinitely so, and a loss of input due to imbalanced completion, after an early period of development, cannot be recovered (Berardi et al., 2000). Similarly, early misregistration of maps cannot be wholly undone. Much current research is focused on genetic, pharmacological, or situational factors that can extend or reinstate plasticity (Hübener and Bonhoeffer, 2014). This thorough description of early organization and critical periods was the great focus and accomplishment of initial work on developing neural systems.

Though the temptation to give these established mechanisms much credit for the basic organization of the nervous system is strong, it is important to remember what remains unexplained. To begin with the most elaborated case, the complex interlocking maps of dimensions of ocular dominance, orientation, motion, and visuotopic location are laid down in the visual cortex in the first several years of human life, and if this process is disturbed, irrecoverable amblyopia can result (Levi et al., 2015). The establishment of these patterns has been described almost exclusively in terms of associative networks, with excitatory and inhibitory components that serve to reduce redundancy in central representations and sharpen intrinsic topographic representations. Interestingly, however, real-world function may impinge on the ongoing success and functionality of such projections (Rodger and Dunlop, 2015). In later development, however, the visual cortex continues to participate in learning-related changes, both long and short term, assembling environmental statistics, and even altering topographic projections if part of an input is silenced (Gilbert and Li, 2012, 2013). The cortical regions corresponding to inactive or lost locations in the retina or inactive digits on the hand will become responsive to adjacent regions of the sensorium in short intervals (Florence et al., 1997). Across the lifespan, the brain is constantly amassing and updating the information that makes Bayesian prediction feasible, from details of early sensory processing to higher-order organization, in both long-term assemblies such as the statistics of images to short-term priming (Burr and Cicchini, 2014). In an extreme case, after only 2 or so weeks of visual deprivation and Braille-learning in adulthood, the visual cortex will become both active in Braille reading and essential in its execution (Burton et al., 2002; Merabet and Pascual-Leone, 2010). Activation of so-called "visual" cortex in multiple nonvisual tasks is commonplace in imaging studies (Anderson et al., 2013). Are the elaborate topographies of retinal topography, interocular alignment, wavelength calculation, and orientation selectivity established in early development employed in some useful fashion in the distributed activation demonstrated in adulthood, or are they circumvented? How does apparent reuse and reassignment dissociate itself from the "critical period" nature of early morphological organization (Anderson, 2014)?

34.2.5.2 Genes, Species-Typical Behaviors, and Cortical Areas

Another possible source of information about the nature of sensitive or critical periods might be found in gene expression networks (Hensch, 2004). The genes expressed in early development across the cortex have now been mapped in some detail, though their "meaning" has hardly begun to be understood. The features that stand out in these examinations are first, the rostral-to-caudal pattern of maturation of gene expression described earlier in relation to the generation and early termination of neurogenesis from rostral to caudal (Miller et al., 2014; Pletikos et al., 2014; Silbereis et al., 2016), and second, the separate status of the primary sensory cortices, in terms of early maturation superimposed upon the early maturational gradient, genes involved in axon-target recognition presumably important for the unique trophic and topographic dependence of thalamus-to-cortex in those areas, and associated neurotransmitter systems (Hamasaki et al., 2004; Yamamori, 2011). Notably absent, at this admittedly early point in the understanding of developmental genetic networks, is any signature of genes associated with a sensitive period, or early or other individuated maturation of a nonprimary cortical area (Pletikos et al., 2014). From the behavioral end, evidence exists for genetic loading of face recognition ability (Shakeshaft and Plomin, 2015) and aspects of language learning (Newbury and Monaco, 2010). Working backward from plasticity-relevant mechanisms such as synapse stabilization, particular genes involved with such mechanisms have been described for some time, as described by Hensch (2004); what is absent is a signature of coordination of the multiple mechanisms required for a temporally specified critical period. Conceivably, we might expect something of this kind if the special circuitry which must be present (somewhere in the brain) for the reliably elicited infant preference for faces was found in or around cortical regions related to face recognition in adults, but nothing as yet has been described. This question has been examined in bird song learning, in particular, where readiness for song learning can be partially decoupled from the bird's age (London et al., 2009; Clayton, 2013). So far, however, it has been difficult to recognize any genetic state associated with a "sensitive period" using a variety of assays of gene expression. So far, a very large number of genes can be identified with this state, so many as to be uninformative about particular mechanistic questions. Similar excess of genes in flux at any point has also been described for the cortex, where the link to a singular change in perceptual or behavioral competence is even less clear.

Parallel examination of the functional properties of the cortex, however, has made it increasingly unlikely that special genetic mechanisms defining the computational properties of a defined location in cortex could be the way a species-specific sensitive period is produced. The best-researched example is human infant's special competence in "recognizing" faces and preferring to look at them (Pascalis and Kelly, 2009). Ignoring at this point whether adult face-sensitive cortex is best described by localized activity in a particular region or a network of regions (Haxby et al., 2000), we may simply ask if the infant pattern of activation resembles the adult configuration, for example, showing an early nucleus of activation in a region also indexed in adults, perhaps later spreading to other linked regions. Nothing of this kind is seen but rather a qualitatively different pattern of activation than the adult, which takes close to 10 years of experience to approximate the adult configuration (Cohen Kadosh et al., 2011; de Haan et al., 2002). Similarly, considering language organization, the immediate and largest deficits and delays in language acquisition result from damage to the right side of the cortex, not the left, opposite to the adult pattern (Bates and Roe, 2001). Early infant preferences for faces and speech must of course have some neural correlate, but the origin of such early preferences may quite likely be found outside of the neocortex, though adult processing competence will later require the cortex.

34.2.5.3 Gradients of Activation, Activity, or Cessation of Activity

Another way of the developing cortex might organize incoming information is by coming "on line" area by area, seeded by initial activity in primary sensory and motor areas, and then entering into computation in a rough caudal-to-rostral hierarchy (Finlay and Uchiyama, 2015). This hypothesis is in line with information on "feed-forward" and "feed-back" cortical projections, and what we know about the gross morphology of the sequence of maturation in cortical areas (Gogtay et al., 2004). The idea of "self-initiating, self-terminating" circuitry for sensitive periods would be consistent with progressive entry of regions, cortical or otherwise, into synaptic integration. A critical problem in the development of learning in distributed systems was identified by Elman (1993) that makes this possibility particularly attractive: if a network is to learn a number of tasks, a critical requirement is to have a way of releasing computational resources gradually, so that

old tasks are not routinely overwritten by new tasks. Is there evidence for any version of this possibility?

With the now often-mentioned exception of primary sensory areas, the evidence for sequential activation of cortical areas is poor (though explicit examination of specified regions in a plausible hierarchy has not actually been examined with this question in mind). By "activation," we might mean the idea that regions of cortex far forward in the hierarchy are simply inactive in early development, or their level of activity is lower by some metric. The most impressive evidence contrary to this argument is the presence of robust activity in the frontal cortex quite early in development while performing "cognitive control" tasks, far before those abilities dependent on frontal cortex have stabilized in adolescence and adulthood (Casey et al., 2005). Overall, the general pattern of learning in development recapitulates what is typically seen in individual bouts of learning in specified tasks: the initial pattern of activation is widespread, and often bilateral, settling down into a more localized pattern of "crystallized" or "mature" activation after extensive learning tasks. Therefore, whatever organizes the roughly hierarchical organization of maturation in the cortex is not expressed in activity by itself, but some consequence of that activity, or the cessation of some kind of activity. One quite interesting hint has emerged in a recent study of ocular dominance plasticity in mice, which argued that the maturation of silent glutaminergic synapses into active synapses on principal neurons marked the end of the critical period (Huang et al., 2015). This maturation, which involved the conversion of as many as 50% of synaptic profiles, could be prevented with the animal's visual responsivity, both behavioral and electrophysiological, essentially normal, although ocular dominance profiles remained malleable.

34.2.5.4 Summary: Deploy With Military Precision, Then (Eagerly) Sit and Wait

In the first section, we reviewed "translating time" research, in which a multitude of developmental events in the basic construction of the brain occurred in precise sequence over numerous species, appropriately scaled to the developmental duration for each species. Though echoes of this scheduling persist, for example, in changes across the cortex in grey and white matter volumes, once constructed, the brain appears to tolerate a great amount of deviation in the onset and the organization of experience. These changes appear as broad gradients in achievement of maximum volume, and the small decrement in volume associated with the end of some sensitive periods. Birth can occur at a variety of times with respect to neural maturation, and the "synaptic surge," an apparent morphological index of the onset of significant experience, is also decoupled from an exact point in neural maturation and may also be decoupled from birth. Most organizational sensory events that have been investigated in detail, such as the formation of ocular dominance columns, do not begin at a specified maturational state independent of experience but are initiated by the experience itself. Over a wide range, the isocortex is as yet embarrassingly free of any gene, activity, or morphological indication of cortical area—specific sensitive periods, in onset or offset. The entire cortex, however, presents an organized processing device to incoming input, privileging primary sensory and motor areas, with those embedded in a generally hierarchically arranged sequence of areas. Perplexingly, it is not the relative onset or any progressive change in amount of activity that is best associated with a cortical area's hierarchical position, but perhaps when its uncommitted synaptic elements are depleted (Huang et al., 2015). The following discussion of social learning and the lesser predictability of life history events will give further rationale for what might appear a perplexing contrast in the lockstep of initial neural development compared with the initiation of learning in the world.

Before turning to that subject, however, we suggest that associative networks and the isocortex have been given unwarranted prominence in understanding the emergence of early behavior. To be sure, a great deal of the initial representation of the body in the brain, such as understanding nearest-neighbor relationships in sensory surfaces, the number of sensors on the face, the size and positioning of the eyes, the length of the limbs, and the patterns of coactivation of muscles are directly dependent on associative learning and are well understood in that context. All of those features of brain—body organization, however, are notably free of any motivational, or even motor context in the sense of directed movement. Attending to any cue, such as the preference to look at simple contrast over blur, or for the mother's voice, a face-like configuration, or beginning the babbling and attention to response necessary for language intrinsically involves a component of motivation and some action by the infant, if only an eye movement (Syal and Finlay, 2011). Circuitry attached to motivation, movement, and reward does not lie in the cortex but in the basal ganglia and basal forebrain and is the more likely place to look for the sources of sequential organization of behavior involving any motor decisions. These regions are also known to have circuitry reflecting the continually changing behavioral disposition of the animals, from suckling, to dispersal, to attending to new infants. These preferences can be manifest in multiple ways, from the innate predispositions to approach, avoid, or learn from certain stimulus configurations that we have just described. However, these preferences may be quite more complex, minimally containing the circuitry to adapt responses to contexts (Graybiel,

1998). They may also change the weighting and linkage of rewards, for example, the adolescent emphasis on positive and discounting of negative outcomes (Cohen et al., 2016), or obligatory linkages, such as the requirement of a particular individual for social reward in monogamous animals (Young and Wang, 2004).

34.3 Life History

34.3.1 Construction of Individual Brains Versus Life History Transactions

"Translating time", as we have described it, is well described by a single goal: to construct the brain in the specific duration allowed for a mammal, appropriately scaled for a particular brain size. The events of "life history", by contrast—birth, weaning, adolescence, dispersal, first parenthood, and so forth—are transactional and social, defined by competing interests and multiple goals. These two contrasting organizational features of maturing systems are not entirely mutually exclusive, of course. We described one tradeoff in one organ system against another within brain development, whether to allocate tissue preferentially to olfactory-limbic structures or to isocortex. Conversely, the timing of life history events will have necessary constraints from neural and somatic maturation. Overall, however, the timing of life history events such as birth or weaning depends not only on the maturational state of the child but also the competing and aligning interests of the child and mother both, in both in individual variation and cross-species contexts (Royle et al., 2012). For another example, the timing of dispersal from a home range is unlikely to depend on a single maturational signal from the nervous system of an individual reaching a criterion age but will more likely also involve seasonality, competition for resources prior to leaving and likely competition consequent to leaving (Royle et al., 2012). In the general context of evo-devo, we will present some preliminary evidence and arguments here that brains have evolved in the context of such contingency and variability and thus possess the circuitry that demonstrates the evolutionary prevalence of such transactions.

34.3.2 Human Birth in Its Primate and Mammalian Contexts

We will discuss here some general considerations involving the timing of birth, and the specific timing of human birth in the context of other primates, and mammals in general. Overall, a long gestation requires more investment of resources from the mother than does a short one. This investment involves not only the transfer of nutrients to the fetus but also a lengthening of the interval to the next conception, and hence a decrease of maternal reproductive rate. However, there are several reasons why it might be adaptive for a mother to invest in a long gestation and produce precocial offspring; these might include the need to avoid predators and the need to move around safely shortly after birth, the latter including the need to avoid drowning in the case of aquatic mammals (Pagel and Harvey, 1988). Because brain mass is a power-law function of developmental duration during the early phases of development that include gestation (Passingham, 1985; Workman et al., 2013; Halley, 2015), gestation length is also tightly coupled to the degree of brain maturation. This means that for species of equal brain size, the later born species will necessarily be neurologically more mature at birth.

Humans are usually viewed as an altricial species among the great apes, often with reference to the large proportion of brain growth that occurs after birth. This is frequently interpreted as a prolongation of a fetal rate of development far beyond birth (Coqueugniot et al., 2004), but this interpretation is based on the erroneous assumption that birth occurs at the same stage in fetal development across different species, and any changes that are observed in development after birth are the result of evolutionary modifications of the developmental schedule itself. The same assumption governs statements about behavioral development such as, "it takes the human newborn a year to reach the stage of motor development equivalent to that of a newborn great ape" (Smith and Tompkins, 1995: p. 269). However, such manifestations of "secondary altriciality" do not necessarily reflect an evolutionary reshaping of the developmental schedule; it could simply be that the position of birth is being moved to an earlier time with respect to a developmental schedule that is conserved across species and elongation of human development that is "allometrically expected." With regard to development of the nervous system, our analysis shows that this is likely to be the case. Humans are born in a neurologically more altricial state than other apes, but this does not mean that humans are an evolutionary anomaly; to the contrary: human altriciality is well within general mammalian norms.

The classic explanation for human altriciality has been the "obstetrical dilemma": In this scenario, the transition to bipedal locomotion in human evolution led to a narrowing of the skeletal structure of the birth canal, conflicting with selection pressure toward larger brains that make passage through the birth canal more difficult, and thus necessitating early birth so that a substantial portion of cranial expansion could take place outside the womb (Schultz, 1949; Rosenberg and Trevathan, 1995). Another explanation is that it may be maternal metabolic constraints rather than cephalopelvic proportion that determine the timing of birth in humans

(Dunsworth et al., 2012). In either interpretation, both reinforce the idea that birth timing is a negotiation between the requirements of the fetus and the mother, not linked to a precise degree of neural development. Evolution can then act as a filter, with surviving individuals possessing overall organizational properties robust to commonly occurring variation. The comparative record shows that variation in the timing of birth with respect to neural maturation is very broad, both in widely separated taxonomic groups, but also in relatively closely related species, such as current chimpanzees and humans. Therefore, we may ask, what features of developmental patterns for neural maturation versus environmental learning will be most robust to this challenge?

34.3.3 Human Weaning in Its Primate and Mammalian Contexts

The transition from suckling to independent feeding constitutes a major shift in the behavioral and cognitive capacities required for survival (Lee, 1996). Knowing the stage of brain maturity at which weaning takes place is therefore of clear benefit for understanding the relationship between development of the nervous system and life history adaptations. And by situating weaning in a developmental plan that is normalized across species, we can make principled comparisons between species or taxa.

Our analysis shows that weaning, such as birth, is an event that is variable and uncoupled from the highly coordinated schedule of brain development (Fig. 34.10). An animal that can no longer depend on its mother for energetic input must be able to supplement its diet with other sources of nutrition, and this requires a degree of cognitive and behavioral competence that allows for the independent acquisition and ingestion of foods. And because many of these forms of competence, particularly the early-developing ones, will be synchronized with brain development, any variability in the timing of weaning for a species is likely to be associated with the duration of developmental time over which a young individual of that species is able to get support from its social group.

In a cross-cultural survey of weaning practices, the median age of weaning across 133 nonindustrial societies was reported to be 29 months postbirth, with a standard deviation of 10 months (Sellen, 2001). In traditional societies, earlier weaning is associated with shorter interbirth intervals. Early weaning should be conducive to higher reproductive output, all other things being held equal. As an illustration of how late weaning can be when initiated by the child, it has been reported in a large sample of US mothers that the average age of child-led weaning is 4.4 years, or 53 months (Dettwyler, 2004), which is much later than

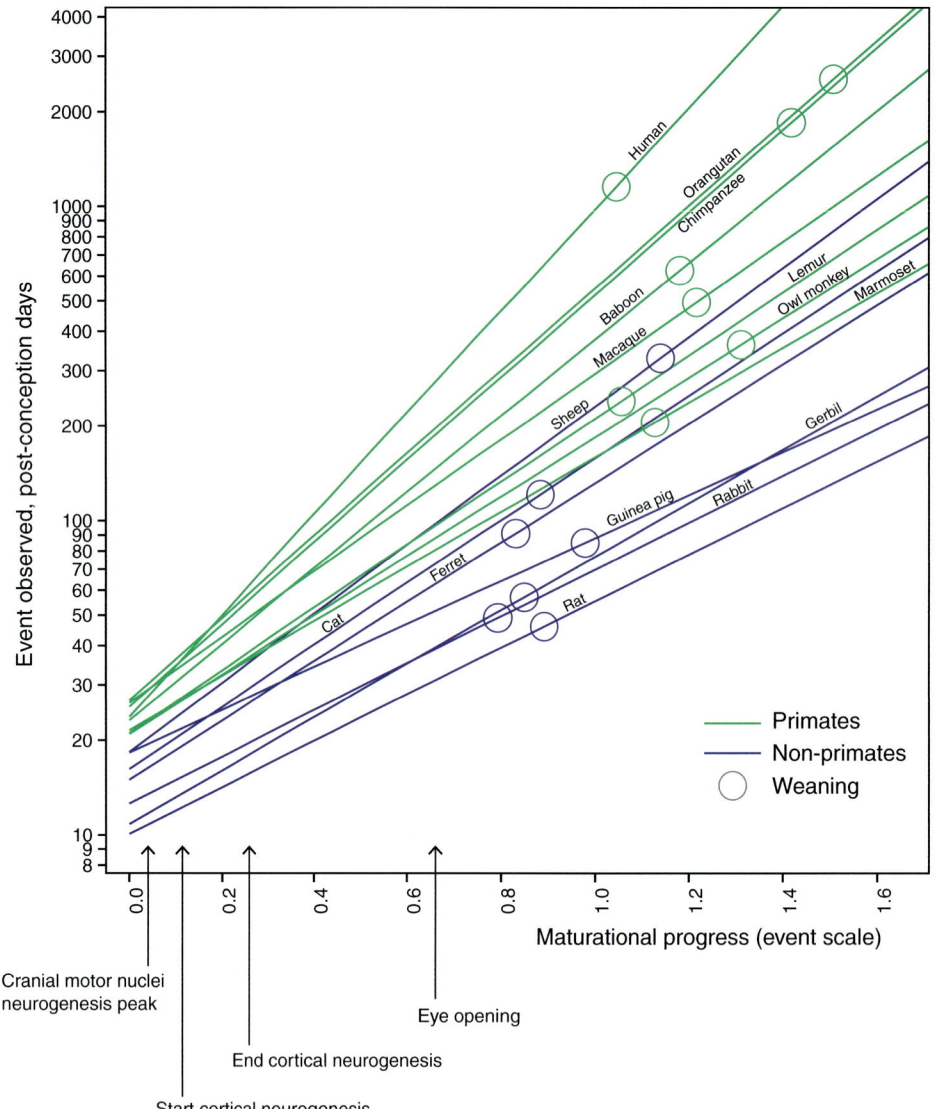

FIGURE 34.10 Variability in weaning compared to neural maturational state. Predicted developmental schedules for eight primate and seven nonprimate species, with observed timing of weaning indicated for each. The event scale on the x-axis has been extended beyond its original range of 0–1, to allow for extrapolation of developmental trajectories into the range in which weaning occurs. Examples of events included in the model are displayed at their respective positions on the event scale, on the x-axis. The y-axis indicates the estimated date of the occurrence of each event in each species. Plotted species include those whose developmental schedules are directly predicted by the model as well as those that are not. The latter consists of all of the primates except for the human and rhesus macaque, which are the two primates included in the current version of model. For the unmodeled primate species, we used adult brain weight to estimate slope and gestation length to estimate intercept, as these variables have been shown to predict the timing of events with reasonable accuracy. Timing of weaning in postconception days and associated references for each species are as follows: baboon (*Papio cynocephalus*), 625.5, Jones et al. (2009); cat (*Felis catus*), 121, Tacutu et al. (2013); chimpanzee (*Pan troglodytes*), 1849, Robson and Wood (2008); ferret (*Mustela putorius*), 91, Isler and van Schaik (2012); gerbil (*Meriones unguiculatus*), 49, Tacutu et al. (2013); guinea pig (*Cavia porcellus*), 85, Jones et al. (2009); hamster (*Mesocricetus auratus*), 34, Jones et al. (2009); human (*Homo sapiens*), 1140, Sellen (2001); lemur (*Lemur catta*), 240, Rowe (1996); macaque (*Macaca mulatta*), 495, Promislow (1991); marmoset (*Callithrix jacchus*), 205.5, Tacutu et al. (2013); Orangutan (*Pongo pygmaeus pygmaeus*), 2540, Barrickman (2008); owl monkey (*Aotus azarae*), 363, Jones et al. (2009); rabbit (*Oryctolagus cuniculus*), 57, Jones et al. (2009); rat (*Rattus norvegicus*), 46, Tacutu et al. (2013); sheep (*Ovis aries*), 329, Tacutu et al. (2013).

the nonindustrial average noted earlier, and significantly later than the median weaning age for US mothers, which is around 7 months (CDC).

If we compare any of these measures, even the latest reported, to reported weaning ages for other apes, such as the Western gorilla at around 3 years and the chimpanzee at 4–5 years, we see that weaning is earlier in humans, as has been emphasized by many researchers studying human life history evolution (Kennedy, 2005; Promislow, 1991; Reichard and Barelli, 2008). But this observation is based on absolute duration. If we examine allometric predictions for these species compared to brain maturation, we see that humans are weaned even earlier than the linear projection would suggest. So, it is the case that humans are weaned significantly earlier than other primates. Considering all mammals, and not only primates, however, humans are weaned at a maturational point well within general mammalian variability (Figs. 34.3 and 34.10).

34.3.4 The Serial Litter and the Social World

One popular anthropological writer went so far as to characterize human reproduction, compared to other primates, as a "serial litter" (Morris, 2012). Having one offspring maturing at a time, with the long duration of development that a big brain seems to absolutely require, as we discussed in the earlier sections, with an 8–9 year interbirth interval is a low-output reproductive strategy that seems risky at best. So, humans have made two alterations in reproduction associated with the "serial litter" strategy, not only the early weaning observed by anthropologists, but also the relatively shortened gestation with respect to neural maturation we describe. Both of these are methods of reducing unique maternal investment to tractable levels, depending on either biparental care, or other alloparenting in the early postnatal period and progressively more as childhood progresses (Hrdy, 2011). One particularly attractive idea about a reliable source of alloparental care is the "grandmother hypothesis," which links late aspects of maternal life history to human altriciality and early weaning together (Hawkes et al., 1998). Female fertility ends at the same (absolute) age in humans as it does in other great apes (Robson et al., 2006; Robbins et al., 2006). Unlike other great ape females, however, women remain productive for decades longer: great apes become frail and rarely survive to menopause, while even in hunter-gatherer populations (where mortality is relatively high) about a third of the women are past the child-bearing years (Blurton Jones et al., 2002; Blurton Jones, 2016; Hawkes, 2003, 2010). The linkage of postmenopausal longevity is combined with weaning ages early compared to other primates may be the critical source of the markedly increased reproductive productivity in humans.

In contrast to this essentially economic rationale, an alternate interpretation of relatively early birth combined with a long developmental duration is that extended development is critical for some aspect of learning or cognition, with language learning the most obvious candidate (Bjorklund, 1997; Locke and Bogin, 2006). While it is possible that this is the case, several arguments can be martialed against any such single-factor causal scenario. Language experience can be quite variable: cross-culturally, while infant-directed speech has several hallmark similarities (Broesch and Bryant, 2015), aspects like the total amount of speech can be widely variable (Bornstein et al., 2015), and the rate of acquisition of markers of language maturity can vary widely by socioeconomic status (Fernald et al., 2013). Finally, large brains and long development are indissociable (Charvet and Finlay, 2012), and language is a necessary outcome of neither.

One of the most profound differences in human development, however, we suggest, is the changed social environment produced by early weaning. A long period of development results, with the brain positioned during the entire time for maximum learning, from early childhood to adolescence, where the infant and child is not being solely provisioned by the mother, but most typically by a larger group, from grandmothers, to the minimal "nuclear family" to much more extended related and nonrelated groups. Humans have an extended childhood where language, custom, and allegiance are being defined by an extended social group, not by the immediate parents. This allegiance to a peer group is often made over the complaints of the parents (Locke and Bogin, 2006). So it may be that an essentially economic decision, to limit maternal investment to a viable amount, may be one foundation of our extensive and very unusual sociality.

This economic decision thrusts the human child, unlike other primates, in the early parts of its "sensitive period" of development of any number of sensory, cognitive, motor, and social abilities out of the small society of mother and child, and into the village of age peers, other relatives, and any number of unrelated others. While much evidence suggests relatively greater attunement of the human child for social interaction, imitation, and cooperation (examples from a large literature: Bullinger et al., 2011; Haun et al., 2014; Tomasello, 1999), it is not only the motivations and preferences of the child that differ from its primate ancestors. Possessed of an exceptionally large brain constructed in a primate-typical schedule, with an allometrically predictable extended period of maturation filtered by evolution to be permissive of variability, the human child exercises those motivations and preferences in social environments more variable in every respect than those of any immediate primate relative. The developmental niche we inhabit is thus a curious mixture of a conserved neurodevelopmental schedule and a specially adapted life history.

References

Aggarwal, M., Gobius, I., Richards, L.J., Mori, S., 2015. Diffusion MR microscopy of cortical development in the mouse embryo. Cereb. Cortex 25, 1970–1980.

Alberts, J.R., Ronca, A.E., 1993. Fetal experience revealed by rats - psychobiological insights. Early Hum. Dev. 35, 153–166.

Allendoerfer, K.L., Shatz, C.J., 1994. The subplate, a transient neocortical structure. Annu. Rev. Neurosci. 17, 185–218.

Allman, J.M., Tetreault, N.A., Hakeem, A.Y., Manaye, K.F., Semendeferi, K., Erwin, J.M., et al., 2010. The von Economo neurons in frontoinsular and anterior cingulate cortex in great apes and humans. Brain Struct. Funct. 214, 495–517.

Altman, J., Bayer, S.A., 1988. Development of the rat thalamus. I. Mosaic organization of the thalamic neuroepithelium. J. Comp. Neurol. 275, 346–377.

Altman, J., Bayer, S.A., 1988. Development of the rat thalamus. II. Time of origin and settling pattern of neurons derived from the anterior lobule of the thalamic neuroepithelium. J. Comp. Neurol. 275, 378–405.

Altman, J., Bayer, S.A., 1988. Development of the rat thalamus. III. Time of origin and settling pattern of neurons of the reticular nucleus. J. Comp. Neurol. 275, 406–428.

Amunts, K., Schleicher, A., Ditterich, A., Zilles, K., 2003. Broca's region: cytoarchitectonic asymmetry and developmental changes. J. Comp. Neurol. 465, 72–89.

Anderson, M.L., 2014. After Phrenology: neural reuse and the interactive brain. MIT Press, Cambridge.

Anderson, M.L., Kinnison, J., Pessoa, L., 2013. Describing functional diversity of brain regions and brain networks. Neuroimage 73, 50–58.

Azadbakht, H., Parkes, L.M., Haroon, H.A., Augath, M., Logothetis, N.K., de Crespigny, A., et al., 2015. Validation of high-resolution tractography against in vivo tracing in the macaque visual cortex. Cereb. Cortex 25, 4299–4309.

Barrickman, N.L., 2008. Evolutionary relationship between life history and brain growth in anthropoid primates. PhD Dissertation. Duke University.

Barrickman, N.L., Bastian, M.L., Isler, K., van Schaik, C.P., 2008. Life history costs and benefits of encephalization: a comparative test using data from long-term studies of primates in the wild. J. Hum. Evol. 54, 568–590.

Barton, R.A., Venditti, C., 2013. Human frontal lobes are not relatively large. Proc. Natl. Acad. Sci. 110, 9001–9006.

Bastian, A.J., 2006. Learning to predict the future: the cerebellum adapts feedforward movement control. Curr. Opin. Neurobiol. 16, 645–649.

Bates, E., Roe, K., 2001. Language Development in Children with Unilateral Brain Injury. In: Nelson, C.A., Luciana, M. (Eds.), Handbook of Developmental Cognitive Neuroscience. MIT Press, Cambridge, MA, pp. 281–307.

Bates, E., Thal, D., Finlay, B.L., Clancy, B., 2002. Early language development and its neural correlates. In: Segalowitz, S.J., Rapin, I. (Eds.), Child Neurology, vol. 8. Elsevier Science B.V., Amsterdam, pp. 109–1076.

Bayer, S.A., Altman, J., 1991. Neocortical Development. Raven Press, New York.

Bear, M.F., 2003. Bidirectional synaptic plasticity: from theory to reality. Philos. Trans. R. Soc. Lond. B Biol. Sci. 358, 649–655.

Beauchemin, M., Gonzalez-Frankenberger, B., Tremblay, J., Vannasing, P., Martinez-Montes, E., Belin, P., et al., 2011. Mother and stranger: an electrophysiological study of voice processing in newborns. Cereb. Cortex 21, 1705–1711.

Ben-Ari, Y., 2002. Excitatory actions of gaba during development: the nature of the nurture. Nat. Rev. Neurosci. 3, 728–739.

Berardi, N., Pizzorusso, T., Maffei, L., 2000. Critical periods during sensory development. Curr. Opin. Neurobiol. 10, 138–145.

Bi, G.-Q., Poo, M.-M., 2001. Synaptic modification by correlated activity: Hebb's postulate revisited. Annu. Rev. Neurosci. 24, 139–166.

Bjorklund, D.F., 1997. The role of immaturity in human development. Psychol. Bull. 122, 153.

Blass, E.M., 1987. Critical events during sensitive periods of social development in rats. In: Bornstein, M.H. (Ed.), Sensitive Periods in Development: Interdisciplinary Perspectives. Psychology Press, New York, pp. 81–91.

Blue, M.E., Parnevelas, J.G., 1983. The formation and maturation of synapses in the visual cortex of the rat: quantitatve analysis. J. Neurocytol. 12, 697–712.

Bluml, S., Wisnowski, J.L., Nelson, M.D., Paquette, L., Gilles, F.H., Kinney, H.C., et al., 2013. Metabolic maturation of the human brain from birth through adolescence: insights from in vivo magnetic resonance spectroscopy. Cereb. Cortex 23, 2944–2955.

Blurton Jones, N.G., 2016. Demography and Evolutionary Ecology of Hadza Hunter-Gatherers. Cambridge University Press, Cambridge.

Blurton Jones, N.G., Hawkes, K., O'Connell, J.F., 2002. Antiquity of post-reproductive life: are there modern impacts on hunter-gatherer post-reproductive life spans? Am. J. Hum. Biol. 14 (2), 184–205.

Bornstein, M.H., Putnick, D.L., Cote, L.R., Haynes, O.M., Suwalsky, J.T.D., 2015. Mother-infant contingent vocalizations in 11 countries. Psychol. Sci. 28, 1272–1284.

Bourgeois, J.P., Goldman-Rakic, P.S., Rakic, P., 1994. Synaptogenesis in the prefrontal cortex of rhesus monkeys. Cereb. Cortex 4, 78–96.

Bourgeois, J.P., Jastreboff, P.J., Rakic, P., 1989. Synaptogenesis in visual cortex of normal and preterm monkey: evidence for intrinsic regulation of synaptic overproduction. Proc. Natl. Acad. Sci. U.S.A. 86, 4297–4301.

Bourgeois, J.P., Rakic, P., 1993. Changes of synaptic density in the primary visual cortex of the macaque monkey from fetal to adult stage. J. Neurosci. 13, 2801–2820.

Brand, S., Rakic, P., 1979. Cytodifferentiation and synaptogenesis in the neostriatum of fetal and neonatal rhesus monkeys. Anat. Embryol. 169, 21–34.

Brauer, J., Anwander, A., Friederici, A.D., 2011. Neuroanatomical prerequisites for language functions in the maturing brain. Cereb. Cortex 21, 459–466.

Broesch, T.L., Bryant, G.A., 2015. Prosody in infant-directed speech is similar across western and traditional cultures. J. Cognition Dev. 16, 31–43.

Brown, K.E., Arends, J.J.A., Wasserstrom, S.P., Zantua, J.B., Jacquin, M.F., Woolsey, T.A., 1995. Developmental transformation of dendritic arbors in mouse whisker thalamus. Dev. Brain Res. 86, 335–339.

Bryant, K.L., Suwyn, C., Reding, K.M., Smiley, J.F., Hackett, T.A., Preuss, T.M., 2012. Evidence for ape and human specializations in geniculostriate projections from VGLUT2 immunohistochemistry. Brain Behav. Evol. 80, 210–221.

Bullinger, A.F., Zimmermann, F., Kaminski, J., Tomasello, M., 2011. Different social motives in the gestural communication of chimpanzees and human children. Dev. Sci. 14, 58–68.

Burr, D., Cicchini, G.M., 2014. Vision: efficient adaptive coding. Curr. Biol. 24, R1096–R1098.

Burton, H., Snyder, A.Z., Conturo, T.E., Akbudak, E., Ollinger, J.M., Raichle, M.E., 2002. Adaptive changes in early and late blind: a fMRI study of Braille reading. J. Neurophysiol 87, 589–607.

Cahalane, D.J., Charvet, C.J., Finlay, B.L., 2012. Systematic, balancing gradients in neuron density and number across the primate isocortex. Front Neuroanat 6, 28.

Cahalane, D.J., Charvet, C.J., Finlay, B.L., 2014. Modeling local and cross-species neuron number variations in the cerebral cortex as arising from a common mechanism. Proc. Natl. Acad. Sci. 111, 17642–17647.

Cancedda, L., Fiumelli, H., Chen, K., Poo, M.M., 2007. Excitatory GABA action Is essential for morphological maturation of cortical neurons in vivo. J. Neurosci. 27, 5224–5235.

Casey, B.J., Tottenham, N., Liston, C., Durston, S., 2005. Imaging the developing brain: what have we learned about cognitive development? Trends Cognit. Sci. 9, 104–110.

Caviness, V.S., Takahashi, T., Nowakowski, R.S., 1995. Numbers, time and neocortical neuronogenesis: a general developmental and evolutionary model. Trends Neurosci. 18, 379–383.

Chalfin, B.P., Cheung, D.T., Muniz, J.A.P.C., Silveira, L.C.L., Finlay, B.L., 2007. Scaling of neuron number and volume of the pulvinar complex in New World primates: comparisons with humans, other primates and mammals. J. Comp. Neurol. 504, 265–274.

Chaplin, T.A., Yu, H.-H., Soares, J.G.M., Gattass, R., Rosa, M.G.P., 2013. A conserved pattern of differential expansion of cortical areas in simian primates. J. Neurosci. 33, 15120–15125.

Charnov, E.L., Downhower, J.F., 1995. A trade-off-invariant life-history rule for optimal offspring size. Nature 376, 418–419.

Charvet, C.J., Cahalane, D.J., Finlay, B.L., 2014. Systematic, cross-cortex variation in neuron numbers in rodents and primates. Cereb. Cortex 25, 147–160.

Charvet, C.J., Finlay, B.L., 2012. Embracing covariation in brain evolution: large brains, extended development, and flexible primate social systems. In: Michel, A.H., Dean, F. (Eds.), Progress in Brain Research, vol. 195. Elsevier, pp. 71–87.

Charvet, C.J., Finlay, B.L., 2014. Evo-Devo and the primate isocortex: the central organizing role of intrinsic gradients of neurogenesis. Brain. Behav. Evol. 84, 81–92.

Choe, M.-S., Ortiz-Mantilla, S., Makris, N., Gregas, M., Bacic, J., Haehn, D., et al., 2013. Regional infant brain development: an MRI-based morphometric analysis in 3 to 13 month olds. Cereb. Cortex 23, 2100–2117.

Clancy, B., Darlington, R.B., Finlay, B.L., 2000. The course of human events: predicting the timing of primate neural development. Dev. Sci. 3, 57–66.

Clancy, B., Darlington, R.B., Finlay, B.L., 2001. Translating developmental time across mammalian species. Neuroscience 105, 7–17.

Clancy, B., Kersh, B., Hyde, J., Anand, K.J.S., Darlington, R.B., Finlay, B.L., 2007. Web-based method for translating neurodevelopment from laboratory species to humans. Neuroinformatics 5, 79–94.

Clayton, D.F., 2013. The genomics of memory and learning in songbirds. Annu. Rev. Genomics Hum. Genet. 14, 45–65.

Cohen, A.O., Breiner, K., Steinberg, L., Bonnie, R.J., Scott, E.S., Taylor-Thompson, K.A., et al., 2016. When is an adolescent an adult? Assessing cognitive control in emotional and nonemotional contexts. Psychol. Sci. 27 (4) (Online).

Cohen Kadosh, K., Cohen Kadosh, R., Dick, F., Johnson, M.H., 2011. Developmental changes in effective connectivity in the emerging core face network. Cereb. Cortex 21, 1389–1394.

Conrad, M.S., Dilger, R.N., Johnson, R.W., 2012. Brain growth of the domestic pig (sus scrofa) from 2 to 24 weeks of age: a longitudinal MRI study. Dev. Neurosci. 34, 291–298.

Coqueugniot, H., Hublin, J.-J., Veillon, F., Houët, F., Jacob, T., 2004. Early brain growth in Homo erectus and implications for cognitive ability. Nature 431, 299–302.

Croteau-Chonka, E.C., Dean, D.C., Remer, J., Dirks, H., O'Muircheartaigh, J., Deoni, S.C.L., 2016. Examining the relationships between cortical maturation and white matter myelination throughout early childhood. Neuroimage 125, 413–421.

Damaraju, E., Caprihan, A., Lowe, J.R., Allen, E.A., Calhoun, V.D., Phillips, J.P., 2014. Functional connectivity in the developing brain: a longitudinal study from 4 to 90 months of age. Neuroimage 84, 169–180.

Darlington, R.B., Dunlop, S.A., Finlay, B.L., 1999. Neural development in metatherian and eutherian mammals: variation and constraint. J. Comp. Neurol. 411, 359–368.

de Haan, M., Humphreys, K., Johnson, M.H., 2002. Developing a brain specialized for face perception: a converging methods approach. Dev. Psychobiol. 40, 200–212.

Dehay, C., Kennedy, H., Kosik, K.S., 2015. The outer subventricular zone and primate-specific cortical complexification. Neuron 85, 683–694.

Dettwyler, K.A., 2004. When to wean: biological versus cultural perspectives. Clin. Obstetr. Gynecol 47, 712–723.

Dick, F., Lee, H.L., Nusbaum, H., Price, C.J., 2011. Auditory-motor expertise alters "speech selectivity" in professional musicians and actors. Cereb. Cortex 21, 938–948.

Donahue, C.J., Sotiropoulos, S.N., Jbabdi, S., Hernandez-Fernandez, M., Behrens, T.E., Dyrby, T.B., Coalson, T., Kennedy, H., Knoblauch, K., Van Essen, D.C., Glasser, M.F., 2016. Using diffusion tractography to predict cortical connection strength and distance: a quantitative comparison with tracers in the monkey. J. Neurosci. 36, 6758–6770.

Doya, K., 2001. Complementary roles of basal ganglia and cerebellum in learning and motor control. Curr. Opin. Neurobiol. 10, 20–58.

Dreher, B., Robinson, S.R., 1988. Development of the retinofugal pathway in birds and mammals: evidence for a common timetable. Brain. Behav. Evol. 31, 369–390.

Dunsworth, H.M., Warrener, A.G., Deacon, T., Ellison, P.T., Pontzer, H., 2012. Metabolic hypothesis for human altriciality. Proc. Natl. Acad. Sci. 109, 15212–15216.

Eckenhoff, M.F., Rakic, P., 1991. A quantitative analysis of synaptogenesis in the molecular layer of the dentate gyrus in the rhesus monkey. Dev. Brain Res. 64, 129–135.

Elman, J.L., 1993. Learning and development in neural networks: the importance of starting small. Cognition 48, 71–99.

Elman, J.L., Bates, E.A., Johnson, M.H., Karmiloff-Smith, A., Parisi, D., Plunkett, K., 1996. Rethinking Innateness: a connectionist perspective on development. MIT Press, Cambridge, MA.

Fernald, A., Marchman, V.A., Weisleder, A., 2013. SES differences in language processing skill and vocabulary are evident at 18 months. Dev. Sci. 16, 234–248.

Fields, R.D., 2015. A new mechanism of nervous system plasticity: activity-dependent myelination. Nat. Rev. Neurosci. 16, 756–767.

Finlay, B.L., 1991. Control of cell number and type in the developing and evolving neocortex. In: Finlay, B.L., Innocenti, G., Scheich, H. (Eds.), The Neocortex: Ontogeny and Phylogeny NATO ASI Series. Plenum Press, New York, pp. 33–41.

Finlay, B.L., 1992. Cell death and the creation of regional differences in neuronal numbers. J. Neurobiol. 23, 1159–1171.

Finlay, B.L., Darlington, R.B., 1995. Linked regularities in the development and evolution of mammalian brains. Science 268, 1578–1584.

Finlay, B.L., Innocenti, G., Scheich, H., 1991. The neocortex: ontogeny and phylogeny. Plenum Press, New York.

Finlay, B.L., Silveira, L.C.L., Reichenbach, A., 2005. Comparative Aspects of Visual System Development. In: Kremers, J. (Ed.), The Structure, Function and Evolution of the Primate Visual System. John Wiley and Sons, New York, pp. 37–72.

Finlay, B.L., Slattery, M., 1983. Local differences in amount of early cell death in neocortex predict adult local specializations. Science 219, 1349–1351.

Finlay, B.L., Uchiyama, R., 2015. Developmental mechanisms channeling cortical evolution. Trends Neurosci. 38, 69–76.

Finlay, B.L., Workman, A.D., 2013. Human exceptionalism. Trends Cognit. Sci. 17, 199–201.

Fitch, W.T., Reby, D., 2001. The descended larynx is not uniquely human. Proc. R. Soc. Lond. B 268, 1669–1675.

Florence, S.L., Jain, N., Kaas, J.H., 1997. Plasticity of somatosensory cortex in primates. Semin. Neurosci. 9, 3–12.

Foster, R.E., Connors, B.W., Waxman, S.G., 1982. Rat optic nerve: electrophysiological, pharmacological and anatomical studies during development. Dev. Brain Res. 3, 371–386.

Gao, W., Alcauter, S., Elton, A., Hernandez-Castillo, C.R., Smith, J.K., Ramirez, J., et al., 2015. Functional network development during the first year: relative sequence and socioeconomic correlations. Cereb. Cortex 25, 2919–2928.

Garwicz, M., Christensson, M., Psouni, E., 2009. A unifying model for timing of walking onset in humans and other mammals. Proc. Natl. Acad. Sci. 106, 21889–21893.

Ge, X., Shi, Y., Li, J., Zhang, Z., Lin, X., Zhan, J., et al., 2015. Development of the human fetal hippocampal formation during early second trimester. Neuroimage 119, 33–43.

Geerligs, L., Renken, R.J., Saliasi, E., Maurits, N.M., Lorist, M.M., 2015. A brain-wide study of age-related changes in functional connectivity. Cereb. Cortex 25, 1987–1999.

Giedd, J.N., Blumenthal, J., Jeffries, N.O., Castellanos, F.X., Liu, H., Zijdenbos, A., et al., 1999. Brain development during childhood and adolescence: a longitudinal MRI study. Nat. Neurosci. 2, 861–863.

Giedd, J.N., Castellanos, F.X., Rajapakse, J.C., Vaituzis, A.C., Rapoport, J.L., 1997. Sexual dimorphism of the developing human brain. Prog. Neuro Psychopharmacol. Biol. Psychiatry 21, 1185–1201.

Giedd, J.N., Rapoport, J.L., 2010. Structural MRI of pediatric brain development: what have we learned and where are we going? Neuron 67, 728–734.

Giedd, J.N., Snell, J.W., Lange, N., Rajapakse, J.C., Casey, B.J., Kozuch, P.L., et al., 1996. Quantitative magnetic resonance imaging of human brain development: ages 4-18. Cereb. Cortex 6, 551–560.

Gilbert, C.D., Li, W., 2012. Adult visual cortical plasticity. Neuron 75, 250–264.

Gilbert, C.D., Li, W., 2013. Top-down influences on visual processing. Nat. Rev. Neurosci. 14, 350–363.

Gilmore, J.H., Shi, F., Woolson, S.L., Knickmeyer, R.C., Short, S.J., Lin, W., et al., 2012. Longitudinal development of cortical and subcortical gray matter from birth to 2 years. Cereb. Cortex 22, 2478–2485.

Gogtay, N., Giedd, J.N., Lusk, L., Hayashi, K.M., Greenstein, D., Vaituzis, A.C., et al., 2004. Dynamic mapping of human cortical development during childhood through early adulthood. Proc. Natl. Acad. Sci. U.S.A. 101, 8174–8179.

Goldman-Rakic, P.S., 1987. Development of cortical circuitry and cognitive function. Child Dev. 58, 601–622.

Goodson, J.L., Thompson, R.R., 2010. Nonapeptide mechanisms of social cognition, behavior and species-specific social systems. Curr. Opin. Neurobiol. 20, 784–794.

Gould, S.J., 1977. Ontogeny and Phylogeny. Harvard University Press, Cambridge, Massachusetts.

Graybiel, A.M., 2008. Habits, rituals, and the evaluative brain. Annu. Rev. Neurosci. 31, 359–387.

Graybiel, A.N., 1998. The basal ganglia and chunking of action repertoires. Neurobiol. Learn. Mem. 70, 119–136.

Greene, D.J., Laumann, T.O., Dubis, J.W., Ihnen, S.K., Neta, M., Power, J.D., et al., 2014. Developmental changes in the organization of functional connections between the basal ganglia and cerebral cortex. J. Neurosci. 34, 5842–5854.

Gu, S., Satterthwaite, T.D., Medaglia, J.D., Yang, M., Gur, R.E., Gur, R.C., et al., 2015. Emergence of system roles in normative neurodevelopment. Proc. Natl. Acad. Sci. 112, 13681–13686.

Gunnar, M., Quevedo, K., 2006. The neurobiology of stress and development. Annu. Rev. Psychol. 58, 145–173.

Habas, P.A., Scott, J.A., Roosta, A., Rajagopalan, V., Kim, K., Rousseau, F., et al., 2012. Early folding patterns and asymmetries of the normal human brain detected from in utero MRI. Cereb. Cortex 22, 13–25.

Halley, A.C., 2015. Meta-analysis of ontogenetic brain/body allometry across mammals: implications for primate encephalization and fetal growth theories of relative brain size. Brain Behav. Evol. 85, 287–293.

Halley, A.C., 2016. Prenatal brain-body allometry in mammals. Brain Behav. Evol. 88 (1), 14–24.

Hamasaki, T., Leingärtner, A., Ringstedt, T., O'Leary, D.D., 2004. EMX2 regulates sizes and positioning of the primary sensory and motor areas in neocortex by direct specification of cortical progenitors. Neuron 43, 359–372.

Hammelrath, L., Škokić, S., Khmelinskii, A., Hess, A., van der Knaap, N., Staring, M., et al., 2016. Morphological maturation of the mouse brain: an in vivo MRI and histology investigation. Neuroimage 125, 144–152.

Haun, D.B.M., Rekers, Y., Tomasello, M., 2014. Children conform to the behavior of peers; Other great apes stick with what they know. Psychol. Sci. 25, 2160–2167.

Hawkes, K., 2003. Grandmothers and the evolution of human longevity. Am. J. Hum. Biol. 15, 380–400.

Hawkes, K., 2010. How grandmother effects plus individual variation in frailty shape fertility and mortality: guidance from human–chimpanzee comparisons. Proc. Nat. Acad. Sci. U.S.A. 107 (Suppl. 2), 8977–8984.

Hawkes, K., O'Connell, J.F., Jones, N.B., Alvarez, H., Charnov, E.L., 1998. Grandmothering, menopause, and the evolution of human life histories. Proc. Natl. Acad. Sci. 95, 1336–1339.

Haxby, J.V., Hoffman, E.A., Gobbini, M.I., 2000. The distributed human neural system for face perception. Trends Cognit. Sci. 4, 223–233.

Hedden, T., Schultz, A.P., Rieckmann, A., Mormino, E.C., Johnson, K.A., Sperling, R.A., et al., 2016. Multiple brain markers are linked to age-related variation in cognition. Cereb. Cortex 26, 1388–1400.

Hensch, T.K., 2004. Critical period regulation. Annu. Rev. Neurosci. 27, 549–579.

Hofman, M.A., 1989. On the evolution and geometry of the brain in mammals. Prog. Neurobiol. 32, 137–158.

Hoskins, S.G., 1990. The metamorphosis of the amphibian eye. J. Neurobiol. 21, 970–989.

Hrdy, S.B., 2011. Mothers and others: the evolutionary origins of mutual understanding. Harvard University Press, Massachusetts.

Huang, H., Jeon, T., Sedmak, G., Pletikos, M., Vasung, L., Xu, X., et al., 2013. Coupling diffusion imaging with histological and gene expression analysis to examine the dynamics of cortical areas across the fetal period of human brain development. Cereb. Cortex 23, 2620–2631.

Huang, X., Stodieck, S.K., Goetze, B., Cui, L., Wong, M.H., Wenzel, C., Hosan, L., Dong, Y., Lowel, S., Schlüter, O.M., 2015. Progressive maturation of silent synapses governs the duration of a critical period. Proc. Natl. Acad. Sci. 112, E3131–E3140.

Hübener, M., Bonhoeffer, T., 2014. Neuronal plasticity: beyond the critical period. Cell 159, 727–737.

Huttenlocher, P.R., Dabholkar, A.S., 1997. Regional differences in synaptogenesis in human cerebral cortex. J. Comp. Neurol. 387, 167–178.

Huttenlocher, P.R., de Courten, C., Garey, L.J., Van der Loos, H., 1982. Synaptogenesis in human visual cortex — evidence for synapse elimination during normal development. Neurosci. Lett. 33, 247–252.

Ingalhalikar, M., Parker, D., Ghanbari, Y., Smith, A., Hua, K., Mori, S., et al., 2015. Connectome and maturation profiles of the developing mouse brain using diffusion tensor imaging. Cereb. Cortex 25, 2696–2706.

Innocenti, G.M., Price, D.J., 2005. Exuberance in the development of cortical networks. Nat. Rev. Neurosci. 6, 955–965.

Isler, K., van Schaik, C.P., 2009. Why are there so few smart mammals (but so many smart birds)? Biol. Lett. 5, 125–129.

Isler, K., van Schaik, C.P., 2012. Allomaternal care, life history and brain size evolution in mammals. J. Hum. Evol. 63, 52–63.

Ivy, G.O., Killackey, H.P., 1981. The ontogeny of the distribution of callosal projection neurons in the rat parietal cortex. J. Comp. Neurol. 195, 367–389.

Janowsky, J.S., Finlay, B.L., 1983. Cell degeneration in the early development of forebrain and cerebellum. Anat. Embryol. 167, 439–447.

Janowsky, J.S., Finlay, B.L., 1986. Normal neuron loss and axon retraction in early development: their significance for the outcome of perinatal brain damage. Dev. Med. Child Neurol. 28, 375–389.

Jbabdi, S., Sotiropoulos, S.N., Haber, S.N., Van Essen, D.C., Behrens, T.E., 2015. Measuring macroscopic brain connections in vivo. Nat. Neurosci. 18, 1546–1555.

Jerison, H.J., 1973. Evolution of the Brain and Intelligence. Academic Press, New York.

Jerison, H.J., 1997. Evolution of prefrontal cortex. In: Krasnegor, N.A., Lyon, G.R., Goldman-Rakic, P.S. (Eds.), Development of the Prefrontal Cortex: Evolution, Neurobiolgy and Behavior. Pall H. Brooks Publishing Co., Baltimore, pp. 9–26.

Johnson, M.H., 2005. Subcortical face processing. Nat. Rev. Neurosci. 6, 766–774.

Jones, K.E., Orme, K.E., Safi, K., Sechrest, W., Boakes, E.H., et al., 2009. PanTHERIA: a species-level database of life history, ecology, and geography of extant and recently extinct mammals. Ecology 90, 2648–2700.

Jusczyk, P.W., Friederici, A.D., Wessels, J.M.I., Svenkerud, V., Jusczyk, A.M., 1993. Infants' sensitivity to the sound pattern of native-language words. J. Mem. Lang. 32, 402–420.

Kaas, J.H., 1997. Topographic maps are fundamental to sensory processing. Brain Res. Bull. 44, 107–112.

Kang, H.J., Kawasawa, Y.I., Cheng, F., Zhu, Y., Xu, X., Li, M., et al., 2011. Spatio-temporal transcriptome of the human brain. Nature 478, 483–489.

Katz, L.C., Crowley, J.C., 2002. Development of cortical circuits: lessons from ocular dominance columns. Nat. Rev. Neurosci. 3, 34–42.

Kennedy, G.E., 2005. From the ape's dilemma to the weanling's dilemma: early weaning and its evolutionary context. J. Hum. Evol. 48, 123–145.

Kiecker, C., Lumsden, A., 2012. The role of organizers in patterning the nervous system. Annu. Rev. Neurosci. 35, 347–367.

Konishi, M., 2003. Coding of auditory space. Annu. Rev. Neurosci. 26, 31–55.

Kuklisova-Murgasova, M., Aljabar, P., Srinivasan, L., Counsell, S.J., Doria, V., Serag, A., et al., 2011. A dynamic 4D probabilistic atlas of the developing brain. Neuroimage 54, 2750–2763.

LaMantia, A.-S., Rakic, P., 1990. Axon overproduction and elimination in the corpus callosum of the developing rhesus monkey. J. Neurosci. 10, 2156–2175.

Lee, P.C., 1996. The meanings of weaning: growth, lactation, and life history. Evol. Anthropol. Issues News Rev. 5, 87–98.

Leroy, F., Glasel, H., Dubois, J., Hertz-Pannier, L., Thirion, B., Mangin, J.-F., et al., 2011. Early maturation of the linguistic dorsal pathway in human infants. J. Neurosci. 31, 1500–1506.

Levi, D.M., Knill, D.C., Bavelier, D., 2015. Stereopsis and amblyopia: a mini-review. Vis. Res. 114, 17–30.

Li, G., Lin, W., Gilmore, J.H., Shen, D., 2015. Spatial patterns, longitudinal development, and hemispheric asymmetries of cortical thickness in infants from birth to 2 years of age. J. Neurosci. 35, 9150–9162.

Lieberman, P., 1984. The Biology and Evolution of Language. Harvard University Press, Cambridge, MA.

Lieven, E., Stoll, S., 2013. Early communicative development in two cultures: a comparison of the communicative environments of children from two cultures. Hum. Dev. 56, 178–206.

Locke, J.L., Bogin, B., 2006. Language and life history: a new perspective on the development and evolution of human language. Behav. Brain Sci. 29, 259–325.

London, S.E., Dong, S., Replogle, K., Clayton, D.F., 2009. Developmental shifts in gene expression in the auditory forebrain during the sensitive period for song learning. Dev. Neurobiol. 69, 437–450.

Lumsden, A., Krumlauf, R., 1996. Patterning the vertebrate neuraxis. Science 274, 1109–1115.

Mahmoudzadeh, M., Dehaene-Lambertz, G., Fournier, M., Kongolo, G., Goudjil, S., Dubois, J., et al., 2013. Syllabic discrimination in premature human infants prior to complete formation of cortical layers. Proc. Natl. Acad. Sci. 110, 4846–4851.

Makropoulos, A., Aljabar, P., Wright, R., Hüning, B., Merchant, N., Arichi, T., et al., 2016. Regional growth and atlasing of the developing human brain. Neuroimage 125, 456–478.

Mampe, B., Friederici, A.D., Christophe, A., Wermke, K., 2009. Newborns' cry melody is shaped by their native language. Curr. Biol. 19, 1994–1997.

May, A., 2011. Experience-dependent structural plasticity in the adult human brain. Trends Cognit. Sci. 15, 475–482.

McClelland, J.L., McNaughton, B.L., O'Reilly, R.C., 1995. Why there are complementary learning systems in the hippocampus and neocortex: insights from successes and failures of connectionist models of learning and memory. Psychol. Bull. 102, 419–457.

McMurray, B., 2007. Defusing the childhood vocabulary explosion. Science 317, 631.

Merabet, L.B., Pascual-Leone, A., 2010. Neural reorganization following sensory loss: the opportunity of change. Nat. Rev. Neurosci. 11, 44–52.

Miller, B., Nagy, D., Finlay, B.L., Chance, B., Kobayashi, A., Nioka, S., 1993. Consequences of reduced cerebral blood flow in brain development: I. Gross morphology, histology, and callosal connectivity. Exp. Neurol. 124, 326–342.

Miller, D.J., Duka, T., Stimpson, C.D., Schapiro, S.J., Baze, W.B., McArthur, M.J., et al., 2012. Prolonged myelination in human neocortical evolution. Proc. Natl. Acad. Sci. 109, 16480–16485.

Miller, J.A., Ding, S.-L., Sunkin, S.M., Smith, K.A., Ng, L., Szafer, A., et al., 2014. Transcriptional landscape of the prenatal human brain. Nature 508, 199–206.

Missler, M., Eins, S., Merker, H.J., Rothe, H., Wolff, J.R., 1993. Prenatal and postnatal development of the primary visual cortex of the common marmoset .1. a changing space for synaptogenesis. J. Comp. Neurol. 333, 41–52.

Moon, C., Lagercrantz, H., Kuhl, P.K., 2013. Language experienced in utero affects vowel perception after birth: a two-country study. Acta Paediatr. 102, 156–160.

Moriceau, S., Roth, T.L., Sullivan, R.M., 2010. Rodent model of infant attachment learning and stress. Dev. Psychobiol. 52, 651–660.

Morris, D., 2012. The Naked Man: A Study of the Male Body. Random House, New York.

Nathanielscz, P.W., 1998. Comparative studies on the initiation of labor. Eur. J. Obstetr. Gynecol. Reprod. Biol. 78, 127–132.

Newbury, D.F., Monaco, A.P., 2010. Genetic advances in the study of speech and language disorders. Neuron 68, 309–320.

Niewenhuys, R., 2012. The insular cortex: a review. In: Hofman, M.A., Falk, D. (Eds.), Evolution of the Primate Brain: From Neuron to Behavior, vol. 195. Elsevier, pp. 123–153.

O'Leary, D.D.M., Koester, S.E., 1993. Development of projection neuron types, axon pathways, and patterned connections of the mammalian cortex. Neuron 10, 991–1006.

Oishi, K., Faria, A.V., Yoshida, S., Chang, L., Mori, S., 2013. Quantitative evaluation of brain development using anatomical MRI and diffusion tensor imaging. Int. J. Dev. Neurosci. 31, 512–524.

Oishi, K., Nakagawa, N., Tachikawa, K., Sasaki, S., Aramaki, M., Hirano, S., et al., 2016. Identity of neocortical layer 4 neurons is specified through correct positioning into the cortex. eLife 5, e10907.

Oppenheim, R.W., 1991. Cell death during development of the nervous system. Annu. Rev. Neurosci. 14, 453–502.

Pagel, M.D., Harvey, P.H., 1988. How mammals produce large-brained offspring. Evolution 101, 948–957.

Pandit, A.S., Robinson, E., Aljabar, P., Ball, G., Gousias, I.S., Wang, Z., et al., 2014. Whole-brain mapping of structural connectivity in infants reveals altered connection strength associated with growth and preterm birth. Cereb. Cortex 24, 2324–2333.

Pascalis, O., Kelly, D.J., 2009. The origins of face processing in humans: phylogeny and ontogeny. Perspect. Psychol. Sci. 4, 200–209.

Passingham, R.E., 1985. Rates of brain development in mammals including man. Brain Behav. Evol. 26, 167–175.

Passingham, R.E., Smaers, J.B., 2014. Is the prefrontal cortex especially enlarged in the human brain allometric relations and remapping factors. Brain Behav. Evol. 84, 156–166.

Petanjek, Z., Judas, M., Simic, G., Rasin, M.R., Uylings, H.B.M., Rakic, P., et al., 2011. Extraordinary neoteny of synaptic spines in the human prefrontal cortex. Proc. Natl. Acad. Sci. 108, 13281–13286.

Pfister, B.J., Iwata, A., Meaney, D.F., Smith, D.H., 2004. Extreme stretch growth of integrated axons. J. Neurosci. 24, 7978–7983.

Pletikos, M., Sousa, A.M.M., Sedmak, G., Meyer, K., Zhu, Y., Cheng, F., et al., 2014. Temporal specification and bilaterality of human neocortical topographic gene expression. Neuron 81, 321–332.

Promislow, D.E., 1991. Senescence in natural populations of mammals: a comparative study. Evolution 1010, 1869–1887.

Pulvermüller, F., Schumann, J.H., 1994. Neurobiological mechanisms of language acquisition. Lang. Learn. 44, 681–734.

Rajagopalan, V., Scott, J., Habas, P.A., Kim, K., Corbett-Detig, J., Rousseau, F., et al., 2011. Local tissue growth patterns underlying normal fetal human brain gyrification quantified in utero. J. Neurosci. 31, 2878–2887.

Rakic, P., 1974. Neurons in the rhesus monkey visual cortex: systematic relation between time of origin and eventual disposition. Science 183, 425–427.

Rakic, P., 1977. Genesis of the dorsal lateral geniculate nucleus in the rhesus monkey: site and time of origin, kinetics of proliferation, routes of migration and pattern of distribution of neurons. J. Comp. Neurol. 176, 23–52.

Rakic, P., 2002. Neurogenesis in adult primate neocortex: an evaluation of the evidence. Nat. Rev. Neurosci. 3, 65–71.

Rakic, P., Bourgeois, J.P., Goldman-Rakic, P.S., 1994. Synaptic development of the cerebral cortex: implications for learning, memory, and mental illness. Self Organ. Brain 102, 227–243.

Rakic, P., Bourgeouis, J.-P., Eckenhoff, M., Zecevic, N., Goldman-Rakic, P.S., 1986. Concurrent overproduction of synapses in diverse regions of the primate cerebral cortex. Science 232, 232–235.

Raznahan, A., Greenstein, D., Lee, N.R., Clasen, L.S., Giedd, J.N., 2012. Prenatal growth in humans and postnatal brain maturation into late adolescence. Proc. Natl. Acad. Sci. 109, 11366–11371.

Raznahan, A., Shaw, P., Lalonde, F., Stockman, M., Wallace, G.L., Greenstein, D., et al., 2011. How does your cortex grow? J. Neurosci. 31, 7174–7177.

Raznahan, A., Shaw, P.W., Lerch, J.P., Clasen, L.S., Greenstein, D., Berman, R., et al., 2014. Longitudinal four-dimensional mapping of subcortical anatomy in human development. Proc. Natl. Acad. Sci. 111, 1592–1597.

Reep, R., Darlington, R.B., Finlay, B.L., 2007. The limbic system in mammalian brain evolution. Brain Behav. Evol. 70, 57–70.

Reichard, U.H., Barelli, C., 2008. Life history and reproductive strategies of Khao Yai Hylobates lar: implications for social evolution in apes. Int. J. Primatol. 29, 823–844.

Robbins, A.M., Robbins, M.M., Gerald-Steklis, N., Steklis, H.D., 2006. Age-related patterns of reproductive success among female mountain gorillas. Am. J. Phys. Anthropol. 131, 511–521.

Robertson, S., Dierker, L., Sorokin, Y., Rosen, M., 1982. Human fetal movement: spontaneous oscillations near one cycle per minute. Science 218, 1327–1330.

Robson, S.L., Wood, B., 2008. Hominin life history: reconstruction and evolution. J. Anat. 212, 394–425.

Robson, S.L., van Schaik, C.P., Hawkes, K., 2006. The Derived Features of Human Life History. In: Hawkes, K., Paine, R.R. (Eds.), The Evolution of Human Life History. SAR Press, Santa Fe, USA, pp. 17–44.

Rodger, J., Dunlop, S.A., 2015. Central nerve regeneration in reptiles. Neural Regen 43, 1–24.

Rosenberg, K., Trevathan, W., 1995. Bipedalism and human birth: the obstetrical dilemma revisited. Evol. Anthropol. Issues News Rev. 4, 161–168.

Rowe, N., 1996. The Pictorial Guide to the Living Primates. Pogonias Press, New York.

Roy, B.C., Frank, M.C., DeCamp, P., Miller, M., Roy, D., 2015. Predicting the birth of a spoken word. Proc. Natl. Acad. Sci. 112, 12663–12668.

Royle, N.J., Smiseth, P.T., Kölliker, M., 2012. The Evolution of Parental Care. Oxford University Press, Oxford.

Saffran, J.R., Aslin, R.N., Newport, E.L., 1996. Statistical learning by 8-month-old infants. Science 274, 1926–1928.

Sakai, T., Matsui, M., Mikami, A., Malkova, L., Hamada, Y., Tomonaga, M., et al., 2013. Developmental patterns of chimpanzee cerebral tissues provide important clues for understanding the remarkable enlargement of the human brain. Proc. R. Soc. Biol. Sci. 280, 753–757.

Sanchez, M.M., Ladd, C.O., Plotsky, P., 2001. Early adverse experience as a developmental risk factor for later psychopathology: evidence from rodent and primate models. Dev. Psychopathol. 13, 419–449.

Sanes, J.R., Lichtman, J.W., 1999. Development of the vertebrate neuro-muscular junction. Annu. Rev. Neurosci. 22, 389–442.

Schall, J.D., Leventhal, A.G., 1987. Relationships between ganglion cell dendritic structure and retinal topography in the cat. J. Comp. Neurol. 257, 149–159.

Scheibel, A.B., 1993. Dendritic Structure and Language Development. In: Boysson-Bardies, B., Schonen, S., Jusczyk, P., McNeilage, P., Morton, J. (Eds.), Developmental Neurocognition: Speech and Face Processing in the First Year of Life. Springer, Dordrecht, Netherlands, pp. 51–62.

Schmidhuber, J., 2015. Deep learning in neural networks: an overview. Neural Netw. 61, 85–117.

Schoenemann, P.T., 2006. Evolution of the size and functional areas of the human brain. Annu. Rev. Anthropol. 35, 379–406.

Schultz, A.H., 1949. Sex differences in the pelves of primates. Am. J. Phys. Anthropol. 7, 401–424.

Schwartz, M.L., Rakic, P., Goldman-Rakic, P.S., 1991. Early phenotype expression of cortical neurons: evidence that a subclass of migrating neurons have callosal axons. Proc. Natl. Acad. Sci. 88, 1354–1358.

Sellen, D.W., 2001. Comparison of infant feeding patterns reported for nonindustrial populations with current recommendations. J. Nutr. 131, 2707–2715.

Semendeferi, K., Lu, A., Schenker, N., Damasio, H., 2002. Humans and great apes share a large frontal cortex. Nat. Neurosci. 5, 272–276.

Senft, S.L., Woolsey, T.A., 1991. Growth of thalamic afferents into mouse barrel cortex. Cereb. Cortex 1, 308–336.

Shakeshaft, N.G., Plomin, R., 2015. Genetic specificity of face recognition. Proc. Natl. Acad. Sci. 112, 12887–12892.

Shaw, P., Greenstein, D., Lerch, J., Clasen, L., Lenroot, R., Gogtay, N., et al., 2006. Intellectual ability and cortical development in children and adolescents. Nature 440, 676–679.

Sherman, D.L., Brophy, P.J., 2005. Mechanisms of axon ensheathment and myelin growth. Nat. Rev. Neurosci. 6, 683–690.

Shiraishi, N., Katayama, A., Nakashima, T., Yamada, S., Uwabe, C., Kose, K., et al., 2015. Morphology and morphometry of the human embryonic brain: a three-dimensional analysis. Neuroimage 115, 96–103.

Silbereis, J.C., Pochareddy, S., Zhu, Y., Li, M., Sestan, N., 2016. The cellular and molecular landscapes of the developing human central nervous system. Neuron 89, 248–268.

Simmonds, D., Hallquist, M.N., Asato, M., Luna, B., 2013. Developmental stages and sex differences of white matter and behavioral development through adolescence: a longitudinal diffusion tensor imaging (DTI) study. Neuroimage 92, 356–368.

Simonds, R.J., Scheibel, A.B., 1989. The postnatal development of the motor speech area: a preliminary study. Brain Lang. 37, 42–58.

Smith, B.H., Tompkins, R.L., 1995. Toward a life history of the. Hominidae. Annu. Rev. Anthropol 20, 257–279.

Stearns, S.C., 1992. The Evolution of Life Histories. Oxford University Press, Oxford.

Stein, B.E., Stanford, T.R., 2008. Multisensory integration: current issues from the perspective of the single neuron. Nat. Rev. Neurosci. 9, 255–266.

Syal, S., Finlay, B.L., 2011. Thinking outside the cortex: social motivation in the evolution and development of language. Dev. Sci. 14, 417–430.

Tacutu, R., Craig, T., Budovsky, A., Wuttke, D., Lehmann, G., et al., 2013. Human ageing genomic resources: integrated databases and tools for the biology and genetics of ageing. Nucleic Acids Res. 41, D1027–D1033.

Takahashi, E., Folkerth, R.D., Galaburda, A.M., Grant, P.E., 2012. Emerging cerebral connectivity in the human fetal brain: an MR tractography study. Cereb. Cortex 22, 455–464.

Takahashi, E., Hayashi, E., Schmahmann, J.D., Ellen Grant, P., 2014. Development of cerebellar connectivity in human fetal brains revealed by high angular resolution diffusion tractography. Neuroimage 96, 326–333.

Tate, M.C., Lindquist, R.A., Nguyen, T., Sanai, N., Barkovich, A.J., Huang, E.J., et al., 2015. Postnatal growth of the human pons: a morphometric and immunohistochemical analysis. J. Comp. Neurol. 523, 449–462.

Tomasello, M., 1999. The Cultural Origins of Human Cognition. Harvard University Press, Harvard.

Trevathan, W., 2010. Ancient Bodies, Modern Lives. Oxford University Press, Oxford.

Udin, S.B., Fawcett, J.W., 1988. The formation of topographic maps. Annu. Rev. Neurosci. 11, 289–327.

Van der Loos, H., Welker, E., 1985. Development and plasticity of somatosensory brain maps. In: Rowe, M., Willis, J.W.D. (Eds.), Development, Organization, and Processing in Somatosensory Pathways; Neurology and Neurobiology, vol. 14. Alan R. Liss, New York, pp. 53–67.

Van Essen, D., 1997. A tension-based theory of morphogenesis and compact wiring in the central nervous system. Nature 385, 313–318.

Walker, L., Chang, L.-C., Nayak, A., Irfanoglu, M.O., Botteron, K.N., McCracken, J., et al., 2016. The diffusion tensor imaging (DTI) component of the NIH MRI study of normal brain development. Neuroimage 124, 1125–1130.

Werker, J.F., Tees, R.C., 1992. The organization and reorganization of human speech perception. Annu. Rev. Neurosci. 15, 377–402.

Wierenga, L.M., Langen, M., Oranje, B., Durston, S., 2013. Unique developmental trajectories of cortical thickness and surface area. NeuroImage 87.

Wiesel, T.N., 1999. Early explorations of the development and plasticity of the visual cortex: a personal view. J. Neurobiol. 41, 7–9.

Wiesel, T.N., Hubel, D.H., 1965. Comparison of the effects of unilateral and bilateral eye closure on cortical unit responses in kittens. J. Neurophysiol 28, 1029–1040.

Windrem, M.S., Jan de Beur, S., Finlay, B.L., 1988. Control of cell number in the developing neocortex: II. Effects of corpus callosum transection. Dev. Brain Res. 43, 13–22.

Wong, R.O.L., 1990. Differential growth and remodelling of ganglion cell dendrites in the postnatal rabbit retina. J. Comp. Neurol. 294, 109–132.

Wong, R.O.L., 1999. Retinal waves and visual system development. Annu. Rev. Neurosci. 22, 29–47.

Wong, R.O.L., Chernjavsky, A., Smith, S.J., Shatz, C.J., 1995. Early functional neural networks in the developing retina. Nature 374, 716–718.

Woo, T.U., Finlay, B.L., 1996. Cortical target depletion and ingrowth of geniculocortical axons: implications for cortical specification. Cereb. Cortex 6, 457–469.

Woo, T.U., Niederer, J.K., Finlay, B.L., 1996. Cortical target depletion and the developing lateral geniculate nucleus: implications for trophic dependence. Cereb. Cortex 6, 446–456.

Workman, A.D., Charvet, C.J., Clancy, B., Darlington, R.B., Finlay, B.L., 2013. Modeling transformations of neurodevelopmental sequences across mammalian species. J. Neurosci. 33, 7368–7383.

Yamamori, T., 2011. Selective gene expression in regions of primate neocortex: implications for cortical specialization. Prog. Neurobiol. 94, 201–222.

Young, L.J., Wang, Z., 2004. The neurobiology of pair bonding. Nat. Neurosci 7, 1048–1054.

Zatorre, R.J., Fields, R.D., Johansen-Berg, H., 2012. Plasticity in gray and white: neuroimaging changes in brain structure during learning. Nat. Neurosci. 15, 528–536.

Zecevic, N., Bourgeois, J.-P., Rakic, P., 1989. Changes in synaptic density in motor cortex of rhesus monkey during fetal and postnatal life. Dev. Brain Res. 50, 11–32.

Zecevic, N., Rakic, P., 1991. Synaptogenesis in monkey somatosensory cortex. Cereb. Cortex 1, 510–523.

Zilles, K., Palomero-Gallagher, N., Amunts, K., 2013. Development of cortical folding during evolution and ontogeny. Trends Neurosci. 36, 275–284.

Relevant Websites

http://neurosciencelibrary.org/ – Brain Museum Source.

www.translatingtime.net – Home Page for Translating Time Utility and Model.

35

Human Association Cortex: Expanded, Untethered, Neotenous, and Plastic

F.M. Krienen, R.L. Buckner

Harvard Medical School, Boston, MA, United States; Broad Institute, Cambridge, MA, United States; Harvard University, Cambridge, MA, United States; Massachusetts General Hospital, Boston, MA, United States

35.1 Association Cortex Is Disproportionately Expanded in Great Apes and Particularly Humans

Compared with other primates, the human neocortex is remarkably enlarged and exhibits distinct patterns of connectivity, gene expression, development, and maturation (Preuss et al., 2004; Vallender et al., 2008; Preuss, 2012; Sherwood and Duka, 2012; Bozek et al., 2014). Certain aspects of this expansion and reorganization can be predicted by scaling rules either specific to primates or applicable to all mammals. For example, the size of extrastriate cortex, primary visual area V1, and other primary sensory and motor areas scale predictably with the total size of the cerebral cortex across mammals. In general, larger brains tend to have proportionately smaller primary sensory and motor areas (Kaskan et al., 2005). Primates are particularly encephalized, that is, they have large brain size relative to body size and also have larger brains in absolute terms (Striedter, 2005). Hence the proportion of the neocortex devoted to conserved primary sensory and motor areas tends to be lower in primates and particularly in the large-brained apes (Krubitzer, 2007; Fig. 35.1). Similarly, proportions devoted to allocortex—three- or four-layered cortex including the parahippocampus, hippocampus, and piriform/olfactory cortex—have also diminished in primates with the expansion of neocortex.

In primates, the relative diminution of primary sensory cortex is accompanied by the expansion of fields devoted to higher-order cortical processing, especially in association cortex. Often, the prefrontal cortex is singled out as being exceptional in humans. While prefrontal cortex is disproportionately enlarged as compared with primary sensory areas, so too are parietal and temporal association regions. Thus, a characteristic feature of primate and particularly human neocortex is that it possesses large, distributed zones of association cortex. Notably, these regions are physically disconnected from one another on the cortical surface (Hill et al., 2010), interposed between the conserved sensory and motor areas. There is ongoing debate about what physical changes to cortex and white matter underlie this expansion (Schoenemann et al., 2005; Barton and Venditti, 2013; Passingham and Smaers, 2014; Smaers, 2013; Smaers et al., 2011), as well as the extent to which the organization of association cortex in humans can be predicted from trends observed in other primates (Herculano-Houzel, 2009).

In addition to these questions, there is the basic issue of how relative enlargement of association regions could have occurred in multiple locations across the cortical mantle. How might coordinated expansion occur across distributed, noncontiguous portions of association cortex? This puzzle is compounded by the observation that association cortex appears to preferentially wire to itself, forming long-range, distributed networks that span considerable distance. What is clear is that such an organization is a departure from the predominantly local corticocortical wiring observed in sensory areas.

Other observations similarly point to a departure in the developmental trajectories of the conserved areas compared with association cortex. Postnatal expansion of association cortex in humans is developmentally protracted, mirroring evolutionary expansion estimated by comparing humans to macaques (Hill et al., 2010). Association regions myelinate late—well into adolescence in humans and other great apes. Projection neurons within association cortex have morphological characteristics that distinguish them from their counterparts in primary sensory areas (Elston et al., 2006; Bianchi et al., 2012).

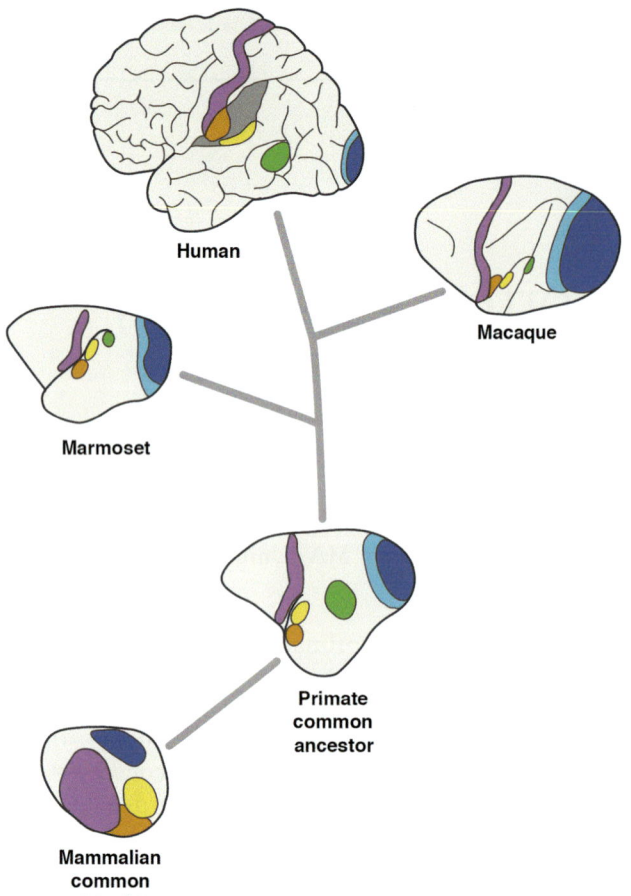

FIGURE 35.1 Relative scaling of conserved sensory and motor areas in primates. Relative to a hypothetical estimation of the placental mammalian common ancestor, extant primates tend to have a greater percentage of the cortical mantle that falls between primary and secondary sensory systems. A hypothetical estimation of the primate common ancestor suggests that this trend was already present in the early divergence of primates. *Adapted from Krubitzer, L., 2007. The magnificent compromise: cortical field evolution in mammals. Neuron 56 (2), 201–208.*

Though the effects are subtle as compared with the relatively high degree of transcriptional conservation across the neocortex, sensory regions also have molecular identities that distinguish them from association regions (Hawrylycz et al., 2012). Functionally, association cortex is implicated in some of our most advanced cognitive behaviors, including complex reasoning, social cognition, and imagination (Cabeza and Nyberg, 1997; Mesulam, 1990, 1998; Fuster, 2004; Buckner et al., 2008; Badre and D'Esposito, 2009; Spreng et al., 2009; Vendetti and Bunge, 2014).

Why does association cortex has these properties, and how did they arise? In this chapter, we explore these peculiarities of association cortex as a way to understand how they impact cortical processing and how they might contribute to the cognitive faculties that are advanced in primates (especially humans). We offer a framework to think about how association cortex comes to take a network organization distinct from conserved sensory regions (Buckner and Krienen, 2013). In short, the rapid expansion of the neocortex over the recent course of hominin evolution may have created distributed expanses of cortex that lie distant from the strong influences of ancient patterning centers that cascade to confer areal fate and specify connectional properties. The "untethering" from the strong constraints of activity-dependent patterning from sensory relays leads to a network profile in which the expanded cortical regions preferentially wire together. This chapter puts particular emphasis on how conserved developmental constraints play out on a massively scaled cortical sheet. It also highlights gaps in our present knowledge of how these constraints, which have been primarily studied in rodents, might translate to humans and other primates.

35.2 Long-Range Projections Connect Distributed Regions of Association Cortex Together

The mammalian neocortex is generally composed of six layers, each associated with a distinct profile of cell types, densities, connectional properties, and developmental histories. The distribution of connections within the six layers of neocortex provides information about their origin and/or destination (eg, corticocortical, corticofugal), though the extent to which connections respect laminar classifications can differ by area (Selemon and Goldman-Rakic, 1988; Felleman and Van Essen, 1991; Barbas and Zikopoulos, 2007). Variation exists across species and across cortical areas. Layers may be absent or further subdivided. In rodents, Layer II and Layer III, known as supragranular layers since they are located above the "granular" Layer IV, are the origin of the majority of corticocortical connections. In mice, neurons in infragranular layers (Layer V and Layer VI) disproportionately project to subcortical targets. Cortically projecting neurons that reside in deeper layers are relatively scarce in sensory areas (Mitchell and Maklis, 2004) but may be more prominent in association areas such as temporal area TE (Zingg et al., 2014). In rodents, most deep layer cortically projecting neurons also maintain a dual projection to subcortical structures, whereas few supragranular neurons have these dual projections. This observation led to the suggestion that the cortically projecting deep-layer neurons may represent a population that derived from the evolutionarily older deep-layer corticofugal projecting neurons (Mitchell and Macklis, 2004; Fame et al., 2011; Custo Greig et al., 2013). In primates, the relative proportion of supra- versus infragranular neurons that project intracortically varies across development (Meissirel et al., 1991) and across areas, and displays a complex relationship to the distance between connected areas (Markov et al., 2013).

In rodents, ipsilateral corticocortical connections tend to be local. With some exceptions (Frostig et al., 2008; Zingg et al., 2014), corticocortical connections tend to favor adjacent fields. One prevalent form of long-range cortiocortical projection, whether in primary sensory or association regions, is to the homotopic region in the opposite hemisphere (Fig. 35.2; Oh et al., 2014). These two types of cortical projections also have different developmental time courses; callosal projections peak in late embryonic and early postnatal ages, while ipsilateral projections peak about a week later; both continue to be pruned and refined into the third postnatal week (Mitchell and Macklis, 2004).

In primates, as in rodents, the densest corticocortical connections tend to occur between neighboring regions. A series of elegant papers by Markov, Kennedy, and colleagues (Markov et al., 2013b,a, 2014, 2011; Ercsey-Ravasz et al., 2013) quantified connection strengths in macaque interareal networks and demonstrated that connection density tends to drop off logarithmically with distance. Given the dominance of local connections, it is the long-range connections that distinguish the connectivity profiles of neighboring regions. In the macaque, a neocortical area may project to dozens of

other cortical locations beyond its immediate neighborhood (Fig. 35.2). One study quantitatively mapped retrograde inputs to 29 cortical areas and found that each area received input from 26 to 87 of the 91 areas examined (Markov et al., 2013a). This work extends a large body of earlier literature demonstrating that cortical areas in primates are interconnected in complex networks that can span considerable distance (Jones and Powell, 1970; Barbas and Mesulam, 1981; Ungerleider and Desimone, 1986; Zeki and Shipp, 1988; Seltzer and Pandya, 1989; Felleman and Van Essen, 1991; Romanski et al., 1999; Saleem et al., 2008). Few studies to date have compared the corticocortical properties of rodents and primates directly using the same metrics. One recent paper to do so made the surprising discovery that a number of network properties, including the overall distribution of local and long-range connections, are similar for mice and primates (Horvat et al., 2016). A key difference is the presence of strong long-range connections between primary sensory areas in the mouse, which are largely absent in primates (Markov et al., 2014). Instead, in primates, long-distance connections are enriched between association areas.

In the 1980s, Patricia Goldman-Rakic and her colleagues offered a new paradigm for thinking about the organization and development of cortical projections in primates (Schwartz and Goldman-Rakic, 1984; Cavada and Goldman-Rakic, 1989a,b; Selemon and Goldman-Rakic, 1988; Goldman-Rakic, 1988; Schwartz and Goldman-Rakic, 1991). Guided by double-labeling experiments which permitted dual injections of tracer conjugated to different fluorescent probes in a single animal, they noted that parietal, prefrontal, and temporal areas as well as cingulate regions tended to be interconnected in what she termed "parallel systems of distributed neural networks." In her synthesis of the literature of the era, Goldman-Rakic (1988) noted that prefrontal cortex should not simply be considered a place for the integration of input streams, as was a prevailing view at the time. Instead, the data indicate that adjacent regions within prefrontal cortex connects to adjacent regions in parietal, temporal, and cingulate regions. This suggested that prefrontal cortex is not just a site of massive convergence, and further that subregions within prefrontal (and other association) zones could support processing of multiple streams of information in parallel through participation in distributed association networks.

More recently, neuroimaging techniques for inferring network organization have been used to probe the question of how association networks are organized in the human brain. Despite differences in methodological and analytical approach, these studies converge on the presence of distributed networks that interdigitate across the cortical mantle (reviewed in Raichle, 2011;

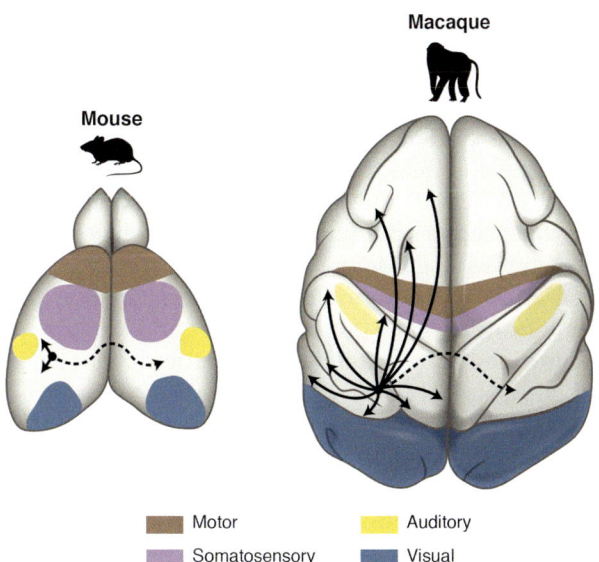

FIGURE 35.2 Long-range corticocortical connections in rodents and primates. A prevalent form of long-range corticocortical connection is to the homologous region in the opposite hemisphere. In both species, connections within ipsilateral cortex are predominantely local. In primates, association cortex in particular is characterized by long-range corticocortical connections that connect distributed regions together. Long-range corticocortical connections within association cortex may be a key building block of the information processing capabilities of the primate brain.

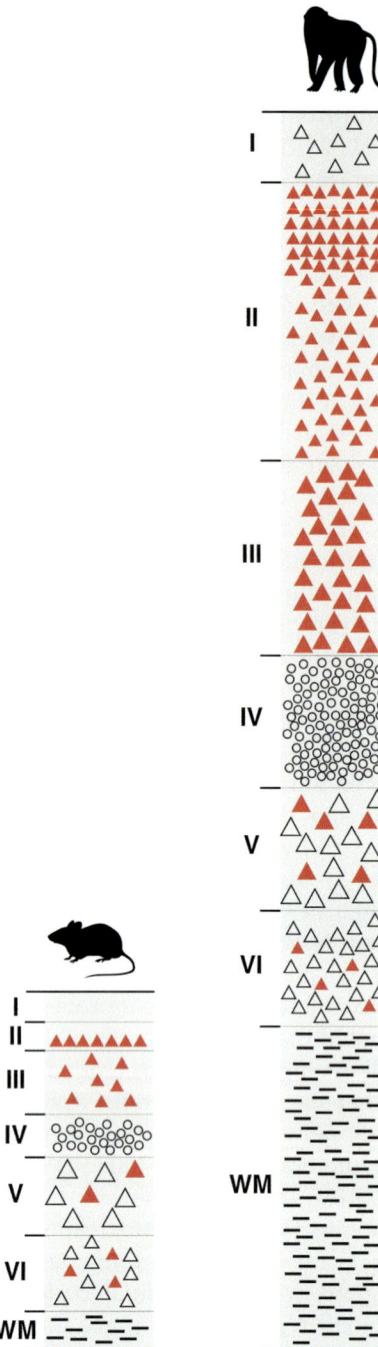

FIGURE 35.3 Evolutionary expansion of upper cortical layers in primates. Relative thickness of cortical layers differs between rodents and primates. The upper layers II and III are most expanded in primates and are also the layers that house the dominant proportion of corticocortically projecting neurons (*red*). *Adapted with permission from MacDonald, J.L., et al., 2013. Specification of cortical projection neurons. In: Rubenstein, J.L.R., Rakic, P. (Eds.), Patterning and Cell Type Specification in the Developing CNS and PNS, first ed. Academic Press, San Diego, CA, pp. 475–502.*

Fox and Raichle, 2007; Deco and Corbetta, 2011; Buckner et al., 2013; Van Essen, 2013). The expanded portions of association cortex are consistently shown to contain

multiple, distributed networks that connect prefrontal, parietal, temporal, and cingulate domains (Fig. 35.4; Yeo et al., 2011). These "association networks" stand out as having proportionately greater distant or long-range corticocortical connectivity compared with sensory and motor zones (Sepulcre et al., 2010) and increased intersubject variability (Mueller et al., 2013).

Core constraints underlying corticocortical connectivity in the mammalian cerebral cortex, such as minimizing path length, preserving modularity, and increasing specificity have been proposed as a way to anticipate how connectivity changes with increasing brain size (Cherniak, 1995; Striedter, 2005; Bullmore and Sporns, 2009; Markov et al., 2013b; Goulas et al., 2015). These general constraints likely go a long way in accounting for the prominence of long-range connections and distributed networks in large-brained primates such as humans. Less clear are the mechanisms that specify how long-range corticocortical networks become established and take their mature forms. Can we discover rules that provide insight into the developmental, molecular, and activity-dependent mechanisms that allow for an emphasis on distributed, long-range corticocortical networks in human association cortex? The next section reviews theories of how the subdivision of the cortex is achieved. Then, we discuss how these developmental rules might play out in the context of rapid expansion of the neocortex over human evolution.

35.3 Theories of Areal Patterning

Cortical patterning emerges early as the cortical plate forms. The protomap hypothesis proposed that areal parcellation is achieved by overlapping gradients of molecules across the proliferative ventricular zone (VZ) that act as a prefiguring map during development. These gradients guide the formation of specific afferent innervation to appropriate destinations in the newly forming cortical plate (Rakic, 1988). Foundational experiments in the mouse illustrate that patterning of neocortical areas is achieved at early embryonic stages by signaling centers that secrete proteins (Bmps, Wnts, Fgfs, and Shh) carrying positional information. Within the VZ, transcription factors (such as *Emx2*, *Pax6*, *Nr2f1*) are expressed as gradients that confer regional fate identities to the developing cortical plate. Postmitotic neurons also express transcription factors that further delineate primary sensory and motor areas. Spontaneous and evoked extrinsic activity further shape cortical fields and their connectivity to the thalamus, subcortical structures, and each other (reviewed in Sur and Rubenstein, 2005).

Of interest, even before patterning centers were discovered and characterized in the mammalian central

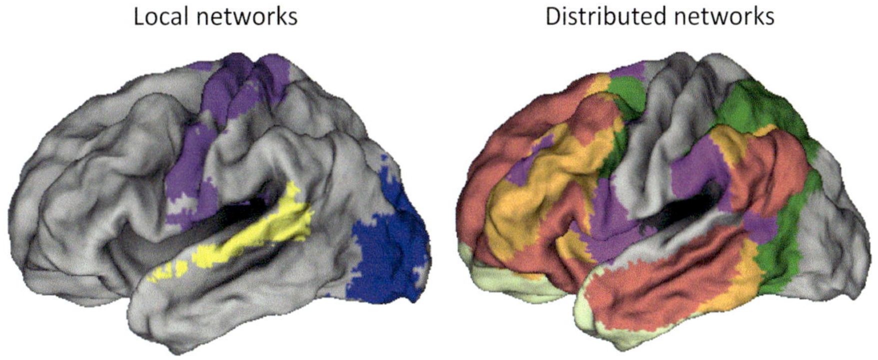

Local networks   Distributed networks

FIGURE 35.4 Balance of local and distant connectivity in human neocortex. Neuroimaging techniques reveal two broad patterns of connectivity in the human neocortex, which were anticipated by anatomical study of the primate. Local networks form in zones that include unimodal primary and secondary sensory regions. Association cortex contains multiple, distributed networks that connect prefrontal, parietal, temporal and cingulate regions. Each color represents an interconnected network. Note how highly interconnected networks, such as the network illustrated in orange that is important to cognitive control, spans widely distributed zones of association cortex (see Yeo et al., 2011).

nervous system, their existence was anticipated as a critical component of the development of discrete cortical areas and of the evolution of new ones. In 1988, a "group report" produced by Kaas, Allman, Rakic, and colleagues (Stryker et al., 1988) laid out several concepts for how primary areas are specified and their borders refined. An early idea was that new areas could be formed by the duplication of existing ones; once doubled, the redundant area would be free to gradually diverge its structure and function over subsequent generations. Allman and Kaas (1971), noting the topographic similarities between visual middle temporal area (MT) in primate extrastriate cortex and primary visual area V1, proposed MT may have been formed as a result of such a duplication event. Some 30 years after this proposal, elegant experimental work showed that shifting or ectopically manipulating the expression of secreted signaling molecule FGF8 causes dramatic shifts in the extent and boundaries, or even duplications, of primary sensory (barrel) cortex in mouse (Fukuchi-Shimogori and Grove, 2001). This provides a proof of concept for how a cortical field duplication event of a primary sensory area might occur.

In addition to the patterning centers that specify the borders of primary areas, the supraareal configuration of areas can provide insight into their development. In primates, the organization of extrastriate visual areas gives an important clue. Rosa (2002) noted that if one flattens primate extrastriate cortex and overlays visuotopic receptive field eccentricity, a supraareal organization is apparent in which eccentricity varies in a gradient that cuts across areal boundaries. This organization is probably specified through cascading events unfolding during early stages of development. Molecular gradients and activity-dependent cascades constrain the development of primary visual cortex, which then influences the patterning of adjacent, developing areas

though activity-dependent interactions (Rosa, 2002). A key concept in this framework is that a small number of cortical maps serve as topographic anchors or "tethers." Once these are specified, adjacent areas inherit their topography by activity-dependent interactions with the anchors.

35.4 The Tethering Hypothesis: Making Sense of the Gaps

The basic arrangement of sensory and motor areas on the cortical plate is strikingly conserved across mammals (Fig. 35.1; Krubitzer, 2007). The consensus is that the first mammals possessed a small cortex with large olfactory (piriform) fields, retrosplenial and cingulate cortices, and a small number of primary and secondary sensory fields that closely abutted one another (Kaas, 2011). Though the relative proportions of these fields vary across species according to sensory specializations, the basic conservation of their spatial arrangement suggests that the signaling gradients that prefigure these areas on the developing cortical plate are also conserved. From this basic plan, large cortices evolved in multiple lineages.

Recently we proposed a general form of the tethering hypothesis (Buckner and Krienen, 2013) as a way to think about how the expansion of association cortex across distributed regions of the cortical mantle in primates may have conferred different properties to these zones relative to the conserved primary sensory areas. The hypothesis has two main assumptions. The first is that the rapid expansion of the association cortex in primates, and particularly the lineage that led to modern humans, produced zones of cortex that lie far from the evolutionarily conserved organizing maps that serve as anchors. The anchors, specified by a few key molecular gradients, initially evolved on a smaller surface area and may exert less influence on the development, organization, and connectivity of distal association regions that emerge as the cortex expands during later stages. A critical detail is that conserved patterning centers arise early in development, when the cortical mantle is still relatively small. At early developmental stages, molecular gradients that prefigure future cortical areas are physically closer together, much as in the rodent. The protracted period of cortical development in humans is likely key to explain how particular regions and layers of cortex become disproportionately expanded, and why they come to have particular network properties. It is possible that a modest number of new patterning centers have arisen from novel transcription factors and other molecular gradients after primates diverged from rodents. However, it is unlikely that many evolved in the last 3 million years during which the human lineage experienced the most rapid cranial expansion (Carroll, 2003).

The second assumption of the tethering hypothesis is that absent the strong constraints from patterning centers characteristic of the developing sensory areas, the higher-order neocortical zones tend to wire to each other in network forms distinct from the prototypical sensory cascades. The protracted developmental schedule of some properties of association cortex, along with activity-dependent sculpting likely contributes to why connectivity between association regions departs from the form observed in conserved sensory regions. However, studying the developmental events that guide connectivity in primates remains challenging, and many open questions remain as to how these differences emerge. The following sections discuss some possible mechanisms and highlight the gaps that remain in our understanding of cortical organization and connectivity.

35.5 Determinants of Cortical Expansion and Patterning

The tethering hypothesis is consistent with the notion that many of the apparently exceptional features of association cortex arose as a consequence of rapid cortical expansion in humans. The extended period of cortical neurogenesis in primates increases surface area and disproportionately increases neuron numbers in the upper cortical layers (Fig. 35.3; Bystron et al., 2008; Charvet et al., 2015; Lewitus et al., 2014). The emergent circuit organization characteristic of association cortex may have arisen because expanded zones lie physically distant from conserved patterning centers and the strong influence of sensory activity cascades. The mechanisms of nonuniform expansion are not yet clear, but the answer

likely includes modifications to proliferative zones early in cortical development.

The major site of proliferation for excitatory neurogenesis in rodents is the VZ. In primates and some other gyrencephalic species, additional germinal structures have emerged as the primary source of neurogenic proliferation. The outer subventricular zone (OSVZ) in particular is substantially expanded in primates and probably accounts for the increase in neurons that populate upper cortical layers (Bystron et al., 2008; Dehay et al., 2015). Exquisite imaging and single-cell transcriptional profiling studies using fetal cortical slices have revealed unique migration and replication dynamics as well as gene expression profiles of progenitor cells that populate the human and nonhuman primate OSVZ (Lui et al., 2011; Pollen et al., 2015; Betizeau et al., 2013; Hansen et al., 2010; Ostrem et al., 2014).

In rodents, excitatory neurons largely find residence in the cortical plate via radial migration from the VZ. In this way, the final cortical destination of an excitatory neuron can be predicted by where its progenitor was located (Rakic, 1995). In gyrencephalic species including ferrets, cats, and humans, regions destined to become gyri in the adult cortex have greater proliferation in the OSVZ than in the other ventricular compartments (Gertz et al., 2014; Reillo et al., 2011). The OSVZ may be a key innovation that allows for greater tangential migration, and thus greater cortical expansion and clonal dispersion in primates (Lui et al., 2011).

In rodents, cortical territories of primary and secondary sensory areas, as well as frontal motor areas, are configured by the expression of transcription factors in progenitor and postmitotic cell populations. For instance, *Lmo4* and *Bhlhb5* are expressed in specific postmitotic neurons and guide thalamic afferentation as well as the development of corticofugal pathways (Kashani et al., 2006; Joshi et al., 2008). In general, areal identity is intimately connected with its connectivity, which in turn is determined by the specific constellation of projection neuron subtypes (Custo Greig et al., 2013). As such, understanding how neurons—and particularly excitatory projection neurons—in different regions of cortex become biased toward different fates will be key to determining how areal identity is acquired (Custo Greig et al., 2013).

How and when do higher-order areas differentiate? In mice, some of the earliest born glutamatergic neurons, the Cajal–Retzius neurons that populate the cell-sparse layer I, appear to play a role in establishing the boundaries of secondary sensory areas as well as parietal association areas in the mouse (Barber et al., 2015). A critical factor appears to be the speed of migration of these early cortical pioneers, which is regulated cell-autonomously by vesicle-associated membrane protein (VAMP)-dependent mechanisms (likely by subfamily VAMP1-

3). VAMPs accumulate in the leading edge of cells and are necessary for the formation of lamellipodia and efficient migration. Since Cajal—Retzius cells are thought to be necessary for anchoring radial glia and for influencing the targeting of thalamocortical axons (Aboitiz and Zamorano, 2013; Tissir and Goffinet, 2003), controlling Cajal—Retzius migration might be one mechanism for how the size of secondary sensory and parietal association regions are delineated in mouse (Barber et al., 2015). Manipulating the speed of migration affects how much cortical territory is covered by a given Cajal—Retzius neuron subtype and has downstream effects on the size of resulting cortical areas.

The *Vamp3−/−* mouse mutants have an unusual areal feature evident in flat mounts of the cortex. The border of area V1 extends rostrally such that the normally continuous border, consisting of areas V2L and V2M, was pinched off. As a result, V2L and V2M no longer abut in this mutant. In flattened maps of the primate brain, area V1 is bordered continuously by V2, but V3 is pinched off such that there is a dorsal V3 that does not share a border with ventral V3. The physical separation of the dorsal and ventral portions of primate V3, coupled with the technical difficulties of accessing ventral V3 for electrophysiological mapping, has caused some debate for years as to whether they were actually part of a single area (Lyon and Kaas, 2002; Zeki, 2003; Lyon and Connolly, 2012). The mechanism demonstrated in Barber et al. (2015) shows how it is possible for subtle changes in migration rates of early pioneer neurons, controlled in this case by VAMPs, to regulate the cortical real estate of both primary (V1) and higher-order areas.

35.6 Activity-Dependent Sculpting Shapes Cortical Territories

Sensory thalamic nuclei, including the lateral and medial geniculate, transmit information from the periphery to the neocortex. Thalamic afferents shape both the boundaries and internal topography of emerging primary sensory areas (Pons et al., 1991). Perturbations of thalamic afferents can shift the territories of immediately adjacent secondary areas. A prime example is that deleting thalamic axons from the lateral geniculate nucleus drives the positioning of differentially expressed genes that distinguish primary V1 and neighboring higher-order visual areas (Chou et al., 2013). In visual cortex, thalamocortical connections bias the selection of which corticocortical afferents will remain from the initial exuberant population (Carić and Price, 1999).

Thalamic association nuclei including the mediodorsal nucleus send afferents to multiple-distributed neocortical association areas, for instance to prefrontal and parietal cortex (Fig. 35.5; Selemon and Goldman-Rakic, 1988). In primates there may be additional association nuclei, such as the pulvinar and center median nucleus, which are not found in rodents (Jones and Rubenstein, 2004). What role does the thalamus play in shaping the boundaries and internal topographies of association areas, and guiding the formation of their corticocortical connections? Unfortunately, it is still not clear how and when thalamic afferents guide the development of higher-order association regions that receive input from the thalamic association nuclei. A clue might be found in the human neuroimaging literature: functional connectivity between the thalamus and neocortex is detectable in somatomotor regions in neonates, whereas connectivity between the thalamus and association networks only becomes robust in the first and second year of life (Alcauter et al., 2014).

It has been proposed that the thalamic association nuclei serve to unify or coordinate the development of corticocortical association networks (Fig. 35.6; Goldman-Rakic, 1988). This idea is compelling, but no developmental mechanism has been described to account for how single thalamic nuclei simultaneously target multiple association zones across the cortex, and how these then come to wire to each other. Ipsilateral corticocortical connectivity of sensory areas can be properly established in absence of thalamic input (Huffman et al., 2004), and sensory deprivation (whisker cauterization) prevents callosal projections to some, but not all, contralateral cortical targets (Suárez et al., 2014). This indicates that even within sensory systems, aberrant thalamic input can selectively affect a subset of

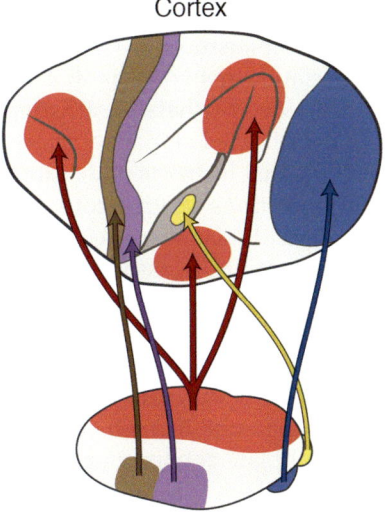

Cortex

Thalamus

FIGURE 35.5 Thalamocortical connections in primates. Sensory thalamic nuclei project to primary sensory areas in the neocortex. Association nuclei such as the mediodorsal nucleus send and receive projections from distributed association regions.

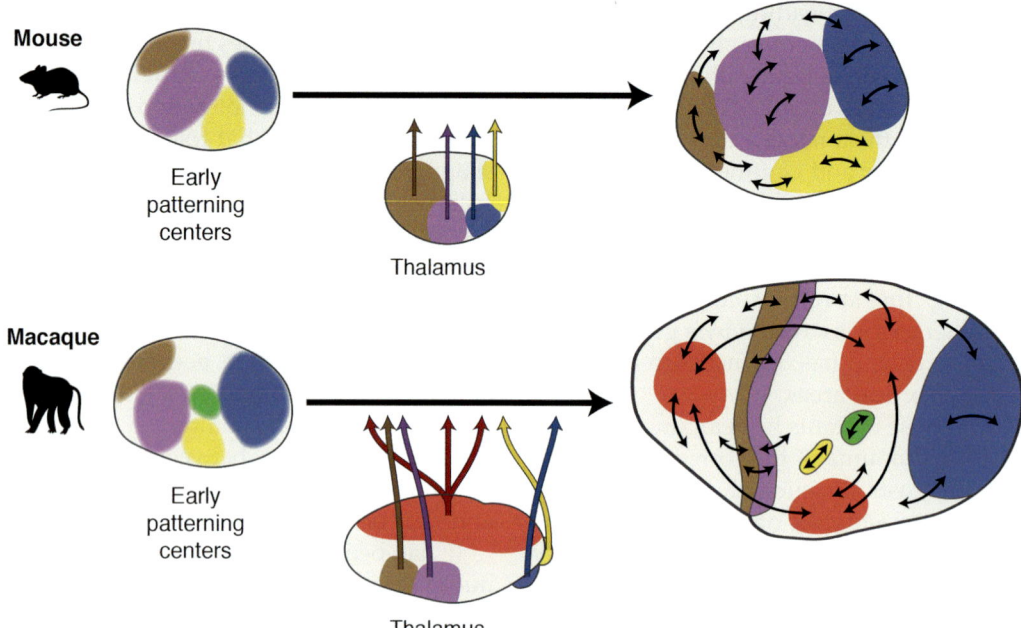

FIGURE 35.6 Early cortical patterning centers and thalamic afferents establish functional zones and connectivity in the developing cortex. In rodents, primary and secondary sensory areas take up a large proportion of the cortical mantle. In primates, association nuclei make up the largest proportion of the thalamus and innervate distributed cortical regions. Later, these regions tend to preferentially wire together forming long-range distributed cortical circuits. A key consideration is that cortical development is protracted in primates relative to rodents. At the earliest stages of cortical plate formation, during the time when patterning centers influence areal patterning, the primate cortical plate is small. The subsequent expansion of the cortical plate causes the conserved sensory areas to move further apart. The interposed regions of association cortex develop long-range connections to each other.

corticocortical projections originating from the same pool of cortical cells.

35.7 Intrinsic Developmental Factors Shape Cortical Territories and Their Connectivity

One possible explanation for the emergence of distributed association networks is that corticocortical connections are guided by thalamic afferents or another mechanism of coordinated signaling across distributed association regions that instructs immature cortical neurons to extend projections to other cortical (association) targets. For instance, thalamic association nuclei might extend afferents to multiple association regions at the same time, forming the basis of a distributed network. Further segregation of corticocortical networks could be achieved if different cohorts of cortical regions received these signals at different times.

Neurogenesis does not happen in all parts of the developing cortical plate at once; one idea is that variation in neuron numbers across the cortex reflects a posterior–anterior gradient in neurogenesis timing (Cahalane et al., 2014). Regional differences in neurogenesis timing entails that cells destined to become cortical neurons experience different extrinsic factors (such as contact with thalamic afferents and secreted patterning

molecules) because they are dividing and migrating at different times. Posterior regions stay proliferatively active longer than anterior ones, which in turn accounts for increased neuron numbers in posterior primary visual cortex (Cahalane et al., 2014; Smart et al., 2002). However, quantification of neuron numbers in a chimpanzee cortex suggests that neuron numbers do not vary in a smooth gradient. Sensory areas have higher neuron densities than adjacent nonprimary regions, as does a region of dorsomedial prefrontal cortex (Collins et al., 2016). Primary visual area V1 in primates is an extreme, with dramatically high density and total neuron numbers (Cahalane et al., 2012; Collins et al., 2016).

Heterochrony in neurogenesis across the cortex could conceivably provide a basis for heterochrony in the establishment of circuits in different areas (Beul et al., 2015). However, if anything, evidence thus far suggests there is a certain amount of synchrony in the development of prefrontal, sensory, and motor synapses. For instance, the early time course of synaptogenesis in rhesus macaques appears to be concurrent for prefrontal, motor, and sensory areas, leading to the notion that there may be a general process by which cortical circuitry is established regardless of functional domain (Rakic et al., 1986; Bourgeois et al., 1994). Still, early synchrony across sensory and association regions does not preclude later asynchrony. This may be relevant since the upper cortical layers are the last to develop. Subtle timing differences in the extension of corticocortical projections across the cortex at later stages of cortical development may yet be revealed.

Other aspects of cortical circuit development more clearly distinguish association cortex from sensory areas. This is evident in the neoteny of synapse production and elimination. Dramatic axon overproduction and elimination occurs over the course of development in primates in corticocortical (commissural and callosal) pathways (LaMantia and Rakic, 1994). In rhesus monkeys, the explosive period of perinatal synaptogenesis in prefrontal cortex largely occurs in supragranular layers (Bourgeois et al., 1994). In humans, maturation, as well as the process of myelination, persists in prefrontal association cortex over a period spanning into the third decade of life (Petanjek et al., 2011; Miller et al., 2012). Similarly, dendritic arborization is delayed in prefrontal cortex relative to sensory and motor regions (Bianchi et al., 2013). Posterior association cortices have not yet been studied, and an open question is whether the neoteny of these features occurs in concert across prefrontal, temporal, and parietal association regions.

Another possible contribution to the development of association networks might be that overproduction of connections between association regions occurs pervasively. The subsequent elimination of projections would prune back and refine corticocortical association pathways over the long period of postnatal development. In this possibility, the multiple parallel, distributed networks evident in the adult emerge from a single or small number of pervasively interconnected association cortical networks. No special mechanism would be needed to guide corticocortical afferents to precise targets; rather, experience and competition could serve to eliminate redundant connections and increase specificity of targets over time. Early evidence of synaptic exuberance was documented with the observation of transient callosal connections between primary visual areas in the cat (Innocenti and Frost, 1979). Since then, several examples of fully transient thalamocortical and corticocortical projections have been described (Innocenti and Price, 2005). However, the general rule appears to be that corticocortical projection elimination achieves refinement by degree rather than kind. For instance, projection elimination between V1 and V2 further refines visual topographic correspondences between these two areas. Though as much as two-thirds of the corticocortical projections are eliminated, transient projections tend to respect a topographical order. This suggests that even exuberant projections followed guidance cues to appropriate targets before being eliminated; in other words, they were more redundant than inappropriate (Bressoud and Innocenti, 1999). In general, projections appear to first invade the cortex at roughly the same location in which they are detected in the adult, suggesting a certain amount of precision in the initial guidance and decision of where to innervate (Dehay et al., 1988; Webster et al., 1991; Meissirel et al., 1991).

35.8 Conserved and Divergent Properties of Neocortical Gene Expression Between Rodents and Primates

The preceding sections highlight conserved properties of cortical development, thalamic innervation, and refinement of circuits between rodents and primates and also point to possible divergences. Species-specific patterns of gene expression can similarly provide important clues as to how cortical diversity is achieved (Belgard et al., 2011). Particularly informative genes are those with mutations or expression patterns that are or expression patterns that are preferential to the lineage leading to humans (Hill and Walsh, 2005; Konopka et al., 2012). In general, however, gene expression is remarkably conserved between mouse and human brains, despite the approximate 3500-fold difference in brain size.

A number of examples highlight the strong conservation in cortical gene expression across taxa. One is that anatomic proximity accounts for a substantial proportion of the variance in transcriptional profiles, with a number of genes displaying pronounced gradients. It has been suggested that molecular similarities of adjacent regions provide a basis for local corticocortical connectivity (Bernard et al., 2012). A fascinating question, which is explored in the next section, is whether genetic similarity can explain patterns of long-range corticocortical connectivity as well. Another example of conserved principles across species is that cortical layers have idiosyncratic gene expression differences that are sustained across areal boundaries (Chen et al., 2011, 2012; Hawrylycz et al., 2010; Bernard et al., 2012).

Subtle differences in spatial and temporal expression of genes that are conserved between species may have important functional consequences. For example, genetically identified callosal projection neurons in mice and macaques share, in common, many of the same markers during comparable stages of development (Fame et al., 2016). Conservation of laminar expression patterns is greatest in the deep layers of cortex, while differences tended to emerge in supragranular layers. For instance, *Gfra2* is expressed exclusively in deep layers in the mouse. In macaque, it is expressed in deep as well as supragranular layers. This could mean that the expansion of the upper layers in primates involved cooption of ancient, projection neuron populations that retain some of the same markers but develop very different projection properties (Fame et al., 2016).

As another example, the developing visual system in marmosets and mice relies on a guidance molecule pathway (Sema3A/Npn1) that is conserved between the two species but has different temporal and spatial profiles (Homman-Ludiye and Bourne, 2014). Sema3A expression is first detected as a ubiquitous embryonic distribution in both species, but only in the marmoset, does it persist in later stages of development (in later-maturing visual areas). The expansion of the marmoset visual processing system may have extended the function of Sema3A. Possibly, this expansion included enabling the Sema3A pathway to contribute to the formation of visual area identity, organization, and connectivity in a species- (or primate-) specific manner (Homman-Ludiye and Bourne, 2014).

The search for area-specific gene expression in the mouse has yielded a number of modules consistent with areal anatomical borders, leading to the idea that a combinatorial code of gene expression can be used to delineate cortical areas in that species (Hawrylycz et al., 2010). A more nuanced picture emerges from gene profiling studies of primates. Regionally specific gene expression tends to respect classical, but broad, cortical subdivisions (sensory, motor, limbic, and association) (Yamamori and Rockland, 2006). One study used microarrays to map transcriptional profiles of 10 cortical regions in the rhesus macaque, including primary sensory areas, higher-order visual areas, and prefrontal cortex (Bernard et al., 2012). The top principal components distinguished V1 and V2 from other cortical samples, a result that was recapitulated in a similar approach using human postmortem cortical samples (Hawrylycz et al., 2012, 2015). In fact, the first principal component in the human cortical dataset largely distinguished primary motor and sensory samples from nonprimary cortical regions (Hawrylycz et al., 2012). With the exception of a small number of primary and secondary sensory and motor areas that have distinct gene expression profiles, the available data so far do not point to a combinatorial code that delineates each cortical area in primates.

A recent study examined laminar expression profiles of 1000 genes that are expressed in the mammalian neocortex and found that 79% have highly conserved laminar expression patterns between mice and humans (Zeng et al., 2012). Several of the remaining 21% exhibited species- or neocortical area-specific distributions. Some had different laminar expression patterns in putatively homologous brain regions across mice and humans, and 19 genes of these were uniquely or disproportionately enriched in the supragranular layers of the neocortex in humans. Zeng et al. (2012) hypothesized that the selective enrichment of these genes may be

a molecular signature of the enhancement of long-range corticocortical projections emanating from Layer III pyramidal neurons in humans. This idea is intriguing because supragranular pyramidal neurons are overrepresented in primates, particularly in humans (Marín-Padilla, 1992). But what is the relationship between these changes to supragranular layers and corticocortical connectivity?

35.9 Gene Expression Topography Links to Human Association Networks

Powerful high-throughput technologies such as microarrays and RNAseq enable a comprehensive picture of which genes are expressed in a given brain region. These technologies allow sensitive read-outs of which genes are being actively transcribed in a sample, which may come from bulk brain tissue, microdissected substructures, or even single disaggregated cells (Macosko et al., 2015; Pollen et al., 2015; Darmanis et al., 2015).

Several studies have harnessed microarray profiling of small cortical brain samples to ask whether variation in transcriptional profiles is associated with variation in connectivity profiles across the human neocortex (Cioli et al., 2014; Richiardi et al., 2015; Hawrylycz et al., 2015; Wang et al., 2015; Krienen et al., 2016). A common theme across these initial observations is that the topography of functional brain networks defined by neuroimaging approaches is recapitulated by the covariation of gene expression in human postmortem brain tissue. Each study used different approaches for candidate gene selection and emphasized different properties of the association, for instance, focusing on enrichment of particular gene families (Cioli et al., 2014; Richiardi et al., 2015; Wang et al., 2015), genes whose relationship to connectivity is conserved across species (Richiardi et al., 2015), or relationships between the implicated genes and human disease (Richiardi et al., 2015; Wang et al., 2015). Richiardi et al. (2015) derived a consensus list of 136 genes whose relative expression levels were consistently correlated across postmortem samples with MRI-based connectivity estimates. They demonstrated that the transcriptional covariance matrix of these genes was associated with structural (axonal) connectivity in the mouse. This suggests that genes in the consensus list may be associated with conserved properties of connectivity between humans and mice.

In addition to conserved properties, this approach can be used to ask whether there are genes whose expression patterns are associated with different forms of connectivity across the neocortex. In the macaque monkey, corticocortical connections in association cortex have distinct neurochemical phenotypes relative to sensory and motor regions (Hof et al., 1995). For example, nonphosphorylated neurofilament protein heavy (NEFH) is expressed in cortical projection neurons in a regional- and laminar-specific way. Critically, the distribution varies across corticocortical pathways validated by systematic tracer injection experiments (Hof et al., 1995). This was an early indication that different corticocortical networks may be distinguished by their molecular profiles. The expression of the gene *NEFH* is enriched in supragranular layers in human neocortex, whereas it is enriched in infragranular layers in mouse (Zeng et al., 2012). One question is whether the differential expression of supragranular-enriched genes in general might relate to the disproportionate expansion of association cortical networks in the human brain.

Using the Allen Institute's human brain transcriptional atlas, we found that transcriptional expression of the human-specific supragranular-enriched (HSE) genes identified in Zeng et al. (2012) reflect distributed corticocortical network organization as assessed by functional connectivity neuroimaging (Krienen et al., 2016). Spatial proximity is likely to be a primary factor constraining gene expression across the neocortex in mammals generally, with many of the most differentially expressed genes exhibiting graded expression patterns that fall along principal axes (Fig. 35.7A; Chen et al., 2011, 2012; Bernard et al., 2012). Beyond these spatial gradients, association, paralimbic, and sensory/motor cortices each showed distinct HSE transcriptional phenotypes. For instance, HSE expression was more similar between pairs of regions in somatosensory and visual

cortex than between either of those and the physically more proximal parietal association cortex. Similarly, prefrontal, parietal, and temporal association cortex had more similar HSE profiles to each other than to the sensory/motor regions (Fig. 35.7B; Krienen et al., 2016).

Taken together, these observations show that variation in the transcriptional landscape across the adult human neocortex associates with broad corticocortical connectivity properties. It does not imply that the expression of these genes evolved to directly produce the particular forms of corticocortical connectivity observed across networks. Rather, regions that share similar connectional phenotypes may tend to express a constellation of genes, including the HSE genes, at similar levels, even though the regions may not themselves directly connect. This variation is also subtle, against a backdrop of a high degree of transcriptional similarity in the neocortex overall relative to subcortical structures and the cerebellum (Hawrylycz et al., 2010, 2012).

Differences in the cortical regional transcriptomes are the most divergent early in development and become more synchronized postnatally (Pletikos et al., 2014). The best opportunity to understand how different cortical fields acquire their corticocortical projection phenotypes will likely come from mapping transcriptional variation at the times in which these specific projections begin to form (Pletikos et al., 2014). New data are suggestive of at least some transient primate- and human-specific gene expression patterns early in

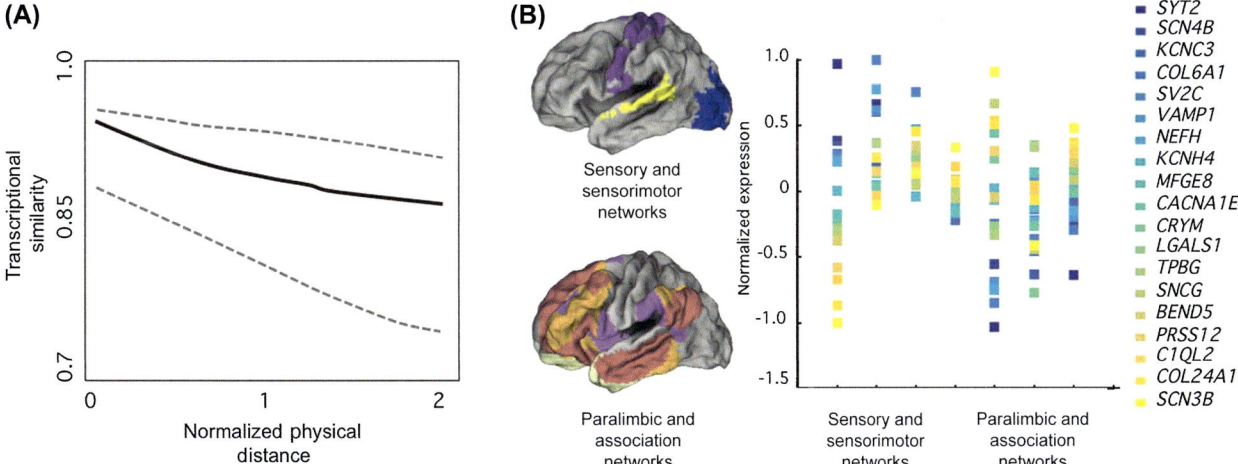

FIGURE 35.7 Transcriptional architecture of human neocortex. (A) In general, similarity in transcriptional profiles decreases with increasing physical distance between cortical regions. (Adapted with permission from Hawrylycz, M.J., et al., 2012. An anatomically comprehensive atlas of the adult human brain transcriptome. Nature 489 (7416), 391–399.). (B) Though physical proximity is a strong predictor, other modes of transcriptional similarity are also apparent (Krienen et al., 2016). In particular, genes preferentially expressed in the supragranular layers in humans but not mice tend to have expression profiles that dissociate cortical subtypes. Regions falling within sensory and sensorimotor networks have similar expression profiles of these genes, whereas regions falling in paralimbic and association networks have more similar expression to each other than to sensory and motor regions. Expression of this particular gene collection is shared across distributed regions of association cortex. As an example, regions in parietal association cortex had more similar gene expression profiles to zones in prefrontal association cortex than to the more proximal somatosensory cortex or visual cortices.

development (Pletikos et al., 2014). For instance, patches in human embryonic frontal and temporal lobes have regionalized gene expression not seen at comparable timepoints in macaques or mice and may be related to language regions that emerge later in development (reviewed in Geschwind and Rakic, 2013).

35.10 Spatial Gradients and Sharp Boundaries

Throughout this chapter, we have suggested that the distributed network architecture of association cortex in humans arises because ancient developmental programs play out on an expanded cortical mantle. Under this view, association cortex is not prefigured with sharp boundaries specified early in development but rather emerges from the cortical zones that lie farthest from the strong influences of conserved patterning centers. Few novel morphogens are expected to have evolved in humans during the last few million years of rapid cranial expansion.

Other possibilities exist. As discussed earlier, one idea is that shifts or duplications in patterning centers provide mechanisms for the evolution of new cortical areas (Allman and Kaas, 1971; Sur and Rubenstein, 2005). Duplications, if they have recently occurred at all in primate evolution, are likely restricted to a relatively small number of areas (eg, MT). Nonetheless, rare duplicative events or other forms of dramatic reorganization could occur and contribute a major shift in how conserved organizational properties play out in the expanding cortical mantle.

Another possibility is that areas are specified by a combinatorial code of gene expression (Hawrylycz et al., 2010). Two recent papers show that a large number of enhancers—short stretches of DNA that can be bound by transcription factors to activate transcription of a gene some distance away—have activity patterns that reflect the combinatorial integration of transcription factors in mouse. These emerge as sharp boundaries as early as embryonic day E11.5—the start of cortical plate development—in mouse progenitor zones (Visel et al., 2013; Pattabiraman et al., 2014). The large number of enhancers whose activity is circumscribed in sharp boundaries across progenitor zones in the mouse could be consistent with the idea that, as a general rule, cortical fields are prefigured with sharp boundaries from the earliest stages of cortical neurogenesis (Pattabiraman et al., 2014).

While this idea is powerful, it is still unclear whether there is prefiguration of higher-order areal boundaries directly (except in the sense that higher-order areas are shifted when primary sensory and motor boundaries shift). Several difficult and unresolved issues remain.

The first is that mouse cortex has relatively few association areas. As such, the enhancer and transcription factor boundaries described in mouse have focused on the boundaries of primary motor and sensory fields— areas that have relatively clear and identifiable boundaries in the adult. The second issue is that factors controlling areal development and the transcription factor landscape have not been extensively characterized in embryonic primates to determine whether cortical arealization is similarly prefigured by gene, transcription factor, and enhancer combinations.

Perhaps a more fundamental issue is the extent to which primate neocortical areas, beyond conserved primary sensory and motor areas, possess sharp boundaries in the first place. The idea that many or most brain areas may lack sharp boundaries is not new. Bonin and Bailey's monograph on cortical areas in the macaque monkey made this point explicitly by using watercolor shading to emphasize gradual transitions in cytoarchitectonic features across the cortical mantle (Bonin and Bailey, 1947). Sanides wrote of the concept of "poleward gradation," noting the gradual changes in the granularity of Layer IV and the appearance of large pyramidal cells in Layer V moving anteriorly from motor cortex to the frontal pole (Sanides, 1970). Work from Barbas and colleagues has emphasized the transition from granular cortex to progressively more agranular zones across the medial prefrontal cortex (Barbas and Rempel-Clower, 1997) and the implications for connectivity between cortical areas that possess different cytoarchitectural properties.

The lack of clear and consistent boundaries has led to debate on how many cortical areas exist and motivated the proposal that multiple criteria must converge for an areal boundary to be drawn (Stryker et al., 1988; Kaas, 1987). Trends such as gradients of gene expression, connectivity, and cytoarchitectonic features suggest the possibility that cortical fields may, with few exceptions, be viewed as gradually transitioning into each other across the cortical surface.

35.11 Conclusions

The rapid expansion of association cortex in great apes and particularly humans may have created large zones of the cortical mantle located far from patterning centers that constrain areal regionalization and connectivity. In the adult, subtle differences in gene expression patterns distinguish association cortex from sensory, motor, and paralimbic cortices and may reflect differences in connectional properties among these broad classes of cortex. Protracted developmental schedules indicate that association networks continue to mature late into development. The question of how and when widely separated regions of association cortex come to wire to each other remains unresolved, as does the role of thalamic association nuclei in this process.

References

Aboitiz, F., Zamorano, F., 2013. Neural progenitors, patterning and ecology in neocortical origins. Front. Neuroanat. 7, 1–15.

Alcauter, S., et al., 2014. Development of thalamocortical connectivity during infancy and its cognitive correlations. J. Neurosci. 34 (27), 9067–9075.

Allman, J.M., Kaas, J.H., 1971. A representation of the visual field in the caudal third of the middle temporal gyrus of the owl monkey (*Aotus trivirgatus*). Brain Res. 31, 85–105.

Badre, D., D'Esposito, M., 2009. Is the rostro-caudal axis of the frontal lobe hierarchical? Nat. Rev. Neurosci. 10 (9), 659–669.

Barbas, H., Mesulam, M.M., 1981. Organization of afferent input to subdivisions of area 8 in the rhesus monkey. J. Comp. Neurol. 200, 407–431.

Barbas, H., Rempel-Clower, N., 1997. Cortical structure predicts the pattern of corticocortical connections. Cereb. Cortex 7 (7), 635–646.

Barbas, H., Zikopoulos, B., 2007. The prefrontal cortex and flexible behavior. Neurosci 13 (5), 532–545.

Barber, M., et al., 2015. Migration speed of Cajal-Retzius cells modulated by vesicular trafficking controls the size of higher-order cortical areas. Curr. Biol. 25 (19), 1–13.

Barton, R.A., Venditti, C., 2013. Human frontal lobes are not relatively large. Proc. Natl. Acad. Sci. U.S.A. 110 (22), 9001–9006.

Belgard, T.G., et al., 2011. A transcriptomic atlas of mouse neocortical layers. Neuron 71 (4), 605–616.

Bernard, A., et al., 2012. Transcriptional architecture of the primate neocortex. Neuron 73 (6), 1083–1099.

Betizeau, M., et al., 2013. Precursor diversity and complexity of lineage relationships in the outer subventricular zone of the primate. Neuron 80 (2), 442–457.

Beul, S.F., Grant, S., Hilgetag, C.C., 2015. A predictive model of the cat cortical connectome based on cytoarchitecture and distance. Brain Struct. Funct. 220, 3167–3184.

Bianchi, S., et al., 2012. Dendritic morphology of pyramidal neurons in the chimpanzee neocortex: regional specializations and comparison to humans. Cereb. Cortex 23 (10), 2429–2436.

Bianchi, S., et al., 2013. Synaptogenesis and development of pyramidal neuron dendritic morphology in the chimpanzee neocortex resembles humans. Proc. Natl. Acad. Sci. U.S.A. 110 (Suppl. 2), 10395–10401.

Bonin, V.,G., Bailey, P., 1947. The Neocortex of Macaca mulatta. University of Illinois Press, Champaign.

Bourgeois, J.-P., Goldman-Rakic, P.S., Rakic, P., 1994. Synaptogenesis in the prefrontal cortex of rhesus monkeys. Cereb. Cortex 4 (1), 78–96.

Bozek, K., et al., 2014. Exceptional evolutionary divergence of human muscle and brain metabolomes parallels human cognitive and physical uniqueness. PLoS Biol. 12 (5), e1001871.

Bressoud, R., Innocenti, G.M., 1999. Typology, early differentiation, and exuberant growth of a set of cortical axons. J. Comp. Neurol. 406 (1), 87–108.

Buckner, R.L., Krienen, F.M., 2013. The evolution of distributed association networks in the human brain. Trends Cogn. Sci. 17 (12), 648–665.

Buckner, R.L., Andrews-Hanna, J.R., Schacter, D.L., 2008. The brain's default network: anatomy, function, and relevance to disease. Ann. N.Y. Acad. Sci. 1124, 1–38.

Buckner, R.L., Krienen, F.M., Yeo, B.T.T., 2013. Opportunities and limitations of intrinsic functional connectivity MRI. Nat. Neurosci. 16 (7), 832–837.

Bullmore, E., Sporns, O., 2009. Complex brain networks: graph theoretical analysis of structural and functional systems. Nat. Rev. Neurosci. 10 (4), 312.

Bystron, I., Blakemore, C., Rakic, P., 2008. Development of the human cerebral cortex: Boulder Committee revisited. Nat. Rev. Neurosci. 9 (2), 110–122.

Cabeza, R., Nyberg, L., 1997. Imaging cognition: an empirical review of pet studies with normal subjects. J. Cogn. Neurosci. 9 (1), 1–26.

Cahalane, D.J., Charvet, C.J., Finlay, B.L., 2014. Modeling local and cross-species neuron number variations in the cerebral cortex as arising from a common mechanism. Proc. Natl. Acad. Sci. U.S.A. 111 (49), 17642–17647.

Cahalane, D.J., Charvet, C.J., Finlay, B.L., 2012. Systematic, balancing gradients in neuron density and number across the primate isocortex. Front. Neuroanat. 6, 1–28.

Carić, D., Price, D.J., 1999. Evidence that the lateral geniculate nucleus regulates the normal development of visual corticocortical projections in the cat. Exp. Neurol. 156 (2), 353–362.

Carroll, S.B., 2003. Genetics and the making of *Homo sapiens*. Nature 422 (6934), 849–857.

Cavada, C., Goldman-Rakic, P.S., 1989. Posterior parietal cortex in rhesus monkey: I. Parcellation of areas based on distinctive limbic and sensory corticocortical connections. J. Comp. Neurol. 287 (4), 393–421.

Cavada, C., Goldman-Rakic, P.S., 1989. Posterior parietal cortex in rhesus monkey: II. Evidence for segregated corticocortical networks linking sensory and limbic areas with the frontal lobe. J. Comp. Neurol. 287 (4), 422–445.

Charvet, C.J., Cahalane, D.J., Finlay, B.L., 2015. Systematic, cross-cortex variation in neuron numbers in rodents and primates. Cereb. Cortex 25 (1), 147–160.

Chen, C.-H., et al., 2011. Genetic influences on cortical regionalization in the human brain. Neuron 72 (4), 537–544.

Chen, C.-H., et al., 2012. Hierarchical genetic organization of human cortical surface area. Science 335 (6076), 1634–1636.

Cherniak, C., 1995. Neural component placement. Trends Neurosci. 18 (12), 522–527.

Chou, S.-J., et al., 2013. Geniculocortical input drives genetic distinctions between primary and higher-order visual areas. Science 340 (6137), 1239–1242.

Cioli, C., et al., 2014. Differences in human cortical gene expression match the temporal properties of large-scale functional networks. PLoS One 9 (12), e115913.

Collins, C.E., et al., 2016. Cortical cell and neuron density estimates in one chimpanzee hemisphere. Proc. Natl. Acad. Sci. U.S.A. 113 (3), 740–745.

Custo Greig, L., et al., 2013. Molecular logic of neocortical projection neuron specification, development and diversity. Nat. Rev. Neurosci. 14 (11), 755–769.

Darmanis, S., et al., 2015. A survey of human brain transcriptome diversity at the single cell level. Proc. Natl. Acad. Sci. U.S.A. 112 (23), 7285–7290.

Deco, G., Corbetta, M., 2011. The dynamical balance of the brain at rest. Neurosci 17 (1), 107–123.

Dehay, C., Kennedy, H., Bullier, J., 1988. Characterization of transient cortical projections from auditory, somatosensory, and motor cortices to visual areas 17, 18, and 19 in the kitten. J. Comp. Neurol. 272 (1), 68–S.

Dehay, C., Kennedy, H., Kosik, K.S., 2015. The outer subventricular zone and primate-specific cortical complexification. Neuron 85 (4), 683–694.

Elston, G.N., et al., 2006. Specializations of the granular prefrontal cortex of primates: implications for cognitive processing. Anat. Rec. Part A Discov. Mol. Cell. Evol. Biol. 288 (1), 26−35.

Ercsey-Ravasz, M., et al., 2013. A predictive network model of cerebral cortical connectivity based on a distance rule. Neuron 80 (1), 184−197.

Fame, R.M., et al., 2016. Subtype-specific genes that characterize subpopulations of callosal projection neurons in mouse identify molecularly homologous populations in macaque cortex. Cereb. Cortex 1−14.

Fame, R.M., MacDonald, J.L., Macklis, J.D., 2011. Development, specification, and diversity of callosal projection neurons. Trends Neurosci. 34 (1), 41−50.

Felleman, D.J., Van Essen, D.C., 1991. Distributed hierarchical processing in the primate cerebral cortex. Cereb. Cortex 1 (1), 1−47.

Fox, M.D., Raichle, M.E., 2007. Spontaneous fluctuations in brain activity observed with functional magnetic resonance imaging. Nat. Rev. Neurosci. 8 (9), 700−711.

Frostig, R.D., et al., 2008. Large-scale organization of rat sensorimotor cortex based on a motif of large activation spreads. J. Neurosci. 28 (49), 13274−13284.

Fukuchi-Shimogori, T., Grove, E.A., 2001. Neocortex patterning by the secreted signaling molecule FGF8. Science 294 (5544), 1071−1074.

Fuster, J.M., 2004. Upper processing stages of the perception-action cycle. Trends Cogn. Sci. 8 (4), 143−145.

Gertz, C.C., et al., 2014. Diverse behaviors of outer radial glia in developing ferret and human cortex. J. Neurosci. 34 (7), 2559−2570.

Geschwind, D.H., Rakic, P., 2013. Cortical evolution: judge the brain by its cover. Neuron 80 (3), 633−647.

Goldman-Rakic, P.S., 1988. Topography of cognition: parallel distributed networks in primate association cortex. Annu. Rev. Neurosci. 11 (1), 137−156.

Goulas, A., Uylings, H.B., Hilgetag, C.C., 2015. Principles of ipsilateral and contralateral cortico-cortical connectivity in the mouse. bioRxiv.

Hansen, D.V., et al., 2010. Neurogenic radial glia in the outer subventricular zone of human neocortex. Nature 464 (7288), 554−561.

Hawrylycz, M.J., et al., 2010. Areal and laminar differentiation in the mouse neocortex using large scale gene expression data. Methods 50 (2), 113−121.

Hawrylycz, M.J., et al., 2015. Canonical genetic signatures of the adult human brain. Nat. Neurosci. 18 (12), 1832−1844.

Hawrylycz, M.J., et al., 2012. An anatomically comprehensive atlas of the adult human brain transcriptome. Nature 489 (7416), 391−399.

Herculano-Houzel, S., 2009. The human brain in numbers: a linearly scaled-up primate brain. Front. Hum. Neurosci. 3, 31.

Hill, J., et al., 2010. Similar patterns of cortical expansion during human development and evolution. Proc. Natl. Acad. Sci. U.S.A. 107 (29), 13135−13140.

Hill, R.S., Walsh, C.A., 2005. Molecular insights into human brain evolution. Nature 437 (7055), 64−67.

Hof, P.R., Nimchinsky, E.A., Morrison, J.H., 1995. Neurochemical phenotype of corticocortical connections in the macaque monkey: quantitative analysis of a subset of neurofilament protein-immunoreactive projection neurons in frontal, parietal, temporal, and cingulate cortices. J. Comp. Neurol. 362 (1), 109−133.

Homman-Ludiye, J., Bourne, J.A., 2014. The guidance molecule Semaphorin3A is differentially involved in the arealization of the mouse and primate neocortex. Cereb. Cortex 24 (11), 2884−2898.

Horvát, S., et al., 2016. Spatial embedding and wiring cost constrain the functional layout of the cortical network of rodents and primates. PLoS Bio 14 (7), e1002512−e1002530.

Huffman, K.J., Garel, S., Rubenstein, J.L.R., 2004. Fgf8 regulates the development of intra-neocortical projections. J. Neurosci. 24 (41), 8917−8923.

Innocenti, G.M., Frost, D.O., 1979. Effects of visual experience on the maturation of the efferent system to the corpus callosum. Nature 280 (5719), 231−234.

Innocenti, G.M., Price, D.J., 2005. Exuberance in the development of cortical networks. Nat. Rev. Neurosci. 6 (12), 955−965.

Jones, E.G., Powell, T.P.S., 1970. An anatomical study of converging sensory pathways within the cerebral cortex of the monkey. Brain 93 (4), 793−820.

Jones, E.G., Rubenstein, J.L.R., 2004. Expression of regulatory genes during differentiation of thalamic nuclei in mouse and monkey. J. Comp. Neurol. 477 (1), 55−80.

Joshi, P.S., et al., 2008. Bhlhb5 regulates the postmitotic acquisition of area identities in layers II-V of the developing neocortex. Neuron 60 (2), 258−272.

Kaas, J.H., 2011. Neocortex in early mammals and its subsequent variations. Ann. N.Y. Acad. Sci. 1225 (1), 28−36.

Kaas, J.H., 1987. The organization of neocortex in mammals: implications for theories of brain function. Annu. Rev. Psychol. 38 (1), 129−151.

Kashani, A.H., et al., 2006. Calcium activation of the LMO4 transcription complex and its role in the patterning of thalamocortical connections. J. Neurosci. 26 (32), 8398−8408.

Kaskan, P.M., et al., 2005. Peripheral variability and central constancy in mammalian visual system evolution. Proc. R. Soc. Biol. Sci. 272 (1558), 91−100.

Konopka, G., et al., 2012. Human-specific transcriptional networks in the brain. Neuron 75 (4), 601−617.

Krienen, F.M., et al., 2016. Transcriptional profiles of supragranular-enriched genes associate with corticocortical network architecture in the human brain. Proc. Natl. Acad. Sci. U.S.A. 113 (4), E469−E478.

Krubitzer, L., 2007. The magnificent compromise: cortical field evolution in mammals. Neuron 56 (2), 201−208.

LaMantia, A.S., Rakic, P., 1994. Axon overproduction and elimination in the anterior commissure of the developing rhesus monkey. J. Comp. Neurol. 340 (3), 328−336.

Lewitus, E., et al., 2014. An adaptive threshold in mammalian neocortical evolution. PLoS Biol. 12 (11), e1002000−e1002015.

Lui, J.H., Hansen, D.V., Kriegstein, A.R., 2011. Development and evolution of the human neocortex. Cell 146 (1), 18−36.

Lyon, D.C., Connolly, J.D., 2012. The case for primate V3. Proc. R. Soc. Biol. Sci. 279, 625−633.

Lyon, D.C., Kaas, J.H., 2002. Evidence for a modified V3 with dorsal and ventral halves in macaque monkeys. Neuron 33 (3), 453−461.

MacDonald, J.L., et al., 2013. Specification of cortical projection neurons. In: Rubenstein, J.L.R., Rakic, P. (Eds.), Patterning and Cell Type Specification in the Developing CNS and PNS, first ed. Academic Press, San Diego, CA, pp. 475−502.

Macosko, E.Z., et al., 2015. Highly parallel genome-wide expression profiling of individual cells using nanoliter droplets. Cell 161 (5), 1202−1214.

Marín-Padilla, M., 1992. Ontogenesis of the pyramidal cell of the mammalian neocortex and developmental cytoarchitectonics: a unifying theory. J. Comp. Neurol. 321 (2), 223−240.

Markov, N.T., et al., 2014. A weighted and directed interareal connectivity matrix for macaque cerebral cortex. Cereb. Cortex 24 (1), 17−36.

Markov, N.T., et al., 2011. Weight consistency specifies regularities of macaque cortical networks. Cereb. Cortex 21 (6), 1254−1272.

Markov, N.T., Ercsey-Ravasz, M., Lamy, C., et al., 2013. The role of long-range connections on the specificity of the macaque interareal cortical network. Proc. Natl. Acad. Sci. U.S.A. 110 (13), 5187−5192.

Markov, N.T., Ercsey-Ravasz, M., Van Essen, D.C., et al., 2013. Cortical high-density counterstream architectures. Science 342 (6158), 1238406.

Markov, N.T., et al., 2013. Anatomy of hierarchy: feedforward and feedback pathways in macaque visual cortex. J. Comp. Neurol. 522 (1), 225−259.

Meissirel, C., et al., 1991. Segregation of callosal and association pathways during development in the visual cortex of the primate. J. Neurosci. 11 (11), 3297–3316.

Mesulam, M.M., 1998. From sensation to cognition. Brain 121, 1013–1052.

Mesulam, M.M., 1990. Large-scale neurocognitive networks and distributed processing for attention, language, and memory. Ann. Neurol. 28 (5), 597–613.

Miller, D.J., et al., 2012. Prolonged myelination in human neocortical evolution. Proc. Natl. Acad. Sci. U.S.A. 109 (41), 16480–16485.

Mitchell, B.D., Macklis, J.D., 2004. Large-scale maintenance of dual projections by callosal and frontal cortical projection neurons in adult mice. J. Comp. Neurol. 482 (1), 17–32.

Mueller, S., et al., 2013. Individual variability in functional connectivity architecture of the human brain. Neuron 77 (3), 586–595.

Oh, S.W., et al., 2014. A mesoscale connectome of the mouse brain. Nature 508 (7495), 207–214.

Ostrem, B.E.L., et al., 2014. Control of outer radial glial stem cell mitosis in the human brain. Cell Rep. 8 (3), 656–664.

Passingham, R.E., Smaers, J.B., 2014. Is the prefrontal cortex especially enlarged in the human brain? Allometric relations and remapping factors. Brain Behav. Evol. 84 (2), 156–166.

Pattabiraman, K., et al., 2014. Transcriptional regulation of enhancers active in protodomains of the developing cerebral cortex. Neuron 82 (5), 989–1003.

Petanjek, Z., et al., 2011. Extraordinary neoteny of synaptic spines in the human prefrontal cortex. Proc. Natl. Acad. Sci. U.S.A. 108 (32), 13281–13286.

Pletikos, M., et al., 2014. Temporal specification and bilaterality of human neocortical topographic gene expression. Neuron 81 (2), 321–332.

Pollen, A.A., et al., 2015. Molecular identity of human outer radial glia during cortical development. Cell 163 (1), 55–67.

Pons, T.P., Garraghty, P.E., Ommaya, A.K., Kaas, J.H., Taub, E., Mishkin, M., 1991. Massive cortical reorganization after sensory deafferentation in adult macaques. Science 252 (5014), 1857–1860.

Preuss, T.M., 2012. Human brain evolution: from gene discovery to phenotype discovery. Proc. Natl. Acad. Sci. U.S.A. 109 (Suppl. 1), 10709–10716.

Preuss, T.M., et al., 2004. Human brain evolution: insights from microarrays. Nat. Rev. Genet. 5 (11), 850–860.

Raichle, M.E., 2011. The restless brain. Brain Connect. 1 (1), 3–12.

Rakic, P., 1995. A small step for the cell, a giant leap for mankind: a hypothesis of neocortical expansion during evolution. Trends Neurosci. 18 (9), 383–388.

Rakic, P., 1988. Specification of cerebral cortical areas. Science 241 (4862), 170–176.

Rakic, P., et al., 1986. Concurrent overproduction of synapses in diverse regions of the primate cerebral cortex. Science 232 (4747), 232–235.

Reillo, I., et al., 2011. A role for intermediate radial glia in the tangential expansion of the mammalian cerebral cortex. Cereb. Cortex 21 (7), 1674–1694.

Richiardi, J., et al., 2015. Correlated gene expression supports synchronous activity in brain networks. Science 348 (6240), 1241–1244.

Romanski, L., Bates, J., Goldman-Rakic, P.S., 1999. Auditory belt and parabelt projections to the prefrontal cortex in the rhesus monkey. J. Comp. Neurol. 403 (2), 141–157.

Rosa, M.G.P., 2002. Visual maps in the adult primate cerebral cortex: some implications for brain development and evolution. Braz. J. Med. Biol. Res. 35 (12), 1485–1498.

Saleem, K.S., Kondo, H., Price, J.L., 2008. Complementary circuits connecting the orbital and medial prefrontal networks with the temporal, insular, and opercular cortex in the macaque monkey. J. Comp. Neurol. 506 (4), 659–693.

Sanides, F., 1970. Functional architecture of motor and sensory cortices in primates in the light of a new concept of neocortex evolution. In: Noback, C.R., Montagne, W. (Eds.), The Primate Brain: Advances in Primatology, pp. 1–72.

Schoenemann, P.T., Sheehan, M.J., Glotzer, L.D., 2005. Prefrontal white matter volume is disproportionately larger in humans than in other primates. Nat. Neurosci. 8 (2), 242–252.

Schwartz, M., Goldman-Rakic, P.S., 1984. Callosal and intrahemispheric connectivity of the prefrontal association cortex in rhesus monkey: relation between intraparietal and principal sulcal cortex. J. Comp. Neurol. 226 (3), 403–420.

Schwartz, M.L., Goldman-Rakic, P.S., 1991. Prenatal specification of callosal connections in rhesus monkey. J. Comp. Neurol. 307 (1), 144–162.

Selemon, L.D., Goldman-Rakic, P.S., 1988. Common cortical and subcortical targets of the dorsolateral prefrontal and posterior parietal cortices in the rhesus monkey: evidence for a distributed neural network subserving spatially guided behavior. J. Neurosci. 8 (11), 4049–4068.

Seltzer, B., Pandya, D.N., 1989. Frontal lobe connections of the superior temporal sulcus in the rhesus monkey. J. Comp. Neurol. 281 (1), 97–113.

Sepulcre, J., et al., 2010. The organization of local and distant functional connectivity in the human brain. PLoS Comput. Biol. 6 (6), e1000808.

Sherwood, C.C., Duka, T., 2012. Now that we've got the map, where are we going? Moving from gene candidate lists to function in studies of brain evolution. Brain Behav. Evol. 80 (3), 167–169.

Smaers, J.B., 2013. How humans stand out in frontal lobe scaling. Proc. Natl. Acad. Sci. U.S.A. 110 (39), E3682.

Smaers, J.B., et al., 2011. Primate prefrontal cortex evolution: human brains are the extreme of a lateralized ape trend. Brain Behav. Evol. 77 (2), 67–78.

Smart, I.H.M., et al., 2002. Unique morphological features of the proliferative zones and postmitotic compartments of the neural epithelium giving rise to striate and extrastriate cortex in the monkey. Cereb. Cortex 12 (1), 37–53.

Spreng, R.N., Mar, R.A., Kim, A.S.N., 2009. The common neural basis of autobiographical memory, prospection, navigation, theory of mind, and the default mode: a quantitative meta-analysis. J. Cogn. Neurosci. 21 (3), 489–510.

Striedter, G.F., 2005. Principles of Brain Evolution, first ed. Sinauer Associates Incorporated, Sunderland, MA.

Stryker, M., et al., 1988. Group report. Principles of cortical self-organization. In: Rakic, P., Singer, W. (Eds.), Neurobiology of Neocortex, pp. 115–136.

Suárez, R., et al., 2014. Balanced interhemispheric cortical activity is required for correct targeting of the corpus callosum. Neuron 82 (6), 1289–1298.

Sur, M., Rubenstein, J.L.R., 2005. Patterning and plasticity of the cerebral cortex. Science 310 (5749), 805–810.

Tissir, F., Goffinet, A.M., 2003. Reelin and brain development. Nat. Rev. Neurosci. 4 (6), 496–505.

Ungerleider, L.G., Desimone, R., 1986. Cortical connections of visual area MT in the macaque. J. Comp. Neurol. 248, 190–222.

Vallender, E.J., Mekel-Bobrov, N., Lahn, B.T., 2008. Genetic basis of human brain evolution. Trends Neurosci. 31 (12), 637–644.

Van Essen, D.C., 2013. Cartography and connectomes. Neuron 80 (3), 775–790.

Veale, K.J., et al., 2010. Recycling endosome membrane incorporation into the leading edge regulates lamellipodia formation and macrophage migration. Traffic 11 (10), 1370–1379.

Vendetti, M.S., Bunge, S.A., 2014. Evolutionary and developmental changes in the lateral frontoparietal network: a little goes a long way for higher-level cognition. Neuron 84 (5), 906–917.

Visel, A., et al., 2013. A high-resolution enhancer atlas of the developing telencephalon. Cell 152 (4), 895–908.

Wang, G.-Z., et al., 2015. Correspondence between resting-state activity and brain gene expression. Neuron 88 (4), 659–666.

Webster, M.J., Ungerleider, L.G., Bachevalier, J., 1991. Connections of inferior temporal areas TE and TEO with medial temporal-lobe structures in infant and adult monkeys. J. Neurosci. 11 (4), 1095–1116.

Yamamori, T., Rockland, K.S., 2006. Neocortical areas, layers, connections, and gene expression. Neurosci. Res. 55 (1), 11–27.

Yeo, B.T.T., et al., 2011. The organization of the human cerebral cortex estimated by intrinsic functional connectivity. J. Neurophysiol. 106 (3), 1125–1165.

Zeki, S., 2003. Improbable areas in the visual brain. Trends Neurosci. 26 (1), 23–26.

Zeki, S., Shipp, S., 1988. The functional logic of cortical connections. Nature 335 (6188), 311–317.

Zeng, H., et al., 2012. Large-scale cellular-resolution gene profiling in human neocortex reveals species-specific molecular signatures. Cell 149 (2), 483–496.

Zingg, B., et al., 2014. Neural networks of the mouse neocortex. Cell 156 (5), 1096–1111.

36

On the Evolution of the Frontal Eye Field: Comparisons of Monkeys, Apes, and Humans

J.D. Schall, W. Zinke, J.D. Cosman, M.S. Schall, M. Paré, P. Pouget

Vanderbilt University, Nashville, TN, United States; Queen's University, Kingston, ON, Canada; Université Pierre et Marie Curie, Institut du Cerveau et de la Moelle épinière, Paris, France

36.1 Overview

Originally considered only an ocular motor area, interest in frontal eye field (FEF) has increased markedly with its recognition as an anatomical hub region, which is critically involved in covert and overt orienting in monkeys and humans (reviewed by Squire et al., 2013; Schall, 2015). Thus, FEF is associated with motor, sensory, and cognitive processes, which are mediated in association with neighboring cortical areas.

Ferrier (1874) described an area in the dorsal portion of the arcuate sulcus of anesthetized monkeys where electric stimulation elicited movements of the eyes and head in the direction opposite the stimulated hemisphere. These are referred to as contraversive movements. This finding has been replicated in multiple species including prosimians, New World monkeys, Old World monkeys, apes, and humans. Förster (1926) described the area from which he could evoke eye movement as the "frontales Augenfelds," which was translated into "frontal eye fields" (Davidoff, 1928; Förster and Penfield, 1930).

Before proceeding, we introduce our nomenclature. This chapter describes the location of the FEF assessed with a variety of methods. For efficiency we use phrases such as "FEF is located …," when we should more clearly state "A region activated during saccades …" or "A region from which saccades were elicited …." We emphasize this here to alert the reader that the location and boundaries of the FEF(s) are uncertain, operational, and have rarely been compared systematically across methods.

Today, the majority of research on FEF is done with macaque monkeys and humans, and this chapter, like previous reviews (Paus, 1996; Tehovnik et al., 2000; Amiez and Petrides, 2009; Vernet et al., 2014; Percheron

et al., 2015) is animated by an apparent discrepancy in the location of FEF between monkeys and human. In both macaque and human the FEF is located in a sulcus rostral to the central sulcus, known either as arcuate or precentral (see Connolly, 1936, 1950). All authors agree that FEF in monkeys is located in the dysgranular region of caudal prefrontal cortex commonly included in Brodmann's area 8; however, most locate FEF in humans in Brodmann's agranular area 6. We address this apparent enigma by highlighting the following observations: (1) particular similarities and differences in eye-, head-, and body-orienting behavior across species, (2) variation of frontal sulcal morphology across species and individuals, (3) the location of FEF at a cytoarchitectonic transition zone between agranular area 6 and granular area 8 in a cortical region with greater than average variation of sulcal pattern, and (4) the embrace of FEF caudally (in premotor cortex) and rostrally (in prefrontal cortex) by areas also contributing to visually guided gaze behavior. These observations are framed by an appreciation of the varying technical limits and spatial resolution of methods used to locate FEF in monkeys, apes, and humans, and of the functional and anatomical position of FEF in a surrounding network of areas contributing to orienting behavior that includes but is not limited to eye movements.

36.2 Gaze Control and Coordination in Prosimians, Monkeys, Apes, and Humans

We briefly review key similarities and major differences in the organization of the eye and head across species. The structure of the neural circuits producing eye movements will naturally vary with differences in the nature of movement execution dictated by the size and configuration of the eye, head, body, and habitat.

A search for homologies among primates in the cortical and subcortical organization of eye movements is encouraged by the many similarities in gaze behavior observed between monkeys, apes, and humans. Members of each species produce each of the different types of eye movements—fixation maintained by gaze-holding vestibular and visual reflexes, which is interrupted by rapid gaze shifts and slow pursuit eye movements to direct vision in three dimensions. In numerous quantitative respects the eye movements of macaque monkeys and humans are more similar than different in execution but can differ in planning and context (eg, Martinez-Conde and Macknik, 2008; Baizer and Bender, 1989; Einhäuser et al., 2006; Berg et al., 2009; Shepherd et al., 2010). In a wide range of testing conditions, the advanced cognitive control of eye movements is indistinguishable between humans and macaque monkeys (eg, Munoz and Everling, 2004; Camalier et al., 2007; Hanes and Schall, 1995; Hanes and Carpenter, 1999; Nelson et al., 2010). Consequently, the macaque has been a faithful source of insights for human clinics of neurology (eg, Leigh and Zee, 2015) and psychiatry (eg, Driscoll and Barr, 2016). Comparisons of the eye movements of chimpanzees, gorillas, and orangutans with those of humans also have noted numerous similarities with particular differences (Kano et al., 2011; Kano and Tomonaga, 2009, 2011). Primates in general are visual creatures with particular ocular, muscular (eg, Blumer et al., 2016), and neural adaptations including devotion of a large proportion of the brain to visual processing.

Primates execute gaze shifts to scrutinize objects with the high acuity visual resolution afforded by the area centralis or fovea. There are qualitative differences in the organization of central retina between nocturnal and diurnal primates related to their gaze behavior that may be reflected in differences in the ocular motor network. With the exception of the nocturnal owl monkey and Strepsirrhini, all Haplorrhini, including the tarsier, possess a fovea (Wolin and Massopust, 1967). The absolute size of the fovea is comparable across diurnal primates, and a higher acuity of vision is achieved by increasing the size of the eye; for example, human eyes are about four times larger than marmoset eyes (Finlay et al., 2008). On the other hand, galagos and owl monkeys have a rudimentary area centralis (DeBruyn et al., 1980; Stone and Johnston, 1981; Wikler and Rakic, 1990, 1990; Webb and Kaas, 1976). Accordingly, the spatial resolution in galagos and owl monkeys ($\sim$5 cycles deg^{-1}) is much poorer than in macaques, apes, or humans ($\sim$50 cycles deg^{-1}) (Langston et al., 1986; Mitchell and Leopold, 2015). To appreciate this difference, note that when held at arm's length, your finger subtends about 1 degree of arc in the visual field. A

human fingerprint is composed of about 50 ridges, which may be visible to you. A galago or owl monkey could resolve only five large strips across a finger.

The distinctive visual abilities of primates are conferred primarily by a retina equipped with a specialized central region of high resolution that is associated with a sophisticated neural system producing large variety, range, and coordination of eye movements to gather visual information. The large oculomotor range that primates possess appears to be evolutionarily recent. The human oculomotor range spans about ±50 degrees of visual angle (Guitton and Volle, 1987; Stahl, 1999) and is nearly matched by that of the macaque monkey (Tomlinson and Bahra, 1986) and baboon (Marchetti et al., 1983). In contrast, the oculomotor range of more distantly related primates such as prosimians (Shepherd and Platt, 2006) or the New World squirrel monkey (McCrea and Gdowski, 2003; Heiney and Blazquez, 2011) and marmoset (Mitchell et al., 2014) is limited to <15 degrees. Limited ocular motility is commonly reported in nonhuman primates with small heads and poorly developed fovea, such as tree shrew (Remple et al., 2006) and cat (Guitton et al., 1984). Larger gaze shifts are accomplished with a combined rotation of the head and eyes. What qualifies as "larger" varies across species. The greater inertial mass of the larger heads of apes and humans makes head movements more energetically costly, slower, and socially revealing. Thus, while smaller prosimians and New World monkeys naturally make gaze shifts of more than about 5 degrees with a head rotation too (eg, McCrea and Gdowski, 2003), humans naturally make gaze shifts of more than about 20 degrees with a head rotation too. Moreover, relative to apes and monkeys, the human eye is positioned further forward in the orbit, which expands the effective visual field with horizontal eye movements achieving larger abductions (Denion et al., 2015a). Such unique orbital morphology is adaptive for bipedal locomotion in a habitat of open planes (Denion et al., 2015b). Horizontal eye movements are more common for terrestrial species compared with arboreal species (Kobayashi and Kohshima, 2001).

The eyes may be a window of the soul for humans, but gaze direction is camouflaged in other primates by coloration of the sclera resembling surrounding skin color. Comparing the external features of eyes from 88 primate species, humans are the only primate with a white sclera and the largest amount of exposed sclera relative to the outline of the eye (Kobayashi and Kohshima, 1997). This morphology affords a clear visual marker of gaze direction that facilitates communication (Kobayashi and Kohshima, 1997, 2001). Gaze following is observed across primate species (Rosati and Hare, 2009; Shepherd, 2010), but most often it utilizes the orientation of the head. The reliance on the direction of the eye seems to be specific to humans (Tomasello et al., 2007). In summary, the specialization of the fovea and the oculomotor range delimit the visual ability of primates and rationalize the amount of cortex dedicated to visual processing.

36.3 Variability of Sulci in the Primate Frontal Cortex

In every primate species including humans FEF is located in a sulcus (or dimple) immediately rostral to the central sulcus. In humans this is referred to as the precentral sulcus, but in monkeys, the arcuate sulcus. Mingazzini (1888) first coined the term arcuate sulcus ("solco arcuato") to describe the inferior precentral sulcus in the Old World monkey, which others had previously called "sillon courbe" by Gromier or "sillon kypsiloïde" by Broca (see Hervé, 1888).

The location of FEF in standardized coordinates for the individual species can vary in proportion to variability of the morphology of this sulcus. The high variability of sulcal patterns in humans is well known, and a variety of patterns can be identified across individuals (Ono et al., 1990). The precentral sulcus is interrupted into superior and inferior segments in over 50% of individuals; three segments are found in about 30% of individuals, and four segments are found in fewer than 5%. About 30% of individuals also exhibit a longitudinally oriented marginal precentral sulcus, reminiscent of the precentral dimple in macaques. The major superior and inferior segments develop at different times. In more than 90% of individuals the superior precentral sulcus connects with the superior frontal sulcus, and the inferior precentral sulcus connects with the inferior frontal sulcus. None of these tendencies are the same across hemispheres.

The first reference to a sulcus praecentralis referred only to the inferior precentral sulcus (IPrCS) (Ecker, 1869). According to Eberstaller (1890), the superior precentral sulcus was first described as a separate sulcal entity by Jensen (1871). Recent studies have further subdivided these sulci—Germann et al. (2005) distinguished three parts of the IPrCS: a dorsal and a ventral part, and a so-called horizontal extension separating the other two parts. Also, the superior precentral sulcus was subdivided into a ventral and a dorsal part, separated by the superior frontal sulcus (Amiez et al., 2006).

Naturally, the location of FEF and surrounding areas will vary as a function of systematic and random variability of sulcal patterns. In fact, as detailed in the following, Amiez et al. (2006) conducted a subject-by-subject analysis of the locus of eye movement–related functional activity revealed in relation to the detailed morphology of the precentral and superior frontal sulci.

A focus of activation associated with saccadic eye movements was located in the ventral branch of the superior precentral sulcus in both hemispheres. A second focus has been found in the dorsal part of the inferior precentral sulcus. Imaging during a hand response selection task revealed activation focused in the dorsal branch of the superior precentral sulcus close to the caudal end of the superior frontal sulcus. Activation in primary motor cortex was focused in the precentral knob (or Broca's *pli de passage moyen*). The relationship of FEF to the bordering premotor cortex is elaborated in the following paragraphs.

Compared to humans, the cerebral sulcal pattern is much less variable across Old World monkeys (Cercopithecidae) (Falk, 1978). The frontal lobe consists of an arcuate sulcus rostral to the central sulcus and a longitudinal principal sulcus (known also as sulcus rectus) rostral to the arcuate sulcus (Connolly, 1936, 1950; Walker, 1940). The arcuate sulcus consists of a superior (horizontal) limb or branch and an inferior (vertical); at the genu or curve, typically located caudal to the end of the principal sulcus, some individuals exhibit a posterior extension of the sulcus referred to as a "spur." Thus, the shape and extent of the arcuate sulcus in monkeys varies. However, no one has systematically characterized the variation of its sulcal morphology. For example, the prevalence of the arcuate spur is unknown in various macaque species. Therefore, we investigated the incidence of an arcuate spur in a sample of 235 hemispheres obtained from nine direct observations before histological processing, 132 magnetic resonance (MR) images, and 109 figures from the literature. This sample consists of 221 left hemispheres and 161 right hemispheres; for 147 brains both hemispheres were available. Samples were obtained from *Macaca fascicularis* ($n = 19$), *Macaca mulatta* ($n = 162$), *Macaca nemestrina* ($n = 7$), and *Macaca radiata* ($n = 21$). We could determine the gender of 54 male and 7 female monkeys. Table 36.1 summarizes the findings. In about one-third of both left and right hemispheres a large spur was present, another one-third exhibited a small spur, and the remaining one-third exhibited no spur. Thus, two-thirds of hemispheres exhibit an arcuate spur. In about two-thirds of hemispheres the spur was symmetric in appearance in both hemispheres.

These values may underestimate the prevalence of an arcuate spur because we found a spur more often in our sample of histological (89%) and MR (86%) images as compared with figures in the literature (57%). We cannot evaluate how faithfully authors portray cortical sulcal patterns. Also, even though the sample of female macaques was small, we found no clear difference in prevalence between females (71%) and males (67%). An arcuate spur was most common in *M. mulatta* (75%) and *M. nemestrina* (71%), only slightly less common in

TABLE 36.1 Prevalence of arcuate spur in macaques

Spur size	Hemisphere	Number	Percentage (%)
Large	Left	65/221	29
	Right	48/161	30
Small	Left	83/221	38
	Right	64/161	40
Absent	Left	73/221	33
	Right	49/161	30
Symmetric	Both	94/147	64

M. radiata (67%), and least common in *M. fascicularis* (53%).

These modest findings have several implications. First, atlases based on averages of macaque brains (eg, Frey et al., 2011; Rohlfing et al., 2012; Calabrese et al., 2015) imply that sulcal patterns are more regular and uniform across individuals than they actually are. Such averaged atlases underestimate the length of the spur observed in many individuals by representing only a small spur (Frey et al., 2011) or none at all (Calabrese et al., 2015; Rohlfing et al., 2012). Likewise, atlases based on a single individual will misrepresent the incidental presence or lack of a spur as standard anatomy (eg, Saleem and Logothetis, 2012). Second, sulcal patterns can serve as reliable landmarks to identify the location of cortical areas. For example, in macaques the location and shape of the arcuate sulcus reliably predicts the location of FEF. Also, the arcuate spur has also been identified as a boundary between functionally and anatomically distinguished dorsal and ventral premotor areas (eg, Gabernet et al., 1999). Third, a variety of mechanisms for cortical folding have been proposed—differential cell proliferation, differential tangential expansion, radial intercalation, axon tension, radial glia, and so on (see for review Striedter et al., 2015; see also Mota and Herculano-Houzel, 2015; Tallinen et al., 2016). These data cannot distinguish among these alternatives, but they offer an opportunity for further exploration. Finally, variation of structure most likely announces variation of function. The common lack of symmetrical occurrences of the spur represents clear evidence for cerebral asymmetry in macaques. While others have reported a rightward bias in the length of the arcuate sulcus in macaques (Sakamoto et al., 2014; Imai et al., 2011), some have reported no asymmetry (Falk et al., 1990; Heilbroner and Holloway, 1989). Whether such asymmetry is systematic and functionally meaningful requires a larger sample. However, behavioral and anatomical evidence for some degree of cerebral asymmetry in macaques has been reported (eg, Hamilton and Vermeire, 1988; Heilbroner and Holloway, 1989; Falk et al., 1990). Perhaps an arcuate spur relates to

individual hand preferences (eg, Lehman, 1980; Mangalam et al., 2014). In any case, when quantified in the MRI atlases of macaque brains, the magnitude of deformation needed to co-register the set of macaque brains varies across the cortex and was noticeably elevated in the region of the spur (Frey et al., 2011; Calabrese et al., 2015). Does this greater variability in the organization of this region provide greater opportunity for natural selection? Further research is needed to determine how the variation of sulcal morphology relates to the location and boundaries of FEF and neighboring areas.

36.4 Location of FEF Across Primate Species

36.4.1 Scandentia

Together with Dermoptera, for which very little is known, Scandentia is the closest mammalian order relative to Primates (Perelman et al., 2011). Most of their visual specializations evolved independently of primates. They inhabit tropical forests of southern Asia. Electrical stimulation mapping of the (small) frontal lobe has not found a discrete FEF, although stimulation of sites in frontal motor areas did elicit eye blinks (Lende, 1970; Remple et al., 2006; Baldwin et al., 2017). This may not be not surprising, because tree shrews have neither a specialized focus of high resolution in the retina nor much range of eye movements. However, they are very visual creatures that orient readily with head and body.

36.4.2 Strepsirrhini: Prosimians

Most prosimians are nocturnal animals. As noted in the earlier section, the retina lacks a clear fovea and instead has only a rudimentary area centralis. Their relatively poor distance vision is compensated by a long snout ending in a moist and touch-sensitive surface, similar to that of dogs. The wet nose provides a conduit for pheromones into the vomeronasal organ.

Prosimians resemble other primates and are distinct from other mammals in the presence of a clear granular prefrontal cortex rostral to the agranular motor cortex. Therefore, prosimians are a good model for the evolution of FEFs, because their frontal cortex is subdivided into a granular and agranular portion, but they do not rely as much on the visual system as simians do.

36.4.2.1 Lorisoidea (Galago)

Galagos are nocturnal primates living in low forests. Arboreal animals, they move by quadrupedal walking and leaping. Galagos have relatively large eyes with a rudimentary fovea supporting relatively low acuity vision (DeBruyn et al., 1980; Stone and Johnston, 1981). Among primates they rely least on vision. Because of the relatively large size of the eyes, galagos have a limited range of eye movements and compensate for this with more head movements to explore novel objects (Rogers et al., 1993; Cantalupo et al., 2002). Although their form of locomotion might suggest that they rely on good eye–hand coordination for their agile leaping, they primarily use olfactory and auditory information. Galagos have large ears with great mobility that are continuously moving. As insectivores, they rely on sound to localize their prey (Charles-Dominique, 1977).

Electrical stimulation of a small region medial to the anterior frontal sulcus evokes contraversive eye movements (Wu et al., 2000; Fig. 36.1). This very restricted representation of eye movements was compensated by a larger representation of ear movements in the cortex caudal to FEF (Fogassi et al., 1994). While the more rostral region had cytoarchitectonic characteristics of FEF seen in other species with a granular layer 4 and medium-to-large densely packed layer 5 pyramidal neurons, the caudal region appeared more dysgranular (Preuss and Goldman-Rakic, 1991). Like FEF in macaques the rostral region is connected with the multiform division of the mediodorsal nucleus and with the intermediate and superficial layers of the superior colliculus (Markowitsch et al., 1980; Preuss and Goldman-Rakic, 1991).

Curiously, a dorsomedial eye field corresponding to the supplementary eye field has not been found (Wu et al., 2000). Research with macaque monkeys and humans has led to the general conclusion that the supplementary eye field contributes to high level, executive control of gaze behavior (eg, Schall and Boucher, 2007). If correct, the absence of a supplementary eye field in galagos could be understood in light of the stimulus-bound simplicity of galago behavior relative to that of macaques and humans. Regardless, this reminds us that the number and organization of cortical areas mediating gaze control need not be equivalent across species.

36.4.3 Haplorrhini: Simians

Simians (anthropoids) are all diurnal with the exception of owl monkeys. They have less reliance on olfaction and are primarily visual animals with a well-developed fovea supporting high acuity vision and a suitably matched ocular motor system supporting the control of eye movements in coordination with head, limb, and body movements.

36.4.3.1 Platyrrhini (New World Monkey)

Platyrrhini are small to medium size monkeys with flat noses, which distinguishes them from Old World monkeys. Other than howler monkeys, they lack the typical trichromatic vision of the Old World monkeys.

36.4.3.1.1 Callitricidae (Marmoset)

Marmosets are the smallest New World monkeys. Those used in research, *Callithrix jacchus*, naturally live in open forest habitats. Marmosets are very agile and active, moving with quadrupedal walking and leaping. Claw-like nails allow them to cling to trees while foraging. The retina of the marmoset has a well-developed fovea comparable to that found in other diurnal primates (Finlay et al., 2008; Franco et al., 2000). However, marmosets have the smallest eyes among the primates (Wolin and Massopust, 1967) affording less spatial acuity (Finlay et al., 2008). With lightweight heads, marmosets can shift gaze with rapid head movements and so have correspondingly smaller oculomotor range (Mitchell et al., 2014). Marmosets also display head-cocking, rapid head rotations along the longitudinal axis, for visual exploration (Kaplan and Rogers, 2006).

With a lissencephalic (smooth) cortex and an opportunity for genetic studies, marmosets have gained new popularity in research. A review by Bakola et al. (2015) summarizes knowledge about the frontal motor system. Although the frontal cortex of marmosets is overall a smooth surface, some individuals exhibit an arcuate dimple. Eye movements with and without head movements are elicited from the rostral frontal lobe with electrical stimulation (Mott et al., 1910; Blum et al., 1982; Fig. 36.2). Such movements were also elicited by stimulation of the rostral frontal lobe in prosimian lemurs (Mott and Halliburton, 1908). Subsequent research with marmosets has verified visual and ocular motor connectivity of these regions (Reser et al., 2013; Lyon and Kaas, 2001; Spatz and Tigges, 1972; Rosa et al., 2009; Krubitzer and Kaas, 1990; Collins et al., 2005). Areas 8aV and 45, which might correspond to the frontal visual (FV) area described by Krubitzer and Kaas (1990), show similar patterns of connections with parietal and extrastriate visual areas. This observation suggests that FEF in the marmoset spans area 8 and perhaps partially area 45 (Reser et al., 2013). The cytoarchitecture of the FEF region is characterized by larger neurons in layer 5, higher myelin density, and a granular layer 4 (Burman et al., 2006; Krubitzer and Kaas, 1990), similar to other species examined. The locally elevated density of myelin was also found with structural MRI techniques (Bock et al., 2009). Information about the neurophysiology of FEF in particular or prefrontal cortex in general in marmosets is lacking. However, an fMRI study reported elevated BOLD signal within the location of area 8aV when visual stimuli were presented (Hung et al., 2015).

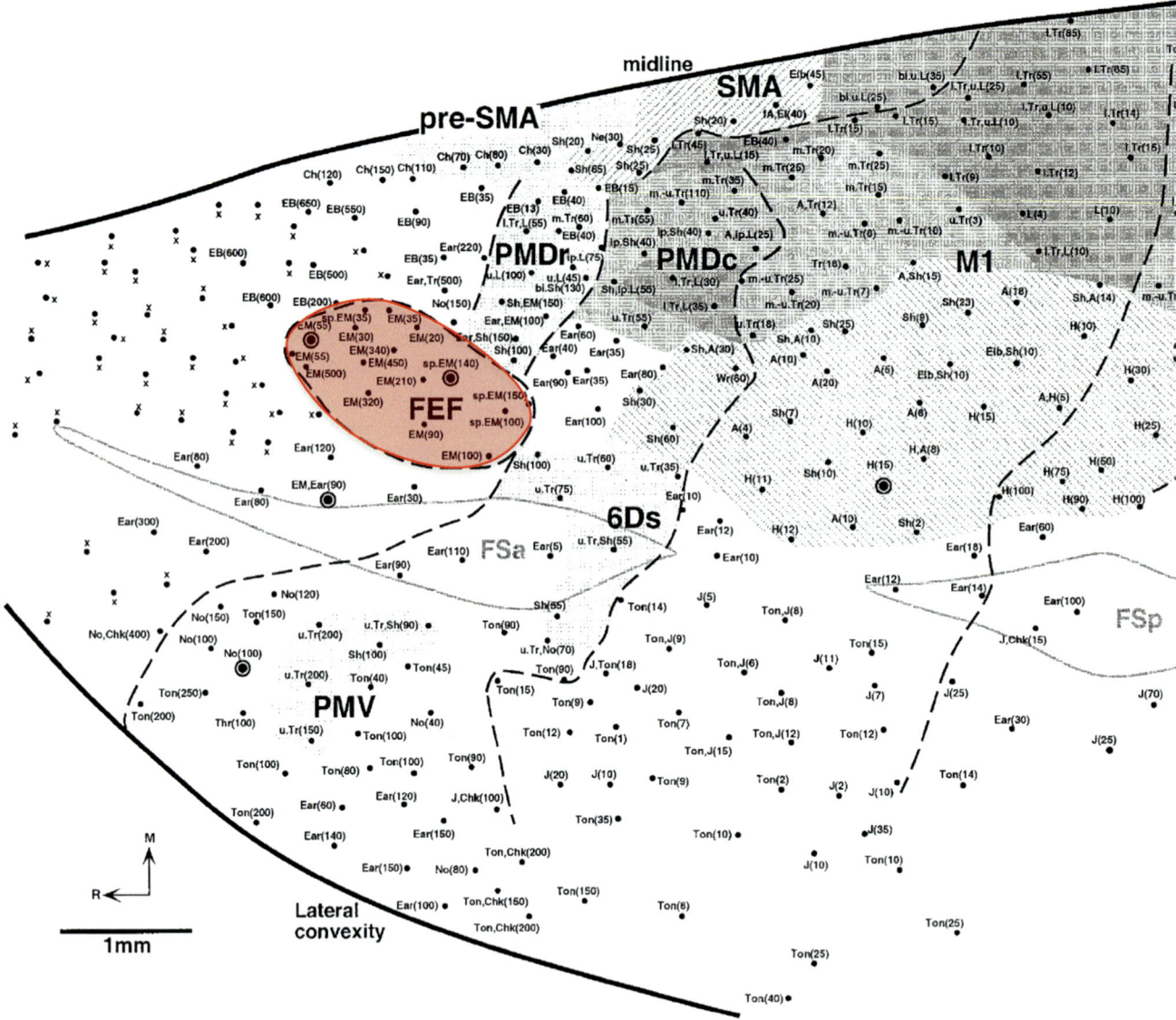

FIGURE 36.1 Map of eye and body movements elicited by intracortical microstimulation in *Galago garnetti*. The frontal eye field (FEF) is *highlighted red*, located rostral to premotor cortex. Smooth pursuit eye movements were evoked at more caudal sites, and saccadic eye movements, more rostral. Curiously, a supplementary eye field was not found. *Reproduction from Wu, C.W., Bichot, N.P., Kaas, J.H., 2000. Converging evidence from microstimulation, architecture, and connections for multiple motor areas in the frontal and cingulate cortex of prosimian primates. J. Comp. Neurol. 423, 140–177.*

Eye movements with and without head movements were also elicited by stimulation caudal to the FEF (Mott et al., 1910); sites eliciting pinna movements were not found, contrasting with prosimian lemurs (Mott and Halliburton, 1908). In marmosets, an additional caudal area 8 (area 8C) was described at the junction between dorsal and ventral premotor cortex (Paxinos et al., 2012; Burman et al., 2014, 2015; Bakola et al., 2015), which is distinguished by denser myelination. This area has a connectivity pattern like area 6Va, but lacks some of the somatosensory connections and instead receives input from area 8aV and is interconnected with visual areas (Bakola et al., 2015; Burman

et al., 2015; Palmer and Rosa, 2006; Rosa et al., 2009). Burman et al. (2015) consider area 8C as part of the premotor network rather than as a prefrontal area because it lacks a distinct layer 4. However, they also suggest that this region might be specialized for visually guided movements and suggest that it might correspond to a region where Preuss et al. (1996) evoked eye and neck movements in the owl monkey. Marmoset area 8C might correspond to the PMvr spur described by Gabernet et al. (1999) for the macaque monkey as is described in the following section. Burman et al. (2015) point out that area 8C might be similar to area 6Va involved in head movements. Because marmosets compensate for

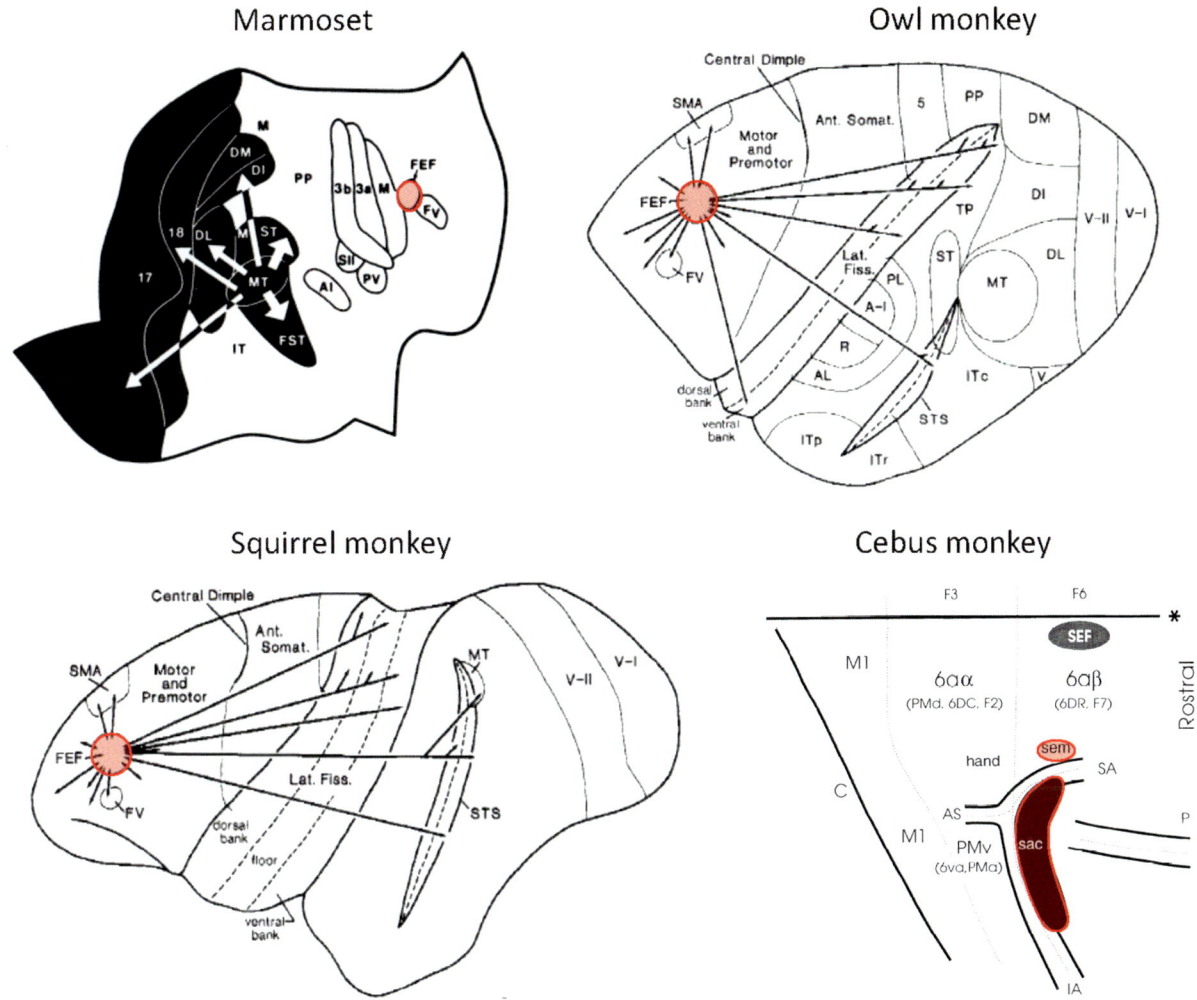

FIGURE 36.2 Location and connectivity of frontal eye field defined by microstimulation and intracortical connectivity in marmoset, owl monkey, and squirrel monkey with map of saccadic and smooth pursuit eye movements elicited by intracortical microstimulation in *Cebus*. *Reproduced from Krubitzer, L., Kaas, J., 1990. The organization and connections of somatosensory cortex in marmosets. J. Neurosci. 10, 952–974; Huerta, M., Krubitzer, L., Kaas, J., 1987. Frontal eye field as defined by intracortical microstimulation in squirrel monkeys, owl monkeys, and macaque monkeys. II. Cortical connections. J. Comp. Neurol. 265, 332–361; Tian, J., Lynch, J., 1996. Corticocortical input to the smooth and saccadic eye movement subregions of the frontal eye field in* Cebus *monkeys. J. Neurophysiol. 76, 2754–2771.*

their limited oculomotor range with rapid head movements, an area linked with FEF that is involved in directed head movements is not surprising.

36.4.3.1.2 Aotidae (Owl Monkey)

Owl monkeys are nocturnal. They live in higher elevations in forests and tropical environments of Central America and northern South America (Groves, 2001). They feed upon small fruit, insects, leaves, and flowers. They are able to snatch flying insects out of the air or off of a branch, unlike foragers or tool users who exploit holes and spaces in the tree bark (Wright, 1989). Even though the nocturnal environment would be a disadvantage for animals dependent upon vision, owl monkeys have been able to thrive in the niche. They possess large eyes without a fovea and a nearly pure

rod retina (Woollard, 1927; Jacobs et al., 1993). These structural features provide for better low-light spatial resolution that improves their night vision, but leaves the owl monkey with monochromatic vision.

Owl monkeys are lissencephalic anterior to the central sulcus, only rarely with an arcuate dimple. FEF was identified with microstimulation at low currents around the arcuate dimple, and because of the smaller prefrontal FEF appears located relatively more rostral (Fig. 36.2; Gould et al., 1986; Huerta et al., 1986, 1987). Caudal to FEF is an expanded region where head movements are represented (Gould et al., 1986). Gould et al. (1986) describe another eye field that is located more dorsal and also has a caudally adjacent region representing head movements and likely corresponds to SEF of macaque monkeys (Preuss et al., 1996). Huerta et al.

(1986, 1987) also characterize an FV field located ventral to FEF. Saccadic eye movements could be evoked with higher currents in area 8B—referred to it as dorsal oculomotor area—as well as in the dorsal premotor cortex (Stepniewska et al., 1993; Preuss et al., 1996). Stimulation in the cortex surrounding FEF at higher currents also leads to ear, eyelid, or vibrissa movements (Huerta et al., 1986).

FEF in owl monkeys shows a connectivity pattern comparable to squirrel monkeys and macaque monkeys with connections to the upper temporal lobe, the dorsomedial visual area, and posterior parietal cortex as well as the superior colliculus and the lateral band of the mediodorsal nucleus (Kaas and Morel, 1993; Weller et al., 1984; Krubitzer and Kaas, 1993; Kaas et al., 1977; Collins et al., 2005; Huerta et al., 1986, 1987). However, the FEF region of owl monkeys apparently lacks strong connections with the middle temporal visual area that are found in diurnal monkeys (Huerta et al., 1987; Weller et al., 1984; but see Cerkevich et al., 2014). A connectivity pattern similar to FEF is shared by the more ventral FV area.

36.4.3.1.3 Saimiriinae (Squirrel Monkey)

Squirrel monkeys live in Central and South America in the midlevel tangled layers of branches, vines, and trees of tropical forests. Squirrel monkeys use all four limbs to move but will occasionally walk short distances on two limbs. Squirrel monkeys predominantly eat fruits and insects. They rarely reach for insects in motion, but prefer finding insects on leaves. They engage in long periods of visual observing behavior (Glickman and Sroges, 1966); however, they tend to show notably less visual investigative behavior relative to other primates (DuMond, 1968; Glickman and Sroges, 1966; but see Haude and Ray, 1974). Male squirrel monkeys are dichromatic, while females can be either dichromatic or trichromatic. The eye movements of squirrel monkeys are in general comparable to macaque monkeys, but they have a much narrower oculomotor range of about 10—15 degrees horizontally (Heiney and Blazquez, 2011; McCrea and Gdowski, 2003). Interestingly, the oculomotor range is larger for the vertical axis (about 25 degrees); this asymmetry contrasts with terrestrial species in which horizontal eye movements are more frequent and of larger average amplitude. Like other monkeys, squirrel monkeys communicate with visual signals and social behaviors, including facial expressions, as well as signaling with their ears.

In squirrel monkeys, gyral variations of the frontal lobe range from no sulcus to a small inferior arcuate dimple to a small arcuate sulcus (Emmers and Akert, 1963; Blum et al., 1982; Huerta et al., 1987). FEF was identified with electric stimulation near the dimple, with considerable variation between individuals (Huerta et al., 1986; Fig. 36.2).

FEF of squirrel monkeys is connected with visual areas in the superior temporal sulcus including area MT in addition to subcortical structures paralleling macaques (Huerta et al., 1987; Tigges et al., 1981).

36.4.3.1.4 Cebidae (Capuchin)

Capuchin monkeys are considered to be the most intelligent of New World monkeys, possessing a relatively large brain relative to their body size. They have coordinated movements between hands, tail, feet, and mouth. They perceive objects, movements, and surfaces in much the same way humans do. They have a large repertoire of visual signals and gestural behaviors to communicate, including submissive grins, raising eyebrows, head tilt, play face, threat face, and lip smacking. Capuchins are found commonly in northern parts of Argentina but are also distributed around North and South America. They live in different regions of trees—some very high and others in the lower regions. They spend most of their waking hours foraging for a variety of types of food. They eat a greater variety of food types than any other type of monkeys. They exhibit rudimentary tool use, for example, using stones to access food.

The capuchin (*Cebus apella*) shows a clearly defined gyral pattern (Connolly, 1936, 1950), resembling that of macaque monkeys with an arcuate sulcus that is usually accompanied by a principal sulcus (Sanides, 1970; Tian and Lynch, 1997).

FEF as identified with microstimulation is found along the anterior bank of the arcuate sulcus (Lynch et al., 1994) at a location comparable to that of macaque monkeys. It shows a comparable connectivity pattern as described for other monkeys, with connections to MT and to the posterior parietal cortex as well as subcortically to the superior colliculus and the mediodorsal nucleus of the thalamus (Tian and Lynch, 1997; Leichnetz and Gonzalo-Ruiz, 1996; Rosa et al., 1993; Lynch et al., 1994).

In addition to FEF in the anterior bank of the arcuate sulcus, two more eye fields have been characterized in *Cebus*. One is located dorsomedially and likely corresponds to the supplementary eye field of macaque monkeys (Tian and Lynch, 1995) and another one is located in the posterior bank of the superior limb of the arcuate sulcus (Tian and Lynch, 1996). In contrast to FEF where microstimulation evokes saccadic eye movements, stimulation at the later more dorsomedial location results in smooth eye movements. This location might correspond to the smooth pursuit zone described for macaque monkeys (Gottlieb et al., 1993, 1994); however, in macaque monkeys it is in the posterior bank of the arcuate sulcus adjacent to FEF and not located dorsomedially as in *Cebus* (Fig. 36.2). This illustrates that a common function can be mediated by cortical areas in different relative locations across species.

36.4.4 Cercopithecidae (Old World Monkey)

36.4.4.1 Macaca (Macaque)

Macaque monkeys are the most common nonhuman primate used in research to gain insights into human brain function (Passingham, 2009). Hence, the most detailed knowledge about the organization and location of the primate FEF has been obtained from various macaque species (*M. mulatta*, *M. fascicularis*, *M. radiata*, *M. nemestrina*, *Macaca sinica*, and *Macaca fuscata*). Given the relatively minor known differences in the brain organization across macaque species, many studies have used more than a single species or some do not specify clearly what macaque species was used.

Macaques are generally acknowledged as terrestrial, but some (eg, *M. fascicularis*) are also arboreal. Macaques are mostly quadrupedal animals; however, they also are very dexterous in using their hands for visually guided reaching and grasping behavior. Macaques use saccadic and smooth pursuit eye movements to explore the visual field, including vergence movements to explore both far and near locations. Macaques are social animals that usually live in hierarchical groups. Gaze direction is a behaviorally relevant cue, and macaques show gaze-following behavior as directed by the head direction. Gaze direction is also an important signal in the social structure; direct staring at an animal higher in the hierarchy is an offensive behavior that could trigger retribution. Accordingly, macaques covertly observe other animals.

Following the groundbreaking experiments of Fritsch and Hitzig (1870) demonstrating that electric stimulation of certain locations in the cerebral cortex evoked specific body movements in a dog, Ferrier, (1874) carried out a thorough mapping to localize motor functions in the cortex of macaque monkeys. Based on his maps of the frontal cortex, Ferrier located a region in which stimulation evoked contraversive movements of the eyes and head as well as other ocular movements such as vergence and blinks (Fig. 36.3). Numerous subsequent investigators mapped locations eliciting orienting movements in macaque monkeys resolving over time the effects of anesthesia, eye movement recording sensitivity, electrode size, and electrical current levels (Horsley and Schäfer, 1888; Mott and Schäfer, 1890; Russell, 1894; Levinsohn, 1909; Smith, 1949; Crosby et al., 1952; Henderson and Crosby, 1952; Wagman et al., 1957, 1958, 1961; Robinson and Fuchs, 1969; Bruce et al., 1985). The evolution of FEF location across these studies has been reviewed elsewhere (Schall, 1997). The current definition of FEF is the region where currents less than 50 μA evoke short latency saccades, which restricts FEF in the rostral bank and fundus around the genu of the arcuate sulcus. Within this region saccade amplitude, but not direction, is topographically organized;

shorter (<2 degrees) saccades are represented ventrolaterally, and progressively longer saccades (15−20 degrees) are represented dorsomedially.

The macaque FEF is well known to be highly visually responsive with pronounced modulation according to the allocation of attention and other cognitive demands (reviewed by Squire et al., 2013; Schall, 2015). Visually responsive neurons are found in cortical areas extending rostral to FEF. Indeed, the map of saccade amplitude is paralleled by a map of visual receptive field eccentricity described in the prearcuate gyrus (Suzuki and Azuma, 1983).

A region related to slow tracking pursuit eye movements is located at the fundus of the arcuate sulcus immediately caudal to the principal sulcus (MacAvoy et al., 1991; Gottlieb et al., 1994; Fukushima et al., 2002; Tanaka and Lisberger, 2002; Fig. 36.3). Low-intensity electrical stimulation of FEF in head-free monkeys also elicits contraversive head movements and natural eye-with-head gaze shifts (Tu and Keating, 2000; Knight and Fuchs, 2007; Monteon et al., 2010, 2013). Medial to FEF in the dorsomedial cortex is an anatomically separate and functionally distinct supplementary eye field (Schlag and Schlag-Rey, 1987; Schall, 1991). Along the superior limb of the arcuate is a region that represents orienting movements of the pinna (Bon and Lucchetti, 1994; Lucchetti et al., 2008; Lanzilotto et al., 2013) with a gradual transition from eye movements and ear movements that also correspond to the eccentricity of localized events (Barbas and Mesulam, 1981; Lanzilotto et al., 2013). Finally, neurons in the rostral convexity of the arcuate sulcus are active during vergence and accommodation for perception of objects at different distances (Gamlin and Yoon, 2000).

The location of FEF has a distinctive cyto- and myeloarchitecture (Walker, 1940; von Bonin and Bailey, 1947; Stanton et al., 1989; Preuss and Goldman-Rakic, 1991). Not unexpectedly, it is distinguished by a concentration of large pyramidal cells in layer 5 higher than surrounding areas. It has a thin granular layer 4 distinct from the agranular area 6 at its caudal boundary but noticeably thinner than the rostrally bordering area referred to now as 8Ar. Area 8Ar should probably be considered as functionally distinct from FEF; it may correspond to the area FV that has been distinguished from the heavily myelinated part of FEF in owl monkeys based on patterns of connectivity with prestriate visual areas (Weller and Kaas, 1987; Krubitzer and Kaas, 1990). The medial aspect of FEF has relatively fewer large pyramidal cells in layer 3 and a loosely organized granular layer and has more connectivity with cortical areas representing the peripheral visual field. The dorsomedial border of FEF transitions into area 8B. The ventral aspect of FEF has a higher density of larger pyramidal cells in layer 3 and has more connectivity with

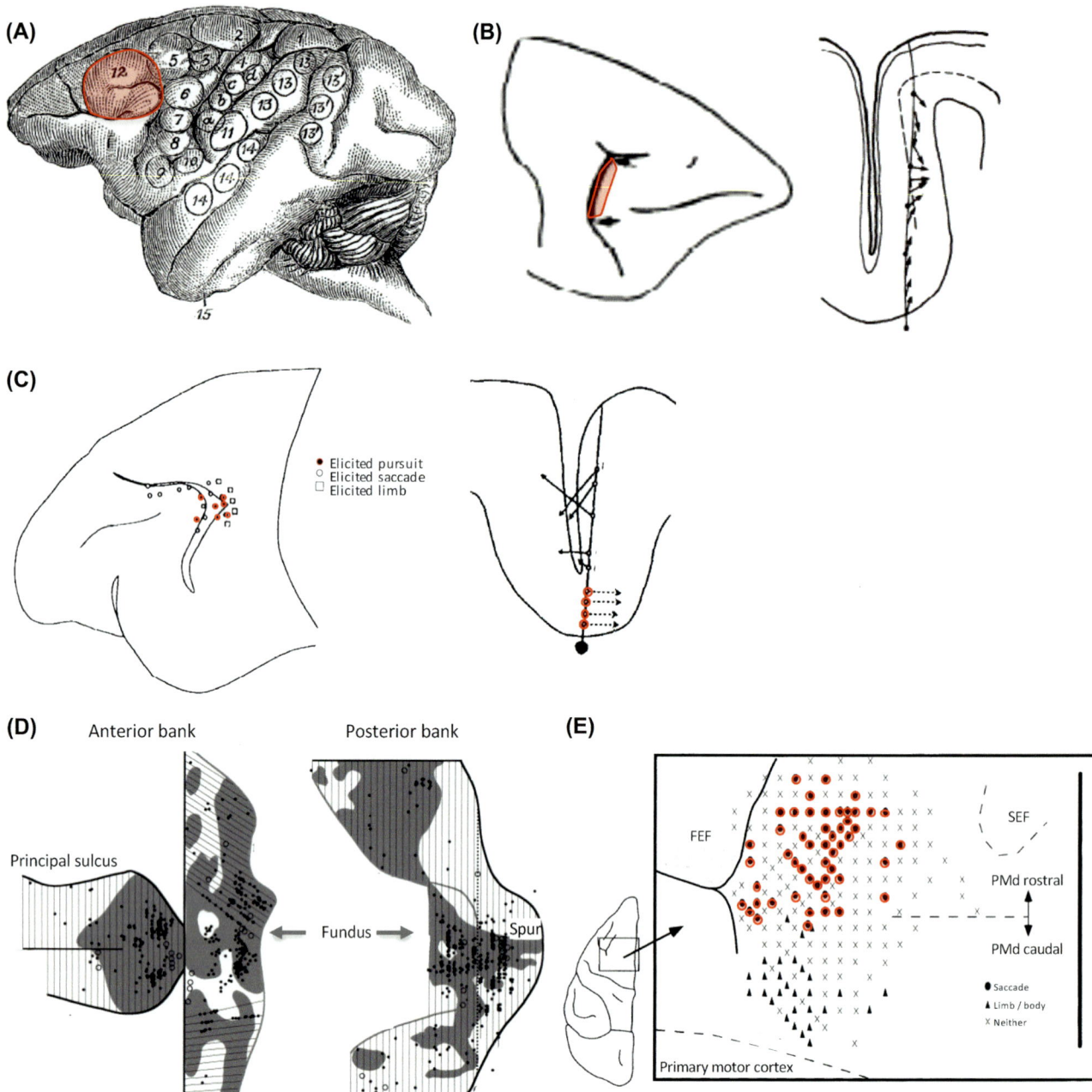

FIGURE 36.3 Representative maps of the frontal eye field (FEF) across eras, methods, and eye movements. (A) Map of body movements elicited by electrical stimulation. Stimulation within the region numbered 12 produced "Elevation of the eyebrows and the upper eyelids, turning of the eyes and head to the opposite side, and great dilatation of both pupils." Reproduced from Ferrier, D., 1874. Experiments on the brain of monkeys. No. I. Proc. R. Soc. Lond. 23, 409–430. (B) Location of FEF in the rostral bank of the arcuate sulcus extending from the convexity to the fundus defined by saccades evoked with the lowest currents of intracortical microstimulation in awake, behaving macaques. Reproduced from Bruce, C.J., Goldberg, M.E., Bushnell, M.C., Stanton, G.B., 1985. Primate frontal eye fields. II. physiological and anatomical correlates of electrically evoked eye movements. J. Neurophysiol. 54, 714–734. (C) Location of pursuit zone in FEF at the fundus of the arcuate sulcus shown in a surface view (left) and cross section (right). Reproduced from MacAvoy, M.G., Gottlieb, J.P., Bruce, C.J., 1991. Smooth-pursuit eye movement representation in the primate frontal eye field. Cereb. Cortex 1, 95–102; Gottlieb, J.P., Bruce, C.J., MacAvoy, M.G., 1993. Smooth eye movements elicited by microstimulation in the primate frontal eye field. J. Neurophysiol. 69, 786–799. (D) Unfolded arcuate and principle sulci showing locations of neurons transneuronally labeled by rabies virus injected into an extraocular muscle (*black dots*) and regions of metabolic activation during saccade production (*gray*). Reproduced from Moschovakis, A.K., Gregoriou, G.G., Ugolini, G., Doldan, M., Graf, W., Guldin, W., Hadjidimitrakis, K., Savaki, H.E. 2004. Oculomotor areas of the primate frontal lobes: A transneuronal transfer of rabies virus and [14C]-2-deoxyglucose functional imaging study. J Neurosci, 24, 5726–5740. (E) Map of body movements evoked by intracortical microstimulation of a cortical region caudal to the arcuate sulcus, showing a region from which saccades were evoked in awake behaving macaques. *Reproduced from Fujii, N., Mushiake, H., Tanji, J., 2000. Rostrocaudal distinction of the dorsal premotor area based on oculomotor involvement. J. Neurophysiol. 83, 1764–1769.*

cortical areas representing the central visual field. The ventrolateral border transitions into area 45. The medial and lateral parts of FEF in the arcuate sulcus are also distinguished by connectivity; they are connected with many common but several distinct cortical areas (Schall et al., 1995; Markov et al., 2014). The lateral segment, which is responsible for generating short saccades, receives visual inputs from the foveal representation in retinotopically organized areas, from areas that represent central vision in inferotemporal cortex and from other areas having no retinotopic order. Meanwhile, the medial segment, which is responsible for generating longer saccades coupled with head and also pinna movements, is innervated by the peripheral representation of retinotopically organized areas, from areas that emphasize peripheral vision or are multimodal and from other areas that have no retinotopic order or are auditory. Convergence from the dorsal and ventral visual processing streams occurs in lateral but not in medial FEF. These differences of connectivity must have corresponding differences of functionality, but they are unknown at this time.

The location of FEF in macaques has also been investigated with a variety of neuroimaging techniques. One approach used 2-deoxyglucose imaging to locate activated areas in both the anterior and posterior banks of the arcuate sulcus, the prearcuate convexity, dorsomedial frontal, caudal principal and periprincipal, anterior cingulate, and even some of the orbitofrontal cortex (Moschovakis et al., 2004; Savaki et al., 2015). The relation of these diverse cortical areas to eye movement production was reinforced by retrograde transneuronal labeling by rabies virus that was injected into the lateral rectus muscle (Moschovakis et al., 2004; Fig. 36.3D).

FEF has also been localized using fMRI with macaques. Early studies identified an involvement of FEF in visual motion processing in awake behaving monkeys (Vanduffel et al., 2001). Subsequent primate fMRI studies characterized involvement of FEF in visually guided saccade tasks (Koyama et al., 2004; Baker et al., 2006; Premereur et al., 2015), memory guided saccade tasks (Kagan et al., 2010), antisaccade tasks (Ford et al., 2009), visual search tasks (Wardak et al., 2010), and spatial attention tasks (Caspari et al., 2015). The connectivity found in resting state fMRI of the macaque FEF corresponds very well to the connectivity pattern found with anatomical tract tracers (Babapoor-Farrokhran et al., 2013).

Finally, the location of FEF in the arcuate sulcus has been confirmed through inactivation (eg, Sommer and Tehovnik, 1997) and lesion studies demonstrating temporary impairments in eye movements (Schiller et al., 1980, 1987; Keating, 1993; Morrow and Sharpe, 1995; Heide et al., 1996) and in eye—head movements (van

der Steen et al., 1986). Inactivation of FEF also impairs visual search (Latto, 1978; Wardak et al., 2006).

While there is no disputing the location of the FEF in the rostral bank of the arcuate sulcus of macaques, multiple studies have described visual- and saccade-related function in areas surrounding the arcuate sulcus. For example, visually responsive, eye movement, and eye position—related neural activity is found in premotor cortex caudal to the arcuate sulcus around the arcuate spur (Boussaoud, 1995; Boussaoud et al., 1993, 1998; Fujii et al., 1998, 2000). Neural activity recorded in this general region is also modulated during tasks requiring visually guided selection of targets for action (eg, Cisek and Kalaska, 2005; Song and McPeek, 2010). Also, measurements of 2-DG uptake during production of saccades of various directions and amplitudes defined two maps, one running between area 44 and ventral area 6, and the other spanning areas 8 and 45 in the anterior bank of the arcuate sulcus and extending onto the prearcuate convexity (Savaki et al., 2015). Finally, fMRI measurements in macaque monkeys reveal saccade-related activation in areas 45A, 45B, and 46 outside the conventional FEF (eg, Premereur et al., 2015).

Thus, numerous studies in macaques agree on the core location of FEF in the rostral bank of the arcuate sulcus, surrounded by areas associated with pursuit and vergence movements. Many other studies demonstrate that neural activity associated with visual responses and eye movement production is found in areas surrounding FEF both caudally in premotor cortex and rostrally in prefrontal cortex. Thus, defining the boundaries of "FEF" in the macaque is a matter of specifying criteria.

36.4.4.2 Papio (Baboons)

The frontal lobe of baboons resembles that of ma caques although with somewhat more complexity in the relationship of the arcuate and principal sulci and the presence of more dimples or partial sulci (Connolly 1936, 1950).

Less is known about the properties of FEF in baboons. Confidence in its location in the rostral bank of the arcuate sulcus, though, has guided studies of the effects of FEF lesions (eg, Pribram, 1955). Some evidence from cytoarchitecture and connectivity reinforces this confidence (Watanabe-Sawaguchi et al., 1991).

36.4.5 Hominoidea

The evolution of visually guided reaching and grasping is closely associated with primate ecology and the need to move through the forest canopy. Apes and monkeys that use brachiation rely upon vision to select, reach for, and secure their grip onto appropriate branches. These actions are often rapid and coordinated with other self-generated movements that strongly affect the visual input, such as the motion of the body, turning of the head, and frequent saccades. Since a miscalculation can have fatal consequences, primates have evolved under a strong selective pressure for accuracy in this domain. The origins of precise manual behavior may stem from specific aspects of the feeding behaviors of early primates. One influential hypothesis holds that early primates were nocturnal predators who were aided by a wide field of stereoscopic vision for catching insects in their hands (Cartmill, 1992). An alternative view is that precision reaching and grasping evolved for the purposes of extracting small fruits from terminal branches of angiosperms (Bloch and Boyer, 2002; Sussman et al., 2013).

36.4.5.1 *Pongo (Orangutan), Gorilla (Gorilla), and Pan (Chimpanzee and Bonobo)*

Relative to the Cercopithecidae brain, major changes have occurred in the cerebral cortex of hominids, especially in the frontal lobe (Connolly, 1936, 1950). Apes have no sulcus resembling the arcuate sulcus of Cercipithecidae (eg, Bogart et al., 2012). The chimpanzee frontal lobe consists of a prominent superior precentral sulcus and an inferior frontal sulcus, both with multiple limbs. The inferior precentral sulcus branches off of the inferior frontal sulcus. Ventrolaterally is found a fronto-orbital sulcus. Differential studies carried out on a large number of chimpanzee brains show considerable individual variation, particularly in the inferior precentral region (Sherwood et al., 2003).

Classic studies used electrical stimulation of the cortical surface under anesthesia to map sites eliciting movements of the body and eyes in chimpanzees (Grünbaum and Sherrington, 1901; Leyton and Sherrington, 1917; Hines, 1940; Dusser de Barenne et al., 1941; Bailey et al., 1950), orangutan (Beevor and Horsley, 1890; Leyton and Sherrington, 1917), and gorilla (Grünbaum and Sherrington, 1901; Leyton and Sherrington, 1917). Leyton and Sherrington (1917) mapped body and eye movements elicited by electrical stimulation of the cortical surface in 22 chimpanzees, 3 orangutans, and 3 gorillas. In chimpanzees, they showed that conjugate contraversive ocular rotations were elicited at a range of sites around the middle and inferior sulci with some sites eliciting eye opening as well (Fig. 36.4). Most of the rotations were horizontal, but some sites elicited rotations with upward or more often downward components. A few sites elicited convergence movements. They noted that higher currents were needed to elicit eye movements as compared to limb movements from the precentral gyrus. They further showed that contraversive ocular rotations were not a primary movement from the precentral gyrus and occurred only associated with head movements.

In a sample of 38 chimpanzees Bailey et al. (1950) located regions from which ocular deviations could be elicited in 11. Their results indicate two general regions from which ocular deviation was evoked with greatest frequency. One was located ventrolaterally straddling the inferior frontal sulcus, and the other was located dorsomedially ventral to and straddling the superior frontal sulcus. In most of the chimpanzees, eye movements were elicited much more rostral than the precentral sulci, and in many cases stimulation of the cortex rostral to the regions from which ocular deviations were elicited resulted in no movements. When explored, stimulation of the cortex caudal to the eye fields often elicited skeletal movements and other eye movements such as eye blinks and pupil size changes. The dorsomedial locations eliciting ocular rotations most commonly were located in area FC, which is described as transitional between agranular and prefrontal cortex having a slight granular layer. The ventrolateral locations appeared most commonly located in the weakly granular area FC and possibly extending into granular areas FDΔ and FDΓ. By the way, note that the summary figure offered by the authors seems not to most accurately portray their findings because it identifies an eye movement region more ventral than was observed (Fig. 36.4C).

In orangutan and gorilla, Leyton and Sherrington (1917) found sites eliciting ocular rotations similarly situated about the middle frontal sulcus (Figs. 36.5 and 36.6). Thus, the site most clearly identified with FEF is located comparably in chimpanzee, orangutan, and gorilla, distant from primary motor cortex and centered rostral to the inferior principal sulcus around the inferior frontal sulcus.

36.4.5.2 *Human*

Readers of this chapter need no introduction to the importance of visually guided eye movements coordinated with ongoing manual, locomotor, and social behavior. They will also know that the human cerebral sulcal pattern is noticeably elaborated from that of apes and appreciate how the significant individual gyral variation, sometimes even between hemispheres, can obscure descriptions of the location of cortical areas. We should also remind the reader that different investigators have arrived at somewhat different descriptions of the organization of cortical areas (Fig. 36.7). In considering which cortical areas in the human are homologous to areas in the brains of nonhuman primates, the FEF provides an interesting test case. The novice is surprised how far caudal FEF is until appreciating how much frontal lobe is in the human brain.

FEF has been located in humans using electrical stimulation in various forms for many years. To summarize, most studies have evoked ocular rotations by

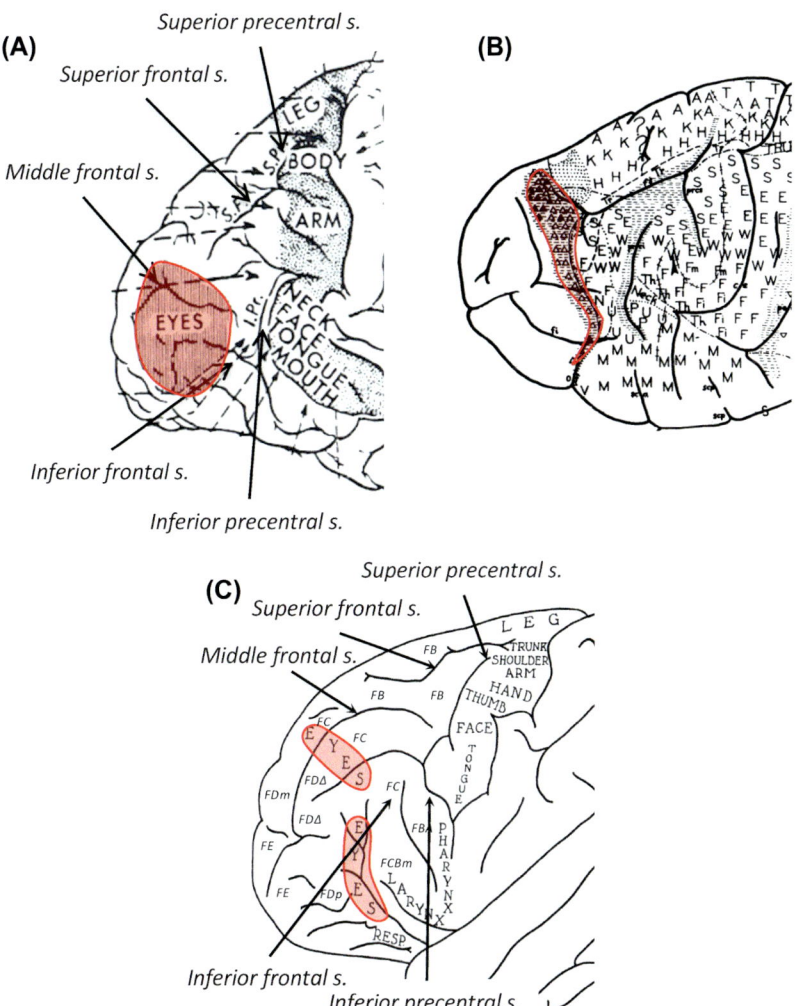

FIGURE 36.4   Location of frontal eye field based on electrical stimulation in the chimpanzee. *Reproduced from (A) Grünbaum, A., Sherrington, C.S., 1901. Observations on the physiology of the cerebral cortex of some of the higher apes. Proc. R. Soc. Lond. 69, 206–209; (B) Dusser de Barenne, J., Garol, H., McCulloch, W., 1941. Functional organization of sensory and adjacent cortex of the monkey. J. Neurophysiol. 4, 287–303; (C) Bailey, P., von Bonin, G., McCulloch, W.S., 1950. The Isocortex of the Chimpanzee. University Illinois Press.*

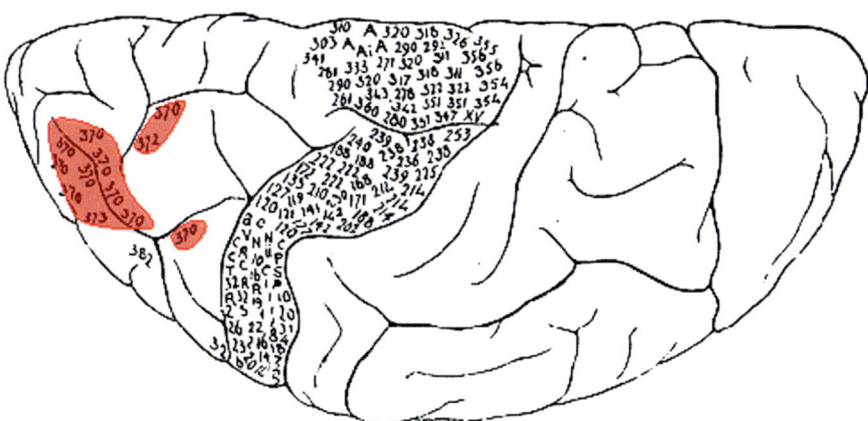

FIGURE 36.5   Location of frontal eye field based on electrical stimulation in the orangutan. *From Leyton, A.S., Sherrington, C.S., 1917. Observations on the excitable cortex of the chimpanzee, orangutan, and gorilla. Q. J. Exp. Physiol. 11, 135–222.*

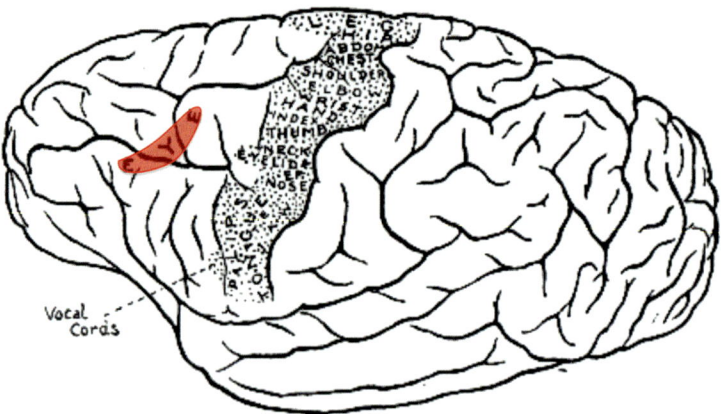

FIGURE 36.6   Location of frontal eye field based on electrical stimulation in the gorilla. *From Leyton, A.S., Sherrington, C.S., 1917. Observations on the excitable cortex of the chimpanzee, orangutan, and gorilla. Q. J. Exp. Physiol. 11, 135—222.*

stimulation of the caudal end of the middle frontal gyrus, but results disagree about the extent of surrounding cortex that also elicits eye movements and whether it extends to the skeletal motor representation in the precentral gyrus. Dejerine and colleagues located a region, which they called "area D," from which conjugate deviation of the head and eyes were evoked (Dejerine and Roussy, 1906; Dejerine, 1914). Förster (1931, 1936) identified FEF as the region from which contraversive ocular rotation was elicited at the caudal end of the middle frontal gyrus (Fig. 36.8) in a region designated 8αβγ by Vogt and Vogt (1926) (Fig. 36.7A). Förster also reported coordinated contraversive eye, head, and trunk rotation following stimulation of a dorsomedial region designated 6αβ, corresponding to Dejerine's area D. Subsequently, Penfield and coworkers evoked eye movements over a wider area of frontal cortex, extending caudally onto the precentral gyrus; however, most

sites were rostral to the precentral sulcus at a location Penfield identifies as Vogts' 8αβγ (Penfield and Boldrey, 1937; Rasmussen and Penfield, 1947). Subsequently, Lemmen et al. (1959) reported conjugate and divergent eye movements from stimulation of the posterior end of the middle frontal gyrus rostral to sites in the precentral gyrus that elicited limb and lip movements (Fig. 36.9); these authors noted a particular absence of eye movements elicited by stimulation of the precentral gyrus.

Later studies used subdural electrode arrays implanted over the frontal lobe of patients and found FEF in a zone rostrally contiguous with the motor cortex representation of head and forelimb (Godoy et al., 1990). More recent work has located FEF in the posterior part of the middle frontal gyrus, just below the intersection of the precentral sulcus and the superior frontal sulcus (Blanke et al., 2000; Blanke and Seeck, 2003; Lobel

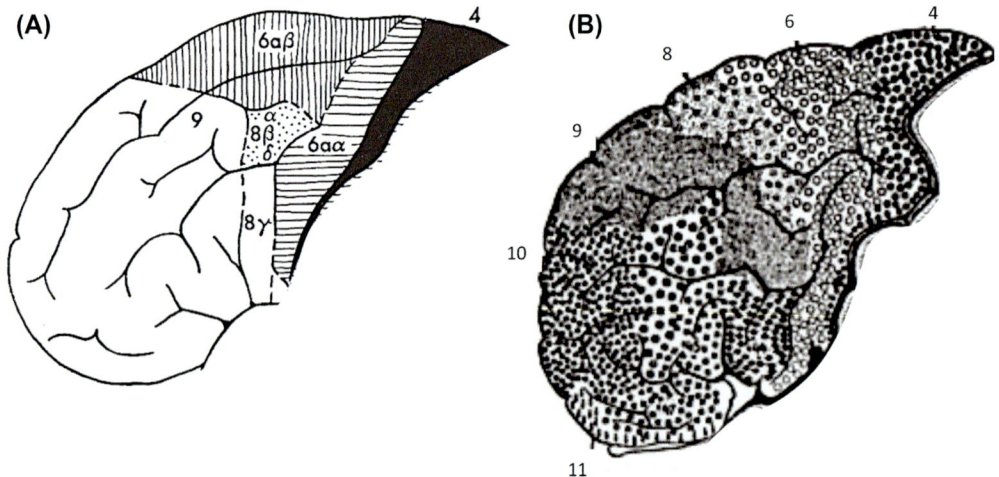

FIGURE 36.7   Frontal lobe areas identified by Vogt and Vogt (1926) (A) and Brodmann (1909) (B). *Reproduced from (A) Vogt, C., Vogt, O., 1926. Die vergleichend-architektonische und die vergleichend-reizphysiologische Felderung der Großhirnrinde unter besonderer Berücksichtigung der menschlichen. Naturwissenschaften 14, 1190—1194; (B) Brodmann, K., 1909. Localisation in the Cerebral Cortex. Springer, US.*

(A)

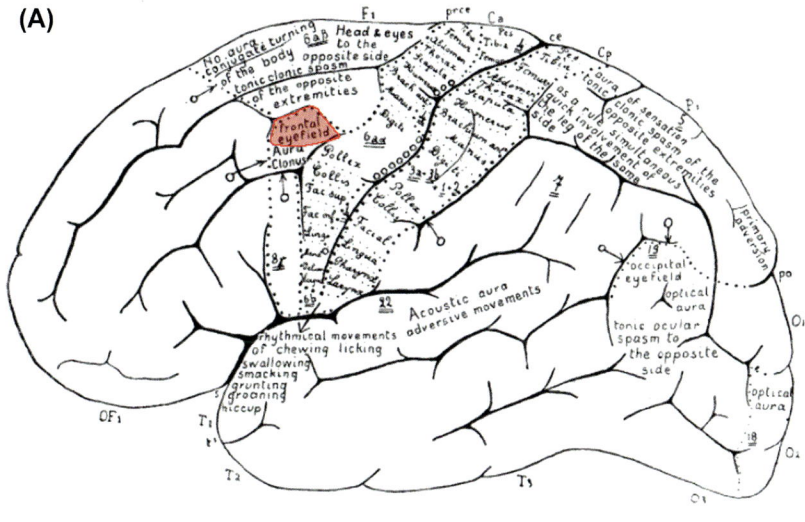

(B)

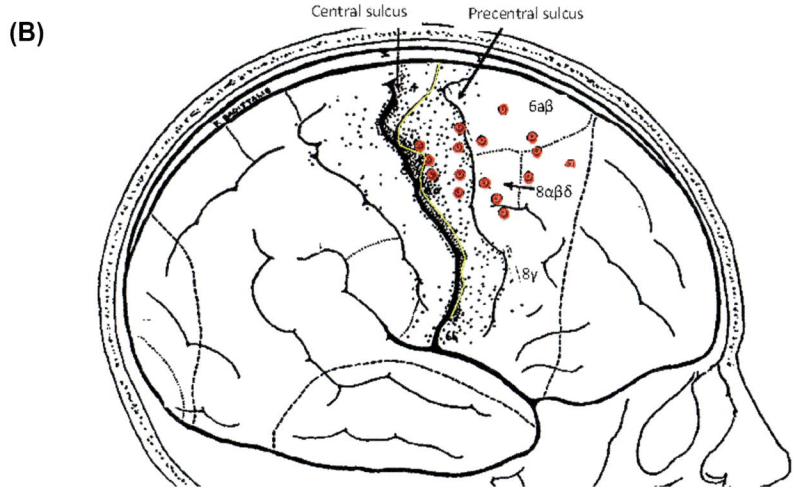

FIGURE 36.8 Location of frontal eye field based on surgical electrical stimulation in the human. (A) *Red* highlights location of FEF rostral to precentral sulcus. Reproduced from Förster, O., 1931. The cerebral cortex in man. Lancet 2, 309–312. (B) Summary of sites eliciting conjugate movements of the eyes (*red*) superimposed on summary map of sites eliciting movements of other parts of the body (*small points*). *Yellow line* marks rostral border of area 4. The remainder of the convexity of the precentral gyrus is made up of area 6aα. *From Penfield, W., Boldrey, E., 1937. Somatic motor and sensory representation in the cerebral cortex of man as studied by electrical stimulation. Brain 60, 389–443.*

et al., 2001; Milea et al., 2002; Yamamoto et al., 2004; Thurtell et al., 2009; Kaiboriboon et al., 2012; Montemurro et al., 2016; Fig. 36.10). Some studies have also located a more lateral region associated with eye movement production close to the surface of the precentral gyrus (Lobel et al., 2001). The reader should note that these electrical stimulation studies typically can explore only the cortical surface not buried in sulci and that spatial localization is difficult to interpret because of current spread influencing remote sites.

FEF has also been localized using transcranial magnetic stimulation (eg, Wessel and Kömpf, 1991; Müri et al., 1991; Olk et al., 2006; Neggers et al., 2007). Transcranial magnetic stimulation does not elicit ocular rotations,

so the influence is inferred through indirect measures of influence on saccade latency or accuracy. Such studies point toward the caudal end of the middle frontal gyrus at the intersection of the superior frontal sulcus and the superior precentral sulcus. Some investigators also locate FEF relative to the motor hand area (eg, Ro et al., 1999), on average 5 cm lateral of the sagittal midline and 3–4 cm rostral of each subject's motor hand area.

Human FEF has been located through specific deficits in eye movements observed after focal lesions (reviewed by Pierrot-Deseilligny, 1994). Early studies noted gaze deficits with damage involving large parts of frontal cortex; however, more recent studies of patients with more restricted lesions have converged on the precentral

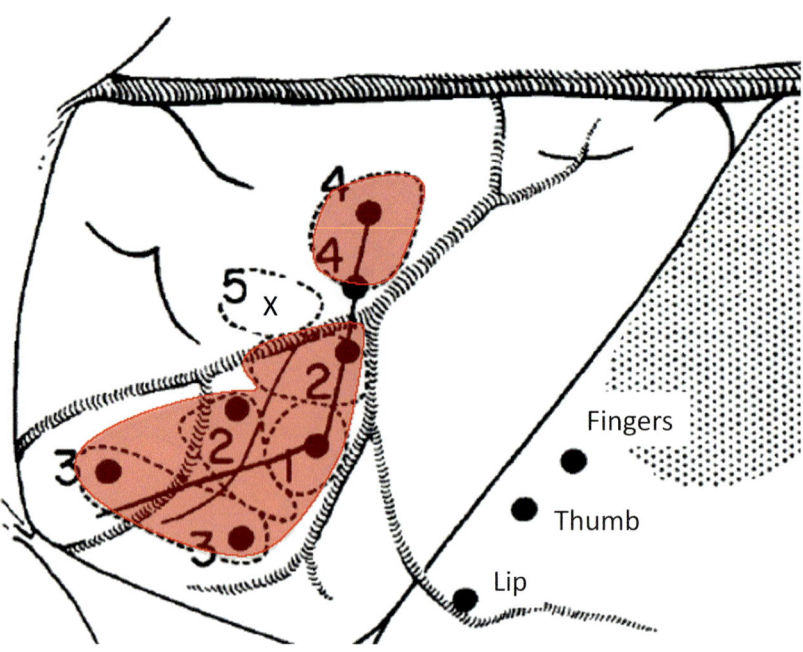

FIGURE 36.9   *Red highlights* sites from which cortical stimulation resulted in conjugate deviation of the eyes. Stimulation at site 5 had no effect. Stimulation at more caudal sites elicited movements of the indicated parts. *From Lemmen, L.J., Davis, J.S., Radnor, L.L., 1959. Observations on stimulation of the human frontal eye field. J. Comp. Neurol. 112, 163–168.*

sulcus ventral to the superior frontal sulcus (Rivaud et al., 1994; Fig. 36.11).

Human FEF has been described in numerous PET and fMRI studies (reviewed by Grosbras et al., 2005; see also

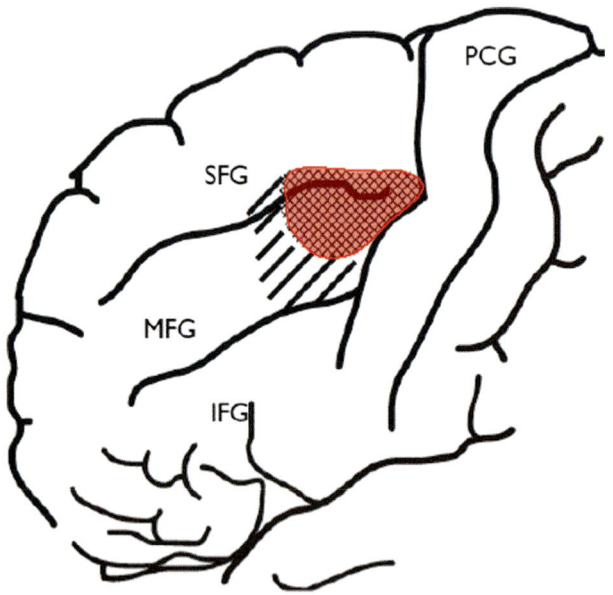

FIGURE 36.10   *Red* highlights location where conjugate eye movements were elicited with the lowest currents using subdural electrode arrays. *From Blanke, O., Spinelli, L., Thut, G., Michel, C.M., Perrig, S., Landis, T., Seeck, M., 2000. Location of the human frontal eye field as defined by electrical cortical stimulation: anatomical, functional and electrophysiological characteristics. Neuroreport 11, 1907–1913.*

Luna et al., 1998; Petit et al., 1997, 1999; Neggers et al., 2012; Ford et al., 2005; Amiez et al., 2006; Kastner et al., 2007; Ikkai and Curtis, 2008; Derrfuss et al., 2012; Thakkar et al., 2014). Execution of saccadic eye movements is accompanied by relatively strong and consistent bilateral activation in the superior precentral sulcus and by relatively weaker and less consistent activation in the superior segment of the inferior precentral sulcus. Pursuit eye movements are accompanied by activation in the superior precentral sulcus occupying a smaller region that tends to be inferior and slightly lateral to that occupied during saccades, closer to the fundus. Within individuals, the activation in both precentral sulci is restricted to the banks and does not extend onto the surface convexity, even though maps of average activation give that appearance. High-resolution imaging locates the most active voxels on the rostral bank of the superior precentral sulcus. The dorsal activation is located at the junction of the precentral sulcus and the superior frontal sulcus.

Amiez et al. (2006) conducted a subject-by-subject analysis of the locus of eye movement–related functional activity revealed in relation to the individual morphology of the precentral and superior frontal sulci (Fig. 36.15). A focus of activation associated with saccadic eye movements was located in the ventral branch of the superior precentral sulcus in both hemispheres. A second focus has found in the dorsal part of the inferior precentral sulcus. Imaging during a hand response selection task revealed activation focused

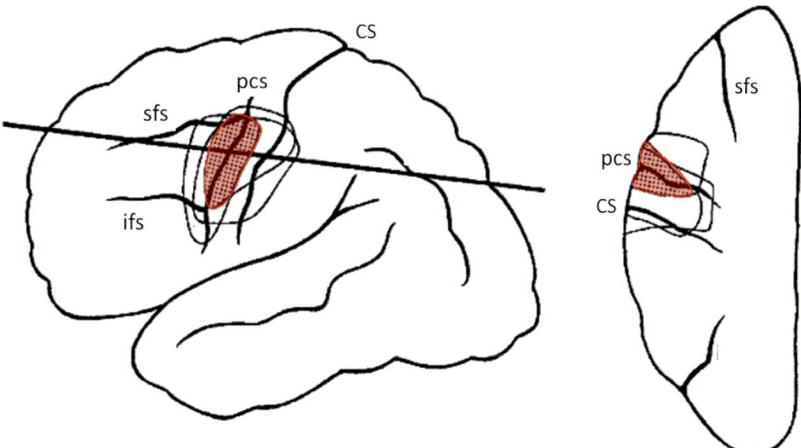

FIGURE 36.11   Lesions of frontal eye field. Lateral (left) and cross-sectional (right) views of the lesions of three patients (*thin outline*) that extend caudally to the central sulcus (cs) intersect (*stippled*) on the precentral sulcus (pcs) ventral to the superior frontal sulcus (sfs) and dorsal to the inferior frontal sulcus (ifs). The patients exhibited a range of deficits in producing saccadic eye movements. *From Rivaud, S., Müri, R., Gaymard, B., Vermersch, A., Pierrot-Deseilligny, C., 1994. Eye movement disorders after frontal eye field lesions in humans. Exp. Brain Res. 102, 110–120.*

in the dorsal branch of the superior precentral sulcus close to the caudal end of the superior frontal sulcus. The relative magnitude of activation in the superior and inferior precentral sulci varies with the nature of the eye movement (eg, saccade or pursuit, in light or dark) and task demands (eg, prosaccade or antisaccade) (Fig. 36.12A; eg, Neggers et al., 2012). Medial and lateral foci are also observed during covert attention tasks as well (Beauchamp et al., 2001). Furthermore, two regions recruited in auditory attention tasks were reported to be interdigitated with these two visual attention regions (Fig. 36.12B; Michalka et al., 2015).

The interpretation of the location of BOLD activation during eye movements can be complicated by other factors. First, eye blinks are commonly associated with saccades (eg, Evinger et al., 1994; Gandhi, 2012). Since the original electrical stimulation mapping it has been clear that cortical regions caudal to the arcuate sulcus in monkeys produce other ocular movements including blinks. Thus, some of the BOLD activation during eye movement tasks could arise from regions producing eye blinks. However, an fMRI study of spontaneous eye blinks in macaque monkeys did not report activation in or around FEF, calling into question whether this is an actual confounding factor for FEF localization (Guipponi et al., 2015). Second, saccades may be associated with neck contractions to rotate the head (eg, Goonetilleke et al., 2015). Thus, some of the BOLD activation during eye movement tasks could arise from premotor cortical areas that contribute to head movements (eg, Preuss et al., 1996). Third, in the macaque areas rostral and caudal to the arcuate sulcus have visually responsive neurons, which are modulated by attention tasks. If the human FEF is similarly surrounded, then BOLD activation would occupy a larger region than FEF

proper. Also, as noted earlier, in the macaque areas caudal to the arcuate sulcus include neurons active during saccade tasks. We suggest alternative hypotheses concerning the relation of the monkey and human FEF and premotor region in the following section.

36.5 Is FEF Located Differently Across Species?

In this section we consider from multiple perspectives whether the FEF can be regarded as homologous across primate species, including humans.

36.5.1 Comparative Neuroimaging

Several studies compare directly macaque and human fMRI data (Koyama et al., 2004; Baker et al., 2006; Ford et al., 2009; Kagan et al., 2010). For example, Koyama et al. (2004) compared fMRI activation patterns in humans and macaques performing visually guided saccades. BOLD activation in macaques was found in the rostral bank of the arcuate sulcus, and also in premotor cortex caudal to the arcuate sulcus. In humans a large domain of activation occupied the banks of the precentral sulcus. The authors propose that the peak of activation near the junction of the precentral and the superior frontal sulci corresponds to the FEF in monkeys. They observed another focus of activation in the inferior precentral sulcus (Fig. 36.14). Resting-state fMRI connectivity of FEF has been compared in macaques and humans (Hutchison et al., 2012; Babapoor-Farrokhran et al., 2013). These studies have found consistent functional connections of the FEF with frontal and parietal cortical areas in humans and macaques (Fig. 36.14). However, the analysis also revealed more lateralization of

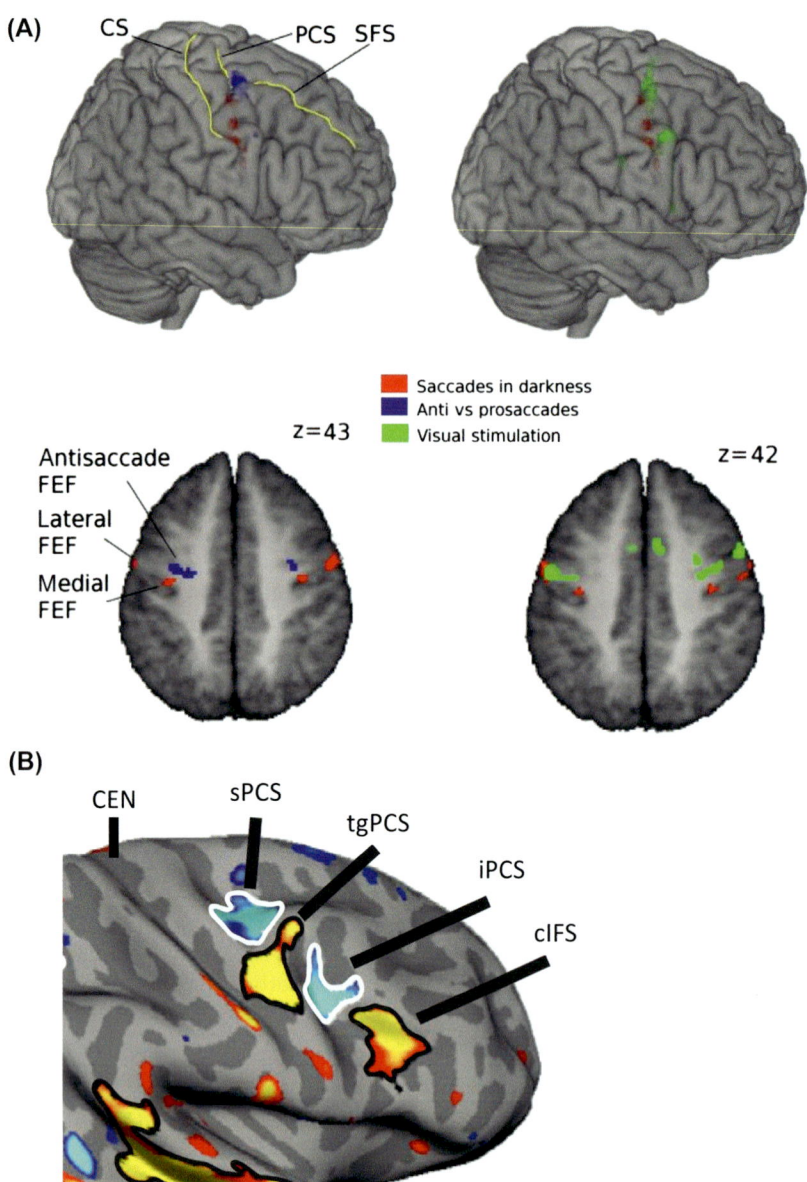

FIGURE 36.12   Functional activation observed in attention and saccade tasks. (A) Activation maps for three tasks. The location of greater activation for saccades in the dark (*red*) is compared with that for antisaccades as compared to prosaccades (*blue*) and that for visual stimulation (*green*) (from Neggers et al., 2012). (B) Map of the right frontal lobe of an individual subject performing a task requiring sustained attention to sequences of visual or auditory stimuli. The *colors* highlight regions with statistically significantly greater BOLD activation during auditory attention (*hot colors*) or visual attention (*cool colors*). The cortex was inflated to reveal interior of sulci (*dark gray*) between gyri (*light gray*). Note the interdigitated pattern of auditory and visual biases in the caudal lateral frontal cortex. *CEN*, central sulcus; *sPCS*, superior precentral sulcus; *tgPCS*, transverse gyrus intersecting the precentral sulcus; *iPCS*, inferior precentral sulcus; *cIFS*, caudal inferior frontal sulcus (from Michalka et al., 2015).   *(A) Reproduced from Neggers, S.F.W., Diepen, R.M., van Zandbelt, B.B., Vink, M., Mandl, R.C.W., Gutteling, T.P., 2012. A functional and structural investigation of the human fronto-basal volitional saccade network. PLoS One 7, e29517. (B) Reproduced from Michalka, S.W., Kong, L., Rosen, M.L., Shinn-Cunningham, B.G., Somers, D.C., 2015. Short-term memory for space and time flexibly recruit complementary sensory-biased frontal lobe attention networks. Neuron. 87, 882–892.*

connectivity of the region identified as FEF in humans than in monkeys.

The homology or at least comparability of the cortical region in and rostral to the arcuate sulcus in macaques and the superior precentral sulcus in humans has also been revealed in quantitative functional connectivity maps (Sallet et al., 2013; see also Goulas et al., 2012). Of note, in macaques a particular hot spot of connectivity with this arcuate region was caudal to the arcuate, dorsal to a spur in the region overlapping that associated with visually guided saccades described previously.

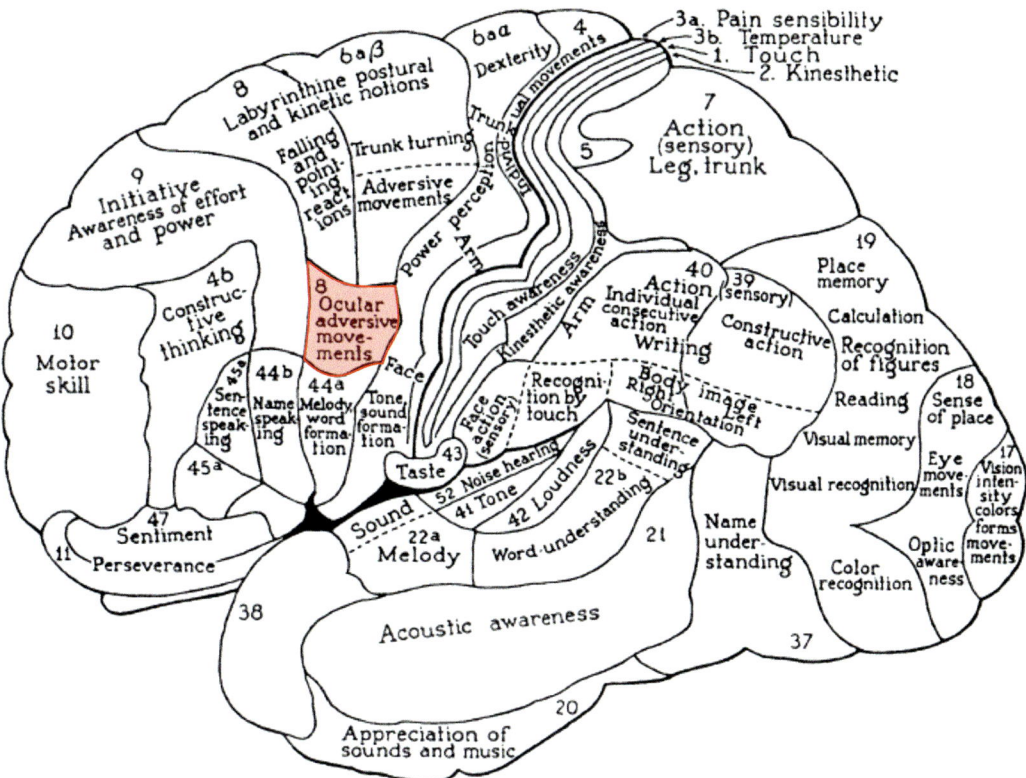

FIGURE 36.13   Lateral view of the human cerebral hemispheres, showing the localization of functions in the cerebral cortex. The numbers list Brodmann's (1909) cytoarchitectonic areas. Note that the domain for "ocular adversive movements" is identified as area 8. *From von Kleist, K., 1934. Gehirnpathologie. J.A. Barth, Leipzig.*

36.5.2 Comparative Architecture

As noted earlier, the FEF in macaque monkeys is centered in the rostral bank of the arcuate sulcus at the caudal end of prefrontal cortex. The cytoarchitecture of the area is included as Brodmann's area 8, generally. Likewise, the FEF mapped in other monkeys is recognized as being located in granular prefrontal cortex. The FEF in apes also appears to be in prefrontal cortex, although we have less specific information about the cytoarchitecture of the regions mapped. However, many authors describe the human FEF as being located in Brodmann's agranular area 6. Here we address this apparent discrepancy.

The literature on the location of FEF in humans has referred exclusively to Brodmann's cytoarchitectonic map (1909). Of course this is not the only or the last description, nor does it correspond to modern descriptions in many respects (Zilles and Amunts, 2010; Nieuwenhuys, 2013). In Brodmann's map area 6 occupies a very large amount of the frontal lobe, but contemporaneous as well subsequent maps by other investigators subdivide Brodmann's area 6 into many more areas. For example, myeloarchitectonic studies have

distinguished the caudal end of the middle frontal gyrus as distinct from surrounding areas (eg, Nieuwenhuys et al., 2015). While other authors locate the caudal end of the middle frontal gyrus in area 6 (Sarkissov et al., 1955), it has also been labeled area FB (von Economo and Koskinas, 1925), area 4s (von Bonin, 1949), the boundary of FA and FB (Bailey and von Bonin, 1951), and 8αβγ (Vogt and Vogt, 1926). Penfield, with Förster, described the majority of stimulation sites eliciting eye movements as being in 8αβγ (Penfield and Rasmussen, 1950), and the exceedingly elaborate map of von Kleist (1934) identifies ocular adversive movements with Brodmann's area 8 (Fig. 36.13).

Of note, the structure of the cortex occupied by FEF has been reevaluated more recently. Human FEF can be distinguished from surrounding areas by MRI myelin mapping (Glasser et al., 2016). In addition, a recent anatomical study reexamined the architecture of this region using modern chemoarchitectonic methods in postmortem tissue from six subjects (Rosano et al., 2003; see also Schmitt et al., 2005). The histological structure of the superior precentral sulcus was distinct from adjacent rostral and caudal regions. A thin granular layer 4 was observed in sections labeled with neuronal nuclear

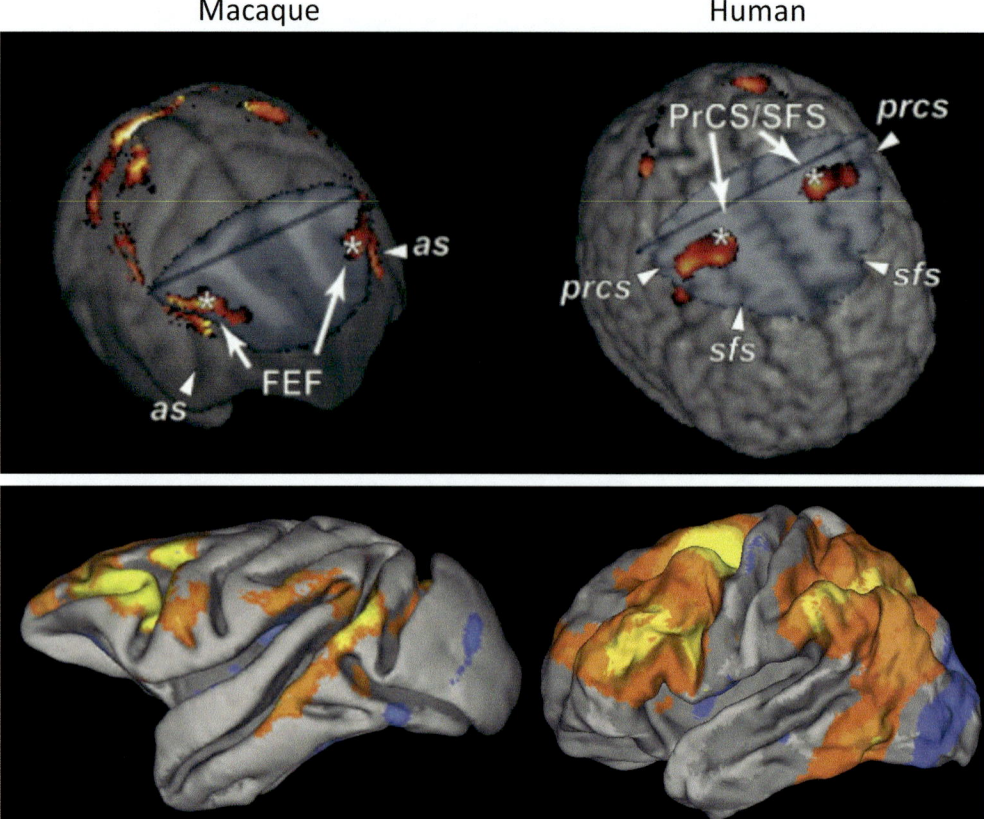

FIGURE 36.14   Direct comparison of monkey and human FEF. Top panel shows fMRI activation in the monkey (left) and human (right) brains during visually guided saccades. From Koyama, M., Hasegawa, I., Osada, T., Adachi, Y., Nakahara, K., Miyashita, Y., 2004. Functional magnetic resonance imaging of macaque monkeys performing visually guided saccade tasks: comparison of cortical eye fields with humans. Neuron 41, 795–807. Bottom panel compares resting state connectivity with a seed in FEF for monkey (left) and human (right). *From Hutchison, R.M., Gallivan, J.P., Culham, J.C., Gati, J.S., Menon, R.S., Everling, S., 2012. Functional connectivity of the frontal eye fields in humans and macaque monkeys investigated with resting-state fMRI. J. Neurophysiol. 107, 2463–2474.*

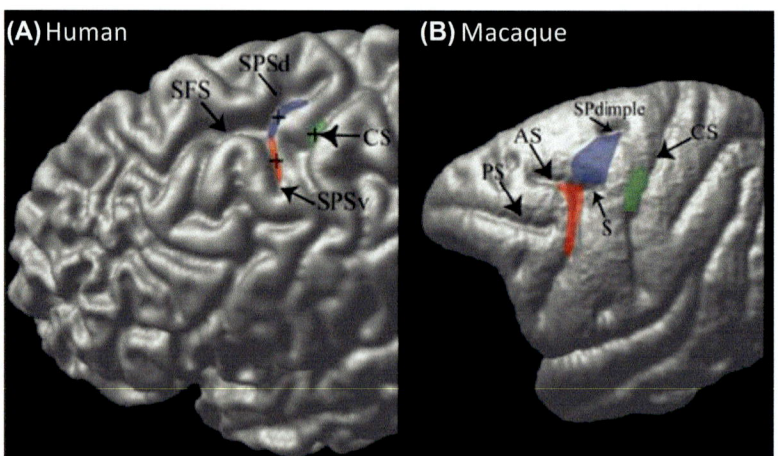

FIGURE 36.15   Location of frontal eye field relative to premotor and motor cortex in human (A) and macaque monkey (B). In humans the FEF is in the ventral limb of the superior precentral sulcus (SPSv) (*red*) relative to premotor cortex in the dorsal limb of the superior precentral sulcus (SPSd) (*blue*) and the hand representation of primary motor cortex (*green*) in the central sulcus (CS). In macaque monkeys the FEF is in the rostral bank of the arcuate sulcus (AS) caudal to the principal sulcus (PS). Dorsal premotor cortex is bounded by the arcuate sulcus and a precentral dimple (SPdimple). The hand representation in primary motor cortex is at the indicated level in the central sulcus. *From Amiez, C., Kostopoulos, P., Champod, A.-S., Petrides, M., 2006. Local morphology predicts functional organization of the dorsal premotor region in the human brain. J. Neurosci. 26, 2724–2731.*

protein (NeuN), and the nonphosphorylated neurofilament triplet protein (NNFP). Also, clusters of large, intensely immunoreactive pyramidal cells were located in deep layers 3 and 5. In sections labeled for calcium-binding proteins, the two walls of the sulcus were characterized by higher density of calretinin-labeled interneurons, lower density of calbindin-labeled pyramidal neurons, higher density of calbindin-labeled interneurons in layers 2–3, and higher density of large parvalbumin-labeled interneurons in deep layer 3. These histological features resemble the macaque FEF more than agranular area 6. These immunohistochemistry methods highlighted distinctions across this cortical region that are obscured in Nissl-stained section. Based on this analysis of cytoarchitectural, myeloarchitectural, and histochemical studies, one can conclude that Brodmann's description is unlikely to be correct because it misplaced the caudal boundary of area 8 of humans rostrally.

This characterization seems to resolve the discrepancy. However, questions remain. As described earlier, BOLD activation is also observed in the inferior precentral sulcus, near areas 44 and 45. In nonhuman primates tested, the FEF in the arcuate sulcus is bordered ventrally by areas 44 and 45. Also, the macaque has a region caudal to the FEF in agranular premotor cortex with visual responsiveness and from which saccadic eye movements can be elicited by intracortical microstimulation. Does this area have a homologue in humans? To motivate future experimental work, we suggest for consideration the hypothesis that the "FEF" in the inferior precentral sulcus is homologous with the "FEF" in the arcuate sulcus, while the "FEF" in the superior precentral sulcus is homologous with the premotor eye movement region caudal to the arcuate sulcus of macaques. We note that this assignment is opposite that featured in a recent comprehensive mapping of human cerebral cortical areas (Glasser et al., 2016; see also Amiez and Petrides, 2009). Is it conceivable that premotor cortex in humans takes over functions that are performed by prefrontal cortex in monkeys? The emergence of language, for example, may entail such a phylogenetic displacement. We find it instructive to consider the finding that cortical control of the larynx in monkeys is located in area 6, while in humans it is located in area 4, presumably to provide for articulated speech (Simonyan, 2014).

Identification of the FEF region in the superior precentral sulcus as a premotor area can provide rationale for the findings of a recent study that compared diffusion tractography of corticostriatal pathways in humans and macaques (Neggers et al., 2015). In macaques a seed region in the rostral bank of the arcuate sulcus had connectivity primarily with the head of the caudate and also the anteromedial putamen, while a seed in primary

motor cortex in the rostral bank of the central sulcus was connected with more posterior sections of caudate and mainly putamen. These observations replicate neuroanatomical tract tracing findings in macaques. In humans, though, the medial region of activation associated with saccadic eye movements was connected primarily to putamen and only a small portion of the caudate. Moreover, the region of striatum connected with the dorsal FEF in humans overlapped substantially the region of striatal connectivity with a point in primary motor cortex in the rostral bank of the central sulcus. We look forward to a future study examining the connectivity of the FEF region in the inferior precentral sulcus.

36.5.3 Comparative Sulcal Patterns

Another way to compare the location of FEF across species involves analyzing the topographic pattern of sulci (Fig. 36.16). The FEF in lissencephalic species is located in the rostral frontal lobe, sometimes around a cortical dimple. In monkeys with an arcuate and principal (previously known as rectus) sulcus, the saccade-related FEF is consistently located in the rostral bank of the arcuate sulcus. Curiously, the smooth pursuit–related region has been found dorsal to the arcuate in *Cebus* and at the fundus of the arcuate in macaques. Another region associated with visually guided saccades has been located immediately caudal to the arcuate sulcus around the arcuate spur (when it exists).

The sulcus pattern in the apes and human is notably more complex than that in monkeys, but particular associations have been described (eg, Hervé, 1888; Owen, 1900; Connolly, 1936, 1950; Rizzolatti and Arbib, 2002; Falk, 2014). First, there is general agreement that the arcuate sulcus, or at least the ventral (vertical) branch, of monkeys corresponds to the inferior precentral sulcus in apes and humans. Second, the superior frontal sulcus and superior precentral sulcus of apes and humans may be derived from the precentral dimple of monkeys. Alternatively, they may correspond to the superior (horizontal) limb of the macaque arcuate sulcus. Third, early authors identified the macaque principal sulcus (also referred to as rectus sulcus) with the hominid inferior frontal sulcus, but later analyses conclude that the inferior frontal sulcus of apes and humans has no counterpart in monkeys and instead the principal sulcus of monkeys corresponds to the middle frontal sulcus of humans. Thus, the ventral frontal lobe of humans has expanded relative to apes and monkeys. Such an expansion could lead to a relatively more dorsal location of the FEF in humans relative to apes and monkeys.

These topological identifications of sulci can guide a perspective on the location of FEF in nonhuman primates and humans. Hence, if the FEF of monkeys is in the arcuate sulcus, and the arcuate sulcus is homologous

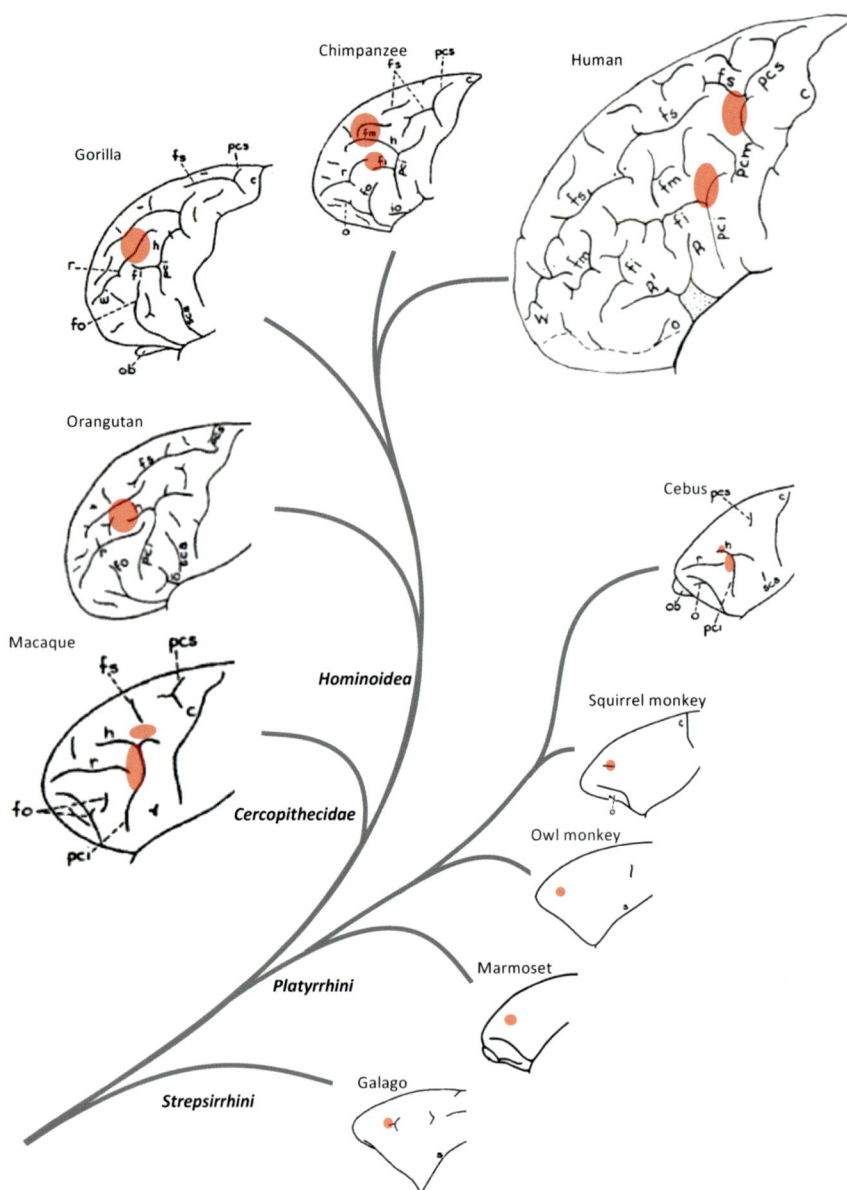

FIGURE 36.16   A frontal cortex cladogram to illustrate the location of FEF in a range of prosimian and primate species. The diagrams are not to scale. Sulcal labels are as follows: *c*, central; *fi*, frontal inferior; *fm*, frontal medius; *fo*, frontal orbital; *fs*, frontal superior; *h*, horizontal ramus of precentral inferior sulcus; *io*, inferior orbital; *o*, orbital; *ob*, olfactory bulb; *pci*, precentral inferior; *pcm*, precentral medius; *pcs*, precentral superior; *r*, rectus; *R*, ramus ascendens fissurae Sylvii; *sca*, subcentral anterior; *W*, frontomarginal of Wernicke. Note the use of the label "rectus" for the macaque and *Cebus* principal sulcus and of the labels "pci" and "h" for the macaque and *Cebus* arcuate sulcus. Two regions are highlighted in *Cebus*, macaque, chimpanzee, and human based on evidence for at least two discrete zones. *Reproduced from Connolly, C.J., 1936. The fissural pattern of the primate brain. Am. J. Phys. Anthropol. 21, 301–422; Connolly, C. J., 1950. External Morphology of the Primate Brain. CC Thomas.*

with the inferior precentral sulcus, then does it follow that the FEF of humans is located in the inferior precentral sulcus? Such reasoning can account for the lateral BOLD activation observed in imaging studies. If this reasoning is sound and valid, then what is the status of the FEF region at the intersection of the superior frontal sulcus and the dorsal precentral sulcus? One possibility is that this dorsal region is homologous to the premotor eye field of macaques. On the other hand, if the dorsal region is correctly understood as the homologue of the macaque FEF, based on evidence reviewed previously, then what is the proper relation of the ventral FEF region? Perhaps it is a premotor eye field as suggested earlier. Confusing all of this, though, is the evidence that the eye movement region in orangutans, gorillas, and chimpanzees is located clearly rostral to the precentral sulci. We note, though, that in spite of the similar appearance of the frontal sulci of humans and chimpanzees the precentral sulcus of chimpanzees does not mark the border between premotor and prefrontal cortex as it may in humans; instead it marks the border between motor and premotor cortex (Preuss, personnel communication). This can explain why the FEF is located on the middle frontal gyrus of the apes, but it cannot help us understand the apparent caudal migration of the FEF in humans. Clearly, many questions remain.

36.6 Conclusion

We end this chapter with more questions and cautions than conclusions. First, should the term "frontal eye field" be singular or plural? Shall we regard "FEF" (now in scare quotes to remind the reader of the uncertainty) as a collection of distinct areas that can be of different sizes (and locations) in different species? Or shall we regard "FEF" as a single area with multiple modules adapted according to the lifestyle and habitat of the respective species? Under either hypothesis the apparent location of "FEF" might appear to differ across individuals and species as one function is emphasized over another. In the macaque, for example, we have described both arcuate and post-arcuate regions from which saccades are elicited with weak electrical stimulation. Interposed between these is a pursuit region. Rostral to the saccade map in the arcuate sulcus is a vergence region. The medial and lateral segments within the arcuate sulcus support very different patterns of cortical connections. More dorsomedially in the macaque is a region concerned with ear movements and orienting to sounds. Evolutionary specialization of any (or all of these) could result in elaboration of at least five or six subregions or areas.

Next, across primate species does the size (and complexity) of "FEF" scale with total neocortical area? With the area of visual cortical areas in the parietal and temporal lobes? Does it scale with changes in the subcortical network mediating orienting such as the mediodorsal thalamus, basal ganglia, and—in particular—the superior colliculus. Does it scale with the degree of elaboration of extrastriate visual areas and other sensory systems? Does it scale with the complexity of the musculature used for orienting and associated visually guided behaviors (eye, head, eye + head, eye + head + hand)? Does it scale with the complexity of cognitive control over gaze behavior (eg, dissociation of gaze from visual salience for social deception tested by an ability to make antisaccades)? We cannot answer these questions because precise boundaries of FEF are uncertain, especially in humans. For this reason, it remains essential that the human imaging studies identify the anatomical landmarks and full extent of any activation interpreted as corresponding to FEF.

The size and location of "FEF" is a matter of operational definition with resolution specified by technique.

Consider, for example, mapping based on the eye movements elicited by electrical stimulation. Results differ according to state of consciousness (anesthetized or alert), electrode size (surface or intracortical), stimulation parameters (brief or long stimulation trains), method of movement inspection (simple observation or quantitative eye tracking), and inclusion of coordinated orienting movements (head free or fixed). Until common methods are applied across all species, including humans, ground truth comparisons cannot be made. Therefore, seeking convergence of criteria across methods seems the only viable approach. Ultimately, a comparative analysis of the location, architecture, and connectivity of "FEF" should be framed by an appreciation of the interaction of gaze control with the body size, posture, locomotion, and reaching behavior, plus the elaborated relevance of gaze for communication in monkeys, apes, and humans living in larger social groups.

Acknowledgments

We thank C. Curtis, J. Kaas, T. Preuss, and J. Sweeney for many helpful comments and suggestions. J.D.S., W.Z., J.D.C., and M.S.S. were supported by the Vanderbilt Center for Integrative and Cognitive Neuroscience and by grants from the National Eye Institute and the National Institute of Mental Health. J.D.S. is also supported by Robin and Richard Patton through the E. Bronson Ingram Chair in Neuroscience. M.P. is supported by the Natural Science and Engineering Council of Canada and the Canadian Institute of Health Research. P.P. is supported by CNRS France. Address of M.P. is Center for Neuroscience Studies, Department of Biomedical and Molecular Sciences, Queen's University, Botterell Hall, Rm 438, Kingston ON K7L 3N6 Canada. Address of P.P. is CM, UMR 7225, UMRS 975, CNRS-INSERM-Université Pierre et Marie Curie Hôpital de la Salpêtrière, 47 boulevard de l'Hôpital, 75651 Paris CEDEX 13 France.

References

Amiez, C., Petrides, M., 2009. Anatomical organization of the eye fields in the human and non-human primate frontal cortex. Prog. Neurobiol. 89, 220—230.

Amiez, C., Kostopoulos, P., Champod, A.-S., Petrides, M., 2006. Local morphology predicts functional organization of the dorsal premotor region in the human brain. J. Neurosci. 26, 2724—2731.

Babapoor-Farrokhran, S., Hutchison, R.M., Gati, J.S., Menon, R.S., Everling, S., 2013. Functional connectivity patterns of medial and lateral macaque frontal eye fields reveal distinct visuomotor networks. J. Neurophysiol. 109, 2560—2570.

Bailey, P., von Bonin, G., 1951. The Isocortex of Man. University of Illinois Press.

Bailey, P., von Bonin, G., McCulloch, W.S., 1950. The Isocortex of the Chimpanzee. University Illinois Press.

Baizer, J., Bender, D., 1989. Comparison of saccadic eye movements in humans and macaques to single-step and double-step target movements. Vis. Res. 29, 485—495.

Baker, J.T., Patel, G.H., Corbetta, M., Snyder, L.H., 2006. Distribution of activity across the monkey cerebral cortical surface, thalamus and midbrain during rapid, visually guided saccades. Cereb. Cortex 16, 447—459.

Bakola, S., Burman, K.J., Rosa, M.G.P., 2015. The cortical motor system of the marmoset monkey (Callithrix jacchus). Neurosci. Res. 93, 72—81.

Baldwin, M.K., Cooke, D.F., Krubitzer, L., 2017. Intracortical microstimulation maps of motor, somatosensory, and posterior parietal cortex in tree shrews (*Tupaia belangeri*) reveal complex movement representations. Cereb 27 (2), 1439–1456.

Barbas, H., Mesulam, M., 1981. Organization of afferent input to subdivisions of area 8 in the rhesus monkey. J. Comp. Neurol. 200, 407–431.

Beauchamp, M.S., Petit, L., Ellmore, T.M., Ingeholm, J., Haxby, J.V., 2001. A parametric fMRI study of overt and covert shifts of visuospatial attention. Neuroimage 14, 310–321.

Beevor, C.E., Horsley, V., 1890. A record of the results obtained by electrical excitation of the so-called motor cortex and internal capsule in an orangoutang (Simia satyrus). Philos. Trans. R. Soc. Lond. B Biol. Sci. 181, 129–158.

Berg, D.J., Boehnke, S.E., Marino, R.A., Munoz, D.P., Itti, L., 2009. Free viewing of dynamic stimuli by humans and monkeys. J. Vis. 9, 19.1–19.15.

Blanke, O., Seeck, M., 2003. Direction of saccadic and smooth eye movements induced by electrical stimulation of the human frontal eye field: effect of orbital position. Exp. Brain Res. 150, 174–183.

Blanke, O., Spinelli, L., Thut, G., Michel, C.M., Perrig, S., Landis, T., Seeck, M., 2000. Location of the human frontal eye field as defined by electrical cortical stimulation: anatomical, functional and electrophysiological characteristics. Neuroreport 11, 1907–1913.

Bloch, J.I., Boyer, D.M., 2002. Grasping primate origins. Science 298, 1606–1610.

Blum, B., Kulikowski, J., Carden, D., Harwood, D., 1982. Eye movements induced by electrical stimulation of the frontal eye fields of marmosets and squirrel monkeys. Brain Behav. Evol. 21, 34–41.

Blumer, R., Maurer-Gesek, B., Gesslbauer, B., Blumer, M., Pechriggl, E., Davis-López de Carrizosa, M.A., Horn, A.K., May, P.J., Streicher, J., de la Cruz, R.R., Pastor, Á.M., 2016. Palisade endings are a constant feature in the extraocular muscles of frontal-eyed, but not lateral-eyed, animals. Invest. Ophthalmol. Vis. Sci. 57, 320–331.

Bock, N.A., Kocharyan, A., Liu, J.V., Silva, A.C., 2009. Visualizing the entire cortical myelination pattern in marmosets with magnetic resonance imaging. J. Neurosci. Methods 185, 15–22.

Bogart, S.L., Mangin, J.-F., Schapiro, S.J., Reamer, L., Bennett, A.J., Pierre, P.J., Hopkins, W.D., 2012. Cortical sulci asymmetries in chimpanzees and macaques: a new look at an old idea. Neuroimage 61, 533–541.

Bon, L., Lucchetti, C., 1994. Ear and eye representation in the frontal cortex, area 8b, of the macaque monkey: an electrophysiological study. Exp. Brain Res. 102, 259–271.

von Bonin, G., 1949. Architecture of the precentral motor cortex and some adjacent areas. In: Bucy, P.C. (Ed.), The Precentral Motor Cortex. University of Illinois Press, pp. 7–82.

von Bonin, G., Bailey, P., 1947. The Neocortex of *Macaca mulatta*. University of Illinois Press.

Borrell, V., Götz, M., 2014. Role of radial glial cells in cerebral cortex folding. Curr. Opin. Neurobiol. 27, 39–46.

Boussaoud, D., 1995. Primate premotor cortex: modulation of preparatory neuronal activity by gaze angle. J. Neurophysiol. 73, 886–890.

Boussaoud, D., Wise, S., 1993. Primate frontal cortex: effects of stimulus and movement. Exp. Brain Res. 95, 28–40.

Boussaoud, D., Barth, T., Wise, S., 1993. Effects of gaze on apparent visual responses of frontal cortex neurons. Exp. Brain Res. 93, 423–434.

Boussaoud, D., Jouffrais, C., Bremmer, F., 1998. Eye position effects on the neuronal activity of dorsal premotor cortex in the macaque monkey. J. Neurophysiol. 80, 1132–1150.

Brodmann, K., 1909. Localisation in the Cerebral Cortex. Springer, US.

Bruce, C.J., Goldberg, M.E., Bushnell, M.C., Stanton, G.B., 1985. Primate frontal eye fields. II. Physiological and anatomical correlates of electrically evoked eye movements. J. Neurophysiol. 54, 714–734.

Burman, K.J., Palmer, S.M., Gamberini, M., Rosa, M.G.P., 2006. Cytoarchitectonic subdivisions of the dorsolateral frontal cortex of the marmoset monkey (*Callithrix jacchus*), and their projections to dorsal visual areas. J. Comp. Neurol. 495, 149–172.

Burman, K.J., Bakola, S., Richardson, K.E., Reser, D.H., Rosa, M.G., 2014. Patterns of afferent input to the caudal and rostral areas of the dorsal premotor cortex (6DC and 6DR) in the marmoset monkey. J. Comp. Neuro. 522, 3683–3716.

Burman, K.J., Bakola, S., Richardson, K.E., Yu, H.-H., Reser, D.H., Rosa, M.G.P., 2015. Cortical and thalamic projections to cytoarchitectural areas 6Va and 8C of the marmoset monkey: connectionally distinct subdivisions of the lateral premotor cortex. J. Comp. Neurol. 523, 1222–1247.

Calabrese, E., Badea, A., Coe, C.L., Lubach, G.R., Shi, Y., Styner, M.A., Johnson, G.A., 2015. A diffusion tensor MRI atlas of the postmortem rhesus macaque brain. Neuroimage 117, 408–416.

Camalier, C., Gotler, A., Murthy, A., Thompson, K., Logan, G., Palmeri, T., Schall, J., 2007. Dynamics of saccade target selection: race model analysis of double step and search step saccade production in human and macaque. Vis. Res. 47, 2187–2211.

Cantalupo, C., McCain, D., Ward, J.P., 2002. Function of head-cocking in Garnett's greater bush baby (*Otolemur garnettii*). Int. J. Primatol. 23, 203–221.

Cartmill, M., 1992. New views on primate origins. Evol. Anthr. 1, 105–111.

Caspari, N., Janssens, T., Mantini, D., Vandenberghe, R., Vanduffel, W., 2015. Covert shifts of spatial attention in the macaque monkey. J. Neurosci. 35, 7695–7714.

Cerkevich, C.M., Collins, C.E., Kaas, J.H., 2014. Cortical inputs to the middle temporal visual area in New World owl monkeys. Eye Brain 2015, 1–15.

Charles-Dominique, P., 1977. Ecology and Behaviour of Nocturnal Primates: Prosimians of Equatorial West Africa. Columbia University Press.

Cisek, P., Kalaska, J.F., 2005. Neural correlates of reaching decisions in dorsal premotor cortex: specification of multiple direction choices and final selection of action. Neuron 45, 801–814.

Collins, C.E., Lyon, D.C., Kaas, J.H., 2005. Distribution across cortical areas of neurons projecting to the superior colliculus in new world monkeys. Anat. Rec. 285, 619–627.

Connolly, C.J., 1936. The fissural pattern of the primate brain. Am. J. Phys. Anthropol. 21, 301–422.

Connolly, C.J., 1950. External Morphology of the Primate Brain. CC Thomas.

Crosby, E.C., Yoss, R.E., Henderson, J.W., 1952. The mammalian midbrain and isthmus regions. Part II. The fiber connections. D. The pattern for eye movements on the frontal eye field and the discharge of specific portions of this field to and through midbrain levels. J. Comp. Neurol. 97, 357–383.

Davidoff, L.M., 1928. A visit to professor Förster's clinic in Breslau with special observation of his treatment of epilepsy. Psychiatric Quart 2, 307–313.

DeBruyn, E., Wise, V., Casagrande, V., 1980. The size and topographic arrangement of retinal ganglion cells in the galago. Vis. Res. 20, 315–327.

Dejerine, J., 1914. Sémiologie Des Affections Du Système Nerveux. Masson et cie.

Dejerine, J., Roussy, G., 1906. Le syndrome thalamique. Rev. Neurol. 14, 521–532.

Denion, E., Hitier, M., Guyader, V., Dugué, A.-E., Mouriaux, F., 2015. Unique human orbital morphology compared with that of apes. Sci. Rep. 5, 11528.

Denion, E., Hitier, M., Levieil, E., Mouriaux, F., 2015. Human rather than ape-like orbital morphology allows much greater lateral visual field expansion with eye abduction. Sci. Rep. 5, 12437.

Derrfuss, J., Vogt, V., Fiebach, C.J., von Cramon, D.Y., Tittgemeyer, M., 2012. Functional organization of the left inferior precentral sulcus: dissociating the inferior frontal eye field and the inferior frontal junction. Neuroimage 59, 3829–3837.

Driscoll, C.A., Barr, C.S., 2016. Studying longitudinal trajectories in animal models of psychiatric illness and their translation to the human condition. Neurosci. Res. 102, 67–77.

DuMond, F.V., 1968. The squirrel monkey in a semi-natural environment. In: Rosenblum, L.A., Cooper, R.W. (Eds.), The Squirrel Monkey. Academic Press, pp. 87–145.

Dusser de Barenne, J., Garol, H., McCulloch, W., 1941. Functional organization of sensory and adjacent cortex of the monkey. J. Neurophysiol. 4, 287–303.

Eberstaller, O., 1890. Das Stirnhirn. Urban & Schwarzenberg.

Ecker, A., 1869. Die Hirnwindungen des Menschen. Vieweg, Braunschweig.

von Economo, C.F., Koskinas, G.N., 1925. Die Cytoarchitektonik der Hirnrinde des erwachsenen Menschen. Springer, Berlin, Wien.

Einhäuser, W., Kruse, W., Hoffmann, K.-P., König, P., 2006. Differences of monkey and human overt attention under natural conditions. Vis. Res. 46, 1194–1209.

Emmers, R., Akert, K., 1963. A Stereotaxic Atlas of the Brain of the Squirrel Monkey Saimiri sciureus. University of Wisconsin Press, Madison.

Van Essen, D., 1997. A tension-based theory of morphogenesis and compact wiring in the central nervous system. Nature 385, 313–318.

Evinger, C., Manning, K., Pellegrini, J., Basso, M., Powers, A., Sibony, P., 1994. Not looking while leaping: the linkage of blinking and saccadic gaze shifts. Exp. Brain Res. 100, 337–344.

Falk, D., 1978. Cerebral asymmetry in old world monkeys. Acta Anat. 101, 334–339.

Falk, D., 2014. Interpreting sulci on hominin endocasts: old hypotheses and new findings. Front. Hum. Neurosci. 8, 134. https://doi.org/10.3389/fnhum.2014.00134.

Falk, D., Hildebolt, C., Cheverud, J., Vannier, M., Helmkamp, R., Konigsberg, L., 1990. Cortical asymmetries in frontal lobes of rhesus monkeys (Macaca mulatta). Brain Res. 512, 40–45.

Ferrier, D., 1874. Experiments on the brain of monkeys. No. I. Proc. R. Soc. Lond. 23, 409–430.

Finlay, B.L., Franco, E.C.S., Yamada, E.S., Crowley, J.C., Parsons, M., Muniz, J.A.P.C., Silveira, L.C.L., 2008. Number and topography of cones, rods and optic nerve axons in New and Old World primates. Vis. Neurosci. 25, 289–299.

Fogassi, L., Gallese, V., Gentilucci, M., Luppino, G., Matelli, M., Rizzolatti, G., 1994. The fronto-parietal cortex of the prosimian Galago: patterns of cytochrome oxidase activity and motor maps. Behav. Brain Res. 60, 91–113.

Ford, K.A., Goltz, H.C., Brown, M.R.G., Everling, S., 2005. Neural processes associated with antisaccade task performance investigated with event-related fMRI. J. Neurophysiol. 94, 429–440.

Ford, K.A., Gati, J.S., Menon, R.S., Everling, S., 2009. BOLD fMRI activation for anti-saccades in nonhuman primates. Neuroimage 45, 470–476.

Förster, O., 1926. Zur operativen Behandlung der Epilepsie. J. Neurol. 89, 137–147.

Förster, O., 1931. The cerebral cortex in man. Lancet 2, 309–312.

Förster, O., 1936. The motor cortex in man in the light of Hughlings Jackson's doctrines. Brain 59, 135–159.

Förster, O., Penfield, W., 1930. The structural basis of traumatic epilepsy and results of radical operation. Brain 53, 99–119.

Franco, E., Finlay, B., Silveira, L., Yamada, E., Crowley, J., 2000. Conservation of absolute foveal area in New World monkeys. a constraint on eye size and conformation. Brain Behav. Evol. 56, 276–286.

Frey, S., Pandya, D.N., Chakravarty, M.M., Bailey, L., Petrides, M., Collins, D.L., 2011. An MRI based average macaque monkey stereotaxic atlas and space (MNI monkey space). Neuroimage 55, 1435–1442.

Fritsch, G., Hitzig, E., 1870. Über die elektrische Erregbarkeit des Grosshirns. archives für Anatomie Physiologie und Wissenshaftlicke Medicin 37, 300–332.

Fritsch, G., Hitzig, E., 2009. Electric excitability of the cerebrum (Über die elektrische Erregbarkeit des Grosshirns). Epilepsy Behav. 15, 123–130.

Fujii, N., Mushiake, H., Tanji, J., 1998. An oculomotor representation area within the ventral premotor cortex. Proc. Natl. Acad. Sci. 95, 12034–12037.

Fujii, N., Mushiake, H., Tanji, J., 2000. Rostrocaudal distinction of the dorsal premotor area based on oculomotor involvement. J. Neurophysiol. 83, 1764–1769.

Fukushima, K., Yamanobe, T., Shinmei, Y., Fukushima, J., Kurkin, S., Peterson, B.W., 2002. Coding of smooth eye movements in three-dimensional space by frontal cortex. Nature 419, 157–162.

Gabernet, L., Meskenaïte, V., Hepp-Reymond, M.C., 1999. Parcellation of the lateral premotor cortex of the macaque monkey based on staining with the neurofilament antibody SMI-32. Exp. Brain Res. 128, 188–193.

Gamlin, P., Yoon, K., 2000. An area for vergence eye movement in primate frontal cortex. Nature 407, 1003–1007.

Gandhi, N.J., 2012. Interactions between gaze-evoked blinks and gaze shifts in monkeys. Exp. Brain Res. 216, 321–339.

Germann, J., Robbins, S., Halsband, U., Petrides, M., 2005. Precentral sulcal complex of the human brain: morphology and statistical probability maps. J. Comp. Neurol. 493, 334–356.

Glasser, M.F., Coalson, T.S., Robinson, E.C., Hacker, C.D., Harwell, J., Yacoub, E., Ugurbil, K., Andersson, J., Beckmann, C.F., Jenkinson, M., Smith, S.M., Van Essen, D.C., 2016. A multi-modal parcellation of human cerebral cortex. Nature 536, 171–178.

Glickman, S.E., Sroges, R.W., 1966. Curiosity in zoo animals. Behaviour 26, 151–187.

Godoy, J., Lüders, H., Dinner, D., Morris, H., Wyllie, E., 1990. Versive eye movements elicited by cortical stimulation of the human brain. Neurology 40, 296–299.

Goonetilleke, S.C., Katz, L., Wood, D.K., Gu, C., Huk, A.C., Corneil, B.D., 2015. Cross-species comparison of anticipatory and stimulus-driven neck muscle activity well before saccadic gaze shifts in humans and nonhuman primates. J. Neurophysiol. 114, 902–913.

Gottlieb, J.P., Bruce, C.J., MacAvoy, M.G., 1993. Smooth eye movements elicited by microstimulation in the primate frontal eye field. J. Neurophysiol. 69, 786–799.

Gottlieb, J.P., MacAvoy, M.G., Bruce, C.J., 1994. Neural responses related to smooth-pursuit eye movements and their correspondence with electrically elicited smooth eye movements in the primate frontal eye field. J. Neurophysiol. 72, 1634–1653.

Goulas, A., Uylings, H.B., Stiers, P., 2012. Unravelling the intrinsic functional organization of the human lateral frontal cortex: a parcellation scheme based on resting state FMRI. J. Neurosci. 32, 10238–10252.

Gould 3rd, H., Cusick, C., Pons, T., Kaas, J., 1986. The relationship of corpus callosum connections to electrical stimulation maps of motor, supplementary motor, and the frontal eye fields in owl monkeys. J. Comp. Neurol. 247, 297–325.

Grosbras, M.-H., Laird, A.R., Paus, T., 2005. Cortical regions involved in eye movements, shifts of attention, and gaze perception. Hum. Brain Mapp. 25, 140–154.

Groves, C., 2001. Primate Taxonomy. Smithsonian Series in Comparative Evolutionary Biology. Smithsonian, Washington, p. 1.

Grünbaum, A., Sherrington, C.S., 1901. Observations on the physiology of the cerebral cortex of some of the higher apes. Proc. R. Soc. Lond. 69, 206–209.

Guipponi, O., Odouard, S., Pinède, S., Wardak, C., Ben Hamed, S., 2015. fMRI cortical correlates of spontaneous eye blinks in the nonhuman primate. Cereb. Cortex 25, 2333–2345.

Guitton, D., Volle, M., 1987. Gaze control in humans: eye-head coordination during orienting movements to targets within and beyond the oculomotor range. J. Neurophysiol. 58, 427–459.

Guitton, D., Douglas, R.M., Volle, M., 1984. Eye-head coordination in cats. J. Neurophysiol. 52, 1030–1050.

Hamilton, C., Vermeire, B., 1988. Complementary hemispheric specialization in monkeys. Science 242, 1691–1694.

Hanes, D., Carpenter, R., 1999. Countermanding saccades in humans. Vis. Res. 39, 2777–2791.

Hanes, D., Schall, J., 1995. Countermanding saccades in macaque. Vis. Neurosci. 12, 929–937.

Haude, R.H., Ray, O.S., 1974. Visual observing behavior in the squirrel monkey. Anim. Learn. Behav. 2, 138–140.

Heide, W., Kurzidim, K., Kömpf, D., 1996. Deficits of smooth pursuit eye movements after frontal and parietal lesions. Brain 119 (Pt 6), 1951–1969.

Heilbroner, P., Holloway, R., 1989. Anatomical brain asymmetry in monkeys: frontal, temporoparietal, and limbic cortex in Macaca. Am. J. Phys. Anthropol. 80, 203–211.

Heiney, S.A., Blazquez, P.M., 2011. Behavioral responses of trained squirrel and rhesus monkeys during oculomotor tasks. Exp. Brain Res. 212, 409–416.

Henderson, J.W., Crosby, E.C., 1952. An experimental study of optokinetic responses. AMA Arch. Ophthalmol. 47, 43–54.

Hervé, G., 1888. La Circonvolution de Broca. A. Delahaye & E. Lecrosnier, Paris.

Hines, M., 1940. Movements elicited from precentral gyrus of adult chimpanzees by stimulation with sine wave currents. J. Neurophysiol. 3, 442–466.

Hopkins, W.D., Meguerditchian, A., Coulon, O., Bogart, S., Mangin, J.-F., Sherwood, C.C., Grabowski, M.W., Bennett, A.J., Pierre, P.J., Fears, S., Woods, R., Hof, P.R., Vauclair, J., 2014. Evolution of the central sulcus morphology in primates. Brain Behav. Evol. 84, 19–30.

Horsley, V., Schäfer, E.A., 1888. A record of experiments upon the functions of the cerebral cortex. Philos. Trans. R. Soc. Lond. B Biol. Sci. 179, 1–45.

Huerta, M.F., Krubitzer, L.A., Kaas, J.H., 1986. Frontal eye field as defined by intracortical microstimulation in squirrel monkeys, owl monkeys, and macaque monkeys: I. subcortical connections. J. Comp. Neurol. 253, 415–439.

Huerta, M., Krubitzer, L., Kaas, J., 1987. Frontal eye field as defined by intracortical microstimulation in squirrel monkeys, owl monkeys, and macaque monkeys. II. Cortical connections. J. Comp. Neurol. 265, 332–361.

Hung, C.-C., Yen, C.C., Ciuchta, J.L., Papoti, D., Bock, N.A., Leopold, D.A., Silva, A.C., 2015. Functional MRI of visual responses in the awake, behaving marmoset. Neuroimage 120, 1–11.

Hutchison, R.M., Gallivan, J.P., Culham, J.C., Gati, J.S., Menon, R.S., Everling, S., 2012. Functional connectivity of the frontal eye fields in humans and macaque monkeys investigated with resting-state fMRI. J. Neurophysiol. 107, 2463–2474.

Ikkai, A., Curtis, C.E., 2008. Cortical activity time locked to the shift and maintenance of spatial attention. Cereb. Cortex 18, 1384–1394.

Imai, N., Sawada, K., Fukunishi, K., Sakata-Haga, H., Fukui, Y., 2011. Sexual dimorphism of sulcal length asymmetry in the cerebrum of adult cynomolgus monkeys (Macaca fascicularis). Congenit. Anom. 51, 161–166.

Jacobs, G.H., Deegan 2nd, J.F., Neitz, J., Crognale, M.A., Neitz, M., 1993. Photopigments and color vision in the nocturnal monkey. Aotus. Vis. Res. 33, 1773–1783.

Jensen, J., 1871. Die Furchen und Windungen der menschlichen Grosshirn-Hemisphären. Allgemeine Zeitschrift für Psychiatrie. Druck und Verlag von Georg Reimer.

Kaas, J., Morel, A., 1993. Connections of visual areas of the upper temporal lobe of owl monkeys: the MT crescent and dorsal and ventral subdivisions of FST. J. Neurosci. 13, 534–546.

Kaas, J., Lin, C., Wagor, E., 1977. Cortical projections of posterior parietal cortex in owl monkeys. J. Comp. Neurol. 72, 387–408.

Kagan, I., Iyer, A., Lindner, A., Andersen, R.A., 2010. Space representation for eye movements is more contralateral in monkeys than in humans. Proc. Natl. Acad. Sci. 107, 7933–7938.

Kaiboriboon, K., Lüders, H.O., Miller, J.P., Leigh, R.J., 2012. Upward gaze and head deviation with frontal eye field stimulation. Epileptic Disord. 14, 64–68.

Kano, F., Tomonaga, M., 2009. How chimpanzees look at pictures: a comparative eye-tracking study. Proc. R. Soc. Lond. 276, 1949–1955.

Kano, F., Tomonaga, M., 2011. Species difference in the timing of gaze movement between chimpanzees and humans. Anim. Cogn. 14, 879–892.

Kano, F., Hirata, S., Call, J., Tomonaga, M., 2011. The visual strategy specific to humans among hominids: a study using the gap-overlap paradigm. Vis. Res. 51, 2348–2355.

Kaplan, G., Rogers, L.J., 2006. Head-cocking as a form of exploration in the common marmoset and its development. Dev. Psychobiol. 48, 551–560.

Kastner, S., DeSimone, K., Konen, C.S., Szczepanski, S.M., Weiner, K.S., Schneider, K.A., 2007. Topographic maps in human frontal cortex revealed in memory-guided saccade and spatial working-memory tasks. J. Neurophysiol. 97, 3494–3507.

Kato, M., Miyauchi, S., 2003. Human precentral cortical activation patterns during saccade tasks: an fMRI comparison with activation during intentional eyeblink tasks. Neuroimage 19, 1260–1272.

Keating, E., 1993. Lesions of the frontal eye field impair pursuit eye movements, but preserve the predictions driving them. Behav. Brain Res. 53, 91–104.

von Kleist, K., 1934. Gehirnpathologie. J.A. Barth, Leipzig.

Knight, T.A., Fuchs, A.F., 2007. Contribution of the frontal eye field to gaze shifts in the head-unrestrained monkey: effects of microstimulation. J. Neurophysiol. 97, 618–634.

Kobayashi, H., Kohshima, S., 1997. Unique morphology of the human eye. Nature 387, 767–768.

Kobayashi, H., Kohshima, S., 2001. Unique morphology of the human eye and its adaptive meaning: comparative studies on external morphology of the primate eye. J. Hum. Evol. 40, 419–435.

Koyama, M., Hasegawa, I., Osada, T., Adachi, Y., Nakahara, K., Miyashita, Y., 2004. Functional magnetic resonance imaging of macaque monkeys performing visually guided saccade tasks: comparison of cortical eye fields with humans. Neuron 41, 795–807.

Krubitzer, L., Kaas, J., 1990. The organization and connections of somatosensory cortex in marmosets. J. Neurosci. 10, 952–974.

Krubitzer, L., Kaas, J., 1993. The dorsomedial visual area of owl monkeys: connections, myeloarchitecture, and homologies in other primates. J. Comp. Neurol. 334, 497–528.

Langston, A., Casagrande, V., Fox, R., 1986. Spatial resolution of the Galago. Vis. Res. 26, 791–796.

Lanzilotto, M., Perciavalle, V., Lucchetti, C., 2013. A new field in monkey's frontal cortex: premotor ear-eye field (PEEF). Neurosci. Biobehav. Rev. 37, 1434–1444.

Latto, R., 1978. The effects of bilateral frontal eye-field, posterior parietal or superior collicular lesions on visual search in the rhesus monkeys. Brain Res. 146, 35–50.

Lehman, R., 1980. Distribution and changes in strength of hand preference of cynomolgus monkeys. Brain Behav. Evol. 17, 209–217.

Leichnetz, G., Gonzalo-Ruiz, A., 1996. Prearcuate cortex in the Cebus monkey has cortical and subcortical connections like the macaque frontal eye field and projects to fastigial-recipient oculomotor-related brainstem nuclei. Brain Res. Bull. 41, 1−29.

Leigh, R.J., Zee, D.S., 2015. The Neurology of Eye Movements. Oxford University Press, New York.

Lemmen, L.J., Davis, J.S., Radnor, L.L., 1959. Observations on stimulation of the human frontal eye field. J. Comp. Neurol. 112, 163−168.

Lende, R., 1970. Cortical localization in the tree shrew (*Tupaia*). Brain Res. 18, 61−75.

Levinsohn, G., 1909. Über die Beziehungen der Grosshirnrinde beim Affen zu den Bewegungen des Auges. Graefes Arch. Clin. Exp. Ophthalmol. 71, 313−378.

Leyton, A.S., Sherrington, C.S., 1917. Observations on the excitable cortex of the chimpanzee, orangutan, and gorilla. Q. J. Exp. Physiol. 11, 135−222.

Lobel, E., Kahane, P., Leonards, U., Grosbras, M.-H., Lehéricy, S., Bihan, D.L., Berthoz, A., 2001. Localization of human frontal eye fields: anatomical and functional findings of functional magnetic resonance imaging and intracerebral electrical stimulation. J. Neurosurg. 95, 804−815.

Love, S.A., Marie, D., Roth, M., Lacoste, R., Nazarian, B., Bertello, A., Coulon, O., Anton, J.-L., Meguerditchian, A., 2016. The average baboon brain: MRI templates and tissue probability maps from 89 individuals. Neuroimage 132, 526−533.

Lucchetti, C., Lanzilotto, M., Bon, L., 2008. Auditory-motor and cognitive aspects in area 8B of macaque monkey's frontal cortex: a premotor ear-eye field (PEEF). Exp. Brain Res. 186, 131−141.

Luna, B., Thulborn, K.R., Strojwas, M.H., McCurtain, B.J., Berman, R.A., Genovese, C.R., Sweeney, J.A., 1998. Dorsal cortical regions subserving visually guided saccades in humans: an fMRI study. Cereb. Cortex 8, 40−47.

Lynch, J., Hoover, J., Strick, P., 1994. Input to the primate frontal eye field from the substantia nigra, superior colliculus, and dentate nucleus demonstrated by transneuronal transport. Exp. Brain Res. 100, 181−186.

Lyon, D., Kaas, J., 2001. Connectional and architectonic evidence for dorsal and ventral V3, and dorsomedial area in marmoset monkeys. J. Neurosci. 21, 249−261.

MacAvoy, M.G., Gottlieb, J.P., Bruce, C.J., 1991. Smooth-pursuit eye movement representation in the primate frontal eye field. Cereb. Cortex 1, 95−102.

Mangalam, M., Desai, N., Singh, M., 2014. Division of labor in hand usage in free-ranging bonnet macaques, *Macaca radiata*. Am. J. Primatol. 76, 576−585.

Marchetti, E., Gauthier, G.M., Pellet, J., 1983. Cerebellar control of eye movements studied with injection of harmaline in the trained baboon. Arch. Ital. Biol. 121, 1−17.

Markov, N.T., Vezoli, J., Chameau, P., Falchier, A., Quilodran, R., Huissoud, C., Lamy, C., Misery, P., Giroud, P., Ullman, S., Barone, P., Dehay, C., Knoblauch, K., Kennedy, H., 2014. Anatomy of hierarchy: feedforward and feedback pathways in macaque visual cortex. J. Comp. Neurol. 522, 225−259.

Markowitsch, H., Pritzel, M., Wilson, M., Divac, I., 1980. The prefrontal cortex of a prosimian (*Galago senegalensis*) defined as the cortical projection area of the thalamic mediodorsal nucleus. Neuroscience 5, 1771−1779.

Martinez-Conde, S., Macknik, S.L., 2008. Fixational eye movements across vertebrates: comparative dynamics, physiology, and perception. J. Vis. 8, 28.1−28.16.

McCrea, R.A., Gdowski, G.T., 2003. Firing behaviour of squirrel monkey eye movement-related vestibular nucleus neurons during gaze saccades. J. Physiol. 546, 207−224.

Michalka, S.W., Kong, L., Rosen, M.L., Shinn-Cunningham, B.G., Somers, D.C., 2015. Short-term memory for space and time flexibly recruit complementary sensory-biased frontal lobe attention networks. Neuron 87, 882−892.

Milea, D., Lobel, E., Lehéricy, S., Duffau, H., Rivaud-Péchoux, S., Berthoz, A., Pierrot-Deseilligny, C., 2002. Intraoperative frontal eye field stimulation elicits ocular deviation and saccade suppression. Neuroreport 13, 1359−1364.

Mingazzini, G., 1888. Intorno ai solchi e le circonvoluzioni cerebrali dei primati e del feto umano (Estratto dagli Atti della R. Accademia medica di Roma, anno XV, vol. IV, serie II). Tipografia Fratelli Centenari, Roma.

Mitchell, J.F., Leopold, D.A., 2015. The marmoset monkey as a model for visual neuroscience. Neurosci. Res. 93, 20−46.

Mitchell, J.F., Reynolds, J.H., Miller, C.T., 2014. Active vision in marmosets: a model system for visual neuroscience. J. Neurosci. 34, 1183−1194.

Montemurro, N., Herbet, G., Duffau, H., 2016. Right cortical and axonal structures eliciting ocular deviation during electrical stimulation mapping in awake patients. Brain Topogr. 29.

Monteon, J.A., Constantin, A.G., Wang, H., Martinez-Trujillo, J., Crawford, J.D., 2010. Electrical stimulation of the frontal eye fields in the head-free macaque evokes kinematically normal 3D gaze shifts. J. Neurophysiol. 104, 3462−3475.

Monteon, J.A., Wang, H., Martinez-Trujillo, J., Crawford, J.D., 2013. Frames of reference for eye-head gaze shifts evoked during frontal eye field stimulation. EJNS 37, 1754−1765.

Morrow, M., Sharpe, J., 1995. Deficits of smooth-pursuit eye movement after unilateral frontal lobe lesions. Ann. Neurol. 37, 443−451.

Moschovakis, A.K., Gregoriou, G.G., Ugolini, G., Doldan, M., Graf, W., Guldin, W., Hadjidimitrakis, K., Savaki, H.E., 2004. Oculomotor areas of the primate frontal lobes: a transneuronal transfer of rabies virus and [14C]-2-deoxyglucose functional imaging study. J. Neurosci. 24, 5726−5740.

Mota, B., Herculano-Houzel, S., 2015. Brain structure. Cortical folding scales universally with surface area and thickness, not number of neurons. Science 349, 74−77.

Mott, F.W., Halliburton, W.D., 1908. Localisation of function in the lemur's brain. Proc. R. Soc. Lond. 80, 136−147.

Mott, F.W., Schäfer, E.A., 1890. On associated eye-movements produced by cortical faradization of the monkey's brain. Brain 13, 165−173.

Mott, F.W., Schuster, E., Halliburton, W.D., 1910. Cortical lamination and localisation in the brain of the marmoset. Proc. R. Soc. Lond. 82, 124−134.

Munoz, D.P., Everling, S., 2004. Look away: the anti-saccade task and the voluntary control of eye movement. Nat. Rev. Neurosci. 5, 218−228.

Müri, R., Hess, C., Meienberg, O., 1991. Transcranial stimulation of the human frontal eye field by magnetic pulses. Exp. Brain Res. 86, 219−223.

Neggers, S.F.W., Huijbers, W., Vrijlandt, C., Vlaskamp, B.N.S., Schutter, D.J.L.G., Kenemans, J., 2007. TMS pulses on the frontal eye fields break coupling between visuospatial attention and eye movements. J. Neurophysiol. 98, 2765−2778.

Neggers, S.F.W., Diepen, R.M., van Zandbelt, B.B., Vink, M., Mandl, R.C.W., Gutteling, T.P., 2012. A functional and structural investigation of the human fronto-basal volitional saccade network. PLoS One 7, e29517.

Neggers, S.F., Zandbelt, B.B., Schall, M.S., Schall, J.D., 2015. Comparative diffusion tractography of corticostriatal motor pathways reveals differences between humans and macaques. J. Neurophysiol. 113, 2164−2172.

Nelson, M.J., Boucher, L., Logan, G.D., Palmeri, T.J., Schall, J.D., 2010. Nonindependent and nonstationary response times in stopping and stepping saccade tasks. Atten. Percept. Psychophys. 72, 1913−1929.

Nieuwenhuys, R., 2013. The myeloarchitectonic studies on the human cerebral cortex of the Vogt-Vogt school, and their significance for the interpretation of functional neuroimaging data. Brain Struct. Funct. 218, 303−352.

Nieuwenhuys, R., Broere, C.A., Cerliani, L., 2015. A new myeloarchitectonic map of the human neocortex based on data from the Vogt—Vogt school. Brain Struct. Funct. 220, 2551—2573.

Olk, B., Chang, E., Kingstone, A., Ro, T., 2006. Modulation of antisaccades by transcranial magnetic stimulation of the human frontal eye field. Cereb. Cortex 16, 76—82.

Ono, M., Kubik, S., Abernathey, C.D., 1990. Atlas of the Cerebral Sulci. Thieme Medical Publishers, New York.

Owen, R., 1900. Descriptive and Illustrated Catalogue of the Physiological Series of Comparative Anatomy Contained in the Museum of the Royal College of Surgeons of England. Taylor & Francis, London.

Palmer, S.M., Rosa, M.G.P., 2006. Quantitative analysis of the corticocortical projections to the middle temporal area in the marmoset monkey: evolutionary and functional implications. Cereb. Cortex 16, 1361—1375.

Passingham, R., 2009. How good is the macaque monkey model of the human brain? Curr. Opin. Neurobiol. 19, 6—11.

Paus, T., 1996. Location and function of the human frontal eye-field: a selective review. Neuropsychologia 34, 475—483.

Paxinos, G., Watson, C., Petrides, M., Rosa, M., Tokuno, H., 2012. The Marmoset Brain in Stereotaxic Coordinates. Elsevier.

Penfield, W., Boldrey, E., 1937. Somatic motor and sensory representation in the cerebral cortex of man as studied by electrical stimulation. Brain 60, 389—443.

Penfield, W., Rasmussen, T., 1950. The Cerebral Cortex of Man; a Clinical Study of Localization of Function.

Percheron, G., François, C., Pouget, P., 2015. What makes a frontal area of primate brain the frontal eye field? Front. Integr. Neurosci. 9, 33.

Perelman, P., Johnson, W.E., Roos, C., Seuánez, H.N., Horvath, J.E., Moreira, M.A., Kessing, B., Pontius, J., Roelke, M., Rumpler, Y., Schneider, M.P., Silva, A., O'Brien, S.J., Pecon-Slattery, J., 2011. A molecular phylogeny of living primates. PLoS Genet. 7 (3), e1001342.

Petit, L., Clark, V.P., Ingeholm, J., Haxby, J.V., 1997. Dissociation of saccade-related and pursuit-related activation in human frontal eye fields as revealed by fMRI. J. Neurophysiol. 77, 3386—3390.

Petit, L., Dubois, S., Tzourio, N., Dejardin, S., Crivello, F., Michel, C., Etard, O., Denise, P., Roucoux, A., Mazoyer, B., 1999. PET study of the human foveal fixation system. Hum. Brain Mapp. 8, 28—43.

Pierrot-Deseilligny, C., 1994. Saccade and smooth-pursuit impairment after cerebral hemispheric lesions. Eur. Neurol. 34, 121—134.

Premereur, E., Janssen, P., Vanduffel, W., 2015. Effector specificity in macaque frontal and parietal cortex. J. Neurosci. 35, 3446—3459.

Preuss, T., Goldman-Rakic, P., 1991. Myelo- and cytoarchitecture of the granular frontal cortex and surrounding regions in the strepsirhine primate Galago and the anthropoid primate Macaca. J. Comp. Neurol. 310, 429—474.

Preuss, T.M., Stepniewska, I., Kaas, J.H., 1996. Movement representation in the dorsal and ventral premotor areas of owl monkeys: a microstimulation study. J. Comp. Neurol. 371, 649—676.

Pribram, K.H., 1955. Lesions of "frontal eye fields" and delayed response of baboons. J. Neurophysiol. 18, 105—112.

Rasmussen, T., Penfield, W., 1947. Further studies of the sensory and motor cerebral cortex of man. Federation Proceedings, pp. 452—460.

Remple, M.S., Reed, J.L., Stepniewska, I., Kaas, J.H., 2006. Organization of frontoparietal cortex in the tree shrew (Tupaia belangeri). I. architecture, microelectrode maps, and corticospinal connections. J. Comp. Neurol. 497, 133—154.

Reser, D.H., Burman, K.J., Yu, H.-H., Chaplin, T.A., Richardson, K.E., Worthy, K.H., Rosa, M.G.P., 2013. Contrasting patterns of cortical input to architectural subdivisions of the area 8 complex: a retrograde tracing study in marmoset monkeys. Cereb. Cortex 23, 1901—1922.

Rivaud, S., Müri, R., Gaymard, B., Vermersch, A., Pierrot-Deseilligny, C., 1994. Eye movement disorders after frontal eye field lesions in humans. Exp. Brain Res. 102, 110—120.

Rizzolatti, G., Arbib, M.A., 2002. Language within our grasp. In: Cacioppo, J.T., et al. (Eds.), Foundations in Social Neuroscience. MIT Press, Cambridge, Mass.

Ro, T., Cheifet, S., Ingle, H., Shoup, R., Rafal, R., 1999. Localization of the human frontal eye fields and motor hand area with transcranial magnetic stimulation and magnetic resonance imaging. Neuropsychologia 37, 225—231.

Robinson, D.A., Fuchs, A.F., 1969. Eye movements evoked by stimulation of frontal eye fields. J. Neurophysiol. 32, 637—648.

Rogers, L., Stafford, D., Ward, J., 1993. Head cocking in galagos. Anim. Behav. 45, 943—952.

Rohlfing, T., Kroenke, C.D., Sullivan, E.V., Dubach, M.F., Bowden, D.M., Grant, K.A., Pfefferbaum, A., 2012. The INIA19 template and neuroMaps atlas for primate brain image parcellation and spatial normalization. Front. Neuroinform. 6, 27.

Rosa, M., Soares, J., Fiorani Jr., M., Gattass, R., 1993. Cortical afferents of visual area MT in the Cebus monkey: possible homologies between New and Old World monkeys. Vis. Neurosci. 10, 827—855.

Rosa, M.G.P., Palmer, S.M., Gamberini, M., Burman, K.J., Yu, H.-H., Reser, D.H., Bourne, J.A., Tweedale, R., Galletti, C., 2009. Connections of the dorsomedial visual area: pathways for early integration of dorsal and ventral streams in extrastriate cortex. J. Neurosci. 29, 4548—4563.

Rosano, C., Krisky, C.M., Welling, J.S., Eddy, W.F., Luna, B., Thulborn, K.R., Sweeney, J.A., 2002. Pursuit and saccadic eye movement subregions in human frontal eye field: a high-resolution fMRI investigation. Cereb. Cortex 12, 107—115.

Rosano, C., Sweeney, J.A., Melchitzky, D.S., Lewis, D.A., 2003. The human precentral sulcus: chemoarchitecture of a region corresponding to the frontal eye fields. Brain Res. 972, 16—30.

Rosati, A.G., Hare, B., 2009. Looking past the model species: diversity in gaze-following skills across primates. Curr. Opin. Neurobiol. 19, 45—51.

Russell, J.R., 1894. An experimental investigation of eye movements. J. Physiol. 17, 1.

Sajad, A., Sadeh, M., Keith, G.P., Yan, X., Wang, H., Crawford, J.D., 2015. Visual-motor transformations within frontal eye fields during head-unrestrained gaze shifts in the monkey. Cereb. Cortex 25, 3932—3952.

Sakamoto, K., Sawada, K., Fukunishi, K., Noritaka, I., Sakata-Haga, H., Yoshihiro, F., 2014. Postnatal change in sulcal length asymmetry in cerebrum of cynomolgus monkeys (Macaca fascicularis). Anat. Rec. 297, 200—207.

Saleem, K.S., Logothetis, N.K., 2012. A Combined MRI and Histology Atlas of the Rhesus Monkey Brain in Stereotaxic Coordinates. Academic Press.

Sallet, J., Mars, R.B., Noonan, M.P., Neubert, F.X., Jbabdi, S., O'Reilly, J.X., Filippini, N., Thomas, A.G., Rushworth, M.F., 2013. The organization of dorsal frontal cortex in humans and macaques. J. Neurosci. 33, 12255—12274.

Sanides, F., 1970. Functional architecture of motor and sensory cortices in primates in the light of a new concept of neocortex evolution. In: Noback, C.R., Montagna, W. (Eds.), The Primate Brain. Appleton-Century-Crofts., New York, pp. 137—208.

Sarkissov, S., Filimonoff, I., Kononowa, E., Preobraschenskaja, I., Kukuew, L., 1955. Atlas of the Cytoarchitectonics of the Human Cerebral Cortex. Medgiz, Moscow.

Savaki, H.E., Gregoriou, G.G., Bakola, S., Moschovakis, A.K., 2015. Topography of visuomotor parameters in the frontal and premotor eye fields. Cereb. Cortex 25, 3095—3106.

Schall, J.D., 1991. Neuronal activity related to visually guided saccadic eye movements in the supplementary motor area of rhesus monkeys. J. Neurophysiol. 66, 530−558.

Schall, J.D., 1997. Visuomotor areas of the frontal lobe. In: Rockland, K.S., Kaas, J.H., Peters, A. (Eds.), Extrastriate Cortex in Primates. Springer, Boston, MA, US, pp. 527−638.

Schall, J.D., 2015. Visuomotor functions in the frontal lobe. Annu. Rev. Vis. Sci. 1, 469−498.

Schall, J., Boucher, L., 2007. Executive control of gaze by the frontal lobes. Cogn. Affect. Behav. Neurosci. 7, 396−412.

Schall, J.D., Morel, A., King, D.J., Bullier, J., 1995. Topography of visual cortex connections with frontal eye field in macaque: convergence and segregation of processing streams. J. Neurosci. 15, 4464−4487.

Schiller, P., True, S., Conway, J., 1980. Deficits in eye movements following frontal eye-field and superior colliculus ablations. J. Neurophysiol. 44, 1175−1189.

Schiller, P., Sandell, J., Maunsell, J., 1987. The effect of frontal eye field and superior colliculus lesions on saccadic latencies in the rhesus monkey. J. Neurophysiol. 57, 1033−1049.

Schlag, J., Schlag-Rey, M., 1987. Evidence for a supplementary eye field. J. Neurophysiol. 57, 179−200.

Schmitt, O., Modersitzki, J., Heldmann, S., Wirtz, S., Hömke, L., Heide, W., Kömpf, D., Wree, A., 2005. Three-dimensional cytoarchitectonic analysis of the posterior bank of the human precentral sulcus. Anat. Embryol. 210, 387−400.

Shepherd, S.V., 2010. Following gaze: gaze-following behavior as a window into social cognition. Front. Integr. Neurosci. 4, 5.

Shepherd, S.V., Platt, M.L., 2006. Noninvasive telemetric gaze tracking in freely moving socially housed prosimian primates. Methods 38, 185−194.

Shepherd, S.V., Steckenfinger, S.A., Hasson, U., Ghazanfar, A.A., 2010. Human-monkey gaze correlations reveal convergent and divergent patterns of movie viewing. Curr. Biol. 20, 649−656.

Sherwood, C.C., Broadfield, D.C., Holloway, R.L., Gannon, P.J., Hof, P.R., 2003. Variability of broca's area homologue in African great apes: implications for language evolution. Anat. Rec. 271, 276−285.

Simonyan, K., 2014. The laryngeal motor cortex: its organization and connectivity. Curr. Opin. Neurobiol. 28, 15−21.

Smith, W.K., 1949. The frontal eye fields. In: Bucy, P.C. (Ed.), The Precentral Motor Cortex. University of Illinois Press.

Sommer, M., Tehovnik, E., 1997. Reversible inactivation of macaque frontal eye field. Exp. Brain Res. 116, 229−249.

Song, J.H., McPeek, R.M., 2010. Roles of narrow- and broad-spiking dorsal premotor area neurons in reach target selection and movement production. J. Neurophysiol. 103, 2124−2138.

Spatz, W., Tigges, J., 1972. Experimental-anatomical studies on the "middle temporal visual area (MT)" in primates. I. Efferent cortico-cortical connections in the marmoset Callithrix jacchus. J. Comp. Neurol. 146, 451−464.

Squire, R.F., Noudoost, B., Schafer, R.J., Moore, T., 2013. Prefrontal contributions to visual selective attention. Annu. Rev. Neurosci. 36, 451−466.

Stahl, J.S., 1999. Amplitude of human head movements associated with horizontal saccades. Exp. Brain Res. 126, 41−54.

Stanton, G.B., Deng, S.Y., Goldberg, M.E., McMullen, N.T., 1989. Cytoarchitectural characteristic of the frontal eye fields in macaque monkeys. J. Comp. Neurol. 282, 415−427.

van der Steen, J., Russell, I., James, G., 1986. Effects of unilateral frontal eye-field lesions on eye-head coordination in monkey. J. Neurophysiol. 55, 696−714.

Stepniewska, I., Preuss, T.M., Kaas, J.H., 1993. Architectonics, somatotopic organization, and ipsilateral cortical connections of the primary motor area (M1) of owl monkeys. J. Comp. Neurol. 330, 238−271.

Stone, J., Johnston, E., 1981. The topography of primate retina: a study of the human, bushbaby, and new- and old-world monkeys. J. Comp. Neurol. 196, 205−223.

Striedter, G.F., Srinivasan, S., Monuki, E.S., 2015. Cortical folding: when, where, how, and why? Annu. Rev. Neurosci. 38, 291−307.

Sun, T., Hevner, R.F., 2014. Growth and folding of the mammalian cerebral cortex: from molecules to malformations. Nat. Rev. Neurosci. 15, 217−232.

Sussman, R.W., Tab Rasmussen, D., Raven, P.H., 2013. Rethinking primate origins again. Am. J. Primatol. 75, 95−106.

Suzuki, H., Azuma, M., 1983. Topographic studies on visual neurons in the dorsolateral prefrontal cortex of the monkey. Exp. Brain Res. 53, 47−58.

Tallinen, T., Chung, J.Y., Rousseau, F., Girard, N., Lefèvre, J., Mahadevan, L., 2016. On the growth and form of cortical convolutions. Nat. Phys. 12.

Tanaka, M., Lisberger, S.G., 2002. Role of arcuate frontal cortex of monkeys in smooth pursuit eye movements. I. basic response properties to retinal image motion and position. J. Neurophysiol. 87, 2684−2699.

Tehovnik, E.J., Sommer, M.A., Chou, I.H., Slocum, W.M., Schiller, P.H., 2000. Eye fields in the frontal lobes of primates. Brain Res. Brain Res. Rev. 32, 413−448.

Thakkar, K.N., van den Heiligenberg, F.M.Z., Kahn, R.S., Neggers, S.F.W., 2014. Frontal-subcortical circuits involved in reactive control and monitoring of gaze. J. Neurosci. 34, 8918−8929.

Thompson, K.G., Bichot, N.P., Sato, T.R., 2005. Frontal eye field activity before visual search errors reveals the integration of bottom-up and top-down salience. J. Neurophysiol. 93, 337−351.

Thurtell, M., Mohamed, A., Lüders, H., Leigh, R., 2009. Evidence for three-dimensional cortical control of gaze from epileptic patients. J. Neurol. Neurosurg. Psychiatry 80, 683−685.

Tian, J., Lynch, J., 1995. Slow and saccadic eye movements evoked by microstimulation in the supplementary eye field of the Cebus monkey. J. Neurophysiol. 74, 2204−2210.

Tian, J., Lynch, J., 1996. Corticocortical input to the smooth and saccadic eye movement subregions of the frontal eye field in Cebus monkeys. J. Neurophysiol. 76, 2754−2771.

Tian, J., Lynch, J., 1997. Subcortical input to the smooth and saccadic eye movement subregions of the frontal eye field in Cebus monkey. J. Neurosci. 17, 9233−9247.

Tigges, J., Tigges, M., Anschel, S., Cross, N., Letbetter, W., McBride, R., 1981. Areal and laminar distribution of neurons interconnecting the central visual cortical areas 17, 18, 19, and MT in squirrel monkey (Saimiri). J. Comp. Neurol. 202, 539−560.

Tomasello, M., Hare, B., Lehmann, H., Call, J., 2007. Reliance on head versus eyes in the gaze following of great apes and human infants: the cooperative eye hypothesis. J. Hum. Evol. 52, 314−320.

Tomlinson, R.D., Bahra, P.S., 1986. Combined eye-head gaze shifts in the primate. I. Metrics. J. Neurophysiol. 56, 1542−1557.

Tu, T.A., Keating, E.G., 2000. Electrical stimulation of the frontal eye field in a monkey produces combined eye and head movements. J. Neurophysiol. 84, 1103−1106.

Vanduffel, W., Fize, D., Mandeville, J., Nelissen, K., Van Hecke, P., Rosen, B., Tootell, R., Orban, G., 2001. Visual motion processing investigated using contrast agent-enhanced fMRI in awake behaving monkeys. Neuron 32, 565−577.

Vernet, M., Quentin, R., Chanes, L., Mitsumasu, A., Valero-Cabré, A., 2014. Frontal eye field, where art thou? Anatomy, function, and non-invasive manipulation of frontal regions involved in eye movements and associated cognitive operations. Front. Integr. Neurosci. 8, 66.

Vogt, C., Vogt, O., 1926. Die vergleichend-architektonische und die vergleichend-reizphysiologische Felderung der Großhirnrinde unter besonderer Berücksichtigung der menschlichen. Naturwissenschaften 14, 1190−1194.

Wagman, I.H., Werman, R., Feldman, D.S., Sugarman, L., Krieger, H.P., 1957. The oculomotor effects of cortical and subcortical stimulation in the monkey. J. Neuropathol. Exp. Neurol. 16, 269−278.

Wagman, I.H., Krieger, H., Bender, M., 1958. Eye movements elicited by surface and depth stimulation of the occipital lobe of *Macaque mulatta*. J. Comp. Neurol. 109, 169−193.

Wagman, I.H., Krieger, H.P., Papatheodorou, C.A., Bender, M.B., 1961. Eye movements elicited by surface and depth stimulation of the frontal lobe of *Macaque mulatta*. J. Comp. Neurol. 117, 179−188.

Walker, A.E., 1940. A cytoarchitectural study of the prefrontal area of the macaque monkey. J. Comp. Neurol. 73, 59−86.

Wardak, C., Ibos, G., Duhamel, J.-R., Olivier, E., 2006. Contribution of the monkey frontal eye field to covert visual attention. J. Neurosci. 26, 4228−4235.

Wardak, C., Vanduffel, W., Orban, G.A., 2010. Searching for a salient target involves frontal regions. Cereb. Cortex 20, 2464−2477.

Watanabe-Sawaguchi, K., Kubota, K., Arikuni, T., 1991. Cytoarchitecture and intrafrontal connections of the frontal cortex of the brain of the hamadryas baboon (*Papio hamadryas*). J. Comp. Neurol. 311, 108−133.

Webb, S., Kaas, J., 1976. The sizes and distribution of ganglion cells in the retina of the owl monkey. *Aotus trivirgatus*. Vis. Res. 16, 1247−1254.

Weller, R.E., Kaas, J.H., 1987. Subdivisions and connections of inferior temporal cortex in owl monkeys. J. Comp. Neurol. 256, 137−172.

Weller, R., Wall, J., Kaas, J., 1984. Cortical connections of the middle temporal visual area (MT) and the superior temporal cortex in owl monkeys. J. Comp. Neurol. 228, 81−104.

Wessel, K., Kömpf, D., 1991. Transcranial magnetic brain stimulation: lack of oculomotor response. Exp. Brain Res. 86, 216−218.

Wikler, K., Rakic, P., 1990. Distribution of photoreceptor subtypes in the retina of diurnal and nocturnal primates. J. Neurosci. 10, 3390−3401.

Wolin, L., Massopust Jr., L., 1967. Characteristics of the ocular fundus in primates. J. Anat. 101, 693−699.

Woollard, H., 1927. The differentiation of the retina in the primates. Proc. Zool. Soc. Lond. 97, 1−18.

Wright, P.C., 1989. The nocturnal primate niche in the New World. J. Hum. Evol. 18, 635−658.

Wu, C., Bichot, N., Kaas, J., 2000. Converging evidence from microstimulation, architecture, and connections for multiple motor areas in the frontal and cingulate cortex of prosimian primates. J. Comp. Neurol. 423, 140−177.

Yamamoto, J., Ikeda, A., Satow, T., Matsuhashi, M., Baba, K., Yamane, F., Miyamoto, S., Mihara, T., Hori, T., Taki, W., Hashimoto, N., Shibasaki, H., 2004. Human eye fields in the frontal lobe as studied by epicortical recording of movement-related cortical potentials. Brain 127, 873−887.

Zilles, K., Amunts, K., 2010. Centenary of Brodmann's map—Conception and fate. Nat. Rev. Neurosci. 11, 139−145.

37

The Evolution of Auditory Cortex in Humans

J.P. Rauschecker

Georgetown University Medical Center, Washington, DC, United States; TUM, Munich, Germany

37.1 Auditory Cortex: Core, Belt, and Parabelt

37.1.1 Macaque

A revival of interest in the organization of primate auditory cortex occurred in the 1990s, after the field lay largely dormant for almost 20 years following the initial discoveries by Merzenich and Brugge (1973) and others. Groundbreaking were the anatomical studies by Kaas et al. (Morel et al., 1993; Hackett et al., 1998, 1999; Kaas and Hackett, 2000) using cytoarchitectonic and histochemical techniques (staining for, eg, acetyl-cholinesterase and parvalbumin), which allowed the partition of auditory cortex into a core region surrounded by a belt both laterally and medially and a parabelt beyond that (Fig. 37.1). Similarly, Jones et al. (1995) also used immunohistochemical techniques for the visualization of calcium-binding protein to characterize subdivisions of auditory cortex in the rhesus monkey.

Neurophysiological studies began around the same time and were able to use the neuroanatomical characterization as guidance for the mapping of functional regions (Rauschecker et al., 1995, 1997). This resulted in the finding of ultimately three core areas (A1, R, and RT) and three lateral belt areas (AL, ML, and CL). The sets of core and belt regions were identified by reversals of best frequencies along an anterior–posterior gradient in rhesus monkeys, whereas core and belt were set apart by different best bandwidths, when tested with pure tones and band-passed noise (BPN) bursts (Rauschecker et al., 1995; Fig. 37.2). A medial belt, postulated originally on neuroanatomical grounds, was verified electrophysiologically later (Kusmierek and Rauschecker, 2009).

37.1.2 Human

Rivier and Clarke (1997) used the same histochemical stains as in monkeys to demonstrate the existence of multiple areas in human auditory cortex. Direct comparison in the macaque, chimpanzee, and human confirmed the existence of similar subdivisions in all three species (Hackett et al., 2001). At the same time, probabilistic measurement of cytoarchitectonic differences made it possible to identify subdivisions of Heschl region (Morosan et al., 2001). The functional partition of auditory cortex into core, belt, and parabelt regions in humans was verified by different sets of studies. Wessinger et al. (2001), using functional MRI (fMRI) [blood-oxygenation-level-dependent (BOLD) imaging], first showed separate regions responding differentially to pure tones and BPN bursts. Later, Chevillet et al. (2011) used tones, BPN bursts, and vowel sounds (spectrally complex harmonic tones) to identify a concentric organization of core and belt regions in human auditory cortex, surrounded along the anterior edge by a "parabelt" region responding best to vowel sounds. Most recently, Leaver and Rauschecker (2016), using again tones and BPN bursts as in the monkey studies (Fig. 37.3), not only identified core and belt but also confirmed tonotopic organization in the same human subjects, allowing direct comparison with the cytoarchitectonic scheme of Morosan et al. (2001). These studies make it appear very likely that human auditory cortex is organized very similarly to the macaque.

37.2 Functional Topography of Auditory Cortex

Tonotopic organization was shown to be a universal feature of auditory cortical organization in animal electrophysiological studies in a variety of species early on (from dogs and cats to monkeys; Merzenich and Brugge, 1973; Tunturi and Barrett, 1977; Reale and Imig, 1980). Given the topographic organization of other sensory systems (visual and somatosensory), mapping their sensory epithelia onto at least primary sensory cortical areas, tonotopy (or more appropriately cochleotopy) in

891

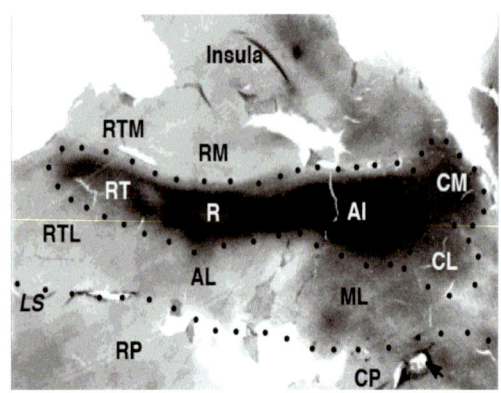

FIGURE 37.1 Histochemical definition of core and belt regions in rhesus monkey auditory cortex (reproduced from Hackett, T.A., Stepniewska, I., Kaas, J.H., 1998. Subdivisions of auditory cortex and ipsilateral cortical connections of the parabelt auditory cortex in macaque monkeys. J. Comp. Neurol. 394, 475–495). Core areas (RT, R, and A1) show darker staining for parvalbumin than surrounding areas in the lateral belt (RTL, AL, ML, and CL) and medial belt (RTM, RM, and CM). Parabelt (RP and CP) and insula show lightest staining.

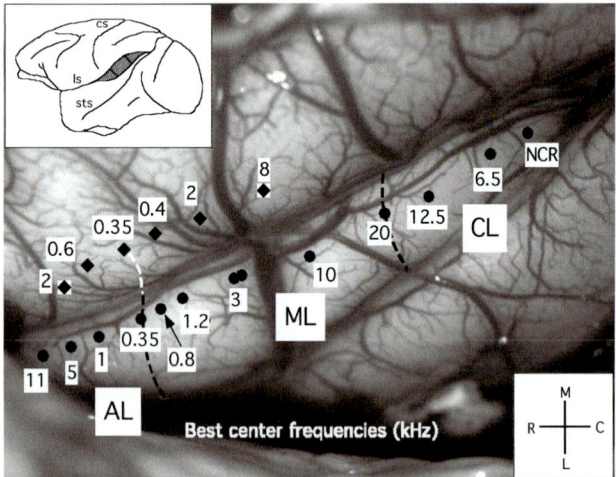

FIGURE 37.2 Cochleotopic organization of lateral belt areas (AL, ML, and CL) in rhesus monkey auditory cortex (reproduced from Rauschecker, J.P., Tian, B., Hauser, M., 1995. Processing of complex sounds in the macaque nonprimary auditory cortex. Science 268, 111–114). *NCR*, no clear response. *Rhomboid* symbols refer to recording sites from core areas R and A1.

primary auditory cortex did not come as a huge surprise. Monkey studies also showed clear cochleotopy in (secondary) lateral belt areas (Rauschecker et al., 1995).

Tonotopy of auditory cortex in humans and its exact organization across multiple areas, however, has proven a more persistently vexing problem than in other species. This may be due to the scarcity of data from direct electrophysiological surface recordings, the limited resolution of noninvasive recording techniques, or the semiquantitative nature of analytic methods, which could

not account for interindividual variability. Most studies agree at least that tonotopic organization of auditory cortex does exist in humans. Having been shown originally on the basis of electrical and neuromagnetic evoked potential studies (Celesia, 1976; Romani et al., 1982), tonotopy was also later verified by functional neuroimaging techniques (Lauter et al., 1985; Talavage et al., 2000, 2004). The existence of multiple representations of the hearing range became clear from these studies, but multiple mirror-symmetric tonotopic maps in human auditory cortex were not established until high-resolution fMRI became available (Formisano et al., 2003; Humphries et al., 2010; Striem-Amit et al., 2011; Saenz and Langers, 2014). However, the axes or gradients of tonotopy reported were different in different studies. Again, this may have been partly due not only to the poor spatial resolution of some imaging techniques such as EEG or PET but also to the limited extent and density of sampling.

Apart from the topographic organization of best frequency, some studies have looked for topographic organization of other parameters, eg, the rate of periodic stimuli. Such "periodotopy" had been reported in the inferior colliculus of cats (Schreiner and Langner, 1988) and monkeys (Baumann et al., 2011), but not in the auditory cortex of primates. One study claimed evidence for periodotopic organization in primary auditory cortex of humans (Barton et al., 2012), but a more recent study, using a more extensive range of frequencies and modulation rates, argued that the former data confused frequency with pitch, which is generated by periodicity (Leaver and Rauschecker, 2016). Indeed, a unified representation of pitch seems to be generated in the anterolateral portion of auditory cortex in marmosets (Bendor and Wang, 2005), and the existence of such a "pitch center" has also been confirmed in humans (Norman-Haignere et al., 2013).

Leaver and Rauschecker (2016) also provided the most extensive set of individual data from tonotopic mapping in human auditory cortex thus far and have argued that the topographic organization of auditory cortex in humans is indeed very similar to that of monkeys, thus concluding one of the most basic quests for evolutionary similarity in two primate species (see Fig. 37.3).

37.2.1 Hierarchical Processing of Complex Sounds

In the early studies of rhesus monkey auditory cortex, it soon became clear that, apart from responding better to BPN bursts than to pure tones, belt neurons in rhesus monkey auditory cortex also responded better to more complex, spectrally broadband sounds, including

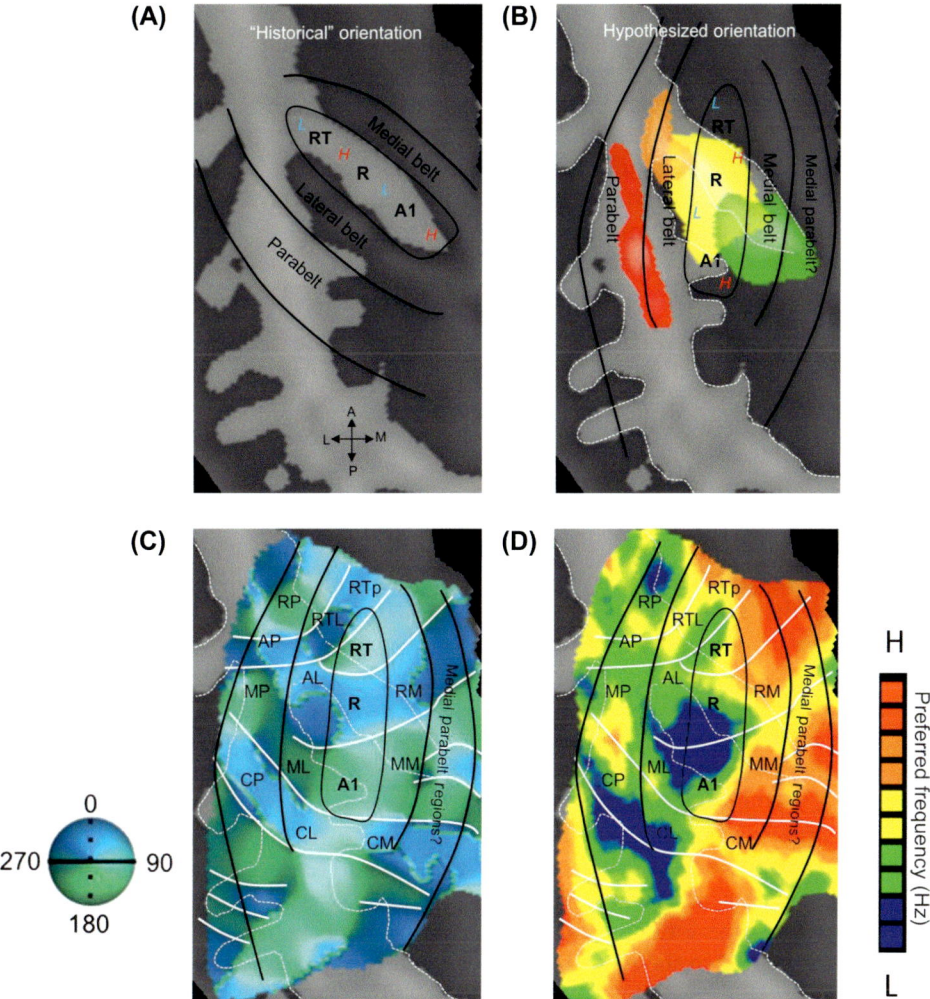

FIGURE 37.3 Tonotopic organization of human auditory cortex (Leaver and Rauschecker, 2016) superimposed on cytoarchitectonic parcellation of Heschl gyrus (HG). Hypothesized positions of auditory cortical regions coincide with probabilistic maps of koniocortical areas in humans (Morosan et al., 2001). (A) Previous neuroimaging research placed the orientation of core auditory fields *along* HG, with high frequencies represented medially (*red* "H") and low frequencies represented laterally (*blue* "L"). (B) The new functional imaging data show an orientation of core regions *oblique* to HG, with high and low frequencies alternating from posterior to anterior. In this scheme, the functional definition of core regions overlaps with the koniocortical/primary region Te1.0, as defined from underlying cytoarchitecture (Rademacher et al., 2001; Morosan et al., 2001), which is shown in *yellow* according to the WFU Pick Atlas (Maldjian et al., 2003). Medial, nonprimary region Te1.1 is shown in *green*, lateral region Te1.2 in orange, and Te3.0 in *red*. (C) A map of frequency-gradient direction is shown, derived from a map of frequency preference independent of stimulus bandwidth (ie, including responses to PT, narrowband noise, and broadband noise together). *White lines* indicate the position of reversals in local frequency gradients. (Reversals between 0/360 and 180 degrees are shown as a switch from *green* to *blue*, and reversals between 90 and 270 degrees are shown as a change from dark to light colors, respectively.) The hypothesized locations of putative core, belt, and parabelt regions are marked by *solid black lines*, along with hypothesized subregions (areas) homologous to those identified by microelectrode mapping in nonhuman primates. (D) A group tonotopic map is displayed, which matches gradient map displayed in C. Data from all stimulus frequencies and bandwidths were used to create this map. Group frequency preference (tonotopic) maps are shown on cortical surfaces; color indicates the center frequency eliciting the greatest blood-oxygenation-level-dependent activity (see also the color scale to the right of the panel). Frequency-gradient reversals that delineate functional areas in C remain in D. All panels display a group-average cortical surface (right hemisphere); *white dashed lines* mark major sulci and gyri. Auditory subfield names are taken from the nonhuman primate literature for convenience and follow these abbreviations: *R*, rostral; *C*, caudal; *M*, medial; *A*, anterior; *L*, lateral belt; *M*, medial belt; *P*, parabelt; *T*, temporal; *p*, pole.

species-specific vocalizations, arguing for a cortical processing hierarchy similar to that in visual cortex (Rauschecker, 1998, 2015; Tian et al., 2013). From an auditory point of view, spoken language also starts with the processing of complex auditory signals. Thus, the recordings in nonhuman primates suggest that neurons as early as the secondary processing stage along the auditory cortical pathway (the lateral belt areas) can show a preference for species-specific communication calls (Rauschecker et al., 1995). It appears as if this response tuning is generated by convergence of input from lower-order neurons that respond to simple sounds such as tones, frequency-modulated sweeps, or BPN bursts (Tian and Rauschecker, 2004; Rauschecker and Tian, 2004). As in other species (Suga, 1988; Margoliash, 1986), neurons are sensitive to highly specific combinations of such inputs, and combining signals in a nonlinear conjunctive AND-logic lead to the existence of neurons that respond specifically to certain types of calls. There is no reason to believe that human auditory cortex does not contain similar neurons with combination sensitivity and a similar hierarchy from rather simple to more complex neurons, whose incidence increases from primary auditory cortex to more anterior regions of the superior temporal lobe (Rauschecker, 1998; Rauschecker and Tian, 2000; Rauschecker and Scott, 2009).

Indeed, as mentioned earlier, early studies of human auditory cortex with fMRI have shown that primary auditory cortex responds best to tones, while at the next stage, the equivalent of the lateral belt in the monkey, BPN bursts are more effective stimuli (Wessinger et al., 2001). Further along the anteroventral pathway, cortical regions are selectively activated by words and intelligible speech sounds (Binder et al., 2000; Scott et al., 2000). This hierarchical organization of the auditory ventral stream with regard to speech-sound processing was later corroborated with more refined techniques (Chevillet et al., 2011). Furthermore, a meta-analysis of more than 100 neuroimaging studies of human speech processing has demonstrated that cortical regions close to primary core areas are sensitive to phonemes; farther afield in anterior superior temporal gurus (STG), words are processed; finally, in the most anterior locations of superior temporal sulcus, short phrases lead to selective activation (DeWitt and Rauschecker, 2012). The search for phoneme selectivity and phonetic feature representations in humans continues with direct recordings from the cortical surface (Mesgarani et al., 2014).

Invariant representation of sounds is another important step toward establishing a usable system for auditory communication, such as speech. There is evidence that invariances are formed along the anteroventral stream (DeWitt and Rauschecker, 2012). However, other reports have found that premotor regions may be involved as well (eg, Chevillet et al., 2013). It appears, therefore, as if invariances are formed in different ways: once on the basis of spectrotemporal information, which is pooled along the frequency domain as an OR-logic within the auditory ventral stream, and independently in the domain of motor gestures, which are formed originally in premotor cortex for speech production, but are invoked downstream (as inverse models) during the processing of speech as well (Rauschecker and Scott, 2009; Rauschecker, 2011). The same is almost inevitably true for the processing of other complex sounds that can be classified into discrete categories (Leaver and Rauschecker, 2010). Such auditory objects

are also represented in anterior regions of the STG, but premotor cortex participates in their encoding as long as they can be produced. Monkeys are naturally handicapped by their less sophisticated vocal apparatus, which limits their vocal repertoire and their capacity to mimic sounds. The involvement of the dorsal pathway (including premotor regions) in the processing and categorization of self-produced sounds will therefore have to be tested by other means (Remedios et al., 2009; Artchakov et al., 2012).

37.2.2 Involvement of the Auditory Dorsal Stream in the Processing of Sound Sequences

The role of the auditory dorsal pathway [caudal belt or posterior superior temporal cortex (STC) to dorsolateral prefrontal regions; Romanski et al., 1999] in the processing of space and motion, by analogy to the visual dorsal pathway, is well documented in monkeys (Rauschecker and Tian, 2000; Miller and Recanzone, 2009) and in humans (Griffiths et al., 1996; Baumgart et al., 1999; Warren et al., 2002; Deouell et al., 2007). This spatial role of the auditory dorsal stream has recently been generalized into one of sensorimotor integration and control (Rauschecker, 2011), which comprises the processing of not only space and motion but also sequences in space and time.

The involvement of the dorsal auditory pathway, including premotor and inferior parietal regions, in the encoding and representation of temporally extended sounds (or sound sequences) became especially evident, when imagery of musical melodies was investigated (Leaver et al., 2009). During the learning of such sequences, the basal ganglia were actively engaged, whereas after these sequences became highly familiar, the same sequences activated more and more prefrontal regions. It appears, therefore, that the basal ganglia are responsible for the concatenation of sequential auditory information or formation of "chunks," which represent information about conditional probabilities for one sound being followed by another. Once the chunks have been formed, they are stored in prefrontal regions. A similar chunking process seems to occur with cued sequences of learned finger movements and involves prefrontal cortex near Broca language region (Koechlin and Jubault, 2006), thus demonstrating that even higher-order "cognitive" functions of language may well have evolved from simpler sensorimotor functions in other species.

Thus, the role of the dorsal stream can be conceptualized into one of sensorimotor integration and control and applies to all kinds of sequential stimuli, even beyond the auditory domain. Specifically for speech, the auditory dorsal processing stream plays a role in speech production as well as categorization of phonemes during online processing of speech (Rauschecker and Scott, 2009; Rauschecker, 2011; Fig. 37.4). The former role conforms to the classical idea of an "efference copy" or feedforward model and allows for fast and efficient online control of speech production. By contrast, the latter function can be formalized as an inverse model during real-time speech processing, creating the affordances of the speech signal in a Gibsonian sense

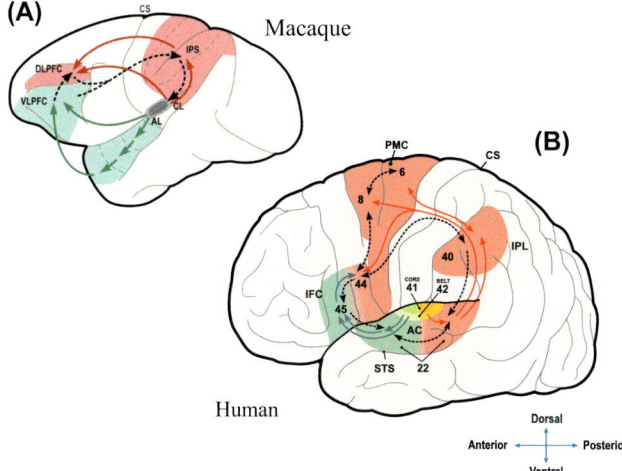

FIGURE 37.4 Ventral and dorsal streams for the processing of complex sounds in the primate brain: (A) in the rhesus monkey (modified from Rauschecker, J.P., Tian, B., 2000. Mechanisms and streams for processing of "what" and "where" in auditory cortex. Proc. Natl. Acad. Sci. U.S.A. 97 (22), 11800–11806); (B) in the human (modified from Rauschecker, J.P., Scott, S.K., 2009. Maps and streams in the auditory cortex: non-human primates illuminate human speech processing. Nat. Neurosci. 12 (6), 718–724; Rauschecker, J.P., 2011. An expanded role for the dorsal auditory pathway in sensorimotor integration and control. Hear. Res. 271, 16–25; Rauschecker, J.P., 2012. Ventral and dorsal streams in the evolution of speech and language. Front. Evol. Neurosci. 4, 7; Rauschecker, J.P., 2015a. Auditory and visual cortex of primates: a comparison of two sensory systems. Eur. J. Neurosci. 41 (5), 579–585; Rauschecker, J.P., 2015b. Evolution of human language. In: Wright, J.D. (Ed.), International Encyclopedia of the Social and Behavioral Sciences, second ed. Elsevier, pp. 323–328. The ventral stream (in *green*) plays a general role in auditory object recognition, including perception of vocalizations and speech. The dorsal stream (in *red*) pivots around inferior/posterior parietal cortex, where a quick sketch of sensory event information is compared with an efference copy of motor plans (*dashed lines*). Thus, the dorsal stream plays a general role in sensorimotor integration and control. In clockwise fashion, starting out from auditory cortex (AC), the processing loop performs as a forward model: object information, such as vocalizations and speech, is decoded in the anteroventral stream all the way to category-invariant inferior frontal cortex (IFC or VLPFC in monkeys) and transformed into articulatory representations (DLPFC or ventral PMC). Frontal activations are transmitted to the IPL and pST, where they are compared with auditory and other sensory information. *AL*, anterolateral area; *CL*, caudolateral area; *STS*, superior temporal sulcus; *DLPFC, VLPFC*, dorsolateral and ventrolateral prefrontal cortex; *PMC*, premotor cortex; *IPL*, inferior parietal lobule; *IPS*, inferior parietal sulcus; *CS*, central sulcus.

(Gibson, 1966; Rauschecker, 2005). Both functions require a (direct or indirect) connection between sensory and motor cortical structures of the brain, whereby subcortical structures (eg, the basal ganglia) provide an additional link setting up transitional probabilities during associative learning of sound sequences.

37.2.3 Brain Connectivity in Monkeys and Humans

Comparing human and monkey brain connectivity, there may be quantitative differences in the strengths of these connections, but not a difference in principle. Similarly, in the ventral stream, the fine-grain organization of cortical areas and the fine-tuning of its neuronal elements may be richer in humans than in monkeys, providing humans with a perceptual network for the detection of more subtle differences in the acoustic signal. The decisive distinction between humans and monkeys may, however, lie in a third component: the prefrontal network. With its own hierarchical organization, it provides the substrate for recursive processing of nested sequences, as they are typical for human syntactic language structures (Friederici, 2004; Bornkessel-Schlesewsky et al., 2015). Again, however, this emergent new ability of humans may be based on a quantitative rather than principal difference in human and monkey brain organization, which ties in the existing strengths of both ventral and dorsal processing streams with frontoparietal networks underlying working memory and the phonological loop (Baddeley, 1992).

To further test the evolutionary similarity of human and monkey ventral and dorsal streams, connectivity studies in both species will have to investigate in greater detail which areas or regions are connected. This will establish a greater amount of homology than other approaches, especially when the same techniques of structural and functional imaging are utilized. While anatomical tracer studies in monkeys will remain the gold standard (Romanski et al., 1999; Petrides and Pandya, 2009; Hackett, 2011), noninvasive fiber tractography using MRI-based technology will gain increasing importance as its resolution improves, because the same approach can then be used in both species. Early attempts using diffusion tensor imaging have had insufficient power to resolve crossing fibers within a single voxel or disentangle fibers with crossing trajectories (Anwander et al., 2007; Catani et al., 2005; Croxson et al., 2005; Rilling et al., 2008). Such studies have, therefore, remained inconclusive with regard to monkey–human homologies in language evolution. High-angular-resolution techniques, such as diffusion spectrum imaging (DSI), have been utilized successfully in humans (eg, Frey et al., 2008) and in monkeys

(Schmahmann et al., 2007). Cross-validation studies of autoradiographic tract tracing and DSI in monkeys have shown a remarkable concordance of results between tracer studies and DSI (Schmahmann et al., 2007). Recent improvements in resolution and reductions in scan time are of monkey diffusion MRI in vivo and ex vivo have resulted in remarkable progress: a direct comparison of macaque and human extreme capsule projections visualized using the same technique revealed surprising similarities across species despite earlier suggestions of a uniquely human extreme capsule projection (Mars et al., 2016a). Further comparison of connectivity fingerprints between cortical areas in the human and nonhuman primate brain can be used not only to identify between-species homologues, but also to illustrate differences between species (Mars et al., 2016b). Functional studies based on BOLD responses are feasible in both species as well (Petkov et al., 2006; Ortiz-Rios et al., 2015) and can elucidate connectivity to a certain extent. However, microstimulation techniques as another approach to analyze connectivity (Petkov et al., 2015) are limited to animal studies.

Acknowledgments

The collection of the underlying data as well as the writing of this chapter was supported by grants from the National Science Foundation (BCS-0519127, OISE-0730255), from the National Institutes of Health (R01NS052494, RC1DC010720), the Academy of Finland (FiDiPro), and the German Excellence Initiative and the European Union Seventh Framework Programme under grant agreement no. 291763 (IAS-TUM). Some of the ideas contained here have also been expressed in prior publications by the author.

References

Anwander, A., Tittgemeyer, M., von Cramon, D.Y., Friederici, A.D., Knösche, T.R., 2007. Connectivity-based parcellation of Broca's area. Cereb. Cortex 17 (4), 816—825.

Artchakov, D., Ortiz, M., Kusmierek, P., Cui, D., VanMeter, I., Jääskeläinen, I., Sams, M., Rauschecker, J.P., 2012. Representation of sound sequences in the auditory dorsal stream after sensorimotor learning in the rhesus monkey. Soc. Neurosci. Abstr. 368, 04.

Baddeley, A., 1992. Working memory. Science 255 (5044), 556—559.

Barton, B., Venezia, J.H., Saberi, K., Hickok, G., Brewer, A.A., 2012. Orthogonal acoustic dimensions define auditory field maps in human cortex. Proc. Natl. Acad. Sci. U.S.A. 109, 20738—20743.

Baumann, S., Griffiths, T.D., Sun, L., Petkov, C.I., Thiele, A., Rees, A., 2011. Orthogonal representation of sound dimensions in the primate midbrain. Nat. Neurosci. 14, 423—425.

Baumgart, F., Gaschler-Markefski, B., Woldorff, M.G., Heinze, H.-J., Scheich, H., 1999. A movement-sensitive area in auditory cortex. Nature 400, 724—726.

Bendor, D., Wang, X., 2005. The neuronal representation of pitch in primate auditory cortex. Nature 436 (7054), 1161—1165.

Binder, J.R., Frost, J.A., Hammeke, T.A., Bellgowan, P.S., Springer, J.A., Kaufman, J.N., Possing, E.T., 2000. Human temporal lobe activation by speech and nonspeech sounds. Cereb. Cortex 10, 512—528.

Bornkessel-Schlesewsky, I., Schlesewsky, M., Small, S.L., Rauschecker, J.P., 2015. Neurobiological roots of language in

primate audition: common computational properties. Trends Cogn. Sci. 19 (3), 142—150. https://doi.org/10.1016/j.tics.2014.12.008.

Catani, M., Jones, D.K., ffytche, D.H., 2005. Perisylvian language networks of the human brain. Ann. Neurol. 57, 8—16.

Celesia, G.G., 1976. Organization of auditory cortical areas in man. Brain 99, 403—414.

Chevillet, M., Riesenhuber, M., Rauschecker, J.P., 2011. Functional localization of the ventral auditory "what" stream hierarchy. J. Neurosci. 31 (25), 9345—9352.

Chevillet, M.A., Jiang, X., Rauschecker, J.P., Riesenhuber, M., 2013. Automatic phoneme category selectivity in the dorsal auditory stream. J. Neurosci. 33 (12), 5208—5215.

Croxson, P.L., Johansen-Berg, H., Behrens, T.E.J., Robson, M.D., Pinsk, M.A., Gross, C.G., Richter, W., Richter, M.C., Kastner, S., Rushworth, M.F.S., 2005. Quantitative investigation of connections of the prefrontal cortex in the human and macaque using probabilistic diffusion tractography. J. Neurosci. 25, 8854—8866.

Deouell, L.Y., Heller, A.S., Malach, R., D'Esposito, M., Knight, R.T., 2007. Cerebral responses to change in spatial location of unattended sounds. Neuron 55, 985—996.

DeWitt, I., Rauschecker, J.P., 2012. Phoneme and word recognition in the auditory ventral stream. Proc. Natl. Acad. Sci. U.S.A. 109 (8), E505—E514.

Formisano, E., et al., 2003. Mirror-symmetric tonotopic maps in human primary auditory cortex. Neuron 40, 859—869.

Frey, S., Campbell, J.S., Pike, G.B., Petrides, M., 2008. Dissociating the human language pathways with high angular resolution diffusion fiber tractography. J. Neurosci. 28, 11435—11444.

Friederici, A.D., 2004. Processing local transitions versus long-distance syntactic hierarchies. Trends Cogn. Sci. 8, 245—247.

Gibson, J.J., 1966. The Senses Considered as Perceptual Systems. Allen and Unwin, London.

Griffiths, T.D., et al., 1996. Evidence for a sound movement area in the human cerebral cortex. Nature 383, 425—427.

Hackett, T.A., Stepniewska, I., Kaas, J.H., 1998. Subdivisions of auditory cortex and ipsilateral cortical connections of the parabelt auditory cortex in macaque monkeys. J. Comp. Neurol. 394, 475—495.

Hackett, T.A., Stepniewska, I., Kaas, J.H., 1999. Prefrontal connections of the parabelt auditory cortex in macaque monkeys. Brain Res. 817, 45—58.

Hackett, T.A., Preuss, T.M., Kaas, J.H., 2001. Architectonic identification of the core region in auditory cortex of macaques, chimpanzees, and humans. J. Comp. Neurol. 441, 197—222.

Hackett, T.A., 2011. Information flow in the auditory cortical network. Hear. Res. 271 (1—2), 133—146.

Humphries, C., Liebenthal, E., Binder, J.R., 2010. Tonotopic organization of human auditory cortex. Neuroimage 50, 1202—1211.

Jones, E.G., Dell'Anna, M.E., Molinari, M., Rausell, E., Hashikawa, T., 1995. Subdivisions of macaque monkey auditory cortex revealed by calcium-binding protein immunoreactivity. J. Comp. Neurol. 362, 153—170.

Kaas, J.H., Hackett, T.A., 2000. Subdivisions of auditory cortex and processing streams in primates. Proc. Natl. Acad. Sci. U.S.A. 97 (22), 11793—11799.

Koechlin, E., Jubault, T., 2006. Broca's area and the hierarchical organization of human behavior. Neuron 50, 963—974.

Kusmierek, P., Rauschecker, J.P., 2009. Responses of rhesus monkey rostral and middle medial belt neurons to tones, noises, and natural sounds. J. Neurophysiol. 102 (3), 1606—1622.

Lauter, J.L., Herscovitch, P., Formby, C., Raichle, M.E., 1985. Tonotopic organization in human auditory cortex revealed by positron emission tomography. Hear. Res. 20 (3), 199—205.

Leaver, A.M., Rauschecker, J.P., 2010. Cortical representation of natural complex sounds: effects of acoustic features and auditory object category. J. Neurosci. 30 (22), 7604—7612.

Leaver, A.M., Rauschecker, J.P., 2016. Functional topography of human auditory cortex. J. Neurosci. 36 (4), 1416–1428.

Leaver, A.M., Van Lare, J.E., Zielinski, B.A., Halpern, A., Rauschecker, J.P., 2009. Brain activation during anticipation of sound sequences. J. Neurosci. 29 (8), 2477–2485.

Maldjian, J.A., Laurienti, P.J., Kraft, R.A., Burdette, J.H., 2003. An automated method for neuroanatomic and cytoarchitectonic atlas-based interrogation of fMRI data sets. Neuroimage 19 (3), 1233–1239.

Margoliash, D., 1986. Preference for autogenous song by auditory neurons in a song system nucleus of the white-crowned sparrow. J. Neurosci. 6, 1643–1661.

Mars, R.B., Foxley, S., Verhagen, L., Jbabdi, S., Sallet, J., Nonnan, M.P., Neubert, F.-X., Andersson, J.L., Croxson, P.L., Dunbar, R.I.M., Khrapitchev, A.A., Sibson, N.R., Miller, K.L., Rushworth, M.F.S., 2016a. The extreme capsule fiber complex in humans and macaque monkeys: a comparative diffusion MRI tractography study. Brain Struct. Funct. 221 (8), 4059–4071.

Mars, R.B., Verhagen, L., Gladwin, T.E., Neubert, F.X., Sallet, J., Rushworth, M.F., 2016b. Comparing brains by matching connectivity profiles. Neurosci. Biobehav. Rev. 60, 90–97.

Merzenich, M.M., Brugge, J.F., 1973. Representation of the cochlear partition of the superior temporal plane of the macaque monkey. Brain Res. 50 (2), 275–296.

Mesgarani, N., Cheung, C., Johnson, K., Chang, E.F., 2014. Phonetic feature encoding in human superior temporal gyrus. Science 343 (6174), 1006–1010.

Miller, L.M., Recanzone, G.H., 2009. Populations of auditory cortical neurons can accurately encode acoustic space across stimulus intensity. Proc. Natl. Acad. Sci. U.S.A. 106, 5931–5935.

Morel, A., Garraghty, P.E., Kaas, J.H., 1993. Tonotopic organization, architectonic fields, and connections of auditory cortex in macaque monkeys. J. Comp. Neurol. 335, 437–459.

Morosan, P., Rademacher, J., Schleicher, A., Amunts, K., Schormann, T., Zilles, K., 2001. Human primary auditory cortex: cytoarchitectonic subdivisions and mapping into a spatial reference system. Neuroimage 13, 684–701.

Norman-Haignere, S., Kanwisher, N., McDermott, J.H., 2013. Cortical pitch regions in humans respond primarily to resolved harmonics and are located in specific tonotopic regions of anterior auditory cortex. J. Neurosci. 33 (50), 19451–19469.

Ortiz-Rios, M., Kusmierek, P., DeWitt, I., Archakov, D.A., Azevedo, F.A.C., Sams, M., Jääskeläinen, I.P., Keliris, G.A., Rauschecker, J.P., 2015. Functional MRI of the vocalization-processing network in the macaque brain. Front. Neurosci. 9, 113. https://doi.org/10.3389/fnins.2015.00113 eCollection 2015.

Petkov, C.I., Kayser, C., Augath, M., Logothetis, N.K., 2006. Functional imaging reveals numerous fields in the monkey auditory cortex. PLoS Biol. 4, e215.

Petkov, C.I., Kikuchi, Y., Milne, A., Mishkin, M., Rauschecker, J.P., Logothetis, N.K., 2015. Different forms of effective connectivity in primate frontotemporal pathways. Nat. Commun. 6, 6000. https://doi.org/10.1038/ncomms7000 published online. January 23, 2015.

Petrides, M., Pandya, D.N., 2009. Distinct parietal and temporal pathways to the homologues of Broca's area in the monkey. PLoS Biol. 7 (8), e1000170. Epub 2009 Aug 11.

Rademacher, J., Bürgel, U., Geyer, S., Schormann, T., Schleicher, A., Freund, H.J., Zilles, K., 2001. Variability and asymmetry in the human precentral motor system. A cytoarchitectonic and myeloarchitectonic brain mapping study. Brain 124 (Pt 11), 2232–2258.

Rauschecker, J.P., 1998. Cortical processing of complex sounds. Curr. Opin. Neurobiol. 8, 516–521.

Rauschecker, J.P., 2005. Vocal gestures and auditory objects. Commentary on target article: from monkey-like action recognition to human language: an evolutionary framework for neurolinguistics by M. A. Arbib. Beh. Brain Sci. 28 (2), 143–144.

Rauschecker, J.P., 2011. An expanded role for the dorsal auditory pathway in sensorimotor integration and control. Hear. Res. 271, 16–25.

Rauschecker, J.P., Tian, B., 2000. Mechanisms and streams for processing of "what" and "where" in auditory cortex. Proc. Natl. Acad. Sci. U.S.A. 97 (22), 11800–11806.

Rauschecker, J.P., Scott, S.K., 2009. Maps and streams in the auditory cortex: non-human primates illuminate human speech processing. Nat. Neurosci. 12 (6), 718–724.

Rauschecker, J.P., Tian, B., Hauser, M., 1995. Processing of complex sounds in the macaque nonprimary auditory cortex. Science 268, 111–114.

Rauschecker, J.P., Tian, B., Pons, T., Mishkin, M., 1997. Serial and parallel processing in rhesus monkey auditory cortex. J. Comp. Neurol. 382, 89–103.

Rauschecker, J.P., Tian, B., 2004. Processing of band-passed noise in the lateral auditory belt cortex of the rhesus monkey. J. Neurophysiol. 91 (6), 2578–2589.

Rauschecker, J.P., 2012. Ventral and dorsal streams in the evolution of speech and language. Front. Evol. Neurosci. 4, 7.

Rauschecker, J.P., 2015. Auditory and visual cortex of primates: a comparison of two sensory systems. Eur. J. Neurosci. 41 (5), 579–585.

Rauschecker, J.P., 2015. Evolution of human language. In: Wright, J.D. (Ed.), International Encyclopedia of the Social and Behavioral Sciences, second ed. Elsevier, pp. 323–328.

Reale, R.A., Imig, T.J., 1980. Tonotopic organization in auditory cortex of the cat. J. Comp. Neurol. 192, 265–291.

Remedios, R., Logothetis, N.K., Kayser, C., 2009. Monkey drumming reveals common networks for perceiving vocal and nonvocal communication sounds. Proc. Natl. Acad. Sci. U.S.A. 106 (42), 18010–18015.

Rilling, J.K., Glasser, M.F., Preuss, T.M., Ma, X., Zhao, T., Hu, X., Behrens, T.E.J., 2008. The evolution of the arcuate fasciculus revealed with comparative DTI. Nat. Neurosci. 11, 426–428.

Rivier, F., Clarke, S., 1997. Cytochrome oxidase, acetylcholinesterase, and NADPH-diaphorase staining in human supratemporal and insular cortex: evidence for multiple auditory areas. Neuroimage 6, 288–304.

Romani, G.L., Williamson, S.J., Kaufman, L., Brenner, D., 1982. Characterization of the human auditory cortex by the neuromagnetic method. Exp. Brain Res. 47, 381–393.

Romanski, L.M., Tian, B., Fritz, J., Mishkin, M., Goldman-Rakic, P.S., Rauschecker, J.P., 1999. Dual streams of auditory afferents target multiple domains in the primate prefrontal cortex. Nat. Neurosci. 2, 1131–1136.

Saenz, M., Langers, D.R.M., 2014. Tonotopic mapping of human auditory cortex. Hear. Res. 307, 42–52.

Schmahmann, J.D., Pandya, D.N., Wang, R., Dai, G., D'Arceuil, H.E., de Crespigny, A.J., Wedeen, V.J., 2007. Association fibre pathways of the brain: parallel observations from diffusion spectrum imaging and autoradiography. Brain 130 (3), 630–653.

Schreiner, C.E., Langner, G., 1988. Periodicity coding in the inferior colliculus of the cat. II. Topographical organization. J. Neurophysiol. 60, 1823–1840.

Scott, S.K., Blank, C.C., Rosen, S., Wise, R.J., 2000. Identification of a pathway for intelligible speech in the left temporal lobe. Brain 123, 2400–2406.

Striem-Amit, E., Hertz, U., Amedi, A., 2011. Extensive cochleotopic mapping of human auditory cortical fields obtained with phase-encoding fMRI. PLoS One 6, e17832.

Suga, N., 1988. Auditory neuroethology and speech processing: complex-sound processing by combination-sensitive neurons. In: Edelman, G.M., Gall, W.E., Cowan, W.M. (Eds.), Auditory Function. Wiley, New York.

Talavage, T.M., Ledden, P.J., Benson, R.R., Rosen, B.R., Melcher, J.R., 2000. Frequency-dependent responses exhibited by multiple regions in human auditory cortex. Hear. Res. 150, 225–244.

Talavage, T.M., Sereno, M.I., Melcher, J.R., Ledden, P.J., Rosen, B.R., Dale, A.M., 2004. Tonotopic organization in human auditory cortex revealed by progressions of frequency sensitivity. J. Neurophysiol. 91, 1282–1296.

Tian, B., Reser, D., Durham, A., Kustov, A., Rauschecker, J.P., 2001. Functional specialization in rhesus monkey auditory cortex. Science 292, 290–293.

Tian, B., Rauschecker, J.P., 2004. Processing of frequency-modulated sounds in the lateral auditory belt cortex of the rhesus monkey. J. Neurophysiol. 92 (5), 2993–3013.

Tian, B., Kusmierek, P., Rauschecker, J.P., 2013. Analogues of simple and complex cells in the auditory cortex of the rhesus monkey. Proc. Natl. Acad. Sci. U.S.A. 110 (19), 7892–7897.

Tunturi, A.R., Barrett, T.W., 1977. Tonotopic pattern for single neurons in dog cortex, using elementary signals. Physiol. Chem. Phys. 9 (1), 81–84.

Warren, J.D., Zielinski, B.A., Green, G.G.R., Rauschecker, J.P., Griffiths, T.D., 2002. Perception of sound-source motion by the human brain. Neuron 34, 139–148.

Wessinger, C.M., Van Meter, J., Tian, B., Van Lare, J., Pekar, J., Rauschecker, J.P., 2001. Hierarchical organization of human auditory cortex revealed by functional magnetic resonance imaging. J. Cogn. Neurosci. 13 (1), 1–7.

38

Language Evolution

C. Boeckx

ICREA, Universitat de Barcelona, Barcelona, Spain

38.1 An Evolving Landscape

There has never been a better time, in my opinion, to work on the topic of language evolution. The field of "evolutionary linguistics," as one may call it, used to be plagued by lack of relevant data and therefore constituted a forum for all kinds of speculation. Such just-so stories were denounced most forcefully in a famous essay by Lewontin (1998). Unfortunately, they continue to this day. However, next to, and slowly supplanting this literature, a body of knowledge is emerging that holds great promise. It is a body of knowledge that is the epitome of interdisciplinarity, bringing together scholars from evolutionary biology, genetics, linguistics, cognitive neuroscience, and developmental psychology. This growing body of knowledge, which is the focus of this chapter, rests on several foundational assumptions:

1. Any evolutionary account of the human language phenotype will have to take into account both the species-specific biological endowment and the powerful structuring forces of culture and social interactions. The human language phenotype is too complex to capture by means of single-level, unidimensional, one-magic-bullet explanations.
2. Any account intended to appeal to biological mechanisms, including cultural accounts that of course must recognize the human biological commitment to culture, will have to center its efforts on the brain. There is simply no alternative to linking genotype and phenotype. Accordingly, cognitive descriptions will have to be linked to neural circuit mechanisms prior to being linked to genes.
3. Any approach aiming to characterize the evolutionary trajectory of human language will have to be Darwinian. This (hopefully) sounds like a no-brainer, but it is worth stressing that for an account to be properly Darwinian, it must at the very least take the notion of "descent with modification" seriously.

This means that it must strive to exploit as broad a comparative database as possible (in the case of humans, focus on the lessons one can derive from other animals) and avoid resorting to "special" explanations (be they in terms of special brain regions, mechanisms, etc.). No matter how modestly "linguistic" nonhuman creatures appear to be, the Darwinian logic of descent requires us to be able to discern an evolutionary path from there (them) to here (us). That is, ultimately, the human language phenotype must be rooted in nonstrictly human and nonstrictly linguistic cognition. "Only us" approaches just will not do. (To be clear: this is not to say that there is no specialization at the brain level. There is. But specialized regions follow principles of the surrounding brain circuits and work in an integrated fashion with less-specialized circuits.)

It seems to me that the field of evolutionary linguistics has matured enough now to demand that points (1)−(3) be present in any study falling under its scope. In particular, it simply will not do to state (as has all too frequently been the case until recently) that "we know so little about the brain," or that "some cognitive process (ultimately implemented in neural hardware) is encoded in our genes," or that "animal studies are not useful to capture the core human language phenotype because language is something unique to humans."

Of the three guidelines just listed, I take (2) to be the most important one: evolutionary linguistics is a brain initiative. As Marcus (2004) put it, genes do not build behavior or cognition. They code for and regulate the expression patterns of proteins that build neural building blocks that assemble into circuits, whose (environment-dependent) activities amount to processing information— that is to say, computing—that we can describe in cognitive terms and take to underlie behavior. There is no way around this long chain of inferential steps from genotypes to phenotypes. Each and every one of these steps is necessary and equally important, and moving from one level of description to the next requires a specific linking hypothesis. (For a very clear discussion of this issue with respect to the first gene causally related to some aspects of language, FOXP2, see Fisher, 2015.)

For several decades, the field of linguistics has been dominated by debates concerning "nature" or "nurture," until at some point advocates of the "rich and domain-specific biological endowment" camp (typically associated with work from the generative grammar tradition most closely connected to Noam Chomsky) realized that they were probably making too many demands on said biological endowment (demands that it could not possibly meet, especially given the assumption that human language evolved recently). They then began cutting down on the cognitive apparatus they relied on. This in turn led them to realize that this trimmed-down apparatus was incapable of capturing all the subtleties of linguistic knowledge. As a result, they began to delegate, explicitly or more often implicitly, some (indeed, much) of the mechanisms supporting adult linguistic knowledge onto the learning task and on the context in which learning takes place. In other words, the learning component came to play a much more important role. In so doing, they converged with advocates of the strong cultural component school (for more detailed discussion, see Boeckx, 2014). Interestingly, some of the strongest results of the latter school emerged in the context of evolutionary studies (see Kirby, 2013). Such studies, which formed the bulk of works presented at all the Evolang meetings (the major forum for studies on language evolution), demonstrate that generations of language learners, equipped with generic learning biases, shape the grammatical profile of the languages they learned. If this is true, then it would be mistaken to insist on building these aspects of grammatical knowledge into the initial substrate/biological endowment. Rather, these attractors populating the grammar space emerge from learning and communicative pressures. As a result, the focus, for the biologically inclined, should not only be on tracing down the origins of these generic biases in other animals, but also on our inclination to engaging in learning and communicating (verbally).

Following authors such as Michael Arbib (2002), I like to distinguish between the evolution of language and the evolution of the language-ready brain. The term "language-ready brain" not only draws attention to the centrality of the brain (cf. point (2) mentioned earlier), but also stresses the distance we ought to expect between what we will find at the neural level and what we can observe in the linguistic phenotype (adult linguistic knowledge).

The distinction between language and language-ready brain should not, I think, be conflated with the distinction between either of these terms and the notion of "protolanguage." The latter, exhaustively reviewed in Fitch (2010), refers to a stage (or multiple stages) in language evolution that is intermediate between what we can reasonably infer about the linguistic capacity of the last ancestor we shared with great apes from the behavior of living great apes and our species-specific linguistic capacity. Some scholars, such as Arbib (2002), take whatever degree of grammatical complexity that may exist between modern language and protolanguage to be a matter of culture. Others, such as myself, while recognizing the importance of cultural learning and transmission, still allow for significant changes at the level of the brain between a protolanguage user and a full-fledged, modern-language user. In other words, just as one distinguishes between language and protolanguage, it may be worth distinguishing between a protolanguage-ready brain and a language-ready brain. The difference between the two may just be a matter of degree of commitment to culture, but it would still be reflected in our biology, and need not be regarded as less significant than the step that led to the emergence of protolanguage.

The position just outlined appears to match the model insightfully sketched by Scott-Phillips and Kirby (2010). They observe that "[w]e can characterise the study of language evolution as being concerned with the emergence of language out of non-language. This involves two main processes of information transmission and change: a biological one (…) and cultural one (…). Prior to the existence of a culturally transmitted communication system, we can consider only the various preadaptations for language (e.g. vocal learning, conceptual structure [etc.]). Once cultural transmission is in place, then it might operate simultaneously with biological evolution in a co-evolutionary process and/or there might be cultural evolution alone [reference omitted]. In either case, we urgently need a better general understanding of how cultural transmission and social coordination shape language if we are to achieve a complete picture of the evolution of language. Once language has emerged, further changes can and do occur. This is the domain of language change and historical linguistics." What this passage makes clear is that the distinction between "phylogeny" (evolution of language; genetic evolution) and "glossogeny" (evolution of languages; cultural evolution), introduced by Hurford (1990), is not as clear-cut as it may be, given the recognized role of culture in adding complexity and structuring the linguistic phenotype.

The passage quoted from Scott-Phillips and Kirby (2010) also allows me to address another, related point.

They talk about "the emergence of language out of nonlanguage," but just to be clear: the Darwinian logic of descent forces us to recognize a continuum, at some level, between these two states. If, as is generally recognized, we must ground aspects of our linguistic capacity into the brains of nonlinguistic creatures, calling these creatures nonlinguistic is a bit misleading. This is exactly the issue that Hauser et al. (2002) addressed in a widely cited article. In an attempt to capture both the species-specific character of language and the roots of aspects of our language faculty in other species, they distinguished between the faculty of language in the narrow sense (FLN, corresponding to what makes us unique) and the faculty of language in the broad sense (FLB). In this way, nonhuman animals could be called broadly linguistic creatures, but not narrowly linguistic creatures, a term that should be reserved to us.

Unfortunately, instead of focusing on the fact that this distinction drew attention to the roots of aspects of language in other species and called for renewed appreciation for the promises of comparative cognition, many scholars, especially linguists, used the FLN-FLB distinction to stress what made humans different. In other words, instead of focusing on FLB, they focused on FLN. Specifically, they tried to provide content to FLN (something which, incidentally, Hauser, Chomsky, and Fitch invited by hypothesizing that FLN consisted of "recursion and its mapping to the sensorimotor and conceptual interfaces") and looked for some unique traits. In so doing, they not only departed from the Darwinian logic of descent but also departed from the consensus in current biology that "phenotypic novelty is largely reorganizational rather than a product of innovative genes" (West-Eberhard, 2003). It seems to me that statements such as West-Eberhard's demand that the idea that what makes human language unique is the way it organizes or collects attributes that individually can be rooted in nonhuman cognition be taken as the default hypothesis. Of course, once collected under a single roof ("language-ready brain"), these traits may give rise to nonlinear, "emergent" effects. Likewise, as Fujita (2016) has stressed, when placed in the context of the human brain, "old" pieces may acquire new roles that transform their nature (the sort of feedback loop familiar in biology). So, in the end, as I already expressed in Boeckx (2013), the FLN-FLB distinction does not strike me as the most felicitous way of capturing the continuum that the Darwinian logic of descent demands between our linguistic capacity and nonhuman cognition. I prefer the perspective advocated in Petkov and Jarvis (2012) and Arriaga and Jarvis (2013).

As a final note on this continuum, let me point out that continuum need not mean "scale." Rather, following François Jacob's well-known tinkering metaphor, we should think of our language-ready brain as a mosaic or patchwork, composed of parts (possibly, multiple protolanguage components) that do not fit neatly a cognitive *scala naturae* but rather display deep homologies with capacities found in a variety of species. Of course, our language-ready brain evolved from a primate brain, but the latent potential of subtle tinkering events, as I will show later, need not always be best understood by examining primate behavior and studying primate cognition (for more on this way of understanding cognitive phylogenies, see Theofanopoulou and Boeckx, 2015).

38.2 Deep Homology

Until recently most language-oriented books on comparative cognition just had one message: animals do not have the cognitive capacities necessary to acquire language (see Anderson, 2004 for a particularly clear example of this). But the tides have changed. De Waal and Ferrari (2010) put it best: "Over the last few decades, comparative cognitive research has focused on the pinnacles of mental evolution, asking all-or-nothing questions such as which animals (if any) possess a theory of mind, culture, linguistic abilities, future planning, and so on. Research programs adopting this top-down perspective have often pitted one taxon against another, resulting in sharp dividing lines. Insight into the underlying mechanisms has lagged behind … A dramatic change in focus now seems to be under way, however, with increased appreciation that the basic building blocks of cognition might be shared across a wide range of species. We argue that this bottom-up perspective, which focuses on the constituent capacities underlying larger cognitive phenomena, is more in line with both neuroscience and evolutionary biology."

In the domain of language this bottom-up perspective is best illustrated in the context of birdsong studies. There is now an extensive literature documenting parallels between birdsong and aspects of human language (not just speech, as I will discuss later) at various levels of description: development, neural basis, and evolution (for a recent state-of-the-art survey, see Bolhuis and Everaert, 2013). Both birdsong and human language are vivid illustrations of what Peter Marler (2004) called the "instinct to learn." Both show strong evidence for the need to have a brain configured in a particular way to engage in vocal learning (Jarvis, 2004) and at the same time highlight the importance of the environment in shaping the structure of the songs. Thus, like humans, songbirds raised in isolation, without any conspecific adult models during the critical period for song learning, never recover. They never move much beyond the babbling stage called "subsong." But Feher et al. (2009) showed that in a colony of songbirds founded by an isolate, normal song structure reemerges over the course of a few generations, in a way strongly reminiscent of what happens with creole languages.

Like language users, then, birds demonstrate that possessing a song-ready brain is not enough. Rather, learning, taking place over several generations, is required for the song to develop completely. The birdsong literature also demonstrates that specific environmental circumstances, such as domestication (crucially, not targeting song structure per se), can lead to a complexification of the song phenotype (Okanoya, 2012), raising the possibility that self-domestication processes often mentioned in the context of recent human evolution (see Thomas, 2014; Benitez-Burraco et al., 2016) may have also played a role in structuring the human language phenotype, an issue I will return to later.

Arguably the biggest advance coming from the birdsong literature concerns the neurobiological basis of vocal learning. The literature of the past 20 years has accumulated robust findings concerning the brain pathways necessary for vocal learning and the molecular underpinnings of said pathways. Vocal learning abilities in birds depend on two main pathways: a nidopallial-striatal-thalamic loop responsible for the acquisition of new vocalizations, known as the anterior pathway, and a posterior pathway directly connecting the arcopallium and the syrinx, which controls intentional vocal production. Translating this to humans, a corticobasal ganglia-thalamic loop, along with a direct connection between the motor cortex and the larynx, appears to be the required neurological substrate for vocal learning (see Jarvis, 2004 and much subsequent work since). Analyzing the rudimentary substrates for vocal learning found in a non–vocal-learning suboscine, the eastern phoebe, Liu et al. (2013) suggest that the posterior pathway is probably the first step in achieving a vocal-learning readiness.

More remarkably, in addition to uncovering these neural pathways, birdsong scholars were capable of canalizing the genomic revolution to uncover deep homologies at the molecular level, so much so that comparisons of brain transcriptomes of song-learning birds and humans relative to vocal nonlearners identified convergent gene expression specializations in specific song and speech brain regions of avian vocal learners and humans (Pfenning et al., 2014). In particular, the forebrain part of the vocal-learning circuit that makes a robust direct connection to brain stem vocal motor neurons in independent lineages of vocal-learning birds (songbird, parrot, and hummingbird) was shown to have specialized regulation of axon guidance genes belonging to the SLIT-ROBO molecular pathway (Wang et al., 2015). Wang et al. showed that, unlike in mice and non–vocal-learning birds, in vocal-learning birds, SLIT1

was differentially downregulated in the motor song output nucleus that provides a direct link between the bird equivalent of the motor cortex and the syrinx, whereas SLIT1's receptor ROBO1 was developmentally upregulated during critical periods for vocal learning. Interestingly, the SLIT-ROBO pathway has been associated with a range of disorders in humans where language is affected (see references in Boeckx and Benítez-Burraco, 2014b). In addition, SLIT1 is a direct target of FOXP2 (Vernes et al., 2007; Konopka et al., 2009). Thus, Wang et al.'s finding is directly related to the remarkable literature on FOXP2 and its role in language that was built on Lai et al.'s (2001) landmark association between a mutation affecting this transcription factor and a development language disorder.

This is not the place to provide an extensive review of the FOXP2 literature (for excellent recent overviews, see Fisher, 2016; Fisher and Vernes, 2015; Graham and Fisher, 2015). Suffice it to say that this literature has provided a unique point of entry into the molecular basis of the neural implementation of aspects of the human language capacity. Especially worthy of note are (1) discoveries pertaining to the neural function of FOXP2; (2) the strong expression profile of FOXP2 in subcorticular circuits; and (3) the recent evolutionary changes affecting FOXP2. Let me touch on each of these briefly.

Following Lai et al.'s discovery that heterozygous mutations of the human FOXP2 gene cause a monogenic speech and language disorder (reviewed in the overviews cited earlier), intensive work has uncovered that reduced functional dosage of the mouse version (Foxp2) causes deficient corticostriatal synaptic plasticity and impairs motor-skill learning and that the songbird orthologue appears to be critically important for vocal learning. Significant research has been made in delineating the gene's interactome, which has led to results such as Vernes et al. (2011), where Foxp2 was shown to regulate gene networks implicated in neurite outgrowth in the developing brain by directly and indirectly regulating mRNAs involved in the development and plasticity of neuronal connections. In another study, Tsui et al. (2013) concluded that FOXP2 regulates genesis of some intermediate progenitors and neurons in the mammalian cortex and suggested that the evolution of the gene may be associated with the expansion of the human cortex.

One of the most robust findings in the FOXP2 literature concerns the gene's strong expression in subcortical structures such as the basal ganglia, the thalamus, and the cerebellum (Vargha-Khadem et al., 2005; Haesler et al., 2004; Teramitsu et al., 2004). This has led to the renewed interest in these structures for language and has led to the reappraisal of models that attribute a key role, both ontogenetically and phylogenetically, to subcortical structures (eg, Jarvis, 2004; Lieberman, 2006; but see also Boeckx, 2013; Boeckx and Benítez-Burraco, 2014a,b; Barton, 2012; among others).

In terms of impact, the evolutionary history of FOXP2 cannot be matched. FOXP2 is a highly conserved gene, with only two amino acids separating the human version from the chimpanzee version, and only three amino acids separating the human version from the mouse version. Needless to say, the discovery of two recent changes in the otherwise remarkable stable history of the gene has led to a lot of speculation concerning language evolution (speculation unfortunately rarely accompanied by detailed linking hypotheses between these mutations and neurobiology). The discovery that the Neanderthal genome contained the two key mutations found in modern humans (Krause et al., 2007) led to the claim that language has a deeper history than most scholars claimed (for the most compelling case in favor of this position, see Dediu and Levinson, 2013). Although hard to prove at this point, independent data suggest that this position is worth taking very seriously (although it is also important to bear in mind the evidence pointing to species-specific regulation of the gene; cf. Maricic et al., 2013). The SLIT-ROBO GTPase activating protein 2 (SRGAP2) is duplicated three times in the human genome compared to the chimpanzee genome, and one of the partial duplicated copies partially suppresses SLIT protein activity, thereby causing slower forebrain dendritic pruning, leaving more and longer dendrites (Charrier et al., 2012; Dennis et al., 2012). Crucially, the relevant duplications are not only specific to our species but are also part of the Neanderthal genome. Given that the SLIT-ROBO pathway appears to be significant in the context of vocal learning, and that the SRGAP2 duplication is not unique to the genome of anatomically modern humans, it is tempting to claim that some of our extinct ancestors were, if not equipped with all aspects of the modern language faculty, at least capable of complex vocal learning.

Independent evidence for this position may come once again from the birdsong literature. Liu et al. (2013) point out in their analysis that "vocal learners share relatively small body size, which may allow these birds (and their ancestors) to better manoeuvre flight and create more ecological niches for foraging (nectar feeding, flying-insect catching) and aerial vocal display. Such elaborate flight manoeuvreing may require a better coordination or reconfiguration of respiratory control from the forebrain. The forebrain respiratory control may subsequently integrate pre-existing motor pathway in the arcopallium for the control of flight, jaw, and vocal movement, and/or auditory relays, and lead to evolution of vocal learning (see a similar view proposed by Janik and Slater (1997) for vocal learning in mammals)." Although, as Liu et al. point out, this "respiratory

control" hypothesis is highly speculative, it is worth bearing in mind, in line with Levinson (2016), that the African variety of *Homo erectus* (*c*.1.6 My) appears to have lacked the breathing control necessary for modern speech. By contrast, the last common ancestor between anatomically modern humans and Neanderthals appears to have had all the physiological prerequisites. Levinson suggests, reasonably, that by that time communication was predominantly vocal, as opposed to gestural.

This reasoning is to be placed in the context of the motor theory of vocal learning put forth by Feenders et al. (2008). This theory states that the "cerebral brain pathways for vocal learning in distantly related animals evolved independently as specializations of a preexisting motor system inherited from their common ancestors" and is based on gene expression in very closely located brain circuits responsible for activities such as hopping and signing. Plausibly, as discussed by Chakraborty and Jarvis (2015), the pathway critical for vocal learning may have evolved by duplication and then divergence from the motor pathways regulating nonvocal motor behaviors.

Chakraborty and Jarvis (2015) have also argued for another duplication event, responsible for the type of advanced vocal imitation found in parrots. By means of baseline gene expression, singing-driven gene expression, and neural connectivity tracing experiments, Chakraborty et al. (2015) demonstrated that the parrot brain contains what looks like a song system within another song system. The inner song system, which they call "core," appears similar to other vocal-learning birds, whereas the outer system, called "shell," appears unique. Chakraborty et al. propose that it is this shell pathway that is responsible for their imitative abilities compared to other vocal-learning bird lineages. As Chakraborty et al. observe, the connectivity of the shells had some significant differences to the core nuclei. Also, the size of some of the shells, contrary to what happens in the case of cores, seems to scale allometrically with brain size and positively correlates with both vocal and cognitive complexity in different parrot species. I find the latter finding of great interest, as Deacon (1989, 1997) suggested that one of the factors allowing for the establishment of the vocal learning circuit is encephalization. Deacon points out that when space is sparse, the innate, noncortically controlled vocalization circuit outcompetes the cortical connections making intentional vocalization possible. But as brains expand, space is created for both systems to be maintained. The vocal-learning literature in birds does not seem to support a strong version of this hypothesis, as vocal learning does not seem to require big brains, although there are two caveats to be made in this context. First, encephalization appears to be a factor for shell systems, and it is precisely in those birds with shell systems that we see a richer semantics associated with vocal output. Second, although absolute size may not be a factor, neuron number may be. That is, it has recently been shown by Olkowicz et al. (2016) that bird brains pack more neurons in the same brain space as mammals, and vocal-learning birds have an even higher packing density than other bird species and twice as many neurons as some nonhuman primate brains bigger than theirs.

In this context, it is also worth pointing out, following Hillert (2015), that the timing of the duplication events (Dennis et al., 2012) of the SRGAP2 gene mentioned earlier coincides with landmarks of neocortical expansion in the transition from *Australopithecus* to *Homo*. This suggests to me that a better understanding of the functional consequences of the peculiar evolution of SRGAP2 may bear directly on the vocal-learning system we possess(ed).

Be that as it may, Chakraborty et al.'s work highlights the relevance of complexity in vocal learning that complements the notion of complexity that emerged from, eg, the work of Okanoya (2012) on the complexification of songs in the context of domestication. [I should note that there may be different kinds of complexity, as discussed in Deacon (2006), just like there are different degrees of novelty/complexification in an organism; cf. Müller (2010).] There appear to be (possibly not unrelated) environmental and neural factors contributing to the range of vocal-learning capacities. In fact, Petkov and Jarvis (2012) and Arriaga and Jarvis (2013) propose that vocal-learning abilities are distributed along a continuum. Thus, instead of thinking of vocal learning as a relative rare trait found only in three distantly related groups of mammals (humans, bats, and cetaceans) and three groups of birds (parrots, hummingbirds, and songbirds), we may conceive of vocal learning as more widespread. According to this view, the degree of voluntary control that an animal has over its vocalizations is primarily determined by the robustness of the cortical—laryngeal (or equivalent) pathway, very weak in mice and nonhuman primates, but very strong in canonical vocal learners. This continuum hypothesis makes it possible to study a wider range of species to reveal the neurobiological substrates of vocal learning and parallels attempts to decompose the human language faculty into more basic building blocks that can be studied in a wide range of species (see, eg, Samuels, 2011 for a detailed attempt in the context of phonology).

The ultimate lesson I want to highlight from the literature on vocal learning in birds (which I suspect will soon be enriched by work on bats; see Rodenas-Cuadrado et al., 2015) is that solid progress and formulation of testable hypotheses in understanding the evolution of aspects of language must go hand in hand with a very precise characterization of the neural basis of these cognitive/behavioral capacities.

38.3 Primate Ancestry

Whereas the previous section provided a survey of substantial results building on more than a decade of sustained work on the neurobiology of vocal learning, this section will focus on an equally important, but less well understood, topic in the domain of language evolution: the cognitive abilities we inherited from our primate ancestry.

Until recently, researchers disregarded nonhuman primates in the context of evolutionary linguistics because unlike vocal-learning birds, the imitative abilities of our closest living relatives, especially their vocal imitative abilities, were far from remarkable. In addition, experimenting with nonhuman primates is far more challenging when it comes to neurobiological issues than it is with birds or mice. But as I hope to convey in this section, we can still gather valuable information concerning our primate ancestry; in fact, in the few cases where investigating the neural basis of their behavior, primates have been shown to shed light on the evolution of human language, so much so that I anticipate significant progress in this area in the near future.

There is very good evidence of a rich cognitive life among our primate relatives. Works such as Cheney and Seyfarth (1992, 2008), Hauser (2001), and, especially Hurford (2007) have documented a wide range of behaviors that strike me as providing solid grounds for trusting in Darwinian descent when it comes to "semantics." This is not to say, of course, that there was no innovation in this domain when full-blown language emerged, but in the domain of semantics, and in fact language as a whole, the gap between "them" and "us" has not infrequently been exaggerated by the fact that our own behavior was described in extremely rich, domain-specific computational terms, but not so for our relatives. When less "intellectualized" descriptions of our behavior are provided, the gap is considerably reduced, and the plausibility of descent reinforced. To my mind the clearest demonstration of the statement just made comes from the work of Moore (2016a,b) on intentional communication and its cognitive prerequisites. Often described in uniquely human terms (semantics/pragmatics), it turns out that more "minimalist/elementary" descriptions can be offered and shown to be adequate.

The reason I will not discuss semantics/pragmatics much here is because even for humans, the neurobiological basis of these abilities is not well understood. It is often said that semantics remains the "last frontier" for both neurolinguistics and evolutionary linguistics, but

I think that it is especially in the context of neurolinguistics that insight is lacking. Evolutionary speaking, one can be fairly confident that a rich conceptual basis, "waiting" to be modified, was in place millions of years ago. Hopefully, the growing appreciation for the cognitive sophistication of parrots and corvid songbirds (Emery and Clayton, 2004) will also provide additional insights into this area, following the logic of deep homology discussed in the previous section.

The question I will focus on in the rest of this section is, how did this rich conceptual structure we have every reason to believe was present in our last common ancestor with great apes, got to be paired with a vocal-learning ability? Put another way, how did we come to take volitional control of our vocalizations?

It is not the case that the auditory processing system of primates is poor. In fact, Bornkessel-Schlesewsky et al. (2015) and Wilson et al. (2015) have revealed substantial homology in this domain, with both a ventral and dorsal pathway linking up the frontal and temporal cortex to organize auditory perception at various scales. Rather, the issue pertains to vocal imitation.

To address these questions, recent work has focused on several areas and primate behaviors. One has been to show that primate calls, though not learned, have a richer internal structure than one may have suspected, with some scholars seeing the seeds of compositional semantics in them (Arnold and Zuberbühler, 2008 et seq.; work which incidentally led to a reexamination of combinatorial meaning in birds' vocal output; Engesser et al., 2016).

Another line of research has focused on the ways in which primates get around the poor control of their vocal apparatus. Consider, for example, the way in which orangutans achieve the desired modification of their calls by manipulating leaves (Hardus et al., 2009; Lameira et al., 2012): by positioning a hand or holding leaves in front of their lips, orangutans lower the maximum frequency but maintain other parameters of the call similar. This is a bit like playing music with a wind instrument. More interestingly still, when orangutans modify their calls in this way, they sound as if their body size is bigger than it actually is, which has obvious functional consequences. Since size exaggeration has also been claimed to be at the heart of the evolution of a descended larynx (Fitch, 2000; Fitch and Reby, 2001), it is tempting to see these modified calls by orangutans as a solution forced upon them by the lack of laryngeal control.

Yet another line of research has identified sophisticated vocal behaviors in primates such as the elaborate songs in gibbons (Koda et al., 2012 and references therein). These have been shown to engage in duets and perform sophisticated movements of the vocal apparatus (very reminiscent of human soprano singers) to propagate their long-distance vocalizations.

These cases aside, studies on primate communication have focused on the fact that primates may be poor vocal learners but far better gesture users. There is a rich literature showing the complexity of primate gestures (see, eg, Roberts et al., 2014), with very clear evidence of intentionality in them (Genty and Zuberbühler, 2014, 2015). This has often been used as evidence for the primacy of a gestural protolanguage (see, eg, Arbib, 2002) and claims that human languages are not always a matter of sounds, but signs (sign languages). In fact, the very term "protolanguage" was first used by Hewes (1973) with a gestural protolanguage in mind. But it seems to me that the question is not so much whether sounds or signs were first, but how sounds and signs were paired to achieve a multimodal signal like human language. This is so because (1) if we set aside sign language for a moment, it is clear that sounds and signs are paired very early in ontogeny in humans (Esteve-Gibert and Prieto, 2014) and are always coupled in adults (speakers gesturing and deaf signers vocalizing) and (2) the motor theory of vocal learning put forth by Jarvis et al. for birds, which I discussed in the previous section, suggests that gestures may well have provided a first evolutionary step, but the crucial leaning step came afterward [inspired by the work on vocal-learning birds, Fitch (2011) suggests that the direct corticolaryngeal connection key to vocal learning exapted from the corticospinal pathway for nonvocal motor production]. One has to bear in mind that speech does not go away as a linguistic medium, except in extreme situations.

It is for these reasons that I find particularly illuminating studies that identify multimodal signals in nonhuman primates. I am certainly not the first to call for a multimodal study of nonhuman primate communication (see Waller et al., 2013; Zuberbühler, 2015), but it is quite likely that many previous studies missed valuable information by focusing on one or the other modality. Recent work (Liebal et al., 2004; Micheletta et al., 2013; Genty et al., 2015; Taglialatela et al., 2015) has tried to correct this bias and has suggested that the activation of the homologous Broca area in chimpanzees in both attention calls and gestures points to a multimodal origin of language (Taglialatela et al., 2011).

As a matter of fact, multimodal communication appears deeply rooted in evolutionary history. Bass and Chagnaud (2012) identified shared developmental and evolutionary origins of neural basis of vocal-acoustic and pectoral-gestural signaling present in vocalizing fishes. In a similar vein, Lawton et al. (2014) has revealed a conserved role of *Drosophila melanogaster* FoxP

(FOXP2's homolog) in motor coordination and courtship song. Male flies with reduced FoxP expression exhibit decreased levels of courtship behavior, altered pulse-song structure, and sex-specific motor impairments in walking and flight. Multimodal communication has also been found in birds (Pika and Bugnyar, 2011; Bostwick and Prum, 2005; Hoepfner and Goller, 2013; Williams, 2001; Soma and Mori, 2015), where wing movement, beak movement, and even feet movement accompany song in a synchronized fashion.

All in all, it seems to me that the study of multimodal communication holds great promise for the study of human language evolving, breaking free of the somewhat sterile debate between musical and gestural protolanguage (incidentally, song and dance might be just another instance of multimodal communication whose evolutionary history may be worth elucidating).

Among the instances of multimodal communication in primates, there is one that I wish to highlight: lip smacking. Lip smacking is an affiliative facial expression observed in many primates, and it has been the focus of intense and very productive research recently (Ghazanfar et al., 2012). The reason researchers were drawn to it is because although it is not "vocal" (with at least one exception to which I return later), it constitutes a case of "silent" vocalization and provides one of the best precursors for aspects of human speech. Lip smacking is made up of regular cycles of vertical jaw movement and is clearly directed toward a conspecific. During the lip-smacking act, the lips, tongue, and hyoid have been shown to move at 4–5 Hz (Ghazanfar et al., 2012), which is exactly within the range of the universal rhythm of speech and distinct from the rhythm of chewing movements in both monkeys and humans. In addition, (aspects of) speech and lip smacking appear to be dependent on homologous cortical circuits (unlike innate calls, they are cortically controlled) (Ghazanfar et al., 2010) and display a remarkably similar developmental trajectory (Ghazanfar et al., 2013; Ghazanfar and Takahashi, 2014). All these parallelisms suggest that aspects of human speech, most specifically the syllabic envelope, evolved from the rhythmic facial expressions of a common ancestor to both humans and macaques (Ghazanfar and Poeppel, 2014a; MacNeilage, 2010). In my view, the lip-smacking studies by Ghazanfar and colleagues offer us the strongest parallelism with human speech in the nonhuman primate domain, only surpassed by work on vocal-learning birds.

Remarkably, gelada baboons have been shown to pair lip smacks and vocalizations, in a way even more similar to human syllables (Bergman, 2013; Richman, 1976). This is not to say, of course that these "syllables" uttered by geladas are identical to syllables produced by humans (on these differences, see Martins and Boeckx, 2014), but it reinforces the idea that some aspects of human vocal behavior can be found in nonhuman primates, exactly as the Darwinian logic of descent leads us to expect.

Pursuing this hypothesis, Lameira et al. (2015) have looked at other possible precursors of aspects of human speech in great apes, beyond lip smacks, and have documented orangutan calls at a speechlike rhythm, coined "clicks" and "faux speech." They suggest that "like voiceless consonants, clicks required no vocal fold action, but did involve independent manoeuvring over lips and tongue." And "in parallel to vowels, faux speech showed harmonic and formant modulations, implying vocal fold and supralaryngeal action." They conclude that great apes may be less respiratorily, articulatorily, or neurologically constrained for the production of consonant- and vowel-like calls at speech rhythm than previously thought. Certainly, as Lameira et al. (2014) point out, there seems to be growing evidence for evolutionary continuity within the great apes as far the control of the supralaryngeal vocal tract. Thus, they argue, control of the vocal folds (that key aspect of vocal learning we understand well from vocal-learning birds) evolved as a subsequent step (see Levinson, 2016 for a similar historical reconstruction based on a comparative study of turn taking).

In a related line of research, Pisanski et al. (2016) offer an impressive review arguing that "the ability of humans to flexibly control the size-related source–filter dimensions of our vocal signals (i.e., vocal control) is likely to predate our ability to articulate the verbal dimensions of speech and therefore may provide an evolutionary pathway from nonhuman primate vocal communication to human speech." Pisanski et al. present a range of evidence, most of it from the past 5 years or so, that nonhuman primates may share our capacity to modulate F0 and formants to perhaps a greater extent than previously thought. As an example, they mention cases of chimpanzees producing "novel and apparently flexible attention-seeking grunts toward humans." They point out that although it does not occur in the wild, this type of vocal behavior demonstrates "a latent capacity to control vocal fold adduction and airflow that is required to produce sustained laryngeal vibration." Thus, among nonhuman primates, manipulation of the larynx and vocal tract may be more flexible than once thought. I personally anticipate more evidence in favor of this position in the near future, especially from bonobos, where evidence for functional flexibility is accumulating rapidly (Clay et al., 2015). Pisanski et al. conclude their review by pointing out that "vocal flexibility in nonhuman primates suggests that other species have greater neuroanatomical elaboration of the direct lateral motor cortical route than previously thought or, alternatively, may be achieving flexibility with older neural structures." This I take to be a crucial issue in the domain of the evolution of the neurobiology of language that hopefully will be the focus of intense research in the years to come.

To conclude this brief survey of our primate ancestry when it comes to vocal behavior, I would like to highlight current work on marmoset and the evolution of turn taking. Levinson (2016) has made a strong case for it being a crucial trait of human verbal communication, one that can be found among our closest living relatives, the bonobos and chimpanzees (Fröhlich et al., 2016). As a highly vocal primate, marmosets offer a rich source of data concerning turn taking. In fact, Takahashi et al. (2013) argue that the turn-taking behavior of marmosets is so similar to what is found in our behavior that this is yet another piece of evidence that our primate ancestry may have been rich enough to provide a foundation from which human cooperative vocal (and not only gestural) communication could have evolved.

Takahashi et al. (2016) investigated the ontogeny of this behavior in marmosets and were able to show how engaging in vocal turn taking with parents at a very early age shapes vocal acoustics in infant marmoset monkeys. Reviewing a related study by Zhang and Ghazanfar (2016), Tchernichovski and Oller (2016) offer a very interesting scenario for the evolution of vocal communication. They suggest that, at first, slow fluctuations in respiratory rate are mirrored in the rate of vocalization. As the channel evolves, according to them, "additional physiological variables start affecting acoustic features until, eventually, fine forebrain control drives differentiation of syllable type, producing vocal sequences, which may carry rich information about behavioral state."

I find this hypothesis interesting, not because it suggests that the cortex may have taken control over emotional vocalizations, but because it means that emotional vocalizations controlled by the limbic system in nonvocal learners, including mice, could provide a rich source of aspects of human linguistic behavior (see Theofanopoulou, 2016 for interesting suggestions in this direction).

As should be obvious, the neurobiology of nonhuman primates (with the possible exception of lip smacking) is nowhere nearly as precisely understood as the neural (and molecular) underpinning of vocal learning in birds, but, as I hope to have shown, recent work has taken the crucial step of showing that nonhuman primates can provide a rich evolutionary substrate for human language, which is an important departure from what was received wisdom only a few years ago.

38.4 Tinkering With Our Inheritance

Up until now I have been keen to stress descent, because it has been minimized for too long, and it offers genuine results and promises for the field as a whole. But the Darwinian logic of descent comes jointly with that of "modification," and this is the issue I want to address in this section. What got modified in our recent history that matters for the evolution of our linguistic capacity? Even granting, as I think we must, the real possibility that our ancestors became vocal learners long ago, were there further modification steps that took place more recently in our lineage that may be cognitively relevant?

Recall from the discussion in Section 38.2 that the vocal learning literature, especially that focused on birds, has identified at least two ways in which vocal-learning capacities can be enhanced. One is via domestication, the other via the elaboration of shell structures on top of the core vocal-learning circuit, which seems to go hand in hand with brain growth.

I believe that both processes were critically involved in the emergence of our species. In particular, the fossil record, enriched by discoveries made possible by the successful retrieval of ancient DNA (Pääbo, 2014), suggests that the brain growth trajectory in our species departed significantly from that our extinct ancestors, early in life. There is every reason to believe that although infants do not produce grammatically complex expressions at that time, the first months after birth constitute a critical period for subsequent language (and indeed, cognitive) development. Although much work that adopted the concept of critical period for language development in the wake of Lenneberg (1967) usually refers to a period from birth to puberty, a closer examination reveals that there are, in fact, several critical periods for different aspects of language development, and many of these periods are much shorter than anticipated. As reviewed by Friedmann and Rusou (2015), the acquisition of some of the most distinctive properties of language such as complex syntax has a critical period that ends during the first year of life, and children who missed this window of opportunity later show severe syntactic impairments. Early postnatal life is also a critical window during which many neurodevelopmental disorders, such as autism, develop (though some of them, such as schizophrenia, may only manifest themselves later in life). As LeBlanc and Fagiolini (2011) write in their review article on whether autism should be seen as a "critical period disorder," brain circuits that are key to language and higher-order cognition are refined by experience during critical periods early in postnatal life, and even slight dysregulations can have large-scale effects.

It is for this reason that I find it extremely interesting that a series of work (Neubauer et al., 2010; Gunz et al., 2010, 2012; Scott et al., 2014; Hublin et al., 2015) has shown on the basis of data from living primates as well as computer-assisted reconstructions of brain developments from Neanderthal skulls that anatomically modern humans, with their globularized braincases, follow an early postnatal brain growth trajectory that sets us apart from our closest extinct and extant relatives. Although Neanderthal brains were significantly bigger than what we can observe from our closest living primate relatives, their brain growth trajectory was essentially a "blown-up" version of a chimpanzee brain development. Ours, by contrast, follows a different course that is reflected in our basketball-like head shape. Interestingly, it is during this very same period (first months postnatally) that the growth pattern of our faces also departs from that of our closest extinct and extant relatives (Lacruz et al., 2015).

I have argued in a series of articles (Boeckx, 2013, 2017a; Boeckx and Benítez-Burraco, 2014a) that this globularization phase must have been significant for understanding the evolution of our cognitive capacities, including language. If the brain grows differently, it wires differently, because growth and wiring are two sides of the same coin. These two units coevolve. And if we believe that different connectivity patterns give rise to distinct circuits with functional consequences, then globularizing a brain at such a critical period must have been cognitively significant.

I recognize that we cannot yet be sure about the precise consequences of globularization, because we do not yet understand fully how different brains can produce different cognitive modes. But the long list of disorders displaying both craniofacial and cognitive/linguistic deficits (Boeckx and Benítez-Burraco, 2014a) can help in this regard. Furthermore, Gunz et al., as well as Bruner (2010) et seq., have provided us with important information: the globularization of the human brain reconfigures the seat of cognition in several ways. It has been claimed to affect most clearly the parietal lobe, the cerebellum, and the frontal pole; plausibly as well, it affects the temporal lobe and the olfactory bulbs. This much can be inferred from fossilized crania. But our knowledge of brain development leads us to suspect that more deeply embedded structures that do not leave obvious marks on bones must have also been affected, be it in terms of volume or wiring diagrams (or both): the thalamus, the hippocampus, and other subcortical structures that strongly interact with regions of the brain most visibly affected by globularization. Many of these regions have not traditionally been associated with neural circuits thought to be responsible for language but in the next section, I will list a series of reasons to doubt the classical neurolinguistic models.

Work on archaic genomes has provided us with lists of genes that were modified in the course of evolution of our species since the separation from the Neanderthal lineage. Several of these changes appear to be related to brain development and function (as well as to craniofacial bone development; see Boeckx and Benítez-Burraco, 2015). In work in progress I am exploring the possibility that quite of few of these changes were in fact connected and caused the globularization of the brain (case) by affecting neural processes that delayed neurogenesis and influence brain growth (affecting most clearly the subventricular zone), generating a brain that matured more slowly (essentially a remarkably neotenous brain), and as a result a brain that was more dependent on the environment. Such a brain had to be a better "language-learning" brain. Accordingly, structures traditionally responsible with learning and memory, both cortical and subcorticular, came to play an even more salient role in our species.

Apart from allowing us to detect a different brain growth trajectory, the fossil record also leads us to suspect that our species underwent an intensification of what amounts to a self-domestication process (as Benitez-Burraco et al., 2016 discuss, globularization and self-domestication may in fact be linked). As reviewed in Thomas (2014), several scholars have long entertained the idea that anatomically modern humans were self-domesticated. Behavioral similarities (as well as neurobiological similarities, see Rilling et al., 2012) between us and bonobos, who have also claimed to be self-domesticated (Hare et al., 2012), reinforce this idea (MacLean, 2016). The idea of self-domestication will be obvious to anyone who recognizes the heavy dependence of ours species on its sociocultural environment. This is the cooperative spirit that seems to make us so special among our closest relatives (Tomasello, 2009).

But behavior is not the only source of evidence, which is a good thing given the notorious difficulty of reconstructing behavior from the fossil record. As stressed in Theofanopoulou and Boeckx (2016a), several anatomical characteristics (as well as emerging molecular data) of our species are reminiscent of what one finds in the context of domestication, especially when contrasted with what we know about Neanderthals (taken as a representative example of archaic *Homo*). As reviewed under the rubric of the "domestication syndrome" by Wilkins et al. (2014), domesticated species display a range of characteristics that set them apart from their wild counterparts: the suite of characteristics that signals domestication includes depigmentation, floppy, reduced ears, shorter muzzles, curly tails, smaller teeth, smaller brain/cranial capacity, neotenous (juvenile) behavior, docility, and reproductive cycle changes (more frequent estrous cycles). Of course, not all of these characteristics are found in all domesticates, but most of them are indeed present in all domesticates. Theofanopoulou and Boeckx (2016a) note that many of these changes match fairly well some of the well-known anatomical differences between anatomically modern humans and Neanderthals. I will review some of their observations here: first, the distinct ontogenetic trajectories discussed earlier under the rubric of globularization (Hublin et al., 2015; Lacruz et al., 2015) result in craniofacial differences that invariably lead to a more "gracile," "juvenile" profile in anatomically modern humans. This profile is sometimes considered "feminized" (Cieri et al., 2014) and is associated with an overall reduction of sexual dimorphism, which also appears to be a trait associated with domestication (Hare et al., 2012). Incidentally, Theofanopoulou and Boeckx point out this process of "feminization" (reduction of androgen levels, rise of estrogen levels; see Cieri et al., 2014 for references) is often associated with a reduced reactivity of the hypothalamus–pituitary–adrenal axis (Trut et al., 2009), a physiological trait also associated with domestication (Kunzl and Sachser, 1999).

Returning to craniofacial considerations, it is well established that prognathism is significantly reduced in our species (Maureille and Bar, 1999; Lacruz et al., 2015). Browridges and nasal projections are smaller in us than in our most closely related (extinct) relatives (Cieri et al., 2014), as is our cranial capacity (Mellars et al., 2007), and our tooth size (Zilberman and Smith, 1992). Differences in other traits associated with domestication may also exist, but there are either confounding factors (eg, geography for pigmentation) involved or the data are subject to more controversial interpretation (eg, Knight et al., 1995 in the case of reproductive cycle changes) than the anatomical record we just reviewed briefly. What is clear is that on balance we appear to bear more anatomical characteristics associated with domestication than Neanderthals.

Surprisingly, proponents of the self-domestication hypothesis have not always been explicit about what species they were comparing anatomically modern humans to in order to make their case. Comparison with more distantly related species, such as chimpanzees (taken as proxy for the last common ancestors they shared with us), strikes me as inherently less compelling, as there are just too many anatomical differences (and thus, confounding factors) involved. In addition, the domestication process is supposed to be rapid and, as such, fits better with a Neanderthal/anatomically modern human comparison. It is also unlikely that the self-domestication process took place after the emergence of anatomically modern humans (contra Cieri et al., 2014), because some of the traits associated with a domestication process are already present in the earliest specimens of our species, although of course it

is very likely the case that this self-domestication process intensified after our species emerged.

What is most important for present purposes is that much like changes at the level of brain growth, domestication and concomitant-relaxed selection have been claimed to lead to an enhancement of complexity in vocal learning (Okanoya, 2012; Deacon, 2010), so that if the anatomical record (as well as the emerging molecular evidence, which I have not reviewed here) is anything to go by, it is quite plausible that early anatomically modern humans were biologically committed to a greater dependence on culture and learning, which naturally led to a more complex grammatical mind (and maybe a greater dependence on activity-dependent genes, which appear to be important in all vocal learners).

38.5 Updating the Neurobiological Model for Human Language

Apart from the fact that many scholars have insisted too much on the species uniqueness of our language capacity, the lack of a comprehensive neurobiological model for said capacity has certainly played a major role in keeping the evolution of human language shrouded in mystery. In the absence of an adequate neurobiological target for evolutionary studies, the temptation to bypass the brain and spin evolutionary tales is just too great.

For over a century, researchers stuck to the classical model derived from the work of Broca and Wernicke. This is not, of course, to say that the classical frontotemporal cortical circuit makes no contribution to high-level linguistic processing. But it is equally clear that language processing recruits a far more widely distributed network (Fedorenko and Thompson-Schill, 2014; Blank et al., 2016; Friederici and Singer, 2015; Hagoort and Indefrey, 2014; Poeppel et al., 2012), drawing on multiple brain rhythms to provide a rich enough oscillatory regime for such a complex task (Theofanopoulou and Boeckx, 2016c; Lam et al., 2016). At the same time, an increasing number of papers on subcortical structures such as the hippocampus or the cerebellum start with statements such as "a growing body of work suggests [x; a subcortical structure] contributes to a variety of cognitive domains beyond its traditional role in [y; a very basic function not part of any core description of language]" (see Jarvis, 2004 on the forebrain broadly, Marien and Manto, 2016 for the cerebellum; Theofanopoulou and Boeckx, 2016b for the thalamus; Kurczek et al., 2013 for the hippocampus).

In the remainder of this section, I would like to highlight a few research paths and trends that strike me as worth pursuing, since they provide a window of opportunity to tackle, productively, very difficult questions about the evolution of human language.

Most of these highlights bear on semantic cognition, rather than vocal learning in the narrow sense (although I repeat that I do not think one should draw a sharp distinction between the two), because I think that this is the domain that we understand the least, precisely because comparative studies are so hard [in the domain of vocal learning, Jarvis (2004) offers a compelling candidate model]. Incidentally, what I call "semantic cognition" likely falls under the rubric of syntax for most researchers. This is because decades of work in theoretical linguistics have made clear that syntax cannot be dissociated from (structural) semantics.

The first trend I would like to discuss is the importance of the temporal lobe. In my experience, when people think of higher-order (linguistic) cognition, they immediately associate it with the frontal lobe and, more specifically, Broca area. No doubt, Broca area is an important node of the language network, but it is certainly not the more central one, at least when the construction of compositionally interpreted structures (ie, syntax/semantics) is concerned. True, as revealed by Schenker et al. (2010), Broca area has expanded sixfold compared to what we find in chimpanzees (Brodmann areas 44 and 45 that make up most of Broca region appear to be the most greatly expanded cortical areas yet identified in humans), but a growing number of experiments have revealed that the temporal lobe, traditionally associated with "memory" and "the lexicon," plays an equally, if not more, important role in cognition (on the possible role of Broca region, see Boeckx et al., 2014; Bornkessel-Schlesewsky et al., 2015; Fitch and Martins, 2014). For instance, work by Pylkkänen and colleagues has focused on the contribution of the left anterior temporal lobe to combinatorial processing. Together with the ventromedial prefrontal cortex, the angular gyrus, and the medial parietal cortex, the left anterior temporal lobe plays a central function in the compositional interpretation of hierarchically structured expressions (Pylkkänen, 2015; Westerlund et al., 2015).

A recent study by Davey et al. (2016) points to the role of another temporal region, the posterior middle temporal gyrus in semantic cognition. The latter appears to function as a hub within a large-scale network that allows the integration of automatic retrieval in the default mode network with demanding, goal-oriented cognition. In terms of connectivity, this temporal region appears to be uniquely placed to link two macrocircuits, the default mode network, thought to be the seat of "self-generated thoughts," and a frontoparietal "multiple-demand"/"executive control"/"dorsal attention circuit." As reviewed in Buckner and Krienen (2013), cortical association regions spanning the frontal and parietal cortices are disproportionately expanded

in humans compared with other primates. They mature late in development and are often disrupted in mental disorders (Sato et al., 2016). They appear to impose order on the self-generated thoughts by the default mode network.

A more deeply embedded temporal structure, the hippocampus is a core member of the default mode network. Recently, and largely thanks to the work of Duff and colleagues, its potential role in linguistic cognition has been reappreciated (see Alamri, 2016 for a more extensive review and discussion). Kurczek et al. (2015) have highlighted the significance of the hippocampus in the neural network that supports a range of abilities including remembering the past, thinking about the future, and introspecting about oneself and others, abilities that are often said to be human specific and shown to break down in a range of mental disorders where language is also affected (eg, schizophrenia, see Wible, 2012). Studies involving individuals with hippocampal amnesia, such as Duff et al. (2013), show that quintessentially human behaviors, such as the creative use of language as evidenced in verbal play, is severely damaged, which serves to highlight the interdependence of language and memory. Like evolution, language appears to make new use of old parts, so memorized units are as important as novel combinations, and in recent years scholars have been calling for a renewed appreciation of memory as an integral component of online information processing (see Hasson et al., 2015). Traditional divisions like declarative versus procedural memory systems may have to make way to a more integrated system that recognizes a strong hippocampal—striatal axis (prototypical declarative vs. procedural memory hubs) in learning, prediction, and goal-directed behavior (Pennartz et al., 2011). In this context, it is worth mentioning that mice endowed with the humanized version of Foxp2 have been claimed to accelerate learning by enhancing transitions from declarative to procedural performance (Schreiweis et al., 2014).

As Rubin et al. (2014) write, the frontal lobes have figured prominently in most studies of flexible or goal-directed behavior, but the hippocampus appears to play an equally important role by "forming and reconstructing relational memory representations that underlie flexible cognition and social behavior." Recently, Ellamil et al. (2016) showed how the hippocampus is recruited in the spontaneous generation of thoughts. As they point out, "[i]n contrast to the connectivity of other cortical areas, where short-distance synapses to nearby neurons predominate and long-distance connections are rare, neurons within a large part of the hippocampus are equally likely to connect to nearby or distant neighbors (Buzsaki, 2006). This highly variable microcircuitry may facilitate the creation of arbitrary or unlikely connections between groups of neurons that otherwise encode distinct memories or experiences (Buzsaki, 2006)." This description fits quite well with the core creativity function often attributed to language (see, eg, Boeckx, 2014). Such a generation of thought would have to be regulated by circuits responsible for dorsal attention, but it may be that the source of novel combinations may not lie in the neocortex. If true, this would be excellent news for evolutionary linguistics, for the hippocampus is one of the most closely scrutinized brain structures in neuroscience. Perhaps, some of our higher-order cognitive functions were indeed exapted from basic circuits associated with memory and spatial cognition. At the very least, this possibility reinforces the need to adopt a much broader conception of the neural circuits responsible for linguistic cognition.

As a final note on the possible hippocampal involvement in language, I would like to comment in the fact that it is usually the temporal cortex that is taken to provide the basis for linguistic memory ("the lexicon"), not the hippocampus. But the two structures clearly interact and may play complementary roles. When Teyler and DiScenna (1986) put forth their "hippocampal memory indexing theory," they suggested that the role of hippocampus may be "to form and retain an index of neocortical areas activated by experiential events," serving to "establish a cortically based memory trace." This role of the hippocampus as a "pointer" to more elaborate conceptual structures fits well with descriptions of word meaning (Pietroski, 2007) and could explain how hippocampal lesions need not completely abolish linguistic output, as the matured cortex could supplement it.

This raises an important issue, I think. Many brain structures have often been discarded as peripheral for language because of lesion studies showing how linguistic cognition can be preserved after seriously damaging them. But this sort of conclusion is often arrived at by studying the language system in adults, as opposed to asking how the system developed in the individual. It may well be that some structures were once critical, but their roles can be transient (and harder to detect, like the Wittgensteinian ladder that got used to climb and was then kicked off). Too often, the frontotemporal cortical circuit is studied once it is fully in place, but this may just reflect a specialization (modularization or routinization) lying at the confluence of broader, more distributed corticosubcortical circuits engaged in more generic functions. Instead of viewing the "core language circuit" in isolation, ontogenetic as well as phylogenetic considerations force us to view it as part and parcel of a broader cognitive basis. (I think this is consistent with the high degree of functional specificity for high-level linguistic processing in the adult human brain described in Fedorenko et al. (2011).)

As a final topic for this section, I would like to touch on the role of the parietal lobe in the context of language. I do so because this is the most conspicuous brain structural change that can be associated with the emergence of anatomically modern humans from the fossil record (Bruner, 2010). As I point out in Boeckx (2017b), there is evidence of significant "parietalization" of certain neural circuits highly relevant for language. For instance, Kumar et al. (2016) suggest that the enhanced laryngeal motor cortex—parietal connections in humans could provide the right neural substrate for more complex synchrony of higher-order sensorimotor coordination, proprioceptive and tactile feedback, and modulation of learned voice for speech production.

I also argued that the expansion of the parietal region played a role on the "semantic" side. Bruner (2010) reviews studies that point to the hub status of the parietal lobe for functions associated with the default mode network, which I already alluded to in the context of the hippocampus. Moreover, there is a range of studies that have documented a human-specific modification of how Broca region and Wernicke region are connected with one another through white matter pathways. In addition to a direct connection that is shared with nonhuman primates, there is also an indirect pathway, through the parietal lobe in humans (Catani and Jones, 2005; Dick and Tremblay, 2012; Hecht et al., 2013; Mendoza and Merchant, 2014; Catani and Bambini, 2014). Although this remains to be confirmed, it is plausible to envisage that this indirect connection may provide the basis for richer representational capacities. For instance, it could be that the formation of a frontoparietotemporal loop allows for the formation of recursive representations, since both frontoparietal and frontotemporal networks are routinely associated with processing sequences, and recursive representations are, at some level of description, sequences of sequences. Several authors have independently stressed the relevance (both ontogenetically and phylogenetically) of an indirect, frontoparietotemporal pathway (Friederici, 2012; de Diego-Balaguer et al., 2016) for language tasks, and so evolutionary studies stand to benefit from future investigations of the role of what is sometimes called Geschwind territory, next to the more familiar Broca and Wernicke regions.

38.6 Conclusion

To summarize the main message of this chapter, like vocal learning in birds, human language is a highly complex, polygenic trait that recruits numerous brain regions, over and above the classical language regions, to provide a computational regime supporting linguistic cognition. Strong adherence to the Darwinian logic of descent, with its emphasis on a rich cognitive life for nonhuman species, offers the hope to shed light on what looks like a very human-specific, domain-specific capacity like language. Such an approach requires studies that embrace the multifactorial (not only genes, not only environment), multidimensional (genome, connectome, "dynome," "cognome") nature of the capacity to master grammatical systems of the kinds we do.

There is no doubt in my mind that a richer data set thanks to the genomic revolution, and a broader comparative basis adopting a bottom-up approach to study primate cognition guarantees that in the very near future a much more elaborate, and more precise picture of language evolution will be available than the one I have offered here. But even the present one strikes me as a far cry from claims not so long ago that we would never find out.

Acknowledgments

For helping me understand some of the issues discussed here, I am indebted to Erich Jarvis, Tecumseh Fitch, Simon Fisher, Sonja Vernes, Simon Kirby, Marc Hauser, David Poeppel, Gary Marcus, Ev Fedorenko, Dick Lewontin, Koji Fujita, Kazuo Okanoya, Bridget Samuels, Saleh Alamri, Pedro Tiago Martins, and Constantina Theofanopoulou. I am particularly grateful for the comments on the prefinal draft I received from Todd Preuss and Erich Jarvis.

This chapter was written while I was a visitor at the UPF-CSIC Institute for Evolutionary Biology. I thank all the members of that institute for providing an environment conducive to research.

I also would like to acknowledge the financial support from the Spanish Ministry of Economy and Competitiveness (grants FFI2013-43823-P and FFI2014-61888-EXP), a Marie Curie International Reintegration Grant from the European Union (PIRG-GA-2009-256413), research funds from the Fundació Bosch i Gimpera, and from the Generalitat de Catalunya (2014-SGR-200).

References

Alamri, S., 2016. The Role of the Hippocampus in Language Processing. PhD. Universitat de Barcelona (in progress).

Anderson, S., 2004. Doctor Dolittle's Delusion. Yale University Press, New Haven.

Arbib, M., 2002. How the Brain Got Language. Oxford University Press, Oxford.

Arnold, K., Zuberbühler, K., 2008. Meaningful call combinations in a non-human primate. Curr. Biol. 18 (5), R202–R203.

Arriaga, G., Jarvis, E.D., 2013. Mouse vocal communication system: are ultrasounds learned or innate? Brain Lang. 124 (1), 96–116.

Barton, R.A., 2012. Embodied cognitive evolution and the cerebellum. Philos. Trans. R. Soc. B 367 (1599), 2097–2107.

Bass, A.H., Chagnaud, B.P., 2012. Shared developmental and evolutionary origins for neural basis of vocal–acoustic and pectoral–gestural signaling. Proc. Natl. Acad. Sci. U.S.A. 109 (Suppl. 1), 10677–10684.

Benitez-Burraco, A., Theofanopoulou, C., Boeckx, C., 2016. Globularization and domestication. Topoi 1–14.

Bergman, T.J., 2013. Speech-like vocalized lip-smacking in geladas. Curr. Biol. 23 (7), R268–R269.

Blank, I., Balewski, Z., Mahowald, K., Fedorenko, E., 2016. Syntactic processing is distributed across the language system. Neuroimage 127, 307–323.

Boeckx, C., Benítez-Burraco, A., 2014. The shape of the human language-ready brain. Front. Psychol. 5, 282. https://doi.org/10.3389/fpsyg.2014.00282.

Boeckx, C., Benítez-Burraco, A., 2014. Globularity and language-readiness: generating new predictions by expanding the set of genes of interest. Front. Psychol. 5, 1324.

Boeckx, C., Benítez-Burraco, A., 2015. Osteogenesis and neurogenesis: a robust link also for language evolution. Front. Cell. Neurosci. 9, 291. https://doi.org/10.3389/fncel.2015.00291.

Boeckx, C., Martinez-Alvarez, A., Leivada, E., 2014. The functional neuroanatomy of serial order in language. J. Neurolinguist. 32, 1–15.

Boeckx, C., 2013. Biolinguistics: forays into human cognitive biology. J. Anthropol. Sci. 91, 63–89.

Boeckx, C., 2014. Elementary Syntactic Structures. Cambridge University Press.

Boeckx, C., 2017a. The language-ready head: evolutionary considerations. Psychon. Bull. Rev. 24 (1), 194–199. https://doi.org/10.3758/s13423-016-1087-5.

Boeckx, C., 2017b. A conjecture about the neural basis of recursion in light of descent with modification. J. Neurolinguist. 43, 193–198.

Bolhuis, J.J., Everaert, M., 2013. Birdsong, Speech, and Language: Exploring the Evolution of Mind and Brain. MIT Press.

Bornkessel-Schlesewsky, I., Schlesewsky, M., Small, S.L., Rauschecker, J.P., 2015. Neurobiological roots of language in primate audition: common computational properties. Trends Cogn. Sci. 19 (3), 142–150.

Bostwick, K.S., Prum, R.O., 2005. Courting bird sings with stridulating wing feathers. Science 309 (5735), 736.

Bruner, E., 2010. Morphological differences in the parietal lobes within the human genus. Curr. Anthropol. 51, S77–S88. https://doi.org/10.1086/650729.

Buckner, R.L., Krienen, F.M., 2013. The evolution of distributed association networks in the human brain. Trends Cogn. Sci. 17 (12), 648–665.

Buzsaki, G., 2006. Rhythms of the Brain. Oxford University Press, Oxford.

Catani, M., Bambini, V., 2014. A model for social communication and language evolution and development (SCALED). Curr. Opin. Neurobiol. 28, 165–171.

Catani, M., Jones, D.K., 2005. Perisylvian language networks of the human brain. Ann. Neurol. 57 (1), 8–16.

Chakraborty, M., Jarvis, E.D., 2015. Brain evolution by brain pathway duplication. Philos. Trans. R. Soc. B 370 (1684), 20150056.

Chakraborty, M., Walløe, S., Nedergaard, S., Fridel, E.E., Dabelsteen, T., Pakkenberg, B., Jarvis, E.D., 2015. Core and shell song systems unique to the parrot brain. PLoS One 10 (6), e0118496.

Charrier, C., Joshi, K., Coutinho-Budd, J., Kim, J.E., Lambert, N., De Marchena, J., Polleux, F., 2012. Inhibition of SRGAP2 function by its human-specific paralogs induces neoteny during spine maturation. Cell 149 (4), 923–935.

Cheney, D.L., Seyfarth, R.M., 1992. How Monkeys See the World: Inside the Mind of Another Species. University of Chicago Press.

Cheney, D.L., Seyfarth, R.M., 2008. Baboon Metaphysics: The Evolution of a Social Mind. University of Chicago Press.

Cieri, R.L., Churchill, S.E., Franciscus, R.G., Tan, J., Hare, B., 2014. Craniofacial feminization, social tolerance, and the origins of behavioral modernity. Curr. Anthropol. 55 (4), 419–443.

Clay, Z., Archbold, J., Zuberbühler, K., 2015. Functional flexibility in wild bonobo vocal behaviour. PeerJ 3, e1124.

Davey, J., Thompson, H.E., Hallam, G., Karapanagiotidis, T., Murphy, C., De Caso, I., Jefferies, E., 2016. Exploring the role of the posterior middle temporal gyrus in semantic cognition: integration of anterior temporal lobe with executive processes. Neuroimage 137.

de Diego-Balaguer, R., Martinez-Alvarez, A., Pons, F., 2016. Temporal attention as a scaffold for language development. Front. Psychol. 7, 44.

De Waal, F.B., Ferrari, P.F., 2010. Towards a bottom-up perspective on animal and human cognition. Trends Cogn. Sci. 14 (5), 201–207.

Deacon, T.W., 1989. The neural circuitry underlying primate calls and human language. Hum. Evol. 4 (5), 367–401.

Deacon, T., 1997. The Symbolic Species. Norton, New York.

Deacon, T.W., 2006. Emergence: the hole at the Wheel's Hub1. In: The Re-emergence of Emergence: The Emergentist Hypothesis from Science to Religion: The Emergentist Hypothesis from Science to Religionvol, vol. 159, p. 111.

Deacon, T.W., 2010. A role for relaxed selection in the evolution of the language capacity. Proc. Natl. Acad. Sci. U.S.A. 107 (Suppl. 2), 9000–9006.

Dediu, D., Levinson, S.C., 2013. On the antiquity of language: the reinterpretation of Neandertal linguistic capacities and its consequences. Front. Psychol. 4, 397.

Dennis, M.Y., Nuttle, X., Sudmant, P.H., Antonacci, F., Graves, T.A., Nefedov, M., Curry, C.J., 2012. Evolution of human-specific neural SRGAP2 genes by incomplete segmental duplication. Cell 149 (4), 912–922.

Dick, A.S., Tremblay, P., 2012. Beyond the arcuate fasciculus: consensus and controversy in the connectional anatomy of language. Brain 135 aws222.

Duff, M.C., Kurczek, J., Rubin, R., Cohen, N.J., Tranel, D., 2013. Hippocampal amnesia disrupts creative thinking. Hippocampus 23 (12), 1143–1149.

Ellamil, M., Fox, K.C., Dixon, M.L., Pritchard, S., Todd, R.M., Thompson, E., Christoff, K., 2016. Dynamics of neural recruitment surrounding the spontaneous arising of thoughts in experienced mindfulness practitioners. Neuroimage 136.

Emery, N.J., Clayton, N.S., 2004. The mentality of crows: convergent evolution of intelligence in corvids and apes. Science 306 (5703), 1903–1907.

Engesser, S., Ridley, A.R., Townsend, S.W., 2016. Meaningful call combinations and compositional processing in the southern pied babbler. Proc. Natl. Acad. Sci. U.S.A. 113, 201600970.

Esteve-Gibert, N., Prieto, P., 2014. Infants temporally coordinate gesture-speech combinations before they produce their first words. Speech Commun. 57, 301–316.

Fedorenko, E., Thompson-Schill, S.L., 2014. Reworking the language network. Trends Cogn. Sci. 18 (3), 120–126.

Fedorenko, E., Behr, M.K., Kanwisher, N., 2011. Functional specificity for high-level linguistic processing in the human brain. Proc. Natl. Acad. Sci. U.S.A. 108 (39), 16428–16433.

Feenders, G., Liedvogel, M., Rivas, M., Zapka, M., Horita, H., Hara, E., et al., 2008. Molecular mapping of movement-associated areas in the avian brain: a motor theory for vocal learning origin. PLoS One 3 (3), e1768.

Feher, O., Wang, H., Saar, S., Mitra, P.P., Tchernichovski, O., 2009. De novo establishment of wild-type song culture in the zebra finch. Nature 459 (7246), 564–568.

Fisher, S.E., Vernes, S.C., 2015. Genetics and the language sciences. Annu. Rev. Linguist. 1, 289–310.

Fisher, S.E., 2015. Translating the genome in human neuroscience. In: Marcus, G., Freeman, J. (Eds.), The Future of the Brain: Essays by the World's Leading Neuroscientists. Princeton University Press, Princeton, NJ, pp. 149–159.

Fisher, S.E., 2016. A molecular genetic perspective on speech and language. In: Hickok, G., Small, S. (Eds.), Neurobiology of Language. Elsevier, Amsterdam, pp. 13–24.

Fitch, W., Martins, M.D., 2014. Hierarchical processing in music, language, and action: Lashley revisited. Ann. N.Y. Acad. Sci. 1316 (1), 87–104.

Fitch, W.T., Reby, D., 2001. The descended larynx is not uniquely human. Proc. R. Soc. Lond. B Biol. Sci. 268 (1477), 1669–1675.

Fitch, W.T., 2000. The evolution of speech: a comparative review. Trends Cogn. Sci. 4 (7), 258–267.

Fitch, W.T., 2010. The Evolution of Language. Cambridge University Press, Cambridge.

Fitch, W.T., 2011. The evolution of syntax: an exaptationist perspective. Front. Evol. Neurosci. 3 (9).

Friederici, A.D., Singer, W., 2015. Grounding language processing on basic neurophysiological principles. Trends Cogn. Sci. 19 (6), 329–338.

Friederici, A.D., 2012. Language development and the ontogeny of the dorsal pathway. Front. Evol. Neurosci. 4 (3).

Friedmann, N., Rusou, D., 2015. Critical period for first language: the crucial role of language input during the first year of life. Curr. Opin. Neurobiol. 35, 27–34.

Fröhlich, M., Kuchenbuch, P., Müller, G., Fruth, B., Furuichi, T., Wittig, R.M., Pika, S., 2016. Unpeeling the layers of language: bonobos and chimpanzees engage in cooperative turn-taking sequences. Sci. Rep. 6.

Fujita, K., 2016. On certain fallacies in evolutionary linguistics and how one can eliminate them. In: Fujita, K., Boeckx, C. (Eds.), Advances in Biolinguistics: The Human Language Faculty and its Biological Basis. Routledge.

Genty, E., Zuberbühler, K., 2014. Spatial reference in a bonobo gesture. Curr. Biol. 24 (14), 1601–1605.

Genty, E., Zuberbühler, K., 2015. Iconic gesturing in bonobos. Commun. Integr. Biol. 8 (1), e992742.

Genty, E., Neumann, C., Zuberbühler, K., 2015. Bonobos modify communication signals according to recipient familiarity. Sci. Rep. 5.

Ghazanfar, A.A., Poeppel, D., 2014. The neurophysiology and evolution of the speech rhythm. In: Gazzaniga, M.S., Mangun, R. (Eds.), The Cognitive Neurosciences V. MIT Press.

Ghazanfar, A.A., Takahashi, D.Y., 2014. The evolution of speech: vision, rhythm, cooperation. Trends Cognitive Sci 18 (10), 543–553.

Ghazanfar, A.A., Chandrasekaran, C., Morrill, R.J., 2010. Dynamic, rhythmic facial expressions and the superior temporal sulcus of macaque monkeys: implications for the evolution of audiovisual speech. Eur. J. Neurosci. 31 (10), 1807–1817.

Ghazanfar, A.A., Takahashi, D.Y., Mathur, N., Fitch, W.T., 2012. Cineradiography of monkey lip-smacking reveals putative precursors of speech dynamics. Curr. Biol. 22 (13), 1176–1182.

Ghazanfar, A.A., Morrill, R.J., Kayser, C., 2013. Monkeys are perceptually tuned to facial expressions that exhibit a theta-like speech rhythm. Proc. Natl. Acad. Sci. U.S.A. 110 (5), 1959–1963.

Graham, S.A., Fisher, S.E., 2015. Understanding language from a genomic perspective. Annu. Rev. Genet. 49, 131–160.

Gunz, P., Neubauer, S., Maureille, B., Hublin, J.-J., 2010. Brain development after birth differs between Neanderthals and modern humans. Curr. Biol. 20, R921–R922. https://doi.org/10.1016/j.cub.2010.10.018.

Gunz, P., Neubauer, S., Golovanova, L., Doronichev, V., Maureille, B., Hublin, J.-J., 2012. A uniquely modern human pattern of endocranial development. Insights from a new cranial reconstruction of the Neandertal newborn from Mezmaiskaya. J. Hum. Evol. 62, 300–313. https://doi.org/10.1016/j.jhevol.2011.11.013.

Haesler, S., Wada, K., Nshdejan, A., Morrisey, E.E., Lints, T., Jarvis, E.D., Scharff, C., 2004. FoxP2 expression in avian vocal learners and non-learners. J. Neurosci. 24 (13), 3164–3175.

Hagoort, P., Indefrey, P., 2014. The neurobiology of language beyond single words. Annu. Rev. Neurosci. 37, 347–362.

Hardus, M.E., Lameira, A.R., Singleton, I., Morrogh-Bernard, H.C., Knott, C.D., Ancrenaz, M., Wich, S.A., 2009. A description of the orangutan's vocal and sound repertoire, with a focus on geographic variation. Orangutans: Geographic Variation in Behavioral Ecology and Conservation, pp. 49–64.

Hare, B., Wobber, V., Wrangham, R., 2012. The self-domestication hypothesis: evolution of bonobo psychology is due to selection against aggression. Anim. Behav. 83 (3), 573–585.

Hasson, U., Chen, J., Honey, C.J., 2015. Hierarchical process memory: memory as an integral component of information processing. Trends Cogn. Sci. 19 (6), 304–313.

Hauser, M.D., Chomsky, N., Tecumseh Fitch, W., November 22, 2002. The faculty of language: what is it, who has it, and how did it evolve? Science 298 (5598), 1569–1579. https://doi.org/10.1126/science.298.5598.1569.

Hauser, M.D., 2001. Wild Minds: What Animals Really Think. Macmillan.

Hecht, E.E., Murphy, L.E., Gutman, D.A., Votaw, J.R., Schuster, D.M., Preuss, T.M., Orban, G.A., Stout, D., Parr, L.A., 2013. Differences in neural activation for object-directed grasping in chimpanzees and humans. J. Neurosci. 33, 14117–14134.

Hewes, G.W., 1973. Primate communication and the gestural origin of language [and comments and reply]. Curr. Anthropol. 5–24.

Hillert, D.G., 2015. On the evolving biology of language. Front. Psychol. 6.

Hoepfner, A.R., Goller, F., 2013. Atypical Song Reveals Spontaneously Developing Coordination between Multi-Modal Signals in Brown-Headed Cowbirds (*Molothrus ater*).

Hublin, J.J., Neubauer, S., Gunz, P., 2015. Brain ontogeny and life history in Pleistocene hominins. Philos. Trans. R. Soc. 370, 20140062. https://doi.org/10.1098/rstb.2014.0062.

Hurford, J., 1990. Nativist and functional explanations in language acquisition. In: Roca, I.M. (Ed.), Logical Issues in Language Acquistion. Foris Publications, Dordrecht, The Netherlands, pp. 85–136.

Hurford, J.R., 2007. The Origin of Meaning. Oxford University Press.

Janik, V.M., Slater, P.B., 1997. Vocal learning in mammals. Adv. Study Behav. 26, 59–99.

Jarvis, E.D., 2004. Learned birdsong and the neurobiology of human language. Ann. N.Y. Acad. Sci. 1016 (1), 749–777.

Kirby, S., 2013. Language, culture and computation: an adaptive systems approach to biolinguistics. In: Boeckx, C., Grohmann, K.K. (Eds.), Cambridge Handbook of Biolinguistics. Cambridge Handbooks in Language and Linguistics. Cambridge University Press.

Knight, C., Power, C., Watts, I., 1995. The human symbolic revolution: a Darwinian account. Camb. Archaeol. J. 5 (01), 75–114.

Koda, H., Nishimura, T., Tokuda, I.T., Oyakawa, C., Nihonmatsu, T., Masataka, N., 2012. Soprano singing in gibbons. Am. J. Phys. Anthropol. 149 (3), 347–355.

Konopka, G., Bomar, J.M., Winden, K., Coppola, G., Jonsson, Z.O., Gao, F., Geschwind, D.H., 2009. Human-specific transcriptional regulation of CNS development genes by FOXP2. Nature 462 (7270), 213–217.

Krause, J., Lalueza-Fox, C., Orlando, L., Enard, W., Green, R.E., Burbano, H.A., Bertranpetit, J., 2007. The derived FOXP2 variant of modern humans was shared with Neandertals. Curr. Biol. 17 (21), 1908–1912.

Kumar, V., Croxson, P.L., Simonyan, K., 2016. Structural organization of the laryngeal motor cortical network and its implication for evolution of speech production. J. Neurosci. 36 (15), 4170–4181.

Kunzl, C., Sachser, N., 1999. The behavioral endocrinology of domestication: a comparison between the domestic Guinea pig (*Cavia aperea* f. *porcellus*) and its wild ancestor, the cavy (*Cavia aperea*). Horm. Behav. 35, 28–37.

Kurczek, J., Brown-Schmidt, S., Duff, M., 2013. Hippocampal contributions to language: evidence of referential processing deficits in amnesia. J. Exp. Psychol. Gen. 142 (4), 1346.

Kurczek, J., Wechsler, E., Ahuja, S., Jensen, U., Cohen, N.J., Tranel, D., Duff, M., 2015. Differential contributions of hippocampus and medial prefrontal cortex to self-projection and self-referential processing. Neuropsychologia 73, 116–126.

Lacruz, R., Bromage, T., O'Higgins, P., Arsuaga, J.L., Stringer, C., Godinho, R., Carbonell, E., 2015. Facial ontogeny in Neanderthals and their ancestors. Nat. Commun. 6.

Lai, C.S., Fisher, S.E., Hurst, J.A., Vargha-Khadem, F., Monaco, A.P., 2001. A forkhead-domain gene is mutated in a severe speech and language disorder. Nature 413 (6855), 519–523.

Lam, N.H., Schoffelen, J.M., Uddén, J., Hultén, A., Hagoort, P., 2016. Neural activity during sentence processing as reflected in theta, alpha, beta, and gamma oscillations. Neuroimage 142, 43–54. https://doi.org/10.1016/j.neuroimage.2016.03.007.

Lameira, A.R., Hardus, M.E., Wich, S.A., 2012. Orangutan instrumental gesture-calls: reconciling acoustic and gestural speech evolution models. Evol. Biol. 39 (3), 415–418.

Lameira, A.R., Maddieson, I., Zuberbühler, K., 2014. Primate feedstock for the evolution of consonants. Trends Cogn. Sci. 18 (2), 60–62.

Lameira, A.R., Hardus, M.E., Bartlett, A.M., Shumaker, R.W., Wich, S.A., Menken, S.B., 2015. Speech-like rhythm in a voiced and voiceless orangutan call. PLoS One 10 (1), e116136.

Lawton, K.J., Wassmer, T.L., Deitcher, D.L., 2014. Conserved role of *Drosophila melanogaster* FoxP in motor coordination and courtship song. Behav. Brain Res. 268, 213–221.

LeBlanc, J.J., Fagiolini, M., 2011. Autism: a "critical period" disorder? Neural Plast. 2011.

Lenneberg, E.H., 1967. Biological Foundations of Language. Wiley, New York.

Levinson, S.C., 2016. Turn-taking in human communication – origins and implications for language processing. Trends Cogn. Sci. 20 (1), 6–14.

Lewontin, R., 1998. Evolution of cognition. In: Scarborough, D., Sternberg, S. (Eds.), Questions We Will Never Answer. An Invitation to Cognitive Science., Methods, Models, and Conceptual Issues, vol. 4 MIT Press.

Liebal, K., Pika, S., Tomasello, M., 2004. Social communication in siamangs (*Symphalangus syndactylus*): use of gestures and facial expressions. Primates 45 (1), 41–57.

Lieberman, P., 2006. Toward an Evolutionary Biology of Language. Harvard University Press.

Liu, W.C., Wada, K., Jarvis, E.D., Nottebohm, F., 2013. Rudimentary substrates for vocal learning in a suboscine. Nat. Commun. 4.

MacLean, E.L., 2016. Unraveling the evolution of uniquely human cognition. Proc. Natl. Acad. Sci. U.S.A. 113 (23), 6348–6354.

MacNeilage, P.F., 2010. The Origin of Speech. Oxford University Press.

Marcus, G., 2004. The Birth of the Mind. Basic Books, New York.

Maricic, T., Günther, V., Georgiev, O., Gehre, S., Ćurlin, M., Schreiweis, C., de la Rasilla, M., 2013. A recent evolutionary change affects a regulatory element in the human FOXP2 gene. Mol. Biol. Evol. 30 (4), 844–852.

Marien, P., Manto, M., 2016. The Linguistic Cerebellum. Academic Press.

Marler, P., 2004. Innateness and the instinct to learn. An. Acad. Bras. Cienc. 76 (2), 189–200.

Martins, P.T., Boeckx, C., 2014. Attention mechanisms and the mosaic evolution of speech. Front. Psychol. 5, 1463.

Maureille, B., Bar, D., 1999. The premaxilla in Neandertal and early modern children: ontogeny and morphology. J. Hum. Evol. 37 (2), 137–152.

Mellars, P., Boyle, K., Bar-Yosef, O., Stringer, C.H.R., 2007. Rethinking the Human Revolution. McDonald Institute Monographs, Cambridge.

Mendoza, G., Merchant, H., 2014. Motor system evolution and the emergence of high cognitive functions. Prog. Neurobiol. 122, 73–93.

Micheletta, J., Engelhardt, A., Matthews, L.E.E., Agil, M., Waller, B.M., 2013. Multicomponent and multimodal lipsmacking in crested macaques (*Macaca nigra*). Am. J. Primatol. 75 (7), 763–773.

Moore, R., 2016. Gricean communication and cognitive development. Philos. Q. (in press).

Moore, R., 2016. Gricean Communication, Joint Action, and the Evolution of Cooperation. Topoi (in press).

Müller, G.B., 2010. Epigenetic Innovation. Evolution: The Extended Synthesis. MIT Press, Cambridge, MA, pp. 307–328.

Neubauer, S., Gunz, P., Hublin, J.-J., 2010. Endocranial shape changes during growth in chimpanzees and humans: a morphometric analysis of unique and shared aspects. J. Hum. Evol. 59, 555–566. https://doi.org/10.1016/j.jhevol.2010.06.011.

Okanoya, K., 2012. Behavioural factors governing song complexity in Bengalese finches. Int. J. Comp. Psychol. 25 (1).

Olkowicz, S., Kocourek, M., Lucan, R.K., Portes, M., Fitch, W.T., Herculano-Houzel, S., Nemec, P., 2016. Birds have primate-like numbers of neurons in the forebrain. Proc. Nat. Acad. Sci. U.S.A. 113 (26), 7255–7260.

Pääbo, S., 2014. The human condition—a molecular approach. Cell 157 (1), 216–226.

Pennartz, C.M.A., Ito, R., Verschure, P.F.M.J., Battaglia, F.P., Robbins, T.W., 2011. The hippocampal–striatal axis in learning, prediction and goal-directed behavior. Trends Neurosci. 34 (10), 548–559.

Petkov, C.I., Jarvis, E.D., 2012. Birds, primates, and spoken language origins: behavioral phenotypes and neurobiological substrates. Front. Evol. Neurosci. 4, 12.

Pfenning, A.R., Hara, E., Whitney, O., Rivas, M.V., Wang, R., Roulhac, P.L., Jarvis, E.D., 2014. Convergent transcriptional specializations in the brains of humans and song-learning birds. Science 346 (6215), 1256846.

Pietroski, P.M., 2007. Systematicity via monadicity. Croat. J. Philos. 7 (3), 343–374.

Pika, S., Bugnyar, T., 2011. The use of referential gestures in ravens (*Corvus corax*) in the wild. Nat. Commun. 2, 560.

Pisanski, K., Cartei, V., McGettigan, C., Raine, J., Reby, D., 2016. Voice modulation: a window into the origins of human vocal control? Trends Cogn. Sci. 20 (4), 304–318.

Poeppel, D., Emmorey, K., Hickok, G., Pylkkänen, L., 2012. Towards a new neurobiology of language. J. Neurosci. 32 (41), 14125–14131.

Pylkkänen, L., 2015. Composition of complex meaning: interdisciplinary perspectives on the left anterior temporal lobe. Neurobiology of Language. Elsevier, New York.

Richman, B., 1976. Some vocal distinctive features used by gelada monkeys. J. Acoust. Soc. Am. 60 (3), 718–724.

Rilling, J.K., Scholz, J., Preuss, T.M., Glasser, M.F., Errangi, B.K., Behrens, T.E., 2012. Differences between chimpanzees and bonobos in neural systems supporting social cognition. Soc. Cogn. Affect. Neurosci. 7, 369–379.

Roberts, A.I., Roberts, S.G.B., Vick, S.J., 2014. The repertoire and intentionality of gestural communication in wild chimpanzees. Anim. Cogn. 17 (2), 317–336.

Rodenas-Cuadrado, P., Chen, X.S., Wiegrebe, L., Firzlaff, U., Vernes, S.C., 2015. A novel approach identifies the first transcriptome networks in bats: a new genetic model for vocal communication. BMC Genom 16 (1), 836.

Rubin, R.D., Watson, P.D., Duff, M.C., Cohen, N.J., 2014. The role of the hippocampus in flexible cognition and social behavior. Front. Hum. Neurosci. 8, 742.

Samuels, B., 2011. Phonological Architecture. Oxford University Press, Oxford.

Sato, J.R., Biazoli, C.E., Salum, G.A., Gadelha, A., Crossley, N., Vieira, G., Anés, M., 2016. Connectome hubs at resting state in children and adolescents: reproducibility and psychopathological correlation. Dev. Cogn. Neurosci. 20, 2–11.

Schenker, N.M., Hopkins, W.D., Spocter, M.A., Garrison, A.R., Stimpson, C.D., Erwin, J.M., Sherwood, C.C., 2010. Broca's area homologue in chimpanzees (*Pan troglodytes*): probabilistic mapping, asymmetry, and comparison to humans. Cereb. Cortex 20 (3), 730–742.

Schreiweis, C., Bornschein, U., Burguière, E., Kerimoglu, C., Schreiter, S., Dannemann, M., Groszer, M., 2014. Humanized Foxp2 accelerates learning by enhancing transitions from declarative to procedural performance. Proc. Natl. Acad. Sci. U.S.A. 111 (39), 14253–14258.

Scott, N., Neubauer, S., Hublin, J.J., Gunz, P., 2014. A shared pattern of postnatal endocranial development in extant hominoids. Evol. Biol. 41 (4), 572–594.

Scott-Phillips, T.C., Kirby, S., 2010. Language evolution in the laboratory. Trends Cogn. Sci. 14, 411–417.

Soma, M., Mori, C., 2015. The songbird as a percussionist: syntactic rules for non-vocal sound and song production in java sparrows. PLoS One 10 (5), e0124876.

Taglialatela, J.P., Russell, J.L., Schaeffer, J.A., Hopkins, W.D., 2011. Chimpanzee vocal signaling points to a multimodal origin of human language. PLoS One 6 (4), e18852.

Taglialatela, J.P., Russell, J.L., Pope, S.M., Morton, T., Bogart, S., Reamer, L.A., Hopkins, W.D., 2015. Multimodal communication in chimpanzees. Am. J. Primatol. 77 (11), 1143–1148.

Takahashi, D.Y., Fenley, A.R., Ghazanfar, A.A., 2016. Early development of turn-taking with parents shapes vocal acoustics in infant marmoset monkeys. Philos. Trans. R. Soc. B 371 (1693), 20150370.

Takahashi, D.Y., Narayanan, D.Z., Ghazanfar, A.A., 2013. Coupled oscillator dynamics of vocal turn-taking in monkeys. Curr. Biol. 23 (21), 2162–2168.

Tchernichovski, O., Oller, D.K., 2016. Vocal development: how marmoset infants express their feelings. Curr. Biol. 26 (10), R422–R424.

Teramitsu, I., Kudo, L.C., London, S.E., Geschwind, D.H., White, S.A., 2004. Parallel FoxP1 and FoxP2 expression in songbird and human brain predicts functional interaction. J. Neurosci. 24 (13), 3152–3163.

Teyler, T.J., DiScenna, P., 1986. The hippocampal memory indexing theory. Behav. Neurosci. 100 (2), 147.

Theofanopoulou, C., Boeckx, C., 2015. Cognitive phylogenies, the Darwinian logic of descent, and the inadequacy of cladistic thinking. Front. Cell Dev. Biol. 3, 64.

Theofanopoulou, C., Boeckx, C., 2016. Molecular Footprints of the Self-Domestication Process in *Homo sapiens* (Submitted MS).

Theofanopoulou, C., Boeckx, C., 2016. The central role of the thalamus in language and cognition. Advances in Biolinguistics: The Human Language Faculty its Biological Basis.

Theofanopoulou, C., Boeckx, C., 2016. (Neural) syntax. In: Martin, R., Gallego, A. (Eds.), Language, Syntax, and the Natural Sciences. Cambridge University Press, Cambridge (to appear).

Theofanopoulou, C., 2016. Implications of oxytocin in human linguistic cognition: from genome to phenome. Front. Neurosci. 10, 271.

Thomas, J., 2014. Self-domestication and Language Evolution. PhD, University of Edinburgh.

Tomasello, M., 2009. The Cultural Origins of Human Cognition. Harvard University Press.

Trut, L., Oskina, I., Kharlamova, A., 2009. Animal evolution during domestication: the domesticated fox as a model. Bioessays 31, 349.

Tsui, D., Vessey, J.P., Tomita, H., Kaplan, D.R., Miller, F.D., 2013. FoxP2 regulates neurogenesis during embryonic cortical development. J. Neurosci. 33 (1), 244–258.

Vargha-Khadem, F., Gadian, D.G., Copp, A., Mishkin, M., 2005. FOXP2 and the neuroanatomy of speech and language. Nat. Rev. Neurosci. 6 (2), 131–138.

Vernes, S.C., Spiteri, E., Nicod, J., Groszer, M., Taylor, J.M., Davies, K.E., Fisher, S.E., 2007. High-throughput analysis of promoter occupancy reveals direct neural targets of FOXP2, a gene mutated in speech and language disorders. Am. J. Hum. Genet. 81 (6), 1232–1250.

Vernes, S.C., Oliver, P.L., Spiteri, E., Lockstone, H.E., Puliyadi, R., Taylor, J.M., Nicod, J., 2011. Foxp2 regulates gene networks implicated in neurite outgrowth in the developing brain. PLoS Genet. 7 (7), e1002145.

Waller, B., Liebal, K., Burrows, A., Slocombe, K., 2013. How can a multimodal approach to primate communication help us understand the evolution of communication? Evol. Psychol. 11 (3), 538–549.

Wang, R., Chen, C.C., Hara, E., Rivas, M.V., Roulhac, P.L., Howard, J.T., Jarvis, E.D., 2015. Convergent differential regulation of SLIT-ROBO axon guidance genes in the brains of vocal learners. J. Comp. Neurol. 523 (6), 892–906.

West-Eberhard, M.J., 2003. Developmental Plasticity and Evolution. Oxford University Press.

Westerlund, M., Kastner, I., Al Kaabi, M., Pylkkänen, L., 2015. The LATL as locus of composition: MEG evidence from English and Arabic. Brain Lang. 141, 124–134.

Wible, C.G., 2012. Hippocampal temporal-parietal junction interaction in the production of psychotic symptoms: a framework for understanding the schizophrenic syndrome. Front. Hum. Neurosci. 6, 180.

Wilkins, A.S., Wrangham, R.W., Fitch, W.T., 2014. The "domestication syndrome" in mammals: a unified explanation based on neural crest cell behavior and genetics. Genetics 197 (3), 795–808.

Williams, H., 2001. Choreography of song, dance and beak movements in the zebra finch (*Taeniopygia guttata*). J. Exp. Biol. 204 (20), 3497–3506.

Wilson, B., Kikuchi, Y., Sun, L., Hunter, D., Dick, F., Smith, K., Petkov, C.I., 2015. Auditory sequence processing reveals evolutionarily conserved regions of frontal cortex in macaques and humans. Nat. Commun. 6.

Zhang, Y.S., Ghazanfar, A.A., 2016. Perinatally influenced autonomic system fluctuations drive infant vocal sequences. Curr. Biol. 26 (10), 1249–1260.

Zilberman, U., Smith, P., 1992. A comparison of tooth structure in Neanderthals and early *Homo sapiens sapiens*: a radiographic study. J. Anat. 180 (Pt 3), 387.

Zuberbühler, K., 2015. Linguistic capacity of non-human animals. Wiley Interdiscip. Rev. Cogn. Sci. 6 (3), 313–321.

Further reading

Bastir, M., Rosas, A., Gunz, P., Peña-Melian, A., Manzi, G., Harvati, K., Hublin, J.J., 2011. Evolution of the base of the brain in highly encephalized human species. Nat. Commun. 2, 588.

Benítez-Burraco, A., Boeckx, C., 2015. Possible functional links among brain- and skull-related genes selected in modern humans. Front. Psychol. 6, 794. https://doi.org/10.3389/fpsyg.2015.00794.

Ghazanfar, A.A., Poeppel, D., 2014. The Cognitive Neurosciences V.

Lewontin, R., 2000. The Triple Helix: Gene, Organism, and Environment. Harvard University Press, Cambridge, MA.

Theofanopoulou, C., 2015. Brain asymmetry in the white matter making and globularity. Front. Psychol. 6, 1355. https://doi.org/10.3389/fpsyg.2015.01355.

39

The Search for Human Cognitive Specializations

H. Clark Barrett

UCLA, Los Angeles, CA, United States

39.1 Introduction

The search for human cognitive specializations is one of the most active frontiers of research in the behavioral sciences, and also one of the most fraught with controversy. One reason is that the entities that this research seeks to identify—human cognitive specializations—are shrouded in epistemological uncertainty akin to the uncertainty surrounding "units of inheritance" in Darwin's time. While Darwin and his contemporaries did not know what the units of inheritance were, there were enough observations at the time to suggest that they were highly likely to exist. Today, a large body of psychological data, including data on cognitive differences between humans and other species, suggests that human cognitive specializations of some kind are likely to exist. But there is as yet no widespread agreement about what human cognitive specializations are, or how we would know one if we saw it.

There are reasons to suspect that satisfying answers to the units of cognition problem will be slower and harder to come by than answers to the units of inheritance problem were. For example, it is widely agreed that whatever human cognitive specializations might be, they are not likely to be strictly "particulate" in the same sense as genes. Whereas genes sometimes do appear to behave like functional Lego pieces than can be snapped in and out of the genome, information-processing specializations in the brain are not likely to be Lego-like. Rather, they are likely to be complex, polygenic traits with fuzzy boundaries that, like everything made of neurons, change dynamically and interactively over multiple timescales (Barrett, 2012; Barrett and Kurzban, 2006). Thus, while Mendel and other early geneticists got lucky in selecting a domain of study with several simple lawlike properties, human cognitive scientists can expect nothing of the sort.

It is not surprising, then, that not a single proposal of a human cognitive specialization has gained widespread acceptance in the scientific community in the manner that Mendel's findings did in the early 20th century. Even for a case as seemingly obvious as human language—obvious in the sense that it is clearly a complex functional trait that is uniquely elaborated in humans—there is little agreement about the underlying cognitive specializations that give rise to the trait (Christiansen and Kirby, 2003).

The reasons for this disappointing state of affairs are many (Barrett, 2015). One insight into the controversy lies in the term "human cognitive specialization" itself. As it turns out, this is a near perfect storm of scientific nomenclature, because each of its components is contentious in its own way. That any given aspect of cognition is unique to humans, that it involves a specialization in a cognitive system, and that it reflects an evolved biological adaptation, are all claims that invite controversy. Thus, identifying an intersection of these terms that a majority of behavioral and brain researchers would agree on is a difficult task indeed.

One way out of this perfect storm would be for researchers and critics to settle on some reasonable ground rules for what these terms mean. First, we should accept that there *might* be some cognitive specializations that explain aspects of human cognition, and that some of these might be cognitive autapomorphies: mechanisms unique to our lineage (though as explored later, even uniquely human cognitive processes can rely on mechanisms that are not uniquely human). We must at least entertain the possibility of uniquely human cognitive specializations to have any chance of finding them. Second, we must recognize that our enterprise is one of discovery and not of a priori definition. If we do not yet know exactly what human cognitive specializations

Evolutionary Neuroscience, Second Edition
https://doi.org/10.1016/B978-0-12-820584-6.00039-8

917

might be, we must leave some parameters free to vary. A problem in the search for human cognitive specializations has been that some have focused too narrowly on crisp philosophical definitions, such as the idea of a cognitive "module," that virtually define the phenomenon out of existence by setting biologically unreasonable criteria (Fodor, 1983). For those interested in the wet reality of biology, such as brain researchers, this literature has largely, and wisely, been ignored.

What is needed instead is a set of biologically informed criteria and methods that would allow the search for human cognitive specializations to have a reasonable possibility of discovering something. Here I will offer a brief sketch of a strategy for searching for human cognitive specializations, focusing in particular on the problem of generating and comparing proposals. Then I will offer a brief, nonexhaustive list of psychological abilities that have been proposed to be unique to humans. I will select three of these—language, mindreading, and culture—as case studies in which there are many competing proposals about the putative specializations underlying these abilities. These case studies show how far we are from being to make satisfying conclusions about the mechanisms underlying human psychological uniqueness, but they also point to the need for a more systematic approach to the problem. I conclude with suggestions for concrete steps we can take toward realizing the goal of mapping human cognitive specializations.

39.2 A Search Strategy for Human Cognitive Specializations

The first challenge for a proper science of human cognitive specializations is theoretical: to establish some theoretical and terminological ground rules that allow the search for human cognitive specializations to proceed without undue bias from conceptual frameworks that prejudge what we expect to see. This has been a major problem in the search for human cognitive specializations. On the evolutionary side, there is substantial controversy over the search for adaptations, with some arguing that theories about adaptations are largely "just-so stories" that are difficult to test, and others arguing that there is systematic bias to see humans as purely blank slates. On the cognitive side, debates have raged for decades regarding how to properly formulate models of cognition and to conceptualize what a cognitive mechanism is. Yet, hypotheses about human cognitive specializations must be, fundamentally, hypotheses about evolved adaptations (specializations, including ontogenetically shaped ones) that carry out cognitive (information-processing or psychological) functions. Thus, the primary source domains for

constructing theories of human cognitive specializations are theoretical biology, including evolutionary theory, genetics, and development, and theoretical psychology, including theoretical neuroscience and other areas of theory concerned with understanding how the brain processes information to guide behavior. In principle, any combination of theoretical tools from these domains can be used to construct hypotheses about cognitive specializations, under constraints, of course, of biological and psychological plausibility.

Even with agreement on those points, a crucial problem arises: in the realm of cognition, what counts as a "mechanism"? On biological grounds, we might expect the neural mechanisms giving rise to cognitive abilities to be diverse, and therefore not to conform to a single narrow model of functional design (Barrett, 2015). Empirically, this implies that we should remain open to multiple models of what the psychological mechanisms underlying a given cognitive skill might be. There are often, if not always, many possible mechanistic models of how a particular aspect of cognition might be carried out, and when that is true, we should be careful not to go beyond what the data can adjudicate. This arises especially where diverse cognitive phenomena appear to be both complex and causally interconnected. For example, spoken/gestural language and mindreading are both often nominated as uniquely human cognitive abilities. As discussed in the following paragraphs, however, these abilities appear causally intertwined as they manifest in psychology and behavior, leaving open various possibilities for how we should differentiate these abilities both in the behavioral phenotype, and for models of the architecture of cognitive mechanisms underlying them. For any given set of behaviorally observed abilities in a species, there are usually multiple possible cognitive architectures that could, in theory, give rise to those abilities. For just two putative "abilities," such as language and mindreading, the mapping functions between underlying mechanisms and these skills could be one-to-one, many-to-one, one-to-many, or many-to-many, as illustrated in Fig. 39.1. Moreover, the question of how to differentiate "abilities," such as language and mindreading, is usually a matter for debate, with some theories treating them as subtypes or products of a more generalized ability (eg, cultural transmission or general-purpose learning mechanisms).

Proper adjudication between models of cognitive architecture that might explain an observed set of abilities requires that all plausible possibilities be on the table; otherwise, data might falsely point us toward a model that happens to suit our prior theoretical biases. Ideally, then, the goal should be to formulate as many mechanistic hypotheses for a given set of observations as are consistent with knowledge from biology and neuroscience, and then to use existing data to choose between

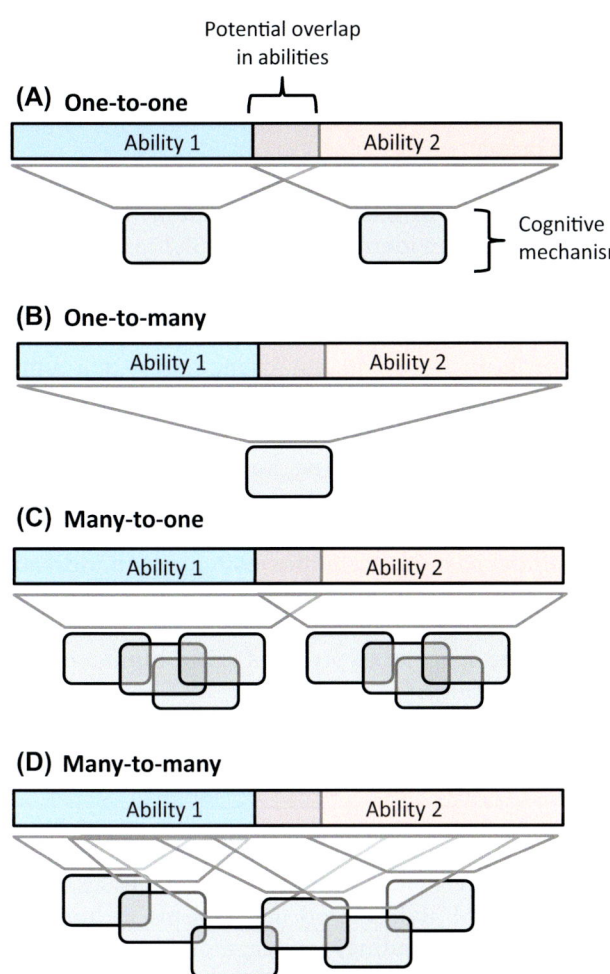

FIGURE 39.1 Potential mapping functions between cognitive specializations (*gray boxes*) and observed behavioral or psychological abilities (*colored bars*).

these hypotheses. Navigating this tradeoff optimally—to generate and entertain all possible mechanistic models that could account for the data in a biologically realistic way, but no more—is clearly very difficult and probably impossible to do perfectly. However, it contrasts rather starkly with current practice, in which theorists propose a small number of high-profile models, leading to a hypothesis set that is almost certainly many orders of magnitude smaller than the real set of possibilities. This makes it difficult to assign levels of confidence to proposals that do exist, or to estimate the probability that we have missed the true biological explanation for some set of observations. Moreover, because many proposals are presented as stark alternatives it is often difficult to see when one proposal is consistent with another, or when they have partly overlapping theoretical predictions. For example, models of uniquely human cognition that attribute our abilities to increased brain size are consistent with a host of more multimechanistic theories, because the properties and

interactions of specialized brain mechanisms may change as overall brain size increases (Allman, 2000; Kaas, 2000; Striedter, 2005).

Based on these considerations, a strategy for searching for human cognitive specializations would consist of three steps (not necessarily in a strict temporal order): (1) collection of data on the cognitive abilities of humans and other species in relatively standardized formats that allow systematic mapping of skills to proposed underlying mechanisms; (2) generation of the full set of hypotheses that could, in principle, account for the observed data; (3) use of techniques such as model comparison, metaanalysis, and data reduction to narrow down the set of hypotheses generated in step (2).

Step 1. Construct systematic maps of the cognitive phenomes of humans and other species. At present there exists a substantial body of data on the cognitive abilities of humans and other species, but typically these data exist in heterogeneous formats that are not easily compared. Data on cognitive abilities are scattered across many journals and academic fields, and often are not collected with an eye toward use beyond the immediate research context in which they are collected. Recently there have begun efforts in neuroscience and other fields to remedy this problem by translating data into standardized formats and assembling them into large databases, and to ask researchers to consider collecting and storing data in ways that allow them to be transported across research contexts. Biologically speaking, the cognitive abilities of humans and other species can be considered part of their cognitive "phenomes," ie, the phenotypic manifestations of evolved cognitive mechanisms (Bilder et al., 2009; Houle et al., 2010). Attention to the systematic assembly of databases on the cognitive phenomes of humans and other species—including data from behavioral studies, brain mapping, and the like—is key to systematic comparison of hypotheses about human cognitive specializations.

Step 2. Develop a large set of computational models of mechanisms that could generate the observed phenomic data. Theoretical neuroscience and computational modeling more generally offers tools for modeling neural mechanisms that give rise to observed patterns of behavior, including responses on experimental tasks and associated patterns of activity as measured through functional brain mapping and other technologies (eg, Eliasmith, 2013). Also possible, though less well-explored, are evolutionary developmental models of the ontogenetic reaction norms that could give rise to these neural structures. To arrive at the correct model of the human cognitive specializations that give rise to the data collected in Step 1, it is important to entertain the full range of theoretical possibilities that might explain the data. Thus, in the theorizing stage, a host of hypotheses should be developed. Moreover, these hypotheses should be formulated to be as specific as possible in terms of their empirical (phenomic) entailments, in order for them to be properly pitted against each other with respect to the data. In Bayesian terms, an important component of each hypothesis is the likelihood (probability) of observing a given pattern of data if the hypothesis were true. The fact that hypotheses may overlap in many regions of predicted data is a reason why the empirical entailments of different hypotheses must be specified as precisely as possible to properly adjust posterior probabilities on the basis of observations. Thus, the (perhaps continually expanding) result of this step would be a large set of models, or hypotheses, H_1, H_2, H_3, ... H_n.

Step 3. Use model comparison approaches to adjudicate between models generated in Step 2 using data from Step 1. Model comparison approaches differ from null hypothesis significance testing approaches in that they entertain, in principle, all possible models and allow existing data to weigh for or against each model. Bayesian approaches, for example, assign posterior probabilities to different hypotheses based on the likelihood of having observed the existing data if each hypothesis were true. Model comparison is an approach that is sometimes used in the literature on human cognitive specializations, but often in a piecemeal way using data from one or a few cognitive tasks, and comparing a small number of models. Ideally, the search for human cognitive specializations should attempt to compare diverse models using multiple data sets, and it should make use of tools such as metaanalysis and data-reduction techniques. At present, such techniques are often impossible due to the heterogeneous and patchy nature of data, which is why an explicit effort to assemble systematic maps of regions of the cognitive phenome will be an important step toward resolving debates about human cognitive specializations.

39.3 A Partial List of Uniquely Human Cognitive Abilities

We are currently far from being able to comprehensively execute the search strategy proposed earlier. We do not as yet have anything like a systematic map of the cognitive phenomes of humans or other species, and despite substantial progress in the development of computational models of cognition, we probably remain far from having explored the full possibility space of such models. Moreover, many if not most proposals of human cognitive specializations have not been adequately modeled from a computational perspective. However, it is possible to assemble a brief survey of uniquely human cognitive abilities that has been proposed, and to consider the range of possibilities of underlying mechanisms that might give rise to these skills.

Table 39.1 provides a (nonexhaustive) list of cognitive abilities that have been proposed to be unique to, or at least uniquely elaborated in, humans—though homologs of most, if not all, of these abilities likely appear in other primates.

A major question for the study of human cognitive specializations is as follows: Does each of the abilities and/or subabilities identified in Table 39.1 (and perhaps others) represent a distinct cognitive specialization? As illustrated in Fig. 39.1, there are usually multiple possibilities for the mapping function between underlying brain architecture and phenotypically observable skills. The literature on uniquely human abilities can be regarded as a source of hypotheses about these mapping functions, which can then be compared with existing data. This task is, of course, enormous, and beyond the scope of this chapter. Instead, I will select three proposed uniquely human abilities—language, mindreading, and cultural transmission—for closer examination as case studies for examining the challenges in narrowing down the space of hypotheses about human cognitive specializations.

39.4 Case Study: Language

Human language is a good example of a complex cognitive ability that almost certainly makes use of a mix of ancestral and derived mechanisms in humans. Similarly to mindreading, language as it manifests in humans is so functionally well organized and apparently seamless in its operation that it is difficult to decompose it into the underlying cognitive mechanisms that give rise to the observable surface skill. Hauser et al. (2002), for example, distinguish between a "broad" and "narrow" sense of the language faculty, suggesting that the faculty of language "narrow" (FLN) consists only of recursion and is the only aspect of language uniquely derived in humans. Additionally, they suggest that the FLN (recursion) might not be an adaptation specifically for language, leaving open the question of what it might be a specialization for (in other words, whether its proper domain is language per se, or whether it gives rise to language as a by-product of selection for some other function). While this is in principle a plausible proposal, it is just one of many possible hypotheses about the nature of the cognitive specializations underlying human language abilities. We might call it $H_{\text{Hauser et al. (2002)}}$. This hypothesis has empirical implications that are, in principle, testable with comparative cognitive phenomic data. For example, if human language turns out to require mechanisms other than recursion that are uniquely derived in humans, or if recursion proves not to be a universal feature of human languages, then at least aspects of $H_{\text{Hauser et al. (2002)}}$ are falsified.

TABLE 39.1   A partial list of proposed uniquely human cognitive abilities and associated subspecializations[a]

Proposed uniquely human cognitive ability	Proposed subskills and/or cognitive specializations supporting the ability
Language	Recursion (Hauser et al., 2002; Premack, 2007) Symbols (Deacon, 1997) Grammatical rules (Pinker and Bloom, 1990) Pragmatics/relevance (Sperber and Wilson, 1995) Mirror neurons (Rizzolatti and Arbib, 1998) Derived neural pathways (Rilling et al., 2008, 2012)
Mindreading	Theory of mind mechanism (Baron-Cohen, 1995; Leslie, 1994; Povinelli and Preuss, 1995) Shared intentionality (Tomasello, 2014) Empathy (Decety and Jackson, 2004; Premack, 2007; Saxe, 2006) Implicit mindreading (Apperly, 2010; Heyes and Frith, 2014)
Cultural transmission	Prestige and conformity bias (Boyd and Richerson, 1985) Pedagogy (Csibra and Gergely, 2011) Imitation (Meltzoff and Decety, 2003) Overimitation (Lyons et al., 2007) Shared intentionality (Tomasello, 2014)
Cooperation and morality	Third-party/altruistic punishment (Fehr and Fischbacher, 2004; Fehr and Gächter, 2002) Moral emotions (Haidt, 2007) Coalitional psychology (Kurzban et al., 2001) Norm acquisition (Boyd and Richerson, 2009) Strong reciprocity (Fehr et al., 2002) Cheater detection (Cosmides, 1989)
Tool use	Pedagogy (Csibra and Gergely, 2011) Function-based categorization (Johnson-Frey, 2004; Kemler Nelson et al., 2000) Tool-specific action knowledge (Johnson-Frey, 2004) Cause–effect inference (Povinelli, 2000; Premack, 2007) Functional fixedness (German and Defeyter, 2000) Overimitation (Lyons et al., 2007) Planning/working memory (Baddeley, 1986; Coolidge and Wynn, 2009; Premack, 2007; Stout et al., 2015)
Higher level cognitive and executive functions	Executive functions (Damasio, 1994; Luria, 1980; Stout, 2010; Stuss and Benson, 1984) Self-awareness (Leary and Buttermore, 2003; Wheeler et al., 1997) Planning/working memory (Baddeley, 1986; Coolidge and Wynn, 2009; Premack, 2007; Stout et al., 2015) Episodic memory/metamemory/mental time travel (Suddendorf and Corballis, 2007; Tulving, 1983; Wheeler et al., 1997) Metacognition (Metcalfe and Shimamura, 1994) Metarepresentation (Cosmides and Tooby, 2000; Perner, 1991; Sperber, 2000) Representation of unobservable states (Povinelli, 2000) Inhibitory and attentional control (Aron et al., 2004; Knight et al., 1995)

[a]Note: this is not intended as an exhaustive list of proposals. References are provided as examples of each proposal and are not comprehensive.

H$_{Hauser\ et\ al.\ (2002)}$ is an example of a rather simple, "one reason" hypothesis that accounts for the uniqueness of human language abilities via a single uniquely derived mechanism. The literature contains others that similarly propose a small number of mechanisms to account for human language abilities. For example, Tomasello's (2014) "shared intentionality" hypothesis roots human language abilities and cultural abilities more generally in two evolutionary steps: first, the evolution of mechanisms of "joint intentionality" (dyadic sharing of intentions and representations), and second, the evolution of "collective intentionality," giving rise to conventions and norms, such as those of human languages. These jointly enable uniquely human language (including pragmatics, proposed to be a unique feature of human language), and other abilities as well, such as large-scale cooperation. Thus, the mechanism of collective intentionality, while cognitively specialized, might not be a specialization for language per se; its proper domain, according to H$_{Tomasello\ (2014)}$, is cooperation, with language arising as a by-product.

Yet another hypothesis that roots the origins of language in a single mechanism is the mirror neuron hypothesis of language, proposed by Rizzolatti and Arbib

(1998). According to $H_{Rizzolatti\ and\ Arbib\ (1998)}$, the mirror neuron system, present in both humans and nonhuman primates, formed the evolutionary substrate from which language evolved, by providing a cognitive link between communicative acts (eg, speech) and action representation. Thus, the mirror neuron system served as the ancestral mechanism or set of mechanisms from which uniquely derived language abilities evolved. Note that each of these hypotheses—$H_{Hauser\ et\ al.\ (2002)}$, $H_{Tomasello\ (2014)}$, and $H_{Rizzolatti\ and\ Arbib\ (1998)}$—has empirical implications for observations we might make about the cognitive phenomes of humans and apes, and that some of these predictions are overlapping (eg, the unique presence of spoken language in humans) while others diverge (eg, the specific relationship between action and language implied by $H_{Rizzolatti\ and\ Arbib\ (1998)}$).

When we consider the set of possible hypotheses that could explain uniquely human language abilities, these are of course only several possibilities from among a much larger set. And it is important to note that the set of published hypotheses must be vastly smaller than the set of possible hypotheses. In addition to relatively simple models that posit one or a few factors to account for uniquely human language, there exist a host of more complex models involving multiple derived differences in the human lineage that give rise to the apparently seamless phenomenon of language. Decades of work on the brain basis of language have shown that it involves complex interactions between many brain regions (Binder et al., 1997; Price, 2000). While comparative brain anatomy so far has not revealed evidence for any entirely new (uniquely derived) brain regions (Petrides and Pandya, 1999), there is evidence for evolutionary *modification* of brain networks that have come to play an important role in human language, including derived cortical asymmetries in language areas in humans (Corballis, 1991; Geschwind and Galaburda, 1984) and modifications to connections between brain regions that are important in the language network, including white matter pathways (Friederici, 2009; Rilling et al., 2008, 2012). Moreover, we are just beginning to grapple with the significance of differences in gene sequences and patterns of gene expression between humans and other apes, which could have important consequences for the functional properties of brain mechanisms of language (Enard et al., 2002a,b; Cáceres et al., 2003). In a sense, each of these changes could imply its own H_i regarding uniquely derived specializations for language in humans. And, given that a proper analysis would require assessing the entire hypothesis space using available phenome data, all of the various combinations of $H_1 \ldots H_n$ should be allowed to compete for their share of explaining the data. Even hybrids of existing proposals, such as, for example, the union of $H_{Hauser\ et\ al.\ (2002)}$ (recursion) and $H_{Corballis\ (1991)}$ (left–

right asymmetry) should be on the table, if they are cogent possibilities.

39.5 Case Study: Mindreading

Mindreading, or "theory of mind" (ToM), refers to the ability to make inferences about, or to be contingently sensitive to, the mental states of others such as their intentions, goals, knowledge, beliefs, emotions, and motivations (Apperly, 2010; Baron-Cohen, 1995; Nichols and Stich, 2003). Like language, this is a complex and potentially multipart ability that appears to take a uniquely elaborated form in humans, though it is likely at least partly homologous with less elaborated versions of the skill in other primates (Call and Tomasello, 2008), and may even have convergently evolved analogs in more distantly related social taxa such as corvids (Clayton et al., 2007) and canids (Hare et al., 2002). As is the case for language, what appears to manifest as a relatively seamless ability at the behavioral level—eg, the ability to guess or respond appropriately to what someone else is thinking—might be the product of diverse underlying mechanisms, with a variety of possibilities for how these are carved up functionally and how they are distributed across species. Moreover, there are multiple possibilities regarding mechanisms that might be uniquely specialized to this domain, as opposed to shared across a broader domain or coopted from other domains. Many theories of language, for example, hold pragmatics to be a key feature of human language, and pragmatics draws heavily on mindreading (Tomasello, 2014; Scott-Phillips, 2014; Sperber and Wilson, 1995).

From the perspective of cognitive phenomics, it would be useful to have a comprehensive map of the full set of mindreading abilities in humans and across many taxa. Unfortunately, as is the case for almost all aspects of cognitive phenomes, the map is patchy in several ways. First, for historical reasons, a large majority of the research on mindreading abilities in humans and other species have focused on variants of a single kind of task, the false-belief task (Wellman et al., 2001). This task tests an individual's ability to use knowledge of another's false belief about the world (eg, knowledge that the individual falsely believes food to be hidden in location A when it is in fact in location B) to anticipate the other's behavior (eg, that the individual will search in location A and not B). Hundreds of replications of the standard version of this task suggested that it develops around the age of 4 years in humans (Wellman et al., 2001), but newer versions that do not require an explicit response on the part of the child suggest that the ability to track false beliefs may be present in infancy across human societies (Onishi and Baillargeon, 2005; Barrett et al., 2013). However, to date, no other species has been found to be able to pass any version of the false-belief task, suggesting that this aspect of ToM may be unique to humans (Call and Tomasello, 2008).

In some circles, false-belief understanding has become synonymous with ToM, but a broader ethological perspective suggests that tracking and using others' mental states may be important across a wide variety of social contexts and mental state types (Barrett, 2015). Indeed, tracking of other kinds of mental states, such as knowledge (true beliefs) is not unique to humans (Call and Tomasello, 2008; Clayton et al., 2007). And there may be unique aspects of human mindreading that go beyond false belief—for example, human-specific forms of empathy (Decety and Jackson, 2004) or the ability to form collective intentions (Tomasello, 2014).

Thus, one problem with our knowledge of the cognitive phenome space of mindreading is that task coverage of the space is patchy, with large amounts of research effort in some areas of the space and little or none in others. Another problem is comparative: it is difficult to draw comprehensive conclusions about what aspects of mindreading are truly unique to humans without a better coverage of the phenomes of closely and distantly related species. Much research effort has been devoted to a small number of taxa including chimpanzees, a few other primates, corvids, and canids. This has led to some intriguing proposals, such as the possibility of convergently evolved mindreading abilities in corvids and canids, but comparative data are as yet too incomplete to make firm conclusions about homology versus convergence in the distribution of mindreading skills across animal taxa.

As was the case with language, there exist a variety of proposals regarding the specific mechanisms, or cognitive specializations, underlying the skill of mindreading. One of the earliest in the neuropsychology literature drew on research with the false-belief task, and in particular, an apparent neural dissociation in patients with autism, who showed large deficits in performance on the false-belief task compared to age-matched controls, but not on comparable nonmentalistic tasks (Baron-Cohen et al., 1985). This led to an early proposed architecture of mindreading composed of several mechanisms, or modules: an eye direction detector (EDD) that forms representations of what another sees, an intentionality detector that forms representations of others' goals and desires, a shared attention mechanism that computes triadic representations of shared attention, and a Theory of Mind Mechanism (ToMM), responsible for tracking beliefs, which was held to be specifically impaired in autism, leading to a selective deficit in false-belief task performance (Baron-Cohen, 1995). Of these mechanisms, the EDD mechanism was proposed to be shared widely across taxa. Although the proposed phylogenetic distribution of the other mechanisms was

less clear, $H_{Baron-Cohen\ (1995)}$, thus proposes a four-part architecture with at least one part, ToMM, unique to humans. There are other proposals of relatively complex architectures. For example, Nichols and Stich (2003) proposed an architecture for "third-person" mindreading that involves at least 12 distinct mechanisms, including mechanisms for representing "possible worlds," a mechanism for detecting discrepant belief attribution, and so on. Not all of the mechanisms proposed in $H_{Nichols\ and\ Stich\ (2003)}$ would necessarily be unique to humans (eg, perception detection mechanisms), but this is an example of a very complex multipart hypothesis that would require detailed computational modeling to pit it against other proposals in terms of their relative likelihoods of observing particular patterns of data.

Other hypotheses attempt specifically to match mechanisms of mindreading to specific brain areas, such as Saxe's proposed components of uniquely human social cognition (Saxe, 2006). $H_{Saxe\ (2006)}$ posits two unique aspects of human mindreading and social cognition: first, "representational" ToM, in which human adults only (but not children or apes) represent what a person believes to be true of an object or state of affairs, and second, representing triadic relations [ie, between you, me, and something we are attending to or talking about; similar to proposals by Tomasello et al. (2005) and Tomasello (2014)]. She suggests that representational ToM is produced by a uniquely derived mechanism in the temporoparietal junction, and that triadic awareness might be produced by a uniquely derived mechanism in dorsal medial prefrontal cortex (with the additional proposal of partially unique human capacity to engage in emotional empathy located in ventral medial prefrontal cortex). $H_{Saxe\ (2006)}$ is precise in matching mechanisms with brain areas, and thus generates predictions about, eg, regional activation in mindreading tasks, that other theories might not.

Some proposals posit a much smaller number of underlying mechanisms. Heyes and Frith (2014) propose a "dual systems" model in which the mechanism responsible for "implicit" mindreading (which is automatic and does not require conscious reasoning) develops early in infancy, and "explicit" mindreading of the sort tested by standard false-belief tasks is a culturally acquired skill (see also Apperly, 2010). Like some of the hypotheses we saw regarding language, $H_{Heyes\ and\ Frith\ (2014)}$ is agnostic whether the implicit mindreading mechanism is an adaptation *for* mindreading per se, as opposed to a developmental by-product of mechanisms specialized for some other domain. However, it suggests that the uniquely human skill lies only in the "start-up kit" given by the uniquely human implicit mindreading mechanism, and that the rest of explicit mindreading is a by-product of human skills of cultural transmission, which are not specialized for mindreading.

As was the case for language, then, there is an abundance of diverse proposals in the literature, and many as yet unproposed possibilities, for the cognitive architecture of mindreading. A problem here, as for other putative human cognitive specializations, is that many existing hypotheses overlap in their predictions to a degree that renders existing data insufficient to adjudicate between them. For example, $H_{Baron-Cohen\ (1995)}$, $H_{Heyes\ and\ Frith\ (2014)}$, and $H_{Saxe\ (2006)}$ all predict that performance on "explicit" or standard false-belief tasks will only occur in human adults and older children but not infants or other apes, but offer different accounts of the underlying mechanistic reasons for this. $H_{Heyes\ and\ Frith\ (2014)}$ and $H_{Saxe\ (2006)}$ overlap at least partly on the involvement of the temporoparietal junction, for example, but differ (potentially) concerning how the adult skill is acquired, whereas $H_{Baron-Cohen\ (1995)}$ is agnostic about neural instantiation and differs starkly from $H_{Heyes\ and\ Frith\ (2014)}$ on the issue of acquisition. Currently it is clear that competence on standard false-belief tasks is not reached until middle childhood, a prediction made by all three accounts (Wellman et al., 2001). There is also evidence for individual and cross-cultural variation in age at which above-chance performance is reached (Heyes and Frith, 2014), but there is not yet sufficient phenomic resolution to adjudicate between different accounts of acquisition (eg, between a more maturation-based hypothesis, $H_{Baron-Cohen\ (1995)}$, and a strictly learning-based hypothesis, $H_{Heyes\ and\ Frith\ (2014)}$). Because the details of ontogeny are, ultimately, necessary features of hypotheses about human cognitive specialization, this is an area where it seems that multiple hypotheses must remain on the table.

An additional problem in the case of mindreading is that, while there seems to be substantial agreement about the uniquely human capacity to track others' false beliefs, there is more ambiguity about which, if any, other aspects of mindreading might be uniquely human. This is due in part to the large imbalance of research on the false-belief task as opposed to other aspects of intersubjective cognition. Another putatively uniquely human aspect of mindreading is the ability to represent triadic relations between self, other, and one-third entity, predicted by $H_{Saxe\ (2006)}$, $H_{Tomasello\ (2014)}$, and $H_{Nichols\ and\ Stich\ (2003)}$. While there is some cognitive phenomic evidence to suggest this might be uniquely human, this is a much less-studied area of mindreading, and its relationship to false-belief tracking—is it part of the same derived trait or an entirely separate evolutionary event?—is not clear. Moreover, there may well be additional aspects of mindreading that are uniquely human, such as a uniquely elaborated form of emotional empathy proposed by Saxe (2006), and others as well.

Finally, there is the question of the independence of mindreading from other domains of cognition. It is plausible that mindreading is causally related, both proximately and ultimately, to other potentially uniquely derived human cognitive abilities such as language, cooperation, cultural transmission, and executive control. For example, Sperber and Wilson (1995) and Scott-Phillips (2014) have proposed that pragmatic understanding is a unique aspect of human language that depends on mindreading. Tomasello (2014) proposes that the ability to form group-level cooperative intentions is uniquely human. Csibra and Gergely (2011) propose that human learning about tools and other artifacts depends on pedagogy, a mindreading—like communicative ability that requires infants to interpret adults' goals. Heyes (2014) suggests that explicit mindreading depends on uniquely human abilities of executive control, as evidenced by experimental data showing that competition from other executive tasks interferes with explicit mindreading (Bull et al., 2008; Perner and Lang, 1999). This is an intriguing proposal, given that executive control is another area where uniquely human abilities have been proposed (Baddeley, 1986; Damasio, 1994; Knight et al., 1995; Luria, 1980; Stout, 2010; Stuss and Benson, 1984; see Table 39.1).

All of these proposals are relevant to the units of cognition question: whether mindreading constitutes a distinct ability; whether it is the by-product of other abilities; whether mindreading and other abilities share architecture in a many-to-many fashion; and so on. At the ultimate (evolutionary) level, they are questions about whether the evolution of mindreading and the evolution of these other abilities represent separate evolutionary events, or whether some of these cases might represent the evolution of seemingly multiple aspects of the cognitive phenome via selection for single effects (eg, pragmatic communication, cooperation). At the proximate (mechanistic) level, they are questions about the mechanistic independence of these aspects of the cognitive architecture. Consider, for example, Hauser et al.'s (2002) proposal to distinguish the "faculty of language broad," shared with other primates, from the "FLN," unique to humans. According to $H_{Sperber\ and\ Wilson\ (1995)}$ and $H_{Scott-Phillips\ (2014)}$, mindreading is both unique to humans and necessary for human language, making it part of FLN, but in that case it is not the same FLN as $H_{Hauser\ et\ al.\ (2002)}$, which includes only recursion. These examples show that there remains a large set of possible hypotheses carving the set of things that humans seem uniquely able to do into the underlying cognitive specializations that give rise to them.

39.6 Case Study: Culture

Of the putatively unique human abilities listed in Table 39.1, culture—and our abilities to transmit and use it— is perhaps the most difficult of all when it comes to the problem of discerning the underlying cognitive specializations that give rise to it. Here again, it seems a near certainty that there must be uniquely derived cognitive mechanisms in humans that allow us, in Klein's (2003) terms, to "wield culture" in ways that no other species appears able to do. As anthropologists have documented with increasing degrees of certainty, human capacities to transmit and use cultural information appear to have been a large and necessary part of the "secret of our success" (Henrich, 2015). But, is culture itself a "thing," a unified domain with its own cognitive specializations that enable it? If we make a distinction, following Hauser et al. (2002), between a faculty of culture "broad", which includes mechanisms shared by humans and nonhuman primates, and a faculty of culture "narrow" (FCN), which includes all and only the cognitive specializations that enable uniquely derived aspects of human culture, then what mechanisms does FCN contain?

Unfortunately, this is yet another domain where the strong and highly plausible intuition that humans must have uniquely derived mechanisms exceeds our ability, at present, to say exactly what those uniquely derived mechanisms are. As in the other domains we have examined—language and mindreading—there is no shortage of proposals regarding the cognitive specializations underlying uniquely human cultural capacities, and anthropologists, psychologists, biologists, and neuroscientists are increasingly turning their attention to them.

Evolutionary theorists of cultural transmission point to a potentially important distinction between the social learning capacities of other animals, and the cultural capacities of humans. Human cultural change is "ratcheting" or "cumulative," meaning that it can produce representations, practices, and products that increase in complexity over generations (Boyd and Richerson, 1985; Tennie et al., 2009). Social learning abilities seen in other species can lead to the spread of behavioral preferences and practices in a population, such as preferences for particular mates (Dugatkin, 1992), food preferences (Galef, 1993), tool use techniques (Whiten et al., 2009), and social rituals (Perry, 2011). However, these preferences and practices do not generally become increasingly complex and elaborate over the course of many generations, as human cultural practices do (eg, techniques of artifact manufacture, architecture, art, clothing, music). Moreover, humans possess uniquely derived means of cultural transmission—most notably, language—that are not present in other apes. Because chimpanzees possess nonratcheting capacities of social learning—for example, transmission of food-processing technologies such as termite fishing—it is assumed, via phylogenetic parsimony, that the common ancestor of humans and chimps did as well, and that these nonratcheting social transmission abilities are the

ancestral state from which derived capacities of cumulative cultural evolution evolved in the hominin lineage.

This raises the question, of course, of what the proper description of the cognitive specializations underlying social learning in the great apes and other nonhuman primates might be, and which of them are or are not shared with humans. As for the other domains reviewed earlier, there exists a broad range of proposals for the architecture of mechanisms underlying social learning, ranging from the very simple to the very complex. Brass and Heyes (2005), for example, propose that all "imitation" (a type of social learning) is carried out by mechanisms of associative learning and action control that are not specific to social learning, and that imitation as a specific form of learning is enabled by learned perceptual-motor links. Thus, according to H$_{\text{Brass and Heyes (2005)}}$, FCN might be an empty set, if mechanisms of social learning are the same as mechanisms of individual learning, and therefore, not specialized for social learning at all. At the more elaborate end of the spectrum of possibilities, Whiten et al. (2004) propose that social learning encompasses at least four types— copying, affordance learning, observational conditioning, and enhancement—some of which are broken down into many more subtypes. Within copying, for example, there is imitation, object movement reenactment, and end-state emulation, the last of which itself comes in two different kinds. Thus, according to H$_{\text{Whiten et al. (2004)}}$, social learning can be broken down into anywhere between half a dozen to a dozen or so subtypes, each presumably making use of different underlying mechanisms, and distributed in different combinations across taxa. According to the search strategy outlined in Section 39.2, then, H$_{\text{Whiten et al. (2004)}}$, would actually be best broken down into many subhypotheses, each postulating a specific combination, across taxa, from this menu of proposed mechanisms.

All of this leaves open the question of the FCN: which mechanisms, if any, are uniquely derived in humans to carry out the function of enabling the putatively unique form of cumulative culture seen in our species? Again, there are many possibilities. One reading of H$_{\text{Brass and Heyes (2005)}}$, for example, is that FCN is empty because imitation and other forms of cultural transmission are all by-products of mechanisms not specific to the domain. Tomasello (2009), in contrast, propose that cumulative cultural evolution is made possible by human-specific features of mindreading (in particular, shared intentionality), along with language and teaching (also both uniquely human), so H$_{\text{Tomasello (2009)}}$ would presumably require at least three underlying cognitive specializations and perhaps more. H$_{\text{Tomasello (2014)}}$ proposes a specific, two-stage process in which the ability to share dyadic intentions later gave rise to the capacity for collective intentionality, with benefits arising from cooperative activities providing the main source of selection. Another proposal is that of Csibra and Gergely (2011), who propose that a specific form of cultural transmission, which they call "natural pedagogy," evolved originally because of the benefits of transmitting category-specific information about tool use (complex tool use is another putatively uniquely derived human cognitive ability, which might or might not be a separate domain from culture more generally; see Table 39.1). According to H$_{Csibra and Gergely (2011)}$, what is unique to humans is neither social learning nor communication per se, but rather, social learning via communication, or natural pedagogy. H$_{Csibra and Gergely (2011)}$ specifies a particular evolutionary time frame for the evolution of the proposed specialization (constrained by archeological data on the appearance of complex tools beginning approximately 2.5 million years ago; see also Coolidge and Wynn, 2009; Stout et al., 2015). It posits the mechanism to be separate from language and mindreading, and makes specific predictions about the cognitive phenomics of the ability, including its cross-cultural universality and appearance early in development. Each of these hypotheses—as well as others—make distinct, though partly overlapping, predictions about what should be observed phenomically. However, as in other domains, there has not yet been systematic comparison of them using phenomic data to compare all the hypotheses simultaneously. Given the partial overlap of predictions, the use of some form of model comparison would be particularly helpful, because it would in theory take into account, in a simultaneous way, where predictions of different hypotheses overlap and diverge. However it is not clear whether existing data are fine grained enough to adjudicate between them.

To add to the complexity, there are many proposals of structuring and submechanisms within the domain of cumulative cultural evolution. According to one tradition, culture–gene coevolution theory, learning mechanisms can be categorized based on the learning biases they create (in this literature, the term "mechanism" typically means a mechanism of cultural evolutionary change, but sometimes this implies the existence of distinct underlying cognitive mechanisms of learning and transmission; Boyd and Richerson, 1985; Mesoudi, 2009). Context biases influence the tendency to acquire variants of learnable information as a function of the context in which those acquired, such as whether they are common in the population (conformity bias) or come from prestigious individuals (prestige bias). Content biases influence the learnability of an item based on what it is about, eg, whether the information is social (Mesoudi et al., 2006) or negative (Fessler et al., 2014), or whether it falls into a particular ancestrally important

category such as animals or artifacts (Barrett and Broesch, 2012; Barrett et al., 2016).

Experimentally, there exists evidence for all of these kinds of bias. Important questions for the study of human cognitive specializations are whether these biases map onto underlying brain mechanisms in a one-to-one fashion, and whether each bias corresponds to a uniquely derived mechanism of cultural transmission in humans. While many such relationships are possible, it is nearly certain that the relationship is not entirely one to one, and that at least some cognitive processes shaping the "learnability landscape" of cultural information might predate the evolution of human culture per se. For example, cross-cultural developmental studies in humans have shown a bias in human children for learning danger information, a bias that is particularly strong in the domain of animals (as opposed to, eg, artifacts; Barrett and Broesch, 2012; Barrett et al., 2016). However, similar biases for learning rapidly about animal danger from conspecifics have been shown in other species of primate (Cook and Mineka, 1989), raising the possibility that this is a bias that is ancestral to the hominin lineage and therefore not a uniquely derived specialization for cultural transmission in humans, and not part of FCN.

The question of whether culture is an independent domain or skill from other putatively uniquely human skills, such as language, mindreading, and cooperation, is an important and difficult one. There are at least two questions here: whether the domain of culture is independent of these other domains at an ultimate evolutionary level (ie, did distinct selection pressures, or the same ones, give rise to them?), and whether they are independent or partly so at the proximate, mechanistic level (ie, are there mechanisms of cultural transmission that are unique to that domain?). There are reasons to suspect that the answers to these questions might be complex. For example, one could make the claim that all linguistic transmission is a form of cultural transmission, and so the capacity for language should be nested within the capacity for culture—raising questions such as which underlying cognitive mechanisms might be special to language, whether language can make use of all mechanisms within the domain of cultural transmission, etc. Similarly, some have claimed that every act of linguistic comprehension (or very nearly so) is also an act of mindreading, because to properly understand a speech act is to understand the meaning the speaker intended (Sperber and Wilson, 1995). While this claim is debated, there is no doubt that much of language relies heavily on mindreading, again raising the question of the independence of these domains, evolutionarily and mechanistically. For culture and the domains it makes contact with or even possibly subsumes, then, questions of underlying architecture remain largely unanswered. What are the distinct parts, or specializations, if any? What is their evolutionary history, and how do they give rise to the heavily overlapping phenomena of language, culture, and social cognition?

39.7 Conclusion: Narrowing the Search

Typically, a review of human cognitive specializations would endorse a particular picture of what those specializations are, based on the state of the evidence. The message of this chapter, however, is that this is usually done too hastily. Such a synthesis might be possible, but I have chosen instead to focus on what we do not yet know, and how we might develop a more explicit and rationally designed search strategy for the pursuit of human cognitive specializations. At present, the search for human cognitive specializations is akin to a kind of free-for-all gold rush, with different parties searching in different places, some striking gold, some falsely thinking they have, and large sections of the landscape left untouched. I use metaphors of maps and landscapes intentionally here, and urge that we explicitly conceptualize the search for human cognitive specializations as an effort to create a map of the human cognitive phenome. Ours, at present, resembles early European maps of the world: incomplete and distorted, with the blank spaces filled in by conjecture.

One problem, then, is the incompleteness of our map of what human cognitive abilities are, what their correlation structure is, and how they vary or do not vary across cultures and across development. Houle et al. (2010) are correct in identifying the mapping of the phenome—including, in this case, the cognitive phenome—as "the next challenge." Adding to this challenge is the comparative challenge of producing comparable maps for other species; without such maps, identifying what is unique about human cognition is a hopeless task.

A second major problem is the inadequacy of current theoretical models of human cognitive specializations. In fact, there are at least two subproblems here. One is that many, if not most, current hypotheses about human cognitive specializations are massively underspecified, in terms of what they would predict, and which patterns of data would or would not be consistent with them. This underspecification means that it is difficult to adjudicate between hypotheses with large areas of predictive overlap. It also means that it is possible, and sometimes even justified, for opposing theoretical camps to claim the same data as consistent with their proprietary hypotheses. This is indeed the state of affairs for all three of the case study domains I surveyed before—language, mindreading, and culture—as well as others I did not survey, such as cooperation, where decades of research has failed to resolve debates about possible explanations for uniquely human

patterns of behavior, such as cultural group selection (Boyd and Richerson, 2010; West et al., 2007).

Another problem for models of human cognitive specializations, in addition to the problem of underspecification, is that there are not enough of them (Barrett, 2015). This might seem like a radical proposal, especially since the theoretical landscape seems to be awash in competing proposals that we are unable to satisfyingly test between. However, that problem, I argue, arises mainly from underspecification (ie, vagueness or the capacity to explain too much), combined with the gold rush mentality of staking out a claim in the vast territory of potential hypotheses and then defending it against all comers. What this can mean is that a hypothesis might seem to be the best in town because it is, indeed, the best among those available: other, perhaps equally plausible hypotheses might exist, but simply have not been proposed. And, large areas of the landscape of possible hypotheses might not have been explored for many reasons, including conformity, which can lead to failure to converge on the best explanation (eg, a new hypothesis that differs only in small details from an existing, well-known hypothesis might be difficult to publish). For these reasons, even if we were to adopt the best possible information—theoretic approach to adjudicating between existing hypotheses, such as a Bayesian approach, our estimates of the posterior probabilities of hypotheses being true would be warped by our inadequate coverage of the hypothesis space and distorted estimates of the priors.

Even if these points are agreed upon, this does not make the task of a broader and more inclusive search an easy one. For example, even if we wanted to move away from the more intuition-based method of generating hypotheses that many evolutionary behavioral scientists currently follow, there do not currently exist algorithms for populating hypotheses space in an unbiased manner. Nor is there an obvious way to map the cognitive phenome in a neutral fashion that prevents researchers from massively overstudying certain aspects of cognition while leaving others unstudied. However, the argument here is that it is better to recognize and face these problems explicitly, as daunting as they might be, than to simply shrug and carry on with the gold rush. The search for human cognitive specializations is likely to produce a much better map of reality if its geographers use the best possible tools for the job.

References

Allman, J.M., 2000. Evolving Brains. Scientific American Library, New York.

Apperly, I., 2010. Mindreaders: The Cognitive Basis of "Theory of Mind". Psychology Press, New York.

Aron, A.R., Robbins, T.W., Poldrack, R.A., 2004. Inhibition and the right inferior frontal cortex. Trends Cogn. Sci. 8, 170−177.

Baddeley, A., 1986. Working Memory. Oxford University Press, New York.

Baron-Cohen, S., 1995. Mindblindness: An Essay on Autism and Theory of Mind. MIT Press, Cambridge, MA.

Baron-Cohen, S., Leslie, A.M., Frith, U., 1985. Does the autistic child have a "theory of mind"? Cognition 21, 37–46.

Barrett, H.C., 2012. A hierarchical model of the evolution of human brain specializations. Proc. Natl. Acad. Sci. 109, 10733–10740.

Barrett, H.C., Broesch, J., 2012. Prepared social learning about dangerous animals in children. Evol. Hum. Behav. 33, 499–508.

Barrett, H.C., Kurzban, R., 2006. Modularity in cognition: framing the debate. Psychol. Rev. 113, 628–647.

Barrett, H.C., Broesch, T., Scott, R.M., He, Z., Baillargeon, R., Wu, D., Bolz, M., Henrich, J., Setoh, P., Wang, J., Laurence, S., 2013. Early false-belief understanding in traditional non-Western societies. Proc. R. Soc. Biol. Sci. 280, 20122654.

Barrett, H.C., Peterson, C.D., Frankenhuis, W.E., 2016. Mapping the cultural learnability landscape of danger. Child Dev. 87, 770–781.

Barrett, H.C., 2015. The Shape of Thought: How Mental Adaptations Evolve. Oxford University Press, New York.

Bilder, R.M., Sabb, F.W., Cannon, T.D., London, E.D., Jentsch, J.D., Parker, D.S., Poldrack, R.A., Evans, C., Freimer, N.B., 2009. Phenomics: the systematic study of phenotypes on a genome-wide scale. Neuroscience 164, 30–42.

Binder, J.R., Frost, J.A., Hammeke, T.A., Cox, R.W., Rao, S.M., Prieto, T., 1997. Human brain language areas identified by functional magnetic resonance imaging. J. Neurosci. 17, 353–362.

Boyd, R., Richerson, P.J., 1985. Culture and the Evolutionary Process. University of Chicago Press, Chicago.

Boyd, R., Richerson, P.J., 2009. Culture and the evolution of human cooperation. Philos. Trans. R. Soc. Lond. B Biol. Sci. 364, 3281–3288.

Boyd, R., Richerson, P.J., 2010. Transmission coupling mechanisms: cultural group selection. Philos. Trans. R. Soc. Lond. B Biol. Sci. 365, 3787–3795.

Brass, M., Heyes, C., 2005. Imitation: is cognitive neuroscience solving the correspondence problem? Trends Cogn. Sci. 9, 489–495.

Bull, R., Phillips, L.H., Conway, C.A., 2008. The role of control functions in mentalizing: dual-task studies of theory of mind and executive function. Cognition 107, 663–672.

Cáceres, M., Lachuer, J., Zapala, M.A., Redmond, J.C., Kudo, L., Geschwind, D.H., Lockhart, D.J., Preuss, T.M., Barlow, C., 2003. Elevated gene expression levels distinguish human from nonhuman primate brains. Proc. Natl. Acad. Sci. 100, 13030–13035.

Call, J., Tomasello, M., 2008. Does the chimpanzee have a theory of mind? 30 years later. Trends Cogn. Sci. 12, 187–192.

Christiansen, M.H., Kirby, S., 2003. Language evolution: consensus and controversies. Trends Cogn. Sci. 7, 300–307.

Clayton, N.S., Dally, J.M., Emery, N.J., 2007. Social cognition by food-caching corvids: the western scrub-jay as a natural psychologist. Philos. Trans. R. Soc. Biol. Sci. 362, 507–522.

Cook, S., Mineka, S., 1989. Observational conditioning of fear to fear-relevant versus fear-irrelevant stimuli in rhesus monkeys. J. Abnorm. Psychol. 98, 448–459.

Coolidge, F.L., Wynn, T., 2009. The Rise of Homo Sapiens: The Evolution of Modern Thinking. John Wiley & Sons, New York.

Corballis, M.C., 1991. The Lopsided Ape: Evolution of the Generative Mind. Oxford University Press, New York.

Cosmides, L., 1989. The logic of social exchange: has natural selection shaped how humans reason? Studies with the Wason selection task. Cognition 31, 187–276.

Cosmides, L., Tooby, J., 2000. Consider the source: the evolution of adaptations for decoupling and metarepresentation. In: Sperber, D. (Ed.), Metarepresentations: A Multidisciplinary Perspective. Oxford University Press, New York, pp. 53–116.

Csibra, G., Gergely, G., 2011. Natural pedagogy as evolutionary adaptation. Philos. Trans. R. Soc. Biol. Sci. 366, 1149–1157.

Damasio, A.R., 1994. Descartes' Error. Putnam, New York.

Deacon, T.W., 1997. The Symbolic Species: The Co-evolution of Language and the Brain. Norton, New York.

Decety, J., Jackson, P.L., 2004. The functional architecture of human empathy. Behav. Cogn. Neurosci. Rev. 3, 71–100.

Dugatkin, L.A., 1992. Sexual selection and imitation: females copy the mate choice of others. Am. Nat. 139, 1384–1389.

Eliasmith, C., 2013. How to Build a Brain: A Neural Architecture for Biological Cognition. Oxford University Press, New York.

Enard, W., Khaitovich, P., Klose, J., Zöllner, S., Heissig, F., Giavalisco, P., Nieselt-Struwe, K., Muchmore, E., Varki, A., Ravid, R., Doxiadis, G.M., 2002. Intra- and interspecific variation in primate gene expression patterns. Science 296, 340–343.

Enard, W., Przeworski, M., Fisher, S.E., Lai, C.S., Wiebe, V., Kitano, T., Monaco, A.P., Pääbo, S., 2002. Molecular evolution of FOXP2, a gene involved in speech and language. Nature 418, 869–872.

Fehr, E., Fischbacher, U., 2004. Third-party punishment and social norms. Evol. Hum. Behav. 25, 63–87.

Fehr, E., Gächter, S., 2002. Altruistic punishment in humans. Nature 415, 137–140.

Fehr, E., Fischbacher, U., Gächter, S., 2002. Strong reciprocity, human cooperation, and the enforcement of social norms. Hum. Nat. 13, 1–25.

Fessler, D.M., Pisor, A.C., Navarrete, C.D., 2014. Negatively-biased credulity and the cultural evolution of beliefs. PLoS One 9, e95167.

Fodor, J.A., 1983. The Modularity of Mind. MIT Press, Cambridge, MA.

Friederici, A.D., 2009. Pathways to language: fiber tracts in the human brain. Trends Cogn. Sci. 13, 175–181.

Galef Jr., B.G., 1993. Functions of social learning about food: a causal analysis of effects of diet novelty on preference transmission. Anim. Behav. 46, 257–265.

German, T.P., Defeyter, M.A., 2000. Immunity to functional fixedness in young children. Psychon. Bull. Rev. 7, 707–712.

Geschwind, N., Galaburda, A.M., 1984. Cerebral Dominance: The Biological Foundations. Harvard University Press, Cambridge, MA.

Haidt, J., 2007. The new synthesis in moral psychology. Science 316, 998–1002.

Hare, B., Brown, M., Williamson, C., Tomasello, M., 2002. The domestication of social cognition in dogs. Science 298, 1634–1636.

Hauser, M.D., Chomsky, N., Fitch, W.T., 2002. The faculty of language: what is it, who has it, and how did it evolve? Science 298, 1569–1579.

Henrich, J., 2015. The Secret of Our Success: How Culture Is Driving Human Evolution, Domesticating Our Species, and Making Us Smarter. Princeton University Press, Princeton, NJ.

Heyes, C., 2014. Submentalizing: I am not really reading your mind. Perspect. Psychol. Sci. 9, 131–143.

Heyes, C.M., Frith, C.D., 2014. The cultural evolution of mindreading. Science 344, 1243091.

Houle, D., Govindaraju, D.R., Omholt, S., 2010. Phenomics: the next challenge. Nat. Rev. Genet. 11, 855–866.

Johnson-Frey, S.H., 2004. The neural bases of complex tool use in humans. Trends Cogn. Sci. 8, 71–78.

Kaas, J.H., 2000. Why is brain size so important: design problems and solutions as neocortex gets bigger or smaller. Brain Mind 1, 7–23.

Kemler Nelson, D.G., Frankenfield, A., Morris, C., Blair, E., 2000. Young children's use of functional information to categorize artifacts: three factors that matter. Cognition 77, 133–168.

Klein, R.G., 2003. Paleoanthropology. Whither the Neanderthals? Science 299, 1525–1527.

Knight, R.T., Grabowecky, M.F., Scabini, D., 1995. Role of human prefrontal cortex in attention control. Adv. Neurol. 66, 21–36.

Kurzban, R., Tooby, J., Cosmides, L., 2001. Can race be erased? Coalitional computation and social categorization. Proc. Natl. Acad. Sci. 98, 15387–15392.

Leary, M.R., Buttermore, N.R., 2003. The evolution of the human self: tracing the natural history of self-awareness. J. Theory Soc. Behav. 33, 365–404.

Leslie, A.M., 1994. ToMM, ToBy, and Agency: core architecture and domain specificity. In: Hirschfeld, L.A., Gelman, S.A. (Eds.), Mapping the Mind: Domain Specificity in Cognition and Culture. Cambridge University Press, New York, pp. 119–148.

Luria, A.R., 1980. Higher Cortical Functions in Man, second ed. Basic Books, New York.

Lyons, D.E., Young, A.G., Keil, F.C., 2007. The hidden structure of overimitation. Proc. Natl. Acad. Sci. 104, 19751–19756.

Meltzoff, A.N., Decety, J., 2003. What imitation tells us about social cognition: a rapprochement between developmental psychology and cognitive neuroscience. Philos. Trans. R. Soc. Biol. Sci. 358, 491–500.

Mesoudi, A., 2009. How cultural evolutionary theory can inform social psychology and vice versa. Psychol. Rev. 116, 929.

Mesoudi, A., Whiten, A., Dunbar, R., 2006. A bias for social information in human cultural transmission. Br. J. Psychol. 97, 405–423.

Metcalfe, J.E., Shimamura, A.P., 1994. Metacognition: Knowing about Knowing. MIT Press, Cambridge, MA.

Nichols, S., Stich, S.P., 2003. Mindreading: An Integrated Account of Pretence, Self-Awareness, and Understanding Other Minds. Oxford University Press, New York.

Onishi, K.H., Baillargeon, R., 2005. Do 15-month-old infants understand false beliefs? Science 308, 255–258.

Perner, J., 1991. Understanding the Representational Mind. MIT Press, Cambridge, MA.

Perner, J., Lang, B., 1999. Development of theory of mind and executive control. Trends Cogn. Sci. 3, 337–344.

Perry, S., 2011. Social traditions and social learning in capuchin monkeys (*Cebus*). Philos. Trans. R. Soc. Biol. Sci. 366, 988–996.

Petrides, M., Pandya, D.N., 1999. Dorsolateral prefrontal cortex: comparative cytoarchitectonic analysis in the human and the macaque brain and corticocortical connection patterns. Eur. J. Neurosci. 11, 1011–1036.

Pinker, S., Bloom, P., 1990. Natural language and natural selection. Behav. Brain Sci. 13, 707–727.

Povinelli, D.J., 2000. Folk Physics for Apes: The Chimpanzee's Theory of How the World Works. Oxford University Press, New York.

Povinelli, D.J., Preuss, T.M., 1995. Theory of mind: evolutionary history of a cognitive specialization. Trends Neurosci. 18, 418–424.

Premack, D., 2007. Human and animal cognition: continuity and discontinuity. Proc. Natl. Acad. Sci. 104, 13861–13867.

Preuss, T.M., 2004. What is it like to be a human? In: Gazzaniga, M.S. (Ed.), The Cognitive Neurosciences, third ed. MIT Press, Cambridge, MA, pp. 5–22.

Price, C.J., 2000. The anatomy of language: contributions from functional neuroimaging. J. Anat. 197, 335–359.

Rilling, J.K., Glasser, M.F., Preuss, T.M., Ma, X., Zhao, T., Hu, X., Behrens, T.E.J., 2008. The evolution of the arcuate fasciculus revealed with comparative DTI. Nat. Neurosci. 11, 426–428.

Rilling, J., Glasser, M.F., Jbabdi, S., Andersson, J., Preuss, T.M., 2012. Continuity, divergence, and the evolution of brain language pathways. Front. Evol. Neurosci. 3, 11.

Rizzolatti, G., Arbib, M.A., 1998. Language within our grasp. Trends Neurosci. 21, 188–194.

Saxe, R., 2006. Uniquely human social cognition. Curr. Opin. Neurobiol. 16, 235–239.

Scott-Phillips, T., 2014. Speaking Our Minds: Why Human Communication Is Different, and How Language Evolved to Make it Special. Palgrave MacMillan, New York.

Sperber, D., 2000. Metarepresentations in an evolutionary perspective. In: Sperber, D. (Ed.), Metarepresentations: A Multidisciplinary Perspective. Oxford University Press, New York, pp. 117–138.

Sperber, D., Wilson, D., 1995. Relevance: Communication and Cognition, second ed. Blackwell, Oxford, UK.

Stout, D., 2010. The evolution of cognitive control. Top. Cogn. Sci. 2, 614–630.

Stout, D., Hecht, E., Khreisheh, N., Bradley, B., Chaminade, T., 2015. Cognitive demands of lower paleolithic toolmaking. PLoS One 10, e0121804.

Striedter, G.F., 2005. Principles of Brain Evolution. Sinauer, Sunderland, MA.

Stuss, D.T., Benson, D.F., 1984. Neuropsychological studies of the frontal lobes. Psychol. Bull. 95, 3–28.

Suddendorf, T., Corballis, M.C., 2007. The evolution of foresight: what is mental time travel, and is it unique to humans? Behav. Brain Sci. 30, 299–313.

Tennie, C., Call, J., Tomasello, M., 2009. Ratcheting up the ratchet: on the evolution of cumulative culture. Philos. Trans. R. Soc. Biol. Sci. 364, 2405–2415.

Tomasello, M., Carpenter, M., Call, J., Behne, T., Moll, H., 2005. Understanding and sharing intentions: the origins of cultural cognition. Behav. Brain Sci. 28, 675–691.

Tomasello, M., 2009. The Cultural Origins of Human Cognition. Harvard University Press, Cambridge, MA.

Tomasello, M., 2014. A Natural History of Human Thinking. Harvard University Press, Cambridge, MA.

Tulving, E., 1983. Elements of Episodic Memory. Clarendon Press, Oxford, UK.

Wellman, H.M., Cross, D., Watson, J., 2001. Meta-analysis of theory-of-mind development: the truth about false belief. Child Dev. 72, 655–684.

West, S.A., Griffin, A.S., Gardner, A., 2007. Social semantics: altruism, cooperation, mutualism, strong reciprocity and group selection. J. Evol. Biol. 20, 415–432.

Wheeler, M.A., Stuss, D.T., Tulving, E., 1997. Toward a theory of episodic memory: the frontal lobes and autonoetic consciousness. Psychol. Bull. 121, 331–354.

Whiten, A., Horner, V., Litchfield, C.A., Marshall-Pescini, S., 2004. How do apes ape? Anim. Learn. Behav. 32, 36–52.

Whiten, A., McGuigan, N., Marshall-Pescini, S., Hopper, L.M., 2009. Emulation, imitation, over-imitation and the scope of culture for child and chimpanzee. Philos. Trans. R. Soc. Biol. Sci. 364, 2417–2428.

Index

Printed in the United States
By Bookmasters